Project Co-ordinator: Gill Dawson **Editorial and Listings Co-ordinator:** Alison Nash
Production Manager: Emma Griffin **Design & Page Layout:** Kathryn Teece, Ian Cooper
Researchers: Natalie Bunegar, Isobel Harris, Max Harris
Advertising Sales: Gill Grimshaw, Debra Greer, Heather Gallimore, Clare Devanney, Jackie Smith
Editorial Contibutors: Michael Holmes, Jason Orme, Natasha Brinsmead, Melanie Griffiths, Alana Clogan
Subscriptions: Alexandra Worthington **Marketing:** Caroline Hawksworth
Technical & Database Support: Stephen Marks & ICO Solutions Ltd
Reprographics: Atelier Data Services Ltd **Publishing Director:** Peter Harris

Additional copies of The Homebuilder's Handbook are available, priced at £25.00 inclusive of UK delivery.
To order please call 01527 834435.

Third edition published in the United Kingdom in December 2005 by Ascent Publishing Ltd copyright © Ascent Publishing 2005

First published in the United Kingdom in December 2003 by Ascent Publishing Ltd copyright © Ascent Publishing 2003

Ascent Publishing Ltd,
2 Sugar Brook Court, Aston Road,
Bromsgrove, Worcestershire, B60 3EX

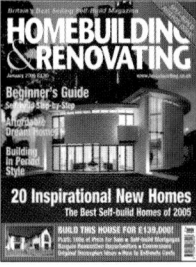

From the publishers of Homebuilding & Renovating magazine and Move or Improve? magazine
www.homebuilding.co.uk, www.moveorimprove.co.uk, www.plotfinder.net, www.propertyfinderfrance.net
www.renovationfrance.net, www.sitefinderireland.com

ISBN 0-9552043-0-5

Printed in Italy

Foreword

My copy of The Homebuilder's Handbook has never been far from my reach over the past few months. I have been researching and building my third self-build project and will shortly embark on my first speculative development, building five new houses and converting a barn and former pub. The Handbook has been a very useful and reliable ally in finding suppliers, saving me considerable time as well as turning up some great new products.

I have to admit to being pretty useless at keeping all of my contacts organised: I tend to write everything down on the back of any scrap of paper to hand and my scribbled notes inevitably get lost or thrown out, leaving me scrabbling desperately through piles of paper, the bins, or worse still, navigating through the labyrinth of 'sponsored entries' on the internet to get hold of companies I already know. Having a copy of the Handbook around has been a lifesaver at times, helping me find a telephone number or an official website quickly and easily.

It is when it comes to finding new products or suppliers that the Handbook comes into its own though. Over the past year the Handbook's pages have yielded suppliers of everything from stone flooring and structural glazing, to a local supplier of the non-setting silicone sealant for my flat roofing.

Not surprisingly then, my copy is now looking more than a bit dog-eared, (I've promised it to one of the builders on site) so it is with great timing that this new Third Edition finds its way onto my desk. The Handbook team have spent the last year verifying every one of the thousands of existing directory entries, and have also scoured the industry to expand the content with the inclusion of several hundred new contacts. To make the Handbook even easier to use, they have also restructured its content, with the addition of several new sub categories.

This Third Edition of The Homebuilder's Handbook brings together fifteen years of contacts from Homebuilding & Renovating magazine. It is the definitive sourcebook for anyone planning to build their own home or undertake a renovation project and will prove as useful to the amateur DIYer as it does to the construction professional.

Of course, in a fast changing market full of new ideas and innovation, no directory can be 100% comprehensive, and so work has already started on the Fourth Edition. In the meantime, if you have any suggestions for improvements, or further contacts you feel should be included in the next edition, please let us know either by writing to us or emailing us as homebuilders.handbook@centaur.co.uk

I hope that your copy of the Handbook proves itself as useful in your project as it has in mine.

Michael

Michael Holmes
Editor in Chief
Homebuilding & Renovating Magazine

PRODUCT INDEX

PRODUCT INDEX

Contents

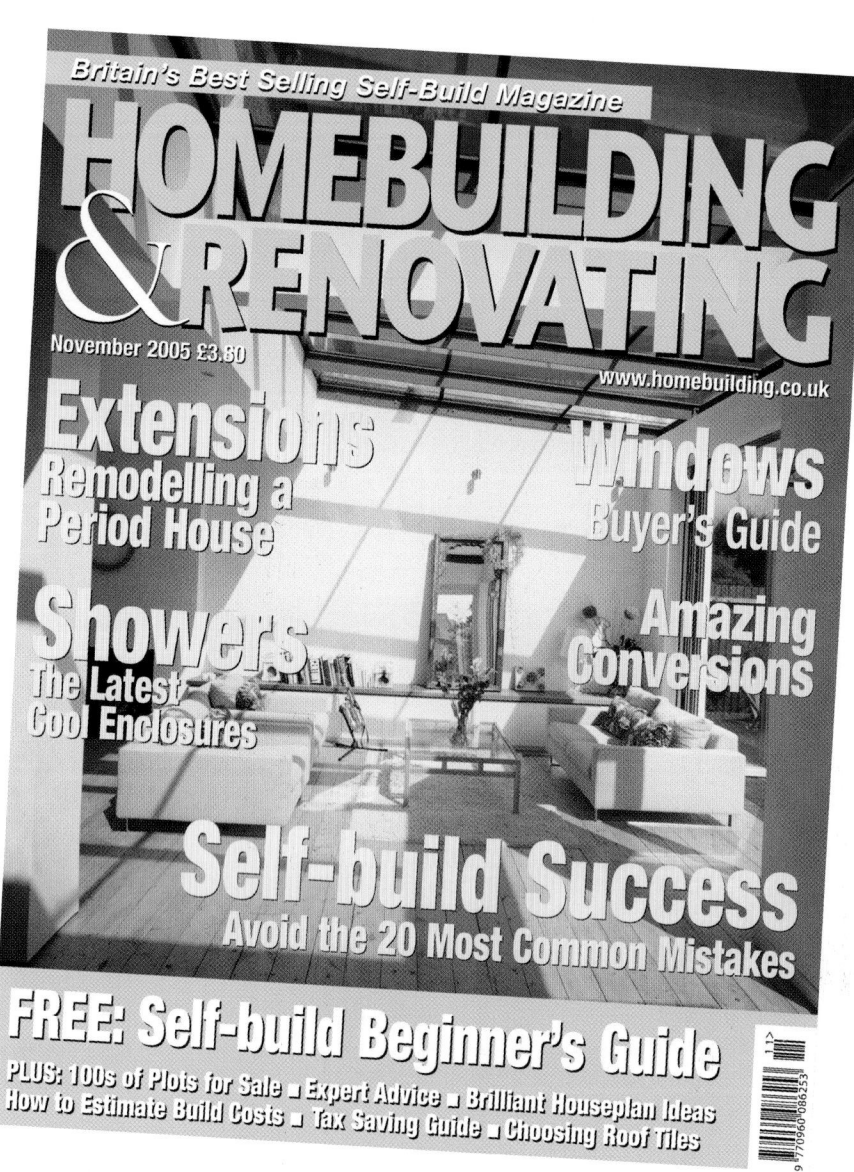

SUBSTRUCTURE & FOUNDATIONS

Groundworks

The groundwork trade is probably the most important trade of all, as the structural stability and integrity of a new home relies on the work having been done properly. It is also the part of the building process where the self-builder can feel out of control, where costs can spiral and where dramatic changes of direction, or even design, can be forced upon you. However, the commencement of groundworks is very exciting: it means that your project is finally up and running and your plans are about to be realised.

In most cases homes built with either timber frame or masonry will share precisely the same foundations, but in a few cases, where loads are critical, the lightweight timber frame or other prefabricated type structures may be able to have lesser foundations.

All foundations should be accurate to the plan but unfortunately that is not always possible. However, if you are building in timber frame, foundations should be level to within 20mm and square to within 12mm.

TYPES OF FOUNDATIONS

Standard strip foundations

For most soil conditions this is the best and cheapest way of constructing foundations. A trench of between 450mm and 600mm in width is dug to a depth of between 1.0 and 1.2m deep beneath all external and loadbearing walls. In the bottom of this a layer of concrete, at least 225mm thick, is placed and upon this the foundation walls are built in blockwork with two skins to external walls and a single skin to internal loadbearing walls.

Trenchfill foundations

Where the water table is high or where the trench sides are unstable, it is often better to revert to this type of foundation. The trenches are essentially the same except that, instead of just putting concrete in the bottom, you fill them almost to the top. This means that you are out of the ground in one day and it also means that the amount of below damp proof course blockwork is reduced.

Deep trenchfill foundations

Where heavy clay is in the presence of trees it is necessary to take the foundations below the level at which the tree's thirst for water is active. This can

mean digging to three metres deep and filling the trenches with massive amounts of concrete. At these levels special care needs to be taken and shoring might be required. Additionally, it is usually necessary to insulate the foundation from any ground movement by lining one or both sides of the trench with a compressible material plus a slip membrane.

At some stage it is no longer economically or physically possible to dig any deeper by conventional means, or the cost of sending spoil away becomes prohibitive. At this point it often becomes more economically viable to switch to a piled foundation, which avoids the need for deep trenches and huge amounts of concrete.

Piled and ringbeam

Piled and ringbeam foundations are also used in situations where the top layers of ground have poor bearing capacity and good bearing can only be found at deeper levels. The piles are driven or bored into the ground to support a reinforced concrete ringbeam (groundbeam) spanning from cap to cap to support the house walls. This ringbeam can be cast in situ or delivered as a prefabricated unit. Systems are also available which combine the ringbeam with the floor as a stable platform upon which the house can be built.

Reinforced concrete raft

Used where the ground has good bearing but is inherently unstable due to geological or mining conditions far below the surface. A large hole is dug and filled with consolidated layers of hardcore. Upon this a specially designed reinforced concrete raft is cast which suports the whole house. It takes a lot of concrete and it needs to be carefully designed, but in areas where such conditions are common, most groundworkers see it as 'standard'.

Images BELOW show Abbey Pynford's Housedeck - a proprietory piled raft system.

BUILDING STRUCTURE & MATERIALS - **Substructure & Foundations** - Piling & Specialist Foundation Systems; Underpinning

BUILDING STRUCTURE & MATERIALS

PILING AND SPECIALIST FOUNDATION SYSTEMS

KEY

PRODUCT TYPES: 1 = Specialist Foundations
2 = Piling 3 = Clayboard

SEE ALSO: BRICKS BLOCKS STONE & CLADDING - Blocks

OTHER: ▽ Reclaimed ⌂ On-line shopping
✍ Bespoke ✋ Hand-made ECO Ecological

AARSLEFF PILING
Hawton Lane, Balderton,
Newark, Nottinghamshire, NG24 3BU
Area of Operation: UK (Excluding Ireland)
Tel: 01636 611140
Fax: 01636 611142
Email: piling@aarsleff.co.uk
Web: www.aarsleff.co.uk
Product Type: 2

ABBEY PYNFORD HOUSE FOUNDATIONS LIMITED
Second Floor, Hille House,
132 St Albans Road, Watford,
Hertfordshire, WD24 4AQ
Area of Operation: UK & Ireland
Tel: 0870 085 8400
Fax: 0870 085 8401
Email: sales@abbeypynford.co.uk
Web: www.abbeypynford.co.uk
Product Type: 1, 2

ACE MINIMIX
Millfields Road, Ettingshall,
Wolverhampton, West Midlands, WV4 6JP
Area of Operation: UK (Excluding Ireland)
Tel: 01902 353522
Fax: 0121 585 5557
Email: info@tarmac.co.uk
Web: www.tarmac.co.uk

ALL FOUNDATIONS LIMITED
Primrose Business Park,
White Lane, Blackwell, Derby,
Derbyshire, DE55 5JR
Area of Operation: UK & Ireland
Tel: 0870 350 2050
Email: mail@allfoundations.co.uk
Web: www.allfoundationsltd.co.uk
Product Type: 1, 2

ANVIL FOUNDATIONS LTD
8 Beaufort Chase, Dean Row,
Wilmslow, Cheshire, SK9 2BZ
Area of Operation: UK (Excluding Ireland)
Tel: 01625 522 800
Fax: 01625 522 850
Email: info@minipiling.co.uk
Web: www.minipiling.co.uk
Product Type: 1, 2

CEMEX READYMIX
CEMEX House, Rugby , Warwickshire, CV21 2DT
Area of Operation: Worldwide
Tel: 0800 667 827
Fax: 01788 564404
Email: info.readymix@cemex.co.uk
Web: www.cemex.co.uk ⌂
Product Type: 1

CENTRAL PILING LTD
Central Park, Colchester Road,
Halstead, Essex, CO9 2EU
Area of Operation: East England,
Greater London, South East England,
South West England and South Wales
Tel: 01787 474000
Fax: 01787 472113
Email: central@piling.uk.com
Web: www.centralpiling.com
Product Type: 2

CITIBUILD MINI PILING LTD
Walnut Tree Farm, Swanton Abbott,
Norfolk, NR10 5DL
Area of Operation: UK (Excluding Ireland)
Tel: 01692 538888
Fax: 01692 538100
Email: mark@citibuild.co.uk
Web: www.citibuild.co.uk
Product Type: 1, 2

CJ O'SHEA & CO LTD
Unit 1 Granard Business Centre, Bunns Lane,
Mill Hill, London, NW7 2DZ
Area of Operation: East England, Greater London,
South East England, South West England & South Wales
Tel: 0208 959 3600
Fax: 0208 959 0184
Email: admin@oshea.co.uk
Web: www.oshea.co.uk

CONCEPT ENGINEERING CONSULTANTS LTD
Unit 8 Warple Mews, Warple Way,
London, Greater London, W3 0RF
Area of Operation: East England, Greater London,
South East England, South West England & South Wales
Tel: 0208 840 3321
Fax: 0208 579 6910
Email: si@conceptconsultants.co.uk
Web: www.conceptconsultants.co.uk
Product Type: 1, 2

FALCON STRUCTURAL REPAIRS
2B Harbour Road, Portishead, Bristol, BS20 7DD
Area of Operation: Europe
Tel: 01275 844889
Fax: 01275 847002
Email: sjoyner@falconstructural.co.uk
Web: www.falconstructural.co.uk
Product Type: 3

FOUNDATION PILING LTD
Unit 21, Area A, Rednal Industrial Estate,
Queens Head, Oswestry, Shropshire, SY11 4HS
Area of Operation: UK (Excluding Ireland)
Tel: 01691 610638
Fax: 01691 610500
Email: info@foundation-piling.com
Web: www.foundation-piling.com
Product Type: 1, 2

FOUNDATIONS EASTERN LTD
Moulsham Mill, Parkway,
Chelmsford, Essex, CM2 7PX
Area of Operation: East England, Greater London,
Midlands & Mid Wales, South East England, South
West England and South Wales
Tel: 01245 226502
Fax: 01245 223967
Email: fel.minipiling@virgin.net
Web: www.minipiling.com
Product Type: 1, 2

FOUNDEX (UK) LTD
Bretby Business Park, Building 1A,
Ashby Road, Burton upon Trent,
Staffordshire, DE15 0YZ
Area of Operation: UK (Excluding Ireland)
Tel: 01283 553240
Fax: 01283 553242
Email: email@foundex.fsnet.co.uk
Web: www.foundexukltd.co.uk
Product Type: 1, 2

GEOBOND
Sheiling House, Invincible Road,
Farnborough, Hampshire, GU14 7QU
Area of Operation: UK (Excluding Ireland)
Tel: 01252 519224
Fax: 01252 378665
Product Type: 1, 2

H+H CELCON LIMITED
Celcon House, Ightham, Sevenoaks, Kent, TN15 9HZ
Area of Operation: UK (Excluding Ireland)
Tel: 01732 880520
Fax: 01732 880531
Email: marketing@celcon.co.uk
Web: www.celcon.co.uk
Product Type: 1

HOLDEN + PARTNERS
26 High Street, Wimbledon,
Greater London, SW19 5BY
Area of Operation: Worldwide
Tel: 0208 946 5502
Fax: 0208 879 0310
Email: arch@holdenpartners.co.uk
Web: www.holdenpartners.co.uk

MAXIT UK
The Heath, Runcorn, Cheshire, WA7 4QX
Area of Operation: UK & Ireland
Tel: 01928 515656
Fax: 01928 576792
Email: sales@maxit-uk.co.uk
Web: www.maxit-uk.co.uk
Product Type: 1

R E DESIGN
97 Lincoln Avenue, Glasgow, G133DH
Area of Operation: Scotland
Tel: 0141 959 1902
Fax: 0141 959 1902
Email: mail@r-e-design.co.uk
Web: www.r-e-design.co.uk
Product Type: 1, 2

SAFEGUARD EUROPE LTD
Redkiln Close, Horsham,
West Sussex, RH13 5QL
Area of Operation: Worldwide
Tel: 01403 210204
Fax: 01403 217529
Email: info@safeguardeurope.com
Web: www.safeguardeurope.com
Product Type: 1

THE BIG BASEMENT COMPANY LIMITED
Trussley Works, Trussley Road,
Shepherds Bush, London, W6 7PR
Area of Operation: Greater London,
South East England
Tel: 0700 2442273
Fax: 0208 7488957
Email: enquiries@bigbasement.co.uk
Web: www.bigbasement.co.uk
Product Type: 1, 2, 3

THE MINI PILING COMPANY
Sandy Lane, Wildmoor,
Bromsgrove, Worcestershire, B61 0QU
Area of Operation: UK (Excluding Ireland)
Tel: 0121 457 9966
Fax: 0121 457 9918
Email: steve@theminipilingco.co.uk
Web: www.theminipilingco.co.uk
Product Type: 1, 2, 3

URETEK (UK) LTD
Peel House, Peel Road,
Skelmersdale, Lancashire, WN8 9PT
Area of Operation: UK & Ireland
Tel: 01695 50525
Fax: 01695 555212
Email: sales@uretek.co.uk
Web: www.uretek.co.uk

VAN ELLE LTD
Kirkby Lane, Pinxton,
Nottinghamshire, NG16 6JA
Area of Operation: UK & Ireland
Tel: 01773 580580
Fax: 01773 862100
Email: vic@van-elle.co.uk
Web: www.van-elle.co.uk
Product Type: 1, 2

VANDEX UK LTD
Redkiln Close, Redkiln Way,
Horsham, Sussex,
West Sussex, RH13 5QL
Area of Operation: UK & Ireland
Tel: 0870 241 6264
Fax: 01403 217529
Email: info@vandex.co.uk
Web: www.vandex.co.uk
Product Type: 1

UNDERPINNING

KEY

OTHER: ▽ Reclaimed ⌂ On-line shopping
✍ Bespoke ✋ Hand-made ECO Ecological

ABBEY PYNFORD HOUSE FOUNDATIONS LIMITED
Second Floor, Hille House,
132 St Albans Road, Watford,
Hertfordshire, WD24 4AQ
Area of Operation: UK & Ireland
Tel: 0870 085 8400
Fax: 0870 085 8401
Email: sales@abbeypynford.co.uk
Web: www.abbeypynford.co.uk
Other Info: ✍

ACE MINIMIX
Millfields Road, Ettingshall,
Wolverhampton, West Midlands, WV4 6JP
Area of Operation: UK (Excluding Ireland)
Tel: 01902 353522
Fax: 0121 585 5557
Email: info@tarmac.co.uk
Web: www.tarmac.co.uk

ALL FOUNDATIONS LIMITED
Primrose Business Park, White Lane,
Blackwell, Derby, Derbyshire, DE55 5JR
Area of Operation: UK & Ireland
Tel: 0870 3502050
Email: mail@allfoundations.co.uk
Web: www.allfoundationsltd.co.uk

ANVIL FOUNDATIONS LTD
8 Beaufort Chase, Dean Row,
Wilmslow, Cheshire, SK9 2BZ
Area of Operation: UK (Excluding Ireland)
Tel: 01625 522 800
Fax: 01625 522 850
Email: info@minipiling.co.uk
Web: www.minipiling.co.uk
Other Info: ✍ ✋

CITIBUILD MINI PILING LTD
Walnut Tree Farm, Swanton Abbott,
Norfolk, NR10 5DL
Area of Operation: UK (Excluding Ireland)
Tel: 01692 538888
Fax: 01692 538100
Email: mark@citibuild.co.uk
Web: www.citibuild.co.uk

CONCEPT ENGINEERING CONSULTANTS LTD
Unit 8 Warple Mews,
Warple Way, London, W3 0RF
Area of Operation: East England,
Greater London, South East England,
South West England and South Wales
Tel: 0208 840 3321
Fax: 0208 579 6910
Email: si@conceptconsultants.co.uk
Web: www.conceptconsultants.co.uk

FOUNDATION PILING LTD
Unit 21, Area A,
Rednal Industrial Estate,
Queens Head, Oswestry,
Shropshire, SY11 4HS
Area of Operation: UK (Excluding Ireland)
Tel: 01691 610638
Fax: 01691 610500
Email: info@foundation-piling.com
Web: www.foundation-piling.com

GEOBOND
Sheiling House, Invincible Road,
Farnborough, Hampshire , GU14 7QU
Area of Operation: UK (Excluding Ireland)
Tel: 01252 519224
Fax: 01252 378665

BASEMENTS
UK

HELICAL SYSTEMS LTD
The Old Police Station, 195 Main Road,
Biggin Hill, Westerham, Kent, TN16 3JU
Area of Operation: UK & Ireland
Tel: 01959 541148 **Fax:** 01959 540841
Email: janecannon@helicalsystems.co.uk
Web: www.helicalsystems.co.uk ✏
Other Info: ✏

R E DESIGN
97 Lincoln Avenue, Glasgow, G133DH
Area of Operation: Scotland
Tel: 0141 959 1902 **Fax:** 0141 959 1902
Email: mail@r-e-design.co.uk
Web: www.r-e-design.co.uk

**THE BIG BASEMENT
COMPANY LIMITED**
Trussley Works, Trussley Road,
Shepherds Bush, Greater London, W6 7PR
Area of Operation: Greater London,
South East England
Tel: 0700 244 2273 **Fax:** 0208 748 8957
Email: enquiries@bigbasement.co.uk
Web: www.bigbasement.co.uk

THE MINI PILING COMPANY
Sandy Lane, Wildmoor,
Bromsgrove, Worcestershire, B61 0QU
Area of Operation: UK (Excluding Ireland)
Tel: 0121 457 9966 **Fax:** 0121 457 9918
Email: steve@theminipilingco.co.uk
Web: www.theminipilingco.co.uk

URETEK (UK) LTD
Peel House, Peel Road,
Skelmersdale, Lancashire, WN8 9PT
Area of Operation: UK & Ireland
Tel: 01695 50525 **Fax:** 01695 555212
Email: sales@uretek.co.uk
Web: www.uretek.co.uk

VAN ELLE LTD
Kirkby Lane, Pinxton, Nottinghamshire, NG16 6JA
Area of Operation: UK & Ireland
Tel: 01773 580580 **Fax:** 01773 862100
Email: vic@van-elle.co.uk
Web: www.van-elle.co.uk

BASEMENTS

KEY

PRODUCT TYPES: 1= Modular Basements
2 = Basement Tanking 3 = Basement
Specialists 4. Basement Membranes
5. Other

OTHER: ▽ Reclaimed ✑ On-line shopping
✏ Bespoke ✋ Hand-made ECO Ecological

**ABBEY PYNFORD HOUSE
FOUNDATIONS LIMITED**
Second Floor, Hille House, 132 St Albans Road,
Watford, Hertfordshire, WD24 4AQ
Area of Operation: UK & Ireland
Tel: 0870 085 8400 **Fax:** 0870 085 8401
Email: sales@abbeypynford.co.uk
Web: www.abbeypynford.co.uk
Product Type: 3

ABTECH (UK) LTD
Sheiling House, Invincible Road, Farnborough,
Hampshire , GU14 7QU
Area of Operation: UK (Excluding Ireland)
Tel: 0800 085 1431 **Fax:** 01252 378665
Email: sales@abtechbasements.co.uk
Web: www.abtechbasements.co.uk
Product Type: 3, 4

AQUATECNIC
211 Heathhall Industrial Estate, Dumfries,
Dumfries & Galloway, DG1 3PH
Area of Operation: UK & Ireland
Tel: 0845 226 8283 **Fax:** 0845 226 8293
Email: info@aquatecnic.net
Web: www.aquatecnic.net ✑
Product Type: 2

BASEMENTS UK
Unit 7 Nations Business Park, Curdridge Lane,
Curdridge, Southampton, Hampshire, SO32 2BH
Area of Operation: East England, Greater London,
Midlands & Mid Wales, North East England, North
West England and North Wales, South East England,
South West England and South Wales
Tel: 0845 060 4488 **Fax:** 01489 786109
Email: info@bukfirst.co.uk
Web: www.basementsuk.co.uk
Product Type: 2, 3, 4

BILCO UK
Park Farm Business Centre, Fornham St Genevieve,
Bury St Edmunds, Suffolk, IP28 6TS
Area of Operation: UK (Excluding Ireland)
Tel: 01284 701696 **Fax:** 01284 702531
Email: bilcouk@bilco.com
Web: www.bilco.com
Product Type: 5

BIOCRAFT LTD
25B Chapel Hill, Reading, Berkshire, RG31 5BT
Area of Operation: East England, Greater London,
South East England, South West England & South Wales
Tel: 0118 945 1144
Email: info@biocraft.co.uk
Web: www.biocraft.co.uk
Product Type: 3, 4

CAVITY TRAYS LTD
New Administration Centre, Boundary Avenue,
Yeovil, Somerset, BA22 8HU
Area of Operation: Worldwide
Tel: 01935 474769
Fax: 01935 428223
Email: enquiries@cavitytrays.co.uk
Web: www.cavitytrays.com ✑
Product Type: 4

CONCEPT ENGINEERING CONSULTANTS LTD
Unit 8 Warple Mews,
Warple Way, London, W3 0RF
Area of Operation: East England, Greater London,
South East England, South West England & South Wales
Tel: 0208 840 3321
Fax: 0208 579 6910
Email: si@conceptconsultants.co.uk
Web: www.conceptconsultants.co.uk
Product Type: 1, 2, 3, 4, 5

DELTA MEMBRANE SYSTEMS LTD
Bassett Business Centre, Hurricane Way,
North Weald, Epping, Essex, CM16 6AA
Area of Operation: UK & Ireland
Tel: 01992 523811
Fax: 01992 524046
Email: info@deltamembranes.com
Web: www.deltamembranes.com
Product Type: 2, 3, 4

DELTA MEMBRANES
Area of Operation: UK & Ireland
Tel: 01992 523811 **Fax:** 01992 524046
Email: info@deltamembranes.com
Web: www.deltamembranes.com

Solutions for Basements
New build and Refurbishment
30 Year Guarantee
Solutions for Damp Problems
Barn conversions, damp walls, damp floors
30 Year Guarantee

FOUNDEX (UK) LTD
Bretby Business Park, Building 1A, Ashby Road,
Burton upon Trent, Staffordshire, DE15 0YZ
Area of Operation: UK (Excluding Ireland)
Tel: 01283 553240
Fax: 01283 553242
Email: email@foundex.fsnet.co.uk
Web: www.foundexukltd.com
Product Type: 1, 2, 4

HOLDEN + PARTNERS
26 High Street, Wimbledon,
London, SW19 5BY
Area of Operation: Worldwide
Tel: 0208 946 5502
Fax: 0208 879 0310
Email: arch@holdenpartners.co.uk
Web: www.holdenpartners.co.uk
Product Type: 5

**INDUSTRIAL TEXTILES
& PLASTICS LTD**
Stillington Road, Easingwold, York,
North Yorkshire, YO61 3FA
Area of Operation: Worldwide
Tel: 01347 825200
Email: info@itpltd.com
Web: www.itpltd.com
Product Type: 4

NEWMIL LTD
17 Arundel Close, New Milton,
Hampshire, BH25 5UH
Area of Operation: UK (Excluding Ireland)
Tel: 0845 090 0109
Fax: 0870 094 1258
Email: enquiries@newmil.co.uk
Web: www.newmil.co.uk
Product Type: 2, 3

PHOENIX (GB) LTD
Chestnut Field House, Chestnut Field,
Rugby, Warwickshire, CU21 2PA
Area of Operation: UK & Ireland
Tel: 01788 571482
Fax: 01788 542245
Email: tonyb@phoenix-gb.com
Web: www.phoenix-ag.com
Product Type: 2, 4

QUAD-LOCK (ENGLAND) LTD
Unit B3.1, Maws Centre,
Jackfield, Telford, Shropshire, TF8 7LS
Area of Operation: UK (Excluding Ireland)
Tel: 0870 443 1901
Fax: 0870 443 1902
Email: k.turner@quadlock.co.uk
Web: www.quadlock.co.uk
Product Type: 5

QUADRIGA CONCEPTS LTD
Gadbrook House, Gadbrook Park,
Rudheath, Northwich, Cheshire, CW9 7RG
Area of Operation: UK & Ireland
Tel: 0808 100 3777
Fax: 01606 330777
Email: info@quadrigaltd.com
Web: www.quadrigaltd.com
Product Type: 2, 4

R E DESIGN
97 Lincoln Avenue,
Glasgow, G133DH
Area of Operation: Scotland
Tel: 0141 959 1902
Fax: 0141 959 1902
Email: mail@r-e-design.co.uk
Web: www.r-e-design.co.uk
Product Type: 5

SAFEGUARD EUROPE LTD
Redkiln Close, Horsham,
West Sussex, RH13 5QL
Area of Operation: Worldwide
Tel: 01403 210204
Fax: 01403 217529
Email: info@safeguardeurope.com
Web: www.safeguardeurope.com
Product Type: 2, 3, 4

SAFEGUARD
Area of Operation: Worldwide
Tel: 01403 210204
Fax: 01403 217529
Email: info@safeguardeurope.com
Web: www.safeguardeurope.com
Product Type: 2, 3, 4

Safeguard supply a choice of systems for
waterproofing and refurbishing existing
basements. CAD drawings can be downloaded
from the company's website.

SPRY PRODUCTS LTD
64 Nottingham Road, Long Eaton,
Nottingham, Nottinghamshire, NG10 2AU
Area of Operation: UK & Ireland
Tel: 0115 9732914
Fax: 0115 9725172
Email: jerspry@aol.com
Web: www.spryproducts.com
Product Type: 2, 3, 4

THE BIG BASEMENT COMPANY LIMITED
Trussley Works, Trussley Road,
Shepherds Bush, London, W6 7PR
Area of Operation: Greater London,
South East England
Tel: 0700 244 2273
Fax: 0208 748 8957
Email: enquiries@bigbasement.co.uk
Web: www.bigbasement.co.uk
Product Type: 1, 2, 3, 4, 5

THERMONEX LTD
Delcon House, 65 Manchester Road,
Bolton, Lancashire, BL2 1ES
Area of Operation: UK (Excluding Ireland)
Tel: 01204 559551
Fax: 01204 559552
Email: salesadmin@thermonex.co.uk
Web: www.thermonex.co.uk

TRACE BASEMENT SYSTEMS
Unit 8, Hurst Mill, Hurst Road,
Glossop, Derbyshire, SK13 7QB
Area of Operation: East England, Greater London,
Midlands & Mid Wales, North East England,
North West England and North Wales
Tel: 01457 865165
Fax: 01457 866253
Email: enquiries@traceremedial.co.uk
Web: www.tracebasementsystems.co.uk
Product Type: 3

TRITON CHEMICAL MANUFACTURING LTD
129 Felixstowe Road,
Abbey Wood, London, SE2 9SG
Area of Operation: Worldwide
Tel: 0208 310 3929
Fax: 0208 312 0349
Email: neil@triton-chemicals.com
Web: www.triton-chemicals.com
Product Type: 2, 3, 4

VANDEX UK LTD
Redkiln Close, Redkiln Way,
Horsham, Sussex, West Sussex, RH13 5QL
Area of Operation: UK & Ireland
Tel: 0870 241 6264 **Fax:** 01403 217529
Email: info@vandex.co.uk
Web: www.vandex.co.uk
Product Type: 2, 3, 4

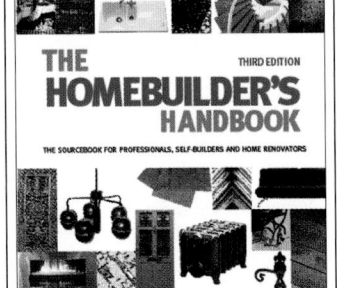

THE HOMEBUILDER'S HANDBOOK
THIRD EDITION

THE SOURCEBOOK FOR PROFESSIONALS, SELF-BUILDERS AND HOME RENOVATORS

1,000S OF ESSENTIAL CONTACTS

Don't forget to let us know about companies you think should be listed in the next edition
email:
customerservice@centaur.co.uk

GROUND INVESTIGATION, STABILISATION AND SOIL TESTING

KEY
OTHER: ▽ Reclaimed ⌐ On-line shopping
✎ Bespoke ✋ Hand-made ECO Ecological

A PROCTOR GROUP LTD
The Haugh, Blairgowrie,
Perth and Kinross, PH10 7ER
Area of Operation: Worldwide
Tel: 01250 872261
Fax: 01250 872727
Email: karrina.andrews@proctorgroup.com
Web: www.proctorgroup.com

ABBEY PYNFORD HOUSE FOUNDATIONS LIMITED
Second Floor, Hille House, 132 St Albans Road,
Watford, Hertfordshire, WD24 4AQ
Area of Operation: UK & Ireland
Tel: 0870 085 8400 **Fax:** 0870 085 8401
Email: sales@abbeypynford.co.uk
Web: www.abbeypynford.co.uk
Other Info: ✎

ACE MINIMIX
Millfields Road, Ettingshall,
Wolverhampton, West Midlands, WV4 6JP
Area of Operation: UK (Excluding Ireland)
Tel: 01902 353522 **Fax:** 0121 5855557
Email: info@tarmac.co.uk
Web: www.tarmac.co.uk

ALL FOUNDATIONS LIMITED
Primrose Business Park, White Lane,
Blackwell, Derby, Derbyshire, DE55 5JR
Area of Operation: UK & Ireland
Tel: 0870 3502050
Email: mail@allfoundations.co.uk
Web: www.allfoundationsltd.co.uk

CONCEPT ENGINEERING CONSULTANTS LTD
Unit 8 Warple Mews,
Warple Way, London, W3 0RF
Area of Operation: East England, Greater London,
South East England, South West England & South Wales
Tel: 0208 840 3321
Fax: 0208 579 6910
Email: si@conceptconsultants.co.uk
Web: www.conceptconsultants.co.uk

GEOBOND
Sheiling House, Invincible Road,
Farnborough, Hampshire, GU14 7QU
Area of Operation: UK (Excluding Ireland)
Tel: 01252 519224 **Fax:** 01252 378665

GEOINVESTIGATE LTD
Units 4-5 Terry Dicken Industrial Estate, Ellerbeck
Way, Stokesley, North Yorkshire, TS9 7AE
Area of Operation: UK (Excluding Ireland)
Tel: 01642 713779 **Fax:** 01642 719923
Email: geoinvestigate@qnetadsl.com

PHI GROUP LTD
Harcourt House, Royal Crescent,
Cheltenham, Gloucestershire, GL50 3DA
Area of Operation: UK (Excluding Ireland)
Tel: 0870 333 4126
Fax: 0870 333 4127
Email: marketing@phigroup.co.uk
Web: www.phigroup.co.uk

R E DESIGN
97 Lincoln Avenue, Glasgow, G133DH
Area of Operation: Scotland
Tel: 0141 959 1902
Fax: 0141 959 1902
Email: mail@r-e-design.co.uk
Web: www.r-e-design.co.uk

SEVENOAKS ENVIRONMENTAL CONSULTANCY LTD
19 Gimble Way, Pembury,
Tunbridge Wells, Kent, TN2 4BX
Area of Operation: East England, Greater London,
Midlands & Mid Wales, North East England, North
West England and North Wales, South East England,
South West England and South Wales
Tel: 01892 822999
Fax: 01892 822992
Email: d.jones@sevenoaksenvironmental.co.uk
Web: www.sevenoaksenvironmental.co.uk

URETEK (UK) LTD
Peel House, Peel Road,
Skelmersdale, Lancashire, WN8 9PT
Area of Operation: UK & Ireland
Tel: 01695 50525
Fax: 01695 555212
Email: sales@uretek.co.uk
Web: www.uretek.co.uk

VAN ELLE LTD
Kirkby Lane, Pinxton,
Nottinghamshire, NG16 6JA
Area of Operation: UK & Ireland
Tel: 01773 580580
Fax: 01773 862100
Email: vic@van-elle.co.uk
Web: www.van-elle.co.uk

GENERAL FOUNDATIONS

KEY
OTHER: ▽ Reclaimed ⌐ On-line shopping
✎ Bespoke ✋ Hand-made ECO Ecological

FAIRCLEAR LTD
16 Sibthorpe Drive, Sudbrooke,
Lincoln, Lincolnshire, LN2 2RQ
Area of Operation: East England
Tel: 01522 595189
Fax: 01522 595189
Email: enquiries@fairclear.co.uk
Web: www.fairclear.co.uk

JOE WILLIAMS GROUNDWORK & DEMOLITION
The Holdings, Aston Bury Farm,
Aston, Stevenage, Hertfordshire, SG2 7EG
Area of Operation: East England, South East England
Tel: 01438 880824
Fax: 01438 880345
Email: jwgroundworks@yahoo.co.uk
Web: www.joe-williams.co.uk

COMPLETE STRUCTURAL SYSTEMS

Image courtesy of Beco Wallform (01724 747576)

SPONSORED BY BECO WALLFORM
Tel 01724 747576 Web www.becowallform.co.uk

Beco **Wallform**

Structural Systems

Choosing which structure is right for you will probably be made even more confusing by some of the claims and myths perpetuated in the building industry - many of which have little if any basis in fact. Bear in mind that whilst the choice of structure is important, the other factors - external cladding, roofing, foundations, doors, windows, heating and interiors - are all decisions that can almost always be made independently from your choice of structure, so don't be fooled into letting these factors influence your decision. Here are some of the areas that can be most misleading:

Image by Border Oak

Appearance: There is no inherent reason why the structural system you choose for your home should influence the way it looks either inside or out - unless you want it to. Bricks, stone, render and other claddings are equally applicable to masonry, timber or steel frame, or any other construction system. It is, however, worth investigating the cost implications of your external cladding in combination with your chosen building system - stone, for example, is usually most economically used on a blockwork structure.

Some structural systems become a design feature in their own right and are chosen for this reason alone. For instance, the heavy oak beams of a traditional post and beam timber frame are difficult to recreate unless a genuine oak frame is used.

Build Speed: With good site conditions a timber frame home can be much faster to build than a masonry home, mainly as much of the shell has been built in a factory some weeks before going on site, but also because the internal trades can start as soon as the shell is weathertight. With timber and steel frames, the manufacturer should be able to erect the frame to a weathertight stage, leaving the self-builder to manage only the finishing trades themselves. This is convenient, but a contractor could do the same for any other construction system.

Cost: Most designs can be built using any structural system and it is possible to obtain quotes and make comparisons. The difference in price in relation to overall build costs is not usually a big factor. However, in some situations, for example on a sloping site, one system may be far more economical than others.

Availability: A frame home manufactured in a factory will usually have to be ordered, and a deposit paid, some weeks in advance of the time it is required on site. Masonry materials are readily available from all merchants.

Alternative Building Materials

Structural Insulated Panels (SIPS) and Insulated Concrete Formwork (ICF) are the leaders in the growing number of alternatives to traditional building materials.

SIPS are sheets of plywood or oriented strand board laminated around a core of insulating board, and can be used for both wall panels and roofing. They can be moved into place quickly and easily, and have excellent insulating properties. They also avoid the need for roof trusses, therefore they are perfect for attic conversions where space is at a premium.

ICF is a system of hollow, interlocking lightweight blocks, which are stacked together into the shape you

want and then filled with concrete. The blocks remain in place permanently as insulation, rather than being removed like conventional poured concrete formwork. The structure can then be rendered, or a brick skin can be added.

Natural Building Materials

It is also possible to build a home out of natural materials, including cob, hemp and straw bales. There are a number of associations and companies who specialise in these types of homes, and there are also many courses around the country where you can learn more about these building methods.

For more information, see the Trade, Regulatory Bodies and Professional Associations chapter, the Courses and Events Diary, and the list of companies within the following Eco Structures section.

TIMBER FRAME

KEY

PRODUCT TYPES: 1= Post and Beam
2 = Open Panel   3 = Closed Panel
4 = Green Oak   5 = Log Homes
6 = Barn Frames   7 = Other

SEE ALSO: MERCHANTS - Timber Merchants

OTHER: ▽ Reclaimed ⌐🖱 On-line shopping
◢ Bespoke 🖐Hand-made ECO Ecological

**ADVANCED TIMBER TECHNOLOGY
BY BENFIELD ATT**
Benfield ATT, Castle Way, Caldicot,
Monmouthshire, NP26 5PR
Area of Operation: Worldwide
Tel: 01291 437050 **Fax:** 01291 437051
Email: info@adtimtec.com
Web: www.adtimtec.com

ALLWOOD BUILDINGS LTD
Talewater Works, Talaton, Exeter, Devon, EX5 2RT
Area of Operation: UK (Excluding Ireland)
Tel: 01404 850977 **Fax:** 01404 850946
Email: frames@allwoodtimber.co.uk
Web: www.allwoodtimber.co.uk
Product Type: 1, 2, 3

**ANIRINA OY (FINLAND) - LOG HOME
SUPPLIERS & BUILDERS UK**
The Gate House, Home Park Terrace, Hampton Court
Road, Hampton Wick, Kingston upon Thames,
Surrey, KT1 4AE
Area of Operation: Europe
Tel: 0208 943 0430 **Fax:** 0208 943 0430
Email: chris.drayson@virgin.net
Product Type: 5

ANTIQUE BUILDINGS LTD
Dunsfold, Godalming,
Surrey, GU8 4NP
Area of Operation: UK & Ireland
Tel: 01483 200477
Email: info@antiquebuildings.com
Web: www.antiquebuildings.com
Product Type: 6
Other Info: ▽
Material Type: A) 2

**ARCADIAN TIMBER
FRAMES LTD**
Wisteria House, May Lane,
Dursley, Gloucestershire, GL11 4JH
Area of Operation: UK (Excluding Ireland)
Tel: 01453 542248
Fax: 01453 549954
Email: apwithers@arcadianframes.com
Web: www.arcadianframes.com

**ARTICHOUSE IRELAND,
LOGART HOMES**
Arva, Tivoli Terrace North,
Dun Laoghaire, Co Dublin, Ireland
Area of Operation: UK & Ireland
Tel: +353 1280 2879
Fax: +353 1280 0955
Email: info@logart.ie
Web: www.logart.ie
Product Type: 5
Other Info: ECO ◢

AXIS TIMBER LTD
9 Chapel Road, Sarisbury Green,
Southampton, Hampshire, SO31 7FB
Area of Operation: UK & Ireland
Tel: 01489 575073
Fax: 01489 571607
Email: douglas@axistimber.com
Web: www.axistimber.co.uk
Product Type: 1, 7
Material Type: B) 1, 2, 5, 7, 8, 9, 10, 12

BEAVER TIMBER COMPANY
Barcaldine, Argyll & Bute, PA37 1SG
Area of Operation: UK (Excluding Ireland)
Tel: 01631 720353
Fax: 01631 720430
Email: info@beavertimber.co.uk
Web: www.beavertimber.co.uk ⌐🖱
Product Type: 5

BENFIELD ATT
Castle Way, Caldicot,
Monmouthshire, NP26 5PR
Area of Operation: UK & Ireland
Tel: 01291 437050
Fax: 01291 437051
Email: info@benfieldatt.co.uk
Web: www.benfieldatt.co.uk ⌐🖱
Product Type: 2, 3

**BORDER OAK DESIGN
& CONSTRUCTION**
Kingsland Sawmills,
Kingsland, Leominster,
Herefordshire, HR6 9SF
Area of Operation: Worldwide
Tel: 01568 708752
Fax: 01568 708295
Email: sales@borderoak.com
Web: www.borderoak.com
Product Type: 4, 6
Other Info: ECO ▽ ◢ 🖐

BROCH LIMITED
Unit 7, Parsons Road,
Manor Trading Estate,
South Benfleet,
Essex, SS7 4PY
Area of Operation: Europe
Tel: 0870 879 3070
Fax: 0870 879 3071
Email: info@brochsolid.com
Web: www.brochsolid.com
Product Type: 1, 2, 3, 4, 5, 6, 7

BROCH LTD

Area of Operation: UK & Europe
Tel: 0870 8793070
Fax: 0870 8793071
Email: info@brochsolid.com
Web: www.brochsolid.com

Broch solid timber frame building: houses to
hotels; conservatories to conference halls

BUYDIRECT.CO.UK LTD
Robjohns, Colam Lane, Little Baddow,
Chelmsford, Essex, CM3 4SY
Area of Operation: UK & Ireland
Tel: 01245 224637 **Fax:** 01245 225384
Email: sales@buydirect.co.uk
Web: www.buydirect.co.uk
Product Type: 5

CANADA WOOD UK
PO Box 1, Farnborough, Hampshire, GU14 6WE
Area of Operation: UK (Excluding Ireland)
Tel: 01252 522545 **Fax:** 01252 522546
Email: office@canadawooduk.org
Web: www.canadawood.info
Product Type: 2, 5 **Material Type:** A) 12

Beco Wallform

For individual design flair and the best standards of living comfort Beco WALLFORM offers service, choice and high performance results in the construction of the ideal energy efficient home.

Contact us:

Tel: 01724 74 75 76
Fax: 01724 74 75 79

Web: www.becowallform.co.uk
Email: info@becowallform.co.uk

DESIGN FLEXIBILITY

The versatility of the WALLFORM system offers maximum potential for any site development.

Tel: 01724 74 75 76

PLANNING AND PERFORMANCE

Building with WALLFORM even in conservation areas maximises energy performance for any style.

Back to the Future -
with Beco WALLFORM

QUICK TO BUILD

For professionals and Selfbuilders alike, the speed and practicality of the construction process reduces build time.

Web: www.becowallform.co.uk

DIFFICULT SITES

Components are light and easy to handle and the practical building method overcomes the problems posed by difficult and inaccessible sites.

BETTER BUILDING
Beco WALLFORM

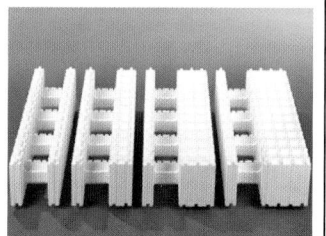

COMPREHENSIVE PRODUCT RANGE

Beco WALLFORM gives a wide range of energy and structural options which will satisfy the most demanding design.

Email: info@becowallform.co.uk

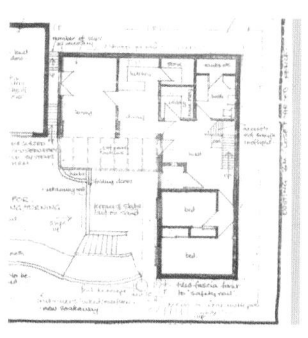

DESIGN FREEDOM

Traditional or contemporary design, straight or curved walls, earth-sheltered house or landmark tower - Beco WALLFORM has the flexibility and performance which gives CUSTOMER SATISFACTION.

Tel: 01724 74 75 76

WYSIWYG -
What You See Is What You Get!

The structual and energy performance of Beco WALLFORM means that the finished building is a true reflection of the design, not the limitations of the building method.

Web: www.becowallform.co.uk

LOW ENERGY CONSTRUCTION.

Low energy homes are quick to build, comfortable to live in and have low running costs.

LOW BILLS -
LOW ENVIRONMENTAL IMPACT

FUNCTIONAL EXTENSIONS

Sun room, basement, gymnasium, swimming pool and pool house - Beco WALLFORM provides comfort, low bills **and no condensation!**

Web: www.becowallform.co.uk

DOWN TOOLS!

The speed and practicality of Beco WALLFORM construction means less downtime and more sundowner time at completion of the project.

BUILDING BETTER
Beco WALLFORM

Beco Wallform

Over thirty years of product development and experience worldwide make BECO WALLFORM the leader in ICF construction.

For performance and service choose Beco WALLFORM.

Contact us:

Tel: 01724 74 75 76
Fax: 01724 74 75 79

Web: www.becowallform.co.uk
Email: info@becowallform.co.uk

CUSTOM HOMES™

Energy efficient homes from the UK's largest self-build package company with offices nationwide

A complete service from design to construction

DESIGN

100's of designs collected over 30 years are readily available for adaptation to suit individual family requirements or one-off specials can be provided at no extra cost.

SERVICE THROUGH LOCAL OFFICES

We believe that no other company offers the same level of service from design to construction with expert guidance available from our offices nationwide.

PROJECT MANAGERS

Custom Homes can assist individuals with no knowledge of the building industry to construct their dream home through their nationwide independent project manager service.

For free guidance and advice at any stage please contact:
Custom Homes, South Suffolk Business Centre, Alexandra Road, Sudbury, Suffolk CO10 2ZX
Tel: 01787 377388 Fax: 01787 377622   E-mail: admin@customhomes.co.uk

This guide is regarded as a must for those contemplating a self build or for those with land wishing to finalise a design or carefully working out their budget. Featuring over 150 home designs, it is packed with information about land, finance, costings and build time scales. To obtain your copy simply send a cheque made payable to Custom Homes for ~~£19.95~~ £14.95 including p&p UK only or phone our

credit card hotline 01787 377388

HBH 2006

Name: Mr/Mrs/Miss:. . . . .

Address: . . . .

. . . .

Tel Evening: . . . .   Day: . . . .

Do you have a plot of land? Yes ☐   No ☐

Guide to Building Your Own
DREAM HOME

CUSTOM
HOMES

30th Edition

HB&R

www.customhomes.co.uk

Photo: Finn Architects

CANADA WOOD

Area of Operation: UK
Tel: 01252 522545 **Fax:** 01252 522546
Email: office@canadawooduk.org
Web: www.canadawood.info
Product Type: 2, 5
Materials Type: A) 12

Your projects need not be this grand to use Canadian wood products! Wood-frame housing to wood floors and much more - start at www.canadawood.info

CARPENTER OAK & WOODLAND
Hall Farm, Thickwood Lane, Colerne, Wilts, SN14 8BE
Area of Operation: Worldwide
Tel: 01225 743089 **Fax:** 01225 744100
Email: enquiries@cowco.biz **Web:** www.cowco.biz

CARPENTER OAK LTD & RODERICK JAMES ARCHITECT LTD
The Framing Yard, East Cornworthy, Totnes, Devon, TQ9 7HF
Area of Operation: Worldwide
Tel: 01803 732900 **Fax:** 01803 732901
Email: enquiries@carpenteroak.com
Web: www.carpenteroak.com
Product Type: 1, 4, 6 **Other Info:** ✐

CEDAR SELF-BUILD HOMES
PO Box 3167, Chester, Cheshire, CH4 0ZH
Area of Operation: Europe
Tel: 01244 661048 **Fax:** 01244 660707
Email: info@cedar-self-build.com
Web: www.cedar-self-build.com
Product Type: 1 **Other Info:** ✐
Material Type: A) 12

CEDAR QUALITY HOMES LTD
Area of Operation: Europe
Tel: 01244 661048 **Fax:** 01244 660707
Email: info@cedar-self-build.com
Web: www.cedar-self-build.com
Product Type: 1 **Materials Type:** A) 12
Other Info: ✐

• Timber frame kit homes • Ecologically friendly • Choice of cedar, pine, brick, render or stone exterior walls • Preliminary design service and advice • Erection service including: Foundations • Kit Erection • Exterior • Turnkey also available

CENTURY HOMES TIMBER FRAME LTD
Lammas Gate, 84 Meadrow, Godalming, Surrey, GU7 3HT
Area of Operation: Worldwide
Tel: 01495 726910 **Fax:** 01495 724758
Email: salesdesk@centuryhomes.co.uk
Web: www.centuryhomes.co.uk ✐

CJ O'SHEA & CO LTD
Unit 1 Granard Business Centre, Bunns Lane, Mill Hill, London, NW7 2DZ
Area of Operation: East England, Greater London, South East England, South West England & South Wales
Tel: 0208 959 3600 **Fax:** 0208 959 0184
Email: admin@oshea.co.uk
Web: www.oshea.co.uk

CLASSICS TIMBERFRAME
Harelow Moor, Greenlaw, Borders, TD10 6XT
Area of Operation: UK (Excluding Ireland)
Tel: 01578 740218 **Fax:** 01578 740218
Email: borderdesign@constructionplus.net
Web: www.borderdesign.co.uk
Product Type: 1, 2

CONCEPT TIMBER
35 Lancaster Road, Bowerhill Industrial Estate, Melksham, Wiltshire, SN12 6SS
Area of Operation: UK (Excluding Ireland)
Tel: 01225 792939 **Fax:** 01225 792949
Email: enquiries@concept-timber.co.uk
Web: www.concept-timber.co.uk

COUNTRY HOMES
The Mill House, Marsh Farm, Cross Keys, Withington, Hereford, Herefordshire, HR1 3NN
Area of Operation: UK (Excluding Ireland)
Tel: 01432 820660 **Fax:** 01432 820404
Email: enquiries@country-homes.org
Web: www.country-homes.org
Product Type: 2
Material Type: B) 12

COWAN JOINERY CONSTRUCTION LTD
3 Clem Attlee Gardens, Larkhall, Lanarkshire, ML9 1HB
Area of Operation: UK (Excluding Ireland)
Tel: 01698 884362 **Fax:** 01698 883132
Email: office@cowanjoinery.co.uk

CRANNOG HOMES
799a Lordswood Lane, Lordswood, Chatham, Kent, ME5 8JP
Area of Operation: UK & Ireland
Tel: 01634 201143 **Fax:** 01634 201143
Email: johnm@crannoghomes.com
Web: www.crannoghomes.com
Product Type: 1, 5
Material Type: A) 12

CREATIVE ESTATES TIMBER FRAME BUILDING
Unit 62, Thornhill Industrial Estate, South Marston, Swindon, Wiltshire, SN3 4TA
Area of Operation: UK & Ireland
Tel: 0870 432 8268 **Fax:** 0870 432 8269
Email: info@creativeestates.co.uk
Web: www.creativeestates.co.uk
Product Type: 2

CUSTOM HOMES
PO Box 267, Horley, Surrey, RH6 9BW
Area of Operation: UK & Ireland
Tel: 01787 377388
Email: admin@customhomes.co.uk
Web: www.customhomes.co.uk
Product Type: 2

Remarkable **timber frame homes**
from concept to completion

Tel: 0871 200 2430
Web: www.fusiontimberframe.com
Email: homes@fusiontimberframe.com

Fusion Timber Frame

CUSTOM HOMES
Area of Operation: UK & Ireland
Tel: 01787 377388 **Fax:** 01787 377622
Email: admin@customhomes.co.uk
Web: www.customhomes.co.uk

Independent project managers offer a full turn-key service together with an individual design and planning service with the benefit of local knowledge. Custom Homes leads where others follow.

CUSTOM HOMES
Area of Operation: UK & Ireland
Tel: 01787 377388 **Fax:** 01787 377622
Email: admin@customhomes.co.uk
Web: www.customhomes.co.uk

Custom Homes is the largest self-build homes package company in the UK with offices nationwide offering a complete service from design to construction and finance to plot search.

CUSTOM HOMES
Area of Operation: UK & Ireland
Tel: 01787 377388 **Fax:** 01787 377622
Email: admin@customhomes.co.uk
Web: www.customhomes.co.uk

Custom Homes has built up a reputation the envy of the industry by placing the highest priority on the requirements of its customers right down to the last detail, providing a level of service others just dream about.

DEESIDE HOMES TIMBERFRAME
Broomhill Road, Spurryhillock Industrial Estate, Stonehaven, Near Aberdeen, Aberdeenshire, AB39 2NH
Area of Operation: UK (Excluding Ireland)
Tel: 01569 767123
Fax: 01569 767766
Email: john.wright@bancon.co.uk
Web: www.bancon.co.uk

DESIGNER HOMES
Pooh Cottage, Minto, Hawick,
Roxburghshire, Borders, TD9 8SB
Area of Operation: UK & Ireland
Tel: 01450 870127
Fax: 01450 870127
Product Type: 2, 3
Other Info:

DGS CONSTRUCTION LTD
The Glebe, Nash Road, Whaddon,
Milton Keynes, Buckinghamshire, MK17 0NQ
Area of Operation: East England, Greater London, Midlands & Mid Wales, South East England, South West England & South Wales
Tel: 01908 503147
Fax: 01908 504995
Email: info@dgsconstruction.co.uk
Web: www.dgsconstruction.co.uk
Product Type: 2, 3, 7

ECO HOUSES
12 Lee Street, Louth, Lincolnshire, LN11 9HJ
Area of Operation: Worldwide
Tel: 0845 345 2049
Email: sd@jones-nash.co.uk
Web: www.eco-houses.co.uk
Product Type: 1, 2, 3, 4, 5

ECO SYSTEMS IRELAND LTD
40 Glenshesk Road, Ballycastle, Co Antrim, BT54 6PH
Area of Operation: UK & Ireland
Tel: 02820 768708 **Fax:** 02820 769781
Email: info@ecosystemsireland.com
Web: www.ecosystemsireland.com
Product Type: 3, 7

ECOHOMES LTD
First Floor, 52 Briggate, Brighouse,
West Yorkshire, HD6 1ES
Area of Operation: UK & Ireland
Tel: 01484 40 20 40 **Fax:** 01484 40 01 01
Email: sales@ecohomes.ltd.uk
Web: www.ecohomes.ltd.uk

ENGLISH HERITAGE BUILDINGS
Coldharbour Farm Estate, Woods Corner,
East Sussex, TN21 9LQ
Area of Operation: Europe
Tel: 01424 838643
Fax: 01424 838606
Email: info@ehbp.com
Web: www.ehbp.com
Product Type: 1, 4, 6

ENVIROZONE
Encon House, 54 Ballycrochan Road,
Bangor, Co Down, BT19 6NF
Area of Operation: UK & Ireland
Tel: 02891 477799
Fax: 02891 452463
Email: info@envirozone.co.uk
Web: www.envirozone.co.uk
Product Type: 3, 7

EXCEL BUILDING SOLUTIONS
Maerdy Industrial Estate,
Rhymney, Powys, NP22 5PY
Area of Operation: Worldwide
Tel: 01685 845200 **Fax:** 01685 844106
Email: sales@excelfibre.com
Web: www.excelfibre.com
Product Type: 3 **Other Info:** ECO
Material Type: K) 3

FFOREST TIMBER ENGINEERING LTD
Kestrel Way, Garngoch Industrial Estate,
Gorseinon, Swansea, SA4 9WF
Area of Operation: Midlands & Mid Wales, South West England and South Wales
Tel: 01792 895620
Fax: 01792 893969
Email: info@fforest.co.uk
Web: www.fforest.co.uk
Product Type: 2

FIRST LEISURE UK LTD
Oakbank House, Kenmore Street,
Aberfeldy, Perth and Kinross, PH15 2BL
Area of Operation: UK (Excluding Ireland)
Tel: 01887 829418
Fax: 01887 829090
Email: simon@firstleisureuk.co.uk
Web: www.log-buildings.co.uk
Product Type: 5

FLEMING HOMES
Station Road, Duns, Berwickshire,
Borders, TD11 3HS
Area of Operation: UK (Excluding Ireland)
Tel: 01361 883785
Fax: 01361 883898
Email: enquiries@fleminghomes.co.uk
Web: www.fleminghomes.co.uk
Other Info:

FLIGHT TIMBER PRODUCTS
Earls Colne Business Park, Earls Colne,
Essex, CO6 2NS
Area of Operation: East England, Greater London, South East England
Tel: 01787 222336
Fax: 01787 222359
Email: sales@flighttimber.com
Web: www.flighttimber.com

FOUR ACRES CONTRUCTION
Rawnoch Road, Johnstone, Renfrewshire, PA5 05P
Area of Operation: UK (Excluding Ireland)
Tel: 01505 337788
Fax: 01505 337788
Product Type: 2, 3, 5

FRAME UK
Jenson House, Cardrew Industrial Estate,
Redruth, Cornwall, TR15 1SS
Area of Operation: UK (Excluding Ireland)
Tel: 01209 310560
Fax: 01209 310561
Email: enquiries@framehomes.co.uk
Web: www.frameuk.com
Product Type: 2, 3

FRAME WISE
Presteigne Industrial Estate, Presteigne,
Powys, LD8 2UF
Area of Operation: UK (Excluding Ireland)
Tel: 01544 260125
Fax: 01544 260707
Email: framewise@framewiseltd.co.uk
Web: www.framewiseltd.co.uk

FUSION TIMBER FRAME LTD
First Floor, 18 Keymer Road, Hassocks,
West Sussex, BN6 8AN
Area of Operation: UK (Excluding Ireland)
Tel: 0871 200 2430
Fax: 0871 200 2431
Email: homes@fusiontimberframe.com
Web: www.fusiontimberframe.com
Product Type: 1, 2, 3

GREEN OAK STRUCTURES
20 Bushy Coombe Gardens, Glastonbury,
Somerset, BA6 8JT
Area of Operation: UK & Ireland
Tel: 01458 833420
Fax: 01458 833420
Email: timberframes@greenoakstructures.co.uk
Web: www.greenoakstructures.co.uk
Product Type: 1, 4, 6
Other Info: ECO

GRIFFNER COILLTE LTD
Forest Park, Mullingar, Co. Westmeath, Ireland
Area of Operation: UK & Ireland
Tel: +353 443 7800 **Fax:** +353 443 7888
Email: info@griffnercoillte.ie
Web: www.griffnercoillte.ie
Product Type: 3

HOBBANS TIMBERWORKS
Hobbans Farm, Bobbingworth,
Ongar, Essex, CM5 0LZ
Area of Operation: UK & Ireland
Tel: 01277 890165 **Fax:** 01277 890165
Email: rupert@hobbanstimberworks.co.uk
Web: www.hobbanstimberworks.co.uk
Product Type: 1, 4, 6

HOMELODGE BUILDINGS LTD
Kingswell Point, Crawley, Winchester,
Hampshire, SO21 2 PU
Area of Operation: UK (Excluding Ireland)
Tel: 01962 881480 **Fax:** 01962 889070
Email: info@homelodge.co.uk
Web: www.homelodge.co.uk
Product Type: 3

HONEYSUCKLE BOTTOM SAWMILL LTD
Honeysuckle Bottom, Green Dene,
East Horsley, Leatherhead, Surrey, KT24 5TD
Area of Operation: Greater London,
South East England
Tel: 01483 282394 **Fax:** 01483 282394
Email: honeysucklemill@aol.com
Web: www.easisites.co.uk/honeysucklebottomsawmill
Product Type: 4

HOUSE - UK
347 Leverington Common, Leverington,
Wisbech, Cambridgeshire, P13 5JR
Area of Operation: UK (Excluding Ireland)
Tel: 01945 410361 **Fax:** 01945 419038
Email: enquiries@house-uk.co.uk
Web: www.house-uk.co.uk
Product Type: 5 **Other Info:** ✏

INNOVAHOUSE
Alltan Donn House, Altonburn Road, Nairn,
Highlands, IV12 5NB
Area of Operation: UK & Ireland
Tel: 01667 452555 **Fax:** 01667 453777
Email: info@innova-house.co.uk
Web: www.info@innova-house.co.uk

INSIDEOUT BUILDINGS LTD
The Green, Over Kellet, Carnforth, Lancashire, LA6 1BU
Area of Operation: UK (Excluding Ireland)
Tel: 01524 737999
Email: lynn@iobuild.co.uk
Web: www.iobuild.co.uk
Product Type: 7
Other Info: ✏

INTERBILD LTD
2a Ainslie Street, West Pitkerro,
Dundee, Angus, DD5 3RR
Area of Operation: UK & Ireland
Tel: 01382 480481 **Fax:** 01382 480482
Email: tpd@interbild.com
Web: www.interbild.com
Product Type: 1, 2

JOINERY & TIMBER BUILDINGS
6 Lower Beech Cottages, Off Manchester Road,
Tytherington, Macclesfield, Cheshire, SK10 2ED
Area of Operation: North West England & North Wales
Tel: 07909 907656 **Fax:** 01625 501655
Email: jim@timberbuildings.fsnet.co.uk
Product Type: 1, 4, 6
Other Info: ✏

KINGSTON TIMBER FRAME
14 Mill Hill Drive, Huntington, York,
North Yorkshire, YO32 9PU
Area of Operation: East England, North East
England, North West England and North Wales
Tel: 01904 762589 **Fax:** 01904 766686
Email: info@kingstontimberframe.co.uk
Web: www.kingstontimberframe.co.uk
Product Type: 3

LAKELAND TIMBER FRAME
Unit 38c, Holme Mills, Holme,
Carnforth, Lancashire, LA6 1RD
Area of Operation: UK (Excluding Ireland)
Tel: 01524 782596
Fax: 01524 784972
Email: tony@lakelandtimberframe.co.uk
Web: www.lakelandtimberframe.co.uk

LAMINATED WOOD LIMITED
Grain Silo Complex, Abbey Road,
Hempsted, Gloucester,
Gloucestershire, GL2 5HU
Area of Operation: UK (Excluding Ireland)
Tel: 01452 418000
Fax: 01452 418333
Email: mail@lamwood.co.uk
Web: www.lamwood.co.uk
Product Type: 1, 7
Other Info: ✏
Material Type: B) 1, 7, 10

LEISURE SPACE LTD
Unit 1, Top Farm,
Rectory Road, Campton Shefford,
Bedfordshire, SG17 5PF
Area of Operation: Europe
Tel: 01462 816147
Fax: 01462 819252
Email: enquiries@leisurespaceltd.co.uk
Web: www.leisurespaceltd.co.uk
Product Type: 5
Other Info: ✏

**LINDISFARNE
TIMBER FRAME LTD**
197 Rosalind Street, Ashington,
Northumberland, NE63 9BB
Area of Operation: UK (Excluding Ireland)
Tel: 01670 810472
Fax: 01670 810472
Email: info@lindisfarnetimberframeltd.co.uk
Web: www.lindisfarnetimberframeltd.co.uk
Product Type: 3

LINWOOD HOMES LTD.
8250 River Road, Delta, BC Canada V4G 1B5
Area of Operation: UK & Ireland
Tel: +1 604 946 5430 ext.146
Fax: +1 604 940 6276
Email: pdauphinee@linwoodhomes.com
Web: www.linwoodhomes.com ✏
Product Type: 1, 2, 5 **Other Info:** ✏

LINWOOD HOMES LTD

Area of Operation: UK & Ireland
Tel: +1 604 946 5430 ext.146
Fax: +1 604 940 6276
Email: pdauphinee@linwoodhomes.com
Web: www.linwoodhomes.com
Product Type: 1, 2, 5

Craftsmanship in cedar post and beam, log or timber frame. Quality building materials and a custom-designed home package. A complete building solution for your primary residence or vacation home.

LLOYDS TIMBER FRAME LIMITED
Glovers Meadow, Oswestry, Shropshire, SY10 8NH
Area of Operation: Europe
Tel: 01691 656511 **Fax:** 01691 656533
Email: info@lloydstimberframes.co.uk
Web: www.lloydstimberframes.co.uk
Product Type: 2, 3

LOG & CEDAR HOMES LTD
10 Birch Court, Doune, Perth and Kinross, FK16 6JD
Area of Operation: North East England, North West England and North Wales, Scotland
Tel: 01786 842216
Fax: 01786 842216
Email: enquiries@logandcedarhomes.co.uk
Web: www.logandcedarhomes.co.uk
Product Type: 1, 3, 5, 6

MAPLE TIMBER FRAME
Tarnacre Hall Business Park, Tarnacre Lane,
St Michaels, Lancashire, PR3 0SZ
Area of Operation: Worldwide
Tel: 01995 679444
Fax: 01995 679769
Email: enquiry@mapletimberframe.com
Web: www.mapletimberframe.com
Product Type: 2

NEATWOOD HOMES LTD
Unit 6, Westwood Industrial Estate,
Pontrilas, Herefordshire, HR2 0EL
Area of Operation: UK (Excluding Ireland)
Tel: 01981 240860
Fax: 01981 240255
Email: sales@neatwoodhomes.co.uk
Web: www.neatwoodhomes.co.uk
Product Type: 2, 3

NEW WORLD TIMBER FRAME
Mitchell Hanger, Audley End Airfield,
Saffron Walden, Essex, CB11 4LH
Area of Operation: UK (Excluding Ireland)
Tel: 01799 513331
Fax: 01799 513341
Email: info@newworldtimberframe.co.uk
Web: www.newworldtimberframe.com
Product Type: 1, 2, 3, 4, 6

NORDIC WOOD
21 Tartar Road, Cobham, Surrey, KT11 2AS
Area of Operation: UK & Ireland
Tel: 01932 576944
Email: info@nordic-wood.co.uk
Web: www.nordic-wood.co.uk
Product Type: 5

OAKMASTERS
The Mill, Isaacs Lane, Haywards Heath,
West Sussex, RH16 4RZ
Area of Operation: UK (Excluding Ireland)
Tel: 01444 455455
Fax: 01444 455333
Email: oak@oakmasters.co.uk
Web: www.oakmasters.co.uk

PINECONE LOG HOMES
26 Jennings Road, St Albans,
Hertfordshire, AL1 4PD
Area of Operation: UK & Ireland
Tel: 0800 169 6327
Fax: 01727 851558
Email: info@pineconeloghomes.co.uk
Web: www.pineconeloghomes.co.uk
Product Type: 5

PINECONE LOG HOMES

Area of Operation: UK & Ireland
Tel: 0800 169 6327 **Fax:** 01727 851558
Email: info@pineconeloghomes.co.uk
Web: www.pineconeloghomes.co.uk
Product Type: 5
Other Info: ECO

PineCone Log Homes are able to supply custom log homes and other log buildings manufacturered by True North Homes of Canada.

POTTON LTD
Wyboston Lakes, Great North Road,
Wyboston, Bedfordshire, MK44 3BA
Area of Operation: UK & Ireland
Tel: 01480 401401
Fax: 01480 401444
Email: sales@potton.co.uk
Web: www.potton.co.uk
Product Type: 1, 2, 6

RAYNE CONSTRUCTION
Rayne North, Inverurie, Aberdeenshire, AB51 5DB
Area of Operation: UK & Ireland
Tel: 01464 851518 **Fax:** 01464 851555
Email: info@rayne-construction-ltd.freeserve.co.uk
Web: www.rayne-construction-ltd.freeserve.co.uk
Product Type: 5
Other Info: ECO

ROB ROY HOMES
Dalchonzie, Comrie, Perthshire, PH6 2LB
Area of Operation: North East England, North West England and North Wales, Scotland
Tel: 01764 670424 **Fax:** 01764 670419
Email: mail@robroyhomes.co.uk
Web: www.robroyhomes.co.uk
Product Type: 2

ROLLALONG LIMITED
Woolsbridge Industrial Park,
Three Legged Cross, Wimborne,
Dorset, BH21 6SF
Area of Operation: UK (Excluding Ireland)
Tel: 01202 824541 **Fax:** 01202 826525
Email: enquiries@rollalong.co.uk
Web: www.rollalong.co.uk
Product Type: 7

SCANDINAVIAN LOG CABINS DIRECT
6 North End, London Road, East Grinstead,
West Sussex, RH19 1QQ
Area of Operation: UK & Ireland
Tel: 01342 311131
Fax: 01342 311131
Email: cabins@slcd.co.uk
Web: www.slcd.co.uk
Product Type: 5

SCOTFRAME TIMBER ENGINEERING LIMITED
18 Aghnatrisk Road, Hillsborough,
Co Down, BT26 6JJ
Area of Operation: Ireland, Scotland & Northern England
Tel: 028 9268 8807
Fax: 028 9268 8809
Email: hillsborough@scotframe.co.uk
Web: www.scotframe.co.uk
Product Type: 2

SCOTFRAME TIMBER ENGINEERING LIMITED
4 Deerdykes Place, Cumbernauld,
Glasgow, G68 9HE
Area of Operation: Scotland, Northern England & Ireland
Tel: 01236 861200
Fax: 01236 861201
Email: cumbernauld@scotframe.co.uk
Web: www.scotframe.co.uk
Product Type: 2

SCOTFRAME TIMBER ENGINEERING LIMITED
Inverurie Business Park,
Souterford Avenue, Inverurie,
Aberdeenshire, AB51 0ZJ
Area of Operation: Scotland, Northern England & Ireland
Tel: 01467 624440
Fax: 01467 624255
Email: inverurie@scotframe.co.uk
Web: www.scotframe.co.uk
Product Type: 2

STRATHCLYDE TIMBER SYSTEMS LTD
Castlecary, Cumbernauld, City of Glasgow, G68 0DT
Area of Operation: UK (Excluding Ireland)
Tel: 01324 840 909 **Fax:** 01324 840 907
Email: sales@strathclydetimbersystems.com
Web: www.strathclydetimbersystems.com
Product Type: 2

T J CRUMP OAKWRIGHTS LIMITED
The Lakes, Swainhill, Hereford,
Herefordshire, HR4 7PU
Area of Operation: Worldwide
Tel: 01432 353353
Fax: 01432 357733
Email: nick@oakwrights.co.uk
Web: www.oakwrights.co.uk
Product Type: 1, 4, 6
Material Type: A) 2

TAYLOR LANE TIMBER FRAME LTD
Chapel Road, Rotherwas Industrial Estate,
Hereford, Herefordshire, HR2 6LD
Area of Operation: UK & Ireland
Tel: 01432 271912 **Fax:** 01432 351064
Email: info@taylor-lane.co.uk
Web: www.taylor-lane.co.uk
Product Type: 2

TECCO SYSTEMS
1 Elm Close, Hove,
East Sussex, BN3 6TG
Area of Operation: UK & Ireland
Tel: 01273 501210
Fax: 0207 837 3070
Email: rabernstein@onetel.com
Web: www.teccosystems.co.uk
Product Type: 1, 7

THE BORDER DESIGN CENTRE
Harelow Moor, Greenlaw,
Borders, TD10 6XT
Area of Operation: UK (Excluding Ireland)
Tel: 01578 740218
Fax: 01578 740218
Email: borderdesign@btconnect.com
Web: www.borderdesign.co.uk
Product Type: 1, 2
Other Info:

THE GREEN OAK CARPENTRY CO LTD
Langley Farm, Langley, Rake, Liss,
Hampshire, GU33 7JW
Area of Operation: UK & Ireland
Tel: 01730 892049
Fax: 01730 895225
Email: enquiries@greenoakcarpentry.co.uk
Web: www.greenoakcarpentry.co.uk

THE GREEN OAK CARPENTRY CO LTD
Area of Operation: UK & Ireland
Tel: 01730 892049 **Fax:** 01730 895225
Email: enquiries@greenoakcarpentry.co.uk
Web: www.greenoakcarpentry.co.uk

The Green Oak Carpentry Company has 15 years experience in the design, engineering and installation of new oak structures such as barns, halls and houses etc. We also restore old timber structures.

SPONSORED BY: BECO WALLFORM   www.becowallform.co.uk

THE LOG CABIN COMPANY
Potash Garden Centre,
9 Main Road, Hockley,
Hawkwell, Essex, SS5 4JN
Area of Operation: UK (Excluding Ireland)
Tel: 01702 206012
Email: sales@thelogcabincompany.co.uk
Web: www.thelogcabincompany.co.uk
Product Type: 5

**THE OAK FRAME
CARPENTRY CO LTD**
Nupend Farm, Nupend, Stonehouse,
Gloucestershire, GL10 3SU
Area of Operation: Midlands & Mid Wales,
South West England and South Wales
Tel: 01453 828788
Fax: 01453 828788
Email: Simon.eeles@btconnect.com
Product Type: 1, 4, 6

THE PATIO ROOM COMPANY
52 Castlemaine Avenue,
South Croydon, Surrey, CR2 7HR
Area of Operation: UK (Excluding Ireland)
Tel: 0208 406 3001
Fax: 0208 686 4928
Email: helpdesk@patioroom.co.uk
Web: www.patioroom.co.uk
Product Type: 1
Other Info: ✎

**THE SWEDISH
HOUSE COMPANY**
Seabridge House, 8 St Johns Road,
Tunbridge Wells, Kent, TN4 9NP
Area of Operation: UK & Ireland
Tel: 0870 770 0760
Fax: 0870 770 0759
Email: sales@swedishhouses.com
Web: www.swedishhouses.com
Product Type: 3
Other Info: ECO ✎
Material Type: B) 1, 2, 8, 9, 10

THE TIMBER FRAME CO LTD
The Framing Yard, 7 Broadway,
Charlton Adam, Somerset, TA11 7BB
Area of Operation: Worldwide
Tel: 01458 224463
Fax: 01458 224571
Email: admin@thetimberframe.co.uk
Web: www.thetimberframe.co.uk
Product Type: 1, 4, 6, 7
Other Info: ECO ✎ ✋
Material Type: A) 2

THOMAS MITCHELL HOMES LTD
Southend, Thornton, Fife, KY1 4ED
Area of Operation: Worldwide
Tel: 01592 774401   **Fax:** 01592 774088
Email: stuart@tmhomes.co.uk
Web: www.thomasmitchellhomes.com
Product Type: 2

TIMBER DEVELOPMENTS
Unit 13, Stafford Park 12,
Telford, Shrops, TF3 3BJ
Area of Operation: Europe
Tel: 0870 774 0949   **Fax:** 0870 777 7520
Email: info@timberdevelopments.com
Web: www.timberdevelopments.com

TIMBERFRAME WALES
Unit 1 & 5 Maesquarre Road,
Ammanford, Carmarthenshire, SA18 2LF
Area of Operation: UK (Excluding Ireland)
Tel: 01269 595255   **Fax:** 01269 595305
Email: info@timberframewales.com
Web: www.timberframewales.co.uk

TRUE NORTH LOG HOMES INC
PO Box 2169, Winhara Road,
Bracebridge, Ontario, Canada, P1L 1W1
Area of Operation: Worldwide
Tel: 0800 169 6327   **Fax:** 01727 851558
Email: info@pineconeloghomes.co.uk
Web: www.pineconeloghomes.co.uk
Product Type: 5   **Other:** ECO

TRUE NORTH LOG HOMES INC.

Area of Operation: Worldwide
Tel: 0800 169 6327   **Fax:** 01727 851558
Email: info@pineconeloghomes.co.uk
Web: www.pineconeloghomes.co.uk
Product Type: 5
Other: ECO

True North Log Homes provide custom log homes
from Canada using very large machined logs and
give a 25 year warranty of 'zero' air infiltration
through the log wall. They hold 14 patents.
Available through PineCone Log Homes Ltd.

TURNER TIMBER FRAME
Leven Road, Brandesburton, Nr. Driffield,
East Riding of Yorks, YO25 8RT
Area of Operation: UK (Excluding Ireland)
Tel: 01964 543535   **Fax:** 01964 543535
Email: turnertimberframe@hotmail.com
Web: www.turnertimberframe.co.uk
Product Type: 1, 2, 5   **Material Type:** B) 2

**VIDAL PREFABRICATED
TIMBER FRAMED HOUSES**
Old Brightmoor Farm, Thornborough, Bucks, MK18 2EA
Area of Operation: UK & Ireland
Tel: 01280 824787   **Fax:** 01280 824288
Email: vidalhomes@aol.com   **Product Type:** 2, 3

W.H. COLT SON & CO LTD
Prestige House, Landews Meadow,
Green Lane, Challock,
Ashford, Kent, TN25 4BL
Area of Operation: UK & Ireland
Tel: 01233 740074
Fax: 01233 740123
Email: mail@colthouses.co.uk
Web: www.colthouses.co.uk
Product Type: 2
Material Type: B) 10

WESTWIND OAK BUILDINGS LTD
Unit 1, Laurel Farm,
Streamcross, Lower Claverham,
Nr. Bristol, BS49 4PZ
Area of Operation: Europe
Tel: 01934 877317
Email: judy@westwindoak.com
Web: www.westwindoak.com
Product Type: 4

WOODCO
Tofts of Tain, Castletown,
Highlands, KW14 8TB
Area of Operation: Worldwide
Tel: 01847 821418
Fax: 01847 821418
Email: woodcoscotland@btconnect.com
Web: www.woodco.clara.net
Product Type: 1, 3, 5

BUILDING STRUCTURE & MATERIALS - **Complete Structural Systems** - Steel Frame; Brick & Block Structures; 'Eco' Stuctures

SPONSORED BY: BECO WALLFORM www.becowallform.co.uk

BUILDING STRUCTURE & MATERIALS

STEEL FRAME

```
KEY

PRODUCT TYPES:  1= Lightweight
2 = Heavyweight

OTHER:  ▽ Reclaimed  🛒 On-line shopping
✎ Bespoke  ✋ Hand-made  ECO Ecological
```

ALL FOUNDATIONS LIMITED
Primrose Business Park, White Lane,
Blackwell, Derby, Derbyshire, DE55 5JR
Area of Operation: UK & Ireland
Tel: 0870 350 2050
Email: mail@allfoundations.co.uk
Web: www.allfoundationsltd.co.uk
Product Type: 1, 2

**ANIRINA OY (FINLAND) - LOG HOME
SUPPLIERS & BUILDERS UK**
The Gate House, Home Park Terrace,
Hampton Court Road, Hampton Wick,
Kingston upon Thames, Surrey, KT1 4AE
Area of Operation: Europe
Tel: 0208 943 0430 **Fax:** 0208 943 0430
Email: chris.drayson@virgin.net
Product Type: 2

AVON MANUFACTURING LIMITED
Avon House, Kineton Road, Southam,
Leamington Spa, Warwickshire, CV47 0DG
Area of Operation: UK (Excluding Ireland)
Tel: 01926 817292
Fax: 01926 814156
Email: sales@avonmanufacturing.co.uk
Web: www.avonmanufacturing.co.uk
Product Type: 1, 2

CONCEPT ENGINEERING CONSULTANTS LTD
Unit 8 Warple Mews, Warple Way, London, W3 0RF
Area of Operation: East England, Greater London,
South East England, South West England & South Wales
Tel: 0208 840 3321
Fax: 0208 579 6910
Email: si@conceptconsultants.co.uk
Web: www.conceptconsultants.co.uk

CORUS FRAMING SOLUTIONS
Swinden Technology Centre, Swinden House,
Moorgate, Rotherham, South Yorkshire, S60 3AR
Area of Operation: UK (Excluding Ireland)
Tel: 01724 405060
Fax: 01724 404224
Email: corusconstruction@corusgroup.com
Web: www.corusconstruction.com
Product Type: 1

JOY STEEL STRUCTURES (LONDON) LTD
London Industrial Park, 1 Whitings Ways,
London, E6 6LR
Area of Operation: UK (Excluding Ireland)
Tel: 0207 474 0550
Fax: 0207 473 0158
Email: joysteel@dial.pipex.com
Web: www.joysteel.co.uk

LINDAB LTD
Unit 7 Block 2, Shenstone Trading Estate,
Bromsgrove Road, Halesowen,
West Midlands, B63 3XB
Area of Operation: Worldwide
Tel: 0121 585 2780
Fax: 0121 585 2782
Email: building.products@lindab.co.uk
Web: www.lindab.com
Product Type: 1
Material Type: C) 4

METEK UK LIMITED
Heighington Lane, Aycliffe Industrial Park,
Newton Aycliffe, Durham, DL5 6QG
Area of Operation: UK (Excluding Ireland)
Tel: 01325 372700
Fax: 01325 370903
Email: mbs@mmpgroup.co.uk
Web: www.metekbuildingsystems.co.uk

ROLLALONG LIMITED
Woolsbridge Industrial Park,
Three Legged Cross, Wimborne,
Dorset, BH21 6SF
Area of Operation: UK (Excluding Ireland)
Tel: 01202 824541
Fax: 01202 826525
Email: enquiries@rollalong.co.uk
Web: www.rollalong.co.uk
Product Type: 1, 2

STEELFRAME BV
Andromedastraat 5, Netherlands, Tilburg, 5015 AV
Area of Operation: Europe
Tel: +31 135 449 859
Fax: +31 135 449 860
Email: info@steelframe.nl
Web: www.steelframe.nl 🛒
Product Type: 1

STEELSMART HOMES UK LTD
Suite 3-5 George Street Chambers,
36 George Street, Birmingham,
West Midlands, B3 1QA
Area of Operation: UK & Ireland
Tel: 0121 200 8300
Fax: 0121 200 8303
Email: info@steelsmarthomes.co.uk
Web: www.steelsmarthomes.co.uk
Product Type: 1

TITAN CONTAINERS (UK) LTD
Suite 1, 1 Cecil Court, London Road, Enfield,
Middlesex, Greater London, EN2 6DE
Area of Operation: UK & Ireland
Tel: 0208 362 1444
Fax: 0208 362 1555
Email: uk@titancontainer.com
Web: www.titancontainer.com
Product Type: 2

BRICK AND BLOCK STRUCTURES

```
KEY

OTHER:  ▽ Reclaimed  🛒 On-line shopping
✎ Bespoke  ✋ Handmade  ECO Ecological
```

BRADSTONE STRUCTURAL
Aggregate Industries UK Ltd, North End,
Ashton Keynes, Swindon, Wiltshire, SN6 3QX
Area of Operation: UK (Excluding Ireland)
Tel: 01285 646884
Fax: 01285 646891
Email: bradstone.structural@aggregate.com
Web: www.bradstone.com

CJ O'SHEA & CO LTD
Unit 1 Granard Business Centre, Bunns Lane,
Mill Hill, London, NW7 2DZ
Area of Operation: East England, Greater London,
South East England, South West England & South Wales
Tel: 0208 959 3600
Fax: 0208 959 0184
Email: admin@oshea.co.uk
Web: www.oshea.co.uk

CONCEPT ENGINEERING CONSULTANTS LTD
Unit 8 Warple Mews, Warple Way, London, W3 0RF
Area of Operation: East England, Greater London,
South East England, South West England & South Wales
Tel: 0208 840 3321
Fax: 0208 579 6910
Email: si@conceptconsultants.co.uk
Web: www.conceptconsultants.co.uk

DESIGN AND MATERIALS LTD
Lawn Road, Carlton in Lindrick,
Nottinghamshire, S81 9LB
Area of Operation: UK & Ireland
Tel: 01909 730333
Fax: 01909 730605
Email: enquiries@designandmaterials.uk.com
Web: www.designandmaterials.uk.com 🛒

DGS CONSTRUCTION LTD
The Glebe, Nash Road,
Whaddon, Milton Keynes,
Buckinghamshire, MK17 0NQ
Area of Operation: East England, Greater London,
Midlands & Mid Wales, South East England, South
West England and South Wales
Tel: 01908 503147 **Fax:** 01908 504995
Email: info@dgsconstruction.co.uk
Web: www.dgsconstruction.co.uk

DISCOVERY CONTRACTORS LTD
Discovery Contractors Ltd, Discovery House,
Joseph Wilson Industrial Estate,
Whitstable, Kent, CT5 3PS
Area of Operation: UK (Excluding Ireland)
Tel: 01227 275559
Fax: 01227 275918
Email: info@dcontracts.com
Web: www.dcontracts.com
Other Info: ECO ▽ ✎

FAIRCLEAR LTD
16 Sibthorpe Drive, Sudbrooke,
Lincoln, Lincolnshire, LN2 2RQ
Area of Operation: East England
Tel: 01522 595189
Fax: 01522 595189
Email: enquiries@fairclear.co.uk
Web: www.fairclear.co.uk

MARUN CONSTRUCTION
62 Crofton Park, Yeovil, Somerset, BA21 4EE
Area of Operation: South West England
and South Wales
Tel: 01935 426947
Email: marun33@hotmail.com

NORBUR MOOR BUILDING SERVICES LTD
Gawthorne, Hazel Grove,
Stockport, Cheshire, SK7 5AB
Area of Operation: North West England
and North Wales
Tel: 0800 093 5785
Fax: 0161 456 1944
Email: paul@norburymoor.co.uk
Web: www.norburymoor.co.uk /
www.norburymoorbuilding.co.uk 🛒

THIN JOINT TECHNOLOGY LTD
3 Albright Road,
Speke Approaches Industrial Estate,
Liverpool, Merseyside, WA8 8FY
Area of Operation: UK & Ireland
Tel: 0151 422 8000
Fax: 0151 422 8001
Email: sales@thinjoint.com
Web: www.thinjoint.com
Material Type: I) 1, 2, 4

'ECO' STRUCTURES

```
KEY

PRODUCT TYPES:  1= Cob and Rammed Earth
2 = Straw Bale    3 = Wattle and Daub
4 = Hemp      5 = Other

SEE ALSO:  COMPLETE STRUCTURAL SYSTEMS
- Structural Insulated Panels

OTHER:  ▽ Reclaimed  🛒 On-line shopping
✎ Bespoke  ✋ Hand-made  ECO Ecological
```

**BORDER OAK DESIGN
& CONSTRUCTION**
Kingsland Sawmills, Kingsland,
Leominster, Herefordshire, HR6 9SF
Area of Operation: Worldwide
Tel: 01568 708752
Fax: 01568 708295
Email: sales@borderoak.com
Web: www.borderoak.com
Product Type: 5
Other Info: ECO ▽ ✎ ✋

CONCEPT TIMBER
35 Lancaster Road,
Bowerhill Industrial Estate,
Melksham, Wiltshire, SN12 6SS
Area of Operation: UK (Excluding Ireland)
Tel: 01225 792939
Fax: 01225 792949
Email: enquiries@concept-timber.co.uk
Web: www.concept-timber.co.uk
Product Type: 5

DGS CONSTRUCTION LTD
The Glebe, Nash Road, Whaddon,
Milton Keynes, Buckinghamshire, MK17 0NQ
Area of Operation: East England,
Greater London, Midlands & Mid Wales,
South East England, South West England
and South Wales
Tel: 01908 503147
Fax: 01908 504995
Email: info@dgsconstruction.co.uk
Web: www.dgsconstruction.co.uk
Product Type: 5

ECO SYSTEMS IRELAND LTD
40 Glenshesk Road, Ballycastle,
Co Antrim, BT54 6PH
Area of Operation: UK & Ireland
Tel: 02820 768708
Fax: 02820 769781
Email: info@ecosystemsireland.com
Web: www.ecosystemsireland.com
Product Type: 5

ENVIROZONE
Encon House, 54 Ballycrochan Road,
Bangor, Co Down, BT19 6NF
Area of Operation: UK & Ireland
Tel: 02891 477799
Fax: 02891 452463
Email: info@envirozone.co.uk
Web: www.envirozone.co.uk

EXCEL BUILDING SOLUTIONS
Maerdy Industrial Estate, Rhymney,
Powys, NP22 5PY
Area of Operation: Worldwide
Tel: 01685 845200
Fax: 01685 844106
Email: sales@excelfibre.com
Web: www.excelfibre.com
Product Type: 5 **Other Info:** ECO
Material Type: K) 3

GREEN OAK STRUCTURES
20 Bushy Coombe Gardens,
Glastonbury, Somerset, BA6 8JT
Area of Operation: UK & Ireland
Tel: 01458 833420 **Fax:** 01458 833420
Email: timberframes@greenoakstructures.co.uk
Web: www.greenoakstructures.co.uk
Product Type: 3 **Other Info:** ECO ✎
Material Type: A) 2

HEMPHAB PRODUCTS
Rusheens, Ballygriffen,
Kenmare, Kerry, Ireland
Area of Operation: Ireland Only
Tel: +353 644 1747
Email: hempbuilding@eircom.net
Web: www.hempbuilding.com
Product Type: 4

LINDISFARNE TIMBER FRAME LTD
197 Rosalind Street, Ashington,
Northumberland, NE63 9BB
Area of Operation: UK (Excluding Ireland)
Tel: 01670 810472 **Fax:** 01670 810472
Email: info@lindisfarnetimberframeltd.co.uk
Web: www.lindisfarnetimberframeltd.co.uk
Product Type: 5

LOW-IMPACT LIVING INITIATIVE
Redfield Community, Buckingham Road,
Winslow, Buckinghamshire, MK18 3LZ
Area of Operation: UK (Excluding Ireland)
Tel: 01296 714184 **Fax:** 01296 714184
Email: lili@lowimpact.org
Web: www.lowimpact.org 🛒

BUILDING STRUCTURE & MATERIALS - **Complete Structural Systems** - Insulated Concrete Formwork; Structural Insulated Panels

SPONSORED BY: BECO WALLFORM www.becowallform.co.uk

BUILDING STRUCTURE & MATERIALS

SAFEGUARD EUROPE LTD
Redkiln Close, Horsham, West Sussex, RH13 5QL
Area of Operation: Worldwide
Tel: 01403 210204 **Fax:** 01403 217529
Email: info@safeguardeurope.com
Web: www.safeguardeurope.com
Product Type: 4

SAFEGUARD
Area of Operation: Worldwide
Tel: 01403 210204
Fax: 01403 217529
Email: info@safeguardeurope.com
Web: www.safeguardeurope.com

Safeguard supply the Oldroyd Xv waterproofing system for turf roofs. Manufactured to an ISO14001 environmental management system.

STRAWBALE-BUILDING COMPANY UK
34 Rosebery Way, Tring, Hertfordshires, HP23 5DS
Area of Operation: Europe
Email: chug@strawbale-building.co.uk
Web: www.strawbale-building.co.uk

WOODCO
Tofts of Tain, Castletown, Highlands, KW14 8TB
Area of Operation: Worldwide
Tel: 01847 821418 **Fax:** 01847 821418
Email: woodcoscotland@btconnect.com
Web: www.woodco.clara.net

INSULATED CONCRETE FORMWORK

KEY
OTHER: ▽ Reclaimed ⌂ On-line shopping
✐ Bespoke ✋Handmade ECO Ecological

BECO PRODUCTS LTD
Beco House, 6 Exmoor Avenue,
Scunthorpe, Lincolnshire, DN15 8NJ
Area of Operation: UK & Ireland
Tel: 01724 747576 **Fax:** 01724 747579
Email: info@becowallform.co.uk
Web: www.becowallform.co.uk
Product Type: 1

CJ O'SHEA & CO LTD
Unit 1 Granard Business Centre, Bunns Lane,
Mill Hill, London, NW7 2DZ
Area of Operation: East England,
Greater London, South East England,
South West England & South Wales
Tel: 0208 959 3600
Fax: 0208 959 0184
Email: admin@oshea.co.uk
Web: www.oshea.co.uk

FALCON STRUCTURAL REPAIRS
2B Harbour Road, Portishead, Bristol, BS20 7DD
Area of Operation: Europe
Tel: 01275 844889 **Fax:** 01275 847002
Email: sjoyner@falconstructural.co.uk
Web: www.falconstructural.co.uk
Product Type: 3

FORMWORKS UK LTD
Pine Barn, Hamsey, Lewes, East Sussex, BN8 5TB
Area of Operation: Worldwide
Tel: 01273 478110 **Fax:** 01273 471419
Email: sales@formworksuk.com
Web: www.formworksuk.com
Product Type: 1

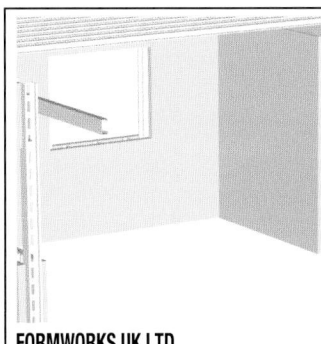

FORMWORKS UK LTD
Area of Operation: Worldwide
Tel: 01273 478110
Fax: 01273 471419
Email: sales@formworksuk.com
Web: www.formworksuk.com

A Modern Method of Construction combining Permanent Insulating Formwork, steel and concrete for the rapid construction of highly insulated, airtight, high mass buildings.

NEWMIL LTD
17 Arundel Close, New Milton, Hampshire, BH25 5UH
Area of Operation: UK (Excluding Ireland)
Tel: 0845 090 0109 **Fax:** 0870 094 1258
Email: enquiries@newmil.co.uk
Web: www.newmil.co.uk **Product Type:** 1

POLARWALL LIMITED
Unit 3 Old Mill Industrial Estate,
Stoke Canon, Exeter, Devon, EX5 4RJ
Area of Operation: Europe
Tel: 01392 841777 **Fax:** 01392 841936
Email: info@polarwall.co.uk
Web: www.polarwall.co.uk
Product Type: 1 **Material Type:** G) 1

POLARWALL
Area of Operation: Europe
Tel: 01392 841777
Fax: 01392 841936
Email: info@polarwall.co.uk
Web: www.polarwall.co.uk

Polarwall represents the lastest generation of ICF technology. There is no other method of construction that can match Polarwall in terms of thermal efficiency, acoustic performance, structural strength, longevity, build strength.

POLYSTEEL UK
Unit 26, Malmesbury Road, Kingsditch Trading Estate, Cheltenham, Gloucestershire, GL51 9PL
Area of Operation: UK & Ireland
Tel: 0870 382 2229 **Fax:** 0870 169 6869
Email: info@polysteel.co.uk
Web: www.polysteel.co.uk
Product Type: 1

QUAD-LOCK (ENGLAND) LTD
Unit B3.1, Maws Centre, Jackfield,
Telford, Shropshire, TF8 7LS
Area of Operation: UK (Excluding Ireland)
Tel: 0870 443 1901 **Fax:** 0870 443 1902
Email: k.turner@quadlock.co.uk
Web: www.quadlock.co.uk **Product Type:** 1

SIPCRETE
Po Box 429, Turvey , MK43 8DT
Area of Operation: Worldwide
Tel: 0870 743 9866 **Fax:** 0870 762 5612
Email: ajp@siptec.com
Web: www.sipcrete.com
Product Type: 1, 2 **Material Type:** H) 6

STYRO STONE
16a High Street, Tenterden, Kent, TN30 6AP
Area of Operation: UK (Excluding Ireland)
Tel: 01580 767707 **Fax:** 01580 767709
Email: infouk@styrostone.com
Web: www.styrostone.com
Product Type: 1

STRUCTURAL INSULATED PANELS

KEY
OTHER: ▽ Reclaimed ⌂ On-line shopping
✐ Bespoke ✋Hand-made ECO Ecological

ACORN SIP BUILDINGS LIMITED
Rye Wharf, Harbour Road,
Rye, East Sussex, TN31 7TE
Area of Operation: South East England
Tel: 01797 224465 **Fax:** 01797 224485
Email: sales@acornsip.com
Web: www.acornsip.com
Product Type: 2

BPAC LTD
Crossway, Donibristle Industrial Estate,
Dalgety Bay, Fife, KY11 9JE
Area of Operation: UK & Ireland
Tel: 01383 823995 **Fax:** 01383 823518
Email: sales@bpac.co.uk
Web: www.bpac.co.uk
Product Type: 2

BUILD EXPRESS (SIPS) LIMITED
15 Salmon Grove, Cottingham Road,
Kingston upon Hull, East Riding of Yorks, HU6 7SX
Area of Operation: UK & Ireland
Tel: 01482 341072 **Fax:** 01482 341072
Email: brianellis@netwise.karoo.co.uk
Product Type: 2

BUILD IT GREEN (UK) LTD
Arena Business Centre, 9 Nimrod Way,
Ferndown, Dorset, BH21 7SH
Area of Operation: UK (Excluding Ireland)
Tel: 01202 862320 **Fax:** 01202 877685
Email: enquiries@buildit-green.co.uk
Web: www.buildit-green.co.uk
Product Type: 2

ECOHOMES LTD
First Floor, 52 Briggate, Brighouse,
West Yorkshire, HD6 1ES
Area of Operation: UK & Ireland
Tel: 01484 402040 **Fax:** 01484 400101
Email: sales@ecohomes.ltd.uk
Web: www.ecohomes.ltd.uk
Product Type: 2

EXCEL BUILDING SOLUTIONS
Maerdy Industrial Estate,
Rhymney, Powys, NP22 5PY
Area of Operation: Worldwide
Tel: 01685 845200 **Fax:** 01685 844106
Email: sales@excelfibre.com
Web: www.excelfibre.com **Material Type:** K) 3
Product Type: 2 **Other Info:** ECO ✐

INSULATED PANEL CONSTRUCTION LTD
The Grove, Chalton Heights,
Chalton, Luton, Bedfordshire, LU4 9UF
Area of Operation: East England,
Midlands & Mid Wales, South East England
Tel: 01525 877322 **Fax:** 01525 877322
Email: bcipc@aol.com **Product Type:** 2

INTERBILD LTD
2a Ainslie Street, West Pitkerro,
Dundee, Angus, DD5 3RR
Area of Operation: UK & Ireland
Tel: 01382 480481 **Fax:** 01382 480482
Email: tpd@interbild.com
Web: www.interbild.com **Product Type:** 2

KINGSPAN TEK
Pembridge, Leominster, Herefordshire, HR6 9LA
Area of Operation: UK & Ireland
Tel: 01544 387308 **Fax:** 0870 850 8666
Email: quotations@tek.kingspan.com
Web: www.tek.kingspan.com **Product Type:** 2

MAXIROOF LTD
36 Lower End, Swaffham Prior, Cambridge,
Cambridgeshire, CB5 0HT
Area of Operation: UK (Excluding Ireland)
Tel: 01638 743380 **Fax:** 01638 743380
Email: info@maxiroof.co.uk
Web: www.maxiroof.co.uk **Product Type:** 2

SAFEGUARD EUROPE LTD
Redkiln Close, Horsham, West Sussex, RH13 5QL
Area of Operation: Worldwide
Tel: 01403 210204 **Fax:** 01403 217529
Email: info@safeguardeurope.com
Web: www.safeguardeurope.com
Product Type: 1, 2, 4

SIP HOME LTD
Unit 1-3 Hadston Industrial Estate,
South Broomhill, Northumberland, NE65 9YG
Area of Operation: UK (Excluding Ireland)
Tel: 01670 760300 **Fax:** 01670 760740
Email: john@siphome.co.uk
Web: www.siphome.co.uk
Product Type: 2 **Material Type:** H) 6

SIPCRETE
Po Box 429, Turvey, MK43 8DT
Area of Operation: Worldwide
Tel: 0870 743 9866 **Fax:** 0870 762 5612
Email: ajp@siptec.com **Web:** www.sipcrete.com
Product Type: 1, 2 **Material Type:** H) 6

SIPIT (SCOTLAND) LTD
66 Strathmore Road,
Balmore Industrial Estate,
Glasgow, Lanarkshire, G22 7DW
Area of Operation: UK & Ireland
Tel: 0141 336 2400 **Fax:** 0141 336 8400
Email: enquiries@sipitscotland.co.uk
Web: www.sipitscotland.co.uk **Product Type:** 2

STRUCTURAL INSULATED PANEL TECHNOLOGY LTD (SIPTEC)
PO Box 429, Turvey, Bedfordshire, MK43 8DT
Area of Operation: Europe
Tel: 0870 743 9866 **Fax:** 0870 762 5612
Email: mail@sips.ws **Web:** www.sips.ws
Product Type: 2 **Material Type:** K) 5, 10, 14

THERMASTRUCTURE
Bankside House, Henfield Road,
Small Dole, West Sussex, BN5 9XQ
Area of Operation: UK (Excluding Ireland)
Tel: 01273 492212 **Fax:** 01273 494328
Email: contact@thermastructure.co.uk
Web: www.thermastructure.co.uk
Product Type: 2, 3 **Material Type:** C) 2

WOODCO
Tofts of Tain, Castletown, Highlands, KW14 8TB
Area of Operation: Worldwide
Tel: 01847 821418 **Fax:** 01847 821418
Email: woodcoscotland@btconnect.com
Web: www.woodco.clara.net
Product Type: 2
Other Info: ✐ ✋

BRICKS, BLOCKS, STONE & CLADDING

Image courtesy of Capital Group Limited (0161 799 7555)

SPONSORED BY CAPITAL GROUP LIMITED
Tel 0161 799 7555 Web www.choosecapital.co.uk

capital
group

Bricks

In the UK, bricks are still by far the most popular choice for the exterior of new homes
— it is estimated that at least 80% of new homes are clad at least in part using brick.

The biggest difficulty in choosing bricks is that there is such a wide variety of products on the market. As well as choosing a brick that you like, you should bear in mind that the planners may influence your choice, especially on sensitive sites: in some situations they may insist on a particular type of brick. Otherwise, factors such as the design of your home, and its vulnerability to environmental conditions, should also play a part in your decision. Remember, your choice of brick, along with the style and colour of mortar and the pattern in which the bricks are laid, can make the difference between having beautiful brickwork or an uninspiring wall.

It is impossible to predict the finished appearance of your masonry by looking at brochures alone. Companies can usually arrange for samples to be sent to you, and ideally the supplier will give you a local address to visit to see an example of your preferred brick in place. Alternatively, it is a good idea to get a bricklayer to make up a sample panel, of around a square metre, on site.

You can order bricks either through a builder's merchants, direct from a specialist brick manufacturer, from a specialist brick merchants or from a salvage

yard. Many suppliers also offer a brick matching service if you are trying to match old or existing brickwork. However, when shopping around for quotes, beware of a system known as "brick registration". In this system, once you have made an enquiry to a merchant regarding a particular brick delivered to a specific site, they will contact the brick company and register the quoted price and details with them. This price will then be the minimum you will be quoted through any other merchant, effectively preventing you from shopping around for the best price, so always be careful what details you give away.

Self-build quotes are generally at a price per thousand — you should also make sure your quote includes delivery fees. To calculate how many bricks you'll need, allow 60 bricks/m2 and remember to add 5-7% extra for wastage. This figure should be increased to 10% when using reclaimed bricks.

Good brickwork requires good bricklayers, and often the best way to find one is through recommendations from family and friends. Not only does this give you some assurance of their competency and dependability, but you can also see examples of their work.

Bricks remain the most popular cladding for homes, and the range available on the market today is huge; from traditional and reclaimed varieties, like those shown RIGHT by Capital Group, to more contemporary styles like these gloss ones ABOVE from Ibstock.

BRICKS

KEY

PRODUCT TYPES: 1 = Facing   2 = Engineering
3 = Brick Slips   4 = Specials   5 = Extruded
6 = Glazed   7 = Wire Cut   8 = Reproduction /
Reclaimed   9 = Sand-faced   10 = Stock
11 = Air Bricks

OTHER: ▽ Reclaimed   ⌂ On-line shopping
✎ Bespoke   ✋ Hand-made   ECO Ecological

ABA BUILDING PRODUCTS LTD
Drakes Lodge, 2 Home Farm Lane, Kirklington,
Newark, Nottinghamshire, NG22 8PE
Area of Operation: UK (Excluding Ireland)
Tel: 01636 815491
Fax: 01636 812869
Email: sales@ababuildingproducts.com
Web: www.ababuildingproducts.com
Product Type: 1, 2, 3, 4, 5, 6, 7, 8, 9, 10, 11
Other Info: ▽ ✎ ✋

ABA BUILDING PRODUCTS

Area of Operation: UK (Excluding Ireland)
Tel: 01636 815491   **Fax:** 01636 812869
Email: sales@ababuildingproducts.com
Web: www.ababuildingproducts.com
Product Type: 1, 2, 3, 4, 5, 6, 7, 8, 9, 10, 11

ABA are one of the leading suppliers of reclaimed and
quality facing bricks to the trade and public in the UK
offering a wide range of colours and sizes for new
build or to match existing brickwork.

Call now and see what ABA can do for you.

ALDERSHAW HANDMADE CLAY TILES LTD
Pokehold Wood, Kent Street,
Sedlescombe, Nr Battle,
East Sussex, TN33 0SD
Area of Operation: Europe
Tel: 01424 756777
Fax: 01424 756888
Email: tiles@aldershaw.co.uk
Web: www.aldershaw.co.uk
Product Type: 1, 3, 4, 11
Other Info: ✋
Material Type: F) 1, 3, 4

BAGGERIDGE BRICK
Fir Street, Sedgley, West Midlands, DY3 4AA
Area of Operation: Worldwide
Tel: 01902 880555
Fax: 01902 880432
Email: marketing@baggeridge.co.uk
Web: www.baggeridge.co.uk
Product Type: 1, 2, 3, 4, 5, 6, 7, 8, 9, 10

BANBURY BRICKS LTD
83a Yorke Street, Mansfield Woodhouse,
Nottinghamshire, NG19 9NH
Area of Operation: UK & Ireland
Tel: 0845 230 0941
Fax: 0845 230 0942
Email: enquires@banburybricks.co.uk
Web: www.banburybricks.co.uk
Product Type: 1, 2, 3, 4, 5, 6, 7, 8, 9, 10, 11
Other Info: ▽ ✎ ✋

BOVINGDON BRICKWORKS LTD
Ley Hill Road, Bovingdon,
Near Hemel Hempstead,
Hertfordshire, HP3 0NW
Area of Operation: UK & Ireland
Tel: 01442 833176
Fax: 01442 834539
Email: info@bovingdonbrick.co.uk
Web: www.bovingdonbrick.co.uk
Product Type: 1, 3, 4, 10

BRICK CENTRE LTD
Pottery Lane East, Brimington Road North,
Chesterfield, Derbyshire, S41 9BH
Area of Operation: UK (Excluding Ireland)
Tel: 01246 260001
Fax: 01246 454597
Email: sales@brickcentre.co.uk
Web: www.brickcentre.co.uk

BRICKABILITY
South Road, Bridgend Industrial Estate,
Bridgend, CF31 3XG
Area of Operation: UK (Excluding Ireland)
Tel: 01656 645222
Fax: 01656 665832
Email: enquiries@brickability.co.uk
Web: www.brickability.co.uk
Product Type: 1, 2, 3, 4, 5, 6, 7, 8, 9, 10, 11

BROAD OAK BUILDING PRODUCTS
Unit R3, Elvington Industrial Estate, Elvington,
York, North Yorkshire, YO41 4AR
Area of Operation: UK (Excluding Ireland)
Tel: 01904 607222
Fax: 01904 607223
Email: york.brick@virgin.net
Web: www.broadoakbuildingproducts.co.uk
Product Type: 1, 2, 3, 4, 5, 7, 8, 9, 10
Other Info: ▽ ✎ ✋

BULMER BRICK CUTTING SERVICES
The Brickfields, Hedingham Road,
Bulmer, Sudbury, Suffolk, CO10 7EF
Area of Operation: UK & Ireland
Tel: 01787 269132
Fax: 01787 269044
Email: info@brickcutters.com
Web: www.brickcutters.com
Product Type: 3, 4

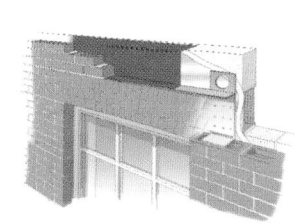

BULMER BRICK CUTTING SERVICES

Area of Operation: UK & Ireland
Tel: 01787 269 132
Fax: 01787 269 044
Email: info@brickcutters.com
Web: www.brickcutters.com

Fully bonded arch lintels, Rubbed & Gauged
Archwork, Hand cutting & Brick Carving.
Cut & Bonded Special Shapes, Refacing,
Herringbone Panels.

CAPITAL GROUP LIMITED
Victoria Mills, Highfield Road,
Little Hulton, Manchester, M38 9ST
Area of Operation: Worldwide
Tel: 0161 799 7555
Fax: 0161 799 7666
Email: leigh@choosecapital.co.uk
Web: www.choosecapital.co.uk ⌂
Product Type: 1, 2, 3, 4, 5, 6, 7, 8, 9, 10

CHESHIRE BRICK & SLATE
Brook House Farm, Salters Bridge,
Tarvin Sands, Tarvin, Cheshire, CH3 8NR
Area of Operation: UK (Excluding Ireland)
Tel: 01829 740883
Fax: 01829 740481
Email: enquiries@cheshirebrickandslate.co.uk
Web: www.cheshirebrickandslate.co.uk
Product Type: 1, 2, 3, 4, 7, 8, 10

COLEFORD BRICK & TILE CO LTD
The Royal Forest of Dean Brickworks,
Cinderford, Gloucestershire, GL14 3JJ
Area of Operation: UK & Ireland
Tel: 01594 822160
Fax: 01594 826655
Email: sales@colefordbrick.co.uk
Web: www.colefordbrick.co.uk
Product Type: 4, 8
Other Info: ✎ ✋

CREST BRICK, SLATE & TILE LTD
Howdenshire Way, Knedlington Road,
Howden, East Riding of Yorks, DN14 7HZ
Area of Operation: UK (Excluding Ireland)
Tel: 0870 241 1398
Fax: 01430 433000
Email: info@crest-bst.co.uk
Web: www.crest-bst.co.uk
Product Type: 1, 2, 3, 4, 8, 9, 10

EUROBRICK SYSTEMS LTD
Unit 7 Wilverley Trading Estate, Bath Road,
Brislington, Bristol, BS4 5NL
Area of Operation: Europe
Tel: 0117 971 7117
Fax: 0117 971 7217
Email: info@eurobrick.co.uk
Web: www.eurobrick.co.uk
Product Type: 3

FASTCLAD CLADDING SOLUTIONS
Granite Close, Enderby, Leicester,
Leicestershire, LE19 4AE
Area of Operation: Europe
Tel: 0116 272 5133
Fax: 0116 272 5131
Email: info@fastclad.com
Web: www.fastclad.com
Product Type: 1, 3, 4, 5, 7, 8, 9, 10
Material Type: E) 1, 2, 4, 7, 11, 13, 14

FURNESS BRICK & TILE CO
Askam in Furness, Cumbria, LA16 7HF
Area of Operation: Worldwide
Tel: 01229 462411
Fax: 01229 462363
Email: peterhubble@mac.com
Web: www.furnessbrick.com
Product Type: 1, 4, 8, 10
Other Info: ✎ ✋
Material Type: F) 1

FYFESTONE
Aquithie Road, Kemnay, Aberdeen,
Aberdeenshire, AB51 5PD
Area of Operation: UK (Excluding Ireland)
Tel: 01467 651000
Fax: 01467 642342
Email: masonry@aggregate.com
Web: www.fyfestone.com
Product Type: 1

HANSON BUILDING PRODUCTS
Stewartby, Bedford, Bedfordshire, MK43 9LZ
Area of Operation: Worldwide
Tel: 08705 258258 **Fax:** 01234 762040
Email: info@hansonplc.com
Web: www.hanson.biz
Product Type: 1, 2, 3, 4, 5, 6, 7, 8, 9, 10

IBSTOCK BRICK LTD
Leicester Road, Ibstock, Leicester,
Leicestershire, LE67 6HS
Area of Operation: UK & Ireland
Tel: 01530 261999 **Fax:** 01530 263478
Email: leicester.sales@ibstock.co.uk
Web: www.ibstock.co.uk
Product Type: 1, 3, 4, 5, 6, 7, 8, 9, 10, 11

INNOVATIVE BUILDING SOLUTIONS
Mill Green House,
48-50 Mill Green Road,
Mitcham, Surrey, CR4 4HY
Area of Operation: UK & Ireland
Tel: 0208 687 2260
Fax: 0208 687 2249
Email: ibs@charterhouseplc.co.uk
Product Type: 1, 2, 3, 4, 5, 6, 7, 8, 9, 10, 11

MANCHESTER BRICK & PRECAST LTD
Haigh Avenue, Whitehill Industrial Estate,
Stockport, Cheshire, SK4 1NU
Area of Operation: UK & Ireland
Tel: 0161 480 2621
Fax: 0161 480 0108
Email: sales@manbrick.co.uk
Web: www.manbrick.co.uk

MIDLANDS SLATE & TILE
Qualcast Road, Horseley Fields,
Wolverhampton, West Midlands, WV1 2QP
Area of Operation: UK & Ireland
Tel: 01902 458780
Fax: 01902 458549
Email: sales@slate-tile-brick.co.uk
Web: www.slate-tile-brick.co.uk
Product Type: 1, 2, 7, 10
Material Type: F) 3

NORTHCOT BRICK LTD
Blockley, Moreton-in-Marsh,
Gloucestershire, GL56 9LH
Area of Operation: UK & Ireland
Tel: 01386 700551
Fax: 01386 700852
Email: info@northcotbrick.co.uk
Web: www.northcotbrick.co.uk
Product Type: 1, 3, 4, 5, 7, 8, 9

NOVABRIK IRELAND LTD
Templemichael, Caherconlish, Limerick
Area of Operation: UK & Ireland
Tel: +353 6135 1400
Fax: +353 6135 1408
Email: ireland@novabrik.com
Web: www.novabrik.com
Product Type: 1

PADIPA LIMITED
Unit East 2, Sway Storage & Workshops,
Sway, Lymington, Hampshire, SO41 6DD
Area of Operation: UK & Ireland
Tel: 01590 681710
Fax: 01590 610336
Email: enquiries@padipa.co.uk
Web: www.padipa.co.uk
Product Type: 3

SELBORNE TILE & BRICK LTD
Honey Lane, Selbourne, Alton,
Hampshire, GU34 3BS
Area of Operation: UK (Excluding Ireland)
Tel: 01420 478 752
Product Type: 1, 4, 7, 9, 10

SOUTH WEST RECLAMATION LTD
Wireworks Estate, Bristol Road,
Bridgwater, Somerset, TA6 4AP
Area of Operation: South West England
and South Wales
Tel: 01278 444141
Fax: 01278 444114
Email: info@southwest-rec.co.uk
Web: www.southwest-rec.co.uk
Other Info: ▽

SURREY RECLAIMED BRICKS
28C The Plantation,
West Park Road, Newchapel,
Lingfield, Surrey, RH7 6HT
Area of Operation: Greater London,
South East England
Tel: 01342 714561
Fax: 01342 714561
Email: surreybricks@aol.com
Web: www.surreyreclaimedbrickwork.co.uk ⌂
Product Type: 1, 7, 10
Other Info: ▽

TAYLOR MAXWELL & COMPANY LIMITED
4 John Oliver Buildings,
53 Wood Street, Barnet,
Hertfordshire, EN5 4BS
Area of Operation: UK (Excluding Ireland)
Tel: 0208 440 0551
Fax: 0208 440 0552
Email: simonepalmer@taylor.maxwell.co.uk
Web: www.taylor.maxwell.co.uk
Product Type: 1, 2, 3, 4, 5, 6, 7, 8, 9, 10

THE MATCHING BRICK COMPANY
Lockes Yard, Hartcliffe Way,
Bedminster, Bristol, BS3 5RJ
Area of Operation: UK (Excluding Ireland)
Tel: 0117 963 7000
Fax: 0117 966 4612
Email: matchingbrick@btconnect.com
Web: www.matchingbrick.co.uk
Product Type: 1, 2, 3, 4, 5, 7, 8, 9, 10

THERMOBRICK
20 Underwood Drive,
Stoney Stanton, Leicester,
Leicestershire, LE9 4TA
Area of Operation: UK & Ireland
Tel: 01455 272860
Fax: 01455 271324
Email: ruttim1@aol.com
Web: www.rutlandtimber.co.uk
Product Type: 1, 3

WETHERBY BUILDING SYSTEMS LIMITED
1 Kidglove Road, Golborne
Enterprise Park, Golborne,
Greater Manchester, WA3 3GS
Area of Operation: UK & Ireland
Tel: 01942 717100
Fax: 01942 717101
Email: info@wbs-ltd.co.uk
Web: www.wbs-ltd.co.uk
Product Type: 3, 7, 9

WIENERBERGER LTD
Wienerberger House, Brooks Drive, Cheadle Royal
Business Park, Cheadle, Cheshire, SK8 3SA
Area of Operation: UK & Ireland
Tel: 0161 491 8200 **Fax:** 0161 488 4827
Email: nicky.webb@wienerberger.com
Web: www.wienerberger.co.uk
Product Type: 1, 2, 4, 5, 7, 8, 9, 10
Other Info: ✏ ✋

YORK HANDMADE BRICK CO LTD
Forest Lane, Alne, York, North Yorkshire, YO61 1TU
Area of Operation: Worldwide
Tel: 01347 838881 **Fax:** 01347 838885
Email: sales@yorkhandmade.co.uk
Web: www.yorkhandmade.co.uk
Product Type: 1, 3, 4, 5 **Other Info:** ✋

YORK HANDMADE BRICK CO LTD

Area of Operation: UK
Tel: 01347 838881
Fax: 01347 838885
Email: sales@yorkhandmade.co.uk
Web: www.yorkhandmade.co.uk
Product Type: 1, 3, 4, 5
Other Info: ✋

York Handmade Brick are the largest
independent manufacturer of genuine handmade
bricks in the UK. Also producers of terracotta
floor tiles and landscape products.

PANELS, SECTIONS, SHEETS AND TILES

KEY
PRODUCT TYPES: 1 = Panels 2 = Sections
3 = Sheets   4 = Tiles   5 = Other

OTHER: ▽ Reclaimed ⌂ On-line shopping
✏ Bespoke ✋ Hand-made ECO Ecological

ALMURA BUILDING PRODUCTS LTD
Cantay House, St George's Place,
Cheltenham, Gloucestershire, GL50 3PN
Area of Operation: Europe
Tel: 01242 262900
Fax: 01242 221333
Email: philipmarsh@almura.co.uk
Web: www.almuracladdings.co.uk
Product Type: 1, 3, 5

BLANC DE BIERGES
Eastrea Road, Whittlesey,
Cambridgeshire, PE7 2AG
Area of Operation: Worldwide
Tel: 01733 202566
Fax: 01733 205405
Email: info@blancdebierges.com
Web: www.blancdebierges.com
Other Info: ✏ ✋

BOARD CENTRAL
Chiltern Business Centre,
Couching Street, Watlington,
Oxfordshire, OX49 5PX
Area of Operation: Greater London,
South East England
Tel: 0845 458 8016
Fax: 01844 354112
Email: howardmorrice@hotmail.com
Product Type: 1, 3

BORDER CONCRETE PRODUCTS
Jedburgh Road, Kelso, Borders, TD5 8JG
Area of Operation: North East England,
North West England and North Wales, Scotland
Tel: 01573 224393 **Fax:** 01573 276360
Email: sales@borderconcrete.co.uk
Web: www.borderconcrete.co.uk

CAVALOK BUILDING PRODUCTS
Green Lane, Newtown, Tewkesbury,
Gloucestershire, GL20 8HD
Area of Operation: UK & Ireland
Tel: 01684 855706 **Fax:** 01684 299124
Email: info@cavalok.co.uk
Web: www.cavalok.com
Product Type: 1, 2, 3
Other Info: ECO ✏

CEMBRIT BLUNN
6 Coleshill Street, Fazeley, Tamworth,
Staffordshire, B78 3XJ
Area of Operation: UK & Ireland
Tel: 01827 288827 **Fax:** 01827 288176
Email: parcher@cembritblunn.co.uk
Web: www.cembritblunn.co.uk
Product Type: 1, 3
Material Type: H) 4

CLANCAST CONTRACTS LTD
48 Shaw Street, Glasgow, G51 3BL
Area of Operation: UK (Excluding Ireland)
Tel: 0141 440 2345 **Fax:** 0141 440 2488
Email: info@clancast.co.uk
Product Type: 1
Material Type: K) 5

CREST BRICK, SLATE & TILE LTD
Howdenshire Way, Knedlington Road, Howden,
East Riding of Yorks, DN14 7HZ
Area of Operation: UK (Excluding Ireland)
Tel: 0870 241 1398 **Fax:** 01430 433000
Email: info@crest-bst.co.uk
Web: www.crest-bst.co.uk
Product Type: 4

SPONSORED BY: CAPITAL BRICK LIMITED www.choosecapital.co.uk

DELTA MEMBRANE SYSTEMS LTD
Bassett Business Centre, Hurricane Way,
North Weald, Epping, Essex, CM16 6AA
Area of Operation: UK & Ireland
Tel: 0870 747 2181
Fax: 0992 524046
Email: info@deltamembranes.com
Web: www.deltamembranes.com
Product Type: 3

EGGER (UK) LIMITED
Anick Grange Road, Hexham,
Northumberland, NE46 4JS
Area of Operation: UK & Ireland
Tel: 01434 602191
Fax: 01434 605103
Email: building.uk@egger.com
Web: www.egger.co.uk
Product Type: 1
Material Type: H) 1, 6

EUROBRICK SYSTEMS LTD
Unit 7 Wilverley Trading Estate, Bath Road,
Brislington, Bristol, BS4 5NL
Area of Operation: Europe
Tel: 0117 971 7117
Fax: 0117 971 7217
Email: info@eurobrick.co.uk
Web: www.eurobrick.co.uk
Product Type: 1

EXTERIOR IMAGE LTD
322 Hursley Road, Chandlers Ford,
Eastleigh, Hampshire, SO53 5PH
Tel: 02380 271777
Fax: 02380 269091
Email: info@exteriorimage.co.uk
Web: www.exteriorimage.co.uk

FASTCLAD CLADDING SOLUTIONS
Granite Close, Enderby, Leicester,
Leicestershire, LE19 4AE
Area of Operation: Europe
Tel: 0116 272 5133
Fax: 0116 272 5131
Email: info@fastclad.com
Web: www.fastclad.com
Product Type: 1, 5
Material Type: E) 1, 2, 4, 5, 11, 12, 13, 14

FRANCIS N. LOWE LTD.
The Marble Works, New Road, Middleton,
Matlock, Derbyshire, DE4 4NA
Area of Operation: Europe
Tel: 01629 822216
Fax: 01629 824348
Email: info@lowesmarble.com
Web: www.lowesmarble.com
Product Type: 4
Material Type: E) 1, 2, 3, 5, 8, 9

HONEYSUCKLE BOTTOM SAWMILL LTD
Honeysuckle Bottom, Green Dene,
East Horsley, Leatherhead, Surrey, KT24 5TD
Area of Operation: Greater London,
South East England
Tel: 01483 282394
Fax: 01483 282394
Email: honeysucklemill@aol.com
Web: www.easisites.co.uk/honeysucklebottomsawmill
Material Type: A) 2, 12

IBSTOCK BRICK LTD
Leicester Road, Ibstock,
Leicester, Leicestershire, LE67 6HS
Area of Operation: UK & Ireland
Tel: 01530 261999
Fax: 01530 263478
Email: leicester.sales@ibstock.co.uk
Web: www.ibstock.co.uk
Product Type: 1, 4

INNOVATIVE BUILDING SOLUTIONS
Mill Green House, 48-50 Mill Green Road,
Mitcham, Surrey, CR4 4HY
Area of Operation: UK & Ireland
Tel: 0208 687 2260
Fax: 0208 687 2249
Email: ibs@charterhouseplc.co.uk
Product Type: 1

KEYMER TILES LTD
Nye Road, Burgess Hill, West Sussex, RH15 0LZ
Area of Operation: East England, Greater London,
Midlands & Mid Wales, South East England, South
West England and South Wales
Tel: 01444 232931 **Fax:** 01444 871852
Email: info@keymer.co.uk
Web: www.keymer.co.uk
Product Type: 4 **Other Info:** ✎ ✋
Material Type: F) 3

MARLEY ETERNIT
Station Road, Coleshill, Birmingham,
West Midlands, B46 1HP
Area of Operation: UK (Excluding Ireland)
Tel: 0870 562 6500 **Fax:** 0870 562 6550
Email: roofingsales@marleyeternit.co.uk
Web: www.marleyeternit.co.uk
Product Type: 1, 3, 4

NATURAL BUILDING TECHNOLOGIES
The Hangar, Worminghall Road, Oakley,
Buckinghamshire, HP18 9UL
Area of Operation: UK & Ireland
Tel: 01844 338338
Fax: 01844 338525
Email: info@natural-building.co.uk
Web: www.natural-building.co.uk
Product Type: 1
Other Info: ECO
Material Type: K) 3, 8, 9

NOISE STOP SYSTEMS
Unit 3 Acaster Industrial Estate,
Cowper Lane, Acaster Malbis, York,
North Yorkshire, YO23 2XB
Area of Operation: UK & Ireland
Tel: 0845 130 6269
Fax: 01904 709857
Email: info@noisestopsystems.co.uk
Web: www.noisestopsystems.co.uk ✋
Product Type: 1, 2, 3

PERMACELL FINESSE LTD
Western Road, Silver End,
Witham, Essex, CM8 3BQ
Area of Operation: UK & Ireland
Tel: 01376 583241
Fax: 01376 583072
Email: sales@pfl.co.uk
Web: www.permacell-finesse.co.uk
Product Type: 1

SKANDA (UK) LTD
64 - 65 Clywedog Road North, Wrexham Industrial
Estate, Wrexham, LL13 9XN
Area of Operation: UK & Ireland
Tel: 01978 664255 **Fax:** 01978 661427
Email: info@skanda-uk.com
Web: www.skanda-uk.com/www.heraklith.co.uk

STANCLIFFE STONE
Grange Mill, Matlock, Derbyshire, DE4 4BW
Area of Operation: Worldwide
Tel: 01629 653000 **Fax:** 01629 650996
Email: sales@stancliffe.com
Web: www.stancliffe.com
Product Type: 1 **Other Info:** ✎
Material Type: E) 4, 5

**STRUCTURAL INSULATED PANEL
TECHNOLOGY LTD (SIPTEC)**
PO Box 429, Turvey, Bedfordshire, MK43 8DT
Area of Operation: Europe
Tel: 0870 743 9866 **Fax:** 0870 762 5612
Email: mail@sips.ws
Web: www.sips.ws
Product Type: 1
Material Type: H) 4, 6

SWISH BUILDING PRODUCTS LTD
Pioneer House, Mariner, Lichfield Road Industrial
Estate, Tamworth, Staffordshire, B79 7TF
Area of Operation: UK & Ireland
Tel: 01827 317200 **Fax:** 01827 317201
Email: info@swishbp.co.uk
Web: www.swishbp.co.uk
Product Type: 1, 2, 3
Material Type: D) 1

TAYLOR MAXWELL & COMPANY LIMITED
4 John Oliver Buildings, 53 Wood Street,
Barnet, Hertfordshire, EN5 4BS
Area of Operation: UK (Excluding Ireland)
Tel: 0208 440 0551 **Fax:** 0208 440 0552
Email: simonepalmer@taylor.maxwell.co.uk
Web: www.taylor.maxwell.co.uk
Product Type: 1, 4, 5

THE EXPANDED METAL COMPANY LIMITED
PO Box 14, Longhill Industrial Estate (North),
Hartlepool, Durham, TS25 1PR
Area of Operation: Worldwide
Tel: 01429 408087 **Fax:** 01429 866795
Email: paulb@expamet.co.uk
Web: www.expandedmetalcompany.co.uk
Material Type: C) 1, 2, 3, 4, 6, 7, 9, 10, 11, 12, 13, 18

THE GRANITE FACTORY
4 Winchester Drive, Peterlee, Durham, SR8 2RJ
Area of Operation: North East England
Tel: 0191 518 3600
Fax: 0191 518 3600
Email: admin@granitefactory.co.uk
Web: www.granitefactory.co.uk ✋
Product Type: 1, 4

TWINFIX
201 Cavendish Place, Birchwood Park,
Birchwood, Warrington, Cheshire, WA3 6WU
Area of Operation: UK & Ireland
Tel: 01925 811311 **Fax:** 01925 852955
Email: enquiries@twinfix.co.uk
Web: www.twinfix.co.uk

VISION ASSOCIATES
Demita House, North Orbital Road,
Denham, Buckinghamshire, UB9 5EY
Area of Operation: UK & Ireland
Tel: 01895 831600 **Fax:** 01895 835323
Email: info@visionassociates.co.uk
Web: www.visionassociates.co.uk
Product Type: 1, 3

WESSEX BUILDING PRODUCTS (MULTITEX)
Dolphin Industrial Estate,
Southampton Road, Salisbury,
Wiltshire, SP1 2NB
Area of Operation: UK & Ireland
Tel: 01722 332139
Fax: 01722 338458
Email: sales@wessexbuildingproducts.co.uk
Web: www.wessexbuildingproducts.co.uk
Product Type: 1
Material Type: D) 6

WETHERBY BUILDING SYSTEMS LIMITED
1 Kidglove Road, Golborne Enterprise Park,
Golborne, Greater Manchester, WA3 3GS
Area of Operation: UK & Ireland
Tel: 01942 717100
Fax: 01942 717101
Email: info@wbs-ltd.co.uk
Web: www.wbs-ltd.co.uk
Product Type: 1, 4

RENDER

KEY
SEE ALSO: FLOOR AND WALL FINISHES -
Paints, Stains and Varnishes (Exterior Paint)

OTHER: ▽ Reclaimed ✎ On-line shopping
✎ Bespoke ✋ Hand-made ECO Ecological

**ALUMASC EXTERIOR
BUILDING PRODUCTS**
White House Works, Bold Road,
Sutton, St Helens, Merseyside, WA9 4JG
Area of Operation: UK & Ireland
Tel: 01744 648400
Fax: 01744 648401
Email: info@alumasc-exteriors.co.uk
Web: www.alumasc-exteriors.co.uk

BRICKABILITY
South Road,
Bridgend Industrial Estate,
Bridgend, CF31 3XG
Area of Operation: UK (Excluding Ireland)
Tel: 01656 645222 **Fax:** 01656 665832
Email: enquiries@brickability.co.uk
Web: www.brickability.co.uk

FIBROCEM
PO Box 4769, Wimborne,
Dorset, BH21 8BA
Area of Operation: UK (Excluding Ireland)
Tel: 01202 8577448
Fax: 01202 824495
Email: info@fibrocem.co.uk
Web: www.fibrocem.co.uk
Material Type: G) 2, 5

LAFARGE CEMENT
Manor Court, Chilton,
Oxfordshire, OX11 0RN
Area of Operation: UK & Ireland
Tel: 01235 448400
Fax: 01235 448600
Email: info@lafargecement.co.uk
Web: www.lafargecement.co.uk
Material Type: G) 2

MID ARGYLL GREEN SHOP
An Tairbeart, Campbeltown Road,
Tarbet, Argyll & Bute, PA29 6SX
Area of Operation: Scotland
Tel: 01880 821212
Email: kate@mackandmags.co.uk
Web: www.mackandmags.co.uk ✎
Other Info: ECO
Material Type: K) 3, 9

NATURAL BUILDING TECHNOLOGIES
The Hangar, Worminghall Road,
Oakley, Buckinghamshire, HP18 9UL
Area of Operation: UK & Ireland
Tel: 01844 338338
Fax: 01844 338525
Email: info@natural-building.co.uk
Web: www.natural-building.co.uk
Material Type: I) 2, 3

SKANDA (UK) LTD
64 - 65 Clywedog Road North,
Wrexham Industrial Estate,
Wrexham, LL13 9XN
Area of Operation: UK & Ireland
Tel: 01978 664255 **Fax:** 01978 661427
Email: info@skanda-uk.com
Web: www.skanda-uk.com/www.heraklith.co.uk

**THE EXPANDED METAL
COMPANY LIMITED**
PO Box 14, Longhill Industrial Estate (North),
Hartlepool, Durham, TS25 1PR
Area of Operation: Worldwide
Tel: 01429 408087
Fax: 01429 866795
Email: paulb@expamet.co.uk
Web: www.expandedmetalcompany.co.uk

**WETHERBY BUILDING
SYSTEMS LIMITED**
1 Kidglove Road, Golborne Enterprise Park,
Golborne, Greater Manchester, WA3 3GS
Area of Operation: UK & Ireland
Tel: 01942 717100
Fax: 01942 717101
Email: info@wbs-ltd.co.uk
Web: www.wbs-ltd.co.uk
Material Type: G) 5

WETHERTEX UK
Bleakhill Way, Mansfield,
Nottinghamshire, NG18 5EZ
Area of Operation: UK (Excluding Ireland)
Tel: 01623 633833
Fax: 01623 635551
Email: sales@wethertex.co.uk
Web: www.wethertex.co.uk

SPONSORED BY: CAPITAL BRICK LIMITED www.choosecapital.co.uk

SHAKES AND SHINGLES

KEY

SEE ALSO: MERCHANTS - Timber Merchants,
ROOFING - Tiles, Shingles and Slates

OTHER: ▽ Reclaimed 🖱 On-line shopping
🗩 Bespoke ✋Hand-made ECO Ecological

ALMURA BUILDING PRODUCTS LTD
Cantay House, St George's Place,
Cheltenham, Gloucestershire, GL50 3PN
Area of Operation: Europe
Tel: 01242 262 900
Fax: 01242 221 333
Email: philipmarsh@almura.co.uk
Web: www.almuracladdings.co.uk

CANADA WOOD UK
PO Box 1, Farnborough, Hampshire, GU14 6WE
Area of Operation: UK (Excluding Ireland)
Tel: 01252 522545
Fax: 01252 522546
Email: office@canadawooduk.org
Web: www.canadawood.info
Material Type: A) 12

JOHN BRASH LTD
The Old Shipyard, Gainsborough,
Lincolnshire, DN21 1NG
Area of Operation: UK & Ireland
Tel: 01427 613858
Fax: 01427 810218
Email: info@johnbrash.co.uk
Web: www.johnbrash.co.uk
Material Type: A) 2, 8, 12

WEATHERBOARDING

KEY

SEE ALSO: MERCHANTS - Timber Merchants

OTHER: ▽ Reclaimed 🖱 On-line shopping
🗩 Bespoke ✋Hand-made ECO Ecological

ALMURA BUILDING PRODUCTS LTD
Cantay House, St George's Place, Cheltenham,
Gloucestershire, GL50 3PN
Area of Operation: Europe
Tel: 01242 262 900
Fax: 01242 221 333
Email: philipmarsh@almura.co.uk
Web: www.almuracladdings.co.uk

ALMURA BUILDING PRODUCTS LTD

Area of Operation: Europe
Tel: 01242 262 900 **Fax:** 01242 221 333
Email: philipmarsh@almura.co.uk
Web: www.almuracladdings.co.uk

Almura Supply the highest quality of cladding of
their type in the world. These include
weatherboarding, stone and brick effect boards
and aggregate faced panels.

BOARD CENTRAL
Chiltern Business Centre, Couching Street,
Watlington, Oxfordshire, OX49 5PX
Area of Operation: Greater London,
South East England
Tel: 0845 458 8016
Fax: 01844 354112
Email: howardmorrice@hotmail.com

BRICKABILITY
South Road, Bridgend Industrial Estate,
Bridgend, CF31 3XG
Area of Operation: UK (Excluding Ireland)
Tel: 01656 645222
Fax: 01656 665832
Email: enquiries@brickability.co.uk
Web: www.brickability.co.uk

CANADA WOOD UK
PO Box 1, Farnborough, Hampshire, GU14 6WE
Area of Operation: UK (Excluding Ireland)
Tel: 01252 522545
Fax: 01252 522546
Email: office@canadawooduk.org
Web: www.canadawood.info
Material Type: A) 12

CAVALOK BUILDING PRODUCTS
Green Lane, Newtown, Tewkesbury,
Gloucestershire, GL20 8HD
Area of Operation: UK & Ireland
Tel: 01684 855706
Fax: 01684 299124
Email: info@cavalok.co.uk
Web: www.cavalok.com

CEMBRIT BLUNN
6 Coleshill Street, Fazeley,
Tamworth, Staffordshire, B78 3XJ
Area of Operation: UK & Ireland
Tel: 01827 288827
Fax: 01827 288176
Email: parcher@cembritblunn.co.uk
Web: www.cembritblunn.co.uk

EGGER (UK) LIMITED
Anick Grange Road, Hexham,
Northumberland, NE46 4JS
Area of Operation: UK & Ireland
Tel: 01434 602191
Fax: 01434 605103
Email: building.uk@egger.com
Web: www.egger.co.uk
Material Type: H) 1

ETERNIT BUILDING MATERIALS
Whaddon Road, Meldreth,
Royston, Hertfordshire, SG8 5RL
Area of Operation: UK (Excluding Ireland)
Tel: 01763 264686
Fax: 01763 262531
Email: marketing@eternit.co.uk
Web: www.eternit.co.uk

JOHN BRASH LTD
The Old Shipyard, Gainsborough,
Lincolnshire, DN21 1NG
Area of Operation: UK & Ireland
Tel: 01427 613858 **Fax:** 01427 810218
Email: info@johnbrash.co.uk
Web: www.johnbrash.co.uk

NATURAL BUILDING TECHNOLOGIES
The Hangar, Worminghall Road, Oakley,
Buckinghamshire, HP18 9UL
Area of Operation: UK & Ireland
Tel: 01844 338338 **Fax:** 01844 338525
Email: info@natural-building.co.uk
Web: www.natural-building.co.uk

NORTHWOOD FORESTRY LTD
Goose Green Lane, Nr Ashington,
Pulborough, West Sussex, RH20 2LW
Area of Operation: South East England
Tel: 01798 813029 **Fax:** 01798 813139
Email: enquiries@northwoodforestry.co.uk
Web: www.northwoodforestry.co.uk
Other Info: 🖱
Material Type: A) 2

PERMACELL FINESSE LTD
Western Road, Silver End,
Witham, Essex, CM8 3BQ
Area of Operation: UK & Ireland
Tel: 01376 583241
Fax: 01376 583072
Email: sales@pfl.co.uk
Web: www.permacell-finesse.co.uk

PLASTIVAN LTD
Unit 4 Bonville Trading Estate,
Bonville Road, Brislington, Bristol, BS5 0EB
Area of Operation: Worldwide
Tel: 0117 300 5625
Fax: 0117 971 5028
Email: sales@plastivan.co.uk
Web: www.plastivan.co.uk

SWISH BUILDING PRODUCTS LTD
Pioneer House, Mariner,
Litchfield Road Industrial Estate,
Tamworth, Staffordshire, B79 7TF
Area of Operation: UK & Ireland
Tel: 01827 317200
Fax: 01827 317201
Email: info@swishbp.co.uk
Web: www.swishbp.co.uk
Material Type: D) 1

TAYLOR MAXWELL & COMPANY LIMITED
4 John Oliver Buildings, 53 Wood Street,
Barnet, Hertfordshire, EN5 4BS
Area of Operation: UK (Excluding Ireland)
Tel: 0208 440 0551
Fax: 0208 440 0552
Email: simonepalmer@taylor.maxwell.co.uk
Web: www.taylor.maxwell.co.uk

VISION ASSOCIATES
Demita House, North Orbital Road,
Denham, Buckinghamshire, UB9 5EY
Area of Operation: UK & Ireland
Tel: 01895 831600
Fax: 01895 835323
Email: info@visionassociates.co.uk
Web: www.visionassociates.co.uk

STONE CLADDING

KEY

PRODUCT TYPES: 1= Ashlar 2 = Cobble
3 = Dressed Stone 4 = Rubble

OTHER: ▽ Reclaimed 🖱 On-line shopping
🗩 Bespoke ✋Hand-made ECO Ecological

ADDSTONE CAST STONE
2 Millers Gate, Stone, Staffordshire, ST15 8ZF
Area of Operation: UK (Excluding Ireland)
Tel: 01785 818810
Fax: 01785 818810
Email: sales@addstone.co.uk
Web: www.addstone.co.uk 🖱
Product Type: 1, 3
Other Info: 🗩 ✋
Material Type: E) 11, 13

ALBION STONE
27-33 Brighton Road, Redhill, Surrey, RH1 6PP
Area of Operation: UK (Excluding Ireland)
Tel: 01737 771772
Fax: 01737 771776
Email: sales@albionstonequarries.com
Web: www.albionstonequarries.com 🖱

ASPECTS OF STONE LTD
Unit 29, Broughton Grounds,
Broughton, Newport Pagnell,
Buckinghamshire, MK16 0HZ
Area of Operation: UK (Excluding Ireland)
Tel: 01908 830061
Fax: 01908 830062
Email: sales@aspectsofstone.co.uk
Web: www.aspectsofstone.co.uk
Product Type: 1, 3

BLACK MOUNTAIN QUARRIES LTD
Howton Court , Pontrilas, Herefordshire, HR2 0BG
Area of Operation: UK & Ireland
Tel: 01981 241541
Email: info@blackmountainquarries.com
Web: www.blackmountainquarries.com 🖱
Product Type: 2, 3, 4 **Other Info:** ECO ✋

BLANC DE BIERGES
Eastrea Road, Whittlesey, Cambridgeshire, PE7 2AG
Area of Operation: Worldwide
Tel: 01733 202566 **Fax:** 01733 205405
Email: info@blancdebierges.com
Web: www.blancdebierges.com

BRADSTONE STRUCTURAL
Aggregate Industries UK Ltd, North End,
Ashton Keynes, Swindon, Wiltshire, SN6 3QX
Area of Operation: UK (Excluding Ireland)
Tel: 01285 646884 **Fax:** 01285 646891
Email: bradstone.structural@aggregate.com
Web: www.bradstone.com
Product Type: 1

BRS YORK STONE
50 High Green Road, Altofts, Normanton,
Lancashire, WF6 2LQ
Area of Operation: UK (Excluding Ireland)
Tel: 01924 220356 **Fax:** 01924 220356
Email: john@york-stone.fsnet.co.uk
Web: www.yorkstonepaving.co.uk

CAWARDEN BRICK CO. LTD
Cawarden Springs Farm, Blithbury Road,
Rugeley, Staffordshire, WS15 3HL
Area of Operation: UK (Excluding Ireland)
Tel: 01889 574066 **Fax:** 01889 575695
Email: home-garden@cawardenreclaim.co.uk
Web: www.cawardenreclaim.co.uk
Product Type: 2

D F FIXINGS
15 Aldham Gardens, Rayleigh, Essex, SS6 9TB
Area of Operation: UK (Excluding Ireland)
Tel: 07956 674673 **Fax:** 01268 655072

FARMINGTON NATURAL STONE
Northleach, Cheltenham, Gloucestershire, GL54 3NZ
Area of Operation: UK (Excluding Ireland)
Tel: 01451 860280 **Fax:** 01451 860115
Email: cotswold.stone@farmington.co.uk
Web: www.farmingtonnaturalstone.co.uk
Product Type: 2, 3

FERNHILL STONE UK LTD
Unit 16 Greencroft Industrial Estate,
Annfield Plain, Stanley, Durham, DH9 7XP
Area of Operation: UK & Ireland
Tel: 01207 521482 **Fax:** 01207 521455
Email: geoffreydriver@btconnect.com
Web: www.fernhillstone.com
Material Type: E) 4, 11

FERNHILL STONE UK LTD

Area of Operation: UK & Ireland
Tel: 01207 521482
Fax: 01207 521455
Email: geoffreydriver@btconnect.com
Web: www.fernhillstone.com

At Fernhill Stone we offer you all the beauty of
natural stone at a fraction of the cost.

FRANCIS N. LOWE LTD.
The Marble Works, New Road,
Middleton, Matlock,
Derbyshire, DE4 4NA
Area of Operation: Europe
Tel: 01629 822216 **Fax:** 01629 824348
Email: info@lowesmarble.com
Web: www.lowesmarble.com
Product Type: 1, 3

FYFESTONE
Aquithie Road, Kemnay,
Aberdeen, Aberdeenshire, AB51 5PD
Area of Operation: UK (Excluding Ireland)
Tel: 01467 651000 **Fax:** 01467 642342
Email: masonry@aggregate.com
Web: www.fyfestone.com
Product Type: 1, 3, 4

HEART OF STONE LTD
114 London Road, Shrewsbury,
Shropshire, SY2 6PP
Area of Operation: UK (Excluding Ireland)
Tel: 01743 247917 **Fax:** 01743 247917
Email: heartofstone@tiscali.co.uk
Web: www.heartofstone.co.uk
Product Type: 2, 3
Material Type: E) 1, 2, 3, 4, 5, 8, 9

INNOVATIVE BUILDING SOLUTIONS
Mill Green House, 48-50 Mill Green Road,
Mitcham, Surrey, CR4 4HY
Area of Operation: UK & Ireland
Tel: 0208 687 2260 **Fax:** 0208 687 2249
Email: ibs@charterhouseplc.co.uk

KIRK NATURAL STONE
Bridgend, Fyvie, Turriff,
Aberdeenshire, AB53 8QB
Area of Operation: Worldwide
Tel: 01651 891891 **Fax:** 01651 891794
Email: info@kirknaturalstone.com
Web: www.kirknaturalstone.com
Product Type: 1, 2, 3, 4

MANCHESTER BRICK & PRECAST LTD
Haigh Avenue, Whitehill Industrial Estate,
Stockport, Cheshire, SK4 1NU
Area of Operation: UK & Ireland
Tel: 0161 480 2621 **Fax:** 0161 480 0108
Email: sales@manbrick.co.uk
Web: www.manbrick.co.uk
Product Type: 3

MEADOWSTONE (DERBYSHIRE) LTD
West Way, Somercotes,
Derbyshire, DE55 4QJ
Area of Operation: UK & Ireland
Tel: 01773 540707 **Fax:** 01773 528119
Email: info@meadowstone.co.uk
Web: www.meadowstone.co.uk
Product Type: 1

**MINCHINHAMPTON ARCHITECTURAL
SALVAGE CO**
Cirencester Road, Aston Downs,
Chalford, Stroud, Gloucestershire, GL6 8PE
Area of Operation: Worldwide
Tel: 01285 760886 **Fax:** 01285 760838
Email: masco@catbrain.com
Web: www.catbrain.com
Product Type: 1, 2, 3

NATIONAL STONE PRODUCTS
Unit 25, M1 Commerce Park,
Markham Lane, Duckmanton,
Chesterfield, Derbyshire, S44 5HS
Area of Operation: UK (Excluding Ireland)
Tel: 01246 240881 **Fax:** 01246 240685
Email: bobhockley-nsp@btconnect.com
Product Type: 3

NICAN STONE LTD
Bank House, School Lane,
Bronington, Shropshire, SY13 3HN
Area of Operation: UK (Excluding Ireland)
Tel: 01948 780670 **Fax:** 01948 780679
Email: enquiries@nicanstone.com
Web: www.nicanstone.com

PADIPA LIMITED
Unit East 2, Sway Storage and Workshops,
Sway, Lymington, Hampshire, SO41 6DD
Area of Operation: UK & Ireland
Tel: 01590 681710
Fax: 01590 610336
Email: enquiries@padipa.co.uk
Web: www.padipa.co.uk

PENNINE STONE LTD
Askern Road, Carcroft, Doncaster,
South Yorkshire, DN6 8DH
Area of Operation: UK & Ireland
Tel: 01302 729277 **Fax:** 01302 729288
Email: neils@penninestone.co.uk
Web: www.penninestone.co.uk

PROCTER CONCRETE PRODUCTS
New Hold, Aberford Road, Garforth, Leeds,
West Yorkshire, LS25 2HG
Area of Operation: UK (Excluding Ireland)
Tel: 0113 286 2586 **Fax:** 0113 286 7376
Email: garry.horsfall@procterbedwas.co.uk
Web: www.caststoneuk.co.uk

ROCK UNIQUE LTD
c/o Select Garden and Pet Centre,
Main Road, Sundridge, Kent, TN14 6ED
Area of Operation: Europe
Tel: 01959 565 608 **Fax:** 01959 569 312
Email: stone@rock-unique.com
Web: www.rock-unique.com
Product Type: 2
Other Info: ▽ ✎ ☝
Material Type: E) 1, 2, 3, 4, 5, 8, 9, 10

SNOWDONIA SLATE & STONE
Glandwr Workshops, Glanypwll Road,
Blaenau Ffestiniog, Gwynedd, LL41 3PG
Area of Operation: UK (Excluding Ireland)
Tel: 01766 832525 **Fax:** 01766 832404
Email: richard@snowdoniaslate.co.uk
Web: www.snowdoniaslate.co.uk
Product Type: 3

SOUTHERN MASONRY LTD
37b Broadway East, West Wilts Trading Estate,
Westbury, Wiltshire, BA13 4AJ
Area of Operation: South East England,
South West England and South Wales
Tel: 07976 613421
Fax: 01749 345170
Email: charlie@stonefiresurrounds.co.uk
Product Type: 1, 2, 3, 4
Other Info: ▽ ✎ ☝

STANCLIFFE STONE
Grange Mill, Matlock,
Derbyshire, DE4 4BW
Area of Operation: Worldwide
Tel: 01629 653000
Fax: 01629 650996
Email: sales@stancliffe.com
Web: www.stancliffe.com
Product Type: 1, 3, 4

STONE ESSENTIALS COMPANY LIMITED
Mount Spring Works, off Burnley Road East,
Waterfoot, Lancashire, BB4 9LA
Area of Operation: UK & Ireland
Tel: 01706 210605 **Fax:** 01706 228707
Email: stoneessentials@btconnect.com
Web: www.stone-essentials.co.uk
Product Type: 1, 3

THE HOME OF STONE LTD
Boot Barn, Newcastle, Monmouth,
Monmouthshire, NP25 5NU
Area of Operation: South West England & South Wales
Tel: 01600 750462 **Fax:** 01600 750462
Material Type: E) 4, 5

THORVERTON STONE COMPANY LTD
Seychelles Farm, Upton Pyne, Exeter, Devon, EX5 5HY
Area of Operation: UK & Ireland
Tel: 01392 851 822 **Fax:** 01392 851833
Email: caststone@thorvertonstone.co.uk
Web: www.thorvertonstone.co.uk
Product Type: 3

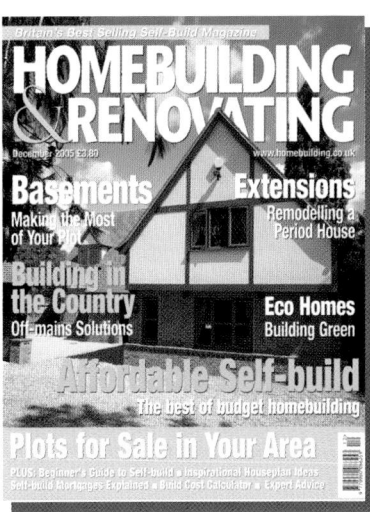

TRADSTOCKS LTD
Dunaverig, Thornhill, Stirling,
Stirlingshire, FK8 3QW
Area of Operation: Scotland
Tel: 01786 850400
Fax: 01786 850404
Email: peter@tradstocks.co.uk
Web: www.tradstocks.co.uk
Product Type: 1, 2, 3, 4

VOBSTER CAST STONE COMPANY LIMITED
Newbury Works, Coleford, Radstock,
Somerset, BA3 5RX
Area of Operation: UK (Excluding Ireland)
Tel: 01373 812514
Fax: 01373 813384
Email: barrie@caststonemasonry.co.uk
Web: www.caststonemasonry.co.uk

WARMSWORTH STONE LIMITED
1-3 Sheffield Road, Warmsworth,
Doncaster, South Yorkshire, DN4 9QH
Area of Operation: UK & Ireland
Tel: 01302 858617
Fax: 01302 855844
Email: info@warmsworth-stone.co.uk
Web: www.warmsworth-stone.co.uk
Product Type: 1, 2, 3, 4
Material Type: E) 1, 2, 4, 5, 8

WELLS CATHEDRAL STONEMASONS
Brunel Stoneworks, Station Road,
Cheddar, Somerset, BS27 3AH
Area of Operation: Worldwide
Tel: 01934 743544
Fax: 01934 744536
Email: wcs@stone-mason.co.uk
Web: www.stone-mason.co.uk
Product Type: 1, 3

BLOCKS

KEY

PRODUCT TYPES: 1= Solid   2 = Dense
3 = Lightweight   4 = Super-lightweight
5 = Fairfaced   6 = Insulant Filled
7 = Glazed   8 = Foundation Blocks

OTHER: ▽ Reclaimed   On-line shopping
Bespoke   Hand-made   ECO Ecological

BORDER CONCRETE PRODUCTS
Jedburgh Road, Kelso, Borders, TD5 8JG
Area of Operation: North East England,
North West England and North Wales, Scotland
Tel: 01573 224393
Fax: 01573 276360
Email: sales@borderconcrete.co.uk
Web: www.borderconcrete.co.uk
Product Type: 5

BRADSTONE STRUCTURAL
Aggregate Industries UK Ltd, North End,
Ashton Keynes, Swindon, Wiltshire, SN6 3QX
Area of Operation: UK (Excluding Ireland)
Tel: 01285 646884 **Fax:** 01285 646891
Email: bradstone.structural@aggregate.com
Web: www.bradstone.com
Product Type: 1

CAPITAL GROUP LIMITED
Victoria Mills, Highfield Road,
Little Hulton, Manchester, M38 9ST
Area of Operation: Worldwide
Tel: 0161 799 7555 **Fax:** 0161 799 7666
Email: leigh@choosecapital.co.uk
Web: www.choosecapital.co.uk
Product Type: 1, 2, 3, 5, 6, 8

CJ O'SHEA & CO LTD
Unit 1 Granard Business Centre,
Bunns Lane, Mill Hill, London, NW7 2DZ
Area of Operation: East England,
Greater London, South East England,
South West England & South Wales
Tel: 0208 959 3600 **Fax:** 0208 959 0184
Email: admin@oshea.co.uk
Web: www.oshea.co.uk

H+H CELCON LIMITED
Celcon House, Ightham, Sevenoaks, Kent, TN15 9HZ
Area of Operation: UK (Excluding Ireland)
Tel: 01732 880520 **Fax:** 01732 880531
Email: marketing@celcon.co.uk
Web: www.celcon.co.uk
Product Type: 3, 4, 8

HANSON BUILDING PRODUCTS
Stewartby, Bedford,
Bedfordshire, MK43 9LZ
Area of Operation: Worldwide
Tel: 08705 258258 **Fax:** 01234 762040
Email: info@hansonplc.com
Web: www.hanson.biz
Product Type: 1, 2, 3, 4, 5, 8

HISTON CONCRETE PRODUCTS LTD
Wisbech Road, Littleport, Cambridgeshire, CB6 1RA
Area of Operation: UK (Excluding Ireland)
Tel: 01353 861416 **Fax:** 01353 862165
Email: sales@histonconcrete.co.uk
Web: www.histonconcrete.co.uk

IBSTOCK BRICK LTD
Leicester Road, Ibstock,
Leicester, Leicestershire, LE67 6HS
Area of Operation: UK & Ireland
Tel: 01530 261999 **Fax:** 01530 263478
Email: leicester.sales@ibstock.co.uk
Web: www.ibstock.co.uk
Product Type: 1, 7

INNOVATIVE BUILDING SOLUTIONS
Mill Green House, 48-50 Mill Green Road,
Mitcham, Surrey, CR4 4HY
Area of Operation: UK & Ireland
Tel: 0208 687 2260 **Fax:** 0208 687 2249
Email: ibs@charterhouseplc.co.uk
Product Type: 3 **Material Type:** G) 1

MARSHALLS
Southowram, Halifax, West Yorkshire, HX3 9TE
Area of Operation: UK (Excluding Ireland)
Tel: 0870 120 7474
Email: marshalls@trilobyte.co.uk
Web: www.marshalls.co.uk

MASTERBLOCK
North End, Ashton Keynes, Wiltshire, SN6 3QX
Area of Operation: UK (Excluding Ireland)
Tel: 01285 646900 **Fax:** 01285 646949
Email: kevin.matthews@aggregate.com
Web: www.masterblock.co.uk
Product Type: 1, 2, 3, 5

**NATURAL BUILDING
TECHNOLOGIES**
The Hangar, Worminghall Road,
Oakley, Buckinghamshire, HP18 9UL
Area of Operation: UK & Ireland
Tel: 01844 338338 **Fax:** 01844 338525
Email: info@natural-building.co.uk
Web: www.natural-building.co.uk
Product Type: 1, 2, 3, 7

PLASMOR
P.O. Box 44, Womersley Road,
Knottingley, West Yorkshire, WF11 0DN
Area of Operation: UK (Excluding Ireland)
Tel: 01977 673221 **Fax:** 01977 607071
Email: knott@plasmor.co.uk
Web: www.plasmor.co.uk
Product Type: 2, 3, 4, 8

POLARLIGHT ACRYLIC BLOCK WINDOWS
W.I.N. Business Park, Canal Quay,
Newry, Co Down, BT35 6PH
Area of Operation: UK & Ireland
Tel: 02830 833888 **Fax:** 02830 833917
Email: info@polarlight.co.uk
Web: www.polarlight.co.uk
Product Type: 7

QUINN-LITE
Derrylin, Co Fermanagh, BT92 9AU
Area of Operation: Europe
Tel: 028 6774 8866 **Fax:** 028 6774 2309
Email: info@quinn-lite.com
Web: www.quinn-lite.com
Product Type: 3, 4

SOUTHERN MASONRY LTD
37b Broadway East, West Wilts Trading Estate,
Westbury, Wiltshire, BA13 4AJ
Area of Operation: South East England,
South West England and South Wales
Tel: 07976 613421 **Fax:** 01749 345170
Email: charlie@stonefiresurrounds.co.uk
Product Type: 2, 3, 4, 5

TARMAC LIMITED
Millfields Road, Ettingshall,
Wolverhampton, West Midlands, WV4 6JP
Area of Operation: Europe
Tel: 01902 353522 **Fax:** 01902 382922
Email: info@tarmac.co.uk
Web: www.tarmac.co.uk
Material Type: G) 1

TARMAC TOPBLOCK LTD
Millfields Road,
Ettingshall, Wolverhampton,
West Midlands, WV4 6JP
Area of Operation: UK (Excluding Ireland)
Tel: 01902 382844
Fax: 01902 382219
Email: wendy.hinett@tarmac.co.uk
Web: www.topblock.co.uk
Product Type: 2, 3, 4, 8

THE MATCHING BRICK COMPANY
Lockes Yard, Hartcliffe Way,
Bedminster, Bristol, BS3 5RJ
Area of Operation: UK (Excluding Ireland)
Tel: 0117 963 7000
Fax: 0117 966 4612
Email: matchingbrick@btconnect.com
Web: www.matchingbrick.co.uk
Product Type: 1

NOTES

Company Name ....
Address ....
email ....
Web ....

Company Name ....
Address ....
email ....
Web ....

Company Name ....
Address ....
email ....
Web ....

Company Name ....
Address ....
email ....
Web ....

Company Name ....
Address ....
email ....
Web ....

BUILDING STRUCTURE & MATERIALS

ROOFING

Image courtesy of Greys Artstone (01484 666400)

Roofing

When it comes to specifying a new house, roof design tends to get rather neglected by self-builders. They often have very definite ideas about the structure of their home or about the external cladding, but the role of the roof in all this is usually relegated to a mere detail to be sorted out, rather like designing the drain runs or the mains wiring. In truth there is often very little choice in the matter — once you have selected a style for your home, both the roof structure and the covering tend to be chosen to blend in with what you have specified, what your budget will allow, or what the planners will accept.

Other than flat roofing systems, which are rarely used outside garages, roof terraces and contemporary home structures, there are two major competing techniques for building roofs. Traditionally, roof timbers were measured, cut and assembled on site — a skilled job involving complex setting-out procedures and cutting lots of obscure angles and notches. However, the rise of the prefabricated roof truss is slowly but surely putting an end to all this. With trussed roofs, the brainwork is done by computer and the cutting by machine. On a simple rectangular box-shaped structure, roof trusses are about three times quicker to erect than traditional roofs and, because of the inherent strength of each individual truss, they use considerably less timber — usually about 30% less by volume.

Whereas traditional cut roofs are built from sawn carcassing, readily purchased from any builder's merchant, trussed roofing tends to get fabricated by specialists. This shouldn't present a problem to self-builders: provided you can present a set of dimensioned plans, you will get a quote back usually within a few days. You can contact specialists or you can take your plans to any builder's merchant who will do the donkey work for you.

ABOVE LEFT: Crested ridges from Eternit.
ABOVE: Handmade clay roof tiles by Midlands Slate & Tile.
RIGHT: Dragon finial by RoofDragon.com

If you choose to build using trusses, bear in mind that prefabricated roof trusses are sensitive things and they perform well only if they are treated well. Care should be taken not to put any twist or undue load on to them, both whilst being handled and when being stored before erection. They should be stored upright on bearers (not standing on their feet). They should never be altered on site. They also cannot be cut around chimneys and openings, so you must get the plan accurately built.

So why doesn't everybody use roof trusses? Mostly they do, but there are some situations where the traditional cut roof holds sway. Roof truss manufacturers sometimes get very busy and cannot deliver for several weeks. Also, complicated roof shapes take longer to build whichever system you use and the difference in erection speeds — which is the trusses' big selling point — is much less marked. Consequently, many builders specify trusses for their main roofs, but prefer to stick with the traditional methods when it comes to odd jobs like building dormer windows, porches or garages.

There is another reason why many builders dislike the trussed roof; it effectively eliminates use of the loft space for anything other than storage. The cross members which make up each truss cannot be removed or altered in any way and this prevents the roof space being opened up at a later date. There is, however, the possibility of using specialised attic trusses which are designed to leave the main loft space open so that any future loft conversion can be arranged with a minimum of fuss and expense. However, whilst the speed of installation is maintained, attic trusses are between two and three times the price of regular ones and this means that they effectively lose their cost advantage over traditionally cut roofs.

TILES, SHINGLES AND SLATES

KEY

PRODUCT TYPES: 1= Profiled Tiles including Pantiles 2 = Plain Tiles including Pegs
3 = Crease Tiles 4 = Interlocking Tiles
5 = Slates 6 = Shingles

SEE ALSO: BRICKS, BLOCKS, STONE AND CLADDING - Shakes and Shingles

OTHER: ▽ Reclaimed ⌐ On-line shopping
✎ Bespoke ✋ Hand-made ECO Ecological

ALDERSHAW HANDMADE CLAY TILES LTD
Pokehold Wood, Kent Street, Sedlescombe,
Nr Battle, East Sussex, TN33 0SD
Area of Operation: Europe
Tel: 01424 756777 **Fax:** 01424 756888
Email: tiles@aldershaw.co.uk
Web: www.aldershaw.co.uk **Product Type:** 2, 3, 5

ALL THINGS STONE LIMITED
Anchor House, 33 Ospringe Street, Ospringe,
Faversham, Kent, ME13 8TW
Area of Operation: UK & Ireland
Tel: 01795 531001 **Fax:** 01795 530742
Email: info@allthingsstone.co.uk
Web: www.allthingsstone.co.uk
Product Type: 5

ASPECT ROOFING CO LTD
The Old Mill, East Harling, Norwich,
Norfolk, NR16 2QW
Area of Operation: UK (Excluding Ireland)
Tel: 01953 717777 **Fax:** 01953 717164
Email: info@raretiles.co.uk
Web: www.raretiles.co.uk
Product Type: 1, 2, 3, 4, 5, 6

BANBURY BRICKS LTD
83a Yorke Street, Mansfield Woodhouse,
Nottinghamshire, NG19 9NH
Area of Operation: UK & Ireland
Tel: 0845 230 0941 **Fax:** 0845 230 0942
Email: enquires@banburybricks.co.uk
Web: www.banburybricks.co.uk

BICESTER ROOFING CO. LTD
Manor Farm, Weston on the Green,
Nr. Bicester, Oxfordshire, OX25 3QL
Area of Operation: South East England
Tel: 0870 264 6454
Fax: 0870 264 6455
Email: sales@bicesterroofing.co.uk
Web: www.bicesterroofing.co.uk

BLACK MOUNTAIN QUARRIES LTD
Howton Court, Pontrilas, Herefordshire, HR2 0BG
Area of Operation: UK & Ireland
Tel: 01981 241541
Email: info@blackmountainquarries.com
Web: www.blackmountainquarries.com
Product Type: 5, 6 **Other Info:** ECO
Material Type: E) 3, 4, 5

BRADSTONE STRUCTURAL
Aggregate Industries UK Ltd, North End, Ashton Keynes, Swindon, Wiltshire, SN6 3QX
Area of Operation: UK (Excluding Ireland)
Tel: 01285 646884
Fax: 01285 646891
Email: bradstone.structural@aggregate.com
Web: www.bradstone.com
Product Type: 5 **Material Type:** E) 13

BRETT LANDSCAPING LTD .
Salt Lane, Cliffe, Rochester, Kent, ME3 7SZ
Area of Operation: Worldwide
Tel: 01634 222188
Fax: 01634 222001
Email: landscapinginfo@brett.co.uk
Web: www.brett.co.uk/landscaping
Product Type: 2

CANADA WOOD UK
PO Box 1, Farnborough, Hampshire, GU14 6WE
Area of Operation: UK (Excluding Ireland)
Tel: 01252 522545 **Fax:** 01252 522546
Email: office@canadawooduk.org
Web: www.canadawood.info
Product Type: 6 **Material Type:** A) 12

CEMBRIT BLUNN
6 Coleshill Street, Fazeley, Tamworth,
Staffordshire, B78 3XJ
Area of Operation: UK & Ireland
Tel: 01827 288827 **Fax:** 01827 288176
Email: parcher@cembritblunn.co.uk
Web: www.cembritblunn.co.uk
Product Type: 5

CEMEX ROOF TILES
Nicolson Way, Wellington Road, Burton On Trent,
Staffordshire, DE14 2AW
Area of Operation: UK & Ireland
Tel: 01283 517070 **Fax:** 01283 516290
Email: enquiries.russellrooftiles@cemex.co.uk
Web: www.russell-rooftiles.co.uk
Product Type: 2

CHESHIRE BRICK & SLATE
Brook House Farm, Salters Bridge, Tarvin Sands,
Tarvin, Cheshire, CH3 8NR
Area of Operation: UK (Excluding Ireland)
Tel: 01829 740883 **Fax:** 01829 740481
Email: enquiries@cheshirebrickandslate.co.uk
Web: www.cheshirebrickandslate.co.uk
Product Type: 1, 4, 5

CHINA SLATE LTD
Wingfield View, Coney Green, Clay Cross,
Chesterfield, Derbyshire, S45 1ZZ
Area of Operation: UK & Ireland
Tel: 01246 865222
Fax: 01246 866622
Email: sales@chinaslate.co.uk
Web: www.chinaslate.co.uk
Product Type: 5

CREST BRICK, SLATE & TILE LTD
Howdenshire Way, Knedlington Road, Howden,
East Riding of Yorks, DN14 7HZ
Area of Operation: UK (Excluding Ireland)
Tel: 0870 241 1398 **Fax:** 01430 433000
Email: info@crest-bst.co.uk
Web: www.crest-bst.co.uk
Product Type: 1, 2, 4, 5

DELABOLE SLATE COMPANY
Pengelly, Delabole, Cornwall, PL33 9AZ
Area of Operation: UK (Excluding Ireland)
Tel: 01840 212242 **Fax:** 01840 212948
Email: sales@delaboleslate.co.uk
Web: www.delaboleslate.co.uk
Product Type: 5

DENBY DALE CAST PRODUCTS LTD
230 Cumberworth Lane, Denby Dale,
Huddersfield, West Yorkshire, HD8 8PR
Area of Operation: UK & Ireland
Tel: 01484 863560 **Fax:** 01484 865597
Email: sales@denbydalecastproducts.co.uk
Web: www.denbydalecastproducts.co.uk
Product Type: 5

DREADNOUGHT WORKS
Dreadnought Road, Pensnett, Brierley Hill,
West Midlands, DY5 4TH
Area of Operation: Europe
Tel: 01384 77405 **Fax:** 01384 74553
Email: office@dreadnought.co.uk
Web: www.dreadnought-tiles.co.uk
Product Type: 2, 3 **Material Type:** F) 1, 3

ETERNIT BUILDING MATERIALS
Whaddon Road, Meldreth, Royston,
Hertfordshire, SG8 5RL
Area of Operation: UK (Excluding Ireland)
Tel: 01763 264686 **Fax:** 01763 262531
Email: marketing@eternit.co.uk
Web: www.eternit.co.uk
Product Type: 1, 2, 5
Other Info: ✋ **Material Type:** E) 3

FORTICRETE ROOFING PRODUCTS
Boss Avenue, Leighton Buzzard,
Bedfordshire, LU7 4SD
Area of Operation: UK & Ireland
Tel: 01525 244900
Fax: 01525 850432
Email: roofing@forticrete.com
Web: www.forticrete.co.uk
Product Type: 1, 4, 5

GREYS ARTSTONE LTD
Burdwell Works, New Mill Road, Brockholes,
Holmfirth, West Yorkshire, HD9 7AZ
Area of Operation: UK (Excluding Ireland)
Tel: 01484 666400
Fax: 01484 662709
Email: info@greysartstone.co.uk
Web: www.greysartstone.co.uk
Product Type: 5
Material Type: E) 12

GREYSLATE & STONE SUPPLIES
Y Maes, The Square, Blaenau Ffestiniog,
Gwynedd, LL41 3UN
Area of Operation: UK & Ireland
Tel: 01766 830521   **Fax:** 01766 831706
Email: greyslate@slateandstone.net
Web: www.slateandstone.net
Product Type: 5   **Material Type:** E) 3

IMERYS ROOFTILES
1 Rue Des Vergers, BP22, Parc d'Activities,
De Limonest, France 69579
Area of Operation: UK (Excluding Ireland)
Tel: 0161 928 4572   **Fax:** 0161 929 8513
Email: enquiries.rooftiles@imerys.com
Web: www.imerys-rooftiles.com
Product Type: 1, 2

INTERLOC BUILDING SOLUTIONS
Forsyth House, Innova Way, Rosyth Europarc,
Rosyth, Fife, KY11 2UU
Area of Operation: Europe
Tel: 01383 428032   **Fax:** 01383 428001
Email: info@interlocbuild.co.uk
Web: www.interlocbuild.co.uk
Product Type: 4, 6   **Material Type:** C) 2, 7, 9

JOHN BRASH LTD
The Old Shipyard, Gainsborough,
Lincolnshire, DN21 1NG
Area of Operation: UK & Ireland
Tel: 01427 613858
Fax: 01427 810218
Email: info@johnbrash.co.uk
Web: www.johnbrash.co.uk
Product Type: 6   **Material Type:** A) 2, 8, 12

KEYMER TILES LTD
Nye Road, Burgess Hill, West Sussex, RH15 0LZ
Area of Operation: East England, Greater London,
Midlands & Mid Wales, South East England, South
West England and South Wales
Tel: 01444 232931   **Fax:** 01444 871852
Email: info@keymer.co.uk
Web: www.keymer.co.uk
Product Type: 2   **Other Info:** ☞ ♨
Material Type: F) 3

MARLEY ETERNIT
Station Road, Coleshill, Birmingham,
West Midlands, B46 1HP
Area of Operation: UK (Excluding Ireland)
Tel: 0870 562 6500
Fax: 0870 562 6550
Email: roofingsales@marleyeternit.co.uk
Web: www.marleyeternit.co.uk
Product Type: 2, 5
Material Type: E) 3

MATTHEW HEBDEN
123 Lonsdale Drive, Enfield,
Greater London, EN2 7LS
Area of Operation: UK (Excluding Ireland)
Tel: 0208 367 6463
Fax: 0208 367 0166
Email: sales@matthewhebden.co.uk
Web: www.matthewhebden.co.uk
Product Type: 6

METROTILE UK LTD
Unit 3 Sheldon Business Park, Sheldon Corner,
Chippenham, Wiltshire, SN14 0RQ
Area of Operation: UK & Ireland
Tel: 01249 658514
Fax: 01249 658453
Email: sales@metrotile.co.uk
Web: www.metrotile.co.uk
Product Type: 1, 2, 6

MIDLANDS SLATE & TILE
Qualcast Road, Horseley Fields,
Wolverhampton, West Midlands, WV1 2QP
Area of Operation: UK & Ireland
Tel: 01902 458780
Fax: 01902 458549
Email: sales@slate-tile-brick.co.uk
Web: www.slate-tile-brick.co.uk
Product Type: 1, 2, 3, 4, 5
Material Type: E) 3, 4

**MINCHINHAMPTON
ARCHITECTURAL SALVAGE CO**
Cirencester Road, Aston Downs, Chalford,
Stroud, Gloucestershire, GL6 8PE
Area of Operation: Worldwide
Tel: 01285 760886
Fax: 01285 760838
Email: masco@catbrain.com
Web: www.catbrain.com
Product Type: 1, 5

**NUMBER 9 STUDIO UK
ARCHITECTURAL CERAMICS**
Mole Cottage Industries, Mole Cottage,
Watertown, Chittlehamholt, Devon, EX37 9HF
Area of Operation: Worldwide
Tel: 01769 540471
Fax: 01769 540471
Email: arch.ceramics@moley.uk.com
Web: www.moley.uk.com

OCTAVEWARD LTD
Balle Street Mill, Balle Street,
Darwen, Lancashire, BB3 2AZ
Area of Operation: Worldwide
Tel: 01254 773300
Fax: 01254 773950
Email: info@octaveward.com
Web: www.octaveward.com

ONDULINE BUILDING PRODUCTS LTD
Eardley House, 182-184 Campden Hill Road,
Kensington, London, W8 7AS
Area of Operation: UK & Ireland
Tel: 0207 727 0533
Fax: 0207 792 1390
Email: ondulineuk@aol.com
Web: www.onduline.net

**RED BANK MANUFACTURING
COMPANY LIMITED**
Measham, Swadlincote, Derbyshire, DE12 7EL
Area of Operation: UK & Ireland
Tel: 01530 270333   **Fax:** 01530 273667
Email: sales@redbankmfg.co.uk
Web: www.redbankmfg.co.uk
Product Type: 1, 5
Other Info: ECO
Material Type: E) 14

ROOFSURE
Metro House, 14-17 Metropolitan Business Park,
Preston New Road, Blackpool, Lancashire, FY3 9LT
Area of Operation: UK (Excluding Ireland)
Tel: 0800 597 2828
Fax: 01253 798193 / 764671
Email: roofsureltd@btconnect.com
Web: www.roofsure.co.uk

RUBEROID BUILDING PRODUCTS
Appley Lane North, Appley Bridge,
Lancashire, WN6 9AB
Area of Operation: UK (Excluding Ireland)
Tel: 08000 285573   **Fax:** 01257 252514
Email: marketing@ruberoid.co.uk
Web: www.ruberoid.co.uk
Product Type: 2
Material Type: M) 2

SANDTOFT ROOF TILES LTD
Belton Road, Sandtoft, Doncaster,
South Yorkshire, DN8 5SY
Area of Operation: UK & Ireland
Tel: 01427 871200   **Fax:** 01427 871222
Email: elaine.liversadge@sandtoft.co.uk
Web: www.sandtoft.com
Product Type: 1, 2, 3, 4, 5
Material Type: E) 3, 12, 14

SLATE WORLD LTD
Westmoreland Road, Kingsbury,
Greater London, NW9 9RN
Area of Operation: Europe
Tel: 0208 204 3444   **Fax:** 0208 204 3311
Email: advertising@americanslate.com
Web: www.slateworld.com
Product Type: 5
Material Type: E) 3

SLATE WORLD LTD

Area of Operation: Europe
Tel: 0208 2043444
Fax: 0208 204 3311
Email: advertising@americanslate.com
Web: www.slateworld.com
Material Type: E) 3

Suppliers of the most comprehensive range of
QUARRY-DIRECT, non-slip and low maintenance
natural slate flooring tiles in the UK.

SNOWDONIA SLATE & STONE
Glandwr Workshops, Glanypwll Road,
Blaenau Ffestiniog, Gwynedd, LL41 3PG
Area of Operation: UK (Excluding Ireland)
Tel: 01766 832525   **Fax:** 01766 832404
Email: richard@snowdoniaslate.co.uk
Web: www.snowdoniaslate.co.uk
Product Type: 5

SOUTH WEST RECLAMATION LTD
Wireworks Estate, Bristol Road,
Bridgwater, Somerset, TA6 4AP
Area of Operation: South West England
and South Wales
Tel: 01278 444141   **Fax:** 01278 444114
Email: info@southwest-rec.co.uk
Web: www.southwest-rec.co.uk
Product Type: 1, 2, 3, 4, 5
Other Info: ▽
Material Type: E) 3

SSQ
301 Elveden Road, Park Royal,
Greater London, NW10 7SS
Area of Operation: Worldwide
Tel: 0208 961 7725
Fax: 0208 965 7013
Email: alain@ssq.co.uk
Web: www.ssq.co.uk
Product Type: 5
Other Info: ECO ☞ ♨

STONE AND SLATE
Coney Green Farm, Lower Market Street,
Claycross, Chesterfield, Derbyshire, S45 9NE
Area of Operation: UK & Ireland
Tel: 01246 250088   **Fax:** 01246 250099
Email: sales@stoneandslate.ltd.uk
Web: www.stoneandslate.co.uk
Product Type: 5

SWALLOWS TILES (CRANLEIGH) LTD
Bookhurst Hill, Cranleigh, Guildford, Surrey, GU6 7DP
Area of Operation: UK (Excluding Ireland)
Tel: 01483 274100
Fax: 01483 267593
Email: info@swallowsrooftiles.co.uk
Web: www.swallowsrooftiles.co.uk

TAYLOR MAXWELL & COMPANY LIMITED
4 John Oliver Buildings, 53 Wood Street,
Barnet, Hertfordshire, EN5 4BS
Area of Operation: UK (Excluding Ireland)
Tel: 0208 440 0551
Fax: 0208 440 0552
Email: simonepalmer@taylor.maxwell.co.uk
Web: www.taylor.maxwell.co.uk
Product Type: 1, 2, 3, 4

TERCA/KORAMIC
Adamson House, Pomona Strand,
Trafford, Manchester, MI6 0BA
Tel: 0161 873 7701
Fax: 0161 876 0736
Email: bricksales@terca.com
Web: www.terra.co.uk
Product Type: 1, 2, 5

THE CLAY ROOF TILE CO. LTD
Tarporley Road, Lower Whitley,
Warrington, Cheshire, WA4 4EZ
Area of Operation: North West England
and North Wales, Scotland
Tel: 01928 796100
Fax: 01928 796101
Email: info@clayrooftile.co.uk
Web: www.clayrooftile.co.uk
Product Type: 1, 2, 3, 4, 5

**THE NATURAL SLATE
COMPANY LTD - DERBYSHIRE**
Unit 2 Armytage Estate, Station Road,
Whittington Moor, Chesterfield,
Derbyshire, S41 9ET
Area of Operation: Worldwide
Tel: 0845 177 5008   **Fax:** 0870 429 9891
Email: sales@theslatecompany.net
Web: www.theslatecompany.net
Product Type: 5
Other Info: ECO ♨
Material Type: E) 3

**THE NATURAL SLATE
COMPANY LTD - LONDON**
161 Ballards Lane, Finchley, London, N3 1LJ
Area of Operation: Worldwide
Tel: 0845 177 5008   **Fax:** 0870 429 9891
Email: sales@theslatecompany.net
Web: www.theslatecompany.net
Product Type: 5
Other Info: ECO ♨
Material Type: E) 3

TRITON CHEMICAL MANUFACTURING LTD
129 Felixstowe Road, Abbey Wood,
London, SE2 9SG
Area of Operation: Worldwide
Tel: 0208 310 3929   **Fax:** 0208 312 0349
Email: neil@triton-chemicals.com
Web: www.triton-chemicals.com
Product Type: 6

TUDOR ROOF TILE CO LTD
Denge Marsh Road, Lydd, Kent, TN29 9JH
Area of Operation: Worldwide
Tel: 01797 320202   **Fax:** 01797 320700
Email: info@tudorrooftiles.co.uk
Web: www.tudorrooftiles.co.uk
Product Type: 2
Other Info: ♨
Material Type: F) 1

WELSH SLATE
Business Design Centre, Unit 205,
52 Upper Street, London, N1 0QH
Area of Operation: Worldwide
Tel: 0207 354 0306   **Fax:** 0207 354 8485
Email: enquiries@welshslate.com
Web: www.welshslate.com
Product Type: 5   **Material Type:** E) 3

CLAY ROOF TILES
Classical...beautiful...timeless...

Beavoise interlocking clay plain tile

...and affordable

HP10   Monopole No. 1   Phalempin   Beauvoise

With almost 200 years of experience and expertise in serving the needs of the self builder we can offer you an extensive and imaginative range of clay roofing tiles to enable you to create your own individual roof of character and distinction – without costing the earth!

Starting with five ranges of traditional clay plain tiles, the unbeatable Imerys Roof Tiles portfolio includes ever popular classical Pantiles, eye-catching single Romans and an array of beautiful, profiled and flat clay interlocking tiles and clay slates.

These affordable and innovative clay roofing products are available in a variety of modular sizes and stunning colours, with certain products suitable for roof pitches as low as 17.5°.

Our quality is second to none and all products are covered by the Imerys Roof Tiles comprehensive 30 year guarantee. All tiles exceed the requirements of BS EN 1304 and many carry certification by the British Board of Agrément. Imerys Tileries also operate quality management systems in compliance with BS EN 9000 (1994).

For product information visit our website www.imerys-rooftiles.com

For your personal copy of the Imerys Roof Tiles Product Selector and details of your local stockist please email us at: enquiries.rooftiles@imerys.com or, if you prefer, please telephone us on

0161 928 4572

...the natural choice

IMERYS
Roof tiles

AFAQ
ISO 9000
VERSION 1994

NBS Plus

NFRC
THE NATIONAL FEDERATION OF
ROOFING CONTRACTORS LIMITED
COLLECTIVE MARK
ASSOCIATE MEMBER

BBA

NF

BS EN 1304
(1998)

WIENERBERGER LTD
Wienerberger House, Brooks Drive,
Cheadle Royal Business Park,
Cheadle, Cheshire, SK8 3SA
Area of Operation: UK & Ireland
Tel: 0161 491 8200
Fax: 0161 488 4827
Email: nicky.webb@wienerberger.com
Web: www.wienerberger.co.uk

WILLIAM BLYTH
Monckton Manor,
Chevet Lane, Notton, Wakefield,
West Yorkshire, DN18 5RD
Area of Operation: UK (Excluding Ireland)
Tel: 01652 632175
Fax: 01227 700350

METAL ROOFING

> ### KEY
>
> **SEE ALSO:** ROOFING - Flat Roofing
>
> **OTHER:** ▽ Reclaimed ⌐ On-line shopping
> ✎ Bespoke ✋ Hand-made ECO Ecological

**ALUMASC EXTERIOR
BUILDING PRODUCTS**
White House Works, Bold Road,
Sutton, St Helens, Merseyside, WA9 4JG
Area of Operation: UK & Ireland
Tel: 01744 648400
Fax: 01744 648401
Email: info@alumasc-exteriors.co.uk
Web: www.alumasc-exteriors.co.uk

B H ROOFING
Oliver Street, Kingsley, Northampton,
Northamptonshire, NN27JJ
Area of Operation: UK (Excluding Ireland)
Tel: 01604 710645
Fax: 01604 710645

INTERLOC BUILDING SOLUTIONS
Forsyth House, Innova Way,
Rosyth Europarc, Rosyth,
Fife, KY11 2UU
Area of Operation: Europe
Tel: 01383 428032
Fax: 01383 428001
Email: info@interlocbuild.co.uk
Web: www.interlocbuild.co.uk

LINDAB LTD
Unit 7 Block 2,
Shenstone Trading Estate,
Bromsgrove Road, Halesowen,
West Midlands, B63 3XB
Area of Operation: Worldwide
Tel: 0121 585 2780
Fax: 0121 585 2782
Email: building.products@lindab.co.uk
Web: www.lindab.co.uk
Material Type: C) 4

LONSDALE METAL COMPANY LTD
Unit 40, Millmead Industrial Centre,
Millmead Road, London, N17 9QU
Area of Operation: UK (Excluding Ireland)
Tel: 0208 801 4221
Fax: 0208 801 1287
Email: info@lonsdalemetal.co.uk
Web: www.roofglazing.co.uk

METROTILE UK LTD
Unit 3 Sheldon Business Park,
Sheldon Corner, Chippenham,
Wiltshire, SN14 0RQ
Area of Operation: UK & Ireland
Tel: 01249 658514
Fax: 01249 658453
Email: sales@metrotile.co.uk
Web: www.metrotile.co.uk
Material Type: C) 2

ONDULINE BUILDING PRODUCTS LTD
Eardley House, 182-184 Campden Hill Road,
Kensington, London, W8 7AS
Area of Operation: UK & Ireland
Tel: 0207 727 0533
Fax: 0207 792 1390
Email: ondulineuk@aol.com
Web: www.onduline.net

BEAMS AND TRUSSES

> ### KEY
>
> **SEE ALSO:** MERCHANTS - Timber Merchants
>
> **OTHER:** ▽ Reclaimed ⌐ On-line shopping
> ✎ Bespoke ✋ Hand-made ECO Ecological

AC ROOF TRUSSES LTD
Severn Farm Industrial Estate,
Welshpool, Powys, SY21 7DF
Area of Operation: UK & Ireland
Tel: 01938 554881
Fax: 01938 556265
Email: info@acrooftrusses.co.uk
Web: www.acrooftrusses.co.uk ⌐
Other Info: ✎

**ALTHAM HARDWOOD
CENTRE LIMITED**
Altham Corn Mill, Burnley Road,
Altham, Accrington,
Lancashire, BB5 5UP
Area of Operation: UK & Ireland
Tel: 01282 771618
Fax: 01282 777932
Email: info@oak-beams.co.uk
Web: www.oak-beams.co.uk

ALTHAM HARDWOOD CENTRE LTD

Area of Operation: UK & Ireland
Tel: 01282 771618
Fax: 01282 777932
Email: info@oak-beams.co.uk
Web: www.oak-beams.co.uk

We supply both hand-cut and bandsawn beams and trusses made from beam quality green oak. Bespoke gazebos, bridges and imaginative structures also available.

ASPECT ROOFING CO LTD
The Old Mill, East Harling,
Norwich, Norfolk, NR16 2QW
Area of Operation: UK (Excluding Ireland)
Tel: 01953 717777
Fax: 01953 717164
Email: info@raretiles.co.uk
Web: www.raretiles.co.uk

BALCAS TIMBER LTD
Laragh, Enniskillen,
Co Fermanagh, BT94 2FQ
Area of Operation: UK & Ireland
Tel: 0286 632 3003
Fax: 0286 632 7924
Email: info@balcas.com
Web: www.balcas.com

COLIN BAKER
Timberyard, Crownhill, Halberton,
Tiverton, Devon, EX16 7AY
Area of Operation: Europe
Tel: 01884 820152
Fax: 01884 820152
Email: colinbaker@colinbakeroak.co.uk
Web: www.colinbakeroak.co.uk
Material Type: A) 2

FFOREST TIMBER ENGINEERING LTD
Kestrel Way, Garngoch Industrial Estate,
Gorseinon, Swansea, SA4 9WF
Area of Operation: Midlands & Mid Wales,
South West England and South Wales
Tel: 01792 895620
Fax: 01792 893969
Email: info@fforest.co.uk
Web: www.fforest.co.uk

FLIGHT TIMBER PRODUCTS
Earls Colne Business Park,
Earls Colne, Essex, CO6 2NS
Area of Operation: East England,
Greater London, South East England
Tel: 01787 222336
Fax: 01787 222359
Email: sales@flighttimber.com
Web: www.flighttimber.com

GREEN OAK STRUCTURES
20 Bushy Coombe Gardens,
Glastonbury, Somerset, BA6 8JT
Area of Operation: UK & Ireland
Tel: 01458 833420
Fax: 01458 833420
Email: timberframes@greenoakstructures.co.uk
Web: www.greenoakstructures.co.uk
Other Info: ECO ✎
Material Type: A) 2

HENRY VENABLES TIMBER LTD
Tollgate Drive, Tollgate Industrial Estate,
Stafford, Staffordshire, ST16 3HS
Area of Operation: UK (Excluding Ireland)
Tel: 01785 270600
Fax: 01785 270626
Email: enquiries@henryvenables.co.uk
Web: www.henryvenables.co.uk

HONEYSUCKLE BOTTOM SAWMILL LTD
Honeysuckle Bottom,
Green Dene, East Horsley,
Leatherhead, Surrey, KT24 5TD
Area of Operation: Greater London,
South East England
Tel: 01483 282394
Fax: 01483 282394
Email: honeysucklemill@aol.com
Web: www.easisites.co.uk/honeysucklebottomsawmill

LAMINATED WOOD LIMITED
Grain Silo Complex, Abbey Road, Hempsted,
Gloucester, Gloucestershire, GL2 5HU
Area of Operation: UK (Excluding Ireland)
Tel: 01452 418000
Fax: 01452 418333
Email: mail@lamwood.co.uk
Web: www.lamwood.co.uk
Material Type: B) 1, 7, 10

LAWSONS
Gorst Lane, Off New Lane, Burscough,
Ormskirk, Lancashire, L40 0RS
Area of Operation: Worldwide
Tel: 01704 893998
Fax: 01704 892526
Email: info@traditionaltimber.co.uk
Web: www.traditionaltimber.co.uk

MILBANK ROOFS
Hargrave Meadow Cottage,
Church Lane, Hargrave, Bury St Edmunds,
Suffolk, IP29 5HH
Area of Operation: UK & Ireland
Tel: 01284 852505
Fax: 01284 850021
Email: info@milbankroofs.com
Web: www.milbankroofs.com

MITEK INDUSTRIES LTD
Mitek House, Grazebrook Industrial Park,
Peartree Lane, Dudley,
West Midlands, DY2 0XW
Area of Operation: UK & Ireland
Tel: 01384 451400
Fax: 01384 451415
Email: roy.troman@mitek.co.uk
Web: www.mitek.co.uk

PERIOD BEAMS
14 Gramby Crescent, Bennethorpe,
Doncaster, South Yorkshire, DN2 6AN
Area of Operation: UK (Excluding Ireland)
Tel: 01302 320638
Fax: 01302 320638

TAYLOR LANE TIMBER FRAME LTD
Chapel Road, Rotherwas Industrial Estate,
Hereford, Herefordshire, HR2 6LD
Area of Operation: UK & Ireland
Tel: 01432 271912
Fax: 01432 351064
Email: info@taylor-lane.co.uk
Web: www.taylor-lane.co.uk

THE GREEN OAK CARPENTRY CO LTD
Langley Farm, Langley, Rake,
Liss, Hampshire, GU33 7JW
Area of Operation: UK & Ireland
Tel: 01730 892049
Fax: 01730 895225
Email: enquiries@greenoakcarpentry.co.uk
Web: www.greenoakcarpentry.co.uk
Other Info: ✎ ✋
Material Type: A) 2

TRADITIONAL OAK & TIMBER COMPANY
P O Stores, Haywards Heath Road,
North Chailey, Nr Lewes, East Sussex, BN8 4EY
Area of Operation: Worldwide
Tel: 01825 723648
Fax: 01825 722215
Email: info@tradoak.co.uk
Web: www.tradoak.com

SARKING

> ### KEY
>
> **OTHER:** ▽ Reclaimed ⌐ On-line shopping
> ✎ Bespoke ✋ Hand-made ECO Ecological

HUNTON FIBER UK LTD
Rockleigh Court, Rock Road,
Finedon, Northamptonshire, NN9 5EL
Area of Operation: UK & Ireland
Tel: 01933 682683
Fax: 01933 680296
Email: admin@huntonfiber.co.uk
Web: www.hunton.no
Other Info: ▽ ECO
Material Type: K) 8

SMARTPLY EUROPE
Hawley Manor, Hawley Road,
Dartford, Kent, DA1 1PX
Area of Operation: Europe
Tel: 01322 424900
Fax: 01322 424920
Email: info@smartply.com
Web: www.smartply.com
Other Info: ECO
Material Type: B) 2, 9

XTRATHERM UK LIMITED
Unit 5 Jensen Court,
Astmoor Industrial Estate,
Runcorn, Cheshire, WA7 1SQ
Area of Operation: UK & Ireland
Tel: 0871 222 1033
Fax: 0871 222 1044
Email: kerry@xtratherm.com
Web: www.xtratherm.com

BUILDING STRUCTURE & MATERIALS

FLAT ROOFING SYSTEMS

KEY

SEE ALSO: ROOFING - Metal Roofing, ROOFING - Green Roofing

OTHER: ▽ Reclaimed ⌐🖱 On-line shopping
✎ Bespoke ✋ Hand-made ECO Ecological

ALUMASC EXTERIOR BUILDING PRODUCTS
White House Works, Bold Road,
Sutton, St Helens, Merseyside, WA9 4JG
Area of Operation: UK & Ireland
Tel: 01744 648400
Fax: 01744 648401
Email: info@alumasc-exteriors.co.uk
Web: www.alumasc-exteriors.co.uk

BRETT MARTIN DAYLIGHT SYSTEMS
Sandford Close, Alderford's Green Industrial Estate,
Coventry, West Midlands, CV2 2QU
Area of Operation: Worldwide
Tel: 02476 602022
Fax: 02476 602744
Email: toridawson@brettmartin.com
Web: www.daylightsystems.com
Other Info: ✎
Material Type: D) 1, 3, 5, 6

BUILD IT GREEN (UK) LTD
Arena Business Centre, 9 Nimrod Way,
Ferndown, Dorset, BH21 7SH
Area of Operation: UK (Excluding Ireland)
Tel: 01202 862320
Fax: 01202 877685
Email: enquiries@buildit-green.co.uk
Web: www.buildit-green.co.uk

DIY ROOFING LTD
Hillcrest House, Featherbed Lane,
Hunt End, Redditch,
Worcestershire, B97 5QL
Area of Operation: UK & Ireland
Tel: 0800 783 4890
Fax: 01527 403483
Email: chris@diyroofing.co.uk
Web: www.diyroofing.co.uk

GLAZING VISION
6 Barns Close, Brandon, Suffolk, IP27 0NY
Area of Operation: Worldwide
Tel: 01842 815581
Fax: 01842 815515
Email: sales@visiongroup.co.uk
Web: www.visiongroup.co.uk

INTERLOC BUILDING SOLUTIONS
Forsyth House, Innova Way, Rosyth Europarc,
Rosyth, Fife, KY11 2UU
Area of Operation: Europe
Tel: 01383 428032 **Fax:** 01383 428001
Email: info@interlocbuild.co.uk
Web: www.interlocbuild.co.uk
Other Info: ECO
Material Type: K) 5, 11, 12

MIDLAND BUTYL LTD
Windmill Farm, Biggin Lane,
Ashbourn, Derbyshire, DE6 3FN
Area of Operation: Worldwide
Tel: 01335 372133 **Fax:** 01335 372199
Email: sales@midland-butyl.co.uk
Web: www.midland-butyl.co.uk

PHOENIX (GB) LTD
Chestnut Field House, Chestnut Field,
Rugby, Warwickshire, CU21 2PA
Area of Operation: UK & Ireland
Tel: 01788 571482 **Fax:** 01788 542245
Email: tonyb@phoenix-gb.com
Web: www.phoenix-ag.com

PMS LTD
Barima House, Springhill Road,
Peebles, Borders, EH45 9ER
Area of Operation: Scotland
Tel: 01721 720917
Email: pmsflatroofing@tiscali.co.uk
Web: www.pmsflatroofing.co.uk

PROTEC SYSTEMS (UK) LTD
93 High Street, Worle,
Weston-super-Mare,
Somerset, BS22 6ET
Area of Operation: UK (Excluding Ireland)
Tel: 01934 524926
Fax: 01934 524903
Email: anne@protecsystems.uk.net
Web: www.protecsystems.uk.net

RENOTHERM
New Street House, New Street,
Petworth, West Sussex, GU 28 0AS
Area of Operation: UK (Excluding Ireland)
Tel: 01798 343658 **Fax:** 01798 344093
Email: sales@renotherm.co.uk
Web: www.renotherm.co.uk
Material Type: D) 4

SMARTPLY EUROPE
Hawley Manor, Hawley Road,
Dartford, Kent, DA1 1PX
Area of Operation: Europe
Tel: 01322 424900 **Fax:** 01322 424920
Email: info@smartply.com
Web: www.smartply.com
Other Info: ECO
Material Type: B) 2, 9

SPARTAN TILES
Slough Lane, Ardleigh, Essex, CO7 7RU
Area of Operation: UK & Ireland
Tel: 01206 230553 **Fax:** 01206 230516
Email: sales@spartantiles.com
Web: www.spartantiles.com
Material Type: G) 1, 5

SPRY PRODUCTS LTD
64 Nottingham Road, Long Eaton,
Nottingham, Nottinghamshire, NG10 2AU
Area of Operation: UK & Ireland
Tel: 0115 973 2914 **Fax:** 0115 972 5172
Email: jerspry@aol.com
Web: www.spryproducts.com
Other Info: ECO
Material Type: D) 1

STRUCTURAL INSULATED PANEL TECHNOLOGY LTD (SIPTEC)
PO Box 429, Turvey, Bedfordshire, MK43 8DT
Area of Operation: Europe
Tel: 0870 743 9866
Fax: 0870 762 5612
Email: mail@sips.ws
Web: www.sips.ws
Material Type: H) 4, 6

SUNSQUARE LIMITED
12L Hardwick Industrial Estate,
Bury St Edmunds, Suffolk, IP33 2QH
Area of Operation: UK & Ireland
Tel: 0845 226 3172
Fax: 0845 226 3173
Email: info@sunsquare.co.uk
Web: www.sunsquare.co.uk
Other Info: ECO ✎

TOPSEAL SYSTEMS LIMITED
Unit 1&5 Hookstone Chase,
Harrogate, North Yorkshire, HG2 7HH
Area of Operation: Europe
Tel: 01423 886495
Fax: 01423 889550
Email: sales@topseal.co.uk
Web: www.topseal.co.uk
Material Type: D) 5, 6

TRITON CHEMICAL MANUFACTURING LTD
129 Felixstowe Road, Abbey Wood, London, SE2 9SG
Area of Operation: Worldwide
Tel: 0208 310 3929
Fax: 0208 312 0349
Email: neil@triton-chemicals.com
Web: www.triton-chemicals.com

TWINFIX
201 Cavendish Place, Birchwood Park,
Birchwood, Warrington, Cheshire, WA3 6WU
Area of Operation: UK & Ireland
Tel: 01925 811311
Fax: 01925 852955
Email: enquiries@twinfix.co.uk
Web: www.twinfix.co.uk

FELTS, MEMBRANES AND WATERPROOFING

KEY

PRODUCT TYPES: 1= Waterproof
2 = Breathable

OTHER: ▽ Reclaimed ⌐🖱 On-line shopping
✎ Bespoke ✋ Hand-made ECO Ecological

ALUMASC EXTERIOR BUILDING PRODUCTS
White House Works, Bold Road,
Sutton, St Helens, Merseyside, WA9 4JG
Area of Operation: UK & Ireland
Tel: 01744 648400 **Fax:** 01744 648401
Email: info@alumasc-exteriors.co.uk
Web: www.alumasc-exteriors.co.uk
Product Type: 1

BICESTER ROOFING CO.LTD
Manor Farm, Weston on the Green,
Nr. Bicester, Oxfordshire, OX25 3QL
Area of Operation: South East England
Tel: 0870 264 6454
Fax: 0870 264 6455
Email: sales@bicesterroofing.co.uk
Web: www.bicesterroofing.co.uk

BRETT MARTIN DAYLIGHT SYSTEMS
Sandford Close, Alderford's Green Industrial Estate,
Coventry, West Midlands, CV2 2QU
Area of Operation: Worldwide
Tel: 02476 602022
Fax: 02476 602744
Email: toridawson@brettmartin.com
Web: www.daylightsystems.com
Material Type: D) 1, 3, 5, 6

DEGUSSA CONSTRUCTION CHEMICALS (UK)
Albany House, Swinton Hall Road,
Swinton, Manchester, M27 4DT
Area of Operation: Worldwide
Tel: 0161 794 7411
Fax: 0161 727 8547
Email: mbtfeb@degussa.com
Web: www.degussa-cc.co.uk
Product Type: 1, 2

DIY ROOFING LTD
Hillcrest House, Featherbed Lane, Hunt End,
Redditch, Worcestershire, B97 5QL
Area of Operation: UK & Ireland
Tel: 0800 7834890
Fax: 01527 403483
Email: chris@diyroofing.co.uk
Web: www.diyroofing.co.uk
Product Type: 1

EPS INSULATION LTD
Unit 10 Amber Close, Amington Industrial Estate,
Tamworth, Staffordshire, B79 8BH
Area of Operation: UK (Excluding Ireland)
Tel: 01827 313951
Fax: 01827 54683
Email: info@eps-systemsltd.com
Web: www.eps-systemsltd.com
Product Type: 2, 3, 4, 5

EPS SYSTEMS

Area of Operation: UK
Tel: 01827 313951
Fax: 01827 54683
Email: info@eps-systemsltd.com
Web: www.eps-systemsltd.com

DIY spray foam kits for that professional finish at a fraction of the cost, consolidate and insulate your roof. 10% off first orders.

INDUSTRIAL TEXTILES & PLASTICS LTD
Stillington Road, Easingwold, York,
North Yorkshire, YO61 3FA
Area of Operation: Worldwide
Tel: 01347 825200
Email: info@itpltd.com
Web: www.itpltd.com
Product Type: 1, 2
Other Info: ECO

INTERLOC BUILDING SOLUTIONS
Forsyth House, Innova Way, Rosyth Europarc,
Rosyth, Fife, KY11 2UU
Area of Operation: Europe
Tel: 01383 428032
Fax: 01383 428001
Email: info@interlocbuild.co.uk
Web: www.interlocbuild.co.uk
Product Type: 1
Material Type: G) 6

KLOBER LTD
Ingleberry Road, Shepshed,
Nr Loughborough,
Leicestershire, LE12 9DE
Area of Operation: UK & Ireland
Tel: 01509 500660
Fax: 01509 600061
Email: support@klober.co.uk
Web: www.klober.co.uk
Product Type: 1, 2

MIDLAND BUTYL LTD
Windmill Farm, Biggin Lane, Ashbourn,
Derbyshire, DE6 3FN
Area of Operation: Worldwide
Tel: 01335 372133
Fax: 01335 372199
Email: sales@midland-butyl.co.uk
Web: www.midland-butyl.co.uk
Product Type: 1

NOISE STOP SYSTEMS
Unit 3 Acaster Industrial Estate, Cowper Lane,
Acaster Malbis, York, North Yorkshire, YO23 2XB
Area of Operation: UK & Ireland
Tel: 0845 130 6269 **Fax:** 01904 709857
Email: info@noisestopsystems.co.uk
Web: www.noisestopsystems.co.uk
Product Type: 1, 2

PHOENIX (GB) LTD
Chestnut Field House, Chestnut Field,
Rugby, Warwickshire, CU21 2PA
Area of Operation: UK & Ireland
Tel: 01788 571482
Fax: 01788 542245
Email: tonyb@phoenix-gb.com
Web: www.phoenix-ag.com
Product Type: 1

RENOTHERM
New Street House, New Street,
Petworth, West Sussex, GU 28 0AS
Area of Operation: UK (Excluding Ireland)
Tel: 01798 343658
Fax: 01798 344093
Email: sales@renotherm.co.uk
Web: www.renotherm.co.uk

ROOFING INSULATION SERVICES
Hilldale House, 9 Hilldale Avenue,
Blackley, Manchester, M9 6PQ
Area of Operation: UK (Excluding Ireland)
Tel: 0800 731 8314
Email: info@roofinginsulationservices.co.uk
Web: http://roofinginsulationservices.co.uk
Product Type: 2, 3, 4

ROOFING INSULATION SERVICES

Area of Operation: UK (Excluding Ireland)
Tel: 0800 7318314
Email: info@roofinginsulationservices.co.uk
Web: http://roofinginsulationservices.co.uk

Polyurethane spray foam offers the hi-tech solution to nearly all roofing problems at a fraction of the cost, and disruption, of a conventional re-roof.

RUBEROID BUILDING PRODUCTS
Appley Lane North, Appley Bridge,
Lancashire, WN6 9AB
Area of Operation: UK (Excluding Ireland)
Tel: 08000 285573
Fax: 01257 252514
Email: marketing@ruberoid.co.uk
Web: www.ruberoid.co.uk

SPRY PRODUCTS LTD
64 Nottingham Road, Long Eaton,
Nottingham, Nottinghamshire, NG10 2AU
Area of Operation: UK & Ireland
Tel: 0115 973 2914
Fax: 0115 972 5172
Email: jerspry@aol.com
Web: www.spryproducts.com
Product Type: 1
Other Info: ECO
Material Type: D) 1

TOPSEAL SYSTEMS LIMITED
Unit 1&5 Hookstone Chase,
Harrogate, North Yorkshire, HG2 7HH
Area of Operation: Europe
Tel: 01423 886495
Fax: 01423 889550
Email: sales@topseal.co.uk
Web: www.topseal.co.uk
Product Type: 1
Material Type: D) 6

TRITON CHEMICAL
MANUFACTURING LTD
129 Felixstowe Road,
Abbey Wood, London, SE2 9SG
Area of Operation: Worldwide
Tel: 0208 310 3929
Fax: 0208 312 0349
Email: neil@triton-chemicals.com
Web: www.triton-chemicals.com
Product Type: 1

VISQUEEN BUILDING PRODUCTS
Maerdy Industrial Estate, Rhymney,
Blaenau, Gwent, NP22 5PY
Area of Operation: UK & Ireland
Tel: 01685 840672
Fax: 01685 842580
Email: enquiries@visqueenbuilding.co.uk
Web: www.visqueenbuilding.co.uk
Product Type: 1, 2

XTRATHERM UK LIMITED
Unit 5 Jensen Court, Astmoor Industrial Estate,
Runcorn, Cheshire, WA7 1SQ
Area of Operation: UK & Ireland
Tel: 0871 222 1033
Fax: 0871 222 1044
Email: kerry@xtratherm.com
Web: www.xtratherm.com

SOAKERS AND FLASHINGS

> **KEY**
>
> **OTHER:** ▽ Reclaimed 🖱 On-line shopping
>
> 🖋 Bespoke ✋ Hand-made ECO Ecological

BICESTER ROOFING CO.LTD
Manor Farm, Weston on the Green, Nr. Bicester,
Oxfordshire, OX25 3QL
Area of Operation: South East England
Tel: 0870 264 6454
Fax: 0870 264 6455
Email: sales@bicesterroofing.co.uk
Web: www.bicesterroofing.co.uk

DIY ROOFING LTD
Hillcrest House, Featherbed Lane, Hunt End,
Redditch, Worcestershire, B97 5QL
Area of Operation: UK & Ireland
Tel: 0800 7834890
Fax: 01527 403483
Email: chris@diyroofing.co.uk
Web: www.diyroofing.co.uk

TOPSEAL SYSTEMS LIMITED
Unit 1&5 Hookstone Chase, Harrogate,
North Yorkshire, HG2 7HH
Area of Operation: Europe
Tel: 01423 886495
Fax: 01423 889550
Email: sales@topseal.co.uk
Web: www.topseal.co.uk

DECORATIVE TRIMS

> **KEY**
>
> **PRODUCT TYPES:** 1= Finials 2 = Ridges
> 3 = Other
>
> **OTHER:** ▽ Reclaimed 🖱 On-line shopping
> 🖋 Bespoke ✋ Hand-made ECO Ecological

ALDERSHAW HANDMADE CLAY TILES LTD
Pokehold Wood, Kent Street, Sedlescombe,
Nr Battle, East Sussex, TN33 0SD
Area of Operation: Europe
Tel: 01424 756777
Fax: 01424 756888
Email: tiles@aldershaw.co.uk
Web: www.aldershaw.co.uk
Product Type: 1

ASPECT ROOFING CO LTD
The Old Mill, East Harling,
Norwich, Norfolk, NR16 2QW
Area of Operation: UK (Excluding Ireland)
Tel: 01953 717777
Fax: 01953 717164
Email: info@raretiles.co.uk
Web: www.raretiles.co.uk
Product Type: 1, 2

ENVIROMAT BY Q LAWNS
Corkway Drove, Hockwold,
Thetford, Norfolk, IP26 4JR
Area of Operation: UK (Excluding Ireland)
Tel: 01842 828266
Fax: 01842 827911
Email: sales@qlawns.co.uk
Web: www.enviromat.co.uk 🖱
Product Type: 3

GAP
Partnership Way, Shadsworth Business Park,
Blackburn, Lancashire, BB1 2QP
Area of Operation: Midlands & Mid Wales, North
East England, North West England and North Wales
Tel: 01254 682888
Web: www.gap.uk.com
Product Type: 1, 2, 3
Material Type: D) 1

NUMBER 9 STUDIO UK
ARCHITECTURAL CERAMICS
Mole Cottage Industries, Mole Cottage,
Watertown, Chittlehamholt, Devon, EX37 9HF
Area of Operation: Worldwide
Tel: 01769 540471
Fax: 01769 540471
Email: arch.ceramics@moley.uk.com
Web: www.moley.uk.com

PERMACELL FINESSE LTD
Western Road, Silver End,
Witham, Essex, CM8 3BQ
Area of Operation: UK & Ireland
Tel: 01376 583241
Fax: 01376 583072
Email: sales@pfl.co.uk
Web: www.permacell-finesse.co.uk
Product Type: 3

PLASTIVAN LTD
Unit 4 Bonville Trading Estate,
Bonville Road, Brislington, Bristol, BS5 0EB
Area of Operation: Worldwide
Tel: 0117 300 5625
Fax: 0117 971 5028
Email: sales@plastivan.co.uk
Web: www.plastivan.co.uk
Product Type: 1

RED BANK MANUFACTURING
COMPANY LIMITED
Measham, Swadlincote,
Derbyshire, DE12 7EL
Area of Operation: UK & Ireland
Tel: 01530 270333
Fax: 01530 273667
Email: sales@redbankmfg.co.uk
Web: www.redbankmfg.co.uk
Product Type: 1, 2
Other Info: ECO ✎
Material Type: F) 3

ROOFDRAGON.COM
63 Moseley Street, Southend-on-Sea,
Essex, SS2 4NL
Area of Operation: UK & Ireland
Tel: 01702 467444
Email: colin@roofdragon.com
Web: www.roofdragon.com 🖱
Product Type: 1, 2

THE EXPANDED METAL COMPANY LIMITED
PO Box 14, Longhill Industrial Estate (North),
Hartlepool, Durham, TS25 1PR
Area of Operation: Worldwide
Tel: 01429 408087
Fax: 01429 866795
Email: paulb@expamet.co.uk
Web: www.expandedmetalcompany.co.uk

TOPSEAL SYSTEMS LIMITED
Unit 1&5 Hookstone Chase, Harrogate,
North Yorkshire, HG2 7HH
Area of Operation: Europe
Tel: 01423 886495 **Fax:** 01423 889550
Email: sales@topseal.co.uk
Web: www.topseal.co.uk
Product Type: 1, 2

BUILDING STRUCTURE & MATERIALS

SOFFITS, FASCIAS & BARGEBOARDS

KEY
OTHER: ▽ Reclaimed 🖱 On-line shopping
✏ Bespoke ✋Hand-made ECO Ecological

ARP LTD
Unit 2 Vitruvius Way, Meridian Business Park,
Braunstone Park, Leicestershire, LE3 2WH
Area of Operation: UK & Ireland
Tel: 0116 289 4400 **Fax:** 0116 289 4433
Email: sales@arp-ltd.com
Web: www.arp-ltd.com

CLEARVIEW WINDOWS & DOORS
Unit 14, Sheddingdean Industrial Estate,
Marchants Way, Burgess Hill,
West Sussex, RH15 8QY
Area of Operation: UK (Excluding Ireland)
Tel: 01444 250111 **Fax:** 01444 250678
Email: sales@clearviewsussex.co.uk
Web: www.clearviewsussex.co.uk

EVERWHITE PLASTICS LTD
Everwhite House, Aberaman Park Industrial Estate,
Aberdare, Mid Glamorgan, CF44 6DA
Area of Operation: UK (Excluding Ireland)
Tel: 01685 882447
Web: www.everwhite.biz 🖱
Material Type: D) 1

EXTERIOR IMAGE LTD
322 Hursley Road, Chandlers Ford,
Eastleigh, Hampshire, SO53 5PH
Tel: 02380 271777
Fax: 02380 269091
Email: info@exteriorimage.co.uk
Web: www.exteriorimage.co.uk

GAP
Partnership Way, Shadsworth Business Park,
Blackburn, Lancashire, BB1 2QP
Area of Operation: Midlands & Mid Wales,
North East England, North West England
and North Wales
Tel: 01254 682888
Web: www.gap.uk.com
Material Type: D) 1

MILBANK ROOFS
Hargrave Meadow Cottage, Church Lane,
Hargrave, Bury St Edmunds, Suffolk, IP29 5HH
Area of Operation: UK & Ireland
Tel: 01284 852 505
Fax: 01284 850 021
Email: info@milbankroofs.com
Web: www.milbankroofs.com

NORTHANTS RAINWATER SYSTEMS
31 Summerfields, Northampton,
Northamptonshire, NN4 9YN
Area of Operation: UK (Excluding Ireland)
Tel: 01604 877775
Fax: 01604 674447
Email: mail@northantsrainwater.co.uk
Web: www.northantsrainwater.co.uk

PERMACELL FINESSE LTD
Western Road, Silver End,
Witham, Essex, CM8 3BQ
Area of Operation: UK & Ireland
Tel: 01376 583241
Fax: 01376 583072
Email: sales@pfl.co.uk
Web: www.permacell-finesse.co.uk

PLASTIVAN LTD
Unit 4 Bonville Trading Estate,
Bonville Road, Brislington, Bristol, BS5 0EB
Area of Operation: Worldwide
Tel: 0117 300 5625 **Fax:** 0117 971 5028
Email: sales@plastivan.co.uk
Web: www.plastivan.co.uk

SUPERIOR FASCIAS
Adelaide House, Portsmouth Road,
Lowford, Southampton, Hampshire, SO31 8EQ
Area of Operation: South East England,
South West England and South Wales
Tel: 0700 596 4603
Fax: 0700 596 4609
Email: info@superiorfascias.co.uk
Web: www.superiorfascias.co.uk

SWISH BUILDING PRODUCTS LTD
Pioneer House, Mariner,
Lichfield Road Industrial Estate,
Tamworth, Staffordshire, B79 7TF
Area of Operation: UK & Ireland
Tel: 01827 317200
Fax: 01827 317201
Email: info@swishbp.co.uk
Web: www.swishbp.co.uk
Material Type: D) 1

SWISH BUILDING PRODUCTS
Area of Operation: UK & Ireland
Tel: 01827 317200
Fax: 01827 317201
Email: info@swishbp.co.uk

Low maintenance, Cellular PVC roofline and cladding systems in standard, coloured and decorative formats. Also Marbrex internal decorative panelling system. All fitted using conventional tools and skills.

THE EXPANDED METAL COMPANY LIMITED
PO Box 14, Longhill Industrial Estate (North),
Hartlepool, Durham, TS25 1PR
Area of Operation: Worldwide
Tel: 01429 408087
Fax: 01429 866795
Email: paulb@expamet.co.uk
Web: www.expandedmetalcompany.co.uk

EAVES CLOSERS

KEY
OTHER: ▽ Reclaimed 🖱 On-line shopping
✏ Bespoke ✋Hand-made ECO Ecological

CLEARVIEW WINDOWS & DOORS
Unit 14, Sheddingdean Industrial Estate,
Marchants Way, Burgess Hill,
West Sussex, RH15 8QY
Area of Operation: UK (Excluding Ireland)
Tel: 01444 250111
Fax: 01444 250678
Email: sales@clearviewsussex.co.uk
Web: www.clearviewsussex.co.uk

KLOBER LTD
Ingleberry Road, Shepshed,
Nr Loughborough, Leicestershire, LE12 9DE
Area of Operation: UK & Ireland
Tel: 01509 500660
Fax: 01509 600061
Email: support@klober.co.uk
Web: www.klober.co.uk
Other Info: ✏

TOPSEAL SYSTEMS LIMITED
Unit 1&5 Hookstone Chase, Harrogate,
North Yorkshire, HG2 7HH
Area of Operation: Europe
Tel: 01423 886495 **Fax:** 01423 889550
Email: sales@topseal.co.uk
Web: www.topseal.co.uk

GREEN ROOFING

KEY
OTHER: ▽ Reclaimed 🖱 On-line shopping
✏ Bespoke ✋Hand-made ECO Ecological

ENVIROMAT BY Q LAWNS
Corkway Drove, Hockwold,
Thetford, Norfolk, IP26 4JR
Area of Operation: UK (Excluding Ireland)
Tel: 01842 828266 **Fax:** 01842 827911
Email: sales@qlawns.co.uk
Web: www.enviromat.co.uk 🖱

TRITON CHEMICAL MANUFACTURING LTD
129 Felixstowe Road, Abbey Wood,
London, SE2 9SG
Area of Operation: Worldwide
Tel: 0208 310 3929 **Fax:** 0208 312 0349
Email: neil@triton-chemicals.com
Web: www.triton-chemicals.com

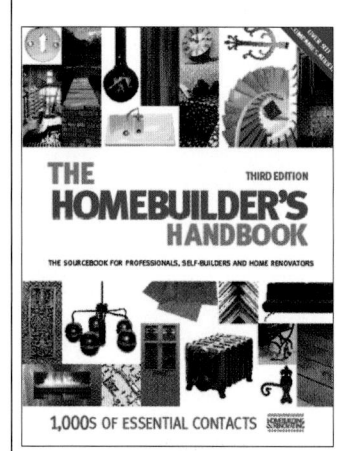

Don't forget to let us know about companies you think should be listed in the next edition
email:
customerservice@centaur.co.uk

NOTES

DOORS, WINDOWS & CONSERVATORIES

Image courtesy of Ultraframe (0500 822340)

SPONSORED BY ULTRAFRAME

Tel 0500 822340   Web www.ultraframe-conservatories.co.uk

Conservatories

Conservatories first became popular in Victorian times when progress in glass manufacturing meant that they became more affordable for the middle classes, and they have since become the first port of call for homeowners looking for a simple add-on to their home.

The design origins of the conservatory lie in the Georgian orangery and the Victorian hot house, but that doesn't mean that today's self-builders and renovators have to adhere to conventional period style. Whilst the vast majority of conservatories sold in the UK are still either Victorian, Edwardian or Georgian in appearance, bespoke designs using largely glass – with a minimal structure usually in steel or aluminium – are becoming ever more affordable and the reinvented, far simpler, contemporary conservatory is now poised to be the latest must-have feature — working equally well on cutting edge and period properties, where the style can form an interesting juxtaposition.

Whilst a well-designed conservatory adds much more in value than it costs to erect, a poor, ill-conceived, condensation-filled add-on that takes light away from the rest of the house is likely to put off potential buyers. As a result, it is essential to plan your purchase carefully.

There are many aspects of design to consider. One of the keys to a successful conservatory is maximising the useable space and, for this reason, the positioning of the entrance from the main house and the exit onto the patio area should be carefully considered. The common scheme is to place the doors in the centre, which can effectively cut the conservatory space in two by creating a corridor effect with furniture arranged on either side to allow clear access to the garden. Having the entrance to the edge of the room will usually create a more impressive sense of space.

In terms of exterior design there is little excuse for not finding something that complements the rest of your home. Approach the design of the conservatory in the same way as you would approach the rest of your new home: individually and with attention to detail. Traditional period style off-the-shelf kits can look terrible on a new Georgian style home; likewise, a contemporary glass structure can look hopelessly out of place on a modest traditional home. Make your conservatory suppliers work hard for their money — a 'one size fits all' solution is unlikely to work.

Sealed double glazing units now mean that there is no excuse for a draughty room, and Part N of the Building Regulations – which all new conservatories have to meet – means that energy efficient double glazing units are effectively compulsory.

Some people find that a simple extension of the heating system with a radiator into the conservatory is an easy way to heat the structure. Where wall space is

A well designed conservatory can transform your home and add considerable value, but remember that interior design (TOP) can be as important as the exterior when selling. Timber conservatory, ABOVE LEFT, by Wickes. PVCu conservatory, ABOVE RIGHT, by Amdega.

limited, underfloor heating is a good solution, especially for a tiled or stone floor. Alternatively, trench radiators sunk into the floor also work well. However, with insulation standards so high, the prevalence of Argon-filled glass and the amount of sun that a south-facing conservatory gets, the main problem for most people will be keeping heat levels down. Ventilation can be brought in through the ridge or eaves in a flow or trickle vent system and is essential in the summer months. The effect of the sun can be diminished by using a 'bronzed' polycarbonate roof, incorporating as many opening windows as possible and, in extreme

cases, the installation of a mechanical ventilation or air conditioning system. Installing blinds can also help to keep a conservatory cool in summer and warm in winter.

Today's conservatory owners also have woken up to the fact that, as with the rest of their home, the internal look of their conservatory needs to be comfortable and stylish. Wooden or slate flooring, accent lighting, neutral shades and leather furniture can make a new conservatory much more 'liveable' — meaning that you will get better use out of it and it will be more attractive to potential buyers.

INTERNAL DOORS

KEY
PRODUCT TYPES: 1= Plain 2 = Panelled
3 = Glazed 4 = French 5 = Folding / Sliding
6 = Fire Doors 7 = Ledged / Braced
8 = Security 9 = Other

OTHER: ▽ Reclaimed On-line shopping
Bespoke Hand-made ECO Ecological

ASTON HOUSE JOINERY
Aston House, Tripontium Business Centre,
Newton Lane, Rugby, Warwickshire, CV23 0TB
Area of Operation: UK & Ireland
Tel: 01788 860032 **Fax:** 01788 860614
Email: mailbox@ovationwindows.co.uk
Web: www.astonhousejoinery.co.uk
Product Type: 1, 2, 3, 4, 5, 6, 7, 8
Other Info: ECO
Material Type: A) 1, 2

BARHAM & SONS
58 Finchley Avenue, Mildenhall,
Bury St. Edmunds, Suffolk, IP28 7BG
Area of Operation: UK (Excluding Ireland)
Tel: 01638 665759 / 711611
Fax: 01638 663913
Email: info@barhamwoodfloors.com
Product Type: 1, 2, 3, 7

BARRON GLASS
Unit 11, Lansdown Industrial Estate,
Cheltenham, Gloucestershire, GL51 8PL
Area of Operation: UK & Ireland
Tel: 01242 22800
Fax: 01242 226555
Email: admin@barronglass.co.uk
Web: www.barronglass.co.uk
Product Type: 1, 2, 3

BECKER SLIDING PARTITIONS LTD
Wemco House, 477 Whippendell Road,
Herefordshire, WD18 7QY
Area of Operation: UK & Ireland
Tel: 01923 236906
Fax: 01923 230 149
Email: Sales@Becker.uk.com
Web: www.becker.uk.com
Product Type: 5

BROADLEAF TIMBER
Llandeilo Road Industrial Estate, Carms,
Carmarthenshire, SA18 3JG
Area of Operation: UK & Ireland
Tel: 01269 851910
Fax: 01269 851911
Email: sales@broadleaftimber.com
Web: www.broadleaftimber.com
Product Type: 2, 3, 4, 6, 7
Other Info:
Material Type: A) 2

BROXWOOD (SCOTLAND) LTD
Inveralmond Way, Inveralmond Industrial Estate,
Perth, Perth and Kinross, PH1 3UQ
Area of Operation: UK & Ireland
Tel: 01738 444456
Fax: 01738 444452
Email: sales@broxwood.com
Web: www.broxwood.com
Product Type: 2, 3, 4, 5, 6
Other Info: ECO
Material Type: A) 1

CENTURION
Westhill Business Centre, Arnhall Business Park,
Westhill, Aberdeen, Aberdeenshire, AB32 6UF
Area of Operation: Scotland
Tel: 01224 744440
Fax: 01224 744819
Email: info@centurion-solutions.co.uk
Web: www.centurion-solutions.co.uk
Product Type: 6

**CHAMBERLAIN & GROVES LTD -
THE DOOR & SECURITY STORE**
101 Boundary Road,
Walthamstow, London, E17 8NQ
Area of Operation: UK (Excluding Ireland)
Tel: 0208 520 6776
Fax: 0208 520 2190
Email: ken@securedoors.co.uk
Web: www.securedoors.co.uk
Product Type: 1, 2, 3, 4, 6, 8, 9
Other Info: ECO
Material Type: A) 1, 2

CHARLES MASON LTD
Unit 11A Brook Street Mill,
Off Goodall Street, Macclesfield,
Cheshire, SK11 7AW
Area of Operation: Worldwide
Tel: 0800 085 3616
Fax: 01625 668789
Email: info@charles-mason.com
Web: www.charles-mason.com
Product Type: 1, 2, 3, 4, 5

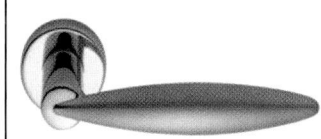

CHARLES MASON LTD
architectural fittings

CHARLES MASON ARCHITECTURAL FITTINGS
Area of Operation: Worldwide
Tel: 0800 085 3616 **Fax:** 01625 668789
Email: info@charles-mason.com
Web: www.charles-mason.com
Product Type: 1, 2, 3, 4, 5

Quality ironmongery and accessories, door handles, knobs, hinges, locks, lighting, switches in steel chrome antique brass and black iron. Buy online or visit our showroom.

CHARLES PEARCE LTD
26 Devonshire Road, London, W4 2HD
Area of Operation: Worldwide
Tel: 0208 995 3333
Email: enquiries@charlespearce.co.uk
Web: www.charlespearce.co.uk
Product Type: 5

CHESHIRE DOOR COMPANY
Paradise Mill, Old Park Lane,
Macclesfield, Cheshire, SK11 6TJ
Area of Operation: UK & Ireland
Tel: 01625 421221 **Fax:** 01625 421422
Email: info@cheshiredoorcompany.co.uk
Web: www.cheshiredoorcompany.co.uk

CHIPPING NORTON GLASS LTD
Units 1 & 2, Station Yard Industrial Estate,
Chipping Norton, Oxfordshire, OX7 5HX
Area of Operation: Midlands & Mid Wales,
South West England and South Wales
Tel: 01608 643261 **Fax:** 01608 641768
Email: gill@cnglass.plus.com
Web: www.chippingnortonglass.co.uk
Product Type: 3, 4, 5, 6
Material Type: D) 1

CONSTRUCTION TECHNICAL SERVICES
Dunedin House, Alexandra Road,
Penzance, Cornwall, TR18 4LZ
Area of Operation: UK & Ireland
Tel: 01736 330303
Fax: 01736 360497
Email: CTecS@aol.com
Web: www.hometown.aol.co.uk/carpinjohn
Product Type: 5

COUNTY HARDWOODS
Creech Mill, Mill Lane, Creech Saint Michael,
Taunton, Somerset, TA3 5PX
Area of Operation: UK & Ireland
Tel: 01823 444217 **Fax:** 01823 443940
Email: hardwood@netcomuk.co.uk
Web: www.countyhardwoods.co.uk
Product Type: 7
Material Type: A) 2

COUNTY JOINERY (SOUTH EAST) LTD
Unit 300, Vinehall Business Centre,
Vinehall Road, Mountfield, Robertsbridge,
East Sussex, TN32 5JW
Area of Operation: Greater London,
South East England
Tel: 01424 871500
Fax: 01424 871550
Email: countyjoinery@aol.com
Web: www.countyjoinery.co.uk
Product Type: 1, 2, 3, 4, 5, 6, 7, 8, 9
Material Type: A) 2, 3, 4, 5, 6, 7, 8, 9, 10, 11, 12, 13, 14

**CROXFORD'S JOINERY
MANUFACTURERS & WOODTURNERS**
Meltham Joinery, Works New Street,
Meltham, Holmfirth, West Yorkshire, HD9 5NT
Area of Operation: UK (Excluding Ireland)
Tel: 01484 850892
Fax: 01484 850969
Email: ralph@croxfords.demon.co.uk
Web: www.croxfords.co.uk
Product Type: 1, 2, 3, 4, 5, 7

CUSTOM WOOD PRODUCTS
Cliffe Road, Easton on the Hill,
Stamford, Lincolnshire, PE9 3NP
Area of Operation: East England
Tel: 01780 755711
Fax: 01780 480834
Email: customwoodprods@aol.com
Web: www.cwpuk.com
Product Type: 1, 2, 3, 4, 5, 7
Other Info: ECO
Material Type: A) 2, 3, 4, 5, 6

DICTATOR DIRECT
Inga House, Northdown Business Park,
Ashford Road, Lenham, Kent, ME17 2DL
Area of Operation: Worldwide
Tel: 01622 854770, **Fax:** 01622 854771
Email: mail@dictatordirect.com
Web: www.dictatordirect.com

DIRECTDOORS.COM
Bay 5 Eastfield Industrial Estate,
Eastfield Drive, Penicuik, Lothian, EH26 8JA
Area of Operation: UK (Excluding Ireland)
Tel: 0131 669 6310 **Fax:** 0131 657 1578
Email: info@directdoors.com
Web: www.directdoors.com
Product Type: 1, 2, 3, 4, 5, 6, 7, 8

DISTINCTIVE COUNTRY FURNITURE LTD
Parrett Works, Martock, Somerset, TA12 6AE
Area of Operation: Worldwide
Tel: 01935 825800 **Fax:** 01935 825800
Email: Dcfltdinteriors@aol.com
Web: www.distinctivecountryfurniture.co.uk
Other Info: ▽
Material Type: A) 2, 3, 4, 7, 14

DISTINCTIVE COUNTRY FURNITURE
Area of Operation: Worldwide
Tel: 01935 825800
Fax: 01935 825800
Email: dcfltdinteriors@aol.com
Web: www.distinctivecountryfurniture.co.uk

Makers of planked doors, frames, linings and architraves. Please call for a free brochure. Quotations available upon request.

DISTINCTIVE DOORS
14 & 15 Chambers Way, Newton Chambers Road,
Chapeltown, Sheffield, South Yorkshire, S35 2PH
Area of Operation: UK & Ireland
Tel: 0114 220 2250 **Fax:** 0114 220 2254
Email: enquiries@distinctivedoors.co.uk
Web: www.distinctivedoors.co.uk
Product Type: 2, 4, 6, 7
Material Type: A) 1, 2, 6

DISTINCTIVE DOORS
Area of Operation: UK & Ireland
Tel: 0114 220 2250 **Fax:** 0114 220 2254
Email: enquiries@distinctivedoors.co.uk
Web: www.distinctivedoors.co.uk
Product Type: 2, 4, 6, 7

Suppliers of High Class External & Internal Doors.
Exclusive Product Range Triple Glazed Doors.
Oak Internal Engineered Doors.
Hardwood Doors, Clear Pine Doors.

**DRUMMONDS ARCHITECTURAL
ANTIQUES LIMITED**
Kirkpatrick Buildings, 25 London Road (A3),
Hindhead, Surrey, GU36 6AB
Area of Operation: Worldwide
Tel: 01428 609444 **Fax:** 01428 609445
Email: davidcox@drummonds-arch.co.uk
Web: www.drummonds-arch.co.uk
Product Type: 1, 2, 3, 4, 7, 9
Other Info: ▽
Material Type: A) 2

DYNASTY DOORS
Unit 3, The Micro Centre, Gillette Way,
Reading, Berkshire, RG2 0LR
Area of Operation: South East England,
South West England and South Wales
Tel: 0118 987 4000 **Fax:** 0118 921 2999
Email: Martin@dynastydoors.co.uk
Web: www.dynastydoors.co.uk
Product Type: 1, 2, 3, 4, 6

Ultraframe – Innovation and expertise in the conservatory market

In today's overcrowded world, light and space are precious and minimal in many of our homes. Adding a conservatory is an ideal way to open up your living space.

Whilst the garden or patio are wonderful places to relax during the Summer, the British weather means that for many months of the year we just

can't enjoy them. Adding a conservatory to your home doesn't just give you extra space it brings the outdoors into your home and creates a unique living space that can be enjoyed all year round.

Ultraframe is the leading designer and manufacturer of conservatory roofs in the UK. The market leader for over 20 years, Ultraframe offers creative solutions for every home, from Litespace, the innovative walk-in bay window, to spectacular showcase conservatories designed to suit all styles of home, all applications and every price point.

Individuality

Ultraframe are passionate about conservatories and seek to ensure that each one is perfectly suited to its owner. As the most configurable roofing system on the market today, Ultraframe offers you the opportunity to create a conservatory that is as individual as you are. Our extensive research and development continues to innovate and

extend the boundaries of design, so when you choose an Ultraframe roof, you can be certain that you're getting the best.

Innovation and Comfort

The position of the conservatory in relation to its surroundings is a key factor in the creation of a comfortable conservatory interior. Good insulation and heating will be required if the conservatory faces North, whilst additional ventilation and shading for South facing conservatories is essential. Ultraframe is the only company that supplies ventilation in every roof as standard. Our patented ventilation systems are second to none and ensure that you can enjoy your conservatory whenever you like, not when the weather dictates. Ultraframe also offer innovative glazing options such as Conservaglass Optimum, which features a unique invisible micro coating to keep solar heat out, yet keep the warmth from heating appliances in.

Expertise

All Ultraframe roofs are engineered for the extremes of the UK climate and are approved by the British Board of Agreement who has declared them to have a life expectancy of at least 25 years. They are the only systems to have passed the USA's BOCA system, the toughest testing regime in the world.

Reassurance

Each Ultraframe roof is supplied with the Ultraframe Certificate of Authenticity - your guarantee that you have a top quality Ultraframe roof. The certificate is a legal document that you can use to reassure future buyers of your home that your conservatory really does have an Ultraframe roof with all of the associated benefits.

Inspiration

FREE Ultra Guide to Conservatories

If you are looking for inspiration when considering extending your home, make sure you have the latest and most comprehensive guide to planning, design and style. The Ultra Guide to Conservatories is packed with inspirational case studies, striking photography and valuable practical advice making it essential reading for anyone planning a conservatory. **Simply call 0500 822 340 or visit www.ultraframe-conservatories.co.uk to order your free copy.**

Choose a conservatory installer with confidence
If you 're considering a conservatory – take a look at Ultraframe's checklist BEFORE you buy!

Consider your positioning
- If you are south facing remember to look at glazing options that help to block the heat of the sun and consider additional ventilation to help keep temperatures pleasant in the summer
- If you are facing north, good heating and insulation is important, as is excellent ventilation

Look at the layout
- Position doors at the side if you want to maximize the useable space in your conservatory, doors placed in the middle will create a corridor through the room

Choosing the Site - Take into account factors such as:
- The room which will provide interior access to the conservatory
- Access from the conservatory into the garden

- Wall, buildings, trees and greenery next to the proposed site

Garden Design
- Having your conservatory built is a great opportunity to re-landscape your garden, so that the conservatory blends in perfectly with its surroundings and you can enjoy them together

Planning permission
- Check whether or not planning permission is required, (In most cases it isn't). If planning permission is necessary you must submit the appropriate forms to your local council before starting work.

Reputation is everything
- Make sure you choose a reputable installer, like those on the Guild Approved Ultra Installer Scheme

---

To ensure homeowners select a company to install their conservatory with confidence, Ultraframe has introduced the Guild Approved Ultra Installer Scheme.

All members of the national scheme have met the standards set by Ultraframe, been vetted by the Guild of Master Craftsmen and have passed thorough inspections carried out by the British Board of Agrément.

By choosing a member of our Scheme to install your conservatory you can be assured of a quality installation and excellent service before, during and after the work is completed.

To find your nearest members of the Guild Approved Ultra Installer Scheme call us on freephone 0500 822340 or visit our website www.ultraframe-conservatories.co.uk Our trained advisors will be able to provide you with the contact details of your local Ultra Installers and can answer any queries you may have about buying a conservatory.

ENGELS WINDOWS & DOORS LTD
1 Kingley Centre, Downs Road, West Stoke,
Chichester, West Sussex, PO18 9HJ
Area of Operation: UK (Excluding Ireland)
Tel: 01243 576633
Fax: 01243 576644
Email: admin@engels.co.uk
Web: www.engels.co.uk

EVEREST LTD
Sopers Road, Cuffley,
Hertfordshire, EN6 4SG
Area of Operation: UK & Ireland
Tel: 0800 010123
Web: www.everest.co.uk
Product Type: 1, 3, 5

FAIRMITRE WINDOWS & JOINERY LTD
2A Cope Road, Banbury, Oxfordshire, OX16 2EH
Area of Operation: UK (Excluding Ireland)
Tel: 01295 268441
Fax: 01295 268468
Email: info@fairmitrewindows.co.uk
Web: www.fairmitrewindows.co.uk
Product Type: 2, 3, 4, 7

FITZROY JOINERY
Garden Close, Langage Industrial Estate,
Plympton, Devon, PL7 5EU
Area of Operation: UK & Ireland
Tel: 0870 428 9110
Fax: 0870 428 9111
Email: admin@fitzroy.co.uk
Web: www.fitzroy.co.uk
Product Type: 1, 2, 3, 4, 5, 6
Other Info: ECO
Material Type: A) 1, 2, 3, 4, 5, 6, 7, 8, 9, 10, 12, 14

FLETCHER JOINERY
261 Whessoe Road, Darlington, Durham, DL3 0YL
Area of Operation: North East England
Tel: 01325 357347
Fax: 01325 357347
Email: enquiries@fletcherjoinery.co.uk
Web: www.fletcherjoinery.co.uk
Product Type: 1, 2, 3, 4, 5, 6, 7, 8, 9

FLOORS AND DOORS DIRECT
Unit 7 Blaydon Trade Park, Toll Bridge Road,
Blaydob, Tyne & Wear, NE21 5TR
Area of Operation: North East England
Tel: 0191 414 5055
Fax: 0191 414 5066
Web: www.floorsanddoorsdirect.co.uk
Product Type: 1

FOLDING SLIDING DOORS LIMITED
FSD Works, Hopbine Avenue,
West Bowling, Bradford,
West Yorkshire, BD5 8ER
Area of Operation: UK (Excluding Ireland)
Tel: 0845 644 6630
Fax: 0845 644 6631
Email: info@foldingslidingdoor.com
Web: www.foldingslidingdoor.com
Product Type: 5

GOODMILL LTD
Ballybay, County Monaghan,
Republic of Ireland,
Area of Operation: UK & Ireland
Tel: +353 429 741 771
Fax: +353 429 741 771
Email: maggsmills@eircom.net
Web: www.goodmillsecurity.com
Product Type: 2, 6, 9
Material Type: C) 2

HAYMANS TIMBER PRODUCTS
Haymans Farm, Hocker Lane,
Over Alderley, Macclesfield,
Cheshire, SK10 4SD
Area of Operation: UK & Ireland
Tel: 01625 590098
Fax: 01625 586174
Email: haymanstimber@aol.com
Web: www.haymanstimber.co.uk
Product Type: 1, 2, 3, 4, 5, 6, 7
Material Type: A) 2, 4, 6, 7

IMAJ STEEL DOORS
45h Leyton Industrial Estate,
Argall Way, London, E8 2NG
Area of Operation: Worldwide
Tel: 0207 275 0466
Fax: 0207 249 0176
Email: sales@imajsteeldoors.com
Web: www.imajsteeldoors.com

IN DOORS
Beechinwood Farm,
Beechinwood Lane, Platt,
Nr. Sevenoaks, Kent, TN15 8QN
Area of Operation: UK (Excluding Ireland)
Tel: 01732 887445
Fax: 01732 887446
Email: info@indoorsltd.co.uk
Web: www.indoorsltd.co.uk
Product Type: 1, 2, 3, 4, 5, 7

INNERDOOR LTD
Royds Enterprise Park, Future Fields,
Bradford, West Yorkshire, BD6 3EW
Area of Operation: Worldwide
Tel: 0845 128 3958
Fax: 0845 128 3959
Email: info@innerdoor.co.uk
Web: www.innerdoor.co.uk
Product Type: 1, 2, 3, 4, 5, 6

INNERDOOR LTD

Area of Operation: Worldwide
Tel: 0845 128 3958 **Fax:** 0845 128 3959
Email: info@innerdoor.co.uk
Web: www.innerdoor.co.uk
Product Type: 1, 2, 3, 4, 5, 6

Innerdoor supply a wide range of doors, hinged wood doors, hinged glass doors, sliding glass doors plus wood door hardware and glass door hardware.

INNOVATION GLASS COMPANY
27-28 Blake Industrial Estate,
Brue Avenue, Bridgwater,
Somerset, TA7 8EQ
Area of Operation: Europe
Tel: 01278 683645
Fax: 01278 450088
Email: magnus@igc-uk.com
Web: www.igc-uk.com
Product Type: 2, 9

J L JOINERY
Cockerton View, Grange Lane,
Preston, Lancashire, PR4 5JE
Area of Operation: UK (Excluding Ireland)
Tel: 01772 616123
Fax: 01772 619182
Email: mail@jljoinery.co.uk
Web: www.jljoinery.co.uk

JB KIND LTD
Shobnall Street, Burton-on-Trent,
Staffordshire, DE14 2HP
Area of Operation: UK & Ireland
Tel: 01283 510210
Fax: 01283 511132
Email: info@jbkind.com
Web: www.jbkind.com
Product Type: 1, 2, 3, 4, 5, 6, 7, 8
Material Type: A) 2, 3, 4, 5, 6, 7, 8, 15

JELD-WEN
Watch House Lane, Doncaster,
South Yorkshire, DN5 9LR
Area of Operation: Worldwide
Tel: 01302 394000 **Fax:** 01302 787383
Email: customer-services@jeld-wen.co.uk
Web: www.jeld-wen.co.uk
Product Type: 1, 2, 3

JOINERY-PLUS
Bentley Hall Barn, Alkmonton,
Ashbourne, Derbyshire, DE6 3DJ
Area of Operation: UK (Excluding Ireland)
Tel: 07931 386233 **Fax:** 01335 330922
Email: info@joinery-plus.co.uk
Web: www.joinery-plus.co.uk
Product Type: 1, 2, 3, 4, 6, 7, 8
Material Type: A) 1, 2, 3, 4, 5, 6, 7, 8, 9, 10, 11, 12, 13, 14

KIVA DOORS LIMITED
Bramingham Business Centre,
Enterprise Way, Luton,
Bedfordshire, LU3 4BU
Area of Operation: UK & Ireland
Tel: 0800 328 5851 **Fax:** 01582 584111
Email: info@kiva-doors.com
Web: www.kiva-doors.com
Product Type: 1, 2, 3, 4, 5, 6, 8
Material Type: A) 2, 3, 4, 5, 6, 7, 9, 12

LAMWOOD LTD
Unit 1, Riverside Works, Station Road,
Cheddleton, Staffordshire, ST13 7EE
Area of Operation: UK & Ireland
Tel: 01538 361888 **Fax:** 01538 361912
Email: sales@lamwoodltd.co.uk
Web: www.lamwoodltd.co.uk
Material Type: A) 2

LAWSONS
Gorst Lane, Off New Lane, Burscough,
Ormskirk, Lancashire, L40 0RS
Area of Operation: Worldwide
Tel: 01704 893998 **Fax:** 01704 892526
Email: info@traditionaltimber.co.uk
Web: www.traditionaltimber.co.uk
Other Info:

MASONITE EUROPE
Jason House, Kerry Hill, Horsforth,
Leeds, West Yorkshire, LS18 4JR
Area of Operation: Europe
Tel: 0113 2587689 **Fax:** 0113 2590015
Email: nbuckle@masonite.com
Web: www.masonite-europe.com
Product Type: 2, 3, 5, 6

MERLIN UK LIMITED
Unit 5 Fence Avenue,
Macclesfield, Cheshire, SK10 1LT
Area of Operation: North West England and North Wales
Tel: 01625 424488 **Fax:** 0871 781 8967
Email: info@merlinuk.net
Web: www.merlindoors.co.uk
Product Type: 1, 2, 3, 4, 5, 6, 7, 9

MIDLANDS SLATE & TILE
Qualcast Road, Horseley Fields,
Wolverhampton, West Midlands, WV1 2QP
Area of Operation: UK & Ireland
Tel: 01902 458780
Fax: 01902 458549
Email: sales@slate-tile-brick.co.uk
Web: www.slate-tile-brick.co.uk
Product Type: 3, 7
Material Type: A) 2

**MINCHINHAMPTON
ARCHITECTURAL SALVAGE CO**
Cirencester Road, Aston Downs,
Chalford, Stroud, Gloucestershire, GL6 8PE
Area of Operation: Worldwide
Tel: 01285 760886
Fax: 01285 760838
Email: masco@catbrain.com
Web: www.catbrain.com
Product Type: 1, 2, 3, 4, 7

NATIONAL DOOR COMPANY
Unit 55 Dinting Vale Business Park, Dinting Vale,
Glossop, Derbyshire, SK13 6JD
Area of Operation: UK (Excluding Ireland)
Tel: 01457 867079 **Fax:** 01457 868795
Email: sales@nationaldoor.co.uk
Web: www.nationaldoor.co.uk
Product Type: 1, 2, 3, 4, 5, 6, 7, 8, 9

NATIONAL DOOR COMPANY

Area of Operation: UK (Excluding Ireland)
Tel: 01457 867079 **Fax:** 01457 868795
Email: sales@nationaldoor.co.uk
Web: www.nationaldoor.co.uk
Product Type: 1, 2, 3, 4, 5, 6, 7, 8, 9

Enhance your surroundings with our exquisite door ranges. Available in many veneers including oak, pine, sapele, ash, beech, etimoe, walnut and cherry. Any quantity, any size, lacquered or unlacquered. Most doors are also available.

NFP JOINERY
Milford Village Hall, Portsmouth Road,
Milford, Godalming, Surrey, GU8 5DS
Area of Operation: UK & Ireland
Tel: 01483 414291 **Fax:** 01483 414831
Email: colin@nfpeurope.co.uk
Web: www.nfpeurope.co.uk
Product Type: 2, 3, 4, 9
Material Type: B) 13

NOISE STOP SYSTEMS
Unit 3 Acaster Industrial Estate,
Cowper Lane, Acaster Malbis, York,
North Yorkshire, YO23 2XB
Area of Operation: UK & Ireland
Tel: 0845 130 6269 **Fax:** 01904 709857
Email: info@noisestopsystems.co.uk
Web: www.noisestopsystems.co.uk
Product Type: 1, 2, 4, 5, 6, 9

**NORBUILD TIMBER FABRICATION
& FINE CARPENTRY LTD**
Marcassie Farm, Rafford,
Forres, Moray, IV36 2RH
Area of Operation: UK & Ireland
Tel: 01309 676865
Fax: 01309 676865
Email: norbuild@marcassie.fsnet.co.uk
Product Type: 2, 3, 5, 9
Other Info: ECO
Material Type: A) 1, 2, 4, 6, 9, 11

NORTH YORKSHIRE TIMBER
Standard House, Thurston Road,
Northallerton Business Park,
Northallerton, North Yorkshire, DL6 2NA
Area of Operation: UK (Excluding Ireland)
Tel: 01609 780777
Fax: 01609 777888
Email: sales@nytimber.co.uk
Web: www.nytimber.co.uk

NORTHWOOD FORESTRY LTD
Goose Green Lane, Nr Ashington,
Pulborough, West Sussex, RH20 2LW
Area of Operation: South East England
Tel: 01798 813029 **Fax:** 01798 813139
Email: enquiries@northwoodforestry.co.uk
Web: www.northwoodforestry.co.uk
Product Type: 7
Material Type: A) 2

SPONSORED BY: ULTRAFRAME www.ultraframe-conservatories.co.uk

))INNERDOOR

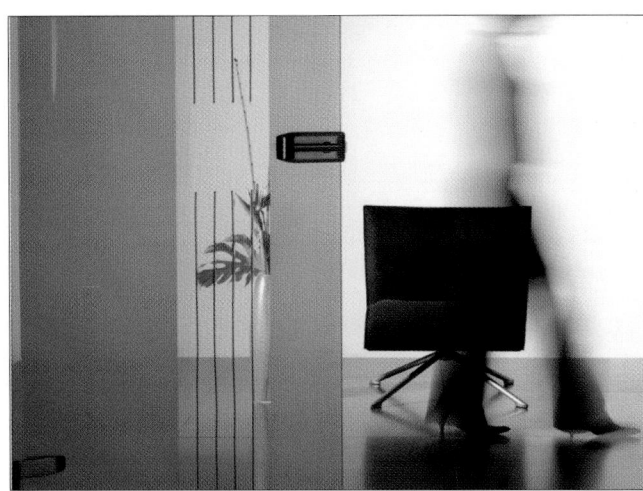

Innerdoor supply a wide range of contemporary internal doors, both hinged and sliding glass and wooden doors plus designer door hardware to compliment the range.

INNERDOOR LTD
www.innerdoor.co.uk
Tel: 0845 128 3958 Fax: 0845 128 3959
Email: info@innerdoor.co.uk

OLDE WORLDE OAK JOINERY LTD
Unit 12, Longford Industrial Estate, Longford Road, Cannock, Staffordshire, WS11 0DG
Area of Operation: Europe
Tel: 01543 469328 **Fax:** 01543 469341
Email: sales@oldeworldeoakjoinery.co.uk
Web: www.oldeworldeoakjoinery.co.uk
Product Type: 1, 2, 6, 7

PARAGON JOINERY
Systems House, Eastbourne Road, Blindley Heath, Surrey, RH7 6JP
Area of Operation: UK (Excluding Ireland)
Tel: 01342 836300 **Fax:** 01342 836364
Email: sales@paragonjoinery.com
Web: www.paragonjoinery.com
Product Type: 1, 2, 3, 4, 5, 6, 7, 8, 9
Other Info: ✐
Material Type: A) 1, 2, 4, 5, 15

PELLFOLD PARTHOS LTD
1 The Quadrant, Howarth Road, Maidenhead, Berkshire, SL6 1AP
Area of Operation: UK & Ireland
Tel: 01628 773353
Fax: 01628 773363
Email: sales@pellfoldparthos.co.uk
Web: www.designs4space.com
Product Type: 5

PERIOD OAK DOORS LTD
Garden House, Prisleilly Lane, Castlemorris, Pembrokeshire, SA62 5EH
Area of Operation: UK (Excluding Ireland)
Tel: 01348 841041
Fax: 01348 840934
Email: perioddoors@yahoo.co.uk
Web: www.perioddoors.com
Product Type: 1, 2, 7

PREMDOR
Gemini House, Hargreaves Road, Groundwell Industrial Estate, Swindon, Wiltshire, SN25 5AJ
Area of Operation: UK & Ireland
Tel: 0870 990 7998
Email: ukmarketing@premdor.com
Web: www.premdor.co.uk
Product Type: 1, 2, 3, 4, 5, 6, 7, 9

REGENCY JOINERY
Unit 18, Leigh Road, Haine Road Industrial Estate, Ramsgate, Kent, CT12 5EU
Area of Operation: East England, Greater London, Midlands & Mid Wales, South East England
Tel: 07782 517392
Email: enquiries@regency-joinery.co.uk
Web: www.regency-joinery.co.uk
Product Type: 2, 3, 4, 7

REYNAERS LTD
Kettles Wood Drive, Birmingham, West Midlands, B32 3DB
Area of Operation: UK & Ireland
Tel: 0121 421 1999
Fax: 0121 421 9797
Email: reynaersltd@reynaers.com
Web: www.reynaers.com
Product Type: 5

RHONES JOINERY (WREXHAM) LTD
Mold Road Industrial Estate, Gwersyllt, Wrexham, LL11 4AQ
Area of Operation: UK (Excluding Ireland)
Tel: 01978 262488
Fax: 01978 262488
Email: info@rhonesjoinery.co.uk
Web: www.rhonesjoinery.co.uk
Product Type: 2, 4, 7
Material Type: A) 2, 4, 5, 6, 12

ROBERT J TURNER & CO
Roe Green, Sandon, Buntingford, Herts, SG9 0QE
Area of Operation: East England, Greater London
Tel: 01763 288371
Fax: 01763 288440
Email: sales@robertjturner.co.uk
Web: www.robertjturner.co.uk
Product Type: 1, 2, 3, 4, 5, 6, 7, 8, 9
Other Info: ✐ ✋

ROBERT MILLS ARCHITECTURAL ANTIQUES
Narroways Road, Eastville, Bristol, BS2 9XB
Area of Operation: Worldwide
Tel: 0117 955 6542
Fax: 0117 955 8146
Email: Info@rmills.co.uk
Web: www.rmills.co.uk
Product Type: 1, 2, 3, 4, 5, 8, 9
Other Info: ▽

SABRINA OAK DOORS LTD
Alma Street, Mountfields, Shrewsbury, Shropshire, SY3 8QL
Area of Operation: UK (Excluding Ireland)
Tel: 01743 357977
Fax: 01743 352233
Email: mail@oakdoors.co.uk
Web: www.oakdoors.co.uk
Product Type: 1, 2, 3, 4, 5, 7, 9
Other Info: ✐ ✋
Material Type: A) 2

SAFEGUARD DOORS LIMITED
Units 6-9 Bridge Business Park, Bridge Street, Wednesbury, West Midlands, WS10 0AW
Area of Operation: Worldwide
Tel: 0121 556 9138
Fax: 0121 556 7275
Email: sales@safeguarddoors.co.uk
Web: www.safeguarddoors.co.uk
Product Type: 1, 2, 3, 6, 9

SANDERSON'S FINE FURNITURE
Unit 5 & 6, The Village Workshop, Four Crosses Business Park, Four Crosses, Powys, SY22 6ST
Area of Operation: UK (Excluding Ireland)
Tel: 01691 830075
Fax: 01691 830075
Email: sales@sandersonsfinefurniture.co.uk
Web: www.sandersonsfinefurniture.co.uk
Product Type: 1, 2, 3, 4, 7, 9
Other Info: ✐ ✋

SASH CRAFT
Unit 1 Mill Farm, Kelston, Bath, Somerset, BA1 9AQ
Area of Operation: South West England & South Wales
Tel: 01225 424434
Fax: 01225 424435
Email: pbale@sashcraftofbath.co.uk
Web: www.sashcraftofbath.co.uk
Product Type: 1, 2, 3, 4, 5, 7
Material Type: B) 8, 12

SCANDINAVIAN WINDOW SYSTEMS LTD
10 Eldon Street, Tuxford, Newark, Nottinghamshire, NG22 0LH
Area of Operation: UK & Ireland
Tel: 01777 871847
Fax: 01777 872650
Email: sue@scandinavian-windows.co.uk
Web: www.scandinavian-windows.co.uk
Product Type: 1, 2, 3, 4, 5, 6, 9
Other Info: ECO
Material Type: A) 1, 2, 3, 4, 5, 6, 7, 8, 9, 10, 12, 14

SCOTTS OF THRAPSTON LTD.
Bridge Street, Thrapston, Northamptonshire, NN14 4LR
Area of Operation: UK & Ireland
Tel: 01832 732366
Fax: 01832 733703
Email: julia@scottsofthrapston.co.uk
Web: www.scottsofthrapston.co.uk
Product Type: 1, 2, 3, 5, 6, 8, 9
Other Info: ✐

SOLARLUX SYSTEMS LTD
Holmfield Lane, Wakefield, West Yorks, WF2 7AD
Area of Operation: Worldwide
Tel: 01924 204444
Fax: 01924 204455
Email: info@solarlux.co.uk
Web: www.solarlux.uk.com
Product Type: 5

SPONSORED BY: ULTRAFRAME www.ultraframe-conservatories.co.uk

SPECTUS SYSTEMS
Queens Avenue, Macclesfield, Cheshire, SK10 2BN
Area of Operation: UK & Ireland
Tel: 01625 420400 **Fax:** 01625 501418
Email: contact@spectus.co.uk
Web: www.spectus.co.uk

STANDARDS GROUP
Bentley Hall Barn, Alkmonton,
Ashbourne, Derbyshire, DE6 3DJ
Area of Operation: UK & Ireland
Tel: 01335 330263 **Fax:** 01335 330922
Email: uk@standardsgroup.com
Web: www.standardsgroup.com
Product Type: 1, 2, 3, 6, 9

STRATHEARN STONE AND TIMBER LTD
Glenearn, Bridge of Earn, Perth,
Perth and Kinross, PH2 9HL
Area of Operation: North East England, Scotland
Tel: 01738 813215 **Fax:** 01738 815946
Email: info@stoneandoak.com
Web: www.stoneandoak.com

STROUDS WOODWORKING COMPANY LTD.
Ashmansworthy, Woolsery, Bideford,
Devon, EX39 5RE
Area of Operation: South West England
and South Wales
Tel: 01409 241624 **Fax:** 01409 241769
Email: enquires@doormakers.co.uk
Web: www.doormakers.co.uk **Other Info:** ✐ ✋
Product Type: 1, 2, 3, 4, 5, 6, 7, 8, 9
Material Type: A) 2, 3, 4, 5, 6, 7, 8, 10

SUN PARADISE UK LTD
Phoenix Wharf, Eel Pie Island,
Twickenham, Middlesex, TW1 3DY
Area of Operation: UK & Ireland
Tel: 0870 240 7604 **Fax:** 0870 240 7614
Email: info@sunparadise.co.uk
Web: www.sunparadise.co.uk **Product Type:** 5

SUNFOLD SYSTEMS
Sunfold House, Wymondham Business Park,
Chestnut Drive, Wymondham, Norfolk, NR18 9SB
Area of Operation: Worldwide
Tel: 01953 423423 **Fax:** 01953 423430
Email: info@sunfoldsystems.co.uk
Web: www.sunfold.com
Product Type: 1, 2, 3, 5, 6 **Other Info:** ✐

THE DISAPPEARING DOOR COMPANY
3 Southern House, Anthony's Way, Medway City
Estate, Strood, Rochester, Kent, ME2 4DN
Area of Operation: UK (Excluding Ireland)
Tel: 01634 720077
Fax: 01634 720111
Email: ddc@johnplanck.co.uk
Web: www.disappearingdoors.co.uk
Product Type: 1, 3, 5, 9 **Material Type:** H) 2

THE REAL DOOR COMPANY
Unit 5, Cadwell Lane, Hitchen,
Hertfordshire, SG4 0SA
Area of Operation: UK & Ireland
Tel: 01462 451230 **Fax:** 01462 440459
Email: sales@realdoor.co.uk
Web: www.realdoor.co.uk
Product Type: 1, 2, 6, 7 **Other Info:** ✐

THE REAL OAK DOOR
AND FLOORING COMPANY
5 Robin Way, Sudbury, Suffolk, CO10 7PF
Area of Operation: UK (Excluding Ireland)
Tel: 01787 310051 **Fax:** 01787 312497
Email: valentinecollins@aol.com
Web: www.realoakdoor.co.uk
Product Type: 7 **Material Type:** A) 2

TIMBER RECLAIMED
PO Box 33727, Greater London, SW3 4XE
Area of Operation: UK & Ireland
Tel: 0207 824 9200
Fax: 0207 824 9207
Email: elaine.barker@dmbarker.com
Web: www.timberreclaimed.com
Product Type: 1, 2, 7 **Other Info:** ▽

TODD DOORS LTD
112-116 Church Rd, Northolt, Greater London, UB5 5AE
Area of Operation: UK (Excluding Ireland)
Tel: 0208 845 2493 **Fax:** 0208 845 7579
Web: www.todd-doors.co.uk **Other Info:** ECO ✐
Product Type: 1, 2, 3, 4, 5, 6, 7, 8, 9
Material Type: A) 1, 2, 3, 4, 5, 6, 7

TOMPKINS LTD
High March Close, Long March Industrial Estate,
Daventry, Northamptonshire, NN11 4EZ
Area of Operation: UK (Excluding Ireland)
Tel: 01327 877187 **Fax:** 01327 310491
Email: info@tompkinswood.co.uk
Web: www.tompkinswood.co.uk **Other Info:** ✐ ✋
Product Type: 1, 2, 3, 4, 5, 6, 7
Material Type: A) 1, 2, 3, 4, 5, 6, 7, 8, 9, 10, 11, 12, 13, 14

TRADITIONAL OAK & TIMBER COMPANY
P O Stores, Haywards Heath Road, North Chailey,
Nr Lewes, East Sussex, BN8 4EY
Area of Operation: Worldwide
Tel: 01825 723648 **Fax:** 01825 722215
Email: info@tradoak.co.uk
Web: www.tradoak.com **Product Type:** 1, 2, 4, 7

TRANIK HOUSE DOORSTORE
63 Cobham Road, Ferndown Industrial Estate,
Wimborne, Dorset, BH21 7QA
Area of Operation: UK (Excluding Ireland)
Tel: 01202 872211
Email: sales@tranik.co.uk
Web: www.tranik.co.uk ⌃

VICAIMA LIMITED
c/o Lundie Marketing, 2 The Square,
Old Town, Swindon, Wiltshire, SN1 3EB
Area of Operation: UK & Ireland
Tel: 01793 488511 **Fax:** 01793 541888
Email: info@lundie.co.uk **Web:** www.vicaima.com
Product Type: 1, 2, 3, 6
Material Type: A) 1, 2, 3, 4, 5, 6, 7

WOODHOUSE TIMBER
Unit 6 Quarry Farm Industrial Estate,
Staplecross Road, Bodiam, East Sussex, TN32 5RA
Area of Operation: UK (Excluding Ireland)
Tel: 01580 831700 **Fax:** 01580 830365
Email: info@woodhousetimber.co.uk
Web: www.woodhousetimber.co.uk
Other Info: ✐ ✋ **Product Type:** 2, 7, 9

WOODLAND PRODUCTS DESIGN LTD
St.Peter's House, 6 Cambridge Road,
Kingston Upon Thames, Surrey, KT1 3JY
Area of Operation: UK (Excluding Ireland)
Tel: 0208 547 2171 **Fax:** 0208 547 1722
Email: enquiries@woodland-products.co.uk
Web: www.woodland-products.co.uk
Product Type: 4

EXTERNAL DOORS

KEY

PRODUCT TYPES: 1= Plain 2 = Panelled
3 = Glazed 4 = French 5 = Folding / Sliding
6 = Ledged / Braced 7 = Stable 8 = Other

SEE ALSO: GARAGES AND OUTHOUSES -
Garage Doors

OTHER: ▽ Reclaimed ⌃ On-line shopping
✐ Bespoke ✋ Hand-made ECO Ecological

AMBASSADOOR WINDOWS & DOORS LTD
18 Bidwell Road, Rackheath Industrial Estate,
Rackheath, Norwich, Norfolk, NR13 6PT
Area of Operation: East England, Greater London,
Midlands & Mid Wales
Tel: 01603 720332
Email: enquiries@ambassadoor.co.uk
Web: www.ambassadoor.co.uk
Product Type: 1, 2, 3, 4, 5, 6 **Material Type:** A) 1

ANDERSEN / BLACK MILLWORK CO.
Andersen House, Dallow Street, Burton on Trent,
Staffordshire, DE14 2PQ
Area of Operation: UK & Ireland
Tel: 01283 511122 **Fax:** 01283 510863
Email: info@blackmillwork.co.uk
Web: www.blackmillwork.co.uk
Product Type: 3, 4, 5
Material Type: A) 2, 5

APROPOS TECTONIC LTD
Greenside House, Richmond Street,
Ashton Under Lyne, Lancashire, OL6 7ES
Area of Operation: UK & Ireland
Tel: 0870 777 0320 **Fax:** 0870 777 0323
Email: enquiries@apropos-tectonic.com
Web: www.apropos-tectonic.com
Product Type: 5

ASTON HOUSE JOINERY
Aston House, Tripontium Business Centre,
Newton Lane, Rugby, Warwickshire, CV23 0TB
Area of Operation: UK & Ireland
Tel: 01788 860032
Fax: 01788 860614
Email: mailbox@ovationwindows.co.uk
Web: www.astonhousejoinery.co.uk
Product Type: 1, 2, 3, 4, 5, 6, 7
Other Info: ECO ✐ ✋
Material Type: A) 1, 2, 12

B & M WINDOW & KITCHEN CENTRE
2-6 Whitworth Drive, Aycliffe Industrial Park,
Durham, DL5 6SZ
Area of Operation: Worldwide
Tel: 01325 308888
Fax: 01325 316002
Email: sales@bandmhomeimprovements.com
Web: www.bandmhomeimprovements.com ⌃
Product Type: 1, 2, 3, 4, 5, 7

BARHAM & SONS
58 Finchley Avenue, Mildenhall,
Bury St. Edmunds, Suffolk, IP28 7BG
Area of Operation: UK (Excluding Ireland)
Tel: 01638 665759 / 711611
Fax: 01638 663913
Email: info@barhamwoodfloors.com
Product Type: 1, 2, 3, 6

BARTHOLOMEW
Rakers Yard, Rake Road, Milland,
West Sussex, GU30 7JS
Area of Operation: Europe
Tel: 01428 742800 **Fax:** 01428 743801
Email: denise@bartholomew-conservatories.co.uk
Web: www.bartholomew-conservatories.co.uk
Product Type: 1, 2, 3, 4, 5, 6, 7
Other Info: ECO
Material Type: A) 2, 10

BDG GROUP LIMITED
5 Wenlock Road, Lurgan, Craigavon,
Co Armagh, BT66 8QR
Area of Operation: Worldwide
Tel: 0283 832 7741 **Fax:** 0283 832 4358
Email: info@bdg.co.uk
Web: www.bdg.co.uk
Product Type: 1, 2, 3, 4, 5, 8
Material Type: D) 1

BETTER LIVING PRODUCTS UK LTD
14 Riverside Business Park,
Stansted, Essex, CM24 8PL
Area of Operation: UK & Ireland
Tel: 01279 812958 **Fax:** 01279 817771
Email: info@dispenser.co.uk
Web: www.dispenser.co.uk ⌃
Product Type: 1 **Material Type:** C) 4

BIRCHDALE GLASS LTD
Unit L, Eskdale Road, Uxbridge,
Greater London, UB8 2RT
Area of Operation: UK & Ireland
Tel: 01895 259111 **Fax:** 01895 810087
Email: info@birchdaleglass.com
Web: www.birchdaleglass.com
Product Type: 1, 2, 3, 4, 5, 7, 8

BIRTLEY BUILDING PRODUCTS LTD
Mary Avenue, Birtley, County Durham, DH3 1JF
Area of Operation: UK & Ireland
Tel: 0191 410 6631 **Fax:** 0191 410 0650
Email: info@birtley-building.co.uk
Web: www.birtley-building.co.uk
Product Type: 3
Material Type: C) 2

BOWATER WINDOWS
NEWBUILD DIVISION
Water Orton Lane, Minworth,
Sutton Coldfield, West Midlands, B76 9BW
Area of Operation: UK (Excluding Ireland)
Tel: 0121 749 3000 **Fax:** 0121 749 8210
Email: info@theprimeconnection.co.uk
Web: www.whs-halo.co.uk
Product Type: 1, 2, 3, 4, 7, 8
Material Type: D) 1, 6

BROADLEAF TIMBER
Llandeilo Road Industrial Estate,
Carms, Carmarthenshire, SA18 3JG
Area of Operation: UK & Ireland
Tel: 01269 851910 **Fax:** 01269 851911
Email: sales@broadleaftimber.com
Web: www.broadleaftimber.com
Product Type: 1, 2, 3, 4, 6, 7
Other Info: ✐ ✋
Material Type: A) 2

BROXWOOD (SCOTLAND) LTD
Inveralmond Way, Inveralmond Industrial Estate,
Perth, Perth and Kinross, PH1 3UQ
Area of Operation: UK & Ireland
Tel: 01738 444456 **Fax:** 01738 444452
Email: sales@broxwood.com
Web: www.broxwood.com
Product Type: 1, 2, 3, 4, 5
Other Info: ECO ✐
Material Type: A) 1, 2, 10

CAREY & FOX LTD
51 Langthwaite Business Park, South Kirkby,
Pontefract, West Yorkshire, WF9 3NR
Area of Operation: UK & Ireland
Tel: 01977 608069 **Fax:** 01977 646791
Email: enquiries@careyandfox.co.uk
Web: www.careyandfox.co.uk
Product Type: 1, 2, 3, 4, 5, 6, 7, 8
Material Type: A) 2

CENTURION
Westhill Business Park, Arnhall Business Park,
Westhill, Aberdeen, Aberdeenshire, AB32 6UF
Area of Operation: Scotland
Tel: 01224 744440 **Fax:** 01224 744819
Email: info@centurion-solutions.co.uk
Web: www.centurion-solutions.co.uk
Product Type: 5

CHAMBERLAIN & GROVES LTD -
THE DOOR & SECURITY STORE
101 Boundary Road,
Walthamstow, London, E17 8NQ
Area of Operation: UK (Excluding Ireland)
Tel: 0208 520 6776 **Fax:** 0208 520 2190
Email: ken@securedoors.co.uk
Web: www.securedoors.co.uk
Product Type: 1, 2, 3, 4, 7, 8
Other Info: ECO ✐ ✋
Material Type: A) 1, 2

CHANTRY HOUSE OAK LTD
Ham Farm, Church Lane, Oving,
Chichester, West Sussex, PO20 2BT
Area of Operation: South East England
Tel: 01243 776811 **Fax:** 01243 839873
Email: sales@chantryhouseoak.co.uk
Web: www.chantryhouseoak.co.uk

CHESHIRE DOOR COMPANY
Paradise Mill, Old Park Lane,
Macclesfield, Cheshire, SK11 6TJ
Area of Operation: UK & Ireland
Tel: 01625 421221 **Fax:** 01625 421422
Email: info@cheshiredoorcompany.co.uk
Web: www.cheshiredoorcompany.co.uk

for those in the home improvement market, here is a new opportunity for you

CHIPPING NORTON GLASS LTD
Units 1 & 2, Station Yard Industrial Estate,
Chipping Norton, Oxfordshire, OX7 5HX
Area of Operation: Midlands & Mid Wales,
South West England and South Wales
Tel: 01608 643261 **Fax:** 01608 641768
Email: gill@cnglass.plus.com
Web: www.chippingnortonglass.co.uk
Product Type: 3, 4, 5, 7
Material Type: C) 1

CLASSIC PVC HOME IMPROVEMENTS
46 Stepney Place, Llanelli,
Carmarthenshire, SA15 1SE
Area of Operation: South West England
and South Wales
Tel: 0800 064 9494 **Fax:** 01554 775086
Email: enquiries@classic.uk.com
Web: www.classic.uk.com
Product Type: 1, 2, 3, 4, 7
Material Type: D) 1, 6

CLEARVIEW WINDOWS & DOORS
Unit 14, Sheddingdean Industrial Estate,
Marchants Way, Burgess Hill,
West Sussex, RH15 8QY
Area of Operation: UK (Excluding Ireland)
Tel: 01444 250111
Fax: 01444 250678
Email: sales@clearviewsussex.co.uk
Web: www.clearviewsussex.co.uk

CLEMENT STEEL WINDOWS
Clement House, Haslemere, Surrey, GU27 1HR
Area of Operation: Worldwide
Tel: 01428 643393
Fax: 01428 644436
Email: info@clementwg.co.uk
Web: www.clementsteelwindows.com
Product Type: 3

CONSERVATORY & WINDOW WORLD LTD
Watling Rd, Bishop Aukland, Durham, DL14 9AU
Area of Operation: North East England
Tel: 01388 458 088 **Fax:** 01388 458 518
Email: paldennis20@aol.com
Web: www.conservatoryandwindowworld.co.uk
Product Type: 2, 3, 4, 5, 7
Material Type: D) 1, 3, 6

CONSERVATORY OUTLET
Unit 8 Headway Business Park, Denby Dale Road,
Wakefield, West Yorkshire, WF2 7AZ
Area of Operation: North East England
Tel: 01924 881920
Web: www.conservatoryoutlet.co.uk
Product Type: 1, 2, 3, 4, 5, 7, 8
Material Type: D) 1

CONSTRUCTION TECHNICAL SERVICES
Dunedin House, Alexandra Road,
Penzance, Cornwall, TR18 4LZ
Area of Operation: UK & Ireland
Tel: 01736 330303 **Fax:** 01736 360497
Email: CTecS@aol.com
Web: www.hometown.aol.co.uk/carpinjohn
Product Type: 5, 8

COUNTY JOINERY (SOUTH EAST) LTD
Unit 300, Vinehall Business Centre, Vinehall Road,
Mountfield, Robertsbridge, East Sussex, TN32 5JW
Area of Operation: Greater London, South East
England
Tel: 01424 871500 **Fax:** 01424 871550
Email: countyjoinery@aol.com
Web: www.countyjoinery.co.uk
Product Type: 1, 2, 3, 4, 5, 6, 7, 8
Material Type: A) 2, 3, 4, 5, 6, 7, 8, 9, 10,
11, 12, 13, 14

CREATE JOINERY
Worldsend, Llanmadoc, Swansea, SA3 1DB
Area of Operation: Greater London, Midlands & Mid
Wales, North West England & North Wales, South
East England, South West England & South Wales
Tel: 01792 386677 **Fax:** 01792 386677
Email: mail@create-joinery.co.uk
Web: www.create-joinery.co.uk
Product Type: 1, 2, 3, 4, 5, 6, 7, 8

CROXFORD'S JOINERY
MANUFACTURERS & WOODTURNERS
Meltham Joinery, Works New Street, Meltham,
Holmfirth, West Yorkshire, HD9 5NT
Area of Operation: UK (Excluding Ireland)
Tel: 01484 850892 **Fax:** 01484 850969
Email: ralph@croxfords.demon.co.uk
Web: www.croxfords.co.uk
Product Type: 1, 2, 3, 4, 5, 6, 7, 8

CUSTOM WOOD PRODUCTS
Cliffe Road, Easton on the Hill, Stamford,
Lincolnshire, PE9 3NP
Area of Operation: East England
Tel: 01780 755711
Fax: 01780 480834
Email: customwoodprods@aol.com
Web: www.cwpuk.co.uk
Product Type: 1, 2, 3, 4, 5, 6, 7
Other Info: ECO ✍ ✋
Material Type: A) 2, 3, 4, 5, 6

DICTATOR DIRECT
Inga House, Northdown Business Park,
Ashford Road, Lenham, Kent, ME17 2DL
Area of Operation: Worldwide
Tel: 01622 854770
Fax: 01622 854771
Email: mail@dictatordirect.com
Web: www.dictatordirect.com

DIRECTDOORS.COM
Bay 5 Eastfield Industrial Estate, Eastfield Drive,
Penicuik, Lothian, EH26 8JA
Area of Operation: UK (Excluding Ireland)
Tel: 0131 669 6310
Fax: 0131 657 1578
Email: info@directdoors.com
Web: www.directdoors.com ✍
Product Type: 1, 2, 3, 4, 5, 6, 7, 8

DISTINCTIVE DOORS
14 & 15 Chambers Way, Newton Chambers Road,
Chapeltown, Sheffield, South Yorkshire, S35 2PH
Area of Operation: UK (Excluding Ireland)
Tel: 0114 220 2250 **Fax:** 0114 220 2254
Email: enquiries@distinctivedoors.co.uk
Web: www.distinctivedoors.co.uk ✍
Product Type: 2, 3, 4, 6, 7
Material Type: A) 1, 2, 6

DISTINCTIVE DOORS

Area of Operation: UK & Ireland
Tel: 0114 220 2250 **Fax:** 0114 220 2254
Email: enquiries@distinctivedoors.co.uk
Web: www.distinctivedoors.co.uk
Product Type: 2, 3, 4, 6, 7

Suppliers of High Class External & Internal Doors.
Exclusive Product Range Triple Glazed Doors.
Oak Internal Engineered Doors.
Hardwood Doors, Clear Pine Doors.

DIY SASH WINDOWS
2 - 6 Whitworth Drive, Aycliffe Industrial Park,
Newton Aycliffe, Durham, DL56SZ
Area of Operation: UK & Ireland
Tel: 01325 308888 **Fax:** 01325 316002
Email: sales@diysashwindows.co.uk
Web: www.diysashwindows.co.uk
Product Type: 1, 2, 3, 4, 5, 7

DORLUXE LIMITED
30 Pinbush Road, Lowestoft,
Suffolk, NR33 7NL
Area of Operation: UK & Ireland
Tel: 01502 567744 **Fax:** 01502 567743
Email: info@dorluxe.co.uk
Web: www.dorluxe.co.uk
Product Type: 3, 7

DRUMMONDS ARCHITECTURAL ANTIQUES LTD
Kirkpatrick Buildings, 25 London Road (A3),
Hindhead, Surrey, GU36 6AB
Area of Operation: Worldwide
Tel: 01428 609444
Fax: 01428 609445
Email: davidcox@drummonds-arch.co.uk
Web: www.drummonds-arch.co.uk
Product Type: 1, 2, 3, 6
Other Info: ▽ **Material Type:** A) 2

DYNASTY DOORS
Unit 3, The Micro Centre, Gillette Way,
Reading, Berkshire, RG2 0LR
Area of Operation: South East England,
South West England and South Wales
Tel: 0118 987 4000 **Fax:** 0118 921 2999
Email: Martin@dynastydoors.co.uk
Web: www.dynastydoors.co.uk
Product Type: 1, 2, 3, 4, 7

ENGELS WINDOWS & DOORS LTD
1 Kingley Centre, Downs Road, West Stoke,
Chichester, West Sussex, PO18 9HJ
Area of Operation: UK (Excluding Ireland)
Tel: 01243 576633 **Fax:** 01243 576644
Email: admin@engels.co.uk
Web: www.engels.co.uk

ENVIROZONE
Encon House, 54 Ballycrochan Road,
Bangor, Co Down, BT19 6NF
Area of Operation: UK & Ireland
Tel: 02891 477799 **Fax:** 02891 452463
Email: info@envirozone.co.uk
Web: www.envirozone.co.uk
Product Type: 1, 2, 3, 4, 5, 6, 7, 8

ESOGRAT LTD
Caldervale Works, River Street, Brighouse,
Huddersfield, West Yorkshire, HD6 1JS
Area of Operation: UK & Ireland
Tel: 01484 716228 **Fax:** 01484 400107
Email: info@esograt.com
Web: www.esograt.com
Product Type: 1, 2, 3, 4, 7, 8

EUROCELL PROFILES LTD
Fairbrook House, Clover Nook Road,
Alfreton, Derby, Derbyshire, DE55 4RF
Area of Operation: UK (Excluding Ireland)
Tel: 01773 842100 **Fax:** 01773 842109
Email: eric.gale@eurocell.co.uk
Web: www.eurocell.co.uk
Product Type: 1, 3, 4, 5, 7, 8
Material Type: D) 1, 6

EVEREST LTD
Sopers Road, Cuffley, Hertfordshire, EN6 4SG
Area of Operation: UK & Ireland
Tel: 0800 010123 **Web:** www.everest.co.uk ✍
Product Type: 1, 3, 5

EVERGREEN DOOR
Unit 1, Oakwell Park Industrial Estate,
Birstall, West Yorkshire, WF17 9LU
Area of Operation: Worldwide
Tel: 01924 423171 **Fax:** 01924 423175
Email: andrewgrogan@evergreendoor.co.uk
Web: www.evergreendoor.co.uk
Product Type: 1, 2, 3, 4, 5, 6, 7
Material Type: C) 2, 4, 10, 11, 13, 14, 16, 18

FAIRMITRE WINDOWS & JOINERY LTD
2A Cope Road, Banbury, Oxfordshire, OX16 2EH
Area of Operation: UK (Excluding Ireland)
Tel: 01295 268441 **Fax:** 01295 268468
Email: info@fairmitrewindows.co.uk
Web: www.fairmitrewindows.co.uk
Product Type: 2, 3, 4, 5, 6, 7

FAIROAK TIMBER PRODUCTS LIMITED
Manor Farm, Chilmark, Salisbury, Wiltshire, SP3 5AG
Area of Operation: UK (Excluding Ireland)
Tel: 01722 716779 **Fax:** 01722 716761
Email: sales@fairoak.co.uk
Web: www.fairoakwindows.co.uk
Product Type: 1, 2, 3, 4, 7
Material Type: A) 2, 12

FAIROAK TIMBER PRODUCTS LTD

Area of Operation: UK
Tel: 01722 716779
Fax: 01722 716761
Email: sales@fairoak.co.uk
Web: www.fairoakwindows.co.uk
Product Type: 1, 2, 3, 4, 7
Material Type: A) 2, 12

High performance timber "Chilmark" doorsets,
manufactured to order in European Oak and
durable hardwoods, supplied complete with
multi-point locks

FITZROY JOINERY
Garden Close, Langage Industrial Estate,
Plympton, Devon, PL7 5EU
Area of Operation: UK & Ireland
Tel: 0870 428 9110
Fax: 0870 428 9111
Email: admin@fitzroy.co.uk
Web: www.fitzroy.co.uk ✍
Product Type: 1, 2, 3, 4, 5, 7
Other Info: ECO ✍
Material Type: A) 1, 2, 3, 4, 5, 6, 7, 8,
9, 10, 12, 14

FLETCHER JOINERY
261 Whessoe Road, Darlington,
Durham, DL3 0YL
Area of Operation: North East England
Tel: 01325 357347
Fax: 01325 357347
Email: enquiries@fletcherjoinery.co.uk
Web: www.fletcherjoinery.co.uk
Product Type: 1, 2, 3, 4, 5, 6, 7, 8

FOLDING SLIDING DOORS LIMITED
FSD Works, Hopbine Avenue,
West Bowling, Bradford,
West Yorkshire, BD5 8ER
Area of Operation: UK (Excluding Ireland)
Tel: 0845 644 6630 **Fax:** 0845 644 6631
Email: info@foldingslidingdoor.com
Web: www.foldingslidingdoor.com ✍
Product Type: 5

FRANKLIN LEEDS LLP
Carlton Works, Cemetery Road,
Yeadon, Leeds, West Yorkshire, LS19 7BD
Area of Operation: UK & Ireland
Tel: 0113 250 2991 **Fax:** 0113 250 0991
Email: david.franklin@franklinwindows.co.uk
Web: www.franklinwindows.co.uk
Product Type: 1, 2, 3, 4, 5, 7, 8

GAP
Partnership Way, Shadsworth Business Park,
Blackburn, Lancashire, BB1 2QP
Area of Operation: Midlands & Mid Wales, North
East England, North West England and North Wales
Tel: 01254 682888
Web: www.gap.uk.com
Product Type: 1, 2, 3, 7 **Material Type:** D) 1

BUILDING STRUCTURE & MATERIALS

GOODMILL LTD
Ballybay, County Monaghan, Republic of Ireland,
Area of Operation: UK & Ireland
Tel: +353 429 741 771
Fax: +353 429 741 771
Email: maggsmills@eircom.net
Web: www.goodmillsecurity.com
Product Type: 2, 8
Material Type: C) 2

GREEN BUILDING STORE
11 Huddersfield Road, Meltham, Holmfirth,
West Yorkshire, HD9 4NJ
Area of Operation: UK & Ireland
Tel: 01484 854898
Fax: 01484 854899
Email: info@greenbuildingstore.co.uk
Web: www.greenbuildingstore.co.uk
Product Type: 1, 2, 3, 4, 5, 7, 8
Other Info: ECO
Material Type: B) 8

HAYMANS TIMBER PRODUCTS
Haymans Farm, Hocker Lane,
Over Alderley, Macclesfield,
Cheshire, SK10 4SD
Area of Operation: UK & Ireland
Tel: 01625 590098
Fax: 01625 586174
Email: haymanstimber@aol.com
Web: www.haymanstimber.co.uk
Product Type: 1, 2, 3, 4, 5, 6, 7
Material Type: A) 2, 4, 6, 7

HOUSE OF BRASS
122 North Sherwood St, Nottingham,Notts, NG1 4EF
Area of Operation: Worldwide
Tel: 0115 947 5430 **Fax:** 0115 947 5430
Email: sales@houseofbrass.co.uk
Web: www.houseofbrass.co.uk
Product Type: 1, 2, 3, 4, 5

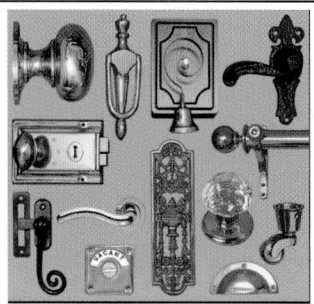

HOUSE OF BRASS LTD
Area of Operation: Worldwide
Tel: 0115 947 5430
Email: sales@houseofbrass.co.uk
Web: www.houseofbrass.co.uk
Product Type: 1, 2, 3, 4, 5 **Other Info:**
Materials Type: C) 5, 11, 14, 16 & 17

Architectural fittings range from classic to contemporary designs in a range of finishes. Also, brass curtain poles and stairrods made to size. Call for a FREE colour catalogue (24 Hrs).

HOWARTH WINDOWS AND DOORS LTD
The Dock, New Holland, Lincolnshire, DN19 7RT
Area of Operation: UK & Ireland
Tel: 01469 530577 **Fax:** 01469 531559
Email: abarker@howarth-timber.co.uk
Web: www.howarth-timber.co.uk

ID SYSTEMS
Sunflex House, Rhombus Business Park,
Diamond Road, Norwich, Norfolk, NR6 6NN
Area of Operation: UK & Ireland
Tel: 01603 408804 **Fax:** 01603 258648
Email: info@i-d-systems.co.uk
Web: www.i-d-systems.co.uk
Product Type: 5

IMAJ STEEL DOORS
45h Leyton Industrial Estate, Argall Way, London, E8 2NG
Area of Operation: Worldwide
Tel: 0207 275 0466 **Fax:** 0207 249 0176
Email: sales@imajsteeldoors.com
Web: www.imajsteeldoors.com **Material Type:** C) 2

IN DOORS
Beechinwood Farm, Beechinwood Lane,
Platt, Nr. Sevenoaks, Kent, TN15 8QN
Area of Operation: UK (Excluding Ireland)
Tel: 01732 887445 **Fax:** 01732 887446
Email: info@indoorsltd.co.uk
Web: www.indoorsltd.co.uk
Product Type: 1, 2, 3, 4, 5, 6, 7

J L JOINERY
Cockerton View, Grange Lane,
Preston, Lancashire, PR4 5JE
Area of Operation: UK (Excluding Ireland)
Tel: 01772 616123 **Fax:** 01772 619182
Email: mail@jljoinery.co.uk
Web: www.jljoinery.co.uk

JOHN LOWRY JOINERY
Area of Operation: UK (Excluding Ireland)
Tel: 01772 616123
Fax: 01772 619182
Email: mail@jljoinery.co.uk
Web: www.jljoinery.co.uk

John Lowry and his staff take pride in their work and attention to detail, which frequently exceeds customer's expectations. Their reputation is built on the high quality of the materials used, expert craftsmanship and quality of finish.

JB KIND LTD
Shobnall Street, Burton-on-Trent,
Staffordshire, DE14 2HP
Area of Operation: UK & Ireland
Tel: 01283 510210 **Fax:** 01283 511132
Email: info@jbkind.com **Web:** www.jbkind.com
Product Type: 1, 2, 3, 4 **Material Type:** A) 1, 2

JELD-WEN
Watch House Lane, Doncaster, South Yorkshire, DN5 9LR
Area of Operation: UK
Tel: 0870 126 0000 **Fax:** 01302 787303
Email: customer-services@jeld-wen.co.uk
Web: www.jeld-wen.co.uk
Product Type: 1, 2, 3, 5 **Material Type:** C) 18

JELD-WEN
Area of Operation: UK
Tel: 0870 126 0000 **Fax:** 01302 787303
Email: customer-services@jeld-wen.co.uk
Web: www.jeld-wen.co.uk

A 2, 3, 4, 5 and 6 leaf door manufactured in attractive hardwood, the Canberra, is JELD-WEN's first slide & fold patio door. Supplied factory glazed and basecoat stained as standard. Factory finishing in Hi-Build stain or paint, available as an option. See JELD-WEN for colour choices. Approved Documents L 2002 and M Compliant.

JOHN FLEMING & CO LTD
Silverburn Place, Bridge of Don,
Aberdeen, Aberdeenshire, AB23 8EG
Area of Operation: Scotland
Tel: 0800 085 8728
Fax: 01224 825377
Email: info@johnfleming.co.uk
Web: www.johnfleming.co.uk

JOINERY-PLUS
Bentley Hall Barn, Alkmonton, Ashbourne,
Derbyshire, DE6 3DJ
Area of Operation: UK (Excluding Ireland)
Tel: 07931 386233 **Fax:** 01335 330922
Email: info@joinery-plus.co.uk
Web: www.joinery-plus.co.uk
Product Type: 1, 2, 3, 4, 6, 7
Material Type: A) 1, 2, 3, 4, 5, 6, 7, 8, 9, 10, 11, 12, 13, 14

KIVA DOORS LIMITED
Bramingham Business Centre,
Enterprise Way, Luton,
Bedfordshire, LU3 4BU
Area of Operation: UK & Ireland
Tel: 0800 328 5851 **Fax:** 01582 584111
Email: info@kiva-doors.com
Web: www.kiva-doors.com
Product Type: 1, 2, 3, 4
Material Type: A) 1, 2, 3, 4, 5, 6, 7, 9, 12, 13

KS PROFILES LTD
Broad March, Long March Industrial Estate,
Daventry, Northamptonshire, NN11 4HE
Area of Operation: UK & Ireland
Tel: 01327 316960 **Fax:** 01327 876412
Email: sales@ksprofiles.com
Web: www.ksprofiles.com
Product Type: 1, 2, 3, 4, 7

LAMWOOD LTD
Unit 1, Riverside Works, Station Road,
Cheddleton, Staffordshire, ST13 7EE
Area of Operation: UK & Ireland
Tel: 01538 361888 **Fax:** 01538 361912
Email: sales@lamwoodltd.co.uk
Web: www.lamwoodltd.co.uk
Material Type: A) 2

LATTICE PERIOD WINDOWS
Unit 85 Northwick Business Centre, Blockley,
Moreton in Marsh, Gloucestershire, GI56 9RF
Area of Operation: UK & Ireland
Tel: 01386 701079 **Fax:** 01386 701114
Email: sales@lattice windows .net
Web: www.latticewindows.net
Product Type: 1, 2, 3, 4, 6
Other Info: ✋
Material Type: A) 2

LAWSONS
Gorst Lane, Off New Lane, Burscough,
Ormskirk, Lancashire, L40 ORS
Area of Operation: Worldwide
Tel: 01704 893998 **Fax:** 01704 892526
Email: info@traditionaltimber.co.uk
Web: www.traditionaltimber.co.uk
Other Info: ▽

MANSE MASTERDOR LIMITED
Hambleton Grove, Knaresborough,
North Yorkshire, HG5 0DB
Area of Operation: UK (Excluding Ireland)
Tel: 01423 866868 **Fax:** 01423 866368
Email: info@masterdor.co.uk
Web: www.masterdor.co.uk
Product Type: 1, 2, 3, 4, 7, 8
Other Info: ✎ **Material Type:** B) 1

MARVIN ARCHITECTURAL
Canal House, Catherine Wheel Road,
Brentford, Middlesex,
Greater London, TW8 8BD
Area of Operation: UK & Ireland
Tel: 0208 569 8222 **Fax:** 0208 560 6374
Email: sales@marvinUK.com
Web: www.marvin-architectural.com
Product Type: 1, 3, 4, 5, 7

MERLIN UK LIMITED
Unit 5 Fence Avenue, Macclesfield,
Cheshire, SK10 1LT
Area of Operation: North West England
and North Wales
Tel: 01625 424488 **Fax:** 0871 781 8967
Email: info@merlinuk.net
Web: www.merlindoors.co.uk
Product Type: 1, 2, 3, 4, 5, 7, 8

**MINCHINHAMPTON
ARCHITECTURAL SALVAGE CO**
Cirencester Road, Aston Downs,
Chalford, Stroud, Gloucestershire, GL6 8PE
Area of Operation: Worldwide
Tel: 01285 760886
Fax: 01285 760838
Email: masco@catbrain.com
Web: www.catbrain.com
Product Type: 1, 2, 3, 6

MULTIWOOD HOME IMPROVEMENTS LTD
1298 Chester Road, Stretford,
Manchester, M32 9AU
Area of Operation: Europe
Tel: 0161 866 9991 **Fax:** 0161 866 9992
Email: sales@multiwoodconservatories.co.uk
Web: www.multiwoodconservatories.co.uk
Product Type: 3, 4, 5

MUMFORD & WOOD
Tower Business Park, Kelvedon Road,
Tiptree, Essex, CO5 0LX
Area of Operation: Worldwide
Tel: 01621 818155
Fax: 01621 818175
Email: sales@mumfordwood.com
Web: www.mumfordwood.com
Product Type: 1, 2, 3, 4

NATIONAL DOOR COMPANY
Unit 55 Dinting Vale Business Park,
Dinting Vale, Glossop, Derbyshire, SK13 6JD
Area of Operation: UK (Excluding Ireland)
Tel: 01457 867079
Fax: 01457 868795
Email: sales@nationaldoor.co.uk
Web: www.nationaldoor.co.uk
Product Type: 1, 2, 3, 4, 6, 7, 8

NOISE STOP SYSTEMS
Unit 3 Acaster Industrial Estate,
Cowper Lane, Acaster Malbis,
York, North Yorkshire, YO23 2XB
Area of Operation: UK & Ireland
Tel: 0845 130 6269
Fax: 01904 709857
Email: info@noisestopsystems.co.uk
Web: www.noisestopsystems.co.uk ✎
Product Type: 1, 2, 4, 5, 7, 8

**NORBUILD TIMBER FABRICATION
& FINE CARPENTRY LTD**
Marcassie Farm, Rafford, Forres, Moray, IV36 2RH
Area of Operation: UK & Ireland
Tel: 01309 676865 **Fax:** 01309 676865
Email: norbuild@marcassie.fsnet.co.uk
Product Type: 2, 3, 8
Other Info: ECO ✎ ✋
Material Type: A) 1, 2, 6, 11

NORTH YORKSHIRE TIMBER
Standard House, Thurston Road,
Northallerton Business Park,
Northallerton, North Yorkshire, DL6 2NA
Area of Operation: UK (Excluding Ireland)
Tel: 01609 780777 **Fax:** 01609 777888
Email: sales@nytimber.co.uk
Web: www.nytimber.co.uk

NORTHWOOD FORESTRY LTD
Goose Green Lane, Nr Ashington,
Pulborough, West Sussex, RH20 2LW
Area of Operation: South East England
Tel: 01798 813029 **Fax:** 01798 813139
Email: enquiries@northwoodforestry.co.uk
Web: www.northwoodforestry.co.uk
Product Type: 6

OCTAVEWARD LTD
Balle Street Mill, Balle Street, Darwen,
Lancashire, BB3 2AZ
Area of Operation: Worldwide
Tel: 01254 773300 **Fax:** 01254 773950
Email: info@octaveward.com
Web: www.octaveward.com
Product Type: 1, 2, 3, 6, 7, 8
Other Info: ECO ▽ ✎ ✋
Material Type: D) 2, 6

OLDE WORLDE OAK JOINERY LTD
Unit 12, Longford Industrial Estate, Longford Road,
Cannock, Staffordshire, WS11 0DG
Area of Operation: Europe
Tel: 01543 469328
Fax: 01543 469341
Email: sales@oldeworldeoakjoinery.co.uk
Web: www.oldeworldeoakjoinery.co.uk
Product Type: 1, 2, 6, 7

ORIGINAL OAK
Ashlands, Burwash, East Sussex, TN19 7HS
Area of Operation: UK (Excluding Ireland)
Tel: 01435 882228
Fax: 01435 882228
Material Type: A) 2

OXFORD SASH WINDOW CO LTD.
Eynsham Park Estate Yard, Cuckoo Lane,
North Leigh, Oxfordshire, OX29 6PW
Area of Operation: Greater London, Midlands
& Mid Wales, South East England, South West
England & South Wales
Tel: 01993 883536
Fax: 01993 883027
Email: oxsash@globalnet.co.uk
Web: www.sashwindow.co.uk
Product Type: 1, 2, 3

PARAGON JOINERY
Systems House, Eastbourne Road,
Blindley Heath, Surrey, RH7 6JP
Area of Operation: UK (Excluding Ireland)
Tel: 01342 836300
Fax: 01342 836364
Email: sales@paragonjoinery.com
Web: www.paragonjoinery.com
Product Type: 1, 2, 3, 4, 5, 6, 7, 8
Other Info: ✎
Material Type: A) 1, 2, 15

PERIOD OAK DOORS LTD
Garden House, Prisleilly Lane,
Castlemorris, Pembrokeshire, SA62 5EH
Area of Operation: UK (Excluding Ireland)
Tel: 01348 841041
Fax: 01348 840934
Email: perioddoors@yahoo.co.uk
Web: www.perioddoors.com
Product Type: 1, 2, 3, 6, 7

PERIOD OAK DOORS LTD.
Area of Operation: UK (Excluding Ireland)
Tel: 01348 841041 **Fax:** 01348 840934
Email: perioddoors@yahoo.co.uk
Web: www.perioddoors.com
Product Type: 1, 2, 3, 6, 7

Quality bespoke doors [made to measure] designed to recreate the cottage style of years gone by, in both pine and beautiful oak, at affordable prices.

PERMACELL FINESSE LTD
Western Road, Silver End,
Witham, Essex, CM8 3BQ
Area of Operation: UK & Ireland
Tel: 01376 583241
Fax: 01376 583072
Email: sales@pfl.co.uk
Web: www.permacell-finesse.co.uk
Product Type: 1, 3, 4

PERMADOOR
Upton-on-Severn, Worcestershire, WR8 0RX
Area of Operation: UK (Excluding Ireland)
Tel: 01684 595200
Fax: 01684 594283
Email: info@permadoor.co.uk
Web: www.permadoor.co.uk
Product Type: 1, 3, 8
Material Type: D) 1, 3, 6

PORTICO GB LTD
Unit 9 Windmill Avenue, Woolpit Business Park,
Woolpit, Bury St Edmunds, Suffolk, IP30 9UP
Area of Operation: Greater London,
South East England
Tel: 01359 244299 **Fax:** 01359 244232
Email: info@portico-gb.co.uk
Web: www.portico-newbuild.co.uk
Product Type: 4, 5

PREMDOR
Gemini House, Hargreaves Road, Groundwell
Industrial Estate, Swindon, Wiltshire, SN25 5AJ
Area of Operation: UK & Ireland
Tel: 0870 990 7998
Email: ukmarketing@premdor.com
Web: www.premdor.co.uk
Product Type: 1, 2, 3, 4, 6, 7, 8

RATIONEL WINDOWS LTD
7 Avonbury Business Park, Howes Lane,
Bicester, Oxfordshire, OX26 2UA
Area of Operation: UK (Excluding Ireland)
Tel: 01869 248181
Fax: 01869 249693
Email: jrh@rationel.dk
Web: www.rationel.com
Product Type: 2, 3, 4, 5, 7
Material Type: B) 2, 13

REGENCY JOINERY
Unit 18, Leigh Road, Haine Road Industrial Estate,
Ramsgate, Kent, CT12 5EU
Area of Operation: East England, Greater London,
Midlands & Mid Wales, South East England
Tel: 07782 517392
Email: enquiries@regency-joinery.co.uk
Web: www.regency-joinery.co.uk
Product Type: 2, 3, 4, 6

SUNPARADISE

Contemporary aluminium
folding/sliding doors and roof systems

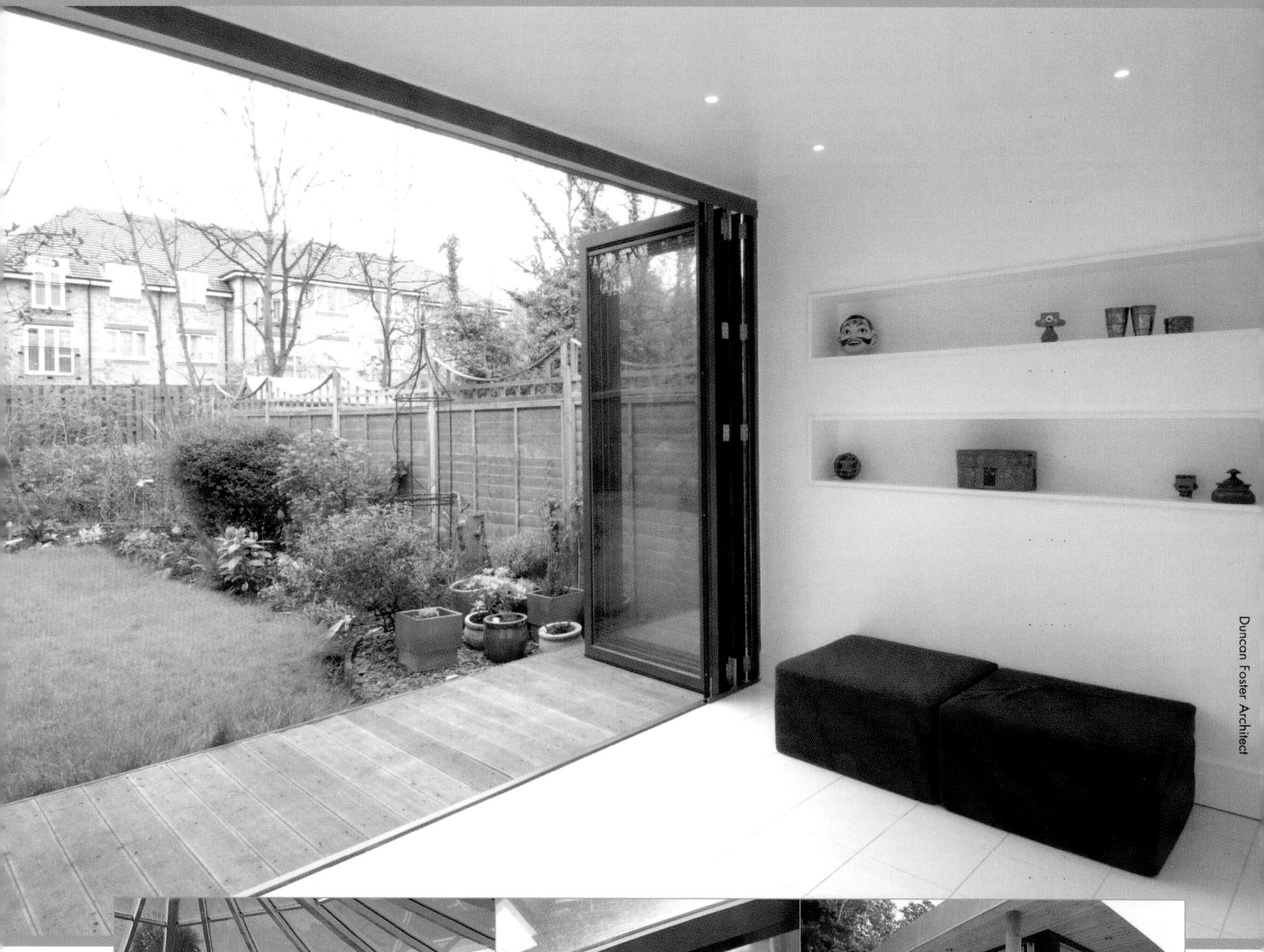

Duncan Foster Architect

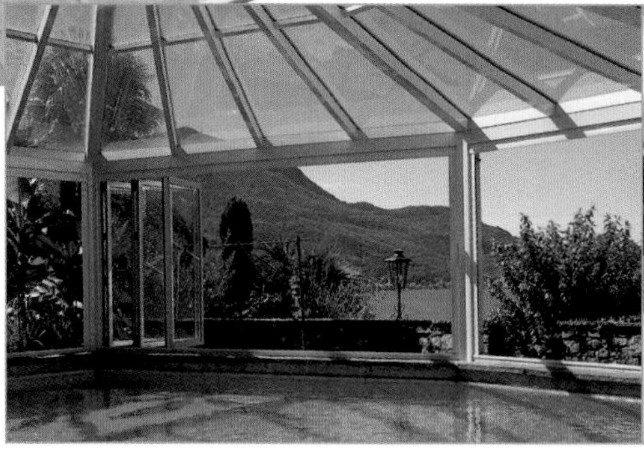

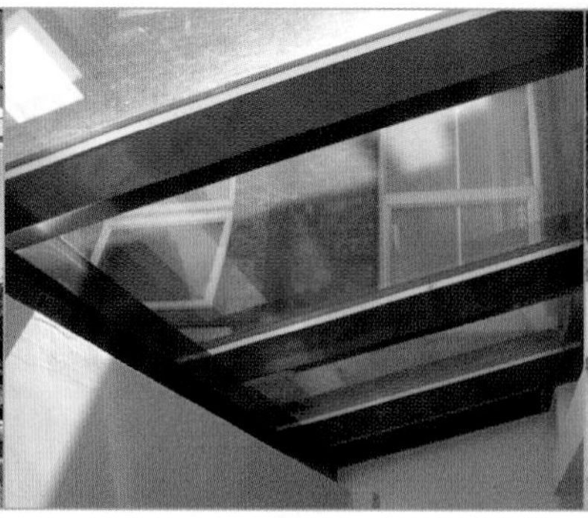

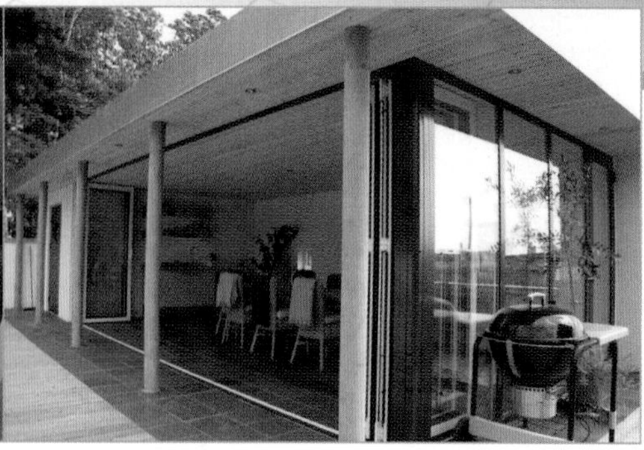

Glazed folding sliding doors are secure, weathertight, thermally efficient and a delight to use, folding right back to bring in the outside world. Designed to compliment our contemporary roof and window systems, unique flexible conservatories can be created with the help of our enthusiastic, knowledgeable staff.

Exceptional quality, technical expertise and level of service from your first phone call to installation and handover are assured. Technical brochures and CAD files can be downloaded from our website or ask for our CD ROM.

REYNAERS LTD
Kettles Wood Drive, Birmingham,
West Midlands, B32 3DB
Area of Operation: UK & Ireland
Tel: 0121 421 1999 **Fax:** 0121 421 9797
Email: reynaersltd@reynaers.com
Web: www.reynaers.com
Product Type: 5

RHONES JOINERY (WREXHAM) LTD
Mold Road Industrial Estate,
Gwersyllt, Wrexham, LL11 4AQ
Area of Operation: UK (Excluding Ireland)
Tel: 01978 262488 **Fax:** 01978 262488
Email: info@rhonesjoinery.co.uk
Web: www.rhonesjoinery.co.uk
Product Type: 2, 4, 6, 7

ROBERT J TURNER & CO
Roe Green, Sandon, Buntingford, Herts, SG9 0QE
Area of Operation: East England, Greater London
Tel: 01763 288371 **Fax:** 01763 288440
Email: sales@robertjturner.co.uk
Web: www.robertjturner.co.uk
Product Type: 1, 2, 3, 4, 5, 6, 7, 8

ROBERT MILLS ARCHITECTURAL ANTIQUES
Narroways Road, Eastville, Bristol, BS2 9XB
Area of Operation: Worldwide
Tel: 0117 955 6542 **Fax:** 0117 955 8146
Email: Info@rmills.co.uk **Web:** www.rmills.co.uk
Product Type: 2, 3, 4, 5, 8 **Other Info:** ▽

RO-DOR LIMITED
Stevens Drove, Houghton,
Stockbridge, Hampshire, SO20 6LP
Area of Operation: UK & Ireland
Tel: 01794 388080 **Fax:** 01794 388090
Email: rdc46@dial.pipex.com
Web: www.ro-dor.co.uk
Product Type: 8 **Other Info:** ✐ ✋

SABRINA OAK DOORS LTD
Alma Street, Mountfields,
Shrewsbury, Shropshire, SY3 8QL
Area of Operation: UK (Excluding Ireland)
Tel: 01743 357977 **Fax:** 01743 352233
Email: mail@oakdoors.co.uk
Web: www.oakdoors.co.uk
Product Type: 1, 2, 3, 4, 5, 6, 7
Other Info: ✐ ✋ **Material Type:** A) 2

SABRINA OAK DOORS LTD
Area of Operation: UK (Excluding Ireland)
Tel: 01743 357977 **Fax:** 01743 352233
Email: mail@oakdoors.co.uk
Web: www.oakdoors.co.uk
Product Type: 1, 2, 3, 4, 6, 7
Materials Type: A) 2

Handcrafted oak doors
- made in the traditional way.
For colour brochure call 01743 357977

SAFEGUARD DOORS LIMITED
Units 6-9 Bridge Business Park, Bridge Street,
Wednesbury, West Midlands, WS10 0AW
Area of Operation: Worldwide
Tel: 0121 556 9138 **Fax:** 0121 556 7275
Email: sales@safeguarddoors.co.uk
Web: www.safeguarddoors.co.uk
Product Type: 1, 2, 3

SANDERSON'S FINE FURNITURE
Unit 5 & 6, The Village Workshop, Four Crosses
Business Park, Four Crosses, Powys, SY22 6ST
Area of Operation: UK (Excluding Ireland)
Tel: 01691 830075 **Fax:** 01691 830075
Email: sales@sandersonsfinefurniture.co.uk
Web: www.sandersonsfinefurniture.co.uk
Product Type: 1, 2, 3, 4, 6, 7, 8 **Other Info:** ✐ ✋

SANDERSONS FINE FURNITURE

Area of Operation: UK
Tel: 01691 830075
Fax: 01691 830075
Email: sales@sandersonsfinefurniture.co.uk
Web: www.sandersonsfinefurniture.co.uk
Product Type: 1, 2, 3, 4, 6, 7, 8

Traditional manufacturers of quality bespoke oak
internal and external doors, frame and filled
cottage, Georgian, Victorian, stable, internal
cottage, skirting, architrave and flooring.

SAPA BUILDING SYSTEMS LIMITED
Alexandra Way, Ashchurch, Tewkesbury,
Gloucestershire, GL20 8NB
Area of Operation: UK & Ireland
Tel: 01684 853500 **Fax:** 01684 851850
Email: nicola.abbey@sapagroup.com
Web: www.sapagroup.com/uk/buildingsystems
Product Type: 1, 2, 3, 4, 5 **Other Info:** ✐
Material Type: C) 1

SASH CRAFT
Unit 1 Mill Farm, Kelston, Bath, Somerset, BA1 9AQ
Area of Operation: South West England & South Wales
Tel: 01225 424434 **Fax:** 01225 424435
Email: pbale@sashcraftofbath.co.uk
Web: www.sashcraftofbath.co.uk
Product Type: 1, 2, 3, 4, 5, 6, 7
Material Type: B) 8, 12

SASH UK LTD
Ferrymoor Way, Park Springs, Grimethorpe,
Barnsley, South Yorkshire, S72 7BN
Area of Operation: Worldwide
Tel: 01226 715619 **Fax:** 01226 719968
Email: mailbox@sashuk.com
Web: www.sashuk.com
Product Type: 1, 3, 7, 8 **Material Type:** D) 1

SCANDINAVIAN WINDOW SYSTEMS LTD
10 Eldon St, Tuxford, Newark, Notts, NG22 0LH
Area of Operation: UK & Ireland
Tel: 01777 871847 **Fax:** 01777 872650
Email: sue@scandinavian-windows.co.uk
Web: www.scandinavian-windows.co.uk
Product Type: 1, 2, 3, 4, 5, 7, 8
Material Type: B) 8, 13

SCHÜCO INTERNATIONAL KG
Whitehall Avenue, Kingston, Milton Keynes,
Buckinghamshire, MK10 0AL
Area of Operation: Worldwide
Tel: 01908 282111 **Fax:** 01908 282124
Email: shamman@schueco.com
Web: www.schueco.co.uk **Material Type:** C) 1

SCOTTS OF THRAPSTON LTD
Bridge St, Thrapston, Northamptonshire, NN14 4LR
Area of Operation: UK & Ireland
Tel: 01832 732366 **Fax:** 01832 733703
Email: julia@scottsofthrapston.co.uk
Web: www.scottsofthrapston.co.uk
Product Type: 1, 2, 3, 7, 8 **Other Info:** ✐

SOLARLUX SYSTEMS LTD
Holmfield Lane, Wakefield, West Yorkshire, WF2 7AD
Area of Operation: Worldwide
Tel: 01924 204444 **Fax:** 01924 204455
Email: info@solarlux.co.uk
Web: www.solarlux.uk.com
Product Type: 5 **Other Info:** ✐

SPECTUS SYSTEMS
Queens Avenue, Macclesfield, Cheshire, SK10 2BN
Area of Operation: UK & Ireland
Tel: 01625 420400 **Fax:** 01625 501418
Email: contact@spectus.co.uk
Web: www.spectus.co.uk

SPS TIMBER WINDOWS
Unit 2, 34 Eveline Road, Mitcham, Surrey, CR4 3LE
Area of Operation: Greater London
Tel: 0208 640 5035 **Fax:** 0208 685 1570
Email: info@spstimberwindows.co.uk
Web: www.spstimberwindows.co.uk
Product Type: 2, 3, 4, 7, 8
Other Info: ✐ ✋ **Material Type:** A) 2, 6

STEEL WINDOW SERVICE AND SUPPLIES LTD
30 Oxford Road, Finsbury Park, London, N4 3EY
Area of Operation: Greater London
Tel: 0207 272 2294 **Fax:** 0207 281 2309
Email: post@steelwindows.co.uk
Web: www.steelwindows.co.uk
Product Type: 1, 3, 4 **Material Type:** C) 2, 4

STRATHEARN STONE AND TIMBER LTD
Glenearn, Bridge of Earn, Perth,
Perth and Kinross, PH2 9HL
Area of Operation: North East England, Scotland
Tel: 01738 813215 **Fax:** 01738 815946
Email: info@stoneandoak.com
Web: www.stoneandoak.com
Product Type: 3, 4, 7 **Other Info:** ✐ ✋

STROUDS WOODWORKING COMPANY LTD.
Ashmansworthy, Woolsery, Bideford, Devon, EX39 5RE
Area of Operation: South West England
and South Wales
Tel: 01409 241624 **Fax:** 01409 241769
Email: enquires@doormakers.co.uk
Web: www.doormakers.co.uk
Product Type: 1, 2, 3, 4, 5, 6, 7
Other Info: ECO ✐ ✋
Material Type: A) 1, 2, 7, 8, 10, 12

SUN PARADISE UK LTD
Phoenix Wharf, Eel Pie Island,
Twickenham, Middlesex, TW1 3DY
Area of Operation: UK & Ireland
Tel: 0870 240 7604 **Fax:** 0870 240 7614
Email: info@sunparadise.co.uk
Web: www.sunparadise.co.uk
Product Type: 5

SUN PARADISE UK LTD

Area of Operation: UK
Tel: 0870 240 7604 **Fax:** 0870 240 7614
Email: info@sunparadise.co.uk
Web: www.sunparadise.co.uk
Product Type: 5

From a simple 'Patio Door' replacement to the glazing of a
complete façade, Sun Paradise sliding doors can be used
in a variety of applications with Aluminium frames or even
'frameless' systems that can slide and stack into
concealed recesses.

SUNFOLD SYSTEMS
Sunfold House, Wymondham Business Park,
Chestnut Drive, Wymondham, Norfolk, NR18 9SB
Area of Operation: Worldwide
Tel: 01953 423423
Fax: 01953 423430
Email: info@sunfoldsystems.co.uk
Web: www.sunfold.co.uk
Product Type: 3, 4, 5 **Other Info:** ✐

**THE ORIGINAL BOX SASH
WINDOW COMPANY**
29/30 The Arches, Alma Road,
Windsor, Berkshire, SL4 1QZ
Area of Operation: UK (Excluding Ireland)
Tel: 01753 858196
Fax: 01753 857827
Email: info@boxsash.com
Web: www.boxsash.com

**THE REAL OAK DOOR
AND FLOORING COMPANY**
5 Robin Way, Sudbury, Suffolk, CO10 7PF
Area of Operation: UK (Excluding Ireland)
Tel: 01787 310051
Fax: 01787 312497
Email: valentinecollins@aol.com
Web: www.realoakdoor.co.uk
Product Type: 6 **Material Type:** A) 2

TODD DOORS LTD
112-116 Church Road, Northolt,
Greater London, UB5 5AE
Area of Operation: UK (Excluding Ireland)
Tel: 0208 845 2493
Fax: 0208 845 7579
Email: info@todd-doors.co.uk
Web: www.todd-doors.co.uk
Product Type: 1, 2, 3, 4, 5, 6, 7, 8
Other Info: ECO ✐ **Material Type:** A) 1, 2, 3, 4, 5, 6, 7

TOMPKINS LTD
High March Close, Long March Industrial Estate,
Daventry, Northamptonshire, NN11 4EZ
Area of Operation: UK (Excluding Ireland)
Tel: 01327 877187
Fax: 01327 310491
Email: info@tompkinswood.co.uk
Web: www.tompkinswood.co.uk
Product Type: 1, 2, 3, 4, 5, 6, 7 **Other Info:** ✐ ✋
Material Type: A) 1, 2, 3, 4, 5, 6, 7, 8, 9, 10, 11, 12, 13, 14

TONY HOOPER
Unit 18 Camelot Court, Bancombe Trading Estate,
Somerton, TA11 6SB
Area of Operation: UK (Excluding Ireland)
Tel: 01458 274221
Fax: 01458 274690
Email: tonyhooper1@aol.com
Web: www.tonyhooper.co.uk

TONY HOOPER

Area of Operation: UK (Excluding Ireland)
Tel: 01458 274221
Fax: 01458 274690
Email: tonyhooper1@aol.com
Web: www.tonyhooper.co.uk

We offer an expert bespoke service for doors,
window frames and joinery of all kinds.

TOTAL GLASS LIMITED
Total Complex, Overbrook Lane,
Knowsley, Merseyside, L34 9FB
Area of Operation: Midlands & Mid Wales, North
East England, North West England and North Wales
Tel: 0151 549 2339 **Fax:** 0151 546 0022
Email: sales@totalglass.com
Web: www.totalglass.com
Product Type: 1, 3, 4, 7, 8

TRADITIONAL OAK & TIMBER COMPANY
P O Stores, Haywards Heath Road,
North Chailey, Nr Lewes, East Sussex, BN8 4EY
Area of Operation: Worldwide
Tel: 01825 723648 **Fax:** 01825 722215
Email: info@tradoak.co.uk
Web: www.tradoak.com **Product Type:** 1, 2, 4, 6, 7

TRANIK HOUSE DOORSTORE
63 Cobham Road, Ferndown Industrial Estate,
Wimborne, Dorset, BH21 7QA
Area of Operation: UK (Excluding Ireland)
Tel: 01202 872211
Email: sales@tranik.co.uk
Web: www.tranik.co.uk

TWC THE WINDOW COMPANY LIMITED
The Drill Hall, 262 Huddersfield Road,
Thongsbridge, Holmfirth, West Yorkshire, HD9 3JQ
Area of Operation: UK & Ireland
Tel: 0800 917 1918 **Fax:** 01484 685210
Email: admin@twcthewindowcompany.net
Web: www.twcthewindowcompany.net
Product Type: 5, 7

UAP
Bank House, 16-18 Bank Street,
Walshaw, Bury, Lancashire, BL8 3AZ
Area of Operation: UK & Ireland
Tel: 0161 763 5290 **Fax:** 0161 763 6726
Email: uap@btconnect.com
Web: www.universal-imports.com
Product Type: 8

URBAN FRONT LTD
Design Studio, 1 Little Hill, Heronsgate,
Rickmansworth, Hertfordshire, WD3 5BX
Area of Operation: UK & Ireland
Tel: 0870 609 1525 **Fax:** 0870 609 3564
Email: elizabeth@urbanfront.co.uk
Web: www.urbanfront.co.uk
Material Type: A) 2, 7, 12

URBAN FRONT
Area of Operation: UK & Ireland
Tel: 0870 609 1525
Fax: 0870 609 3564
Email: elizabeth@urbanfront.co.uk
Web: www.urbanfront.co.uk

Urban Front design and manufacture quality
contemporary exterior wood doors. A matching
internal veneer range is also available.

VENTROLLA LTD
11 Hornbeam Square South,
Harrogate, North Yorkshire, HG2 8NB
Area of Operation: UK & Ireland
Tel: 0800 378278 **Fax:** 01423 859321
Email: info@ventrolla.co.uk
Web: www.ventrolla.co.uk
Product Type: 2, 3, 4

WOODLAND PRODUCTS DESIGN LTD
St.Peter's House, 6 Cambridge Road,
Kingston Upon Thames, Surrey, KT1 3JY
Area of Operation: UK (Excluding Ireland)
Tel: 0208 547 2171
Fax: 0208 547 1722
Email: enquiries@woodland-products.co.uk
Web: www.woodland-products.co.uk
Product Type: 4

WINDOWS

KEY
PRODUCT TYPES: 1= Sash 2 = Casement
3 = Bow / Bay   4 = Oriel   5 = Tilt and Turn
6 = High Performance 7 = PVC Clad Timber
8 = Aluminium Clad Timber

SEE ALSO: DOORS, WINDOWS AND
CONSERVATORIES - Glass and Glazing, DOORS,
WINDOWS AND CONSERVATORIES - Roof
Windows and Light Pipes, DOORS, WINDOWS
AND CONSERVATORIES - Stained Glass

OTHER: ▽ Reclaimed On-line shopping
 Bespoke Hand-made   ECO Ecological

ALUPLAST UK/PLUSPLAN
Leicester Road, Lutterworth, Leics, LE17 4HE
Area of Operation: UK & Ireland
Tel: 01455 556771 **Fax:** 01455 555323
Email: info@aluplastuk.com
Web: www.aluplastuk.com
Product Type: 1, 2, 3, 5

AMBASSADOOR
WINDOWS & DOORS LTD
18 Bidwell Road, Rackheath Industrial Estate,
Rackheath, Norwich, Norfolk, NR13 6PT
Area of Operation: East England,
Greater London, Midlands & Mid Wales
Tel: 01603 720332
Email: enquiries@ambassadoor.co.uk
Web: www.ambassadoor.co.uk
Product Type: 1, 2, 3, 5, 6
Material Type: A) 1

ANDERSEN / BLACK MILLWORK CO.
Andersen House, Dallow Street,
Burton on Trent, Staffordshire, DE14 2PQ
Area of Operation: UK & Ireland
Tel: 01283 511122 **Fax:** 01283 510863
Email: info@blackmillwork.co.uk
Web: www.blackmillwork.co.uk
Product Type: 1, 2, 3, 4, 6, 8
Material Type: A) 2, 5

ANDERSEN WINDOWS/BLACK MILLWORK
Area of Operation: UK & Ireland
Tel: 01283 511 122 **Fax:** 01283 510 863
Email: info@blackmillwork.co.uk
Web: www.blackmillwork.co.uk

Andersen's new Woodwright™ made to measure
sliding sash window incorporates Fibrex® - a
revolutionary new material with the strength of
natural wood and the low maintenance qualities
of PVC-u.

ARDEN SOUTHERN LTD
Arden House, Spark Brook Street,
Coventry, West Midlands, CV1 56T
Area of Operation: South West England
and South Wales
Tel: 08707 890160 **Fax:** 08707 890161
Email: enquires@ardenwindows.net
Web: www.ardenwindows.net
Product Type: 1, 2 **Other Info:**

ASTON HOUSE JOINERY
Aston House, Tripontium Business Centre,
Newton Lane, Rugby, Warwickshire, CV23 0TB
Area of Operation: UK & Ireland
Tel: 01788 860032 **Fax:** 01788 860614
Email: mailbox@ovationwindows.co.uk
Web: www.astonhousejoinery.co.uk
Product Type: 1, 2, 3, 4, 6
Other Info: ECO
Material Type: A) 1, 2, 12

B & M WINDOW & KITCHEN CENTRE
2-6 Whitworth Drive, Aycliffe Industrial Park,
Durham, DL5 6SZ
Area of Operation: Worldwide
Tel: 01325 308888 **Fax:** 01325 316002
Email: sales@bandmhomeimprovements.com
Web: www.bandmhomeimprovements.com
Product Type: 1, 2, 3, 5

BARTHOLOMEW
Rakers Yard, Rake Road, Milland,
West Sussex, GU30 7JS
Area of Operation: Europe
Tel: 01428 742800 **Fax:** 01428 743801
Email: denise@bartholomew-conservatories.co.uk
Web: www.bartholomew-conservatories.co.uk
Product Type: 1, 2, 3, 4, 5, 6
Other Info: ECO **Material Type:** A) 2, 10

BDG GROUP LIMITED
5 Wenlock Road, Lurgan, Craigavon,
Co Armagh, BT66 8QR
Area of Operation: Worldwide
Tel: 0283 832 7741
Fax: 0283 832 4358
Email: info@bdg.co.uk **Web:** www.bdg.co.uk
Product Type: 2, 5 **Material Type:** D) 1

BIRCHDALE GLASS LTD
Unit L, Eskdale Road, Uxbridge,
Greater London, UB8 2RT
Area of Operation: UK & Ireland
Tel: 01895 259111
Fax: 01895 810087
Email: info@birchdaleglass.com
Web: www.birchdaleglass.com
Product Type: 1, 2, 3, 4, 5, 6

BOWATER WINDOWS NEWBUILD DIVISION
Water Orton Lane, Minworth, Sutton Coldfield,
West Midlands, B76 9BW
Area of Operation: UK (Excluding Ireland)
Tel: 0121 749 3000
Fax: 0121 749 8210
Email: info@theprimeconnection.co.uk
Web: www.whs-halo.co.uk
Product Type: 1, 2, 3, 4, 5, 6
Material Type: D) 1

BROXWOOD (SCOTLAND) LTD
Inveralmond Way, Inveralmond Industrial Estate,
Perth, Perth and Kinross, PH1 3UQ
Area of Operation: UK & Ireland
Tel: 01738 444456
Fax: 01738 444452
Email: sales@broxwood.com
Web: www.broxwood.com
Product Type: 1, 2, 3, 5, 6, 8
Other Info: ECO **Material Type:** A) 1, 2, 15

CAREY & FOX LTD
51 Langthwaite Business Park, South Kirkby,
Pontefract, West Yorkshire, WF9 3NR
Area of Operation: UK & Ireland
Tel: 01977 608069 **Fax:** 01977 646791
Email: enquiries@careyandfox.co.uk
Web: www.careyandfox.co.uk
Product Type: 1, 2, 3, 4, 5, 6 **Material Type:** A) 2

CHIPPING NORTON GLASS LTD
Units 1 & 2, Station Yard Industrial Estate,
Chipping Norton, Oxfordshire, OX7 5HX
Area of Operation: Midlands & Mid Wales,
South West England and South Wales
Tel: 01608 643261 **Fax:** 01608 641768
Email: gill@cnglass.plus.com
Web: www.chippingnortonglass.co.uk
Product Type: 1, 2, 3, 4, 5, 6 **Material Type:** C) 1

CLASSIC PVC HOME IMPROVEMENTS
46 Stepney Place, Llanelli, Carms, SA15 1SE
Area of Operation: South West England & South Wales
Tel: 0800 064 9494 **Fax:** 01554 775086
Email: enquiries@classic.uk.com
Web: www.classic.uk.com **Product Type:** 1, 2, 3, 4, 5, 6

CLEARVIEW WINDOWS
Northfields Industrial Estate, Blenheim Way,
Market Deeping, Peterborough, PE6 8LD
Area of Operation: UK (Including Ireland)
Tel: 01778 347147 **Fax:** 01778 341363
Email: sales@clearviewgroup.co.uk
Web: www.clearviewgroup.co.uk

CLEARVIEW WINDOWS LTD
Area of Operation: UK including Ireland
Tel: 01778 347 147 **Fax:** 01778 341 363
Email: sales@clearviewgroup.co.uk
Web: www.clearviewgroup.co.uk

Established nearly 50 years, we are a family run
company, supplying bespoke sash windows
nationwide. We use Aluminium in any colour or
Upvc profiles in white, oak or Mahogany.

CLEARVIEW WINDOWS LTD
Area of Operation: UK including Ireland
Tel: 01778 347 147 **Fax:** 01778 341 363
Email: sales@clearviewgroup.co.uk
Web: www.clearviewgroup.co.uk

Established nearly 50 years, we are a family run
company, supplying bespoke sash windows
nationwide. We use Aluminium in any colour or
Upvc profiles in white, oak or Mahogany.

CLEARVIEW WINDOWS & DOORS
Unit 14, Sheddingdean Industrial Estate, Marchants
Way, Burgess Hill, West Sussex, RH15 8QY
Area of Operation: UK (Excluding Ireland)
Tel: 01444 250111 **Fax:** 01444 250678
Email: sales@clearviewsussex.com
Web: www.clearviewsussex.com

CLEMENT STEEL WINDOWS
Clement House, Haslemere, Surrey, GU27 1HR
Area of Operation: Worldwide
Tel: 01428 643393 **Fax:** 01428 644436
Email: info@clementwg.co.uk
Web: www.clementsteelwindows.com
Product Type: 2, 5, 6

CONSERVATORY & WINDOW WORLD LTD
Watling Road, Bishop Aukland, Durham, DL14 9AU
Area of Operation: North East England
Tel: 01388 458 088 **Fax:** 01388 458 518
Email: paldennis20@aol.com
Web: www.conservatoryandwindowworld.co.uk
Product Type: 1, 2, 3, 4, 5, 6 **Material Type:** D) 1, 3, 6

CONSERVATORY OUTLET
Unit 8 Headway Business Park, Denby Dale Road,
Wakefield, West Yorkshire, WF2 7AZ
Area of Operation: North East England
Tel: 01924 881920
Web: www.conservatoryoutlet.co.uk
Product Type: 2, 3, 5 **Material Type:** D) 1

COUNTY JOINERY (SOUTH EAST) LTD
Unit 300, Vinehall Business Centre, Vinehall Road,
Mountfield, Robertsbridge, East Sussex, TN32 5JW
Area of Operation: Greater London,
South East England
Tel: 01424 871500 **Fax:** 01424 871550
Email: countyjoinery@aol.com
Web: www.countyjoinery.co.uk

CREATE JOINERY
Worldsend, Llanmadoc, Swansea, SA3 1DB
Area of Operation: Greater London, Midlands & Mid
Wales, North West England and North Wales, South
East England, South West England and South Wales
Tel: 01792 386677 **Fax:** 01792 386677
Email: mail@create-joinery.co.uk
Web: www.create-joinery.co.uk
Product Type: 1, 2, 3, 5, 6

CROXFORD'S JOINERY MANUFACTURERS & WOODTURNERS
Meltham Joinery, Works New Street, Meltham,
Holmfirth, West Yorkshire, HD9 5NT
Area of Operation: UK (Excluding Ireland)
Tel: 01484 850892 **Fax:** 01484 850969
Email: ralph@croxfords.demon.co.uk
Web: www.croxfords.co.uk
Product Type: 1, 2, 3, 5, 6 **Material Type:** A) 1, 2

CUSTOM WOOD PRODUCTS
Cliffe Road, Easton on the Hill,
Stamford, Lincolnshire, PE9 3NP
Area of Operation: East England
Tel: 01780 755711 **Fax:** 01780 480834
Email: customwoodprods@aol.com
Web: www.cwpuk.com **Product Type:** 1, 2, 3, 4, 6
Other Info: ECO ✍ ✋ **Material Type:** A) 2

DISTINCTIVE DOORS
14 & 15 Chambers Way, Newton Chambers Road,
Chapeltown, Sheffield, South Yorkshire, S35 2PH
Area of Operation: UK (Excluding Ireland)
Tel: 0114 220 2250 **Fax:** 0114 220 2254
Email: enquiries@distinctivedoors.co.uk
Web: www.distinctivedoors.co.uk ✍
Product Type: 1, 2, 3

DIY SASH WINDOWS
2 - 6 Whitworth Drive, Aycliffe Industrial Park,
Newton Aycliffe, Durham, DL56SZ
Area of Operation: UK & Ireland
Tel: 01325 308888 **Fax:** 01325 316002
Email: sales@diysashwindows.co.uk
Web: www.diysashwindows.co.uk
Product Type: 1, 2, 3, 5

DRUMMONDS ARCHITECTURAL ANTIQUES LIMITED
Kirkpatrick Buildings, 25 London Road (A3),
Hindhead, Surrey, GU36 6AB
Area of Operation: Worldwide
Tel: 01428 609444 **Fax:** 01428 609445
Email: davidcox@drummonds-arch.co.uk
Web: www.drummonds-arch.co.uk **Other Info:** ▽
Product Type: 1, 2 **Material Type:** A) 2

DUPLUS DOMES LTD
370 Melton Road, Leicester,
Leicestershire, LE4 7SL
Area of Operation: Worldwide
Tel: 0116 261 0710
Fax: 0116 261 0539
Email: sales@duplus.co.uk
Web: www.duplus.co.uk

EDGETECH IG INC
Unit 3, Swallow Gate Business Park,
Holbrook Lane, Coventry, CV6 4BL
Area of Operation: UK (Excluding Ireland)
Tel: 02476 705570
Fax: 02476 705510
Email: aoconnor@edgetechig.com
Web: www.edgetechig.com
Product Type: 1, 2, 5, 6

ENGELS WINDOWS & DOORS LTD
1 Kingley Centre, Downs Road, West Stoke,
Chichester, West Sussex, PO18 9HJ
Area of Operation: UK (Excluding Ireland)
Tel: 01243 576633
Fax: 01243 576644
Email: admin@engels.co.uk
Web: www.engels.co.uk

ENVIROZONE
Encon House, 54 Ballycrochan Road,
Bangor, Co Down, BT19 6NF
Area of Operation: UK & Ireland
Tel: 02891 477799
Fax: 02891 452463
Email: info@envirozone.co.uk
Web: www.envirozone.co.uk
Product Type: 5, 6

ESOGRAT LTD
Caldervale Works, River Street,
Brighouse, Huddersfield,
West Yorkshire, HD6 1JS
Area of Operation: UK & Ireland
Tel: 01484 716228
Fax: 01484 400107
Email: info@esograt.com
Web: www.esograt.com
Product Type: 2, 3, 5, 6

EUROCELL PROFILES LTD
Fairbrook House, Clover Nook Road,
Alfreton, Derby, Derbyshire, DE55 4RF
Area of Operation: UK (Excluding Ireland)
Tel: 01773 842100
Fax: 01773 842109
Email: eric.gale@eurocell.co.uk
Web: www.eurocell.co.uk
Product Type: 2, 3, 5
Material Type: D) 1

EVEREST LTD
Sopers Road, Cuffley,
Hertfordshire, EN6 4SG
Area of Operation: UK & Ireland
Tel: 0800 010123
Web: www.everest.co.uk ✍
Product Type: 8

FAIRMITRE WINDOWS & JOINERY LTD
2A Cope Road, Banbury,
Oxfordshire, OX16 2EH
Area of Operation: UK (Excluding Ireland)
Tel: 01295 268441
Fax: 01295 268468
Email: info@fairmitrewindows.co.uk
Web: www.fairmitrewindows.co.uk
Product Type: 1, 2, 3, 5, 6, 8

FAIROAK TIMBER PRODUCTS LTD
Manor Farm, Chilmark,
Salisbury, Wiltshire, SP3 5AG
Area of Operation: UK (Excluding Ireland)
Tel: 01722 716779
Fax: 01722 716761
Email: sales@fairoak.co.uk
Web: www.fairoakwindows.co.uk
Product Type: 2, 3, 4, 6
Material Type: A) 2, 12

FAIROAK TIMBER PRODUCTS LTD
Area of Operation: UK (Excluding Ireland)
Tel: 01722 716779
Fax: 01722 716761
Email: sales@fairoak.co.uk
Web: www.fairoakwindows.co.uk
Product Type: 2, 3, 4, 6
Material Type: A) 2, 12

High performance severe weather tested timber
"Amesbury" casement windows, manufactured to
order in European Oak, durable hardwoods, and
softwoods.

FITZROY JOINERY
Garden Close, Langage Industrial Estate,
Plympton, Devon, PL7 5EU
Area of Operation: UK & Ireland
Tel: 0870 428 9110
Fax: 0870 428 9111
Email: admin@fitzroy.co.uk
Web: www.fitzroy.co.uk ✍
Product Type: 1, 2, 3, 4, 5, 6, 8
Other Info: ECO ✍
Material Type: A) 1, 2, 3, 4, 5, 6, 7, 15

FLETCHER JOINERY
261 Whessoe Road, Darlington, Durham, DL3 0YL
Area of Operation: North East England
Tel: 01325 357347
Fax: 01325 357347
Email: enquiries@fletcherjoinery.co.uk
Web: www.fletcherjoinery.co.uk
Product Type: 1, 2, 3, 4, 6

FRANKLIN LEEDS LLP
Carlton Works, Cemetery Road,
Yeadon, Leeds, West Yorkshire, LS19 7BD
Area of Operation: UK & Ireland
Tel: 0113 250 2991
Fax: 0113 250 0991
Email: david.franklin@franklinwindows.co.uk
Web: www.franklinwindows.co.uk
Product Type: 1, 2, 3, 4, 5, 6, 8

FRANKLIN LEEDS LLP
Area of Operation: UK & Ireland
Tel: 0113 250 2991
Fax: 0113 250 0991
Email: davidfranklin@franklinwindows.co.uk
Web: www.franklinwindows.co.uk

At last a real alternative to timber, metal and
Upvc windows. Enjoy higher security, longer life,
choice of colours, larger glass area. UK made.

G2K JOINERY LTD
Trench Farm, Tilley Green,
Wem, Shropshire, SY4 5PJ
Area of Operation: UK (Excluding Ireland)
Tel: 01939 236640 **Fax:** 01939 236650
Email: graham@g2kjoinery.co.uk
Web: www.g2kjoinery.co.uk

GREEN BUILDING STORE
11 Huddersfield Road, Meltham,
Holmfirth, West Yorkshire, HD9 4NJ
Area of Operation: UK & Ireland
Tel: 01484 854898 **Fax:** 01484 854899
Email: info@greenbuildingstore.co.uk
Web: www.greenbuildingstore.co.uk ✍
Product Type: 1, 2, 3, 4, 5, 6, 8
Other Info: ECO ✍ **Material Type:** B) 8

HAYMANS TIMBER PRODUCTS
Haymans Farm, Hocker Lane, Over Alderley,
Macclesfield, Cheshire, SK10 4SD
Area of Operation: UK & Ireland
Tel: 01625 590098 **Fax:** 01625 586174
Email: haymanstimber@aol.com
Web: www.haymanstimber.co.uk
Product Type: 1, 3

HOLDSWORTH WINDOWS LTD
Darlingscote Road, Shipston-on-Stour,
Warwickshire, CV36 4PR
Area of Operation: UK & Ireland
Tel: 01608 661883 **Fax:** 01608 661008
Email: info@holdsworthwindows.co.uk
Web: www.holdsworthwindows.co.uk
Product Type: 2 **Material Type:** C) 2, 4

HOWARTH WINDOWS AND DOORS LTD
The Dock, New Holland, Lincolnshire, DN19 7RT
Area of Operation: UK & Ireland
Tel: 01469 530577 **Fax:** 01469 531559
Email: abarker@howarth-timber.co.uk
Web: www.howarth-timber.co.uk
Product Type: 1, 2, 3, 5 **Material Type:** B) 1, 8

J L JOINERY
Cockerton View, Grange Lane, Preston,
Lancashire, PR4 5JE
Area of Operation: UK (Excluding Ireland)
Tel: 01772 616123 **Fax:** 01772 619182
Email: mail@jljoinery.co.uk
Web: www.jljoinery.co.uk

JOHN LOWRY JOINERY
Area of Operation: UK (Excluding Ireland)
Tel: 01772 616123
Fax: 01772 619182
Email: mail@jljoinery.co.uk
Web: www.jljoinery.co.uk

John Lowry and his staff take pride in their work and
attention to detail, which frequently exceeds
customer's expectations. Their reputation is built on
the high quality of the materials used, expert
craftsmanship and quality of finish.

JELD-WEN
Watch House Lane, Doncaster, South Yorks, DN5 9LR
Area of Operation: Worldwide
Tel: 01302 394000
Fax: 01302 787383
Email: customer-services@jeld-wen.co.uk
Web: www.jeld-wen.co.uk
Product Type: 1, 2, 3 **Material Type:** D) 1

JOHN FLEMING & CO LTD
Silverburn Place, Bridge of Don,
Aberdeen, Aberdeenshire, AB23 8EG
Area of Operation: Scotland
Tel: 0800 085 8728 **Fax:** 01224 825377
Email: info@johnfleming.co.uk
Web: www.johnfleming.co.uk

JOINERY-PLUS
Bentley Hall Barn, Alkmonton,
Ashbourne, Derbyshire, DE6 3DJ
Area of Operation: UK (Excluding Ireland)
Tel: 07931 386233 **Fax:** 01335 330922
Email: info@joinery-plus.co.uk
Web: www.joinery-plus.co.uk
Product Type: 1, 2, 3, 4, 5, 6, 8 **Material Type:** A) 2

KS PROFILES LTD
Broad March, Long March Industrial Estate,
Daventry, Northamptonshire, NN11 4HE
Area of Operation: UK & Ireland
Tel: 01327 316960 **Fax:** 01327 876412
Email: sales@ksprofiles.com
Web: www.ksprofiles.com
Product Type: 2, 3, 4, 5, 6

LAMWOOD LTD
Unit 1, Riverside Works, Station Road, Cheddleton,
Staffordshire, ST13 7EE
Area of Operation: UK & Ireland
Tel: 01538 361888 **Fax:** 01538 361912
Email: sales@lamwoodltd.co.uk
Web: www.lamwoodltd.co.uk **Material Type:** A) 2

LATTICE PERIOD WINDOWS
Unit 85 Northwick Business Centre,
Blockley, Moreton in Marsh,
Gloucestershire, Gl56 9RF
Area of Operation: UK & Ireland
Tel: 01386 701079 **Fax:** 01386 701114
Email: sales@lattice windows .net
Web: www.latticewindows.net
Product Type: 2, 3, 5
Material Type: A) 2

M. H. JOINERY SERVICES
25b Camwal Road, Harlgate, North Yorks, HG1 4PT
Area of Operation: North East England
Tel: 01423 888856 **Fax:** 01432 888856
Email: info@mhjoineryservices.co.uk
Web: www.mhjoineryservices.co.uk
Product Type: 1
Other Info: ✍ ✋

MARVIN ARCHITECTURAL
Canal House, Catherine Wheel Road, Brentford,
Middlesex, Greater London, TW8 8BD
Area of Operation: UK & Ireland
Tel: 0208 569 8222 **Fax:** 0208 560 6374
Email: sales@marvinuk.com
Web: www.marvin-architectural.com
Product Type: 1, 2, 3, 5, 6, 8
Material Type: A) 2, 3, 15

MASTERFRAME WINDOWS LTD
4 Crittall Road, Witham, Essex, CM8 3DR
Area of Operation: UK (Excluding Ireland)
Tel: 0800 591854
Web: www.thebygonecollection.co.uk
Product Type: 1, 6 **Other Info:** ✍
Material Type: D) 1

MERLIN UK LIMITED
Unit 5 Fence Avenue,
Macclesfield, Cheshire, SK10 1LT
Area of Operation: North West England
and North Wales
Tel: 01625 424488 **Fax:** 0871 781 8967
Email: info@merlinuk.net
Web: www.merlindoors.co.uk
Product Type: 1, 2, 3, 5

MIGHTON PRODUCTS LTD
PO Box 1, Saffron Walden, Essex, CB10 1QJ
Area of Operation: UK (Excluding Ireland)
Tel: 0800 0560471
Email: sales@mightonproducts.com
Web: www.mightonproducts.com ✍
Product Type: 1, 5

**MINCHINHAMPTON ARCHITECTURAL
SALVAGE COMPANY**
Cirencester Road, Aston Downs,
Chalford, Stroud, Gloucestershire, GL6 8PE
Area of Operation: Worldwide
Tel: 01285 760886 **Fax:** 01285 760838
Email: masco@catbrain.com
Web: www.catbrain.com **Product Type:** 1

MULTIWOOD HOME IMPROVEMENTS LTD
1298 Chester Road, Stretford, Manchester, M32 9AU
Area of Operation: Europe
Tel: 0161 866 9991 **Fax:** 0161 866 9992
Email: sales@multiwoodconservatories.co.uk
Web: www.multiwoodconservatories.co.uk
Product Type: 1, 2, 5

MUMFORD & WOOD
Tower Business Park, Kelvedon Road,
Tiptree, Essex, CO5 0LX
Area of Operation: Worldwide
Tel: 01621 818155 **Fax:** 01621 818175
Email: sales@mumfordwood.com
Web: www.mumfordwood.com
Product Type: 1, 2, 3, 4, 5, 6, 8

**NORBUILD TIMBER FABRICATION
& FINE CARPENTRY LTD**
Marcassie Farm, Rafford, Forres, Moray, IV36 2RH
Area of Operation: UK & Ireland
Tel: 01309 676865 **Fax:** 01309 676865
Email: norbuild@marcassie.fsnet.co.uk
Product Type: 2, 6
Other Info: ECO ✍ ✋
Material Type: A) 1, 2, 6

NORTH YORKSHIRE TIMBER
Standard House, Thurston Road,
Northallerton Business Park,
Northallerton, North Yorkshire, DL6 2NA
Area of Operation: UK (Excluding Ireland)
Tel: 01609 780777 **Fax:** 01609 777888
Email: sales@nytimber.co.uk
Web: www.nytimber.co.uk

NSB CASEMENTS LTD
Steele Road, Park Royal, London, NW10 7AR
Area of Operation: UK (Excluding Ireland)
Tel: 0208 961 3090 **Fax:** 0208 961 3050
Email: info@nsbcasements.co.uk
Product Type: 2

NULITE LTD
41 Hutton Close, Crowther Industrial Estate,
Washington, Tyne & Wear, NE38 0AH
Area of Operation: UK & Ireland
Tel: 0191 419 1111 **Fax:** 0191 419 1123
Email: sales@nulite-ltd.co.uk
Web: www.nulite-ltd.co.uk
Other Info: ✍
Material Type: C) 1, 2, 4

OCTAVEWARD LTD
Balle Street Mill, Balle Street,
Darwen, Lancashire, BB3 2AZ
Area of Operation: Worldwide
Tel: 01254 773300 **Fax:** 01254 773950
Email: info@octaveward.com
Web: www.octaveward.com
Product Type: 1, 2, 5, 6 **Other Info:** ECO ✍
Material Type: D) 6

OLD MANOR COTTAGES LTD
Turnpike Lane, Ickleford, Hitchin,
Hertfordshire, SG5 3UZ
Area of Operation: UK (Excluding Ireland)
Tel: 01462 456033 **Fax:** 01462 456033
Email: oldmanorco@aol.com

OXFORD SASH WINDOW CO LTD.
Eynsham Park Estate Yard, Cuckoo Lane,
North Leigh, Oxfordshire, OX29 6PW
Area of Operation: Greater London, Midlands
& Mid Wales, South East England, South West
England & South Wales
Tel: 01993 883536 **Fax:** 01993 883027
Email: oxsash@globalnet.com
Web: www.sashwindow.co.uk
Product Type: 1, 2, 3, 4, 5

SPONSORED BY: ULTRAFRAME www.ultraframe-conservatories.co.uk

BUILDING STRUCTURE & MATERIALS

PARAGON JOINERY
Systems House, Eastbourne Road,
Blindley Heath, Surrey, RH7 6JP
Area of Operation: UK (Excluding Ireland)
Tel: 01342 836300 **Fax:** 01342 836364
Email: sales@paragonjoinery.com
Web: www.paragonjoinery.com
Product Type: 1, 2, 3, 4, 5, 6, 8
Other Info:

PERMACELL FINESSE LTD
Western Road, Silver End,
Witham, Essex, CM8 3BQ
Area of Operation: UK & Ireland
Tel: 01376 583241 **Fax:** 01376 583072
Email: sales@pfl.co.uk
Web: www.permacell-finesse.co.uk
Product Type: 1, 3, 5

PORTICO GB LTD
Unit 9 Windmill Avenue, Woolpit Business Park,
Woolpit, Bury St Edmunds, Suffolk, IP30 9UP
Area of Operation: Greater London,
South East England
Tel: 01359 244299 **Fax:** 01359 244232
Email: info@portico-gb.co.uk
Web: www.portico-newbuild.co.uk
Product Type: 1, 2, 5

PREMDOR
Gemini House, Hargreaves Road,
Groundwell Industrial Estate,
Swindon, Wiltshire, SN25 5AJ
Area of Operation: UK & Ireland
Tel: 0870 990 7998
Email: ukmarketing@premdor.com
Web: www.premdor.co.uk
Product Type: 2, 3, 4, 6

PRITCHARDS
Unit 22 Brookend Lane, Kempsey,
Worcester, Worcestershire, WR5 3LF
Area of Operation: UK (Excluding Ireland)
Tel: 07721 398074 **Fax:** 01905 356507
Email: info@pritchards.tv
Web: www.pritchards.tv

QUICKSLIDE LTD
St Thomas Road, Longroyd Bridge,
Huddersfield, West Yorkshire, HD1 3LF
Area of Operation: UK & Ireland
Tel: 01484 487000 **Fax:** 01484 487001
Email: sales@quickslide.co.uk
Web: www.quickslide.co.uk
Product Type: 1

RATIONEL WINDOWS LTD
7 Avonbury Business Park, Howes Lane,
Bicester, Oxfordshire, OX26 2UA
Area of Operation: UK (Excluding Ireland)
Tel: 01869 248181 **Fax:** 01869 249693
Email: jrh@rationel.dk
Web: www.rationel.com
Product Type: 2, 3, 4, 6, 8
Material Type: B) 2, 13

REDDISEALS
The Furlong, Berry Hill Industrial Estate,
Droitwich, Worcestershire, WR9 9BG
Area of Operation: UK & Ireland
Tel: 0845 165 9507 **Fax:** 01905 791877
Email: reddiseals@reddiplex.co.uk
Web: www.reddiseals.com **Product Type:** 1, 2

REGENCY JOINERY
Unit 18, Leigh Road, Haine Road Industrial Estate,
Ramsgate, Kent, CT12 5EU
Area of Operation: East England, Greater London,
Midlands & Mid Wales, South East England
Tel: 07782 517392
Email: enquiries@regency-joinery.co.uk
Web: www.regency-joinery.co.uk **Product Type:** 1, 2, 3

REYNAERS LTD
Kettles Wood Drive, Birmingham, West Mids, B32 3DB
Area of Operation: UK & Ireland
Tel: 0121 421 1999 **Fax:** 0121 421 9797
Email: reynaersltd@reynaers.com
Web: www.reynaers.com
Product Type: 2, 5 **Material Type:** C) 1

RHONES JOINERY (WREXHAM) LTD
Mold Road Industrial Estate,
Gwersyllt, Wrexham, LL11 4AQ
Area of Operation: UK (Excluding Ireland)
Tel: 01978 262488 **Fax:** 01978 262488
Email: info@rhonesjoinery.co.uk
Web: www.rhonesjoinery.co.uk
Product Type: 1, 2, 3 **Other Info:**

ROBERT J TURNER & CO
Roe Green, Sandon, Buntingford, Hertfordshire, SG9 0QE
Area of Operation: East England, Greater London
Tel: 01763 288371 **Fax:** 01763 288440
Email: sales@robertjturner.co.uk
Web: www.robertjturner.co.uk
Product Type: 1, 2, 3, 6

ROBERT MILLS ARCHITECTURAL ANTIQUES
Narroways Road, Eastville, Bristol, BS2 9XB
Area of Operation: Worldwide
Tel: 0117 955 6542 **Fax:** 0117 955 8146
Email: info@rmills.co.uk **Web:** www.rmills.co.uk
Product Type: 3 **Other Info:**

SAPA BUILDING SYSTEMS LIMITED
Alexandra Way, Ashchurch, Tewkesbury,
Gloucestershire, GL20 8NB
Area of Operation: UK & Ireland
Tel: 01684 853500 **Fax:** 01684 851850
Email: nicola.abbey@sapagroup.com
Web: www.sapagroup.com/uk/buildingsystems
Product Type: 1, 2, 3, 4, 5, 6
Other Info: **Material Type:** C) 1

SASH CRAFT
Unit 1 Mill Farm, Kelston, Bath, Somerset, BA1 9AQ
Area of Operation: South West England & South Wales
Tel: 01225 424434 **Fax:** 01225 424435
Email: pbale@sashcraftofbath.co.uk
Web: www.sashcraftofbath.co.uk
Product Type: 1, 2, 3, 4 **Material Type:** B) 8, 12

**SASH RESTORATION
COMPANY (HEREFORD) LTD**
Pigeon House Farm, Lower Breinton,
Hereford, Herefordshire, HR4 7PG
Area of Operation: UK & Ireland
Tel: 01432 359562 **Fax:** 01432 269749
Email: sales@sash-restoration.co.uk
Web: www.sash-restoration.co.uk
Product Type: 1, 3, 6 **Other Info:**
Material Type: A) 2

SASH UK LTD
Ferrymoor Way, Park Springs, Grimethorpe,
Barnsley, South Yorkshire, S72 7BN
Area of Operation: Worldwide
Tel: 01226 715619 **Fax:** 01226 719968
Email: mailbox@sashuk.com
Web: www.sashuk.com
Product Type: 2, 3, 5, 6
Material Type: D) 1

SASHPRO
51a Norbury Road, Thornton Heath, Surrey, CR7 8JP
Area of Operation: UK & Ireland
Tel: 0208 653 6477 **Fax:** 0208 771 5956
Email: enquiries@sashpro.co.uk
Web: www.sashpro.co.uk
Product Type: 1, 6

SCANDINAVIAN WINDOW SYSTEMS LTD
10 Eldon Street, Tuxford, Newark,
Nottinghamshire, NG22 0LH
Area of Operation: UK & Ireland
Tel: 01777 871847 **Fax:** 01777 872650
Email: enquiries@scandinavian-windows.co.uk
Web: www.scandinavian-windows.co.uk
Product Type: 2, 5, 6, 8 **Other Info:** ECO
Material Type: B) 1, 8, 13

SCANDINAVIAN WINDOW SYSTEMS
Area of Operation: UK & Ireland
Tel: 01777 871847 **Fax:** 01777 872650
Email: enquiries@scandinavian-windows.co.uk
Web: www.scandinavian-windows.co.uk
Product Type: 2, 5, 6, 8
Materials Type: B) 1, 8, 13

Suppliers of made to measure high performance windows, doors
and sliding doors from Norway. Options include factory finished
& factory glazed frames in all timber or aluminium and timber
composite. Sunflex folding sliding doors also available in all
timber, all aluminium or aluminium clad timber.

SCHUCO INTERNATIONAL KG
Whitehall Avenue, Kingston,
Milton Keynes, Buckinghamshire, MK10 0AL
Area of Operation: Worldwide
Tel: 01908 282111 **Fax:** 01908 282124
Email: shamman@schueco.com
Web: www.schueco.co.uk **Material Type:** C) 1

SLIDING SASH WINDOWS LTD
The Workshop, Rope Walk, Wotton-Under-Edge,
Gloucestershire, GL12 7DN
Area of Operation: UK (Excluding Ireland)
Tel: 01453 844877 **Fax:** 01453 844877
Email: mark@slidingsashwindows.biz
Web: slidingsashwindows.com
Product Type: 1, 2, 3, 4, 6 **Material Type:** B) 8, 12

SPECTUS SYSTEMS
Queens Avenue, Macclesfield, Cheshire, SK10 2BN
Area of Operation: UK & Ireland
Tel: 01625 420400 **Fax:** 01625 501418
Email: contact@spectus.co.uk
Web: www.spectus.co.uk
Product Type: 1, 2, 3, 5, 8
Material Type: D) 1

SPS TIMBER WINDOWS
Unit 2, 34 Eveline Road,
Mitcham, Surrey, CR4 3LE
Area of Operation: Greater London
Tel: 0208 640 5035 **Fax:** 0208 685 1570
Email: info@spstimberwindows.co.uk
Web: www.spstimberwindows.co.uk
Product Type: 1, 2, 3, 4, 5, 6
Other Info: ECO
Material Type: B) 2, 8, 12

STEEL WINDOW SERVICE & SUPPLIES LTD
30 Oxford Road, Finsbury Park, London, N4 3EY
Area of Operation: Greater London
Tel: 0207 272 2294 **Fax:** 0207 281 2309
Email: post@steelwindows.co.uk
Web: www.steelwindows.co.uk
Product Type: 2, 3, 6 **Material Type:** C) 2, 4

STORM WINDOWS LIMITED
Unit 7, James Scott Road, Off Park Lane,
Halesowen, West Midlands, B63 2QT
Area of Operation: UK (Excluding Ireland)
Tel: 01384 636365 **Fax:** 01384 410307
Email: sandra.lamb@virgin.net
Web: www.stormwindows.co.uk
Product Type: 1, 2, 3
Material Type: J) 4

**STROUDS WOODWORKING
COMPANY LIMITED.**
Ashmansworthy, Woolsery,
Bideford, Devon, EX39 5RE
Area of Operation: South West England
& South Wales
Tel: 01409 241624 **Fax:** 01409 241769
Email: enquires@doormakers.co.uk
Web: www.doormakers.co.uk
Product Type: 1, 2, 3, 4, 5, 6
Other Info: ECO **Material Type:** A) 1, 2, 8, 10, 12

SUN PARADISE UK LTD
Phoenix Wharf, Eel Pie Island,
Twickenham, Middlesex, TW1 3DY
Area of Operation: UK & Ireland
Tel: 0870 240 7604 **Fax:** 0870 240 7614
Email: info@sunparadise.co.uk
Web: www.sunparadise.co.uk
Product Type: 5

SUN PARADISE UK LTD
Area of Operation: UK
Tel: 0870 240 7604 **Fax:** 0870 240 7614
Email: info@sunparadise.co.uk
Web: www.sunparadise.co.uk
Product Type: 5

From a simple 'Patio Door' replacement to the glazing of a complete façade, Sun Paradise sliding doors can be used in a variety of applications with Aluminium frames or even 'frameless' systems that can slide and stack into concealed recesses.

SUNFOLD SYSTEMS
Sunfold House, Wymondham Business Park,
Chestnut Drive, Wymondham, Norfolk, NR18 9SB
Area of Operation: Worldwide
Tel: 01953 423423 **Fax:** 01953 423430
Email: info@sunfoldsystems.co.uk
Web: www.sunfold.com
Product Type: 5, 8 **Other Info:** ✎

THE ORIGINAL BOX SASH WINDOW COMPANY
29/30 The Arches, Alma Road,
Windsor, Berkshire, SL4 1QZ
Area of Operation: UK (Excluding Ireland)
Tel: 01753 858196 **Fax:** 01753 857827
Email: info@boxsash.com
Web: www.boxsash.com
Product Type: 1 **Other Info:** ✎

TOMPKINS LTD
High March Close, Long March Industrial Estate,
Daventry, Northamptonshire, NN11 4EZ
Area of Operation: UK (Excluding Ireland)
Tel: 01327 877187 **Fax:** 01327 310491
Email: info@tompkinswood.co.uk
Web: www.tompkinswood.co.uk
Product Type: 1, 2, 3, 4, 5 **Other Info:** ✎ ✋
Material Type: A) 1, 2, 3, 4, 5, 6, 7, 8, 9, 10, 11, 12, 13, 14

TONY HOOPER
Unit 18 Camelot Court, Bancombe Trading Estate,
Somerton, TA11 6SB
Area of Operation: UK (Excluding Ireland)
Tel: 01458 274221 **Fax:** 01458 274690
Email: tonyhooper1@aol.com
Web: www.tonyhooper.co.uk

TOTAL GLASS LIMITED
Total Complex, Overbrook Lane,
Knowsley, Merseyside, L34 9FB
Area of Operation: Midlands & Mid Wales,
North East England, North West England
and North Wales
Tel: 0151 549 2339 **Fax:** 0151 546 0022
Email: sales@totalglass.com
Web: www.totalglass.com
Product Type: 1, 2, 3, 5, 6

TWC THE WINDOW COMPANY LIMITED
The Drill Hall, 262 Huddersfield Road,
Thongsbridge, Holmfirth, West Yorkshire, HD9 3JQ
Area of Operation: UK & Ireland
Tel: 0800 917 1918 **Fax:** 01484 685210
Email: admin@twcthewindowcompany.net
Web: www.twcthewindowcompany.net
Product Type: 1, 2, 5 **Material Type:** D) 1

UNIVERSAL ARCHES LTD
103 Peasley Cross Lane, Peasley Cross,
St Helens, Merseyside, WA11 7NW
Area of Operation: UK & Ireland
Tel: 01744 612844 **Fax:** 01744 694250
Email: paula@universalarches.com
Web: www.universalarches.com
Other Info: ✎ ✋ **Material Type:** D) 1

VALE GARDEN HOUSES LTD
Londonthorpe Road, Grantham, Lincolnshire, NG31 9SJ
Area of Operation: UK & Ireland
Tel: 01476 564433 **Fax:** 01476 578555
Email: ken@valegardenhouses.com
Web: www.bronzecasements.com
Product Type: 2 **Material Type:** C) 12

VELFAC
Kettering Parkway, Wellingborough Road,
Kettering, Northamptonshire, NN15 6XR
Area of Operation: UK (Excluding Ireland)
Tel: 01223 897100 **Fax:** 01223 897101
Email: ac@velfac.co.uk **Web:** www.velfac.co.uk
Product Type: 2, 5, 6, 8 **Other Info:** ✎

VELUX WINDOWS
Woodside Way, Glenrothes East, Fife, KY7 4ND
Area of Operation: Worldwide
Tel: 0870 240 0617 **Fax:** 0870 401 4951
Web: www.velux.co.uk ✐

VENTROLLA LTD
11 Hornbeam Square South,
Harrogate, North Yorkshire, HG2 8NB
Area of Operation: UK & Ireland
Tel: 0800 378278 **Fax:** 01423 859321
Email: info@ventrolla.co.uk
Web: www.ventrolla.co.uk **Product Type:** 1, 2, 3

VENTROLLA LTD
Area of Operation: UK & Ireland
Tel: 0800 378 278 **Fax:** 01423 859321
Email: info@ventrolla.co.uk
Web: www.ventrolla.co.uk
Product Type: 1,2,3

The UK's market leader in the repair, renovation and upgrade of original sliding sash windows. Recognised for use in listed buildings and properties within conservation areas.

VICTORIAN SLIDERS LTD
Unit B Greenfields Business Centre,
Kidwelly, Carmarthenshire, SA17 4PT
Area of Operation: UK & Ireland
Tel: 0845 1700 810 **Fax:** 0845 1700 820
Email: info@victoriansliders.co.uk
Web: www.victorianslidersltd.co.uk

WOODLAND PRODUCTS DESIGN LTD
St.Peter's House, 6 Cambridge Road,
Kingston Upon Thames, Surrey, KT1 3JY
Area of Operation: UK (Excluding Ireland)
Tel: 0208 547 2171 **Fax:** 0208 547 1722
Email: enquiries@woodland-products.co.uk
Web: www.woodland-products.co.uk

GLASS & GLAZING

KEY
PRODUCT TYPES: 1= Standard   2 = Double Glazing   3 = Toughened   4 = Secondary Glazing   5 = Decorative   6 = Leaded Windows   7 = Fibreglass   8 = Other
SEE ALSO: DOORS, WINDOWS AND CONSERVATORIES - Windows, DOORS, WINDOWS AND CONSERVATORIES - Roof Windows and Light Pipes, DOORS, WINDOWS AND CONSERVATORIES - Stained Glass

OTHER: ▽ Reclaimed ✐ On-line shopping ✎ Bespoke ✋ Hand-made ECO Ecological

ADDISON DESIGN SYSTEMS
Unit 23, 106A Bedford Road, Wootton,
Bedfordshire, MK43 9JB
Area of Operation: UK (Excluding Ireland)
Tel: 01234 767721 **Fax:** 01234 767781
Email: marketing.styring@addison-design.co.uk
Web: www.addison-design.co.uk

AMBASSADOOR WINDOWS & DOORS LTD
18 Bidwell Road, Rackheath Industrial Estate,
Rackheath, Norwich, Norfolk, NR13 6PT
Area of Operation: East England, Greater London,
Midlands & Mid Wales
Tel: 01603 720332
Email: enquiries@ambassadoor.co.uk
Web: www.ambassadoor.co.uk **Product Type:** 1, 2, 3, 5

APROPOS TECTONIC LTD
Greenside House, Richmond Street,
Ashton Under Lyne, Lancashire, OL6 7ES
Area of Operation: UK & Ireland
Tel: 0870 777 0320 **Fax:** 0870 777 0323
Email: enquiries@apropos-tectonic.com
Web: www.apropos-tectonic.com **Product Type:** 2

B & M WINDOW & KITCHEN CENTRE
2-6 Whitworth Drive, Aycliffe Industrial Park,
Durham, DL5 6SZ
Area of Operation: Worldwide
Tel: 01325 308888 **Fax:** 01325 316002
Email: sales@bandmhomeimprovements.com
Web: www.bandmhomeimprovements.com ✐
Product Type: 1, 2, 3, 4, 5, 6

BARRON GLASS
Unit 11, Lansdown Industrial Estate,
Cheltenham, Gloucestershire, GL51 8PL
Area of Operation: UK & Ireland
Tel: 01242 22800 **Fax:** 01242 226555
Email: admin@barronglass.co.uk
Web: www.barronglass.co.uk **Product Type:** 1, 2, 5, 6, 8

BIRCHDALE GLASS LTD
Unit L, Eskdale Road, Uxbridge,
Greater London, UB8 2RT
Area of Operation: UK & Ireland
Tel: 01895 259111 **Fax:** 01895 810087
Email: info@birchdaleglass.com
Web: www.birchdaleglass.com
Product Type: 1, 2, 3, 4, 5, 6, 8

CHAMBERLAIN & GROVES LTD THE DOOR & SECURITY STORE
101 Boundary Road, Walthamstow, London, E17 8NQ
Area of Operation: UK (Excluding Ireland)
Tel: 0208 520 6776 **Fax:** 0208 520 2190
Email: ken@securedoors.co.uk
Web: www.securedoors.co.uk
Product Type: 1, 2, 3, 5, 6, 8

CHARLES HENSHAW & SONS LTD
Russell Road, Edinburgh, Lothian, EH11 2LS
Area of Operation: UK (Excluding Ireland)
Tel: 0131 337 4204 **Fax:** 0131 346 2441
Email: admin@charles-henshaw.co.uk
Web: www.charles-henshaw.com

CHIPPING NORTON GLASS LTD
Units 1 & 2, Station Yard Industrial Estate,
Chipping Norton, Oxfordshire, OX7 5HX
Area of Operation: Midlands & Mid Wales,
South West England and South Wales
Tel: 01608 643261
Fax: 01608 641768
Email: gill@cnglass.plus.com
Web: www.chippingnortonglass.co.uk
Product Type: 1, 2, 3, 4, 5, 6
Material Type: J) 1, 2, 4, 5, 6, 7, 8, 9, 10

CLASSIC PVC HOME IMPROVEMENTS
46 Stepney Place, Llanelli,
Carmarthenshire, SA15 1SE
Area of Operation: South West England
and South Wales
Tel: 0800 064 9494   **Fax:** 01554 775086
Email: enquiries@classic.uk.com
Web: www.classic.uk.com
Product Type: 1, 2, 3, 4, 5, 6, 7
Material Type: D) 1

DUPLUS DOMES LTD
370 Melton Road, Leicester,
Leicestershire, LE4 7SL
Area of Operation: Worldwide
Tel: 0116 261 0710
Fax: 0116 261 0539
Email: sales@duplus.co.uk
Web: www.duplus.co.uk
Material Type: D) 3

DURABUILD GLAZED STRUCTURES LTD
Carlton Road, Coventry,
West Midlands, CV6 7FL
Area of Operation: Worldwide
Tel: 02476 669169   **Fax:** 02476 669170
Email: enquiries@durabuild.co.uk
Web: www.durabuild.co.uk
Product Type: 2, 3, 5, 6, 8
Other Info: ECO
Material Type: J) 2, 4, 5, 10

EDGETECH IG INC
Unit 3, Swallow Gate Business Park,
Holbrook Lane, Coventry, CV6 4BL
Area of Operation: UK (Excluding Ireland)
Tel: 02476 705570   **Fax:** 02476 705510
Email: aoconnor@edgetechig.com
Web: www.edgetechig.com
Product Type: 1, 2, 4

ENVIROZONE
Encon House, 54 Ballycrochan Road,
Bangor, Co Down, BT19 6NF
Area of Operation: UK & Ireland
Tel: 02891 477799   **Fax:** 02891 452463
Email: info@envirozone.co.uk
Web: www.envirozone.co.uk
Product Type: 2, 3

EVEREST LTD
Sopers Road, Cuffley, Hertfordshire, EN6 4SG
Area of Operation: UK & Ireland
Tel: 0800 010123
Web: www.everest.co.uk

EVERGREEN DOOR
Unit 1, Oakwell Park Industrial Estate,
Birstall, West Yorkshire, WF17 9LU
Area of Operation: Worldwide
Tel: 01924 423171
Fax: 01924 423175
Email: andrewgrogan@evergreendoor.co.uk
Web: www.evergreendoor.co.uk
Product Type: 3, 5, 6, 7

FITZROY JOINERY
Garden Close, Langage Industrial Estate,
Plympton, Devon, PL7 5EU
Area of Operation: UK & Ireland
Tel: 0870 428 9110
Fax: 0870 428 9111
Email: admin@fitzroy.co.uk
Web: www.fitzroy.co.uk
Product Type: 1, 2, 3, 4, 5, 6
Other Info: ECO
Material Type: J) 2, 4, 5, 6, 7, 8, 9, 10

FOREST OF DEAN STAINED GLASS
Units 18-20, Harts Barn, Monmouth Road,
Longhope, Gloucestershire, GL17 0QD
Area of Operation: UK (Excluding Ireland)
Tel: 01452 830994   **Fax:** 01452 830994
Email: mike@forestglass.co.uk
Web: www.forestglass.co.uk   **Product Type:** 5, 6, 8

FRANKLIN LEEDS LLP
Carlton Works, Cemetery Road, Yeadon,
Leeds, West Yorkshire, LS19 7BD
Area of Operation: UK & Ireland
Tel: 0113 250 2991   **Fax:** 0113 250 0991
Email: david.franklin@franklinwindows.co.uk
Web: www.franklinwindows.co.uk
Product Type: 1, 2, 3, 5, 6, 8

GLASS INSERTS LTD
Unit 6, Wood Road, Kingswood, Bristol, BS15 8NN
Area of Operation: UK (Excluding Ireland)
Tel: 0117 961 9953
Email: paulrichards@glassinsertsltd.com
Product Type: 5

GLAVERBEL (UK) LTD.
Chestnut Field, Regent Place,
Rugby, Warwickshire, CV21 2TL
Area of Operation: Europe
Tel: 01788 535353   **Fax:** 01788 560853
Email: gvb.uk@glaverbel.com
Web: www.myglaverbel.com
Material Type: J) 1, 2, 4, 5, 6, 7, 9, 10

GREEN BUILDING STORE
11 Huddersfield Road, Meltham,
Holmfirth, West Yorkshire, HD9 4NJ
Area of Operation: UK & Ireland
Tel: 01484 854898
Fax: 01484 854899
Email: info@greenbuildingstore.co.uk
Web: www.greenbuildingstore.co.uk
Product Type: 2, 3, 8
Other Info: ECO   **Material Type:** J) 2, 4, 10

HARAN GLASS
Southpoint, 15 Lawmoor Road,
Dixon Blazes, City of Glasgow
Area of Operation: UK (Excluding Ireland)
Tel: 0141 418 4510   **Fax:** 0141 429 8655
Web: www.haranglass.com

HOLDSWORTH WINDOWS LTD
Darlingscote Road, Shipston-on-Stour,
Warwickshire, CV36 4PR
Area of Operation: UK & Ireland
Tel: 01608 661883   **Fax:** 01608 661008
Email: info@holdsworthwindows.co.uk
Web: www.holdsworthwindows.co.uk
Product Type: 6   **Other Info:**
Material Type: C) 2, 4

HOT GLASS DESIGN
Unit 24 Crosby Yard Industrial Estate,
Bridgend, Mid Glamorgan, CF31 1JZ
Area of Operation: Europe
Tel: 01656 659884   **Fax:** 01656 659884
Email: info@hotglassdesign.co.uk
Web: www.hotglassdesign.co.uk
Product Type: 3, 5, 8
Material Type: J) 1, 2, 4, 5, 6

INNOVATION GLASS COMPANY
27-28 Blake Industrial Estate, Brue Avenue,
Bridgwater, Somerset, TA7 8EQ
Area of Operation: Europe
Tel: 01278 683645   **Fax:** 01278 450088
Email: magnus@igc-uk.com
Web: www.igc-uk.com
Product Type: 3, 5   **Other Info:**

LATTICE PERIOD WINDOWS
Unit 85 Northwick Business Centre, Blockley,
Moreton in Marsh, Gloucestershire, GI56 9RF
Area of Operation: UK & Ireland
Tel: 01386 701079   **Fax:** 01386 701114
Email: sales@lattice windows .net
Web: www.latticewindows.net
Product Type: 1, 2, 3, 4, 5, 6
Material Type: J) 1, 2, 3, 4, 5, 6, 7, 8, 9, 10

SPONSORED BY: ULTRAFRAME www.ultraframe-conservatories.co.uk

M. H. JOINERY SERVICES
25b Camwal Road, Harlgate, North Yorkshire, HG1 4PT
Area of Operation: North East England
Tel: 01423 888856
Fax: 01432 888856
Email: info@mhjoineryservices.co.uk
Web: www.mhjoineryservices.co.uk
Product Type: 1, 2, 3, 6

NULITE LTD
41 Hutton Close, Crowther Industrial Estate,
Washington, Tyne & Wear, NE38 OAH
Area of Operation: UK & Ireland
Tel: 0191 419 1111   **Fax:** 0191 419 1123
Email: sales@nulite-ltd.co.uk
Web: www.nulite-ltd.co.uk
Product Type: 2, 3, 7, 8
Other Info: ✎   **Material Type:** C) 1, 2, 4

OXFORD SASH WINDOW CO LTD.
Eynsham Park Estate Yard, Cuckoo Lane,
North Leigh, Oxfordshire, OX29 6PW
Area of Operation: Greater London, Midlands & Mid
Wales, South East England, South West England &
South Wales
Tel: 01993 883536
Fax: 01993 883027
Email: oxsash@globalnet.co.uk
Web: www.sashwindow.co.uk
Product Type: 1, 2, 3, 8

PERMACELL FINESSE LTD
Western Road, Silver End, Witham, Essex, CM8 3BQ
Area of Operation: UK & Ireland
Tel: 01376 583241
Fax: 01376 583072
Email: sales@pfl.co.uk
Web: www.permacell-finesse.co.uk

REDDISEALS
The Furlong, Berry Hill Industrial Estate,
Droitwich, Worcestershire, WR9 9BG
Area of Operation: UK & Ireland
Tel: 0845 165 9507
Fax: 01905 791877
Email: reddiseals@reddiplex.co.uk
Web: www.reddiseals.com ✎
Product Type: 8

ROBERT J TURNER & CO
Roe Green, Sandon, Buntingford,
Hertfordshire, SG9 0QE
Area of Operation: East England, Greater London
Tel: 01763 288371
Fax: 01763 288440
Email: sales@robertjturner.co.uk
Web: www.robertjturner.co.uk
Product Type: 1, 2, 3, 5, 6

SASH UK LTD
Ferrymoor Way, Park Springs, Grimethorpe,
Barnsley, South Yorkshire, S72 7BN
Area of Operation: Worldwide
Tel: 01226 715619
Fax: 01226 719968
Email: mailbox@sashuk.com
Web: www.sashuk.com
Product Type: 1, 2, 3, 5, 6
Material Type: D) 1

SCHUCO INTERNATIONAL KG
Whitehall Avenue , Kingston, Milton Keynes,
Buckinghamshire, MK10 0AL
Area of Operation: Worldwide
Tel: 01908 282111
Fax: 01908 282124
Email: shamman@schueco.com
Web: www.schueco.co.uk

SPS TIMBER WINDOWS
Unit 2, 34 Eveline Road, Mitcham, Surrey, CR4 3LE
Area of Operation: Greater London
Tel: 0208 640 5035
Fax: 0208 685 1570
Email: info@spstimberwindows.co.uk
Web: www.spstimberwindows.co.uk
Product Type: 1, 2, 3, 5, 6
Other Info: ✎

STANDARDS GROUP
Bentley Hall Barn, Alkmonton,
Ashbourne, Derbyshire, DE6 3DJ
Area of Operation: UK & Ireland
Tel: 01335 330263
Fax: 01335 330922
Email: uk@standardsgroup.com
Web: www.standardsgroup.com
Product Type: 1, 3, 5, 6

STORM WINDOWS LIMITED
Unit 7, James Scott Road, Off Park Lane,
Halesowen, West Midlands, B63 2QT
Area of Operation: UK (Excluding Ireland)
Tel: 01384 636365
Fax: 01384 410307
Email: sandra.lamb@virgin.net
Web: www.stormwindows.co.uk
Product Type: 4

STUART OWEN NORTON GLASS & SIGN LTD
Unit 15 Dukes Court, Prudhoe,
Northumberland, NE42 6DA
Area of Operation: Worldwide
Tel: 01661 833227   **Fax:** 01661 833732
Email: brilliant.cutter@btinternet.com
Web: www.glass-and-sign.com
Product Type: 1, 2, 3, 5, 6
Other Info: ✎ ✋   **Material Type:** J) 4, 5, 6

STUART NORTON GLASS & SIGN LTD

Area of Operation: Worldwide
Tel: 01661 833227   **Fax:** 01661 833732
Email: brilliant.cutter@btinternet.com
Web: www.glass-and-sign.com

• Brilliant cutting • Victorian Door Glass •
• Dull Cutting • Sandblasting • Acid Etched •
• Stained Glass • Leaded Panels and much,
much more.

SUNFOLD SYSTEMS
Sunfold House, Wymondham Business Park,
Chestnut Drive, Wymondham, Norfolk, NR18 9SB
Area of Operation: Worldwide
Tel: 01953 423423
Fax: 01953 423430
Email: info@sunfoldsystems.co.uk
Web: www.sunfold.co.uk
Product Type: 1, 2, 3

SUNRISE STAINED GLASS
58-60 Middle Street, Southsea,
Portsmouth, Hampshire, PO5 4BP
Area of Operation: UK (Excluding Ireland)
Tel: 02392 750512
Fax: 02392 875488
Email: sunrise@stained-windows.co.uk
Web: www.stained-windows.co.uk
Product Type: 5, 6

SUNSQUARE LIMITED
12L Hardwick Industrial Estate,
Bury St Edmunds, Suffolk, IP33 2QH
Area of Operation: UK & Ireland
Tel: 0845 226 3172
Fax: 0845 226 3173
Email: info@sunsquare.co.uk
Web: www.sunsquare.co.uk
Product Type: 1, 2, 3
Other Info: ECO ✎

THE STAINED GLASS STUDIO
Unit 5, Brewery Arts, Brewery Court,
Cirencester, Gloucestershire, GL7 1JH
Area of Operation: Worldwide
Tel: 01285 644430
Fax: 0800 454627
Email: daniella@wilson-dunne.wanadoo.co.uk
Product Type: 5

TOTAL GLASS LIMITED
Total Complex, Overbrook Lane,
Knowsley, Merseyside, L34 9FB
Area of Operation: Midlands & Mid Wales,
North East England, North West England
and North Wales
Tel: 0151 549 2339
Fax: 0151 546 0022
Email: sales@totalglass.com
Web: www.totalglass.com
Product Type: 1, 2, 3, 5, 6, 7, 8

TWINFIX
201 Cavendish Place, Birchwood Park,
Birchwood, Warrington, Cheshire, WA3 6WU
Area of Operation: UK & Ireland
Tel: 01925 811311   **Fax:** 01925 852955
Email: enquiries@twinfix.co.uk
Web: www.twinfix.co.uk   **Product Type:** 8

ULTRAFRAME UK
Salthill Road, Clitheroe, Lancashire, BB7 1PE
Area of Operation: UK & Ireland
Tel: 0500 822340   **Fax:** 0870 414 1020
Email: sales@ultraframe.com
Web: www.ultraframe.com
Product Type: 1, 3, 8   **Material Type:** D) 1, 3

ROOF WINDOWS & LIGHT PIPES

KEY

OTHER: ▽ Reclaimed   ✎ On-line shopping
✎ Bespoke   ✋ Hand-made   ECO Ecological

AUTOMATED CONTROL SERVICES LTD
Unit 16, Hightown Industrial Estate,
Crow Arch Lane, Ringwood, Dorset, BH24 1ND
Area of Operation: UK (Excluding Ireland)
Tel: 01425 461008   **Fax:** 01425 461009
Email: sales@automatedcontrolservices.co.uk
Web: www.automatedcontrolservices.co.uk

E. RICHARDS
PO Box 1115, Winscombe, Somerset, BS25 1WA
Area of Operation: UK & Ireland
Tel: 0845 330 8859   **Fax:** 0845 330 7260
Email: paultrace@e-richards.co.uk
Web: www.e-richards.co.uk ✎

E RICHARDS
Area of Operation: UK & Ireland
Tel: 0845 330 8859
Web: www.e-richards.co.uk
Product Type: 7
Materials Type: C) 1   **Other Info:** ✎

Since 1893 Richards Brackets have been manufacturing and supplying
cast iron and steel products to the building and plumbing industry.
Included in the range is the "Cast" conservation rooflight. 7 sizes to suit
all applications conservation are available, including an escape rooflight.

GLAZING VISION
6 Barns Close, Brandon, Suffolk, IP27 0NY
Area of Operation: Worldwide
Tel: 01842 815581   **Fax:** 01842 815515
Email: sales@visiongroup.co.uk
Web: www.visiongroup.co.uk

JOULESAVE EMES LTD
27 Water Lane, South Witham,
Grantham, Lincolnshire, NG33 5PH
Area of Operation: UK & Ireland
Tel: 01572 768362   **Fax:** 01572 767146
Email: sales@joulesave.co.uk
Web: www.joulesave.co.uk

KEYLITE ROOF WINDOWS
Derryloran Industrial Estate, Sandholes Road,
Cookstown, Co Tyrone, BT80 9LU
Area of Operation: UK & Ireland
Tel: 028 8675 8921   **Fax:** 028 8675 8923
Email: info@keylite.co.uk   **Web:** www.keylite.co.uk

MONODRAUGHT LTD
Halifax House, Cressex Business Park,
High Wycombe, Buckinghamshire, HP12 3SE
Area of Operation: UK (Excluding Ireland)
Tel: 01494 897700   **Fax:** 01494 532465
Email: info@monodraught.com
Web: www.sunpipe.co.uk ✎

NULITE LTD
41 Hutton Close, Crowther Industrial Estate,
Washington, Tyne & Wear, NE38 OAH
Area of Operation: UK & Ireland
Tel: 0191 419 1111   **Fax:** 0191 419 1123
Email: sales@nulite-ltd.co.uk
Web: www.nulite-ltd.co.uk

ROTO FRANK LTD
Swift Point, Rugby, Warwickshire, CV21 1QH
Area of Operation: Worldwide
Tel: 01788 558600   **Fax:** 01788 558605
Email: uksales@roto-frank.co.uk
Web: www.roto-frank.co.uk

RPL SOLATUBE IN SCOTLAND
4 Blythbank Cottages, West Linton,
Lothian, EH46 7DF
Area of Operation: Scotland
Tel: 0845 601 5785   **Fax:** 01721 752624
Email: info@solabright.co.uk
Web: www.solatubescotland.co.uk

SOLA SKYLIGHTS UK LTD
12 Furnace Industrial Estate, Shildon, Durham, DL4 1QB
Area of Operation: UK & Ireland
Tel: 01388 778445   **Fax:** 01388 778216
Email: info@solaskylights.com
Web: www.solaskylights.com   **Other Info:** ECO

SOLALIGHTING
17 High Street, Olney, Buckinghamshire, MK46 4EB
Area of Operation: UK (Excluding Ireland)
Tel: 01234 241466   **Fax:** 01234 241766
Email: sales@solalighting.com
Web: www.solalighting.com

SUNSQUARE LIMITED
12L Hardwick Industrial Estate,
Bury St Edmunds, Suffolk, IP33 2QH
Area of Operation: UK & Ireland
Tel: 0845 226 3172   **Fax:** 0845 226 3173
Email: info@sunsquare.co.uk
Web: www.sunsquare.co.uk
Other Info: ECO ✎

THE GREEN SHOP
Cheltenham Road, Bisley, Nr Stroud,
Gloucestershire, GL6 7BX
Area of Operation: UK & Ireland
Tel: 01452 770629   **Fax:** 01452 770104
Email: paint@greenshop.co.uk
Web: www.greenshop.co.uk ✎

THE LIGHTPIPE CO. LTD
116b High Street, Cranfield, Bedfordshire, MK43 0DG
Area of Operation: UK (Excluding Ireland)
Tel: 08702 416680   **Fax:** 01234 751144
Email: sales@lightpipe.org.uk
Web: www.lightpipe.org.uk

BUILDING STRUCTURE & MATERIALS - **Doors, Windows & Conservatories** - Roof Windows & LIght Pipes; Curtains, Blinds, Shutters & Accessories

SPONSORED BY: ULTRAFRAME www.ultraframe-conservatories.co.uk

BUILDING STRUCTURE & MATERIALS

VISIONARY DESIGN - INNOVATIVE SOLUTIONS

At Glazing Vision we apply our philosophy of streamlined design to our range of opening skylights and roofs. Unobstructed daylight and access areas have become prerequisites to our core design concepts, along with minimising framework & concealing all fixings and opening equipment.

For more information please contact us for an info pack or visit our website where you will find more examples of all our products & also useful tools, such as the quote generator, to assist you.

GLAZING VISION

Part of the Vision Group of companies
www.visiongroup.co.uk
tel: 01842 815 581

LIGHTPIPE COMPANY LTD

Area of Operation: UK
Tel: 08702 41 66 80
Email: sales@lightpipe.org.uk
Web: www.lightpipe.org.uk

Being the only independent Lightpipe retailer in the UK we can offer advice on the most suitable Lightpipe for you or your clients project from the widest choice available.

THE LOFT SHOP LTD

Eldon Way, Littlehampton, West Sussex, BN17 7HE
Area of Operation: UK (Excluding Ireland)
Tel: 0870 604 0404 **Fax:** 0870 603 9075
Email: marjory.kay@loftshop.co.uk
Web: www.loftshop.co.uk

THE ROOFLIGHT COMPANY

Unit 8, Wychwood Business Centre,
Shipton Under Wychwood, Oxfordshire, OX7 6XU
Area of Operation: UK (Excluding Ireland)
Tel: 01993 830613 **Fax:** 01993 831066
Email: info@therooflightcompany.co.uk
Web: www.therooflightcompany.co.uk

THE ROOFLIGHT COMPANY

Area of Operation: UK (Excluding Ireland)
Tel: 01993 830613
Fax: 01993 831066
Email: info@therooflightcompany.co.uk
Web: www.therooflightcompany.co.uk

Conservation Rooflights® are chosen not just because they are favoured by planners, but also because they look so good. With their slim lines and low profile it is easy to see why!

WESSEX BUILDING PRODUCTS (MULTITEX)

Dolphin Industrial Estate, Southampton Road,
Salisbury, Wiltshire, SP1 2NB
Area of Operation: UK & Ireland
Tel: 01722 332139 **Fax:** 01722 338458
Email: sales@wessexbuildingproducts.co.uk
Web: www.wessexbuildingproducts.co.uk

PLEASE MENTION
THE HOMEBUILDER'S HANDBOOK
WHEN YOU CALL

CURTAINS, BLINDS, SHUTTERS AND ACCESSORIES

KEY

OTHER: ▽ Reclaimed On-line shopping
 Bespoke Hand-made ECO Ecological

A W PROTECTION FILMS

P.O Box 62, Hedge End, Southampton,
Hampshire, SO30 3ZJ
Area of Operation: Worldwide
Tel: 02380 477550 **Fax:** 02380 477886
Email: sales@andywrap.co.uk
Web: www.maskingfilm.co.uk
Other Info:

AMBASSADOOR WINDOWS & DOORS LTD

18 Bidwell Road, Rackheath Industrial Estate,
Rackheath, Norwich, Norfolk, NR13 6PT
Area of Operation: East England, Greater London, Midlands & Mid Wales
Tel: 01603 720332
Email: enquiries@ambassadoor.co.uk
Web: www.ambassadoor.co.uk
Material Type: A) 1

AMDEGA CONSERVATORIES

Woodside, Church Lane, Bursledon,
Southampton, Hampshire, SO31 8AB
Area of Operation: Worldwide
Tel: 0800 591523
Web: www.amdega-conservatories.co.uk

ARC-COMP

Friedrichstr. 3, 12205 Berlin-Lichterfelde, Germany
Area of Operation: Europe
Tel: +49 (0)308 430 9956
Fax: +49 (0)308 430 9957
Email: jvs@arc-comp.com
Web: www.arc-com.com **Material Type:** C) 1

ARC-COMP (IRISH BRANCH)

Whitefield Cottage, Lugduff,
Tinahely, Co. Wicklow, Republic of Ireland
Area of Operation: Europe
Tel: +353 (0)868 729 945
Fax: +353 (0)402 28900
Email: jvs@arc-comp.com
Web: www.arc-comp.com **Material Type:** C) 1

B ROURKE & CO LTD

Vulcan Works, Accrington Road,
Burnley, Lancashire, BB11 5QD
Area of Operation: Worldwide
Tel: 01282 422841
Fax: 01282 458901
Email: info@rourkes.co.uk
Web: www.rourkes.co.uk
Material Type: C) 5, 11, 14

BALMORAL BLINDS

12 Beresford Close, Frimley Green,
Camberley, Surrey, GU16 6LB
Area of Operation: Greater London, South East England
Tel: 01252 674172
Fax: 0870 132 7683
Email: sales@balmoralblinds.co.uk
Web: www.balmoralblinds.co.uk

BLINDS.CO.UK

17 - 19 Gt. Eastern Street, London , EC2A 3EJ
Area of Operation: UK (Excluding Ireland)
Tel: 0207 375 1053
Email: sales@blinds.co.uk
Web: www.blinds.co.uk

BRITISH BLIND AND SHUTTER ASSOCIATION

42 Heath Street, Tamworth, Staffordshire, B79 7JH
Area of Operation: UK & Ireland
Tel: 01827 52337
Fax: 01827 310827
Email: info@bbsa.org.uk
Web: www.bbsa.org.uk

BUILDING STRUCTURE & MATERIALS - **Doors, Windows & Conservatories** - Curtains, Blinds, Shutters & Accessories; Conservatories

SPONSORED BY: ULTRAFRAME www.ultraframe-conservatories.co.uk

BUILDING STRUCTURE & MATERIALS

CONSERVATORY BLINDS LTD
8-10 Ruxley Lane, Ewell, Epsom, Surrey, KT19 0JD
Area of Operation: East England, Greater London,
Midlands & Mid Wales, North East England, North
West England and North Wales, South East England,
South West England and South Wales
Tel: 0800 071 8888 **Fax:** 0208 394 0022
Email: info@conservatoryblinds.co.uk
Web: www.conservatoryblinds.co.uk

ELERO UK LIMITED
Unit 4, Foundry Lane, Halebank,
Widnes, Cheshire, WA8 8TZ
Area of Operation: UK & Ireland
Tel: 0870 240 4219 **Fax:** 0870 240 4086
Email: sales@elerouk.co.uk
Web: www.elerouk.co.uk

GRANT MERCER
PO Box 246, Banstead, Surrey, SM7 3AF
Area of Operation: Europe
Tel: 01737 357957 **Fax:** 01737 373003
Email: grantmercer@shuttersuk.co.uk
Web: www.shuttersuk.com

KESTREL SHUTTERS
9 East Race Street, Stowe,
Pennsylvania, 19464, USA
Area of Operation: Worldwide
Tel: +1 610 326 6679 **Fax:** +1 610 326 6779
Email: sales@diyshutters.com
Web: www.diyshutters.com

LEVOLUX LTD
1 Forward Drive, Harrow, Middlesex,
Greater London, HA3 8NT
Area of Operation: Worldwide
Tel: 0208 863 9111 **Fax:** 0208 863 8760
Email: info@levolux.com **Web:** www.levolux.com

MEL-TEC LTD
1 Boundary Road, Buckingham Road Industrial
Estate, Brackley, Northamptonshire, NN13 7ES
Area of Operation: UK & Ireland
Tel: 01280 705323 **Fax:** 01280 702258
Email: sales@meltec.co.uk
Web: www.meltec.co.uk
Other Info: **Material Type:** C) 1

MERRICK & DAY
Redbourne Hall, Redbourne,
Gainsborough, Lincolnshire, DN21 4JG
Tel: 0870 757 0980
Fax: 0870 757 0985
Email: sales@merrick-day.com
Web: www.merrick-day.com

NIGEL TYAS HANDCRAFTED IRONWORK
Green Works,18 Green Road,
Penistone, Sheffield,
South Yorkshire, S36 6BE
Area of Operation: Worldwide
Tel: 01226 766618
Email: sales@nigeltyas.co.uk
Web: www.nigeltyas.co.uk
Material Type: C) 2, 4, 5, 6

OPEN N SHUT
Forum House, Stirling Road,
Chichester, West Sussex, PO19 7DN
Area of Operation: UK (Excluding Ireland)
Tel: 01243 774888 **Fax:** 01243 774333
Email: shutters@opennshut.co.uk
Web: www.opennshut.co.uk

PLANTATION SHUTTERS
131 Putney Bridge Road, London, SW15 2PA
Area of Operation: UK & Ireland
Tel: 0208 871 9222 / 9333 **Fax:** 0208 871 0041
Email: sales@plantation-shutters.co.uk
Web: www.plantation-shutters.co.uk

RUFFLETTE
Sharston Road, Manchester, M22 4TH
Area of Operation: UK & Ireland
Tel: 0161 998 1811 **Fax:** 0161 945 9468
Email: customer-care@rufflette.com
Web: www.rufflette.com

SURFACEMATERIALDESIGN
17 Skiffington Close, London, SW2 3UL
Area of Operation: Worldwide
Tel: 0208 671 3383
Email: info@surfacematerialdesign.co.uk
Web: www.surfacematerialdesign.co.uk

THE BRADLEY COLLECTION
Lion Barn, Maitland Road,
Needham Market, Suffolk, IP6 8NS
Area of Operation: Worldwide
Tel: 01449 722724 **Fax:** 01449 722728
Email: claus.fortmann@bradleycollection.co.uk
Web: www.bradleycollection.co.uk

VELUX WINDOWS
Woodside Way, Glenrothes East, Fife, KY7 4ND
Area of Operation: Worldwide
Tel: 0870 240 0617 **Fax:** 0870 401 4951
Web: www.velux.co.uk

WINDOWSCREENS UK
P.O.Box 181, Upminster, Essex, RM14 1GX
Area of Operation: UK & Ireland
Tel: 01708 222273 **Fax:** 01708 641898
Email: info@flyscreensuk.co.uk
Web: www.flyscreensuk.co.uk

CONSERVATORIES

KEY

PRODUCT TYPES: 1= Victorian
2 = Edwardian   3 = Lean-to
4 = Georgian   5 = P-Shaped
6 = L-Shaped   7 = Double Height

OTHER: ▽ Reclaimed   On-line shopping
 Bespoke   Hand-made   ECO Ecological

ALUPLAST UK/PLUSPLAN
Leicester Road, Lutterworth,
Leicestershire, LE17 4HE
Area of Operation: UK & Ireland
Tel: 01455 556771 **Fax:** 01455 555323
Email: info@aluplastuk.com
Web: www.aluplastuk.com
Product Type: 1, 2, 3, 4, 5, 6
Material Type: D) 1

AMDEGA CONSERVATORIES
Woodside, Church Lane, Bursledon,
Southampton, Hampshire, SO31 8AB
Area of Operation: Worldwide
Tel: 0800 591523
Web: www.amdega-conservatories.co.uk

ANDERSEN / BLACK MILLWORK CO.
Andersen House, Dallow Street,
Burton on Trent, Staffordshire, DE14 2PQ
Area of Operation: UK & Ireland
Tel: 01283 511122
Fax: 01283 510863
Email: info@blackmillwork.co.uk
Web: www.blackmillwork.co.uk
Product Type: 1, 2, 3, 4, 5, 6
Material Type: A) 2, 5

APROPOS TECTONIC LTD
Greenside House, Richmond Street,
Ashton Under Lyne, Lancashire, OL6 7ES
Area of Operation: UK & Ireland
Tel: 0870 777 0320 **Fax:** 0870 777 0323
Email: enquiries@apropos-tectonic.com
Web: www.apropos-tectonic.com
Other Info: **Material Type:** C) 1

ARC-COMP
Friedrichstr. 3, 12205 Berlin-Lichterfelde, Germany
Area of Operation: Europe
Tel: +49 (0)308 430 9956
Fax: +49 (0)308 430 9957
Email: jvs@arc-comp.com
Web: www.arc-com.com
Product Type: 3 **Material Type:** C) 1

ARC-COMP (IRISH BRANCH)
Whitefield Cottage, Lugduff, Tinahely,
Co. Wicklow, Republic of Ireland
Area of Operation: Europe
Tel: +353 (0)868 729 945
Fax: +353 (0)402 28900
Email: jvs@arc-comp.com
Web: www.arc-comp.com
Product Type: 3 **Material Type:** C) 1

ARDEN SOUTHERN LTD
Arden House, Spark Brook Street,
Coventry, West Midlands, CV1 56T
Area of Operation: South West England
and South Wales
Tel: 0870 789 0160 **Fax:** 0870 789 0161
Email: enquires@ardenwindows.net
Web: www.ardenwindows.net
Product Type: 1, 2

ASTON HOUSE JOINERY
Aston House, Tripontium Business Centre,
Newton Lane, Rugby, Warwickshire, CV23 0TB
Area of Operation: UK & Ireland
Tel: 01788 860032 **Fax:** 01788 860614
Email: mailbox@ovationwindows.co.uk
Web: www.astonhousejoinery.co.uk
Product Type: 1, 2, 3, 4, 5, 6, 7
Other Info: ECO   **Material Type:** A) 1, 2, 12

AURORA CONSERVATORIES
The Old Station, Naburn, York,
North Yorkshire, YO19 4RW
Area of Operation: UK (Excluding Ireland)
Tel: 0800 026 4212 **Fax:** 01904 610318
Email: louiseaurora@btconnect.com
Web: www.auroraconservatories.co.uk
Product Type: 1, 2, 3, 4, 5, 6
Other Info: **Material Type:** J) 4, 10

B & M WINDOW & KITCHEN CENTRE
2-6 Whitworth Drive, Aycliffe Industrial Park,
Durham, DL5 6SZ
Area of Operation: Worldwide
Tel: 01325 308888 **Fax:** 01325 316002
Email: sales@bandmhomeimprovements.com
Web: www.bandmhomeimprovements.com
Product Type: 1, 2, 3, 4, 5, 6

BALTIC PINE CONSERVATORIES LTD
Baltic House, Longrock,
Penzance, Cornwall, TR20 8HX
Area of Operation: UK & Ireland
Tel: 01736 332200 **Fax:** 01736 332219
Email: enquiries@balticpine.co.uk
Web: www.balticpine.co.uk
Product Type: 1, 2, 3, 4, 5, 6 **Other Info:**

BARTHOLOMEW
Rakers Yard, Rake Road,
Milland, West Sussex, GU30 7JS
Area of Operation: Europe
Tel: 01428 742800 **Fax:** 01428 743801
Email: denise@bartholomew-conservatories.co.uk
Web: www.bartholomew-conservatories.co.uk
Product Type: 1, 2, 3, 4, 5, 6
Other Info: ECO   **Material Type:** A) 2, 10

BDG GROUP LIMITED
5 Wenlock Road, Lurgan, Craigavon,
Co Armagh, BT66 8QR
Area of Operation: Worldwide
Tel: 0283 832 7741 **Fax:** 0283 832 4358
Email: info@bdg.co.uk
Web: www.bdg.co.uk
Product Type: 1, 2, 3, 4, 5, 6
Other Info: **Material Type:** C) 1

BIRCHDALE GLASS LTD
Unit L, Eskdale Road, Uxbridge,
Greater London, UB8 2RT
Area of Operation: UK & Ireland
Tel: 01895 259111
Fax: 01895 810087
Email: info@birchdaleglass.com
Web: www.birchdaleglass.com

BOWATER WINDOWS NEWBUILD DIVISION
Water Orton Lane, Minworth, Sutton Coldfield,
West Midlands, B76 9BW
Area of Operation: UK (Excluding Ireland)
Tel: 0121 749 3000
Fax: 0121 749 8210
Email: info@theprimeconnection.co.uk
Web: www.whs-halo.co.uk
Product Type: 1, 2, 3, 4, 5, 6
Material Type: D) 1

BRECKENRIDGE CONSERVATORIES
Unit 10 Papyrus Road, Werrington,
Peterborough, Cambridgeshire, PE4 5BH
Area of Operation: East England
Tel: 01733 575750
Fax: 01733 292875
Email: sales@breckenridgeconservatories.co.uk
Web: www.breckenridgeconservatories.co.uk
Material Type: D) 1

CAREY & FOX LTD
51 Langthwaite Business Park, South Kirkby,
Pontefract, West Yorkshire, WF9 3NR
Area of Operation: UK & Ireland
Tel: 01977 608069 **Fax:** 01977 646791
Email: enquiries@careyandfox.co.uk
Web: www.careyandfox.co.uk
Product Type: 1, 2, 3, 4, 5, 6
Material Type: A) 2

**CARPENTER OAK LTD & RODERICK
JAMES ARCHITECT LTD**
The Framing Yard, East Cornworthy,
Totnes, Devon, TQ9 7HF
Area of Operation: Worldwide
Tel: 01803 732900 **Fax:** 01803 732901
Email: enquiries@carpenteroak.com
Web: www.carpenteroak.com
Material Type: A) 2, 8

CHARTERHOUSE CONSERVATORIES
Park Street, Gosport, Hampshire , PO12 4UH
Area of Operation: South East England
Tel: 02392 504006 **Fax:** 02392 513765
Email: info@charterhouse.nu
Web: www.charterhouse.nu

CHIPPING NORTON GLASS LTD
Units 1 & 2, Station Yard Industrial Estate,
Chipping Norton, Oxfordshire, OX7 5HX
Area of Operation: Midlands & Mid Wales,
South West England and South Wales
Tel: 01608 643261 **Fax:** 01608 641768
Email: gill@cnglass.plus.com
Web: www.chippingnortonglass.co.uk
Product Type: 1, 2, 3, 4, 5, 6, 7
Material Type: C) 1

CLASSIC PVC HOME IMPROVEMENTS
46 Stepney Place, Llanelli,
Carmarthenshire, SA15 1SE
Area of Operation: South West England
and South Wales
Tel: 0800 064 9494
Fax: 01554 775086
Email: enquiries@classic.uk.com
Web: www.classic.uk.com
Product Type: 1, 2, 3, 4, 5, 6 **Material Type:** D) 1

CLEARVIEW WINDOWS & DOORS
Unit 14, Sheddingdean Industrial Estate, Marchants
Way, Burgess Hill, West Sussex, RH15 8QY
Area of Operation: UK (Excluding Ireland)
Tel: 01444 250111
Fax: 01444 250678
Email: sales@clearviewsussex.co.uk
Web: www.clearviewsussex.co.uk

CONSERVATORIES 2 YOU
16 Teal Close, Quedgeley,
Gloucester, Gloucestershire, GL2 4GR
Area of Operation: UK (Excluding Ireland)
Tel: 01452 533221
Fax: 01452 55322
Email: info@conservatories2you.co.uk
Web: www.conservatories2you.co.uk
Product Type: 1, 2, 3, 4, 5, 6

CONSERVATORY & WINDOW WORLD LTD
Watling Rd, Bishop Aukland, Durham, DL14 9AU
Area of Operation: North East England
Tel: 01388 458088 **Fax:** 01388 458518
Email: paldennis20@aol.com
Web: www.conservatoryandwindowworld.co.uk
Product Type: 1, 2, 3, 4, 5, 6
Material Type: D) 1, 3, 6

CONSERVATORY OUTLET
Unit 8 Headway Business Park, Denby Dale Road,
Wakefield, West Yorkshire, WF2 7AZ
Area of Operation: North East England
Tel: 01924 881920
Web: www.conservatoryoutlet.co.uk
Product Type: 1, 2, 3, 4, 5, 6, 7
Other Info: ✎ **Material Type:** D) 1

CUSTOM WOOD PRODUCTS
Cliffe Road, Easton on the Hill,
Stamford, Lincolnshire, PE9 3NP
Area of Operation: East England
Tel: 01780 755711 **Fax:** 01780 480834
Email: customwoodprods@aol.com
Web: www.cwpuk.com
Product Type: 1, 2, 3, 4, 5, 6
Other Info: ECO ✎ ✋ **Material Type:** A) 2

DAVID SALISBURY CONSERVATORIES
Bennett Road, Isleport Business Park,
Highbridge, Somerset, TA9 4PW
Area of Operation: UK (Excluding Ireland)
Tel: 01278 764444 **Fax:** 01278 764422
Email: clarewinstanley@davidsalisbury.com
Web: www.davidsalisbury.com
Product Type: 1, 2, 3, 4, 5, 6
Other Info: ✎ **Material Type:** A) 15

DESIGN IN WOOD
73 Nannymar Road, Darfield,
South Yorkshire, S73 9AW
Area of Operation: UK (Excluding Ireland)
Tel: 01226 750527 **Fax:** 01226 341790

DIRECT CONSERVATORIES 4 U
Suite 27, Silk House, Park Green,
Macclesfield, SK11 7QJ
Area of Operation: UK & Ireland
Tel: 0800 279 3928 **Fax:** 0845 058 6002
Email: info@directconservatories4u.co.uk
Web: www.directconservatories4u.co.uk ✋
Product Type: 1, 2, 3, 4, 5, 6, 7
Material Type: D) 1

DIY SASH WINDOWS
2 - 6 Whitworth Drive, Aycliffe Industrial Park,
Newton Aycliffe, Durham, DL56SZ
Area of Operation: UK & Ireland
Tel: 01325 308888 **Fax:** 01325 316002
Email: sales@diysashwindows.co.uk
Web: www.diysashwindows.co.uk
Product Type: 1, 2, 3, 4, 5, 6, 7

DRUMMONDS ARCHITECTURAL ANTIQUES LTD
Kirkpatrick Buildings, 25 London Road (A3),
Hindhead, Surrey, GU36 6AB
Area of Operation: Worldwide
Tel: 01428 609444 **Fax:** 01428 609445
Email: davidcox@drummonds-arch.co.uk
Web: www.drummonds-arch.co.uk
Product Type: 1, 3 **Other Info:** ✎ ✋
Material Type: C) 2, 5

DURABUILD GLAZED STRUCTURES LTD
Carlton Road, Coventry, West Midlands, CV6 7FL
Area of Operation: Worldwide
Tel: 02476 669169 **Fax:** 02476 669170
Email: enquiries@durabuild.co.uk
Web: www.durabuild.co.uk
Product Type: 1, 2, 3, 4, 5, 6
Other Info: ECO ✎ ✋ **Material Type:** A) 12, 15

ELAM
Llettyreos, Llanfinhanges, Llanfyllin,
Powys, SY22 5JF
Area of Operation: UK (Excluding Ireland)
Tel: 01691 648495 **Fax:** 01691 648495
Email: info@elam-timber.co.uk
Web: www.elam-timber.co.uk

ENGELS WINDOWS & DOORS LTD
1 Kingley Centre, Downs Road, West Stoke,
Chichester, West Sussex, PO18 9HJ
Area of Operation: UK (Excluding Ireland)
Tel: 01243 576633 **Fax:** 01243 576644
Email: admin@engels.co.uk
Web: www.engels.co.uk

ESOGRAT LTD
Caldervale Works, River Street, Brighouse,
Huddersfield, West Yorkshire, HD6 1JS
Area of Operation: UK & Ireland
Tel: 01484 716228 **Fax:** 01484 400107
Email: info@esograt.com **Web:** www.esograt.com
Product Type: 1, 2, 3, 4, 5, 6

EUROCELL PROFILES LTD
Fairbrook House, Clover Nook Road,
Alfreton, Derby, Derbyshire, DE55 4RF
Area of Operation: UK (Excluding Ireland)
Tel: 01773 842100 **Fax:** 01773 842109
Email: eric.gale@eurocell.co.uk
Web: www.eurocell.co.uk
Product Type: 1, 2, 3, 4, 5, 6
Material Type: D) 1, 3

EVEREST LTD
Sopers Road, Cuffley, Hertfordshire, EN6 4SG
Area of Operation: UK & Ireland
Tel: 0800 010123
Web: www.everest.co.uk ✋
Product Type: 1, 2, 3, 4, 5, 6

FAIRMITRE WINDOWS & JOINERY LTD
2A Cope Road, Banbury, Oxfordshire, OX16 2EH
Area of Operation: UK (Excluding Ireland)
Tel: 01295 268441 **Fax:** 01295 268468
Email: info@fairmitrewindows.co.uk
Web: www.fairmitrewindows.co.uk
Product Type: 1, 2, 3, 4, 5, 6, 7

FITZROY JOINERY
Garden Close, Langage Industrial Estate,
Plympton, Devon, PL7 5EU
Area of Operation: UK & Ireland
Tel: 0870 428 9110 **Fax:** 0870 428 9111
Email: admin@fitzroy.co.uk
Web: www.fitzroy.co.uk ✋
Product Type: 1, 2, 3, 4, 5, 6
Other Info: ECO ✎
Material Type: A) 1, 2, 3, 4, 5, 6, 7, 8

FLETCHER JOINERY
261 Whessoe Road,
Darlington, Durham, DL3 0YL
Area of Operation: North East England
Tel: 01325 357347 **Fax:** 01325 357347
Email: enquiries@fletcherjoinery.co.uk
Web: www.fletcherjoinery.co.uk
Product Type: 1, 2, 3, 4, 5, 6
Other Info: ✎ ✋ **Material Type:** A) 2

FRANKLIN LEEDS LLP
Carlton Works, Cemetery Road, Yeadon,
Leeds, West Yorkshire, LS19 7BD
Area of Operation: UK & Ireland
Tel: 0113 250 2991 **Fax:** 0113 250 0991
Email: david.franklin@franklinwindows.co.uk
Web: www.franklinwindows.co.uk
Product Type: 1, 2, 3, 4, 5, 6, 7

**FROST CONSERVATORIES
AND GARDEN BUILDINGS LTD**
The Old Forge, Tempsford,
Sandy, Bedfordshire, SG19 2AG
Area of Operation: UK (Excluding Ireland)
Tel: 01767 640808 **Fax:** 01767 640561
Email: sales@frostconservatories.co.uk
Web: www.frostconservatories.co.uk

G MIDDLETON LTD
Cross Croft Industrial Estate,
Appleby, Cumbria, CA16 6HX
Area of Operation: Europe
Tel: 01768 352060
Fax: 01768 353228
Email: g.middleton@timber75.freeserve.co.uk
Web: www.graham-middleton.co.uk ✋
Material Type: A) 2

G2K JOINERY LTD
Trench Farm, Tilley Green,
Wem, Shropshire, SY4 5PJ
Area of Operation: UK (Excluding Ireland)
Tel: 01939 236640
Fax: 01939 236650
Email: graham@g2kjoinery.co.uk
Web: www.g2kjoinery.co.uk

GREEN BUILDING STORE
11 Huddersfield Road, Meltham,
Holmfirth, West Yorkshire, HD9 4NJ
Area of Operation: UK & Ireland
Tel: 01484 854898
Fax: 01484 854899
Email: info@greenbuildingstore.co.uk
Web: www.greenbuildingstore.co.uk ✋
Product Type: 1, 2, 3, 5, 6
Other Info: ECO
Material Type: B) 8

JAMES HARCOURT
Hockley Court, Stratford Road,
Hockley Heath, Solihull,
West Midlands, B94 6NW
Area of Operation: UK (Excluding Ireland)
Tel: 0870 241 6337
Web: www.jamesharcourt.co.uk
Product Type: 1, 2, 3, 4, 5, 6

KIRK NATURAL STONE
Bridgend, Fyvie, Turriff,
Aberdeenshire, AB53 8QB
Area of Operation: Worldwide
Tel: 01651 891891
Fax: 01651 891794
Email: info@kirknaturalstone.com
Web: www.kirknaturalstone.com
Product Type: 1, 2, 3, 4

KS PROFILES LTD
Broad March, Long March Industrial Estate,
Daventry, Northamptonshire, NN11 4HE
Area of Operation: UK & Ireland
Tel: 01327 316960
Fax: 01327 876412
Email: sales@ksprofiles.com
Web: www.ksprofiles.com
Product Type: 1, 2, 3, 4, 5, 6

LAMWOOD LTD
Unit 1, Riverside Works, Station Road,
Cheddleton, Staffordshire, ST13 7EE
Area of Operation: UK & Ireland
Tel: 01538 361888
Fax: 01538 361912
Email: sales@lamwoodltd.co.uk
Web: www.lamwoodltd.co.uk
Material Type: A) 2

LATTICE PERIOD WINDOWS
Unit 85 Northwick Business Centre, Blockley,
Moreton in Marsh, Gloucestershire, Gl56 9RF
Area of Operation: UK & Ireland
Tel: 01386 701079 **Fax:** 01386 701114
Email: sales@lattice windows .net
Web: www.latticewindows.net
Product Type: 1, 2, 3, 4, 6
Material Type: A) 2

LEISURE SPACE LTD
Unit 1, Top Farm, Rectory Road,
Campton Shefford, Bedfordshire, SG17 5PF
Area of Operation: Europe
Tel: 01462 816147
Fax: 01462 819252
Email: enquiries@leisurespaceltd.com
Web: www.leisurespaceltd.co.uk
Other Info: ✎

LLOYD CHRISTIE
15 Langton Street, London, SW10 0JL
Area of Operation: Worldwide
Tel: 0207 351 2108
Fax: 0207 376 5867
Email: info@lloydchristie.com
Web: www.lloydchristie.com
Other Info: ✎

M AND N MANUFACTURING LTD
M and N House, Barleyfield Industrial Estate,
Brynmawr, Blaenau Gwent, NP23 4LU
Area of Operation: Worldwide
Tel: 01495 313700
Fax: 01495 313716
Email: info@mngroup.co.uk
Web: www.mngroup.co.uk
Product Type: 1, 2, 3, 4, 5, 6
Other Info: ✐
Material Type: C) 1

MERLIN UK LIMITED
Unit 5 Fence Avenue, Macclesfield,
Cheshire, SK10 1LT
Area of Operation: North West England
and North Wales
Tel: 01625 424488
Fax: 0871 781 8967
Email: info@merlinuk.net
Web: www.merlindoors.co.uk
Product Type: 1, 2, 3, 4, 5, 6, 7

MULTIWOOD HOME IMPROVEMENTS LTD
1298 Chester Road , Stretford,
Manchester, M32 9AU
Area of Operation: Europe
Tel: 0161 866 9991
Fax: 0161 866 9992
Email: sales@multiwoodconservatories.co.uk
Web: www.multiwoodconservatories.co.uk

OLDE WORLDE OAK JOINERY LTD
Unit 12, Longford Industrial Estate,
Longford Road, Cannock,
Staffordshire, WS11 0DG
Area of Operation: Europe
Tel: 01543 469328
Fax: 01543 469341
Email: sales@oldeworldeoakjoinery.co.uk
Web: www.oldeworldeoakjoinery.co.uk

OPTIMUM CONSERVATORY KITS LTD
2A Hakkiwell Mill, Raglan Street, Bolton,
Greater Manchester, BL1 8AG
Area of Operation: UK (Excluding Ireland)
Tel: 01204 555920
Fax: 01204 385111
Email: sales@optimum.co.uk
Web: www.optimumconservatorykits.com
Product Type: 1, 2, 3, 4, 5, 6
Material Type: D) 1

PARAGON JOINERY
Systems House, Eastbourne Road,
Blindley Heath, Surrey, RH7 6JP
Area of Operation: UK (Excluding Ireland)
Tel: 01342 836300
Fax: 01342 836364
Email: sales@paragonjoinery.com
Web: www.paragonjoinery.com

PERMACELL FINESSE LTD
Western Road, Silver End, Witham, Essex, CM8 3BQ
Area of Operation: UK & Ireland
Tel: 01376 583241
Fax: 01376 583072
Email: sales@pfl.co.uk
Web: www.permacell-finesse.co.uk

PORTICO GB LTD
Unit 9 Windmill Avenue, Woolpit Business Park,
Woolpit, Bury St Edmunds, Suffolk, IP30 9UP
Area of Operation: Greater London, South East
England
Tel: 01359 244299
Fax: 01359 244232
Email: info@portico-gb.co.uk
Web: www.portico-newbuild.co.uk

PRITCHARDS
Unit 22 Brookend Lane, Kempsey, Worcester,
Worcestershire, WR5 3LF
Area of Operation: UK (Excluding Ireland)
Tel: 07721 398074
Fax: 01905 356507
Email: info@pritchards.tv
Web: www.pritchards.tv

ROBERT J TURNER & CO
Roe Green, Sandon, Buntingford,
Hertfordshire, SG9 0QE
Area of Operation: East England, Greater London
Tel: 01763 288371
Fax: 01763 288440
Email: sales@robertjturner.co.uk
Web: www.robertjturner.co.uk
Product Type: 1, 2, 3, 4, 5, 6

SAPA BUILDING SYSTEMS LIMITED
Alexandra Way, Ashchurch, Tewkesbury,
Gloucestershire, GL20 8NB
Area of Operation: UK & Ireland
Tel: 01684 853500
Fax: 01684 851850
Email: nicola.abbey@sapagroup.com
Web: www.sapagroup.com/uk/buildingsystems
Product Type: 1, 2, 3, 4, 5, 6
Other Info: ✐ ✋
Material Type: C) 1

SASH CRAFT
Unit 1 Mill Farm, Kelston,
Bath, Somerset, BA1 9AQ
Area of Operation: South West England
and South Wales
Tel: 01225 424434
Fax: 01225 424435
Email: pbale@sashcraftofbath.co.uk
Web: www.sashcraftofbath.co.uk
Product Type: 1, 2, 3, 4, 5, 6

SASH UK LTD
Ferrymoor Way, Park Springs, Grimethorpe,
Barnsley, South Yorkshire, S72 7BN
Area of Operation: Worldwide
Tel: 01226 715619
Fax: 01226 719968
Email: mailbox@sashuk.com
Web: www.sashuk.com
Product Type: 1, 2, 3, 4, 5, 6
Material Type: D) 1

SCHUCO INTERNATIONAL KG
Whitehall Avenue, Kingston, Milton Keynes,
Buckinghamshire, MK10 0AL
Area of Operation: Worldwide
Tel: 01908 282111
Fax: 01908 282124
Email: shamman@schueco.com
Web: www.schueco.co.uk

SPECTUS SYSTEMS
Queens Avenue, Macclesfield,
Cheshire, SK10 2BN
Area of Operation: UK & Ireland
Tel: 01625 420400
Fax: 01625 501418
Email: contact@spectus.co.uk
Web: www.spectus.co.uk
Material Type: D) 1

STUDLEY CONSERVATORIES
Shawbank House, Shawbank Road, Lakeside,
Redditch, Warwickshire, B98 8YN
Area of Operation: Greater London, Midlands
& Mid Wales, South East England, South West
England & South Wales
Tel: 01527 854811
Fax: 01527 854106
Email: sales@studleyconservatoryvillage.com
Web: www.studleyconservatoryvillage.com

SUN PARADISE UK LTD
Phoenix Wharf, Eel Pie Island,
Twickenham, Middlesex, TW1 3DY
Area of Operation: UK & Ireland
Tel: 0870 240 7604
Fax: 0870 240 7614
Email: info@sunparadise.co.uk
Web: www.sunparadise.co.uk
Product Type: 3, 4, 5, 6
Material Type: C) 1

SUN PARADISE UK LTD
Area of Operation: UK
Tel: 0870 240 7604 **Fax:** 0870 240 7614
Email: info@sunparadise.co.uk
Web: www.sunparadise.co.uk
Product Type: 3, 4, 5, 6

A range of contemporarily styled, thermally broken Aluminium roof systems are available for construction of bespoke Conservatories, Sunrooms and extensions. Aluminium frames are great looking, easy maintenance, slimmer than PVCu and won't twist, warp, rust or discolour over time. They are polyester powder coated to one of 14 standard colours or to your own specification and are glazed with Low E safety glass to be compliant with Building regulations Document L.

SUNFOLD SYSTEMS
Sunfold House, Wymondham Business Park,
Chestnut Drive, Wymondham, Norfolk, NR18 9SB
Area of Operation: Worldwide
Tel: 01953 423423 **Fax:** 01953 423430
Email: info@sunfoldsystems.co.uk
Web: www.sunfold.com
Other Info: ✐

THE GREEN OAK CARPENTRY CO LTD
Langley Farm, Langley, Rake, Liss,
Hampshire, GU33 7JW
Area of Operation: UK & Ireland
Tel: 01730 892049 **Fax:** 01730 895225
Email: enquiries@greenoakcarpentry.co.uk
Web: www.greenoakcarpentry.co.uk
Material Type: A) 2

THE PATIO ROOM COMPANY
52 Castlemaine Avenue, South Croydon,
Surrey, CR2 7HR
Area of Operation: UK (Excluding Ireland)
Tel: 0208 406 3001 **Fax:** 0208 686 4928
Email: helpdesk@patioroom.co.uk
Web: www.patioroom.co.uk
Other Info: ✐

TOMPKINS LTD
High March Close, Long March Industrial Estate,
Daventry, Northamptonshire, NN11 4EZ
Area of Operation: UK (Excluding Ireland)
Tel: 01327 877187
Fax: 01327 310491
Email: info@tompkinswood.co.uk
Web: www.tompkinswood.co.uk
Product Type: 1, 2, 3, 4, 5, 6, 7
Other Info: ✐ ✋
Material Type: A) 1, 2, 3, 4, 5, 6, 7, 8, 9,
10, 11, 12, 13, 14

TOTAL GLASS LIMITED
Total Complex, Overbrook Lane,
Knowsley, Merseyside, L34 9FB
Area of Operation: Midlands & Mid Wales,
North East England, North West England
and North Wales
Tel: 0151 549 2339
Fax: 0151 546 0022
Email: sales@totalglass.com
Web: www.totalglass.com
Product Type: 1, 2, 3, 4, 5, 6, 7

TRADITIONAL CONSERVATORIES LTD
486 Warwick Road, Tyseley, Birmingham,
West Midlands, B11 2HP
Area of Operation: Midlands & Mid Wales
Tel: 0121 706 0102 **Fax:** 0121 628 3033
Email: sales@traditionalconservatories.com
Web: www.traditionalconservatories.co.uk
Product Type: 1, 2, 3, 4, 5, 6, 7

TWC THE WINDOW COMPANY LIMITED
The Drill Hall, 262 Huddersfield Road, Thongsbridge,
Holmfirth, West Yorkshire, HD9 3JQ
Area of Operation: UK & Ireland
Tel: 0800 917 1918
Fax: 01484 685210
Email: admin@twcthewindowcompany.net
Web: www.twcthewindowcompany.net
Product Type: 1, 2, 3, 5
Material Type: D) 1

ULTRAFRAME UK
Salthill Road, Clitheroe, Lancashire, BB7 1PE
Area of Operation: UK & Ireland
Tel: 0500 822340 **Fax:** 0870 414 1020
Email: sales@ultraframe.com
Web: www.ultraframe.com
Product Type: 1, 2, 3, 4, 5, 6, 7
Material Type: D) 1, 3

WICKES
Wickes Customer Services, Wickes House,
120-138 Station Road, Harrow, Middlesex,
Greater London, HA1 2QB
Area of Operation: UK (Excluding Ireland)
Tel: 0870 608 9001
Fax: 0208 863 6225
Web: www.wickes.co.uk

STAINED GLASS

KEY

SEE ALSO: DOORS, WINDOWS AND
CONSERVATORIES - Glass and Glazing

OTHER: ▽ Reclaimed ⌂ On-line shopping
✐ Bespoke ✋ Hand-made ECO Ecological

BARRON GLASS
Unit 11, Lansdown Industrial Estate,
Cheltenham, Gloucestershire, GL51 8PL
Area of Operation: UK & Ireland
Tel: 01242 22800
Fax: 01242 226555
Email: admin@barronglass.co.uk
Web: www.barronglass.co.uk

BIRCHDALE GLASS LTD
Unit L, Eskdale Road, Uxbridge,
Greater London, UB8 2RT
Area of Operation: UK & Ireland
Tel: 01895 259111
Fax: 01895 810087
Email: info@birchdaleglass.com
Web: www.birchdaleglass.com

BOURNEMOUTH STAINED GLASS
790 Wimborne Road, Moordown,
Bournemouth, Dorset, BH9 2DX
Area of Operation: Worldwide
Tel: 01202 514734
Fax: 01202 250239
Email: shop@stainedglass.co.uk
Web: www.stainedglass.co.uk

CHIPPING NORTON GLASS LTD
Units 1 & 2, Station Yard Industrial Estate,
Chipping Norton, Oxfordshire, OX7 5HX
Area of Operation: Midlands & Mid Wales,
South West England and South Wales
Tel: 01608 643261
Fax: 01608 641768
Email: gill@cnglass.plus.com
Web: www.chippingnortonglass.co.uk

FOREST OF DEAN STAINED GLASS
Units 18-20, Harts Barn, Monmouth Road,
Longhope, Gloucestershire, GL17 0QD
Area of Operation: UK (Excluding Ireland)
Tel: 01452 830994
Fax: 01452 830994
Email: mike@forestglass.co.uk
Web: www.forestglass.co.uk

SPONSORED BY: ULTRAFRAME www.ultraframe-conservatories.co.uk

FRANKLIN LEEDS LLP
Carlton Works, Cemetery Road,
Yeadon, Leeds, West Yorkshire, LS19 7BD
Area of Operation: UK & Ireland
Tel: 0113 250 2991
Fax: 0113 250 0991
Email: david.franklin@franklinwindows.co.uk
Web: www.franklinwindows.co.uk

HOLDSWORTH WINDOWS LTD
Darlingscote Road, Shipston-on-Stour,
Warwickshire, cv36 4pr
Area of Operation: UK & Ireland
Tel: 01608 661883
Fax: 01608 661008
Email: info@holdsworthwindows.co.uk
Web: www.holdsworthwindows.co.uk

IT'S A BLAST!
27a Westfield Road, Bishops Stortford,
Hertfordshire, CM23 2RE
Area of Operation: UK (Excluding Ireland)
Tel: 01279 656133
Email: carole@itsablast.co.uk
Web: www.itsablast.co.uk

PEELS OF LONDON
PO Box 160, Richmond,
Surrey, TW10 7XL
Area of Operation: Worldwide
Tel: 0208 948 0689
Fax: 0208 948 0689
Email: info@e-peels.co.uk
Web: www.e-peels.co.uk

**ROBERT MILLS
ARCHITECTURAL ANTIQUES**
Narroways Road, Eastville, Bristol, BS2 9XB
Area of Operation: Worldwide
Tel: 0117 955 6542
Fax: 0117 955 8146
Email: Info@rmills.co.uk
Web: www.rmills.co.uk

SIMPLY STAINED
7 Farm Mews, Farm Road,
Brighton & Hove, East Sussex, BN3 1GH
Area of Operation: UK (Excluding Ireland)
Tel: 01273 220030
Email: david@simply-stained.co.uk
Web: www.simply-stained.co.uk

**STUART OWEN NORTON
GLASS AND SIGN LTD**
Unit 15 Dukes Court, Prudhoe,
Northumberland, NE42 6DA
Area of Operation: Worldwide
Tel: 01661 833227
Fax: 01661 833732
Email: brilliant.cutter@btinternet.com
Web: www.glass-and-sign.com

SUNRISE STAINED GLASS
58-60 Middle Street, Southsea,
Portsmouth, Hampshire, PO5 4BP
Area of Operation: UK (Excluding Ireland)
Tel: 02392 750512
Fax: 02392 875488
Email: sunrise@stained-windows.co.uk
Web: www.stained-windows.co.uk

THE STAINED GLASS STUDIO
Unit 5, Brewery Arts, Brewery Court,
Cirencester, Gloucestershire, GL7 1JH
Area of Operation: Worldwide
Tel: 01285 644430
Fax: 0800 454627
Email: daniella@wilson-dunne.wanadoo.co.uk

TOTAL GLASS LIMITED
Total Complex, Overbrook Lane,
Knowsley, Merseyside, L34 9FB
Area of Operation: Midlands & Mid Wales, North
East England, North West England & North Wales
Tel: 0151 549 2339
Fax: 0151 546 0022
Email: sales@totalglass.com
Web: www.totalglass.com

UAP
Bank House, 16-18 Bank Street, Walshaw,
Bury, Lancashire, BL8 3AZ
Area of Operation: UK & Ireland
Tel: 0161 763 5290 **Fax:** 0161 763 6726
Email: uap@btconnect.com
Web: www.universal-imports.com

LINTELS AND CILLS

KEY

PRODUCT TYPES: 1= Sleeper Lintels
2 = Cavity Closing Lintels   3 = Cills

OTHER: ▽ Reclaimed   ⌐ On-line shopping
✎ Bespoke   ✋ Hand-made   ECO Ecological

ACE STONE BUILDING PRODUCTS
98/99 Reddal Hill Road, Cradley Heath,
West Midlands, B64 5JT
Area of Operation: UK & Ireland
Tel: 01384 638076
Fax: 01384 566179
Email: info@acenu-look.co.uk
Web: www.acenu-look.co.uk
Product Type: 3

ADDSTONE CAST STONE
2 Millers Gate, Stone, Staffordshire, ST15 8ZF
Area of Operation: UK (Excluding Ireland)
Tel: 01785 818810   **Fax:** 01785 818810
Email: sales@addstone.co.uk
Web: www.addstone.co.uk ⌐
Product Type: 2, 3
Other Info: ✎ ✋
Material Type: E) 11, 13

BIRTLEY BUILDING PRODUCTS LTD
Mary Avenue, Birtley, County Durham, DH3 1JF
Area of Operation: UK & Ireland
Tel: 0191 410 6631
Fax: 0191 410 0650
Email: info@birtley-building.co.uk
Web: www.birtley-building.co.uk
Product Type: 2
Material Type: C) 4

BLACK MOUNTAIN QUARRIES LTD
Howton Court, Pontrilas,
Herefordshire, HR2 0BG
Area of Operation: UK & Ireland
Tel: 01981 241541
Email: info@blackmountainquarries.com
Web: www.blackmountainquarries.com ⌐

BORDER CONCRETE PRODUCTS
Jedburgh Road, Kelso, Borders, TD5 8JG
Area of Operation: North East England,
North West England and North Wales, Scotland
Tel: 01573 224393
Fax: 01573 276360
Email: sales@borderconcrete.co.uk
Web: www.borderconcrete.co.uk
Product Type: 3

BULMER BRICK CUTTING SERVICES
The Brickfields, Hedingham Road,
Bulmer, Sudbury, Suffolk, CO10 7EF
Area of Operation: UK (Excluding Ireland)
Tel: 01787 269132
Fax: 01787 269044
Email: info@brickcutters.com
Web: www.brickcutters.com
Other Info: ✎

CAWARDEN BRICK CO. LTD
Cawarden Springs Farm, Blithbury Road,
Rugeley, Staffordshire, WS15 3HL
Area of Operation: UK (Excluding Ireland)
Tel: 01889 574066
Fax: 01889 575695
Email: home-garden@cawardenreclaim.co.uk
Web: www.cawardenreclaim.co.uk
Product Type: 1, 3

CJ O'SHEA & CO LTD
Unit 1 Granard Business Centre,
Bunns Lane, Mill Hill, London, NW7 2DZ
Area of Operation: East England,
Greater London, South East England,
South West England & South Wales
Tel: 0208 959 3600
Fax: 0208 959 0184
Email: admin@oshea.co.uk
Web: www.oshea.co.uk

EXCEL BUILDING SOLUTIONS
Maerdy Industrial Estate, Rhymney,
Powys, NP22 5PY
Area of Operation: Worldwide
Tel: 01685 845200 **Fax:** 01685 844106
Email: sales@excelfibre.com
Web: www.excelfibre.com

FLIGHT TIMBER PRODUCTS
Earls Colne Business Park, Earls Colne,
Essex, CO6 2NS
Area of Operation: East England,
Greater London, South East England
Tel: 01787 222336
Fax: 01787 222359
Email: sales@flighttimber.com
Web: www.flighttimber.com

GREYSLATE & STONE SUPPLIES
Y Maes, The Square, Blaenau Ffestiniog,
Gwynedd, LL41 3UN
Area of Operation: UK & Ireland
Tel: 01766 830521
Fax: 01766 831706
Email: greyslate@slateandstone.net
Web: www.slateandstone.net
Product Type: 3

I G LTD (STEEL LINTELS)
Avondale Road, Cumbran,
Gwent, Torfaen, NP44 1XY
Area of Operation: UK & Ireland
Tel: 01633 486486
Fax: 01633 486492
Email: gloria.stephens@igltd.co.uk
Web: www.igltd.co.uk
Product Type: 1, 2

KEYSTONE LINTELS LIMITED
Ballyreagh Industrial Estate, Sandholes Road,
Cookstown, Co Tyrone, BT80 9DG
Area of Operation: UK & Ireland
Tel: 028 8676 2184
Fax: 028 8676 1011
Email: info@keystonelintels.co.uk
Web: www.keystonelintels.co.uk
Product Type: 1, 2

MANCHESTER BRICK & PRECAST LTD
Haigh Avenue, Whitehill Industrial Estate,
Stockport, Cheshire, SK4 1NU
Area of Operation: UK & Ireland
Tel: 0161 480 2621
Fax: 0161 480 0108
Email: sales@manbrick.co.uk
Web: www.manbrick.co.uk
Product Type: 2, 3

**MINCHINHAMPTON
ARCHITECTURAL SALVAGE CO**
Cirencester Road, Aston Downs,
Chalford, Stroud, Gloucestershire, GL6 8PE
Area of Operation: Worldwide
Tel: 01285 760886
Fax: 01285 760838
Email: masco@catbrain.com
Web: www.catbrain.com
Product Type: 1, 3

SNOWDONIA SLATE & STONE
Glandwr Workshops, Glanypwll Road,
Blaenau Ffestiniog, Gwynedd, LL41 3PG
Area of Operation: UK (Excluding Ireland)
Tel: 01766 832525
Fax: 01766 832404
Email: richard@snowdoniaslate.co.uk
Web: www.snowdoniaslate.co.uk
Product Type: 1, 3

TAYLOR LANE TIMBER FRAME LTD
Chapel Road, Rotherwas Industrial Estate,
Hereford, Herefordshire, HR2 6LD
Area of Operation: UK & Ireland
Tel: 01432 271912 **Fax:** 01432 351064
Email: info@taylor-lane.co.uk
Web: www.taylor-lane.co.uk
Product Type: 2

THE GREEN OAK CARPENTRY CO LTD
Langley Farm, Langley, Rake,
Liss, Hampshire, GU33 7JW
Area of Operation: UK & Ireland
Tel: 01730 892049 **Fax:** 01730 895225
Email: enquiries@greenoakcarpentry.co.uk
Web: www.greenoakcarpentry.co.uk
Product Type: 2

**TRADITIONAL OAK
AND TIMBER COMPANY**
P O Stores, Haywards Heath Road,
North Chailey, Nr Lewes,
East Sussex, BN8 4EY
Area of Operation: Worldwide
Tel: 01825 723648   **Fax:** 01825 722215
Email: info@tradoak.co.uk
Web: www.tradoak.com
Product Type: 1

DOOR FURNITURE AND
ARCHITECTURAL IRONMONGERY

KEY

PRODUCT TYPES: 1 = Door Handles and Knobs
2 = Door Closers and Locks   3 = Letter Boxes
4 = Finger Plates   5 = Other

OTHER: ▽ Reclaimed   ⌐ On-line shopping
✎ Bespoke   ✋ Hand-made   ECO Ecological

A & H BRASS
201-203 Edgware Road, London, W2 1ES
Area of Operation: Worldwide
Tel: 0207 402 1854 **Fax:** 0207 402 0110
Email: ahbrass@btinternet.com
Web: www.aandhbrass.co.uk ⌐
Product Type: 1, 2, 3, 4, 5
Material Type: C) 3, 11, 12, 13, 14

A C LEIGH
61-67 Benedicts Street,
Norwich, Norfolk, NR2 4PD
Area of Operation: UK (Excluding Ireland)
Tel: 01603 216500 **Fax:** 01603 760707
Email: marketing@acleigh.co.uk
Web: www.acleigh-handles.co.uk ⌐
Product Type: 1, 2, 3, 4, 5

ALCESTER LOCKS LTD
30 High Street, Alcester,
Warwickshire, B49 5AB
Area of Operation: Midlands & Mid Wales
Tel: 01527 401011 **Fax:** 01527 457056
Email: sales@alcesterlocks.co.uk
Web: www.alcesterlocks.co.uk
Product Type: 2

ANY OLD IRON
PO Box 198, Ashford, Kent, TN26 3SE
Area of Operation: Worldwide
Tel: 01622 685336
Fax: 01622 685336
Email: anyoldiron@aol.com
Web: www.anyoldiron.co.uk
Product Type: 1, 2, 3, 4

APPART
72-75 Warren Street, London, W1T 5PE
Area of Operation: Worldwide
Tel: 0870 7521054
Fax: 0207 255 9356
Email: info@appart.co.uk
Web: www.appart.co.uk
Product Type: 1, 2, 3, 4, 5

ARCHITECTURAL COMPONENTS LTD
4-8 Exhibition Road, South Kensington,
Greater London, SW7 2HF
Area of Operation: Worldwide
Tel: 0207 581 2401 **Fax:** 0207 589 4928
Email: sales@knobs.co.uk
Web: www.doorhandles.co.uk
Product Type: 1, 2, 3, 4, 5

ARCHITECTURAL IRONMONGERY LTD
28 Kyrle Street, Ross-on-Wye,
Herefordshire, HR9 7DB
Area of Operation: Worldwide
Tel: 01989 567946
Fax: 01989 567946
Email: info@arciron.co.uk
Web: www.arciron.com
Product Type: 1, 2, 3, 4, 5

ASHFIELD TRADITIONAL
119 High Street, Needham Market,
Ipswich, Suffolk, IP6 8DQ
Area of Operation: Europe
Tel: 01449 723601 **Fax:** 01449 723602
Email: mail@limelightgb.com
Web: www.limelightgb.com
Product Type: 1, 2, 3

B ROURKE & CO LTD
Vulcan Works, Accrington Road,
Burnley, Lancashire, BB11 5QD
Area of Operation: Worldwide
Tel: 01282 422841 **Fax:** 01282 458901
Email: info@rourkes.co.uk
Web: www.rourkes.co.uk
Product Type: 1, 2, 3, 4, 5

BASTA PARSONS
Alma St, Wolverhampton,
West Midlands, WV10 0EY
Area of Operation: UK & Ireland
Tel: 01902 877770
Fax: 01902 877771
Email: sjohnson@bastaparsonsgb.com
Web: www.bastaparsons.com
Product Type: 1, 2

BENNETTS (IRONGATE) LIMITED
222 Mansfield Road, Derby,
Derbyshire, DE1 3RB
Area of Operation: UK (Excluding Ireland)
Tel: 01332 346521
Fax: 01332 293453
Email: mark-bennetts@btconnect.com
Product Type: 1, 2, 3, 4, 5
Material Type: C) 1, 2, 3, 4, 5, 6, 7, 8, 9, 10,
11, 12, 13, 14, 17, 18

BOYALLS ARCHITECTURAL IRONMONGERS
187 High Street, Hampton Hill,
Greater London, TW12 1NL
Area of Operation: UK (Excluding Ireland)
Tel: 0208 941 0880
Fax: 0208 941 3718
Email: sales@boyalls.com
Web: www.boyalls.com

BRAMAH SECURITY EQUIPMENT LTD
31 Oldbury Place, London, W1U 5PT
Area of Operation: Worldwide
Tel: 0207 486 1739 **Fax:** 0207 935 2779
Email: lock.sales@bramah.co.uk
Web: www.bramah.co.uk **Product Type:** 2

BRASS ART
Unit L1, Lockside, Anchor Brook
Industrial Estate, Aldridge, Walsall,
West Midlands, WS9 8EG
Area of Operation: UK (Excluding Ireland)
Tel: 01922 740512 **Fax:** 01922 740510
Email: sales@brassart.co.uk
Web: www.brassart.co.uk
Product Type: 1, 2, 4

BRASS FOUNDRY CASTINGS
PO Box 151, Westerham, Kent, TN16 1YF
Area of Operation: Worldwide
Tel: 01959 563863 **Fax:** 01959 561262
Email: info@brasscastings.co.uk
Web: www.brasscastings.co.uk
Product Type: 1, 3, 4, 5

BROADLEAF TIMBER
Llandeilo Road Industrial Estate,
Carms, Carmarthenshire, SA18 3JG
Area of Operation: UK & Ireland
Tel: 01269 851910 **Fax:** 01269 851911
Email: sales@broadleaftimber.com
Web: www.broadleaftimber.com **Product Type:** 1, 2

BROUGHTONS OF LEICESTER
The Old Cinema, 69 Cropston Road,
Anstey, Leicester, Leicestershire, LE7 7BP
Area of Operation: Worldwide
Tel: 0116 235 2555 **Fax:** 0116 234 1188
Email: sale@broughtons.com
Web: www.broughtons.com

**C & R ARCHITECTURAL
HARDWARE LIMITED**
Unit 6, Scotshawbrook Ind. Est., Branch Road, Lower
Darwen, Darwen, Lancashire, BB30PR
Area of Operation: UK & Ireland
Tel: 01254 278757 **Fax:** 01254 278767
Email: info@theaishop.com
Web: www.theaishop.com

CARLISLE BRASS
Park House Road, Carlisle, Cumbria, CA3 0JU
Area of Operation: UK & Ireland
Tel: 01228 511770 **Fax:** 01228 511885
Email: enquiries@carlislebrass.co.uk
Web: www.carlislebrass.co.uk
Product Type: 1, 2, 3, 4, 5

**CHAMBERLAIN & GROVES LTD -
THE DOOR & SECURITY STORE**
101 Boundary Road, Walthamstow,
London, E17 8NQ
Area of Operation: UK (Excluding Ireland)
Tel: 0208 520 6776 **Fax:** 0208 520 2190
Email: ken@securedoors.co.uk
Web: www.securedoors.co.uk **Product Type:** 1

CHARLES MASON LTD
Unit 11A Brook Street Mill, Off Goodall Street,
Macclesfield, Cheshire, SK11 7AW
Area of Operation: Worldwide
Tel: 0800 085 3616 **Fax:** 01625 668789
Email: info@charles-mason.com
Web: www.charles-mason.com
Product Type: 1, 2, 3, 4, 5

CHARTERHOUSE CONSERVATORIES
Park Street, Gosport, Hampshire, PO12 4UH
Area of Operation: South East England
Tel: 02392 504006 **Fax:** 02392 513765
Email: info@charterhouse.nu
Web: www.charterhouse.nu

CIFIAL UK LTD
7 Faraday Court, Park Farm Industrial Estate,
Wellingborough, Northamptonshire, NN8 6XY
Area of Operation: UK & Ireland
Tel: 01933 402008 **Fax:** 01933 402063
Email: sales@cifial.co.uk **Web:** www.cifial.co.uk

CIFIAL UK
Area of Operation: UK & Ireland
Tel: 01933 402008 **Fax:** 01933 402063
Email: sales@cifial.co.uk
Web: www.cifial.co.uk

Add a touch of glamour to your home with luxury
door furniture from Cifial. Their new designer
collection, EuroDesign, comprises of 6 stylish
handle options.

COBALT BLACKSMITHS
The Forge, English Farm, English Lane,
Nuffield, Oxfordshire, RG9 5TH
Tel: 01491 641990
Fax: 01491 640909
Email: enquiries@cobalt-blacksmiths.co.uk
Web: www.cobalt-blacksmiths.co.uk
Product Type: 1, 2

CORE AND ORE LTD
16 Portland Street, Clifton, Bristol, BS8 4JH
Area of Operation: UK (Excluding Ireland)
Tel: 01179 042408
Fax: 01179 658057
Email: sales@coreandore.com
Web: www.coreandore.com
Product Type: 1

D & E ARCHITECTURAL HARDWARE LTD
17 Royce Road, Carr Road Industrial Estate,
Fengate, Peterborough, Cambridgeshire, PE1 5YB
Area of Operation: UK & Ireland
Tel: 01733 896123
Fax: 01733 894466
Email: wan@DandE.co.uk
Web: www.DandE.co.uk
Product Type: 1, 2, 3, 4, 5

DANICO BRASS LTD
31-35 Winchester Road,
Swiss Cottage, London, NW3 3NR
Area of Operation: Worldwide
Tel: 0207 483 4477
Fax: 0207 722 7992
Email: sales@danico.co.uk
Product Type: 1, 2, 3, 4, 5
Other Info:

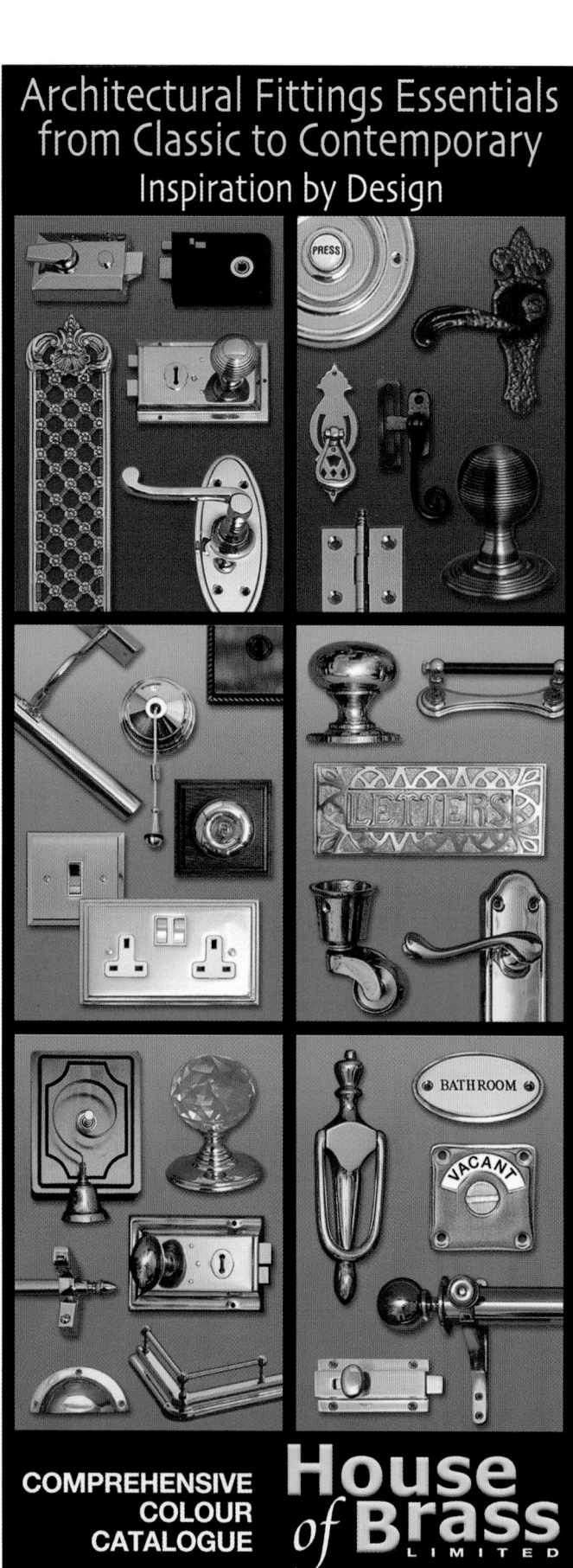

DARTINGTON STEEL DESIGN
Units 3-4 Webbers Yard, Dartington
Industrial Estate, Totnes, Devon, TQ9 6JY
Area of Operation: Worldwide
Tel: 01803 868671
Fax: 01803 868665
Email: sales@dartington.com
Web: www.dartington.com
Product Type: 1, 3, 4
Other Info: ✋

DARTINGTON STEEL DESIGN

Area of Operation: Worldwide
Tel: 01803 868671
Fax: 01803 868665
Email: sales@dartington.com
Web: www.dartington.com
Product Type: 1, 3, 4

Contemporary door, cabinet and window furniture from Dartington Steel Design, also specialists in restoration ironmongery.

DAVID K COOPER ARCHITECTURAL IRONMONGERY
Unit 27 Maybrook Industrial Estate,
Maybrook Road, Walsall Wood,
Walsall, West Midlands, WS8 7DG
Area of Operation: UK & Ireland
Tel: 01543 454479 **Fax:** 01543 453707
Email: sales@dkcooper.co.uk
Web: www.dkcooper.co.uk
Product Type: 1, 2, 3, 4, 5

DESA UK LTD
11 Beech House, Padgate Business Park,
Green Lane, Warrington, Cheshire, WA1 4JN
Area of Operation: UK (Excluding Ireland)
Tel: 01925 828854
Fax: 01925 284124
Email: sholmes@desauk.co.uk
Web: www.desauk.co.uk
Product Type: 5

DICTATOR DIRECT
Inga House, Northdown Business Park,
Ashford Road, Lenham, Kent, ME17 2DL
Area of Operation: Worldwide
Tel: 01622 854770
Fax: 01622 854771
Email: mail@dictatordirect.com
Web: www.dictatordirect.com
Product Type: 2

DIRECTDOORS.COM
Bay 5 Eastfield Industrial Estate,
Eastfield Drive, Penicuik, Lothian, EH26 8JA
Area of Operation: UK (Excluding Ireland)
Tel: 0131 669 6310 **Fax:** 0131 657 1578
Email: info@directdoors.com
Web: www.directdoors.com ✐
Product Type: 1, 2, 3, 4, 5

DISTINCTIVE DOORS
14 & 15 Chambers Way, Newton Chambers Road,
Chapeltown, Sheffield, South Yorkshire, S35 2PH
Area of Operation: UK (Excluding Ireland)
Tel: 0114 220 2250
Fax: 0114 220 2254
Email: enquiries@distinctivedoors.co.uk
Web: www.distinctivedoors.co.uk ✐
Product Type: 1, 2, 3, 4, 5 **Material Type:** A) 2, 3

DRUMMONDS ARCHITECTURAL ANTIQUES LIMITED
Kirkpatrick Buildings, 25 London Road (A3),
Hindhead, Surrey, GU36 6AB
Area of Operation: Worldwide
Tel: 01428 609444 **Fax:** 01428 609445
Email: davidcox@drummonds-arch.co.uk
Web: www.drummonds-arch.co.uk
Product Type: 1, 3, 4, 5

EVERGREEN DOOR
Unit 1, Oakwell Park Industrial Estate, Birstall,
West Yorkshire, WF17 9LU
Area of Operation: Worldwide
Tel: 01924 423171
Fax: 01924 423175
Email: andrewgrogan@evergreendoor.co.uk
Web: www.evergreendoor.co.uk

FORGERIES
The Loft, 108 Brassey Road, Winchester,
Hampshire, SO22 6SA
Area of Operation: UK (Excluding Ireland)
Tel: 01962 842822
Fax: 01962 842822
Email: penny@forgeriesonline.co.uk
Web: www.forgeriesonline.co.uk
Product Type: 1, 2

GATESTUFF.COM
17-19 Old Woking Road, West Byfleet,
Surrey, KT14 6LW
Area of Operation: UK & Ireland
Tel: 01932 344434
Email: sales@gatestuff.com
Web: www.gatestuff.com ✐
Product Type: 2, 5

HARBRINE LTD
27-31 Payne Road, London, E3 2SP
Area of Operation: Worldwide
Tel: 0208 980 8000 **Fax:** 0208 980 6050
Email: info@harbrine.co.uk
Web: www.harbrine.co.uk
Product Type: 1, 2

HAYMANS TIMBER PRODUCTS
Haymans Farm, Hocker Lane, Over Alderley,
Macclesfield, Cheshire, SK10 4SD
Area of Operation: UK & Ireland
Tel: 01625 590098 **Fax:** 01625 586174
Email: haymanstimber@aol.com
Web: www.haymanstimber.co.uk

HOLDEN + PARTNERS
26 High Street, Wimbledon,
Greater London, SW19 5BY
Area of Operation: Worldwide
Tel: 0208 946 5502 **Fax:** 0208 879 0310
Email: arch@holdenpartners.co.uk
Web: www.holdenpartners.co.uk
Product Type: 1, 3, 5

HOUSE OF BRASS
122 North Sherwood Street, Nottingham,
Nottinghamshire, NG1 4EF
Area of Operation: Worldwide
Tel: 0115 947 5430 **Fax:** 0115 947 5430
Email: sales@houseofbrass.co.uk
Web: www.houseofbrass.co.uk ✐
Product Type: 1, 2, 3, 4, 5

IN DOORS
Beechinwood Farm, Beechinwood Lane,
Platt, Nr. Sevenoaks, Kent, TN15 8QN
Area of Operation: UK (Excluding Ireland)
Tel: 01732 887445 **Fax:** 01732 887446
Email: info@indoorsltd.co.uk
Web: www.indoorsltd.co.uk
Product Type: 1, 3

INTERIOR ASSOCIATES
3 Highfield Road, Windsor,
Berkshire, SL4 4DN
Area of Operation: UK & Ireland
Tel: 01753 865339 **Fax:** 01753 865339
Email: sales@interiorassociates.fsnet.co.uk
Web: www.interiorassociates.co.uk
Product Type: 5

IRONMONGERY DIRECT
Unit 2-3 Eldon Way Trading Estate,
Eldon Way, Hockley, Essex, SS5 4AD
Area of Operation: Worldwide
Tel: 01702 562770 **Fax:** 01702 562799
Email: sales@ironmongerydirect.com
Web: www.ironmongerydirect.com
Product Type: 1, 2, 3, 4

ISAAC LORD LTD
West End Court, Suffield Road, High Wycombe,
Buckinghamshire, HP11 2JY
Area of Operation: East England, Greater London,
North East England, South East England, South West
England and South Wales
Tel: 01494 462121 **Fax:** 01494 461376
Email: info@isaaclord.co.uk
Web: www.isaaclord.co.uk

ITFITZ
PO Box 960, Maidenhead, Berkshire, SL6 9FU
Area of Operation: UK (Excluding Ireland)
Tel: 01628 890432 **Fax:** 0870 133 7955
Email: sales@itfitz.co.uk
Web: www.itfitz.co.uk
Product Type: 1, 2, 3, 4, 5

JAMES GIBBONS FORMAT LTD
Vulcan Road, Bilston, Wolverhampton,
West Midlands, WV14 7JG
Area of Operation: Worldwide
Tel: 01902 405500 **Fax:** 01902 385915
Email: info@jgf.co.uk **Web:** www.jgf.co.uk
Product Type: 1, 2, 3, 4

JOHN PLANCK LTD
Southern House, Anthonys Way,
Medway City Estate,
Rochester, Kent, ME2 4DN
Area of Operation: UK (Excluding Ireland)
Tel: 01634 720077 **Fax:** 01634 720111
Email: john@johnplanck.co.uk
Web: www.johnplanck.co.uk
Product Type: 1, 2, 3, 4, 5

KASPAR SWANKEY
405, Goldhawk Road, Hammersmith,
West London, W6 0SA
Area of Operation: Worldwide
Tel: 0208 746 3586 **Fax:** 0208 746 3586
Email: kaspar@swankeypankey.com
Web: www.swankeypankey.com
Product Type: 1, 5

LEOHARDWARE.COM
Unit.3, Scotshawbrook Industrial Estate,
Branch Road, Lower Darwen, Darwen,
Lancashire, BB3 0PR
Area of Operation: UK & Ireland
Tel: 01254 278757 **Fax:** 01254 278767
Email: info@leohardware.com
Web: www.leohardware.com
Product Type: 1, 2, 3, 4, 5
Material Type: A) 4, 7, 15

LEVOLUX LTD
1 Forward Drive, Harrow, Middlesex,
Greater London, HA3 8NT
Area of Operation: Worldwide
Tel: 0208 863 9111 **Fax:** 0208 863 8760
Email: info@levolux.com
Web: www.levolux.com **Product Type:** 5

MBL
55 High St, Biggleswade, Bedfordshire, SG18 0JH
Area of Operation: UK (Excluding Ireland)
Tel: 01767 318695 **Fax:** 01767 318834
Email: info@mblai.co.uk
Web: www.mblai.co.uk
Product Type: 1, 2, 3, 4, 5

MERLIN UK LIMITED
Unit 5 Fence Avenue, Macclesfield,
Cheshire, SK10 1LT
Area of Operation: North West England and North Wales
Tel: 01625 424488 **Fax:** 0871 781 8967
Email: info@merlinuk.net
Web: www.merlindoors.co.uk
Product Type: 1, 2, 3, 4, 5

MIGHTON PRODUCTS LTD
PO Box 1, Saffron Walden, Essex, CB10 1QJ
Area of Operation: UK (Excluding Ireland)
Tel: 0800 0560471
Email: sales@mightonproducts.com
Web: www.mightonproducts.com

**MINCHINHAMPTON ARCHITECTURAL
SALVAGE CO**
Cirencester Road, Aston Downs, Chalford,
Stroud, Gloucestershire, GL6 8PE
Area of Operation: Worldwide
Tel: 01285 760886 **Fax:** 01285 760838
Email: masco@catbrain.com
Web: www.catbrain.com
Product Type: 1, 2, 3, 4

NIGEL TYAS HANDCRAFTED IRONWORK
Green Works, 18 Green Road, Penistone,
Sheffield, South Yorkshire, S36 6BE
Area of Operation: Worldwide
Tel: 01226 766618
Email: sales@nigeltyas.co.uk
Web: www.nigeltyas.co.uk
Product Type: 1, 2, 3

PEPLOW ROBERTS LIMITED
Unit 11 Eden Way, Pages Industrial Park,
Billington Road, Leighton Buzzard,
Bedfordshire, LU7 4TZ
Area of Operation: Worldwide
Tel: 01525 375118
Fax: 01525 852130
Email: info@paperstream.net
Web: www.paperstream.net
Product Type: 5

PLASTIC MIRRORS
18 Blundell Lane, Preston, Lancashire, PR1 0EA
Area of Operation: Worldwide
Tel: 01772 746769 **Fax:** 01772 746869
Email: enquiries@plasticmirrors.co.uk
Web: www.plasticmirrors.co.uk
Product Type: 5

REDDISEALS
The Furlong, Berry Hill Industrial Estate,
Droitwich, Worcestershire, WR9 9BG
Area of Operation: UK & Ireland
Tel: 01905 791876 **Fax:** 01905 791877
Email: Reddiseals@Reddiseals.com
Web: www.reddiseals.com
Product Type: 2, 5

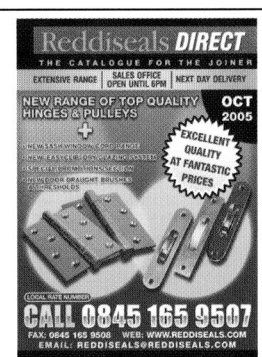

RYELUND LTD
Sawmill Lane, Helmsley, North Yorkshire, YO62 5DQ
Area of Operation: Worldwide
Tel: 01439 772802 **Fax:** 01439 771002
Email: info@ryelund.co.uk
Web: www.ryelund.co.uk

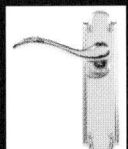

SERAMICO LTD
9 Billers Chase, Beaulieu Park, Chelmsford, Essex, HP4 3LF
Area of Operation: UK & Ireland
Tel: 01245 464 490 **Fax:** 01245 464 490
Email: seramicoltd@aol.com
Web: www.seramico.co.uk **Product Type:** 1, 2

STANDARDS GROUP
Bentley Hall Barn, Alkmonton,
Ashbourne, Derbyshire, DE6 3DJ
Area of Operation: UK & Ireland
Tel: 01335 330263 **Fax:** 01335 330922
Email: uk@standardsgroup.com
Web: www.standardsgroup.com **Product Type:** 1

STRATHEARN STONE AND TIMBER LTD
Glenearn, Bridge of Earn, Perth,
Perth and Kinross, PH2 9HL
Area of Operation: North East England, Scotland
Tel: 01738 813215 **Fax:** 01738 815946
Email: info@stoneandoak.com
Web: www.stoneandoak.com **Product Type:** 1, 5
Other Info: **Material Type:** C) 5

THE BRADLEY COLLECTION
Lion Barn, Maitland Road, Needham Market, Suffolk, IP6 8NS
Area of Operation: Worldwide
Tel: 01449 722724 **Fax:** 01449 722728
Email: claus.fortmann@bradleycollection.co.uk
Web: www.bradleycollection.co.uk **Product Type:** 1

THE CAST IRON RECLAMATION COMPANY
The Barn, Preston Farm Court,
Little Bookham, Surrey, KT23 4EF
Area of Operation: Worldwide
Tel: 0208 977 5977 **Fax:** 0208 786 6690
Email: enquiries@perfect-irony.com
Web: www.perfect-irony.com
Product Type: 1, 2, 3, 7, 12, 16, 20

THE EXPANDED METAL COMPANY LIMITED
PO Box 14, Longhill Industrial Estate (North),
Hartlepool, Durham, TS25 1PR
Area of Operation: Worldwide
Tel: 01429 408087 **Fax:** 01429 866795
Email: paulb@expamet.co.uk
Web: www.expandedmetalcompany.co.uk

TRAPEX
26 Pindor Road, Hoddesdon, Hertfordshire, EN11 0DE
Area of Operation: Worldwide
Tel: 01992 462150 **Fax:** 01992 446736
Email: info@trapex.com
Web: www.trapex.com **Product Type:** 1, 2, 3, 4, 5

URBAN FRONT LTD
Design Studio, 1 Little Hill, Heronsgate,
Rickmansworth, Hertfordshire, WD3 5BX
Area of Operation: UK & Ireland
Tel: 0870 609 1525 **Fax:** 0870 609 3564
Email: elizabeth@urbanfront.co.uk
Web: www.urbanfront.co.uk
Product Type: 1, 2

WAGNER (GB) LTD
VBH House, Bailey Drive, Gillingham Business Park,
Gillingham, Kent, ME8 0WG
Area of Operation: Worldwide
Tel: 01634 263263 **Fax:** 01634 263504
Email: sales@wagnergb.com
Web: www.wagnergb.com **Product Type:** 2

WMC ANTIQUES
141 Baldocks Lane, Melton Mowbray,
Leicester, Leicestershire, LE13 1EP
Area of Operation: Worldwide
Tel: 01664 851488
Email: willcoo1@ntlworld.com
Web: www.wmcantiques.co.uk
Product Type: 1, 2, 3, 4, 5

ZERO SEAL SYSTEMS LTD
Unit 6, Ladford Covert, Seighford,
Stafford, Staffordshire, ST18 9QG
Area of Operation: Europe
Tel: 01785 282910 **Fax:** 01785 282498
Email: sales@zeroplus.co.uk
Web: www.zeroplus.co.uk
Product Type: 5

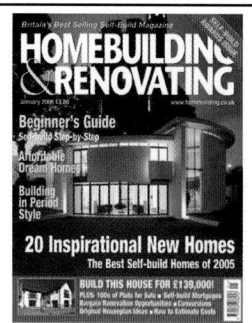

HOMEBUILDING & RENOVATING MAGAZINE
Area of Operation: UK & Ireland
Tel: 01527 834435 **Fax:** 01527 837810
Email: alex.worthington@centaur.co.uk
Web: www.homebuilding.co.uk

Homebuilding & Renovating, Britain's best selling self-build magazine is an essential read for anyone extending, renovating, converting or building their own home, providing practical advice and inspirational ideas.

MOVE OR IMPROVE? MAGAZINE
Area of Operation: UK & Ireland
Tel: 01527 834435 **Fax:** 01527 837810
Email: alex.worthington@centaur.co.uk
Web: www.moveorimprove.co.uk

Move or Improve? magazine for people adding adding space and value to their homes. Includes design guides, practical advice and inspiration on extensions, loft and basement conversions and improving your current home.

WWW.PLOTFINDER.NET
Area of Operation: UK & Ireland
Tel: 01527 834436 **Fax:** 01527 837810
Email: laura.cassidy@centaur.co.uk
Web: www.plotfinder.net

Plotfinder is an online database which holds details of almost 6,000 building plots and properties in need of renovation or conversion currently for sale throughout the UK.

WWW.PROPERTYFINDERFRANCE.NET
Area of Operation: UK & France
Tel: 01527 834435 **Fax:** 01527 837810
Email: alex.worthington@centaur.co.uk
Web: www.propertyfinderfrance.net

Looking for a property in France, this website has around 50,000 properties throughout France, you can search by type of property, size, price, area and number of bedrooms. To get a weeks free trial visit www.propertyfinderfrance.net.

WWW.SITEFINDERIRELAND.COM
Area of Operation: Ireland
Tel: 01527 834435 **Fax:** 01527 837810
Email: laura.cassidy@centaur.co.uk
Web: www.sitefinderireland.com

Sites for sale. Whether you are looking for a building site, a property that need renovating or a building to convert www.sitefinderireland.com can make your search easier; it lists over 2,000 currently for sale throughout Ireland.

WWW.HOMEBUILDING.CO.UK
Area of Operation: UK and Ireland
Tel: 01527 834435 **Fax:** 01527 837810
Email: customerservice@centaur.co.uk
Web: www.homebuilding.co.uk

Includes vast amounts of self-build and renovation information, a huge directory of products and suppliers, houseplans, readers' homes, various advice such as finance and planning plus The Beginners Guide.

NOTES

Company Name ....
Address ....
email ....
Web ....

Company Name ....
Address ....
email ....
Web ....

Company Name ....
Address ....
email ....
Web ....

Company Name ....
Address ....
email ....
Web ....

Company Name ....
Address ....
email ....
Web ....

INSULATION & PROOFING MATERIALS

Image courtesy of Kingspan Insulation Ltd (0870 850 8555)

SPONSORED BY KINGSPAN INSULATION LTD
Tel 0870 850 8555 Web www.insulation.kingspan.com

Kingspan
Kingspan Insulation Ltd

Insulation

Today's building regulations have strict guidelines on thermal efficiency, which has had a dramatic effect on the energy efficiency of our homes. A centrally heated house, built without insulation in the early 1970s, (before the imposition of the relevant building regulations) would consume five times more fuel than a modern equivalent.

ABOVE: Pipe insulation from Wickes.

All materials insulate to an extent: they have a certain level of thermal conductivity (often referred to as a lambda value). However, the insulating properties of traditional building materials cannot compare with the range of dedicated insulating materials available today. Realistically, you cannot build a house – or an extension to an existing house – without specifying some form of dedicated insulation in your walls, roof and floors.

Glass and Mineral Fibres

Perhaps the most easily understood types of insulation are wools, whether synthetic or natural. They work by trapping air between the fibres, giving lambda values of around 0.04. Synthetic wool is one of the cheapest materials to work with, although natural wool can cost up to five times more. Both excel in applications where they can be laid flat and where compression is not an issue, such as in the floors of uninhabited lofts, They are incredibly easy to work and are quick to install, but it is easy to leave holes in the overall insulating blanket, thus greatly reducing the overall effectiveness.

Rigid Boards

The most common alternative to insulating wool is to use a rigid board material. For underfloor insulation this is really the only option, but it also makes a lot of sense in walls and roofing applications. The cheapest and thickest board is made of expanded polystyrene, which has a similar lambda value as mineral wools and thus needs to be used in similar thicknesses in order to get sufficiently low U-values.

Extruded polystyrene is tougher and stronger than the expanded version, but the lambda value is significantly better (usually around 0.03) and it is suitable for areas that get a lot of heavy foot traffic. However, this comes at a cost: it is two to three times more expensive than expanded polystyrene.

The polyurethane family of chemicals produce the thinnest insulating boards, with lambda values around half that of mineral wool and expanded polystyrene (between 0.02 and 0.025) which makes them approximately twice as effective in terms of insulation capabilities for a given thickness. In turn they are two to three times the price of expanded polystyrene.

Spray-in or Blow-in Solutions

Applying insulation in liquid or semi-liquid form means that you are going to get an excellent fit, something which all of the pre-packaged options struggle with. Research has shown that the effect of even very small gaps in the insulated envelope has a dramatic impact on the overall U-values of the structure. Blowing or spraying is one way of ensuring this doesn't happen.

Cellulose (recycled newsprint), Polyiso foam, and expanded polystyrene in bead form, are examples of materials used in this method. All spray-in or blow-in solutions need a void in which to sit; this may be a masonry cavity or a timber stud wall or rafter space. In order to get the assembly down to an acceptable U-value, it may be necessary to combine with another insulating material, such as a rigid board.

Reflective Foils

Reflective foils aim to reflect radiant heat, similar to the way that low-e glass works, but need a 25mm air gap on at least one side — without this gap, they have hardly any insulation value at all. This limits their application, but they do promise good insulation levels from a minimum of space.

Structural Insulation

Structural insulated panels (SIPs) incorporate a section of insulation between two sheets of board, and as such enjoy inherently high levels of energy efficiency. Those wishing to incorporate high levels of insulation may wish to consider this form of construction as an alternative to the traditional masonry route, especially as the construction of SIPs is inherently more airtight than other methods — and much of the heat in a house is lost through imperfect seals.

BELOW LEFT: Reflective foils can be tacked to the underside of your rafters to drastically reduce heat loss through the roof. Image from A Proctor Group. BELOW: Full-fill cavity wall insulation from Knauf Insulation.

THERMAL INSULATION

KEY

PRODUCT TYPES: 1 = Wall Cavity 2 = Floor
3 = Roof 4 = Internal Wall 5 = External Wall
6 = Pipe

OTHER: ▽ Reclaimed ⌂ On-line shopping
✎ Bespoke ✋ Hand-made ECO Ecological

A PROCTOR GROUP LTD
The Haugh, Blairgowrie, Perth & Kinross, PH10 7ER
Area of Operation: Worldwide
Tel: 01250 872261
Fax: 01250 872727
Email: karrina.andrews@proctorgroup.com
Web: www.proctorgroup.com
Product Type: 1, 3, 4

APOLLO INSULATION (UK) LTD
PO Box 200, Horley, Surrey, RH6 7FU
Area of Operation: Worldwide
Tel: 01293 776974
Fax: 01293 776975
Email: information@apollo-energy.com
Web: www.apollo-energy.com
Product Type: 1, 2, 3, 5

ARMACELL (UK) LTD
Mars Street, Oldham, Lancashire, OL9 6LY
Area of Operation: Worldwide
Tel: 0161 287 7100
Fax: 0161 633 2685
Email: info.uk@armacell.com
Web: www.armacell.com
Product Type: 6

KINGSPAN KOOLTHERM® K12 FRAMING BOARD
Area of Operation: Worldwide
Tel: 0870 850 8555
Fax: 0870 850 8666
Email: info.uk@insulation.kingspan.com
Web: www.insulation.kingspan.com

Timber and Steel framing insulation solution.
Literature contains design considerations,
technical data, sitework details, thermal
performance and U-value tables.

BUILD IT GREEN (UK) LTD
Arena Business Centre, 9 Nimrod Way,
Ferndown, Dorset, BH21 7SH
Area of Operation: UK (Excluding Ireland)
Tel: 01202 862320
Fax: 01202 877685
Email: enquiries@buildit-green.co.uk
Web: www.buildit-green.co.uk
Product Type: 5

CELOTEX LIMITED
Lady Lane Industrial Estate, Hadleigh,
Ipswich, Suffolk, IP7 6BA
Area of Operation: UK & Ireland
Tel: 01473 822093
Fax: 01473 820880
Email: info@celotex.co.uk
Web: www.celotex.co.uk
Product Type: 1, 2, 3, 4, 5
Material Type: K) 14

CELOTEX
Area of Operation: UK & Ireland
Tel: 01473 822093
Fax: 01473 820880
Email: info@celotex.co.uk
Web: www.celotex.co.uk
Product Type: 1, 2, 3, 4, 5
Material Type: K) 14

UK building insulation specialist Celotex has
commenced roll-out of a new generation of rigid
polyisocyanurate (PIR) foam board products using
blowing agents with Zero Ozone Depletion
Potential (Zero ODP).

DACATIE BUILDING SOLUTIONS
Quantum House, Salmon Fields, Royton,
Oldham, Lancashire, OL2 6JG
Area of Operation: UK & Ireland
Tel: 0161 627 4222
Fax: 0161 627 4333
Email: info@dacatie.co.uk
Web: www.dacatie.co.uk
Product Type: 1

DOW CONSTRUCTION PRODUCTS
2 Heathrow Boulevard, 284 Bath Road,
West Drayton, Middlesex, Greater London, UB7 0DQ
Area of Operation: Worldwide
Tel: 0208 917 5050
Email: styrofoam-uk@dow.com
Web: www.styrofoameurope.com
Product Type: 1, 2, 3, 4, 6

ECOSHOP
Unit 1, Glen of the Downs Garden Centre,
Kilmacanogue, Co Wicklow, Republic of Ireland
Area of Operation: Ireland Only
Tel: +353 01 287 2914
Fax: +353 01 201 6480
Email: info@ecoshop.ie
Web: www.ecoshop.ie
Product Type: 1, 2, 3, 4, 5

EDULAN UK LTD
Unit M, Northstage, Broadway, Salford,
Greater Manchester, M50 2UW
Area of Operation: Worldwide
Tel: 0161 876 8040
Fax: 0161 876 8041
Email: peter@polyurethane.uk.com
Web: www.edulan.com
Product Type: 1, 3, 5
Material Type: K) 5, 14

ELLIOTTS INSULATION AND DRYLINING
Unit 8 Goodwood Road, Boyatt Wood Industrial
Estate, Eastleigh, Hampshire, SO50 4NT
Area of Operation: South East England,
South West England and South Wales
Tel: 02380 623960
Fax: 02380 623965
Email: insulation@elliott-brothers.co.uk
Web: www.elliotts.uk.com
Product Type: 1, 2, 3, 4, 5

ENCON INSULATION
Brunswick House, 1 Deighton Close,
Wetherby, West Yorkshire, LS22 7GZ
Area of Operation: UK (Excluding Ireland)
Tel: 01937 524200
Fax: 01937 524280
Email: s.roy@encon.co.uk
Web: www.encon.co.uk
Product Type: 1, 3, 5

EPS INSULATION LTD
Unit 10 Amber Close, Amington Industrial Estate,
Tamworth, Staffordshire, B79 8BH
Area of Operation: UK (Excluding Ireland)
Tel: 01827 313951 **Fax:** 01827 54683
Email: info@eps-systemsltd.com
Web: www.eps-systemsltd.com
Product Type: 2, 3, 4, 5

EXCEL BUILDING SOLUTIONS
Maerdy Industrial Estate,
Rhymney, Powys, NP22 5PY
Area of Operation: Worldwide
Tel: 01685 845200 **Fax:** 01685 844106
Email: sales@excelfibre.com
Web: www.excelfibre.com
Product Type: 1, 2, 3, 4, 5
Other Info: ▽ ECO
Material Type: K) 3

FILLCRETE LTD
Maple House, 5 Over Minnis,
New Ash Green, Longfield, Kent, DA3 8JA
Area of Operation: UK (Excluding Ireland)
Tel: 01474 872444 **Fax:** 01474 872426
Email: timfolkes@fillcrete.com
Web: www.fillcrete.com
Product Type: 3, 4, 5

FOAMSEAL LTD
New Street House, New Street,
Petworth, West Sussex, GU28 0AS
Area of Operation: Worldwide
Tel: 01798 345400 **Fax:** 01798 345410
Email: info@foamseal.co.uk
Web: www.foamseal.co.uk
Product Type: 1, 2, 3, 4, 5, 6
Material Type: K) 5

GREEN BUILDING STORE
11 Huddersfield Road, Meltham,
Holmfirth, West Yorkshire, HD9 4NJ
Area of Operation: UK & Ireland
Tel: 01484 854898 **Fax:** 01484 854899
Email: info@greenbuildingstore.co.uk
Web: www.greenbuildingstore.co.uk ⌂
Product Type: 2, 3, 4
Other Info: ECO

H+H CELCON LIMITED
Celcon House, Ightham,
Sevenoaks, Kent, TN15 9HZ
Area of Operation: UK (Excluding Ireland)
Tel: 01732 880520 **Fax:** 01732 880531
Email: marketing@celcon.co.uk
Web: www.celcon.co.uk
Product Type: 1, 4, 5

HUNTON FIBER UK LTD
Rockleigh Court, Rock Road, Finedon,
Northamptonshire, NN9 5EL
Area of Operation: UK & Ireland
Tel: 01933 682683
Fax: 01933 680296
Email: admin@huntonfiber.co.uk
Web: www.hunton.no
Product Type: 1, 3
Other Info: ▽ ECO

JOULESAVE EMES LTD
27 Water Lane, South Witham,
Grantham, Lincolnshire, NG33 5PH
Area of Operation: UK & Ireland
Tel: 01572 768362
Fax: 01572 767146
Email: sales@joulesave.co.uk
Web: www.joulesave.co.uk
Product Type: 3

KINGSPAN INSULATION LTD
Pembridge, Nr Leominster,
Herefordshire, HR6 9LA
Area of Operation: Worldwide
Tel: 0870 850 8555 **Fax:** 0870 850 8666
Email: info.uk@insulation.kingspan.com
Web: www.insulation.kingspan.com
Product Type: 1, 2, 3, 4, 5, 6
Other Info: ECO
Material Type: K) 5, 14

KINGSPAN KOOLTHERM® K17 INSULATED DRY-LINING BOARD
Area of Operation: Worldwide
Tel: 0870 850 8555
Fax: 0870 850 8666
Email: info.uk@insulation.kingspan.com
Web: www.insulation.kingspan.com

Insulated plasterboard solution for plaster-dab or
mechanical fixing. Literature contains design
considerations, technical data, sitework details, thermal
performance and U-value tables.

KINGSPAN TEK
Pembridge, Leominster,
Herefordshire, HR6 9LA
Area of Operation: UK & Ireland
Tel: 01544 387308
Fax: 0870 850 8666
Email: quotations@tek.kingspan.com
Web: www.tek.kingspan.com
Product Type: 1, 3, 4, 5

KNAUF DIY
PO Box 732, Maidstone,
Kent, ME15 6ST
Area of Operation: UK (Excluding Ireland)
Tel: 0845 601 1763
Fax: 0845 601 1762
Email: knaufdiy@knauf.co.uk
Web: www.teachmediy.co.uk

KNAUF INSULATION
PO Box 10, Stafford Road,
St Helens, Merseyside, WA10 3NS
Area of Operation: UK & Ireland
Tel: 01270 824024 (Enquiries Hotline)
Fax: 01270 824025
Email: info@knaufinsulation.com
Web: www.knaufinsulation.co.uk
Product Type: 1, 2, 3, 4, 5, 6

LAFARGE GYVLON LTD
221 Europa Boulevard,
Westbrook, Warrington,
Cheshire, WA5 7TN
Area of Operation: UK & Ireland
Tel: 01925 428780
Fax: 01925 428788
Email: sales@gyvlon-floors.co.uk
Web: www.gyvlon-floors.co.uk
Product Type: 2

M. H. JOINERY SERVICES
25b Camwal Road, Harlgate,
North Yorkshire, HG1 4PT
Area of Operation: North East England
Tel: 01423 888856
Fax: 01432 888856
Email: info@mhjoineryservices.co.uk
Web: www.mhjoineryservices.co.uk

MAXIROOF LTD
36 Lower End, Swaffham Prior,
Cambridge, Cambridgeshire, CB5 0HT
Area of Operation: UK (Excluding Ireland)
Tel: 01638 743380
Fax: 01638 743380
Email: info@maxiroof.co.uk
Web: www.maxiroof.co.uk
Product Type: 3

Bellissim**O**

Whether you are comparing the fire & smoke, thermal or longevity performance the premium performance *Kingspan* **Kool**therm K-range is best in its class

Fire & Smoke
The CFC/HCFC-free phenolic core of the *Kingspan* **Kool**therm K-range achieves a Class **O** rating to the Building Regulations / Low Risk to Technical Standards and achieves the best possible rating of less than 5% smoke emission when tested to BS 5111: Part 1: 1974.

These properties are critical when looking at the design of buildings. If the worst happens and there is a fire, the building should be able to provide some resistance to structural damage and also should not compromise the ability of any inhabitants to escape from the building.

Thermal
When considering thermal conductivity (λ-value) the *Kingspan* **Kool**therm K-range is in a class of its own with a league topping 0.022 – 0.024 W/m.K. This means that less thickness of insulation is needed to achieve the required Building Regulations/Standard. Thinner insulation allows the specifier to reduce the impact of the structure on the usable area of the building.

Longevity
The closed cell structure of the *Kingspan* **Kool**therm K-range means that its superior thermal conductivity is unaffected by moisture (H_2**O**) and air movement. This consistency of performance results in the design standard of the element being achieved for the lifetime of the building. This characteristic is essential if we are to benefit from the current drive towards reducing the CO_2 emissions from our buildings.

Kooltherm K3 Floorboard
INSULATION FOR SOLID CONCRETE AND SUSPENDED GROUND FLOORS

Kooltherm K5 EWB
INSULATION FOR USE BEHIND TRADITIONAL AND LIGHTWEIGHT POLYMER MODIFIED RENDERS AND TIMBER OR TILE RAINSCREENS

Kooltherm K7 Pitched Roof Board
RAFTER LEVEL INSULATION FOR TILED OR SLATED PITCHED WARM ROOF SPACES

Kooltherm K8 Cavity Board
PARTIAL FILL CAVITY WALL INSULATION

Kooltherm K10 Soffit Board
INSULATION FOR STRUCTURAL CEILINGS (SOFFITS)

Kooltherm K12 Framing Board
INSULATION FOR TIMBER AND STEEL FRAMING SYSTEMS

Kooltherm K15 Rainscreen Board
INSULATION FOR USE BEHIND RAINSCREEN CLADDING SYSTEMS

Kooltherm K17 Drylining Board
INSULATION FOR PLASTER-DAB/ADHESIVE BONDING

Kooltherm K18 Drylining Board
INSULATION FOR MECHANICAL FIXING

Further information on the *Kingspan* **Kool**therm K-range of rigid phenolic insulation products for roofs, walls and floors is available from Kingspan Insulation on:

Telephone: 0870 733 8333 (UK)
email: literature.uk@insulation.kingspan.com

Telephone: 042 97 95038 (Ireland)
email: literature.ie@insulation.kingspan.com

www.insulation.kingspan.com

® Kingspan, Kooltherm and the Lion device are Registered Trademarks of the Kingspan Group plc

Kingspan ®

Kingspan Insulation Ltd
Pembridge, Leominster, Herefordshire HR6 9LA, UK
Castleblayney, County Monaghan, Ireland

MAXIROOF LTD
Area of Operation: UK (Excluding Ireland)
Tel: 01638 743380 Fax: 01638 743380
Email: info@maxiroof.co.uk
Web: www.maxiroof.co.uk

FAST. SIMPLE. ECONOMICAL.
Build rooms in the roof the intelligent way with
MAXIROOF airtight SIPS.
There is no better roof.

MGC LTD
McMillan House, 54-56 Cheam Common Road,
Worcester Park, Surrey, KT4 8RH
Area of Operation: UK & Ireland
Tel: 0208 337 0731 Fax: 0208 337 3739
Email: info@mgcltd.co.uk
Web: www.mgcltd.co.uk

MID ARGYLL GREEN SHOP
An Tairbeart, Campbeltown Road, Tarbet,
Argyll & Bute, PA29 6SX
Area of Operation: Scotland
Tel: 01880 821212
Email: kate@mackandmags.co.uk
Web: www.mackandmags.co.uk
Product Type: 1, 2, 3, 4, 5
Other Info: ECO
Material Type: K) 3, 8, 9

MIKE WYE & ASSOCIATES
Buckland Filleigh Sawmills, Buckland Filleigh,
Beaworthy, Devon, EX21 5RN
Area of Operation: Worldwide
Tel: 01409 281644 Fax: 01409 281669
Email: sales@mikewye.co.uk
Web: www.mikewye.co.uk
Other Info: ECO
Material Type: K) 9

MIRATEX LIMITED
Unit 10, Park Hall Farm, Brookhouse Road,
Cheadle, Staffordshire, ST10 2NJ
Area of Operation: UK & Ireland
Tel: 01538 750923 Fax: 01538 752078
Email: info@miratex.co.uk
Web: www.miratex.co.uk
Product Type: 5

NATURAL BUILDING TECHNOLOGIES
The Hangar, Worminghall Road,
Oakley, Buckinghamshire, HP18 9UL
Area of Operation: UK & Ireland
Tel: 01844 338338 Fax: 01844 338525
Email: info@natural-building.co.uk
Web: www.natural-building.co.uk
Product Type: 1, 2, 3, 4, 5
Other Info: ECO Material Type: K) 3, 8, 9

NOISE STOP SYSTEMS
Unit 3 Acaster Industrial Estate, Cowper Lane,
Acaster Malbis, York, North Yorkshire, YO23 2XB
Area of Operation: UK & Ireland
Tel: 0845 130 6269 Fax: 01904 709857
Email: info@noisestopsystems.co.uk
Web: www.noisestopsystems.co.uk
Product Type: 1, 2, 3, 4, 5

KINGSPAN KOOLTHERM® K8 CAVITY BOARD
Area of Operation: Worldwide
Tel: 0870 850 8555
Fax: 0870 850 8666
Email: info.uk@insulation.kingspan.com
Web: www.insulation.kingspan.com

Partial fill cavity wall insulation solution.
Literature contains design considerations,
technical data, sitework details, thermal
performance and U-value tables.

OLD HOUSE STORE LTD
Hampstead Farm, Binfield Heath,
Henley on Thames, Oxfordshire, RG9 4LG
Area of Operation: Worldwide
Tel: 0118 969 7711 Fax: 0118 969 8822
Email: info@oldhousestore.co.uk
Web: www.oldhousestore.co.uk
Product Type: 1, 2, 3, 4
Other Info: ECO Material Type: K) 3, 9

QUAD-LOCK (ENGLAND) LTD
Unit B3.1, Maws Centre, Jackfield,
Telford, Shropshire, TF8 7LS
Area of Operation: UK (Excluding Ireland)
Tel: 0870 443 1901 Fax: 0870 443 1902
Email: k.turner@quadlock.co.uk
Web: www.quadlock.co.uk
Product Type: 4, 5

RENOTHERM
New Street House, New Street,
Petworth, West Sussex, GU 28 0AS
Area of Operation: UK (Excluding Ireland)
Tel: 01798 343658
Fax: 01798 344093
Email: sales@renotherm.co.uk
Web: www.renotherm.co.uk
Product Type: 1, 2, 3
Material Type: K) 5, 14

ROOFING INSULATION SERVICES
Hilldale House, 9 Hilldale Avenue,
Blackley, Manchester, M9 6PQ
Area of Operation: UK (Excluding Ireland)
Tel: 0800 731 8314
Email: info@roofinginsulationservices.co.uk
Web: http://roofinginsulationservices.co.uk
Product Type: 2, 3, 4

ROOFING INSULATION SERVICES
Area of Operation: UK (Excluding Ireland)
Tel: 0800 7318314
Email: info@roofinginsulationservices.co.uk
Web: http://roofinginsulationservices.co.uk
Product Type: 2, 3, 4

Spray foam insulation provides a seamless
envelope of thermal insulation, delivering much
greater energy efficiency and lower heating
costs, as well as eliminating condensation.

ROOFSURE
Metro House, 14-17 Metropolitan Business Park,
Preston New Road, Blackpool, Lancashire, FY3 9LT
Area of Operation: UK (Excluding Ireland)
Tel: 0800 597 2828 Fax: 01253 798193 / 764671
Email: roofsureltd@btconnect.com
Web: www.roofsure.co.uk

SECOND NATURE UK LTD
Soulands Gate, Soulby, Dacre,
Penrith, Cumbria, CA11 0JF
Area of Operation: Worldwide
Tel: 01768 486285 Fax: 01768 486825
Email: info@secondnatureuk.com
Web: www.secondnatureuk.com
Product Type: 2, 3, 4 Other Info: ECO
Material Type: K) 9

SECONDS & CO
Industrial Estate, Presteigne, Powys, LD8 2UF
Area of Operation: UK (Excluding Ireland)
Tel: 01544 260501 Fax: 01544 260525
Email: secondsandco@aol.com
Web: www.secondsandco.co.uk
Product Type: 1, 2, 3, 4

KINGSPAN KOOLTHERM® K7 PITCHED ROOF BOARD
Area of Operation: Worldwide
Tel: 0870 850 8555
Fax: 0870 850 8666
Email: info.uk@insulation.kingspan.com
Web: www.insulation.kingspan.com

Pitched roof insulation solution. Literature
contains design considerations, technical data,
sitework details, thermal performance and
U-value tables.

SIP HOME LTD
Unit 1-3 Hadston Industrial Estate,
South Broomhill, Northumberland, NE65 9YG
Area of Operation: UK (Excluding Ireland)
Tel: 01670 760300 Fax: 01670 760740
Email: john@siphome.co.uk
Web: www.siphome.co.uk

SKANDA (UK) LTD
64 - 65 Clywedog Road North,
Wrexham Industrial Estate,
Wrexham, LL13 9XN
Area of Operation: UK & Ireland
Tel: 01978 664255
Fax: 01978 661427
Email: info@skanda-uk.com
Web: www.skanda-uk.com / www.heraklith.co.uk

SPRAYSEAL CONTRACTS LTD
Bollin House, Blakeley Lane,
Mobberley, Cheshire, WA16 7LX
Area of Operation: East England, Greater London,
Midlands & Mid Wales, North East England, North
West England and North Wales, South East England,
South West England and South Wales
Tel: 01565 872303 Fax: 01565 872599
Email: info@sprayseal.co.uk
Web: www.sprayseal.co.uk
Product Type: 3

SPONSORED BY: KINGSPAN INSULATION LTD   www.insulation.kingspan.com

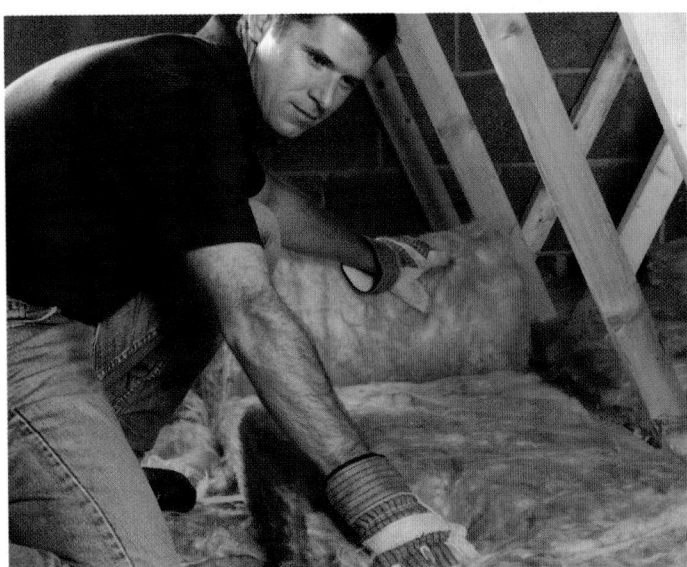

level: a 100 mm layer between the joists with a second layer, about 200 mm thick, across them. If the loft is needed for storage, **Polyfoam® Supadeck** offers an easy solution: a layer of **Polyfoam® Insulation** replaces the top layer of insulation, its strength providing support for the wooden decking surface.

Insulation at rafter level allows the interior of the roof's pitch to be used as living space. Installing rigid foam boards, **Polyfoam® Roofboard** and **Polyfoam® Raftersqueeze** for example, above, below or between the rafters being the usual method. An increasingly popular alternative is **Crown® Rafter Roll 32**, a high-performance glass mineral wool specifically developed for use between rafters. Combined with

Knauf Breatheline, a vapour permeable underlay, it creates a roof that can 'breathe' which is possibly the easiest and most cost-effective to construct.

As the only company in the UK to manufacture both mineral wool and foam board insulation, Knauf Insulation can provide an insulation solution just for you, no matter how unique your needs are. Basic advice and guidance are available on the website at www.knaufinsulation.co.uk, a wide range of application-specific brochures are available free of charge and, if you've still got a question, the Technical Advisory Centre on 01744 766666 will be delighted to help you.

Please mention ID GA42805 when replying.

What Insulation *Where?*

Thermal insulation makes a huge contribution to the comfort and cost of running a home and, if that's not enough, it's also an essential element required by the Building Regulations! Stephen Wise, Knauf Insulation's Market Development Manager, provides a quick tour.

Mineral wool and foam board are the most widely used building insulation materials, their chalk and cheese characteristics dictating where they are best used. Mineral wool is made from molten sand or rock spun into a mat and, 'as standard', has excellent thermal, acoustic and fire insulation properties. In fact it's specified by many of the Robust Details used to meet Part E of the Building Regulations and has a Euroclass A1 fire classification; the highest rating possible. Rigid foam insulation boards include expanded polystyrene (EPS), polyisocyanurates (PIR and PUR) and extruded polystyrene (XPS) with the latter capable of having exceptional compressive strength.

Self-build homes tend to be unique in design and usually need a similarly

unique insulation solution, but a brief look at the most frequently used types of construction gives a valuable insight into what's used where.

If the ground floor is concrete then it's vital the insulation has sufficient strength to withstand dynamic loading. Extruded polystyrene (XPS) is the only real choice and **Polyfoam® Floorboard 'Domestic'**, with a compressive strength of 165 kPa, is ideal for homes. For suspended timber floors the best option is a mineral wool, **Rocksilk® Flexible Slab** for example, installed between the floor joists.

Full-fill cavity wall insulation is the easiest, least expensive, method of insulating walls. Mineral wool – both **Crown®** and **Rocksilk® DriTherm® Cavity Slabs** are designed for the task – giving the most reliable performance. If the wall has a complex shape, or the cavity is particularly wide, glass mineral wool such as **Crown® Supafil™ Cavity Wall Insulation** can be injected into the cavity through either leaf.

The Building Regulations specify that timber frame external walls should have a half-hour fire resistance and this is easy to achieve using mineral wool – **Crown® FrameTherm, Timber Roll and Timber Slab Insulation** being specifically designed for timber frame construction.

Pitched roofs are the most popular type of roof construction. **Crown® Loft Roll**, a lightweight glass mineral wool blanket, is ideal for use at ceiling joist

Concrete ground floor with floating screed

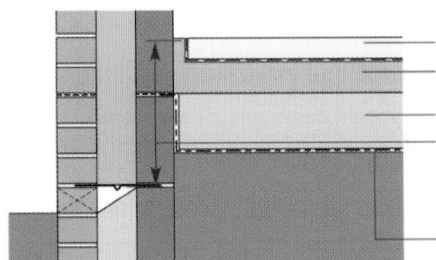

- Floating screed on slip sheet/VCL
- Polyfoam® Floorboard
- Concrete slab
- Continue wall insulation at least 150mm below top of perimeter insulation and support on a row of ties.
- Damp proof membrane

LIFE CYCLE ASSESSMENT

Area of Operation: Worldwide
Tel: 0870 850 8555
Fax: 0870 850 8666
Email: info.uk@insulation.kingspan.com
Web: www.insulation.kingspan.com

First insulation manufacturer to commission and openly publish BRE certified Life Cycle Assessment. Paper shows manufacturing impact on the environment.

SPRINGVALE EPS LTD
Dinting Vale Business Park, Dinting Vale, Glossop, Derbyshire, SK13 6LG
Area of Operation: UK & Ireland
Tel: 01457 863211 **Fax:** 01457 869269
Email: salesg@springvale.com
Web: www.springvale.com
Product Type: 1, 2, 3, 5
Material Type: K) 11

STOPGAP LTD
PO Box 2389, Cardiff, CF23 5WJ
Area of Operation: Worldwide
Tel: 02920 213736 **Fax:** 02920 213736
Email: info@stopg-p.com
Web: www.stopg-p.com ⌐
Product Type: 2

STRUCTURAL INSULATED PANEL TECHNOLOGY LTD (SIPTEC)
PO Box 429, Turvey, Bedfordshire, MK43 8DT
Area of Operation: Europe
Tel: 0870 743 9866 **Fax:** 0870 762 5612
Email: mail@sips.ws
Web: www.sips.ws
Product Type: 2, 3, 4, 5
Material Type: H) 4, 6

THE GREEN SHOP
Cheltenham Road, Bisley, Nr Stroud, Gloucestershire, GL6 7BX
Area of Operation: UK & Ireland
Tel: 01452 770629 **Fax:** 01452 770104
Email: paint@greenshop.co.uk
Web: www.greenshop.co.uk ⌐
Product Type: 1, 2, 3, 4 **Other Info:** ▽ ECO
Material Type: K) 3, 9

THERMILATE EUROPE
The Media Centre, 7 Northumberland Street, Huddersfield, West Yorkshire, HD1 1RL
Area of Operation: Worldwide
Tel: 0870 744 1759
Fax: 0870 744 1760
Email: info@thermilate.com
Web: www.thermilate.com ⌐
Product Type: 2, 3, 4, 5, 6

THERMOBRICK
20 Underwood Drive, Stoney Stanton, Leicester, Leicestershire, LE9 4TA
Area of Operation: UK & Ireland
Tel: 01455 272860
Fax: 01455 271324
Email: ruttim1@aol.com
Web: www.rutlandtimber.co.uk
Product Type: 5

WEBER BUILDING SOLUTIONS
Dickens House, Enterprise Way, Maulden Road, Flitwick, Bedford, Bedfordshire, MK45 5BY
Area of Operation: Worldwide
Tel: 0870 333 0070 **Fax:** 01525 718988
Email: info@weberbuildingsolutions.co.uk
Web: www.weberbuildingsolutions.co.uk
Product Type: 5

WEBSTERS INSULATION LTD
Crow Tree Farm, Crow Tree Bank, Thorne Levels, Doncaster, South Yorkshire, DN8 5TF
Area of Operation: UK (Excluding Ireland)
Tel: 01405 812682 **Fax:** 01405 817201
Email: info@webstersinsulation.com
Web: www.webstersinsulation.com
Product Type: 1, 2, 3, 4, 5

WEBSTERS INSULATION

Area of Operation: UK (Excluding Ireland)
Tel: 01405 812682 **Fax:** 01405 817201
Email: info@webstersinsulation.com
Web: www.webstersinsulation.com
Product Type: 1, 2, 3, 4, 5

Websters have 25 years experience in Polyurethane Spray Foam Insulation. Our workmanship and materials are of the highest quality, and our prices are highly competitive, which is why we are the market leaders in our industry. Websters have 25 years experience in Polyurethane Spray Foam Insulation. Our workmanship and materials are of the highest quality, and our prices are highly competitive, which is why we are the market leaders in our industry.

WETHERBY BUILDING SYSTEMS LTD
1 Kidglove Road, Golborne Enterprise Park, Golborne, Greater Manchester, WA3 3GS
Area of Operation: UK & Ireland
Tel: 01942 717100
Fax: 01942 717101
Email: info@wbs-ltd.co.uk
Web: www.wbs-ltd.co.uk
Product Type: 4, 5

XTRATHERM UK LIMITED
Unit 5 Jensen Court, Astmoor Industrial Estate, Runcorn, Cheshire, WA7 1SQ
Area of Operation: UK & Ireland
Tel: 0871 222 1033
Fax: 0871 222 1044
Email: kerry@xtratherm.com
Web: www.xtratherm.com
Product Type: 1, 2, 3, 4
Material Type: K) 14

ACOUSTIC INSULATION

KEY

PRODUCT TYPES: 1= Door and Window
2 = Wall   3 = Floor

OTHER: ▽ Reclaimed ⌐ On-line shopping
✎ Bespoke 🖐 Hand-made ECO Ecological

A PROCTOR GROUP LTD
The Haugh, Blairgowrie, Perth & Kinross, PH10 7ER
Area of Operation: Worldwide
Tel: 01250 872261
Fax: 01250 872727
Email: karrina.andrews@proctorgroup.com
Web: www.proctorgroup.com
Product Type: 2, 3

AUDIO AGENCY EUROPE
PO Box 4601, Kiln Farm, Milton Keynes, Buckinghamshire, MK19 7ZN
Area of Operation: Europe
Tel: 01908 510123
Fax: 01908 511123
Email: info@audioagencyeurope.com
Web: www.audioagencyeurope.com
Product Type: 2

CUSTOM AUDIO DESIGNS LTD
Unit 2 Amey Industrial Estate, Petersfield, Hampshire, GU32 3LN
Area of Operation: UK & Ireland
Tel: 0870 747 5511
Fax: 0870 747 9878
Email: sales@customaudiodesigns.co.uk
Web: www.customaudiodesigns.co.uk
Product Type: 1, 2, 3
Other Info: ECO ✎
Material Type: K) 2, 4, 10

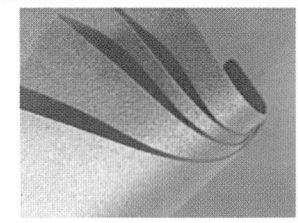
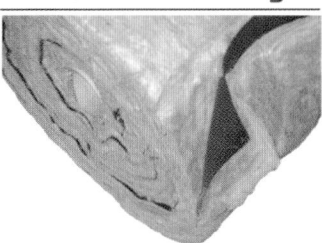

BUILDING STRUCTURE & MATERIALS

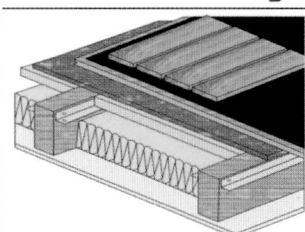

Custom Audio Designs

Impact Reduction Mats for uprating the acoustic performance of separating floors. Easy to handle & work with.
A cost effective noise reduction product offering high degrees of impact & airborne noise reduction at minimal thickness

Sales: 0870 747 5432
www.nonoise.co.uk

Custom Audio Designs

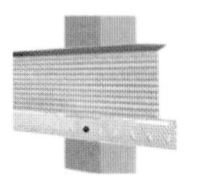

Acoustic Resilient Bars for uprating the acoustic performance of studwall partitions and ceilings. Simple & easy to use. A cost effective noise reduction product offering high degrees of impact & airborne noise reduction at minimal cost

Sales: 0870 747 5432
www.nonoise.co.uk

EDULAN UK LTD
Unit M, Northstage, Broadway, Salford,
Greater Manchester, M50 2UW
Area of Operation: Worldwide
Tel: 0161 876 8040 **Fax:** 0161 876 8041
Email: peter@polyurethane.uk.com
Web: www.edulan.com
Product Type: 2, 3
Material Type: K) 5

ELLIOTTS INSULATION AND DRYLINING
Unit 8 Goodwood Road, Boyatt Wood Industrial
Estate, Eastleigh, Hampshire, SO50 4NT
Area of Operation: South East England,
South West England and South Wales
Tel: 02380 623960 **Fax:** 02380 623965
Email: insulation@elliott-brothers.co.uk
Web: www.elliotts.uk.com
Product Type: 2, 3

EXCEL BUILDING SOLUTIONS
Maerdy Industrial Estate, Rhymney,
Powys, NP22 5PY
Area of Operation: Worldwide
Tel: 01685 845200 **Fax:** 01685 844106
Email: sales@excelfibre.com
Web: www.excelfibre.com
Product Type: 2, 3
Other Info: ▽ ECO
Material Type: K) 3

FOAMSEAL LTD
New Street House, New Street, Petworth,
West Sussex, GU28 0AS
Area of Operation: Worldwide
Tel: 01798 345400
Fax: 01798 345410
Email: info@foamseal.co.uk
Web: www.foamseal.co.uk
Product Type: 1, 2, 3
Material Type: K) 5

KINGSPAN KOOLTHERM® K3 FLOORBOARD

Area of Operation: Worldwide
Tel: 0870 850 8555
Fax: 0870 850 8666
Email: info.uk@insulation.kingspan.com
Web: www.insulation.kingspan.com

Suspended and solid floor insulation solution.
Literature contains design considerations, technical data, sitework details, thermal performance and U-value tables.

H+H CELCON LIMITED
Celcon House, Ightham, Sevenoaks, Kent, TN15 9HZ
Area of Operation: UK (Excluding Ireland)
Tel: 01732 880520
Fax: 01732 880531
Email: marketing@celcon.co.uk
Web: www.celcon.co.uk
Product Type: 2

HUNTON FIBER UK LTD
Rockleigh Court, Rock Road, Finedon,
Northamptonshire, NN9 5EL
Area of Operation: UK & Ireland
Tel: 01933 682683
Fax: 01933 680296
Email: admin@huntonfiber.co.uk
Web: www.hunton.no
Product Type: 2
Other Info: ▽ ECO

INSTACOUSTIC LIMITED
Insta House, Ivanhoe Road, Hogwood Business Park,
Finchampstead, Wokingham, Berkshire, RG40 4PZ
Area of Operation: UK & Ireland
Tel: 0118 932 8811
Fax: 0118 973 9547
Email: instacoustic@instagroup.co.uk
Web: www.instagroup.co.uk
Product Type: 2, 3

KNAUF INSULATION
PO Box 10, Stafford Road, St Helens,
Merseyside, WA10 3NS
Area of Operation: UK & Ireland
Tel: 01270 824024 (Enquiries Hotline)
Fax: 01270 824025
Email: info@knaufinsulation.com
Web: www.knaufinsulation.co.uk
Product Type: 2, 3

LAFARGE GYVLON LTD
Lafarge Gyvlon, 221 Europa Boulevard,
Westbrook, Warrington, Cheshire, WA5 7TN
Area of Operation: UK & Ireland
Tel: 01925 428780
Fax: 01925 428788
Email: sales@gyvlon-floors.co.uk
Web: www.gyvlon-floors.co.uk
Product Type: 3

LORIENT POLYPRODUCTS LTD
Fairfax Road, Heathfield Industrial Estate,
Newton Abbot, Devon, TQ12 6UD
Area of Operation: Worldwide
Tel: 01626 834252 **Fax:** 01626 833166
Email: mktg@lorientuk.com
Web: www.lorientgroup.com
Product Type: 1

MGC LTD
McMillan House, 54-56 Cheam Common Road,
Worcester Park, Surrey, KT4 8RH
Area of Operation: UK & Ireland
Tel: 0208 337 0731 **Fax:** 0208 337 3739
Email: info@mgcltd.co.uk
Web: www.mgcltd.co.uk

MINELCO SPECIALITIES LIMITED
Raynesway, Derby, Derbyshire, DE21 7BE
Area of Operation: Worldwide
Tel: 01332 545224 **Fax:** 01332 677590
Email: minelco.specialities@minelco.com
Web: www.minelco.com
Product Type: 2, 3

THE MINELCO GROUP

Area of Operation: Product Featured UK Only
Tel: 01332 545224 **Fax:** 01332 677590
Email: minelco.specialities@minelco.com
Web: www.minelco.com
Product Type: 2, 3

Micafil, a vermiculite product, is a versatile material used in thermal and acoustic insulation, great for cavity walls & lofts, also for chimney linings and screeds.

NATURAL BUILDING TECHNOLOGIES
The Hangar, Worminghall Road,
Oakley, Buckinghamshire, HP18 9UL
Area of Operation: UK & Ireland
Tel: 01844 338338 **Fax:** 01844 338525
Email: info@natural-building.co.uk
Web: www.natural-building.co.uk
Other Info: ECO
Material Type: K) 3, 8, 9

KINGSPAN NILVENT®

Area of Operation: Worldwide
Tel: 0870 850 8555
Fax: 0870 850 8666
Email: info.uk@insulation.kingspan.com
Web: www.insulation.kingspan.com

Premium performance next generation breathable membrane – airtight, 100% waterproof, strong, durable and self-sealing.

KINGSPAN INSULATION QUICK GUIDE

Area of Operation: Worldwide
Tel: 0870 850 8555
Fax: 0870 850 8666
Email: info.uk@insulation.kingspan.com
Web: www.insulation.kingspan.com

Comprehensive information on premium and high performance insulation solutions for pitched and flat roofs, walls and floors.

NOISE STOP SYSTEMS
Unit 3 Acaster Industrial Estate, Cowper Lane,
Acaster Malbis, York, North Yorkshire, YO23 2XB
Area of Operation: UK & Ireland
Tel: 0845 130 6269 **Fax:** 01904 709857
Email: info@noisestopsystems.co.uk
Web: www.noisestopsystems.co.uk ⌁
Product Type: 1, 2, 3

QUAD-LOCK (ENGLAND) LTD
Unit B3.1, Maws Centre, Jackfield,
Telford, Shropshire, TF8 7LS
Area of Operation: UK (Excluding Ireland)
Tel: 0870 443 1901 **Fax:** 0870 443 1902
Email: k.turner@quadlock.co.uk
Web: www.quadlock.co.uk
Product Type: 2

SECOND NATURE UK LTD
Soulands Gate, Soulby, Dacre,
Penrith, Cumbria, CA11 0JF
Area of Operation: Worldwide
Tel: 01768 486285 **Fax:** 01768 486825
Email: info@secondnatureuk.com
Web: www.secondnatureuk.com
Product Type: 2, 3

SIP HOME LTD
Unit 1-3 Hadston Industrial Estate,
South Broomhill, Northumberland, NE65 9YG
Area of Operation: UK (Excluding Ireland)
Tel: 01670 760300 **Fax:** 01670 760740
Email: john@siphome.co.uk
Web: www.siphome.co.uk

SKANDA (UK) LTD
64 - 65 Clywedog Road North,
Wrexham Industrial Estate, Wrexham, LL13 9XN
Area of Operation: UK & Ireland
Tel: 01978 664255 **Fax:** 01978 661427
Email: info@skanda-uk.com
Web: www.skanda-uk.com / www.heraklith.co.uk

SOUND REDUCTION SYSTEMS LTD
Adam Street, Bolton, Lancashire, BL3 2AP
Area of Operation: UK & Ireland
Tel: 01204 380074 **Fax:** 01204 380957
Email: info@soundreduction.co.uk
Web: www.soundreduction.co.uk
Product Type: 2, 3

STRUCTURAL INSULATED PANEL TECHNOLOGY LTD (SIPTEC)
PO Box 429, Turvey, Bedfordshire, MK43 8DT
Area of Operation: Europe
Tel: 0870 743 9866 **Fax:** 0870 762 5612
Email: mail@sips.ws
Web: www.sips.ws
Product Type: 2, 3
Material Type: H) 4, 6

BUILDING STRUCTURE & MATERIALS - **Insulation & Proofing Materials** - Acoustic Insulation; Fire Proofing; Damp Proofing

SPONSORED BY: KINGSPAN INSULATION LTD www.insulation.kingspan.com

BUILDING STRUCTURE & MATERIALS

THE EXPANDED METAL COMPANY LIMITED
PO Box 14, Longhill Industrial Estate (North),
Hartlepool, Durham, TS25 1PR
Area of Operation: Worldwide
Tel: 01429 408087
Fax: 01429 866795
Email: paulb@expamet.co.uk
Web: www.expandedmetalcompany.co.uk

WEBSTERS INSULATION LTD
Crow Tree Farm, Crow Tree Bank, Thorne Levels,
Doncaster, South Yorkshire, DN8 5TF
Area of Operation: UK (Excluding Ireland)
Tel: 01405 812682
Fax: 01405 817201
Email: info@webstersinsulation.com
Web: www.webstersinsulation.com
Product Type: 2, 3

WINTUN
Wintun Works, Millerston, Paisley,
Renfrewshire, PA1 2XR
Area of Operation: UK & Ireland
Tel: 0141 889 5969
Fax: 0141 887 8907
Email: mail@wintun.co.uk
Web: www.wintun.co.uk
Product Type: 1

XETAL CONSULTANTS LIMITED
Unit 28 Cryant Business Park, Crynant,
Neath, Swansea, SA10 8PA
Area of Operation: Worldwide
Tel: 01639 751056
Fax: 01639 751058
Email: jdickson@xetal.co.uk
Web: www.xetal.co.uk
Product Type: 2, 3
Material Type: K) 2, 3, 4, 7, 8, 9

FIRE PROOFING

KEY

PRODUCT TYPES: 1= Barriers and Stops
2 = Boards and Sheets   3 = Coatings
4 = Membranes   5 = Sealants

SEE ALSO: SMART HOMES AND SECURITY -
Fire Protection

OTHER: ▽ Reclaimed   On-line shopping
Bespoke   Hand-made   ECO Ecological

CAVITY TRAYS LTD
New Administration Centre, Boundary Avenue,
Yeovil, Somerset, BA22 8HU
Area of Operation: Worldwide
Tel: 01935 474769
Fax: 01935 428223
Email: enquiries@cavitytrays.co.uk
Web: www.cavitytrays.com
Product Type: 1

DACATIE BUILDING SOLUTIONS
Quantum House, Salmon Fields, Royton,
Oldham, Lancashire, OL2 6JG
Area of Operation: UK & Ireland
Tel: 0161 627 4222
Fax: 0161 627 4333
Email: info@dacatie.co.uk
Web: www.dacatie.co.uk
Product Type: 1

ELLIOTTS INSULATION AND DRYLINING
Unit 8 Goodwood Road, Boyatt Wood Industrial
Estate, Eastleigh, Hampshire, SO50 4NT
Area of Operation: South East England,
South West England and South Wales
Tel: 02380 623960
Fax: 02380 623965
Email: insulation@elliott-brothers.co.uk
Web: www.elliotts.uk.com
Product Type: 1, 2, 5

SUSTAINABILITY APPRAISAL
Area of Operation: Worldwide
Tel: 0870 850 8555
Fax: 0870 850 8666
Email: info.uk@insulation.kingspan.com
Web: www.insulation.kingspan.com

The first study of its kind within the UK's construction product industry looking at every aspect of Kingspan Insulation's Pembridge production facility in Herefordshire. The detailed findings have been published in full in a report called Sustainability Appraisal.

H+H CELCON LIMITED
Celcon House, Ightham, Sevenoaks, Kent, TN15 9HZ
Area of Operation: UK (Excluding Ireland)
Tel: 01732 880520 **Fax:** 01732 880531
Email: marketing@celcon.co.uk
Web: www.celcon.co.uk
Product Type: 1

KNAUF INSULATION
PO Box 10, Stafford Road, St Helens,
Merseyside, WA10 3NS
Area of Operation: UK & Ireland
Tel: 01270 824024 (Enquiries Hotline)
Fax: 01270 824025
Email: info@knaufinsulation.com
Web: www.knaufinsulation.co.uk
Product Type: 1, 2

LORIENT POLYPRODUCTS LTD
Fairfax Road, Heathfield Industrial Estate,
Newton Abbot, Devon, TQ12 6UD
Area of Operation: Worldwide
Tel: 01626 834252 **Fax:** 01626 833166
Email: mktg@lorientuk.com
Web: www.lorientgroup.com
Product Type: 1, 5

ROOFSURE
Metro House, 14-17 Metropolitan Business Park,
Preston New Road, Blackpool, Lancashire, FY3 9LT
Area of Operation: UK (Excluding Ireland)
Tel: 0800 597 2828 **Fax:** 01253 798193 / 764671
Email: roofsureltd@btconnect.com
Web: www.roofsure.co.uk

WINTUN
Wintun Works, Millerston, Paisley,
Renfrewshire, PA1 2XR
Area of Operation: UK & Ireland
Tel: 0141 889 5969 **Fax:** 0141 887 8907
Email: mail@wintun.co.uk
Web: www.wintun.co.uk
Product Type: 5

DAMP PROOFING

KEY

PRODUCT TYPES: 1= Cavity Closers
2 = Boards and Sheets   3 = Backing Boards
4 = Other

SEE ALSO: SUBSTRUCTURE AND
FOUNDATIONS - Basements

OTHER: ▽ Reclaimed   On-line shopping
Bespoke   Hand-made   ECO Ecological

A PROCTOR GROUP LTD
The Haugh, Blairgowrie,
Perth & Kinross, PH10 7ER
Area of Operation: Worldwide
Tel: 01250 872261
Fax: 01250 872727
Email: karrina.andrews@proctorgroup.com
Web: www.proctorgroup.com
Product Type: 2, 3

ABTECH (UK) LTD
Sheiling House, Invincible Road,
Farnborough, Hampshire, GU14 7QU
Area of Operation: UK (Excluding Ireland)
Tel: 0800 085 1431
Fax: 01252 378665
Email: sales@abtechbasements.co.uk
Web: www.abtechbasements.co.uk

AQUAPANEL
PO Box 732, Maidstone, Kent, ME15 6ST
Area of Operation: UK (Excluding Ireland)
Tel: 0800 169 6545
Fax: 0845 6011762
Email: knaufdiy@knauf.co.uk
Web: www.aquapanel.co.uk
Product Type: 3

AQUATECNIC
211 Heathhall Industrial Estate, Dumfries,
Dumfries & Galloway, DG1 3PH
Area of Operation: UK & Ireland
Tel: 0845 226 8283
Fax: 0845 226 8293
Email: info@aquatecnic.net
Web: www.aquatecnic.net
Product Type: 2, 4

ARDEX UK LIMITED
Homefield Road, Haverhill,
Suffolk, CB9 8QP
Area of Operation: UK (Excluding Ireland)
Tel: 01440 714939
Fax: 01440 716660
Email: info@ardex.co.uk
Web: www.ardex.co.uk
Product Type: 4

CAVITY TRAYS LTD
New Administration Centre,
Boundary Avenue,
Yeovil, Somerset, BA22 8HU
Area of Operation: Worldwide
Tel: 01935 474769
Fax: 01935 428223
Email: enquiries@cavitytrays.co.uk
Web: www.cavitytrays.com
Product Type: 1

DACATIE BUILDING SOLUTIONS
Quantum House, Salmon Fields,
Royton, Oldham, Lancashire, OL2 6JG
Area of Operation: UK & Ireland
Tel: 0161 627 4222
Fax: 0161 627 4333
Email: info@dacatie.co.uk
Web: www.dacatie.co.uk
Product Type: 1

DELTA MEMBRANE SYSTEMS LTD
Bassett Business Centre,
Hurricane Way, North Weald,
Epping, Essex, CM16 6AA
Area of Operation: UK & Ireland
Tel: 0870 747 2181
Fax: 01992 524046
Email: info@deltamembranes.com
Web: www.deltamembranes.com
Product Type: 2, 3

DIY ROOFING LTD
Hillcrest House, Featherbed Lane,
Hunt End, Redditch, Worcestershire, B97 5QL
Area of Operation: UK & Ireland
Tel: 0870 783 4890
Fax: 01527 403483
Email: chris@diyroofing.co.uk
Web: www.diyroofing.co.uk
Product Type: 2

INSULATION FOR SUSTAINABILITY

Area of Operation: Worldwide
Tel: 0870 850 8555
Fax: 0870 850 8666
Email: info.uk@insulation.kingspan.com
Web: www.insulation.kingspan.com

Ground breaking widely respected report showing how CO2 emissions can be reduced by designing well insulated energy efficient buildings.

FOAMSEAL LTD
New Street House, New Street,
Petworth, West Sussex, GU28 0AS
Area of Operation: Worldwide
Tel: 01798 345400
Fax: 01798 345410
Email: info@foamseal.co.uk
Web: www.foamseal.co.uk

HIGHTEX-COATINGS LTD
Unit 14 Chapel Farm, Hanslope Road,
Hartwell, Northamptonshire, NN7 2EU
Area of Operation: UK (Excluding Ireland)
Tel: 01604 861250 **Fax:** 01604 871116
Email: bob@hightexcoatings.co.uk
Web: www.hightexcoatings.co.uk
Product Type: 4

INDOOR CLIMATE SOLUTIONS
10 Cryersoak Close, Monkspath,
Solihull, West Midlands, B90 4UW
Area of Operation: Midlands & Mid Wales
Tel: 0870 011 5432 **Fax:** 0870 011 5432
Email: info@ics-aircon.com
Web: www.ics-aircon.com
Product Type: 4

KILTOX CONTRACTS LIMITED
Unit 6 Chiltonian Industrial Estate,
203 Manor Lane, Lee, London, SE12 0TX
Area of Operation: Worldwide
Tel: 0845 166 2040 **Fax:** 0845 166 2050
Email: info@kiltox.co.uk
Web: www.kiltox.co.uk

MGC LTD
McMillan House, 54-56 Cheam Common Road,
Worcester Park, Surrey, KT4 8RH
Area of Operation: Worldwide
Tel: 0208 337 0731 **Fax:** 0208 337 3739
Email: info@mgcltd.co.uk
Web: www.mgcltd.co.uk

MUNTERS LTD
Blackstone Road, Huntingdon,
Cambridgeshire, PE29 6EE
Area of Operation: UK & Ireland
Tel: 01480 442327 **Fax:** 01480 458333
Email: denice.smith@munters.co.uk
Web: www.munters.co.uk
Product Type: 4

NEVER PAINT AGAIN INTERNATIONAL
Claro Court Business Centre, Claro Road,
Harrogate, West Yorkshire, HG1 4BA
Area of Operation: Worldwide
Tel: 0870 067 0643 **Fax:** 0870 067 0643
Email: info@neverpaintagain.co.uk
Web: www.neverpaintagain.co.uk
Product Type: 4

Rely on Aquapanel for tiling success

aquapanel

THE UK'S NUMBER ONE TILE BACKER RANGE

- **Ideal tile backing board for wet & humid areas (including showers)**
- **Will not rot or warp unlike conventional backing boards**
- **Easy to cut and fix**
- **Will support 50kg/m² of tiles**

Available from B&Q Warehouse, Screwfix, Wickes, quality builders' merchants and tile outlets.

The tile backing boards that are not afraid of water.

aquapanel thermal

- **Moisture resistant floor tile backer with 'Polyfoam' thermal insulation**
- **Ideal for use with undertile heating systems**
- **Easy to install**

For further information call Aquapanel now on Freephone 0800 1696545

KNAUFDIY

www.aquapanel.com

To help make every Aquapanel project a success,
use Knauf DIY fixings and accessories

THE TILE ASSOCIATION
serving the tile industry and its customers

Ceramic coated screws

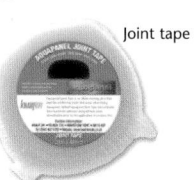
Joint tape

Thermal fixings

PHOENIX (GB) LTD
Chestnut Field House, Chestnut Field,
Rugby, Warwickshire, CU21 2PA
Area of Operation: UK & Ireland
Tel: 01788 571482 **Fax:** 01788 542245
Email: tonyb@phoenix-gb.com
Web: www.phoenix-ag.com
Product Type: 4

KINGSPAN THERMAFLOOR TF73

Area of Operation: Worldwide
Tel: 0870 850 8555
Fax: 0870 850 8666
Email: info.uk@insulation.kingspan.com
Web: www.insulation.kingspan.com

Insulated chipboard solution for suspended
and solid floors. Literature contains design
considerations, technical data, sitework details,
thermal performance and U-value tables.

RUBEROID BUILDING PRODUCTS
Appley Lane North, Appley Bridge, Lancs, WN6 9AB
Area of Operation: UK (Excluding Ireland)
Tel: 08000 285573 **Fax:** 01257 252514
Email: marketing@ruberoid.co.uk
Web: www.ruberoid.co.uk
Product Type: 4

SAFEGUARD EUROPE LTD
Redkiln Close, Horsham, West Sussex, RH13 5QL
Area of Operation: Worldwide
Tel: 01403 210204 **Fax:** 01403 217529
Email: info@safeguardeurope.com
Web: www.safeguardeurope.com
Product Type: 4

SAFEGUARD

Area of Operation: Worldwide
Tel: 01403 210204
Fax: 01403 217529
Email: info@safeguardchem.com
Web: www.safeguardchem.com
Product Type: 4

Safeguard supply a wide range of damp-proofing
products including the easy-to-use Dryzone
damp-proofing cream (pictured).

SPRY PRODUCTS LTD
64 Nottingham Road, Long Eaton,
Nottingham, Nottinghamshire, NG10 2AU
Area of Operation: UK & Ireland
Tel: 0115 9732914 **Fax:** 0115 9725172
Email: jerspry@aol.com
Web: www.spryproducts.com
Product Type: 2, 4 **Other Info:** ECO

**TRITON CHEMICAL
MANUFACTURING LIMITED**
129 Felixstowe Road,
Abbey Wood, London, SE2 9SG
Area of Operation: Worldwide
Tel: 0208 310 3929
Fax: 0208 312 0349
Email: neil@triton-chemicals.com
Web: www.triton-chemicals.com
Product Type: 2, 4

VANDEX UK LTD
Redkiln Close, Redkiln Way,
Horsham, Sussex,
West Sussex, RH13 5QL
Area of Operation: UK & Ireland
Tel: 0870 241 6264
Fax: 01403 217529
Email: info@vandex.co.uk
Web: www.vandex.co.uk

VISQUEEN BUILDING PRODUCTS
Maerdy Industrial Estate,
Rhymney, Blaenau,
Gwent, NP22 5PY
Area of Operation: UK & Ireland
Tel: 01685 840672
Fax: 01685 842580
Email: enquiries@visqueenbuilding.co.uk
Web: www.visqueenbuilding.co.uk
Product Type: 1, 2, 4
Other Info: ▽ ECO
Material Type: D) 1

WIGOPOL UK
Old Icknield Way, Benson,
Oxfordshire, OX10 6PW
Area of Operation: UK (Excluding Ireland)
Tel: 01491 836847
Fax: 01491 826020
Email: enquiries@wigopol.co.uk
Web: www.wigopol.co.uk
Product Type: 4

BUILDING REGULATIONS / BUILDING STANDARDS

Area of Operation: Worldwide
Tel: 0870 850 8555
Fax: 0870 850 8666
Email: info.uk@insulation.kingspan.com
Web: www.insulation.kingspan.com

Literature detailing insulation solutions for walls, roofs and
floors in accordance with the new Building Regulations:
Approved Documents L1A/B and L2A/B in England and
Wales and the Technical Standards Part J in Scotland.

DRAINAGE & RAINWATER GOODS

Image courtesy of Tuscan Foundry Products (01403 860040)

Drainage

Aiming for as few downpipes on your home as possible, located where they will have minimal visual impact, is a sure way to guarantee your rainwater goods don't dominate the main elevations. With careful planning and the right product, however, they can become design features in their own right, playing an important part in defining the building's character.

TOP: Copper guttering with green patina by Good Directions.
ABOVE: Guttering and downpipe by Lindab.
RIGHT: Copper guttering and downpipe by Klober.

Calculations to determine the number of downpipes you will need should be based on the area of your roof — the bigger the roof area, the more downpipes you require. To minimise the number of downpipes, consider installing deeper guttering sections which hold more water and thus require less outlets. If in any doubt consult the manufacturers, many of whom will work out your requirements and produce a design scheme and layout for your project as part of a quote.

PVCu accounts for around 90% of new rainwater systems in the UK. It is a low cost option, virtually maintenance free and self-coloured so requires no finishing. It is also easy to cut and can be joined using simple rubber gaskets, something that most DIYers should be able to carry out.

On the downside, PVCu has a relatively short lifespan, estimated at around 10 years. It does not cope well with sunlight, although high gloss PVCu is available to reflect the sunlight away and minimise damage.

Cast iron generally works out at four times the price of PVCu per linear metre, but is the material traditionally used for rainwater goods; therefore it is

still widely specified on high end traditional style properties. It is durable and long lasting — with correct maintenance cast iron should last as long, if not longer, than the house itself.

Cast aluminium is lighter and cheaper than cast iron and available in both traditional and modern designs. It is sometimes accepted as an alternative to cast iron on listed buildings or in Conservation Areas.

If you are fixing rainwater goods to unstable masonry, glass reinforced plastic (GRP) may be the best option, being lighter and easier to fix than cast iron but with a longer lifespan then PVCu. GRP comes in many finishes including cast iron and lead effects, but it is an expensive option.

Copper develops a lovely green protective patina over time, or alternatively many specialist companies can carry out an artificial ageing process to give the patina within days.

You could also consider glass, or cheaper clear polycarbonate, downpipes — a great option for contemporary style homes.

For something different, try a rain chain or raincup system descending from the guttering instead of a downpipe — the water will run either straight down the chain, or will flow from cup to cup, into a soakaway below. This looks stylish and contemporary with metal systems, especially copper, and gives a modern, subtle alternative.

Tips

- Decide on your rainwater system as early as possible - preferably in the design stages. The system you choose will help determine below ground drainage as well as the width of the tile overhang required at the eaves.

- Make sure your chosen system complements your design - it is surprising how much an unsuitable system will stand out.

- Make sure you have sufficient down pipes.

- Check your ground porosity is suitable for soakaways - building control will need to know.

PIPES AND PUMPS

KEY

PRODUCT TYPES: 1= Below Ground Pipes
2 = Pumps   3 = Other

SEE ALSO: MERCHANTS, TOOLS AND
EQUIPMENT - Plumbers Merchants, PLUMBING
- Plumbing, Pipes and Fittings

OTHER: ▽ Reclaimed ⏁ On-line shopping
✐ Bespoke ✋ Hand-made ECO Ecological

ALLERTON DRAINAGE
Woodbridge Road, Sleaford, Lincolnshire, NG34 7EW
Area of Operation: UK (Excluding Ireland)
Tel: 01529 305757 **Fax:** 01529 414232
Email: sales@allertonuk.com
Web: www.allertonuk.com ⏁
Product Type: 2

ALTON PUMPS
Redwood Lane, Medstead,
Alton, Hampshire, GU34 5PE
Area of Operation: Worldwide
Tel: 01420 561661 **Fax:** 01420 561661
Email: sales@altonpumps.com
Web: www.altonpumps.com
Product Type: 2

CONDER PRODUCTS LTD
2 Whitehouse Way, South West Industrial Estate,
Peterlee, Durham, SR8 2HZ
Area of Operation: Worldwide
Tel: 0870 264 0004 **Fax:** 0870 264 0005
Email: sales@conderproducts.com
Web: www.conderproducts.com
Product Type: 2

DRAINSTORE.COM
Units 1 & 2, Heanor Gate Road, Heanor Gate
Industrial Estate, Heanor, Derbyshire, DE75 7RJ
Area of Operation: Europe
Tel: 01773 767611 **Fax:** 01773 767613
Email: adrian@drainstore.com
Web: www.drainstore.com ⏁
Product Type: 1, 2 **Material Type:** D) 1

E. RICHARDS
PO Box 1115, Winscombe, Somerset, BS25 1WA
Area of Operation: UK & Ireland
Tel: 0845 330 8859 **Fax:** 0845 330 7260
Email: paultrace@e-richards.co.uk
Web: www.e-richards.co.uk ⏁
Product Type: 1

EDINCARE
Unit 10 Avebury Court, Mark Road, Hemel
Hempstead, Hertfordshire, HP2 7TA
Area of Operation: UK (Excluding Ireland)
Tel: 01442 211554 **Fax:** 01442 211553
Email: info@edincare.com
Web: www.edincare.com
Product Type: 2

EUROPEAN PIPE SUPPLIERS LTD
Unit 6 Severnlink Distribution Centre, off Newhouse
Farm Estate, Chepstow, Monmouthshire, NP16 6UN
Area of Operation: Europe
Tel: 01291 622215 **Fax:** 01291 620381
Email: sales@epsonline.co.uk
Web: www.epsonline.co.uk
Product Type: 2 **Material Type:** C) 5

GEBERIT LTD
New Hythe Business Park,
New Hythe Lane, Aylesford,
Kent, ME20 7PJ
Area of Operation: UK & Ireland
Tel: 01622 717811 **Fax:** 01622 716 920
Web: www.geberit.co.uk
Product Type: 1, 3

GREEN BUILDING STORE
11 Huddersfield Road, Meltham,
Holmfirth, West Yorkshire, HD9 4NJ
Area of Operation: UK & Ireland
Tel: 01484 854898 **Fax:** 01484 854899
Email: info@greenbuildingstore.co.uk
Web: www.greenbuildingstore.co.uk ⏁
Product Type: 3
Other Info: ECO
Material Type: C) 4, 7

GRUNDFOS PUMPS LTD
Grovebury Road, Leighton Buzzard,
Bedfordshire, LU7 4TL
Area of Operation: Worldwide
Tel: 01525 850000 **Fax:** 01525 850011
Email: uk_sales@grundfos.com
Web: www.grundfos.com
Product Type: 2, 3

GRUNDFOS PUMPS LTD
Area of Operation: Worldwide
Tel: 01525 850000
Fax: 01525 850011
Email: uk_sales@grundfos.com
Web: www.grundfos.com
Product Type: 2, 3

Grundfos Pump plan – 2 heads are better than 1.
The Pump plan provides independent pumping for
heating and hot water in a compact, integrated unit.
Pump plan is compatible with all types of domestic
controls and modern boilers (which incorporate high
efficiency heat exchangers) and includes a unique
wiring centre.

GRUNDFOS PUMPS LTD
Area of Operation: Worldwide
Tel: 01525 850000
Fax: 01525 850011
Email: uk_sales@grundfos.com
Web: www.grundfos.com
Product Type: 2, 3

The new Grundfos Alpha+ - One pump fits all.
The latest innovation to emerge from the Grundfos
pumps design team is the Alpha+, the domestic
circulator with added value. The new Grundfos Alpha+
features both automatic and fixed speed operation,
which means that the Alpha+ is suitable for all domestic
heating systems.

GRUNDFOS
Area of Operation: Worldwide
Tel: 01525 775402 **Fax:** 01525 775236
Email: uk_sales@grundfos.com
Web: www.grundfos.co.uk
Product Type: 2, 3

The MQ is a complete, whole house domestic booster
pump suitable for pumping potable water, rain water or
other clean, thin, non-aggressive liquids containing no
solid particles or fibres. This makes it Ideal for use in
private homes, summer houses and weekend cottages.

GRUNDFOS
Area of Operation: Worldwide
Tel: 01525 775402 **Fax:** 01525 775236
Email: uk_sales@grundfos.com
Web: www.grundfos.co.uk
Product Type: 2, 3

The Grundfos Unilift KP/AP is a portable drainage
pump designed for the removal of wastewater from
washing machines, swimming pools and ponds
among many other general transfer duties. These
Grundfos Pumps are ideal for use in flooded areas,
for example: cellars.

HARGREAVES FOUNDRY LTD
Water Lane, South Parade, Halifax,
West Yorkshire, HX3 9HG
Area of Operation: UK & Ireland
Tel: 01422 330607
Fax: 01422 320349
Email: info@hargreavesfoundry.co.uk
Web: www.hargreavesfoundry.co.uk
Product Type: 1

HUNTER PLASTICS
Nathan Way, London, SE28 0AE
Area of Operation: Worldwide
Tel: 0208 855 9851
Fax: 0208 317 7764
Email: john.morris@hunterplastics.co.uk
Web: www.hunterplastics.co.uk
Product Type: 1

HUTCHINSON DRAINAGE LTD
White Wall Nook, Wark, Hexham,
Northumberland, NE48 3PX
Area of Operation: UK (Excluding Ireland)
Tel: 01434 220508
Fax: 01434 220745
Email: enquiries@hutchinson-drainage.co.uk
Web: www.hutchinson-drainage.co.uk
Product Type: 2

INTERFLOW UK
Leighton, Shrewsbury, Shropshire, SY5 6SQ
Area of Operation: UK & Ireland
Tel: 01952 510050
Fax: 01952 510967
Email: villiers@interflow.co.uk
Web: www.interflow.co.uk
Product Type: 1

KLARGESTER ENVIRONMENTAL
College Road North, Aston Clinton,
Aylesbury, Buckinghamshire, HP22 5EW
Area of Operation: Worldwide
Tel: 01296 633000
Fax: 01296 633001
Email: uksales@klargester.co.uk
Web: www.klargester.com
Product Type: 2

MARLEY PLUMBING & DRAINAGE
Lenham, Maidstone, Kent, ME17 2DE
Area of Operation: Europe
Tel: 01622 858888 **Fax:** 01622 858725
Email: marketing@marleyext.com
Web: www.marley.co.uk
Product Type: 1, 3
Material Type: D) 1

PUMPMASTER UK LIMITED
Manor House Offices, Malvern Road,
Lower Wick, Worcester,
Worcestershire, WR2 4BS
Area of Operation: UK (Excluding Ireland)
Tel: 01905 420170 **Fax:** 01905 749419
Email: pumpsales@pumpmaster.co.uk
Web: www.pumpmaster.co.uk
Product Type: 2

RAINHARVESTING SYSTEMS LTD
Units 23 & 24, Merretts Mill Industrial Centre,
Bath Road, Woodchester, Stroud,
Gloucestershire, GL5 5EX
Area of Operation: UK & Ireland
Tel: 01452 772000 **Fax:** 01452 770115
Email: sales@rainharvesting.co.uk
Web: www.rainharvesting.co.uk
Product Type: 2, 3

ROCKBOURNE ENVIRONMENTAL
6 Silver Business Park, Airfield Way,
Christchurch, Dorset, BH23 3TA
Area of Operation: UK & Ireland
Tel: 01202 480980 **Fax:** 01202 490590
Email: info@rockbourne.net
Web: www.rockbourne.net
Product Type: 2

SAINT-GOBAIN PIPELINES
Lows Lane, Stanton-by-Dale,
Ilkeston, Derbyshire, DE7 4QU
Area of Operation: UK & Ireland
Tel: 0115 930 5000 **Fax:** 0115 932 9513
Email: mike.rawlings@saint-gobain.com
Web: www.saint-gobain-pipelines.co.uk
Product Type: 1
Material Type: C) 5

SANIFLO LTD
Howard House, The Runway,
South Ruislip, Middlesex, HA4 6SE
Area of Operation: UK (Excluding Ireland)
Tel: 0208 842 4040 **Fax:** 0208 842 1671
Email: andrews@saniflo.co.uk
Web: www.saniflo.co.uk
Product Type: 2

UPONOR HOUSING SOLUTIONS LTD
Snapethorpe House, Rugby Road,
Lutterworth, Leicestershire, LE17 4HN
Area of Operation: UK (Excluding Ireland)
Tel: 01455 550355
Fax: 01455 550366
Email: hsenquiries@uponor.co.uk
Web: www.uponorhousingsolutions.co.uk
Product Type: 1

GROUNDWATER MANAGEMENT

KEY

PRODUCT TYPES: 1= Culverts 2 = Gullies
and Gratings 3 = Land Drains 4 = Soakaways
5 = Surface Water Drainage

OTHER: ▽ Reclaimed ⌒ On-line shopping
✎ Bespoke ✋ Hand-made ECO Ecological

DRAINSTORE.COM
Units 1 & 2, Heanor Gate Road, Heanor Gate
Industrial Estate, Heanor, Derbyshire, DE75 7RJ
Area of Operation: Europe
Tel: 01773 767611
Fax: 01773 767613
Email: adrian@drainstore.com
Web: www.drainstore.com ⌒
Product Type: 4
Material Type: D) 1

GRAMM ENVIRONMENTAL LIMITED
17-19 Hight Street, Ditchling, Hassocks,
West Sussex, BN6 8SY
Area of Operation: Worldwide
Tel: 01892 506935
Fax: 01892 506926
Email: sales@grammenvironmental.com
Web: www.grammenvironmental.com
Product Type: 4, 5

HIBERNIA RAINHARVESTING
Unit 530 Stanstead Distribution Centre, Start Hill,
Bishops Stortford, Hertfordshire, CM22 7DG
Area of Operation: UK & Ireland
Tel: 02890 249954
Fax: 02890 249964
Email: rain@hiberniaeth.com
Web: www.hiberniaeth.com
Product Type: 5

KLARGESTER ENVIRONMENTAL
College Road North, Aston Clinton, Aylesbury,
Buckinghamshire, HP22 5EW
Area of Operation: Worldwide
Tel: 01296 633000
Fax: 01296 633001
Email: sales@klargester.co.uk
Web: www.klargester.com
Product Type: 5

MAYFIELD (MANUFACTURING) LTD
Wenden House, New End, Hemingby,
Horncastle, Lincolnshire, LN9 5QQ
Area of Operation: Worldwide
Tel: 01507 578630 **Fax:** 01507 578609
Email: info@mayfieldmanufacturing.co.uk
Web: www.aludrain.co.uk
Product Type: 2

ROCKBOURNE ENVIRONMENTAL
6 Silver Business Park, Airfield Way,
Christchurch, Dorset, BH23 3TA
Area of Operation: UK & Ireland
Tel: 01202 480980 **Fax:** 01202 490590
Email: info@rockbourne.net
Web: www.rockbourne.net
Product Type: 1, 3, 4, 5

SPRY PRODUCTS LTD
64 Nottingham Road, Long Eaton,
Nottingham, Nottinghamshire, NG10 2AU
Area of Operation: UK & Ireland
Tel: 0115 9732914 **Fax:** 0115 9725172
Email: jerspry@aol.com
Web: www.spryproducts.com
Product Type: 3, 5 **Other Info:** ✋
Material Type: D) 1

TERRAIN AERATION SERVICES
Aeration House, 20 Mill Fields, Haughley,
Stowmarket, Suffolk, IP14 3PU
Area of Operation: Europe
Tel: 01449 673783 **Fax:** 01449 614564
Email: terrainaeration@aol.com
Web: www.terrainaeration.co.uk
Product Type: 3

WATLING HOPE (INSTALLATIONS) LTD
1 Goldicote Business Park, Banbury Road,
Stratford Upon Avon, Warwickshire, CV37 7NB
Area of Operation: UK & Ireland
Tel: 01789 740757 **Fax:** 01789 740 404
Email: enquiries@watling-hope.co.uk
Web: www.watling-hope.co.uk
Product Type: 3, 4, 5

RAINWATER MANAGEMENT

KEY

PRODUCT TYPES: 1= Rainwater Goods
2 = Rainwater Storage 3 = Water Recycling

OTHER: ▽ Reclaimed ⌒ On-line shopping
✎ Bespoke ✋ Hand-made ECO Ecological

ACO TECHNOLOGIES PLC
ACO Business Park, Hitchin Road,
Shefford, Bedfordshire, SG17 5TE
Area of Operation: UK & Ireland
Tel: 01462 816666 **Fax:** 01462 851490
Email: marketing@aco.co.uk
Web: www.aco.co.uk
Product Type: 1 **Material Type:** D) 1

ACORN ENVIRONMENTAL SYSTEMS LIMITED
Somerset Bridge, Bridgwater, Somerset, TA6 6LL
Area of Operation: Europe
Tel: 01278 439325
Fax: 01278 439324
Email: info@acornsystems.com
Web: www.acornsystems.com
Product Type: 3

ALUMASC EXTERIOR BUILDING PRODUCTS
White House Works, Bold Road, Sutton, St Helens,
Merseyside, WA9 4JG
Area of Operation: UK & Ireland
Tel: 01744 648400
Fax: 01744 648401
Email: info@alumasc-exteriors.co.uk
Web: www.alumasc-exteriors.co.uk
Product Type: 1

ALUMINIUM ROOFLINE PRODUCTS
Unit 2 Vitruvius Way, Meridian Business Park,
Braunstone, Leicester, Leicestershire, LE19 1WA
Area of Operation: UK (Excluding Ireland)
Tel: 0116 289 4400
Email: jim.muddimer@arp-ltd.com
Web: www.arp-ltd.com
Product Type: 1

ARP LTD
Unit 2 Vitruvius Way, Meridian Business Park,
Braunstone Park, Leicestershire, LE3 2WH
Area of Operation: UK & Ireland
Tel: 0116 289 4400
Fax: 0116 289 4433
Email: sales@arp-ltd.com
Web: www.arp-ltd.com
Product Type: 1

BALMORAL TANKS
Balmoral Park, Loirston, Aberdeen,
Aberdeenshire, AB12 3GY
Area of Operation: Worldwide
Tel: 01224 859100
Fax: 01224 859123
Email: s.gibb@balmoral.co.uk
Web: www.balmoraltanks.com
Product Type: 2

CAVALOK BUILDING PRODUCTS
Green Lane, Newtown, Tewkesbury,
Gloucestershire, GL20 8HD
Area of Operation: UK & Ireland
Tel: 01684 855706
Fax: 01684 299124
Email: info@cavalok.co.uk
Web: www.cavalok.com
Product Type: 1

COBURG GUTTER GRID
Little Gunnerby, Hatcliffe, Grimsby,
Lincolnshire, DN37 0SP
Area of Operation: UK & Ireland
Tel: 01472 371406
Fax: 01469 560435
Email: sue@guttergrid.com
Web: www.guttergrid.com
Product Type: 1

CONDER PRODUCTS LTD
2 Whitehouse Way,
South West Industrial Estate,
Peterlee, Durham, SR8 2HZ
Area of Operation: Worldwide
Tel: 0870 264 0004
Fax: 0870 264 0005
Email: sales@conderproducts.com
Web: www.conderproducts.com
Product Type: 2, 3

COPPAGUTTA
8 Bottings Industrial Estate,
Hilltons Road, Botley, Southampton,
Hampshire, SO30 2DY
Area of Operation: UK & Ireland
Tel: 01489 797774
Fax: 01489 797774
Email: office@good-directions.com
Web: www.coppagutta.com
Product Type: 1
Material Type: C) 7

CRESS WATER LTD
61 Woodstock Road, Worcester,
Worcestershire, WR2 5ND
Area of Operation: Europe
Tel: 01905 422707
Fax: 01905 422744
Email: info@cresswater.co.uk
Web: www.cresswater.co.uk
Product Type: 3
Other Info: ECO

DIY ROOFING LTD
Hillcrest House, Featherbed Lane,
Hunt End, Redditch,
Worcestershire, B97 5QL
Area of Operation: UK & Ireland
Tel: 0800 783 4890
Fax: 01527 403483
Email: chris@diyroofing.co.uk
Web: www.diyroofing.co.uk
Product Type: 1

DRAINSMART
5 Hursley Road,
Chandlers Ford, Eastleigh,
Hampshire, SO53 2FW
Area of Operation: Worldwide
Tel: 02380 269091
Fax: 02380 269091
Email: info@drainsmart.co.uk
Web: www.drainsmart.co.uk.
Product Type: 1
Other Info: ✎

E. RICHARDS
PO Box 1115, Winscombe,
Somerset, BS25 1WA
Area of Operation: UK & Ireland
Tel: 0845 330 8859
Fax: 0845 330 7260
Email: paultrace@e-richards.co.uk
Web: www.e-richards.co.uk ✎
Product Type: 1

ECO HOUSES
12 Lee Street, Louth, Lincolnshire, LN11 9HJ
Area of Operation: Worldwide
Tel: 0845 345 2049
Email: sd@jones-nash.co.uk
Web: www.eco-houses.co.uk
Product Type: 2, 3

ECOSHOP
Unit 1, Glen of the Downs Garden Centre,
Kilmacanogue, Co Wicklow, Republic of Ireland
Area of Operation: Ireland Only
Tel: +353 01 287 2914
Fax: +353 01 201 6480
Email: info@ecoshop.ie
Web: www.ecoshop.ie
Product Type: 1, 2, 3

EUROPEAN PIPE SUPPLIERS LTD
Unit 6 Severnlink Distribution Centre, off Newhouse
Farm Estate, Chepstow, Monmouthshire, NP16 6UN
Area of Operation: Europe
Tel: 01291 622215 **Fax:** 01291 620381
Email: sales@epsonline.co.uk
Web: www.epsonline.co.uk
Product Type: 1
Material Type: C) 1, 2, 3, 4, 5, 6, 7

EXTERIOR IMAGE LTD
322 Hursley Road, Chandlers Ford,
Eastleigh, Hampshire, SO53 5PH
Tel: 02380 271777
Fax: 02380 269091
Email: info@exteriorimage.co.uk
Web: www.exteriorimage.co.uk
Product Type: 1

GAP
Partnership Way, Shadsworth Business Park,
Blackburn, Lancashire, BB1 2QP
Area of Operation: Midlands & Mid Wales, North
East England, North West England and North Wales
Tel: 01254 682888
Web: www.gap.uk.com
Product Type: 1
Material Type: D) 1

GOOD DIRECTIONS LTD
8 Bottings Industrial Estate, Hilltons Road,
Botley, Southampton, Hampshire, SO30 2DY
Area of Operation: UK & Ireland
Tel: 01489 797773 **Fax:** 01489 796700
Email: office@good-directions.co.uk
Web: www.good-directions.com
Product Type: 1
Other Info: ECO
Material Type: C) 7

GRAMM ENVIRONMENTAL LIMITED
17-19 Hight Street, Ditchling,
Hassocks, West Sussex, BN6 8SY
Area of Operation: Worldwide
Tel: 01892 506935 **Fax:** 01892 506926
Email: sales@grammenvironmental.com
Web: www.grammenvironmental.com
Product Type: 1, 2, 3

HARGREAVES FOUNDRY LTD
Water Lane, South Parade, Halifax,
West Yorkshire, HX3 9HG
Area of Operation: UK & Ireland
Tel: 01422 330607 **Fax:** 01422 320349
Email: info@hargreavesfoundry.co.uk
Web: www.hargreavesfoundry.co.uk
Product Type: 1

HARRISON THOMPSON & CO. LTD.
(YEOMAN RAINGUARD)
Yeoman House, Whitehall Estate,
Whitehall Road, Leeds,
West Yorkshire, LS12 5JB
Area of Operation: UK & Ireland
Tel: 0113 279 5854 **Fax:** 0113 231 0406
Email: info@rainguard.co.uk
Web: www.rainguard.co.uk
Product Type: 1

HIBERNIA RAINHARVESTING
Unit 530 Stanstead Distribution Centre,
Start Hill, Bishops Stortford,
Hertfordshire, CM22 7DG
Area of Operation: UK & Ireland
Tel: 02890 249954
Fax: 02890 249964
Email: rain@hiberniaeth.com
Web: www.hiberniaeth.com
Product Type: 1, 2, 3

HIGHWATER (SCOTLAND) LTD
Winewell, Grantown Road, Nairn,
Highlands, IV12 5QN
Area of Operation: Scotland
Tel: 01667 451009 **Fax:** 01667 451009
Email: highwater@btinternet.com
Web: www.highwatertechnologies.co.uk
Product Type: 2

HUNTER PLASTICS
Nathan Way, London, SE28 0AE
Area of Operation: Worldwide
Tel: 0208 855 9851 **Fax:** 0208 317 7764
Email: john.morris@hunterplastics.co.uk
Web: www.hunterplastics.co.uk
Product Type: 1

HYDRO INTERNATIONAL
Shearwater House, Clevedon Hall Estate,
Victoria Road, Clevedon, BS21 7RD
Area of Operation: UK & Ireland
Tel: 01275 878371 **Fax:** 01275 874979
Email: enquiries@hydro-international.co.uk
Web: www.hydro-international.biz
Product Type: 1, 2
Other Info: ✎

KLARGESTER ENVIRONMENTAL
College Road North, Aston Clinton,
Aylesbury, Buckinghamshire, HP22 5EW
Area of Operation: Worldwide
Tel: 01296 633000 **Fax:** 01296 633001
Email: uksales@klargester.co.uk
Web: www.klargester.com
Product Type: 2, 3

LINDAB LTD
Unit 7 Block 2, Shenstone Trading Estate,
Bromsgrove Road, Halesowen,
West Midlands, B63 3XB
Area of Operation: Worldwide
Tel: 0121 585 2780 **Fax:** 0121 585 2782
Email: building.products@lindab.co.uk
Web: www.lindab.com
Product Type: 1
Material Type: C) 4, 7, 17

MARLEY PLUMBING & DRAINAGE
Lenham, Maidstone, Kent, ME17 2DE
Area of Operation: Europe
Tel: 01622 858888 **Fax:** 01622 858725
Email: marketing@marleyext.com
Web: www.marley.co.uk
Product Type: 1
Material Type: C) 1

MAYFIELD (MANUFACTURING) LTD
Wenden House, New End, Hemingby,
Horncastle, Lincolnshire, LN9 5QQ
Area of Operation: Worldwide
Tel: 01507 578630 **Fax:** 01507 578609
Email: info@mayfieldmanufacturing.co.uk
Web: www.aludrain.co.uk
Product Type: 1

MID ARGYLL GREEN SHOP
An Tairbeart, Campbeltown Road,
Tarbet, Argyll & Bute, PA29 6SX
Area of Operation: Scotland
Tel: 01880 821212
Email: kate@mackandmags.co.uk
Web: www.mackandmags.co.uk ✎
Product Type: 1 **Other Info:** ECO
Material Type: C) 4

NORTHANTS RAINWATER SYSTEMS
31 Summerfields, Northampton,
Northamptonshire, NN4 9YN
Area of Operation: UK (Excluding Ireland)
Tel: 01604 877775 **Fax:** 01604 674447
Email: mail@northantsrainwater.co.uk
Web: www.northantsrainwater.co.uk
Product Type: 1

RAINHARVESTING SYSTEMS LTD
Units 23 & 24, Merretts Mill Industrial Centre,
Bath Road, Woodchester, Stroud,
Gloucestershire, GL5 5EX
Area of Operation: UK & Ireland
Tel: 01452 772000 **Fax:** 01452 770115
Email: sales@rainharvesting.co.uk
Web: www.rainharvesting.co.uk
Product Type: 1, 2, 3

**RAINWATER MANAGEMENT
SYSTEMS (UK) LTD**
15 Mill Walk, Whitwell, Worksop,
Nottinghamshire, S80 4SH
Area of Operation: UK (Excluding Ireland)
Tel: 01909 723297 **Fax:** 01909 721482
Email: inforainwater@aol.com
Product Type: 2, 3

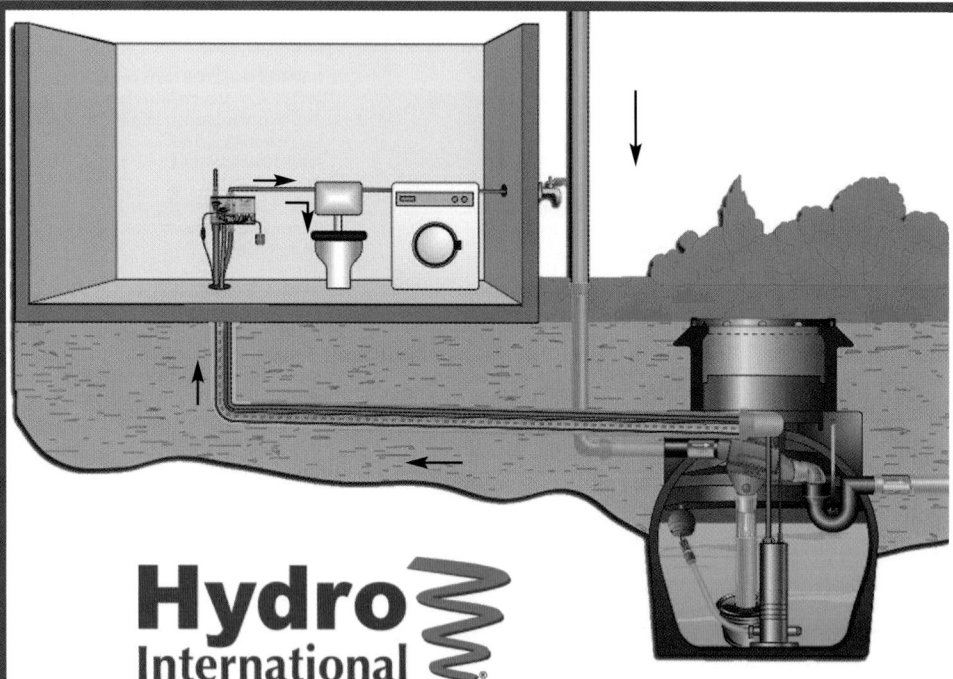

ROCKBOURNE ENVIRONMENTAL
6 Silver Business Park, Airfield Way,
Christchurch, Dorset, BH23 3TA
Area of Operation: UK & Ireland
Tel: 01202 480980
Fax: 01202 490590
Email: info@rockbourne.net
Web: www.rockbourne.net
Product Type: 2, 3

SAINT-GOBAIN PIPELINES
Lows Lane, Stanton-by-Dale, Ilkeston,
Derbyshire, DE7 4QU
Area of Operation: UK & Ireland
Tel: 0115 930 5000
Fax: 0115 932 9513
Email: mike.rawlings@saint-gobain.com
Web: www.saint-gobain-pipelines.co.uk 🔾
Product Type: 1
Material Type: C) 5

SURAFLOW
Horsebridge Mill, Kings Somborne,
Stockbridge, Hampshire, SO20 6PX
Area of Operation: UK & Ireland
Tel: 01794 389589
Fax: 01794 389597
Email: enquiries@suraflow.co.uk
Web: www.suraflow.co.uk
Product Type: 1
Material Type: C) 1

TITAN POLLUTION CONTROL
West Portway Industrial Estate, Andover,
Hampshire, SP10 3LF
Area of Operation: UK (Excluding Ireland)
Tel: 01264 353222
Fax: 01264 366446
Email: info@titanpc.co.uk
Web: www.titanpc.co.uk
Product Type: 2, 3

TUSCAN FOUNDRY PRODUCTS
Units C1-C3, Oakendene Industrial Estate,
Bolney Road, Cowfold, West Sussex, RH13 8AZ
Area of Operation: Worldwide
Tel: 01403 860040
Fax: 0845 345 0215
Email: enquiries@tuscanfoundry.co.uk
Web: www.tuscanfoundry.co.uk 🔾
Product Type: 1

WATER SUPPORT SERVICES
Norway Business Centre, Compton Road,
Yeovil, Somerset, BA21 5BU
Area of Operation: East England,
Midlands & Mid Wales, South East England,
South West England & South Wales
Tel: 01935 382490
Fax: 01935 412371
Email: info@water-support.co.uk
Web: www.water-support.co.uk
Product Type: 2, 3

WATERBANK LTD
PO Box 6446, Towcester,
Northamptonshire, NN12 8WY
Area of Operation: UK (Excluding Ireland)
Tel: 01327 831351
Fax: 01327 831242
Email: sales@waterbank.co.uk
Web: www.waterbank.co.uk
Product Type: 2, 3

SEWAGE MANAGEMENT

KEY
PRODUCT TYPES: 1= Treatment Systems
2 = Cess Pools 3 = Septic Tanks 4 = Reed
Beds 5 = Access Fittings 6 = Pumping
Equipment 7 = Other

OTHER: ▽ Reclaimed 🔾 On-line shopping
✐ Bespoke ✋Hand-made ECO Ecological

ACORN ENVIRONMENTAL SYSTEMS LIMITED
Somerset Bridge, Bridgwater,
Somerset, TA6 6LL
Area of Operation: Europe
Tel: 01278 439325
Fax: 01278 439324
Email: info@acornsystems.com
Web: www.acornsystems.com
Product Type: 1, 2, 3, 6

ALLERTON DRAINAGE
Woodbridge Road, Sleaford,
Lincolnshire, NG34 7EW
Area of Operation: UK (Excluding Ireland)
Tel: 01529 305757
Fax: 01529 414232
Email: sales@alltertonuk.com
Web: www.allertonuk.com 🔾
Product Type: 1

ALTON PUMPS
Redwood Lane, Medstead,
Alton, Hampshire, GU34 5PE
Area of Operation: Worldwide
Tel: 01420 561661
Fax: 01420 561661
Email: sales@altonpumps.com
Web: www.altonpumps.com
Product Type: 6

BALMORAL TANKS
Balmoral Park, Loirston, Aberdeen,
Aberdeenshire, AB12 3GY
Area of Operation: Worldwide
Tel: 01224 859100
Fax: 01224 859123
Email: s.gibb@balmoral.co.uk
Web: www.balmoraltanks.com
Product Type: 1, 3, 5

BIO BUBBLE LTD
Emsworth Yacht Harbour, Thorney Road,
Emsworth, Hampshire, PO10 8BW
Area of Operation: UK & Ireland
Tel: 01243 370100
Fax: 01243 370090
Email: sales@bio-bubble.com
Web: www.bio-bubble.com
Product Type: 1
Other Info: ECO ✐ ✋

BIOCLERE TECHNOLOGY
Bioclere House, Moons Hill,
Frensham, Surrey, GU103AW
Area of Operation: Worldwide
Tel: 01252 792688
Fax: 01252 795636
Email: sales@bioclere.co.uk
Web: www.bioclere.co.uk

BIODIGESTER
27 Brightstowe Road, Burnham On Sea,
Somerset, TA8 2HW
Area of Operation: UK (Excluding Ireland)
Tel: 01278 786104
Fax: 01278 793380
Email: sales@biodigester.co.uk
Web: www.biodigester.co.uk
Product Type: 1, 3

CLEANWATER SOUTH WEST LTD
Foxfield, Welcombe, Bideford, Devon, EX39 6HF
Area of Operation: South East England,
South West England and South Wales
Tel: 01288 331561
Fax: 01288 331561
Email: sales@cleanwatersw.co.uk
Web: www.cleanwatersw.co.uk
Product Type: 1, 2, 3, 4, 6, 7

CONDER PRODUCTS LTD
2 Whitehouse Way, South West Industrial Estate,
Peterlee, Durham, SR8 2HZ
Area of Operation: Worldwide
Tel: 0870 264 0004 **Fax:** 0870 264 0005
Email: sales@conderproducts.com
Web: www.conderproducts.com
Product Type: 1, 2, 3, 6

CONDER PRODUCTS LTD
Area of Operation: Worldwide
Tel: 0870 264 0004 **Fax:** 0870 264 0005
Email: sales@conderproducts.com
Web: www.conderproducts.com
Product Type: 1, 2, 3, 6

Conder Products Ltd is an environmental solutions
provider, manufacturing package products for
sewage treatment, pump stations, oil/water
separators, attenuation, above and below ground
storage vessels and rainwater harvesting.

CRESS WATER LTD
61 Woodstock Road, Worcester,
Worcestershire, WR2 5ND
Area of Operation: Europe
Tel: 01905 422707
Fax: 01905 422744
Email: info@cresswater.co.uk
Web: www.cresswater.co.uk
Product Type: 1, 3, 4, 6
Other Info: ECO

DRAINSTORE.COM
Units 1 & 2, Heanor Gate Road,
Heanor Gate Industrial Estate,
Heanor, Derbyshire, DE75 7RJ
Area of Operation: Europe
Tel: 01773 767611
Fax: 01773 767613
Email: adrian@drainstore.com
Web: www.drainstore.com 🔾
Product Type: 1, 3, 6

ECO HOUSES
12 Lee Street, Louth, Lincolnshire, LN11 9HJ
Area of Operation: Worldwide
Tel: 0845 345 2049
Email: sd@jones-nash.co.uk
Web: www.eco-houses.co.uk
Product Type: 1, 4, 6

ECOSHOP
Unit 1, Glen of the Downs Garden Centre,
Kilmacanogue, Co Wicklow, Republic of Ireland
Area of Operation: Ireland Only
Tel: +353 01 287 2914
Fax: +353 01 201 6480
Email: info@ecoshop.ie
Web: www.ecoshop.ie
Product Type: 1, 4

EDINCARE
Unit 10 Avebury Court,
Mark Road, Hemel Hempstead,
Hertfordshire, HP2 7TA
Area of Operation: UK (Excluding Ireland)
Tel: 01442 211554
Fax: 01442 211553
Email: info@edincare.com
Web: www.edincare.com
Product Type: 1, 6, 7

GRAMM ENVIRONMENTAL LIMITED
17-19 Hight Street, Ditchling,
Hassocks, West Sussex, BN6 8SY
Area of Operation: Worldwide
Tel: 01892 506935
Fax: 01892 506926
Email: sales@grammenvironmental.com
Web: www.grammenvironmental.com
Product Type: 2, 3

GREEN ROCK
3 Elmhurst Road, Harwich, Essex, CO12 3SA
Area of Operation: UK (Excluding Ireland)
Tel: 01255 554055
Fax: 01255 554055
Web: www.greenrock.fi

GRUNDFOS PUMPS LTD
Grovebury Road, Leighton Buzzard,
Bedfordshire, LU7 4TL
Area of Operation: Worldwide
Tel: 01525 850000
Fax: 01525 850011
Email: uk_sales@grundfos.com
Web: www.grundfos.com

GRUNDFOS
Area of Operation: Worldwide
Tel: 01525 850000
Fax: 01525 850011
Email: uk_sales@grundfos.com
Web: www.grundfos.co.uk

For sewage and wastewater removal, these compact,
easy to install Grundfos Products are ideal for many
refurbishment situations – from attics to basements,
where toilet bathroom, kitchens and laundry facilities
are required.

GRUNDFOS PUMPS LTD
Area of Operation: Worldwide
Tel: 01525 850000
Fax: 01525 850011
Email: uk_sales@grundfos.com
Web: www.grundfos.com
Product Type: 2, 3

Grundfos Pump plan – 2 heads are better than 1.
The Pump plan provides independent pumping for
heating and hot water in a compact, integrated unit.
Pump plan is compatible with all types of domestic
controls and modern boilers (which incorporate high
efficiency heat exchangers) and includes a unique
wiring centre.

HUNTER PLASTICS
Nathan Way, London, SE28 0AE
Area of Operation: Worldwide
Tel: 0208 855 9851 **Fax:** 0208 317 7764
Email: john.morris@hunterplastics.co.uk
Web: www.hunterplastics.co.uk
Product Type: 5

HUTCHINSON DRAINAGE LTD
White Wall Nook, Wark, Hexham,
Northumberland, NE48 3PX
Area of Operation: UK (Excluding Ireland)
Tel: 01434 220508 **Fax:** 01434 220745
Email: enquiries@hutchinson-drainage.co.uk
Web: www.hutchinson-drainage.co.uk
Product Type: 1, 3, 4, 6

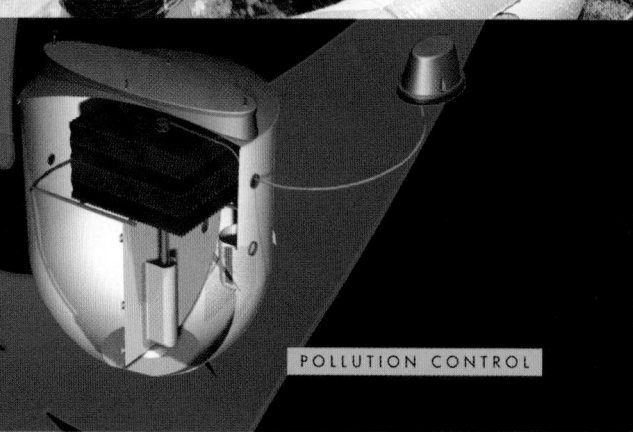

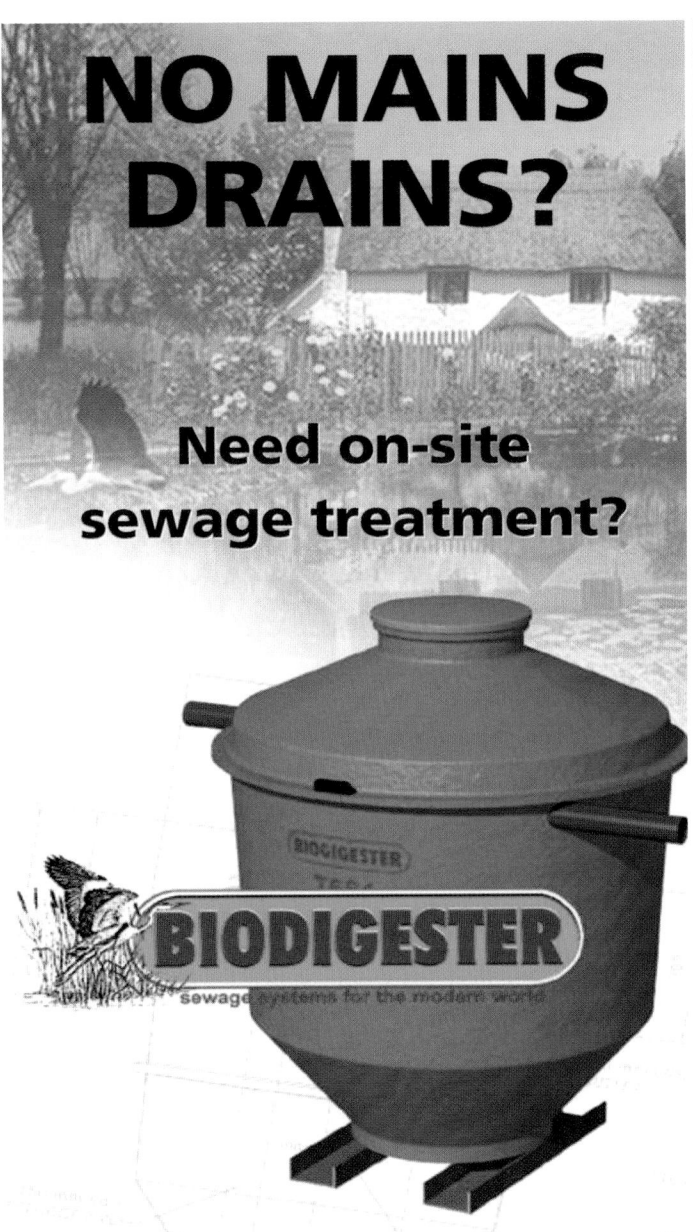

NO MAINS DRAINS?

Need on-site sewage treatment?

BIODIGESTER
sewage systems for the modern world

- Simple process
- Simple installation
- Simple maintenance

.. AT LOW COST

BIODIGESTER LIMITED
27 Brightstowe Road, Burnham on Sea
Somerset TA8 2HW
Tel: **01278 786104**
Fax: 01278 793380
Email: **sales@biodigester.co.uk**
www.biodigester.co.uk

KLARGESTER ENVIRONMENTAL
College Road North,
Aston Clinton, Aylesbury,
Buckinghamshire, HP22 5EW
Area of Operation: Worldwide
Tel: 01296 633000
Fax: 01296 633001
Email: sales@klargester.co.uk
Web: www.klargester.com
Product Type: 1, 2, 3, 4, 6, 7

RAINWATER MANAGEMENT SYSTEMS (UK) LTD
15 Mill Walk, Whitwell, Worksop,
Nottinghamshire, S80 4SH
Area of Operation: UK (Excluding Ireland)
Tel: 01909 723297
Fax: 01909 721482
Email: inforainwater@aol.com
Product Type: 1, 2, 3, 4, 6

RIVERSIDE WATER TECHNOLOGIES
Pipe House Wharf,
Morfa Road, Swansea, SA1 1TD
Area of Operation: UK & Ireland
Tel: 01792 655968
Fax: 01792 644461
Email: sales@riverside-water.co.uk
Web: www.riverside-water.co.uk
Product Type: 1

ROCKBOURNE ENVIRONMENTAL
6 Silver Business Park,
Airfield Way, Christchurch,
Dorset, BH23 3TA
Area of Operation: UK & Ireland
Tel: 01202 480980
Fax: 01202 490590
Email: info@rockbourne.net
Web: www.rockbourne.net
Product Type: 1, 2, 3, 4, 6

SERIOUS WASTE MANAGEMENT LIMITED
58-60 Wetmore Road,
Burton-upon-Trent,
Staffordshire, DE14 1SN
Area of Operation: UK (Excluding Ireland)
Tel: 01283 562382
Fax: 01283 562312
Email: info@weareserious.co.uk
Web: www.weareserious.co.uk
Product Type: 1, 2, 3, 4, 5, 6

TITAN POLLUTION CONTROL
West Portway Industrial Estate,
Andover, Hampshire, SP10 3LF
Area of Operation: UK (Excluding Ireland)
Tel: 01264 353222
Fax: 01264 366446
Email: info@titanpc.co.uk
Web: www.titanpc.co.uk
Product Type: 1, 2, 3, 4, 6, 7

WATLING HOPE (INSTALLATIONS) LIMITED
1 Goldicote Business Park,
Banbury Road, Stratford Upon Avon,
Warwickshire, CV37 7NB
Area of Operation: UK & Ireland
Tel: 01789 740757
Fax: 01789 740404
Email: enquiries@watling-hope.co.uk
Web: www.watling-hope.co.uk
Product Type: 1, 3, 5, 6

WPL LTD
Units 1 & 2 Aston Road,
Waterlooville, Hampshire, PO7 7UX
Area of Operation: Europe
Tel: 0845 450 4818
Fax: 02392 242624
Email: sales@wpl.co.uk
Web: www.wpl.co.uk
Product Type: 1
Material Type: D) 6

WPL LTD

Area of Operation: Europe
Tel: 0845 450 4818 **Fax:** 02392 242624
Email: sales@wpl.co.uk
Web: www.wpl.co.uk
Product Type: 1

WPL's Diamond provides discreet, cost-effective wastewater treatment. Fast to install and 'odour-free', the eco-system operates with extended desludging intervals of 3/5 years and requires minimal maintenance. A range of plants, up to 55pe.

WPL LTD

Area of Operation: Europe
Tel: 0845 450 4818 **Fax:** 02392 242624
Email: sales@wpl.co.uk
Web: www.wpl.co.uk
Product Type: 1

WPL's popular Diamond S-Range wastewater treament plant is now available with an extended five-year warranty. This covers all hardware and includes three years' supply of all servicing consumables. A range of plants, up to 55pe.

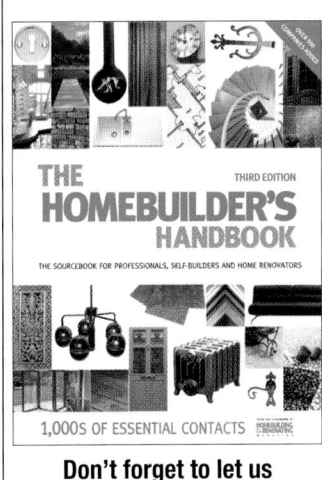
Don't forget to let us know about companies you think should be listed in the next edition
email:
customerservice@centaur.co.uk

FIXINGS, ADHESIVES AND TREATMENTS

Screws

Woodscrews probably evolved from ridged or threaded nails that were just hammered home. Then a bright spark thought of slotting the head — and DIY was born.

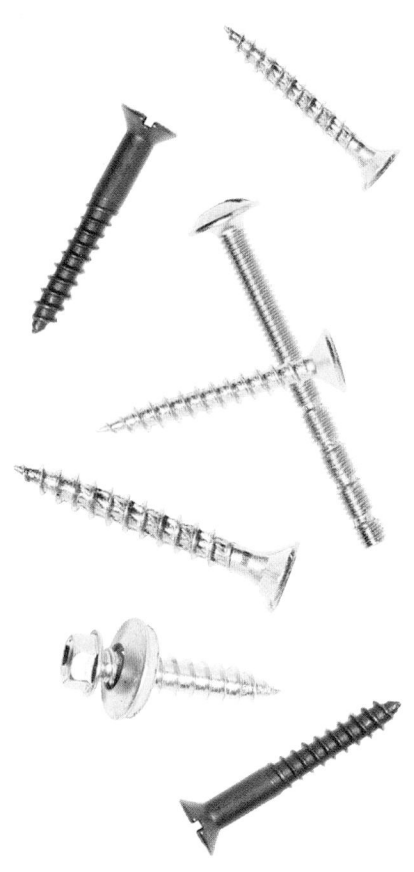

We can thank the Ancient Greeks for coming up with the idea of the screw but it took hundreds of years before anyone made the first woodscrews. No one knows for sure, but it seems that the first ones were made, filed painstakingly by hand, perhaps 500 years ago.

The real revolution in woodscrew design happened in the 1970s when automatic assembly in UK furniture manufacture required a screw that did not need a pilot hole. GKN produced the Supascrew which was heat-treatment hardened, had a twin-thread, enabling the screw to be inserted with twice the speed, had a cross-recessed drive in the head (a development of Pozidrive) and was bright zinc plated. Supascrew was enhanced and became the SupaXpress and this – now widely copied – is still the screw most often used for general building work.

The next big development was the Deck-Tite screw designed for decking, where screws tend to unscrew over time. The core of the screw, instead of being round in section, is tri-lobular — roughly speaking, like a rounded triangle. Once the screw is driven in, the surrounding material relaxes around the tri-lobular shank and holds the screw tight.

Screws work by forcing their way into the material into which they are being screwed – which is why the threaded section is tapered – then pulling the materials together as the screw is turned. This is why a pilot hole is normally drilled first, followed by a clearance hole for the non-threaded shank on the screw. Modern screws without a shank reduce the need for a pilot hole; however, without one there is an increased risk of a split near the end of timber. With a pilot hole, you'll also find it easier to control the positioning and angle of the screw.

ADHESIVES

KEY

PRODUCT TYPES: 1= Carpet 2 = Ceramics
3 = Concrete 4 = Cork 5 = Rubber
6 = Timber 7 = Linoleum 8 = Plastics
9 = Other

OTHER: ▽ Reclaimed ⌐ On-line shopping
✎ Bespoke ✋ Hand-made ECO Ecological

ARDEX UK LIMITED
Homefield Road, Haverhill,
Suffolk, CB9 8QP
Area of Operation: UK (Excluding Ireland)
Tel: 01440 714939
Fax: 01440 716660
Email: info@ardex.co.uk
Web: www.ardex.co.uk
Product Type: 1, 2, 4, 5, 7

BATHROOM CITY
Seeleys Road, Tyseley Industrial Estate,
Birmingham, West Midlands, B11 2LQ
Area of Operation: UK & Ireland
Tel: 0121 753 0700
Fax: 0121 753 1110
Email: sales@bathroomcity.com
Web: www.bathroomcity.com ⌐
Product Type: 2

BUILDING ADHESIVES LTD
Longton Road, Trentham,
Stoke on Trent, Staffordshire, ST4 8JB
Area of Operation: Worldwide
Tel: 01782 591100
Fax: 01782 591193
Email: info@building-adhesives.com
Web: www.building-adhesives.com

DEGUSSA CONSTRUCTION CHEMICALS (UK)
Albany House, Swinton Hall Road,
Swinton, Manchester, M27 4DT
Area of Operation: Worldwide
Tel: 0161 794 7411 **Fax:** 0161 727 8547
Email: mbtfeb@degussa.com
Web: www.degussa-cc.co.uk
Product Type: 2, 3, 6

GEOCEL LIMITED
Western Wood Way, Langage Science Park,
Plympton, Plymouth, Devon, PL7 5BG
Area of Operation: UK & Ireland
Tel: 01752 334350 **Fax:** 01752 202065
Email: info@geocel.co.uk
Web: www.geocel.co.uk
Product Type: 2, 5, 6, 8, 9

HAFIXS INDUSTRIAL PRODUCTS
Park Royal House, 23 Park Royal Road,
London, NW10 7JH
Area of Operation: Europe
Tel: 0208 969 3034 **Fax:** 0871 443 8060
Email: sales@hafixs.co.uk
Web: www.hafixs.co.uk ⌐
Product Type: 1, 2, 3, 4, 5, 6, 7, 8, 9

HENKEL CONSUMER ADHESIVES
Apollo Court, 2 Bishop Square Business Park,
Hatfield, Herefordshire, AL10 9EY
Area of Operation: Worldwide
Tel: 01606 593933 **Fax:** 01707 289048
Web: www.henkel.co.uk ⌐
Product Type: 3, 9

MAPEI UK LTD
Mapei House, Steel Park Road,
Halesowen, West Midlands, B62 8HD
Area of Operation: UK & Ireland
Tel: 0121 508 6970 **Fax:** 0121 508 6960
Email: info@mapei.co.uk
Web: www.mapei.co.uk
Product Type: 1, 2, 3, 4, 5, 6, 7, 8, 9
Other Info: ECO

MASKING FILM COMPANY
P.O.Box 62, Hedge End,
Southampton, Hampshire, SO30 3ZJ
Area of Operation: Worldwide
Tel: 02380 477550
Fax: 02380 477886
Email: mckphil@lowtackfilm.co.uk
Web: www.lowtackfilm.co.uk ⌐
Product Type: 1

NATURAL WOOD FLOORING COMPANY LTD
20 Smugglers Way, Wandsworth,
London, SW18 IEG
Area of Operation: Worldwide
Tel: 0208 871 9771 **Fax:** 0208 877 0273
Email: sales@naturalwoodfloor.co.uk
Web: www.naturalwoodfloor.co.uk
Product Type: 6

RESIN BONDED LTD
Unit 7 Ashdown Court, Vernon Road,
Uckfield, East Sussex, TN22 5DX
Area of Operation: UK & Ireland
Tel: 01825 766186
Fax: 01825 766186
Email: info@resinbonded.co.uk
Web: www.resinbonded.co.uk ⌐
Product Type: 3

THE INDUSTRIAL TAPE COMPANY
8 Glovers, Great Leighs,
Chelmsford, Essex, CM3 1PY
Area of Operation: UK & Ireland
Tel: 01245 361074 **Fax:** 01245 361074
Email: industrial.tapecompany@virgin.net

TOPSEAL SYSTEMS LIMITED
Unit 1&5 Hookstone Chase, Harrogate,
North Yorkshire, HG2 7HH
Area of Operation: Europe
Tel: 01423 886495 **Fax:** 01423 889550
Email: sales@topseal.co.uk
Web: www.topseal.co.uk
Product Type: 8, 9

UNIFIX LTD
St Georges House, Grove Lane,
Smethwick, Birmingham,
West Midlands, B66 2QT
Area of Operation: Europe
Tel: 0800 096 1110
Fax: 0800 096 1115
Email: sales@unifix.com
Web: www.unifix-online.co.uk ⌐
Product Type: 6, 8, 9

WEBER BUILDING SOLUTIONS
Dickens House, Enterprise Way,
Maulden Road, Flitwick, Bedford,
Bedfordshire, MK45 5BY
Area of Operation: Worldwide
Tel: 0870 333 0070
Fax: 01525 718988
Email: info@weberbuildingsolutions.co.uk
Web: www.weberbuildingsolutions.co.uk
Product Type: 2, 9

BUILDING STRUCTURE & MATERIALS - **Fixings, Adhesives and Treatments** - Treatments, Preservatives & Stains; Fixings

BUILDING STRUCTURE & MATERIALS

TREATMENTS, PRESERVATIONS & STAINS

KEY

PRODUCT TYPES: 1= Dry Rot 2 = Wet Rot
3 = Pre-paint Repair 4 = Woodworm
5 = Coloured Stains 6 = Rust Proofing
7 = Other

OTHER: ▽ Reclaimed ⏚ On-line shopping
✎ Bespoke ✋ Hand-made ECO Ecological

AQUAFIRE SYSTEMS
1 Newhaven Main Street,
Edinburgh, EH6 4LJ
Area of Operation: UK & Ireland
Tel: 0131 551 6551
Fax: 0131 551 6575
Email: sales@aquafire.co.uk
Web: www.aquafire.co.uk
Product Type: 5, 7

BEHLEN LTD
15 Huss's Lane, Main Street,
Long Eaton, Nottinghamshire, NG10 1GS
Area of Operation: Europe
Tel: 0871 910 0900
Fax: 0871 271 0451
Email: enquiries@behlen.co.uk
Web: www.behlen.co.uk ⏚
Product Type: 3, 5, 7

ECO SOLUTIONS LIMITED
Summerleaze House, Church Road,
Winscombe, Somerset, BS25 1BH
Area of Operation: UK (Excluding Ireland)
Tel: 01934 844484
Fax: 01934 844119
Email: info@ecosolutions.co.uk
Web: www.strip-paint.com
Product Type: 7

GEOCEL LIMITED
Western Wood Way,
Langage Science Park, Plympton,
Plymouth, Devon, PL7 5BG
Area of Operation: UK & Ireland
Tel: 01752 334350
Fax: 01752 202065
Email: info@geocel.co.uk
Web: www.geocel.co.uk
Product Type: 7 **Other Info:** ECO

GREEN BUILDING STORE
11 Huddersfield Road, Meltham,
Holmfirth, West Yorkshire, HD9 4NJ
Area of Operation: UK & Ireland
Tel: 01484 854898
Fax: 01484 854899
Email: info@greenbuildingstore.co.uk
Web: www.greenbuildingstore.co.uk ⏚
Product Type: 3, 4, 5, 7
Other Info: ECO

INDOOR CLIMATE SOLUTIONS
10 Cryersoak Close, Monkspath,
Solihull, West Midlands, B90 4UW
Area of Operation: Midlands & Mid Wales
Tel: 0870 011 5432
Fax: 0870 011 5432
Email: info@ics-aircon.com
Web: www.ics-aircon.com
Product Type: 1, 2, 4

MGC LTD
McMillan House, 54-56 Cheam Common Rd,
Worcester Park, Surrey, KT4 8RH
Area of Operation: UK & Ireland
Tel: 0208 337 0731
Fax: 0208 337 3739
Email: info@mgcltd.co.uk
Web: www.mgcltd.co.uk
Product Type: 7

NEW MARKETS LTD
Theocsbury House,
18-20 Barton Street, Tewkesbury,
Gloucestershire, GL20 5PP
Area of Operation: UK & Ireland
Tel: 01684 291544
Fax: 01684 291545
Email: info@newmarkets.co.uk
Web: www.newmarkets.co.uk
Product Type: 6

NUTSHELL NATURAL PAINTS
PO Box 72, South Brent,
Devon, TQ10 9YR
Area of Operation: Worldwide
Tel: 0870 033 1140
Fax: 01364 73801
Email: info@nutshellpaints.com
Web: www.nutshellpaints.com ⏚
Product Type: 5

OSMO UK LTD
Unit 2 Pembroke Road,
Stocklake Industrial Estate, Aylesbury,
Buckinghamshire, HP20 1DB
Area of Operation: UK & Ireland
Tel: 01296 481220
Fax: 01296 424090
Email: info@osmouk.com
Web: www.osmouk.com
Product Type: 5

RUSTINS LTD
Waterloo Road, London, NW2 7TX
Area of Operation: Worldwide
Tel: 0208 450 4666
Fax: 0208 452 2008
Email: rustins@rustins.co.uk
Web: www.rustins.co.uk

SAFEGUARD EUROPE LTD
Redkiln Close, Horsham,
West Sussex, RH13 5QL
Area of Operation: Worldwide
Tel: 01403 210204
Fax: 01403 217529
Email: info@safeguardeurope.com
Web: www.safeguardeurope.com
Product Type: 1, 2, 4

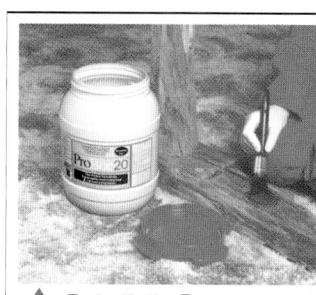

SAFEGUARD
Area of Operation: Worldwide
Tel: 01403 210204
Fax: 01403 217529
Email: info@safeguardeurope.com
Web: www.safeguardeurope.com
Product Type: 1, 2, 4

Safeguard's ProBor range of timber treatments are specifically designed to treat woodworm, dry rot and wet rot in refurbishment projects.

SAFEGUARD
Area of Operation: Worldwide
Tel: 01403 210204
Fax: 01403 217529
Email: info@safeguardeurope.com
Web: www.safeguardeurope.com
Product Type: 2

Protect expensive wooden flooring from damp & contaminated floor-slabs with Oldroyd Xs membrane – available from Safeguard.

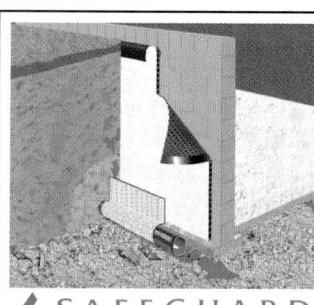

SAFEGUARD
Area of Operation: Worldwide
Tel: 01403 210204
Fax: 01403 217529
Email: info@safeguardeurope.com
Web: www.safeguardeurope.com
Product Type: 2, 3, 4

Safeguard supply a choice of systems for waterproofing new-build and existing basements. CAD drawings can be downloaded from the company's website.

THE STILL ROOM SOAPS & POTIONS
The Green Business Centre,
Stanford Bridge, Worcester,
Worcestershire, WR6 6SA
Area of Operation: UK & Ireland
Tel: 01299 896514
Fax: 01584 881012
Email: info@thestillroom.co.uk
Web: www.thestillroom.co.uk ⏚

WD-40 COMPANY LTD
PO Box 440, Kiln Farm,
Milton Keynes, MK11 3LF
Area of Operation: Worldwide
Tel: 01908 555400
Fax: 01908 266900
Email: info@wd40.co.uk
Web: www.wd40.co.uk
Product Type: 6, 7

WINDOW CARE SYSTEMS LTD
Unit E, Sawtry Business Park,
Glatton Road, Sawtry, Huntingdon,
Cambridgeshire, PE28 5GQ
Area of Operation: UK & Ireland
Tel: 01487 830311
Fax: 01487 832876
Email: sales@window-care.net
Web: www.window-care.com
Product Type: 2, 3

WINN & COALES (DENSO) LTD
Denso House, Chapel Road,
London, SE27 0TR
Area of Operation: Worldwide
Tel: 0208 670 7511 **Fax:** 0208 761 2456
Email: mail@denso.net
Web: www.denso.net
Product Type: 7

FIXINGS

KEY

PRODUCT TYPES: 1= Mechanical Fixings
2 = Threaded Fixings

SEE ALSO: MERCHANTS - Builders Merchants,
FIXINGS, ADHESIVES AND TREATMENTS - Wall Ties

OTHER: ▽ Reclaimed ⏚ On-line shopping
✎ Bespoke ✋ Hand-made ECO Ecological

ANCON BUILDING PRODUCTS
President Way, President Park,
Sheffield, South Yorkshire, S4 7UR
Area of Operation: Worldwide
Tel: 0114 238 1238 **Fax:** 0114 276 8543
Email: info@ancon.co.uk
Web: www.ancon.co.uk
Product Type: 1, 2

ARTEX-RAWLPLUG
Pasture Lane, Ruddington,
Nottingham, Nottinghamshire, NG11 6AE
Area of Operation: UK (Excluding Ireland)
Tel: 0115 984 5679 **Fax:** 0115 940 5240
Email: info@bpb.com
Web: www.artex-rawlplug.co.uk

BATHROOM CITY
Seeleys Road, Tyseley Industrial Estate,
Birmingham, West Midlands, B11 2LQ
Area of Operation: UK & Ireland
Tel: 0121 753 0700 **Fax:** 0121 753 1110
Email: sales@bathroomcity.com
Web: www.bathroomcity.com ⏚
Product Type: 1

BLIND BOLT COMPANY
Tollgate Industrial Estate,
Stafford, Staffordshire, ST16 3HS
Area of Operation: Worldwide
Tel: 01785 270629
Email: enquiries@blindbolt.co.uk
Web: www.blindbolt.co.uk
Product Type: 1

BLUEBIRD FIXINGS LIMITED
Westminster Road Industrial Estate,
Station Road, North Hykehem,
Lincolnshire, LN6 3QY
Area of Operation: UK & Ireland
Tel: 01522 697776 **Fax:** 01522 697771
Email: info@bluebird-fixings.ltd.uk
Web: www.bluebird-fixings.ltd.uk
Product Type: 1

BPC BUILDING PRODUCTS LTD.
Flanshaw Way, Wakefield,
West Yorkshire, WF2 9LP
Area of Operation: UK & Ireland
Tel: 01924 364794 **Fax:** 01924 373846
Email: gareth@bpcfixings.com
Web: www.bpcfixings.com ⏚
Product Type: 1

HELICAL SYSTEMS LTD
The Old Police Station, 195 Main Road,
Biggin Hill, Westerham, Kent, TN16 3JU
Area of Operation: UK & Ireland
Tel: 01959 541148
Fax: 01959 540841
Email: janecannon@helicalsystems.co.uk
Web: www.helicalsystems.co.uk ⏚
Product Type: 1

BUILDING STRUCTURE & MATERIALS

HENKEL CONSUMER ADHESIVES
Apollo Court, 2 Bishop Square Business Park,
Hatfield, Herefordshire, AL10 9EY
Area of Operation: Worldwide
Tel: 01606 593 933
Fax: 01707 289048
Web: www.henkel.co.uk
Product Type: 1

ORAC (UK) LIMITED
Unit 5, Hewitts Estate,
Elmbridge Road, Cranleigh,
Surrey, GU6 8LW
Area of Operation: Worldwide
Tel: 01483 271211
Fax: 01483 278317
Email: stewart@oracdecor.com
Web: www.oracdecor.com

SEAC LTD
46 Chesterfield Road, Leicester,
Leicestershire, LE5 5LP
Area of Operation: Worldwide
Tel: 0116 273 9501 **Fax:** 0116 273 8373
Email: enquiries@seac.uk.com
Web: www.seac.uk.com
Product Type: 1, 2

T I MIDWOOD & CO LIMITED
Green Lane, Wardle, Nantwich,
Cheshire, CW5 6BJ
Area of Operation: Europe
Tel: 01829 261111 **Fax:** 01829 261102
Email: simon@timcouk.com
Web: www.timcouk.com
Product Type: 2

THIN JOINT TECHNOLOGY LTD
3 Albright Road,
Speke Approaches Industrial Estate,
Liverpool, Merseyside, WA8 8FY
Area of Operation: UK & Ireland
Tel: 0151 422 8000 **Fax:** 0151 422 8001
Email: sales@thinjoint.com
Web: www.thinjoint.com
Product Type: 1, 2

UNIFIX LTD
St Georges House, Grove Lane,
Smethwick, Birmingham,
West Midlands, B66 2QT
Area of Operation: Europe
Tel: 0800 096 1110 **Fax:** 0800 096 1115
Email: sales@unifix.com
Web: www.unifix-online.co.uk
Product Type: 1, 2

WETHERBY BUILDING SYSTEMS LIMITED
1 Kidglove Road, Golborne Enterprise Park,
Golborne, Greater Manchester, WA3 3GS
Area of Operation: UK & Ireland
Tel: 01942 717100 **Fax:** 01942 717101
Email: info@wbs-ltd.co.uk
Web: www.wbs-ltd.co.uk
Product Type: 2

WALL TIES

KEY

PRODUCT TYPES: 1= Brick-to-Block Ties
2 = Brick-to-Timber Ties 3 = Brick-to-Steel
Ties 4 = Wall Starter Systems

OTHER: ▽ Reclaimed On-line shopping
Bespoke Hand-made ECO Ecological

ANCON BUILDING PRODUCTS
President Way, President Park,
Sheffield, South Yorkshire, S4 7UR
Area of Operation: Worldwide
Tel: 0114 238 1238 **Fax:** 0114 276 8543
Email: info@ancon.co.uk
Web: www.ancon.co.uk
Product Type: 1, 2, 3, 4

BLUEBIRD FIXINGS LIMITED
Westminster Road Industrial Estate,
Station Road, North Hykehem,
Lincolnshire, LN6 3QY
Area of Operation: UK & Ireland
Tel: 01522 697776
Fax: 01522 697771
Email: info@bluebird-fixings.ltd.uk
Web: www.bluebird-fixings.ltd.uk

BPC BUILDING PRODUCTS LTD.
Flanshaw Way, Wakefield,
West Yorkshire, WF2 9LP
Area of Operation: UK & Ireland
Tel: 01924 364794
Fax: 01924 373846
Email: gareth@bpcfixings.com
Web: www.bpcfixings.com
Product Type: 1, 2

HELICAL SYSTEMS LTD
The Old Police Station,
195 Main Road, Biggin Hill,
Westerham, Kent, TN16 3JU
Area of Operation: UK & Ireland
Tel: 01959 541148
Fax: 01959 540841
Email: janecannon@helicalsystems.co.uk
Web: www.helicalsystems.co.uk
Product Type: 2

SAFEGUARD EUROPE LTD
Redkiln Close, Horsham,
West Sussex, RH13 5QL
Area of Operation: Worldwide
Tel: 01403 210204
Fax: 01403 217529
Email: info@safeguardeurope.com
Web: www.safeguardeurope.com
Product Type: 2

THIN JOINT TECHNOLOGY LTD
3 Albright Road,
Speke Approaches Industrial Estate,
Liverpool, Merseyside, WA8 8FY
Area of Operation: UK & Ireland
Tel: 0151 422 8000
Fax: 0151 422 8001
Email: sales@thinjoint.com
Web: www.thinjoint.com
Product Type: 1, 2

TRITON CHEMICAL MANUFACTURING LTD
129 Felixstowe Road,
Abbey Wood, London, SE2 9SG
Area of Operation: Worldwide
Tel: 0208 310 3929 **Fax:** 0208 312 0349
Email: neil@triton-chemicals.com
Web: www.triton-chemicals.com

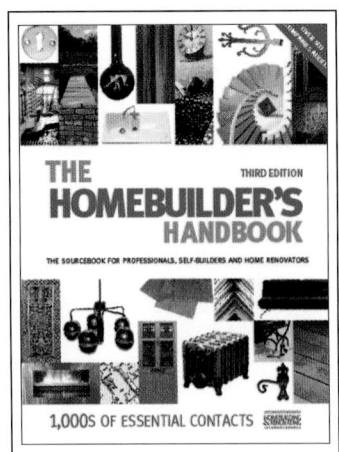

THIRD EDITION
THE HOMEBUILDER'S HANDBOOK
THE SOURCEBOOK FOR PROFESSIONALS, SELF-BUILDERS AND HOME RENOVATORS

1,000s OF ESSENTIAL CONTACTS

**Don't forget to let us
know about companies
you think should be listed
in the next edition**
email:
customerservice@centaur.co.uk

CONCRETE, SAND, MORTARS & AGGREGATES

Image courtesy of Hewden (0845 60 70 111)

Concrete and Mortars

Of all the major elements that go into building a house, concrete gets the least attention: we tend to stick it in the ground and forget about it. Yet concrete occupies a unique niche in the world of construction. For a start, it usually arrives on site as a sloppy liquid and has to be shaped there and then. But once this is done – and providing it's done correctly – it provides a rock-like platform. There is really nothing else quite like it and it's a measure of its success that it is taken for granted the way it is. But in taking it for granted, we risk closing our minds to many of the unusual uses for concrete.

Techniques for making concrete have moved on since the material's emergence in classical times (the word concrete comes from the Latin concretus, meaning grown together or compounded), but the basic principal remains the same. Pieces of an inert hard material such as gravel or crushed rock are mixed together with a paste of cement and water, and this is left to harden over time. Early versions of concrete used lime as a binding agent – you sometimes hear this referred to as a lime concrete – but this has now been superseded by Portland cements.

Despite the standardisation of its constituent parts, concrete exists in many different formats. By changing the type of aggregate, concrete can be made so light it can float, or so heavy that it is almost twice its usual density. It can be made waterproof or porous. The surface can be as smooth as glass or rough textured and patterned, with the option now available to sandblast or stencil concrete, or even imprint photographs into it. And it doesn't have to be dull grey – a variety of pigments are available which can transform the look and feel of concrete.

Some environmental campaigners have attacked the use of concrete in construction, as cement production is a potentially polluting process. However, a close analysis of this impact shows that it is arguably one of the most benign materials used and production is very well controlled. Also, more recently, there has been a move towards blending Portland cement with other products which are the processed wastes of other industries, and these have been found effective. Concrete technology is constantly developing new innovations that make the material perform better or recycle more waste materials.

What is also not generally realised is that concrete itself is readily recyclable. Although correctly specified it will last for centuries, it can also be readily broken up and reused, usually as hardcore in foundations but increasingly in place of aggregates in new concrete mixes.

Mortars

You might think that knocking up a few batches of 5:1 mortar for the brickwork you intend to carry out on your self-build home is easy. Think again – or at least think hard before you choose the mix. Mortars are complex in the way they work, and a little understanding of this will help the self-builder combat problems before they occur.

For a start it is a good idea to know the precise function of mortar. In a brick wall, its many tasks include a gap filling and adhesive function. As well as this it helps to remove irregularities and spread loads evenly, thus giving the building stability and solidity. It is, therefore, wise to select a mortar for your task that has the durability and strength to suit the application.

Many people make the mistake of using sand that is either too soft, with too many silts and fine clay particles, which can lead to unacceptable shrinkage movement in the mortar once set, or too sharp. Sharp sands are the ones that lack finer particles, which means that water retention in the working mix is poor and the mortar is harsh and unmanageable. A well-graded sand, suitable for mortar, will have an even distribution of particle sizes from the finest to the coarsest, and the void – the labyrinth of holes – should be one third of the total volume. It is the void that is filled by the binder – cement or lime or air, if it is encapsulated form, or a combination of the three. The normal binder for mortars up to the early part of this century was lime: however the building trade found that limes varied greatly in different parts of the country, and between the wars Portland cement, which comprises limestone burned to a high temperature in combination with clay, became almost universally adopted as a binder.

Cement mortars are ideal for today's thin-walled, highly-stressed buildings with brick skins that are constructed very rapidly and require a quick set.

However the practice of using dry hydrate of lime (bagged lime from your builders' merchant) as a binder and adding a pinch of cement to give it setting power has become popular with self-builders who wish to match old brickwork, although this is not generally advocated by professionals. If you want to have the benefit of lime as a plasticiser – something that promotes bonding in the mortar in your building – a 1:1:6 mix of cement, bagged lime and sand respectively is most commonly used. If you intend to use lime to give a white mortar, or to match the mortar in an older section of their house, a standard mix with weaker lime mortar for pointing is an alternative, but

you should ensure that the bagged lime you use is mixed dry with sand, then wet and kept moist for several days if not weeks before use. The other option is to buy lime putty made from quicklime, but this is usually extremely expensive on a new build project.

If you are using blockwork on the inner leaf of your house, it is important to use a weaker mix because blocks are made with an excess of water and take time to dry out, shrinking in the process. A very hard mortar will tend to crack the block if there is any movement in the wall. A softer mortar with a high proportion of lime as a plasticiser, or an air entraining agent, which performs the same function, will help to counteract the shrinkage and ensure that any cracking is in the joints rather than the blocks.

With stone, the mortar should not normally be harder or denser than the stone employed. This usually means a lime-rich mix, unless you are working with a very hard stone like granite, so that the stone is allowed to expand and contract within the wall.

CONCRETE

KEY

PRODUCT TYPES: 1= Concrete Ready Mix
2 = Bagged Concrete   3 = Mini Loads
4 = Cement and Concrete Mixers

OTHER: ▽ Reclaimed   🛒 On-line shopping
✏ Bespoke ✋ Hand-made   ECO Ecological

ACE MINIMIX
Millfields Road, Ettingshall, Wolverhampton,
West Midlands, WV4 6JP
Area of Operation: UK (Excluding Ireland)
Tel: 01902 353522 **Fax:** 0121 5855557
Email: info@tarmac.co.uk
Web: www.tarmac.co.uk
Product Type: 1, 3, 4

ARDEX UK LIMITED
Homefield Road, Haverhill, Suffolk, CB9 8QP
Area of Operation: UK (Excluding Ireland)
Tel: 01440 714939 **Fax:** 01440 716660
Email: info@ardex.co.uk **Web:** www.ardex.co.uk
Product Type: 1, 2, 4

BORDER CONCRETE PRODUCTS
Jedburgh Road, Kelso, Borders, TD5 8JG
Area of Operation: North East England,
North West England and North Wales, Scotland
Tel: 01573 224393 **Fax:** 01573 276360
Email: sales@borderconcrete.co.uk
Web: www.borderconcrete.co.uk
Other Info: ✏ ✋

CASTLE CEMENT LIMITED
3160 Solihull Parkway, Birmingham Business Park,
Birmingham, West Midlands, B37 7YN
Area of Operation: UK (Excluding Ireland)
Tel: 0845 600 1616 **Fax:** 0121 606 1436
Email: customer.services@castlecement.co.uk
Web: www.castlecement.co.uk
Material type: G)2

CEMEX READYMIX
CEMEX House, Rugby, Warwickshire, CV21 2DT
Area of Operation: Worldwide
Tel: 0800 667 827 **Fax:** 01788 564404
Email: readymix@cemex.co.uk
Web: www.cemex.co.uk 🛒
Product Type: 1, 3

H+H CELCON LIMITED
Celcon House, Ightham, Sevenoaks, Kent, TN15 9HZ
Area of Operation: UK (Excluding Ireland)
Tel: 01732 880520 **Fax:** 01732 880531
Email: marketing@celcon.co.uk
Web: www.celcon.co.uk

HANSON BUILDING PRODUCTS
Stewartby, Bedford, Bedfordshire, MK43 9LZ
Area of Operation: Worldwide
Tel: 08705 258258 **Fax:** 01234 762040
Email: info@hansonplc.com
Web: www.hanson.biz
Product Type: 1, 2, 4

NATURAL CEMENT DISTRIBUTION LTD
Unit 12, Redbrook Business Park,
Wilthorpe Road, Barnsley,
South Yorkshire, S75 1JN
Area of Operation: Worldwide
Tel: 01226 299333
Email: phil@naturalcement.co.uk
Web: www.naturalcement.co.uk

RMC CONCRETE PRODUCTS (UK) LTD
RMC House, Evreux Way, Rugby,
Warwickshire, CV21 2DT
Area of Operation: UK (Excluding Ireland)
Tel: 01788 542111 **Fax:** 01788 514747
Email: andrew.norgett@rmc.co.uk
Web: www.rmc.co.uk
Product Type: 1

TARMAC LIMITED
Millfields Road, Ettingshall, Wolverhampton,
West Midlands, WV4 6JP
Area of Operation: Europe
Tel: 01902 353522 **Fax:** 01902 382922
Email: info@tarmac.co.uk
Web: www.tarmac.co.uk

AGGREGATES

KEY

OTHER: ▽ Reclaimed   🛒 On-line shopping
✏ Bespoke ✋ Hand-made   ECO Ecological

ACE MINIMIX
Millfields Road, Ettingshall, Wolverhampton,
West Midlands, WV4 6JP
Area of Operation: UK (Excluding Ireland)
Tel: 01902 353522
Fax: 0121 5855557
Email: info@tarmac.co.uk
Web: www.tarmac.co.uk

BARDON AGGREGATES
Hulland Ward, Ashbourne, Derbyshire, DE6 3ET
Area of Operation: Worldwide
Tel: 0845 600 0860 **Fax:** 01335 372485
Web: www.bardon-aggregates.com

BRETT AGGREGATES LIMITED
Brett House, Bysing Wood Road,
Faversham, Kent, ME13 7UD
Area of Operation: South East England
Tel: 0800 028 5980
Fax: 0800 028 5979
Web: www.brett.co.uk

HANSON BUILDING PRODUCTS
Stewartby, Bedford, Bedfordshire, MK43 9LZ
Area of Operation: Worldwide
Tel: 08705 258258 **Fax:** 01234 762040
Email: info@hansonplc.com
Web: www.hanson.biz

KIRK NATURAL STONE
Bridgend, Fyvie, Turriff, Aberdeenshire, AB53 8QB
Area of Operation: Worldwide
Tel: 01651 891891 **Fax:** 01651 891794
Email: info@kirknaturalstone.com
Web: www.kirknaturalstone.com

LAFARGE CEMENT
Manor Court, Chilton, Oxfordshire, OX11 0RN
Area of Operation: UK & Ireland
Tel: 01235 448400
Fax: 01235 448600
Email: info@lafargecement.co.uk
Web: www.lafargecement.co.uk

LIMEBASE PRODUCTS LTD
Walronds Park, Isle Brewers, Taunton,
Somerset, TA3 6QP
Area of Operation: UK & Ireland
Tel: 01460 281921
Fax: 01460 281100
Email: info@limebase.co.uk
Web: www.limebase.co.uk

MIRATEX LIMITED
Unit 10, Park Hall Farm, Brookhouse Road,
Cheadle, Staffordshire, ST10 2NJ
Area of Operation: UK & Ireland
Tel: 01538 750923
Fax: 01538 752078
Email: info@miratex.co.uk
Web: www.miratex.co.uk

RIVAR SAND & GRAVEL LTD
Pinchington Lane, Newbury, Berkshire, RG19 8SR
Area of Operation: South East England
Tel: 01635 523524 **Fax:** 01635 521621
Email: sales@rivarsandandgravel.co.uk
Web: www.rivarsandandgravel.co.uk
Other Info: ✏

RMC CONCRETE PRODUCTS (UK) LTD
RMC House, Evreux Way, Rugby,
Warwickshire, CV21 2DT
Area of Operation: UK (Excluding Ireland)
Tel: 01788 542111 **Fax:** 01788 514747
Email: andrew.norgett@rmc.co.uk
Web: www.rmc.co.uk

ROSE OF JERICHO
Horchester Farm, Holywell,
Dorchester, Dorset, DT2 0LL
Area of Operation: UK (Excluding Ireland)
Tel: 01935 83676 **Fax:** 01935 83903
Email: info@rose-of-jericho.demon.co.uk
Web: www.rose-of-jericho.demon.co.uk

SCOTTISH LIME CENTRE TRUST
The School House, Rocks Road, Charlestown,
Nr Dunfermline, Fife, KY11 3EN
Area of Operation: Scotland
Tel: 01383 872722 **Fax:** 01383 872744
Email: info@scotlime.org **Web:** www.scotlime.org

TARMAC LIMITED
Millfields Road, Ettingshall, Wolverhampton,
West Midlands, WV4 6JP
Area of Operation: Europe
Tel: 01902 353522 **Fax:** 01902 382922
Email: info@tarmac.co.uk **Web:** www.tarmac.co.uk

THE CORNISH LIME COMPANY LTD
Brims Park, Old Callywith Road,
Bodmin, Cornwall, PL31 2DZ
Area of Operation: East England, Greater London,
Midlands & Mid Wales, South East England,
South West England and South Wales
Tel: 01208 79779 **Fax:** 01208 73744
Email: phil@cornishlime.co.uk
Web: www.cornishlime.co.uk 🛒

WENLOCK LIME
Coates Kiln, Stretton Road, Much Wenlock,
Shropshire, TF13 6DG
Area of Operation: UK (Excluding Ireland)
Tel: 01952 728611 **Fax:** 01952 728361
Email: enquire@wenlocklime.co.uk
Web: www.wenlocklime.co.uk

SAND

KEY

OTHER: ▽ Reclaimed   🛒 On-line shopping
✏ Bespoke ✋ Hand-made   ECO Ecological

ACE MINIMIX
Millfields Road, Ettingshall, Wolverhampton,
West Midlands, WV4 6JP
Area of Operation: UK (Excluding Ireland)
Tel: 01902 353522 **Fax:** 0121 5855557
Email: info@tarmac.co.uk
Web: www.tarmac.co.uk

LIMEBASE PRODUCTS LTD
Walronds Park, Isle Brewers, Taunton,
Somerset, TA3 6QP
Area of Operation: UK & Ireland
Tel: 01460 281921 **Fax:** 01460 281100
Email: info@limebase.co.uk
Web: www.limebase.co.uk

RIVAR SAND & GRAVEL LTD
Pinchington Lane, Newbury, Berkshire, RG19 8SR
Area of Operation: South East England
Tel: 01635 523524 **Fax:** 01635 521621
Email: sales@rivarsandandgravel.co.uk
Web: www.rivarsandandgravel.co.uk

ROSE OF JERICHO
Horchester Farm, Holywell,
Dorchester, Dorset, DT2 0LL
Area of Operation: UK (Excluding Ireland)
Tel: 01935 83676 **Fax:** 01935 83903
Email: info@rose-of-jericho.demon.co.uk
Web: www.rose-of-jericho.demon.co.uk

SCOTTISH LIME CENTRE TRUST
The School House, Rocks Road, Charlestown,
Nr Dunfermline, Fife, KY11 3EN
Area of Operation: Scotland
Tel: 01383 872722 **Fax:** 01383 872744
Email: info@scotlime.org **Web:** www.scotlime.org

TARMAC LIMITED
Millfields Road, Ettingshall, Wolverhampton,
West Midlands, WV4 6JP
Area of Operation: Europe
Tel: 01902 353522 **Fax:** 01902 382922
Email: info@tarmac.co.uk **Web:** www.tarmac.co.uk

THE CORNISH LIME COMPANY LTD
Brims Park, Old Callywith Road,
Bodmin, Cornwall, PL31 2DZ
Area of Operation: East England, Greater London,
Midlands & Mid Wales, South East England,
South West England and South Wales
Tel: 01208 79779 **Fax:** 01208 73744
Email: phil@cornishlime.co.uk
Web: www.cornishlime.co.uk 🛒

WENLOCK LIME
Coates Kiln, Stretton Road, Much Wenlock,
Shropshire, TF13 6DG
Area of Operation: UK (Excluding Ireland)
Tel: 01952 728611 **Fax:** 01952 728361
Email: enquire@wenlocklime.co.uk
Web: www.wenlocklime.co.uk

LIMES

KEY

PRODUCT TYPES: 1= Bagged Limes
2 = Lime Putty   3 = Hydrated Lime
4 = Unhydrated Lime   5 = Lime Cement Mixes
6 = Pre-mixed Lime Renders, Lime Paints &
Washes

OTHER: ▽ Reclaimed   🛒 On-line shopping
✏ Bespoke ✋ Hand-made   ECO Ecological

ANGLIA LIME COMPANY
Fishers Farm, Belchamp Walter,
Sudbury, Suffolk, CO10 7AP
Area of Operation: East England, Greater London
Tel: 01787 313974 **Fax:** 01787 313944
Email: info@anglialime.com
Web: www.anglialime.com **Product Type:** 2, 6

BUXTON LIME INDUSTRIES
Tunstead House, Buxton, Derbyshire, SK17 8TG
Area of Operation: UK (Excluding Ireland)
Tel: 01298 768444 **Fax:** 01298 72195
Email: buxton.sales@buxtonlime.co.uk
Web: www.buxtonlime.co.uk **Product Type:** 3, 4

HYDRAULIC LIMES
The Lime Loft, Priestlands Lane,
Sherborne, Dorset, DT9 4RS
Area of Operation: Worldwide
Tel: 01935 815290 **Fax:** 01935 815290
Email: info@hydrauliclimes.co.uk
Web: www.hydrauliclimes.co.uk
Product Type: 1, 2, 3, 4, 6 **Other Info:** ECO

J&J SHARPE
Woodside, Merton, Okehampton, Devon, EX20 3EG
Area of Operation: UK (Excluding Ireland)
Tel: 01805 603587 **Fax:** 01805 603179
Email: mail@jjsharpe.co.uk
Web: www.jjsharpe.co.uk
Product Type: 2, 4, 6

KEIM MINERAL PAINTS LTD
Muckley Cross, Morville, Nr Bridgnorth,
Shropshire, WV16 4RR
Area of Operation: Worldwide
Tel: 01746 714543 **Fax:** 01746 714526
Email: sales@keimpaints.co.uk
Web: www.keimpaints.co.uk
Product Type: 6 **Other Info:** ECO

LIMEBASE PRODUCTS LTD
Walronds Park, Isle Brewers, Taunton,
Somerset, TA3 6QP
Area of Operation: UK & Ireland
Tel: 01460 281921
Fax: 01460 281100
Email: info@limebase.co.uk
Web: www.limebase.co.uk
Product Type: 1, 2, 6

M CARRINGTON
20 High Street, Somersham, Huntingdon,
Cambridgeshire, PE28 3JA
Area of Operation: UK (Excluding Ireland)
Tel: 01487 840305 **Fax:** 01487 840305
Email: malcolm@mcarrington.com
Web: www.mcarrington.com
Product Type: 1, 2, 3, 4, 5, 6

MIKE WYE & ASSOCIATES
Buckland Filleigh Sawmills, Buckland Filleigh,
Beaworthy, Devon, EX21 5RN
Area of Operation: Worldwide
Tel: 01409 281644 **Fax:** 01409 281669
Email: sales@mikewye.co.uk
Web: www.mikewye.co.uk
Product Type: 2, 6 **Other Info:** ECO

NATURAL BUILDING TECHNOLOGIES
The Hangar, Worminghall Road, Oakley,
Buckinghamshire, HP18 9UL
Area of Operation: UK & Ireland
Tel: 01844 338338
Fax: 01844 338525
Email: info@natural-building.co.uk
Web: www.natural-building.co.uk
Product Type: 5 **Other Info:** ECO

ROSE OF JERICHO
Horchester Farm, Holywell, Dorchester,
Dorset, DT2 0LL
Area of Operation: UK (Excluding Ireland)
Tel: 01935 83676
Fax: 01935 83903
Email: info@rose-of-jericho.demon.co.uk
Web: www.rose-of-jericho.demon.co.uk
Product Type: 2, 6

SCOTTISH LIME CENTRE TRUST
The School House, Rocks Road, Charlestown,
Nr Dunfermline, Fife, KY11 3EN
Area of Operation: Scotland
Tel: 01383 872722
Fax: 01383 872744
Email: info@scotlime.org
Web: www.scotlime.org
Product Type: 1, 2, 4, 6

TARMAC LIMITED
Millfields Road, Ettingshall, Wolverhampton,
West Midlands, WV4 6JP
Area of Operation: Europe
Tel: 01902 353522
Fax: 01902 382922
Email: info@tarmac.co.uk
Web: www.tarmac.co.uk

THE CORNISH LIME COMPANY LTD
Brims Park, Old Callywith Road,
Bodmin, Cornwall, PL31 2DZ
Area of Operation: East England, Greater London,
Midlands & Mid Wales, South East England,
South West England and South Wales
Tel: 01208 79779 **Fax:** 01208 73744
Email: phil@cornishlime.co.uk
Web: www.cornishlime.co.uk ⌐
Product Type: 2, 3, 4, 5, 6 **Other Info:** ECO ✎ ✋

WENLOCK LIME
Coates Kiln, Stretton Road, Much Wenlock,
Shropshire, TF13 6DG
Area of Operation: UK (Excluding Ireland)
Tel: 01952 728611 **Fax:** 01952 728361
Email: enquire@wenlocklime.co.uk
Web: www.wenlocklime.co.uk
Product Type: 1, 2, 3, 4, 6

MORTARS

KEY

PRODUCT TYPES: 1= Cement Mortar
2 = Mortar Additives 3 = Quick-setting Mortars
4 = Specialist Mortars

OTHER: ▽ Reclaimed ⌐ On-line shopping
✎ Bespoke ✋ Hand-made ECO Ecological

ANGLIA LIME COMPANY
Fishers Farm, Belchamp Walter, Sudbury,
Suffolk, CO10 7AP
Area of Operation: East England, Greater London
Tel: 01787 313974 **Fax:** 01787 313944
Email: info@anglialime.com
Web: www.anglialime.com
Product Type: 4

ARDEX UK LIMITED
Homefield Road, Haverhill, Suffolk, CB9 8QP
Area of Operation: UK (Excluding Ireland)
Tel: 01440 714939 **Fax:** 01440 716660
Email: info@ardex.co.uk **Web:** www.ardex.co.uk
Product Type: 1, 2, 3, 4

BUXTON LIME INDUSTRIES
Tunstead House, Buxton, Derbyshire, SK17 8TG
Area of Operation: UK (Excluding Ireland)
Tel: 01298 768444 **Fax:** 01298 72195
Email: buxton.sales@buxtonline.co.uk
Web: www.buxtonline.co.uk
Product Type: 1

CEMEX READYMIX
CEMEX House, Rugby, Warwickshire, CV21 2DT
Area of Operation: Worldwide
Tel: 0800 667 827 **Fax:** 01788 564404
Email: info.readymix@cemex.co.uk
Web: www.cemex.co.uk ⌐
Product Type: 1, 2, 3, 4

EASIPOINT MARKETING LIMITED
Restoration House, Drumhead Road, Chorley North
Industrial Estate, Chorley, Lancashire, PR6 7DE
Area of Operation: UK & Ireland
Tel: 01257 224900
Fax: 01257 224901
Email: enquiries@easipoint.co.uk
Web: www.easipoint.co.uk
Product Type: 4

HANSON BUILDING PRODUCTS
Stewartby, Bedford, Bedfordshire, MK43 9LZ
Area of Operation: Worldwide
Tel: 08705 258258
Fax: 01234 762040
Email: info@hansonplc.com
Web: www.hanson.biz
Product Type: 1, 2, 3, 4

HYDRAULIC LIMES
The Lime Loft, Priestlands Lane,
Sherborne, Dorset, DT9 4RS
Area of Operation: Worldwide
Tel: 01935 815290 **Fax:** 01935 815290
Email: info@hydrauliclimes.co.uk
Web: www.hydrauliclimes.co.uk
Product Type: 4 **Other Info:** ECO

KEIM MINERAL PAINTS LTD
Muckley Cross, Morville, Nr Bridgnorth,
Shropshire, WV16 4RR
Area of Operation: Worldwide
Tel: 01746 714543
Fax: 01746 714526
Email: sales@keimpaints.co.uk
Web: www.keimpaints.co.uk
Product Type: 1

LIMEBASE PRODUCTS LTD
Walronds Park, Isle Brewers, Taunton,
Somerset, TA3 6QP
Area of Operation: UK & Ireland
Tel: 01460 281921
Fax: 01460 281100
Email: info@limebase.co.uk
Web: www.limebase.co.uk
Product Type: 4

MIRATEX LIMITED
Unit 10, Park Hall Farm, Brookhouse Road,
Cheadle, Staffordshire, ST10 2NJ
Area of Operation: UK & Ireland
Tel: 01538 750923
Fax: 01538 752078
Email: info@miratex.co.uk
Web: www.miratex.co.uk
Product Type: 4

NATURAL CEMENT DISTRIBUTION LTD
Unit 12, Redbrook Business Park, Wilthorpe Road,
Barnsley, South Yorkshire, S75 1JN
Area of Operation: Worldwide
Tel: 01226 299333
Email: phil@naturalcement.co.uk
Web: www.naturalcement.co.uk
Product Type: 1, 3, 4 **Other Info:** ECO ✎

ROSE OF JERICHO
Horchester Farm, Holywell, Dorchester,
Dorset, DT2 0LL
Area of Operation: UK (Excluding Ireland)
Tel: 01935 83676 **Fax:** 01935 83903
Email: info@rose-of-jericho.demon.co.uk
Web: www.rose-of-jericho.demon.co.uk
Product Type: 2, 4

SCOTTISH LIME CENTRE TRUST
The School House, Rocks Road,
Charlestown, Nr Dunfermline, Fife, KY11 3EN
Area of Operation: Scotland
Tel: 01383 872722 **Fax:** 01383 872744
Email: info@scotlime.org
Web: www.scotlime.org
Product Type: 4

TARMAC LIMITED
Millfields Road, Ettingshall, Wolverhampton,
West Midlands, WV4 6JP
Area of Operation: Europe
Tel: 01902 353522 **Fax:** 01902 382922
Email: info@tarmac.co.uk **Web:** www.tarmac.co.uk

THE CORNISH LIME COMPANY LTD
Brims Park, Old Callywith Road, Bodmin,
Cornwall, PL31 2DZ
Area of Operation: East England, Greater London,
Midlands & Mid Wales, South East England, South
West England and South Wales
Tel: 01208 79779
Fax: 01208 73744
Email: phil@cornishlime.co.uk
Web: www.cornishlime.co.uk ⌐
Product Type: 3, 4

THIN JOINT TECHNOLOGY LTD
3 Albright Road, Speke Approaches Industrial Estate,
Liverpool, Merseyside, WA8 8FY
Area of Operation: UK & Ireland
Tel: 0151 422 8000
Fax: 0151 422 8001
Email: sales@thinjoint.com
Web: www.thinjoint.com
Product Type: 4

WEBER BUILDING SOLUTIONS
Dickens House, Enterprise Way, Maulden Road,
Flitwick, Bedford, Bedfordshire, MK45 5BY
Area of Operation: Worldwide
Tel: 0870 333 0070
Fax: 01525 718988
Email: info@weberbuildingsolutions.co.uk
Web: www.weberbuildingsolutions.co.uk
Product Type: 4

WENLOCK LIME
Coates Kiln, Stretton Road, Much Wenlock,
Shropshire, TF13 6DG
Area of Operation: UK (Excluding Ireland)
Tel: 01952 728611
Fax: 01952 728361
Email: enquire@wenlocklime.co.uk
Web: www.wenlocklime.co.uk
Product Type: 2, 3, 4

MERCHANTS

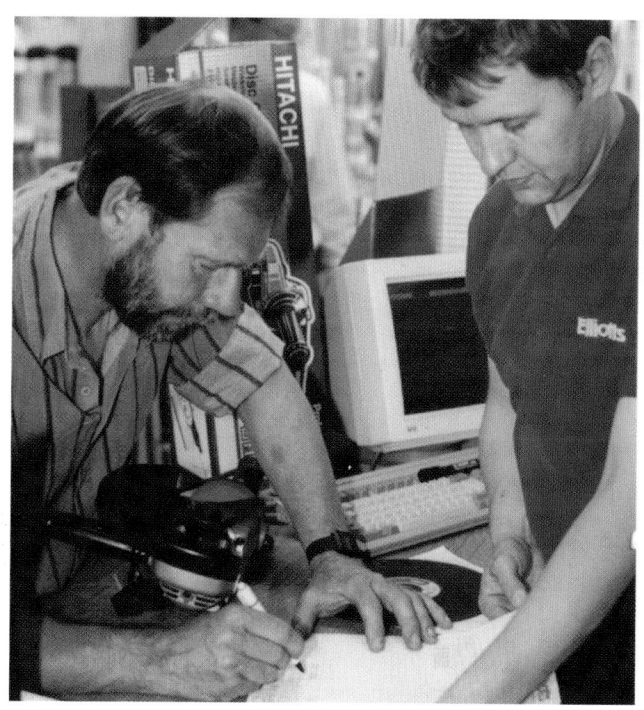

Image courtesy of Independent Builders' Merchants (02380 226852)

MERCHANTS, TOOLS & EQUIPMENT

SPONSORED BY INDEPENDENT BUILDERS' MERCHANTS
Tel 02380 226852. Web www.cbagroup.co.uk/homebuild

Salvage

Many salvage items can be far cheaper than their new equivalent, offering a real saving as well as the advantages of age and character. There is also a strong ecological argument for recycling salvaged goods. However, if you are building or renovating your home, what you really need to know is whether you can buy salvage that is reasonably priced, and relatively easy to work with, which will add character to your house without overstretching the budget.

You may be concerned about the condition of previously used architectural elements — and you would be right to worry, especially with materials such as reclaimed bricks, that are bought ready-wrapped on a pallet. Obviously with salvage there is bound to be a certain amount of wastage, so when purchasing pre-wrapped products you could try to negotiate an arrangement with the dealer before you buy, in case more than 15% of the goods are broken.

Beware, too, of functional salvage that may be missing vital parts. If you find the stove or radiator of your dreams in a general salvage yard, ask whether it is complete and in working order, and make sure the dealer will take it back if you discover otherwise. Generally speaking, it is best to buy functional products from experts, who recondition and test their vintage products thoroughly.

Vintage bathroom fittings need extra scrutiny before you buy. Look carefully at any loos and basins that have been stored outside, as frost can make all-but-invisible cracks in the porcelain. Baths, sinks and loos made in the days of imperial measurements and non-standard sizes can easily be converted to modern metric use, but check that your supplier has fitted, or offers, proper conversion connections. If the dealer doesn't offer a fitting service, check that your plumber understands the challenges of the task.

Before buying a reclaimed bathroom suite, bear in mind that the bathroom is a hot spot that can enhance or lower the value of your property, and estate agents advise that fixtures must be pristine, and preferably white. Similarly, in the kitchen, freshly re-enamelled Agas and sparkling Belfast sinks are very desirable touches that will encourage purchasers, but, aside from these iconic pieces, estate agents advise that kitchens must be clean and contemporary.

Dealers often display rather grand staircases or handsome iron spirals, but unless they were built for a house very similar to yours, they will only be usable if your design is partly, and perhaps expensively, adapted around them. Salvaged staircases are often more trouble than they are worth: finials and newel posts, decorative tread ends and strings can all be put to use more easily.

Salvage bargains sometimes cost you extra money in unexpected ways. If you are buying either doors or windows, check what your insurance policy says about security. Is it possible to fulfil physical security requirements with your salvaged treasures, or will they cause a hike in your premium? Be warned that old windows will rarely comply with modern insulation requirements: but if you are living in a listed building

and have a responsibility to replace 'like for like', Part 'L' of the Building Regulations need not trouble you.

As well as hidden costs, there may be hidden dangers to reclamation. Old electrical appliances, for instance, may be hazardous. However, you can purchase beautiful antique lighting quite safely as long as you ask a professional to check the wiring, replace the bulb holder and flex, if necessary, and earth the fitting.

Unless you have your own flatbed truck, you will be asking your dealer to deliver the larger purchases, and this cost can seriously eat into your budget. Clarify costs beforehand so that you can factor them into your budget, and always have the cheek to ask whether the dealer might throw in delivery costs if you place a large order

BUILDERS MERCHANTS

KEY
OTHER: ▽ Reclaimed   ⌐ On-line shopping
✎ Bespoke   ✋ Hand-made   ECO Ecological

ALPHA LIGHTWEIGHT BUILDING PRODUCTS
Horsehay, Telford, Shropshire, TF4 2PA
Area of Operation: UK (Excluding Ireland)
Tel: 01952 504082   **Fax:** 01952 505190
Email: info@alphalbp.co.uk

AQUALUX
Universal Point, Steelmans Road, off Park Lane,
Wednesbury, West Midlands, WS10 9UZ
Area of Operation: Europe
Tel: 0870 241 6131
Fax: 0870 241 6133
Email: david.hall@bhdgroup.co.uk
Web: www.aqualux.co.uk

ASHMEAD BUILDING SUPPLIES LIMITED
Portview Road, Avonmouth, Bristol, BS11 9LD
Area of Operation: South West England
and South Wales
Tel: 0117 982 8281   **Fax:** 0117 982 0135
Email: avon@ashmead.co.uk
Web: www.ashmead.co.uk

B & Q PLC
Portswood House, 1 Hampshire Corporate Park,
Chandlers Ford, Eastleigh, Hampshire, SO5 3YX
Area of Operation: UK & Ireland
Tel: 02380 256256
Fax: 02380 256020
Email: melanie.james@b-and-q.co.uk
Web: www.diy.com ⌐

BRADFORDS BUILDING SUPPLIES
96 Hendford Hill, Yeovil, Somerset, BA20 2QT
Area of Operation: South West England
and South Wales
Tel: 01935 845245   **Fax:** 01935 845242
Email: marketing@bradfords.co.uk
Web: www.bradfords.co.uk

BUILD CENTER
Trowel House, Kettering Parkway,
Kettering, Northamptonshire, NN15 6XR
Area of Operation: UK (Excluding Ireland)
Tel: 01536 311300
Fax: 01536 311345
Email: selfbuild@centers.co.uk
Web: www.buildercenter.co.uk ⌐

BUILDBASE LTD
Watlington Road, Cowley, Oxford, Oxfordshire, OX4 6LN
Area of Operation: UK (Excluding Ireland)
Tel: 0800 107 2255
Fax: 01865 747594
Email: tony.newcombe@buildbase.co.uk
Web: www.buildbase.co.uk

BUILDERS MERCHANTS FEDERATION
15 Soho Square, London, W1D 3HL
Area of Operation: UK (Excluding Ireland)
Tel: 0870 901 3380   **Fax:** 0207 734 2766
Email: info@bmf.org.uk
Web: www.bmf.org.uk

CARE DESIGN
Moorgate, Ormskirk, Lancashire, L39 4RX
Area of Operation: UK & Ireland
Tel: 01695 579061
Fax: 01695 570489
Email: caredesign@clara.net

COOPER CLARKE GROUP LTD
Bloomfield Road, Farnworth, Bolton,
Lancashire, BL4 9LP
Area of Operation: UK (Excluding Ireland)
Tel: 01204 862222
Fax: 01204 795296
Email: bolton@cooperclarke.co.uk
Web: www.heitonuk.com

ELLIOTT BROTHERS LTD
Millbank Wharf, Northam, Southampton,
Hampshire, SO14 5AG
Area of Operation: South East England,
South West England and South Wales
Tel: 02380 226852
Fax: 02380 638780
Email: laurenh@elliott-brothers.co.uk
Web: www.elliotts.uk.com

FIRST STOP BUILDERS MERCHANTS
Queens Drive, Kilmarnock, Ayrshire, KA1 3XA
Area of Operation: Scotland
Tel: 01563 534818   **Fax:** 01563 537848
Email: admin@firststopbm.co.uk
Web: www.firststopbm.co.uk

GIBBS AND DANDY PLC
226 Dallow Road, Luton, Bedfordshire, LU1 1YB
Area of Operation: South East England
Tel: 01582 798798   **Fax:** 01582 798799
Email: luton@gibbsanddandy.com
Web: www.gibbsanddandy.com

GO FIX IT LTD
Unit 10 Castle Industrial Estate, Louvain Street,
Off Oldham Road, Failsworth, Manchester, M35 0HB
Area of Operation: UK (Excluding Ireland)
Tel: 0161 681 4109   **Fax:** 0161 681 8169
Email: admin@gofixit.co.uk
Web: www.gofixit.co.uk ⌐

GRANT & STONE LTD
Head Office, Unit 1, Blenheim Road, Cressex,
High Wycombe, Buckinghamshire, HP12 3RS
Area of Operation: UK (Excluding Ireland)
Tel: 01494 441191
Fax: 01494 536543
Email: info@grantandstone.co.uk
Web: www.grantandstone.co.uk

GREYSLATE & STONE SUPPLIES
Y Maes, The Square, Blaenau Ffestiniog,
Gwynedd, LL41 3UN
Area of Operation: UK & Ireland
Tel: 01766 830521   **Fax:** 01766 831706
Email: greyslate@slateandstone.net
Web: www.slateandstone.net

JACKSON BUILDING CENTRES
Pelham House, Canwick Road,
Lincoln, Lincolnshire, LN5 8HG
Area of Operation: UK (Excluding Ireland)
Tel: 01522 511115
Fax: 01522 559156
Web: www.jacksonbc.co.uk

JAMES BURRELL LTD
Head Office, Deptford Road,
Gateshead, Tyne & Wear, NE8 2BR
Area of Operation: North East England
Tel: 0191 477 2249   **Fax:** 0191 477 4816
Email: jamesburrell@compuserve.com
Web: www.jamesburrell.com

LEGEND INDUSTRIES LTD
Wembley Point, One Harrow Road,
Wembley, Greater London, HA9 6DE
Area of Operation: UK & Ireland
Tel: 0208 903 3344   **Fax:** 0208 900 2120
Email: sales@theproductzone.com
Web: www.theproductzone.com

MACKENZIE DEAN LTD
Satinstown Farm Business Centre, Burwash Road,
Broad Oak, Heathfield, East Sussex, TN21 8RU
Area of Operation: UK & Ireland
Tel: 01435 862244
Fax: 01435 867781
Email: info@mackenziedean.co.uk
Web: www.mackenziedean.co.uk

MERRITT & FRYERS LTD
Firth Street Works, Skipton,
North Yorkshire, BD23 2PX
Area of Operation: North East England,
North West England and North Wales
Tel: 01756 792485
Fax: 01756 700391

N & C BUILDING PRODUCTS LTD
41-51 Freshwater Road, Chadwell Heath,
Romford, Essex, RM8 1SP
Area of Operation: Worldwide
Tel: 0208 586 4600   **Fax:** 0208 586 4646
Email: mark.spradbery@nichollsandclarke.com
Web: www.ncdirect.co.uk ⌐

OLD HOUSE STORE LTD
Hampstead Farm, Binfield Heath,
Henley on Thames, Oxfordshire, RG9 4LG
Area of Operation: Worldwide
Tel: 0118 969 7711   **Fax:** 0118 969 8822
Email: info@oldhousestore.co.uk
Web: www.oldhousestore.co.uk ⌐

ROBERT PRICE BUILDERS' MERCHANTS
Park Road, Abergavenny, Monmouthshire, NP7 5PF
Area of Operation: South Wales
and Bordering Counties
Tel: 01873 858585   **Fax:** 01873 856854
Email: info@robert-price.co.uk
Web: www.robert-price.co.uk

ROBERT PRICE BUILDERS' MERCHANTS
Area of Operation: South Wales & Bordering Counties
Tel: 01873 858585
Fax: 01873 856854
Email: info@robert-price.co.uk
Web: www.robert-price.co.uk

Robert Price Builders' Merchants have over 40
years experience helping self-builders in South
Wales build their dream homes. Self Build guide
available.

THE BUILDING CENTRE
26 Store Street, off Tottenham
Court Road, London, WC1E 7BT
Area of Operation: Worldwide
Tel: 0207 692 4000
Fax: 0207 631 0329
Email: information@buildingcentre.co.uk
Web: www.buildingcentre.co.uk ⌐

THE HOMEBUILDER'S HANDBOOK
THIRD EDITION
THE SOURCEBOOK FOR PROFESSIONALS, SELF-BUILDERS AND HOME RENOVATORS
1,000S OF ESSENTIAL CONTACTS

**Don't forget to let us
know about companies
you think should be listed
in the next edition**
email: customerservice@centaur.co.uk

TILES UK LTD
1-13 Montford Street, Off South Langworthy Road,
Salford, Greater Manchester, M50 2XD
Area of Operation: UK (Excluding Ireland)
Tel: 0161 872 5155
Fax: 0161 848 7948
Email: info@tilesuk.com
Web: www.tilesuk.com ⌐

TRAVIS PERKINS PLC
Head Office, Lodge Way House, Lodge Way,
Harlestone Road, Northampton,
Northamptonshire, NN5 7UG
Area of Operation: UK (Excluding Ireland)
Tel: 01604 752424
Fax: 01604 591180
Email: nicola.mcgonagle@travisperkins.co.uk
Web: www.travisperkins.co.uk ⌐

UK BUILDING MATERIALS.COM
5 Belmont Drive, Belmont Drive,
Taunton, Somerset, TA1 4QB
Area of Operation: South West England
and South Wales
Tel: 01823 333364
Fax: 01823 333381
Email: sales@ukbuildingmaterials.com
Web: www.ukbuildingmaterials.com

VALUE DIY LIMITED
Unit 4, Phillips House, Chapel Lane,
Emley, West Yorkshire, HD8 9ST
Area of Operation: UK (Excluding Ireland)
Tel: 0845 644 2306
Fax: 01924 844606
Email: sales@valuediy.co.uk
Web: www.valuediy.co.uk ⌐

PLUMBERS MERCHANTS

KEY
OTHER: ▽ Reclaimed   ⌐ On-line shopping
✎ Bespoke   ✋ Hand-made   ECO Ecological

ANDERSON FLOORWARMING LIMITED
IPPEC Scottish System House,
Unit 119, Atlas Express Industrial Estate,
1 Rutherglen Road, Glasgow, G73 1SX
Area of Operation: Scotland
Tel: 0141 647 6716   **Fax:** 0141 647 6751
Email: scottishbranch@ippec.co.uk
Web: www.ippec.co.uk ⌐

ANDY PLUMB
Hathern House, Hathern Ware Industrial Estate,
Rempstone Road, Sutton Bonnington,
Leicestershire, LE12 5EH
Area of Operation: Worldwide
Tel: 0871 425 1672   **Fax:** 01509 844102
Email: sales@andyplumb.co.uk
Web: www.andyplumb.co.uk ⌐

AQUALUX
Universal Point, Steelmans Road, off Park Lane,
Wednesbury, West Midlands, WS10 9UZ
Area of Operation: Europe
Tel: 0870 241 6131   **Fax:** 0870 241 6133
Email: david.hall@bhdgroup.co.uk
Web: www.aqualux.co.uk

B & Q PLC
Portswood House, 1 Hampshire Corporate Park,
Chandlers Ford, Eastleigh, Hampshire, SO5 3YX
Area of Operation: UK & Ireland
Tel: 02380 256256   **Fax:** 02380 256020
Email: melanie.james@b-and-q.co.uk
Web: www.diy.com ⌐

BATHROOM STUDIOS
139 Old Road, Clacton-On-Sea, Essex, CO15 3AX
Area of Operation: South East England
Tel: 01255 434435
Email: sales@bathroomstudios.co.uk
Web: www.bathroomstudios.co.uk

MERCHANTS, TOOLS & EQUIPMENT

INDEPENDENT BUILDERS' MERCHANTS
helping homebuilders locally

Tippers
Europa Way
Lichfield Staffordshire
WS14 9TZ
Tel 01543 440000
info@tippersbm.co.uk

ROBERT PRICE BUILDERS' MERCHANTS
Park Road
Abergavenny
South Wales NP7 5PF
Tel 01873 858585
www.robert-price.co.uk

Bradfords
96 Hendford Hill Yeovil
Somerset
BA20 2QT
Tel 01935 845245
www.bradfords.co.uk

RGB Building Supplies
Rolles Quay
Barnstaple Devon
EX31 1JD
Tel 01271 375501
www.rgbltd.co.uk

Merchants that understand Self Build

Our aim is to help you turn your dreams of a new home into reality. By choosing an Independent Builders Merchant you will be dealing with people who understand Self Build.
Our knowledge of your local area gives us the ability to help and guide you to ensure your project runs smoothly and efficiently.

JOHN A. STEPHENS LTD.
John A Stephens Ltd
Castle Meadow Road,
Nottingham, NG2 1AG
Tel 0115 9412861
www.johnastephens.co

Typical services for Self Builders

- Help and advice on finding land
- Free Quantity Take off service provided on receipt of full working drawings
- Special Events tailored for Self Build - including Club Meetings, Newsletters, Building Seminars and much more
- Leading Manufacturers trust us to distribute their materials Nationwide
- Friendly experienced staff
- Accounts tailored for Self Build
- Specialist Product areas - Brick Libraries, Kitchen & Bathroom Showrooms, Drainage, Plumbing, Heating, Windows, Doors, Roofing etc.
- Credit Accounts for Self Builders
- Regular visits from trained Sales Representatives
- Free crane offload deliveries from Merchants branches
- Provision of SAP calculations
- Assistance with VAT claim back
- All this and competitive pricing

Elliotts
Millbank Wharf
Northam Southampton
SO14 5AG
Tel 023 8022 6852
www.elliotts.uk.com

Beesley & Fildes Ltd
Wilson Road
Huyton Liverpool
L36 6AF
Tel 0151 480 8304
www.beesleyandfildes.co.

gibbs & dandy
PO Box 17
226 Dallow Road
Luton LU1 1JG
Tel 01582 798798
www.gibbsanddandy.cor

RIDGEONS TIMBER & BUILDERS MERCHANTS
Nuffield Road
Trinity Hall Ind Est
Cambridge CB4 1TS
Tel 01223 466000
www.ridgeons.co.uk

COVERS
Sussex House
Quarry Lane
Chichester PO19 2PE
Tel 01243 785141
www.dcover.co.uk

HOMEBUILDING & RENOVATING SHOW

For information on Homebuilding and Renovating Shows please contact your local merchant who will be pleased to let you know dates for the National and Regional Events.

Contact your local Independent Builders Merchant for more information about the services they can offer and also to request a copy our FREE 'Homebuilders Guide'.

Full details are on the Independent Builders Merchants Website

www.cbagroup.co.uk/homebuild

BEESLEY & FILDES LTD

"Established since 1820, an independent family run business with branches throughout the North West of England, offering a complete package of building products, timber, roofing, plumbing and bathrooms. We provide a prompt delivery service and a comprehensive sales support."

Head Office
Beesley & Fildes Ltd
Wilson Road
Huyton
Liverpool
L36 6AF
Tel: 0151 480 8304 Fax: 0151 481 0248
Email: jstanton@beesleyandfildes.co.uk
Web: www.beesleyandfildes.co.uk

BRADFORDS BUILDING SUPPLIES

Bradfords has been supplying building materials for over 200 years and is one of the UK's largest independent builders merchants with 26 branches throughout the South West, Herefordshire and Worcestershire. We have developed a renowned specialist expertise in home-build, having been the principal supplier to thousands of projects during the last decade.

Bradfords Building Supplies
96 Hendford Hill,
Yeovil,
Somerset,
BA20 2QR
Tel: 01935 845245
Fax: 01935 845242
Web: www.bradfords.co.uk

COVERS LTD

Covers is one of the largest privately owned timber and builders merchants on the south coast, with a total of ten depots situated in East and West Sussex, Hampshire and the Channel Islands. Covers stock a complete range of timber and building materials for the construction and allied trades, with comprehensive flooring, bathroom, and kitchen showrooms.

For more information, or to order a catalogue contact Covers on 01243 785141.

Covers Ltd
Sussex House,
Quarry Lane,
Chichester,
West Sussex
PO19 2PE
Tel: 01243 785141
Fax: 01243 531151
Web: www.dcover.co.uk

ELLIOTT BROTHERS LIMITED

Elliotts are Hampshire's leading independent Builders Merchant ideally located for self builders across Hampshire and Dorset. Established in 1842 and still family run with 10 branches across the county. To join our Self Build Club contact Lauren Haines on 023 8038 5305.

Elliott Brothers Limited
Head Office
Millbank Wharf
Northam
Southampton   SO14 5AG
Web: www.elliotts.uk.com

GIBBS & DANDY PLC

Gibbs & Dandy has everything for the home builder: building materials, timber, plumbing and heating supplies, sanitaryware, electrical fittings, paint and glass. Branches branches in Luton, Bedford, St Ives (Cambs), St Neots, Northampton, Slough, Maidenhead, Henley, Brackley and Kettering.

Gibbs and Dandy plc
PO Box 17
226 Dallow Road
Luton
LU1 1JG
Tel: 01582 798798 Fax: 01582 798799
Email: mail@gibbsanddandy.com
Web: www.gibbsanddandy.com

JOHN A. STEPHENS LTD.

JOHN A STEPHENS

"For over 35 years John A Stephens has provided a consistently high quality service to the building and construction industry from its five acre site in central Nottingham and is regarded as one of the largest and most comprehensively stocked depots in the region."

John A Stephens
Castle Meadow Road,
Nottingham
NG2 1AG
Tel: 0115 9412861
Web: www.johnastephens.co.uk

RAWLE GAMMON & BAKER

Rawle Gammon & Baker started in 1850 in Barnstaple, Devon as timber importers with premises beside the river and has gone from strength to strength over the generations, expanding gradually to sell the whole range of building materials. Still family owned, the Company now has ten branches, with branch number eleven due to open in Tiverton, Devon in 2006. RGB covers Devon, East Cornwall and West Somerset, with its own sawmills at Chapelton, North Devon. Under the Managing Directorship of Giles Isaac, RGB is proud to offer a wide range of building materials at competitive prices and our helpful and knowledgeable staff is very well placed to aid all builders in achieving their goals and avoiding pitfalls.

RGB Holdings Ltd,
Gammon House,
Riverside Road,
Barnstaple,
Devon,
EX31 1QN
Tel: 01271 313000
Fax: 01271 329982
E-mail: rgb@rgbltd.co.uk

RIDGEONS

"With 18 trading branches based in East Anglia, the Ridgeon Group is the largest family owned independent timber and builders merchant in the United Kingdom".

For your nearest contact visit www.ridgeons.co.uk

ROBERT PRICE BUILDERS' MERCHANTS

Robert Price Builders' Merchants have over 40 years experience helping self-builders in South Wales build their dream homes. Self Build guide available.

Margaret Roy
Marketing Manager
Park Road
Abergavenny
NP7 5PF
01873 858585
01873 856854
07966 117952

TIPPERS

Midlands based Builders Merchant offering a full range of building materials including bricks, blocks, timber & joinery, landscaping materials, plumbing & heating and kitchens & bathrooms.

Tel: 01543 440000
Email: info@tippersbm.co.uk

MERCHANTS, TOOLS & EQUIPMENT

Beautiful bathrooms from B&Q

At B&Q we've been creating **beautiful bathrooms** for over 30 years – and now our vast range is **inspirational** and more **desirable** than ever! You'll be **surprised** what you can find.

View our **extensive range** in store or online at **www.diy.com**

B&Q

BUILDBASE LTD
Watlington Road, Cowley, Oxford,
Oxfordshire, OX4 6LN
Area of Operation: UK (Excluding Ireland)
Tel: 0800 107 2255
Fax: 01865 747594
Email: tony.newcombe@buildbase.co.uk
Web: www.buildbase.co.uk

CARE DESIGN
Moorgate, Ormskirk, Lancashire, L39 4RX
Area of Operation: UK & Ireland
Tel: 01695 579061 **Fax:** 01695 570489
Email: caredesign@clara.net

COMFORT ZONE SYSTEMS LTD
Suite 5 Caxton House, 143 South Coast Road,
Peacehaven, East Sussex, BN10 8NN
Area of Operation: Europe
Tel: 01273 580888
Fax: 01273 580848
Email: clive@czsystems.co.uk
Web: www.czsystems.co.uk

DISCOUNTED HEATING
57 Faringdon Road, Plymouth, Devon, PL4 9ER
Area of Operation: UK (Excluding Ireland)
Tel: 0870 042 8884 **Fax:** 0870 330 5815
Email: marc@discountedheating.co.uk
Web: www.discountedheating.co.uk

GIBBS AND DANDY PLC
226 Dallow Road, Luton, Bedfordshire, LU1 1YB
Area of Operation: South East England
Tel: 01582 798798
Fax: 01582 798799
Email: luton@gibbsanddandy.com
Web: www.gibbsanddandy.com

GO FIX IT LTD
Unit 10 Castle Industrial Estate, Louvain Street,
Off Oldham Road, Failsworth,
Manchester, M35 0HB
Area of Operation: UK (Excluding Ireland)
Tel: 0161 681 4109
Fax: 0161 681 8169
Email: admin@gofixit.co.uk
Web: www.gofixit.co.uk

GRAHAM
4 Binley Street, Kirkstall Road, Leeds,
West Yorkshire, LS3 1LU
Area of Operation: UK (Excluding Ireland)
Tel: 0800 539766
Fax: 0113 243 0102
Email: customer.services@graham-group.co.uk.
Web: www.graham-group.co.uk

HEPWORTH PLUMBING PRODUCTS
Edlington Lane, Edlington, Doncaster,
South Yorkshire, DN12 1BY
Area of Operation: Worldwide
Tel: 01709 856300
Fax: 01709 856301
Email: info@hepworthplumbing.co.uk
Web: www.hepworthplumbing.co.uk

LEGEND INDUSTRIES LTD
Wembley Point, One Harrow Road,
Wembley, Greater London, HA9 6DE
Area of Operation: UK & Ireland
Tel: 0208 903 3344
Fax: 0208 900 2120
Email: sales@theproductzone.com
Web: www.theproductzone.com

PLUMB CENTER
Trowel House, Kettering Parkway,
Kettering, Northamptonshire, NN15 6XR
Area of Operation: UK (Excluding Ireland)
Tel: 0800 622557
Email: customerservices@wolseley.co.uk
Web: www.plumbcenter.co.uk

PUMP WORLD LTD
Unit 11, Woodside Road,
South Marston Business Park, Swindon,
Wiltshire, SN3 4WA
Area of Operation: UK & Ireland
Tel: 01793 820142
Fax: 01793 823800
Email: enquiries@pumpworld.co.uk
Web: www.pumpworld.co.uk

**RICHMONDS PLUMBING & HEATING
MERCHANTS LTD**
15-25 Carnoustie Place, Scotland Street,
Glasgow, City of Glasgow, G5 8PA
Area of Operation: Scotland
Tel: 0141 429 7441
Fax: 0141 420 1406
Email: sales@richmonds-phm.co.uk
Web: www.richmonds-phm.co.uk

ROOFING MERCHANTS

KEY
OTHER: ▽ Reclaimed   ◌ On-line shopping
✎ Bespoke ✋ Hand-made   ECO Ecological

ELLIOTT BROTHERS LTD
Millbank Wharf, Northam, Southampton,
Hampshire , SO14 5AG
Area of Operation: South East England,
South West England and South Wales
Tel: 02380 226852
Fax: 02380 638780
Email: laurenh@elliott-brothers.co.uk
Web: www.elliotts.uk.com

GREYSLATE & STONE SUPPLIES
Y Maes, The Square, Blaenau Ffestiniog,
Gwynedd, LL41 3UN
Area of Operation: UK & Ireland
Tel: 01766 830521
Fax: 01766 831706
Email: greyslate@slateandstone.net
Web: www.slateandstone.net

MILBANK ROOFS
Hargrave Meadow Cottage, Church Lane,
Hargrave, Bury St Edmunds, Suffolk, IP29 5HH
Area of Operation: UK & Ireland
Tel: 01284 852 505
Fax: 01284 850 021
Email: info@milbankroofs.com
Web: www.milbankroofs.com

**ROBERT PRICE TIMBER
AND ROOFING MERCHANTS**
The Wood Yard, Forest Road, Taffs Well,
Cardiff, CF15 7YE
Area of Operation: South West England
and South Wales
Tel: 02920 811681
Fax: 02920 813605
Email: sales@robert-price.co.uk
Web: www.robert-price.co.uk

SIG ROOFING SUPPLIES GROUP
Harding Way, St. Ives, Cambridge,
Cambridgeshire, PE27 3YJ
Area of Operation: UK (Excluding Ireland)
Tel: 01480 466777
Fax: 01480 302576
Email: danielwaters@sigroofing.co.uk
Web: www.sigplc.co.uk

TIMBER MERCHANTS

KEY
OTHER: ▽ Reclaimed   ◌ On-line shopping
✎ Bespoke ✋ Hand-made   ECO Ecological

B & Q PLC
Portswood House, 1 Hampshire Corporate Park,
Chandlers Ford, Eastleigh, Hampshire , SO5 3YX
Area of Operation: UK & Ireland
Tel: 02380 256256
Fax: 02380 256020
Email: melanie.james@b-and-q.co.uk
Web: www.diy.com

BOARD CENTRAL
Chiltern Business Centre, Couching Street,
Watlington, Oxfordshire, OX49 5PX
Area of Operation: Greater London,
South East England
Tel: 0845 4588016
Fax: 01844 354112
Email: howardmorrice@hotmail.com

BSW TIMBER
Cargo, Carlisle, Cumbria, CA6 4BA
Area of Operation: UK (Excluding Ireland)
Tel: 01228 673366
Fax: 01228 673353
Email: marketing@bsw.co.uk
Web: www.bsw.co.uk

CARE DESIGN
Moorgate, Ormskirk, Lancashire, L39 4RX
Area of Operation: UK & Ireland
Tel: 01695 579061
Fax: 01695 570489
Email: caredesign@clara.net

DEVON HARDWOODS LTD
Dotton, Colaton Raleigh, Sidmouth, Devon, EX10 0JH
Area of Operation: South West England
and South Wales
Tel: 01395 568991
Fax: 01395 567881
Email: sales@devonhardwoods.ltd.uk

ELLIOTT BROTHERS LTD
Millbank Wharf, Northam, Southampton,
Hampshire, SO14 5AG
Area of Operation: South East England,
South West England and South Wales
Tel: 02380 226852
Fax: 02380 638780
Email: laurenh@elliott-brothers.co.uk
Web: www.elliotts.uk.com

GIBBS AND DANDY PLC
226 Dallow Road, Luton, Bedfordshire, LU1 1YB
Area of Operation: South East England
Tel: 01582 798798
Fax: 01582 798799
Email: luton@gibbsanddandy.com
Web: www.gibbsanddandy.com

HARDWOOD WAREHOUSE
Unit 9c Haig Business Park, Markinch, Fife, KY7 6AQ
Area of Operation: UK & Ireland
Tel: 01592 769100
Fax: 01592 769399
Email: enquiries@hardwoodwarehouse.co.uk
Web: www.hardwoodwarehouse.co.uk

HOEBEEK (UK) LTD
Hoebeek House, Castlefields Lane,
Bingley, West Yorkshire, BD16 2AB
Area of Operation: UK & Ireland
Tel: 01274 561999   **Fax:** 01274 562999
Email: hoebeekukltd@msn.com
Web: www.hoebeek.be

HONEYSUCKLE BOTTOM SAWMILL LTD
Honeysuckle Bottom, Green Dene, East Horsley,
Leatherhead, Surrey, KT24 5TD
Area of Operation: Greater London,
South East England
Tel: 01483 282394
Fax: 01483 282394
Email: honeysucklemill@aol.com
Web: www.easisites.co.uk/honeysucklebottomsawmill

JOHN BODDY TIMBER LTD
Riverside Sawmills, Boroughbridge,
North Yorkshire, YO51 9LJ
Area of Operation: UK (Excluding Ireland)
Tel: 01423 322370 **Fax:** 01423 324334
Email: info@john-boddy-timber.ltd.uk
Web: www.john-boddy-timber.ltd.uk

LANARKSHIRE HARDWOODS
Girdwoodend Farm, Auchengray, Carnwath,
Lanark, Lanarkshire, ML11 8LL
Area of Operation: Scotland
Tel: 01501 785460
Email: patrickbaxter@girdwoodend.wanadoo.co.uk
Web: www.lanarkshirehardwoods.co.uk

MCKAY FLOORING LTD
8 Harmony Square, Govan, City of Glasgow, G51 3LW
Area of Operation: UK & Ireland
Tel: 0141 440 1586   **Fax:** 0141 425 1020
Email: enquiries@mckayflooring.co.uk
Web: www.mckayflooring.co.uk

N.P. TIMBER CO. LTD
Welham Lane, Great Bowden,
Market Harborough, Leicestershire, LE16 7HS
Area of Operation: Worldwide
Tel: 01858 468064   **Fax:** 01858 469408
Email: sales@nptimber.co.uk
Web: www.nptimber.co.uk

NORTH YORKSHIRE TIMBER
Standard House, Thurston Road, Northallerton
Business Park, Northallerton, North Yorkshire, DL6 2NA
Area of Operation: UK (Excluding Ireland)
Tel: 01609 780777   **Fax:** 01609 777888
Email: sales@nytimber.co.uk
Web: www.nytimber.co.uk

NORTHWOOD FORESTRY LTD
Goose Green Lane, Nr Ashington, Pulborough,
West Sussex, RH20 2LW
Area of Operation: South East England
Tel: 01798 813029   **Fax:** 01798 813139
Email: enquiries@northwoodforestry.co.uk
Web: www.northwoodforestry.co.uk

PANEL AGENCY LIMITED
Maple House, 5 Over Minnis, New Ash Green,
Longfield, Kent, DA3 8JA
Area of Operation: UK & Ireland
Tel: 01474 872578   **Fax:** 01474 872426
Email: sales@panelagency.com
Web: www.panelagency.com

PINE SUPPLIES
Lower Tongs Farm, Longshaw Ford Road,
Smithills, Bolton, Lancashire, BL1 7PP
Area of Operation: UK (Excluding Ireland)
Tel: 01204 841416   **Fax:** 01204 845814
Email: pine-info@telinco.co.uk
Web: www.pine-supplies.co.uk

**ROBERT PRICE TIMBER AND ROOFING
MERCHANTS**
The Wood Yard, Forest Road, Taffs Well,
Cardiff, CF15 7YE
Area of Operation: South West England & South Wales
Tel: 02920 811681   **Fax:** 02920 813605
Email: sales@robert-price.co.uk
Web: www.robert-price.co.uk

SCOTTISH WOOD
Inzievar Woods, Oakley, Dunfermline, Fife, KY12 8HB
Area of Operation: Scotland
Tel: 01383 851328   **Fax:** 01383 851339
Email: enquiries@scottishwood.co.uk
Web: www.scottishwood.co.uk

T.BREWER & CO
Timber Mill Way, Gauden Road,
Clapham, London, SW4 6LY
Area of Operation: Greater London,
South East England
Tel: 0207 720 9494
Fax: 0207 622 0426
Email: clapham@tbrewer.co.uk
Web: www.tbrewer.co.uk

THE CARPENTRY INSIDER - AIRCOMDIRECT
1 Castleton Crescent, Skegness,
Lincolnshire, PE25 2TJ
Area of Operation: Worldwide
Tel: 01754 767163
Email: aircom8@hotmail.com
Web: www.carpentry.tk

THE ROUNDWOOD TIMBER COMPANY LTD
The Roundwood Timber Company Ltd,
Roundwood, Newick Lane, Mayfield,
East Sussex, TN20 6RG
Area of Operation: UK & Ireland
Tel: 01435 867072
Fax: 01435 864708
Email: sales@roundwoodtimber.com
Web: www.roundwoodtimber.com

THE SPA & WARWICK TIMBER CO LTD
Harriott Drive, Heathcote Industrial Estate,
Warwick, Warwickshire, CV34 6TJ
Area of Operation: Midlands & Mid Wales
Tel: 01926 883876
Fax: 01926 450831
Email: sales@spa-warwick.co.uk

THOROGOOD TIMBER PLC
Colchester Road, Ardleigh,
Colchester, Essex, C07 7PQ
Area of Operation: East England
Tel: 01206 233100
Fax: 01206 233115
Email: barry@thorogood.co.uk
Web: www.thorogood.co.uk

TREESPANNER TIMBER
East Cottage, Dry Hill Farm, Moons Lane,
Dormansland, Surrey, RH7 6PD
Area of Operation: South East England
Tel: 01342 871529
Email: charles.willment@virgin.net
Web: www.treespanner.co.uk

VINCENT TIMBER LTD
8 Montgomery Street, Birmingham,
West Midlands, B11 1DU
Area of Operation: UK & Ireland
Tel: 0121 772 5511
Fax: 0121 766 6002
Email: enquires@vincenttimber.co.uk
Web: www.vincenttimber.co.uk

WOODHOUSE TIMBER
Unit 6 Quarry Farm Industrial Estate,
Staplecross Rd, Bodiam,
East Sussex, TN32 5RA
Area of Operation: UK (Excluding Ireland)
Tel: 01580 831700
Fax: 01580 830365
Email: info@woodhousetimber.co.uk
Web: www.woodhousetimber.co.uk

ARCHITECTURAL ANTIQUES & SALVAGE YARDS

KEY

PRODUCT TYPES: 1= Bathroom & Accessories
2 = Chimney Pieces, Fireplaces and Grates
3 = Church Salvage   4 = Doors and Door
Furniture   5 = Garden   6 = Furniture & Mirrors
7 = Kitchen and Accessories   8 = Lighting
9 = Staircases   10 = Statuary
11 = Windows & Window Furniture
12 = Architectural Metalwork
13 = Architectural Stone and Terracotta
14 = Architectural Woodwork and Panelling
15 = Bricks   16 = Flagstones and Floor Tiles
17 = Roof Slates and Tiles   18 = Stone
19 = Timber   20 = Other

OTHER: ▽ Reclaimed   ✋ On-line shopping
✍ Bespoke   ✋ Hand-made   ECO Ecological

ACE RECLAMATION
Pineview, Barrack Road, West Parley,
Ferndown, Dorset, BH22 8UB
Area of Operation: Worldwide
Tel: 01202 579222   **Fax:** 01202 582043
Email: info@acereclamation.com
Web: www.acereclamation.com
Product Type: 1, 2, 4, 5, 10, 13, 14, 15, 16, 17, 18, 19

ANTIQUE BUILDINGS LTD
Dunsfold, Godalming, Surrey, GU8 4NP
Area of Operation: UK & Ireland
Tel: 01483 200477
Email: info@antiquebuildings.com
Web: www.antiquebuildings.com
Product Type: 14, 15, 16, 19

ARCHITECTURAL ANTIQUES
351 King Street, Hammersmith, London, W6 9NH
Area of Operation: UK (Excluding Ireland)
Tel: 0208 741 7883   **Fax:** 0208 741 1109
Email: info@aa-fireplaces.co.uk
Web: www.aa-fireplaces.co.uk
Product Type: 1, 2, 6, 8

ARCSAL.COM
3 Lower Street, Rode, Somerset, BA11 6PU
Area of Operation: Worldwide
Tel: 07966 416 745
Email: info@arcsal.com
Web: www.arcsal.com
Product Type: 1, 3, 4, 5, 6, 7, 8, 9, 10, 11, 12, 13, 14, 15, 16, 17, 18, 19

ATC (MONMOUTHSHIRE) LTD
Unit 2, Mayhill Industrial Estate,
Monmouth, Monmouthshire, NP25 3LX
Area of Operation: Worldwide
Tel: 01600 713036   **Fax:** 01600 715512
Email: info@floorsanddecking.com
Web: www.floorsanddecking.com
Product Type: 14

BCA MATERIAUX ANCIENS S.A.
Route de Craon, L'Hotellerie-de-Flee,
Maine et Loire, France, 49500
Area of Operation: Worldwide
Tel: +33 233 947 400   **Fax:** +33 233 944 656
Email: enquiries@bca-materiauxanciens.com
Web: www.bca-antiquematerials.com

BINGLEY ANTIQUES
Springfield Farm Estate, Haworth,
West Yorkshire, BD21 5PT
Area of Operation: UK (Excluding Ireland)
Tel: 01535 646666   **Fax:** 01535 648527
Email: john@bingleyantiques.com
Web: www.bingleyantiques.com
Product Type: 2, 3, 4, 5, 6, 9, 11, 12, 13, 14, 18

CAWARDEN BRICK CO. LTD
Cawarden Springs Farm, Blithbury Road,
Rugeley, Staffordshire, WS15 3HL
Area of Operation: UK (Excluding Ireland)
Tel: 01889 574066   **Fax:** 01889 575695
Email: home-garden@cawardenreclaim.co.uk
Web: www.cawardenreclaim.co.uk

CHESHIRE BRICK & SLATE
Brook House Farm, Salters Bridge,
Tarvin Sands, Tarvin, Cheshire, CH3 8NR
Area of Operation: UK (Excluding Ireland)
Tel: 01829 740883   **Fax:** 01829 740481
Email: enquiries@cheshirebrickandslate.co.uk
Web: www.cheshirebrickandslate.co.uk
Product Type: 1, 2, 3, 4, 5, 6, 8, 10, 11, 13, 14, 15, 16, 17, 18, 19, 20

CONSERVATION BUILDING PRODUCTS
Forge Lane, Cradley Heath, Warley,
West Midlands, B64 5AL
Area of Operation: Midlands & Mid Wales
Tel: 01384 569551   **Fax:** 01384 410625
Email: conservationbuildingproducts@ukonline.co.uk
Web: www.conservationbuildingproducts.co.uk
Product Type: 1, 2, 3, 4, 5, 6, 7, 8, 9, 10, 11, 12, 13, 14, 15, 16, 17, 18, 19, 20

COX'S ARCHITECTURAL SALVAGE YARD LTD.
10 Fosseway Business Park, Moreton in Marsh,
Gloucestershire, GL56 9NQ
Area of Operation: UK & Ireland
Tel: 01608 652505
Fax: 01608 652881
Email: info@coxsarchitectural.co.uk
Web: www.coxsarchitectural.co.uk
Product Type: 1, 2, 3, 4, 5, 6, 8, 9, 11, 13, 14, 16, 18, 19, 20

COX'S ARCHITECTURAL SALVAGE YARD LTD
Area of Operation: UK & Ireland
Tel: 01608 652505   **Fax:** 01608 652881
Email: info@coxsarchitectural.co.uk
Web: www.coxsarchitectural.co.uk
Product Type: 1, 2, 3, 4, 5, 6, 8, 9, 11, 13, 14, 16, 18, 19, 20

Cox's Architectural Salvage Yard Ltd. (CASY) has been trading in Moreton in Marsh since 1992. With 12,500 sq. ft. of covered warehouse and 1/2 acre of outside yard, CASY offers one of the largest and most varied stocks of reclaimed building materials and architectural antiques in the country.

**CRONINS RECLAMATION
& SOLID WOOD FLOORING**
Preston Farm Court, Lower Road,
Little Bookham, Surrey, KT23 4EF
Area of Operation: Worldwide
Tel: 0208 614 4370   **Fax:** 01932 241918
Email: dfc1@supanet.com
Web: www.croninsreclamation.co.uk
Product Type: 1, 2, 4, 5, 6, 11, 14, 15, 16, 17, 19, 20

DISMANTLE AND DEAL DIRECT
108, London Road, Aston Clinton,
Buckinghamshire, HP22 5HS
Area of Operation: Worldwide
Tel: 01296 632 300
Email: info@ddd-uk.com
Web: www.ddd-uk.com
Product Type: 2, 4, 5, 8, 9, 10, 11, 12, 13, 14, 15, 16, 18, 19, 20

DRUMMONDS ARCHITECTURAL ANTIQUES LTD
Kirkpatrick Buildings, 25 London Road (A3),
Hindhead, Surrey, GU36 6AB
Area of Operation: Worldwide
Tel: 01428 609444
Fax: 01428 609445
Email: davidcox@drummonds-arch.co.uk
Web: www.drummonds-arch.co.uk
Product Type: 1, 2, 3, 4, 5, 6, 7, 8, 9, 10, 11, 12, 13, 14, 15, 16, 17, 18, 19, 20

GREYSLATE & STONE SUPPLIES
Y Maes, The Square, Blaenau Ffestiniog,
Gwynedd, LL41 3UN
Area of Operation: UK & Ireland
Tel: 01766 830521
Fax: 01766 831706
Email: greyslate@slateandstone.net
Web: www.slateandstone.net

HOLYROOD ARCHITECTURAL SALVAGE
Holyrood Business Park, 146 Duddingston Road
West, Edinborough, Lothian, EH16 4AP
Area of Operation: UK & Ireland
Tel: 0131 661 9305
Fax: 0131 656 9404
Email: holyroodsalvage@btconnect.com
Web: www.holyroodarchitecturalsalvage.com
Product Type: 1, 2, 3, 4, 5, 6, 10, 20

HOUSE OF CLOCKS
98 Dunstable Street, Ampthill,
Bedford, Bedfordshire, MK45 2JP
Area of Operation: Worldwide
Tel: 01525 403136
Email: houseofclocks@tiscali.co.uk
Web: www.houseofclocks.co.uk
Product Type: 6

LASSCO
Brunswick House, 30 Wandsworth Road,
Vauxhall, London, SW8 2LG
Area of Operation: Worldwide
Tel: 0207 394 2100
Fax: 0207 5017797
Email: brunswick@lassco.co.uk
Web: www.lassco.co.uk
Product Type: 1, 2, 4, 5, 6, 7, 8, 9, 10, 11, 12, 14

LAWSONS
Gorst Lane, Off New Lane,
Burscough, Ormskirk,
Lancashire, L40 0RS
Area of Operation: Worldwide
Tel: 01704 893998
Fax: 01704 892526
Email: info@traditionaltimber.co.uk
Web: www.traditionaltimber.co.uk
Product Type: 1, 2, 3, 4, 5, 6, 7, 9, 10, 11, 12, 13, 14, 15, 16, 17, 18, 19, 20

MINCHINHAMPTON ARCHITECTURAL SALVAGE CO
Cirencester Road, Aston Downs,
Chalford, Stroud, Gloucestershire, GL6 8PE
Area of Operation: Worldwide
Tel: 01285 760886
Fax: 01285 760838
Email: masco@catbrain.com
Web: www.catbrain.com
Product Type: 1, 2, 4, 5, 6, 9, 10, 11, 12, 13, 14, 16, 17, 18, 19

NATIONWIDE RECLAIM
Goosey Lodge Farm, Wyminton Lane,
Wymington, Bedfordshire, NN10 9LU
Area of Operation: UK (Excluding Ireland)
Tel: 01933 313121 **Fax:** 01933 623089
Email: info@nationwidereclaim.co.uk
Web: www.nationwidereclaim.co.uk
Product Type: 1, 2, 3, 4, 5, 6, 7, 8, 9, 10, 11, 12, 13, 14, 15, 16, 17, 18, 19

NOTTINGHAM RECLAIMS
St Albans Works, 181 Hartley Road,
Nottinghamshire, NG7 3DW
Area of Operation: UK (Excluding Ireland)
Tel: 0115 9790666 **Fax:** 0115 9791607
Email: nottm.aar@ntlworld.com
Web: www.naar.co.uk
Product Type: 1, 2, 4, 5, 6, 7, 11, 13, 14, 16, 17, 19, 20

OAKBEAMS.COM
Hunterswood Farm, Alfold Road,
Dunsfold, Godalming, Surrey, GU8 4NP
Area of Operation: Worldwide
Tel: 01483 200477 **Email:** info@oakbeams.com
Web: www.oakbeams.com
Product Type: 14, 15, 16, 17, 19

PATTISONS ARCHITECTURAL ANTIQUES
108 London Road, Aston Clinton,
Buckinghamshire, HP22 5HS
Area of Operation: Worldwide
Tel: 01926 632300 **Fax:** 01926 631329
Email: info@ddd-uk.com **Web:** www.ddd-uk.com
Product Type: 20

PRIORS RECLAMATION LTD
Ditton Priors Industrial Estate, Station Road,
Ditton Priors, Bridgnorth, Shropshire, WV16 6SS
Area of Operation: Worldwide
Tel: 01746 712450 **Fax:** 01746 712450
Email: vicki@priorsrec.co.uk
Web: www.priorsrec.co.uk
Product Type: 4, 19

RANSFORDS
Drayton Way, Drayton Fields Industrial Estate,
Daventry, Northamptonshire, NN11 5XW
Area of Operation: Worldwide
Tel: 01327 705310 **Fax:** 01327 706831
Email: sales@ransfords.com
Web: www.ransfords.com /
www.stoneflooringandpaving.com
Product Type: 1, 2, 3, 4, 5, 6, 7, 8, 9, 10, 11, 12, 13, 14, 15, 16, 17, 18, 19, 20

ROBERT MILLS ARCHITECTURAL ANTIQUES
Narroways Road, Eastville, Bristol, BS2 9XB
Area of Operation: Worldwide
Tel: 0117 955 6542 **Fax:** 0117 955 8146
Email: Info@rmills.co.uk
Web: www.rmills.co.uk
Product Type: 2, 3, 4, 6, 8, 9, 11, 13, 20

SOUTH WEST RECLAMATION LTD
Wireworks Estate, Bristol Road,
Bridgwater, Somerset, TA6 4AP
Area of Operation: South West England and South Wales
Tel: 01278 444141
Fax: 01278 444114
Email: info@southwest-rec.co.uk
Web: www.southwest-rec.co.uk
Product Type: 1, 2, 3, 4, 5, 9, 10, 11, 18

STRIP IT LTD
109 Pope Street, Birmingham,
West Midlands, B1 3AG
Area of Operation: Midlands & Mid Wales
Tel: 0121 243 4000
Email: sales@stripit.biz **Web:** www.stripit.biz
Product Type: 2, 3, 4, 6, 11, 14

THE CAST IRON RECLAMATION COMPANY
The Barn, Preston Farm Court,
Little Bookham, Surrey, KT23 4EF
Area of Operation: Worldwide
Tel: 0208 977 5977
Fax: 0208 786 6690
Email: enquiries@perfect-irony.com
Web: www.perfect-irony.com
Product Type: 1, 2, 3, 7, 12, 16, 20

THE HOUSE HOSPITAL
14a Winders Road, Battersea, London, SW11 3HE
Area of Operation: UK (Excluding Ireland)
Tel: 0207 223 3179
Email: info@thehousehospital.com
Web: www.thehousehospital.com
Product Type: 2, 4, 6, 7, 12, 20

VIKING RECLAMATION LTD
Old Brick Yard, Cow House Lane,
Armthorpe Industrial Estate, Armthorpe,
Doncaster, South Yorkshire, DN3 3EE
Area of Operation: UK (Excluding Ireland)
Tel: 01302 835449 **Fax:** 01302 835449
Email: info@reclaimed.co.uk
Web: www.reclaimed.co.uk
Product Type: 1, 2, 3, 4, 5, 6, 9, 10, 13, 14, 15, 16, 17, 18, 19, 20

WALCOT RECLAMATION
108 Walcot Street, Bath, Somerset, BA1 5BG
Area of Operation: Worldwide
Tel: 01225 444404 **Fax:** 01225 448163
Email: rick@walcot.com **Web:** www.walcot.com
Product Type: 1, 2, 3, 4, 5, 6, 9, 10, 12, 13, 15, 16, 17, 18, 19

WOODSIDE RECLAMATION
Woodside, Scremerston, Berwick upon Tweed,
Northumberland, TD15 2SY
Area of Operation: Worldwide
Tel: 01289 331211
Email: info@redbaths.co.uk
Web: www.redbaths.co.uk
Product Type: 1, 2, 3, 4, 6, 9, 11, 12, 13, 14, 15, 16, 17, 18, 19, 20

WYE VALLEY RECLAMATION LTD
Fordshill Road, Rotherwas, Hereford,
Herefordshire, HR2 6NS
Area of Operation: UK (Excluding Ireland)
Tel: 01432 353606
Fax: 01432 340020
Email: enquiries@valley-reclamation.co.uk
Web: www.wye-valley-reclamation.co.uk

PLEASE MENTION
THE HOMEBUILDER'S HANDBOOK
WHEN YOU CALL

FEIN MultiMaster
renovation system.

FEIN MultiMaster – the perfect tool for professionals and DIY hobbyists.
Its astonishing suitability for a wealth of different applications makes the
FEIN MultiMaster one of the most versatile and therefore most sensible special tools
on the market. Any application will produce professional results: for renovating all
around the house (windows, floor coverings, interior fittings, furniture, tiles, etc.),
for car repairs, boat maintenance and model building.
You will find the FEIN MultiMaster in specialised trades.
Information under: **www.multimaster.info/gb**

Powered by innovation

FEIN Power Tools Ltd.
Telephone Number: 01327 308730 · Fax Number: 01327 308739 · email: sales@fein-uk.co.uk

Image courtesy of Screwfix Direct Ltd (0800 096 6226)

MERCHANTS, TOOLS & EQUIPMENT

SPONSORED BY SCREWFIX DIRECT LTD
Tel 0800 096 6226   Web www.screwfix.com

Hire Equipment

In many cases, hiring tools and equipment for your project has numerous advantages over buying.

With developments in modern technology, equipment is constantly being changed and models are becoming outdated very quickly. By hiring equipment, you have access to the latest models without having to continually pay the purchase price to update the version you own. Ownership can be extremely costly, especially on larger items which are used infrequently and often need rented storage space to keep them in, so hiring these pieces will keep your capital free for other uses.

Of course all equipment, however new, is subject to breakdown from time to time, but when the apparatus is hired it can be replaced quickly and easily at no extra cost. It is also often delivered to your site, saving you time as well as money.

Trained and qualified staff are always on hand in hire outlets to advise on the best tools and machinery for any specified job, and will show you how to use them safely and effectively. In many cases you will also receive a safety guide, which will advise on safe operation and any additional protective equipment which may be required.

It is worthwhile to maintain a good relationship with the hire company, as not only will you then be able to borrow more goods from them in the future, but you may also be able to strike a deal on prices you pay. Therefore you should always clean tools before you return them, and treat them as if they were your own during use.

However, in situations where your rental period may be long and prone to overrunning deadlines, it may be worth buying equipment and selling it on after completion. There is always a market for second hand equipment in good condition, and the launch of auction sites such as Ebay has only strengthened a seller's position; making it easy to reach buyers throughout the world without having to leave your home.

Image courtesy of Hewden.

TOOLS & EQUIPMENT

KEY

SEE ALSO: TOOLS AND EQUIPMENT - Hire
Companies

OTHER: ▽ Reclaimed ⬠ On-line shopping
✍ Bespoke 🖐Hand-made ECO Ecological

AXMINSTER POWER TOOL CENTRE LTD
Unit 10, Weycroft Avenue, Axminster,
Devon, EX13 5PH
Area of Operation: Worldwide
Tel: 0800 371822 **Fax:** 01297 35242
Email: email@axminster.co.uk
Web: www.axminster.co.uk

B & Q PLC
Portswood House, 1 Hampshire Corporate Park,
Chandlers Ford, Eastleigh, Hampshire, SO5 3YX
Area of Operation: UK & Ireland
Tel: 02380 256256 **Fax:** 02380 256020
Email: melanie.james@b-and-q.co.uk
Web: www.diy.com ⬠

CARL KAMMERLING
INTERNATIONAL LIMITED
Glanydon, Pwllheli, Gwynedd, LL53 5LH
Area of Operation: Worldwide
Tel: 01758 701070 **Fax:** 01758 704777
Email: sales@ck-tools.com
Web: www.ck-tools.com

DUMPERLAND
No.1 Venn Place, Joiners Square, Hanley,
Stoke-on-Trent, Staffordshire, ST1 3HP
Area of Operation: Worldwide
Tel: 07812 217647
Fax: 01782 744401
Email: info@dumperland.com
Web: www.dumperland.com

ELLIOTTS TOOL WAREHOUSE
Unit 10 Winchester Trade Park, Easton Lane,
Winchester, Hampshire, SO23 7FA
Area of Operation: UK (Excluding Ireland)
Tel: 01962 827610 **Fax:** 01962 827611
Email: tools@elliott-brothers.co.uk
Web: www.elliotts4tools.com ⬠

EUROFLEX
Portman House, Millbrook Road East,
Southampton, Hampshire, SO15 1HN
Area of Operation: UK & Ireland
Tel: 02380 635131
Fax: 02380 635303
Email: sales@euroflex.co.uk
Web: www.euroflex.co.uk

EXPRESS TOOLS LTD
7 Cooper Dean Drive, Bournemouth,
Dorset, BH8 9LN
Area of Operation: UK & Ireland
Tel: 01202 395481
Fax: 01202 395481
Email: sales@expresstools.co.uk
Web: www.expresstools.co.uk ⬠

FEIN POWER TOOLS
4 Badby Park, Heartlands Business Park,
Daventry, Northamptonshire, NN11 5YT
Area of Operation: UK (Excluding Ireland)
Tel: 01327 308730
Fax: 01327 308739
Email: sales@fein-uk.co.uk
Web: www.fein.com

GE PROTIMETER C/O PANAMETRICS
Shannon Industrial Estate,
Shannon, Co Clare, Dublin, Ireland
Area of Operation: Worldwide
Tel: +353 61 470 200
Fax: +353 61 471 359
Email: protimeter@ge.com
Web: www.protimeter.com ⬠

GIBBS AND DANDY PLC
226 Dallow Road, Luton, Bedfordshire, LU1 1YB
Area of Operation: South East England
Tel: 01582 798798 **Fax:** 01582 798799
Email: luton@gibbsanddandy.com
Web: www.gibbsanddandy.com

HC SLINGSBY PLC
Preston Street, Bradford,
West Yorkshire, BD7 1JF
Area of Operation: Worldwide
Tel: 01274 721591
Fax: 01274 723044
Email: sales@slingsby.com
Web: www.slingsby.com ⬠

ICF-IT LIMITED
Unit 26, Malmesbury Road,
Kingsditch Trading Estate, Cheltenham,
Gloucestershire, GL51 9PL
Area of Operation: UK & Ireland
Tel: 0870 169 6869
Fax: 0870 169 6869
Email: info@icf-it.com

IRWIN INDUSTRIAL
TOOL COMPANY LTD
Parkway Works, Kettlebridge Road,
Sheffield, South Yorkshire, S9 3BL
Area of Operation: Worldwide
Tel: 0114 251 9101
Fax: 0114 243 4302
Email: uksales@irwin.co.uk
Web: www.irwin.co.uk

KUBOTA (UK) LTD
Dormer Road, Thame,
Oxfordshire, OX9 3UN
Area of Operation: UK & Ireland
Tel: 01844 214500
Fax: 01844 216685
Email: richardh@kubota.co.uk
Web: www.kubota.co.uk

LOADMASTER ENGINEERING LTD
Andover House, Fairmile, Henley on Thames,
Oxfordshire, RG9 2LA
Area of Operation: Worldwide
Tel: 0870 386 7377 **Fax:** 01491 414224
Email: sales@loadrunner.co.uk
Web: www.loadrunner.co.uk

LOADMASTER ENGINEERING LTD

Area of Operation: Worldwide
Tel: 0870 386 7377 **Fax:** 01491 414224
Email: sales@loadrunner.co.uk
Web: www.loadrunner.co.uk

The LoadRunner Micro~Dumper is designed to
be the most efficient power barrow on the
market. With a 1/3ton (350kg) payload, the
machine delivers tremendous productivity and is
easy to use.

OLD HOUSE STORE LTD
Hampstead Farm, Binfield Heath,
Henley on Thames, Oxfordshire, RG9 4LG
Area of Operation: Worldwide
Tel: 0118 969 7711
Fax: 0118 969 8822
Email: info@oldhousestore.co.uk
Web: www.oldhousestore.co.uk ▽⬠

PAUL HELPS LANDSCAPING
Catalpa, Ham Street, Baltonsborough,
Nr Glastonbury, Somerset, BA6 8QQ
Area of Operation: UK & Ireland
Tel: 01458 850084 **Fax:** 01458 850853
Email: info@paulhelpslandscaping.co.uk
Web: www.paulhelpsmucktrucksales.co.uk

POWERMECH
Battersea Road,
Heaton Mersey Industrial Estate,
Stockport, Cheshire, SK4 3EA
Area of Operation: UK (Excluding Ireland)
Tel: 0161 432 1999

RIGHT LINES
Waverley House, Waverley Road,
Huddersfield, West Yorkshire, HD1 5NA
Area of Operation: UK (Excluding Ireland)
Tel: 01484 544111 **Fax:** 01484 549111
Email: enquiries@rightlines.ltd.uk
Web: www.rightlines.ltd.uk ⬠

SCAFFOLDING SUPPLIES LTD
15 Wybers Way, Grimsby, Lincolnshire, DN37 9QR
Area of Operation: UK (Excluding Ireland)
Tel: 07712 322636 **Fax:** 01472 501022
Email: scaffolding@hotmail.com
Web: www.scaffoldingsupplies.com

SCAFFOLDING SUPPLIES LTD

Area of Operation: UK (Excluding Ireland)
Tel: 0771 232 2636
Fax: 01472 501022
Email: scaffolding@hotmail.com
Web: www.scaffoldingsupplies.com

New & Used Scaffolding For Sale
• Free Design • Guaranteed Buy-Back •
Nationwide Delivery •

SCREWFIX DIRECT LTD
Houndstone Business Park, Mead Avenue,
Yeovil, Somerset, BA22 8RT
Area of Operation: UK (Excluding Ireland)
Tel: 0800 096 6226 **Fax:** 01935 401665
Email: online@screwfix.com
Web: www.screwfix.com ⬠

SOUTHERN PLANT & TOOL HIRE
Centenary Business Park,
Station Road, Henley on Thames,
Oxfordshire, RG9 1DS
Area of Operation: UK (Excluding Ireland)
Tel: 01491 576063 **Fax:** 01491 410596
Email: southernplant@supanet.com
Web: www.southernplant.co.uk

STRIPPERS PAINT REMOVERS
PO Box 6, Sudbury, Suffolk, CO10 6TW
Area of Operation: UK & Ireland
Tel: 01787 371524 **Fax:** 01787 313944
Email: david.crawte@strripperspaintremovers.com
Web: www.strripperspaintremovers.com

T.B DAVIES (CARDIFF) LTD
Penarth Road, Cardiff, CF11 8TD
Area of Operation: UK (Excluding Ireland)
Tel: 02920 713000 **Fax:** 02920 702386
Email: sales@tbdavies.co.uk
Web: www.ladders-online.com ⬠

THE DRYWALL EMPORIUM LTD
3a The Maltings, Station Road, Sawbridgeworth,
Hertfordshire, CM21 9JX
Area of Operation: Worldwide
Tel: 01279 722282 **Fax:** 01279 722286
Email: info@drywall-emporium.com
Web: www.drywall-emporium.com ⬠

THE DRYWALL EMPORIUM LTD

Area of Operation: Worldwide
Tel: 01279 722282 **Fax:** 01279 722286
Email: info@drywall-emporium.com
Web: www.drywall-emporium.com

A unique range of brick building / repointing
and plasterboard tools / tape supplies for
the "hands on" self-builder & renovator.
Try us – we offer a service second to none!

THE MAX DISTRIBUTION CO LTD
The Old Bakery, 39 High Street,
East Malling, Kent, ME19 6AJ
Area of Operation: Europe
Tel: 01732 840845 **Fax:** 01732 841845
Email: sales@the-max.co.uk
Web: www.the-max.co.uk

TOOLSTATION
Express Park, Bridgwater, Somerset, TA6 4RN
Area of Operation: UK (Excluding Ireland)
Tel: 0808 100 7211
Fax: 0808 100 7210
Email: info@toolstation.com
Web: www.toolstation.com ⬠

WD-40 COMPANY LTD
PO Box 440, Kiln Farm,
Milton Keynes, MK11 3LF
Area of Operation: Worldwide
Tel: 01908 555400
Fax: 01908 266900
Email: info@wd40.co.uk
Web: www.wd40.co.uk

SCREWFIX DIRECT LIMITED

Mead Avenue,
Houndstone Business Park,
Yeovil
BA22 8RT
Tel: 0500 41 41 41

www.screwfix.com
Email: online@screwfix.com

18V DRILL/DRIVER

Great performance and an ergonomic design. Features 13mm keyless Jacobs chuck, run-out brake, two 2.0 Ah Ni-Cd batteries, spirit bubble, magnetic tray and carry case.

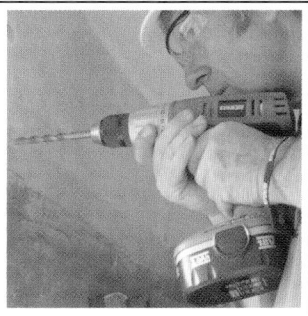

24V COMBI DRILL

Enhanced hammer action and versatility combined with 13mm keyless Jacobs 500 series chuck and run-out brake. Includes spirit bubble, magnetic tray and two 2.o Ah Ni-Cd batteries.

3" BELT SANDER

Variable speed Belt Sander offers excellent control ensuring quick and efficient sanding of large surfaces. Flush front handle and soft grip handles for increased comfort when using over a long time.

650W JIGSAW

Perfectly balanced design for enhanced functionality. Features tool-less blade changes, dust extraction facility and 4 pendulum settings.

6KG SDS PLUS HAMMER DRILL

Robust design delivers outstanding performance when drilling, hammering or chiselling. Features a rotary stop, with 12 chisal position for precise working. Safety clutch and 1100W for great performance & power.

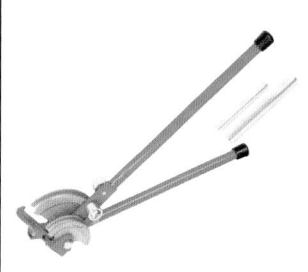

HEAVY DUTY PIPE BENDER

Pipe Bender for 15 and 22mm copper tube or 15mm stainless tube. Includes formers and guides.

SPREADER CLAMPS

Easy to use with push button, quick release and padded jaws.

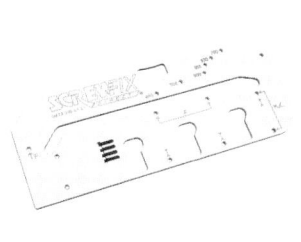

WORKTOP JIGS

Made exclusively for Screwfix Direct by one of Europe's leading manufacturers of worktop jigs. Incredible low price made possible by bulk purchases!
For kitchen, bedroom and bathroom fitters. Cuts 500, 600, 650 and 700mm width worktops. 4 nylon aligning pegs included. Made of 13mm Tufnol, guaranteed never to warp.

POZI 2 CONTRACTORS BITS BOX

Every contractor should have one in the tool box.

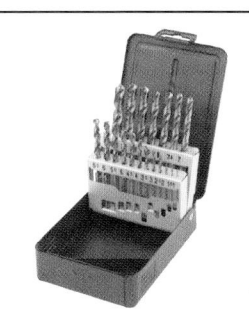

ERBAUER GROUND HSS SET

19 Piece Set.
High quality ground HSS Drills from Erbauer with 135° split point and fully ground flute.

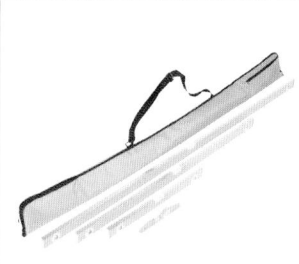

PROFESSIONAL LEVEL SET

4 Piece Set.
Comprises 24", 48", 72" spirit levels and 10" magnetic scaffolders' level. Accurate to 0.5mm/m (0.029°). Milled on both edges. Supplied in padded storage bag.

MERCHANTS, TOOLS & EQUIPMENT

MERCHANTS, TOOLS & EQUIPMENT *(vertical sidebar)*

HIRE COMPANIES

KEY

SEE ALSO: TOOLS AND EQUIPMENT - Tools and Equipment

OTHER: ▽ Reclaimed   🖱 On-line shopping   ✏ Bespoke   ✋ Hand-made   ECO Ecological

COMPACT HOISTS
Unit 8B, Blackbrook Business Park,
Narrowboat Way, Dudley,
West Midlands, DY2 0XQ
Area of Operation: Midlands & Mid Wales
Tel: 01384 240400
Fax: 01384 240300
Email: compactft@aol.com
Web: www.thecompactgroup.co.uk

FRAME WISE
Presteigne Industrial Estate,
Presteigne, Powys, LD8 2UF
Area of Operation: UK (Excluding Ireland)
Tel: 01544 260125
Fax: 01544 260707
Email: framewise@framewiseltd.co.uk
Web: www.framewiseltd.co.uk

HEWDEN
Trafford House, Chester Road,
Stretford, Manchester, M32 0RL
Area of Operation: UK (Excluding Ireland)
Tel: 0845 60 70 111 **Fax:** 0247 666 6988
Email: jeff.schofield@hewden.co.uk
Web: www.hewden.co.uk

HIRE CENTER
Trowel House, Kettering Parkway,
Kettering, Northamptonshire, NN15 6XR
Tel: 0800 529 529
Email: centerline@centers.co.uk
Web: www.hirecenter.co.uk

ICF-IT LIMITED
Unit 26, Malmesbury Road,
Kingsditch Trading Estate,
Cheltenham, Gloucestershire, GL51 9PL
Area of Operation: UK & Ireland
Tel: 0870 169 6869 **Fax:** 0870 169 6869
Email: info@icf-it.com

OUTSOURCE SITE SERVICES
UK Control Centre, Bradford Street,
Shifnal, Shropshire, TF11 8AU
Area of Operation: UK & Ireland
Tel: 01952 277763 **Fax:** 01952 277764
Email: roger@out-source.biz
Web: www.out-source.biz

SOUTHERN PLANT AND TOOL HIRE
Centenary Business Park,
Station Road, Henley on Thames,
Oxfordshire, RG9 1DS
Area of Operation: UK (Excluding Ireland)
Tel: 01491 576063 **Fax:** 01491 410596
Email: southernplant@supanet.com
Web: www.southernplant.co.uk

TOPSKIPS.COM
Baxall Business Centre,
Adswood Road, Stockport,
Cheshire, SK3 8LF
Area of Operation: UK & Ireland
Tel: 0800 019 2410 **Fax:** 0870 054 0078
Email: mark.attwood@topskips.com
Web: www.topskips.com 🖱

NOTES

Company Name ....
....
Address ....
....
email ....
Web ....

Company Name ....
....
Address ....
....
email ....
Web ....

Company Name ....
....
Address ....
....
email ....
Web ....

Company Name ....
....
Address ....
....
email ....
Web ....

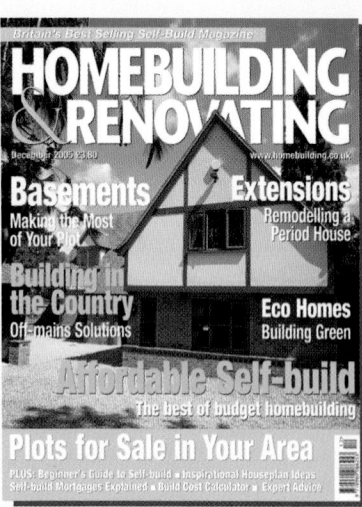

HEATING

Image courtesy of CVO Fire (01325 327221)

HEATING, PLUMBING & ELECTRICAL

SPONSORED BY CVO FIRE LTD
Tel 01325 327221 Web www.cvo.co.uk

CVO FIRE
Creators of the world's most desirable fireplaces

Radiators

ABOVE: Whilst it looks contemporary, this model by Feature Radiators was actually inspired by 1950s aircraft technology.
BELOW: In a twist on the traditional radiator, Bisque Radiators offer this model in over 1000 colour options, meaning that it can be fitted in to even the most contemporary of homes.

ABOVE: Many modern radiators, like this example from Tuscan Foundry Products, are sectional, so can be built to your specific size requirements.

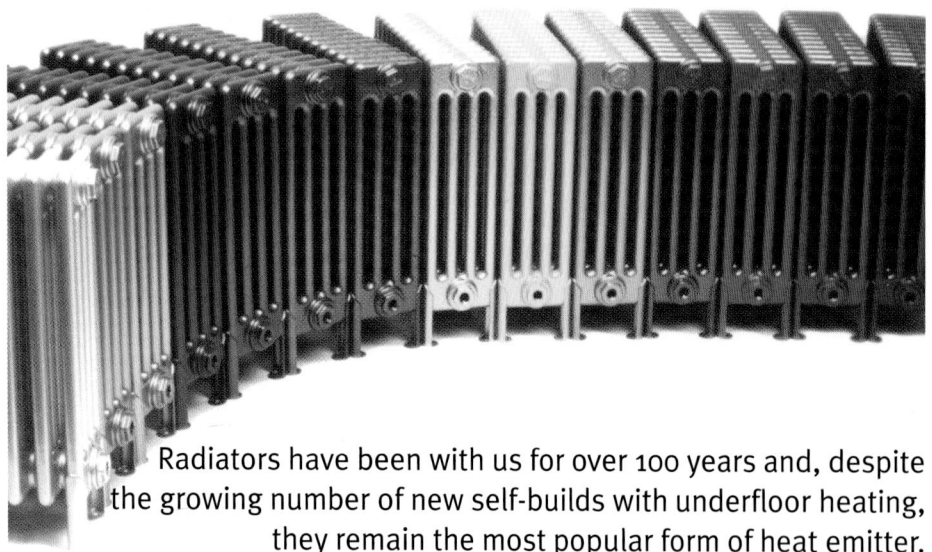

Radiators have been with us for over 100 years and, despite the growing number of new self-builds with underfloor heating, they remain the most popular form of heat emitter.

Radiators have their drawbacks — their location can often restrict the positioning of furniture in a room. They also tend to emit heat unevenly, creating warm and cold spots and, because they convect heat rather than radiate it, they tend to cause draughts. In their favour however, they are cheap, easy to get hold of, simple to install, convenient for drying clothes and towels and are easily controlled.

Pressed steel panel radiators remain the most cost effective option for home heating. However, not all radiators are cheap — if you opt for traditional cast iron models or some of the modern designer examples, you could spend a small fortune.

Cast iron column radiators look fantastic in period style homes, and even in contemporary homes they can add a distinctive touch. Many manufacturers produce accurate period reproductions with either Victorian or Edwardian styling, but for real authenticity you should check out the many reclamation yards which now offer reconditioned radiators. When using a reclaimed radiator, make sure you get a guarantee that is is safe and functional, and remember to allow space for 'bushing' down the old size pipe fittings.

Modern style radiators are available in a dizzying myriad of forms, shapes and colours from curved ones, designed to follow the contours of circular rooms, to spirals, cubes, and ones which look more like modern art sculptures than heat emitters.

Some manufacturers offer as many as 150 different colours so there's sure to be something to suit your room. But unless your room is monochrome and you intend it to stay that way for some time, be wary of some of very bright colours.

Low surface temperature (LST) radiators are ideal if you have children or care for elderly relatives. Casings cover the hot surface but don't generally affect heat output. Ideally, surface temperatures shouldn't be above 43° and for even greater safety, look for models with lockable casings and grilles that cover controls and pipework.

Thermostatic radiator valves (TVRs) can easily be added to radiators and provide accurate and economical fuel saving temperature control in individual rooms. Look for designs with a quick response time and for the latest remote sensing versions (these work best in draughty or sunny locations which can otherwise give false readings). Models with click stop settings, which cannot be accidentally knocked and with an energy saving frost position, are worth considering. Also, new systems can be fitted with zone controls, allowing upstairs and downstairs to be run on different programmes.

If you're on such a tight budget that you can't afford to replace ugly radiators in an existing house, play down their visual impact by painting them the same colour as your walls, or alternatively you could fit simple radiator covers.

UNDERFLOOR HEATING

ADVANCED HEATING TECHNOLOGIES LTD
26 Stanley Avenue, Minster, Sheerness, Kent, ME12 2EY
Area of Operation: Worldwide
Tel: 0781 393 7360 **Fax:** 01795 877232
Email: wilson.mark@tinyworld.co.uk
Web: www.aht-heating.com
Product Type: 1

ANDERSON FLOORWARMING LTD
IPPEC Scottish System House,
Unit 119, Atlas Express Industrial Estate,
1 Rutherglen Road, Glasgow, G73 1SX
Area of Operation: Scotland
Tel: 0141 647 6716 **Fax:** 0141 647 6751
Email: scottishbranch@ippec.co.uk
Web: www.ippec.co.uk 🖱
Product Type: 2

APOLLO INSULATION (UK) LTD
PO Box 200, Horley, Surrey, RH6 7FU
Area of Operation: Worldwide
Tel: 01293 776974 **Fax:** 01293 776975
Email: information@apollo-energy.com
Web: www.apollo-energy.com
Product Type: 1, 2

APPLIED HEATING SERVICES LTD
17, Wilden Road, Pattinson South Industrial Estate, Washington, Tyne & Wear, NE38 8QB
Area of Operation: North East England
Tel: 0191 4177604 **Fax:** 0191 4171549
Email: georgecossey1@btconnect.com
Web: www.appliedheat.co.uk
Product Type: 1, 2 **Other Info:** ECO ✎

ASTRA CEILING FANS & LIGHTING
Unit 1, Pilsworth Way, Pilsworth, Bury, Lancashire, BL9 8RE
Area of Operation: Europe
Tel: 0161 766 9090 **Fax:** 0161 766 9191
Email: support@astra247.com
Web: www.astra247.com 🖱
Product Type: 1

ATB AIR CONDITIONING & HEATING
67 Melloway Road, Rushden, Northamptonshire, NN10 6XX
Area of Operation: East England, Greater London, Midlands & Mid Wales, South East England
Tel: 0870 260 1650 **Fax:** 01933 411731
Email: enquiries@atbairconditioning.co.uk
Web: www.atbairconditioning.co.uk
Product Type: 2

BEGETUBE UK LTD
8 Carsegate Road South, Inverness, Highlands, IV3 8LL
Area of Operation: UK & Ireland
Tel: 01463 246600 **Fax:** 01463 246624
Email: rory@begetube.co.uk
Web: www.begetube-uk.co.uk 🖱
Product Type: 2

BESPOKE UNDERFLOOR HEATING
Unit 28, Rotheram Close, Norwood Ind Estate, Sheffield, South Yorkshire, S21 2 JU
Area of Operation: UK (Excluding Ireland)
Tel: 0114 2483396 **Fax:** 0114 2486146
Email: bespoke_ufh@btopenworld.com
Web: www.bespokeunderfloorheating.co.uk 🖱
Product Type: 1

BORDERS UNDERFLOOR HEATING
26 Coopersknowe Crescent, Galashiels, Borders, TD1 2DS
Area of Operation: UK & Ireland
Tel: 01896 668667
Fax: 01896 668678
Email: underfloor@btinternet.com
Web: www.bordersunderfloor.co.uk
Product Type: 2

BUY DIRECT HEATING SUPPLIES
Unit 7, Commerce Business Centre, Commerce Close, Westbury, Wiltshire, BA13 4LS
Area of Operation: UK (Excluding Ireland)
Tel: 01373 301360
Fax: 01373 865101
Email: info@buydirect.co.uk
Web: www.buydirecttheatingsupplies.co.uk 🖱
Product Type: 1, 2

CHELMER HEATING SERVICES LIMITED
Unit 12A, Baddow Park, West Hanningfield Road, Chelmsford, Essex, CM2 7SY
Area of Operation: UK (Excluding Ireland)
Tel: 01245 471111
Fax: 01245 471117
Email: sales@chelmerheating.co.uk
Web: www.chelmerheating.co.uk
Product Type: 2
Other Info: ✎

COMFOOT FLOORING
Walnut Tree, Redgrave Road, South Lopham, Diss, Norfolk, IP22 2HN
Area of Operation: Europe
Tel: 01379 688516
Fax: 01379 688517
Email: sales@comfoot.com
Web: www.comfoot.com 🖱
Product Type: 1

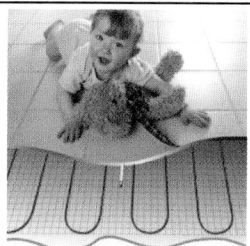

COMFOOT FLOORING
Area of Operation: Europe
Tel: 01379 688516
Fax: 01379 688517
Email: sales@comfoot.com
Web: www.comfoot.com
Product Type: 1

Cold feet, wet floors? Not anymore. Suppliers and installers of electric under floor heating used anywhere. Clean... warm... and soft. Looking for inspiration? You will when you see the new Wiltex range of laminates. Some even have real wood incorporated. Fed up with that dusty garage? Go on treat yourself to an epoxy coating that will outlast normal floor paints.

COMFORT ZONE SYSTEMS LTD
Suite 5 Caxton House, 143 South Coast Road, Peacehaven, East Sussex, BN10 8NN
Area of Operation: Europe
Tel: 01273 580888 **Fax:** 01273 580848
Email: clive@czsystems.co.uk
Web: www.czsystems.co.uk **Product Type:** 2

CONSERVATION ENGINEERING LTD
The Street, Troston, Bury St Edmunds, Suffolk, IP31 1EW
Area of Operation: UK & Ireland
Tel: 01359 269360 **Fax:** 01359 268340
Email: anne@conservation-engineering.co.uk
Web: www.heating-designs.co.uk
Product Type: 2

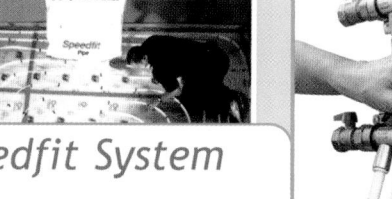
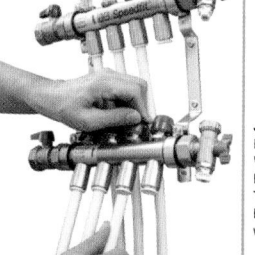

CVO **FIRE**

Creators of the world's most desirable fireplaces

"I particularly admired your professionalism an

"We were impressed with the level of knowledge that CVO had i

BORDERS UNDERFLOOR HEATING

Quality comes as standard

BEFORE

AFTER

Has somebody told you that you can't have underfloor heating...

- with a suspended timber floor
- on an upper floor
- with carpet
- with a hardwood floor
- with different temperatures in different rooms
- in separate automatically controlled zones
- with lots of glazing
- with high ceilings
- because you can't afford it?

We've got good news... YOU CAN!

Admittedly underfloor heating costs a bit more to install than a radiator system, but it's worth it. Today's rising fuel prices mean you'll save more money on running costs than ever before. Combine underfloor heating with a heat pump and you could save even more. Apart from this, the floor space taken up by radiators costs you a lot of money for the space they take up before you even buy them to hang on your walls.

Underfloor heating lasts the lifetime of the building and costs a fraction of the price of a fitted kitchen that you'll probably want to change in a few years' time.

So don't believe everything you read in the papers, ask us first. Underfloor heating is all we do.

AFTER

BEFORE

Advice, brochures, and quotes are free.

Borders Underfloor Heating   Tel: **01896 668667**
26 Coopersknowe Crescent, GALASHIELS. TD1 2DS.
www.bordersunderfloor.co.uk

We leave our clients with a warm feeling.

Robbens Versatility Perfect for *"Living Room"* Extension

Robbens' ability to work with all kinds of floor constructions can reap dividends in bringing the advantages of top quality "designed" underfloor heating to the even smaller project.

All the benefits of Robbens' versatile technology are featured in the recent extension to the Townsend family home at Hurstpierpoint near Brighton.

Alan Townsend and his wife Lynda certainly needed more space for their fast growing family - two boys and two girls aged between nine and two. Reluctant to leave their home in the quiet Sussex village - which is also ideally located for Alan's work as a director of a seafood importing company - they opted to "add on".

Plans took a major step forward when they discovered that Jock Heller, a local friend for some time, was also the man who had built their house 25 years earlier!

Jock agreed to act as project manager for the new scheme and work was soon under way.

Impressed

The decision to incorporate Robbens underfloor heating was made at an early stage.

"We'd been impressed by the obvious advantages of underfloor heating in regard to safety and space saving," says Alan Townsend "and Robbens were recommended as the system of choice."

The extension comprises a major enlargement of the kitchen, which has been completely refitted, with adjoining study and utility room.

A suspended floor system was used for the main kitchen area. During installation the spaces between the joists were first filled with insulation material to prevent downward heat transfer **(see www.underfloorheating.co.uk)**

The loops of Robbens multi-layer aluminium and p-EX pipework could then be clipped in position below special aluminium conducting sheets with the loop ends connected to a manifold on the primary heating circuit, supplied by a new gas-fired boiler.

When the system is in operation the aluminium sheets ensure that the heat is conducted right across the entire floor surface, providing mainly radiant energy into the room, rapidly bringing it up to the desired temperature.

Solid floors

The new study and utility room feature solid floors. Loops of Robbens multi-layer pipework were fixed in position above a layer of insulation.

After pressure testing, a normal sand and cement screed was laid to integrate the heating system into the floor structure and provide a smooth, even surface for the oak-veneered, engineering grade plywood, which is the final floor finish.

The ends results have been spectacular. After just one heating season with the new system, Alan Townsend wishes he could convert the whole house to Robbens underfloor heating!

> "We'd been impressed by the obvious advantages of underfloor heating in regard to safety and space saving,"

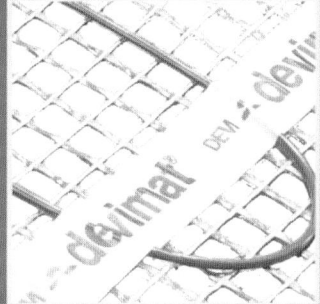
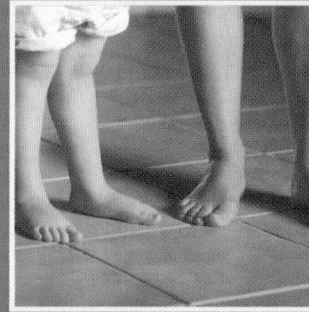

CONTINENTAL UNDER FLOOR HEATING
Continental House, Kings Hill,
Bude, Cornwall, EX23 0LU
Area of Operation: UK & Ireland
Tel: 0128 8357880 **Fax:** 0845 108 1205
Email: info@continental-ufh.co.uk
Web: www.continental-ufh.co.uk
Product Type: 1, 2 **Other Info:** ECO

COSY ROOMS (COSY-HEATING.CO.UK)
17 Chiltern Way, North Hykeham,
Lincoln, Lincolnshire, LN6 9SY
Area of Operation: UK (Excluding Ireland)
Tel: 01522 696002
Fax: 01522 696002
Email: keith@cosy-rooms.com
Web: www.cosy-heating.co.uk
Product Type: 1

DANFOSS RANDALL LTD
Ampthill Road, Bedford,
Bedfordshire, MK42 9ER
Area of Operation: UK & Ireland
Tel: 0845 121 7400 **Fax:** 0845 121 7515
Email: danfossrandall@danfoss.com
Web: www.danfoss-randall.co.uk
Product Type: 1, 2

DCD SYSTEMS LTD
43 Howards Thicket, Gerrards Cross,
Buckinghamshire, SL9 7NU
Area of Operation: UK & Ireland
Tel: 01753 882028 **Fax:** 01753 882029
Email: peter@dcd.co.uk **Web:** www.dcd.co.uk
Product Type: 1, 2

DEVI
Unit 4, Brickfields Business Park,
Woolpit, Suffolk, IP30 9QS
Area of Operation: UK & Ireland
Tel: 01359 242400 **Fax:** 01359 242525
Email: uk@de-vi.co.uk
Web: www.devi.co.uk
Product Type: 1

DIFFUSION ENVIRONMENTAL SYSTEMS
47 Central Avenue, West Molesey, Surrey, KT8 2QZ
Area of Operation: UK (Excluding Ireland)
Tel: 0208 783 0033 **Fax:** 0208 783 0140
Email: diffusion@etenv.co.uk
Web: www.energytechniqueplc.co.uk

DISCOUNT FLOOR HEATING
16 Ashgrove Close, Sebastopol,
Pontypool, Torfaen, NP4 5DA
Area of Operation: UK (Excluding Ireland)
Tel: 0845 658 1511 **Fax:** 0871 661 3557
Email: sales@discountfloorheating.co.uk
Web: www.discountfloorheating.com
Product Type: 1

EBECO UNDERFLOOR HEATING UK LTD
Unit N, Kingsfield Business Centre,
Philanthropic Road, Redhill, Surrey, RH1 4DP
Area of Operation: Europe
Tel: 01737 761767 **Fax:** 01737 507907
Email: uksales@ebeco.com
Web: www.ebeco.com
Product Type: 1

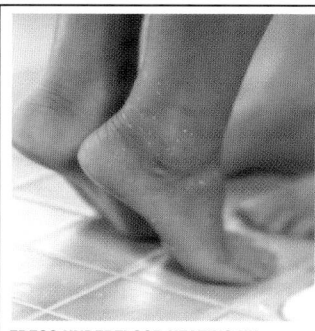

ECO HOMETEC UK LTD
Unit 11E, Carcroft Enterprise Park, Carcroft,
Doncaster, South Yorkshire, DN6 8DD
Area of Operation: Europe
Tel: 01302 722266
Fax: 01302 728634
Email: Stephen@eco-hometec.co.uk
Web: www.eco-hometec.co.uk
Product Type: 2

ECO SYSTEMS IRELAND LTD
40 Glenshesk Road, Ballycastle, Co Antrim, BT54 6PH
Area of Operation: UK & Ireland
Tel: 02820 768708
Fax: 02820 769781
Email: info@ecosystemsireland.com
Web: www.ecosystemsireland.com
Product Type: 1

ELEKTRA (UK) LTD
19 Manning Road, Felixstowe, Suffolk, IP11 2AY
Area of Operation: UK & Ireland
Tel: 0845 226 8142 **Fax:** 0845 225 8143
Email: info@elektra-uk.com
Web: www.elektra-uk.com
Product Type: 1

ENERFOIL MAGNUM LTD
Kenmore Road, Comrie Bridge, Kenmore,
Aberfeldy, Perthshire, PH15 2LS
Area of Operation: Europe
Tel: 01887 822999 **Fax:** 01887 822954
Email: sales@enerfoil.com
Web: www.enerfoil.com
Product Type: 1

ENERGY & ENVIRONMENT LTD
91 Claude Road, Chorlton, Manchester, M21 8DE
Area of Operation: UK & Ireland
Tel: 0161 881 1383
Email: mail@energyenv.co.uk
Web: www.energyenv.co.uk
Product Type: 2

ENERGY MASTER
Keltic Business Park, Unit 1 Clieveragh Industrial
Estate, Listowel, Ireland
Area of Operation: Ireland Only
Tel: +353 68 24300 **Fax:** +353 68 24533
Email: info@energymaster.ie
Web: www.energymaster.ie
Product Type: 2

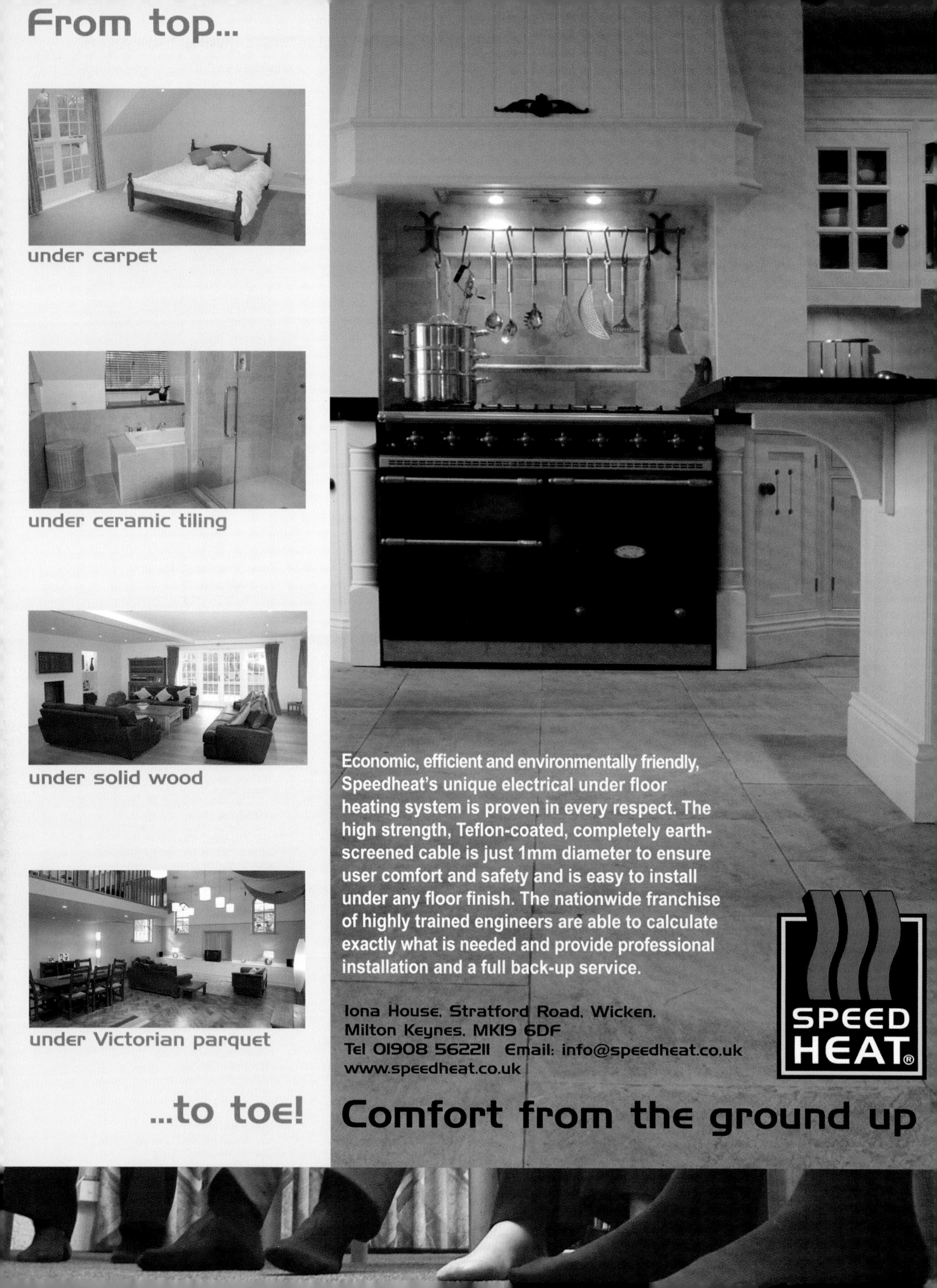
From top...
under carpet
under ceramic tiling
under solid wood
under Victorian parquet
...to toe!
Economic, efficient and environmentally friendly, Speedheat's unique electrical under floor heating system is proven in every respect. The high strength, Teflon-coated, completely earth-screened cable is just 1mm diameter to ensure user comfort and safety and is easy to install under any floor finish. The nationwide franchise of highly trained engineers are able to calculate exactly what is needed and provide professional installation and a full back-up service.
Iona House, Stratford Road, Wicken,
Milton Keynes, MK19 6DF
Tel 01908 562211 Email: info@speedheat.co.uk
www.speedheat.co.uk
SPEED HEAT®
Comfort from the ground up

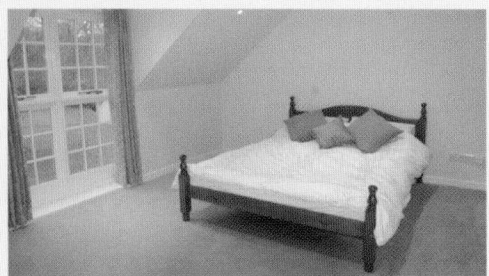

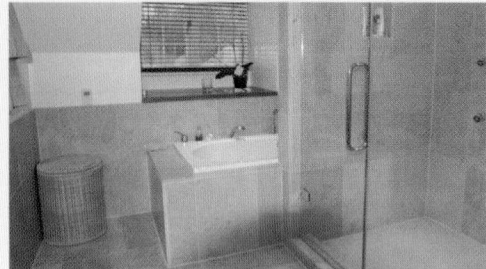

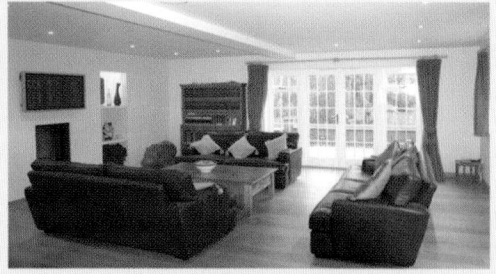

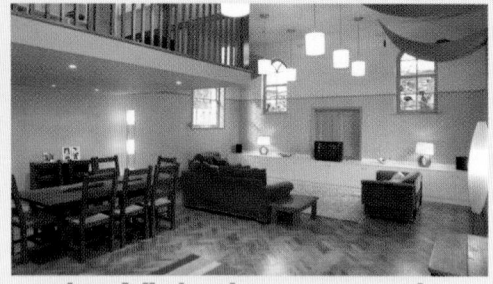

ENVIROZONE
Encon House, 54 Ballycrochan Road,
Bangor, Co Down, BT19 6NF
Area of Operation: UK & Ireland
Tel: 02891 477799 **Fax:** 02891 452463
Email: info@envirozone.co.uk
Web: www.envirozone.co.uk
Product Type: 2

EU SOLUTIONS
Maghull Business Centre,
1 Liverpool Road North, Maghull,
Liverpool, L31 2HB
Area of Operation: UK & Ireland
Tel: 0870 160 1660 **Fax:** 0151 526 8849
Email: ths@blueyonder.co.uk
Web: www.totalhomesolutions.co.uk
Product Type: 1, 2

EURO BATHROOMS
102 Annareagh Road, Richhill, Co Armagh, BT61 9JY
Area of Operation: Ireland Only
Tel: 028 3887 9996 **Fax:** 028 3887 9996
Email: paul@euro-bathrooms.co.uk
Web: www.euro-bathrooms.co.uk
Product Type: 2

EVEN-HEAT
Unit 10, Thorpe Way, Thorpe Way Industrial Estate,
Banbury, Oxfordshire, OX16 4SP
Area of Operation: UK (Excluding Ireland)
Tel: 01295 277881 **Fax:** 01295 277556
Email: enquiries@even-heat.co.uk
Web: www.even-heat.co.uk
Product Type: 1, 2

FLEXELEC (JIMI HEAT)
Unit C, 200 Rickmansworth Road,
Watford, Hertfordshire, WD18 7JS
Area of Operation: Worldwide
Tel: 01923 234477 **Fax:** 01923 240264
Email: sales@flexelec.co.uk
Web: www.flexelec.com
Product Type: 1

FLOOR HEATING SYSTEMS
Unit G12 Imex Enterprise Park, Wigwam Lane,
Hucknall, Nottinghamshire, NG15 7SZ
Area of Operation: Worldwide
Tel: 0115 963 2314 **Fax:** 0115 963 2317
Email: sales@floorheatingsystems.com
Web: www.floorheatingsystems.com
Product Type: 1

FLOORHEATECH LTD
Bowen House, Bredgar Road,
Gillingham, Kent, ME8 6PL
Area of Operation: UK & Ireland
Tel: 0800 814 4328 **Fax:** 0870 131 6520
Email: tracey@floorheatech.co.uk
Web: www.floorheatech.co.uk
Product Type: 1

FLOORWARMING (UK) LTD
Warwick Mill, Warwick Bridge,
Carlisle, Cumbria, CA4 8RR
Area of Operation: UK (Excluding Ireland)
Tel: 01228 631300 **Fax:** 01228 631333
Email: michael@floorwarming.co.uk
Web: www.floorwarming.co.uk
Product Type: 1, 2

FLORAD
Unit 1, Horshoe Business Park,
Lye Lane, Hertfordshire, AL2 3TA
Area of Operation: Europe
Tel: 01923 893025 **Fax:** 01923 670723
Email: info@florad.co.uk
Web: www.florad.co.uk **Product Type:** 2

GEOTHERMAL LTD
4 Imex Centre, Broadleys Business Park,
Stirling, Stirlingshire, FK7 7LQ
Area of Operation: Scotland
Tel: 01786 473666 **Fax:** 01786 475599
Email: geothermal@geoheat.co.uk
Web: www.geoheat.co.uk
Product Type: 2

GLEN DIMPLEX UK LTD
Millbrook House, Grange Drive, Hedge End,
Southampton, Hampshire, SO30 2DF
Area of Operation: UK & Ireland
Tel: 0870 077 7117 **Fax:** 0870 727 0114
Email: marketing@glendimplex.com
Web: www.dimplex.co.uk
Product Type: 1

GREENSHOP SOLAR LTD
Units 23/24, Merretts Mill Industrial Centre,
Woodchester, Stroud, Gloucestershire, GL5 5EX
Area of Operation: UK (Excluding Ireland)
Tel: 01452 772030 **Fax:** 01452 770115
Email: eddie@greenshop.co.uk
Web: www.greenshop-solar.co.uk
Product Type: 2 **Other Info:** ECO

HEATSAFE CABLE SYSTEMS LIMITED
Cromwell Road, Bredbury,
Stockport, Cheshire, SK6 2RF
Area of Operation: Worldwide
Tel: 0161 430 8333 **Fax:** 0161 430 8654
Email: webenquiry@heat-trace.com
Web: www.heatsafe.com
Product Type: 1

HEPWORTH PLUMBING PRODUCTS
Edlington Lane, Edlington, Doncaster,
South Yorkshire, DN12 1BY
Area of Operation: Worldwide
Tel: 01709 856300 **Fax:** 01709 856301
Email: info@hepworthplumbing.co.uk
Web: www.hepworthplumbing.co.uk
Product Type: 2 **Other Info:**

HILTON CROFT LTD
53-55 Gatwick Road, Manor Royal,
Crawley, West Sussex, RH10 9RD
Area of Operation: UK (Excluding Ireland)
Tel: 01293 452657 **Fax:** 01293 426561
Email: info@hilton-croft.co.uk
Web: www.hilton-croft.co.uk
Product Type: 1, 2

GLEN DIMPLEX UK LTD — (INVISIBLE HEATING SYSTEMS)
INVISIBLE HEATING SYSTEMS
IHS Design Centre, Morefield Industrial Estate,
Ullapool, Highlands, IV26 2SX
Area of Operation: UK & Ireland
Tel: 01854 613161 **Fax:** 01854 613160
Email: sales@invisibleheating.co.uk
Web: www.invisibleheating.com
Product Type: 1, 2

IPPEC SYSTEMS LTD
66 Rea Street South, Birmingham,
West Midlands, B5 6LB
Area of Operation: UK & Ireland
Tel: 0121 622 4333 **Fax:** 0121 622 5768
Email: info@ippec.co.uk
Web: www.ippec.co.uk
Product Type: 1, 2

JOHN GUEST SPEEDFIT LTD
Horton Road, West Drayton, Middlesex, UB7 8JL
Area of Operation: Worldwide
Tel: 01895 449233
Fax: 01895 425314
Email: info@johnguest.co.uk
Web: www.speedfit.co.uk
Product Type: 2

KILSPINDIE
Hill Studios, Kilspindie,
Perth and Kinross, PH2 7RX
Area of Operation: UK (Excluding Ireland)
Tel: 01821 670260
Fax: 01821 670268
Email: kuh@emler.co.uk

KV RADIATORS
6 Postle Close, Kilsby, Rugby,
Warwickshire, CV23 8YG
Area of Operation: UK & Ireland
Tel: 01788 823286
Fax: 01788 823 002
Email: solutions@kvradiators.com
Web: www.kvradiators.com
Product Type: 1, 2

HEATING, PLUMBING & ELECTRICAL

Electric Underfloor Heating by WarmFloors Ltd

Warmfloors are specialist suppliers of electric underfloor heating system for all floor types including tiles, wood & laminate.

WarmFloors supply electric underfloor heating direct to the public, conservatory companies, builders, developers & retail tile outlets - all at discount prices.

A revolutionary carbon under floor heating system that is so thin it hardly makes an impression on floor height, ideal for when every millimetre counts!

With wood and laminate flooring becoming evermore popular this under floor heating system is the perfect solution.

The under floor carbon offers an attractive alternative to other heat sources such as radiators, as it is hidden from view & can free up much needed wall space, giving you the freedom to design your room how you want. With the product installed directly below your wood or laminate flooring, providing a fast acting heat system which in most cases can replace radiators and other conventional forms of heating. It's unique double laminated design means that the element itself is less than half a millimetre thick and yet is extremely durable and robust which is backed by a TEN YEAR GUARANTEE.

For further information please call: 0800 0433 195

Visit us on-line at: www.warmfloorsonline.com
Fax: 01535 631196 Email: sales@warmfloorsonline.com

Unit 1, Aire Street, Cross Hills, Keighley, West Yorkshire BD20 7RT

LAFARGE GYVLON LTD
Lafarge Gyvlon, 221 Europa Boulevard,
Westbrook, Warrington, Cheshire, WA5 7TN
Area of Operation: UK & Ireland
Tel: 01925 428780 **Fax:** 01925 428788
Email: sales@gyvlon-floors.co.uk
Web: www.gyvlon-floors.co.uk **Product Type:** 1, 2

LEEMICK LTD
79 Windermere Drive, Rainham, Kent, ME8 9DX
Area of Operation: East England, Greater London,
Midlands & Mid Wales, South East England,
South West England and South Wales
Tel: 01634 351666 **Fax:** 01634 351666
Email: sales@leemick.co.uk **Web:** www.leemick.co.uk
Product Type: 1, 2

LeeMick Ltd
UNDERFLOOR HEATING

LEEMICK LTD
Area of Operation: UK
Tel: 01634 351666
Fax: 01634 351666
Email: sales@leemick.co.uk
Web: www.leemick.co.uk

Design, Supply and installation of Water based
underfloor heating for a conservatory extension or
large detached house. Underfloor heating, the
modern efficient answer.

MYSON
Emlyn Street, Farnworth, Bolton, Lancashire, BL4 7EB
Area of Operation: Worldwide
Tel: 01204 863200 **Fax:** 01204 863229
Email: salesmtw@myson.co.uk
Web: www.myson.co.uk
Product Type: 1, 2

NEOHEAT
Smallmead Gate, Pingemead Business Centre,
Reading, Berkshire, RG30 3UR
Area of Operation: Worldwide
Tel: 0845 108 0361 **Fax:** 0845 108 1295
Email: info@neoheat.com
Web: www.neoheat.com
Product Type: 1

NU-HEAT UK LIMITED
Heathpark House, Devonshire Road, Heathpark
Industrial Estate, Honiton, Devon, EX14 1SD
Area of Operation: UK & Ireland
Tel: 0800 731 1976 **Fax:** 01404 549771
Email: ufh@nu-heat.co.uk
Web: www.nu-heat.co.uk
Product Type: 2

**PARAGON SYSTEMS
(SCOTLAND) LIMITED**
The Office, Corbie Cottage, Maryculter, Aberdeen,
Aberdeenshire, AB12 5FT
Area of Operation: Scotland
Tel: 01224 735536 **Fax:** 01224 735537
Email: info@paragon-systems.co.uk
Web: www.paragon-systems.co.uk
Product Type: 2

PARKSIDE TILES
49-51 Highmeres Road, Thurmeston,
Leicester, Leicestershire, LE4 9LZ
Area of Operation: UK (Excluding Ireland)
Tel: 0116 276 2532 **Fax:** 0116 246 0649
Email: parkside@tiles2.wanadoo.co.uk
Web: www.parksidetiles.co.uk
Product Type: 1

PEDARSON HEATING
Elektrek House, 19 Manning Road,
Felixstowe, Suffolk, IP11 2AY
Area of Operation: UK & Ireland
Tel: 01394 270777 **Fax:** 01394 670189
Email: info@pedarson.com
Web: www.pedarsonheating.co.uk
Product Type: 1

R&D MARKETING (DEMISTA) LTD
Land House, Anyards Road,
Cobham, Surrey, KT11 2LW
Area of Operation: UK (Excluding Ireland)
Tel: 01932 866600 **Fax:** 01932 866 688
Email: rd@demista.co.uk
Web: www.demista.co.uk
Product Type: 1

RADIANT HEATING SOLUTIONS LTD
Mill Farm, Hougham, Grantham,
Lincolnshire, NG32 2HZ
Tel: 01400 250572 **Fax:** 01400 251264
Email: sales@heating-solutions.biz
Web: www.heating-solutions.biz
Product Type: 1, 2

RAYOTEC LTD
Unit 3, Brooklands Close,
Sunbury on Thames, Surrey, TW16 7DX
Area of Operation: UK & Ireland
Tel: 01932 784848 **Fax:** 01932 784848
Email: info@rayotec.com
Web: www.rayotec.com
Product Type: 1, 2

REID UNDERFLOOR HEATING (SCOTLAND) LTD
8 Hillside Grove, Barrhead, Glasgow,
Renfrewshire, G78 1HB
Area of Operation: East England, Midlands
& Mid Wales, North East England, North West
England & North Wales, Scotland
Tel: 0141 880 4443 **Fax:** 0141 880 4442
Email: sales@ruhs.co.uk **Web:** www.ruhs.co.uk
Product Type: 2

ROBBENS SYSTEMS UNDERFLOOR HEATING
69 Castleham Road, St Leonards On Sea,
East Sussex, TN38 9NU
Area of Operation: UK & Ireland
Tel: 01424 851111 **Fax:** 01424 851135
Email: robbens@underfloorheating.co.uk
Web: www.underfloorheating.co.uk
Product Type: 1, 2
Other Info: ✎

RUH(S) LTD
8 Hillside Grove, Barrhead, Glasgow, G78 1HB
Area of Operation: UK (Excluding Ireland)
Tel: 0141 880 4443 **Fax:** 0141 880 4442
Email: office@ruhs.co.uk
Web: www.ruhs.co.uk
Product Type: 2

**SCANDINAVIAN UNDERFLOOR
HEATING CO. LTD.**
314 Croydon Road, Beckenham, Kent, BR3 4HR
Area of Operation: UK & Ireland
Tel: 0208 6636171 **Fax:** 0208 6503556
Email: info@heatingunderfloor.co.uk
Web: www.heatingunderfloor.co.uk
Product Type: 2

SHIRES TECHNICAL SERVICES
57-63 Lea Road, Northampton,
Northamptonshire, NN1 4PE
Area of Operation: UK (Excluding Ireland)
Tel: 01604 472525 **Fax:** 01604 473837
Email: enquiries@shireservices.com
Web: www.shireservices.com
Product Type: 2

SOLAR TWIN
2nd Floor, 50 Watergate Street,
Chester, Cheshire, CH1 2LA
Area of Operation: UK & Ireland
Tel: 01244 403407 **Fax:** 01244 403654
Email: oliver@solartwin.com
Web: www.solartwin.com
Product Type: 2

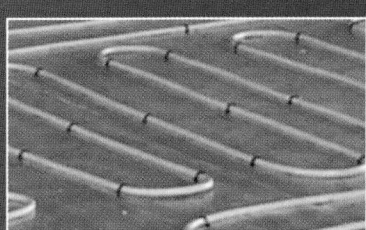

HEATING, PLUMBING & ELECTRICAL

Feel the warmth

Warmup
UNDERTILE HEATING

As the UK's leading manufacturer of undertile heating systems, we design our systems to be both simple and flexible to install.

Systems for an average size bathroom floor (2.5 to 3.4m²) start at just £181.62 *including* the VAT and programmable thermostat. Running costs for such a room are less than 2p per hour.

To give you extra confidence, our heating elements come with both our LIFETIME Guarantee* and *Safety Net Guarantee*™*.

There are plenty of good reasons to choose Warmup:

- Simple and flexible installation
- So thin it won't raise floor levels
- 24 hour customer helpline
- Lifetime Guarantee
- Safety Net Installation Guarantee
- UK's leading supplier for over 10 years

All Warmup Undertile Heating products are designed for easy installation. As such, should you accidentally damage your heater during installation, we will provide you with a new one, free of charge.*

*Terms & conditions apply.

*For a **FREE DVD** & brochure, call 0845 345 2288 or visit **warmup.com***

SPEEDHEAT UK
Iona House, Stratford Road, Wicken,
Milton Keynes, Buckinghamshire, MK19 6DF
Area of Operation: UK & Ireland
Tel: 0800 783 5831 **Fax:** 01908 562205
Email: info@speedheat.co.uk
Web: www.speedheat.co.uk
Product Type: 1

STEP WARMFLOOR UK LTD
Fir Bank, 400 Tottington Rd, Bury,
Lancashire, BL8 1TU
Area of Operation: UK & Ireland
Tel: 0161 763 4077 **Fax:** 0161 763 4078
Email: ann@altech.co.uk
Web: www.stepwarmfloor.co.uk
Product Type: 1

SUPAWARM UNDERFLOOR HEATING
Rose Brae, Toll Bar, Distington,
Cumbria, CA14 4PD
Area of Operation: UK & Ireland
Tel: 01946 832984 **Fax:** 01946 833588
Email: alanc348@aol.com
Web: www.supawarmunderfloorheating.co.uk
Product Type: 2

THERMALFLOOR
Unit 3, Nether Friarton, Perth,
Perth & Kinross, PH2 8DF
Area of Operation: UK & Ireland
Tel: 08450 620400 **Fax:** 08450 620401
Email: heat@thermalfloor.sol.co.uk
Web: www.thermalfloor-heating.co.uk
Product Type: 2

THERMO-FLOOR (GB) LTD
Unit 1 Babsham Farm, Chichester Road,
Bognor Regis, West Sussex, PO21 5EL
Area of Operation: UK (Excluding Ireland)
Tel: 01243 822058 **Fax:** 01243 860379
Email: sales@thermo-floor.co.uk
Web: www.thermo-floor.co.uk
Product Type: 1, 2 **Other Info:** ECO

TOG SYSTEMS
Unit 2a Wildmere Road Industrial Estate,
Banbury, Oxfordshire, OX16 3JU
Area of Operation: UK & Ireland
Tel: 01295 277600
Fax: 01295 279402
Email: kevin@togsystems.net
Web: www.togsystems.net
Product Type: 1, 2

TRIANCO LIMITED
Thorncliffe, Chapeltown, Sheffield,
South Yorkshire, S35 2PH
Area of Operation: UK & Ireland
Tel: 0114 257 2300
Fax: 0114 257 1419
Email: info@trianco.co.uk
Web: www.trianco.co.uk
Product Type: 1

UFH DESIGN LTD
Alexander Villas, Shrubbery Road,
Worcester, Worcestershire, WR1 1QR
Area of Operation: UK (Excluding Ireland)
Tel: 01905 726516
Fax: 01905 726516

UNDERFLOOR DIRECT LTD
Unit1 Lisburn Enterprise Centre,
Ballinderry Road, Lisburn,
Co Antrim, BT28 2BP
Area of Operation: UK & Ireland
Tel: 02892 634068 **Fax:** 02892 669667
Email: info@keeheating.co.uk
Web: www.keeheating.co.uk /
www.underfloordirect.co.uk
Product Type: 2 **Other Info:**

UNDERFLOOR HEATING SERVICES
Bakers Cottage, Pitt Hill Lane,
Moorlynch, Somerset, TA7 9BT
Area of Operation: UK (Excluding Ireland)
Tel: 07808 328135 **Fax:** 01278 427272
Web: www.underfloorheatingservices-sw.co.uk

UNDERFLOOR HEATING SYSTEMS LTD
Unit 1, 79 Friar Street, Worcester,
Worcestershire, WR1 2NT
Area of Operation: Europe
Tel: 01905 616928 **Fax:** 01905 611240
Email: info@underfloorheatingsystems.co.uk
Web: www.underfloorheatingsystems.co.uk
Product Type: 2

UNDERFLOOR HEATING UK
Norris House, Elton Park Business Centre,
Hadleigh Road, Ipswich, Suffolk, IP2 0HU
Area of Operation: Europe
Tel: 01473 280444 **Fax:** 01473 231850
Email: sales@cjelectrical.co.uk
Web: www.cjelectrical.co.uk
Product Type: 1

UPONOR HOUSING SOLUTIONS LTD
Snapethorpe House, Rugby Road,
Lutterworth, Leicestershire, LE17 4HN
Area of Operation: UK (Excluding Ireland)
Tel: 01455 550355 **Fax:** 01455 550366
Email: hsenquiries@uponor.co.uk
Web: www.uponorhousingsolutions.co.uk

**VELTA - THE UNDERFLOOR
HEATING COMPANY**
Unit 1B Denby Dale Industrial Park,
Wakefield Road, Denby Dale, Huddersfield,
West Yorkshire, HD8 8QH
Area of Operation: Worldwide
Tel: 01484 860811 **Fax:** 01484 865775
Email: info@velta-uk.com
Web: www.u-h-c.co.uk
Product Type: 2

VIESSMANN LTD
Hortonwood 30, Telford, Shropshire, TF1 7YP
Area of Operation: UK & Ireland
Tel: 01952 675000 **Fax:** 01952 675040
Email: info@viessmann.co.uk
Web: www.viessmann.co.uk
Product Type: 2

WARM TILES LTD
18 Ernleigh Road, Ipswich, Suffolk, IP4 5LU
Area of Operation: UK & Ireland
Tel: 01473 725743 **Fax:** 01473 725743
Email: barry@warmtiles.co.uk
Web: www.warmtiles.co.uk
Product Type: 1

WARMAFLOOR
42 Botley Road, Park Gate,
Southampton, Hampshire, SO31 1AJ
Area of Operation: East England, Greater London,
Midlands & Mid Wales, South East England,
South West England & South Wales
Tel: 01489 581787 **Fax:** 01489 576 444
Email: Sales@warmafloor.co.uk
Web: www.warmafloor.co.uk
Product Type: 1, 2

WARMALUX MANUFACTURING
PO Box 1333, Huddersfield, West Yorkshire, HD1 9WB
Area of Operation: UK & Ireland
Tel: 01422 374801 **Fax:** 01422 370681
Email: sales@warmalux.co.uk
Web: www.warmalux.co.uk
Product Type: 1

WARMFLOOR SOLUTIONS LTD
Business and Innovation Centre, Wearfield,
Sunderland, Tyne & Wear, SR5 2TA
Area of Operation: UK & Ireland
Tel: 0191 516 6289 **Fax:** 0191 516 6287
Email: sales@warmfloor-solutions.com
Web: www.warmfloor-solutions.com
Product Type: 1

WARMFLOORS ONLINE
Unit1 Aire Street, Cross Hills,
Keighley, West Yorkshire, BD20 7RT
Area of Operation: UK & Ireland
Tel: 01535 631195 **Fax:** 01535 631196
Email: sales@warmfloorsonline.com
Web: www.warmfloorsonline.com
Product Type: 1

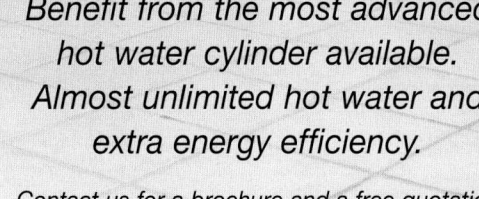
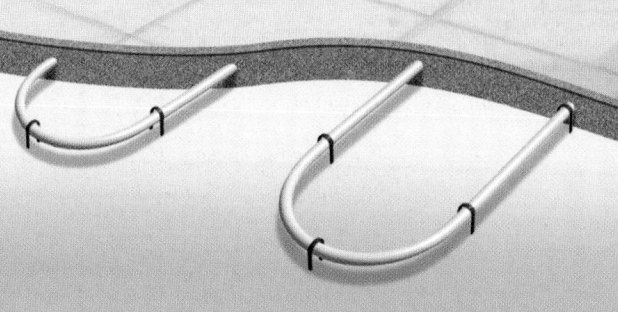
HEATING, PLUMBING & ELECTRICAL

WARMUP PLC
702 Tudor Estate, Abbey Road, London, NW10 7UW
Area of Operation: UK (Excluding Ireland)
Tel: 0845 345 2288 **Fax:** 0845 345 2299
Email: uk@warmup.com
Web: www.warmup.com
Product Type: 1

WARMUP PLC

Area of Operation: UK
Tel: 0845 345 2288
Fax: 0845 345 2299
Email: uk@warmup.com
Web: www.warmup.com

Warmup is the UK's leading supplier of electric undertile heating systems. The product comes with a Lifetime and Safety Net Installation Guarantee, and 24 hour customer helpline. Flexible and simple installation in any tiled room.

XETAL CONSULTANTS LIMITED
Unit 28 Crynant Business Park,
Crynant, Neath, Swansea, SA10 8PA
Area of Operation: Worldwide
Tel: 01639 751056 **Fax:** 01639 751058
Email: jdickson@xetal.co.uk
Web: www.xetal.co.uk
Product Type: 1

YEOVIL PLUMBING SUPPLIES
Unit 1, Bartlett Park, Linx Trading Estate,
Yeovil, Somerset, BA20 2PJ
Area of Operation: UK (Excluding Ireland)
Tel: 01935 474780 **Fax:** 01935 432405

STOVES

KEY

PRODUCT TYPES: 1= Antique
2 = Contemporary 3 = Glass Fronted
4 = Open Fronted 5 = Integral Boilers 6 =
Low Canopy 7 = High Canopy 8 = Flat Top
9 = Top Loading 10 = Double Sided

ENERGY SOURCES: ★ Gas ○ Oil
□ Coal ● Wood ✪ Multi-Fuel
✿ Electric ◗ Gel

OTHER: ▽ Reclaimed ✍ On-line shopping
✏ Bespoke ✋ Hand-made ECO Ecological

AARROW FIRES
The Fireworks, Bridport,
Dorset, DT6 3BE
Area of Operation: UK (Excluding Ireland)
Tel: 01308 427234 **Fax:** 01308 423441
Email: markbrettell@aarrowfires.com
Web: www.aarrowfires.com

ACANTHA LIFESTYLE LTD
32-34 Park Royal Road, Park Royal,
Greater London, NW10 7LN
Area of Operation: Worldwide
Tel: 0208 453 1537
Fax: 0208 453 1538
Email: sales@acanthalifestyle.com
Web: www.acanthalifestyle.co.uk

AGA-RAYBURN
Station Road, Ketley, Telford, Shropshire, TF1 5AQ
Area of Operation: Worldwide
Tel: 01952 642000 **Fax:** 01952 243 138
Email: jkingsbury-webber@aga-web.co.uk
Web: www.aga-web.co.uk

ANGLIA FIREPLACES & DESIGN LTD
Anglia House, Kendal Court, Cambridge Road,
Impington, Cambridgeshire, CB4 9YS
Area of Operation: UK & Ireland
Tel: 01223 234713
Fax: 01223 235116
Email: info@fireplaces.co.uk
Web: www.fireplaces.co.uk
Product Type: 2, 3, 5, 6, 7, 8, 10 ★ □ ● ✪

BAXI FIRES DIVISION
Wood Lane, Erdington, Birmingham,
West Midlands, B24 9QP
Area of Operation: UK (Excluding Ireland)
Tel: 0121 373 8111 **Fax:** 0121 373 8181
Email: cgarner@valor.co.uk
Web: www.firesandstoves.co.uk
Product Type: 1, 2, 3, 6, 7, 8

BD BROOKS FIREPLACES
109 Halifax Road, Ripponden, Halifax,
West Yorkshire, HX6 4DA
Area of Operation: UK (Excluding Ireland)
Tel: 01422 822220 **Fax:** 01422 822220
Email: bdbrooks2000@yahoo.com
Web: www.bdbrooksfireplaces.com
Product Type: 3, 4, 6, 7, 8, 10

BROSELEY FIRES
Knights Way, Battlefield Enterprise Park,
Shrewsbury, Shropshire, SY1 3AB
Area of Operation: UK & Ireland
Tel: 01743 461444 **Fax:** 01743 461446
Email: sales@broseleyfires.com
Web: www.broseleyfires.com
Product Type: 3, 4, 5, 6, 8

BURLEY APPLIANCES LIMITED
Lands End Way, Oakham, Leicestershire, LE15 6RB
Area of Operation: Worldwide
Tel: 01572 756956 **Fax:** 01572 724390
Email: info@burley.co.uk
Web: www.burley.co.uk

CHARNWOOD STOVES & FIRES
Bishops Way, Newport, Isle of Wight, PO30 5WS
Area of Operation: Worldwide
Tel: 01983 537780 **Fax:** 01983 537788
Email: charnwood@ajwells.co.uk
Web: www.charnwood.com ✍
Product Type: 1, 2, 3, 5, 6, 7, 8 ★ □ ● ✪ ✿
Other Info: ECO ✋

CHARNWOOD STOVES

Area of Operation: Worldwide
Tel: 01983 537780 **Fax:** 01983 537788
Email: charnwood@ajwells.co.uk
Web: www.charwood.com
Product Type: 1, 2, 3, 5, 6, 7, 8

Britain's original designers and manufacturers of clean-burn multi-fuel, wood burning, Gas and Electric stoves, fires & boilers. Various models, choice of colours and accessories to suit every situation.

CLEARVIEW STOVES
More Works, Squilver Hill,
Bishops Castle, Shropshire, SY9 5HH
Area of Operation: Worldwide
Tel: 01588 650401
Fax: 01588 650493
Email: mail@clearviewstoves.com
Web: www.clearviewstoves.com
Product Type: 5, 6, 7, 8 ■ ● ❄

CONTINENTAL FIRES LIMITED
Laundry Bank, Church Stretton,
Shropshire, SY6 6PH
Area of Operation: UK & Ireland
Tel: 01694 724199 **Fax:** 01694 720100
Email: sales@continentalfires.com
Web: www.continentalfires.com
Product Type: 1, 2, 3, 4, 10

COSY ROOMS (COSY-HEATING.CO.UK)
17 Chiltern Way, North Hykeham,
Lincoln, Lincolnshire, LN6 9SY
Area of Operation: UK (Excluding Ireland)
Tel: 01522 696002 **Fax:** 01522 696002
Email: keith@cosy-rooms.com
Web: www.cosy-heating.co.uk ✌
Product Type: 3, 4

COUNTRY STYLE COOKERS
Unit 8, Oakleys Yard, Gatehouse Rd, Rotherwas Ind
Est, Hereford, Herefordshire, HR2 6RQ
Area of Operation: UK & Ireland
Tel: 01432 342351 **Fax:** 01432 371331
Email: sales@countrystyle-cookers.com
Web: www.countrystyle-cookers.com
Product Type: 1 ★ ○ □ ● ✪
Other Info: ▽

DCD SYSTEMS LTD
43 Howards Thicket, Gerrards Cross,
Buckinghamshire, SL9 7NU
Area of Operation: UK & Ireland
Tel: 01753 882028
Fax: 01753 882029
Email: peter@dcd.co.uk
Web: www.dcd.co.uk
Product Type: 5

DD HEATING LTD
16-19 The Manton Centre, Manton Lane,
Bedford, Bedfordshire, MK41 7PX
Area of Operation: Worldwide
Tel: 0870 777 8323
Fax: 0870 777 8320
Email: info@heatline.co.uk
Web: www.heatline.co.uk ✌
Product Type: 1, 2, 3, 4, 5

DOVRE
Falcon Road, Sowton Industrial Estate,
Exeter, Devon, EX2 7LF
Area of Operation: UK & Ireland
Tel: 01392 474057 **Fax:** 01392 219932
Email: enquiries@dovre.co.uk
Web: www.dovre.co.uk
Product Type: 1, 2, 3, 4, 5, 6, 7, 8, 9, 10 ★ ● ✪

DOWLING STOVES
Unit 3, Bladnoch Bridge Estate,
Newton Stewart,
Dumfries & Galloway, DG8 9AB
Area of Operation: UK & Ireland
Tel: 01988 402 666
Fax: 01988 402 666
Email: enquiries@dowlingstoves.co.uk
Web: www.dowlingstoves.co.uk
Product Type: 2, 3, 4, 5, 6, 7, 8, 10 ■ ● ❄

DUNSLEY HEAT LTD
Bridge Mills, Huddersfield Rd,
Holmfirth, Huddersfield,
West Yorkshire, HD9 3TW
Area of Operation: UK & Ireland
Tel: 01484 682635
Fax: 01484 688428
Email: sales@dunsleyheat.co.uk
Web: www.dunsleyheat.co.uk
Product Type: 3, 5, 6, 8, 10 ■ ● ❄

ECO SYSTEMS IRELAND LIMITED
40 Glenshesk Road,
Ballycastle, Co Antrim, BT54 6PH
Area of Operation: UK & Ireland
Tel: 02820 768708
Fax: 02820 769781
Email: info@ecosystemsireland.com
Web: www.ecosystemsireland.com
Product Type: 2, 3, 5

EMSWORTH FIREPLACES LTD
Unit 3, Station Approach,
Emsworth, Hampshire, PO10 7PW
Area of Operation: Worldwide
Tel: 01243 373431
Fax: 01243 371023
Email: sales@emsworth.co.uk
Web: www.emsworth.co.uk ✌
Product Type: 2, 3, 4, 6, 7, 8, 10 ★ □ ● ✪

ESSE
Ouzledale Foundry,
Long Ing, Barnoldswick,
Lancashire, BB18 6BN
Area of Operation: Worldwide
Tel: 01282 813235
Fax: 01282 816876
Email: esse@ouzledale.co.uk
Web: www.esse.com
Product Type: 1, 2, 3, 5

EUROHEAT LTD
Court Farm Business Park,
Bishops Frome,
Worcestershire, WR6 5AY
Area of Operation: UK & Ireland
Tel: 01885 491100
Fax: 01885 491101
Email: sales@euroheat.co.uk
Web: www.euroheat.co.uk
Product Type: 2, 3, 5, 6 ★ ○ ● ✪

EUROHEAT
Area of Operation: UK & Ireland
Tel: 01885 491100 **Fax:** 01885 491101
Email: sales@euroheat.co.uk
Web: www.euroheat.co.uk
Product Type: 2, 3

Euroheat heating stoves are at the forefront of
style, innovation and efficiency. The stoves come
in traditional and contemporary styles in
multi-fuel, gas or oil models.

FIREBELLY WOODSTOVES
27 Cuerden Close, Bamber Bridge,
Preston, Lancashire, PR5 6BX
Area of Operation: UK & Ireland
Tel: 0161 408 1710
Email: mail@firebellystoves.com
Web: www.firebellystoves.com ✌
Product Type: 2

FIREPLACE & TIMBER PRODUCTS
Unit 2 Holyrood Drive,
Skippingdale Industrial Estate,
Scunthorpe, Lincolnshire, DN15 8NN
Area of Operation: UK & Ireland
Tel: 01724 852888
Fax: 01724 277255
Email: ftprdcts@yahoo.co.uk
Product Type: 3, 4, 6, 8

FIREPLACE CONSULTANTS LTD
The Studio, The Old Rothschild Arms,
Buckland Rd, Buckland, Aylesbury,
Buckinghamshire, HP22 5LP
Area of Operation: Greater London,
Midlands & Mid Wales, South East England
Tel: 01296 632287
Fax: 01296 632287
Email: info@fireplaceconsultants.com
Web: www.fireplaceconsultants.com
Product Type: 2, 3, 4, 6, 7, 8, 9, 10

FLAMEWAVE FIRES - DK UK LTD
PO Box 611, Folkestone, Kent, CT18 7WY
Area of Operation: UK (Excluding Ireland)
Tel: 0845 257 5028
Fax: 0845 257 5038
Web: www.flamewavefires.co.uk

FRANCO BELGE
Unit 1 Weston Works, Weston Lane, Tyseley,
Birmingham, West Midlands, B11 3RP
Area of Operation: UK (Excluding Ireland)
Tel: 0121 706 8266
Fax: 0121 706 9182
Email: vicky@franco-belge.co.uk
Web: www.franco-belge.co.uk
Product Type: 1, 3, 4, 5 ★ ○ □ ● ✪

GAZCO LTD
Osprey Road, Sowton Industrial Estate,
Exeter, Devon, EX2 7JG
Area of Operation: Europe
Tel: 01392 261999
Fax: 01392 444148
Email: info@gazco.com
Web: www.gazco.com
Product Type: 2, 3, 4, 8 ★ ❀

GLEN DIMPLEX UK LTD
Millbrook House, Grange Drive, Hedge End,
Southampton, Hampshire, SO30 2DF
Area of Operation: UK & Ireland
Tel: 0870 077 7117
Fax: 0870 727 0114
Email: marketing@glendimplex.com
Web: www.dimplex.co.uk

GRENADIER FIRELIGHTERS LIMITED
Unit 3C, Barrowmore Enterprise Estate,
Great Barrow, Chester, Cheshire, CH3 7JS
Area of Operation: UK & Ireland
Tel: 01829 741649
Fax: 01829 741659
Email: enquiries@grenadier.uk.com
Web: www.grenadier.uk.com ✌
Product Type: 4
Other Info: ✎

HETA UK
The Street, Hatfield Peverel, Chelmsford,
Essex, CN3 2DY
Area of Operation: UK (Excluding Ireland)
Tel: 01245 381247
Fax: 01245 381606
Email: hetauk@woodstoves.co.uk
Web: www.woodstoves.co.uk
Product Type: 1, 2, 3, 6, 7, 8, 9, 10

JOTUL (UK) LTD
1 The IO Centre, Nash Road, Park Farm North,
Redditch, Worcestershire, B98 7AS
Area of Operation: UK & Ireland
Tel: 01527 506010 **Fax:** 01527 528181
Email: sales@jotuluk.com **Web:** www.jotul.com
Product Type: 1, 2, 3, 4, 7, 10 ★ ● ✪

KINGSWORTHY FOUNDRY CO LTD
London Road, Kingsworthy, Winchester,
Hampshire, SO23 7QG
Area of Operation: UK & Ireland
Tel: 01962 883776 **Fax:** 01962 882925
Email: kwf@fsbdial.co.uk
Web: www.kingsworthyfoundry.co.uk
Product Type: 3, 5, 6, 7, 8, 10 □ ● ✪ ❀

HEATING, PLUMBING & ELECTRICAL

LEL FIREPLACES
Tre-Ifan Farmhouse, Caergeiliog,
Holyhead, Anglesey, LL65 3HP
Area of Operation: UK & Ireland
Tel: 01407 742240 **Fax:** 01407 742262
Email: sales@lel-fireplaces.com
Web: www.lel-fireplaces.com
Product Type: 1, 2, 3, 4, 5, 6, 7, 8, 10
★ ○ □ ● ✪ ❄

LIVINGSTYLE.CO.UK
Bridge Street, Shotton, Flintshire, CH5 1DU
Area of Operation: UK (Excluding Ireland)
Tel: 0800 2989190
Email: info@livingstyle.co.uk
Web: www.livingstyle.co.uk
Product Type: 2, 3, 4, 5, 6, 7, 8, 9, 10

MARK RIPLEY FORGE & FIREPLACES
Robertsbridge, Bridge Bungalow,
East Sussex, TN32 5NY
Tel: 01580 880324 **Fax:** 01580 881927
Email: ripleym@gxn.co.uk
Web: www.ripleyfireplaces.co.uk
Product Type: 2, 3, 4, 6, 7, 8, 9, 10

METAL DEVELOPMENTS/OILWARM
The Workshop, Wheatcroft Farm,
Cullompton, Devon, EX15 1RA
Area of Operation: Europe
Tel: 01884 35806 **Fax:** 01884 35505
Email: sales@metaldev.demon.co.uk
Web: www.metaldev.demon.co.uk
Product Type: 2, 3, 5, 6, 7, 8, 10 ○ □ ● ✪
Other Info: ✋

OLDE ENGLANDE REPRODUCTIONS
Fireplace Works, Normacot Road, Longton,
Stoke-on-Trent, Staffordshire, ST3 1PN
Area of Operation: UK (Excluding Ireland)
Tel: 01782 319350 **Fax:** 01782 593479
Email: sales@oerfireplaces.com
Web: www.oerfireplaces.com
Product Type: 1, 2, 3, 4, 6, 8, 9

OPIES' THE STOVE SHOP
The Stove Shop, The Street, Hatfield Peverel,
Chelmsford, Essex, CM3 2DY
Area of Operation: East England,
Greater London, South East England
Tel: 01245 380471
Email: enquiries@opie-woodstoves.co.uk
Web: www.opie-woodstoves.co.uk
Product Type: 1, 2, 3, 6, 7, 10 ● ○

PERCY DOUGHTY & CO
Imperial Point, Express Trading Estate, Stonehill
Road, Farnworth, Bolton, Lancashire, M38 9ST
Area of Operation: Midlands & Mid Wales,
North West England and North Wales
Tel: 01204 868550
Fax: 01204 868551
Email: sales@percydoughty.com
Web: www.percydoughty.co.uk
Product Type: 2, 3, 4, 5, 6, 7, 8, 9, 10

R W KNIGHT & SON LTD
Castle Farm, Marshfield,
Chippenham, Wiltshire, SN14 8HU
Area of Operation: Midlands & Mid Wales,
South East England, South West England
and South Wales
Tel: 01225 891469 **Fax:** 01225 892369
Email: Enquiries@knight-stoves.co.uk
Web: www.knight-stoves.co.uk
Product Type: 2, 3, 4, 5, 6, 7, 8, 9, 10
★ ○ □ ● ✪ ❄
Other Info: ECO

ROBEYS
Old School House, Green Lane,
Belper, Derbyshire, DE56 1BY
Area of Operation: UK & Ireland
Tel: 01773 820940 **Fax:** 01773 821652
Email: info@robeys.co.uk
Web: www.robeys.co.uk
Product Type: 1, 2, 3, 4, 5, 6, 7, 8, 9, 10
Other Info: ECO ✎ ✋

RUDLOE STONEWORKS LTD
Leafield Stoneyard, Potley Lane,
Corsham, Wiltshire, SN13 9RS
Area of Operation: UK & Ireland
Tel: 01225 816400 **Fax:** 01225 811343
Email: paul@rudloe-stone.com
Web: www.rudloe-stone.com
Product Type: 2, 3, 4, 6, 8, 9, 10

SCAN OF DENMARK
28 Darmonds Green, West Kirby,
Wirral, Merseyside, CH48 5DU
Area of Operation: UK & Ireland
Tel: 0151 625 0504 **Fax:** 0151 625 0501
Email: info@scanstoves.net
Web: www.krog-iversen.dk
Product Type: 2, 3

SKANTHERM
Stovax Ltd, Falcon Road, Sowton Industrial Estate,
Exeter, Devon, EX2 7LF
Area of Operation: UK & Ireland
Tel: 01392474060 **Fax:** 01392 219932
Email: skan@stovax.com
Web: www.stovax.com
Product Type: 2 ●

STOVAX LIMITED
Falcon Road, Sowton Industrial Estate,
Exeter, Devon, EX2 7LF
Area of Operation: UK & Ireland
Tel: 01392 474011 **Fax:** 01392 219932
Email: info@stovax.com
Web: www.stovax.com
Product Type: 1, 2, 3, 4, 5, 6, 7, 8
★ ○ □ ● ✪ ❄

TAYLOR AND PORTWAY
52 Broton Drive Industrial Estate,
Halstead, Essex, CO9 1HB
Area of Operation: Europe
Tel: 01787 47255 **Fax:** 01787 476589
Email: sales@portwayfires.com
Web: www.portwayfires.com

THE CERAMIC STOVE CO.
4 Earl Street, Oxford, Oxfordshire, OX2 0JA
Area of Operation: Worldwide
Tel: 01865 245077 **Fax:** 01865 245077
Email: info@ceramicstove.com
Web: www.ceramicstove.com
Product Type: 1, 2, 3, 10 ★ ●
Other Info: ▽ ✎ ✋ ECO

THE CERAMIC STOVE COMPANY

Area of Operation: Worldwide
Tel: 01865 245077 **Fax:** 01865 245077
Email: info@ceramicstove.com
Web: www.ceramicstove.com
Product Type: 1, 2, 3, 10
Other Info: ECO

The Ceramic Stove Company's portfolio continues to
expand into areas of astonishing diversity. Recent
commissions range from a hand-built terracotta stove for
the National Trust in Hebden Bridge, through an Osier in a
strawbale house in Wales, to a Gustavian stove glazed in
dark green and with a gas-fired firebox, for designer
Thierry Despont's client in St Moritz, all built to
specification.

THE FIREPLACE GALLERY (UK) LTD
Clarence Road, Worksops, Nottinghamshire, S80 1QA
Area of Operation: UK & Ireland
Tel: 01909 500802 **Fax:** 01909 500810
Email: fireplacegallery@btinternet.com
Web: www.fireplacegallery.co.uk
Product Type: 2, 3, 4

HEATING, PLUMBING & ELECTRICAL

THE FIREPLACE MARKETING COMPANY LIMITED
The Old Coach House, Southern Road,
Thame, Oxfordshire, OX9 2ED
Area of Operation: Worldwide
Tel: 01844 260960 **Fax:** 01844 260267
Email: david@fireplacemarketing.co.uk
Web: www.fireplace.co.uk
Product Type: 1, 2, 3, 4, 5, 6, 7, 8, 9, 10
★ ○ □ ● ✪ ✿ ◆

THE ORGANIC ENERGY COMPANY
Severn Road, Welshpool, Powys, SY21 7AZ
Area of Operation: UK (Excluding Ireland)
Tel: 0845 458 4076
Fax: 01938 559 222
Email: hbenq@organicenergy.co.uk
Web: www.organicenergy.co.uk
Other Info: ECO

THEALE FIREPLACES RDG LTD
Milehouse Farm, Bath Road,
Theale, Berkshire, RG7 5HJ
Area of Operation: UK (Excluding Ireland)
Tel: 0118 930 2232
Fax: 0118 932 3344
Email: mail@theale-fireplaces.co.uk
Web: www.theale-fireplaces.co.uk

THORSTOVES
Canada Hill, East Ogwell,
Newton Abbot, Devon, TQ12 6AF
Area of Operation: UK (Excluding Ireland)
Tel: 01626 363 507
Email: info@thorstoves.com
Web: www.thorstoves.com
Product Type: 1
Other Info: ▽

TOWN AND COUNTRY FIRES
1 Enterprise Way, Thornton Road Industrial Estate,
Pickering, North Yorkshire, YO18 7NA
Area of Operation: Europe
Tel: 01751 474803
Fax: 01751 475205
Email: sales@townandcountryfires.co.uk
Web: www.townandcountryfires.co.uk

VERINE
52 Broton Drive, Halstead, Essex, CO9 1HB
Area of Operation: UK & Ireland
Tel: 01787 472551
Fax: 01787 476589
Email: sales@verine.co.uk
Web: www.verine.co.uk
Product Type: 3, 8 ★

VILLAGER STOVES
Millwey Industrial Estate,
Axminster, Dorset, EX13 5HU
Area of Operation: UK & Ireland
Tel: 0870 160 2202
Fax: 01297 35900
Email: stoves@villager.co.uk
Web: www.villager.co.uk
Product Type: 2, 5, 6, 7, 8, 10 ★ ● ✪ ✿

VILLAGER STOVES
Area of Operation: UK & Ireland
Tel: 0870 160 2202 **Fax:** 01297 35900
Email: stoves@villager.co.uk
Web: www.villager.co.uk
Product Type: 2, 5, 6, 7, 8, 10

Quality........
Over 40 model options of British made wood burning, multi-fuel, gas fired and electric stoves. ....assured

WWW.FIREPLACE.CO.UK
Old Coach House, Southern Road,
Thame, Oxfordshire, OX9 2ED
Area of Operation: UK (Excluding Ireland)
Tel: 01844 260960
Fax: 01844 260267
Web: www.fireplace.co.uk

YEOMAN STOVES
Falcon Road, Sowton Industrial Estate,
Devon, EX2 7LF
Area of Operation: UK (Excluding Ireland)
Tel: 01395 474011
Fax: 01395 219932
Email: yeoman@stovax.co.uk
Web: www.yeoman-stoves.co.uk

BOILERS

KEY

PRODUCT TYPES: 1= Standard / System
2 = Condensing   3 = Dual Output Condensing
4 = Combination   5 = Kitchen Ranges with Boilers   6 = Other

ENERGY SOURCES: ★ Gas   ○ Oil
□ Coal   ● Wood   ✪ Multi-Fuel
✿ Electric   ◆ Gel

OTHER: ▽ Reclaimed   ⌂ On-line shopping
✎ Bespoke   ✋ Hand-made   ECO Ecological

ACV UK LTD
St Davids Business Park,
Dalgety Bay, Fife, KY11 9PF
Area of Operation: UK & Ireland
Tel: 01383 820100
Fax: 01383 820180
Email: information@acv-uk.com
Web: www.acv-uk.com
Product Type: 1, 2, 3, 4, 6 ★ ○ ✿

AEL
4 Berkeley Court, Manor Park,
Runcorn, Cheshire, WA7 1TQ
Area of Operation: UK & Ireland
Tel: 01928 579068
Fax: 01928 579523
Email: ael@totalise.co.uk
Web: www.aelheating.com
Product Type: 2

AGA-RAYBURN
Station Road, Ketley, Telford,
Shropshire, TF1 5AQ
Area of Operation: Worldwide
Tel: 01952 642000 **Fax:** 01952 243 138
Email: jkingsbury-webber@aga-web.co.uk
Web: www.aga-web.co.uk

ALPHA BOILERS
Nepicar House, London Road,
Wrotham Heath, Kent, TN15 7RS
Area of Operation: UK (Excluding Ireland)
Tel: 01732 783000 **Fax:** 01732 783080
Email: info@alphatherm.co.uk
Web: www.alpha-boilers.com
Product Type: 1, 2, 4

APPLIED HEATING SERVICES LTD
17, Wilden Road, Pattinson South Industrial Estate,
Washington, Tyne & Wear, NE38 8QB
Area of Operation: North East England
Tel: 0191 4177604 **Fax:** 0191 4171549
Email: georgecossey1@btconnect.com
Web: www.appliedheat.co.uk
Product Type: 1, 2, 3, 4 ★ ○ ✿

ARCHIE KIDD (THERMAL) LTD
Poulshot, Devizes, Wiltshire, SN10 1RT
Area of Operation: UK & Ireland
Tel: 01380 828123 **Fax:** 01380 828186

BAXI POTTERTON
Brownedge Road, Bamber Bridge,
Preston, Lancashire, PR5 6SN
Area of Operation: UK & Ireland
Tel: 0870 606 0780 **Fax:** 01926 410006
Web: www.baxipotterton.co.uk
Product Type: 1, 2, 3, 4, 6

BLAZES FIREPLACE CENTRES
23 Standish Street, Burnley, Lancashire, BB11 1AP
Area of Operation: UK (Excluding Ireland)
Tel: 01777 838554 **Fax:** 01777 839370
Email: info@blazes.co.uk
Web: www.blazes.co.uk
Product Type: 2, 4

BORDERS UNDERFLOOR HEATING
26 Coopersknowe Crescent,
Galashiels, Borders, TD1 2DS
Area of Operation: UK & Ireland
Tel: 01896 668667 **Fax:** 01896 668678
Email: underfloor@btinternet.com
Web: www.bordersunderfloor.co.uk
Product Type: 1, 2, 4, 6 ○

BTU HEATING
38 Weyside Road, Guildford, Surrey, GU1 1JB
Area of Operation: South East England
Tel: 01483 590600 **Fax:** 01483 590601
Email: enquiries@btu-heating.com
Web: www.btu-group.com
Product Type: 1, 2, 3, 4 ★ ○ ✪ ✿
Other Info: ECO ✎

BUY DIRECT HEATING SUPPLIES
Unit 7, Commerce Business Centre,
Commerce Close, Westbury,
Wiltshire, BA13 4LS
Area of Operation: UK (Excluding Ireland)
Tel: 01373 301360
Fax: 01373 865101
Email: info@buydirectuk.co.uk
Web: www.buydirectheatingsupplies.co.uk ⌂
Product Type: 1, 2, 3, 4, 5, 6

CHARNWOOD STOVES & FIRES
Bishops Way, Newport, Isle of Wight, PO30 5WS
Area of Operation: Worldwide
Tel: 01983 537780 **Fax:** 01983 537788
Email: charnwood@ajwells.co.uk
Web: www.charnwood.com ⌂
Product Type: 1 □ ● ✪
Other Info: ECO ✋

CHARNWOOD STOVES
Area of Operation: Worldwide
Tel: 01983 537780 **Fax:** 01983 537788
Email: charnwood@ajwells.co.uk
Web: www.charnwood.com
Product Type: 1

Britain's original designers and manufacturers of clean-burn multi-fuel and wood burning fires & boilers. Ultra efficient and eco-friendly. Various models, choice of colours and accessories to suit every situation.

HEATING, PLUMBING & ELECTRICAL

Dualstream
WATER CONTROL
setting the standards

The Ultimate Mains Pressure System for your House

Dualstream systems increase flow rates at mains pressure to both hot and cold supplies within the house, delivering a truly outstanding performance. They have been designed to work on poor mains supplies in towns or rural locations with flow rates as low as 9 litre. With most house designs now incorporating more than one bathroom or ensuite Dualstream Systems are the ideal choice for multi bathroom dwellings.

When buying or building your new home consider the importance of mains pressure and flow rates and the design of a system that can meet your demands. Then call GAH hating Products a company that offers a range of products at affordable prices to suit all house types.

Full design service and technical support available.

GAH (HEATING PRODUCTS) LIMITED

MELTON ROAD•MELTON•WOODBRIDGE•SUFFOLK IP12 1NH
Tel: 01394 386699 • Fax: 01394 386609
Email:dcooper@gah.co.uk • www.gah.co.uk

Thermecon
OIL BOILERS

CLEARVIEW STOVES
More Works, Squilver Hill, Bishops Castle,
Shropshire, SY9 5HH
Area of Operation: Worldwide
Tel: 01588 650401 **Fax:** 01588 650493
Email: mail@clearviewstoves.com
Web: www.clearviewstoves.com
Product Type: 1 □ ● ✪

COPPERJOB LTD
25 Wain Park, Plympton, Plymouth, Devon, PL7 2HX
Area of Operation: UK & Ireland
Tel: 01752 339234 **Email:** john@copperjob.com
Web: www.centralheating.co.uk ✑
Product Type: 1, 2, 4

COSY ROOMS (COSY-HEATING.CO.UK)
17 Chiltern Way, North Hykeham,
Lincoln, Lincolnshire, LN6 9SY
Area of Operation: UK (Excluding Ireland)
Tel: 01522 696002 **Fax:** 01522 696002
Email: keith@cosy-rooms.com
Web: www.cosy-heating.co.uk ✑
Product Type: 2, 4

DCD SYSTEMS LTD
43 Howards Thicket, Gerrards Cross,
Buckinghamshire, SL9 7NU
Area of Operation: UK & Ireland
Tel: 01753 882028 **Fax:** 01753 882029
Email: peter@dcd.co.uk
Web: www.dcd.co.uk
Product Type: 1, 2, 3, 4, 5, 6 ★ ○ □ ● ✪ ❄

DD HEATING LTD
16-19 The Manton Centre, Manton Lane,
Bedford, Bedfordshire, MK41 7PX
Area of Operation: Worldwide
Tel: 0870 777 8323 **Fax:** 0870 777 8320
Email: info@heatline.co.uk
Web: www.heatline.co.uk ✑
Product Type: 1, 2, 3, 4

DUNSLEY HEAT LTD
Bridge Mills, Huddersfield Road, Holmfirth,
Huddersfield, West Yorkshire, HD9 3TW
Area of Operation: UK & Ireland
Tel: 01484 682635 **Fax:** 01484 688428
Email: sales@dunsleyheat.co.uk
Web: www.dunsleyheat.co.uk
Product Type: 5 ○

ECO HOMETEC UK LTD
Unit 11E, Carcroft Enterprise Park, Carcroft,
Doncaster, South Yorkshire, DN6 8DD
Area of Operation: Europe
Tel: 01302 722266
Fax: 01302 728634
Email: Stephen@eco-hometec.co.uk
Web: www.eco-hometec.co.uk
Product Type: 2, 3, 4, 6

ENERGY AND ENVIRONMENT LTD
91 Claude Road, Chorlton, Manchester, M21 8DE
Area of Operation: UK & Ireland
Tel: 0161 881 1383
Email: mail@energyenv.co.uk
Web: www.energyenv.co.uk ✑
Product Type: 2

ENERGY MASTER
Keltic Business Park,
Unit 1 Clieveragh Industrial Estate,
Listowel, Ireland,
Area of Operation: Ireland Only
Tel: +353 68 24300
Fax: +353 68 24533
Email: info@energymaster.ie
Web: www.energymaster.ie
Product Type: 2, 3, 4, 6

EURO BATHROOMS
102 Annareagh Road, Richhill, Co Armagh, BT61 9JY
Area of Operation: Ireland Only
Tel: 028 3887 9996
Fax: 028 3887 9996
Email: paul@euro-bathrooms.co.uk
Web: www.euro-bathrooms.co.uk
Product Type: 1, 2, 3

FERROLI UK
Lichfield Road, Branston Industrial Estate,
Burton Upon Trent, Staffordshire, DE14 3HD
Area of Operation: UK (Excluding Ireland)
Tel: 08707 282 882
Fax: 08707 282 883
Email: sales@ferroli.co.uk
Web: www.ferroli.co.uk
Product Type: 1, 2, 3, 4, 6 ★

GAH HEATING PRODUCTS
Melton Road, Melton,
Woodbridge, Suffolk, IP12 1NH
Area of Operation: UK (Excluding Ireland)
Tel: 01394 386606 **Fax:** 01394 386609
Email: info@dualstream.co.uk
Web: www.gah.co.uk

GEMINOX UK
Blenheim House, 1 Blenheim Road,
Epsom, Surrey, KT19 9AP
Area of Operation: UK & Ireland
Tel: 01372 722277 **Fax:** 01372 744477
Email: sales@geminox-uk.com
Web: www.geminox-uk.com
Product Type: 2, 3, 4 ★ ○ ✪

GLEDHILL WATER STORAGE LIMITED
Sycamore Estate, Squires Gate,
Blackpool, Lancashire, FY4 3RL
Area of Operation: UK (Excluding Ireland)
Tel: 01253 474444 **Fax:** 01253 474445
Email: sales@gledhill.net
Web: www.gledhill.net
Product Type: 2

GRANT ENGINEERING LTD
Crinkle, Birr, Co. Offaly, Ireland
Area of Operation: Ireland Only
Tel: +353 509 20089 **Fax:** +353 509 21060
Email: info@grantengineering.ie
Web: www.grantengineering.ie
Product Type: 1, 2, 4, 6 ○

GRANT UK
Hopton House, Hopton Industrial Estate,
Devizes, Wiltshire, SN10 2EU
Area of Operation: UK (Excluding Ireland)
Tel: 0870 777 5553 **Fax:** 0870 777 5559
Email: sales@grantuk.com
Web: www.grantuk.com
Product Type: 1, 2, 4 ○

HRM BOILERS LTD
Haverscroft Ind. Estate, Attleborough, Norfolk, NR17 1YE
Area of Operation: UK (Excluding Ireland)
Tel: 01953 455400 **Fax:** 01953 454483
Email: info@hrmboilers.co.uk
Web: www.hrmboilers.co.uk
Product Type: 1, 4 ○

HRM BOILERS

Area of Operation: UK
Tel: 01953 455400
Fax: 01953 454483
Email: info@hrmboilers.co.uk
Web: www.hrmboilers.co.uk
Product Type: 1, 4 ○

HRM manufactures the unique Wallstar range of domestic wall mounted oil fired boilers. Regular, system and combination models are available.

HEATING, PLUMBING & ELECTRICAL

POWERGEN

A company of **e·on**

Generate your own energy

and save £120* each year
with WhisperGen

WhisperGen – the revolutionary new central heating system from Powergen.

Whether you're renovating your home or building a new one, now's the time to install WhisperGen, our new Micro Combined Heat and Power central heating system.

Not only does WhisperGen provide your heating and hot water, it also generates electricity for your home without burning any extra gas, saving you around £120* a year on your energy bill compared to the most efficient A–rated gas condensing boiler.

WhisperGen also produces around 1.5 tonnes* less carbon dioxide every year than an A–rated boiler, helping protect the environment for future generations too.

Call now to find out more about installing WhisperGen in your home

→ **0800 068 6515**

or email whispergen@powergen.co.uk

* Subject to Terms and Conditions. Taken from the EA Technology report for the Energy Saving Trust based on a three to four bed house saving £120 per year.

Positive Energy

JOHNSON & STARLEY
Rhosili Road, Brackmills, Northampton,
Northamptonshire, NN4 7LZ
Area of Operation: UK & Ireland
Tel: 01604 762881 **Fax:** 01604 767408
Email: marketing@johnsonandstarleyltd.co.uk
Web: www.johnsonandstarley.co.uk
Product Type: 4

KESTON BOILERS
34 West Common Road,
Hayes, Bromley, Kent, BR2 7BX
Area of Operation: UK & Ireland
Tel: 0208 462 0262 **Fax:** 0208 462 4459
Email: info@keston.co.uk
Web: www.keston.co.uk
Product Type: 2, 3, 4 ★
Other Info: ✎

KILTOX CONTRACTS LIMITED
Unit 6 Chiltonian Industrial Estate,
203 Manor Lane, Lee, London, SE12 0TX
Area of Operation: Worldwide
Tel: 0845 166 2040 **Fax:** 0845 166 2050
Email: info@kiltox.co.uk
Web: www.kiltox.co.uk ⌐
Product Type: 6 ❀ **Other Info:** ECO

LIONHEART HEATING SERVICES
PO Box 741, Harworth Park, Doncaster,
South Yorkshire, DN11 8WY
Area of Operation: UK (Excluding Ireland)
Tel: 01302 755200 **Fax:** 01302 750155
Email: lionheart@microgroup.ltd.uk
Web: www.lionheartheating.co.uk ⌐

MALVERN BOILERS LTD
Spring Lane North, Malvern,
Worcestershire, WR14 1BW
Area of Operation: UK (Excluding Ireland)
Tel: 01684 893777 **Fax:** 01684 893776
Email: danar@malvernboilers.co.uk
Web: www.malvernboilers.co.uk
Product Type: 1, 2, 4

MHS RADIATORS & BOILERS
35 Nobel Square, Burnt Mills Industrial Estate,
Basildon, Essex, SS13 1LT
Area of Operation: UK & Ireland
Tel: 01268 591010 **Fax:** 01268 728202
Email: sales@modular-heating-group.co.uk
Web: www.mhsradiators.com
Product Type: 1, 2, 3, 4 ★ ○ ❀

OSO HOTWATER (UK) LTD
E15 Marquis Court, Team Valley Trading Estate,
Gateshead, Tyne & Wear, NE11 0RU
Area of Operation: UK (Excluding Ireland)
Tel: 0191 482 0800 **Fax:** 0191 491 3655
Email: sales.uk@oso-hotwater.com
Web: www.oso-hotwater.com
Product Type: 1, 2, 3, 4

PARAGON SYSTEMS (SCOTLAND) LIMITED
The Office, Corbie Cottage, Maryculter,
Aberdeen, Aberdeenshire, AB12 5FT
Area of Operation: Scotland
Tel: 01224 735536 **Fax:** 01224 735537
Email: info@paragon-systems.co.uk
Web: www.paragon-systems.co.uk
Product Type: 2, 4, 6 ★ ○ ❀

PERCY DOUGHTY & CO
Imperial Point, Express Trading Estate, Stonehill
Road, Farnworth, Bolton, Lancashire, M38 9ST
Area of Operation: Midlands & Mid Wales,
North West England and North Wales
Tel: 01204 868550 **Fax:** 01204 868551
Email: sales@percydoughty.com
Web: www.percydoughty.co.uk **Product Type:** 6

PHIL GREEN & SON
Unit 7, Maylite Trading Estate, Berrow Green Road,
Martley, Worcester, Worcestershire, WR6 6PQ
Area of Operation: UK (Excluding Ireland)
Tel: 01885 488936 **Fax:** 01885 488936
Email: info@philgreenandson.co.uk
Web: www.philgreenandson.co.uk
Product Type: 5 ★ ○ □

POWERGEN
Newstead Court, Sherwood Park, Little Oak Drive,
Annesley, Nottinghamshire, NG15 0DR
Area of Operation: UK (Excluding Ireland)
Tel: 0800 068 6515
Email: whisper@powergen.co.uk
Web: www.powergen.co.uk
Product Type: 1, 4 ★

R W KNIGHT & SON LTD
Castle Farm, Marshfield, Chippenham,
Wiltshire, SN14 8HU
Area of Operation: Midlands & Mid Wales, South
East England, South West England & South Wales
Tel: 01225 891469 **Fax:** 01225 892369
Email: Enquiries@knight-stoves.co.uk
Web: www.knight-stoves.co.uk
Product Type: 5 ★ ○ □ ● ❖

RAVENHEAT MANUFACTURING LTD
Chartists Way, Morley, Leeds,
West Yorkshire, LS27 9ET
Area of Operation: Europe
Tel: 0113 252 7007 **Fax:** 0113 238 0229
Email: sales@ravenheat.co.uk
Web: www.ravenheat.co.uk
Product Type: 2, 4 ★

SCALGON
115 Park Lane, Reading, Berkshire, RG31 4DR
Area of Operation: UK (Excluding Ireland)
Tel: 0118 9424981
Email: postmaster@scalgon.co.uk
Web: www.scalgon.co.uk
Product Type: 2, 3, 4, 6

SOUTHERN SOLAR
Unit 6, Allington Farm, Allington Lane,
Offham, Lewes, East Sussex, BN7 3QL
Area of Operation: Greater London, South East
England, South West England & South Wales
Tel: 0845 456 9474 **Fax:** 01273 483928
Email: info@southernsolar.co.uk
Web: www.southernsolar.co.uk
Product Type: 1, 2 ★

STREBEL LTD
1F Albany Park Industrial Estate, Frimley Road,
Camberley, Surrey, GU16 7PB
Area of Operation: UK & Ireland
Tel: 01276 685422 **Fax:** 01276 685405
Email: info@strebel.co.uk
Web: www.strebel.co.uk
Product Type: 1, 2, 3, 4 ★ ○ ● ❖

THE ORGANIC ENERGY COMPANY
Severn Road, Welshpool, Powys, SY21 7AZ
Area of Operation: UK (Excluding Ireland)
Tel: 0845 4584076 **Fax:** 01938 559 222
Email: hbenq@organicenergy.co.uk
Web: www.organicenergy.co.uk
Product Type: 6 ●
Other Info: ECO

THERMALFLOOR
Unit 3, Nether Friarton, Perth,
Perth & Kinross, PH2 8DF
Area of Operation: UK & Ireland
Tel: 0845 062 0400 **Fax:** 0845 062 0401
Email: heat@thermalfloor.sol.co.uk
Web: www.thermalfloor-heating.co.uk
Product Type: 1, 2, 3, 4 ★ ○ ❀

TOWN AND COUNTRY FIRES
1 Enterprise Way, Thornton Road Industrial Estate,
Pickering, North Yorkshire, YO18 7NA
Area of Operation: Europe
Tel: 01751 474803 **Fax:** 01751 475205
Email: sales@townandcountryfires.co.uk
Web: www.townandcountryfires.co.uk

TRIANCO LIMITED
Thorncliffe, Chapeltown, Sheffield,
South Yorkshire, S35 2PH
Area of Operation: UK & Ireland
Tel: 0114 257 2300 **Fax:** 0114 257 1419
Email: info@trianco.co.uk
Web: www.trianco.co.uk
Product Type: 1, 2, 4, 5 ★ ○ ❀

HEATING, PLUMBING & ELECTRICAL

VIESSMANN LTD
Hortonwood 30, Telford, Shropshire, TF1 7YP
Area of Operation: UK & Ireland
Tel: 01952 675000 **Fax:** 01952 675040
Email: info@viessmann.co.uk
Web: www.viessmann.co.uk
Product Type: 1, 2, 3, 4 ★ ○
Other Info: ECO

VOKERA LTD
4th Floor, Catherine House, Boundary Way,
Hemel Hempstead, Hertfordshire, HP2 7RP
Area of Operation: UK & Ireland
Tel: 0870 333 0220 **Fax:** 01442 281460
Email: enquiries@vokera.co.uk
Web: www.vokera.co.uk
Product Type: 1, 2, 4 ✳

WORCESTER BOSCH GROUP
Cotswold Way, Warndon, Worcester,
Worcestershire, WR4 9SW
Area of Operation: UK & Ireland
Tel: 01905 754624 **Fax:** 01905 753103
Email: general.worcester@uk.bosch.com
Web: www.worcester-bosch.co.uk
Product Type: 1, 2, 4, 5

RADIATORS

KEY

SEE ALSO: HEATING - Convectors,
BATHROOMS - Heated Towel Rails
PRODUCT TYPES: 1= Column   2 = Panel
3 = Skirting   4 = Trench   5 = Low Surface
Temperature   6 = Radiator Covers   7 = Other

ENERGY SOURCES: ★ Gas   ○ Oil
☐ Coal   ● Wood   ✪ Multi-Fuel
✿ Electric   ◗ Gel

OTHER: ▽ Reclaimed   ⌐ On-line shopping
✍ Bespoke   ✋ Hand-made   ECO Ecological

ACOVA RADIATORS (UK) LTD
B15 Armstrong Mall, Southwood Business Park,
Farnborough, Hampshire , GU14 0NR
Area of Operation: UK & Ireland
Tel: 01252 531207
Fax: 01252 531201
Email: pam.hay@zehnder.co.uk
Web: www.acova.co.uk
Product Type: 1, 2, 7 ✪
Material Type: C) 1, 2, 11

AEL
4 Berkeley Court, Manor Park,
Runcorn, Cheshire, WA7 1TQ
Area of Operation: UK & Ireland
Tel: 01928 579068
Fax: 01928 579523
Email: ael@totalise.co.uk
Web: www.aelheating.com
Product Type: 1, 2, 3, 5, 6, 7
Material Type: C) 5

AESTUS
Unit 5 Strawberry Lane Ind. Estate,
Strawberry Lane, Willenhall,
West Midlands, WV13 3RS
Area of Operation: UK & Ireland
Tel: 0870 403 0115
Fax: 0870 403 0116
Email: melissa@publicityengineers.com
Product Type: 1, 2, 6

AGA-RAYBURN
Station Road, Ketley, Telford,
Shropshire, TF1 5AQ
Area of Operation: Worldwide
Tel: 01952 642000
Fax: 01952 243138
Email: jkingsbury-webber@aga-web.co.uk
Web: www.aga-web.co.uk

ALURAD RADIATORS
Gazelle House, Old Wickford Road, South Woodham
Ferrers, Chelmsford, Essex, CM3 5QX
Area of Operation: UK & Ireland
Tel: 01245 322990 **Fax:** 01245 321055
Email: enquiries@alurad.co.uk
Web: www.alurad.co.uk
Product Type: 1, 2, 3

ARC-COMP
Friedrichstr. 3, 12205 Berlin-Lichterfelde, Germany
Area of Operation: Europe
Tel: +49 (0)308 430 9956
Fax: +49 (0)308 430 9957
Email: jvs@arc-comp.com
Web: www.arc-comp.com
Product Type: 2 ✿
Material Type: J) 5, 6

ARC-COMP (IRISH BRANCH)
Whitefield Cottage, Lugduff, Tinahely,
Co. Wicklow, Republic of Ireland, .
Area of Operation: Europe
Tel: +353 (0)868 729 945
Fax: +353 (0)402 28900
Email: jvs@arc-comp.com
Web: www.arc-comp.com
Product Type: 2 ✿
Material Type: J) 5, 6

AUTRON PRODUCTS LTD
Unit 17 Second Avenue, Bluebridge Ind. Estate,
Halstead, Essex, C09 2SU
Area of Operation: UK & Ireland
Tel: 01787 473964 **Fax:** 01787 474061
Email: sales@autron.co.uk
Web: www.autron.co.uk
Product Type: 1, 4, 5, 6, 7

BATHROOMSTUFF.CO.UK
40 Evelegh Road, Farlington,
Portsmouth, Hampshire, PO6 1DL
Area of Operation: UK (Excluding Ireland)
Tel: 08450 580 540 **Fax:** 023 9221 5695
Email: sales@bathroomstuff.co.uk
Web: www.bathroomstuff.co.uk ⌐
Product Type: 1, 2, 3, 7

BISQUE LTD
23 Queen Square, Bath, Somerset, BA1 2HX
Area of Operation: Worldwide
Tel: 01225 478500 **Fax:** 01225 478586
Email: marketing@bisque.co.uk
Web: www.bisque.co.uk
Product Type: 1, 2, 3, 4, 5

BLAZES FIREPLACE CENTRES
23 Standish Street, Burnley, Lancashire, BB11 1AP
Area of Operation: UK (Excluding Ireland)
Tel: 01777 838554 **Fax:** 01777 839370
Email: info@blazes.co.uk
Web: www.blazes.co.uk

BRASS GRILLES UK
Unit 174, 78 Marylebone High Street,
London, W1U 5AP
Area of Operation: UK (Excluding Ireland)
Tel: 07905 292101 / 01923 451600
Fax: 01923 451600
Email: sales@brass-grilles..co.uk
Web: www.brass-grilles.co.uk ⌐
Product Type: 6

BUY DIRECT HEATING SUPPLIES
Unit 7, Commerce Business Centre,
Commerce Close, Westbury, Wiltshire, BA13 4LS
Area of Operation: UK (Excluding Ireland)
Tel: 01373 301360 **Fax:** 01373 865101
Email: info@buydirectuk.co.uk
Web: www.buydirectheatingsupplies.co.uk ⌐
Product Type: 1, 2, 3, 4, 5, 6, 7

CHATSWORTH HEATING PRODUCTS LTD
Unit B Watchmoor Point, Camberley, Surrey, GU15 3EX
Area of Operation: UK (Excluding Ireland)
Tel: 01276 605880 **Fax:** 01276 605881
Email: enquiries@chatsworth-heating.co.uk
Web: www.chatsworth-heating.co.uk
Product Type: 2

CLASSIC WARMTH DIRECT LTD
Unit 8, Brunel Workshops, Ashburton Industrial
Estate, Ross-on-Wye, Herefordshire, HR9 7DX
Area of Operation: UK & Ireland
Tel: 01989 565555 **Fax:** 01989 561058
Email: patbur@tiscali.co.uk
Product Type: 2

CLYDE COMBUSTIONS LTD
Unit 10, Lion Park Avenue,
Chessington, Surrey, KT9 1ST
Area of Operation: UK & Ireland
Tel: 0208 391 2020 **Fax:** 0208 397 4598
Email: info@clyde4heat.co.uk
Web: www.columnradiators.com
Product Type: 1

CLYDE COLUMN RADIATORS
Area of Operation: UK
Tel: 0208 391 2020
Fax: 0208 397 4598
Email: info@clyde4heat.co.uk
Web: www.columnradiators.com
Product Type: 1

Discover cast iron radiators, multi column tubular
radiators, single column tubular radiators, formed
steel column radiators, kitchen and bathroom
radiators at website, www.columnradiators.com.

CORNER FRIDGE COMPANY
Unit 6 Harworth Enterprise Park,
Brunel Industrial Estate, Harworth,
Doncaster, South Yorkshire, DN11 8SG
Area of Operation: UK & Ireland
Tel: 01302 759308 **Fax:** 01302 751233
Email: info@cornerfridge.com
Web: www.cornerfridge.com
Product Type: 2

COSY ROOMS (COSY-HEATING.CO.UK)
17 Chiltern Way, North Hykeham,
Lincoln, Lincolnshire, LN6 9SY
Area of Operation: UK (Excluding Ireland)
Tel: 01522 696002 **Fax:** 01522 696002
Email: keith@cosy-rooms.com
Web: www.cosy-heating.co.uk ⌐
Product Type: 2

COSY ROOMS (DESIGNER-WARMTH.CO.UK)
17 Chiltern Way, North Hykeham,
Lincoln, Lincolnshire, LN6 9SY
Area of Operation: UK (Excluding Ireland)
Tel: 01522 696002 **Fax:** 01522 696002
Email: keith@cosy-rooms.com
Web: www.designer-warmth.co.uk ⌐
Product Type: 1, 3, 5, 7

COVERSCREEN
Unit 174, 78 Marylebone High Street,
London, W1U 5AP
Area of Operation: UK (Excluding Ireland)
Tel: 01923 451600 **Fax:** 01923 451600
Email: sales@coverscreen.co.uk
Web: www.coverscreen.co.uk ⌐
Product Type: 6

DCD SYSTEMS LTD
43 Howards Thicket, Gerrards Cross,
Buckinghamshire, SL9 7NU
Area of Operation: UK & Ireland
Tel: 01753 882028 **Fax:** 01753 882029
Email: peter@dcd.co.uk
Web: www.dcd.co.uk

DD HEATING LTD
16-19 The Manton Centre,
Manton Lane, Bedford,
Bedfordshire, MK41 7PX
Area of Operation: Worldwide
Tel: 0870 777 8323
Fax: 0870 777 8320
Email: info@heatline.co.uk
Web: www.heatline.co.uk ⌐
Product Type: 1, 2, 4

DEEP BLUE SHOWROOM
299-313 Lewisham High Street,
Lewisham, London, SE13 6NW
Area of Operation: UK (Excluding Ireland)
Tel: 0208 690 3401 **Fax:** 0208 690 1408

DIFFUSION ENVIRONMENTAL SYSTEMS
47 Central Avenue, West Molesey, Surrey, KT8 2QZ
Area of Operation: UK (Excluding Ireland)
Tel: 0208 783 0033 **Fax:** 0208 783 0140
Email: diffusion@etenv.co.uk
Web: www.energytechniqueplc.co.uk

DRAYTON PROPERTY
5 Lake Crescent, Daventry,
Northamptonshire, NN11 9EB
Area of Operation: UK (Excluding Ireland)
Tel: 01327 300249 **Fax:** 01327 300249
Email: draytonproperty@aol.com
Web: www.draytonproperty.com ⌐
Product Type: 7
Material Type: C) 5

E. RICHARDS
PO Box 1115, Winscombe,
Somerset, BS25 1WA
Area of Operation: UK & Ireland
Tel: 0845 330 8859 **Fax:** 0845 330 7260
Email: info@e-richards.co.uk
Web: www.e-richards.co.uk ⌐

E RICHARDS
Area of Operation: UK & Ireland
Tel: 0845 330 8859
Web: www.e-richards.co.uk
Product Type: 1 ★
Materials Type: C) 5   **Other Info:** ⌐

Since 1893 Richards have been supplying cast iron and steel products to
the building and plumbing industry, included in the range is the "Victorian"
cast iron radiator. Copied from a 19th century pattern, the "Victorian" is an
ideal choice for both period and contemporary settings.

ECOLEC
Sharrocks Street, Wolverhampton,
West Midlands, WV1 3RP
Area of Operation: UK & Ireland
Tel: 01902 457575
Fax: 01902 457797
Email: sales@ecolec.co.uk
Web: www.ecolec.co.uk ⌐
Product Type: 1, 2, 3, 5, 7

ENERFOIL MAGNUM LTD
Kenmore Road, Comrie Bridge, Kenmore,
Aberfeldy, Perthshire, PH15 2LS
Area of Operation: Europe
Tel: 01887 822999
Fax: 01887 822954
Email: sales@enerfoil.com
Web: www.enerfoil.com ⌐

ENERGY MASTER
Keltic Business Park, Unit 1 Clieveragh Industrial
Estate, Listowel, Ireland
Area of Operation: Ireland Only
Tel: +353 68 24300 **Fax:** +353 68 24533
Email: info@energymaster.ie
Web: www.energymaster.ie
Product Type: 2, 5

ESKIMO DESIGN LTD
25 Horsell Road, London, N5 1XL
Area of Operation: Worldwide
Tel: 020 7 117 0110
Email: ed@eskimodesign.co.uk
Web: www.eskimodesign.co.uk
Product Type: 1, 2, 3, 5, 6, 7 ★ ☼
Other Info: ✐
Material Type: C) 1, 2, 3, 7, 11, 14, 15, 16

EURO BATHROOMS
102 Annareagh Road, Richhill,
Co Armagh, BT61 9JY
Area of Operation: Ireland Only
Tel: 028 3887 9996 **Fax:** 028 3887 9996
Email: paul@euro-bathrooms.co.uk
Web: www.euro-bathrooms.co.uk
Product Type: 1, 2

FEATURE RADIATORS
Bingley Railway Station, Wellington Street,
Bingley, West Yorkshire, BD16 2NB
Area of Operation: UK (Excluding Ireland)
Tel: 01274 567789
Fax: 01274 561183
Email: contactus@featureradiators.com
Web: www.featureradiators.com
Product Type: 1, 2, 3, 4, 5, 7

FERROLI UK
Lichfield Road, Branston Industrial Estate,
Burton Upon Trent, Staffordshire, DE14 3HD
Area of Operation: UK (Excluding Ireland)
Tel: 0870 728 2882
Fax: 0870 728 2883
Email: sales@ferroli.co.uk
Web: www.ferroli.co.uk
Product Type: 5, 7

GEA
GEA Air Treatment Division, Sudstrasse 48,
D-44625, Hern, Germany
Area of Operation: Europe
Tel: 0800 783 0073
Email: info@gea-acqua.com
Web: www.gea-acqua.com
Product Type: 7

GLEN DIMPLEX UK LTD
Millbrook House, Grange Drive, Hedge End,
Southampton, Hampshire, SO30 2DF
Area of Operation: UK & Ireland
Tel: 0870 077 7117 **Fax:** 0870 727 0114
Email: marketing@glendimplex.com
Web: www.dimplex.co.uk
Product Type: 1, 2 ☼

HEATPROFILE LTD
Horizon House, Wey Court,
Mary Road, Guildford,
Surrey, GU1 4QU
Area of Operation: UK & Ireland
Tel: 01483 537000
Fax: 01483 537500
Email: sales@heatprofile.co.uk
Web: www.heatprofile.co.uk
Product Type: 3, 4

HUDSON REED
Rylands Street, Burnley,
Lancashire, BB10 1RH
Area of Operation: Europe
Tel: 01282 418000
Fax: 01282 428915
Email: info@ultra-group.com
Web: www.hudsonreed.com ✓☐
Product Type: 7

INTERESTING INTERIORS LTD
37-39 Queen Street, Aberystwyth,
Ceredigion, SY23 1PU
Area of Operation: UK (Excluding Ireland)
Tel: 01970 626162
Fax: 0870 051 7959
Email: enquiries@interestinginteriors.com
Web: www.interiors-ltd.demon.co.uk
Product Type: 1, 2, 3, 4, 5, 6, 7

**JAGA HEATING
PRODUCTS (UK) LIMITED**
Jaga House, Orchard Business Park,
Bromyard Road, Ledbury,
Herefordshire, HR8 1LG
Area of Operation: Worldwide
Tel: 01531 631533
Fax: 01531 631534
Email: Jaga@jaga.co.uk
Web: www.theradiatorfactory.com
Product Type: 1, 2, 3, 4, 5, 6
Other Info: ECO

JALI LTD
Albion Works, Church Lane,
Barham, Kent, CT4 6QS
Area of Operation: UK (Excluding Ireland)
Tel: 01227 833333
Fax: 01227 831950
Email: sales@jali.co.uk
Web: www.jali.co.uk ✓☐
Product Type: 6
Material Type: H) 2

**JIS EUROPE
(SUSSEX RANGE)**
Warehouse 2, Nash Lane,
Scaynes Hill, Haywards Heath,
West Sussex, RH17 7NJ
Area of Operation: Europe
Tel: 01444 831200
Fax: 01444 831900
Email: info@jiseurope.co.uk
Product Type: 1, 7

JOULESAVE EMES LTD
27 Water Lane, South Witham,
Grantham, Lincolnshire, NG33 5PH
Area of Operation: UK & Ireland
Tel: 01572 768362 **Fax:** 01572 767146
Email: sales@joulesave.co.uk
Web: www.joulesave.co.uk
Product Type: 6, 7

KEELING HEATING PRODUCTS
Cranbourne Road, Gosport, Hampshire, PO12 1RJ
Area of Operation: UK & Ireland
Tel: 02392 796633 **Fax:** 02392 425028
Email: sales@keeling.co.uk
Web: www.keeling.co.uk
Product Type: 7

KERMI (UK) LTD
7 Brunel Road, Corby, Northamptonshire, NN17 4JW
Area of Operation: UK & Ireland
Tel: 01536 400004 **Fax:** 01536 446614
Email: c.radcliff@kermi.co.uk
Web: www.kermi.co.uk
Product Type: 1, 2

KV RADIATORS
6 Postle Close, Kilsby, Rugby, Warwickshire, CV23 8YG
Area of Operation: UK & Ireland
Tel: 01788 823286 **Fax:** 01788 823 002
Email: solutions@kvradiators.com
Web: www.kvradiators.com
Product Type: 1, 4, 5, 7

LIONHEART HEATING SERVICES
PO Box 741, Harworth Park, Doncaster,
South Yorkshire, DN11 8WY
Area of Operation: UK (Excluding Ireland)
Tel: 01302 755200 **Fax:** 01302 750155
Email: lionheart@microgroup.ltd.uk
Web: www.lionheartheating.co.uk ✓☐

LOBLITE ELECTRIC LTD
Third Avenue, Team Valley Trading Estate,
Gateshead, Tyne & Wear, NE11 0QQ
Area of Operation: Europe
Tel: 0191 487 8103 **Fax:** 0191 491 5541
Email: sales@loblite.co.uk
Web: www.heatec-rads.com
Product Type: 1, 2, 7 ☼ **Other Info:** ▽ ✐
Material Type: C) 1, 2, 5, 14

M & O BATHROOM CENTRE
174-176 Goswell Road, Clarkenwell,
London, EC1V 7DT
Area of Operation: East England,
Greater London, South East England
Tel: 0207 608 0111 **Fax:** 0207 490 3083
Email: mando@lineone.net

M&T (UK) LTD
PO Box 382, Grimsby, Lincolnshire, DN37 9XB
Area of Operation: Europe
Tel: 01472 886155 **Fax:** 01472 590887
Email: meinertzinuk@tiscali.co.uk
Web: www.meinertz.com
Product Type: 7 ☼

MARBLE HEATING CO LTD
139 Kennington Park Road, London, SE11 4JJ
Area of Operation: UK & Ireland
Tel: 0845 230 0877 **Fax:** 0845 230 0878
Email: sales@marbleheating.co.uk
Web: www.marbleheating.co.uk
Product Type: 1, 2 ☼ **Other Info:** ECO
Material Type: E) 1, 2, 5

MFT (UK) LTD.
P.O. Box 382, Grimsby, Lincolnshire, DN37 9SW
Area of Operation: UK & Ireland
Tel: 01472 886155 **Fax:** 01472 590887
Email: meinertzinuk@tiscali.co.uk
Web: www.meinertz.com
Product Type: 1, 2, 3, 4, 7

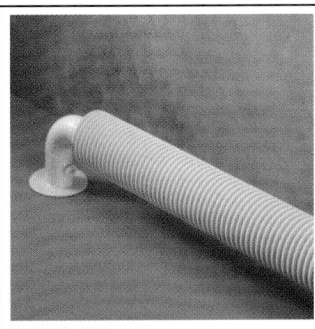

MFT (UK) LTD

Area of Operation: UK & Ireland
Tel: 01472 886155 **Fax:** 01472 590887
Email: meinertzinuk@tiscali.co.uk
Web: www.meinertz.com
Product Type: 1, 2, 3, 4, 7

The MFT range of stylish finned tube radiators,
convectors and convection grilles reflect their
Scandinavian origins, both in their appearance
and efficiency.

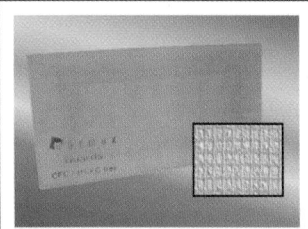

ASTRA247.COM
Area of Operation: Worldwide
Tel: 0161 797 3222 **Fax:** 0161 797 3444
Email: info@astra247.com
Web: www.astra247.com

Underfloor Heating Insulation
Marmox insulation board is ideal for laying beneath underfloor heating mats, accelerates warm up time and inceases efficiency.

'easy online ordering'

ASTRA247.COM
Area of Operation: Worldwide
Tel: 0161 797 3222 **Fax:** 0161 797 3444
Email: info@astra247.com
Web: www.astra247.com

Underfloor Heating Mats
Electric underfloor heating mats in all sizes for conservatories, kitchens, bathrooms and most other rooms.

'easy online ordering'

ASTRA247.COM
Area of Operation: Worldwide
Tel: 0161 797 3222 **Fax:** 0161 797 3444
Email: info@astra247.com
Web: www.astra247.com

Marble Heating
Range of slimline electric heaters available in several different marble finishes.

'easy online ordering'

ASTRA247.COM
Area of Operation: Worldwide
Tel: 0161 797 3222 **Fax:** 0161 797 3444
Email: info@astra247.com
Web: www.astra247.com

Electric Stoves
Wide range of electric stoves from Dimplex, Delonghi and many other top brands.

'easy online ordering'

ASTRA247.COM
Area of Operation: Worldwide
Tel: 0161 797 3222 **Fax:** 0161 797 3444
Email: info@astra247.com
Web: www.astra247.com

Smart Electric Heating Systems
Energy efficient and intelligent complete electric heating systems designed and supplied throughout the UK.

'easy online ordering'

ASTRA247.COM
Area of Operation: Worldwide
Tel: 0161 797 3222 **Fax:** 0161 797 3444
Email: info@astra247.com
Web: www.astra247.com

Electric Radiators
Wide range of electric heating solutions: intelligent electronic complete systems, oil filled radiators, stainless steel designer panel heaters.

'easy online ordering'

ASTRA247.COM
Area of Operation: Worldwide
Tel: 0161 797 3222 **Fax:** 0161 797 3444
Email: info@astra247.com
Web: www.astra247.com

Electric Panel Heaters
Wide range of electric heating solutions: intelligent electronic complete systems, oil filled radiators, stainless steel designer panel heaters.

'easy online ordering'

ASTRA247.COM
Area of Operation: Worldwide
Tel: 0161 797 3222 **Fax:** 0161 797 3444
Email: info@astra247.com
Web: www.astra247.com

Electric Fires
Wide range of electric fires and stoves from Dimplex, Delonghi and other top brands.

'easy online ordering'

ASTRA247.COM
Area of Operation: Worldwide
Tel: 0161 797 3222 **Fax:** 0161 797 3444
Email: info@astra247.com
Web: www.astra247.com

Underfloor Heating Kits
Complete electric underfloor heating kits designed and supplied direct to your door, helpful, friendly and free advice.

'easy online ordering'

ASTRA247.COM
Area of Operation: Worldwide
Tel: 0161 797 3222 **Fax:** 0161 797 3444
Email: info@astra247.com
Web: www.astra247.com

Heating Controls
Range of central heating controls including: digital controls and timers, room and cylinder stats etc.

'easy online ordering'

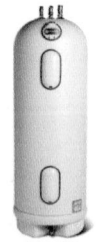

ASTRA247.COM
Area of Operation: Worldwide
Tel: 0161 797 3222 **Fax:** 0161 797 3444
Email: info@astra247.com
Web: www.astra247.com

Electric Water Heating
Electric water heating products to suit all domestic needs. Unvented, instantaneous, combination and point of use. Contact us with your requirements.

'easy online ordering'

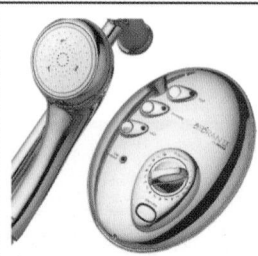

ASTRA247.COM
Area of Operation: Worldwide
Tel: 0161 797 3222 **Fax:** 0161 797 3444
Email: info@astra247.com
Web: www.astra247.com

Electric Showers
We offer a selected range of high power, well specified, quality electric showers. Each with technical specifications available on our website.

'easy online ordering'

HEATING, PLUMBING & ELECTRICAL

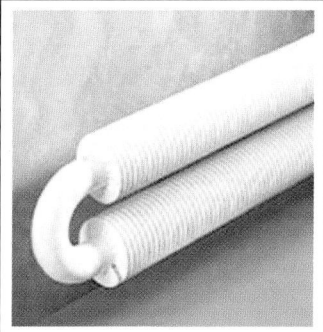

MFT (UK) LTD
Area of Operation: UK & Ireland
Tel: 01472 886155 Fax: 01472 590887
Email: meinertzinuk@tiscali.co.uk
Web: www.meinertz.com
Product Type: 1, 2, 3, 4, 7

The MFT range of stylish finned tube radiators, convectors and convection grills reflect their Scandinavian origins in both appearance and efficiency.

MHS RADIATORS & BOILERS
35 Nobel Square, Burnt Mills Industrial Estate, Basildon, Essex, SS13 1LT
Area of Operation: UK & Ireland
Tel: 01268 591010 Fax: 01268 728202
Email: sales@modular-heating-group.co.uk
Web: www.mhsradiators.com
Product Type: 1, 2, 7 Material Type: C) 1, 5, 6

MYSON
Emlyn Street, Farnworth, Bolton, Lancashire, BL4 7EB
Area of Operation: Worldwide
Tel: 01204 863200 Fax: 01204 863229
Email: salesmtw@myson.co.uk
Web: www.myson.co.uk
Product Type: 1, 2

NU TECH RADIANT LTD
Peppercombe Farm, Newham Lane, Steyning, West Sussex, BN44 3LR
Area of Operation: UK (Excluding Ireland)
Tel: 01903 810033 Fax: 01903 810033
Email: mki.radiant@aol.com
Web: www.conservatoryheat.co.uk

PARAGON SYSTEMS (SCOTLAND) LIMITED
The Office, Corbie Cottage, Maryculter, Aberdeen, Aberdeenshire, AB12 5FT
Area of Operation: Scotland
Tel: 01224 735536 Fax: 01224 735537
Email: info@paragon-systems.co.uk
Web: www.paragon-systems.co.uk

PURMO HEATING PRODUCTS LTD
Rettig Park, Drum Lane, Birtley, Durham, DH2 1AB
Area of Operation: UK (Excluding Ireland)
Tel: 0845 070 1090 Fax: 0845 070 1080
Email: kgunn@rettigheating.co.uk
Web: www.purmo.com
Product Type: 2, 5, 7

QUINN RADIATORS
Spinning Jenny Way, Leigh, Lancashire, WN7 4PE
Area of Operation: Europe
Tel: 01942 671105 Fax: 01942 261801
Email: marketing@quinn-radiators.co.uk
Web: www.quinn-radiators.co.uk
Product Type: 1, 2, 5, 6

RADIATING ELEGANCE
32 Main Street, Orton-on-the-Hill, Warwickshire, CV9 3NN
Area of Operation: UK (Excluding Ireland)
Tel: 0800 028 0921
Email: info@radiatingelegance.co.uk
Web: www.radiatingelegance.co.uk
Product Type: 6

RADIATING STYLE
Unit 15 Thompon Road, Hounslow, Middlesex, Greater London, TW3 3UH
Area of Operation: UK (Excluding Ireland)
Tel: 0870 072 3428 Fax: 0208 577 9222
Email: sales@radiatingstyle.com
Web: www.radiatingstyle.com
Product Type: 1, 2

SIMPLY RADIATORS
Sandycroft House, 20 Station Road, Woburn Sands, Buckinghamshire, MK17 8RW
Area of Operation: UK & Ireland
Tel: 0870 991 3648 Fax: 0870 432 5213
Email: tony.trott@simplyradiators.co.uk
Web: www.simplyradiators.co.uk
Product Type: 1, 2, 3, 4, 5, 6, 7

STREBEL LTD
1F Albany Park Industrial Estate, Frimley Road, Camberley, Surrey, GU16 7PB
Area of Operation: UK & Ireland
Tel: 01276 685422 Fax: 01276 685405
Email: info@strebel.co.uk
Web: www.strebel.co.uk
Product Type: 1, 2, 3, 4, 5 ★ ○ ● ✪

TASKWORTHY
The Old Brickyard, Pontrilas, Herefordshire, HR2 0DJ
Area of Operation: UK (Excluding Ireland)
Tel: 01981 242900 Fax: 01981 242901
Email: peter@taskworthy.co.uk
Web: www.taskworthy.co.uk
Product Type: 6
Other Info: ✍ ✋

THE CAST IRON RECLAMATION COMPANY
The Barn, Preston Farm Court, Little Bookham, Surrey, KT23 4EF
Area of Operation: Worldwide
Tel: 0208 977 5977 Fax: 0208 786 6690
Email: enquiries@perfect-irony.com
Web: www.perfect-irony.com
Product Type: 1, 2, 3, 7, 12, 16, 20

THE RADIATOR COMPANY LIMITED
Elan House, Charlwoods Road, East Grinstead, West Sussex, RH19 2HG
Area of Operation: UK & Ireland
Tel: 01342 302250 Fax: 01342 305561
Email: ian@theradiatorcompany.co.uk
Web: www.theradiatorcompany.co.uk
Product Type: 1, 2, 3, 4, 5, 7

THE RADIATOR COVER COMPANY
29 Croft Avenue, Kidlington, Oxford, Oxfordshire, OX5 2HT
Area of Operation: UK & Ireland
Tel: 01865 849399 Fax: 01865 849399
Email: enquiries@theradiatorcoverco.co.uk
Web: www.theradiatorcoverco.co.uk
Product Type: 6

TUBES RADIATORS UK
341 Kings Road, London, SW3 5ES
Area of Operation: Europe
Tel: 0207 351 1988
Fax: 0207 351 3507
Email: salesdept@tubesradiatori.com
Web: www.tubesradiatori.com
Product Type: 1 ★ ✪ ✤
Material Type: C) 2, 14

TUBISM
27 Cauldwell Road, Linton, Swadlincote, Derbyshire, DE12 6RX
Area of Operation: UK (Excluding Ireland)
Tel: 01283 761477
Fax: 01283 763852
Email: warm@tubism.co.uk
Web: www.tubism.co.uk
Product Type: 7 Other Info: ✍ ✋
Material Type: C) 3

TUSCAN FOUNDRY PRODUCTS
Units C1-C3, Oakendene Industrial Estate, Bolney Road, Cowfold, West Sussex, RH13 8AZ
Area of Operation: Worldwide
Tel: 01403 860040
Fax: 0845 345 0215
Email: enquiries@tuscanfoundry.co.uk
Web: www.tuscanfoundry.co.uk ✍
Product Type: 1, 2

URBIS (GB) LTD.
55 Lidgate Crescent, Langthwaite Grange Industrial Estate, South Kirkby, West Yorkshire, WF9 3NR
Area of Operation: UK & Ireland
Tel: 01977 659829 Fax: 01977 659808
Email: david@urbis-gb.co.uk
Web: www.urbis-gb.co.uk ✍
Product Type: 1, 7

'Special Reader offer 20% off all website prices'*

URBIS (GB) LTD

Area of Operation: UK & Ireland
Tel: 01977 659829 Fax: 01977 659808
Email: david@urbis-gb.co.uk
Web: www.urbis-gb.co.uk
Product Type: 1, 7

SAVE POUNDS WITH URBIS!
The home of urban lifestyle products at unique, unbeatable prices! Limited stock of Chronus stainless steel radiators 600 x 1200 (pictured).

Special reader offer 'On-line' price from £149.83 inc. VAT. Only available on-line at: www.urbis-gb.co.uk while stocks last! Please quote ref: HBH06 when ordering!

VIESSMANN LTD
Hortonwood 30, Telford, Shropshire, TF1 7YP
Area of Operation: UK & Ireland
Tel: 01952 675000
Fax: 01952 675040
Email: info@viessmann.co.uk
Web: www.viessmann.co.uk
Product Type: 2

VOGUE UK
Units 6-10, Strawberry Lane, Industrial Estate, Willenhall, West Midlands, WV13 3RS
Area of Operation: UK & Ireland
Tel: 0870 403 0101
Fax: 0870 403 0102
Email: sales@vogue-uk.com
Web: www.vogue-uk.com

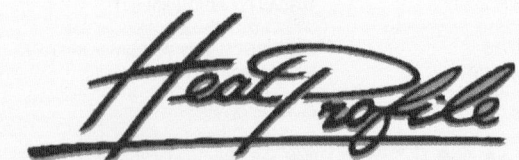

Full central heating without radiators?

If you are looking for a full central heating system that provides all round warmth without radiators or under floor heating, then *HeatProfile Skirting Heating Systems* are the answer.

HeatProfile Skirting Systems are designed to meet the most exacting requirements and, as there are no radiators, they are space saving, visually attractive, energy saving and virtually maintenance free. Unlike a radiator which works by 80% convection and 20% radiation, *HeatProfile* skirting heating is 80% radiant and 20% convected, which means far less air turbulence, resulting in less allergen and dust in the air, giving a cleaner living environment.

HeatProfile Skirting Heating generates *radiant* warmth at low level, therefore providing a higher level of all round comfort with no cold spots or wasted energy.

Skirting heating laid flat under floor to ceiling glazing.

The systems, either electric or water, are extremely easy and quick to install and have major environmental advantages. The heat emitter of the system is a very attractive skirting feature, which, of course, dispenses with the need for wooden skirting boards, and as there are no radiators, you can place the furniture *where you want it* and not where radiators dictate. The Systems are also very quick to respond to temperature demand. Another hidden benefit is that the bracket that fixes the heat emitter to the wall has a 25mm adjustment allowing the skirting to be moved up and down so that speaker cables etc can be hidden from view and fitted carpeting is literally "wall to wall".

These are just some of the many added value benefits that HeatProfile skirting heating systems offer, but there are many more, so if you would like to find out what they are, you can either telephone Ian Mitchell on **01483 537000** or visit our website

www.heatprofile.co.uk

HeatProfile Ltd Horizon House, 4 Wey Court, Mary Road, Guildford GU1 4QU
Tel: 01483 537000 Fax: 01483 537500
Email: sales@heatprofile.co.uk

WARMROOMS
24 Corncroft Lane, St Leonards Park,
Gloucester, Gloucestershire, GL4 6XU
Area of Operation: UK (Excluding Ireland)
Tel: 01452 304460 **Fax:** 01452 304460
Email: sales@warmrooms.co.uk
Web: www.warmrooms.co.uk ⌁
Product Type: 1, 2, 3, 4, 5, 6, 7 ★ ✪ ❋

WINTHER BROWNE
75 Bilton Way, Enfield, London, EN3 7ER
Area of Operation: UK (Excluding Ireland)
Tel: 0208 3449050 **Fax:** 0845 612 1894
Email: sales@wintherbrowne.co.uk
Web: www.wintherbrowne.co.uk
Product Type: 6 **Material Type:** B) 2

WWW.DESIGNER-RADIATORS.COM
Regent Street, Colne, Lancashire, BB8 8LD
Area of Operation: UK & Ireland
Tel: 01282 862509 **Fax:** 01282 871192
Web: www.designer-radiators.com ⌁
Product Type: 1, 2, 3, 4, 5, 6, 7

ZEHNDER LTD
B15 Armstrong Mall, Southwood Business Park,
Farnborough, Hampshire , GU14 0NR
Area of Operation: UK & Ireland
Tel: 01252 515151 **Fax:** 01252 522528
Email: sales@zehnder.co.uk
Web: www.zehnder.co.uk
Product Type: 1, 2, 3, 4, 5, 7
Other Info: ✎ **Material Type:** C) 2

FIRES

..FIRES4U..
PO Box 6843, Swadlincote, Derbyshire, DE12 7XX
Area of Operation: Worldwide
Tel: 0845 612 0001 **Email:** sales@fires4u.co.uk
Web: www.fires4u.co.uk ⌁
Product Type: 1, 2, 3, 4, 5, 6, 8
★ ○ □ ● ✪ ❋ ◆

FIRES4U

Area of Operation: Worldwide
Tel: 0845 612 0001
Email: sales@fires4u.co.uk
Web: www.fires4u.co.uk
Product Type: 1, 2, 3, 4, 5, 6, 8

An extensive online website of Fires, Fireplaces & Stoves with a multitude of options and at low prices. Free delivery to most parts of the UK. Price match promise. Good source of information and specification.

A & M ENERGY FIRES
Pool House, Huntley,
Gloucestershire, GL19 3DZ
Area of Operation: Europe
Tel: 01452 830662 **Fax:** 01452 830891
Email: am@energyfires.co.uk
Web: www.energyfires.co.uk
Product Type: 9

ACANTHA LIFESTYLE LTD
32-34 Park Royal Road, Park Royal,
London, NW10 7LN
Area of Operation: Worldwide
Tel: 0208 453 1537 **Fax:** 0208 453 1538
Email: sales@acanthalifesyle.com
Web: www.acanthalifestyle.co.uk
Product Type: 1, 2, 3, 4, 5, 6, 7, 8
Material Type: E) 1, 2, 5, 8

AGA-RAYBURN
Station Road, Ketley, Telford,
Shropshire, TF1 5AQ
Area of Operation: Worldwide
Tel: 01952 642000 **Fax:** 01952 243138
Email: jkingsbury-webber@aga-web.co.uk
Web: www.aga-web.co.uk

ANGLIA FIREPLACES & DESIGN LTD
Anglia House, Kendal Court, Cambridge Road,
Impington, Cambridgeshire, CB4 9YS
Area of Operation: UK & Ireland
Tel: 01223 234713 **Fax:** 01223 235116
Email: info@fireplaces.co.uk
Web: www.fireplaces.co.uk
Product Type: 1, 8 ★ □ ● ✪

BAXI FIRES DIVISION
Baxi Fire Division, Wood Lane, Erdington,
Birmingham, West Midlands, B24 9QP
Area of Operation: UK (Excluding Ireland)
Tel: 0121 373 8111 **Fax:** 0121 373 8181
Email: cgarner@valor.co.uk
Web: www.firesandstoves.co.uk
Product Type: 1, 2, 3, 4, 5, 6, 8

BD BROOKS FIREPLACES
109 Halifax Road, Ripponden, Halifax,
West Yorkshire, HX6 4DA
Area of Operation: UK (Excluding Ireland)
Tel: 01422 822220 **Fax:** 01422 822220
Email: bdbrooks2000@yahoo.com
Web: www.bdbrooksfireplaces.com
Product Type: 1, 2, 3, 4, 5, 6, 7, 8

BE MODERN GROUP LTD
Western Approach, South Shields,
Tyne & Wear, NE33 5QZ
Area of Operation: UK & Ireland
Tel: 0191 455 3571 **Fax:** 0191 456 5556
Web: www.bemodern.co.uk
Product Type: 1, 6, 8, 9 ★ ❋
Material Type: C) 2, 3, 4, 5, 11, 12, 14

BLAZES FIREPLACE CENTRES
23 Standish Street, Burnley, Lancashire, BB11 1AP
Area of Operation: UK (Excluding Ireland)
Tel: 01777 838554 **Fax:** 01777 839370
Email: info@blazes.co.uk
Web: www.blazes.co.uk

BRILLIANT FIRES
Thwaites Close, Shadsworth Business Park,
Blackburn, Lancashire, BB1 2QQ
Area of Operation: UK & Ireland
Tel: 01254 682384
Fax: 01254 672 647
Email: info@brilliantfires.co.uk
Web: www.brilliantfires.co.uk
Product Type: 1, 2, 3, 4, 5, 8, 9 ★ ❋
Material Type: C) 1, 2, 3, 11, 14, 16

BROSELEY FIRES
Knights Way, Battlefield Enterprise Park,
Shrewsbury, Shropshire, SY1 3AB
Area of Operation: UK & Ireland
Tel: 01743 461444
Fax: 01743 461446
Email: sales@broseleyfires.com
Web: www.broseleyfires.com

BURLEY APPLIANCES LIMITED
Lands End Way, Oakham, Leicestershire, LE15 6RB
Area of Operation: Worldwide
Tel: 01572 756956
Fax: 01572 724390
Email: info@burley.co.uk
Web: www.burley.co.uk
Product Type: 1, 8

CHARLTON & JENRICK LTD
Units G1 & G2, Halesfield 5, Telford,
Shropshire, TF7 4QJ
Area of Operation: UK & Ireland
Tel: 01952 278020 **Fax:** 01952 278043
Email: hiedi@charltonandjenrick.co.uk
Web: www.charltonandjenrick
Product Type: 1, 8, 9 ★ ❄

CHARNWOOD STOVES & FIRES
Bishops Way, Newport, Isle of Wight, PO30 5WS
Area of Operation: Worldwide
Tel: 01983 537780 **Fax:** 01983 537788
Email: charnwood@ajwells.co.uk
Web: www.charnwood.com ✎
Product Type: 1, 8 ★ □ ● ✿ ❄
Other Info: ECO ✋
Material Type: C) 2, 5

CHARNWOOD STOVES

Area of Operation: Worldwide
Tel: 01983 537780 **Fax:** 01983 537788
Email: charnwood@ajwells.co.uk
Web: www.charwood.com
Product Type: 1, 8

Britain's original designers and manufacturers of
clean-burn multi-fuel and wood burning fires &
boilers. Ultra efficient and eco-friendly. Various
models, choice of colours and accessories to suit
every situation.

CHISWELL FIREPLACES LTD
Fireplace Showroom, 192 Watford Road,
Chiswell Green, St Albans, Hertfordshire, AL2 3EB
Area of Operation: South East England
Tel: 01727 859512
Email: sales@chiswellfireplaces.com
Web: www.chiswellfireplaces.com
Product Type: 1, 2, 3, 4, 5, 6, 7, 8, 9 ★ □ ● ❄
Material Type: C) 2, 3, 5, 11, 13, 14

CONTINENTAL FIRES LIMITED
Laundry Bank, Church Stretton, Shropshire, SY6 6PH
Area of Operation: UK & Ireland
Tel: 01694 724199 **Fax:** 01694 720100
Email: sales@continentalfires.com
Web: www.continentalfires.com

COSY ROOMS (COSY-HEATING.CO.UK)
17 Chiltern Way, North Hykeham,
Lincoln, Lincolnshire, LN6 9SY
Area of Operation: UK (Excluding Ireland)
Tel: 01522 696002 **Fax:** 01522 696002
Email: keith@cosy-rooms.com
Web: www.cosy-heating.co.uk ✎
Product Type: 1, 9

CVO FIRE LTD
4 Beaumont Square, Durham Way South, Aycliffe
Industrial Park, Newton Aycliffe, Durham, DL5 6SW
Area of Operation: UK & Ireland
Tel: 01325 327221 **Fax:** 01325 327292
Web: www.cvo.co.uk

DRAYTON PROPERTY
5 Lake Crescent, Daventry, Northants, NN11 9EB
Area of Operation: UK (Excluding Ireland)
Tel: 01327 300249 **Fax:** 01327 300249
Email: draytonproperty@aol.com
Web: www.draytonproperty.com ✎
Product Type: 2, 3, 4, 8

DUNSLEY HEAT LTD
Bridge Mills, Huddersfield Road, Holmfirth,
Huddersfield, West Yorkshire, HD9 3TW
Area of Operation: UK & Ireland
Tel: 01484 682635 **Fax:** 01484 688428
Email: sales@dunsleyheat.co.uk
Web: www.dunsleyheat.co.uk
Product Type: 9 □ ●

ELGIN & HALL
Adelphi House, Hunton, Bedale, North Yorks, DL8 1LY
Area of Operation: UK (Excluding Ireland)
Tel: 01677 450 100 **Fax:** 01677 450 713
Email: info@elgin.co.uk
Web: www.elgin.co.uk

EMSWORTH FIREPLACES LTD
Unit 3, Station Approach, Emsworth,
Hampshire, PO10 7PW
Area of Operation: Worldwide
Tel: 01243 373431 **Fax:** 01243 371023
Email: sales@emsworth.co.uk
Web: www.emsworth.co.uk ✎
Product Type: 1, 2, 4, 5, 6, 8
Material Type: A) 2, 3, 4, 5, 8

ESSE
Ouzledale Foundry, Long Ing, Barnoldswick,
Lancashire, BB18 6BN
Area of Operation: Worldwide
Tel: 01282 813235 **Fax:** 01282 816876
Email: esse@ouzledale.co.uk **Web:** www.esse.com
Product Type: 1, 9

EUROHEAT LTD
Court Farm Business Park, Bishops Frome,
Worcestershire, WR6 5AY
Area of Operation: UK & Ireland
Tel: 01885 491100 **Fax:** 01885 491101
Email: sales@euroheat.co.uk
Web: www.euroheat.co.uk
Product Type: 1, 8 ★ ○ ● ✿
Material Type: C) 2, 5, 11

FABER FIRES
Touchstone House, 82 High Street, Measham,
Derbyshire, DE12 7JB
Area of Operation: UK (Excluding Ireland)
Tel: 0845 130 1862 **Fax:** 01530 274271
Email: sales@faber-fires.co.uk
Web: www.faber-fires.co.uk
Product Type: 1, 9 ★ □ ●

FABER FIRES

Area of Operation: UK (Excluding Ireland)
Tel: 0845 130 1862
Fax: 01530 274 271
Email: sales@faber-fires.co.uk
Web: www.faber-fires.co.uk
Product Type: 1,9

Faber Fires from the outstanding European
manufacturer available in the UK. From classic
traditional solid stone, to the ultra modern
minimalist designs.

FIREPLACE & TIMBER PRODUCTS
Unit 2 Holyrood Drive,
Skippingdale Industrial Estate,
Scunthorpe, Lincolnshire, DN15 8NN
Area of Operation: UK & Ireland
Tel: 01724 852888 **Fax:** 01724 277255
Email: ftprdcts@yahoo.co.uk
Product Type: 1, 2, 3, 4, 5, 6, 8, 9

FIREPLACE CONSULTANTS LTD
The Studio, The Old Rothschild Arms,
Buckland Road, Buckland, Aylesbury,
Buckinghamshire, HP22 5LP
Area of Operation: Greater London,
Midlands & Mid Wales, South East England
Tel: 01296 632287 **Fax:** 01296 632287
Email: info@fireplaceconsultants.com
Web: www.fireplaceconsultants.com
Product Type: 1, 2, 3, 4, 5, 6, 7, 8
Other Info: ✐ ✋

FLAMERITE FIRES LTD
Greenhough Road, Lichfield,
West Midlands, WS13 7AU
Area of Operation: Worldwide
Tel: 01543 251122
Fax: 01543 251133
Email: info@flameritefires.com
Web: www.flameritefires.com
Product Type: 1, 5, 6, 8, 9 ❄
Other Info: ✋
Material Type: C) 1, 2, 3, 4, 5, 6, 11, 14

FLAMERIRE FIRES

Area of Operation: Worldwide
Tel: 01543 251 122 **Fax:** 01543 251 133
Email: info@flameritefires.com
Web: www.flameritefires.com
Energy Sources: ❄

Innovative, award-winning, designs create a
dramatic focal point for any style of room
instantly and cost-effectively. Manufactured in
the UK to the highest quality standards.

FLAMEWAVE FIRES - DK UK LTD
PO Box 611, Folkestone, Kent, CT18 7WY
Area of Operation: UK (Excluding Ireland)
Tel: 0845 257 5028
Fax: 0845 257 5038
Web: www.flamewavefires.co.uk

GAZCO LTD
Osprey Road, Sowton Industrial Estate,
Exeter, Devon, EX2 7JG
Area of Operation: Europe
Tel: 01392 261999
Fax: 01392 444148
Email: info@gazco.com
Web: www.gazco.com
Product Type: 1, 2, 3, 4, 5, 6, 8 ★ □ ● ✿ ❄
Other Info: ✐
Material Type: C) 2, 3, 5, 11

GLEN DIMPLEX UK LTD
Millbrook House, Grange Drive, Hedge End,
Southampton, Hampshire, SO30 2DF
Area of Operation: UK & Ireland
Tel: 0870 077 7117 **Fax:** 0870 727 0114
Email: marketing@glendimplex.com
Web: www.dimplex.co.uk
Product Type: 1, 9 ❄

GRENADIER FIRELIGHTERS LIMITED
Unit 3C, Barrowmore Enterprise Estate,
Great Barrow, Chester, Cheshire, CH3 7JS
Area of Operation: UK & Ireland
Tel: 01829 741649 **Fax:** 01829 741659
Email: enquiries@grenadier.com
Web: www.grenadier.uk.com ✎
Product Type: 9 **Other Info:** ✋

JETMASTER FIRES LTD
Unit 2 Peacock Trading Estate, Goodwood Road,
Chandlers Ford, Eastleigh, Hampshire, SO50 4NT
Area of Operation: Europe
Tel: 0870 727 0105 **Fax:** 0870 727 0106
Email: jetmastersales@aol.com
Web: www.jetmaster.co.uk
Product Type: 1, 2, 3, 4, 5, 6, 7, 8, 9
Other Info: ✋

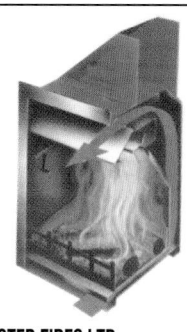

JETMASTER FIRES LTD

Area of Operation: Europe
Tel: 0870 727 0105 **Fax:** 0870 727 0106
Email: jetmastersales@aol.com
Web: www.jetmaster.co.uk
Product Type: 1,2,3,4,5,6,7,8,9

Jetmaster's original two way heating system with
radiant and convective heat, warms the whole
room right into the corners, while the flames and
embers make a natural and effective focal draw.
There's nothing like a real fire!

JETMASTER FIRES LTD

Area of Operation: Europe
Tel: 0870 727 0105 **Fax:** 0870 727 0106
Email: jetmastersales@aol.com
Web: www.jetmaster.co.uk
Product Type: 1, 2, 3, 4, 5, 6 ,7, 8, 9

Jetmaster has been building stylish and efficient,
smoke free fireboxes for wood and coal or gas
since 1951, real fires! real quality! A welcoming
warmth to enhance every home.

HEATING, PLUMBING & ELECTRICAL

HEATING, PLUMBING & ELECTRICAL

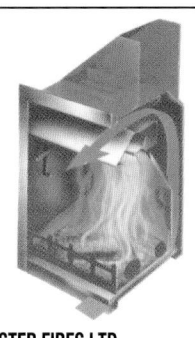

JETMASTER FIRES LTD
Area of Operation: Europe
Tel: 0870 727 0105 **Fax:** 0870 727 0106
Email: jetmastersales@aol.com
Web: www.jetmaster.co.uk
Product Type: 1,2,3,4,5,6,7,8,9

Jetmaster gas fires are often mistaken for real coal fires. Flame variation flicker and ember colour changes combine to produce a completely realistic effect. Also available with remote control, that ignites the pilot and controls the fire.

KINGSWORTHY FOUNDRY CO LTD
London Road, Kingsworthy, Winchester, Hampshire, SO23 7QG
Area of Operation: UK & Ireland
Tel: 01962 883776 **Fax:** 01962 882925
Email: kwf@fsbdial.co.uk
Web: www.kingsworthyfoundry.co.uk
Product Type: 9

LEL FIREPLACES
Tre-Ifan Farmhouse, Caergeiliog, Holyhead, Anglesey, LL65 3HP
Area of Operation: UK & Ireland
Tel: 01407 742240 **Fax:** 01407 742262
Email: sales@lel-fireplaces.com
Web: www.lel-fireplaces.com ⌐
Product Type: 1, 2, 3, 4, 5, 6, 7, 8, 9 ★ □ ● ✪ ❀

LIVINGSTYLE.CO.UK
Bridge Street, Shotton, Flintshire, CH5 1DU
Area of Operation: UK (Excluding Ireland)
Tel: 0800 298 9190
Email: info@livingstyle.co.uk
Web: www.livingstyle.co.uk ⌐
Product Type: 1, 2, 3, 4, 5, 6, 7, 8, 9

MAGIGLO LTD
7 Lysander Close, Pysons Road Industrial Estate, Broadstairs, Kent, CT10 2YJ
Area of Operation: UK & Ireland
Tel: 01843 602863 **Fax:** 01843 860108
Email: info@magiglo.co.uk
Web: www.magiglo.co.uk
Product Type: 1, 2, 4, 5, 6, 8 ★
Other Info: ✐

MARK RIPLEY FORGE & FIREPLACES
Robertsbridge, Bridge Bungalow, East Sussex, TN32 5NY
Tel: 01580 880324 **Fax:** 01580 881927
Email: ripleym@gxn.co.uk
Web: www.ripleyfireplaces.co.uk
Product Type: 2, 4, 6

OLDE ENGLANDE REPRODUCTIONS
Fireplace Works, Normacot Road, Longton, Stoke-on-Trent, Staffordshire, ST3 1PN
Area of Operation: UK (Excluding Ireland)
Tel: 01782 319350 **Fax:** 01782 593479
Email: sales@oerfireplaces.com
Web: www.oerfireplaces.com
Product Type: 1, 2, 3, 4, 5, 6, 8

PARAGON FIRES
Unit G1 & G2, Halesfield, Telford, Shropshire, TF7 4QJ
Area of Operation: UK & Ireland
Tel: 01952 278020 **Fax:** 01952 278043
Email: info@charltonandjenrick.co.uk
Web: www.paragonfires.co.uk
Product Type: 1, 8 ★ ❀

PENDRAGON FIREPLACES.COM
12 Market St, Stourbridge, West Midlands, DY8 1AD
Area of Operation: UK & Ireland
Tel: 01384 376441 **Fax:** 01384 376441
Email: sales@pendragonfireplaces.co.uk
Web: www.pendragonfireplaces.com ⌐
Product Type: 2, 3, 4, 5, 6, 7

PERCY DOUGHTY & CO
Imperial Point, Express Trading Estate, Stonehill Road, Farnworth, Bolton, Lancashire, M38 9ST
Area of Operation: Midlands & Mid Wales, North West England and North Wales
Tel: 01204 868550 **Fax:** 01204 868551
Email: sales@percydoughty.com
Web: www.percydoughty.co.uk
Product Type: 1, 4, 5, 6, 8

PETRA HELLAS
Toll Bar Business Park, Newchurch Road, Stacksteads, Bacup, Lancashire, OI13 0NA
Area of Operation: Europe
Tel: 01706 876102 **Fax:** 01706 876194
Email: info@petrahellas.co.uk
Web: www.petrahellas.co.uk
Product Type: 1, 8

R W KNIGHT & SON LTD
Castle Farm, Marshfield, Chippenham, Wiltshire, SN14 8HU
Area of Operation: Midlands & Mid Wales, South East England, South West England and South Wales
Tel: 01225 891469 **Fax:** 01225 892369
Email: Enquiries@knight-stoves.co.uk
Web: www.knight-stoves.co.uk
Product Type: 1, 2, 3, 4, 5 ★ □ ● ✪
Material Type: C) 2, 3, 5, 11, 14

REAL FLAME
80 New Kings Rd, Fulham, Greater London, SW6 4LT
Area of Operation: Worldwide
Tel: 0207 731 5025 **Fax:** 0207 736 4625
Email: info@realflame.co.uk
Web: www.realflame.co.uk ⌐
Product Type: 1, 2, 3, 4, 5, 6, 7, 8 ★
Other Info: ✐ ✋

ROBEYS
Old School House, Green Lane, Belper, Derbyshire, DE56 1BY
Area of Operation: UK & Ireland
Tel: 01773 820940 **Fax:** 01773 821652
Email: info@robeys.co.uk **Web:** www.robeys.co.uk

ROCAL FIRES
Touchstone House, Measham, Derbyshire, DE12 7JB
Area of Operation: UK (Excluding Ireland)
Tel: 0845 130 1862 **Fax:** 01530 274271
Email: sales@rocal-fires.co.uk
Web: www.rocal-fires.co.uk
Product Type: 1, 9 ★ □ ●

RUDLOE STONEWORKS LTD
Leafield Stoneyard, Potley Lane, Corsham, Wiltshire, SN13 9RS
Area of Operation: UK & Ireland
Tel: 01225 816400
Fax: 01225 811343
Email: paul@rudloe-stone.com
Web: www.rudloe-stone.com
Product Type: 1, 2, 5, 7

SOLAR FIRES & FIREPLACES LTD
Alyn Works, Mostyn Road, Holywell, Flintshire, CH8 9DT
Area of Operation: UK (Excluding Ireland)
Tel: 01745 561685
Fax: 01745 580 987
Email: sales@solarfiresandfireplaces.co.uk
Web: www.solarfiresandfireplaces.co.uk

STOVAX LIMITED
Falcon Road, Sowton Industrial Estate, Exeter, Devon, EX2 7LF
Area of Operation: UK & Ireland
Tel: 01392 474011
Fax: 01392 219932
Email: info@stovax.com
Web: www.stovax.com
Product Type: 1, 2, 3, 4, 5, 6, 7, 8, 9 □ ● ✪
Other Info: ✐ ✋

SUPERIOR FIRES
Touchstone House, 82 High Street, Measham, Derbyshire, DE12 7JB
Area of Operation: UK (Excluding Ireland)
Tel: 0845 130 1862
Fax: 01530 274271
Email: sales@superior-fires.co.uk
Web: www.superior-fires.co.uk
Product Type: 1, 9 ★

SUPERIOR FIRES
Area of Operation: UK (Excluding Ireland)
Tel: 0845 130 1862 **Fax:** 01530 274 271
Email: sales@superior-fires.co.uk
Web: www.superior-fires.co.uk
Product Type: 1,9

Superior Fires - The Ultimate In No Compromise Design. A range of five stunning modern designs that are wall mounted. These fires all use the latest in technology and do not require any form of Flu or Chimney. Ideal for all main rooms and not just limited to the more traditional living room fireplace. Easy to install.

TAYLOR AND PORTWAY
52 Broton Drive Industrial Estate, Halstead, Essex, CO9 1HB
Area of Operation: Europe
Tel: 01787 472551
Fax: 01787 476589
Email: sales@portwayfires.com
Web: www.portwayfires.com
Product Type: 1, 2, 3, 4, 6, 8 ★ ❀

THE EDWARDIAN FIREPLACE COMPANY
Former All Saints Church, Armoury Way, Wandsworth, London, SW18 1HZ
Area of Operation: Greater London, South East England
Tel: 0208 870 0167
Fax: 0208 877 9069
Email: mike@edwardianfires.com
Web: www.edwardianfires.com
Product Type: 1, 2, 3, 4, 5, 6, 7, 8, 9

THE FIREPLACE GALLERY (UK) LTD
Clarence Road, Worksop, Nottinghamshire, S80 1QA
Area of Operation: UK & Ireland
Tel: 01909 500802 **Fax:** 01909 500810
Email: fireplacegallery@btinternet.com
Web: www.fireplacegallery.co.uk
Product Type: 1, 8

THE FIREPLACE MARKETING COMPANY LIMITED
The Old Coach House, Southern Road, Thame, Oxfordshire, OX9 2ED
Area of Operation: Worldwide
Tel: 01844 260960 **Fax:** 01844 260267
Email: david@fireplacemarketing.co.uk
Web: www.fireplace.co.uk ⌐
Product Type: 1, 2, 3, 4, 5, 6, 7, 8

THE JRG GROUP
3 Crompton Way, North Newmoor Industrial Estate, Irvine, Ayrshire, KA11 4HU
Area of Operation: Europe
Tel: 0871 200 8080 **Fax:** 01294 211222
Email: chris@jrgfiresurrounds.com
Web: www.jrggroup.com
Product Type: 1, 8

THEALE FIREPLACES RDG LTD
Milehouse Farm, Bath Road, Theale, Berkshire, RG7 5HJ
Area of Operation: UK (Excluding Ireland)
Tel: 0118 930 2232 **Fax:** 0118 932 3344
Email: mail@theale-fireplaces.co.uk
Web: www.theale-fireplaces.co.uk
Product Type: 1, 2, 3, 4, 5

TOWN AND COUNTRY FIRES
1 Enterprise Way, Thornton Road Industrial Estate, Pickering, North Yorkshire, YO18 7NA
Area of Operation: Europe
Tel: 01751 474803 **Fax:** 01751 475205
Email: sales@townandcountryfires.co.uk
Web: www.townandcountryfires.co.uk

VERINE
52 Broton Drive, Halstead, Essex, CO9 1HB
Area of Operation: UK & Ireland
Tel: 01787 472551 **Fax:** 01787 476589
Email: sales@verine.co.uk
Web: www.verine.co.uk
Product Type: 1, 2, 3, 4, 5, 6, 8 ★ ❀

WWW.FIREPLACE.CO.UK
Old Coach House, Southern Road, Thame, Oxfordshire, OX9 2ED
Area of Operation: UK (Excluding Ireland)
Tel: 01844 260960 **Fax:** 01844 260267
Web: www.fireplace.co.uk

FIREPLACES

KEY
PRODUCT TYPES: 1= Contemporary
2 = Edwardian 3 = Georgian 4 = Victorian
5 = Art Deco 6 = Art Nouveau 7 = Arts & Crafts 8 = Classical 9 = Other

ENERGY SOURCES: ★ Gas ○ Oil
□ Coal ● Wood ✪ Multi-Fuel
❀ Electric ◆ Gel

OTHER: ▽ Reclaimed ⌐ On-line shopping
✐ Bespoke ✋ Hand-made ECO Ecological

DELLSTONE - FIRES, FIREPLACES & STOVES
Touchstone House, 82 High Street, Measham, Derbyshire, DE12 7JB
Area of Operation: UK (Excluding Ireland)
Tel: 0845 130 1862 **Fax:** 01530 274271
Email: sales@dellstone.com
Web: www.dellstone.com
Product Type: 1, 2, 3, 4, 5, 6, 8, 9
★ ○ □ ● ✪ ❀ ◆
Other Info: ✐ ✋ **Material Type:** C) 2

HEATING, PLUMBING & ELECTRICAL

CVO FIRE

CVO FIRE

4 Beaumont Square,
Durham Way South,
Aycliffe Industrial Park,
Newton Aycliffe,
Durham DL5 6SW
Enquiry Hotline: 01325 301020
Tel: 01325 327221
Fax: 01325 327292
Website: www.cvo.co.uk
Area of operation: UK & Ireland

FLUELESS FIRE LINE

A flueless gas fire that functions without the use of a chimney or flue that can virtually go anywhere. Shown in Rosal Limestone, available in granite, madre perle, stainless steel, mirror and powder coated steel.

FLUELESS FIRE LINE

The cleanest burning flueless fire in the world that doesn't require a catalytic converter or glass fronted guard to function safely.

FLUELESS FIRE LINE

Beautiful realistic flames, easy to install and 100% heat efficient. Hand made in the UK utilizing the highest grade of materials.

FIREBOWL

The first and original of its kind. Winner of the Prince of Wales award Available in white, black concrete and handcast aluminium and bronze finishes.

FIREBOWL

Incorporates pioneering gas technology with the finest of materials. Ceramat enables the flames to dance across the firebowl in a very realistic and stunning effect.

FIRE RIBBON

Our most popular hole in the wall fire that incorporates a beautiful simplistic design with the worlds most powerful standard burner. Two sizes available.

FIRE RIBBON

This big beautiful fire delivers a magnificent continuous ribbon of flame and is available in limestone, madre perle, granite, stainless steel, and powder coated steel.

FIRE BLOCK

A beautiful hole-in-the-wall fire combing a ribbon effect flame that can be used with or without pebbles. Limestone fascia shown.

RIPPLE

A larger piece formed in concrete with a textured surface and a small concrete firebowl interior. A convex iron guard slides up and down.

SEED

A stunning piece with a hand-applied, soft-skin finish echoing the cracks and fractures of a seed pod.

NY BOLECTION

A classic CVO sandstone mantel shown with a contemporary white firebowl gas burner. Other classic mantels include marble, and american black walnut.

AARROW FIRES
The Fireworks, Bridport, Dorset, DT6 3BE
Area of Operation: UK (Excluding Ireland)
Tel: 01308 427234 **Fax:** 01308 423441
Email: markbrettell@aarrowfires.com
Web: www.aarrowfires.com

ACANTHA LIFESTYLE LTD
32-34 Park Royal Road, Park Royal,
London, NW10 7LN
Area of Operation: Worldwide
Tel: 0208 453 1537 **Fax:** 0208 453 1538
Email: sales@acanthalifesyle.com
Web: www.acanthalifestyle.co.uk
Product Type: 1, 2, 3, 4, 5, 6, 7, 8

ALL THINGS STONE LIMITED
Anchor House, 33 Ospringe Street,
Ospringe, Faversham, Kent, ME13 8TW
Area of Operation: UK & Ireland
Tel: 01795 531001 **Fax:** 01795 530742
Email: info@allthingsstone.co.uk
Web: www.allthingsstone.co.uk

ANGLIA FIREPLACES & DESIGN LTD
Anglia House, Kendal Court, Cambridge Road,
Impington, Cambridgeshire, CB4 9YS
Area of Operation: UK & Ireland
Tel: 01223 234713 **Fax:** 01223 235116
Email: info@fireplaces.co.uk
Web: www.fireplaces.co.uk
Product Type: 1, 2, 3, 4, 6, 8, 9

ARO MARBLE
18 Minerva Road, London, NW10 6HJ
Area of Operation: UK (Excluding Ireland)
Tel: 0208 965 1144 **Fax:** 0208 965 1818
Email: info@aromarble.com
Web: www.aromarble.com
Product Type: 1, 2, 3, 4, 5, 6, 7, 8, 9
Material Type: E) 1, 2

BD BROOKS FIREPLACES
109 Halifax Road, Ripponden,
Halifax, West Yorkshire, HX6 4DA
Area of Operation: UK (Excluding Ireland)
Tel: 01422 822220 **Fax:** 01422 822220
Email: bdbrooks2000@yahoo.com
Web: www.bdbrooksfireplaces.com
Product Type: 1, 2, 3, 4, 5, 6, 7, 8, 9

BE MODERN GROUP LTD
Western Approach, South Shields,
Tyne & Wear, NE33 5QZ
Area of Operation: UK & Ireland
Tel: 0191 455 3571 **Fax:** 0191 456 5556
Web: www.bemodern.co.uk
Product Type: 1, 5, 6, 8 ★ ❀
Material Type: A) 2, 5

BLAZES FIREPLACE CENTRES
23 Standish Street, Burnley, Lancashire, BB11 1AP
Area of Operation: UK (Excluding Ireland)
Tel: 01777 838554 **Fax:** 01777 839370
Email: info@blazes.co.uk **Web:** www.blazes.co.uk

BRILLIANT FIRES
Thwaites Close, Shadsworth Business Park,
Blackburn, Lancashire, BB1 2QQ
Area of Operation: UK & Ireland
Tel: 01254 682384 **Fax:** 01254 672 647
Email: info@brilliantfires.co.uk
Web: www.brilliantfires.co.uk
Product Type: 1, 8 **Other Info:** ✏ ✋
Material Type: C) 3

BURLEY APPLIANCES LIMITED
Lands End Way, Oakham, Leicestershire, LE15 6RB
Area of Operation: Worldwide
Tel: 01572 756956 **Fax:** 01572 724390
Email: info@burley.co.uk **Web:** www.burley.co.uk

CHARLES PEARCE LTD
26 Devonshire Road, London, W4 2HD
Area of Operation: Worldwide
Tel: 0208 995 3333
Email: enquiries@charlespearce.co.uk
Web: www.charlespearce.co.uk
Material Type: M) 4

CHARLTON & JENRICK LTD
Units G1 & G2, Halesfield 5, Telford, Shrops, TF7 4QJ
Area of Operation: UK & Ireland
Tel: 01952 278020 **Fax:** 01952 278043
Email: hiedi@charltonandjenrick.co.uk
Web: www.charltonandjenrick
Product Type: 1, 2, 8 **Other Info:** ECO
Material Type: C) 1, 2, 3, 5, 11, 14

CHARNWOOD STOVES & FIRES
Bishops Way, Newport, Isle of Wight, PO30 5WS
Area of Operation: Worldwide
Tel: 01983 537780 **Fax:** 01983 537788
Email: charnwood@ajwells.co.uk
Web: www.charnwood.com ✋
Product Type: 1, 8, 9 **Other Info:** ECO ✋
Material Type: C) 2, 5

CHISWELL FIREPLACES LTD
Fireplace Showroom, 192 Watford Road,
Chiswell Green, St Albans, Hertfordshire, AL2 3EB
Area of Operation: South East England
Tel: 01727 859512
Email: sales@chiswellfireplaces.com
Web: www.chiswellfireplaces.com
Product Type: 1, 2, 3, 4, 5, 6, 7, 8, 9 ★ □ ● ❀
Material Type: A) 1, 2, 4, 5, 7, 10

CLARKES ANTIQUE FIREPLACES
Old Forge, 32 Fore Street, Buckfastleigh,
Devon, TQ11 0AA
Area of Operation: UK (Excluding Ireland)
Tel: 01364 643060
Product Type: 2, 3, 4, 8, 9 **Other Info:** ▽

CVO FIRE LTD
4 Beaumont Square, Durham Way South, Aycliffe
Industrial Park, Newton Aycliffe, Durham, DL5 6SW
Area of Operation: UK & Ireland
Tel: 01325 327221 **Fax:** 01325 327292
Web: www.cvo.co.uk

DRAYTON PROPERTY
5 Lake Crescent, Daventry, Northants, NN11 9EB
Area of Operation: UK (Excluding Ireland)
Tel: 01327 300249 **Fax:** 01327 300249
Email: draytonproperty@aol.com
Web: www.draytonproperty.com ✋
Product Type: 2, 3, 4, 5, 8

EMSWORTH FIREPLACES LTD
Unit 3, Station Approach, Emsworth,
Hampshire, PO10 7PW
Area of Operation: Worldwide
Tel: 0845 230 5200 **Fax:** 01243 371023
Email: sales@emsworth.co.uk
Web: www.emsworth.co.uk ✋
Product Type: 1, 2, 3, 4, 5, 6, 7, 8
Material Type: A) 2, 3, 4, 5, 8

EMSWORTH
Area of Operation: Worldwide
Tel: 0845 2305200 **Fax:** 01243 371023
Email: sales@emsworth.co.uk
Web: www.emsworth.co.uk
Product Type: 1, 2, 3, 4, 5, 6, 7, 8
Materials Type: A) 2, 3, 4, 5, 8 **Other Info:** ✋

Manufacturers of contemporary and traditional
fireplaces in marble, limestone, steel and timber
since 1934. Bespoke work a speciality, contact us
for a brochure. Nationwide delivery service.

ENGLISH FIREPLACES
Old Firs, Hillbrow, Liss, Hampshire, GU33 7QE
Area of Operation: Worldwide
Tel: 01730 890218
Email: mjostephens@btinternet.com
Web: www.englishfireplaces.co.uk ✋
Product Type: 2, 3, 4, 6, 7, 8 ★ □ ● ❀
Material Type: E) 1, 2, 3, 4, 5

FINESSE FIREPLACES
Finesse Fireplaces, Unit 3,
The Tannery Industrial Estate, Holt,
Trowbridge, Wiltshire, BA14 6BB
Area of Operation: UK (Excluding Ireland)
Tel: 01225 783558
Fax: 01225 783558
Email: neil@finesse-stone.fsnet.co.uk
Web: www.finessefireplaces.com
Other Info: ✏
Material Type: E) 5

FIREPLACE & TIMBER PRODUCTS
Unit 2 Holyrood Drive, Skippingdale Industrial Estate,
Scunthorpe, Lincolnshire, DN15 8NN
Area of Operation: UK & Ireland
Tel: 01724 852888
Fax: 01724 277255
Email: ftprdcts@yahoo.co.uk
Product Type: 1, 2, 3, 4, 5, 6, 7, 8, 9

FIREPLACE CONSULTANTS LTD
The Studio, The Old Rothschild Arms,
Buckland Road, Buckland, Aylesbury,
Buckinghamshire, HP22 5LP
Area of Operation: Greater London,
Midlands & Mid Wales, South East England
Tel: 01296 632287
Fax: 01296 632287
Email: info@fireplaceconsultants.com
Web: www.fireplaceconsultants.com
Product Type: 1, 2, 3, 4, 5, 6, 7, 8

FIREPLACE DESIGN CONSULTANCY
Stansley Wood Farm, Dapple Heath,
Rugeley, Staffordshire, WS15 3PH
Area of Operation: UK (Excluding Ireland)
Tel: 01889 500500
Fax: 01889 500500
Email: info@inglenooks.co.uk
Web: www.inglenooks.co.uk

FIREPLACE WAREHOUSE GROUP LTD
Design House, Walnut Tree Close,
Guildford, Surrey, GU1 4UQ
Area of Operation: UK (Excluding Ireland)
Tel: 01483 568777
Fax: 01483 570013
Email: info@designfireplaces.co.uk
Web: www.fireplacewarehouse.co.uk
Product Type: 1, 2, 3, 4, 5, 6, 8

EMSWORTH
Area of Operation: Worldwide
Tel: 0845 2305200 **Fax:** 01243 371023
Email: sales@emsworth.co.uk
Web: www.emsworth.co.uk
Product Type: 1, 2, 3, 4, 5, 6, 7, 8
Materials Type: A) 2, 3, 4, 5, 8 **Other Info:** ✋

Manufacturers of contemporary and traditional
fireplaces in marble, limestone, steel and timber
since 1934. Bespoke work a speciality, contact us
for a brochure. Nationwide delivery service.

FLAMERITE FIRES LTD
Greenhough Road, Lichfield,
West Midlands, WS13 7AU
Area of Operation: Europe
Tel: 01543 251122
Fax: 01543 251133
Email: info@flameritefires.com
Web: www.flamritefires.com
Product Type: 1, 5, 6, 8, 9 ❀
Other Info: ✋
Material Type: A) 1, 2, 3

FRANCIS N. LOWE LTD.
The Marble Works, New Road, Middleton,
Matlock, Derbyshire, DE4 4NA
Area of Operation: Europe
Tel: 01629 822216
Fax: 01629 824348
Email: info@lowesmarble.com
Web: www.lowesmarble.com

GAZCO LTD
Osprey Road, Sowton Industrial Estate,
Exeter, Devon, EX2 7JG
Area of Operation: Europe
Tel: 01392 261999
Fax: 01392 444148
Email: info@gazco.com
Web: www.gazco.com
Product Type: 1 ★ ❀
Material Type: C) 2, 3

GLEN DIMPLEX UK LTD
Millbrook House, Grange Drive, Hedge End,
Southampton, Hampshire, SO30 2DF
Area of Operation: UK & Ireland
Tel: 0870 077 7117
Fax: 0870 727 0114
Email: marketing@glendimplex.com
Web: www.dimplex.co.uk
Product Type: 1, 9 ❀

HADDONSTONE LTD
The Forge House, East Haddon,
Northampton, Northamptonshire, NN6 8DB
Area of Operation: Worldwide
Tel: 01604 770711
Fax: 01604 770027
Email: info@haddonstone.co.uk
Web: www.haddonstone.co.uk

HADDONSTONE
Area of Operation: Worldwide
Tel: 01604 770711 **Fax:** 01604 770027
Email: info@haddonstone.co.uk
Web: www.haddonstone.co.uk
Product Type: 1,3,5,8,9
Other Info: ✋ ✋

Haddonstone manufacture cast stone fireplaces
from the ornate and traditional to the stylish and
contemporary. Extensive range of standard
designs. Custom made also available.

KINGSWORTHY FOUNDRY CO LTD
London Road, Kingsworthy,
Winchester, Hampshire, SO23 7QG
Area of Operation: UK & Ireland
Tel: 01962 883776
Fax: 01962 882925
Email: kwf@fsbdial.co.uk
Web: www.kingsworthyfoundry.co.uk
Product Type: 9

LANCASHIRE PLASTER PRODUCTS
167 Chorley New Road, Horwich,
Bolton, Lancashire, BL6 5QE
Area of Operation: UK (Excluding Ireland)
Tel: 01204 693900
Email: plasterproducts@gahope.com
Web: www.gahope.com

LEL FIREPLACES
Tre-Ifan Farmhouse, Caergeiliog,
Holyhead, Anglesey, LL65 3HP
Area of Operation: UK & Ireland
Tel: 01407 742240 **Fax:** 01407 742262
Email: sales@lel-fireplaces.com
Web: www.lel-fireplaces.com ⌂
Product Type: 1, 2, 3, 4, 5, 6, 7, 8
★ ○ □ ● ✪ ✿ ♦
Other Info: ✌ **Material Type:** C) 5

LIVINGSTYLE.CO.UK
Bridge Street, Shotton, Flintshire, CH5 1DU
Area of Operation: UK (Excluding Ireland)
Tel: 0800 298 9190
Email: info@livingstyle.co.uk
Web: www.livingstyle.co.uk ⌂
Product Type: 1, 2, 3, 4, 5, 6, 7, 8, 9

MARBLE HILL FIREPLACES
70-72 Richmond Road, Twickenham,
Greater London, TW1 3BE
Area of Operation: Greater London,
South West England and South Wales
Tel: 0208 892 1488 **Fax:** 0208 8916591
Email: sales@marblehill.co.uk
Web: www.marblehill.co.uk
Other Info: ▽ ✎ ✌ **Material Type:** E) 2, 4, 5, 11

MARK RIPLEY FORGE & FIREPLACES
Robertsbridge, Bridge Bungalow,
East Sussex, TN32 5NY
Tel: 01580 880324 **Fax:** 01580 881927
Email: ripleym@gxn.co.uk
Web: www.ripleyfireplaces.co.uk
Product Type: 9

NUNNA UUNI LTD
Karelia House, Kenmore Road,
Comrie Bridge By Aberfeldy,
Perth and Kinross, PH15 2PF
Area of Operation: UK & Ireland
Tel: 01887 822 025 **Fax:** 01887 822954
Email: sales@enerfoil.com
Web: www.nunnauuni.com
Product Type: 1, 9
Material Type: E) 10

OLD FLAMES
30 Long Street, Easingwold, York,
North Yorkshire, YO61 3HT
Area of Operation: UK (Excluding Ireland)
Tel: 01347 821188 **Fax:** 01347 821188
Email: philiplynas@aol.com
Web: www.oldflames.co.uk

OLDE ENGLANDE REPRODUCTIONS
Fireplace Works, Normacot Road, Longton,
Stoke-on-Trent, Staffordshire, ST3 1PN
Area of Operation: UK (Excluding Ireland)
Tel: 01782 319350 **Fax:** 01782 593479
Email: sales@oerfireplaces.com
Web: www.oerfireplaces.com
Product Type: 1, 2, 3, 4, 5, 6, 8

ORE
Fireplace Works, Normacot Road, Longton,
Stoke-On-Trent, Staffordshire, ST3 1PN
Area of Operation: UK (Excluding Ireland)
Tel: 01782 319350 **Fax:** 01782 593479
Web: www.orefireplaces.com
Product Type: 1, 2, 3, 4, 5, 6, 8

PENDRAGON FIREPLACES.COM
12 Market Street, Stourbridge,
West Midlands, DY8 1AD
Area of Operation: UK & Ireland
Tel: 01384 376441 **Fax:** 01384 376441
Email: sales@pendragonfireplaces.co.uk
Web: www.pendragonfireplaces.com ⌂
Product Type: 2, 3, 4, 5, 6, 7

PEPPERS FIREPLACE
72 Avenue Road, Bexley Heath, Kent, DA7 4EG
Area of Operation: UK (Excluding Ireland)
Tel: 0208 303 7318 **Fax:** 0208 301 1012
Email: sales@peppers.uk.com
Web: www.peppers.uk.com

PERCY DOUGHTY & CO
Imperial Point, Express Trading Estate, Stonehill Rd,
Farnworth, Bolton, Lancashire, M38 9ST
Area of Operation: Midlands & Mid Wales,
North West England and North Wales
Tel: 01204 868550 **Fax:** 01204 868551
Email: sales@percydoughty.com
Web: www.percydoughty.co.uk
Product Type: 1, 2, 3, 4, 5, 6, 8

PETRA HELLAS
Toll Bar Business Park, Newchurch Road,
Stacksteads, Bacup, Lancashire, OI13 0NA
Area of Operation: Europe
Tel: 01706 876102 **Fax:** 01706 876194
Email: info@petrahellas.co.uk
Web: www.petrahellas.co.uk
Product Type: 1, 5, 8

R W KNIGHT & SON LTD
Castle Farm, Marshfield, Chippenham, Wilts, SN14 8HU
Area of Operation: Midlands & Mid Wales, South
East England, South West England & South Wales
Tel: 01225 891469 **Fax:** 01225 892369
Email: enquiries@knight-stoves.co.uk
Web: www.knight-stoves.co.uk
Product Type: 1, 2, 3, 4, 5
Material Type: A) 2, 4, 5

REAL FLAME
80 New Kings Rd, Fulham, Greater London, SW6 4LT
Area of Operation: Worldwide
Tel: 0207 731 5025 **Fax:** 0207 736 4625
Email: info@realflame.co.uk
Web: www.realflame.co.uk ⌂
Product Type: 1, 2, 3, 4, 5, 6, 7, 8, 9 ★
Other Info: ✎ ✌

RENAISSANCE MOULDINGS
262 Handsworth Road, Handsworth,
Sheffield, South Yorkshire, S13 9BS
Area of Operation: UK (Excluding Ireland)
Tel: 0114 244 6622 **Fax:** 0114 261 0472
Email: enq@rfppm.com
Web: www.rfppm.co.uk

ROBERT AAGAARD & CO
Frogmire House, Stockwell Road,
Knaresborough, North Yorkshire, HG5 0JP
Area of Operation: UK (Excluding Ireland)
Tel: 01423 864805 **Fax:** 01423 869356
Email: info@robertaagaard.co.uk
Web: www.robertaagaard.co.uk
Product Type: 1, 2, 3, 4, 5, 6, 7, 8
Material Type: E) 2, 11, 13

ROBEYS
Old School House, Green Lane,
Belper, Derbyshire, DE56 1BY
Area of Operation: UK & Ireland
Tel: 01773 820940 **Fax:** 01773 821652
Email: info@robeys.co.uk
Web: www.robeys.co.uk
Product Type: 1, 2, 3, 4, 5, 6, 8

RUDLOE STONEWORKS LTD
Leafield Stoneyard, Potley Lane,
Corsham, Wiltshire, SN13 9RS
Area of Operation: UK & Ireland
Tel: 01225 816400 **Fax:** 01225 811343
Email: paul@rudloe-stone.com
Web: www.rudloe-stone.com
Product Type: 1, 2, 3, 4, 7, 8

SIMALI STONE FLOORING
40E Wincombe Business Park,
Shaftesbury, Dorset, SP7 9QJ
Area of Operation: Europe
Tel: 01747 852557 **Fax:** 01747 852557
Email: stonefloors@aol.com
Web: www.stoneflooringonline.com

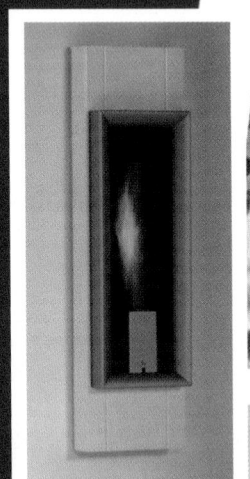

SOLAR FIRES & FIREPLACES LTD
Alyn Works, Mostyn Road,
Holywell, Flintshire, CH8 9DT
Area of Operation: UK (Excluding Ireland)
Tel: 01745 561685 **Fax:** 01745 580 987
Email: sales@solarfiresandfireplaces.co.uk
Web: www.solarfiresandfireplaces.co.uk

STOVAX LIMITED
Falcon Road, Sowton Industrial Estate,
Exeter, Devon, EX2 7LF
Area of Operation: UK & Ireland
Tel: 01392 474011 **Fax:** 01392 219932
Email: info@stovax.com
Web: www.stovax.com
Product Type: 1, 2, 3, 4, 5, 6, 7, 8, 9 □ ● ✿

TAYLOR AND PORTWAY
52 Broton Drive Industrial Estate,
Halstead, Essex, CO9 1HB
Area of Operation: Europe
Tel: 01787 472551 **Fax:** 01787 476589
Email: sales@portwayfires.com
Web: www.portwayfires.com
Product Type: 1, 3, 8 ★ □ ● ✿

TEMPLESTONE
Station Wharf, Castle Cary, Somerset, BA7 7PE
Area of Operation: Worldwide
Tel: 01963 350242 **Fax:** 01963 350258
Email: sales@templestone.co.uk
Web: www.templestone.co.uk
Other Info: ✋
Material Type: E) 5, 13

THE EDWARDIAN FIREPLACE COMPANY
Former All Saints Church, Armoury Way,
Wandsworth, London, SW18 1HZ
Area of Operation: Greater London,
South East England
Tel: 0208 870 0167 **Fax:** 0208 877 9069
Email: mike@edwardianfires.com
Web: www.edwardianfires.com
Product Type: 1, 2, 3, 4, 5, 6, 7, 8, 9

THE FIREPLACE GALLERY (UK) LTD
Clarence Road, Worksop,
Nottinghamshire, S80 1QA
Area of Operation: UK & Ireland
Tel: 01909 500802 **Fax:** 01909 500810
Email: fireplacegallery@btinternet.com
Web: www.fireplacegallery.co.uk
Product Type: 1, 2, 3, 4, 5, 6, 7, 8, 9 ★ □ ● ✿
Other Info: ECO ✍ ✋
Material Type: A) 1, 2, 3, 4, 5, 7, 11

**THE FIREPLACE MARKETING
COMPANY LIMITED**
The Old Coach House, Southern Road,
Thame, Oxfordshire, OX9 2ED
Area of Operation: Worldwide
Tel: 01844 260960
Fax: 01844 260267
Email: david@fireplacemarketing.co.uk
Web: www.fireplace.co.uk ✍
Product Type: 1, 2, 3, 4, 5, 6, 7, 8, 9

THE GRANITE FACTORY
4 Winchester Drive, Peterlee, Durham, SR8 2RJ
Area of Operation: North East England
Tel: 0191 518 3600 **Fax:** 0191 518 3600
Email: admin@granitefactory.co.uk
Web: www.granitefactory.co.uk ✍

THE JRG GROUP
3 Crompton Way, North Newmoor
Industrial Estate, Irvine, Ayrshire, KA11 4HU
Area of Operation: Europe
Tel: 0871 2008080 **Fax:** 01294 211222
Email: chris@jrgfiresurrounds.com
Web: www.jrggroup.com
Product Type: 1, 8

THE MASON'S YARD
Penhenllan, Cusop, Hay on Wye,
Herefordshire, HR3 5TE
Area of Operation: UK (Excluding Ireland)
Tel: 01497 821333
Email: hugh@themasonsyard.co.uk
Web: www.themasonsyard.co.uk
Product Type: 1, 2, 3, 4, 5, 6, 7, 8, 9

THEALE FIREPLACES RDG LTD
Milehouse Farm, Bath Road, Theale, Berks, RG7 5HJ
Area of Operation: UK (Excluding Ireland)
Tel: 0118 930 2232 **Fax:** 0118 932 3344
Email: mail@theale-fireplaces.co.uk
Web: www.theale-fireplaces.co.uk
Product Type: 1, 2, 3, 4, 5, 8

THORNHILL GALLERIES
No. 3, 19 Osiers Road, London, SW18 1NL
Area of Operation: Worldwide
Tel: 0208 874 2101 **Fax:** 0208 877 0313
Email: sales@thornhillgalleries.co.uk
Web: www.thornhillgalleries.co.uk
Product Type: 3, 5, 6, 7, 8

VERINE
52 Broton Drive, Halstead, Essex, CO9 1HB
Area of Operation: UK & Ireland
Tel: 01787 472551 **Fax:** 01787 476589
Email: sales@verine.co.uk **Web:** www.verine.co.uk
Product Type: 1 ★ ✿
Material Type: C) 1, 2, 3, 5, 11, 14

WARMSWORTH STONE FIREPLACES
Sheffield Road, Warmsworth, Doncaster,
South Yorkshire, DN4 9QH
Area of Operation: UK (Excluding Ireland)
Tel: 01302 858617 **Fax:** 01302 855844
Email: info@warmsworth-stone.co.uk
Web: www.warmsworth-stone.co.uk

WESSEX STONE FIREPLACES
Ilsom Farm, Cirencester Road, Tetbury,
Gloucestershire, GL8 8RX
Area of Operation: UK (Excluding Ireland)
Tel: 01666 504658 **Fax:** 01666 502285
Email: info@wells-group.co.uk
Web: www.wells-group.co.uk
Product Type: 1, 8 ★ □ ● ✿
Other Info: ✍ ✋ **Material Type:** E) 5, 13

WINTHER BROWNE
75 Bilton Way, Enfield, London, EN3 7ER
Area of Operation: UK (Excluding Ireland)
Tel: 0208 344 9050
Fax: 0845 612 1894
Email: sales@wintherbrowne.co.uk
Web: www.wintherbrowne.co.uk
Product Type: 1, 2, 3, 4, 5, 6, 7, 8
Material Type: A) 2, 4, 9

WWW.FIREPLACE.CO.UK
Old Coach House, Southern Road,
Thame, Oxfordshire, OX9 2ED
Area of Operation: UK (Excluding Ireland)
Tel: 01844 260960 **Fax:** 01844 260267
Web: www.fireplace.co.uk

YEOMAN STOVES
Falcon Rd, Sowton Industrial Estate,
Devon, EX2 7LF
Area of Operation: UK (Excluding Ireland)
Tel: 01395 474011 **Fax:** 01395 219932
Email: yeoman@stovax.co.uk
Web: www.yeoman-stoves.co.uk

FIREPLACE ACCESSORIES

KEY

PRODUCT TYPES: 1= Mantel Pieces
2 = Fenders   3 = Fire Curtains   4 = Fire
Screens and Canopies   5 = Fire Backs
6 = Fire Baskets   7 = Fireplace Doors
8 = Dampers   9 = Other

OTHER: ▽ Reclaimed   ✍ On-line shopping
✍ Bespoke   ✋ Hand-made   ECO Ecological

A & M ENERGY FIRES
Pool House, Huntley, Gloucestershire, GL19 3DZ
Area of Operation: Europe
Tel: 01452 830662 **Fax:** 01452 830891
Email: am@energyfires.co.uk
Web: www.energyfires.co.uk
Product Type: 3, 6, 7

ACANTHA LIFESTYLE LTD
32-34 Park Royal Rd, Park Royal,
London, NW10 7LN
Area of Operation: Worldwide
Tel: 0208 453 1537 **Fax:** 0208 453 1538
Email: sales@acanthalifesyle.com
Web: www.acanthalifestyle.co.uk
Product Type: 1

ANGLIA FIREPLACES & DESIGN LTD
Anglia House, Kendal Court, Cambridge Road,
Impington, Cambridgeshire, CB4 9YS
Area of Operation: UK & Ireland
Tel: 01223 234713 **Fax:** 01223 235116
Email: info@fireplaces.co.uk
Web: www.fireplaces.co.uk
Product Type: 1

ARCHITECTURAL HERITAGE
Taddington Manor, Taddington, Nr Cutsdean,
Cheltenham, Gloucestershire, GL54 5RY
Area of Operation: Worldwide
Tel: 01386 584414 **Fax:** 01386 584236
Email: puddy@architectural-heritage.co.uk
Web: www.architectural-heritage.co.uk
Product Type: 5

BD BROOKS FIREPLACES
109 Halifax Road, Ripponden,
Halifax, West Yorkshire, HX6 4DA
Area of Operation: UK (Excluding Ireland)
Tel: 01422 822220 **Fax:** 01422 822220
Email: bdbrooks2000@yahoo.com
Web: www.bdbrooksfireplaces.com
Product Type: 1, 2, 3, 4, 5, 6, 7, 8, 9

BRILLIANT FIRES
Thwaites Close, Shadsworth Business Park,
Blackburn, Lancashire, BB1 2QQ
Area of Operation: UK & Ireland
Tel: 01254 682384 **Fax:** 01254 672647
Email: info@brilliantfires.co.uk
Web: www.brilliantfires.co.uk
Product Type: 1

BROSELEY FIRES
Knights Way, Battlefield Enterprise Park,
Shrewsbury, Shropshire, SY1 3AB
Area of Operation: UK & Ireland
Tel: 01743 461444
Fax: 01743 461446
Email: sales@broseleyfires.com
Web: www.broseleyfires.com
Product Type: 4

COBALT BLACKSMITHS
The Forge, English Farm, English Lane,
Nuffield, Oxfordshire, RG9 5TH
Tel: 01491 641990
Fax: 01491 640909
Email: enquiries@cobalt-blacksmiths.co.uk
Web: www.cobalt-blacksmiths.co.uk
Product Type: 4, 6

DRAYTON PROPERTY
5 Lake Crescent, Daventry,
Northamptonshire, NN11 9EB
Area of Operation: UK (Excluding Ireland)
Tel: 01327 300249
Fax: 01327 300249
Email: draytonproperty@aol.com
Web: www.draytonproperty.com ✍
Product Type: 1, 5, 6

EMSWORTH FIREPLACES LTD
Unit 3, Station Approach,
Emsworth, Hampshire, PO10 7PW
Area of Operation: Worldwide
Tel: 0845 2305200
Fax: 01243 371023
Email: sales@emsworth.co.uk
Web: www.emsworth.co.uk ✍
Product Type: 1, 2, 5, 6

HEATING, PLUMBING & ELECTRICAL

EMSWORTH
Area of Operation: Worldwide
Tel: 0845 2305200 Fax: 01243 371023
Email: sales@emsworth.co.uk
Web: www.emsworth.co.uk
Product Type: 1, 2, 5, 6

Manufacturers of contemporary and traditional fireplaces in marble, limestone, steel and timber since 1934. Bespoke work a speciality, contact us for a brochure. Nationwide delivery service.

ENGLISH FIREPLACES
Old Firs, Hillbrow, Liss,
Hampshire, GU33 7QE
Area of Operation: Worldwide
Tel: 01730 890218
Email: mjostephens@btinternet.com
Web: www.englishfireplaces.co.uk
Product Type: 1, 5, 6

ESSE
Ouzledale Foundry, Long Ing,
Barnoldswick, Lancashire, BB18 6BN
Area of Operation: Worldwide
Tel: 01282 813235
Fax: 01282 816876
Email: esse@ouzledale.co.uk
Web: www.esse.com
Product Type: 4, 6, 8, 9

FIREPLACE & TIMBER PRODUCTS
Unit 2 Holyrood Drive, Skippingdale Ind. Est.,
Scunthorpe, Lincolnshire, DN15 8NN
Area of Operation: UK & Ireland
Tel: 01724 852888
Fax: 01724 277255
Email: ftprdcts@yahoo.co.uk
Product Type: 1, 2, 3, 4, 5, 6, 7, 8, 9

FIREPLACE CONSULTANTS LTD
The Studio, The Old Rothschild Arms,
Buckland Road, Buckland, Aylesbury,
Buckinghamshire, HP22 5LP
Area of Operation: Greater London, Midlands
& Mid Wales, South East England
Tel: 01296 632287 Fax: 01296 632287
Email: info@fireplaceconsultants.com
Web: www.fireplaceconsultants.com
Product Type: 1, 2, 3, 4, 5, 6, 7, 8

GLEN DIMPLEX UK LTD
Millbrook House, Grange Drive, Hedge End,
Southampton, Hampshire, SO30 2DF
Area of Operation: UK & Ireland
Tel: 0870 077 7117
Fax: 0870 727 0114
Email: marketing@glendimplex.com
Web: www.dimplex.co.uk
Product Type: 9

GRENADIER FIRELIGHTERS LIMITED
Unit 3C, Barrowmore Enterprise Estate,
Great Barrow, Chester, Cheshire, CH3 7JS
Area of Operation: UK & Ireland
Tel: 01829 741649
Fax: 01829 741659
Email: enquiries@grenadier.uk.com
Web: www.grenadier.uk.com
Product Type: 9
Other Info:

KINGSWORTHY FOUNDRY CO LTD
London Road, Kingsworthy, Winchester,
Hampshire , SO23 7QG
Area of Operation: UK & Ireland
Tel: 01962 883776
Fax: 01962882925
Email: kwf@fsbdial.co.uk
Web: www.kingsworthyfoundry.co.uk
Product Type: 2, 4, 5, 6

LEL FIREPLACES
Tre-Ifan Farmhouse, Caergeiliog,
Holyhead, Anglesey, LL65 3HP
Area of Operation: UK & Ireland
Tel: 01407 742240
Fax: 01407 742262
Email: sales@lel-fireplaces.com
Web: www.lel-fireplaces.com
Product Type: 1, 2, 4, 5, 6, 7, 8, 9
Other Info: ECO
Material Type: A) 1, 2, 3, 4, 5, 6, 7, 10, 15

LIVINGSTYLE.CO.UK
Bridge Street, Shotton, Flintshire, CH5 1DU
Area of Operation: UK (Excluding Ireland)
Tel: 0800 298 9190
Email: info@livingstyle.co.uk
Web: www.livingstyle.co.uk
Product Type: 1, 2, 3, 4, 5, 6, 7, 8, 9

MARBLE HILL FIREPLACES
70-72 Richmond Road, Twickenham,
Greater London, TW1 3BE
Area of Operation: Greater London,
South West England and South Wales
Tel: 0208 892 1488
Fax: 0208 891 6591
Email: sales@marblehill.co.uk
Web: www.marblehill.co.uk
Product Type: 1, 6

MARK RIPLEY FORGE & FIREPLACES
Robertsbridge, Bridge Bungalow,
East Sussex, TN32 5NY
Tel: 01580 880324 Fax: 01580 881927
Email: ripleym@gxn.co.uk
Web: www.ripleyfireplaces.co.uk
Product Type: 1, 2, 5, 6

OLDE ENGLANDE REPRODUCTIONS
Fireplace Works, Normacot Road, Longton,
Stoke-on-Trent, Staffordshire, ST3 1PN
Area of Operation: UK (Excluding Ireland)
Tel: 01782 319350 Fax: 01782 593479
Email: sales@oerfireplaces.com
Web: www.oerfireplaces.com
Product Type: 1, 2, 4, 6

PENDRAGON FIREPLACES.COM
12 Market Street, Stourbridge,
West Midlands, DY8 1AD
Area of Operation: UK & Ireland
Tel: 01384 376441 Fax: 01384 376441
Email: sales@pendragonfireplaces.co.uk
Web: www.pendragonfireplaces.com
Product Type: 1, 6

PERCY DOUGHTY & CO
Imperial Point, Express Trading Estate, Stonehill
Road, Farnworth, Bolton, Lancashire, M38 9ST
Area of Operation: Midlands & Mid Wales, North
West England and North Wales
Tel: 01204 868550 Fax: 01204 868551
Email: sales@percydoughty.com
Web: www.percydoughty.co.uk
Product Type: 1, 2, 4, 5, 6, 8

R W KNIGHT & SON LTD
Castle Farm, Marshfield,
Chippenham, Wiltshire, SN14 8HU
Area of Operation: Midlands & Mid Wales,
South East England, South West England
and South Wales
Tel: 01225 891469
Fax: 01225 892369
Email: Enquiries@knight-stoves.co.uk
Web: www.knight-stoves.co.uk
Product Type: 1, 2, 4, 5, 6, 8
Material Type: B) 2, 13

REAL FLAME
80 New Kings Road, Fulham,
Greater London, SW6 4LT
Area of Operation: Worldwide
Tel: 0207 731 5025
Fax: 0207 736 4625
Email: info@realflame.co.uk
Web: www.realflame.co.uk
Product Type: 1, 2, 4, 5, 6, 8
Other Info:

ROCKINGHAM FENDER SEATS
Grange Farm, Thorney, Peterborough,
Cambridgeshire, PE6 0PJ
Area of Operation: Worldwide
Tel: 01733 270233
Fax: 01733 270512
Email: clubfenders@rockingham-fenderseats.com
Web: www.rockingham-fenderseats.com
Product Type: 2, 4
Material Type: A) 2

RUDLOE STONEWORKS LTD
Leafield Stoneyard, Potley Lane,
Corsham, Wiltshire, SN13 9RS
Area of Operation: UK & Ireland
Tel: 01225 816400
Fax: 01225 811343
Email: paul@rudloe-stone.com
Web: www.rudloe-stone.com
Product Type: 1, 2, 3, 4, 5, 6, 7

SCHIEDEL-ISOKERN CHIMNEY SYSTEMS
14 Haviland Road, Ferndown Industrial Estate,
Wimborne, Dorset, BH21 7RF
Area of Operation: UK & Ireland
Tel: 01202 861650 Fax: 01202 861632
Email: sales@isokern.co.uk
Web: www.isokern.co.uk
Product Type: 9

STOVAX LIMITED
Falcon Road, Sowton Industrial Estate,
Exeter, Devon, EX2 7LF
Area of Operation: UK & Ireland
Tel: 01392 474011 Fax: 01392 219932
Email: info@stovax.com Web: www.stovax.com
Product Type: 1, 2, 4, 5, 6, 7, 8, 9,

THE EDWARDIAN FIREPLACE COMPANY
Former All Saints Church, Armoury Way,
Wandsworth, London, SW18 1HZ
Area of Operation: Greater London,
South East England
Tel: 0208 870 0167
Fax: 0208 877 9069
Email: mike@edwardianfires.com
Web: www.edwardianfires.com
Product Type: 1, 2, 3, 4, 5, 6, 7, 8, 9

THE FIREPLACE DOOR COMPANY
106 Alfreton Road, Sutton in Ashfield,
Nottinghamshire, NG17 1FQ
Area of Operation: Worldwide
Tel: 01623 477435
Fax: 01623 456734
Email: fdc@ntlworld.com
Web: www.fireplacedoorcompany.co.uk
Product Type: 7

THE FIREPLACE GALLERY (UK) LTD
Clarence Road, Worksops,
Nottinghamshire, S80 1QA
Area of Operation: UK & Ireland
Tel: 01909 500802 Fax: 01909 500810
Email: fireplacegallery@btinternet.com
Web: www.fireplacegallery.co.uk
Product Type: 1, 2, 6
Other Info: ECO
Material Type: A) 1, 2, 3, 4, 5, 7, 11

**THE FIREPLACE MARKETING
COMPANY LIMITED**
The Old Coach House, Southern Road,
Thame, Oxfordshire, OX9 2ED
Area of Operation: Worldwide
Tel: 01844 260960 Fax: 01844 260267
Email: david@fireplacemarketing.co.uk
Web: www.fireplace.co.uk
Product Type: 1, 2, 3, 4, 5, 6, 7, 8

THE MASON'S YARD
Penhenllan, Cusop, Hay on Wye,
Herefordshire, HR3 5TE
Area of Operation: UK (Excluding Ireland)
Tel: 01497 821333
Email: hugh@themasonsyard.co.uk
Web: www.themasonsyard.co.uk
Product Type: 1

THEALE FIREPLACES RDG LTD
Milehouse Farm, Bath Road, Theale,
Berkshire, RG7 5HJ
Area of Operation: UK (Excluding Ireland)
Tel: 0118 930 2232 Fax: 0118 932 3344
Email: mail@theale-fireplaces.co.uk
Web: www.theale-fireplaces.co.uk
Product Type: 1, 2, 3, 4, 5, 6, 8

THORNHILL GALLERIES
No. 3, 19 Osiers Road, London, SW18 1NL
Area of Operation: Worldwide
Tel: 020 8874 2101 Fax: 020 8877 0313
Email: sales@thornhillgalleries.co.uk
Web: www.thornhillgalleries.co.uk
Product Type: 1, 2, 4, 5, 6

TRADITIONAL OAK & TIMBER COMPANY
P O Stores, Haywards Heath Road, North Chailey,
Nr Lewes, East Sussex, BN8 4EY
Area of Operation: Worldwide
Tel: 01825 723648 Fax: 01825 722215
Email: info@tradoak.co.uk
Web: www.tradoak.com
Product Type: 1

WWW.FIREPLACE.CO.UK
Old Coach House, Southern Road, Thame,
Oxfordshire, OX9 2ED
Area of Operation: UK (Excluding Ireland)
Tel: 01844 260960
Fax: 01844 260267
Web: www.fireplace.co.uk

'ECO' HEATING

KEY
PRODUCT TYPES: 1= Solar Powered
2 = Geothermal   3 = Wind Powered
4 = Other

ENERGY SOURCES: ★ Gas   ○ Oil
□ Coal   ● Wood   ✪ Multi-Fuel
✲ Electric   ◆ Gel

OTHER: ▽ Reclaimed   On-line shopping
Bespoke   Hand-made   ECO Ecological

AERODYN-SHOREPOWER
16 Popes Lane, Rockwell Green,
Wellington, Somerset, TA21 9DQ
Area of Operation: UK (Excluding Ireland)
Tel: 01823 666177
Fax: 01823 666177
Email: shorepower@ukonline.co.uk
Product Type: 1, 3

AICO LTD
Mile End Business Park, Maesbury Road,
Oswestry, Shropshire, SY10 8NN
Area of Operation: UK (Excluding Ireland)
Tel: 0870 758 4000
Fax: 0870 758 4010
Email: sales@aico.co.uk
Web: www.aico.co.uk
Product Type: 4

BORDERS UNDERFLOOR HEATING
26 Coopersknowe Crescent,
Galashiels, Borders, TD1 2DS
Area of Operation: UK & Ireland
Tel: 01896 668667
Fax: 01896 668678
Email: underfloor@btinternet.com
Web: www.bordersunderfloor.co.uk
Product Type: 1, 2, 4

BRISTOL & SOMERSET RENEWABLE ENERGY ADVICE SERVICE
The CREATE Centre,
Smeaton Rd, Bristol, BS1 6XN
Area of Operation: South West England & South Wales
Tel: 0800 512012 **Fax:** 0117 929 9114
Email: reas@cse.org.uk
Web: www.cse.org.uk/renewables
Product Type: 1, 2, 4

BTU HEATING
38 Weyside Road, Guildford,
Surrey, GU1 1JB
Area of Operation: South East England
Tel: 01483 590600 **Fax:** 01483 590601
Email: enquiries@btu-heating.com
Web: www.btu-group.com

CHELMER HEATING SERVICES LTD
Unit 12A, Baddow Park, West Hanningfield Rd,
Chelmsford, Essex, CM2 7SY
Area of Operation: UK (Excluding Ireland)
Tel: 01245 471111 **Fax:** 01245 471117
Email: sales@chelmerheating.co.uk
Web: www.chelmerheating.co.uk
Product Type: 1, 2, 4

COMFORT AIR CONDITIONING LTD
Comfort Works, Newchapel Road,
Lingfield, Surrey, RH7 6LE
Area of Operation: East England,
Greater London, Midlands & Mid Wales,
South East England,
South West England and South Wales
Tel: 01342 830600 **Fax:** 01342 830605
Email: info@comfort.uk.com
Web: www.comfort.uk.com
Product Type: 2, 4 **Other Info:** ✎

CONSERVATION ENGINEERING LTD
The Street, Troston, Bury St Edmunds,
Suffolk, IP31 1EW
Area of Operation: UK & Ireland
Tel: 01359 269360 **Fax:** 01359 268340
Email: anne@conservation-engineering.co.uk
Web: www.heating-designs.co.uk
Product Type: 1, 2

EARTHWISE SCOTLAND LTD
9a Netherton Business Centre, Kemnay,
Aberdeenshire, AB51 5LX
Area of Operation: UK (Excluding Ireland)
Tel: 01467 641640
Email: admin@earthwisescotland.co.uk
Web: www.earthwisescotland.co.uk
Product Type: 2

ECO HEAT PUMPS
Sheffield Technology Park, 60 Shirland Lane,
Sheffield, South Yorkshire, S9 3SP
Area of Operation: UK (Excluding Ireland)
Tel: 0114 296 2227 **Fax:** 0114 296 2229
Email: info@ecoheatpumps.co.uk
Web: www.ecoheatpumps.co.uk
Product Type: 2

ECO HOMETEC UK LTD
Unit 11E, Carcroft Enterprise Park,
Carcroft, Doncaster,
South Yorkshire, DN6 8DD
Area of Operation: Europe
Tel: 01302 722266
Fax: 01302 728634
Email: Stephen@eco-hometec.co.uk
Web: www.eco-hometec.co.uk
Product Type: 1, 2

ECO HOUSES
12 Lee Street, Louth,
Lincolnshire, LN11 9HJ
Area of Operation: Worldwide
Tel: 0845 345 2049
Email: sd@jones-nash.co.uk
Web: www.eco-houses.co.uk

ECO SYSTEMS IRELAND LTD
40 Glenshesk Rd, Ballycastle,
Co Antrim, BT54 6PH
Area of Operation: UK & Ireland
Tel: 02820 768708
Fax: 02820 769781
Email: info@ecosystemsireland.com
Web: www.ecosystemsireland.com
Product Type: 4

ECOLEC
Sharrocks Street, Wolverhampton,
West Midlands, WV1 3RP
Area of Operation: UK & Ireland
Tel: 01902 457575
Fax: 01902 457797
Email: sales@ecolec.co.uk
Web: www.ecolec.co.uk ✎
Product Type: 4

ECOSHOP
Unit 1, Glen of the Downs Garden Centre,
Kilmacanogue, Co Wicklow, Republic of Ireland
Area of Operation: Ireland Only
Tel: +353 01 287 2914
Fax: +353 01 201 6480
Email: info@ecoshop.ie
Web: www.ecoshop.ie

ECOWARM SOLAR LTD
23 St.Marys Avenue, Haughley,
Suffolk, IP14 3NZ
Area of Operation: East England
Tel: 01449 771130
Email: enquiries@ecowarm-solar.co.uk
Web: www.ecowarm-solar.co.uk
Product Type: 1

ENERGY AND ENVIRONMENT LTD
91 Claude Road, Chorlton,
Manchester, M21 8DE
Area of Operation: UK & Ireland
Tel: 0161 881 1383
Email: mail@energyenv.co.uk
Web: www.energyenv.co.uk ✎
Product Type: 1, 2

ENERGY MASTER
Keltic Business Park, Unit 1 Clieveragh
Industrial Estate, Listowel, Ireland
Area of Operation: Ireland Only
Tel: +353 68 24300
Fax: +353 68 24533
Email: info@energymaster.ie
Web: www.energymaster.ie
Product Type: 1, 2

ENVIROZONE
Encon House, 54 Ballycrochan Road, Bangor,
Co Down, BT19 6NF
Area of Operation: UK & Ireland
Tel: 02891 477799
Fax: 02891 452463
Email: info@envirozone.co.uk
Web: www.envirozone.co.uk
Product Type: 1, 2, 3

FERROLI UK
Lichfield Road, Branston Industrial Estate,
Burton Upon Trent, Staffordshire, DE14 3HD
Area of Operation: UK (Excluding Ireland)
Tel: 08707 282882
Fax: 08707 282883
Email: sales@ferroli.co.uk
Web: www.ferroli.co.uk
Product Type: 4

FIRSTLIGHT ENERGY LIMITED
Riverside Business Centre, River Lawn Road,
Tonbridge, Kent, TN9 1EP
Area of Operation: UK (Excluding Ireland)
Tel: 01424 753235
Email: tony@firstlightenergy.co.uk
Web: www.firstlightenergy.co.uk
Product Type: 1

FOUNDATION FIREWOOD
39B Park Farm Industrial Estate,
Buntingford, Hertfordshire, SG9 9AZ
Area of Operation: UK (Excluding Ireland)
Tel: 01763 271271 **Fax:** 01763 271270
Email: info@fbcgroup.co.uk
Web: www.fbcgroup.co.uk
Product Type: 4 ●

GEOSCIENCE LTD
Falmouth Business Park,
Bickland Water Road,
Falmouth, Cornwall, TR11 4SZ
Area of Operation: UK (Excluding Ireland)
Tel: 01326 211 070 **Fax:** 01326 211 071
Email: earthenergy@geoscience.co.uk
Web: www.earthenergy.co.uk
Product Type: 2

GEOTHERMAL INTERNATIONAL LTD
Spencer Court, 143 Albany Road,
Coventry, Warwickshire, CV5 6ND
Area of Operation: Europe
Tel: 02476 673131 **Fax:** 02476 679999
Email: info@geoheat.co.uk
Web: www.geoheat.co.uk
Product Type: 2

GEOTHERMAL LTD
4 Imex Centre, Broadleys Business Park,
Stirling, Stirlingshire, FK7 7LQ
Area of Operation: Scotland
Tel: 01786 473666 **Fax:** 01786 475599
Email: geothermal@geoheat.co.uk
Web: www.geoheat.co.uk
Product Type: 2

GREENSHOP SOLAR LTD
Units 23/24, Merretts Mill Industrial Centre,
Woodchester, Stroud, Gloucestershire, GL5 5EX
Area of Operation: UK (Excluding Ireland)
Tel: 01452 772030 **Fax:** 01452 770115
Email: eddie@greenshop.co.uk
Web: www.greenshop-solar.co.uk
Product Type: 1 **Other Info:** ECO

ICE ENERGY
Unit 4 Oakfields House,
Oakfields Industrial Estate, Eynsham,
Oxford, Oxfordshire, OX29 4TR
Area of Operation: UK (Excluding Ireland)
Tel: 01865 882202 **Fax:** 01865 882539
Email: info@iceenergy.co.uk
Web: www.iceenergy.co.uk
Product Type: 2

INVISIBLE HEATING SYSTEMS
IHS Design Centre,
Morefield Industrial Estate,
Ullapool, Highlands, IV26 2SX
Area of Operation: UK (Excluding Ireland)
Tel: 01854 613161 **Fax:** 01854 613160
Email: sales@invisibleheating.co.uk
Web: www.invisibleheating.com
Product Type: 1, 2, 4
Other Info: ✎

IPPEC SYSTEMS LTD
66 Rea Street South, Birmingham,
West Midlands, B5 6LB
Area of Operation: UK & Ireland
Tel: 0121 622 4333 **Fax:** 0121 622 5768
Email: info@ippec.co.uk
Web: www.ippec.co.uk
Product Type: 1

KILTOX CONTRACTS LIMITED
Unit 6 Chiltonian Industrial Estate,
203 Manor Lane, Lee, London, SE12 0TX
Area of Operation: Worldwide
Tel: 0845 166 2040 **Fax:** 0845 166 2050
Email: info@kiltox.co.uk
Web: www.kiltox.co.uk ✎
Product Type: 4
Other Info: ECO

POWERGEN
Newstead Court, Sherwood Park, Little Oak Drive,
Annesley, Nottinghamshire, NG15 0DR
Area of Operation: UK (Excluding Ireland)
Tel: 0800 068 6515
Email: whispergen@powergen.co.uk
Web: www.powergen.co.uk
Product Type: 4 **Other Info:** ECO

HEATING, PLUMBING & ELECTRICAL

SPONSORED BY: CVO FIRE LTD www.cvo.co.uk

POWERTECH SOLAR LTD
21 Haviland Road, Forndown Industrial Estate,
Wimborne, Dorset, BH21 7RZ
Area of Operation: UK (Excluding Ireland)
Tel: 08707 300111
Fax: 08707 300222
Email: sales@solar.org.uk
Web: www.solar.org.uk
Product Type: 1, 2, 4

RADIANT HEATING SOLUTIONS LTD
Mill Farm, Hougham, Grantham,
Lincolnshire, NG32 2HZ
Tel: 01400 250572
Fax: 01400 251264
Email: sales@heating-solutions.biz
Web: www.heating-solutions.biz
Product Type: 2

RAYOTEC LTD
Unit 3, Brooklands Close,
Sunbury on Thames, Surrey, TW16 7DX
Area of Operation: UK & Ireland
Tel: 01932 784848
Fax: 01932 784848
Email: info@rayotec.com
Web: www.rayotec.com
Product Type: 1

RIOMAY LTD
1 Birch Road, Eastbourne,
East Sussex, BN23 6PL
Area of Operation: UK & Ireland
Tel: 01323 648641
Fax: 01323 720682
Email: tonybook@riomay.com
Web: www.riomay.com
Product Type: 1
Other Info: ECO ✎

SECON SOLAR
50 Business & Innovation Centre, Wearfield,
Sunderland, Tyne & Wear, SR5 2TA
Area of Operation: UK & Ireland
Tel: 0191 516 6554
Fax: 0191 516 6558
Email: info@seconsolar.com
Web: www.seconsolar.com
Product Type: 1

SOLAR FIRES & FIREPLACES LTD
Alyn Works, Mostyn Road, Holywell,
Flintshire, CH8 9DT
Area of Operation: UK (Excluding Ireland)
Tel: 01745 561685
Fax: 01745 580 987
Email: sales@solarfiresandfireplaces.co.uk
Web: www.solarfiresandfireplaces.co.uk
Product Type: 1

SOLAR SENSE
Energy Parc, Sandy Lane, Pennard,
Swansea, SA3 2EN
Area of Operation: Worldwide
Tel: 0845 458 3141
Fax: 0870 163 8620
Email: info@solarsense.co.uk
Web: www.solarsense.co.uk ✎
Product Type: 1, 3
Other Info: ECO ✎

SOLARIS SOLAR ENERGY SYSTEMS
Toames, Macroom, Co. Cork
Area of Operation: UK & Ireland
Tel: +353 264 6312
Fax: +353 264 6313
Email: solaris@eircom.net
Web: www.solaris-energy.com
Product Type: 1

SOLARUK
Crabtree, The Warren, Crowborough,
East Sussex, TN6 1UB
Area of Operation: Worldwide
Tel: 01892 667320
Fax: 01892 667622
Email: info@solaruk.net
Web: www.solaruk.net
Product Type: 1

SOUTHERN SOLAR
Unit 6, Allington Farm, Allington Lane,
Offham, Lewes, East Sussex, BN7 3QL
Area of Operation: Greater London, South East
England, South West England and South Wales
Tel: 0845 456 9474 **Fax:** 01273 483928
Email: info@southernsolar.co.uk
Web: www.southernsolar.co.uk
Product Type: 1, 2
Other Info: ✎

STREBEL LTD
1F Albany Park Industrial Estate, Frimley Road,
Camberley, Surrey, GU16 7PB
Area of Operation: UK & Ireland
Tel: 01276 685422 **Fax:** 01276 685405
Email: info@strebel.co.uk
Web: www.strebel.co.uk
Product Type: 1 ★ ○ ● ✿

SUNDWEL SOLAR LTD
Unit 1, Tower Road, Washington,
Sunderland, Tyne & Wear, NE37 2SH
Area of Operation: UK (Excluding Ireland)
Tel: 0191 416 3001
Fax: 0191 415 4297
Email: solar@sundwel.com
Web: www.sundwel.com
Product Type: 1

SUNUSER LTD
157 Buslingthorpe Lane, Leeds,
West Yorkshire, LS7 2DQ
Area of Operation: UK (Excluding Ireland)
Tel: 0113 262 0261
Fax: 0113 262 3970
Email: solar@sunuser.co.uk
Web: www.sunuser.com
Product Type: 1
Other Info: ECO ✎

THE GREEN SHOP
Cheltenham Road, Bisley, Nr Stroud,
Gloucestershire, GL6 7BX
Area of Operation: UK & Ireland
Tel: 01452 770629
Fax: 01452 770104
Email: paint@greenshop.co.uk
Web: www.greenshop.co.uk ✎
Product Type: 1, 3

THE HEAT PUMP COMPANY UK LIMITED
29 Claymoor Park, Booker, Nr. Marlow,
Buckinghamshire, SL7 3DL
Area of Operation: UK (Excluding Ireland)
Tel: 01494 450154
Fax: 01494 458159
Email: sales@npsair.com
Web: www.npsair.com
Product Type: 2

THE ORGANIC ENERGY COMPANY
Severn Road, Welshpool, Powys, SY21 7AZ
Area of Operation: UK (Excluding Ireland)
Tel: 0845 4584076
Fax: 01938 559 222
Email: hbenq@organicenergy.co.uk
Web: www.organicenergy.co.uk
Product Type: 1
Other Info: ECO

THERMALFLOOR
Unit 3, Nether Friarton, Perth,
Perth and Kinross, PH2 8DF
Area of Operation: UK & Ireland
Tel: 08450 620400 **Fax:** 08450 620401
Email: heat@thermalfloor.sol.co.uk
Web: www.thermalfloor-heating.co.uk
Product Type: 1, 2 ★ ○ ✿

VIESSMANN LTD
Hortonwood 30, Telford,
Shropshire, TF1 7YP
Area of Operation: UK & Ireland
Tel: 01952 675000 **Fax:** 01952 675040
Email: info@viessmann.co.uk
Web: www.viessmann.co.uk
Product Type: 1, 2 ★ ○
Other Info: ECO

WORCESTER BOSCH GROUP
Cotswold Way, Warndon, Worcester,
Worcestershire, WR4 9SW
Area of Operation: UK & Ireland
Tel: 01905 754624
Fax: 01905 753103
Email: general.worcester@uk.bosch.com
Web: www.worcester-bosch.co.uk
Product Type: 1, 2

HEATING CONTROLS AND VALVES

KEY

PRODUCT TYPES: 1= Thermostats
2 = Time Switches   3 = TRVs
4 = UFH Controls   5 = Zone Controls
6 = Weather Compensation / Optimisation
7 = Other

OTHER: ▽ Reclaimed ✎ On-line shopping
✎ Bespoke ✋Hand-made   ECO Ecological

ACM INSTALLATIONS LTD
71 Western Gailes Way, Hull,
East Riding of Yorks, HU8 9EQ
Area of Operation: North East England
Tel: 0870 242 3285
Fax: 01482 377530
Email: enquiries@acminstallations.com
Web: www.acminstallations.com
Product Type: 2, 5
Other Info: ECO ✎

AESTUS
Unit 5 Strawberry Lane Industrial Estate,
Strawberry Lane, Willenhall,
West Midlands, WV13 3RS
Area of Operation: UK & Ireland
Tel: 0870 403 0115
Fax: 0870 403 0116
Email: melissa@publicityengineers.com

AUTRON PRODUCTS LTD
Unit 17 Second Avenue,
Bluebridge Industrial Estate,
Halstead, Essex, CO9 2SU
Area of Operation: UK & Ireland
Tel: 01787 473964
Fax: 01787 474061
Email: sales@autron.co.uk
Web: wwww.autron.co.uk
Product Type: 3

BATHROOMSTUFF.CO.UK
40 Evelegh Road, Farlington,
Portsmouth, Hampshire, PO6 1DL
Area of Operation: UK (Excluding Ireland)
Tel: 0845 058 0540
Fax: 02392 215695
Email: sales@bathroomstuff.co.uk
Web: www.bathroomstuff.co.uk ✎
Product Type: 3

BAXI POTTERTON
Brownedge Road, Bamber Bridge,
Preston, Lancashire, PR5 6SN
Area of Operation: UK & Ireland
Tel: 0870 606 0780
Fax: 01926 410006
Web: www.baxipotterton.co.uk
Product Type: 1, 2, 5

BEGETUBE UK LTD
8 Carsegate Road South,
Inverness, Highlands, IV3 8LL
Area of Operation: UK & Ireland
Tel: 01463 246600
Fax: 01463 246624
Email: rory@begetube.co.uk
Web: www.begetube-uk.co.uk ✎
Product Type: 1, 2, 3, 4, 5, 6

BORDERS UNDERFLOOR HEATING
26 Coopersknowe Crescent,
Galashiels, Borders, TD1 2DS
Area of Operation: UK & Ireland
Tel: 01896 668667 **Fax:** 01896 668678
Email: underfloor@btinternet.com
Web: www.bordersunderfloor.co.uk
Product Type: 1, 2, 4, 5, 6

BTU HEATING
38 Weyside Road, Guildford, Surrey, GU1 1JB
Area of Operation: South East England
Tel: 01483 590600 **Fax:** 01483 590601
Email: enquiries@btu-heating.com
Web: www.btu-group.com

BUY DIRECT HEATING SUPPLIES
Unit 7, Commerce Business Centre, Commerce
Close, Westbury, Wiltshire, BA13 4LS
Area of Operation: UK (Excluding Ireland)
Tel: 01373 301360 **Fax:** 01373 865101
Email: info@buydirectuk.com
Web: www.buydirectheatingsupplies.co.uk ✎
Product Type: 1, 2, 3, 4, 5, 6, 7

CHATSWORTH HEATING PRODUCTS LTD
Unit B Watchmoor Point, Camberley,
Surrey, GU15 3EX
Area of Operation: UK (Excluding Ireland)
Tel: 01276 605880 **Fax:** 01276 605881
Email: enquiries@chatsworth-heating.co.uk
Web: www.chatsworth-heating.co.uk

CONSERVATION ENGINEERING LTD
The Street, Troston, Bury St Edmunds,
Suffolk, IP31 1EW
Area of Operation: UK & Ireland
Tel: 01359 269360 **Fax:** 01359 268340
Email: anne@conservation-engineering.co.uk
Web: www.heating-designs.co.uk

CONTINENTAL UNDER FLOOR HEATING
Continental House, Kings Hill, Bude,
Cornwall, EX23 0LU
Area of Operation: UK & Ireland
Tel: 01288 357880
Fax: 0845 108 1205
Email: info@continental-ufh.co.uk
Web: www.continental-ufh.co.uk ✎
Product Type: 1, 4, 5, 6, 7
Other Info: ECO ✎

COPPERJOB LTD
25 Wain Park, Plympton,
Plymouth, Devon, PL7 2HX
Area of Operation: UK & Ireland
Tel: 0870 460 7793
Email: john@copperjob.com
Web: www.centralheatingrepair.co.uk ✎
Product Type: 1, 2, 3, 5, 6

COPPER JOB
Area of Operation: UK & Ireland
Tel: 0870 460 7793
Email: john@copperjob.com
Web: www.centralheatingrepair.co.uk
Product Type: 1, 2, 3, 5, 6

The Industry's newest pbublication designed to
aid installers and engineers of all experience
levels. This superb manual helps you to
understand domestic Central Heating and Hot
Water Systems, quickly trace faults and effect
professional repairs. Available online at:
www.centralheatingrepair.co.uk

HEATING, PLUMBING & ELECTRICAL

**COSY ROOMS
(COSY-HEATING.CO.UK)**
17 Chiltern Way, North Hykeham,
Lincoln, Lincolnshire, LN6 9SY
Area of Operation: UK (Excluding Ireland)
Tel: 01522 696002
Fax: 01522 696002
Email: keith@cosy-rooms.com
Web: www.cosy-heating.co.uk
Product Type: 3

DANFOSS RANDALL LTD
Ampthill Road, Bedford,
Bedfordshire, MK42 9ER
Area of Operation: UK & Ireland
Tel: 0845 121 7400
Fax: 0845 121 7515
Email: danfossrandall@danfoss.com
Web: www.danfoss-randall.co.uk
Product Type: 1, 2, 3, 4, 5, 6

DCD SYSTEMS LTD
43 Howards Thicket,
Gerrards Cross,
Buckinghamshire, SL9 7NU
Area of Operation: UK & Ireland
Tel: 01753 882028
Fax: 01753 882029
Email: peter@dcd.co.uk
Web: www.dcd.co.uk
Product Type: 1, 2, 4, 5, 6, 7

E. RICHARDS
PO Box 1115, Winscombe,
Somerset, BS25 1WA
Area of Operation: UK & Ireland
Tel: 0845 330 8859
Fax: 0845 330 7260
Email: paultrace@e-richards.co.uk
Web: www.e-richards.co.uk

ECO HOMETEC UK LTD
Unit 11E, Carcroft Enterprise Park,
Carcroft, Doncaster, South Yorkshire, DN6 8DD
Area of Operation: Europe
Tel: 01302 722266
Fax: 01302 728634
Email: Stephen@eco-hometec.co.uk
Web: www.eco-hometec.co.uk
Product Type: 1, 2, 4, 5, 6

ECOLEC
Sharrocks Street, Wolverhampton,
West Midlands, WV1 3RP
Area of Operation: UK & Ireland
Tel: 01902 457575
Fax: 01902 457797
Email: sales@ecolec.co.uk
Web: www.ecolec.co.uk
Product Type: 1, 2, 5, 7

ESKIMO DESIGN LTD.
25 Horsell Road, London, N5 1XL
Area of Operation: Worldwide
Tel: 020 7117 0110
Email: ed@eskimodesign.co.uk
Web: www.eskimodesign.co.uk
Product Type: 3

EURO BATHROOMS
102 Annareagh Road, Richhill,
Co Armagh, BT61 9JY
Area of Operation: Ireland Only
Tel: 028 3887 9996
Fax: 028 3887 9996
Email: paul@euro-bathrooms.co.uk
Web: www.euro-bathrooms.co.uk
Product Type: 1, 2, 4, 5

FEATURE RADIATORS
Bingley Railway Station,
Wellington Street, Bingley,
West Yorkshire, BD16 4BX
Area of Operation: UK (Excluding Ireland)
Tel: 01274 567789
Fax: 01274 561183
Email: contactus@featureradiators.com
Web: www.featureradiators.com
Product Type: 3

FLOORWARMING (UK) LTD
Warwick Mill, Warwick Bridge,
Carlisle, Cumbria, CA4 8RR
Area of Operation: UK (Excluding Ireland)
Tel: 01228 631300
Fax: 01228 631333
Email: michael@floorwarming.co.uk
Web: www.floorwarming.co.uk
Product Type: 1, 2, 3, 4, 5, 6, 7

GEMINOX UK
Blenheim House, 1 Blenheim Road,
Epsom, Surrey, KT19 9AP
Area of Operation: UK & Ireland
Tel: 01372 722277
Fax: 01372 744477
Email: sales@geminox-uk.com
Web: www.geminox-uk.com
Product Type: 4, 6

GRANT ENGINEERING LTD
Crinkle, Birr, Co. Offaly, Ireland
Area of Operation: Ireland Only
Tel: +353 509 20089
Fax: +353 509 21060
Email: info@grantengineering.ie
Web: www.grantengineering.ie
Product Type: 1

GRUNDFOS PUMPS LTD
Grovebury Road, Leighton Buzzard,
Bedfordshire, LU7 4TL
Area of Operation: Worldwide
Tel: 01525 850000
Fax: 01525 850011
Email: uk-sales@grundfos.com
Web: www.grundfos.com
Product Type: 7

HEPWORTH PLUMBING PRODUCTS
Edlington Lane, Edlington, Doncaster,
South Yorkshire, DN12 1BY
Area of Operation: Worldwide
Tel: 01709 856300
Fax: 01709 856301
Email: info@hepworthplumbing.co.uk
Web: www.hepworthplumbing.co.uk
Product Type: 4
Other Info: ✎

HONEYWELL CONTROL SYSTEMS LTD
Honeywell House,
Arlington Business Park,
Bracknell, Berkshire, RG12 1EB
Area of Operation: Worldwide
Tel: 01344 656000
Fax: 01344 656054
Email: literature@honeywell.com
Web: www.honeywelluk.com
Product Type: 1, 2, 3, 4, 5, 6, 7

INTERACTIVE HOMES LTD
Lorton House, Conifer Crest,
Newbury, Berkshire, RG14 6RS
Area of Operation: UK (Excluding Ireland)
Tel: 01635 49111
Fax: 01635 40735
Email: sales@interactivehomes.co.uk
Web: www.interactivehomes.co.uk
Product Type: 1, 2, 3, 4, 5, 6, 7

**JIS EUROPE
(SUSSEX RANGE)**
Warehouse 2, Nash Lane,
Scaynes Hill, Haywards Heath,
West Sussex, RH17 7NJ
Area of Operation: Europe
Tel: 01444 831200 **Fax:** 01444 831900
Email: info@jiseurope.co.uk
Product Type: 1, 3

KV RADIATORS
6 Postle Close, Kilsby, Rugby,
Warwickshire, CV23 8YG
Area of Operation: UK & Ireland
Tel: 01788 823286 **Fax:** 01788 823 002
Email: solutions@kvradiators.com
Web: www.kvradiators.com
Product Type: 3, 7

LEEMICK LTD
79 Windermere Drive,
Rainham, Kent, ME8 9DX
Area of Operation: East England,
Greater London, Midlands & Mid Wales,
South East England, South West England
and South Wales
Tel: 01634 351666 **Fax:** 01634 351666
Email: sales@leemick.co.uk
Web: www.leemick.co.uk
Product Type: 1, 2, 4, 5, 6, 7
Other Info: ✎

LIONHEART HEATING SERVICES
PO Box 741, Harworth Park, Doncaster,
South Yorkshire, DN11 8WY
Area of Operation: UK (Excluding Ireland)
Tel: 01302 755200 **Fax:** 01302 750155
Email: lionheart@microgroup.ltd.uk
Web: www.lionheartheating.co.uk

LOBLITE ELECTRIC LTD
Third Avenue, Team Valley Trading Estate,
Gateshead, Tyne & Wear, NE11 0QQ
Area of Operation: Europe
Tel: 0191 487 8103
Fax: 0191 491 5541
Email: sales@loblite.co.uk
Web: www.heatec-rads.com
Product Type: 1, 5
Other Info: ECO

PARAGON SYSTEMS (SCOTLAND) LIMITED
The Office, Corbie Cottage,
Maryculter, Aberdeen,
Aberdeenshire, AB12 5FT
Area of Operation: Scotland
Tel: 01224 735536
Fax: 01224 735537
Email: info@paragon-systems.co.uk
Web: www.paragon-systems.co.uk
Product Type: 1, 4, 5, 6

PARK SYSTEMS INTEGRATION LTD
Unit 3 Queen's Park, Earlsway,
Team Valley Trading Estate,
Gateshead, Tyne & Wear, NE11 0QD
Area of Operation: UK & Ireland
Tel: 0191 497 0770
Fax: 0191 497 0772
Email: sales@parksystemsintegration.co.uk
Web: www.parksystemsintegration.co.uk
Product Type: 1, 2, 5, 6

SOUTHERN SOLAR
Unit 6, Allington Farm, Allington Lane,
Offham, Lewes, East Sussex, BN7 3QL
Area of Operation: Greater London,
South East England, South West England
and South Wales
Tel: 0845 456 9474
Fax: 01273 483928
Email: info@southernsolar.co.uk
Web: www.southernsolar.co.uk
Product Type: 1, 3, 5

SYSTEMAIR LTD
Pharoah House, Arnolde Close,
Medway City Estate, Rochester,
Kent, ME2 4SP
Area of Operation: Worldwide
Tel: 01634 735 000
Fax: 01634 735 001
Email: sales@systemair.co.uk
Web: www.systemair.co.uk
Product Type: 1, 2, 3, 4, 5, 6, 7

THE RADIATOR COMPANY LIMITED
Elan House, Charlwoods Road,
East Grinstead, West Sussex, RH19 2HG
Area of Operation: UK & Ireland
Tel: 01342 302250
Fax: 01342 305561
Email: ian@theradiatorcompany.co.uk
Web: www.theradiatorcompany.co.uk
Product Type: 3, 7

THERMO-FLOOR (GB) LTD
Unit 1 Babsham Farm,
Chichester Road, Bognor Regis,
West Sussex, PO21 5EL
Area of Operation: UK (Excluding Ireland)
Tel: 01243 822058
Fax: 01243 860379
Email: sales@thermo-floor.co.uk
Web: www.thermo-floor.co.uk
Product Type: 1, 2, 4, 5, 6, 7
Other Info: ✎

UNIFIX LTD
St Georges House, Grove Lane,
Smethwick, Birmingham,
West Midlands, B66 2QT
Area of Operation: Europe
Tel: 0800 096 1110
Fax: 0800 096 1115
Email: sales@unifix.com
Web: www.unifix-online.co.uk
Product Type: 1, 2, 3

**VELTA - THE UNDERFLOOR
HEATING COMPANY**
Unit 1B, Denby Dale Industrial Park,
Wakefield Road, Denby Dale,
Huddersfield, West Yorkshire, HD8 8QH
Area of Operation: Worldwide
Tel: 01484 860811
Fax: 01484 865775
Email: info@velta-uk.com
Web: www.u-h-c.co.uk
Product Type: 1, 2, 4, 5, 6, 7

HEATING, PLUMBING & ELECTRICAL

VOKERA LTD
4th Floor, Catherine House,
Boundary Way, Hemel Hempstead,
Hertfordshire, HP2 7RP
Area of Operation: UK & Ireland
Tel: 0870 333 0220 **Fax:** 01442 281460
Email: enquiries@vokera.co.uk
Web: www.vokera.co.uk
Product Type: 1, 2, 6

WARMROOMS
24 Corncroft Lane, St Leonards Park,
Gloucester, Gloucestershire, GL4 6XU
Area of Operation: UK (Excluding Ireland)
Tel: 01452 304460
Fax: 01452 304460
Email: sales@warmrooms.co.uk
Web: www.warmrooms.co.uk ⌐🖰
Product Type: 3

WORCESTER BOSCH GROUP
Cotswold Way, Warndon, Worcester,
Worcestershire, WR4 9SW
Area of Operation: UK & Ireland
Tel: 01905 754624 **Fax:** 01905 753103
Email: general.worcester@uk.bosch.com
Web: www.worcester-bosch.co.uk
Product Type: 1, 6

YORKSHIRE FITTINGS
PO Box 166, Leeds,
West Yorkshire, LS10 1NA
Tel: 0113 270 1104 **Fax:** 0113 271 5275
Email: info@yorkshirefittings.co.uk
Web: www.yorkshirefittings.co.uk
Product Type: 3

FLUES AND HEATING ACCESSORIES

KEY

PRODUCT TYPES: 1= Flue Liners
2 = Fan Flues   3 = Boiler Flues
4 = Flue Pipes   5 = Other

OTHER: ▽ Reclaimed ⌐🖰 On-line shopping
✍ Bespoke 🖐 Hand-made   ECO Ecological

ALPHA BOILERS
Nepicar House, London Road,
Wrotham Heath, Kent, TN15 7RS
Area of Operation: UK (Excluding Ireland)
Tel: 01732 783000 **Fax:** 01732 783080
Email: info@alphatherm.co.uk
Web: www.alpha-boilers.com
Product Type: 3, 4

ANKI CHIMNEY SYSTEMS
Bishops Way, Newport,
Isle of Wight, PO30 5WS
Area of Operation: UK (Excluding Ireland)
Tel: 01983 527997 **Fax:** 01983 537788
Email: anki@ajwells.co.uk
Web: www.anki.co.uk
Product Type: 1

APPLIED HEATING SERVICES LTD
17 Wilden Road, Pattinson South Industrial Estate,
Washington, Tyne & Wear, NE38 8QB
Area of Operation: North East England
Tel: 0191 417 7604 **Fax:** 0191 417 1549
Email: georgecossey1@btconnect.com
Web: www.appliedheat.co.uk
Product Type: 3

CHARNWOOD STOVES & FIRES
Bishops Way, Newport, Isle of Wight, PO30 5WS
Area of Operation: Worldwide
Tel: 01983 537780
Fax: 01983 537788
Email: charnwood@ajwells.co.uk
Web: www.charnwood.com ⌐🖰
Product Type: 1, 3, 4, 5
Other Info: ECO

CHARNWOOD STOVES & ANKI CHIMNEY SYSTEMS

Area of Operation: Worldwide
Tel: 01983 537780 **Fax:** 01983 537788
Email: charnwood@ajwells.co.uk
Web: www.charwood.com or www.anki.co.uk
Product Type: 1,3, 4, 5

Exceptional wood burning stoves and pumice chimney systems for the perfect combination in highly efficient home heating. Ideal for new build and installation into existing properties.

CLEARVIEW STOVES
More Works, Squilver Hill,
Bishops Castle, Shropshire, SY9 5HH
Area of Operation: Worldwide
Tel: 01588 650401 **Fax:** 01588 650493
Email: mail@clearviewstoves.com
Web: www.clearviewstoves.com
Product Type: 1, 2, 4

DCD SYSTEMS LTD
43 Howards Thicket, Gerrards Cross,
Buckinghamshire, SL9 7NU
Area of Operation: UK & Ireland
Tel: 01753 882028
Fax: 01753 882029
Email: peter@dcd.co.uk
Web: www.dcd.co.uk
Product Type: 5

ECO HOMETEC UK LTD
Unit 11E, Carcroft Enterprise Park, Carcroft,
Doncaster, South Yorkshire, DN6 8DD
Area of Operation: Europe
Tel: 01302 722266
Fax: 01302 728634
Email: Stephen@eco-hometec.co.uk
Web: www.eco-hometec.co.uk
Product Type: 2, 3, 4

EXHAUSTO
Unit 3 Lancaster Court, Coronation Road,
Cressex Business Park, High Wycombe,
Buckinghamshire, HP12 3TD
Area of Operation: UK & Ireland
Tel: 01494 465166
Fax: 01494 465163
Email: info@exhausto.co.uk
Web: www.exhausto.co.uk
Product Type: 2, 3, 4

FERROLI UK
Lichfield Road, Branston Industrial Estate,
Burton Upon Trent, Staffordshire, DE14 3HD
Area of Operation: UK (Excluding Ireland)
Tel: 08707 282882
Fax: 08707 282883
Email: sales@ferroli.co.uk
Web: www.ferroli.co.uk
Product Type: 3

FIREPLACE CONSULTANTS LTD
The Studio, The Old Rothschild Arms,
Buckland Road, Buckland, Aylesbury,
Buckinghamshire, HP22 5LP
Area of Operation: Greater London,
Midlands & Mid Wales, South East England
Tel: 01296 632287 **Fax:** 01296 632287
Email: info@fireplaceconsultants.com
Web: www.fireplaceconsultants.com
Product Type: 1, 2, 4

GEMINOX UK
Blenheim House, 1 Blenheim Road,
Epsom, Surrey, KT19 9AP
Area of Operation: UK & Ireland
Tel: 01372 722277
Fax: 01372 744477
Email: sales@geminox-uk.com
Web: www.geminox-uk.com
Product Type: 1, 3, 4

GRANT ENGINEERING LTD
Crinkle, Birr, Co. Offaly, Ireland
Area of Operation: Ireland Only
Tel: +353 509 20089
Fax: +353 509 21060
Email: info@grantengineering.ie
Web: www.grantengineering.ie
Product Type: 3

GRANT UK
Hopton House, Hopton Industrial Estate,
Devizes, Wiltshire, SN10 2EU
Area of Operation: UK (Excluding Ireland)
Tel: 0870 777 5553
Fax: 0870 777 5559
Email: sales@grantuk.com
Web: www.grantuk.com
Product Type: 3

LIONHEART HEATING SERVICES
PO Box 741, Harworth Park, Doncaster,
South Yorkshire, DN11 8WY
Area of Operation: UK (Excluding Ireland)
Tel: 01302 755200
Fax: 01302 750155
Email: lionheart@microgroup.ltd.uk
Web: www.lionheartheating.co.uk ⌐🖰

MHS RADIATORS & BOILERS
35 Nobel Square, Burnt Mills Industrial Estate,
Basildon, Essex, SS13 1LT
Area of Operation: UK & Ireland
Tel: 01268 591010
Fax: 01268 728202
Email: sales@modular-heating-group.co.uk
Web: www.mhsradiators.com
Product Type: 3

PARAGON SYSTEMS (SCOTLAND) LIMITED
The Office, Corbie Cottage, Maryculter,
Aberdeen, Aberdeenshire, AB12 5FT
Area of Operation: Scotland
Tel: 01224 735536
Fax: 01224 735537
Email: info@paragon-systems.co.uk
Web: www.paragon-systems.co.uk
Product Type: 3, 4

PERCY DOUGHTY & CO
Imperial Point, Express Trading Estate, Stonehill
Road, Farnworth, Bolton, Lancashire, M38 9ST
Area of Operation: Midlands & Mid Wales, North
West England and North Wales
Tel: 01204 868550
Fax: 01204 868551
Email: sales@percydoughty.com
Web: www.percydoughty.co.uk
Product Type: 4

R W KNIGHT & SON LTD
Castle Farm, Marshfield,
Chippenham, Wiltshire, SN14 8HU
Area of Operation: Midlands & Mid Wales, South
East England, South West England and South Wales
Tel: 01225 891469
Fax: 01225 892369
Email: Enquiries@knight-stoves.co.uk
Web: www.knight-stoves.co.uk
Product Type: 1, 2, 3, 4

RED BANK MANUFACTURING COMPANY LIMITED
Measham, Swadlincote, Derbyshire, DE12 7EL
Area of Operation: UK & Ireland
Tel: 01530 270333
Fax: 01530 273667
Email: sales@redbankmfg.co.uk
Web: www.redbankmfg.co.uk
Product Type: 1

SCALGON
115 Park Lane, Reading,
Berkshire, RG31 4DR
Area of Operation: UK (Excluding Ireland)
Tel: 0118 9424981
Email: postmaster@scalgon.co.uk
Web: www.scalgon.co.uk

SCHIEDEL-ISOKERN CHIMNEY SYSTEMS
14 Haviland Road, Ferndown Industrial Estate,
Wimborne, Dorset, BH21 7RF
Area of Operation: UK & Ireland
Tel: 01202 861650
Fax: 01202 861632
Email: sales@isokern.co.uk
Web: www.isokern.co.uk
Product Type: 1, 4

SOUTHERN SOLAR
Unit 6, Allington Farm, Allington Lane,
Offham, Lewes, East Sussex, BN7 3QL
Area of Operation: Greater London,
South East England, South West England
and South Wales
Tel: 0845 456 9474
Fax: 01273 483928
Email: info@southernsolar.co.uk
Web: www.southernsolar.co.uk
Product Type: 3

THE FIREPLACE MARKETING COMPANY LIMITED
The Old Coach House, Southern Road,
Thame, Oxfordshire, OX9 2ED
Area of Operation: Worldwide
Tel: 01844 260960
Fax: 01844 260267
Email: david@fireplacemarketing.co.uk
Web: www.fireplace.co.uk ⌐🖰
Product Type: 1, 2, 4

THEALE FIREPLACES RDG LTD
Milehouse Farm, Bath Road,
Theale, Berkshire, RG7 5HJ
Area of Operation: UK (Excluding Ireland)
Tel: 0118 930 2232
Fax: 0118 932 3344
Email: mail@theale-fireplaces.co.uk
Web: www.theale-fireplaces.co.uk
Product Type: 1

THERMOCRETE
Mortimer Street, Bradford,
West Yorkshire, BD8 9RL
Area of Operation: Worldwide
Tel: 01274 544442
Fax: 01274 484448
Email: info@thermocrete.com
Web: www.thermocrete.com
Product Type: 1, 3

TRIANCO LIMITED
Thorncliffe, Chapeltown, Sheffield,
South Yorkshire, S35 2PH
Area of Operation: UK & Ireland
Tel: 0114 257 2300
Fax: 0114 257 1419
Email: info@trianco.co.uk
Web: www.trianco.co.uk

VOKERA LTD
4th Floor, Catherine House, Boundary Way,
Hemel Hempstead, Hertfordshire, HP2 7RP
Area of Operation: UK & Ireland
Tel: 0870 333 0220
Fax: 01442 281460
Email: enquiries@vokera.co.uk
Web: www.vokera.co.uk
Product Type: 2, 3, 4

WORCESTER BOSCH GROUP
Cotswold Way, Warndon, Worcester,
Worcestershire, WR4 9SW
Area of Operation: UK & Ireland
Tel: 01905 754624
Fax: 01905 753103
Email: general.worcester@uk.bosch.com
Web: www.worcester-bosch.co.uk
Product Type: 2, 3

FUEL SUPPLIERS AND FUEL TANKS

KEY

ENERGY SOURCES: ★ Gas   ○ Oil
□ Coal   ● Wood   ✪ Multi-Fuel
✿ Electric   ◗ Gel

OTHER: ▽ Reclaimed   ☝ On-line shopping
✏ Bespoke   ✋ Hand-made   ECO Ecological

BALMORAL TANKS
Balmoral Park, Loirston, Aberdeen,
Aberdeenshire, AB12 3GY
Area of Operation: Worldwide
Tel: 01224 859100
Fax: 01224 859123
Email: s.gibb@balmoral.co.uk
Web: www.balmoraltanks.com

BORDERS UNDERFLOOR HEATING
26 Coopersknowe Crescent, Galashiels,
Borders, TD1 2DS
Area of Operation: UK & Ireland
Tel: 01896 668667
Fax: 01896 668678
Email: underfloor@btinternet.com
Web: www.bordersunderfloor.co.uk

BP LPG
1 Cambuslang Way, Cambuslang,
Glasgow, G32 8ND
Area of Operation: UK (Excluding Ireland)
Tel: 0845 300 0038
Fax: 0141 307 4869
Email: bplpg@bp.com
Web: www.bplpg.co.uk

BP LPG

Area of Operation: UK (Excluding Ireland)
Tel: 0845 300 0038
Fax: 0141 307 4869
Email: bplpg@bp.com
Web: www.bplpg.co.uk

BP LPG supplies propane and butane gas to
homes and businesses.
Contact us to find out how we can meet your
energy needs.

BUY DIRECT HEATING SUPPLIES
Unit 7, Commerce Business Centre,
Commerce Close, Westbury,
Wiltshire, BA13 4LS
Area of Operation: UK (Excluding Ireland)
Tel: 01373 301360
Fax: 01373 865101
Email: info@buydirectuk.co.uk
Web: www.buydirectheatingsupplies.co.uk ☝

CALOR GAS LTD
Athena Drive, Tachbrook Park,
Warwick, Warwickshire, CV34 6RL
Area of Operation: UK (Excluding Ireland)
Tel: 01926 330088
Fax: 01926 318718
Email: enquiry@calor.co.uk
Web: www.calor.co.uk ☝

CLEARVIEW STOVES
More Works, Squilver Hill,
Bishops Castle, Shropshire, SY9 5HH
Area of Operation: Worldwide
Tel: 01588 650401 **Fax:** 01588 650493
Email: mail@clearviewstoves.com
Web: www.clearviewstoves.com

COUNTRYWIDE ENERGY
Defford Mill, Earls Croome, Worcestershire, WR8 9DF
Area of Operation: Midlands & Mid Wales,
South West England and South Wales
Tel: 01386 757333 **Fax:** 01386 757341
Email: juliejones-ford@countrywidefarmers.co.uk
Web: www.countrywidefarmers.co.uk

CROWN OIL
The Oil Centre, Bury New Road, Heap Bridge,
Bury, Greater Manchester, BL9 7HY
Area of Operation: UK (Excluding Ireland)
Tel: 0161 764 6622 **Fax:** 0161 762 7685
Email: sales@crownoil.co.uk
Web: www.crownoil.co.uk

OPIES' THE STOVE SHOP
The Stove Shop, The Street, Hatfield Peverel,
Chelmsford, Essex, CM3 2DY
Area of Operation: East England,
Greater London, South East England
Tel: 01245 380471
Email: enquiries@opie-woodstoves.co.uk
Web: www.opie-woodstoves.co.uk

SHELL GAS LTD
PO Box 1100, Chesterfield, S44 5YQ
Area of Operation: Worldwide
Tel: 0845 128 4541
Fax: 0870 128 4541
Email: enquiries@shell.com
Web: www.shellgas.co.uk

SOLID FUEL ASSOCIATION LTD (SFA)
7 Swanwick Court, Alfreton, Derbyshire, DE55 7AS
Area of Operation: UK (Excluding Ireland)
Tel: 01773 835400
Fax: 01773 834351
Email: sfa@solidfuel.co.uk
Web: www.solidfuel.co.uk ☝

CONVECTORS

KEY

SEE ALSO: HEATING - Radiators
ENERGY SOURCES: ★ Gas   ○ Oil
□ Coal   ● Wood   ✪ Multi-Fuel
✿ Electric   ◗ Gel

OTHER: ▽ Reclaimed   ☝ On-line shopping
✏ Bespoke   ✋ Hand-made   ECO Ecological

BUY DIRECT HEATING SUPPLIES
Unit 7, Commerce Business Centre,
Commerce Close, Westbury, Wiltshire, BA13 4LS
Area of Operation: UK (Excluding Ireland)
Tel: 01373 301360
Fax: 01373 865101
Email: info@buydirectuk.co.uk
Web: www.buydirectheatingsupplies.co.uk ☝

CONSORT EQUIPMENT PRODUCTS LTD
Thornton Industrial Estate, Milford Haven,
Pembrokeshire, SA73 2RT
Area of Operation: Worldwide
Tel: 01646 692172
Fax: 01646 695195
Email: enquiries@consortepl.com
Web: www.consortepl.com

COSY ROOMS (COSY-HEATING.CO.UK)
17 Chiltern Way, North Hykeham, Lincoln,
Lincolnshire, LN6 9SY
Area of Operation: UK (Excluding Ireland)
Tel: 01522 696002 **Fax:** 01522 696002
Email: keith@cosy-rooms.com
Web: www.cosy-heating.co.uk ☝

COUNTRY STYLE COOKERS
Unit 8, Oakleys Yard, Gatehouse Road,
Rotherwas Industrial Estate, Hereford,
Herefordshire, HR2 6RQ
Area of Operation: UK & Ireland
Tel: 01432 342351 **Fax:** 01432 371331
Email: sales@countrystyle-cookers.com
Web: www.countrystyle-cookers.com
Other Info: ▽ ○

COUNTRYWIDE ENERGY
Defford Mill, Earls Croome,
Worcestershire, WR8 9DF
Area of Operation: Midlands & Mid Wales,
South West England and South Wales
Tel: 01386 757333 **Fax:** 01386 757341
Email: juliejones-ford@countrywidefarmers.co.uk
Web: www.countrywidefarmers.co.uk

CROWN OIL
The Oil Centre, Bury New Road, Heap Bridge,
Bury, Greater Manchester, BL9 7HY
Area of Operation: UK (Excluding Ireland)
Tel: 0161 764 6622 **Fax:** 0161 762 7685
Email: sales@crownoil.co.uk
Web: www.crownoil.co.uk

ELECTRICALSHOP.NET
81 Kinson Road, Wallisdown,
Bournemouth, Dorset, BH10 4DG
Area of Operation: UK (Excluding Ireland)
Tel: 0870 027 3730 **Fax:** 0870 027 3731
Email: sales@electricalshop.net
Web: www.electricalshop.net ☝

EURO BATHROOMS
102 Annareagh Rd, Richhill, Co Armagh, BT61 9JY
Area of Operation: Ireland Only
Tel: 028 3887 9996 **Fax:** 028 3887 9996
Email: paul@euro-bathrooms.co.uk
Web: www.euro-bathrooms.co.uk

JETMASTER FIRES LTD
Unit 2 Peacock Trading Estate, Goodwood Road,
Chandlers Ford, Eastleigh, Hampshire, SO50 4NT
Area of Operation: Europe
Tel: 0870 727 0105 **Fax:** 0870 727 0106
Email: jetmastersales@aol.com
Web: www.jetmaster.co.uk

LIONHEART HEATING SERVICES
PO Box 741, Harworth Park, Doncaster,
South Yorkshire, DN11 8WY
Area of Operation: UK (Excluding Ireland)
Tel: 01302 755200 **Fax:** 01302 750155
Email: lionheart@microgroup.ltd.uk
Web: www.lionheartheating.co.uk ☝

M&T (UK) LTD
PO Box 382, Grimsby,
Lincolnshire, DN37 9XB
Area of Operation: Europe
Tel: 01472 886155 **Fax:** 01472 590887
Email: meinertzinuk@tiscali.co.uk
Web: www.meinertz.com

MYSON
Emlyn Street, Farnworth,
Bolton, Lancashire, BL4 7EB
Area of Operation: Worldwide
Tel: 01204 863200 **Fax:** 01204 863229
Email: salesmtw@myson.co.uk
Web: www.myson.co.uk

TITAN PLASTECH LIMITED
Barbot Hall Industrial Estate, Mangham Road,
Rotherham, South Yorkshire, S61 4RJ
Area of Operation: Europe
Tel: 01709 538300 **Fax:** 01709 538301
Email: tony.soper@titantanks.co.uk
Web: www.titanplastech.co.uk

VORTICE LTD
Beeches House, Eastern Avenue,
Burton on Trent, Staffordshire, DE13 0BB
Area of Operation: Worldwide
Tel: 01283 492949 **Fax:** 01283 544121
Email: sales@vortice.ltd.uk
Web: www.vortice.ltd.uk ☝

NOTES

Company Name ....
Address ....
....
email ....
Web ....

Company Name ....
Address ....
....
email ....
Web ....

Company Name ....
Address ....
....
email ....
Web ....

Company Name ....
Address ....
....
email ....
Web ....

Company Name ....
Address ....
....
email ....
Web ....

HEATING, PLUMBING & ELECTRICAL

For all your bulk domestic and commercial LPG requirements, call;

0845 076 55 44

MOVE OR IMPROVE MAGAZINE

Move or Improve? magazine provides the facts that anyone determined to add space and value to their home needs to make the right decisions.

Features:

Design Masterclass: Making efficient use of space

The Design Doctor: Design expert Michael Holmes helps two readers find solutions to their improving dilemas

The Guide: A complete guide to renovating and extending your home, with advice on planning, building regulations, finance, etc

Renovation Price Book: Full estimated costings for different aspects of remodelling

Special Features, for example: Loft conversion guide; How to buy replacement windows; Build, buy or renovate?; Wooden floors — complete design guide

Real Life Projects: Inspiring and practical readers' studies including cellar conversions, luxurious new sunrooms, kitchen extensions and perfect side extensions

Plus much more

On sale now in all good newsagents and supermarkets

Ascent Publishing Limited
Unit 2 Sugar Brook Court
Aston Road
Bromsgrove
Worcestershire
B60 3EX

SUBSCRIBE TODAY!

Tel: 01527 834435
Website: www.moveorimprove.co.uk

VISIT: www.moveorimprove.co.uk

PLUMBING

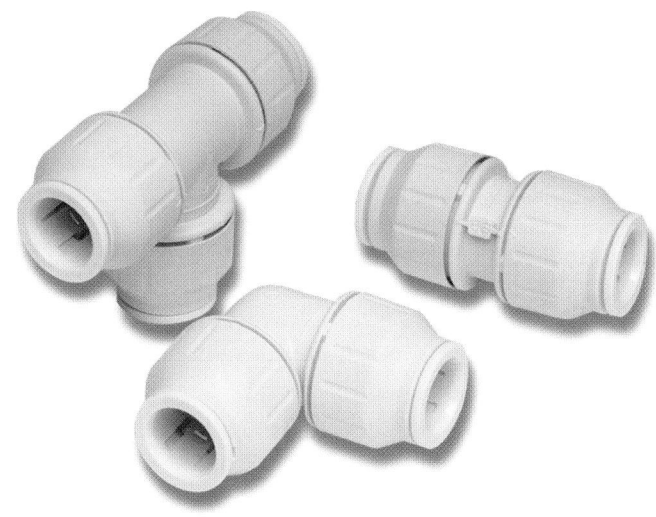

Image courtesy of John Guest Speedfit Ltd (01895 449233)

HEATING, PLUMBING & ELECTRICAL

SPONSORED BY JOHN GUEST SPEEDFIT LTD
Tel 01895 449233 Web www.speedfit.co.uk

SPONSORED BY: JOHN GUEST SPEEDFIT LTD www.speedfit.co.uk

Domestic Hot Water

The majority of smaller homes now have a combination boiler which provides hot water on demand but, as anyone who has lived with a combi that is underpowered for its job will know only too well, they can have their limitations. With the exception of one or two models, even high-powered combination boilers will still find it difficult to meet demand from two bathrooms simultaneously.

The most popular way of providing sufficient domestic hot water to meet demand is to have a store of some kind. The choice for stored hot water is between a traditional hot water cylinder (a vented system) and a pressurized cylinder (unvented).

Standard Cylinders: The standard hot water cylinder uses gravity to provide the pressure, and so needs to be placed above the highest hot water tap in order to work. The cylinder will usually be fed by the cold water storage tank in the loft. The flow rate will be limited at this pressure and so showers will need to be pumped (i.e. power showers).

Unvented Cylinders: A mains pressure cylinder stores hot water at around 3 bar with the pressure in the system provided by the mains — the cylinder is fed directly by the mains in the road rather than via a cold

water storage tank in the loft, which has the advantage of better flow rates, eliminating the need for pumps on showers. It also means that hot and cold pressure is balanced, thereby preventing fluctuations in temperature of showers and mixer taps. As with a combi boiler, an unvented hot water cylinder eliminates the need for a header tank in the attic, thereby freeing up valuable space.

To cope with any potential malfunctions, an unvented hot water cylinder is fitted with a number of mandatory safety features, including a pressure relief valve and temperature cut-off devices to avoid overheating. All these controls, together with the necessity to make the cylinder strong enough to withstand the full pressure of mains water, make unvented hot water more expensive than a

ABOVE: Image by Aquabeau.

conventional cylinder, but the extra cost must be judged against the savings created by eliminating the need for loft work and noisy pumps for showers.

Quick Recovery Cylinders: The quick recovery cylinder incorporates a vastly increased heat exchanger to transfer heat more swiftly from the primary heating circuit to the stored water. The rapid recovery speed means that the tank can be smaller than for a standard cylinder, allowing for easy placement and greater economy. This sort of cylinder can be vented or unvented.

Multi-coil Cylinders: Conventional hot water cylinders have a single heat exchanger coil built into them, within which the primary flow from the boiler is fed. Cylinders are available with a second or even third coil designed to allow multiple heat sources to power the cylinder, including solar panels, open fires and kitchen ranges. Thermostatic valves are available that will control the heat sources to the cylinder to optimise energy efficiency.

A conventional vented cylinder can also be fitted with a coil at the top (the hottest part) in which pressurized water from the mains is fed and heated directly to give a limited supply of mains pressure hot water for showers (it operates in the same way as a thermal store).

Solar Water Heating: Active solar panels make use of free energy from the sun to provide hot water which can be used to heat the hot water cylinder or a thermal store which can provide space heating. Solar panels cost from £700 upwards installed. A 3-4m2 solar panel will provide around 1,000kWh of hot water per year, which is just over half the typical annual household requirement, so some other form of heating is required to provide domestic hot water in the winter months. The payback for a solar panel will depend on the cost of the fuel it is replacing, but it is long – at least ten years – however it will help to reduce CO_2 emissions from day one.

Image by Ideal Standard

PLUMBING, PIPES & FITTINGS

KEY

SEE ALSO: MERCHANTS - Plumbers
Merchants, DRAINAGE - Pipes and Pumps

OTHER: ▽ Reclaimed 🖱 On-line shopping
✎ Bespoke ✋ Hand-made ECO Ecological

AL CHALLIS LTD
Europower House, Lower Road,
Cookham, Maidenhead,
Berkshire, SL6 9EH
Area of Operation: UK & Ireland
Tel: 01628 529024
Fax: 0870 458 0577
Email: chris@alchallis.com
Web: www.alchallis.com
Other Info: ECO ✎

ALTON PUMPS
Redwood Lane, Medstead,
Alton, Hampshire, GU34 5PE
Area of Operation: Worldwide
Tel: 01420 561661
Fax: 01420 561661
Email: sales@altonpumps.com
Web: www.altonpumps.com

AQUALISA PRODUCTS LTD
The Flyers Way, Westerham,
Kent, TN16 1DE
Area of Operation: UK & Ireland
Tel: 01959 560000
Fax: 01959 560030
Email: sue.anderson@aqualisa.co.uk
Web: www.aqualisa.co.uk

AUTRON PRODUCTS LTD
Unit 17 Second Avenue,
Bluebridge Industrial Estate,
Halstead, Essex, C09 2SU
Area of Operation: UK & Ireland
Tel: 01787 473964
Fax: 01787 474061
Email: sales@autron.co.uk
Web: wwww.autron.co.uk

BTU HEATING
38 Weyside Road,
Guildford, Surrey, GU1 1JB
Area of Operation: South East England
Tel: 01483 590600
Fax: 01483 590601
Email: enquiries@btu-heating.com
Web: www.btu-group.com

COBURG GUTTER GRID
Little Gunnerby, Hatcliffe, Grimsby,
Lincolnshire, DN37 0SP
Area of Operation: UK & Ireland
Tel: 01472 371406
Fax: 01469 560435
Email: sue@guttergrid.com
Web: www.guttergrid.com

CONTINENTAL UNDER FLOOR HEATING
Continental House, Kings Hill,
Bude, Cornwall, EX23 0LU
Area of Operation: UK & Ireland
Tel: 01288 357880
Fax: 0845 108 1205
Email: info@continental-ufh.co.uk
Web: www.continental-ufh.co.uk 🖱
Other Info: ECO ✎

COPPERJOB LTD
25 Wain Park, Plympton,
Plymouth, Devon, PL7 2HX
Area of Operation: UK & Ireland
Tel: 01752 339234
Email: john@copperjob.com
Web: www.centralheating.co.uk 🖱

DALLMER LTD
4, Norman Way, Lavenham,
Sudbury, Suffolk, CO10 9PY
Area of Operation: Worldwide
Tel: 01787 248244 **Fax:** 01787 248246
Email: sales@dallmer.com **Web:** www.dallmer.com

DCD SYSTEMS LTD
43 Howards Thicket, Gerrards Cross,
Buckinghamshire, SL9 7NU
Area of Operation: UK & Ireland
Tel: 01753 882028 **Fax:** 01753 882029
Email: peter@dcd.co.uk
Web: www.dcd.co.uk **Other Info:** ECO

DECORMASTER LTD
Unit 16, Waterside Industrial Estate,
Wolverhampton, West Midlands, WV2 2RH
Area of Operation: Worldwide
Fax: 01902 353126
Email: sales@oldcolours.co.uk
Web: www.oldcolours.co.uk

DRAINSTORE.COM
Units 1 & 2, Heanor Gate Road, Heanor Gate
Industrial Estate, Heanor, Derbyshire, DE75 7RJ
Area of Operation: Europe
Tel: 01773 767611 **Fax:** 01773 767613
Email: adrian@drainstore.com
Web: www.drainstore.com 🖱

ECP GROUP
103 Burrell Road, Ipswich, Suffolk, IP2 8AD
Area of Operation: East England,
South East England
Tel: 01473 400101 **Fax:** 01473 400103
Email: sales@ecpgroup.com
Web: www.ecpgroup.com

EURO BATHROOMS
102 Annareagh Road, Richhill, Co Armagh, BT61 9JY
Area of Operation: Ireland Only
Tel: 028 3887 9996 **Fax:** 028 3887 9996
Email: paul@euro-bathrooms.co.uk
Web: www.euro-bathrooms.co.uk

EUROCARE SHOWERS LTD
Unit 19, Doncaster Industry Park, Watch House Lane,
Bentley, Doncaster, South Yorkshire, DN5 9LZ
Area of Operation: Worldwide
Tel: 01302 788684 **Fax:** 01302 780010
Email: sales@eurocare-showers.com
Web: www.eurocare-showers.com
Other Info: ✎

FREERAIN
Millennium Green Business Centre,
Rio Drive, Collingham,
Nottinghamshire, NG23 7NB
Area of Operation: UK & Ireland
Tel: 01636 894900 **Fax:** 01636 894909
Email: info@freerain.co.uk
Web: www.freerain.co.uk

GRUNDFOS PUMPS LTD
Grovebury Road, Leighton Buzzard,
Bedfordshire, LU7 4TL
Area of Operation: Worldwide
Tel: 01525 850000 **Fax:** 01525 850011
Email: uk-sales@grundfos.com
Web: www.grundfos.com

HARGREAVES FOUNDRY LTD
Water Lane, South Parade, Halifax,
West Yorkshire, HX3 9HG
Area of Operation: UK & Ireland
Tel: 01422 330607 **Fax:** 01422 320349
Email: info@hargreavesfoundry.co.uk
Web: www.hargreavesfoundry.co.uk

HEPWORTH PLUMBING PRODUCTS
Edlington Lane, Edlington, Doncaster,
South Yorkshire, DN12 1BY
Area of Operation: Worldwide
Tel: 01709 856300
Fax: 01709 856301
Email: info@hepworthplumbing.co.uk
Web: www.hepworthplumbing.co.uk
Other Info: ✎

HIBERNIA RAINHARVESTING
Unit 530 Stanstead Distribution Centre, Start Hill,
Bishops Stortford, Hertfordshire, CM22 7DG
Area of Operation: UK & Ireland
Tel: 02890 249954 **Fax:** 02890 249964
Email: rain@hiberniaeth.com
Web: www.hiberniaeth.com

HIGHWATER (SCOTLAND) LTD
Winewell, Grantown Road,
Nairn, Highlands, IV12 5QN
Area of Operation: Scotland
Tel: 01667 451009 **Fax:** 01667 451009
Email: highwater@btinternet.com
Web: www.highwatertechnologies.co.uk
Other Info: ✎

HUNTER PLASTICS
Nathan Way, London, SE28 0AE
Area of Operation: Worldwide
Tel: 0208 855 9851 **Fax:** 0208 317 7764
Email: john.morris@hunterplastics.co.uk
Web: www.hunterplastics.co.uk

HYDRAQUIP
Unit 2 Raleigh Court,
Priestley Way, Crawley,
West Sussex, RH10 9PD
Area of Operation: Worldwide
Tel: 01293 615166
Fax: 01293 614965
Email: sales@hydraquip.co.uk
Web: www.hydraquip.co.uk 🖱

INTERFLOW UK
Leighton, Shrewsbury, Shropshire, SY5 6SQ
Area of Operation: UK & Ireland
Tel: 01952 510050
Fax: 01952 510967
Email: villiers@interflow.co.uk
Web: www.interflow.co.uk

IPPEC SYSTEMS LTD
66 Rea Street South, Birmingham,
West Midlands, B5 6LB
Area of Operation: UK & Ireland
Tel: 0121 622 4333
Fax: 0121 622 5768
Email: info@ippec.co.uk
Web: www.ippec.co.uk

JOHN GUEST SPEEDFIT LTD
Horton Road, West Drayton,
Middlesex, UB7 8JL
Area of Operation: Worldwide
Tel: 01895 449233
Fax: 01895 425314
Email: info@johnguest.co.uk
Web: www.speedfit.co.uk

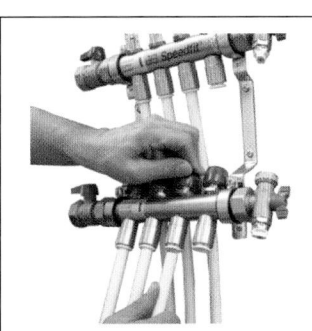

JG SPEEDFIT LTD
Area of Operation: Worldwide
Tel: 01895 449233
Fax: 01895 425314
Email: info@johnguest.co.uk
Web: www.speedfit.co.uk

The Push fit Solution for Plumbing, Heating and
Underfloor Heating.

LOBLITE ELECTRIC LTD
Third Avenue, Team Valley
Trading Estate, Gateshead,
Tyne & Wear, NE11 0QQ
Area of Operation: Europe
Tel: 0191 487 8103
Fax: 0191 491 5541
Email: sales@loblite.co.uk
Web: www.heatec-rads.com

MARLEY PLUMBING & DRAINAGE
Lenham, Maidstone, Kent, ME17 2DE
Area of Operation: Europe
Tel: 01622 858888 **Fax:** 01622 858725
Email: marketing@marleyext.com
Web: www.marley.co.uk

MAYFIELD (MANUFACTURING) LTD
Wenden House, New End, Hemingby,
Horncastle, Lincolnshire, LN9 5QQ
Area of Operation: Worldwide
Tel: 01507 578630
Fax: 01507 578609
Email: info@mayfieldmanufacturing.co.uk
Web: www.aludrain.co.uk

MEKON PRODUCTS
25 Bessemer Park, Milkwood Road,
London, SE24 0HG
Area of Operation: Greater London
Tel: 0207 733 8011 **Fax:** 0207 737 0840
Email: info@mekon.net
Web: www.mekon.net

MISCELLANEA DISCONTINUED BATHROOMWARE
Churt Place Nurseries, Tilford Road,
Churt, Farnham, Surrey, GU10 2LN
Area of Operation: UK (Excluding Ireland)
Tel: 01428 608164
Fax: 01428 608165
Email: email@brokenbog.com
Web: www.brokenbog.com

NR PLUMBING SERVICES
15 Augustine Close, Durham, DH1 5FE
Area of Operation: North East England
Tel: 0191 384 5893
Fax: 0191 384 5893
Email: nrplumbing@aol.com
Web: www.nrplumbing.co.uk

OPELLA LTD
Twyford Road, Rotherwas Industrial Estate,
Hereford, Herefordshire, HR2 6JR
Area of Operation: Worldwide
Tel: 01432 357331
Fax: 01432 264014
Email: sales@opella.co.uk
Web: www.opella.co.uk

SPONSORED BY: JOHN GUEST SPEEDFIT LTD www.speedfit.co.uk

PARAGON SYSTEMS (SCOTLAND) LIMITED
The Office, Corbie Cottage, Maryculter,
Aberdeen, Aberdeenshire, AB12 5FT
Area of Operation: Scotland
Tel: 01224 735536
Fax: 01224 735537
Email: info@paragon-systems.co.uk
Web: www.paragon-systems.co.uk

PLUMBING IMPORTS
Dalton Airfield, Dalton, Thirsk,
North Yorkshire, YO7 3HE
Area of Operation: UK (Excluding Ireland)
Tel: 0845 310 8059
Fax: 01845 577838
Email: sales@plumbingimports.co.uk
Web: www.plumbingimports.co.uk

POLYPIPE KITCHENS & BATHROOMS LTD
Warmsworth, Halt Industrial Estate,
South Yorkshire, DN4 9LS
Area of Operation: Worldwide
Tel: 01709 770990
Fax: 01302 310602
Email: davery@ppbp.co.uk
Web: www.polypipe.com/bk

PULSAR DIRECT LTD
70 High Park Drive,
Mill Park Industrial Estate,
Old Wolverton, Milton Keynes,
Buckinghamshire, MK12 5TT
Area of Operation: UK (Excluding Ireland)
Tel: 0800 298 8701 **Fax:** 0800 298 8702
Email: s.hogg@pulsardirect.co.uk
Web: www.pulsardirect.co.uk

PUMP WORLD LTD
Unit 11, Woodside Road,
South Marston Business Park,
Swindon, Wiltshire, SN3 4WA
Area of Operation: UK & Ireland
Tel: 01793 820142 **Fax:** 01793 823800
Email: enquiries@pumpworld.co.uk
Web: www.pumpworld.co.uk

RADIANT HEATING SOLUTIONS LTD
Mill Farm, Hougham, Grantham,
Lincolnshire, NG32 2HZ
Tel: 01400 250572
Fax: 01400 251264
Email: sales@heating-solutions.biz
Web: www.heating-solutions.biz

RANGEMASTER
Clarence Street, Royal Leamington Spa,
Warwickshire, CV31 2AD
Area of Operation: Worldwide
Tel: 01926 457400
Fax: 01926 450526
Email: consumers@rangemaster.co.uk
Web: www.rangemaster.co.uk

ROCKBOURNE ENVIRONMENTAL
6 Silver Business Park, Airfield Way,
Christchurch, Dorset, BH23 3TA
Area of Operation: UK & Ireland
Tel: 01202 480980
Fax: 01202 490590
Email: info@rockbourne.net
Web: www.rockbourne.net

RYTONS BUILDING PRODUCTS LTD
Design House, Kettering Business Park,
Kettering, Northamptonshire, NN15 6NL
Area of Operation: UK & Ireland
Tel: 01536 511874
Fax: 01536 310455
Email: lit@rytons.com
Web: www.vents.co.uk

SANIFLO
I D A Industrial Estate, Edenderry,
Co Offaly, Edenderry
Area of Operation: Ireland Only
Tel: 046 973 3077
Fax: 046 973 3078
Email: robin@sanirish.com
Web: www.saniflo.ie

SANIFLO LTD
Howard House, The Runway,
South Ruislip, Middlesex, HA4 6SE
Area of Operation: UK (Excluding Ireland)
Tel: 0208 842 4040 **Fax:** 0208 842 1671
Email: andrews@saniflo.co.uk
Web: www.saniflo.co.uk

SMART SHOWERS LTD
Unit 11, Woodside Road, South Marston Business
Park, Swindon, Wiltshire, SN3 4WA
Area of Operation: UK & Ireland
Tel: 01793 822775 **Fax:** 01793 823800
Email: louise@smartshowers.co.uk
Web: www.smartshowers.co.uk

SOLAR TWIN
2nd Floor, 50 Watergate Street,
Chester, Cheshire, CH1 2LA
Area of Operation: UK & Ireland
Tel: 01244 403407
Fax: 01244 403654
Email: oliver@solartwin.com
Web: www.solartwin.com

SURE GB LTD
Unit 3, Century Park, Starley Way,
Bickenhill, Solihull,
West Midlands, B37 7HF
Area of Operation: Worldwide
Tel: 0121 782 5666
Fax: 0121 782 4304
Email: sales@surestop.co.uk
Web: www.surestop.co.uk

TAPS SHOP
35 Bridge Street, Witney,
Oxfordshire, OX28 1DA
Area of Operation: Worldwide
Tel: 0845 430 3035
Fax: 01993 779653
Email: info@tapsshop.co.uk
Web: www.tapsshop.co.uk

TECHFLOW PRODUCTS LTD
Unit 7 Sovereign Business Park,
Albert Drive, Victoria Industrial Estate,
Burgess Hill, West Sussex, RH15 9TY
Area of Operation: UK & Ireland
Tel: 01444 258003
Fax: 01444 258004
Email: rod@techflow.co.uk
Web: www.techflow.co.uk
Other Info: ECO

THE WISEMAN GROUP
PO Box 58, Ingatestone, Essex, CM4 9DL
Area of Operation: Worldwide
Tel: 01277 633200
Fax: 01277 632700
Email: sales@wisemangroup.co.uk
Web: www.wisemangroup.co.uk
Other Info: ECO

TOOLSTATION
Express Park, Bridgwater,
Somerset, TA6 4RN
Area of Operation: UK (Excluding Ireland)
Tel: 0808 100 7211
Fax: 0808 100 7210
Email: info@toolstation.com
Web: www.toolstation.com

UK COPPER BOARD
5 Grovelands Business Centre, Boundary Way,
Hemel Hempstead, Hertfordshire, HP2 7TE
Area of Operation: UK & Ireland
Tel: 01442 275700 **Fax:** 01442 275716
Email: copperboard@copperdev.co.uk
Web: www.ukcopperboard.co.uk

UNDERFLOOR DIRECT LTD
Unit 1 Lisburn Enterprise Centre,
Ballinderry Road, Lisburn, Co Antrim, BT28 2BP
Area of Operation: UK & Ireland
Tel: 02892 634068 **Fax:** 02892 669667
Email: info@keeheating.co.uk
Web: www.underfloordirect.co.uk
Other Info: ✎

UNIFIX LTD
St Georges House, Grove Lane,
Smethwick, Birmingham,
West Midlands, B66 2QT
Area of Operation: Europe
Tel: 0800 096 1110 **Fax:** 0800 096 1115
Email: sales@unifix.co.uk
Web: www.unifix-online.co.uk

**UPONOR HOUSING
SOLUTIONS LTD**
Snapethorpe House, Rugby Road,
Lutterworth, Leicestershire, LE17 4HN
Area of Operation: UK (Excluding Ireland)
Tel: 01455 550355
Fax: 01455 550366
Email: hsenquiries@uponor.co.uk
Web: www.uponorhousingsolutions.co.uk

WATERMILL PRODUCTS LTD
Watermill House,
Fairview Industrial Estate,
Holland Road, Hurst Green,
Oxted, Surrey, RH8 9BD
Area of Operation: UK & Ireland
Tel: 01883 715425
Fax: 01883 716422
Email: sales@watermillshowers.co.uk
Web: www.watermillshowers.co.uk

YORKSHIRE FITTINGS
PO Box 166, Leeds,
West Yorkshire, LS10 1NA
Tel: 0113 270 1104
Fax: 0113 271 5275
Email: info@yorkshirefittings.co.uk
Web: www.yorkshirefittings.co.uk

WATER TANKS & CYLINDERS

KEY

PRODUCT TYPES: 1= Vented Hot Water
Cylinders 2 = Unvented Hot Water Cylinders
3 = Tanks 4 = Thermal Stores 5 = Other

OTHER: ▽ Reclaimed On-line shopping
Bespoke Hand-made ECO Ecological

ACV UK LTD
St Davids Business Park,
Dalgety Bay, Fife, KY11 9PF
Area of Operation: UK & Ireland
Tel: 01383 820100
Fax: 01383 820180
Email: information@acv-uk.com
Web: www.acv-uk.com
Product Type: 1, 2, 4

ALBION WATER HEATERS
Shelah Road, Halesowen, West Midlands, B63 3PG
Area of Operation: UK & Ireland
Tel: 0121 585 5151
Fax: 0121 501 3826
Email: geoff.egginton@albionwaterheaters.com
Web: www.albionwaterheaters.com
Product Type: 1, 2, 4

BALMORAL TANKS
Balmoral Park, Loirston, Aberdeen,
Aberdeenshire, AB12 3GY
Area of Operation: Worldwide
Tel: 01224 859100
Fax: 01224 859123
Email: s.gibb@balmoral.co.uk
Web: www.balmoraltanks.com
Product Type: 3

BTU HEATING
38 Weyside Road, Guildford, Surrey, GU1 1JB
Area of Operation: South East England
Tel: 01483 590600
Fax: 01483 590601
Email: enquiries@btu-heating.com
Web: www.btu-group.com
Product Type: 1, 2, 3, 4

COPPERJOB LTD
25 Wain Park, Plympton, Plymouth, Devon, PL7 2HX
Area of Operation: UK & Ireland
Tel: 01752 339234
Email: john@copperjob.com
Web: www.centralheating.co.uk
Product Type: 1, 2, 3, 4

COSY ROOMS (COSY-HEATING.CO.UK)
17 Chiltern Way, North Hykeham, Lincoln,
Lincolnshire, LN6 9SY
Area of Operation: UK (Excluding Ireland)
Tel: 01522 696002
Fax: 01522 696002
Email: keith@cosy-rooms.com
Web: www.cosy-heating.co.uk
Product Type: 1, 3

ECO HOMETEC UK LTD
Unit 11E, Carcroft Enterprise Park, Carcroft,
Doncaster, South Yorkshire, DN6 8DD
Area of Operation: Europe
Tel: 01302 722266
Fax: 01302 728634
Email: Stephen@eco-hometec.co.uk
Web: www.eco-hometec.co.uk
Product Type: 1, 2, 4

ENERGY MASTER
Keltic Business Park, Unit 1
Clieveragh Industrial Estate, Listowel, Ireland
Area of Operation: Ireland Only
Tel: +353 68 24300
Fax: +353 68 24533
Email: info@energymaster.ie
Web: www.energymaster.ie
Product Type: 1, 2, 3, 5

EURO BATHROOMS
102 Annareagh Road, Richhill,
Co Armagh, BT61 9JY
Area of Operation: Ireland Only
Tel: 028 3887 9996
Fax: 028 3887 9996
Email: paul@euro-bathrooms.co.uk
Web: www.euro-bathrooms.co.uk
Product Type: 1, 3, 4

FABDEC
Grange Road, Ellesmere,
Shropshire, SY12 9DG
Area of Operation: UK & Ireland
Tel: 01691 627200
Fax: 01691 627222
Email: gavin.watson@fabdec.com
Web: www.fabdec.com
Product Type: 2

FREERAIN
Millennium Green Business Centre, Rio Drive,
Collingham, Nottinghamshire, NG23 7NB
Area of Operation: UK & Ireland
Tel: 01636 894900
Fax: 01636 894909
Email: info@freerain.co.uk
Web: www.freerain.co.uk
Product Type: 3

GAH HEATING PRODUCTS
Melton Road, Melton, Woodbridge,
Suffolk, IP12 1NH
Area of Operation: UK (Excluding Ireland)
Tel: 01394 386606
Fax: 01394 386609
Email: info@dualstream.co.uk
Web: www.gah.co.uk
Product Type: 2

GREENSHOP SOLAR LTD
Units 23/24, Merretts Mill Industrial Centre,
Woodchester, Stroud, Gloucestershire, GL5 5EX
Area of Operation: UK (Excluding Ireland)
Tel: 01452 772030
Fax: 01452 770115
Email: eddie@greenshop.co.uk
Web: www.greenshop-solar.co.uk
Product Type: 4
Other Info: ECO

HEATRAE SADIA HEATING
Hurricane Way, Norwich, Norfolk, NR6 6EA
Area of Operation: Worldwide
Tel: 01603 420110 **Fax:** 01603 420149
Email: sales@heatraesadia.com
Web: www.heatraesadia.com **Product Type:** 2

HRM BOILERS LTD
Haverscroft Industrial Estate,
Attleborough, Norfolk, NR17 1YE
Area of Operation: UK (Excluding Ireland)
Tel: 01953 455400 **Fax:** 01953 454483
Email: info@hrmboilers.co.uk
Web: www.hrmboilers.co.uk
Product Type: 1

JOHNSON & STARLEY
Rhosili Road, Brackmills, Northampton,
Northamptonshire, NN4 7LZ
Area of Operation: UK & Ireland
Tel: 01604 762881 **Fax:** 01604 767408
Email: marketing@johnsonandstarleyltd.co.uk
Web: www.johnsonandstarley.co.uk
Product Type: 2

KESTON BOILERS
34 West Common Road, Hayes,
Bromley, Kent, BR2 7BX
Area of Operation: UK & Ireland
Tel: 0208 462 0262
Fax: 0208 462 4459
Email: info@keston.co.uk
Web: www.keston.co.uk **Product Type:** 2

POLYPIPE KITCHENS & BATHROOMS LTD
Warmsworth, Halt Industrial Estate,
South Yorkshire, DN4 9LS
Area of Operation: Worldwide
Tel: 01709 770990
Fax: 01302 310602
Email: davery@ppbp.co.uk
Web: www.polypipe.com/bk
Product Type: 3

POLYTANK LTD
Naze Lane East, Freckleton,
Preston, Lancashire, PR4 1UN
Area of Operation: UK & Ireland
Tel: 01772 632850 **Fax:** 01772 679615
Email: sales@polytank.co.uk
Web: www.polytank.co.uk
Product Type: 3

PUMP WORLD LTD
Unit 11, Woodside Road, South Marston Business
Park, Swindon, Wiltshire, SN3 4WA
Area of Operation: UK & Ireland
Tel: 01793 820142 **Fax:** 01793 823800
Email: enquiries@pumpworld.co.uk
Web: www.pumpworld.co.uk
Product Type: 2, 3

ROCKBOURNE ENVIRONMENTAL
6 Silver Business Park, Airfield Way,
Christchurch, Dorset, BH23 3TA
Area of Operation: UK & Ireland
Tel: 01202 480980 **Fax:** 01202 490590
Email: info@rockbourne.net
Web: www.rockbourne.net
Product Type: 3

SOUTHERN SOLAR
Unit 6, Allington Farm, Allington Lane,
Offham, Lewes, East Sussex, BN7 3QL
Area of Operation: Greater London, South East
England, South East England and South Wales
Tel: 0845 456 9474 **Fax:** 01273 483928
Email: info@southernsolar.co.uk
Web: www.southernsolar.co.uk
Product Type: 1, 2, 3, 4

TITAN PLASTECH LIMITED
Barbot Hall Industrial Estate, Mangham Road,
Rotherham, South Yorkshire, S61 4RJ
Area of Operation: Europe
Tel: 01709 538300 **Fax:** 01709 538301
Email: tony.soper@titantanks.co.uk
Web: www.titanplastech.co.uk
Product Type: 2, 3

WATERMILL PRODUCTS LTD
Watermill House, Fairview Industrial Estate,
Holland Road, Hurst Green, Oxted, Surrey, RH8 9BD
Area of Operation: UK & Ireland
Tel: 01883 715425
Fax: 01883 716422
Email: sales@watermillshowers.co.uk
Web: www.watermillshowers.co.uk
Product Type: 5

WHIRLPOOL EXPRESS (UK) LTD
61-62 Lower Dock Street, Kingsway,
Newport, NP20 1EF
Area of Operation: Europe
Tel: 01633 244555
Fax: 01633 244555
Email: reception@whirlpoolexpress.co.uk
Web: www.whirlpoolexpress.co.uk ⟨⌃⟩

WATER TREATMENT

KEY

PRODUCT TYPES: 1= Chlorination
2 = De-alkalisation   3 = Demineralisation
4 = Descaling   5 = Filtration
6 = Water Softening 7 = Other

OTHER: ▽ Reclaimed ⟨⌃⟩ On-line shopping
✎ Bespoke ✋ Hand-made ECO Ecological

AQUA CLEAR WATER TREATMENT
Ballinahinch, Knocklong, Co.Limerick, Ireland
Area of Operation: Ireland Only
Tel: +353 (0)62 53482
Fax: +353-(0)62-53482
Email: aquacq@eircom.net
Web: www.aquaclear.ie
Product Type: 1, 2, 3, 4, 5, 6, 7

AQUA CURE PLC
Aqua Cure House, Hall Street,
Southport, Merseyside, PR9 0SE
Area of Operation: UK (Excluding Ireland)
Tel: 01704 516916
Fax: 01704 544916
Email: sales@aquacure.plc.uk
Web: www.aquacure.co.uk
Product Type: 1, 2, 3, 5, 6, 7
Other Info: ✎

**BROOK-WATER DESIGNER
BATHROOMS & KITCHENS**
The Downs, Woodhouse Hill, Uplyme,
Lyme Regis, Dorset, DT7 3SL
Area of Operation: Worldwide
Tel: 01235 201256
Email: sales@brookwater.co.uk
Web: www.brookwater.co.uk ⟨⌃⟩
Product Type: 6

CISTERMISER LTD
Unit 1 Woodley Park Estate, 59-69 Reading Road,
Woodley, Reading, Berkshire, RG5 3AN
Area of Operation: UK & Ireland
Tel: 0118 969 1611 **Fax:** 0118 944 1426
Email: sales@cistermiser.co.uk
Web: www.cistermiser.co.uk
Product Type: 4

CLEANWATER SOUTH WEST LTD
Foxfield, Welcombe, Bideford, Devon, EX39 6HF
Area of Operation: South East England,
South West England and South Wales
Tel: 01288 331561 **Fax:** 01288 331561
Email: sales@cleanwatersw.co.uk
Web: www.cleanwatersw.co.uk
Product Type: 7

COPPERJOB LTD
25 Wain Park, Plympton, Plymouth, Devon, PL7 2HX
Area of Operation: UK & Ireland
Tel: 01752 339234
Email: john@copperjob.com
Web: www.centralheating.co.uk ⟨⌃⟩
Product Type: 4

ECP GROUP
103 Burrell Road, Ipswich, Suffolk, IP2 8AD
Area of Operation: East England,
South East England
Tel: 01473 400101 **Fax:** 01473 400103
Email: sales@ecpgroup.com
Web: www.ecpgroup.com
Product Type: 1, 5, 6, 7

FAST SYSTEMS LTD
Dalton House, Newtown Road, Henley on Thames,
Oxfordshire, RG9 1HG
Area of Operation: East England, Greater London,
South East England, South West England
and South Wales
Tel: 01491 419200
Fax: 01491 419201
Email: paul@fsltd.co.uk
Web: www.scalewatcher.co.uk ⟨⌃⟩
Product Type: 4

GAH HEATING PRODUCTS
Melton Road, Melton, Woodbridge, Suffolk, IP12 1NH
Area of Operation: UK (Excluding Ireland)
Tel: 01394 386606 **Fax:** 01394 386609
Email: info@dualstream.co.uk
Web: www.gah.co.uk
Product Type: 6

GREEN ROCK
3 Elmhurst Road, Harwich, Essex, CO12 3SA
Area of Operation: UK (Excluding Ireland)
Tel: 01255 554055 **Fax:** 01255 554055
Web: www.greenrock.fi

H2ONICS
Anode House, Unit 14 Berkeley Crescent,
Frimley, Camberley, Surrey, GU16 8YN
Area of Operation: Worldwide
Tel: 0800 298 5031 **Fax:** 07000 353683
Email: enquiries@h2onics.co.uk
Web: www.h2onics.co.uk ⟨⌃⟩
Product Type: 4, 5, 6, 7

HIGHWATER (SCOTLAND) LTD
Winewell, Grantown Rd, Nairn, Highlands, IV12 5QN
Area of Operation: Scotland
Tel: 01667 451000 **Fax:** 01667 451009
Email: highwater@btinternet.com
Web: www.highwatertechnologies.co.uk
Product Type: 1, 3, 5, 6
Other Info: ✎

KINETICO UK LTD
Bridge House, Park Gate Business Centre,
Park Gate, Hampshire, SO31 1FQ
Area of Operation: UK (Excluding Ireland)
Tel: 01489 566970 **Fax:** 01489 566976
Email: info@kinetico.co.uk
Web: www.kinetico.co.uk
Product Type: 2, 3, 4, 5, 6

LIFESCIENCE PRODUCTS LTD
185 Milton Park, Abingdon, Oxfordshire, OX20 1BY
Area of Operation: Worldwide
Tel: 01235 832111
Fax: 01235 832129
Email: sales@lifescience.co.uk
Web: www.lifescience.co.uk ⟨⌃⟩
Product Type: 4, 5, 6

SALAMANDER ENGINEERING LTD
24 Reddicap Trading Estate, Sutton Coldfield,
West Midlands, B75 7BU
Area of Operation: Worldwide
Tel: 0121 378 0952 **Fax:** 0121 3111521
Email: enquiries@salamander-engineering.co.uk
Web: www.salamander-engineering.co.uk
Product Type: 4, 5, 6, 7

THE SOFT WATER SHOP
54 Westfield Road, Harpenden,
Hertfordshire, AL5 4HW
Area of Operation: UK (Excluding Ireland)
Tel: 01582 461313 **Fax:** 01582 461313
Email: softwatershop@ntlworld.com
Web: www.softwatershop.co.uk
Product Type: 1, 2, 3, 4, 5, 6, 7

THE WISEMAN GROUP
PO Box 58, Ingatestone, Essex, CM4 9DL
Area of Operation: Worldwide
Tel: 01277 633200 **Fax:** 01277 632700
Email: sales@wisemangroup.co.uk
Web: www.wisemangroup.co.uk
Product Type: 2, 4, 6

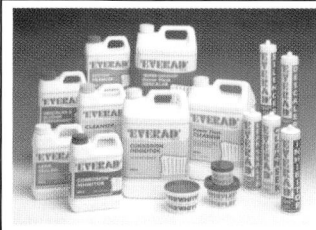

THE WISEMAN GROUP

Area of Operation: Worldwide
Tel: 01277 633200 **Fax:** 01277 632700
Email: sales@wisemangroup.co.uk
Web: www.wisemangroup.co.uk
Product Type: 2,4,6

The Wiseman group manufacture and supply a
large range of chemicals specifically aimed at the
homebuilder. Whether you need fire-rated
expanding foam or a corrosion inhibitor for your
heating system view the whole range at
www.wisemangroup.co.uk or
call 01277 633200 for more information.

NOTES

Company Name ....
....
Address ....
....
....
email ....
Web ....

Company Name ....
....
Address ....
....
....
email ....
Web ....

Company Name ....
....
Address ....
....
....
email ....
Web ....

HEATING, PLUMBING & ELECTRICAL

Essential ideas and information for your self-build or renovation

The entire archives of Homebuilding & Renovating issues from 2002, 03, 04 or 05 are now available on easy-to-use double CDs.

All pages and pictures appear just as they do in the magazine (to view as pdfs through Acrobat Reader). Easy to navigate and fully text searchable. All supplier information and contacts are available as web and email links. Clickable contents pages.

£24.99 each

22 Brilliant Barn Conversions

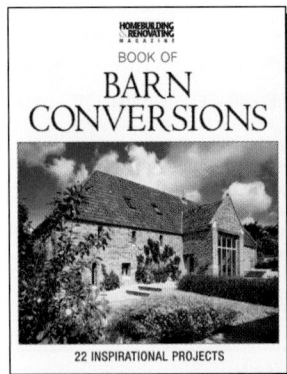

Project to suit all budgets, from £15,000 upwards. Fully costed and detailed project descriptions. Plus contact details for suppliers and craftsmen for each project.

"Beautiful barn restorations, plus the nitty gritty of how they were done. Makes you want to have a go."
Paul Carslake, Author,
How to Renovate a House in France

ONLY £15

24 High-quality, Low-budget Homes

A collection of inspiring self-built homes from £32,000 up to £150,000. They all show how it is possible for you to achieve a spacious family home in any style on a budget. How to maximise usable floorspace and at the same time create a unique family home.

ONLY £14.95

19 Inspirational Individually Designed Homes

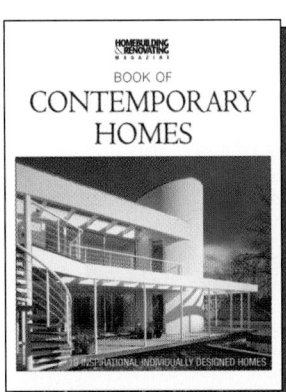

An invaluable source of inspiration for anyone planning to build. Remarkable projects built from £60,000 to £1 million plus.

"If you dream of designing and building a contemporary style home then this book is definitely for you."
Michael Holmes, Editor-in-Chief
Homebuilding & Renovating Magazine

ONLY £14.95

How to Renovate a House in France

An essential guide to help you turn an old rural property in France into a beautiful home. 'How to Renovate' covers the whole process, from assessing and buying a property, through all jobs, large and small, required to get it into shape.

ONLY £25

NEW - How to Create a Jardin Paysan

Written by Louise Ranck, an expert on traditional French rural gardens. With plant lists, tips and techniques going back generations, it shows you how to create an authentic, natural garden and the perfect backdrop for a home in the country.

ONLY £25

BUY How to Renovate a House in France and How to Create a Jardin Paysan at the same time **FOR JUST £40 SAVE £10** (RRP £25 each)

ORDER YOUR COPY:
VISIT: www.homebuilding.co.uk or CALL: 01527 834435

£5 OFF any of the above books
when you subscribe to *Homebuilding & Renovating* magazine

Postage and packing charges apply to all of the above, call or see website for more details. Details and prices are correct at time of going to press but may be subject to change.

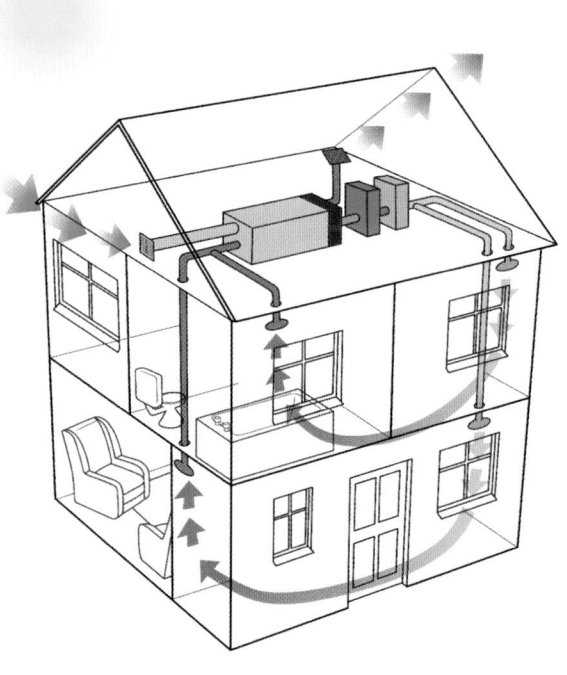

Image courtesy of Villavent Ltd (01993 778481)

Villavent® Comfort Conditioning™
Villavent's revolutionary new package embracing energy efficiency, cooling and clean living.

HEATING, PLUMBING & ELECTRICAL

SPONSORED BY VILLAVENT LTD
Tel 01993 778481. Web www.villavent.co.uk

Ventilation

Adequate ventilation, the replacement of stale indoor air with fresh outside air, is required under the statutory Building Regulations. It is seen as essential for controlling moisture levels and indoor air pollutants, such as the droppings from house dust mites, tobacco smoke and volatile organic compounds (VOCs).

ABOVE: Image from Fantasia Ceiling Fans

Which solution is right for your project will depend on how much you are willing to spend, the system maintenance regime acceptable to you, the relative importance of energy efficiency and ecology, aesthetics, and whether or not you want to filter, pre-warm or cool the incoming fresh air. With new build, it will also depend on how energy efficient and airtight a house you are building, as this will set the effectiveness of features such as heat recovery.

If you are building a new house to sell in the near future, or are on a tight budget, the best option is likely to be the cheapest and simplest and this will usually mean individual extractor fans in wet areas and passive vents, or a positive input ventilation system (PIV). If you are building an ecological house with a very low heat requirement, then your choice is likely to be between a passive stack system that uses no direct electricity, a PIV system, and a heat recovery ventilation system (HRV). A new option is solar panels, which heat air rather than water, to pre-warm incoming fresh air, diverting to heat the hot water cylinder when heating is not required. If you are building or renovating a high value property, then you might consider a system that combines air conditioning with ventilation. If you do not want trickle vents in windows for aesthetic reasons, then you should consider a positive input ventilation system, or a mechanical whole-house system that operates continuously. If you require air filtration, for instance to control allergens, a mechanical system with controlled air intake is the best option, although positive input ventilation systems can also filter incoming air.

For a typical small renovation project with a single bathroom, the most cost-effective option will be to go for a single wall or ceiling mounted extract fan. For a larger renovation, a single centrally located positive input ventilation system is likely to be more cost-effective than two, or more, individual extract fans, as this will also do away with the need for, and cost of, trickle vents or airbricks other than those required under Part J for combustion air supply.

Whichever system you opt for, you must include rapid ventilation such as opening windows to all habitable rooms. These must be of a minimum size dictated by the Building Regulations. You may also decide to provide direct extraction above the kitchen hob. This can be combined into a whole-house system, but it is considered best to keep this independent from the rest of the house ducting due to the high levels of grease and moisture. It is also still necessary to ventilate the roof structure and any voids beneath suspended floors and ceilings under Part C of the Building Regulations, and to provide ventilation for any open fires or stoves under Part J.

Image by Velfac

AIR CONDITIONING

KEY

OTHER: ▽ Reclaimed 🖱 On-line shopping

✏ Bespoke ✋ Hand-made ECO Ecological

ADM
Ling Fields, Gargrave Road, Skipton,
North Yorkshire, BD23 1UX
Area of Operation: Europe
Tel: 01756 701051 **Fax:** 01756 701076
Email: info@admsystems.co.uk
Web: www.admsystems.co.uk
Other Info: ✏

AIRCONWAREHOUSE.COM
Unit 2 Chichester Road, Romiley, Cheshire, SK6 4BL
Area of Operation: UK (Excluding Ireland)
Tel: 0161 430 7878 **Fax:** 0161 430 7979
Email: info@airconwarehouse.com
Web: www.airconwarehouse.com
Other Info: ECO ✏

ALLERGYPLUS LIMITED
65 Stowe Drive, Southam, Warwickshire, CV47 1NZ
Area of Operation: UK & Ireland
Tel: 0870 190 0022 **Fax:** 0870 190 0044
Email: info@allergyplus.co.uk
Web: www.allergyplus.co.uk 🖱

ATB AIR CONDITIONING AND HEATING
67 Melloway Road, Rushden, Northants, NN10 6XX
Area of Operation: East England, Greater London,
Midlands & Mid Wales, South East England
Tel: 0870 260 1650 **Fax:** 01933 411731
Email: enquiries@atbairconditioning.co.uk
Web: www.atbairconditioning.co.uk

BIDDLE AIR SYSTEMS
St Marys Road, Nuneaton, Warwickshire, CV11 5AU
Area of Operation: Worldwide
Tel: 024 7638 4233 **Fax:** 024 7637 3621
Email: biddle@biddle-air.co.uk
Web: www.biddle-air.co.uk

BTU HEATING
38 Weyside Road, Guildford, Surrey, GU1 1JB
Area of Operation: South East England
Tel: 01483 590600 **Fax:** 01483 590601
Email: enquiries@btu-heating.com
Web: www.btu-group.com

CELSIUS AIR CONDITIONING
1 Well Street, Heywood, Lancashire, OL10 1NT
Area of Operation: UK & Ireland
Tel: 01706 367500 **Fax:** 01706 367355
Email: sales@celsiusair.co.uk
Web: www.celsiusair.co.uk

COMFORT AIR CONDITIONING LTD
Comfort Works, Newchapel Road,
Lingfield, Surrey, RH7 6LE
Area of Operation: East England, Greater London,
Midlands & Mid Wales, South East England, South
West England and South Wales
Tel: 01342 830600 **Fax:** 01342 830605
Email: info@comfort.uk.com
Web: www.comfort.uk.com
Other Info: ECO ✏

**CRYOTEC REFRIGERATION
& AIR CONDITIONING**
Unit 4, Wolf Business Park, Alton Road,
Ross on Wye, Herefordshire, HR9 5NB
Area of Operation: South West England
and South Wales
Tel: 0800 389 2369 **Fax:** 01989 764401
Email: info@cryotec.co.uk
Web: www.cryotec.co.uk

DANFOSS RANDALL LTD
Ampthill Road, Bedford, Bedfordshire, MK42 9ER
Area of Operation: UK & Ireland
Tel: 0845 121 7400 **Fax:** 0845 121 7515
Email: danfossrandall@danfoss.com
Web: www.danfoss-randall.co.uk

DD HEATING LTD
16-19 The Manton Centre, Manton Lane,
Bedford, Bedfordshire, MK41 7PX
Area of Operation: Worldwide
Tel: 0870 777 8323 **Fax:** 0870 777 8320
Email: info@heatline.co.uk
Web: www.heatline.co.uk 🖱

DIFFUSION ENVIRONMENTAL SYSTEMS
47 Central Avenue, West Molesey, Surrey, KT8 2QZ
Area of Operation: UK (Excluding Ireland)
Tel: 0208 783 0033
Fax: 0208 783 0140
Email: diffusion@etenv.co.uk
Web: www.energytechniqueplc.co.uk

DRY-IT-OUT LIMITED
The Cwm, Churchstoke, Montgomery,
Powys, SY15 6TJ
Area of Operation: Europe
Tel: 0870 011 7987
Fax: 01588 620145
Email: enquiries@dry-it-out.com
Web: www.dry-it-out.com 🖱

EU SOLUTIONS
Maghull Business Centre, 1 Liverpool Road North,
Maghull, Liverpool, L31 2HB
Area of Operation: UK & Ireland
Tel: 0870 160 1660 **Fax:** 0151 526 8849
Email: ths@blueyonder.co.uk
Web: www.totalhomesolutions.co.uk
Other Info: ECO

GEA
GEA Air Treatment Division, Sudstrasse 48,
D-44625, Hern, Germany
Area of Operation: Europe
Tel: 0800 783 0073
Email: info@gea-acqua.com
Web: www.gea-acqua.com

HAMPTON
PO Box 6074, Newbury, Berkshire, RG14 9AF
Area of Operation: UK & Ireland
Tel: 01635 25440
Fax: 01635 255762
Email: fieldenj@freeuk.co.uk /
john.fielden@btconnect.com
Web: www.hamptonventilation.com
Other Info: ✏

HONEYWELL CONTROL SYSTEMS LTD
Honeywell House, Arlington Business Park,
Bracknell, Berkshire, RG12 1EB
Area of Operation: Worldwide
Tel: 01344 656000
Fax: 01344 656054
Email: literature@honeywell.com
Web: www.honeywelluk.com

INDOOR CLIMATE SOLUTIONS
10 Cryersoak Close, Monkspath,
Solihull, West Midlands, B90 4UW
Area of Operation: Midlands & Mid Wales
Tel: 0870 011 5432 **Fax:** 0870 011 5432
Email: info@ics-aircon.com
Web: www.ics-aircon.com

INTERACTIVE AIR LTD
Wolfelands, High Street,
Westerham, Kent, TN16 1RQ
Area of Operation: Worldwide
Tel: 01959 565959 **Fax:** 01959 569933
Email: info@interavtiveair.com
Web: www.interactiveair.com
Other Info: ✏

JOHNSON & STARLEY
Rhosili Road, Brackmills, Northampton,
Northamptonshire, NN4 7LZ
Area of Operation: UK & Ireland
Tel: 01604 762881
Fax: 01604 767408
Email: marketing@johnsonandstarleyltd.co.uk
Web: www.johnsonandstarley.co.uk

SHIRES TECHNICAL SERVICES
57-63 Lea Road, Northampton,
Northamptonshire, NN1 4PE
Area of Operation: UK (Excluding Ireland)
Tel: 01604 472525
Fax: 01604 473837
Email: enquiries@shireservices.com
Web: www.shireservices.com

SPACE AIR SOLUTIONS LTD
Willway Court, 1 Opus Park, Moorfield Road,
Guildford, Surrey, GU1 1SZ
Area of Operation: UK (Excluding Ireland)
Tel: 01483 504883
Fax: 01483 574835
Email: solutions@spaceair.co.uk
Web: www.homeairsolutions.com

THE AIR CONDITIONING COMPANY
Summit House, 40 Highgate West Hill,
Greater London, N6 6LS
Area of Operation: Europe
Tel: 0208 340 8000
Fax: 0208 341 0365
Email: info@airconco.com
Web: www.airconco.com 🖱
Other Info: ✏

THE HEAT PUMP COMPANY UK LIMITED
29 Claymoor Park, Booker, Nr. Marlow,
Buckinghamshire, SL7 3DL
Area of Operation: UK (Excluding Ireland)
Tel: 01494 450154 **Fax:** 01494 458159
Email: sales@npsair.com
Web: www.npsair.com
Other Info: ✏

TLC ELECTRICAL WHOLESALERS
TLC Building, Off Fleming Way, Crawley,
West Sussex, RH10 9JY
Area of Operation: Worldwide
Tel: 01293 565630
Fax: 01293 425234
Email: sales@tlc-direct.co.uk
Web: www.tlc-direct.co.uk 🖱

UNICO SYSTEM INTERNATIONAL LIMITED
Unit 3, Ynyshir Industrial Estate, Llanwonno Road,
Porth, Rhondda Cynon Taff, CF39 0HU
Area of Operation: Worldwide
Tel: 01443 684828
Fax: 01443 684838
Email: scott@unicosystem.com
Web: www.unicosystem.co.uk

VENT AXIA
Fleming Way, Crawley,
West Sussex, RH10 9YX
Area of Operation: UK (Excluding Ireland)
Tel: 01293 526062
Fax: 01293 552375
Email: sales@vent-axia.com
Web: www.vent-axia.com

VORTICE LTD
Beeches House, Eastern Avenue,
Burton on Trent, Staffordshire, DE13 0BB
Area of Operation: Worldwide
Tel: 01283 492949 **Fax:** 01283 544121
Email: sales@vortice.ltd.uk
Web: www.vortice.ltd.uk 🖱
Other Info: ECO

WAGNER (GB) LTD
VBH House, Bailey Drive, Gillingham
Business Park, Gillingham, Kent, ME8 0WG
Area of Operation: Worldwide
Tel: 01634 263263 **Fax:** 01634 263504
Email: sales@wagnergb.com
Web: www.wagnergb.com 🖱

VENTILATION & HEAT RECOVERY

KEY

PRODUCT TYPES: 1= Mechanical 2 = Passive

OTHER: ▽ Reclaimed 🖱 On-line shopping

✏ Bespoke ✋ Hand-made ECO Ecological

ADM
Ling Fields, Gargrave Road, Skipton,
North Yorkshire, BD23 1UX
Area of Operation: Europe
Tel: 01756 701051 **Fax:** 01756 701076
Email: info@admsystems.co.uk
Web: www.admsystems.co.uk
Product Type: 1, 2 **Other Info:** ▽ ✏ ECO

AIRCONWAREHOUSE.COM
Unit 2 Chichester Rd, Romiley, Cheshire, SK6 4BL
Area of Operation: UK (Excluding Ireland)
Tel: 0161 430 7878 **Fax:** 0161 430 7979
Email: info@airconwarehouse.com
Web: www.airconwarehouse.com
Product Type: 1 **Other Info:** ECO ✏

AIRFLOW DEVELOPMENTS LTD
Lancaster Road, Cressex Business Park,
High Wycombe, Buckinghamshire, HP12 3QP
Area of Operation: Worldwide
Tel: 01494 525252 **Fax:** 01494 461073
Email: info@airflow.co.uk
Web: www.iconfan.co.uk 🖱

ALLERGYPLUS LIMITED
65 Stowe Drive, Southam, Warwickshire, CV47 1NZ
Area of Operation: UK & Ireland
Tel: 0870 190 0022 **Fax:** 0870 190 0044
Email: info@allergyplus.co.uk
Web: www.allergyplus.co.uk 🖱
Product Type: 1, 2 **Other Info:** ✏

HEATING, PLUMBING & ELECTRICAL

Villavent Ltd are the brand leaders in whole house environmental solutions. With over 40 years experience in Scandinavia and 20 here in the UK we have both the product quality and expertise to provide an array of services and products to provide a 'one stop' shop, bespoke service to the marketplace.

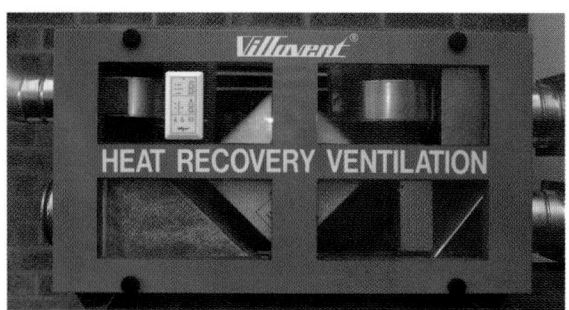

With 20 years experience here in the UK, Villavent strive to provide what we all expect – 'value for money'. It is our experience with what is generally regarded as the quality product in the marketplace and we back this up with an installation service, expert advice, good levels of guarantee, extended warranty packages and a clear informative website. As we move forward we can already boost the National Installation Service – a first direct from a manufacturer and a DIY installation booklet for those individuals keen to do it themselves.

Villavent are part of a large European network within the ventilation industry ensuring we remain strong and competitive within this field in a growing international market, and can call upon manufacture from Germany, Norway, Sweden and Canada.

Our product quality is second to none with manufacture to ISO 9002 and the environmental standard 140001.

As the systems are generally a whole house installation complete with piping and ducts the design is of paramount importance with any installation.

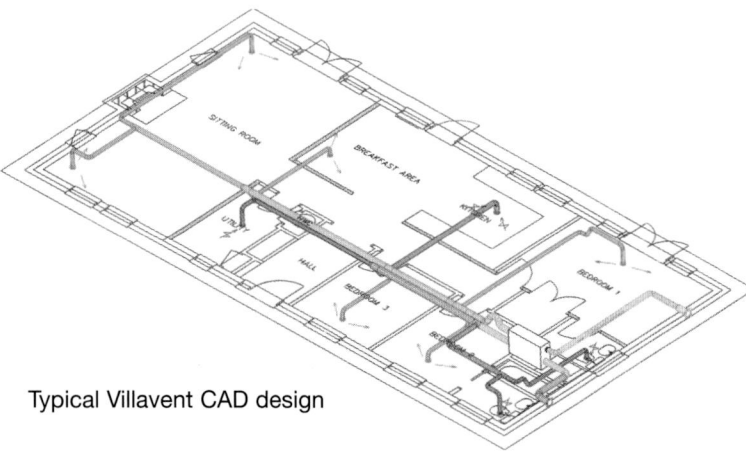

Typical Villavent CAD design

Villavent operate a CAD system, which is drawn in 4, colours and is very clear in its imaging. Generally we provide 3 issues of drawings: For Comment, For Approval and For Construction once all comments and notes have been discussed and agreed. An accurate materials list can then be called off to ensure a smooth installation.

Villavent offer a 2 year guarantee date of invoice and covers all parts. Villavent offer a further 3 year extended warranty, which can be taken out within 12 months of the original invoice. This can be paid monthly, annually by cheque or credit card or up front for 3 years with 5% discount.

Villavent has its own service and maintenance engineers and access to a national network of service engineers. All service queries should be directed to 01993 772270.

So how can we be of assistance to you?

As detailed on the following pages we specialise in:-

- **Central Extract Ventilation**
- **Heat Recovery Ventilation**
- **Comfort Cooling**
- **Villavent® Comfort Conditioning™**
- **Central Vacuum**

Villavent Ltd
Avenue 2, Station Lane Industrial Est,
Witney, Oxon OX28 4YL
Tel: 01993 778481
Fax: 01993 779962
Email: sales@villavent.co.uk
Web: www.villavent.co.uk

on products supplied with LM12-85/EU7 filter

Central Extract

Whole house ventilation from one fan unit providing quiet unobtrusive ventilation with only one external grill and one electrical point.

Villavent
Tel: 01993 778481
Fax: 01993 779962
Email: sales@villavent.co.uk
Web: www.villavent.co.uk

Comfort Cooling

Simplified unit to provide 4kw of cooling at a very modest cost, to the whole house. Can also be retrofitted to the ventilation.

Villavent
Tel: 01993 778481
Fax: 01993 779962
Email: sales@villavent.co.uk
Web: www.villavent.co.uk

Central Vac

One, stationary unit enabling easier more efficient cleaning whilst healthier and no unit noise in the room being cleaned.

Villavent
Tel: 01993 778481
Fax: 01993 779962
Email: sales@villavent.co.uk
Web: www.villavent.co.uk.

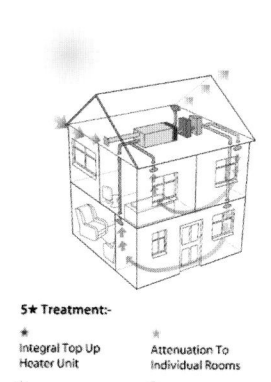

5★ Treatment:-

★ Integral Top Up Heater Unit
★ Attenuation To Individual Rooms
★ Electrostatic Filtration
★ 80% and 90% Heat Recovery Efficiency
★ Comfort Cooling

Villavent® Comfort Conditioning™

The ability to choose the level of comfort within your home.

Heat Recovery – VX400EV

Loft mounted **70% efficient** Heat Recovery unit.

Villavent
Tel: 01993 778481
Fax: 01993 779962
Email: sales@villavent.co.uk
Web: www.villavent.co.uk

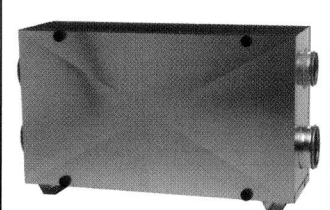

Heat Recovery – VX400E

Wall mounted **70% efficient** Heat Recovery unit.

Villavent
Tel: 01993 778481
Fax: 01993 779962
Email: sales@villavent.co.uk
Web: www.villavent.co.uk

Heat Recovery – VM400

Wall or loft mounted **90% efficient** Heat Recovery unit.

Villavent
Tel: 01993 778481
Fax: 01993 779962
Email: sales@villavent.co.uk
Web: www.villavent.co.uk

Heat Recovery – VR400

Wall, ceiling or loft mounted **80% efficient** Heat Recovery unit with rotary wheel and automatic summer by-pass.

Villavent
Tel: 01993 778481
Fax: 01993 779962
Email: sales@villavent.co.uk
Web: www.villavent.co.uk

Retirement Housing

Heat Recovery ventilation to each luxury flat.

Villavent
Tel: 01993 778481
Fax: 01993 779962
Email: sales@villavent.co.uk
Web: www.villavent.co.uk

Health Care

Abingdon Health Care
Heat Recovery ventilation to all areas.

Villavent
Tel: 01993 778481
Fax: 01993 779962
Email: sales@villavent.co.uk
Web: www.villavent.co.uk.

Student Accomodation

Magdalen College - Oxford

Villavent
Tel: 01993 778481
Fax: 01993 779962
Email: sales@villavent.co.uk
Web: www.villavent.co.uk

Residential Development

Hargreaves Homes - West Sussex.
Family house development - South coast.

Villavent
Tel: 01993 778481
Fax: 01993 779962
Email: sales@villavent.co.uk
Web: www.villavent.co.uk

So what's the best thing about REGAVENT - the big noise in HRV?

You can't <u>hear</u> it!

Send your plans for FREE DESIGN & QUOTE!

It's True! Rega's unique WhisperFlow™ technology ensures that even when operating at full 'Boost' mode (during the morning bathroom rush hour!) the system remains whisper quiet, unobtrusively, yet effectively, venting stale and damp air and replacing it with air that's gently warmed and filtered.

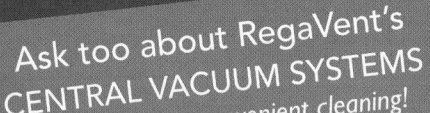

- Powerful AC or DC electric motors
- Fully programmable operation
- Acoustically insulated ducting
- Filters reduce effects of respiratory allergies
- Optional switch systems
- Reduces heating bills
- Can promote conditions for improved health
- Easy to install for DIYer or professional alike
- Conforms to Building regulations

RegaVent HRV systems have been at the forefront of the market for more than 20 years. With thousands of troublefree installations, nationwide, you can be sure of fast, efficient, delivery and installation.

Ask too about RegaVent's CENTRAL VACUUM SYSTEMS - the ultimate in convenient cleaning!

Fresh thinking in the air...

www.rega-uk.com

rega vent

Rega Ventilation Limited 21/22 Eldon Way, Biggleswade, Beds SG18 8NH
Telephone: 01767 600499   fax: 01767 600487   email: sales@rega-uk.com

ATB AIR CONDITIONING & HEATING
67 Melloway Road, Rushden,
Northamptonshire, NN10 6XX
Area of Operation: East England, Greater London,
Midlands & Mid Wales, South East England
Tel: 0870 260 1650 **Fax:** 01933 411731
Email: enquiries@atbairconditioning.co.uk
Web: www.atbairconditioning.co.uk
Product Type: 1

AUTOMATED CONTROL SERVICES LTD
Unit 16, Hightown Ind Est, Crow Arch Lane,
Ringwood, Dorset, BH24 1ND
Area of Operation: UK (Excluding Ireland)
Tel: 01425 461008 **Fax:** 01425 461009
Email: sales@automatedcontrolservices.co.uk
Web: www.automatedcontrolservices.co.uk
Product Type: 1, 2

BEAM CENTRAL VACUUMS (UK) LTD
Unit 2 Campden Business Park, Station Road,
Chipping Camden, Gloucestershire, GL55 6JX
Area of Operation: UK (Excluding Ireland)
Tel: 0845 260 0123 **Fax:** 01386 849002
Email: info@beamvac.co.uk
Web: www.beamvac.co.uk
Product Type: 1

BROOK DESIGN HARDWARE LTD
Brook House, Dunmurry Industrial Estate,
Dunmurry, Belfast, Co Antrim, BT17 9HU
Area of Operation: UK (Excluding Ireland)
Tel: 028 9061 6505
Fax: 028 9061 6518 / 9061 9379
Email: admin@brookvent.co.uk
Web: www.brookvent.co.uk

COMFORT AIR CONDITIONING LTD
Comfort Works, Newchapel Road,
Lingfield, Surrey, RH7 6LE
Area of Operation: East England, Greater London,
Midlands & Mid Wales, South East England, South
West England and South Wales
Tel: 01342 830600 **Fax:** 01342 830605
Email: info@comfort.uk.com
Web: www.comfort.uk.com
Product Type: 1 **Other Info:** ECO

COMFYAIR LTD
3 Parkside Works, Parkwood Street,
Keighley, West Yorkshire, BD21 4PJ
Area of Operation: UK & Ireland
Tel: 01535 611333 **Fax:** 01535 611334
Email: contact@comfyair.co.uk
Web: www.comfyair.co.uk

CONSERVATION ENGINEERING LTD
The Street, Troston, Bury St Edmunds,
Suffolk, IP31 1EW
Area of Operation: UK & Ireland
Tel: 01359 269360 **Fax:** 01359 268340
Email: anne@conservation-engineering.co.uk
Web: www.heating-designs.co.uk
Product Type: 1

DD HEATING LTD
16-19 The Manton Centre, Manton Lane,
Bedford, Bedfordshire, MK41 7PX
Area of Operation: Worldwide
Tel: 0870 777 8323 **Fax:** 0870 777 8320
Email: info@heatline.co.uk
Web: www.heatline.co.uk
Product Type: 1

DIFFUSION ENVIRONMENTAL SYSTEMS
47 Central Avenue, West Molesey, Surrey, KT8 2QZ
Area of Operation: UK (Excluding Ireland)
Tel: 0208 783 0033 **Fax:** 0208 783 0140
Email: diffusion@etenv.co.uk
Web: www.energytechniqueplc.co.uk

DOMUS VENTILATION LTD
Bearwarden House, Bearwarden Business Park,
Royston Road, Wendens Ambo, Essex, CB11 4JX
Area of Operation: Worldwide
Tel: 01799 540602 **Fax:** 01799 541143
Email: info@domusventilation.com
Web: www.domusventilation.com
Product Type: 1

DRY-IT-OUT LIMITED
The Cwm, Churchstoke,
Montgomery, Powys, SY15 6TJ
Area of Operation: Europe
Tel: 0870 011 7987 **Fax:** 01588 620145
Email: enquiries@dry-it-out.com
Web: www.dry-it-out.com
Product Type: 1

DYER ENVIRONMENTAL CONTROLS LTD
Unit 10, Lawnhurst Trading Estate, Cheadle Heath,
Stockport, Cheshire, SK3 0SD
Area of Operation: UK (Excluding Ireland)
Tel: 0161 491 4840 **Fax:** 0161 491 4841
Email: enquiry@dyerenvironmental.co.uk
Web: www.dyerenvironmental.co.uk
Product Type: 2

ECO SYSTEMS IRELAND LTD
40 Glenshesk Rd, Ballycastle, Co Antrim, BT54 6PH
Area of Operation: UK & Ireland
Tel: 02820 768708 **Fax:** 02820 769781
Email: info@ecosystemsireland.com
Web: www.ecosystemsireland.com
Product Type: 1

ENVIROZONE
Encon House, 54 Ballycrochan Road,
Bangor, Co Down, BT19 6NF
Area of Operation: UK & Ireland
Tel: 02891 477799 **Fax:** 02891 452463
Email: info@envirozone.co.uk
Web: www.envirozone.co.uk
Product Type: 1

EU SOLUTIONS
Maghull Business Centre, 1 Liverpool Road North,
Maghull, Liverpool, L31 2HB
Area of Operation: UK & Ireland
Tel: 0870 160 1660 **Fax:** 0151 526 8849
Email: ths@blueyonder.co.uk
Web: www.totalhomesolutions.co.uk
Product Type: 1

EXHAUSTO
Unit 3 Lancaster Court, Coronation Road,
Cressex Business Park, High Wycombe,
Buckinghamshire, HP12 3TD
Area of Operation: UK & Ireland
Tel: 01494 465166 **Fax:** 01494 465163
Email: info@exhausto.co.uk
Web: www.exhausto.co.uk
Product Type: 1

HAMPTON
PO Box 6074, Newbury, Berkshire, RG14 9AF
Area of Operation: UK & Ireland
Tel: 01635 25440 **Fax:** 01635 255762
Email: fieldenj@freeuk.co.uk /
john.fielden@btconnect.com
Web: www.hamptonventilation.com
Product Type: 1 **Other Info:**

HONEYWELL CONTROL SYSTEMS LTD
Honeywell House, Arlington Business Park,
Bracknell, Berkshire, RG12 1EB
Area of Operation: Worldwide
Tel: 01344 656000 **Fax:** 01344 656054
Email: literature@honeywell.com
Web: www.honeywelluk.com
Product Type: 1, 2

JOHNSON & STARLEY
Rhosili Road, Brackmills, Northampton,
Northamptonshire, NN4 7LZ
Area of Operation: UK & Ireland
Tel: 01604 762881 **Fax:** 01604 767408
Email: marketing@johnsonandstarleyltd.co.uk
Web: www.johnsonandstarley.co.uk
Product Type: 1, 2

KAIR VENTILATION LTD
6 Chiltonian Industrial Estate, 203 Manor Lane,
Lee, London, SE12 0TX
Area of Operation: Worldwide
Tel: 0845 166 2240 **Fax:** 0845 166 2050
Email: info@kair.info
Web: www.kair.co.uk
Product Type: 1

HEATING, PLUMBING & ELECTRICAL

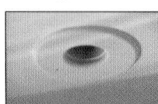

HEATING, PLUMBING & ELECTRICAL

KILTOX CONTRACTS LIMITED
6 Chiltonian Ind. Est, 203 Manor Lane,
Lee, London, Greater London, SE12 0TX
Area of Operation: Worldwide
Tel: 0845 166 2040 **Fax:** 0845 166 2050
Email: info@kiltox.co.uk **Web:** www.kiltox.co.uk
Product Type: 1 **Other Info:** ECO

PARAGON SYSTEMS (SCOTLAND) LIMITED
The Office, Corbie Cottage, Maryculter,
Aberdeen, Aberdeenshire, AB12 5FT
Area of Operation: Scotland
Tel: 01224 735536 **Fax:** 01224 735537
Email: info@paragon-systems.co.uk
Web: www.paragon-systems.co.uk
Product Type: 1

PASSIVENT LIMITED
2 Brooklands Road, Sale, Cheshire, M33 3SS
Area of Operation: UK & Ireland
Tel: 0161 962 7113 **Fax:** 0161 905 2085
Email: davidp@passivent.com
Web: www.passiventselfbuild.com
Product Type: 2

REGA VENTILATION LTD
21-22 Eldon Way, Biggleswade,
Bedfordshire, SG18 8NH
Area of Operation: Europe
Tel: 01767 600499 **Fax:** 01767 600487
Email: sales@rega-uk.com
Web: www.rega-uk.com
Product Type: 1, 2

RIDGE VENTILATION SYSTEMS
The Old Forge, Tempsford, Sandy,
Bedfordshire, SG19 2AG
Area of Operation: UK & Ireland
Tel: 01767 640808 **Fax:** 01767 640561
Email: sales@ridgeventilationsystems.co.uk
Web: www.ridgeventilationsystems.co.uk
Product Type: 1

RYTONS BUILDING PRODUCTS LTD
Design House, Kettering Business Park,
Kettering, Northamptonshire, NN15 6NL
Area of Operation: UK & Ireland
Tel: 01536 511874 **Fax:** 01536 310455
Email: lit@rytons.com **Web:** www.vents.co.uk
Product Type: 2

SOLARIS SOLAR ENERGY SYSTEMS
Toames, Macroom, Co. Cork
Area of Operation: UK & Ireland
Tel: +353 264 6312 **Fax:** +353 264 6313
Email: solaris@eircom.net
Web: www.solaris-energy.com
Product Type: 1

SPACE AIR SOLUTIONS LTD
Willway Court, 1 Opus Park, Moorfield Road,
Guildford, Surrey, GU1 1SZ
Area of Operation: UK (Excluding Ireland)
Tel: 01483 504883 **Fax:** 01483 574835
Email: solutions@spaceair.co.uk
Web: www.homeairsolutions.com
Product Type: 1

STARKEY SYSTEMS UK LTD
4a St Martins House, St Martins Gate,
Worcester, Worcestershire, WR1 2DT
Area of Operation: UK (Excluding Ireland)
Tel: 01905 611041
Fax: 01905 27462
Email: info@centralvacuums.co.uk
Web: www.centralvacuums.co.uk
Product Type: 1

SYSTEMAIR LTD
Pharoah House, Arnolde Close,
Medway City Estate, Rochester, Kent, ME2 4SP
Area of Operation: Worldwide
Tel: 01634 735000
Fax: 01634 735001
Email: sales@systemair.co.uk
Web: www.systemair.co.uk

THE HEAT PUMP COMPANY UK LIMITED
29 Claymoor Park, Booker, Nr. Marlow,
Buckinghamshire, SL7 3DL
Area of Operation: UK (Excluding Ireland)
Tel: 01494 450154
Fax: 01494 458159
Email: sales@npsair.com
Web: www.npsair.com
Product Type: 1, 2

TITON
International House, Peartree Road,
Stanway, Colchester, Essex, CO3 0JL
Area of Operation: Worldwide
Tel: 01206 713800
Fax: 01206 543126
Email: sales@titon.co.uk
Web: www.titon.co.uk
Product Type: 1, 2

TOTAL HOME ENVIRONMENT LTD
Unit 2 Campden Business Park, Station Road,
Chipping Campden, Gloucestershire, GL55 6JX
Area of Operation: UK (Excluding Ireland)
Tel: 0845 260 0123 **Fax:** 01386 849002
Email: info@beamvac.co.uk
Web: www.beamvac.co.uk
Product Type: 1
Other Info: ECO

UBBINK (UK) LTD
Borough Rd, Brackley,
Northamptonshire, NN13 7TB
Area of Operation: Europe
Tel: 01280 700211
Fax: 01280 705332
Email: info@ubbink.co.uk
Web: www.ubbink.co.uk
Product Type: 1, 2

UNICO SYSTEM INTERNATIONAL LIMITED
Unit 3, Ynyshir Industrial Estate, Llanwonno Road,
Porth, Rhondda Cynon Taff, CF39 0HU
Area of Operation: Worldwide
Tel: 01443 684828
Fax: 01443 684838
Email: scott@unicosystem.com
Web: www.unicosystem.co.uk

UNIFIX LTD
St Georges House, Grove Lane, Smethwick,
Birmingham, West Midlands, B66 2QT
Area of Operation: Europe
Tel: 0800 096 1110
Fax: 0800 096 1115
Email: sales@unifix.com
Web: www.unifix-online.co.uk
Product Type: 1

VECTAIRE LTD
Lincoln Road, Cressex Business Park,
High Wycombe, Buckinghamshire, HP12 3RH
Area of Operation: Worldwide
Tel: 01494 522333
Fax: 01494 522337
Email: sales@vectaire.co.uk
Web: www.vectaire.co.uk

VECTAIRE

Area of Operation: Worldwide
Tel: 01494 522333
Fax: 01494 522337
Email: sales@vectaire.co.uk
Web: www.vectaire.co.uk

2 Speed fans in 3 sizes for near silent whole house ventilation. Installation into roof or ceiling void for balanced extract system. Vectaire can assist on design and selection.

VENT AXIA
Fleming Way, Crawley,
West Sussex, RH10 9YX
Area of Operation: UK (Excluding Ireland)
Tel: 01293 526062
Fax: 01293 552375
Email: sales@vent-axia.com
Web: www.vent-axia.com

VILLAVENT LTD
Avenue 2, Station Lane Industrial Estate,
Witney, Oxfordshire, OX28 4YL
Area of Operation: UK & Ireland
Tel: 01993 778481
Fax: 01993 779962
Email: sales@villavent.co.uk
Web: www.villavent.co.uk
Product Type: 1

VORTICE LTD
Beeches House, Eastern Avenue, Burton on Trent,
Staffordshire, DE13 0BB
Area of Operation: Worldwide
Tel: 01283 492949 **Fax:** 01283 544121
Email: sales@vortice.ltd.uk
Web: www.vortice.ltd.uk
Product Type: 1 **Other Info:** ECO

EXTRACTOR FANS

KEY

PRODUCT TYPES: 1= Centrifugal Fans
2 = Propeller Fans

OTHER: ▽ Reclaimed On-line shopping
Bespoke Hand-made ECO Ecological

ADM
Ling Fields, Gargrave Road,
Skipton, North Yorkshire, BD23 1UX
Area of Operation: Europe
Tel: 01756 701051 **Fax:** 01756 701076
Email: info@admsystems.co.uk
Web: www.admsystems.co.uk
Product Type: 1 **Other Info:** ▽ ECO

ALLERGYPLUS LIMITED
65 Stowe Drive, Southam, Warwickshire, CV47 1NZ
Area of Operation: UK & Ireland
Tel: 0870 190 0022 **Fax:** 0870 190 0044
Email: info@allergyplus.co.uk
Web: www.allergyplus.co.uk
Product Type: 1, 2 **Other Info:**

DOMUS VENTILATION LTD
Bearwarden House, Bearwarden Business Park,
Royston Road, Wendens Ambo, Essex, CB11 4JX
Area of Operation: Worldwide
Tel: 01799 540602 **Fax:** 01799 541143
Email: info@domusventilation.com
Web: www.domusventilation.com
Product Type: 1

ELECTRICALSHOP.NET
81 Kinson Road, Wallisdown,
Bournemouth, Dorset, BH10 4DG
Area of Operation: UK (Excluding Ireland)
Tel: 0870 027 3730 **Fax:** 0870 027 3731
Email: sales@electricalshop.net
Web: www.electricalshop.net
Product Type: 1, 2

EU SOLUTIONS
Maghull Business Centre, 1 Liverpool Road North,
Maghull, Liverpool, L31 2HB
Area of Operation: UK & Ireland
Tel: 0870 160 1660 **Fax:** 0151 526 8849
Email: ths@blueyonder.co.uk
Web: www.totalhomesolutions.co.uk
Product Type: 1

HEATING, PLUMBING & ELECTRICAL

SPONSORED BY: VILLAVENT LTD www.villavent.co.uk

EXHAUSTO
Unit 3 Lancaster Court, Coronation Rd, Cressex Business Park, High Wycombe, Bucks, HP12 3TD
Area of Operation: UK & Ireland
Tel: 01494 465166 **Fax:** 01494 465163
Email: info@exhausto.co.uk
Web: www.exhausto.co.uk

GET PLC
Key Point, 3-17 High St, Potters Bar, Herts, EN6 5AJ
Area of Operation: Worldwide
Tel: 01707 601601 **Fax:** 01707 601701
Email: info@getplc.co.uk
Web: www.getplc.com
Product Type: 1, 2

GET PLC
Area of Operation: Worldwide
Tel: 01707 601601 **Fax:** 01707 601701
Email: info@getplc.co.uk
Web: www.getplc.com
Product Type: 1, 2

Our new slimline 100mm extractor fan has been designed to provide an innovative approach that complements modern styling and fittings and is available in a range of finishes.

HAMPTON
PO Box 6074, Newbury, Berkshire, RG14 9AF
Area of Operation: UK & Ireland
Tel: 01635 25440 **Fax:** 01635 255762
Email: fieldenj@freeuk.co.uk /
john.fielden@btconnect.com
Web: www.hamptonventilation.com
Product Type: 1

KAIR VENTILATION LTD
6 Chiltonian Industrial Estate,
203 Manor Lane, Lee, London, SE12 0TX
Area of Operation: Worldwide
Tel: 08451 662240 **Fax:** 08451 662050
Email: info@kair.info
Web: www.kair.co.uk
Product Type: 1, 2

KILTOX CONTRACTS LIMITED
Unit 6 Chiltonian Industrial Estate,
203 Manor Lane, Lee, London, SE12 0TX
Area of Operation: Worldwide
Tel: 0845 166 2040 **Fax:** 0845 166 2050
Email: info@kiltox.co.uk
Web: www.kiltox.co.uk
Product Type: 2 **Other Info:** ECO

PARAGON SYSTEMS (SCOTLAND) LIMITED
The Office, Corbie Cottage, Maryculter,
Aberdeen, Aberdeenshire, AB12 5FT
Area of Operation: Scotland
Tel: 01224 735536 **Fax:** 01224 735537
Email: info@paragon-systems.co.uk
Web: www.paragon-systems.co.uk
Product Type: 1

RIDGE VENTILATION SYSTEMS
The Old Forge, Tempsford, Sandy,
Bedfordshire, SG19 2AG
Area of Operation: UK & Ireland
Tel: 01767 640808 **Fax:** 01767 640561
Email: sales@ridgeventilationsystems.co.uk
Web: www.ridgeventilationsystems.co.uk
Product Type: 1 **Other Info:** ECO

STARKEY SYSTEMS UK LTD
4a St Martins House, St Martins Gate,
Worcester, Worcestershire, WR1 2DT
Area of Operation: UK (Excluding Ireland)
Tel: 01905 611041
Fax: 01905 27462
Email: info@centralvacuums.co.uk
Web: www.centralvacuums.co.uk
Product Type: 2

SYSTEMAIR LTD
Pharoah House, Arnolde Close, Medway City Estate,
Rochester, Kent, ME2 4SP
Area of Operation: Worldwide
Tel: 01634 735 000 **Fax:** 01634 735 001
Email: sales@systemair.co.uk
Web: www.systemair.co.uk
Product Type: 1, 2

TLC ELECTRICAL WHOLESALERS
TLC Building, Off Fleming Way,
Crawley, West Sussex, RH10 9JY
Area of Operation: Worldwide
Tel: 01293 565630 **Fax:** 01293 425234
Email: sales@tlc-direct.co.uk
Web: www.tlc-direct.co.uk
Product Type: 1, 2

UBBINK (UK) LTD
Borough Road, Brackley,
Northamptonshire, NN13 7TB
Area of Operation: Europe
Tel: 01280 700211 **Fax:** 01280 705332
Email: info@ubbink.co.uk
Web: www.ubbink.co.uk
Product Type: 1, 2

UNIFIX LTD
St Georges House, Grove Lane, Smethwick,
Birmingham, West Midlands, B66 2QT
Area of Operation: Europe
Tel: 0800 096 1110 **Fax:** 0800 096 1115
Email: sales@unifix.com
Web: www.unifix-online.co.uk
Product Type: 2

VECTAIRE LTD
Lincoln Road, Cressex Business Park,
High Wycombe, Buckinghamshire, HP12 3RH
Area of Operation: Worldwide
Tel: 01494 522333 **Fax:** 01494 522337
Email: sales@vectaire.co.uk
Web: www.vectaire.co.uk

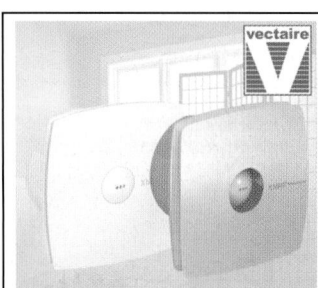

VECTAIRE
Area of Operation: Worldwide
Tel: 01494 522333
Fax: 01494 522337
Email: sales@vectaire.co.uk
Web: www.vectaire.co.uk

X-Mart deluxe low profile fans are finished in stainless steel or white; 3 sizes and extract capacities; splashproof to IPX4; timer and humidity control; standard or automatic shutters.

VENT AXIA
Fleming Way, Crawley, West Sussex, RH10 9YX
Area of Operation: UK (Excluding Ireland)
Tel: 01293 526062 **Fax:** 01293 552375
Email: sales@vent-axia.com
Web: www.vent-axia.com

VILLAVENT LTD
Avenue 2, Station Lane
Industrial Estate, Witney,
Oxfordshire, OX28 4YL
Area of Operation: UK & Ireland
Tel: 01993 778481
Fax: 01993 779962
Email: sales@villavent.co.uk
Web: www.villavent.co.uk
Product Type: 1

VORTICE LTD
Beeches House, Eastern Avenue,
Burton on Trent, Staffordshire, DE13 0BB
Area of Operation: Worldwide
Tel: 01283 492949
Fax: 01283 544121
Email: sales@vortice.ltd.uk
Web: www.vortice.ltd.uk
Product Type: 1, 2
Other Info: ECO

CEILING FANS

KEY
OTHER: ▽ Reclaimed ✑ On-line shopping
✑ Bespoke ✋ Hand-made ECO Ecological

AIRFLOW DEVELOPMENTS LTD
Lancaster Road, Cressex Business Park,
High Wycombe, Buckinghamshire, HP12 3QP
Area of Operation: Worldwide
Tel: 01494 525252
Fax: 01494 461073
Email: info@airflow.co.uk
Web: www.iconfan.co.uk

ASTRA CEILING FANS AND LIGHTING
Unit 1, Pilsworth Way, Pilsworth,
Bury, Lancashire, BL9 8RE
Area of Operation: Europe
Tel: 0161 766 9090
Fax: 0161 766 9191
Email: support@astra247.com
Web: www.astra247.com

BROUGHTONS OF LEICESTER
The Old Cinema, 69 Cropston Road,
Anstey, Leicester, Leicestershire, LE7 7BP
Area of Operation: Worldwide
Tel: 0116 235 2555
Fax: 0116 234 1188
Email: sale@broughtons.com
Web: www.broughtons.com

COOLANDWARM.COM
25 Waterside Point, Anhalt Road,
London, SW11 4PD
Area of Operation: Europe
Tel: 0207 228 7107
Fax: 0207 228 7121
Email: info@coolandwarm.com
Web: www.coolandwarm.com

DOMUS VENTILATION LTD
Bearwarden House, Bearwarden Business Park,
Royston Road, Wendens Ambo, Essex, CB11 4JX
Area of Operation: Worldwide
Tel: 01799 540602
Fax: 01799 541143
Email: info@domusventilation.com
Web: www.domusventilation.com

EXHAUSTO
Unit 3 Lancaster Court, Coronation Road,
Cressex Business Park, High Wycombe,
Buckinghamshire, HP12 3TD
Area of Operation: UK & Ireland
Tel: 01494 465166
Fax: 01494 465163
Email: info@exhausto.co.uk
Web: www.exhausto.co.uk

FANTASIA CEILING FANS
Unit B, The Flyers Way,
Westerham, Kent, TN16 1DE
Area of Operation: UK (Excluding Ireland)
Tel: 01959 564440
Fax: 01959 564829
Email: info@fantasiaceilingfans.com
Web: www.fantasiaceilingfans.com

FANTASIA CEILNG FANS
Area of Operation: UK (Excluding Ireland)
Tel: 01959 564440
Fax: 01959 564829
Email: info@fantasiaceilingfans.com
Web: www.fantasiaceilingfans.com

The Viper Ceiling from Fantasia, 137cm diameter, 100watt halogen light, remote controlled. Just one of many stunning ceiling fans available from Fantasia.

TLC ELECTRICAL WHOLESALERS
TLC Building, Off Fleming Way, Crawley,
West Sussex, RH10 9JY
Area of Operation: Worldwide
Tel: 01293 565630
Fax: 01293 425234
Email: sales@tlc-direct.co.uk
Web: www.tlc-direct.co.uk

VECTAIRE LTD
Lincoln Road, Cressex Business Park,
High Wycombe, Buckinghamshire, HP12 3RH
Area of Operation: Worldwide
Tel: 01494 522333 **Fax:** 01494 522337
Email: sales@vectaire.co.uk
Web: www.vectaire.co.uk

VORTICE LTD
Beeches House, Eastern Avenue,
Burton on Trent, Staffordshire, DE13 0BB
Area of Operation: Worldwide
Tel: 01283 492949 **Fax:** 01283 544121
Email: sales@vortice.ltd.uk
Web: www.vortice.ltd.uk
Other Info: ECO

ASTRA247.COM
Area of Operation: Worldwide
Tel: 0161 797 3222 **Fax:** 0161 797 3444
Email: info@astra247.com
Web: www.astra247.com

Conservatory Air Conditioning
We offer a range of quiet, efficient and energy saving conservatory air conditioning units.

'easy online ordering'

ASTRA247.COM
Area of Operation: Worldwide
Tel: 0161 797 3222 **Fax:** 0161 797 3444
Email: info@astra247.com
Web: www.astra247.com

Split Unit Air Conditioning
We offer a range of quiet, efficient and energy saving split unit air conditioning units.

'easy online ordering'

ASTRA247.COM
Area of Operation: Worldwide
Tel: 0161 797 3222 **Fax:** 0161 797 3444
Email: info@astra247.com
Web: www.astra247.com

Portable Air Conditioning
We offer a range of quiet, efficient and energy saving portable air conditioning units suitable for domestic use.

'easy online ordering'

ASTRA247.COM
Area of Operation: Worldwide
Tel: 0161 797 3222 **Fax:** 0161 797 3444
Email: info@astra247.com
Web: www.astra247.com

Conservatory Fans
Astra offer a wide choice of ceiling fans for use in your conservatory. Free, helpful advice and a full range of accessories and controls.

'easy online ordering'

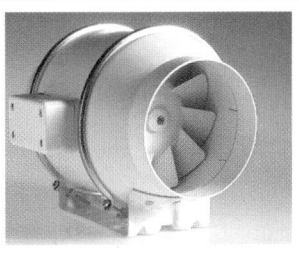

ASTRA247.COM
Area of Operation: Worldwide
Tel: 0161 797 3222 **Fax:** 0161 797 3444
Email: info@astra247.com
Web: www.astra247.com

Domestic Ventilation
Adequate ventilation is an important consideration when building or renovating your home. We offer a wide range of quality ventilation products on our website.
'easy online ordering'

ASTRA247.COM
Area of Operation: Worldwide
Tel: 0161 797 3222 **Fax:** 0161 797 3444
Email: info@astra247.com
Web: www.astra247.com

Extractor Fans
Adequate ventilation is an important consideration when building or renovating your home. We offer a wide range of quality extractor fans and related products.
'easy online ordering'

ASTRA247.COM
Area of Operation: Worldwide
Tel: 0161 797 3222 **Fax:** 0161 797 3444
Email: info@astra247.com
Web: www.astra247.com

Complete Extract Systems
We offer a wide range of quality extractor fans and related products, as components or as complete kits.

'easy online ordering'

ASTRA247.COM
Area of Operation: Worldwide
Tel: 0161 797 3222 **Fax:** 0161 797 3444
Email: info@astra247.com
Web: www.astra247.com

Designer Ceiling Fans
With over 20 designer models available from our extensive range of 150+ ceiling fans online, you will be spoilt for choice. Friendly, helpful advice available.
'easy online ordering'

ASTRA247.COM
Area of Operation: Worldwide
Tel: 0161 797 3222 **Fax:** 0161 797 3444
Email: info@astra247.com
Web: www.astra247.com

Industrial Ceiling Fans
A quality range of industrial ceiling fans suitable for domestic installation. We have over 150 fans online to choose from with free, technical advice available.
'easy online ordering'

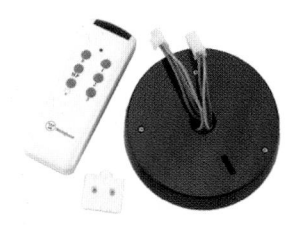

ASTRA247.COM
Area of Operation: Worldwide
Tel: 0161 797 3222 **Fax:** 0161 797 3444
Email: info@astra247.com
Web: www.astra247.com

Ceiling Fan Remote Controls
We offer a selection of ceiling fan remote controls to control fan speeds and light functions for use with all well known brands.

'easy online ordering'

ASTRA247.COM
Area of Operation: Worldwide
Tel: 0161 797 3222 **Fax:** 0161 797 3444
Email: info@astra247.com
Web: www.astra247.com

Smoke and Heat Detection
We offer a full range of smoke and heat detectors, interlinked and with battery back-up. Order directly from our website.

'easy online ordering'

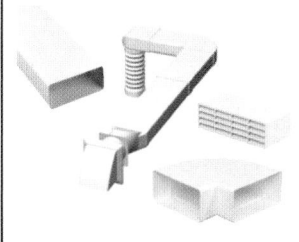

ASTRA247.COM
Area of Operation: Worldwide
Tel: 0161 797 3222 **Fax:** 0161 797 3444
Email: info@astra247.com
Web: www.astra247.com

Ducting Systems
We offer a full range of ducting products on our website including a good selection of flat ducting components for your domestic ventilation project.

'easy online ordering'

HUGE RANGE OF LIGHTING

Why buy from Astra?
We offer low prices, a wide choice of products, friendly knowledgeable staff and a quality service.

Recent customer comments
'Great service with fast delivery. Many thanks.'
'Brilliant item, fast delivery, would highly recommend!'

FREEPHONE 0800 195 4495
SHOP ONLINE www.astra247.com

ELECTRICAL

Image courtesy of Hamilton Litestat (01747 860088)

HEATING, PLUMBING & ELECTRICAL

Eco Electricity

We are all familiar with photovoltaic panels being used on a small scale, in the form of solar powered calculators or roadside signs, but increasingly they are being used on rooftops where they can convert enough energy from the sun's rays to power the home.

Photovoltaic cells are made from silicon semi-conductor cells which produce a current when light hits them. The cells are arranged in panels which typically produce 10-100 watts of electricity in bright sunlight. Ideally panels should be mounted at an angle of 10-35_ from the horizontal, pointing south (or south-east or south-west), and be unshaded. They can form part of the roof covering and so save on slates/tiles.

Depending on the type, 7-10m2 of panels will produce 1kW of electricity in bright sunlight, or about 25% of an average house's annual electricity requirement.

Over the last 50 years the efficiency of PVs has improved significantly and costs have come down. A 2.5kW system will currently cost around £13,000 installed, and produce 1,800kW hours of electricity each year. Grants are available from the Energy Saving Trust for £2,500/kW, and many major manufacturers now guarantee the output of their panels for 20 years.

Photovoltaic panels will convert your house into a miniature power station. When it is sunny you will probably produce more electricity than you use, which means surplus electricity will be exported to the National Grid. Savings are likely to be around £150-£200 per year for the £13,000 system. Without the grants taken into account, this gives a payback period of some 65 years — which for many will mean that the benefits have to be more than purely financial. The

system will also save around 750kg of carbon dioxide emissions each year.

It can also be a good idea to combine photovoltaic panels with wind generators to take advantage of all weather conditions. Wind generators work on the same principle as a bicycle dynamo, with a magnet and a coil of wire inside the body of a generator. The force of the wind causes the wire to move, and because it is within a magnetic field, this movement generates an electrical current in the wire. In the UK we currently produce less than 1% of our energy from the wind, but have the potential to produce much more.

Obviously, the planning office will have some say in whether you can use either system but, unless you are in a conservation area or similar, a small turbine or PV panel shouldn't pose a problem. The use of green energy is seen as a positive thing, and there are grants available for the installation of such systems - visit www.clear-skies.org for more information..

ELECTRICAL FITTINGS

KEY

SEE ALSO: LIGHTING - Lighting Controls and Accessories

PRODUCT TYPES: 1= Sockets 2 = Switches 3 = Plugs 4 = Terminal Connectors 5 = Fuse Boxes 6 = Other

OTHER: ▽ Reclaimed 🖱 On-line shopping ✏ Bespoke ✋ Hand-made ECO Ecological

A ALEXANDER & SON (ELECTRICAL) LTD
9 Cathkinview Road, Mount Florida, Glasgow, Lanarkshire, G42 9EH
Area of Operation: North East England, Scotland
Tel: 0141 632 0868 **Fax:** 0141 636 0020
Email: christine@alexander-electrical.co.uk
Web: www.alexander-electrical.co.uk
Product Type: 1, 2, 3, 4, 5, 6

AICO LTD
Mile End Business Park, Maesbury Road, Oswestry, Shropshire, SY10 8NN
Area of Operation: UK (Excluding Ireland)
Tel: 0870 758 4000 **Fax:** 0870 758 4010
Email: sales@aico.co.uk **Web:** www.aico.co.uk
Product Type: 6

ARCHITECTURAL IRONMONGERY LTD
28 Kyrle Street, Ross-on-Wye, Herefordshire, HR9 7DB
Area of Operation: Worldwide
Tel: 01989 567946 **Fax:** 01989 567946
Email: info@arciron.co.uk
Web: www.arciron.com 🖱
Product Type: 1, 2, 6

ASTRA CEILING FANS & LIGHTING
Unit 1, Pilsworth Way, Pilsworth, Bury, Lancashire, BL9 8RE
Area of Operation: Europe
Tel: 0161 766 9090 **Fax:** 0161 766 9191
Email: support@astra247.com
Web: www.astra247.com 🖱
Product Type:

BLUE BEACON LIGHTING
Intermail Plc - Horizon West, Canal View Road, Newbury, Berkshire, RG14 5XF
Area of Operation: UK & Ireland
Tel: 0870 241 3992 **Fax:** 01635 41678
Email: michaelm@crescent.co.uk
Web: www.bluebeacon.co.uk 🖱

BLUE RIDGE ELECTRICAL SYSTEMS LTD
18 Ridge Drive, Rugby, Warwickshire, CV25 3FE
Area of Operation: UK (Excluding Ireland)
Tel: 01788 561701 **Fax:** 01788 536242
Email: sales@blueridge-electrical.com
Web: www.blueridge-electrical.com
Product Type: 1, 2, 3, 4, 5

BROMLEIGHS
Unit 12, Goudhurst Road, Marden, Kent, TN12 9NW
Area of Operation: UK (Excluding Ireland)
Tel: 0800 018 3993 **Product Type:** 1, 2, 3

BROUGHTONS OF LEICESTER
The Old Cinema, 69 Cropston Road, Anstey, Leicester, Leicestershire, LE7 7BP
Area of Operation: Worldwide
Tel: 0116 235 2555 **Fax:** 0116 234 1188
Email: sale@broughtons.com
Web: www.broughtons.com 🖱

BYRON
Byron House, 34 Sherwood Road, Aston Fields Industrial Estate, Bromsgrove, West Midlands, B60 3DR
Area of Operation: Worldwide
Tel: 01527 557700 **Fax:** 01527 557701
Email: info@chbyron.com
Web: www.chbyron.com
Product Type: 1, 2, 6

CONTACTUM LTD
Victoria Works, Edgeware Road, Cricklewood, London, NW2 6LF
Area of Operation: Worldwide
Tel: 0208 452 6366 **Fax:** 0208 208 3340
Email: binum@contactum.co.uk
Web: www.contactum.co.uk
Product Type: 1, 2, 5, 6

CRESCENT LIGHTING LTD
8 Rivermead, Pipers Lane, Thatcham, Berkshire, RG19 4EP
Area of Operation: Europe
Tel: 01635 878888 **Fax:** 01635 873888
Email: sales@crescent.co.uk
Web: www.crescent.co.uk
Product Type: 6

DANLERS LIMITED
Vincients Road, Bumpers Farm Industrial Estate, Chippenham, Wiltshire, SN14 6NQ
Area of Operation: Europe
Tel: 01249 443377 **Fax:** 01249 443388
Email: sales@danlers.co.uk
Web: www.danlers.co.uk
Product Type: 2

DCD SYSTEMS LTD
43 Howards Thicket, Gerrards Cross, Buckinghamshire, SL9 7NU
Area of Operation: UK & Ireland
Tel: 01753 882028 **Fax:** 01753 882029
Email: peter@dcd.co.uk
Web: www.dcd.co.uk
Product Type: 2
Other Info: ECO

ELECTRICALSHOP.NET
81 Kinson Road, Wallisdown, Bournemouth, Dorset, BH10 4DG
Area of Operation: UK (Excluding Ireland)
Tel: 0870 027 3730 **Fax:** 0870 027 3731
Email: sales@electricalshop.net
Web: www.electricalshop.net 🖱
Product Type: 1, 2, 3, 4, 5, 6

ELECTROPRESS LTD
13 Hall Walk, Coleshill, Birmingham, West Midlands, B46 3ES
Area of Operation: UK & Ireland
Tel: 01675 463984
Email: jimmy@alfra.uk.com
Product Type: 6

EURO BATHROOMS
102 Annareagh Road, Richhill, Co Armagh, BT61 9JY
Area of Operation: Ireland Only
Tel: 028 3887 9996
Fax: 028 3887 9996
Email: paul@euro-bathrooms.co.uk
Web: www.euro-bathrooms.co.uk
Product Type: 1, 2, 3, 4, 5, 6

FLEXELEC (JIMI HEAT)
Unit C, 200 Rickmansworth Road, Watford, Hertfordshire, WD18 7JS
Area of Operation: Worldwide
Tel: 01923 234477 **Fax:** 01923 240264
Email: sales@flexelec.co.uk
Web: www.flexelec.com
Product Type: 6

FOCUS SB
Napier Road, Castleham Industrial Estate, St-Leonards-on-Sea, East Sussex, TN38 9NY
Area of Operation: Worldwide
Tel: 01424 440734 **Fax:** 01424 853862
Email: sales@focus-sb.co.uk
Web: www.focus-sb.co.uk
Product Type: 1, 2
Other Info: ✏

GET PLC
Key Point, 3-17 High St, Potters Bar, Herts, EN6 5AJ
Area of Operation: Worldwide
Tel: 01707 601601 **Fax:** 01707 601701
Email: info@getplc.co.uk **Web:** www.getplc.com
Product Type: 1, 2, 5, 6

GET PLC
Area of Operation: Worldwide
Tel: 01707 601601 **Fax:** 01707 601701
Email: info@getplc.co.uk
Web: www.getplc.com 🖱
Product Type: 1, 2, 5, 6

Available in three high quality finishes the ergonomic design of the Rocca range is characterised by broad, smooth lines. Concealed fixing screws also meet consumer demand for streamlined, screwless appearance.

GIBBS AND DANDY PLC
226 Dallow Road, Luton, Bedfordshire, LU1 1YB
Area of Operation: South East England
Tel: 01582 798798
Fax: 01582 798799
Email: luton@gibbsanddandy.com
Web: www.gibbsanddandy.com
Product Type: 1, 2, 3, 4, 5, 6

HAANI CABLES LTD
Tofts Farm Industrial Estate, Hartlepool, Durham, TS25 2BS
Area of Operation: Worldwide
Tel: 01429 221184
Fax: 01429 272714
Email: cables@haanicables.co.uk
Web: www.haanicables.co.uk
Product Type: 6
Other Info: ✏

HAF DESIGNS LTD
HAF House, Mead Lane, Hertford, Hertfordshire, SG13 7AP
Area of Operation: UK & Ireland
Tel: 0800 389 8821
Fax: 01992 505705
Email: info@hafltd.co.uk
Web: www.hafdesigns.co.uk 🖱
Product Type: 1, 2

HAMILTON LITESTAT
Quarry Industrial Estate, Mere, Wiltshire, BA12 6LA
Area of Operation: Worldwide
Tel: 01747 860088
Fax: 01747 861032
Email: info@hamilton-litestat.com
Web: www.hamilton-litestat.com
Product Type: 1, 2

HOUSE OF BRASS
122 North Sherwood Street, Nottingham, Nottinghamshire, NG1 4EF
Area of Operation: Worldwide
Tel: 0115 947 5430
Fax: 0115 947 5430
Email: sales@houseofbrass.co.uk
Web: www.houseofbrass.co.uk 🖱 **Product Type:** 1, 2

HOUSE OF BRASS LTD
Area of Operation: Worldwide
Tel: 0115 947 5430
Email: sales@houseofbrass.co.uk
Web: www.houseofbrass.co.uk
Product Type: 1, 2, 6
Other Info: 🖱

Extensive range of luxury electrical accessories from classic to contemporary designs in a wide range of plates and finishes delivered direct to you. Call for a FREE colour catalogue (24 Hrs).

INTELLIGENT HOUSE
Langham House, Suite 401, 302 Regent Street, London, W1B 3HH
Area of Operation: UK & Ireland
Tel: 0207 394 9344
Fax: 0207 252 3879
Email: info@intelligenthouse.net
Web: www.intelligenthouse.net
Product Type: 1, 2, 6
Other Info: ✏

INTERACTIVE HOMES LTD
Lorton House, Conifer Crest, Newbury, Berkshire, RG14 6RS
Area of Operation: UK (Excluding Ireland)
Tel: 01635 49111
Fax: 01635 40735
Email: sales@interactivehomes.co.uk
Web: www.interactivehomes.co.uk
Product Type: 2

IRONMONGERY DIRECT
Unit 2-3 Eldon Way Trading Estate, Eldon Way, Hockley, Essex, SS5 4AD
Area of Operation: Worldwide
Tel: 01702 562770
Fax: 01702 562799
Email: sales@ironmongerydirect.com
Web: www.ironmongerydirect.com 🖱
Product Type: 1, 2

JOHN CLAYTON LIGHTING LTD
Worthingham House, Deep Lane, Hagworthingham, Lincolnshire, PE23 4LZ
Area of Operation: Europe
Tel: 0800 389 6395
Fax: 0870 240 6417
Email: rachel@jclighting.com
Web: www.flexidim.com
Product Type: 2, 6

LAMPS & LIGHTING LTD
Bridgewater Court, Network 65 Business Park, Burnley, Lancashire, BB11 5ST
Area of Operation: UK (Excluding Ireland)
Tel: 01282 448666
Fax: 01282 417703
Email: sales@lampslighting.co.uk
Web: www.lampslighting.co.uk 🖱
Product Type: 1, 2

ASTRA247.COM
Area of Operation: Worldwide
Tel: 0161 797 3222 **Fax:** 0161 797 3444
Email: info@astra247.com
Web: www.astra247.com

White Accessories
We offer a full range of white accessories from a choice of over 1000 sockets and switches in various styles and finishes to suit your interior décor.

'easy online ordering'

ASTRA247.COM
Area of Operation: Worldwide
Tel: 0161 797 3222 **Fax:** 0161 797 3444
Email: info@astra247.com
Web: www.astra247.com

Flat Plate Sockets and Switches
We offer a full range of flat plate accessories from a choice of over 1000 sockets and switches in various styles and finishes to suit your interior décor.

'easy online ordering'

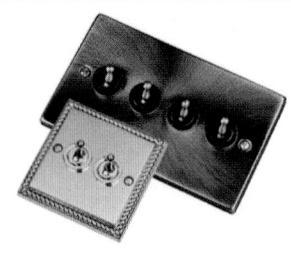

ASTRA247.COM
Area of Operation: Worldwide
Tel: 0161 797 3222 **Fax:** 0161 797 3444
Email: info@astra247.com
Web: www.astra247.com

Toggle style Light Switches
Brass, chrome, black nickel and other quality finishes available from one gang to four gang.

'easy online ordering'

ASTRA247.COM
Area of Operation: Worldwide
Tel: 0161 797 3222 **Fax:** 0161 797 3444
Email: info@astra247.com
Web: www.astra247.com

Chrome Sockets and Switches
Extensive selection of Polished Chrome and Satin Chrome wiring accessories for all rooms of the house. Similar flat plate versions also available.

'easy online ordering'

ASTRA247.COM
Area of Operation: Worldwide
Tel: 0161 797 3222 **Fax:** 0161 797 3444
Email: info@astra247.com
Web: www.astra247.com

Brass Sockets and Switches
Beautiful selection of 100s of sockets and switches finished in antique, polished and satin brass to suit your interior fixings and decor.

'easy online ordering'

ASTRA247.COM
Area of Operation: Worldwide
Tel: 0161 797 3222 **Fax:** 0161 797 3444
Email: info@astra247.com
Web: www.astra247.com

Black Nickel Sockets and Switches
Stylish selection of sockets and switches finished in a modern Black Nickel colour.

'easy online ordering'

ASTRA247.COM
Area of Operation: Worldwide
Tel: 0161 797 3222 **Fax:** 0161 797 3444
Email: info@astra247.com
Web: www.astra247.com

Transparent Plate
We offer a stylish new range of transparent flat place accessories. All available on our website.

'easy online ordering'

ASTRA247.COM
Area of Operation: Worldwide
Tel: 0161 797 3222 **Fax:** 0161 797 3444
Email: info@astra247.com
Web: www.astra247.com

Round Pin Sockets
As well as our extensive selection of 13A sockets we stock a range of round pin sockets in 5A and 2A in chrome, brass and black nickel finishes.

'easy online ordering'

ASTRA247.COM
Area of Operation: Worldwide
Tel: 0161 797 3222 **Fax:** 0161 797 3444
Email: info@astra247.com
Web: www.astra247.com

Outdoor Sockets & Switches
A comprehensive range of IP rated outdoor sockets and switches all available to order directly from our website.

'easy online ordering'

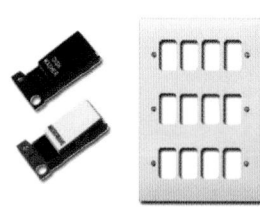

ASTRA247.COM
Area of Operation: Worldwide
Tel: 0161 797 3222 **Fax:** 0161 797 3444
Email: info@astra247.com
Web: www.astra247.com

Modular Grid Switches
We have a complete range of modular grid switches available for control of appliances and fittings in your home.

'easy online ordering'

ASTRA247.COM
Area of Operation: Worldwide
Tel: 0161 797 3222 **Fax:** 0161 797 3444
Email: info@astra247.com
Web: www.astra247.com

TV, Telephone and Data Sockets
Data sockets for telephone, ethernet, satellite and television. All available in a range of black nickel, brass and chrome finishes.

'easy online ordering'

ASTRA247.COM
Area of Operation: Worldwide
Tel: 0161 797 3222 **Fax:** 0161 797 3444
Email: info@astra247.com
Web: www.astra247.com

Remote and Touch control dimmers
Stylish touch control and remote control dimmers suitable for mains lighting and electronic transformers including those requiring "trailing-edge" control.

'easy online ordering'

HEATING, PLUMBING & ELECTRICAL

LIGHT RIGHT LTD
SBC House, Restmor Way, Wallington,
Surrey, SM6 7AH
Area of Operation: Europe
Tel: 0208 255 2022 **Fax:** 0208 286 1900
Email: enquiries@lightright.co.uk
Product Type: 6

LIGHTING DIRECT 2U LIMITED
Venture Court, Broadlands, Wolverhampton,
West Midlands, WV10 6TB
Area of Operation: UK & Ireland
Tel: 0870 600 0076
Email: sales@lightingdirect2u.co.uk
Web: www.lightingdirect2u.co.uk
Product Type: 1, 2

MAYFIELD (MANUFACTURING) LTD
Wenden House, New End, Hemingby,
Horncastle, Lincolnshire, LN9 5QQ
Area of Operation: Worldwide
Tel: 01507 578630
Fax: 01507 578609
Email: info@mayfieldmanufacturing.co.uk
Web: www.aludrain.co.uk **Product Type:** 6

MUSICAL APPROACH
111 Wolverhampton Road, Stafford,
Staffordshire, ST17 4BG
Area of Operation: UK (Excluding Ireland)
Tel: 01785 255154
Fax: 01785 243623
Email: info@musicalapproach.co.uk
Web: www.musicalapproach.co.uk
Product Type: 2, 6
Other Info:

OXFORD LIGHTING & ELECTRICAL SOLUTIONS
Unit 117, Culham Site No 1, Station Road,
Culham, Abingdon, Oxfordshire, OX14 3DA
Area of Operation: UK (Excluding Ireland)
Tel: 01865 408522
Fax: 01865 408522
Email: olessales@tiscali.co.uk
Web: www.oles.co.uk
Product Type: 2

PESTWEST ELECTRONICS
Denholme Drive, Ossett, West Yorkshire, WF5 9NB
Area of Operation: Worldwide
Tel: 01924 268500
Fax: 01924 273591
Email: info@pestwest.com
Web: www.pestwest.com
Product Type: 6

PITACS LTD
7 Grovebury Road, Leighton Buzzard,
Bedfordshire, LU7 4SR
Area of Operation: Worldwide
Tel: 01525 379 505
Fax: 01525 379 170
Email: info@pitacs.com
Web: www.pitacs.com
Product Type: 6

**PREMIER ELECTRICAL
& SECURITY SERVICES**
Unit D10, The Seedbed Centre,
Langston Rd, Loughton, Essex, IG10 3TQ
Area of Operation: East England,
Greater London, Midlands & Mid Wales
Tel: 0208 787 7063
Fax: 0208 508 3249
Email: info@premierelectrical.com
Web: www.premierelectrical.com
Product Type: 1, 2, 3, 4, 5, 6

PYRAMID ELECTRICAL & ALARMS
1 Morland Drive, Hinckley, Leicestershire,
Leicestershire, LE10 0GG
Area of Operation: East England, Greater London,
Midlands & Mid Wales, North East England,
North West England and North Wales
Tel: 01455 458325
Fax: 01455 458 325
Email: pyramid@dotdotnetdot.net
Web: www.pyramidelectrical.net
Product Type: 1, 2, 3, 4, 5, 6

QVS ELECTRICAL
4C The Birches Industrial Estate,
Imberhorne Lane, East Grinstead,
West Sussex, RH19 1XZ
Area of Operation: Worldwide
Tel: 0800 197 6565
Fax: 0800 197 6566
Email: sales@qvsdirect.co.uk
Web: www.qvsdirect.co.uk
Product Type: 1, 2, 6

SCHNEIDER ELECTRIC
Stafford Park 5, Telford, Shropshire, TF3 3BL
Area of Operation: UK (Excluding Ireland)
Tel: 0870 608 8608
Fax: 0870 608 8606
Email: sean.jordan@gb.schneider-electric.com
Web: www.squared.co.uk
Product Type: 1, 2, 5, 6

SCOLMORE INTERNATIONAL LIMITED
1 Scolmore Park, Landsberg, Lichfield Road
Industrial Estate, Tamworth, Staffordshire, B79 7XB
Area of Operation: UK & Ireland
Tel: 01827 63454 **Fax:** 01827 63362
Email: sales@scolmore.com
Web: www.scolmore.com
Product Type: 1, 2, 3, 4, 5, 6

SUSSEX BRASSWARE
Napier Road, Castleham Industrial Estate, St
Leonards-on-Sea, East Sussex, TN38 9NY
Area of Operation: Worldwide
Tel: 01424 857913
Fax: 01424 853862
Email: sales@sussexbrassware.co.uk
Web: www.sussexbrassware.co.uk
Product Type: 1, 2, 6

SWITCH TO WOOD
Unit 4, Firsland Park Estate, Henfield Road,
Albourne, West Sussex, BN6 9JJ
Area of Operation: Worldwide
Tel: 01273 495999
Fax: 01273 495019
Email: sales@switchtowood.co.uk
Web: www.switchtowood.co.uk
Product Type: 1, 2

TLC ELECTRICAL WHOLESALERS
TLC Building, Off Fleming Way,
Crawley, West Sussex, RH10 9JY
Area of Operation: Worldwide
Tel: 01293 565630
Fax: 01293 425234
Email: sales@tlc-direct.co.uk
Web: www.tlc-direct.co.uk
Product Type: 1, 2, 3, 4, 5, 6

TOOLSTATION
Express Park, Bridgwater,
Somerset, TA6 4RN
Area of Operation: UK (Excluding Ireland)
Tel: 0808 100 7211
Fax: 0808 100 7210
Email: info@toolstation.com
Web: www.toolstation.com
Product Type: 1, 2, 3, 4, 5, 6

UNIFIX LTD
St Georges House, Grove Lane,
Smethwick, Birmingham,
West Midlands, B66 2QT
Area of Operation: Europe
Tel: 0800 096 1110
Fax: 0800 096 1115
Email: sales@unifix.com
Web: www.unifix-online.co.uk
Product Type: 1, 2, 3, 4, 5, 6

WANDSWORTH GROUP LTD
Albert Drive, Sheerwater,
Woking, Surrey, GU21 5SE
Area of Operation: UK (Excluding Ireland)
Tel: 01483 740740
Fax: 01483 740384
Email: info@wandsworthgroup.com
Web: www.wandsworthgroup.com
Product Type: 1, 2, 3

ALTERNATIVE POWER SOURCES

BYRON
Byron House, 34 Sherwood Road,
Aston Fields Industrial Estate,
Bromsgrove, West Midlands, B60 3DR
Area of Operation: Worldwide
Tel: 01527 557700
Fax: 01527 557701
Email: info@chbyron.com
Web: www.chbyron.com

ENERGY AND ENVIRONMENT LTD
91 Claude Road, Chorlton, Manchester, M21 8DE
Area of Operation: UK & Ireland
Tel: 0161 881 1383
Email: mail@energyenv.co.uk
Web: www.energyenv.co.uk

GENERATORS NOW
The Birches , Megg Lane, Chipperfield,
Hertfordshire, WD49JW
Area of Operation: UK (Excluding Ireland)
Tel: 01923 262818
Fax: 01923 262882
Email: jon@generatorsnow.co.uk
Web: www.generatorsnow.co.uk

LEAX LIGHTING CONTROLS
11 Mandeville Courtyard,
142 Battersea Park Rd, London, SW11 4NB
Area of Operation: Europe
Tel: 0207 501 0880
Fax: 0207 501 0890
Email: sales@leax.co.uk
Web: www.leax.co.uk
Other Info:

POWERTECH SOLAR LTD
21 Haviland Road, Forndown Industrial Estate,
Wimborne, Dorset, BH21 7RZ
Area of Operation: UK (Excluding Ireland)
Tel: 08707 300111
Fax: 08707 300222
Email: sales@solar.org.uk
Web: www.solar.org.uk

SIP HOME LTD
Unit 1-3 Hadston Industrial Estate,
South Broomhill, Northumberland, NE65 9YG
Area of Operation: UK (Excluding Ireland)
Tel: 01670 760300
Fax: 01670 760740
Email: john@siphome.co.uk
Web: www.siphome.co.uk

SOLARUK
Crabtree, The Warren, Crowborough,
East Sussex, TN6 1UB
Area of Operation: Worldwide
Tel: 01892 667320
Fax: 01892 667622
Email: info@solaruk.net
Web: www.solaruk.net

THE GREEN SHOP
Cheltenham Road, Bisley, Nr Stroud,
Gloucestershire, GL6 7BX
Area of Operation: UK & Ireland
Tel: 01452 770629
Fax: 01452 770104
Email: paint@greenshop.co.uk
Web: www.greenshop.co.uk

WARMFLOORS ONLINE
Unit1 Aire Street, Cross Hills, Keighley,
West Yorkshire, BD20 7RT
Area of Operation: UK & Ireland
Tel: 01535 631195
Fax: 01535 631196
Email: sales@warmfloorsonline.com
Web: www.warmfloorsonline.com

NOTES

Company Name ....
....
Address ....
....
....
email ....
Web ....

Company Name ....
....
Address ....
....
....
email ....
Web ....

Company Name ....
....
Address ....
....
....
email ....
Web ....

Company Name ....
....
Address ....
....
....
email ....
Web ....

Company Name ....
....
Address ....
....
....
email ....
Web ....

HEATING, PLUMBING & ELECTRICAL

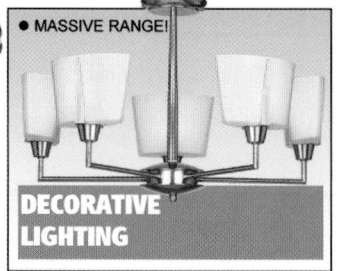

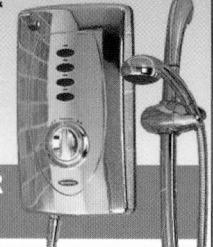

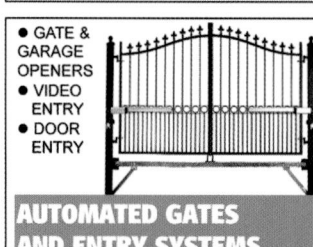

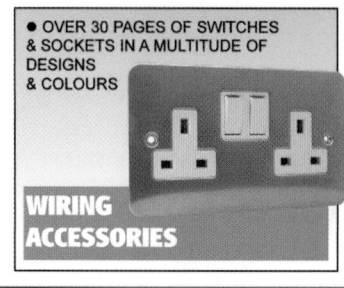

HEATING, PLUMBING & ELECTRICAL

SMART HOMES & SECURITY

Image courtesy of Sharp Electronics (UK) Ltd (0161 204 2333)

HEATING, PLUMBING & ELECTRICAL

Automation

The ability to control all of the lighting within a room with the touch of a single button offers a level of convenience not available with traditional switching. Groups of lights can be controlled together at preset levels for different activities, and a large room can be divided into independently controlled 'zones'. Different settings or 'moods' can change the atmosphere within a room or zone instantly.

With automated lighting you'll never have to get up to dim the lights for a movie again. Image ABOVE by Helvar.

Automated Systems can now be incorporated into any room of your home, including "wet" areas like the kitchen and bathroom. Image LEFT courtesy of Smartkontrols. Image ABOVE courtesy of CEDIA.

The number of lighting zones or circuits in a room can be limited by the sheer quantity of conventional switches or dimmers on the wall. For example, a lounge area may have four circuits including main and wall lights, spotlights and table lamps, making conventional and dimmer switches seem inflexible and cumbersome.

By contrast, lighting control units can be hidden in electrical cupboards and keypads in a number of styles can replace traditional switches. Each keypad has a number of scene selection buttons, programmed to hold a scene and mood for the room.

Installation differs from conventional lighting as each zone or circuit is wired directly to the controlling system. Wall keypads are not wired with electrical cable, but by either a low-voltage data control cable wired from the keypad directly to the controlling system, or are entirely wireless. This allows keypads to be located in bathrooms or outside, where conventional wall switches are not allowed.

Specialist systems can also reduce costs — dimmable lighting prolongs the life of light bulbs and minimum/low settings can save electricity and energy. It is also easy to add automation for security and safety to this equation; after all, a lighting control system, a smart climate control system, and a security system all-in-one will afford convenience, comfort and luxury to any home. United under a common control system, independent subsystems can alter the atmosphere of the entire house with one command.

For example, at the press of a single button on a home control keypad, you could brighten certain lights, increase the thermostats, disarm the security system and select a suitable ambience from the whole-house music system. Accomplishing this same scenario without the help of a home control system would require at least half a dozen button presses and take a great deal longer.

Keyboards are a relatively cheap choice of control, but they are perhaps not suitable for all the family,

whilst touch-screens are user-friendly but very expensive. Web tablets, remote controls and even telephones can also be used to control home-based systems, so seek advice from the experts and don't proceed with the installation until these issues are settled to your satisfaction

Ideally, the decision to install a controlling system will be made at the first fix stage, as the installer can then confidently plan and install the complex wiring required, hiding the lot behind walls, ceilings and floors. If not, installing a control system can become a little more invasive. If you're lucky, you may have room in the ceiling or floor areas to hold wiring, which can be completed with planned and structured installation procedures to limit the amount of inconvenience to you and your family. This option might not be available for older houses, however. Hence, a wireless system might be a viable alternative. Again, detailed and thorough meetings with your chosen installer will be critical at the planning stage.

HOME AUTOMATION AND WIRELESS SYSTEMS

KEY

SEE ALSO: LIGHTING - Lighting Controls and Accessories (Smart Lighting Systems)

PRODUCT TYPES: 1= Specialists
2 = Systems 3 = Hubs 4 = Controls

OTHER: ▽ Reclaimed ⌂ On-line shopping
✎ Bespoke ✋ Hand-made ECO Ecological

A.C.E - AUTOMATION CONTRACTORS & ENGINEERS

64 Wannock Avenue, Willingdon,
East Sussex, BN20 9RH
Area of Operation: UK (Excluding Ireland)
Tel: 01323 489798
Fax: 07884 599411
Email: sales@ace-automation.co.uk
Web: www.ace-automation.co.uk
Product Type: 1, 2, 4

ABITANA NV

4 Myson Way, Raynham Road Industrial Estate,
Bishop's Stortford, Hertfordshire, CM23 5JZ
Area of Operation: Europe
Tel: 01279 757775
Fax: 01279 653535
Email: sminns@minitran.co.uk
Web: www.minitran.co.uk ⌂

ABSOLUTE ELECTRICS

47 Tanfield Road, Croydon,
Surrey, CR0 1AN
Area of Operation: Greater London,
South East England
Tel: 0208 681 7835
Fax: 0208 667 9836
Email: mark@absolute-electrics.co.uk
Web: www.absolute-electrics.com
Product Type: 1, 2, 3, 4

ACA APEX LIMITED

Apex House, Ivinghoe Way, Edlesborough,
Dunstable, Bedfordshire, LU6 2EL
Area of Operation: UK & Ireland
Tel: 01525 220782
Fax: 01525 220782
Email: sales@aca-apex.com
Web: www.aca-apex.com
Product Type: 1, 2, 3

ACM INSTALLATIONS LTD

71 Western Gailes Way, Hull,
East Riding of Yorks, HU8 9EQ
Area of Operation: North East England
Tel: 0870 242 3285
Fax: 01482 377530
Email: enquiries@acminstallations.com
Web: www.acminstallations.com
Product Type: 1, 2, 3, 4
Other Info: ✎

ADV AUDIO VISUAL INSTALLATIONS LTD

12 York Place, Leeds,
West Yorkshire, LS1 2DS
Area of Operation: North East England,
North West England and North Wales
Tel: 0870 199 5755
Email: info@adv-installs.co.uk
Web: www.adv-installs.co.uk
Product Type: 4

ALDOUS SYSTEMS (EUROPE) LTD:

28 Bishopstone, Aylesbury,
Buckinghamshire, HP17 8SF
Area of Operation: Europe
Tel: 0870 240 1162
Fax: 0207 691 7844
Email: sales@aldoussystems.co.uk
Web: www.aldoussystems.co.uk
Product Type: 2

ANDROMEDA TELEMATICS LTD

Tec. 6, Byfleet Technical Centre,
Canada Road, Byfleet, Surrey, KT14 7JX
Area of Operation: Europe
Tel: 01932 341200 **Fax:** 01932 331980
Email: sales@andromeda-telematics.com
Web: www.andromeda-telematics.com
Product Type: 1, 2, 3, 4

ARCHANGEL CUSTOM INSTALLATION SPECIALISTS (JERSEY) LIMITED

Mont St Aubyn, La Rue Du Carrefour Au Clercq,
Grouville, Jersey, JE3 9US
Area of Operation: Europe
Tel: 0845 838 2636
Email: david.fell@archangelmultimedia.co.uk
Web: www.archangelmultimedia.co.uk
Product Type: 1, 2, 3, 4

AUDIO DESIGNS CUSTOM INSTALL

7/9 Park Place, Horsham, West Sussex, RH12 1DF
Area of Operation: South East England
Tel: 01403 252255
Fax: 01342 328065
Email: adci@audiodesigns.co.uk
Web: www.audiodesigns.co.uk
Product Type: 2, 3, 4

AUDIO VENUE

36 Queen Street, Maidenhead, Berkshire, SL6 1HZ
Area of Operation: Greater London, South East
England, South West England and South Wales
Tel: 01628 633995
Fax: 01628 633654
Email: info@audiovenue.com
Web: www.audiovenue.co.uk
Product Type: 1, 2, 4

AUDIOFILE

27 Hockerill Street, Bishops Stortford,
Hertfordshire, CM23 2DH
Area of Operation: UK & Ireland
Tel: 01279 506576
Fax: 01279 506638
Email: info@audiofile.co.uk
Web: www.audiofile.co.uk

AUDIOVISION

46 Market Square, St Neots,
Cambridgeshire, PE19 2AF
Area of Operation: Europe
Tel: 01480 471202
Fax: 01480 471115
Email: sales@audiovisiononline.co.uk
Web: www.audiovisiononline.co.uk ⌂
Product Type: 1, 2, 4

AUTO AUDIO HOME DIVISION

32-34 Park Royal Road, London, NW10 7LN
Area of Operation: Worldwide
Tel: 0208 838 8838
Email: nick@autoaudio.uk.com
Product Type: 1, 2, 3, 4

AUTOMATED CONTROL SERVICES LTD

Unit 16, Hightown Ind Est, Crow Arch Lane,
Ringwood, Dorset, BH24 1ND
Area of Operation: UK (Excluding Ireland)
Tel: 01425 461008 **Fax:** 01425 461009
Email: sales@automatedcontrolservices.co.uk
Web: www.automatedcontrolservices.co.uk
Product Type: 1

AUTOMATIC BUILDINGS LTD

Unit 12b Gorsey Lane, Coleshill,
Warwickshire, B46 1JU
Area of Operation: UK & Ireland
Tel: 01675 466881 **Fax:** 01675 464138
Email: info@automaticbuildings.co.uk
Web: www.automaticbuildings.co.uk
Product Type: 1, 2, 4

AVM

Le Panorama, 456, Chemin de Carimai,
Mougins, France, 06250
Area of Operation: Europe
Tel: +33 492 181 015 **Fax:** +33 492 181 015
Email: phavy@wanadoo.fr
Product Type: 4

AVNEX LTD

3 Hartfield Close, Kents Hill, Milton Keynes,
Buckinghamshire, MK7 6HN
Area of Operation: UK & Ireland
Tel: 01908 200585
Fax: 01908 200586
Email: susan@avnex.co.uk
Web: www.avnex.co.uk
Product Type: 2

AVNEX LTD

Area of Operation: UK
Tel: 01908 200585
Fax: 01908 200586
Email: susan@avnex.co.uk
Web: www.avnex.co.uk

The revolutionary NEW system that distributes
audio, video, control, telephony and data around
your home.

BCW DESIGN

1 Church Farm Court, Neston Road,
Willaston, Wirral, CH64 2XP
Area of Operation: UK (Excluding Ireland)
Tel: 0870 757 9690
Fax: 0151 327 4307
Email: info@bcwdesign.com
Web: www.bcwdesign.com

BESPOKE INSTALLATIONS

1 Rogers Close, Tiverton, Devon, EX16 6UW
Area of Operation: UK (Excluding Ireland)
Tel: 01884 243497 **Fax:** 01884 243497
Email: michael@bespokeinstallations.com
Web: www.bespoke.biz
Product Type: 1, 2, 3, 4
Other Info: ✎

BEYOND THE INVISIBLE LIMITED

162-164 Arthur Road, London, SW19 8AQ
Area of Operation: Greater London
Tel: 0870 740 5859 **Fax:** 0870 740 5860
Email: info@beyondtheinvisible.com
Web: www.beyondtheinvisible.com

CCTV (UK) LTD

109-111 Pope Street, Birmingham,
West Midlands, B1 3AG
Area of Operation: Worldwide
Tel: 0121 200 1031
Email: sales@cctvgroup.com
Web: www.cctvgroup.com
Product Type: 1, 2, 3, 4

CENTURION

Westhill Business Park, Arnhall Business Park,
Westhill, Aberdeen, Aberdeenshire, AB32 6UF
Area of Operation: Scotland
Tel: 01224 744440 **Fax:** 01224 744819
Email: info@centurion-solutions.co.uk
Web: www.centurion-solutions.co.uk
Product Type: 2

CHOICE HI-FI

Denehurst Gardens, Richmond, Surrey, TW10
Area of Operation: UK (Excluding Ireland)
Tel: 0208 392 1959
Email: info@choice-hifi.com
Web: www.choice-hifi.com
Product Type: 1, 2, 3, 4

CLOUD9 SYSTEMS

87 Bishops Park Road, London, SW6 6DY
Area of Operation: Greater London,
South East England
Tel: 0870 420 5495 **Fax:** 0870 402 0121
Email: info@cloud9systems.co.uk
Web: www.cloud9systems.co.uk
Product Type: 1, 2, 4

COASTAL ACOUSTICS LTD

16 Trinity Church Yard, Guildford, Surrey, GU1 3RR
Area of Operation: UK (Excluding Ireland)
Tel: 01483 885670 **Fax:** 01483 885677
Email: info@coastalacoustics.com

COASTAL ACOUSTICS

Area of Operation: UK (Excluding Ireland)
Tel: 01483 885670
Fax: 01483 885677
Email: info@coastalacoustics.com

Acoustic design, multi-room A/V systems, IT
networks, telephone systems, lighting and
security for residential, marine and professional
installations. Cedia Members.

COHERE LTD

10 Singleton Scarp, London, N12 7AR
Area of Operation: Europe
Tel: 0845 456 0695
Email: info@wirefreeliving.com
Web: www.wirefreeliving.com
Product Type: 1, 2, 4

COMFORT HOME CONTROLS

Carlton House, Carlton Avenue, Chester,
Cheshire, CH4 8UE
Area of Operation: UK & Ireland
Tel: 08707 605528 **Fax:** 01244 671455
Email: sales@home-control.co.uk
Web: www.home-control.co.uk ⌂
Product Type: 2, 4 **Other Info:** ✎

CONTROLWISE LTD

Unit 7, St. Davids Industrial Estate, Pengam,
Blackwood, Caerphilly, NP12 3SW
Area of Operation: South West England
and South Wales
Tel: 01443 836836 **Fax:** 01443 836502
Email: stuart@controlwise.co.uk
Web: www.controlwise.co.uk
Product Type: 1, 2, 3, 4

CREO DESIGN

62 North Street, Leeds, West Yorkshire, LS2 7PN
Area of Operation: UK (Excluding Ireland)
Tel: 0113 246 7373 **Fax:** 0113 242 5114
Email: info@creo-designs.com
Web: www.creo-designs.co.uk
Product Type: 1, 2, 3, 4
Other Info: ✎

CUSTOM INSTALLATIONS

Custom House, 51 Baymead Lane, North Petherton,
Bridgwater, Somerset, TA6 6RN
Area of Operation: South West England
and South Wales
Tel: 01278 662555 **Fax:** 01278 662975
Email: roger@custom-installations.co.uk
Web: www.custom-installations.co.uk
Product Type: 1, 2, 3, 4

HEATING, PLUMBING & ELECTRICAL

DARTMOOR DIGITAL DESIGN
Bigadon Farm, Buckfastleigh,
Devon, TQ11 0ND
Area of Operation: UK & Ireland
Tel: 01364 644300
Fax: 01364 644308
Email: sales@iiid.co.uk
Web: www.iiid.co.uk
Product Type: 2, 3

DCD SYSTEMS LTD
43 Howards Thicket, Gerrards Cross,
Buckinghamshire, SL9 7NU
Area of Operation: UK & Ireland
Tel: 01753 882028
Fax: 01753 882029
Email: peter@dcd.co.uk
Web: www.dcd.co.uk
Product Type: 4
Other Info: ECO

DECKORUM LIMITED
4c Royal Oak Lane, Pirton,
Hitchin, Hertfordshire, CG5 3QT
Area of Operation: Greater London,
South East England, South West England
and South Wales
Tel: 0845 020 4360
Fax: 0845 020 4361
Email: simon@deckorum.com
Web: www.deckorum.com
Product Type: 1, 2, 3, 4

DESIGN 2 AUTOMATE
Unit 14 Deanfield Court,
Link 59 Business Park,
Clitheroe, Lancashire, BB7 1QS
Area of Operation: UK & Ireland
Tel: 01200 444 356
Fax: 01200 444 359
Email: info@d2a.co.uk
Web: www.d2a.co.uk
Product Type: 1, 2, 3, 4

DIGITAL DECOR
Sonas House, Button End,
Harston, Cambridge,
Cambridgeshire, CB2 5NX
Area of Operation: East England,
Greater London
Tel: 01223 870935
Fax: 01223 870935
Email: seamus@digital-decor.co.uk
Web: www.digital-decor.co.uk
Product Type: 1, 2, 3, 4
Other Info: ✐

DIGITAL PLUMBERS
84 The Chase, London, SW4 0NF
Area of Operation: Greater London
Tel: 0207 819 1730
Fax: 0207 819 1731
Email: help@digitalplumbers.com
Web: www.digitalplumbers.com
Product Type: 1, 2, 3, 4

DISCOVERY SYSTEMS LTD
1 Corn Mill Close, The Spindles, Ashton in
Makerfield, Wigan, Lancashire, WN4 0PX
Area of Operation: North West England
and North Wales
Tel: 01942 723756
Email: mike@discoverysystems.co.uk
Web: www.discoverysystems.co.uk
Product Type: 2

EAST TECHNOLOGY INTEGRATORS
22 Seer Mead, Seer Green, Beaconsfield,
Buckinghamshire, HP9 2QL
Area of Operation: Greater London,
South East England
Tel: 0845 056 0245
Email: sales@east-ti.co.uk
Web: www.east-ti.co.uk
Product Type: 1, 2, 4

EASYCOMM
1 George House, Church Street,
Buntingford, Hertfordshire, SG9 9AS
Area of Operation: UK (Excluding Ireland)
Tel: 0800 389 9459
Email: steve@ezcomm.co.uk
Web: www.ezcomm.co.uk
Product Type: 1, 2, 4

EASYGATES
75 Long Lane, Halesowen, Birmingham,
West Midlands, B62 9DJ
Area of Operation: Europe
Tel: 0870 7606 536 **Fax:** 0121 561 3395
Email: info@easygates.co.uk
Web: www.easygates.co.uk
Product Type: 1, 2, 4

EASYLIFE AUTOMATION LIMITED
Kullan House, 12 Meadowfield Road,
Stocksfield, Northumberland, NE43 7QX
Area of Operation: UK (Excluding Ireland)
Tel: 01661 844159
Fax: 01661 842594
Email: sales@easylifeautomation.com
Web: www.easylifeautomation.com
Product Type: 1, 2, 3, 4

EU SOLUTIONS
Maghull Business Centre, 1 Liverpool Road North,
Maghull, Liverpool, L31 2HB
Area of Operation: UK & Ireland
Tel: 0870 160 1660 **Fax:** 0151 526 8849
Email: ths@blueyonder.co.uk
Web: www.totalhomesolutions.co.uk
Product Type: 2

FLAMINGBOX
Perry Farm, Maiden Bradley, Wiltshire, BA12 7JQ
Area of Operation: UK (Excluding Ireland)
Tel: 01985 845440
Fax: 01985 845448
Email: james.ratcliffe@flamingbox.com
Web: www.flamingbox.com
Product Type: 1, 2, 4
Other Info: ✐

FUSION GROUP
Unit 4 Winston Avenue, Croft,
Leicester, Leicestershire, LE9 3GQ
Area of Operation: UK (Excluding Ireland)
Tel: 01455 285 252
Fax: 0116 260 3805
Email: gary.mills@fusiongroup.uk.com
Web: www.fusiongroup.uk.com
Product Type: 1, 2, 3, 4

GATESTUFF.COM
17-19 Old Woking Road, West Bysleet,
Surrey, KT14 6LW
Area of Operation: UK & Ireland
Tel: 01932 344434
Email: sales@gatestuff.com
Web: www.gatestuff.com ✐
Product Type: 4

GOLDSYSTEM LTD
46 Nightingale Lane, Bromley, Kent, BR1 2SB
Area of Operation: Greater London,
South East England
Tel: 0208 313 0485 **Fax:** 0208 313 0665
Email: info@goldsystem.co.uk
Web: www.goldsystem.co.uk
Product Type: 1, 2, 3, 4
Other Info: ✐

HELLERMANN TYTON
43-45 Salthouse Road, Brackmills,
Northampton, Northamptonshire, NN4 7EX
Area of Operation: Worldwide
Tel: 01604 706633 **Fax:** 01604 705454
Email: sales@htdata.co.uk
Web: www.homenetworksciences.com
Product Type: 1, 2, 3

HELLERMANNTYTON
Area of Operation: Worldwide
Tel: 01604 706633
Fax: 01604 705454
Email: sales@htdata.co.uk
Web: www.homenetworksciences.com ✐
Product Type: 1, 2, 3

Hellermann Tyton Home Network Sciences provide a
networking infrastructure to support home
communications and entertainment. The modular
system allows integration of Telephony, Data, TV,
CCTV and Audio Solutions.

HOMETECH INTEGRATION LTD
Earlsgate House 35, St. Ninian's Road,
Stirling, Stirlingshire, FK8 2HE
Area of Operation: UK (Excluding Ireland)
Tel: 0870 766 1060 **Fax:** 0870 766 1070
Email: info@hometechintegration.com
Web: www.hometechintegration.com
Product Type: 1, 2, 3, 4

HONEYWELL CONTROL SYSTEMS LIMITED
Honeywell House, Arlington Business Park,
Bracknell, Berkshire, RG12 1EB
Area of Operation: Worldwide
Tel: 01344 656000 **Fax:** 01344 656054
Email: literature@honeywell.com
Web: www.honeywelluk.com
Product Type: 1, 2, 4

HOUSE ELECTRONIC LTD
97 Francisco Close, Hollyfields Park,
Chafford Hundred, Essex, RM16 6YE
Area of Operation: UK & Ireland
Tel: 01375 483595 **Fax:** 01375 483595
Email: info@houselectronic.co.uk
Web: www.houselectronic.co.uk
Product Type: 1, 2, 3, 4

HOUSEHOLD AUTOMATION
Foxways, Pinkhurst Lane, Slinfold,
Horsham, West Sussex, RH13 0QR
Area of Operation: UK & Ireland
Tel: 0870 330 0071
Fax: 0870 330 0072
Email: ellis-andrew@btconnect.com
Web: www.household-automation.co.uk
Product Type: 1, 2, 3, 4

INTELLIGENT HOUSE
Langham House, Suite 401,
302 Regent Street, London, W1B 3HH
Area of Operation: UK & Ireland
Tel: 0207 394 9344
Fax: 0207 252 3879
Email: info@intelligenthouse.net
Web: www.intelligenthouse.net
Product Type: 1, 2, 3, 4

INTERACTIVE HOMES LTD
Lorton House, Conifer Crest,
Newbury, Berkshire, RG14 6RS
Area of Operation: UK (Excluding Ireland)
Tel: 01635 49111
Fax: 01635 40735
Email: sales@interactivehomes.co.uk
Web: www.interactivehomes.co.uk
Product Type: 1, 2, 3, 4

INTERCONNECTION
12 Chamberlayne Road,
Moreton Hall Industrial Estate,
Bury St Edmunds, Suffolk, IP32 7EY
Area of Operation: East England
Tel: 01284 768676 **Fax:** 01284 767161
Email: info@interconnectionltd.co.uk
Web: www.interconnectionltd.co.uk
Product Type: 2

HEATING, PLUMBING & ELECTRICAL *(side tab)*

IPHOMENET.COM
1045 Stratford Road,
Hall Green, Birmingham,
West Midlands, B28 8AS
Area of Operation: UK & Ireland
Tel: 0121 778 3003
Fax: 0121 778 1117
Email: info@iphomenet.com
Web: www.iphomenet.com
Product Type: 2

J K AUDIO VISUAL
Unit 7 Newport Business Park,
Audley Avenue, Newport,
Shropshire, TF10 7DP
Area of Operation: UK (Excluding Ireland)
Tel: 01952 825088
Fax: 01952 814884
Email: sales@jk-audiovisual.co.uk
Product Type: 1, 2, 4

JELLYBEAN CTRL LTD
Eagle House, Passfield Business Centre,
Lynchborough Road, Passfield,
Hampshire, GU30 7SB
Area of Operation: UK (Excluding Ireland)
Tel: 01428 751729
Fax: 01428 751772
Email: david@jellybeanctrl.com
Web: www.jellybeanctrl.com
Product Type: 1, 2, 3, 4

JOHN CLAYTON LIGHTING LTD
Worthingham House,
Deep Lane, Hagworthingham,
Lincolnshire, PE23 4LZ
Area of Operation: Europe
Tel: 0800 389 6395
Fax: 0870 240 6417
Email: rachel@jclighting.com
Web: www.flexidim.com
Product Type: 1, 2, 4

LAMBDA TECHNOLOGIES LTD
Stonebridge Court, Wakefield Rd,
Horbury, Wakefield,
West Yorkshire, WF4 5HQ
Area of Operation: East England,
Greater London, Midlands & Mid Wales,
North East England, North West England and
North Wales, Scotland, South East England,
South West England and South Wales
Tel: 01924 202100
Fax: 01924 299283
Email: john.brown@lambdatechnologies.com
Web: www.lambdatechnologies.com
Product Type: 1, 2, 3, 4

LAUREL HOME NETWORKS
Unit 11a, Brook Street Mill,
Macclesfield, Cheshire, SK11 7AW
Area of Operation: UK (Excluding Ireland)
Tel: 01625 668678
Email: tom.mattocks@laurel.uk.com
Web: www.laurel.uk.com
Product Type: 1, 2, 3, 4

LAUREL HOME NETWORKS
Area of Operation: UK (Excluding Ireland)
Tel: 01625 668678
Email: tom.mattocks@laurel.uk.com
Web: www.laurel.uk.com
Product Type: 1, 2, 3, 4

Specialist company offering consultation,
installation and ongoing support for today's smart
homes. For structured cabling throughout your
development be prepared at first fix.

LIFESTYLE ELECTRONICS
Woodview, Castlebridge, County Wexford
Area of Operation: Europe
Tel: +353 535 9880
Fax: +353 535 9844
Email: lifestyle123@eircom.net
Product Type: 1, 2, 3, 4
Other Info: ✐

LINK MEDIA SYSTEMS
68 St John Street, London, EC1M 4DT
Area of Operation: UK (Excluding Ireland)
Tel: 0207 251 2638
Fax: 0207 251 2487
Email: sohan@linkmediasystems.com
Web: www.linkmediasystems.com
Product Type: 1, 2, 3, 4

LUTRON EA LTD
Lutron House, 6 Sovereign Close,
Wapping, Greater London, E1W 3JF
Area of Operation: Worldwide
Tel: 0207 702 0657
Fax: 0207 480 6899
Email: ddanby@lutron.com
Web: www.lutron.com/europe
Product Type: 2, 4

MARATA VISION
20 Greenhill Crescent, Watford Business Park,
Watford, Herefordshire, WD18 8JA
Area of Operation: UK & Ireland
Tel: 01923 495595
Fax: 01923 495599
Email: info@marata.co.uk
Web: www.marata.co.uk
Product Type: 4

MARQUEE HOME LIMITED
Unit 6, Eversley Way, Thorpe Industrial Estate,
Egham, Surrey, TW20 8RF
Area of Operation: Greater London,
South East England
Tel: 07004 567888
Fax: 07004 567788
Email: paulendersby@marqueehome.co.uk
Web: www.marqueehome.co.uk
Product Type: 1, 2, 3, 4

MARTINS HI-FI
85/87 Ber Street, Norwich, Norfolk, NR1 3EY
Area of Operation: East England
Tel: 01603 627010
Fax: 01603 878019
Email: info@martinshifi.co.uk
Web: www.martinshifi.co.uk
Product Type: 1, 2, 3, 4

MICHAEL BLAKE BESPOKE
1 Rogers Close, Tiverton, Devon, EX16 6UW
Area of Operation: UK (Excluding Ireland)
Tel: 07968 105807 **Fax:** 01884 243497
Email: michael@bespoke.biz
Web: www.bespoke.biz
Product Type: 1

MODE LIGHTING (UK) LTD
The Maltings, 63 High Street, Ware,
Hertfordshire, SG12 9AD
Area of Operation: Worldwide
Tel: 01920 462121
Fax: 01920 466881
Email: james.king@modelighting.co.uk
Web: www.modelighting.co.uk
Product Type: 2, 4

MUSICAL APPROACH
111 Wolverhampton Road, Stafford,
Staffordshire, ST17 4BG
Area of Operation: UK (Excluding Ireland)
Tel: 01785 255154 **Fax:** 01785 243623
Email: info@musicalapproach.co.uk
Web: www.musicalapproach.co.uk
Product Type: 1, 2, 3, 4
Other Info: ✐

NDR SYSTEMS UK
Kendlehouse, 236 Nantwich Road,
Crewe, Cheshire, CW2 6BP
Area of Operation: UK & Ireland
Tel: 01270 219575
Email: daverehman@ndrsystemsuk.com
Web: www.ndrsystemsuk.com
Product Type: 1, 2, 3, 4

ORANGES & LEMONS
61/63 Webbs Road, Battersea,
London, SW11 6RX
Area of Operation: Greater London,
South East England
Tel: 0207 924 2040 **Fax:** 0207 924 3665
Email: oranges.lemons@virgin.net
Web: www.oandlhifi.co.uk
Product Type: 1, 2, 4

PARK SYSTEMS
INTEGRATION LIMITED
Unit 3 Queen's Park, Earlsway,
Team Valley Trading Estate,
Gateshead, Tyne & Wear, NE11 0QD
Area of Operation: UK & Ireland
Tel: 0191 497 0770
Fax: 0191 497 0772
Email: sales@parksystemsintegration.co.uk
Web: www.parksystemsintegration.co.uk
Product Type: 1, 2, 3, 4

PHILLSON LTD
144 Rickerscote Road, Stafford,
Staffordshire, ST17 4HE
Area of Operation: UK & Ireland
Tel: 0845 612 0128
Fax: 01477 535090
Email: info@phillson.co.uk
Web: www.phillson.co.uk
Product Type: 1, 2, 3, 4

PREMIER ELECTRICAL
& SECURITY SERVICES
Unit D10, The Seedbed Centre,
Langston Road, Loughton, Essex, IG10 3TQ
Area of Operation: East England,
Greater London, Midlands & Mid Wales
Tel: 0208 787 7063
Fax: 0208 508 3249
Email: info@premierelectrical.com
Web: www.premierelectrical.com
Product Type: 1, 2, 3, 4

PRESTIGE AUDIO
12 High Street, Rickmansworth,
Hertfordshire, WD3 1ER
Area of Operation: Worldwide
Tel: 01923 711113
Fax: 01923 776606
Email: info@prestigeaudio.co.uk
Web: www.prestigeaudio.co.uk
Product Type: 1, 2, 3, 4
Other Info: ✐

PW TECHNOLOGIES
Windy House, Hutton Roof,
Carnforth, Lancashire, LA6 2PE
Area of Operation: UK (Excluding Ireland)
Tel: 01524 272400
Fax: 01524 272402
Email: info@pwtechnologies.co.uk
Web: www.pwtechnologies.co.uk

PYRAMID ELECTRICAL & ALARMS
1 Morland Drive, Hinckley,
Leicestershire, LE10 0GG
Area of Operation: East England,
Greater London, Midlands & Mid Wales,
North East England, North West England
and North Wales
Tel: 01455 458325
Fax: 01455 458 325
Email: pyramid@dotdotnetdot.net
Web: www.pyramidelectrical.net
Product Type: 2, 3, 4

HEATING, PLUMBING & ELECTRICAL

The complete home entertainment facility

Enjoy your entertainment sources right around the home with LexCom Home from Square D.

- Distributed digital television
- Telephone
- Broadband
- Audio

Listen to your favourite music from wherever you choose using discreet ceiling mounted speakers, or create a home cinema experience with 5.1 surround sound.

Elite solutions for prestige homes

The ultimate in intelligent operation C-Bus from Clipsal always ensures your home is in tune with your needs.

Control and operation of lighting, audio and visual as well as home utilities is brought together in one easy to use system.

User friendly interfaces like glass switches and elegant touch screens add an extra degree of genuine style.

An integrated approach to the home

Intelligent Home Control (IHC) from Square D is unobtrusive technology that brings real benefits

- Greater safety
- Greater security
- Greater convenience

Tailored to meet your particular needs, the system will give new levels of control over your electrical services making sure your new home is a smarter place to live.

Electrical solutions to rely on

Whatever your electrical needs Schneider Electric will provide the solution

- Intelligent Home Systems
- Consumer units
- Wiring accessories
- Decorative accessories
- Outdoor accessories

For further information please call
0870 608 8 608 and ask to speak
to a member of the residential team
or visit our website
www.schneider.co.uk

Schneider Electric

ROCOCO SYSTEMS & DESIGN
26 Danbury Street, London, N1 8JU
Area of Operation: Europe
Tel: 0207 454 1234
Fax: 0207 870 0604
Email: mary@rococosystems.com
Web: www.rococosystems.com
Product Type: 1, 2, 3, 4

SCAN AUDIO
1 Honeycrock Lane, Salford,
Redhill, Surrey, RH1 5DG
Area of Operation: UK (Excluding Ireland)
Tel: 01737 778620
Fax: 01737 778620
Email: sales@scanaudio.co.uk
Web: www.scanaudio.co.uk
Product Type: 2
Other Info: ✐

SCHNEIDER ELECTRIC
Stafford Park 5, Telford, Shropshire, TF3 3BL
Area of Operation: UK (Excluding Ireland)
Tel: 0870 608 8608
Fax: 0870 608 8606
Email: sean.jordan@gb.schneider-electric.com
Web: www.squared.co.uk
Product Type: 2, 3, 4

SEISMIC INTERAUDIO
3 Maypole Drive, Kings Hill,
West Malling, Kent, ME19 4BP
Area of Operation: Europe
Tel: 0870 073 4764
Fax: 0870 073 4765
Email: info@seismic.co.uk
Web: www.seismic.co.uk
Product Type: 1, 2, 3, 4

SENSORY INTERNATIONAL LIMITED
48A London Road, Alderley Edge,
Cheshire, SK9 7DZ
Area of Operation: Worldwide
Tel: 0870 350 2244
Fax: 0870 350 2255
Email: garyc@sensoryinternational.com
Web: www.sensoryinternational.com
Product Type: 1, 2, 3, 4

SEVENOAKS SOUND & VISION (LEEDS)
62 North Street, Leeds, West Yorkshire, LS2 7PN
Area of Operation: Midlands & Mid Wales, North
East England, North West England and North Wales
Tel: 0113 245 2775
Fax: 0113 242 5114
Email: leeds@sevenoakssoundandvision.co.uk
Web: www.sevenoakssoundandvision.co.uk
Product Type: 1, 2, 4

SEVENOAKS SOUND & VISION (OXFORD)
41 St Clements Street, Oxford,
Oxfordshire, OX4 1AG
Area of Operation: UK (Excluding Ireland)
Tel: 01865 241773
Fax: 01865 794904
Email: oxford@sevenoakssoundandvision.co.uk
Web: www.sevenoakssoundandvision.co.uk

SG SYSTEMS LTD
Unit 38 Elderpark Workspace,
100 Elderpark Street, Glasgow,
Renfrewshire, G51 3TR
Area of Operation: Europe
Tel: 0141 445 4125
Fax: 0141 433 9014
Email: sales@sgsystems.ltd.uk
Web: www.sgsystems.ltd.uk
Product Type: 1, 2, 3, 4
Other Info: ✐ ✋

SIEMENS AUTOMATION & DRIVES
Sir William Siemens House, Princess Road,
Manchester, M20 2UR
Area of Operation: UK & Ireland
Tel: 0845 770 5070
Fax: 01625 665111
Email: adukmarketing@siemens.com
Web: www.siemens.co.uk\smarthomes
Product Type: 2

SILVERTEAM LIMITED
14 Brinell Way, Harfreys Road Industrial Estate,
Great Yarmouth, Norfolk, NR31 0LY
Area of Operation: Worldwide
Tel: 01493 669879 **Fax:** 01493 669647
Email: stuart@silverteam.co.uk
Web: www.silverteam.co.uk
Product Type: 1, 2, 3, 4

SINCLAIR YOUNGS LTD
Basingstoke Business Centre,
Winchester Road, Basingstoke,
Hampshire, RG22 4AU
Area of Operation: South East England
Tel: 01256 355015 **Fax:** 01256 818344
Email: david.wilson@sinclairyoungs.com
Web: www.sinclairyoungs.com

SMART HOUSE SOLUTIONS
7 Erik Road, Bowers Gifford, Essex, S13 2HY
Area of Operation: UK & Ireland
Tel: 01277 264369 **Fax:** 01277 262143
Web: www.smarthousesolutions.co.uk
Product Type: 1, 2, 3, 4

SMART KONTROLS LTD
8-9 Horsted Square, Bell Lane Business Park,
Uckfield, East Sussex, TN22 1QG
Area of Operation: Europe
Tel: 01825 769812
Fax: 01825 769813
Web: www.smartkontrols.co.uk
Product Type: 1, 2, 3, 4

SMARTCOMM LTD
45 Cressex Enterprise Centre, Lincoln Road,
Cressex Business Park, High Wycombe,
Buckinghamshire, HP12 3RL
Area of Operation: Worldwide
Tel: 01494 471912
Fax: 01494 472464
Email: info@smartcomm.co.uk
Web: www.smartcomm.co.uk

SMARTCOMM
Area of Operation: Worldwide
Tel: 01494 471 912 **Fax:** 01494 472 464
Email: info@smartcomm.co.uk
Web: www.smartcomm.co.uk

Independent installation specialists in quality Home
automation systems.
We design, install and program systems for; Entertainment,
Communications, Lighting, Music, Security and
Heating/AC. Contact us for your initial FREE
consultation.

SQUARE D
Stafford Park 5, Telford, Shropshire, TF3 3BL
Area of Operation: Worldwide
Tel: 0870 608 8608 **Fax:** 0870 608 8606
Email: john_robb@schneider.co.uk
Web: www.schneider.co.uk
Product Type: 2

STONEAUDIO UK LIMITED
Holmead Walk, Poundbury, Dorchester,
Dorset, DT1 3GE
Area of Operation: Worldwide
Tel: 01305 257555 **Fax:** 01305 257666
Email: info@stoneaudio.co.uk
Web: www.stoneaudio.co.uk ✐
Product Type: 2, 4

STUART WESTMORELAND (HOLDINGS) LTD
33 Cattle Market, Loughborough,
Leicestershire, LE11 3DL
Area of Operation: East England,
Midlands & Mid Wales
Tel: 01509 230465
Fax: 01509 267478
Email: loughborough@stuartwestmoreland.co.uk
Web: www.custom-install.co.uk
Product Type: 1, 2, 4

SUB SYSTEMS INTEGRATION LTD
202 A Dawes Road, London, SW6 7RQ
Area of Operation: Greater London
Tel: 0207 381 1177
Fax: 0207 381 8833
Email: info@sub.eu.com
Web: www.sub.eu.com
Product Type: 1, 2, 3, 4
Other Info: ✐

TEC 4 HOME
10 Carver Street, Hockley,
Birmingham, West Midlands, B1 3AS
Area of Operation: UK (Excluding Ireland)
Tel: 0121 693 9292
Fax: 0121 693 9293
Email: sammy@tec4home.com
Web: www.tec4home.com

TECCHO
Unit 19W, Kilroot Business Park,
Carrickfergus, Co Antrim, BT38 7PR
Area of Operation: UK & Ireland
Tel: 0845 890 1150
Fax: 0870 063 4120
Email: enquiries@teccho.net
Web: www.teccho.net ✐
Product Type: 1, 2, 3, 4

TELE-CONTROLS LTD
Unit 21, Three Point Business Park,
Haslingden Rossendale, Lancashire, BB4 5EH
Area of Operation: Worldwide
Tel: 01706 226333
Fax: 01706 226444
Email: info@tele-control.co.uk
Web: www.tele-power-net.com
Product Type: 4

**THE BIG PICTURE
AV & HOME CINEMA LIMITED**
51 Sutton Road, Walsall,
West Midlands, WS1 2PQ
Area of Operation: UK & Ireland
Tel: 01922 623000
Email: info@getthebigpicture.co.uk
Web: www.getthebigpicture.co.uk
Product Type: 1, 2

THE EDGE
90-92 Norwich Road, Ipswich,
Suffolk, IP1 2NL
Area of Operation: UK (Excluding Ireland)
Tel: 01473 288211
Fax: 01473 288255
Email: nick@theedge.eu.com
Web: www.theedge.eu.com
Product Type: 1, 2

THE IVY HOUSE
Ivy House, Bottom Green,
Upper Broughton, Melton Mowbray,
Leicestershire, LE14 3BA
Area of Operation: East England,
Midlands & Mid Wales
Tel: 01664 822628
Email: enquiries@the-ivy-house.com
Web: www.the-ivy-house.com

THE LITTLE CINEMA COMPANY
72 New Bond Street, London, W1S 1RR
Area of Operation: Worldwide
Tel: 0207 385 5521
Fax: 0207 385 5524
Email: sales@littlecinema.co.uk
Web: www.littlecinema.co.uk
Product Type: 1, 2, 3, 4
Other Info: ✐

THE MAJIK HOUSE COMPANY LTD
Unit J Mainline Industrial Estate, Crooklands Rd,
Milnthorpe, Cumbria, LA7 7LR
Area of Operation: North East England,
North West England and North Wales
Tel: 0870 240 8350
Fax: 0870 240 8350
Email: tim@majikhouse.com
Web: www.majikhouse.com
Product Type: 1, 2, 3, 4

THE MAX DISTRIBUTION CO LTD
The Old Bakery, 39 High Street,
East Malling, Kent, ME19 6AJ
Area of Operation: Europe
Tel: 01732 840845
Fax: 01732 841845
Email: sales@the-max.co.uk
Web: www.the-max.co.uk

THE MULTI ROOM COMPANY LTD
4 Churchill House, Churchill Road,
Cheltenham, Gloucestershire, GL53 7EG
Area of Operation: UK & Ireland
Tel: 01242 539100
Fax: 01242 539300
Email: info@multi-room.com
Web: www.multi-room.com
Product Type: 1, 2, 3, 4
Other Info: ✐

THE THINKING HOME
The White House, Wilderspool Park, Greenalls
Avenue, Warrington, Cheshire, WA4 6HL
Area of Operation: UK & Ireland
Tel: 0800 881 8319
Fax: 0800 881 8329
Email: info@thethinkinghome.com
Web: www.thethinkinghome.com
Product Type: 1, 2, 3, 4
Other Info: ✐

THINKINGBRICKS LTD
6 High Street, West Wickham, Cambridge,
Cambridgeshire, CB1 6RY
Area of Operation: UK & Ireland
Tel: 01223 290886
Email: ian@thinkingbricks.co.uk
Web: www.thinkingbricks.co.uk
Product Type: 1, 2, 3, 4

TONY TAYLOR ELECTRICAL CONTRACTING
34 Goodrest Walk, Worcester,
Worcestershire, WR3 8LG
Area of Operation: Europe
Tel: 01905 723301
Email: tony@purpleink.wanadoo.co.uk
Web: www.ttecltd.com
Product Type: 1, 2, 3, 4

TRENTA UK LIMITED
PO Box 24445, Ealing, Greater London, W5 4ZT
Area of Operation: UK & Ireland
Tel: 0208 840 6988
Fax: 0208 567 2043
Email: gg@domotics.uk.com
Web: www.domotics.uk.com
Product Type: 1, 2, 3, 4

TRI CUSTOM
Unit 38 Elderpark Workspace,
100 Elderpark Street, Govan,
Glasgow, Lanarkshire, G51 3TR
Area of Operation: UK & Ireland
Tel: 0141 4454195
Email: Info@tricustom.co.uk
Web: www.tricustom.co.uk
Product Type: 1, 2, 3, 4
Other Info: ✐

TRICAD
Woodgrove, Crampshaw Lane,
Ashtead, Surrey, KT21 2TX
Area of Operation: UK & Ireland
Tel: 07778 282675
Fax: 01372 272223
Email: darren.sparks@tricad.co.uk
Web: www.tricad.co.uk
Product Type: 1

HEATING, PLUMBING & ELECTRICAL

TRUFI LIMITED
Unit 14, Horizon Business Village, 1 Brooklands Road, Weybridge, Surrey, KT13 0TJ
Area of Operation: South East England
Tel: 01932 340750
Fax: 01932 345459
Email: info@trufi.co.uk
Web: www.trufi.co.uk
Product Type: 1, 2, 4

UNITY CONTROL
Pinkhurst Lane, Slinfold, Horsham, West Sussex, RH13 0QR
Area of Operation: Worldwide
Tel: 01403 791305
Email: info@unitycontrol.co.uk
Web: www.unitycontrol.co.uk
Product Type: 1, 2, 3, 4

UNLIMITED DREAM HOME SYSTEMS
The Long Barn, Wakefield Road, Hampole, Doncaster, South Yorkshire, DN6 7EU
Area of Operation: UK (Excluding Ireland)
Tel: 01302 725550
Fax: 01302 727274
Email: info@thehifistudios.freeserve.co.uk
Web: www.unlimiteddream.co.uk 🖱
Product Type: 1, 2, 3, 4

VALTEK
Ardykeohane, Bruff, Co Limerick, Ireland
Area of Operation: UK & Ireland
Tel: +353 6138 2116 **Fax:** +353 6138 2161
Email: aaal@eircom.net
Web: www.valtek.biz
Product Type: 1, 2, 3, 4

VISPEC
PO Box 191, Ware, Hertfordshire, SG12 0ZJ
Area of Operation: South East England
Tel: 01920 421061
Email: info@vispec.co.uk
Web: www.vispec.co.uk
Product Type: 1, 2, 3, 4

WEBSTRACT LTD
2-4 Place Farm, Wheathampstead, Hertfordshire, AL4 8SB
Area of Operation: UK (Excluding Ireland)
Tel: 0870 446 0146 **Fax:** 0870 762 6431
Email: info@webstract.co.uk
Web: www.webstract.co.uk
Product Type: 2, 4

WIRED FOR LIVING
Systems House, 11 Lomax Street, Great Harwood, Blackburn, Lancashire, BB6 7DJ
Area of Operation: UK (Excluding Ireland)
Tel: 01254 880288 **Fax:** 01254 880289
Email: info@wiredforliving.co.uk
Web: www.wiredforliving.co.uk
Product Type: 4

WIRED-AVD
9 Addison Square, Ringwood, Hampshire, BH24 1NY
Area of Operation: South East England, South West England and South Wales
Tel: 01425 479014 **Fax:** 01425 471733
Email: sales@wired-avd.co.uk
Web: www.wired-avd.co.uk
Product Type: 1, 2, 3, 4

WISELAN LIMITED
27 Old Gloucester Street, London, Greater London, WC1N 3XX
Area of Operation: UK & Ireland
Tel: 0870 787 2144 **Fax:** 0870 787 8823
Email: sales@wiselan.com
Web: www.wiselan.com 🖱
Product Type: 1, 2, 3, 4

X-HOME
Unit W Williamsons Holdings, Uphall, Lothian, EH52 6PA
Area of Operation: UK (Excluding Ireland)
Tel: 0845 130 1091 **Fax:** 0845 130 1092
Email: sales@xhome.biz
Web: www.xhome.biz
Product Type: 1, 2, 3, 4

ZEN CUSTOM SOLUTIONS LTD
25 Story Street, Hull, East Riding of Yorks, HU1 3SA
Area of Operation: UK (Excluding Ireland)
Tel: 01482 587397
Fax: 01482 326533
Email: mick@zenaudioandvision.co.uk
Web: www.zenaudioandvision.co.uk
Product Type: 1, 2, 3, 4

TELECOMMUNICATIONS

KEY
OTHER: ▽ Reclaimed 🖱 On-line shopping
🖋 Bespoke 🖐 Hand-made ECO Ecological

ABITANA NV
4 Myson Way, Raynham Road Industrial Estate, Bishop's Stortford, Hertfordshire, CM23 5JZ
Area of Operation: Europe
Tel: 01279 757775
Fax: 01279 653535
Email: sminns@minitran.co.uk
Web: www.minitran.co.uk 🖱

ABSOLUTE ELECTRICS
47 Tanfield Road, Croydon, Surrey, CR0 1AN
Area of Operation: Greater London, South East England
Tel: 0208 681 7835 **Fax:** 0208 667 9836
Email: mark@absolute-electrics.co.uk
Web: www.absolute-electrics.com

ACA APEX LIMITED
Apex House, Ivinghoe Way, Edlesborough, Dunstable, Bedfordshire, LU6 2EL
Area of Operation: UK & Ireland
Tel: 01525 220782 **Fax:** 01525 220782
Email: sales@aca-apex.com
Web: www.aca-apex.com

ALDOUS SYSTEMS (EUROPE) LTD:
28 Bishopstone, Aylesbury, Buckinghamshire, HP17 8SF
Area of Operation: Europe
Tel: 0870 240 1162
Fax: 0207 691 7844
Email: sales@aldoussystems.co.uk
Web: www.aldoussystems.co.uk

ARCHANGEL CUSTOM INSTALLATION SPECIALISTS (JERSEY) LIMITED
Mont St Aubyn, La Rue Du Carrefour Au Clercq, Grouville, Jersey, JE3 9US
Area of Operation: Europe
Tel: 0845 838 2636
Email: david.fell@archangelmultimedia.co.uk
Web: www.archangelmultimedia.co.uk

AUDIOFILE
27 Hockerill Street, Bishops Stortford, Hertfordshire, CM23 2DH
Area of Operation: UK & Ireland
Tel: 01279 506576
Fax: 01279 506638
Email: info@audiofile.co.uk
Web: www.audiofile.co.uk

AUTO AUDIO HOME DIVISION
32-34 Park Royal Road, London, NW10 7LN
Area of Operation: Worldwide
Tel: 0208 838 8838
Email: nick@autoaudio.uk.com

BCW DESIGN
1 Church Farm Court, Neston Road, Willaston, Wirral, CH64 2XP
Area of Operation: UK (Excluding Ireland)
Tel: 0870 757 9690
Fax: 0151 327 4307
Email: info@bcwdesign.com
Web: www.bcwdesign.com

BESPOKE INSTALLATIONS
1 Rogers Close, Tiverton,
Devon, EX16 6UW
Area of Operation: UK (Excluding Ireland)
Tel: 01884 243497 **Fax:** 01884 243497
Email: michael@bespokeinstallations.com
Web: www.bespoke.biz

BEYOND THE INVISIBLE LIMITED
162-164 Arthur Road, London, SW19 8AQ
Area of Operation: Greater London
Tel: 0870 740 5859 **Fax:** 0870 740 5860
Email: info@beyondtheinvisible.com
Web: www.beyondtheinvisible.com

CHOICE HI-FI
Denehurst Gardens, Richmond, Surrey, TW10
Area of Operation: UK (Excluding Ireland)
Tel: 0208 392 1959
Email: info@choice-hifi.com
Web: www.choice-hifi.com

CLOUD9 SYSTEMS
87 Bishops Park Road, London, SW6 6DY
Area of Operation: Greater London,
South East England
Tel: 0870 420 5495 **Fax:** 0870 402 0121
Email: info@cloud9systems.co.uk
Web: www.cloud9systems.co.uk

COASTAL ACOUSTICS LTD
16 Trinity Church Yard, Guildford, Surrey, GU1 3RR
Area of Operation: UK (Excluding Ireland)
Tel: 01483 885670 **Fax:** 01483 885677
Email: info@coastalacoustics.com

CREO DESIGN
62 North Street, Leeds, West Yorkshire, LS2 7PN
Area of Operation: UK (Excluding Ireland)
Tel: 0113 246 7373
Fax: 0113 242 5114
Email: info@creo-designs.co.uk
Web: www.creo-designs.co.uk

CUSTOM INSTALLATIONS
Custom House, 51 Baymead Lane, North Petherton,
Bridgwater, Somerset, TA6 6RN
Area of Operation: South West England and South
Wales
Tel: 01278 662555
Fax: 01278 662975
Email: roger@custom-installations.co.uk
Web: www.custom-installations.co.uk

DARTMOOR DIGITAL DESIGN
Bigadon Farm, Buckfastleigh, Devon, TQ11 0ND
Area of Operation: UK & Ireland
Tel: 01364 644300
Fax: 01364 644308
Email: sales@iiid.co.uk
Web: www.iiid.co.uk

DECKORUM LIMITED
4c Royal Oak Lane, Pirton, Hitchin,
Hertfordshire, CG5 3QT
Area of Operation: Greater London, South East
England, South West England and South Wales
Tel: 0845 020 4360
Fax: 0845 020 4361
Email: simon@deckorum.com
Web: www.deckorum.com

DIGITAL DECOR
Sonas House, Button End, Harston,
Cambridge, Cambridgeshire, CB2 5NX
Area of Operation: East England, Greater London
Tel: 01223 870935
Fax: 01223 870935
Email: seamus@digital-decor.co.uk
Web: www.digital-decor.co.uk
Other Info: ✍

DIGITAL PLUMBERS
Digital Plumbers, 84 The Chase, London, SW4 0NF
Area of Operation: Greater London
Tel: 0207 819 1730
Fax: 0207 819 1731
Email: help@digitalplumbers.com
Web: www.digitalplumbers.com

DISCOVERY SYSTEMS LTD
1 Corn Mill Close, The Spindles, Ashton in
Makerfield, Wigan, Lancashire, WN4 0PX
Area of Operation: North West England
and North Wales
Tel: 01942 723756
Email: mike@discoverysystems.co.uk
Web: www.discoverysystems.co.uk

DORO UK LTD
22 Walkers Road, North Moons Moat,
Redditch, Worcestershire, B98 9HE
Area of Operation: Europe
Tel: 01527 583800
Fax: 01527 583801
Email: joanne.wyke@doro-uk.com
Web: www.doro-uk.com

EASYCOMM
1 George House, Church Street,
Buntingford, Hertfordshire, SG9 9AS
Area of Operation: UK (Excluding Ireland)
Tel: 0800 389 9459
Email: steve@ezcomm.co.uk
Web: www.ezcomm.co.uk

FLAMINGBOX
Perry Farm, Maiden Bradley, Wiltshire, BA12 7JQ
Area of Operation: UK (Excluding Ireland)
Tel: 01985 845440 **Fax:** 01985 845448
Email: james.ratcliffe@flamingbox.com
Web: www.flamingbox.com
Other Info: ✍

FUSION GROUP
Unit 4 Winston Avenue, Croft, Leicester,
Leicestershire, LE9 3GQ
Area of Operation: UK (Excluding Ireland)
Tel: 01455 285252
Fax: 0116 260 3805
Email: gary.mills@fusiongroup.uk.com
Web: www.fusiongroup.uk.com

GOLDSYSTEM LTD
46 Nightingale Lane, Bromley, Kent, BR1 2SB
Area of Operation: Greater London, South East
England
Tel: 0208 313 0485
Fax: 0208 313 0665
Email: info@goldsystem.co.uk
Web: www.goldsystem.co.uk
Other Info: ✍

HELLERMANN TYTON
43-45 Salthouse Road, Brackmills,
Northampton, Northamptonshire, NN4 7EX
Area of Operation: Worldwide
Tel: 01604 706633
Fax: 01604 705454
Email: sales@htdata.co.uk
Web: www.homenetworksciences.com

HOME NETWORKING ONLINE
The Chestnuts, 28 Bishopstone, Aylesbury,
Buckinghamshire, HP17 8SF
Area of Operation: Europe
Tel: 0870 240 1162
Fax: 0207 691 7844
Email: sales@home-networking-online.co.uk
Web: www.home-networking-online.co.uk ✍

HOMETECH INTEGRATION LTD
Earlsgate House 35, St. Ninian's Road,
Stirling, Stirlingshire, FK8 2HE
Area of Operation: UK (Excluding Ireland)
Tel: 0870 766 1060
Fax: 0870 766 1070
Email: info@hometechintegration.com
Web: www.hometechintegration.com

HOUSE ELECTRONIC LTD
97 Francisco Close, Hollyfields Park,
Chafford Hundred, Essex, RM16 6YE
Area of Operation: UK & Ireland
Tel: 01375 483595
Fax: 01375 483595
Email: info@houseelectronic.co.uk
Web: www.houseelectronic.co.uk
Other Info: ✍

HOUSEHOLD AUTOMATION
Foxways, Pinkhurst Lane, Slinfold,
Horsham, West Sussex, RH13 0QR
Area of Operation: UK & Ireland
Tel: 0870 330 0071
Fax: 0870 330 0072
Email: ellis-andrew@btconnect.com
Web: www.household-automation.co.uk

INTELLIGENT HOUSE
Langham House, Suite 401,
302 Regent Street, London, W1B 3HH
Area of Operation: UK & Ireland
Tel: 0207 394 9344
Fax: 0207 252 3879
Email: info@intelligenthouse.net
Web: www.intelligenthouse.net

INTERACTIVE HOMES LTD
Lorton House, Conifer Crest,
Newbury, Berkshire, RG14 6RS
Area of Operation: UK (Excluding Ireland)
Tel: 01635 49111
Fax: 01635 40735
Email: sales@interactivehomes.co.uk
Web: www.interactivehomes.co.uk

IPHOMENET.COM
1045 Stratford Road, Hall Green,
Birmingham, West Midlands, B28 8AS
Area of Operation: UK & Ireland
Tel: 0121 778 3003 **Fax:** 0121 778 1117
Email: info@iphomenet.com
Web: www.iphomenet.com

J K AUDIO VISUAL
Unit 7 Newport Business Park, Audley Avenue,
Newport, Shropshire, TF10 7DP
Area of Operation: UK (Excluding Ireland)
Tel: 01952 825088 **Fax:** 01952 814884
Email: sales@jk-audiovisual.co.uk

JELLYBEAN CTRL LTD
Eagle House, Passfield Business Centre,
Lynchborough Road, Passfield, Hampshire, GU30 7SB
Area of Operation: UK (Excluding Ireland)
Tel: 01428 751729 **Fax:** 01428 751772
Email: david@jellybeanctrl.com
Web: www.jellybeanctrl.com

LAMBDA TECHNOLOGIES LTD
Stonebridge Court, Wakefield Road, Horbury,
Wakefield, West Yorkshire, WF4 5HQ
Area of Operation: UK (Excluding Ireland)
Tel: 01924 202100 **Fax:** 01924 299283
Email: john.brown@lambdatechnologies.com
Web: www.lambdatechnologies.com

LAUREL HOME NETWORKS
Unit 11a, Brook Street Mill, Macclesfield,
Cheshire, SK11 7AW
Area of Operation: UK (Excluding Ireland)
Tel: 01625 668678
Email: tom.mattocks@laurel.uk.com
Web: www.laurel.uk.com

LINK MEDIA SYSTEMS
68 St John Street, London, EC1M 4DT
Area of Operation: UK (Excluding Ireland)
Tel: 0207 251 2638 **Fax:** 0207 251 2487
Email: sohan@linkmediasystems.com
Web: www.linkmediasystems.com

MARQUEE HOME LIMITED
Unit 6, Eversley Way, Thorpe Industrial Estate,
Egham, Surrey, TW20 8RF
Area of Operation: Greater London,
South East England
Tel: 07004 567888 **Fax:** 07004 567788
Email: paulendersby@marqueehome.co.uk
Web: www.marqueehome.co.uk

MARTINS HI-FI
85/87 Ber Street, Norwich, Norfolk, NR1 3EY
Area of Operation: East England
Tel: 01603 627010
Fax: 01603 878019
Email: info@martinshifi.co.uk
Web: www.martinshifi.co.uk

MUSICAL APPROACH
111 Wolverhampton Road, Stafford,
Staffordshire, ST17 4BG
Area of Operation: UK (Excluding Ireland)
Tel: 01785 255154
Fax: 01785 243623
Email: info@musicalapproach.co.uk
Web: www.musicalapproach.co.uk

NDR SYSTEMS UK
Kendlehouse, 236 Nantwich Road, Crewe,
Cheshire, CW2 6BP
Area of Operation: UK & Ireland
Tel: 01270 219575
Email: daverehman@ndrsystemsuk.com
Web: www.ndrsystemsuk.com

PARK SYSTEMS INTEGRATION LTD
Unit 3 Queen's Park, Earlsway, Team Valley
Trading Estate, Gateshead, Tyne & Wear, NE11 0QD
Area of Operation: UK & Ireland
Tel: 0191 497 0770
Fax: 0191 497 0772
Email: sales@parksystemsintegration.co.uk
Web: www.parksystemsintegration.co.uk

PHILLSON LTD
144 Rickerscote Road, Stafford,
Staffordshire, ST17 4HE
Area of Operation: UK & Ireland
Tel: 0845 612 0128
Fax: 01477 535090
Email: info@phillson.co.uk
Web: www.phillson.co.uk

PREMIER ELECTRICAL & SECURITY SERVICES
Unit D10, The Seedbed Centre, Langston Road,
Loughton, Essex, IG10 3TQ
Area of Operation: East England,
Greater London, Midlands & Mid Wales
Tel: 0208 787 7063
Fax: 0208 508 3249
Email: info@premierelectrical.com
Web: www.premierelectrical.com

PW TECHNOLOGIES
Windy House, Hutton Roof, Carnforth,
Lancashire, LA6 2PE
Area of Operation: UK (Excluding Ireland)
Tel: 01524 272400
Fax: 01524 272402
Email: info@pwtechnologies.co.uk
Web: www.pwtechnologies.co.uk

ROCOCO SYSTEMS & DESIGN
26 Danbury Street, London, N1 8JU
Area of Operation: Europe
Tel: 0207 454 1234
Fax: 0207 870 0604
Email: mary@rococosystems.com
Web: www.rococosystems.com

SEVENOAKS SOUND & VISION (LEEDS)
62 North Street, Leeds, West Yorkshire, LS2 7PN
Area of Operation: Midlands & Mid Wales,
North East England, North West England
and North Wales
Tel: 0113 245 2775
Fax: 0113 242 5114
Email: leeds@sevenoakssoundandvision.co.uk
Web: www.sevenoakssoundandvision.co.uk

SEVENOAKS SOUND & VISION (OXFORD)
41 St Clements Street, Oxford,
Oxfordshire, OX4 1AG
Area of Operation: UK (Excluding Ireland)
Tel: 01865 241773 **Fax:** 01865 794904
Email: oxford@sevenoakssoundandvision.co.uk
Web: www.sevenoakssoundandvision.co.uk

SG SYSTEMS LTD
Unit 38 Elderpark Workspace,
100 Elderpark Street, Glasgow,
Renfrewshire, G51 3TR
Area of Operation: Europe
Tel: 0141 445 4125
Fax: 0141 433 9014
Email: info@sgsystems.ltd.uk
Web: www.sgsystems.ltd.uk

SMART HOUSE SOLUTIONS
7 Erik Road, Bowers Gifford, Essex, S13 2HY
Area of Operation: UK & Ireland
Tel: 01277 264369 **Fax:** 01277 262143
Web: www.smarthousesolutions.co.uk

SMART KONTROLS LTD
8-9 Horsted Square, Bell Lane Business Park,
Uckfield, East Sussex, TN22 1QG
Area of Operation: Europe
Tel: 01825 769812 **Fax:** 01825 769813
Web: www.smartkontrols.co.uk

SMARTCOMM LTD
45 Cressex Enterprise Centre,
Lincoln Road, Cressex Business Park,
High Wycombe, Buckinghamshire, HP12 3RL
Area of Operation: Worldwide
Tel: 01494 471912 **Fax:** 01494 472464
Email: info@smartcomm.co.uk
Web: www.smartcomm.co.uk

SUB SYSTEMS INTEGRATION LTD
202a Dawes Road, London, SW6 7RQ
Area of Operation: Greater London
Tel: 0207 381 1177
Fax: 0207 381 8833
Email: info@sub.eu.com
Web: www.sub.eu.com
Other Info: ✏

TECCHO
Unit 19W, Kilroot Business Park,
Carrickfergus, Co Antrim, BT38 7PR
Area of Operation: UK & Ireland
Tel: 0845 890 1150
Fax: 0870 063 4120
Email: enquiries@teccho.net
Web: www.teccho.net ✒

TECHNOTREND LTD
Unit B5 Armstrong Mall,
Southwood Business Park,
Farnborough, Hampshire, GU14 0NR
Area of Operation: UK & Ireland
Tel: 01252 513346
Fax: 01252 547498
Email: david.roberts@technotrend.co.uk
Web: www.technotrend.co.uk ✒

TELESTIAL LTD
109-111 Pope Street, Birmingham,
West Midlands, B1 3AG
Area of Operation: UK (Excluding Ireland)
Tel: 0870 855 0010
Email: sales@telestial.biz
Web: www.telestial.biz

THE IVY HOUSE
Ivy House, Bottom Green, Upper Broughton,
Melton Mowbray, Leicestershire, LE14 3BA
Area of Operation: East England,
Midlands & Mid Wales
Tel: 01664 822628
Email: enquiries@the-ivy-house.com
Web: www.the-ivy-house.com

THE MAJIK HOUSE COMPANY LTD
Unit J Mainline Industrial Estate, Crooklands Road,
Milnthorpe, Cumbria, LA7 7LR
Area of Operation: North East England,
North West England and North Wales
Tel: 0870 240 8350 **Fax:** 0870 240 8350
Email: tim@majikhouse.com
Web: www.majikhouse.com

THE MULTI ROOM COMPANY LTD
4 Churchill House, Churchill Road, Cheltenham,
Gloucestershire, GL53 7EG
Area of Operation: UK & Ireland
Tel: 01242 539100 **Fax:** 01242 539300
Email: info@multi-room.com
Web: www.multi-room.com
Other Info: ✏

THE THINKING HOME
The White House, Wilderspool Park, Greenalls
Avenue, Warrington, Cheshire, WA4 6HL
Area of Operation: UK & Ireland
Tel: 0800 881 8319 **Fax:** 0800 881 8329
Email: info@thethinkinghome.com
Web: www.thethinkinghome.com

THINKINGBRICKS LTD
6 High Street, West Wickham, Cambridge,
Cambridgeshire, CB1 6RY
Area of Operation: UK & Ireland
Tel: 01223 290886
Email: ian@thinkingbricks.co.uk
Web: www.thinkingbricks.co.uk
Other Info: ✏

TONY TAYLOR ELECTRICAL CONTRACTING
34 Goodrest Walk, Worcester,
Worcestershire, WR3 8LG
Area of Operation: Europe
Tel: 01905 723301
Email: tony@purpleink.wanadoo.co.uk
Web: www.ttecltd.com
Other Info: ✏

TRICAD
Woodgrove, Crampshaw Lane, Ashtead,
Surrey, KT21 2TX
Area of Operation: UK & Ireland
Tel: 07778 282675 **Fax:** 01372 272223
Email: darren.sparks@tricad.co.uk
Web: www.tricad.co.uk

TRUFI LIMITED
Unit 14, Horizon Business Village,
1 Brooklands Road, Weybridge, Surrey, KT13 0TJ
Area of Operation: South East England
Tel: 01932 340750 **Fax:** 01932 345459
Email: info@trufi.co.uk
Web: www.trufi.co.uk

VISPEC
PO Box 191, Ware, Hertfordshire, SG12 0ZJ
Area of Operation: South East England
Tel: 01920 421061 **Email:** info@vispec.co.uk
Web: www.vispec.co.uk
Other Info: ✏

WEBSTRACT LTD
28 Lattimore Road, Wheathampstead,
Hertfordshire, AL4 8QE
Area of Operation: UK (Excluding Ireland)
Tel: 0870 446 0146 **Fax:** 0870 762 6431
Email: info@webstract.co.uk
Web: www.webstract.co.uk

WIRED-AVD
9 Addison Square, Ringwood,
Hampshire, BH24 1NY
Area of Operation: South East England,
South West England and South Wales
Tel: 01425 479014 **Fax:** 01425 471733
Email: sales@wired-avd.co.uk
Web: www.wired-avd.co.uk

WISELAN LIMITED
27 Old Gloucester Street, London, WC1N 3XX
Area of Operation: UK & Ireland
Tel: 0870 787 2144 **Fax:** 0870 787 8823
Email: sales@wiselan.com
Web: www.wiselan.com ✒

X-HOME
Unit W Williamsons Holdings, Uphall,
Lothian, EH52 6PA
Area of Operation: UK (Excluding Ireland)
Tel: 0845 130 1091
Fax: 0845 130 1092
Email: sales@xhome.biz
Web: www.xhome.biz

ZEN CUSTOM SOLUTIONS LTD
25 Story Street, Hull, East Riding of Yorks, HU1 3SA
Area of Operation: UK (Excluding Ireland)
Tel: 01482 587397
Fax: 01482 326533
Email: mick@zenaudioandvision.co.uk
Web: www.zenaudioandvision.co.uk

HOME ENTERTAINMENT

KEY

SEE ALSO: LIGHTING - Lighting Controls and
Accessories (Smart Lighting Systems)
PRODUCT TYPES: 1= Music Systems 2 =
Home Theatre Systems   3 = Entertainment
Equipment   4 = PC Entertainment Hubs
OTHER: ▽ Reclaimed ✒ On-line shopping
✏ Bespoke ✋ Hand-made   ECO Ecological

ABITANA NV
4 Myson Way, Raynham Road Industrial Estate,
Bishop's Stortford, Hertfordshire, CM23 5JZ
Area of Operation: Europe
Tel: 01279 757775
Fax: 01279 653535
Email: sminns@minitran.co.uk
Web: www.minitran.co.uk ✒

ABSOLUTE ELECTRICS
47 Tanfield Road, Croydon, Surrey, CR0 1AN
Area of Operation: Greater London,
South East England
Tel: 0208 681 7835 **Fax:** 0208 667 9836
Email: mark@absolute-electrics.co.uk
Web: www.absolute-electrics.com
Product Type: 1, 2, 3, 4

ACA APEX LIMITED
Apex House, Ivinghoe Way, Edlesborough,
Dunstable, Bedfordshire, LU6 2EL
Area of Operation: UK & Ireland
Tel: 01525 220782 **Fax:** 01525 220782
Email: sales@aca-apex.com
Web: www.aca-apex.com
Product Type: 1, 2

ACA-APEX

Area of Operation: UK & Ireland
Tel: 01525 220782 **Fax:** 01525 220782
Email: sales@aca-apex.com
Web: www.aca-apex.com
Product Type: 1, 2

ACA-APEX Limited is a design and manufacturing
company dedicated to producing innovative
networking solutions for the residential and small
office market.

ACM INSTALLATIONS LTD
71 Western Gailes Way, Hull,
East Riding of Yorks, HU8 9EQ
Area of Operation: North East England
Tel: 0870 242 3285
Fax: 01482 377530
Email: enquiries@acminstallations.com
Web: www.acminstallations.com
Product Type: 1, 2, 3, 4
Other Info: ✏

ADV AUDIO VISUAL INSTALLATIONS LTD
12 York Place, Leeds, West Yorkshire, LS1 2DS
Area of Operation: North East England,
North West England and North Wales
Tel: 0870 199 5755
Email: info@adv-installs.co.uk
Web: www.adv-installs.co.uk
Product Type: 1, 2, 3

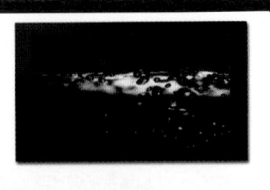

ALDOUS SYSTEMS (EUROPE) LTD:
28 Bishopstone, Aylesbury,
Buckinghamshire, HP17 8SF
Area of Operation: Europe
Tel: 0870 240 1162
Fax: 0207 691 7844
Email: sales@aldoussystems.co.uk
Web: www.aldoussystems.co.uk
Product Type: 4

AMINA TECHNOLOGIES
Cirrus House, Glebe Road, Huntingdon,
Cambridgeshire, PE29 7DX
Area of Operation: UK (Excluding Ireland)
Tel: 01480 354390
Fax: 01480 356564
Email: richard@amina.co.uk
Web: www.amina.co.uk
Product Type: 1
Other Info: ✐

**ARCHANGEL CUSTOM INSTALLATION
SPECIALISTS (JERSEY) LIMITED**
Mont St Aubyn, La Rue Du Carrefour Au Clercq,
Grouville, Jersey, JE3 9US
Area of Operation: Europe
Tel: 0845 838 2636
Email: david.fell@archangelmultimedia.co.uk
Web: www.archangelmultimedia.co.uk
Product Type: 1, 2, 3, 4

ARMOUR HOME ELECTRONICS
Unit B3, Kings Wey Business Park,
Foresyth Road, Woking, Surrey, GU21 5CA
Area of Operation: Worldwide
Tel: 01279 501111
Fax: 01279 501080
Email: info@armourhe.co.uk
Web: www.armourhe.co.uk
Product Type: 1, 3

AUDIO & CINEMA EXPERIENCE
Trenoweth, Barbican Hill, East Looe,
Cornwall, PL13 1BG
Area of Operation: UK (Excluding Ireland)
Tel: 0870 458 4438
Email: kay@audio-cinema.co.uk
Web: www.audio-cinema.co.uk
Product Type: 1, 2, 3, 4

AUDIO DESIGNS CUSTOM INSTALL
7/9 Park Place, Horsham, West Sussex, RH12 1DF
Area of Operation: South East England
Tel: 01403 252255
Fax: 01403 328065
Email: adci@audiodesigns.co.uk
Web: www.audiodesigns.co.uk
Product Type: 1, 2, 3, 4

AUDIO T CUSTOM INSTALLATIONS
15 Upper High Street, Epsom, Surrey, KT17 4QY
Area of Operation: Greater London, South East
England, South West England and South Wales
Tel: 01372 748888
Fax: 01372 747451
Email: daveadams@audio-t.co.uk
Web: www.audio-t.co.uk/custom
Product Type: 1, 2, 3, 4

AUDIO VENUE
36 Queen Street, Maidenhead,
Berkshire, SL6 1HZ
Area of Operation: Greater London,
South East England, South West England
and South Wales
Tel: 01628 633995
Fax: 01628 633654
Email: info@audiovenue.com
Web: www.audiovenue.co.uk
Product Type: 1, 2, 3

AUDIOFILE
27 Hockerill Street, Bishops Stortford,
Hertfordshire, CM23 2DH
Area of Operation: UK & Ireland
Tel: 01279 506576
Fax: 01279 506638
Email: info@audiofile.co.uk
Web: www.audiofile.co.uk

AUTO AUDIO HOME DIVISION
32-34 Park Royal Road, London, NW10 7LN
Area of Operation: Worldwide
Tel: 0208 838 8838
Email: nick@autoaudio.uk.com
Product Type: 1, 2, 3, 4

AUTOMATIC BUILDINGS LTD
Unit 12b Gorsey Lane, Coleshill,
Warwickshire, B46 1JU
Area of Operation: UK & Ireland
Tel: 01675 466881
Fax: 01675 464138
Email: info@automaticbuildings.co.uk
Web: www.automaticbuildings.co.uk

AVM
Le Panorama, 456, Chemin de Carimai,
Mougins, France, 06250
Area of Operation: Europe
Tel: +33 492 181 015
Fax: +33 492 181 015
Email: phavy@wanadoo.fr
Product Type: 1, 2, 3, 4

BCW DESIGN
1 Church Farm Court, Neston Road,
Willaston, Wirral, CH64 2XP
Area of Operation: UK (Excluding Ireland)
Tel: 0870 757 9690
Fax: 0151 327 4307
Email: info@bcwdesign.com
Web: www.bcwdesign.com

BESPOKE INSTALLATIONS
1 Rogers Close, Tiverton, Devon, EX16 6UW
Area of Operation: UK (Excluding Ireland)
Tel: 01884 243497
Fax: 01884 243497
Email: michael@bespokeinstallations.com
Web: www.bespoke.biz
Product Type: 1, 2, 3, 4

BEYOND THE INVISIBLE LIMITED
162-164 Arthur Road, London, SW19 8AQ
Area of Operation: Greater London
Tel: 0870 740 5859
Fax: 0870 740 5860
Email: info@beyondtheinvisible.com
Web: www.beyondtheinvisible.com

BRETTCOM LTD
AV House, Colne Vale Business Park,
Colne Vale Road, Milnsbridge, Huddersfield,
North Yorkshire, HD3 4NX
Area of Operation: UK (Excluding Ireland)
Tel: 01484 649905
Fax: 01484 646623
Email: sales@brettcom.co.uk
Web: www.brettcom.co.uk
Product Type: 1, 3, 4

CCTV (UK) LTD
109-111 Pope Street, Birmingham,
West Midlands, B1 3AG
Area of Operation: Worldwide
Tel: 0121 200 1031
Email: sales@cctvgroup.com
Web: www.cctvgroup.com
Product Type: 1, 2, 3, 4

CHOICE HI-FI
Denehurst Gardens, Richmond, Surrey, TW10
Area of Operation: UK (Excluding Ireland)
Tel: 0208 392 1959
Email: info@choice-hifi.com
Web: www.choice-hifi.com
Product Type: 1, 2, 3, 4
Other Info: ✐

CLOUD9 SYSTEMS
87 Bishops Park Road, London, SW6 6DY
Area of Operation: Greater London,
South East England
Tel: 0870 420 5495
Fax: 0870 402 0121
Email: info@cloud9systems.co.uk
Web: www.cloud9systems.co.uk
Product Type: 1, 2, 3, 4

COMFORT HOME CONTROLS
Carlton House, Carlton Avenue, Chester,
Cheshire, CH4 8UE
Area of Operation: UK & Ireland
Tel: 08707 605528
Fax: 01244 671455
Email: sales@home-control.co.uk
Web: www.home-control.co.uk ✐
Product Type: 1, 2, 4

CONTROLWISE LTD
Unit 7, St. Davids Industrial Estate, Pengam,
Blackwood, Caerphilly, NP12 3SW
Area of Operation: South West England
and South Wales
Tel: 01443 836836
Fax: 01443 836502
Email: stuart@controlwise.co.uk
Web: www.controlwise.co.uk
Product Type: 1, 2, 3, 4

CREO DESIGN
62 North Street, Leeds, West Yorkshire, LS2 7PN
Area of Operation: UK (Excluding Ireland)
Tel: 0113 246 7373
Fax: 0113 242 5114
Email: info@creo-designs.co.uk
Web: www.creo-designs.co.uk
Product Type: 1, 2, 3, 4

CUSTOM INSTALLATIONS
Custom House, 51 Baymead Lane, North Petherton,
Bridgwater, Somerset, TA6 6RN
Area of Operation: South West England
and South Wales
Tel: 01278 662555
Fax: 01278 662975
Email: roger@custom-installations.co.uk
Web: www.custom-installations.co.uk
Product Type: 1, 2, 3, 4

D&T ELECTRONICS
Unit 9a Cranborne Industrial Estate, Cranborne Road,
Potters Bar, Hertfordshire, EN6 3JN
Area of Operation: UK & Ireland
Tel: 0870 241 5891
Fax: 01707 653570
Email: info@dandt.co.uk
Web: www.dandt.co.uk
Product Type: 1, 2, 3

DARTMOOR DIGITAL DESIGN
Bigadon Farm, Buckfastleigh, Devon, TQ11 0ND
Area of Operation: UK & Ireland
Tel: 01364 644300
Fax: 01364 644308
Email: sales@iiid.co.uk
Web: www.iiid.co.uk
Product Type: 2

DECKORUM LIMITED
4c Royal Oak Lane, Pirton, Hitchin,
Hertfordshire, CG5 3QT
Area of Operation: Greater London, South East
England, South West England and South Wales
Tel: 0845 020 4360
Fax: 0845 020 4361
Email: simon@deckorum.com
Web: www.deckorum.com
Product Type: 1, 2, 3

DECO LEISURE
10 Carver Street, Birmingham,
West Midlands, B1 3AS
Area of Operation: Worldwide
Tel: 0121 693 9292
Fax: 0121 693 9293
Email: paul.dunkley@btinternet.com
Web: www.tec4home.com
Product Type: 3

DESIGN 2 AUTOMATE
Unit 14 Deanfield Court, Link 59 Business Park,
Clitheroe, Lancashire, BB7 1QS
Area of Operation: UK & Ireland
Tel: 01200 444356 **Fax:** 01200 444359
Email: info@d2a.co.uk
Web: www.d2a.co.uk
Product Type: 1, 2, 3, 4

DIGITAL DECOR
Sonas House, Button End, Harston,
Cambridge, Cambridgeshire, CB2 5NX
Area of Operation: East England, Greater London
Tel: 01223 870935
Fax: 01223 870935
Email: seamus@digital-decor.co.uk
Web: www.digital-decor.co.uk

DIGITAL PLUMBERS
Digital Plumbers, 84 The Chase, London, SW4 0NF
Area of Operation: Greater London
Tel: 0207 819 1730
Fax: 0207 819 1731
Email: help@digitalplumbers.com
Web: www.digitalplumbers.com
Product Type: 1, 2, 3, 4

DISCOVERY SYSTEMS LTD
1 Corn Mill Close, The Spindles, Ashton in
Makerfield, Wigan, Lancashire, WN4 0PX
Area of Operation: North West England
and North Wales
Tel: 01942 723756
Email: mike@discoverysystems.co.uk
Web: www.discoverysystems.co.uk
Product Type: 3

EAST TECHNOLOGY INTEGRATORS
22 Seer Mead, Seer Green, Beaconsfield,
Buckinghamshire, HP9 2QL
Area of Operation: Greater London,
South East England
Tel: 0845 056 0245
Email: sales@east-ti.co.uk
Web: www.east-ti.co.uk
Product Type: 1, 2

EASYCOMM
1 George House, Church Street,
Buntingford, Hertfordshire, SG9 9AS
Area of Operation: UK (Excluding Ireland)
Tel: 0800 389 9459
Email: steve@ezcomm.co.uk
Web: www.ezcomm.co.uk
Product Type: 1, 2

EASYLIFE AUTOMATION LIMITED
Kullan House, 12 Meadowfield Road,
Stocksfield, Northumberland, NE43 7QX
Area of Operation: UK (Excluding Ireland)
Tel: 01661 844159
Fax: 01661 842594
Email: sales@easylifeautomation.com
Web: www.easylifeautomation.com
Product Type: 1, 2, 3

**ECT PROJECTION SCREENS
AND ACCESSORIES**
PO Box 4020, Pangbourne, Berkshire, RG8 8TX
Area of Operation: UK & Ireland
Tel: 0118 984 1141
Fax: 0118 984 1847
Email: lucinda@ect-av.com
Web: www.ect-av.com
Product Type: 3

FLAMINGBOX
Perry Farm, Maiden Bradley,
Wiltshire, BA12 7JQ
Area of Operation: UK (Excluding Ireland)
Tel: 01985 845440
Fax: 01985 845448
Email: james.ratcliffe@flamingbox.com
Web: www.flamingbox.com
Product Type: 1, 2
Other Info: ✐

**FOCUS 21 VISUAL
COMMUNICATIONS LTD**
Tims Boatyard, Timsway, Staines,
Surrey, TW18 3JY
Area of Operation: UK & Ireland
Tel: 01784 441153
Fax: 01784 225840
Email: info@focus21.co.uk
Web: www.focus21.co.uk
Product Type: 1, 2, 3
Other Info: ✐

FUSION GROUP
Unit 4 Winston Avenue, Croft, Leicester,
Leicestershire, LE9 3GQ
Area of Operation: UK (Excluding Ireland)
Tel: 01455 285 252
Fax: 0116 260 3805
Email: gary.mills@fusiongroup.uk.com
Web: www.fusiongroup.uk.com
Product Type: 1, 2, 3, 4

GOLDSYSTEM LTD
46 Nightingale Lane, Bromley, Kent, BR1 2SB
Area of Operation: Greater London,
South East England
Tel: 0208 313 0485
Fax: 0208 313 0665
Email: info@goldsystem.co.uk
Web: www.goldsystem.co.uk
Product Type: 1, 2, 3, 4
Other Info: ✎

GRAHAMS HI-FI LTD
Canonbury Yard, 190a New North Road,
London, N1 7BS
Area of Operation: Europe
Tel: 020 7226 5500
Email: enq@grahams.co.uk
Web: www.grahams.co.uk ✎
Product Type: 1, 2, 3

HELLERMANN TYTON
43-45 Salthouse Road, Brackmills,
Northampton, Northamptonshire, NN4 7EX
Area of Operation: Worldwide
Tel: 01604 706633
Fax: 01604 705454
Email: sales@htdata.co.uk
Web: www.homenetworksciences.com
Product Type: 1

HIFI CINEMA
1 Mars House, Calleva Park,
Aldermaston, Berkshire, RG7 8LA
Area of Operation: Greater London,
South East England
Tel: 0118 982 0402
Fax: 0118 977 3535
Email: experience@hificinema.co.uk
Web: www.hificinema.co.uk
Product Type: 2

HOME NETWORKING ONLINE
The Chestnuts, 28 Bishopstone,
Aylesbury, Buckinghamshire, HP17 8SF
Area of Operation: Europe
Tel: 0870 240 1162
Fax: 0207 691 7844
Email: sales@home-networking-online.co.uk
Web: www.home-networking-online.co.uk ✎
Product Type: 3, 4

HOMETECH INTEGRATION LTD
Earlsgate House 35, St. Ninian's Road,
Stirling, Stirlingshire, FK8 2HE
Area of Operation: UK (Excluding Ireland)
Tel: 0870 766 1060
Fax: 0870 766 1070
Email: info@hometechintegration.com
Web: www.hometechintegration.com
Product Type: 1, 2, 3, 4

HOUSE ELECTRONIC LTD
97 Francisco Close, Hollyfields Park,
Chafford Hundred, Essex, RM16 6YE
Area of Operation: UK & Ireland
Tel: 01375 483595
Fax: 01375 483595
Email: info@houselectronic.co.uk
Web: www.houselectronic.co.uk
Product Type: 1, 2, 3, 4

HOUSEHOLD AUTOMATION
Foxways, Pinkhurst Lane, Slinfold,
Horsham, West Sussex, RH13 0QR
Area of Operation: UK & Ireland
Tel: 0870 330 0071 **Fax:** 0870 330 0072
Email: ellis-andrew@btconnect.com
Web: www.household-automation.co.uk
Product Type: 2, 3, 4

IKURE (AUDIO VISUAL) LTD
Brackendale House, Crowsley Road,
Shiplake, Berkshire, RG9 3JT
Area of Operation: Worldwide
Tel: 0118 940 2160
Fax: 0118 940 6845
Email: timv@ikure.co.uk
Web: www.ikure.co.uk
Product Type: 2

INTERACTIVE HOMES LTD
Lorton House, Conifer Crest, Newbury,
Berkshire, RG14 6RS
Area of Operation: UK (Excluding Ireland)
Tel: 01635 49111
Fax: 01635 40735
Email: sales@interactivehomes.co.uk
Web: www.interactivehomes.co.uk
Product Type: 1, 2, 3, 4

INTERCONNECTION
12 Chamberlayne Road,
Moreton Hall Industrial Estate,
Bury St Edmunds, Suffolk, IP32 7EY
Area of Operation: East England
Tel: 01284 768676
Fax: 01284 767161
Email: info@interconnectionltd.co.uk
Web: www.interconnectionltd.co.uk
Product Type: 1, 2, 3

IPHOMENET.COM
1045 Stratford Road, Hall Green,
Birmingham, West Midlands, B28 8AS
Area of Operation: UK & Ireland
Tel: 0121 778 3003
Fax: 0121 778 1117
Email: info@iphomenet.com
Web: www.iphomenet.com
Product Type: 3

J K AUDIO VISUAL
Unit 7 Newport Business Park, Audley Avenue,
Newport, Shropshire, TF10 7DP
Area of Operation: UK (Excluding Ireland)
Tel: 01952 825088
Fax: 01952 814884
Email: sales@jk-audiovisual.co.uk
Product Type: 2, 3

JELLYBEAN CTRL LTD
Eagle House, Passfield Business Centre,
Lynchborough Road, Passfield,
Hampshire, GU30 7SB
Area of Operation: UK (Excluding Ireland)
Tel: 01428 751729
Fax: 01428 751772
Email: david@jellybeanctrl.com
Web: www.jellybeanctrl.com
Product Type: 1, 2, 3, 4

LAMBDA TECHNOLOGIES LTD
Stonebridge Court, Wakefield Road, Horbury,
Wakefield, West Yorkshire, WF4 5HQ
Area of Operation: UK (Excluding Ireland)
Tel: 01924 202100
Fax: 01924 299283
Email: john.brown@lambdatechnologies.com
Web: www.lambdatechnologies.com
Product Type: 1, 2, 3, 4

LAUREL HOME NETWORKS
Unit 11a, Brook Street Mill,
Macclesfield, Cheshire, SK11 7AW
Area of Operation: UK (Excluding Ireland)
Tel: 01625 668678
Email: tom.mattocks@laurel.uk.com
Web: www.laurel.uk.com
Product Type: 4

LIFESTYLE ELECTRONICS
Woodview, Castlebridge, County Wexford
Area of Operation: Europe
Tel: +353 535 9880
Fax: +353 535 9844
Email: lifestyle123@eircom.net
Product Type: 1, 2, 3, 4
Other Info: ✎

LINK MEDIA SYSTEMS
68 St John Street, London, EC1M 4DT
Area of Operation: UK (Excluding Ireland)
Tel: 0207 251 2638
Fax: 0207 251 2487
Email: sohan@linkmediasystems.com
Web: www.linkmediasystems.com
Product Type: 1, 2, 3, 4

**MALCOLM J. LATHAM
CUSTOM INSTALLATION B.V.**
Belsebaan 3a, 5131 PH Alphen (NB), Netherlands
Area of Operation: Europe
Tel: +31 135 082 411
Fax: +31 135 082 617
Email: info@latham-ci.com
Web: www.latham-ci.com
Product Type: 1, 2, 3

MARANTZ HIFI UK
Kingsbridge House, Padbury Oaks,
575-583 Bath Road, Longford,
Middlesex, UB7 0EH
Area of Operation: Worldwide
Tel: 01753 680868
Fax: 01753 680428
Email: joe.thurston@marantz.co.uk
Web: www.marantz.com
Product Type: 1, 2, 3

MARATA VISION
20 Greenhill Crescent, Watford Business Park,
Watford, Herefordshire, WD18 8JA
Area of Operation: UK & Ireland
Tel: 01923 495595
Fax: 01923 495599
Email: info@marata.co.uk
Web: www.marata.co.uk
Product Type: 2, 3

MARQUEE HOME LIMITED
Unit 6, Eversley Way, Thorpe Industrial Estate,
Egham, Surrey, TW20 8RF
Area of Operation: Greater London,
South East England
Tel: 07004 567888
Fax: 07004 567788
Email: paulendersby@marqueehome.co.uk
Web: www.marqueehome.co.uk
Product Type: 1, 2, 3, 4

MARTINS HI-FI
85/87 Ber Street, Norwich, Norfolk, NR1 3EY
Area of Operation: East England
Tel: 01603 627010
Fax: 01603 878019
Email: info@martinshifi.co.uk
Web: www.martinshifi.co.uk
Product Type: 1, 2, 3, 4

MJ ACOUSTICS
9 Venture Court, Boleness Road,
Wisbech, Cambridgeshire, PE13 2XQ
Area of Operation: Worldwide
Tel: 01945 467770
Fax: 01945 467778
Email: phill@mjacoustics.co.uk
Web: www.mjacoustics.co.uk
Product Type: 1, 2, 3

MODE LIGHTING (UK) LTD
The Maltings, 63 High Street, Ware,
Hertfordshire, SG12 9AD
Area of Operation: Worldwide
Tel: 01920 462121
Fax: 01920 466881
Email: james.king@modelighting.co.uk
Web: www.modelighting.co.uk
Product Type: 3

MUSICAL APPROACH
111 Wolverhampton Road, Stafford,
Staffordshire, ST17 4BG
Area of Operation: UK (Excluding Ireland)
Tel: 01785 255154 **Fax:** 01785 243623
Email: info@musicalapproach.co.uk
Web: www.musicalapproach.co.uk
Product Type: 1, 2, 3, 4
Other Info: ✎

NDR SYSTEMS UK
Kendlehouse, 236 Nantwich Road,
Crewe, Cheshire, CW2 6BP
Area of Operation: UK & Ireland
Tel: 01270 219575
Email: daverehman@ndrsystemsuk.com
Web: www.ndrsystemsuk.com
Product Type: 1, 2, 3, 4

NEVILLE JOHNSON
Broadoak Business Park, Ashburton Road West,
Trafford Park, Manchester, Greater
Manchester, M17 1RW
Tel: 0161 873 8333
Fax: 0161 873 8335
Email: sales@nevillejohnson.co.uk
Web: www.nevillejohnson.co.uk
Product Type: 2
Other Info: ✎

OPUS GB
Gallery Court, Pilgrmage St, London, SE1 4LL
Area of Operation: Worldwide
Tel: 0207 940 2205
Fax: 0207 940 2206
Email: info@opus-technologies.co.uk
Web: www.opus-technologies.co.uk
Product Type: 1, 3

ORANGES & LEMONS
61/63 Webbs Rd, Battersea, London, SW11 6RX
Area of Operation: Greater London,
South East England
Tel: 0207 924 2040
Fax: 0207 924 3665
Email: oranges.lemons@virgin.net
Web: www.oandlhifi.co.uk
Product Type: 1, 2, 3

PARK SYSTEMS INTEGRATION LTD
Unit 3 Queen's Park, Earlsway,
Team Valley Trading Estate,
Gateshead, Tyne & Wear, NE11 0QD
Area of Operation: UK & Ireland
Tel: 0191 497 0770
Fax: 0191 497 0772
Email: sales@parksystemsintegration.co.uk
Web: www.parksystemsintegration.co.uk
Product Type: 1, 2, 3, 4

PHILLSON LTD
144 Rickerscote Road, Stafford,
Staffordshire, ST17 4HE
Area of Operation: UK & Ireland
Tel: 0845 612 0128
Fax: 01477 535090
Email: info@phillson.co.uk
Web: www.phillson.co.uk
Product Type: 1, 2, 3, 4

PLASMATVINFO.COM LIMITED
Dorisima House, Ipswich Road, Brantham,
Manningtree, Suffolk, CO11 1NR
Area of Operation: Worldwide
Tel: 01206 391001
Fax: 01206 391096
Email: info@plasmatvinfo.com
Web: www.plasmatvinfo.com
Product Type: 3

**PREMIER ELECTRICAL
AND SECURITY SERVICES**
Unit D10, The Seedbed Centre,
Langston Road, Loughton, Essex, IG10 3TQ
Area of Operation: East England,
Greater London, Midlands & Mid Wales
Tel: 0208 787 7063 **Fax:** 0208 508 3249
Email: info@premierelectrical.com
Web: www.premierelectrical.com
Product Type: 1, 2, 3, 4

PRESTIGE AUDIO
12 High Street, Rickmansworth, Herts, WD3 1ER
Area of Operation: Worldwide
Tel: 01923 711113 **Fax:** 01923 776606
Email: info@prestigeaudio.co.uk
Web: www.prestigeaudio.co.uk
Product Type: 1, 2, 3, 4
Other Info: ✎

PW TECHNOLOGIES
Windy House, Hutton Roof,
Carnforth, Lancashire, LA6 2PE
Area of Operation: UK (Excluding Ireland)
Tel: 01524 272400
Fax: 01524 272402
Email: info@pwtechnologies.co.uk
Web: www.pwtechnologies.co.uk

ROCOCO SYSTEMS & DESIGN
26 Danbury Street, London, N1 8JU
Area of Operation: Europe
Tel: 0207 454 1234
Fax: 0207 870 0604
Email: mary@rococosystems.com
Web: www.rococosystems.com
Product Type: 1, 2, 3, 4

SCAN AUDIO
1 Honeycrock Lane, Salford,
Redhill, Surrey, RH1 5DG
Area of Operation: UK (Excluding Ireland)
Tel: 01737 778620
Fax: 01737 778620
Email: sales@scanaudio.co.uk
Web: www.scanaudio.co.uk
Product Type: 1, 2, 3, 4
Other Info: ✏

SCHNEIDER ELECTRIC
Stafford Park 5, Telford, Shropshire, TF3 3BL
Area of Operation: UK (Excluding Ireland)
Tel: 0870 608 8608
Fax: 0870 608 8606
Email: sean.jordan@gb.schneider-electric.com
Web: www.squared.co.uk
Product Type: 1, 2

SEISMIC INTERAUDIO
3 Maypole Drive, Kings Hill,
West Malling, Kent, ME19 4BP
Area of Operation: Europe
Tel: 0870 073 4764
Fax: 0870 073 4765
Email: info@seismic.co.uk
Web: www.seismic.co.uk
Product Type: 1, 2, 3

SENSORY INTERNATIONAL LIMITED
48A London Road, Alderley Edge,
Cheshire, SK9 7DZ
Area of Operation: Worldwide
Tel: 0870 350 2244
Fax: 0870 350 2255
Email: garyc@sensoryinternational.com
Web: www.sensoryinternational.com
Product Type: 3, 4

SEVENOAKS SOUND & VISION (LEEDS)
62 North Street, Leeds, West Yorkshire, LS2 7PN
Area of Operation: Midlands & Mid Wales,
North East England, North West England
and North Wales
Tel: 0113 245 2775
Fax: 0113 242 5114
Email: leeds@sevenoakssoundandvision.co.uk
Web: www.sevenoakssoundandvision.co.uk
Product Type: 1, 2, 3, 4

SEVENOAKS SOUND & VISION (OXFORD)
41 St Clements Street, Oxford,
Oxfordshire, OX4 1AG
Area of Operation: UK (Excluding Ireland)
Tel: 01865 241773
Fax: 01865 794904
Email: oxford@sevenoakssoundandvision.co.uk
Web: www.sevenoakssoundandvision.co.uk

SG SYSTEMS LTD
Unit 38 Elderpark Workspace,
100 Elderpark Street, Glasgow,
Renfrewshire, G51 3TR
Area of Operation: Europe
Tel: 0141 445 4125
Fax: 0141 433 9014
Email: info@sgsystems.ltd.uk
Web: www.sgsystems.ltd.uk
Product Type: 1, 2, 3, 4

SHARP ELECTRONICS (UK) LTD.
Sharp House, Thorp Road,
Newton Heath, Manchester, M40 5BE
Area of Operation: UK & Ireland
Tel: 0161 204 2333
Fax: 0161 204 2579
Email: mike.gabriel@sharp-uk.co.uk
Web: www.sharp.co.uk ✏
Product Type: 1, 3

SINCLAIR YOUNGS LTD
Basingstoke Business Centre,
Winchester Road, Basingstoke,
Hampshire, RG22 4AU
Area of Operation: South East England
Tel: 01256 355015
Fax: 01256 818344
Email: david.wilson@sinclairyoungs.com
Web: www.sinclairyoungs.com
Product Type: 3

SMART HOUSE SOLUTIONS
7 Erik Road, Bowers Gifford,
Essex, S13 2HY
Area of Operation: UK & Ireland
Tel: 01277 264369
Fax: 01277 262143
Web: www.smarthousesolutions.co.uk
Product Type: 1, 2, 3, 4

SMART KONTROLS LTD
8-9 Horsted Square, Bell Lane Business Park,
Uckfield, East Sussex, TN22 1QG
Area of Operation: Europe
Tel: 01825 769812
Fax: 01825 769813
Web: www.smartkontrols.co.uk
Product Type: 1, 2, 3, 4

SMARTCOMM LTD
45 Cressex Enterprise Centre,
Lincoln Road, Cressex Business Park,
High Wycombe, Buckinghamshire, HP12 3RL
Area of Operation: Worldwide
Tel: 01494 471912
Fax: 01494 472464
Email: info@smartcomm.co.uk
Web: www.smartcomm.co.uk

SQUARE D
Stafford Park 5, Telford,
Shropshire, TF3 3BL
Area of Operation: Worldwide
Tel: 0870 608 8608
Fax: 0870 608 8606
Email: john_robb@schneider.co.uk
Web: www.schneider.co.uk

STONEAUDIO UK LIMITED
Holmead Walk, Poundbury,
Dorchester, Dorset, DT1 3GE
Area of Operation: Worldwide
Tel: 01305 257555
Fax: 01305 257666
Email: info@stoneaudio.co.uk
Web: www.stoneaudio.co.uk ✏
Product Type: 1, 2, 3, 4

**STUART WESTMORELAND
(HOLDINGS) LIMITED**
33 Cattle Market, Loughborough,
Leicestershire, LE11 3DL
Area of Operation: East England,
Midlands & Mid Wales
Tel: 01509 230465
Fax: 01509 267478
Email: loughborough@stuartwestmoreland.co.uk
Web: www.custom-install.co.uk
Product Type: 1, 2, 3

SUB SYSTEMS INTEGRATION LTD
202a Dawes Road, London, SW6 7RQ
Area of Operation: Greater London
Tel: 0207 381 1177
Fax: 0207 381 8833
Email: info@sub.eu.com
Web: www.sub.eu.com
Product Type: 1, 2, 3, 4
Other Info: ✏

TEC 4 HOME
10 Carver Street, Hockley,
Birmingham, West Midlands, B1 3AS
Area of Operation: UK (Excluding Ireland)
Tel: 0121 693 9292
Fax: 0121 693 9293
Email: info@tec4home.com
Web: www.tec4home.com

TECCHO
Unit 19W, Kilroot Business Park,
Carrickfergus, Co Antrim, BT38 7PR
Area of Operation: UK & Ireland
Tel: 0845 890 1150
Fax: 0870 063 4120
Email: enquiries@teccho.net
Web: www.teccho.net ✏
Product Type: 1, 2, 3, 4

TELESTIAL LTD
109-111 Pope Street, Birmingham,
West Midlands, B1 3AG
Area of Operation: UK (Excluding Ireland)
Tel: 0870 855 0010
Email: sales@telestial.biz
Web: www.telestial.biz
Product Type: 1, 3

**THE BIG PICTURE -
AV & HOME CINEMA LIMITED**
51 Sutton Road, Walsall,
West Midlands, WS1 2PQ
Area of Operation: UK & Ireland
Tel: 01922 623000
Email: info@getthebigpicture.co.uk
Web: www.getthebigpicture.co.uk
Product Type: 1, 2, 3

THE EDGE
90-92 Norwich Road,
Ipswich, Suffolk, IP1 2NL
Area of Operation: UK (Excluding Ireland)
Tel: 01473 288211
Fax: 01473 288255
Email: nick@theedge.eu.com
Web: www.theedge.eu.com
Product Type: 2, 3

THE IVY HOUSE
Ivy House, Bottom Green,
Upper Broughton, Melton Mowbray,
Leicestershire, LE14 3BA
Area of Operation: East England,
Midlands & Mid Wales
Tel: 01664 822628
Email: enquiries@the-ivy-house.com
Web: www.the-ivy-house.com

THE LITTLE CINEMA COMPANY
72 New Bond Street, London, W1S 1RR
Area of Operation: Worldwide
Tel: 0207 385 5521
Fax: 0207 385 5524
Email: sales@littlecinema.co.uk
Web: www.littlecinema.co.uk
Product Type: 1, 2, 3
Other Info: ✏

THE MAJIK HOUSE COMPANY LTD
Unit J Mainline Industrial Estate,
Crooklands Road, Milnthorpe,
Cumbria, LA7 7LR
Area of Operation: North East England,
North West England and North Wales
Tel: 0870 240 8350
Fax: 0870 240 8350
Email: tim@majikhouse.com
Web: www.majikhouse.com
Product Type: 1, 2, 3

**THE MAX DISTRIBUTION
COMPANY LIMITED**
The Old Bakery, 39 High Street,
East Malling, Kent, ME19 6AJ
Area of Operation: Europe
Tel: 01732 840845
Fax: 01732 841845
Email: sales@the-max.co.uk
Web: www.the-max.co.uk

THE MULTI ROOM COMPANY LTD
4 Churchill House, Churchill Road,
Cheltenham, Gloucestershire, GL53 7EG
Area of Operation: UK & Ireland
Tel: 01242 539100
Fax: 01242 539300
Email: info@multi-room.com
Web: www.multi-room.com
Product Type: 1, 2, 3
Other Info: ✏

THE THINKING HOME
The White House, Wilderspool Park,
Greenalls Avenue, Warrington,
Cheshire, WA4 6HL
Area of Operation: UK & Ireland
Tel: 0800 881 8319
Fax: 0800 881 8329
Email: info@thethinkinghome.com
Web: www.thethinkinghome.com
Product Type: 1, 2, 3, 4

THINKINGBRICKS LTD
6 High Street, West Wickham,
Cambridge, Cambridgeshire, CB1 6RY
Area of Operation: UK & Ireland
Tel: 01223 290886
Email: ian@thinkingbricks.co.uk
Web: www.thinkingbricks.co.uk
Product Type: 1, 2, 3, 4

**TONY TAYLOR
ELECTRICAL CONTRACTING**
34 Goodrest Walk, Worcester,
Worcestershire, WR3 8LG
Area of Operation: Europe
Tel: 01905 723301
Email: tony@purpleink.wanadoo.co.uk
Web: www.ttecltd.com
Product Type: 1, 2, 3, 4

TRI CUSTOM
Unit 38 Elderpark Workspace,
100 Elderpark Street, Govan,
Glasgow, Lanarkshire, G51 3TR
Area of Operation: UK & Ireland
Tel: 0141 4454195
Email: Info@tricustom.co.uk
Web: www.tricustom.co.uk
Product Type: 1, 2, 3
Other Info: ✏

TRICAD
Woodgrove, Crampshaw Lane,
Ashtead, Surrey, KT21 2TX
Area of Operation: UK & Ireland
Tel: 07778 282675
Fax: 01372 272223
Email: darren.sparks@tricad.co.uk
Web: www.tricad.co.uk
Product Type: 2, 3

TRUFI LIMITED
Unit 14, Horizon Business Village,
1 Brooklands Road, Weybridge,
Surrey, KT13 0TJ
Area of Operation: South East England
Tel: 01932 340750 **Fax:** 01932 345459
Email: info@trufi.co.uk
Web: www.trufi.co.uk
Product Type: 1, 2, 3

UNLIMITED DREAM HOME SYSTEMS
The Long Barn, Wakefield Road, Hampole,
Doncaster, South Yorkshire, DN6 7EU
Area of Operation: UK (Excluding Ireland)
Tel: 01302 725550 **Fax:** 01302 727274
Email: info@thehifistudios.freeserve.co.uk
Web: www.unlimiteddream.co.uk ✏
Product Type: 1, 2, 3, 4

VALTEK
Ardykeohane, Bruff, Co Limerick, Ireland
Area of Operation: UK & Ireland
Tel: +353 6138 2116
Fax: +353 6138 2161
Email: aaal@eircom.net
Web: www.valtek.biz
Product Type: 1, 2, 3, 4

HEATING, PLUMBING & ELECTRICAL

VISPEC
PO Box 191, Ware, Hertfordshire, SG12 02J
Area of Operation: South East England
Tel: 01920 421061
Email: info@vispec.co.uk
Web: www.vispec.co.uk
Product Type: 1, 2, 3, 4
Other Info:

WEBSTRACT LTD
28 Lattimore Road, Wheathampstead,
Hertfordshire, AL4 8QE
Area of Operation: UK (Excluding Ireland)
Tel: 0870 446 0146
Fax: 0870 762 6431
Email: info@webstract.co.uk
Web: www.webstract.co.uk
Product Type: 1, 2, 3

WIRED FOR LIVING
Systems House, 11 Lomax Street,
Great Harwood, Blackburn, Lancashire, BB6 7DJ
Area of Operation: UK (Excluding Ireland)
Tel: 01254 880288
Fax: 01254 880289
Email: info@wiredforliving.co.uk
Web: www.wiredforliving.co.uk
Product Type: 1, 2, 3

WIRED-AVD
9 Addison Square, Ringwood,
Hampshire, BH24 1NY
Area of Operation: South East England,
South West England and South Wales
Tel: 01425 479014 **Fax:** 01425 471733
Email: sales@wired-avd.co.uk
Web: www.wired-avd.co.uk
Product Type: 1, 2, 3, 4

WISELAN LIMITED
27 Old Gloucester Street, London, WC1N 3XX
Area of Operation: UK & Ireland
Tel: 0870 787 2144
Fax: 0870 787 8823
Email: sales@wiselan.com
Web: www.wiselan.com
Product Type: 1, 2, 3, 4

X-HOME
Unit W Williamsons Holdings,
Uphall, Lothian, EH52 6PA
Area of Operation: UK (Excluding Ireland)
Tel: 0845 130 1091
Fax: 0845 130 1092
Email: sales@xhome.biz
Web: www.xhome.biz
Product Type: 1, 2, 3, 4

ZEN CUSTOM SOLUTIONS LTD
25 Story Street, Hull, East Riding of Yorks, HU1 3SA
Area of Operation: UK (Excluding Ireland)
Tel: 01482 587397
Fax: 01482 326533
Email: mick@zenaudioandvision.co.uk
Web: www.zenaudioandvision.co.uk
Product Type: 1, 2, 3, 4

BUILT-IN VACUUMS

KEY
OTHER: ▽ Reclaimed On-line shopping
Bespoke Hand-made ECO Ecological

ADM
Ling Fields, Gargrave Road, Skipton,
North Yorkshire, BD23 1UX
Area of Operation: Europe
Tel: 01756 701051
Fax: 01756 701076
Email: info@admsystems.co.uk
Web: www.admsystems.co.uk
Other Info:

ALLERGYPLUS LIMITED
65 Stowe Drive, Southam, Warwickshire, CV47 1NZ
Area of Operation: UK & Ireland
Tel: 0870 190 0022
Fax: 0870 190 0044
Email: info@allergyplus.co.uk
Web: www.allergyplus.co.uk

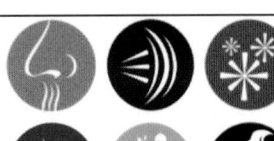

AllergyPlus
Specialists in Environmental Solutions

ALLERGYPLUS LTD
Area of Operation: UK & Ireland
Tel: 08701 9000 22 **Fax:** 08701 9000 44
Email: info@allergyplus.co.uk
Web: www.allergyplus.co.uk

Central Vacuums

Get the best for less- for the informed choice for
central vacuum systems and effective ventilation,
heat recovery, comfort cooling, climate
conditioning systems.

Say goodbye to bags and poor suction.

BEAM CENTRAL VACUUMS (UK) LTD
Unit 2 Campden Business Park, Station Road,
Chipping Camden, Gloucestershire, GL55 6JX
Area of Operation: UK (Excluding Ireland)
Tel: 0845 260 0123
Fax: 01386 849002
Email: info@beamvac.co.uk
Web: www.beamvac.co.uk

BEAM VACUUM SYSTEMS LTD
65 Deerpark Road, Castledawson, BT45 8BS
Area of Operation: UK & Ireland
Tel: 028 7938 6307 **Fax:** 028 7938 6869
Email: enquiries@beamvacuums.ie
Web: www.beamvacuums.ie

BEL-AIR CENTRAL VACUUM SYSTEMS
1 Kirkhill Gate, Newton Mearns,
East Renfrewshire, G77 5RH
Area of Operation: UK (Excluding Ireland)
Tel: 0141 639 7327 **Fax:** 0141 639 7827
Email: gordon@centralvacuumsystems.co.uk
Web: www.centralvacuumsystems.co.uk

BESPOKE INSTALLATIONS
1 Rogers Close, Tiverton, Devon, EX16 6UW
Area of Operation: UK (Excluding Ireland)
Tel: 01884 243497 **Fax:** 01884 243497
Email: michael@bespokeinstallations.com
Web: www.bespoke.biz **Other Info:**

CENTRAVAC LTD
Unit 6, Mills Hill Trading Estate, Mills Hill Road,
Middleton, Greater Manchester, M24 2ES
Area of Operation: UK & Ireland
Tel: 08000 13 15 17 **Fax:** 0161 6433377
Email: enquiries@centravac.co.uk
Web: www.centravac.co.uk

CVC DIRECT LIMITED
Longways House, High Road, Brightwell cum Sotwell,
Wallingford, Oxfordshire, OX10 0QT
Area of Operation: Europe
Tel: 01491 836666 **Fax:** 01491 838086
Email: info@cvcdirect.co.uk
Web: www.cvcdirect.co.uk

CVC DIRECT LIMITED
Area of Operation: Europe
Tel: 01491 836666 **Fax:** 01491 838086
Email: info@cvcdirect.co.uk
Web: www.cvcdirect.co.uk
Other Info:

CVC Direct Ltd supply a wide range of easy to
install, competitively priced central vacuum
systems of high technical quality with full EU
certification.

DUOVAC UK
21 Ellingham Way, Ashford, Kent, TN23 6NF
Area of Operation: UK (Excluding Ireland)
Tel: 01233 664244
Email: info@duovac.co.uk
Web: www.duovac.co.uk

EASYLIFE AUTOMATION LIMITED
Kullan House, 12 Meadowfield Road,
Stocksfield, Northumberland, NE43 7QX
Area of Operation: UK (Excluding Ireland)
Tel: 01661 844159
Fax: 01661 842594
Email: sales@easylifeautomation.com
Web: www.easylifeautomation.com

EU SOLUTIONS
Maghull Business Centre, 1 Liverpool Road North,
Maghull, Liverpool, L31 2HB
Area of Operation: UK & Ireland
Tel: 0870 160 1660 **Fax:** 0151 526 8849
Email: ths@blueyonder.co.uk
Web: www.totalhomesolutions.co.uk
Other Info: ECO

FUSION GROUP
Unit 4 Winston Avenue, Croft,
Leicester, Leicestershire, LE9 3GQ
Area of Operation: UK (Excluding Ireland)
Tel: 01455 285 252 **Fax:** 0116 260 3805
Email: gary.mills@fusiongroup.uk.com
Web: www.fusiongroup.uk.com

HAMPTON
PO Box 6074, Newbury, Berkshire, RG14 9AF
Area of Operation: UK & Ireland
Tel: 01635 25440 **Fax:** 01635 255762
Email: fieldenj@freeuk.com /
john.fielden@btconnect.com
Web: www.hamptonventilation.com

HOUSEHOLD AUTOMATION
Foxways, Pinkhurst Lane, Slinfold,
Horsham, West Sussex, RH13 0QR
Area of Operation: UK & Ireland
Tel: 0870 330 0071 **Fax:** 0870 330 0072
Email: ellis-andrew@btconnect.com
Web: www.household-automation.co.uk

NUTONE PRODUCTS UK
30 Harmer Street, Gravesend,
Kent, DA12 2AX
Area of Operation: Europe
Tel: 01474 352264
Fax: 01474 334438
Email: marionmoney@tiscali.co.uk
Other Info:

PARAGON SYSTEMS (SCOTLAND) LTD
The Office, Corbie Cottage, Maryculter,
Aberdeen, Aberdeenshire, AB12 5FT
Area of Operation: Scotland
Tel: 01224 735536 **Fax:** 01224 735537
Email: info@paragon-systems.co.uk
Web: www.paragon-systems.co.uk
Other Info: ECO

REGA VENTILATION LTD
21-22 Eldon Way, Biggleswade,
Bedfordshire, SG18 8NH
Area of Operation: Europe
Tel: 01767 600499
Fax: 01767 600487
Email: sales@rega-uk.com
Web: www.rega-uk.com

SMART CENTRAL VACUUMS LTD
Unit 3a, Whitestone Business Park,
Whitestone, Herefordshire, HR1 3SE
Area of Operation: UK (Excluding Ireland)
Tel: 01432 853019
Fax: 01432 853070
Email: info@smartvacuums.co.uk
Web: www.smartvacuums.co.uk

STARKEY SYSTEMS UK LTD
4a St Martins House, St Martins Gate,
Worcester, Worcestershire, WR1 2DT
Area of Operation: UK (Excluding Ireland)
Tel: 01905 611041
Fax: 01905 27462
Email: info@centralvacuums.co.uk
Web: www.centralvacuums.co.uk

TOTAL HOME ENVIRONMENT LTD
Unit 2 Campden Business Park, Station Road,
Chipping Campden, Gloucestershire, GL55 6JX
Area of Operation: UK (Excluding Ireland)
Tel: 0845 260 0123 **Fax:** 01386 849002
Email: info@beamvac.co.uk
Web: www.beamvac.co.uk
Other Info: ECO

BEAM CENTRAL VACUUMS / T.H.E LTD

Area of Operation: UK (Excluding Ireland)
Tel: 0845 260 0123
Fax: 01386 849002
Email: info@beamvac.co.uk
Web: www.beamvac.co.uk

Vacuum quickly, quietly, with more suction but
no re-circulated dust and nothing to lug around.
10 years Heat Recovery Ventilation experience
also. Both systems benefit Asthma and allergy
sufferers.

VACUDUCT LTD
Unit 3, Brynberth Enterprise Park,
Rhayader, Powys, LD6 5EN
Area of Operation: Europe
Tel: 0800 783 6264 **Fax:** 01597 810755
Email: info@vacuduct.co.uk
Web: www.vacuduct.co.uk
Other Info:

ACCESS SYSTEMS

KEY
OTHER: ▽ Reclaimed   On-line shopping
  Bespoke   Hand-made   ECO Ecological

4N PRODUCTS LIMITED
23 Mill Lane, Kegworth, Derby, Derbyshire, DE74 2FX
Area of Operation: UK & Ireland
Tel: 0870 241 8468 **Fax:** 01509 670029
Email: sales@4nproducts.co.uk
Web: www.4nproducts.co.uk

**A.C.E - AUTOMATION
CONTRACTORS & ENGINEERS**
64 Wannock Avenue, Willingdon,
East Sussex, BN20 9RH
Area of Operation: UK (Excluding Ireland)
Tel: 01323 489798
Fax: 07884 599411
Email: sales@ace-automation.co.uk
Web: www.ace-automation.co.uk

ABSOLUTE ELECTRICS
47 Tanfield Road, Croydon, Surrey, CR0 1AN
Area of Operation: Greater London,
South East England
Tel: 0208 681 7835
Fax: 0208 667 9836
Email: mark@absolute-electrics.co.uk
Web: www.absolute-electrics.com

ACM INSTALLATIONS LTD
71 Western Gailes Way, Hull,
East Riding of Yorks, HU8 9EQ
Area of Operation: North East England
Tel: 0870 242 3285
Fax: 01482 377530
Email: enquiries@acminstallations.com
Web: www.acminstallations.com
Other Info:

AFM SECURITY
Head Office, The Stables, Foxes Loke,
Loddon, Norfolk, NR14 6UL
Area of Operation: East England, Greater London
Tel: 01508 522078 **Fax:** 01508 522079
Email: sales@afmsecurity.co.uk
Web: www.afmsecurity.co.uk

**ARCHANGEL CUSTOM INSTALLATION
SPECIALISTS (JERSEY) LIMITED**
Mont St Aubyn, La Rue Du Carrefour Au Clercq,
Grouville, Jersey, JE3 9US
Area of Operation: Europe
Tel: 0845 838 2636
Email: david.fell@archangelmultimedia.co.uk
Web: www.archangelmultimedia.co.uk

ATG ACCESS LTD.
Automation House, Lowton Business Park,
Newton Road, Lowton, Cheshire, WA3 2AP
Area of Operation: Worldwide
Tel: 01942 685522
Fax: 01942 269676
Email: marketing@atgaccess.com
Web: www.atgaccess.com

ATLAS GROUP
Design House, 27 Salt Hill Way,
Slough, Berkshire, SL1 3TR
Area of Operation: UK (Excluding Ireland)
Tel: 01753 573573
Fax: 01753 552424
Email: info@atlasgroup.co.uk
Web: www.atlasgroup.co.uk

AUTOMATED CONTROL SERVICES LTD
Unit 16, Hightown Ind Est, Crow Arch Lane,
Ringwood, Dorset, BH24 1ND
Area of Operation: UK (Excluding Ireland)
Tel: 01425 461008
Fax: 01425 461009
Email: sales@automatedcontrolservices.co.uk
Web: www.automatedcontrolservices.co.uk

AUTOMATIC BUILDINGS LTD
Unit 12b Gorsey Lane, Coleshill,
Warwickshire, B46 1JU
Area of Operation: UK & Ireland
Tel: 01675 466881
Fax: 01675 464138
Email: info@automaticbuildings.co.uk
Web: www.automaticbuildings.co.uk

AUTOPA LIMITED
Cottage Leap, Rugby, Warwickshire, CV21 3XP
Area of Operation: Worldwide
Tel: 01788 550556
Fax: 01788 550265
Email: info@autopa.co.uk
Web: www.autopa.co.uk

BCW DESIGN
1 Church Farm Court, Neston Road,
Willaston, Wirral, CH64 2XP
Area of Operation: UK (Excluding Ireland)
Tel: 0870 757 9690
Fax: 0151 327 4307
Email: info@bcwdesign.com
Web: www.bcwdesign.com

BESPOKE INSTALLATIONS
1 Rogers Close, Tiverton, Devon, EX16 6UW
Area of Operation: UK (Excluding Ireland)
Tel: 01884 243497 **Fax:** 01884 243497
Email: michael@bespokeinstallations.com
Web: www.bespoke.biz

BEYOND THE INVISIBLE LIMITED
162-164 Arthur Road, London, SW19 8AQ
Area of Operation: Greater London
Tel: 0870 740 5859 **Fax:** 0870 740 5860
Email: info@beyondtheinvisible.com
Web: www.beyondtheinvisible.com

BPT AUTOMATION LTD
Unit 16, Sovereign Park, Cleveland Way,
Hemel Hempstead, Hertfordshire, HP2 7DA
Area of Operation: UK & Ireland
Tel: 01442 235355 **Fax:** 01442 244729
Email: sales@bpt.co.uk
Web: www.bptautomation.co.uk

BYRON
Byron House, 34 Sherwood Road,
Aston Fields Industrial Estate, Bromsgrove,
West Midlands, B60 3DR
Area of Operation: Worldwide
Tel: 01527 557700 **Fax:** 01527 557701
Email: info@chbyron.com
Web: www.chbyron.com

CCTV (UK) LTD
109-111 Pope Street, Birmingham,
West Midlands, B1 3AG
Area of Operation: Worldwide
Tel: 0121 200 1031
Email: sales@cctvgroup.com
Web: www.cctvgroup.com

CENTURION
Westhill Business Park, Arnhall Business Park,
Westhill, Aberdeen, Aberdeenshire, AB32 6UF
Area of Operation: Scotland
Tel: 01224 744440
Fax: 01224 744819
Email: info@centurion-solutions.co.uk
Web: www.centurion-solutions.co.uk

CLOUD9 SYSTEMS
87 Bishops Park Road, London, SW6 6DY
Area of Operation: Greater London,
South East England
Tel: 0870 420 5495
Fax: 0870 402 0121
Email: info@cloud9systems.co.uk
Web: www.cloud9systems.co.uk

COASTAL ACOUSTICS LTD
16 Trinity Church Yard, Guildford, Surrey, GU1 3RR
Area of Operation: UK (Excluding Ireland)
Tel: 01483 885670
Fax: 01483 885677
Email: info@coastalacoustics.com

COASTFORM SYSTEMS LTD.
18, Dinnington Business Centre,
Outgang Lane, Dinnington, Sheffield,
South Yorkshire, S25 3QX
Area of Operation: Worldwide
Tel: 01909 561470 **Fax:** 08700 516793
Email: sales@coastform.co.uk
Web: www.coastform.co.uk

COMFORT HOME CONTROLS
Carlton House, Carlton Avenue, Chester,
Cheshire, CH4 8UE
Area of Operation: UK & Ireland
Tel: 08707 605528 **Fax:** 01244 671455
Email: sales@home-control.co.uk
Web: www.home-control.co.uk

CONTROLWISE LTD
Unit 7, St. Davids Industrial Estate,
Pengam, Blackwood, Caerphilly, NP12 3SW
Area of Operation: South West England
and South Wales
Tel: 01443 836836
Fax: 01443 836502
Email: stuart@controlwise.co.uk
Web: www.controlwise.co.uk

DECKORUM LIMITED
4c Royal Oak Lane, Pirton, Hitchin,
Hertfordshire, CG5 3QT
Area of Operation: Greater London,
South East England, South West England
and South Wales
Tel: 0845 020 4360
Fax: 0845 020 4361
Email: simon@deckorum.com
Web: www.deckorum.com

DIGITAL DECOR
Sonas House, Button End, Harston,
Cambridge, Cambridgeshire, CB2 5NX
Area of Operation: East England, Greater London
Tel: 01223 870935
Fax: 01223 870935
Email: seamus@digital-decor.co.uk
Web: www.digital-decor.co.uk
Other Info:

DISCOVERY SYSTEMS LTD
1 Corn Mill Close, The Spindles, Ashton in
Makerfield, Wigan, Lancashire, WN4 0PX
Area of Operation: North West England
and North Wales
Tel: 01942 723756
Email: mike@discoverysystems.co.uk
Web: www.discoverysystems.co.uk

DORCUM LTD
11 Lyndhurst Road, Hove, East Sussex, BN3 6FA
Area of Operation: UK & Ireland
Tel: 01273 230023
Fax: 01273 220108
Email: sales@teleporter.co.uk
Web: www.teleporter.co.uk

EAST TECHNOLOGY INTEGRATORS
22 Seer Mead, Seer Green, Beaconsfield,
Buckinghamshire, HP9 2QL
Area of Operation: Greater London,
South East England
Tel: 0845 056 0245
Email: sales@east-ti.co.uk
Web: www.east-ti.co.uk

EASYGATES
75 Long Lane, Halesowen, Birmingham,
West Midlands, B62 9DJ
Area of Operation: Europe
Tel: 0870 7606 536
Fax: 0121 561 3395
Email: info@easygates.co.uk
Web: www.easygates.co.uk

FARFISA SECURITY PRODUCTS LTD
The Brows, Farnham Road,
Liss, Hampshire, GU33 6JG
Area of Operation: UK (Excluding Ireland)
Tel: 01730 893839 **Fax:** 01730 893472
Email: info@farfisasecurity.co.uk
Web: www.farfisasecurity.co.uk

FUSION GROUP
Unit 4 Winston Avenue, Croft,
Leicester, Leicestershire, LE9 3GQ
Area of Operation: UK (Excluding Ireland)
Tel: 01455 285252 **Fax:** 0116 260 3805
Email: gary.mills@fusiongroup.uk.com
Web: www.fusiongroup.uk.com

GATESTUFF.COM
17-19 Old Woking Road,
West Bysleet, Surrey, KT14 6LW
Area of Operation: UK & Ireland
Tel: 01932 344434
Email: sales@gatestuff.com
Web: www.gatestuff.com

HEATING, PLUMBING & ELECTRICAL

HOMETECH INTEGRATION LTD
Earlsgate House 35, St. Ninian's Road,
Stirling, Stirlingshire, FK8 2HE
Area of Operation: UK (Excluding Ireland)
Tel: 0870 766 1060
Fax: 0870 766 1070
Email: info@hometechintegration.com
Web: www.hometechintegration.com

HOUSE ELECTRONIC LTD
97 Francisco Close, Hollyfields Park,
Chafford Hundred, Essex, RM16 6YE
Area of Operation: UK & Ireland
Tel: 01375 483595
Fax: 01375 483595
Email: info@houselectronic.co.uk
Web: www.houselectronic.co.uk

INTERACTIVE HOMES LTD
Lorton House, Conifer Crest,
Newbury, Berkshire, RG14 6RS
Area of Operation: UK (Excluding Ireland)
Tel: 01635 49111
Fax: 01635 40735
Email: sales@interactivehomes.co.uk
Web: www.interactivehomes.co.uk

JOHN PLANCK LTD
Southern House, Anthonys Way,
Medway City Estate, Rochester,
Kent, ME2 4DN
Area of Operation: UK (Excluding Ireland)
Tel: 01634 720077
Fax: 01634 720111
Email: john@johnplanck.co.uk
Web: www.johnplanck.co.uk

LAMBDA TECHNOLOGIES LTD
Stonebridge Court, Wakefield Road, Horbury,
Wakefield, West Yorkshire, WF4 5HQ
Area of Operation: UK (Excluding Ireland)
Tel: 01924 202100
Fax: 01924 299283
Email: john.brown@lambdatechnologies.com
Web: www.lambdatechnologies.com

LINK MEDIA SYSTEMS
68 St John Street, London, EC1M 4DT
Area of Operation: UK (Excluding Ireland)
Tel: 0207 251 2638
Fax: 0207 251 2487
Email: sohan@linkmediasystems.com
Web: www.linkmediasystems.com

MYGARD SAFE & SECURE PLC
Suite 10 Lloyd Berkeley Place, Pebble Lane,
Aylesbury, Buckinghamshire, HP20 2JH
Area of Operation: UK (Excluding Ireland)
Tel: 01296 433746
Email: businessdevelopment@mygard.com
Web: www.mygard.com

PARK SYSTEMS INTEGRATION LTD
Unit 3 Queen's Park, Earlsway,
Team Valley Trading Estate,
Gateshead, Tyne & Wear, NE11 0QD
Area of Operation: UK & Ireland
Tel: 0191 497 0770
Fax: 0191 497 0772
Email: sales@parksystemsintegration.co.uk
Web: www.parksystemsintegration.co.uk

PHILLSON LTD
144 Rickerscote Road, Stafford,
Staffordshire, ST17 4HE
Area of Operation: UK & Ireland
Tel: 0845 612 0128
Fax: 01477 535090
Email: info@phillson.co.uk
Web: www.phillson.co.uk

PREMIER ELECTRICAL & SECURITY SERVICES
Unit D10, The Seedbed Centre, Langston Road,
Loughton, Essex, IG10 3TQ
Area of Operation: East England,
Greater London, Midlands & Mid Wales
Tel: 0208 787 7063 **Fax:** 0208 508 3249
Email: info@premierelectrical.com
Web: www.premierelectrical.com

PRESTIGE AUDIO
12 High Street, Rickmansworth,
Hertfordshire, WD3 1ER
Area of Operation: Worldwide
Tel: 01923 711113
Fax: 01923 776606
Email: info@prestigeaudio.co.uk
Web: www.prestigeaudio.co.uk
Other Info:

PW TECHNOLOGIES
Windy House, Hutton Roof, Carnforth,
Lancashire, LA6 2PE
Area of Operation: UK (Excluding Ireland)
Tel: 01524 272400
Fax: 01524 272402
Email: info@pwtechnologies.co.uk
Web: www.pwtechnologies.co.uk

PYRAMID ELECTRICAL & ALARMS
1 Morland Drive, Hinckley,
Leicestershire, Leicestershire, LE10 0GG
Area of Operation: East England, Greater London,
Midlands & Mid Wales, North East England,
North West England and North Wales
Tel: 01455 458325
Fax: 01455 458 325
Email: pyramid@dotdotnetdot.net
Web: www.pyramidelectrical.net

ROCOCO SYSTEMS & DESIGN
26 Danbury Street, London, N1 8JU
Area of Operation: Europe
Tel: 0207 454 1234
Fax: 0207 870 0604
Email: mary@rococosystems.com
Web: www.rococosystems.com

SECURIKEY LTD
P O Box 18, Aldershot, Hampshire, GU12 4SL
Area of Operation: Worldwide
Tel: 01252 311888
Fax: 01252 343950
Email: enquiries@securikey.co.uk
Web: www.securikey.co.uk

SENSORY INTERNATIONAL LIMITED
48A London Road, Alderley Edge, Cheshire, SK9 7DZ
Area of Operation: Worldwide
Tel: 0870 350 2244
Fax: 0870 350 2255
Email: garyc@sensoryinternational.com
Web: www.sensoryinternational.com

SEVENOAKS SOUND & VISION (LEEDS)
62 North Street, Leeds, West Yorkshire, LS2 7PN
Area of Operation: Midlands & Mid Wales, North
East England, North West England and North Wales
Tel: 0113 245 2775
Fax: 0113 242 5114
Email: leeds@sevenoakssoundandvision.co.uk
Web: www.sevenoakssoundandvision.co.uk

SEVENOAKS SOUND & VISION (OXFORD)
41 St Clements Street, Oxford, Oxfordshire, OX4 1AG
Area of Operation: UK (Excluding Ireland)
Tel: 01865 241773
Fax: 01865 794904
Email: oxford@sevenoakssoundandvision.co.uk
Web: www.sevenoakssoundandvision.co.uk

SMARTCOMM LTD
45 Cressex Enterprise Centre, Lincoln Road,
Cressex Business Park, High Wycombe,
Buckinghamshire, HP12 3RL
Area of Operation: Worldwide
Tel: 01494 471912
Fax: 01494 472464
Email: info@smartcomm.co.uk
Web: www.smartcomm.co.uk

TELGUARD
Ashcombe House, London Road,
Dorking, Surrey, RH4 1TA
Area of Operation: UK & Ireland
Tel: 01306 877889
Fax: 01306 877990
Email: sales@telguard.co.uk
Web: www.doorentry.co.uk

THE MULTI ROOM COMPANY LTD
4 Churchill House, Churchill Road,
Cheltenham, Gloucestershire, GL53 7EG
Area of Operation: UK & Ireland
Tel: 01242 539100
Fax: 01242 539300
Email: info@multi-room.com
Web: www.multi-room.com

THE THINKING HOME
The White House, Wilderspool Park, Greenalls
Avenue, Warrington, Cheshire, WA4 6HL
Area of Operation: UK & Ireland
Tel: 0800 881 8319
Fax: 0800 881 8329
Email: info@thethinkinghome.com
Web: www.thethinkinghome.com
Other Info:

TLC ELECTRICAL WHOLESALERS
TLC Building, Off Fleming Way,
Crawley, West Sussex, RH10 9JY
Area of Operation: Worldwide
Tel: 01293 565630
Fax: 01293 425234
Email: sales@tlc-direct.co.uk
Web: www.tlc-direct.co.uk

DISCOUNTED PRICES FROM TLC ELECTRICAL!

Area of Operation: Worldwide
Tel: 01293 565630
Fax: 01293 425234
Email: sales@tlc-direct.co.uk
Web: www.tlc-direct.co.uk

TLC Electrical Wholesalers offer ' below trade' prices
on over 10,000 products.
22 Warehouses nationwide.
Online website: www.tlc-direct.co.uk
Free Catalogue (24hr Catalogue line) 01293 42 00 00

TRUFI LIMITED
Unit 14, Horizon Business Village,
1 Brooklands Road, Weybridge,
Surrey, KT13 0TJ
Area of Operation: South East England
Tel: 01932 340750 **Fax:** 01932 345459
Email: info@trufi.co.uk
Web: www.trufi.co.uk
Other Info:

UNLIMITED DREAM HOME SYSTEMS
The Long Barn, Wakefield Road, Hampole,
Doncaster, South Yorkshire, DN6 7EU
Area of Operation: UK (Excluding Ireland)
Tel: 01302 725550 **Fax:** 01302 727274
Email: info@thehifistudios.freeserve.co.uk
Web: www.unlimiteddream.co.uk

VEERMOUNT TECHNOLOGY LTD
15 Ancaster Crescent,
New Malden, Surrey, KT3 6BD
Area of Operation: Worldwide
Tel: 0208 241 6161 **Fax:** 0208 241 6515
Email: jmoore@veermounttechnology.co.uk
Web: www.veermounttechnology.co.uk

VISPEC
PO Box 191, Ware, Hertfordshire, SG12 0ZJ
Area of Operation: South East England
Tel: 01920 421061
Email: info@vispec.co.uk
Web: www.vispec.co.uk

WANDSWORTH GROUP LTD
Albert Drive, Sheerwater, Woking, Surrey, GU21 5SE
Area of Operation: UK (Excluding Ireland)
Tel: 01483 740740
Fax: 01483 740384
Email: info@wandsworthgroup.com
Web: www.wandsworthgroup.com

WIRED-AVD
9 Addison Square, Ringwood,
Hampshire, BH24 1NY
Area of Operation: South East England,
South West England and South Wales
Tel: 01425 479014 **Fax:** 01425 471733
Email: sales@wired-avd.co.uk
Web: www.wired-avd.co.uk

X-HOME
Unit W Williamsons Holdings, Uphall,
Lothian, EH52 6PA
Area of Operation: UK (Excluding Ireland)
Tel: 0845 130 1091 **Fax:** 0845 130 1092
Email: sales@xhome.biz
Web: www.xhome.biz

ALARMS

KEY
OTHER: ▽ Reclaimed   ⌐ On-line shopping
Bespoke   Hand-made   ECO Ecological

ABSOLUTE ELECTRICS
47 Tanfield Road, Croydon, Surrey, CR0 1AN
Area of Operation: Greater London,
South East England
Tel: 0208 681 7835
Fax: 0208 667 9836
Email: mark@absolute-electrics.co.uk
Web: www.absolute-electrics.com

ACM INSTALLATIONS LTD
71 Western Gailes Way, Hull,
East Riding of Yorks, HU8 9EQ
Area of Operation: North East England
Tel: 0870 242 3285
Fax: 01482 377530
Email: enquiries@acminstallations.com
Web: www.acminstallations.com
Other Info:

AFM SECURITY
Head Office, The Stables, Foxes Loke,
Loddon, Norfolk, NR14 6UL
Area of Operation: East England, Greater London
Tel: 01508 522078
Fax: 01508 522079
Email: sales@afmsecurity.co.uk
Web: www.afmsecurity.co.uk

ARCHANGEL CUSTOM INSTALLATION
SPECIALISTS (JERSEY) LIMITED
Mont St Aubyn, La Rue Du Carrefour Au Clercq,
Grouville, Jersey, JE3 9US
Area of Operation: Europe
Tel: 0845 838 2636
Email: david.fell@archangelmultimedia.co.uk
Web: www.archangelmultimedia.co.uk

AUTOMATIC BUILDINGS LTD
Unit 12b Gorsey Lane, Coleshill,
Warwickshire, B46 1JU
Area of Operation: UK & Ireland
Tel: 01675 466881
Fax: 01675 464138
Email: info@automaticbuildings.co.uk
Web: www.automaticbuildings.co.uk

BCW DESIGN
1 Church Farm Court, Neston Road,
Willaston, Wirral, CH64 2XP
Area of Operation: UK (Excluding Ireland)
Tel: 0870 757 9690
Fax: 0151 327 4307
Email: info@bcwdesign.com
Web: www.bcwdesign.com

BESPOKE INSTALLATIONS
1 Rogers Close, Tiverton, Devon, EX16 6UW
Area of Operation: UK (Excluding Ireland)
Tel: 01884 243497 **Fax:** 01884 243497
Email: michael@bespokeinstallations.com
Web: www.bespoke.biz

BYRON
Byron House, 34 Sherwood Road,
Aston Fields Industrial Estate, Bromsgrove,
West Midlands, B60 3DR
Area of Operation: Worldwide
Tel: 01527 557700
Fax: 01527 557701
Email: info@chbyron.com
Web: www.chbyron.com

CCTV (UK) LTD
109-111 Pope Street, Birmingham,
West Midlands, B1 3AG
Area of Operation: Worldwide
Tel: 0121 200 1031
Email: sales@cctvgroup.com
Web: www.cctvgroup.com

CENTURION
Westhill Business Park, Arnhall Business Park,
Westhill, Aberdeen, Aberdeenshire, AB32 6UF
Area of Operation: Scotland
Tel: 01224 744440
Fax: 01224 744819
Email: info@centurion-solutions.co.uk
Web: www.centurion-solutions.co.uk

CHARNOCK ENTERPRISES
Disklok House, Preston Road, Charnock Richard,
Chorley, Lancashire, PR7 5HH
Area of Operation: UK (Excluding Ireland)
Tel: 01257 795100
Fax: 01257 795101
Email: sales@disklokuk.co.uk
Web: www.charnockenterprises.co.uk

COMFORT HOME CONTROLS
Carlton House, Carlton Avenue, Chester,
Cheshire, CH4 8UE
Area of Operation: UK & Ireland
Tel: 08707 605528
Fax: 01244 671455
Email: sales@home-control.co.uk
Web: www.home-control.co.uk ⌐⊟

CONTROLWISE LTD
Unit 7, St. Davids Industrial Estate, Pengam,
Blackwood, Caerphilly, NP12 3SW
Area of Operation: South West England
and South Wales
Tel: 01443 836836
Fax: 01443 836502
Email: stuart@controlwise.co.uk
Web: www.controlwise.co.uk

DIGITAL DECOR
Sonas House, Button End, Harston,
Cambridge, Cambridgeshire, CB2 5NX
Area of Operation: East England, Greater London
Tel: 01223 870935
Fax: 01223 870935
Email: seamus@digital-decor.co.uk
Web: www.digital-decor.co.uk
Other Info: ✎

DISCOVERY SYSTEMS LTD
1 Corn Mill Close, The Spindles, Ashton in
Makerfield, Wigan, Lancashire, WN4 0PX
Area of Operation: North West England
and North Wales
Tel: 01942 723756
Email: mike@discoverysystems.co.uk
Web: www.discoverysystems.co.uk

ELECTRICALSHOP.NET
81 Kinson Road, Wallisdown,
Bournemouth, Dorset, BH10 4DG
Area of Operation: UK (Excluding Ireland)
Tel: 0870 027 3730
Fax: 0870 027 3731
Email: sales@electricalshop.net
Web: www.electricalshop.net ⌐⊟

FIREANGEL LIMITED
The TechnoCentre, Puma Way,
Coventry, Warwickshire, CV1 2TT
Area of Operation: Worldwide
Tel: 02476 236600
Email: info@fireangel.co.uk
Web: www.fireangel.co.uk ⌐⊟

GOLDSYSTEM LTD
46 Nightingale Lane, Bromley, Kent, BR1 2SB
Area of Operation: Greater London,
South East England
Tel: 0208 313 0485
Fax: 0208 313 0665
Email: info@goldsystem.co.uk
Web: www.goldsystem.co.uk
Other Info: ✎

HOUSE ELECTRONIC LTD
97 Francisco Close, Hollyfields Park,
Chafford Hundred, Essex, RM16 6YE
Area of Operation: UK & Ireland
Tel: 01375 483595
Fax: 01375 483595
Email: info@houselectronic.co.uk
Web: www.houselectronic.co.uk

LINK MEDIA SYSTEMS
68 St John Street, London, EC1M 4DT
Area of Operation: UK (Excluding Ireland)
Tel: 0207 251 2638
Fax: 0207 251 2487
Email: sohan@linkmediasystems.com
Web: www.linkmediasystems.com

MYGARD SAFE & SECURE PLC
Suite 10 Lloyd Berkeley Place, Pebble Lane,
Aylesbury, Buckinghamshire, HP20 2JH
Area of Operation: UK (Excluding Ireland)
Tel: 01296 433746
Email: businessdevelopment@mygard.com
Web: www.mygard.com ⌐⊟

PARK SYSTEMS INTEGRATION LTD
Unit 3 Queen's Park, Earlsway, Team Valley Trading
Estate, Gateshead, Tyne & Wear, NE11 0QD
Area of Operation: UK & Ireland
Tel: 0191 497 0770
Fax: 0191 497 0772
Email: sales@parksystemsintegration.co.uk
Web: www.parksystemsintegration.co.uk

PHILLSON LTD
144 Rickerscote Road, Stafford,
Staffordshire, ST17 4HE
Area of Operation: UK & Ireland
Tel: 0845 612 0128
Fax: 01477 535090
Email: info@phillson.co.uk
Web: www.phillson.co.uk

PREMIER ELECTRICAL & SECURITY SERVICES
Unit D10, The Seedbed Centre,
Langston Road, Loughton, Essex, IG10 3TQ
Area of Operation: East England,
Greater London, Midlands & Mid Wales
Tel: 0208 787 7063
Fax: 0208 508 3249
Email: info@premierelectrical.com
Web: www.premierelectrical.com

PYRAMID ELECTRICAL & ALARMS
1 Morland Drive, Hinckley, Leicestershire,
Leicestershire, LE10 0GG
Area of Operation: East England, Greater London,
Midlands & Mid Wales, North East England, North
West England and North Wales
Tel: 01455 458325
Fax: 01455 458 325
Email: pyramid@dotdotnetdot.net
Web: www.pyramidelectrical.net

SCHNEIDER ELECTRIC
Stafford Park 5, Telford, Shropshire, TF3 3BL
Area of Operation: UK (Excluding Ireland)
Tel: 0870 608 8608
Fax: 0870 608 8606
Email: sean.jordan@gb.schneider-electric.com
Web: www.squared.co.uk

SECURIKEY LTD
P O Box 18, Aldershot, Hampshire, GU12 4SL
Area of Operation: Worldwide
Tel: 01252 311888 **Fax:** 01252 343950
Email: enquiries@securikey.co.uk
Web: www.securikey.co.uk

TECHNOTREND LTD
Unit B5 Armstrong Mall, Southwood Business Park,
Farnborough, Hampshire, GU14 0NR
Area of Operation: UK & Ireland
Tel: 01252 513346 **Fax:** 01252 547498
Email: david.roberts@technotrend.co.uk
Web: www.technotrend.co.uk ⌐⊟

THE MULTI ROOM COMPANY LTD
4 Churchill House, Churchill Road, Cheltenham,
Gloucestershire, GL53 7EG
Area of Operation: UK & Ireland
Tel: 01242 539100 **Fax:** 01242 539300
Email: info@multi-room.com
Web: www.multi-room.com

TLC ELECTRICAL WHOLESALERS
TLC Building, Off Fleming Way, Crawley,
West Sussex, RH10 9JY
Area of Operation: Worldwide
Tel: 01293 565630 **Fax:** 01293 425234
Email: sales@tlc-direct.co.uk
Web: www.tlc-direct.co.uk ⌐⊟

TRUFI LIMITED
Unit 14, Horizon Business Village, 1 Brooklands
Road, Weybridge, Surrey, KT13 0TJ
Area of Operation: South East England
Tel: 01932 340750 **Fax:** 01932 345459
Email: info@trufi.co.uk
Web: www.trufi.co.uk

VISPEC
PO Box 191, Ware, Hertfordshire, SG12 O2J
Area of Operation: South East England
Tel: 01920 421061
Email: info@vispec.co.uk
Web: www.vispec.co.uk

WANDSWORTH GROUP LTD
Albert Drive, Sheerwater, Woking, Surrey, GU21 5SE
Area of Operation: UK (Excluding Ireland)
Tel: 01483 740740 **Fax:** 01483 740384
Email: info@wandsworthgroup.com
Web: www.wandsworthgroup.com

WEBSTRACT LTD
28 Lattimore Road, Wheathampstead,
Hertfordshire, AL4 8QE
Area of Operation: UK (Excluding Ireland)
Tel: 0870 446 0146 **Fax:** 0870 762 6431
Email: info@webstract.co.uk
Web: www.webstract.co.uk
Other Info: ✎

X-HOME
Unit W Williamsons Holdings, Uphall,
Lothian, EH52 6PA
Area of Operation: UK (Excluding Ireland)
Tel: 0845 130 1091 **Fax:** 0845 130 1092
Email: sales@xhome.biz
Web: www.xhome.biz

CCTV & SURVEILLANCE SYSTEMS

KEY
OTHER: ▽ Reclaimed ⌐⊟ On-line shopping
✎ Bespoke ✋ Hand-made ECO Ecological

A.C.E - AUTOMATION CONTRACTORS & ENGINEERS
64 Wannock Avenue, Willingdon,
East Sussex, BN20 9RH
Area of Operation: UK (Excluding Ireland)
Tel: 01323 489798 **Fax:** 07884 599411
Email: sales@ace-automation.co.uk
Web: www.ace-automation.co.uk

ABSOLUTE ELECTRICS
47 Tanfield Road, Croydon, Surrey, CR0 1AN
Area of Operation: Greater London,
South East England
Tel: 0208 681 7835
Fax: 0208 667 9836
Email: mark@absolute-electrics.co.uk
Web: www.absolute-electrics.com

ACM INSTALLATIONS LTD
71 Western Gailes Way, Hull,
East Riding of Yorks, HU8 9EQ
Area of Operation: North East England
Tel: 0870 242 3285
Fax: 01482 377530
Email: enquiries@acminstallations.com
Web: www.acminstallations.com
Other Info: ✎

AFM SECURITY
Head Office, The Stables, Foxes Loke,
Loddon, Norfolk, NR14 6UL
Area of Operation: East England,
Greater London
Tel: 01508 522078
Fax: 01508 522079
Email: sales@afmsecurity.co.uk
Web: www.afmsecurity.co.uk

ALDOUS SYSTEMS (EUROPE) LTD:
28 Bishopstone, Aylesbury,
Buckinghamshire, HP17 8SF
Area of Operation: Europe
Tel: 0870 240 1162
Fax: 0207 691 7844
Email: sales@aldoussystems.co.uk
Web: www.aldoussystems.co.uk

ARCHANGEL CUSTOM INSTALLATION SPECIALISTS (JERSEY) LIMITED
Mont St Aubyn, La Rue Du Carrefour Au Clercq,
Grouville, Jersey, JE3 9US
Area of Operation: Europe
Tel: 0845 838 2636
Email: david.fell@archangelmultimedia.co.uk
Web: www.archangelmultimedia.co.uk

BCW DESIGN
1 Church Farm Court, Neston Road,
Willaston, Wirral, CH64 2XP
Area of Operation: UK (Excluding Ireland)
Tel: 0870 757 9690
Fax: 0151 327 4307
Email: info@bcwdesign.com
Web: www.bcwdesign.com

BESPOKE INSTALLATIONS
1 Rogers Close, Tiverton, Devon, EX16 6UW
Area of Operation: UK (Excluding Ireland)
Tel: 01884 243497
Fax: 01884 243497
Email: michael@bespokeinstallations.com
Web: www.bespoke.biz

BEYOND THE INVISIBLE LIMITED
162-164 Arthur Road, London, SW19 8AQ
Area of Operation: Greater London
Tel: 0870 740 5859
Fax: 0870 740 5860
Email: info@beyondtheinvisible.com
Web: www.beyondtheinvisible.com

CCTV (UK) LTD
109-111 Pope Street, Birmingham,
West Midlands, B1 3AG
Area of Operation: Worldwide
Tel: 0121 200 1031
Email: sales@cctvgroup.com
Web: www.cctvgroup.com

CENTURION
Westhill Business Park,
Arnhall Business Park, Westhill,
Aberdeen, Aberdeenshire, AB32 6UF
Area of Operation: Scotland
Tel: 01224 744440
Fax: 01224 744819
Email: info@centurion-solutions.co.uk
Web: www.centurion-solutions.co.uk

CHOICE HI-FI
Denehurst Gardens, Richmond, Surrey, TW10
Area of Operation: UK (Excluding Ireland)
Tel: 0208 392 1959
Email: info@choice-hifi.com
Web: www.choice-hifi.com

CLOUD9 SYSTEMS
87 Bishops Park Road, London, SW6 6DY
Area of Operation: Greater London,
South East England
Tel: 0870 420 5495 **Fax:** 0870 402 0121
Email: info@cloud9systems.co.uk
Web: www.cloud9systems.co.uk

COASTAL ACOUSTICS LTD
16 Trinity Church Yard, Guildford, Surrey, GU1 3RR
Area of Operation: UK (Excluding Ireland)
Tel: 01483 885670 **Fax:** 01483 885677
Email: info@coastalacoustics.com

COMFORT HOME CONTROLS
Carlton House, Carlton Avenue,
Chester, Cheshire, CH4 8UE
Area of Operation: UK & Ireland
Tel: 08707 605528 **Fax:** 01244 671455
Email: sales@home-control.co.uk
Web: www.home-control.co.uk

CONTROLWISE LTD
Unit 7, St. Davids Industrial Estate, Pengam,
Blackwood, Caerphilly, NP12 3SW
Area of Operation: South West England
& South Wales
Tel: 01443 836836 **Fax:** 01443 836502
Email: stuart@controlwise.co.uk
Web: www.controlwise.co.uk

CREO DESIGN
62 North Street, Leeds, West Yorkshire, LS2 7PN
Area of Operation: UK (Excluding Ireland)
Tel: 0113 246 7373 **Fax:** 0113 242 5114
Email: info@creo-designs.co.uk
Web: www.creo-designs.co.uk
Other Info: ✐

CUSTOM INSTALLATIONS
Custom House, 51 Baymead Lane, North Petherton,
Bridgwater, Somerset, TA6 6RN
Area of Operation: South West England
and South Wales
Tel: 01278 662555 **Fax:** 01278 662975
Email: roger@custom-installations.co.uk
Web: www.custom-installations.co.uk

DARTMOOR DIGITAL DESIGN
Bigadon Farm, Buckfastleigh, Devon, TQ11 0ND
Area of Operation: UK & Ireland
Tel: 01364 644300 **Fax:** 01364 644308
Email: sales@iiid.co.uk
Web: www.iiid.co.uk

DECKORUM LIMITED
4c Royal Oak Lane, Pirton, Hitchin,
Hertfordshire, CG5 3QT
Area of Operation: Greater London, South East
England, South West England and South Wales
Tel: 0845 020 4360 **Fax:** 0845 020 4361
Email: simon@deckorum.com
Web: www.deckorum.com

DIGITAL DECOR
Sonas House, Button End, Harston,
Cambridge, Cambridgeshire, CB2 5NX
Area of Operation: East England, Greater London
Tel: 01223 870935 **Fax:** 01223 870935
Email: seamus@digital-decor.co.uk
Web: www.digital-decor.co.uk
Other Info: ✐

DISCOVERY SYSTEMS LTD
1 Corn Mill Close, The Spindles,
Ashton in Makerfield, Wigan,
Lancashire, WN4 0PX
Area of Operation: North West England
and North Wales
Tel: 01942 723756
Email: mike@discoverysystems.co.uk
Web: www.discoverysystems.co.uk

DORCUM LTD
11 Lyndhurst Road, Hove, East Sussex, BN3 6FA
Area of Operation: UK & Ireland
Tel: 01273 230023 **Fax:** 01273 220108
Email: sales@teleporter.co.uk
Web: www.teleporter.co.uk

EAST TECHNOLOGY INTEGRATORS
22 Seer Mead, Seer Green,
Beaconsfield, Buckinghamshire, HP9 2QL
Area of Operation: Greater London,
South East England
Tel: 0845 056 0245
Email: sales@east-ti.co.uk
Web: www.east-ti.co.uk

EASYGATES
75 Long Lane, Halesowen,
Birmingham, West Midlands, B62 9DJ
Area of Operation: Europe
Tel: 0870 7606 536 **Fax:** 0121 561 3395
Email: info@easygates.co.uk
Web: www.easygates.co.uk

EASYLIFE AUTOMATION LIMITED
Kullan House, 12 Meadowfield Road,
Stocksfield, Northumberland, NE43 7QX
Area of Operation: UK (Excluding Ireland)
Tel: 01661 844159 **Fax:** 01661 842594
Email: sales@easylifeautomation.com
Web: www.easylifeautomation.com

FLAMINGBOX
Perry Farm, Maiden Bradley, Wiltshire, BA12 7JQ
Area of Operation: UK (Excluding Ireland)
Tel: 01985 845440 **Fax:** 01985 845448
Email: james.ratcliffe@flamingbox.com
Web: www.flamingbox.com
Other Info: ✐

FUSION GROUP
Unit 4 Winston Avenue, Croft, Leicester,
Leicestershire, LE9 3GQ
Area of Operation: UK (Excluding Ireland)
Tel: 01455 285 252 **Fax:** 0116 260 3805
Email: gary.mills@fusiongroup.uk.com
Web: www.fusiongroup.uk.com

GATESTUFF.COM
17-19 Old Woking Road, West Bysleet,
Surrey, KT14 6LW
Area of Operation: UK & Ireland
Tel: 01932 344434
Email: sales@gatestuff.com
Web: www.gatestuff.com ✐

GET PLC
Key Point, 3-17 High Street, Potters Bar,
Hertfordshire, EN6 5AJ
Area of Operation: Worldwide
Tel: 01707 601601 **Fax:** 01707 601701
Email: info@getplc.com
Web: www.getplc.com

GJD MANUFACTURING LTD
Unit 2, Birch Industrial Estate, Whittle Lane,
Heywood, Lancashire, OL10 2SX
Area of Operation: Worldwide
Tel: 01706 363998 **Fax:** 01706 363991
Email: info@gjd.co.uk
Web: www.gjd.co.uk

GOLDSYSTEM LTD
46 Nightingale Lane, Bromley, Kent, BR1 2SB
Area of Operation: Greater London,
South East England
Tel: 0208 313 0485 **Fax:** 0208 313 0665
Email: info@goldsystem.co.uk
Web: www.goldsystem.co.uk
Other Info: ✐

HELLERMANN TYTON
43-45 Salthouse Road, Brackmills,
Northampton, Northamptonshire, NN4 7EX
Area of Operation: Worldwide
Tel: 01604 706633
Fax: 01604 705454
Email: sales@htdata.co.uk
Web: www.homenetworksciences.com

HOUSE ELECTRONIC LTD
97 Francisco Close, Hollyfields Park,
Chafford Hundred, Essex, RM16 6YE
Area of Operation: UK & Ireland
Tel: 01375 483595 **Fax:** 01375 483595
Email: info@houselectronic.co.uk
Web: www.houselectronic.co.uk

HOUSEHOLD AUTOMATION
Foxways, Pinkhurst Lane, Slinfold,
Horsham, West Sussex, RH13 0QR
Area of Operation: UK & Ireland
Tel: 0870 330 0071 **Fax:** 0870 330 0072
Email: ellis-andrew@btconnect.com
Web: www.household-automation.co.uk

INTERACTIVE HOMES LTD
Lorton House, Conifer Crest,
Newbury, Berkshire, RG14 6RS
Area of Operation: UK (Excluding Ireland)
Tel: 01635 49111 **Fax:** 01635 40735
Email: sales@interactivehomes.co.uk
Web: www.interactivehomes.co.uk

LAMBDA TECHNOLOGIES LTD
Stonebridge Court, Wakefield Road,
Horbury, Wakefield, West Yorkshire, WF4 5HQ
Area of Operation: UK (Excluding Ireland)
Tel: 01924 202100 **Fax:** 01924 299283
Email: john.brown@lambdatechnologies.com
Web: www.lambdatechnologies.com

LINK MEDIA SYSTEMS
68 St John Street, London, EC1M 4DT
Area of Operation: UK (Excluding Ireland)
Tel: 0207 251 2638 **Fax:** 0207 251 2487
Email: sohan@linkmediasystems.com
Web: www.linkmediasystems.com

MARTINS HI-FI
85/87 Ber Street, Norwich,
Norfolk, NR1 3EY
Area of Operation: East England
Tel: 01603 627010 **Fax:** 01603 878019
Email: info@martinshifi.com
Web: www.martinshifi.co.uk

MUSICAL APPROACH
111 Wolverhampton Road, Stafford,
Staffordshire, ST17 4BG
Area of Operation: UK (Excluding Ireland)
Tel: 01785 255154 **Fax:** 01785 243623
Email: info@musicalapproach.co.uk
Web: www.musicalapproach.co.uk

MYGARD SAFE & SECURE PLC
Suite 10 Lloyd Berkeley Place, Pebble Lane,
Aylesbury, Buckinghamshire, HP20 2JH
Area of Operation: UK (Excluding Ireland)
Tel: 01296 433746
Email: businessdevelopment@mygard.com
Web: www.mygard.com ✐

NDR SYSTEMS UK
Kendlehouse, 236 Nantwich Road,
Crewe, Cheshire, CW2 6BP
Area of Operation: UK & Ireland
Tel: 01270 219575
Email: daverehman@ndrsystemsuk.com
Web: www.ndrsystemsuk.com

PARK SYSTEMS INTEGRATION LTD
Unit 3 Queen's Park, Earlsway, Team Valley
Trading Estate, Gateshead, Tyne & Wear, NE11 0QD
Area of Operation: UK & Ireland
Tel: 0191 497 0770
Fax: 0191 497 0772
Email: sales@parksystemsintegration.co.uk
Web: www.parksystemsintegration.co.uk

PHILLSON LTD
144 Rickerscote Road, Stafford,
Staffordshire, ST17 4HE
Area of Operation: UK & Ireland
Tel: 0845 612 0128
Fax: 01477 535090
Email: info@phillson.co.uk
Web: www.phillson.co.uk

PREMIER ELECTRICAL & SECURITY SERVICES
Unit D10, The Seedbed Centre,
Langston Road, Loughton, Essex, IG10 3TQ
Area of Operation: East England,
Greater London, Midlands & Mid Wales
Tel: 0208 787 7063 **Fax:** 0208 508 3249
Email: info@premierelectrical.com
Web: www.premierelectrical.com

PW TECHNOLOGIES
Windy House, Hutton Roof,
Carnforth, Lancashire, LA6 2PE
Area of Operation: UK (Excluding Ireland)
Tel: 01524 272400 **Fax:** 01524 272402
Email: info@pwtechnologies.co.uk
Web: www.pwtechnologies.co.uk

PYRAMID ELECTRICAL & ALARMS
1 Morland Drive, Hinckley,
Leicestershire, Leicester, LE10 0GG
Area of Operation: East England, Greater London,
Midlands & Mid Wales, North East England,
North West England and North Wales
Tel: 01455 458325 **Fax:** 01455 458 325
Email: pyramid@dotdotnetdot.net
Web: www.pyramidelectrical.net

SECURIKEY LTD
P O Box 18, Aldershot, Hampshire, GU12 4SL
Area of Operation: Worldwide
Tel: 01252 311888 **Fax:** 01252 343950
Email: enquiries@securikey.co.uk
Web: www.securikey.co.uk

SECURITY VISION INTERNATIONAL LTD
Berwick Courtyard, Berwick St Leonard,
Salisbury, Wiltshire, SP3 5SN
Area of Operation: UK & Ireland
Tel: 01747 820820 **Fax:** 01747 820821
Email: enquiries@securityvision.co.uk
Web: www.securityvision.co.uk

SENSORY INTERNATIONAL LIMITED
48A London Road, Alderley Edge, Cheshire, SK9 7DZ
Area of Operation: Worldwide
Tel: 0870 350 2244 **Fax:** 0870 350 2255
Email: garyc@sensoryinternational.com
Web: www.sensoryinternational.com

SEVENOAKS SOUND & VISION (LEEDS)
62 North Street, Leeds, West Yorkshire, LS2 7PN
Area of Operation: Midlands & Mid Wales, North
East England, North West England and North Wales
Tel: 0113 245 2775 **Fax:** 0113 242 5114
Email: leeds@sevenoakssoundandvision.co.uk
Web: www.sevenoakssoundandvision.co.uk

SEVENOAKS SOUND & VISION (OXFORD)
41 St Clements Street, Oxford, Oxfordshire, OX4 1AG
Area of Operation: UK (Excluding Ireland)
Tel: 01865 241773 **Fax:** 01865 794904
Email: oxford@sevenoakssoundandvision.co.uk
Web: www.sevenoakssoundandvision.co.uk

SMARTCOMM LTD
45 Cressex Enterprise Centre, Lincoln Road,
Cressex Business Park, High Wycombe,
Buckinghamshire, HP12 3RL
Area of Operation: Worldwide
Tel: 01494 471912 **Fax:** 01494 472464
Email: info@smartcomm.co.uk
Web: www.smartcomm.co.uk

STONEAUDIO UK LIMITED
Holmead Walk, Poundbury,
Dorchester, Dorset, DT1 3GE
Area of Operation: Worldwide
Tel: 01305 257555 **Fax:** 01305 257666
Email: info@stoneaudio.co.uk
Web: www.stoneaudio.co.uk ✐

SUB SYSTEMS INTEGRATION LTD
202a Dawes Road, London, SW6 7RQ
Area of Operation: Greater London
Tel: 0207 381 1177
Fax: 0207 381 8833
Email: info@sub.eu.com
Web: www.sub.eu.com
Other Info: ✐

THE BIG PICTURE
AV & HOME CINEMA LIMITED
51 Sutton Road, Walsall,
West Midlands, WS1 2PQ
Area of Operation: UK & Ireland
Tel: 01922 623000
Email: info@getthebigpicture.co.uk
Web: www.getthebigpicture.co.uk

THE EDGE
90-92 Norwich Road, Ipswich, Suffolk, IP1 2NL
Area of Operation: UK (Excluding Ireland)
Tel: 01473 288211
Fax: 01473 288255
Email: nick@theedge.eu.com
Web: www.theedge.eu.com

THE MAJIK HOUSE COMPANY LTD
Unit J Mainline Industrial Estate,
Crooklands Road, Milnthorpe, Cumbria, LA7 7LR
Area of Operation: North East England,
North West England and North Wales
Tel: 0870 240 8350
Fax: 0870 240 8350
Email: tim@majikhouse.com
Web: www.majikhouse.com

THE MULTI ROOM COMPANY LTD
4 Churchill House, Churchill Road,
Cheltenham, Gloucestershire, GL53 7EG
Area of Operation: UK & Ireland
Tel: 01242 539100 **Fax:** 01242 539300
Email: info@multi-room.com
Web: www.multi-room.com

TLC ELECTRICAL WHOLESALERS
TLC Building, Off Fleming Way,
Crawley, West Sussex, RH10 9JY
Area of Operation: Worldwide
Tel: 01293 565630
 Fax: 01293 425234
Email: sales@tlc-direct.co.uk
Web: www.tlc-direct.co.uk

TRUFI LIMITED
Unit 14, Horizon Business Village,
1 Brooklands Road, Weybridge, Surrey, KT13 0TJ
Area of Operation: South East England
Tel: 01932 340750 **Fax:** 01932 345459
Email: info@trufi.co.uk
Web: www.trufi.co.uk

UNLIMITED DREAM HOME SYSTEMS
The Long Barn, Wakefield Road, Hampole,
Doncaster, South Yorkshire, DN6 7EU
Area of Operation: UK (Excluding Ireland)
Tel: 01302 725550
Fax: 01302 727274
Email: info@thehifistudios.freeserve.co.uk
Web: www.unlimiteddream.co.uk

VISPEC
PO Box 191, Ware, Hertfordshire, SG12 02J
Area of Operation: South East England
Tel: 01920 421061
Email: info@vispec.co.uk
Web: www.vispec.co.uk

WEBSTRACT LTD
28 Lattimore Road, Wheathampstead,
Hertfordshire, AL4 8QE
Area of Operation: UK (Excluding Ireland)
Tel: 0870 446 0146 **Fax:** 0870 762 6431
Email: info@webstract.co.uk
Web: www.webstract.co.uk

WIRED-AVD
9 Addison Square, Ringwood,
Hampshire, BH24 1NY
Area of Operation: South East England,
South West England and South Wales
Tel: 01425 479014 **Fax:** 01425 471733
Email: sales@wired-avd.co.uk
Web: www.wired-avd.co.uk

WISELAN LIMITED
27 Old Gloucester Street, London,
Greater London, WC1N 3XX
Area of Operation: UK & Ireland
Tel: 0870 787 2144
Fax: 0870 787 8823
Email: sales@wiselan.com
Web: www.wiselan.com

X-HOME
Unit W Williamsons Holdings,
Uphall, Lothian, EH52 6PA
Area of Operation: UK (Excluding Ireland)
Tel: 0845 130 1091
Fax: 0845 130 1092
Email: sales@xhome.biz
Web: www.xhome.biz

FIRE PROTECTION

KEY

SEE ALSO: DOORS, WINDOWS AND
CONSERVATORIES - Fire Doors, INSULATION
AND PROOFING MATERIALS - Fire Proofing
OTHER: ▽ Reclaimed 🏷 On-line shopping
🖎 Bespoke 🖐 Hand-made ECO Ecological

ABSOLUTE ELECTRICS
47 Tanfield Road, Croydon, Surrey, CR0 1AN
Area of Operation: Greater London,
South East England
Tel: 0208 681 7835
Fax: 0208 667 9836
Email: mark@absolute-electrics.co.uk
Web: www.absolute-electrics.com

AFM SECURITY
Head Office, The Stables, Foxes Loke,
Loddon, Norfolk, NR14 6UL
Area of Operation: East England,
Greater London
Tel: 01508 522078
Fax: 01508 522079
Email: sales@afmsecurity.co.uk
Web: www.afmsecurity.co.uk

AUTOMATED CONTROL SERVICES LTD
Unit 16, Hightown Ind Est, Crow Arch Lane,
Ringwood, Dorset, BH24 1ND
Area of Operation: UK (Excluding Ireland)
Tel: 01425 461008
Fax: 01425 461009
Email: sales@automatedcontrolservices.co.uk
Web: www.automatedcontrolservices.co.uk

AUTOMATIC BUILDINGS LTD
Unit 12b Gorsey Lane, Coleshill,
Warwickshire, B46 1JU
Area of Operation: UK & Ireland

Tel: 01675 466881 **Fax:** 01675 464138
Email: info@automaticbuildings.co.uk
Web: www.automaticbuildings.co.uk

CENTURION
Westhill Business Park, Arnhall Business Park,
Westhill, Aberdeen, Aberdeenshire, AB32 6UF
Area of Operation: Scotland
Tel: 01224 744440
Fax: 01224 744819
Email: info@centurion-solutions.co.uk
Web: www.centurion-solutions.co.uk

DOMESTIC SPRINKLERS PLC
6 Kent CLose, Weymouth, Dorset, DT4 9TF
Area of Operation: UK & Ireland
Tel: 01305 765763
Fax: 01305 777700
Email: email@domesticsprinklers.co.uk
Web: www.domesticsprinklers.co.uk

FIREANGEL LIMITED
The TechnoCentre, Puma Way,
Coventry, Warwickshire, CV1 2TT
Area of Operation: Worldwide
Tel: 02476 236600
Email: info@fireangel.co.uk
Web: www.fireangel.co.uk

FIREFIGHTER LTD
18 Albert Road, West Kirby,
Wirral, Merseyside, CH48 0RS
Area of Operation: UK & Ireland
Tel: 0151 625 4133
Email: mail@firefighter.ltd.uk
Web: www.firefighter.ltd.uk

MYGARD SAFE & SECURE PLC
Suite 10 Lloyd Berkeley Place, Pebble Lane,
Aylesbury, Buckinghamshire, HP20 2JH
Area of Operation: UK (Excluding Ireland)
Tel: 01296 433746
Email: businessdevelopment@mygard.com
Web: www.mygard.com

NDR SYSTEMS UK
Kendlehouse, 236 Nantwich Road,
Crewe, Cheshire, CW2 6BP
Area of Operation: UK & Ireland
Tel: 01270 219575
Email: daverehman@ndrsystemsuk.com
Web: www.ndrsystemsuk.com

PARK SYSTEMS INTEGRATION LTD
Unit 3 Queen's Park, Earlsway, Team Valley Trading
Estate, Gateshead, Tyne & Wear, NE11 0QD
Area of Operation: UK & Ireland
Tel: 0191 497 0770
Fax: 0191 497 0772
Email: sales@parksystemsintegration.co.uk
Web: www.parksystemsintegration.co.uk

PREMIER ELECTRICAL & SECURITY SERVICES
Unit D10, The Seedbed Centre,
Langston Road, Loughton, Essex, IG10 3TQ
Area of Operation: East England,
Greater London, Midlands & Mid Wales
Tel: 0208 787 7063
Fax: 0208 508 3249
Email: info@premierelectrical.com
Web: www.premierelectrical.com

PYRAMID ELECTRICAL & ALARMS
1 Morland Drive, Hinckley, Leicestershire,
Leicestershire, LE10 0GG
Area of Operation: East England, Greater London,
Midlands & Mid Wales, North East England,
North West England and North Wales
Tel: 01455 458325 **Fax:** 01455 458 325
Email: pyramid@dotdotnetdot.net
Web: www.pyramidelectrical.net

SAFELINCS
9 West Street, Alford, Lincolnshire, LN13 9DG
Area of Operation: UK & Ireland
Tel: 01507 462176 **Fax:** 01507 463288
Email: service@safelincs.co.uk
Web: www.safelincs.co.uk

SECURIKEY LTD
P O Box 18, Aldershot, Hampshire, GU12 4SL
Area of Operation: Worldwide
Tel: 01252 311888 **Fax:** 01252 343950
Email: enquiries@securikey.co.uk
Web: www.securikey.co.uk

THE THINKING HOME
The White House, Wilderspool Park, Greenalls
Avenue, Warrington, Cheshire, WA4 6HL
Area of Operation: UK & Ireland
Tel: 0800 881 8319 **Fax:** 0800 881 8329
Email: info@thethinkinghome.com
Web: www.thethinkinghome.com

TLC ELECTRICAL WHOLESALERS
TLC Building, Off Fleming Way,
Crawley, West Sussex, RH10 9JY
Area of Operation: Worldwide
Tel: 01293 565630
Fax: 01293 425234
Email: sales@tlc-direct.co.uk
Web: www.tlc-direct.co.uk

VISPEC
PO Box 191, Ware,
Hertfordshire, SG12 02J
Area of Operation: South East England
Tel: 01920 421061
Email: info@vispec.co.uk
Web: www.vispec.co.uk

NOTES

Company Name ....

....

Address ....

....

email ....
Web ....

Company Name ....

....

Address ....

....

email ....
Web ....

Company Name ....

....

Address ....

....

email ....
Web ....

Company Name ....

....

Address ....

....

email ....
Web ....

Company Name ....

....

Address ....

....

email ....
Web ....

Company Name ....

....

Address ....

....

email ....
Web ....

Company Name ....

....

Address ....

....

email ....
Web ....

Company Name ....

....

Address ....

....

email ....
Web ....

Company Name ....

....

Address ....

....

email ....
Web ....

Company Name ....

....

Address ....

....

email ....
Web ....

HEATING, PLUMBING & ELECTRICAL

INTERIOR STRUCTURE

Lofts

If you want a bigger house, but don't want the upheaval of moving, converting your loft can make a lot of sense. Conversion can yield up to 30% more usable space in an average two-storey house, and it is cost effective: metre for metre, a loft conversion is less expensive than building a ground level extension because there are no foundations to worry about. The price depends on location, the existing roof structure, and final specification. But wherever you live, converting the loft is almost always cheaper than moving to a bigger house.

Image by Velux

Loft conversions are also built quickly – typically it takes five to eight weeks – and disruption is minimal because materials go in through the roof. Planning approval is usually straightforward because the new structure is built within the volume of the existing building.

Few people would feel competent to undertake a complete loft conversion without some professional help. Depending on your budget, design ambitions and the type of roof you have, this help can come from either an architect, a contractor with design-and-build services, or a specialist loft conversion company.

There are also several areas of the Building Regulations to consider. For example, your roof insulation will need to be rearranged, and will be checked to ensure it meets the regulated levels. Also, if you are altering the timbers in a roof space, you need to prove that your design will be strong enough to bear the new loads, and rigid enough to stop the roof shape deflecting. The building inspector will want to see that calculations have been carried out to ensure that these aims have been met.

Usually your existing ceiling joists will not be strong enough to act as floor joists for the loft conversion, and there are various ways of reinforcing them. The normal method is to leave the existing joists in place so minimum disruption is caused to the ceiling below, and place larger joists by their side to take the added weight of the new floor.

The roof carpentry will also have to be carefully designed to ensure that it stays in place whilst structural alterations are undertaken, and is then adequate for the open roof space you need, which may require the insertion of steel beams to support the new loadings. In some instances, where property prices justify the expense, it can pay to replace the existing roof structure altogether, creating a new one that gives more usable living space.

However, perhaps the biggest hurdle faced by loft converters is meeting the fire safety regulations. When you create habitable space more than 4.5m above ground level, you must allow for the fact that people will not be able to climb out of a window to escape a fire below. You have to provide at least one egress window – a window from which you can escape easily – plus there needs to be a fire-proofed route from the loft down to the front door. Adequate fire-proofing is not difficult to achieve, but will require additional work to be carried out on the doorways (and possibly the walls) leading onto your escape route – typically the bedroom doors opening onto a landing. These doors will have to be treated either with fire resistant paint or some fireproof board, or be replaced with fire doors. If your existing stairwell exits into a living room rather than an enclosed hallway, this will be deemed as unacceptable and you might have to build an enclosure around the staircase, or find another route down to the ground.

FLOORING STRUCTURE

KEY

PRODUCT TYPES: 1= Beam and Block
2 = Floating 3 = Floor Beams 4 = Decking
5 = Structural Floors 6 = Other
OTHER: ▽ Reclaimed ⊕ On-line shopping
✍ Bespoke ✋Hand-made ECO Ecological

AC ROOF TRUSSES LTD
Severn Farm Industrial Estate,
Welshpool, Powys, SY21 7DF
Area of Operation: UK & Ireland
Tel: 01938 554881 **Fax:** 01938 556265
Email: info@acrooftrusses.co.uk
Web: www.acrooftrusses.co.uk ⊕
Product Type: 3, 4, 6

BOARD CENTRAL
Chiltern Business Centre,
Couching Street, Watlington,
Oxfordshire, OX49 5PX
Area of Operation: Greater London,
South East England
Tel: 0845 458 8016 **Fax:** 01844 354112
Email: howardmorrice@hotmail.com
Product Type: 6

FFOREST TIMBER ENGINEERING LTD
Kestrel Way, Garngoch Industrial Estate,
Gorseinon, Swansea, SA4 9WF
Area of Operation: Midlands & Mid Wales,
South West England and South Wales
Tel: 01792 895620 **Fax:** 01792 893969
Email: info@fforest.co.uk
Web: www.fforest.co.uk
Product Type: 3

MERSEYBEAMS LTD
Riverbank Road, Bromborough,
Wirral, Merseyside, CH62 3LQ
Area of Operation: UK (Excluding Ireland)
Tel: 0151 334 7346 **Fax:** 0151 334 0600
Email: info@merseybeams.com
Web: www.merseybeams.com
Product Type: 3

MERSEYBEAMS LTD
Area of Operation: UK (Excluding Ireland)
Tel: 0151 334 7346 **Fax:** 0151 334 0600
Email: info@merseybeams.com
Web: www.merseybeams.com
Product Type: 3

We manufacture 150mm & 225mm Deep Beam
& Block along with 100mm,150mm,200mm and
225mm Deep Wideslab systems suitable for both
ground and upper floors, specialising in self build
projects, completing over 100 self build houses
each year.

MIDLANDS CEILINGS & PARTITIONS
201 High Street, Harborne, Birmingham,
West Midlands, B17 9QG
Area of Operation: UK (Excluding Ireland)
Tel: 0121 694 8258 **Fax:** 0121 694 3258
Email: deanoneill100@msn.com
Product Type: 2

MILBANK FLOORS
Earls Colne Business Park, Earls Colne,
Colchester, Essex, CO6 2NS
Area of Operation: UK (Excluding Ireland)
Tel: 01787 223931 **Fax:** 01787 220522
Email: floors@milbank.co.uk
Web: www.milbank-floors.co.uk
Product Type: 1, 5

MITEK INDUSTRIES LTD
Mitek House, Grazebrook Industrial Park,
Peartree Lane, Dudley, West Midlands, DY2 0XW
Area of Operation: UK & Ireland
Tel: 01384 451400 **Fax:** 01384 451415
Email: roy.troman@mitek.co.uk
Web: www.mitek.co.uk

OAKBEAMS.COM
Hunterswood Farm, Alfold Road, Dunsfold,
Godalming, Surrey, GU8 4NP
Area of Operation: Worldwide
Tel: 01483 200477
Email: info@oakbeams.com
Web: www.oakbeams.com
Product Type: 3 **Other Info:** ▽

S.L. HARDWOODS
390 Sydenham Road, Croydon, Surrey, CR0 2EA
Area of Operation: UK (Excluding Ireland)
Tel: 0208 683 0292 **Fax:** 0208 683 0404
Email: info@slhardwoods.co.uk
Web: www.slhardwoods.co.uk
Product Type: 5

SCHLUTER SYSTEMS LTD
Units 4-5, Bardon 22 Industrial Park,
Beveridge Lane, Coalville, Leicestershire, LE67 1TE
Area of Operation: UK & Ireland
Tel: 01530 813396 **Fax:** 01530 813376
Email: admin@schluter.co.uk
Web: www.schluter.co.uk

SPANCAST CONCRETE FLOORS
Stephenson Way, Barrington Industrial Estate,
Bedlington, Northumberland, NE22 7DQ
Area of Operation: East England,
North East England, North West England
and North Wales, Scotland
Tel: 01670 531160 **Fax:** 01670 531170
Email: sales@spancast.co.uk
Web: www.spancast.co.uk
Product Type: 5 **Material Type:** G) 1

THE CARPENTRY INSIDER - AIRCOMDIRECT
1 Castleton Crescent, Skegness, Lincolnshire, PE25 2TJ
Area of Operation: Worldwide
Tel: 01754 767163
Email: aircom8@hotmail.com
Web: www.carpentry.tk ⊕ **Product Type:** 4, 6

THE EXPANDED METAL COMPANY LIMITED
PO Box 14, Longhill Industrial Estate (North),
Hartlepool, Durham, TS25 1PR
Area of Operation: Worldwide
Tel: 01429 408087 **Fax:** 01429 866795
Email: paulb@expamet.co.uk
Web: www.expandedmetalcompany.co.uk

TOMPKINS LTD
High March Close, Long March Industrial Estate,
Daventry, Northamptonshire, NN11 4EZ
Area of Operation: UK (Excluding Ireland)
Tel: 01327 877187 **Fax:** 01327 310491
Email: info@tompkinswood.co.uk
Web: www.tompkinswood.co.uk
Product Type: 4, 5 **Other Info:** ✍ ✋

LOFTS AND LOFT CONVERSIONS

KEY

PRODUCT TYPES: 1= Loft Installation 2 = Loft
Conversion 3 = Loft Doors 4 = Loft Ladders
OTHER: ▽ Reclaimed ⊕ On-line shopping
✍ Bespoke ✋Hand-made ECO Ecological

AA LOFT SERVICES
36 Summerley Lane, Bognor Regis,
West Sussex, PO22 7HX
Area of Operation: South East England
Tel: 01243 855696
Product Type: 2, 4

ABOVE IT ALL
56 Hardman Avenue, Rawtenstall,
Rossendale, Lancashire, BB4 6BB
Area of Operation: North West England
and North Wales
Tel: 0800 505 3344
Email: info@above-it-all.co.uk
Web: www.above-it-all.co.uk
Product Type: 2
Material Type: B) 2

ANDREW WILSON
184 Upper Shoreham Road,
Shoreham by the Sea, West Sussex, BN43 6BG
Area of Operation: South East England
Tel: 01273 880257
Fax: 01273 880257
Email: barbandy3@hotmail.com
Product Type: 1, 2

ASSET LOFT CONVERSIONS
32 Saltwell View, Gateshead,
Tyne & Wear, NE8 4NT
Area of Operation: North East England
Tel: 0191 477 9057
Fax: 0191 420 7057
Web: www.assetlofts.co.uk

ECONOLOFT LTD
Unit 5, Kingfisher Court,
South Lancs Industrial Estate, Bryn,
Ashton in Makerfield, Lancashire, WN4 8DY
Area of Operation: UK (Excluding Ireland)
Tel: 01942 722754
Email: sales@econoloft.co.uk
Web: www.econoloft.co.uk
Product Type: 2

EPS INSULATION LTD
Unit 10 Amber Close, Amington Industrial Estate,
Tamworth, Staffordshire, B79 8BH
Area of Operation: UK (Excluding Ireland)
Tel: 01827 313951 **Fax:** 01827 54683
Email: info@eps-systemsltd.com
Web: www.eps-systemsltd.com

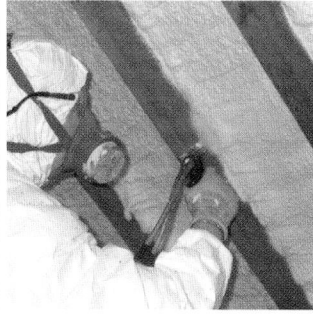

EPS SYSTEMS
Area of Operation: UK
Tel: 01827 313951
Fax: 01827 54683
Email: info@eps-systemsltd.com
Web: www.eps-systemsltd.com

DIY spray foam kits for that professional finish at
a fraction of the cost, consolidate and insulate
your roof. 10% off first orders.

HALLS STAIRS & LANDING
The Triangle, Paddock, Huddersfield,
West Yorkshire, HD1 4RN
Area of Operation: UK & Ireland
Tel: 01484 451485
Email: info@loftaccess.com
Web: www.loftaccess.com
Product Type: 4

LADDERS-ONLINE
Penarth Road, Cardiff, CF11 8TD
Area of Operation: Worldwide
Tel: 08450 647647 **Fax:** 02920 702386
Email: sales@tbdavies.co.uk
Web: www.ladders-online.com ⊕
Product Type: 4

LEE WOODWARD
Unit 1, Alpha Business Centre, 60 South Grove,
Walthamstow, London, E17 7NX
Tel: 0207 692 4967 **Fax:** 0208 503 7966
Product Type: 2

LOFT CENTRE PRODUCTS
Thicket Lane, Halnaker, Nr Chichester,
West Sussex, PO18 0QS
Area of Operation: UK & Ireland
Tel: 01243 785246 **Fax:** 01243 533184
Email: sales@loftcentreproducts.co.uk
Web: www.loftcentreproducts.co.uk
Product Type: 4

LOFT CENTRE PRODUCTS
Area of Operation: UK
Tel: 01243 785246 **Fax:** 01243 533184
Email: sales@loftcentreproducts.co.uk
Web: www.loftcentreproducts.co.uk

Loft Centre Products comprehensive range of
spirals include all metal, all timber, metal and
timber, round and square to suit most internal
and external applications.

Posi-Joist™
Supporting a cleaner Lifestyle

MiTek

The Clear advantages of Posi-Joist"

Open web design for easy installation of services

Greater design flexibility

A fixing surface that provides a truly quiet floor

Light, strong and easy to install

Added profit through savings in labour, time and materials

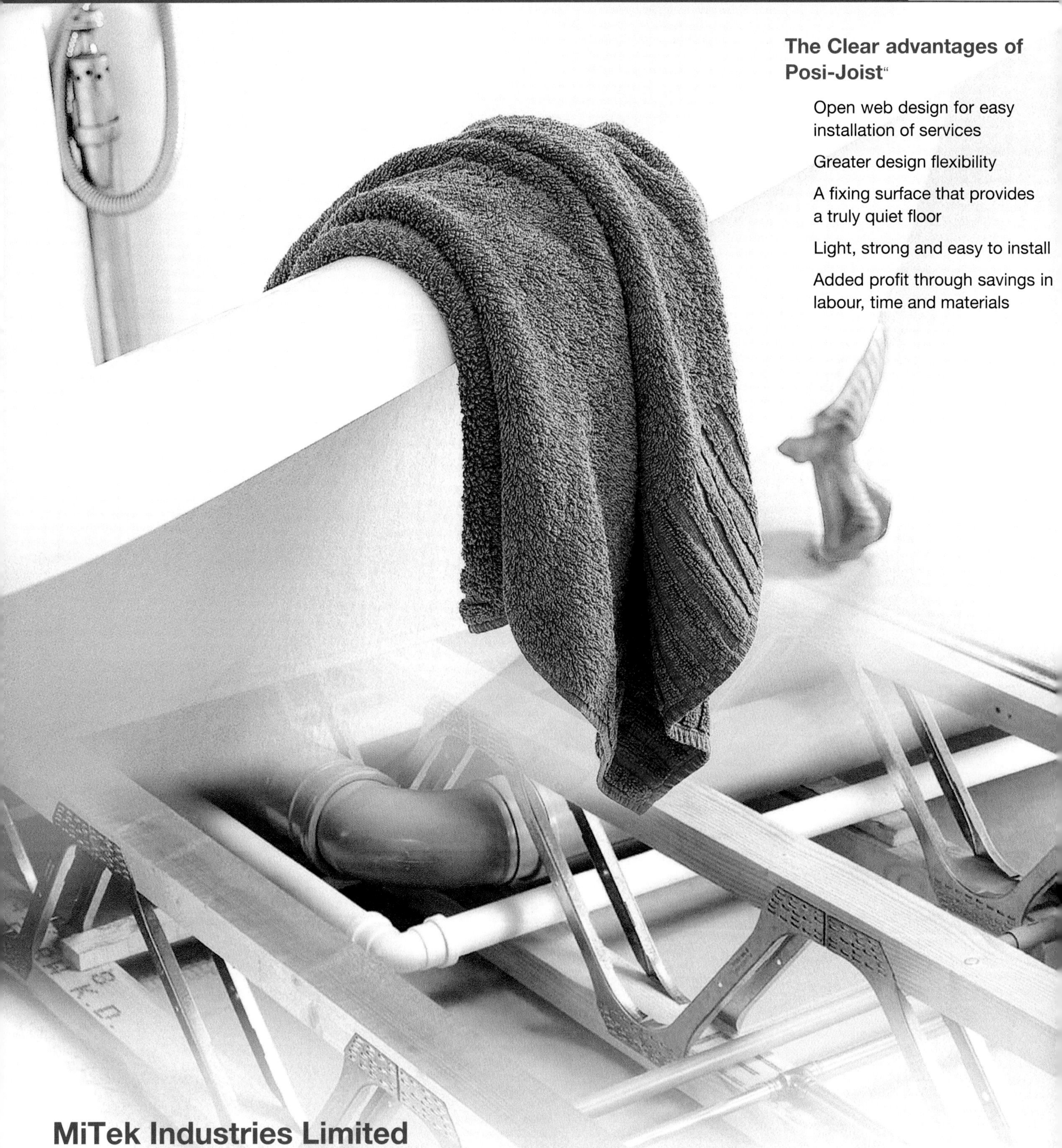

MiTek Industries Limited
MiTek House • Grazebrook Industrial Park • Peartree Lane • Dudley DY2 0XW
Telephone: 01384 451400 • Facsimile: 01384 451411 • www.mitek.co.uk

LOFT CONVERSION CENTRE
27 Old Shoreham Road, Brighton,
East Sussex, BN1 6SB
Area of Operation: UK (Excluding Ireland)
Tel: 01273 733333
Web: www.loftconversioncentre.co.uk

LOFT CONVERSION WAREHOUSE LTD
9 Motcombe Street, London, SW1X 8LA
Area of Operation: UK (Excluding Ireland)
Tel: 0207 245 1150
Fax: 0207 201 2569
Email: enquiries@loftconversionwarehouse.com
Web: www.loftconversionwarehouse.com
Product Type: 1, 2, 4

LOFT MASTERS
795 Great Cambridge Rd,
Enfield, London, EN1 3PN
Area of Operation: UK (Excluding Ireland)
Tel: 0800 917 7532
Email: sales@loftmasters.co.uk
Web: www.loftmasters.co.uk
Product Type: 1, 2

MIDLANDS CEILINGS & PARTITIONS
201 High Street, Harborne,
Birmingham, West Midlands, B17 9QG
Area of Operation: UK (Excluding Ireland)
Tel: 0121 694 8258
Fax: 0121 694 3258
Email: deanoneill100@msn.com
Product Type: 2

ROTO FRANK LTD
Swift Point, Rugby,
Warwickshire, CV21 1QH
Area of Operation: Worldwide
Tel: 01788 558600
Fax: 01788 558605
Email: uksales@roto-frank.co.uk
Web: www.roto-frank.co.uk
Product Type: 4

SPACE LOFT CONVERSIONS
Halton, 156 Bures Road,
Sudbury, Suffolk, CO10 0JG
Area of Operation: East England
Tel: 01787 373570
Email: info@loftconversionscolchester.co.uk
Web: www.loftconversionscolchester.co.uk
Product Type: 2

SPACEX UK LTD
Suite 3, The Red House, 53 High Street,
Lymington, Hampshire, SO41 9AH
Area of Operation: UK (Excluding Ireland)
Tel: 01590 610380 **Fax:** 01590 688864
Email: enquiries@spacexloftconversions.com
Web: www.spacexuk.com
Product Type: 2

STIRA FOLDING ATTIC STAIRS
Dunmore, Co Galway, Ireland
Area of Operation: UK & Ireland
Tel: 0800 973111 **Fax:** +353 933 8428
Email: stira@stira.ie
Web: www.stira.co.uk ⊕
Product Type: 4

SYSTEMATTIC LTD
4 Springfield Road, Altrincham, Cheshire, WA14 1HE
Area of Operation: North West England
and North Wales
Tel: 0161 928 0034 **Fax:** 0161 928 0048
Email: contact_us@systemattic.ltd.uk
Web: www.systemattic.ltd.uk
Product Type: 1, 2

T.B DAVIES (CARDIFF) LTD
Penarth Road, Cardiff, CF11 8TD
Area of Operation: UK (Excluding Ireland)
Tel: 02920 713000
Fax: 02920 702386
Email: sales@tbdavies.co.uk
Web: www.ladders-online.com ⊕
Product Type: 4

TEAM TECHNOLOGY
32 High Street, Guilden Morden,
Nr Royston, Hertfordshire, SG8 0JR
Area of Operation: UK (Excluding Ireland)
Tel: 01763 853369
Fax: 01763 853164
Email: info@teamtechnology.co.uk
Web: www.teamtechnologyltd.co.uk
Product Type: 4

THE LOFT SHOP LTD
Eldon Way, Littlehampton,
West Sussex, BN17 7HE
Area of Operation: UK (Excluding Ireland)
Tel: 0870 604 0404
Fax: 0870 603 9075
Email: marjory.kay@loftshop.co.uk
Web: www.loftshop.co.uk ⊕
Product Type: 1, 4

THE TELESCOPIC LADDER COMPANY
Canal Wharf, Horsenden Lane North,
Greenford, Greater London, UB6 7PH
Area of Operation: Worldwide
Tel: 0208 900 1902
Fax: 0208 900 1906
Email: sales@telescopicladders.co.uk
Web: www.telescopicladders.co.uk ⊕
Product Type: 4

TRUSS LOFT CONVERSIONS LTD
Bellwood Farm, Harrogate Road, Ripon,
North Yorkshire, HG4 3AA
Area of Operation: UK (Excluding Ireland)
Tel: 0800 195 3855
 Fax: 01765 692 189
Email: sales@trussloft.co.uk
Web: www.trussloft.co.uk
Product Type: 2

CEILINGS

KEY
OTHER: ▽ Reclaimed ⊕ On-line shopping
✏ Bespoke 🖐 Hand-made ECO Ecological

CREATIVE CEILINGS LTD
West Midlands House, Gipsy Lane, Willenhall,
West Midlands, WV13 2HA
Area of Operation: UK (Excluding Ireland)
Tel: 0870 755 7830
Fax: 0121 609 7001
Email: info@creativeceilings.co.uk
Web: www.creativeceilings.co.uk

MIDLANDS CEILINGS & PARTITIONS
201 High Street, Harborne, Birmingham,
West Midlands, B17 9QG
Area of Operation: UK (Excluding Ireland)
Tel: 0121 694 8258
Fax: 0121 694 3258
Email: deanoneill100@msn.com

SOUND REDUCTION SYSTEMS LTD
Adam Street, Bolton, Lancashire, BL3 2AP
Area of Operation: UK & Ireland
Tel: 01204 380074
Fax: 01204 380957
Email: info@soundreduction.co.uk
Web: www.soundreduction.co.uk

PLEASE MENTION

THE HOMEBUILDER'S HANDBOOK

WHEN YOU CALL

INTERIORS, FIXTURES & FINISHES

HOMEBUILDING & RENOVATING ARCHIVE CDS

12 ISSUES OF HOMEBUILDING & RENOVATING ON 2 CDS ONLY £24.99!

- All pages and pictures appear exactly as they do in the magazine.

- Easy to navigate and fully searchable.

- All supplier information and contacts are linked to email and web addresses.

- Clickable contents page.

NEW FOR 2006

INTERIORS, FIXTURES & FINISHES

STAIRCASES, STAIRPARTS & WALKWAYS

Image courtesy of Blanc De Bierges (01733 202566)

INTERIORS, FIXTURES & FINISHES

Stairs

The staircase matters enormously, both on a functional level and as a design statement. It is often the first thing visitors see on entering your house, and is about as dominant a part of the interior as the windows are to the outside. It should be far more than just a means of getting you up to the next floor. So what are the first things to consider when deciding on a staircase?

ABOVE: Staircase by Bisca

Making efficient use of space often involves a staircase that turns, rather than a single straight flight. Adding turns increases the cost of the staircase, but it can also make it look more interesting. Staircases can turn using landings that connect shorter straight flights: in many cases these are available off-the-shelf from most national joinery suppliers, together with half flights which can be used to configure all sorts of staircase designs inexpensively.

Another potential problem for the self-builder is that today's building regulations do not permit, on grounds of safety, staircases that in the past would have been considered perfectly safe. Large numbers of cottages and farmhouses, often altered many times over the years, have staircases that have attractive features such as tiny flights of combined straight steps and winders tucked in wherever suitable space could be found, often beside the fireplace. They may be charming, but they are often considered hazardous by the standards of today, and replicating them in new cottage-style homes can prove extremely difficult. Self-builders often give up because of the irregularity of the walls, the difficulty of finding someone capable of constructing them, and because of the problems posed by building regulations. Nevertheless it is sometimes possible to placate the building control officer by building a staircase of this sort if you are prepared to build a second staircase of a more open, contemporary design in a different section of the building. This is a solution that can also help overcome the problem of getting large pieces of furniture upstairs.

You should bear in mind that the staircase is a fixed feature, and thus will be hard to replace if you change your mind after installation, therefore allow the money to get it right first time. Ingenuity and good ideas do not necessarily cost money, however, and even simple flights can be 'dressed up' very effectively. The industry standard wooden staircase with open treads is a flight of 13 steps and a total rise of 2,565-2,700mm. Standard staircases like this can be purchased for as little as £170 incl. VAT. Remember, don't be tempted to insert a staircase that is too 'grand' for the style of building; simple and understated can be just as effective in the right surroundings.

Constructing staircases poses many problems, while calculating the space and headroom required for the staircase is one of the greatest challenges for any self-builder. It is all a question of style and subtlety — and it may mean a long search for a joinery you consider capable of the job.

The width and configuration of the staircase is critical in making the most of space. Where the stairs start from and land defines how much circulation space – dead space – will be required on each floor to provide access to all the other rooms. Where space is restricted, the stair design can dictate whether or not an attic conversion is practical.

Image by Richard Burbidge

STAIRCASES, GENERAL

KEY

OTHER: ▽ Reclaimed ⌃ On-line shopping
✎ Bespoke 🖑 Hand-made ECO Ecological

ARC ANGEL
Angel Works, Bendish Farm,
Whitwell, Hitchin, Hertfordshire, SG4 8JD
Area of Operation: Europe
Tel: 01438 871100 **Fax:** 01438 871100
Email: angels@arcangelmetalwork.co.uk
Web: www.arcangelmetalwork.co.uk
Other Info: ✎ 🖑

AVON MANUFACTURING LIMITED
Avon House, Kineton Road, Southam,
Leamington Spa, Warwickshire, CV47 0DG
Area of Operation: UK (Excluding Ireland)
Tel: 01926 817292 **Fax:** 01926 814156
Email: sales@avonmanufacturing.co.uk
Web: www.avonmanufacturing.co.uk

B ROURKE & CO LTD
Vulcan Works, Accrington Road,
Burnley, Lancashire, BB11 5QD
Area of Operation: Worldwide
Tel: 01282 422841
Fax: 01282 458901
Email: info@rourkes.co.uk
Web: www.rourkes.co.uk
Other Info: ✎
Material Type: C) 2, 5, 6

BAYFIELD STAIR CO
Unit 4, Praed Road, Trafford Park,
Manchester, M17 1PQ
Area of Operation: Worldwide
Tel: 0161 848 0700 **Fax:** 0161 872 2230
Email: sales@bayfieldstairs.co.uk
Web: www.bayfieldstairs.co.uk

BISCA
Sawmill Lane, Helmsley,
North Yorkshire, YO62 5DQ
Area of Operation: UK (Excluding Ireland)
Tel: 01439 771702 **Fax:** 01439 771002
Email: info@bisca.co.uk
Web: www.bisca.co.uk

BISCA
Area of Operation: UK (Excluding Ireland)
Tel: 01439 771 702 **Fax:** 01439 771 002
Email: info@bisca.co.uk
Web: www.bisca.co.uk

Bisca, (an acronym for brass, iron, steel, copper, aluminium,) specialise in bold, sculptural, bespoke staircases. They have, indeed, turned the humble staircase into a work of art and have tackled everything from curved, cantilevered, spiral and suspended staircases in materials ranging from oak to stainless steel and glass.

BISCA
Area of Operation: UK (Excluding Ireland)
Tel: 01439 771 702
Fax: 01439 771 002
Email: info@bisca.co.uk
Web: www.bisca.co.uk

Curved, cantilevered, spiral or suspended – Bisca designs and builds one-off staircases which are works of art...dramatic constructions, built to take centre stage.

BISCA
Area of Operation: UK (Excluding Ireland)
Tel: 01439 771 702
Fax: 01439 771 002
Email: info@bisca.co.uk
Web: www.bisca.co.uk

A Bisca staircase is a work of art – making imaginative use of materials such as glass, timber and steel, and utilising modern technical know-how.

BLANC DE BIERGES
Eastrea Road, Whittlesey, Cambridgeshire, PE7 2AG
Area of Operation: Worldwide
Tel: 01733 202566 **Fax:** 01733 205405
Email: info@blancdebierges.com
Web: www.blancdebierges.com   **Other Info:** ✎ 🖑

BURBEARY JOINERY LTD
Units 2 - 4, 47 Robin Lane, Beighton,
Sheffield, South Yorkshire, S20 1BB
Area of Operation: UK (Excluding Ireland)
Tel: 0114 247 5003 **Fax:** 0114 247 5007
Email: info@burbearyjoinery.co.uk
Web: www.burbearyjoinery.co.uk ⌃
Other Info: ECO ✎ 🖑
Material Type: A) 2, 3, 4, 5, 6, 15

CAMBRIDGE STRUCTURES
2 Huntingdon Street, St. Neots,
Cambridgeshire, PE19 1BG
Area of Operation: Worldwide
Tel: 01480 477700 **Fax:** 01480 477766
Email: info@cambridgestructures.com
Web: www.cambridgestructures.com
Other Info: ECO ✎
Material Type: A) 1, 2, 3, 4, 5, 6, 7, 8, 9, 10, 11, 12, 13, 14, 15

CHRIS TOPP & COMPANY WROUGHT IRONWORKS
Lyndhurst, Carlton Husthwaite,
Thirsk, North Yorkshire, YO7 2BJ
Area of Operation: Worldwide
Tel: 01845 501415
Fax: 01845 501072
Email: enquiry@christopp.co.uk
Web: www.christopp.co.uk
Other Info: ✎ 🖑
Material Type: C) 1, 2, 3, 4, 5, 6, 7, 11, 12, 14, 18

COMPLETE STAIR SYSTEMS LTD
Unit 10/A, Acorn Business Centre,
Milton Street, Maidstone, Kent, ME16 8LL
Area of Operation: UK & Ireland
Tel: 0845 838 1622
Fax: 01622 721951
Email: info@completestairsystems.co.uk
Web: www.completestairsystems.co.uk
Other Info: ✎ 🖑
Material Type: A) 1, 2, 3, 4, 5, 6, 7, 8, 9, 10, 11, 12, 13, 14, 15

CONSCULPT SPIRAL STAIRCASES LTD
4 Morrab Road, Penzance, Cornwall, TR18 4EL
Area of Operation: Worldwide
Tel: 01736 364477
Fax: 01736 364477
Email: info@spiralstairs.uk.com
Web: www.spiralstairs.uk.com
Other Info: ✎ 🖑
Material Type: A) 1, 15

COTTAGE CRAFT SPIRALS
The Barn, Gorsley Low Farm,
The Wash, Chapel-En-Le-Frith,
Derbyshire, SK23 0QL
Area of Operation: Worldwide
Tel: 01663 750716
Fax: 01663 751093
Email: sales@castspiralstairs.com
Web: www.castspiralstairs.com
Other Info: ✎ 🖑
Material Type: A) 2, 3, 4, 5, 6, 7, 8, 9, 10, 11, 12, 13, 14

COUNTY JOINERY (SOUTH EAST) LTD
Unit 300, Vinehall Business Centre,
Vinehall Road, Mountfield, Robertsbridge,
East Sussex, TN32 5JW
Area of Operation: Greater London,
South East England
Tel: 01424 871500 **Fax:** 01424 871550
Email: countyjoinery@aol.com
Web: www.countyjoinery.co.uk

CREATE JOINERY
Worldsend, Llanmadoc, Swansea, SA3 1DB
Area of Operation: Greater London, Midlands & Mid Wales, North West England and North Wales, South East England, South West England and South Wales
Tel: 01792 386677 **Fax:** 01792 386677
Email: mail@create-joinery.co.uk
Web: www.create-joinery.co.uk

INTERIORS, FIXTURES & FINISHES

CROXFORD'S JOINERY MANUFACTURERS & WOODTURNERS
Meltham Joinery, Works New Street,
Meltham, Holmfirth, West Yorkshire, HD9 5NT
Area of Operation: UK (Excluding Ireland)
Tel: 01484 850892
Fax: 01484 850969
Email: ralph@croxfords.demon.co.uk
Web: www.croxfords.co.uk
Material Type: A) 2, 3, 6

CUSTOM WOOD PRODUCTS
Cliffe Road, Easton on the Hill,
Stamford, Lincolnshire, PE9 3NP
Area of Operation: East England
Tel: 01780 755711
Fax: 01780 480834
Email: customwoodprods@aol.com
Web: www.cwpuk.com
Other Info: ECO

DRESSER MOULDINGS LTD
Lostock Industrial Estate West, Cranfield Road,
Lostock, Bolton, Lancashire, BL6 4SB
Area of Operation: UK (Excluding Ireland)
Tel: 01204 667667
Fax: 01204 667600
Email: mac@dresser.uk.com
Web: www.dresser.uk.com

E.A. HIGGINSON & CO LTD
Unit 1, Carlisle Road, London, NW9 0HD
Area of Operation: UK (Excluding Ireland)
Tel: 0208 200 4848
Fax: 0208 200 8249
Email: sales@higginson.co.uk
Web: www.higginson.co.uk
Material Type: A) 1, 2, 3, 4, 5, 6, 7, 10

ERMINE ENGINEERING COMPANY LTD
Francis House, Silver Birch Park, Great Northern
Terrace, Lincoln, Lincolnshire, LN5 8LG
Area of Operation: UK (Excluding Ireland)
Tel: 01522 510977
Fax: 01522 510929
Email: info@ermineengineering.co.uk
Web: www.ermineengineering.co.uk
Other Info:
Material Type: A) 1, 2, 3, 4, 5, 6, 15

FAHSTONE LTD
Michaels Stud Farm, Meer End, Kenilworth,
Warwickshire, CV8 1PU
Area of Operation: UK (Excluding Ireland)
Tel: 01676 533679
Fax: 01676 532224
Email: sales@meer-end.co.uk
Web: www.meer-end.co.uk

FAIRMITRE WINDOWS & JOINERY LTD
2A Cope Road, Banbury, Oxfordshire, OX16 2EH
Area of Operation: UK (Excluding Ireland)
Tel: 01295 268441
Fax: 01295 268468
Email: info@fairmitrewindows.co.uk
Web: www.fairmitrewindows.co.uk

FITZROY JOINERY
Garden Close, Langage Industrial Estate,
Plympton, Devon, PL7 5EU
Area of Operation: UK & Ireland
Tel: 0870 4289 110
Fax: 0870 4289 111
Email: admin@fitzroy.co.uk
Web: www.fitzroy.co.uk
Other Info: ECO
Material Type: A) 1, 2, 3, 4, 5, 6, 7, 8, 9, 10, 11

FLETCHER JOINERY
261 Whessoe Road, Darlington, Durham, DL3 0YL
Area of Operation: North East England
Tel: 01325 357347
Fax: 01325 357347
Email: enquiries@fletcherjoinery.co.uk
Web: www.fletcherjoinery.co.uk

FLIGHT TIMBER PRODUCTS
Earls Colne Business Park,
Earls Colne, Essex, CO6 2NS
Area of Operation: East England,
Greater London, South East England
Tel: 01787 222336 **Fax:** 01787 222359
Email: sales@flighttimber.com
Web: www.flighttimber.com
Other Info:

G MIDDLETON LTD
Cross Croft Industrial Estate,
Appleby, Cumbria, CA16 6HX
Area of Operation: Europe
Tel: 01768 352067 **Fax:** 01768 353228
Email: g.middleton@timber75.freeserve.co.uk
Web: www.graham-middleton.co.uk
Material Type: A) 2

HALDANE UK LTD
Blackwood Way, Bankwood Industrial Estate,
Glenrothes, Fife, KY7 6JF
Area of Operation: UK & Ireland
Tel: 01592 775656 **Fax:** 01592 775757
Email: sales@haldaneuk.com
Web: www.haldaneuk.com
Other Info:
Material Type: A) 2, 3, 4, 5, 6, 7, 8, 9, 10, 11, 12, 13, 14

HALLS STAIRS & LANDING
The Triangle, Paddock, Huddersfield,
West Yorkshire, HD1 4RN
Area of Operation: UK & Ireland
Tel: 01484 451485
Email: info@loftaccess.com
Web: www.loftaccess.com
Material Type: A) 2, 6

INTERIOR ASSOCIATES
3 Highfield Road, Windsor, Berkshire, SL4 4DN
Area of Operation: UK & Ireland
Tel: 01753 865339 **Fax:** 01753 865339
Email: sales@interiorassociates.fsnet.co.uk
Web: www.interiorassociates.co.uk
Other Info:
Material Type: C) 2, 3, 14, 17

JAIC LTD
Pattern House, Southwell Business Park,
Portland, Dorset, DT5 2NR
Area of Operation: UK (Excluding Ireland)
Tel: 01305 826991
Fax: 01305 823535
Email: info@jaic.co.uk
Other Info:
Material Type: A) 2, 4, 5

JELD-WEN
Watch House Lane, Doncaster,
South Yorkshire, DN5 9LR
Area of Operation: Worldwide
Tel: 01302 394000
Fax: 01302 787383
Email: customer-services@jeld-wen.co.uk
Web: www.jeld-wen.co.uk

JIMPEX LTD
1 Castle Farm, Cholmondeley,
Malpas, Cheshire, SY14 8AQ
Area of Operation: UK & Ireland
Tel: 01829 720433
Fax: 01829720553
Email: jan@jimpex.co.uk
Web: www.jimpex.co.uk

JOINERY-PLUS
Bentley Hall Barn,
Alkmonton, Ashbourne,
Derbyshire, DE6 3DJ
Area of Operation: UK (Excluding Ireland)
Tel: 07931 386233
Fax: 01335 330922
Email: info@joinery-plus.co.uk
Web: www.joinery-plus.co.uk
Material Type: A) 2, 6, 7

KASPAR SWANKEY
405, Goldhawk Road, Hammersmith,
West London, W6 0SA
Area of Operation: Worldwide
Tel: 0208 746 3586
Fax: 0208 746 3586
Email: kaspar@swankeypankey.com
Web: www.swankeypankey.com

LADDERS-ONLINE
Penarth Road, Cardiff, CF11 8TD
Area of Operation: Worldwide
Tel: 08450 647647
Fax: 02920 702386
Email: sales@tbdavies.co.uk
Web: www.ladders-online.com

LAPPIPORRAS
Tresparrett Farm Villa,
Tresparrett, Camelford,
Cornwall, PL32 9ST
Area of Operation: UK & Europe
Tel: 01840 261415
Fax: 01840 261415
Email: sales@finnishwoodproducts.com
Web: www.finnishwoodproducts.com
Material Type: A) 9

LAPPIPORRAS STAIRS
Area of Operation: UK & Europe
Tel: 01840 261415
Fax: 01840 261415
Email: sales@finnishwoodproducts.co.uk
Web: www.finnishwoodproducts.com

Stairs from Finland, manufactured in high-quality Finnish Redwood or Birch.
Factory lacquered in clear of coloured finishes. Various styles and components to create individualism. Made to your measurements.

LOFT CENTRE PRODUCTS
Thicket Lane, Halnaker, Nr Chichester,
West Sussex, PO18 0QS
Area of Operation: UK & Ireland
Tel: 01243 785246
Fax: 01243 533184
Email: sales@loftcentreproducts.co.uk
Web: www.loftcentreproducts.co.uk

LOFT CENTRE PRODUCTS
Area of Operation: UK
Tel: 01243 785246
Fax: 01243 533184
Email: sales@loftcentreproducts.co.uk
Web: www.loftcentreproducts.co.uk

Loft Centre Products have developed a comprehensive range of staircases both standard and bespoke, offering a wide range of styles and materials to match most applications.

METALCRAFT [TOTTENHAM]
6-40 Durnford Street, Tottenham, London, N15 5NQ
Area of Operation: UK (Excluding Ireland)
Tel: 0208 802 1715
Fax: 0208 802 1258
Email: sales@makingmetalwork.com
Web: www.makingmetalwork.com
Material Type: C) 2, 3, 4, 5

NIGEL TYAS HANDCRAFTED IRONWORK
Green Works, 18 Green Road,
Penistone, Sheffield, South Yorkshire, S36 6BE
Area of Operation: Worldwide
Tel: 01226 766618
Email: sales@nigeltyas.co.uk
Web: www.nigeltyas.co.uk
Other Info: ✎ ✋
Material Type: C) 2, 4, 5, 6

OAKLEAF INDUSTRIES LTD
D5 Flightway Business Park,
Dunkeswell, Honiton, Devon, EX14 4RD
Area of Operation: UK & Ireland
Tel: 01404 891902
Fax: 01404 891912
Email: info@stairsolutions.co.uk
Web: www.stairsolutions.co.uk
Material Type: A) 1, 2, 4, 5, 6, 9

PARAGON JOINERY
Systems House, Eastbourne Road,
Blindley Heath, Surrey, RH7 6JP
Area of Operation: UK (Excluding Ireland)
Tel: 01342 836300
Fax: 01342 836364
Email: sales@paragonjoinery.com
Web: www.paragonjoinery.com

ROBERT J TURNER & CO
Roe Green, Sandon, Buntingford,
Hertfordshire, SG9 0QE
Area of Operation: East England, Greater London
Tel: 01763 288371
Fax: 01763 288440
Email: sales@robertjturner.co.uk
Web: www.robertjturner.co.uk

SANDERSON'S FINE FURNITURE
Unit 5 & 6, The Village Workshop, Four Crosses
Business Park, Four Crosses, Powys, SY22 6ST
Area of Operation: UK (Excluding Ireland)
Tel: 01691 830075
Fax: 01691 830075
Email: sales@sandersonsfinefurniture.co.uk
Web: www.sandersonsfinefurniture.co.uk

SPIRAL STAIRS LTD
Unit 3, Annington Commercial Centre, Annington
Road, Steyning, West Sussex, BN44 3WA
Area of Operation: Europe
Tel: 01903 812310
Fax: 01903 812306
Email: sales@spiralstairs.org
Web: www.spiralstairs.org
Other Info: ✎
Material Type: A) 1, 2, 3, 4, 5, 6, 9, 15

STAIRFLIGHT LTD
Unit 17, Landford Common Industrial Estate,
New Road, Landford, Salisbury, Wiltshire, SP5 2AZ
Area of Operation: UK (Excluding Ireland)
Tel: 01794 324150
Fax: 01794 324151
Email: sales@prestige-staircases.com
Web: www.prestige-staircases.com
Other Info: ✎ ✋
Material Type: A) 1, 2, 3, 4, 5, 6

STAIRPLAN LTD
Unit C4, Stafford Park 4, Telford,
Shropshire, TF3 3BA
Area of Operation: UK (Excluding Ireland)
Tel: 01952 216000
Fax: 01952 216021
Email: sales@stairplan.com
Web: www.stairplan.com
Material Type: A) 1, 2, 4, 6, 9

STAIRRODS (UK) LTD
Unit 6, Park Road North Industrial Estate,
Blackhill, Consett, Durham, DH8 5UN
Area of Operation: Worldwide
Tel: 01207 591176
Fax: 01207 591911
Email: sales@stairrods.co.uk
Web: www.stairrods.co.uk
Other Info: ✎

STIRA FOLDING ATTIC STAIRS
Dunmore, Co Galway, Ireland
Area of Operation: UK & Ireland
Tel: 0800 973111
Fax: +353 933 8428
Email: stira@stira.ie
Web: www.stira.co.uk ✑
Other Info: ✎   **Material Type:** B) 2

STUART INTERIORS
Barrington Court, Barrington,
Ilminster, Somerset, TA19 0NQ
Area of Operation: Worldwide
Tel: 01460 240349
Fax: 01460 242069
Email: design@stuartinteriors.com
Web: www.stuartinteriors.com

T.B DAVIES (CARDIFF) LTD
Penarth Road, Cardiff, CF11 8TD
Area of Operation: UK (Excluding Ireland)
Tel: 02920 713000
Fax: 02920 702386
Email: sales@tbdavies.co.uk
Web: www.ladders-online.com ✑

TEAM TECHNOLOGY
32 High Street, Guilden Morden,
Nr Royston, Hertfordshire, SG8 0JR
Area of Operation: UK (Excluding Ireland)
Tel: 01763 853369
Fax: 01763 853164
Email: info@teamtechnology.co.uk
Web: www.teamtechnologyltd.co.uk
Material Type: A) 2, 3, 4, 5, 6, 9, 14

THE GRANITE FACTORY
4 Winchester Drive, Peterlee,
Durham, SR8 2RJ
Area of Operation: North East England
Tel: 0191 518 3600
 Fax: 0191 518 3600
Email: admin@granitefactory.co.uk
Web: www.granitefactory.co.uk ✑

THE LOFT SHOP LTD
Eldon Way, Littlehampton,
West Sussex, BN17 7HE
Area of Operation: UK (Excluding Ireland)
Tel: 0870 6040404
Fax: 0870 6039075
Email: marjory.kay@loftshop.co.uk
Web: www.loftshop.co.uk ✑

THE SWEDISH WINDOW COMPANY
Old Maltings House, Hall Street,
Long Melford, Suffolk, CO10 9JB
Area of Operation: UK & Ireland
Tel: 01787 467297
 Fax: 01787 319982
Email: info@swedishwindows.com
Web: www.swedishwindows.com

THE WOODEN HILL COMPANY LTD
The Lodge, Old House, The Street,
Betchworth, Surrey, RH3 7DJ
Area of Operation: Europe
Tel: 0845 456 1088
Fax: 01932 264 693
Email: scott@the-wooden-hill-company.co.uk
Web: www.the-wooden-hill-company.co.uk
Other Info: ✎ ✋
Material Type: A) 1, 2, 3, 4, 7, 15

THE WOODEN HILL COMPANY

Area of Operation: Europe
Tel: 0845 456 1088
Fax: 01932 264693
Email: scott@the-wooden-hill-company.co.uk
Web: www.the-wooden-hill-company.co.uk

Building on 25 years of experience TWHC offer a comprehensive staircase range. With over 100 different models ranging from kits for as little as £600 moving up to purpose built staircases for all domestic and commercial applications. So please keep in touch via our website or call us and let us know if you would like to be added to our mailing list.

TOMPKINS LTD
High March Close, Long March Industrial Estate,
Daventry, Northamptonshire, NN11 4EZ
Area of Operation: UK (Excluding Ireland)
Tel: 01327 877187
Fax: 01327 310491
Email: info@tompkinswood.co.uk
Web: www.tompkinswood.co.uk
Other Info: ✎

TONY HOOPER
Unit 18 Camelot Court, Bancombe Trading Estate,
Somerton, TA11 6SB
Area of Operation: UK (Excluding Ireland)
Tel: 01458 274221
Fax: 01458 274690
Email: tonyhooper1@aol.com
Web: www.tonyhooper.co.uk

WOODCHESTER KITCHENS & INTERIORS
Unit 18a Chalford Industrial Estate,
Chalford, Gloucestershire, GL6 8NT
Area of Operation: UK (Excluding Ireland)
Tel: 01453 886411
Fax: 01453 886411
Email: enquires@woodchesterkitchens.co.uk
Web: www.woodchesterkitchens.co.uk
Other Info: ✎ ✋

WOODSIDE JOINERY
40 Llantarnam Park, Cwmbran,
Torfaen, NP44 3AW
Area of Operation: UK & Ireland
Tel: 01633 875232
Fax: 01633 482718
Email: woodsidejoinery@talk21.com

SPIRAL & HELICAL STAIRCASES

KEY
OTHER: ▽ Reclaimed ✑ On-line shopping
✎ Bespoke ✋ Hand-made ECO Ecological

B ROURKE & CO LTD
Vulcan Works, Accrington Road, Burnley,
Lancashire, BB11 5QD
Area of Operation: Worldwide
Tel: 01282 422841
Fax: 01282 458901
Email: info@rourkes.co.uk
Web: www.rourkes.co.uk
Material Type: C) 2, 5, 6

BAYFIELD STAIR CO
Unit 4, Praed Road, Trafford Park,
Manchester, M17 1PQ
Area of Operation: Worldwide
Tel: 0161 848 0700
Fax: 0161 872 2230
Email: sales@bayfieldstairs.co.uk
Web: www.bayfieldstairs.co.uk

BISCA
Sawmill Lane, Helmsley,
North Yorkshire, YO62 5DQ
Area of Operation: UK (Excluding Ireland)
Tel: 01439 771702
Fax: 01439 771002
Email: info@bisca.co.uk
Web: www.bisca.co.uk

BLANC DE BIERGES
Eastrea Road, Whittlesey,
Cambridgeshire, PE7 2AG
Area of Operation: Worldwide
Tel: 01733 202566
Fax: 01733 205405
Email: info@blancdebierges.com
Web: www.blancdebierges.com
Other Info: ✎ ✋

BURBEARY JOINERY LTD
Units 2 - 4, 47 Robin Lane,
Beighton, Sheffield,
South Yorkshire, S20 1BB
Area of Operation: UK (Excluding Ireland)
Tel: 0114 247 5003
Fax: 0114 247 5007
Email: info@burbearyjoinery.co.uk
Web: www.burbearyjoinery.co.uk ✍
Other Info: ECO ✎ ✋
Material Type: A) 1, 2, 4, 5, 6, 15

CAMBRIDGE STRUCTURES
2 Huntingdon Street, St. Neots,
Cambridgeshire, PE19 1BG
Area of Operation: Worldwide
Tel: 01480 477700
Fax: 01480 477766
Email: info@cambridgestructures.com
Web: www.cambridgestructures.com
Other Info: ECO ✎ ✋
Material Type: A) 1, 2, 3, 4, 5, 6, 7, 8, 9,
10, 11, 12, 13, 14, 15

**CHRIS TOPP & COMPANY
WROUGHT IRONWORKS**
Lyndhurst, Carlton Husthwaite,
Thirsk, North Yorkshire, YO7 2BJ
Area of Operation: Worldwide
Tel: 01845 501415
Fax: 01845 501072
Email: enquiry@christopp.co.uk
Web: www.christopp.co.uk
Other Info: ✎ ✋
Material Type: C) 2, 3, 4, 5, 6, 7, 11, 12, 17

COMPLETE STAIR SYSTEMS LTD
Unit 10/A, Acorn Business Centre,
Milton Street, Maidstone, Kent, ME16 8LL
Area of Operation: UK & Ireland
Tel: 0845 838 1622
Fax: 01622 721951
Email: info@completestairsystems.co.uk
Web: www.completestairsystems.co.uk
Other Info: ✎ ✋
Material Type: A) 1, 2, 3, 4, 5, 6, 7, 8, 9,
10, 11, 12, 13, 14, 15

CONSCULPT SPIRAL STAIRCASES LTD
4 Morrab Road, Penzance, Cornwall, TR18 4EL
Area of Operation: Worldwide
Tel: 01736 364477
Fax: 01736 364477
Email: info@spiralstairs.uk.com
Web: www.spiralstairs.uk.com

COTTAGE CRAFT SPIRALS
The Barn, Gorsley Low Farm, The Wash,
Chapel-En-Le-Frith, Derbyshire, SK23 0QL
Area of Operation: Worldwide
Tel: 01663 750716 **Fax:** 01663 751093
Email: sales@castspiralstairs.com
Web: www.castspiralstairs.com
Other Info: ✎ ✋
Material Type: A) 2, 3, 4, 5, 6, 7, 8, 9,
10, 11, 12, 13, 14

CREATE JOINERY
Worldsend, Llanmadoc, Swansea, SA3 1DB
Area of Operation: Greater London, Midlands & Mid
Wales, North West England and North Wales, South
East England, South West England and South Wales
Tel: 01792 386677
Fax: 01792 386677
Email: mail@create-joinery.co.uk
Web: www.create-joinery.co.uk

DRESSER MOULDINGS LTD
Lostock Industrial Estate West, Cranfield Road,
Lostock, Bolton, Lancashire, BL6 4SB
Area of Operation: UK (Excluding Ireland)
Tel: 01204 667667 **Fax:** 01204 667600
Email: mac@dresser.uk.com
Web: www.dresser.uk.com

DRUMMONDS ARCHITECTURAL ANTIQUES LTD
Kirkpatrick Buildings, 25 London Road (A3),
Hindhead, Surrey, GU36 6AB
Area of Operation: Worldwide
Tel: 01428 609444
Fax: 01428 609445
Email: davidcox@drummonds-arch.co.uk
Web: www.drummonds-arch.co.uk
Other Info: ▽
Material Type: A) 2

E.A. HIGGINSON & CO LTD
Unit 1, Carlisle Road, London, NW9 0HD
Area of Operation: UK (Excluding Ireland)
Tel: 0208 200 4848 **Fax:** 0208 200 8249
Email: sales@higginson.co.uk
Web: www.higginson.co.uk
Material Type: A) 2, 4, 5

ERMINE ENGINEERING COMPANY LTD
Francis House, Silver Birch Park, Great Northern
Terrace, Lincoln, Lincolnshire, LN5 8LG
Area of Operation: UK (Excluding Ireland)
Tel: 01522 510977 **Fax:** 01522 510929
Email: davidcox@ermineengineering.co.uk
Web: www.ermineengineering.co.uk
Other Info: ✎ ✋
Material Type: A) 1, 2, 3, 4, 5, 6, 15

EUROPEAN PIPE SUPPLIERS LTD
Unit 6 Severnlink Distribution Centre,
off Newhouse Farm Estate, Chepstow,
Monmouthshire, NP16 6UN
Area of Operation: Europe
Tel: 01291 622215
Fax: 01291 620381
Email: sales@epsonline.co.uk
Web: www.epsonline.co.uk
Material Type: C) 1, 2, 3, 4, 5, 6

FAHSTONE LTD
Michaels Stud Farm, Meer End,
Kenilworth, Warwickshire, CV8 1PU
Area of Operation: UK (Excluding Ireland)
Tel: 01676 533679
Fax: 01676 532224
Email: sales@meer-end.co.uk
Web: www.meer-end.co.uk

FITZROY JOINERY
Garden Close, Langage Industrial Estate,
Plympton, Devon, PL7 5EU
Area of Operation: UK & Ireland
Tel: 0870 428 9110
Fax: 0870 428 9111
Email: admin@fitzroy.co.uk
Web: www.fitzroy.co.uk ✍
Other Info: ECO ✎
Material Type: A) 1, 2, 3, 4, 5, 6, 7, 8, 9, 10, 11, 12, 14

HALDANE UK LTD
Blackwood Way, Bankwood Industrial Estate,
Glenrothes, Fife, KY7 6JF
Area of Operation: UK & Ireland
Tel: 01592 775656
Fax: 01592 775757
Email: sales@haldaneuk.com
Web: www.haldaneuk.com
Other Info: ✎
Material Type: A) 2, 3, 4, 5, 6, 7, 8, 9, 10, 11, 12, 13, 14

HALLS STAIRS & LANDING
The Triangle, Paddock, Huddersfield,
West Yorkshire, HD1 4RN
Area of Operation: UK & Ireland
Tel: 01484 451485
Email: info@loftaccess.com
Web: www.loftaccess.com
Material Type: A) 4

JAIC LTD
Pattern House, Southwell Business Park,
Portland, Dorset, DT5 2NR
Area of Operation: UK (Excluding Ireland)
Tel: 01305 826991
Fax: 01305 823535
Email: info@jaic.co.uk

LADDERS-ONLINE
Penarth Road, Cardiff, CF11 8TD
Area of Operation: Worldwide
Tel: 08450 647647
Fax: 02920 702386
Email: sales@tbdavies.co.uk
Web: www.ladders-online.com ✍

LOFT CENTRE PRODUCTS
Thicket Lane, Halnaker, Nr Chichester,
West Sussex, PO18 0QS
Area of Operation: UK & Ireland
Tel: 01243 785246
Fax: 01243 533184
Email: sales@loftcentreproducts.co.uk
Web: www.loftcentreproducts.co.uk

LOFT CENTRE PRODUCTS
Area of Operation: UK
Tel: 01243 785246 **Fax:** 01243 533184
Email: sales@loftcentreproducts.co.uk
Web: www.loftcentreproducts.co.uk

Loft Centre Products comprehensive range of
spirals include all metal, all timber, metal and
timber, round and square to suit most internal
and external applications.

METALCRAFT [TOTTENHAM]
6-40 Durnford Street,
Tottenham, London, N15 5NQ
Area of Operation: UK (Excluding Ireland)
Tel: 0208 802 1715
Fax: 0208 802 1258
Email: sales@makingmetalwork.com
Web: www.makingmetalwork.com

OAKLEAF INDUSTRIES LTD
D5 Flightway Business Park,
Dunkeswell, Honiton,
Devon, EX14 4RD
Area of Operation: UK & Ireland
Tel: 01404 891902
Fax: 01404 891912
Email: info@stairsolutions.co.uk
Web: www.stairsolutions.co.uk
Material Type: A) 1, 2, 3, 4, 5, 6, 9

RICHARD BURBIDGE LTD
Whittington Road, Oswestry,
Shropshire, SY11 1HZ
Area of Operation: Worldwide
Tel: 01691 655131
Fax: 01691 657694
Email: info@richardburbidge.co.uk
Web: www.richardburbidge.co.uk

SPIRAL CONSTRUCTION LTD
Water-Ma-Trout Industrial Estate,
Helston, Cornwall, TR13 0LW
Area of Operation: UK & Ireland
Tel: 01326 574497
Fax: 01326 574760
Email: enquiries@spiral.uk.com
Web: www.spiral.uk.com
Material Type: A) 2, 3, 4, 5, 6, 7, 9

SPIRAL STAIRCASE SYSTEMS
The Mill, Glynde, Near Lewes,
East Sussex, BN8 6SS
Area of Operation: Worldwide
Tel: 01273 858341
Fax: 01273 858200
Email: sales@spiralstairs.co.uk
Web: www.spiralstairs.co.uk
Material Type: A) 1, 2, 3, 4, 5, 6, 7, 8, 9, 10, 12

SPIRAL STAIRS LTD
Unit 3, Annington Commercial Centre,
Annington Road, Steyning,
West Sussex, BN44 3WA
Area of Operation: Europe
Tel: 01903 812310
Fax: 01903 812306
Email: sales@spiralstairs.org
Web: www.spiralstairs.org
Other Info: ✎
Material Type: A) 1, 2, 3, 4, 5, 6, 9, 15

STAIRFLIGHT LTD
Unit 17, Landford Common Industrial Estate,
New Road, Landford, Salisbury, Wiltshire, SP5 2AZ
Area of Operation: UK (Excluding Ireland)
Tel: 01794 324150
Fax: 01794 324151
Email: sales@prestige-staircases.com
Web: www.prestige-staircases.com
Other Info: ✎
Material Type: A) 1, 2, 3, 4, 5, 6

T.B DAVIES (CARDIFF) LTD
Penarth Road, Cardiff, CF11 8TD
Area of Operation: UK (Excluding Ireland)
Tel: 02920 713000
Fax: 02920 702386
Email: sales@tbdavies.co.uk
Web: www.ladders-online.com ✎

TEAM TECHNOLOGY
32 High Street, Guilden Morden,
nr Royston, Hertfordshire, SG8 0JR
Area of Operation: UK (Excluding Ireland)
Tel: 01763 853369
Fax: 01763 853164
Email: info@teamtechnology.co.uk
Web: www.teamtechnologyltd.co.uk
Material Type: A) 2, 3, 4, 5, 6, 9, 14

THE GRANITE FACTORY
4 Winchester Drive, Peterlee, Durham, SR8 2RJ
Area of Operation: North East England
Tel: 0191 518 3600
Fax: 0191 518 3600
Email: admin@granitefactory.co.uk
Web: www.granitefactory.co.uk ✎
Material Type: E) 1, 2, 3, 4, 5, 7, 8, 9, 11

THE LOFT SHOP LTD
Eldon Way, Littlehampton,
West Sussex, BN17 7HE
Area of Operation: UK (Excluding Ireland)
Tel: 0870 6040404
Fax: 0870 6039075
Email: marjory.kay@loftshop.co.uk
Web: www.loftshop.co.uk ✎

THE WOODEN HILL COMPANY LTD
The Lodge, Old House, The Street,
Betchworth, Surrey, RH3 7DJ
Area of Operation: Europe
Tel: 0845 456 1088
Fax: 01932 264 693
Email: scott@the-wooden-hill-company.co.uk
Web: www.the-wooden-hill-company.co.uk
Other Info: ✎ ✋
Material Type: A) 1, 2, 3, 4, 7, 15

TOMPKINS LTD
High March Close, Long March Industrial Estate,
Daventry, Northamptonshire, NN11 4EZ
Area of Operation: UK (Excluding Ireland)
Tel: 01327 877187
Fax: 01327 310491
Email: info@tompkinswood.co.uk
Web: www.tompkinswood.co.uk
Other Info: ✎ ✋

WOODSIDE JOINERY
40 Llantarnam Park, Cwmbran, Torfaen, NP44 3AW
Area of Operation: UK & Ireland
Tel: 01633 875232 **Fax:** 01633 482718
Email: woodsidejoinery@talk21.com

STAIRPARTS

KEY
OTHER: ▽ Reclaimed 🛒 On-line shopping
✎ Bespoke ✋ Hand-made ECO Ecological

ARC ANGEL
Angel Works, Bendish Farm, Whitwell,
Hitchin, Hertfordshire, SG4 8JD
Area of Operation: Europe
Tel: 01438 871100
Fax: 01438 871100
Email: angels@arcangelmetalwork.co.uk
Web: www.arcangelmetalwork.co.uk

AVON MANUFACTURING LIMITED
Avon House, Kineton Road, Southam,
Leamington Spa, Warwickshire, CV47 0DG
Area of Operation: UK (Excluding Ireland)
Tel: 01926 817292
Fax: 01926 814156
Email: sales@avonmanufacturing.co.uk
Web: www.avonmanufacturing.co.uk

B ROURKE & CO LTD
Vulcan Works, Accrington Road,
Burnley, Lancashire, BB11 5QD
Area of Operation: Worldwide
Tel: 01282 422841
Fax: 01282 458901
Email: info@rourkes.co.uk
Web: www.rourkes.co.uk
Material Type: C) 2, 4, 5, 6

BALCAS TIMBER LTD
Laragh, Enniskillen, Co Fermanagh, BT94 2FQ
Area of Operation: UK & Ireland
Tel: 0286 632 3003
Fax: 0286 632 7924
Email: info@balcas.com
Web: www.balcas.com

BAYFIELD STAIR CO
Unit 4, Praed Road, Trafford Park,
Manchester, M17 1PQ
Area of Operation: Worldwide
Tel: 0161 848 0700
Fax: 0161 872 2230
Email: sales@bayfieldstairs.co.uk
Web: www.bayfieldstairs.co.uk

BISCA
Sawmill Lane, Helmsley, North Yorkshire, YO62 5DQ
Area of Operation: UK (Excluding Ireland)
Tel: 01439 771702
Fax: 01439 771002
Email: info@bisca.co.uk
Web: www.bisca.co.uk

BOB LANE WOODTURNERS
Unit 1, White House Workshop,
Old London Road, Swinfen,
Lichfield, Staffordshire, WS14 9QW
Area of Operation: UK (Excluding Ireland)
Tel: 01543 483148
Fax: 01543 481245
Email: sales@theturner.co.uk
Web: www.theturner.co.uk

BURBEARY JOINERY LTD
Units 2 - 4, 47 Robin Lane, Beighton,
Sheffield, South Yorkshire, S20 1BB
Area of Operation: UK (Excluding Ireland)
Tel: 0114 247 5003
Fax: 0114 247 5007
Email: info@burbearyjoinery.co.uk
Web: www.burbearyjoinery.co.uk ✎
Other Info: ECO ✎
Material Type: A) 2, 3, 4, 5, 6, 7, 9

**CHRIS TOPP & COMPANY
WROUGHT IRONWORKS**
Lyndhurst, Carlton Husthwaite, Thirsk,
North Yorkshire, YO7 2BJ
Area of Operation: Worldwide
Tel: 01845 501415
Fax: 01845 501072
Email: enquiry@christopp.co.uk
Web: www.christopp.co.uk
Other Info: ✎ ✋
Material Type: C) 2, 3, 4, 5, 6, 7, 11, 12, 17

CHRISTIE TIMBER SERVICES LTD
New Victoria Sawmills, Bridgeness Road,
Bo'ness, Falkirk, EH51 9LG
Area of Operation: North East England, Scotland
Tel: 01506 828222
Fax: 01506 828226
Email: sales@christie-timber.co.uk
Web: www.christie-timber.co.uk ✎
Material Type: A) 1, 2, 3, 5, 6, 12

COTTAGE CRAFT SPIRALS
The Barn, Gorsley Low Farm, The Wash,
Chapel-En-Le-Frith, Derbyshire, SK23 0QL
Area of Operation: Worldwide
Tel: 01663 750716
Fax: 01663 751093
Email: sales@castspiralstairs.com
Web: www.castspiralstairs.com
Other Info: ✎
Material Type: A) 2, 3, 4, 5, 6, 7, 8, 9, 10, 11, 12, 13, 14

COUNTY JOINERY (SOUTH EAST) LTD
Unit 300, Vinehall Business Centre,
Vinehall Road, Mountfield, Robertsbridge,
East Sussex, TN32 5JW
Area of Operation: Greater London, South East England
Tel: 01424 871500
Fax: 01424 871550
Email: countyjoinery@aol.com
Web: www.countyjoinery.co.uk

**CROXFORD'S JOINERY
MANUFACTURERS & WOODTURNERS**
Meltham Joinery, Works New Street,
Meltham, Holmfirth, West Yorkshire, HD9 5NT
Area of Operation: UK (Excluding Ireland)
Tel: 01484 850892
Fax: 01484 850969
Email: ralph@croxfords.demon.co.uk
Web: www.croxfords.co.uk

CUSTOM WOOD PRODUCTS
Cliffe Road, Easton on the Hill,
Stamford, Lincolnshire, PE9 3NP
Area of Operation: East England
Tel: 01780 755711 **Fax:** 01780 480834
Email: customwoodprods@aol.com
Web: www.cwpuk.com
Other Info: ECO ✎ ✋

DESIGNS BY DAVID
84 Merlin Avenue, Nuneaton,
Warwickshire, CV10 9JZ
Area of Operation: UK & Ireland
Tel: 02476 744580
Email: david@designsbydavid.co.uk
Web: www.designsbydavid.co.uk
Other Info: ✋

DRESSER MOULDINGS LTD
Lostock Industrial Estate West, Cranfield Road,
Lostock, Bolton, Lancashire, BL6 4SB
Area of Operation: UK (Excluding Ireland)
Tel: 01204 667667 **Fax:** 01204 667600
Email: mac@dresser.uk.com
Web: www.dresser.uk.com

ESL HEALTHCARE LTD
Potts Marsh Industrial Estate, Eastbourne Road,
Westham, East Sussex, BN24 5NH
Area of Operation: UK & Ireland
Tel: 01323 465800
Fax: 01323 460248
Email: sales@eslindustries.co.uk
Web: www.eslindustries

EUROPEAN PIPE SUPPLIERS LTD
Unit 6 Severnlink Distribution Centre, off Newhouse
Farm Estate, Chepstow, Monmouthshire, NP16 6UN
Area of Operation: Europe
Tel: 01291 622215 **Fax:** 01291 620381
Email: sales@epsonline.co.uk
Web: www.epsonline.co.uk
Material Type: C) 5, 6

FAHSTONE LTD
Michaels Stud Farm, Meer End,
Kenilworth, Warwickshire, CV8 1PU
Area of Operation: UK (Excluding Ireland)
Tel: 01676 533679
Fax: 01676 532224
Email: sales@meer-end.co.uk
Web: www.meer-end.co.uk

HALDANE UK LTD
Blackwood Way, Bankwood Industrial Estate,
Glenrothes, Fife, KY7 6JF
Area of Operation: UK & Ireland
Tel: 01592 775656
Fax: 01592 775757
Email: sales@haldaneuk.com
Web: www.haldaneuk.com
Other Info: ✎
Material Type: A) 2, 3, 4, 5, 6, 7, 8, 9, 10, 11, 12, 13, 14

HALLS STAIRS & LANDING
The Triangle, Paddock, Huddersfield,
West Yorkshire, HD1 4RN
Area of Operation: UK & Ireland
Tel: 01484 451485
Email: info@loftaccess.com
Web: www.loftaccess.com
Material Type: A) 1, 2, 6

HERITAGE WOOD TURNING
Unit 3 The Old Brickworks, Bakestonedale Road,
Pott Shrigley, Macclesfield, Cheshire, SK10 5RX
Area of Operation: UK (Excluding Ireland)
Tel: 01625 560655 **Fax:** 01625 560697
Email: sales@heritageoak.co.uk
Web: www.heritageoak.co.uk
Other Info: ✎

HOUSE OF BRASS
122 North Sherwood Street,
Nottingham, Nottinghamshire, NG1 4EF
Area of Operation: Worldwide
Tel: 0115 947 5430 **Fax:** 0115 947 5430
Email: sales@houseofbrass.co.uk
Web: www.houseofbrass.co.uk ✎

ION GLASS
Unit 7 Grange Industrial Estate, Albion Street,
Southwick, Brighton, West Sussex, BN42 4EN
Area of Operation: UK (Excluding Ireland)
Tel: 0845 658 9988 **Fax:** 0845 658 9989
Email: sales@ionglass.co.uk
Web: www.ionglass.co.uk
Other Info: ✎
Material Type: J) 4, 5

JAIC LTD
Pattern House, Southwell Business Park,
Portland, Dorset, DT5 2NR
Area of Operation: UK (Excluding Ireland)
Tel: 01305 826991 **Fax:** 01305 823535
Email: info@jaic.co.uk

JOINERY-PLUS
Bentley Hall Barn, Alkmonton,
Ashbourne, Derbyshire, DE6 3DJ
Area of Operation: UK (Excluding Ireland)
Tel: 07931 386233 **Fax:** 01335 330922
Email: info@joinery-plus.co.uk
Web: www.joinery-plus.co.uk
Material Type: A) 2, 6, 7

M.B.L
55 High Street, Biggleswade,
Bedfordshire, SG18 0JH
Area of Operation: UK (Excluding Ireland)
Tel: 01767 318695
Fax: 01767 318695
Email: info@mblai.co.uk
Web: www.mblai.co.uk
Other Info: ▽ ✎ ✋

METALCRAFT [TOTTENHAM]
6-40 Durnford Street, Tottenham,
London, N15 5NQ
Area of Operation: UK (Excluding Ireland)
Tel: 0208 802 1715
Fax: 0208 802 1258
Email: sales@makingmetalwork.com
Web: www.makingmetalwork.com
Other Info: ✎
Material Type: C) 2, 3, 4, 5

NIGEL TYAS HANDCRAFTED IRONWORK
Green Works, 18 Green Road, Penistone,
Sheffield, South Yorkshire, S36 6BE
Area of Operation: Worldwide
Tel: 01226 766618
Email: sales@nigeltyas.co.uk
Web: www.nigeltyas.co.uk
Other Info: ✎ ✋
Material Type: C) 2, 5, 6

RICHARD BURBIDGE LTD
Whittington Road,
Oswestry, Shropshire, SY11 1HZ
Area of Operation: Worldwide
Tel: 01691 655131
Fax: 01691 657694
Email: info@richardburbidge.co.uk
Web: www.richardburbidge.co.uk

SCHLUTER SYSTEMS LTD
Units 4-5, Bardon 22 Industrial Park,
Beveridge Lane, Coalville,
Leicestershire, LE67 1TE
Area of Operation: UK & Ireland
Tel: 01530 813396
Fax: 01530 813376
Email: admin@schluter.co.uk
Web: www.schluter.co.uk
Other Info: ✎
Material Type: C) 1, 2, 3, 11

STAIRFLIGHT LTD
Unit 17, Landford Common Industrial Estate,
New Road, Landford, Salisbury, Wiltshire, SP5 2AZ
Area of Operation: UK (Excluding Ireland)
Tel: 01794 324150
Fax: 01794 324151
Email: sales@prestige-staircases.com
Web: www.prestige-staircases.com
Material Type: A) 1, 2, 3, 4, 5, 6

STAIRPLAN LTD
Unit C4, Stafford Park 4, Telford,
Shropshire, TF3 3BA
Area of Operation: UK (Excluding Ireland)
Tel: 01952 216000
Fax: 01952 216021
Email: sales@stairplan.com
Web: www.stairplan.com
Material Type: A) 1, 2, 4, 6

STAIRRODS (UK) LTD
Unit 6, Park Road North Industrial Estate,
Blackhill, Consett, Durham, DH8 5UN
Area of Operation: Worldwide
Tel: 01207 591176 **Fax:** 01207 591911
Email: sales@stairrods.co.uk
Web: www.stairrods.co.uk
Other Info: ✎ ✋

SUSSEX BRASSWARE
Napier Road, Castleham Industrial Estate,
St Leonards-on-Sea, East Sussex, TN38 9NY
Area of Operation: Worldwide
Tel: 01424 857913 **Fax:** 01424 853862
Email: sales@sussexbrassware.co.uk
Web: www.sussexbrassware.co.uk ✎
Other Info: ✎ ✋

T.B DAVIES (CARDIFF) LTD
Penarth Road, Cardiff, CF11 8TD
Area of Operation: UK (Excluding Ireland)
Tel: 02920 713000
Fax: 02920 702386
Email: sales@tbdavies.co.uk
Web: www.ladders-online.com ✎

THE EXPANDED METAL COMPANY LIMITED
PO Box 14, Longhill Industrial Estate (North),
Hartlepool, Durham, TS25 1PR
Area of Operation: Worldwide
Tel: 01429 408087
Fax: 01429 866795
Email: paulb@expamet.co.uk
Web: www.expandedmetalcompany.co.uk

THE SPA & WARWICK TIMBER CO LTD
Harriott Drive, Heathcote Industrial Estate,
Warwick, Warwickshire, CV34 6TJ
Area of Operation: Midlands & Mid Wales
Tel: 01926 883876 **Fax:** 01926 450831
Email: sales@spa-warwick.co.uk

WINTHER BROWNE
75 Bilton Way, Enfield, London, EN3 7ER
Area of Operation: UK (Excluding Ireland)
Tel: 0208 344 9050
Fax: 0845 612 1894
Email: sales@wintherbrowne.co.uk
Web: www.wintherbrowne.co.uk
Material Type: A) 2, 4

WOODCHESTER KITCHENS & INTERIORS
Unit 18a Chalford Industrial Estate, Chalford,
Gloucestershire, GL6 8NT
Area of Operation: UK (Excluding Ireland)
Tel: 01453 886411
Fax: 01453 886411
Email: enquires@woodchesterkitchens.co.uk
Web: www.woodchesterkitchens.co.uk
Material Type: A) 1, 2, 3, 4, 5, 6, 7, 8, 9, 10, 11, 12

WOODHOUSE TIMBER
Unit 6 Quarry Farm Industrial Estate,
Staplecross Road, Bodiam, East Sussex, TN32 5RA
Area of Operation: UK (Excluding Ireland)
Tel: 01580 831700
Fax: 01580 830365
Email: info@woodhousetimber.co.uk
Web: www.woodhousetimber.co.uk
Other Info: ✎
Material Type: A) 2

WOODSIDE JOINERY
40 Llantarnam Park, Cwmbran, Torfaen, NP44 3AW
Area of Operation: UK & Ireland
Tel: 01633 875232
Fax: 01633 482718
Email: woodsidejoinery@talk21.com

ZOROUFY UK LTD
Unit A, 8 Capel Hendre Industrial Estate, Capel Hendre,
Ammanford, Carms, Carmarthenshire, SA18 3SJ
Area of Operation: Europe
Tel: 01269 832244
Email: enquiries@zoroufy.co.uk
Web: www.zoroufy.com ✎
Material Type: C) 11, 14, 18

WALKWAYS

KEY		
OTHER: ▽ Reclaimed	✎ On-line shopping	
✎ Bespoke	✋ Hand-made	ECO Ecological

B ROURKE & CO LTD
Vulcan Works, Accrington Road,
Burnley, Lancashire, BB11 5QD
Area of Operation: Worldwide
Tel: 01282 422841 **Fax:** 01282 458901
Email: info@rourkes.co.uk
Web: www.rourkes.co.uk
Material Type: C) 2, 4, 5, 6

BAYFIELD STAIR CO
Unit 4, Praed Road, Trafford Park,
Manchester, M17 1PQ
Area of Operation: Worldwide
Tel: 0161 848 0700
Fax: 0161 872 2230
Email: sales@bayfieldstairs.co.uk
Web: www.bayfieldstairs.co.uk

BISCA
Sawmill Lane, Helmsley,
North Yorkshire, YO62 5DQ
Area of Operation: UK (Excluding Ireland)
Tel: 01439 771702
Fax: 01439 771002
Email: info@bisca.co.uk
Web: www.bisca.co.uk

BURBEARY JOINERY LTD
Units 2 - 4, 47 Robin Lane,
Beighton, Sheffield,
South Yorkshire, S20 1BB
Area of Operation: UK (Excluding Ireland)
Tel: 0114 247 5003
Fax: 0114 247 5007
Email: info@burbearyjoinery.co.uk
Web: www.burbearyjoinery.co.uk ✎

CAMBRIDGE STRUCTURES
2 Huntingdon Street, St. Neots,
Cambridgeshire, PE19 1BG
Area of Operation: Worldwide
Tel: 01480 477700
Fax: 01480 477766
Email: info@cambridgestructures.com
Web: www.cambridgestructures.com
Material Type: A) 1, 2, 3, 4, 5, 6, 7, 8, 9, 10, 11, 12, 13, 14, 15

ERMINE ENGINEERING COMPANY LTD
Francis House, Silver Birch Park,
Great Northern Terrace, Lincoln,
Lincolnshire, LN5 8LG
Area of Operation: UK (Excluding Ireland)
Tel: 01522 510977
Fax: 01522 510929
Email: info@ermineengineering.co.uk
Web: www.ermineengineering.co.uk
Other Info: ✎ ✋
Material Type: C) 1, 2, 3, 4

FAHSTONE LTD
Michaels Stud Farm,
Meer End, Kenilworth,
Warwickshire, CV8 1PU
Area of Operation: UK (Excluding Ireland)
Tel: 01676 533679
Fax: 01676 532224
Email: sales@meer-end.co.uk
Web: www.meer-end.co.uk

FLOORS GALORE
Unit 5, Centurion Park, Kendal Road,
Shrewsbury, Shropshire, SY1 4EH
Area of Operation: UK & Ireland
Tel: 0800 083 0623
Email: info@floors-galore.co.uk
Web: www.floors-galore.co.uk ✎
Material Type: A) 2, 4, 5, 6, 7, 9

ION GLASS
Unit 7 Grange Industrial Estate,
Albion Street, Southwick, Brighton,
West Sussex, BN42 4EN
Area of Operation: UK (Excluding Ireland)
Tel: 0845 658 9988
Fax: 0845 658 9989
Email: sales@ionglass.co.uk
Web: www.ionglass.co.uk
Other Info: ✎
Material Type: J) 4, 5

JAIC LTD
Pattern House, Southwell Business Park,
Portland, Dorset, DT5 2NR
Area of Operation: UK (Excluding Ireland)
Tel: 01305 826991
Fax: 01305 823535
Email: info@jaic.co.uk

JIMPEX LTD
1 Castle Farm, Cholmondeley,
Malpas, Cheshire, SY14 8AQ
Area of Operation: UK & Ireland
Tel: 01829 720433
Fax: 01829 720553
Email: jan@jimpex.co.uk
Web: www.jimpex.co.uk
Material Type: A) 2, 3, 4, 5, 6, 7, 9, 10

KIRK NATURAL STONE
Bridgend, Fyvie, Turriff,
Aberdeenshire, AB53 8QB
Area of Operation: Worldwide
Tel: 01651 891891
Fax: 01651 891794
Email: info@kirknaturalstone.com
Web: www.kirknaturalstone.com
Material Type: E) 1, 2, 3, 4, 5

METALCRAFT [TOTTENHAM]
6-40 Durnford Street,
Tottenham, London, N15 5NQ
Area of Operation: UK (Excluding Ireland)
Tel: 0208 802 1715
Fax: 0208 802 1258
Email: sales@makingmetalwork.com
Web: www.makingmetalwork.com
Other Info: ✎
Material Type: C) 2, 3, 4, 5

THE EXPANDED METAL COMPANY LIMITED
PO Box 14, Longhill Industrial Estate (North),
Hartlepool, Durham, TS25 1PR
Area of Operation: Worldwide
Tel: 01429 408087
Fax: 01429 866795
Email: paulb@expamet.co.uk
Web: www.expandedmetalcompany.co.uk

TOMPKINS LTD
High March Close, Long March Industrial Estate,
Daventry, Northamptonshire, NN11 4EZ
Area of Operation: UK (Excluding Ireland)
Tel: 01327 877187
Fax: 01327 310491
Email: info@tompkinswood.co.uk
Web: www.tompkinswood.co.uk
Other Info: ✋

STAIRLIFTS

KEY		
OTHER: ▽ Reclaimed	✎ On-line shopping	
✎ Bespoke	✋ Hand-made	ECO Ecological

BROOKS STAIRLIFTS
Spring Mills, Norwood Avenue,
Shipley, West Yorkshire, BD18 2AX
Area of Operation: Worldwide
Tel: 01274 717766
Fax: 01274 533129
Email: info@stairlifts.co.uk
Web: www.stairlifts.co.uk

ESL HEALTHCARE LTD
Potts Marsh Industrial Estate,
Eastbourne Road, Westham,
East Sussex, BN24 5NH
Area of Operation: UK & Ireland
Tel: 01323 465800
Fax: 01323 460248
Email: sales@eslindustries.co.uk
Web: www.eslindustries.com

HOUSE ELECTRONIC LTD
97 Francisco Close, Hollyfields Park,
Chafford Hundred, Essex, RM16 6YE
Area of Operation: UK & Ireland
Tel: 01375 483595
Fax: 01375 483595
Email: info@houselectronic.co.uk
Web: www.houselectronic.co.uk

LIGHTING

Image courtesy of Astra Ceiling Fans & Lighting (0161 766 9090)

SPONSORED BY ASTRA CEILING FANS & LIGHTING
Tel 0161 766 9090 Web www.astra247.com

INTERIORS, FIXTURES & FINISHES

Lighting

Lighting need not be viewed in a purely functional capacity when creating your ideal home. There are endless exciting possibilities and effects which can be achieved simply by choosing the right lighting fixtures for your living space. A growing trend for more elaborate lighting designs has meant a shift away from traditional central pendant or ceiling lighting, meaning that creating the right lighting environment really is a chance to be creative and, with so many products around, may give you the individual stamp on your home you are looking for. As such, lighting should be designed at the earliest possible stage, ensuring that you can get the most out of this ornamental opportunity.

Interior Lighting

The positioning of each of your light fittings will dictate the look you are aiming to achieve. Questions to ask are; what effect am I looking for? Functional lighting for a specific task or safety function, or a more atmospheric and artistic feel? Areas such as the kitchen or study will demand a high level of functional lighting but this can be embellished with the aid of task lighting which directs light to the areas where it is most needed. Downlighters with adjustable heads are useful and consequently, a popular choice here. Other living areas throughout the home afford a more extravagant selection of fittings. Ambient or background lighting serves as a base upon which to build. Additions may take the form of the more subtle use of accent lighting, consisting of picture lights, spotlights and low-voltage strip lights which can highlight the colour and contours of your favourite art works within your home to great effect. An interesting idea might be to layer this with a lighting design feature using fibre optics for an individual, quirky feel to your home. As well as the design aspects, it is also important to consider the automating possibilities with lighting; pre-programming your light settings can increase comfort in your home as well as adding a crucial security dimension.

ABOVE: Teir lights by All Weather Lighting are perfect for lighting up paths.

ABOVE: Fibre optic starscape by Blue Beacon Lighting.

Exterior Lighting

Lighting the outside of your home also carries security benefits, but this is by no means the limit of exterior lighting's capabilities. For example, lighting your paths, steps and decking can serve as a safety, as well as an aesthetic, function. More ornamental highlighters can be used to enhance the most attractive features of your garden. They include: spotlighting, silhouetting and water lighting for accentuating sculptures, trees and water features. The wide range of designs on the market will enable you to match the style of your exterior lighting to that of your interior, with the two working in harmony to strike the right tone for your ideal home. There is also the more traditional option of flood lighting to keep your premises secure, again all with the additional option of automated models. Finally, it is worth considering the energy saving options open to you such as solar powered lighting and energy saving bulbs, which can help to ensure that you get your ideal lighting scheme at a more ideal price.

ABOVE: Images show a room set with two different lighting scenes to demonstrate a lighting control system. Images supplied by John Clayton Lighting Ltd.

LIGHTING

KEY

PRODUCT TYPES: 1= Downlights
2 = Spotlights 3 = Chandeliers 4 = Wall
Lights 5 = Emergency 6 = Security
7 = External 8 = Low Voltage
9 = Waterproof 10 = Fibre Optic
11 = Solar 12 = Fluorescent

OTHER: ▽ Reclaimed 🖱 On-line shopping
✏ Bespoke ✋ Hand-made ECO Ecological

A & H BRASS
201-203 Edgware Road, London, W2 1ES
Area of Operation: Worldwide
Tel: 0207 402 1854 **Fax:** 0207 402 0110
Email: ahbrass@btinternet.com
Web: www.aandhbrass.co.uk 🖱
Product Type: 3, 7, 8

ADVANCED LEDS LTD
Unit 14, Bow Court, Fletchworth Gate,
Burnsall Road, Coventry, West Midlands, CV5 6SP
Area of Operation: Europe
Tel: 02476 716151 **Fax:** 02476 712161
Email: sales@advanced-led.com
Web: www.advanced-led.com
Product Type: 1, 4, 12

ALL WEATHER LIGHTING LTD
Shrubbery Court, Cross Bank, Bewdley,
Worcestershire, DY12 2XF
Area of Operation: Europe
Tel: 01299 269246 **Fax:** 01299 269246
Email: chris@allweatherlighting.co.uk
Web: www.allweatherlighting.co.uk
Product Type: 3, 7, 9

ANDY THORNTON LTD
Ainleys Industrial Estate, Elland,
West Yorkshire, HX5 9JP
Area of Operation: Worldwide
Tel: 01422 375 595 **Fax:** 01422 377 455
Email: email@ataa.co.uk
Web: www.andythornton.com
Product Type: 3, 7

ARC-COMP
Friedrichstr. 3, 12205 Berlin-Lichterfelde, Germany
Area of Operation: Europe
Tel: +49 (0)308 430 9956
Fax: +49 (0)308 430 9957
Email: jvs@arc-comp.com
Web: www.arc-com.com
Product Type: 7

ARC-COMP (IRISH BRANCH)
Whitefield Cottage, Lugduff, Tinahely,
Co. Wicklow, Republic of Ireland
Area of Operation: Europe
Tel: +353 (0)868 729 945
Fax: +353 (0)402 28900
Email: jvs@arc-comp.com
Web: www.arc-comp.com
Product Type: 7

ASHFIELD TRADITIONAL
119 High Street, Needham Market,
Ipswich, Suffolk, IP6 8DQ
Area of Operation: Europe
Tel: 01449 723601 **Fax:** 01449 723602
Email: mail@limelightgb.com
Web: www.limelightgb.com 🖱
Product Type: 1, 3

ASTRA CEILING FANS & LIGHTING
Unit 1, Pilsworth Way, Pilsworth, Bury,
Lancashire, BL9 8RE
Area of Operation: Europe
Tel: 0161 766 9090 **Fax:** 0161 766 9191
Email: support@astra247.com
Web: www.astra247.com 🖱
Product Type: 3, 5, 6, 7, 8, 9, 10

B ROURKE & CO LTD
Vulcan Works, Accrington Road,
Burnley, Lancashire, BB11 5QD
Area of Operation: Worldwide
Tel: 01282 422841
Fax: 01282 458901
Email: info@rourkes.co.uk
Web: www.rourkes.co.uk
Product Type: 1, 2, 3, 4, 6, 7
Material Type: C) 1, 2, 3, 4, 5, 6, 7, 11, 14

BATHROOM CITY
Seeleys Road, Tyseley Industrial Estate,
Birmingham, West Midlands, B11 2LQ
Area of Operation: UK & Ireland
Tel: 0121 753 0700
Fax: 0121 753 1110
Email: sales@bathroomcity.com
Web: www.bathroomcity.com 🖱
Product Type: 3, 8, 9

BEACON LINK LTD
Beacon House, Melton Road,
Barrow upon Soar, Leicester,
Leicestershire, LE11 1NW
Area of Operation: Midlands & Mid Wales
Tel: 01509 620606
Fax: 01509 621688
Email: anthonyp@mood-lighting.co.uk
Web: www.mood-lighting.co.uk

BESPOKE INSTALLATIONS
1 Rogers Close, Tiverton, Devon, EX16 6UW
Area of Operation: UK (Excluding Ireland)
Tel: 01884 243497
Fax: 01884 243497
Email: michael@bespokeinstallations.com
Web: www.bespoke.biz
Product Type: 2, 3, 4, 6, 7, 8, 10

BESSELINK & JONES
1.04 Chelsea Harbour Design Centre,
Chelsea Harbour, London, SW10 0XE
Area of Operation: UK & Ireland
Tel: 0207 351 4669
Fax: 0207 352 3898
Email: enquiry@besselink.com
Web: www.besselink.com 🖱
Product Type: 1, 2, 3, 4, 5, 8, 9
Material Type: A) 2, 3, 4, 5, 6, 7, 8, 9, 10, 11, 12, 14

BLOWZONE HOT STUDIO
The Ruskin Glass Centre, Wollaston Road,
Amblecote, West Midlands, DY8 4HF
Area of Operation: Worldwide
Tel: 01384 399464
Fax: 01384 377746
Email: sales@blowzone.co.uk
Web: www.blowzone.co.uk
Product Type: 10

BLUE BEACON LIGHTING
Intermail Plc, Horizon West, Canal View Road,
Newbury, Berkshire, RG14 5XF
Area of Operation: UK & Ireland
Tel: 0870 241 3992
Fax: 01635 41678
Email: michaelm@crescent.co.uk
Web: www.bluebeacon.co.uk 🖱
Product Type: 1, 2, 3, 4, 6, 7, 8, 9, 10

BROUGHTONS OF LEICESTER
The Old Cinema, 69 Cropston Road, Anstey,
Leicester, Leicestershire, LE7 7BP
Area of Operation: Worldwide
Tel: 0116 235 2555
Fax: 0116 234 1188
Email: sale@broughtons.com
Web: www.broughtons.com 🖱
Product Type: 3

BURWOOD LIGHTING COMPANY LTD
Market Street, Exeter, Devon, EX1 1BW
Area of Operation: UK & Ireland
Tel: 01392 259367
Fax: 01392 210239
Email: sales@burwoodlighting.co.uk
Web: www.burwoodlighting.co.uk
Product Type: 1, 2, 3, 4, 5, 6, 7, 8, 9, 10

CAMERON PETERS LTD
The Old Dairy, Home Farm, Ardington,
Wantage, Oxfordshire, OX12 8PD
Area of Operation: Worldwide
Tel: 01235 835000
Fax: 01235 835005
Email: info@cameronpeters.co.uk
Web: www.cameronpeters.co.uk
Product Type: 1, 2, 3, 4, 7, 8, 10

CANDELA LTD
47 Spaces Business Centre, Ingate Place,
London, SW8 3NS
Area of Operation: Worldwide
Tel: 0207 720 4480
Fax: 0207 498 0026
Email: mail@candela.ltd.uk
Web: www.candela.ltd.uk
Product Type: 1, 4, 5, 7, 8, 9

CHARLES MASON LTD
Unit 11A Brook Street Mill, Off Goodall Street,
Macclesfield, Cheshire, SK11 7AW
Area of Operation: Worldwide
Tel: 0800 085 3616
Fax: 01625 668789
Email: info@charles-mason.com
Web: www.charles-mason.com 🖱
Product Type: 1, 2, 3, 4, 7
Material Type: C) 2, 3, 4

COMMERCIAL ILLUMINATION LTD
Thurleston Hall, Ipswich, Suffolk, IP1 6TD
Area of Operation: Worldwide
Tel: 01473 257813
Email: sales@period-lighting.co.uk
Web: www.period-lighting.co.uk
Product Type: 1, 3, 7, 8, 9
Other Info: ✏ ✋

EASYLIGHTING
c/o Aladdins Lighting, The Street, Long Stratton,
Norwich, Norfolk, NR15 2XJ
Area of Operation: UK (Excluding Ireland)
Tel: 01508 532528
Fax: 01508 532528
Email: info@easylighting.co.uk
Web: www.easylighting.co.uk

ELECTRICALSHOP.NET
81 Kinson Road, Wallisdown, Bournemouth,
Dorset, BH10 4DG
Area of Operation: UK (Excluding Ireland)
Tel: 0870 027 3730
Fax: 0870 027 3731
Email: sales@electricalshop.net
Web: www.electricalshop.net 🖱
Product Type: 1, 2, 5, 7, 8, 13

ELECTRO TECHNIK
Bordesley Hall, Alvechurch, Birmingham,
West Midlands, B48 7QA
Area of Operation: Worldwide
Tel: 01527 595349
Fax: 01527 595092
Email: alan@flpatents.co.uk
Web: www.flpatents.co.uk
Product Type: 1, 8

F W LIGHTING LTD
Unit 19, The Lays Business Centre, Charlton Road,
Keynsham, Bristol, Somerset, BS31 2SE
Area of Operation: UK & Ireland
Tel: 0117 986 7500
Fax: 0117 986 7600
Email: info@fw-lighting.co.uk
Web: www.fw-lighting.co.uk
Product Type: 1, 7, 8, 9
Other Info: ✏ ✋

FORMFOLLOWS
27A Russell Square, Brighton,
East Sussex, BN1 2EE
Area of Operation: Worldwide
Tel: 01273 550180
Fax: 01273 823995
Email: sk@formfollows.co.uk
Web: www.formfollows.co.uk 🖱
Other Info: ✋

FUTURISTIC FIBRE OPTICS LTD
The Innovation Centre, Brunswick Street,
Nelson, Lancashire, BB9 0PQ
Area of Operation: UK (Excluding Ireland)
Tel: 01282 877177
Fax: 01282 877178
Email: ejl@f2olighting.com
Web: www.f2olighting.com
Product Type: 10

GET PLC
Key Point, 3-17 High Street,
Potters Bar, Hertfordshire, EN6 5AJ
Area of Operation: Worldwide
Tel: 01707 601601
Fax: 01707 601701
Email: info@getplc.co.uk
Web: www.getplc.com
Product Type: 1, 2, 4, 6, 7, 9, 12, 13

GET PLC
Area of Operation: Worldwide
Tel: 01707 601601 **Fax:** 01707 601701
Email: info@getplc.co.uk
Web: www.getplc.com 🖱
Product Type: 1, 2, 4, 6, 7, 9, 12, 13

The new screwless range will add the perfect
finishing touch to a domestic or commercial
environment by offering an ultra slim high quality
alternative.

GJD MANUFACTURING LTD
Unit 2, Birch Industrial Estate, Whittle Lane,
Heywood, Lancashire, OL10 2SX
Area of Operation: Worldwide
Tel: 01706 363998
Fax: 01706 363991
Email: info@gjd.co.uk
Web: www.gjd.co.uk
Product Type: 6

GOBBSMAK DESIGN LTD
Drill Hall, Meadow Place, Crieff,
Perth and Kinross, PH7 4DU
Area of Operation: Worldwide
Tel: 01764 655392
Fax: 01764 654300
Email: catrina@gobbsmak.co.uk
Web: www.gobbsmak.co.uk
Other Info: ECO ✏ ✋

GREEN ISLAND LTD
The Lighthouse,
Unit 7 Eastwood Park Industrial Estate,
Penryn, Cornwall, TR10 8LA
Area of Operation: Europe
Tel: 01326 377775
Fax: 01326 377773
Email: sales@greenisland.co.uk
Web: www.greenisland.co.uk
Product Type: 1, 2, 3, 4, 7, 8, 9, 10

HOUSE ELECTRONIC LTD
97 Francisco Close, Hollyfields Park,
Chafford Hundred, Essex, RM16 6YE
Area of Operation: UK & Ireland
Tel: 01375 483595
Fax: 01375 483595
Email: info@houselectronic.co.uk
Web: www.houselectronic.co.uk
Product Type: 1, 3, 6, 7, 8, 9, 10

INTERIORS, FIXTURES & FINISHES

ASTRA247.COM
Area of Operation: Worldwide
Tel: 0161 797 3222 **Fax:** 0161 797 3444
Email: info@astra247.com
Web: www.astra247.com

Remote and Touch control dimmers
Stylish touch control and remote control dimmers suitable for mains lighting and electronic transformers including those requiring "trailing-edge" control.

'easy online ordering'

ASTRA247.COM
Area of Operation: Worldwide
Tel: 0161 797 3222 **Fax:** 0161 797 3444
Email: info@astra247.com
Web: www.astra247.com

Ceiling Fan Lights
A wide choice of over 150 quality ceiling fans and ceiling fan lights as well as accessories from all the major brands. Free advice, quick delivery.

'easy online ordering'

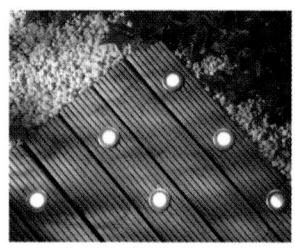

ASTRA247.COM
Area of Operation: Worldwide
Tel: 0161 797 3222 **Fax:** 0161 797 3444
Email: info@astra247.com
Web: www.astra247.com

Outdoor Lighting
We have a large selection of outdoor lighting products on our website including walkway, pond, security, LED and more.

'easy online ordering'

ASTRA247.COM
Area of Operation: Worldwide
Tel: 0161 797 3222 **Fax:** 0161 797 3444
Email: info@astra247.com
Web: www.astra247.com

Decorative Lighting
Astra offer an extensive selection of beautfiul, stylish lighting products to suit all interior designs.

'easy online ordering'

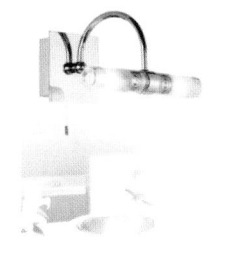

ASTRA247.COM
Area of Operation: Worldwide
Tel: 0161 797 3222 **Fax:** 0161 797 3444
Email: info@astra247.com
Web: www.astra247.com

Bathroom Lighting
Astra offer an range of stylish bathroom lighting and bathroom mirror lights, all available to order directly from our website.

'easy online ordering'

ASTRA247.COM
Area of Operation: Worldwide
Tel: 0161 797 3222 **Fax:** 0161 797 3444
Email: info@astra247.com
Web: www.astra247.com

Spotlights and Recessed
We have a wide range of spotlights and recessed lighting in all finishes and styles. All available to order directly from our website.

'easy online ordering'

ASTRA247.COM
Area of Operation: Worldwide
Tel: 0161 797 3222 **Fax:** 0161 797 3444
Email: info@astra247.com
Web: www.astra247.com

LED Lighting
A complete range of Indoor and outdoor LED lighting products all available to order directly from our website.

'easy online ordering'

ASTRA247.COM
Area of Operation: Worldwide
Tel: 0161 797 3222 **Fax:** 0161 797 3444
Email: info@astra247.com
Web: www.astra247.com

Conservatory Fans
Astra offer a wide choice of ceiling fans for use in your conservatory. Free, helpful advice and a full range of accessories and controls.

'easy online ordering'

ASTRA247.COM
Area of Operation: Worldwide
Tel: 0161 797 3222 **Fax:** 0161 797 3444
Email: info@astra247.com
Web: www.astra247.com

Security Lighting
Complete range of modern and traditional security lighting products available direct from our website.

'easy online ordering'

ASTRA247.COM
Area of Operation: Worldwide
Tel: 0161 797 3222 **Fax:** 0161 797 3444
Email: info@astra247.com
Web: www.astra247.com

Fire Proof Downlighters
Superb line of Flame Proof downlighters, available as IP65 & Compact Flourescent, low voltage and mains voltage. Full 90 minutes fire rating.

'easy online ordering'

ASTRA247.COM
Area of Operation: Worldwide
Tel: 0161 797 3222 **Fax:** 0161 797 3444
Email: info@astra247.com
Web: www.astra247.com

Table Lamps
Large selection of table lamps in all styles and finishes available to suit any room.

'easy online ordering'

ASTRA247.COM
Area of Operation: Worldwide
Tel: 0161 797 3222 **Fax:** 0161 797 3444
Email: info@astra247.com
Web: www.astra247.com

Track Lighting Systems
Complete range of modular track lighting systems utilising low voltage, halogen and dichroic lamps.

'easy online ordering'

Timeless Quality, Elegance and Excellence...

Hartland, Sheer and Cheriton - superb quality electrical wiring accessories available in Polished, Satin and Antique Brass, Bright and Satin Chrome, Bright and Satin Stainless Steel, Black Nickel and Pearl Oyster.

making connections beautifully...

ICA LIGHTING
10 Newton Place, Glasgow, G3 7PR
Area of Operation: Worldwide
Tel: 0141 331 2366 **Fax:** 0141 353 3588
Email: info@icalighting.com
Web: www.icalighting.com
Product Type: 1, 2, 3, 4, 5, 6, 7, 8, 9, 10, 12, 13

IMPACT LIGHTING SERVICUS LTD
1 Thatched Cottage, Herd Lane,
Corringham, Essex, SS17 9BH
Area of Operation: East England,
Greater London, South East England
Tel: 01375 361391 **Fax:** 01375 361392
Email: ray@impact-lighting.freeserve.co.uk
Product Type: 1, 2, 3, 4, 5, 6, 7, 8, 9, 10

INTERACTIVE HOMES LTD
Lorton House, Conifer Crest,
Newbury, Berkshire, RG14 6RS
Area of Operation: UK (Excluding Ireland)
Tel: 01635 49111 **Fax:** 01635 40735
Email: sales@interactivehomes.co.uk
Web: www.interactivehomes.co.uk
Product Type: 1, 3, 4, 6, 7, 8, 9
Material Type: C) 2, 3, 7, 9, 11, 12, 13, 14, 15, 16

ISAAC LORD LTD
West End Court, Suffield Road,
High Wycombe, Buckinghamshire, HP11 2JY
Area of Operation: East England,
Greater London, North East England, South East
England, South West England and South Wales
Tel: 01494 462121 **Fax:** 01494 461376
Email: info@isaaclord.co.uk
Web: www.isaaclord.co.uk
Product Type: 2, 8

JIM LAWRENCE LTD
Scotland Hall Farm, Stoke by Nayland,
Colchester, Essex, CO6 4QG
Area of Operation: UK (Excluding Ireland)
Tel: 01206 263459 **Fax:** 01206 262166
Email: sales@jim-lawrence.co.uk
Web: www.jim-lawrence.co.uk
Product Type: 1, 3, 7, 9

JOHN GIBBONS
Montrose House, Willersey Road,
Badsey, Worcestershire, WR11 7HD
Area of Operation: Worldwide
Tel: 01386 830216 **Fax:** 01386 830216
Email: johngibbons@btconnect.com
Web: www.johngibbonslighting.com
Product Type: 3

KASPAR SWANKEY
405, Goldhawk Road, Hammersmith,
West London, W6 0SA
Area of Operation: Worldwide
Tel: 0208 746 3586 **Fax:** 0208 746 3586
Email: kaspar@swankeypankey.com
Web: www.swankeypankey.com

KITCHEN SUPPLIES
East Chesters, North Way, Hillend Industrial Estate,
Dalgety Bay, Fife, KY11 9JA
Area of Operation: UK (Excluding Ireland)
Tel: 01383 824729
Email: sales@kitchensupplies.co.uk
Web: www.kitchensupplies.co.uk
Product Type: 1, 4, 8

KNIGHT DESIGN LIGHTING
PO Box 15, Brackley,
Northamptonshire, NN13 5YN
Area of Operation: UK & Ireland
Tel: 01280 851092 **Fax:** 01280 851093
Email: knightdesign@btconnect.com
Web: www.knightdesignlighting.co.uk
Product Type: 1, 2, 3, 4, 5, 7, 8, 9

LAMPHOLDER 2000 PLC
Unit 3 TU House, Thorpe Underwood,
Northampton, NN6 9PA
Area of Operation: UK (Excluding Ireland)
Tel: 01536 713 642
Fax: 01536 713 994
Email: bc@lampholder.co.uk
Web: www.lampholder.co.uk
Product Type: 1, 2, 4, 8, 13

LAMPS & LIGHTING LTD
Bridgewater Court, Network 65 Business Park,
Burnley, Lancashire, BB11 5ST
Area of Operation: UK (Excluding Ireland)
Tel: 01282 448666
Fax: 01282 417703
Email: sales@lampslighting.co.uk
Web: www.lampslighting.co.uk
Product Type: 1, 2, 3, 4, 5, 6, 7, 8, 9, 10

LEAX LIGHTING CONTROLS
11 Mandeville Courtyard,
142 Battersea Park Road, London, SW11 4NB
Area of Operation: Europe
Tel: 0207 501 0880
Fax: 0207 501 0890
Email: sales@leax.co.uk
Web: www.leax.co.uk
Product Type: 1, 7, 8, 10

LEIGH LIGHTING
1591-93 London Road,
Leigh On Sea, Essex, SS9 2SG
Area of Operation: Worldwide
Tel: 01702 477633 **Fax:** 01702 470112
Email: leigh.lighting@virgin.net
Web: www.leighlighting.com
Product Type: 1, 3, 7, 8

LIGHT IQ LTD
Carpenters Yard, 27 Gironde Road,
London, SW6 7DY
Area of Operation: Europe
Tel: 0207 386 3949 **Fax:** 0207 386 3950
Email: philip@lightiq.com
Web: www.lightiq.com

LIGHT RIGHT LTD
SBC House, Restmor Way,
Wallington, Surrey, SM6 7AH
Area of Operation: Europe
Tel: 0208 255 2022 **Fax:** 0208 286 1900
Email: enquiries@lightright.co.uk
Product Type: 1, 3, 7, 8, 9

LIGHTING DIRECT 2U LIMITED
Venture Court, Broadlands, Wolverhampton,
West Midlands, WV10 6TB
Area of Operation: UK & Ireland
Tel: 0870 600 0076
Email: sales@lightingdirect2u.co.uk
Web: www.lightingdirect2u.co.uk
Product Type: 3, 6, 7, 8, 9

LIGHTING FOR GARDENS LTD
20 Furmston Court, Letchworth Garden City,
Hertfordshire, SG6 1UJ
Area of Operation: UK (Excluding Ireland)
Tel: 01462 486777
Fax: 01462 480344
Email: sales@lightingforgardens.com
Web: www.lightingforgardens.com
Product Type: 7

LIGHTINGFORYOU
Grange Farm, Pinley Green,
Claverdon, Warwickshire, CV35 8NA
Area of Operation: UK (Excluding Ireland)
Tel: 01926 842738
Fax: 01926 840114
Email: sales@lightingforyou.co.uk
Web: www.lightingforyou.co.uk

LINOLITE:SYLVANIA
Avis Way, Newhaven, East Sussex, BN9 0ED
Area of Operation: Europe
Tel: 0870 606 2030 **Fax:** 0870 241 0803
Email: tom.anderson@sylvania-lighting.com
Web: www.sylvania-lighting.com
Product Type: 1, 4, 8, 13

LOW ENERGY WORLD
38 Carter Street, Utoxeter, Staffordshire, ST14 8EU
Area of Operation: UK (Excluding Ireland)
Tel: 0845 060 0222 **Fax:** 0845 060 0444
Email: sales@lowenergyworld.com
Web: www.lowenergyworld.com
Product Type: 1, 2, 3, 4, 6, 7, 8, 9, 13
Other Info: ECO

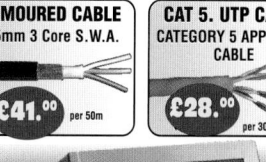

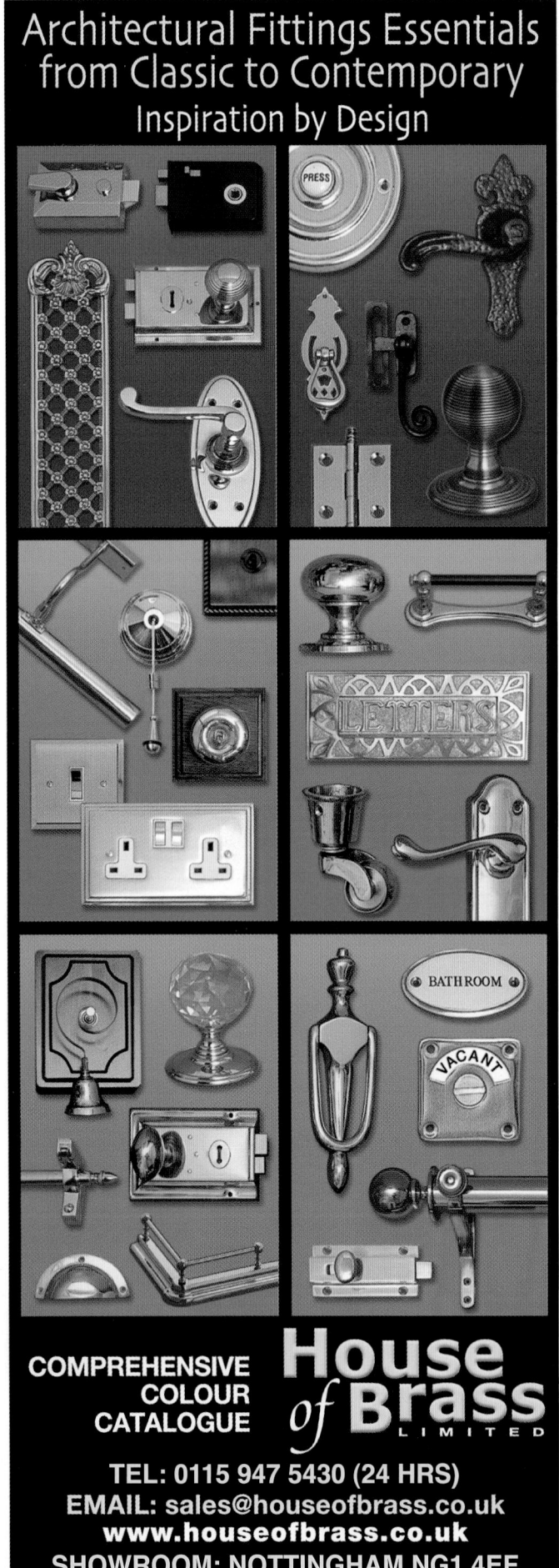

METCRAFT LIGHTING LIMITED
13a Gatesway Crescent, Broadway Industrial Estate, Chadderton, Oldham, Lancashire, OL9 9LB
Area of Operation: Europe
Tel: 0161 683 4298
Fax: 0161 688 8004
Email: sales@metcraftlighting.com
Web: www.metcraftlighting.com

MOOD LIGHTING
PO Box 111, Loughborough, Leicestershire, LE12 8ZS
Area of Operation: UK (Excluding Ireland)
Tel: 01509 621681
Fax: 01509 621688
Email: sales@mood-lighting.co.uk
Web: www.mood-lighting.co.uk
Product Type: 3, 4

MYFLOATINGWORLD.COM LIMITED
3 Hartlepool Court, Royal Docks, Galleons Lock, Royal Docks, London, E16 2RL
Area of Operation: Europe
Tel: 0870 777 0728
Fax: 0870 777 0729
Email: info@myfloatingworld.com
Web: www.myfloatingworld.com
Product Type: 1, 2, 3, 4, 8
Other Info:

NIGEL TYAS HANDCRAFTED IRONWORK
Green Works, 18 Green Road, Penistone, Sheffield, South Yorkshire, S36 6BE
Area of Operation: Worldwide
Tel: 01226 766618
Email: sales@nigeltyas.co.uk
Web: www.nigeltyas.co.uk
Product Type: 1, 3, 8
Other Info:
Material Type: C) 2, 6, 11

OLD FLAMES
30 Long Street, Easingwold, York, North Yorkshire, YO61 3HT
Area of Operation: UK (Excluding Ireland)
Tel: 01347 821188
Fax: 01347 821188
Email: philiplynas@aol.com
Web: www.oldflames.co.uk
Product Type: 3

OPTIC LIGHTING LTD
The Innovation Centre, Brunswick Street, Nelson, Lancashire, BB9 0PQ
Area of Operation: UK & Ireland
Tel: 01282 877171
Fax: 01282 877178
Email: info@opticlighting.co.uk
Web: www.opticlighting.co.uk
Product Type: 10

OXFORD LIGHTING & ELECTRICAL SOLUTIONS
Unit 117, Culham Site No 1, Station Road, Culham, Abingdon, Oxfordshire, OX14 3DA
Area of Operation: UK (Excluding Ireland)
Tel: 01865 408522
Fax: 01865 408522
Email: olessales@tiscali.co.uk
Web: www.oles.co.uk

PESTWEST ELECTRONICS
Denholme Drive, Ossett, West Yorkshire, WF5 9NB
Area of Operation: Worldwide
Tel: 01924 268500
Fax: 01924 273591
Email: info@pestwest.com
Web: www.pestwest.com
Product Type: 13

POLARON PLC
26 Greenhill Crescent, Watford Business Park, Watford, Hertfordshire, WD18 8XG
Area of Operation: UK (Excluding Ireland)
Tel: 01923 495495
Fax: 01923 228796
Email: arichards@polaron.co.uk
Web: www.polaron.com
Product Type: 1, 2, 3, 4, 6, 8

PS INTERIORS
11 Cecil Road, Hale, Altrincham, Cheshire, WA15 9NY
Area of Operation: Europe
Tel: 0161 926 9398
Fax: 0161 929 0363
Email: sales@ps-interiors.co.uk
Web: www.ps4interiors.co.uk
Product Type: 2, 3, 4, 7, 8, 9

QVS ELECTRICAL
4C The Birches Industrial Estate, Imberhorne Lane, East Grinstead, West Sussex, RH19 1XZ
Area of Operation: Worldwide
Tel: 0800 197 6565
Fax: 0800 197 6566
Email: sales@qvsdirect.co.uk
Web: www.qvsdirect.co.uk
Product Type: 1, 2, 3, 4, 7, 8, 13

RAYLIGHT LTD
1 Cherry Trees, Stanbridge Road Terrace, Leighton Buzzard, Bedfordshire, LU7 4QU
Area of Operation: Europe
Tel: 01525 385511
Fax: 01525 372255
Email: info@raylight.co.uk
Web: www.raylight.co.uk
Product Type: 1, 2, 4, 7, 8, 9, 13

RE:DESIGN
11 Bognor Road, Chichester, West Sussex, PO19 7TF
Area of Operation: UK & Ireland
Tel: 07787 987652
Email: barry@redesign.uk.com
Web: www.redesign.uk.com
Product Type: 1, 4, 7, 8, 9, 13

RELICS OF WITNEY LTD
1 Tristram Road, Ducklington, Witney, Oxfordshire, OX29 7HX
Area of Operation: Worldwide
Tel: 0845 430 3035
Fax: 01993 779653
Email: sales@lightsshop.co.uk
Web: www.lightsshop.co.uk
Product Type: 1, 3, 7, 8, 9
Material Type: C) 1, 2, 3, 11, 14

SHELDON COONEY
Chapel Glassworks, Leek Road, Cellarhead, Werrington, Staffordshire, ST9 0DQ
Area of Operation: UK & Ireland
Tel: 01782 551699
Email: info@sheldoncooney.com
Web: www.sheldoncooney.com

SIMPLY STAINED
7 Farm Mews, Farm Road, Brighton & Hove, East Sussex, BN3 1GH
Area of Operation: UK (Excluding Ireland)
Tel: 01273 220030
Email: david@simply-stained.co.uk
Web: www.simply-stained.co.uk
Product Type: 1, 3, 4, 7
Other Info:
Material Type: E) 9

SMART HOUSE SOLUTIONS
7 Erik Road, Bowers Gifford, Essex, S13 2HY
Area of Operation: UK & Ireland
Tel: 01277 264369
Fax: 01277 262143
Web: www.smarthousesolutions.co.uk
Product Type: 6

SMITHBROOK LIGHTING
Manfield Park, Cranleigh, Surrey, GU6 8PT
Area of Operation: UK & Ireland
Tel: 01483 272744
Fax: 01483 267863
Email: sales@smithbrooklighting.co.uk
Web: www.smithbrooklighting.co.uk
Product Type: 3
Material Type: C) 5, 6, 11

STARSCAPE STAR CEILINGS
7 Main Street, Lowick,
Berwick upon Tweed,
Northumberland, TD15 2UD
Area of Operation: Europe
Tel: 01289 388399
Email: enquiries@starceiling.co.uk
Web: www.starceiling.co.uk
Product Type: 10

STUART INTERIORS
Barrington Court, Barrington,
Ilminster, Somerset, TA19 0NQ
Area of Operation: Worldwide
Tel: 01460 240349
Fax: 01460 242069
Email: design@stuartinteriors.com
Web: www.stuartinteriors.com
Product Type: 3, 4

SYLVANIA LIGHTING INTERNATIONAL
Avis Way, Newhaven, East Sussex, BN9 0ED
Area of Operation: Europe
Tel: 0870 606 2030
Fax: 0870 241 0803
Email: sales.office@sylvania-lighting.com
Web: www.sylvania-lighting.com
Product Type: 1, 3, 4, 5, 7, 8, 9

THE BRADLEY COLLECTION
Lion Barn, Maitland Road, Needham Market,
Suffolk, IP6 8NS
Area of Operation: Worldwide
Tel: 01449 722724
Fax: 01449 722728
Email: claus.fortmann@bradleycollection.co.uk
Web: www.bradleycollection.co.uk
Product Type: 3, 4, 13
Material Type: C) 2, 3, 13

THE GREEN SHOP
Cheltenham Road, Bisley, Nr Stroud,
Gloucestershire, GL6 7BX
Area of Operation: UK & Ireland
Tel: 01452 770629
Fax: 01452 770104
Email: paint@greenshop.co.uk
Web: www.greenshop.co.uk
Product Type: 7, 8, 12
Other Info: ECO

TLC ELECTRICAL WHOLESALERS
TLC Building, Off Fleming Way,
Crawley, West Sussex, RH10 9JY
Area of Operation: Worldwide
Tel: 01293 565630
Fax: 01293 425234
Email: sales@tlc-direct.co.uk
Web: www.tlc-direct.co.uk
Product Type: 1, 2, 3, 5, 6, 7, 8, 9, 10

VICTORIA HAMMOND INTERIORS
Bury Farm, Church Street,
Bovingdon, Hemel Hempstead,
Hertfordshire, HP3 0LU
Area of Operation: UK (Excluding Ireland)
Tel: 01442 831641
Fax: 01442 831641
Email: victoria@victoriahammond.com
Web: www.victoriahammond.com
Product Type: 1, 2, 3, 4, 8

LIGHTING CONTROLS AND ACCESSORIES

ACM INSTALLATIONS LTD
71 Western Gailes Way, Hull,
East Riding of Yorks, HU8 9EQ
Area of Operation: North East England
Tel: 0870 242 3285
Fax: 01482 377530
Email: enquiries@acminstallations.com
Web: www.acminstallations.com
Product Type: 5
Other Info:

ADV AUDIO VISUAL INSTALLATIONS LTD
12 York Place, Leeds, West Yorkshire, LS1 2DS
Area of Operation: North East England,
North West England and North Wales
Tel: 0870 199 5755
Email: info@adv-installs.co.uk
Web: www.adv-installs.co.uk
Product Type: 5

ALL WEATHER LIGHTING LTD
Shrubbery Court, Cross Bank,
Bewdley, Worcestershire, DY12 2XF
Area of Operation: Europe
Tel: 01299 269246
Fax: 01299 269246
Email: chris@allweatherlighting.co.uk
Web: www.allweatherlighting.co.uk
Product Type: 5

ARCHITECTURAL IRONMONGERY LTD
28 Kyrle Street, Ross-on-Wye,
Herefordshire, HR9 7DB
Area of Operation: Worldwide
Tel: 01989 567946
Fax: 01989 567946
Email: info@arciron.co.uk
Web: www.arciron.com
Product Type: 1, 2, 3, 4

AUDIO VENUE
36 Queen Street, Maidenhead,
Berkshire, SL6 1HZ
Area of Operation: Greater London,
South East England, South West England
and South Wales
Tel: 01628 633995
Fax: 01628 633654
Email: info@audiovenue.com
Web: www.audiovenue.co.uk
Product Type: 3, 5

AUDIOFILE
27 Hockerill Street, Bishops Stortford,
Hertfordshire, CM23 2DH
Area of Operation: UK & Ireland
Tel: 01279 506576
Fax: 01279 506638
Email: info@audiofile.co.uk
Web: www.audiofile.co.uk
Product Type: 5

AVM
Le Panorama, 456, Chemin de Carimai,
Mougins, France, 06250
Area of Operation: Europe
Tel: +33 492 181 015
Fax: +33 492 181 015
Email: phavy@wanadoo.fr
Product Type: 5

B ROURKE & CO LTD
Vulcan Works, Accrington Road,
Burnley, Lancashire, BB11 5QD
Area of Operation: Worldwide
Tel: 01282 422841
Fax: 01282 458901
Email: info@rourkes.co.uk
Web: www.rourkes.co.uk

BEACON LINK LTD
Beacon House, Melton Road, Barrow upon Soar,
Leicester, Leicestershire, LE11 1NW
Area of Operation: Midlands & Mid Wales
Tel: 01509 620606
Fax: 01509 621688
Email: anthonyp@mood-lighting.co.uk
Web: www.mood-lighting.co.uk
Product Type: 3, 4, 5, 6

BESPOKE INSTALLATIONS
1 Rogers Close, Tiverton,
Devon, EX16 6UW
Area of Operation: UK (Excluding Ireland)
Tel: 01884 243497
Fax: 01884 243497
Email: michael@bespokeinstallations.com
Web: www.bespoke.biz
Product Type: 1, 2, 3, 4, 5

BEYOND THE INVISIBLE LIMITED
162-164 Arthur Road, London, SW19 8AQ
Area of Operation: Greater London
Tel: 0870 740 5859
Fax: 0870 740 5860
Email: info@beyondtheinvisible.com
Web: www.beyondtheinvisible.com
Product Type: 5

BLUE BEACON LIGHTING
Intermail Plc, Horizon West,
Canal View Road, Newbury,
Berkshire, RG14 5XF
Area of Operation: UK & Ireland
Tel: 0870 241 3992
Fax: 01635 41678
Email: michaelm@crescent.co.uk
Web: www.bluebeacon.co.uk
Product Type: 5

BURWOOD LIGHTING COMPANY LTD
Market Street, Exeter, Devon, EX1 1BW
Area of Operation: UK & Ireland
Tel: 01392 259367
Fax: 01392 210239
Email: sales@burwoodlighting.co.uk
Web: www.burwoodlighting.co.uk
Product Type: 1, 2, 3, 4, 5

CANDELA LTD
47 Spaces Business Centre,
Ingate Place, London, SW8 3NS
Area of Operation: Worldwide
Tel: 0207 720 4480
Fax: 0207 498 0026
Email: mail@candela.ltd.uk
Web: www.candela.ltd.uk
Product Type: 3, 5
Other Info:

CHARLES MASON LTD
Unit 11A Brook Street Mill,
Off Goodall Street, Macclesfield,
Cheshire, SK11 7AW
Area of Operation: Worldwide
Tel: 0800 085 3616
Fax: 01625 668789
Email: info@charles-mason.com
Web: www.charles-mason.com
Product Type: 1, 2, 3, 4

CLOUD9 SYSTEMS
87 Bishops Park Road, London, SW6 6DY
Area of Operation: Greater London,
South East England
Tel: 0870 420 5495
Fax: 0870 402 0121
Email: info@cloud9systems.co.uk
Web: www.cloud9systems.co.uk
Product Type: 5

COHERE LTD
10 Singleton Scarp, London, N12 7AR
Area of Operation: Europe
Tel: 0845 456 0695
Email: info@wirefreeliving.com
Web: www.wirefreeliving.com
Product Type: 5

D&T ELECTRONICS
Unit 9a Cranborne Industrial Estate,
Cranborne Road, Potters Bar,
Hertfordshire, EN6 3JN
Area of Operation: UK & Ireland
Tel: 0870 241 5891
Fax: 01707 653570
Email: info@dandt.co.uk
Web: www.dandt.co.uk
Product Type: 5

DANICO BRASS LTD
31-35 Winchester Road,
Swiss Cottage, London, NW3 3NR
Area of Operation: Worldwide
Tel: 0207 483 4477 **Fax:** 0207 722 7992
Email: sales@danico.co.uk
Product Type: 1, 2, 3, 4, 6, 7
Other Info:

DANLERS LIMITED
Vincients Road, Bumpers Farm Industrial Estate,
Chippenham, Wiltshire, SN14 6NQ
Area of Operation: Europe
Tel: 01249 443377 **Fax:** 01249 443388
Email: sales@danlers.co.uk
Web: www.danlers.co.uk
Product Type: 3, 5

DECKORUM LIMITED
4c Royal Oak Lane, Pirton, Hitchin,
Hertfordshire, CG5 3QT
Area of Operation: Greater London,
South East England, South West England
and South Wales
Tel: 0845 020 4360 **Fax:** 0845 020 4361
Email: simon@deckorum.com
Web: www.deckorum.com
Product Type: 5, 6

DIGITAL DECOR
Sonas House, Button End, Harston,
Cambridge, Cambridgeshire, CB2 5NX
Area of Operation: East England, Greater London
Tel: 01223 870935 **Fax:** 01223 870935
Email: seamus@digital-decor.co.uk
Web: www.digital-decor.co.uk
Product Type: 5, 6

EASYLIGHTING
c/o Aladdins Lighting, The Street,
Long Stratton, Norwich, Norfolk, NR15 2XJ
Area of Operation: UK (Excluding Ireland)
Tel: 01508 532528
Fax: 01508 532528
Email: info@easylighting.co.uk
Web: www.easylighting.co.uk

F W LIGHTING LTD
Unit 19, The Lays Business Centre,
Charlton Road, Keynsham, Bristol,
Somerset, BS31 2SE
Area of Operation: UK & Ireland
Tel: 0117 986 7500
Fax: 0117 986 7600
Email: info@fw-lighting.co.uk
Web: www.fw-lighting.co.uk
Product Type: 1, 2, 3, 4, 5, 6
Other Info:

FLAMINGBOX
Perry Farm, Maiden Bradley,
Wiltshire, BA12 7JQ
Area of Operation: UK (Excluding Ireland)
Tel: 01985 845440
Fax: 01985 845448
Email: james.ratcliffe@flamingbox.com
Web: www.flamingbox.com
Product Type: 4
Other Info:

FUTRONIX
Futronix House, 143 Croydon Road,
Caterham, Surrey, CR3 6PF
Area of Operation: Worldwide
Tel: 01883 373333
Fax: 01883 373335
Email: sales@futronix.info
Web: www.futronix.info
Product Type: 3, 5

GET PLC
Key Point, 3-17 High Street,
Potters Bar, Hertfordshire, EN6 5AJ
Area of Operation: Worldwide
Tel: 01707 601601
Fax: 01707 601701
Email: info@getplc.com
Web: www.getplc.com
Product Type: 3

SPONSORED BY: ASTRA CEILING FANS & LIGHTING www.astra247.co.uk

GOBBSMAK DESIGN LTD
Drill Hall, Meadow Place, Crieff,
Perth and Kinross, PH7 4DU
Area of Operation: Worldwide
Tel: 01764 655392
Fax: 01764 654300
Email: catrina@gobbsmak.co.uk
Web: www.gobbsmak.co.uk
Product Type: 5
Other Info: ECO ✎

HAF DESIGNS LTD
HAF House, Mead Lane, Hertford,
Hertfordshire, SG13 7AP
Area of Operation: UK & Ireland
Tel: 0800 389 8821
Fax: 01992 505705
Email: info@hafltd.co.uk
Web: www.hafdesigns.co.uk ✎
Product Type: 2, 3, 4

HELVAR
Hawley Mill, Hawley Road,
Dartford, Kent, DA2 7SY
Area of Operation: Worldwide
Tel: 01322 222211
Fax: 01322 282216
Email: gary.brown@helvar.com
Web: www.helvar.co.uk
Product Type: 5

HOMETECH INTEGRATION LTD
Earlsgate House 35, St. Ninian's Road,
Stirling, Stirlingshire, FK8 2HE
Area of Operation: UK (Excluding Ireland)
Tel: 0870 766 1060
Fax: 0870 766 1070
Email: info@hometechintegration.com
Web: www.hometechintegration.com
Product Type: 5

HOUSE ELECTRONIC LTD
97 Francisco Close, Hollyfields Park,
Chafford Hundred, Essex, RM16 6YE
Area of Operation: UK & Ireland
Tel: 01375 483595
Fax: 01375 483595
Email: info@houselectronic.co.uk
Web: www.houselectronic.co.uk
Product Type: 3, 4, 5
Other Info: ✎

HOUSE OF BRASS
122 North Sherwood Street, Nottingham,
Nottinghamshire, NG1 4EF
Area of Operation: Worldwide
Tel: 0115 947 5430
Fax: 0115 947 5430
Email: sales@houseofbrass.co.uk
Web: www.houseofbrass.co.uk ✎
Product Type: 1, 3, 4

HOUSEHOLD AUTOMATION
Foxways, Pinkhurst Lane, Slinfold,
Horsham, West Sussex, RH13 0QR
Area of Operation: UK & Ireland
Tel: 0870 330 0071
Fax: 0870 330 0072
Email: ellis-andrew@btconnect.com
Web: www.household-automation.co.uk
Product Type: 5

ICA LIIGHTING
10 Newton Place, Glasgow, G3 7PR
Area of Operation: Worldwide
Tel: 0141 331 2366
Fax: 0141 353 3588
Email: info@icalighting.com
Web: www.icalighting.com
Product Type: 1, 2, 3, 4, 5, 6, 7

IMPACT LIGHTING SERVICES LTD
1 Thatched Cottage, Herd Lane,
Corringham, Essex, SS17 9BH
Area of Operation: East England,
Greater London, South East England
Tel: 01375 361391 **Fax:** 01375 361392
Email: ray@impact-lighting.freeserve.co.uk
Product Type: 3, 4, 5

INTELLIGENT HOUSE
Langham House, Suite 401,
302 Regent Street, London, W1B 3HH
Area of Operation: UK & Ireland
Tel: 0207 394 9344
Fax: 0207 252 3879
Email: info@intelligenthouse.net
Web: www.intelligenthouse.net
Product Type: 2, 3, 5, 6

INTERACTIVE HOMES LTD
Lorton House, Conifer Crest, Newbury,
Berkshire, RG14 6RS
Area of Operation: UK (Excluding Ireland)
Tel: 01635 49111
Fax: 01635 40735
Email: sales@interactivehomes.co.uk
Web: www.interactivehomes.co.uk
Product Type: 3, 5

INTERCONNECTION
12 Chamberlayne Road, Moreton Hall Industrial
Estate, Bury St Edmunds, Suffolk, IP32 7EY
Area of Operation: East England
Tel: 01284 768676
Fax: 01284 767161
Email: info@interconnectionltd.co.uk
Web: www.interconnectionltd.co.uk
Product Type: 5

IRONMONGERY DIRECT
Unit 2-3 Eldon Way Trading Estate, Eldon Way,
Hockley, Essex, SS5 4AD
Area of Operation: Worldwide
Tel: 01702 562770
Fax: 01702 562799
Email: sales@ironmongerydirect.com
Web: www.ironmongerydirect.com ✎
Product Type: 5

J K AUDIO VISUAL
Unit 7 Newport Business Park,
Audley Avenue, Newport,
Shropshire, TF10 7DP
Area of Operation: UK (Excluding Ireland)
Tel: 01952 825088
Fax: 01952 814884
Email: sales@jk-audiovisual.co.uk
Product Type: 5

JELLYBEAN CTRL LTD
Eagle House, Passfield Business Centre,
Lynchborough Road, Passfield,
Hampshire GU30 7SB
Area of Operation: UK (Excluding Ireland)
Tel: 01428 751729
Fax: 01428 751772
Email: david@jellybeanctrl.com
Web: www.jellybeanctrl.com
Product Type: 5

JOHN CLAYTON LIGHTING LTD
Worthingham House, Deep Lane,
Hagworthingham, Lincolnshire, PE23 4LZ
Area of Operation: Europe
Tel: 0800 389 6395
Fax: 0870 240 6417
Email: rachel@jclighting.com
Web: www.flexidim.com
Product Type: 5

KITCHEN SUPPLIES
East Chesters, North Way,
Hillend Industrial Estate,
Dalgety Bay, Fife, KY11 9JA
Area of Operation: UK (Excluding Ireland)
Tel: 01383 824729
Email: sales@kitchensupplies.co.uk
Web: www.kitchensupplies.co.uk ✎
Product Type: 6

KNIGHT DESIGN LIGHTING
PO Box 15, Brackley,
Northamptonshire, NN13 5YN
Area of Operation: UK & Ireland
Tel: 01280 851092
Fax: 01280 851093
Email: knightdesign@btconnect.com
Web: www.knightdesignlighting.co.uk

LAMPS & LIGHTING LTD
Bridgewater Court,
Network 65 Business Park,
Burnley, Lancashire, BB11 5ST
Area of Operation: UK (Excluding Ireland)
Tel: 01282 448666
Fax: 01282 417703
Email: sales@lampslighting.co.uk
Web: www.lampslighting.co.uk ✎
Product Type: 3, 5

LEAX LIGHTING CONTROLS
11 Mandeville Courtyard,
142 Battersea Park Road, London, SW11 4NB
Area of Operation: Europe
Tel: 0207 501 0880
Fax: 0207 501 0890
Email: sales@leax.co.uk
Web: www.leax.co.uk
Product Type: 3, 5, 6

LIFESTYLE ELECTRONICS
Woodview, Castlebridge, County Wexford
Area of Operation: Europe
Tel: +353 535 9880
Fax: +353 535 9844
Email: lifestyle123@eircom.net
Product Type: 5
Other Info: ✎

LINK MEDIA SYSTEMS
68 St John Street, London, EC1M 4DT
Area of Operation: UK (Excluding Ireland)
Tel: 0207 251 2638
Fax: 0207 251 2487
Email: sohan@linkmediasystems.com
Web: www.linkmediasystems.com
Product Type: 5

LOW ENERGY WORLD
38 Carter Street, Utoxeter,
Staffordshire, ST14 8EU
Area of Operation: UK (Excluding Ireland)
Tel: 0845 060 0222
Fax: 0845 060 0444
Email: sales@lowenergyworld.com
Web: www.lowenergyworld.com ✎
Product Type: 7
Other Info: ECO

LUTRON EA LTD
Lutron House, 6 Sovereign Close,
Wapping, Greater London, E1W 3JF
Area of Operation: Worldwide
Tel: 0207 702 0657
Fax: 0207 480 6899
Email: ddanby@lutron.com
Web: www.lutron.com/europe
Product Type: 3, 5

MARQUEE HOME LIMITED
Unit 6, Eversley Way, Thorpe Industrial Estate,
Egham, Surrey, TW20 8RF
Area of Operation: Greater London,
South East England
Tel: 07004 567888
Fax: 07004 567788
Email: paulendersby@marqueehome.co.uk
Web: www.marqueehome.co.uk
Product Type: 2, 3, 4, 5, 6

MODE LIGHTING (UK) LTD
The Maltings, 63 High Street, Ware,
Hertfordshire, SG12 9AD
Area of Operation: Worldwide
Tel: 01920 462121
Fax: 01920 466881
Email: james.king@modelighting.co.uk
Web: www.modelighting.co.uk
Product Type: 3, 5

MOOD LIGHTING
PO Box 111, Loughborough,
Leicestershire, LE12 8ZS
Area of Operation: UK (Excluding Ireland)
Tel: 01509 621681
Fax: 01509 621688
Email: sales@mood-lighting.co.uk
Web: www.mood-lighting.co.uk ✎

MUSICAL APPROACH
111 Wolverhampton Road,
Stafford, Staffordshire, ST17 4BG
Area of Operation: UK (Excluding Ireland)
Tel: 01785 255154
Fax: 01785 243623
Email: info@musicalapproach.co.uk
Web: www.musicalapproach.co.uk
Product Type: 4, 5, 6

NDR SYSTEMS UK
Kendlehouse, 236 Nantwich Road,
Crewe, Cheshire, CW2 6BP
Area of Operation: UK & Ireland
Tel: 01270 219575
Email: daverehman@ndrsystemsuk.com
Web: www.ndrsystemsuk.com
Product Type: 1, 2, 3, 4, 5, 6, 7

**OXFORD LIGHTING
& ELECTRICAL SOLUTIONS**
Unit 117, Culham Site No 1,
Station Road, Culham,
Abingdon, Oxfordshire, OX14 3DA
Area of Operation: UK (Excluding Ireland)
Tel: 01865 408522
Fax: 01865 408522
Email: olessales@tiscali.co.uk
Web: www.oles.co.uk ✎

PHILLSON LTD
144 Rickerscote Road,
Stafford, Staffordshire, ST17 4HE
Area of Operation: UK & Ireland
Tel: 0845 612 0128
Fax: 01477 535090
Email: info@phillson.co.uk
Web: www.phillson.co.uk
Product Type: 5

POLARON PLC
26 Greenhill Crescent,
Watford Business Park, Watford,
Hertfordshire, WD18 8XG
Area of Operation: UK (Excluding Ireland)
Tel: 01923 495495
Fax: 01923 228796
Email: arichards@polaron.co.uk
Web: www.polaron.co.uk
Product Type: 5

PW TECHNOLOGIES
Windy House, Hutton Roof,
Carnforth, Lancashire, LA6 2PE
Area of Operation: UK (Excluding Ireland)
Tel: 01524 272400
Fax: 01524 272402
Email: info@pwtechnologies.co.uk
Web: www.pwtechnologies.co.uk

ROCOCO SYSTEMS & DESIGN
26 Danbury Street, London, N1 8JU
Area of Operation: Europe
Tel: 0207 454 1234
Fax: 0207 870 0604
Email: mary@rococosystems.com
Web: www.rococosystems.com
Product Type: 5

SEISMIC INTERAUDIO
3 Maypole Drive, Kings Hill,
West Malling, Kent, ME19 4BP
Area of Operation: Europe
Tel: 0870 073 4764
Fax: 0870 073 4765
Email: info@seismic.co.uk
Web: www.seismic.co.uk
Product Type: 5

SENSORY INTERNATIONAL LIMITED
48A London Road, Alderley Edge,
Cheshire, SK9 7DZ
Area of Operation: Worldwide
Tel: 0870 350 2244
Fax: 0870 350 2255
Email: garyc@sensoryinternational.com
Web: www.sensoryinternational.com
Product Type: 5

SMART HOUSE SOLUTIONS
7 Erik Road, Bowers Gifford, Essex, S13 2HY
Area of Operation: UK & Ireland
Tel: 01277 264369
Fax: 01277 262143
Web: www.smarthousesolutions.co.uk
Product Type: 5

SMARTCOMM LTD
45 Cressex Enterprise Centre,
Lincoln Road, Cressex Business Park,
High Wycombe, Buckinghamshire, HP12 3RL
Area of Operation: Worldwide
Tel: 01494 471912
Fax: 01494 472464
Email: info@smartcomm.co.uk
Web: www.smartcomm.co.uk

TECCHO
Unit 19W, Kilroot Business Park,
Carrickfergus, Co Antrim, BT38 7PR
Area of Operation: UK & Ireland
Tel: 0845 890 1150
Fax: 0870 063 4120
Email: enquiries@teccho.net
Web: www.teccho.net
Product Type: 3, 5

THE BIG PICTURE
AV & HOME CINEMA LIMITED
51 Sutton Road, Walsall,
West Midlands, WS1 2PQ
Area of Operation: UK & Ireland
Tel: 01922 623000
Email: info@getthebigpicture.co.uk
Web: www.getthebigpicture.co.uk
Product Type: 5

THE EDGE
90-92 Norwich Road,
Ipswich, Suffolk, IP1 2NL
Area of Operation: UK (Excluding Ireland)
Tel: 01473 288211
Fax: 01473 288255
Email: nick@theedge.eu.com
Web: www.theedge.eu.com
Product Type: 5

THE IVY HOUSE
Ivy House, Bottom Green,
Upper Broughton, Melton Mowbray,
Leicestershire, LE14 3BA
Area of Operation: East England,
Midlands & Mid Wales
Tel: 01664 822628
Email: enquiries@the-ivy-house.com
Web: www.the-ivy-house.com
Product Type: 5

THE MAJIK HOUSE COMPANY LTD
Unit J Mainline Industrial Estate,
Crooklands Road, Milnthorpe,
Cumbria, LA7 7LR
Area of Operation: North East England,
North West England and North Wales
Tel: 0870 240 8350
Fax: 0870 240 8350
Email: tim@majikhouse.com
Web: www.majikhouse.com
Product Type: 5, 6

THE MAX DISTRIBUTION CO LTD
The Old Bakery, 39 High Street,
East Malling, Kent, ME19 6AJ
Area of Operation: Europe
Tel: 01732 840845
Fax: 01732 841845
Email: sales@the-max.co.uk
Web: www.the-max.co.uk
Product Type: 5

THE MULTI ROOM COMPANY LTD
4 Churchill House, Churchill Road,
Cheltenham, Gloucestershire, GL53 7EG
Area of Operation: UK & Ireland
Tel: 01242 539100 **Fax:** 01242 539300
Email: info@multi-room.com
Web: www.multi-room.com
Product Type: 5

THE THINKING HOME
The White House,
Wilderspool Park, Greenalls Avenue,
Warrington, Cheshire, WA4 6HL
Area of Operation: UK & Ireland
Tel: 0800 881 8319
Fax: 0800 881 8329
Email: info@thethinkinghome.com
Web: www.thethinkinghome.com
Product Type: 5

THINKINGBRICKS LTD
6 High Street, West Wickham,
Cambridge, Cambridgeshire, CB1 6RY
Area of Operation: UK & Ireland
Tel: 01223 290886
Email: ian@thinkingbricks.co.uk
Web: www.thinkingbricks.co.uk
Product Type: 5

TLC ELECTRICAL WHOLESALERS
TLC Building, Off Fleming Way,
Crawley, West Sussex, RH10 9JY
Area of Operation: Worldwide
Tel: 01293 565630 **Fax:** 01293 425234
Email: sales@tlc-direct.co.uk
Web: www.tlc-direct.co.uk
Product Type: 1, 2, 3, 4, 5, 6

TONY TAYLOR
ELECTRICAL CONTRACTING
34 Goodrest Walk, Worcester,
Worcestershire, WR3 8LG
Area of Operation: Europe
Tel: 01905 723301
Email: tony@purpleink.wanadoo.co.uk
Web: www.ttecltd.com
Product Type: 5

TRUFI LIMITED
Unit 14, Horizon Business Village,
1 Brooklands Road, Weybridge,
Surrey, KT13 0TJ
Area of Operation: South East England
Tel: 01932 340750 **Fax:** 01932 345459
Email: info@trufi.co.uk
Web: www.trufi.co.uk
Product Type: 5

WANDSWORTH GROUP LTD
Albert Drive, Sheerwater,
Woking, Surrey, GU21 5SE
Area of Operation: UK (Excluding Ireland)
Tel: 01483 740740 **Fax:** 01483 740384
Email: info@wandsworthgroup.com
Web: www.wandsworthgroup.com
Product Type: 1, 2, 3, 4, 5, 6

WEBSTRACT LTD
28 Lattimore Road,
Wheathampstead,
Hertfordshire, AL4 8QE
Area of Operation: UK (Excluding Ireland)
Tel: 0870 446 0146 **Fax:** 0870 762 6431
Email: info@webstract.co.uk
Web: www.webstract.co.uk
Product Type: 3, 5

WIRED FOR LIVING
Systems House, 11 Lomax Street,
Great Harwood, Blackburn,
Lancashire, BB6 7DJ
Area of Operation: UK (Excluding Ireland)
Tel: 01254 880288
Fax: 01254 880289
Email: info@wiredforliving.co.uk
Web: www.wiredforliving.co.uk
Product Type: 5

WIRED-AVD
9 Addison Square, Ringwood,
Hampshire, BH24 1NY
Area of Operation: South East England,
South West England and South Wales
Tel: 01425 479014
Fax: 01425 471733
Email: sales@wired-avd.co.uk
Web: www.wired-avd.co.uk
Product Type: 5

NOTES

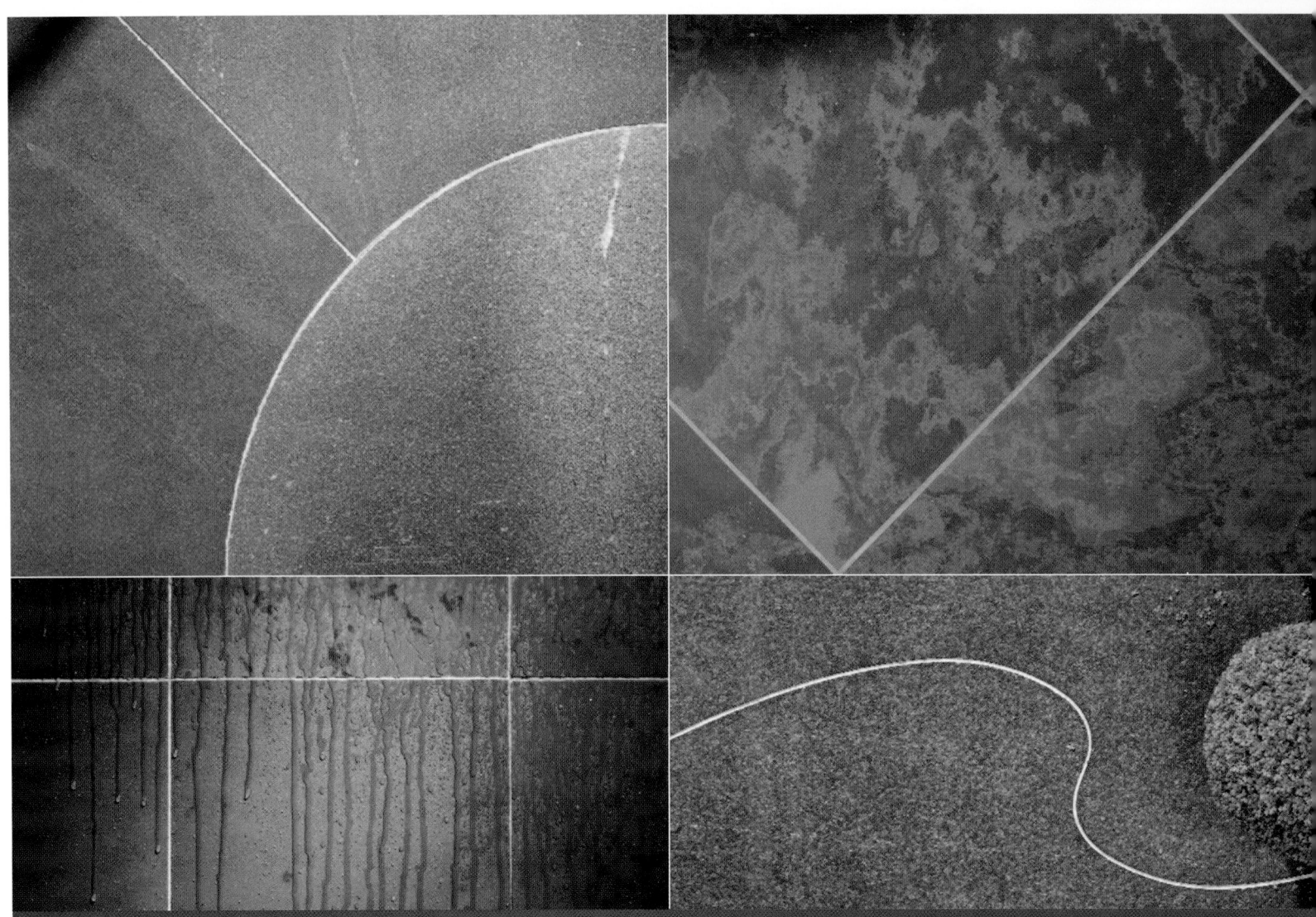

Inspiration in stone

Floor tiles
Wall tiles
Kitchen worktops
Bathroom finishes
Basins and shower trays
Slate
Travertine
Limestone
Granite

FLOOR & WALL FINISHES

Image courtesy of Pergo Ltd (01235 556300)

SPONSORED BY PERGO LTD
Tel 01235 556300 Web www.pergo.com

Designed for Generations

INTERIORS, FIXTURES & FINISHES

SPONSORED BY: PERGO LTD www.pergo.com

Flooring

Carpets, laminates, timbers and tiles; the vast choice of wall and floor finishes available today reflects the ever changing demands of the modern home. As the functions of living areas have expanded, so too have our demands on the finishes we choose.

With such a huge amount of choice on the market, we can now tailor floor and wall finishes to specific areas of the home, and selecting these finishes can be as exciting and diverse as choosing furniture and fittings to suit individual rooms.

Carpet tends to be viewed as an old-fashioned choice, but for some homeowners the alternatives just cannot beat it for a luxurious feel. Laying carpet in the sitting room and bedrooms can give a warm, cosy feeling underfoot in these living spaces, whilst leaving you free to use more durable materials where they are most required. Wool carpet is the traditional choice as it is particularly soft and maintains its appearance longest, but consequently it is the most expensive option. Whilst synthetic materials do offer a cheaper alternative, the expense is still supplemented by the fact that carpet does not qualify for VAT relief in new builds and conversions, as it is regarded as a 'moveable item'. The other downside of carpets is that they harbour dust mites and other allergens, which can trigger asthma and allergies, and are often difficult to remove with everyday cleaning methods.

Natural flooring options such as coir and seagrass matting are becoming increasingly popular as an inexpensive alternative to carpet. As well as being relatively cheap, they offer the additional bonus of being a renewable natural resource. However these mattings, like carpet, can also harbour allergens.

An ecological alternative is linoleum, which is completely biodegradable. Linoleum is a favourite for kitchens, bathrooms, playrooms and nurseries as it contains linseed oil, a natural antiseptic, and is also very hardwearing, cost-effective, and is easy to install.

Solid hardwood may be an expensive option but, when fitted well in the right areas, will need little-or-no further upkeep and will look very impressive. However, natural woods can be affected by moisture levels in areas such as kitchens and bathrooms, making it perhaps most suitable for living room areas. Using reclaimed boards can be a good move; they can be less expensive than using new boards, and will also have worn-in character, which works particularly well in older houses for a look of authenticity. However, you should note that reclaimed boards may need work such as sanding and de-nailing before they are laid. Alternatively, laminate flooring can provide a realistic wood-look finish at a cheaper price than the real thing, and whilst it may need to be replaced after excessive wear, this is not an expensive process.

Stone and ceramic tiles, suitable for both floor and wall finishes, are best suited to kitchen and bathroom areas which capitalise on their hardwearing, waterproof properties and hygienic, easy to clean surfaces. Although stone appears an expensive option at first glance, it is worth remembering that these tiles are free of VAT for new builds and renovations, significantly reducing their price tag, and that mass-produced tiles lower the price further.

Perhaps the best course of action is to look at each room individually considering which type of floor and wall finish they demand. Think of the home in terms of its separate parts, as opposed to as a whole; using different materials throughout your house will give it a modern feel which will reflect the careful thought which went into each room's design.

PVCu or vinyl flooring which comes as tiles or sheet material, is the cheapest smooth floor covering and is widely used in kitchens and bathrooms. But it is a major source of toxic substances in the environment and can 'off gas' a cocktail of synthetic chemicals which are very rarely tested for their effects on human health. A much more attractive and durable option – though more expensive – is to use real linoleum, made with wood flour, linseed oil and pine resin on a jute backing.

LEFT: Image courtesy of York Handmade Brick.
TOP: Wooden floor by Richwood Flooring.

WOODEN FLOORING

KEY

PRODUCT TYPES: 1= Planks 2 = Boards
3 = Parquet 4 = Laminate 5 = Glue Free
Systems 6 = Other

OTHER: ▽ Reclaimed ᵀᴴ On-line shopping
✎ Bespoke ✋Hand-made ECO Ecological

1926 TRADING COMPANY LTD
2 Daimler Close, Royal Oak,
Daventry, Northamptonshire, NN11 8QJ
Area of Operation: UK & Ireland
Tel: 0800 587 2027
Fax: 01327 310123
Email: sales@1926woodflooring.co.uk
Web: www.1926woodflooring.co.uk
Product Type: 1, 2, 3, 4, 5, 6
Other Info: ECO ✎ ✋
Material Type: A) 1, 2, 3, 4, 5, 6, 7, 8, 9, 10, 14

AARDVARK WHOLESALE LTD
PO Box 3733, Dronfield,
Derbyshire, S18 9AD
Area of Operation: UK (Excluding Ireland)
Tel: 0800 279 0486
Email: rawwoodfloors@zicom.net
Product Type: 1, 2, 4, 5, 6
Other Info: ✎

ATC (MONMOUTHSHIRE) LTD
Unit 2, Mayhill Industrial Estate,
Monmouth, Monmouthshire, NP25 3LX
Area of Operation: Worldwide
Tel: 01600 713036
Fax: 01600 715512
Email: info@floorsanddecking.com
Web: www.floorsanddecking.com ᵀᴴ
Product Type: 1, 2, 3, 6
Material Type: A) 2, 3, 5, 6, 7, 8, 9, 10, 12, 13

ATKINSON & KIRBY
2 Burscough Road, Ormskirk,
Lancashire, L39 2XG
Area of Operation: UK & Ireland
Tel: 01695 573234
Fax: 01695 586902
Email: sales@akirby.co.uk
Web: www.akirby.co.uk
Product Type: 1, 2, 3, 5
Material Type: A) 1, 2, 3, 4, 5, 6, 7, 9, 10

BARHAM & SONS
58 Finchley Avenue,
Mildenhall, Bury St. Edmunds,
Suffolk, IP28 7BG
Area of Operation: UK (Excluding Ireland)
Tel: 01638 665759 / 711611
Fax: 01638 663913
Email: info@barhamwoodfloors.com
Product Type: 1, 2

BATONS TO BEAMS LTD
Unit 4, Pool Bank Park, Tarvin,
Chester, Cheshire, CH3 8JH
Area of Operation: UK & Ireland
Tel: 01829 741900
Fax: 01829 741101
Email: batonstobeams@yahoo.co.uk
Web: www.batonstobeams.co.uk
Product Type: 1, 2, 3, 5, 6
Other Info: ▽ ECO ✎ ✋

BAYFIELD STAIR CO
Unit 4, Praed Rd, Trafford Park,
Manchester, M17 1PQ
Area of Operation: Worldwide
Tel: 0161 848 0700 **Fax:** 0161 872 2230
Email: sales@bayfieldstairs.co.uk
Web: www.bayfieldstairs.co.uk
Product Type: 1, 2, 3, 5, 6

BHK (UK) LTD
Davy Drive North West Industrial Estate,
Peterlee, Durham, SR8 2JF
Area of Operation: UK & Ireland
Tel: 0191 518 6538
Fax: 0191 518 6536
Email: eleanor.smith@peterlee.bh.de
Web: www.bhkonline.com
Product Type: 4, 5

BLACK MOUNTAIN QUARRIES LTD
Howton Court, Pontrilas,
Herefordshire, HR2 0BG
Area of Operation: UK & Ireland
Tel: 01981 241541
Email: info@blackmountainquarries.com
Web: www.blackmountainquarries.com ᵀᴴ

BONAKEMI LIMITED
1 Radian Court, Davy Avenue, Knowlhill,
Milton Keynes, Buckinghamshire, MK5 8PJ
Area of Operation: UK & Ireland
Tel: 01908 399740 **Fax:** 01908 232722
Email: info.uk@bona.com
Web: www.bona.com **Product Type:** 6

BROADLEAF TIMBER
Llandeilo Road Industrial Estate,
Carms, Carmarthenshire, SA18 3JG
Area of Operation: UK & Ireland
Tel: 01269 851910
Fax: 01269 851911
Email: sales@broadleaftimber.com
Web: www.broadleaftimber.com
Product Type: 1, 2, 3
Material Type: A) 2, 5, 7, 8, 10

CALLAGHAN WOOD FLOORS LTD
Unit 4 Inwood Business Park,
Whitton Road, Hounslow, TW3 2EB
Area of Operation: UK (Excluding Ireland)
Tel: 0208 577 1100 **Fax:** 0208 577 0400
Email: info@solid-wood-flooring.co.uk
Web: www.woodenfloors.net
Product Type: 1, 2, 3, 5, 6

CAMPBELL MARSON & COMPANY LTD
34 Wimbledon Business Centre,
Riverside Road, London, SW17 0BA
Area of Operation: UK & Ireland
Tel: 0208 879 1909
Fax: 0208 946 9395
Email: sales@campbellmarson.com
Web: www.campbellmarson.com ᵀᴴ
Product Type: 1, 2, 3, 5

CANADA WOOD UK
PO Box 1, Farnborough,
Hampshire, GU14 6WE
Area of Operation: UK (Excluding Ireland)
Tel: 01252 522545
Fax: 01252 522546
Email: office@canadawooduk.org
Web: www.canadawood.info
Product Type: 1, 2
Material Type: A) 2, 3, 4, 5, 6, 7, 9

CHRISTIE TIMBER SERVICES LTD
New Victoria Sawmills,
Bridgeness Road, Bo'ness,
Falkirk, EH51 9LG
Area of Operation: North East England, Scotland
Tel: 01506 828222
Fax: 01506 828226
Email: sales@christie-timber.co.uk
Web: www.christie-timber.co.uk ᵀᴴ
Product Type: 1, 2
Other Info: ✎
Material Type: A) 2, 3, 4, 5, 6, 12

CITY BATHROOMS & KITCHENS
158 Longford Road, Longford,
Coventry, West Midlands, CV6 6DR
Area of Operation: UK & Ireland
Tel: 02476 365877
Fax: 02476 644992
Email: citybathrooms@hotmail.com
Web: www.citybathrooms.co.uk

COMFOOT FLOORING
Walnut Tree, Redgrave Rd,
South Lopham,
Diss, Norfolk, IP22 2HN
Area of Operation: Europe
Tel: 01379 688516
Fax: 01379 688517
Email: sales@comfoot.com
Web: www.comfoot.com ᵀᴴ

COMFOOT FLOORING
Area of Operation: Europe
Tel: 01379 688516
Fax: 01379 688517
Email: sales@comfoot.com
Web: www.comfoot.com ᵀᴴ
Material Type: H) 4, 5

Day after day, the floor in your kid's room takes
the beating of it's life, but who wants to get the
mop out every night?
Comfoot Flooring suppliers of Witex Products

COUNTY HARDWOODS
Creech Mill, Mill Lane, Creech Saint Michael,
Taunton, Somerset, TA3 5PX
Area of Operation: UK & Ireland
Tel: 01823 444217 **Fax:** 01823 443940
Email: hardwood@netcomuk.co.uk
Web: www.countyhardwoods.co.uk ᵀᴴ
Product Type: 1, 2, 3, 5
Material Type: A) 2

COURTENAY JOHN BOTTERILL
24 Orwell Road, Clacton on Sea, Essex, CO15 1PP
Area of Operation: Greater London,
South East England
Tel: 01255 428837
Email: courtenay1@aol.com
Product Type: 1, 2, 3
Material Type: A) 2, 3, 5, 7

CRONINS RECLAMATION
& SOLID WOOD FLOORING
Preston Farm Court, Lower Road,
Little Bookham, Surrey, KT23 4EF
Area of Operation: Worldwide
Tel: 0208 614 4370 **Fax:** 01932 241918
Email: dfc1@supanet.com
Web: www.croninsreclamation.co.uk
Product Type: 2

CSM CARPETS & FLOORING LTD
Brickmakers Arms Lane, Doddington,
Cambridgeshire, PE15 0TR
Area of Operation: East England
Tel: 01354 740727 **Fax:** 01354 740078
Email: alan.csm@virgin.net
Web: www.csm-flooring.co.uk
Product Type: 2, 4

EXPRESSION
COLLECTION

A BRAND NEW EXPRESSION
With bevelled edges for a brand new look

For a brochure on flooring ideas
call 0800 374771

www.pergo.com

Designed for Generations

PERGO®
Designed for Generations

The original laminate flooring from Sweden with real guarantee.

For a brochure on flooring ideas and stockist list, call 0800 374771.

www.pergo.com

PERGO Expression

Floors with a bevelled edge for a new and modern expression .

PERGO Country

A modern rustic with bevelled edges.

PERGO Naturaltouch

A wooden floor feeling without the worries.

PERGO Vintage

A floor with an antique textured surface.

PERGO Modern Tile

Mix, match and design your own floor.

PERGO Exotic

Exotic designs with an exclusive ploished finish,

PERGO Original

Pure and elegant floors with the spirit of Scandinavian design

PERGO Classic Plus

The essential elegance of classic wood designs

4-in-1 Molding

4 different flooring transitions in a single package; T-moldings, hard surafce reducer, acrpet transition amd end moldings.
No glue necessary. Just snap parts together and push the assembly into the track.

simple
SOLUTIONS
FLOORING ACCESSORIES

PERGO quality goes beyond the installed floor. With our complete accessory assortment we offer that perfect look you can expect from a PERGO floor.

INTERIORS, FIXTURES & FINISHES

E&B QUALITY FLOORING SUPPLIES
Lynn Hill House, Yaxham Road, Norfolk, NR19 1HA
Area of Operation: East England
Tel: 01362 695081 **Fax:** 01362 695088
Email: info@laminateafloor.co.uk
Web: www.laminateafloor.co.uk
Product Type: 4

ECO IMPACT LTD
50a Kew Green, Richmond,
Greater London, TW9 3BB
Area of Operation: UK & Ireland
Tel: 0208 940 7072
Fax: 0208 332 1218
Email: sales@ecoimpact.co.uk
Web: www.ecoimpact.co.uk
Product Type: 2, 3
Other Info: ECO

EGGER (UK) LIMITED
Anick Grange Road, Hexham,
Northumberland, NE46 4JS
Area of Operation: UK & Ireland
Tel: 01434 602191
Fax: 01434 605103
Email: building.uk@egger.com
Web: www.egger.co.uk
Product Type: 6
Material Type: H) 1, 6

ENGLISH TIMBERS LTD
1A Main Street, Kirkburn, Driffield,
East Riding of Yorks, YO25 8NT
Area of Operation: UK & Ireland
Tel: 01377 229301
Fax: 01377 229303
Email: info@englishtimbers.co.uk
Web: www.englishtimbers.co.uk
Product Type: 1, 2
Material Type: A) 2, 3, 5, 6, 7

EUROPEAN HERITAGE
48-54 Dawes Road, Fulham, London, SW6 7EN
Area of Operation: Worldwide
Tel: 0207 381 6063
Fax: 0207 381 9534
Email: fulham@europeanheritage.co.uk
Web: www.europeanheritage.co.uk
Product Type: 1, 2, 5, 6
Material Type: A) 2, 3, 4, 5, 6, 9, 10

EVANS FLOORING
Pontyclerc, Penybanc Road, Ammanford,
Carmarthenshire, SA18 3HP
Area of Operation: UK & Ireland
Tel: 01269 591600
Fax: 01269 596116
Email: enquiries@evanshardwoodflooring.co.uk
Web: www.evanshardwoodflooring.co.uk
Product Type: 1, 2, 3, 5, 6

EXAKT PRECISION TOOLS LTD
Unit 9 Urquhart Trade Centre,
109 Urquhart Road, Aberdeen,
Aberdeenshire, AB24 5NH
Area of Operation: Europe
Tel: 01224 643434 **Fax:** 01224 643232
Email: info@exaktpt.com
Web: www.exaktpt.com
Product Type: 3, 4, 6

FINE OAK FLOORING
30 Folly Lane, St Albans, Hertfordshire, AL3 5JP
Area of Operation: UK (Excluding Ireland)
Tel: 01727 730680 **Fax:** 01727 730680
Email: erik@fineoakflooring.co.uk
Web: www.fineoakflooring.co.uk
Product Type: 1

FITZROY JOINERY
Garden Close, Langage Industrial Estate,
Plympton, Devon, PL7 5EU
Area of Operation: UK & Ireland
Tel: 0870 428 9110 **Fax:** 0870 428 9111
Email: admin@fitzroy.co.uk
Web: www.fitzroy.co.uk
Product Type: 4
Other Info: ECO
Material Type: A) 2

FLOORBRAND
Cavendish Business Centre,
High Street Green, Sible Hedingham,
Essex, CO9 1LH
Area of Operation: Europe
Tel: 0800 612 0101
Fax: 0845 166 8002
Email: sales@floorbrand.com
Web: www.floorbrand.com
Product Type: 3, 4, 5

FLOORCO
69-71 Washbrook Road, Rushden,
Northamptonshire, NN10 6UR
Area of Operation: UK (Excluding Ireland)
Tel: 01933 418899

FLOORS AND DOORS DIRECT
Unit 7 Blaydon Trade Park,
Toll Bridge Road, Blaydob,
Tyne & Wear, NE21 5TR
Area of Operation: North East England
Tel: 0191 414 5055
Fax: 0191 414 5066
Web: www.floorsanddoorsdirect.co.uk
Product Type: 1, 2, 4

FLOORS GALORE
Unit 5, Centurion Park,
Kendal Road, Shrewsbury,
Shropshire, SY1 4EH
Area of Operation: UK & Ireland
Tel: 0800 083 0623
Email: info@floors-galore.co.uk
Web: www.floors-galore.co.uk
Product Type: 1, 2, 3, 4, 5, 6
Material Type: A) 1, 2, 3, 4, 5, 6, 7, 8, 9, 10, 11, 12, 13, 14

FLOORSDIRECT
Unit 6, Fountain Drive, Hertford,
Hertfordshire, SG13 7UB
Area of Operation: UK (Excluding Ireland)
Tel: 01992 552447
Fax: 01992 558760
Email: jenny@floorsdirect.co.uk
Web: www.floorsdirect.co.uk
Product Type: 1, 4, 5, 6
Material Type: A) 1, 2, 3, 4, 5, 6, 7, 9

FLOORTECK LTD
8 Kingsdown Road, Northfield,
Birmingham, West Midlands, B31 1AH
Area of Operation: UK (Excluding Ireland)
Tel: 0121 476 6271
Fax: 0121 476 6271
Email: sarahcollier@go.com
Product Type: 1, 2, 3, 4, 5, 6

FLOORZ.CO.UK
Unit 13 Spa Industrial Estate,
Longfield Road, North Farm Estate,
Tunbridge Wells, Kent, TN2 3EY
Area of Operation: UK & Ireland
Tel: 01892 678855
Fax: 01892 678856
Email: info@floorz.co.uk
Web: www.floorz.co.uk
Product Type: 1, 2, 6
Other Info:

FOREST INSIGHT
East Farm, Knook, Warminster,
Wiltshire, BA12 0JG
Area of Operation: UK (Excluding Ireland)
Tel: 01985 850 088
Fax: 01985 850185
Email: nick@forest-insight.fsnet.co.uk

FRENCH PARQUET DIRECT LTD.
Woodhead House, Sorn, Ayrshire, KA5 6JA
Area of Operation: Europe
Tel: 01290 559028
Fax: 01290 559188
Email: enquiries@frenchparquet.co.uk
Web: www.frenchparquet.com
Product Type: 1, 2, 3
Other Info: ECO
Material Type: A) 2, 8

HARDWOOD FLOOR STORE
North Way, Hillend Industrial Estate,
Dalgety Bay, Fife, KY11 9JA
Area of Operation: UK (Excluding Ireland)
Tel: 01383 824729
Email: sales@hardwoodfloorstore.co.uk
Web: www.hardwoodfloorstore.co.uk
Product Type: 1, 2, 3, 4

HENRY VENABLES TIMBER LTD
Tollgate Drive, Tollgate Industrial Estate,
Stafford, Staffordshire, ST16 3HS
Area of Operation: UK (Excluding Ireland)
Tel: 01785 270600
Fax: 01785 270626
Email: enquiries@henryvenables.co.uk
Web: www.henryvenables.co.uk
Product Type: 1, 2

HITT OAK LTD
13a Northview Crescent, London, NW10 1RD
Area of Operation: UK (Excluding Ireland)
Tel: 0208 450 3821
Fax: 0208 911 0299
Email: zhitas@yahoo.co.uk
Product Type: 1, 2, 3, 4

HONEYSUCKLE BOTTOM SAWMILL LTD
Honeysuckle Bottom, Green Dene,
East Horsley, Leatherhead, Surrey, KT24 5TD
Area of Operation: Greater London, South East England
Tel: 01483 282394
Fax: 01483 282394
Email: honeysucklemill@aol.com
Web: www.easisites.co.uk/honeysucklebottomsawmill
Product Type: 2

HYPERION TILES
67 High Street, Ascot, Berkshire, SL5 7HP
Area of Operation: Greater London, South East England
Tel: 01344 620211
Fax: 01344 620100
Email: graham@hyperiontiles.com
Web: www.hyperiontiles.com
Product Type: 1, 2, 3, 4, 5, 6

INTERESTING INTERIORS LTD
37-39 Queen Street, Aberystwyth,
Ceredigion, SY23 1PU
Area of Operation: UK (Excluding Ireland)
Tel: 01970 626162
Fax: 0870 051 7959
Email: enquiries@interestinginteriors.com
Web: www.interiors-ltd.demon.co.uk
Product Type: 1, 3

JIMPEX LTD
1 Castle Farm, Cholmondeley,
Malpas, Cheshire, SY14 8AQ
Area of Operation: UK & Ireland
Tel: 01829 720433
Fax: 01829720553
Email: jan@jimpex.co.uk
Web: www.jimpex.co.uk
Product Type: 1, 3, 5

JOHN FLEMING & CO LTD
Silverburn Place, Bridge of Don,
Aberdeen, Aberdeenshire, AB23 8EG
Area of Operation: Scotland
Tel: 0800 085 8728
Fax: 01224 825377
Email: info@johnfleming.co.uk
Web: www.johnfleming.co.uk
Product Type: 2, 4

JUNCKERS LTD
Unit A,1 Wheaton Road,
Witham, Essex, CM8 3UJ
Area of Operation: UK & Ireland
Tel: 01376 517512
Fax: 01376 514401
Email: sales@junckers.co.uk
Web: www.junckers.com
Product Type: 1, 2, 3, 5
Other Info: ECO
Material Type: A) 2, 4, 6, 11

KAHRS (UK) LTD
Unit 2 West, 68 Bognor Road,
Chichester, West Sussex, P019 8NS
Area of Operation: UK & Ireland
Tel: 01243 778747 **Fax:** 01243 531237
Email: sales@kahrs.co.uk
Web: www.kahrs.co.uk
Product Type: 1, 2, 3, 5
Other Info: ECO
Material Type: A) 2, 3, 4, 5, 6, 7, 9, 14

KARELIA WOOD FLOORING
Havenhurst Mill, Bexhill Road,
St Leonards-on-Sea, East Sussex, TN38 0AJ
Area of Operation: UK (Excluding Ireland)
Tel: 01424 456805 **Fax:** 01424 440505
Email: enquiries@kareliawoodflooring-uk.com
Web: www.kareliaparketti.com
Product Type: 1, 2, 3, 4, 5

KITCHEN SUPPLIES
East Chesters, North Way,
Hillend Industrial Estate,
Dalgety Bay, Fife, KY11 9JA
Area of Operation: UK (Excluding Ireland)
Tel: 01383 824729
Email: sales@kitchensupplies.co.uk
Web: www.kitchensupplies.co.uk
Product Type: 1, 2, 3, 4, 5

LAMBETH DIXON
Unit 5, Drury Lane, Wood Hall Business Park,
Sudbury, Suffolk, CO10 1WH
Area of Operation: UK (Excluding Ireland)
Tel: 01787 379311 **Fax:** 01787 379344
Email: lambethdixon@btconnect.com
Product Type: 1, 2, 3, 4

LAWSONS
Gorst Lane, Off New Lane, Burscough,
Ormskirk, Lancashire, L40 0RS
Area of Operation: Worldwide
Tel: 01704 893998
Fax: 01704 892526
Email: info@traditionaltimber.co.uk
Web: www.traditionaltimber.co.uk

MCKAY FLOORING LTD
8 Harmony Square, Govan,
City of Glasgow, G51 3LW
Area of Operation: UK & Ireland
Tel: 0141 440 1586
Fax: 0141 425 1020
Email: enquiries@mckayflooring.co.uk
Web: www.mckayflooring.co.uk
Product Type: 1, 2, 3, 4, 5, 6
Other Info: ECO
Material Type: A) 1, 2, 3, 4, 5, 6, 7, 8, 9, 10, 11, 12, 13, 14, 15

MERCIA FLOORING
59 The Square, Dunchurch,
Rugby, Warwickshire, CV22 6NU
Area of Operation: UK (Excluding Ireland)
Tel: 01788 522168
Fax: 01788 811847
Email: sales@merciaflooring.co.uk
Web: www.merciaflooring.co.uk
Product Type: 1, 2, 3, 4, 5

MIDLANDS SLATE & TILE
Qualcast Road, Horseley Fields,
Wolverhampton, West Midlands, WV1 2QP
Area of Operation: UK & Ireland
Tel: 01902 458780
Fax: 01902 458549
Email: sales@slate-tile-brick.co.uk
Web: www.slate-tile-brick.co.uk
Product Type: 1, 2, 3, 4
Material Type: A) 1, 2, 5, 7

N&C NICOBOND
41-51 Freshwater Road, Chadwell Heath,
Romford, Essex, RM8 1SP
Area of Operation: Worldwide
Tel: 0208 586 4600
Fax: 0208 586 4646
Email: info@nichollsandclarke.com
Web: www.ncdirect.co.uk

INTERIORS, FIXTURES & FINISHES

NATURAL FLOORING
152-154 Wellingborough Road,
Northampton, Northamptonshire, NN1 4DT
Area of Operation: UK (Excluding Ireland)
Tel: 01604 239238/7 **Fax:** 01604 627799
Web: www.naturalflooring.co.uk

NATURAL IMAGE (GRANITE, MARBLE, STONE)
Spelmonden Estate, Goudhurst Road,
Goudhurst, Kent, TN17 1HE
Area of Operation: UK & Ireland
Tel: 01580 212222
Fax: 01580 211841
Web: www.naturalstonefloors.org
Product Type: 1, 2

NATURAL WOOD FLOORING COMPANY LTD
20 Smugglers Way,
Wandsworth, London, SW18 IEG
Area of Operation: Worldwide
Tel: 0208 871 9771 **Fax:** 0208 877 0273
Email: sales@naturalwoodfloor.co.uk
Web: www.naturalwoodfloor.co.uk
Product Type: 1, 2, 3, 4, 5

NORTH YORKSHIRE TIMBER
Standard House, Thurston Road,
Northallerton Business Park, Northallerton,
North Yorkshire, DL6 2NA
Area of Operation: UK (Excluding Ireland)
Tel: 01609 780777 **Fax:** 01609 777888
Email: sales@nytimber.co.uk
Web: www.nytimber.co.uk

OAKBEAMS.COM
Hunterswood Farm, Alfold Road,
Dunsfold, Godalming, Surrey, GU8 4NP
Area of Operation: Worldwide
Tel: 01483 200477
Email: info@oakbeams.com
Web: www.oakbeams.com
Product Type: 2

ORIGINAL OAK
Ashlands, Burwash, East Sussex, TN19 7HS
Area of Operation: UK (Excluding Ireland)
Tel: 01435 882228
Fax: 01435 882228
Material Type: A) 2

OSMO UK LTD
Unit 2 Pembroke Road, Stocklake Industrial Estate,
Aylesbury, Buckinghamshire, HP20 1DB
Area of Operation: UK & Ireland
Tel: 01296 481220
Fax: 01296 424090
Email: info@osmouk.com
Web: www.osmouk.com
Product Type: 1, 2, 3, 4, 5

PARAGON JOINERY
Systems House, Eastbourne Road,
Blindley Heath, Surrey, RH7 6JP
Area of Operation: UK (Excluding Ireland)
Tel: 01342 836300
Fax: 01342 836364
Email: sales@paragonjoinery.com
Web: www.paragonjoinery.com

PARQUET & GENERAL FLOORING CO. LTD.
Grange Lane, Winsford, Cheshire, CW7 2PS
Area of Operation: UK (Excluding Ireland)
Tel: 01606 861442
Fax: 01606 861445
Email: floors@wideboards.com
Web: www.wideboards.com
Product Type: 1, 2, 3, 5, 6
Material Type: A) 2, 3, 4, 5, 6, 7, 8, 9, 10,
11, 12, 13, 14

PERGO LTD
19 Blacklands Way, Abingdon Business Park,
Abingdon, Oxfordshire, OX11 9PD
Area of Operation: Worldwide
Tel: 01235 556300 **Fax:** 01235 556350
Email: vicky.sweet@pergo.com
Web: www.pergo.com
Product Type: 1, 4, 5, 6
Other Info: ECO

PETERSONS NATURAL FLOORINGS
Unit 10/11 Woodlands Park Industrial Estate,
Short Thorn Road, Stratton Strawless, Norwich,
Norfolk, NR10 5NU
Area of Operation: UK & Ireland
Tel: 01603 755511
Fax: 01603 755019
Email: office@petersons-natural-floorings.co.uk
Web: www.petersons-natural-floorings.co.uk
Product Type: 2
Other Info: ECO
Material Type: A) 2, 4, 6, 8

PILKINGTON'S TILES GROUP
PO Box 4, Clifton Junction, Manchester, M27 8LP
Area of Operation: UK (Excluding Ireland)
Tel: 0161 727 1000
Fax: 0161 727 1122
Email: sales@pilkingtons.com
Web: www.pilkingtons.com
Other Info: ✎

RED ROSE PLASTICS (BURNLEY) LTD
Parliament Street, Burnley, Lancashire, BB11 3JT
Area of Operation: East England, North East
England, North West England and North Wales
Tel: 01282 724600 **Fax:** 01282 724644
Email: info@redroseplastics.co.uk
Web: www.redroseplastics.co.uk
Product Type: 1, 4, 5
Other Info: ECO

REDLAM TIMBERS LTD
PO Box 1908, Meriden, Coventry,
Warwickshire, CV7 7YR
Area of Operation: UK (Excluding Ireland)
Tel: 01676 521222 **Fax:** 01676 521221
Email: redlamtimbers@aol.com
Web: www.redlam-flooring.co.uk
Other Info: ECO
Material Type: A) 2, 4, 5

RICHARD BURBIDGE LTD
Whittington Road, Oswestry,
Shropshire, SY11 1HZ
Area of Operation: Worldwide
Tel: 01691 655131 **Fax:** 01691 657694
Email: info@richardburbidge.co.uk
Web: www.richardburbidge.co.uk
Product Type: 2, 4

RICHWOOD FLOORING
Transpoint, Doncaster Road, Kirk Sandall,
Doncaster, South Yorkshire, DN3 1HT
Area of Operation: UK & Ireland
Tel: 01302 888800 **Fax:** 01302 888899
Email: info@richwoodflooring.com
Web: www.richwoodflooring.com
Product Type: 1, 5
Material Type: A) 1, 2, 3, 5, 7

S.L. HARDWOODS
390 Sydenham Road, Croydon, Surrey, CR0 2EA
Area of Operation: UK (Excluding Ireland)
Tel: 0208 683 0292 **Fax:** 0208 683 0404
Email: info@slhardwoods.co.uk
Web: www.slhardwoods.co.uk

SCANDAFLOOR LIMITED
208 St Annes Road East,
Lytham St Annes, Lancashire, FY8 3HT
Area of Operation: UK (Excluding Ireland)
Tel: 01253 714907 **Fax:** 01253 729348
Email: info@scandafloor.co.uk
Web: www.scandafloor.co.uk
Product Type: 1, 2, 3

SCOTTSTYLE
140 Castlemain Avenue, Southbourne, Dorset, BH6 5ER
Area of Operation: UK (Excluding Ireland)
Tel: 01202 427352
Email: nick@scottstyle.com
Web: www.scottstyle.com/flooring ⌂
Product Type: 1

SOLID FLOOR
53 Pembridge Road, Notting Hill, London, W11 3HG
Area of Operation: Greater London,
Scotland, South East England
Tel: 0207 221 9166 **Fax:** 0207 221 8193
Email: nottinghill@solidfloor.co.uk
Web: www.solidfloor.co.uk
Product Type: 1, 2, 3, 6 **Other Info:** ▽ ✎ ✋
Material Type: A) 1, 2, 3, 5, 6, 7, 8, 9, 10

SOUTH WESTERN FLOORING SERVICES
145-147 Park Lane, Frampton Cotterell,
Bristol, BS36 2ES
Area of Operation: UK & Ireland
Tel: 01454 880982 **Fax:** 01454 880982
Email: mikeflanders@blueyonder.co.uk
Web: www.southwesternflooring.com
Product Type: 1, 2, 3, 5
Material Type: A) 1, 2, 3, 4, 5, 6, 7, 8, 9, 10, 11, 12,
13, 14

STRATHEARN STONE AND TIMBER LTD
Glenearn, Bridge of Earn, Perth,
Perth and Kinross, PH2 9HL
Area of Operation: North East England, Scotland
Tel: 01738 813215 **Fax:** 01738 815946
Email: info@stoneandoak.com
Web: www.stoneandoak.com
Product Type: 2
Material Type: A) 2, 10

SWIFTWOOD IMPORTS LTD
Quay House, Nene Parade, Wisbech,
Cambridgeshire, PE13 3BY
Area of Operation: UK (Excluding Ireland)
Tel: 01945 587000 **Fax:** 01945 581203
Email: timber.floors@swiftwood.com
Web: www.basecofloors.co.uk
Product Type: 1, 2
Material Type: B) 2, 8, 10

THE CARPENTRY INSIDER - AIRCOMDIRECT
1 Castleton Crescent, Skegness,
Lincolnshire, PE25 2TJ
Area of Operation: Worldwide
Tel: 01754 767163
Email: aircom8@hotmail.com
Web: www.carpentry.tk ⌂
Product Type: 1, 6

THE NATURAL FLOORING WAREHOUSE
East Building, Former All Saints Church, Armoury
Way, Wandsworth, London, SW18 1HZ
Area of Operation: UK (Excluding Ireland)
Tel: 0208 870 5555 **Fax:** 0208 877 2847
Email: natalie@edwardianfires.com
Product Type: 1, 2, 3, 5, 6

THE REAL DOOR COMPANY
Unit 5, Cadwell Lane, Hitchen,
Hertfordshire, SG4 0SA
Area of Operation: UK & Ireland
Tel: 01462 451230 **Fax:** 01462 440459
Email: sales@realdoor.co.uk
Web: www.realdoor.co.uk

**THE REAL OAK DOOR
& FLOORING COMPANY**
5 Robin Way, Sudbury, Suffolk, CO10 7PF
Area of Operation: UK (Excluding Ireland)
Tel: 01787 310051 **Fax:** 01787 312497
Email: valentinecollins@aol.com
Web: www.realoakdoor.co.uk
Material Type: A) 2

THE SPA & WARWICK TIMBER CO LTD
Harriott Drive, Heathcote Industrial Estate,
Warwick, Warwickshire, CV34 6TJ
Area of Operation: Midlands & Mid Wales
Tel: 01926 883876 **Fax:** 01926 450831
Email: sales@spa-warwick.co.uk
Product Type: 1, 2, 5

THE WORLDWIDE WOOD COMPANY LTD
154 Colney Hatch Lane, Muswell Hill,
London, N10 1ER
Area of Operation: Worldwide
Tel: 0800 458 3366 **Fax:** 0208 365 3965
Email: info@solidwoodflooring.com
Web: www.solidwoodflooring.com
Product Type: 1, 2, 3, 5, 6

THOROGOOD TIMBER PLC
Colchester Road, Ardleigh, Colchester,
Essex, CO7 7PQ
Area of Operation: East England
Tel: 01206 233100 **Fax:** 01206 233115
Email: barry@thorogood.co.uk
Web: www.thorogood.co.uk
Product Type: 1, 2

TIMBER RECLAIMED
PO Box 33727, London, SW3 4XE
Area of Operation: UK & Ireland
Tel: 0207 824 9200
Fax: 0207 824 9207
Email: elaine.barker@dmbarker.com
Web: www.timberreclaimed.com
Product Type: 1, 2
Other Info: ▽

TONGLING BAMBOO FLOORING
6 Camellia Drive, Priorslee, Telford, Shrops, TF2 9UA
Area of Operation: UK & Ireland
Tel: 01952 200032 **Fax:** 01952 291938
Email: sales@tlflooring.co.uk
Web: www.tlflooring.co.uk
Product Type: 1, 2 **Material Type:** M) 6

TOPPS TILES
Thorpe Way, Grove Park,
Enderby, Leicestershire, LE19 1SU
Area of Operation: UK (Excluding Ireland)
Tel: 0116 282 8000 **Fax:** 0116 282 8100
Email: mlever@toppstiles.co.uk
Web: www.toppstiles.co.uk

TOPPS TILES

Area of Operation: UK (Excluding Ireland)
Tel: 0116 282 8000 **Fax:** 0116 282 8100
Email: mlever@toppstiles.co.uk
Web: www.toppstiles.co.uk

Topps Tiles is Britain's biggest tile and wood flooring specialist, with over 200 stores nationwide. For details of your nearest store or free brochure call 0800 138 1673 www.toppstiles.co.uk.

INTERIORS, FIXTURES & FINISHES

TRADITIONAL OAK & TIMBER COMPANY
P O Stores, Haywards Heath Road,
North Chailey, Nr Lewes,
East Sussex, BN8 4EY
Area of Operation: Worldwide
Tel: 01825 723648
Fax: 01825 722215
Email: info@tradoak.co.uk
Web: www.tradoak.com
Product Type: 2
Other Info: ▽
Material Type: A) 2

TREEWORK FLOORING LTD
Cheston Combe, Church Town, Backwell,
Bristol, BS48 3JQ
Area of Operation: UK (Excluding Ireland)
Tel: 01275 790049
Fax: 01275 463078
Email: johnemery@treeworkflooring.co.uk
Web: www.treeworkflooring.co.uk
Product Type: 1, 2, 3

UK HARDWOODS LTD
Wade Mill, Molland, South Molton, Devon, EX36 3NL
Area of Operation: UK (Excluding Ireland)
Tel: 01769 550526
Product Type: 1, 2, 3

UK WOOD FLOORS LTD
Unit 8 Arrow Industrial Estate,
Farnborough, Hampshire, GU14 7QH
Area of Operation: UK (Excluding Ireland)
Tel: 01252 520520
Fax: 01252 520440
Email: info@ukwoodfloors.co.uk
Web: www.ukwoodfloors.co.uk
Product Type: 1, 2, 3, 5
Other Info: ECO
Material Type: A) 2, 3, 4, 5, 6, 7, 9

VICTORIA HAMMOND INTERIORS
Bury Farm, Church Street, Bovingdon,
Hemel Hempstead, Hertfordshire, HP3 0LU
Area of Operation: UK (Excluding Ireland)
Tel: 01442 831641
Fax: 01442 831641
Email: victoria@victoriahammond.com
Web: www.victoriahammond.com
Product Type: 2, 3, 5, 6
Other Info: ▽ ✎

VICTORIAN WOOD WORKS LTD
54 River Road, Creekmouth,
Barking, Essex, IG11 0DW
Area of Operation: Worldwide
Tel: 0208 534 1000
Fax: 0208 534 2000
Email: sales@victorianwoodworks.co.uk
Web: www.victorianwoodworks.co.uk
Product Type: 1, 2, 3, 6
Other Info: ▽ ✎ ✋

VISION ASSOCIATES
Demita House, North Orbital Road,
Denham, Buckinghamshire, UB9 5EY
Area of Operation: UK & Ireland
Tel: 01895 831600
Fax: 01895 835323
Email: info@visionassociates.co.uk
Web: www.visionassociates.co.uk
Product Type: 1, 2, 3

WAXMAN CERAMICS LTD
Grove Mills, Elland, West Yorkshire, HX5 9DZ
Tel: 01422 311331
Fax: 01422 310654
Email: sales@waxmanceramics.co.uk
Web: www.waxmanceramics.co.uk

WOOD YOU LIKE
School Road, Charing, Ashford, Kent, TN27 0JN
Area of Operation: East England
Tel: 0845 1661190
Email: info@wood-you-like.co.uk
Web: www.wood-you-like.co.uk ✎
Product Type: 1, 2, 3, 5, 6
Other Info: ECO ✎
Material Type: A) 1, 2, 3, 4, 5, 6, 7, 9, 10

WOOD2U
26 Waring Way, Dunchurch, Rugby,
Warwickshire, CV22 6PH
Area of Operation: UK (Excluding Ireland)
Tel: 0870 241 8847
Email: sales@wood2u.com
Web: www.wood2u.co.uk ✎
Product Type: 1, 2, 3, 4, 5, 6

WOODHOUSE TIMBER
Unit 6 Quarry Farm Industrial Estate,
Staplecross Road, Bodiam, East Sussex, TN32 5RA
Area of Operation: UK (Excluding Ireland)
Tel: 01580 831700
Fax: 01580 830365
Email: info@woodhousetimber.co.uk
Web: www.woodhousetimber.co.uk
Product Type: 1, 2
Material Type: A) 2, 6

WOODLINE FLOORS LTD
Unit 3, Brook Farm, Horsham Road,
Cowfold, Horsham, West Sussex, RH13 8AH
Area of Operation: UK & Ireland
Tel: 0870 840 8484
Fax: 0870 840 0040
Email: sales@woodlinefloors.co.uk
Web: www.woodlinefloors.co.uk ✎
Product Type: 1, 5
Material Type: A) 1, 2, 3, 4, 5, 6, 7, 9

WOODLINE FLOORS

Area of Operation: UK & Ireland
Tel: 0870 840 8484 **Fax:** 0870 840 0040
Email: sales@woodlinefloors.co.uk
Web: www.woodlinefloors.co.uk
Product Type: 1, 5

Woodline Floors specialises in supplying a range of quality real wood floors direct from Finland at trade prices offering a variety of styles and finishes suitable for both contemporary and traditional interiors.

Contact us for:
• Free Brochures • Free Samples • No Obligation Quotations

CERAMIC FLOORING, INCLUDING MOSAICS

KEY

OTHER: ▽ Reclaimed ✎ On-line shopping
✎ Bespoke ✋ Hand-made ECO Ecological

ALISTAIR MACKINTOSH LTD
Bannerley Road, Garretts Green,
Birmingham, West Midlands, B33 0SL
Area of Operation: UK & Ireland
Tel: 0121 784 6800
Fax: 0121 789 7068
Email: info@alistairmackintosh.co.uk
Web: www.alistairmackintosh.co.uk

BRITISH CERAMIC TILE
Heathfield, Newton Abbot, Devon, TQ12 6RF
Area of Operation: UK & Ireland
Tel: 01626 831480
Fax: 01626 831465
Email: sales@bctltd.co.uk
Web: www.candytiles.com
Material Type: F) 1, 2, 4

CAPITAL MARBLE DESIGN
Unit 1 Pall Mall Deposit,
124-128 Barlby Road, London, W10 6BL
Area of Operation: UK & Ireland
Tel: 0208 968 5340
Fax: 0208 968 8827
Email: stonegallery@capitalmarble.co.uk
Web: www.capitalmarble.co.uk
Other Info: ECO ✎
Material Type: J) 5, 6

CERAMIQUE INTERNATIONALE LTD
Unit 1 Royds Lane, Lower Wortley Ring Road,
Leeds, West Yorkshire, LS12 6DU
Area of Operation: UK & Ireland
Tel: 0113 231 0218
Fax: 0113 231 0353
Email: cameron@ceramiqueinternationale.co.uk
Web: www.ceramiqueinternationale.co.uk
Other Info: ✎

CITY BATHROOMS & KITCHENS
158 Longford Road, Longford,
Coventry, West Midlands, CV6 6DR
Area of Operation: UK & Ireland
Tel: 02476 365877
Fax: 02476 644992
Email: citybathrooms@hotmail.com
Web: www.citybathrooms.co.uk

CONCEPT TILING LIMITED
Unit 3, Jones Court, Jones Square,
Stockport, Cheshire, SK1 4LJ
Area of Operation: UK & Ireland
Tel: 0161 480 0994
Fax: 0161 480 0911
Email: enquiries@concept-tiles.com
Web: www.concept-tiles.com
Material Type: E) 1, 2, 3, 4, 5, 8, 9, 11, 13

CTD SCOTLAND
72 Hydepark Street, Glasgow,
Dunbartonshire, G3 8BW
Area of Operation: Scotland
Tel: 0141 221 4591
Fax: 0141 221 8442
Email: info@ctdscotland.co.uk
Web: www.ctdscotland.co.uk

DAR INTERIORS
Arch 11, Miles Street, London , SW8 1RZ
Area of Operation: Worldwide
Tel: 0207 720 9678
Fax: 0207 627 5129
Email: enquiries@darinteriors.com
Web: www.darinteriors.com

EUROPEAN HERITAGE
48-54 Dawes Road, Fulham,
London, SW6 7EN
Area of Operation: Worldwide
Tel: 0207 381 6063 **Fax:** 0207 381 9534
Email: fulham@europeanheritage.co.uk
Web: www.europeanheritage.co.uk
Material Type: F) 1, 2, 3

EXAKT PRECISION TOOLS LTD
Unit 9 Urquhart Trade Centre, 109 Urquhart Road,
Aberdeen, Aberdeenshire, AB24 5NH
Area of Operation: Europe
Tel: 01224 643434 **Fax:** 01224 643232
Email: info@exaktpt.com
Web: www.exaktpt.com

FIRED EARTH INTERIORS
3 Twyford Mill, Oxford Road, Adderbury,
Banbury, Oxfordshire, OX17 3SX
Area of Operation: Worldwide
Tel: 01295 814300 **Fax:** 01295 810832
Email: enquiries@firedearth.com.
Web: www.firedearth.com

FLOORBRAND
Cavendish Business Centre, High Street Green,
Sible Hedingham, Essex, CO9 3LH
Area of Operation: Europe
Tel: 0800 612 0101 **Fax:** 0845 166 8002
Email: sales@floorbrand.com
Web: www.floorbrand.com ✎

FLOORTECK LTD
8 Kingsdown Road, Northfield,
Birmingham, West Midlands, B31 1AH
Area of Operation: UK (Excluding Ireland)
Tel: 0121 476 6271
Fax: 0121 476 6271
Email: sarahcollier@go.com

H & R JOHNSON TILES LTD
Harewood Street, Tunstall, Stoke on Trent,
Staffordshire, ST6 5JZ
Area of Operation: Worldwide
Tel: 01782 575575
Fax: 01782 524138
Email: techsales@johnson-tiles.com
Web: www.johnson-tiles.com

HERITAGE TILE CONSERVATION LTD
The Studio, 2 Harris Green, Broseley,
Shropshire, TF12 5HJ
Area of Operation: UK & Ireland
Tel: 01952 881039
Fax: 01952 881039
Email: enquiries@heritagetile.co.uk
Web: www.heritagetile.co.uk
Material Type: F) 1, 2, 3, 4, 5

HITT OAK LTD
13a Northview Crescent, London, NW10 1RD
Area of Operation: UK (Excluding Ireland)
Tel: 0208 450 3821
Fax: 0208 911 0299
Email: zhitas@yahoo.co.uk

HYPERION TILES
67 High Street, Ascot, Berkshire, SL5 7HP
Area of Operation: Greater London,
South East England
Tel: 01344 620211
Fax: 01344 620100
Email: graham@hyperiontiles.com
Web: www.hyperiontiles.com

INNOVATION GLASS COMPANY
27-28 Blake Industrial Estate,
Brue Avenue, Bridgwater, Somerset, TA7 8EQ
Area of Operation: Europe
Tel: 01278 683645 **Fax:** 01278 450088
Email: magnus@igc-uk.com
Web: www.igc-uk.com

LAWLEY CERAMICS
8 Stourbridge Road, Bromsgrove,
Worcestershire, B61 0AB
Area of Operation: UK (Excluding Ireland)
Tel: 01527 570455 **Fax:** 01527 570455
Email: lawleyceramics@hotmail.com

MAESTROTILE
Unit 7 Waterloo Park, Waterloo Road Industrial
Estate, Bidford on Avon, Warwickshire, B50 4JH
Area of Operation: UK & Ireland
Tel: 01789 778700 **Fax:** 01789 778852
Email: malcolm@maestrotile.com
Web: www.maestrotile.com

MARIA STARLING MOSAICS
40 Strand House, Merbury Close,
London, SE28 0LU
Area of Operation: UK (Excluding Ireland)
Tel: 07775 517409
Email: mosaics@mariastarling.com
Web: www.mariastarling.com
Other Info: ✎ ✋

MARLBOROUGH TILES LTD.
Elcot Lane, Marlborough, Wiltshire, SN8 2AY
Area of Operation: UK & Ireland
Tel: 01672 512422 **Fax:** 01672 515791
Email: admin@marlboroughtiles.com
Web: www.marlboroughtiles.com

MERCIA FLOORING
59 The Square, Dunchurch, Rugby,
Warwickshire, CV22 6NU
Area of Operation: UK (Excluding Ireland)
Tel: 01788 522168 **Fax:** 01788 811847
Email: sales@merciaflooring.co.uk
Web: www.merciaflooring.co.uk

INTERIORS, FIXTURES & FINISHES

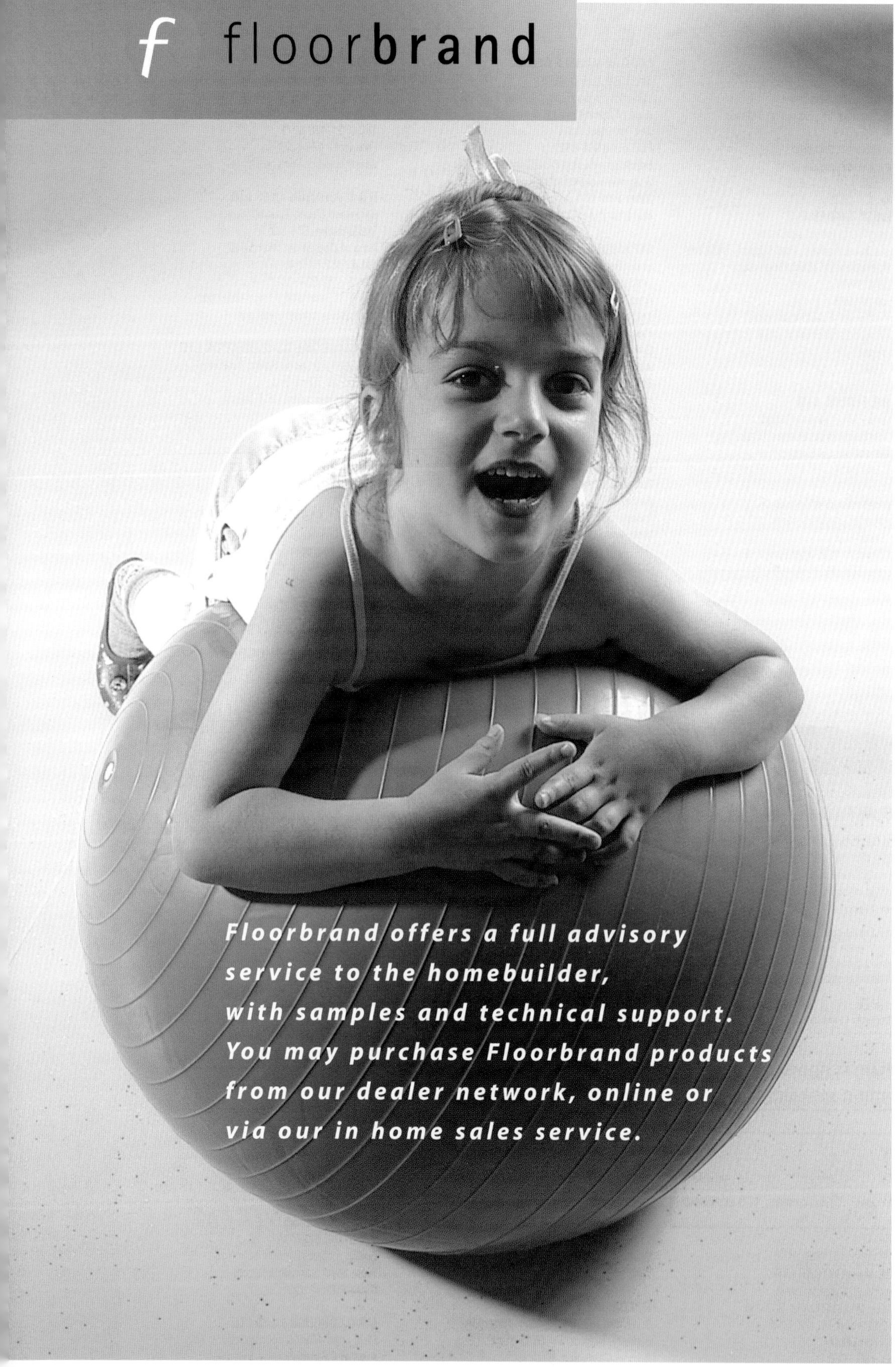

Purline, for those looking for striking colour and design!

Purline is manufactured from 100% polyurethane, which is fully recyclable. This extremely hard wearing product, suitable for the heaviest commercial usage, also looks stunning in modern home interiors, where pure colour is required. Available in tiles, rolls or interlocking panels for rapid installation. Purline is so easy to keep clean and offers a very hygienic flooring solution to kitchens, bathrooms, conservatories or anywhere in the home.

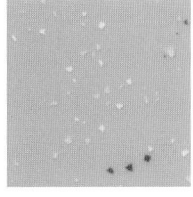   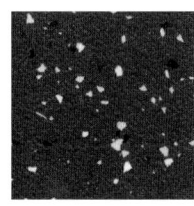

Available in 16 solid shades or with flecks

Cristalite engineered granite flooring, walls and surfaces

Granite flooring has never been so accessible. Our engineered granite takes 85% quartz, the main constituent of granite, and adds 15% acrylic to it, making it more flexible and able to be produced at 6mm thickness for easier installation and incredible large format tiles.

You can now match your floors, walls and work surfaces all in Cristalite engineered granite.

  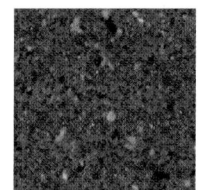 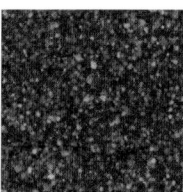

Available in 25+ colours

Duraceramic is the original ceramic tile alternative

Duraceramic is the first product on the market that offers the genuine grouted finish of a ceramic tile, with the warmth and comfort of vinyl. With its Life Time Wear Warranty, it offers total peace of mind and fantastic looks for many areas of the house. Because you can fit and grout in the same day, Duraceramic reduces the cost of installation compared with traditional ceramics. If grout is not for you, then install the product edge to edge without grout for a really clean modern look.

Available in 41 colours

Our laminate flooring selection includes over 70 designs

www.floorbrand.com • Call 0800 612 0101

SPONSORED BY: PERGO LTD www.pergo.com

N&C NICOBOND
41-51 Freshwater Road, Chadwell Heath,
Romford, Essex, RM8 1SP
Area of Operation: Worldwide
Tel: 0208 586 4600
Fax: 0208 586 4646
Email: info@nichollsandclarke.com
Web: www.ncdirect.co.uk

NATURAL FLOORING
152-154 Wellingborough Road,
Northampton, Northamptonshire, NN1 4DT
Area of Operation: UK (Excluding Ireland)
Tel: 01604 239238/7
Fax: 01604 627799
Web: www.naturalflooring.co.uk

ORIGINAL OAK
Ashlands, Burwash, East Sussex, TN19 7HS
Area of Operation: UK (Excluding Ireland)
Tel: 01435 882228
Fax: 01435 882228
Material Type: F) 3

ORIGINAL STYLE
Falcon Road, Sowton Industrial Estate,
Exeter, Devon, EX2 7LF
Area of Operation: Worldwide
Tel: 01392 474011
Fax: 01392 219932
Email: vmattison@originalstyle.com
Web: www.originalstyle.com
Other Info: 🖑
Material Type: E) 2, 11

PARKSIDE TILES
49-51 Highmeres Road, Thurmeston,
Leicester, Leicestershire, LE4 9LZ
Area of Operation: UK (Excluding Ireland)
Tel: 0116 276 2532
Fax: 0116 246 0649
Email: parkside@tiles2.wanadoo.co.uk
Web: www.parksidetiles.co.uk

PILKINGTON'S TILES GROUP
PO Box 4, Clifton Junction,
Manchester, M27 8LP
Area of Operation: UK (Excluding Ireland)
Tel: 0161 727 1000
Fax: 0161 727 1122
Email: sales@pilkingtons.com
Web: www.pilkingtons.com

RUSTICA LTD
154c Milton Park, Oxfordshire, OX14 4SD
Area of Operation: UK & Ireland
Tel: 01235 834192
Fax: 01235 835162
Email: sales@rustica.co.uk
Web: www.rustica.co.uk
Other Info: 🖑
Material Type: E) 2, 3, 4, 5, 8, 9

**SHARON JONES HANDMADE
ARCHITECTURAL CERAMIC TILES**
Trevillian Cottage, Barrington,
Nr Ilminster, Somerset, TA19 0JB
Area of Operation: UK (Excluding Ireland)
Tel: 01460 259074
Email: info@handmadearchitecturaltiles.co.uk
Web: www.handmadearchitecturaltiles.co.uk

STONE AND SLATE
Coney Green Farm, Lower Market Street,
Claycross, Chesterfield, Derbyshire, S45 9NE
Area of Operation: UK & Ireland
Tel: 01246 250088
Fax: 01246 250099
Email: sales@stoneandslate.ltd.uk
Web: www.stoneandslate.co.uk

THE CANDY TILE COMPANY
Heathfield, Newton Abbot, Devon, TQ12 6RF
Area of Operation: UK (Excluding Ireland)
Tel: 01626 834774
Fax: 01626 834775
Email: sales@candytiles.com
Web: www.candytiles.com
Material Type: F) 1

THE NATURAL FLOORING WAREHOUSE
East Building, Former All Saints Church,
Armoury Way, Wandsworth, London, SW18 1HZ
Area of Operation: UK (Excluding Ireland)
Tel: 0208 870 5555
Fax: 0208 877 2847
Email: natalie@edwardianfires.com

THE ORIGINAL TILE COMPANY
23A Howe Street, Lothian, EH3 6TF
Area of Operation: UK & Ireland
Tel: 0131 556 2013
Fax: 0131 558 3172
Email: info@originaltile.freeserve.co.uk
Other Info: 🖉 🖑
Material Type: E) 1, 2, 3, 4, 5, 7, 8, 12

**THE REALLY SAFE FLOORING COMPANY
LIMITED**
Unit 29 Wivenhoe Business Centre,
Book Street, Colchester, Essex, CO7 9DP
Area of Operation: UK & Ireland
Tel: 01206 827870
Fax: 01206 827881
Email: sales@realsafe.co.uk
Web: www.realsafe.co.uk

TILECO
3 Molesey Business Centre, Central Avenue,
West Molesey, Surrey, KT8 2QZ
Area of Operation: UK (Excluding Ireland)
Tel: 0208 481 9500 **Fax:** 0208 481 9501
Email: info@tileco.co.uk
Web: www.tileco.co.uk
Other Info: 🖉
Material Type: C) 2, 3

TILES UK LTD
1-13 Montford Street, Off South Langworthy Road,
Salford, Greater Manchester, M50 2XD
Area of Operation: UK (Excluding Ireland)
Tel: 0161 872 5155 **Fax:** 0161 848 7948
Email: info@tilesuk.com
Web: www.tilesuk.com 🖑

TOPPS TILES
Thorpe Way, Grove Park, Enderby,
Leicestershire, LE19 1SU
Area of Operation: UK (Excluding Ireland)
Tel: 0116 282 8000 **Fax:** 0116 282 8100
Email: mlever@toppstiles.co.uk
Web: www.toppstiles.co.uk

TOPPS TILES

Area of Operation: UK (Excluding Ireland)
Tel: 0116 282 8000 **Fax:** 0116 282 8100
Email: mlever@toppstiles.co.uk
Web: www.toppstiles.co.uk

Topps Tiles is Britain's biggest tile and wood
flooring specialist, with over 200 stores
nationwide. For details of your nearest store
or free brochure call 0800 138 1673
www.toppstiles.co.uk.

TRADEMARK TILES
Airfield Industrial Estate, Warboys,
Huntingdon, Cambridgeshire, PE28 2SH
Area of Operation: East England, South East England
Tel: 01487 825300 **Fax:** 01487 825305
Email: info@trademarktiles.co.uk

WAXMAN CERAMICS LTD
Grove Mills, Elland, West Yorkshire, HX5 9DZ
Tel: 01422 311331 **Fax:** 01422 310654
Email: sales@waxmanceramics.co.uk
Web: www.waxmanceramics.co.uk

WILTON STUDIOS
Cleethorpes Business Centre, Wilton Road Industrial
Estate, Grimsby, Lincolnshire, DN36 4AS
Area of Operation: UK (Excluding Ireland)
Tel: 01472 210820 **Fax:** 01472 812602
Email: postbox@wiltonstudios.co.uk
Web: www.wiltonstudios.co.uk
Other Info: 🖉 🖑
Material Type: C) 1, 2, 3, 4, 12, 14, 15, 16

WORLD'S END TILES
Silverthorne Road, Battersea, SW8 3HE
Area of Operation: UK (Excluding Ireland)
Tel: 0207 819 2100
Fax: 0207 819 2101
Email: info@worldsendtiles.co.uk
Web: www.worldsendtiles.co.uk
Material Type: F) 1, 2, 3, 4, 5

STONE & SLATE FLOORING, INCLUDING MARBLE, GRANITE, TRAVERTINE & TERRAZZO

KEY

OTHER: ▽ Reclaimed 🖑 On-line shopping
🖉 Bespoke 🖑 Hand-made ECO Ecological

AFFORDABLE GRANITE LTD
Unit 5 Charlwood Place,
Norwood Hill Road, Surrey, RH6 0EB
Area of Operation: UK (Excluding Ireland)
Tel: 0845 330 1692
Fax: 0845 330 1260
Email: sales@affordablegranite.co.uk
Web: www.affordablegranite.co.uk
Material Type: E) 1

ALICANTE STONE
Damaso Navarro, 6 Bajo, 03610 Petrer (Alicante),
P.O. Box 372, Spain
Area of Operation: Europe
Tel: +34 966 31 96 97
Fax: +34 966 31 96 98
Email: info@alicantestone.com
Web: www.alicantestone.com
Material Type: E) 1, 2, 4, 5, 8

ALISTAIR MACKINTOSH LTD
Bannerley Road, Garretts Green,
Birmingham, West Midlands, B33 0SL
Area of Operation: UK & Ireland
Tel: 0121 784 6800
Fax: 0121 789 7068
Email: info@alistairmackintosh.co.uk
Web: www.alistairmackintosh.co.uk
Material Type: E) 1, 2, 5, 8

ALL THINGS STONE LIMITED
Anchor House, 33 Ospringe Street,
Ospringe, Faversham, Kent, ME13 8TW
Area of Operation: UK & Ireland
Tel: 01795 531001
Fax: 01795 530742
Email: info@allthingsstone.co.uk
Web: www.allthingsstone.co.uk
Material Type: E) 1, 2, 3, 5, 8

ASHLAR MASON
Manor Building, Manor Road, London, W13 0JB
Area of Operation: East England, Greater London,
Midlands & Mid Wales, South East England,
South West England and South Wales
Tel: 0208 997 0002
Fax: 0208 997 0008
Email: enquiries@ashlarmason.co.uk
Web: www.ashlarmason.co.uk
Material Type: E) 1, 2, 3, 4, 5, 8, 9, 11, 13

BANBURY BRICKS LTD
83a Yorke Street, Mansfield Woodhouse,
Nottinghamshire, NG19 9NH
Area of Operation: UK & Ireland
Tel: 0845 230 0941
Fax: 0845 230 0942
Email: enquires@banburybricks.co.uk
Web: www.banburybricks.co.uk
Material Type: E) 3

BLACK MOUNTAIN QUARRIES LTD
Howton Court, Pontrilas, Herefordshire, HR2 0BG
Area of Operation: UK & Ireland
Tel: 01981 241541
Email: info@blackmountainquarries.com
Web: www.blackmountainquarries.com 🖑

BLANC DE BIERGES
Eastrea Road, Whittlesey, Cambridgeshire, PE7 2AG
Area of Operation: Worldwide
Tel: 01733 202566
Fax: 01733 205405
Email: info@blancdebierges.com
Web: www.blancdebierges.com
Other Info: 🖉 🖑

BRETT LANDSCAPING LTD .
Salt Lane, Cliffe, Rochester, Kent, ME3 7SZ
Area of Operation: Worldwide
Tel: 01634 222188
Fax: 01634 222001
Email: landscapinginfo@brett.co.uk
Web: www.brett.co.uk/landscaping

BRITISH CERAMIC TILE
Heathfield, Newton Abbot, Devon, TQ12 6RF
Area of Operation: UK & Ireland
Tel: 01626 831480
Fax: 01626 831465
Email: sales@bctltd.co.uk
Web: www.candytiles.com
Material Type: F) 1, 2, 4

CAPITAL GROUP LIMITED
Victoria Mills, Highfield Road,
Little Hulton, Manchester, M38 9ST
Area of Operation: Worldwide
Tel: 0161 799 7555
Fax: 0161 799 7666
Email: leigh@choosecapital.co.uk
Web: www.choosecapital.co.uk 🖑

CAPITAL MARBLE DESIGN
Unit 1 Pall Mall Deposit, 124-128
Barlby Road, London, W10 6BL
Area of Operation: UK & Ireland
Tel: 0208 968 5340
Fax: 0208 968 8827
Email: stonegallery@capitalmarble.co.uk
Web: www.capitalmarble.co.uk
Other Info: ECO 🖉

CERAMIQUE INTERNATIONALE LTD
Unit 1 Royds Lane, Lower Wortley Ring Road,
Leeds, West Yorkshire, LS12 6DU
Area of Operation: UK & Ireland
Tel: 0113 231 0218
Fax: 0113 231 0353
Email: cameron@ceramiqueinternationale.co.uk
Web: www.ceramiqueinternationale.co.uk

CITY BATHROOMS & KITCHENS
158 Longford Road, Longford, Coventry,
West Midlands, CV6 6DR
Area of Operation: UK & Ireland
Tel: 02476 365877
Fax: 02476 644992
Email: citybathrooms@hotmail.com
Web: www.citybathrooms.co.uk

CLASSICAL FLAGSTONES LTD
Lower Ledge Farm, Dyrham, Wiltshire, FN14 8EY
Area of Operation: UK & Ireland
Tel: 0117 937 1960
Fax: 0117 303 9088
Email: info@classical-flagstones.com
Web: www.classical-flagstones.com
Material Type: E) 11

CONCEPT TILING LIMITED
Unit 3, Jones Court, Jones Square,
Stockport, Cheshire, SK1 4LJ
Area of Operation: UK & Ireland
Tel: 0161 480 0994
Fax: 0161 480 0911
Email: enquiries@concept-tiles.com
Web: www.concept-tiles.com
Material Type: E) 1, 2, 3, 4, 5, 8, 9, 11, 13

CORE AND ORE LTD
16 Portland Street,
Clifton, Bristol, BS8 4JH
Area of Operation: UK (Excluding Ireland)
Tel: 01179 042408
Fax: 01179 658057
Email: sales@coreandore.com
Web: www.coreandore.com

COUNTRY FLOORING
16 Julian Road, Orpington, Kent, BR6 6HU
Area of Operation: UK & Ireland
Tel: 01689 619044
Email: naturalstone2000@aol.com
Web: www.naturalstone2000.co.uk
Material Type: E) 2, 3, 4, 5, 8, 9

COUNTY GRANITE AND MARBLE
Mill Lane, Creech Saint Michael,
Taunton, Somerset, TA3 5PX
Area of Operation: UK (Excluding Ireland)
Tel: 01823 444554
Fax: 01823 445013
Email: granites@netcomuk.co.uk
Web: www.countygranite.co.uk

CTD SCOTLAND
72 Hydepark Street, Glasgow,
Dunbartonshire, G3 8BW
Area of Operation: Scotland
Tel: 0141 221 4591
Fax: 0141 221 8442
Email: info@ctdscotland.co.uk
Web: www.ctdscotland.co.uk

D F FIXINGS
15 Aldham Gardens, Rayleigh, Essex, SS6 9TB
Area of Operation: UK (Excluding Ireland)
Tel: 07956 674673
Fax: 01268 655072

DELABOLE SLATE COMPANY
Pengelly, Delabole, Cornwall, PL33 9AZ
Area of Operation: UK (Excluding Ireland)
Tel: 01840 212242
Fax: 01840 212948
Email: sales@delaboleslate.co.uk
Web: www.delaboleslate.co.uk

DEVON STONE LTD
38 Station Road, Budleigh Salterton,
Devon, EX9 6RT
Area of Operation: UK (Excluding Ireland)
Tel: 01395 446841
Email: amy@devonstone.com
Material Type: E) 1, 2, 3, 4, 5, 8

ELON
12 Silver Road, London, W12 7SG
Area of Operation: UK & Ireland
Tel: 0208 932 3000
Fax: 0208 932 3001
Email: marketing@elon.co.uk
Web: www.elon.co.uk
Material Type: E) 2, 3

EUROPEAN HERITAGE
48-54 Dawes Road, Fulham, London, SW6 7EN
Area of Operation: Worldwide
Tel: 0207 381 6063
Fax: 0207 381 9534
Email: fulham@europeanheritage.co.uk
Web: www.europeanheritage.co.uk
Material Type: E) 1, 2, 3, 4, 5, 7, 8

EXAKT PRECISION TOOLS LTD
Unit 9 Urquhart Trade Centre,
109 Urquhart Road, Aberdeen,
Aberdeenshire, AB24 5NH
Area of Operation: Europe
Tel: 01224 643434
Fax: 01224 643232
Email: info@exaktpt.com
Web: www.exaktpt.com

FIRED EARTH INTERIORS
3 Twyford Mill, Oxford Road, Adderbury,
Banbury, Oxfordshire, OX17 3SX
Area of Operation: Worldwide
Tel: 01295 814300
Fax: 01295 810832
Email: enquiries@firedearth.com.
Web: www.firedearth.com

FLOORTECK LTD
8 Kingsdown Road, Northfield,
Birmingham, West Midlands, B31 1AH
Area of Operation: UK (Excluding Ireland)
Tel: 0121 476 6271
Fax: 0121 476 6271
Email: sarahcollier@go.com

GREYSLATE & STONE SUPPLIES
Y Maes, The Square,
Blaenau Ffestiniog, Gwynedd, LL41 3UN
Area of Operation: UK & Ireland
Tel: 01766 830521
Fax: 01766 831706
Email: greyslate@slateandstone.net
Web: www.slateandstone.net
Material Type: E) 3

HARD ROCK FLOORING
Fleet Marston Farm,
Fleet Marston, Aylesbury,
Buckinghamshire, HP18 0PZ
Area of Operation: Europe
Tel: 01296 658755
Fax: 01296 655735
Email: showroom@hardrockflooring.co.uk
Web: www.hardrockflooring.co.uk
Material Type: E) 2, 3, 4, 5, 8, 9

HERITAGE FLAGSTONES
The Sanctuary, Newchurch, Hoar Cross,
Burton upon Trent, Derbyshire, DE13 8RQ
Area of Operation: UK (Excluding Ireland)
Tel: 01283 575822
Fax: 01283 575180
Email: info@heritageflagstones.co.uk
Web: www.heritageflagstones.co.uk

HYPERION TILES
67 High Street, Ascot, Berkshire, SL5 7HP
Area of Operation: Greater London,
South East England
Tel: 01344 620211
Fax: 01344 620100
Email: graham@hyperiontiles.com
Web: www.hyperiontiles.com

INDIGENOUS LTD
Cheltenham Road, Burford,
Oxfordshire, OX18 4JA
Area of Operation: Worldwide
Tel: 01993 824200
Fax: 01993 824300
Email: enquiries@indigenoustiles.com
Web: www.indigenoustiles.com
Other Info: ✍ ✋
Material Type: A) 2, 4, 6

J & R MARBLE COMPANY LTD
Unit 9, Period Works, Lammas Road,
Leyton, London, E10 7QT
Area of Operation: UK (Excluding Ireland)
Tel: 0208 539 6471
Fax: 0208 539 9264
Email: sales@jrmarble.co.uk
Web: www.jrmarble.co.uk
Other Info: ✍
Material Type: E) 1, 2, 3, 4, 5, 8, 9, 13, 14

KEYSTONE NATURAL STONE FLOORING
204 Duggins Lane, Tile Hill,
Coventry, West Midlands, CV4 9GP
Area of Operation: Worldwide
Tel: 02476 422580
Fax: 02476 695794

KIRK NATURAL STONE
Bridgend, Fyvie, Turriff,
Aberdeenshire, AB53 8QB
Area of Operation: Worldwide
Tel: 01651 891891
Fax: 01651 891794
Email: info@kirknaturalstone.com
Web: www.kirknaturalstone.com
Material Type: E) 1, 2, 3, 4, 5

KIRKSTONE QUARRIES LIMITED
Skelwith Bridge, Ambleside,
Cumbria, LA22 9NN
Area of Operation: UK & Ireland
Tel: 01539 433296
Fax: 01539 434006
Email: info@kirkstone.com
Web: www.kirkstone.com
Material Type: E) 1, 2, 3, 5, 8, 9

LAPICIDA
Killinghall Stone Quarry, Ripon Road, Killinghall,
Harrogate, North Yorkshire, HG3 2BA
Area of Operation: UK (Excluding Ireland)
Tel: 01423 560262
Fax: 01423 529517
Email: sales@lapicida.co.uk
Web: www.lapicida.co.uk

MAESTROTILE
Unit 7 Waterloo Park, Waterloo Road Industrial
Estate, Bidford on Avon, Warwickshire, B50 4JH
Area of Operation: UK & Ireland
Tel: 01789 778700 **Fax:** 01789 778852
Email: malcolm@maestrotile.com
Web: www.maestrotile.com

MANDARIN
Unit 1 Wonastow Road Industrial Estate,
Monmouth, Monmouthshire, NP25 5JB
Area of Operation: Europe
Tel: 01600 715444 **Fax:** 01600 715494
Email: info@mandarinstone.com
Web: www.mandarinstone.com

MARBLE CLASSICS
Unit 3, Station Approach,
Emsworth, Hampshire, PO10 7PW
Area of Operation: UK & Ireland
Tel: 01243 370011 **Fax:** 01243 371023
Email: info@marbleclassics.co.uk
Web: www.marbleclassics.co.uk
Material Type: E) 1, 2, 3, 5, 8, 13

MARIA STARLING MOSAICS
40 Strand House, Merbury Close,
Greater London, SE28 0LU
Area of Operation: UK (Excluding Ireland)
Tel: 07775 517409
Email: mosaics@mariastarling.com
Web: www.mariastarling.com

MARLBOROUGH TILES LTD.
Elcot Lane, Marlborough, Wiltshire, SN8 2AY
Area of Operation: UK & Ireland
Tel: 01672 512422
Fax: 01672 515791
Email: admin@marlboroughtiles.com
Web: www.marlboroughtiles.com

MGLW
44 Linford Street, London, SW8 4UN
Area of Operation: Greater London
Tel: 0207 720 9944
Email: contact@naturalstonefloor.com
Web: www.naturalstonefloor.com
Material Type: E) 1, 2, 3, 4, 5, 6, 7, 8, 9, 10, 11, 12, 13, 14

MIDLANDS SLATE & TILE
Qualcast Road, Horseley Fields,
Wolverhampton, West Midlands, WV1 2QP
Area of Operation: UK & Ireland
Tel: 01902 458780 **Fax:** 01902 458549
Email: sales@slate-tile-brick.co.uk
Web: www.slate-tile-brick.co.uk
Material Type: E) 3, 4, 5, 8

INTERIORS, FIXTURES & FINISHES

INTERIORS, FIXTURES & FINISHES - **Floor & Wall Finishes** - Stone & Slate Flooring including Marble, Granite, Travertine & Terrazzo

SPONSORED BY: PERGO LTD www.pergo.com

THE TIMELESS MASTERPIECE

KeyStone

NATURAL STONE FLOORING & PAVING

204 DUGGINS LANE
TILE HILL
COVENTRY CV4 9GP

TEL: 024 7642 2580
FAX: 024 7669 5794

- Guaranteed personal service
- 100's of stones for flooring, paving & cladding
- Samples available on request
- Each order is hand-picked and delivered to your door
- Over 10 years experience successfully supplying natural stone to customers nationwide
- Full colour brochure or CD available on request

N&C NICOBOND
41-51 Freshwater Road, Chadwell Heath,
Romford, Essex, RM8 1SP
Area of Operation: Worldwide
Tel: 0208 586 4600
Fax: 0208 586 4646
Email: info@nichollsandclarke.com
Web: www.ncdirect.co.uk

**NATURAL IMAGE
(GRANITE, MARBLE, STONE)**
Spelmonden Estate, Goudhurst Road,
Goudhurst, Kent, TN17 1HE
Area of Operation: UK & Ireland
Tel: 01580 212222
Fax: 01580 211841
Web: www.naturalstonefloors.org

NICAN STONE LTD
Bank House, School Lane,
Bronington, Shropshire, SY13 3HN
Area of Operation: UK (Excluding Ireland)
Tel: 01948 780670
Fax: 01948 780679
Email: enquiries@nicanstone.com
Web: www.nicanstone.com

ORIGINAL OAK
Ashlands, Burwash, East Sussex, TN19 7HS
Area of Operation: UK (Excluding Ireland)
Tel: 01435 882228
Fax: 01435 882228
Other Info: ▽ ✋
Material Type: E) 1, 2

ORIGINAL STYLE
Falcon Road, Sowton Industrial Estate,
Exeter, Devon, EX2 7LF
Area of Operation: Worldwide
Tel: 01392 474011
Fax: 01392 219932
Email: vmattison@originalstyle.com
Web: www.originalstyle.com
Material Type: E) 2, 3, 5, 8, 11

PILKINGTON'S TILES GROUP
PO Box 4, Clifton Junction,
Manchester, M27 8LP
Area of Operation: UK (Excluding Ireland)
Tel: 0161 727 1000
Fax: 0161 727 1122
Email: sales@pilkingtons.com
Web: www.pilkingtons.com
Material Type: G) 1, 5

RANSFORDS
Drayton Way, Drayton Fields
Industrial Estate, Daventry,
Northamptonshire, NN11 5XW
Area of Operation: Worldwide
Tel: 01327 705310
Fax: 01327 706831
Email: sales@ransfords.com
Web: www.ransfords.com /
www.stoneflooringandpaving.com

RAPID GRANITE LIMITED
Tollgate Farm, Golford Road,
Cranbrook, Kent, TN17 3NX
Area of Operation: UK (Excluding Ireland)
Tel: 01580 720900
Fax: 01580 720901
Email: info@rapidgranite.co.uk
Web: www.rapidgranite.co.uk ✋
Material Type: E) 1, 2, 7, 8

**RED ROSE PLASTICS
(BURNLEY) LTD**
Parliament Street, Burnley,
Lancashire, BB11 3JT
Area of Operation: East England,
North East England, North West England
and North Wales
Tel: 01282 724600
Fax: 01282 724644
Email: info@redroseplastics.co.uk
Web: www.redroseplastics.co.uk

RIVERSTONE LTD
301 Elveden Road, Park Royal,
London, NW10 7SS
Area of Operation: Worldwide
Tel: 0208 961 7725 **Fax:** 0208 965 7013
Email: alain@ssq.co.uk
Web: www.ssq.co.uk
Other Info: ✎
Material Type: E) 3

ROCK UNIQUE LTD
c/o Select Garden and Pet Centre,
Main Road, Sundridge, Kent, TN14 6ED
Area of Operation: Europe
Tel: 01959 565608 **Fax:** 01959 569312
Email: stone@rock-unique.com
Web: www.rock-unique.com

RUSTICA LTD
154c Milton Park, Oxfordshire, OX14 4SD
Area of Operation: UK & Ireland
Tel: 01235 834192 **Fax:** 01235 835162
Email: sales@rustica.co.uk
Web: www.rustica.co.uk
Material Type: E) 2, 3, 4, 5, 8, 9

SIMALI STONE FLOORING
40E Wincombe Business Park,
Shaftesbury, Dorset, SP7 9QJ
Area of Operation: Europe
Tel: 01747 852557 **Fax:** 01747 852557
Email: stonefloors@aol.com
Web: www.stoneflooringonline.com
Material Type: E) 1, 2, 3, 4, 5, 6, 7, 8, 9

SLATE WORLD LTD
Westmoreland Road, Kingsbury,
Greater London, NW9 9RN
Area of Operation: Europe
Tel: 0208 204 3444
Fax: 0208 204 3311
Email: advertising@americanslate.com
Web: www.slateworld.com
Material Type: E) 3, 4, 5, 9

SLATE WORLD LTD

Area of Operation: Europe
Tel: 0208 2043444
Fax: 0208 204 3311
Email: advertising@americanslate.com
Web: www.slateworld.com
Material Type: E) 3

Suppliers of the most comprehensive range of
QUARRY-DIRECT, non-slip and low maintenance
natural slate flooring tiles in the UK.

SOUTH WESTERN FLOORING SERVICES
145-147 Park Lane, Frampton Cotterell,
Bristol, BS36 2ES
Area of Operation: UK & Ireland
Tel: 01454 880982 **Fax:** 01454 880982
Email: mikeflanders@blueyonder.co.uk
Web: www.southwesternflooring.com
Other Info: ▽ **Material Type:** E) 1, 2, 3, 4, 5, 7, 8

STONE AND SLATE
Coney Green Farm, Lower Market Street,
Claycross, Chesterfield, Derbyshire, S45 9NE
Area of Operation: UK & Ireland
Tel: 01246 250088 **Fax:** 01246 250099
Email: sales@stoneandslate.ltd.uk
Web: www.stoneandslate.co.uk

INTERIORS, FIXTURES & FINISHES

INTERIORS, FIXTURES & FINISHES - Floor & Wall Finishes - Stone & Slate Flooring including Marble, Granite, Travertine & Terrazzo

SPONSORED BY: PERGO LTD www.pergo.com

STONEHOUSE TILES (LONDON) LTD
Unit 33 Enterprise Industrial Estate,
Bolima Road, London, SE16 3LF
Area of Operation: UK (Excluding Ireland)
Tel: 0207 237 5375
Fax: 0207 231 7597
Email: info@stonehousetiles.co.uk
Web: www.stonehousetiles.co.uk

STONELL DIRECT
100-105 Victoria Crescent, Burton upon Trent,
Staffordshire, DE14 2QF
Area of Operation: Worldwide
Tel: 08000 832283
Fax: 01283 501098
Email: liz.may@stonelldirect.com
Web: www.stonelldirect.com

STONEVILLE (UK) LTD
Unit 12, Set Star Estate, Transport Avenue, Great
West Road, Brentford, Greater London, TW8 9HF
Area of Operation: Europe
Tel: 0208 560 1000
Fax: 0208 560 4060
Email: info@stoneville.co.uk
Web: www.stoneville.co.uk
Material Type: E) 1, 2, 3, 5, 8

STONEWAYS LTD
Railside Works, Marlbrook,
Leominster, Herefordshire, HR6 0PH
Area of Operation: UK (Excluding Ireland)
Tel: 01568 616818 **Fax:** 01568 620085
Email: stoneways@msn.com
Web: www.stoneways.co.uk

STRATHEARN STONE AND TIMBER LTD
Glenearn, Bridge of Earn, Perth,
Perth and Kinross, PH2 9HL
Area of Operation: North East England, Scotland
Tel: 01738 813215 **Fax:** 01738 815946
Email: info@stoneandoak.com
Web: www.stoneandoak.com
Material Type: E) 3, 5

STUDIO STONE
The Stone Yard, Alton Lane, Four Marks,
Hampshire, GU34 5AJ
Area of Operation: UK (Excluding Ireland)
Tel: 01420 562500
Fax: 01420 563192
Email: sales@studiostone.co.uk
Web: www.studiostone.co.uk
Other Info: ✎
Material Type: E) 1, 2, 3, 4, 5, 8

THE GRANITE FACTORY
4 Winchester Drive, Peterlee,
Durham, SR8 2RJ
Area of Operation: North East England
Tel: 0191 518 3600 **Fax:** 0191 518 3600
Email: admin@granitefactory.co.uk
Web: www.granitefactory.co.uk ✁

THE HOME OF STONE LTD
Boot Barn, Newcastle, Monmouth,
Monmouthshire, NP25 5NU
Area of Operation: South West England
and South Wales
Tel: 01600 750462 **Fax:** 01600 750462

THE NATURAL FLOORING WAREHOUSE
East Building, Former All Saints Church,
Armoury Way, Wandsworth, London, SW18 1HZ
Area of Operation: UK (Excluding Ireland)
Tel: 0208 870 5555 **Fax:** 0208 877 2847
Email: natalie@edwardianfires.com

**THE NATURAL SLATE COMPANY LTD -
DERBYSHIRE**
Unit 2 Armytage Estate,
Station Road, Whittington Moor,
Chesterfield, Derbyshire, S41 9ET
Area of Operation: Worldwide
Tel: 0845 177 5008 **Fax:** 0870 429 9891
Email: sales@theslatecompany.net
Web: www.theslatecompany.net
Other Info: ECO ✋
Material Type: E) 3, 8

**THE NATURAL SLATE
COMPANY LTD - LONDON**
161 Ballards Lane, Finchley, London, N3 1LJ
Area of Operation: Worldwide
Tel: 0845 177 5008 **Fax:** 0870 429 9891
Email: sales@theslatecompany.net
Web: www.theslatecompany.net
Other Info: ECO ✋
Material Type: E) 3, 8

THE ORIGINAL TILE COMPANY
23A Howe Street, Lothian, EH3 6TF
Area of Operation: UK & Ireland
Tel: 0131 556 2013 **Fax:** 0131 558 3172
Email: info@originaltile.freeserve.co.uk
Other Info: ✋

**THE REALLY SAFE FLOORING
COMPANY LIMITED**
Unit 29 Wivenhoe Business Centre,
Book Street, Colchester, Essex, CO7 9DP
Area of Operation: UK & Ireland
Tel: 01206 827870 **Fax:** 01206 827881
Email: sales@realsafe.co.uk
Web: www.realsafe.co.uk

TILECO
3 Molesey Business Centre,
Central Avenue, West Molesey,
Surrey, KT8 2QZ
Area of Operation: UK (Excluding Ireland)
Tel: 0208 481 9500
Fax: 0208 481 9501
Email: info@tileco.co.uk
Web: www.tileco.co.uk

TILES UK LTD
1-13 Montford Street,
Off South Langworthy Road,
Salford, Greater Manchester, M50 2XD
Area of Operation: UK (Excluding Ireland)
Tel: 0161 872 5155 **Fax:** 0161 848 7948
Email: info@tilesuk.com
Web: www.tilesuk.com ✁

TOPPS TILES
Thorpe Way, Grove Park, Enderby,
Leicestershire, LE19 1SU
Area of Operation: UK (Excluding Ireland)
Tel: 0116 282 8000 **Fax:** 0116 282 8100
Email: mlever@toppstiles.co.uk
Web: www.toppstiles.co.uk

TOPPS TILES

Area of Operation: UK (Excluding Ireland)
Tel: 0116 282 8000 **Fax:** 0116 282 8100
Email: mlever@toppstiles.co.uk
Web: www.toppstiles.co.uk

Topps Tiles is Britain's biggest tile and wood
flooring specialist, with over 200 stores
nationwide. For details of your nearest store
or free brochure call 0800 138 1673
www.toppstiles.co.uk.

TRADEMARK TILES
Airfield Industrial Estate, Warboys,
Huntingdon, Cambridgeshire, PE28 2SH
Area of Operation: East England,
South East England
Tel: 01487 825300 **Fax:** 01487 825305
Email: info@trademarktiles.co.uk

INTERIORS, FIXTURES & FINISHES

WALLS AND FLOORS LTD
Wilson Terrace, Kettering, Northants, NN19 9RT
Area of Operation: UK (Excluding Ireland)
Tel: 01536 410484
Email: sales@wallsandfloors.uk.com
Web: www.wallsandfloors.uk.com

WELSH SLATE
Business Design Centre, Unit 205, 52
Upper Street, London, N1 0QH
Area of Operation: Worldwide
Tel: 0207 354 0306
Fax: 0207 354 8485
Email: enquiries@welshslate.com
Web: www.welshslate.com
Material Type: E) 3

WILTON STUDIOS
Cleethorpes Business Centre,
Wilton Road Industrial Estate,
Grimsby, Lincolnshire, DN36 4AS
Area of Operation: UK (Excluding Ireland)
Tel: 01472 210820 **Fax:** 01472 812602
Email: postbox@wiltonstudios.co.uk
Web: www.wiltonstudios.co.uk
Other Info: ✋

ZARKA MARBLE
41A Belsize Lane, Hampstead, London, NW3 5AU
Area of Operation: South East England
Tel: 0207 431 3042 **Fax:** 0207 431 3879
Email: enquiries@zarkamarble.co.uk
Web: www.zarkamarble.co.uk
Material Type: E) 1, 2, 5

CLAY & QUARRY TILES, INCLUDING TERRACOTTA

KEY

OTHER: ▽ Reclaimed ✓ On-line shopping
✎ Bespoke ✋ Hand-made ECO Ecological

ALDERSHAW HANDMADE CLAY TILES LTD
Pokehold Wood, Kent Street, Sedlescombe,
Nr Battle, East Sussex, TN33 0SD
Area of Operation: Europe
Tel: 01424 756777
Fax: 01424 756888
Email: tiles@aldershaw.co.uk
Web: www.aldershaw.co.uk
Other Info: ✎ ✋
Material Type: F) 1, 3, 4, 5

ASHLAR MASON
Manor Building, Manor Road, London, W13 0JB
Area of Operation: East England, Greater London,
Midlands & Mid Wales, South East England,
South West England and South Wales
Tel: 0208 997 0002 **Fax:** 0208 997 0008
Email: enquiries@ashlarmason.co.uk
Web: www.ashlarmason.co.uk

BLACK MOUNTAIN QUARRIES LTD
Howton Court, Pontrilas, Herefordshire, HR2 0BG
Area of Operation: UK & Ireland
Tel: 01981 241541
Email: info@blackmountainquarries.com
Web: www.blackmountainquarries.com ✓

BODJ FAIR TRADE FLOORING
Bishops Way, Newport, Isle of Wight, PO30 5WS
Area of Operation: Europe
Tel: 01983 537760 **Fax:** 01983 537788
Email: bodj@ajwells.co.uk
Web: www.bodj.co.uk
Other Info: ECO ✋
Material Type: F) 3

CAPITAL MARBLE DESIGN
Unit 1 Pall Mall Deposit,
124-128 Barlby Road, London, W10 6BL
Area of Operation: UK & Ireland
Tel: 0208 968 5340 **Fax:** 0208 968 8827
Email: stonegallery@capitalmarble.co.uk
Web: www.capitalmarble.co.uk

CERAMIQUE INTERNATIONALE LTD
Unit 1 Royds Lane,
Lower Wortley Ring Road,
Leeds, West Yorkshire, LS12 6DU
Area of Operation: UK & Ireland
Tel: 0113 231 0218 **Fax:** 0113 231 0353
Email: cameron@ceramiqueinternationale.co.uk
Web: www.ceramiqueinternationale.co.uk

CITY BATHROOMS & KITCHENS
158 Longford Road, Longford,
Coventry, West Midlands, CV6 6DR
Area of Operation: UK & Ireland
Tel: 02476 365877
Fax: 02476 644992
Email: citybathrooms@hotmail.com
Web: www.citybathrooms.co.uk

CRAVEN DUNNILL JACKFIELD LTD
Jackfield Tile Museum,
Ironbridge Gorge, Telford,
Shropshire, TF8 7LJ
Area of Operation: Worldwide
Tel: 01952 884124
Fax: 01952 884487
Email: gemma@cravendunnill-jackfield.co.uk
Web: www.cravendunnill-jackfield.co.uk

DAR INTERIORS
Arch 11, Miles Street, London, SW8 1RZ
Area of Operation: Worldwide
Tel: 0207 720 9678
Fax: 0207 627 5129
Email: enquiries@darinteriors.com
Web: www.darinteriors.com

DEVON STONE LTD
38 Station Road, Budleigh Salterton,
Devon, EX9 6RT
Area of Operation: UK (Excluding Ireland)
Tel: 01395 446841
Email: amy@devonstone.com
Material Type: F) 3

ELON
12 Silver Road, London, W12 7SG
Area of Operation: UK & Ireland
Tel: 0208 932 3000
Fax: 0208 932 3001
Email: marketing@elon.co.uk
Web: www.elon.co.uk ✓
Material Type: F) 3

EUROPEAN HERITAGE
48-54 Dawes Road, Fulham, London, SW6 7EN
Area of Operation: Worldwide
Tel: 0207 381 6063
Fax: 0207 381 9534
Email: fulham@europeanheritage.co.uk
Web: www.europeanheritage.co.uk
Other Info: ✋
Material Type: F) 1, 2, 3

FIRED EARTH INTERIORS
3 Twyford Mill, Oxford Road, Adderbury,
Banbury, Oxfordshire, OX17 3SX
Area of Operation: Worldwide
Tel: 01295 814300
Fax: 01295 810832
Email: enquiries@firedearth.com.
Web: www.firedearth.com

FLOORTECK LTD
8 Kingsdown Road, Northfield,
Birmingham, West Midlands, B31 1AH
Area of Operation: UK (Excluding Ireland)
Tel: 0121 476 6271
Fax: 0121 476 6271
Email: sarahcollier@go.com

FRANCIS N. LOWE LTD.
The Marble Works,
New Road, Middleton,
Matlock, Derbyshire, DE4 4NA
Area of Operation: Europe
Tel: 01629 822216
Fax: 01629 824348
Email: info@lowesmarble.com
Web: www.lowesmarble.com

H & R JOHNSON TILES LTD
Harewood Street, Tunstall, Stoke on Trent,
Staffordshire, ST6 5JZ
Area of Operation: Worldwide
Tel: 01782 575 575
Fax: 01782 524138
Email: techsales@johnson-tiles.com
Web: www.johnson-tiles.com

HARD ROCK FLOORING
Fleet Marston Farm, Fleet Marston,
Aylesbury, Buckinghamshire, HP18 0PZ
Area of Operation: Europe
Tel: 01296 658755 **Fax:** 01296 655735
Email: showroom@hardrockflooring.co.uk
Web: www.hardrockflooring.co.uk

HERITAGE TILE CONSERVATION LTD
The Studio, 2 Harris Green, Broseley,
Shropshire, TF12 5HJ
Area of Operation: UK & Ireland
Tel: 01952 881039 **Fax:** 01952 881039
Email: enquiries@heritagetile.co.uk
Web: www.heritagetile.co.uk

HYPERION TILES
67 High Street, Ascot, Berkshire, SL5 7HP
Area of Operation: Greater London,
South East England
Tel: 01344 620211 **Fax:** 01344 620100
Email: graham@hyperiontiles.com
Web: www.hyperiontiles.com

INDIGENOUS LTD
Cheltenham Road, Burford, Oxfordshire, OX18 4JA
Area of Operation: Worldwide
Tel: 01993 824200 **Fax:** 01993 824300
Email: enquiries@indigenoustiles.com
Web: www.indigenoustiles.com
Other Info: ✎ ✋

LAWLEY CERAMICS
8 Stourbridge Road, Bromsgrove,
Worcestershire, B61 0AB
Area of Operation: UK (Excluding Ireland)
Tel: 01527 570455 **Fax:** 01527 570455
Email: lawleyceramics@hotmail.com

MARLBOROUGH TILES LTD.
Elcot Lane, Marlborough, Wiltshire, SN8 2AY
Area of Operation: UK & Ireland
Tel: 01672 512422 **Fax:** 01672 515791
Email: admin@marlboroughtiles.com
Web: www.marlboroughtiles.com

MIDLANDS SLATE & TILE
Qualcast Road, Horseley Fields,
Wolverhampton, West Midlands, WV1 2QP
Area of Operation: UK & Ireland
Tel: 01902 458780 **Fax:** 01902 458549
Email: sales@slate-tile-brick.co.uk
Web: www.slate-tile-brick.co.uk
Material Type: F) 3

NATURAL FLOORING
152-154 Wellingborough Road,
Northampton, Northamptonshire, NN1 4DT
Area of Operation: UK (Excluding Ireland)
Tel: 01604 239238/7 **Fax:** 01604 627799
Web: www.naturalflooring.co.uk

ORIGINAL STYLE
Falcon Road, Sowton Industrial Estate,
Exeter, Devon, EX2 7LF
Area of Operation: Worldwide
Tel: 01392 474011 **Fax:** 01392 219932
Email: vmattison@originalstyle.com
Web: www.originalstyle.com
Material Type: F) 3

RUSTICA LTD
154c Milton Park,
Oxfordshire, OX14 4SD
Area of Operation: UK & Ireland
Tel: 01235 834192 **Fax:** 01235 835162
Email: sales@rustica.co.uk
Web: www.rustica.co.uk
Other Info: ▽ ✋
Material Type: F) 3

SIMALI STONE FLOORING
40E Wincombe Business Park,
Shaftesbury, Dorset, SP7 9QJ
Area of Operation: Europe
Tel: 01747 852557 **Fax:** 01747 852557
Email: stonefloors@aol.com
Web: www.stoneflooringonline.com

SOUTH WESTERN FLOORING SERVICES
145-147 Park Lane, Frampton Cotterell,
Bristol, BS36 2ES
Area of Operation: UK & Ireland
Tel: 01454 880982 **Fax:** 01454 880982
Email: mikeflanders@blueyonder.co.uk
Web: www.southwesternflooring.com
Material Type: E) 1, 2, 3, 4, 5, 7, 8

STONE AND SLATE
Coney Green Farm, Lower Market Street,
Claycross, Chesterfield, Derbyshire, S45 9NE
Area of Operation: UK & Ireland
Tel: 01246 250088 **Fax:** 01246 250099
Email: sales@stoneandslate.ltd.uk
Web: www.stoneandslate.ltd.uk

THE ORIGINAL TILE COMPANY
23A Howe Street, Lothian, EH3 6TF
Area of Operation: UK & Ireland
Tel: 0131 556 2013
Fax: 0131 558 3172
Email: info@originaltile.freeserve.co.uk
Material Type: F) 3, 5

THE REALLY SAFE FLOORING COMPANY LIMITED
Unit 29 Wivenhoe Business Centre,
Book Street, Colchester, Essex, CO7 9DP
Area of Operation: UK & Ireland
Tel: 01206 827870
Fax: 01206 827881
Email: sales@realsafe.co.uk
Web: www.realsafe.co.uk

TILES UK LTD
1-13 Montford Street,
Off South Langworthy Road,
Salford, Greater Manchester, M50 2XD
Area of Operation: UK (Excluding Ireland)
Tel: 0161 872 5155
Fax: 0161 848 7948
Email: info@tilesuk.com
Web: www.tilesuk.com ✓

TOPPS TILES
Thorpe Way, Grove Park, Enderby,
Leicestershire, LE19 1SU
Area of Operation: UK (Excluding Ireland)
Tel: 0116 282 8000 **Fax:** 0116 282 8100
Email: mlever@toppstiles.co.uk
Web: www.toppstiles.co.uk

TOPPS TILES

Area of Operation: UK (Excluding Ireland)
Tel: 0116 282 8000 **Fax:** 0116 282 8100
Email: mlever@toppstiles.co.uk
Web: www.toppstiles.co.uk

Topps Tiles is Britain's biggest tile and wood flooring specialist, with over 200 stores nationwide. For details of your nearest store or free brochure call 0800 138 1673 www.toppstiles.co.uk.

TRADEMARK TILES
Airfield Industrial Estate, Warboys,
Huntingdon, Cambridgeshire, PE28 2SH
Area of Operation: East England,
South East England
Tel: 01487 825300
Fax: 01487 825305
Email: info@trademarktiles.co.uk

VICTORIAN WOOD WORKS LTD
54 River Road, Creekmouth,
Barking, Essex, IG11 0DW
Area of Operation: Worldwide
Tel: 0208 534 1000
Fax: 0208 534 2000
Email: sales@victorianwoodworks.co.uk
Web: www.victorianwoodworks.co.uk
Other Info: ▽

WALLS AND FLOORS LTD
Wilson Terrace, Kettering,
Northamptonshire, NN19 9RT
Area of Operation: UK (Excluding Ireland)
Tel: 01536 410484
Email: sales@wallsandfloors.uk.com
Web: www.wallsandfloors.uk.com ⌐

WAXMAN CERAMICS LTD
Grove Mills, Elland, West Yorkshire, HX5 9DZ
Tel: 01422 311331
Fax: 01422 310654
Email: sales@waxmanceramics.co.uk
Web: www.waxmanceramics.co.uk

WILTON STUDIOS
Cleethorpes Business Centre,
Wilton Road Industrial Estate,
Grimsby, Lincolnshire, DN36 4AS
Area of Operation: UK (Excluding Ireland)
Tel: 01472 210820
Fax: 01472 812602
Email: postbox@wiltonstudios.co.uk
Web: www.wiltonstudios.co.uk

WORLD'S END TILES
Silverthorne Road, Battesea, London, SW8 3HE
Area of Operation: UK (Excluding Ireland)
Tel: 0207 819 2100
Fax: 0207 819 2101
Email: info@worldsendtiles.co.uk
Web: www.worldsendtiles.co.uk
Material Type: E) 1, 2, 3, 5

YORK HANDMADE BRICK CO LTD
Forest Lane, Alne, York, North Yorkshire, YO61 1TU
Area of Operation: Worldwide
Tel: 01347 838881
Fax: 01347 838885
Email: sales@yorkhandmade.co.uk
Web: www.yorkhandmade.co.uk
Material Type: F) 3

CONCRETE FLOORING

KEY
OTHER: ▽ Reclaimed   ⌐ On-line shopping
✎ Bespoke   ✋ Hand-made   ECO Ecological

DEGUSSA CONSTRUCTION CHEMICALS (UK)
Albany House, Swinton Hall Road,
Swinton, Manchester, M27 4DT
Area of Operation: Worldwide
Tel: 0161 794 7411
Fax: 0161 727 8547
Email: mbtfeb@degussa.com
Web: www.degussa-cc.co.uk

FLOORTECK LTD
8 Kingsdown Road, Northfield,
Birmingham, West Midlands, B31 1AH
Area of Operation: UK (Excluding Ireland)
Tel: 0121 476 6271 **Fax:** 0121 476 6271
Email: sarahcollier@go.com

INSTARMAC GROUP PLC
Kingsbury Link, Trinity Road,
Tamworth, Staffordshire, B78 2EX
Area of Operation: Europe
Tel: 01827 872244 **Fax:** 01827 874466
Email: email@instarmac.co.uk
Web: www.instarmac.co.uk

NATURAL FLOORING
152-154 Wellingborough Road,
Northampton, Northamptonshire, NN1 4DT
Area of Operation: UK (Excluding Ireland)
Tel: 01604 239238/7 **Fax:** 01604 627799
Web: www.naturalflooring.co.uk

PERMANENT FLOORING LTD
Britannia House, High St,
Bagillt, Flintshire, CH6 6HE
Area of Operation: UK & Ireland
Tel: 01352 714869 **Fax:** 01352 713666
Email: info@permanentflooring.com
Web: www.permanentflooringltd.com

SPANCAST CONCRETE FLOORS
Stephenson Way, Barrington Industrial Estate,
Bedlington, Northumberland, NE22 7DQ
Area of Operation: East England, North East
England, North West England and North Wales,
Scotland
Tel: 01670 531160 **Fax:** 01670 531170
Email: sales@spancast.co.uk
Web: www.spancast.co.uk

LINOLEUM, VINYL & RUBBER FLOORING

KEY
OTHER: ▽ Reclaimed   ⌐ On-line shopping
✎ Bespoke   ✋ Hand-made   ECO Ecological

AARDVARK WHOLESALE LTD
PO Box 3733, Dronfield, Derbyshire, S18 9AD
Area of Operation: UK (Excluding Ireland)
Tel: 0800 279 0486
Email: rawwoodfloors@zicom.net

AMTICO INTERNATIONAL
Solar Park, Southside, Solihull,
West Midlands, B90 4SH
Area of Operation: Worldwide
Tel: 0121 745 0800
Fax: 0121 745 0888
Email: philip.lusher@amtico.com
Web: www.amtico.com
Other Info: ECO ✎ ✋

CSM CARPETS & FLOORING LTD
Brickmakers Arms Lane, Doddington,
Cambridgeshire, PE15 0TR
Area of Operation: East England
Tel: 01354 740727
Fax: 01354 740078
Email: alan.csm@virgin.net
Web: www.csm-flooring.co.uk

DALSOUPLE
Unit 1 Showground Road,
Bridgwater, Somerset, TA6 6AJ
Area of Operation: Worldwide
Tel: 01278 727733
Fax: 01278 727766
Email: info@dalsouple.com
Web: www.dalsouple.com
Other Info: ECO ✎

FIRST FLOOR(FULHAM) LTD
174 Wandsworth Bridge Road,
Fulham, London, SW6 2UQ
Area of Operation: UK (Excluding Ireland)
Tel: 0207 736 1123
Fax: 0207 371 9812
Email: annie@firstfloor.uk.com
Web: www.firstfloor.uk.com

INTERIORS, FIXTURES & FINISHES

FLOORBRAND
Cavendish Business Centre, High Street Green,
Sible Hedingham, Essex, CO9 3LH
Area of Operation: Europe
Tel: 0800 612 0101
Fax: 0845 166 8002
Email: sales@floorbrand.com
Web: www.floorbrand.com

FLOORSDIRECT
Unit 6, Fountain Drive, Hertford,
Hertfordshire, SG13 7UB
Area of Operation: UK (Excluding Ireland)
Tel: 01992 552447
Fax: 01992 558760
Email: jenny@floorsdirect.co.uk
Web: www.floorsdirect.co.uk
Material Type: D) 1, 4

FLOORTECK LTD
8 Kingsdown Road,
Northfield, Birmingham,
West Midlands, B31 1AH
Area of Operation: UK (Excluding Ireland)
Tel: 0121 476 6271
Fax: 0121 476 6271
Email: sarahcollier@go.com

FORBO FLOORING
P O Box 1, Den Road,
Kirkclady, Fife, KY1 2SB
Area of Operation: Worldwide
Tel: 01592 643111
Fax: 01592 643999
Email: info.uk@forbo.com
Web: www.forbo-flooring.co.uk
Other Info: ECO

HARVEY MARIA LTD
17 Riverside Business Park, Lyon Road,
Wimbledon, London, SW19 2RL
Area of Operation: UK & Ireland
Tel: 0208 542 0088
Fax: 0208 542 0099
Email: info@harveymaria.co.uk
Web: www.harveymaria.co.uk
Material Type: D) 4

INTERESTING INTERIORS LTD
37-39 Queen Street, Aberystwyth,
Ceredigion, SY23 1PU
Area of Operation: UK (Excluding Ireland)
Tel: 01970 626162
Fax: 0870 051 7959
Email: enquiries@interestinginteriors.com
Web: www.interiors-ltd.demon.co.uk

JIMPEX LTD
1 Castle Farm, Cholmondeley,
Malpas, Cheshire, SY14 8AQ
Area of Operation: UK & Ireland
Tel: 01829 720433
Fax: 01829720553
Email: jan@jimpex.co.uk
Web: www.jimpex.co.uk

NATURAL FLOORING
152-154 Wellingborough Road,
Northampton, Northamptonshire, NN1 4DT
Area of Operation: UK (Excluding Ireland)
Tel: 01604 239238/7 **Fax:** 01604 627799
Web: www.naturalflooring.co.uk

THE NATURAL FLOORING WAREHOUSE
East Building, Former All Saints Church,
Armoury Way, Wandsworth,
London, SW18 1HZ
Area of Operation: UK (Excluding Ireland)
Tel: 0208 870 5555 **Fax:** 0208 877 2847
Email: natalie@edwardianfires.com

THEISSEN GARDINEN-RAUMAUSSTATTUNG
Virchowstraße 18, Essen, Germany, 45147
Area of Operation: Europe
Tel: +49 201 734 011
Fax: +49 201 701 871
Email: naturcom-theissen@t-online.de
Web: www.gardinen-theissen.de
Other Info: ECO

UNNATURAL FLOORING
389 King Street, London, W6 9NJ
Area of Operation: UK & Ireland
Tel: 0870 766 1088 **Fax:** 0208 846 9552
Email: info@unnaturalflooring.com
Web: www.unnaturalflooring.com

SOFT FLOOR COVERINGS

KEY
PRODUCT TYPES: 1= Carpets
2 = Rugs 3 = Natural Matting
4 = Carpet Accessories and Underlays
5 = Oriental Carpets   6 = Other
OTHER: ▽ Reclaimed   On-line shopping
  Bespoke Hand-made ECO Ecological

AARDVARK WHOLESALE LTD
PO Box 3733, Dronfield, Derbyshire, S18 9AD
Area of Operation: UK (Excluding Ireland)
Tel: 0800 279 0486
Email: rawwoodfloors@zicom.net
Product Type: 1, 4

ANNETTE NIX
150a Camden Street, London, NW1 9PA
Area of Operation: Worldwide
Tel: 07956 451719
Email: annette.n@virgin.net
Web: www.annettenix.com
Product Type: 1, 2

AXMINSTER CARPETS
Axminster, Devon, EX13 5PQ
Area of Operation: UK & Ireland
Tel: 01297 630650 **Fax:** 01297 35241
Email: sales@axminster-carpets.co.uk
Web: www.axminster-carpets.co.uk
Product Type: 1

BIRCH INTERNATIONAL CARPETS
Hazel Park Mills, 318 Coleford Road,
Sheffield, South Yorkshire, S9 5PH
Area of Operation: UK & Ireland
Tel: 0114 243 1230 **Fax:** 0114 243 5118
Email: info@birchinternational.co.uk
Web: www.birchinternational.co.uk
Product Type: 1

BLENHEIM CARPETS
41 Pimlico Road, London, SW1W 8NE
Area of Operation: Worldwide
Tel: 0207 823 6333 **Fax:** 0207 823 5210
Email: admin@blenheim-carpets.com
Web: www.blenheim-carpets.com
Product Type: 1, 2, 3, 4

BRINTONS LTD
PO Box 16, Exchange Street,
Kidderminster, Worcestershire, DY10 1AG
Area of Operation: UK & Ireland
Tel: 01562 820000
Fax: 01562 634523
Email: selfbuild@brintons.co.uk
Web: www.brintons.net
Product Type: 1, 2

BRONTE CARPETS
Bankfield Mill, Greenfield Road,
Colne, Lancashire, BB8 9PD
Area of Operation: Europe
Tel: 01282 862736
Fax: 01282 868307
Email: office@brontecarpets.co.uk
Web: www.brontecarpets.co.uk
Product Type: 1, 2

CAVALIER CARPETS
Thompson Street Industrial Estate,
Blackburn, Lancashire, BB2 1TX
Area of Operation: UK & Ireland
Tel: 01254 268053
Email: sales@cavalier-carpets.co.uk
Web: www.cavaliercarpets.co.uk
Product Type: 1

CORE AND ORE LTD
16 Portland Street, Clifton, Bristol, BS8 4JH
Area of Operation: UK (Excluding Ireland)
Tel: 01179 042408
Fax: 01179 658057
Email: sales@coreandore.com
Web: www.coreandore.com
Product Type: 6
Material Type: M) 4

CRUCIAL TRADING
PO Box 10469, Birmingham,
West Midlands, B46 1WB
Area of Operation: Worldwide
Tel: 01562 743747
Email: sales@crucial-trading.com
Web: www.crucial-trading.com
Product Type: 1, 2, 3
Other Info:

CSM CARPETS & FLOORING LTD
Brickmakers Arms Lane,
Doddington, Cambridgeshire, PE15 0TR
Area of Operation: East England
Tel: 01354 740727
Fax: 01354 740078
Email: alan.csm@virgin.net
Web: www.csm-flooring.co.uk
Product Type: 1

FIRST FLOOR (FULHAM) LTD
174 Wandsworth Bridge Road,
Fulham, London, SW6 2UQ
Area of Operation: UK (Excluding Ireland)
Tel: 0207 736 1123
Fax: 0207 371 9812
Email: annie@firstfloor.uk.com
Web: www.firstfloor.uk.com
Product Type: 1

FLOORSDIRECT
Unit 6, Fountain Drive, Hertford,
Hertfordshire, SG13 7UB
Area of Operation: UK (Excluding Ireland)
Tel: 01992 552447
Fax: 01992 558760
Email: jenny@floorsdirect.co.uk
Web: www.floorsdirect.co.uk
Product Type: 1, 3, 4, 6

FLOORTECK LTD
8 Kingsdown Road,
Northfield, Birmingham,
West Midlands, B31 1AH
Area of Operation: UK (Excluding Ireland)
Tel: 0121 476 6271
Fax: 0121 476 6271
Email: sarahcollier@go.com
Product Type: 1, 3, 4, 5, 6

GIANO DESIGNS LTD
7 Blakedown Road,
Halesowen, Birmingham,
West Midlands, B634NE
Area of Operation: UK (Excluding Ireland)
Tel: 0121 550 2071
Email: creative@giano.co.uk
Web: www.giano.co.uk
Product Type: 2

NATURAL FLOORING
152-154 Wellingborough Road,
Northampton, Northamptonshire, NN1 4DT
Area of Operation: UK (Excluding Ireland)
Tel: 01604 239238/7
Fax: 01604 627799
Web: www.naturalflooring.co.uk

OLLERTON HALL DECOR
Ollerton Hall, Ollerton,
Knutsford, Cheshire, WA16 8SF
Area of Operation: UK & Ireland
Tel: 01565 650222
Fax: 01565 754411
Email: sales@ollertonhalldiscountcarpets.co.uk
Web: www.ollertonhalldiscountcarpets.co.uk
Product Type: 1, 2, 3, 4

POWNALL CARPETS
Spenbrook Mill,
Newchurch-in-Pendle, Fence,
Burnley, Lancashire, BB12 9JH
Area of Operation: UK & Ireland
Tel: 01282 611711
Email: carpet@pownallcarpets.com
Web: www.pownallcarpets.com
Product Type: 1

RENEWAL CARPET TILE LTD
P.O.Box 428, Blackburn,
Lancashire, BB2 2WQ
Area of Operation: Europe
Tel: 0870 350 2602
Email: timwright@renewalcarpettiles.com
Web: www.renewalcarpettiles.com
Product Type: 1
Other Info: ECO

RUG DESIGN CO
220 Tantallon Road,
Glasgow, Lanarkshire, G41 3JP
Area of Operation: Worldwide
Tel: 0845 345 1744
Fax: 0871 989 5421
Email: info@rugdesign.co.uk
Web: www.rugdesign.co.uk
Product Type: 1, 2

SCHLUTER SYSTEMS LTD
Units 4-5, Bardon 22 Industrial Park,
Beveridge Lane, Coalville,
Leicestershire, LE67 1TE
Area of Operation: UK & Ireland
Tel: 01530 813396
Fax: 01530 813376
Email: admin@schluter.co.uk
Web: www.schluter.co.uk
Product Type: 4

SELECT FIRST
4 Ridgmount Street, London, WC1E 7AA
Area of Operation: UK & Ireland
Tel: 0207 580 6960
Fax: 0207 580 4173
Email: myles@selectfirst.com
Web: www.carpetinfo.co.uk
Product Type: 1

STAIRRODS (UK) LTD
Unit 6, Park Road North Industrial Estate,
Blackhill, Consett, Durham, DH8 5UN
Area of Operation: Worldwide
Tel: 01207 591176
Fax: 01207 591911
Email: sales@stairrods.co.uk
Web: www.stairrods.co.uk
Product Type: 4

SWIFTEC
Pennine House, Tilson Road,
Roundthorn Estate,
Manchester, M23 9GF
Area of Operation: UK & Ireland
Tel: 0800 074 4145
Fax: 0800 074 0005
Product Type: 6

**THE ALTERNATIVE
FLOORING COMPANY**
3b Stephenson Close, East Portway,
Andover, Hampshire, SP10 3RU
Area of Operation: UK & Ireland
Tel: 01264 335111
Fax: 01264 336445
Email: sales@alternativeflooring.com
Web: www.alternativeflooring.com
Product Type: 2, 3, 6
Other Info: ECO

THE CANE STORE
Wash Dyke Cottage, No1 Witham Road,
Long Bennington, Lincolnshire, NG23 5DS
Area of Operation: Worldwide
Tel: 01400 282271 **Fax:** 01400 281103
Email: jaki@canestore.co.uk
Web: www.canestore.co.uk
Product Type: 3

SPONSORED BY: PERGO LTD www.pergo.com

THE CARPET FOUNDATION
MCF Complex, 60 New Road,
Kidderminster, Worcestershire, DY10 1AQ
Area of Operation: UK (Excluding Ireland)
Tel: 01562 755 568
Fax: 01562 865405
Email: info@carpetfoundation.com
Web: www.comebacktocarpet.com
Product Type: 1, 2

THE NATURAL FLOOR COMPANY
389 King Street, London, W6 9NJ
Area of Operation: UK & Ireland
Tel: 0208 741 4451
Fax: 0208 846 9552
Email: barnabygreen@onetel.net.uk
Web: www.natfloorco.com
Product Type: 1, 2, 3, 4, 6

THEISSEN GARDINEN-RAUMAUSSTATTUNG
Virchowstraße 18, Essen, Germany, 45147
Area of Operation: Europe
Tel: +49 201 734 011
Fax: +49 201 701 871
Email: naturcom-theissen@t-online.de
Web: www.gardinen-theissen.de
Product Type: 1, 3, 4, 5, 6

UNNATURAL FLOORING
389 King Street, London, W6 9NJ
Area of Operation: UK & Ireland
Tel: 0870 766 1088
Fax: 0208 846 9552
Email: info@unnaturalflooring.com
Web: www.unnaturalflooring.com
Product Type: 3

VICTORIA CARPETS LIMITED
Worcester Road, Kidderminster,
Worcestershire, DY10 1HL
Area of Operation: Europe
Tel: 01562 749300
Fax: 01562 749349
Email: sales@victoriacarpets.com
Web: www.victoriacarpets.com
Product Type: 1

VICTORIA HAMMOND INTERIORS
Bury Farm, Church Street,
Bovingdon, Hemel Hempstead,
Hertfordshire, HP3 0LU
Area of Operation: UK (Excluding Ireland)
Tel: 01442 831641 **Fax:** 01442 831641
Email: victoria@victoriahammond.com
Web: www.victoriahammond.com
Product Type: 1, 2, 3, 4, 6

WILLIAM POWNALL & SONS LTD
Spenbrook Mill, Newchurch in Pendle,
Burnley, Lancashire, BB12 9JH
Area of Operation: Worldwide
Tel: 01282 611711 **Fax:** 01282 614510
Email: carpet@pownallcarpets.co.uk
Web: www.pownallcarpets.com
Product Type: 1

WOOLS OF NEW ZEALAND
International Design Centre, Little Lane, Ilkley,
West Yorkshire, LS29 8UG
Area of Operation: UK & Ireland
Tel: 01943 603888 **Fax:** 01943 817083
Email: info.uk@canesis.com
Web: www.woolcarpet.com **Product Type:** 1, 2

WALL TILES & BACKING BOARDS

KEY
PRODUCT TYPES: 1= Hand Painted Tiles
2 = Mirrored Tiles 3 = Matt Tiles
4 = Satin Tiles 5 = High Gloss Tiles
6 = Backing Boards
OTHER: ▽ Reclaimed ⌐🖰 On-line shopping
✏ Bespoke 🖐Hand-made ECO Ecological

ALICANTE STONE
Damaso Navarro, 6 Bajo, 03610 Petrer
(Alicante), P.O. Box 372, Spain
Area of Operation: Europe
Tel: +34 966 31 96 97 **Fax:** +34 966 31 96 98
Email: info@alicantestone.com
Web: www.alicantestone.com
Product Type: 2, 3, 4
Material Type: E) 2, 4, 5, 8

ALISTAIR MACKINTOSH LTD
Bannerley Road, Garretts Green,
Birmingham, West Midlands, B33 0SL
Area of Operation: UK & Ireland
Tel: 0121 784 6800 **Fax:** 0121 789 7068
Email: info@alistairmackintosh.co.uk
Web: www.alistairmackintosh.co.uk

AMABIS TILES
Ubique Park , March Way,
Battlefield Enterprise Park,
Shrewsbury, Shropshire, SY1 3JE
Area of Operation: Europe
Tel: 01743 440860 **Fax:** 01743 462440
Email: sales@amabis.co.uk
Web: www.amabis.co.uk
Product Type: 5 **Other Info:** 🖐
Material Type: F) 1

ART ON TILES
2 Orchard Terrace, The Street, Walberton,
Arundel, West Sussex, BN18 0PH
Area of Operation: Worldwide
Tel: 01243 552346
Email: info@artontiles.co.uk
Web: www.artontiles.co.uk
Product Type: 1
Other Info: ✏

ASHLAR MASON
Manor Building, Manor Road, London, W13 0JB
Area of Operation: East England, Greater London,
Midlands & Mid Wales, South East England, South
West England and South Wales
Tel: 0208 997 0002 **Fax:** 0208 997 0008
Email: enquiries@ashlarmason.co.uk
Web: www.ashlarmason.co.uk
Product Type: 3

AZTEC METAL TILES
Bowl Road, Charing, Kent, TN27 0HB
Area of Operation: UK & Ireland
Tel: 01233 712332 **Fax:** 01233 714994
Email: enquiries@aztecmetaltiles.co.uk
Web: www.aztecmetaltiles.co.uk
Product Type: 2, 3, 4

BATHROOM CITY
Seeleys Road, Tyseley Industrial Estate,
Birmingham, West Midlands, B11 2LQ
Area of Operation: UK & Ireland
Tel: 0121 753 0700 **Fax:** 0121 753 1110
Email: sales@bathroomcity.com
Web: www.bathroomcity.com ⌐🖰

BRITISH CERAMIC TILE
Heathfield, Newton Abbot, Devon, TQ12 6RF
Area of Operation: UK & Ireland
Tel: 01626 831480 **Fax:** 01626 831465
Email: sales@bctltd.co.uk
Web: www.candytiles.co.uk
Product Type: 3, 4, 5

CAPITAL MARBLE DESIGN
Unit 1 Pall Mall Deposit,
124-128 Barlby Road, London, W10 6BL
Area of Operation: UK & Ireland
Tel: 0208 968 5340 **Fax:** 0208 968 8827
Email: stonegallery@capitalmarble.co.uk
Web: www.capitalmarble.co.uk
Product Type: 1, 2, 3, 4, 5
Other Info: ECO ✏

CARE DESIGN
Moorgate, Ormskirk, Lancashire, L39 4RX
Area of Operation: UK & Ireland
Tel: 01695 579061
Fax: 01695 570489
Email: caredesign@clara.net

CERAMIQUE INTERNATIONALE LTD
Unit 1 Royds Lane,
Lower Wortley Ring Road,
Leeds, West Yorkshire, LS12 6DU
Area of Operation: UK & Ireland
Tel: 0113 231 0218
Fax: 0113 231 0353
Email: cameron@ceramiqueinternationale.co.uk
Web: www.ceramiqueinternationale.co.uk

CITY BATHROOMS & KITCHENS
158 Longford Road, Longford,
Coventry, West Midlands, CV6 6DR
Area of Operation: UK & Ireland
Tel: 02476 365877
Fax: 02476 644992
Email: citybathrooms@hotmail.com
Web: www.citybathrooms.co.uk

CONCEPT TILING LIMITED
Unit 3, Jones Court, Jones Square,
Stockport, Cheshire, SK1 4LJ
Area of Operation: UK & Ireland
Tel: 0161 480 0994 **Fax:** 0161 480 0911
Email: enquiries@concept-tiles.com
Web: www.concept-tiles.com

CRAVEN DUNNILL JACKFIELD LTD
Jackfield Tile Museum,
Ironbridge Gorge,
Telford, Shropshire, TF8 7LJ
Area of Operation: Worldwide
Tel: 01952 884124
Fax: 01952 884487
Email: gemma@cravendunnill-jackfield.co.uk
Web: www.cravendunnill-jackfield.co.uk
Product Type: 1, 5

CTD SCOTLAND
72 Hydepark Street, Glasgow,
Dunbartonshire, G3 8BW
Area of Operation: Scotland
Tel: 0141 221 4591
Fax: 0141 221 8442
Email: info@ctdscotland.co.uk
Web: www.ctdscotland.co.uk
Product Type: 1, 2, 3, 4, 5
Material Type: E) 1, 2, 3, 7, 8

D F FIXINGS
15 Aldham Gardens,
Rayleigh, Essex, SS6 9TB
Area of Operation: UK (Excluding Ireland)
Tel: 07956 674673 **Fax:** 01268 655072

DAR INTERIORS
Arch 11, Miles Street, London, SW8 1RZ
Area of Operation: Worldwide
Tel: 0207 720 9678 **Fax:** 0207 627 5129
Email: enquiries@darinteriors.com
Web: www.darinteriors.com
Product Type: 1, 3, 4
Other Info: ✏ 🖐

DEGUSSA CONSTRUCTION CHEMICALS (UK)
Albany House, Swinton Hall Road,
Swinton, Manchester, M27 4DT
Area of Operation: Worldwide
Tel: 0161 794 7411 **Fax:** 0161 727 8547
Email: mbtfeb@degussa.com
Web: www.degussa-cc.co.uk
Product Type: 2

DEVON STONE LTD
38 Station Road, Budleigh
Salterton, Devon, EX9 6RT
Area of Operation: UK (Excluding Ireland)
Tel: 01395 446841
Email: amy@devonstone.com

ELON
12 Silver Road, London, W12 7SG
Area of Operation: UK & Ireland
Tel: 0208 932 3000
Fax: 0208 932 3001
Email: marketing@elon.co.uk
Web: www.elon.co.uk ⌐🖰
Product Type: 1

EUROPEAN HERITAGE
48-54 Dawes Road,
Fulham, London, SW6 7EN
Area of Operation: Worldwide
Tel: 0207 381 6063
Fax: 0207 381 9534
Email: fulham@europeanheritage.co.uk
Web: www.europeanheritage.co.uk
Product Type: 2, 3, 4, 5

EXAKT PRECISION TOOLS LTD
Unit 9 Urquhart Trade Centre,
109 Urquhart Road, Aberdeen,
Aberdeenshire, AB24 5NH
Area of Operation: Europe
Tel: 01224 643434 **Fax:** 01224 643232
Email: info@exaktpt.com
Web: www.exaktpt.com

FIRED EARTH INTERIORS
3 Twyford Mill, Oxford Road, Adderbury,
Banbury, Oxfordshire, OX17 3SX
Area of Operation: Worldwide
Tel: 01295 814300
Fax: 01295 810832
Email: enquiries@firedearth.com.
Web: www.firedearth.

FLORIAN TILES
PO Box 4684, Sturminster Neton, Dorset, DT10 2SX
Area of Operation: UK (Excluding Ireland)
Tel: 01963 251025
Fax: 01963 251025
Email: enquiries@floriantiles.co.uk
Web: www.floriantiles.co.uk
Product Type: 1

H & R JOHNSON TILES LTD
Harewood Street, Tunstall,
Stoke on Trent, Staffordshire, ST6 5JZ
Area of Operation: Worldwide
Tel: 01782 575 575
Fax: 01782 524138
Email: techsales@johnson-tiles.com
Web: www.johnson-tiles.com
Product Type: 4, 5

HERITAGE TILE CONSERVATION LTD
The Studio, 2 Harris Green,
Broseley, Shropshire, TF12 5HJ
Area of Operation: UK & Ireland
Tel: 01952 881039 **Fax:** 01952 881039
Email: enquiries@heritagetile.co.uk
Web: www.heritagetile.co.uk
Product Type: 1, 3, 5

HITT OAK LTD
13a Northview Crescent, London, NW10 1RD
Area of Operation: UK (Excluding Ireland)
Tel: 0208 450 3821 **Fax:** 0208 911 0299
Email: zhitas@yahoo.co.uk

HOT GLASS DESIGN
Unit 24 Crosby Yard Industrial Estate,
Bridgend, Mid Glamorgan, CF31 1JZ
Area of Operation: Europe
Tel: 01656 659884
Fax: 01656 659884
Email: info@hotglassdesign.co.uk
Web: www.hotglassdesign.co.uk
Material Type: J) 1, 2, 4, 5, 6

HUNTER ART & TILE STUDIO
Craft Courtyard, Harestanes, Ancrum,
Jedburgh, Borders, TD8 6UQ
Area of Operation: UK & Ireland
Tel: 01835 830 328
Email: enquiries@hunterartandtilestudio.co.uk
Web: www.hunterartandtilestudio.co.uk
Product Type: 1

HYPERION TILES
67 High Street, Ascot, Berkshire, SL5 7HP
Area of Operation: Greater London,
South East England
Tel: 01344 620211 **Fax:** 01344 620100
Email: graham@hyperiontiles.com
Web: www.hyperiontiles.com
Product Type: 1, 2, 3, 4, 5

INDIGENOUS LTD
Cheltenham Road, Burford, Oxfordshire, OX18 4JA
Area of Operation: Worldwide
Tel: 01993 824200 **Fax:** 01993 824300
Email: enquiries@indigenoustiles.com
Web: www.indigenoustiles.com
Product Type: 1

INNOVATION GLASS COMPANY
27-28 Blake Industrial Estate, Brue Avenue,
Bridgwater, Somerset, TA7 8EQ
Area of Operation: Europe
Tel: 01278 683645
Fax: 01278 450088
Email: magnus@igc-uk.com
Web: www.igc-uk.com
Product Type: 2, 3, 5
Other Info: ✎
Material Type: J) 4, 5, 6

KIRK NATURAL STONE
Bridgend, Fyvie, Turriff,
Aberdeenshire, AB53 8QB
Area of Operation: Worldwide
Tel: 01651 891891 **Fax:** 01651 891794
Email: info@kirknaturalstone.com
Web: www.kirknaturalstone.com

LAWLEY CERAMICS
8 Stourbridge Road, Bromsgrove,
Worcestershire, B61 0AB
Area of Operation: UK (Excluding Ireland)
Tel: 01527 570455 **Fax:** 01527 570455
Email: lawleyceramics@hotmail.com

MAESTROTILE
Unit 7 Waterloo Park,
Waterloo Road Industrial Estate,
Bidford on Avon, Warwickshire, B50 4JH
Area of Operation: UK & Ireland
Tel: 01789 778700
Fax: 01789 778852
Email: malcolm@maestrotile.com
Web: www.maestrotile.com

MAGGIE JONES
HAND PAINTED TILES
19, Langland Road, Mumbles,
Swansea, Swansea, SA3 4ND
Area of Operation: Europe
Tel: 01792 360551
Email: maggie@maggiejonestiles.co.uk
Web: www.maggiejonestiles.co.uk ✎
Product Type: 1

MANDARIN
Unit 1 Wonastow Road Industrial Estate,
Monmouth, Monmouthshire, NP25 5JB
Area of Operation: Europe
Tel: 01600 715444
Fax: 01600 715494
Email: info@mandarinstone.com
Web: www.mandarinstone.com ✎

MARIA STARLING MOSAICS
40 Strand House, Merbury Close, SE28 0LU
Area of Operation: UK (Excluding Ireland)
Tel: 07775 517409
Email: mosaics@mariastarling.com
Web: www.mariastarling.com

MARLBOROUGH TILES LTD.
Elcot Lane, Marlborough, Wiltshire, SN8 2AY
Area of Operation: UK & Ireland
Tel: 01672 512422
Fax: 01672 515791
Email: admin@marlboroughtiles.com
Web: www.marlboroughtiles.com
Product Type: 1, 3, 4

METAL TILES LTD
6c Waterloo Works, Gorsey Mount Street,
Stockport, Cheshire, SK1 3BU
Area of Operation: UK (Excluding Ireland)
Tel: 0161 480 1166
Fax: 0161 480 2838
Email: info@metaltiles.ltd.uk
Web: www.metaltiles.co.uk
Material Type: C) 3, 7, 11

N&C NICOBOND
41-51 Freshwater Road, Chadwell Heath,
Romford, Essex, RM8 1SP
Area of Operation: Worldwide
Tel: 0208 586 4600
Fax: 0208 586 4646
Email: info@nichollsandclarke.com
Web: www.ncdirect.co.uk

£££SAVE£££
BUY DIRECT
DEPOTS NATIONWIDE

**Manufacturers and suppliers
of wall and floor tiles
together with a full range
of adhesives and grouts**

N&C
Nicobond

41-51 Freshwater Road, Chadwell Heath
Romford, Essex RM8 1SP
Tel: 020 8586 4600
Fax: 020 8586 4646
e-mail: info@nichollsandclarke.com
www.ncdirect.co.uk

NORTON TILE COMPANY LIMITED
The Coach House, Norton Lane,
Norton, Chichester, West Sussex, PO20 3NH
Area of Operation: UK (Excluding Ireland)
Tel: 01243 544224
Fax: 01243 544033
Email: information@norton-tile.co.uk
Web: www.norton-tile.co.uk
Product Type: 1

ORIGINAL STYLE
Falcon Road, Sowton Industrial Estate,
Exeter, Devon, EX2 7LF
Area of Operation: Worldwide
Tel: 01392 474011
Fax: 01392 219932
Email: vmattison@originalstyle.com
Web: www.originalstyle.com
Product Type: 1, 3, 4, 5
Other Info: ✋
Material Type: F) 1, 3

ORIGINAL STYLE LTD
Falcon Road, Sowton Industrial Estate,
Exeter, Devon, EX2 7LF
Area of Operation: Worldwide
Tel: 01392 473000
Fax: 01392 219932
Email: info@originalstyle.com
Web: www.originalstyle.com
Product Type: 1, 3, 4, 5

PARKSIDE TILES
49-51 Highmeres Road, Thurmeston,
Leicester, Leicestershire, LE4 9LZ
Area of Operation: UK (Excluding Ireland)
Tel: 0116 276 2532 **Fax:** 0116 246 0649
Email: parkside@tiles2.wanadoo.co.uk
Web: www.parksidetiles.co.uk
Product Type: 1, 2, 3, 4, 5
Material Type: F) 1, 2, 4

PHOENIX TILE STUDIO
Unit 3 Winkhill Mill, Swan Street,
Stoke on Trent, Staffordshire, ST4 7RH
Area of Operation: Worldwide
Tel: 01782 745599 **Fax:** 01782 745599
Email: office@phoenixtilestudio.fsbusiness.co.uk
Web: www.phoenixtilestudio.co.uk
Product Type: 1, 3, 4, 5
Material Type: F) 1

PILKINGTON'S TILES GROUP
PO Box 4, Clifton Junction, Manchester, M27 8LP
Area of Operation: UK (Excluding Ireland)
Tel: 0161 727 1000 **Fax:** 0161 727 1122
Email: sales@pilkingtons.com
Web: www.pilkingtons.com
Product Type: 3, 4, 5

RAPID GRANITE LIMITED
Tollgate Farm, Golford Road,
Cranbrook, Kent, TN17 3NX
Area of Operation: UK (Excluding Ireland)
Tel: 01580 720900 **Fax:** 01580 720901
Email: info@rapidgranite.co.uk
Web: www.rapidgranite.co.uk ✎

ROCK UNIQUE LTD
c/o Select Garden and Pet Centre,
Main Road, Sundridge, Kent, TN14 6ED
Area of Operation: Europe
Tel: 01959 565608 **Fax:** 01959 569312
Email: stone@rock-unique.com
Web: www.rock-unique.com
Product Type: 3, 4, 5
Other Info: ECO ✎ ✋
Material Type: E) 1, 2, 3, 4, 5, 9

RUPERT SCOTT LTD
The Glass Studio, Mytton Mill, Montford Bridge,
Shrewsbury, Shropshire, SY4 1HA
Area of Operation: UK (Excluding Ireland)
Tel: 01743 851393 **Fax:** 01743 851393
Email: info@rupertscott.com
Web: www.rupertscott.com
Other Info: ✋ **Material Type:** J) 5

RUSTICA LTD
154c Milton Park, Oxfordshire, OX14 4SD
Area of Operation: UK & Ireland
Tel: 01235 834192 **Fax:** 01235 835162
Email: sales@rustica.co.uk
Web: www.rustica.co.uk
Product Type: 1, 3, 4, 5

**SHARON JONES HANDMADE
ARCHITECTURAL CERAMIC TILES**
Trevillian Cottage, Barrington,
Nr Ilminster, Somerset, TA19 0JB
Area of Operation: UK (Excluding Ireland)
Tel: 01460 259074
Email: info@handmadearchitecturaltiles.co.uk
Web: www.handmadearchitecturaltiles.co.uk
Product Type: 1, 3, 4, 5
Other Info: ✎ ✋

SIESTA CORK TILE CO
Unit 21, Tait Road, Gloucester Road,
Croydon, Surrey, CR0 2DP
Area of Operation: UK (Excluding Ireland)
Tel: 0208 683 4055 **Fax:** 0208 663 4480
Email: siestacork@aol.com
Web: www.siestacorktiles.co.uk
Material Type: M) 1

SIMALI STONE FLOORING
40E Wincombe Business Park,
Shaftesbury, Dorset, SP7 9QJ
Area of Operation: Europe
Tel: 01747 852557 **Fax:** 01747 852557
Email: stonefloors@aol.com
Web: www.stoneflooringonline.com

STONE AND SLATE
Coney Green Farm, Lower Market Street,
Claycross, Chesterfield, Derbyshire, S45 9NE
Area of Operation: UK & Ireland
Tel: 01246 250088
Fax: 01246 250099
Email: sales@stoneandslate.ltd.uk
Web: www.stoneandslate.co.uk
Product Type: 1, 2, 4, 5

STONELL DIRECT
100-105 Victoria Crescent,
Burton upon Trent, Staffordshire, DE14 2QF
Area of Operation: Worldwide
Tel: 08000 832283 **Fax:** 01283 501098
Email: liz.may@stonelldirect.com
Web: www.stonelldirect.com

STONEVILLE (UK) LTD
Unit 12, Set Star Estate, Transport Avenue, Great
West Road, Brentford, Greater London, TW8 9HF
Area of Operation: Europe
Tel: 0208 560 1000 **Fax:** 0208 560 4060
Email: info@stoneville.co.uk
Web: www.stoneville.co.uk
Material Type: E) 1, 2, 5, 8

STUDIO STONE
The Stone Yard, Alton Lane, Four Marks,
Hampshire, GU34 5AJ
Area of Operation: UK (Excluding Ireland)
Tel: 01420 562500 **Fax:** 01420 563192
Email: sales@studiostone.co.uk
Web: www.studiostone.co.uk

THE CANDY TILE COMPANY
Heathfield, Newton Abbot, Devon, TQ12 6RF
Area of Operation: UK (Excluding Ireland)
Tel: 01626 834774 **Fax:** 01626 834775
Email: sales@candytiles.com
Web: www.candytiles.com **Product Type:** 3, 4, 5

THE GRANITE FACTORY
4 Winchester Drive, Peterlee, Durham, SR8 2RJ
Area of Operation: North East England
Tel: 0191 518 3600 **Fax:** 0191 518 3600
Email: admin@granitefactory.co.uk
Web: www.granitefactory.co.uk ✎

THE ORIGINAL TILE COMPANY
23A Howe Street, Lothian, EH3 6TF
Area of Operation: UK & Ireland
Tel: 0131 556 2013 **Fax:** 0131 558 3172
Email: info@originaltile.freeserve.co.uk
Product Type: 1, 2, 3, 4, 5 **Other Info:** ✎ ✋

TILECO
3 Molesey Business Centre, Central Avenue,
West Molesey, Surrey, KT8 2QZ
Area of Operation: UK (Excluding Ireland)
Tel: 0208 481 9500 **Fax:** 0208 481 9501
Email: info@tileco.co.uk **Web:** www.tileco.co.uk

TILES UK LTD
1-13 Montford Street, Off South Langworthy Road,
Salford, Greater Manchester, M50 2XD
Area of Operation: UK (Excluding Ireland)
Tel: 0161 872 5155 **Fax:** 0161 848 7948
Email: info@tilesuk.com **Web:** www.tilesuk.com ✎
Product Type: 1, 2, 3, 4, 5

TOPPS TILES
Thorpe Way, Grove Park, Enderby,
Leicestershire, LE19 1SU
Area of Operation: UK (Excluding Ireland)
Tel: 0116 282 8000 **Fax:** 0116 282 8100
Email: mlever@toppstiles.co.uk
Web: www.toppstiles.co.uk

INTERIORS, FIXTURES & FINISHES

TRADEMARK TILES
Airfield Industrial Estate, Warboys, Huntingdon,
Cambridgeshire, PE28 2SH
Area of Operation: East England,
South East England
Tel: 01487 825300 **Fax:** 01487 825305
Email: info@trademarktiles.co.uk
Product Type: 1, 3, 4, 5

WALLS AND FLOORS LTD
Wilson Terrace, Kettering, Northamptonshire, NN19 9RT
Area of Operation: UK (Excluding Ireland)
Tel: 01536 410484
Email: sales@wallsandfloors.uk.com
Web: www.wallsandfloors.uk.com 🖰

WAXMAN CERAMICS LTD
Grove Mills, Elland, West Yorkshire, HX5 9DZ
Tel: 01422 311331 **Fax:** 01422 310654
Email: sales@waxmanceramics.co.uk
Web: www.waxmanceramics.co.uk

WILTON STUDIOS
Cleethorpes Business Centre,
Wilton Road Industrial Estate,
Grimsby, Lincolnshire, DN36 4AS
Area of Operation: UK (Excluding Ireland)
Tel: 01472 210820 **Fax:** 01472 812602
Email: postbox@wiltonstudios.co.uk
Web: www.wiltonstudios.co.uk
Product Type: 1, 2, 3, 4, 5 **Other Info:** 🖉 🖐

WORLD'S END TILES
Silverthorne Road, Battesea,
London, SW8 3HE
Area of Operation: UK (Excluding Ireland)
Tel: 0207 819 2100 **Fax:** 0207 819 2101
Email: info@worldsendtiles.co.uk
Web: www.worldsendtiles.co.uk **Other Info:** 🖐
Product Type: 1, 3, 4, 5   **Material Type:** E) 2

WALLPAPER

KEY
PRODUCT TYPES: 1= Flock 2 = Paper
3 = Paperweave 4 = Stretch Fabric
5 = Suede Effect 6 = Textile 7 = Other
OTHER: ▽ Reclaimed 🖰 On-line shopping
🖉 Bespoke 🖐 Hand-made ECO Ecological

HAMILTON WESTON WALLPAPERS
18 St Marys Grove, Richmond, Surrey, TW9 1UY
Area of Operation: UK & Ireland
Tel: 0208 940 4850
Fax: 0208 332 0296
Email: info@hamiltonweston.com
Web: www.hamiltonweston.com **Product Type:** 2

RAY MUNN LTD
861-863 Fulham Road, London, SW6 5HP
Area of Operation: UK (Excluding Ireland)
Tel: 0207 736 9876
Email: rishi@raymunn.com
Web: www.raymunn.com
Product Type: 1, 2, 3, 5, 7

ROOMS FOR KIDS LTD
Unit 2 Gordon Street (Off Craven Street),
Birkenhead, Wirral, Cheshire, CH41 4AB
Area of Operation: Worldwide
Tel: 0151 650 2401
Fax: 0151 650 2403
Email: roger@friezeframe.com
Web: www.friezeframe.com 🖰
Product Type: 2, 7

THE WORLDWIDE WOOD COMPANY LIMITED
154 Colney Hatch Lane, Muswell Hill,
London, N10 1ER
Area of Operation: Worldwide
Tel: 0800 458 3366
Fax: 0208 365 3965
Email: info@solidwoodflooring.com
Web: www.solidwoodflooring.com
Product Type: 7

THEISSEN GARDINEN-RAUMAUSSTATTUNG
Virchowstraße 18, Essen, Germany, 45147
Area of Operation: Europe
Tel: +49 201 734 011
Fax: +49 201 701 871
Email: naturcom-theissen@t-online.de
Web: www.gardinen-theissen.de
Product Type: 2, 6

VICTORIA HAMMOND INTERIORS
Bury Farm, Church Street, Bovingdon,
Hemel Hempstead, Hertfordshire, HP3 0LU
Area of Operation: UK (Excluding Ireland)
Tel: 01442 831641
Fax: 01442 831641
Email: victoria@victoriahammond.com
Web: www.victoriahammond.com
Product Type: 1, 2, 3, 4, 5, 6, 7
Other Info: 🖉

WALL PANELLING

KEY
PRODUCT TYPES: 1= Laminated 2 = Wood
Veneered 3 = Timber Effect 4 = Other
OTHER: ▽ Reclaimed 🖰 On-line shopping
🖉 Bespoke 🖐 Hand-made ECO Ecological

ABP-TBS PARTNERSHIP LTD.
Martens Road, Northbank Industrial Park,
Irlam, Manchester, M44 5AX.
Area of Operation: UK & Ireland
Tel: 0161 775 1871.
Fax: 0161 775 8929.
Email: info@tbs-fabrications.co.uk
Web: www.abp-tbswashrooms.co.uk
Product Type: 1, 2, 3, 4
Material Type: C) 1, 3

ARCHITECTURAL HERITAGE
Taddington Manor, Taddington, Nr Cutsdean,
Cheltenham, Gloucestershire, GL54 5RY
Area of Operation: Worldwide
Tel: 01386 584414
Fax: 01386 584236
Email: puddy@architectural-heritage.co.uk
Web: www.architectural-heritage.co.uk
Other Info: ▽ 🖉 🖐

BHK (UK) LTD
Davy Drive North West Industrial Estate,
Peterlee, Durham, SR8 2JF
Area of Operation: UK & Ireland
Tel: 0191 518 6538 **Fax:** 0191 518 6536
Email: eleanor.smith@peterlee.bh.de
Web: www.bhkonline.com
Product Type: 1

CANADA WOOD UK
PO Box 1, Farnborough, Hampshire, GU14 6WE
Area of Operation: UK (Excluding Ireland)
Tel: 01252 522545
Fax: 01252 522546
Email: office@canadawooduk.org
Web: www.canadawood.info
Product Type: 4 **Material Type:** A) 12

CITY BATHROOMS & KITCHENS
158 Longford Road, Longford, Coventry,
West Midlands, CV6 6DR
Area of Operation: UK & Ireland
Tel: 02476 365877 **Fax:** 02476 644992
Email: citybathrooms@hotmail.com
Web: www.citybathrooms.co.uk

CORE AND ORE LTD
16 Portland Street, Clifton, Bristol, BS8 4JH
Area of Operation: UK (Excluding Ireland)
Tel: 01179 042408
Fax: 01179 658057
Email: sales@coreandore.com
Web: www.coreandore.com
Product Type: 4
Material Type: M) 4

COURTENAY JOHN BOTTERILL
24 Orwell Road, Clacton on Sea, Essex, CO15 1PP
Area of Operation: Greater London,
South East England
Tel: 01255 428837
Email: courtenay1@aol.com
Product Type: 4 **Material Type:** A) 2, 3, 4, 5, 7, 11

DISTINCTIVE COUNTRY FURNITURE LIMITED
Parrett Works, Martock, Somerset, TA12 6AE
Area of Operation: Worldwide
Tel: 01935 825800
Fax: 01935 825800
Email: Dcfltdinteriors@aol.com
Web: www.distinctivecountryfurniture.co.uk
Other Info: ▽ 🖉 🖐
Material Type: A) 2, 3, 4, 7, 14

DISTINCTIVE COUNTRY FURNITURE

Area of Operation: Worldwide
Tel: 01935 825800
Fax: 01935 825800
Email: dcfltdinteriors@aol.com
Web: www.distinctivecountryfurniture.co.uk

Makers of 16th and 17th Century furniture and
interiors. Staircases, kitchens, panelling, oak
flooring, doors and furniture. Please call for a free
brochure. Quotations available upon request.

ECO IMPACT LTD
50a Kew Green, Richmond, London, TW9 3BB
Area of Operation: UK & Ireland
Tel: 0208 940 7072 **Fax:** 0208 332 1218
Email: sales@ecoimpact.co.uk
Web: www.ecoimpact.co.uk **Product Type:** 4

FORMICA LIMITED
Coast Road, North Shields,
Tyne & Wear, NE29 8RE
Area of Operation: UK & Ireland
Tel: 0191259 3000 **Fax:** 0191259 2719
Email: info@formica.co.uk
Web: www.formica.co.uk **Product Type:** 1

HANNAH WHITE
Area of Operation: Worldwide
Tel: 01372 806703
Fax: 01372 806703
Email: hannah@hannahwhite.co.uk
Web: www.hannahwhite.co.uk
Product Type: 4 **Other Info:** 🖉 🖐
Material Type: M) 4

HERITAGE WOOD TURNING
Unit 3 The Old Brickworks,
Bakestonedale Road, Pott Shrigley,
Macclesfield, Cheshire, SK10 5RX
Area of Operation: UK (Excluding Ireland)
Tel: 01625 560655
Fax: 01625 560697
Email: sales@heritageoak.co.uk
Web: www.heritageoak.co.uk

INNOVATION GLASS COMPANY
27-28 Blake Industrial Estate,
Brue Avenue, Bridgwater, Somerset, TA7 8EQ
Area of Operation: Europe
Tel: 01278 683645
Fax: 01278 450088
Email: magnus@igc-uk.com
Web: www.igc-uk.com
Product Type: 4 **Material Type:** J) 4, 5, 6, 7, 8

INTERESTING INTERIORS LTD
37-39 Queen Street, Aberystwyth,
Ceredigion, SY23 1PU
Area of Operation: UK (Excluding Ireland)
Tel: 01970 626162 **Fax:** 0870 051 7959
Email: enquiries@interestinginteriors.com
Web: www.interiors-ltd.demon.co.uk
Product Type: 1

ION GLASS
Unit 7 Grange Ind Est, Albion Street, Southwick,
Brighton, West Sussex, BN42 4EN
Area of Operation: UK (Excluding Ireland)
Tel: 0845 658 9988 **Fax:** 0845 658 9989
Email: sales@ionglass.co.uk
Web: www.ionglass.co.uk
Product Type: 4 **Other Info:** 🖉
Material Type: J) 1, 4, 5, 6

LAWSONS
Gorst Lane, Off New Lane, Burscough,
Ormskirk, Lancashire, L40 0RS
Area of Operation: Worldwide
Tel: 01704 893998
Fax: 01704 892526
Email: info@traditionaltimber.co.uk
Web: www.traditionaltimber.co.uk

NATURAL BUILDING TECHNOLOGIES
The Hangar, Worminghall Road,
Oakley, Buckinghamshire, HP18 9UL
Area of Operation: UK & Ireland
Tel: 01844 338338
Fax: 01844 338525
Email: info@natural-building.co.uk
Web: www.natural-building.co.uk

NORSKE INTERIORS
Estate Road One, South Humberside Industrial
Estate, Grimsby, Lincolnshire, DN31 2TA
Area of Operation: UK (Excluding Ireland)
Tel: 01472 240832 **Fax:** 01472 360112
Email: sales@norske-int.co.uk
Web: www.norske-int.co.uk
Product Type: 4

PANEL MASTER UK LIMITED
Unit 7, Spring Vale Mill, Waterside Road,
Haslingden, Rossendale, Lancashire, BB4 5EZ
Area of Operation: UK & Ireland
Tel: 01706 219196 **Fax:** 01706 222173
Email: info@panelmaster.co.uk
Web: www.panelmaster.co.uk 🖰
Product Type: 1, 2, 3, 4

PANELITDIRECT.COM
Unit 1, Oldhall Industrial Estate,
Bromborough, Wirral, Cheshire, CH62 3QA
Area of Operation: UK & Ireland
Tel: 0845 466 0123 **Fax:** 0870 170 9870
Email: info@panelitdirect.com
Web: www.panelitdirect.com
Product Type: 4 **Material Type:** D) 1

PERMACELL FINESSE LTD
Western Road, Silver End, Witham, Essex, CM8 3BQ
Area of Operation: UK & Ireland
Tel: 01376 583241 **Fax:** 01376 583072
Email: sales@pfl.co.uk
Web: www.permacell-finesse.co.uk
Product Type: 4

PLASTIVAN LTD
Unit 4 Bonville Trading Estate,
Bonville Road, Brislington, Bristol, BS5 0EB
Area of Operation: Worldwide
Tel: 0117 300 5625 **Fax:** 0117 971 5028
Email: sales@plastivan.co.uk
Web: www.plastivan.co.uk
Product Type: 1, 3, 4 **Other Info:** ▽

SIESTA CORK TILE CO
Unit 21, Tait Road, Gloucester Road,
Croydon, Surrey, CR0 2DP
Area of Operation: UK (Excluding Ireland)
Tel: 0208 683 4055 **Fax:** 0208 663 4480
Email: siestacork@aol.com
Web: www.siestacorktiles.co.uk
Product Type: 4 **Material Type:** M) 1

STATELY HOMES PANELLING
81 Hursley Road, Chandlers Ford,
Eastleigh, Hampshire, SO53 2FS
Area of Operation: UK (Excluding Ireland)
Tel: 02380 255765   **Material Type:** A) 2

STATELY HOMES PANELLING

Area of Operation: UK
Tel: 02380 255765
Email: lambradle@aol.com

Traditional waxed real wood panelling, made the old fashioned way.
Panelling made in any shape, size or type to any height in most timbers.

SWISH BUILDING PRODUCTS LTD
Pioneer House, Mariner, Litchfield Road Industrial Estate, Tamworth, Staffordshire, B79 7TF
Area of Operation: UK & Ireland
Tel: 01827 317200   **Fax:** 01827 317201
Email: info@swishbp.co.uk
Web: www.swishbp.co.uk
Material Type: D) 1

**THE CARPENTRY INSIDER
- AIRCOMDIRECT**
1 Castleton Crescent, Skegness,
Lincolnshire, PE25 2TJ
Area of Operation: Worldwide
Tel: 01754 767163
Email: aircom8@hotmail.com
Web: www.carpentry.tk ⌂
Product Type: 3, 4

THE GRANITE FACTORY
4 Winchester Drive, Peterlee,
Durham, SR8 2RJ
Area of Operation: North East England
Tel: 0191 518 3600   **Fax:** 0191 518 3600
Email: admin@granitefactory.co.uk
Web: www.granitefactory.co.uk ⌂

**THE WORLDWIDE WOOD
COMPANY LIMITED**
154 Colney Hatch Lane, Muswell Hill,
London, N10 1ER
Area of Operation: Worldwide
Tel: 0800 458 3366
Fax: 0208 365 3965
Email: info@solidwoodflooring.com
Web: www.solidwoodflooring.com
Product Type: 2

TIMBER RECLAIMED
PO Box 33727, London, SW3 4XE
Area of Operation: UK & Ireland
Tel: 0207 824 9200 **Fax:** 0207 824 9207
Email: elaine.barker@dmbarker.com
Web: www.timberreclaimed.com
Product Type: 1 **Other Info:** ▽

VICTORIAN WOOD WORKS LTD
54 River Road, Creekmouth,
Barking, Essex, IG11 0DW
Area of Operation: Worldwide
Tel: 0208 534 1000   **Fax:** 0208 534 2000
Email: sales@victorianwoodworks.co.uk
Web: www.victorianwoodworks.co.uk

VISION ASSOCIATES
Demita House, North Orbital Road,
Denham, Buckinghamshire, UB9 5EY
Area of Operation: UK & Ireland
Tel: 01895 831600 **Fax:** 01895 835323
Email: info@visionassociates.co.uk
Web: www.visionassociates.co.uk **Product Type:** 2

WOODHOUSE TIMBER
Unit 6 Quarry Farm Industrial Estate,
Staplecross Road, Bodiam, East Sussex, TN32 5RA
Area of Operation: UK (Excluding Ireland)
Tel: 01580 831700 **Fax:** 01580 830365
Email: info@woodhousetimber.co.uk
Web: www.woodhousetimber.co.uk
Material Type: A) 2

PAINTS, STAINS AND VARNISHES

KEY

PRODUCT TYPES: 1= Interior Paint 2 = Exterior Paint 3 = Stains and Varnishes
OTHER: ▽ Reclaimed ⌂ On-line shopping
✏ Bespoke ✋ Hand-made   ECO Ecological

BEDEC PRODUCT LTD.
Units 1 & 2 Poplars Farm, Aythorpe Roding,
Dunmow, Essex, CM6 1RY
Area of Operation: Worldwide
Tel: 01279 876657   **Fax:** 01279 876008
Email: info@bedec.co.uk   **Web:** www.bedec.co.uk

BONAKEMI LIMITED
1 Radian Court, Davy Avenue, Knowlhill,
Milton Keynes, Buckinghamshire, MK5 8PJ
Area of Operation: UK & Ireland
Tel: 01908 399740 **Fax:** 01908 232722
Email: info.uk@bona.com
Web: www.bona.com
Product Type: 3

CENTRE FOR ALTERNATIVE TECHNOLOGY
Llwyngweren Quarry, Machynlleth, Powys, SY20 9AZ
Area of Operation: Worldwide
Tel: 01654 705950   **Fax:** 01654 702782
Email: lucy.stone@cat.org.uk
Web: www.cat.org.uk ⌂   **Other Info:** ECO

COMFOOT FLOORING
Walnut Tree, Redgrave Rd, South Lopham,
Diss, Norfolk, IP22 2HN
Area of Operation: Europe
Tel: 01379 688516   **Fax:** 01379 688517
Email: sales@comfoot.com
Web: www.comfoot.com ⌂

COMFOOT FLOORING
Area of Operation: Europe
Tel: 01379 688516
Fax: 01379 688517
Email: sales@comfoot.com
Web: www.comfoot.com ⌂
Material Type: H) 4, 5

Day after day, the floor in your kid's room takes the beating of it's life, but who wants to get the mop out every night?
Comfoot Flooring suppliers of Witex Products

ECOSHOP
Unit 1, Glen of the Downs Garden Centre,
Kilmacanogue, Co Wicklow, Republic of Ireland
Area of Operation: Ireland Only
Tel: +353 01 287 2914   **Fax:** +353 01 201 6480
Email: info@ecoshop.ie **Web:** www.ecoshop.ie

FLEXI COATINGS UK
Meridion House, Heron Way,
Truro, Cornwall, TR1 2XN
Area of Operation: UK (Excluding Ireland)
Tel: 0870 080 2313   **Fax:** 01872 260066
Email: info@flexicoatings.com
Web: www.flexicoatings.com   **Product Type:** 2

HOLKHAM LINSEED PAINTS
The Clock Tower, Longlands, Holkham Park,
Wells-next-the-Sea, Norfolk, NR23 1RU
Area of Operation: UK & Ireland
Tel: 01328 711348   **Fax:** 01328 710368
Email: linseedpaint@holkham.co.uk
Web: www.holkhamlinseedpaints.co.uk ⌂
Product Type: 1, 2   **Other Info:** ECO

INTERNATIONAL PAINTS
Meadow Lane, St Ives,
Cambridgeshire, PE27 4UY
Area of Operation: UK & Ireland
Tel: 01480 484284
Web: www.international-paints.co.uk
Product Type: 1, 2

INTERNATIONAL PAINTS
Area of Operation: UK & Ireland
Tel: 01480 484284
Web: www.international-paints.co.uk
Product Type: 1,2
International's range of radiator paints provide tough, decorative protection against wear & tear. Paints are available in a choice of finishes & colours including chrome & stainless steel, & for the ultimate in flexible colour choice there is International's Radiator Clearcoat. (Prices start from rrp £4.98 for Quick Drying Radiator Enamel)

KEIM MINERAL PAINTS LTD
Muckley Cross, Morville, Nr Bridgnorth,
Shropshire, WV16 4RR
Area of Operation: Worldwide
Tel: 01746 714543 **Fax:** 01746 714526
Email: Sales@Keimpaints.co.uk
Web: www.keimpaints.co.uk
Product Type: 1, 2 **Other Info:** ECO

MALABAR PAINTWORKS
31-33 The South Bank Business Centre,
Ponton Road, London, SW8 4B
Area of Operation: Europe
Tel: 0207 5014 200   **Fax:** 0207 501 4210
Email: info@malabar.co.uk
Web: www.malabar.co.uk ⌂
Product Type: 1 **Other Info:** ✏

MID ARGYLL GREEN SHOP
An Tairbeart, Campbeltown Road, Tarbet,
Argyll & Bute, PA29 6SX
Area of Operation: Scotland
Tel: 01880 821212
Email: kate@mackandmags.co.uk
Web: www.mackandmags.co.uk ⌂
Product Type: 1, 2, 3
Other Info: ECO

MIKE WYE & ASSOCIATES
Buckland Filleigh Sawmills, Buckland Filleigh,
Beaworthy, Devon, EX21 5RN
Area of Operation: Worldwide
Tel: 01409 281644
Fax: 01409 281669
Email: sales@mikewye.co.uk
Web: www.mikewye.co.uk
Product Type: 1, 3

NATURAL BUILDING TECHNOLOGIES
The Hangar, Worminghall Road, Oakley,
Buckinghamshire, HP18 9UL
Area of Operation: UK & Ireland
Tel: 01844 338338 **Fax:** 01844 338525
Email: info@natural-building.co.uk
Web: www.natural-building.co.uk **Other Info:** ECO

NEVER PAINT AGAIN INTERNATIONAL
Claro Court Business Centre, Claro Road,
Harrogate, West Yorkshire, HG1 4BA
Area of Operation: Worldwide
Tel: 0870 067 0643 **Fax:** 0870 067 0643
Email: info@neverpaintagain.co.uk
Web: www.neverpaintagain.co.uk
Product Type: 1, 2, 3

NUTSHELL NATURAL PAINTS
PO Box 72, South Brent, Devon, TQ10 9YR
Area of Operation: Worldwide
Tel: 0870 033 1140
Fax: 01364 73801
Email: info@nutshellpaints.com
Web: www.nutshellpaints.com ⌂
Product Type: 1, 2, 3

OLD HOUSE STORE LTD
Hampstead Farm, Binfield Heath,
Henley on Thames,
Oxfordshire, RG9 4LG
Area of Operation: Worldwide
Tel: 0118 969 7711
Fax: 0118 969 8822
Email: info@oldhousestore.co.uk
Web: www.oldhousestore.co.uk ⌂
Other Info: ECO

ORIGINAL STYLE
Falcon Road, Sowton Industrial Estate,
Exeter, Devon, EX2 7LF
Area of Operation: Worldwide
Tel: 01392 474011
Fax: 01392 219932
Email: vmattison@originalstyle.com
Web: www.originalstyle.com **Product Type:** 1

OSMO UK LTD
Unit 2 Pembroke Road, Stocklake Industrial Estate,
Aylesbury, Buckinghamshire, HP20 1DB
Area of Operation: UK & Ireland
Tel: 01296 481220 **Fax:** 01296 424090
Email: info@osmouk.com **Web:** www.osmouk.com

RAY MUNN LTD
861-863 Fulham Road, London, SW6 5HP
Area of Operation: UK (Excluding Ireland)
Tel: 0207 736 9876
Email: rishi@raymunn.com
Web: www.raymunn.com
Product Type: 1, 2, 3 **Other Info:** ECO

SANDED FLOORS
7 Pleydell Avenue, Upper Norwood,
London, SE19 2LN
Area of Operation: South East England
Tel: 0209 653 1326
Email: peter-weller@sandedfloors.co.uk
Web: www.sandedfloors.co.uk ⌂
Product Type: 3

INTERIORS, FIXTURES & FINISHES - **Floor & Wall Finishes** - Access Covers; Tile Adhesives & Grout; Plaster, Plasterboard & Dry Lining

SPONSORED BY: PERGO LTD www.pergo.com

THE GREEN SHOP
Cheltenham Road, Bisley, Nr Stroud,
Gloucestershire, GL6 7BX
Area of Operation: UK & Ireland
Tel: 01452 770629 **Fax:** 01452 770104
Email: paint@greenshop.co.uk
Web: www.greenshop.co.uk ⌐
Product Type: 3 **Other Info:** ECO ✋

WILTON STUDIOS
Cleethorpes Business Centre,
Wilton Road Industrial Estate,
Grimsby, Lincolnshire, DN36 4AS
Area of Operation: UK (Excluding Ireland)
Tel: 01472 210820   **Fax:** 01472 812602
Email: postbox@wiltonstudios.co.uk
Web: www.wiltonstudios.co.uk

ACCESS COVERS

KEY
OTHER: ▽ Reclaimed ⌐ On-line shopping
✎ Bespoke ✋ Hand-made   ECO Ecological

EGGER (UK) LIMITED
Anick Grange Road, Hexham,
Northumberland, NE46 4JS
Area of Operation: UK & Ireland
Tel: 01434 602191
Fax: 01434 605103
Email: building.uk@egger.com
Web: www.egger.co.uk

INTERFLOW UK
Leighton, Shrewsbury, Shropshire, SY5 6SQ
Area of Operation: UK & Ireland
Tel: 01952 510050
Fax: 01952 510967
Email: villiers@interflow.co.uk
Web: www.interflow.co.uk

TILE ADHESIVES AND GROUT

KEY
OTHER: ▽ Reclaimed ⌐ On-line shopping
✎ Bespoke ✋ Hand-made   ECO Ecological

BUILDING ADHESIVES LTD
Longton Road, Trentham, Stoke on Trent,
Staffordshire, ST4 8JB
Area of Operation: Worldwide
Tel: 01782 591100
Fax: 01782 591193
Email: info@building-adhesives.com
Web: www.building-adhesives.com

CERAMIQUE INTERNATIONALE LTD
Unit 1 Royds Lane, Lower Wortley Ring Road,
Leeds, West Yorkshire, LS12 6DU
Area of Operation: UK & Ireland
Tel: 0113 231 0218
Fax: 0113 231 0353
Email: cameron@ceramiqueinternationale.co.uk
Web: www.ceramiqueinternationale.co.uk

**DEGUSSA CONSTRUCTION
CHEMICALS (UK)**
Albany House, Swinton Hall Road,
Swinton, Manchester, M27 4DT
Area of Operation: Worldwide
Tel: 0161 794 7411
Fax: 0161 727 8547
Email: mbtfeb@degussa.com
Web: www.degussa-cc.co.uk

DEVON STONE LTD
38 Station Road, Budleigh
Salterton, Devon, EX9 6RT
Area of Operation: UK (Excluding Ireland)
Tel: 01395 446841
Email: amy@devonstone.co.uk

EUROPEAN HERITAGE
48-54 Dawes Road, Fulham, London, SW6 7EN
Area of Operation: Worldwide
Tel: 0207 381 6063 **Fax:** 0207 381 9534
Email: fulham@europeanheritage.co.uk
Web: www.europeanheritage.co.uk

H & R JOHNSON TILES LTD
Harewood Street, Tunstall,
Stoke on Trent, Staffordshire, ST6 5JZ
Area of Operation: Worldwide
Tel: 01782 575 575 **Fax:** 01782 524138
Email: techsales@johnson-tiles.com
Web: www.johnson-tiles.com

INDIGENOUS LTD
Cheltenham Road, Burford, Oxfordshire, OX18 4JA
Area of Operation: Worldwide
Tel: 01993 824200 **Fax:** 01993 824300
Email: enquiries@indigenoustiles.com
Web: www.indigenoustiles.com

MAESTROTILE
Unit 7 Waterloo Park, Waterloo Road Industrial
Estate, Bidford on Avon, Warwickshire, B50 4JH
Area of Operation: UK & Ireland
Tel: 01789 778700 **Fax:** 01789 778852
Email: malcolm@maestrotile.com
Web: www.maestrotile.com

N&C NICOBOND
41-51 Freshwater Road, Chadwell Heath,
Romford, Essex, RM8 1SP
Area of Operation: Worldwide
Tel: 0208 586 4600   **Fax:** 0208 586 4646
Email: info@nichollsandclarke.com
Web: www.ncdirect.co.uk

ROCK UNIQUE LTD
c/o Select Garden and Pet Centre,
Main Road, Sundridge, Kent, TN14 6ED
Area of Operation: Europe
Tel: 01959 565 608 **Fax:** 01959 569 312
Email: stone@rock-unique.com
Web: www.rock-unique.com

STONELL DIRECT
100-105 Victoria Crescent,
Burton upon Trent, Staffordshire, DE14 2QF
Area of Operation: Worldwide
Tel: 0800 083 2283 **Fax:** 01283 501098
Email: liz.may@stonelldirect.com
Web: www.stonelldirect.com

TOPPS TILES
Thorpe Way, Grove Park, Enderby, Leicestershire, LE19 1SU
Area of Operation: UK (Excluding Ireland)
Tel: 0116 282 8000 **Fax:** 0116 282 8100
Email: mlever@toppstiles.co.uk
Web: www.toppstiles.co.uk

TOPPS TILES

Area of Operation: UK (Excluding Ireland)
Tel: 0116 282 8000 **Fax:** 0116 282 8100
Email: mlever@toppstiles.co.uk
Web: www.toppstiles.co.uk

Topps Tiles is Britain's biggest tile and wood
flooring specialist, with over 200 stores
nationwide. For details of your nearest store
or free brochure call 0800 138 1673
www.toppstiles.co.uk.

PLASTER, PLASTERBOARD AND DRY LINING

KEY
PRODUCT TYPES: 1= Resin Bonded Plaster
2 = Gypsum Based Plaster
3 = Lightweight Plaster 4 = Textured Coatings
5 = Plaster Bonding 6 = Other Plasters
7 = Square-edged Plasterboard
8 = Feather-edged Plasterboard
9 = Fibre-reinforced Plasterboard
10 = Insulation-backed Plasterboard
OTHER: ▽ Reclaimed ⌐ On-line shopping
✎ Bespoke ✋ Hand-made   ECO Ecological

**ALLTEK UK (INTERNATIONAL
COATING PRODUCTS LTD)**
The Old Railway Buildings, The Silk Road,
Macclesfield, Cheshire, SK10 1JD
Area of Operation: UK & Ireland
Tel: 01625 434043
Fax: 01625 619070
Email: info@alltekuk.com
Web: www.alltekuk.com & www.icp-alltek.com
Product Type: 4, 6

ARTEX-RAWLPLUG
Pasture Lane, Ruddington, Nottingham,
Nottinghamshire, NG11 6AE
Area of Operation: UK (Excluding Ireland)
Tel: 0115 984 5679
Fax: 0115 940 5240
Email: info@bpb.com
Web: www.artex-rawlplug.co.uk

B.WILLIAMSON & DAUGHTERS
Copse Cottage, Ford Manor Road,
Dormansland, Lingfield, Surrey, RH7 6NZ
Area of Operation: Greater London,
South East England
Tel: 01342 834829
Fax: 01342 870390
Email: bryan.williamson@btclick.com
Web: www.specialistcleaning4me.co.uk
Product Type: 1, 3, 4, 5, 6

BRITISH GYPSUM LTD
East Leake, Loughborough, Leicestershire, LE12 6HX
Area of Operation: UK (Excluding Ireland)
Tel: 0115 945 1000
Fax: 0115 945 1901
Web: www.british-gypsum.com

CENTREPIECE MOULDINGS
The Plaster Shop, C6 Bersham
Enterprise Park, Wrexham, LL14 4EG
Area of Operation: North West England
and North Wales
Tel: 01978 363923
Email: plasterware@yahoo.co.uk
Web: www.centrepiecemouldings.co.uk
Product Type: 2

DEESIDE HOMES TIMBERFRAME
Broomhill Road, Spurryhillock Industrial Estate,
Stonehaven, Near Aberdeen, Aberdeenshire, AB39 2NH
Area of Operation: UK (Excluding Ireland)
Tel: 01569 767123
Fax: 01569 767766
Email: john.wright@bancon.co.uk
Web: www.bancon.co.uk

ELLIOTTS INSULATION AND DRYLINING
Unit 8 Goodwood Road, Boyatt Wood Industrial
Estate, Eastleigh, Hampshire, SO50 4NT
Area of Operation: South East England,
South West England and South Wales
Tel: 02380 623960
Fax: 02380 623965
Email: insulation@elliott-brothers.co.uk
Web: www.elliotts.uk.com
Product Type: 2, 3, 4, 5, 7, 8, 9, 10

HEATHFIELD COMBINED SERVICES LTD
Creative House, 81d Crayford Way,
Crayford, Kent, DA1 4JY
Area of Operation: Europe
Tel: 01322 552255 **Fax:** 01322 524488
Email: des@plasterer.co.uk
Web: www.plasterer.co.uk ⌐
Product Type: 1, 2, 3, 4, 5, 6

J & A PLASTERING
44 Cobden Avenue, Southampton, Hampshire, SO18 1FW
Area of Operation: South East England
Tel: 02380 552180
Email: crosbykriss@aol.com **Product Type:** 2, 3, 5, 6

KNAUF DIY
PO Box 732, Maidstone, Kent, ME15 6ST
Area of Operation: UK (Excluding Ireland)
Tel: 0845 601 1763 **Fax:** 0845 601 1762
Email: knaufdiy@knauf.co.uk
Web: www.teachmediy.co.uk

MID ARGYLL GREEN SHOP
An Tairbeart, Campbeltown Road,
Tarbet, Argyll & Bute, PA29 6SX
Area of Operation: Scotland
Tel: 01880 821212
Email: kate@mackandmags.co.uk
Web: www.mackandmags.co.uk ⌐
Product Type: 6 **Other Info:** ECO

OLD HOUSE STORE LTD
Hampstead Farm, Binfield Heath,
Henley on Thames, Oxfordshire, RG9 4LG
Area of Operation: Worldwide
Tel: 0118 969 7711 **Fax:** 0118 969 8822
Email: info@oldhousestore.co.uk
Web: www.oldhousestore.co.uk ⌐

POLISHED PLASTER
1 Hardgate Lane, Cross Roads,
Keighley, West Yorkshire, BD21 5PS
Area of Operation: UK & Ireland
Tel: 07789 861315 **Fax:** 0870 762 3738
Email: info@polishedplaster.co.uk
Web: www.polishedplaster.co.uk
Product Type: 4, 6 **Other Info:** ✎ ✋

RAY MUNN LTD
861-863 Fulham Road, London, SW6 5HP
Area of Operation: UK (Excluding Ireland)
Tel: 0207 736 9876
Email: rishi@raymunn.com
Web: www.raymunn.com
Product Type: 4, 6 **Other Info:** ✎

THE ARTFUL PLASTERER
Unit 1 Lovedere Farm, Goathurst,
Bridgewater, Somerset, TA5 2DD
Area of Operation: Worldwide
Tel: 0870 333 6335 **Fax:** 0870 333 6339
Email: colour@theartfulplasterer.co.uk
Web: www.theartfulplasterer.co.uk

MOULDINGS AND ARCHITRAVE

KEY
PRODUCT TYPES: 1= Ceiling Centres
2 = Beams 3 = Cornices 4 = Coving
5 = Dado Rails 6 = Edge Trims 7 = Panel
Mouldings 8 = Picture Rails 9 = Skirting
10 = Window Boards 11 = Other
OTHER: ▽ Reclaimed ⌐ On-line shopping
✎ Bespoke ✋ Hand-made   ECO Ecological

ANDY THORNTON LTD
Ainleys Industrial Estate, Elland,
West Yorkshire, HX5 9JP
Area of Operation: Worldwide
Tel: 01422 375 595 **Fax:** 01422 377 455
Email: email@ataa.co.uk
Web: www.andythornton.com
Product Type: 3, 5, 7

INTERIORS, FIXTURES & FINISHES

BALMORAL MOULDINGS
Balmoral Park, Lourston, Aberdeenshire, AB12 3GY
Area of Operation: UK (Excluding Ireland)
Tel: 01224 859100
Fax: 01224 859123
Email: group@balmoral.co.uk
Web: www.balmoral-group.com

BOB LANE WOODTURNERS
Unit 1, White House Workshop, Old London Road,
Swinfen, Lichfield, Staffordshire, WS14 9QW
Area of Operation: UK (Excluding Ireland)
Tel: 01543 483148
Fax: 01543 481245
Email: sales@theturner.co.uk
Web: www.theturner.co.uk

BRIGHTON MOULDINGS LTD
12 Preston Road, Brighton, East Sussex, BN1 4QF
Area of Operation: East England, Greater London,
South East England, South West England & South Wales
Tel: 01273 622230 **Fax:** 01273 622240
Email: info@brightonmouldings.com
Web: www.brightonmouldings.com
Product Type: 1, 3, 4, 5, 6, 7, 8, 9

CENTREPIECE MOULDINGS
The Plaster Shop, C6 Bersham Enterprise Park,
Wrexham, LL14 4EG
Area of Operation: North West England
and North Wales
Tel: 01978 363923
Email: plasterware@yahoo.co.uk
Web: www.centrepiecemouldings.co.uk
Product Type: 1, 3, 4, 5, 7, 8

COPLEY DECOR LTD
Leyburn Business Park, Leyburn,
North Yorkshire, DL8 5QA
Area of Operation: UK & Ireland
Tel: 01969 623410 **Fax:** 01969 624398
Email: mouldings@copleydecor.co.uk
Web: www.copleydecor.co.uk
Product Type: 1, 3, 4, 5, 7, 11

CORNICES CENTRE
2 Slade Court, 230 Walm Lane, London, NW2 3BT
Area of Operation: Greater London
Tel: 0208 452 3310 **Fax:** 0208 452 3310
Email: info@cornicescentre.co.uk
Web: www.cornicescentre.co.uk
Product Type: 1, 3, 4, 5, 7, 8, 9, 11

DAVUKA GROUP LTD
2C The Wend, Coulsdon, Surrey, CR5 2AX
Area of Operation: UK (Excluding Ireland)
Tel: 0208 660 2854 **Fax:** 0208 645 2556
Email: info@davuka.co.uk
Web: www.davuka.co.uk
Product Type: 1, 3, 4, 5, 6, 7, 8, 9

DAVUKA GROUP LTD

Area of Operation: UK (Excluding Ireland)
Tel: 0208 660 2854 **Fax:** 0208 645 2556
Email: info@davuka.co.uk
Web: www.decorative-coving.co.uk
Product Type: 1,3,4,5,6,7,8,9
Complete range of high quality decorative
mouldings, internal & external application.
Profiles viewable in 3D on our website:
www.decorative-coving.co.uk or call for
brochure: 020 8660 2854 (Davuka GRP Ltd)

DRESSER MOULDINGS LTD
Lostock Industrial Estate West, Cranfield Road,
Lostock, Bolton, Lancashire, BL6 4SB
Area of Operation: UK (Excluding Ireland)
Tel: 01204 667667
Fax: 01204 667600
Email: mac@dresser.uk.com
Web: www.dresser.uk.com
Product Type: 3, 5, 7, 8, 9

HAYLES & HOWE LTD
25 Picton Street, Montpelier, Bristol, BS6 5PZ
Area of Operation: Worldwide
Tel: 0117 924 6673
Fax: 0117 924 3928
Email: info@haylesandhowe.co.uk
Web: www.haylesandhowe.co.uk
Product Type: 1, 3, 4, 5, 6, 7, 8, 9, 11

J & A PLASTERING
44 Cobden Avenue, Southampton,
Hampshire, SO18 1FW
Area of Operation: South East England
Tel: 02380 552180
Email: crosbykriss@aol.com
Product Type: 1, 3, 4, 5, 6, 7, 8, 9

LANCASHIRE PLASTER PRODUCTS
167 Chorley New Road, Horwich, Bolton,
Lancashire, BL6 5QE
Area of Operation: UK (Excluding Ireland)
Tel: 01204 693900
Email: plasterproducts@gahope.com
Web: www.gahope.com
Product Type: 1, 3, 5, 6, 7, 11

LEEWAY HOME IMPROVEMENTS
19 The Hurst, Moseley, Birmingham,
West Midlands, B13 0DA
Area of Operation: UK & Ireland
Tel: 0121 777 5460
Email: info@lee-hath.dircon.co.uk
Web: www.ukhomeinteriors.co.uk
Product Type: 1, 3, 4, 5, 6, 7, 9, 10

LOCKER & RILEY
(FIBROUS PLASTERING) LIMITED
Capital House, Hawk Hill, Battlebridge,
Wickford, Essex, SS11 7RJ
Area of Operation: UK (Excluding Ireland)
Tel: 01268 574100 **Fax:** 01268 574101
Email: jonriley@lockerandriley.com
Web: www.lockerandriley.com
Product Type: 1, 2, 3, 4, 5, 6, 7, 8, 9

LONDON FINE ART PLASTER
Unit 3, Romeo Business Centre,
Juliet Way, Purfleet, Essex, RM15 4YD
Area of Operation: UK (Excluding Ireland)
Tel: 01708 252400 **Fax:** 01708 252401
Email: enquiry@londonfineartplaster.com
Web: www.londonfineartplaster.com
Product Type: 1, 2, 3, 4, 5, 6, 7, 8, 9, 11

NATIONAL DOOR COMPANY
Unit 55 Dinting Vale Business Park,
Dinting Vale, Glossop, Derbyshire, SK13 6JD
Area of Operation: UK (Excluding Ireland)
Tel: 01457 867079 **Fax:** 01457 868795
Email: sales@nationaldoor.co.uk
Web: www.nationaldoor.co.uk
Product Type: 9

NMC (UK) LTD
Tafarnaubach Industrial Estate,
Tredegar, Blaenau Gwent, NP22 3AA
Area of Operation: UK & Ireland
Tel: 01495 713266 **Fax:** 01495 713277
Email: enquiries@nmc-uk.com
Web: www.nmc-uk.com
Product Type: 1, 2, 3, 4, 5

nmc (uk) Ltd
Area of Operation: UK & Ireland
Tel: 01495 713266 **Fax:** 01495 713277
Email: enquiries@nmc-uk.com
Web: www.nmc-uk.com
Product Type: 1, 2, 3, 4, 5

A quality range of lightweight, high performance and high definition ceiling centres, covings, dado rails and skirting boards. Easy to cut, shape and install.

OAKLEAF (GLASGOW)
Birch Court, Doune, Stirlingshire, FK16 6JD
Area of Operation: Scotland
Tel: 01786 842216
Fax: 01786 842216
Email: sales@oakleaf.co.uk
Web: www.oakleaf.co.uk
Product Type: 1, 2, 3, 4, 5, 6, 7, 8, 9

OAKLEAF REPRODUCTIONS LTD
Main Street, Wilsden, West Yorkshire, BD15 0JP
Area of Operation: Worldwide
Tel: 01535 272878
Fax: 01535 275748
Email: sales@oakleaf.co.uk
Web: www.oakleaf.co.uk
Product Type: 2

**OAKLEY BESPOKE
REPRODUCTION BEAMS**
31Fort, Picklecombe Maker,
Torpoint, Cornwall, PL10 1JB
Area of Operation: Worldwide
Tel: 01752 829299
Fax: 01752 829299
Email: shaun@repro-beams.freeserve.co.uk
Web: www.reproduction-beams.co.uk
Product Type: 2

OAKMASTERS
The Mill, Isaacs Lane, Haywards Heath,
West Sussex, RH16 4RZ
Area of Operation: UK (Excluding Ireland)
Tel: 01444 455455 **Fax:** 01444 455333
Email: oak@oakmasters.co.uk
Web: www.oakmasters.co.uk
Product Type: 2

OCTAVEWARD LTD
Balle Street Mill, Balle Street,
Darwen, Lancashire, BB3 2AZ
Area of Operation: Worldwide
Tel: 01254 773300
Fax: 01254 773950
Email: info@octaveward.com
Web: www.octaveward.com

ORAC (UK) LIMITED
Unit 5, Hewitts Estate, Elmbridge Road,
Cranleigh, Surrey, GU6 8LW
Area of Operation: Worldwide
Tel: 01483 271211
Fax: 01483 278317
Email: stewart@oracdecor.com
Web: www.oracdecor.com
Product Type: 1, 3, 4, 5, 6, 7, 9, 11

RENAISSANCE MOULDINGS
262 Handsworth Road, Handsworth,
Sheffield, South Yorkshire, S13 9BS
Area of Operation: UK (Excluding Ireland)
Tel: 01142 446622
Fax: 01142 610472
Email: enq@rfppm.com
Web: www.rfppm.co.uk

REVIVAL DECORATIVE MOULDINGS
2 Park Street, Ampthill, Bedfordshire, MK45 2LR
Area of Operation: Worldwide
Tel: 01525 406690
Fax: 01525 406690
Email: sales@revivalplaster.co.uk
Web: www.revivalplaster.co.uk
Product Type: 1, 2, 3, 4, 5, 7, 8, 9, 11
Other Info: ✍ ✎

RICHARD BURBIDGE LTD
Whittington Road, Oswestry,
Shropshire, SY11 1HZ
Area of Operation: Worldwide
Tel: 01691 655131
Fax: 01691 657694
Email: info@richardburbidge.co.uk
Web: www.richardburbidge.co.uk
Product Type: 5, 6, 7, 8, 9

SCULPTURE GRAIN LIMITED
Unit 8 Warren Court,
Knockholt Road, Halstead,
Sevenoaks, Kent, TN14 7ER
Area of Operation: UK & Ireland
Tel: 01959 534060
Fax: 01959 532696
Email: mick@sculpturegrain.freeserve.co.uk
Product Type: 2, 9, 11

STEVENSONS OF NORWICH
Roundtree Way, Norwich, Norfolk, NR7 8SQ
Area of Operation: Worldwide
Tel: 01603 400824
Fax: 01603 405113
Email: info@stevensons-of-norwich.co.uk
Web: www.stevensons-of-norwich.co.uk
Product Type: 1, 3, 4, 5, 7, 8, 11

THE SPA & WARWICK TIMBER CO LTD
Harriott Drive, Heathcote Industrial Estate,
Warwick, Warwickshire, CV34 6TJ
Area of Operation: Midlands & Mid Wales
Tel: 01926 883876
Fax: 01926 450831
Email: sales@spa-warwick.co.uk
Product Type: 5, 8, 9, 10, 11

WINTHER BROWNE
75 Bilton Way, Enfield, London, EN3 7ER
Area of Operation: UK (Excluding Ireland)
Tel: 0208 344 9050
Fax: 0845 612 1894
Email: sales@wintherbrowne.co.uk
Web: www.wintherbrowne.co.uk

WOODHOUSE TIMBER
Unit 6 Quarry Farm Industrial Estate,
Staplecross Road, Bodiam,
East Sussex, TN32 5RA
Area of Operation: UK (Excluding Ireland)
Tel: 01580 831700
Fax: 01580 830365
Email: info@woodhousetimber.co.uk
Web: www.woodhousetimber.co.uk
Product Type: 9

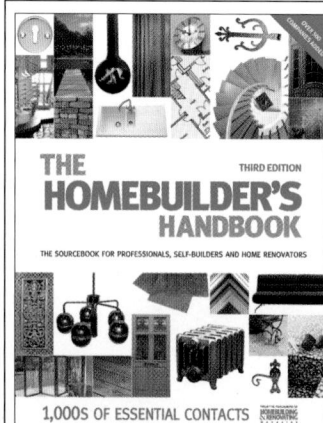
NOTES

Company Name ....
....
Address ....
....
....
email ....
Web ....

Company Name ....
....
Address ....
....
....
email ....
Web ....

Company Name ....
....
Address ....
....
....
email ....
Web ....

Company Name ....
....
Address ....
....
....
email ....
Web ....

INTERIORS, FIXTURES & FINISHES

BATHROOMS

Image courtesy of Pegler Limited (0870 120 0284)

SPONSORED BY PEGLER LIMITED
Tel 0870 120 0284 Web www.pegler.co.uk/francis

Pegler Limited
QUALITY • RELIABILITY • INNOVATION

INTERIORS, FIXTURES & FINISHES

Bathrooms

ABOVE: The key to a successful wetroom is effective water-proofing, drainage and ventilation. Tiles shown by Fired Earth

ABOVE: The luxury Riverbath by Kohler.
LEFT: Traditional baths can look stunning in modern bathrooms, as shown in this example by Alexanders.

Tailoring a bathroom to meet your ideal requirements can be the ultimate luxury. However, it is worth keeping in mind the practicality that any future buyers will be looking for when working on your bathroom, as this is one of the best places to add value to your home. A second bathroom can add around ten per cent to your house value, and a large bathroom will provide a hugely attractive selling point.

The first thing to prioritise is what will fit into your budget. An en suite and a downstairs WC can be excellent investments if you can afford them on top of a main bathroom, but where the budget is tight you need to decide what bathroom space will work best for you. There are ways of maximising even the smallest of main bathrooms; with the use of careful design, coupled with good quality space saving fixtures and fittings, you can ensure that your bathroom achieves its maximum potential.

Shower units are an obvious means of saving space, as well as offering ecological and economic benefits through the drop in water consumption. If space is holding back your design ambitions rather than budget, you can create a more luxurious shower area with the addition of a music system or steam release system.

A modern alternative to baths and showers is building a wet room, which involves fully waterproofing your bathroom and installing a shower and a large drain into the floor. The advantage here is that there is no

step or cill into the shower, which is especially useful for the elderly or those with young children. This design also creates more floor space, making the room appear more spacious, clean and stream-lined. Utilising all wall space, by fitting shelving and furniture to walls where possible, will reinforce the effect.

If, however, you are set on installing a bath - which, of course is advisable as a selling point for the future - a great way of exploiting all available space and capitalising on unusual room dimensions is to fit a corner bath. Installing a shower screen on your corner bath will increase utility even further, as will installing space-saving compact lavatories and basins, which can compensate for having a large bath, or will enable you to create an en-suite or WC out of the smallest space.

Installing mass produced fitted bathroom furniture is the cheapest way to create storage and minimise fuss. Saving money here will also free up some of your budget to spend on wall and floor finishes. Stone,

ceramic and quarry tiles are the hardest and most durable materials and a fully tiled bathroom will look impressive. However, tiled floors are cold underfoot so it is worth considering underfloor heating before the building work begins. Other accessories, such as heated mirrors or towel rails, will ensure that your bathroom is comfortable with a luxurious feel. When it comes to these bathroom accessories it is best to spend more on quality items in brass or chrome finishes, which will cope well with the steamy bathroom environment.

Adding an interesting tap design is the ideal way to finish off your bathroom and sourcing these, in antique shops and salvage yards for traditional designs, or in specialist outlets for more contemporary items, will again give your bathroom an individual stamp. Keep in mind that planning carefully and spending more on your bathroom at the design stage will reap two-fold benefits; providing comfort now, and increased profits when you come to sell.

BASINS

KEY

OTHER: ▽ Reclaimed 🖱 On-line shopping
✎ Bespoke ✋Hand-made ECO Ecological

A & H BRASS
201-203 Edgware Road, London, W2 1ES
Area of Operation: Worldwide
Tel: 0207 402 1854
Fax: 0207 402 0110
Email: ahbrass@btinternet.com
Web: www.aandhbrass.co.uk 🖱

ALEXANDERS
Unit 3, 1/1A Spilsby Road, Harold Hill,
Romford, Essex, RM3 8SB
Area of Operation: UK & Ireland
Tel: 01708 384574
Fax: 01708 384089
Email: sales@alexanders-uk.net
Web: www.alexanders-uk.net
Material Type: J) 5

ALICANTE STONE
Damaso Navarro, 6 Bajo, 03610 Petrer
(Alicante), P.O. Box 372, Spain
Area of Operation: Europe
Tel: +34 966 319 697
Fax: +34 966 319 698
Email: info@alicantestone.com
Web: www.alicantestone.com
Material Type: E) 2, 5, 8

AMBIANCE BAIN
Cumberland House,
80 Scrubs Lane, London, NW10 6RF
Area of Operation: UK & Ireland
Tel: 0870 902 1313
Fax: 0870 902 1312
Email: sghirardello@mimea.com
Web: www.ambiancebain.com

ANTIQUE BATHS OF IVYBRIDGE LTD
Emebridge Works, Ermington Road,
Ivybridge, Devon, PL21 9DE
Area of Operation: UK (Excluding Ireland)
Tel: 01752 698250
Fax: 01752 698266
Email: sales@antiquebaths.com
Web: www.antiquebaths.com 🖱

AQUAPLUS SOLUTIONS
Unit 106, Indiana Building, London, SE13 7QD
Area of Operation: UK & Ireland
Tel: 0870 201 1915
Fax: 0870 201 1916
Email: info@aquaplussolutions.com
Web: www.aquaplussolutions.com

BARRHEAD SANITARYWARE PLC
15-17 Nasmyth Road, Hillington Industrial Estate,
Hillington, Glasgow, Renfrewshire, G52 4RG
Area of Operation: UK & Ireland
Tel: 0141 883 0066
Fax: 0141 883 0077
Email: sales@barrhead.co.uk

BATHROOM BARN
Uplands, Wyton, Huntingdon,
Cambridgeshire, PE28 2JZ
Area of Operation: UK (Excluding Ireland)
Tel: 01480 458900
Email: sales@bathroombarn.co.uk
Web: www.bathroombarn.co.uk

BATHROOM CITY
Seeleys Road, Tyseley Industrial Estate,
Birmingham, West Midlands, B11 2LQ
Area of Operation: UK & Ireland
Tel: 0121 753 0700
Fax: 0121 753 1110
Email: sales@bathroomcity.com
Web: www.bathroomcity.com 🖱

BATHROOM DISCOUNT CENTRE
297 Munster Road, Fulham,
Greater London, SW6 6BW
Area of Operation: UK (Excluding Ireland)
Tel: 0207 381 4222
Fax: 0207 381 6792
Web: www.bathroomdiscount.co.uk

BATHROOM OPTIONS
394 Ringwood Road, Ferndown, Dorset, BH22 9AU
Area of Operation: South East England
Tel: 01202 873379
Fax: 01202 892252
Email: bathroom@options394.fsnet.co.uk
Web: www.bathroomoptions.co.uk

BATHROOMS PLUS LTD
222 Malmesbury Park Road,
Bournemouth, Dorset, BH8 8PR
Area of Operation: South West England
and South Wales
Tel: 01202 294417
Fax: 01202 316425
Email: info@bathrooms-plus.co.uk
Web: www.bathrooms-plus.co.uk
Material Type: C) 3, 13, 14, 16, 18

BATHROOMSTUFF.CO.UK
40 Evelegh Road , Farlington,
Portsmouth, Hampshire, PO6 1DL
Area of Operation: UK (Excluding Ireland)
Tel: 0845 058 0540
Fax: 02392 215695
Email: sales@bathroomstuff.co.uk
Web: www.bathroomstuff.co.uk 🖱

**BROOK-WATER DESIGNER
BATHROOMS & KITCHENS**
The Downs, Woodhouse Hill, Uplyme,
Lyme Regis, Dorset, DT7 3SL
Area of Operation: Worldwide
Tel: 01235 201256
Email: sales@brookwater.co.uk
Web: www.brookwater.co.uk 🖱
Other Info: ✎

CAPITAL MARBLE DESIGN
Unit 1 Pall Mall Deposit,
124-128 Barlby Road, London, W10 6BL
Area of Operation: UK & Ireland
Tel: 0208 968 5340
Fax: 0208 968 8827
Email: stonegallery@capitalmarble.co.uk
Web: www.capitalmarble.co.uk
Other Info: ✎ ✋

CAPLE
Fourth Way, Avonmouth, Bristol, BS11 8DW
Area of Operation: UK (Excluding Ireland)
Tel: 0870 606 9606
Fax: 0117 938 7449
Email: amandalowe@mlay.co.uk
Web: www.mlay.co.uk

CIFIAL UK LTD
7 Faraday Court, Park Farm Industrial Estate,
Wellingborough, Northamptonshire, NN8 6XY
Area of Operation: UK & Ireland
Tel: 01933 402008
Fax: 01933 402063
Email: sales@cifial.co.uk
Web: www.cifial.co.uk

CITY BATHROOMS & KITCHENS
158 Longford Road, Longford,
Coventry, West Midlands, CV6 6DR
Area of Operation: UK & Ireland
Tel: 02476 365877
Fax: 02476 644992
Email: citybathrooms@hotmail.com
Web: www.citybathrooms.co.uk

COSY ROOMS (THE-BATHROOM-SHOP.CO.UK)
17 Chiltern Way, North Hykeham,
Lincoln, Lincolnshire, LN6 9SY
Area of Operation: UK (Excluding Ireland)
Tel: 01522 696002 **Fax:** 01522 696002
Email: keith@cosy-rooms.com
Web: www.the-bathroom-shop.co.uk 🖱

CP HART GROUP
Newnham Terrace,
Hercules Road, London, SE1 7DR
Area of Operation: Greater London
Tel: 0207 902 5250
Fax: 0207 902 1030
Email: paulr@cphart.co.uk
Web: www.cphart.co.uk

CROSSWATER
Unit 5 Butterly Avenue,
Questor, Dartford, Kent, DA1 1JG
Area of Operation: UK & Ireland
Tel: 01322 628270
Fax: 01322 628280
Email: sales@crosswater.co.uk
Web: www.crosswater.co.uk
Material Type: J) 5

CYMRU KITCHENS
63 Caerleon Road, St. Julians,
Newport, NP19 7BX
Area of Operation: UK (Excluding Ireland)
Tel: 01633 676767 **Fax:** 01633 212512
Email: sales@cymrukitchens.com
Web: www.cymrukitchens.com

DART VALLEY SYSTEMS LIMITED
Kemmings Close, Long Road,
Paignton, Devon, TQ4 7TW
Area of Operation: UK & Ireland
Tel: 01803 529021
Fax: 01803 559016
Email: sales@dartvalley.co.uk
Web: www.dartvalley.co.uk

DECORLUX LTD
18 Ghyll Industrial Estate, Ghyll Road,
Heathfield, East Sussex, TN21 8AW
Area of Operation: UK (Excluding Ireland)
Tel: 01435 866638 **Fax:** 01435 866641
Email: info@decorlux.co.uk
Other Info: ▽ ✋

DECORMASTER LTD
Unit 16, Waterside Industrial Estate,
Wolverhampton, West Midlands, WV2 2RH
Area of Operation: Worldwide
Fax: 01902 353126
Email: sales@oldcolours.co.uk
Web: www.oldcolours.co.uk

DEEP BLUE SHOWROOM
299-313 Lewisham High Street,
Lewisham, London, SE13 6NW
Area of Operation: UK (Excluding Ireland)
Tel: 0208 690 3401 **Fax:** 0208 690 1408

DESIGNER BATHROOMS
140-142 Pogmoor Road, Pogmoor,
Barnsley, North Yorkshire, S75 2DX
Area of Operation: UK (Excluding Ireland)
Tel: 01226 280200 **Fax:** 01226 733273
Email: mike@designer-bathrooms.com
Web: www.designer-bathrooms.com

DEVON STONE LTD
38 Station Road, Budleigh Salterton, Devon, EX9 6RT
Area of Operation: UK (Excluding Ireland)
Tel: 01395 446841
Email: amy@devonstone.com
Other Info: ✎ ✋
Material Type: E) 1, 2, 3, 4, 5, 8

DOLPHIN BATHROOMS
Bromwich Road, Worcester,
Worcestershire, WR2 4BD
Area of Operation: UK (Excluding Ireland)
Tel: 0800 626717
Web: www.dolphinbathrooms.co.uk

DUPONT CORIAN & ZODIAQ
10 Quarry Court, Pitstone Green Business Park,
Pitstone, nr. Tring, Buckinghamshire, LU7 9GW
Area of Operation: UK & Ireland
Tel: 01296 663555
Fax: 01296 663599
Email: sales@corian.co.uk
Web: www.corian.co.uk

DURAVIT
Unit 7, Stratus Park, Brudenell Drive, Brinklow,
Milton Keynes, Buckinghamshire, MK10 0DE
Area of Operation: UK & Ireland
Tel: 0870 730 7787
Fax: 0870 730 7786
Email: info@uk.duravit.com
Web: www.duravit.co.uk

GOODWOOD BATHROOMS LTD
Church Road, North Mundham, Chichester,
West Sussex, PO20 1JU
Area of Operation: UK (Excluding Ireland)
Tel: 01243 532121
Fax: 01243 533423
Email: sales@goodwoodbathroom.co.uk
Web: www.goodwoodbathrooms.co.uk

HERITAGE BATHROOMS
Princess Street, Bedminster, Bristol, BS3 4AG
Area of Operation: UK & Ireland
Tel: 0117 963 3333
Email: marketing@heritagebathrooms.com
Web: www.heritagebathrooms.com

IDEAL STANDARD
The Bathroom Works, National Avenue, Kingston
Upon Hull, East Riding of Yorks, HU5 4HS
Area of Operation: Worldwide
Tel: 0800 590311
Fax: 01482 445886
Email: ideal-standard@aseur.com
Web: www.ideal-standard.co.uk

INNOVATION GLASS COMPANY
27-28 Blake Industrial Estate, Brue Avenue,
Bridgwater, Somerset, TA7 8EQ
Area of Operation: Europe
Tel: 01278 683645
Fax: 01278 450088
Email: magnus@igc-uk.com
Web: www.igc-uk.com

INSPIRED BATHROOMS
Unit R4, Innsworth Technology Park, Innsworth Lane,
Gloucester, Gloucestershire, GL3 1DL
Area of Operation: UK & Ireland
Tel: 01452 559121
Fax: 01452 530908
Email: sales@inspired-bathrooms.co.uk
Web: www.inspired-bathrooms.co.uk 🖱

ION GLASS
Unit 7 Grange Industrial Estate, Albion Street,
Southwick, Brighton, West Sussex, BN42 4EN
Area of Operation: UK (Excluding Ireland)
Tel: 0845 658 9988
Fax: 0845 658 9989
Email: sales@ionglass.co.uk
Web: www.ionglass.co.uk

J & R MARBLE COMPANY LTD
Unit 9,Period Works, Lammas Road,
Leyton, London, E10 7QT
Area of Operation: UK (Excluding Ireland)
Tel: 0208 539 6471
Fax: 0208 539 9264
Email: sales@jrmarble.co.uk
Web: www.jrmarble.co.uk
Other Info: ✎
Material Type: E) 1, 2, 3, 4, 5, 8, 9, 13, 14

JACUZZI UK
Silverdale Road, Newcastle under Lyme,
Staffordshire, ST5 6EL
Area of Operation: UK & Ireland
Tel: 01782 717175 **Fax:** 01782 717166
Email: jacuzzisalesdesk@jacuzziuk.com
Web: www.jacuzzi.co.uk
Material Type: F) 1

JACUZZI UK - BC SANITAN
Silverdale Road, Newcastle under Lyme,
Staffordshire, ST5 6EL
Area of Operation: UK & Ireland
Tel: 01782 717175 **Fax:** 01782 717166
Email: jacuzzisalesdesk@jacuzziuk.com
Web: www.jacuzziuk.com
Material Type: F) 1

bespoke

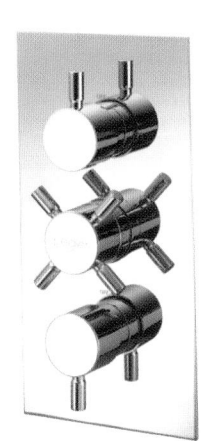

design

Francis Pegler matching thermostatic showers, taps and mixers give bathrooms a prestigious bespoke look. These superb creations redefine the art of bathroom design and offer an aesthetically rewarding visual and sensual experience.

For the complete range of Francis Pegler bathroom products, visit the Pegler website or call the number below for your free copy of the sumptuous new Francis Pegler brochures.

francis pegler

Pegler Limited
QUALITY • RELIABILITY • INNOVATION

St. Catherine's Avenue, Doncaster, South Yorkshire, DN4 8DF.
Fax: 01302 560109   E-mail: uk.sales@pegler.co.uk   Tel: **0870 1200284**

5 YEAR GUARANTEE
Showers

10 YEAR GUARANTEE
Taps & Mixers

an Aalberts Industries company

KENSINGTON STUDIO
13c/d Kensington Road, Earlsdon,
Coventry, West Midlands, CV5 6GG
Area of Operation: UK (Excluding Ireland)
Tel: 02476 713326
Fax: 02476 713136
Email: sales@kensingtonstudio.com
Web: www.kensingtonstudio.com

KERAMAG WAVENEY LTD
London Road, Beccles, Suffolk, NR34 8TS
Area of Operation: UK & Ireland
Tel: 01502 716600
Fax: 01502 717767
Email: info@keramagwaveney.co.uk
Web: www.keramag.com

KIRKSTONE QUARRIES LIMITED
Skelwith Bridge, Ambleside, Cumbria, LA22 9NN
Area of Operation: UK & Ireland
Tel: 01539 433296
Fax: 01539 434006
Email: info@kirkstone.com
Web: www.kirkstone.com

LECICO PLC
Unit 47a Hobbs Industrial Estate,
Newchapel, Nr Lingfield, Surrey, RH7 6HN
Area of Operation: UK & Ireland
Tel: 01342 834777
Fax: 01342 834783
Email: info@lecico.co.uk
Web: www.lecico.co.uk

M & O BATHROOM CENTRE
174-176 Goswell Road,
Clarkenwell, London, EC1V 7DT
Area of Operation: East England,
Greater London, South East England
Tel: 0207 608 0111
Fax: 0207 490 3083
Email: mando@lineone.net

MARBLE CLASSICS
Unit 3, Station Approach, Emsworth,
Hampshire, PO10 7PW
Area of Operation: UK & Ireland
Tel: 01243 370011
Fax: 01243 371023
Email: info@marbleclassics.co.uk
Web: www.marbleclassics.co.uk
Material Type: E) 1, 2, 3, 5, 8

MEKON PRODUCTS
25 Bessemer Park,
Milkwood Road, London, SE24 0HG
Area of Operation: Greater London
Tel: 0207 733 8011
Fax: 0207 737 0840
Email: info@mekon.net
Web: www.mekon.net

MEREWAY CONTRACTS
413 Warwick Road, Birmingham,
West Midlands, B11 2JR
Area of Operation: UK (Excluding Ireland)
Tel: 0121 707 3288
Fax: 0121 707 3871
Email: info@merewaycontracts.co.uk
Web: www.merewaycontracts.co.uk

MIKE WALKER DISTRIBUTION
Clutton Hill Estate, King Lane,
Clutton Hill, Bristol, BS39 5QQ
Area of Operation: Midlands & Mid Wales,
South West England and South Wales
Tel: 01761 453838
Fax: 01761 453060
Email: office.mwd@zoom.co.uk
Web: www.mikewalker.co.uk

**MISCELLANEA DISCONTINUED
BATHROOMWARE**
Churt Place Nurseries, Tilford Road,
Churt, Farnham, Surrey, GU10 2LN
Area of Operation: UK (Excluding Ireland)
Tel: 01428 608164 **Fax:** 01428 608165
Email: email@brokenbog.com
Web: www.brokenbog.com

Broken Basin?

www.brokenbog.com

MISCELLANEA
DISCONTINUED BATHROOMWARE
The Largest Collection Of
Bathroomware In The World
Specialists in Discontinued Colours
From replacement toilet seats
to complete bathroom suites
Over 100 colours in stock
See our main advert on p. 288 for details

NICHOLAS ANTHONY LTD
42-44 Wigmore Street, London, W1U 2RX
Area of Operation: East England,
Greater London, South East England
Tel: 0800 068 3603
Fax: 01206 762698
Email: info@nicholas-anthony.co.uk
Web: www.nicholas-anthony.co.uk

NO CODE
Larkwhistle Farm Road, Micheldever,
Hampshire, SO21 3BG
Area of Operation: UK & Ireland
Tel: 01962 870078
Fax: 01962 870077
Email: sales@nocode.co.uk
Web: www.nocode.co.uk

OLD FASHIONED BATHROOMS
Foresters Hall, 52 High Street, Debenham,
Stowmarket, Suffolk, IP14 6QW
Area of Operation: East England,
Greater London, South East England
Tel: 01728 860926
Fax: 01728 860446
Email: ofbrooms@lycos.co.uk
Web: www.oldfashionedbathrooms.co.uk
Other Info: ▽
Material Type: F) 1, 2, 4

QUALCERAM SHIRES PLC
Unit 9/10, Campus 5,
Off Third Avenue, Letchworth Garden City,
Hertfordshire, SG6 2JF
Area of Operation: UK & Ireland
Tel: 01462 676710
Fax: 01462 676701
Email: info@qualceram-shires.com
Web: www.qualceram-shires.com
Other Info: ✋

REGINOX UK LTD
Radnor Park Trading Estate,
Congleton, Cheshire, CW12 4XJ
Area of Operation: UK & Ireland
Tel: 01260 280033
Fax: 01260 298889
Email: sales@reginoxuk.co.uk
Web: www.reginox.com

**ROOTS KITCHENS,
BEDROOMS & BATHROOMS**
Vine Farm, Stockers Hill,
Boughton-under-Blean,
Faversham, Kent, ME13 9AB
Area of Operation: UK (Excluding Ireland)
Tel: 01227 751130
Fax: 01227 750033
Email: showroom@rootskitchens.co.uk
Web: www.rootskitchens.co.uk

ROSCO COLLECTIONS
Stone Allerton, Axbridge,
Somerset, BS26 2NS
Area of Operation: UK & Ireland
Tel: 01934 712299
Fax: 01934 713222
Email: jonathan@roscobathrooms.demon.co.uk

SHOWERLUX UK LTD
Sibree Road, Coventry,
West Midlands, CV3 4FD
Area of Operation: UK & Ireland
Tel: 02476 882515
Fax: 02476 305457
Email: sales@showerlux.co.uk
Web: www.showerlux.com

SMART SHOWERS LTD
Unit 11, Woodside Road,
South Marston Business Park,
Swindon, Wiltshire, SN3 4WA
Area of Operation: UK & Ireland
Tel: 01793 822775
Fax: 01793 823800
Email: louise@smartshowers.co.uk
Web: www.smartshowers.co.uk

SOGA LTD
41 Mayfield Street, Hull,
East Riding of Yorks, HU3 1NT
Area of Operation: Worldwide
Tel: 01482 327025
Email: info@soga.co.uk
Web: www.soga.co.uk

SPLASH DISTRIBUTION
113 High Street, Cuckfield,
West Sussex, RH17 5JX
Area of Operation: UK & Ireland
Tel: 01444 473355
Fax: 01444 473366
Email: stuart@splashdistribution.co.uk

STIFFKEY BATHROOMS
89 Upper Saint Giles Street,
Norwich, Norfolk, NR2 1AB
Area of Operation: Worldwide
Tel: 01603 627850
Fax: 01603 619775
Email: stiffkeybathrooms.norwich@virgin.net
Web: www.stiffkeybathrooms.com

STONELL DIRECT
100-105 Victoria Crescent,
Burton upon Trent, Staffordshire, DE14 2QF
Area of Operation: Worldwide
Tel: 08000 832283
Fax: 01283 501098
Email: liz.may@stonelldirect.com
Web: www.stonelldirect.com

STONEVILLE (UK) LTD
Unit 12, Set Star Estate,
Transport Avenue, Great West Road,
Brentford, Greater London, TW8 9HF
Area of Operation: Europe
Tel: 0208 560 1000
Fax: 0208 560 4060
Email: info@stoneville.co.uk
Web: www.stoneville.co.uk

STUDIO STONE
The Stone Yard, Alton Lane,
Four Marks, Hampshire, GU34 5AJ
Area of Operation: UK (Excluding Ireland)
Tel: 01420 562500
Fax: 01420 563192
Email: sales@studiostone.co.uk
Web: www.studiostone.co.uk
Material Type: E) 1, 2, 3, 5, 8

SVEDBERGS
London House, 100 New Kings Rd,
London, SW6 4LX
Area of Operation: UK & Ireland
Tel: 0207 348 6107 **Fax:** 0207 348 6108
Email: info@svedbergs.co.uk
Web: www.svedbergs.co.uk
Material Type: F) 1, 2

TAPS SHOP
35 Bridge Street, Witney, Oxfordshire, OX28 1DA
Area of Operation: Worldwide
Tel: 0845 430 3035
Fax: 01993 779653
Email: info@tapsshop.co.uk
Web: www.tapsshop.co.uk

THE DORSET CUPBOARD COMPANY
2 Mayo Farm, Cann, Shaftesbury, Dorset, SP7 0EF
Area of Operation: Greater London, South East
England, South West England and South Wales
Tel: 01747 855044 **Fax:** 01747 855045
Email: nchturner@aol.com

THE KITCHEN & BATHROOM COLLECTION
Nelson House, Nelson Road,
Salisbury, Wiltshire, SP1 3LT
Area of Operation: South West England
and South Wales
Tel: 01722 334800
Fax: 01722 412252
Email: info@kbc.co.uk
Web: www.kbc.co.uk
Other Info: ✎

THE WATER MONOPOLY
16/18 Lonsdale Road, London, NW6 6RD
Area of Operation: UK (Excluding Ireland)
Tel: 0207 624 2636
Fax: 0207 624 2631
Email: enquiries@watermonopoly.com
Web: www.watermonopoly.com

TRADEMARK TILES
Airfield Industrial Estate, Warboys,
Huntingdon, Cambridgeshire, PE28 2SH
Area of Operation: East England,
South East England
Tel: 01487 825300
Fax: 01487 825305
Email: info@trademarktiles.co.uk

VICTORIAN BATHROOMS
Ingsmill Complex, Dale Street,
Ossett, West Yorkshire, WF5 9HQ
Area of Operation: UK (Excluding Ireland)
Tel: 0845 130 6911
Fax: 01924 261223
Email: sales@victorianbathrooms.co.uk
Web: www.victorianbathrooms.co.uk

WALTON BATHROOMS
The Hersham Centre, The Green,
Molesey Road, Hersham, Walton on Thames,
Surrey, KT12 4HL
Area of Operation: UK (Excluding Ireland)
Tel: 01932 224784 **Fax:** 01932 253447
Email: sales@waltonbathrooms.co.uk
Web: www.waltonbathrooms.co.uk

WATER FRONT
All Worcester Buildings, Birmingham Road,
Redditch, Worcestershire, B97 6DY
Area of Operation: UK & Ireland
Tel: 01527 584244
Fax: 01527 61127
Email: waterfront@btconnect.com
Web: www.waterfrontbathrooms.com
Other Info: ✎
Material Type: E) 2, 5

WHIRLPOOL EXPRESS (UK) LTD
61-62 Lower Dock Street, Kingsway,
Newport, NP20 1EF
Area of Operation: Europe
Tel: 01633 244555
Fax: 01633 244555
Email: reception@whirlpoolexpress.co.uk
Web: www.whirlpoolexpress.co.uk

WILLIAM GARVEY
Leyhill, Payhembury, Honiton, Devon, EX14 3JG
Area of Operation: Worldwide
Tel: 01404 841430 **Fax:** 01404 841626
Email: webquery@williamgarvey.co.uk
Web: www.williamgarvey.co.uk
Other Info: ✎ ✋
Material Type: A) 10

WWW.BOUNDARYBATHROOMS.CO.UK
Regent Street, Colne, Lancashire, BB8 8LD
Area of Operation: UK & Ireland
Tel: 01282 862509
Fax: 01282 871192
Web: www.boundarybathrooms.co.uk

TAPS

KEY

SEE ALSO: MERCHANTS - Plumbers
Merchants; KITCHENS- Taps
OTHER: ▽ Reclaimed On-line shopping
 Bespoke Hand-made ECO Ecological

A & H BRASS
201-203 Edgware Road, London, W2 1ES
Area of Operation: Worldwide
Tel: 0207 402 1854
Fax: 0207 402 0110
Email: ahbrass@btinternet.com
Web: www.aandhbrass.co.uk
Material Type: C) 14, 16

A&J GUMMERS LIMITED
Unit H, Redfern Park Way, Tyseley,
Birmingham, West Midlands, B11 2DN
Area of Operation: Worldwide
Tel: 0121 706 2241
Fax: 0121 706 2960
Email: sales@gummers.com
Web: www.gummers.com

AL CHALLIS LTD
Europower House, Lower Road, Cookham,
Maidenhead, Berkshire, SL6 9EH
Area of Operation: UK & Ireland
Tel: 01628 529024
Fax: 0870 458 0577
Email: chris@alchallis.com
Web: www.alchallis.com

ALEXANDERS
Unit 3, 1/1A Spilsby Road, Harold Hill,
Romford, Essex, RM3 8SB
Area of Operation: UK & Ireland
Tel: 01708 384574
Fax: 01708 384089
Email: sales@alexanders-uk.net
Web: www.alexanders-uk.net

AMBIANCE BAIN
Cumberland House, 80 Scrubs Lane,
London, NW10 6RF
Area of Operation: UK & Ireland
Tel: 0870 902 1313
Fax: 0870 902 1312
Email: sghirardello@mimea.com
Web: www.ambiancebain.com

ANTIQUE BATHS OF IVYBRIDGE LTD
Emebridge Works, Ermington Road,
Ivybridge, Devon, PL21 9DE
Area of Operation: UK (Excluding Ireland)
Tel: 01752 698250
Fax: 01752 698266
Email: sales@antiquebaths.com
Web: www.antiquebaths.com

AQUALISA PRODUCTS LTD
The Flyers Way, Westerham, Kent, TN16 1DE
Area of Operation: UK & Ireland
Tel: 01959 560000
Fax: 01959 560030
Email: sue.anderson@aqualisa.co.uk
Web: www.aqualisa.co.uk

AQUAPLUS SOLUTIONS
Unit 106, Indiana Building, London, SE13 7QD
Area of Operation: UK & Ireland
Tel: 0870 201 1915
Fax: 0870 201 1916
Email: info@aquaplussolutions.com
Web: www.aquaplussolutions.com

BARRHEAD SANITARYWARE PLC
15-17 Nasmyth Road,
Hillington Industrial Estate, Hillington,
Glasgow, Renfrewshire, G52 4RG
Area of Operation: UK & Ireland
Tel: 0141 883 0066
Fax: 0141 883 0077
Email: sales@barrhead.co.uk

BATHROOM BARN
Uplands, Wyton, Huntingdon,
Cambridgeshire, PE28 2JZ
Area of Operation: UK (Excluding Ireland)
Tel: 01480 458900
Email: sales@bathroombarn.co.uk
Web: www.bathroombarn.co.uk

BATHROOM CITY
Seeleys Road, Tyseley Industrial Estate,
Birmingham, West Midlands, B11 2LQ
Area of Operation: UK & Ireland
Tel: 0121 753 0700
Fax: 0121 753 1110
Email: sales@bathroomcity.com
Web: www.bathroomcity.com

BATHROOM DISCOUNT CENTRE
297 Munster Road, Fulham,
Greater London, SW6 6BW
Area of Operation: UK (Excluding Ireland)
Tel: 0207 381 4222
Fax: 0207 381 6792
Web: www.bathroomdiscount.co.uk

BATHROOM EXPRESS
61-62 Lower Dock Street,
Kingsway, Newport, NP20 1EF
Area of Operation: UK & Ireland
Tel: 0845 130 2000
Fax: 01633 244881
Email: sales@bathroomexpress.co.uk
Web: www.bathroomexpress.co.uk

BATHROOM HEAVEN
25 Eccleston Square,
London, Sw1V 1NS
Area of Operation: UK & Ireland
Tel: 0845 121 6700
Fax: 0207 233 6074
Email: james@bathroomheaven.com
Web: www.bathroomheaven.com

BATHROOM OPTIONS
394 Ringwood Road,
Ferndown, Dorset, BH22 9AU
Area of Operation: South East England
Tel: 01202 873379
Fax: 01202 892252
Email: bathroom@options394.fsnet.co.uk
Web: www.bathroomoptions.co.uk

BATHROOMS PLUS LTD
222 Malmesbury Park Road,
Bournemouth, Dorset, BH8 8PR
Area of Operation: South West England
and South Wales
Tel: 01202 294417
Fax: 01202 316425
Email: info@bathrooms-plus.co.uk
Web: www.bathrooms-plus.co.uk

BATHROOMSTUFF.CO.UK
40 Evelegh Road, Farlington,
Portsmouth, Hampshire, PO6 1DL
Area of Operation: UK (Excluding Ireland)
Tel: 0845 058 0540
Fax: 023 9221 5695
Email: sales@bathroomstuff.co.uk
Web: www.bathroomstuff.co.uk

BROOK-WATER DESIGNER
BATHROOMS & KITCHENS
The Downs, Woodhouse Hill,
Uplyme, Lyme Regis, Dorset, DT7 3SL
Area of Operation: Worldwide
Tel: 01235 201256
Email: sales@brookwater.co.uk
Web: www.brookwater.co.uk

CAPLE
Fourth Way, Avonmouth, Bristol, BS11 8DW
Area of Operation: UK (Excluding Ireland)
Tel: 0870 606 9606
Fax: 0117 938 7449
Email: amandalowe@mlay.co.uk
Web: www.mlay.co.uk

CHARLES MASON LTD
Unit 11A Brook Street Mill, Off Goodall Street,
Macclesfield, Cheshire, SK11 7AW
Area of Operation: Worldwide
Tel: 0800 085 3616
Fax: 01625 668789
Email: info@charles-mason.com
Web: www.charles-mason.com
Material Type: C) 2, 3, 14

CIFIAL UK LTD
7 Faraday Court, Park Farm Industrial Estate,
Wellingborough, Northamptonshire, NN8 6XY
Area of Operation: UK & Ireland
Tel: 01933 402008
Fax: 01933 402063
Email: sales@cifial.co.uk
Web: www.cifial.co.uk

CIFIAL UK
Area of Operation: UK & Ireland
Tel: 01933 402008
Fax: 01933 402063
Email: sales@cifial.co.uk
Web: www.cifial.co.uk

Cifial's new contemporary TECHNOvation 35 range
offers solutions for both high & low water pressure.
Designed exclusively for Cifial by international award
winning designer, Carlos Aquiar.

CITY BATHROOMS & KITCHENS
158 Longford Road, Longford, Coventry,
West Midlands, CV6 6DR
Area of Operation: UK & Ireland
Tel: 02476 365877
Fax: 02476 644992
Email: citybathrooms@hotmail.com
Web: www.citybathrooms.co.uk

COSY ROOMS (THE-BATHROOM-SHOP.CO.UK)
17 Chiltern Way, North Hykeham,
Lincoln, Lincolnshire, LN6 9SY
Area of Operation: UK (Excluding Ireland)
Tel: 01522 696002
Fax: 01522 696002
Email: keith@cosy-rooms.com
Web: www.the-bathroom-shop.co.uk

CROSSWATER
Unit 5 Butterly Avenue, Questor,
Dartford, Kent, DA1 1JG
Area of Operation: UK & Ireland
Tel: 01322 628270
Fax: 01322 628280
Email: sales@crosswater.co.uk
Web: www.crosswater.co.uk

CYMRU KITCHENS
63 Caerleon Road, St. Julians, Newport, NP19 7BX
Area of Operation: UK (Excluding Ireland)
Tel: 01633 676767 **Fax:** 01633 212512
Email: sales@cymrukitchens.com
Web: www.cymrukitchens.com

DANICO BRASS LTD
31-35 Winchester Road, Swiss Cottage,
London, NW3 3NR
Area of Operation: Worldwide
Tel: 0207 483 4477
Fax: 0207 722 7992
Email: sales@danico.co.uk

DART VALLEY SYSTEMS LIMITED
Kemmings Close, Long Road,
Paignton, Devon, TQ4 7TW
Area of Operation: UK & Ireland
Tel: 01803 529021
Fax: 01803 559016
Email: sales@dartvalley.co.uk
Web: www.dartvalley.co.uk

DECORLUX LTD
18 Ghyll Industrial Estate, Ghyll Road,
Heathfield, East Sussex, TN21 8AW
Area of Operation: UK (Excluding Ireland)
Tel: 01435 866638
Fax: 01435 866641
Email: info@decorlux.co.uk

DEEP BLUE SHOWROOM
299-313 Lewisham High Street, Lewisham,
London, SE13 6NW
Area of Operation: UK (Excluding Ireland)
Tel: 0208 690 3401
Fax: 0208 690 1408

DESIGNER BATHROOMS
140-142 Pogmoor Road, Pogmoor,
Barnsley, North Yorkshire, S75 2DX
Area of Operation: UK (Excluding Ireland)
Tel: 01226 280200
Fax: 01226 733273
Email: mike@designer-bathrooms.com
Web: www.designer-bathrooms.com

DOLPHIN BATHROOMS
Bromwich Road, Worcester,
Worcestershire, WR2 4BD
Area of Operation: UK (Excluding Ireland)
Tei: 0800 626717
Web: www.dolphinbathrooms.co.uk

EUROBATH INTERNATIONAL LTD
Eurobath House, Wedmore Road,
Cheddar, Somerset, BS27 3EB
Area of Operation: Worldwide
Tel: 01934 744466
Fax: 01934 744345
Email: sales@eurobath.co.uk
Web: www.eurobath.co.uk

GEBERIT LTD
New Hythe Business Park, New Hythe Lane,
Aylesford, Kent, ME20 7PJ
Area of Operation: UK & Ireland
Tel: 01622 717811
Fax: 01622 716920
Web: www.geberit.co.uk

GROHE UK
1 River Road, Barking, Essex, IG11 0HD
Area of Operation: Worldwide
Tel: 0208 594 7292`
Fax: 0208 594 8898
Email: info@grohe.co.uk
Web: www.grohe.co.uk

HANSGROHE
Units D1 and D2, Sandown Park Trading Estate,
Royal Mills, Esher, Surrey, KT10 8BL
Area of Operation: Worldwide
Tel: 0870 770 1972
Fax: 0870 770 1973
Email: sales@hansgrohe.co.uk
Web: www.hansgrohe.co.uk
Material Type: C) 1, 2, 3, 11, 14

HERITAGE BATHROOMS
Princess Street, Bedminster, Bristol, BS3 4AG
Area of Operation: UK & Ireland
Tel: 0117 963 3333
Email: marketing@heritagebathrooms.com
Web: www.heritagebathrooms.com

INTERIORS, FIXTURES & FINISHES

Pegler Limited
QUALITY • RELIABILITY • INNOVATION

PEGLER LIMITED

St Catherine's Avenue,
Doncaster,
South Yorkshire
DN4 8DF

Tel: 0870 1200 284
Fax: 01302 580109
E-mail: uk.sales@pegler.co.uk
Website: www.pegler.co.uk

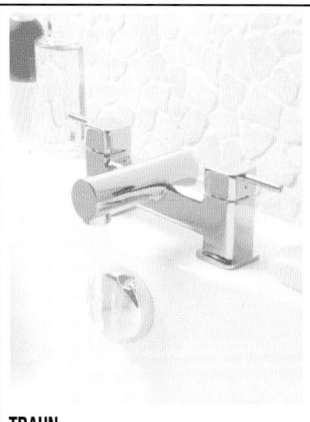

TRAUN

An avant garde minimalist design with a 'cubist' theme that is sure to enhance even the most up to the minute designer wet room.

LUSSO

Lusso's smoothly sculptured organic curves and delicate, leaf motif creates a natural theme which is perfect for eco-friendly designer homes.

VISIO

Visio's striking hardened chrome tubular design is a contemporary classic.

XIA

Xia period styling with a modern twist, swan neck mixers for the basin, bath and bidet makes this the complete range.

TIKO

The smooth organic shape of Tiko bathroom fittings are ergonomically designed to fit your hand perfectly and give an indefinable pleasure every time you use them.

SIGNIA

All models have a graceful curved lever control, which allows smooth, single handed adjustment of both water temperature and flow.

SLIQUE

The cylindrical theme of this Slique bath filler is perfect for today's designer wet rooms. The controls on all models in this minimalist designer range are ergonomically sculpted to allow finger touch water flow control.

OPAL

'Twist & turn' tap heads provide precise control at a touch on all models except the mono basin mixer which has an easy to use 'tilt head' lever control for simultaneous adjustment of water temperature and flow.

RIVEAU

Riveau's period styling is deceptively simple and recalls the timeless elegance of an earlier age. Beneath the flawless surface, Pegler advanced water control technology ensures superb performance and trouble free operation for years to come.

APPLAUSE

Smooth, organic curves and a deceptively understanding style make Applause perfect for today's bathrooms.

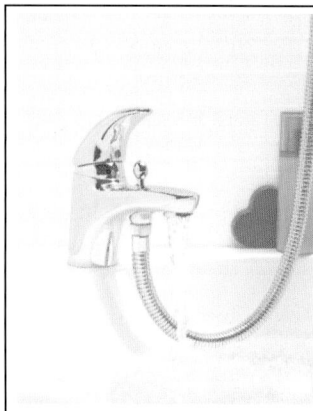

OVATION

Ovation's distinctive style is designed to blend beautifully with bathrooms of all kinds.

INTERIORS, FIXTURES & FINISHES

HUDSON REED
Rylands Street, Burnley, Lancashire, BB10 1RH
Area of Operation: Europe
Tel: 01282 418000
Fax: 01282 428915
Email: info@ultra-group.com
Web: www.hudsonreed.com 〜🖰
Material Type: C) 11, 14, 16, 17

IDEAL STANDARD
The Bathroom Works, National Avenue,
Kingston Upon Hull, East Riding of Yorks, HU5 4HS
Area of Operation: Worldwide
Tel: 0800 590311
Fax: 01482 445886
Email: ideal-standard@aseur.com
Web: www.ideal-standard.co.uk

INNOVATION GLASS COMPANY
27-28 Blake Industrial Estate,
Brue Avenue, Bridgwater,
Somerset, TA7 8EQ
Area of Operation: Europe
Tel: 01278 683645
Fax: 01278 450088
Email: magnus@igc-uk.com
Web: www.igc-uk.com

INSPIRED BATHROOMS
Unit R4, Innsworth Technology Park,
Innsworth Lane, Gloucester,
Gloucestershire, GL3 1DL
Area of Operation: UK & Ireland
Tel: 01452 559121
Fax: 01452 530908
Email: sales@inspired-bathrooms.co.uk
Web: www.inspired-bathrooms.co.uk 〜🖰

ITFITZ
PO Box 960, Maidenhead, Berkshire, SL6 9FU
Area of Operation: UK (Excluding Ireland)
Tel: 01628 890432
Fax: 0870 133 7955
Email: sales@itfitz.co.uk
Web: www.itfitz.co.uk 〜🖰

JACUZZI UK
Silverdale Road, Newcastle under Lyme,
Staffordshire, ST5 6EL
Area of Operation: UK & Ireland
Tel: 01782 717175
Fax: 01782 717166
Email: jacuzzisalesdesk@jacuzziuk.com
Web: www.jacuzzi.co.uk

JACUZZI UK - BC SANITAN
Silverdale Road, Newcastle under Lyme,
Staffordshire, ST5 6EL
Area of Operation: UK & Ireland
Tel: 01782 717175
Fax: 01782 717166
Email: jacuzzisalesdesk@jacuzziuk.com
Web: www.jacuzziuk.com

JOHN SYDNEY
Lagrange, Tamworth, Staffordshire, B79 7XD
Area of Operation: Worldwide
Tel: 01827 304000
Fax: 01827 68553
Email: enquiries@johnsydney.com
Web: www.johnsydney.com
Material Type: C) 11, 14

KENSINGTON STUDIO
13c/d Kensington Road, Earlsdon,
Coventry, West Midlands, CV5 6GG
Area of Operation: UK (Excluding Ireland)
Tel: 02476 713326
Fax: 02476 713136
Email: sales@kensingtonstudio.com
Web: www.kensingtonstudio.com 〜🖰

KERAMAG WAVENEY LTD
London Road, Beccles, Suffolk, NR34 8TS
Area of Operation: UK & Ireland
Tel: 01502 716600
Fax: 01502 717767
Email: info@keramagwaveney.co.uk
Web: www.keramag.com

KEUCO (UK)
2 Claridge Court,
Lower Kings Road, Berkhamsted,
Hertfordshire, HP42AF
Area of Operation: Worldwide
Tel: 01442 865220
Fax: 01442 865260
Email: klaus@keuco.co.uk
Web: www.keuco.de
Material Type: C) 3, 14, 17

KOHLER MIRA LIMITED
Gloucester Road, Cheltenham,
Gloucestershire, GL51 8TP
Area of Operation: UK & Ireland
Tel: 0870 850 5551
Fax: 0870 850 5552
Email: info@kohleruk.com
Web: www.kohleruk.com

LECICO PLC
Unit 47a Hobbs Industrial Estate,
Newchapel, Nr Lingfield,
Surrey, RH7 6HN
Area of Operation: UK & Ireland
Tel: 01342 834777
Fax: 01342 834783
Email: info@lecico.co.uk
Web: www.lecico.co.uk
Material Type: C) 3, 14, 16

M & O BATHROOM CENTRE
174-176 Goswell Road,
Clarkenwell, London, EC1V 7DT
Area of Operation: East England,
Greater London, South East England
Tel: 0207 608 0111
Fax: 0207 490 3083
Email: mando@lineone.net

MEKON PRODUCTS
25 Bessemer Park,
Milkwood Road,
London, SE24 0HG
Area of Operation: Greater London
Tel: 0207 733 8011
Fax: 0207 737 0840
Email: info@mekon.net
Web: www.mekon.net

MEREWAY CONTRACTS
413 Warwick Road, Birmingham,
West Midlands, B11 2JR
Area of Operation: UK (Excluding Ireland)
Tel: 0121 707 3288
Fax: 0121 707 3871
Email: info@merewaycontracts.co.uk
Web: www.merewaycontracts.co.uk

MIKE WALKER DISTRIBUTION
Clutton Hill Estate, King Lane,
Clutton Hill, Bristol, BS39 5QQ
Area of Operation: Midlands & Mid Wales,
South West England and South Wales
Tel: 01761 453838
Fax: 01761 453060
Email: office.mwd@zoom.co.uk
Web: www.mikewalker.co.uk

MISCELLANEA DISCONTINUED BATHROOMWARE
Churt Place Nurseries,
Tilford Road, Churt,
Farnham, Surrey, GU10 2LN
Area of Operation: UK (Excluding Ireland)
Tel: 01428 608164
Fax: 01428 608165
Email: email@brokenbog.com
Web: www.brokenbog.com

NICHOLAS ANTHONY LTD
42-44 Wigmore Street,
London, W1U 2RX
Area of Operation: East England,
Greater London, South East England
Tel: 0800 068 3603
Fax: 01206 762698
Email: info@nicholas-anthony.co.uk
Web: www.nicholas-anthony.co.uk

NO CODE
Larkwhistle Farm Road,
Micheldever, Hampshire, SO21 3BG
Area of Operation: UK & Ireland
Tel: 01962 870078
Fax: 01962 870077
Email: sales@nocode.co.uk
Web: www.nocode.co.uk

OLD FASHIONED BATHROOMS
Foresters Hall, 52 High Street,
Debenham, Stowmarket, Suffolk, IP14 6QW
Area of Operation: East England,
Greater London, South East England
Tel: 01728 860926
Fax: 01728 860446
Email: ofbrooms@lycos.co.uk
Web: www.oldfashionedbathrooms.co.uk
Other Info: ▽

PEGLER LIMITED
St Catherine's Avenue,
Doncaster, South Yorkshire, DN4 8DF
Area of Operation: Worldwide
Tel: 0870 120 0284
Fax: 01302 560109
Email: uk.sales@pegler.co.uk
Web: www.pegler.co.uk/francis

POLYPIPE KITCHENS & BATHROOMS LTD
Warmsworth, Halt Industrial Estate,
South Yorkshire, DN4 9LS
Area of Operation: Worldwide
Tel: 01709 770990
Fax: 01302 310602
Email: davery@ppbp.co.uk
Web: www.polypipe.com/bk

QUALCERAM SHIRES PLC
Unit 9/10, Campus 5, Off Third Avenue,
Letchworth Garden City, Hertfordshire, SG6 2JF
Area of Operation: UK & Ireland
Tel: 01462 676710
Fax: 01462 676701
Email: info@qualceram-shires.com
Web: www.qualceram-shires.com

REGINOX UK LTD
Radnor Park Trading Estate,
Congleton, Cheshire, CW12 4XJ
Area of Operation: UK & Ireland
Tel: 01260 280033
Fax: 01260 298889
Email: sales@reginoxuk.co.uk
Web: www.reginox.com 〜🖰

ROOTS KITCHENS, BEDROOMS & BATHROOMS
Vine Farm, Stockers Hill,
Boughton-under-Blean,
Faversham, Kent, ME13 9AB
Area of Operation: UK (Excluding Ireland)
Tel: 01227 751130 **Fax:** 01227 750033
Email: showroom@rootskitchens.co.uk
Web: www.rootskitchens.co.uk 〜🖰

ROSCO COLLECTIONS
Stone Allerton, Axbridge,
Somerset, BS26 2NS
Area of Operation: UK & Ireland
Tel: 01934 712299 **Fax:** 01934 713222
Email: jonathan@roscobathrooms.demon.co.uk

SAMUEL HEATH
Leopold Street, Birmingham,
West Midlands, B12 0UJ
Area of Operation: Worldwide
Tel: 0121 772 2303 **Fax:** 0121 772 3334
Email: lloyd@samuel-heath.com
Web: www.samuel-heath.com
Material Type: C) 11, 13, 14, 16

SMART SHOWERS LTD
Unit 11, Woodside Road, South Marston Business
Park, Swindon, Wiltshire, SN3 4WA
Area of Operation: UK & Ireland
Tel: 01793 822775 **Fax:** 01793 823800
Email: louise@smartshowers.co.uk
Web: www.smartshowers.co.uk 〜🖰

SOGA LTD
41 Mayfield Street, Hull,
East Riding of Yorks, HU3 1NT
Area of Operation: Worldwide
Tel: 01482 327025
Email: info@soga.co.uk
Web: www.soga.co.uk

SPLASH DISTRIBUTION
113 High Street, Cuckfield,
West Sussex, RH17 5JX
Area of Operation: UK & Ireland
Tel: 01444 473355
Fax: 01444 473366
Email: stuart@splashdistribution.co.uk

STIFFKEY BATHROOMS
89 Upper Saint Giles Street,
Norwich, Norfolk, NR2 1AB
Area of Operation: Worldwide
Tel: 01603 627850
Fax: 01603 619775
Email: stiffkeybathrooms.norwich@virgin.net
Web: www.stiffkeybathrooms.com

SVEDBERGS
London House,
100 New Kings Road,
London, SW6 4LX
Area of Operation: UK & Ireland
Tel: 0207 348 6107
Fax: 0207 348 6108
Email: info@svedbergs.co.uk
Web: www.svedbergs.co.uk

SWADLING BRASSWARE - A MATKI COMPANY
Churchward Road, Yate, Bristol, BS37 5PL
Area of Operation: Europe
Tel: 01454 322888
Fax: 01454 315284
Email: enquiries@swadlingbrassware.com
Web: www.swadlingbrassware.com

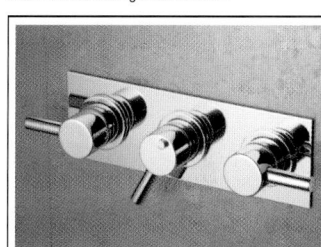

SWADLING BRASSWARE A MATKI COMPANY

Area of Operation: Europe
Tel: 01454 322 888 **Fax:** 01454 315 284
Email: enquiries@swadlingbrassware.com
Web: www.swadlingbrassware.com

Swadling thermostatic showers and taps
combine state-of-the-art technology with lasting
quality. In addition to the mid-priced Précis
range, Swadling also has two exclusive
colllections aimed at the upper end of the
market.

TAPS SHOP
35 Bridge Street, Witney,
Oxfordshire, OX28 1DA
Area of Operation: Worldwide
Tel: 0845 430 3035
Fax: 01993 779653
Email: info@tapsshop.co.uk
Web: www.tapsshop.co.uk 〜🖰

THE DORSET CUPBOARD COMPANY
2 Mayo Farm, Cann, Shaftesbury, Dorset, SP7 0EF
Area of Operation: Greater London, South East
England, South West England and South Wales
Tel: 01747 855044
Fax: 01747 855045
Email: nchturner@aol.com

INTERIORS, FIXTURES & FINISHES

new passions in the bathroom

If, like us, you're passionate about design, you'll love the three new tap ranges from Francis Pegler. Giving a sensuous visual and tactile experience each time they are used. The robust design brings the latest water control technology and a touch of finesse to today's designer bathrooms. See the complete range on our website, or call for a copy of our latest brochure.

10 YEAR GUARANTEE

St. Catherine's Avenue, Doncaster, South Yorkshire, DN4 8DF
Tel: 0870 1200284   Fax: 01302 560109
Email:uk.sales@pegler.co.uk   **www.pegler.co.uk/francis**

Pegler Limited
QUALITY • RELIABILITY • INNOVATION

THE KITCHEN AND BATHROOM COLLECTION
Nelson House, Nelson Road,
Salisbury, Wiltshire, SP1 3LT
Area of Operation: South West England
and South Wales
Tel: 01722 334800 **Fax:** 01722 412252
Email: info@kbc.co.uk
Web: www.kbc.co.uk
Other Info: ✏

THE WATER MONOPOLY
16/18 Lonsdale Road, London, NW6 6RD
Area of Operation: UK (Excluding Ireland)
Tel: 0207 624 2636 **Fax:** 0207 624 2631
Email: enquiries@watermonopoly.com
Web: www.watermonopoly.com

TRADEMARK TILES
Airfield Industrial Estate, Warboys,
Huntingdon, Cambridgeshire, PE28 2SH
Area of Operation: East England,
South East England
Tel: 01487 825300 **Fax:** 01487 825305
Email: info@trademarktiles.co.uk

TRITON PLC
Triton Road, Shepperton Business Park,
Caldwell Road, Nuneaton,
Warwickshire, CV11 4NR
Area of Operation: UK & Ireland
Tel: 02476 344441 **Fax:** 02476 349828
Web: www.tritonshowers.co.uk

URBIS (GB) LTD.
55 Lidgate Crescent,
Langthwaite Grange Industrial Estate,
South Kirkby, West Yorkshire, WF9 3NR
Area of Operation: UK & Ireland
Tel: 01977 659829 **Fax:** 01977 659808
Email: david@urbis-gb.co.uk
Web: www.urbis-gb.co.uk ✏

WALTON BATHROOMS
The Hersham Centre, The Green, Molesey Road,
Hersham, Walton on Thames, Surrey, KT12 4HL
Area of Operation: UK (Excluding Ireland)
Tel: 01932 224784 **Fax:** 01932 253447
Email: sales@waltonbathrooms.co.uk
Web: www.waltonbathrooms.co.uk

WATER FRONT
All Worcester Buildings, Birmingham Road,
Redditch, Worcestershire, B97 6DY
Area of Operation: UK & Ireland
Tel: 01527 584244 **Fax:** 01527 61127
Email: waterfront@btconnect.com
Web: www.waterfrontbathrooms.com

WHIRLPOOL EXPRESS (UK) LTD
61-62 Lower Dock Street,
Kingsway, Newport, NP20 1EF
Area of Operation: Europe
Tel: 01633 244555
Fax: 01633 244555
Email: reception@whirlpoolexpress.co.uk
Web: www.whirlpoolexpress.co.uk ✏

WWW.BOUNDARYBATHROOMS.CO.UK
Regent Street, Colne, Lancashire, BB8 8LD
Area of Operation: UK & Ireland
Tel: 01282 862 509
Fax: 01282 871 192
Web: www.boundarybathrooms.co.uk

BATHS & BATHROOM SUITES

KEY

PRODUCT TYPES: 1= Standard Baths
2 = Roll Top Baths   3 = Whirlpool Baths
4 = Corner Baths   5 = Double Baths
6 = Shower Baths   7 = Inset Baths
8 = Round Baths   9 = Other
OTHER: ▽ Reclaimed   ✏ On-line shopping
✏ Bespoke   ✋ Hand-made   ECO Ecological

A & H BRASS
201-203 Edgware Road, London, W2 1ES
Area of Operation: Worldwide
Tel: 0207 402 1854
Fax: 0207 402 0110
Email: ahbrass@btinternet.com
Web: www.aandhbrass.co.uk ✏
Product Type: 1

AIRBATH
Aquabeau Limited, Swinnow Lane,
Leeds, West Yorkshire, LS13 4TY
Area of Operation: Worldwide
Tel: 0113 386 9150
Fax: 0113 239 3406
Email: sales@aquabeau.com
Web: www.airbath.co.uk
Product Type: 3

ALEXANDERS
Unit 3, 1/1A Spilsby Road,
Harold Hill, Romford, Essex, RM3 8SB
Area of Operation: UK & Ireland
Tel: 01708 384574
Fax: 01708 384089
Email: sales@alexanders-uk.net
Web: www.alexanders-uk.net
Product Type: 1, 3, 4, 5, 7, 8, 9

ALTHEA UK LTD
Concept House, Blanche Street,
Bradford, West Yorkshire, BD4 8DA
Area of Operation: UK (Excluding Ireland)
Tel: 01274 660770
Fax: 01274 667929
Email: sales@altheauk.com
Web: www.periodbathrooms.co.uk

AMBIANCE BAIN
Cumberland House,
80 Scrubs Lane, London, NW10 6RF
Area of Operation: UK & Ireland
Tel: 0870 902 1313
Fax: 0870 902 1312
Email: sghirardello@mimea.com
Web: www.ambiancebain.com

**ANTIQUE BATHS
OF IVYBRIDGE LTD**
Emebridge Works,
Ermington Road, Ivybridge,
Devon, PL21 9DE
Area of Operation: UK (Excluding Ireland)
Tel: 01752 698250
Fax: 01752 698266
Email: sales@antiquebaths.com
Web: www.antiquebaths.com ✏
Product Type: 1, 2, 3, 5, 6, 7

APPOLLO
Aquabeau Limited,
Swinnow Lane, Leeds,
West Yorkshire, LS13 4TY
Area of Operation: Worldwide
Tel: 0113 386 9150
Fax: 0113 239 3406
Email: sales@aquabeau.com
Web: www.appollobathing.co.uk
Product Type: 1, 6, 9

AQUABEAU LIMITED
Swinnow Lane, Leeds,
West Yorkshire, LS13 4TY
Area of Operation: Worldwide
Tel: 0113 386 9150
Fax: 0113 239 3406
Email: sales@aquabeau.com
Web: www.aquabeau.com
Product Type: 1, 3, 4, 5, 6, 7, 8

AQUACRAFT
Aquabeau Limited, Swinnow Lane,
Leeds, West Yorkshire, LS13 4TY
Area of Operation: Worldwide
Tel: 0113 386 9150
Fax: 0113 239 3406
Email: sales@aquabeau.com
Web: www.aquacraft.co.uk
Product Type: 1, 4, 5, 6, 7, 8

BARRHEAD SANITARYWARE PLC
15-17 Nasmyth Road,
Hillington Industrial Estate, Hillington,
Glasgow, Renfrewshire, G52 4RG
Area of Operation: UK & Ireland
Tel: 0141 883 0066
Fax: 0141 883 0077
Email: sales@barrhead.co.uk
Product Type: 1, 4

BATHROOM BARN
Uplands, Wyton, Huntingdon,
Cambridgeshire, PE28 2JZ
Area of Operation: UK (Excluding Ireland)
Tel: 01480 458900
Email: sales@bathroombarn.co.uk
Web: www.bathroombarn.co.uk

BATHROOM CITY
Seeleys Road, Tyseley Industrial Estate,
Birmingham, West Midlands, B11 2LQ
Area of Operation: UK & Ireland
Tel: 0121 753 0700
Fax: 0121 753 1110
Email: sales@bathroomcity.com
Web: www.bathroomcity.com
Product Type: 1, 2, 3, 4, 5, 6, 7, 8, 9

BATHROOM DISCOUNT CENTRE
297 Munster Road, Fulham,
Greater London, SW6 6BW
Area of Operation: UK (Excluding Ireland)
Tel: 0207 381 4222
Fax: 0207 381 6792
Web: www.bathroomdiscount.co.uk

BATHROOM EXPRESS
61-62 Lower Dock Street,
Kingsway, Newport, NP20 1EF
Area of Operation: UK & Ireland
Tel: 0845 130 2000
Fax: 01633 244881
Email: sales@bathroomexpress.co.uk
Web: www.bathroomexpress.co.uk
Product Type: 1, 2, 3, 4, 5, 6, 7, 8

BATHROOM HEAVEN
25 Eccleston Square,
London, Sw1V 1NS
Area of Operation: UK & Ireland
Tel: 0845 121 6700
Fax: 0207 233 6074
Email: james@bathroomheaven.com
Web: www.bathroomheaven.com
Product Type: 1, 2, 4, 9
Material Type: D) 5

BATHROOM OPTIONS
394 Ringwood Road,
Ferndown, Dorset, BH22 9AU
Area of Operation: South East England
Tel: 01202 873379
Fax: 01202 892252
Email: bathroom@options394.fsnet.co.uk
Web: www.bathroomoptions.co.uk
Product Type: 1, 2, 3, 4, 5, 6, 7, 8, 9

BATHROOMS PLUS LTD
222 Malmesbury Park Road,
Bournemouth, Dorset, BH8 8PR
Area of Operation: South West England
and South Wales
Tel: 01202 294417 **Fax:** 01202 316425
Email: info@bathrooms-plus.co.uk
Web: www.bathrooms-plus.co.uk
Product Type: 1, 2, 3, 4, 5, 6, 7, 8, 9
Material Type: C) 2

BATHROOMSTUFF.CO.UK
40 Evelegh Road , Farlington,
Portsmouth, Hampshire, PO6 1DL
Area of Operation: UK (Excluding Ireland)
Tel: 0845 058 0540 **Fax:** 02392 215695
Email: sales@bathroomstuff.co.uk
Web: www.bathroomstuff.co.uk
Product Type: 1, 2, 4, 5, 6, 7, 8, 9

BRONTE WHIRLPOOLS
Unit 10, Ryefield Way, Silsden,
West Yorkshire, BD20 0EF
Area of Operation: UK (Excluding Ireland)
Tel: 01535 656524 **Fax:** 01535 658823
Email: info@brontewhirlpools.co.uk
Web: www.brontewhirlpools.co.uk
Product Type: 3

**BROOK-WATER DESIGNER
BATHROOMS & KITCHENS**
The Downs, Woodhouse Hill, Uplyme,
Lyme Regis, Dorset, DT7 3SL
Area of Operation: Worldwide
Tel: 01235 201256
Email: sales@brookwater.co.uk
Web: www.brookwater.co.uk

CABUCHON BATHFORMS
Whitegate, White Lund Estate,
Lancaster, Lancashire, LA3 3BT
Area of Operation: Worldwide
Tel: 01524 66022 **Fax:** 01524 844927
Email: info@cabuchon.com
Web: www.cabuchon.com
Product Type: 1, 4, 7, 9
Other Info: ✎

CAPITAL MARBLE DESIGN
Unit 1 Pall Mall Deposit,
124-128 Barlby Road, London, W10 6BL
Area of Operation: UK & Ireland
Tel: 0208 968 5340
Fax: 0208 968 8827
Email: stonegallery@capitalmarble.co.uk
Web: www.capitalmarble.co.uk
Product Type: 1, 8

CAPLE
Fourth Way, Avonmouth, Bristol, BS11 8DW
Area of Operation: UK (Excluding Ireland)
Tel: 0870 606 9606
Fax: 0117 938 7449
Email: amandalowe@mlay.co.uk
Web: www.mlay.co.uk
Product Type: 2, 4, 6, 7

CITY BATHROOMS & KITCHENS
158 Longford Road,
Longford, Coventry,
West Midlands, CV6 6DR
Area of Operation: UK & Ireland
Tel: 02476 365877
Fax: 02476 644992
Email: citybathrooms@hotmail.com
Web: www.citybathrooms.co.uk

CLEARWATER COLLECTION
Enterprise House, Iron Works Park,
Bowling Back Lane, Bradford,
West Yorkshire, BD4 8SX
Area of Operation: UK (Excluding Ireland)
Tel: 01274 738140
Fax: 01274 732461
Email: enquiries@clearwater-collection.com
Web: www.clearwater-collection.com
Product Type: 2

**COSY ROOMS
(THE-BATHROOM-SHOP.CO.UK)**
17 Chiltern Way, North Hykeham,
Lincoln, Lincolnshire, LN6 9SY
Area of Operation: UK (Excluding Ireland)
Tel: 01522 696002
Fax: 01522 696002
Email: keith@cosy-rooms.com
Web: www.the-bathroom-shop.co.uk ✎
Product Type: 1, 2, 6

CP HART GROUP
Newnham Terrace,
Hercules Road, London, SE1 7DR
Area of Operation: Greater London
Tel: 0207 902 5250
Fax: 0207 902 1030
Email: paulr@cphart.co.uk
Web: www.cphart.co.uk

CYMRU KITCHENS
63 Caerleon Road, St. Julians,
Newport, NP19 7BX
Area of Operation: UK (Excluding Ireland)
Tel: 01633 676767
Fax: 01633 212512
Email: sales@cymrukitchens.com
Web: www.cymrukitchens.com

DECORMASTER LTD
Unit 16, Waterside Industrial Estate,
Wolverhampton, West Midlands, WV2 2RH
Area of Operation: Worldwide
Fax: 01902 353126
Email: sales@oldcolours.co.uk
Web: www.oldcolours.co.uk
Product Type: 1, 2, 3, 4, 5, 6, 7, 8

DEEP BLUE SHOWROOM
299-313 Lewisham High Street,
Lewisham, London, SE13 6NW
Area of Operation: UK (Excluding Ireland)
Tel: 0208 690 3401
Fax: 0208 690 1408

DESIGNER BATHROOMS
140-142 Pogmoor Road, Pogmoor,
Barnsley, North Yorkshire, S75 2DX
Area of Operation: UK (Excluding Ireland)
Tel: 01226 280200
Fax: 01226 733273
Email: mike@designer-bathrooms.com
Web: www.designer-bathrooms.com
Product Type: 1, 2, 3, 4, 5, 6, 7, 8, 9

DIY WETROOMS
PO Box 124, Poulton le Fylde, Lancashire, FY6 8WU
Area of Operation: UK (Excluding Ireland)
Tel: 0871 200 2206
Email: info@diywetrooms.com
Web: www.diywetrooms.com
Product Type: 1, 6

DOLPHIN BATHROOMS
Bromwich Road , Worcester,
Worcestershire, WR2 4BD
Area of Operation: UK (Excluding Ireland)
Tel: 0800 626717
Web: www.dolphinbathrooms.co.uk
Product Type: 2, 3, 4, 6, 7

DURAVIT
Unit 7, Stratus Park, Brudenell Drive, Brinklow,
Milton Keynes, Buckinghamshire, MK10 0DE
Area of Operation: UK & Ireland
Tel: 0870 730 7787
Fax: 0870 730 7786
Email: info@uk.duravit.com
Web: www.duravit.co.uk
Product Type: 1, 3, 4, 5, 7
Material Type: D) 5

ELLIOTT BROTHERS LTD
Millbank Wharf, Northam, Southampton,
Hampshire , SO14 5AG
Area of Operation: South East England,
South West England and South Wales
Tel: 02380 226852
Fax: 02380 638780
Email: laurenh@elliott-brothers.co.uk
Web: www.elliotts.uk.com
Product Type: 1, 2, 3, 4, 5, 6, 8

EUROCARE SHOWERS LTD
Unit 19, Doncaster Industry Park, Watch House Lane,
Bentley, Doncaster, South Yorkshire, DN5 9LZ
Area of Operation: Worldwide
Tel: 01302 788684 **Fax:** 01302 780010
Email: sales@eurocare-showers.com
Web: www.eurocare-showers.com
Product Type: 3

HAROLD MOORE & SON LTD
Rawson Spring Road, Hillsborough,
Sheffield, South Yorkshire, S6 1PD
Area of Operation: UK & Ireland
Tel: 0114 233 6161
Fax: 0114 232 6375
Email: admin@haroldmoorebaths.co.uk
Web: www.haroldmoorebaths.co.uk
Product Type: 1, 3, 4, 8

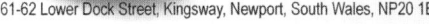

INTERIORS, FIXTURES & FINISHES

HERITAGE BATHROOMS
Princess Street, Bedminster, Bristol, BS3 4AG
Area of Operation: UK & Ireland
Tel: 0117 963 3333
Email: marketing@heritagebathrooms.com
Web: www.heritagebathrooms.com
Product Type: 1, 2, 3, 4, 5, 6, 7, 8

IDEAL STANDARD
The Bathroom Works, National Avenue,
Kingston Upon Hull, East Riding of Yorks, HU5 4HS
Area of Operation: Worldwide
Tel: 0800 590311
Fax: 01482 445886
Email: ideal-standard@aseur.com
Web: www.ideal-standard.co.uk

INSPIRED BATHROOMS
Unit R4, Innsworth Technology Park, Innsworth Lane,
Gloucester, Gloucestershire, GL3 1DL
Area of Operation: UK & Ireland
Tel: 01452 559121
Fax: 01452 530908
Email: sales@inspired-bathrooms.co.uk
Web: www.inspired-bathrooms.co.uk 🖱
Product Type: 1, 2, 3, 4, 5, 6, 7, 8

INTERESTING INTERIORS LTD
37-39 Queen Street, Aberystwyth,
Ceredigion, SY23 1PU
Area of Operation: UK (Excluding Ireland)
Tel: 01970 626162
Fax: 01970 051 7959
Email: enquiries@interestinginteriors.com
Web: www.interiors-ltd.demon.co.uk
Product Type: 1, 2, 3, 4, 5, 6, 7, 8, 9
Other Info: ▽ ECO ✎ 👍

JACUZZI UK
Silverdale Road, Newcastle under Lyme,
Staffordshire, ST5 6EL
Area of Operation: UK & Ireland
Tel: 01782 717175
Fax: 01782 717166
Email: jacuzzisalesdesk@jacuzziuk.com
Web: www.jacuzziuk.co.uk
Product Type: 3, 4, 6, 7, 9

JACUZZI UK - BC SANITAN
Silverdale Road, Newcastle under Lyme,
Staffordshire, ST5 6EL
Area of Operation: UK & Ireland
Tel: 01782 717175
Fax: 01782 717166
Email: jacuzzisalesdesk@jacuzziuk.com
Web: www.jacuzziuk.com
Product Type: 2

JOHN FLEMING & CO LTD
Silverburn Place, Bridge of Don,
Aberdeen, Aberdeenshire, AB23 8EG
Area of Operation: Scotland
Tel: 0800 085 8728
Fax: 01224 825377
Email: info@johnfleming.co.uk
Web: www.johnfleming.co.uk

JUST ADD WATER
202 - 228 York Way , Kings Cross,
London, N7 9AZ
Area of Operation: Greater London
Tel: 0207 697 3161
Fax: 0207 697 3162
Email: jawsales@justaddwater.co.uk
Web: www.justaddwater.co.uk
Product Type: 1

KALDEWEI
Unit 7, Sundial Court, Tolworth Rise South,
Surbiton, Surrey, KT5 9RN
Area of Operation: Worldwide
Tel: 0870 777 2223
Fax: 0870 777 2225
Email: info@kaldewei-uk.com
Web: www.kaldewei.com
Product Type: 1, 3, 4, 5, 6, 7, 8, 9

KALDEWEI
Area of Operation: Worldwide
Tel: 0870 777 2223 **Fax:** 0870 777 2225
Email: info@kaldewei-uk.com
Web: www.kaldewei.com
Product Type: 1, 3, 4, 5, 6, 7, 8, 9
Materials Type: C) 2

Kaldewei is Europe's largest producer of bath and shower trays made from beautiful and durable enamelled steel. Whirlpool systems are also now available.

KENSINGTON STUDIO
13 c & d Kensington Road, Earlsdon,
Coventry, West Midlands, CV5 6GG
Area of Operation: UK (Excluding Ireland)
Tel: 02476 713326
Fax: 02476 713136
Email: sales@kensingtonstudio.com
Web: www.kensingtonstudio.com 🖱
Product Type: 1, 2, 3, 4, 5, 6, 7, 8, 9

KERAMAG WAVENEY LTD
London Road, Beccles, Suffolk, NR34 8TS
Area of Operation: UK & Ireland
Tel: 01502 716600 **Fax:** 01502 717767
Email: info@keramagwaveney.co.uk
Web: www.keramag.com
Product Type: 1, 2, 3, 4, 5, 6, 7, 9

KIRKSTONE QUARRIES LIMITED
Skelwith Bridge, Ambleside, Cumbria, LA22 9NN
Area of Operation: UK & Ireland
Tel: 01539 433296 **Fax:** 01539 434006
Email: info@kirkstone.com
Web: www.kirkstone.com
Product Type: 9

KOHLER MIRA LIMITED
Gloucester Road, Cheltenham,
Gloucestershire, GL51 8TP
Area of Operation: UK & Ireland
Tel: 0870 850 5551
Fax: 0870 850 5552
Email: info@kohleruk.com
Web: www.kohleruk.com

LECICO PLC
Unit 47a Hobbs Industrial Estate,
Newchapel, Nr Lingfield, Surrey, RH7 6HN
Area of Operation: UK & Ireland
Tel: 01342 834777
Fax: 01342 834783
Email: info@lecico.co.uk
Web: www.lecico.co.uk
Product Type: 1

M & O BATHROOM CENTRE
174-176 Goswell Road, Clarkenwell,
London, EC1V 7DT
Area of Operation: East England,
Greater London, South East England
Tel: 0207 608 0111
Fax: 0207 490 3083
Email: mando@lineone.net
Product Type: 1, 2, 3, 4, 5, 6, 7

MARBLE CLASSICS
Unit 3, Station Approach, Emsworth,
Hampshire, PO10 7PW
Area of Operation: UK & Ireland
Tel: 01243 370011
Fax: 01243 371023
Email: info@marbleclassics.co.uk
Web: www.marbleclassics.co.uk
Product Type: 9
Material Type: E) 2, 5, 8

MEREWAY CONTRACTS
413 Warwick Road, Birmingham,
West Midlands, B11 2JR
Area of Operation: UK (Excluding Ireland)
Tel: 0121 707 3288
Fax: 0121 707 3871
Email: info@merewaycontracts.co.uk
Web: www.merewaycontracts.co.uk

MISCELLANEA DISCONTINUED BATHROOMWARE
Churt Place Nurseries, Tilford Road,
Churt, Farnham, Surrey, GU10 2LN
Area of Operation: UK (Excluding Ireland)
Tel: 01428 608164 **Fax:** 01428 608165
Email: email@brokenbog.com
Web: www.brokenbog.com
Product Type: 1, 2, 3, 4, 5, 6, 7, 8, 9

MULTIJET HYDROTHERAPY
Unit A Wanborough Business Park, West Flexford
Lane, Wanborough, Guildford, Surrey, GU3 2JS
Area of Operation: UK & Ireland
Tel: 01483 813181
Fax: 01483 813182
Web: www.multijet-hydrotherapy.com
Product Type: 1, 3, 4, 5, 7, 8, 9

N&C NICOBOND
41-51 Freshwater Road, Chadwell Heath,
Romford, Essex, RM8 1SP
Area of Operation: Worldwide
Tel: 0208 586 4600
Fax: 0208 586 4646
Email: info@nichollsandclarke.com
Web: www.ncdirect.com

NICHOLAS ANTHONY LTD
42-44 Wigmore Street, London, W1U 2RX
Area of Operation: East England,
Greater London, South East England
Tel: 0800 068 3603
Fax: 01206 762698
Email: info@nicholas-anthony.co.uk
Web: www.nicholas-anthony.co.uk
Product Type: 1, 3, 4, 6, 7

NO CODE
Larkwhistle Farm Road, Micheldever,
Hampshire, SO21 3BG
Area of Operation: UK & Ireland
Tel: 01962 870078 **Fax:** 01962 870077
Email: sales@nocode.co.uk
Web: www.nocode.co.uk
Product Type: 1, 2, 3, 4, 5, 7

NORDIC
Unit 5, Trading Estate, Holland Road,
Oxted, Surrey, RH8 9BZ
Area of Operation: UK (Excluding Ireland)
Tel: 01883 732400 **Fax:** 01883 716970
Email: info@nordic.co.uk
Web: www.nordic.co.uk
Product Type: 3

OLD FASHIONED BATHROOMS
Foresters Hall, 52 High Street,
Debenham, Stowmarket, Suffolk, IP14 6QW
Area of Operation: East England,
Greater London, South East England
Tel: 01728 860926 **Fax:** 01728 860446
Email: ofbrooms@lycos.co.uk
Web: www.oldfashionedbathrooms.co.uk
Product Type: 1, 2, 3, 5, 6, 7, 9
Other Info: ▽
Material Type: C) 1, 2, 5

OMNITUB
The Bothy, Lays Lane, Blagdon,
Somerset, BS40 7RQ
Area of Operation: UK & Ireland
Tel: 01761 462 641 **Fax:** 01761 462 641
Email: info@omnitub.co.uk
Web: www.omnitub.co.uk
Product Type: 6, 9

INTERIORS, FIXTURES & FINISHES

PRETTY SWIFT LTD
Units 7&8, 51 Chancery Lane,
Debenham, Suffolk, IP14 6PJ
Area of Operation: UK (Excluding Ireland)
Tel: 01728 861818
Fax: 01728 861919
Email: prettyswiftltd@aol.com

QUALCERAM SHIRES PLC
Unit 9/10, Campus 5, Off Third Avenue,
Letchworth Garden City, Hertfordshire, SG6 2JF
Area of Operation: UK & Ireland
Tel: 01462 676710
Fax: 01462 676701
Email: info@qualceram-shires.com
Web: www.qualceram-shires.com
Product Type: 1, 2, 3, 4, 5, 6, 7, 8, 9

RELAXAIR
Tiber House, Hall Lane, Off Lostock Lane,
Lostock, Bolton, Lancashire, BL6 4BR
Area of Operation: UK & Ireland
Tel: 01204 675804
Fax: 01204 675809
Email: info@relaxair.co.uk
Web: www.relaxair.co.uk ⊸🖰
Product Type: 3

ROOTS KITCHENS,
BEDROOMS & BATHROOMS
Vine Farm, Stockers Hill,
Boughton-under-Blean,
Faversham, Kent, ME13 9AB
Area of Operation: UK (Excluding Ireland)
Tel: 01227 751130
Fax: 01227 750033
Email: showroom@rootskitchens.co.uk
Web: www.rootskitchens.co.uk ⊸🖰
Product Type: 1, 2, 3, 4, 5, 6, 7

ROSCO COLLECTIONS
Stone Allerton, Axbridge,
Somerset, BS26 2NS
Area of Operation: UK & Ireland
Tel: 01934 712299
Fax: 01934 713222
Email: jonathan@roscobathrooms.demon.co.uk
Product Type: 1, 2, 3, 4, 5, 6, 8

SHOWERLUX UK LTD
Sibree Road, Coventry,
West Midlands, CV3 4FD
Area of Operation: UK & Ireland
Tel: 02476 882515
Fax: 02476 305457
Email: sales@showerlux.co.uk
Web: www.showerlux.com
Product Type: 1, 3, 4, 5, 6

SMART SHOWERS LTD
Unit 11, Woodside Road,
South Marston Business Park,
Swindon, Wiltshire, SN3 4WA
Area of Operation: UK & Ireland
Tel: 01793 822775
Fax: 01793 823800
Email: louise@smartshowers.co.uk
Web: www.smartshowers.co.uk ⊸🖰
Product Type: 3

SOGA LTD
41 Mayfield Street, Hull,
East Riding of Yorks, HU3 1NT
Area of Operation: Worldwide
Tel: 01482 327025
Email: info@soga.co.uk
Web: www.soga.co.uk
Product Type: 2, 3, 4, 6, 8

STIFFKEY BATHROOMS
89 Upper Saint Giles Street,
Norwich, Norfolk, NR2 1AB
Area of Operation: Worldwide
Tel: 01603 627850
Fax: 01603 619775
Email: stiffkeybathrooms.norwich@virgin.net
Web: www.stiffkeybathrooms.com
Product Type: 2, 5, 6
Other Info: ▽

SURFACE MAGIC
102 Tomswood Hill, Barkingside,
Ilford, Essex, IG6 2QJ
Area of Operation: South East England
Tel: 0800 027 1951 **Fax:** 0208 501 1230
Email: info@surfacemagic.co.uk
Web: www.surfacemagic.co.uk
Product Type: 1, 2, 3, 4, 5, 6, 7, 8, 9

SVEDBERGS
London House, 100 New Kings Rd,
London, SW6 4LX
Area of Operation: UK & Ireland
Tel: 0207 348 6107 **Fax:** 0207 348 6108
Email: info@svedbergs.co.uk
Web: www.svedbergs.co.uk
Product Type: 1, 2, 3, 4, 5, 8, 9

THE GRANITE FACTORY
4 Winchester Drive,
Peterlee, Durham, SR8 2RJ
Area of Operation: North East England
Tel: 0191 518 3600 **Fax:** 0191 518 3600
Email: admin@granitefactory.co.uk
Web: www.granitefactory.co.uk ⊸🖰
Product Type: 7

THE KITCHEN & BATHROOM COLLECTION
Nelson House, Nelson Road,
Salisbury, Wiltshire, SP1 3LT
Area of Operation: South West England
and South Wales
Tel: 01722 334800
Fax: 01722 412252
Email: info@kbc.co.uk
Web: www.kbc.co.uk
Product Type: 1, 2, 3, 4, 5, 6, 7

THE MANTALEDA BATHROOM CO. LTD
Thurston Road, Northallerton Business Park,
Northallerton, North Yorkshire, DL6 2NA
Area of Operation: Worldwide
Tel: 01609 771211
Fax: 01609 760100
Email: baths@mantaleda.fsnet.co.uk
Web: www.mantaleda.fsnet.co.uk
Product Type: 3, 4, 5, 6, 7, 8

THE WATER MONOPOLY
16/18 Lonsdale Road, London, NW6 6RD
Area of Operation: UK (Excluding Ireland)
Tel: 0207 624 2636
Fax: 0207 624 2631
Email: enquiries@watermonopoly.com
Web: www.watermonopoly.com

TRADEMARK TILES
Airfield Industrial Estate, Warboys,
Huntingdon, Cambridgeshire, PE28 2SH
Area of Operation: East England,
South East England
Tel: 01487 825300 **Fax:** 01487 825305
Email: info@trademarktiles.co.uk
Product Type: 1, 2, 3, 4, 5, 6, 7, 8

VASCO
Clitheroe Works, Clitheroe Street,
Skipton, North Yorkshire, BD23 1SU
Area of Operation: UK (Excluding Ireland)
Tel: 0870 027 4528 **Fax:** 0870 027 4531
Email: vascobathrooms@yahoo.co.uk

VICTORIA & ALBERT BATHS
Unit 14, Stafford Park 12,
Telford, Shropshire, TF3 3BJ
Area of Operation: Worldwide
Tel: 01952 210814 **Fax:** 01952 210810
Email: info@vandabaths.com
Web: www.vandabaths.com
Product Type: 1, 2, 5, 7

VICTORIAN BATHROOMS
Ingsmill Complex, Dale Street,
Ossett, West Yorkshire, WF5 9HQ
Area of Operation: UK (Excluding Ireland)
Tel: 0845 130 6911 **Fax:** 01924 261223
Email: sales@victorianbathrooms.co.uk
Web: www.victorianbathrooms.co.uk
Product Type: 1, 2

WALTON BATHROOMS
The Hersham Centre, The Green,
Molesey Road, Hersham,
Walton on Thames, Surrey, KT12 4HL
Area of Operation: UK (Excluding Ireland)
Tel: 01932 224784
Fax: 01932 253447
Email: sales@waltonbathrooms.co.uk
Web: www.waltonbathrooms.co.uk
Product Type: 1, 2, 3, 4, 6

WHIRLPOOL EXPRESS (UK) LTD
61-62 Lower Dock Street,
Kingsway, Newport, NP20 1EF
Area of Operation: Europe
Tel: 01633 244555
Fax: 01633 244555
Email: reception@whirlpoolexpress.co.uk
Web: www.whirlpoolexpress.co.uk ⊸🖰
Product Type: 1, 2, 3, 4, 5, 6, 7, 8

WICKES
Wickes Customer Services, Wickes House,
120-138 Station Road, Harrow, Middlesex,
Greater London, HA1 2QB
Area of Operation: UK (Excluding Ireland)
Tel: 0870 608 9001
Fax: 0208 863 6225
Web: www.wickes.co.uk

WWW.BOUNDARYBATHROOMS.CO.UK
Regent Street, Colne, Lancashire, BB8 8LD
Area of Operation: UK & Ireland
Tel: 01282 862 509
Fax: 01282 871 192
Web: www.boundarybathrooms.co.uk
Product Type: 1, 2, 3, 4, 5, 6, 7, 8, 9

LAVATORIES

┌───┐
│ **KEY**   │
│ **OTHER:** ▽ Reclaimed ⊸🖰 On-line shopping │
│ ✎ Bespoke ✋ Hand-made ECO Ecological   │
└───┘

ALEXANDERS
Unit 3, 1/1A Spilsby Road, Harold Hill,
Romford, Essex, RM3 8SB
Area of Operation: UK & Ireland
Tel: 01708 384574
Fax: 01708 384089
Email: sales@alexanders-uk.net
Web: www.alexanders-uk.net

ANTIQUE BATHS OF IVYBRIDGE LTD
Emebridge Works, Ermington Road,
Ivybridge, Devon, PL21 9DE
Area of Operation: UK (Excluding Ireland)
Tel: 01752 698250
Fax: 01752 698266
Email: sales@antiquebaths.com
Web: www.antiquebaths.com ⊸🖰

BARRHEAD SANITARYWARE PLC
15-17 Nasmyth Road, Hillington Industrial Estate,
Hillington, Glasgow, Renfrewshire, G52 4RG
Area of Operation: UK & Ireland
Tel: 0141 883 0066 **Fax:** 0141 883 0077
Email: sales@barrhead.co.uk

BATHROOM BARN
Uplands, Wyton, Huntingdon,
Cambridgeshire, PE28 2JZ
Area of Operation: UK (Excluding Ireland)
Tel: 01480 458900
Email: sales@bathroombarn.co.uk
Web: www.bathroombarn.co.uk

BATHROOM CITY
Seeleys Road, Tyseley Industrial Estate,
Birmingham, West Midlands, B11 2LQ
Area of Operation: UK & Ireland
Tel: 0121 753 0700 **Fax:** 0121 753 1110
Email: sales@bathroomcity.com
Web: www.bathroomcity.com ⊸🖰

BATHROOM DISCOUNT CENTRE
297 Munster Road, Fulham,
Greater London, SW6 6BW
Area of Operation: UK (Excluding Ireland)
Tel: 0207 381 4222
Fax: 0207 381 6792
Web: www.bathroomdiscount.co.uk

BATHROOM OPTIONS
394 Ringwood Road,
Ferndown, Dorset, BH22 9AU
Area of Operation: South East England
Tel: 01202 873379
Fax: 01202 892252
Email: bathroom@options394.fsnet.co.uk
Web: www.bathroomoptions.co.uk

BATHROOMS PLUS LTD
222 Malmesbury Park Road,
Bournemouth, Dorset, BH8 8PR
Area of Operation: South West England
and South Wales
Tel: 01202 294417
Fax: 01202 316425
Email: info@bathrooms-plus.co.uk
Web: www.bathrooms-plus.co.uk

BATHROOMSTUFF.CO.UK
40 Evelegh Road, Farlington,
Portsmouth, Hampshire, PO6 1DL
Area of Operation: UK (Excluding Ireland)
Tel: 0845 058 0540
Fax: 02392 215695
Email: sales@bathroomstuff.co.uk
Web: www.bathroomstuff.co.uk ⊸🖰

BROOK-WATER DESIGNER
BATHROOMS & KITCHENS
The Downs, Woodhouse Hill,
Uplyme, Lyme Regis, Dorset, DT7 3SL
Area of Operation: Worldwide
Tel: 01235 201256
Email: sales@brookwater.co.uk
Web: www.brookwater.co.uk ⊸🖰

CAPLE
Fourth Way, Avonmouth, Bristol, BS11 8DW
Area of Operation: UK (Excluding Ireland)
Tel: 0870 606 9606
Fax: 0117 938 7449
Email: amandalowe@mlay.co.uk
Web: www.mlay.co.uk

CIFIAL UK LTD
7 Faraday Court, Park Farm Industrial Estate,
Wellingborough, Northamptonshire, NN8 6XY
Area of Operation: UK & Ireland
Tel: 01933 402008 **Fax:** 01933 402063
Email: sales@cifial.co.uk
Web: www.cifial.co.uk

CIFIAL UK
Area of Operation: UK & Ireland
Tel: 01933 402008
Fax: 01933 402063
Email: sales@cifial.co.uk
Web: www.cifial.co.uk

Cifial's new TECHNOvation C1 & C4 collection of
pure white designer ceramics has strong, chunky
styling yet is space efficient suitable for large &
small bathrooms.

CISTERMISER LTD
Unit 1 Woodley Park Estate, 59-69 Reading Road,
Woodley, Reading, Berkshire, RG5 3AN
Area of Operation: UK & Ireland
Tel: 0118 969 1611 **Fax:** 0118 944 1426
Email: sales@cistermiser.co.uk
Web: www.cistermiser.co.uk

CITY BATHROOMS & KITCHENS
158 Longford Road, Longford, Coventry,
West Midlands, CV6 6DR
Area of Operation: UK & Ireland
Tel: 02476 365877 **Fax:** 02476 644992
Email: citybathrooms@hotmail.com
Web: www.citybathrooms.co.uk

COSY ROOMS (THE-BATHROOM-SHOP.CO.UK)
17 Chiltern Way, North Hykeham,
Lincoln, Lincolnshire, LN6 9SY
Area of Operation: UK (Excluding Ireland)
Tel: 01522 696002 **Fax:** 01522 696002
Email: keith@cosy-rooms.com
Web: www.the-bathroom-shop.co.uk ✏

CYMRU KITCHENS
63 Caerleon Road, St. Julians, Newport, NP19 7BX
Area of Operation: UK (Excluding Ireland)
Tel: 01633 676767 **Fax:** 01633 212512
Email: sales@cymrukitchens.com
Web: www.cymrukitchens.com

DART VALLEY SYSTEMS LIMITED
Kemmings Close, Long Road, Paignton,
Devon, TQ4 7TW
Area of Operation: UK & Ireland
Tel: 01803 529021
Fax: 01803 559016
Email: sales@dartvalley.co.uk
Web: www.dartvalley.co.uk

DECORMASTER LTD
Unit 16, Waterside Industrial Estate,
Wolverhampton, West Midlands, WV2 2RH
Area of Operation: Worldwide
Fax: 01902 353126
Email: sales@oldcolours.co.uk
Web: www.oldcolours.co.uk

DEEP BLUE SHOWROOM
299-313 Lewisham High Street, Lewisham,
London, SE13 6NW
Area of Operation: UK (Excluding Ireland)
Tel: 0208 690 3401
Fax: 0208 690 1408

DESIGNER BATHROOMS
140-142 Pogmoor Road, Pogmoor, Barnsley,
North Yorkshire, S75 2DX
Area of Operation: UK (Excluding Ireland)
Tel: 01226 280200
Fax: 01226 733273
Email: mike@designer-bathrooms.com
Web: www.designer-bathrooms.com

DOLPHIN BATHROOMS
Bromwich Road, Worcester,
Worcestershire, WR2 4BD
Area of Operation: UK (Excluding Ireland)
Tel: 0800 626717
Web: www.dolphinbathrooms.co.uk

DURAVIT
Unit 7, Stratus Park, Brudenell Drive, Brinklow,
Milton Keynes, Buckinghamshire, MK10 0DE
Area of Operation: UK & Ireland
Tel: 0870 730 7787
Fax: 0870 730 7786
Email: info@uk.duravit.com
Web: www.duravit.co.uk

ESL HEALTHCARE LTD
Potts Marsh Industrial Estate, Eastbourne Road,
Westham, East Sussex, BN24 5NH
Area of Operation: UK & Ireland
Tel: 01323 465800
Fax: 01323 460248
Email: sales@eslindustries.co.uk
Web: www.eslindustries.com
Material Type: D) 3, 6

GEBERIT LTD
New Hythe Business Park, New Hythe Lane,
Aylesford, Kent, ME20 7PJ
Area of Operation: UK & Ireland
Tel: 01622 717811
Fax: 01622 716920
Web: www.geberit.co.uk

GOODWOOD BATHROOMS LTD
Church Road, North Mundham,
Chichester, West Sussex, PO20 1JU
Area of Operation: UK (Excluding Ireland)
Tel: 01243 532121
Fax: 01243 533423
Email: sales@goodwoodbathroom.co.uk
Web: www.goodwoodbathrooms.co.uk

HERITAGE BATHROOMS
Princess Street, Bedminster, Bristol, BS3 4AG
Area of Operation: UK & Ireland
Tel: 0117 963 3333
Email: marketing@heritagebathrooms.com
Web: www.heritagebathrooms.com

IDEAL STANDARD
The Bathroom Works, National Avenue,
Kingston Upon Hull, East Riding of Yorks, HU5 4HS
Area of Operation: Worldwide
Tel: 0800 590311
Fax: 01482 445886
Email: ideal-standard@aseur.com
Web: www.ideal-standard.co.uk

INSPIRED BATHROOMS
Unit R4, Innsworth Technology Park,
Innsworth Lane, Gloucester,
Gloucestershire, GL3 1DL
Area of Operation: UK & Ireland
Tel: 01452 559121
Fax: 01452 530908
Email: sales@inspired-bathrooms.co.uk
Web: www.inspired-bathrooms.co.uk ✏

INTERESTING INTERIORS LTD
37-39 Queen Street, Aberystwyth,
Ceredigion, SY23 1PU
Area of Operation: UK (Excluding Ireland)
Tel: 01970 626162
Fax: 0870 051 7959
Email: enquiries@interestinginteriors.com
Web: www.interiors-ltd.demon.co.uk

JACUZZI UK
Silverdale Road, Newcastle under Lyme,
Staffordshire, ST5 6EL
Area of Operation: UK & Ireland
Tel: 01782 717175
Fax: 01782 717166
Email: jacuzzisalesdesk@jacuzziuk.com
Web: www.jacuzzi.co.uk
Material Type: F) 1

JACUZZI UK - BC SANITAN
Silverdale Road,
Newcastle under Lyme,
Staffordshire, ST5 6EL
Area of Operation: UK & Ireland
Tel: 01782 717175
Fax: 01782 717166
Email: jacuzzisalesdesk@jacuzziuk.com
Web: www.jacuzziuk.com
Material Type: F) 1

KENSINGTON STUDIO
13 c & d Kensington Road, Earlsdon,
Coventry, West Midlands, CV5 6GG
Area of Operation: UK (Excluding Ireland)
Tel: 02476 713326
Fax: 02476 713136
Email: sales@kensingtonstudio.com
Web: www.kensingtonstudio.com ✏

KERAMAG WAVENEY LTD
London Road, Beccles, Suffolk, NR34 8TS
Area of Operation: UK & Ireland
Tel: 01502 716600
Fax: 01502 717767
Email: info@keramagwaveney.co.uk
Web: www.keramag.com

LECICO PLC
Unit 47a Hobbs Industrial Estate,
Newchapel, Nr Lingfield, Surrey, RH7 6HN
Area of Operation: UK & Ireland
Tel: 01342 834777 **Fax:** 01342 834783
Email: info@lecico.co.uk
Web: www.lecico.co.uk

M & O BATHROOM CENTRE
174-176 Goswell Road,
Clarkenwell, London, EC1V 7DT
Area of Operation: East England,
Greater London, South East England
Tel: 0207 608 0111 **Fax:** 0207 490 3083
Email: mando@lineone.net

MEREWAY CONTRACTS
413 Warwick Road, Birmingham,
West Midlands, B11 2JR
Area of Operation: UK (Excluding Ireland)
Tel: 0121 707 3288 **Fax:** 0121 707 3871
Email: info@merewaycontracts.co.uk
Web: www.merewaycontracts.co.uk

**MISCELLANEA DISCONTINUED
BATHROOMWARE**
Churt Place Nurseries,
Tilford Road, Churt,
Farnham, Surrey, GU10 2LN
Area of Operation: UK (Excluding Ireland)
Tel: 01428 608164 **Fax:** 01428 608165
Email: email@brokenbog.com
Web: www.brokenbog.com

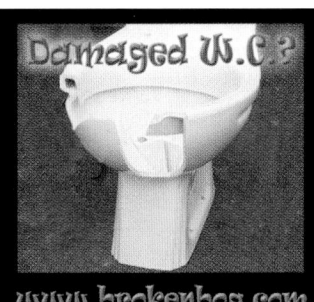

NICHOLAS ANTHONY LTD
42-44 Wigmore Street, London, W1U 2RX
Area of Operation: East England,
Greater London, South East England
Tel: 0800 068 3603 **Fax:** 01206 762698
Email: info@nicholas-anthony.co.uk
Web: www.nicholas-anthony.co.uk

NO CODE
Larkwhistle Farm Road,
Micheldever, Hampshire, SO21 3BG
Area of Operation: UK & Ireland
Tel: 01962 870078 **Fax:** 01962 870077
Email: sales@nocode.co.uk
Web: www.nocode.co.uk

OLD FASHIONED BATHROOMS
Foresters Hall, 52 High Street, Debenham,
Stowmarket, Suffolk, IP14 6QW
Area of Operation: East England,
Greater London, South East England
Tel: 01728 860926 **Fax:** 01728 860446
Email: ofbrooms@lycos.co.uk
Web: www.oldfashionedbathrooms.co.uk
Other Info: ▽
Material Type: F) 1, 2, 4

QUALCERAM SHIRES PLC
Unit 9/10, Campus 5, Off Third Avenue,
Letchworth Garden City, Hertfordshire, SG6 2JF
Area of Operation: UK & Ireland
Tel: 01462 676710
Fax: 01462 676701
Email: info@qualceram-shires.com
Web: www.qualceram-shires.com

ROOTS KITCHENS, BEDROOMS & BATHROOMS
Vine Farm, Stockers Hill, Boughton-under-Blean,
Faversham, Kent, ME13 9AB
Area of Operation: UK (Excluding Ireland)
Tel: 01227 751130
Fax: 01227 750033
Email: showroom@rootskitchens.co.uk
Web: www.rootskitchens.co.uk ✏

ROSCO COLLECTIONS
Stone Allerton, Axbridge, Somerset, BS26 2NS
Area of Operation: UK & Ireland
Tel: 01934 712299
Fax: 01934 713222
Email: jonathan@roscobathrooms.demon.co.uk

SANIFLO LTD
Howard House, The Runway,
South Ruislip, Middlesex, HA4 6SE
Area of Operation: UK (Excluding Ireland)
Tel: 0208 842 4040
Fax: 0208 842 1671
Email: andrews@saniflo.co.uk
Web: www.saniflo.co.uk

STIFFKEY BATHROOMS
89 Upper Saint Giles Street,
Norwich, Norfolk, NR2 1AB
Area of Operation: Worldwide
Tel: 01603 627850
Fax: 01603 619775
Email: stiffkeybathrooms.norwich@virgin.net
Web: www.stiffkeybathrooms.com

SVEDBERGS
London House, 100 New Kings Road,
London, SW6 4LX
Area of Operation: UK & Ireland
Tel: 0207 348 6107
Fax: 0207 348 6108
Email: info@svedbergs.co.uk
Web: www.svedbergs.co.uk
Other Info: ECO

TAPS SHOP
35 Bridge Street, Witney,
Oxfordshire, OX28 1DA
Area of Operation: Worldwide
Tel: 0845 430 3035
Fax: 01993 779653
Email: info@tapsshop.co.uk
Web: www.tapsshop.co.uk ✏

THE KITCHEN & BATHROOM COLLECTION
Nelson House, Nelson Road,
Salisbury, Wiltshire, SP1 3LT
Area of Operation: South West England
and South Wales
Tel: 01722 334800
Fax: 01722 412252
Email: info@kbc.co.uk
Web: www.kbc.co.uk
Other Info: ✍

THE WATER MONOPOLY
16/18 Lonsdale Road, London, NW6 6RD
Area of Operation: UK (Excluding Ireland)
Tel: 0207 624 2636
Fax: 0207 624 2631
Email: enquiries@watermonopoly.com
Web: www.watermonopoly.com

TRADEMARK TILES
Airfield Industrial Estate, Warboys,
Huntingdon, Cambridgeshire, PE28 2SH
Area of Operation: East England,
South East England
Tel: 01487 825300
Fax: 01487 825305
Email: info@trademarktiles.co.uk

SPONSORED BY: PEGLER LIMITED www.pegler.co.uk/francis

WALTON BATHROOMS
The Hersham Centre, The Green,
Molesey Road, Hersham,
Walton on Thames, Surrey, KT12 4HL
Area of Operation: UK (Excluding Ireland)
Tel: 01932 224784
Fax: 01932 253447
Email: sales@waltonbathrooms.co.uk
Web: www.waltonbathrooms.co.uk

WHIRLPOOL EXPRESS (UK) LTD
61-62 Lower Dock Street,
Kingsway, Newport, NP20 1EF
Area of Operation: Europe
Tel: 01633 244555
Fax: 01633 244555
Email: reception@whirlpoolexpress.co.uk
Web: www.whirlpoolexpress.co.uk

WWW.BOUNDARYBATHROOMS.CO.UK
Regent Street, Colne, Lancashire, BB8 8LD
Area of Operation: UK & Ireland
Tel: 01282 862 509
Fax: 01282 871 192
Web: www.boundarybathrooms.co.uk

SHOWERS

KEY

PRODUCT TYPES: 1= Enclosures 2 = Trays
3 = Heads 4 = Shower Pumps 5 = Other
OTHER: ▽ Reclaimed On-line shopping
Bespoke Hand-made ECO Ecological

A & H BRASS
201-203 Edgware Road, London, W2 1ES
Area of Operation: Worldwide
Tel: 0207 402 1854
Fax: 0207 402 0110
Email: ahbrass@btinternet.com
Web: www.aandhbrass.co.uk
Product Type: 1, 2, 3, 4

A&J GUMMERS LIMITED
Unit H, Redfern Park Way,
Tyseley, Birmingham,
West Midlands, B11 2DN
Area of Operation: Worldwide
Tel: 0121 706 2241
Fax: 0121 706 2960
Email: sales@gummers.com
Web: www.gummers.com
Product Type: 1, 2, 3, 4, 5

ADVANCED SHOWERS
32 Frederick Sanger Road,
Surrey Research Park,
Guildford, Surrey, GU2 7YD
Area of Operation: UK & Ireland
Tel: 01483 295930
Fax: 01483 295935
Email: sales@advanced-showers.com
Web: www.advanced-showers.com
Product Type: 1

AL CHALLIS LTD
Europower House, Lower Road,
Cookham, Maidenhead, Berkshire, SL6 9EH
Area of Operation: UK & Ireland
Tel: 01628 529024
Fax: 0870 458 0577
Email: chris@alchallis.com
Web: www.alchallis.com
Product Type: 3

ALEXANDERS
Unit 3, 1/1A Spilsby Road,
Harold Hill, Romford, Essex, RM3 8SB
Area of Operation: UK & Ireland
Tel: 01708 384574
Fax: 01708 384089
Email: sales@alexanders-uk.net
Web: www.alexanders-uk.net
Product Type: 1, 2, 3, 4, 5

AMBIANCE BAIN
Cumberland House, 80 Scrubs Lane,
London, NW10 6RF
Area of Operation: UK & Ireland
Tel: 0870 902 1313
Fax: 0870 902 1312
Email: sghirardello@mimea.com
Web: www.ambiancebain.com
Product Type: 1

ANTIQUE BATHS OF IVYBRIDGE LTD
Emebridge Works, Ermington Road,
Ivybridge, Devon, PL21 9DE
Area of Operation: UK (Excluding Ireland)
Tel: 01752 698250
Fax: 01752 698266
Email: sales@antiquebaths.com
Web: www.antiquebaths.com
Product Type: 1, 2, 3, 4, 5

APPOLLO
Aquabeau Limited, Swinnow Lane,
Leeds, West Yorkshire, LS13 4TY
Area of Operation: Worldwide
Tel: 0113 386 9150
Fax: 0113 239 3406
Email: sales@aquabeau.com
Web: www.appollobathing.co.uk
Product Type: 1, 5

AQATA SHOWER ENCLOSURES
Brookfield, Harrowbrook Industrial Estate,
Hinckley, Leicestershire, LE10 3DU
Area of Operation: UK & Ireland
Tel: 01455 896500
Fax: 01455 896501
Email: sales@aqata.co.uk
Web: www.aqata.co.uk
Product Type: 1, 2
Other Info:
Material Type: C) 1

AQUABEAU LIMITED
Swinnow Lane, Leeds,
West Yorkshire, LS13 4TY
Area of Operation: Worldwide
Tel: 0113 386 9150
Fax: 0113 239 3406
Email: sales@aquabeau.com
Web: www.aquabeau.com
Product Type: 1, 2

AQUALISA PRODUCTS LTD
The Flyers Way, Westerham, Kent, TN16 1DE
Area of Operation: UK & Ireland
Tel: 01959 560000
Fax: 01959 560030
Email: sue.anderson@aqualisa.co.uk
Web: www.aqualisa.co.uk
Product Type: 3, 4

AQUALUX
Universal Point, Steelmans Road,
off Park Lane, Wednesbury,
West Midlands, WS10 9UZ
Area of Operation: Europe
Tel: 0870 241 6131
Fax: 0870 241 6133
Email: david.hall@bhdgroup.co.uk
Web: www.aqualux.co.uk
Product Type: 1, 2, 5

AQUAPLUS SOLUTIONS
Unit 106, Indiana Building,
London, SE13 7QD
Area of Operation: UK & Ireland
Tel: 0870 201 1915
Fax: 0870 201 1916
Email: info@aquaplussolutions.com
Web: www.aquaplussolutions.com
Product Type: 1, 3

BATHROOM BARN
Uplands, Wyton, Huntingdon,
Cambridgeshire, PE28 2JZ
Area of Operation: UK (Excluding Ireland)
Tel: 01480 458900
Email: sales@bathroombarn.co.uk
Web: www.bathroombarn.co.uk

BATHROOM CITY
Seeleys Road, Tyseley Industrial Estate,
Birmingham, West Midlands, B11 2LQ
Area of Operation: UK & Ireland
Tel: 0121 753 0700
Fax: 0121 753 1110
Email: sales@bathroomcity.com
Web: www.bathroomcity.com
Product Type: 1, 2, 3, 4, 5

BATHROOM DISCOUNT CENTRE
297 Munster Road, Fulham,
Greater London, SW6 6BW
Area of Operation: UK (Excluding Ireland)
Tel: 0207 381 4222
Fax: 0207 381 6792
Web: www.bathroomdiscount.co.uk

BATHROOM EXPRESS
61-62 Lower Dock Street,
Kingsway, Newport, NP20 1EF
Area of Operation: UK & Ireland
Tel: 0845 130 2000
Fax: 01633 244881
Email: sales@bathroomexpress.co.uk
Web: www.bathroomexpress.co.uk
Product Type: 1, 3, 4

BATHROOM HEAVEN
25 Eccleston Square,
London, Sw1V 1NS
Area of Operation: UK & Ireland
Tel: 0845 121 6700
Fax: 0207 233 6074
Email: james@bathroomheaven.com
Web: www.bathroomheaven.com
Product Type: 1, 2, 3, 4, 5

BATHROOM OPTIONS
394 Ringwood Road,
Ferndown, Dorset, BH22 9AU
Area of Operation: South East England
Tel: 01202 873379
Fax: 01202 892252
Email: bathroom@options394.fsnet.co.uk
Web: www.bathroomoptions.co.uk
Product Type: 1, 2, 3, 4, 5

BATHROOMS PLUS LTD
222 Malmesbury Park Road,
Bournemouth, Dorset, BH8 8PR
Area of Operation: South West England
and South Wales
Tel: 01202 294417
Fax: 01202 316425
Email: info@bathrooms-plus.co.uk
Web: www.bathrooms-plus.co.uk
Product Type: 1, 2, 3, 4, 5

BATHROOMSTUFF.CO.UK
40 Evelegh Road,
Farlington, Portsmouth,
Hampshire, PO6 1DL
Area of Operation: UK (Excluding Ireland)
Tel: 08450 580 540
Fax: 023 9221 5695
Email: sales@bathroomstuff.co.uk
Web: www.bathroomstuff.co.uk
Product Type: 1, 2, 3, 5

**BROOK-WATER DESIGNER
BATHROOMS & KITCHENS**
The Downs, Woodhouse Hill,
Uplyme, Lyme Regis, Dorset, DT7 3SL
Area of Operation: Worldwide
Tel: 01235 201256
Email: sales@brookwater.co.uk
Web: www.brookwater.co.uk

CAPLE
Fourth Way, Avonmouth,
Bristol, BS11 8DW
Area of Operation: UK (Excluding Ireland)
Tel: 0870 606 9606
Fax: 0117 938 7449
Email: amandalowe@mlay.co.uk
Web: www.mlay.co.uk
Product Type: 1, 2, 3

CHARLES MASON LTD
Unit 11A Brook Street Mill,
Off Goodall Street, Macclesfield,
Cheshire, SK11 7AW
Area of Operation: Worldwide
Tel: 0800 085 3616
Fax: 01625 668789
Email: info@charles-mason.com
Web: www.charles-mason.com
Product Type: 3, 4
Material Type: C) 2, 3, 14

CITY BATHROOMS & KITCHENS
158 Longford Road, Longford,
Coventry, West Midlands, CV6 6DR
Area of Operation: UK & Ireland
Tel: 02476 365877
Fax: 02476 644992
Email: citybathrooms@hotmail.com
Web: www.citybathrooms.co.uk

CORAM SHOWERS LIMITED
Stanmore Industrial Estate,
Bridgnorth, Shropshire, WV15 5HP
Area of Operation: UK (Excluding Ireland)
Tel: 01746 766466
Fax: 01746 764140
Email: sales@coram.co.uk
Web: www.coram.co.uk
Product Type: 1, 2, 5

CORE AND ORE LTD
16 Portland Street, Clifton, Bristol, BS8 4JH
Area of Operation: UK (Excluding Ireland)
Tel: 0117 904 2408
Fax: 0117 965 8057
Email: sales@coreandore.com
Web: www.coreandore.com
Product Type: 2

**COSY ROOMS
(THE-BATHROOM-SHOP.CO.UK)**
17 Chiltern Way,
North Hykeham, Lincoln,
Lincolnshire, LN6 9SY
Area of Operation: UK (Excluding Ireland)
Tel: 01522 696002
Fax: 01522 696002
Email: keith@cosy-rooms.com
Web: www.the-bathroom-shop.co.uk
Product Type: 1, 4

CROSSWATER
Unit 5 Butterly Avenue,
Questor, Dartford, Kent, DA1 1JG
Area of Operation: UK & Ireland
Tel: 01322 628270
Fax: 01322 628280
Email: sales@crosswater.co.uk
Web: www.crosswater.co.uk
Product Type: 3

CYMRU KITCHENS
63 Caerleon Road, St. Julians,
Newport, NP19 7BX
Area of Operation: UK (Excluding Ireland)
Tel: 01633 676767
Fax: 01633 212512
Email: sales@cymrukitchens.com
Web: www.cymrukitchens.com

DART VALLEY SYSTEMS LIMITED
Kemmings Close, Long Road,
Paignton, Devon, TQ4 7TW
Area of Operation: UK & Ireland
Tel: 01803 529021
Fax: 01803 559016
Email: sales@dartvalley.co.uk
Web: www.dartvalley.co.uk
Product Type: 3, 5

DECORMASTER LTD
Unit 16, Waterside Industrial Estate,
Wolverhampton, West Midlands, WV2 2RH
Area of Operation: Worldwide
Fax: 01902 353126
Email: sales@oldcolours.co.uk
Web: www.oldcolours.co.uk
Product Type: 1, 2, 4

DEEP BLUE SHOWROOM
299-313 Lewisham High Street,
Lewisham, London, SE13 6NW
Area of Operation: UK (Excluding Ireland)
Tel: 0208 690 3401
Fax: 0208 690 1408

DESIGNER BATHROOMS
140-142 Pogmoor Road, Pogmoor, Barnsley,
North Yorkshire, S75 2DX
Area of Operation: UK (Excluding Ireland)
Tel: 01226 280200
Fax: 01226 733273
Email: mike@designer-bathrooms.com
Web: www.designer-bathrooms.com
Product Type: 1, 2, 3, 4
Other Info: ✎

DOLPHIN BATHROOMS
Bromwich Road, Worcester, Worcestershire, WR2 4BD
Area of Operation: UK (Excluding Ireland)
Tel: 0800 626717
Web: www.dolphinbathrooms.co.uk
Product Type: 1

DORMA UK LTD
Wilbury Way, Hitchin, Hertfordshire, SG4 0AB
Area of Operation: UK & Ireland
Tel: 01462 477600
Fax: 01462 477601
Email: info@dorma-uk.co.uk
Web: www.dorma-uk.co.uk
Product Type: 1
Material Type: J) 4

DUPONT CORIAN & ZODIAQ
10 Quarry Court, Pitstone Green Business Park,
Pitstone, Nr Tring, Buckinghamshire, LU7 9GW
Area of Operation: UK & Ireland
Tel: 01296 663555
Fax: 01296 663599
Email: sales@corian.co.uk
Web: www.corian.co.uk
Product Type: 2
Other Info: ✎

EUROBATH INTERNATIONAL LTD
Eurobath House, Wedmore Road,
Cheddar, Somerset, BS27 3EB
Area of Operation: Worldwide
Tel: 01934 744466
Fax: 01934 744345
Email: sales@eurobath.co.uk
Web: www.eurobath.co.uk
Product Type: 3, 4, 5

EUROCARE SHOWERS LTD
Unit 19, Doncaster Industry Park,
Watch House Lane, Bentley,
Doncaster, South Yorkshire, DN5 9LZ
Area of Operation: Worldwide
Tel: 01302 788684 **Fax:** 01302 780010
Email: sales@eurocare-showers.com
Web: www.eurocare-showers.com
Product Type: 1, 2, 4

FLAIR INTERNATIONAL
Bailieborough, Co.Cavan, Ireland
Area of Operation: Europe
Tel: 01344 467342 **Fax:** +353 429 665 516
Email: flairshowers@aol.com
Web: www.flairshowers.com
Product Type: 1
Material Type: C) 1

FLAIR INTERNATIONAL
Area of Operation: Europe
Tel: 01344 467342 **Fax:** 00 353 429 665516
Email: flairshowers@aol.com
Web: www.flairshowers.com
Product Type: 6, 9

Ireland's leading manufacturer of quality glass
power-shower proof shower enclosures and bath
screens. Wide choice of shapes and styles
suitable for every space.

GROHE UK
1 River Road, Barking, Essex, IG11 0HD
Area of Operation: Worldwide
Tel: 0208 594 7292 **Fax:** 0208 594 8898
Email: info@grohe.co.uk
Web: www.grohe.co.uk
Product Type: 3

HANSGROHE
Units D1 and D2, Sandown Park Trading Estate,
Royal Mills, Esher, Surrey, KT10 8BL
Area of Operation: Worldwide
Tel: 0870 770 1972 **Fax:** 0870 770 1973
Email: sales@hansgrohe.co.uk
Web: www.hansgrohe.co.uk
Product Type: 1, 3, 4
Other Info: ✎
Material Type: C) 2, 3, 11, 14

HANSGROHE
Area of Operation: UK & Ireland
Tel: 0870 7701972 **Fax:** 0870 7701973
Email: sales@hansgrohe.co.uk
Web: www.hansgrohe.co.uk
Product Type: 1, 3, 4
Materials Type: C) 2, 3, 11, 14 **Other Info:** ✎

Europe's largest manufacturer of design-led
showers, mixer taps, thermostatic valves and
accessories, plus a range of wellbeing products
like steam cabins.

HAROLD MOORE & SON LTD
Rawson Spring Road, Hillsborough,
Sheffield, South Yorkshire, S6 1PD
Area of Operation: UK & Ireland
Tel: 0114 233 6161
Fax: 0114 232 6375
Email: admin@haroldmoorebaths.co.uk
Web: www.haroldmoorebaths.co.uk
Product Type: 2

HEATRAE SADIA HEATING
Hurricane Way, Norwich, Norfolk, NR6 6EA
Area of Operation: Worldwide
Tel: 01603 420110
Fax: 01603 420149
Email: sales@heatraesadia.com
Web: www.heatraesadia.com
Product Type: 4

HERITAGE BATHROOMS
Princess Street, Bedminster, Bristol, BS3 4AG
Area of Operation: UK & Ireland
Tel: 0117 963 3333
Email: marketing@heritagebathrooms.com
Web: www.heritagebathrooms.com
Product Type: 1, 2, 3

HOT GLASS DESIGN
Unit 24 Crosby Yard Industrial Estate,
Bridgend, Mid Glamorgan, CF31 1JZ
Area of Operation: Europe
Tel: 01656 659884
Fax: 01656 659884
Email: info@hotglassdesign.co.uk
Web: www.hotglassdesign.co.uk
Product Type: 1
Material Type: J) 1, 2, 4, 5, 6

HUDSON REED
Rylands Street, Burnley, Lancashire, BB10 1RH
Area of Operation: Europe
Tel: 01282 418000 **Fax:** 01282 428915
Email: info@ultra-group.com
Web: www.hudsonreed.com ✎
Product Type: 3, 5
Material Type: C) 11, 14, 16, 17

IDEAL STANDARD
The Bathroom Works, National Avenue, Kingston
Upon Hull, East Riding of Yorks, HU5 4HS
Area of Operation: Worldwide
Tel: 0800 590311
Fax: 01482 445886
Email: ideal-standard@aseur.com
Web: www.ideal-standard.co.uk

INNOVATION GLASS COMPANY
27-28 Blake Industrial Estate, Brue Avenue,
Bridgwater, Somerset, TA7 8EQ
Area of Operation: Europe
Tel: 01278 683645 **Fax:** 01278 450088
Email: magnus@igc-uk.com
Web: www.igc-uk.com
Product Type: 1, 2
Material Type: C) 3

INSPIRED BATHROOMS
Unit R4, Innsworth Technology Park,
Innsworth Lane, Gloucester,
Gloucestershire, GL3 1DL
Area of Operation: UK & Ireland
Tel: 01452 559121
Fax: 01452 530908
Email: sales@inspired-bathrooms.co.uk
Web: www.inspired-bathrooms.co.uk ✎
Product Type: 1, 2, 3, 4, 5

ION GLASS
Unit 7 Grange Industrial Estate,
Albion Street, Southwick, Brighton,
West Sussex, BN42 4EN
Area of Operation: UK (Excluding Ireland)
Tel: 0845 658 9988 **Fax:** 0845 658 9989
Email: sales@ionglass.co.uk
Web: www.ionglass.co.uk
Product Type: 1, 2
Other Info: ✎

ITFITZ
PO Box 960, Maidenhead, Berkshire, SL6 9FU
Area of Operation: UK (Excluding Ireland)
Tel: 01628 890432 **Fax:** 0870 133 7955
Email: sales@itfitz.co.uk
Web: www.itfitz.co.uk ✎
Product Type: 3

JACUZZI UK
Silverdale Road, Newcastle under Lyme,
Staffordshire, ST5 6EL
Area of Operation: UK & Ireland
Tel: 01782 717175 **Fax:** 01782 717166
Email: jacuzzisalesdesk@jacuzziuk.com
Web: www.jacuzzi.co.uk
Product Type: 1, 2, 3, 5

JOHN SYDNEY
Lagrange, Tamworth,
Staffordshire, B79 7XD
Area of Operation: Worldwide
Tel: 01827 304000
Fax: 01827 68553
Email: enquiries@johnsydney.com
Web: www.johnsydney.com
Product Type: 3, 4, 5
Material Type: C) 11, 14

KALDEWEI
Unit 7, Sundial Court,
Tolworth Rise South,
Surbiton, Surrey, KT5 9RN
Area of Operation: Worldwide
Tel: 0870 777 2223
Fax: 0870 777 2225
Email: info@kaldewei-uk.com
Web: www.kaldewei.com
Product Type: 2

KENSINGTON STUDIO
13c/d Kensington Road, Earlsdon,
Coventry, West Midlands, CV5 6GG
Area of Operation: UK (Excluding Ireland)
Tel: 02476 713326
Fax: 02476 713136
Email: sales@kensingtonstudio.com
Web: www.kensingtonstudio.com
Product Type: 1, 2, 3, 4, 5

KERAMAG WAVENEY LTD
London Road, Beccles, Suffolk, NR34 8TS
Area of Operation: UK & Ireland
Tel: 01502 716600
Fax: 01502 717767
Email: info@keramagwaveney.co.uk
Web: www.keramag.com
Product Type: 1, 2, 3, 4

KERMI (UK) LTD
7 Brunel Road, Corby,
Northamptonshire, NN17 4JW
Area of Operation: UK & Ireland
Tel: 01536 400004
Fax: 01536 446614
Email: c.radcliff@kermi.co.uk
Web: www.kermi.co.uk
Product Type: 1

KEUCO (UK)
2 Claridge Court, Lower Kings Road,
Berkhamsted, Hertfordshire, HP42AF
Area of Operation: Worldwide
Tel: 01442 865220
Fax: 01442 865260
Email: klaus@keuco.co.uk
Web: www.keuco.de
Product Type: 5
Material Type: C) 3, 14, 17

KOHLER MIRA LIMITED
Gloucester Road, Cheltenham,
Gloucestershire, GL51 8TP
Area of Operation: UK & Ireland
Tel: 0870 850 5551
Fax: 0870 850 5552
Email: info@kohleruk.com
Web: www.kohleruk.com

M & O BATHROOM CENTRE
174-176 Goswell Road,
Clarkenwell, London, EC1V 7DT
Area of Operation: East England,
Greater London, South East England
Tel: 0207 608 0111
Fax: 0207 490 3083
Email: mando@lineone.net
Product Type: 1, 2, 3, 4

MANHATTAN SHOWERS
Marsden Mill, Brunswick Street,
Nelson, Lancashire, BB9 0LY
Area of Operation: UK & Ireland
Tel: 01282 605000
Fax: 01282 604762
Web: www.manhattanshowers.co.uk
Product Type: 1, 2

MARBLE CLASSICS
Unit 3, Station Approach,
Emsworth, Hampshire, PO10 7PW
Area of Operation: UK & Ireland
Tel: 01243 370011
Fax: 01243 371023
Email: info@marbleclassics.co.uk
Web: www.marbleclassics.co.uk
Product Type: 2
Material Type: E) 1, 2, 5

MATKI PLC
Churchward Road, Yate,
Bristol, BS37 5PL
Area of Operation: UK & Ireland
Tel: 01454 322888
Fax: 01454 315284
Email: helpline@matki.co.uk
Web: www.matki.co.uk
Product Type: 1, 2, 3

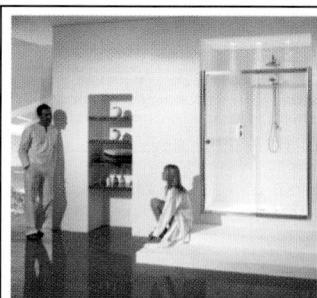

MATKI SHOWERING

Area of Operation: UK
Tel: 01454 322888
Fax: 01454 315284
Email: helpline@matki.co.uk
Web: www.matki.co.uk
Product Type: 1, 2, 3

Matki has been a major innovator in the high quality shower enclosure market for the past 30 years. Their shower doors and trays have a 10 year guarantee.

MEKON PRODUCTS
25 Bessemer Park,
Milkwood Road, London, SE24 0HG
Area of Operation: Greater London
Tel: 0207 733 8011
Fax: 0207 737 0840
Email: info@mekon.net
Web: www.mekon.net
Product Type: 1, 2, 3

MEREWAY CONTRACTS
413 Warwick Road, Birmingham,
West Midlands, B11 2JR
Area of Operation: UK (Excluding Ireland)
Tel: 0121 707 3288
Fax: 0121 707 3871
Email: info@merewaycontracts.co.uk
Web: www.merewaycontracts.co.uk

METROPOLITAN SHOWERS
Marsden Mill, Brunswick Street,
Nelson, Lancashire, BB9 0LY
Area of Operation: UK & Ireland
Tel: 01282 606070
Fax: 01282 606065
Web: www.metropolitanshowers.co.uk
Product Type: 1, 2

**MISCELLANEA DISCONTINUED
BATHROOMWARE**
Churt Place Nurseries,
Tilford Road, Churt, Farnham,
Surrey, GU10 2LN
Area of Operation: UK (Excluding Ireland)
Tel: 01428 608164
Fax: 01428 608165
Email: email@brokenbog.com
Web: www.brokenbog.com
Product Type: 1, 2, 3, 4, 5

N&C NICOBOND
41-51 Freshwater Road, Chadwell Heath,
Romford, Essex, RM8 1SP
Area of Operation: Worldwide
Tel: 0208 586 4600
Fax: 0208 586 4646
Email: info@nichollsandclarke.com
Web: www.ncdirect.co.uk

NICHOLAS ANTHONY LTD
42-44 Wigmore Street, London, W1U 2RX
Area of Operation: East England,
Greater London, South East England
Tel: 0800 068 3603
Fax: 01206 762698
Email: info@nicholas-anthony.co.uk
Web: www.nicholas-anthony.co.uk
Product Type: 1, 2, 3, 4

NO CODE
Larkwhistle Farm Road,
Micheldever, Hampshire, SO21 3BG
Area of Operation: UK & Ireland
Tel: 01962 870078
Fax: 01962 870077
Email: sales@nocode.co.uk
Web: www.nocode.co.uk
Product Type: 1, 2, 3

NORDIC
Unit 5, Trading Estate, Holland Road,
Oxted, Surrey, RH8 9BZ
Area of Operation: UK (Excluding Ireland)
Tel: 01883 732400
Fax: 01883 716970
Email: info@nordic.co.uk
Web: www.nordic.co.uk
Product Type: 1, 2

OLD FASHIONED BATHROOMS
Foresters Hall, 52 High Street, Debenham,
Stowmarket, Suffolk, IP14 6QW
Area of Operation: East England,
Greater London, South East England
Tel: 01728 860926
Fax: 01728 860446
Email: ofbrooms@lycos.co.uk
Web: www.oldfashionedbathrooms.co.uk
Product Type: 1, 2, 3, 4
Other Info: ▽ ✎
Material Type: C) 7, 13, 14, 16

PINE GARDEN
Unit 2E, Smith Green Depot, Stoney Lane,
Galgate, Lancaster, Lancashire, LA2 0PX
Area of Operation: UK & Ireland
Tel: 01524 752882 **Fax:** 01524 751744
Email: info@pinegarden.co.uk
Web: www.pinegarden.co.uk
Product Type: 1 **Other Info:** ✎
Material Type: A) 9

POLYPIPE KITCHENS & BATHROOMS LTD
Warmsworth, Halt Industrial Estate,
South Yorkshire, DN4 9LS
Area of Operation: Worldwide
Tel: 01709 770990 **Fax:** 01302 310602
Email: davery@ppbp.com
Web: www.polypipe.com/bk
Product Type: 2

PUMP WORLD LTD
Unit 11, Woodside Road,
South Marston Business Park,
Swindon, Wiltshire, SN3 4WA
Area of Operation: UK & Ireland
Tel: 01793 820142 **Fax:** 01793 823800
Email: enquiries@pumpworld.co.uk
Web: www.pumpworld.co.uk
Product Type: 3, 4, 5

QUALCERAM SHIRES PLC
Unit 9/10, Campus 5, Off Third Avenue,
Letchworth Garden City, Hertfordshire, SG6 2JF
Area of Operation: UK & Ireland
Tel: 01462 676710 **Fax:** 01462 676701
Email: info@qualceram-shires.com
Web: www.qualceram-shires.com
Product Type: 1, 2, 3, 4, 5

ROMAN LIMITED
Whitworth Avenue,
Aycliffe Industrial Park,
Durham, DL5 6YN
Area of Operation: UK & Ireland
Tel: 01325 311318
Fax: 01325 319889
Email: info@roman-showers.com
Web: www.roman-showers.com
Product Type: 1, 2

**ROOTS KITCHENS,
BEDROOMS & BATHROOMS**
Vine Farm, Stockers Hill,
Boughton-under-Blean,
Faversham, Kent, ME13 9AB
Area of Operation: UK (Excluding Ireland)
Tel: 01227 751130
Fax: 01227 750033
Email: showroom@rootskitchens.co.uk
Web: www.rootskitchens.co.uk
Product Type: 1, 2, 3

ROSCO COLLECTIONS
Stone Allerton, Axbridge,
Somerset, BS26 2NS
Area of Operation: UK & Ireland
Tel: 01934 712299
Fax: 01934 713222
Email: jonathan@roscobathrooms.demon.co.uk
Product Type: 1, 2

RSJ ASSOCIATES LTD
Unit 5, Greenfield Road,
Greenfield Farm Industrial Estate,
Congleton, Cheshire, CW12 4TR
Area of Operation: UK & Ireland
Tel: 01260 276188
Fax: 01260 280889
Email: info@rsjassociates.co.uk
Web: www.hueppe.com
Product Type: 1, 2, 3

SAMUEL HEATH
Leopold Street, Birmingham,
West Midlands, B12 0UJ
Area of Operation: Worldwide
Tel: 0121 772 2303 **Fax:** 0121 772 3334
Email: lloyd@samuel-heath.com
Web: www.samuel-heath.com
Product Type: 3, 5
Material Type: C) 11, 13, 14, 16

SANIFLO LTD
Howard House, The Runway,
South Ruislip, Middlesex, HA4 6SE
Area of Operation: UK (Excluding Ireland)
Tel: 0208 842 4040 **Fax:** 0208 842 1671
Email: andrews@saniflo.co.uk
Web: www.saniflo.co.uk
Product Type: 4

SHOWERLUX UK LTD
Sibree Road, Coventry,
West Midlands, CV3 4FD
Area of Operation: UK & Ireland
Tel: 02476 882515 **Fax:** 02476 305457
Email: sales@showerlux.com
Web: www.showerlux.com
Product Type: 1, 2

SMART SHOWERS LTD
Unit 11, Woodside Road,
South Marston Business Park,
Swindon, Wiltshire, SN3 4WA
Area of Operation: UK & Ireland
Tel: 01793 822775 **Fax:** 01793 823800
Email: louise@smartshowers.co.uk
Web: www.smartshowers.co.uk
Product Type: 1, 2, 3, 4, 5

SOGA LTD
41 Mayfield Street, Hull,
East Riding of Yorks, HU3 1NT
Area of Operation: Worldwide
Tel: 01482 327025
Email: info@soga.co.uk
Web: www.soga.co.uk
Product Type: 1

INTERIORS, FIXTURES & FINISHES

SPLASH DISTRIBUTION
113 High Street, Cuckfield,
West Sussex, RH17 5JX
Area of Operation: UK & Ireland
Tel: 01444 473355
Fax: 01444 473366
Email: stuart@splashdistribution.co.uk

STEAM DIRECT
12 Reeves Way, South Woodham Ferrers,
Chelmsford, Essex, CM3 5XF
Area of Operation: UK & Ireland
Tel: 01245 324577
Fax: 01245 325665
Email: steamdirect@aol.com
Web: www.steam-direct.co.uk
Product Type: 1

STIFFKEY BATHROOMS
89 Upper Saint Giles Street,
Norwich, Norfolk, NR2 1AB
Area of Operation: Worldwide
Tel: 01603 627850
Fax: 01603 619775
Email: stiffkeybathrooms.norwich@virgin.net
Web: www.stiffkeybathrooms.com
Product Type: 3, 5
Other Info: ▽

STUDIO STONE
The Stone Yard, Alton Lane,
Four Marks, Hampshire, GU34 5AJ
Area of Operation: UK (Excluding Ireland)
Tel: 01420 562500
Fax: 01420 563192
Email: sales@studiostone.co.uk
Web: www.studiostone.co.uk
Product Type: 1, 2
Other Info: ✎
Material Type: E) 1, 2, 3, 5

SVEDBERGS
London House, 100 New Kings Road,
London, SW6 4LX
Area of Operation: UK & Ireland
Tel: 0207 348 6107
Fax: 0207 348 6108
Email: info@svedbergs.co.uk
Web: www.svedbergs.co.uk
Product Type: 1, 3, 5

**SWADLING BRASSWARE
- A MATKI COMPANY**
Churchward Road,
Yate, Bristol, BS37 5PL
Area of Operation: Europe
Tel: 01454 322888 **Fax:** 01454 315284
Email: enquiries@swadlingbrassware.com
Web: www.swadlingbrassware.com

TAB UK
7. Kingsway House, Kingsway,
Team Valley, Gateshead,
Tyne & Wear, NE11 0HW
Area of Operation: UK & Ireland
Tel: 0191 491 5188 **Fax:** 0191 491 5189
Email: sales@tab-uk.com
Web: www.tab-uk.com
Product Type: 1
Material Type: C) 1, 11, 14

TAPS SHOP
35 Bridge Street, Witney,
Oxfordshire, OX28 1DA
Area of Operation: Worldwide
Tel: 0845 430 3035 **Fax:** 01993 779653
Email: info@tapsshop.co.uk
Web: www.tapsshop.co.uk ⌐⊕
Product Type: 3, 4, 5

TECHFLOW PRODUCTS LTD
Unit 7 Sovereign Business Park, Albert Drive,
Victoria Industrial Estate, Burgess Hill,
West Sussex, RH15 9TY
Area of Operation: UK & Ireland
Tel: 01444 258003 **Fax:** 01444 258004
Email: rod@techflow.co.uk
Web: www.techflow.co.uk
Product Type: 4

THE KITCHEN AND BATHROOM COLLECTION
Nelson House, Nelson Road,
Salisbury, Wiltshire, SP1 3LT
Area of Operation: South West England
and South Wales
Tel: 01722 334800
Fax: 01722 412252
Email: info@kbc.co.uk
Web: www.kbc.co.uk
Product Type: 1, 2, 3, 4, 5
Other Info: ✎
Material Type: E) 1, 2, 5

THE WATER MONOPOLY
16/18 Lonsdale Road, London, NW6 6RD
Area of Operation: UK (Excluding Ireland)
Tel: 0207 624 2636
Fax: 0207 624 2631
Email: enquiries@watermonopoly.com
Web: www.watermonopoly.com

TRADEMARK TILES
Airfield Industrial Estate, Warboys,
Huntingdon, Cambridgeshire, PE28 2SH
Area of Operation: East England, South East
England
Tel: 01487 825300
Fax: 01487 825305
Email: info@trademarktiles.co.uk
Product Type: 1, 2, 3, 4

TRITON PLC
Triton Road, Shepperton Business Park,
Caldwell Road, Nuneaton, Warwickshire, CV11 4NR
Area of Operation: UK & Ireland
Tel: 02476 344441
Fax: 02476 349828
Web: www.tritonshowers.co.uk
Product Type: 3, 4

WALTON BATHROOMS
The Hersham Centre, The Green, Molesey Road,
Hersham, Walton on Thames, Surrey, KT12 4HL
Area of Operation: UK (Excluding Ireland)
Tel: 01932 224784
Fax: 01932 253447
Email: sales@waltonbathrooms.co.uk
Web: www.waltonbathrooms.co.uk
Product Type: 1

WATER FRONT
All Worcester Buildings, Birmingham Road,
Redditch, Worcestershire, B97 6DY
Area of Operation: UK & Ireland
Tel: 01527 584244
Fax: 01527 61127
Email: waterfront@btconnect.com
Web: www.waterfrontbathrooms.com
Product Type: 3, 5
Material Type: C) 11, 14

WATERMILL PRODUCTS LTD
Watermill House, Fairview Industrial Estate,
Holland Road, Hurst Green,
Oxted, Surrey, RH8 9BD
Area of Operation: UK & Ireland
Tel: 01883 715425
Fax: 01883 716422
Email: sales@watermillshowers.co.uk
Web: www.watermillshowers.co.uk
Product Type: 4

WHIRLPOOL EXPRESS (UK) LTD
61-62 Lower Dock Street, Kingsway,
Newport, NP20 1EF
Area of Operation: Europe
Tel: 01633 244555
Fax: 01633 244555
Email: reception@whirlpoolexpress.co.uk
Web: www.whirlpoolexpress.co.uk ⌐⊕
Product Type: 1, 2, 3, 4, 5

WWW.BOUNDARYBATHROOMS.CO.UK
Regent Street, Colne, Lancashire, BB8 8LD
Area of Operation: UK & Ireland
Tel: 01282 862509
Fax: 01282 871192
Web: www.boundarybathrooms.co.uk
Product Type: 1, 2, 3, 4, 5

HEATED TOWEL RAILS

KEY

SEE ALSO: HEATING - Radiators
OTHER: ▽ Reclaimed ⌐⊕ On-line shopping
 ✎ Bespoke ✋ Hand-made ECO Ecological

A&J GUMMERS LIMITED
Unit H, Redfern Park Way, Tyseley,
Birmingham, West Midlands, B11 2DN
Area of Operation: Worldwide
Tel: 0121 706 2241
Fax: 0121 706 2960
Email: sales@gummers.com
Web: www.gummers.com

AESTUS
Unit 5 Strawberry Lane Industrial Estate,
Strawberry Lane, Willenhall,
West Midlands, WV13 3RS
Area of Operation: UK & Ireland
Tel: 0870 403 0115
Fax: 0870 403 0116
Email: melissa@publicityengineers.com

AL CHALLIS LTD
Europower House, Lower Road,
Cookham, Maidenhead, Berkshire, SL6 9EH
Area of Operation: UK & Ireland
Tel: 01628 529024
Fax: 0870 458 0577
Email: chris@alchallis.com
Web: www.alchallis.com

ALEXANDERS
Unit 3, 1/1A Spilsby Road,
Harold Hill, Romford, Essex, RM3 8SB
Area of Operation: UK & Ireland
Tel: 01708 384574
Fax: 01708 384089
Email: sales@alexanders-uk.net
Web: www.alexanders-uk.net

ANTIQUE BATHS OF IVYBRIDGE LTD
Emebridge Works, Ermington Road,
Ivybridge, Devon, PL21 9DE
Area of Operation: UK (Excluding Ireland)
Tel: 01752 698250
Fax: 01752 698266
Email: sales@antiquebaths.com
Web: www.antiquebaths.com ⌐⊕

BATHROOM CITY
Seeleys Road, Tyseley Industrial Estate,
Birmingham, West Midlands, B11 2LQ
Area of Operation: UK & Ireland
Tel: 0121 753 0700
Fax: 0121 753 1110
Email: sales@bathroomcity.com
Web: www.bathroomcity.com ⌐⊕

BATHROOM DISCOUNT CENTRE
297 Munster Road, Fulham,
Greater London, SW6 6BW
Area of Operation: UK (Excluding Ireland)
Tel: 0207 381 4222
Fax: 0207 381 6792
Web: www.bathroomdiscount.co.uk

BATHROOM EXPRESS
61-62 Lower Dock Street,
Kingsway, Newport, NP20 1EF
Area of Operation: UK & Ireland
Tel: 0845 130 2000
Fax: 01633 244881
Email: sales@bathroomexpress.co.uk
Web: www.bathroomexpress.co.uk ⌐⊕

BATHROOM HEAVEN
25 Eccleston Square, London, SW1V 1NS
Area of Operation: UK & Ireland
Tel: 0845 121 6700
Fax: 0207 233 6074
Email: james@bathroomheaven.com
Web: www.bathroomheaven.com

BATHROOM OPTIONS
394 Ringwood Road, Ferndown, Dorset, BH22 9AU
Area of Operation: South East England
Tel: 01202 873379
Fax: 01202 892252
Email: bathroom@options394.fsnet.co.uk
Web: www.bathroomoptions.co.uk

BATHROOMS PLUS LTD
222 Malmesbury Park Road,
Bournemouth, Dorset, BH8 8PR
Area of Operation: South West England
and South Wales
Tel: 01202 294417
Fax: 01202 316425
Email: info@bathrooms-plus.co.uk
Web: www.bathrooms-plus.co.uk

BATHROOMSTUFF.CO.UK
40 Evelegh Road, Farlington, Portsmouth,
Hampshire, PO6 1DL
Area of Operation: UK (Excluding Ireland)
Tel: 08450 580540
Fax: 02392 215695
Email: sales@bathroomstuff.co.uk
Web: www.bathroomstuff.co.uk ⌐⊕

CHATSWORTH HEATING PRODUCTS LTD
Unit B Watchmoor Point, Camberley,
Surrey, GU15 3EX
Area of Operation: UK (Excluding Ireland)
Tel: 01276 605880
Fax: 01276 605881
Email: enquiries@chatsworth-heating.co.uk
Web: www.chatsworth-heating.co.uk

CITY BATHROOMS & KITCHENS
158 Longford Road, Longford,
Coventry, West Midlands, CV6 6DR
Area of Operation: UK & Ireland
Tel: 02476 365877
Fax: 02476 644992
Email: citybathrooms@hotmail.com
Web: www.citybathrooms.co.uk

CLASSIC WARMTH DIRECT LTD
Unit 8, Brunel Workshops, Ashburton Industrial
Estate, Ross-on-Wye, Herefordshire, HR9 7DX
Area of Operation: UK & Ireland
Tel: 01989 565555
Fax: 01989 561058
Email: patbur@tiscali.co.uk

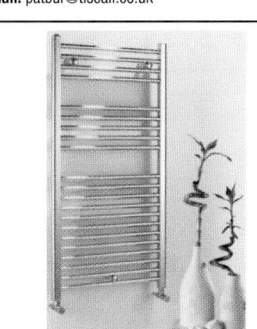

CLASSIC WARMTH DIRECT LTD

Area of Operation: UK & Ireland
Tel: 01989 565555 **Fax:** 01989 561058
Email: patbur@tiscali.co.uk

Classic Warmth's range of tubular ladder towel
rails meet the requirement for most bathroom
and kitchen applications with options for chrome,
white; straight, curved; hot water, electric or duel
fuel supply.

COMFOOT FLOORING
Walnut Tree, Redgrave Road,
South Lopham, Diss, Norfolk, IP22 2HN
Area of Operation: Europe
Tel: 01379 688516 **Fax:** 01379 688517
Email: sales@comfoot.com
Web: www.comfoot.com ⌐⊕

MIRAGE CRYSTAL

A unique award winning electric heated glass Radiator/Towel warmer design. Designed with your safety in mind it can withstand the hardest of knocks and will not splinter.

GAVOTTE MULTI RAIL

A unique contemporary elegant and artistic multi rail design from Myson.

BESPOKE MYSON PRODUCT

We can design and manufacture your unique designer radiator/towel warmer. Send your sketch detailing sizes and we will bring your design to life.

FANTASIA

A contemporary square multi rail design from Myson that will not only keep your towels warm but will also add style to your bathroom.

MYSON ELECTRIC UNDERFLOOR

A headache free design solution for your kitchen, bathroom or conservatory renovation project. Space saving, easy to install and cheap to run. ALL YOU NEED IN ONE BOX.

the first name in designer warmth

Emlyn Street
Off Egerton Street
Farnworth
Bolton BL4 7EB
Tel: 01204 863200
Fax: 01204 863229
Email: salesmtw@myson.co.uk
www.myson.co.uk

PLEASE QUOTE: HANDBOOKMYS06

MIRAGE MIRROR

A unique award winning electric heated glass design from Myson. Very safe and easy to install, this product is guaranteed to provide a focal point in your home.

RHAPSODY

A unique example of engineering disguised as art that is guaranteed to provide a focal point in your room.

ACAPPELLA

The curvaceous Acappella combines beauty, warmth and the simple function of a mirror as well as adding a bold design statement to your bedroom, hallway or bathroom.

SONATA

The height of contemporary design which combines perfect symmetry tubing to create a dramatic focal point for any wall in your home.

NOCTURNE

A masterpiece combining beauty, elegance, style and warmth to bring ambience to any room.

INTERIORS, FIXTURES & FINISHES

CONSORT EQUIPMENT PRODUCTS LTD
Thornton Industrial Estate, Milford Haven,
Pembrokeshire, SA73 2RT
Area of Operation: Worldwide
Tel: 01646 692172 **Fax:** 01646 695195
Email: enquiries@consortepl.com
Web: www.consortepl.com

**COSY ROOMS
(THE-BATHROOM-SHOP.CO.UK)**
17 Chiltern Way, North Hykeham,
Lincoln, Lincolnshire, LN6 9SY
Area of Operation: UK (Excluding Ireland)
Tel: 01522 696002 **Fax:** 01522 696002
Email: keith@cosy-rooms.com
Web: www.the-bathroom-shop.co.uk

CYMRU KITCHENS
63 Caerleon Road, St. Julians,
Newport, NP19 7BX
Area of Operation: UK (Excluding Ireland)
Tel: 01633 676767 **Fax:** 01633 212512
Email: sales@cymrukitchens.com
Web: www.cymrukitchens.com

DEEP BLUE SHOWROOM
299-313 Lewisham High Street,
Lewisham, London, SE13 6NW
Area of Operation: UK (Excluding Ireland)
Tel: 0208 690 3401 **Fax:** 0208 690 1408

DOLPHIN BATHROOMS
Bromwich Road, Worcester,
Worcestershire, WR2 4BD
Area of Operation: UK (Excluding Ireland)
Tel: 0800 626717
Web: www.dolphinbathrooms.co.uk

ECOLEC
Sharrocks Street, Wolverhampton,
West Midlands, WV1 3RP
Area of Operation: UK & Ireland
Tel: 01902 457575
Fax: 01902 457797
Email: sales@ecolec.co.uk
Web: www.ecolec.co.uk
Other Info:

ESKIMO DESIGN LTD.
25 Horsell Road, London, N5 1XL
Area of Operation: Worldwide
Tel: 0207 117 0110
Email: ed@eskimodesign.co.uk
Web: www.eskimodesign.co.uk
Other Info:
Material Type: C) 1, 2, 3, 7, 11, 14, 15, 16

HERITAGE BATHROOMS
Princess Street, Bedminster,
Bristol, BS3 4AG
Area of Operation: UK & Ireland
Tel: 0117 963 3333
Email: marketing@heritagebathrooms.com
Web: www.heritagebathrooms.com

HUDSON REED
Rylands Street, Burnley,
Lancashire, BB10 1RH
Area of Operation: Europe
Tel: 01282 418000
Fax: 01282 428915
Email: info@ultra-group.com
Web: www.hudsonreed.com
Material Type: C) 11, 14, 16

INSPIRED BATHROOMS
Unit R4, Innsworth Technology Park,
Innsworth Lane, Gloucester,
Gloucestershire, GL3 1DL
Area of Operation: UK & Ireland
Tel: 01452 559121
Fax: 01452 530908
Email: sales@inspired-bathrooms.co.uk
Web: www.inspired-bathrooms.co.uk

JIS EUROPE (SUSSEX RANGE)
Warehouse 2, Nash Lane,
Scaynes Hill, Haywards Heath,
West Sussex, RH17 7NJ
Area of Operation: Europe
Tel: 01444 831200
Fax: 01444 831900
Email: info@jiseurope.co.uk

KEELING HEATING PRODUCTS
Cranbourne Road, Gosport,
Hampshire, PO12 1RJ
Area of Operation: UK & Ireland
Tel: 02392 796633
Fax: 02392 425028
Email: sales@keeling.co.uk
Web: www.keeling.co.uk
Material Type: C) 2, 7, 11, 13, 14, 16

KENSINGTON STUDIO
13 c & d Kensington Road,
Earlsdon, Coventry,
West Midlands, CV5 6GG
Area of Operation: UK (Excluding Ireland)
Tel: 02476 713326
Fax: 02476 713136
Email: sales@kensingtonstudio.com
Web: www.kensingtonstudio.com

KERMI (UK) LTD
7 Brunel Road, Corby,
Northamptonshire, NN17 4JW
Area of Operation: UK & Ireland
Tel: 01536 400004
Fax: 01536 446614
Email: c.radcliff@kermi.co.uk
Web: www.kermi.co.uk

M & O BATHROOM CENTRE
174-176 Goswell Road,
Clarkenwell, London, EC1V 7DT
Area of Operation: East England,
Greater London, South East England
Tel: 0207 608 0111
Fax: 0207 490 3083
Email: mando@lineone.net

MEKON PRODUCTS
25 Bessemer Park, Milkwood Road,
London, SE24 0HG
Area of Operation: Greater London
Tel: 0207 733 8011
Fax: 0207 737 0840
Email: info@mekon.net
Web: www.mekon.net

MEREWAY CONTRACTS
413 Warwick Road, Birmingham,
West Midlands, B11 2JR
Area of Operation: UK (Excluding Ireland)
Tel: 0121 707 3288
Fax: 0121 707 3871
Email: info@merewaycontracts.co.uk
Web: www.merewaycontracts.co.uk

**MISCELLANEA DISCONTINUED
BATHROOMWARE**
Churt Place Nurseries,
Tilford Road, Churt, Farnham,
Surrey, GU10 2LN
Area of Operation: UK (Excluding Ireland)
Tel: 01428 608164
Fax: 01428 608165
Email: email@brokenbog.com
Web: www.brokenbog.com

MYSON
Emlyn Street, Farnworth,
Bolton, Lancashire, BL4 7EB
Area of Operation: Worldwide
Tel: 01204 863200
Fax: 01204 863229
Email: salesmtw@myson.co.uk
Web: www.myson.co.uk

NICHOLAS ANTHONY LTD
42-44 Wigmore Street, London, W1U 2RX
Area of Operation: East England,
Greater London, South East England
Tel: 0800 068 3603
Fax: 01206 762698
Email: info@nicholas-anthony.co.uk
Web: www.nicholas-anthony.co.uk

NO CODE
Larkwhistle Farm Road,
Micheldever, Hampshire, SO21 3BG
Area of Operation: UK & Ireland
Tel: 01962 870078
Fax: 01962 870077
Email: sales@nocode.co.uk
Web: www.nocode.co.uk

OLD FASHIONED BATHROOMS
Foresters Hall, 52 High Street,
Debenham, Stowmarket, Suffolk, IP14 6QW
Area of Operation: East England,
Greater London, South East England
Tel: 01728 860926
Fax: 01728 860446
Email: ofbrooms@lycos.co.uk
Web: www.oldfashionedbathrooms.co.uk
Material Type: C) 14, 16

QUALCERAM SHIRES PLC
Unit 9/10, Campus 5, Off Third Avenue,
Letchworth Garden City, Hertfordshire, SG6 2JF
Area of Operation: UK & Ireland
Tel: 01462 676710
Fax: 01462 676701
Email: info@qualceram-shires.com
Web: www.qualceram-shires.com

RADIATING STYLE
Unit 15 Thompon Road, Hounslow,
Middlesex, Greater London, TW3 3UH
Area of Operation: UK (Excluding Ireland)
Tel: 0870 072 3428
Fax: 0208 577 9222
Email: sales@radiatingstyle.com
Web: www.radiatingstyle.com

**ROOTS KITCHENS,
BEDROOMS & BATHROOMS**
Vine Farm, Stockers Hill,
Boughton-under-Blean,
Faversham, Kent, ME13 9AB
Area of Operation: UK (Excluding Ireland)
Tel: 01227 751130
Fax: 01227 750033
Email: showroom@rootskitchens.co.uk
Web: www.rootskitchens.co.uk

STIFFKEY BATHROOMS
89 Upper Saint Giles Street,
Norwich, Norfolk, NR2 1AB
Area of Operation: Worldwide
Tel: 01603 627850
Fax: 01603 619775
Email: stiffkeybathrooms.norwich@virgin.net
Web: www.stiffkeybathrooms.com

INTERIORS, FIXTURES & FINISHES

SVEDBERGS
London House, 100 New Kings Road,
London, SW6 4LX
Area of Operation: UK & Ireland
Tel: 0207 348 6107 **Fax:** 0207 348 6108
Email: info@svedbergs.co.uk
Web: www.svedbergs.co.uk

THE KITCHEN & BATHROOM COLLECTION
Nelson House, Nelson Road,
Salisbury, Wiltshire, SP1 3LT
Area of Operation: South West England
and South Wales
Tel: 01722 334800 **Fax:** 01722 412252
Email: info@kbc.co.uk
Web: www.kbc.co.uk
Other Info: ✎

THE RADIATOR COMPANY LIMITED
Elan House, Charlwoods Road,
East Grinstead, West Sussex, RH19 2HG
Area of Operation: UK & Ireland
Tel: 01342 302250 **Fax:** 01342 305561
Email: ian@theradiatorcompany.co.uk
Web: www.theradiatorcompany.co.uk

TRADEMARK TILES
Airfield Industrial Estate, Warboys,
Huntingdon, Cambridgeshire, PE28 2SH
Area of Operation: East England,
South East England
Tel: 01487 825300 **Fax:** 01487 825305
Email: info@trademarktiles.co.uk

URBIS (GB) LTD.
55 Lidgate Crescent,
Langthwaite Grange Industrial Estate,
South Kirkby, West Yorkshire, WF9 3NR
Area of Operation: UK & Ireland
Tel: 01977 659829
Fax: 01977 659808
Email: david@urbis-gb.co.uk
Web: www.urbis-gb.co.uk ⌂

VOGUE UK
Units 6-10, Strawberry Lane, Industrial Estate,
Willenhall, West Midlands, WV13 3RS
Area of Operation: UK & Ireland
Tel: 0870 403 0101 **Fax:** 0870 403 0102
Email: sales@vogue-uk.com
Web: www.vogue-uk.com

WALTON BATHROOMS
The Hersham Centre, The Green, Molesey Road,
Hersham, Walton on Thames, Surrey, KT12 4HL
Area of Operation: UK (Excluding Ireland)
Tel: 01932 224784 **Fax:** 01932 253447
Email: sales@waltonbathrooms.co.uk
Web: www.waltonbathrooms.co.uk

WARMROOMS
24 Corncroft Lane, St Leonards Park,
Gloucester, Gloucestershire, GL4 6XU
Area of Operation: UK (Excluding Ireland)
Tel: 01452 304460 **Fax:** 01452 304460
Email: sales@warmrooms.co.uk
Web: www.warmrooms.co.uk ⌂

WHIRLPOOL EXPRESS (UK) LTD
61-62 Lower Dock Street, Kingsway,
Newport, NP20 1EF
Area of Operation: Europe
Tel: 01633 244555 **Fax:** 01633 244555
Email: reception@whirlpoolexpress.co.uk
Web: www.whirlpoolexpress.co.uk ⌂

WWW.BOUNDARYBATHROOMS.CO.UK
Regent Street, Colne, Lancashire, BB8 8LD
Area of Operation: UK & Ireland
Tel: 01282 862509 **Fax:** 01282 871192
Web: www.boundarybathrooms.co.uk

WWW.DESIGNER-RADIATORS.COM
Regent Street, Colne, Lancashire, BB8 8LD
Area of Operation: UK & Ireland
Tel: 01282 862509
Fax: 01282 871192
Web: www.designer-radiators.com ⌂

BATHROOM FURNITURE

KEY
PRODUCT TYPES: 1= Vanity Units 2 =
Cabinets 3 = Washstands 4 = Other
OTHER: ▽ Reclaimed ⌂ On-line shopping
✎ Bespoke ✋ Hand-made ECO Ecological

ABP-TBS PARTNERSHIP LTD.
Martens Road, Northbank Industrial Park,
Irlam, Manchester, M44 5AX.
Area of Operation: UK & Ireland
Tel: 0161 775 1871.
Fax: 0161 775 8929.
Email: info@tbs-fabrications.co.uk
Web: www.abp-tbswashrooms.co.uk
Product Type: 1, 2

ALEXANDERS
Unit 3, 1/1A Spilsby Road, Harold Hill,
Romford, Essex, RM3 8SB
Area of Operation: UK & Ireland
Tel: 01708 384574
Fax: 01708 384089
Email: sales@alexanders-uk.net
Web: www.alexanders-uk.net
Product Type: 1, 2, 3, 4

ALICANTE STONE
Damaso Navarro, 6 Bajo, 03610 Petrer
(Alicante), P.O. Box 372, Spain
Area of Operation: Europe
Tel: +34 966 319 697
Fax: +34 966 319 698
Email: info@alicantestone.com
Web: www.alicantestone.com
Product Type: 1
Material Type: E) 1, 2, 5, 8

ALLMILMO UK
Unit 17 Plantaganet House,
Kingsclere Business Park,
Kingsclere, Newbury, Berkshire, RG20 4SW
Area of Operation: UK & Ireland
Tel: 01635 868181
Fax: 01635 869693
Email: allmilmo@aol.com
Web: www.allmilmo.com
Product Type: 1, 2, 3, 4

ALTHEA UK LTD
Concept House, Blanche Street,
Bradford, West Yorkshire, BD4 8DA
Area of Operation: UK (Excluding Ireland)
Tel: 01274 660770
Fax: 01274 667929
Email: sales@altheauk.com
Web: www.periodbathrooms.co.uk

AMBIANCE BAIN
Cumberland House,
80 Scrubs Lane, London, NW10 6RF
Area of Operation: UK & Ireland
Tel: 0870 902 1313
Fax: 0870 902 1312
Email: sghirardello@mimea.com
Web: www.ambiancebain.com
Product Type: 1, 2, 3

ANTIQUE BATHS OF IVYBRIDGE LTD
Emebridge Works, Ermington Road,
Ivybridge, Devon, PL21 9DE
Area of Operation: UK (Excluding Ireland)
Tel: 01752 698250 **Fax:** 01752 698266
Email: sales@antiquebaths.com
Web: www.antiquebaths.com ⌂
Product Type: 1, 3

AQUAPLUS SOLUTIONS
Unit 106, Indiana Building, London, SE13 7QD
Area of Operation: UK & Ireland
Tel: 0870 201 1915 **Fax:** 0870 201 1916
Email: info@aquaplussolutions.com
Web: www.aquaplussolutions.com

**ARTICHOKE
(KITCHEN AND CABINET MAKERS)**
Hortswood, Long Lane,
Wrington, Somerset, BS40 5SP
Area of Operation: UK & Ireland
Tel: 01934 863840
Fax: 01934 863841
Email: mail@artichoke-ltd.com
Web: www.artichoke-ltd.com
Product Type: 1, 2, 3

B ROURKE & CO LTD
Vulcan Works, Accrington Road,
Burnley, Lancashire, BB11 5QD
Area of Operation: Worldwide
Tel: 01282 422841
Fax: 01282 458901
Email: info@rourkes.co.uk
Web: www.rourkes.co.uk
Product Type: 3, 4
Material Type: C) 2, 4, 5, 6

BARRHEAD SANITARYWARE PLC
15-17 Nasmyth Road,
Hillington Industrial Estate,
Hillington, Glasgow,
Renfrewshire, G52 4RG
Area of Operation: UK & Ireland
Tel: 0141 883 0066
Fax: 0141 883 0077
Email: sales@barrhead.co.uk
Product Type: 1, 2

BATHROOM BARN
Uplands, Wyton, Huntingdon,
Cambridgeshire, PE28 2JZ
Area of Operation: UK (Excluding Ireland)
Tel: 01480 458900
Email: sales@bathroombarn.co.uk
Web: www.bathroombarn.co.uk

BATHROOM CITY
Seeleys Road, Tyseley Industrial Estate,
Birmingham, West Midlands, B11 2LQ
Area of Operation: UK & Ireland
Tel: 0121 753 0700
Fax: 0121 753 1110
Email: sales@bathroomcity.com
Web: www.bathroomcity.com ⌂
Product Type: 1, 2, 3, 4

BATHROOM OPTIONS
394 Ringwood Road,
Ferndown, Dorset, BH22 9AU
Area of Operation: South East England
Tel: 01202 873379
Fax: 01202 892252
Email: bathroom@options394.fsnet.co.uk
Web: www.bathroomoptions.co.uk
Product Type: 1, 2, 3, 4

BATHROOMS PLUS LTD
222 Malmesbury Park Road,
Bournemouth, Dorset, BH8 8PR
Area of Operation: South West England
and South Wales
Tel: 01202 294417
Fax: 01202 316425
Email: info@bathrooms-plus.co.uk
Web: www.bathrooms-plus.co.uk
Product Type: 1, 2, 3, 4
Material Type: A) 1, 2, 4, 5, 9

**BROOK-WATER DESIGNER
BATHROOMS & KITCHENS**
The Downs, Woodhouse Hill,
Uplyme, Lyme Regis, Dorset, DT7 3SL
Area of Operation: Worldwide
Tel: 01235 201256
Email: sales@brookwater.co.uk
Web: www.brookwater.co.uk ⌂

BUSHBOARD LTD
9-29 Rixon Road, Wellingborough,
Northamptonshire, NN8 4BA
Area of Operation: UK & Ireland
Tel: 01933 232200
Email: washrooms@bushboard.co.uk
Web: www.bushboard.co.uk

BUYDESIGN
c/o Woodschool Ltd, Monteviot Nurseries,
Ancrum, Jedburgh, Borders, TD8 6TU
Area of Operation: North East England,
North West England & North Wales, Scotland
Tel: 01835 830740
Fax: 01835 830750
Email: info@buydesign-furniture.com
Web: www.buydesign-furniture.com
Product Type: 2, 4
Other Info: ECO ✎ ✋

CAMERON PETERS LTD
The Old Dairy, Home Farm, Ardington,
Wantage, Oxfordshire, OX12 8PD
Area of Operation: Worldwide
Tel: 01235 835000
Fax: 01235 835005
Email: info@cameronpeters.co.uk
Web: www.cameronpeters.co.uk
Product Type: 4

CAPITAL MARBLE DESIGN
Unit 1 Pall Mall Deposit, 124-128 Barlby Road,
London, W10 6BL
Area of Operation: UK & Ireland
Tel: 0208 968 5340
Fax: 0208 968 8827
Email: stonegallery@capitalmarble.co.uk
Web: www.capitalmarble.co.uk
Product Type: 1, 2, 3

CITY BATHROOMS & KITCHENS
158 Longford Road, Longford, Coventry,
West Midlands, CV6 6DR
Area of Operation: UK & Ireland
Tel: 02476 365877
Fax: 02476 644992
Email: citybathrooms@hotmail.com
Web: www.citybathrooms.co.uk

COSY ROOMS (THE-BATHROOM-SHOP.CO.UK)
17 Chiltern Way, North Hykeham,
Lincoln, Lincolnshire, LN6 9SY
Area of Operation: UK (Excluding Ireland)
Tel: 01522 696002
Fax: 01522 696002
Email: keith@cosy-rooms.com
Web: www.the-bathroom-shop.co.uk ⌂
Product Type: 1, 2, 3

COTTAGE FARM ANTIQUES
Stratford Road, Aston Sub Edge,
Chipping Campden, Gloucestershire, GL55 6PZ
Area of Operation: Worldwide
Tel: 01386 438263
Fax: 01386 438263
Email: info@cottagefarmantiques.co.uk
Web: www.cottagefarmantiques.co.uk ⌂
Product Type: 2, 4
Other Info: ▽ ECO ✎ ✋
Material Type: B) 2

CYMRU KITCHENS
63 Caerleon Road, St. Julians, Newport, NP19 7BX
Area of Operation: UK (Excluding Ireland)
Tel: 01633 676767 **Fax:** 01633 212512
Email: sales@cymrukitchens.com
Web: www.cymrukitchens.com

D.A FURNITURE LTD
Woodview Workshop, Pitway Lane,
Farrington Gurney, Bristol, BS39 6TX
Area of Operation: South West England
and South Wales
Tel: 01761 453117
Email: enquiries@dafurniture.co.uk
Web: www.dafurniture.co.uk
Product Type: 1, 2, 3 **Other Info:** ✎ ✋
Material Type: A) 2, 3, 4, 5, 6, 7, 8, 9, 10, 11, 12,
13, 14, 15

DECORLUX LTD
18 Ghyll Industrial Estate, Ghyll Road,
Heathfield, East Sussex, TN21 8AW
Area of Operation: UK (Excluding Ireland)
Tel: 01435 866638 **Fax:** 01435 866641
Email: info@decorlux.co.uk
Product Type: 1, 2, 3

INTERIORS, FIXTURES & FINISHES

DECORMASTER LTD
Unit 16, Waterside Industrial Estate,
Wolverhampton, West Midlands, WV2 2RH
Area of Operation: Worldwide
Fax: 01902 353126
Email: sales@oldcolours.co.uk
Web: www.oldcolours.co.uk
Product Type: 1, 2, 3

DEEP BLUE SHOWROOM
299-313 Lewisham High Street,
Lewisham, London, SE13 6NW
Area of Operation: UK (Excluding Ireland)
Tel: 0208 690 3401
Fax: 0208 690 1408
Product Type: 1, 2, 3

DESIGNER BATHROOMS
140-142 Pogmoor Road, Pogmoor,
Barnsley, North Yorkshire, S75 2DX
Area of Operation: UK (Excluding Ireland)
Tel: 01226 280200
Fax: 01226 733273
Email: mike@designer-bathrooms.com
Web: www.designer-bathrooms.com
Product Type: 1, 2, 3, 4

DOLPHIN BATHROOMS
Bromwich Road, Worcester,
Worcestershire, WR2 4BD
Area of Operation: UK (Excluding Ireland)
Tel: 0800 626717
Web: www.dolphinbathrooms.co.uk
Product Type: 2, 3

DUPONT CORIAN & ZODIAQ
10 Quarry Court,
Pitstone Green Business Park,
Pitstone, nr Tring,
Buckinghamshire, LU7 9GW
Area of Operation: UK & Ireland
Tel: 01296 663555
Fax: 01296 663599
Email: sales@corian.co.uk
Web: www.corian.co.uk
Product Type: 1
Other Info: ✎

DURAVIT
Unit 7, Stratus Park, Brudenell Drive,
Brinklow, Milton Keynes,
Buckinghamshire, MK10 0DE
Area of Operation: UK & Ireland
Tel: 0870 730 7787
Fax: 0870 730 7786
Email: info@uk.duravit.com
Web: www.duravit.co.uk
Product Type: 1, 2, 3, 4
Material Type: A) 1, 2, 3, 4, 5, 6, 7, 9

EMCO
Centre Point Distribution Ltd,
Unit A5 Regent Park, Booth Drive,
Park Farm Industrial Estate, Wellingborough,
Northamptonshire, NN8 6GR
Area of Operation: UK & Ireland
Tel: 01933 403786 **Fax:** 01933 403789
Email: k.pedrick@mailcity.com
Web: www.emco-bath.com
Product Type: 1, 2, 3, 4

ESKIMO DESIGN LTD.
25 Horsell Road, London, N5 1XL
Area of Operation: Worldwide
Tel: 020 7117 0110
Email: ed@eskimodesign.co.uk
Web: www.eskimodesign.co.uk
Product Type: 1, 2
Other Info: ♨
Material Type: A) 1, 6, 7, 12

FRANCIS N. LOWE LTD.
The Marble Works, New Road,
Middleton, Matlock, Derbyshire, DE4 4NA
Area of Operation: Europe
Tel: 01629 822216 **Fax:** 01629 824348
Email: info@lowesmarble.com
Web: www.lowesmarble.com
Product Type: 1

G MIDDLETON LTD
Cross Croft Industrial Estate,
Appleby, Cumbria, CA16 6HX
Area of Operation: Europe
Tel: 01768 352067
Fax: 01768 353228
Email: g.middleton@timber75.freeserve.co.uk
Web: www.graham-middleton.co.uk ✎
Product Type: 1, 2
Material Type: A) 2, 3, 4, 5, 9

GOODWOOD BATHROOMS LTD
Church Road, North Mundham,
Chichester, West Sussex, PO20 1JU
Area of Operation: UK (Excluding Ireland)
Tel: 01243 532121
Fax: 01243 533423
Email: sales@goodwoodbathroom.co.uk
Web: www.goodwoodbathrooms.co.uk
Product Type: 1, 2, 3

HERITAGE BATHROOMS
Princess Street, Bedminster, Bristol, BS3 4AG
Area of Operation: UK & Ireland
Tel: 0117 963 3333
Email: marketing@heritagebathrooms.com
Web: www.heritagebathrooms.com
Product Type: 1, 2, 3

HYGROVE
152/154 Merton Road,
Wimbledon, London, SW19 1EH
Area of Operation: South East England
Tel: 0208 543 1200
Fax: 0208 543 6521
Email: sales@hygrove.fsnet.co.uk
Web: www.hygrovefurniture.co.uk
Product Type: 1, 2, 3, 4

IDEAL STANDARD
The Bathroom Works,
National Avenue, Kingston Upon Hull,
East Riding of Yorks, HU5 4HS
Area of Operation: Worldwide
Tel: 0800 590311
Fax: 01482 445886
Email: ideal-standard@aseur.com
Web: www.ideal-standard.co.uk

INSPIRED BATHROOMS
Unit R4, Innsworth Technology Park,
Innsworth Lane, Gloucester,
Gloucestershire, GL3 1DL
Area of Operation: UK & Ireland
Tel: 01452 559121
Fax: 01452 530908
Email: sales@inspired-bathrooms.co.uk
Web: www.inspired-bathrooms.co.uk ✎
Product Type: 1, 2, 3

INTERESTING INTERIORS LTD
37-39 Queen Street, Aberystwyth,
Ceredigion, SY23 1PU
Area of Operation: UK (Excluding Ireland)
Tel: 01970 626162
Fax: 0870 051 7959
Email: enquiries@interestinginteriors.com
Web: www.interiors-ltd.demon.co.uk
Product Type: 1, 2, 3, 4

JACUZZI UK
Silverdale Road, Newcastle under Lyme,
Staffordshire, ST5 6EL
Area of Operation: UK & Ireland
Tel: 01782 717175
Fax: 01782 717166
Email: jacuzzisalesdesk@jacuzziuk.com
Web: www.jacuzzi.co.uk
Product Type: 1, 2, 4

JACUZZI UK - BC SANITAN
Silverdale Road, Newcastle under Lyme,
Staffordshire, ST5 6EL
Area of Operation: UK & Ireland
Tel: 01782 717175
Fax: 01782 717166
Email: jacuzzisalesdesk@jacuzziuk.com
Web: www.jacuzziuk.com
Product Type: 1, 2, 4

KENSINGTON STUDIO
13 c & d Kensington Road,
Earlsdon, Coventry, West Midlands, CV5 6GG
Area of Operation: UK (Excluding Ireland)
Tel: 02476 713326
Fax: 02476 713136
Email: sales@kensingtonstudio.com
Web: www.kensingtonstudio.com ✎
Product Type: 1, 2, 3, 4

KERAMAG WAVENEY LTD
London Road, Beccles,
Suffolk, NR34 8TS
Area of Operation: UK & Ireland
Tel: 01502 716600
Fax: 01502 717767
Email: info@keramagwaveney.co.uk
Web: www.keramag.com
Product Type: 1, 2, 3, 4

KEUCO (UK)
2 Claridge Court,
Lower Kings Road, Berkhamsted,
Hertfordshire, HP42AF
Area of Operation: Worldwide
Tel: 01442 865220
Fax: 01442 865260
Email: klaus@keuco.co.uk
Web: www.keuco.de
Product Type: 1, 2
Material Type: H) 1, 2, 5

KOHLER MIRA LIMITED
Gloucester Road, Cheltenham,
Gloucestershire, GL51 8TP
Area of Operation: UK & Ireland
Tel: 0870 850 5551
Fax: 0870 850 5552
Email: info@kohleruk.com
Web: www.kohleruk.com

LECICO PLC
Unit 47a Hobbs Industrial Estate,
Newchapel, Nr Lingfield, Surrey, RH7 6HN
Area of Operation: UK & Ireland
Tel: 01342 834777
Fax: 01342 834783
Email: info@lecico.co.uk
Web: www.lecico.co.uk
Product Type: 1

M & O BATHROOM CENTRE
174-176 Goswell Road, Clarkenwell,
London, EC1V 7DT
Area of Operation: East England,
Greater London, South East England
Tel: 0207 608 0111
Fax: 0207 490 3083
Email: mando@lineone.net
Product Type: 1, 2, 3

MEKON PRODUCTS
25 Bessemer Park, Milkwood Road,
London, SE24 0HG
Area of Operation: Greater London
Tel: 0207 733 8011
Fax: 0207 737 0840
Email: info@mekon.net
Web: www.mekon.net
Product Type: 1, 2, 3

MEREWAY CONTRACTS
413 Warwick Road, Birmingham,
West Midlands, B11 2JR
Area of Operation: UK (Excluding Ireland)
Tel: 0121 707 3288
Fax: 0121 707 3871
Email: info@merewaycontracts.co.uk
Web: www.merewaycontracts.co.uk

MIKE WALKER DISTRIBUTION
Clutton Hill Estate, King Lane,
Clutton Hill, Bristol, BS39 5QQ
Area of Operation: Midlands & Mid Wales,
South West England and South Wales
Tel: 01761 453838
Fax: 01761 453060
Email: office.mwd@zoom.co.uk
Web: www.mikewalker.co.uk

**MISCELLANEA DISCONTINUED
BATHROOMWARE**
Churt Place Nurseries, Tilford Road,
Churt, Farnham, Surrey, GU10 2LN
Area of Operation: UK (Excluding Ireland)
Tel: 01428 608164
Fax: 01428 608165
Email: email@brokenbog.com
Web: www.brokenbog.com
Product Type: 4

NICHOLAS ANTHONY LTD
42-44 Wigmore Street, London, W1U 2RX
Area of Operation: East England,
Greater London, South East England
Tel: 0800 068 3603
Fax: 01206 762698
Email: info@nicholas-anthony.co.uk
Web: www.nicholas-anthony.co.uk
Product Type: 1, 2, 3

NO CODE
Larkwhistle Farm Road, Micheldever,
Hampshire, SO21 3BG
Area of Operation: UK & Ireland
Tel: 01962 870078
Fax: 01962 870077
Email: sales@nocode.co.uk
Web: www.nocode.co.uk
Product Type: 1, 2, 3

OLD FASHIONED BATHROOMS
Foresters Hall, 52 High Street, Debenham,
Stowmarket, Suffolk, IP14 6QW
Area of Operation: East England,
Greater London, South East England
Tel: 01728 860926
Fax: 01728 860446
Email: ofbrooms@lycos.co.uk
Web: www.oldfashionedbathrooms.co.uk
Product Type: 1, 2, 3
Other Info: ✍
Material Type: B) 2

PANELS PLUS
22 Mill Place, Kingston Upon Thames,
Surrey, KT1 2RJ
Area of Operation: Greater London
Tel: 0208 399 6343
Fax: 0208 399 6343
Email: sales@panelsplusltd.com
Web: www.panelsplusltd.com ✎
Product Type: 1, 2

PARLOUR FARM
Unit 12b, Wilkinson Road, Love Lane Industrial
Estate, Cirencester, Gloucestershire, GL7 1YT
Area of Operation: Europe
Tel: 01285 885336
Fax: 01285 643189
Email: info@parlourfarm.com
Web: www.parlourfarm.com

PRETTY SWIFT LTD
Units 7&8, 51 Chancery Lane,
Debenham, Suffolk, IP14 6PJ
Area of Operation: UK (Excluding Ireland)
Tel: 01728 861818
Fax: 01728 861919
Email: prettyswiftltd@aol.com

QUALCERAM SHIRES PLC
Unit 9/10, Campus 5, Off Third Avenue,
Letchworth Garden City, Hertfordshire, SG6 2JF
Area of Operation: UK & Ireland
Tel: 01462 676710
Fax: 01462 676701
Email: info@qualceram-shires.com
Web: www.qualceram-shires.com
Product Type: 1, 2, 3, 4

R&D MARKETING (DEMISTA) LTD
Land House, Anyards Road,
Cobham, Surrey, KT11 2LW
Area of Operation: UK (Excluding Ireland)
Tel: 01932 866600 **Fax:** 01932 866 688
Email: rd@demista.co.uk
Web: www.demista.co.uk
Product Type: 2

INTERIORS, FIXTURES & FINISHES

RICHARD BAKER FURNITURE LTD
Wimbledon Studios, 257 Burlington Road,
New Malden, Surrey, KT3 4NE
Area of Operation: Europe
Tel: 0208 336 1777 **Fax:** 0208 336 1666
Email: sales@richardbakerfurniture.co.uk
Web: www.richardbakerfurniture.co.uk
Product Type: 1, 2

**ROOTS KITCHENS,
BEDROOMS & BATHROOMS**
Vine Farm, Stockers Hill,
Boughton-under-Blean,
Faversham, Kent, ME13 9AB
Area of Operation: UK (Excluding Ireland)
Tel: 01227 751130 **Fax:** 01227 750033
Email: showroom@rootskitchens.co.uk
Web: www.rootskitchens.co.uk 🖑
Product Type: 1, 2, 3

ROSCO COLLECTIONS
Stone Allerton, Axbridge, Somerset, BS26 2NS
Area of Operation: UK & Ireland
Tel: 01934 712299 **Fax:** 01934 713222
Email: jonathan@roscobathrooms.demon.co.uk
Product Type: 1, 2, 3

SAMUEL HEATH
Leopold Street, Birmingham,
West Midlands, B12 0UJ
Area of Operation: Worldwide
Tel: 0121 772 2303 **Fax:** 0121 772 3334
Email: lloyd@samuel-heath.com
Web: www.samuel-heath.com
Product Type: 4

SHOWERLUX UK LTD
Sibree Road, Coventry, West Midlands, CV3 4FD
Area of Operation: UK & Ireland
Tel: 02476 882515 **Fax:** 02476 305457
Email: sales@showerlux.co.uk
Web: www.showerlux.com
Product Type: 1, 2

SMART SHOWERS LTD
Unit 11, Woodside Road,
South Marston Business Park,
Swindon, Wiltshire, SN3 4WA
Area of Operation: UK & Ireland
Tel: 01793 822775 **Fax:** 01793 823800
Email: louise@smartshowers.co.uk
Web: www.smartshowers.co.uk 🖑
Product Type: 1, 2

SOGA LTD
41 Mayfield Street, Hull,
East Riding of Yorks, HU3 1NT
Area of Operation: Worldwide
Tel: 01482 327025
Email: info@soga.co.uk
Web: www.soga.co.uk
Product Type: 1, 3

SPLASH DISTRIBUTION
113 High Street, Cuckfield,
West Sussex, RH17 5JX
Area of Operation: UK & Ireland
Tel: 01444 473355 **Fax:** 01444 473366
Email: stuart@splashdistribution.co.uk
Product Type: 1, 2, 3

STIFFKEY BATHROOMS
89 Upper Saint Giles Street, Norwich,
Norfolk, NR2 1AB
Area of Operation: Worldwide
Tel: 01603 627850 **Fax:** 01603 619775
Email: stiffkeybathrooms.norwich@virgin.net
Web: www.stiffkeybathrooms.com
Product Type: 3

SVEDBERGS
London House, 100 New Kings Road,
London, SW6 4LX
Area of Operation: UK & Ireland
Tel: 0207 348 6107 **Fax:** 0207 348 6108
Email: info@svedbergs.co.uk
Web: www.svedbergs.co.uk
Product Type: 1, 2, 3, 4
Material Type: A) 1, 2, 9

TAB UK
7. Kingsway House, Kingsway, Team Valley,
Gateshead, Tyne & Wear, NE11 0HW
Area of Operation: UK & Ireland
Tel: 0191 491 5188
Fax: 0191 491 5189
Email: sales@tab-uk.com
Web: www.tab-uk.com
Product Type: 1, 2
Material Type: D) 5

TAPS SHOP
35 Bridge Street, Witney,
Oxfordshire, OX28 1DA
Area of Operation: Worldwide
Tel: 0845 430 3035
Fax: 01993 779653
Email: info@tapsshop.co.uk
Web: www.tapsshop.co.uk 🖑
Product Type: 2, 3

THE DORSET CUPBOARD COMPANY
2 Mayo Farm, Cann, Shaftesbury, Dorset, SP7 0EF
Area of Operation: Greater London,
South East England, South West England
and South Wales
Tel: 01747 855044
Fax: 01747 855045
Email: nchturner@aol.com
Product Type: 1, 2, 3
Other Info: ✎

THE KITCHEN & BATHROOM COLLECTION
Nelson House, Nelson Road,
Salisbury, Wiltshire, SP1 3LT
Area of Operation: South West England
and South Wales
Tel: 01722 334800
Fax: 01722 412252
Email: info@kbc.co.uk
Web: www.kbc.co.uk
Product Type: 1, 2, 3
Other Info: ✎

THE WATER MONOPOLY
16/18 Lonsdale Road, London, NW6 6RD
Area of Operation: UK (Excluding Ireland)
Tel: 0207 624 2636
Fax: 0207 624 2631
Email: enquiries@watermonopoly.com
Web: www.watermonopoly.com

TOUCHSTONE - MARBLE SURFACES
Touchstone House, 82 High Street,
Measham, Derbyshire, DE12 7JB
Area of Operation: UK (Excluding Ireland)
Tel: 0845 130 1862
Fax: 01530 274271
Email: sales@touchstone-uk.com
Web: www.touchstone-uk.com

TOUCHSTONE - MARBLE SURFACES
Area of Operation: UK (Excluding Ireland)
Tel: 0845 130 1862
Fax: 01530 274271
Email: sales@touchstone-uk.com
Web: www.touchstone-uk.com

Add that distinctive finishing touch to your
bath/wash room with beautiful marble panelling,
vanity tops, window cills etc.. Produced to your
own specification from any natural marble or
micro marble. Friendly help and assistance
provided.

TRADEMARK TILES
Airfield Industrial Estate, Warboys,
Huntingdon, Cambridgeshire, PE28 2SH
Area of Operation: East England,
South East England
Tel: 01487 825300
Fax: 01487 825305
Email: info@trademarktiles.co.uk
Product Type: 1, 2, 3

WALTON BATHROOMS
The Hersham Centre, The Green, Molesey Road,
Hersham, Walton on Thames, Surrey, KT12 4HL
Area of Operation: UK (Excluding Ireland)
Tel: 01932 224784
Fax: 01932 253447
Email: sales@waltonbathrooms.co.uk
Web: www.waltonbathrooms.co.uk
Product Type: 2

WICKES
Wickes Customer Services, Wickes House,
120-138 Station Road, Harrow, Middlesex,
Greater London, HA1 2QB
Area of Operation: UK (Excluding Ireland)
Tel: 0870 608 9001
Fax: 0208 863 6225
Web: www.wickes.co.uk

WILLIAM GARVEY
Leyhill, Payhembury, Honiton, Devon, EX14 3JG
Area of Operation: Worldwide
Tel: 01404 841430
Fax: 01404 841626
Email: webquery@williamgarvey.co.uk
Web: www.williamgarvey.co.uk
Product Type: 1, 2, 3, 4
Other Info: ✎ 🖑

WWW.BOUNDARYBATHROOMS.CO.UK
Regent Street, Colne, Lancashire, BB8 8LD
Area of Operation: UK & Ireland
Tel: 01282 862 509
Fax: 01282 871 192
Web: www.boundarybathrooms.co.uk
Product Type: 1, 2, 3, 4

BATHROOM ACCESSORIES

KEY
PRODUCT TYPES: 1= Mirrors 2 = Heated
Mirrors   3 = Containers and Organisers
4 = Shelving   5 = Toilet Seats   6 = Other
OTHER: ▽ Reclaimed 🖑 On-line shopping
✎ Bespoke 🖐 Hand-made   ECO Ecological

A & H BRASS
201-203 Edgware Road, London, W2 1ES
Area of Operation: Worldwide
Tel: 0207 402 1854
Fax: 0207 402 0110
Email: ahbrass@btinternet.com
Web: www.aandhbrass.co.uk 🖑
Product Type: 1, 4, 5
Material Type: C) 14, 16

ALEXANDERS
Unit 3, 1/1A Spilsby Road, Harold Hill,
Romford, Essex, RM3 8SB
Area of Operation: UK & Ireland
Tel: 01708 384574
Fax: 01708 384089
Email: sales@alexanders-uk.net
Web: www.alexanders-uk.net
Product Type: 1, 3, 4, 5, 6

ALTHEA UK LTD
Concept House, Blanche Street,
Bradford, West Yorkshire, BD4 8DA
Area of Operation: UK (Excluding Ireland)
Tel: 01274 660770
Fax: 01274 667929
Email: sales@altheauk.com
Web: www.periodbathrooms.co.uk

AMBIANCE BAIN
Cumberland House, 80 Scrubs Lane,
London, NW10 6RF
Area of Operation: UK & Ireland
Tel: 0870 902 1313
Fax: 0870 902 1312
Email: sghirardello@mimea.com
Web: www.ambiancebain.com
Product Type: 1, 3, 4

ANTIQUE BATHS OF IVYBRIDGE LTD
Emebridge Works, Ermington Road,
Ivybridge, Devon, PL21 9DE
Area of Operation: UK (Excluding Ireland)
Tel: 01752 698250
Fax: 01752 698266
Email: sales@antiquebaths.com
Web: www.antiquebaths.com 🖑
Product Type: 1, 3, 5, 6

AQUACREST UK LTD
Haven Light Industrial Estate, Gilbey Road,
Grimsby, Lincolnshire, DN31 2SJ
Area of Operation: UK & Ireland
Tel: 01472 241233
Fax: 01472 241233
Email: aquatop@aquatop.co.uk
Web: www.aquatop.co.uk
Product Type: 6

ARCHITECTURAL COMPONENTS LTD
4-8 Exhibition Road, South Kensington,
London, SW7 2HF
Area of Operation: Worldwide
Tel: 0207 581 2401
Fax: 0207 589 4928
Email: sales@knobs.co.uk
Web: www.doorhandles.co.uk 🖑
Product Type: 3, 4, 5, 6
Material Type: C) 3, 5, 11, 12, 14, 16, 17

ARCHITECTURAL IRONMONGERY LTD
28 Kyrle Street, Ross-on-Wye,
Herefordshire, HR9 7DB
Area of Operation: Worldwide
Tel: 01989 567946
Fax: 01989 567946
Email: info@arciron.co.uk
Web: www.arciron.com 🖑
Product Type: 6

B ROURKE & CO LTD
Vulcan Works, Accrington Road,
Burnley, Lancashire, BB11 5QD
Area of Operation: Worldwide
Tel: 01282 422841
Fax: 01282 458901
Email: info@rourkes.co.uk
Web: www.rourkes.co.uk
Product Type: 1, 4, 6
Material Type: C) 2

BARRHEAD SANITARYWARE PLC
15-17 Nasmyth Road, Hillington Industrial Estate,
Hillington, Glasgow, Renfrewshire, G52 4RG
Area of Operation: UK & Ireland
Tel: 0141 883 0066
Fax: 0141 883 0077
Email: sales@barrhead.co.uk
Product Type: 5

BATHROOM CITY
Seeleys Road, Tyseley Industrial Estate,
Birmingham, West Midlands, B11 2LQ
Area of Operation: UK & Ireland
Tel: 0121 753 0700
Fax: 0121 753 1110
Email: sales@bathroomcity.com
Web: www.bathroomcity.com 🖑
Product Type: 1, 5, 6

BATHROOM EXPRESS
61-62 Lower Dock Street,
Kingsway, Newport, NP20 1EF
Area of Operation: UK & Ireland
Tel: 0845 130 2000 **Fax:** 01633 244881
Email: sales@bathroomexpress.co.uk
Web: www.bathroomexpress.co.uk 🖑
Product Type: 1, 3, 4, 5, 6

BATHROOM OPTIONS
394 Ringwood Road, Ferndown, Dorset, BH22 9AU
Area of Operation: South East England
Tel: 01202 873379
Fax: 01202 892252
Email: bathroom@options394.fsnet.co.uk
Web: www.bathroomoptions.co.uk
Product Type: 1, 5, 6

BATHROOMS PLUS LTD
222 Malmesbury Park Road,
Bournemouth, Dorset, BH8 8PR
Area of Operation: South West England
and South Wales
Tel: 01202 294417
Fax: 01202 316425
Email: info@bathrooms-plus.co.uk
Web: www.bathrooms-plus.co.uk
Product Type: 1, 2, 5

BETTER LIVING PRODUCTS UK LTD
14 Riverside Business Park,
Stansted, Essex, CM24 8PL
Area of Operation: UK & Ireland
Tel: 01279 812958
Fax: 01279 817771
Email: info@dispenser.co.uk
Web: www.dispenser.co.uk
Product Type: 1, 3, 4, 6
Material Type: C) 3

**BROOK-WATER DESIGNER
BATHROOMS & KITCHENS**
The Downs, Woodhouse Hill, Uplyme,
Lyme Regis, Dorset, DT7 3SL
Area of Operation: Worldwide
Tel: 01235 201256
Email: sales@brookwater.co.uk
Web: www.brookwater.co.uk

CAMERON PETERS LTD
The Old Dairy, Home Farm, Ardington,
Wantage, Oxfordshire, OX12 8PD
Area of Operation: Worldwide
Tel: 01235 835000
Fax: 01235 835005
Email: info@cameronpeters.co.uk
Web: www.cameronpeters.co.uk
Product Type: 6

CAPITAL MARBLE DESIGN
Unit 1 Pall Mall Deposit,
124-128 Barlby Road, London, W10 6BL
Area of Operation: UK & Ireland
Tel: 0208 968 5340
Fax: 0208 968 8827
Email: stonegallery@capitalmarble.co.uk
Web: www.capitalmarble.co.uk
Product Type: 1

CARLISLE BRASS
Park House Road,
Carlisle, Cumbria, CA3 0JU
Area of Operation: UK & Ireland
Tel: 01228 511770
Fax: 01228 511885
Email: enquiries@carlislebrass.co.uk
Web: www.carlislebrass.co.uk

CHARLES MASON LTD
Unit 11A Brook Street Mill,
Off Goodall Street, Macclesfield,
Cheshire, SK11 7AW
Area of Operation: Worldwide
Tel: 0800 085 3616
Fax: 01625 668789
Email: info@charles-mason.com
Web: www.charles-mason.com
Product Type: 3, 4, 6
Material Type: C) 2, 3, 13, 14

CITY BATHROOMS & KITCHENS
158 Longford Road, Longford,
Coventry, West Midlands, CV6 6DR
Area of Operation: UK & Ireland
Tel: 02476 365877
Fax: 02476 644992
Email: citybathrooms@hotmail.com
Web: www.citybathrooms.co.uk

**COSY ROOMS
(THE-BATHROOM-SHOP.CO.UK)**
17 Chiltern Way, North Hykeham,
Lincoln, Lincolnshire, LN6 9SY
Area of Operation: UK (Excluding Ireland)
Tel: 01522 696002 **Fax:** 01522 696002
Email: keith@cosy-rooms.com
Web: www.the-bathroom-shop.co.uk
Product Type: 3, 4, 5, 6

CROSSWATER
Unit 5 Butterly Avenue, Questor,
Dartford, Kent, DA1 1JG
Area of Operation: UK & Ireland
Tel: 01322 628270 **Fax:** 01322 628280
Email: sales@crosswater.co.uk
Web: www.crosswater.co.uk
Product Type: 1, 3, 4

CYMRU KITCHENS
63 Caerleon Road, St. Julians,
Newport, NP19 7BX
Area of Operation: UK (Excluding Ireland)
Tel: 01633 676767 **Fax:** 01633 212512
Email: sales@cymrukitchens.com
Web: www.cymrukitchens.com

DANICO BRASS LTD
31-35 Winchester Road,
Swiss Cottage, London, NW3 3NR
Area of Operation: Worldwide
Tel: 0207 483 4477 **Fax:** 0207 722 7992
Email: sales@danico.co.uk
Product Type: 1, 3, 4

DART VALLEY SYSTEMS LTD
Kemmings Close, Long Road,
Paignton, Devon, TQ4 7TW
Area of Operation: UK & Ireland
Tel: 01803 529021 **Fax:** 01803 559016
Email: sales@dartvalley.co.uk
Web: www.dartvalley.co.uk
Product Type: 6

DECORLUX LTD
18 Ghyll Industrial Estate, Ghyll Road,
Heathfield, East Sussex, TN21 8AW
Area of Operation: UK (Excluding Ireland)
Tel: 01435 866638 **Fax:** 01435 866641
Email: info@decorlux.co.uk
Product Type: 1, 3, 4

DECORMASTER LTD
Unit 16, Waterside Industrial Estate,
Wolverhampton, West Midlands, WV2 2RH
Area of Operation: Worldwide
Fax: 01902 353126
Email: sales@oldcolours.co.uk
Web: www.oldcolours.co.uk
Product Type: 5

DESIGNER BATHROOMS
140-142 Pogmoor Road, Pogmoor,
Barnsley, North Yorkshire, S75 2DX
Area of Operation: UK (Excluding Ireland)
Tel: 01226 280200 **Fax:** 01226 733273
Email: mike@designer-bathrooms.com
Web: www.designer-bathrooms.com
Product Type: 1, 2, 3, 4, 5, 6

DOLPHIN BATHROOMS
Bromwich Road, Worcester,
Worcestershire, WR2 4BD
Area of Operation: UK (Excluding Ireland)
Tel: 0800 626717
Web: www.dolphinbathrooms.co.uk
Product Type: 1

EMCO
Centre Point Distribution Ltd,
Unit A5 Regent Park, Booth Drive,
Park Farm Industrial Estate, Wellingborough,
Northamptonshire, NN8 6GR
Area of Operation: UK & Ireland
Tel: 01933 403786 **Fax:** 01933 403789
Email: k.pedrick@mailcity.com
Web: www.emco-bath.com
Product Type: 1, 2, 3, 4, 6
Material Type: C) 11

ENERFOIL MAGNUM LTD
Kenmore Road, Comrie Bridge, Kenmore,
Aberfeldy, Perthshire, PH15 2LS
Area of Operation: Europe
Tel: 01887 822999
Fax: 01887 822954
Email: sales@enerfoil.com
Web: www.enerfoil.com
Product Type: 2

EUROBATH INTERNATIONAL LTD
Eurobath House, Wedmore Road,
Cheddar, Somerset, BS27 3EB
Area of Operation: Worldwide
Tel: 01934 744466
Fax: 01934 744345
Email: sales@eurobath.co.uk
Web: www.eurobath.co.uk
Product Type: 1, 3, 4, 6

HAF DESIGNS LTD
HAF House, Mead Lane, Hertford,
Hertfordshire, SG13 7AP
Area of Operation: UK & Ireland
Tel: 0800 389 8821
Fax: 01992 505705
Email: info@hafltd.co.uk
Web: www.hafdesigns.co.uk
Product Type: 3, 4

HANSGROHE
Units D1 and D2, Sandown Park Trading Estate,
Royal Mills, Esher, Surrey, KT10 8BL
Area of Operation: Worldwide
Tel: 0870 770 1972 **Fax:** 0870 770 1973
Email: sales@hansgrohe.co.uk
Web: www.hansgrohe.co.uk
Product Type: 1, 3, 4, 6
Material Type: C) 2, 3, 11, 14

HARBRINE LTD
27-31 Payne Road, London, E3 2SP
Area of Operation: Worldwide
Tel: 0208 980 8000 **Fax:** 0208 980 6050
Email: info@harbrine.co.uk
Web: www.harbrine.co.uk

HERITAGE BATHROOMS
Princess Street, Bedminster, Bristol, BS3 4AG
Area of Operation: UK & Ireland
Tel: 0117 963 3333
Email: marketing@heritagebathrooms.com
Web: www.heritagebathrooms.com
Product Type: 1, 2, 3, 4, 5

HOMESTYLE DIRECT LTD
Unit 21 Hainault Works, Hainault Road,
Little Heath, Romford, Essex, RM6 5SS
Area of Operation: UK (Excluding Ireland)
Tel: 0208 599 8080 **Fax:** 0208 599 7070
Email: sofia@homestyle-bathrooms.co.uk
Web: www.homestyle-bathrooms.co.uk
Product Type: 1, 3, 4, 5, 6

HUDSON REED
Rylands Street, Burnley, Lancashire, BB10 1RH
Area of Operation: Europe
Tel: 01282 418000 **Fax:** 01282 428915
Email: info@ultra-group.com
Web: www.hudsonreed.com
Product Type: 6

IDEAL STANDARD
The Bathroom Works, National Avenue,
Kingston Upon Hull, East Riding of Yorks, HU5 4HS
Area of Operation: Worldwide
Tel: 0800 590311 **Fax:** 01482 445886
Email: ideal-standard@aseur.com
Web: www.ideal-standard.co.uk

INNOVATION GLASS COMPANY
27-28 Blake Industrial Estate, Brue Avenue,
Bridgwater, Somerset, TA7 8EQ
Area of Operation: Europe
Tel: 01278 683645 **Fax:** 01278 450088
Email: magnus@igc-uk.com
Web: www.igc-uk.com
Product Type: 1, 4, 6
Material Type: J) 4, 5, 6

INSPIRED BATHROOMS
Unit R4, Innsworth Technology Park,
Innsworth Lane, Gloucester,
Gloucestershire, GL3 1DL
Area of Operation: UK & Ireland
Tel: 01452 559121
Fax: 01452 530908
Email: sales@inspired-bathrooms.co.uk
Web: www.inspired-bathrooms.co.uk
Product Type: 1, 2, 3, 4, 5

ION GLASS
Unit 7 Grange Industrial Estate,
Albion Street, Southwick, Brighton,
West Sussex, BN42 4EN
Area of Operation: UK (Excluding Ireland)
Tel: 0845 658 9988
Fax: 0845 658 9989
Email: sales@ionglass.co.uk
Web: www.ionglass.co.uk

ITFITZ
PO Box 960, Maidenhead, Berkshire, SL6 9FU
Area of Operation: UK (Excluding Ireland)
Tel: 01628 890432
Fax: 0870 133 7955
Email: sales@itfitz.co.uk
Web: www.itfitz.co.uk
Product Type: 1, 6

JACUZZI UK
Silverdale Road, Newcastle under Lyme,
Staffordshire, ST5 6EL
Area of Operation: UK & Ireland
Tel: 01782 717175
Fax: 01782 717166
Email: jacuzzisalesdesk@jacuzziuk.com
Web: www.jacuzzi.co.uk
Product Type: 1, 4, 5, 6

JACUZZI UK - BC SANITAN
Silverdale Road, Newcastle under Lyme,
Staffordshire, ST5 6EL
Area of Operation: UK & Ireland
Tel: 01782 717175
Fax: 01782 717166
Email: jacuzzisalesdesk@jacuzziuk.com
Web: www.jacuzziuk.com
Product Type: 1, 4, 5, 6

JOHN SYDNEY
Lagrange, Tamworth, Staffordshire, B79 7XD
Area of Operation: Worldwide
Tel: 01827 304000
Fax: 01827 68553
Email: enquiries@johnsydney.com
Web: www.johnsydney.com
Product Type: 3, 4, 6
Material Type: C) 11, 14

KENSINGTON STUDIO
13c/d Kensington Road, Earlsdon,
Coventry, West Midlands, CV5 6GG
Area of Operation: UK (Excluding Ireland)
Tel: 02476 713326
Fax: 02476 713136
Email: sales@kensingtonstudio.com
Web: www.kensingtonstudio.com
Product Type: 1, 2, 3, 4, 5, 6

KEUCO (UK)
2 Claridge Court, Lower Kings Road,
Berkhamsted, Hertfordshire, HP42AF
Area of Operation: Worldwide
Tel: 01442 865220
Fax: 01442 865260
Email: klaus@keuco.co.uk
Web: www.keuco.de
Product Type: 1, 3, 4, 6
Material Type: C) 1, 3, 11, 14, 16

KOHLER MIRA LIMITED
Gloucester Road, Cheltenham,
Gloucestershire, GL51 8TP
Area of Operation: UK & Ireland
Tel: 0870 850 5551
Fax: 0870 850 5552
Email: info@kohleruk.com
Web: www.kohleruk.com

INTERIORS, FIXTURES & FINISHES

LECICO PLC
Unit 47a Hobbs Industrial Estate, Newchapel,
Nr Lingfield, Surrey, RH7 6HN
Area of Operation: UK & Ireland
Tel: 01342 834777 **Fax:** 01342 834783
Email: info@lecico.co.uk
Web: www.lecico.co.uk
Product Type: 1, 3, 4, 5

M & O BATHROOM CENTRE
174-176 Goswell Rd, Clarkenwell, London, EC1V 7DT
Area of Operation: East England,
Greater London, South East England
Tel: 0207 608 0111 **Fax:** 0207 490 3083
Email: mando@lineone.net
Product Type: 1, 2, 3, 4, 5

MEKON PRODUCTS
25 Bessemer Park, Milkwood Rd, London, SE24 0HG
Area of Operation: Greater London
Tel: 0207 733 8011 **Fax:** 0207 737 0840
Email: info@mekon.net **Web:** www.mekon.net

MEREWAY CONTRACTS
413 Warwick Rd, Birmingham, West Mids, B11 2JR
Area of Operation: UK (Excluding Ireland)
Tel: 0121 707 3288 **Fax:** 0121 707 3871
Email: info@merewaycontracts.co.uk
Web: www.merewaycontracts.co.uk

**MISCELLANEA DISCONTINUED
BATHROOMWARE**
Churt Place Nurseries, Tilford Road,
Churt, Farnham, Surrey, GU10 2LN
Area of Operation: UK (Excluding Ireland)
Tel: 01428 608164 **Fax:** 01428 608165
Email: email@brokenbog.com
Web: www.brokenbog.com
Product Type: 5, 6

MISTER MIRRORS
Imperial House, Redlands,
Coulsdon, Surrey, CR5 2HT
Area of Operation: UK & Ireland
Tel: 0208 668 7016
Fax: 0208 660 2384
Email: beourguest@clara.co.uk
Web: www.mister-mirrors.co.uk
Product Type: 2

NICHOLAS ANTHONY LTD
42-44 Wigmore Street, London, W1U 2RX
Area of Operation: East England,
Greater London, South East England
Tel: 0800 068 3603
Fax: 01206 762698
Email: info@nicholas-anthony.co.uk
Web: www.nicholas-anthony.co.uk
Product Type: 1, 3

NO CODE
Larkwhistle Farm Road,
Micheldever, Hampshire, SO21 3BG
Area of Operation: UK & Ireland
Tel: 01962 870078
Fax: 01962 870077
Email: sales@nocode.co.uk
Web: www.nocode.co.uk
Product Type: 1, 3, 4, 5

NORSKE INTERIORS
Estate Road One,
South Humberside Industrial Estate,
Grimsby, Lincolnshire, DN31 2TA
Area of Operation: UK (Excluding Ireland)
Tel: 01472 240832
Fax: 01472 360112
Email: sales@norske-int.co.uk
Web: www.norske-int.co.uk
Product Type: 6

OLD FASHIONED BATHROOMS
Foresters Hall, 52 High Street,
Debenham, Stowmarket,
Suffolk, IP14 6QW
Area of Operation: East England,
Greater London, South East England
Tel: 01728 860926
Fax: 01728 860446
Email: ofbrooms@lycos.co.uk
Web: www.oldfashionedbathrooms.co.uk
Product Type: 1, 5, 6
Other Info: ▽ ☝
Material Type: A) 10

PARKSIDE TILES
49-51 Highmeres Road,
Thurmeston, Leicester,
Leicestershire, LE4 9LZ
Area of Operation: UK (Excluding Ireland)
Tel: 0116 276 2532
Fax: 0116 246 0649
Email: parkside@tiles2.wanadoo.co.uk
Web: www.parksidetiles.co.uk
Product Type: 1

PEPLOW ROBERTS LIMITED
Unit 11 Eden Way, Pages Industrial Park, Billington
Road, Leighton Buzzard, Bedfordshire, LU7 4TZ
Area of Operation: Worldwide
Tel: 01525 375118
Fax: 01525 852130
Email: info@paperstream.net
Web: www.paperstream.net
Product Type: 3, 6

PLASTIC MIRRORS
18 Blundell Lane, Preston, Lancashire, PR1 0EA
Area of Operation: Worldwide
Tel: 01772 746869
Fax: 01772 746869
Email: enquiries@plasticmirrors.co.uk
Web: www.plasticmirrors.co.uk
Product Type: 1
Material Type: D) 5

POLYPIPE KITCHENS & BATHROOMS LTD
Warmsworth, Halt Industrial Estate,
South Yorkshire, DN4 9LS
Area of Operation: Worldwide
Tel: 01709 770990
Fax: 01302 310602
Email: davery@ppbp.co.uk
Web: www.polypipe.com/bk
Product Type: 5

PRETTY SWIFT LTD
Units 7&8, 51 Chancery Lane,
Debenham, Suffolk, IP14 6PJ
Area of Operation: UK (Excluding Ireland)
Tel: 01728 861818
Fax: 01728 861919
Email: prettyswiftltd@aol.com

QUALCERAM SHIRES PLC
Unit 9/10, Campus 5, Off Third Avenue,
Letchworth Garden City, Hertfordshire, SG6 2JF
Area of Operation: UK & Ireland
Tel: 01462 676710
Fax: 01462 676701
Email: info@qualceram-shires.com
Web: www.qualceram-shires.com
Product Type: 1, 2, 5

R&D MARKETING (DEMISTA) LTD
Land House, Anyards Road,
Cobham, Surrey, KT11 2LW
Area of Operation: UK (Excluding Ireland)
Tel: 01932 866600
Fax: 01932 866 688
Email: rd@demista.co.uk
Web: www.demista.co.uk
Product Type: 2

R&D MARKETING (DEMISTA™) LTD

Area of Operation: UK
Tel: 01932 866600 **Fax:** 01932 866688
Email: rd@demista.co.uk
Web: www.demista.co.uk
Product Type: 2

Eliminate condensation from your mirrors forever
with the **Original demista™** heated mirror pads
as spcified and fitted by hotels and developments
worldwide.

• *Low Cost* • *Simple To Fit* • *Maintenance Free*
• *International Approvals* • *Many Sizes*
• *Saves on cleaning materials and time*

COSYFLOOR™

Area of Operation: UK
Tel: 01932 866600
Fax: 01932 866688
Email: rd@demista.co.uk
Web: www.demista.co.uk
Product Type: 2

Transform cold but beautiful surfaces into inviting
warm floors with **Cosyfloor™** underfloor heating
systems. DIY or professional installation.

INTERIORS, FIXTURES & FINISHES

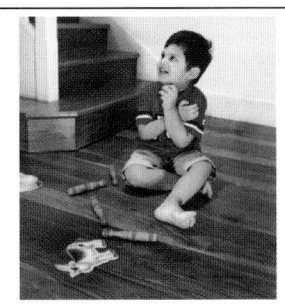

ECOMAT™

Area of Operation: UK
Tel: 01932 866600 **Fax:** 01932 866688
Email: rd@demista.co.uk
Web: www.demista.co.uk
Product Type: 2

ECOMAT™ ultra thin heating elements are ideal beneath laminate and wood floor surfaces. Advantages include: even distribution giving gentle overall warmth; simple installation; maintenance free; 10 year gurantee.

REGINOX UK LTD
Radnor Park Trading Estate,
Congleton, Cheshire, CW12 4XJ
Area of Operation: UK & Ireland
Tel: 01260 280033
Fax: 01260 298889
Email: sales@reginoxuk.co.uk
Web: www.reginox.com
Product Type: 6

**ROOTS KITCHENS,
BEDROOMS & BATHROOMS**
Vine Farm, Stockers Hill,
Boughton-under-Blean,
Faversham, Kent, ME13 9AB
Area of Operation: UK (Excluding Ireland)
Tel: 01227 751130
Fax: 01227 750033
Email: showroom@rootskitchens.co.uk
Web: www.rootskitchens.co.uk
Product Type: 1, 5

ROSCO COLLECTIONS
Stone Allerton, Axbridge, Somerset, BS26 2NS
Area of Operation: UK & Ireland
Tel: 01934 712299
Fax: 01934 713222
Email: jonathan@roscobathrooms.demon.co.uk
Product Type: 1, 2, 3, 4, 5

SAMUEL HEATH
Leopold Street, Birmingham,
West Midlands, B12 0UJ
Area of Operation: Worldwide
Tel: 0121 772 2303
Fax: 0121 772 3334
Email: lloyd@samuel-heath.com
Web: www.samuel-heath.com
Product Type: 1, 3, 4, 6
Material Type: C) 11, 13, 14, 16

SATANA INTERNATIONAL LTD
Unit E Winford Rural Workshops,
Higher Halstock Leigh, Yeovil,
Somerset, BA22 9QX
Area of Operation: UK (Excluding Ireland)
Tel: 01935 891888
Fax: 01935 891819
Email: satanaltd@aol.com
Web: www.heatedmirrors.co.uk
Product Type: 2

SHOWERLUX UK LTD
Sibree Road, Coventry, West Midlands, CV3 4FD
Area of Operation: UK & Ireland
Tel: 02476 882515
Fax: 02476 305457
Email: sales@showerlux.co.uk
Web: www.showerlux.com
Product Type: 1, 3, 4

SHUTTER FRONTIER LTD
2 Rosemary Farmhouse, Rosemary Lane,
Flimwell, Wadhurst, East Sussex, TN5 7PT
Area of Operation: Worldwide
Tel: 01580 878137
Fax: 01580 878137
Email: jane@shutterfrontier.co.uk
Web: www.shutterfrontier.co.uk

SMART SHOWERS LTD
Unit 11, Woodside Road,
South Marston Business Park,
Swindon, Wiltshire, SN3 4WA
Area of Operation: UK & Ireland
Tel: 01793 822775
Fax: 01793 823800
Email: louise@smartshowers.co.uk
Web: www.smartshowers.co.uk
Product Type: 1, 4, 5
Material Type: D) 5, 6

STIFFKEY BATHROOMS
89 Upper Saint Giles Street,
Norwich, Norfolk, NR2 1AB
Area of Operation: Worldwide
Tel: 01603 627850
Fax: 01603 619775
Email: stiffkeybathrooms.norwich@virgin.net
Web: www.stiffkeybathrooms.com
Product Type: 1, 4, 5
Other Info: ▽

TAPS SHOP
35 Bridge Street, Witney, Oxfordshire, OX28 1DA
Area of Operation: Worldwide
Tel: 0845 430 3035
Fax: 01993 779653
Email: info@tapsshop.co.uk
Web: www.tapsshop.co.uk
Product Type: 1, 3, 6

THE HEATED MIRROR COMPANY LTD
Sherston, Wiltshire, SN16 0LW
Area of Operation: Worldwide
Tel: 01666 840003
Fax: 01666 840856
Email: heated.mirror@virgin.net
Web: www.heated-mirrors.com
Product Type: 2

THE KITCHEN AND BATHROOM COLLECTION
Nelson House, Nelson Road,
Salisbury, Wiltshire, SP1 3LT
Area of Operation: South West England
and South Wales
Tel: 01722 334800
Fax: 01722 412252
Email: info@kbc.co.uk
Web: www.kbc.co.uk
Product Type: 1, 2, 3, 4, 5
Other Info: ✐

THE WATER MONOPOLY
16/18 Lonsdale Road, London, NW6 6RD
Area of Operation: UK (Excluding Ireland)
Tel: 0207 624 2636
Fax: 0207 624 2631
Email: enquiries@watermonopoly.com
Web: www.watermonopoly.com

TRADEMARK TILES
Airfield Industrial Estate, Warboys,
Huntingdon, Cambridgeshire, PE28 2SH
Area of Operation: East England,
South East England
Tel: 01487 825300
Fax: 01487 825305
Email: info@trademarktiles.co.uk
Product Type: 1, 2, 3, 4, 5

TRITON PLC
Triton Road, Shepperton Business Park,
Caldwell Road, Nuneaton,
Warwickshire, CV11 4NR
Area of Operation: UK & Ireland
Tel: 02476 344441
Fax: 02476 349828
Web: www.tritonshowers.co.uk
Product Type: 1, 3, 4

URBIS (GB) LTD.
55 Lidgate Crescent, Langthwaite Grange Industrial
Estate, South Kirkby, West Yorkshire, WF9 3NR
Area of Operation: UK & Ireland
Tel: 01977 659829
Fax: 01977 659808
Email: david@urbis-gb.co.uk
Web: www.urbis-gb.co.uk
Product Type: 3, 4

WALTON BATHROOMS
The Hersham Centre, The Green, Molesey Road,
Hersham, Walton on Thames, Surrey, KT12 4HL
Area of Operation: UK (Excluding Ireland)
Tel: 01932 224784
Fax: 01932 253447
Email: sales@waltonbathrooms.co.uk
Web: www.waltonbathrooms.co.uk
Product Type: 1

WARMROOMS
24 Corncroft Lane, St Leonards Park,
Gloucester, Gloucestershire, GL4 6XU
Area of Operation: UK (Excluding Ireland)
Tel: 01452 304460
Fax: 01452 304460
Email: sales@warmrooms.co.uk
Web: www.warmrooms.co.uk

WATER FRONT
All Worcester Buildings, Birmingham Road,
Redditch, Worcestershire, B97 6DY
Area of Operation: UK & Ireland
Tel: 01527 584244
Fax: 01527 61127
Email: waterfront@btconnect.com
Web: www.waterfrontbathrooms.com
Product Type: 1, 3, 4
Material Type: C) 11, 14, 16

WHIRLPOOL EXPRESS (UK) LTD
61-62 Lower Dock Street, Kingsway,
Newport, NP20 1EF
Area of Operation: Europe
Tel: 01633 244555
Fax: 01633 244555
Email: reception@whirlpoolexpress.co.uk
Web: www.whirlpoolexpress.co.uk
Product Type: 1, 3, 4, 6

WILLIAM GARVEY
Leyhill, Payhembury, Honiton, Devon, EX14 3JG
Area of Operation: Worldwide
Tel: 01404 841430
Fax: 01404 841626
Email: webquery@williamgarvey.co.uk
Web: www.williamgarvey.co.uk
Product Type: 1, 4, 5
Other Info: ✐ ✋
Material Type: A) 2, 10

WWW.BOUNDARYBATHROOMS.CO.UK
Regent Street, Colne, Lancashire, BB8 8LD
Area of Operation: UK & Ireland
Tel: 01282 862509
Fax: 01282 871192
Web: www.boundarybathrooms.co.uk
Product Type: 1, 2, 3, 4, 5, 6

SPECIALIST BATHROOM EQUIPMENT

KEY

OTHER: ▽ Reclaimed ⌂ On-line shopping
✐ Bespoke ✋ Hand-made ECO Ecological

AL CHALLIS LTD
Europower House, Lower Road, Cookham,
Maidenhead, Berkshire, SL6 9EH
Area of Operation: UK & Ireland
Tel: 01628 529024
Fax: 0870 458 0577
Email: chris@alchallis.com
Web: www.alchallis.com

APPOLLO
Aquabeau Limited, Swinnow Lane,
Leeds, West Yorkshire, LS13 4TY
Area of Operation: Worldwide
Tel: 0113 386 9150
Fax: 0113 239 3406
Email: sales@aquabeau.com
Web: www.appollobathing.co.uk

AQUABEAU LIMITED
Swinnow Lane, Leeds, West Yorkshire, LS13 4TY
Area of Operation: Worldwide
Tel: 0113 386 9150
Fax: 0113 239 3406
Email: sales@aquabeau.com
Web: www.aquabeau.com

BATHROOMS PLUS LTD
222 Malmesbury Park Road,
Bournemouth, Dorset, BH8 8PR
Area of Operation: South West England
and South Wales
Tel: 01202 294417
Fax: 01202 316425
Email: info@bathrooms-plus.co.uk
Web: www.bathrooms-plus.co.uk

**BROOK-WATER DESIGNER
BATHROOMS & KITCHENS**
The Downs, Woodhouse Hill, Uplyme,
Lyme Regis, Dorset, DT7 3SL
Area of Operation: Worldwide
Tel: 01235 201256
Email: sales@brookwater.co.uk
Web: www.brookwater.co.uk

**CHAMBERLAIN & GROVES LTD
- THE DOOR & SECURITY STORE**
101 Boundary Road, Walthamstow,
London, E17 8NQ
Area of Operation: UK (Excluding Ireland)
Tel: 0208 520 6776
Fax: 0208 520 2190
Email: ken@securedoors.co.uk
Web: www.securedoors.co.uk

CHILTERN INVADEX LTD
Chiltern House, 6 Wedgwood Road,
Bicester, Oxfordshire, OX26 4UL
Area of Operation: Worldwide
Tel: 01869 246470
Fax: 01869 365552
Email: marketing@chilterninvadex.co.uk
Web: www.chilterninvadex.co.uk

CISTERMISER LTD
Unit 1 Woodley Park Estate, 59-69 Reading Road,
Woodley, Reading, Berkshire, RG5 3AN
Area of Operation: UK & Ireland
Tel: 0118 969 1611
Fax: 0118 944 1426
Email: sales@cistermiser.co.uk
Web: www.cistermiser.co.uk

CITY BATHROOMS & KITCHENS
158 Longford Road, Longford,
Coventry, West Midlands, CV6 6DR
Area of Operation: UK & Ireland
Tel: 02476 365877
Fax: 02476 644992
Email: citybathrooms@hotmail.com
Web: www.citybathrooms.co.uk

CROSSWATER
Unit 5 Butterly Avenue, Questor,
Dartford, Kent, DA1 1JG
Area of Operation: UK & Ireland
Tel: 01322 628270
Fax: 01322 628280
Email: sales@crosswater.co.uk
Web: www.crosswater.co.uk

CYMRU KITCHENS
63 Caerleon Road, St. Julians, Newport, NP19 7BX
Area of Operation: UK (Excluding Ireland)
Tel: 01633 676767
Fax: 01633 212512
Email: sales@cymrukitchens.com
Web: www.cymrukitchens.com

INTERIORS, FIXTURES & FINISHES

SPONSORED BY: PEGLER LIMITED www.pegler.co.uk/francis

DART VALLEY SYSTEMS LIMITED
Kemmings Close, Long Road,
Paignton, Devon, TQ4 7TW
Area of Operation: UK & Ireland
Tel: 01803 529021
Fax: 01803 559016
Email: sales@dartvalley.co.uk
Web: www.dartvalley.co.uk

DECORLUX LTD
18 Ghyll Industrial Estate,
Ghyll Road, Heathfield,
East Sussex, TN21 8AW
Area of Operation: UK (Excluding Ireland)
Tel: 01435 866638
Fax: 01435 866641
Email: info@decorlux.co.uk

DESIGNER BATHROOMS
140-142 Pogmoor Road,
Pogmoor, Barnsley,
North Yorkshire, S75 2DX
Area of Operation: UK (Excluding Ireland)
Tel: 01226 280200
Fax: 01226 733273
Email: mike@designer-bathrooms.com
Web: www.designer-bathrooms.com

D-LINE
Buckingham Court, Brackley,
Northamptonshire, NN13 7EU
Area of Operation: UK & Ireland
Tel: 01280 841200
Fax: 01280 845130
Email: uk@dline.com
Web: www.dline.com

DOLPHIN BATHROOMS
Bromwich Road, Worcester,
Worcestershire, WR2 4BD
Area of Operation: UK (Excluding Ireland)
Tel: 0800 626717
Web: www.dolphinbathrooms.co.uk

ECOLEC
Sharrocks Street, Wolverhampton,
West Midlands, WV1 3RP
Area of Operation: UK & Ireland
Tel: 01902 457575
Fax: 01902 457797
Email: sales@ecolec.co.uk
Web: www.ecolec.co.uk

ESL HEALTHCARE LTD
Potts Marsh Industrial Estate, Eastbourne Road,
Westham, East Sussex, BN24 5NH
Area of Operation: UK & Ireland
Tel: 01323 465800
Fax: 01323 460248
Email: sales@eslindustries.co.uk
Web: www.eslindustries.com

EUROCARE SHOWERS LTD
Unit 19, Doncaster Industry Park,
Watch House Lane, Bentley,
Doncaster, South Yorkshire, DN5 9LZ
Area of Operation: Worldwide
Tel: 01302 788684
Fax: 01302 780010
Email: sales@eurocare-showers.com
Web: www.eurocare-showers.com

GEBERIT LTD
New Hythe Business Park,
New Hythe Lane, Aylesford,
Kent, ME20 7PJ
Area of Operation: UK & Ireland
Tel: 01622 717811
Fax: 01622 716920
Web: www.geberit.co.uk

GOODWOOD BATHROOMS LTD
Church Road, North Mundham,
Chichester, West Sussex, PO20 1JU
Area of Operation: UK (Excluding Ireland)
Tel: 01243 532121
Fax: 01243 533423
Email: sales@goodwoodbathroom.co.uk
Web: www.goodwoodbathrooms.co.uk

GROHE UK
1 River Road, Barking, Essex, IG11 0HD
Area of Operation: Worldwide
Tel: 0208 594 7292
Fax: 0208 594 8898
Email: info@grohe.co.uk
Web: www.grohe.co.uk

HAROLD MOORE & SON LTD
Rawson Spring Road, Hillsborough,
Sheffield, South Yorkshire, S6 1PD
Area of Operation: UK & Ireland
Tel: 0114 233 6161
Fax: 0114 232 6375
Email: admin@haroldmoorebaths.co.uk
Web: www.haroldmoorebaths.co.uk

HERITAGE BATHROOMS
Princess Street, Bedminster, Bristol, BS3 4AG
Area of Operation: UK & Ireland
Tel: 0117 963 3333
Email: marketing@heritagebathrooms.com
Web: www.heritagebathrooms.com

HOUSE ELECTRONIC LTD
97 Francisco Close, Hollyfields Park,
Chafford Hundred, Essex, RM16 6YE
Area of Operation: UK & Ireland
Tel: 01375 483595
Fax: 01375 483595
Email: info@houselectronic.co.uk
Web: www.houselectronic.co.uk

INSPIRED BATHROOMS
Unit R4, Innsworth Technology Park,
Innsworth Lane, Gloucester,
Gloucestershire, GL3 1DL
Area of Operation: UK & Ireland
Tel: 01452 559121 **Fax:** 01452 530908
Email: sales@inspired-bathrooms.co.uk
Web: www.inspired-bathrooms.co.uk

JAMES GIBBONS FORMAT LTD
Vulcan Road, Bilston, Wolverhampton,
West Midlands, WV14 7JG
Area of Operation: Worldwide
Tel: 01902 405500
Fax: 01902 385915
Email: info@jgf.co.uk
Web: www.jgf.co.uk

KENSINGTON STUDIO
13 c & d Kensington Road, Earlsdon,
Coventry, West Midlands, CV5 6GG
Area of Operation: UK (Excluding Ireland)
Tel: 02476 713326
Fax: 02476 713136
Email: sales@kensingtonstudio.com
Web: www.kensingtonstudio.com

KERAMAG WAVENEY LTD
London Road, Beccles,
Suffolk, NR34 8TS
Area of Operation: UK & Ireland
Tel: 01502 716600
Fax: 01502 717767
Email: info@keramagwaveney.co.uk
Web: www.keramag.com

KINGKRAFT LTD
26D Orgreave Crescent,
Dore House Industrial Estate,
Sheffield, South Yorkshire, S13 9NQ
Area of Operation: Europe
Tel: 0114 269 0697
Fax: 0114 269 5145
Email: info@kingkraft.co.uk
Web: www.kingkraft.co.uk
Other Info: ✋

LECICO PLC
Unit 47a Hobbs Industrial Estate,
Newchapel, Nr Lingfield,
Surrey, RH7 6HN
Area of Operation: UK & Ireland
Tel: 01342 834777
Fax: 01342 834783
Email: info@lecico.co.uk
Web: www.lecico.co.uk

MEREWAY CONTRACTS
413 Warwick Road, Birmingham,
West Midlands, B11 2JR
Area of Operation: UK (Excluding Ireland)
Tel: 0121 707 3288 **Fax:** 0121 707 3871
Email: info@merewaycontracts.co.uk
Web: www.merewaycontracts.co.uk

**MISCELLANEA DISCONTINUED
BATHROOMWARE**
Churt Place Nurseries, Tilford Road,
Churt, Farnham, Surrey, GU10 2LN
Area of Operation: UK (Excluding Ireland)
Tel: 01428 608164 **Fax:** 01428 608165
Email: email@brokenbog.com
Web: www.brokenbog.com

MOBALPA
4 Cornflower Close, Willand, Devon, EX15 2TT
Area of Operation: Worldwide
Tel: 07740 633672 **Fax:** 01884 820828
Email: ploftus@mobalpa.com
Web: www.mobalpa.com

OLD FASHIONED BATHROOMS
Foresters Hall, 52 High Street, Debenham,
Stowmarket, Suffolk, IP14 6QW
Area of Operation: East England,
Greater London, South East England
Tel: 01728 860926 **Fax:** 01728 860446
Email: ofbrooms@lycos.co.uk
Web: www.oldfashionedbathrooms.co.uk
Material Type: C) 11, 13, 14, 16

PADIPA LIMITED
Unit East 2, Sway Storage & Workshops,
Sway, Lymington, Hampshire, SO41 6DD
Area of Operation: UK & Ireland
Tel: 01590 681710 **Fax:** 01590 610336
Email: enquiries@padipa.co.uk
Web: www.padipa.co.uk

**POLYPIPE KITCHENS
& BATHROOMS LTD**
Warmsworth, Halt Industrial Estate,
South Yorkshire, DN9 9LS
Area of Operation: Worldwide
Tel: 01709 770990
Fax: 01302 310602
Email: davery@ppbp.co.uk
Web: www.polypipe.com/bk

QUALCERAM SHIRES PLC
Unit 9/10, Campus 5, Off Third Avenue,
Letchworth Garden City, Hertfordshire, SG6 2JF
Area of Operation: UK & Ireland
Tel: 01462 676710
Fax: 01462 676701
Email: info@qualceram-shires.com
Web: www.qualceram-shires.com

RSJ ASSOCIATES LTD
Unit 5, Greenfield Road,
Greenfield Farm Industrial Estate,
Congleton, Cheshire, CW12 4TR
Area of Operation: UK & Ireland
Tel: 01260 276188
Fax: 01260 280889
Email: info@rsjassociates.co.uk
Web: www.hueppe.com

STEAM DIRECT
12 Reeves Way, South Woodham Ferrers,
Chelmsford, Essex, CM3 5XF
Area of Operation: UK & Ireland
Tel: 01245 324577
Fax: 01245 325665
Email: steamdirect@aol.com
Web: www.steam-direct.co.uk

STIFFKEY BATHROOMS
89 Upper Saint Giles Street,
Norwich, Norfolk, NR2 1AB
Area of Operation: Worldwide
Tel: 01603 627850
Fax: 01603 619775
Email: stiffkeybathrooms.norwich@virgin.net
Web: www.stiffkeybathrooms.com

STUDIO STONE
The Stone Yard, Alton Lane,
Four Marks, Hampshire, GU34 5AJ
Area of Operation: UK (Excluding Ireland)
Tel: 01420 562500
Fax: 01420 563192
Email: sales@studiostone.co.uk
Web: www.studiostone.co.uk
Other Info: ✏

THE KITCHEN & BATHROOM COLLECTION
Nelson House, Nelson Road,
Salisbury, Wiltshire, SP1 3LT
Area of Operation: South West England
and South Wales
Tel: 01722 334800
Fax: 01722 412252
Email: info@kbc.co.uk
Web: www.kbc.co.uk
Other Info: ✏

WHIRLPOOL EXPRESS (UK) LTD
61-62 Lower Dock Street,
Kingsway, Newport, NP20 1EF
Area of Operation: Europe
Tel: 01633 244555
Fax: 01633 244555
Email: reception@whirlpoolexpress.co.uk
Web: www.whirlpoolexpress.co.uk

WWW.BOUNDARYBATHROOMS.CO.UK
Regent Street, Colne, Lancashire, BB8 8LD
Area of Operation: UK & Ireland
Tel: 01282 862509 **Fax:** 01282 871192
Web: www.boundarybathrooms.co.uk

NOTES

Company Name . . . .
. . . .
Address . . . .
. . . .
. . . .
email . . . .
Web . . . .

Company Name . . . .
. . . .
Address . . . .
. . . .
. . . .
email . . . .
Web . . . .

Company Name . . . .
. . . .
Address . . . .
. . . .
. . . .
email . . . .
Web . . . .

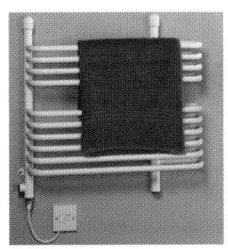

ASTRA247.COM
Area of Operation: Worldwide
Tel: 0161 797 3222 **Fax:** 0161 797 3444
Email: info@astra247.com
Web: www.astra247.com

Dual Fuel Towel Rails
Astra offer a selection of heated towel rails that can run on both your central heating system and electricity.

'easy online ordering'

ASTRA247.COM
Area of Operation: Worldwide
Tel: 0161 797 3222 **Fax:** 0161 797 3444
Email: info@astra247.com
Web: www.astra247.com

Heated Towel Rails
Complete range of heated towel rails in all shapes and sizes. Low surface temperature and Dual fuel.

'easy online ordering'

ASTRA247.COM
Area of Operation: Worldwide
Tel: 0161 797 3222 **Fax:** 0161 797 3444
Email: info@astra247.com
Web: www.astra247.com

Cooker Hoods
Range of stylish, quality, efficient cooker hoods in various finishes and styles to suit.

'easy online ordering'

ASTRA247.COM
Area of Operation: Worldwide
Tel: 0161 797 3222 **Fax:** 0161 797 3444
Email: info@astra247.com
Web: www.astra247.com

Bathroom Mirrors
We offer a selected range of non-heated and heated bathroom mirrors in various styles and finishes to compliment the look of your room.

'easy online ordering'

ASTRA247.COM
Area of Operation: Worldwide
Tel: 0161 797 3222 **Fax:** 0161 797 3444
Email: info@astra247.com
Web: www.astra247.com

Shaver Sockets Dual Voltage
High quality dual voltage shaver sockets. All available in a range of black nickel, brass and chrome finishes.

'easy online ordering'

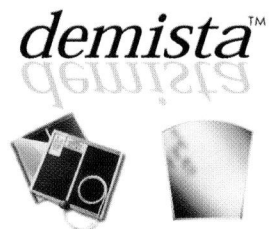

ASTRA247.COM
Area of Operation: Worldwide
Tel: 0161 797 3222 **Fax:** 0161 797 3444
Email: info@astra247.com
Web: www.astra247.com

Demista - Anti misting pads for mirrors
Steamed up bathroom mirrors have always been accepted as an everyday problem that could not be solved, until now. Once fitted, the area will never steam up.
'easy online ordering'

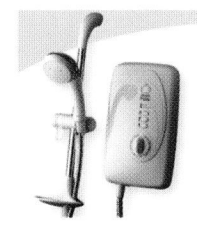

ASTRA247.COM
Area of Operation: Worldwide
Tel: 0161 797 3222 **Fax:** 0161 797 3444
Email: info@astra247.com
Web: www.astra247.com

Electric Showers
We offer a selected range of high power, well specified, quality electric showers. Each with technical specifications available on our website.

'easy online ordering'

ASTRA247.COM
Area of Operation: Worldwide
Tel: 0161 797 3222 **Fax:** 0161 797 3444
Email: info@astra247.com
Web: www.astra247.com

Electric Power Showers
We offer a selected range of well specified, quality electric power showers. Each with technical specifications available on our website.

'easy online ordering'

ASTRA247.COM
Area of Operation: Worldwide
Tel: 0161 797 3222 **Fax:** 0161 797 3444
Email: info@astra247.com
Web: www.astra247.com

Plint Heaters
Astra have a selected range of Plinth heaters available in finishes such as white, black, stainless steel, brown and gilt. Available as 2kW or 3kW.

'easy online ordering'

ASTRA247.COM
Area of Operation: Worldwide
Tel: 0161 797 3222 **Fax:** 0161 797 3444
Email: info@astra247.com
Web: www.astra247.com

Waste Disposal Units
Selection of fast performance waste disposal units for domestic installation.

'easy online ordering'

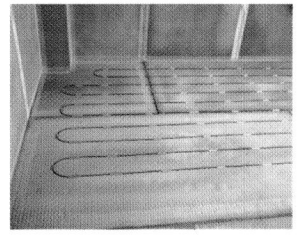

ASTRA247.COM
Area of Operation: Worldwide
Tel: 0161 797 3222 **Fax:** 0161 797 3444
Email: info@astra247.com
Web: www.astra247.com

Marmox Insulation Board
Marmox insulation board is ideal for laying beneath underfloor heating mats, and lining wet areas prior to tiling.

'easy online ordering'

ASTRA247.COM
Area of Operation: Worldwide
Tel: 0161 797 3222 **Fax:** 0161 797 3444
Email: info@astra247.com
Web: www.astra247.com

Tile Backer Board
Marmox insulation board is ideal as a tile backer board and is suitable for all wet areas. It can be shaped and cut to any size.

'easy online ordering'

INTERIORS, FIXTURES & FINISHES

Move or Improve? the new magazine for people adding space and value to their home.

FEATURES:

Design Masterclass: Making efficient use of space

The Design Doctor: Design expert Michael Holmes helps two readers find solutions to their improving dilemas

The Guide: A complete guide to renovating and extending your home, with advice on planning, building regulations, finance, etc

Renovation Price Book: Full estimated costings for different aspects of remodelling

Special Features, for example:
Loft conversion guide
How to buy replacement windows
Build, buy or renovate?
Wooden floors — complete design guide

Real Life Projects: Inspiring and practical readers' studies including cellar conversions, luxurious new sunrooms, kitchen extensions and perfect side extensions

PLUS MUCH MORE

ON SALE NOW in WHSmiths

INTERIORS, FIXTURES & FINISHES

KITCHENS

Image courtesy of William Ball (01375 375151)

SPONSORED BY WILLIAM BALL
Tel 01375 375151 Web www.wball.co.uk

INTERIORS, FIXTURES & FINISHES

Kitchens

Often described as the heart of the home, the kitchen is a hub of activity, which needs to be versatile enough to meet the demands of a busy, modern family. Catering for a range of needs, whilst still maintaining a spacious, calming environment that you will enjoy spending time in, should be priority. This requires in depth planning; not only is it a question of space and budget, but also entails a close look at the way you use your kitchen, as your ideal requirements may differ from the traditional concept of a kitchen fulfilling purely culinary demands.

**TOP : Shaker style kitchen by Harvey Jones.
ABOVE LEFT :Dark wood and light floors provide a striking contrast. Image by Second Nature.
ABOVE: This quartz wortop by Touchstone incorporates Microban, an antibacterial agent.**

A good starting point may be to make a list of items which must be stored in the kitchen, as basic functions should be catered for first. Once this has been decided, you will be free to find space for additions such as dining tables and breakfast bars, as such large furniture can cut into valuable space which might be better used alternatively. Therefore forward planning will enable you to save up space for less conventional items you really want in the room.

The addition of a versatile island unit to your kitchen will prove multi-functional; serving as a workspace, creating a focal point for the room, and providing all-important extra storage. This will also create an interesting and contemporary take on the "work triangle" kitchen layout (an invisible triangle which joins the sink, cooker and refrigerator, the busiest points in the kitchen) by providing an intermediate fourth point.

Hardwearing, practical materials are the best choice for your kitchen furniture and surfaces, but think hard about the style you choose - it should be classic enough for you to enjoy for years to come. For example, fusing stainless steel with wood (or the cheaper laminate wood finish) will be simultaneously practical and stylish, but make sure the design is one which will not go out of style quickly.

If budget is a major factor in your choice, think about combining cheaper units with luxury finishes such as handles and taps. Reflective materials, such as chrome or stainless steel, are a good idea as they help to maximise light in the kitchen, which in turn will make the space appear larger.

Another factor to remember is ventilation. In kitchens, where there is likely to be larger amounts of moisture present from cooking, some form of extract ventilation may be required. This may be in the form of

a cooker hood or extractor fan. However, you should be aware that there are minimum requirements for mechanical extraction rates in the building regulations, so always ensure that your apparatus will comply.

The most important things to keep in mind when designing your kitchen are comfort and purpose. It is crucial to maximise space potential, but at the same time be aware of maintaining a calming environment which is not over-crowded. The overall effect of your kitchen may come down to the colours you use and how you accessorize once most of the work is done. However, whilst homely touches can make your kitchen, remember to avoid clutter; think comfort, but above all, think purpose.

COOKER HOODS AND EXTRACTORS

KEY

SEE ALSO: VENTILATION - Extractor Fans
OTHER: ▽ Reclaimed 🖱 On-line shopping
✎ Bespoke ✋Hand-made ECO Ecological

AMBER KITCHENS OF BISHOP'S STORTFORD
Clarklands House, Parsonage Lane,
Sawbridgeworth, Hertfordshire, CM21 0NG
Area of Operation: East England, Greater London
Tel: 01279 600030
Fax: 01279 721528
Email: info@amberkitchens.com
Web: www.amberkitchens.com

APPLIANCE WORLD UK LTD.
Equity Trade Centre, Hobley Drive,
Swindon, Wiltshire, SN3 4NS
Area of Operation: UK (Excluding Ireland)
Tel: 0870 757 2424
Fax: 0870 757 2475
Email: nickki@applianceworld.co.uk
Web: www.applianceworld.co.uk 🖱

ASHFORD & BROOKS
The Old Workshop, Ashtree Barn, Caters Road,
Bredfield, Woodbridge, Suffolk, IP13 6BE
Area of Operation: UK (Excluding Ireland)
Tel: 01473 737764
Fax: 01473 277176
Email: ashfordbrooks@mailbox.co.uk
Web: www.ashfordandbrooks.co.uk

BATH KITCHEN COMPANY
22 Hensley Road, Bloomfield,
Bath, Somerset, BA2 2DR
Area of Operation: South West England
and South Wales
Tel: 01225 312003
Fax: 01225 312003
Email: david@bathkitchencompany.co.uk
Web: www.bathkitchencompany.co.uk

BAUMATIC LTD
Baumatic Buildings, 6 Bennet Road,
Berkshire, RG2 0QX
Area of Operation: Europe
Tel: 0118 933 6900
Fax: 0118 931 0035
Email: sales@baumatic.co.uk
Web: www.baumatic.com

BEAU-PORT LTD
15 Old Aylesfield Buildings, Froyle Road,
Alton, Hampshire, GU34 4BY
Area of Operation: UK & Ireland
Tel: 0870 350 0133
Fax: 0870 350 0134
Email: sales@beau-port.co.uk
Web: www.beau-port.co.uk 🖱

BELLING
Stoney Lane, Prescot, Merseyside, L35 2XW
Area of Operation: UK & Ireland
Tel: 0870 458 9663
Fax: 0870 458 9693
Web: www.belling.co.uk

BRADSHAW APPLIANCES
Kenn Road, Clevedon, Bristol,
Somerset, BS21 6LH
Area of Operation: UK & Ireland
Tel: 01275 343000 **Fax:** 01275 343454
Email: info@bradshaw.co.uk
Web: www.bradshaw.co.uk

BRITANNIA LIVING LTD
Britannia House, 281 Bristol Avenue,
Blackpool, Lancashire, FY2 0JF
Area of Operation: UK & Ireland
Tel: 01253 471111 **Fax:** 01253 471136
Email: enquiry@britannialiving.co.uk
Web: www.britannialiving.co.uk

CITY BATHROOMS & KITCHENS
158 Longford Road, Longford,
Coventry, West Midlands, CV6 6DR
Area of Operation: UK & Ireland
Tel: 02476 365877
Fax: 02476 644992
Email: citybathrooms@hotmail.com
Web: www.citybathrooms.co.uk

CONNAUGHT KITCHENS
2 Porchester Place, London, W2 2BS
Area of Operation: Greater London
Tel: 0207 706 2210
Fax: 0207 706 2209
Email: design@connaughtkitchens.co.uk
Web: www.connaughtkitchens.co.uk

COUNTRY KITCHENS
The Old Farm House,
Birmingham Road, Blackminster,
Evesham, Worcestershire, WR11 7TD
Area of Operation: UK (Excluding Ireland)
Tel: 01386 831705
Fax: 01386 834051
Email: steve@floridafunbreaks.com
Web: www.handcraftedkitchens.com
Other Info: ▽✎

CRABTREE KITCHENS
17 Station Road, Barnes, London, SW13 0LF
Area of Operation: Greater London,
South West England and South Wales
Tel: 0208 392 6955
Fax: 0208 392 6944
Email: design@crabtreekitchens.co.uk
Web: www.crabtreekitchens.co.uk

CYMRU KITCHENS
63 Caerleon Road, St. Julians,
Newport, NP19 7BX
Area of Operation: UK (Excluding Ireland)
Tel: 01633 676767
Fax: 01633 212512
Email: sales@cymrukitchens.co.uk
Web: www.cymrukitchens.com

DESIGNER KITCHENS
37 High Street, Hertfordshire, EN6 5AJ
Area of Operation: Greater London
Tel: 01707 650565
Fax: 01707 663050
Email: info@designer-kitchens.co.uk
Web: www.designer-kitchens.co.uk

DR COOKERHOODS
2 Alpha Road, Aldershot, Hampshire, GU12 4RG
Area of Operation: UK (Excluding Ireland)
Tel: 01252 351111
Fax: 01252 311608
Email: lynn@DRCookerhoods.co.uk
Web: www.elica.co.uk

ELITE TRADE KITCHENS LTD
90 Willesden Lane, Kilburn,
London, Greater London, NW6 7TA
Area of Operation: UK (Excluding Ireland)
Tel: 0207 328 1234
Fax: 0207 328 1243
Email: Sales@elitekitchens.co.uk
Web: www.elitekitchens.co.uk

FALCON APPLIANCES
Clarence Street, Royal Leamington Spa,
Warwickshire, CV31 2AD
Area of Operation: Worldwide
Tel: 0845 634 0070
Fax: 01926 450526
Email: jparkinson@falconappliances.co.uk
Web: www.falconappliances.co.uk

FLAWFORD TRADE KITCHENS
Sleaford Road, Coddington, Newark,
Nottinghamshire, NG24 2QY
Area of Operation: UK & Ireland
Tel: 01636 626362
Fax: 01636 627971
Email: sales@mailorderkitchens.co.uk
Web: www.mailorderkitchens.co.uk 🖱
Other Info: ▽ ✎ ✋

FOURNEAUX DE FRANCE LTD (FDF)
Unit 3, Albion Close, Newtown Business Park,
Poole, Dorset, BH12 3LL
Area of Operation: Worldwide
Tel: 01202 733011
Fax: 01202 733499
Email: sales@fdef.co.uk
Web: www.fdef.co.uk 🖱

HATT KITCHENS
Hartlebury Trading Estate,
Hartlebury, Worcestershire, DY10 4JB
Area of Operation: UK (Excluding Ireland)
Tel: 01299 251320
Fax: 01299 251579
Email: design@hatt.co.uk
Web: www.hatt.co.uk

HOOVER CANDY GROUP BUILT-IN DIVISION
New Chester Road, Bromborough,
Wirral, CH62 3PE
Area of Operation: Europe
Tel: 0151 334 2781
Fax: 0151 334 9056
Web: www.candy-domestic.co.uk

INDESIT
Morley Way, Peterborough,
Cambridgeshire, PE2 9JB
Area of Operation: UK & Ireland
Tel: 01733 282800
Fax: 01733 341 783
Email: info@indesitcomapny.com
Web: www.indesitcompany.co.uk

INDESIT
Morley Way, Peterborough,
Cambridgeshire, PE2 9JB
Area of Operation: UK & Ireland
Tel: 01733 282800
Email: info@indesitcompany.com
Web: www.indesit.com

IN-TOTO
Shaw Cross Court, Shaw Cross Business Park,
Dewsbury, West Yorkshire, WF12 7RF
Area of Operation: UK (Excluding Ireland)
Tel: 01924 487900
Fax: 01924 437305
Email: graham.russell@intoto.co.uk
Web: www.intoto.co.uk

JOHN LEWIS OF HUNGERFORD
Grove Technology Park, Downsview Road,
Wantage, Oxfordshire, OX12 9FA
Area of Operation: UK (Excluding Ireland)
Tel: 01235 774300
Fax: 01235 769031
Email: park.street@john-lewis.co.uk
Web: www.john-lewis.co.uk
Other Info: 🖱

KENSINGTON STUDIO
13c/d Kensington Road, Earlsdon,
Coventry, West Midlands, CV5 6GG
Area of Operation: UK (Excluding Ireland)
Tel: 02476 713326
Fax: 02476 713136
Email: sales@kensingtonstudio.com
Web: www.kensingtonstudio.com 🖱

MARK LEIGH KITCHENS LTD
11 Common Garden Street,
Lancaster, Lancashire, LA1 1XD
Area of Operation: North West England
and North Wales
Tel: 01524 63273
Fax: 01524 62352
Email: mark@markleigh.co.uk
Web: www.markleigh.co.uk

MAYTAG
2 St. Annes Boulevard, Foxboro Business Park,
Redhill, Surrey, RH1 1AX
Area of Operation: UK & Ireland
Tel: 01737 231000 **Fax:** 01737 778822
Email: ukquery@maytag.com
Web: www.maytag.co.uk

MERCURY APPLIANCES
Whisby Road, Lincoln, Lincolnshire, LN6 3QZ
Area of Operation: Europe
Tel: 01522 881717
Fax: 01522 880220
Email: sales@mercury-appliances.co.uk
Web: www.mercury-appliances.co.uk

MEREWAY CONTRACTS
413 Warwick Road, Birmingham,
West Midlands, B11 2JR
Area of Operation: UK (Excluding Ireland)
Tel: 0121 707 3288
Fax: 0121 707 3871
Email: info@merewaycontracts.co.uk
Web: www.merewaycontracts.co.uk

MOBALPA
4 Cornflower Close, Willand, Devon, EX15 2TT
Area of Operation: Worldwide
Tel: 07740 633672
Fax: 01884 820828
Email: ploftus@mobalpa.com
Web: www.mobalpa.com

MONTANA KITCHENS LTD
BIC 1, Studio 2/3 Innova Business Park,
Mollison Avenue, Enfield, Middlesex, EN3 7XU
Area of Operation: Greater London,
South East England
Tel: 0800 58 75 628
Email: angie@montanakitchens.co.uk
Web: www.montanakitchens.co.uk
Other Info: ✎

NEFF UK
Grand Union House, Old Wolverton Road,
Wolverton, Milton Keynes,
Buckinghamshire, MK12 5PT
Area of Operation: UK (Excluding Ireland)
Tel: 01908 328300
Fax: 01908 328560
Email: info@neff.co.uk
Web: www.neff.co.uk

NEW WORLD
Stoney Lane, Prescot,
Merseyside, L35 2XW
Area of Operation: UK & Ireland
Tel: 0870 458 9663
Fax: 0870 458 9693
Web: www.newworldappliances.co.uk

NICHOLAS ANTHONY LTD
42-44 Wigmore Street, London, W1U 2RX
Area of Operation: East England,
Greater London, South East England
Tel: 0800 068 3603
Fax: 01206 762698
Email: info@nicholas-anthony.co.uk
Web: www.nicholas-anthony.co.uk

NORTHERN TIMBER PRODUCTS LTD
Unit 20/1, West Bowling Green Street,
Edinburgh, EH6 5PE
Area of Operation: Scotland
Tel: 0131 554 8787
Fax: 0131 554 9191
Email: sales@ntp-kitchens.co.uk
Web: www.ntp-kitchens.co.uk

OAKWOOD BESPOKE
80 High Street, Camberley,
Surrey, GU46 7RN
Area of Operation: Greater London,
South East England, South West England
and South Wales
Tel: 01276 708630
Email: enquiries@oakwoodbespokecamberley.co.uk
Web: www.oakwoodbespokecamberley.co.uk

PRICE KITCHENS
11 Imperial Way,
Croydon, Surrey, CR0 4RR
Area of Operation: South East England
Tel: 0208 686 9006
Fax: 0208 686 5958
Email: info@pricekitchens.co.uk
Web: www.pricekitchens.co.uk

INTERIORS, FIXTURES & FINISHES

William Ball Est 1963
Manufacturers of exclusive fitted furniture

Since 1963 the William Ball family has manufactured quality fitted furniture. The third generation are now on the board of directors and the hands on approach by the family ensures you receive professionalism, commitment to quality and attention to detail. William Ball products are made to be enjoyed made to be admired and above all made to last.

Shown here is just a small selection from the full range of 40 kitchens, 15 bedrooms, 10 studies & Open Plan Living.

For more information call us on 01375 375151 or visit our website: www.wball.co.uk

Sorbus
20mm thick solid ash timber with T&G effect centre panel finished in ivory ash with antiquing effect.

William Ball
Contemporary & Traditional Kitchens, Bedrooms, Studies & Open Plan Living & Bathrooms. Showrooms Nationwide
Tel: 01375 375151
Email: marketing@wball.co.uk
Web: www.wball.co.uk

Bianca
18mm thick MDF white high gloss vinyl pressed slab style door with tight radius edges.

William Ball
Contemporary & Traditional Kitchens, Bedrooms, Studies & Open Plan Living & Bathrooms. Showrooms Nationwide
Tel: 01375 375151
Email: marketing@wball.co.uk
Web: www.wball.co.uk

Firefly
22mm thick MDF fox maple vinyl pressed.

William Ball
Contemporary & Traditional Kitchens, Bedrooms, Studies & Open Plan Living & Bathrooms. Showrooms Nationwide
Tel: 01375 375151
Email: marketing@wball.co.uk
Web: www.wball.co.uk

Chalice
23mm thick solid oak timber with T&G effect centre panel finished in light oak with antiquing effect.

William Ball
Contemporary & Traditional Kitchens, Bedrooms, Studies & Open Plan Living & Bathrooms. Showrooms Nationwide
Tel: 01375 375151
Email: marketing@wball.co.uk
Web: www.wball.co.uk

Pallida
20mm thick solid linden wood frame and veneered centre panel stained to pearwood colour.

William Ball
Contemporary & Traditional Kitchens, Bedrooms, Studies & Open Plan Living & Bathrooms. Showrooms Nationwide
Tel: 01375 375151
Email: marketing@wball.co.uk
Web: www.wball.co.uk

Sophia
22mm thick MDF walnut vinyl pressed.

William Ball
Contemporary & Traditional Kitchens, Bedrooms, Studies & Open Plan Living & Bathrooms. Showrooms Nationwide
Tel: 01375 375151
Email: marketing@wball.co.uk
Web: www.wball.co.uk

Merlin
20mm thick solid oak timber with solid centre panel finished in light oak with antiquing effect.

William Ball
Contemporary & Traditional Kitchens, Bedrooms, Studies & Open Plan Living & Bathrooms. Showrooms Nationwide
Tel: 01375 375151
Email: marketing@wball.co.uk
Web: www.wball.co.uk

Citadel
20mm thick shaker style solid alder frame with veneered centre panel stained to fox maple.

William Ball
Contemporary & Traditional Kitchens, Bedrooms, Studies & Open Plan Living & Bathrooms. Showrooms Nationwide
Tel: 01375 375151
Email: marketing@wball.co.uk
Web: www.wball.co.uk

Nigella Open Plan Living
22mm thick MDF ivory vinyl pressed.

William Ball
Contemporary & Traditional Kitchens, Bedrooms, Studies & Open Plan Living & Bathrooms. Showrooms Nationwide
Tel: 01375 375151
Email: marketing@wball.co.uk
Web: www.wball.co.uk

Nigella Bedroom
22mm thick MDF ivory vinyl pressed.

William Ball
Contemporary & Traditional Kitchens, Bedrooms, Studies & Open Plan Living & Bathrooms. Showrooms Nationwide
Tel: 01375 375151
Email: marketing@wball.co.uk
Web: www.wball.co.uk

Thyme Study
20mm thick solid oak frame and centre panel stained to medium oak colour.

William Ball
Contemporary & Traditional Kitchens, Bedrooms, Studies & Open Plan Living & Bathrooms. Showrooms Nationwide
Tel: 01375 375151
Email: marketing@wball.co.uk
Web: www.wball.co.uk

RANGEMASTER
Clarence Street, Royal Leamington Spa,
Warwickshire, CV31 2AD
Area of Operation: Worldwide
Tel: 01926 457400 **Fax:** 01926 450526
Email: consumers@rangemaster.co.uk
Web: www.rangemaster.co.uk

ROBEYS
Old School House, Green Lane,
Belper, Derbyshire, DE56 1BY
Area of Operation: UK & Ireland
Tel: 01773 820940
Fax: 01773 821652
Email: info@robeys.co.uk
Web: www.robeys.co.uk

ROOTS KITCHENS, BEDROOMS & BATHROOMS
Vine Farm, Stockers Hill, Boughton-under-Blean,
Faversham, Kent, ME13 9AB
Area of Operation: UK (Excluding Ireland)
Tel: 01227 751130 **Fax:** 01227 750033
Email: showroom@rootskitchens.co.uk
Web: www.rootskitchens.co.uk

ROUNDEL DESIGN (UK) LTD
Flishinghurst Orchards, Chalk Lane,
Cranbrook, Kent, TN17 2QA
Area of Operation: South East England
Tel: 01580 712666
Email: homebuild@roundeldesign.co.uk
Web: www.roundeldesign.co.uk

SMEG UK
3 Milton Park, Abingdon, Oxfordshire, OX14 4RY
Area of Operation: UK & Ireland
Tel: 0870 990 9908 **Fax:** 01235 861120
Email: sales@smeguk.com
Web: www.smeguk.com

SPILLERS OF CHARD LTD.
The Aga Cooker Centre, Chard Business Park,
Chard, Somerset, TA20 1FA
Area of Operation: South West England
and South Wales
Tel: 01460 67878 **Fax:** 01460 65252
Email: info@cookercentre.com
Web: www.cookercentre.com

STOVES
Stoney Lane, Prescot, Merseyside, L35 2XW
Area of Operation: UK & Ireland
Tel: 0870 458 9663 **Fax:** 0870 458 9693
Web: www.stoves.co.uk

THE AMERICAN APPLIANCE CENTRE
17-19 Mill Lane, Woodford Green, Essex, IG8 0UN
Area of Operation: UK (Excluding Ireland)
Tel: 0208 5055616 **Fax:** 0208 5058700
Email: sales@american-appliance.co.uk
Web: www.american-appliance.co.uk

THE CDA GROUP LTD
Harby Road, Langar, Nottingham,
Nottinghamshire, NG13 9HY
Area of Operation: UK & Ireland
Tel: 01949 862000
Fax: 01949 862401
Email: sales@cda-europe.com
Web: www.cda-europe.com

THE KITCHEN & BATHROOM COLLECTION
Nelson House, Nelson Road,
Salisbury, Wiltshire, SP1 3LT
Area of Operation: South West England
and South Wales
Tel: 01722 334800
Fax: 01722 412252
Email: info@kbc.co.uk
Web: www.kbc.co.uk
Other Info: ✎

THE ORIGINAL KITCHEN COMPANY
4 Main Street, Breaston, Derby,
Derbyshire, DE72 3DX
Area of Operation: UK & Ireland
Tel: 01332 873746
Fax: 01332 873731
Email: originalkitchens@aol.com

UKAPPLIANCES.CO.UK
Castle Mead House, Castle Mead Gardens,
Hertford, Hertfordshire, SG14 1JZ
Area of Operation: UK (Excluding Ireland)
Tel: 0870 760 6600 **Fax:** 0870 760 6600
Email: info@ukappliances.co.uk
Web: www.ukappliances.co.uk ✎

VECTAIRE LTD
Lincoln Road, Cressex Business Park,
High Wycombe, Buckinghamshire, HP12 3RH
Area of Operation: Worldwide
Tel: 01494 522333 **Fax:** 01494 522337
Email: sales@vectaire.co.uk
Web: www.vectaire.co.uk

VECTAIRE

Area of Operation: Worldwide
Tel: 01494 522333
Fax: 01494 522337
Email: sales@vectaire.co.uk
Web: www.vectaire.co.uk

Within Vectaire's exciting range are sizes to fit any hob, up to 1200mm wide, in chimney and island styles; models in stainless steel, glass, black, white and grey.

VENT AXIA
Fleming Way, Crawley, West Sussex, RH10 9YX
Area of Operation: UK (Excluding Ireland)
Tel: 01293 526062
Fax: 01293 552375
Email: sales@vent-axia.com
Web: www.vent-axia.com

W.S. WESTIN LTD
Phoenix Mills, Leeds Road, Huddersfield,
West Yorkshire, HD1 6NG
Area of Operation: UK (Excluding Ireland)
Tel: 01484 421585
Fax: 01484 432420
Email: enquiry@westin.co.uk
Web: www.westin.co.uk

WS WESTIN LTD

Area of Operation: UK (Excluding Ireland)
Tel: 01484 421585
Fax: 01484 432420
Email: enquiry@westin.co.uk
Web: www.westin.co.uk
Product Type: 2

Westin manufacture and supply a range of stainless steel products: bespoke extraction systems; cooker hoods; worktops; splash-backs; doors; coffee machines; all with after sales service.

WINNING DESIGNS
Dyke Farm, West Chiltington Road,
Pulborough, West Sussex, RH20 2EE
Area of Operation: UK & Ireland
Tel: 0870 754 4446
Fax: 0700 032 9946
Email: info@winning-designs-uk.com
Web: www.winning-designs-uk.com
Other Info: ▽ ECO ✎ ✋

WOODEN HEART WAREHOUSE
Laburnum Road, Chertsey, Surrey, KT16 8BY
Area of Operation: Greater London,
South East England
Tel: 01932 568684
Fax: 01932 568685
Email: whw@btclick.com

COOKERS AND RANGES

KEY

PRODUCT TYPES: 1= Cookers 2 = Ranges
3 = Range Style Cookers 4 = Separate Grills
5 = Separate Hobs
OTHER: ▽ Reclaimed ✎ On-line shopping
✎ Bespoke ✋ Hand-made ECO Ecological

AGA-RAYBURN
Station Road , Ketley, Telford,
Shropshire, TF1 5AQ
Area of Operation: Worldwide
Tel: 01952 642000
Fax: 01952 243 138
Email: jkingsbury-webber@aga-web.co.uk
Web: www.aga-web.co.uk
Product Type: 2, 3

AMBER KITCHENS OF BISHOP'S STORTFORD
Clarklands House, Parsonage Lane,
Sawbridgeworth, Hertfordshire, CM21 0NG
Area of Operation: East England, Greater London
Tel: 01279 600030 **Fax:** 01279 721528
Email: info@amberkitchens.com
Web: www.amberkitchens.com
Product Type: 1, 2, 3, 4, 5

APPLIANCE WORLD UK LTD.
Equity Trade Centre, Hobley Drive,
Swindon, Wiltshire, SN3 4NS
Area of Operation: UK (Excluding Ireland)
Tel: 0870 757 2424 **Fax:** 0870 757 2475
Email: nickki@applianceworld.co.uk
Web: www.applianceworld.co.uk ✎
Product Type: 1, 2, 3, 4, 5

ASHFORD & BROOKS
The Old Workshop, Ashtree Barn,
Caters Road, Bredfield,
Woodbridge, Suffolk, IP13 6BE
Area of Operation: UK (Excluding Ireland)
Tel: 01473 737764 **Fax:** 01473 277176
Email: ashfordbrooks@mailbox.co.uk
Web: www.ashfordandbrooks.co.uk
Product Type: 1, 2, 3, 4, 5

BATH KITCHEN COMPANY
22 Hensley Road, Bloomfield,
Bath, Somerset, BA2 2DR
Area of Operation: South West England
and South Wales
Tel: 01225 312003
Fax: 01225 312003
Email: david@bathkitchencompany.co.uk
Web: www.bathkitchencompany.co.uk

BAUMATIC LTD
Baumatic Buildings,
6 Bennet Road, Berkshire, RG2 0QX
Area of Operation: Europe
Tel: 0118 933 6900 **Fax:** 0118 931 0035
Email: sales@baumatic.co.uk
Web: www.baumatic.com
Product Type: 1, 3, 5

BEAU-PORT LTD
15 Old Aylesfield Buildings, Froyle Road,
Alton, Hampshire, GU34 4BY
Area of Operation: UK & Ireland
Tel: 0870 350 0133
Fax: 0870 350 0134
Email: sales@beau-port.co.uk
Web: www.beau-port.co.uk ✎
Product Type: 1, 2, 3, 4, 5

BELLING
Stoney Lane, Prescot, Merseyside, L35 2XW
Area of Operation: UK & Ireland
Tel: 0870 458 9663
Fax: 0870 458 9693
Web: www.belling.co.uk
Product Type: 1, 3, 5

BELLING

Area of Operation: UK & Ireland
Tel: 0870 458 9663 **Fax:** 0870 458 9693
Web: www.belling.co.uk

Belling is an innovative British cooking family brand that offers high performance appliances designed with practical functions in mind. It's wide range of cooking appliances offer style combined with real benefits in terms of time, energy and convenience. For further information visit www.belling.co.uk or call 0870 458 9663.

BRADSHAW APPLIANCES
Kenn Road, Clevedon, Bristol, Somerset, BS21 6LH
Area of Operation: UK & Ireland
Tel: 01275 343000
Fax: 01275 343454
Email: info@bradshaw.co.uk
Web: www.bradshaw.co.uk
Product Type: 1, 2, 3, 5

BRITANNIA LIVING LTD
Britannia House, 281 Bristol Avenue,
Blackpool, Lancashire, FY2 0JF
Area of Operation: UK & Ireland
Tel: 01253 471111
Fax: 01253 471136
Email: enquiry@britannialiving.co.uk
Web: www.britannialiving.co.uk
Product Type: 1, 2, 3, 5

CAPLE
Fourth Way, Avonmouth, Bristol, BS11 8DW
Area of Operation: UK (Excluding Ireland)
Tel: 0870 606 9606 **Fax:** 0117 938 7449
Email: amandalowe@mlay.co.uk
Web: www.mlay.co.uk
Product Type: 1, 3, 5

CHRISTOPHER PETERS ORIGINAL UNFITTED KITCHENS
28 West Street, Warwick, Warwickshire, CV34 6AN
Area of Operation: Europe
Tel: 01926 494106 **Fax:** 02476 303300
Email: enquiries@christopherpetersantiques.co.uk
Web: www.christopherpetersantiques.co.uk

CITY BATHROOMS & KITCHENS
158 Longford Road, Longford,
Coventry, West Midlands, CV6 6DR
Area of Operation: UK & Ireland
Tel: 02476 365877 **Fax:** 02476 644992
Email: citybathrooms@hotmail.com
Web: www.citybathrooms.co.uk

NEFF BUILT-IN KITCHEN APPLIANCES
THE BUILT IN SPECIALIST

Tel: 01908 328300 Fax: 01908 328560
Email: info@neff.co.uk
Website: www.neff.co.uk

Neff is the UK's foremost brand of built-in appliances designed specifically for building in to kitchen furniture. The company is most well known for its award-winning collection of single and double ovens featuring the famous Circotherm system of fan ducted cooking, much imitated but never equalled.

Also available are matching hobs, extractor hoods and microwaves as well as dishwashers, refrigeration and laundry products to complete the coordinated line-up for the modern fitted kitchen.

For a Neff brochure & stockist list, call 0870 513 3090 or visit the website at www.neff.co.uk

D8250

At 120cm wide, this is the largest cooker hood in the Neff range. Specifically designed for wall mounting and siting over an extra wide hob.

D99T7

This stylish new 90cm wide hood boasts electronic control at the touch of a fingertip plus push buttons for 3 speeds.

D96M5

For a stunning impact at the heart of your kitchen, this 60cm wide model combines futuristic looks with state-of-the-art technology.

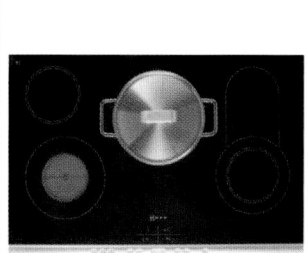

T1683

There are four cooking zones to choose from on this impressive new Piezo ceramic hob available in both 60cm and 90cm wide versions.

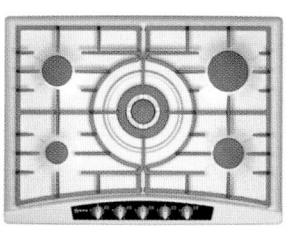

T2766

With its elegant design and central wok burner, this 70cm wide model is one of the most popular gas hobs in the Neff range.

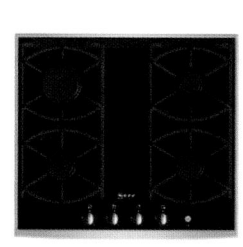

T2646

Electronic control means multi faceted safety benefits are offered by this eye catching 60cm wide gas hob with clever flame failure device.

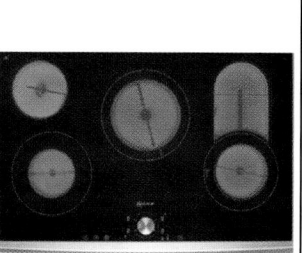

T1583

' Point & Twist' is the name of this new ceramic hob featuring magnetic control via a removable knob. In 60cm & 80cm versions.

U1644

This top-of-the-range double oven boasts electronic control with clear text for a host of impressive cooking and easy cleaning options.

B1664

Scroll down to see what's cooking on this top of the range single oven available in stainless steel to coordinate with other Neff appliances.

B15H4

Unique to Neff is this superb single oven with innovative slide-away door to provide safer and easier access for cooking and cleaning.

U1722

Designed to fit under the worktop, this compact double oven is the answer to all your space requirements. Available in three colourways.

Taste is everything.

A Britannia cooker is a stunning addition to any kitchen and with our tempting selection of styles, widths and colours, from classic cream to contemporary stainless steel, there is sure to be a Britannia cooker to suit every taste.

There's far more to a Britannia than its good looks. Innovative features, including Quickstart - our fast pre-heat system, Cheftop - providing a healthy way of cooking, and the ergonomic telescopic shelf system, ensure the beauty of a Britannia cooker is much more than skin deep.

To see how a Britannia will transform your kitchen order a brochure by calling:

FREEPHONE
0800 073 1003

or visit us at:
www.britannialiving.co.uk

Taste is everything

CONNAUGHT KITCHENS
2 Porchester Place, London, W2 2BS
Area of Operation: Greater London
Tel: 0207 706 2210
Fax: 0207 706 2209
Email: design@connaughtkitchens.co.uk
Web: www.connaughtkitchens.co.uk
Product Type: 1, 2, 3, 4, 5

COUNTRY KITCHENS
The Old Farm House, Birmingham Road,
Blackminster, Evesham, Worcestershire, WR11 7TD
Area of Operation: UK (Excluding Ireland)
Tel: 01386 831705
Fax: 01386 834051
Email: steve@floridafunbreaks.com
Web: www.handcraftedkitchens.com
Product Type: 1, 2, 3, 4, 5

CRABTREE KITCHENS
17 Station Road, Barnes,
Greater London, SW13 0LF
Area of Operation: Greater London,
South West England and South Wales
Tel: 0208 392 6955
Fax: 0208 392 6944
Email: design@crabtreekitchens.co.uk
Web: www.crabtreekitchens.co.uk
Product Type: 1, 2, 3, 4, 5

CYMRU KITCHENS
63 Caerleon Road, St. Julians,
Newport, NP19 7BX
Area of Operation: UK (Excluding Ireland)
Tel: 01633 676767
Fax: 01633 212512
Email: sales@cymrukitchens.com
Web: www.cymrukitchens.com
Product Type: 1, 2, 3, 4, 5

DESIGNER KITCHENS
37 High Street, Hertfordshire, EN6 5AJ
Area of Operation: Greater London
Tel: 01707 650565
Fax: 01707 663050
Email: info@designer-kitchens.co.uk
Web: www.designer-kitchens.co.uk
Product Type: 1, 2, 3, 5

DISCOUNT APPLIANCE CENTRE
Cook House, Brunel Drive,
Newark, Nottingham, Nottinghamshire, NG24 2FB
Area of Operation: Worldwide
Tel: 0870 067 1420
Fax: 01636 707 737
Email: info@thedac.co.uk
Web: www.thedac.co.uk
Product Type: 1, 2, 3, 4, 5

DUNSLEY HEAT LTD
Bridge Mills, Huddersfield Road,
Holmfirth, Huddersfield,
West Yorkshire, HD9 3TW
Area of Operation: UK & Ireland
Tel: 01484 682635
Fax: 01484 688428
Email: sales@dunsleyheat.co.uk
Web: www.dunsleyheat.co.uk
Product Type: 2

ELITE TRADE KITCHENS LTD
90 Willesden Lane, Kilburn, London, NW6 7TA
Area of Operation: UK (Excluding Ireland)
Tel: 0207 328 1234
Fax: 0207 328 1243
Email: Sales@elitekitchens.co.uk
Web: www.elitekitchens.co.uk
Product Type: 1, 2, 3, 5

EVERHOT COOKERS
Coaley Mill, Coaley, Nr Dursley,
Gloucestershire, GL11 5DS
Area of Operation: UK & Ireland
Tel: 01453 890018
Fax: 01453 890958
Email: sales@everhot.co.uk
Web: www.everhot.co.uk
Product Type: 3
Other Info: ✋

FALCON APPLIANCES
Clarence Street, Royal Leamington Spa,
Warwickshire, CV31 2AD
Area of Operation: Worldwide
Tel: 0845 634 0070
Fax: 01926 450526
Email: jparkinson@falconappliances.co.uk
Web: www.falconappliances.co.uk
Product Type: 1, 2, 3

FLAWFORD TRADE KITCHENS
Sleaford Road, Coddington, Newark,
Nottinghamshire, NG24 2QY
Area of Operation: UK & Ireland
Tel: 01636 626362
Fax: 01636 627971
Email: sales@mailorderkitchens.co.uk
Web: www.mailorderkitchens.co.uk ✋
Product Type: 1, 2, 3, 4, 5

FOURNEAUX DE FRANCE LTD (FDF)
Unit 3, Albion Close,
Newtown Business Park,
Poole, Dorset, BH12 3LL
Area of Operation: Worldwide
Tel: 01202 733011
Fax: 01202 733499
Email: sales@fdef.co.uk
Web: www.fdef.co.uk ✋
Product Type: 1, 2, 3, 4, 5

HATT KITCHENS
Hartlebury Trading Estate, Hartlebury,
Worcestershire, DY10 4JB
Area of Operation: UK (Excluding Ireland)
Tel: 01299 251320
Fax: 01299 251579
Email: design@hatt.co.uk
Web: www.hatt.co.uk
Product Type: 1, 2, 3, 4, 5

HERITAGE RANGE COOKERS
2/3 Miller Business Park, Station Road,
Liskeard, Cornwall, PL14 4DA
Area of Operation: UK & Ireland
Tel: 01579 345680
Fax: 01579 346439
Email: enquiries@heritagecookers.co.uk
Web: www.heritagecookers.co.uk
Product Type: 2, 3

HOOVER CANDY GROUP BUILT-IN DIVISION
New Chester Road, Bromborough, Wirral, CH62 3PE
Area of Operation: Europe
Tel: 0151 334 2781
Fax: 0151 334 9056
Web: www.candy-domestic.co.uk
Product Type: 1, 2, 5

HUNTER STOVES LTD
Unit 6, Old Mill Industrial Estate,
Stoke Canon, Exeter, Devon, EX5 4RJ
Area of Operation: UK (Excluding Ireland)
Tel: 01392 841744
Fax: 01392 841382
Email: info@hunterstoves.co.uk
Web: www.hunterstoves.co.uk
Product Type: 1, 2, 3

INDESIT
Morley Way, Peterborough,
Cambridgeshire, PE2 9JB
Area of Operation: UK & Ireland
Tel: 01733 282800
Fax: 01733 341783
Email: info@indesitcompany.com
Web: www.indesit.com
Product Type: 1, 2, 3, 5

IN-TOTO
Shaw Cross Court,
Shaw Cross Business Park,
Dewsbury, West Yorkshire, WF12 7RF
Area of Operation: UK (Excluding Ireland)
Tel: 01924 487900
Fax: 01924 437305
Email: graham.russell@intoto.co.uk
Web: www.intoto.co.uk
Product Type: 1, 2, 3, 5

JOHN LEWIS OF HUNGERFORD
Grove Technology Park, Downsview Road,
Wantage, Oxfordshire, OX12 9FA
Area of Operation: UK (Excluding Ireland)
Tel: 01235 774300
Fax: 01235 769031
Email: park.street@john-lewis.co.uk
Web: www.john-lewis.co.uk
Product Type: 1, 2, 3, 5

KENSINGTON STUDIO
13c/d Kensington Road, Earlsdon,
Coventry, West Midlands, CV5 6GG
Area of Operation: UK (Excluding Ireland)
Tel: 02476 713326
Fax: 02476 713136
Email: sales@kensingtonstudio.com
Web: www.kensingtonstudio.com ✋
Product Type: 1, 2, 3, 4, 5

MARK LEIGH KITCHENS LTD
11 Common Garden Street, Lancaster,
Lancashire, LA1 1XD
Area of Operation: North West England
and North Wales
Tel: 01524 63273
Fax: 01524 62352
Email: mark@markleigh.co.uk
Web: www.markleigh.co.uk
Product Type: 1, 2, 3, 4, 5
Other Info: ✍

MAYTAG
2 St. Annes Boulevard,
Foxboro Business Park,
Redhill, Surrey, RH1 1AX
Area of Operation: UK & Ireland
Tel: 01737 231000
Fax: 01737 778822
Email: ukquery@maytag.com
Web: www.maytag.co.uk
Product Type: 3

MERCURY APPLIANCES
Whisby Road, Lincoln, Lincolnshire, LN6 3QZ
Area of Operation: Europe
Tel: 01522 881717
Fax: 01522 880220
Email: sales@mercury-appliances.co.uk
Web: www.mercury-appliances.co.uk
Product Type: 1, 2, 3, 4, 5

MERCURY APPLIANCES
Area of Operation: Europe
Tel: 01522 881717 **Fax:** 01522 880220
Email: sales@mercury-appliances.co.uk
Web: www.mercury-appliances.co.uk
Product Type: 1, 2, 3, 4, 5

Mercury Appliances, manufacturer of exclusive kitchen appliances, has celebrated its fifth birthday with the launch of an accessories range for their award-winning range cooker

MEREWAY CONTRACTS
413 Warwick Road, Birmingham,
West Midlands, B11 2JR
Area of Operation: UK (Excluding Ireland)
Tel: 0121 707 3288
Fax: 0121 707 3871
Email: info@merewaycontracts.co.uk
Web: www.merewaycontracts.co.uk

MONTANA KITCHENS LTD
BIC 1, Studio 2/3 Innova Business Park,
Mollison Avenue, Enfield, Middlesex, EN3 7XU
Area of Operation: Greater London,
South East England
Tel: 0800 58 75 628
Email: angie@montanakitchens.co.uk
Web: www.montanakitchens.co.uk
Product Type: 1, 2, 3, 4, 5

NEFF UK
Grand Union House,
Old Wolverton Road,
Wolverton, Milton Keynes,
Buckinghamshire, MK12 5PT
Area of Operation: UK (Excluding Ireland)
Tel: 01908 328300
Fax: 01908 328560
Email: info@neff.co.uk
Web: www.neff.co.uk
Product Type: 1, 3, 5

NEW WORLD
Stoney Lane, Prescot,
Merseyside, L35 2XW
Area of Operation: UK & Ireland
Tel: 0870 458 9663
Fax: 0870 458 9693
Web: www.newworldappliances.co.uk
Product Type: 1, 2, 3

NEW WORLD

Area of Operation: UK & Ireland
Tel: 0870 458 9663 **Fax:** 0870 458 9693
Web: www.newworldappliances.co.uk

New World offers a selection of modern appliances that provide good looks, essential functions and outstanding performance at value for money prices. For further information visit the website at www.newworldappliances.co.uk or call 0870 458 9663.

NICHOLAS ANTHONY LTD
42-44 Wigmore Street, London, W1U 2RX
Area of Operation: East England,
Greater London, South East England
Tel: 0800 068 3603
Fax: 01206 762698
Email: info@nicholas-anthony.co.uk
Web: www.nicholas-anthony.co.uk
Product Type: 1, 3, 5

NORTHERN TIMBER PRODUCTS LTD
Unit 20/1, West Bowling Green Street,
Edinburgh, EH6 5PE
Area of Operation: Scotland
Tel: 0131 554 8787
Fax: 0131 554 9191
Email: sales@ntp-kitchens.co.uk
Web: www.ntp-kitchens.co.uk
Product Type: 1, 2, 3, 4, 5

OAKWOOD BESPOKE
80 High Street, Camberley, Surrey, GU46 7RN
Area of Operation: Greater London,
South East England, South West England
and South Wales
Tel: 01276 708630
Email: enquiries@oakwoodbespokecamberley.co.uk
Web: www.oakwoodbespokecamberley.co.uk
Product Type: 1, 2, 3, 4, 5

INTERIORS, FIXTURES & FINISHES

PHIL GREEN & SON
Unit 7, Maylite Trading Estate, Berrow Green Road, Martley, Worcester, Worcestershire, WR6 6PQ
Area of Operation: UK (Excluding Ireland)
Tel: 01885 488936
Fax: 01885 488936
Email: info@philgreenandson.co.uk
Web: www.philgreenandson.co.uk
Product Type: 2

PRICE KITCHENS
11 Imperial Way, Croydon, Surrey, CR0 4RR
Area of Operation: South East England
Tel: 0208 686 9006
Fax: 0208 686 5958
Email: info@pricekitchens.co.uk
Web: www.pricekitchens.co.uk
Product Type: 1, 2, 3, 5

RANGEMASTER
Clarence Street, Royal Leamington Spa, Warwickshire, CV31 2AD
Area of Operation: Worldwide
Tel: 01926 457400
Fax: 01926 450526
Email: consumers@rangemaster.co.uk
Web: www.rangemaster.co.uk
Product Type: 1, 3, 5

REDFYRE COOKERS
Osprey Road, Sowton Industrial Estate, Exeter, Devon, EX2 7JG
Area of Operation: UK & Ireland
Tel: 01392 474061
Fax: 01392 444804
Email: redfyre@gazco.com
Web: www.redfyrecookers.co.uk
Product Type: 2

ROBEYS
Old School House, Green Lane, Belper, Derbyshire, DE56 1BY
Area of Operation: UK & Ireland
Tel: 01773 820940
Fax: 01773 821652
Email: info@robeys.co.uk
Web: www.robeys.co.uk
Product Type: 1, 2, 3

ROOTS KITCHENS, BEDROOMS & BATHROOMS
Vine Farm, Stockers Hill, Boughton-under-Blean, Faversham, Kent, ME13 9AB
Area of Operation: UK (Excluding Ireland)
Tel: 01227 751130
Fax: 01227 750033
Email: showroom@rootskitchens.co.uk
Web: www.rootskitchens.co.uk
Product Type: 1, 2, 3, 4

ROUNDEL DESIGN (UK) LTD
Flishinghurst Orchards, Chalk Lane, Cranbrook, Kent, TN17 2QA
Area of Operation: South East England
Tel: 01580 712666
Email: homebuild@roundeldesign.co.uk
Web: www.roundeldesign.co.uk
Product Type: 1, 2, 3, 5

SMEG UK
3 Milton Park, Abingdon, Oxfordshire, OX14 4RY
Area of Operation: UK & Ireland
Tel: 0870 990 9908
Fax: 01235 861120
Email: sales@smeguk.com
Web: www.smeguk.com

SPILLERS OF CHARD LTD.
The Aga Cooker Centre, Chard Business Park, Chard, Somerset, TA20 1FA
Area of Operation: South West England and South Wales
Tel: 01460 67878
Fax: 01460 65252
Email: info@cookercentre.com
Web: www.cookercentre.com
Product Type: 1, 2, 3, 4, 5

STOVES
Stoney Lane, Prescot, Merseyside, L35 2XW
Area of Operation: UK & Ireland
Tel: 0870 458 9663
Fax: 0870 458 9693
Web: www.stoves.co.uk
Product Type: 1, 3, 5

STOVES
Area of Operation: UK & Ireland
Tel: 0870 458 9663
Fax: 0870 458 9693
Web: www.stoves.co.uk

Stoves built-in, range and freestanding cookers combine technological innovation with design flair - of which its high speed Genus built-in oven is the latest. Other innovations include Rotostar, Stoves' unique fanned gas oven, and unique Pristine® enamel. For further information visit the website at www.stoves.co.uk or call 0870 458 9663.

STOVES ON LINE LTD
Capton, Dartmouth, Devon, TQ6 0JE
Area of Operation: UK (Excluding Ireland)
Tel: 0845 226 5754
Fax: 0870 220 0920
Email: info@stovesonline.com
Web: www.stovesonline.co.uk
Product Type: 1

THE AMERICAN APPLIANCE CENTRE
17-19 Mill Lane, Woodford Green, Essex, IG8 0UN
Area of Operation: UK (Excluding Ireland)
Tel: 0208 505 5616 **Fax:** 0208 505 8700
Email: sales@american-appliance.co.uk
Web: www.american-appliance.co.uk
Product Type: 1, 2, 3, 4, 5

THE CDA GROUP LTD
Harby Road, Langar, Nottingham, Nottinghamshire, NG13 9HY
Area of Operation: UK & Ireland
Tel: 01949 862000 **Fax:** 01949 862401
Email: sales@cda-europe.com
Web: www.cda-europe.com
Product Type: 1, 2, 3, 5

THE KITCHEN AND BATHROOM COLLECTION
Nelson House, Nelson Road, Salisbury, Wiltshire, SP1 3LT
Area of Operation: South West England and South Wales
Tel: 01722 334800 **Fax:** 01722 412252
Email: info@kbc.co.uk
Web: www.kbc.co.uk
Product Type: 1, 2, 3, 4, 5
Other Info: ✎

THE ORIGINAL KITCHEN COMPANY
4 Main Street, Breaston, Derby, Derbyshire, DE72 3DX
Area of Operation: UK & Ireland
Tel: 01332 873746 **Fax:** 01332 873731
Email: originalkitchens@aol.com

THE RANGE COOKER CENTRE
5 Market Street, Whittlesey, Peterborough, Cambridgeshire, PE7 1BA
Area of Operation: UK (Excluding Ireland)
Tel: 01733 202600 **Fax:** 01733 205091
Email: info@rangecookercentre.co.uk
Web: www.rangecookercentre.co.uk
Product Type: 2, 3

THE RANGE COOKER COMPANY
Britannia House, 281 Bristol Avenue,
Blackpool, Lancashire, FY2 0JF
Area of Operation: UK & Ireland
Tel: 01253 471111
Fax: 01253 471136
Email: enquiry@rangecooker.co.uk
Web: www.rangecooker.co.uk
Product Type: 2, 3

THE WESTYE GROUP EUROPE LTD
6B Imprimo Park, Lenthall Road,
Debden, Essex, IG10 3UF
Area of Operation: Europe
Tel: 0208 418 3800
Email: gkott@sub-zero.eu.com
Web: www.westye.eu.com
Product Type: 1, 2, 5

TWYFORD COOKERS LIMITED
Twyford House, 31 & 32 Three Elms
Trading Estate, Bakers Lane,
Hereford, Herefordshire, HR4 9PU
Area of Operation: Worldwide
Tel: 01432 355924
Fax: 01432 272664
Email: sales@twyford-cookers.com
Web: www.twyford-cookers.com ⌐⏁
Product Type: 3

UKAPPLIANCES.CO.UK
Castle Mead House,
Castle Mead Gardens,
Hertford, Hertfordshire, SG14 1JZ
Area of Operation: UK (Excluding Ireland)
Tel: 0870 760 6600
Fax: 0870 760 6600
Email: info@ukappliances.co.uk
Web: www.ukappliances.co.uk ⌐⏁
Product Type: 1, 2, 3, 4, 5

WINNING DESIGNS
Dyke Farm, West Chiltington Road,
Pulborough, West Sussex, RH20 2EE
Area of Operation: UK & Ireland
Tel: 0870 754 4446
Fax: 0700 032 9946
Email: info@winning-designs-uk.com
Web: www.winning-designs-uk.com
Product Type: 1, 2, 3, 4, 5

APPLIANCES & WHITE GOODS

KEY

PRODUCT TYPES: 1= Fridges   2 = Freezers
3 = Fridge Freezers   4 = Dishwashers
5 = Waste Disposal Units   6 = Washing
Machines 7 = Tumble Dryers 8 = Microwaves
OTHER: ▽ Reclaimed ⌐⏁ On-line shopping
✐ Bespoke ⚕ Hand-made   ECO Ecological

ADMIRAL
2 St Annes Boulevard,
Foxboro Business Park,
Redhill, Surrey, RH1 1AX
Area of Operation: UK & Ireland
Tel: 01737 231000
Fax: 01737 778822
Email: ukquery@maytag.co.uk
Web: www.maytag.co.uk
Product Type: 1, 2, 3, 6, 7

AMANA
2 St Annes Boulevard,
Foxboro Business Park,
Redhill, Surrey, RH1 1AX
Area of Operation: UK & Ireland
Tel: 01737 231000
Fax: 01737 778822
Email: info.uk@amana.com
Web: www.amana.co.uk
Product Type: 3, 4

AMBER KITCHENS OF BISHOP'S STORTFORD
Clarklands House, Parsonage Lane,
Sawbridgeworth, Hertfordshire, CM21 0NG
Area of Operation: East England, Greater London
Tel: 01279 600030
Fax: 01279 721528
Email: info@amberkitchens.com
Web: www.amberkitchens.com
Product Type: 1, 2, 3, 4, 5, 6, 7, 8
Other Info: ✐

APPLIANCE WORLD UK LTD.
Equity Trade Centre, Hobley Drive,
Swindon, Wiltshire, SN3 4NS
Area of Operation: UK (Excluding Ireland)
Tel: 0870 757 2424
Fax: 0870 757 2475
Email: nickki@applianceworld.co.uk
Web: www.applianceworld.co.uk ⌐⏁
Product Type: 1, 2, 3, 4, 5, 6, 7, 8

ASHFORD & BROOKS
The Old Workshop, Ashtree Barn,
Caters Road, Bredfield, Woodbridge,
Suffolk, IP13 6BE
Area of Operation: UK (Excluding Ireland)
Tel: 01473 737764
Fax: 01473 277176
Email: ashfordbrooks@mailbox.co.uk
Web: www.ashfordandbrooks.co.uk
Product Type: 1, 2, 3, 4, 5, 6, 7, 8

BATH KITCHEN COMPANY
22 Hensley Road, Bloomfield,
Bath, Somerset, BA2 2DR
Area of Operation: South West England
and South Wales
Tel: 01225 312003
Fax: 01225 312003
Email: david@bathkitchencompany.co.uk
Web: www.bathkitchencompany.co.uk

BAUMATIC LTD
Baumatic Buildings, 6 Bennet Road,
Berkshire, RG2 0QX
Area of Operation: Europe
Tel: 0118 933 6900
Fax: 0118 931 0035
Email: sales@baumatic.com
Web: www.baumatic.com
Product Type: 1, 3, 4, 6

BEAU-PORT LTD
15 Old Aylesfield Buildings,
Froyle Road, Alton, Hampshire, GU34 4BY
Area of Operation: UK & Ireland
Tel: 0870 350 0133
Fax: 0870 350 0134
Email: sales@beau-port.co.uk
Web: www.beau-port.co.uk ⌐⏁
Product Type: 1, 2, 3, 4, 5, 6, 7, 8

BELLING
Stoney Lane, Prescot,
Merseyside, L35 2XW
Area of Operation: UK & Ireland
Tel: 0870 458 9663
Fax: 0870 458 9693
Web: www.belling.co.uk
Product Type: 1, 2, 3, 4, 6, 8

BRADSHAW APPLIANCES
Kenn Road, Clevedon, Bristol,
Somerset, BS21 6LH
Area of Operation: UK & Ireland
Tel: 01275 343000
Fax: 01275 343454
Email: info@bradshaw.co.uk
Web: www.bradshaw.co.uk
Product Type: 1, 2, 3, 4, 8

CAPLE
Fourth Way, Avonmouth, Bristol, BS11 8DW
Area of Operation: UK (Excluding Ireland)
Tel: 0870 606 9606
Fax: 0117 938 7449
Email: amandalowe@mlay.co.uk
Web: www.mlay.co.uk
Product Type: 1, 3, 4, 6, 7, 8

CITY BATHROOMS & KITCHENS
158 Longford Road, Longford, Coventry,
West Midlands, CV6 6DR
Area of Operation: UK & Ireland
Tel: 02476 365877
Fax: 02476 644992
Email: citybathrooms@hotmail.com
Web: www.citybathrooms.co.uk

CONNAUGHT KITCHENS
2 Porchester Place, London, W2 2BS
Area of Operation: Greater London
Tel: 0207 706 2210
Fax: 0207 706 2209
Email: design@connaughtkitchens.co.uk
Web: www.connaughtkitchens.co.uk
Product Type: 1, 2, 3, 4, 5, 6, 7, 8

CORNER FRIDGE COMPANY
Unit 6 Harworth Enterprise Park,
Brunel Industrial Esate, Harworth,
Doncaster, South Yorkshire, DN11 8SG
Area of Operation: UK & Ireland
Tel: 01302 759308
Fax: 01302 751233
Email: info@cornerfridge.com
Web: www.cornerfridge.com
Product Type: 1

COUNTRY KITCHENS
The Old Farm House,
Birmingham Road, Blackminster,
Evesham, Worcestershire, WR11 7TD
Area of Operation: UK (Excluding Ireland)
Tel: 01386 831705
Fax: 01386 834051
Email: steve@floridafunbreaks.com
Web: www.handcraftedkitchens.com
Product Type: 1, 2, 3, 4, 5, 6, 7, 8

CRABTREE KITCHENS
17 Station Road, Barnes,
Greater London, SW13 0LF
Area of Operation: Greater London,
South West England and South Wales
Tel: 0208 392 6955
Fax: 0208 392 6944
Email: design@crabtreekitchens.co.uk
Web: www.crabtreekitchens.co.uk
Product Type: 1, 2, 3, 4, 5, 6, 7, 8

CYMRU KITCHENS
63 Caerleon Road, St. Julians,
Newport, NP19 7BX
Area of Operation: UK (Excluding Ireland)
Tel: 01633 676767
Fax: 01633 212512
Email: sales@cymrukitchens.com
Web: www.cymrukitchens.com
Product Type: 1, 2, 3, 4, 5, 6, 7, 8

DESIGNER KITCHENS
37 High Street, Hertfordshire, EN6 5AJ
Area of Operation: Greater London
Tel: 01707 650565
Fax: 01707 663050
Email: info@designer-kitchens.co.uk
Web: www.designer-kitchens.co.uk
Product Type: 1, 2, 3, 4, 5, 6, 7, 8

DISCOUNT APPLIANCE CENTRE
Cook House, Brunel Drive,
Newark, Nottingham,
Nottinghamshire, NG24 2FB
Area of Operation: Worldwide
Tel: 0870 067 1420
Fax: 01636 707 737
Email: info@thedac.co.uk
Web: www.thedac.co.uk ⌐⏁
Product Type: 1, 2, 3, 4, 5, 6, 7, 8

ELITE TRADE KITCHENS LTD
90 Willesden Lane, Kilburn, London, NW6 7TA
Area of Operation: UK (Excluding Ireland)
Tel: 0207 328 1234
Fax: 0207 328 1243
Email: sales@elitekitchens.co.uk
Web: www.elitekitchens.co.uk
Product Type: 1, 2, 3, 4, 6, 8

FALCON APPLIANCES
Clarence Street, Royal Leamington Spa,
Warwickshire, CV31 2AD
Area of Operation: Worldwide
Tel: 0845 634 0070
Fax: 01926 450526
Email: jparkinson@falconappliances.co.uk
Web: www.falconappliances.co.uk
Product Type: 1, 2

FLAWFORD TRADE KITCHENS
Sleaford Road, Coddington,
Newark, Nottinghamshire, NG24 2QY
Area of Operation: UK & Ireland
Tel: 01636 626362
Fax: 01636 627971
Email: sales@mailorderkitchens.co.uk
Web: www.mailorderkitchens.co.uk ⌐⏁
Product Type: 1, 2, 3, 4, 6, 7, 8

FOURNEAUX DE FRANCE LTD (FDF)
Unit 3, Albion Close, Newtown Business Park,
Poole, Dorset, BH12 3LL
Area of Operation: Worldwide
Tel: 01202 733011
Fax: 01202 733499
Email: sales@fdef.co.uk
Web: www.fdef.co.uk ⌐⏁
Product Type: 1, 2, 3

HATT KITCHENS
Hartlebury Trading Estate, Hartlebury,
Worcestershire, DY10 4JB
Area of Operation: UK (Excluding Ireland)
Tel: 01299 251320
Fax: 01299 251579
Email: design@hatt.co.uk
Web: www.hatt.co.uk
Product Type: 1, 2, 3, 4, 5, 6, 7, 8

HOOVER CANDY GROUP BUILT-IN DIVISION
New Chester Road, Bromborough, Wirral, CH62 3PE
Area of Operation: Europe
Tel: 0151 334 2781 **Fax:** 0151 334 9056
Web: www.candy-domestic.co.uk
Product Type: 1, 2, 3, 4, 6, 8

INDESIT
Morley Way, Peterborough,
Cambridgeshire, PE2 9JB
Area of Operation: UK & Ireland
Tel: 01733 282800 **Fax:** 01733 341783
Email: info@indesitcompany.com
Web: www.indesit.com
Product Type: 1, 2, 3, 4, 6, 7

INSINKERATOR
Emerson Electric UK Ltd, Chelmsford Road,
Chelmsford, Essex, CM6 1LP
Area of Operation: Worldwide
Tel: 01371 873073
Email: insinkerator@insinkerator.com
Web: www.insinkerator.com
Product Type: 5

IN-TOTO
Shaw Cross Court, Shaw Cross Business Park,
Dewsbury, West Yorkshire, WF12 7RF
Area of Operation: UK (Excluding Ireland)
Tel: 01924 487900 **Fax:** 01924 437305
Email: graham.russell@intoto.co.uk
Web: www.intoto.co.uk
Product Type: 1, 2, 3, 4, 5, 6, 7, 8

J&J ORMEROD PLC
Colonial House, Bacup, Lancashire, OL13 0EA
Area of Operation: UK & Ireland
Tel: 01706 877877 **Fax:** 01706 879827
Email: npeters@jjoplc.com
Web: www.jjoplc.com

JOHN LEWIS OF HUNGERFORD
Grove Technology Park, Downsview Road,
Wantage, Oxfordshire, OX12 9FA
Area of Operation: UK (Excluding Ireland)
Tel: 01235 774300 **Fax:** 01235 769031
Email: park.street@john-lewis.co.uk
Web: www.john-lewis.co.uk
Product Type: 1, 2, 3, 4, 5, 6, 7, 8

KENLEY KITCHENS
24-26 Godstone Road,
Kenley, Surrey, CR8 5JE
Area of Operation: UK (Excluding Ireland)
Tel: 0208 668 7000

KENSINGTON STUDIO
13c/d Kensington Road,
Earlsdon, Coventry,
West Midlands, CV5 6GG
Area of Operation: UK (Excluding Ireland)
Tel: 02476 713326
Fax: 02476 713136
Email: sales@kensingtonstudio.com
Web: www.kensingtonstudio.com ⌂
Product Type: 1, 2, 3, 4, 5, 6, 7, 8

LEC REFRIGERATION
Glen Dimplex Home Appliances,
Stoney Lane, Prescot,
Merseyside, L35 2XW
Area of Operation: UK & Ireland
Tel: 0870 458 9663
Fax: 0870 458 9693
Web: www.lec.co.uk
Product Type: 1, 2, 3

MARK LEIGH KITCHENS LTD
11 Common Garden Street,
Lancaster, Lancashire, LA1 1XD
Area of Operation: North West England
and North Wales
Tel: 01524 63273
Fax: 01524 62352
Email: mark@markleigh.co.uk
Web: www.markleigh.co.uk
Product Type: 1, 2, 3, 4, 5, 6, 7, 8
Other Info: ✎

MAYTAG
2 St. Annes Boulevard,
Foxboro Business Park,
Redhill, Surrey, RH1 1AX
Area of Operation: UK & Ireland
Tel: 01737 231000
Fax: 01737 778822
Email: ukquery@maytag.com
Web: www.maytag.co.uk
Product Type: 3, 4, 6, 7

MEREWAY CONTRACTS
413 Warwick Road, Birmingham,
West Midlands, B11 2JR
Area of Operation: UK (Excluding Ireland)
Tel: 0121 707 3288
Fax: 0121 707 3871
Email: info@merewaycontracts.co.uk
Web: www.merewaycontracts.co.uk

MIKE WALKER DISTRIBUTION
Clutton Hill Estate, King Lane,
Clutton Hill, Bristol, BS39 5QQ
Area of Operation: Midlands & Mid Wales,
South West England and South Wales
Tel: 01761 453838
Fax: 01761 453060
Email: office.mwd@zoom.co.uk
Web: www.mikewalker.co.uk

MONTANA KITCHENS LTD
BIC 1, Studio 2/3 Innova Business Park,
Mollison Avenue, Enfield, Middlesex, EN3 7XU
Area of Operation: Greater London,
South East England
Tel: 0800 58 75 628
Email: angie@montanakitchens.co.uk
Web: www.montanakitchens.co.uk
Product Type: 1, 2, 3, 4, 5, 6, 7, 8

NEFF UK
Grand Union House, Old Wolverton Road,
Wolverton, Milton Keynes,
Buckinghamshire, MK12 5PT
Area of Operation: UK (Excluding Ireland)
Tel: 01908 328300
Fax: 01908 328560
Email: info@neff.co.uk
Web: www.neff.co.uk
Product Type: 1, 2, 3, 6, 7

NEW WORLD
Stoney Lane, Prescot,
Merseyside, L35 2XW
Area of Operation: UK & Ireland
Tel: 0870 458 9663
Fax: 0870 458 9693
Web: www.newworldappliances.co.uk
Product Type: 1, 2, 3, 4, 6, 8

NICHOLAS ANTHONY LTD
42-44 Wigmore Street, London, W1U 2RX
Area of Operation: East England,
Greater London, South East England
Tel: 0800 068 3603
Fax: 01206 762698
Email: info@nicholas-anthony.co.uk
Web: www.nicholas-anthony.co.uk
Product Type: 1, 2, 3, 4, 5, 6, 7, 8

NORTHERN TIMBER PRODUCTS LTD
Unit 20/1, West Bowling Green Street,
Edinburgh, EH6 5PE
Area of Operation: Scotland
Tel: 0131 554 8787
Fax: 0131 554 9191
Email: sales@ntp-kitchens.co.uk
Web: www.ntp-kitchens.co.uk
Product Type: 1, 2, 3, 4, 5, 6, 7, 8

OAKWOOD BESPOKE
80 High Street, Camberley, Surrey, GU46 7RN
Area of Operation: Greater London, South East
England, South West England and South Wales
Tel: 01276 708630
Email: enquiries@oakwoodbespokecamberley.co.uk
Web: www.oakwoodbespokecamberley.co.uk
Product Type: 1, 2, 3, 4, 5, 6, 7, 8

PRICE KITCHENS
11 Imperial Way, Croydon, Surrey, CR0 4RR
Area of Operation: South East England
Tel: 0208 686 9006
Fax: 0208 686 5958
Email: info@pricekitchens.co.uk
Web: www.pricekitchens.co.uk
Product Type: 1, 2, 3, 4, 5, 6, 7, 8

ROBEYS
Old School House, Green Lane,
Belper, Derbyshire, DE56 1BY
Area of Operation: UK & Ireland
Tel: 01773 820940
Fax: 01773 821652
Email: info@robeys.co.uk
Web: www.robeys.co.uk
Product Type: 1, 3

**ROOTS KITCHENS,
BEDROOMS & BATHROOMS**
Vine Farm, Stockers Hill,
Boughton-under-Blean,
Faversham, Kent, ME13 9AB
Area of Operation: UK (Excluding Ireland)
Tel: 01227 751130
Fax: 01227 750033
Email: showroom@rootskitchens.co.uk
Web: www.rootskitchens.co.uk ⌂
Product Type: 1, 3, 5, 7

ROUNDEL DESIGN (UK) LTD
Flishinghurst Orchards, Chalk Lane,
Cranbrook, Kent, TN17 2QA
Area of Operation: South East England
Tel: 01580 712666
Email: homebuild@roundeldesign.co.uk
Web: www.roundeldesign.co.uk
Product Type: 1, 2, 3, 4, 5, 6, 7, 8

ROY WARING SOUTH LIMITED
Unit , Lodge Lane, Tuxford, Newark,
Nottinghamshire, NG22 0HZ
Area of Operation: East England,
Greater London, Midlands & Mid Wales,
North East England
Tel: 01777 872082
Fax: 01777 871563
Email: james@roywaring.co.uk
Web: www.roywaring.co.uk

SINCLAIR YOUNGS LTD
Basingstoke Business Centre,
Winchester Road, Basingstoke,
Hampshire, RG22 4AU
Area of Operation: South East England
Tel: 01256 355015 **Fax:** 01256 818344
Email: david.wilson@sinclairyoungs.com
Web: www.sinclairyoungs.com

SPILLERS OF CHARD LTD.
The Aga Cooker Centre,
Chard Business Park,
Chard, Somerset, TA20 1FA
Area of Operation: South West England
and South Wales
Tel: 01460 67878
Fax: 01460 65252
Email: info@cookercentre.com
Web: www.cookercentre.com
Product Type: 1, 2, 3, 4, 5, 6, 7, 8

STOVES
Stoney Lane, Prescot,
Merseyside, L35 2XW
Area of Operation: UK & Ireland
Tel: 0870 458 9663
Fax: 0870 458 9693
Web: www.stoves.co.uk
Product Type: 1, 2, 3, 8

THE AMERICAN APPLIANCE CENTRE
17-19 Mill Lane, Woodford Green, Essex, IG8 0UN
Area of Operation: UK (Excluding Ireland)
Tel: 0208 505 5616
Fax: 0208 505 8700
Email: sales@american-appliance.co.uk
Web: www.american-appliance.co.uk
Product Type: 1, 2, 3, 4, 5, 6, 7, 8

THE CDA GROUP LTD
Harby Road, Langar, Nottingham,
Nottinghamshire, NG13 9HY
Area of Operation: UK & Ireland
Tel: 01949 862000
Fax: 01949 862401
Email: sales@cda-europe.com
Web: www.cda-europe.com
Product Type: 1, 2, 3, 4, 6, 7, 8

**THE HAIGH TWEENY
COMPANY LIMITED**
Haigh Industrial Estate,
Alton Road, Ross on Wye,
Herefordshire, HR9 5LA
Area of Operation: UK (Excluding Ireland)
Tel: 0700 489 3369
Fax: 01424 751444
Email: sales@tweeny.co.uk
Web: www.tweeny.co.uk
Product Type: 5

THE KITCHEN & BATHROOM COLLECTION
Nelson House, Nelson Road,
Salisbury, Wiltshire, SP1 3LT
Area of Operation: South West England
and South Wales
Tel: 01722 334800
Fax: 01722 412252
Email: info@kbc.co.uk
Web: www.kbc.co.uk
Product Type: 1, 2, 3, 4, 5
Other Info: ✎

THE ORIGINAL KITCHEN COMPANY
4 Main Street, Breaston,
Derby, Derbyshire, DE72 3DX
Area of Operation: UK & Ireland
Tel: 01332 873746
Fax: 01332 873731
Email: originalkitchens@aol.com

THE WESTYE GROUP EUROPE LTD
6B Imprimo Park, Lenthall Road,
Debden, Essex, IG10 3UF
Area of Operation: Europe
Tel: 0208 418 3800
Email: gkott@sub-zero.eu.com
Web: www.westye.eu.com
Product Type: 1, 2, 3

UKAPPLIANCES.CO.UK
Castle Mead House, Castle Mead Gardens,
Hertford, Hertfordshire, SG14 1JZ
Area of Operation: UK (Excluding Ireland)
Tel: 0870 760 6600
Fax: 0870 760 6600
Email: info@ukappliances.co.uk
Web: www.ukappliances.co.uk ⌂
Product Type: 1, 2, 3, 4, 5, 6, 7, 8

WINNING DESIGNS
Dyke Farm, West Chiltington Road,
Pulborough, West Sussex, RH20 2EE
Area of Operation: UK & Ireland
Tel: 0870 754 4446 **Fax:** 0700 032 9946
Email: info@winning-designs.com
Web: www.winning-designs-uk.com
Product Type: 1, 2, 3, 4, 5, 6, 7, 8

WOODEN HEART WAREHOUSE
Laburnum Road, Chertsey, Surrey, KT16 8BY
Area of Operation: Greater London,
South East England
Tel: 01932 568684 **Fax:** 01932 568685
Email: whw@btclick.com
Product Type: 1, 2, 3, 4, 5, 6, 7, 8

KITCHEN FURNITURE, FITTED

KEY

PRODUCT TYPES: 1= Shaker
2 = Contemporary   3 = Farmhouse
4 = Industrial   5 = Long Island   6 = New
England 7 = Period 8 = Minimalist   9 = Other
OTHER: ▽ Reclaimed ⌂ On-line shopping
✎ Bespoke ✋ Hand-made ECO Ecological

ALLMILMO UK
Unit 17 Plantaganet House, Kingsclere Business
Park, Kingsclere, Newbury, Berkshire, RG20 4SW
Area of Operation: UK & Ireland
Tel: 01635 868181 **Fax:** 01635 869693
Email: allmilmo@aol.com
Web: www.allmilmo.com
Product Type: 1, 2, 5, 9

AMBER KITCHENS OF BISHOP'S STORTFORD
Clarklands House, Parsonage Lane,
Sawbridgeworth, Hertfordshire, CM21 0NG
Area of Operation: East England, Greater London
Tel: 01279 600030 **Fax:** 01279 721528
Email: info@amberkitchens.com
Web: www.amberkitchens.com
Product Type: 1, 2, 3, 4, 5, 6, 7, 8, 9, 10

ARTICHOKE (KITCHEN AND CABINET MAKERS)
Hortswood, Long Lane, Wrington,
Somerset, BS40 5SP
Area of Operation: UK & Ireland
Tel: 01934 863840 **Fax:** 01934 863841
Email: mail@artichoke-ltd.com
Web: www.artichoke-ltd.com
Product Type: 1, 2, 3, 4, 5, 6, 7, 8, 9

ASHFORD & BROOKS
The Old Workshop, Ashtree Barn, Caters Road,
Bredfield, Woodbridge, Suffolk, IP13 6BE
Area of Operation: UK (Excluding Ireland)
Tel: 01473 737764 **Fax:** 01473 277176
Email: ashfordbrooks@mailbox.co.uk
Web: www.ashfordandbrooks.co.uk
Product Type: 1, 2, 3, 5, 6, 7, 8, 9, 10

BATH KITCHEN COMPANY
22 Hensley Road, Bloomfield,
Bath, Somerset, BA2 2DR
Area of Operation: South West England
and South Wales
Tel: 01225 312003 **Fax:** 01225 312003
Email: david@bathkitchencompany.co.uk
Web: www.bathkitchencompany.co.uk
Product Type: 1, 2, 3, 4, 5, 6, 7, 8, 9
Other Info: ✎

BAYFIELD STAIR CO
Unit 4, Praed Road, Trafford Park, Manchester, M17 1PQ
Area of Operation: Worldwide
Tel: 0161 848 0700
Fax: 0161 872 2230
Email: sales@bayfieldstairs.co.uk
Web: www.bayfieldstairs.co.uk
Product Type: 1, 2, 3, 4, 5, 6, 7, 8, 9, 10

BROOKMANS ENGLISH FURNITURE CO LTD
Portland Business Park, Richmond Park Road, Sheffield, South Yorkshire, S13 8HT
Area of Operation: Worldwide
Tel: 0114 280 0970
Fax: 0114 256 1840
Email: jim.brookman@brookmans.co.uk
Web: www.brookmans.co.uk
Product Type: 1, 3, 7, 8

BUYDESIGN
c/o Woodschool Ltd, Monteviot Nurseries, Ancrum, Jedburgh, Borders, TD8 6TU
Area of Operation: North East England, North West England and North Wales, Scotland
Tel: 01835 830740
Fax: 01835 830750
Email: info@buydesign-furniture.com
Web: www.buydesign-furniture.com
Product Type: 2, 3, 10
Other Info: ECO ✎ ✋

CAMERON PETERS LTD
The Old Dairy, Home Farm, Ardington, Wantage, Oxfordshire, OX12 8PD
Area of Operation: Worldwide
Tel: 01235 835000
Fax: 01235 835005
Email: info@cameronpeters.co.uk
Web: www.cameronpeters.co.uk

CAPLE
Fourth Way, Avonmouth, Bristol, BS11 8DW
Area of Operation: UK (Excluding Ireland)
Tel: 0870 606 9606
Fax: 0117 938 7449
Email: amandalowe@mlay.co.uk
Web: www.mlay.co.uk
Product Type: 1, 2, 3, 4, 7, 8, 9

CHURCHWOOD DESIGN
Unit 2, Tideswell Business Park, Meveril Road, Tideswell, Buxton, Derbyshire, SK17 8NY
Area of Operation: UK (Excluding Ireland)
Tel: 01298 872422
Fax: 01298 873068
Email: info@churchwood.co.uk
Web: www.churchwood.co.uk
Product Type: 1, 3, 5
Other Info: ▽ ECO ✎ ✋

CITY BATHROOMS & KITCHENS
158 Longford Road, Longford, Coventry, West Midlands, CV6 6DR
Area of Operation: UK & Ireland
Tel: 02476 365877
Fax: 02476 644992
Email: citybathrooms@hotmail.com
Web: www.citybathrooms.co.uk

CONNAUGHT KITCHENS
2 Porchester Place, London, W2 2BS
Area of Operation: Greater London
Tel: 0207 706 2210
Fax: 0207 706 2209
Email: design@connaughtkitchens.co.uk
Web: www.connaughtkitchens.co.uk
Product Type: 2

COTTAGE FARM ANTIQUES
Stratford Road, Aston Sub Edge, Chipping Campden, Gloucestershire, GL55 6PZ
Area of Operation: Worldwide
Tel: 01386 438263
Fax: 01386 438263
Email: info@cottagefarmantiques.co.uk
Web: www.cottagefarmantiques.co.uk ✎
Product Type: 8

COUNTRY KITCHENS
The Old Farm House, Birmingham Road, Blackminster, Evesham, Worcestershire, WR11 7TD
Area of Operation: UK (Excluding Ireland)
Tel: 01386 831705
Fax: 01386 834051
Email: sales@floridafunbreaks.com
Web: www.handcraftedkitchens.com
Product Type: 1, 3, 8
Other Info: ▽ ✎ ✋

CP HART GROUP
Newnham Terrace, Hercules Road, London, SE1 7DR
Area of Operation: Greater London
Tel: 0207 902 5250
Fax: 0207 902 1030
Email: paulr@cphart.co.uk
Web: www.cphart.co.uk

CRABTREE KITCHENS
17 Station Road, Barnes, Greater London, SW13 0LF
Area of Operation: Greater London, South West England and South Wales
Tel: 0208 392 6955
Fax: 0208 392 6944
Email: design@crabtreekitchens.co.uk
Web: www.crabtreekitchens.co.uk
Product Type: 1, 2, 4, 5, 7, 8, 9, 10

CYMRU KITCHENS
63 Caerleon Road, St. Julians, Newport, NP19 7BX
Area of Operation: UK (Excluding Ireland)
Tel: 01633 676767
Fax: 01633 212512
Email: sales@cymrukitchens.com
Web: www.cymrukitchens.com
Product Type: 1, 2, 3, 4, 5, 6, 7, 8, 9, 10

D.A FURNITURE LTD
Woodview Workshop, Pitway Lane, Farrington Gurney, Bristol, BS39 6TX
Area of Operation: South West England and South Wales
Tel: 01761 453117
Email: enquiries@dafurniture.co.uk
Web: www.dafurniture.co.uk
Product Type: 1, 2, 3, 5
Other Info: ✎ ✋

DESIGNER KITCHENS
37 High Street, Hertfordshire, EN6 5AJ
Area of Operation: Greater London
Tel: 01707 650565
Fax: 01707 663050
Email: info@designer-kitchens.co.uk
Web: www.designer-kitchens.co.uk
Product Type: 1, 2, 3, 5, 6, 7, 8, 9, 10

DUKE CHRISTIE
Hillockhead Farmhouse, Dallas, Forres, Moray, IV36 2SD
Area of Operation: Worldwide
Tel: 01343 890347
Fax: 01343 890347
Email: enquiries@dukechristie.com
Web: www.dukechristie.com
Product Type: 2, 3, 8
Other Info: ✎ ✋

ELITE TRADE KITCHENS LTD
90 Willesden Lane, Kilburn, London, NW6 7TA
Area of Operation: UK (Excluding Ireland)
Tel: 0207 328 1234
Fax: 0207 328 1243
Email: sales@elitekitchens.co.uk
Web: www.elitekitchens.co.uk
Product Type: 1, 2, 3, 4, 5, 7, 8, 9, 10

ELLIOTT BROTHERS LTD
Millbank Wharf, Northam, Southampton, Hampshire, SO14 5AG
Area of Operation: South East England, South West England and South Wales
Tel: 02380 226852
Fax: 02380 638780
Email: laurenh@elliott-brothers.co.uk
Web: www.elliotts.uk.com
Product Type: 1, 2, 3, 5, 7, 8, 9

FLAWFORD TRADE KITCHENS
Sleaford Road, Coddington, Newark, Nottinghamshire, NG24 2QY
Area of Operation: UK & Ireland
Tel: 01636 626362
Fax: 01636 627971
Email: sales@mailorderkitchens.co.uk
Web: www.mailorderkitchens.co.uk ✎
Product Type: 1, 2, 3, 4, 5, 6, 7, 8, 9, 10

G MIDDLETON LTD
Cross Croft Industrial Estate, Appleby, Cumbria, CA16 6HX
Area of Operation: Europe
Tel: 01768 352067
Fax: 01768 353228
Email: g.middleton@timber75.freeserve.co.uk
Web: www.graham-middleton.co.uk ✎
Product Type: 1, 3, 8

GK DESIGN
Greensbury Farm, Thurleigh Road, Bolnhurst, Bedfordshire, MK44 2ET
Area of Operation: UK (Excluding Ireland)
Tel: 01234 376990
Fax: 01234 376991
Email: studio@gkdesign.co.uk
Web: www.gkdesign.co.uk
Product Type: 10
Other Info: ✎

HARVEY JONES
57 New Kings Road, Fulham, London, SW6 4SE
Area of Operation: UK (Excluding Ireland)
Tel: 0800 917 2340
Fax: 0207 371 0735
Web: www.harveyjones.com

HATT KITCHENS
Hartlebury Trading Estate, Hartlebury, Worcestershire, DY10 4JB
Area of Operation: UK (Excluding Ireland)
Tel: 01299 251320
Fax: 01299 251579
Email: design@hatt.co.uk
Web: www.hatt.co.uk
Product Type: 1, 2, 3, 4, 5, 6, 7, 8, 9, 10

HERITAGE WOOD TURNING
Unit 3 The Old Brickworks, Bakestonedale Road, Pott Shrigley, Macclesfield, Cheshire, SK10 5RX
Area of Operation: UK (Excluding Ireland)
Tel: 01625 560655 **Fax:** 01625 560697
Email: sales@heritageoak.co.uk
Web: www.heritageoak.co.uk
Other Info: ✎

HITT OAK LTD
13a Northview Crescent, London, NW10 1RD
Area of Operation: UK (Excluding Ireland)
Tel: 0208 450 3821 **Fax:** 0208 911 0299
Email: zhitas@yahoo.co.uk
Product Type: 3, 5, 7, 8

HOLME TREE LTD
Units 2 and 3 Machins Business Centre, 29 Wood Street, Asby-de-la-Zouch, Leicestershire, LE65 1EZ
Area of Operation: UK (Excluding Ireland)
Tel: 01530 564561 **Fax:** 01530 417986
Email: info@holmetree.co.uk
Web: www.holmetree.co.uk

HYGROVE
152/154 Merton Road, Wimbledon, London, SW19 1EH
Area of Operation: South East England
Tel: 0208 543 1200 **Fax:** 0208 543 6521
Email: sales@hygrove.fsnet.co.uk
Web: www.hygrovefurniture.co.uk

INTERESTING INTERIORS LTD
37-39 Queen Street, Aberystwyth, Ceredigion, SY23 1PU
Area of Operation: UK (Excluding Ireland)
Tel: 01970 626162 **Fax:** 0870 051 7959
Email: enquiries@interestinginteriors.com
Web: www.interiors-ltd.demon.co.uk
Product Type: 1, 2, 3, 4, 5, 6, 7, 8, 9, 10

IN-TOTO
Shaw Cross Court, Shaw Cross Business Park, Dewsbury, West Yorkshire, WF12 7RF
Area of Operation: UK (Excluding Ireland)
Tel: 01924 487900 **Fax:** 01924 437305
Email: graham.russell@intoto.co.uk
Web: www.intoto.co.uk
Product Type: 1, 2, 3, 4, 5

J&J ORMEROD PLC
Colonial House, Bacup, Lancashire, OL13 0EA
Area of Operation: UK & Ireland
Tel: 01706 877877 **Fax:** 01706 879827
Email: npeters@jjoplc.com
Web: www.jjoplc.com
Product Type: 1, 2, 3, 5, 6, 7, 8, 9, 10

JAIC LTD
Pattern House, Southwell Business Park, Portland, Dorset, DT5 2NR
Area of Operation: UK (Excluding Ireland)
Tel: 01305 826991 **Fax:** 01305 823535
Email: info@jaic.co.uk
Product Type: 1, 3, 5

JOHN FLEMING & CO LTD
Silverburn Place, Bridge of Don, Aberdeen, Aberdeenshire, AB23 8EG
Area of Operation: Scotland
Tel: 0800 085 8728 **Fax:** 01224 825377
Email: info@johnfleming.co.uk
Web: www.johnfleming.co.uk

JOHN LADBURY & CO
Unit 11, Alpha Business Park, Travellers Close, Welham Green, Hertfordshire, AL9 7NT
Area of Operation: Greater London
Tel: 01707 262966 **Fax:** 01707 265400
Email: johnladburyandco@tiscali.co.uk
Web: www.johnladbury.co.uk
Product Type: 1, 2, 3, 5, 7, 8, 9
Other Info: ✎ ✋

JOHN LEWIS OF HUNGERFORD
Grove Technology Park, Downsview Road, Wantage, Oxfordshire, OX12 9FA
Area of Operation: UK (Excluding Ireland)
Tel: 01235 774300 **Fax:** 01235 769031
Email: park.street@john-lewis.co.uk
Web: www.john-lewis.co.uk
Product Type: 1, 3, 8

JOHNNY GREY
Fyning Copse, Fyning Lane, Rogate, Petersfield, Hampshire, GU31 5DH
Area of Operation: Worldwide
Tel: 01730 821424
Email: miles@johnnygrey.co.uk
Web: www.johnnygrey.co.uk
Product Type: 2, 5, 10

KENSINGTON STUDIO
13c/ d Kensington Road, Earlsdon, Coventry, West Midlands, CV5 6GG
Area of Operation: UK (Excluding Ireland)
Tel: 02476 713326 **Fax:** 02476 713136
Email: sales@kensingtonstudio.com
Web: www.kensingtonstudio.com ✎
Product Type: 1, 2, 3, 4, 5, 6, 7, 8, 9, 10

KITCHENS BY JULIAN ENGLISH
Ideal Home, Blackbird Road, Leicester, Leicestershire, LE4 0AM
Area of Operation: Midlands & Mid Wales
Tel: 0116 251 9999 **Fax:** 0116 251 9999
Email: philiphalifax@aol.com
Web: www.kitchens-by-julianenglish.co.uk

MARK LEIGH KITCHENS LTD
11 Common Garden Street, Lancaster, Lancashire, LA1 1XD
Area of Operation: North West England and North Wales
Tel: 01524 63273
Fax: 01524 62352
Email: mark@markleigh.co.uk
Web: www.markleigh.co.uk
Product Type: 1, 2, 5, 9
Other Info: ✎

MASTER LTD
30 Station Road, Heacham,
Norfolk, PE31 7EX
Area of Operation: Europe
Tel: 01485 572032 **Fax:** 01485 571675
Email: ron@masterlimited.co.uk
Web: www.masterlimited.co.uk
Product Type: 1, 3, 4

MEREWAY CONTRACTS
413 Warwick Road, Birmingham,
West Midlands, B11 2JR
Area of Operation: UK (Excluding Ireland)
Tel: 0121 707 3288 **Fax:** 0121 707 3871
Email: info@merewaycontracts.co.uk
Web: www.merewaycontracts.co.uk

MOBALPA
4 Cornflower Close, Willand, Devon, EX15 2TT
Area of Operation: Worldwide
Tel: 07740 633672 **Fax:** 01884 820828
Email: ploftus@mobalpa.com
Web: www.mobalpa.com
Product Type: 1, 2, 3, 4, 5, 6, 7, 9, 10

MONTANA KITCHENS LTD
BIC 1, Studio 2/3 Innova Business Park,
Mollison Avenue, Enfield, Middlesex, EN3 7XU
Area of Operation: Greater London,
South East England
Tel: 0800 58 75 628
Email: angie@montanakitchens.co.uk
Web: www.montanakitchens.co.uk
Product Type: 1, 2, 3, 4, 5, 6, 7, 9, 10

MONTANA KITCHENS

Area of Operation: Greater London, South East
England
Tel: 0800 58 75 628
Email: angie@montanakitchens.co.uk
Web: www.montanakitchens.co.uk

Montana Kitchens is a family run business that
offers a personalised, complete all-inclusive
service to reflect your lifestyle and allocate your
budget effectively.

NICHOLAS ANTHONY LTD
42-44 Wigmore Street, London, W1U 2RX
Area of Operation: East England,
Greater London, South East England
Tel: 0800 068 3603 **Fax:** 01206 762698
Email: info@nicholas-anthony.co.uk
Web: www.nicholas-anthony.co.uk
Product Type: 2, 5

NOLTE KITCHENS
41 London Road, Sawbridgeworth,
Hertfordshire, CM21 9EH
Area of Operation: UK & Ireland
Tel: 01279 868500 **Fax:** 01279 868802
Email: email@noltekitchens.co.uk
Web: www.nolte-kuechen.de
Product Type: 2

NORTHERN TIMBER PRODUCTS LTD
Unit 20/1, West Bowling Green Street,
Edinburgh, EH6 5PE
Area of Operation: Scotland
Tel: 0131 554 8787 **Fax:** 0131 554 9191
Email: sales@ntp-kitchens.co.uk
Web: www.ntp-kitchens.co.uk
Product Type: 1, 2, 3, 5, 6, 7, 8, 9, 10

NOW GROUP PLC (CONTRACTS DIVISION)
Red Scar Business Park, Longridge Road,
Preston, Lancashire, PR2 5NA
Area of Operation: UK & Ireland
Tel: 01772 703838 **Fax:** 01772 705788
Email: sales@nowkitchens.co.uk
Web: www.nowkitchens.co.uk
Product Type: 1, 2, 3, 5, 7, 9

OAKWOOD BESPOKE
80 High Street, Camberley, Surrey, GU46 7RN
Area of Operation: Greater London, South East
England, South West England and South Wales
Tel: 01276 708630
Email: enquiries@oakwoodbespokecamberley.co.uk
Web: www.oakwoodbespokecamberley.co.uk
Product Type: 1, 2, 3, 5, 6, 7, 10

PANELS PLUS
22 Mill Place, Kingston Upon Thames,
Surrey, KT1 2RJ
Area of Operation: Greater London
Tel: 0208 399 6343 **Fax:** 0208 399 6343
Email: sales@panelsplusltd.com
Web: www.panelsplusltd.com
Product Type: 1, 2, 3, 4, 5, 6, 7, 10

PARAGON JOINERY
Systems House, Eastbourne Road,
Blindley Heath, Surrey, RH7 6JP
Area of Operation: UK (Excluding Ireland)
Tel: 01342 836300 **Fax:** 01342 836364
Email: sales@paragonjoinery.com
Web: www.paragonjoinery.com

PARLOUR FARM
Unit 12b, Wilkinson Road, Love Lane Industrial
Estate, Cirencester, Gloucestershire, GL7 1YT
Area of Operation: Europe
Tel: 01285 885336 **Fax:** 01285 643189
Email: info@parlourfarm.com
Web: www.parlourfarm.com

PINELAND FURNITURE LTD
Unit 5 Cleobury Trading Estate,
Cleobury Mortimer, Shropshire, DY14 8DP
Area of Operation: UK (Excluding Ireland)
Tel: 01299 271143
Fax: 01299 271166
Email: pineland@onetel.com
Web: www.pineland.co.uk

PINELAND
Area of Operation: UK (Excluding Ireland)
Tel: 01299 271143
Fax: 01299 271166
Email: pineland@onetel.com
Web: www.pineland.co.uk

Manufacturing a full range of traditionally styled
pine furniture using the finest quality kiln dried
timbers (no plywood or chipboard) and employing
proper mortice and tension and dovetail joining
wherever possible.

PLAIN ENGLISH
Stowupland Hall, Stowupland, Suffolk, IP14 4BE
Area of Operation: Worldwide
Tel: 01449 774028
Fax: 01449 613519
Email: info@plainenglishdesign.co.uk
Web: www.plainenglishdesign.co.uk
Product Type: 1, 2, 8

POGGENPOHL
477-481 Finchley Road, London, NW3 6HS
Area of Operation: UK & Ireland
Tel: 0800 298 1098
Fax: 0207 794 6251
Email: kitchens@poggenpohl-group.co.uk
Web: www.poggenpohl.co.uk
Product Type: 5, 8, 9

POSH PANTRIES
81 Kelvin Road North, Lenziemill,
Cumbernauld, Lanarkshire, G67 6BD
Area of Operation: UK (Excluding Ireland)
Tel: 01236 453556
Fax: 01236 728068
Email: office@poshpantries.co.uk
Web: www.poshpantries.co.uk

PRICE KITCHENS
11 Imperial Way, Croydon, Surrey, CR0 4RR
Area of Operation: South East England
Tel: 0208 686 9006
Fax: 0208 686 5958
Email: info@pricekitchens.co.uk
Web: www.pricekitchens.co.uk
Product Type: 1, 2, 3, 4, 5, 6, 7, 8, 9, 10

RAY PEARS JOINERY LTD
42 Mill Hill Road, Hinckley, Leicestershire, LE10 0AX
Area of Operation: UK (Excluding Ireland)
Tel: 01455 616279
Fax: 01455 891533
Email: pglyn@hotmail.com
Product Type: 1, 2, 3, 4, 5, 6, 7, 8, 9

RICHARD BAKER FURNITURE LTD
Wimbledon Studios, 257 Burlington Road,
New Malden, Surrey, KT3 4NE
Area of Operation: Europe
Tel: 0208 336 1777 **Fax:** 0208 336 1666
Email: Sales@richardbakerfurniture.co.uk
Web: www.richardbakerfurniture.co.uk
Product Type: 1, 2, 5, 8, 9

ROBERT J TURNER & CO
Roe Green, Sandon, Buntingford,
Hertfordshire, SG9 0QE
Area of Operation: East England,
Greater London
Tel: 01763 288371 **Fax:** 01763 288440
Email: sales@robertjturner.co.uk
Web: www.robertjturner.co.uk
Product Type: 1, 2, 3, 4, 5, 6, 7, 8, 9, 10

ROBINSON AND CORNISH LTD
St George's House, St George's Road,
Barnstaple, Devon, EX32 7AS
Area of Operation: Europe
Tel: 01271 329300 **Fax:** 01271 328277
Email: richard.martin@robinsonandcornish.co.uk
Web: www.robinsonandcornish.co.uk
Product Type: 2, 3, 8, 10
Other Info: ✎

ROOTS KITCHENS, BEDROOMS & BATHROOMS
Vine Farm, Stockers Hill, Boughton-under-Blean,
Faversham, Kent, ME13 9AB
Area of Operation: UK (Excluding Ireland)
Tel: 01227 751130 **Fax:** 01227 750033
Email: showroom@rootskitchens.co.uk
Web: www.rootskitchens.co.uk
Product Type: 1, 2, 3, 5

ROUNDEL DESIGN (UK) LTD
Flishinghurst Orchards, Chalk Lane,
Cranbrook, Kent, TN17 2QA
Area of Operation: South East England
Tel: 01580 712666
Email: homebuild@roundeldesign.co.uk
Web: www.roundeldesign.co.uk
Product Type: 1, 2, 3, 5, 7, 8, 9

SANDERSON'S FINE FURNITURE
Unit 5 & 6, The Village Workshop, Four Crosses
Business Park, Four Crosses, Powys, SY22 6ST
Area of Operation: UK (Excluding Ireland)
Tel: 01691 830075 **Fax:** 01691 830075
Email: sales@sandersonsfinefurniture.co.uk
Web: www.sandersonsfinefurniture.co.uk

SECOND NATURE
PO Box 20, Station Road,
Newton Aycliffe, Durham, DL5 6XJ
Area of Operation: UK & Ireland
Tel: 01325 505539
Email: mail@sncollection.co.uk
Web: www.sncollection.co.uk

SECOND NATURE

Area of Operation: UK & Ireland
Tel: 01325 505539
Email: mail@sncollection.co.uk
Web: www.sncollection.co.uk

The cool and contemporary Avant Ivory is just
one option from the extensive range of modern
kitchens from the Second Nature Collection.

SHUTTER FRONTIER LTD
2 Rosemary Farmhouse, Rosemary Lane,
Flimwell, Wadhurst, East Sussex, TN5 7PT
Area of Operation: Worldwide
Tel: 01580 878137 **Fax:** 01580 878137
Email: jane@shutterfrontier.co.uk
Web: www.shutterfrontier.co.uk
Product Type: 7

SIEMATIC
Osprey House, Primett Road,
Stevenage, Hertfordshire, SG1 3EE
Area of Operation: UK & Ireland
Tel: 01438 369251
Fax: 01438 368920
Email: sales@siematic.co.uk
Web: www.siematic.co.uk

SMEG UK
3 Milton Park, Abingdon,
Oxfordshire, OX14 4RY
Area of Operation: UK & Ireland
Tel: 0870 990 9908 **Fax:** 01235 861120
Email: sales@smeguk.com
Web: www.smeguk.com

SPILLERS OF CHARD LTD.
The Aga Cooker Centre, Chard Business Park,
Chard, Somerset, TA20 1FA
Area of Operation: South West England
and South Wales
Tel: 01460 67878 **Fax:** 01460 65252
Email: info@cookercentre.com
Web: www.cookercentre.com
Product Type: 1, 2, 3, 5, 9, 10

SURFACE MAGIC
102 Tomswood Hill, Barkingside,
Ilford, Essex, IG6 2QJ
Area of Operation: South East England
Tel: 0800 027 1951 **Fax:** 0208 501 1230
Email: info@surfacemagic.co.uk
Web: www.surfacemagic.co.uk
Product Type: 1, 2, 3, 5, 6, 7, 8, 9, 10

TASKWORTHY
The Old Brickyard, Pontrilas,
Herefordshire, HR2 0DJ
Area of Operation: UK (Excluding Ireland)
Tel: 01981 242900 **Fax:** 01981 242901
Email: peter@taskworthy.co.uk
Web: www.taskworthy.co.uk
Product Type: 1, 3, 4, 5, 7, 8, 9, 10

THE DESIGN STUDIO
39 High Street, Reigate, Surrey, RH2 9AE
Area of Operation: South East England
Tel: 01737 248228
 Fax: 01737 224180
Email: enq@the-design-studio.co.uk
Web: www.the-design-studio.co.uk
Product Type: 1, 2, 3, 5, 9

THE DORSET CUPBOARD COMPANY
2 Mayo Farm, Cann, Shaftesbury, Dorset, SP7 0EF
Area of Operation: Greater London, South East
England, South West England and South Wales
Tel: 01747 855044 **Fax:** 01747 855045
Email: nchturner@aol.com
Product Type: 1, 3, 5, 7, 9 **Other Info:**

THE KITCHEN AND BATHROOM COLLECTION
Nelson House, Nelson Road, Salisbury,
Wiltshire, SP1 3LT
Area of Operation: South West England
and South Wales
Tel: 01722 334800
Fax: 01722 412252
Email: info@kbc.co.uk
Web: www.kbc.co.uk
Product Type: 2, 3, 4, 5, 6, 7, 9
Other Info:

THE MODERN ROOM LTD
54 St.Pauls Road, Luton,
Bedfordshire, LU1 3RX
Area of Operation: UK (Excluding Ireland)
Tel: 01582 612070
Fax: 01582 612070
Email: laura@themodernroom.com
Web: www.themodernroom.com
Product Type: 2, 5, 9
Other Info:

THE OLD PINE STORE
Coxons Yard, Off Union Street,
Ashbourne, Derbyshire, DE6 1FG
Area of Operation: Europe
Tel: 01335 344112
Fax: 01335 344112
Email: martin@old-pine.co.uk
Web: www.old-pine.co.uk
Product Type: 1, 3, 5
Other Info: ▽ ECO

THE ORIGINAL KITCHEN COMPANY
4 Main Street, Breaston,
Derby, Derbyshire, DE72 3DX
Area of Operation: UK & Ireland
Tel: 01332 873746
Fax: 01332 873731
Email: originalkitchens@aol.com

TOMPKINS LTD
High March Close, Long March Industrial Estate,
Daventry, Northamptonshire, NN11 4EZ
Area of Operation: UK (Excluding Ireland)
Tel: 01327 877187
Fax: 01327 310491
Email: info@tompkinswood.co.uk
Web: www.tompkinswood.co.uk
Product Type: 1, 2, 3, 4, 5, 6, 7, 8, 9, 10
Other Info:

**TOUCHSTONE - GRANITE,
MARBLE & SILESTONE WORKTOPS**
Touchstone House, 82 High Street,
Measham, Derbyshire, DE12 7JB
Area of Operation: UK (Excluding Ireland)
Tel: 0845 130 1862
Fax: 01530 274271
Email: sales@touchstone-uk.com
Web: www.touchstone-uk.com
Product Type: 1, 2, 3, 4, 6, 7, 9

TRADESTYLE CABIINETS LTD
Carmaben Road, Easter Quennslie Industrial Estate,
Glasgow, Lothian, G33 4UN
Area of Operation: UK (Excluding Ireland)
Tel: 0141 781 6800
Fax: 0141 781 6801
Email: info@tradestylecabinets.com
Web: www.tradestylecabinets.com

WATTS & WRIGHT (THE JOINERY SHOP)
Watts & Wright, PO Box 4251,
Walsall, West Midlands, WS5 3WY
Area of Operation: Worldwide
Tel: 01922 610800 / 0207 043 7619
Fax: 0870 762 6387
Email: sales@wattsandwright.com
Web: www.wattsandwright.com
Product Type: 1, 2, 3, 4, 5, 7, 8, 10

WICKES
Wickes Customer Services,
Wickes House, 120-138 Station Road,
Harrow, Middlesex,
Greater London, HA1 2QB
Area of Operation: UK (Excluding Ireland)
Tel: 0870 608 9001
Fax: 0208 863 6225
Web: www.wickes.co.uk

WILLIAM BALL
London Road, Grays,
Essex, RM20 4WB
Area of Operation: Worldwide
Tel: 01375 375151
Fax: 01375 379033
Email: marketing@wball.co.uk
Web: www.wball.co.uk

WILLIAM BALL
Area of Operation: Worldwide
Tel: 01375 375151
Fax: 01375 379033
Email: marketing@wball.co.uk
Web: www.wball.co.uk

The FIRA Gold award has been awarded for the
excellence of William Ball products. Over 100
carcase sizes in 12 finishes and over 40 door
designs. Honesty shown here 600mm base unit
with full height door £205 + VAT.

WILLIAM GARVEY
Leyhill, Payhembury, Honiton, Devon, EX14 3JG
Area of Operation: Worldwide
Tel: 01404 841430 **Fax:** 01404 841626
Email: webquery@williamgarvey.co.uk
Web: www.williamgarvey.co.uk
Product Type: 2 **Other Info:**

WINNING DESIGNS
Dyke Farm, West Chiltington Road,
Pulborough, West Sussex, RH20 2EE
Area of Operation: UK & Ireland
Tel: 0870 754 4446 **Fax:** 0700 032 9946
Email: info@winning-designs-uk.com
Web: www.winning-designs-uk.com
Product Type: 1, 2, 3, 4, 5, 6, 7, 8, 9, 10
Other Info: ▽ ECO

WINNING DESIGNS
Area of Operation: UK & Ireland
Tel: 08707 544446 **Fax:** 07000 329946
Email: info@winning-designs-uk.com
Web: www.winning-designs-uk.com
Product Type: 1, 2, 3, 4, 5, 6, 7, 8, 9, 10
Other Info: ▽ ECO

Top quality bespoke furniture designed and built
by highly experienced designers and craftsmen.
Free design and advice, meetings by appointment
only.

WOODCHESTER KITCHENS & INTERIORS
Unit 18a Chalford Industrial Estate,
Chalford, Gloucestershire, GL6 8NT
Area of Operation: UK (Excluding Ireland)
Tel: 01453 886411
Fax: 01453 886411
Email: enquires@woodchesterkitchens.co.uk
Web: www.woodchesterkitchens.co.uk
Product Type: 1, 2, 3, 4, 5, 6, 7, 8, 9, 10

WOODEN HEART WAREHOUSE
Laburnum Road, Chertsey,
Surrey, KT16 8BY
Area of Operation: Greater London,
South East England
Tel: 01932 568684
Fax: 01932 568685
Email: whw@btclick.com
Product Type: 1, 3, 5, 6, 7, 9

WOODSTOCK FURNITURE
17 Park Drive, East Sheen,
London, SW14 8RB
Area of Operation: Worldwide
Tel: 0208 876 0131
Fax: 0207 245 9981
Email: ask@woodstockfurniture.co.uk
Web: www.woodstockfurniture.co.uk
Product Type: 1, 2, 3, 5, 6, 7, 8, 9, 10
Other Info:

WYCLIFFE OF WARWICKSHIRE
14-24 North Street, Upper Stoke, Coventry,
West Midlands, CV2 3FW
Area of Operation: UK (Excluding Ireland)
Tel: 02476 635151

ZIMMER KUCHEN
Jenna House, North Crawley Road,
Newport Pagnell, Buckinghamshire, MK16 9QA
Area of Operation: UK & Ireland
Tel: 01908 219797
Fax: 0800 585531
Email: info@zimmerkuchen.co.uk
Web: www.zimmerkuchen.co.uk
 Product Type: 1, 3, 5, 9

ZIMMER KÜCHEN
Area of Operation: UK & Ireland
Tel: 01908 219797
Fax: 0800 585531
Email: info@zimmerkuchen.co.uk
Web: www.zimmerkuchen.co.uk

Zimmer Küchen bespoke kitchens are designed to
your individual requirements. There is a Zimmer
look whatever your homestyle and a Zimmer
storage plan whatever your lifestyle.

KITCHEN FURNITURE, UNFITTED

KEY
PRODUCT TYPES: 1= Shaker
2 = Contemporary 3 = Farmhouse
4 = Industrial 5 = Long Island
6 = New England 7 = Period
8 = Minimalist 9 = Other
OTHER: ▽ Reclaimed On-line shopping
 Bespoke Hand-made ECO Ecological

ALLMILMO UK
Unit 17 Plantaganet House,
Kingsclere Business Park, Kingsclere,
Newbury, Berkshire, RG20 4SW
Area of Operation: UK & Ireland
Tel: 01635 868181
Fax: 01635 869693
Email: allmilmo@aol.com
Web: www.allmilmo.com
Product Type: 1, 2, 5, 9

ASHFORD & BROOKS
The Old Workshop, Ashtree Barn,
Caters Road, Bredfield, Woodbridge,
Suffolk, IP13 6BE
Area of Operation: UK (Excluding Ireland)
Tel: 01473 737764
Fax: 01473 277176
Email: ashfordbrooks@mailbox.co.uk
Web: www.ashfordandbrooks.co.uk
Product Type: 1, 2, 3, 5, 6, 7, 8, 9, 10

B ROURKE & CO LTD
Vulcan Works, Accrington Road,
Burnley, Lancashire, BB11 5QD
Area of Operation: Worldwide
Tel: 01282 422841
Fax: 01282 458901
Email: info@rourkes.co.uk
Web: www.rourkes.co.uk
Product Type: 10

BATH KITCHEN COMPANY
22 Hensley Road, Bloomfield, Bath,
Somerset, BA2 2DR
Area of Operation: South West England
and South Wales
Tel: 01225 312003
Fax: 01225 312003
Email: david@bathkitchencompany.co.uk
Web: www.bathkitchencompany.co.uk
Product Type: 1, 2, 3, 4, 5, 6, 7, 8, 9
Other Info:

INTERIORS, FIXTURES & FINISHES

BORDERCRAFT
Old Forge, Peterchurch, Hereford,
Herefordshire, HR2 0SD
Area of Operation: UK (Excluding Ireland)
Tel: 01981 550251
Fax: 01981 550552
Email: sales@bordercraft.co.uk
Web: www.bordercraft.co.uk
Other Info: ✐ ✋

BUYDESIGN
c/o Woodschool Ltd, Monteviot Nurseries,
Ancrum, Jedburgh, Borders, TD8 6TU
Area of Operation: North East England,
North West England and North Wales, Scotland
Tel: 01835 830740
Fax: 01835 830750
Email: info@buydesign-furniture.com
Web: www.buydesign-furniture.com
Product Type: 2, 10

CAMERON PETERS LTD
The Old Dairy, Home Farm, Ardington,
Wantage, Oxfordshire, OX12 8PD
Area of Operation: Worldwide
Tel: 01235 835000
Fax: 01235 835005
Email: info@cameronpeters.co.uk
Web: www.cameronpeters.co.uk

CAPLE
Fourth Way, Avonmouth, Bristol, BS11 8DW
Area of Operation: UK (Excluding Ireland)
Tel: 0870 606 9606
Fax: 0117 938 7449
Email: amandalowe@mlay.co.uk
Web: www.mlay.co.uk
Product Type: 2

CHANTRY HOUSE OAK LTD
Ham Farm, Church Lane, Oving,
Chichester, West Sussex, PO20 2BT
Area of Operation: South East England
Tel: 01243 776811 **Fax:** 01243 839873
Email: sales@chantryhouseoak.co.uk
Web: www.chantryhouseoak.co.uk
Product Type: 1, 3, 5, 7, 9

**CHRISTOPHER PETERS
ORIGINAL UNFITTED KITCHENS**
28, West Street, Warwick, Warwickshire, CV34 6AN
Area of Operation: Europe
Tel: 01926 494106
Fax: 02476 303300
Email: enquiries@christopherpetersantiques.co.uk
Web: www.christopherpetersantiques.co.uk
Product Type: 1, 3, 7, 8, 10
Other Info: ✐

CHURCHWOOD DESIGN
Unit 2, Tideswell Business Park,
Meveril Road, Tideswell, Buxton,
Derbyshire, SK17 8NY
Area of Operation: UK (Excluding Ireland)
Tel: 01298 872422
Fax: 01298 873068
Email: info@churchwood.co.uk
Web: www.churchwood.co.uk
Product Type: 1, 3, 5
Other Info: ▽ ECO ✐ ✋

CITY BATHROOMS & KITCHENS
158 Longford Road, Longford,
Coventry, West Midlands, CV6 6DR
Area of Operation: UK & Ireland
Tel: 02476 365877
Fax: 02476 644992
Email: citybathrooms@hotmail.com
Web: www.citybathrooms.co.uk

CONNAUGHT KITCHENS
2 Porchester Place, London, W2 2BS
Area of Operation: Greater London
Tel: 0207 706 2210
Fax: 0207 706 2209
Email: design@connaughtkitchens.co.uk
Web: www.connaughtkitchens.co.uk
Product Type: 2

COTTAGE FARM ANTIQUES
Stratford Road, Aston Sub Edge,
Chipping Campden, Gloucestershire, GL55 6PZ
Area of Operation: Worldwide
Tel: 01386 438263
Fax: 01386 438263
Email: info@cottagefarmantiques.co.uk
Web: www.cottagefarmantiques.co.uk ✋
Product Type: 3, 8, 10

COUNTRY KITCHENS
The Old Farm House, Birmingham Road,
Blackminster, Evesham, Worcestershire, WR11 7TD
Area of Operation: UK (Excluding Ireland)
Tel: 01386 831705
Fax: 01386 834051
Email: steve@floridafunbreaks.com
Web: www.handcraftedkitchens.com
Product Type: 1, 3, 8

CRABTREE KITCHENS
17 Station Road, Barnes, Greater London, SW13 0LF
Area of Operation: Greater London,
South West England and South Wales
Tel: 0208 392 6955 **Fax:** 0208 392 6944
Email: design@crabtreekitchens.co.uk
Web: www.crabtreekitchens.co.uk
Product Type: 1, 2, 4, 5, 7, 8, 9, 10

CYMRU KITCHENS
63 Caerleon Road, St. Julians, Newport, NP19 7BX
Area of Operation: UK (Excluding Ireland)
Tel: 01633 676767 **Fax:** 01633 212512
Email: sales@cymrukitchens.com
Web: www.cymrukitchens.com **Other Info:** ✐ ✋

D.A FURNITURE LTD
Woodview Workshop, Pitway Lane,
Farrington Gurney, Bristol, BS39 6TX
Area of Operation: South West England & South Wales
Tel: 01761 453117
Email: enquiries@dafurniture.co.uk
Web: www.dafurniture.co.uk
Product Type: 1, 2, 3, 5 **Other Info:** ✐ ✋

DESIGNER KITCHENS
37 High Street, Hertfordshire, EN6 5AJ
Area of Operation: Greater London
Tel: 01707 650565 **Fax:** 01707 663050
Email: info@designer-kitchens.co.uk
Web: www.designer-kitchens.co.uk
Product Type: 1, 2, 3, 5, 6, 7, 8, 9, 10

DISTINCTIVE COUNTRY FURNITURE LIMITED
Parrett Works, Martock, Somerset, TA12 6AE
Area of Operation: Worldwide
Tel: 01935 825800 **Fax:** 01935 825800
Email: Dcfltdinteriors@aol.com
Web: www.distinctivecountryfurniture.co.uk
Other Info: ▽ ✐ ✋   **Material Type:** A) 2, 3, 4, 7, 14

DISTINCTIVE COUNTRY FURNITURE

Area of Operation: Worldwide
Tel: 01935 825800
Fax: 01935 825800
Email: dcfltdinteriors@aol.com
Web: www.distinctivecountryfurniture.co.uk

Makers of kitchens and 16th & 17th Century
furniture and interiors. Staircases, panelling, oak
flooring, doors and furniture. Please call for a free
brochure. Quotations available upon request.

DORLUXE LIMITED
30 Pinbush Road, Lowestoft, Suffolk, NR33 7NL
Area of Operation: UK & Ireland
Tel: 01502 567744 **Fax:** 01502 567743
Email: info@dorluxe.co.uk
Web: www.dorluxe.co.uk
Product Type: 2, 3, 5, 8

DUKE CHRISTIE
Hillockhead Farmhouse, Dallas,
Forres, Moray, IV36 2SD
Area of Operation: Worldwide
Tel: 01343 890347 **Fax:** 01343 890347
Email: enquiries@dukechristie.com
Web: www.dukechristie.com
Product Type: 2, 3, 8
Other Info: ✐ ✋

ELITE TRADE KITCHENS LTD
90 Willesden Lane, Kilburn,
London, NW6 7TA
Area of Operation: UK (Excluding Ireland)
Tel: 0207 328 1234 **Fax:** 0207 328 1243
Email: sales@elitekitchens.co.uk
Web: www.elitekitchens.co.uk

FLAWFORD TRADE KITCHENS
Sleaford Road, Coddington, Newark,
Nottinghamshire, NG24 2QY
Area of Operation: UK & Ireland
Tel: 01636 626362 **Fax:** 01636 627971
Email: sales@mailorderkitchens.co.uk
Web: www.mailorderkitchens.co.uk ✋
Product Type: 1, 2, 3, 4, 5, 6, 7, 8, 9, 10

FURNITURE BY JONATHAN ELWELL
Bryn Teg Workshop & Tan y Bryn Cottage,
Tanrallt Road, Gwespyr, Flintshire, CH8 9JT
Area of Operation: UK (Excluding Ireland)
Tel: 01745 887766
Email: jonathanelwell@ukonline.co.uk
Web: www.solidwoodfurniture.co.uk

G MIDDLETON LTD
Cross Croft Industrial Estate,
Appleby, Cumbria, CA16 6HX
Area of Operation: Europe
Tel: 01768 352067 **Fax:** 01768 353228
Email: g.middleton@timber75.freeserve.co.uk
Web: www.graham-middleton.co.uk ✋
Product Type: 1, 3, 8

GK DESIGN
Greensbury Farm, Thurleigh Road,
Bolnhurst, Bedfordshire, MK44 2ET
Area of Operation: Europe
Tel: 01234 376990
Fax: 01234 376991
Email: studio@gkdesign.co.uk
Web: www.gkdesign.co.uk
Product Type: 10
Other Info: ✐

HARVEY JONES
57 New Kings Road, Fulham, London, SW6 4SE
Area of Operation: UK (Excluding Ireland)
Tel: 0800 917 2340
Fax: 0207 371 0735
Web: www.harveyjones.com

HATT KITCHENS
Hartlebury Trading Estate, Hartlebury,
Worcestershire, DY10 4JB
Area of Operation: UK (Excluding Ireland)
Tel: 01299 251320
Fax: 01299 251579
Email: design@hatt.co.uk
Web: www.hatt.co.uk
Product Type: 1, 2, 3, 4, 5, 6, 7, 8, 9, 10

HYGROVE
152/154 Merton Road, Wimbledon,
London, SW19 1EH
Area of Operation: South East England
Tel: 0208 543 1200
Fax: 0208 543 6521
Email: sales@hygrove.fsnet.co.uk
Web: www.hygrovefurniture.co.uk

IMPORTANT ROOMS LTD
62 High Street, Wargrave, Berkshire,
Reading, Berkshire, RG10 8BY
Area of Operation: Greater London, South East
England, South West England and South Wales
Tel: 0118 940 1266 **Fax:** 0118 940 1667
Email: steve@importantrooms.co.uk
Web: www.importantrooms.co.uk
Product Type: 1, 3, 8

INNOVATION GLASS COMPANY
27-28 Blake Industrial Estate, Brue Avenue,
Bridgwater, Somerset, TA7 8EQ
Area of Operation: Europe
Tel: 01278 683645 **Fax:** 01278 450088
Email: magnus@igc-uk.com
Web: www.igc-uk.com **Product Type:** 2, 5, 9

INTERESTING INTERIORS LTD
37-39 Queen Street, Aberystwyth,
Ceredigion, SY23 1PU
Area of Operation: UK (Excluding Ireland)
Tel: 01970 626162 **Fax:** 0870 051 7959
Email: enquiries@interestinginteriors.com
Web: www.interiors-ltd.demon.co.uk
Product Type: 1, 2, 3, 4, 5, 6, 7, 8, 9, 10

J&J ORMEROD PLC
Colonial House, Bacup, Lancashire, OL13 0EA
Area of Operation: UK & Ireland
Tel: 01706 877877 **Fax:** 01706 879827
Email: npeters@jjoplc.com
Web: www.jjoplc.com

JAIC LTD
Pattern House, Southwell Business Park,
Portland, Dorset, DT5 2NR
Area of Operation: UK (Excluding Ireland)
Tel: 01305 826991 **Fax:** 01305 823535
Email: info@jaic.co.uk
Product Type: 1, 3, 5

JOHN LADBURY & CO
Unit 11, Alpha Business Park, Travellers Close,
Welham Green, Hertfordshire, AL9 7NT
Area of Operation: Greater London
Tel: 01707 262966 **Fax:** 01707 265400
Email: johnladburyandco@tiscali.co.uk
Web: www.johnladbury.co.uk **Other Info:** ✐ ✋

JOHN LEWIS OF HUNGERFORD
Grove Technology Park,
Downsview Road, Wantage,
Oxfordshire, OX12 9FA
Area of Operation: UK (Excluding Ireland)
Tel: 01235 774300
Fax: 01235 769031
Email: park.street@john-lewis.co.uk
Web: www.john-lewis.co.uk
Product Type: 1, 3, 8

JOHN STRAND (MK) LTD
12-22, Herga Road, Wealdstone,
Harrow, Middlesex, HA3 5AS
Area of Operation: UK & Ireland
Tel: 0208 930 6006
Fax: 0208 930 6008
Email: enquiry@johnstrand-mk.co.uk
Web: www.johnstrand-mk.co.uk
Product Type: 5, 10

JOHNNY GREY
Fyning Copse, Fyning Lane,
Rogate, Petersfield,
Hampshire, GU31 5DH
Area of Operation: Worldwide
Tel: 01730 821424
Email: miles@johnnygrey.co.uk
Web: www.johnnygrey.co.uk
Product Type: 2, 5, 10

KENSINGTON STUDIO
13 c & d Kensington Road, Earlsdon,
Coventry, West Midlands, CV5 6GG
Area of Operation: UK (Excluding Ireland)
Tel: 02476 713326 **Fax:** 02476 713136
Email: sales@kensingtonstudio.com
Web: www.kensingtonstudio.com ✋
Product Type: 1, 2, 3, 4, 5, 6, 7, 8, 9, 10

INTERIORS, FIXTURES & FINISHES

KITCHEN DOOR SHOP
123 Oldham Road, Middleton,
Greater Manchester, M24 1AU
Area of Operation: UK (Excluding Ireland)
Tel: 0845 634 6444 **Fax:** 0870 300 2003
Email: kitchendoors@clara.net
Web: www.kitchendoorsltd.com

KITCHENS BY JULIAN ENGLISH
Ideal Home, Blackbird Road,
Leicester, Leicestershire, LE4 0AM
Area of Operation: Midlands & Mid Wales
Tel: 0116 251 9999
Fax: 0116 251 9999
Email: philiphalifax@aol.com
Web: www.kitchens-by-julianenglish.co.uk

MASTER LTD
30 Station Road, Heacham, Norfolk, PE31 7EX
Area of Operation: Europe
Tel: 01485 572032 **Fax:** 01485 571675
Email: ron@masterlimited.co.uk
Web: www.masterlimited.co.uk
Product Type: 1, 3, 4

MEREWAY CONTRACTS
413 Warwick Road, Birmingham,
West Midlands, B11 2JR
Area of Operation: UK (Excluding Ireland)
Tel: 0121 707 3288
Fax: 0121 707 3871
Email: info@merewaycontracts.co.uk
Web: www.merewaycontracts.co.uk

MOBALPA
4 Cornflower Close, Willand, Devon, EX15 2TT
Area of Operation: Worldwide
Tel: 07740 633672
Fax: 01884 820828
Email: ploftus@mobalpa.com
Web: www.mobalpa.com
Product Type: 1, 2, 3, 4, 5, 6, 7, 9, 10

MONTANA KITCHENS LTD
BIC 1, Studio 2/3 Innova Business Park,
Mollison Avenue, Enfield, Middlesex, EN3 7XU
Area of Operation: Greater London,
South East England
Tel: 0800 58 75 628
Email: angie@montanakitchens.co.uk
Web: www.montanakitchens.co.uk
Product Type: 1, 2, 3, 4, 5, 6, 7, 9, 10

NICHOLAS ANTHONY LTD
42-44 Wigmore Street, London, W1U 2RX
Area of Operation: East England,
Greater London, South East England
Tel: 0800 068 3603 **Fax:** 01206 762698
Email: info@nicholas-anthony.co.uk
Web: www.nicholas-anthony.co.uk
Product Type: 2, 5

NORTHERN TIMBER PRODUCTS LTD
Unit 20/1, West Bowling Green Street,
Edinburgh, EH6 5PE
Area of Operation: Scotland
Tel: 0131 554 8787
Fax: 0131 554 9191
Email: sales@ntp-kitchens.co.uk
Web: www.ntp-kitchens.co.uk
Product Type: 1, 2, 3, 5, 6, 7, 8, 9, 10

NOW GROUP PLC (CONTRACTS DIVISION)
Red Scar Business Park, Longridge Road,
Preston, Lancashire, PR2 5NA
Area of Operation: UK & Ireland
Tel: 01772 703838
Fax: 01772 705788
Email: sales@nowkitchens.co.uk
Web: www.nowkitchens.co.uk
Product Type: 1, 2, 3, 5, 7, 9

OAKWOOD BESPOKE
80 High Street, Camberley, Surrey, GU46 7RN
Area of Operation: Greater London, South East
England, South West England and South Wales
Tel: 01276 708630
Email: enquiries@oakwoodbespokecamberley.co.uk
Web: www.oakwoodbespokecamberley.co.uk

PANELS PLUS
22 Mill Place, Kingston Upon Thames,
Surrey, KT1 2RJ
Area of Operation: Greater London
Tel: 0208 399 6343
Fax: 0208 399 6343
Email: sales@panelsplusltd.com
Web: www.panelsplusltd.com
Product Type: 1, 2, 3, 5, 6, 10

PARAGON JOINERY
Systems House, Eastbourne Road,
Blindley Heath, Surrey, RH7 6JP
Area of Operation: UK (Excluding Ireland)
Tel: 01342 836300 **Fax:** 01342 836364
Email: sales@paragonjoinery.com
Web: www.paragonjoinery.com

PARLOUR FARM
Unit 12b, Wilkinson Road, Love Lane Industrial
Estate, Cirencester, Gloucestershire, GL7 1YT
Area of Operation: Europe
Tel: 01285 885336 **Fax:** 01285 643189
Email: info@parlourfarm.com
Web: www.parlourfarm.com

PLAIN ENGLISH
Stowupland Hall, Stowupland, Suffolk, IP14 4BE
Area of Operation: Worldwide
Tel: 01449 774028 **Fax:** 01449 613519
Email: info@plainenglishdesign.co.uk
Web: www.plainenglishdesign.co.uk
Product Type: 1, 2, 8

PINELAND FURNITURE LTD
Unit 5 Cleobury Trading Estate,
Cleobury Mortimer, Shropshire, DY14 8DP
Area of Operation: UK (Excluding Ireland)
Tel: 01299 271143 **Fax:** 01299 271166
Email: pineland@onetel.com
Web: www.pineland.co.uk

PINELAND
Area of Operation: UK (Excluding Ireland)
Tel: 01299 271143
Fax: 01299 271166
Email: pineland@onetel.com
Web: www.pineland.co.uk

Manufacturing a full range of traditionally styled
pine furniture using the finest quality kiln dried
timbers (no plywood or chipboard) and employing
proper mortice and tension and dovetail joining
wherever possible.

POSH PANTRIES
81 Kelvin Road North,
Lenziemill, Cumbernauld,
Lanarkshire, G67 6BD
Area of Operation: UK (Excluding Ireland)
Tel: 01236 453556
Fax: 01236 728068
Email: office@poshpantries.co.uk
Web: www.poshpantries.co.uk

PRICE KITCHENS
11 Imperial Way, Croydon, Surrey, CR0 4RR
Area of Operation: South East England
Tel: 0208 686 9006
Fax: 0208 686 5958
Email: info@pricekitchens.co.uk
Web: www.pricekitchens.co.uk

RICHARD BAKER FURNITURE LTD
Wimbledon Studios, 257 Burlington Road,
New Malden, Surrey, KT3 4NE
Area of Operation: Europe
Tel: 0208 336 1777
Fax: 0208 336 1666
Email: sales@richardbakerfurniture.co.uk
Web: www.richardbakerfurniture.co.uk
Product Type: 1, 2, 5, 8

ROBERT J TURNER & CO
Roe Green, Sandon, Buntingford,
Hertfordshire, SG9 0QE
Area of Operation: East England,
Greater London
Tel: 01763 288371
Fax: 01763 288440
Email: sales@robertjturner.co.uk
Web: www.robertjturner.co.uk
Product Type: 1, 2, 3, 4, 5, 6, 7, 8, 9, 10

ROBINSON AND CORNISH LTD
St George's House, St George's Road,
Barnstaple, Devon, EX32 7AS
Area of Operation: Europe
Tel: 01271 329300
Fax: 01271 328277
Email: richard.martin@robinsonandcornish.co.uk
Web: www.robinsonandcornish.co.uk
Product Type: 2, 3, 8, 10
Other Info: ✎

ROOTS KITCHENS,
BEDROOMS & BATHROOMS
Vine Farm, Stockers Hill,
Boughton-under-Blean,
Faversham, Kent, ME13 9AB
Area of Operation: UK (Excluding Ireland)
Tel: 01227 751130
Fax: 01227 750033
Email: showroom@rootskitchens.co.uk
Web: www.rootskitchens.co.uk ✎

ROUNDEL DESIGN (UK) LTD
Flishinghurst Orchards, Chalk Lane,
Cranbrook, Kent, TN17 2QA
Area of Operation: South East England
Tel: 01580 712666
Email: homebuild@roundeldesign.co.uk
Web: www.roundeldesign.co.uk
Product Type: 1, 2, 3, 5, 7, 8, 9

SAMUEL SCOTT PARTNERSHIP
Surgery House, The Square, Skillington,
Lincolnshire, NG33 5HB
Area of Operation: Europe
Tel: 01476 861806
Fax: 01476 861806
Email: samuel_scott@tiscali.co.uk
Product Type: 1, 2, 3, 5, 6, 7, 8
Other Info: ▽ ECO ✎ ✋

SANDERSON'S FINE FURNITURE
Unit 5 & 6, The Village Workshop,
Four Crosses Business Park, Four Crosses,
Powys, SY22 6ST
Area of Operation: UK (Excluding Ireland)
Tel: 01691 830075 **Fax:** 01691 830075
Email: sales@sandersonsfinefurniture.co.uk
Web: www.sandersonsfinefurniture.co.uk

SECOND NATURE
PO Box 20, Station Road, Newton Aycliffe,
Durham, DL5 6XJ
Area of Operation: UK & Ireland
Tel: 01325 505539
Email: mail@sncollection.co.uk
Web: www.sncollection.co.uk

SECOND NATURE

Area of Operation: UK & Ireland
Tel: 01325 505539
Email: mail@sncollection.co.uk
Web: www.sncollection.co.uk

Second Nature offer an extensive choice of smart
storage solutions including magic corners,
pull-out larders and the unique Le Mans corner
system.

SECOND NATURE

Area of Operation: UK & Ireland
Tel: 01325 505539
Email: mail@sncollection.co.uk
Web: www.sncollection.co.uk

An extensive range of classically styled kitchens
is available from Second Nature. The Bede range
is featured here with solid granite worktops.

SIEMATIC
Osprey House, Primett Road,
Stevenage, Hertfordshire, SG1 3EE
Area of Operation: UK & Ireland
Tel: 01438 369251 **Fax:** 01438 368920
Email: sales@siematic.co.uk
Web: www.siematic.co.uk

SMEG UK
3 Milton Park, Abingdon, Oxfordshire, OX14 4RY
Area of Operation: UK & Ireland
Tel: 0870 990 9908 **Fax:** 01235 861120
Email: sales@smeguk.com
Web: www.smeguk.com

SPILLERS OF CHARD LTD.
The Aga Cooker Centre, Chard Business Park,
Chard, Somerset, TA20 1FA
Area of Operation: South West England
and South Wales
Tel: 01460 67878 **Fax:** 01460 65252
Email: info@cookercentre.com
Web: www.cookercentre.com
Product Type: 1, 2, 3, 5, 10

SURFACE MAGIC
102 Tomswood Hill, Barkingside,
Ilford, Essex, IG6 2QJ
Area of Operation: South East England
Tel: 0800 027 1951
Fax: 0208 501 1230
Email: info@surfacemagic.co.uk
Web: www.surfacemagic.co.uk
Product Type: 1, 2, 3, 5, 6, 7, 8, 9, 10

TASKWORTHY
The Old Brickyard, Pontrilas,
Herefordshire, HR2 0DJ
Area of Operation: UK (Excluding Ireland)
Tel: 01981 242900
Fax: 01981 242901
Email: peter@taskworthy.co.uk
Web: www.taskworthy.co.uk
Product Type: 1, 2, 3, 4, 5, 7, 8, 9, 10

THE DESIGN STUDIO
39 High Street, Reigate, Surrey, RH2 9AE
Area of Operation: South East England
Tel: 01737 248228
Fax: 01737 224180
Email: enq@the-design-studio.co.uk
Web: www.the-design-studio.co.uk
Product Type: 1, 2, 3, 5, 9

THE DORSET CUPBOARD COMPANY
2 Mayo Farm, Cann,
Shaftesbury, Dorset, SP7 0EF
Area of Operation: Greater London,
South East England, South West England
and South Wales
Tel: 01747 855044
Fax: 01747 855045
Email: nchturner@aol.com
Product Type: 1, 3, 5, 7, 9

**THE KITCHEN AND
BATHROOM COLLECTION**
Nelson House, Nelson Road,
Salisbury, Wiltshire, SP1 3LT
Area of Operation: South West England
and South Wales
Tel: 01722 334800
Fax: 01722 412252
Email: info@kbc.co.uk
Web: www.kbc.co.uk
Product Type: 2, 3, 4, 5, 6, 7, 9
Other Info: ✏ ✋

THE MODERN ROOM LTD
54 St.Pauls Road, Luton,
Bedfordshire, LU1 3RX
Area of Operation: UK (Excluding Ireland)
Tel: 01582 612070
Fax: 01582 612070
Email: laura@themodernroom.com
Web: www.themodernroom.com
Product Type: 2, 5, 9
Other Info: ✏ ✋

THE OLD PINE STORE
Coxons Yard, Off Union Street,
Ashbourne, Derbyshire, DE6 1FG
Area of Operation: Europe
Tel: 01335 344112
Fax: 01335 344112
Email: martin@old-pine.co.uk
Web: www.old-pine.co.uk
Product Type: 3, 5

THE ORIGINAL KITCHEN COMPANY
4 Main Street, Breaston,
Derby, Derbyshire, DE72 3DX
Area of Operation: UK & Ireland
Tel: 01332 873746
Fax: 01332 873731
Email: originalkitchens@aol.com

TIM WOOD LIMITED
1 Burland Road, London, SW11 6SA
Area of Operation: Worldwide
Tel: 07041 380030
Fax: 0870 054 8645
Email: homeb@timwood.com
Web: www.timwood.com

TOMPKINS LTD
High March Close, Long March Industrial Estate,
Daventry, Northamptonshire, NN11 4EZ
Area of Operation: UK (Excluding Ireland)
Tel: 01327 877187 **Fax:** 01327 310491
Email: info@tompkinswood.co.uk
Web: www.tompkinswood.co.uk
Product Type: 1, 2, 3, 4, 5, 6, 7, 8, 9, 10
Other Info: ✏ ✋

TRADESTYLE CABIINETS LTD
Carmaben Road, Easter Quennslie Industrial Estate,
Glasgow, Lothian, G33 4UN
Area of Operation: UK (Excluding Ireland)
Tel: 0141 781 6800 **Fax:** 0141 781 6801
Email: info@tradestylecabinets.com
Web: www.tradestylecabinets.com

UNPAINTED KITCHENS LTD
258 Battersea Park Road, London, SW11 3BP
Area of Operation: UK (Excluding Ireland)
Tel: 0207 223 2017 **Fax:** 01604 722573
Email: unpaintedks@aol.com
Web: www.unpaintedkitchens.com
Product Type: 1

UNPAINTED KITCHENS
Area of Operation: UK (Excluding Ireland)
Tel: 020 7223 2017 **Fax:** 01604 722573
Email: unpaintedks@aol.com
Web: www.unpaintedkitchens.com
Product Type: 1

At Unpainted Kitchens, we believe in giving value for money.
Why pay astronomical prices for a painted kitchen when you
can create a truly unique look from our range, giving yourself
a high quality kitchen that will almost certainly beat a
pre-fabricated one costing thousands of pounds?
If you can draw it, we can build it. Please contact us for
advice on designing your own units.

WATTS & WRIGHT (THE JOINERY SHOP)
Watts & Wright, PO Box 4251,
Walsall, West Midlands, WS5 3WY
Area of Operation: Worldwide
Tel: 01922 610800 / 0207 043 7619
Fax: 0870 762 6387
Email: sales@wattsandwright.com
Web: www.wattsandwright.com
Product Type: 1, 2, 3, 4, 5, 7, 8, 10

WICKES
Wickes Customer Services, Wickes House,
120-138 Station Road, Harrow, Middlesex,
Greater London, HA1 2QB
Area of Operation: UK (Excluding Ireland)
Tel: 0870 608 9001 **Fax:** 0208 863 6225
Web: www.wickes.co.uk

WILLIAM GARVEY
Leyhill, Payhembury, Honiton, Devon, EX14 3JG
Area of Operation: Worldwide
Tel: 01404 841430 **Fax:** 01404 841626
Email: webquery@williamgarvey.co.uk
Web: www.williamgarvey.co.uk
Product Type: 2 **Other Info:** ✏ ✋

WINE CORNER LTD
Unit 6 Harworth Enterprise Park,
Brunel Industrial Estate, Harworth,
Doncaster, South Yorkshire, DN11 8SG
Area of Operation: Worldwide
Tel: 01302 757047 **Fax:** 01302 751233
Email: elaine@winecorner.co.uk
Web: www.winecorner.co.uk ✏🛒

WINNING DESIGNS
Dyke Farm, West Chiltington Road,
Pulborough, West Sussex, RH20 2EE
Area of Operation: UK & Ireland
Tel: 0870 754 4446
Fax: 0700 032 9946
Email: info@winning-designs-uk.com
Web: www.winning-designs-uk.com
Product Type: 1, 2, 3, 4, 5, 6, 7, 8, 9, 10
Other Info: ▽ ECO ✏ ✋

WINNING DESIGNS
Area of Operation: UK & Ireland
Tel: 08707 544446 **Fax:** 07000 329946
Email: info@winning-designs-uk.com
Web: www.winning-designs-uk.com
Product Type: 1, 2, 3, 4, 5, 6, 7, 8, 9, 10
Other Info: ▽ ECO ✏ ✋

Top quality bespoke furniture designed and built
by highly experienced designers and craftsmen.
Free design and advice, meetings by appointment
only.

WOODCHESTER KITCHENS & INTERIORS
Unit 18a Chalford Industrial Estate,
Chalford, Gloucestershire, GL6 8NT
Area of Operation: UK (Excluding Ireland)
Tel: 01453 886411
Fax: 01453 886411
Email: enquires@woodchesterkitchens.co.uk
Web: www.woodchesterkitchens.co.uk
Product Type: 1, 2, 3, 4, 5, 6, 7, 8, 9, 10

WOODSTOCK FURNITURE
17 Park Drive, East Sheen, London, SW14 8RB
Area of Operation: Worldwide
Tel: 0208 876 0131
Fax: 0207 245 9981
Email: ask@woodstockfurniture.co.uk
Web: www.woodstockfurniture.co.uk
Product Type: 1, 2, 3, 4, 5, 6, 7, 8, 9, 10
Other Info: ✏ ✋

WORKTOPS

AFFORDABLE GRANITE LTD
Unit 5 John Lory Farm, Charlwood Place,
Norwood Hill Road, Charlwwod, Surrey, RH6 0EB
Area of Operation: UK (Excluding Ireland)
Tel: 0845 330 1692
Email: sales@affordablegranite.co.uk
Web: www.affordablegranite.co.uk
Material Type: E) 1

AFFORDABLE GRANITE LTD
Unit 5 Charlwood Place, Norwood Hill Road,
Surrey, RH6 0EB
Area of Operation: UK (Excluding Ireland)
Tel: 0845 330 1692
Fax: 0845 330 1260
Email: sales@affordablegranite.co.uk
Web: www.affordablegranite.co.uk
Material Type: E) 1

ALICANTE STONE
Damaso Navarro, 6 Bajo, 03610 Petrer
(Alicante), P.O. Box 372, Spain
Area of Operation: Europe
Tel: +34 966 31 96 97
Fax: +34 966 31 96 98
Email: info@alicantestone.com
Web: www.alicantestone.com
Material Type: E) 1, 2, 5

ALISTAIR MACKINTOSH LTD
Bannerley Road, Garretts Green,
Birmingham, West Midlands, B33 0SL
Area of Operation: UK & Ireland
Tel: 0121 784 6800
Fax: 0121 789 7068
Email: info@alistairmackintosh.co.uk
Web: www.alistairmackintosh.co.uk
Material Type: E) 1

ALL THINGS STONE LIMITED
Anchor House, 33 Ospringe Street,
Ospringe, Faversham, Kent, ME13 8TW
Area of Operation: UK & Ireland
Tel: 01795 531001
Fax: 01795 530742
Email: info@allthingsstone.co.uk
Web: www.allthingsstone.co.uk
Material Type: E) 1, 2, 3, 5

ALLMILMO UK
Unit 17 Plantaganet House, Kingsclere Business
Park, Kingsclere, Newbury, Berkshire, RG20 4SW
Area of Operation: UK & Ireland
Tel: 01635 868181
Fax: 01635 869693
Email: allmilmo@aol.com
Web: www.allmilmo.com
Material Type: E) 1

AMBER KITCHENS OF BISHOP'S STORTFORD
Clarklands House, Parsonage Lane,
Sawbridgeworth, Hertfordshire, CM21 0NG
Area of Operation: East England, Greater London
Tel: 01279 600030
Fax: 01279 721528
Email: info@amberkitchens.com
Web: www.amberkitchens.com
Material Type: A) 1, 2, 3, 4, 5, 6, 7, 8, 9, 10, 11, 12,
13, 14, 15

ASHFORD & BROOKS
The Old Workshop, Ashtree Barn, Caters Road,
Bredfield, Woodbridge, Suffolk, IP13 6BE
Area of Operation: UK (Excluding Ireland)
Tel: 01473 737764
Fax: 01473 277176
Email: ashfordbrooks@mailbox.co.uk
Web: www.ashfordandbrooks.co.uk
Material Type: A) 2, 3, 4, 5, 6, 7, 8, 9, 10, 11, 12, 14

ASHLAR MASON
Manor Building, Manor Road, London, W13 0JB
Area of Operation: East England, Greater London,
Midlands & Mid Wales, South East England, South
West England and South Wales
Tel: 0208 997 0002
Fax: 0208 997 0008
Email: enquiries@ashlarmason.co.uk
Web: www.ashlarmason.co.uk
Material Type: E) 1, 2, 3, 4, 5, 8, 9, 11, 12, 13

ASTRACAST PLC
Holden Ing Way, Birstall, West Yorkshire, WF17 9AE
Area of Operation: UK & Ireland
Tel: 01924 477466
Fax: 01924 351297
Email: brochures@astracast.co.uk
Web: www.astracast.co.uk 🛒

AXIOM BY FORMICA
Formica Ltd, Coast Road, North Shields,
Tyne & Wear, NE29 8RE
Area of Operation: UK & Ireland
Tel: 0191 259 3478
Fax: 0191 258 2719
Email: axiom.info@formica.co.uk
Web: www.axiomworktops.com
Material Type: A) 1, 2, 3, 4, 5, 7

AZTEC METAL TILES
Bowl Road, Charing, Kent, TN27 0HB
Area of Operation: UK & Ireland
Tel: 01233 712332 **Fax:** 01233 714994
Email: enquiries@aztecmetaltiles.co.uk
Web: www.aztecmetaltiles.co.uk
Material Type: C) 1, 2, 3, 4, 6, 7, 11, 12

BARNCREST HARDWOOD KITCHEN WORKTOPS
Unit 9, Tregoniggie Industrial Estate,
Falmouth, Cornwall, TR11 4SN
Area of Operation: UK & Ireland
Tel: 01326 375982
Email: enquiries@barncrest.co.uk
Web: www.barncrest.co.uk
Material Type: A) 2, 3, 4, 7

BATH KITCHEN COMPANY
22 Hensley Road, Bloomfield,
Bath, Somerset, BA2 2DR
Area of Operation: South West England
and South Wales
Tel: 01225 312003
Fax: 01225 312003
Email: david@bathkitchencompany.co.uk
Web: www.bathkitchencompany.co.uk
Material Type: A) 1, 2, 3, 4, 5, 6, 7, 8, 9, 10, 11, 12, 13, 14, 15

BAYFIELD STAIR CO
Unit 4, Praed Road, Trafford Park,
Manchester, M17 1PQ
Area of Operation: Worldwide
Tel: 0161 848 0700
Fax: 0161 872 2230
Email: sales@bayfieldstairs.co.uk
Web: www.bayfieldstairs.co.uk

BERWYN SLATE QUARRY LTD
The Horseshoe Pass, Llangollen,
Denbighshire, LL20 8DP
Area of Operation: UK (Excluding Ireland)
Tel: 01978 861897
Fax: 01978 869292
Web: www.berwynslate.com
Material Type: E) 3

BORDERCRAFT
Old Forge, Peterchurch, Hereford,
Herefordshire, HR2 0SD
Area of Operation: UK (Excluding Ireland)
Tel: 01981 550251
Fax: 01981 550552
Email: sales@bordercraft.co.uk
Web: www.bordercraft.co.uk
Other Info: ECO
Material Type: A) 2, 3, 4, 5, 6, 7

BUSHBOARD LTD
9-29 Rixon Road, Wellingborough,
Northamptonshire, NN8 4BA
Area of Operation: UK & Ireland
Tel: 01933 232200
Email: washrooms@bushboard.co.uk
Web: www.bushboard.co.uk

CAPLE
Fourth Way, Avonmouth, Bristol, BS11 8DW
Area of Operation: UK (Excluding Ireland)
Tel: 0870 606 9606 **Fax:** 0117 938 7449
Email: amandalowe@mlay.co.uk
Web: www.mlay.co.uk

**CHRISTOPHER PETERS
ORIGINAL UNFITTED KITCHENS**
28, West Street, Warwick, Warwickshire, CV34 6AN
Area of Operation: Europe
Tel: 01926 494106 **Fax:** 02476 303300
Email: enquiries@christopherpetersantiques.co.uk
Web: www.christopherpetersantiques.co.uk

CITY BATHROOMS & KITCHENS
158 Longford Road, Longford,
Coventry, West Midlands, CV6 6DR
Area of Operation: UK & Ireland
Tel: 02476 365877 **Fax:** 02476 644992
Email: citybathrooms@hotmail.com
Web: www.citybathrooms.co.uk

CONNAUGHT KITCHENS
2 Porchester Place, London, W2 2BS
Area of Operation: Greater London
Tel: 0207 706 2210 **Fax:** 0207 706 2209
Email: design@connaughtkitchens.co.uk
Web: www.connaughtkitchens.co.uk

COTTAGE FARM ANTIQUES
Stratford Road, Aston Sub Edge,
Chipping Campden, Gloucestershire, GL55 6PZ
Area of Operation: Worldwide
Tel: 01386 438263 **Fax:** 01386 438263
Email: info@cottagefarmantiques.co.uk
Web: www.cottagefarmantiques.co.uk
Material Type: A) 4, 5, 6

COUNTRY KITCHENS
The Old Farm House, Birmingham Road,
Blackminster, Evesham, Worcestershire, WR11 7TD
Area of Operation: UK (Excluding Ireland)
Tel: 01386 831705 **Fax:** 01386 834051
Email: steve@floridafunbreaks.com
Web: www.handcraftedkitchens.com
Material Type: A) 2, 4, 5, 6

COUNTY GRANITE AND MARBLE
Mill Lane, Creech Saint Michael,
Taunton, Somerset, TA3 5PX
Area of Operation: UK (Excluding Ireland)
Tel: 01823 444554 **Fax:** 01823 445013
Email: granites@netcomuk.co.uk
Web: www.countygranite.co.uk
Material Type: E) 1, 2, 3, 5

COUNTY HARDWOODS
Creech Mill, Mill Lane, Creech Saint Michael,
Taunton, Somerset, TA3 5PX
Area of Operation: UK & Ireland
Tel: 01823 444217 **Fax:** 01823 443940
Email: hardwood@netcomuk.co.uk
Web: www.countyhardwoods.co.uk,
www.countygranite.co.uk
Material Type: A) 2, 5

CRABTREE KITCHENS
17 Station Road, Barnes, Greater London, SW13 0LF
Area of Operation: Greater London, South West
England and South Wales
Tel: 0208 392 6955 **Fax:** 0208 392 6944
Email: design@crabtreekitchens.co.uk
Web: www.crabtreekitchens.co.uk
Other Info:

CRAFTSHIP WOODENTOPS LTD
The Barn, Park Farm, Hundred Acre Lane,
Wivlesfield Green, East Sussex, RH17 7RU
Area of Operation: UK (Excluding Ireland)
Tel: 01273 891891 **Fax:** 01273 890044
Email: sales@woodentops.co.uk
Web: www.woodentops.co.uk

WOODENTOPS
Area of Operation: UK (Excluding Ireland)
Tel: 01273 891891
Fax: 01273 890044
Email: sales@woodentops.co.uk
Web: www.woodentops.co.uk

We specialise in the manufacture and supply of hardwood worksurfaces for domestic and commercial use in kitchens, bathrooms and offices. Off-the shelf or fully bespoke solutions for all your natural worktop requirements.

CYMRU KITCHENS
63 Caerleon Road, St. Julians, Newport, NP19 7BX
Area of Operation: UK (Excluding Ireland)
Tel: 01633 676767 **Fax:** 01633 212512
Email: sales@cymrukitchens.com
Web: www.cymrukitchens.com

DESIGNER KITCHENS
37 High Street, Hertfordshire, EN6 5AJ
Area of Operation: Greater London
Tel: 01707 650565 **Fax:** 01707 663050
Email: info@designer-kitchens.co.uk
Web: www.designer-kitchens.co.uk

DEVON STONE LTD
38 Station Road, Budleigh Salterton, Devon, EX9 6RT
Area of Operation: UK (Excluding Ireland)
Tel: 01395 446841
Email: amy@devonstone.com
Other Info: **Material Type:** E) 1, 2, 3

DUPONT CORIAN & ZODIAQ
10 Quarry Court, Pitstone Green Business Park,
Pitstone, nr. Tring, Buckinghamshire, LU7 9GW
Area of Operation: UK & Ireland
Tel: 01296 663555 **Fax:** 01296 663599
Email: sales@corian.co.uk
Web: www.corian.co.uk

ECO IMPACT LTD
50a Kew Green, Richmond, TW9 3BB
Area of Operation: UK & Ireland
Tel: 0208 940 7072 **Fax:** 0208 332 1218
Email: sales@ecoimpact.co.uk
Web: www.ecoimpact.co.uk
Other Info: ECO

EGGER (UK) LIMITED
Anick Grange Road, Hexham,
Northumberland, NE46 4JS
Area of Operation: UK & Ireland
Tel: 01434 602191
Fax: 01434 605103
Email: building.uk@egger.com
Web: www.egger.co.uk
Material Type: H) 1, 2, 6

ELITE TRADE KITCHENS LTD
90 Willesden Lane,
Kilburn, London, NW6 7TA
Area of Operation: UK (Excluding Ireland)
Tel: 0207 328 1234
Fax: 0207 328 1243
Email: Sales@elitekitchens.co.uk
Web: www.elitekitchens.co.uk

FLAWFORD TRADE KITCHENS
Sleaford Road, Coddington, Newark,
Nottinghamshire, NG24 2QY
Area of Operation: UK & Ireland
Tel: 01636 626362
Fax: 01636 627971
Email: sales@mailorderkitchens.co.uk
Web: www.mailorderkitchens.co.uk
Material Type: A) 1, 2, 3, 4, 5, 6, 7, 8, 9, 10, 11, 12, 13, 14, 15

FRANCIS N. LOWE LTD.
The Marble Works, New Road,
Middleton, Matlock,
Derbyshire, DE4 4NA
Area of Operation: Europe
Tel: 01629 822216
Fax: 01629 824348
Email: info@lowesmarble.com
Web: www.lowesmarble.com
Material Type: E) 1, 2, 3, 5, 8

G MIDDLETON LTD
Cross Croft Industrial Estate,
Appleby, Cumbria, CA16 6HX
Area of Operation: Europe
Tel: 01768 352067
Fax: 01768 353228
Email: g.middleton@timber75.freeserve.co.uk
Web: www.graham-middleton.co.uk

HATT KITCHENS
Hartlebury Trading Estate,
Hartlebury, Worcestershire, DY10 4JB
Area of Operation: UK (Excluding Ireland)
Tel: 01299 251320
Fax: 01299 251579
Email: design@hatt.co.uk
Web: www.hatt.co.uk

HITT OAK LTD
13a Northview Crescent,
London, NW10 1RD
Area of Operation: UK (Excluding Ireland)
Tel: 0208 450 3821
Fax: 0208 911 0299
Email: zhitas@yahoo.co.uk

HYGROVE
152/154 Merton Road, Wimbledon,
London, SW19 1EH
Area of Operation: South East England
Tel: 0208 543 1200
Fax: 0208 543 6521
Email: sales@hygrove.fsnet.co.uk
Web: www.hygrovefurniture.co.uk
Material Type: A) 1, 2, 3, 4, 5, 6, 7, 8, 9,
10, 11, 12, 14

IMPORTANT ROOMS LTD
62 High Street, Wargrave, Berkshire,
Reading, Berkshire, RG10 8BY
Area of Operation: Greater London, South East
England, South West England and South Wales
Tel: 0118 940 1266
Fax: 0118 940 1667
Email: Steve@importantrooms.co.uk
Web: www.importantrooms.co.uk
Material Type: A) 10

INNOVATION GLASS COMPANY
27-28 Blake Industrial Estate, Brue Avenue,
Bridgwater, Somerset, TA7 8EQ
Area of Operation: Europe
Tel: 01278 683645
 Fax: 01278 450088
Email: magnus@igc-uk.com
Web: www.igc-uk.com
Material Type: J) 4, 5, 6

IN-TOTO
Shaw Cross Court, Shaw Cross Business Park,
Dewsbury, West Yorkshire, WF12 7RF
Area of Operation: UK (Excluding Ireland)
Tel: 01924 487900 **Fax:** 01924 437305
Email: graham.russell@intoto.co.uk
Web: www.intoto.co.uk

ISAAC LORD LTD
West End Court, Suffield Road, High Wycombe,
Buckinghamshire, HP11 2JY
Area of Operation: East England, Greater London,
North East England, South East England, South West
England and South Wales
Tel: 01494 462121
Fax: 01494 461376
Email: info@isaaclord.co.uk
Web: www.isaaclord.co.uk

J & R MARBLE COMPANY LTD
Unit 9,Period Works, Lammas Road,
Leyton, London, E10 7QT
Area of Operation: UK (Excluding Ireland)
Tel: 0208 539 6471
Fax: 0208 539 9264
Email: sales@jrmarble.co.uk
Web: www.jrmarble.co.uk
Other Info:
Material Type: E) 1, 2, 3, 4, 5, 8, 9, 13, 14

JOHN LEWIS OF HUNGERFORD
Grove Technology Park, Downsview Road,
Wantage, Oxfordshire, OX12 9FA
Area of Operation: UK (Excluding Ireland)
Tel: 01235 774300
Fax: 01235 769031
Email: park.street@john-lewis.co.uk
Web: www.john-lewis.co.uk
Material Type: F) 2

JOHN STRAND (MK) LTD
12-22, Herga Road, Wealdstone,
Harrow, Middlesex, HA3 5AS
Area of Operation: UK & Ireland
Tel: 0208 930 6006 **Fax:** 0208 930 6008
Email: enquiry@johnstrand-mk.co.uk
Web: www.johnstrand-mk.co.uk

KENLEY KITCHENS
24-26 Godstone Road, Kenley, Surrey, CR8 5JE
Area of Operation: UK (Excluding Ireland)
Tel: 0208 668 7000

KENSINGTON STUDIO
13c/d Kensington Road, Earlsdon,
Coventry, West Midlands, CV5 6GG
Area of Operation: UK (Excluding Ireland)
Tel: 02476 713326 **Fax:** 02476 713136
Email: sales@kensingtonstudio.com
Web: www.kensingtonstudio.com

KIRK NATURAL STONE
Bridgend, Fyvie, Turriff, Aberdeenshire, AB53 8QB
Area of Operation: Worldwide
Tel: 01651 891891 **Fax:** 01651 891794
Email: info@kirknaturalstone.com
Web: www.kirknaturalstone.com

KIRKSTONE QUARRIES LIMITED
Skelwith Bridge, Ambleside, Cumbria, LA22 9NN
Area of Operation: UK & Ireland
Tel: 01539 433296 **Fax:** 01539 434006
Email: info@kirkstone.com
Web: www.kirkstone.com
Material Type: E) 1, 3, 5

KITCHEN SUPPLIES
East Chesters, North Way, Hillend Industrial Estate,
Dalgety Bay, Fife, KY11 9JA
Area of Operation: UK (Excluding Ireland)
Tel: 01383 824729
Email: sales@kitchensupplies.co.uk
Web: www.kitchensupplies.co.uk

KITCHEN WORKTOPS DIRECT
Kingdom House, MacDonald's Rest, 19 Gleniffer
Road, Paisley, Renfrewshire, PA2 8LP
Area of Operation: UK & Ireland
Tel: 0845 330 7642
Email: dmk@unicom-direct.com
Web: www.kitchen-worktops-direct.co.uk
Other Info:
Material Type: A) 1, 2, 3, 4, 5, 6, 7, 9, 10

M D STAINLESS DESIGNS
79 Verity Cresent, Poole, Dorset, BH17 8TT
Area of Operation: UK (Excluding Ireland)
Tel: 01202 684998
Fax: 01202 684998
Email: stainlessdesigns@onetel.com

MANDARIN
Unit 1 Wonastow Road Industrial Estate,
Monmouth, Monmouthshire, NP25 5JB
Area of Operation: Europe
Tel: 01600 715444 **Fax:** 01600 715494
Email: info@mandarinstone.com
Web: www.mandarinstone.com

MARBLE CLASSICS
Unit 3, Station Approach, Emsworth,
Hampshire, PO10 7PW
Area of Operation: UK & Ireland
Tel: 01243 370011 **Fax:** 01243 371023
Email: info@marbleclassics.co.uk
Web: www.marbleclassics.co.uk
Material Type: E) 1, 2, 3

MARK LEIGH KITCHENS LTD
11 Common Garden Street, Lancaster,
Lancashire, LA1 1XD
Area of Operation: North West England
and North Wales
Tel: 01524 63273 **Fax:** 01524 62352
Email: mark@markleigh.co.uk
Web: www.markleigh.co.uk
Other Info:
Material Type: A) 1, 2, 3, 4, 5, 6, 7, 8, 9,
10, 11, 12, 13, 14

MASTER LTD
30 Station Road, Heacham, Norfolk, PE31 7EX
Area of Operation: Europe
Tel: 01485 572032
Fax: 01485 571675
Email: ron@masterlimited.co.uk
Web: www.masterlimited.co.uk
Material Type: A) 2, 3, 4, 5, 11, 15

MEDUSA CREATIONS UK LTD
Unit 2B Lancaster Road, Carnaby Industrial Estate,
Bridlington, East Riding of Yorks, YO15 3QY
Area of Operation: UK (Excluding Ireland)
Tel: 01262 605222
Fax: 01262 605654
Email: info@medusacreationscommercial.co.uk
Web: www.medusacreationscommercial.co.uk

MEREWAY CONTRACTS
413 Warwick Road, Birmingham,
West Midlands, B11 2JR
Area of Operation: UK (Excluding Ireland)
Tel: 0121 707 3288
Fax: 0121 707 3871
Email: info@merewaycontracts.co.uk
Web: www.merewaycontracts.co.uk

MOBALPA
4 Cornflower Close, Willand, Devon, EX15 2TT
Area of Operation: Worldwide
Tel: 07740 633672
Fax: 01884 820828
Email: ploftus@mobalpa.com
Web: www.mobalpa.com

MONTANA KITCHENS LTD
BIC 1, Studio 2/3 Innova Business Park,
Mollison Avenue, Enfield, Middlesex, EN3 7XU
Area of Operation: Greater London,
South East England
Tel: 0800 58 75 628
Email: angie@montanakitchens.co.uk
Web: www.montanakitchens.co.uk
Material Type: A) 1, 2, 3, 4, 5, 6, 7, 8, 9, 10, 11, 12,
13, 14

NATURAL IMAGE (GRANITE, MARBLE, STONE)
Spelmonden Estate, Goudhurst Road,
Goudhurst, Kent, TN17 1HE
Area of Operation: UK & Ireland
Tel: 01580 212222
Fax: 01580 211841
Web: www.naturalstonefloors.org

NICHOLAS ANTHONY LTD
42-44 Wigmore Street, London, W1U 2RX
Area of Operation: East England,
Greater London, South East England
Tel: 0800 068 3603
Fax: 01206 762698
Email: info@nicholas-anthony.co.uk
Web: www.nicholas-anthony.co.uk

NORTHERN TIMBER PRODUCTS LTD
Unit 20/1, West Bowling Green Street,
Edinburgh, EH6 5PE
Area of Operation: Scotland
Tel: 0131 554 8787
Fax: 0131 554 9191
Email: sales@ntp-kitchens.co.uk
Web: www.ntp-kitchens.co.uk
Material Type: A) 2, 4

NOW GROUP PLC (CONTRACTS DIVISION)
Red Scar Business Park, Longridge Road,
Preston, Lancashire, PR2 5NA
Area of Operation: UK & Ireland
Tel: 01772 703838
Fax: 01772 705798
Email: sales@nowkitchens.co.uk
Web: www.nowkitchens.co.uk

OAKWOOD BESPOKE
80 High Street, Camberley, Surrey, GU46 7RN
Area of Operation: Greater London, South East
England, South West England and South Wales
Tel: 01276 708630
Email: enquiries@oakwoodbespokecamberley.co.uk
Web: www.oakwoodbespokecamberley.co.uk

ORAMA
Azalea Close, Clover Nook Industrial Estate,
Somercotes, Derbyshire, DE55 4QX
Area of Operation: UK & Ireland
Tel: 01773 520560
Fax: 01773 520319
Email: pholt@orama.co.uk
Web: www.orama.co.uk

ORAMA
Area of Operation: UK & Ireland
Tel: 01773 520560
Fax: 01773 520319
Email: pholt@orama.co.uk
Web: www.orama.co.uk

Cappuccino, from Orama's innovative Maia solid
surface range. Orama is one of the UK's leading
manufacturers and suppliers of quality
worksurfaces including laminate, timber and
solid surfaces. Product ranges include Maia and
Ardesco.

PANELS PLUS
22 Mill Place, Kingston Upon Thames,
Surrey, KT1 2RJ
Area of Operation: Greater London
Tel: 0208 399 6343 **Fax:** 0208 399 6343
Email: sales@panelsplusltd.com
Web: www.panelsplusltd.com
Material Type: H) 2, 5

PFLEIDERER INDUSTRIES LIMITED
Oakfield House, Springwood Way, Tytherington
Business Park, Macclesfield, Cheshire, SK10 2XA
Area of Operation: Worldwide
Tel: 01625 660410 **Fax:** 01625 617301
Email: info@pfleiderer.co.uk
Web: www.pfleiderer.co.uk

PLAIN ENGLISH
Stowupland Hall, Stowupland, Suffolk, IP14 4BE
Area of Operation: Worldwide
Tel: 01449 774028
 Fax: 01449 613519
Email: info@plainenglishdesign.co.uk
Web: www.plainenglishdesign.co.uk
Material Type: A) 2, 3, 4, 5, 6, 9

POGGENPOHL
477-481 Finchley Road, London, NW3 6HS
Area of Operation: UK & Ireland
Tel: 0800 298 1098 **Fax:** 0207 794 6251
Email: kitchens@poggenpohl-group.co.uk
Web: www.poggenpohl.co.uk
Material Type: A) 1, 2, 3, 4, 5, 7

PRICE KITCHENS
11 Imperial Way, Croydon, Surrey, CR0 4RR
Area of Operation: South East England
Tel: 0208 686 9006 **Fax:** 0208 686 5958
Email: info@pricekitchens.co.uk
Web: www.pricekitchens.co.uk
Other Info: **Material Type:** A) 1, 2, 3, 4, 5, 7, 9

QUINCE STONEWORKS
3 Potton Road, Biggleswade,
Bedfordshire, SG18 ODU
Area of Operation: UK (Excluding Ireland)
Tel: 01767 314180 **Fax:** 01767 600872
Email: sales@qstoneworks.co.uk
Web: www.qstoneworks.co.uk
Other Info: **Material Type:** E) 1, 2, 3, 4, 5

RAPID GRANITE LIMITED
Tollgate Farm, Golford Road,
Cranbrook, Kent, TN17 3NX
Area of Operation: UK (Excluding Ireland)
Tel: 01580 720900 **Fax:** 01580 720901
Email: info@rapidgranite.co.uk
Web: www.rapidgranite.co.uk
Material Type: E) 1, 2, 8, 10

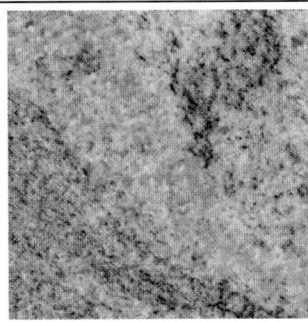

RAPID GRANITE LTD

Area of Operation: UK (Excluding Ireland)
Tel: 01580 720900
Fax: 01580 720901
Email: info@rapidgranite.co.uk
Web: www.rapidgranite.co.uk

Rapid Granite Ltd provides its customers with top quality (Grade A) granite surfaces at extremely low prices which can be used for work tops, splashbacks and tiles.

RED ROSE PLASTICS (BURNLEY) LTD
Parliament Street, Burnley, Lancashire, BB11 3JT
Area of Operation: East England, North East England, North West England and North Wales
Tel: 01282 724600
Fax: 01282 724644
Email: info@redroseplastics.co.uk
Web: www.redroseplastics.co.uk

RICHARD BAKER FURNITURE LTD
Wimbledon Studios, 257 Burlington Road,
New Malden, Surrey, KT3 4NE
Area of Operation: Europe
Tel: 0208 336 1777
Fax: 0208 336 1666
Email: Sales@richardbakerfurniture.co.uk
Web: www.richardbakerfurniture.co.uk

ROBERT J TURNER & CO
Roe Green, Sandon, Buntingford,
Hertfordshire, SG9 0QE
Area of Operation: East England, Greater London
Tel: 01763 288371
Fax: 01763 288440
Email: sales@robertjturner.co.uk
Web: www.robertjturner.co.uk

ROOTS KITCHENS, BEDROOMS & BATHROOMS
Vine Farm, Stockers Hill, Boughton-under-Blean,
Faversham, Kent, ME13 9AB
Area of Operation: UK (Excluding Ireland)
Tel: 01227 751130
Fax: 01227 750033
Email: showroom@rootskitchens.co.uk
Web: www.rootskitchens.co.uk

ROUNDEL DESIGN (UK) LTD
Flishinghurst Orchards, Chalk Lane,
Cranbrook, Kent, TN17 2QA
Area of Operation: South East England
Tel: 01580 712666
Email: homebuild@roundeldesign.co.uk
Web: www.roundeldesign.co.uk
Material Type: A) 2, 3, 5, 6, 7, 8

S.L. HARDWOODS
390 Sydenham Road, Croydon, Surrey, CR0 2EA
Area of Operation: UK (Excluding Ireland)
Tel: 0208 683 0292 **Fax:** 0208 683 0404
Email: info@slhardwoods.co.uk
Web: www.slhardwoods.co.uk

SAMUEL SCOTT PARTNERSHIP
Surgery House, The Square,
Skillington, Lincolnshire, NG33 5HB
Area of Operation: Europe
Tel: 01476 861806 **Fax:** 01476 861806
Email: samuel_scott@tiscali.co.uk
Other Info: ▽ ECO

SCHOCK UK LTD
Unit 444 Walton Summit Centre,
Bamber Bridge, Lancashire, PR5 8AT
Area of Operation: UK (Excluding Ireland)
Tel: 01772 332710 **Fax:** 01772 332717
Email: sales@schock.co.uk
Web: www.schock.de **Other Info:**

SECOND NATURE
PO Box 20, Station Road, Newton Aycliffe, Durham, DL5 6XJ
Area of Operation: UK & Ireland
Tel: 01325 505539
Email: mail@sncollection.co.uk
Web: www.sncollection.co.uk

SECOND NATURE

Area of Operation: UK & Ireland
Tel: 01325 505539
Email: mail@sncollection.co.uk
Web: www.sncollection.co.uk

Second Nature offer an extensive selection of bespoke timber worksurfaces, from rubberwood beech and walnut to rich wenge and exotic bamboo.

SIEMATIC
Osprey House, Primett Road, Stevenage,
Hertfordshire, SG1 3EE
Area of Operation: UK & Ireland
Tel: 01438 369251 **Fax:** 01438 368920
Email: sales@siematic.co.uk
Web: www.siematic.co.uk

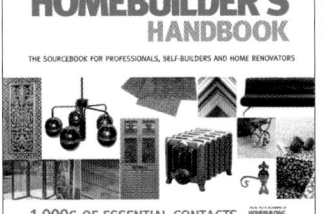
SIMALI STONE FLOORING
40E Wincombe Business Park, Shaftesbury,
Dorset, SP7 9QJ
Area of Operation: Europe
Tel: 01747 852557 **Fax:** 01747 852557
Email: stonefloors@aol.com
Web: www.stoneflooringonline.com
Material Type: E) 1, 2, 3, 5, 7, 8

SLATE WORLD LTD
Westmoreland Road, Kingsbury, London, NW9 9RN
Area of Operation: Europe
Tel: 0208 204 3444 **Fax:** 0208 204 3311
Email: advertising@americanslate.com
Web: www.slateworld.com **Material Type:** E) 3, 5

SPILLERS OF CHARD LTD.
The Aga Cooker Centre,
Chard Business Park,
Chard, Somerset, TA20 1FA
Area of Operation: South West England
and South Wales
Tel: 01460 67878 **Fax:** 01460 65252
Email: info@cookercentre.com
Web: www.cookercentre.com

STONELL DIRECT
100-105 Victoria Crescent,
Burton upon Trent,
Staffordshire, DE14 2QF
Area of Operation: Worldwide
Tel: 08000 832283
Fax: 01283 501098
Email: liz.may@stonelldirect.com
Web: www.stonelldirect.com

STONEVILLE (UK) LTD
Unit 12, Set Star Estate,
Transport Avenue, Great West Road,
Brentford, Greater London, TW8 9HF
Area of Operation: Europe
Tel: 0208 560 1000
Fax: 0208 560 4060
Email: info@stoneville.co.uk
Web: www.stoneville.co.uk
Material Type: E) 1, 2, 3, 5, 8

STUDIO STONE
The Stone Yard, Alton Lane,
Four Marks, Hampshire, GU34 5AJ
Area of Operation: UK (Excluding Ireland)
Tel: 01420 562500
Fax: 01420 563192
Email: sales@studiostone.co.uk
Web: www.studiostone.co.uk
Other Info: **Material Type:** E) 1, 2, 3, 5, 8

TAUROSTONE
Priors Haw Road,
North Weldon Industrial Estate,
Corby, Northamptonshire, NN17 5JG
Area of Operation: UK (Excluding Ireland)
Tel: 01536 740000 **Fax:** 01536 740001
Email: info@taurostone.com
Web: www.taurostone.com

THE CARPENTRY INSIDER - AIRCOMDIRECT
1 Castleton Crescent, Skegness,
Lincolnshire, PE25 2TJ
Area of Operation: Worldwide
Tel: 01754 767163
Email: aircom8@hotmail.com
Web: www.carpentry.tk

THE DESIGN STUDIO
39 High Street, Reigate, Surrey, RH2 9AE
Area of Operation: South East England
Tel: 01737 248228 **Fax:** 01737 224180
Email: enq@the-design-studio.co.uk
Web: www.the-design-studio.co.uk

THE DORSET CUPBOARD COMPANY
2 Mayo Farm, Cann, Shaftesbury, Dorset, SP7 0EF
Area of Operation: Greater London,
South East England, South West England
and South Wales
Tel: 01747 855044 **Fax:** 01747 855045
Email: nchturner@aol.com
Other Info: **Material Type:** A) 2, 3, 4, 5, 6

THE GRANITE FACTORY
4 Winchester Drive, Peterlee, Durham, SR8 2RJ
Area of Operation: North East England
Tel: 0191 518 3600 **Fax:** 0191 518 3600
Email: admin@granitefactory.co.uk
Web: www.granitefactory.co.uk
Material Type: E) 1, 2, 3, 4, 5, 7, 8, 9, 11, 13

THE KITCHEN & BATHROOM COLLECTION
Nelson House, Nelson Road,
Salisbury, Wiltshire, SP1 3LT
Area of Operation: South West England
and South Wales
Tel: 01722 334800 **Fax:** 01722 412252
Email: info@kbc.co.uk **Web:** www.kbc.co.uk
Other Info: **Material Type:** A) 1

THE MODERN ROOM LTD
54 St.Pauls Road, Luton,
Bedfordshire, LU1 3RX
Area of Operation: UK (Excluding Ireland)
Tel: 01582 612070
Fax: 01582 612070
Email: laura@themodernroom.com
Web: www.themodernroom.com
Other Info:

THE ORIGINAL KITCHEN COMPANY
4 Main Street, Breaston, Derby,
Derbyshire, DE72 3DX
Area of Operation: UK & Ireland
Tel: 01332 873746
Fax: 01332 873731
Email: originalkitchens@aol.com

WARMSWORTH STONE LTD
1-3 Sheffield Road, Warmsworth,
Doncaster, South Yorkshire, DN4 9QH
Area of Operation: UK & Ireland
Tel: 01302 858617
Fax: 01302 855844
Email: info@warmsworth-stone.co.uk
Web: www.warmsworth-stone.co.uk
Other Info: **Material Type:** E) 1, 2, 4, 5, 8

WILLIAM GARVEY
Leyhill, Payhembury,
Honiton, Devon, EX14 3JG
Area of Operation: Worldwide
Tel: 01404 841430
Fax: 01404 841626
Email: webquery@williamgarvey.co.uk
Web: www.williamgarvey.co.uk
Other Info: **Material Type:** A) 2, 3, 4, 10

WINNING DESIGNS
Dyke Farm, West Chiltington Road,
Pulborough, West Sussex, RH20 2EE
Area of Operation: UK & Ireland
Tel: 0870 754 4446 **Fax:** 0700 032 9946
Email: info@winning-designs-uk.com
Web: www.winning-designs-uk.com
Other Info: ▽ ECO
Material Type: A) 1, 2, 3, 4, 5, 6, 7, 8, 9, 10, 11, 12, 13, 14, 15

INTERIORS, FIXTURES & FINISHES

WINNING DESIGNS
Area of Operation: UK & Ireland
Tel: 08707 544446 **Fax:** 07000 329946
Email: info@winning-designs-uk.com
Web: www.winning-designs-uk.com
Material Type: A) 1, 2, 3, 4, 5, 6, 7, 8, 9, 10, 11, 12, 13, 14, 15 **Other Info:** ▽ ECO ✐ ✋

All types of worksurface supplied: granite, lavastone, solid wood, corian, stainless steel, laminate and other materials. Free design and advice, meetings by appointment only.

WOODEN HEART WAREHOUSE
Laburnum Road, Chertsey, Surrey, KT16 8BY
Area of Operation: Greater London, South East England
Tel: 01932 568684
Fax: 01932 568685
Email: whw@btclick.com
Material Type: A) 2, 4, 5

WOODENTOPS
The Barn, Park Farm, Hundred Acre Lane, Wivelsfield Green, Haywards Heath, East Sussex, RH17 7RU
Area of Operation: UK (Excluding Ireland)
Tel: 01273 891891
Fax: 01273 890044
Email: sales@woodentops.co.uk
Web: www.woodentops.co.uk
Material Type: A) 2, 3, 4, 5, 6, 7, 8, 10, 11

WOODSTOCK FURNITURE
17 Park Drive, East Sheen, London, SW14 8RB
Area of Operation: Worldwide
Tel: 0208 876 0131
Fax: 0207 245 9981
Email: ask@woodstockfurniture.co.uk
Web: www.woodstockfurniture.co.uk

ZARKA MARBLE
41A Belsize Lane, Hampstead, London, NW3 5AU
Area of Operation: South East England
Tel: 0207 431 3042
Fax: 0207 431 3879
Email: enquiries@zarkamarble.co.uk
Web: www.zarkamarble.co.uk
Material Type: E) 1, 2, 3, 5, 8

SINKS

KEY
PRODUCT TYPES: 1= Contemporary
2 = Traditional 3 = Belfast 4 = Double
5 = Other
OTHER: ▽ Reclaimed ✄ On-line shopping
✐ Bespoke ✋ Hand-made ECO Ecological

AMBER KITCHENS OF BISHOP'S STORTFORD
Clarklands House, Parsonage Lane, Sawbridgeworth, Hertfordshire, CM21 0NG
Area of Operation: East England, Greater London
Tel: 01279 600030
Fax: 01279 721528
Email: info@amberkitchens.com
Web: www.amberkitchens.com
Product Type: 1, 2, 3, 4, 5

APPLIANCE WORLD UK LTD.
Equity Trade Centre, Hobley Drive, Swindon, Wiltshire, SN3 4NS
Area of Operation: UK (Excluding Ireland)
Tel: 0870 757 2424 **Fax:** 0870 757 2475
Email: nickki@applianceworld.co.uk
Web: www.applianceworld.co.uk ✄
Product Type: 1, 2, 3, 4, 5

ASHFORD & BROOKS
The Old Workshop, Ashtree Barn, Caters Road, Bredfield, Woodbridge, Suffolk, IP13 6BE
Area of Operation: UK (Excluding Ireland)
Tel: 01473 737764
Fax: 01473 277176
Email: ashfordbrooks@mailbox.co.uk
Web: www.ashfordandbrooks.co.uk
Product Type: 1, 2, 3, 4, 5

ASHLAR MASON
Manor Building, Manor Road, London, W13 0JB
Area of Operation: East England, Greater London, Midlands & Mid Wales, South East England, South West England and South Wales
Tel: 0208 997 0002
Fax: 0208 997 0008
Email: enquiries@ashlarmason.co.uk
Web: www.ashlarmason.co.uk
Product Type: 1, 2, 3, 4, 5

ASTRACAST PLC
Holden Ing Way, Birstall, West Yorkshire, WF17 9AE
Area of Operation: UK & Ireland
Tel: 01924 477466
Fax: 01924 351297
Email: brochures@astracast.co.uk
Web: www.astracast.co.uk ✄
Product Type: 1, 2, 3, 4, 5

BATH KITCHEN COMPANY
22 Hensley Road, Bloomfield, Bath, Somerset, BA2 2DR
Area of Operation: South West England and South Wales
Tel: 01225 312003
Fax: 01225 312003
Email: david@bathkitchencompany.co.uk
Web: www.bathkitchencompany.co.uk
Product Type: 1, 2, 3, 4

BAUMATIC LTD
Baumatic Buildings, 6 Bennet Road, Berkshire, RG2 0QX
Area of Operation: Europe
Tel: 0118 933 6900
Fax: 0118 931 0035
Email: sales@baumatic.co.uk
Web: www.baumatic.com
Product Type: 1

BEAU-PORT LTD
15 Old Aylesfield Buildings, Froyle Road, Alton, Hampshire, GU34 4BY
Area of Operation: UK & Ireland
Tel: 0870 350 0133
Fax: 0870 350 0134
Email: sales@beau-port.co.uk
Web: www.beau-port.co.uk ✄
Product Type: 1, 2, 3, 4, 5

BROOK-WATER DESIGNER BATHROOMS & KITCHENS
The Downs, Woodhouse Hill, Uplyme, Lyme Regis, Dorset, DT7 3SL
Area of Operation: Worldwide
Tel: 01235 201256
Email: sales@brookwater.co.uk
Web: www.brookwater.co.uk ✄

CAPLE
Fourth Way, Avonmouth, Bristol, BS11 8DW
Area of Operation: UK (Excluding Ireland)
Tel: 0870 606 9606
Fax: 0117 938 7449
Email: amandalowe@mlay.co.uk
Web: www.mlay.co.uk
Product Type: 1, 2, 3, 4

CARRON PHOENIX
Carron Works, Stenhouse Road, Falkirk, FK2 8DW
Area of Operation: UK & Ireland
Tel: 01324 638321
Fax: 01324 620978
Email: fgp-sales@carron.com
Web: www.carron.com
Product Type: 1, 2, 3, 4, 5

CARRON PHOENIX
Area of Operation: UK & Ireland
Tel: 01324 638 321 **Fax:** 01324 620 978
Email: fgp-sales@carron.com
Web: www.carron.com
Product Type: 1, 2, 3, 4, 5

Exceptional quality. Exceptional value. An extensive range of sinks in granite, ceramic and stainless steel to suit the diversity of today's kitchen styling.

CHRISTOPHER PETERS ORIGINAL UNFITTED KITCHENS
28, West Street, Warwick, Warwickshire, CV34 6AN
Area of Operation: Europe
Tel: 01926 494106
Fax: 02476 303300
Email: enquiries@christopherpetersantiques.co.uk
Web: www.christopherpetersantiques.co.uk
Product Type: 2, 3, 4

CITY BATHROOMS & KITCHENS
158 Longford Road, Longford, Coventry, West Midlands, CV6 6DR
Area of Operation: UK & Ireland
Tel: 02476 365877
Fax: 02476 644992
Email: citybathrooms@hotmail.com
Web: www.citybathrooms.co.uk

CONNAUGHT KITCHENS
2 Porchester Place, London, W2 2BS
Area of Operation: Greater London
Tel: 0207 706 2210
Fax: 0207 706 2209
Email: design@connaughtkitchens.co.uk
Web: www.connaughtkitchens.co.uk
Product Type: 1, 2, 3, 4

COTTAGE FARM ANTIQUES
Stratford Road, Aston Sub Edge, Chipping Campden, Gloucestershire, GL55 6PZ
Area of Operation: Worldwide
Tel: 01386 438263
Fax: 01386 438263
Email: info@cottagefarmantiques.co.uk
Web: www.cottagefarmantiques.co.uk ✄
Product Type: 2, 3, 5

COUNTRY KITCHENS
The Old Farm House, Birmingham Road, Blackminster, Evesham, Worcestershire, WR11 7TD
Area of Operation: UK (Excluding Ireland)
Tel: 01386 831705
Fax: 01386 834051
Email: steve@floridafunbreaks.com
Web: www.handcraftedkitchens.com
Product Type: 2, 3, 4

CRABTREE KITCHENS
17 Station Road, Barnes, Greater London, SW13 0LF
Area of Operation: Greater London, South West England and South Wales
Tel: 0208 392 6955
Fax: 0208 392 6944
Email: design@crabtreekitchens.co.uk
Web: www.crabtreekitchens.co.uk
Product Type: 1, 2, 3, 4, 5

CYMRU KITCHENS
63 Caerleon Road, St. Julians, Newport, NP19 7BX
Area of Operation: UK (Excluding Ireland)
Tel: 01633 676767
Fax: 01633 212512
Email: sales@cymrukitchens.com
Web: www.cymrukitchens.com
Product Type: 1, 2, 3, 4, 5

DESIGNER KITCHENS
37 High Street, Hertfordshire, EN6 5AJ
Area of Operation: Greater London
Tel: 01707 650565
Fax: 01707 663050
Email: info@designer-kitchens.co.uk
Web: www.designer-kitchens.co.uk
Product Type: 1, 2, 3, 4, 5

DUPONT CORIAN & ZODIAQ
10 Quarry Court, Pitstone Green Business Park, Pitstone, nr Tring, Buckinghamshire, LU7 9GW
Area of Operation: UK & Ireland
Tel: 01296 663555
Fax: 01296 663599
Email: sales@corian.co.uk
Web: www.corian.co.uk
Product Type: 1, 2, 4

ELITE TRADE KITCHENS LTD
90 Willesden Lane, Kilburn, London, NW6 7TA
Area of Operation: UK (Excluding Ireland)
Tel: 0207 328 1234
Fax: 0207 328 1243
Email: Sales@elitekitchens.co.uk
Web: www.elitekitchens.co.uk
Product Type: 1, 2, 3, 4, 5

ELON
12 Silver Road, London, W12 7SG
Area of Operation: UK & Ireland
Tel: 0208 932 3000
Fax: 0208 932 3001
Email: marketing@elon.co.uk
Web: www.elon.co.uk ✄
Product Type: 1, 2, 3, 4

FALCON APPLIANCES
Clarence Street, Royal Leamington Spa, Warwickshire, CV31 2AD
Area of Operation: Worldwide
Tel: 0845 634 0070
Fax: 01926 450526
Email: jparkinson@falconappliances.co.uk
Web: www.falconappliances.co.uk
Product Type: 1, 2, 4

FLAWFORD TRADE KITCHENS
Sleaford Road, Coddington, Newark, Nottinghamshire, NG24 2QY
Area of Operation: UK & Ireland
Tel: 01636 626362
Fax: 01636 627971
Email: sales@mailorderkitchens.co.uk
Web: www.mailorderkitchens.co.uk ✄
Product Type: 1, 2, 3, 4, 5

FRANKE UK LTD
West Park, Manchester International Office, Styal Road, Manchester, M22 5WB
Area of Operation: UK & Ireland
Tel: 0161 436 6280
Fax: 0161 436 2180
Email: info.uk@franke.com
Web: www.franke.co.uk
Product Type: 1, 2, 3, 5

HATT KITCHENS
Hartlebury Trading Estate, Hartlebury,
Worcestershire, DY10 4JB
Area of Operation: UK (Excluding Ireland)
Tel: 01299 251320 **Fax:** 01299 251579
Email: design@hatt.co.uk
Web: www.hatt.co.uk
Product Type: 1, 2, 3, 4, 5

IN-TOTO
Shaw Cross Court, Shaw Cross Business Park,
Dewsbury, West Yorkshire, WF12 7RF
Area of Operation: UK (Excluding Ireland)
Tel: 01924 487900
Fax: 01924 437305
Email: graham.russell@intoto.co.uk
Web: www.intoto.co.uk
Product Type: 1, 3, 4

J&J ORMEROD PLC
Colonial House, Bacup, Lancashire, OL13 0EA
Area of Operation: UK & Ireland
Tel: 01706 877877
Fax: 01706 879827
Email: npeters@jjoplc.com
Web: www.jjoplc.com
Product Type: 1, 2, 3, 4, 5

JOHN LEWIS OF HUNGERFORD
Grove Technology Park, Downsview Road,
Wantage, Oxfordshire, OX12 9FA
Area of Operation: UK (Excluding Ireland)
Tel: 01235 774300
Fax: 01235 769031
Email: park.street@john-lewis.co.uk
Web: www.john-lewis.co.uk
Product Type: 2, 3, 4

JOHN STRAND (MK) LTD
12-22, Herga Road, Wealdstone,
Harrow, Middlesex, HA3 5AS
Area of Operation: UK & Ireland
Tel: 0208 930 6006
Fax: 0208 930 6008
Email: enquiry@johnstrand-mk.co.uk
Web: www.johnstrand-mk.co.uk
Product Type: 1

KENSINGTON STUDIO
13c/d Kensington Road, Earlsdon,
Coventry, West Midlands, CV5 6GG
Area of Operation: UK (Excluding Ireland)
Tel: 02476 713326
Fax: 02476 713136
Email: sales@kensingtonstudio.com
Web: www.kensingtonstudio.com
Product Type: 1, 2, 3, 4, 5

KITCHEN SUPPLIES
East Chesters, North Way, Hillend Industrial Estate,
Dalgety Bay, Fife, KY11 9JA
Area of Operation: UK (Excluding Ireland)
Tel: 01383 824729
Email: sales@kitchensupplies.co.uk
Web: www.kitchensupplies.co.uk
Product Type: 1, 2, 3, 4

MARBLE CLASSICS
Unit 3, Station Approach, Emsworth,
Hampshire, PO10 7PW
Area of Operation: UK & Ireland
Tel: 01243 370011
Fax: 01243 371023
Email: info@marbleclassics.co.uk
Web: www.marbleclassics.co.uk
Product Type: 1
Other Info:

MARK LEIGH KITCHENS LTD
11 Common Garden Street, Lancaster,
Lancashire, LA1 1XD
Area of Operation: North West England
and North Wales
Tel: 01524 63273
Fax: 01524 62352
Email: mark@markleigh.co.uk
Web: www.markleigh.co.uk
Product Type: 1, 2, 3, 4
Other Info:

MASTER LTD
30 Station Road, Heacham, Norfolk, PE31 7EX
Area of Operation: Europe
Tel: 01485 572032
Fax: 01485 571675
Email: ron@masterlimited.co.uk
Web: www.masterlimited.co.uk
Product Type: 1, 2, 3, 4, 5

MEREWAY CONTRACTS
413 Warwick Road, Birmingham,
West Midlands, B11 2JR
Area of Operation: UK (Excluding Ireland)
Tel: 0121 707 3288
Fax: 0121 707 3871
Email: info@merewaycontracts.co.uk
Web: www.merewaycontracts.co.uk

MIKE WALKER DISTRIBUTION
Clutton Hill Estate, King Lane,
Clutton Hill, Bristol, BS39 5QQ
Area of Operation: Midlands & Mid Wales,
South West England and South Wales
Tel: 01761 453838 **Fax:** 01761 453060
Email: office.mwd@zoom.co.uk
Web: www.mikewalker.co.uk

MOBALPA
4 Cornflower Close,
Willand, Devon, EX15 2TT
Area of Operation: Worldwide
Tel: 07740 633672 **Fax:** 01884 820828
Email: ploftus@mobalpa.com
Web: www.mobalpa.com
Product Type: 1, 2, 3, 4, 5

MONTANA KITCHENS LTD
BIC 1, Studio 2/3 Innova Business Park,
Mollison Avenue, Enfield, Middlesex, EN3 7XU
Area of Operation: Greater London,
South East England
Tel: 0800 58 75 628
Email: angie@montanakitchens.co.uk
Web: www.montanakitchens.co.uk
Product Type: 1, 2, 3, 4, 5

NICHOLAS ANTHONY LTD
42-44 Wigmore Street, London, W1U 2RX
Area of Operation: East England,
Greater London, South East England
Tel: 0800 068 3603 **Fax:** 01206 762698
Email: info@nicholas-anthony.co.uk
Web: www.nicholas-anthony.co.uk
Product Type: 1

NORTHERN TIMBER PRODUCTS LTD
Unit 20/1, West Bowling Green Street,
Edinburgh, EH6 5PE
Area of Operation: Scotland
Tel: 0131 554 8787 **Fax:** 0131 554 9191
Email: sales@ntp-kitchens.co.uk
Web: www.ntp-kitchens.co.uk
Product Type: 1, 2, 3, 4

OAKWOOD BESPOKE
80 High Street, Camberley, Surrey, GU46 7RN
Area of Operation: Greater London,
South East England, South West England
and South Wales
Tel: 01276 708630
Email: enquiries@oakwoodbespokecamberley.co.uk
Web: www.oakwoodbespokecamberley.co.uk
Product Type: 1, 2, 3, 4, 5

POGGENPOHL
477-481 Finchley Road, London, NW3 6HS
Area of Operation: UK & Ireland
Tel: 0800 298 1098 **Fax:** 0207 794 6251
Email: kitchens@poggenpohl-group.co.uk
Web: www.poggenpohl.co.uk
Product Type: 1

PRICE KITCHENS
11 Imperial Way, Croydon, Surrey, CR0 4RR
Area of Operation: South East England
Tel: 0208 686 9006 **Fax:** 0208 686 5958
Email: info@pricekitchens.co.uk
Web: www.pricekitchens.co.uk
Product Type: 1, 2, 3, 4, 5

RANGEMASTER
Clarence St, Royal Leamington Spa, Warks, CV31 2AD
Area of Operation: Worldwide
Tel: 01926 457400 **Fax:** 01926 450526
Email: consumers@rangemaster.co.uk
Web: www.rangemaster.co.uk
Product Type: 1, 2, 3, 4

**ROOTS KITCHENS,
BEDROOMS & BATHROOMS**
Vine Farm, Stockers Hill, Boughton-under-Blean,
Faversham, Kent, ME13 9AB
Area of Operation: UK (Excluding Ireland)
Tel: 01227 751130 **Fax:** 01227 750033
Email: showroom@rootskitchens.co.uk
Web: www.rootskitchens.co.uk
Product Type: 1, 2, 3, 4

ROUNDEL DESIGN (UK) LTD
Flishinghurst Orchards, Chalk Lane,
Cranbrook, Kent, TN17 2QA
Area of Operation: South East England
Tel: 01580 712666
Email: homebuild@roundeldesign.co.uk
Web: www.roundeldesign.co.uk
Product Type: 1, 2, 3, 4

SAMUEL SCOTT PARTNERSHIP
Surgery House, The Square,
Skillington, Lincolnshire, NG33 5HB
Area of Operation: Europe
Tel: 01476 861806 **Fax:** 01476 861806
Email: samuel_scott@tiscali.co.uk
Product Type: 1, 2, 3, 4, 5 **Other Info:** ▽ ECO

SCHOCK UK LTD
Unit 444 Walton Summit Centre,
Bamber Bridge, Lancashire, PR5 8AT
Area of Operation: UK (Excluding Ireland)
Tel: 01772 332710 **Fax:** 01772 332717
Email: sales@schock.co.uk
Web: www.schock.de **Product Type:** 1

SECOND NATURE
PO Box 20, Station Road,
Newton Aycliffe, Durham, DL5 6XJ
Area of Operation: UK & Ireland
Tel: 01325 505539
Email: mail@sncollection.co.uk
Web: www.sncollection.co.uk

SINKCITY.CO.UK
Castle Mead House, Castle Mead Gardens,
Hertford, Hertfordshire, SG14 1JZ
Area of Operation: UK (Excluding Ireland)
Tel: 0870 760 6218 **Fax:** 0870 760 6218
Email: info@sinkcity.co.uk
Web: www.sinkcity.co.uk
Product Type: 1, 2, 3, 4, 5

SMEG UK
3 Milton Park, Abingdon,
Oxfordshire, OX14 4RY
Area of Operation: UK & Ireland
Tel: 0870 990 9908
Fax: 01235 861120
Email: sales@smeguk.com
Web: www.smeguk.com

SPILLERS OF CHARD LTD.
The Aga Cooker Centre,
Chard Business Park,
Chard, Somerset, TA20 1FA
Area of Operation: South West England
and South Wales
Tel: 01460 67878
Fax: 01460 65252
Email: info@cookercentre.com
Web: www.cookercentre.com
Product Type: 1, 2, 3, 4

THE CDA GROUP LTD
Harby Road, Langar, Nottingham,
Nottinghamshire, NG13 9HY
Area of Operation: UK & Ireland
Tel: 01949 862000
Fax: 01949 862401
Email: sales@cda-europe.com
Web: www.cda-europe.com
Product Type: 1, 3, 4

THE DESIGN STUDIO
39 High Street, Reigate, Surrey, RH2 9AE
Area of Operation: South East England
Tel: 01737 248228 **Fax:** 01737 224180
Email: enq@the-design-studio.co.uk
Web: www.the-design-studio.co.uk
Product Type: 1, 2, 3, 4

THE DORSET CUPBOARD COMPANY
2 Mayo Farm, Cann, Shaftesbury, Dorset, SP7 0EF
Area of Operation: Greater London, South East
England, South West England and South Wales
Tel: 01747 855044 **Fax:** 01747 855045
Email: nchturner@aol.com **Product Type:** 1, 2, 3, 4

THE KITCHEN & BATHROOM COLLECTION
Nelson House, Nelson Road, Salisbury, Wiltshire, SP1 3LT
Area of Operation: South West England & South Wales
Tel: 01722 334800 **Fax:** 01722 412252
Email: info@kbc.co.uk **Web:** www.kbc.co.uk
Product Type: 1, 2, 3, 4, 5 **Other Info:**

THE KITCHEN SINK COMPANY
Unit 10, Evans Place,
South Bersted Industrial Estate,
Bognor Regis, West Sussex, PO22 9RY
Area of Operation: UK (Excluding Ireland)
Tel: 01243 841332 **Fax:** 01243 837294
Email: sales@kitchensinkco.co.uk
Web: www.kitchensinkco.com
Product Type: 1, 2, 3, 4, 5

THE ORIGINAL KITCHEN COMPANY
4 Main Street, Breaston, Derby,
Derbyshire, DE72 3DX
Area of Operation: UK & Ireland
Tel: 01332 873746 **Fax:** 01332 873731
Email: originalkitchens@aol.com

UKAPPLIANCES.CO.UK
Castle Mead House, Castle Mead Gardens,
Hertford, Hertfordshire, SG14 1JZ
Area of Operation: UK (Excluding Ireland)
Tel: 0870 760 6600 **Fax:** 0870 760 6600
Email: info@ukappliances.co.uk
Web: www.ukappliances.co.uk
Product Type: 1, 2, 3, 4, 5

WILLIAM GARVEY
Leyhill, Payhembury,
Honiton, Devon, EX14 3JG
Area of Operation: Worldwide
Tel: 01404 841430
Fax: 01404 841626
Email: webquery@williamgarvey.co.uk
Web: www.williamgarvey.co.uk
Product Type: 1, 3, 4 **Other Info:**

WINNING DESIGNS
Dyke Farm, West Chiltington Road,
Pulborough, West Sussex, RH20 2EE
Area of Operation: UK & Ireland
Tel: 0870 754 4446 **Fax:** 0700 032 9946
Email: info@winning-designs-uk.com
Web: www.winning-designs-uk.com
Product Type: 1, 2, 3, 4, 5
Other Info: ▽ ECO

TAPS

KEY
SEE ALSO: MERCHANTS - Plumbers
Merchants; BATHROOMS - Taps
OTHER: ▽ Reclaimed ⌂ On-line shopping
Bespoke Hand-made ECO Ecological

AL CHALLIS LTD
Europower House, Lower Road, Cookham,
Maidenhead, Berkshire, SL6 9EH
Area of Operation: UK & Ireland
Tel: 01628 529024 **Fax:** 0870 458 0577
Email: chris@alchallis.com
Web: www.alchallis.com

INTERIORS, FIXTURES & FINISHES - Kitchens - Taps

SPONSORED BY: WILLIAM BALL www.wball.co.uk

AMBER KITCHENS OF BISHOP'S STORTFORD
Clarklands House, Parsonage Lane,
Sawbridgeworth, Hertfordshire, CM21 0NG
Area of Operation: East England, Greater London
Tel: 01279 600030
Fax: 01279 721528
Email: info@amberkitchens.com
Web: www.amberkitchens.com

APPLIANCE WORLD UK LTD.
Equity Trade Centre, Hobley Drive,
Swindon, Wiltshire, SN3 4NS
Area of Operation: UK (Excluding Ireland)
Tel: 0870 757 2424
Fax: 0870 757 2475
Email: nickki@applianceworld.co.uk
Web: www.applianceworld.co.uk

ASHFORD & BROOKS
The Old Workshop, Ashtree Barn, Caters Road,
Bredfield, Woodbridge, Suffolk, IP13 6BE
Area of Operation: UK (Excluding Ireland)
Tel: 01473 737764
Fax: 01473 277176
Email: ashfordbrooks@mailbox.co.uk
Web: www.ashfordandbrooks.co.uk

ASHLAR MASON
Manor Building, Manor Road, London, W13 0JB
Area of Operation: East England, Greater London,
Midlands & Mid Wales, South East England,
South West England and South Wales
Tel: 0208 997 0002
Fax: 0208 997 0008
Email: enquiries@ashlarmason.co.uk
Web: www.ashlarmason.co.uk

ASTRACAST PLC
Holden Ing Way, Birstall,
West Yorkshire, WF17 9AE
Area of Operation: UK & Ireland
Tel: 01924 477466
Fax: 01924 351297
Email: brochures@astracast.co.uk
Web: www.astracast.co.uk

BATH KITCHEN COMPANY
22 Hensley Road, Bloomfield, Bath,
Somerset, BA2 2DR
Area of Operation: South West England
and South Wales
Tel: 01225 312003 Fax: 01225 312003
Email: david@bathkitchencompany.co.uk
Web: www.bathkitchencompany.co.uk

BAUMATIC LTD
Baumatic Buildings, 6 Bennet Road,
Berkshire, RG2 0QX
Area of Operation: Europe
Tel: 0118 933 6900 Fax: 0118 931 0035
Email: sales@baumatic.co.uk
Web: www.baumatic.com

BEAU-PORT LTD
15 Old Aylesfield Buildings, Froyle Road,
Alton, Hampshire, GU34 4BY
Area of Operation: UK & Ireland
Tel: 0870 350 0133
Fax: 0870 350 0134
Email: sales@beau-port.co.uk
Web: www.beau-port.co.uk
Other Info:

**BROOK-WATER DESIGNER
BATHROOMS & KITCHENS**
The Downs, Woodhouse Hill, Uplyme,
Lyme Regis, Dorset, DT7 3SL
Area of Operation: Worldwide
Tel: 01235 201256
Email: sales@brookwater.co.uk
Web: www.brookwater.co.uk

CAPLE
Fourth Way, Avonmouth, Bristol, BS11 8DW
Area of Operation: UK (Excluding Ireland)
Tel: 0870 606 9606
Fax: 0117 938 7449
Email: amandalowe@mlay.co.uk
Web: www.mlay.co.uk

CARRON PHOENIX
Carron Works, Stenhouse Road, Falkirk, FK2 8DW
Area of Operation: UK & Ireland
Tel: 01324 638321 Fax: 01324 620978
Email: fgp-sales@carron.com
Web: www.carron.com

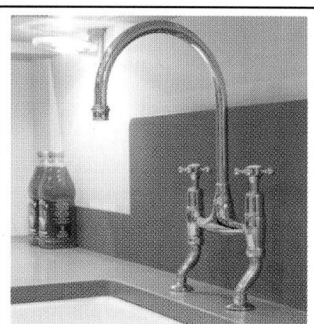

CARRON PHOENIX

Area of Operation: UK & Ireland
Tel: 01324 638 321
Fax: 01324 620 978
Email: fgp-sales@carron.com
Web: www.carron.com

Exceptional quality. Exceptional value. Tap styles
to complement the sink collection including
single lever, dual flow and filtration in a choice of
stunning metal or granite finishes.

**CHRISTOPHER PETERS
ORIGINAL UNFITTED KITCHENS**
28, West Street, Warwick, Warwickshire, CV34 6AN
Area of Operation: Europe
Tel: 01926 494106
Fax: 02476 303300
Email: enquiries@christopherpetersantiques.co.uk
Web: www.christopherpetersantiques.co.uk

CITY BATHROOMS & KITCHENS
158 Longford Road, Longford,
Coventry, West Midlands, CV6 6DR
Area of Operation: UK & Ireland
Tel: 02476 365877
Fax: 02476 644992
Email: citybathrooms@hotmail.com
Web: www.citybathrooms.co.uk

CONNAUGHT KITCHENS
2 Porchester Place, London, W2 2BS
Area of Operation: Greater London
Tel: 0207 706 2210
Fax: 0207 706 2209
Email: design@connaughtkitchens.co.uk
Web: www.connaughtkitchens.co.uk

COTTAGE FARM ANTIQUES
Stratford Road, Aston Sub Edge,
Chipping Campden, Gloucestershire, GL55 6PZ
Area of Operation: Worldwide
Tel: 01386 438263
Fax: 01386 438263
Email: info@cottagefarmantiques.co.uk
Web: www.cottagefarmantiques.co.uk

COUNTRY KITCHENS
The Old Farm House, Birmingham Road,
Blackminster, Evesham, Worcestershire, WR11 7TD
Area of Operation: UK (Excluding Ireland)
Tel: 01386 831705
Fax: 01386 834051
Email: steve@floridafunbreaks.com
Web: www.handcraftedkitchens.com

CRABTREE KITCHENS
17 Station Road, Barnes, Greater London, SW13 0LF
Area of Operation: Greater London, South West
England and South Wales
Tel: 0208 392 6955
Fax: 0208 392 6944
Email: design@crabtreekitchens.co.uk
Web: www.crabtreekitchens.co.uk
Other Info:

CROSSWATER
Unit 5 Butterly Avenue, Questor,
Dartford, Kent, DA1 1JG
Area of Operation: UK & Ireland
Tel: 01322 628270
Fax: 01322 628280
Email: sales@crosswater.co.uk
Web: www.crosswater.co.uk

CYMRU KITCHENS
63 Caerleon Road, St. Julians, Newport, NP19 7BX
Area of Operation: UK (Excluding Ireland)
Tel: 01633 676767 Fax: 01633 212512
Email: sales@cymrukitchens.com
Web: www.cymrukitchens.com

DANICO BRASS LTD
31-35 Winchester Road, Swiss Cottage,
London, NW3 3NR
Area of Operation: Worldwide
Tel: 0207 483 4477 Fax: 0207 722 7992
Email: sales@danico.co.uk

DESIGNER KITCHENS
37 High Street, Hertfordshire, EN6 5AJ
Area of Operation: Greater London
Tel: 01707 650565 Fax: 01707 663050
Email: info@designer-kitchens.co.uk
Web: www.designer-kitchens.co.uk

ELITE TRADE KITCHENS LTD
90 Willesden Lane,
Kilburn, London, NW6 7TA
Area of Operation: UK (Excluding Ireland)
Tel: 0207 328 1234 Fax: 0207 328 1243
Email: sales@elitekitchens.co.uk
Web: www.elitekitchens.co.uk

ELON
12 Silver Road, London, W12 7SG
Area of Operation: UK & Ireland
Tel: 0208 932 3000 Fax: 0208 932 3001
Email: marketing@elon.co.uk
Web: www.elon.co.uk

FALCON APPLIANCES
Clarence Street, Royal Leamington Spa,
Warwickshire, CV31 2AD
Area of Operation: Worldwide
Tel: 0845 634 0070
Fax: 01926 450526
Email: jparkinson@falconappliances.co.uk
Web: www.falconappliances.co.uk

FLAWFORD TRADE KITCHENS
Sleaford Road, Coddington, Newark,
Nottinghamshire, NG24 2QY
Area of Operation: UK & Ireland
Tel: 01636 626362
Fax: 01636 627971
Email: sales@mailorderkitchens.co.uk
Web: www.mailorderkitchens.co.uk

FRANKE UK LTD
West Park, Manchester International Office,
Styal Road, Manchester, M22 5WB
Area of Operation: UK & Ireland
Tel: 0161 436 6280
Fax: 0161 436 2180
Email: info.uk@franke.com
Web: www.franke.co.uk

GROHE UK
1 River Road, Barking, Essex, IG11 0HD
Area of Operation: Worldwide
Tel: 0208 594 7292
Fax: 0208 594 8898
Email: info@grohe.co.uk
Web: www.grohe.co.uk

HATT KITCHENS
Hartlebury Trading Estate,
Hartlebury, Worcestershire, DY10 4JB
Area of Operation: UK (Excluding Ireland)
Tel: 01299 251320
Fax: 01299 251579
Email: design@hatt.co.uk
Web: www.hatt.co.uk

HUDSON REED
Rylands Street, Burnley, Lancashire, BB10 1RH
Area of Operation: Europe
Tel: 01282 418000 Fax: 01282 428915
Email: info@ultra-group.com
Web: www.hudsonreed.co.uk

IN-TOTO
Shaw Cross Court, Shaw Cross Business Park,
Dewsbury, West Yorkshire, WF12 7RF
Area of Operation: UK (Excluding Ireland)
Tel: 01924 487900 Fax: 01924 437305
Email: graham.russell@intoto.co.uk
Web: www.intoto.co.uk

ITFITZ
PO Box 960, Maidenhead, Berkshire, SL6 9FU
Area of Operation: UK (Excluding Ireland)
Tel: 01628 890432 Fax: 0870 133 7955
Email: sales@itfitz.co.uk
Web: www.itfitz.co.uk

JOHN LEWIS OF HUNGERFORD
Grove Technology Park, Downsview Road,
Wantage, Oxfordshire, OX12 9FA
Area of Operation: UK (Excluding Ireland)
Tel: 01235 774300 Fax: 01235 769031
Email: park.street@john-lewis.co.uk
Web: www.john-lewis.co.uk

JOHN STRAND (MK) LTD
12-22, Herga Road, Wealdstone,
Harrow, Middlesex, HA3 5AS
Area of Operation: UK & Ireland
Tel: 0208 930 6006 Fax: 0208 930 6008
Email: enquiry@johnstrand-mk.co.uk
Web: www.johnstrand-mk.co.uk

KITCHEN SUPPLIES
East Chesters, North Way, Hillend
Industrial Estate, Dalgety Bay, Fife, KY11 9JA
Area of Operation: UK (Excluding Ireland)
Tel: 01383 824729
Email: sales@kitchensupplies.co.uk
Web: www.kitchensupplies.co.uk

MARK LEIGH KITCHENS LTD
11 Common Garden Street,
Lancaster, Lancashire, LA1 1XD
Area of Operation: North West England
and North Wales
Tel: 01524 63273 Fax: 01524 62352
Email: mark@markleigh.co.uk
Web: www.markleigh.co.uk Other Info:

MASTER LTD
30 Station Road, Heacham, Norfolk, PE31 7EX
Area of Operation: Europe
Tel: 01485 572032 Fax: 01485 571675
Email: ron@masterlimited.co.uk
Web: www.masterlimited.co.uk
Other Info:

MEREWAY CONTRACTS
413 Warwick Road, Birmingham,
West Midlands, B11 2JR
Area of Operation: UK (Excluding Ireland)
Tel: 0121 707 3288
Fax: 0121 707 3871
Email: info@merewaycontracts.co.uk
Web: www.merewaycontracts.co.uk

MIKE WALKER DISTRIBUTION
Clutton Hill Estate, King Lane, Clutton Hill,
Bristol, BS39 5QQ
Area of Operation: Midlands & Mid Wales,
South West England and South Wales
Tel: 01761 453838
Fax: 01761 453060
Email: office.mwd@zoom.co.uk
Web: www.mikewalker.co.uk

MOBALPA
4 Cornflower Close, Willand, Devon, EX15 2TT
Area of Operation: Worldwide
Tel: 07740 633672
Fax: 01884 820828
Email: ploftus@mobalpa.com
Web: www.mobalpa.com

MONTANA KITCHENS LTD
BIC 1, Studio 2/3 Innova Business Park,
Mollison Avenue, Enfield, Middlesex, EN3 7XU
Area of Operation: Greater London,
South East England
Tel: 0800 58 75 628
Email: angie@montanakitchens.co.uk
Web: www.montanakitchens.co.uk

NICHOLAS ANTHONY LTD
42-44 Wigmore Street, London, W1U 2RX
Area of Operation: East England,
Greater London, South East England
Tel: 0800 068 3603 **Fax:** 01206 762698
Email: info@nicholas-anthony.co.uk
Web: www.nicholas-anthony.co.uk

NORTHERN TIMBER PRODUCTS LTD
Unit 20/1, West Bowling Green Street,
Edinburgh, EH6 5PE
Area of Operation: Scotland
Tel: 0131 554 8787 **Fax:** 0131 554 9191
Email: sales@ntp-kitchens.co.uk
Web: www.ntp-kitchens.co.uk
Other Info: ECO

OAKWOOD BESPOKE
80 High Street, Camberley, Surrey, GU46 7RN
Area of Operation: Greater London, South
England, South West England and South Wales
Tel: 01276 708630
Email: enquiries@oakwoodbespokecamberley.co.uk
Web: www.oakwoodbespokecamberley.co.uk
Other Info:

PEGLER LIMITED
St Catherine's Avenue, Doncaster,
South Yorkshire, DN4 8DF
Area of Operation: Worldwide
Tel: 0870 120 0284 **Fax:** 01302 560109
Email: uk.sales@pegler.co.uk
Web: www.pegler.co.uk/francis

PRICE KITCHENS
11 Imperial Way, Croydon, Surrey, CR0 4RR
Area of Operation: South East England
Tel: 0208 686 9006 **Fax:** 0208 686 5958
Email: info@pricekitchens.co.uk
Web: www.pricekitchens.co.uk

RANGEMASTER
Clarence Street, Royal Leamington Spa,
Warwickshire, CV31 2AD
Area of Operation: Worldwide
Tel: 01926 457400 **Fax:** 01926 450526
Email: consumers@rangemaster.co.uk
Web: www.rangemaster.co.uk

ROBINSON AND CORNISH LTD
St George's House, St George's Road,
Barnstaple, Devon, EX32 7AS
Area of Operation: Europe
Tel: 01271 329300 **Fax:** 01271 328277
Email: richard.martin@robinsonandcornish.co.uk
Web: www.robinsonandcornish.co.uk

**ROOTS KITCHENS,
BEDROOMS & BATHROOMS**
Vine Farm, Stockers Hill, Boughton-under-Blean,
Faversham, Kent, ME13 9AB
Area of Operation: UK (Excluding Ireland)
Tel: 01227 751130 **Fax:** 01227 750033
Email: showroom@rootskitchens.co.uk
Web: www.rootskitchens.co.uk

ROUNDEL DESIGN (UK) LTD
Flishinghurst Orchards, Chalk Lane,
Cranbrook, Kent, TN17 2QA
Area of Operation: South East England
Tel: 01580 712666
Email: homebuild@roundeldesign.co.uk
Web: www.roundeldesign.co.uk

SAMUEL SCOTT PARTNERSHIP
Surgery House, The Square, Skillington,
Lincolnshire, NG33 5HB
Area of Operation: Europe
Tel: 01476 861806 **Fax:** 01476 861806
Email: samuel_scott@tiscali.co.uk **Other Info:**

SCHOCK UK LTD
Unit 444 Walton Summit Centre,
Bamber Bridge, Lancashire, PR5 8AT
Area of Operation: UK (Excluding Ireland)
Tel: 01772 332710 **Fax:** 01772 332717
Email: sales@schock.co.uk
Web: www.schock.de

SECOND NATURE
PO Box 20, Station Road, Newton Aycliffe,
Durham, DL5 6XJ
Area of Operation: UK & Ireland
Tel: 01325 505539
Email: mail@sncollection.co.uk
Web: www.sncollection.co.uk

SINKCITY.CO.UK
Castle Mead House, Castle Mead Gardens,
Hertford, Hertfordshire, SG14 1JZ
Area of Operation: UK (Excluding Ireland)
Tel: 0870 760 6218 **Fax:** 0870 760 6218
Email: info@sinkcity.co.uk
Web: www.sinkcity.co.uk

SMEG UK
3 Milton Park, Abingdon, Oxfordshire, OX14 4RY
Area of Operation: UK & Ireland
Tel: 0870 990 9908 **Fax:** 01235 861120
Email: sales@smeguk.com
Web: www.smeguk.com

SOGA LTD
41 Mayfield St, Hull, East Riding of Yorks, HU3 1NT
Area of Operation: Worldwide
Tel: 01482 327025
Email: info@soga.co.uk **Web:** www.soga.co.uk

SPILLERS OF CHARD LTD.
The Aga Cooker Centre, Chard Business Park,
Chard, Somerset, TA20 1FA
Area of Operation: South West England & South Wales
Tel: 01460 67878 **Fax:** 01460 65252
Email: info@cookercentre.com
Web: www.cookercentre.com

THE CDA GROUP LTD
Harby Road, Langar, Nottingham,
Nottinghamshire, NG13 9HY
Area of Operation: UK & Ireland
Tel: 01949 862000 **Fax:** 01949 862401
Email: sales@cda-europe.com
Web: www.cda-europe.com

THE DESIGN STUDIO
39 High Street, Reigate, Surrey, RH2 9AE
Area of Operation: South East England
Tel: 01737 248228 **Fax:** 01737 224180
Email: enq@the-design-studio.co.uk
Web: www.the-design-studio.co.uk

THE DORSET CUPBOARD COMPANY
2 Mayo Farm, Cann, Shaftesbury, Dorset, SP7 0EF
Area of Operation: Greater London, South East
England, South West England and South Wales
Tel: 01747 855044 **Fax:** 01747 855045
Email: nchturner@aol.com

THE KITCHEN AND BATHROOM COLLECTION
Nelson House, Nelson Road, Salisbury,
Wiltshire, SP1 3LT
Area of Operation: South West England & South Wales
Tel: 01722 334800 **Fax:** 01722 412252
Email: info@kbc.co.uk **Web:** www.kbc.co.uk
Other Info:

THE KITCHEN SINK COMPANY
Unit 10, Evans Place, South Bersted Industrial Estate,
Bognor Regis, West Sussex, PO22 9RY
Area of Operation: UK (Excluding Ireland)
Tel: 01243 841332 **Fax:** 01243 837294
Email: sales@kitchensinkco.com
Web: www.kitchensinkco.com

THE ORIGINAL KITCHEN COMPANY
4 Main Street, Breaston, Derby,
Derbyshire, DE72 3DX
Area of Operation: UK & Ireland
Tel: 01332 873746 **Fax:** 01332 873731
Email: originalkitchens@aol.com

UKAPPLIANCES.CO.UK
Castle Mead House, Castle Mead Gardens,
Hertford, Hertfordshire, SG14 1JZ
Area of Operation: UK (Excluding Ireland)
Tel: 0870 760 6600 **Fax:** 0870 760 6600
Email: info@ukappliances.co.uk
Web: www.ukappliances.co.uk

URBIS (GB) LTD.
55 Lidgate Crescent,
Langthwaite Grange Industrial Estate,
South Kirkby, West Yorkshire, WF9 3NR
Area of Operation: UK & Ireland
Tel: 01977 659829 **Fax:** 01977 659808
Email: david@urbis-gb.co.uk
Web: www.urbis-gb.co.uk

WATER FRONT
All Worcester Buildings, Birmingham Road, Redditch,
Worcestershire, B97 6DY
Area of Operation: UK & Ireland
Tel: 01527 584244 **Fax:** 01527 61127
Email: waterfront@btconnect.com
Web: www.waterfrontbathrooms.com

WHIRLPOOL EXPRESS (UK) LTD
61-62 Lower Dock Street, Kingsway,
Newport, NP20 1EF
Area of Operation: Europe
Tel: 01633 244555 **Fax:** 01633 244555
Email: reception@whirlpoolexpress.co.uk
Web: www.whirlpoolexpress.co.uk

WINNING DESIGNS
Dyke Farm, West Chiltington Road,
Pulborough, West Sussex, RH20 2EE
Area of Operation: UK & Ireland
Tel: 0870 754 4446 **Fax:** 0700 032 9946
Email: info@winning-designs-uk.com
Web: www.winning-designs-uk.com

NOTES

Company Name ....

Address ....

email ....

Web ....

Company Name ....

Address ....

email ....

Web ....

Company Name ....

Address ....

email ....

Web ....

Company Name ....

Address ....

email ....

Web ....

Company Name ....

Address ....

email ....

Web ....

INTERIORS, FIXTURES & FINISHES

INTERIORS, FIXTURES & FINISHES

FURNITURE & FURNISHINGS

Image courtesy of Marvin Architectural (0208 569 8222)

INTERIORS, FIXTURES & FINISHES

Sofas

If you have decided to remodel your home you have the opportunity to choose your sofa with a certain freedom not possible when buying a house – you can design your living room around this significant piece of furniture.

If you are working with a fairly small living room, calculate how much seating you need before rushing out and buying a tiny sofa. Don't rule out a modular sofa – although an oversized model may be impractical, more compact designs, and in particular L-shaped sofas, can squeeze the most seating out of a compact room or wasted corner. If you have the space, consider a U-shaped modular sofa, as if you regularly entertain it is likely that you will want to create a sociable

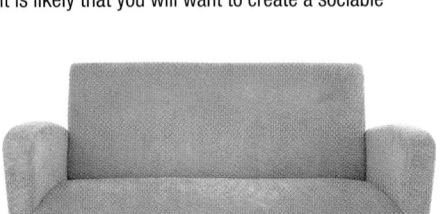

atmosphere in the living room. These can also be added to or reconfigured.

Alternatively, it is almost always better in small or awkward spaces to opt for two smaller sofas than one oversized design which has to be crammed in, leaving no potential for rearrangement should the need arise.

Do not let the television dominate your seating plan – instead place the sofa where it looks and feels right, taking into account windows, doors and the fireplace. Bear in mind that having your back to a window is not a good idea; not only will it block light but your comfort could be compromised by drafts.

In open plan living areas, sofas can work well when doubled up as room dividers, serving to create eating, cooking and seating zones. Low level sofas are a good idea for homes with low ceilings, as they will immediately make the room look higher and feel more spacious. Similarly, sofas with space beneath

are particularly good for smaller rooms, as they will let the light move beneath the furniture, giving the illusion of more space.

You should plan power points in such a way as to accommodate any lighting you wish to have near seating areas for reading or atmosphere. By planning the location of sockets with the sofa in mind, you will avoid having to wedge your arm down the back every time you wish to turn a socket on and off.

Sofa, LEFT, and chaise, ABOVE, both by John Lewis. BELOW Image by Velux.

FURNITURE

KEY

SEE ALSO: BATHROOMS - Bathroom Furniture;
KITCHENS - Kitchen Furniture, Fitted; KITCHENS
- Kitchen Furniture, Unfitted; GARDEN
ACCESSORIES AND DECORATIVE FEATURES -
Garden and Patio Furniture

PRODUCT TYPES: 1= General 2 = Bedroom
3 = Home Office   4 = Other

OTHER: ▽ Reclaimed ⌂ On-line shopping
✐ Bespoke ✋ Hand-made ECO Ecological

ANDY THORNTON LTD
Ainleys Industrial Estate, Elland,
West Yorkshire, HX5 9JP
Area of Operation: Worldwide
Tel: 01422 375595
Fax: 01422 377455
Email: email@ataa.co.uk
Web: www.andythornton.com
Product Type: 1

ARTICHOKE (KITCHEN & CABINET MAKERS)
Hortswood, Long Lane, Wrington,
Somerset, BS40 5SP
Area of Operation: UK & Ireland
Tel: 01934 863840
Fax: 01934 863841
Email: mail@artichoke-ltd.com
Web: www.artichoke-ltd.com
Product Type: 1, 2, 3

B ROURKE & CO LTD
Vulcan Works, Accrington Road,
Burnley, Lancashire, BB11 5QD
Area of Operation: Worldwide
Tel: 01282 422841
Fax: 01282 458901
Email: info@rourkes.co.uk
Web: www.rourkes.co.uk
Product Type: 1
Material Type: C) 1, 4, 5, 6

BADMAN & BADMAN LTD
The Drill Hall, Langford Road,
Weston Super Mare, Somerset, BS23 3PQ
Area of Operation: UK (Excluding Ireland)
Tel: 01934 644122
Fax: 01934 628189
Email: info@badmanandbadman.fsnet.co.uk
Web: www.badmans.co.uk
Product Type: 1, 2, 3

BATHROOM CITY
Seeleys Road, Tyseley Industrial Estate,
Birmingham, West Midlands, B11 2LQ
Area of Operation: UK & Ireland
Tel: 0121 753 0700
Fax: 0121 753 1110
Email: sales@bathroomcity.com
Web: www.bathroomcity.com ⌂
Product Type: 4

BAYFIELD STAIR CO
Unit 4, Praed Road, Trafford Park,
Manchester, M17 1PQ
Area of Operation: Worldwide
Tel: 0161 848 0700
Fax: 0161 872 2230
Email: sales@bayfieldstairs.co.uk
Web: www.bayfieldstairs.co.uk
Product Type: 1, 2, 3, 4

BORDERCRAFT
Old Forge, Peterchurch, Hereford,
Herefordshire, HR2 0SD
Area of Operation: UK (Excluding Ireland)
Tel: 01981 550251
Fax: 01981 550552
Email: info@bordercraft.co.uk
Web: www.bordercraft.co.uk
Other Info: ✐ ✋

BRASS FOUNDRY CASTINGS
PO Box 151, Westerham, Kent, TN16 1YF
Area of Operation: Worldwide
Tel: 01959 563863
Fax: 01959 561262
Email: enquiries@brasscastings.co.uk
Web: www.brasscastings.co.uk ⌂
Product Type: 1
Other Info: ✋
Material Type: C) 7, 11, 12, 13, 14, 15, 16

BROOKMANS ENGLISH FURNITURE CO LTD
Portland Business Park,, Richmond Park Road,,
Sheffield, South Yorkshire, S13 8HT
Area of Operation: Worldwide
Tel: 0114 280 0970
Fax: 0114 256 1840
Email: jim.brookman@brookmans.co.uk
Web: www.brookmans.co.uk
Product Type: 1

BRYN HALL
Knolton Bryn, Overton On Dee, Wrexham, LL13 0LF
Area of Operation: UK (Excluding Ireland)
Tel: 01978 710317
Fax: 01978 710027
Email: office@brynhall.co.uk
Web: www.brynhall.co.uk
Product Type: 1, 2, 3

BUYDESIGN
c/o Woodschool Ltd, Monteviot Nurseries,
Ancrum, Jedburgh, Borders, TD8 6TU
Area of Operation: North East England, North West
England and North Wales, Scotland
Tel: 01835 830740
Fax: 01835 830750
Email: info@buydesign-furniture.com
Web: www.buydesign-furniture.com
Product Type: 1, 2, 3, 4

BYLAW (ROSS) LIMITED
The Workshop, Norwich Road, Lenwade,
Norwich, Norfolk, NR9 5SH
Area of Operation: Worldwide
Tel: 01989 562356
Fax: 01603 872122
Email: graham@bylaw.co.uk
Web: www.bylaw.co.uk
Product Type: 1, 2, 3

CADIRA
233 Sandycomb Road, Kew,
Richmond, London, TW9 2EW
Area of Operation: UK (Excluding Ireland)
Tel: 0870 041 4180
Fax: 01932 829883
Email: info@cadira.co.uk
Web: www.cadira.co.uk ⌂
Product Type: 1

CAMERON PETERS LTD
The Old Dairy, Home Farm, Ardington,
Wantage, Oxfordshire, OX12 8PD
Area of Operation: Worldwide
Tel: 01235 835000
Fax: 01235 835005
Email: info@cameronpeters.co.uk
Web: www.cameronpeters.co.uk
Product Type: 1, 2, 3, 4

CAPLE
Fourth Way, Avonmouth, Bristol, BS11 8DW
Area of Operation: UK (Excluding Ireland)
Tel: 0870 606 9606
Fax: 0117 938 7449
Email: amandalowe@mlay.co.uk
Web: www.mlay.co.uk
Product Type: 2

CLOSETMAID UK
Unit 218, 36 Blenheim Grove,
London, SE15 4QL
Area of Operation: Worldwide
Tel: 0870 225 7002
Fax: 0870 225 7003
Email: ianpurdy@closetmaiduk.com
Web: www.closetmaiduk.com
Product Type: 2

COBALT BLACKSMITHS
The Forge, English Farm, English Lane,
Nuffield, Oxfordshire, RG9 5TH
Tel: 01491 641990
Fax: 01491 640909
Email: enquiries@cobalt-blacksmiths.co.uk
Web: www.cobalt-blacksmiths.co.uk
Product Type: 1
Other Info: ✐ ✋

COTTAGE FARM ANTIQUES
Stratford Road, Aston Sub Edge,
Chipping Campden, Gloucestershire, GL55 6PZ
Area of Operation: Worldwide
Tel: 01386 438263
Fax: 01386 438263
Email: info@cottagefarmantiques.co.uk
Web: www.cottagefarmantiques.co.uk ⌂
Product Type: 1, 2, 3, 4

CUSTOM MADE SHUTTERS
CMS Ltd, Red House Lodge, Limpsfield Common,
Oxted, Surrey, RH8 0QZ
Area of Operation: UK & Ireland
Tel: 01883 722148
Fax: 01883 717173
Email: info@cms.gb.com
Web: www.cms.gb.com
Product Type: 1, 2, 3, 4
Other Info: ✐ ✋
Material Type: A) 1, 2, 3, 4, 5, 6, 7, 9, 10, 12

D.A FURNITURE LTD
Woodview Workshop,
Pitway Lane, Farrington Gurney,
Bristol, BS39 6TX
Area of Operation: South West England
and South Wales
Tel: 01761 453117
Email: enquiries@dafurniture.co.uk
Web: www.dafurniture.co.uk
Product Type: 1, 2, 3, 4
Other Info: ✐ ✋
Material Type: A) 1, 2, 3, 4, 5, 6, 7, 8, 9,
10, 11, 12, 13, 14, 15

DAR INTERIORS
Arch 11, Miles Street,
London, SW8 1RZ
Area of Operation: Worldwide
Tel: 0207 720 9678
Fax: 0207 627 5129
Email: enquiries@darinteriors.com
Web: www.darinteriors.com
Product Type: 1
Other Info: ✐ ✋
Material Type: A) 7, 12

**DISTINCTIVE COUNTRY
FURNITURE LIMITED**
Parrett Works, Martock,
Somerset, TA12 6AE
Area of Operation: Worldwide
Tel: 01935 825800
Fax: 01935 825800
Email: Dcfltdinteriors@aol.com
Web: www.distinctivecountryfurniture.co.uk
Other Info: ▽ ✐ ✋
Material Type: A) 2, 3, 4, 7, 14

D-LINE
Buckingham Court, Brackley,
Northamptonshire, NN13 7EU
Area of Operation: UK & Ireland
Tel: 01280 841200
Fax: 01280 845130
Email: uk@dline.com
Web: www.dline.com

DOVETAIL ENTERPRISES
Dunsinane Avenue,
Dunsinane Industrial Estate,
Dundee, Angus, DD2 3QN
Tel: 01382 833890
Fax: 01382 814816
Email: sales@dovetailenterprises.co.uk
Web: www.dovetailenterprises.co.uk
Product Type: 1, 2
Material Type: A) 1, 2, 3, 4, 6

DUKE CHRISTIE
Hillockhead Farmhouse, Dallas,
Forres, Moray, IV36 2SD
Area of Operation: Worldwide
Tel: 01343 890347
Fax: 01343 890347
Email: enquiries@dukechristie.com
Web: www.dukechristie.com
Product Type: 1, 2, 4
Other Info: ✐ ✋
Material Type: A) 2, 3, 4, 5, 6, 7, 8, 9,
11, 12, 13, 14

FORM FURNITURE
Unit 33, Smithbrook Kilns,
Cranleigh, Surrey, GU6 8JJ
Area of Operation: UK (Excluding Ireland)
Tel: 01428 664430
Fax: 01428 664430
Email: mail@bennewick.com
Web: www.formfurniture.com
Product Type: 1

FORMFOLLOWS
27A Russell Square, Brighton,
East Sussex, BN1 2EE
Area of Operation: Worldwide
Tel: 01273 550180
Fax: 01273 823995
Email: sk@formfollows.co.uk
Web: www.formfollows.co.uk ⌂
Product Type: 1, 2
Other Info: ✐ ✋

FURNITURE BY JONATHAN ELWELL
Bryn Teg Workshop & Tan y Bryn Cottage,
Tanrallt Road, Gwespyr, Flintshire, CH8 9JT
Area of Operation: UK (Excluding Ireland)
Tel: 01745 887766
Email: jonathanelwell@ukonline.co.uk
Web: www.solidwoodfurniture.co.uk
Product Type: 1

FURNITURE123.CO.UK
Sandway Business Centre,
Shannon Street, Leeds,
West Yorkshire, LS9 8SS
Area of Operation: UK & Ireland
Tel: 0113 248 2233
Fax: 0113 248 2266
Email: p.haddock@furniture123.co.uk
Web: www.furniture123.co.uk ⌂
Product Type: 1, 2, 3, 4

GRANGEWOOD
Unit 29, Fairways,
New River Trading Estate
(Brookfield Centre), Cheshunt,
Waltham Cross, Hertfordshire, EN8 0NL
Area of Operation: East England,
Greater London, South East England
Tel: 01992 623933
Fax: 01992 623944
Email: info@grangewood.net
Web: www.grangewood.net
Product Type: 1, 4
Other Info: ✐ ✋
Material Type: A) 4

HAMMONDS FURNITURE LTD
Fleming Road,
Harrowbrook Industrial Estate,
Hinckley, Leicestershire, LE10 3DU
Area of Operation: UK (Excluding Ireland)
Tel: 0800 251505
Fax: 01455 623356
Email: info@hammonds-uk.com
Web: www.hammonds-uk.com
Product Type: 2, 3

HELBENT DESIGNS
Stonehaven, Wet Lane, Draycott,
Cheddar, Somerset, BS27 3TG
Area of Operation: Worldwide
Tel: 01934 744551
Email: enquiries@helbentdesigns.co.uk
Web: www.helbentdesigns.co.uk ⌂
Product Type: 2, 4

HOLME TREE LTD
Units 2 and 3 Machins Business Centre,
29 Wood Street, Asby-de-la-Zouch,
Leicestershire, LE65 1EZ
Area of Operation: UK (Excluding Ireland)
Tel: 01530 564561
Fax: 01530 417986
Email: info@holmetree.co.uk
Web: www.holmetree.co.uk
Product Type: 2, 3

HYGROVE
152/154 Merton Road,
Wimbledon, London, SW19 1EH
Area of Operation: South East England
Tel: 0208 543 1200
Fax: 0208 543 6521
Email: sales@hygrove.fsnet.co.uk
Web: www.hygrovefurniture.co.uk
Product Type: 1, 2, 3

IMPORTANT ROOMS LTD
62 High Street, Wargrave,
Berkshire, Reading,
Berkshire, RG10 8BY
Area of Operation: Greater London,
South East England, South West England
and South Wales
Tel: 0118 940 1266
Fax: 0118 940 1667
Email: Steve@importantrooms.co.uk
Web: www.importantrooms.co.uk
Product Type: 1, 3
Material Type: A) 10

INNOVATION GLASS COMPANY
27-28 Blake Industrial Estate,
Brue Avenue, Bridgwater,
Somerset, TA7 8EQ
Area of Operation: Europe
Tel: 01278 683645
Fax: 01278 450088
Email: magnus@igc-uk.com
Web: www.igc-uk.com
Product Type: 1, 2, 3
Material Type: A) 2, 3

INTERIOR DOOR SYSTEMS LTD
28 Tatton Court, Kingsland Grange,
Warrington, Cheshire, WA1 4RR
Area of Operation: UK & Ireland
Tel: 01925 813100
Fax: 01925 814300
Email: cb@interiordoorsystems.co.uk
Web: www.interiordoorsystems.co.uk
Product Type: 2
Material Type: J) 6

J W JENNINGS ORIENTAL RUGS
10 Church Street, Tewkesbury,
Gloucestershire, GL20 5PA
Area of Operation: Midlands & Mid Wales,
South West England and South Wales
Tel: 01684 292033
Fax: 01684 290292
Email: jwjennings@btconnect.com
Web: www.jenningsrugs.co.uk

J&J ORMEROD PLC
Colonial House, Bacup,
Lancashire, OL13 0EA
Area of Operation: UK & Ireland
Tel: 01706 877877
Fax: 01706 879827
Email: npeters@jjoplc.com
Web: www.jjoplc.com
Product Type: 2

JALI LTD
Albion Works, Church Lane,
Barham, Kent, CT4 6QS
Area of Operation: UK (Excluding Ireland)
Tel: 01227 833333
Fax: 01227 831950
Email: sales@jali.co.uk
Web: www.jali.co.uk
Product Type: 1
Other Info:

JARABOSKY
Old Station Road, Exley Lane, Elland,
West Yorkshire, HX5 0SW
Area of Operation: UK (Excluding Ireland)
Tel: 01422 311922
Fax: 01422 374053
Email: sales@jarabosky.co.uk
Web: www.jarabosky.co.uk
Product Type: 1, 4
Other Info: ▽

JARABOSKY

Area of Operation: UK (Excluding Ireland)
Tel: 01422 311922
Fax: 01422 374053
Email: sales@jarabosky.co.uk
Web: www.jarabosky.co.uk
Product Type: 1, 4   **Other Info:** ▽

The timeless beauty and elegance of wood is
reflected no better than in the unique creations
from Jarabosky. Recycled in a conscious effort
to help preserve our enviroment, each a living
masterpiece; no two pieces can be identical.

JOHN NETHERCOTT & CO
147 Corve Street, Ludlow, Shropshire, SY8 2NN
Area of Operation: UK (Excluding Ireland)
Tel: 01589 877044
Fax: 01589 560255
Email: showroom@johnnethercott.com
Web: www.johnnethercott.com
Product Type: 1, 2, 3, 4
Other Info: ✎ ✋
Material Type: A) 2, 3, 4, 5, 6, 7, 8, 9, 11

KASPAR SWANKEY
405, Goldhawk Road, Hammersmith,
West London, W6 0SA
Area of Operation: Worldwide
Tel: 020 8746 3586
Fax: 020 8746 3586
Email: kaspar@swankeypankey.com
Web: www.swankeypankey.com
Product Type: 1, 4

LANARKSHIRE HARDWOODS
Girdwoodend Farm, Auchengray,
Carnwath, Lanark, Lanarkshire, ML11 8LL
Area of Operation: Scotland
Tel: 01501 785460
Email: patrickbaxter@girdwoodend.wanadoo.co.uk
Web: www.lanarkshirehardwoods.co.uk
Product Type: 1
Other Info: ECO ✎ ✋
Material Type: A) 2, 3, 4, 5, 6, 7, 8, 9, 11

**MINCHINHAMPTON
ARCHITECTURAL SALVAGE CO**
Cirencester Road, Aston Downs,
Chalford, Stroud, Gloucestershire, GL6 8PE
Area of Operation: Worldwide
Tel: 01285 760886
Fax: 01285 760838
Email: masco@catbrain.com
Web: www.catbrain.com
Product Type: 1

MOBALPA
4 Cornflower Close,
Willand, Devon, EX15 2TT
Area of Operation: Worldwide
Tel: 07740 633672
Fax: 01884 820828
Email: ploftus@mobalpa.com
Web: www.mobalpa.com
Product Type: 1, 2, 4

MOSS FURNITURE & INTERIORS LTD
40-42 Old Town, London, SW4 0LB
Area of Operation: UK (Excluding Ireland)
Tel: 0207 622 5999
Fax: 0208 761 4400
Email: clapham@mossfurniture.co.uk
Web: www.mossfurniture.co.uk
Product Type: 1, 2, 3, 4

MYFLOATINGWORLD.COM LIMITED
3 Hartlepool Court, Royal Docks,
Galleons Lock, Royal Docks, London, E16 2RL
Area of Operation: Europe
Tel: 0870 777 0728
Fax: 0870 777 0729
Email: info@myfloatingworld.com
Web: www.myfloatingworld.com ✋
Product Type: 1

NEVILLE JOHNSON
Broadoak Business Park,
Ashburton Road West,
Trafford Park, Manchester, M17 1RW
Tel: 0161 873 8333
Fax: 0161 873 8335
Email: sales@nevillejohnson.co.uk
Web: www.nevillejohnson.co.uk
Product Type: 1, 2, 3, 4

NIGEL TYAS HANDCRAFTED IRONWORK
Green Works, 18 Green Road, Penistone,
Sheffield, South Yorkshire, S36 6BE
Area of Operation: Worldwide
Tel: 01226 766618
Email: sales@nigeltyas.co.uk
Web: www.nigeltyas.co.uk
Product Type: 1
Other Info: ✎ ✋
Material Type: A) 2, 3, 4, 5, 6

OAKWOOD BESPOKE
80 High Street, Camberley, Surrey, GU46 7RN
Area of Operation: Greater London, South East
England, South West England and South Wales
Tel: 01276 708630
Email: enquiries@oakwoodbespokecamberley.co.uk
Web: www.oakwoodbespokecamberley.co.uk
Product Type: 2, 3

PARAGON JOINERY
Systems House, Eastbourne Road,
Blindley Heath, Surrey, RH7 6JP
Area of Operation: UK (Excluding Ireland)
Tel: 01342 836300
Fax: 01342 836364
Email: sales@paragonjoinery.com
Web: www.paragonjoinery.com
Product Type: 1, 2

PARLOUR FARM
Unit 12b, Wilkinson Road,
Love Lane Industrial Estate,
Cirencester, Gloucestershire, GL7 1YT
Area of Operation: Europe
Tel: 01285 885336
Fax: 01285 643189
Email: info@parlourfarm.com
Web: www.parlourfarm.com

PINELAND FURNITURE LTD
Unit 5 Cleobury Trading Estate,
Cleobury Mortimer,
Shropshire, DY14 8DP
Area of Operation: UK (Excluding Ireland)
Tel: 01299 271143
Fax: 01299 271166
Email: pineland@onetel.com
Web: www.pineland.co.uk
Product Type: 1

PINELAND
Area of Operation: UK (Excluding Ireland)
Tel: 01299 271143
Fax: 01299 271166
Email: pineland@onetel.com
Web: www.pineland.co.uk

Manufacturing a full range of traditionally styled
pine furniture using the finest quality kiln dried
timbers (no plywood or chipboard) and employing
proper mortice and tension and dovetail joining
wherever possible.

POETSTYLE
Unit 1, Bayford Street Industrial Centre,
Hackney, Greater London, E8 3SE
Area of Operation: UK (Excluding Ireland)
Tel: 0208 533 0915
Fax: 0208 985 2953
Email: sofachairs@aol.com
Web: www.sofachairs.co.uk
Product Type: 1

PS INTERIORS
11 Cecil Road, Hale,
Altrincham, Cheshire, WA15 9NY
Area of Operation: Europe
Tel: 0161 926 9398
Fax: 0161 929 0363
Email: sales@ps-interiors.co.uk
Web: www.ps4interiors.co.uk
Product Type: 1

RADIATING ELEGANCE
32 Main Street, Orton-on-the-Hill,
Warwickshire, CV9 3NN
Area of Operation: UK (Excluding Ireland)
Tel: 08000 280 921
Email: info@radiatingelegance.co.uk
Web: www.radiatingelegance.co.uk
Product Type: 1

RB UK LTD
Element House, Napier Road,
Bedford, Bedfordshire, MK41 0QS
Area of Operation: UK & Ireland
Tel: 01234 272717
Fax: 01234 270202
Email: info@rbuk.co.uk
Web: www.rbuk.co.uk
Product Type: 4
Material Type: C) 2

RE:DESIGN
11 Bognor Road, Chichester,
West Sussex, PO19 7TF
Area of Operation: UK & Ireland
Tel: 07787 987652
Email: barry@redesign.uk.com
Web: www.redesign.uk.com
Product Type: 1
Material Type: A) 2, 4, 6, 9, 12, 15

RICHARD BAKER FURNITURE LTD
Wimbledon Studios, 257 Burlington Road,
New Malden, Surrey, KT3 4NE
Area of Operation: Europe
Tel: 0208 336 1777
Fax: 0208 336 1666
Email: Sales@richardbakerfurniture.co.uk
Web: www.richardbakerfurniture.co.uk
Product Type: 1, 2, 3, 4

ROOMS FOR KIDS LTD
Unit 2 Gordon Street (Off Craven Street),
Birkenhead, Wirral, Cheshire, CH41 4AB
Area of Operation: Worldwide
Tel: 0151 650 2401
Fax: 0151 650 2403
Email: roger@friezeframe.com
Web: www.friezeframe.com
Product Type: 4

SANDERSON'S FINE FURNITURE
Unit 5 & 6, The Village Workshop, Four Crosses
Business Park, Four Crosses, Powys, SY22 6ST
Area of Operation: UK (Excluding Ireland)
Tel: 01691 830075
Fax: 01691 830075
Email: sales@sandersonsfinefurniture.co.uk
Web: www.sandersonsfinefurniture.co.uk
Product Type: 1, 2, 3

SANDERSONS FINE FURNITURE
Area of Operation: UK (Excluding Ireland)
Tel: 01691 830075
Fax: 01691 830075
Email: sales@sandersonsfinefurniture.co.uk
Web: www.sandersonsfinefurniture.co.uk
Product Type: 1, 2, 3

Traditional manufacturers of quality bespoke oak
furniture, internal and external doors, frame and
filled cottage, Georgian, Victorian, stable,
internal cottage, skirting, architrave and flooring.

SHARPS BEDROOMS
Albany Park, Camberley,
Surrey, GU16 7PU
Area of Operation: UK (Excluding Ireland)
Tel: 01276 802000
Fax: 01276 802030
Email: enquiries@sharps.co.uk
Web: www.sharps.co.uk
Product Type: 2, 3
Other Info: ✎
Material Type: A) 2, 4, 5, 9

SOGA LTD
41 Mayfield Street, Hull,
East Riding of Yorks, HU3 1NT
Area of Operation: Worldwide
Tel: 01482 327025
Email: info@soga.co.uk
Web: www.soga.co.uk
Product Type: 4

STUART INTERIORS
Barrington Court, Barrington,
Ilminster, Somerset, TA19 0NQ
Area of Operation: Worldwide
Tel: 01460 240349
Fax: 01460 242069
Email: design@stuartinteriors.com
Web: www.stuartinteriors.com
Product Type: 1, 2, 3, 4

TASKWORTHY
The Old Brickyard, Pontrilas, Herefordshire, HR2 0DJ
Area of Operation: UK (Excluding Ireland)
Tel: 01981 242900
Fax: 01981 242901
Email: peter@taskworthy.co.uk
Web: www.taskworthy.co.uk
Product Type: 1, 2, 3, 4

TEMPLESTONE
Station Wharf, Castle Cary, Somerset, BA7 7PE
Area of Operation: Worldwide
Tel: 01963 350242
Fax: 01963 350258
Email: sales@templestone.co.uk
Web: www.templestone.co.uk
Product Type: 1
Other Info: ✎ ✋
Material Type: E) 5, 13

THE DESIGN STUDIO
39 High Street, Reigate, Surrey, RH2 9AE
Area of Operation: South East England
Tel: 01737 248228
Fax: 01737 224180
Email: enq@the-design-studio.co.uk
Web: www.the-design-studio.co.uk
Product Type: 1

THE EXPANDED METAL COMPANY LIMITED
PO Box 14, Longhill Industrial Estate (North),
Hartlepool, Durham, TS25 1PR
Area of Operation: Worldwide
Tel: 01429 408087
Fax: 01429 866795
Email: paulb@expamet.co.uk
Web: www.expandedmetalcompany.co.uk

THE FIREPLACE GALLERY (UK) LTD
Clarence Road, Worksop,
Nottinghamshire, S80 1QA
Area of Operation: UK & Ireland
Tel: 01909 500802
Fax: 01909 500810
Email: fireplacegallery@btinternet.com
Web: www.fireplacegallery.co.uk
Product Type: 1, 4
Other Info: ECO ✎ ✋
Material Type: A) 1, 2, 3, 4, 7, 11

THE MODERN ROOM LTD
54 St.Pauls Road, Luton, Bedfordshire, LU1 3RX
Area of Operation: UK (Excluding Ireland)
Tel: 01582 612070
Fax: 01582 612070
Email: laura@themodernroom.com
Web: www.themodernroom.com
Product Type: 1, 2, 3, 4
Other Info: ✎ ✋
Material Type: C) 1, 3, 14

THE OLD PINE STORE
Coxons Yard, Off Union Street,
Ashbourne, Derbyshire, DE6 1FG
Area of Operation: Europe
Tel: 01335 344112
Fax: 01335 344112
Email: martin@old-pine.co.uk
Web: www.old-pine.co.uk
Product Type: 1, 2, 3, 4
Other Info: ▽ ECO ✎ ✋
Material Type: A) 2, 4, 5

THE REAL WOOD FURNITURE CO
London House, 16 Oxford Street,
Woodstock, Oxfordshire, OX20 1TS
Area of Operation: Worldwide
Tel: 01993 813887
Email: info@rwfco.com
Web: www.rwfco.com
Product Type: 1

TIM WOOD LIMITED
1 Burland Road, London, SW11 6SA
Area of Operation: Worldwide
Tel: 07041 380030 **Fax:** 0870 054 8645
Email: homeb@timwood.com
Web: www.timwood.com
Product Type: 1, 2, 3, 4

TOMPKINS LTD
High March Close, Long March Industrial Estate,
Daventry, Northamptonshire, NN11 4EZ
Area of Operation: UK (Excluding Ireland)
Tel: 01327 877187
Fax: 01327 310491
Email: info@tompkinswood.co.uk
Web: www.tompkinswood.co.uk
Product Type: 1, 2, 3
Other Info: ✎ ✋

VICTORIA HAMMOND INTERIORS
Bury Farm, Church Street, Bovingdon,
Hemel Hempstead, Hertfordshire, HP3 0LU
Area of Operation: UK (Excluding Ireland)
Tel: 01442 831641
Fax: 01442 831641
Email: victoria@victoriahammond.com
Web: www.victoriahammond.com
Product Type: 1, 2, 3, 4

WICKES
Wickes Customer Services, Wickes House,
120-138 Station Road, Harrow, Middlesex,
Greater London, HA1 2QB
Area of Operation: UK (Excluding Ireland)
Tel: 0870 608 9001
Fax: 0208 863 6225
Web: www.wickes.co.uk
Product Type: 2

WILLIAM GARVEY
Leyhill, Payhembury, Honiton, Devon, EX14 3JG
Area of Operation: Worldwide
Tel: 01404 841430
Fax: 01404 841626
Email: webquery@williamgarvey.co.uk
Web: www.williamgarvey.co.uk
Product Type: 1, 2, 3, 4
Other Info: ✎ ✋
Material Type: A) 2, 3, 4, 6, 7, 8, 9, 10, 11, 12, 15

WOODCHESTER KITCHENS & INTERIORS
Unit 18a Chalford Industrial Estate,
Chalford, Gloucestershire, GL6 8NT
Area of Operation: UK (Excluding Ireland)
Tel: 01453 886411
Fax: 01453 886411
Email: enquires@woodchesterkitchens.co.uk
Web: www.woodchesterkitchens.co.uk
Product Type: 1, 2, 3, 4

WOODSTOCK FURNITURE
17 Park Drive, East Sheen, London, SW14 8RB
Area of Operation: Worldwide
Tel: 0208 876 0131
Fax: 0207 245 9981
Email: ask@woodstockfurniture.co.uk
Web: www.woodstockfurniture.co.uk
Product Type: 1, 2, 3, 4

ZOKI UK
Zoki Works, 44 Alcester Street,
Birmingham, West Midlands, B12 0PH
Area of Operation: UK & Ireland
Tel: 0121 766 7888 **Fax:** 0121 766 7962
Email: zokiuk@btconnect.com
Web: www.zokiuk.co.uk
Product Type: 1

SCREENS AND ROOM DIVIDERS

KEY
OTHER: ▽ Reclaimed ✎ On-line shopping
✎ Bespoke ✋ Hand-made ECO Ecological

A.S.K. SUPPLIES LIMITED
5 Stretton Close, Bridgnorth,
Shropshire, WV16 5DB
Area of Operation: UK (Excluding Ireland)
Tel: 01746 768164
Fax: 01746 766835
Email: sales@asksupplies.co.uk
Web: www.asksupplies.co.uk

ANY OLD IRON
PO Box 198, Ashford, Kent, TN26 3SE
Area of Operation: Worldwide
Tel: 01622 685336 **Fax:** 01622 685336
Email: anyoldiron@aol.com
Web: www.anyoldiron.co.uk
Other Info: ✎ ✋

ARCHITECTURAL COMPONENTS LTD
4-8 Exhibition Road, South Kensington,
London, SW7 2HF
Area of Operation: Worldwide
Tel: 0207 581 2401 **Fax:** 0207 589 4928
Email: sales@knobs.co.uk
Web: www.doorhandles.co.uk ✎

ARCHITECTURAL IRONMONGERY LTD
28 Kyrle Street, Ross-on-Wye,
Herefordshire, HR9 7DB
Area of Operation: Worldwide
Tel: 01989 567946 **Fax:** 01989 567946
Email: info@arciron.co.uk
Web: www.arciron.com ✎

ASHFIELD TRADITIONAL
119 High Street, Needham Market,
Ipswich, Suffolk, IP6 8DQ
Area of Operation: Europe
Tel: 01449 723601 **Fax:** 01449 723602
Email: mail@limelightgb.com
Web: www.limelightgb.com ✎

B ROURKE & CO LTD
Vulcan Works, Accrington Road,
Burnley, Lancashire, BB11 5QD
Area of Operation: Worldwide
Tel: 01282 422841 **Fax:** 01282 458901
Email: info@rourkes.co.uk
Web: www.rourkes.co.uk

BALCAS TIMBER LTD
Laragh, Enniskillen,
Co Fermanagh, BT94 2FQ
Area of Operation: UK & Ireland
Tel: 0286 632 3003 **Fax:** 0286 632 7924
Email: info@balcas.com
Web: www.balcas.com

BRASS GRILLES UK
Unit 174, 78 Marylebone High Street,
London, W1U 5AP
Area of Operation: UK (Excluding Ireland)
Tel: 07905 292101 / 01923 451600
Fax: 01923 451600
Email: sales@brass-grilles..co.uk
Web: www.brass-grilles.co.uk ✎

BROCKHOUSE MODERNFOLD LTD
Kay One, 23 The Tything,
Worcester, Worcestershire, WR1 1HD
Area of Operation: UK & Ireland
Tel: 01905 330055 **Fax:** 01905 330234
Email: sales@brockhouse.net
Web: www.brockhouse.net

CHARLES PEARCE LTD
26 Devonshire Road, London, W4 2HD
Area of Operation: Worldwide
Tel: 0208 995 3333
Email: enquiries@charlespearce.co.uk
Web: www.charlespearce.co.uk

COVERSCREEN
Unit 174, 78 Marylebone High Street,
London, W1U 5AP
Area of Operation: UK (Excluding Ireland)
Tel: 01923 451600 **Fax:** 01923 451600
Email: sales@coverscreen.co.uk
Web: www.coverscreen.co.uk ✎

CUSTOM MADE SHUTTERS
CMS Ltd, Red House Lodge, Limpsfield Common,
Oxted, Surrey, RH8 0QZ
Area of Operation: UK & Ireland
Tel: 01883 722148 **Fax:** 01883 717173
Email: info@cms.gb.com
Web: www.cms.gb.com
Other Info: ✎ ✋
Material Type: A) 1, 2, 3, 4, 5, 6, 7, 9, 10, 12

D & E ARCHITECTURAL HARDWARE LTD
17 Royce Road, Carr Road Industrial Estate,
Fengate, Peterborough, Cambridgeshire, PE1 5YB
Area of Operation: UK & Ireland
Tel: 01733 896123
Fax: 01733 894466
Email: wan@DandE.co.uk
Web: www.DandE.co.uk

DAR INTERIORS
Arch 11, Miles Street, London , SW8 1RZ
Area of Operation: Worldwide
Tel: 0207 720 9678
Fax: 0207 627 5129
Email: enquiries@darinteriors.com
Web: www.darinteriors.com

DARTINGTON STEEL DESIGN
Units 3-4 Webbers Yard, Dartington Industrial
Estate, Totnes, Devon, TQ9 6JY
Area of Operation: Worldwide
Tel: 01803 868671 **Fax:** 01803 868665
Email: sales@dartington.com
Web: www.dartington.com

D-LINE
Buckingham Court, Brackley,
Northamptonshire, NN13 7EU
Area of Operation: UK & Ireland
Tel: 01280 841200 **Fax:** 01280 845130
Email: uk@dline.com
Web: www.dline.com

ECLIPSE BLIND SYSTEMS LIMITED
Inchinnan Business Park, Inchinnan,
Renfrew, Renfrewshire, PA4 9RE
Area of Operation: UK & Ireland
Tel: 0141 812 3322
Fax: 0141 812 5253
Email: orrd@eclipseblinds.co.uk
Web: www.eclipse-blinds.com

EDEN HOUSE LIMITED
Elveden, Kennel Lane,
Windlesham, Surrey, GU20 6AA
Area of Operation: Greater London,
South East England
Tel: 01276 470192 **Fax:** 01276 489689
Email: info@internalshutters.co.uk
Web: www.internalshutters.co.uk /
www.externalshutters.co.uk
Other Info: ✎

HAF DESIGNS LTD
HAF House, Mead Lane, Hertford,
Hertfordshire, SG13 7AP
Area of Operation: UK & Ireland
Tel: 0800 389 8821 **Fax:** 01992 505705
Email: info@hafdesigns.co.uk
Web: www.hafdesigns.co.uk ⌐🖱

HAF DESIGNS LTD

Area of Operation: UK
Tel: 0800 389 8821
Fax: 01992 505 705
Email: info@hafdesigns.co.uk
Web: www.hafdesigns.co.uk

Bespoke glass sliding systems featuring a mix of
fixed and sliding panels. Installation can be
arranged. Also available, door handles, light
switches and bathroom accessories.

HANNAH WHITE
Area of Operation: Worldwide
Tel: 01372 806703 **Fax:** 01372 806703
Email: hannah@hannahwhite.co.uk
Web: www.hannahwhite.co.uk

HOT GLASS DESIGN
Unit 24 Crosby Yard Industrial Estate,
Bridgend, Mid Glamorgan, CF31 1JZ
Area of Operation: Europe
Tel: 01656 659884
Fax: 01656 659884
Email: info@hotglassdesign.co.uk
Web: www.hotglassdesign.co.uk

IN DOORS
Beechinwood Farm, Beechinwood Lane,
Platt, Nr. Sevenoaks, Kent, TN15 8QN
Area of Operation: UK (Excluding Ireland)
Tel: 01732 887445
Fax: 01732 887446
Email: info@indoorsltd.co.uk
Web: www.indoorsltd.co.uk

INTERIOR ASSOCIATES
3 Highfield Road, Windsor, Berkshire, SL4 4DN
Area of Operation: UK & Ireland
Tel: 01753 865339
Fax: 01753 865339
Email: sales@interiorassociates.fsnet.co.uk
Web: www.interiorassociates.co.uk
Other Info: ✎

INTERIOR DOOR SYSTEMS LTD
28 Tatton Court, Kingsland Grange,
Warrington, Cheshire, WA1 4RR
Area of Operation: UK & Ireland
Tel: 01925 813100
Fax: 01925 814300
Email: cb@interiordoorsystems.co.uk
Web: www.interiordoorsystems.co.uk

IRONMONGERY DIRECT
Unit 2-3 Eldon Way Trading Estate,
Eldon Way, Hockley, Essex, SS5 4AD
Area of Operation: Worldwide
Tel: 01702 562770
Fax: 01702 562799
Email: sales@ironmongerydirect.com
Web: www.ironmongerydirect.com ⌐🖱

ISAAC LORD LTD
West End Court, Suffield Road,
High Wycombe, Buckinghamshire, HP11 2JY
Area of Operation: East England, Greater London,
North East England, South East England,
South West England and South Wales
Tel: 01494 462121
Fax: 01494 461376
Email: info@isaaclord.co.uk
Web: www.isaaclord.co.uk ⌐🖱

JIM LAWRENCE LTD
Scotland Hall Farm, Stoke by Nayland,
Colchester, Essex, CO6 4QG
Area of Operation: UK (Excluding Ireland)
Tel: 01206 263459
Fax: 01206 262166
Email: sales@jim-lawrence.co.uk
Web: www.jim-lawrence.co.uk ⌐🖱

LEVOLUX LTD
1 Forward Drive, Harrow, Middlesex,
Greater London, HA3 8NT
Area of Operation: Worldwide
Tel: 0208 863 9111
Fax: 0208 863 8760
Email: info@levolux.com
Web: www.levolux.com

MBL
55 High Street, Biggleswade,
Bedfordshire, SG18 0JH
Area of Operation: UK (Excluding Ireland)
Tel: 01767 318695
Fax: 01767 318834
Email: info@mblai.co.uk
Web: www.mblai.co.uk
Other Info: ✎

NIGEL TYAS HANDCRAFTED IRONWORK
Green Works, 18 Green Road, Penistone,
Sheffield, South Yorkshire, S36 6BE
Area of Operation: Worldwide
Tel: 01226 766618
Email: sales@nigeltyas.co.uk
Web: www.nigeltyas.co.uk
Other Info: ✎ 🖐
Material Type: C) 2, 6

OAKLEAF (GLASGOW)
Birch Court, Doune, Stirlingshire, FK16 6JD
Area of Operation: Scotland
Tel: 01786 842216
Fax: 01786 842216
Email: sales@oakleaf.co.uk
Web: www.oakleaf.co.uk

OPEN N SHUT
Forum House, Stirling Road, Chichester,
West Sussex, PO19 7DN
Area of Operation: UK (Excluding Ireland)
Tel: 01243 774888
Fax: 01243 774333
Email: shutters@opennshut.co.uk
Web: www.opennshut.co.uk 🖱

PLANTATION SHUTTERS
131 Putney Bridge Road,
London, SW15 2PA
Area of Operation: UK & Ireland
Tel: 0208 871 9222 / 9333
Fax: 0208 871 0041
Email: sales@plantation-shutters.co.uk
Web: www.plantation-shutters.co.uk

**POLARLIGHT ACRYLIC
BLOCK WINDOWS**
W.I.N. Business Park,
Canal Quay, Newry, Co Down, BT35 6PH
Area of Operation: UK & Ireland
Tel: 02830 833888
Fax: 02830 833917
Email: info@polarlight.co.uk
Web: www.polarlight.co.uk

SERAMICO LTD
9 Billers Chase, Beaulieu Park,
Chelmsford, Essex, HP4 3LF
Area of Operation: UK & Ireland
Tel: 01245 464490
Fax: 01245 464490
Email: seramicoltd@aol.com
Web: www.seramico.co.uk 🖱

SHUTTER FRONTIER LTD
2 Rosemary Farmhouse, Rosemary Lane,
Flimwell, Wadhurst, East Sussex, TN5 7PT
Area of Operation: Worldwide
Tel: 01580 878137
Fax: 01580 878137
Email: jane@shutterfrontier.co.uk
Web: www.shutterfrontier.co.uk

SURFACEMATERIALDESIGN
17 Skiffington Close, London, SW2 3UL
Area of Operation: Worldwide
Tel: 0208 671 3383
Email: info@surfacematerialdesign.co.uk
Web: www.surfacematerialdesign.co.uk 🖱

THE CANE STORE
Wash Dyke Cottage, No1 Witham Road,
Long Bennington, Lincolnshire, NG23 5DS
Area of Operation: Worldwide
Tel: 01400 282271
Fax: 01400 281103
Email: jaki@canestore.co.uk
Web: www.canestore.co.uk

**THE EXPANDED METAL
COMPANY LIMITED**
PO Box 14, Longhill Industrial Estate (North),
Hartlepool, Durham, TS25 1PR
Area of Operation: Worldwide
Tel: 01429 408087
Fax: 01429 866795
Email: paulb@expamet.co.uk
Web: www.expandedmetalcompany.co.uk

TOMPKINS LTD
High March Close, Long March Industrial Estate,
Daventry, Northamptonshire, NN11 4EZ
Area of Operation: UK (Excluding Ireland)
Tel: 01327 877187
Fax: 01327 310491
Email: info@tompkinswood.co.uk
Web: www.tompkinswood.co.uk

VISTAMATIC LIMITED
51-55 Fowler Road,
Hainault Industrial Estate,
Hainault, Essex, IG6 3XE
Area of Operation: Worldwide
Tel: 0208 500 2200
Fax: 0208 559 8584
Email: sales@vistamatic.com
Web: www.vistamatic.com

ZERO SEAL SYSTEMS LTD
Unit 6, Ladford Covert,
Seighford, Stafford,
Staffordshire, ST18 9QG
Area of Operation: Europe
Tel: 01785 282910
Fax: 01785 282498
Email: sales@zeroplus.co.uk
Web: www.zeroplus.co.uk

ANTIQUES

KEY

SEE ALSO: MERCHANTS - Architectural
Antiques and Salvage Yards
OTHER: ▽ Reclaimed 🖱 On-line shopping
✎ Bespoke 🖐 Hand-made ECO Ecological

BINGLEY ANTIQUES
Springfield Farm Estate, Haworth,
West Yorkshire, BD21 5PT
Area of Operation: UK (Excluding Ireland)
Tel: 01535 646666
Fax: 01535 648527
Email: john@bingleyantiques.com
Web: www.bingleyantiques.com 🖱

DUKE CHRISTIE
Hillockhead Farmhouse,
Dallas, Forres, Moray, IV36 2SD
Area of Operation: Worldwide
Tel: 01343 890347
Fax: 01343 890347
Email: enquiries@dukechristie.com
Web: www.dukechristie.com

JAIL ANTIQUES
40 Mill Road, Watlington,
King's Lynn, Norfolk, PE33 0HJ
Area of Operation: UK (Excluding Ireland)
Tel: 07773 380880
Email: meljoe43@aol.com
Web: www.jailantiques.com

STUART INTERIORS
Barrington Court, Barrington,
Ilminster, Somerset, TA19 0NQ
Area of Operation: Worldwide
Tel: 01460 240349 **Fax:** 01460 242069
Email: design@stuartinteriors.com
Web: www.stuartinteriors.com

THE REAL WOOD FURNITURE CO
London House, 16 Oxford Street,
Woodstock, Oxfordshire, OX20 1TS
Area of Operation: Worldwide
Tel: 01993 813887
Email: info@rwfco.com
Web: www.rwfco.com

PAINTINGS, PHOTOGRAPHY AND SCULPTURE

> **KEY**
>
> **OTHER:** ▽ Reclaimed ⌐🖰 On-line shopping
> ✐ Bespoke ✋Hand-made ECO Ecological

ALUMINIUM ARTWORKS
Persistence Works, 21 Brown Street,
Sheffield, S1 2BS
Area of Operation: Worldwide
Tel: 0114 249 4748
Email: info@aluminiumartworks.co.uk
Web: www.aluminiumartworks.co.uk

ANNETTE NIX
150a Camden Street, London, NW1 9PA
Area of Operation: Worldwide
Tel: 07956 451719
Email: annette.n@virgin.net
Web: www.annettenix.com ⌐🖰

BLINK RED CONTEMPORARY ART
40 Maritime Street, Edinburgh, Lothian, EH6 6SA
Area of Operation: UK & Ireland
Tel: 0131 625 0192
Fax: 0131 467 7995
Email: customerservices@blinkred.com
Web: www.blinkred.com ⌐🖰

HANNAH WHITE
Area of Operation: Worldwide
Tel: 01372 806703
Fax: 01372 806703
Email: hannah@hannahwhite.co.uk
Web: www.hannahwhite.co.uk

MYFLOATINGWORLD.COM LIMITED
3 Hartlepool Court, Royal Docks, Galleons Lock,
Royal Docks, London, E16 2RL
Area of Operation: Europe
Tel: 0870 777 0728
Fax: 0870 777 0729
Email: info@myfloatingworld.com
Web: www.myfloatingworld.com ⌐🖰

RE:DESIGN
11 Bognor Road, Chichester,
West Sussex, PO19 7TF
Area of Operation: UK & Ireland
Tel: 07787 987652
Email: barry@redesign.uk.com
Web: www.redesign.uk.com

RICEDESIGN
The Barn, Shawbury Lane, Shustoke,
Nr Coleshill, B46 2RR
Area of Operation: Worldwide
Tel: 01675 481183
Email: ricedesigns@hotmail.com

THE REAL WOOD FURNITURE CO
London House, 16 Oxford Street,
Woodstock, Oxfordshire, OX20 1TS
Area of Operation: Worldwide
Tel: 01993 813887
Email: info@rwfco.com
Web: www.rwfco.com

LUXURY FITTINGS

> **KEY**
>
> **PRODUCT TYPES:** 1= Dumb Waiters
> 2 = Bars 3 = Gym Equipment 4 = Other
> **OTHER:** ▽ Reclaimed ⌐🖰 On-line shopping
> ✐ Bespoke ✋Hand-made ECO Ecological

1ST STEP: THE HOME FITNESS EQUIPMENT WAREHOUSE
The Panoramic, 152 Grosvenor Road,
London, SW1V 3JL
Area of Operation: Worldwide
Tel: 0207 932 0833
Fax: 0207 828 4709
Email: info@1ststepgyms.com
Web: www.1ststepgyms.com
Product Type: 3

ACADEMY BILLIARD COMPANY
5 Camphill Industrial Estate, Camphill Road,
West Byfleet, Surrey, KT14 6EW
Area of Operation: Worldwide
Tel: 01932 352067
Fax: 01932 353904
Email: academygames@fsbdial.co.uk
Web: www.games-room.com
Product Type: 2, 3

ACHIEVE FITNESS EQUIPMENT LTD
14 Blackwell Business Park, Blackwell, Near
Shipston-On-Stour, Warwickshire, CV32 6NY
Area of Operation: UK (Excluding Ireland)
Tel: 01608 682335
Fax: 01608 682556
Email: info@achieve-fitness.co.uk
Web: www.achieve-fitness.co.uk ⌐🖰
Product Type: 3

AQATA SHOWER ENCLOSURES
Brookfield, Harrowbrook Industrial Estate,
Hinckley, Leicestershire, LE10 3DU
Area of Operation: UK & Ireland
Tel: 01455 896500
Fax: 01455 896501
Email: sales@aqata.co.uk
Web: www.aqata.co.uk
Product Type: 3

AQUAPLUS SOLUTIONS
Unit 106, Indiana Building, London, SE13 7QD
Area of Operation: UK & Ireland
Tel: 0870 201 1915
Fax: 0870 201 1916
Email: info@aquaplussolutions.com
Web: www.aquaplussolutions.com

B ROURKE & CO LTD
Vulcan Works, Accrington Road,
Burnley, Lancashire, BB11 5QD
Area of Operation: Worldwide
Tel: 01282 422841
Fax: 01282 458901
Email: info@rourkes.co.uk
Web: www.rourkes.co.uk
Product Type: 2, 3

BATHROOM CITY
Seeleys Road, Tyseley Industrial Estate,
Birmingham, West Midlands, B11 2LQ
Area of Operation: UK & Ireland
Tel: 0121 753 0700
Fax: 0121 753 1110
Email: sales@bathroomcity.com
Web: www.bathroomcity.com ⌐🖰
Product Type: 3

BATHROOM OPTIONS
394 Ringwood Road, Ferndown, Dorset, BH22 9AU
Area of Operation: South East England
Tel: 01202 873379
Fax: 01202 892252
Email: bathroom@options394.fsnet.co.uk
Web: www.bathroomoptions.co.uk

BROADWAY SPORTS
Units 17/18 HQ Building,
237 Union Street, Stonehouse,
Plymouth, Devon, PL1 3HQ
Area of Operation: UK (Excluding Ireland)
Tel: 01752 601400
Fax: 01752 601401
Email: info@thefitnessstore.co.uk
Web: www.thefitnessstore.co.uk
Product Type: 3

CAMERON PETERS LTD
The Old Dairy, Home Farm,
Ardington, Wantage,
Oxfordshire, OX12 8PD
Area of Operation: Worldwide
Tel: 01235 835000
Fax: 01235 835005
Email: info@cameronpeters.co.uk
Web: www.cameronpeters.co.uk
Product Type: 3

COTTAGE FARM ANTIQUES
Stratford Road, Aston Sub Edge,
Chipping Campden,
Gloucestershire, GL55 6PZ
Area of Operation: Worldwide
Tel: 01386 438263
Fax: 01386 438263
Email: info@cottagefarmantiques.co.uk
Web: www.cottagefarmantiques.co.uk ⌐🖰
Product Type: 3

DEEP BLUE SHOWROOM
299-313 Lewisham High Street,
Lewisham, London, SE13 6NW
Area of Operation: UK (Excluding Ireland)
Tel: 0208 690 3401
Fax: 0208 690 1408

ENERFOIL MAGNUM LTD
Kenmore Road, Comrie Bridge,
Kenmore, Aberfeldy,
Perthshire, PH15 2LS
Area of Operation: Europe
Tel: 01887 822999
Fax: 01887 822954
Email: sales@enerfoil.com
Web: www.enerfoil.com ⌐🖰

FIREPLACE WAREHOUSE GROUP LTD
Design House, Walnut Tree Close,
Guildford, Surrey, GU1 4UQ
Area of Operation: UK (Excluding Ireland)
Tel: 01483 568777
Fax: 01483 570013
Email: info@designfireplaces.co.uk
Web: www.fireplacewarehouse.co.uk
Product Type: 3

GK DESIGN
Greensbury Farm, Thurleigh Road,
Bolnhurst, Bedfordshire, MK44 2ET
Area of Operation: UK (Excluding Ireland)
Tel: 01234 376990
Fax: 01234 376991
Email: studio@gkdesign.co.uk
Web: www.gkdesign.co.uk
Product Type: 3

HARD ROCK FLOORING
Fleet Marston Farm, Fleet Marston,
Aylesbury, Buckinghamshire, HP18 0PZ
Area of Operation: Europe
Tel: 01296 658755
Fax: 01296 655735
Email: showroom@hardrockflooring.co.uk
Web: www.hardrockflooring.co.uk
Product Type: 2

HARVEY MARIA LTD
17 Riverside Business Park, Lyon Road,
Wimbledon, London, SW19 2RL
Area of Operation: UK & Ireland
Tel: 0208 542 0088
Fax: 0208 542 0099
Email: info@harveymaria.co.uk
Web: www.harveymaria.co.uk ⌐🖰
Product Type: 3

INTERACTIVE HOMES LTD
Lorton House, Conifer Crest,
Newbury, Berkshire, RG14 6RS
Area of Operation: UK (Excluding Ireland)
Tel: 01635 49111
Fax: 01635 40735
Email: sales@interactivehomes.co.uk
Web: www.interactivehomes.co.uk

IRONART OF BATH
Upper Lambridge Street,
Larkhall, Bath,
Somerset, BA1 6RY
Area of Operation: UK (Excluding Ireland)
Tel: 01225 311273
Fax: 01225 443060
Email: ironart@btinternet.com
Web: www.ironart.co.uk
Product Type: 3

JAMES GIBBONS FORMAT LTD
Vulcan Road, Bilston,
Wolverhampton,
West Midlands, WV14 7JG
Area of Operation: Worldwide
Tel: 01902 405500
Fax: 01902 385915
Email: info@jgf.co.uk
Web: www.jgf.co.uk
Product Type: 3

JIM LAWRENCE LTD
Scotland Hall Farm,
Stoke by Nayland,
Colchester, Essex, CO6 4QG
Area of Operation: UK (Excluding Ireland)
Tel: 01206 263459
Fax: 01206 262166
Email: sales@jim-lawrence.co.uk
Web: www.jim-lawrence.co.uk ⌐🖰

KASPAR SWANKEY
405, Goldhawk Road, Hammersmith,
West London, W6 0SA
Area of Operation: Worldwide
Tel: 0208 746 3586
Fax: 0208 746 3586
Email: kaspar@swankeypankey.com
Web: www.swankeypankey.com
Product Type: 3

MOBALPA
4 Cornflower Close, Willand,
Devon, EX15 2TT
Area of Operation: Worldwide
Tel: 07740 633672
Fax: 01884 820828
Email: ploftus@mobalpa.com
Web: www.mobalpa.com
Product Type: 3

MYSTICLINE LTD T/A NEW CITY
PO Box 123, Priorslee, Telford,
Shropshire, TF2 9FF
Area of Operation: UK (Excluding Ireland)
Tel: 01952 200642
Fax: 01952 200738
Email: laura@newcity.biz
Web: www.newcity.biz

NEVILLE JOHNSON
Broadoak Business Park, Ashburton
Road West, Trafford Park,
Manchester, M17 1RW
Area of Operation: Worldwide
Tel: 0161 873 8333
Fax: 0161 873 8335
Email: sales@nevillejohnson.co.uk
Web: www.nevillejohnson.co.uk
Product Type: 4

OPUS GB
Gallery Court, Pilgrmage Street,
London, SE1 4LL
Area of Operation: Worldwide
Tel: 0207 940 2205
Fax: 0207 940 2206
Email: info@opus-technologies.co.uk
Web: www.opus-technologies.co.uk
Product Type: 3

QUENCH! HOME BARS
The Studio, 29 Woodham Way,
Woking, Surrey, GU21 5SJ
Area of Operation: Worldwide
Tel: 01483 740455 **Fax:** 01483 740244
Email: bars@quench.info
Web: www.quench.info
Product Type: 2

QUENCH! HOME BARS

Area of Operation: Worldwide
Tel: 01483 740455
Fax: 01483 740244
Email: bars@quench.info
Web: www.quench.info
Product Type: 2

Quench manufacture a contemporary range of home bars using materials such as stainless steel, brushed aluminium, glass and acrylic.

SECURIKEY LTD
P O Box 18, Aldershot, Hampshire, GU12 4SL
Area of Operation: Worldwide
Tel: 01252 311888 **Fax:** 01252 343950
Email: enquiries@securikey.co.uk
Web: www.securikey.co.uk
Product Type: 3

SERAMICO LTD
9 Billers Chase, Beaulieu Park,
Chelmsford, Essex, HP4 3LF
Area of Operation: UK & Ireland
Tel: 01245 464490
Fax: 01245 464490
Email: seramicoltd@aol.com
Web: www.seramico.co.uk ⌐

TEMPLESTONE
Station Wharf, Castle Cary, Somerset, BA7 7PE
Area of Operation: Worldwide
Tel: 01963 350242 **Fax:** 01963 350258
Email: sales@templestone.co.uk
Web: www.templestone.co.uk
Product Type: 3

VACUDUCT LTD
Unit 3, Brynberth Enterprise Park,
Rhayader, Powys, LD6 5EN
Area of Operation: Europe
Tel: 0800 783 6264
Fax: 01597 810755
Email: info@vacuduct.co.uk
Web: www.vacuduct.co.uk ⌐
Product Type: 3

WILLIAM GARVEY
Leyhill, Payhembury, Honiton, Devon, EX14 3JG
Area of Operation: Worldwide
Tel: 01404 841430
Fax: 01404 841626
Email: webquery@williamgarvey.co.uk
Web: www.williamgarvey.co.uk
Product Type: 2, 4

ZOKI UK
Zoki Works, 44 Alcester Street,
Birmingham, West Midlands, B12 0PH
Area of Operation: UK & Ireland
Tel: 0121 766 7888
Fax: 0121 766 7962
Email: zokiuk@btconnect.com
Web: www.zokiuk.co.uk
Product Type: 3

NOTES

Company Name ....

....

Address ....

....

....

email ....

Web ....

Company Name ....

....

Address ....

....

....

email ....

Web ....

Company Name ....

Address ....

....

....

email ....

Web ....

Company Name ....

Address ....

....

....

email ....

Web ....

EXTERIOR PRODUCTS

EXTERIOR ARCHITECTURAL DETAIL

Image courtesy of Bradstone (01285 646884)

SPONSORED BY BRADSTONE
Tel 01285 646884 Web www.bradstone.com

BRADSTONE

Porches, Porticos & Canopies

What could be worse than returning home in the rain and having to fumble for your keys unprotected from the elements? Porches, canopies and porticos are practical, in terms of providing a sheltered area in which to wait, and can create a stunning focal point to an otherwise dull facade.

ABOVE: Lightweight canopy, distributed in the UK by Interland Trading.

The Georgians were proud of their front doors, which were usually set within a porch under a projecting upper storey or within a moulded projecting lintel, and the Edwardians loved them too. Elaborately carved wooden 'sunray' canopies appeared along with more functional brick arches, flush with the brickwork on the front of the house. At other times the arch was brought forward, sometimes with round pillars on either side to create a larger porch area.

In the 19th Century, porticos — covered entrances supported by columns — were a frequent feature of houses and, if you are constructing a large facade, then an arrangement with columns or pilasters (shallow piers) on either side of the door and an architrave, frieze or cornice over the top will be more in proportion than a smaller entrance. Columns come in a wide choice of styles, sizes and materials and it is often possible to combine standard individual elements to create a completely new design or to commission a custom-made version.

Adjustments to GRP columns can be made on site, often by cutting the square plinth at the base using a hacksaw, and these columns may be left hollow, or filled with concrete if extra load bearing quantities are required. If stone columns are filled with reinforced concrete then the column core must be lined with polystyrene or similar to allow for expansion. With porticos, the contractor will need to know early on the size and material of your chosen design, especially for heavy porticos, where concrete ringbeams or other means of fixing will need to be discussed with the supplier.

Fitting a canopy gives immediate visual impact and a homely, welcoming atmosphere as you approach your front door, with several companies specialising in replicating patterns created during the last century. Alternatively, carpenters will be able to produce something to your own design. Remember to ensure that the tiles or slates match, or at least complement, those on the main roof.

Usually, canopies are delivered part assembled and timber may require priming and painting or staining. A good seal must be made between the roof and the wall with either a mortar filler or flashing to avoid visitors being dripped on from above.

Although most double glazing companies still offer fully enclosed walk-in porches to match their windows, many home owners are now opting for a more traditional approach or, conversely, something high tech or futuristic.

Wood, stone, concrete, reconstituted stone, iron, lead, glass, glass reinforced polyester (GRP) and thatch are just some of the materials used to create canopies, porches and porticos. Glass fibre or GRP, which has become a popular choice, is a tough, smooth, relatively maintenance-free material which will not rot or warp — making it ideal for outdoor use. The versatility of GRP allows for practically any shape or design to be replicated, although many people, and especially planners, prefer the real thing. Those favouring natural stone believe that, as well as longevity, stone becomes more attractive as it weathers, although the weight of stone creates the necessity for greater care during the construction process.

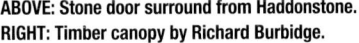

ABOVE: Stone door surround from Haddonstone.
RIGHT: Timber canopy by Richard Burbidge.

EXTERIOR PRODUCTS - **Exterior Architectural Detail** - Weathervanes, Clocks & Clock Towers; Turrets & Cupolas; Porches, Porticos, Canopies & Awnings

EXTERIOR PRODUCTS

SPONSORED BY: BRADSTONE www.bradstone.com

WEATHERVANES, CLOCKS & CLOCK TOWERS

KEY

PRODUCT TYPES: 1= Weathervanes
2 = Clocks 3 = Clock Towers
OTHER: ▽ Reclaimed 🖰 On-line shopping
✎ Bespoke 🖐Hand-made ECO Ecological

ARCHITECTURAL HERITAGE
Taddington Manor, Taddington,
Nr Cutsdean, Cheltenham,
Gloucestershire, GL54 5RY
Area of Operation: Worldwide
Tel: 01386 584414
Fax: 01386 584236
Email: puddy@architectural-heritage.co.uk
Web: www.architectural-heritage.co.uk

B ROURKE & CO LTD
Vulcan Works, Accrington Road,
Burnley, Lancashire, BB11 5QD
Area of Operation: Worldwide
Tel: 01282 422841
Fax: 01282 458901
Email: info@rourkes.co.uk
Web: www.rourkes.co.uk
Product Type: 1, 3

CHRIS TOPP & COMPANY WROUGHT IRONWORKS
Lyndhurst, Carlton Husthwaite,
Thirsk, North Yorkshire, YO7 2BJ
Area of Operation: Worldwide
Tel: 01845 501415
Fax: 01845 501072
Email: enquiry@christopp.co.uk
Web: www.christopp.co.uk
Product Type: 1, 3
Other Info: ✎ 🖐
Material Type: C) 2, 3, 4, 5, 6, 7, 11, 12, 17

GOOD DIRECTIONS LTD
8 Bottings Industrial Estate, Hilltons Road,
Botley, Southampton, Hampshire, SO30 2DY
Area of Operation: UK & Ireland
Tel: 01489 797773
Fax: 01489 796700
Email: office@good-directions.co.uk
Web: www.good-directions.com
Product Type: 1, 2, 3
Material Type: C) 3, 7

HARRISON THOMPSON & CO. LTD. (YEOMAN RAINGUARD)
Yeoman House, Whitehall Estate,
Whitehall Road, Leeds,
West Yorkshire, LS12 5JB
Area of Operation: UK & Ireland
Tel: 0113 279 5854
Fax: 0113 231 0406
Email: info@rainguard.co.uk
Web: www.rainguard.co.uk
Product Type: 3

HAWKINS CLOCK COMPANY
PO Box 39, Market Deeping,
Peterborough, Cambridgeshire, PE6 8XQ
Area of Operation: Worldwide
Tel: 01733 330222
Fax: 01733 333700
Email: sales@hawkinsclocks.co.uk
Web: www.hawkinsclocks.co.uk
Product Type: 1, 2, 3

HAWKINS CLOCK CO

Area of Operation: Worldwide
Tel: 01733 330222 **Fax:** 01733 333700
Email: sales@hawkinsclocks.co.uk
Web: www.hawkinsclocks.co.uk
Product Type: 1, 2, 3 **Other Info:** 🖰

Bespoke or standard clocks, clock towers and
pillar clocks for any building. Design, installation
and renovation service available.

J G S METALWORK
Unit 6 Broomstick Estate, High Street,
Edlesborough, Dunstable, Bedfordshire, LU6 2HS
Area of Operation: East England
Tel: 01525 220360
Fax: 01525 222786
Email: enquiries@weathervanes.org.uk
Web: www.weathervanes.org.uk
Product Type: 1

RENAISSANCE MOULDINGS
262 Handsworth Road, Handsworth,
Sheffield, South Yorkshire, S13 9BS
Area of Operation: UK (Excluding Ireland)
Tel: 0114 244 6622
Fax: 0114 261 0472
Email: enq@rfppm.com
Web: www.rfppm.co.uk
Product Type: 1, 2, 3

ST. GILES WEATHERVANES
Station Approach, Melksham, Wiltshire, SN12 8DB
Area of Operation: UK & Ireland
Tel: 01225 707466
Fax: 01225 704241
Email: rphillips@novacast.co.uk
Web: www.stgilesweathervanes.com 🖰
Product Type: 1

WESSEX BUILDING PRODUCTS (MULTITEX)
Dolphin Industrial Estate, Southampton Road,
Salisbury, Wiltshire, SP1 2NB
Area of Operation: UK & Ireland
Tel: 01722 332139
Fax: 01722 338458
Email: sales@wessexbuildingproducts.co.uk
Web: www.wessexbuildingproducts.co.uk
Product Type: 1, 2, 3

TURRETS & CUPOLAS

KEY

PRODUCT TYPES: 1= Turrets 2 = Cupolas
OTHER: ▽ Reclaimed 🖰 On-line shopping
✎ Bespoke 🖐Hand-made ECO Ecological

CHILSTONE
Victoria Park, Fordcombe Road,
Langton Green, Kent, TN3 0RD
Area of Operation: Worldwide
Tel: 01892 740866
Fax: 01892 740249
Email: ornaments@chilstone.com
Web: www.chilstone.com 🖰
Material Type: E) 13

FERNHILL STONE UK LTD
Unit 16 Greencroft Industrial Estate,
Annfield Plain, Stanley, Durham, DH9 7XP
Area of Operation: UK & Ireland
Tel: 01207 521482
Fax: 01207 521455
Email: geoffreydriver@btconnect.com
Web: www.fernhillstone.com
Material Type: E) 4, 11

GOOD DIRECTIONS LTD
8 Bottings Industrial Estate, Hilltons Road,
Botley, Southampton, Hampshire, SO30 2DY
Area of Operation: UK & Ireland
Tel: 01489 797773
Fax: 01489 796700
Email: office@good-directions.co.uk
Web: www.good-directions.com
Product Type: 1, 2
Material Type: D) 6

HARRISON THOMPSON & CO. LTD. (YEOMAN RAINGUARD)
Yeoman House, Whitehall Estate, Whitehall Road,
Leeds, West Yorkshire, LS12 5JB
Area of Operation: UK & Ireland
Tel: 0113 279 5854
Fax: 0113 231 0406
Email: info@rainguard.co.uk
Web: www.rainguard.co.uk
Product Type: 1

HAWKINS CLOCK COMPANY
PO Box 39, Market Deeping,
Peterborough, Cambridgeshire, PE6 8XQ
Area of Operation: Worldwide
Tel: 01733 330222
Fax: 01733 333700
Email: sales@hawkinsclocks.co.uk
Web: www.hawkinsclocks.co.uk
Product Type: 1, 2

KIRK NATURAL STONE
Bridgend, Fyvie, Turriff, Aberdeenshire, AB53 8QB
Area of Operation: Worldwide
Tel: 01651 891891
Fax: 01651 891794
Email: info@kirknaturalstone.com
Web: www.kirknaturalstone.com

RENAISSANCE MOULDINGS
262 Handsworth Road, Handsworth,
Sheffield, South Yorkshire, S13 9BS
Area of Operation: UK (Excluding Ireland)
Tel: 0114 244 6622
Fax: 0114 261 0472
Email: enq@rfppm.com
Web: www.rfppm.co.uk
Product Type: 1, 2

WESSEX BUILDING PRODUCTS (MULTITEX)
Dolphin Industrial Estate, Southampton Road,
Salisbury, Wiltshire, SP1 2NB
Area of Operation: UK & Ireland
Tel: 01722 332139
Fax: 01722 338458
Email: sales@wessexbuildingproducts.co.uk
Web: www.wessexbuildingproducts.co.uk
Product Type: 1, 2

PORCHES, PORTICOS, CANOPIES & AWNINGS

KEY

OTHER: ▽ Reclaimed 🖰 On-line shopping
✎ Bespoke 🖐Hand-made ECO Ecological

ARC-COMP
Friedrichstr. 3, 12205 Berlin-Lichterfelde, Germany
Area of Operation: Europe
Tel: +49 (0)308 430 9956
Fax: +49 (0)308 430 9957
Email: jvs@arc-com.com
Web: www.arc-com.com
Material Type: C) 1, 3

ARC-COMP (IRISH BRANCH)
Whitefield Cottage, Lugduff, Tinahely,
Co. Wicklow, Republic of Ireland
Area of Operation: Europe
Tel: +353 (0)868 729 945
Fax: +353 (0)402 28900
Email: jvs@arc-comp.com
Web: www.arc-comp.com
Material Type: C) 1, 3

BALMORAL BLINDS
12 Beresford Close, Frimley Green,
Camberley, Surrey, GU16 6LB
Area of Operation: Greater London,
South East England
Tel: 01252 674172
Fax: 0870 132 7683
Email: sales@balmoralblinds.co.uk
Web: www.balmoralblinds.co.uk

BORDER CONCRETE PRODUCTS
Jedburgh Road, Kelso, Borders, TD5 8JG
Area of Operation: North East England,
North West England and North Wales, Scotland
Tel: 01573 224393
Fax: 01573 276360
Email: sales@borderconcrete.co.uk
Web: www.borderconcrete.co.uk
Other Info: ✎ 🖐

BRADSTONE STRUCTURAL
Aggregate Industries UK Ltd,
North End, Ashton Keynes,
Swindon, Wiltshire, SN6 3QX
Area of Operation: UK (Excluding Ireland)
Tel: 01285 646884
Fax: 01285 646891
Email: bradstone.structural@aggregate.com
Web: www.bradstone.com

CHILSTONE
Victoria Park, Fordcombe Road,
Langton Green, Kent, TN3 0RD
Area of Operation: Worldwide
Tel: 01892 740866
Fax: 01892 740249
Email: ornaments@chilstone.com
Web: www.chilstone.com 🖰
Material Type: E) 13

CHISHOLM & COMPANY
Ty Gwyn Farm, Llandrindod Wells,
Powys, LD1 5NY
Area of Operation: Worldwide
Tel: 01597 824404
Fax: 01597 824123
Email: info@chisholmcompany.co.uk
Web: www.chisholmcompany.co.uk

CORNICES CENTRE
2 Slade Court, 230 Walm Lane,
London, NW2 3BT
Area of Operation: Greater London
Tel: 0208 452 3310
Fax: 0208 452 3310
Email: info@cornicescentre.co.uk
Web: www.cornicescentre.co.uk
Material Type: I) 1, 3, 4

DAR INTERIORS
Arch 11, Miles Street,
London, SW8 1RZ
Area of Operation: Worldwide
Tel: 0207 720 9678
Fax: 0207 627 5129
Email: enquiries@darinteriors.com
Web: www.darinteriors.com

ERMINE ENGINEERING COMPANY LTD
Francis House, Silver Birch Park,
Great Northern Terrace, Lincoln,
Lincolnshire, LN5 8LG
Area of Operation: UK (Excluding Ireland)
Tel: 01522 510977
Fax: 01522 510929
Email: info@ermineengineering.co.uk
Web: www.ermineengineering.co.uk
Other Info: ✎ 🖐
Material Type: C) 1, 2, 3, 4, 18

EXTERIOR PRODUCTS *(side tab)*

" Art lasts long, life is short. "

Hippocrates 460-377BC

Classic

Choose Haddonstone stonework, and you can be sure it will make a lasting impression. Whether it's a column, a balustrade, a door surround or a quoin it will be both elegant and permanent.

For details of our extensive cast stone range, contact us for our 172 page catalogue. We even offer a CD Rom which contains our full catalogue, technical specification sheets, assembly advice, CAD drawings and video (£5 each, refundable with your purchase).

We also pride ourselves on our custom-made designs. If you can imagine it, we can make it ~ and it will always be an enduring classic.

Visit us at www.haddonstone.co.uk

Haddonstone Ltd, The Forge House, East Haddon, Northampton NN6 8DB
Tel: 01604 770711 info@haddonstone.co.uk

Offices also in: California, Colorado and New Jersey.

FERNHILL STONE UK LTD
Unit 16 Greencroft Industrial Estate, Annfield Plain, Stanley, Durham, DH9 7XP
Area of Operation: UK & Ireland
Tel: 01207 521482 **Fax:** 01207 521455
Email: geoffreydriver@btconnect.com
Web: www.fernhillstone.com
Material Type: E) 4, 11

GEORGE WOODS
Unit C3b, Westleigh Willows, Feniton, Honiton, Devon, EX14 3BN
Area of Operation: Worldwide
Tel: 01404 851336 **Fax:** 01404 851372
Email: georgewoods@tiscali.co.uk
Web: www.georgewoods.co.uk

GRANT MERCER
PO Box 246, Banstead, Surrey, SM7 3AF
Area of Operation: Europe
Tel: 01737 357957 **Fax:** 01737 373003
Email: grantmercer@shuttersuk.co.uk
Web: www.shuttersuk.com

40 YEAR WARRANTY **SIMPLE INSTALLATION**

Completely Decorate the exterior of your home in superb maintenance free material.

Shutters, Dentils, Door-Surrounds Porticos and much more.

GRANT MERCER
Area of Operation: UK & Europe
Tel: 01737 357957
Fax: 01737 373003
Email: grantmercer@shuttersuk.com
Web: www.shuttersuk.com

*WHOLE HOUSE SYSTEM in MAINTENANCE FREE MATERIAL
SHUTTERS - DOOR SURROUNDS - DENTILS PORTICOS - and MUCH MORE
EASY TO INSTALL - 40 YEAR GUARANTEE*

HADDONSTONE LTD
The Forge House, East Haddon, Northampton, Northamptonshire, NN6 8DB
Area of Operation: Worldwide
Tel: 01604 770711
Fax: 01604 770027
Email: info@haddonstone.co.uk
Web: www.haddonstone.co.uk

HADDONSTONE
Area of Operation: Worldwide
Tel: 01604 770711 **Fax:** 01604 770027
Email: info@haddonstone.co.uk
Web: www.haddonstone.co.uk
Other Info:

Haddonstone has an extensive range of columns, half columns, pilasters and entablatures to create porticos, door surrounds and porches to meet individual design requirements.

INTERLAND TRADING LTD
19 Limetree Close, Cambridge, Cambridgeshire, CB1 8PF
Area of Operation: UK & Ireland
Tel: 01223 246242 **Fax:** 01223 414944
Email: mail@interland-trading.co.uk
Web: www.interland-trading.co.uk

KENT BALUSTERS
1 Gravesend Road, Strood, Kent, ME2 3PH
Area of Operation: UK (Excluding Ireland)
Tel: 01634 711617
Fax: 01634 714644
Email: info@kentbalusters.co.uk
Web: www.kentbalusters.co.uk
Other Info:
Material Type: E) 13

KIRK NATURAL STONE
Bridgend, Fyvie, Turriff, Aberdeenshire, AB53 8QB
Area of Operation: Worldwide
Tel: 01651 891891
Fax: 01651 891794
Email: info@kirknaturalstone.com
Web: www.kirknaturalstone.com

MEADOWSTONE (DERBYSHIRE) LTD
West Way, Somercotes, Derbyshire, DE55 4QJ
Area of Operation: UK & Ireland
Tel: 01773 540707
Fax: 01773 528119
Email: info@meadowstone.co.uk
Web: www.meadowstone.co.uk
Other Info:

OLDE WORLDE OAK JOINERY LTD
Unit 12, Longford Industrial Estate, Longford Road, Cannock, Staffordshire, WS11 0DG
Area of Operation: Europe
Tel: 01543 469328
Fax: 01543 469341
Email: sales@oldeworldeoakjoinery.co.uk
Web: www.oldeworldeoakjoinery.co.uk

RENAISSANCE MOULDINGS
262 Handsworth Road, Handsworth, Sheffield, South Yorkshire, S13 9BS
Area of Operation: UK (Excluding Ireland)
Tel: 0114 244 6622
Fax: 0114 261 0472
Email: enq@rfppm.com
Web: www.rfppm.co.uk
Other Info:

RICHARD BURBIDGE LTD
Whittington Road, Oswestry, Shropshire, SY11 1HZ
Area of Operation: Worldwide
Tel: 01691 655131
Fax: 01691 657694
Email: info@richardburbidge.co.uk
Web: www.richardburbidge.co.uk

SCULPTURE GRAIN LIMITED
Unit 8 Warren Court, Knockholt Road, Halstead, Sevenoaks, Kent, TN14 7ER
Area of Operation: UK & Ireland
Tel: 01959 534060
Fax: 01959 532696
Email: mick@sculpturegrain.freeserve.co.uk
Material Type: A) 2

SEDGEMOOR STONE
Pen Mill Station Yard, Yeovil, Somerset, BA21 5DD
Area of Operation: UK (Excluding Ireland)
Tel: 01935 429797
Fax: 01935 432392
Email: info@sedgemoorstone.co.uk
Web: www.sedgemoorstone.co.uk
Material Type: E) 11, 13

SUPERIOR FASCIAS
Adelaide House, Portsmouth Road, Lowford, Southampton, Hampshire, SO31 8EQ
Area of Operation: South East England, South West England and South Wales
Tel: 0700 596 4603
Fax: 0700 596 4609
Email: info@superiorfascias.co.uk
Web: www.superiorfascias.co.uk

THE PAST RE-CAST LIMITED
Mere House, Harleston Road,
Dickleburgh, Diss, Kent, IP21 4PD
Area of Operation: UK & Ireland
Tel: 01379 741381 **Fax:** 01379 741534
Email: sales@thepastrecast.com
Web: www.thepastrecast.com

TRANIK HOUSE DOORSTORE
63 Cobham Road, Ferndown Industrial Estate,
Wimborne, Dorset, BH21 7QA
Area of Operation: UK (Excluding Ireland)
Tel: 01202 872211
Email: sales@tranik.co.uk
Web: www.tranik.co.uk ⏻

UK AWNINGS
8-10 Ruxley Lane, Ewell, Epsom, Surrey, KT19 0JD
Area of Operation: East England, Greater London,
Midlands & Mid Wales, North East England, North
West England and North Wales, South East England,
South West England and South Wales
Tel: 0800 071 8888
Email: info@conservatoryblinds.co.uk
Web: www.uk-awnings.com

VOUSTONE DESIGNS LIMITED.
Kingdom Cottage, Tibbs Court Lane,
Brenchley, Nr. Tonbridge, Kent, TN12 7AH
Area of Operation: UK (Excluding Ireland)
Tel: 01892 722449 **Fax:** 01892 722573
Email: pauleen@voustone.co.uk
Web: www.voustone.co.uk
Material Type: E) 13

WARMSWORTH STONE LTD
1-3 Sheffield Road, Warmsworth,
Doncaster, South Yorkshire, DN4 9QH
Area of Operation: UK & Ireland
Tel: 01302 858617 **Fax:** 01302 855844
Email: info@warmsworth-stone.co.uk
Web: www.warmsworth-stone.co.uk
Other Info: ✎ ✋ **Material Type:** E) 4, 5

WESSEX BUILDING PRODUCTS (MULTITEX)
Dolphin Industrial Estate, Southampton Road,
Salisbury, Wiltshire, SP1 2NB
Area of Operation: UK & Ireland
Tel: 01722 332139 **Fax:** 01722 338458
Email: sales@wessexbuildingproducts.co.uk
Web: www.wessexbuildingproducts.co.uk

CHIMNEYS

KEY
OTHER: ▽ Reclaimed ⏻ On-line shopping
✎ Bespoke ✋ Hand-made ECO Ecological

ANKI CHIMNEY SYSTEMS
Bishops Way, Newport, Isle of Wight, PO30 5WS
Area of Operation: UK (Excluding Ireland)
Tel: 01983 527997 **Fax:** 01983 537788
Email: anki@ajwells.co.uk **Web:** www.anki.co.uk

ASSOCIATED PLASTIC COMPONENTS
Unit 5, Kingston International Business Park,
Hedon Road, Hull, East Riding of Yorks, HU9 5PE
Area of Operation: UK (Excluding Ireland)
Tel: 01482 783631
Fax: 01482 783292
Email: sales@apcmouldings.co.uk
Web: www.apcmouldings.co.uk

BORDER CONCRETE PRODUCTS
Jedburgh Road, Kelso, Borders, TD5 8JG
Area of Operation: North East England,
North West England and North Wales, Scotland
Tel: 01573 224393
Fax: 01573 276360
Email: sales@borderconcrete.co.uk
Web: www.borderconcrete.co.uk
Other Info: ✎ ✋

CAWARDEN BRICK CO. LTD
Cawarden Springs Farm, Blithbury Road,
Rugeley, Staffordshire, WS15 3HL
Area of Operation: UK (Excluding Ireland)
Tel: 01889 574066
Fax: 01889 575695
Email: home-garden@cawardenreclaim.co.uk
Web: www.cawardenreclaim.co.uk

DUNBRIK LTD
Ferry Lane, Stanley Ferry, Wakefield,
West Yorkshire, WF3 4LT
Area of Operation: UK & Ireland
Tel: 01924 373694
Fax: 01924 383459
Email: tech@dunbrik.co.uk
Web: www.dunbrik.co.uk

HADDONSTONE LTD
The Forge House, East Haddon,
Northampton, Northamptonshire, NN6 8DB
Area of Operation: Worldwide
Tel: 01604 770711
Fax: 01604 770027
Email: info@haddonstone.co.uk
Web: www.haddonstone.co.uk

HLD LTD
The Old Shipyard, Gainsborough,
Lincolnshire, DN21 ING
Area of Operation: UK & Ireland
Tel: 01427 611800
Fax: 01427 612867
Email: technical@hld.co.uk
Web: www.hld.co.uk
Other Info: ✎

NIGEL TYAS HANDCRAFTED IRONWORK
Green Works, 18 Green Road,
Penistone, Sheffield,
South Yorkshire, S36 6BE
Area of Operation: Worldwide
Tel: 01226 766618
Email: sales@nigeltyas.co.uk
Web: www.nigeltyas.co.uk
Other Info: ✎ ✋
Material Type: A) 2, 4

NUMBER 9 STUDIO UK ARCHITECTURAL CERAMICS
Mole Cottage Industries,
Mole Cottage, Watertown,
Chittlehamholt, Devon, EX37 9HF
Area of Operation: Worldwide
Tel: 01769 540471
Fax: 01769 540471
Email: arch.ceramics@moley.uk.com
Web: www.moley.uk.com

PLASTITRIM
Unit 2, New Road Business Estate,
Ditton, Maidstone, Kent, ME20 6AF
Area of Operation: South East England
Tel: 01732 873808
Fax: 01732 874344
Email: info@plastitrim.co.uk
Web: www.plastitrim.co.uk

RED BANK MANUFACTURING COMPANY LIMITED
Measham, Swadlincote,
Derbyshire, DE12 7EL
Area of Operation: UK & Ireland
Tel: 01530 270333
Fax: 01530 273667
Email: sales@redbankmfg.co.uk
Web: www.redbankmfg.co.uk

RENAISSANCE MOULDINGS
262 Handsworth Road,
Handsworth, Sheffield,
South Yorkshire, S13 9BS
Area of Operation: UK (Excluding Ireland)
Tel: 01142 446622
Fax: 01142 610472
Email: enq@rfppm.com
Web: www.rfppm.co.uk

ROOFDRAGON.COM
63 Moseley Street,
Southend-on-Sea, Essex, SS2 4NL
Area of Operation: UK & Ireland
Tel: 01702 467444
Email: colin@roofdragon.com
Web: www.roofdragon.com ⏻
Other Info: ▽

SCHIEDEL-ISOKERN CHIMNEY SYSTEMS
14 Haviland Road, Ferndown Industrial Estate,
Wimborne, Dorset, BH21 7RF
Area of Operation: UK & Ireland
Tel: 01202 861650
Fax: 01202 861632
Email: sales@isokern.co.uk
Web: www.isokern.co.uk

THE CHIMNEY POT SHOP
Unit 41 Birch Road East Industrial Estate,
Witton, Birmingham, West Midlands, B6 7DA
Area of Operation: Worldwide
Tel: 0121 3277776
Email: info@thechimneypotshop.com
Web: www.thechimneypotshop.com

THE EXPANDED METAL COMPANY LIMITED
PO Box 14, Longhill Industrial Estate (North),
Hartlepool, Durham, TS25 1PR
Area of Operation: Worldwide
Tel: 01429 408087
Fax: 01429 866795
Email: paulb@expamet.co.uk
Web: www.expandedmetalcompany.co.uk

WESSEX BUILDING PRODUCTS (MULTITEX)
Dolphin Industrial Estate,
Southampton Road, Salisbury,
Wiltshire, SP1 2NB
Area of Operation: UK & Ireland
Tel: 01722 332139
Fax: 01722 338458
Email: sales@wessexbuildingproducts.co.uk
Web: www.wessexbuildingproducts.co.uk

NOTES

Company Name ....

Address ....

email ....

Web ....

Company Name ....

Address ....

email ....

Web ....

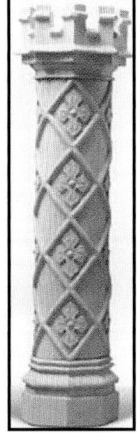
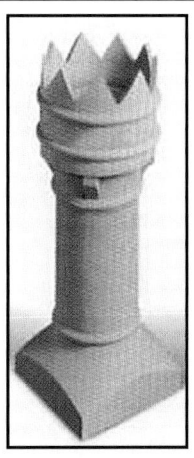

Image courtesy of Gliderol Garage & Industrial Doors (0191 518 0455)

SPONSORED BY GLIDEROL GARAGE & INDUSTRIAL DOORS LTD
Tel 0191 518 0455 Web www.gliderol.co.uk

GLIDEROL
GARAGE &
INDUSTRIAL
DOORS LTD

EXTERIOR PRODUCTS

Garages & Outbuildings

In designing your overall scheme, you need to consider what other structures are required besides the main house. Garaging for the car would be top of most people's wish lists and provision of a garage is now universal in larger homes. Whilst it is a considerable expense, with the structure costing almost as much per unit area as the house does, a secure garage can add substantially to the value of your home.

Provided the details of the garage are included with the original planning consent, then you can choose to build the structure in your own good time (although bear in mind that you will not be able to reclaim VAT if you wait until the house is completed). In fact, many self-builders actually choose to build their garage first to provide a secure storage area from which to manage the rest of the build.

The same planning principle applies to other structures you might wish to build in the longer term - if you want any of these features and wish to postpone their construction until a later date, you should still include them with your original planning application. However, with structures such as sheds and summer houses you can usually afford to take a more relaxed approach. The planning regulations allow you to erect any sort of structure without permission, providing it doesn't constitute a separate dwelling which:

- is no higher than 4m (3m for a flat roof)
- is less than 50% of the garden area

- is further away from the highway than the main building (or at least 20m away).

Note however, that special rules apply within conservation areas, in the setting of Listed Buildings, or if your Permitted Development Rights have been expressly removed by the planners for some reason.

Because garages are generally located adjacent to the front of your house, it is important to match it to the style of your home. Similarly, whilst your choice of garage door may seem unimportant, it is worth remembering that it can account for 15–20% of the front elevation. It is, therefore, important that you choose a style which suits your house, street scene and locality. It is also important to consider glazing and insulation, as garages do not just have to be somewhere to keep the car. Many people also choose to use them for storage or as a workshop or playroom. If you intend to do this then it may be a good idea to install either rooflights or partially glazed doors to save on electricity bills.

ABOVE: A side hung door can make a garage less prominent. Example by Garador.
BELOW: Retractable double door by Jeld-Wen

GARAGES AND CARPORTS

KEY

OTHER: ▽ Reclaimed 🖱 On-line shopping
✎ Bespoke ✋ Hand-made ECO Ecological

BEAVER TIMBER COMPANY
Barcaldine, Argyll & Bute, PA37 1SG
Area of Operation: UK (Excluding Ireland)
Tel: 01631 720353 **Fax:** 01631 720430
Email: info@beavertimber.co.uk
Web: www.beavertimber.co.uk 🖱

CEDAR SELF-BUILD HOMES
PO Box 3167, Chester, Cheshire, CH4 0ZH
Area of Operation: Europe
Tel: 01244 661048 **Fax:** 01244 660707
Email: info@cedar-self-build.com
Web: www.cedar-self-build.com

CHISHOLM & COMPANY
Ty Gwyn Farm, Llandrindod Wells, Powys, LD1 5NY
Area of Operation: Worldwide
Tel: 01597 824404 **Fax:** 01597 824123
Email: info@chisholmcompany.co.uk
Web: www.chisholmcompany.co.uk

COMPTON BUILDINGS LTD
Station Works, Fenny Compton,
Southam, Warwickshire, CV47 2XB
Area of Operation: UK & Ireland
Tel: 0800 975 8860 **Fax:** 01295 770748
Email: sales@compton-buildings.co.uk
Web: www.comptonbuildings.co.uk **Other Info:** ✎

DENCROFT GARAGES LTD
230 Bradford Road, Batley,
West Yorkshire, WF17 6JD
Area of Operation: UK (Excluding Ireland)
Tel: 01924 461996 **Fax:** 01924 465157
Email: phil.denton@btclick.com
Web: www.dencroftgarages.co.uk

EASYGATES
75 Long Lane, Halesowen, Birmingham,
West Midlands, B62 9DJ
Area of Operation: Europe
Tel: 0870 760 6536 **Fax:** 0121 561 3395
Email: info@easygates.co.uk
Web: www.easygates.co.uk

ENGLISH HERITAGE BUILDINGS
Coldharbour Farm Estate, Woods Corner,
East Sussex, TN21 9LQ
Area of Operation: Europe
Tel: 01424 838643 **Fax:** 01424 838606
Email: info@ehbp.com **Web:** www.ehbp.com

LINDAB LTD
Unit 7 Block 2, Shenstone Trading Estate,
Bromsgrove Road, Halesowen, West Midlands,
B63 3XB
Area of Operation: Worldwide
Tel: 0121 585 2780 **Fax:** 0121 585 2782
Email: building.products@lindab.co.uk
Web: www.lindab.com

OSMO UK LTD
Unit 2 Pembroke Road, Stocklake Industrial Estate,
Aylesbury, Buckinghamshire, HP20 1DB
Area of Operation: UK & Ireland
Tel: 01296 481220
Fax: 01296 424090
Email: info@osmouk.com
Web: www.osmouk.com

ROOF-TEK LTD
19 Lynx Crescent, Weston-Super-Mare,
Somerset, BS24 9DJ
Area of Operation: East England, Greater London,
Midlands & Mid Wales, South East England, South
West England and South Wales
Tel: 01934 642929 **Fax:** 01934 644290
Email: roof.tekltd@btinternet.com
Web: www.rooftek.co.uk

THE LOG CABIN COMPANY
Potash Garden Centre, 9 Main Road, Hockley,
Hawkwell, Essex, SS5 4JN
Area of Operation: UK (Excluding Ireland)
Tel: 01702 206012
Email: sales@thelogcabincompany.co.uk
Web: www.thelogcabincompany.co.uk

GARAGE DOORS

KEY

PRODUCT TYPES: 1= Automatic 2 = Roller
3 = Sectional 4 = Barn Style 5 = Other
OTHER: ▽ Reclaimed 🖱 On-line shopping
✎ Bespoke ✋ Hand-made ECO Ecological

AMOURELLE PRODUCTS LTD
8 Ambleside Close, Woodley, Reading,
Berkshire, RG5 4JJ
Area of Operation: UK & Ireland
Tel: 0118 969 4657
 Fax: 0118 962 8682
Email: nick@amourelle.co.uk
Web: www.garage-door-automation.co.uk 🖱
Product Type: 1

ARRIDGE GARAGE DOORS
68-70 Roft Street, Oswestry,
Shropshire, SY11 2EP
Area of Operation: UK (Excluding Ireland)
Tel: 01691 670394
Fax: 01691 670425
Email: info@garagedoor.uk.com
Web: www.discount-garage-doors.co.uk

ARRIDGE GARAGE DOORS LTD

Area of Operation: UK (Excluding Ireland)
Tel: 01691 670394 or 01691 671264
Email: info@garagedoor.uk.com
Web: www.discount-garage-doors.co.uk

Arridge Garage Doors Ltd., Est.1989 specialises
in the supply only of top brand garage doors and
remote controls at big discounts. Nationwide
delivery. Clear fitting instructions with all doors
and remote control units.

B ROURKE & CO LTD
Vulcan Works, Accrington Road,
Burnley, Lancashire, BB11 5QD
Area of Operation: Worldwide
Tel: 01282 422841
Fax: 01282 458901
Email: info@rourkes.co.uk
Web: www.rourkes.co.uk
 Product Type: 1, 2

CARDALE DOORS LTD
Farm Road, Buckingham Road Industrial Estate,
Brackley, Northamptonshire, NN13 7EA
Area of Operation: UK & Ireland
Tel: 01280 703022 / 0800 581082
Fax: 01280 844544
Email: info@cardale.co.uk
Web: www.cardale.co.uk
Product Type: 1, 2, 3, 4, 5
Other Info: ✎ ✋

CHANTRY HOUSE OAK LTD
Ham Farm, Church Lane, Oving, Chichester,
West Sussex, PO20 2BT
Area of Operation: South East England
Tel: 01243 776811 **Fax:** 01243 839873
Email: sales@chantryhouseoak.co.uk
Web: www.chantryhouseoak.co.uk
Product Type: 4, 5

CLEVELAND UP & OVER DOOR CO.
7 Metcalfe Road, Skippers Lane Industrial Estate,
Middlesbrough, North Yorkshire, TS6 6PT
Area of Operation: North East England
Tel: 01642 440920 **Fax:** 01642 456106
Email: clevelandupover@btconnect.com
Web: www.clevelandupover.co.uk
Product Type: 1, 2, 3, 5

DISCOUNT GARAGE DOORS LTD
68-70 Roft Street, Oswestry, Shropshire, SY11 2EP
Area of Operation: UK (Excluding Ireland)
Tel: 01691 670394
Email: info@garagedoor.uk.com
Web: www.discount-garage-doors.co.uk 🖱
Product Type: 1, 2, 3, 4, 5

DYNASTY DOORS
Unit 3, The Micro Centre, Gillette Way,
Reading, Berkshire, RG2 0LR
Area of Operation: South East England,
South West England and South Wales
Tel: 0118 987 4000 **Fax:** 0118 9212 999
Email: martin@dynastydoors.co.uk
Web: www.dynastydoors.co.uk
Product Type: 1, 2, 3, 4

EASYGATES
75 Long Lane, Halesowen, Birmingham,
West Midlands, B62 9DJ
Area of Operation: Europe
Tel: 0870 760 6536 **Fax:** 0121 561 3395
Email: info@easygates.co.uk
Web: www.easygates.co.uk
Product Type: 1

ENGELS WINDOWS & DOORS LTD
1 Kingley Centre, Downs Road, West Stoke,
Chichester, West Sussex, PO18 9HJ
Area of Operation: UK (Excluding Ireland)
Tel: 01243 576633 **Fax:** 01243 576644
Email: admin@engels.co.uk
Web: www.engels.co.uk

EVEREST LTD
Sopers Road, Cuffley, Hertfordshire, EN6 4SG
Area of Operation: UK & Ireland
Tel: 0800 010123
Web: www.everest.co.uk 🖱
Product Type: 1, 2

GARADOR LTD
Bunford Lane, Yeovil, Somerset, BA20 2YA
Area of Operation: UK & Ireland
Tel: 01935 443700 **Fax:** 01935 443744
Email: sally.howell@garador.co.uk
Web: www.garador.co.uk
Product Type: 1, 3, 4

GARAGE DOORS (NORTHERN) LTD
Aspden Street, Bamber Bridge,
Preston, Lancashire, PR5 6TL
Area of Operation: North West England
and North Wales
Tel: 01772 334828
Fax: 01772 627877
Email: post@garagedoorslancs.co.uk
Web: www.garagedoorslancs.co.uk
Product Type: 1, 2, 3, 4, 5

**GLIDEROL GARAGE
& INDUSTRIAL DOORS LIMITED**
Davy Drive, North West Industrial Estate,
Peterlee, Durham, SR8 2JF
Area of Operation: UK (Excluding Ireland)
Tel: 0191 518 0455
Fax: 0191 518 0548
Email: info@gliderol.co.uk
Web: www.gliderol.co.uk
Product Type: 1, 2

**GLIDEROL GARAGE & INDUSTRIAL
DOORS LTD**
Area of Operation: UK (Excluding Ireland)
Tel: 0800 834 250 **Fax:** 0191 518 0548
Email: info@gliderol.co.uk
Web: www.gliderol.co.uk
Product Type: 1, 2

Easy to use, space saving, reliable, quiet, secure,
low maintenance, manual and automatic garage
doors available in a wide range of colours.
Smartens your home and your driveway.

HENDERSON GARAGE DOORS
Durham Road, Bowburn, Durham, DH6 5NG
Area of Operation: UK & Ireland
Tel: 0191 377 0701 **Fax:** 0191 377 0855
Email: marketing@pchenderson.com
Web: www.pchenderson.com
Product Type: 1, 2, 3

HORMANN (UK) LTD
Gee Road, Coalville, Leicestershire, LE67 4JW
Area of Operation: Europe
Tel: 01530 513055 **Fax:** 01530 513051
Email: info@hormann.co.uk
Web: www.hormann.co.uk
Product Type: 1, 2, 3

ILC GARAGE DOORS LTD
South Street, Ossett, Wakefield,
West Yorkshire, WF5 8LE
Area of Operation: UK & Ireland
Tel: 01924 280809 **Fax:** 01924 277453
Email: pat@ilcgaragedoors.co.uk
Web: www.silvelox.com

JELD-WEN
Watch House Lane, Doncaster,
South Yorkshire, DN5 9LR
Area of Operation: Worldwide
Tel: 01302 394000 **Fax:** 01302 787383
Email: customer-services@jeld-wen.co.uk
Web: www.jeld-wen.co.uk

MERIDIAN GARAGE DOORS
Building 237, Bournemouth International Airport,
Christchurch, Dorset, BH23 6NE
Area of Operation: UK (Excluding Ireland)
Tel: 01202 577440 **Fax:** 01202 577449
Email: info@meridian-garage-doors.co.uk
Web: www.meridian-garage-doors.co.uk
Product Type: 1, 3, 5

PREMIER GARAGE DOORS & GATES
Unit 7 Sparrowhall Business Park,
Leighton Road, Edlesborough,
Dunstable, Bedfordshire, LU6 2ES
Area of Operation: East & South East England
Tel: 01525 220212 **Fax:** 01525 222201
Email: sales@premiergaragedoors.co.uk
Web: www.premiergaragedoors.co.uk

PROGRESSIVE SYSTEMS UK
The White House, Downfield,
Stroud, Gloucestershire, GL5 4HJ
Area of Operation: UK & Ireland
Tel: 01453 762262
Fax: 01453 764693
Email: info@progressive-systems.co.uk
Web: progressive-systems.co.uk
Product Type: 1, 2
Other Info: ✎

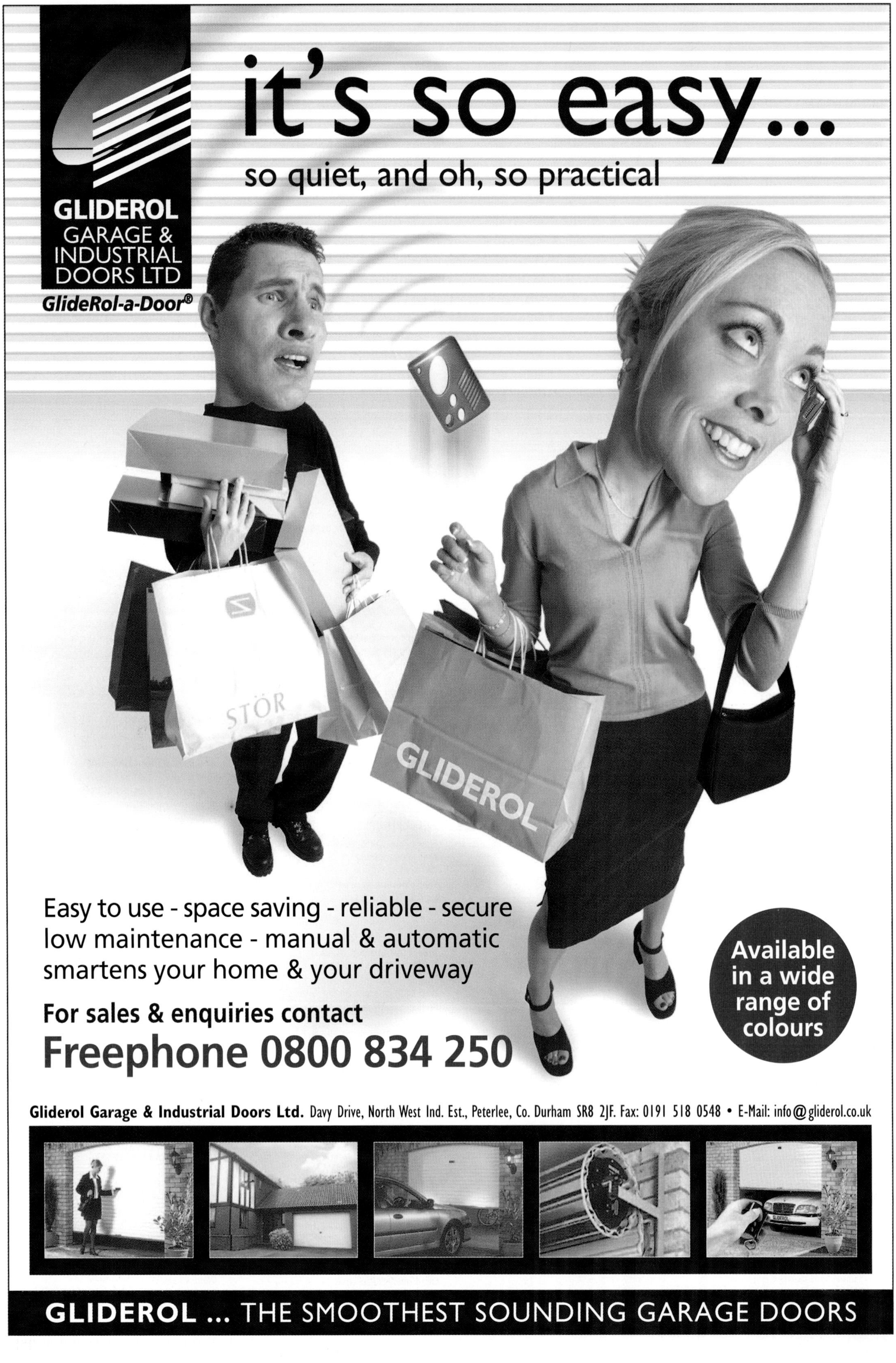

it's so easy...
so quiet, and oh, so practical

GLIDEROL
GARAGE &
INDUSTRIAL
DOORS LTD
GlideRol-a-Door®

Easy to use - space saving - reliable - secure
low maintenance - manual & automatic
smartens your home & your driveway

For sales & enquiries contact
Freephone 0800 834 250

Available in a wide range of colours

Gliderol Garage & Industrial Doors Ltd. Davy Drive, North West Ind. Est., Peterlee, Co. Durham SR8 2JF. Fax: 0191 518 0548 • E-Mail: info@gliderol.co.uk

RO-DOR LIMITED
Stevens Drove, Houghton,
Stockbridge, Hampshire, SO20 6LP
Area of Operation: UK & Ireland
Tel: 01794 388080
Fax: 01794 388090
Email: rdc46@dial.pipex.com
Web: www.ro-dor.co.uk
Product Type: 1, 2, 3, 5
Other Info: ✏ ✋

RUNDUM MEIR
1 Trouthbeck Road, Liverpool,
Merseyside, L18 3LF
Area of Operation: UK (Excluding Ireland)
Tel: 0151 280 6626
Fax: 0151 737 2504
Email: info@rundum.co.uk
Web: www.rundum.co.uk
Product Type: 1, 2, 3

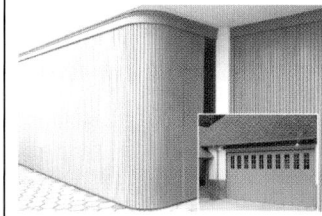

RUNDUM MEIR

Area of Operation: UK (Excluding Ireland)
Tel: 0151 280 6626
Fax: 0151 737 2504
Email: info@rundum.co.uk
Web: www.rundum.co.uk
Product Type: 1, 2, 3

Rundum Meir has been fulfiling customers garage
door requirements since 1968. Bespoke or standard
specification for round the corner, overhead and side
sectional doors.
We manufacture the door and sliding gear as one
package. Manufactured from quality timber,
aluminium or steel. Manual or unique remote
operation. Contact us for a FREE brochure.

SWS UK
Claughton, Lancaster, Lancashire, LA2 9LA
Area of Operation: UK & Ireland
Tel: 01524 772400
Fax: 01524 772411
Email: info@swsuk.co.uk
Web: www.swsuk.co.uk ⌂
Product Type: 1

URBAN FRONT LTD
Design Studio, 1 Little Hill, Heronsgate,
Rickmansworth, Hertfordshire, WD3 5BX
Area of Operation: UK & Ireland
Tel: 0870 609 1525
Fax: 0870 609 3564
Email: elizabeth@urbanfront.co.uk
Web: www.urbanfront.co.uk

WAYNE-DALTON UK
Unit 17, Redbank Court, Redbank,
Manchester, M4 4HF
Area of Operation: UK & Ireland
Tel: 0870 754 1660
Fax: 0870 754 1661
Email: eric@waynedalton.co.uk
Web: www.wayne-dalton.com
Product Type: 1, 3, 5
Other Info: ECO ✏ ✋

WESSEX GARAGE DOORS LTD.
Bessemer Close, Ebblake Industrial Estate,
Verwood, Dorset, BH31 6AZ
Area of Operation: UK (Excluding Ireland)
Tel: 01202 825451
Fax: 01202 823242
Email: info@wessexdoors.co.uk
Web: www.wessexdoors.co.uk
Product Type: 1, 2, 3, 5
Other Info: ✏ ✋

OUTHOUSES

KEY
PRODUCT TYPES: 1= Gazebos
2 = Pergolas 3 = Summerhouses
OTHER: ▽ Reclaimed ⌂ On-line shopping
✏ Bespoke ✋ Hand-made ECO Ecological

ALTHAM HARDWOOD CENTRE LTD
Altham Corn Mill, Burnley Road, Altham,
Accrington, Lancashire, BB5 5UP
Area of Operation: UK & Ireland
Tel: 01282 771618 **Fax:** 01282 777932
Email: info@oak-beams.co.uk
Web: www.oak-beams.co.uk
Product Type: 1, 2
Other Info: ECO ✏ ✋

AMDEGA CONSERVATORIES
Woodside, Church Lane, Bursledon,
Southampton, Hampshire, SO31 8AB
Area of Operation: Worldwide
Tel: 0800 591523
Web: www.amdega-conservatories.co.uk
Product Type: 3

BEAVER TIMBER COMPANY
Barcaldine, Argyll & Bute, PA37 1SG
Area of Operation: UK (Excluding Ireland)
Tel: 01631 720353 **Fax:** 01631 720430
Email: info@beavertimber.co.uk
Web: www.beavertimber.co.uk ⌂
Product Type: 3

CHARLES PEARCE LTD
26 Devonshire Road, London, W4 2HD
Area of Operation: Worldwide
Tel: 0208 995 3333
Email: enquiries@charlespearce.co.uk
Web: www.charlespearce.co.uk

CHISHOLM & COMPANY
Ty Gwyn Farm, Llandrindod Wells, Powys, LD1 5NY
Area of Operation: Worldwide
Tel: 01597 824404 **Fax:** 01597 824123
Email: info@chisholmcompany.co.uk
Web: www.chisholmcompany.co.uk
Product Type: 1

COMPTON BUILDINGS LTD
Station Works, Fenny Compton,
Southam, Warwickshire, CV47 2XB
Area of Operation: UK & Ireland
Tel: 0800 975 8860 **Fax:** 01295 770748
Email: sales@compton-buildings.co.uk
Web: www.comptonbuildings.co.uk
Product Type: 3

DANDF GARDEN PRODUCTS LTD
Unit 6 Onward Business Park, Ackworth,
West Yorkshire, WF7 7BE
Area of Operation: UK (Excluding Ireland)
Tel: 01977 624200 **Fax:** 01977 624201
Email: info@dandf.co.uk
Web: www.dandf.co.uk
Product Type: 1, 2, 3

ENGLISH HERITAGE BUILDINGS
Coldharbour Farm Estate, Woods Corner,
East Sussex, TN21 9LQ
Area of Operation: Europe
Tel: 01424 838643
Fax: 01424 838606
Email: info@ehbp.com
Web: www.ehbp.com
Product Type: 1, 3

FAIROAK TIMBER PRODUCTS LIMITED
Manor Farm, Chilmark, Salisbury,
Wiltshire, SP3 5AG
Area of Operation: UK (Excluding Ireland)
Tel: 01722 716779
Fax: 01722 716761
Email: enquiries@fairoakwindows.co.uk
Web: www.fairoakwindows.co.uk

FAIROAK TIMBER PRODUCTS LTD

Area of Operation: UK (Excluding Ireland)
Tel: 01722 716779
Fax: 01722 716761
Email: enquiries@fairoakwindows.co.uk
Web: www.fairoakwindows.co.uk

Timber garden offices based on the company's
tried and tested sections with insulated walls
and floor, and cedar shingle roof.

FOREST GARDEN
Units 291 & 296, Hartlebury Trading Estate,
Hartlebury, Worcestershire, DY10 4JB
Area of Operation: UK (Excluding Ireland)
Tel: 0870 300 9809
Fax: 0870 191 9888
Email: info@forestgarden.co.uk
Web: www.forestgarden.co.uk
Product Type: 1, 2

GEORGE BARKER & SONS
Backbarrow, Nr Ulverston, Cumbria, LA12 8TA
Area of Operation: UK (Excluding Ireland)
Tel: 01539 531236 **Fax:** 01539 530801
Web: www.gbs-ltd.co.uk

HONEYSUCKLE BOTTOM SAWMILL LTD
Honeysuckle Bottom, Green Dene, East Horsley,
Leatherhead, Surrey, KT24 5TD
Area of Operation: Greater London, South East England
Tel: 01483 282394 **Fax:** 01483 282394
Email: honeysucklemill@aol.com
Web: www.easisites.co.uk/honeysucklebottomsawmill
Product Type: 1, 2

INSIDEOUT BUILDINGS LTD
The Green, Over Kellet, Carnforth, Lancashire, LA6 1BU
Area of Operation: UK (Excluding Ireland)
Tel: 01524 737999
Email: lynn@iobuild.co.uk
Web: www.iobuild.co.uk

JACKSONS FENCING
Stowting Common, Ashford, Kent, TN25 6BN
Area of Operation: UK & Ireland
Tel: 01233 750393 **Fax:** 01233 750403
Email: sales@jacksons-fencing.co.uk
Web: www.jacksons-fencing.co.uk
Product Type: 2

JANE FOLLIS GARDEN DESIGN
71 Eythrope Road, Stone,
Buckinghamshire, HP17 8PH
Area of Operation: South East England
Tel: 01296 747775 **Fax:** 01296 747775
Email: jfgdndesign@aol.com
Web: www.jfgardendesign.co.uk
Product Type: 1, 2, 3
Other Info: ▽ ECO ✏ ✋

LEISURE SPACE LTD
Unit 1, Top Farm, Rectory Road,
Campton Shefford, Bedfordshire, SG17 5PF
Area of Operation: Europe
Tel: 01462 816147
Fax: 01462 819252
Email: enquiries@leisurespaceltd.com
Web: www.leisurespaceltd.co.uk
Product Type: 3
Other Info: ✏

LLOYD CHRISTIE
15 Langton Street, London, SW10 0JL
Area of Operation: Worldwide
Tel: 0207 351 2108 **Fax:** 0207 376 5867
Email: info@lloydchristie.com
Web: www.lloydchristie.com
Product Type: 1, 3

OSMO UK LTD
Unit 2 Pembroke Road, Stocklake Industrial Estate,
Aylesbury, Buckinghamshire, HP20 1DB
Area of Operation: UK & Ireland
Tel: 01296 481220 **Fax:** 01296 424090
Email: info@osmouk.com
Web: www.osmouk.com **Product Type:** 1, 2, 3

PINE GARDEN
Unit 2E, Smith Green Depot , Stoney Lane,
Galgate, Lancaster, Lancashire, LA2 0PX
Area of Operation: UK & Ireland
Tel: 01524 752882 **Fax:** 01524 751744
Email: info@pinegarden.co.uk
Web: www.pinegarden.co.uk
Product Type: 3 **Other Info:** ECO ✏

PINELOG LTD
Riverside Business Park, Bakewell,
Derbyshire, DE45 1GS
Area of Operation: UK (Excluding Ireland)
Tel: 01629 814481 **Fax:** 01629 814634
Email: admin@pinelog.co.uk
Web: www.pinelog.co.uk **Product Type:** 3

RIVERSIDE DECKING COMPANY
4 Chauntry Mews, Chauntry Road,
Maidenhead, Berkshire, SL6 1TT
Area of Operation: Greater London & South East England
Tel: 01628 626545
Email: mklewis3@ukonline.co.uk
Web: www.riversidedeckingcompany.co.uk
Product Type: 2

ROOF-TEK LTD
19 Lynx Crescent, Weston-Super-Mare,
Somerset, BS24 9DJ
Area of Operation: East England, Greater London,
Midlands & Mid Wales, South East England, South
West England and South Wales
Tel: 01934 642929 **Fax:** 01934 644290
Email: roof.tekltd@btinternet.com
Web: www.rooftek.co.uk

SCOTTS OF THRAPSTON LTD.
Bridge Street, Thrapston,
Northamptonshire, NN14 4LR
Area of Operation: UK & Ireland
Tel: 01832 732366 **Fax:** 01832 733703
Email: julia@scottsofthrapston.co.uk
Web: www.scottsofthrapston.co.uk
Product Type: 1, 3
Other Info: ✏

THE LOG CABIN COMPANY
Potash Garden Centre, 9 Main Road,
Hockley, Hawkwell, Essex, SS5 4JN
Area of Operation: UK (Excluding Ireland)
Tel: 01702 206012
Email: sales@thelogcabincompany.co.uk
Web: www.thelogcabincompany.co.uk
Product Type: 3

THE OUTDOOR DECK COMPANY
Mortimer House, 46 Sheen Lane,
London, SW14 8LP
Area of Operation: UK (Excluding Ireland)
Tel: 0208 876 8464 **Fax:** 0208 878 8687
Email: sales@outdoordeck.co.uk
Web: www.outdoordeck.co.uk
Product Type: 1

THE TIMBER FRAME CO LTD
The Framing Yard, 7 Broadway,
Charlton Adam, Somerset, TA11 7BB
Area of Operation: Worldwide
Tel: 01458 224463 **Fax:** 01458 224571
Email: admin@thetimberframe.co.uk
Web: www.thetimberframe.co.uk
Product Type: 1, 2, 3
Other Info: ECO ✏ ✋

Hörmann Makes Renovating Child's Play

New 'Golden Oak' finish for panelled and sectional steel garage doors. An authentic new wood grain effect, offering the appearance of real oak.

"Eco" sectional door in steel

Berry up-and-over garage doors: steel, timber or GRP

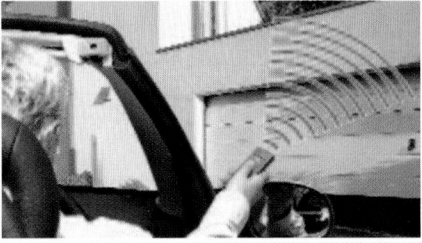

A Hörmann operator offers you greater comfort and convenience

The Hörmann-Programme:
- Sectional doors
- Up-and-over doors
- Side doors
- Garage door and entrance gate operators
- Internal doors
- Security doors

Choose the renowned Hörmann quality

Increased safety to new EN Standard 12604

Hörmann garage doors: increased comfort and convenience, improved safety and aesthetics

Hörmann is Europe's No. 1 garage door manufacturer and with more than 30 door leaf versions in steel, timber and GRP offers you a wide variety to choose from. Whether for a newly built property or renovating an old one, you're sure to find the door that not only best matches the architectural style of your home, but above all offers you permanently good looks coupled with maximum convenience, safety and security.

Hörmann doors comply with the new European Safety Standard EN 12604: finger-trap protection, side trap guards, anti-fall safeguard, with further safety features included as standard at no extra cost!

And of course every Hörmann garage door can also be equipped with an operator. You can operate your garage door using a hand transmitter which allows you to drive straight into your garage without having to leave the comfort of your car!

Don't settle for less!

HÖRMANN

Garage and Industrial Doors

www.hormann.co.uk

For further information, contact:
Tel: 08000 199 333
Hörmann (UK) Limited
Gee Road, Coalville,
Leicestershire, LE67 4JW
e-mail: info@hormann.co.uk

HARD LANDSCAPING

Image courtesy of Wickes (0870 608 9001)

Hard Landscaping

ABOVE: Old railway or log sleepers make an excellent alternative to traditional decking. Image from Bradstone.

Your choice of paving or decking, whether for a driveway, patio or path, will inevitably play a big part in the overall finished appearance of your home and so it is vital to put some real time and thought into what you specify.

You should check that your choice of paving will be able to withstand the relevant amount of traffic; materials used for driveways will obviously need to be stronger than those required for paths.

Obviously it is important that the paving you choose complements the colour and style of your house but this does not mean that it has to match the colour of the brickwork and it is most often the case that a contrast in materials works better. Consider combining colours, shapes, textures and sizes of paving stone to add interesting features, create boundaries, or to break up large areas of monotonous paving. Do not be afraid of mixing different products, for example, paving slabs and gravel, although do ensure that you vary the thickness of the bedding accordingly.

When choosing paving for a path or patio, a much wider choice of materials and patterns are available, such as mosaics, fans and swirls and curved, circular styles. If you want to create a neat, completed finish, consider using edging blocks, corner post edging and pillar caps.

Deciding between natural or artificial paving can be confusing due to the variety of materials available and the differing advice handed out by manufacturers and suppliers. There are three main materials to choose from — concrete, natural stone and reconstituted stone (a mixture of natural raw materials and concrete).

ABOVE: Slate looks as good laid outside as it does on your internal floors. These slate effect tiles are from Wickes.

The biggest difference between these is cost: natural stone is invariably more expensive than concrete or reconstituted stone

You can reclaim VAT on any exterior paving providing you purchase it at the same time as the rest of the materials for your build. Whichever paving you choose, be sure to look at examples that are a couple of years old rather than glossy publicity shots or new paving. It is also worth checking out how the paving looks when wet, as this can dramatically change its appearance.

ABOVE: Bradstone's "Carpet Stones" look like individually laid stones, but are in in fact attached, creating a flexible mat.

Most competent DIYers should be able to take on the task of laying paving and manufacturers will advise on laying techniques. Bear in mind that many types of stone, for instance reclaimed pavers and natural York stone, are more difficult to lay than slabs of an even thickness, and do not forget to slope your paving towards drains in order to disperse water quickly and efficiently.

The alternative option to having a paved area is to choose decking. A deck offers a practical link between the home and garden — ideal for lounging, partying and a safe place for kids to play.

If you have a steeply sloping garden, decking can overcome the problem of forming terraces or retaining walls. The deck boards can be supported on a post and bearer frame, at door level, with wood balustrades used to form a safe barrier. There is also a flexibility of design that is sometimes difficult to achieve with 'hard' materials such as stone flags, as curves, stepped areas and cut-outs for established trees can easily be incorporated into the design.

Bear in mind that deck boards at right angles to the house wall will draw your eye into the garden beyond, whilst boards placed parallel to the wall tend to emphasise the width of the deck, and diagonally placed boards will make a dynamic, contemporary statement.

DECKING

KEY
OTHER: ▽ Reclaimed 🖥 On-line shopping
✏ Bespoke ✋ Hand-made ECO Ecological

ARBORDECK
Lincoln Castle, Lincoln Castle Way,
New Holland, Barrow-upon-Humber,
Lincolnshire, DN19 7RR
Area of Operation: UK (Excluding Ireland)
Tel: 0800 169 5275
Fax: 01469 535526
Email: enquiries@arbordeck.co.uk
Web: www.arbordeck.co.uk

BEAVER TIMBER COMPANY
Barcaldine, Argyll & Bute, PA37 1SG
Area of Operation: UK (Excluding Ireland)
Tel: 01631 720353
Fax: 01631 720430
Email: info@beavertimber.co.uk
Web: www.beavertimber.co.uk 🖥

DALSOUPLE
Unit 1 Showground Road,
Bridgwater, Somerset, TA6 6AJ
Area of Operation: Worldwide
Tel: 01278 727733
Fax: 01278 727766
Email: info@dalsouple.com
Web: www.dalsouple.com
Other Info: ▽ ECO ✏

DECKMASTERS UK LIMITED
The Outdoor Room, 266 Selsdon Road,
South Croydon, Surrey, CR2 7AA
Area of Operation: East England,
Greater London, South East England
Tel: 0800 032 3325
Fax: 0208 681 4090
Email: deckmasters_limited@hotmail.com
Web: www.deckmasters.co.uk
Other Info: ✏

FOREST GARDEN
Units 291 & 296, Hartlebury Trading Estate,
Hartlebury, Worcestershire, DY10 4JB
Area of Operation: UK (Excluding Ireland)
Tel: 0870 300 9809
Fax: 0870 191 9888
Email: info@forestgarden.co.uk
Web: www.forestgarden.co.uk

HLD LTD
The Old Shipyard, Gainsborough,
Lincolnshire, DN21 ING
Area of Operation: UK & Ireland
Tel: 01427 611800
Fax: 01427 612 867
Email: technical@hld.co.uk
Web: www.hld.co.uk

HONEYSUCKLE BOTTOM SAWMILL LTD
Honeysuckle Bottom, Green Dene,
East Horsley, Leatherhead, Surrey, KT24 5TD
Area of Operation: Greater London,
South East England
Tel: 01483 282394 Fax: 01483 282394
Email: honeysucklemill@aol.com
Web: www.easisites.co.uk/honeysucklebottomsawmill

JACKSONS FENCING
Stowting Common, Ashford, Kent, TN25 6BN
Area of Operation: UK & Ireland
Tel: 01233 750393 Fax: 01233 750403
Email: sales@jacksons-fencing.co.uk
Web: www.jacksons-fencing.co.uk

JANE FOLLIS GARDEN DESIGN
71 Eythrope Road, Stone,
Buckinghamshire, HP17 8PH
Area of Operation: South East England
Tel: 01296 747775
Fax: 01296 747775
Email: jfgdndesign@aol.com
Web: www.jfgardendesign.co.uk

JIMPEX LTD
1 Castle Farm, Cholmondeley,
Malpas, Cheshire, SY14 8AQ
Area of Operation: UK & Ireland
Tel: 01829 720433
Fax: 01829720553
Email: jan@jimpex.co.uk
Web: www.jimpex.co.uk
Material Type: A) 2, 3, 7, 10, 13

KIRK NATURAL STONE
Bridgend, Fyvie, Turriff, Aberdeenshire, AB53 8QB
Area of Operation: Worldwide
Tel: 01651 891891
Fax: 01651 891794
Email: info@kirknaturalstone.com
Web: www.kirknaturalstone.com

LAND SKILL PROPERTY
33 Bevin Crescent, Outwood,
West Yorkshire, WF1 3ER
Tel: 01924 826836
Fax: 01924 835174
Email: andrew@landskill.co.uk
Material Type: B) 10

LLOYD CHRISTIE
15 Langton Street, London, SW10 0JL
Area of Operation: Worldwide
Tel: 0207 351 2108
Fax: 0207 376 5867
Email: info@lloydchristie.com
Web: www.lloydchristie.com

RICHARD BURBIDGE LTD
Whittington Road, Oswestry, Shropshire, SY11 1HZ
Area of Operation: Worldwide
Tel: 01691 655131
Fax: 01691 657694
Email: info@richardburbidge.co.uk
Web: www.richardburbidge.co.uk

RIVERSIDE DECKING COMPANY
4 Chauntry Mews, Chauntry Road,
Maidenhead, Berkshire, SL6 1TT
Area of Operation: Greater London,
South East England
Tel: 01628 626545
Email: mklewis3@ukonline.co.uk
Web: www.riversidedeckingcompany.co.uk

THE OUTDOOR DECK COMPANY
Mortimer House, 46 Sheen Lane,
London, SW14 8LP
Area of Operation: UK (Excluding Ireland)
Tel: 0208 876 8464 Fax: 0208 878 8687
Email: sales@outdoordeck.co.uk
Web: www.outdoordeck.co.uk

TIMBER DECKING ASSOCIATION
CIRCE Building, Wheldon Road,
Castleford, West Yorkshire, WF10 2JT
Area of Operation: Europe
Tel: 01977 712718 Fax: 01977 712713
Email: info@tda.org.uk
Web: www.tda.org.uk

WICKES
Wickes Customer Services,
Wickes House, 120-138 Station Road,
Harrow, Middlesex,
Greater London, HA1 2QB
Area of Operation: UK (Excluding Ireland)
Tel: 0870 608 9001
Fax: 0208 863 6225
Web: www.wickes.co.uk

PAVING AND DRIVEWAYS

KEY
PRODUCT TYPES: 1= Paviers 2 = Slabs
3 = Cobbles 4 = Impressed Paving
5 = Gravel 6 = Edgings
OTHER: ▽ Reclaimed 🖥 On-line shopping
✏ Bespoke ✋Hand-made ECO Ecological

ACE MINIMIX
Millfields Road, Ettingshall,
Wolverhampton, West Midlands, WV4 6JP
Area of Operation: UK (Excluding Ireland)
Tel: 01902 353522 Fax: 0121 585 5557
Email: info@tarmac.co.uk
Web: www.tarmac.co.uk

ALISTAIR MACKINTOSH LTD
Bannerley Road, Garretts Green,
Birmingham, West Midlands, B33 0SL
Area of Operation: UK & Ireland
Tel: 0121 784 6800
Fax: 0121 789 7068
Email: info@alistairmackintosh.co.uk
Web: www.alistairmackintosh.co.uk
Product Type: 3, 6
Material Type: E) 5

ALL THINGS STONE LIMITED
Anchor House, 33 Ospringe Street,
Ospringe, Faversham, Kent, ME13 8TW
Area of Operation: UK & Ireland
Tel: 01795 531001
Fax: 01795 530742
Email: info@allthingsstone.co.uk
Web: www.allthingsstone.co.uk
Product Type: 2, 3, 6
Material Type: E) 1, 3, 5

ART OUTSIDE
PO Box 513, Aylesbury,
Buckinghamshire, HP22 6WJ
Area of Operation: Worldwide
Tel: 01296 582004
Email: emma@art-outside.com
Web: www.art-outside.com 🖥
Other Info: ✏

BAGGERIDGE BRICK
Fir Street, Sedgley, West Midlands, DY3 4AA
Area of Operation: Worldwide
Tel: 01902 880555 Fax: 01902 880432
Email: marketing@baggeridge.co.uk
Web: www.baggeridge.co.uk
Product Type: 1, 3, 6

BLACK MOUNTAIN QUARRIES LTD
Howton Court, Pontrilas,
Herefordshire, HR2 0BG
Area of Operation: UK & Ireland
Tel: 01981 241541
Email: info@blackmountainquarries.com
Web: www.blackmountainquarries.com 🖥
Product Type: 1, 2, 3, 6
Other Info: ECO ✋
Material Type: E) 1, 3, 4, 5, 8

BLANC DE BIERGES
Eastrea Road, Whittlesey,
Cambridgeshire, PE7 2AG
Area of Operation: Worldwide
Tel: 01733 202566 Fax: 01733 205405
Email: info@blancdebierges.com
Web: www.blancdebierges.com
Product Type: 1, 2, 3, 6
Other Info: ✏ ✋

BRADSTONE GARDEN
Aggregate Industries UK Ltd, Hulland Ward,
Ashbourne, Derbyshire, DE6 3ET
Area of Operation: UK (Excluding Ireland)
Tel: 01335 372 222 Fax: 01335 370 973
Email: bradstone.structural@aggregate.com
Web: www.bradstone.com
Product Type: 1, 2, 3, 4, 5, 6
Material Type: E) 4, 13

BRETT LANDSCAPING LTD .
Salt Lane, Cliffe, Rochester, Kent, ME3 7SZ
Area of Operation: Worldwide
Tel: 01634 222188 Fax: 01634 222001
Email: landscapinginfo@brett.co.uk
Web: www.brett.co.uk/landscaping
Product Type: 1, 2, 3, 4, 5, 6

BRS YORK STONE
50 High Green Road, Altofts,
Normanton, Lancashire, WF6 2LQ
Area of Operation: UK (Excluding Ireland)
Tel: 01924 220356 **Fax:** 01924 220356
Email: john@york-stone.fsnet.co.uk
Web: www.yorkstonepaving.co.uk

BRS YORK STONE
Area of Operation: UK (Excluding Ireland)
Tel: 01924 220356
Fax: 01924 220356
Email: john@york-stone.fsnet.co.uk
Web: www.yorkstonepaving.co.uk

Our Natural York Stone Paving has been supplied to all
parts of the country, and it's reputation both for quality and
durability is unrivalled. The stone comes direct from the
quarry therefore we can guarantee the best price and
delivery cost.

CAPITAL GROUP LIMITED
Victoria Mills, Highfield Road,
Little Hulton, Manchester, M38 9ST
Area of Operation: Worldwide
Tel: 0161 799 7555 **Fax:** 0161 799 7666
Email: leigh@choosecapital.co.uk
Web: www.choosecapital.co.uk
Product Type: 1, 2, 3

CHESHIRE BRICK & SLATE
Brook House Farm, Salters Bridge,
Tarvin Sands, Tarvin, Cheshire, CH3 8NR
Area of Operation: UK (Excluding Ireland)
Tel: 01829 740883 **Fax:** 01829 740481
Email: enquiries@cheshirebrickandslate.co.uk
Web: www.cheshirebrickandslate.co.uk
Product Type: 1, 2, 3, 5, 6

COLEFORD BRICK & TILE CO LTD
The Royal Forest of Dean Brickworks,
Cinderford, Gloucestershire, GL14 3JJ
Area of Operation: UK & Ireland
Tel: 01594 822160 **Fax:** 01594 826655
Email: sales@colefordbrick.co.uk
Web: www.colefordbrick.co.uk
Product Type: 1, 6

COMER LANDSCAPES
99 London Road, Northwich, Cheshire, CW9 5HQ
Area of Operation: UK (Excluding Ireland)
Tel: 01606 43737 **Fax:** 01606 43736
Email: info@comerlandscapes.co.uk
Web: www.cheshirelawns.co.uk
Product Type: 5

COTTOROSATO
Laterartigiana, Via Della Fornace, 1, Loc. Ponticelli,
06062 citta della pieve, Perugia, Italy
Area of Operation: Europe
Tel: +39 578 248 266 **Fax:** +39 578 248 266
Email: info@cottorosato.it
Web: www.anticocottopievese.it
Product Type: 2
Material Type: F) 3

DALSOUPLE
Unit 1 Showground Road,
Bridgwater, Somerset, TA6 6AJ
Area of Operation: Worldwide
Tel: 01278 727733 **Fax:** 01278 727766
Email: info@dalsouple.com
Web: www.dalsouple.com
Product Type: 1 **Other Info:** ▽ ✐ ECO

DENBY DALE CAST PRODUCTS LTD
230 Cumberworth Lane, Denby Dale,
Huddersfield, West Yorkshire, HD8 8PR
Area of Operation: UK & Ireland
Tel: 01484 863560
Fax: 01484 865597
Email: sales@denbydalecastproducts.co.uk
Web: www.denbydalecastproducts.co.uk
Product Type: 4, 6

DEVON STONE LTD
38 Station Road, Budleigh Salterton, Devon, EX9 6RT
Area of Operation: UK (Excluding Ireland)
Tel: 01395 446841
Email: amy@devonstone.com
Product Type: 1, 2, 3, 6
Other Info: ▽ ✐
Material Type: E) 1, 3, 4

FARMINGTON NATURAL STONE
Northleach, Cheltenham, Gloucestershire, GL54 3NZ
Area of Operation: UK (Excluding Ireland)
Tel: 01451 860280
Fax: 01451 860115
Email: cotswold.stone@farmington.co.uk
Web: www.farmingtonnaturalstone.co.uk
Product Type: 1, 2, 3, 6

FERNHILL STONE UK LTD
Unit 16 Greencroft Industrial Estate,
Annfield Plain, Stanley, Durham, DH9 7XP
Area of Operation: UK & Ireland
Tel: 01207 521482 **Fax:** 01207 521455
Email: geoffreydriver@btconnect.com
Web: www.fernhillstone.com
Product Type: 1
Material Type: E) 4, 11

FOREST WALLING STONE & FLAGSTONES
Boot Barn, Newcastle, Monmouth,
Monmouthshire, NP25 5NU
Area of Operation: UK (Excluding Ireland)
Tel: 01600 750462 **Fax:** 01600 750462
Product Type: 1, 2

FRANCIS N. LOWE LTD.
The Marble Works, New Road, Middleton,
Matlock, Derbyshire, DE4 4NA
Area of Operation: Europe
Tel: 01629 822216
Fax: 01629 824348
Email: info@lowesmarble.com
Web: www.lowesmarble.com

FURNESS BRICK & TILE CO
Askam in Furness, Cumbria, LA16 7HF
Area of Operation: Worldwide
Tel: 01229 462411
Fax: 01229 462363
Email: peterhubble@mac.com
Web: www.furnessbrick.com
Product Type: 1
Material Type: F) 1

HERITAGE FLAGSTONES
The Sanctuary, Newchurch, Hoar Cross,
Burton upon Trent, Derbyshire, DE13 8RQ
Area of Operation: UK (Excluding Ireland)
Tel: 01283 575822 **Fax:** 01283 575180
Email: info@heritageflagstones.co.uk
Web: www.heritageflagstones.co.uk
Product Type: 2

HYPERION TILES
67 High Street, Ascot, Berkshire, SL5 7HP
Area of Operation: Greater London,
South East England
Tel: 01344 620211 **Fax:** 01344 620100
Email: graham@hyperiontiles.com
Web: www.hyperiontiles.com

IBSTOCK BRICK LTD
Leicester Road, Ibstock, Leicester,
Leicestershire, LE67 6HS
Area of Operation: UK & Ireland
Tel: 01530 261999 **Fax:** 01530 263478
Email: leicester.sales@ibstock.co.uk
Web: www.ibstock.co.uk
Product Type: 1

INDIGENOUS LTD
Cheltenham Road, Burford,
Oxfordshire, OX18 4JA
Area of Operation: Worldwide
Tel: 01993 824200
Fax: 01993 824300
Email: enquiries@indigenoustiles.com
Web: www.indigenoustiles.com
Product Type: 2
Material Type: E) 3, 4, 5, 8

JANE FOLLIS GARDEN DESIGN
71 Eythrope Road, Stone,
Buckinghamshire, HP17 8PH
Area of Operation: South East England
Tel: 01296 747775
Fax: 01296 747775
Email: jfgdndesign@aol.com
Web: www.jfgardendesign.co.uk
Product Type: 1, 2, 3, 5, 6

KEYSTONE NATURAL STONE FLOORING
204 Duggins Lane, Tile Hill,
Coventry, West Midlands, CV4 9GP
Area of Operation: Worldwide
Tel: 02476 422580
Fax: 02476 695794

KIRK NATURAL STONE
Bridgend, Fyvie, Turriff,
Aberdeenshire, AB53 8QB
Area of Operation: Worldwide
Tel: 01651 891891
Fax: 01651 891794
Email: info@kirknaturalstone.com
Web: www.kirknaturalstone.com
Product Type: 1, 2, 3, 5, 6

LAND SKILL PROPERTY
33 Bevin Crescent, Outwood,
West Yorkshire, WF1 3ER
Tel: 01924 826836
Fax: 01924 835174
Email: andrew@landskill.co.uk
Product Type: 1, 2, 3, 5, 6
Material Type: E) 4

LLOYD CHRISTIE
15 Langton Street, London, SW10 0JL
Area of Operation: Worldwide
Tel: 0207 351 2108
Fax: 0207 376 5867
Email: info@lloydchristie.com
Web: www.lloydchristie.com

MARSHALLS
Southowram, Halifax, West Yorkshire, HX3 9TE
Area of Operation: UK (Excluding Ireland)
Tel: 0870 120 7474
Email: marshalls@trilobyte.co.uk
Web: www.marshalls.co.uk

MIDLANDS SLATE & TILE
Qualcast Road, Horseley Fields,
Wolverhampton, West Midlands, WV1 2QP
Area of Operation: UK & Ireland
Tel: 01902 458780
Fax: 01902 458549
Email: sales@slate-tile-brick.co.uk
Web: www.slate-tile-brick.co.uk
Product Type: 1, 2, 3
Material Type: E) 1, 3, 4, 5

NATURAL IMAGE (GRANITE, MARBLE, STONE)
Spelmonden Estate, Goudhurst Road,
Goudhurst, Kent, TN17 1HE
Area of Operation: UK & Ireland
Tel: 01580 212222
Fax: 01580 211841
Web: www.naturalstonefloors.org
Product Type: 1, 2

ORIGINAL STONE PAVING CO
The Oaks, Oak Road, Bowling Bank,
Wrexham, LL13 9RG
Area of Operation: UK (Excluding Ireland)
Tel: 01978 661000 **Fax:** 01978 661000
Email: sales@the-original-stone-paving-company.co.uk
Web: www.the-original-stone-paving-company.co.uk

PLASMOR
P.O. Box 44, Womersley Road,
Knottingley, West Yorkshire, WF11 0DN
Area of Operation: UK (Excluding Ireland)
Tel: 01977 673221
Fax: 01977 607071
Email: knott@plasmor.co.uk
Web: www.plasmor.co.uk
Material Type: G) 1

RANSFORDS
Drayton Way, Drayton Fields Industrial Estate,
Daventry, Northamptonshire, NN11 5XW
Area of Operation: Worldwide
Tel: 01327 705310
Fax: 01327 706831
Email: sales@ransfords.com
Web: www.ransfords.com /
www.stoneflooringandpaving.com
Product Type: 2

RESIN BONDED LTD
Unit 7 Ashdown Court, Vernon Road,
Uckfield, East Sussex, TN22 5DX
Area of Operation: UK & Ireland
Tel: 01825 766186
Fax: 01825 766186
Email: info@resinbonded.co.uk
Web: www.resinbonded.co.uk ⏱
Product Type: 5

RIVAR SAND & GRAVEL LTD
Pinchington Lane, Newbury, Berkshire, RG19 8SR
Area of Operation: South East England
Tel: 01635 523524
Fax: 01635 521621
Email: sales@rivarsandandgravel.co.uk
Web: www.rivarsandandgravel.co.uk
Product Type: 1, 2, 3, 5, 6

ROCK UNIQUE LTD
c/o Select Garden and Pet Centre,
Main Road, Sundridge, Kent, TN14 6ED
Area of Operation: Europe
Tel: 01959 565608
Fax: 01959 569312
Email: stone@rock-unique.com
Web: www.rock-unique.com
Product Type: 1, 2, 3, 6
Other Info: ▽ ECO 🖐 🖐
Material Type: E) 1, 2, 3, 4, 5, 8, 9, 10

RUSTICA LTD
154c Milton Park, Oxfordshire, OX14 4SD
Area of Operation: UK & Ireland
Tel: 01235 834192
Fax: 01235 835162
Email: sales@rustica.co.uk
Web: www.rustica.co.uk
Product Type: 2

SPARTAN TILES
Slough Lane, Ardleigh, Essex, CO7 7RU
Area of Operation: UK & Ireland
Tel: 01206 230553
Fax: 01206 230516
Email: sales@spartantiles.com
Web: www.spartantiles.com
Product Type: 2
Material Type: G) 1, 5

STONE IT LTD
Unit 13, Burma Drive, Hull,
East Riding of Yorks, HU9 5SD
Area of Operation: UK & Ireland
Tel: 01482 701442 **Fax:** 01482 702300
Email: info@stoneit.co.uk
Web: www.stoneit.co.uk
Product Type: 2
Material Type: E) 4

STONEMARKET LTD
Oxford Road, Ryton on Dunsmore,
Warwickshire, CV8 3EJ
Area of Operation: UK (Excluding Ireland)
Tel: 02476 518700 **Fax:** 02476 518777
Email: sales@stonemarket.co.uk
Web: www.stonemarket.co.uk
Product Type: 1, 2, 3, 6

STONEWAYS LTD
Railside Works, Marlbrook,
Leominster, Herefordshire, HR6 0PH
Area of Operation: UK (Excluding Ireland)
Tel: 01568 616818
Fax: 01568 620085
Email: stoneways@msn.com
Web: www.stoneways.co.uk
Product Type: 1

SURESET UK LTD
Unit 32 , Deverill Road Trading Estate,
Sutton Veny, Warminster, Wiltshire, BA12 7BZ
Area of Operation: Europe
Tel: 01985 841180
Fax: 01985 841260
Email: mail@sureset.co.uk
Web: www.sureset.co.uk
Material Type: E) 1, 2, 3, 5, 6, 9

TAYLOR MAXWELL & COMPANY LIMITED
4 John Oliver Buildings, 53 Wood Street,
Barnet, Hertfordshire, EN5 4BS
Area of Operation: UK (Excluding Ireland)
Tel: 0208 440 0551
Fax: 0208 440 0552
Email: simonepalmer@taylor.maxwell.co.uk
Web: www.taylor.maxwell.co.uk
Product Type: 1

THE HOME OF STONE LTD
Boot Barn, Newcastle, Monmouth,
Monmouthshire, NP25 5NU
Area of Operation: South West England
and South Wales
Tel: 01600 750462
Fax: 01600 750462
Material Type: E) 4

THE MATCHING BRICK COMPANY
Lockes Yard, Hartcliffe Way,
Bedminster, Bristol, BS3 5RJ
Area of Operation: UK (Excluding Ireland)
Tel: 0117 963 7000
Fax: 0117 966 4612
Email: matchingbrick@btconnect.com
Web: www.matchingbrick.co.uk
Product Type: 1, 3, 6

**THE NATURAL SLATE
COMPANY LTD - DERBYSHIRE**
Unit 2 Armytage Estate, Station Road,
Whittington Moor, Chesterfield,
Derbyshire, S41 9ET
Area of Operation: Worldwide
Tel: 0845 177 5008
Fax: 0870 429 9891
Email: sales@theslatecompany.net
Web: www.theslatecompany.net
Material Type: E) 3, 4

**THE NATURAL SLATE
COMPANY LTD - LONDON**
161 Ballards Lane, Finchley, London, N3 1LJ
Area of Operation: Worldwide
Tel: 0845 177 5008
Fax: 0870 429 9891
Email: sales@theslatecompany.net
Web: www.theslatecompany.net
Material Type: E) 3, 4

TRADSTOCKS LTD
Dunaverig, Thornhill, Stirling,
Stirlingshire, FK8 3QW
Area of Operation: Scotland
Tel: 01786 850400
Fax: 01786 850404
Email: peter@tradstocks.co.uk
Web: www.tradstocks.co.uk
Product Type: 2, 3, 6

VOBSTER CAST STONE CO LTD
Newbury Works, Coleford,
Radstock, Somerset, BA3 5RX
Area of Operation: UK (Excluding Ireland)
Tel: 01373 812514
Fax: 01373 813384
Email: barrie@caststonemasonry.co.uk
Web: www.caststonemasonry.co.uk

WELSH SLATE
Business Design Centre, Unit 205,
52 Upper Street, London, N1 0QH
Area of Operation: Worldwide
Tel: 0207 354 0306
Fax: 0207 354 8485
Email: enquiries@welshslate.com
Web: www.welshslate.com
Material Type: E) 3

YORK HANDMADE BRICK CO LTD
Forest Lane, Alne, York,
North Yorkshire, YO61 1TU
Area of Operation: Worldwide
Tel: 01347 838881
Fax: 01347 838885
Email: sales@yorkhandmade.co.uk
Web: www.yorkhandmade.co.uk
Product Type: 1, 3, 6

ZARKA MARBLE
41A Belsize Lane, Hampstead,
London, NW3 5AU
Area of Operation: South East England
Tel: 0207 431 3042
Fax: 0207 431 3879
Email: enquiries@zarkamarble.co.uk
Web: www.zarkamarble.co.uk

NOTES

Company Name ....

....

Address ....

....

....

email ....

Web ....

Company Name ....

....

Address ....

....

....

email ....

Web ....

Company Name ....

....

Address ....

....

....

email ....

Web ....

BOOKS FROM
HOMEBUILDING & RENOVATING

Each book contains around 20 inspirational projects, drawn from amongst the best featured in Homebuilding & Renovating Magazine over the past five years. These outstanding examples show what can be achieved with the right approach.

- Fully costed case histories
- Projects to suit all budgets
- A variety of styles designed to inspire you
- Design and style tips
- Contact details of suppliers and craftsmen for each project
- Read about real life solutions to some exciting building challenges!

Britain's Best Selling Self-Build Magazine

HOMEBUILDING & RENOVATING

Ascent Publishing Limited
Unit 2 Sugar Brook Court
Aston Road
Bromsgrove
Worcestershire
B60 3EX

ORDER TODAY!

Tel: 01527 834435
Web: www.homebuilding.co.uk

ONLY £14.95

ONLY £14.95

ONLY £15.00

GATES, FENCES & ARCHWAYS

EXTERIOR PRODUCTS

Fencing

Most new building plots require some sort of boundary detailing, generally because you need something in place for security, for privacy, or to keep children or pets in and unwelcome visitors out. However, it's the sort of detail that tends to get left as an afterthought and, to avoid it getting overlooked altogether, many plot vendors make it a condition of sale that a particular boundary fence should be erected within a set time limit — typically three months from sale or before construction work on the house starts. This is understandable — after all, the plot vendor is very often a neighbour and they don't want to have to live with a construction site over-spilling onto their land for months or possibly years.

If you just want to mark a boundary and are not too bothered about privacy or security, then the cheapest permanent option is the timber post-and-rail fence. This arrangement is fairly attractive, is inexpensive and is easily maintained. Alternatively, a chain link fence will cost a similar amount and, though less attractive, is more secure and should stop dogs and children straying. 1.2m-high chestnut palings are another cheap option which, being vertical, are much harder to get over. They are easily fixed, but have a temporary air

Picket fences are excellent for creating unobtrusive boundaries. Example BELOW by Wickes.

about them. The picket fence is similar in design but altogether more permanent in appearance.

If you require privacy and security then you will have to go for a solid or near-solid fence with a height of 1.8m or more (above head height). The traditional way of doing this is to erect something similar to a post-and-rail fence and then to cover it with vertically fixed, featheredge boarding. This is known as a close-boarded fence. It is a little cheaper to use ready-made panels of the type you see in garden centres, but the result is very flimsy in comparison.

Between variations in style and colour, you can create a stunning effect very simply. For example, instead of having a basic boarded fence, why not set the boards horizontally or diagonally, or alternate the boards between the inside and the outside of the fence posts (called hit and miss fencing)? Or instead of using traditional boards, why not try bamboo or willow poles as screening? When experimenting with colour, why not try blues and reds instead of the usual earth colours of green and brown? These will look wonderful when shown amongst lush green foliage, and will appear warm and bright on even the darkest day.

ABOVE: Diagonal boarding provides a modern edge to the traditional fence panel (BELOW). Both examples by Jacksons Fine Fencing.

FENCING AND RAILINGS

KEY

PRODUCT TYPES: 1= Balustrading
2 = Fencing 3 = Railings
OTHER: ▽ Reclaimed   On-line shopping
 Bespoke   Hand-made   ECO Ecological

ADDSTONE CAST STONE
2 Millers Gate, Stone, Staffordshire, ST15 8ZF
Area of Operation: UK (Excluding Ireland)
Tel: 01785 818810 **Fax:** 01785 818810
Email: sales@addstone.co.uk
Web: www.addstone.co.uk
Product Type: 1 **Other Info:**
Material Type: E) 11, 13

ARBORDECK
Lincoln Castle, Lincoln Castle Way, New Holland,
Barrow-upon-Humber, Lincolnshire, DN19 7RR
Area of Operation: UK (Excluding Ireland)
Tel: 0800 169 5275 **Fax:** 01469 535526
Email: enquiries@arbordeck.co.uk
Web: www.arbordeck.co.uk
Product Type: 1

ARCHITECTURAL GATES
(callers by appointment only), Mallard, Hoopers Pool,
Southwick, Trowbridge, Wiltshire, BA14 9NG
Area of Operation: Worldwide
Tel: 01225 766944
Email: architectural_gates@yahoo.co.uk
Web: www.architectural-gates.com
Product Type: 1, 2, 3

ATG ACCESS LTD.
Automation House, Lowton Business Park,
Newton Road, Lowton, Cheshire, WA3 2AP
Area of Operation: Worldwide
Tel: 01942 685 522 **Fax:** 01942 269 676
Email: marketing@atgaccess.com
Web: www.atgaccess.com
Product Type: 3 **Other Info:**

B ROURKE & CO LTD
Vulcan Works, Accrington Road,
Burnley, Lancashire, BB11 5QD
Area of Operation: Worldwide
Tel: 01282 422841 **Fax:** 01282 458901
Email: info@rourkes.co.uk
Web: www.rourkes.co.uk
Product Type: 1, 2

BALCAS TIMBER LTD
Laragh, Enniskillen, Co Fermanagh, BT94 2FQ
Area of Operation: UK & Ireland
Tel: 0286 632 3003
Fax: 0286 632 7924
Email: info@balcas.com
Web: www.balcas.com
Product Type: 1, 2

BLANC DE BIERGES
Eastrea Road, Whittlesey, Cambridgeshire, PE7 2AG
Area of Operation: Worldwide
Tel: 01733 202566
Fax: 01733 205405
Email: info@blancdebierges.com
Web: www.blancdebierges.com
Product Type: 2
Other Info:

CANNOCK GATES (UK) LTD.
Hawks Green, Martindale, Cannock,
Staffordshire, WS11 7XT
Area of Operation: UK (Excluding Ireland)
Tel: 08707 541813
Email: sales@cannockgates.co.uk
Web: www.cannockgates.co.uk

CHAIRWORKS
47 Weir Road, London, SW19 8UG
Area of Operation: UK (Excluding Ireland)
Tel: 0208 247 3700
Fax: 0208 247 3800
Email: info@chairworks.info
Web: www.chairworks.info
Product Type: 2

DICTATOR DIRECT
Inga House, Northdown Business Park,
Ashford Road, Lenham, Kent, ME17 2DL
Area of Operation: Worldwide
Tel: 01622 854770
Fax: 01622 854771
Email: mail@dictatordirect.com
Web: www.dictatordirect.com

EASYGATES
75 Long Lane, Halesowen, Birmingham,
West Midlands, B62 9DJ
Area of Operation: Europe
Tel: 0870 760 6536 **Fax:** 0121 561 3395
Email: info@easygates.co.uk
Web: www.easygates.co.uk
Product Type: 1, 2

ERMINE ENGINEERING COMPANY LTD
Francis House, Silver Birch Park, Great Northern
Terrace, Lincoln, Lincolnshire, LN5 8LG
Area of Operation: UK (Excluding Ireland)
Tel: 01522 510977
Fax: 01522 510929
Email: info@ermineengineering.co.uk
Web: www.ermineengineering.co.uk
Product Type: 1, 2
Other Info:
Material Type: C) 1, 2, 3, 4, 18

F.P. IRONWORK
Unit 44, Oswin Road, Leicester,
Leicestershire, LE3 1HR
Area of Operation: UK (Excluding Ireland)
Tel: 0116 255 0455 **Fax:** 0116 255 6096
Email: sales@fpironwork.com
Web: www.fpironwork.com
Product Type: 1, 2, 3 **Other Info:**
Material Type: C) 2, 4, 6

FOREST GARDEN
Units 291 & 296, Hartlebury Trading Estate,
Hartlebury, Worcestershire, DY10 4JB
Area of Operation: UK (Excluding Ireland)
Tel: 0870 300 9809
 Fax: 0870 191 9888
Email: info@forestgarden.co.uk
Web: www.forestgarden.co.uk
Product Type: 2

GATE-A-MATION LTD
Unit 8 Boundary Business Centre,
Boundary Way, Woking, Surrey, GU21 5DH
Area of Operation: East England, Greater London,
South East England, South West England & South Wales
Tel: 01483 747373
Fax: 01483 776688
Email: sales@gate-a-mation.com
Web: www.gate-a-mation.com
Product Type: 2 **Other Info:**

GEORGE BARKER & SONS
Backbarrow, Nr Ulverston, Cumbria, LA12 8TA
Area of Operation: UK (Excluding Ireland)
Tel: 01539 531236
Fax: 01539 530801
Web: www.gbs-ltd.co.uk

HADDONSTONE LTD
The Forge House, East Haddon,
Northampton, Northamptonshire, NN6 8DB
Area of Operation: Worldwide
Tel: 01604 770711
Fax: 01604 770027
Email: info@haddonstone.co.uk
Web: www.haddonstone.co.uk
Product Type: 1

HADDONSTONE
Area of Operation: Worldwide
Tel: 01604 770711 **Fax:** 01604 770027
Email: info@haddonstone.co.uk
Web: www.haddonstone.co.uk
Other Info:

Haddonstone manufacture garden ornaments
and architectural cast stone from porticos and
balustrading to fountains and fireplaces.
Extensive standard range. Custom made also
available.

HONEYSUCKLE BOTTOM SAWMILL LTD
Honeysuckle Bottom, Green Dene, East Horsley,
Leatherhead, Surrey, KT24 5TD
Area of Operation: Greater London,
South East England
Tel: 01483 282394 **Fax:** 01483 282394
Email: honeysucklemill@aol.com
Web: www.easisites.co.uk/honeysucklebottomsawmill
Product Type: 2

IRONCRAFT
92 High Street, Earl Shilton, Leicester,
Leicestershire, LE9 7DG
Area of Operation: UK (Excluding Ireland)
Tel: 01455 847548 **Fax:** 01455 842422
Email: office@irongraft.co.uk
Web: www.ironcraft.co.uk
Product Type: 2

J.G.S WEATHERVANES
Unit 6, Broomstick Estate, High Street,
Edlesborough, Dunstable, Bedfordshire, LU6 2HS
Area of Operation: UK (Excluding Ireland)
Tel: 01525 220360
Fax: 01525 222786
Email: sales@weathervanes.org.uk
Web: www.weathervanes.org.uk
Product Type: 1, 2
Other Info:

JACKSONS FENCING
Stowting Common, Ashford, Kent, TN25 6BN
Area of Operation: UK & Ireland
Tel: 01233 750393
Fax: 01233 750403
Email: sales@jacksons-fencing.co.uk
Web: www.jacksons-fencing.co.uk
Product Type: 2

JANE FOLLIS GARDEN DESIGN
71 Eythrope Road, Stone,
Buckinghamshire, HP17 8PH
Area of Operation: South East England
Tel: 01296 747775
Fax: 01296 747775
Email: jfgdndesign@aol.com
Web: www.jfgardendesign.co.uk
Product Type: 1, 2, 3

JGS METALWORK
Unit 6 Broomstick Estate, High Street,
Edlesborough, Dunstable, Bedfordshire, LU6 2HS
Area of Operation: East England
Tel: 01525 220360
Fax: 01525 222786
Email: enquiries@weathervanes.org.uk
Web: www.weathervanes.org.uk
Product Type: 1, 2, 3

KEE KLAMP LTD
1 Boulton Road, Reading, Berkshire, RG2 0NH
Area of Operation: Worldwide
Tel: 0118 931 1022
Email: jthelwell@keeklamp.com
Web: www.keeklamp.com
Product Type: 1, 2

KIRK NATURAL STONE
Bridgend, Fyvie, Turriff,
Aberdeenshire, AB53 8QB
Area of Operation: Worldwide
Tel: 01651 891891
Fax: 01651 891794
Email: info@kirknaturalstone.com
Web: www.kirknaturalstone.com
Product Type: 2

LAND SKILL PROPERTY
33 Bevin Crescent, Outwood,
West Yorkshire, WF1 3ER
Tel: 01924 826836
Fax: 01924 835174
Email: andrew@landskill.co.uk
Product Type: 1, 2

LLOYD CHRISTIE
15 Langton Street, London, SW10 OJL
Area of Operation: Worldwide
Tel: 0207 351 2108
Fax: 0207 376 5867
Email: info@lloydchristie.com
Web: www.lloydchristie.com

METALCRAFT [TOTTENHAM]
6-40 Durnford Street, Tottenham,
London, N15 5NQ
Area of Operation: UK (Excluding Ireland)
Tel: 0208 802 1715
Fax: 0208 802 1258
Email: sales@makingmetalwork.com
Web: www.makingmetalwork.com
Product Type: 1, 2
Material Type: C) 2, 3, 4, 5

OSMO UK LTD
Unit 2 Pembroke Road,
Stocklake Industrial Estate, Aylesbury,
Buckinghamshire, HP20 1DB
Area of Operation: UK & Ireland
Tel: 01296 481220
Fax: 01296 424090
Email: info@osmouk.com
Web: www.osmouk.com
Product Type: 2

PREFAB STEEL CO. LTD
114 Brighton Road, Shoreham,
West Sussex, BN43 6RH
Area of Operation: Greater London,
South East England
Tel: 01273 597733
Fax: 01273 597774
Email: sales@prefabsteel.co.uk
Web: www.prefabsteel.co.uk
Product Type: 1, 3
Material Type: C) 2

RIVERSIDE DECKING COMPANY
4 Chauntry Mews, Chauntry Road,
Maidenhead, Berkshire, SL6 1TT
Area of Operation: Greater London,
South East England
Tel: 01628 626545
Email: mklewis3@ukonline.co.uk
Web: www.riversidedeckingcompany.co.uk
Product Type: 1, 2

RUTLAND TIMBER LTD
20 Underwood Drive, Stoney Stanton,
Leicester, Leicestershire, LE9 4TA
Area of Operation: UK & Ireland
Tel: 01455 272 860
Fax: 01455 271 324
Email: Ruttim1@aol.com
Web: www.rutlandtimber.co.uk
Product Type: 2

THE CANE STORE
Wash Dyke Cottage, No1 Witham Road,
Long Bennington, Lincolnshire, NG23 5DS
Area of Operation: Worldwide
Tel: 01400 282271
Fax: 01400 281103
Email: jaki@canestore.co.uk
Web: www.canestore.co.uk
Product Type: 2

THE EXPANDED METAL COMPANY LIMITED
PO Box 14, Longhill Industrial Estate (North),
Hartlepool, Durham, TS25 1PR
Area of Operation: Worldwide
Tel: 01429 408087 **Fax:** 01429 866795
Email: paulb@expamet.com
Web: www.expandedmetalcompany.co.uk

GATES

KEY

PRODUCT TYPES: 1= Gate Closers
2 = Automatic Gates   3 = Other
OTHER: ▽ Reclaimed   🖱 On-line shopping
🖋 Bespoke   ✋ Hand-made   ECO Ecological

4N PRODUCTS LIMITED
23 Mill Lane, Kegworth, Derby,
Derbyshire, DE74 2FX
Area of Operation: UK & Ireland
Tel: 0870 241 8468
Fax: 01509 670029
Email: sales@4nproducts.co.uk
Web: www.4nproducts.co.uk 🖱
Product Type: 2

ARCHITECTURAL GATES
(callers by appointment only),
Mallard, Hoopers Pool, Southwick,
Trowbridge, Wiltshire, BA14 9NG
Area of Operation: Worldwide
Tel: 01225 766944
Email: architectural_gates@yahoo.co.uk
Web: www.architectural-gates.com
Product Type: 1, 2, 3
Material Type: C) 1, 2, 4, 5, 6, 17, 18

ATLAS GROUP
Design House, 27 Salt Hill Way,
Slough, Berkshire, SL1 3TR
Area of Operation: UK (Excluding Ireland)
Tel: 01753 573573 **Fax:** 01753 552424
Email: info@atlasgroup.co.uk
Web: www.atlasgroup.co.uk
Product Type: 1, 2, 3
Other Info: 🖋
Material Type: C) 1, 2, 4, 5, 6

AUTOPA LIMITED
Cottage Leap, Rugby, Warwickshire, CV21 3XP
Area of Operation: Worldwide
Tel: 01788 550556 **Fax:** 01788 550265
Email: info@autopa.co.uk
Web: www.autopa.co.uk
Product Type: 3

B ROURKE & CO LTD
Vulcan Works, Accrington Road,
Burnley, Lancashire, BB11 5QD
Area of Operation: Worldwide
Tel: 01282 422841
Fax: 01282 458901
Email: info@rourkes.co.uk
Web: www.rourkes.co.uk
Product Type: 1, 2, 3

BAYFIELD STAIR CO
Unit 4, Praed Road, Trafford Park,
Manchester, Greater Manchester, M17 1PQ
Area of Operation: Worldwide
Tel: 0161 848 0700
Fax: 0161 872 2230
Email: sales@bayfieldstairs.co.uk
Web: www.bayfieldstairs.co.uk
Product Type: 1, 2, 3

BPT AUTOMATION LTD
Unit 16, Sovereign Park, Cleveland Way,
Hemel Hempstead, Hertfordshire, HP2 7DA
Area of Operation: UK & Ireland
Tel: 01442 235355 **Fax:** 01442 244729
Email: sales@bpt.co.uk
Web: www.bptautomation.co.uk
Product Type: 1, 2

CANNOCK GATES (UK) LTD.
Hawks Green, Martindale, Cannock,
Staffordshire, WS11 7XT
Area of Operation: UK (Excluding Ireland)
Tel: 08707 541813
Email: sales@cannockgates.co.uk
Web: www.cannockgates.co.uk 🖱

CENTURION
Westhill Business Park, Arnhall Business Park,
Westhill, Aberdeen, Aberdeenshire, AB32 6UF
Area of Operation: Scotland
Tel: 01224 744440
Fax: 01224 744819
Email: info@centurion-solutions.co.uk
Web: www.centurion-solutions.co.uk
Product Type: 1, 2, 3

COEL AUTOMATE
Starkbridge Farm, Syke House, Nr Goole,
East Riding of Yorks, DN14 9AZ
Area of Operation: UK (Excluding Ireland)
Tel: 01405 785656
Fax: 01405 785656
Email: info@theelectricalgateshop.co.uk
Web: www.automaticgatekits.co.uk
Product Type: 2

DICTATOR DIRECT
Inga House, Northdown Business Park,
Ashford Road, Lenham, Kent, ME17 2DL
Area of Operation: Worldwide
Tel: 01622 854770
Fax: 01622 854771
Email: mail@dictatordirect.com
Web: www.dictatordirect.com
Product Type: 1

EASYGATES
75 Long Lane, Halesowen,
Birmingham, West Midlands, B62 9DJ
Area of Operation: Europe
Tel: 0870 760 6536
Fax: 0121 561 3395
Email: info@easygates.co.uk
Web: www.easygates.co.uk
Product Type: 1, 2

ERMINE ENGINEERING COMPANY LTD
Francis House, Silver Birch Park, Great Northern
Terrace, Lincoln, Lincolnshire, LN5 8LG
Area of Operation: UK (Excluding Ireland)
Tel: 01522 510977
Fax: 01522 510929
Email: info@ermineengineering.co.uk
Web: www.ermineengineering.co.uk
Product Type: 2, 3
Other Info: 🖋 ✋
Material Type: C) 1, 2, 3, 4, 18

F.P. IRONWORK
Unit 44, Oswin Road, Leicester,
Leicestershire, LE3 1HR
Area of Operation: UK (Excluding Ireland)
Tel: 0116 255 0455
Fax: 0116 255 6096
Email: sales@fpironwork.com
Web: www.fpironwork.com
Product Type: 1, 2, 3
Other Info: 🖋 ✋
Material Type: C) 2, 4, 6

FOREST GARDEN
Units 291 & 296, Hartlebury Trading Estate,
Hartlebury, Worcestershire, DY10 4JB
Area of Operation: UK (Excluding Ireland)
Tel: 0870 300 9809 **Fax:** 0870 191 9888
Email: info@forestgarden.co.uk
Web: www.forestgarden.co.uk
Product Type: 3

FSL LTD
Sandholes Road, Cookstown,
Co Tyrone, BT80 9AR
Area of Operation: UK & Ireland
Tel: 028 8676 6131 **Fax:** 028 8676 2414
Email: gerard_meenan@fsl.ltd.uk
Web: www.fslelectronics.com 🖱
Product Type: 2

GATE-A-MATION LTD
Unit 8 Boundary Business Centre,
Boundary Way, Woking, Surrey, GU21 5DH
Area of Operation: East England, Greater London,
South East England, South West England
and South Wales
Tel: 01483 747373
Fax: 01483 776688
Email: sales@gate-a-mation.com
Web: www.gate-a-mation.com
Product Type: 1, 2
Other Info: 🖋 ✋
Material Type: A) 1, 2, 3, 4, 5, 6, 7, 8, 9, 10, 11, 12,
13, 14, 15

GATESTUFF.COM
17-19 Old Woking Road, West Bysleet,
Surrey, KT14 6LW
Area of Operation: UK & Ireland
Tel: 01932 344434
Email: sales@gatestuff.com
Web: www.gatestuff.com 🖱
Product Type: 1, 2
Other Info: 🖋

GEORGE BARKER & SONS
Backbarrow, Nr Ulverston, Cumbria, LA12 8TA
Area of Operation: UK (Excluding Ireland)
Tel: 01539 531236
Fax: 01539 530801
Web: www.gbs-ltd.co.uk

IRONCRAFT
92 High Street, Earl Shilton, Leicester,
Leicestershire, LE9 7DG
Area of Operation: UK (Excluding Ireland)
Tel: 01455 847548
Fax: 01455 842422
Email: office@irongraft.co.uk
Web: www.ironcraft.co.uk
Product Type: 1

J.G.S WEATHERVANES
Unit 6, Broomstick Estate, High Street, Edlesborough,
Dunstable, Bedfordshire, LU6 2HS
Area of Operation: UK (Excluding Ireland)
Tel: 01525 220360
Fax: 01525 222786
Email: sales@weathervanes.org.uk
Web: www.weathervanes.org.uk
Product Type: 2, 3
Other Info: 🖋 ✋

JACKSONS FENCING
Stowting Common, Ashford, Kent, TN25 6BN
Area of Operation: UK & Ireland
Tel: 01233 750393 **Fax:** 01233 750403
Email: sales@jacksons-fencing.co.uk
Web: www.jacksons-fencing.co.uk
Product Type: 2, 3

JANE FOLLIS GARDEN DESIGN
71 Eythrope Road, Stone,
Buckinghamshire, HP17 8PH
Area of Operation: South East England
Tel: 01296 747775 **Fax:** 01296 747775
Email: jfgdndesign@aol.com
Web: www.jfgardendesign.co.uk
Product Type: 2

JGS METALWORK
Unit 6 Broomstick Estate, High Street,
Edlesborough, Dunstable, Bedfordshire, LU6 2HS
Area of Operation: East England
Tel: 01525 220360 **Fax:** 01525 222786
Email: enquiries@weathervanes.org.uk
Web: www.weathervanes.org.uk
Product Type: 1, 2, 3
Material Type: C) 2, 4

JLC AUTOMATION LTD
The Sussex Barn, New Barn Farm,
Hailsham Rd, Stone Cross, East Sussex, BN24 5BT
Area of Operation: East England,
Greater London, South East England
Tel: 01323 741199 **Fax:** 01323 741150
Email: tina@jlcautomation.co.uk
Web: www.jlcautomation.co.uk
Product Type: 2

KIRK NATURAL STONE
Bridgend, Fyvie, Turriff, Aberdeenshire, AB53 8QB
Area of Operation: Worldwide
Tel: 01651 891891
Fax: 01651 891794
Email: info@kirknaturalstone.com
Web: www.kirknaturalstone.com

LAND SKILL PROPERTY
33 Bevin Crescent, Outwood,
West Yorkshire, WF1 3ER
Tel: 01924 826836
Fax: 01924 835174
Email: andrew@landskill.co.uk
Product Type: 1, 3

METALCRAFT [TOTTENHAM]
6-40 Durnford Street, Tottenham, London, N15 5NQ
Area of Operation: UK (Excluding Ireland)
Tel: 0208 802 1715
Fax: 0208 802 1258
Email: sales@makingmetalwork.com
Web: www.makingmetalwork.com
Product Type: 2
Material Type: C) 2, 3, 4, 5

NIGEL TYAS HANDCRAFTED IRONWORK
Green Works, 18 Green Road,
Penistone, Sheffield, South Yorkshire, S36 6BE
Area of Operation: Worldwide
Tel: 01226 766618
Email: sales@nigeltyas.co.uk
Web: www.nigeltyas.co.uk
Other Info: ✎ ✋
Material Type: A) 2, 4

OSMO UK LTD
Unit 2 Pembroke Road,
Stocklake Industrial Estate,
Aylesbury, Buckinghamshire, HP20 1DB
Area of Operation: UK & Ireland
Tel: 01296 481220
Fax: 01296 424090
Email: info@osmouk.com
Web: www.osmouk.com
Product Type: 1, 2

PARAGON JOINERY
Systems House, Eastbourne Road,
Blindley Heath, Surrey, RH7 6JP
Area of Operation: UK (Excluding Ireland)
Tel: 01342 836300
Fax: 01342 836364
Email: sales@paragonjoinery.com
Web: www.paragonjoinery.com

PATERSON'S AUTOMATION
25 South Road, Bisley,
Woking, Surrey, GU24 9ES
Area of Operation: South East England
Tel: 01483 728276
Email: info@patersonsautomation.co.uk
Web: www.patersonsautomation.co.uk
Product Type: 2
Material Type: A) 2

PREFAB STEEL CO. LTD
114 Brighton Road, Shoreham,
West Sussex, BN43 6RH
Area of Operation: Greater London,
South East England
Tel: 01273 597733
Fax: 01273 597774
Email: sales@prefabsteel.co.uk
Web: www.prefabsteel.co.uk
Product Type: 3
Material Type: C) 2

PREMIER GARAGE DOORS AND GATES
Unit 7 Sparrowhall Business Park,
Leighton Road, Edlesborough,
Dunstable, Bedfordshire, LU6 2ES
Area of Operation: East England,
South East England
Tel: 01525 220212
Fax: 01525 222201
Email: sales@premiergaragedoors.co.uk
Web: www.premiergaragedoors.co.uk
Product Type: 1, 2

SWS UK
Claughton, Lancaster, Lancashire, LA2 9LA
Area of Operation: UK & Ireland
Tel: 01524 772400 **Fax:** 01524 772411
Email: info@swsuk.co.uk
Web: www.swsuk.co.uk ⬆
Product Type: 2

THE ELECTRIC GATE SHOP
Elm Tree Farm, Sykehouse, Nr. Goole,
East Riding of Yorks, DN14 9AE
Area of Operation: UK (Excluding Ireland)
Tel: 01405 785656
Email: info@theelectricgateshop.co.uk
Web: www.theelectricgateshop.co.uk
Product Type: 2

THE EXPANDED METAL COMPANY LIMITED
PO Box 14, Longhill Industrial Estate (North),
Hartlepool, Durham, TS25 1PR
Area of Operation: Worldwide
Tel: 01429 408087
Fax: 01429 866795
Email: paulb@expamet.co.uk
Web: www.expandedmetalcompany.co.uk

TONY HOOPER
Unit 18 Camelot Court, Bancombe Trading Estate,
Somerton, TA11 6SB
Area of Operation: UK (Excluding Ireland)
Tel: 01458 274221
Fax: 01458 274690
Email: tonyhooper1@aol.com
Web: www.tonyhooper.co.uk

ARCHWAYS

KEY

OTHER: ▽ Reclaimed ⬆ On-line shopping
✎ Bespoke ✋ Hand-made ECO Ecological

ARCHITECTURAL GATES
(callers by appointment only), Mallard, Hoopers Pool,
Southwick, Trowbridge, Wiltshire, BA14 9NG
Area of Operation: Worldwide
Tel: 01225 766944
Email: architectural_gates@yahoo.co.uk
Web: www.architectural-gates.com

B ROURKE & CO LTD
Vulcan Works, Accrington Road,
Burnley, Lancashire, BB11 5QD
Area of Operation: Worldwide
Tel: 01282 422841
Fax: 01282 458901
Email: info@rourkes.co.uk
Web: www.rourkes.co.uk
Material Type: C) 2, 4, 6

BAYFIELD STAIR CO
Unit 4, Praed Road, Trafford Park,
Manchester, M17 1PQ
Area of Operation: Worldwide
Tel: 0161 848 0700
Fax: 0161 872 2230
Email: sales@bayfieldstairs.co.uk
Web: www.bayfieldstairs.co.uk

CANNOCK GATES (UK) LTD.
Hawks Green, Martindale, Cannock,
Staffordshire, WS11 7XT
Area of Operation: UK (Excluding Ireland)
Tel: 08707 541813
Email: sales@cannockgates.co.uk
Web: www.cannockgates.co.uk ⬆

DANDF GARDEN PRODUCTS LTD
Unit 6 Onward Business Park,
Ackworth, West Yorkshire, WF7 7BE
Area of Operation: UK (Excluding Ireland)
Tel: 01977 624200
Fax: 01977 624201
Email: sales@dandf.co.uk
Web: www.dandf.co.uk

DENBY DALE CAST PRODUCTS LTD
230 Cumberworth Lane, Denby Dale,
Huddersfield, West Yorkshire, HD8 8PR
Area of Operation: UK & Ireland
Tel: 01484 863560
Fax: 01484 865597
Email: sales@denbydalecastproducts.co.uk
Web: www.denbydalecastproducts.co.uk

EAST OF EDEN PLANTS
38 St Andrews Street, Millbrook,
Torpoint, Devon, PL10 1BE
Area of Operation: UK & Ireland
Tel: 01752 822782
Email: info@eastofedenplants.co.uk
Web: www.eastofedenplants.co.uk ▽⬆

ERMINE ENGINEERING COMPANY LTD
Francis House, Silver Birch Park, Great Northern
Terrace, Lincoln, Lincolnshire, LN5 8LG
Area of Operation: UK (Excluding Ireland)
Tel: 01522 510977
Fax: 01522 510929
Email: info@ermineengineering.co.uk
Web: www.ermineengineering.co.uk
Other Info: ✎ ✋
Material Type: C) 1, 2, 3, 4

F.P. IRONWORK
Unit 44, Oswin Road, Leicester,
Leicestershire, LE3 1HR
Area of Operation: UK (Excluding Ireland)
Tel: 0116 255 0455
Fax: 0116 255 6096
Email: sales@fpironwork.com
Web: www.fpironwork.com
Material Type: C) 2, 4, 6

FOREST GARDEN
Units 291 & 296, Hartlebury Trading Estate,
Hartlebury, Worcestershire, DY10 4JB
Area of Operation: UK (Excluding Ireland)
Tel: 0870 300 9809 **Fax:** 0870 191 9888
Email: info@forestgarden.co.uk
Web: www.forestgarden.co.uk

GEORGE BARKER & SONS
Backbarrow, Nr Ulverston, Cumbria, LA12 8TA
Area of Operation: UK (Excluding Ireland)
Tel: 01539 531236 **Fax:** 01539 530801
Web: www.gbs-ltd.co.uk

IRONCRAFT
92 High Street, Earl Shilton, Leicester,
Leicestershire, LE9 7DG
Area of Operation: UK (Excluding Ireland)
Tel: 01455 847548 **Fax:** 01455 842422
Email: office@irongraft.co.uk
Web: www.ironcraft.co.uk

JACKSONS FENCING
Stowting Common, Ashford, Kent, TN25 6BN
Area of Operation: UK & Ireland
Tel: 01233 750393 **Fax:** 01233 750403
Email: sales@jacksons-fencing.co.uk
Web: www.jacksons-fencing.co.uk

JANE FOLLIS GARDEN DESIGN
71 Eythrope Road, Stone,
Buckinghamshire, HP17 8PH
Area of Operation: South East England
Tel: 01296 747775 **Fax:** 01296 747775
Email: jfgdndesign@aol.com
Web: www.jfgardendesign.co.uk

KIRK NATURAL STONE
Bridgend, Fyvie, Turriff, Aberdeenshire, AB53 8QB
Area of Operation: Worldwide
Tel: 01651 891891 **Fax:** 01651 891794
Email: info@kirknaturalstone.com
Web: www.kirknaturalstone.com

LLOYD CHRISTIE
15 Langton Street, London, SW10 0JL
Area of Operation: Worldwide
Tel: 0207 351 2108
Fax: 0207 376 5867
Email: info@lloydchristie.com
Web: www.lloydchristie.com

DENBY DALE CAST PRODUCTS LTD

METALCRAFT [TOTTENHAM]
6-40 Durnford Street, Tottenham,
London, N15 5NQ
Area of Operation: UK (Excluding Ireland)
Tel: 0208 802 1715 **Fax:** 0208 802 1258
Email: sales@makingmetalwork.com
Web: www.makingmetalwork.com

NIGEL TYAS HANDCRAFTED IRONWORK
Green Works, 18 Green Road, Penistone,
Sheffield, South Yorkshire, S36 6BE
Area of Operation: Worldwide
Tel: 01226 766618
Email: sales@nigeltyas.co.uk
Web: www.nigeltyas.co.uk
Other Info: ✎ ✋
Material Type: C) 2, 4, 5, 6

OSMO UK LTD
Unit 2 Pembroke Road, Stocklake Industrial Estate,
Aylesbury, Buckinghamshire, HP20 1DB
Area of Operation: UK & Ireland
Tel: 01296 481220 **Fax:** 01296 424090
Email: info@osmouk.com
Web: www.osmouk.com

RIVERSIDE DECKING COMPANY
4 Chauntry Mews, Chauntry Road,
Maidenhead, Berkshire, SL6 1TT
Area of Operation: Greater London,
South East England
Tel: 01628 626545
Email: mklewis3@ukonline.co.uk
Web: www.riversidedeckingcompany.co.uk

THE EXPANDED METAL COMPANY LIMITED
PO Box 14, Longhill Industrial Estate (North),
Hartlepool, Durham, TS25 1PR
Area of Operation: Worldwide
Tel: 01429 408087 **Fax:** 01429 866795
Email: paulb@expamet.co.uk
Web: www.expandedmetalcompany.co.uk

TRADSTOCKS LTD
Dunaverig, Thornhill, Stirling,
Stirlingshire, FK8 3QW
Area of Operation: Scotland
Tel: 01786 850400 **Fax:** 01786 850404
Email: peter@tradstocks.co.uk
Web: www.tradstocks.co.uk

NOTES

Company Name ....
....
Address ....
....
....
email ....
Web ....

Company Name ....
....
Address ....
....
....
email ....
Web ....

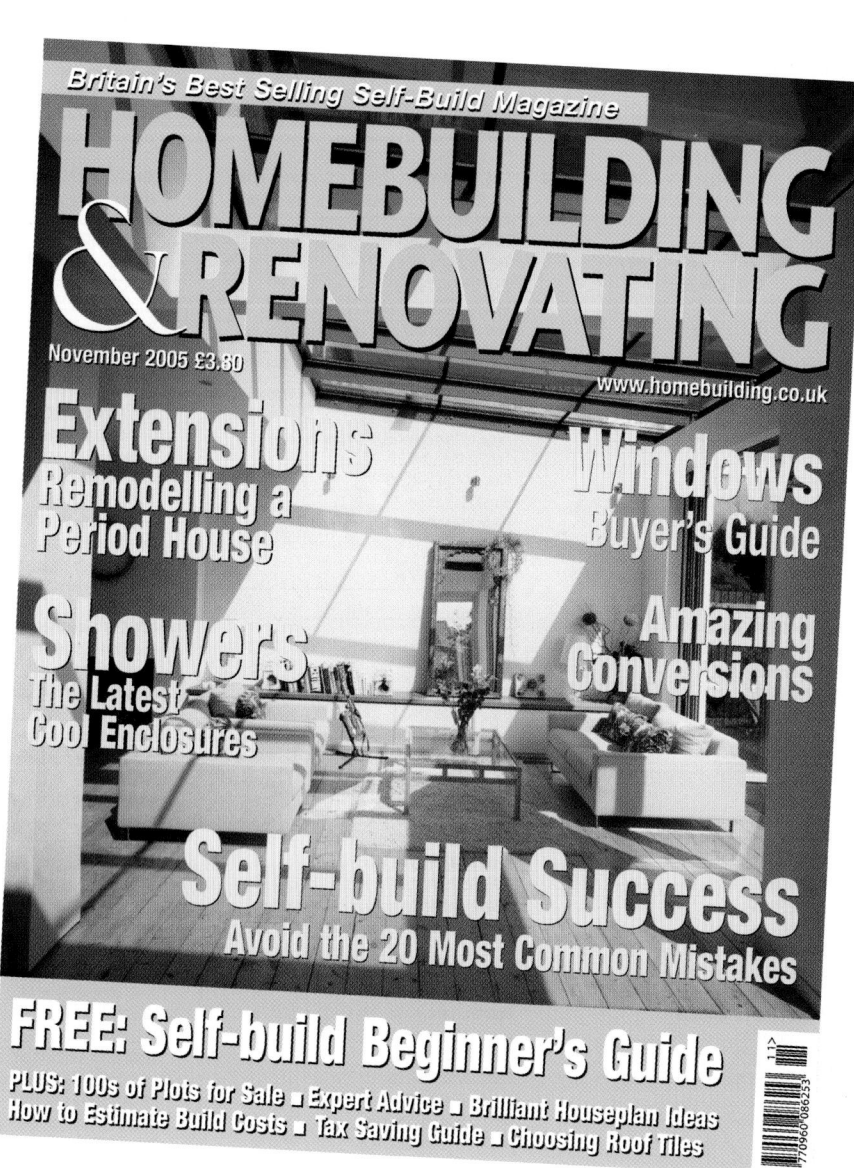

HORTICULTURE & WATER GARDENING

Image courtesy of Marshalls (0870 120 7474)

Landscaping

Image by Marshalls

Image by QLawns

Image by Rolawn

The garden is now much more than a grassy area outside your home; it is more of an external multi-functional living space; a play area for the kids, a dining area for al fresco eating, and somewhere to entertain friends and family.

Therefore a well-designed garden should provide much more than just a decorative surrounding to your home; it should also take into account your family's needs. For example, do you need tall trees or bushes to give added privacy, or smaller shrubs to ensure natural light isn't blocked? Is a pond or water feature a good idea, or will it be a danger to small children? Do you need a large open space where you can easily monitor your children, or could you have a secluded area for dining? By answering such questions you will discover what functions you need your garden to have, and therefore what facilities you will need.

However, you will always need to factor in the natural limitations of your site. For example, a sloping site will need a lot of groundwork doing before it can have any real function; this will include moving large amounts of soil and possibly building retaining walls. Similarly, if you plan to have a pond or water feature, you will need to plot its location in respect to water and electricity sources. Therefore, even if you have no intention of landscaping your garden for some time after completion of your project, it still helps enormously to have some idea of just how you would like it to look when complete. You will be unable to get

heavy plants like JCBs to the back of the house once you have completed the excavations, let alone the building, so it is advisable to plan in the preliminary stage groundworks early in your project.

If you wish to build rockeries or raised features on your site, you may be able to save to save a considerable amount of money by reusing your foundation excavations; in fact the government actively encourages you to do this, by penalising you for removing spoil from site.

Whatever size or shape your plot, most UK homeowners agree that the most essential feature in a beautiful garden is a lush, weed-free lawn. The British climate fortunately provides ideal conditions for propagating a lawn, and as such they are relatively easy to maintain. Cut short, grass has a fine surface texture and, if well maintained, will provide a stunning finish to your garden. If left slightly longer, the grass will be more hardwearing, making it a perfect surface for children to play on.

Trees, shrubs and flowers will add the finishing touches to a natural garden, but always be aware of the aspect of your garden, the type of soil you have, and the proximity to the house when planting. Tree roots can cause considerable damage to your foundations if allowed to, so always plant them a safe distance away. Many plants also need a particular type of soil, or a minimum amount of sunlight per day, in order to flourish, so always check their requirements before purchasing.

LAWNS & TURF

KEY

OTHER: ▽ Reclaimed ⌐ On-line shopping
✐ Bespoke 🖐 Hand-made ECO Ecological

ACCESS IRRIGATION
Yelvertoft Road, Crick, Northampton,
Northamptonshire, NN6 7XS
Area of Operation: UK & Ireland
Tel: 01788 823811
Fax: 01788 82456
Email: sales@access-irrigation.co.uk
Web: www.access-irrigation.co.uk ⌐

COMER LANDSCAPES
99 London Road, Northwich, Cheshire, CW9 5HQ
Area of Operation: UK (Excluding Ireland)
Tel: 01606 43737
Fax: 01606 43736
Email: info@comerlandscapes.co.uk
Web: www.cheshirelawns.co.uk

DEACONS NURSERY
Moorview, Godshill, Isle of Wight, PO38 3HW
Area of Operation: UK & Ireland
Tel: 01983 522243
Fax: 01983 523575
Email: deacons.nursery@btopenworld.com
Web: www.deaconnurseryfruits.co.uk ⌐

EAST OF EDEN PLANTS
38 St Andrews Street, Millbrook,
Torpoint, Devon, PL10 1BE
Area of Operation: UK & Ireland
Tel: 01752 822782
Email: info@eastofedenplants.co.uk
Web: www.eastofedenplants.co.uk ⌐

EASY LAWN
Thingehill Court, Withington,
Herefordshire, HR1 3QG
Area of Operation: UK (Excluding Ireland)
Tel: 01432 850850
Fax: 01432 850064
Email: mail@easylawn.co.uk
Web: www.easylawn.co.uk

ENVIROMAT BY Q LAWNS
Corkway Drove, Hockwold,
Thetford, Norfolk, IP26 4JR
Area of Operation: UK (Excluding Ireland)
Tel: 01842 828266
Fax: 01842 827911
Email: sales@qlawns.co.uk
Web: www.enviromat.co.uk ⌐

GREENSCENE
The Nursery, Gunnersbury Park,
London, W3 8LQ
Area of Operation: UK (Excluding Ireland)
Tel: 0845 345 9808
Fax: 0845 345 9809
Email: enquiries@greenscene.com
Web: www.greenscene.com

HILLIER LANDSCAPES
Ampfield House, Ampfield, Romsey,
Hampshire, SO51 9PA
Area of Operation: South East England,
South West England and South Wales
Tel: 01794 368855
Fax: 01794 368866
Email: hillierlandscapes@btinternet.com
Web: www.hillier-landscapes.co.uk
Other Info: ✐

JANE FOLLIS GARDEN DESIGN
71 Eythrope Road, Stone,
Buckinghamshire, HP17 8PH
Area of Operation: South East England
Tel: 01296 747775
Fax: 01296 747775
Email: jfgdndesign@aol.com
Web: www.jfgardendesign.co.uk

Q LAWNS
Corkway Drove, Hockwold,
Thetford, Norfolk, IP26 4JR
Area of Operation: UK (Excluding Ireland)
Tel: 01842 828266
Fax: 01842 827911
Email: sales@qlawns.co.uk
Web: www.qlawns.co.uk

RIVAR SAND & GRAVEL LTD
Pinchington Lane, Newbury, Berkshire, RG19 8SR
Area of Operation: South East England
Tel: 01635 523524
Fax: 01635 521621
Email: sales@rivarsandandgravel.co.uk
Web: www.rivarsandandgravel.co.uk

ROLAWN LIMITED
Elvington, York, North Yorkshire, YO41 4XR
Area of Operation: UK & Ireland
Tel: 01904 608661
Fax: 01904 608272
Email: info@rolawn.co.uk
Web: www.rolawn.co.uk/www.topsoil.co.uk

TEAL TURF CO.LTD
Teal Trading Ltd, Wadborough, Worcester,
Worcestershire, WR8 9HJ
Area of Operation: UK (Excluding Ireland)
Tel: 01905 840279
Fax: 01905 841460
Email: enquirires@tealturf.co.uk
Web: www.tealturf.co.uk

TURF CENTRE
Ham Barn Farm, Farnham Road, Liss,
Hampshire, GU33 6LG
Area of Operation: South East England,
South West England and South Wales
Tel: 01420 538188
Fax: 01420 538208
Email: turfcentre@btinternet.com
Web: www.turfcentre.co.uk

TREES & PLANTS

KEY

OTHER: ▽ Reclaimed ⌐ On-line shopping
✐ Bespoke 🖐 Hand-made ECO Ecological

DEACONS NURSERY
Moorview, Godshill, Isle of Wight, PO38 3HW
Area of Operation: UK & Ireland
Tel: 01983 522243 **Fax:** 01983 523575
Email: deacons.nursery@btopenworld.com
Web: www.deaconnurseryfruits.co.uk ⌐

EAST OF EDEN PLANTS
38 St Andrews Street, Millbrook,
Torpoint, Devon, PL10 1BE
Area of Operation: UK & Ireland
Tel: 01752 822782
Email: info@eastofedenplants.co.uk
Web: www.eastofedenplants.co.uk ⌐

GREENSCENE
The Nursery, Gunnersbury Park,
London, W3 8LQ
Area of Operation: UK (Excluding Ireland)
Tel: 0845 345 9808
Fax: 0845 345 9809
Email: enquiries@greenscene.com
Web: www.greenscene.com

HILLIER LANDSCAPES
Ampfield House, Ampfield,
Romsey, Hampshire, SO51 9PA
Area of Operation: South East England,
South West England and South Wales
Tel: 01794 368855
Fax: 01794 368866
Email: hillierlandscapes@btinternet.com
Web: www.hillier-landscapes.co.uk
Other Info: ✐

JANE FOLLIS GARDEN DESIGN
71 Eythrope Road, Stone,
Buckinghamshire, HP17 8PH
Area of Operation: South East England
Tel: 01296 747775
Fax: 01296 747775
Email: jfgdndesign@aol.com
Web: www.jfgardendesign.co.uk

WYVALE HAWKINS
Thingehill Court, Withington, Herefordshire, HR1 3QG
Area of Operation: UK (Excluding Ireland)
Tel: 01432 850433
Fax: 01432 850762
Email: sales@wyvale-hawkins.co.uk
Web: www.wyvale-hawkins.co.uk

WATER FEATURES & AQUATICS

KEY

SEE ALSO: GARDEN ACCESSORIES AND
DECORATIVE FEATURES - Ornamental Stonework
PRODUCT TYPES: 1= Fountains 2 = Statuary
3 = Pond Liners 4 = Pumps
OTHER: ▽ Reclaimed ⌐ On-line shopping
✐ Bespoke 🖐 Hand-made ECO Ecological

ARCHITECTURAL HERITAGE
Taddington Manor, Taddington, Nr Cutsdean,
Cheltenham, Gloucestershire, GL54 5RY
Area of Operation: Worldwide
Tel: 01386 584414 **Fax:** 01386 584236
Email: puddy@architectural-heritage.co.uk
Web: www.architectural-heritage.co.uk
Product Type: 1

ART OUTSIDE
PO Box 513, Aylesbury,
Buckinghamshire, HP22 6WJ
Area of Operation: Worldwide
Tel: 01296 582004
Email: emma@art-outside.com
Web: www.art-outside.com ⌐
Product Type: 1, 2
Other Info: ✐

BLOWZONE HOT STUDIO
The Ruskin Glass Centre, Wollaston Road,
Amblecote, West Midlands, DY8 4HF
Area of Operation: Worldwide
Tel: 01384 399464
Fax: 01384 377746
Email: sales@blowzone.co.uk
Web: www.blowzone.co.uk
Product Type: 1

EAST OF EDEN PLANTS
38 St Andrews Street, Millbrook,
Torpoint, Devon, PL10 1BE
Area of Operation: UK & Ireland
Tel: 01752 822782
Email: info@eastofedenplants.co.uk
Web: www.eastofedenplants.co.uk ⌐

GREENSCENE
The Nursery, Gunnersbury Park,
London, W3 8LQ
Area of Operation: UK (Excluding Ireland)
Tel: 0845 345 9808
Fax: 0845 345 9809
Email: enquiries@greenscene.com
Web: www.greenscene.com

HILLIER LANDSCAPES
Ampfield House, Ampfield,
Romsey, Hampshire, SO51 9PA
Area of Operation: South East England,
South West England and South Wales
Tel: 01794 368855
Fax: 01794 368866
Email: hillierlandscapes@btinternet.com
Web: www.hillier-landscapes.co.uk
Other Info: ✐

JANE FOLLIS GARDEN DESIGN
71 Eythrope Road, Stone,
Buckinghamshire, HP17 8PH
Area of Operation: South East England
Tel: 01296 747775
Fax: 01296 747775
Email: jfgdndesign@aol.com
Web: www.jfgardendesign.co.uk

WYVALE HAWKINS
Thingehill Court, Withington, Herefordshire, HR1 3QG
Area of Operation: UK (Excluding Ireland)
Tel: 01432 850433
Fax: 01432 850762
Email: sales@wyvale-hawkins.co.uk
Web: www.wyvale-hawkins.co.uk

CHILSTONE
Victoria Park, Fordcombe Road,
Langton Green, Kent, TN3 0RD
Area of Operation: Worldwide
Tel: 01892 740866
Fax: 01892 740249
Email: ornaments@chilstone.com
Web: www.chilstone.com ⌐
Product Type: 1, 2
Material Type: E) 13

CRESS WATER LTD
61 Woodstock Road, Worcester,
Worcestershire, WR2 5ND
Area of Operation: Europe
Tel: 01905 422707
Fax: 01905 422744
Email: info@cresswater.co.uk
Web: www.cresswater.co.uk

HADDONSTONE LTD
The Forge House, East Haddon,
Northampton, Northamptonshire, NN6 8DB
Area of Operation: Worldwide
Tel: 01604 770711 **Fax:** 01604 770027
Email: info@haddonstone.co.uk
Web: www.haddonstone.co.uk
Product Type: 1, 2

HADDONSTONE
Area of Operation: Worldwide
Tel: 01604 770711 **Fax:** 01604 770027
Email: info@haddonstone.co.uk
Web: www.haddonstone.co.uk
Product Type: 1, 2
Other Info: ⌐ 🖐

Haddonstone has an extensive range of cast stone pool surrounds and fountains including multi-tiered, figured and self-contained. Traditional, classical and contemporary styles.

HILLIER LANDSCAPES
Ampfield House, Ampfield,
Romsey, Hampshire, SO51 9PA
Area of Operation: South East England,
South West England and South Wales
Tel: 01794 368855 **Fax:** 01794 368866
Email: hillierlandscapes@btinternet.com
Web: www.hillier-landscapes.co.uk
Other Info: 📎

IRIS WATER & DESIGN
Langburn Bank, Castleton,
Whitby, North Yorkshire, YO21 2EU
Area of Operation: Europe
Tel: 01287 660002 **Fax:** 01287 660004
Email: info@iriswater.com
Web: www.iriswater.com
Product Type: 1, 2, 3, 4

JANE FOLLIS GARDEN DESIGN
71 Eythrope Road, Stone,
Buckinghamshire, HP17 8PH
Area of Operation: South East England
Tel: 01296 747775 **Fax:** 01296 747775
Email: jfgdndesign@aol.com
Web: www.jfgardendesign.co.uk
Product Type: 1, 3, 4

KIRK NATURAL STONE
Bridgend, Fyvie, Turriff,
Aberdeenshire, AB53 8QB
Area of Operation: Worldwide
Tel: 01651 891891 **Fax:** 01651 891794
Email: info@kirknaturalstone.com
Web: www.kirknaturalstone.com
Product Type: 1, 2

POND GUARD/DAVREN ENTERPRISES
Units 1-1b, Central Industrial Estate,
Rear 138/146 Bolton Road, Atherton,
Greater Manchester, M46 9LF
Area of Operation: UK (Excluding Ireland)
Tel: 01942 888601 **Fax:** 01942 888601
Email: info@davren.co.uk
Web: www.pondguardonline.co.uk

RE:DESIGN
11 Bognor Road, Chichester,
West Sussex, PO19 7TF
Area of Operation: UK & Ireland
Tel: 07787 987652
Email: barry@redesign.uk.com
Web: www.redesign.uk.com
Product Type: 1

RIVAR SAND & GRAVEL LTD
Pinchington Lane, Newbury,
Berkshire, RG19 8SR
Area of Operation: South East England
Tel: 01635 523524 **Fax:** 01635 521621
Email: sales@rivarsandandgravel.co.uk
Web: www.rivarsandandgravel.co.uk
Product Type: 1

THE CANE STORE
Wash Dyke Cottage, No1 Witham Road,
Long Bennington, Lincolnshire, NG23 5DS
Area of Operation: Worldwide
Tel: 01400 282271 **Fax:** 01400 281103
Email: jaki@canestore.co.uk
Web: www.canestore.co.uk
Product Type: 2

TITAN PLASTECH LIMITED
Barbot Hall Industrial Estate,
Mangham Road, Rotherham,
South Yorkshire, S61 4RJ
Area of Operation: Europe
Tel: 01709 538300 **Fax:** 01709 538301
Email: tony.soper@titantanks.co.uk
Web: www.titanplastech.co.uk

NOTES

Company Name ....

Address ....

....

....

email ....

Web ....

Company Name ....

Address ....

....

....

email ....

Web ....

Company Name ....

Address ....

....

....

email ....

Web ....

Company Name ....

Address ....

....

....

email ....

Web ....

GARDEN ACCESSORIES & DECORATIVE FEATURES

Image courtesy of Haddonstone (01604 770711)

Garden Accessories and Decorative Features

When the landscaping of your garden is complete and all plants, trees and shrubs are in place, it is time to concentrate on the things that are going to add the finishing touches to your outdoor space. With the hard work over, selecting your decorative garden features gives you the opportunity to look forward to relaxing in your garden when it is finally complete. Furniture and some well-chosen decorative items are just the things to give your garden the homely touches you desire.

Statues, sundials, urns and clock towers work well in formally styled gardens, giving a classic feel and focusing the eye on a centre piece in your garden which, in turn, can increase the sense of perspective, enlarging your garden space. Similarly, the installation of straight paths, which play with vanishing points to seemingly extend your garden's boundaries, or having cleverly placed reflective mirrors to reflect specific areas, will also seemingly increase the size of your outdoor space.

Garden and patio furniture will also create a focal point for your garden, as well as adding the multi-functional dimension that many designers now advise. A seating or dining area can maximise the spots in your garden which receive the most direct sunlight, and the vast range of products on the market ensures that there is a style to suit every garden. Timber products tend to compliment natural gardens, whilst metalwork offsets traditional gardens to perfection.

The addition of a barbeque will open up the option of al-fresco dining, whilst a patio heater will ensure that you can make the most of your garden into the evenings and at colder times of the year. However, you should check what fuel your burner will use; a gel or wood burner will have a lesser environmental impact than one run on gas.

Finishing touches, such as signs for the front of your house, can extend the look you are aiming to achieve

TOP: Firebowls make excellent patio heaters and, being gel burners, they give off no smoke or fumes. Example by N&C Nicobond. RIGHT: Having garden furniture means that your outdoor space can effectively be used as an extra room in warmer months. Image courtesy of Bradstone.

ORNAMENTAL STONEWORK

KEY
OTHER: ▽ Reclaimed 🖱 On-line shopping
✏ Bespoke 🖐Hand-made ECO Ecological

AD CALVERT ARCHITECTURAL STONE SUPPLIES LTD
Smithy Lane, Grove Square, Leyburn,
North Yorkshire, DL8 5DZ
Area of Operation: UK & Ireland
Tel: 01969 622515
Fax: 01969 624345
Email: stone@calverts.co.uk
Web: www.calverts.co.uk
Other Info: ✏

ALL THINGS STONE LIMITED
Anchor House, 33 Ospringe Street,
Ospringe, Faversham, Kent, ME13 8TW
Area of Operation: UK & Ireland
Tel: 01795 531001
Fax: 01795 530742
Email: info@allthingsstone.co.uk
Web: www.allthingsstone.co.uk
Other Info: ✏

ARCHITECTURAL HERITAGE
Taddington Manor, Taddington, Nr Cutsdean,
Cheltenham, Gloucestershire, GL54 5RY
Area of Operation: Worldwide
Tel: 01386 584414
Fax: 01386 584236
Email: puddy@architectural-heritage.co.uk
Web: www.architectural-heritage.co.uk
Other Info: ▽

ART OUTSIDE
PO Box 513, Aylesbury, Buckinghamshire, HP22 6WJ
Area of Operation: Worldwide
Tel: 01296 582004
Email: emma@art-outside.com
Web: www.art-outside.com 🖱

BRADSTONE GARDEN
Aggregate Industries UK Ltd, Hulland Ward,
Ashbourne, Derbyshire, DE6 3ET
Area of Operation: UK (Excluding Ireland)
Tel: 01335 372 222
Fax: 01335 370 973
Email: bardstone.structural@aggregate.com
Web: www.bradstone.com

BRETT LANDSCAPING LTD
Salt Lane, Cliffe, Rochester, Kent, ME3 7SZ
Area of Operation: Worldwide
Tel: 01634 222188
Fax: 01634 222001
Email: landscapinginfo@brett.co.uk
Web: www.brett.co.uk/landscaping

CHILSTONE
Victoria Park, Fordcombe Road,
Langton Green, Kent, TN3 0RD
Area of Operation: Worldwide
Tel: 01892 740866
Fax: 01892 740249
Email: ornaments@chilstone.com
Web: www.chilstone.com 🖱

DAR INTERIORS
Arch 11, Miles Street, London, SW8 1RZ
Area of Operation: Worldwide
Tel: 0207 720 9678
Fax: 0207 627 5129
Email: enquiries@darinteriors.com
Web: www.darinteriors.com

DENBY DALE CAST PRODUCTS LTD
230 Cumberworth Lane, Denby Dale,
Huddersfield, West Yorkshire, HD8 8PR
Area of Operation: UK & Ireland
Tel: 01484 863560
Fax: 01484 865597
Email: sales@denbydalecastproducts.co.uk
Web: www.denbydalecastproducts.co.uk

DEVON STONE LTD
38 Station Road, Budleigh Salterton, Devon, EX9 6RT
Area of Operation: UK (Excluding Ireland)
Tel: 01395 446841
Email: amy@devonstone.com
Other Info: ▽ ✏ 🖐

EUROPEAN PIPE SUPPLIERS LTD
Unit 6 Severnlink Distribution Centre, off Newhouse
Farm Estate, Chepstow, Monmouthshire, NP16 6UN
Area of Operation: Europe
Tel: 01291 622215
Fax: 01291 620381
Email: sales@epsonline.co.uk
Web: www.epsonline.co.uk
Other Info: ✏

HADDONSTONE LTD
The Forge House, East Haddon,
Northampton, Northamptonshire, NN6 8DB
Area of Operation: Worldwide
Tel: 01604 770711 **Fax:** 01604 770027
Email: info@haddonstone.co.uk
Web: www.haddonstone.co.uk

HADDONSTONE
Area of Operation: Worldwide
Tel: 01604 770711 **Fax:** 01604 770027
Email: info@haddonstone.co.uk
Web: www.haddonstone.co.uk
Other Info: 🖱 🖐

Haddonstone manufacture custom cast stone
designs for restoration projects. The company
also has an extensive range of standard designs
including balustrading, columns and window
surrounds.

JANE FOLLIS GARDEN DESIGN
71 Eythrope Road, Stone,
Buckinghamshire, HP17 8PH
Area of Operation: South East England
Tel: 01296 747775
Fax: 01296 747775
Email: jfgdndesign@aol.com
Web: www.jfgardendesign.co.uk

KENT BALUSTERS
1 Gravesend Road, Strood, Kent, ME2 3PH
Area of Operation: UK (Excluding Ireland)
Tel: 01634 711617
Fax: 01634 714644
Email: info@kentbalusters.co.uk
Web: www.kentbalusters.co.uk
Other Info: ✏ 🖐

KIRK NATURAL STONE
Bridgend, Fyvie, Turriff,
Aberdeenshire, AB53 8QB
Area of Operation: Worldwide
Tel: 01651 891891
Fax: 01651 891794
Email: info@kirknaturalstone.com
Web: www.kirknaturalstone.com

MEADOWSTONE (DERBYSHIRE) LTD
West Way, Somercotes, Derbyshire, DE55 4QJ
Area of Operation: UK & Ireland
Tel: 01773 540707
Fax: 01773 528119
Email: info@meadowstone.co.uk
Web: www.meadowstone.co.uk 🖱
Other Info: ✏ 🖐

ROCK UNIQUE LTD
c/o Select Garden and Pet Centre, Main Road,
Sundridge, Kent, TN14 6ED
Area of Operation: Europe
Tel: 01959 565608
Fax: 01959 569312
Email: stone@rock-unique.com
Web: www.rock-unique.com
Other Info: ▽ ✏ 🖐

THE MASON'S YARD
Penhenllan, Cusop, Hay on Wye, Herefordshire, HR3 5TE
Area of Operation: UK (Excluding Ireland)
Tel: 01497 821333
Email: hugh@themasonsyard.co.uk
Web: www.themasonsyard.co.uk

VOUSTONE DESIGNS LIMITED.
Kingdom Cottage, Tibbs Court Lane,
Brenchley, Nr. Tonbridge, Kent, TN12 7AH
Area of Operation: UK (Excluding Ireland)
Tel: 01892 722449
Fax: 01892 722573
Email: pauleen@voustone.co.uk
Web: www.voustone.co.uk
Other Info: ✏ 🖐

WARMSWORTH STONE LTD
1-3 Sheffield Road, Warmsworth,
Doncaster, South Yorkshire, DN4 9QH
Area of Operation: UK & Ireland
Tel: 01302 858617
Fax: 01302 855844
Email: info@warmsworth-stone.co.uk
Web: www.warmsworth-stone.co.uk
Other Info: ✏ 🖐

WELLS CATHEDRAL STONEMASONS
Brunel Stoneworks, Station Road,
Cheddar, Somerset, BS27 3AH
Area of Operation: Worldwide
Tel: 01934 743544 **Fax:** 01934 744536
Email: wcs@stone-mason.co.uk
Web: www.stone-mason.co.uk

GARDEN & PATIO FURNITURE

KEY
OTHER: ▽ Reclaimed 🖱 On-line shopping
✏ Bespoke 🖐Hand-made ECO Ecological

ALTHAM HARDWOOD CENTRE LTD
Altham Corn Mill, Burnley Road, Altham,
Accrington, Lancashire, BB5 5UP
Area of Operation: UK & Ireland
Tel: 01282 771618 **Fax:** 01282 777932
Email: info@oak-beams.co.uk
Web: www.oak-beams.co.uk
Other Info: ECO ✏ 🖐

ART OUTSIDE
PO Box 513, Aylesbury,
Buckinghamshire, HP22 6WJ
Area of Operation: Worldwide
Tel: 01296 582004
Email: emma@art-outside.com
Web: www.art-outside.com 🖱
Other Info: ▽ ✏ 🖐

B ROURKE & CO LTD
Vulcan Works, Accrington Road,
Burnley, Lancashire, BB11 5QD
Area of Operation: Worldwide
Tel: 01282 422841 **Fax:** 01282 458901
Email: info@rourkes.co.uk
Web: www.rourkes.co.uk

BEAVER TIMBER COMPANY
Barcaldine, Argyll & Bute, PA37 1SG
Area of Operation: UK (Excluding Ireland)
Tel: 01631 720353 **Fax:** 01631 720430
Email: info@beavertimber.co.uk
Web: www.beavertimber.co.uk 🖱

C&V CARMICHAEL LTD
Fabrication Facility, Mossmorran,
Cowdenbeath, Fife, KY4 8EP
Area of Operation: UK (Excluding Ireland)
Tel: 01383 510469
Fax: 01383 610515
Email: cvcarmichael@cvcarmichael.com
Web: www.cvcarmichael.com

CHAIRWORKS
47 Weir Road, London, SW19 8UG
Area of Operation: UK (Excluding Ireland)
Tel: 0208 247 3700
Fax: 0208 247 3800
Email: info@chairworks.info
Web: www.chairworks.info

DAR INTERIORS
Arch 11, Miles Street, London, SW8 1RZ
Area of Operation: Worldwide
Tel: 0207 720 9678
Fax: 0207 627 5129
Email: enquiries@darinteriors.com
Web: www.darinteriors.com
Other Info: ✏ 🖐

DD HEATING LTD
16-19 The Manton Centre, Manton Lane,
Bedford, Bedfordshire, MK41 7PX
Area of Operation: Worldwide
Tel: 0870 777 8323
Fax: 0870 777 8320
Email: info@heatline.co.uk
Web: www.heatline.co.uk 🖱

FURNITURE123.CO.UK
Sandway Business Centre, Shannon Street,
Leeds, West Yorkshire, LS9 8SS
Area of Operation: UK & Ireland
Tel: 0113 248 2233
Fax: 0113 248 2266
Email: p.haddock@furniture123.co.uk
Web: www.furniture123.co.uk 🖱

GEORGE BARKER & SONS
Backbarrow, Nr Ulverston,
Cumbria, LA12 8TA
Area of Operation: UK (Excluding Ireland)
Tel: 01539 531236
Fax: 01539 530801
Web: www.gbs-ltd.co.uk

JANE FOLLIS GARDEN DESIGN
71 Eythrope Road, Stone,
Buckinghamshire, HP17 8PH
Area of Operation: South East England
Tel: 01296 747775
Fax: 01296 747775
Email: jfgdndesign@aol.com
Web: www.jfgardendesign.co.uk
Other Info: ECO ✏ 🖐

LLOYD CHRISTIE
15 Langton Street, London, SW10 0JL
Area of Operation: Worldwide
Tel: 0207 351 2108
Fax: 0207 376 5867
Email: info@lloydchristie.com
Web: www.lloydchristie.com

MADE ON EARTH LTD
Unit A & B, The Coach Works,
Kingsfield Lane, Longwell Green,
Bristol, BS30 6DL
Area of Operation: Europe
Tel: 0845 095 6162
Fax: 0845 095 6162
Email: sales@made-on-earth.co.uk
Web: www.made-on-earth.co.uk

NIGEL TYAS HANDCRAFTED IRONWORK
Green Works, 18 Green Road,
Penistone, Sheffield,
South Yorkshire, S36 6BE
Area of Operation: Worldwide
Tel: 01226 766618
Email: sales@nigeltyas.co.uk
Web: www.nigeltyas.co.uk
Other Info: ✏ 🖐

NUMBER 9 STUDIO UK
ARCHITECTURAL CERAMICS
Mole Cottage Industries, Mole Cottage,
Watertown, Chittlehamholt, Devon, EX37 9HF
Area of Operation: Worldwide
Tel: 01769 540471 **Fax:** 01769 540471
Email: arch.ceramics@moley.uk.com
Web: www.moley.uk.com

OLLERTON LTD
Samlesbury Bottoms,
Preston, Lancashire, PR5 0RN
Area of Operation: Worldwide
Tel: 01254 852127 **Fax:** 01254 854383
Email: sales@ollerton.u-net.com
Web: www.ollerton.co.uk
Other Info: ✎

ONEWAY GARDEN LEISURE
18/19 John Samuel Building, Arthur Drive,
Hoo Farm Industrial Estate, Kidderminster,
Worcestershire, DY11 7RA
Area of Operation: UK (Excluding Ireland)
Tel: 01562 750222 **Fax:** 01562 750222
Email: maureenr18@tiscali.co.uk

OSMO UK LTD
Unit 2 Pembroke Road, Stocklake Industrial Estate,
Aylesbury, Buckinghamshire, HP20 1DB
Area of Operation: UK & Ireland
Tel: 01296 481220 **Fax:** 01296 424090
Email: info@osmouk.com
Web: www.osmouk.com

PARLOUR FARM
Unit 12b, Wilkinson Road, Love Lane Industrial
Estate, Cirencester, Gloucestershire, GL7 1YT
Area of Operation: Europe
Tel: 01285 885336
Fax: 01285 643189
Email: info@parlourfarm.com
Web: www.parlourfarm.com

PATIOHEATERS4U LIMITED
Thor Industrial Estate, Braydon, Cricklade,
Swindon, Wiltshire, SN6 6HQ
Area of Operation: Europe
Tel: 01793 613900
Email: sales@patioheaters4u.com
Web: www.patioheaters4u.com ✎

THE GREEN SHOP
Cheltenham Road, Bisley, Nr Stroud,
Gloucestershire, GL6 7BX
Area of Operation: UK & Ireland
Tel: 01452 770629
Fax: 01452 770104
Email: paint@greenshop.co.uk
Web: www.greenshop.co.uk ✎

TITAN PLASTECH LIMITED
Barbot Hall Industrial Estate, Mangham Road,
Rotherham, South Yorkshire, S61 4RJ
Area of Operation: Europe
Tel: 01709 538300
Fax: 01709 538301
Email: tony.soper@titantanks.co.uk
Web: www.titanplastech.co.uk

SIGNS & MAILBOXES

KEY
OTHER: ▽ Reclaimed 🖱 On-line shopping
✎ Bespoke 🖐 Hand-made ECO Ecological

.TOUCHSTONE UK.
Touchstone House, 82 High Street, Measham,
Derbyshire, DE12 7JB
Area of Operation: UK (Excluding Ireland)
Tel: 0845 130 1862
Fax: 01530 274271
Email: sales@touchstone-uk.com
Web: www.touchstone-uk.com
Other Info: ✎ 🖐

TOUCHSTONE UK

Area of Operation: UK (Excluding Ireland)
Tel: 0845 130 1862
Fax: 01530 274 271
Email: sales@touchstone-uk.com
Web: www.touchstone-uk.com

Stone, Marble, Slate & Granite Signs made to
your own requirements. Individual or Batches

ART OUTSIDE
PO Box 513, Aylesbury, Buckinghamshire, HP22 6WJ
Area of Operation: Worldwide
Tel: 01296 582004
Email: emma@art-outside.com
Web: www.art-outside.com ✎
Other Info: ✎ 🖐

B ROURKE & CO LTD
Vulcan Works, Accrington Road, Burnley,
Lancashire, BB11 5QD
Area of Operation: Worldwide
Tel: 01282 422841
Fax: 01282 458901
Email: info@rourkes.co.uk
Web: www.rourkes.co.uk

EMERY ETCHINGS LTD
C7 Laser Quay, Medway City Estate,
Rochester, Kent, ME2 4 HU
Area of Operation: UK (Excluding Ireland)
Tel: 01634 719396
Fax: 01634 720887
Email: sales@emeryetchings.com
Web: www.signs4houses.co.uk ✎

GIRAFFE MARKETING
10 Duncan Grove, East Acton, London, W3 7NN
Area of Operation: UK (Excluding Ireland)
Tel: 0208 743 0233
Email: charles@giraffemarketing.co.uk
Web: www.hippo-box.co.uk

GLENCALL INTERNATIONAL
One Lomond House, GlenLomond,
Kinross-shire, KY13 9HF
Area of Operation: UK & Ireland
Tel: 01592 840853 **Fax:** 01592 840005
Email: enquiries@glencall.co.uk
Web: www.glencall.co.uk/1gi/pillh.html ✎

HOUSE SIGNS BY TRUDY
32 Rose Gardens, Weston-super-Mare,
Somerset, BS22 7PX
Area of Operation: UK & Ireland
Tel: 01934 517327
Email: mail@trudysilcox.co.uk
Web: www.trudysilcox.co.uk
Other Info: ✎ 🖐

HOUSE SIGNS BY TRUDY

Area of Operation: UK & Ireland
Tel: 01934 517327
Email: mail@trudysilcox.co.uk
Web: www.trudysilcox.co.uk

Stunning bespoke signs each one completely
designed and hand painted.Any design or subject
matter you require.Detailed painting to bring your
subject to life.

JGS METALWORK
Unit 6 Broomstick Estate, High Street,
Edlesborough, Dunstable, Bedfordshire, LU6 2HS
Area of Operation: East England
Tel: 01525 220360
Fax: 01525 222786
Email: enquiries@weathervanes.org.uk
Web: www.weathervanes.org.uk
Other Info: ✎ 🖐

KIRK NATURAL STONE
Bridgend, Fyvie, Turriff, Aberdeenshire, AB53 8QB
Area of Operation: Worldwide
Tel: 01651 891891
Fax: 01651 891794
Email: info@kirknaturalstone.com
Web: www.kirknaturalstone.com

MAILBOXES DIRECT
Unit 4, Canalside, North Bridge Road,
Berkhamsted, Hertfordshire, HP4 1EG
Area of Operation: Worldwide
Tel: 01442 878440
Fax: 01442 871472
Email: info@mailboxesdirect.co.uk
Web: www.mailboxesdirect.co.uk ✎

SIGNS & LABELS LTD
Douglas Bruce House, Corrie Way, Bredbury
Industrial Park, Stockport, Cheshire, SK6 2RR
Area of Operation: Worldwide
Tel: 0800 132323
Fax: 0800 389 5311
Email: sales@safetyshop.co.uk
Web: www.safetyshop.com ✎

SIGNS OF THE TIMES
Wingfield Road, Tebworth,
Leighton Buzzard, Bedfordshire, LU7 9QG
Area of Operation: Worldwide
Tel: 01525 874 185 **Fax:** 01525 875 746
Email: enquiries@sott.co.uk
Web: www.sott.co.uk ✎
Other Info: ✎ 🖐

STOCKSIGNS LTD
43 Ormside Way, Redhill, Surrey, RH1 2LG
Area of Operation: Worldwide
Tel: 01737 764764 **Fax:** 01737 763763
Email: info@stocksigns.co.uk
Web: www.stocksigns.co.uk ✎

VILLAGE GREEN SIGNS
Unit 7 Watford Enterprise Centre,
25 Greenhill Crescent , Watford,
Hertfordshire, WD18 8XU
Area of Operation: Worldwide
Tel: 01923 243777
Fax: 01923 243775
Email: vgsigns@tiscali.co.uk
Web: www.villagegreensigns.co.uk ✎
Other Info: ✎ 🖐

BARBEQUES & PATIO HEATERS

KEY
OTHER: ▽ Reclaimed 🖱 On-line shopping
✎ Bespoke 🖐 Hand-made ECO Ecological

COMPLETE CHIMENEA COMPANY
Casa Dali, 31 Upper East Hayes,
Bath, Somerset, BA1 6LP
Area of Operation: UK (Excluding Ireland)
Tel: 01225 465280
Fax: 01225 471581
Email: da@glo-art.co.uk
Web: www.glo-art.co.uk ✎
Other Info: 🖐

GRENADIER FIRELIGHTERS LIMITED
Unit 3C, Barrowmore Enterprise Estate,
Great Barrow, Chester, Cheshire, CH3 7JS
Area of Operation: UK & Ireland
Tel: 01829 741649
Fax: 01829 741659
Email: enquiries@grenadier.uk.com
Web: www.grenadier.uk.com ✎
Other Info: 🖐

JANE FOLLIS GARDEN DESIGN
71 Eythrope Road, Stone,
Buckinghamshire, HP17 8PH
Area of Operation: South East England
Tel: 01296 747775
Fax: 01296 747775
Email: jfgdndesign@aol.com
Web: www.jfgardendesign.co.uk
Other Info: ECO

ONEWAY GARDEN LEISURE
18/19 John Samuel Building,
Arthur Drive, Hoo Farm Industrial Estate,
Kidderminster, Worcestershire, DY11 7RA
Area of Operation: UK (Excluding Ireland)
Tel: 01562 750222
Fax: 01562 750222
Email: maureenr18@tiscali.co.uk

PATIOHEATERS4U LIMITED
Thor Industrial Estate,
Braydon, Cricklade, Swindon,
Wiltshire, SN6 6HQ
Area of Operation: Europe
Tel: 01793 613900
Email: sales@patioheaters4u.com
Web: www.patioheaters4u.com ✎

SAPPHIRE SPAS HOT TUBS
A1 Rowood Estate,
Murdock Road, Bicester,
Nr Oxford, Oxfordshire, OX26 4PP
Area of Operation: Worldwide
Tel: 01869 327698
Fax: 01869 369552
Email: info@sapphirespas.co.uk
Web: www.sapphirespas.co.uk

TANSUN LIMITED
Spectrum House,
Unit 1 Ridgacre Road, West Bromwich,
West Midlands, B71 1BW
Area of Operation: Worldwide
Tel: 0121 580 6200
Fax: 0121 580 6222
Email: quartzinfo@tansun.co.uk
Web: www.quartzheat.com

LEISURE

Image courtesy of Sundance Pools (01296 715071)

Leisure

It is received wisdom in development circles that, while an indoor pool certainly adds value to a property, an outdoor pool can deter potential buyers who might view them as expensive to maintain, unusable for most of the year and a waste of garden space. Your insurance against this – if you decide that an outdoor pool is something you can't live without – is a design proportionate to the size of house and garden and stylistically in keeping.

Image by Sunfold Systems

A well designed indoor pool will add considerable value to the right property and can be used on even the coldest days of the year. They are easier to keep clean than outdoor pools, as they do not attract the same dirt, and are also cheaper to heat. However, building an indoor pool comes with a hefty price tag and is not a feasible option for every budget.

A quickly assembled outdoor pool in a small garden is likely to decrease the property's value, but if designed with care, can be an attractive focal point. To ensure the pool looks good, make sure you position and landscape it well. Try to plant the surrounding area so it blends in well with the garden, and don't build too close to the house.

Investing in an enclosure for an outdoor pool will enable you to enjoy the water at almost any time of the year. There are several options available, ranging from basic inflatable covers to conservatory-like structures. Telescopic enclosures are an increasingly popular option, as they can be removed in summer if not needed, but when in place they allow the heat of the sun through the cover, which in turn warms the pool.

Generally you do not need planning permission for an outdoor pool, unless you are building in an Area of Outstanding Natural Beauty or your property is listed. An indoor pool is more likely to require planning permission, and you will always need to apply for Building Regulations Approval. Always check with your local authority before commencing work to find out if you need planning permission.

If you employ a pool installer, make sure they are suitably experienced by checking that they are a Spatashield member of the Swimming Pool and Allied Trade Association (SPATA). The members trade under a code of ethics and will automatically have Spatashield Bond and Warranty insurance cover, which covers a new pool up to the builder's limit (maximum £150,000). The bond protects against a SPATA member being unable to complete a contract and the warranty guarantees that the pool will conform to SPATA construction and installation standards.

One of the often-overlooked aspects of owning a pool is its maintenance. Running costs alone are around £500-£1000 a year for heating and chemicals, and you will also have to pay for any damages that may occur. Check whether your supplier offers a maintenance service.

As with a self-built house, it is possible to build a swimming pool largely free of VAT. To do this, it must be constructed at the same time as a new build, and be attached to the property. However, many items of swimming pool equipment, such as diving boards, slides, floating covers and rollers are subject to VAT at the standard rate. Similarly, if you are planning to build either an indoor or outdoor pool as an addition to an existing property, you will be subject to VAT.

SAUNAS

KEY

OTHER: ▽ Reclaimed ⌐ On-line shopping
 Bespoke Hand-made ECO Ecological

DAR INTERIORS
Arch 11, Miles Street, London, SW8 1RZ
Area of Operation: Worldwide
Tel: 0207 720 9678
Fax: 0207 627 5129
Email: enquiries@darinteriors.com
Web: www.darinteriors.com

DREAM LEISURE LTD
Squires Garden Centre, Sixth Cross Road,
Twickenham, Middlesex, TW2 5PA
Area of Operation: UK (Excluding Ireland)
Tel: 0208 977 9900
Fax: 0208 977 9933
Email: info@dream-leisure.co.uk
Web: www.dream-leisure.co.uk
Other Info: ECO

ENERFOIL MAGNUM LTD
Kenmore Road, Comrie Bridge, Kenmore,
Aberfeldy, Perthshire, PH15 2LS
Area of Operation: Europe
Tel: 01887 822999
Fax: 01887 822954
Email: sales@enerfoil.com
Web: www.enerfoil.com

GEORGE BARKER & SONS
Backbarrow, Nr Ulverston, Cumbria, LA12 8TA
Area of Operation: UK (Excluding Ireland)
Tel: 01539 531236
Fax: 01539 530801
Web: www.gbs-ltd.co.uk

GILLINGHAM POOLS
Portinfer Coast Road, Vale, Guernsey, GY6 8LG
Area of Operation: Europe
Tel: 01481 255026
Fax: 01481 253626
Email: gillpools@aol.com
Web: www.gillinghampools.co.uk

GOLDEN COAST LTD
Fishleigh Road, Roundswell Business Park West,
Barnstaple, Devon, EX31 3UA
Area of Operation: UK & Ireland
Tel: 01271 378100
Fax: 01271 371699
Email: swimmer@goldenc.com
Web: www.goldenc.com

HIGH TECH HEALTH
PO Box 235, 2 Forest Court, Egham,
Surrey, TW20 9SH
Area of Operation: Worldwide
Tel: 0845 225 5610
Fax: 0845 225 5612
Email: info@hightechhealth.net
Web: www.hightechhealth.net

HOT TUBS UK LTD
Office 102, Rangefield Court, Farnham Trading
Estate, Farnham, Surrey, GU99NP
Area of Operation: UK (Excluding Ireland)
Tel: 01252 716213
Email: hottubs@uknet.fsnet.co.uk
Web: hottubsuk.net
Other Info:

J W GREEN SWIMMING POOLS LIMITED
Regency House, 88A Great Brickkiln Street,
Graisley, Wolverhampton,
West Midlands, WV3 0PU
Area of Operation: Midlands & Mid Wales
Tel: 01902 427709
Email: info@jwgswimming.co.uk
Web: www.jwgswimming.co.uk
Other Info:

MAGMED LTD.- PHYSIOTHERM INFRARED SAUNAS
3 Willetts Court, Pottergate,
Norwich, Norfolk, NR2 1DG
Area of Operation: UK & Ireland
Tel: 0845 22 5 5008
Fax: 0870 432 0406
Email: physiotherm@magmed.com
Web: www.magmed.com

NATIONAL LEISURE
Suite 179, Maritime House, Southwell
Business Park, Portland, Dorset, DT5 2NB
Area of Operation: UK (Excluding Ireland)
Tel: 01305 824610
Fax: 01305 824611
Email: pools@national-leisure.com
Web: www.national-leisure.com

NORDIC
Unit 5, Trading Estate, Holland Road,
Oxted, Surrey, RH8 9BZ
Area of Operation: UK (Excluding Ireland)
Tel: 01883 732400
Fax: 01883 716970
Email: info@nordic.co.uk
Web: www.nordic.co.uk

PINE GARDEN
Unit 2E, Smith Green Depot , Stoney Lane,
Galgate, Lancaster, Lancashire, LA2 0PX
Area of Operation: UK & Ireland
Tel: 01524 752882
Fax: 01524 751744
Email: info@pinegarden.co.uk
Web: www.pinegarden.co.uk
Other Info: ECO

SAUNASHOP.COM
1 Station Road, Kelly Bray, Callington,
Cornwall, PL17 8ES
Area of Operation: UK & Ireland
Tel: 0500 432132
Fax: 01579 384333
Email: info@saunashop.com
Web: www.saunashop.co.uk

SUMMIT LEISURE LTD
Unit 2, Garlands Trading Estate,
Cadley Road, Collingbourne Ducis,
Marlborough, Wiltshire, SN8 3EB
Area of Operation: UK & Ireland
Tel: 01264 790888
Fax: 01264 790199
Email: office@summitleisure.co.uk
Web: www.summitleisure.co.uk

THE HOME SAUNA COMPANY LTD
37 Sandhurst Close,
Church Hill, Redditch,
Worcestershire, B98 9JY
Area of Operation: UK & Ireland
Tel: 0845 430 3123
Fax: 01527 60072
Email: info@homesauna.co.uk
Web: www.homesauna.co.uk

SWIMMING POOLS

KEY

PRODUCT TYPES: 1= Swimming Pools
2 = Pool Covers and Enclosures
OTHER: ▽ Reclaimed ⌐ On-line shopping
 Bespoke Hand-made ECO Ecological

AQUAFLEX
1 Edison Road, Churchfields Industrial Estate,
Salisbury, Wiltshire, SP2 7NU
Area of Operation: UK & Ireland
Tel: 01722 328873
Fax: 01722 413068
Email: info@aquaflex.co.uk
Web: www.aquaflex.co.uk
Other Info:

DAR INTERIORS
Arch 11, Miles Street, London, SW8 1RZ
Area of Operation: Worldwide
Tel: 0207 720 9678
Fax: 0207 627 5129
Email: enquiries@darinteriors.com
Web: www.darinteriors.com
Product Type: 1

DOLPHIN LEISURE SUPPLIES
Unit 26 Mountney Business Park, Eastbourne Road,
Westham, Pevensey, East Sussex, BN24 5NJ
Area of Operation: UK (Excluding Ireland)
Tel: 01323 766600
Fax: 01323 766600
Email: sales@dolphinpools.co.uk
Web: www.dolphinpools.co.uk

DRIPOOL
Unit 3, Westwood Court,
Brunel Road, Totton, Southampton,
Hampshire, SO40 3WX
Area of Operation: Worldwide
Tel: 02380 663131
Fax: 02380 663232
Email: sales@dripool.co.uk
Web: www.dripool.co.uk
Product Type: 2
Other Info:

ENDLESS POOLS
200 East Dutton Mill Road,
Aston, Pennsylvania, USA, PA 19014
Area of Operation: Worldwide
Tel: 0800 0281056
Fax: +1 610 497 9328
Email: swim@endlesspools.com
Web: www.endlesspools.co.uk/3590

GILLINGHAM POOLS
Portinfer Coast Road, Vale,
Guernsey, GY6 8LG
Area of Operation: Europe
Tel: 01481 255026
Fax: 01481 253626
Email: gillpools@aol.com
Web: www.gillinghampools.co.uk

GOLDEN COAST LTD
Fishleigh Road,
Roundswell Business Park West,
Barnstaple, Devon, EX31 3UA
Area of Operation: UK & Ireland
Tel: 01271 378100
Fax: 01271 371699
Email: swimmer@goldenc.com
Web: www.goldenc.com
Product Type: 1

GOWER POOL ENCLOSURES
Factory 3, Barley Field Industrial Estate,
Brynmwr, Gwynedd, NP23 4LU
Area of Operation: Worldwide
Tel: 01495 313800
Fax: 01495 313716
Email: info@poolenclosures.co.uk
Web: www.poolenclosures.co.uk
Product Type: 2
Other Info:

IAN LEWIS DESIGN
Chapleton, Tulliemet, Pitlochry,
Perth & Kinross, PH9 0PA
Area of Operation: UK (Excluding Ireland)
Tel: 0870 240 6356
Fax: 01796 482733
Email: irl@pooldesign.co.uk
Web: www.selfbuildpools.co.uk

J W GREEN SWIMMING POOLS LIMITED
Regency House, 88A Great Brickkiln Street,
Graisley, Wolverhampton,
West Midlands, WV3 0PU
Area of Operation: Midlands & Mid Wales
Tel: 01902 427709
Email: info@jwgswimming.co.uk
Web: www.jwgswimming.co.uk
Product Type: 1, 2
Other Info:

DAR INTERIORS
Arch 11, Miles Street, London, SW8 1RZ
Area of Operation: Worldwide
Tel: 0207 720 9678
Fax: 0207 627 5129
Email: enquiries@darinteriors.com
Web: www.darinteriors.com
Product Type: 1

DOLPHIN LEISURE SUPPLIES
Unit 26 Mountney Business Park, Eastbourne Road,
Westham, Pevensey, East Sussex, BN24 5NJ
Area of Operation: UK (Excluding Ireland)
Tel: 01323 766600
Fax: 01323 766600
Email: sales@dolphinpools.co.uk
Web: www.dolphinpools.co.uk

NATIONAL LEISURE
Suite 179, Maritime House,
Southwell Business Park,
Portland, Dorset, DT5 2NB
Area of Operation: UK (Excluding Ireland)
Tel: 01305 824610
Fax: 01305 824611
Email: pools@national-leisure.com
Web: www.national-leisure.com

PINELOG LTD
Riverside Business Park, Bakewell,
Derbyshire, DE45 1GS
Area of Operation: UK (Excluding Ireland)
Tel: 01629 814481
Fax: 01629 814634
Email: admin@pinelog.co.uk
Web: www.pinelog.co.uk
Product Type: 1, 2

SELFBUILDPOOLS.CO.UK
Chapelton, Tulliemet, Pitlochry,
Perth and Kinross, PH9 0PA
Area of Operation: UK & Ireland
Tel: 0870 240 6356
Fax: 01796 482733
Email: irl@selfbuildpools.co.uk
Web: www.selfbuildpools.co.uk
Product Type: 1

SUMMIT LEISURE LTD
Unit 2, Garlands Trading Estate,
Cadley Road, Collingbourne Ducis,
Marlborough, Wiltshire, SN8 3EB
Area of Operation: UK & Ireland
Tel: 01264 790888
Fax: 01264 790199
Email: office@summitleisure.co.uk
Web: www.summitleisure.co.uk
Product Type: 1, 2
Other Info:

SUNDANCE POOLS UK LTD
P O Box 284, Milton Keynes,
Buckinghamshire, MK17 0QD
Area of Operation: UK & Ireland
Tel: 01296 715071
Fax: 01296 714991
Email: enquiries@sundancepools.com
Web: www.sundancepools.com
Product Type: 1, 2

TELESCOPIC POOL ENCLOSURES LTD
Unit 13 Wynford Industrial Park, Belbins,
Romsey, Hampshire, SO51 0PW
Area of Operation: UK & Ireland
Tel: 0800 074 0872
Email: info@telescopicpoolenclosures.com
Web: www.telescopicpoolenclosures.com
Product Type: 2
Other Info:

SPAS AND HOT TUBS

KEY

OTHER: ▽ Reclaimed ⌐ On-line shopping
 Bespoke Hand-made ECO Ecological

AEGEAN SPAS & HOT TUBS
The Aegean National Spa & Hot Tub Centre, 2 Hale
Lane, Mill Hill, London, NW7 3NX
Area of Operation: UK (Excluding Ireland)
Tel: 0208 959 1529
Fax: 0208 906 0511
Email: david@aegeanspas.co.uk
Web: www.aegeanspas.co.uk

DAR INTERIORS
Arch 11, Miles Street, London, SW8 1RZ
Area of Operation: Worldwide
Tel: 0207 720 9678
Fax: 0207 627 5129
Email: enquiries@darinteriors.com
Web: www.darinteriors.com

DOLPHIN LEISURE SUPPLIES
Unit 26 Mountney Business Park, Eastbourne Road,
Westham, Pevensey, East Sussex, BN24 5NJ
Area of Operation: UK (Excluding Ireland)
Tel: 01323 766600
Fax: 01323 766600
Email: sales@dolphinpools.co.uk
Web: www.dolphinpools.co.uk

DREAM LEISURE LTD
Squires Garden Centre, Sixth Cross Road,
Twickenham, Middlesex, TW2 5PA
Area of Operation: UK (Excluding Ireland)
Tel: 0208 977 9900
Fax: 0208 977 9933
Email: info@dream-leisure.co.uk
Web: www.dream-leisure.co.uk
Other Info: ECO

GEORGE BARKER & SONS
Backbarrow, Nr Ulverston,
Cumbria, LA12 8TA
Area of Operation: UK (Excluding Ireland)
Tel: 01539 531236
Fax: 01539 530801
Web: www.gbs-ltd.co.uk

GILLINGHAM POOLS
Portinfer Coast Road, Vale,
Guernsey, GY6 8LG
Area of Operation: Europe
Tel: 01481 255026
Fax: 01481 253626
Email: gillpools@aol.com
Web: www.gillinghampools.co.uk

GOLDEN COAST LTD
Fishleigh Road, Roundswell Business Park West,
Barnstaple, Devon, EX31 3UA
Area of Operation: UK & Ireland
Tel: 01271 378100
Fax: 01271 371699
Email: swimmer@goldenc.com
Web: www.goldenc.com

HOT TUBS UK LTD
Office 102, Rangefield Court, Farnham Trading
Estate, Farnham, Surrey, GU99NP
Area of Operation: UK (Excluding Ireland)
Tel: 01252 716213
Email: hottubs@uknet.fsnet.co.uk
Web: hottubsuk.net
Other Info:

J W GREEN SWIMMING POOLS LIMITED
Regency House,
88A Great Brickkiln Street, Graisley,
Wolverhampton, West Midlands, WV3 0PU
Area of Operation: Midlands & Mid Wales
Tel: 01902 427709
Email: info@jwgswimming.co.uk
Web: www.jwgswimming.co.uk

NATIONAL LEISURE
Suite 179, Maritime House,
Southwell Business Park,
Portland, Dorset, DT5 2NB
Area of Operation: UK (Excluding Ireland)
Tel: 01305 824610
Fax: 01305 824611
Email: pools@national-leisure.com
Web: www.national-leisure.com

ONEWAY GARDEN LEISURE
18/19 John Samuel Building,
Arthur Drive, Hoo Farm Industrial Estate,
Kidderminster, Worcestershire, DY11 7RA
Area of Operation: UK (Excluding Ireland)
Tel: 01562 750222
Fax: 01562 750222
Email: maureenr18@tiscali.co.uk

OZTUBS
35 Gladeside, Croydon,
Greater London, CR0 7RL
Area of Operation: UK (Excluding Ireland)
Tel: 0845 124 9531
Email: sales@oztubs.com
Web: www.oztubs.com

SAPPHIRE SPAS HOT TUBS
A1 Rowood Estate, Murdock Road,
Bicester, Nr Oxford, Oxfordshire, OX26 4PP
Area of Operation: Worldwide
Tel: 01869 327698
Fax: 01869 369552
Email: info@sapphirespas.co.uk
Web: www.sapphirespas.co.uk

SPAFORM LIMITED
Spa House, Walton Road, Farlington,
Portsmouth, Hampshire, PO6 1TB
Area of Operation: Worldwide
Tel: 02392 313131
Fax: 02392 377 597
Email: enquiries@spaform.co.uk
Web: www.spaform.co.uk

SUMMIT LEISURE LTD
Unit 2, Garlands Trading Estate,
Cadley Road, Collingbourne Ducis,
Marlborough, Wiltshire, SN8 3EB
Area of Operation: UK & Ireland
Tel: 01264 790888
Fax: 01264 790199
Email: office@summitleisure.co.uk
Web: www.summitleisure.co.uk

SUNDANCE SPAS UK LTD
Unit 8, The Markham Centre, Station Road,
Theal, Reading, Berkshire, RG7 4PE
Area of Operation: UK (Excluding Ireland)
Tel: 0800 146 106 **Fax:** 01952 811797
Email: sales@sundance-spas.co.uk
Web: www.sundance-spas.co.uk

SUSSEX SPAS
The Old Garage, High Street, Handcross,
Haywards Heath, West Sussex, RH17 6BJ
Area of Operation: Greater London,
South East England
Tel: 01444 401861 **Fax:** 01444 401861
Email: info@sussexspas.co.uk
Web: www.sussexspas.co.uk

Image courtesy of Buildstore (0870 870 9991)

CONSULTANTS, LABOUR & FINANCE

SPONSORED BY BUILDSTORE
Tel 0870 870 9991. Web www.buildstore.co.uk

Finance

Providing you have a regular income and a reasonable credit history, you should have no difficulty in obtaining a loan to build your own home or to undertake a renovation or conversion project. It is also possible to finance your project using a bridging loan secured against equity in your current home, either in conjunction with, or instead of, a stage payment mortgage secured against the building project. This kind of funding, often charged at a small premium above mortgage rates, is offered primarily by banks. Lending is not usually restricted by the standard income multipliers that apply to mortgages, but such a facility usually carries an arrangement fee, typically 1-1.5% of the advance. As bridging finance is more expensive than mortgage finance, it is best suited to those who do not want or need a mortgage once they have sold their current home, or those who want to raise funds in addition to taking out a stage payment mortgage to help fund the project without having to sell their current home.

How Much Can You Borrow?

The maximum amount you can borrow on a conventional mortgage is usually calculated using income multipliers to assess afford-ability. These are typically 2.5 x joint income, 3 x a higher income plus 1 x a second income or, for sole earners, 3-4 x income. Existing commitments, e.g. mortgage payments on your current home, may be taken into account when assessing affordability. Some lenders may consider offering a stage payment mortgage alongside an existing home loan, allowing you to remain in your current home during construction.

As part of your application, some proof of income will be required, typically in the form of three months' payslips and your latest P60. Self-employed applicants will usually have to provide two years' audited accounts or approach lenders, some of whom will consider offering funding on a self-certified basis.

Borrowing to Buy Land

Advances on land, or existing buildings for renovation/conversion, are available up to a maximum of 95% of valuation or purchase price, whichever is the lower. In most cases some form of planning consent must be in place for the development of a plot or conversion opportunity before a lender will release funds.

Borrowing for Construction

Funding for construction is usually released in arrears on completion of key stages in the building work, after it has been signed off by either the lender's valuer or a supervising professional. Some specialist self-build mortgage arrangers offer an indemnity policy, which allows some lenders to release funds in advance of the build stages.

This can help cashflow and eliminate the need for bridging finance. The stages, and the percentage of the total advance released, will vary according to the number of storeys being built and the type of construction employed. Exact stage payment details should always be discussed and agreed with your lender. You need to make certain you have sufficient funds to cover the initial fees.

Raising a Deposit

Capital of at least 10-15% of the total project cost is usually needed to get a homebuilding project going. Although it is possible to get a self-build mortgage without selling your current home, most people choose to sell up in order to release capital. If you choose this route, do not forget to budget for somewhere to live and for storage _ bear in mind that rental can be at least as expensive as a mortgage. The most common option for temporary accommodation is rental, although many people stay with family or in a mobile home on site.

Borrowing Costs

Fees vary from lender to lender. Lenders may charge a mortgage application fee, typically £2-300. There will also be a valuation fee payable, which will vary according to the value of your plot, typically £160-400. There may also be a separate fee for specialist products such as a fixed rate or capped rate mortgage. Further fees are usually payable for the re-inspection of building work prior to the release of each stage payment. This charge is typically £30-50 for each of four or more stage payments _ check with your lender for further details. This role can also be undertaken by a supervising professional, such as a chartered surveyor, architect or approved warranty inspector, in which case there will not be an additional fee. Interest will be charged at the agreed interest rate - usually the lender's current variable rate - from the moment funds are released, but only on the amount that has been borrowed/drawn down to date. Interest payments therefore start low and increase towards completion.

Table, BELOW, correct at time of printing
(created October 2005)

Mortgage cost calculator

Monthly payment per £1,000 of borrowing at various rates of interest

% Rate	25 Yr Repayment	Interest Only	% Rate	25 Yr Repayment	Interest Only
0.75	£ 3.67	£ 0.63	3.75	£ 5.19	£ 3.13
1.00	£ 3.78	£ 0.83	4.00	£ 5.33	£ 3.33
1.25	£ 3.90	£ 1.04	4.25	£ 5.48	£ 3.54
1.50	£ 4.02	£ 1.25	4.50	£ 5.62	£ 3.75
1.75	£ 4.14	£ 1.46	4.75	£ 5.77	£ 3.96
2.00	£ 4.27	£ 1.67	5.00	£ 5.91	£ 4.17
2.25	£ 4.39	£ 1.88	5.25	£ 6.06	£ 4.38
2.50	£ 4.52	£ 2.08	5.50	£ 6.21	£ 4.58
2.75	£ 4.65	£ 2.29	5.75	£ 6.36	£ 4.79
3.00	£ 4.79	£ 2.50	6.00	£ 6.52	£ 5.00
3.25	£ 4.92	£ 2.71	6.25	£ 6.67	£ 5.21
3.50	£ 5.06	£ 2.92	6.50	£ 6.83	£ 5.42

INSURANCE

KEY
OTHER: ▽ Reclaimed   On-line shopping
Bespoke   Hand-made   ECO Ecological

BUILDSTORE
Unit 1 Kingsthorne Park, Houstoun Industrial Estate, Livingston, Lothian, EH54 5DB
Area of Operation: UK (Excluding Ireland)
Tel: 0870 870 9991   Fax: 0870 870 9992
Email: enquiries@buildstore.co.uk
Web: www.buildstore.co.uk

CARTEL MARKETING LTD
5th Floor, Building 7, Salford Quays, Exchange Quay, Manchester, M5 3EP
Area of Operation: Worldwide
Tel: 0161 836 4343   Fax: 0161 836 4241
Email: elaine.morley@cartelgroupholdings.plc.uk
Web: www.cartelgroupholdings.plc.uk

DMS SERVICES LTD
Orchard House, Sheffield Road, Blyth, Nr. Worksop, Nottinghamshire, S81 8HF
Area of Operation: UK (Excluding Ireland)
Tel: 01909 591652   Fax: 01909 591031
Email: insurance@armor.co.uk
Web: www.selfbuildonline.co.uk

FLEXIBLE MORTGAGE.NET
Commerce Way, Edenbridge, Kent, TN8 6ED
Area of Operation: UK (Excluding Ireland)
Tel: 01732 866007   Fax: 01732 866155
Email: Davidc@flexible-mortgage.net
Web: www.flexible-mortgage.net

INTELLIGENT MORTGAGE SERVICES
208 High Road, Leytonstone, London, E11 3HU
Area of Operation: UK & Ireland
Tel: 0208 279 0719
Fax: 0208 279 0720
Email: info@intelligentmortgageservices.co.uk
Web: www.intelligentmortgageservices.co.uk

LIBERTY SYNDICATES
One Minster Court, 5th Floor, Mincing Lane, London, EC3R 7AA
Area of Operation: Worldwide
Tel: 0207 895 0011
Fax: 0207 860 8573
Web: www.libertysyndicates.com

NHBC
Buildmark House, Chiltern Avenue, Amersham, Buckinghamshire, HP6 5AP
Area of Operation: Worldwide
Tel: 01494 735363
Web: www.nhbc.co.uk

PROJECT BUILDER INSURANCE FACILITY
Tower Gate House, St. Edwards Court, London Road, Romford, Essex, RM7 9QD
Area of Operation: UK (Excluding Ireland)
Tel: 01708 777402
Fax: 01708 777737
Email: projectbuilder@towergate.co.uk
Web: www.siteinsurance.net

SELF-BUILD ZONE
London House, 77 High Street, Sevenoaks, Kent, TN13 1LD
Area of Operation: UK & Ireland
Tel: 0845 230 9874
Fax: 01732 740994
Email: sales@selfbuildzone.com
Web: www.selfbuildzone.com

SELF-BUILDER.COM
Belmont International, Becket House, Vestry Road, Sevenoaks, Kent, TN14 5EL
Area of Operation: UK & Ireland
Tel: 0800 018 7660   Fax: 01732 744729
Email: sales@self-builder.com
Web: www.self-builder.com

STERLING HAMILTON WRIGHT
Towergate House, St Edwards Court, London Road, Romford, Essex, RM7 6QD
Area of Operation: UK (Excluding Ireland)
Tel: 0870 333 3810   Fax: 0207 515 0350
Email: projectbuilder@towergate.co.uk
Web: www.siteinsurance.net

THE MORTGAGE SHOP ONLINE
231 Grimsby Road, Cleethorpes, Lincolnshire, DN35 7HE
Area of Operation: East England, North East England, North West England and North Wales, South East England, South West England & South Wales
Tel: 01472 200664   Fax: 01472 200664
Email: mail@mortgageshop-online.co.uk

WISEMONEY.COM LTD
24 Charlton Drive, Cheltenham, Gloucestershire, GL53 8ES
Area of Operation: Worldwide
Email: sdye@wisemoney.com
Web: www.wisemoney.com

ZURICH CUSTOMBUILD
Southwood Crescent, Farnborough, Hampshire, GU14 0NJ
Area of Operation: UK (Excluding Ireland)
Tel: 01252 522000
Web: www.zurich.co.uk

MORTGAGE PROVIDERS

KEY
OTHER: ▽ Reclaimed   On-line shopping
Bespoke   Hand-made   ECO Ecological

ADVANCED FLEXIBLE SELF BUILD MORTGAGE LIMITED
Unit 61-62, Alloa Business Centre, Alloa Business Park, Whins Road, Alloa, Clackmannanshire, FK10 3SA
Area of Operation: UK (Excluding Ireland)
Tel: 01259 726650   Fax: 01259 726651
Email: info@afsbm.co.uk
Web: www.afsbm.co.uk

BISHOPSGATE FUNDING LTD
Tower Business Centre, Portland Tower, Portland Street, Manchester, M1 3LF
Area of Operation: UK (Excluding Ireland)
Tel: 0845 601 2654   Fax: 0845 601 2657
Email: enquiries@bishopsgatefunding.com
Web: www.bishopsgatefunding.com

CONSULTANTS, LABOUR & FINANCE

BUILDSTORE
TRADE CARD

BUILDSTORE
TRADE CARD

SELF BUILD • RENOVATION • HOME IMPROVEMENT

0123 5467 9120 4567

MR J SMITH

Start your building with a bit of card

Not quite as daft as it sounds.

With the free Buildstore Tradecard, you'll get access to trade prices on more than 220,000 product lines from some of the country's top suppliers. They include materials, plant and tool hire, roofing, insulation, white goods, flooring – everything you'll need to complete your home.

With the Buildstore Tradecard, you'll also get higher credit limits with our merchant partners. That means increased spending power and improved cashflow.

And we'll give you expert advice throughout your project. A dedicated account manager and a trade-card co-ordinator will be available to you, and we'll give you the comprehensive Tradecard Manual – a wealth of technical information for self builders.

Apply now for your Buildstore Tradecard.
Phone 0870 870 9497 or visit our website – www.buildstore.co.uk

SELF BUILD • RENOVATION • HOME IMPROVEMENT

BUILDSTORE
Unit 1 Kingsthorne Park, Houstoun Industrial Estate,
Livingston, Lothian, EH54 5DB
Area of Operation: UK (Excluding Ireland)
Tel: 0870 870 9991
Fax: 0870 870 9992
Email: enquiries@buildstore.co.uk
Web: www.buildstore.co.uk

CARTEL MARKETING LTD
5th Floor, Building 7, Salford Quays,
Exchange Quay, Manchester, M5 3EP
Area of Operation: Worldwide
Tel: 0161 836 4343
Fax: 0161 836 4241
Email: elaine.morley@cartelgroupholdings.plc.uk
Web: www.cartelgroupholdings.plc.uk

CREDIT & MERCANTILE PLC
Mercantile House, Lingfield, Surrey, RH7 6NG
Area of Operation: UK (Excluding Ireland)
Tel: 01342 837111
Fax: 01342 837901
Email: info@creditmercantile.co.uk
Web: www.creditmercantile.co.uk

ECOLOGY BUILDING SOCIETY
7 Belton Road, Silsden, Keighley,
West Yorkshire, BD20 0EE
Area of Operation: UK (Excluding Ireland)
Tel: 01535 650770
Fax: 01535 650790
Email: loans@ecology.co.uk
Web: www.ecology.co.uk

FIRST PROPERTY FINANCE PLC
Maple House, High Street, Potters Bar,
Hertfordshire, EN6 5BS
Area of Operation: UK (Excluding Ireland)
Tel: 01707 828 705
Fax: 01707 828 087
Email: info@firstpropertyfinance.com
Web: www.firstpropertyfinance.com

FLEXIBLE MORTGAGE.NET
Commerce Way, Edenbridge, Kent, TN8 6ED
Area of Operation: UK (Excluding Ireland)
Tel: 01732 866007
Fax: 01732 866155
Email: davidc@flexible-mortgage.net
Web: www.flexible-mortgage.net

INTELLIGENT MORTGAGE SERVICES
208 High Road, Leytonstone, London, E11 3HU
Area of Operation: UK & Ireland
Tel: 0208 279 0719 **Fax:** 0208 279 0720
Email: info@intelligentmortgageservices.co.uk
Web: www.intelligentmortgageservices.co.uk

KENT RELIANCE BUILDING SOCIETY
Reliance House, Sun Pier,
Chatham, Kent, ME4 4ET
Area of Operation: UK & Ireland
Tel: 01634 848 944 **Fax:** 01634 830912
Email: sales@krbs.com
Web: www.krbs.co.uk

LANCASHIRE MORTGAGE CORPORATION
6th Floor, Bracken House, Charles Street,
Manchester, M1 7BD
Area of Operation: UK (Excluding Ireland)
Tel: 0161 276 2476 **Fax:** 0161 276 2477
Email: webenquirey@financeyourproperty.co.uk
Web: www.financeyourproperty.co.uk

LLOYDS TSB SCOTLAND
Llolyds TSB Scotland Plc Registered Office,
Henry Duncan House, 120 George Street,
Edinburgh, EH2 4LH
Area of Operation: Scotland
Tel: 0131 225 4555
Fax: 0131 260 0881
Web: www.lloydstsb.co.uk

MORTGAGE GUARANTEE PLC
26 Church Road, Rainford, St Helens,
Merseyside, WA11 8HE
Area of Operation: UK (Excluding Ireland)
Tel: 01744 886884
Fax: 01744 886865

PHONE A LOAN
1st Floor, Bracken House, Charles Street,
Manchester, M1 7BD
Area of Operation: UK (Excluding Ireland)
Tel: 0870 112 5011
Fax: 0870 112 5012
Email: brokerageunit@phone-a-loan.co.uk
Web: www.phone-a-loan.co.uk

REGENTSMEAD LIMITED
Russell House, 140 High Street, Edgware,
Middlesex, HA8 7LW
Area of Operation: Greater London, Midlands,
South East England, South West England
Tel: 0208 952 1414
Fax: 0208 952 2424
Email: info@regentsmead.com
Web: www.regentsmead.com

ROWANBANK MORTGAGES
6 Summer Place, Edinburgh, EH3 5NR
Area of Operation: UK (Excluding Ireland)
Tel: 0131 557 3909 **Fax:** 0131 558 3601
Email: enquiries@rowanbankmortgages.co.uk
Web: ww.rowanbankmortgages.co.uk

THE MORTGAGE SHOP ONLINE
231 Grimsby Road, Cleethorpes,
Lincolnshire, DN35 7HE
Area of Operation: East England, North East
England, North West England and North Wales,
South East England, South West England
and South Wales
Tel: 01472 200664 **Fax:** 01472 200664
Email: mail@mortgageshop-online.co.uk

WISEMONEY.COM LTD
24 Charlton Drive, Cheltenham, Gloucestershire,
GL53 8ES
Area of Operation: Worldwide
Email: sdye@wisemoney.com
Web: www.wisemoney.com

WARRANTY PROVIDERS

KEY
OTHER: ▽ Reclaimed ⌐ On-line shopping
📖 Bespoke ✋ Hand-made ECO Ecological

BUILDSTORE
Unit 1 Kingsthorne Park,
Houstoun Industrial Estate,
Livingston, Lothian, EH54 5DB
Area of Operation: UK (Excluding Ireland)
Tel: 0870 870 9991
Fax: 0870 870 9992
Email: enquiries@buildstore.co.uk
Web: www.buildstore.co.uk

NHBC
Buildmark House, Chiltern Avenue,
Amersham, Buckinghamshire, HP6 5AP
Area of Operation: Worldwide
Tel: 01494 735363
Web: www.nhbc.co.uk

PROJECT BUILDER INSURANCE FACILITY
Tower Gate House, St. Edwards Court,
London Road, Romford, Essex, RM7 9QD
Area of Operation: UK (Excluding Ireland)
Tel: 01708 777402
Fax: 01708 777737
Email: projectbuilder@towergate.co.uk
Web: www.siteinsurance.net

SELF-BUILD ZONE
London House, 77 High Street,
Sevenoaks, Kent, TN13 1LD
Area of Operation: UK & Ireland
Tel: 0845 230 9874
Fax: 01732 740994
Email: sales@selfbuildzone.com
Web: www.selfbuildzone.com

ZURICH CUSTOMBUILD
Southwood Crescent, Farnborough,
Hampshire, GU14 0NJ
Area of Operation: UK (Excluding Ireland)
Tel: 01252 522000
Web: www.zurich.co.uk

NOTES

Company Name . . . .
. . . .
Address . . . .
. . . .
. . . .
email . . . .
Web . . . .

Company Name . . . .
. . . .
Address . . . .
. . . .
. . . .
email . . . .
Web . . . .

Company Name . . . .
. . . .
Address . . . .
. . . .
. . . .
email . . . .
Web . . . .

CONSULTANTS, LABOUR & FINANCE

You first ![Lloyds TSB logo] Lloyds TSB Scotland

Safe hands

By moving day our Lloyds TSB Scotland mortgage had taken care of all the big problems, leaving us plenty of time to take care of all the small ones.

Call **0800 056 0156*** for more information or log on to **www.lloydstsb.com/scotland**

YOUR HOME MAY BE REPOSSESSED IF YOU DO NOT KEEP UP REPAYMENTS ON YOUR MORTGAGE.

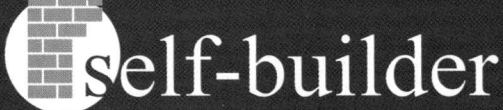

self-builder

Converting, renovating, extending or building your home?

Self-builder insurance provides cover for;

- Contract works - 125% works value

- Site huts and caravans - £30,000 and their contents £5,000

- Owned or hired tools, plant & equipment - £30,000 for each

- Employees' personal tools and effects - £2,000

- Existing structures - 110% of reinstatement value

- Employers Liability - £10 million

- Public Liability - £5 million

- Personal Accident - £20,000

- Legal Expenses - £25,000

- JCT clause 21.2.1 - optional

For instant cover, call **0800 018 7660** or visit **www.self-builder.com**
(10% discount with on-line quotes)

Underwritten by Royal & SunAlliance, one of the world's largest and oldest multinational quoted insurance groups. With a 300 year heritage, R&SA is still transacting business under its own name. Self-builder is a trading style of Belmont International Limited. Belmont International Limited is a wholly owned subsidiary of Belmont Insurance Holdings, which is a company registered in England & Wales under company registration number 02217475. The registered office is located at Becket House, Vestry Road, Otford, Sevenoaks, Kent TN14 5EL. Belmont International Limited is authorised and regulated by the Financial Services Authority.

ROYAL & SUNALLIANCE

Project Builder

Site Insurance
Home Insurance
Warranty

www.siteinsurance.net

Let Project Builder take care of you while you build your home

Building Works Cover Includes:
Employers Liability
Public Liability
Hired in Plant
Personal Accident
Legal Expenses
Free 24/7 Health & Safety Helpline

Optional Covers:
Existing Structures
The Premier Guarantee
(10 year Warranty)

Completed Housing Warranty
If you have already built your house and now need a Warranty

Quote Line
0870 3333 810

Towergate Schemes, Towergate House, St Edwards Court London Road, Romford, Essex, RM7 9QD

Towergate Schemes is a trading name of Edgar Hamilton Limited authorised & regulated by the Financial Services Authority (FSA)

MOVE OR IMPROVE? MAGAZINE

Area of Operation: UK & Ireland
Tel: 01527 834435 **Fax:** 01527 837810
Email: alex.worthington@centaur.co.uk
Web: www.moveorimprove.co.uk

Move or Improve? magazine for people adding adding space and value to their homes. Includes design guides, practical advice and inspiration on extensions, loft and basement conversions and improving your current home.

WWW.PLOTFINDER.NET

Area of Operation: UK & Ireland
Tel: 01527 834436 **Fax:** 01527 837810
Email: laura.cassidy@centaur.co.uk
Web: www.plotfinder.net

Plotfinder is an online database which holds details of almost 6,000 building plots and properties in need of renovation or conversion currently for sale throughout the UK.

WWW.PROPERTYFINDERFRANCE.NET

Area of Operation: UK & France
Tel: 01527 834435 **Fax:** 01527 837810
Email: alex.worthington@centaur.co.uk
Web: www.propertyfinderfrance.net

Looking for a property in France, this website has around 50,000 properties throughout France, you can search by type of property, size, price, area and number of bedrooms. To get a weeks free trial visit www.propertyfinderfrance.net.

HOMEBUILDING & RENOVATING MAGAZINE

Area of Operation: UK & Ireland
Tel: 01527 834435 **Fax:** 01527 837810
Email: alex.worthington@centaur.co.uk
Web: www.homebuilding.co.uk

Homebuilding & Renovating, Britain's best selling self-build magazine is an essential read for anyone extending, renovating, converting or building their own home, providing practical advice and inspirational ideas.

WWW.HOMEBUILDING.CO.UK

Area of Operation: UK and Ireland
Tel: 01527 834435 **Fax:** 01527 837810
Email: customerservice@centaur.co.uk
Web: www.homebuilding.co.uk

Includes vast amounts of self-build and renovation information, a huge directory of products and suppliers, houseplans, readers' homes, various advice such as finance and planning plus The Beginners Guide.

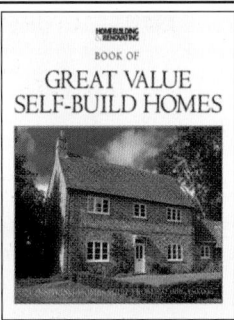

BOOK OF GREAT VALUE SELF-BUILD HOMES

Area of Operation: UK & Ireland
Tel: 01527 834435 **Fax:** 01527 837810
Email: alex.worthington@centaur.co.uk
Web: www.homebuilding.co.uk

Twenty-four homes built for between £32,000 and £150,000. The features show how it is possible to use floor space without sacrificing unique features, and how to achieve maximum style without spending a fortune.

12 ISSUES OF HOMEBUILDING & RENOVATING

Area of Operation: UK and Ireland
Tel: 01527 834435 **Fax:** 01527 837810
Email: alex.worthington@centaur.co.uk
Web: www.homebuilding.co.uk

The entire archive of Homebuilding & Renovating magazine issues from 2002, '03, '04 and '05 are available on easy-to-use double CDs. All pages and pictures appear as they do in the magazine (to view as PDFs through Acrobat Reader).

WWW.SITEFINDERIRELAND.COM

Area of Operation: Ireland
Tel: 01527 834435 **Fax:** 01527 837810
Email: laura.cassidy@centaur.co.uk
Web: www.sitefinderireland.com

Sites for sale. Whether you are looking for a building site, a property that need renovating or a building to convert www.sitefinderireland.com can make your search easier; it lists over 2,000 currently for sale throughout Ireland.

HOW TO RENOVATE A HOUSE IN FRANCE

Area of Operation: UK & France
Tel: 01527 834435 **Fax:** 01527 837810
Email: alex.worthington@centaur.co.uk
Web: www.propertyfinderfrance.net

An essential guide to help you turn and old rural property in France into a beautiful home. How to Renovate covers the whole process, from assessing and buying a property, through all jobs, large and small, required to get it into shape.

BOOK OF CONTEMPORARY HOMES

Area of Operation: UK & Ireland
Tel: 01527 834435 **Fax:** 01527 837810
Email: alex.worthington@centaur.co.uk
Web: www.homebuilding.co.uk

Nineteen individually designed, contemporary-styled homes, from urban homes to country houses. Each case study includes its floorplan and layout, inspirational pictures, costs for the build and a list of useful contacts.

BOOK OF BARN CONVERSIONS

Area of Operation: UK & Ireland
Tel: 01527 834435 **Fax:** 01527 837810
Email: alex.worthington@centaur.co.uk
Web: www.homebuilding.co.uk

Containing 22 inspirational case studies, ranging from rustic to contemporary, this book is a must for anyone contemplating a barn conversion or restoration, whatever their budget. Each project is fully costed with contact details for suppliers.

HOW TO CREATE A JARDIN PAYSAN

Area of Operation: UK & France
Tel: 01527 834435 **Fax:** 01527 837810
Email: alex.worthington@centaur.co.uk
Web: www.propertyfinderfrance.net

How to create a traditional, rural French-style garden which will have a timeless charm. With plant lists, tips and techniques going back generations, this book shows you how to create an authentic, natural garden.

Image courtesy of Move or Improve? magazine (01527 834435)

CONSULTANTS, LABOUR & FINANCE

Thatch

Thatching is, arguably, the oldest building craft still used in Britain today, and the fact that thatch has survived as a roofing material for so long proves that is not only attractive, but is also extremely effective and eco-friendly. The quintessentially British ideal of the "chocolate box" cottage also means that a home with a thatched roof can command premium prices in the property market.

The lifespan of thatch can vary greatly depending upon a number of factors, including the thatch material used and its quality, the skill of the craftsman, and the environmental conditions the roof is exposed to. Also important is how well the roof is cared for. For example, preventing birds from nesting in your thatch can ensure a thatched roof's longevity, as can making minor strategic repairs as and when needed. In fact, some thatchers have maintenance schemes whereby they periodically check your roof and make minor repairs as necessary. This can save a homeowner thousands of pounds in the long run.

The main materials used for thatched roofs are long straw and reeds. Long straw is less durable, and therefore has a lesser expected lifespan of between ten and twenty years. However, aesthetically the material is stunning, as its flexibility allows the thatcher to round corners. It is also cheaper to use than reed, which is a much stronger material. Many thatchers also prefer to use reed as the canes are knocked into shape as they are laid, meaning that they do not require shearing afterwards.

Contrary to popular opinion, it is not hard to find a good, reliable thatcher – visit the National Society of Master Thatchers' website at www.nsmt.co.uk to find details of their members in your area. If you decide to employ a thatcher who doesn't come recommended by a relevant association, there are a few precautions you can take to protect yourself and your property. Firstly, see examples of their previous work. Check that it has lasted for at least 15 years, and get references from previous employers. Secondly, ask for a written estimate and make sure their work comes with a guarantee; any reputable thatcher should willingly return if you find fault with their work within reasonable time (usually up to ten years). Finally, ask to see their public liability certificate. This should show cover for at least £2 million, but is of minimal cost to a contractor, therefore not having one is a suspicious circumstance.

KEY
OTHER: ▽ Reclaimed ⌐ On-line shopping
✎ Bespoke ✋ Hand-made ECO Ecological

BLACKSMITHING AND METAL CRAFTS

ALUMINIUM ARTWORKS
Persistence Works, 21 Brown Street,
Sheffield, S1 2BS
Area of Operation: Worldwide
Tel: 0114 249 4748
Email: info@aluminiumartworks.co.uk
Web: www.aluminiumartworks.co.uk

ARCHITECTURAL METALWORK CONSERVATION
Unit 19, Hoddesdon Industrial Centre,
Pindar Road, Hoddesdon, Hertfordshire, EN11 0DD
Area of Operation: East England
Tel: 01992 443132 **Fax:** 01992 443132

BRITISH ARTIST BLACKSMITHS ASSOCIATION
Anwick Forge, 62 Main Road, Anwick,
Fleaford, Lincolnshire, NG34 9SU
Area of Operation: UK & Ireland
Tel: 01526 830303
Email: babasecretary@anwickforge.co.uk
Web: www.baba.org.uk

CHRIS TOPP & COMPANY WROUGHT IRONWORKS
Lyndhurst, Carlton Husthwaite,
Thirsk, North Yorkshire, YO7 2BJ
Area of Operation: Worldwide
Tel: 01845 501415 **Fax:** 01845 501072
Email: enquiry@christopp.co.uk
Web: www.christopp.co.uk

COBALT BLACKSMITHS
The Forge, English Farm, English Lane,
Nuffield, Oxfordshire, RG9 5TH
Tel: 01491 641990 **Fax:** 01491 640909
Email: enquiries@cobalt-blacksmiths.co.uk
Web: www.cobalt-blacksmiths.co.uk

JGS METALWORK
Unit 6 Broomstick Estate,
High Street, Edlesborough,
Dunstable, Bedfordshire, LU6 2HS
Area of Operation: East England
Tel: 01525 220360 **Fax:** 01525 222786
Email: enquiries@weathervanes.org.uk
Web: www.weathervanes.org.uk

LARBERRY PASTURES
Larberry Pastures, Longnewton,
Stockton on Tees, Aberdeenshire, TS21 1BN
Area of Operation: UK (Excluding Ireland)
Tel: 01642 582000 **Fax:** 01642 582000
Email: info@aferites.com
Web: www.aferites.com

DRYSTONE WALLING

ANDREW LOUDON - TRADITIONAL & DECORATIVE STONEWORK
Bowmanstead Cottage, Bowmanstead,
Coniston, Cumbria, LA21 8HB
Area of Operation: UK (Excluding Ireland)
Tel: 01539 441985
Email: info@drystone-walling.co.uk
Web: www.drystone-walling.co.uk

ARTISAN STONE WORK
1 Isis Close, Abingdon, Oxfordshire, OX14 3TA
Area of Operation: Europe
Tel: 07789 952007
Email: piersconway@yahoo.com
Web: www.artisanstonework.com

DONALD GUNN
2 East Cottage, Rayheugh Farm, Chathill,
Northumberland, NE67 5HL
Area of Operation: Worldwide
Tel: 01668 219818
Email: dgunnwalls@hotmail.com
Web: www.drystone-walls.com

HISTORIC BUILDING CONSERVATION
Brabazon House, Scalford Road, Eastwell,
Melton Mowbray, Leicestershire, LE14 4EF
Area of Operation: UK (Excluding Ireland)
Tel: 01949 861333
Fax: 01949 861331
Email: info@historic-buildings.net
Web: www.historic-buildings.net

OLDBUILDERS COMPANY
Mount Palmer, Birr Demesne, Birr, Co. Offaly
Area of Operation: Ireland Only
Tel: 0509 21133
Email: info@oldbuilders.com
Web: www.oldbuilders.com

LIME PLASTERING & RENDERING

CONSERVATRIX
17 Culkerton, Tetbury, Gloucestershire, GL8 8SS
Area of Operation: UK (Excluding Ireland)
Tel: 01285 841540
Email: david@conservatrix.co.uk
Web: www.conservatrix.co.uk

HISTORIC BUILDING CONSERVATION
Brabazon House, Scalford Road, Eastwell,
Melton Mowbray, Leicestershire, LE14 4EF
Area of Operation: UK (Excluding Ireland)
Tel: 01949 861333 **Fax:** 01949 861331
Email: info@historic-buildings.net
Web: www.historic-buildings.net

J&J SHARPE
Woodside, Merton, Okehampton, Devon, EX20 3EG
Area of Operation: UK (Excluding Ireland)
Tel: 01805 603587 **Fax:** 01805 603179
Email: mail@jjsharpe.co.uk
Web: www.jjsharpe.co.uk

OLDBUILDERS COMPANY
Mount Palmer, Birr Demesne, Birr, Co. Offaly
Area of Operation: Ireland Only
Tel: 0509 21133
Email: info@oldbuilders.com
Web: www.oldbuilders.com

SCOTTISH LIME CENTRE TRUST
The School House, Rocks Road,
Charlestown, Nr Dunfermline, Fife, KY11 3EN
Area of Operation: Scotland
Tel: 01383 872722
Fax: 01383 872744
Email: info@scotlime.org
Web: www.scotlime.org

THATCHING

INTHATCH
Higher Whatley, Otterford,
Chard, Somerset, TA20 3QL
Area of Operation: Worldwide
Tel: 01460 234477
Fax: 01460 234127
Email: sales@inthatch.co.uk
Web: www.inthatch.co.uk

MASTER THATCHERS (NORTH) LIMITED
8 Thorsby Road, Timperley, Altrincham,
Cheshire, WA15 7QP
Area of Operation: UK & Ireland
Tel: 0161 941 1986
Email: peterbrugge@thatching.net
Web: www.thatching.net

SPECIALIST JOINERY MANUFACTURERS

ALTHAM HARDWOOD CENTRE LTD
Altham Corn Mill, Burnley Road,
Altham, Accrington, Lancashire, BB5 5UP
Area of Operation: UK & Ireland
Tel: 01282 771618
Fax: 01282 777932
Email: info@oak-beams.co.uk
Web: www.oak-beams.co.uk

BADMAN & BADMAN LTD
The Drill Hall, Langford Road,
Weston Super Mare, Somerset, BS23 3PQ
Area of Operation: UK (Excluding Ireland)
Tel: 01934 644122
Fax: 01934 628189
Email: info@badmanandbadman.fsnet.co.uk
Web: www.badmans.co.uk

BENCHMARK JOINERY GUISBOROUGH LTD
Unit 3 Morgan Drive, Guisborough,
Cleveland, TS14 7DH
Area of Operation: North East England
Tel: 01287 203317
Email: alan@custom-joinery.com
Web: www.custom-joinery.com

BROADLEAF TIMBER
Llandeilo Road Industrial Estate,
Carms, Carmarthenshire, SA18 3JG
Area of Operation: UK & Ireland
Tel: 01269 851910
Fax: 01269 851911
Email: sales@broadleaftimber.com
Web: www.broadleaftimber.com

COLIN BAKER
Timberyard, Crownhill, Halberton,
Tiverton, Devon, EX16 7AY
Area of Operation: Europe
Tel: 01884 820152
Fax: 01884 820152
Email: colinbaker@colinbakeroak.co.uk
Web: www.colinbakeroak.co.uk

COUNTY JOINERY (SOUTH EAST) LTD
Unit 300, Vinehall Business Centre,
Vinehall Road, Mountfield, Robertsbridge,
East Sussex, TN32 5JW
Area of Operation: Greater London,
South East England
Tel: 01424 871500
Fax: 01424 871550
Email: countyjoinery@aol.com
Web: www.countyjoinery.co.uk

CREATE JOINERY
Worldsend, Llanmadoc, Swansea, SA3 1DB
Area of Operation: Greater London,
Midlands & Mid Wales, North West England
and North Wales, South East England,
South West England and South Wales
Tel: 01792 386677
Fax: 01792 386677
Email: mail@create-joinery.co.uk
Web: www.create-joinery.co.uk

DOVETAIL COMMERCIAL JOINERY
Gamston Industrial Estate, Gamston,
Retford, Nottinghamshire, DN22 0QL
Area of Operation: UK (Excluding Ireland)
Tel: 01777 838138
Fax: 01777 839132
Email: psavill@aol.com
Web: www.dovetailcommercial.co.uk

FLETCHER JOINERY
261 Whessoe Road,
Darlington, Durham, DL3 0YL
Area of Operation: North East England
Tel: 01325 357347
Fax: 01325 357347
Email: enquiries@fletcherjoinery.co.uk
Web: www.fletcherjoinery.co.uk

J L JOINERY
Cockerton View, Grange Lane,
Preston, Lancashire, PR4 5JE
Area of Operation: UK (Excluding Ireland)
Tel: 01772 616123
Fax: 01772 619182
Email: mail@jljoinery.co.uk
Web: www.jljoinery.co.uk

CONSULTANTS, LABOUR & FINANCE - **Specialist Crafts & Services** - Building Restoration; Decorative Interior Features Restoration; Joinery Restoration

SPONSORED BY: MOVE OR IMPROVE? www.moveorimprove.co.uk

JOHN NETHERCOTT & CO
147 Corve Street, Ludlow, Shropshire, SY8 2NN
Area of Operation: UK (Excluding Ireland)
Tel: 01589 877044 **Fax:** 01589 560255
Email: showroom@johnnethercott.com
Web: www.johnnethercott.com

SCOTTS OF THRAPSTON LTD.
Bridge Street, Thrapston,
Northamptonshire, NN14 4LR
Area of Operation: UK & Ireland
Tel: 01832 732366 **Fax:** 01832 733703
Email: julia@scottsofthrapston.co.uk
Web: www.scottsofthrapston.co.uk

TASKWORTHY
The Old Brickyard, Pontrilas,
Herefordshire, HR2 0DJ
Area of Operation: UK (Excluding Ireland)
Tel: 01981 242900 **Fax:** 01981 242901
Email: peter@taskworthy.co.uk
Web: www.taskworthy.co.uk

TIM WOOD LIMITED
1 Burland Road, London, SW11 6SA
Area of Operation: Worldwide
Tel: 07041 380 0030 **Fax:** 0870 054 8645
Email: homeb@timwood.com
Web: www.timwood.com

TONY HOOPER
Unit 18 Camelot Court, Bancombe Trading Estate,
Somerton, TA11 6SB
Area of Operation: UK (Excluding Ireland)
Tel: 01458 274221 **Fax:** 01458 274690
Email: tonyhooper1@aol.com
Web: www.tonyhooper.co.uk

BUILDING RESTORATION

B.WILLIAMSON & DAUGHTERS
Copse Cottage, Ford Manor Road,
Dormansland, Lingfield, Surrey, RH7 6NZ
Area of Operation: Greater London,
South East England
Tel: 01342 834829 **Fax:** 01342 870390
Email: bryan.williamson@btclick.com
Web: www.specialistcleaning4me.co.uk

BELLCREST REFURBISHMENT (FRANCE)
13 Leicester Drive, Tunbridge Wells,
Kent, TN2 5PH
Area of Operation: Europe
Tel: 07979 220861
Email: paul@bellcrestrefurbishment.com
Web: www.bellcrestrefurbishment.com

CONSERVATRIX
17 Culkerton, Tetbury, Gloucestershire, GL8 8SS
Area of Operation: UK (Excluding Ireland)
Tel: 01285 841540
Email: david@conservatrix.co.uk
Web: www.conservatrix.co.uk

FALCON STRUCTURAL REPAIRS
2B Harbour Road, Portishead, Bristol, BS20 7DD
Area of Operation: Europe
Tel: 01275 844889 **Fax:** 01275 847002
Email: sjoyner@falconstructural.co.uk
Web: www.falconstructural.co.uk
Product Type: 3

HIGHTEX-COATINGS LTD
Unit 14 Chapel Farm, Hanslope Road,
Hartwell, Northamptonshire, NN7 2EU
Area of Operation: UK (Excluding Ireland)
Tel: 01604 861250 **Fax:** 01604 871116
Email: bob@hightexcoatings.co.uk
Web: www.hightexcoatings.co.uk

HISTORIC BUILDING CONSERVATION
Brabazon House, Scalford Road,
Eastwell, Melton Mowbray,
Leicestershire, LE14 4EF
Area of Operation: UK (Excluding Ireland)
Tel: 01949 861333 **Fax:** 01949 861331
Email: info@historic-buildings.net
Web: www.historic-buildings.net

J&J SHARPE
Woodside, Merton,
Okehampton, Devon, EX20 3EG
Area of Operation: UK (Excluding Ireland)
Tel: 01805 603587 **Fax:** 01805 603179
Email: mail@jjsharpe.co.uk
Web: www.jjsharpe.co.uk

KEN NEGUS LTD
90 Garfield Road, Wimbledon, London, SW19 8SB
Area of Operation: Greater London,
South East England
Tel: 0208 543 9266
Fax: 0208 543 9100
Email: graham@kennegus.co.uk
Web: www.kennegus.co.uk

M.B.L
55 High Street, Biggleswade,
Bedfordshire, SG18 0JH
Area of Operation: UK (Excluding Ireland)
Tel: 01767 318695
Fax: 01767 318695
Email: info@mblai.co.uk
Web: www.mblai.co.uk

**MALCOLM SMITH -
POWER WASHING & GRAFFITI REMOVAL**
45 Roundponds, Melksham, Wiltshire, SN12 8DW
Area of Operation: Midlands & Mid Wales, South
East England, South West England and South Wales
Tel: 01225 707200
Email: powerwashinguk@aol.com
Web: www.powerwashinguk.co.uk

NIMBUS CONSERVATION LTD
Eastgate, Christchurch Street East,
Frome, Somerset, BA11 1QD
Area of Operation: UK (Excluding Ireland)
Tel: 01373 474646
Fax: 01373 474648
Email: enquiries@nimbusconservation.com
Web: www.nimbusconservation.com

NORBUR MOOR BUILDING SERVICES LTD
Gawthorne, Hazel Grove,
Stockport, Cheshire, SK7 5AB
Area of Operation: North West England
and North Wales
Tel: 0800 093 5785
Fax: 0161 456 1944
Email: paul@norbury moor.co.uk
Web: www.norburymoor.co.uk /
www.norburymoorbuilding.co.uk

OAKBEAMS.COM
Hunterswood Farm, Alfold Road,
Dunsfold, Godalming, Surrey, GU8 4NP
Area of Operation: Worldwide
Tel: 01483 200477
Email: info@oakbeams.com
Web: www.oakbeams.com

OLDBUILDERS COMPANY
Mount Palmer, Birr Demesne,
Birr, Co. Offaly
Area of Operation: Ireland Only
Tel: 0509 21133
Email: info@oldbuilders.com
Web: www.oldbuilders.com

ONE-CALL PROPERTY MAINTENANCE
PO Box 4413, Salisbury House,
Ringwood, Hampshire, BH24 1YR
Area of Operation: Greater London,
South East England
Tel: 01202 828715
Fax: 01202 828715
Email: paul@one-callgroup.co.uk
Web: www.one-callgroup.co.uk

QUADRIGA CONCEPTS LTD
Gadbrook House, Gadbrook Park,
Rudheath, Northwich, Cheshire, CW9 7RG
Area of Operation: UK & Ireland
Tel: 0808 100 3777
Fax: 01606 330777
Email: info@quadrigaltd.com
Web: www.quadrigaltd.co.uk

RENAISSANCE ARCHITECTURAL LTD
15 Gay Street, Bath, BA1 2PH
Area of Operation: East England, Greater London,
Midlands, North East England, North West England,
South East England, South West England
Tel: 01761 479605
Fax: 01761 479605
Email: renaissance@ren.uk.com
Web: www.ren.uk.com

STONEHEALTH LTD
Bowers Court, Broadwell, Dursley,
Gloucestershire, GL11 4JE
Area of Operation: Worldwide
Tel: 01453 540600
Fax: 01453 540609
Email: info@stonehealth.com
Web: www.stonehealth.com

SUSTAINABLE PROPERTY GROUP LTD.
Courtyard Studio, Cowley Farm, Aylesbury Road,
Cuddington, Buckinghamshire, HP18 0AD
Area of Operation: Greater London, Midlands
& Mid Wales, South East England
Tel: 01296 747121
Fax: 01296 747703
Email: info@sustainable.uk.com
Web: www.sustainable.uk.com

**TOUCHSTONE - GRANITE, MARBLE,
SLATE & STONE**
Touchstone House, 82 High Street,
Measham, Derbyshire, DE12 7JB
Area of Operation: UK (Excluding Ireland)
Tel: 0845 130 1862
Fax: 01530 274271
Email: sales@touchstone-uk.com
Web: www.touchstone-uk.com

TOUCHSTONE - GRANITE, MARBLE, SLATE & STONE

Area of Operation: UK (Excluding Ireland)
Tel: 0845 130 1862
Fax: 01530 274271
Email: sales@touchstone-uk.com
Web: www.touchstone-uk.com

Architectural natural stone production. Utilising the
latest in computer based machinery in conjunction
with traditional skilled craftsmen we produce all
items of architectural solid natural stone. e.g.
window casements, bay windows, window cills, door
frames, steps, balustrade etc.. Single item pieces, to
full projects undertaken.

WISE PROPERTY CARE
8 Muriel Street, Barrhead,
Glasgow, G78 1QB
Area of Operation: Scotland
Tel: 0141 876 0300
Fax: 0141 876 0301
Email: les@wisepropertycare.com
Web: www.wisepropertycare.com

DECORATIVE INTERIOR FEATURES RESTORATION

HERITAGE TILE CONSERVATION LTD
The Studio, 2 Harris Green, Broseley,
Shropshire, TF12 5HJ
Area of Operation: UK & Ireland
Tel: 01952 881039
Fax: 01952 881039
Email: enquiries@heritagetile.co.uk
Web: www.heritagetile.co.uk

HERITAGE TILING DESIGN & RESTORATION CO.
P.O. Box 18 Seaforth Vale, Seaforth, Liverpool,
Merseyside, L21 0EQ
Area of Operation: Worldwide
Tel: 0151 920 7349
Fax: 0151 920 7349
Email: info@heritagetiling.com
Web: www.tiling.co.uk

ONE-CALL PROPERTY MAINTENANCE
PO Box 4413, Salisbury House, Ringwood,
Hampshire, BH24 1YR
Area of Operation: Greater London,
South East England
Tel: 01202 828715
Fax: 01202 828715
Email: paul@one-callgroup.co.uk
Web: www.one-callgroup.co.uk

PAXTON RESTORATION LTD
130 Shaftesbury Avenue, London, W1D 5EU
Area of Operation: Greater London
Tel: 0870 027 8424
Fax: 0870 127 7642
Email: richard@paxtonrestoration.co.uk
Web: www.paxtonrestoration.co.uk

POETSTYLE
Unit 1, Bayford Street Industrial Centre,
Hackney, Greater London, E8 3SE
Area of Operation: UK (Excluding Ireland)
Tel: 0208 533 0915
Fax: 0208 985 2953
Email: sofachairs@aol.com
Web: www.sofachairs.co.uk

SANDED FLOORS
7 Pleydell Avenue, Upper Norwood,
London, SE19 2LN
Area of Operation: South East England
Tel: 0209 653 1326
Email: peter-weller@sandedfloors.co.uk
Web: www.sandedfloors.co.uk

SOUTH WESTERN FLOORING SERVICES
145-147 Park Lane, Frampton Cotterell,
Bristol, BS36 2ES
Area of Operation: UK & Ireland
Tel: 01454 880982
Fax: 01454 880982
Email: mikeflanders@blueyonder.co.uk
Web: www.southwesternflooring.com

STONEHEALTH LTD
Bowers Court, Broadwell, Dursley,
Gloucestershire, GL11 4JE
Area of Operation: Worldwide
Tel: 01453 540600
Fax: 01453 540609
Email: info@stonehealth.com
Web: www.stonehealth.com

SUNRISE STAINED GLASS
58-60 Middle Street, Southsea,
Portsmouth, Hampshire, PO5 4BP
Area of Operation: UK (Excluding Ireland)
Tel: 02392 750512
Fax: 02392 875488
Email: sunrise@stained-windows.co.uk
Web: www.stained-windows.co.uk

THE PAINT PRACTICE
18 Hallam Chase, Sandygate, Sheffield,
South Yorkshire, S10 5SW
Area of Operation: Europe
Tel: 0114 230 6828
Fax: 0114 230 6828

JOINERY RESTORATION

B G H JOINERY CO LTD
Unicorn Business Centre, The Ridgeway,
Chiseldon, Swindon, Wiltshire, SN4 0HT
Area of Operation: UK (Excluding Ireland)
Tel: 01793 741330
Fax: 01793 741310
Email: sales@bghjoinery.co.uk
Web: www.bghjoinery.co.uk

CONSERVATRIX
17 Culkerton, Tetbury,
Gloucestershire, GL8 8SS
Area of Operation: UK (Excluding Ireland)
Tel: 01285 841540
Email: david@conservatrix.co.uk
Web: www.conservatrix.co.uk

HISTORIC BUILDING CONSERVATION
Brabazon House, Scalford Road,
Eastwell, Melton Mowbray,
Leicestershire, LE14 4EF
Area of Operation: UK (Excluding Ireland)
Tel: 01949 861333
Fax: 01949 861331
Email: info@historic-buildings.net
Web: www.historic-buildings.net

JOHN NETHERCOTT & CO
147 Corve Street, Ludlow,
Shropshire, SY8 2NN
Area of Operation: UK (Excluding Ireland)
Tel: 01589 877044
Fax: 01589 560255
Email: showroom@johnnethercott.com
Web: www.johnnethercott.com

M. H. JOINERY SERVICES
25b Camwal Road, Harlgate,
North Yorkshire, HG1 4PT
Area of Operation: North East England
Tel: 01423 888856
Fax: 01432 888856
Email: info@mhjoineryservices.co.uk
Web: www.mhjoineryservices.co.uk

NORBUR MOOR BUILDING SERVICES LTD
Gawthorne, Hazel Grove, Stockport,
Cheshire, SK7 5AB
Area of Operation: North West England
and North Wales
Tel: 0800 093 5785
Fax: 0161 456 1944
Email: paul@norburymoor.co.uk
Web: www.norburymoor.co.uk /
www.norburymoorbuilding.co.uk

OLDBUILDERS COMPANY
Mount Palmer, Birr Demesne, Birr, Co. Offaly
Area of Operation: Ireland Only
Tel: 0509 21133
Email: info@oldbuilders.com
Web: www.oldbuilders.com

PAXTON RESTORATION LTD
130 Shaftesbury Avenue, London, W1D 5EU
Area of Operation: Greater London
Tel: 0870 027 8424
Fax: 0870 127 7642
Email: richard@paxtonrestoration.co.uk
Web: www.paxtonrestoration.co.uk

QUADRIGA CONCEPTS LTD
Gadbrook House, Gadbrook Park,
Rudheath, Northwich, Cheshire, CW9 7RG
Area of Operation: UK & Ireland
Tel: 0808 100 3777
Fax: 01606 330777
Email: info@quadrigaltd.com
Web: www.quadrigaltd.com

SANDERSON'S FINE FURNITURE
Unit 5 & 6, The Village Workshop,
Four Crosses Business Park,
Four Crosses, Powys, SY22 6ST
Area of Operation: UK (Excluding Ireland)
Tel: 01691 830075
Fax: 01691 830075
Email: sales@sandersonsfinefurniture.co.uk
Web: www.sandersonsfinefurniture.co.uk

VENTROLLA LTD
11 Hornbeam Square South,
Harrogate, North Yorkshire, HG2 8NB
Area of Operation: UK & Ireland
Tel: 0800 378278
Fax: 01423 859321
Email: info@ventrolla.co.uk
Web: www.ventrolla.co.uk

WINDOW CARE SYSTEMS LTD
Unit E, Sawtry Business Park, Glatton Road,
Sawtry, Huntingdon, Cambridgeshire, PE28 5GQ
Area of Operation: UK & Ireland
Tel: 01487 830311 **Fax:** 01487 832876
Email: sales@window-care.net
Web: www.window-care.com

WISE PROPERTY CARE
8 Muriel Street, Barrhead, Glasgow, G78 1QB
Area of Operation: Scotland
Tel: 0141 876 0300 **Fax:** 0141 876 0301
Email: les@wisepropertycare.com
Web: www.wisepropertycare.com

METALWORK RESTORATION

ARCHITECTURAL METALWORK CONSERVATION
Unit 19, Hoddesdon Industrial Centre,
Pindar Road, Hoddesdon, Hertfordshire, EN11 0DD
Area of Operation: East England
Tel: 01992 443132
Fax: 01992 443132

B.WILLIAMSON & DAUGHTERS
Copse Cottage, Ford Manor Road,
Dormansland, Lingfield, Surrey, RH7 6NZ
Area of Operation: Greater London,
South East England
Tel: 01342 834829
Fax: 01342 870390
Email: bryan.williamson@btclick.com
Web: www.specialistcleaning4me.co.uk

CHRIS TOPP & COMPANY WROUGHT IRONWORKS
Lyndhurst, Carlton Husthwaite,
Thirsk, North Yorkshire, YO7 2BJ
Area of Operation: Worldwide
Tel: 01845 501415
Fax: 01845 501072
Email: enquiry@christopp.co.uk
Web: www.christopp.co.uk

COBALT BLACKSMITHS
The Forge, English Farm, English Lane,
Nuffield, Oxfordshire, RG9 5TH
Tel: 01491 641990
Fax: 01491 640909
Email: enquiries@cobalt-blacksmiths.co.uk
Web: www.cobalt-blacksmiths.co.uk

MBL
55 High Street, Biggleswade,
Bedfordshire, SG18 0JH
Area of Operation: UK (Excluding Ireland)
Tel: 01767 318695
Fax: 01767 318834
Email: info@mblai.co.uk
Web: www.mblai.co.uk

STEEL WINDOW SERVICE & SUPPLIES LTD
30 Oxford Road, Finsbury Park,
London, N4 3EY
Area of Operation: Greater London
Tel: 0207 272 2294
Fax: 0207 281 2309
Email: post@steelwindows.co.uk
Web: www.steelwindows.co.uk

PLEASE MENTION

THE
HOMEBUILDER'S
HANDBOOK

WHEN YOU CALL

STONEWORK RESTORATION

ABBEY MASONRY & RESTORATION
Plot 4 Cross Hands Business Park,
Cross Hands, Carmarthenshire, SA67 8DD
Area of Operation: South West England
and South Wales
Tel: 01269 845084
Fax: 01269 831774
Email: info@abbeymasonry.com
Web: Abbey Masonry & Restoration

AD CALVERT ARCHITECTURAL STONE SUPPLIES LTD
Smithy Lane, Grove Square, Leyburn,
North Yorkshire, DL8 5DZ
Area of Operation: UK & Ireland
Tel: 01969 622515
Fax: 01969 624345
Email: stone@calverts.co.uk
Web: www.calverts.co.uk

ALBA MASONRY
17 Borestone Crescent,
Stirling, Stirlingshire, FK7 9BQ
Area of Operation: Scotland
Tel: 01786 450459
Web: www.albamasonry.co.uk

ARTISAN STONE WORK
1 Isis Close, Abingdon,
Oxfordshire, OX14 3TA
Area of Operation: Europe
Tel: 07789 952007
Email: piersconway@yahoo.com
Web: www.artisanstonework.com

B.WILLIAMSON & DAUGHTERS
Copse Cottage, Ford Manor Road,
Dormansland, Lingfield,
Surrey, RH7 6NZ
Area of Operation: Greater London,
South East England
Tel: 01342 834829
Fax: 01342 870390
Email: bryan.williamson@btclick.com
Web: www.specialistcleaning4me.co.uk

BURLEIGH STONE CLEANING & RESTORATION COMPANY LTD
The Old Stables, 56 Balliol Road,
Bootle, Merseyside, L20 7EJ
Area of Operation: UK (Excluding Ireland)
Tel: 0151 922 3366
Fax: 0151 922 3377
Email: info@burleighstone.co.uk
Web: www.burleighstone.co.uk

CONSERVATRIX
17 Culkerton, Tetbury,
Gloucestershire, GL8 8SS
Area of Operation: UK (Excluding Ireland)
Tel: 01285 841540
Email: david@conservatrix.co.uk
Web: www.conservatrix.co.uk

D F FIXINGS
15 Aldham Gardens,
Rayleigh, Essex, SS6 9TB
Area of Operation: UK (Excluding Ireland)
Tel: 07956 674673
Fax: 01268 655072

HERITAGE TILE CONSERVATION LTD
The Studio, 2 Harris Green,
Broseley, Shropshire, TF12 5HJ
Area of Operation: UK & Ireland
Tel: 01952 881039 **Fax:** 01952 881039
Email: enquiries@heritagetile.co.uk
Web: www.heritagetile.co.uk

HISTORIC BUILDING CONSERVATION
Brabazon House, Scalford Road, Eastwell,
Melton Mowbray, Leicestershire, LE14 4EF
Area of Operation: UK (Excluding Ireland)
Tel: 01949 861333
Fax: 01949 861331
Email: info@historic-buildings.net
Web: www.historic-buildings.net

J & R MARBLE COMPANY LTD
Unit 9,Period Works, Lammas Road,
Leyton, London, E10 7QT
Area of Operation: UK (Excluding Ireland)
Tel: 0208 539 6471
Fax: 0208 539 9264
Email: sales@jrmarble.co.uk
Web: www.jrmarble.co.uk

J&J SHARPE
Woodside, Merton,
Okehampton, Devon, EX20 3EG
Area of Operation: UK (Excluding Ireland)
Tel: 01805 603587
Fax: 01805 603179
Email: mail@jjsharpe.co.uk
Web: www.jjsharpe.co.uk

KEN NEGUS LTD
90 Garfield Road, Wimbledon,
London, SW19 8SB
Area of Operation: Greater London,
South East England
Tel: 0208 543 9266
Fax: 0208 543 9100
Email: graham@kennegus.co.uk
Web: www.kennegus.co.uk

NIMBUS CONSERVATION LTD
Eastgate, Christchurch Street East,
Frome, Somerset, BA11 1QD
Area of Operation: UK (Excluding Ireland)
Tel: 01373 474646
Fax: 01373 474648
Email: enquiries@nimbusconservation.com
Web: www.nimbusconservation.com

NT BLASTING
Penny Three, Park Lane, Lane End,
High Wycombe, Buckinghamshire, HP14 3DE
Area of Operation: UK (Excluding Ireland)
Tel: 01494 883022
Email: info@ntpartners.co.uk

OLDBUILDERS COMPANY
Mount Palmer, Birr Demesne, Birr, Co. Offaly
Area of Operation: Ireland Only
Tel: 0509 21133
Email: info@oldbuilders.com
Web: www.oldbuilders.com

ONE-CALL PROPERTY MAINTENANCE
PO Box 4413, Salisbury House,
Ringwood, Hampshire, BH24 1YR
Area of Operation: Greater London,
South East England
Tel: 01202 828715
Fax: 01202 828715
Email: paul@one-callgroup.co.uk
Web: www.one-callgroup.co.uk

QUADRIGA CONCEPTS LTD
Gadbrook House, Gadbrook Park,
Rudheath, Northwich,
Cheshire, CW9 7RG
Area of Operation: UK & Ireland
Tel: 0808 100 3777
Fax: 01606 330777
Email: info@quadrigaltd.com
Web: www.quadrigaltd.com

RESTORE BRICK & STONE CLEANING LTD
6 Greenoak Way, Wimbledon Village,
London, SW19 5EN
Area of Operation: Greater London,
South East England
Tel: 0208 286 3579
Fax: 0208 947 2622
Email: info@restorebrick.co.uk
Web: www.restorebrick.co.uk

STONEHEALTH LTD
Bowers Court, Broadwell,
Dursley, Gloucestershire, GL11 4JE
Area of Operation: Worldwide
Tel: 01453 540600 **Fax:** 01453 540609
Email: info@stonehealth.com
Web: www.stonehealth.com

CONSULTANTS, LABOUR & FINANCE

CONSULTANTS, LABOUR & FINANCE - **Specialist Crafts & Services** - Sandblasting; French Polishers; Frescos, Murals & Trompe L'oeil; Aerial Photography

SPONSORED BY: MOVE OR IMPROVE? www.moveorimprove.co.uk

SANDBLASTING & CLEANING SERVICES

MALCOLM SMITH - POWER WASHING & GRAFFITI REMOVAL
45 Roundponds, Melksham, Wiltshire, SN12 8DW
Area of Operation: Midlands & Mid Wales, South East England, South West England and South Wales
Tel: 01225 707200
Email: powerwashinguk@aol.com
Web: www.powerwashinguk.co.uk

OLDBUILDERS COMPANY
Mount Palmer, Birr Demesne, Birr, Co. Offaly
Area of Operation: Ireland Only
Tel: 0509 21133
Email: info@oldbuilders.com
Web: www.oldbuilders.com

RESTORE BRICK + STONE CLEANING LTD
6 Greenoak Way, Wimbledon Village, London, SW19 5EN
Area of Operation: Greater London, South East England
Tel: 0208 286 3579 **Fax:** 0208 947 2622
Email: info@restorebrick.co.uk
Web: www.restorebrick.co.uk

SANDED FLOORS
7 Pleydell Ave, Upper Norwood, London, SE19 2LN
Area of Operation: South East England
Tel: 0209 653 1326
Email: peter-weller@sandedfloors.co.uk
Web: www.sandedfloors.co.uk

PLEASE MENTION

THE HOMEBUILDER'S HANDBOOK

WHEN YOU CALL

FRENCH POLISHERS

BROWNS FRENCH POLISHING CO LTD
Unit A2 Pixmore Estate, Pixmore Avenue, Letchworth Garden City, Hertfordshire, SG6 1JJ
Area of Operation: East England, Greater London
Tel: 01462 680241 **Fax:** 01462 482999
Email: info@brownsfrenchpolishing.co.uk
Web: www.brownsfrenchpolishing.co.uk

FRESCOS, MURALS & TROMPE L'OEIL

BLINK RED CONTEMPORARY ART
40 Maritime Street, Edinburgh, Lothian, EH6 6SA
Area of Operation: UK & Ireland
Tel: 0131 625 0192 **Fax:** 0131 467 7995
Email: customerservices@blinkred.com
Web: www.blinkred.com

HUNTER ART & TILE STUDIO
Craft Courtyard, Harestanes, Ancrum, Jedburgh, Borders, TD8 6UQ
Area of Operation: UK & Ireland
Tel: 01835 830328
Email: enquiries@hunterartandtilestudio.co.uk
Web: www.hunterartandtilestudio.co.uk

RICEDESIGN
The Barn, Shawbury Lane, Shustoke, Nr Coleshill, B46 2RR
Area of Operation: Worldwide
Tel: 01675 481183
Email: ricedesigns@hotmail.com

AERIAL PHOTOGRAPHY

AEROLENS
Unit C16 St.George's Business Park, Castle Road, Sittingbourne, Kent, ME10 3TB
Area of Operation: Greater London, South East England
Tel: 0845 838 1764
Email: sales@aerolens.co.uk
Web: www.aerolens.co.uk

NOTES

Company Name ....
....
Address ....
....
....
email ....
Web ....

Company Name ....
....
Address ....
....
....
email ....
Web ....

Company Name ....
....
Address ....
....
....
email ....
Web ....

Company Name ....
....
Address ....
....
....
email ....
Web ....

Inspiration and Practical Advice for Self-builders & Renovators...

Britain's best selling self-build and renovation magazine, Homebuilding & Renovating is an essential read for anyone extending, converting, renovating or building their own home.

● **How to get started**; organising finance, how to save money, estimating build costs, insurance advice and getting planning persmission.

● **Build, renovate or extend?** – how to assess building plots, spot renovation or conversion bargains, and things you need to know before extending.

● **Find your ideal building plot or property** for renovation with www.plotfinder.net.

● **Inspiration and ideas** from reader case studies.

● **Practical advice** – how to manage your subcontractors so that your build completes on time and within budget.

● **Ask the Experts** – have your problems solved by our experts.

SUBSCRIBE TODAY

Visit **www.homebuilding.co.uk** or call **01527 834435**

PROFESSIONAL SERVICES

CONSULTANTS, LABOUR & FINANCE

Architects

The term "architect" is all to often used generically to describe anyone who designs or supervises the construction of buildings. In fact the title is protected by law, with a person only allowed to call themselves an architect if professionally qualified and registered with the Architects Registration Board (ARB). The ARB is the authority established by statute to maintain a register of architects and to regulate the conduct of the architect profession. Anyone calling themselves an architectural consultant or architectural designer has not completed this training and may have few qualifications at all.

ABOVE: This contemporary barn conversion was the overall winner of the Daily Telegraph H&R Awards 2004, and was featured as a case study in the January 2005 issue of Homebuilding and Renovating magazine.

When planning any project, a knowledgable professional is always going to be an asset, but never more so that when you are converting, or planning extensive changes to, a period property. Here it is particularly important to tread carefully, as period buildings can be all too easily damaged or devalued by unsympathetic work. Therefore it makes sense to employ someone with experience of older properties who understand the special challenges they pose. A professional with particular knowledge of traditional buildings can provide insight into their structure and materials, and will appreciate the design customs of period structures. In addition, such specialists should have a better understanding of the legal requirements of working on a historic building and the complexities of the relevant VAT legislation.

Finding an architect with an interest in conservation has been made easy by the RIBA Client's Advisory Service. They offer a tailored search, whereby you list your requirements and they send you a list of registered practices with the right skills and resources for your project.

However, it is also possible to check for suitable architects yourself, as a Register of Practices was set up in 1996 to clarify the status of each architectural practice, its expertise and the services it offers. RIBA has a website, www.ribafind.org, where you can search for a practice using various criteria — including location, type of service and specialist expertise. This has the added advantage of providing easy access to individual company websites, many of which show images and details of their past projects. The style of such websites will also give you a feel for the character of that particular practice.

Architects with a particular interest in conservation are likely to have gained extra qualifications, so look for membership of IHBC (Institute of Historic Building Conservation) or AABC (Register of Architects Accredited in Building Conservation).

All architects in the UK are registered with the Architects Registration Board, to show that they have achieved recognised educational qualifications and work experience. Inclusion in the RIBA Register of Practices means that they also have management procedures in place to ensure they comply with RIBA's continuing professional development obligations, and that the architects they employ adhere to the RIBA Code of Conduct.

ARCHITECTS

3S ARCHITECTS LLP
47 High Street, Kingston Upon Thames,
Surrey, KT1 1LQ
Area of Operation: Europe
Tel: 0208 549 2000
Fax: 0208 549 3636
Email: info@3sarchitects.com
Web: www.3sArchitects.com

A G N HUGHES
30 Wildwood Court, Hawkhirst Road,
Kenley, Surrey, CR8 5DL
Area of Operation: UK (Excluding Ireland)
Tel: 0208 660 2637
Fax: 0208 660 2642
Email: agn.hughes@tiscali.co.uk

A&S DESIGN SERVICES
Cornlands, Sampford Peverell,
Tiverton, Devon, EX16 7UA
Area of Operation: UK (Excluding Ireland)
Tel: 01884 829285
Fax: 01884 829285
Email: ascad@aol.com
Web: www.asdesignservices.co.uk

ADS UK LTD
30 Wild Wood Court, Hawkhirst Rd,
Kenley, Surrey, CR8 5DL
Area of Operation: Worldwide
Tel: 0208 660 2637
Fax: 0208 660 2642
Email: agn.hughes@virgin.net

ANDREW SMITH ASSOCIATES
5 Mount View, Billericay, Essex, CM11 1HB
Area of Operation: East England,
Greater London, South East England
Tel: 01277 630310 **Fax:** 01277 651833
Email: architect@andrewsmithassociates.freeserve.co.uk

ARCHITECTURE VERTE LIMITED
41 Devonshire Buildings, Bath, Somerset, BA2 4SU
Area of Operation: Greater London, Midlands &
Mid Wales, South East England, South West England
& South Wales
Tel: 01225 484835
Email: paul@verte.co.uk

ASBA ARCHITECTS LTD
37 Lewis Way, Peterchurch, Herefordshire, HR2 0SE
Area of Operation: UK & Ireland
Tel: 0800 387310
Email: asba@asba-architects.org
Web: www.asba-architects.org

ASBA SCOTLAND
10 Lynedoch Cresent, Glasgow, G3 6EQ
Area of Operation: Scotland
Tel: 0800 731 3405
Fax: 0141 331 2751
Email: architects@design-practice.com

BRENNAN & WILSON ARCHITECTS
Belhaven Villa, Edinburgh Road, Belhaven,
Dunbar, Lothian, EH42 1PA
Area of Operation: North East England, Scotland
Tel: 01368 860897
Fax: 01368 860897
Email: mail@bwarchitects.co.uk
Web: www.bwarchitects.co.uk

BURNS ASSOCIATES
32 Market Place, Swaffham, Norfolk, PO37 7QH
Area of Operation: UK (Excluding Ireland)
Tel: 01760 722254 **Fax:** 01760 724424
Email: info@burnsassociates-architects.co.uk
Web: www.burnsassociates-architects.co.uk

CA SUSTAINABLE ARCHITECTURE
83 Old Newtown Road, Newbury,
Berkshire, RG14 7DE
Area of Operation: UK (Excluding Ireland)
Tel: 01635 48363
Email: isabel.carmona@ca-sa.co.uk
Web: www.ca-sa.co.uk

CHARTER
St. Mary's House, 15 Cardington Road,
Bedford, Bedfordshire, MK42 0BP
Area of Operation: UK & Ireland
Tel: 01234 342551
Fax: 01234 360055
Email: bedford@charter-architects.com
Web: www.charter.eu.com

**CHRISTOPHER MAGUIRE ARCHITECT
AND GARDEN DESIGNER**
15 Harston Road, Newton, Cambridge,
Cambridgeshire, CB2 5PA
Area of Operation: East England
Tel: 01223 872800
Web: www.christophermaguire.co.uk

CONSTRUCTIVE INDIVIDUALS
Trinity Buoy Wharf, 64 Orchard Place,
London, E14 0JW
Area of Operation: UK (Excluding Ireland)
Tel: 0207 515 9299
Fax: 0207 515 9737
Email: info@constructiveindividuals.com
Web: www.constructiveindividuals.com

CORK TOFT PARTNERSHIP LTD
Greenbank, Howick Cross Lane,
Penwortham, Preston, Lancashire, PR1 0NS
Area of Operation: North West England
and North Wales
Tel: 01772 749014 /5
Fax: 01772 749034
Email: mail@corktoft.com

CORLECO MANAGEMENT
PO Box 255, Douglas,
Isle of Man, IM99 1XS
Area of Operation: UK (Excluding Ireland)
Tel: 01624 853314
Fax: 01624 853315
Email: ajc@corleco.com
Web: www.poundperfoot.co.uk

CURTIS WOOD ARCHITECTS LIMITED
23-28 Penn Street, London, N1 5DL
Area of Operation: UK & Ireland
Tel: 0207 684 1400
Fax: 0207 729 1411
Email: andrew@curtiswoodarchitects.co.uk
Web: www.curtiswoodarchitects.co.uk

CUTLER ARCHITECTS
43 St. Mary's Street, Wallingford,
Oxfordshire, OX10 0EU
Area of Operation: Greater London,
South East England
Tel: 01491 838130
Fax: 01491 836504
Email: mail@cutlerarch.com
Web: www.cutlerarch.com

DAVID NEILL
Ridge Croft, 37 Dalby Avenue,
Bushby, Leicestershire, LE7 9RE
Area of Operation: East England, Midlands
& Mid Wales
Tel: 0116 243 2236
Fax: 0116 243 3580
Email: mail@davidneill.net

DAVID RANDELL ARCHITECTS LTD
The Studio, 7 William Street,
Tiverton, Devon, EX16 6BJ
Area of Operation: South West England
and South Wales
Tel: 01884 254465
Fax: 01884 243451
Email: davidrandell@dra-architects.co.uk
Web: www.davidrandellarchitects.co.uk

DESIGN AND MATERIALS LTD
Lawn Road, Carlton in Lindrick,
Nottinghamshire, S81 9LB
Area of Operation: UK & Ireland
Tel: 01909 730333
Fax: 01909 730605
Email: enquiries@designandmaterials.uk.com
Web: www.designandmaterials.uk.com 🖱

DESIGN FOR HOMES
The Building Centre,
26 Store Street, London, WC1E 7BT
Area of Operation: UK (Excluding Ireland)
Tel: 0870 416 3378
Fax: 0207 436 0573
Email: richard@designforhomes.org
Web: www.designforhomes.org

DIL GREEN ARCHITECT
206 Lyham Road, Brixton, London, SW2 5NR
Area of Operation: UK (Excluding Ireland)
Tel: 0208 671 2242
Fax: 0208 674 9757
Email: w_info@dilgreenarchitect.co.uk
Web: www.dilgreenarchitect.co.uk

ENGLISHAUS CHARTERED ARCHITECTS LTD
30 Lawrence Road, Hampton, Middlesex, TW12 2RJ
Area of Operation: Greater London, South East
England, South West England and South Wales
Tel: 0208 255 0595
Fax: 0208 287 3441
Email: enquiries@englishaus.co.uk
Web: www.englishaus.co.uk

FOUR SQUARE DESIGN LTD
The Old Surgery, Crowle Road, Lambourne,
Hungerford, Berkshire, RG17 8NR
Area of Operation: UK (Excluding Ireland)
Tel: 01488 71384
Fax: 01488 73207
Email: mathewsonwhittaker@compuserve.com
Web: www.foursquaredesign.co.uk

G.M.MOORE & ASSOCIATES
Old Laughton Sawmills, Park Lane,
Laughton, East Sussex, BN8 6BP
Area of Operation: South East England
Tel: 01323 811689 **Fax:** 01323 811325
Email: sales@gmassociates.co.uk
Web: www.gmassociates.co.uk

GEOFFREY G FRY + ASSOCIATES
'North Hangleton', 63 Langbury Lane,
Ferring, Worthing, West Sussex, BN12 6QA
Area of Operation: UK (Excluding Ireland)
Tel: 01903 816619 **Fax:** 01903 816619
Email: geoffrey.g.fry@btinternet.com

GEORGE BLACK ARCHITECT
12 Kingsknowe Crescent, Edinburgh,
Lothian, EH14 2JZ
Area of Operation: Scotland
Tel: 0131 443 9898
Email: george@gbarch.abel.co.uk
Web: www.gbarch.abel.co.uk

GRAEME OGG (ARCHITECTURE & DESIGN)
Keepers Cottage, Derculich, Strathtay,
Pitlochry, Perth and Kinross, PH9 0LR
Area of Operation: Scotland
Tel: 01887 840354 **Fax:** 01887 840354
Email: graemeogg@mac.com

GRINDEY CONSULTING
5 Granville Street, Monton, Manchester, M30 9PX
Area of Operation: North West England
and North Wales
Tel: 07773 371488
Email: ian.grindey@scottwilson.com
Web: www.grindeyconsulting.co.uk

HINTON COOK ARCHITECTS
214 Upper Fifth Street, Milton Keynes,
Buckinghamshire, MK9 2HR
Area of Operation: South East England
Tel: 01908 235544 **Fax:** 01908 236688
Email: hintoncook@btconnect.com
Web: www.hintoncook.co.uk

HOLDEN + PARTNERS
26 High Street, Wimbledon, London, SW19 5BY
Area of Operation: Worldwide
Tel: 0208 946 5502
Fax: 0208 879 0310
Email: arch@holdenpartners.co.uk
Web: www.holdenpartners.co.uk

HOT ARCHITECTURE
2 Minshull Street, Knutsford, Cheshire, WA16 6HG
Area of Operation: North West England
and North Wales
Tel: 01565 650401
Email: rosie@hotarchitecture.com
Web: www.hotarchitecture.com

IAIN FREARSON
68 Cavendish Road, Cambridge,
Cambridgeshire, CB1 3AF
Area of Operation: UK (Excluding Ireland)
Tel: 01223 473997
Fax: 01223 473997
Email: iain.frearson@ntlworld.com

IBI DESIGN ASSOCIATES
55 Chase Way, Southgate, London, N14 5EA
Area of Operation: Greater London, South East
England
Tel: 0208 361 2542
Fax: 0208 361 7062
Email: ibides@aol.com

J G HOOD & ASSOCIATES LTD
50 High Street, Bruton, Somerset, BA10 0AN
Area of Operation: Midlands & Mid Wales, South
East England, South West England and South Wales
Tel: 01749 812139
Fax: 0870 133 3776
Email: mail@jghood-assoc.co.uk
Web: www.jghood-assoc.co.uk

JIM MORRISON ARCHITECTS
31 Cricklewood Park, Belfast, Co Antrim, BT9 5GW
Area of Operation: UK & Ireland
Tel: 02890 660017
Fax: 02890 201710
Email: jim.morrison@dnet.co.uk
Web: www.jimmorrisonarchitects.co.uk

JNA ARCHITECTS
9 East Park, Southgate, West Sussex, RH10 6AN
Area of Operation: South East England
Tel: 01293 439323
Fax: 01293 530160
Email: james@jna-architects.co.uk
Web: www.jna-architects.co.uk

JOHN C. ANGELL
25 Whinfield Lane, Preston, Lancashire, PR2 1NQ
Area of Operation: North West England
and North Wales
Tel: 01772 725308
Email: architects@angell.org.uk
Web: www.angell.org.uk

JOHN PEATE ARCHITECTURAL SERVICES LTD
6 Newport Rd, Shifnal, Shropshire, TF11 8BP
Area of Operation: Midlands & Mid Wales,
North West England and North Wales
Tel: 01952 460175
Fax: 01952 460175
Email: ray.peate@btinternet.com
Web: www.johnpeate.org.uk

JOHN SHORE
16 Popes Lane, Rockwell Green,
Wellington, Somerset, TA21 9DQ
Area of Operation: UK & Ireland
Tel: 01823 666177
Fax: 01823 666177
Email: shorepower@ukonline.co.uk

JOHN SOLOMON DESIGN LIMITED
48 Ham Street, Richmond, Surrey, TW10 7HT
Area of Operation: Worldwide
Tel: 0208 940 2444
Fax: 0208 940 1188
Email: mail@jsajsd.com
Web: www.jsajsd.com

CONSULTANTS, LABOUR & FINANCE

JULIAN OWEN ASSOCIATES ARCHITECTS
6 Cumberland Avenue, Beeston,
Nottingham, Nottinghamshire, NG9 4DH
Area of Operation: Midlands & Mid Wales
Tel: 0115 922 9831
Email: julian@julianowen.co.uk
Web: www.julianowen.co.uk

LEEDS ENVIRONMENTAL DESIGN ASSOCIATES LTD
Micklethwaite House, 70 Cross Green Lane,
Leeds, West Yorkshire, LS9 0DG
Area of Operation: North East England
Tel: 0113 200 9380
Fax: 0113 200 9381
Email: office@leda.org.uk
Web: www.leda.org.uk

LINT DESIGN
2a Bury Lane, Codicote, Hertfordshire, SG4 8XT
Area of Operation: East England,
South East England
Tel: 01438 822064
Fax: 0870 432 0685
Email: mail@lintdesign.co.uk
Web: www.lintdesign.co.uk

LOREN DESIGN LTD
Unit 6, 51 Derbyshire Street, London, E2 6JQ
Area of Operation: East England,
Greater London, South East England
Tel: 0207 729 4878
Fax: 0207 729 6033
Email: lorendes@aol.com

MARMOT ASSOCIATES
Higher Tor Farm, Poundsgate,
Newton Abbot, Devon, TQ13 7PD
Area of Operation: South West England
and South Wales
Tel: 01364 631566
Fax: 01364 631556
Email: marmot-tor@zen.co.uk
Web: www.marmot-tor.com

MICHAEL RIGBY ASSOCIATES
15 Market Street,
Wotton Under Edge,
Gloucestershire, GL12 7AE
Area of Operation: UK & Ireland (Excluding Ireland)
Tel: 01453 521621
Fax: 01453 521681
Email: paula@521621.com

MJW ARCHITECTS
The Old Chapel, Mendip Road,
Stoke St Michael, Somerset, BA3 5JU
Area of Operation: UK & Ireland
Tel: 01749 840180
Fax: 01749 841380
Email: info@mjwarchitects.com
Web: www.mjwarchitects.com

MOLE ARCHITECTS
The Black House, Kingdon Avenue,
Prickwillow, Ely, Cambridgeshire, CB7 4UL
Area of Operation: East England
Tel: 01353 688287
Fax: 01353 688287
Email: studio@molearchitects.co.uk
Web: www.molearchitects.co.uk

NICHOLAS RAY ASSOCIATES
13-15 Convert Garden, Cambridge,
Cambridgeshire, CB1 2HS
Area of Operation: UK & Ireland
Tel: 01223 464455
Email: design@nray-arch.co.uk
Web: www.nray-arch.co.uk

NICOLAS TYE ARCHITECTS
The Long Barn Studio,
Limbersey Lane,
Bedfordshire, MK45 2EA
Area of Operation: UK (Excluding Ireland)
Tel: 01525 406677
Fax: 01525 406688
Email: info@nicolastyearchitects.co.uk
Web: www.nicolastyearchitects.co.uk

NINETTE EDWARDS ARCHITECT
12 Alnside, Whittingham, Northumberland, NE66 4SJ
Area of Operation: North East England
Tel: 01655 574733
Fax: 01491 680384
Email: ninette@alnsideassociates.co.uk
Web: www.alnsideassociates.co.uk

NORDSTROM ASSOCIATES
32 Oswald Road, St Albans, Hertfordshire, AL1 3AQ
Area of Operation: Greater London
Tel: 01727 831971
Fax: 01727 752981
Email: gunnar.nordstrom@ntlworld.com

NTARCHITECTS
9 Cumberland Lodge, Cumberland Road,
Brighton, East Sussex, BN1 6ST
Area of Operation: Greater London, South East
England, South West England and South Wales
Tel: 01273 267184
Fax: 01273 267184
Email: nt@nicolathomas.co.uk
Web: www.nicolathomas.co.uk

PATRICIA HEPPLE ARCHITECT
Rectory Cottage, Railway Hill, Starston,
Harleston, Norfolk, IP20 9NJ
Area of Operation: East England
Tel: 01379 853871

PAUL A STOWELL
'Sorriso', Farleigh Road, Cliddesden,
Basingstoke, Hampshire , RG25 2JL
Area of Operation: Greater London, South East
England, South West England and South Wales
Tel: 01256 320470
Fax: 01256 320470
Email: paulstowell@constructionplans.co.uk
Web: www.constructionplans.co.uk

PETER BLOCKLEY - CHARTERED ARCHITECT
26 High Street, Worton, Devizes, Wiltshire, SN10 5RU
Area of Operation: Greater London, South West
England and South Wales
Tel: 01380 739394
Fax: 01380 739395
Email: peter@pbca.co.uk
Web: www.pbca.co.uk

R E DESIGN
97 Lincoln Avenue, Glasgow, G133DH
Area of Operation: Scotland
Tel: 0141 959 1902
Fax: 0141 959 1902
Email: mail@r-e-design.co.uk
Web: www.r-e-design.co.uk

ROMAN PROJECTS LTD
Roman Heights, Llanfair Hill, Llandovery,
Carmarthenshire, SA20 0YF
Area of Operation: UK (Excluding Ireland)
Tel: 01550 720533 **Fax:** 01550 720 533
Email: romanprojects@aol.com

RUMBALL SEDGWICK
Abbotts House, 198 Lower High Street,
Watford, Hertfordshire, WD17 2FF
Area of Operation: Greater London,
South East England
Tel: 01923 224275 **Fax:** 01923 255005
Email: shaun@w.rumballsedgwick.co.uk
Web: www.rumballsedgwick.co.uk

SARAH ROBERTS ARCHITECTS LTD
3 Church Lane, Bressingham, Diss, Norfolk, IP22 2AE
Area of Operation: UK (Excluding Ireland)
Tel: 01379 688135 **Fax:** 01379 687642
Email: sarah@sr-architects.co.uk
Web: www.sarahrobertsarchitects.co.uk

SIERRA DESIGNS, ARCHITECTURAL AND BUILDING CONSULTANTS
Carfeld House, Pinfold Lane, Kirk Smeaton,
North Yorkshire, WF8 3JT
Area of Operation: UK & Ireland
Tel: 01977 621360 **Fax:** 01977 621365
Email: info@sierradesigns.co.uk
Web: www.sierradesigns.co.uk

SIMON J CUSHING CHARTERED ARCHITECT
102 Woodchurch Road, Birkenhead,
Wirral, Merseyside, CH42 9LP
Area of Operation: North West England
and North Wales
Tel: 0151 653 9900
Fax: 0151 653 9797
Email: simonjcushing@btconnect.com

SNP ASSOCIATES
248 Kingston Road, New Malden, Surrey, KT3 3RN
Area of Operation: Worldwide
Tel: 0208 942 6238
Email: snp@snpassociates.co.uk
Web: www.snpassociates.co.uk

SOUTHPOINT
45 The Dell, Westbury-on-Trym, Bristol, BS9 3UF
Area of Operation: UK (Excluding Ireland)
Tel: 0845 644 6639
Fax: 0870 706 1866
Email: mail@southpoint.co.uk
Web: www.southpoint.co.uk

SUSTAINABLE PROPERTY GROUP LTD.
Courtyard Studio, Cowley Farm, Aylesbury Road,
Cuddington, Buckinghamshire, HP18 0AD
Area of Operation: Greater London,
Midlands & Mid Wales, South East England
Tel: 01296 747121
Fax: 01296 747703
Email: info@sustainable.uk.com
Web: www.sustainable.uk.com

THE BILLINGTON CONSULTANCY LTD
Unit 2A, Station Road, Sandbach,
Cheshire, CW11 3JG
Area of Operation: UK (Excluding Ireland)
Tel: 0800 716703
Web: www.billington.co.uk

TRADA TECHNOLOGY
Stocking Lane, Hughenden Valley,
High Wycombe, Buckinghamshire, HP14 4ND
Area of Operation: Worldwide
Tel: 01494 569600
Fax: 01494 565487
Email: rscott@trada.co.uk
Web: www.trada.co.uk

ARCHITECTURAL TECHNOLOGISTS

A1 MASTERPLANS LTD
Dulford Business Park, Dulford,
Cullompton, Devon, EX15 2DY
Area of Operation: UK (Excluding Ireland)
Tel: 01884 266800
Fax: 01884 266882
Email: contact@masterplansltd.co.uk
Web: www.masterplansltd.co.uk

ACROPOLIS DESIGN LTD
Ideas House, 98 Bradford Road, East Ardsley,
Wakefield, West Yorkshire, WF3 2JL
Area of Operation: UK (Excluding Ireland)
Tel: 01924 870880
Fax: 0871 733 4738
Email: info@acropolisdesign.co.uk
Web: www.acropolisdesign.co.uk

ARCHITECTURE PLUS
5 Dunkery Road, Weston Super Mare,
Somerset, BS23 2TD
Area of Operation: UK (Excluding Ireland)
Tel: 01934 416416
Fax: 01934 622583
Email: office@architecture-plus.co.uk
Web: www.architectureplus.co.uk

ARCHITECTURE VERTE LIMITED
41 Devonshire Buildings,
Bath, Somerset, BA2 4SU
Area of Operation: Greater London, Midlands
& Mid Wales, South East England, South West
England and South Wales
Tel: 01225 484835
Email: paul@verte.co.uk

ASAP- ARCHITECTURAL SERVICES & PLANNING
Howard Buildings, 69-71, Burpham Lane,
Guildford, Surrey, GU4 7LX
Area of Operation: South East England
Tel: 01483 457922
Email: david.haines4@virgin.net
Web: www.asaparchitectural.co.uk

COLIN WILLIAMS BUILDING CONSULTANCY
Courtyard Studio, 52A Dereham Road,
Mattishall, Dereham, Norfolk, NR20 3NS
Area of Operation: East England
Tel: 01362 850171 **Fax:** 01362 850171
Email: design@cooptel.net

CO-ORDINATED CONSTRUCTION SERVICES
18 Kiln Road, Crawley Down,
West Sussex, RH10 4JY
Area of Operation: South East England
Tel: 01342 714511 **Fax:** 01342 714511
Email: info@coordsvcs.com
Web: www.coordsvcs.com

DESIGNER HOMES
Pooh Cottage, Minto, Hawick,
Roxburghshire, Borders, TD9 8SB
Area of Operation: UK & Ireland
Tel: 01450 870127 **Fax:** 01450 870127

ECLIPSE DESIGN
Staunton Harold Hall, Melbourne Road,
Staunton Harold, Leicestershire, LE65 1RT
Area of Operation: UK (Excluding Ireland)
Tel: 0870 460 4758
Email: valton@valton.freeserve.co.uk
Web: www.eclipsedesign.copperstream.co.uk

ENCRAFT
46 Northumberland Road, Leamington Spa,
Warwickshire, CV32 6HB
Area of Operation: UK & Ireland
Tel: 01926 312159 **Fax:** 01926 772480
Email: enquiries@encraft.co.uk
Web: www.encrafthome.co.uk

ENGLISHAUS CHARTERED ARCHITECTS LTD
30 Lawrence Road, Hampton, Middlesex, TW12 2RJ
Area of Operation: Greater London, South East
England, South West England and South Wales
Tel: 0208 255 0595
Fax: 0208 287 3441
Email: enquiries@englishaus.co.uk
Web: www.englishaus.co.uk

FG DESIGN
6 Denmark Drive, Sedbury, Chepstow,
Monmouthshire, NT16 7BD
Area of Operation: South West England
and South Wales
Tel: 01291 624366
Fax: 01291 624366
Email: frank.fgdesign@virgin.net

J G HOOD & ASSOCIATES LTD
50 High Street, Bruton, Somerset, BA10 0AN
Area of Operation: Midlands & Mid Wales, South
East England, South West England and South Wales
Tel: 01749 812139
Fax: 0870 133 3776
Email: mail@jghood-assoc.co.uk
Web: www.jghood-assoc.co.uk

JNA ARCHITECTS
9 East Park, Southgate, West Sussex, RH10 6AN
Area of Operation: South East England
Tel: 01293 439323
Fax: 01293 530160
Email: james@jna-architects.co.uk
Web: www.jna-architects.co.uk

JOHN PEATE ARCHITECTURAL SERVICES LTD
6 Newport Rd, Shifnal, Shropshire, TF11 8BP
Area of Operation: Midlands & Mid Wales,
North West England and North Wales
Tel: 01952 460175
Fax: 01952 460175
Email: ray.peate@btinternet.com
Web: www.johnpeate.org.uk

LARNER SING
29 Lower Street, Rode, Frome,
Somerset, BA11 6PU
Area of Operation: South East England,
South West England and South Wales
Tel: 01373 830527
Fax: 01373 830527
Email: ian@larner-sing.co.uk
Web: www.larner-sing.com

M J BRAIN
Kamala House, North Lane,
Weston On The Green, Bicester,
Oxfordshire, OX25 3RG
Area of Operation: UK (Excluding Ireland)
Tel: 01869 350771
Fax: 01869 351445
Email: malcolmbrain@mjbrain.freeserve.co.uk

MICHAEL D HALL
Studio A, 339 London Road, Bexhill on Sea,
East Sussex, TN39 4AJ
Area of Operation: South East England
Tel: 01424 214541
Fax: 01424 731555
Email: bds@michaeldhall.co.uk
Web: www.michaeldhall.co.uk

MORNINGTIDE DEVELOPMENTS LTD
Beauvale, Loamy Hill, Tolleshunt Major,
Essex, CM9 8LS
Area of Operation: UK (Excluding Ireland)
Tel: 01621 815485
Fax: 01621 819511
Email: morningtide@fsnet.co.uk
Web: www.morningtide.fsnet.co.uk

NICHOLAS RAY ASSOCIATES
13-15 Convert Garden, Cambridge,
Cambridgeshire, CB1 2HS
Area of Operation: UK & Ireland
Tel: 01223 464455
Email: design@nray-arch.co.uk
Web: www.nray-arch.co.uk

NICHOLSON DESIGN CONSULTANTS
1B Knightsbrook, 10 Grassington Road,
Eastbourne, East Sussex, BN20 7BP
Area of Operation: Europe
Tel: 01323 734230
Fax: 01323 731283
Email: nicholson.design@dial.pipex.com

R E DESIGN
97 Lincoln Avenue, Glasgow, G133DH
Area of Operation: Scotland
Tel: 0141 959 1902
Fax: 0141 959 1902
Email: mail@r-e-design.co.uk
Web: www.r-e-design.co.uk

**SIERRA DESIGNS, ARCHITECTURAL
& BUILDING CONSULTANTS**
Carfeld House, Pinfold Lane, Kirk Smeaton,
North Yorkshire, WF8 3JT
Area of Operation: UK & Ireland
Tel: 01977 621360
Fax: 01977 621365
Email: info@sierradesigns.co.uk
Web: www.sierradesigns.co.uk

THE STEVEN BARLOW PARTNERSHIP
81 Manor Road, Kingston,
Portsmouth, Hampshire, PO1 5LB
Area of Operation: UK (Excluding Ireland)
Tel: 07786 577416
Email: svbarlow@hotmail.com

BUILDING SURVEYORS

A1 MASTERPLANS LTD
Dulford Business Park, Dulford,
Cullompton, Devon, EX15 2DY
Area of Operation: UK (Excluding Ireland)
Tel: 01884 266800
Fax: 01884 266882
Email: contact@masterplansltd.co.uk
Web: www.masterplansltd.co.uk

BUILDING PLANS LTD
Unit 10 Beech Avenue, Taverham, Norwich,
Norfolk, NR8 6HW
Area of Operation: UK (Excluding Ireland)
Tel: 01603 868377
Fax: 01603 868412
Email: john@constructionhelp.co.uk
Web: www.constructionhelp.co.uk

CONSTRUCTION COST SOLUTIONS
September Cottage, 1A Wigwam Close,
Poynton, Stockport, Cheshire, SK12 1XF
Area of Operation: Midlands & Mid Wales, North
East England, North West England and North Wales
Tel: 01625 875488
Fax: 01625 878143
Email: david@constructioncostsolutions.co.uk
Web: www.constructioncostsolutions.co.uk

DESIGN PLUS (KENT) LTD
59 Marshall Crescent, Broadstairs, Kent, CT10 2HR
Area of Operation: UK (Excluding Ireland)
Tel: 01843 602218
Email: designpluskent@btinternet.com
Web: www.designpluskent.co.uk

FG DESIGN
6 Denmark Drive, Sedbury, Chepstow,
Monmouthshire, NT16 7BD
Area of Operation: South West England
and South Wales
Tel: 01291 624366
Fax: 01291 624366
Email: frank.fgdesign@virgin.net

GRAHAM G BISHOP - CHARTERED SURVEYORS
9 Church Lane Drive, Coulsdon, Surrey, CR5 3RG
Area of Operation: Greater London, South East
England
Tel: 01737 558473
Fax: 01737 558473
Email: graham@grahamgbishop.co.uk
Web: www.grahamgbishop.co.uk

GRINDEY CONSULTING
5 Granville Street, Monton,
Manchester, M30 9PX
Area of Operation: North West England
and North Wales
Tel: 07773 371488
Email: ian.grindey@scottwilson.com
Web: www.grindeyconsulting.co.uk

**HUTTON+ROSTRON ENVIRONMENTAL
INVESTIGATIONS LTD**
Netley House, Gomshall,
Guildford, Surrey, GU5 9QA
Area of Operation: UK & Ireland
Tel: 01483 203221
Fax: 01483 202911
Email: ei@handr.co.uk
Web: www.handr.co.uk

J.S. BUILDING CONSULTANCY
53 Hawthorn Road, Yeadon, Leeds,
West Yorkshire, LS19 7UT
Area of Operation: UK (Excluding Ireland)
Tel: 0113 250 1303
Email: jsharples@ricsonline.org
Web: www.ukbuildingconsultancy.co.uk

LARNER SING
29 Lower Street, Rode, Frome,
Somerset, BA11 6PU
Area of Operation: South East England,
South West England and South Wales
Tel: 01373 830527
Fax: 01373 830527
Email: ian@larner-sing.co.uk
Web: www.larner-sing.com

M J BRAIN
Kamala House, North Lane,
Weston On The Green, Bicester,
Oxfordshire, OX25 3RG
Area of Operation: UK (Excluding Ireland)
Tel: 01869 350771
Fax: 01869 351445
Email: malcolmbrain@mjbrain.freeserve.co.uk

PAUL A STOWELL
'Sorriso', Farleigh Road, Cliddesden,
Basingstoke, Hampshire, RG25 2JL
Area of Operation: Greater London, South East
England, South West England and South Wales
Tel: 01256 320470
Fax: 01256 320470
Email: paulstowell@constructionplans.co.uk
Web: www.constructionplans.co.uk

PROPERTYPORTAL.COM
Brookfield House, Ford Lane, Frilford,
Abingdon, Oxfordshire, OX13 5NT
Area of Operation: UK (Excluding Ireland)
Tel: 0870 380 0520
Fax: 0870 380 0521
Email: info@propertyportal.com
Web: www.propertyportal.com

RENAISSANCE ARCHITECTURAL LTD
15 Gay Street, Bath, BA1 2PH
Area of Operation: East England,
Greater London, Midlands, North East England,
North West England, South East England,
South West England
Tel: 01761 479605
Fax: 01761 479605
Email: renaissance@ren.uk.com
Web: www.ren.uk.com

RUMBALL SEDGWICK
Abbotts House, 198 Lower High Street,
Watford, Hertfordshire, WD17 2FF
Area of Operation: Greater London,
South East England
Tel: 01923 224275
Fax: 01923 255005
Email: shaun@w.rumballsedgwick.co.uk
Web: www.rumballsedgwick.co.uk

SELF BUILD PRO
Belmont Business Centre, Brook Lane,
Endon, Staffordshire, ST9 9EZ
Area of Operation: UK & Ireland
Tel: 01782 505127
Fax: 01782 505127
Email: enquiries@self-build-pro.co.uk
Web: www.self-build-pro.co.uk

**SIERRA DESIGNS, ARCHITECTURAL
AND BUILDING CONSULTANTS**
Carfeld House, Pinfold Lane, Kirk Smeaton,
North Yorkshire, WF8 3JT
Area of Operation: UK & Ireland
Tel: 01977 621360
Fax: 01977 621365
Email: info@sierradesigns.co.uk
Web: www.sierradesigns.co.uk

TRADA TECHNOLOGY
Stocking Lane, Hughenden Valley,
High Wycombe, Buckinghamshire, HP14 4ND
Area of Operation: Worldwide
Tel: 01494 569600
Fax: 01494 565487
Email: rscott@trada.co.uk
Web: www.trada.co.uk

WISE PROPERTY CARE
8 Muriel Street, Barrhead, Glasgow, G78 1QB
Area of Operation: Scotland
Tel: 0141 876 0300
Fax: 0141 876 0301
Email: les@wisepropertycare.com
Web: www.wisepropertycare.com

STRUCTURAL
AND CIVIL ENGINEERS

A1 MASTERPLANS LTD
Dulford Business Park, Dulford,
Cullompton, Devon, EX15 2DY
Area of Operation: UK (Excluding Ireland)
Tel: 01884 266800
Fax: 01884 266882
Email: contact@masterplansltd.co.uk
Web: www.masterplansltd.co.uk

BILLINGTON CONSULTANCY LTD
Unit 2a, Station Yard, Station Road,
Sandbach, Cheshire, CW11 3JG
Area of Operation: UK (Excluding Ireland)
Tel: 0800 716 703
Email: chrisb@billington.co.uk
Web: www.billington.co.uk

CONCEPT ENGINEERING CONSULTANTS LTD
Unit 8 Warple Mews,
Warple Way, London, W3 0RF
Area of Operation: East England, Greater London,
South East England, South West England
and South Wales
Tel: 0208 840 3321
Fax: 0208 579 6910
Email: si@conceptconsultants.co.uk
Web: www.conceptconsultants.co.uk

CO-ORDINATED CONSTRUCTION SERVICES
18 Kiln Road, Crawley Down,
West Sussex, RH10 4JY
Area of Operation: South East England
Tel: 01342 714511
Fax: 01342 714511
Email: info@coordsvcs.com
Web: www.coordsvcs.com

EDGBASTON DESIGN LIMITED
24 Stirling Road, Birmingham,
West Midlands, B16 9BG
Area of Operation: Midlands & Mid Wales
Tel: 07775 580404
Email: rannett@constructionplus.net

FINDANENGINEER.COM
11 Upper Belgrave Street, London, SW1X 8BH
Area of Operation: UK (Excluding Ireland)
Tel: 0207 235 4535
Fax: 0207 235 4294
Email: mail@istructe.org.uk
Web: www.findanengineer.com

FINDANENGINEER.COM

Area of Operation: UK
Tel: 020 7235 4535
Email: mail@istructe.org.uk
Web: www.Findanengineer.com

Findanengineer.com contains details of structural engineering companies across the UK. All companies on Findanengineer.com have at least one professional member of the Institution of Structural Engineers working for them, which means you can trust the reputation and quality of the companies your search generates.

GRINDEY CONSULTING
5 Granville Street, Monton,
Manchester, M30 9PX
Area of Operation: North West England
and North Wales
Tel: 07773 371488
Email: ian.grindey@scottwilson.com
Web: www.grindeyconsulting.co.uk

JNA ARCHITECTS
9 East Park, Southgate,
West Sussex, RH10 6AN
Area of Operation: South East England
Tel: 01293 439323
Fax: 01293 530160
Email: james@jna-architects.co.uk
Web: www.jna-architects.co.uk

CONSULTANTS, LABOUR & FINANCE

R E DESIGN
97 Lincoln Avenue, Glasgow, G133DH
Area of Operation: Scotland
Tel: 0141 959 1902 **Fax:** 0141 959 1902
Email: mail@r-e-design.co.uk
Web: www.r-e-design.co.uk

ROCKBOURNE ENVIRONMENTAL
6 Silver Business Park, Airfield Way,
Christchurch, Dorset, BH23 3TA
Area of Operation: UK & Ireland
Tel: 01202 480980 **Fax:** 01202 490590
Email: info@rockbourne.net
Web: www.rockbourne.net

SELF BUILD PRO
Belmont Business Centre, Brook Lane,
Endon, Staffordshire, ST9 9EZ
Area of Operation: UK & Ireland
Tel: 01782 505127
Fax: 01782 505127
Email: enquiries@self-build-pro.co.uk
Web: www.self-build-pro.co.uk

TRADA TECHNOLOGY
Stocking Lane, Hughenden Valley,
High Wycombe, Buckinghamshire, HP14 4ND
Area of Operation: Worldwide
Tel: 01494 569600 **Fax:** 01494 565487
Email: rscott@trada.co.uk
Web: www.trada.co.uk

ENVIRONMENTAL CONSULTANTS

BRENNAN & WILSON ARCHITECTS
Belhaven Villa, Edinburgh Road, Belhaven,
Dunbar, Lothian, EH42 1PA
Area of Operation: North East England, Scotland
Tel: 01368 860897
Fax: 01368 860897
Email: mail@bwarchitects.co.uk
Web: www.bwarchitects.co.uk

CA SUSTAINABLE ARCHITECTURE
83 Old Newtown Road, Newbury,
Berkshire, RG14 7DE
Area of Operation: UK (Excluding Ireland)
Tel: 01635 48363
Email: isabel.carmona@ca-sa.co.uk
Web: www.ca-sa.co.uk

CENTRE FOR ALTERNATIVE TECHNOLOGY
Llwyngweren Quarry, Machynlleth,
Powys, SY20 9AZ
Area of Operation: Worldwide
Tel: 01654 705950
Fax: 01654 702782
Email: lucy.stone@cat.org.uk
Web: www.cat.org.uk

CONCEPT ENGINEERING CONSULTANTS LTD
Unit 8 Warple Mews, Warple Way, London, W3 0RF
Area of Operation: East England, Greater London,
South East England, South West England
and South Wales
Tel: 0208 840 3321
Fax: 0208 579 6910
Email: si@conceptconsultants.co.uk
Web: www.conceptconsultants.co.uk

CONSERVATION ENGINEERING LTD
The Street, Troston, Bury St Edmunds,
Suffolk, IP31 1EW
Area of Operation: UK & Ireland
Tel: 01359 269360
Fax: 01359 268340
Email: anne@conservation-engineering.co.uk
Web: www.heating-designs.co.uk

ECOSHOP
Unit 1, Glen of the Downs Garden Centre,
Kilmacanogue, Co Wicklow, Republic of Ireland
Area of Operation: Ireland Only
Tel: +353 012 872 914
Fax: +353 012 016 480
Email: info@ecoshop.ie
Web: www.ecoshop.ie

ENCRAFT
46 Northumberland Road,
Leamington Spa, Warwickshire, CV32 6HB
Area of Operation: UK & Ireland
Tel: 01926 312159
Fax: 01926 772480
Email: enquiries@encraft.co.uk
Web: www.encrafthome.co.uk

ENERGY CHECK
16 Morgan Street, Blaenavon, Torfaen, NP4 9ER
Area of Operation: UK (Excluding Ireland)
Tel: 01495 791406
Fax: 01495 791406
Email: office@sap-rating.com
Web: www.sap-rating.com

ENERGY MASTER
Keltic Business Park,
Unit 1 Clieveragh Industrial Estate,
Listowel, Ireland
Area of Operation: Ireland Only
Tel: +353 682 4300
Fax: +353 682 4533
Email: info@energymaster.ie
Web: www.energymaster.ie

FREERAIN
Millennium Green Business Centre,
Rio Drive, Collingham,
Nottinghamshire, NG23 7NB
Area of Operation: UK & Ireland
Tel: 01636 894900
Fax: 01636 894909
Email: info@freerain.co.uk
Web: www.freerain.co.uk

GEO THERM LTD
PO Box 16, Lowestoft, Suffolk, NR33 8RU
Area of Operation: Europe
Tel: 01502 515707
Fax: 01502 530275
Email: info@geo-centric.net
Web: www.geo-centric.net

**HUTTON+ROSTRON
ENVIRONMENTAL INVESTIGATIONS LTD**
Netley House, Gomshall,
Guildford, Surrey, GU5 9QA
Area of Operation: UK & Ireland
Tel: 01483 203221
Fax: 01483 202911
Email: ei@handr.co.uk
Web: www.handr.co.uk

IAIN FREARSON
68 Cavendish Road, Cambridge,
Cambridgeshire, CB1 3AF
Area of Operation: UK (Excluding Ireland)
Tel: 01223 473997
Fax: 01223 473997
Email: iain.frearson@ntlworld.com

**LEEDS ENVIRONMENTAL
DESIGN ASSOCIATES LTD**
Micklethwaite House, 70 Cross Green Lane,
Leeds, West Yorkshire, LS9 0DG
Area of Operation: North East England
Tel: 0113 200 9380
Fax: 0113 200 9381
Email: office@leda.org.uk
Web: www.leda.org.uk

LOW-IMPACT LIVING INITIATIVE
Redfield Community, Buckingham Road,
Winslow, Buckinghamshire, MK18 3LZ
Area of Operation: UK (Excluding Ireland)
Tel: 01296 714184
Fax: 01296 714184
Email: lili@lowimpact.org
Web: www.lowimpact.org

MINI SOIL SURVEYS LTD
Viking House, Manchester Road,
Bolton, Lancashire, BL2 1DU
Area of Operation: UK & Ireland
Tel: 01204 386661 **Fax:** 01204 386611
Email: peter@minisoils.co.uk
Web: www.minisoils.co.uk

ROCKBOURNE ENVIRONMENTAL
6 Silver Business Park, Airfield Way,
Christchurch, Dorset, BH23 3TA
Area of Operation: UK & Ireland
Tel: 01202 480980
Fax: 01202 490590
Email: info@rockbourne.net
Web: www.rockbourne.net

**SEVENOAKS ENVIRONMENTAL
CONSULTANCY LTD**
19 Gimble Way, Pembury,
Tunbridge Wells, Kent, TN2 4BX
Area of Operation: East England,
Greater London, Midlands & Mid Wales,
North East England, North West England
and North Wales, South East England,
South West England and South Wales
Tel: 01892 822999
Fax: 01892 822992
Email: d.jones@sevenoaksenvironmental.co.uk
Web: www.sevenoaksenvironmental.co.uk

SUSTAINABLE PROPERTY GROUP LTD.
Courtyard Studio, Cowley Farm, Aylesbury Road,
Cuddington, Buckinghamshire, HP18 0AD
Area of Operation: Greater London, Midlands
& Mid Wales, South East England
Tel: 01296 747121
Fax: 01296 747703
Email: info@sustainable.uk.com
Web: www.sustainable.uk.com

TERRAIN AERATION SERVICES
Aeration House, 20 Mill Fields, Haughley,
Stowmarket, Suffolk, IP14 3PU
Area of Operation: Europe
Tel: 01449 673783
Fax: 01449 614564
Email: terrainaeration@aol.com
Web: www.terrainaeration.co.uk

THE NATIONAL ENERGY FOUNDATION
The National Energy Centre, Davy Avenue, Knowlhill,
Milton Keynes, Buckinghamshire, MK5 8NG
Area of Operation: UK (Excluding Ireland)
Tel: 01908 665555
Fax: 01908 665577
Email: info@nef.org.uk
Web: www.nef.org.uk

WOOD FOR GOOD
211 High Road, London, N2 8AN
Area of Operation: UK (Excluding Ireland)
Tel: 0800 279 0016
Fax: 0208 883 6700
Email: info@woodforgood.com
Web: www.woodforgood.com

FINANCIAL ADVISORS & VAT SPECIALISTS

BNB TAX CONSULTANTS
Union Chambers, 63 Temple Row,
Birmingham, West Midlands, B2 5LS
Area of Operation: Europe
Tel: 0121 483 6850
Fax: 0121 483 6851
Email: steve.botham@bnbtax.com
Web: www.bnbtax.com

MICHAEL J. FLINT
16 Paynsbridge Way, Horam, Heathfield,
East Sussex, TN21 0HQ
Area of Operation: UK (Excluding Ireland)
Tel: 01435 813360
Fax: 01435 813279
Email: mjfvat@btinternet.com
Web: www.mjfvat.btinternet.co.uk

NCB ASSOCIATES LTD
Gate House, Fretheren Road,
Welwyn Garden City, Hertfordshire, AL8 6NS
Area of Operation: UK (Excluding Ireland)
Tel: 0845 230 9897
Email: enquiry@ncbassociates.co.uk
Web: www.ncbassociates.co.uk

THE VAT CONSULTANCY
Laurel House, Station Approach,
Alresford, Hampshire, SO24 9JH
Area of Operation: UK (Excluding Ireland)
Tel: 01962 735350
Fax: 01962 735352
Email: vat@thevatconsultancy.com
Web: www.thevatconsultancy.com

HOUSE DESIGNERS

A1 MASTERPLANS LTD
Dulford Business Park, Dulford,
Cullompton, Devon, EX15 2DY
Area of Operation: UK (Excluding Ireland)
Tel: 01884 266800
Fax: 01884 266882
Email: contact@masterplansltd.co.uk
Web: www.masterplansltd.co.uk

ACROPOLIS DESIGN LTD
Ideas House, 98 Bradford Road, East Ardsley,
Wakefield, West Yorkshire, WF3 2JL
Area of Operation: UK (Excluding Ireland)
Tel: 01924 870880
Fax: 0871 733 4738
Email: info@acropolisdesign.co.uk
Web: www.acropolisdesign.co.uk

ADS UK LTD
30 Wild Wood Court, Hawkhirst Road,
Kenley, Surrey, CR8 5DL
Area of Operation: Worldwide
Tel: 0208 660 2637
Fax: 0208 660 2642
Email: agn.hughes@virgin.net

ARCHITECTURE PLUS
5 Dunkery Road, Weston Super Mare,
Somerset, BS23 2TD
Area of Operation: UK (Excluding Ireland)
Tel: 01934 416416
Fax: 01934 622485
Email: office@architectureplus.co.uk
Web: www.architectureplus.co.uk

ARCHITECTURE VERTE LIMITED
41 Devonshire Buildings, Bath,
Somerset, BA2 4SU
Area of Operation: Greater London,
Midlands & Mid Wales, South East England,
South West England and South Wales
Tel: 01225 484835
Email: paul@verte.co.uk

**ASAP- ARCHITECTURAL SERVICES
& PLANNING**
Howard Buildings, 69-71, Burpham Lane,
Guildford, Surrey, GU4 7LX
Area of Operation: South East England
Tel: 01483 457922
Email: david.haines4@virgin.net
Web: www.asaparchitectural.co.uk

ASBA SCOTLAND
10 Lynedoch Cresent, Glasgow, G3 6EQ
Area of Operation: Scotland
Tel: 0800 731 3405
Fax: 0141 331 2751
Email: architects@design-practice.com

BACK TO FRONT EXTERIOR DESIGN
37 West Street, Farnham, Surrey, GU9 7DR
Area of Operation: UK & Ireland
Tel: 01252 820984
Fax: 01252 821907
Email: design@backtofrontexteriordesign.com
Web: www.backtofrontexteriordesign.com

BARLEY WEST LIMITED
22 Virginia Place, Cobham,
Surrey, KT11 1AE
Area of Operation: Greater London,
South East England, South West England
and South Wales
Tel: 01932 865459
Email: info@barleywest.co.uk
Web: www.barleywest.co.uk

BILLINGTON CONSULTANCY LTD
Unit 2a, Station Yard, Station Road,
Sandbach, Cheshire, CW11 3JG
Area of Operation: UK (Excluding Ireland)
Tel: 0800 716703
Email: chrisb@billington.co.uk
Web: www.billington.co.uk

BRENNAN & WILSON ARCHITECTS
Belhaven Villa, Edinburgh Road, Belhaven,
Dunbar, Lothian, EH42 1PA
Area of Operation: North East England, Scotland
Tel: 01368 860897
Fax: 01368 860897
Email: mail@bwarchitects.co.uk
Web: www.bwarchitects.co.uk

BUILDING PLANS LTD
Unit 10 Beech Avenue, Taverham,
Norwich, Norfolk, NR8 6HW
Area of Operation: UK (Excluding Ireland)
Tel: 01603 868377
Fax: 01603 868412
Email: john@constructionhelp.co.uk
Web: www.constructionhelp.co.uk

CA SUSTAINABLE ARCHITECTURE
83 Old Newtown Road, Newbury,
Berkshire, RG14 7DE
Area of Operation: UK (Excluding Ireland)
Tel: 01635 48363
Email: isabel.carmona@ca-sa.co.uk
Web: www.ca-sa.co.uk

CENTRE FOR ALTERNATIVE TECHNOLOGY
Llwyngweren Quarry, Machynlleth,
Powys, SY20 9AZ
Area of Operation: Worldwide
Tel: 01654 705950
Fax: 01654 702782
Email: lucy.stone@cat.org.uk
Web: www.cat.org.uk

**CHRISTOPHER MAGUIRE ARCHITECT
AND GARDEN DESIGNER**
15 Harston Road, Newton,
Cambridge, Cambridgeshire, CB2 5PA
Area of Operation: East England
Tel: 01223 872800
Web: www.christophermaguire.co.uk

COLIN WILLIAMS BUILDING CONSULTANCY
Courtyard Studio, 52A Dereham Road,
Mattishall, Dereham, Norfolk, NR20 3NS
Area of Operation: East England
Tel: 01362 850171
Fax: 01362 850171
Email: design@cooptel.net

CONSTRUCTIVE INDIVIDUALS
Trinity Buoy Wharf, 64 Orchard Place,
London, E14 0JW
Area of Operation: UK (Excluding Ireland)
Tel: 0207 515 9299 **Fax:** 0207 515 9737
Email: info@constructiveindividuals.com
Web: www.constructiveindividuals.com

CO-ORDINATED CONSTRUCTION SERVICES
18 Kiln Road, Crawley Down,
West Sussex, RH10 4JY
Area of Operation: South East England
Tel: 01342 714511 **Fax:** 01342 714511
Email: info@coordsvcs.com
Web: www.coordsvcs.com

DESIGN AND MATERIALS LTD
Lawn Road, Carlton in Lindrick,
Nottinghamshire, S81 9LB
Area of Operation: UK & Ireland
Tel: 01909 730333 **Fax:** 01909 730605
Email: enquiries@designandmaterials.uk.com
Web: www.designandmaterials.uk.com

DESIGN PLUS (KENT) LTD
59 Marshall Crescent, Broadstairs, Kent, CT10 2HR
Area of Operation: UK (Excluding Ireland)
Tel: 01843 602218
Email: designpluskent@btinternet.com
Web: www.designpluskent.co.uk

DESIGNER HOMES
Pooh Cottage, Minto, Hawick,
Roxburghshire, Borders, TD9 8SB
Area of Operation: UK & Ireland
Tel: 01450 870127
Fax: 01450 870127

ECLIPSE DESIGN
Staunton Harold Hall, Melbourne Road,
Staunton Harold, Leicestershire, LE65 1RT
Area of Operation: UK (Excluding Ireland)
Tel: 0870 4604758
Email: valton@valton.freeserve.co.uk
Web: www.eclipsedesign.copperstream.co.uk

ECO HOUSES
12 Lee Street, Louth, Lincolnshire, LN11 9HJ
Area of Operation: Worldwide
Tel: 0845 345 2049
Email: sd@jones-nash.co.uk
Web: www.eco-houses.co.uk

ECOLOGIC DESIGN
16 Popes Lane, Rockwell Green,
Wellington, Somerset, TA21 9DQ
Area of Operation: UK (Excluding Ireland)
Tel: 01823 666177
Fax: 01823 666177

ENGLISHAUS CHARTERED ARCHITECTS LTD
30 Lawrence Road, Hampton, Middlesex, TW12 2RJ
Area of Operation: Greater London, South East
England, South West England and South Wales
Tel: 0208 255 0595
Fax: 0208 287 3441
Email: enquiries@englishaus.co.uk
Web: www.englishaus.co.uk

FG DESIGN
6 Denmark Drive, Sedbury, Chepstow,
Monmouthshire, NT16 7BD
Area of Operation: South West England
and South Wales
Tel: 01291 624366
Fax: 01291 624366
Email: frank.fgdesign@virgin.net

GEORGE BLACK ARCHITECT
12 Kingsknowe Crescent, Edinburgh,
Lothian, EH14 2JZ
Area of Operation: Scotland
Tel: 0131 443 9898
Email: george@gbarch.abel.co.uk
Web: www.gbarch.abel.co.uk

**GRAHAM G BISHOP
- CHARTERED SURVEYORS**
9 Church Lane Drive, Coulsdon, Surrey, CR5 3RG
Area of Operation: Greater London,
South East England
Tel: 01737 558473
Fax: 01737 558473
Email: graham@grahamgbishop.co.uk
Web: www.grahamgbishop.co.uk

GRINDEY CONSULTING
5 Granville Street, Monton,
Manchester, M30 9PX
Area of Operation: North West England
and North Wales
Tel: 07773 371488
Email: ian.grindey@scottwilson.com
Web: www.grindeyconsulting.co.uk

HEMPHAB PRODUCTS
Rusheens, Ballygriffen,
Kenmare, Kerry, Ireland
Area of Operation: Ireland Only
Tel: +353 644 1747
Email: hempbuilding@eircom.net
Web: www.hempbuilding.com

IAIN FREARSON
68 Cavendish Road, Cambridge,
Cambridgeshire, CB1 3AF
Area of Operation: UK (Excluding Ireland)
Tel: 01223 473997
Fax: 01223 473997
Email: iain.frearson@ntlworld.com

IAN MONTGOMERY
Rolles Court, Church Whitfield,
Dover, Kent, CT16 3HY
Area of Operation: South East England
Tel: 01304 827487
Email: ian@homearranger.net
Web: www.homearranger.net

J G HOOD & ASSOCIATES LTD
50 High Street, Bruton, Somerset, BA10 0AN
Area of Operation: Midlands & Mid Wales, South
East England, South West England and South Wales
Tel: 01749 812139
Fax: 0870 133 3776
Email: mail@jghood-assoc.co.uk
Web: www.jghood-assoc.co.uk

J.S. BUILDING CONSULTANCY
53 Hawthorn Road, Yeadon, Leeds,
West Yorkshire, LS19 7UT
Area of Operation: UK (Excluding Ireland)
Tel: 0113 250 1303
Email: jsharples@ricsonline.org
Web: www.ukbuildingconsultancy.co.uk

JEREMY RAWLINGS PERIOD HOMES
Coombe Lee, Blackborough,
Cullompton, Devon, EX15 2HJ
Area of Operation: UK & Ireland
Tel: 01884 266444
Fax: 01884 266758
Email: jeremy.rawlings@btinternet.com
Web: www.periodhome.net

JNA ARCHITECTS
9 East Park, Southgate, West Sussex, RH10 6AN
Area of Operation: South East England
Tel: 01293 439323
Fax: 01293 530160
Email: james@jna-architects.co.uk
Web: www.jna-architects.co.uk

JOHN PEATE ARCHITECTURAL SERVICES LTD
6 Newport Rd, Shifnal, Shropshire, TF11 8BP
Area of Operation: Midlands & Mid Wales,
North West England and North Wales
Tel: 01952 460175 **Fax:** 01952 460175
Email: ray.peate@btinternet.com
Web: www.johnpeate.org.uk

JONATHAN PENTON DESIGN LTD
3 Church Road, Spratton, Northampton,
Northamptonshire, NN6 8HR
Area of Operation: UK (Excluding Ireland)
Tel: 01604 820824 **Fax:** 01604 820823
Email: jon@creative-team.com
Web: www.jonathanpentondesign.com

LARNER SING
29 Lower Street, Rode, Frome, Somerset, BA11 6PU
Area of Operation: South East England, South West
England and South Wales
Tel: 01373 830527
Fax: 01373 830527
Email: ian@larner-sing.co.uk
Web: www.larner-sing.com

M J BRAIN
Kamala House, North Lane, Weston On The Green,
Bicester, Oxfordshire, OX25 3RG
Area of Operation: UK (Excluding Ireland)
Tel: 01869 350771
Fax: 01869 351445
Email: malcolmbrain@mjbrain.freeserve.co.uk

MICHAEL D HALL
Studio A, 339 London Road,
Bexhill on Sea, East Sussex, TN39 4AJ
Area of Operation: South East England
Tel: 01424 214541
Fax: 01424 731555
Email: bds@michaeldhall.co.uk
Web: www.michaeldhall.co.uk

NICHOLAS RAY ASSOCIATES
13-15 Convert Garden, Cambridge,
Cambridgeshire, CB1 2HS
Area of Operation: UK & Ireland
Tel: 01223 464455
Email: design@nray-arch.co.uk
Web: www.nray-arch.co.uk

NICOLAS TYE ARCHITECTS
The Long Barn Studio, Limbersey Lane,
Bedfordshire, MK45 2EA
Area of Operation: UK (Excluding Ireland)
Tel: 01525 406677
Fax: 01525 406688
Email: info@nicolastyearchitects.co.uk
Web: www.nicolastyearchitects.co.uk

NORDSTROM ASSOCIATES
32 Oswald Road, St Albans, Hertfordshire, AL1 3AQ
Area of Operation: Greater London
Tel: 01727 831971
Fax: 01727 752981
Email: gunnar.nordstrom@ntlworld.com

**O'EVE DON INTERIORS
& THE KITCHEN PLANNER**
9 Maple Drive, Castle Park, Mallow, Co Cork
Area of Operation: Ireland Only
Tel: 022 20858
Fax: 022 20858
Email: oevedoninteriors@eircom.net

PAUL A STOWELL
'Sorriso', Farleigh Road, Cliddesden,
Basingstoke, Hampshire , RG25 2JL
Area of Operation: Greater London, South East
England, South West England and South Wales
Tel: 01256 320470
Fax: 01256 320470
Email: paulstowell@constructionplans.co.uk
Web: www.constructionplans.co.uk

R E DESIGN
97 Lincoln Avenue, Glasgow, G133DH
Area of Operation: Scotland
Tel: 0141 959 1902
Fax: 0141 959 1902
Email: mail@r-e-design.co.uk
Web: www.r-e-design.co.uk

RENAISSANCE ARCHITECTURAL LTD
15 Gay Street, Bath, BA1 2PH
Area of Operation: East England, Greater London,
Midlands, North East England, North West England,
South East England, South West England
Tel: 01761 479605
Fax: 01761 479605
Email: renaissance@ren.uk.com
Web: www.ren.uk.com

RUMBALL SEDGWICK
Abbotts House, 198 Lower High Street,
Watford, Hertfordshire, WD17 2FF
Area of Operation: Greater London,
South East England
Tel: 01923 224275
Fax: 01923 255005
Email: shaun@w.rumballsedgwick.co.uk
Web: www.rumballsedgwick.co.uk

SELF BUILD PRO
Belmont Business Centre, Brook Lane,
Endon, Staffordshire, ST9 9EZ
Area of Operation: UK & Ireland
Tel: 01782 505127
Fax: 01782 505127
Email: enquiries@self-build-pro.co.uk
Web: www.self-build-pro.co.uk

**SIERRA DESIGNS, ARCHITECTURAL
AND BUILDING CONSULTANTS**
Carfeld House, Pinfold Lane, Kirk Smeaton,
North Yorkshire, WF8 3JT
Area of Operation: UK & Ireland
Tel: 01977 621360
Fax: 01977 621365
Email: info@sierradesigns.co.uk
Web: www.sierradesigns.co.uk

SNP ASSOCIATES
248 Kingston Road,
New Malden, Surrey, KT3 3RN
Area of Operation: Worldwide
Tel: 0208 942 6238
Email: snp@snpassociates.co.uk
Web: www.snpassociates.co.uk

ST DESIGN CONSULTANTS
Othona House, Waterside, Bradwell on Sea,
Southminster, Essex, CM0 7QT
Area of Operation: UK (Excluding Ireland)
Tel: 01621 776736
Fax: 01621 776736
Email: info@stdesignconsultants.co.uk
Web: www.stdesignconsultants.co.uk

SUSTAINABLE PROPERTY GROUP LTD.
Courtyard Studio, Cowley Farm,
Aylesbury Road, Cuddington,
Buckinghamshire, HP18 0AD
Area of Operation: Greater London, Midlands
& Mid Wales, South East England
Tel: 01296 747121
Fax: 01296 747703
Email: info@sustainable.uk.com
Web: www.sustainable.uk.com

THE BILLINGTON CONSULTANCY LTD
Unit 2A, Station Road, Sandbach,
Cheshire, CW11 3JG
Area of Operation: UK (Excluding Ireland)
Tel: 0800 716703
Web: www.billington.co.uk

THE BORDER DESIGN CENTRE
Harelow Moor, Greenlaw, Borders, TD10 6XT
Area of Operation: UK (Excluding Ireland)
Tel: 01578 740218
Fax: 01578 740218
Email: borderdesign@btconnect.com
Web: www.borderdesign.co.uk

**TROTMAN & TAYLOR ARCHITECTURAL
CONSULTANTS**
40 Deer Park, Ivybridge, Devon, PL21 0HY
Area of Operation: UK (Excluding Ireland)
Tel: 01752 698410
Fax: 01752 698410
Email: clayton.taylor@btinternet.com
Web: www.trotmantaylor.com

VICTORIA HAMMOND INTERIORS
Bury Farm, Church Street, Bovingdon, Hemel
Hempstead, Hertfordshire, HP3 0LU
Area of Operation: UK (Excluding Ireland)
Tel: 01442 831641
Fax: 01442 831641
Email: victoria@victoriahammond.com
Web: www.victoriahammond.com

VIRTUAL-LIVING
11 Mornington Road, Cheadle, Cheshire, SK8 1NJ
Area of Operation: Worldwide
Tel: 01614911162
Email: info@virtual-living.co.uk
Web: www.virtual-living.co.uk

WM DESIGN PARTNERSHIP
First Floor, 14 Bridge Street, Menai Bridge,
Anglesey, LL59 5DN
Area of Operation: UK (Excluding Ireland)
Tel: 01248 717230 **Fax:** 01248 714930
Email: info@wmdesign.co.uk
Web: www.wmdesign.co.uk

XSPACE
99 Woodlands Avenue, Poole, Dorset, BH15 4EG
Area of Operation: South East England,
South West England
Tel: 01202 665387 **Fax:** 01202 380235
Email: design@xspace.biz
Web: www.xspace.biz

INTERIOR DESIGNERS

ACROPOLIS DESIGN LTD
Ideas House, 98 Bradford Road, East Ardsley,
Wakefield, West Yorkshire, WF3 2JL
Area of Operation: UK (Excluding Ireland)
Tel: 01924 870880 **Fax:** 0871 733 4738
Email: info@acropolisdesign.co.uk
Web: www.acropolisdesign.co.uk

ADRIENNE CHINN DESIGN COMPANY LTD
C216 Trident Business Centre,
89 Bickersteth Road, London, SW17 9SH
Area of Operation: UK (Excluding Ireland)
Tel: 0208 516 7783 **Fax:** 0208 516 7785
Email: info@adriennechinn.co.uk
Web: www.adriennechinn.co.uk

ARCHITECTURE VERTE LIMITED
41 Devonshire Buildings, Bath, Somerset, BA2 4SU
Area of Operation: Greater London,
Midlands & Mid Wales, South East England,
South West England and South Wales
Tel: 01225 484835
Email: paul@verte.co.uk

ASBA SCOTLAND
10 Lynedoch Cresent, Glasgow, G3 6EQ
Area of Operation: Scotland
Tel: 0800 731 3405
Fax: 0141 331 2751
Email: architects@design-practice.com

CA SUSTAINABLE ARCHITECTURE
83 Old Newtown Road, Newbury,
Berkshire, RG14 7DE
Area of Operation: UK (Excluding Ireland)
Tel: 01635 48363
Email: isabel.carmona@ca-sa.co.uk
Web: www.ca-sa.co.uk

CANNING & SHERIDAN INTERIORS
718 The Alaska Buildings, 61 Grange Road,
London, SE1 3BD
Area of Operation: Greater London
Tel: 0207 740 2117
Email: info@canning-sheridan.co.uk
Web: www.canning-sheridan.co.uk

CHRISTOPHER COOK DESIGNS LTD
29-33 Creek Road, Hampton Court,
East Molesey, Surrey, KT8 9BE
Area of Operation: Worldwide
Tel: 0208 941 9135
Fax: 0208 941 7282
Email: contact@christophercook.co.uk
Web: www.christophercook.co.uk

CO-ORDINATED CONSTRUCTION SERVICES
18 Kiln Road, Crawley Down,
West Sussex, RH10 4JY
Area of Operation: South East England
Tel: 01342 714511
Fax: 01342 714511
Email: info@coordsvcs.com
Web: www.coordsvcs.com

DEBBIE NEAL INTERIORS
32 Clifton Road, Crouch End, London, N8 8JA
Area of Operation: Greater London
Tel: 0208 340 0046
Email: info@debbienealinteriors.co.uk
Web: www.debbienealinteriors.co.uk

DNA DESIGN
43 View Road, Cliffe Woods,
Rochester, Kent, ME3 8UE
Area of Operation: South East England
Tel: 01634 222266
Fax: 01634 222868
Email: interiors@dna-design.co.uk
Web: dna-design.co.uk

ENGLISHAUS CHARTERED ARCHITECTS LTD
30 Lawrence Road, Hampton, Middlesex, TW12 2RJ
Area of Operation: Greater London, South East
England, South West England and South Wales
Tel: 0208 255 0595
Fax: 0208 287 3441
Email: enquiries@englishaus.co.uk
Web: www.englishaus.co.uk

JNA ARCHITECTS
9 East Park, Southgate, West Sussex, RH10 6AN
Area of Operation: South East England
Tel: 01293 439323
Fax: 01293 530160
Email: james@jna-architects.co.uk
Web: www.jna-architects.co.uk

JOHN SOLOMON DESIGN LIMITED
48 Ham Street, Richmond, Surrey, TW10 7HT
Area of Operation: Worldwide
Tel: 020 8940 2444
Fax: 020 8940 1188
Email: mail@jsajsd.com
Web: www.jsajsd.com

JONATHAN PENTON DESIGN LTD
3 Church Road, Spratton, Northampton,
Northamptonshire, NN6 8HR
Area of Operation: UK (Excluding Ireland)
Tel: 01604 820824
Fax: 01604 820823
Email: jon@creative-team.com
Web: www.jonathanpentondesign.com

LIGHT IQ LTD
Carpenters Yard, 27 Gironde Road,
London, SW6 7DY
Area of Operation: Europe
Tel: 0207 386 3949
Fax: 0207 386 3950
Email: philip@lightiq.com
Web: www.lightiq.com

MARMOT ASSOCIATES
Higher Tor Farm, Poundsgate, Newton Abbot,
Devon, TQ13 7PD
Area of Operation: South West England
and South Wales
Tel: 01364 631566
Fax: 01364 631556
Email: marmot-tor@zen.co.uk
Web: www.marmot-tor.com

NEVILLE JOHNSON
Broadoak Business Park, Ashburton Road
West, Trafford Park, Manchester, M17 1RW
Tel: 0161 873 8333 **Fax:** 0161 873 8335
Email: sales@nevillejohnson.co.uk
Web: www.nevillejohnson.co.uk

NICOLAS TYE ARCHITECTS
The Long Barn Studio, Limbersey Lane,
Bedfordshire, MK45 2EA
Area of Operation: UK (Excluding Ireland)
Tel: 01525 406677 **Fax:** 01525 406688
Email: info@nicolastyearchitects.co.uk
Web: www.nicolastyearchitects.co.uk

**O'EVE DON INTERIORS
& THE KITCHEN PLANNER**
9 Maple Drive, Castle Park, Mallow, Co Cork, .
Area of Operation: Ireland Only
Tel: 022 20858 **Fax:** 022 20858
Email: oevedoninteriors@eircom.net

PROPERTYPORTAL.COM
Brookfield House, Ford Lane, Frilford,
Abingdon, Oxfordshire, OX13 5NT
Area of Operation: UK (Excluding Ireland)
Tel: 0870 380 0520 **Fax:** 0870 380 0521
Email: info@propertyportal.com
Web: www.propertyportal.com

SALLY DERNIE INTERIOR DESIGN
4/29 Sisters Avenue, London, SW11 5SR
Area of Operation: Worldwide
Tel: 0207 738 1628
Fax: 0207 738 9981
Email: info@sallydernie.com
Web: www.sallydernie.com

THE DESIGN STUDIO
39 High Street, Reigate, Surrey, RH2 9AE
Area of Operation: South East England
Tel: 01737 248228
Fax: 01737 224180
Email: enq@the-design-studio.co.uk
Web: www.the-design-studio.co.uk ⌐

THE FINISHING SCHOOL
12 Wallis Road, Bournemouth, Dorset, BH10 4AG
Area of Operation: South East England,
South West England and South Wales
Tel: 01202 387986
Email: sam@thefinishingschool.org.uk
Web: www.thefinishingschool.org.uk

VICTORIA HAMMOND INTERIORS
Bury Farm, Church Street, Bovingdon,
Hemel Hempstead, Hertfordshire, HP3 0LU
Area of Operation: UK (Excluding Ireland)
Tel: 01442 831641
Fax: 01442 831641
Email: victoria@victoriahammond.com
Web: www.victoriahammond.com

VIRTUAL-LIVING
11 Mornington Road, Cheadle, Cheshire, SK8 1NJ
Area of Operation: Worldwide
Tel: 0161491 1162
Email: info@virtual-living.co.uk
Web: www.virtual-living.co.uk

WOOLS OF NEW ZEALAND
International Design Centre, Little Lane,
Ilkley, West Yorkshire, LS29 8UG
Area of Operation: UK & Ireland
Tel: 01943 603888
Fax: 01943 817083
Email: info.uk@canesis.co.uk
Web: www.woolcarpet.com

PLANNING CONSULTANTS

**ASAP- ARCHITECTURAL
SERVICES & PLANNING**
Howard Buildings, 69-71, Burpham Lane,
Guildford, Surrey, GU4 7LX
Area of Operation: South East England
Tel: 01483 457922
Email: david.haines4@virgin.net
Web: www.asaparchitectural.co.uk

BILLINGTON CONSULTANCY LTD
Unit 2a, Station Yard, Station Road,
Sandbach, Cheshire, CW11 3JG
Area of Operation: UK (Excluding Ireland)
Tel: 0800 716 703
Email: chrisb@billington.co.uk
Web: www.billington.co.uk

BUILDING PLANS LTD
Unit 10 Beech Avenue, Taverham,
Norwich, Norfolk, NR8 6HW
Area of Operation: UK (Excluding Ireland)
Tel: 01603 868377
Fax: 01603 868412
Email: john@constructionhelp.co.uk
Web: www.constructionhelp.co.uk

CO-ORDINATED CONSTRUCTION SERVICES
18 Kiln Road, Crawley Down,
West Sussex, RH10 4JY
Area of Operation: South East England
Tel: 01342 714511
Fax: 01342 714511
Email: info@coordsvcs.com
Web: www.coordsvcs.com

CORK TOFT PARTNERSHIP LTD
Greenbank, Howick Cross Lane,
Penwortham, Preston, Lancashire, PR1 0NS
Area of Operation: North West England
and North Wales
Tel: 01772 749014 /5
Fax: 01772 749034
Email: mail@corktoft.com

DESIGN PLUS (KENT) LTD
59 Marshall Crescent,
Broadstairs, Kent, CT10 2HR
Area of Operation: UK (Excluding Ireland)
Tel: 01843 602218
Email: designpluskent@btinternet.com
Web: www.designpluskent.co.uk

**ENGLISHAUS CHARTERED
ARCHITECTS LTD**
30 Lawrence Road,
Hampton, Middlesex, TW12 2RJ
Area of Operation: Greater London, South East
England, South West England and South Wales
Tel: 0208 255 0595
Fax: 0208 287 3441
Email: enquiries@englishaus.co.uk
Web: www.englishaus.co.uk

FG DESIGN
6 Denmark Drive, Sedbury, Chepstow,
Monmouthshire, NT16 7BD
Area of Operation: South West England
and South Wales
Tel: 01291 624366
Fax: 01291 624366
Email: frank.fgdesign@virgin.net

G.M.MOORE & ASSOCIATES
Old Laughton Sawmills, Park Lane,
Laughton, East Sussex, BN8 6BP
Area of Operation: South East England
Tel: 01323 811689
Fax: 01323 811325
Email: sales@gmassociates.co.uk
Web: www.gmassociates.co.uk

GRINDEY CONSULTING
5 Granville Street, Monton, Manchester, M30 9PX
Area of Operation: North West England
and North Wales
Tel: 07773 371488
Email: ian.grindey@scottwilson.com
Web: www.grindeyconsulting.co.uk

J G HOOD & ASSOCIATES LTD
50 High Street, Bruton, Somerset, BA10 0AN
Area of Operation: Midlands & Mid Wales,
South East England, South West England
and South Wales
Tel: 01749 812139
Fax: 0870 133 3776
Email: mail@jghood-assoc.co.uk
Web: www.jghood-assoc.co.uk

J.S. BUILDING CONSULTANCY
53 Hawthorn Road, Yeadon, Leeds,
West Yorkshire, LS19 7UT
Area of Operation: UK (Excluding Ireland)
Tel: 0113 250 1303
Email: jsharples@ricsonline.org
Web: www.ukbuildingconsultancy.co.uk

JNA ARCHITECTS
9 East Park, Southgate, West Sussex, RH10 6AN
Area of Operation: South East England
Tel: 01293 439323
Fax: 01293 530160
Email: james@jna-architects.co.uk
Web: www.jna-architects.co.uk

JOSEPH R NIXON PROJECT MANAGEMENT
Brookview 52a Foxcotte Road,
Charlton, Andover, Hampshire, SP10 4AT
Area of Operation: Greater London,
South East England, South West England
and South Wales
Tel: 01264 364232
Fax: 01264 364232
Email: jrn-pm@tiscali.co.uk

LARNER SING
29 Lower Street, Rode,
Frome, Somerset, BA11 6PU
Area of Operation: South East England,
South West England and South Wales
Tel: 01373 830527
Fax: 01373 830527
Email: ian@larner-sing.co.uk
Web: www.larner-sing.com

NCB ASSOCIATES LTD
Gate House, Fretheren Road, Welwyn
Garden City, Hertfordshire, AL8 6NS
Area of Operation: UK (Excluding Ireland)
Tel: 0845 230 9897
Email: enquiry@ncbassociates.co.uk
Web: www.ncbassociates.co.uk

PAUL A STOWELL
'Sorriso', Farleigh Road, Cliddesden,
Basingstoke, Hampshire, RG25 2JL
Area of Operation: Greater London, South East
England, South West England and South Wales
Tel: 01256 320470
Fax: 01256 320470
Email: paulstowell@constructionplans.co.uk
Web: www.constructionplans.co.uk

PLANNING & DEVELOPMENT SOLUTIONS LTD.
Aspect Court, 47 Park Square East, Leeds,
West Yorkshire, LS1 2NL
Area of Operation: UK (Excluding Ireland)
Tel: 0113 383 3735
Fax: 0113 383 3746
Email: info@pds.uk.com
Web: www.pds.uk.com

PROPERTYPORTAL.COM
Brookfield House, Ford Lane, Frilford,
Abingdon, Oxfordshire, OX13 5NT
Area of Operation: UK (Excluding Ireland)
Tel: 0870 380 0520
Fax: 0870 380 0521
Email: info@propertyportal.com
Web: www.propertyportal.com

RENAISSANCE ARCHITECTURAL LTD
15 Gay Street, Bath, BA1 2PH
Area of Operation: East England, Greater London,
Midlands, North East England, North West England,
South East England, South West England
Tel: 01761 479605
Fax: 01761 479605
Email: renaissance@ren.uk.com
Web: www.ren.uk.com

RUMBALL SEDGWICK
Abbotts House, 198 Lower High Street,
Watford, Hertfordshire, WD17 2FF
Area of Operation: Greater London,
South East England
Tel: 01923 224275
Fax: 01923 255005
Email: shaun@w.rumballsedgwick.co.uk
Web: www.rumballsedgwick.co.uk

**SIERRA DESIGNS, ARCHITECTURAL
& BUILDING CONSULTANTS**
Carfeld House, Pinfold Lane,
Kirk Smeaton, North Yorkshire, WF8 3JT
Area of Operation: UK & Ireland
Tel: 01977 621360
Fax: 01977 621365
Email: info@sierradesigns.co.uk
Web: www.sierradesigns.co.uk

SNP ASSOCIATES
248 Kingston Road, New Malden, Surrey, KT3 3RN
Area of Operation: Worldwide
Tel: 0208 942 6238
Email: snp@snpassociates.co.uk
Web: www.snpassociates.co.uk

SPEER DADE PLANNING CONSULTANTS
10 Stonepound Rd, Hassocks, West Sussex, BN6 8PP
Area of Operation: UK (Excluding Ireland)
Tel: 01273 843737
Fax: 01273 842155
Email: Roy@stonepound.co.uk
Web: www.stonepound.co.uk

SUSTAINABLE PROPERTY GROUP LTD.
Courtyard Studio, Cowley Farm, Aylesbury Road,
Cuddington, Buckinghamshire, HP18 0AD
Area of Operation: Greater London, Midlands
& Mid Wales, South East England
Tel: 01296 747121
Fax: 01296 747703
Email: info@sustainable.uk.com
Web: www.sustainable.uk.com

**TROTMAN & TAYLOR
ARCHITECTURAL CONSULTANTS**
40 Deer Park, Ivybridge, Devon, PL21 0HY
Area of Operation: UK (Excluding Ireland)
Tel: 01752 698410
Fax: 01752 698410
Email: clayton.taylor@btinternet.com
Web: www.trotmantaylor.com

WHEATMAN PLANNING LIMITED
The Gables, Church Lane, Haddiscoe,
Norwich, Norfolk, NR14 6PB
Area of Operation: East England
Tel: 01502 677636 **Fax:** 01502 677636
Email: simon@wheatmanplanning.co.uk
Web: www.wheatmanplanning.co.uk

PROJECT MANAGERS

A1 MASTERPLANS LTD
Dulford Business Park, Dulford,
Cullompton, Devon, EX15 2DY
Area of Operation: UK (Excluding Ireland)
Tel: 01884 266800
Fax: 01884 266882
Email: contact@masterplansltd.co.uk
Web: www.masterplansltd.co.uk

ARCHITECTURE VERTE LIMITED
41 Devonshire Buildings, Bath, Somerset, BA2 4SU
Area of Operation: Greater London, Midlands & Mid
Wales, South East England, South West England
& South Wales
Tel: 01225 484835
Email: paul@verte.co.uk

BARLEY WEST LIMITED
22 Virginia Place, Cobham, Surrey, KT11 1AE
Area of Operation: Greater London, South East
England, South West England and South Wales
Tel: 01932 865459
Email: info@barleywest.co.uk
Web: www.barleywest.co.uk

BILLINGTON CONSULTANCY LTD
Unit 2a, Station Yard, Station Road, Sandbach,
Cheshire, CW11 3JG
Area of Operation: UK (Excluding Ireland)
Tel: 0800 716 703
Email: chrisb@billington.co.uk
Web: www.billington.co.uk

CDL PROJECT SERVICES LTD
The Studio, Mill Cottage, Sedgeberrow,
Worcester, Worcestershire, WR11 7UA
Area of Operation: UK (Excluding Ireland)
Tel: 0870 062 0018
Fax: 01386 882217

CONCEPT ENGINEERING CONSULTANTS LTD
Unit 8 Warple Mews, Warple Way, London, W3 0RF
Area of Operation: East England, Greater London, South
East England, South West England and South Wales
Tel: 0208 840 3321
Fax: 0208 579 6910
Email: si@conceptconsultants.co.uk
Web: www.conceptconsultants.co.uk

CO-ORDINATED CONSTRUCTION SERVICES
18 Kiln Road, Crawley Down,
West Sussex, RH10 4JY
Area of Operation: South East England
Tel: 01342 714511
Fax: 01342 714511
Email: info@coordsvcs.com
Web: www.coordsvcs.com

DESIGN PLUS (KENT) LTD
59 Marshall Crescent, Broadstairs, Kent, CT10 2HR
Area of Operation: UK (Excluding Ireland)
Tel: 01843 602218
Email: designpluskent@btinternet.com
Web: www.designpluskent.co.uk

DIL GREEN ARCHITECT
206 Lyham Road, Brixton,
Greater London, SW2 5NR
Area of Operation: UK (Excluding Ireland)
Tel: 0208 671 2242 **Fax:** 0208 674 9757
Email: w_info@dilgreenarchitect.co.uk
Web: www.dilgreenarchitect.co.uk

ECO HOUSES
12 Lee Street, Louth, Lincolnshire, LN11 9HJ
Area of Operation: Worldwide
Tel: 0845 345 2049
Email: sd@jones-nash.co.uk
Web: www.eco-houses.co.uk

ENCRAFT
46 Northumberland Road, Leamington Spa,
Warwickshire, CV32 6HB
Area of Operation: UK & Ireland
Tel: 01926 312159 **Fax:** 01926 772480
Email: enquiries@encraft.co.uk
Web: www.encrafthome.com

G.M.MOORE & ASSOCIATES
Old Laughton Sawmills, Park Lane,
Laughton, East Sussex, BN8 6BP
Area of Operation: South East England
Tel: 01323 811689 **Fax:** 01323 811325
Email: sales@gmassociates.co.uk
Web: www.gmassociates.co.uk

GEORGE BLACK ARCHITECT
12 Kingsknowe Crescent, Edinburgh,
Lothian, EH14 2JZ
Area of Operation: Scotland
Tel: 0131 443 9898
Email: george@gbarch.abel.co.uk
Web: www.gbarch.abel.co.uk

GRAHAM G BISHOP - CHARTERED SURVEYORS
9 Church Lane Drive, Coulsdon, Surrey, CR5 3RG
Area of Operation: Greater London,
South East England
Tel: 01737 558473 **Fax:** 01737 558473
Email: graham@grahambishop.co.uk
Web: www.grahambishop.co.uk

GRINDEY CONSULTING
5 Granville Street, Monton, Manchester, M30 9PX
Area of Operation: North West England
and North Wales
Tel: 07773 371488
Email: ian.grindey@scottwilson.com
Web: www.grindeyconsulting.co.uk

HCT CONSTRUCTION CONSULTANTS LTD
Mercury House, The Court Yard, Roman Way,
Coleshill, Warwickshire, B46 1HQ
Area of Operation: UK (Excluding Ireland)
Tel: 01675 466010
Fax: 01675 464543
Email: jamie.timmins@hctcc.co.uk
Web: www.hctcc.co.uk

HORIZON CONSTRUCTION MANAGEMENT
Lodge House, Lodge Lane, Langham, Essex, CO4 5NE
Area of Operation: UK (Excluding Ireland)
Tel: 01206 231531
Fax: 01206 231631
Email: info@horizonconstruction.co.uk
Web: www.horizonconstruction.co.uk

IAN MONTGOMERY
Rolles Court, Church Whitfield, Dover, Kent, CT16 3HY
Area of Operation: South East England
Tel: 01304 827487
Email: ian@homearranger.net
Web: www.homearranger.net

J G HOOD & ASSOCIATES LTD
50 High Street, Bruton, Somerset, BA10 0AN
Area of Operation: Midlands & Mid Wales, South
East England, South West England and South Wales
Tel: 01749 812139
Fax: 0870 133 3776
Email: mail@jghood-assoc.co.uk
Web: www.jghood-assoc.co.uk

J.S. BUILDING CONSULTANCY
53 Hawthorn Road, Yeadon,
Leeds, West Yorkshire, LS19 7UT
Area of Operation: UK (Excluding Ireland)
Tel: 0113 250 1303
Email: jsharples@ricsonline.org
Web: www.ukbuildingconsultancy.co.uk

JNA ARCHITECTS
9 East Park, Southgate, West Sussex, RH10 6AN
Area of Operation: South East England
Tel: 01293 439323
Fax: 01293 530160
Email: james@jna-architects.co.uk
Web: www.jna-architects.co.uk

JOSEPH R NIXON PROJECT MANAGEMENT
Brookview 52a Foxcotte Road, Charlton,
Andover, Hampshire, SP10 4AT
Area of Operation: Greater London, South East
England, South West England and South Wales
Tel: 01264 364232
Fax: 01264 364232
Email: jrn-pm@tiscali.net

LARNER SING
29 Lower Street, Rode, Frome, Somerset, BA11 6PU
Area of Operation: South East England, South West
England and South Wales
Tel: 01373 830527
Fax: 01373 830527
Email: ian@larner-sing.co.uk
Web: www.larner-sing.com

M D PROJECTS (NORTH WEST) LTD
Abbey Works, Back King Street, Whalley,
Nr Clitheroe, Lancashire, BB7 9SP
Area of Operation: North West England
and North Wales
Tel: 01254 823738
Fax: 01254 823830
Email: peterdowns@mdprojects.co.uk

MARMOT ASSOCIATES
Higher Tor Farm, Poundsgate,
Newton Abbot, Devon, TQ13 7PD
Area of Operation: South West England
and South Wales
Tel: 01364 631566
Fax: 01364 631556
Email: marmot-tor@zen.co.uk
Web: www.marmot-tor.com

NCB ASSOCIATES LTD
Gate House, Fretheren Road, Welwyn Garden City,
Hertfordshire, AL8 6NS
Area of Operation: UK (Excluding Ireland)
Tel: 0845 230 9897
Email: enquiry@ncbassociates.co.uk
Web: www.ncbassociates.co.uk

NICHOLAS RAY ASSOCIATES
13-15 Convert Garden, Cambridge,
Cambridgeshire, CB1 2HS
Area of Operation: UK & Ireland
Tel: 01223 464455
Email: design@nray-arch.co.uk
Web: www.nray-arch.co.uk

**O'EVE DON INTERIORS & THE KITCHEN
PLANNER**
9 Maple Drive, Castle Park, Mallow, Co Cork
Area of Operation: Ireland Only
Tel: 022 20858
Fax: 022 20858
Email: oevedoninteriors@eircom.net

OTT PROJECTS LTD
2 Whitney Drive, Stevenage, Hertfordshire, SG1 4BG
Area of Operation: East England, Greater London,
South East England
Tel: 01438 223333
Fax: 01438 359594
Email: sales@ott.demon.co.uk
Web: www.ott-projects.co.uk

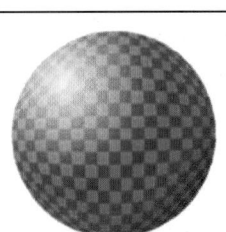

OTT PROJECTS LTD

Area of Operation: UK
Tel: 01438 223333
Fax: 01438 359 594
Email: sales@ott.demon.co.uk
Web: www.ott-project.co.uk
Additional Services: Free Quote, CAD, Planning
Specialism: Renovation, Listed, Barn, Other

Do you need help with your Self Build/Renovation Project?
Talk to the Professionals
Some of the services we offer:
Construction and Project Management, Specification writing
Cost Planning, Competitive Tendering, For more information phone
OTT PROJECTS LTD

PROPERTYPORTAL.COM
Brookfield House, Ford Lane, Frilford,
Abingdon, Oxfordshire, OX13 5NT
Area of Operation: UK (Excluding Ireland)
Tel: 0870 380 0520
Fax: 0870 380 0521
Email: info@propertyportal.com
Web: www.propertyportal.com

SELF BUILD PRO
Belmont Business Centre, Brook Lane,
Endon, Staffordshire, ST9 9EZ
Area of Operation: UK & Ireland
Tel: 01782 505127
Fax: 01782 505127
Email: enquiries@self-build-pro.co.uk
Web: www.self-build-pro.co.uk

**SIERRA DESIGNS, ARCHITECTURAL AND
BUILDING CONSULTANTS**
Carfeld House, Pinfold Lane, Kirk Smeaton,
North Yorkshire, WF8 3JT
Area of Operation: UK & Ireland
Tel: 01977 621360
Fax: 01977 621365
Email: info@sierradesigns.co.uk
Web: www.sierradesigns.co.uk

SIMON J CUSHING CHARTERED ARCHITECT
102 Woodchurch Road, Birkenhead,
Wirral, Merseyside, CH42 9LP
Area of Operation: North West England
and North Wales
Tel: 0151 653 9900
Fax: 0151 653 9797
Email: simonjcushing@btconnect.com

SNP ASSOCIATES
248 Kingston Road, New Malden, Surrey, KT3 3RN
Area of Operation: Worldwide
Tel: 0208 942 6238
Email: snp@snpassociates.co.uk
Web: www.snpassociates.co.uk

ST DESIGN CONSULTANTS
Othona House, Waterside, Bradwell on Sea,
Southminster, Essex, CM0 7QT
Area of Operation: UK (Excluding Ireland)
Tel: 01621 776736
Fax: 01621 776736
Email: info@stdesignconsultants.co.uk
Web: www.stdesignconsultants.co.uk

SUSTAINABLE PROPERTY GROUP LTD.
Courtyard Studio, Cowley Farm, Aylesbury Road,
Cuddington, Buckinghamshire, HP18 0AD
Area of Operation: Greater London, Midlands
& Mid Wales, South East England
Tel: 01296 747121
Fax: 01296 747703
Email: info@sustainable.uk.com
Web: www.sustainable.uk.com

THE BILLINGTON CONSULTANCY LTD
Unit 2A, Station Road, Sandbach,
Cheshire, CW11 3JG
Area of Operation: UK (Excluding Ireland)
Tel: 0800 716703
Web: www.billington.co.uk

XSPACE
99 Woodlands Avenue, Poole, Dorset, BH15 4EG
Area of Operation: South East England,
South West England
Tel: 01202 665387
Fax: 01202 380235
Email: design@xspace.biz
Web: www.xspace.biz

QUANTITY SURVEYORS

A1 MASTERPLANS LTD
Dulford Business Park, Dulford,
Cullompton, Devon, EX15 2DY
Area of Operation: UK (Excluding Ireland)
Tel: 01884 266800 **Fax:** 01884 266882
Email: contact@masterplansltd.co.uk
Web: www.masterplansltd.co.uk

BILLINGTON CONSULTANCY LTD
Unit 2a, Station Yard, Station Road,
Sandbach, Cheshire, CW11 3JG
Area of Operation: UK (Excluding Ireland)
Tel: 0800 716703
Email: chrisb@billington.co.uk
Web: www.billington.co.uk

CONSTRUCTION COST SOLUTIONS
September Cottage, 1A Wigwam Close,
Poynton, Stockport, Cheshire, SK12 1XF
Area of Operation: Midlands & Mid Wales, North
East England, North West England and North Wales
Tel: 01625 875488 **Fax:** 01625 878143
Email: david@constructioncostsolutions.co.uk
Web: www.constructioncostsolutions.co.uk

DEREK GOUGH ASSOCIATES
Toft Smithy, Toft Road, Knutsford,
Cheshire, WA16 9PA
Area of Operation: UK (Excluding Ireland)
Tel: 01565 751500
Fax: 01565 751330
Email: derekgough@dgough.co.uk
Web: www.dgough.co.uk

HCT CONSTRUCTION CONSULTANTS LTD
Mercury House, The Court Yard, Roman Way,
Coleshill, Warwickshire, B46 1HQ
Area of Operation: UK (Excluding Ireland)
Tel: 01675 466010
Fax: 01675 464543
Email: jamie.timmins@hctcc.co.uk
Web: www.hctcc.co.uk

JOSEPH R NIXON PROJECT MANAGEMENT
Brookview 52a Foxcotte Road, Charlton,
Andover, Hampshire, SP10 4AT
Area of Operation: Greater London, South East
England, South West England and South Wales
Tel: 01264 364232
Fax: 01264 364232
Email: jrn-pm@tiscali.co.uk

MARMOT ASSOCIATES
Higher Tor Farm, Poundsgate,
Newton Abbot, Devon, TQ13 7PD
Area of Operation: South West England
and South Wales
Tel: 01364 631566 **Fax:** 01364 631556
Email: marmot-tor@zen.co.uk
Web: www.marmot-tor.com

PAUL A STOWELL
'Sorriso', Farleigh Road, Cliddesden,
Basingstoke, Hampshire, RG25 2JL
Area of Operation: Greater London, South East
England, South West England and South Wales
Tel: 01256 320470
Fax: 01256 320470
Email: paulstowell@constructionplans.co.uk
Web: www.constructionplans.co.uk

QUANTI-QUOTE
Trewint, Main Street, Shap, Penrith,
Cumbria, CA10 3NH
Area of Operation: UK & Ireland
Tel: 01931 716810
Email: peter@Quantiquote.co.uk
Web: www.quantiquote.co.uk

SELF BUILD PRO
Belmont Business Centre, Brook Lane,
Endon, Staffordshire, ST9 9EZ
Area of Operation: UK & Ireland
Tel: 01782 505127
Fax: 01782 505127
Email: enquiries@self-build-pro.co.uk
Web: www.self-build-pro.co.uk

**SIERRA DESIGNS, ARCHITECTURAL AND
BUILDING CONSULTANTS**
Carfeld House, Pinfold Lane, Kirk Smeaton,
North Yorkshire, WF8 3JT
Area of Operation: UK & Ireland
Tel: 01977 621360 **Fax:** 01977 621365
Email: info@sierradesigns.co.uk
Web: www.sierradesigns.co.uk

LANDSCAPE DESIGNERS

ACRES WILD LTD
1 Helm Cottages, Nuthurst,
Horsham, West Sussex, RH13 6RG
Area of Operation: UK & Ireland
Tel: 01403 891084 **Fax:** 01403 891084
Email: enquiries@acreswild.co.uk
Web: www.acreswild.co.uk

**ALICE BOWE - ENGLISH LANDSCAPE
& GARDEN DESIGN (LONDON)**
9a Welldon Crescent, London, HA1 1QU
Area of Operation: UK (Excluding Ireland)
Tel: 0845 838 2649
Email: alice@alicebowe.co.uk
Web: www.alicebowe.co.uk

**ALICE BOWE - ENGLISH LANDSCAPE
& GARDEN DESIGN (NOTTINGHAM)**
1 Briar Gate, Nottingham,
Nottinghamshire, NG10 4BN
Area of Operation: Greater London,
Midlands & Mid Wales, South East England
Tel: 0845 838 2649
Email: alice@alicebowe.co.uk
Web: www.alicebowe.co.uk

BRADSTONE GARDEN
Aggregate Industries UK Ltd,
Hulland Ward, Ashbourne,
Derbyshire, DE6 3ET
Area of Operation: UK (Excluding Ireland)
Tel: 01335 372222
Fax: 01335 370973
Email: bradstone.structural@aggregate.com
Web: www.bradstone.com

CATHERINE HEATHERINGTON DESIGNS
9 Cecil Road, London, N10 2BU
Area of Operation: UK (Excluding Ireland)
Tel: 0208 374 2321
Email: gardens@chdesigns.co.uk
Web: www.chdesigns.co.uk

**CATHERINE THOMAS LANDSCAPE
AND GARDEN DESIGN**
Fisherton Mill, 108 Fisherton Street,
Salisbury, Wiltshire, SP2 7QY
Area of Operation: UK & Ireland
Tel: 01722 339936
Fax: 01722 339936
Email: catherines_gardens@hotmail.com
Web: www.catherinethomas.co.uk

CHARLESWORTH DESIGN
21 Derbyshire Road, Sale,
Cheshire, M33 3EB
Area of Operation: North West England
and North Wales
Tel: 0161 905 3871
Email: robertsfrier@hotmail.com
Web: www.charlesworthdesign.com

CHERRY MILLS GARDEN DESIGN
Flora Cottage, The Drive,
Godalming, Surrey, GU7 1PH
Area of Operation: South East England
Tel: 01483 421499
Fax: 01483 418678
Email: cmills@cmgardendesign.com
Web: www.cmgardendesign.com

**CHRISTOPHER MAGUIRE ARCHITECT
AND GARDEN DESIGNER**
15 Harston Road, Newton, Cambridge,
Cambridgeshire, CB2 5PA
Area of Operation: East England
Tel: 01223 872800
Web: www.christophermaguire.co.uk

**SIERRA DESIGNS, ARCHITECTURAL AND
BUILDING CONSULTANTS**
Carfeld House, Pinfold Lane, Kirk Smeaton,
North Yorkshire, WF8 3JT
Area of Operation: UK & Ireland
Tel: 01977 621360 **Fax:** 01977 621365
Email: info@sierradesigns.co.uk
Web: www.sierradesigns.co.uk

CORK TOFT PARTNERSHIP LTD
Greenbank, Howick Cross Lane, Penwortham,
Preston, Lancashire, PR1 0NS
Area of Operation: North West England
and North Wales
Tel: 01772 749014 /5 **Fax:** 01772 749034
Email: mail@corktoft.com

COURTYARD GARDEN DESIGN
The Workshop, 32 Broadway Avenue,
East Twickenham, Greater London, TW1 1RH
Area of Operation: Worldwide
Tel: 0208 892 0118
Fax: 0208 892 0118/0024
Email: sally.cgd@btconnect.com
Web: www.courtyardgardendesign.co.uk

DIZZY SHOEMARK (UK) LIMITED
Lingmell Crescent, Seascale, Cumbria, CA20 1JX
Area of Operation: North East England, North
West England and North Wales
Tel: 01946 721767
Fax: 01946 721618
Email: aspects@dizzyshoemark.com
Web: www.dizzyshoemark.com

EAST OF EDEN PLANTS
38 St Andrews Street, Millbrook,
Torpoint, Devon, PL10 1BE
Area of Operation: UK & Ireland
Tel: 01752 822782
Email: info@eastofedenplants.co.uk
Web: www.eastofedenplants.co.uk

ENGLISH GARDEN DESIGN ASSOCIATES LTD
The Annexe, Ponchydown Farm, Blackborough ,
Cullompton, Devon, Exeter, Devon, EX15 2HQ
Area of Operation: South West England
and South Wales
Tel: 01884 266188
Fax: 01884 266188
Email: hugh.oconnell@btopenworld.com
Web: www.ukgardendesigner.com

FISHER TOMLIN
74 Sydney Road, Wimbledon, London, SW20 8EF
Area of Operation: Europe
Tel: 0208 542 0683
Email: info@fishertomlin.com
Web: www.fishertomlin.com

GILLIAN TEMPLE ASSOCIATES
Capel House, Capel Road, Orlestone,
Hamstreet, Ashford, Kent, TN26 2EH
Area of Operation: UK (Excluding Ireland)
Tel: 01233 733073
Email: gillian@gilliantemple.co.uk
Web: www.gilliantemple.co.uk

HILLIER LANDSCAPES
Ampfield House, Ampfield,
Romsey, Hampshire, SO51 9PA
Area of Operation: South East England,
South West England and South Wales
Tel: 01794 368855
Fax: 01794 368866
Email: hillierlandscapes@btinternet.com
Web: www.hillier-landscapes.co.uk

HONLEY GARDEN DESIGN LTD
8 Well Hill, Honley, Holmfirth,
West Yorkshire, HD9 6JF
Area of Operation: Europe
Tel: 01484 660783
Email: barrykellington@aol.com
Web: www.lifestylegardens.co.uk

JANO WILLIAMS GARDEN DESIGN
54 Berkeley Road, Westbury Park, Bristol, BS6 7PL
Area of Operation: Worldwide
Tel: 0117 914 1078
Email: janowilliams@blueyonder.co.uk
Web: www.janowilliams.com

JILL FENWICK
53 Hampstead Road, Dorking,
Surrey, RH4 3AE
Area of Operation: Greater London,
Midlands & Mid Wales, South East England
Tel: 01306 889465
Fax: 01306 889465
Email: privategardendesign@tiscali.co.uk
Web: www.privategardendesign.co.uk

JOHN NASH ASSOCIATES
19 Cannon Street, St. Albans,
Hertfordshire, AL3 5JR
Area of Operation: Greater London,
South East England
Tel: 01727 869989
Fax: 01727 869491
Email: jnassoc@globalnet.co.uk
Web: www.johnnashassociates.co.uk

JOHN SOLOMON DESIGN LIMITED
48 Ham Street, Richmond, Surrey, TW10 7HT
Area of Operation: Worldwide
Tel: 0208 940 2444 **Fax:** 0208 940 1188
Email: mail@jsajsd.com
Web: www.jsajsd.com

JONATHAN PRINGLE GARDEN DESIGN
Two Bell House Yard, Hare Street,
Near Buntingford, Hertfordshire, SG9 0DZ
Area of Operation: East England,
Greater London, Midlands & Mid Wales
Tel: 01279 303367
Fax: 01279 303367
Email: jonathan.pringle@ntlworld.com

**JULIET SARGEANT GARDENS
& PRIVATE LANDSCAPES**
39 Falmer Road, Rottingdean,
Brighton, East Sussex, BN2 7DA
Area of Operation: UK & Ireland
Tel: 01273 300587
Email: julietdesigns@supanet.com
Web: www.julietdesigns.co.uk

KEITH PULLAN GARDEN DESIGN
1 Amotherby Close, Amotherby,
Malton, North Yorkshire, YO17 6TG
Area of Operation: Midlands & Mid Wales,
North East England, North West England
& North Wales
Tel: 01653 693885
Email: keithpullan@fastmail.fm
Web: www.keithpullan.co.uk

**LAURENCE MAUNDER GARDEN
DESIGN & CONSULTANCY**
Newton Cottage, Fitzhead,
Taunton, Somerset, TA4 3JW
Area of Operation: South West England
and South Wales
Tel: 01823 401208 **Fax:** 01823 401208
Email: info@laurencemaunder.co.uk
Web: www.laurencemaunder.co.uk

MICHAEL DAY GARDEN DESIGN
The Chalet, Marston Meysey,
Swindon, Wiltshire, SN6 6LQ
Area of Operation: Midlands & Mid Wales,
South West England and South Wales
Tel: 01285 810486 **Fax:** 01285 810970
Email: info@michaeldaygardendesign.co.uk
Web: www.michaeldaygardendesign.co.uk

MIRIAM BOOK GARDEN DESIGNS
50 Hill Drive, Hove, East Sussex, BN3 6QL
Area of Operation: Greater London,
South East England
Tel: 01273 541600 **Fax:** 01273 541600
Email: miriam@gardenbook.co.uk
Web: www.gardenbook.co.uk

**PETER MATTHEWS GARDEN
& LANDSCAPE DESIGN**
The Cart Lodge, Gore Farm, Upchurch,
Sittingbourne, Kent, ME9 7BE
Area of Operation: South East England
Tel: 01634 235564
Email: peter@petermatthews.co.uk
Web: www.petermatthews.co.uk

**PETER THOMAS
ASSOCIATES GARDEN DESIGN**
113 High Street, Codicote,
Hertfordshire, SG4 8UA
Area of Operation: East England,
Greater London, Midlands & Mid Wales,
South East England,
South West England & South Wales
Tel: 01438 821408
Email: info@ptadesign.com
Web: www.ptadesign.com

PICKARD GARDEN & LANDSCAPE DESIGN
45 Business Village, Dyson Way,
Staffordshire Technology Park,
Beaconside, Stafford, ST218 0TW
Area of Operation: Europe
Tel: 01785 850240
Email: admin@pickardsgd.com
Web: www.pickardsgd.com

SAM MCGOWAN DESIGN
26 Dunrobin Place, Edinburgh, EH3 5HZ
Area of Operation: Scotland
Tel: 0131 343 6536
Email: sammcg.edinburgh@virgin.net
Web: www.sam-mcgowan.co.uk

THE PARSONS GARDEN
The Garden Room, 30 Gladwin Road,
Colchester, Essex, CO2 7HS
Area of Operation: UK (Excluding Ireland)
Tel: 01206 570440 **Fax:** 01206 561091
Email: design@theparsonsgarden.co.uk
Web: www.theparsonsgarden.co.uk

UP THE GARDEN PATH
10 Paget Drive, Burntwood,
Staffordshire, WS7 1HP
Area of Operation: Midlands & Mid Wales,
North West England and North Wales
Tel: 01543 670342
Fax: 01543 670342
Email: david@upthegardenpath.net
Web: www.upthegardenpath.net

LOOKING FOR A BUILDING PLOT?

Plotfinder.net is an online database which holds details of almost 6,000 building plots and properties currently for sale throughout the whole of the UK.

www.plotfinder.net
THE UK'S LAND AND RENOVATION DATABASE

ONE YEAR SUBSCRIPTION ONLY £40

BENEFITS OF
www.plotfinder.net

- Instant access to the database (updated daily).
- FREE email alert when new plots are added to your chosen counties.
- 100s of private sales which you may not find anywhere else.
- Unlimited access to any five counties.
- Save favourite plots into a separate folder.
- Mark viewed plots as read.

Current Subscriber Comments:

'This is a fantastic service - I use it all the time - we restore small residential properties.' Emma Nickols

"Your web site is easy to use, flexible, containing a large amount of very useful information and leads - well worth the very modest yearly subscription fee." Jay McDonagh

Image courtesy of www.plotfinder.net

USEFUL INFORMATION

Trade Associations

Derek Vaughan, Managing Director of The UK Trades Confederation, explains how a trade association operates and the benefits of using regulatory bodies for both members and consumers.

There are a number of different types of trade associations, which vary from ones that are specific to a trade, to ones that encourage a number of different trades. However, they all have a common objective to provide services to promote and protect their members' business, reduce overheads and look after consumers.

Trade Associations separate reputable tradesmen from the 'rogue traders' by providing members with a trademark logo, which proves they have abided by a working code of practice. This helps consumers to identify reputable tradesmen if the trademark logo is used on business stationery or marketing material.

Protecting a company name is imperative, which is why a trade association offers its members free advice and continually informs them about new legislation that may affect them, such as different building regulations and employment laws. This in turn aids the consumer, as they can be confident that their tradesman will be up to speed on law changes, industry policies and the work carried out will meet the applicable building regulations.

One of the main pressures on a small business is keeping afloat financially due to huge overheads and late payments. This is where belonging to a trade association is hugely beneficial, as they have more buying power to negotiate better deals on a number of services such as telephone systems, vehicle hire, office equipment and loan schemes. Trade associations often form alliances with other organisations, which open up even more opportunities to join financial, self-certification or business development schemes at lower costs. This means that consumers can be confident that the tradesman's price is competitive, that they can afford sufficient insurance and they have good quality materials and assets such as reliable equipment and machinery.

It is also far better for consumers looking for advice or trades people to contact trade associations which employ a thorough vetting procedure: this makes sure they are recommended a credible trades person and not a 'rogue'. Many trade bodies offer a free online search facility, whereby you can find one of their members quickly and easily. The UKTC is no exception - their free service, www.easyquoteuk.com, is a site where consumers can submit their contact details and a description of the work they would like carried out, and then an accredited UKTC trades person will contact them. Having a directory of vetted businesses is invaluable for the trade association's reputation, and also the homeowner's peace of mind that they are dealing with a reputable company.

Some trade associations also offer homeowners other benefits. These can include consumer advice, a comparison of different homeowner loans, checklists and local planning office contact details for guidance about planning permission, party wall awards and listed buildings.

The greatest benefit to consumers of selecting a tradesperson through a trade association is the protection of the project for both parties. As a homeowner your tradesperson could present you with a final bill for several hundred pounds more than was quoted, or as a tradesperson your client could deny that work was authorised, or was not as stated before work commenced. Therefore, it is important that a written contract exists between a tradesperson and customer to protect both parties in the event that a dispute should arise. The UKTC eliminates this uncertainty by providing homeowners with a free consumer pack containing a basic Homeowner Contract and a handy Consumer Check List to ensure their chosen tradesperson meets all the appropriate requirements.

For more information about The UK Trades Confederation please call on 020 8842 4442 or email mail@uktc.org or visit www.uktc.org

LEFT: House by Potton.

TRADE, REGULATORY BODIES & ASSOCIATIONS

KEY

OTHER: ▽ Reclaimed   ✏ On-line shopping
✐ Bespoke   ✋ Hand-made   ECO Ecological

ASSOCIATION OF BUILDING ENGINEERS

Lutyens House, Billing Brook Road,
Northampton, Northamptonshire, NN3 8NW
Area of Operation: Worldwide
Tel: 01604 404121
Fax: 01604 784220
Email: building.engineers@abe.org.uk
Web: www.abe.org.uk ✏

ASSOCIATION OF PLUMBING & HEATING CONTRACTORS (APHC)

14 Ensign House, Ensign Business Centre,
Westwood Way, Coventry, West Midlands, CV4 8JA
Area of Operation: UK (Excluding Ireland)
Tel: 02476 470626
Fax: 02476 470942
Email: enquiries@aphc.co.uk
Web: www.aphc.co.uk

ASSOCIATION OF SCOTTISH HARDWOOD SAWMILLERS (ASHS)

Marcassie Farm, Rafford, Forres, Moray, IV36 2RH
Area of Operation: UK (Excluding Ireland)
Email: coordinator@ashs.co.uk
Web: www.ashs.co.uk

BASEMENT INFORMATION CENTRE

Riverside House, 4 Meadows Business Park,
Station Approach, Blackwater,
Camberley, Surrey, GU17 9AB
Area of Operation: UK (Excluding Ireland)
Tel: 01276 33155
Fax: 01276 606801
Email: info@tbic.org.uk
Web: www.tbic.org.uk / www.basements.org.uk

BATHROOM MANUFACTURERS ASSOCIATION

Federation House, Station Road,
Stoke on Trent, Staffordshire, ST4 2RT
Area of Operation: UK & Ireland
Tel: 01782 747123
Fax: 01782 747161
Email: yvonne.orgill@bathroom-association.org.uk
Web: www.bathroom-association.org.uk

BRE CERTIFICATION LIMITED

Bucknalls Lane, Garston, Watford,
Hertfordshire, WD25 9XX
Area of Operation: Worldwide
Tel: 01923 664100
Fax: 01923 664603
Email: enquiries@brecertification.co.uk
Web: www.brecertification.co.uk

BRISTOL & SOMERSET RENEWABLE ENERGY ADVICE SERVICE

The CREATE Centre,
Smeaton Road, Bristol, BS1 6XN
Area of Operation: South West England
and South Wales
Tel: 0800 512012
Fax: 0117 929 9114
Email: reas@cse.org.uk
Web: www.cse.org.uk/renewables

BRITISH ARTIST BLACKSMITHS ASSOCIATION

Anwick Forge, 62 Main Road,
Anwick, Sleaford, Lincolnshire, NG34 9SU
Area of Operation: UK & Ireland
Tel: 01526 830303
Email: babasecretary@anwickforge.co.uk
Web: www.baba.org.uk

BRITISH BLIND & SHUTTER ASSOCIATION

42 Heath Street, Tamworth,
Staffordshire, B79 7JH
Area of Operation: UK & Ireland
Tel: 01827 52337
Fax: 01827 310827
Email: info@bbsa.org.uk
Web: www.bbsa.org.uk

BRITISH CEMENT ASSOCIATION

The Concrete Centre, Riverside House,
4 Meadows Business Park,
Station Approach, Blackwater,
Camberley, Surrey, GU17 9AB
Area of Operation: UK (Excluding Ireland)
Tel: 01276 608700
Fax: 01276 608701
Email: enquiries@concretecentre.com
Web: www.concretecentre.com

BRITISH WOODWORKING FEDERATION

55 Tufton Street, London, SW1P 3QL
Area of Operation: UK & Ireland
Tel: 0870 458 6939
Fax: 0870 458 6949
Email: bwf@bwf.org.uk
Web: www.bwf.org.uk ✏

THE BRITISH WOODWORKING FEDERATION

Area of Operation: UK & Ireland
Tel: 0870 458 6939   **Fax:** 0870 458 6949
Email: bwf@bwf.org.uk
Web: www.bwf.org.uk

As the UK woodworking and joinery industry's leading representative body, our 500+ members produce a variety of timber products and comply with a comprehensive Code of Conduct.

BWF - CERTIFIRE FIRE DOOR & DOORSET SCHEME

Area of Operation: UK & Ireland
Tel: 0870 458 6939   **Fax:** 0870 458 6949
Email: firedoors@bwf.org.uk
Web: www.bwf.org.uk

The BWF-CERTIFIRE Fire Door & Doorset Scheme can help you find the right fire door and correct, compatible components to suit your needs. Contact for members' details.

USEFUL INFORMATION

SPONSORED BY: SITEFINDERIRELAND.COM www.sitefinderireland.com

TWA SCHEME

Area of Operation: UK & Ireland
Tel: 0870 458 6939 **Fax:** 0870 458 6949
Email: windows@bwf.org.uk
Web: www.bwf.org.uk

In partnership with the British Standards Institute, TWA Scheme members' high quality timber windows & doorsets are rigorously tested and audited. Contact for members' details.

BUILDERS MERCHANTS FEDERATION
15 Soho Square, London, W1D 3HL
Area of Operation: UK (Excluding Ireland)
Tel: 0870 901 3380 **Fax:** 0207 734 2766
Email: info@bmf.org.uk
Web: www.bmf.org.uk

BUILDING PRODUCTS INDEX LTD
Acorn Centre, 30 Gorst Road, London, NW10 6LE
Area of Operation: Worldwide
Tel: 0208 838 1904
Fax: 0208 838 1905
Email: info@bpindex.co.uk
Web: www.bpindex.co.uk

CEDIA UK LTD
The Pixmore Centre, Pixmore Avenue, Letchworth, Hertfordshire, SG6 1JG
Area of Operation: Worldwide
Tel: 01462 627377
Fax: 01462 620429
Email: info@cedia.co.uk
Web: www.cedia.co.uk

CERAM
Queens Road, Penkhull, Stoke-on-Trent, Staffordshire, ST4 7LQ
Area of Operation: Worldwide
Tel: 01782 764444
Fax: 01782 412331
Email: enquiries@ceram.com
Web: www.ceram.com

CONSERVATION REGISTER
c/o Institute of Conservation, 3rd Floor, Downstream Building, 1 London Bridge, London, SE1 9BG
Area of Operation: UK & Ireland
Tel: 0207 785 3804
Email: info@conservationregister.com
Web: www.conservationregister.com

CORGI
1 Elmwood, Chineham Park, Crockford Lane, Chineham, Hampshire, RG24 8WG
Area of Operation: UK & Ireland
Tel: 0870 401 2200
Email: enquiries@corgi-group.com
Web: www.corgi-group.com

ELECSA LTD
44-48 Borough High Street, London, SE1 1XB
Area of Operation: East England, Greater London, Midlands & Mid Wales, North East England, North West England and North Wales, South East England, South West England and South Wales
Tel: 0870 749 0080
Fax: 0870 749 0085
Email: enquiries@elecsa.org.uk
Web: www.elecsa.org.uk

FAIRTRADES
Quadrant House, The Quadrant, Hoylake, Wirral, CH47 2EE
Area of Operation: UK & Ireland
Tel: 0870 738 4858
Fax: 0870 738 4868
Web: www.fairtrades.co.uk

FEDERATION OF MASTER BUILDERS
Gordon Fisher House, 14-15 Great James Street, London, WC1N 3DP
Area of Operation: UK & Ireland
Tel: 0207 242 7583
Fax: 0207 404 0296
Email: central@fmb.org.uk
Web: www.findabuilder.co.uk

FENSA LTD
44 - 48 Borough High Street, London, Greater London, SE1 1XB
Area of Operation: East England, Greater London, Midlands & Mid Wales, North East England, North West England and North Wales, South East England, South West England and South Wales
Tel: 0870 780 2028
Fax: 0870 780 2029
Email: enquiries@fensa.org.uk
Web: www.fensa.co.uk

FOREST STEWARDSHIP COUNCIL UK WORKING GROUP
11-13 Great Oak Street, Llanidloes, Powys, SY18 6BU
Area of Operation: UK (Excluding Ireland)
Tel: 01686 413916 **Fax:** 01686 412176
Email: info@fsc-uk.org
Web: www.fsc-uk.org

GLASS AND GLAZING FEDERATION
44-48 Borough High Street, London, SE1 1XB
Area of Operation: UK & Ireland
Tel: 0870 042 4255 **Fax:** 0870 042 4266
Email: info@ggf.org.uk
Web: www.ggf.org.uk

HISTORIC SCOTLAND
Longmore House, Salisbury Place, Edinburgh, Lothian, EH9 1SH
Area of Operation: Worldwide
Tel: 0131 668 8668
Fax: 0131 668 8669
Email: hs.conservation.bureau@scotland.gsi.gov.uk
Web: www.historic-scotland.gov.uk

HOMEPRO LTD
Quadrant House, The Quadrant, Hoylake, Wirral, CH47 2EE
Area of Operation: UK & Ireland
Tel: 0870 738 4858
Fax: 0870 738 4868
Web: www.homepro.com

ISTRUCTE
Institution of Structural Engineers, 11 Upper Belgrave Street, London, SW1X 8BH
Area of Operation: UK (Excluding Ireland)
Tel: 0207 201 9104
Fax: 0207 201 9148
Email: pile@istructe.org.uk
Web: www.istructe.org.uk

KBSA
12 Top Barn Business Centre, Holt Heath, Worcester, Worcestershire, WR6 6NH
Area of Operation: UK (Excluding Ireland)
Tel: 01905 621787
Fax: 01905 621887
Email: info@kbsa.co.uk
Web: www.kbsa.co.uk

KELLY-WOODWARD & ASSOCIATES
Technology House, Lissadel Street, Manchester, M6 6AP
Area of Operation: Midlands & Mid Wales, North West England and North Wales
Tel: 0161 278 2684/5
Fax: 0161 278 2686
Email: debbiekelly@kwapr.co.uk
Web: www.kwapr.co.uk

USEFUL INFORMATION

NATIONAL FIREPLACE ASSOCIATION
6th Floor, McLaren Building, 35 Dale End,
Birmingham, West Midlands, B4 7LN
Area of Operation: UK & Ireland
Tel: 0121 200 1310
Fax: 0121 200 1306
Email: enquiries@nfa.org.uk
Web: www.nfa.org.uk

NATIONAL FIREPLACE ASSOCIATION
Area of Operation: UK & Ireland
Tel: 0121 200 1310
Fax: 0121 200 1306
Email: enquiries@nfa.org.uk
Web: www.nfa.org.uk

The NFA is the trade association of the fire & fireplace industry. Its primary objective is to promote stoves, fires and fireplaces for all fuels. Many of the better, most established and experienced Showrooms and Manufacturers in the UK are members of the Association. When purchasing a fire, fireplace or stove, look for the logo that proves a Showroom knows its business! Full details on our newly updated web site www.nfa.org.uk

**NATIONAL HOME
IMPROVEMENT COUNCIL**
Carlye House, 235 Vauxhall Bridge Road,
London, SW1V 1EJ
Area of Operation: UK (Excluding Ireland)
Tel: 0207828 8230
Fax: 0207828 0667
Email: info@nhic.org.uk

NATIONAL PLANT HIRE GUIDE
140 Wales Farm Road, London, W3 SUG
Area of Operation: UK (Excluding Ireland)
Tel: 0870 737 4040
Fax: 0870 737 6060
Email: info@planthireguide.co.uk
Web: www.planthireguide.co.uk

PRECAST FLOORING FEDERATION
60 Charles Street, Leicester,
Leicestershire, LE1 1FB
Area of Operation: UK (Excluding Ireland)
Tel: 0116 253 6161
Fax: 0116 251 4568
Email: info@precastfloors.info
Web: www.pff.org.uk

SOCIETY OF GARDEN DESIGNERS
Katepwa, Ashfield Park Avenue, Ross on Wye,
Herefordshire, HR9 5AX
Area of Operation: Worldwide
Tel: 01989 566695
Fax: 01989 567676
Email: info@sgd.org.uk
Web: www.sgd.org.uk

SOLID FUEL ASSOCIATION LTD (SFA)
7 Swanwick Court, Alfreton,
Derbyshire, DE55 7AS
Area of Operation: UK (Excluding Ireland)
Tel: 01773 835400
Fax: 01773 834351
Email: sfa@solidfuel.co.uk
Web: www.solidfuel.co.uk

STRAW BALE BUILDING ASSOCIATION
Holinroyd Farm, Butts Lane, Todmorden,
West Yorkshire, OL14 8RJ
Area of Operation: UK & Ireland
Tel: 01706 814696
Email: info@strawbalebuildingassociation.co.uk
Web: www.strawbalebuildingassociation.co.uk

**SWIMMING POOL
& ALLIED TRADES ASSOCIATION**
1a Junction Road, Andover,
Hampshire, SP10 3QT
Area of Operation: Worldwide
Tel: 01264 356210
Fax: 01264 332628
Email: admin@spata.co.uk
Web: www.spata.co.uk

**THE ASSOCIATION
FOR PROJECT SAFETY**
16 Rutland Square, Edinburgh,
Lothian, EH1 2BB
Area of Operation: UK & Ireland
Tel: 0131 221 9959
Fax: 0131 221 0061
Email: info@aps.org.uk
Web: www.aps.org.uk

**THE BRICK DEVELOPMENT
ASSOCIATION LTD.**
Woodside House, Winkfield,
Windsor, Berkshire, SL4 2DX
Area of Operation: UK (Excluding Ireland)
Tel: 01344 885651
Fax: 01344 890129
Email: brick@brick.org.uk
Web: www.brick.org.uk

THE BRITISH PHOTOVOLTAIC ASSOCIATION
The National Energy Centre,
Davy Avenue, Knowlhill, Milton Keynes,
Buckinghamshire, MK5 8NG
Area of Operation: UK (Excluding Ireland)
Tel: 01908 442291
Fax: 0870 052 9193
Email: enquiries@pv-uk.org.uk
Web: www.pv-uk.org.uk

**THE CHARTERED INSTITUTE OF
ARCHITECTURAL TECHNOLOGISTS**
397 City Road, London,
Greater London, EC1V 1NH
Area of Operation: Worldwide
Tel: 0207 278 2206
Fax: 0207 837 3194
Email: info@ciat.org.uk
Web: www.ciat.org.uk

THE GUILD OF BUILDERS AND CONTRACTORS
Office 7 Epic House, 128 Fulwell Road,
Teddington, Middlesex,
Greater London, GW11 ORQ
Area of Operation: Worldwide
Tel: 0208 977 1105
Fax: 0208 943 3151
Email: info@buildersguild.co.uk
Web: www.buildersguild.co.uk

THE LISTED PROPERTY OWNERS CLUB
Lower Dane, Hartlip,
Sittingbourne, Kent, ME9 7TE
Area of Operation: Worldwide
Tel: 01795 844939
Fax: 01795 844862
Email: info@lpoc.co.uk
Web: www.lpoc.co.uk

THE MORTGAGE SHOP ONLINE
231 Grimsby Road, Cleethorpes,
Lincolnshire, DN35 7HE
Area of Operation: East England,
North East England, North West England
and North Wales, South East England,
South West England and South Wales
Tel: 01472 200664
Fax: 01472 200664
Email: mail@mortgageshop-online.co.uk

THE NATIONAL ENERGY FOUNDATION
The National Energy Centre,
Davy Avenue, Knowlhill, Milton Keynes,
Buckinghamshire, MK5 8NG
Area of Operation: UK (Excluding Ireland)
Tel: 01908 665555
Fax: 01908 665577
Email: info@nef.org.uk
Web: www.nef.org.uk

**THE NATIONAL FEDERATION
OF ROOFING CONTRACTORS**
24 Weymouth Street,
London, W1G 7LX
Area of Operation: UK (Excluding Ireland)
Tel: 0207 436 0387
Fax: 0207 637 5215
Email: info@nfrc.co.uk
Web: www.nfrc.co.uk

THE SOLAR TRADE ASSOCIATION
The National Energy Centre,
Davy Avenue, Knowlhill, Milton Keynes,
Buckinghamshire, MK5 8NG
Area of Operation: UK (Excluding Ireland)
Tel: 01908 442290
Fax: 0870 052 9194
Email: enquiries@solartradeassociation.org.uk
Web: www.solartradeassociation.org.uk

THE TILE ASSOCIATION
Forum Court, 83 Copers Cope Road,
Beckenham, Kent, BR3 1NR
Area of Operation: UK & Ireland
Tel: 0208 663 0946
Fax: 0208 663 0949
Email: info@tiles.org.uk
Web: www.tiles.org.uk

THE UK TRADES CONFEDERATION
1st Floor Braintree House,
Braintree Road, Ruisliip, Middlesex,
Greater London, HA4 0EJ
Area of Operation: UK & Ireland
Tel: 0870 922 0442
Fax: 0870 922 0441
Email: mail@uktc.org
Web: www.uktc.org

THE UK TRADES CONFEDERATION
Area of Operation: UK & Ireland
Tel: 0870 922 0442 **Fax:** 0870 922 0441
Email: mail@uktc.org
Web: www.uktc.org

The UK Trades Confederation exists to help members increase their business, protect their company and reduce overheads through an extensive range of benefits and services.

THE WELSH TIMBER FORUM
Boughrood House, 97 The Struet,
Brecon, Powys, LD3 7LS
Area of Operation: Midlands & Mid Wales,
North West England and North Wales, South
West England and South Wales
Tel: 0845 456 0342
Fax: 01874 625965
Email: welshtimber@lineone.net
Web: www.welshtimberforum.co.uk

THE WOOD SHOP LIMITED
15 Spinney Way,
Needingworth, St Ives,
Cambridgeshire, PE27 4SR
Area of Operation: Worldwide
Tel: 01480 469367
Fax: 01480 469366
Email: rgalpin@thewoodshop.biz
Web: www.thewoodshop.biz

TIMBER DECKING ASSOCIATION
CIRCE Building, Wheldon Road,
Castleford, West Yorkshire, WF10 2JT
Area of Operation: Europe
Tel: 01977 712718
Fax: 01977 712713
Email: info@tda.org.uk
Web: www.tda.org.uk

TRADITIONAL HOUSING BUREAU
4th Floor, 60 Charles Street,
Leicester, Leicestershire, LE1 1FB
Area of Operation: UK (Excluding Ireland)
Tel: 0116 253 6161
Fax: 0116 251 4568
Email: info@housebuilder.org
Web: www.housebuilder.org.uk

UK CAST STONE ASSOCIATION
15 Stone Hill Court, The Arbours,
Northampton, Northamptonshire, NN3 3RA
Area of Operation: UK & Ireland
Tel: 01604 405666 **Fax:** 01604 405666
Email: info@ukcsa.co.uk
Web: www.ukcsa.co.uk

UK TIMBER FRAME ASSOCIATION
The E Centre, Cooperage Way Business Village,
Alloa, Clackmannanshire, FK10 3LP
Area of Operation: UK (Excluding Ireland)
Tel: 01259 272140
Fax: 01259 272141
Email: info@timber-frame.org
Web: www.timber-frame.org

WWW.FIREPLACE.CO.UK
Old Coach House, Southern Road,
Thame, Oxfordshire, OX9 2ED
Area of Operation: UK (Excluding Ireland)
Tel: 01844 260960
Fax: 01844 260267
Web: www.fireplace.co.uk

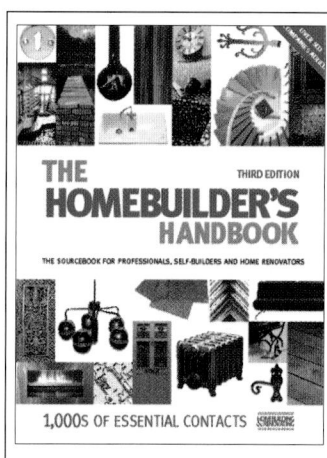

Find a Plot

In any assessment of a potential plot it's important to look beyond the immediate boundaries of the property you've come to see. Don't just start looking at things as you arrive at the site - as you get near the plot from either direction, watch out for changes in the social character of the area. Also look out for 'For Sale' signs and ring the agents to check on values in the area. If there are loads of properties up for sale, ask yourself why. The value of our homes depends to a large degree on those around and about us sharing those values.

Also study the architectural vernacular - in particular, watch out for houses that have recently been built to give you some indication of the types and designs of homes that the planners have approved. Look out for older houses as well and pay attention to how they have faired over the years. Keep an eye open for cracks in the walling, areas that have been re-pointed or rendered - they may be of substandard construction, or the area may suffer from bad ground conditions.

Good plots with planning permission are difficult to come by, and there is great competition for the few that are available. This is why "bungalow eating", the process of buying a house to demolish and rebuilding on the land (with planning permission for a 'replacement dwelling'), is becoming increasingly popular. It also means that the plot will have the advantages of existing highway access and connections to key sevices.

When looking for a renovation opportunity, always use your imagination and look for the potential in the property. Look at the existing space you have available, including basement or attic space to expand into. Also check whether there are partition walls which can be knocked down to increase room sizes, and see how much outside space would be available for an extension or conservatory. In period houses, it is always a good idea to look for any original features, as these can add a great deal of character and value to the finished house.

Alternatively, there is the possibility of a conversion project. Barns are the favourite option, but other structures can also offer unique opportunities. It is also often easier to get planning permission for a conversion than it is for a new build on a fresh plot; The Government is encouraging more houses on brown-field sites, and this includes the residential re-use of existing buildings. However, before planning permission is granted, a full schedule of works may need to be submitted, so make sure you see this before agreeing the property price. Always buy subject to planning permission, using an Option Agreement or a Conditional Contract.

5 things to check before you buy a plot

- Assess the area – approach the site from all directions and look for facilities, changes in the social character etc.
- Check planning permission is in place and look for any conditions associated to it.
- If no planning permission is in place, check whether it is likely to be granted with the Local Planning Authority.
- Look for possible problems, like the possibility of flooding; positioning of access roads; any trees with preservation orders; or connection to drains and services.
- Watch for legal problems such as restrictive covenants, ransom strips, or the land being unregistered.

5 things to check before you buy a renovation/conversion opportunity:

- Find out the maximum end value of houses in the same street / area.
- Assess the cost of basic works, and compare with the estimated sell on value.
- Check Permitted Development Rights are in place.
- Check the extent of any structural damage.
- Investigate the quality of local amenities; look for good transport links, schools and entertainment facilities, and check for local regeneration plans.

USEFUL INFORMATION

USEFUL INFORMATION - Building Plots, Renovation & Conversion Opportunities; Area Checks; Caravans, Static Homes & Site Accommodation

SPONSORED BY: PLOTFINDER.NET www.plotfinder.net

Area Checks

When buying a building plot or property, it is always a good idea to look thoroughly at the area you will be moving to. It is all too easy to buy an excellent property or plot, only to discover later that the quiet pub down the road releases gangs of noisy drunks into your street at closing time.

Before you buy, there are a number of factors you should take into account.

■ Look further afield than the boundaries of your property. Check out the street you'll be living in and the surrounding area thoroughly before you proceed.

■ Visit at different times of the day – it may seem like a quiet area in mid-afternoon, but both traffic and parking may be difficult at school times and rush hour. It is also worth visiting late at night to see whether the street becomes noisy or dangerous when the pubs and clubs close.

■ Buy local newspapers to see what sort of news and events may effect you in your new home.

■ Check crime figures and employment levels in the area.

■ Look how far away amenities such as schools, shops, bars and restaurants are, and also check travel links for convenience.

Of course, it is always best to visit the area yourself to get a feel for it's atmosphere, but finding out the facts and figures can be time consuming work. This is where area check services can prove useful; as they are usually online, you can get most of this information within seconds.

Many of these companies will also be able to provide information on the environmental condition of the area, including air pollution levels, soil type, risk of flooding and subsidence etc., as well as what the history of the site. This can be important, as if it has previously been used for industrial or landfill purposes then you may wish to have a ground survey done prior to purchase.

BUILDING PLOTS, RENOVATION & CONVERSION OPPORTUNITIES

KEY
OTHER: ▽ Reclaimed ⌃ On-line shopping
  Bespoke Hand-made ECO Ecological

BRITISH BOOKSHOPS
Sussex Stationers, 55/56 East Street,
Brighton, East Sussex, BN1 2JT
Area of Operation: South East England
Tel: 01273 220967
Fax: 01273 202116
Email: maproom@britishbookshops.co.uk
Web: www.britishbookshops.co.uk

FRENCH PROPERTY SHOP
Elwick Club, Church Road,
Ashford, Kent, TN23 1QG
Area of Operation: Europe
Tel: 01233 666902
Fax: 01233 666903
Email: nicholas@frenchpropertyshop.com
Web: www.frenchpropertyshop.com

LANDBANK SERVICES
P O Box 2035, Reading, Berkshire, RG6 7FJ
Area of Operation: UK (Excluding Ireland)
Tel: 0118 962 6022 **Fax:** 0118 962 6023
Email: info@landbank.co.uk
Web: www.landbank.co.uk ⌃

PLOTFINDER
2 Sugar Brook Court, Aston Road,
Bromsgrove, Worcestershire, B60 3EX
Area of Operation: UK & Ireland
Tel: 01527 834436 **Fax:** 01527 837810
Email: laura.cassidy@centaur.co.uk
Web: www.plotfinder.net ⌃

PROPERTYPORTAL.COM
Brookfield House, Ford Lane, Frilford,
Abingdon, Oxfordshire, OX13 5NT
Area of Operation: UK (Excluding Ireland)
Tel: 0870 380 0520
Fax: 0870 380 0521
Email: info@propertyportal.com
Web: www.propertyportal.com

WWW.PROPERTYFINDERFRANCE.NET
2 Sugar Brook Court, Aston Road,
Bromsgrove, Worcestershire, B60 3EX
Area of Operation: Worldwide
Tel: 01527 834435
Fax: 01527 837810
Email: alex.worthington@centaur.co.uk
Web: www.propertyfinderfrance.net ⌃

AREA CHECKS AND ADVICE

KEY
PRODUCT TYPES: 1= Environmental Information 2 = Facilities Information 3 = Other
OTHER: ▽ Reclaimed ⌃ On-line shopping
  Bespoke Hand-made ECO Ecological

NATIONAL MAP CENTRE
11 Hertfordshire Business Centre,
Alexander Road, London Colney,
St Albans, Hertfordshire, AL2 1JG
Area of Operation: UK (Excluding Ireland)
Tel: 0845 606 1060
Fax: 01727 827 256
Email: enquiries@mapsnmc.co.uk
Web: www.planningmaps.co.uk ⌃
Product Type: 3

SEVENOAKS ENVIRONMENTAL CONSULTANCY LTD
19 Gimble Way, Pembury,
Tunbridge Wells, Kent, TN2 4BX
Area of Operation: East England, Greater London, Midlands & Mid Wales, North East England, North West England and North Wales, South East England, South West England and South Wales
Tel: 01892 822999
Fax: 01892 822992
Email: d.jones@sevenoaksenvironmental.co.uk
Web: www.sevenoaksenvironmental.co.uk
Product Type: 1

THE GLASGOW MAP CENTRE
TISO Glasgow,
50 Couper Street,
Glasgow, Lanarkshire, G4 0DL
Area of Operation: UK (Excluding Ireland)
Tel: 0141 552 7722
Fax: 0141 552 3345
Email: info@glasgowmapcentre.com
Web: www.glasgowmapcentre.com ⌃
Product Type: 3

UKVILLAGES LTD
37 Church Street, Harston,
Cambridgeshire, CB2 5NP
Area of Operation: UK & Ireland
Tel: 01223 874500
Email: info@ukvillages.co.uk
Web: www.ukvillages.co.uk

UP MY STREET
10th Floor,
Portland House, Stag Place,
London, SW1E 5BH
Area of Operation: UK & Ireland
Tel: 0207 802 2992
Fax: 0207233 5933
Email: pr@upmystreet.com
Web: www.upmystreet.com

CARAVANS, STATIC HOMES & SITE ACCOMMODATION

KEY
PRODUCT TYPES: 1= Caravans 2 = Static Homes 3 = Modular Buildings
OTHER: ▽ Reclaimed ⌃ On-line shopping
  Bespoke Hand-made ECO Ecological

AMBER LEISURE
Bentley Country Park, Flag Hill,
Great Bentley, Colchester, Essex, CO7 8RF
Area of Operation: South East England
Tel: 01255 821817
Fax: 01255 821909
Email: bentley@amberleisure.com
Web: www.amberleisure.com
Product Type: 1

ATLAS CARAVAN COMPANY LTD
Wykeland Industrial Estate, Wiltshire Road,
Hull, East Riding of Yorks, HU4 6PH
Area of Operation: North East England
Tel: 01482 562101
Fax: 01482 566033
Email: info@atlas-caravans.co.uk
Web: www.atlas-caravans.co.uk
Product Type: 1

C. JENKIN & SON
East Mascalls Farm,
East Mascalls Lane,
Lindfield, Haywards Heath,
West Sussex, RH16 2QN
Area of Operation: UK (Excluding Ireland)
Tel: 01444 482333
Fax: 01444 484580
Email: info@jenkinmobilehomes.com
Web: www.jenkinmobilehomes.com
Product Type: 1, 2, 3

USEFUL INFORMATION

Removals

Moving house is generally a stressful experience, but this can be lessened somewhat by choosing a good removals company to help you through the process.

Image supplied by White and Company

It is advisable that you contact a number of firms initially, in both your local area and the area you are moving to. Obviously it is important to get quotes, but don't choose a company simply based on price. Speak to as many companies as possible in person to get a feel for the company and ask any questions you may have to see how they deal with your queries: their staff will be coming into your home, so it is important that you feel comfortable with them. Also check with friends and family for any recommendations they may have.

Firstly you should prepare an inventory of your belongings to show to company representatives. By showing all companies the same details you will be guaranteed accurate quotes for the same amount of work. Also, be honest about what you will be moving: this will ensure that the companies can not only give you an accurate costing, but can also allocate the right sized van and enough staff for the job.

Don't forget to include less obvious items in your inventory, such as items in the garage, attic or garden, as these are often overlooked. This can also incur extra costs as you may need additional van space to transport them later.

If you expect any difficulties with moving large or bulky pieces of furniture, you should tell your removals company in advance; that way they can prepare

themselves with extra staff or equipment to aid the situation. Also, if necessary you should check in advance whether your workmen will be able to do minor jobs as they go along or whether you will be expected to do it: this should include things like lifting carpets, disconnecting white goods, dismantling furniture etc. Be aware that there may be an additional cost for such services.

Finally, you should prepare a list of questions to ask representatives before you make a final decision. Some examples of good questions would be

- ■ Do you have staff and vans available on my set moving date?
- ■ Will you take a provisional booking for this date?
- ■ Is there a cancellation fee if I have to change my moving date?
- ■ Do you offer insurance? If so, what is covered and what are the conditions? (Check with your own household contents insurance, as you may already be covered for some circumstances)
- ■ What is the expected timeframe, from arrival through to completion?
- ■ Are you a member of any trade association or governing body?
- ■ Do you have references from past clients?
- ■ What are the terms and conditions if I book with you?

Software

With a little bit of patience, even the most technophobic self-builder can make effective use of some, if not all, CAD packages, and will benefit greatly from doing so.

It's unlikely that anybody other than a professional will be able to produce a full set of working construction drawings of their dream home to take to a builder and say, "over to you." What is likely, however, is that by using CAD software you'll be able to try out different designs fully, taking into account the size and shape of your plot, to produce a very fine replication of what you're hoping for. Self-builders are full of ideas, some of them good and some of them best left unsaid, and domputers are fantastic at allowing ideas to be tried out quickly and cheaply. If you don't like something you can simply click 'delete' and your folly is consigned to memory.

For the self-builder choosing home design software, there are two main types to pick from. First, and most basic, are the simple floorplan packages which have limited usage and retail often for under a tenner. The second are the self-build 3D packages. These tend to pick up on the more accessible aspects of the pricey professional packages and give basic 3D images of your dream home without entering unnecessary detail, and at £30–50, are good value. Professional packages

are available, with price tag to reflect their intended market. It is highly unlikely that self-builders will find the initial outlay worthwhile, or find the time to really master the package, so in most cases they are best left to the professionals.

Computers are not just useful to the self-builder as a design tool, but can also help manage the build process. There are a range of software packages designed to assist in preparing simple budget estimations through to fully detailed construction costings. These programmes can also be used to develop build schedules and cashflow projections which will prove invaluable to the DIY project manager.

A lot of these programs come with in-built cost estimates of materials and labour and although these obviously quickly go out of date, they can be re-entered to reflect inflation, in some cases via the Internet. Some packages are beginning to merge project management tools with 3D design features — the perfect combination for the self-builder. Design your home, and then let the computer estimate how much it is likely to cost to build.

See how your future home will look without even laying a brick: modern CAD software allows you to view both the outside and inside of your home.
ABOVE: Exterior CAD image by Arcon Visual Architecture.
BELOW: CAD kitchen design by In-toto.

CARA-SALES
Newspace Site,
Catfoss Lane Industrial Estate,
Brandesburton, Hull,
East Riding of Yorks, YO25 8EJ
Area of Operation: UK (Excluding Ireland)
Tel: 01964 542266
Fax: 01964 542277
Email: info@cara-sales.com
Product Type: 1, 2

CARAVAN HIRE UK
Glanyrafon Industrial Estate,
Aberystwyth, Ceredigion, SY23 3JQ
Area of Operation: UK (Excluding Ireland)
Tel: 01970 626920
Fax: 01970 611922
Email: info@caravanhireuk.co.uk
Web: www.caravanhireuk.co.uk
Product Type: 1, 2

IAN JAMES CARAVANS
Yellowsands Holiday Park,
Coast Road, Brean, Somerset, TA8 2RH
Area of Operation: Europe
Tel: 01278 751349
Fax: 01278 751666
Email: icj@breanbeach.co.uk
Web: www.ianjamescaravans.co.uk
Product Type: 1, 2

OUTSOURCE SITE SERVICES
UK Control Centre, Bradford Street,
Shifnal, Shropshire, TF11 8AU
Area of Operation: UK & Ireland
Tel: 01952 277763
Fax: 01952 277764
Email: roger@out-source.biz
Web: www.out-source.biz
Product Type: 3

ROLLALONG LIMITED
Woolsbridge Industrial Park,
Three Legged Cross,
Wimborne, Dorset, BH21 6SF
Area of Operation: UK (Excluding Ireland)
Tel: 01202 824541
Fax: 01202 826525
Email: enquiries@rollalong.co.uk
Web: www.rollalong.co.uk
Product Type: 3

ROUNDSTONE CARAVANS
Worthing Road, Southwater,
Horsham, West Sussex, RH13 9JG
Area of Operation: UK (Excluding Ireland)
Tel: 01403 730218
Fax: 01403 732828
Email: sales@roundstonecaravans.com
Web: www.roundstonecaravans.com
Product Type: 1

SAMBECK CARAVANS LTD
Woodlands Business Park, Tenpenny Hill,
Thorrington, Essex, CO7 8JD
Area of Operation: UK (Excluding Ireland)
Tel: 01206 255223
Fax: 01206 257391
Email: sambeckcaravans@btconnect.com
Web: www.sambeckcaravans.co.uk
Product Type: 1, 2

SQUARE DEAL CARAVANS
Gundrymor Trading Estate,
Collingwood Road, West Moors,
Dorset, BH21 6QW
Area of Operation: South East England,
South West England and South Wales
Tel: 01202 892710
Fax: 01202 894511
Email: sqdealcaravans@Aol.com
Web: www.squaredealcaravans.com
Product Type: 1, 2

SURF BAY LEISURE
The Airfield, Winkleigh,
Devon, EX19 8DW
Area of Operation: UK & Ireland
Tel: 01837 680100
Fax: 01837 680200
Email: info@surfbay.dircon.co.uk
Web: www.surfbayleisure.co.uk
Product Type: 1, 2

WORCESTERSHIRE CARAVAN SALES
Nelson Road, Sandy Lane Industrial Estate,
Stourport-on-Severn,
Worcestershire, DY13 9QB
Area of Operation: UK (Excluding Ireland)
Tel: 01299 878872
Fax: 01299 879988
Email: wcaravan@aol.com
Web: www.worcestershire-caravan-sales.com
Product Type: 1

REMOVALS, STORAGE, AND CHANGE OF ADDRESS SERVICES

> **KEY**
>
> **PRODUCT TYPES:** 1= Removal Services
> 2 = Storage Facilities   3 = Change of Address
> Services
> **OTHER:** ▽ Reclaimed ⌂ On-line shopping
> ✎ Bespoke ✋ Hand-made   ECO Ecological

A & L REMOVALS
151 West Barnes Lane,
New Malden, Surrey, KT3 6HR
Area of Operation: Greater London,
South East England
Tel: 0208 949 8286
Fax: 0208 949 8286
Email: sales@removals.ws
Web: www.removals.ws
Product Type: 1

**ALLIANCE REMOVALS
& STORAGE LTD**
Unit 1, 57 North Street,
Portslade, Brighton,
East Sussex, BN41 1EH
Area of Operation: Europe
Tel: 01273 771741
Email: info@allianceremovals.co.uk
Web: www.allianceremovals.co.uk
Product Type: 1, 2

CAPITAL REMOVAL SERVICE
3 Bowland Road,
Woodford Green, Essex, IG8 7LX
Area of Operation: UK (Excluding Ireland)
Tel: 0208 505 7287
Email: capitalremovalservice@btinternet.com
Web: www.capitalremovalservice.co.uk
Product Type: 1

EASTEND REMOVALS
20 Salisbury Road, London, E10 5RG
Area of Operation: UK (Excluding Ireland)
Tel: 0845 838 1528
Email: eastendremovals@hotmail.com
Web: www.eastendremovals.co.uk ⌂
Product Type: 1, 2

HELP I AM MOVING.COM
Unit 3a, Engine Shed Lane,
Skipton, North Yorkshire, BD23 1UP
Area of Operation: Worldwide
Web: www.helpiammoving.com
Product Type: 1, 2

I AM MOVING.COM
Tel: 0845 0900198
Email: customersupport@iammoving.com
Web: www.iammoving.com

LO-COST MOVES
5 Newlands Avenue, Rochdale,
Lancashire, OL12 0BN
Area of Operation: UK (Excluding Ireland)
Tel: 0800 043 9014 **Fax:** 01706 673869
Email: pete@lo-costremovals.co.uk
Web: www.lo-costremovals.co.uk
Product Type: 1, 2

PAUL MATTHEWS AND SON LTD
Unit 21, Heol Ffaldau,
Brackla Industrial Estate,
Bridgend, CF31 2AJ
Area of Operation: East England,
Greater London, Midlands & Mid Wales,
North East England, Scotland, South East England,
South West England & South Wales
Tel: 01656 655466 **Fax:** 01656 655466
Email: paul@matthewsremovals.wanadoo.co.uk
Product Type: 1, 2

SPACES PERSONAL STORAGE
88 Bushey Road, Raynes Park, London, SW20 0JH
Area of Operation: UK (Excluding Ireland)
Tel: 0800 622244 **Fax:** 0208 946 8446
Email: enquiries@spacestore.co.uk
Web: www.spacestore.co.uk
Product Type: 2

WHITE AND COMPANY
International House, Britannia Road, London, EN8 7PF
Area of Operation: Worldwide
Tel: 0800 801999 **Fax:** 0208 441 0206
Email: melvyn.neal@whiteandcompanyyes.co.uk
Web: www.whiteandcompany.co.uk
Product Type: 1, 2

COMPUTER SOFTWARE

> **KEY**
>
> **PRODUCT TYPES:** 1= CAD Software
> 2 = Estimating Software
> 3 = Project Management Software
> **OTHER:** ▽ Reclaimed ⌂ On-line shopping
> ✎ Bespoke ✋ Hand-made   ECO Ecological

**ADVANCED COMPUTER
SOLUTIONS (EUROPE) LIMITED**
Unit 1 Home Farm Business Centre,
Cardington, Bedfordshire, MK44 3SN
Area of Operation: Worldwide
Tel: 01234 834920 **Fax:** 01234 832601
Email: derekb@caddie.co.uk
Web: www.caddiesoftware.com
Product Type: 1

ARCON VISUAL ARCHITECTURE
Online Warehouse Ltd, Unit 2, Bentley Stores,
Bentley, Farnham, Surrey, GU10 5HY
Area of Operation: Europe
Tel: 01420 520023 **Fax:** 01420 520077
Email: sales@3darchitect.co.uk
Web: www.3darchitect.co.uk
Product Type: 1

AUTODESK
Unit 8, Thame 40,
Jane Morbey Road, Thame,
Oxfordshire, OX9 3RR
Area of Operation: UK (Excluding Ireland)
Tel: 01844 261872
Fax: 01844 216737
Email: sales@manandmachine.co.uk
Web: www.manandmachine.co.uk
Product Type: 1

AVANQUEST UK
Sheridan House, 40-43 Jewry Street,
Winchester, Hampshire, SO33 8RY
Area of Operation: Worldwide
Tel: 01962 835000
Email: info@turbocad.co.uk
Web: www.avanquest.co.uk ⌂
Product Type: 1

EASY PRICE PRO LTD
The Software Centre,
Wattisfield Hall, Wattisfirld,
Suffolk, IP22 1NX
Area of Operation: UK & Ireland
Tel: 0845 612 4747
Fax: 0845 612 4748
Email: info@easypricepro.com
Web: www.easypricepro.com ⌂
Product Type: 2, 3

ESTIMATORS LIMITED
12A High Street, Cheadle,
Stockport, Cheshire, SK8 1AL
Area of Operation: UK (Excluding Ireland)
Tel: 0161 286 8601
Fax: 0161 428 5788
Email: mail@estimators-online.com
Web: www.estimators-online.com
Product Type: 2

HBXL
Citypoint, Temple Gate,
Bristol, BS1 6PL
Area of Operation: UK & Ireland
Tel: 0870 850 2444
Fax: 0870 850 2555
Email: sales@hbxl.co.uk
Web: www.hbxl.co.uk ⌂
Product Type: 2

MAC-3D ADVANCED VISUALS
4 Carlton Terrace, London Road,
Stranraer, Dumfries & Galloway, DG9 8AG
Area of Operation: UK & Ireland
Tel: 07941 071575
Email: info@mac-3d.co.uk
Web: www.mac-3d.co.uk
Product Type: 1

WEBINSPIRED LTD
6 Providence Court, Pynes Hill,
Exeter, Devon, EX2 5JL
Area of Operation: UK (Excluding Ireland)
Tel: 0870 922 0765
Fax: 0870 922 0765
Email: info@webinspired.net
Web: www.webinspired.net

USEFUL INFORMATION

BOOKS AND OTHER LITERATURE

KEY

PRODUCT TYPES: 1= Books
2 = Magazines   3 = Other
OTHER: ▽ Reclaimed   On-line shopping
Bespoke Hand-made ECO Ecological

BASEMENT INFORMATION CENTRE
Riverside House,
4 Meadows Business Park,
Station Approach,
Blackwater, Camberley,
Surrey, GU17 9AB
Area of Operation: UK (Excluding Ireland)
Tel: 01276 33155
Fax: 01276 606801
Email: info@tbic.org.uk
Web: www.tbic.org.uk / www.basements.org.uk
Product Type: 3

BLACKBERRY BOOKS
8 Newport Road, Godshill,
Isle of Wight, PO38 3HR
Area of Operation: UK & Ireland
Tel: 01983 840310
Fax: 01983 840310
Email: info@blackberry-books.co.uk
Web: www.blackberry-books.co.uk
Product Type: 1

BRE CERTIFICATION LIMITED
Bucknalls Lane, Garston,
Watford, Hertfordshire, WD25 9XX
Area of Operation: Worldwide
Tel: 01923 664100
Fax: 01923 664603
Email: enquiries@brecertification.co.uk
Web: www.brecertification.co.uk
Product Type: 3

CENTRE FOR ALTERNATIVE TECHNOLOGY
Llwyngwren Quarry,
Machynlleth, Powys, SY20 9AZ
Area of Operation: Worldwide
Tel: 01654 705950
Fax: 01654 702782
Email: lucy.stone@cat.org.uk
Web: www.cat.org.uk
Product Type: 1, 2

DIYDOCTOR
FTSC, Badgers Hill,
Frome, Somerset, BA11 2EH
Area of Operation: UK & Ireland
Tel: 01373 303930
Fax: 01373 301438
Email: office@diydoctor.org.uk
Web: www.diydoctor.org.uk
Product Type: 1, 3

HIDDENWIRES.CO.UK
Gipsy Road, London, SE27 9RB
Area of Operation: Europe
Tel: 0208 761 1042
Email: info@hiddenwires.co.uk
Web: www.hiddenwires.co.uk
Product Type: 3

HOMEBUILDING & RENOVATING BOOK OF BARN CONVERSIONS
Ascent Publishing Ltd, 2 Sugar Brook Court,
Aston Road, Bromsgrove, Worcestershire, B60 3EX
Area of Operation: Worldwide
Tel: 01527 834435
Fax: 01527 837810
Email: customerservice@centaur.co.uk
Web: www.homebuilding.co.uk
Product Type: 1

HOMEBUILDING & RENOVATING BOOK OF CONTEMPORARY HOMES
Ascent Publishing Ltd, 2 Sugar Brook Court,
Aston Road, Bromsgrove, Worcestershire, B60 3EX
Area of Operation: Worldwide
Tel: 01527 834435
Fax: 01527 837810
Email: customerservice@centaur.co.uk
Web: www.homebuilding.co.uk
Product Type: 1

HOMEBUILDING & RENOVATING BOOK OF GREAT VALUE SELF-BUILD HOMES
Ascent Publishing Ltd, 2 Sugar Brook Court,
Aston Road, Bromsgrove, Worcestershire, B60 3EX
Area of Operation: Worldwide
Tel: 01527 834435
Fax: 01527 837810
Email: customerservice@centaur.co.uk
Web: www.homebuilding.co.uk
Product Type: 1

HOMEBUILDING & RENOVATING MAGAZINE
Ascent Publishing Ltd, 2 Sugar Brook Court,
Aston Road, Bromsgrove, Worcestershire, B60 3EX
Area of Operation: Worldwide
Tel: 01527 834400
Fax: 01527 834499
Email: customerservice@centaur.co.uk
Web: www.homebuilding.co.uk
Product Type: 2

HOMEBUILDING & RENOVATING MAGAZINE ARCHIVE CD ROM
Ascent Publishing Ltd,
2 Sugar Brook Court, Aston Road,
Bromsgrove, Worcestershire, B60 3EX
Area of Operation: Worldwide
Tel: 01527 834435
Fax: 01527 837810
Email: customerservice@centaur.co.uk
Web: www.homebuilding.co.uk
Product Type: 3

HOUSE MOUSE
Hills Farm, Lawshall,
Bury St. Edmunds, Suffolk, IP29 4PJ
Area of Operation: UK (Excluding Ireland)
Tel: 01284 830492
Fax: 01284 830495
Email: bpharber@aol.com
Product Type: 1

HOW TO CREATE A JARDIN PAYSIN
Ascent Publishing Ltd,
2 Sugar Brook Court, Aston Road,
Bromsgrove, Worcestershire, B60 3EX
Area of Operation: Worldwide
Tel: 01527 834435
Fax: 01527 837810
Email: customerservice@centaur.co.uk
Web: www.homebuilding.co.uk
Product Type: 1

HOW TO RENOVATE A HOUSE IN FRANCE
Ascent Publishing Ltd,
2 Sugar Brook Court, Aston Road,
Bromsgrove, Worcestershire, B60 3EX
Area of Operation: Worldwide
Tel: 01527 834435
Fax: 01527 837810
Email: customerservice@centaur.co.uk
Web: www.homebuilding.co.uk
Product Type: 1

LOW-IMPACT LIVING INITIATIVE
Redfield Community, Buckingham Road,
Winslow, Buckinghamshire, MK18 3LZ
Area of Operation: UK (Excluding Ireland)
Tel: 01296 714184
Fax: 01296 714184
Email: lili@lowimpact.org
Web: www.lowimpact.org
Product Type: 3

OVOLO PUBLISHING LTD
1 The Granary, Brook Farm,
Ellington, Cambridgeshire, PE28 0AE
Area of Operation: Worldwide
Tel: 01480 891595
Fax: 01480 891777
Email: info@ovolopublishing.co.uk
Web: www.ovolopublishing.co.uk
Product Type: 1

RIBA BOOKSHOPS
15 Bondhill Street,
London, EC2P 2EA
Area of Operation: Worldwide
Tel: 0207 256 7222
Fax: 0207 374 2737
Email: marketing@ribabooks.com
Web: www.ribabookshops.com
Product Type: 1, 2, 3

SCOTTISH LIME CENTRE TRUST
The School House,
Rocks Road, Charlestown,
Nr Dunfermline, Fife, KY11 3EN
Area of Operation: Scotland
Tel: 01383 872722
Fax: 01383 872744
Email: info@scotlime.org
Web: www.scotlime.org
Product Type: 1

SELF BUILD IRELAND LTD
96 Lisburn Road, Saintfield,
Co Down, BT24 7BP
Area of Operation: Ireland Only
Tel: 028 9751 0570
Fax: 028 9751 0576
Email: info@selfbuild.ie
Web: www.selfbuild.ie
Product Type: 2

SELF BUILD PRO
Belmont Business Centre,
Brook Lane, Endon,
Staffordshire, ST9 9EZ
Area of Operation: UK & Ireland
Tel: 01782 505127
Fax: 01782 505127
Email: enquiries@self-build-pro.co.uk
Web: www.self-build-pro.co.uk
Product Type: 1

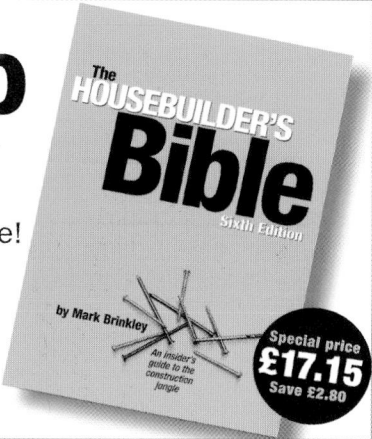

USEFUL INFORMATION

Books and Other Literature

Books are the traditional and most commonly used way to find out answers to your self-build or renovating questions. Even with the rise of the Internet and the expanse of varied information it provides, books have remained popular resource tools, with more and more subject specific texts, from roofing structure to interior decoration, being published every year. However, a relatively new addition to the market is the e-book: either an online version of a traditional text, or completely fresh material which often incorporates extras, like spreadsheet tools, web links and directories.

Ascent Publishing, the company behind Homebuilding & Renovating magazine, currently produces six such books on the subject of self-build (including this Handbook)

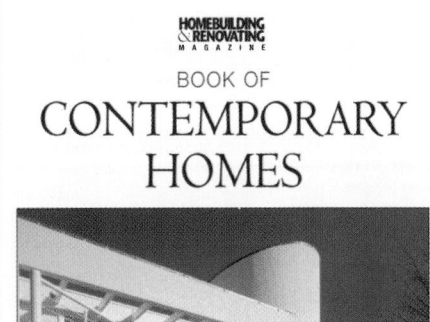

BOOK OF
CONTEMPORARY HOMES

19 INSPIRATIONAL INDIVIDUALLY DESIGNED HOMES

The H&R Book of Great Value Self-Builds profiles 24 homes featured in Homebuilding & Renovating Magazine, which were recently built for between £32,000 and £150,000. These features show how it is possible to best use floor space without sacrificing unique features, and how to achieve maximum style without spending a fortune. It is an inspirational guide to how you can build a smart, elegant, spacious family home on a modest budget, and get more house for your money.

The H&R Book of Barn Conversions contains 22 fully-costed case studies featured in Homebuilding & Renovating Magazine. The projects vary in design from rustic to contemporary conversions, and costs start at just £15,000. The book also contains an introduction to converting a barn, from planning permission through to renovations, as well as tips and style guides on how to make the most of your original features and incorporate more modern designs, both inside and out.

The H&R Book of Contemporary Homes comprises 19 individually designed, contemporary-styled homes previously featured in Homebuilding & Renovating magazine. With a huge variety of locations, from urban homes to country houses, this book aims to show how contemporary homes should not be limited to fashionable or commercial areas, and gives inspiration on how they can be sympathetic to their surrounding environment. Each case study includes inspirational pictures of both the exterior and interior of the property, its floorplan and layout, cost outlines for the build, and a list of useful contacts.

HOW TO RENOVATE A HOUSE IN FRANCE

DAVID ACKERS
JÉRÔME AUMONT
PAUL CARSLAKE

A good number of Brits are choosing to up sticks and move to France, and who can blame them, with good weather, picturesque countryside, vibrant cities and charming culture to look forward to? With flights being fast, cheap and freely available, a home in France can be just minutes away from friends and family left in the UK and, of course, the main perk of building in France is that the price of land and property is considerably cheaper than in Britain. Also, whereas the location of builds in the UK is often dictated by the proximity of school or work locations, buying in France is not usually quite so geographically tied.

How to Renovate a House in France is an essential book for anyone looking to refurbish a property across the Channel, containing advice on a wide variety of subjects, including buying an old house, getting planning permission, hunting for contractors and products, and using traditional or reclaimed materials. It also contains a mini directory of contact details, and an English-French glossary of useful words and phrases.

How to create a
JARDIN PAYSAN

How to Create a Jardin Paysin explains the tradition of French rural gardens, and takes you through the process of creating and maintaining a natural garden using traditional methods. Its author, Louise Ranck, is a renowned expert on rural life style, and the book is filled with beautiful photography, practical illustrations and useful hints and tips.

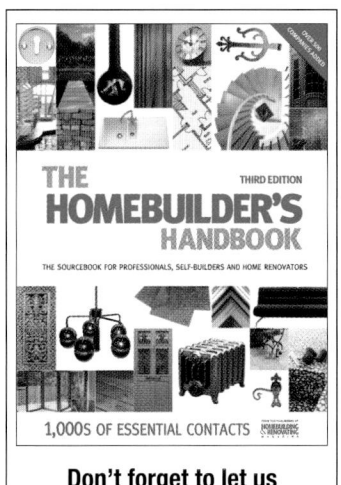

THE HOMEBUILDER'S HANDBOOK

THIRD EDITION

THE SOURCEBOOK FOR PROFESSIONALS, SELF-BUILDERS AND HOME RENOVATORS

1,000S OF ESSENTIAL CONTACTS

Don't forget to let us know about companies you think should be listed in the next edition
email:
customerservice@centaur.co.uk

SPEER DADE PLANNING CONSULTANTS
10 Stonepound Road, Hassocks, West Sussex, BN6 8PP
Area of Operation: UK (Excluding Ireland)
Tel: 01273 843737
Fax: 01273 842155
Email: roy@stonepound.co.uk
Web: www.stonepound.co.uk
Product Type: 1

STOBART DAVIES LIMITED
Stobart House, Pontyclerc, Penybanc Road,
Ammanford, Carmarthenshire, SA18 3HP
Area of Operation: Worldwide
Tel: 01269 593100
Fax: 01269 596116
Email: sales@stobartdavies.com
Web: www.stobartdavies.com
Product Type: 1

THE BUILDING CENTRE
26 Store Street, off Tottenham Court Road,
London, WC1E 7BT
Area of Operation: Worldwide
Tel: 0207 692 4000
Fax: 0207 631 0329
Email: information@buildingcentre.co.uk
Web: www.buildingcentre.co.uk
Product Type: 1, 2, 3

THE HOUSEBUILDER'S BIBLE 6TH EDITION.
Ovolo Publishing Ltd, 1 The Granary, Brook Farm,
Ellington, Cambridge, Cambridgeshire, PE28 0AE
Area of Operation: Worldwide
Tel: 01480 891595
Fax: 01480 893836
Email: info@ovolopublishing.co.uk
Web: www.ovolopublishing.co.uk
Product Type: 1

TRADITIONAL HOUSING BUREAU
4th Floor, 60 Charles Street, Leicester,
Leicestershire, LE1 1FB
Area of Operation: UK (Excluding Ireland)
Tel: 0116 253 6161
Fax: 0116 251 4568
Email: info@housebuilder.org
Web: www.housebuilder.org.uk
Product Type: 2, 3

WWW.RENOVATIONFRANCE.NET
2 Sugarbrook Court, Aston Road,
Bromsgrove, Worcestershire, B60 3EX
Area of Operation: Worldwide
Tel: 01527 834435 **Fax:** 01527 837810
Email: alex.worthington@centaur.co.uk
Web: www.renovationfrance.net
Product Type: 3

NOTES

Company Name ....
....
Address ....
....
....
email ....
Web ....

Company Name ....
....
Address ....
....
....
email ....
Web ....

Company Name ....
....
Address ....
....
....
email ....
Web ....

Company Name ....
....
Address ....
....
....
email ....
Web ....

Company Name ....
....
Address ....
....
....
email ....
Web ....

Company Name ....
....
Address ....
....
....
email ....
Web ....

Company Name ....
....
Address ....
....
....
email ....
Web ....

Company Name ....
....
Address ....
....
....
email ....
Web ....

Company Name ....
....
Address ....
....
....
email ....
Web ....

Company Name ....
....
Address ....
....
....
email ....
Web ....

Company Name ....
....
Address ....
....
....
email ....
Web ....

Company Name ....
....
Address ....
....
....
email ....
Web ....

USEFUL INFORMATION

Land for Sale

Find your perfect plot of land, renovation or conversion project:

www.plotfinder.net
ONLINE LAND AND RENOVATION DATABASE

You too could get to this stage in your self-build by subscribing to www.plotfinder.net to find your ideal building plot.

Searching for your perfect plot of land — be it one with a fantastic view, one in a country hideaway or one close to a buzzing city — is often an uphill struggle. Endless effort and time can be spent trawling through papers, calling estate agents and going to auctions. Which is where plotfinder.net can help you.

www.plotfinder.net is a database which holds details on over 5,500 building plots and properties for renovation currently for sale in the UK. All the time and effort required searching for plots is done for you — which means you will have more time to enjoy planning and designing your dream home. So, to find your perfect plot, start your search by subscribing to www.plotfinder.net.

A one year subscription to www.plotfinder.net is only £40* for unlimited access to five counties of your choice. The website gives instant access to plot and renovation details — these are updated daily. Details of the vendor are on the website for you to contact them directly if you are interested in buying their plot.

Plotfinder.net also has additional features that can help to save you more time, including the option of receiving email alerts when new plots are added to your chosen counties so you don't have to search the website every day. You can save plots or renovation projects into a separate favourites folder and mark plots as read.

It is entirely free to include land or properties in need of work on www.plotfinder.net — to do so, email laura.cassidy@centaur.co.uk with details of your opportunity.

Find your dream building plot with www.plotfinder.net

If you do not have access to the internet, you can access the plotfinder.net database by mail. To receive a listing of development opportunities in the county of your choice by mail, call the following premium rate telephone number: 0906 557 5400. (Calls cost £1 per minute at all times and last approx. three minutes.)

Additional methods:

Fed up with chasing a dream that never seems to materialise? A constant remark from those seeking to buy a plot is that there are none out there, or that those that are get snapped up by builders or developers before they even reach the market.

On the former point they are quite definitely wrong, on the latter point possibly right. But the notion that self-builders stand no chance is incorrect. Estimates indicate that between 15-20,000 self-build projects happen each year, meaning that there are at least 15-20,000 fresh opportunities every 12 months.

Many others complain that they live in an area where plots are never available. Again this is not true. If all the plots that www.plotfinder.net have on the database were converted to pins on a map, the whole of the UK would be covered.

As well as subscribing to plotfinder.net try:

Estate Agents
Estate agents selling a piece of land get a small fee for doing so. What interests them more is the much higher fee for selling the resulting house, so their first thought is for it to be sold to a builder or developer who will engage them for the sale of whatever gets built on the land. Try representing yourself as a 'Private Housebuilder' or give the impression that you're going to sell the house when it's finished.

Local Newspapers
Get papers from surrounding areas, as there is quite often a large overlap. Take out a 'building plots wanted' advertisement.

Planning Register
Whilst you're talking to the planners, ask to see the Planning Register and pay attention to the outline applications. Make a note of the applicant's name and address and write to the owners to enquire if they would be interested in selling their land to you.

Other Self-builders
Talk to other self-builders. They are a mine of information and may well have heard of other plots. If you see a mobile home on a site, you can bet it's a self-builder, or go to discussion forums on sites such as www.homebuilding.co.uk.

Plotfinder.net also includes properties in need of renovation or conversion.

Opportunities in Eire and France:

To find building sites or properties in need of renovation for sale throughout Ireland visit www.sitefinderireland.com.

For properties in need of renovation for sale in France visit www.propertyfinderfrance.net.

USEFUL INFORMATION

Image courtesy of Homebuilding & Renovating Shows (0207 970 4249)

SPONSORED BY HOMEBUILDING & RENOVATING SHOWS
Tel 0207 970 4249 Web www.homebuildingshow.co.uk

USEFUL INFORMATION

SPONSORED BY: HOMEBUILDING & RENOVATING SHOWS www.homebuildingshow.co.uk

MARCH

2 – 5 March
The National Homebuilding & Renovating Show
£8 in advance, £12 on door, free to subscribers
NEC Birmingham, Centaur Exhibitions
Tel: 0207 970 4249 **Fax:** 0207 970 6740
Email: showenquiries@centaur.co.uk
Web: www.homebuildingshow.co.uk
Whether you are building from scratch, converting or extending a home or renovating a property, you will find relevent products and services all under one roof. The show is full of attractions that will help you build a dream home or complete your renovation project. See thousands of dream home designs, choose from hundreds of plots and renovation opportunities, and find out more about finance options. You can also visit seminar sessions, and get experts from the Homebuilding & Renovating Magazine panel to help with your questions and problems. You might just find that inspiration, idea or advice you need to turn your dream home into reality!

2 – 5 March
The Smart Home Show
£8 in advance, £12 on door
(free to subscribers, or with a H&R National Show ticket)
NEC Birmingham, Centaur Exhibitions
Tel: 0207 970 4249 **Fax:** 0207 970 6740
Email: showenquiries@centaur.co.uk
Web: www.smarthomeshow.co.uk
The Smart Home Show is the UK's only exhibition dedicated to bringing together the latest trends in smart home technology. Visitors will be able to experience state of the art products and technology and compare the best intelligent technological solutions available on the market. Discover the latest imaginative solutions and products for home networking, home automation, multi-room entertainment, custom installation, communication and security systems, intelligent heating, lighting and more.

2 – 5 March
Creating Murals for Interiors
Residential £360. Non-residential £210
West Dean College, West Dean, Chichester, West Sussex, PO18 0QZ.
Tel: 01243 818219/811301 **Fax:** 01243 811343
Email: short.courses@westdean.org.uk
Web: www.westdean.org.uk
The theme of the course is the gardens of West Dean, on which a large-scale mural design will be based. Students learn to transfer initial designs on to boards using projection, drawing and painting. Tricks include approach to colour, perspective and fine detailing.

3 – 5 March
The Whole House
£300 high-waged; £225 waged; £160 student
Centre for Alternative Technology, Machynlleth, Powys, SY20 9AZ.
Tel: 01654 705981 **Fax:** 01654 703605
Email: courses@cat.org.uk **Web:** www.cat.org.uk/courses
Pat Borer and Cindy Harris, authors of the best-selling "The Whole House Book", will run this topical course. Course covers ecological building design for low energy consumption; green building technologies; healthier habits; the use of environmentally friendly building materials; water efficiency, conservation and sanitation.

6 – 8 March
Project Management and Risk Assessment £275
Dept of Archaeology, University of York, The King's Manor, York, YO1 7EP **Tel:** 01904 433963
Email: pab11@york.ac.uk **Web:** www.york.ac.uk/depts/arch
Introduces and illustrates the application of project management techniques in conservation projects, considering the place of funding. plus the range of risks involved in conservation work, and strategies for their mitigation.

6 – 10 March
Wind Power
Part 1&2: £575 high-waged; £450 waged; £260 student/unwaged (Part 1 or 2: £300/ £225/ £160)
Centre for Alternative Technology, Machynlleth, Powys, SY20 9AZ.
Tel: 01654 705981 **Fax:** 01654 703605
Email: courses@cat.org.uk **Web:** www.cat.org.uk/courses
Part 1: Demonstrates basic requirements for choosing / installing wind energy systems, including a visit to a wind farm. Suitable for those thinking of buying or building their own system or those with a general interest. Part 2: Workshops on wind machine design.

10 – 12 March
Contemporary Decorative Finishes
Residential £245, Non-residential £145
West Dean College, West Dean, Chichester, West Sussex, PO18 0QZ.
Tel: 01243 818219 / 811301 **Fax:** 01243 811343
Email: short.courses@westdean.org.uk
Web: www.westdean.org.uk
Modern media - irridescents, lustres and spirit dyes with glazes, varnishes and waxes - will be employed with traditional methods to produce samples and finished projects. Students will produce contemporary images and textures for a variety of surfaces.

13 March
Wattle and Daub £100
Weald and Downland Open Air Museum, Singleton, Chichester, West Sussex, PO18 0EU. **Tel:** 01243 811348
Email: courses@wealddown.co.uk **Web:** www.wealddown.co.uk
Course for surveyors, architects, craftsmen, and anyone with an interest in building conservation. Insights into the historic use of wattle and daub, and its repair and conservation today. A morning touring the Museum's examples is followed by an afternoon of 'hands on' practical spent applying wattles and daubing them.

13 - 16 March
Specifying Construction Works
Residential £515. Non-residential £410
West Dean College, West Dean, Chichester, West Sussex, PO18 0QZ.
Tel: 01243 818219/811301 **Fax:** 01243 811343
Email: bcm@westdean.org.uk **Web:** www.westdean.org.uk
Course covers specifying conservation works, from inception to final account and post-contract debrief. It offers coverage of the subject from different professional angles, using discussion topics and consideration of case studies, focussing on what is required to 'make it work in practice': pre-contract trials; assessing quality of specified works; role and requirements in specifying conservation contractors; and preparation for and compilation of an unambiguous tender document that minimises risk.

24 – 27 March
Designing Your Own Garden
Residential £360 Non-residential £210
West Dean College, West Dean, Chichester, West Sussex, PO18 0QZ.
Tel: 01243 818219/811301 **Fax:** 01243 811343
Email: short.courses@westdean.org.uk
Web: www.westdean.org.uk
A course for those who have a new garden with nothing in it, or an established garden they would like to change. Covers the basics of site evaluation, and creating a functional layout and planting plan.

24 March – 2 April
Timber Frame Self Build Parts 1&2: £750 high-waged; £575 waged; £350 student/unwaged (part 1 or 2: £575 / £450 / £260)
Centre for Alternative Technology, Machynlleth, Powys, SY20 9AZ.
Tel: 01654 705981 **Fax:** 01654 703605
Email: courses@cat.org.uk **Web:** www.cat.org.uk/courses
Part 1: Theorectical sessions covering the timber frame method, ecological building methods, design for low energy sustainable housing and appropriate services for such buildings, plus practical frame building. Part 2 follows frame raising through to completion of a building designed to house inverters from CAT's PV roof.

APRIL

7 – 9 April
Straw Bale Building
£180 high-waged; £150 waged; £120 student/unwaged
Low Impact Living Initiative, Redfield Community, Winslow, Buckinghamshire, MK18 3LZ. **Tel/Fax:** 01296 714184
Email: lili@lowimpact.org **Web:** www.lowimpact.org
This is a practical building project, aiming to leave a permanent structure on the site. This will involve some carpentry and general building as well as handling straw bales. We will try to avoid power tools and people of all abilities are welcome. You will also learn about the environmental and practical benefits of constructing buildings using straw bales. The course will be run by Chris 'Chug' Tugby, who is one of the leading straw-bale builders in Britain today.

24 – 26 April
The Historic Interior: Commissioning and Managing Conservation Research
Residential £345. Non-residential £275
West Dean College, West Dean, Chichester, West Sussex, PO18 0QZ
Tel: 01243 818219/811301 **Fax:** 01243 811343
Email: bcm@westdean.org.uk **Web:** www.westdean.org.uk
Primarily designed as a course for project conservators, property managers, curators, historic building inspectors, conservation officers and conservation architects who wish to undertake or commission research in historic interiors to gain an understanding of their development. The course will outline the different methods of analysis commonly undertaken as part of the investigation of historic interiors with reference to completed case studies.

24 – 28 April
Timber Framing From Scratch £450
Weald and Downland Open Air Museum, Singleton, Chichester, West Sussex, PO18 0EU. **Tel:** 01243 811348
Email: courses@wealddown.co.uk **Web:** www.wealddown.co.uk
Course for professionals and anyone with an interest in building conservation. A great opportunity to gain hands on experience of timber framing, this practical course introduces students to the historic use of structural oak framing, tools and techniques.

28 – 30 April
Roofing
£180 high-waged; £150 waged; £120 student/unwaged
Low Impact Living Initiative, Redfield Community, Winslow, Buckinghamshire, MK18 3LZ.
Tel/Fax: 01296 714184
Email: lili@lowimpact.org **Web:** www.lowimpact.org
This course is for anyone considering building a house or extension, employing a builder, repairing a roof, or who wants to understand more about how their house works. Course topics include roofing options and materials, ventilation, fixings and details, loft conversions and solar panels.

28 – 30 April
Basic Blacksmithing
Residential £245 Non-residential £145
West Dean College, West Dean, Chichester, West Sussex, PO18 0QZ.
Tel: 01243 818219/811301 **Fax:** 01243 811343
Email: short.courses@westdean.org.uk
Web: www.westdean.org.uk
In addition to introducing the basic skills of the blacksmith, this course shows students the part these techniques play in developing successful designs. Students may bring draft ideas with them and individual tuition will be given to help realise these.

28 – 30 April
Sustainable Water and Sewage
£180 high-waged; £150 waged; £120 student/unwaged
Low Impact Living Initiative, Redfield Community, Winslow, Buckinghamshire, MK18 3LZ.
Tel/Fax: 01296 714184
Email: lili@lowimpact.org **Web:** www.lowimpact.org
This course is about the collection, conservation and recycling of water, as well as sustainable ways of dealing with sewage. Water shortages are becoming a serious problem all over the world, and there are financial implications too, now that water meters are becoming the norm. Course covers subjects such as compost toilets, water efficiency, reed beds, rainwater harvesting & greywater recycling, and how to find the most appropriate sewage treatment system for your application.

MAY

2 – 5 May
Conservation and Repair of Masonry Ruins
Residential £515. Non-residential £410
West Dean College, West Dean, Chichester, West Sussex, PO18 0QZ.
Tel: 01243 818219/811301 **Fax:** 01243 811343
Email: bcm@westdean.org.uk **Web:** www.westdean.org.uk
Primarily designed for those concerned with the conservation of ruined structures, (although principles are also applicable to roofed buildings). Assessment of the structural and surface condition of masonry walls, identification of traditional materials, fault diagnosis, removal and control of organic growth, consolidation and treatment of wall tops, specification of mortars and grouts and the preparation and placing of mortar, with practical exercises using the ruinette.

3 May
Traditional Timber-Frame Construction £90
Weald and Downland Open Air Museum, Singleton, Chichester, West Sussex, PO18 0EU.
Tel: 01243 811348
Email: courses@wealddown.co.uk **Web:** www.wealddown.co.uk
Course for surveyors, architects, craftsmen, and anyone with a keen interest in building conservation. A course on the traditional systems of timber framing, including demonstrations on timber conversion, principals of layout, scribing method, pegs and assembly.

5 – 7 May
Water Treatment, Conservation & Recycling
£300 high-waged; £225 waged; £160 student, + 10% discount if booked together with the Water Treatment, Conservation & Recycling Course (May 8 - 12)
Centre for Alternative Technology, Machynlleth, Powys, SY20 9AZ.
Tel: 01654 705981 **Fax:** 01654 703605
Email: courses@cat.org.uk **Web:** www.cat.org.uk/courses
A course for people with mains water supplies who are interested in supplementing it from other sources, and for people who either have a private water supply or are considering installing one. It starts by considering water efficiency, then moves on to finding, cleaning and moving water. The appropriate use of rainwater and greywater will be included, and CAT's own water supply and treatment systems will be examined. Practical activities include water testing and building simple treatment systems.

6 – 7 May

The Scottish Homebuilding & Renovating Show
£5 in advance, £8 on door, free to subscribers
SECC Glasgow, Centaur Exhibitions
Tel: 0207 970 4249   **Fax:** 0207 970 6740
Email: showenquiries@centaur.co.uk
Web: www.homebuildingshow.co.uk
Whether you are building from scratch, converting or extending a home or renovating a property, you will find the relevent products and services under one roof. The show is full of attractions that will help you build a dream home or complete your renovation project. See thousands of dream home designs, choose from hundreds of plots and renovation opportunities, and find out more about finance options. You can also visit seminar sessions, and get experts from the Homebuilding & Renovating Magazine panel to help with your questions and problems. You might just find that inspiration, idea or advice you need to turn your dream home into reality!

8 – 12 May

Alternative Sewage Systems
Part 1&2: £575 high-waged; £450 waged; £260 student/unwaged (part 1 or 2: £300 / £225 / £160) + 10% discount if booked together with the Water Treatment, Conservation & Recycling course (May 5 - 7)
Centre for Alternative Technology, Machynlleth, Powys, SY20 9AZ.
Tel: 01654 705981   **Fax:** 01654 703605
Émail: courses@cat.org.uk   **Web:** www.cat.org.uk/courses
This course is suitable for those looking for an alternative to a septic tank. Part 1 covers dry sewage systems and examines theoretical and practical information needed to install a compost toilet system you would be happy to have in your bathroom. Part 2 looks at reed bed systems / aquatic plant treatment, and how plants can be used to absorb nutrients and purify polluted water without using chemicals, machinery or extra sources of energy.

9 – 13 May

The Repair of Old Buildings £650
Society for the Protection of Ancient Buildings, 37 Spital Square, London, E1 6DY
Tel: 020 7377 1644   **Fax:** 0207 247 5296
Email: info@spab.org.uk   **Web:** www.spab.org.uk
Course for professionals only. Demonstrates through lectures by experienced professionals, and practical examples, the manner in which the conservative repair of old buildings can be achieved.

12 May

Oak and Iron £90
Weald and Downland Open Air Museum, Singleton, Chichester, West Sussex, PO18 0EU.   **Tel:** 01243 811348
Email: courses@wealddown.co.uk   **Web:** www.wealddown.co.uk
Course for professionals and anyone with an interest in building conservation, looking at the science craft of using oak for joinery purposes. The use and interaction of iron elements such as bolts, straps, nails, screws, hinges etc is also demonstrated.

15 – 18 May

Conservation and Repair of Plasters and Renders
Residential £515. Non-residential £410
West Dean College, West Dean, Chichester, West Sussex, PO18 0QZ.
Tel: 01243 818219/811301   **Fax:** 01243 811343
Email: bcm@westdean.org.uk   **Web:** www.westdean.org.uk
This course covers history, documentation, repair options, condition survey, specifications, execution and quality control of remedial works to lime, gypsum and cement-based wall plasters and renders. Including practical workshop exercises.

15 – 19 May

Intermediate Timber Framing – Roof Framing £450
Weald and Downland Open Air Museum, Singleton, Chichester, West Sussex, PO18 0EU.
Tel: 01243 811348
Email: courses@wealddown.co.uk   **Web:** www.wealddown.co.uk
Practical course for surveyors, architects, craftsmen, and anyone with an interest in building conservation, who attended the "Timber Framing From Scratch" course. The common principle, hip and jack rafters are marked, cut and fitted to the timber frame made previously. Roof members are pitched on the final afternoon.

19 – 21 May

Self Build Show £10
Ballybrit Racecourse, Galway
SelfBuild Ireland Ltd, 96 Lisburn Rd, Saintfield, Co.Down BT24 7BP
Tel: (UK) +44 (028) 9751 0570, (ROI) (048) 9751 0570
Fax: (UK) +44 (028) 9751 0576, (ROI) (048) 9751 0576
Email: info@selfbuild.ie   **Web:** www.selfbuild.ie
Exhibition dedicated to self-building in Ireland, with high quality companies present to offer information and advice on nearly every aspect of building your own home.

22 – 26 May

Renewable Energy Systems £575 high-waged; £450 waged; £260 student/unwaged
Centre for Alternative Technology, Machynlleth, Powys, SY20 9AZ.
Tel: 01654 705981   **Fax:** 01654 703605
Email: courses@cat.org.uk   **Web:** www.cat.org.uk/courses
This course will look at the potential for generating your own electricity from wind, water and solar power, and at the possibilities for reducing energy consumption. Includes practical sessions in solar, wind and water, and tours of CAT's renewable energy systems.

25 May

Repair of Timber-Framed Buildings £90
Weald and Downland Open Air Museum, Singleton, Chichester, West Sussex, PO18 0EU.
Tel: 01243 811348
Email: courses@wealddown.co.uk   **Web:** www.wealddown.co.uk
Course for surveyors, architects, craftsmen, and anyone with an interest in building conservation. Includes a lecture on the repair of timber framed buildings, a workshop session, and a critical examination of repairs executed at the Museum over 30 years.

26 – 28 May

Cob Building £180 high-waged; £150 waged; £120 student/unwaged
Low Impact Living Initiative, Redfield Community, Winslow, Buckinghamshire, MK18 3LZ   **Tel/Fax:** 01296 714184
Email: lili@lowimpact.org   **Web:** www.lowimpact.org
The course will aim to introduce methods of earth / cob building, and will include theory and practice with short demonstrations; however, the emphasis will be on hands-on learning. Participants should leave feeling confident in their ability to build with cob, and to make and apply earthen plaster.

26 – 31 May

Creative Blacksmithing
Residential £245 Non-residential £145
West Dean College, West Dean, Chichester, West Sussex, PO18 0QZ.
Tel: 01243 818219/811301   **Fax:** 01243 811343
Email: short.courses@westdean.org.uk
Web: www.westdean.org.uk
Aimed at a mixed ability group, this course introduces beginners to the skills of the smith, while enabling those with experience to progress further. Students are encouraged to bring ideas for discussion (considering suitability to medium and available time).

JUNE

5 June

Joinery by Hand: Sash Windows £90
Weald and Downland Open Air Museum, Singleton, Chichester, West Sussex, PO18 0EU.   **Tel:** 01243 811348
Email: courses@wealddown.co.uk   **Web:** www.wealddown.co.uk
Course for professionals and anyone with an interest in building conservation. The historical development of sash windows with practical demonstrations of traditional joinery processes. We will examine original examples from the Brooking Collection.

7 June

Cob Walling – History, Theory and Practice £90
Weald and Downland Open Air Museum, Singleton, Chichester, West Sussex, PO18 0EU.   **Tel:** 01243 811348
Email: courses@wealddown.co.uk   **Web:** www.wealddown.co.uk
Course for surveyors, architects, craftsmen, and anyone with a keen interest in building conservation. We will explore the various types and methods of cob wall construction in the region, and will examine causes of failure, repair strategies and problems relating to alterations to cob structures. Includes some hands-on practice.

9 – 11 June

Solar Electric Systems
£300 high-waged; £225 waged; £160 student/unwaged
Centre for Alternative Technology, Machynlleth, Powys, SY20 9AZ.
Tel: 01654 705981   **Fax:** 01654 703605
Email: courses@cat.org.uk   **Web:** www.cat.org.uk/courses
During the course participants will learn about the different types of photovoltaic modules and how to select, install and maintain the most appropriate systems. It will also cover developments in PV roofing systems, using CAT's 12 kW roof as an example. Both stand-alone and grid connected systems will be outlined.

12 June

Flint Buildings – History, Repair and Restoration £90
Weald and Downland Open Air Museum, Singleton, Chichester, West Sussex, PO18 0EU.   **Tel:** 01243 811348
Email: courses@wealddown.co.uk   **Web:** www.wealddown.co.uk
Course for surveyors, architects, craftsmen, and anyone with an interest in building conservation. Demonstrations and lectures exploring this plentiful, but difficult to use, local building material, will encourage sensitive, authentic repairs using local craft skills.

12 – 15 June

The Ecological Management of Historic Buildings and Sites   Residential £515. Non-residential £410
West Dean College, West Dean, Chichester, West Sussex, PO18 0QZ
Tel: 01243 818219/811301   **Fax:** 01243 811343
Email: short.courses@westdean.org.uk   www.westdean.org.uk
The last decade has seen a renewed interest in the ecological importance of many historic sites and buildings, and there has been a growing acceptance of the need to integrate care of plants and animals into the conservation of historic fabric, sites and landscapes. The course looks at the key habitats involved and their management, with particular reference to some recent examples, including the implications of wildlife legislation.

19 – 21 June

Repair of Traditionally Constructed Brickwork £270
Weald and Downland Open Air Museum, Singleton, Chichester, West Sussex, PO18 0EU.   **Tel:** 01243 811348
Email: courses@wealddown.co.uk   **Web:** www.wealddown.co.uk
Course for professionals and anyone with an interest in building conservation, covering causes of failure and decay, and selection of methods of repair. Practical sessions include cutting out bricks, taking out defective joints, stitch repairs and reinforcement, and patch pointing using lime mortars.

USEFUL INFORMATION

19 – 23 June
Intermediate Timber Framing – Wall Framing £450
Weald and Downland Open Air Museum, Singleton, Chichester, West Sussex, PO18 0EU.   **Tel:** 01243 811348
Email: courses@wealddown.co.uk   **Web:** www.wealddown.co.uk
Practical course for surveyors, architects, craftsmen, and anyone with an interest in building conservation, who attended the "Timber Framing From Scratch" course. The studs and braces of the walls are marked, cut and fitted into the timber frame made on the previous part of the course, and the completed work is erected on the last afternoon.

19 – 23 June
Straw Bale Building
£575 high-waged; £450 waged; £260 student/unwaged
Centre for Alternative Technology, Machynlleth, Powys, SY20 9AZ.
Tel: 01654 705981   **Fax:** 01654 703605
Email: courses@cat.org.uk   **Web:** www.cat.org.uk/courses
Led by Strawbale Futures experts, this course will look at the history and use of straw bale buildings, different techniques, planning considerations and how to build. There will be practical workshops on the load bearing method, plaster and render preparation, and lime and earth finishes.

22 June
Lime Mortars for Traditional Brickwork £90
Weald and Downland Open Air Museum, Singleton, Chichester, West Sussex, PO18 0EU.   **Tel:** 01243 811348
Email: courses@wealddown.co.uk   **Web:** www.wealddown.co.uk
Course for surveyors, architects, craftsmen, and anyone with an interest in building conservation. Lectures and practical demonstrations on the traditional preparation and use of limes and lime mortars and the modern misconceptions about them.

26 - 27 June
Traditional Lime Plasters and Renders £180
Weald and Downland Open Air Museum, Singleton, Chichester, West Sussex, PO18 0EU.   **Tel:** 01243 811348
Email: courses@wealddown.co.uk   **Web:** www.wealddown.co.uk
Course for surveyors, architects, craftsmen, and anyone with an interest in building conservation. A practical based course covering the fundamentals of lime plastering, from simple renders to fine ornamental work. Lectures are followed by hands-on experience, practical demonstrations, and opportunity for discussion.

26 – 29 June
Cleaning Masonry Buildings
Residential £515. Non-residential £410
West Dean College, West Dean, Chichester, West Sussex, PO18 0QZ.
Tel: 01243 818219/811301   **Fax:** 01243 811343
Email: bcm@westdean.org.uk   **Web:** www.westdean.org.uk
An intensive and comprehensive course for practitioners and specifiers, covering the complex aesthetic, technical, practical and health and safety issues involved in the cleaning of stone, brick and terracotta buildings. With opportunities to try out a full range of equipment and techniques, this is an important course to accompany the new BS Code of Practice for Cleaning and Surface Repair of Buildings, Parts I and II.

30 June – 2 July
Solar Water Heating Systems
£300 high-waged; £225 waged; £160 student/unwaged
Centre for Alternative Technology, Machynlleth, Powys, SY20 9AZ.
Tel: 01654 705981   **Fax:** 01654 703605
Email: courses@cat.org.uk   **Web:** www.cat.org.uk/courses
A course for those who want to design or install a solar water heating system, or purchase a system making an informed choice. Sessions cover types of collector, energy storage, plumbing and controls. There will be instruction on the design of solar water heating systems, and a practical demonstration on the construction of a collector.

JULY

1 – 2 July
The Eastern Homebuilding & Renovating Show
£5 in advance, £8 on door, free to subscribers
East of England Showground, Peterborough, Centaur Exhibitions
Tel: 0207 970 4249   **Fax** 0207 970 6740
Email: showenquiries@centaur.co.uk
Web: www.homebuildingshow.co.uk
Whether you are building from scratch, converting or extending a home or renovating a property, you will find the relevent products and services under one roof. The show is full of attractions that will help you build a dream home or complete your renovation project. See thousands of dream home designs, choose from hundreds of plots and renovation opportunities, and find out more about finance options. You can also visit seminar sessions, and get experts from the Homebuilding & Renovating Magazine panel to help with your questions and problems. You might just find that inspiration, idea or advice you need to turn your dream home into reality!

3 – 7 July
Traditional Roofing Methods
£90 per day (all 5 days £400)
Weald and Downland Open Air Museum, Singleton, Chichester, West Sussex, PO18 0EU.   **Tel:** 01243 811348
Email: courses@wealddown.co.uk   **Web:** www.wealddown.co.uk
Course for professionals and anyone with an interest in building conservation. Five linked days exploring the traditions, methods and materials used in the roofing industries. Day 1: The Roofing Square; Day 2: Thatch; Day 3: Tile; Day 4: Slate; Day 5: Leadwork.

7 – 9 July
Straw Bale Building
£180 high-waged; £150 waged; £120 student/unwaged
Low Impact Living Initiative, Redfield Community, Winslow, Buckinghamshire, MK18 3LZ   **Tel/Fax:** 01296 714184
Email: lili@lowimpact.org   **Web:** www.lowimpact.org
A practical building project, aiming to leave a permanent structure on the site, which will involve some carpentry and general building plus handling straw bales. We will try to avoid power tools but people of all abilities are welcome. You will also learn about the environmental and practical benefits of constructing buildings using straw bales. The course will be run by Chris 'Chug' Tugby, one of Britain's leading straw-bale builders.

17 – 21 July
Alternative Building Methods
£575 high-waged; £450 waged; £260 student/unwaged
Centre for Alternative Technology, Machynlleth, Powys, SY20 9AZ.
Tel: 01654 705981   **Fax:** 01654 703605
Email: courses@cat.org.uk   **Web:** www.cat.org.uk/courses
A series of lectures led by Maurice Mitchell (architect and lecturer at CAT for over 20 years) covering earth building, ferro-cement, building with a range of timbers, and emergency shelters. Much of the course will involve the design and construction of a small structure using locally available materials. This course is suitable for building / architecture students, those going to work overseas and anyone wanting to take a broader approach.

21 – 23 July
Self-Build Solar Hot Water
£1695 including system, £150 excluding system
Low Impact Living Initiative, Redfield Community, Winslow, Buckinghamshire, MK18 3LZ   **Tel/Fax:** 01296 714184
Email: lili@lowimpact.org   **Web:** www.lowimpact.org
We think that this is the best value for money solar hot water system that you'll find in the UK! Participants leave with their entire system, tailored to their needs, plus a manual explaining how the system works, and how to install it. Participants can then install themselves, or they can ask a professional installer to install for them (it is also possible to attend the course without purchasing a system). The course begins with a theory session explaining solar water heating and our system in detail. You will then build and pressure test your own pump & control set, expansion vessel kit and air separator, plus more sessions on installing and maintaining your system, including fixing the panels to your roof, and how to replace a cylinder. No specialist skills are needed to take part in this course.

AUGUST

11 – 13 August
Natural Paints and Limes
£180 high-waged; £150 waged; £120 student/unwaged
Low Impact Living Initiative, Redfield Community, Winslow, Buckinghamshire, MK18 3LZ   **Tel/Fax:** 01296 714184
Email: lili@lowimpact.org   **Web:** www.lowimpact.org
How to maintain your property, combining the best of traditional and modern techniques and materials to ensure long-lasting protection, and simultaneously safeguarding your health and that of the environment. Includes theory on the environmental effects of paints, lime and cements, and practical sessions using various paints and finishes, lime washing, rendering and slaking lime. There will be plenty of opportunities to discuss your projects.

18 – 20 August
Green Woodworking
£180 high-waged; £150 waged; £120 student/unwaged
Low Impact Living Initiative, Redfield Community, Winslow, Buckinghamshire, MK18 3LZ   **Tel/Fax:** 01296 714184
Email: lili@lowimpact.org   **Web:** www.lowimpact.org
The techniques of Green woodworking cross many disciplines - from boat and house building to the expressive arts and home crafts. 'Green' in this context refers to unseasoned wood. With a shave-horse and pole-lathe (both of which you can make yourself) and a basic collection of hand tools you can easily start making presents, kitchenware, toys, furniture, garden structures and general fixings for the home. Come along for an introduction to the techniques of green woodworking and learn the origin of DIY. No experience necessary.

18 – 20 August
Rammed Earth Building
£180 high-waged; £150 waged; £120 student/unwaged
Low Impact Living Initiative, Redfield Community, Winslow, Buckinghamshire, MK18 3LZ   **Tel/Fax:** 01296 714184
Email: lili@lowimpact.org   **Web:** www.lowimpact.org
Course includes background information on rammed earth, its benefits as a building material, choosing the right type, construction method and design etc, plus practical elements.

SEPTEMBER

1 – 3 September
Self Build Show £10
Dublin area, contact for details
SelfBuild Ireland Ltd, 96 Lisburn Rd, Saintfield, Co.Down BT24 7BP
Tel: (UK) +44 (028) 9751 0570, (ROI) (048) 9751 0570
Fax: (UK) +44 (028) 9751 0576, (ROI) (048) 9751 0576
Email: info@selfbuild.ie   **Web:** www.selfbuild.ie
Exhibition dedicated to self-building in Ireland, with high quality companies present to offer information and advice on nearly every aspect of building your own home.

1 – 3 September
Straw Bale Building
£180 high-waged; £150 waged; £120 student/unwaged
Low Impact Living Initiative, Redfield Community, Winslow, Buckinghamshire, MK18 3LZ   **Tel/Fax:** 01296 714184
Email: lili@lowimpact.org   **Web:** www.lowimpact.org
A practical building project, aiming to leave a permanent structure on the site. This will involve some carpentry and general building as well as handling straw bales. We will try to avoid power tools and people of all abilities are welcome. You will also learn about the environmental and practical benefits of constructing buildings using straw bales. The course will be run by Chris 'Chug' Tugby, who is one of the leading straw-bale builders in Britain today.

8 – 10 September
The London Homebuilding & Renovating Show
£8 in advance, £12 on door, free to subscribers
ExCeL Centaur Exhibitions
Tel: 0207 970 4249   **Fax** 0207 970 6740
Email: showenquiries@centaur.co.uk
Web: www.homebuildingshow.co.uk
Whether you are building from scratch, converting or extending a home or renovating a property, you will find the relevent products and services under one roof. The show is full of attractions that will help you build a dream home or complete your renovation project. See thousands of dream home designs, choose from hundreds of plots and renovation opportunities, and find out more about finance options. You can also visit seminar sessions, and get experts from the Homebuilding & Renovating Magazine panel to help with your questions and problems. You might just find that inspiration, idea or advice you need to turn your dream home into reality!

8 – 10 September
Cob Building
£300 high-waged; £225 waged; £160 student/unwaged
Centre for Alternative Technology, Machynlleth, Powys, SY20 9AZ.
Tel: 01654 705981   **Fax:** 01654 703605
Email: courses@cat.org.uk   **Web:** www.cat.org.uk/courses
This course covers the practical skills of mixing and building with cob, one of the most sustainable construction materials available today. Participants are shown the principles of sourcing sub soil, mixing cob by foot and making cob blocks. There will be an illustrated talk showing examples of cob buildings and sculptures, plus images from Africa, Arizona and New Mexico.

15 – 17 September
Building with Earth
£300 high-waged; £225 waged; £160 student/unwaged
Centre for Alternative Technology, Machynlleth, Powys, SY20 9AZ.
Tel: 01654 705981   **Fax:** 01654 703605
Email: courses@cat.org.uk   **Web:** www.cat.org.uk/courses
Course covers basic methods and construction techniques, and soil analysis. Practical in block making / rammed earth techniques.

18 – 22 September
Timber Framing From Scratch £450
Weald and Downland Open Air Museum, Singleton, Chichester, West Sussex, PO18 0EU.   **Tel:** 01243 811348
Email: courses@wealddown.co.uk   **Web:** www.wealddown.co.uk
Course for professionals and anyone interested in building conservation. This course introduces students to the historic use of structural oak framing, tools and techniques. Posts, cills, plates and tie beams for a 10' square timber frame are prefabricated during the course, and the frame is erected on the last afternoon.

25 – 29 September
Eco Design and Construction
£575 high-waged; £450 waged; £260 student/unwaged
Centre for Alternative Technology, Machynlleth, Powys, SY20 9AZ.
Tel: 01654 705981 **Fax:** 01654 703605
Email: courses@cat.org.uk **Web:** www.cat.org.uk/courses
A course for builders, architects and those who want to investigate aspects of eco-design and construction that are outside the current mainstream. A holistic approach to eco-design will be outlined, aimed at producing low energy, long-life, sustainable buildings with a minimal environmental footprint. Attention will be given to working out details associated with, for example, "breathing wall construction" and "superinsulation". Material specification will therefore be examined with a view to minimising the long-term environmental impact of their use in buildings. There will be sessions on water treatment and waste management, transport issues, promoting biodiversity, legislation and standards.

OCTOBER

2 – 6 October
The Rural Eco Home
£575 high-waged; £450 waged; £260 student/unwaged
Centre for Alternative Technology, Machynlleth, Powys, SY20 9AZ.
Tel: 01654 705981 **Fax:** 01654 703605
Email: courses@cat.org.uk **Web:** www.cat.org.uk/courses
This course will examine the various eco solutions to building, renovating and living sustainably in a non-urban situation. The potential for energy from renewable sources (both electricity and heat) will be covered, as well as passive solar design, eco building materials, energy efficiency, and water and sanitation systems.

2 – 7 October
The Repair of Old Buildings £650
The Society for the Protection of Ancient Buildings, 37 Spital Square, London, E1 6DY
Tel: 020 7377 1644 **Fax:** 0207 247 5296
Email: info@spab.org.uk **Web:** www.spab.org.uk
Course for architects, surveyors, structural engineers, planners and conservation officers, builders and craftsmen. Demonstrates through lectures by experienced professionals, and practical examples, the manner in which the conservative repair of old buildings can be achieved.

9 – 13 October
Timber Frame Self Build
£575 high-waged; £450 waged; £260 student/unwaged
Centre for Alternative Technology, Machynlleth, Powys, SY20 9AZ.
Tel: 01654 705981 **Fax:** 01654 703605
Email: courses@cat.org.uk **Web:** www.cat.org.uk/courses
A series of frames will be made and erected during the practical sessions, and classroom sessions will cover the timber frame method, ecological building methods, design for low energy sustainable housing and appropriate services for such buildings.

13 – 15 October
Sustainable Energy For Homeowners
£180 high-waged; £150 waged; £120 student/unwaged
Low Impact Living Initiative, Redfield Community, Winslow, Buckinghamshire, MK18 3LZ **Tel/Fax:** 01296 714184
Email: lili@lowimpact.org **Web:** www.lowimpact.org
Renewable energy systems combined with low-energy building design can result in eco-homes with minimal energy costs and low carbon emissions. This course helps you assess which options are appropriate for your home: which technologies to use; typical system sizes; capital and running costs; and resulting carbon emissions. The course focus is domestic energy use, so is relevant to homeowners, prospective homeowners and self-builders.

13 – 15 October
Natural Rendering – Clay Plaster
£300 high-waged; £225 waged; £160 student/unwaged
Centre for Alternative Technology, Machynlleth, Powys, SY20 9AZ.
Tel: 01654 705981 **Fax:** 01654 703605
Email: courses@cat.org.uk **Web:** www.cat.org.uk/courses
This course will examine the use of clay as a natural rendering material. In practical and theoretical sessions, participants will look at preparation of plaster bases, producing mixtures, colour finishes, decorative coats, and external and internal finishes.

16 – 20 October
Timber Framing From Scratch £450
Weald and Downland Open Air Museum, Singleton, Chichester, West Sussex, PO18 0EU. **Tel:** 01243 811348
Email: courses@wealddown.co.uk **Web:** www.wealddown.co.uk
Course for professionals and anyone with an interest in building conservation. A superb opportunity to gain hands on experience of timber framing, this practical course introduces students to the historic use of structural oak framing, tools and techniques. The posts, cills, plates and tie beams or a 10' square timber frame are prefabricated during the course using traditional tools and techniques, and the frame is erected on the last afternoon.

27 – 29 October
DIY For Beginners
£180 high-waged; £150 waged; £120 student/unwaged
Low Impact Living Initiative, Redfield Community, Winslow, Buckinghamshire, MK18 3LZ **Tel/Fax:** 01296 714184
Email: lili@lowimpact.org **Web:** www.lowimpact.org
If you want to install photovoltaics, solar hot water, compost toilets or any of LILI's environmentally-friendly facilities, you will need to know some basic carpentry, plumbing, electrics and brickwork. This course de-mystifies many practical skills, explains environmentally-friendly ways to carry out many common DIY tasks, and covers suppliers / buying materials, fixings and timber.

30 October – 3 November
Renewable Heating Systems
£575 high-waged; £450 waged; £260 student/unwaged
Centre for Alternative Technology, Machynlleth, Powys, SY20 9AZ.
Tel: 01654 705981 **Fax:** 01654 703605
Email: courses@cat.org.uk **Web:** www.cat.org.uk/courses
Various renewable technologies for domestic heating will be examined in this course, including; solar water heating, biomass, heat pumps, domestic CHP and gas condensing boilers. Advice will be given on making an informed choice, costs, environmental benefits, installation and ongoing maintenance.

NOVEMBER

1 – 3 September
Self Build Show £10
Millstreet, Co Cork
SelfBuild Ireland Ltd, 96 Lisburn Rd, Saintfield, Co.Down BT24 7BP
Tel: (UK) +44 (028) 9751 0570, (ROI) (048) 9751 0570
Fax: (UK) +44 (028) 9751 0576, (ROI) (048) 9751 0576
Email: info@selfbuild.ie **Web:** www.selfbuild.ie
Exhibition dedicated to self-building in Ireland, with high quality companies present to offer information and advice on nearly every aspect of building your own home.

6 – 10 November
Green Sanitation & Organic Waste Management
£575 high-waged; £450 waged; £260 student/unwaged
Centre for Alternative Technology, Machynlleth, Powys, SY20 9AZ.
Tel: 01654 705981 **Fax:** 01654 703605
Email: courses@cat.org.uk **Web:** www.cat.org.uk/courses
This course covers two distinct areas: sewage systems and water treatment, and recycling and conservation. We will examine alternatives to conventional sewage treatment works, off mains systems, best environmental practice and legal requirements, followed by an examination of private water supplies: finding and storing water, grey water and rainwater recycling systems, and reducing domestic and commercial water use. Participants work in groups to evaluate and produce solutions to real life case studies, and you will receive advice and guidance as to which systems suit situations and how to size systems. The course is suitable for professionals and those wishing to develop their own systems.

10 – 12 November
The Northern Homebuilding & Renovating Show
£5 in advance, £8 on door, free to subscribers
Harrogate International Centre, Centaur Exhibitions
Tel: 0207 970 4249 **Fax** 0207 970 6740
Email: showenquiries@centaur.co.uk
Web: www.homebuildingshow.co.uk
Whether you are building from scratch, converting or extending a home or renovating a property, you will find the relevent products and services under one roof. The show is full of attractions that will help you build a dream home or complete your renovation project. See thousands of dream home designs, choose from hundreds of plots and renovation opportunities, and find out more about finance options. You can also visit seminar sessions, and get experts from the Homebuilding & Renovating Magazine panel to help with your questions and problems. You might just find that inspiration, idea or advice you need to turn your dream home into reality!

10 – 12 November
Wind and Solar Electricity £180 high-waged; £150 waged; £120 student/unwaged
Low Impact Living Initiative, Redfield Community, Winslow, Buckinghamshire, MK18 3LZ **Tel/Fax:** 01296 714184
Email: lili@lowimpact.org **Web:** www.lowimpact.org
This course provides an overview of the basic principles and the technology of solar and wind electrical systems, offering participants the theoretical knowledge and practical experience required to design and install small renewable energy systems. It is aimed at the general public and those in business, non-profit, public and academic sectors who wish to get an introduction to renewable energy electrical technology in general, as well as those wishing to install renewable energy systems in both urban or rural settings. The emphasis will be on how things work and what it is practicable to do. Participants will have the opportunity to discuss their own projects. No previous electrical knowledge necessary.

10 – 12 November
Wind Power Systems
£300 high-waged; £225 waged; £160 student/unwaged
Centre for Alternative Technology, Machynlleth, Powys, SY20 9AZ.
Tel: 01654 705981 **Fax:** 01654 703605
Email: courses@cat.org.uk **Web:** www.cat.org.uk/courses
This course is suitable for anyone interested in buying and installing an aerogenerator, as well as those with a general interest in the subject. Sessions will cover wind energy, stand-alone systems, siting of machines and grid-linked systems.

20 – 24 November
Intermediate Timber Framing – Roof Framing £450
Weald and Downland Open Air Museum, Singleton, Chichester, West Sussex, PO18 0EU. **Tel:** 01243 811348
Email: courses@wealddown.co.uk **Web:** www.wealddown.co.uk
Course for surveyors, architects, craftsmen, and anyone with an interest in building conservation. A practical course for students who attended the "Timber Framing From Scratch" course. The common principle, hip and jack rafters are marked, cut and fitted to the timber frame made on previous course. All roof members are pitched on the final afternoon.

25 – 26 November
The South West Homebuilding & Renovating Show
£5 in advance, £8 on door, free to subscribers
Bath and West Showground, Shepton Mallet, Somerset, Centaur Exhibitions
Tel: 0207 970 4249 **Fax** 0207 970 6740
Email: showenquiries@centaur.co.uk
Web: www.homebuildingshow.co.uk
Whether you are building from scratch, converting or extending a home or renovating a property, you will find the relevent products and services under one roof. The show is full of attractions that will help you build a dream home or complete your renovation project. See thousands of dream home designs, choose from hundreds of plots and renovation opportunities, and find out more about finance options. You can also visit seminar sessions, and get experts from the Homebuilding & Renovating Magazine panel to help with your questions and problems. You might just find that inspiration, idea or advice you need to turn your dream home into reality!

DECEMBER

1 – 3 December
Self-Build Solar Hot Water
£1695 including system, £150 excluding system
Low Impact Living Initiative, Redfield Community, Winslow, Buckinghamshire, MK18 3LZ **Tel/Fax:** 01296 714184
Email: lili@lowimpact.org **Web:** www.lowimpact.org
We think that this is the best value for money solar hot water system that you'll find in the UK! Participants leave with their entire system, tailored to their needs, plus a manual explaining how the system works, and how to install it. Participants can then install themselves, or they can ask a professional installer to install for them (it is also possible to attend the course without purchasing a system). The course begins with a theory session explaining solar water heating and our system in detail. You will then build and pressure test your own pump & control set, expansion vessel kit and air separator, plus more sessions on installing and maintaining your system, including fixing the panels to your roof, and how to replace a cylinder. No specialist skills are needed to take part in this course.

HOMEBUILDING & RENOVATING SHOW

2 - 5 March: The National Homebuilding & Renovating Show / The Smart Home Show *NEC Birmingham*
6 - 7 May: The Scottish Homebuilding & Renovating Show *SECC Glasgow*
1 - 2 July: The Eastern Homebuilding & Renovating Show *East of England Showground, Peterborough*
8 - 10 Sept: The London Homebuilding & Renovating Show *ExCeL, London*
10 - 12 Nov: The Northern Homebuilding & Renovating Show *Harrogate International Centre*
25 - 26 Nov: The South West Homebuilding & Renovating Show *Bath & West Showground, Shepton Mallet, Somerset*

smarthome show 2-5 March 2006 NEC, Birmingham

When's the next Homebuilding & Renovating Show?

• **National**	NEC	March / April
• **Glasgow**	SECC	May
• **Peterborough**	East of England Showground	July
• **London**	ExCeL	September
• **Harrogate**	Harrogate International Centre	November
• **Somerset**	The Bath & West Showground	November

HOMEBUILDING & RENOVATING SHOW

The UK's leading self-build and renovating shows

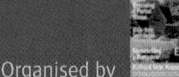

Organised by

For more information and to save £'s on advance tickets

What's on at the Shows and how will a visit benefit me?

- Meet hundreds of exhibitors covering all aspects of self-build and renovation – your opportunity to do all your research under one roof.

- Attend FREE seminars. Self-build experts cover all areas of self-build and renovation in our unique, FREE daily seminars.

- Find your perfect plot, conversion or renovation opportunity. Plotfinder and other land services will help kick start your search.

- Finance your dream. From buying land through to project completion, financial experts will offer you advice to suit your project and your pocket.

- Ask the Experts. Whatever your question, experts, including those from the Homebuilding & Renovating Magazine, will be on hand with the answers!

- Self-build and renovation Masterclasses. Unique daily masterclasses giving you practical demonstrations and a chance to have a 1-1 with key experts from the industry.

- Plan your day at the National Show in advance. The Show planner allows you to select from many product categories relevant to the stage of your project. Visit **www.homebuilding-online.co.uk/national** to find out more.

Is the Show for me?

The Homebuilding & Renovating Show is a must visit event if you are:
- Taking on a self-build, renovation or conversion project
- Wishing to extend or improve your home
- Searching for specific products or services to complete and add finishing touches to your home
- Seeking advice from experts on any aspect of renovation, self-build or home improvement
- Simply looking into the possibilities of creating your dream home

Photos kindly supplied by:
Nigel Rigden, Jeremy Phillips, Border Oak, Andrew Lee,

F

G

WHO WOULD YOU RECOMMEND?

THE HOMEBUILDER'S HANDBOOK

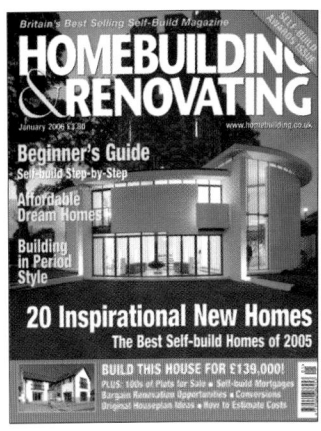

The Homebuilder's Handbook is a regular publication, packed with listings and useful information for the self-builder and renovator. If you have used a company or supplier and would like to recommend them to others (particularly if they offer a service which is above and beyond the usual), please let us know their contact details and their trade so that they can have a **FREE** listing in the Fourth Edition of The Homebuilder's Handbook.

From the publishers of Homebuilding and Renovating Magazine and organisers of The Homebuilding and Renovating Shows.

POSTAL ADDRESS:

2 SUGARBROOK COURT, ASTON ROAD, BROMSGROVE, WORCESTERSHIRE B60 3EX

FAX BACK ON 01527 834486

Contact Name: . . . .

Company Name: . . . .

Company Address: . . . .

. . . .

. . . .

. . . .

County: . . . .

Post Code: . . . .

Company Tel No: . . . .

Fax: . . . .

Website: . . . .

Direct Tel No: . . . .

Direct Fax: . . . .

Direct Email Address: . . . .

Nature of Business: . . . .

YOU CAN RECOMMEND AS MANY COMPANIES AS YOU LIKE
JUST PHOTOCOPY THIS PAGE AND RETURN

Networking:
The Complete Reference

Networking:
The Complete Reference

Craig Zacker

Osborne/**McGraw-Hill**

New York   Chicago   San Francisco
Lisbon   London   Madrid   Mexico City
Milan   New Delhi   San Juan   Seoul
Singapore   Sydney   Toronto

Osborne/**McGraw-Hill**
2600 Tenth Street
Berkeley, California 94710
U.S.A.

To arrange bulk purchase discounts for sales promotions, premiums, or fund-raisers, please contact Osborne/**McGraw-Hill** at the above address. For information on translations or book distributors outside the U.S.A., please see the International Contact Information page immediately following the index of this book.

Networking: The Complete Reference

1234567890 DOC DOC 019876543210

ISBN 0-07-219277-1

Publisher	**Project Manager**	**Indexer**
Brandon A. Nordin	Deidre Dolce	Jack Lewis
Vice President & Associate Publisher	**Freelance Project Manager**	**Computer Designers**
Scott Rogers	Laurie Stewart	Maureen Forys and Kate Kaminski, Happenstance Type-O-Rama
Editorial Director	**Technical Editors**	
Tracy Dunkelberger	Nalneesh Gaur	
	Bill Bruns	**Illustrator**
Acquisitions Editor		Michael Mueller
Jane Brownlow	**Copy Editor**	
	Lunaea Weatherstone	**Series Design**
Acquisitions Coordinator		Peter F. Hancik
Emma Acker	**Proofreader**	
	Sarah Kaminker	

This book was composed with QuarkXPress 4.11 on a Macintosh G4.

For LJ, with love

About the Author

Craig Zacker is a writer, editor, and networker whose computing experience began in the halcyon days of teletypes and paper tape. After making the move from minicomputers to PCs, he worked as an administrator of Novell NetWare networks and as a PC support technician, while operating a freelance desktop publishing business. After earning a Masters degree in English and American literature from NYU, Craig worked extensively on the integration of Windows NT into existing NetWare internetworks, and was employed as a technical writer, content provider, and Webmaster for the online services group of a large software company. Since devoting himself to writing and editing full time, Craig has authored or contributed to many books on networking and operating systems, developed and authored online and printed training materials, and published articles with top industry publications.

Contents at a Glance

Contents

Part I

Network Basics

Part II

Network Hardware

Part III

Network Protocols

Part IV

Network Operating Systems

Part V

Network Connection Services

Part VI

Network Services

Part VII

Network Administration

Introduction

*N*etworking: The Complete Reference is divided into parts, which are as follows:

- Part I outlines networking basics. Chapter 1 explains essential networking terminology and introduces some of the fundamental concepts that you will build on in later chapters. You'll learn about some of the primary components that make up a network and how the technology came about. Chapter 2 introduces the OSI reference model, a theoretical tool that compartmentalizes the networking functionality of a computer into seven discrete layers. These layers work together to enable the computer to communicate effectively with the other computers on the network. In this chapter, you'll also get your first glimpses of the technologies that will be examined in much greater detail later in the book.

- Part II explains the hardware used to construct a network. Chapter 3 examines network interface cards, which enable you to connect a computer to a network. This chapter will discuss the various types of cards you can buy to suit your computer's hardware configuration, your network type, and the role of your system on the network. In Chapter 4, you'll learn about the cables used to connect computers together to form a network. The type of cable you choose determines how easy or difficult network installation and maintenance will be, how long the cable can run, and how well your network will perform. This chapter also examines the standards that should guide your cable installation and some of the tools you'll need to do the job. Chapter 5 discusses wireless LAN technologies, and Chapters 6 and 7 examine the technologies used to construct local area networks, internetworks, and wide area networks. Chapter 8 discusses some of the hardware used to construct network servers, and Chapter 9 explains how to bring all of this hardware together into an effective network design.

■ Part III is about the protocols used at various layers of the OSI reference model. Chapters 10 and 11 examine Ethernet, the most popular data link–layer protocol used in the world today. Chapter 12 addresses other LAN protocols, such as Token Ring and FDDI. Chapters 13, 14, and 15 discuss the three main suites of protocols used at the network and transport layers, TCP/IP, IPX, and NetBEUI.

■ Part IV examines the networking elements of the operating systems most commonly used on today's networks, such as Windows 2000 and NT (Chapter 16), Novell NetWare (Chapter 19), and Unix (Chapter 21). Chapters 17, 18, and 20 discuss the various types of directory services used to store information about networks, their users, and their applications. These include Active Directory, the long-awaited enterprise directory service from Microsoft, which is included with Windows 2000; the domains used to organize Windows NT networks, and Novell Directory Services (NDS), the first commercially successful enterprise directory service. Chapter 22 explains the client capabilities that workstations need in order to access resources hosted by various other operating systems. For example, this chapter describes how to connect Macintosh and Unix systems to Windows and NetWare networks.

■ Part V discusses some of the most important administrative services used on today's networks. Chapter 23 covers the Dynamic Host Configuration Protocol (DHCP), which you can use to automatically configure TCP/IP clients on your network, and Chapter 24 covers the Windows Internet Naming Service (WINS). Windows NT networks use WINS to resolve the NetBIOS names Windows systems are known by into IP addresses necessary to communicate on a TCP/IP network. Chapter 25 examines the Domain Name System (DNS), which is used on the Internet and on private TCP/IP networks to resolve host names into IP addresses.

■ Part VI shows you how to improve your network by adding some of the most useful services available, such as World Wide Web, FTP, and e-mail servers (Chapter 26), network printing (Chapter 27), and access to the Internet (Chapter 28). Chapter 29 examines the security mechanisms you can use to protect your network from unauthorized access.

■ Part VII covers the tools and techniques you can use to administer your network. Chapter 30 outlines Windows-specific network administration techniques, Chapter 31 discusses network troubleshooting and management tools, and Chapter 32 discusses the hardware and software used to create network backups.

The book is intended for beginners, as well as experienced network professionals. I hope that, in addition to the uses mentioned above, it will lead you to learn more about the computer systems you use every day.

The Complete Reference

Networking

Part I

Network Basics

The
Complete
Reference

Networking

What Is a Network?

S imply put, a *network* is a group of computers connected together by cables or some other medium, but the networking process is anything but simple. When computers are able to communicate, they can work together in a variety of ways: by sharing their resources with each other, by distributing the workload of a particular task, or by exchanging messages. This book examines in detail how computers on a network communicate, what functions they perform, and how to go about building, operating, and maintaining them.

The original paradigm for collaborative computing was to have a single large computer connected to a series of terminals, each of which would service a different user. This is called *time-sharing*, because the computer divides its processor clock cycles among the terminals. Time-sharing is the basis for mainframe computing. In this arrangement, the terminals are simply communications devices; they accept input from users through a keyboard and send it to the computer. When the computer returns a result, the terminal displays it on a screen or prints it out on paper. This type of terminal is sometimes called a *dumb terminal*, because it doesn't perform any calculations of its own. The communications between the terminals and the computer are relatively simple on this type of network. Each terminal can only communicate with one device, the computer. Terminals never communicate with each other.

Local Area Networks (LANs)

As time passed and technology progressed, engineers began to connect computers together so that they could communicate. At the same time, computers were becoming smaller and less expensive, giving rise to mini- and microcomputers. The first computer networks used individual links, such as telephone connections, to connect two systems together. Soon after the first IBM PCs hit the market in the 1980s and rapidly became accepted as a business tool, the advantages of connecting these small computers together became obvious. Rather than supplying every computer with its own printer, a network of computers could share a single one. When one user needed to give a file to another user, a network eliminated the need to swap floppy disks. The problem, however, was that connecting a dozen computers in an office with individual point-to-point links between all of them was not practical. The eventual solution to this problem was the *local area network (LAN)*.

A LAN is a group of computers connected by a shared medium, usually a cable. By sharing a single cable, each computer requires only one connection and can conceivably communicate with any other computer on the network. A LAN is limited to a local area by the electrical properties of the cables used to construct them and by the relatively small number of computers that can share a single network medium. LANs are generally restricted to operation within a single building or, at most, a campus of adjacent buildings. Some technologies, such as fiber optics, have extended the range of LANs to several kilometers, but it isn't possible to use a LAN to connect computers in distant cities, for example. This is the province of the *wide area network (WAN)*, as discussed later in this chapter.

In most cases, a LAN is a baseband, packet-switching network. An understanding of the terms *baseband* and *packet switching*, which are examined in the following sections, is necessary to understand how data networks operate, because these terms define how computers transmit data over the network medium.

Baseband versus Broadband

A *baseband* network is one in which the cable or other network medium can carry only a single signal at any one time. A *broadband* network, on the other hand, can carry multiple signals simultaneously, using a discrete part of the cable's bandwidth for each signal. As an example of a broadband network, consider the cable television service that you probably have in your home. Although only one cable runs to your TV, it supplies you with dozens of channels of programming at the same time. If you have more than one television connected to the cable service, the installer probably used a splitter (a coaxial fitting with one connector for the incoming signals and two connectors for outgoing signals) to run the single cable entering your house to two different rooms. The fact that the TVs can be tuned to different programs at the same time while connected to the same cable proves that the cable is providing a separate signal for each channel at all times.

A baseband network uses pulses applied directly to the network medium to create a single signal that carries binary data in encoded form. Compared to broadband technologies, baseband networks span relatively short distances, because they are subject to degradation caused by electrical interference and other factors. The maximum length of a baseband network cable segment diminishes as its transmission rate increases. This is why local area networking protocols such as Ethernet have strict guidelines for cable installations.

Packet Switching versus Circuit Switching

LANs are called *packet-switching networks* because their computers divide their data into small, discrete units called *packets* before transmitting it. There is also a similar technique called *cell switching*, which differs from packet switching only in that cells are always a consistent, uniform size, whereas the size of packets is variable. Most LAN technologies, such as Ethernet, Token Ring, and Fiber Distributed Data Interface (FDDI), use packet switching. Asynchronous Transfer Mode (ATM) is the only cell-switching LAN protocol in common use.

Segmenting the data in this way is necessary because the computers on a LAN share a single cable, and a computer transmitting a single unbroken stream of data would monopolize the network for too long. When you examine the data being transmitted over a packet-switching network, you can see that the data stream consists of packets generated by many different systems, intermixed on the cable. It is normal on this type of network for packets that are part of the same message to take different routes to their destination and even to arrive at the destination in a different order than they were transmitted. The receiving system, therefore, must have a mechanism for reassembling the packets into the correct order and recognizing the absence of packets that may have been lost or damaged in transit.

The opposite of packet switching is *circuit switching*, in which one system establishes a dedicated communication channel to another system before any data is transmitted. In the data networking industry, circuit switching is used for certain types of wide area networking technologies, such as Integrated Services Digital Network (ISDN) and frame relay. The classic example of a circuit-switching network is the public telephone system. When you place a call to another person, a physical circuit is established between your telephone and theirs. This circuit remains active for the entire duration of the call, and no one else can use it, even when it is not carrying any data (that is, when no one is talking). In the early days of the telephone system, every phone was connected to a central office with a dedicated cable, and operators using switchboards manually connected a circuit between the two phones for every call. Today, the process is automated and the telephone system transmits many signals over a single cable, but the underlying principle is the same.

Networks and Internetworks

LANs were originally designed to connect a small number of computers together into what later came to be called a *workgroup*. Rather than investing a huge amount of money into a mainframe computer and the support system needed to run it, business owners came to realize that they could purchase a few computers, cable them together, and perform most of the computing tasks they needed. As the capabilities of personal computers and applications grew, so did the networks, and the technology used to build them progressed as well.

Cables and Topologies

Most LANs are built around copper cables that use standard electrical currents to relay their signals. Originally, most LANs consisted of computers connected with coaxial cables, but eventually, the twisted-pair cabling used for telephone systems became more popular. Another alternative is fiber-optic cable, which doesn't use electrical signals at all, but instead uses pulses of light to encode binary data. Other types of network infrastructures eliminate cables entirely and transmit signals using what is known as *unbounded media,* such as radio waves, infrared, and microwaves.

Note   *For more information about the various types of cables used in data networking, see Chapter 4.*

LANs connect computers using various types of cabling patterns called *topologies* (see Figure 1-1), which depend on the type of cable used and the protocols running on the computers. The most common topologies are as follows:

Bus   A bus topology takes the form of a cable that runs from one computer to the next one in a daisy-chain fashion, much like a string of Christmas tree lights. All of the signals transmitted by the computers on the network travel along the bus in both directions to all of the other computers. The two ends of the bus must be terminated with electrical resistors that nullify the voltages reaching them, so that the signals do not reflect back in the other direction. The primary drawback of the bus topology is that, like the string of Christmas lights it resembles, a fault in the cable anywhere along its length splits the network in two and prevents systems on opposite sides of the break from communicating. In addition, the lack of termination at either half can prevent computers that are still connected from communicating properly. As with Christmas lights, finding a single faulty connection in a large bus network can be troublesome and time-consuming. Most coaxial cable networks, such as the original Ethernet LANs, use a bus topology.

Star   A star topology uses a separate cable for each computer that runs to a central cabling nexus called a *hub* or *concentrator.* The hub propagates the signals entering through any one of its ports out through all of the other ports, so that the signals transmitted by each computer reach all of the other computers. Hubs also amplify the signals as they process them, enabling them to travel longer distances without degrading. A star network is more fault-tolerant than a bus, because a break in a cable affects only the device to which that cable is connected, not the entire network. Most of the networking protocols that call for twisted-pair cable, such as 10Base-T and 100Base-T Ethernet, use the star topology.

Star bus   A star bus topology is one method for expanding the size of a LAN beyond a single star. In this topology, a number of star networks are joined together using a separate bus cable segment to connect their hubs. Each computer can still communicate with any other computer on the network, because each of the hubs transmits its incoming traffic out through the bus port as well as the other star ports. Designed to expand 10Base-T Ethernet networks, the star bus is rarely seen today because of the speed limitations of coaxial bus networks, which can function as a bottleneck that degrades the performance of faster star network technologies such as Fast Ethernet.

Hierarchical star   The hierarchical star topology is the most common method for expanding a star network beyond the capacity of its original hub. When a hub's ports are all filled and you have more computers to connect to the network, you can connect the original hub to a second hub using a cable plugged into a special port designated for this purpose. Traffic arriving at either hub is then propagated to the other hub as well as to the connected computers. The number of hubs that a single LAN can support is dependent on the protocol it uses.

Ring   A ring topology is functionally equivalent to a bus topology with the two ends connected together so that signals travel from one computer to the next in an endless circular fashion. However, the communications ring is only a logical construct, not a physical one. The physical network is actually cabled using a star topology, and a special hub called a *multistation access unit (MAU)* implements the logical ring by taking each incoming signal and transmitting it out through the next downstream port only (instead of through all of the other ports, like a star hub). Each computer, upon receiving an incoming signal, processes it (if necessary) and sends it right back to the hub for transmission to the next station on the ring. Because of this arrangement, systems that transmit signals onto the network must also remove the signals after they have traversed the entire ring. Networks configured in a ring topology can use several different types of cable. Token Ring networks, for example, use twisted-pair cables, while FDDI networks use the ring topology with fiber-optic cable.

Media Access Control

When multiple computers are connected to the same baseband network medium, there must be a *media access control (MAC)* mechanism that arbitrates access to the network, to prevent systems from transmitting data at the same time. A MAC mechanism is a fundamental part of all local area networking protocols that use a shared network medium. The two most common MAC mechanisms are Carrier Sense Multiple Access with Collision Detection (CSMA/CD), which is used by Ethernet networks, and token passing, which is used by Token Ring, FDDI, and other protocols. These two mechanisms are fundamentally different, but they accomplish the same task by providing each system on the network with an equal opportunity to transmit its data. For more information about these MAC mechanisms, see Chapter 10 for CSMA/CD and Chapter 12 for token passing.

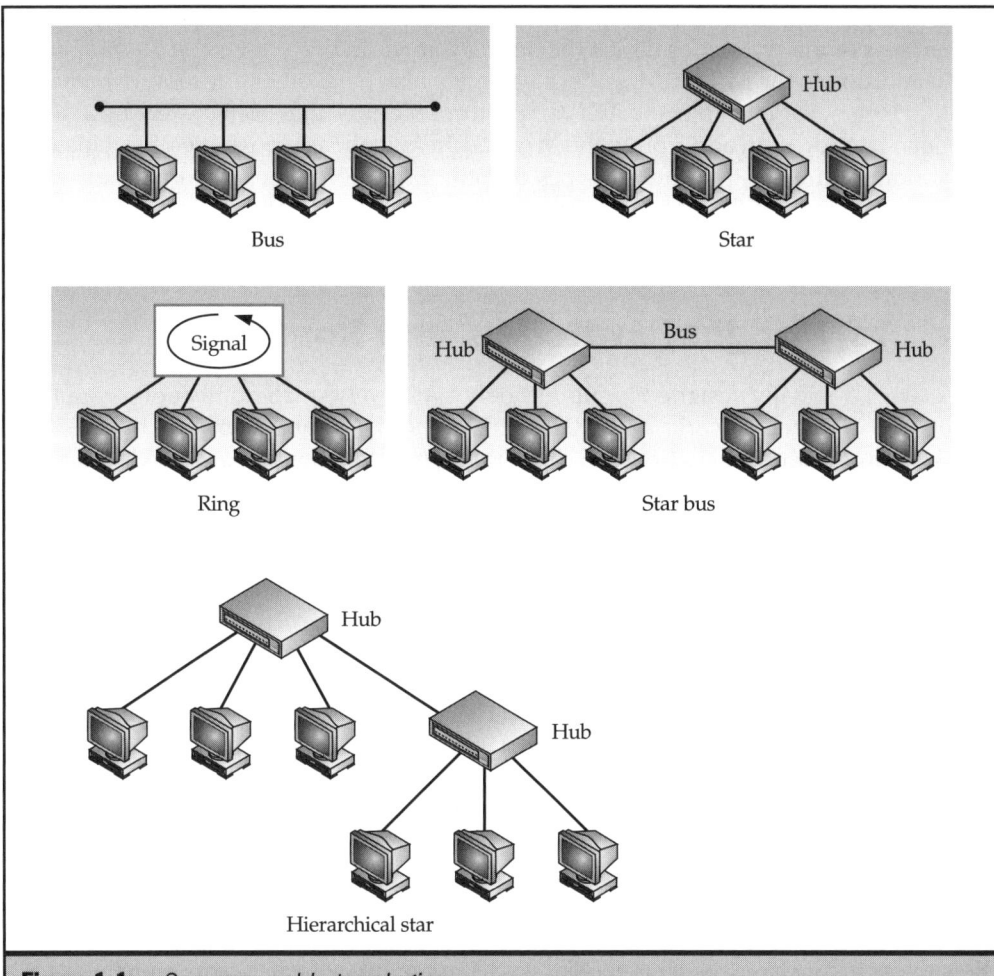

Figure 1-1:   *Common cable topologies*

Addressing

In order for systems on a shared network medium to communicate effectively, they must have some means of identifying each other, usually some form of numerical address. In most cases, the network interface card (NIC) installed into each computer has an address hardcoded into it at the factory, called its *MAC address* or *hardware address,* which uniquely identifies that card among all others. Every packet that each computer transmits over the network contains the address of the sending computer and the address of the system for which the packet is intended.

In addition to the MAC address, systems may also have other addresses operating at other layers. For example, the TCP/IP protocol requires that each system be assigned a unique IP address in addition to the MAC address it already possesses. Systems use the various addresses for different types of communications. For more information on MAC addressing, see Chapter 3; for more information on IP addressing, see Chapter 13.

Repeaters, Bridges, Switches, and Routers

LANs were originally designed to support only a relatively small number of computers—30 for Thin Ethernet networks and 100 for Thick Ethernet—but the needs of businesses quickly outgrew these limitations. To support larger installations, engineers developed products that enabled administrators to connect two or more LANs together into what is known as an *internetwork,* which is essentially a network of networks that enables the computers on one network to communicate with those on another.

Don't confuse the generic term internetwork with the Internet. The Internet is an example of an extremely large internetwork, but any installation that consists of two or more LANs connected together is also called an internetwork.

This terminology is confusing, because it is so often misused. Sometimes what users mean when they refer to a network is actually an internetwork, and at other times, what may seem to be an internetwork is actually a single LAN. Strictly speaking, a LAN or a network segment is a group of computers that share a network cable so that a broadcast message transmitted by one system reaches all of the other systems, even if that segment is actually composed of many pieces of cable. For example, on a typical 10Base-T Ethernet LAN, all of the computers are connected to a hub using individual lengths of cable. Regardless of that fact, this arrangement is still an example of a network segment or LAN.

Individual LANs can be connected together using several different types of devices, some of which simply extend the LAN while another creates an internetwork. These devices are as follows:

Repeaters   A repeater is a purely electrical device that extends the maximum distance a LAN cable can span by amplifying the signals passing through it. The hubs used on star networks are sometimes called *multiport repeaters* because they have signal amplification capabilities integrated into the unit. Standalone repeaters are also available for use on coaxial networks, to extend them over longer distances. Using a repeater to expand a network segment does not divide it into two LANs or create an internetwork.

Bridges   A bridge provides the amplification function of a repeater, along with the ability to selectively filter packets based on their addresses. Packets that originate on one side of the bridge are propagated to the other side only if they are addressed to a system that exists there. Because bridges do not prevent broadcast messages from being propagated across the connected cable segments, they, too, do not create multiple LANs or transform a network into an internetwork.

Switches   Switches are revolutionary devices that in many cases eliminate the shared network medium entirely. A switch is essentially a multiport repeater, like a hub, except that instead of operating at a purely electrical level, the switch reads the destination address in each incoming packet and transmits it out only through the port to which the destination system is connected.

Routers   A router is a device that connects two LANs together to form an internetwork. Like a bridge, a router only forwards traffic that is destined for the connected segment, but unlike repeaters and bridges, routers do not forward broadcast messages. Routers can also connect different types of networks together (such as Ethernet and Token Ring), whereas bridges and repeaters can only connect segments of the same type.

Note   *For more information on repeaters, bridges, routers, and switches, see Chapter 6.*

Wide Area Networks (WANs)

Internetworking enables an organization to build a network infrastructure of almost unlimited size. In addition to connecting multiple LANs together in the same building or campus, an internetwork can also connect LANs at distant locations, through the use of wide area network (WAN) links. A WAN is a collection of LANs, some or all of which are connected using point-to-point links that span relatively long distances. A typical WAN connection consists of two routers, one at each LAN site, connected using a long-distance link such as a leased telephone line. Any computer on one of the LANs can communicate with the other LAN by directing its traffic to the local router, which relays it over the WAN link to the other site.

WAN links differ from LANs in that they do not use a shared network medium and they can span much longer distances. Because the link connects only two systems, there is no need for media access control or a shared network medium. An organization with offices located throughout the world can build an internetwork that provides users with instantaneous access to network resources at any location. The WAN links themselves can use technologies ranging from telephone lines to public data networks to satellite systems. Unlike a LAN, which is nearly always privately owned and operated, an outside service provider (such as a telephone company) is nearly always involved in a WAN connection, because private organizations don't usually own the technologies needed to carry signals over such long distances. Generally speaking, WAN connections are slower and more expensive than LANs, sometimes much slower and much more expensive. As a result, one of the goals of the network administrator is to maximize the efficiency of WAN traffic by eliminating unnecessary communications and choosing the best type of link for the application.

Note   *For more information on WAN technologies, see Chapter 7.*

Protocols and Standards

Communications between computers on a network are defined by *protocols,* standardized languages that the software programs on the computers have in common. These protocols define every part of the communications process, from the signals transmitted over network cables to the query languages that enable applications on different machines to exchange messages. Networked computers run a series of protocols, called a *protocol stack,* that spans from the application user interface at the top to the physical network interface at the bottom. The stack is traditionally split into seven layers. The Open Systems Interconnection (OSI) reference model defines the functions of each layer and how the layers work together to provide network communications.

Note   *For more information on the OSI reference model, see Chapter 2.*

Early networking products tended to be proprietary solutions created by a single manufacturer, but as time passed, interoperability became a greater priority and organizations were formed to develop and ratify networking protocol standards. Most of these bodies are responsible for large numbers of technical and manufacturing standards in many different disciplines. Today, most of the protocols in common use are standardized by these bodies, some of which are as follows:

Institute of Electrical and Electronic Engineers (IEEE)   A U.S.-based society responsible for the publication of the IEEE 802 working group, which includes the standards that define the protocols commonly known as Ethernet and Token Ring, as well as many others.

International Organization for Standardization (ISO)   A worldwide federation of standards bodies from more than 100 countries, responsible for the publication of the OSI reference model document.

Internet Engineering Task Force (IETF)   An ad hoc group of contributors and consultants who collaborate to develop and publish standards for Internet technologies, including the TCP/IP protocols.

Telecommunications Industry Association/Electronic Industry Association (TIA/EIA)   Two organizations that have joined together to develop and publish the Commercial Building Telecommunications Wiring Standards, which define how the cables for data networks should be installed.

Clients and Servers

Local area networking is based on the client/server principle, in which the processes needed to accomplish a particular task are divided between computers functioning as clients and servers. This is in direct contrast to the mainframe model, in which the central computer did all of the processing and simply transmitted the results to a user at a remote terminal. A *server* is a computer running a process that provides a service to other computers when they request it. A *client* is the computer running a program that requests the service from a server.

For example, a LAN-based database application stores its data on a server, which stands by, waiting for clients to request information from it. Users at workstation computers run a database client program in which they generate queries that request specific information in the database and transmit those queries to the server. The server responds to the queries with the requested information and transmits it to the workstations, which format it for display to the users. In this case, the workstations are responsible for providing a user interface and translating the user input into a query language understood by the server. They are also responsible for taking the raw data from the server and displaying it in a comprehensible form to the user. The server may have to service dozens or hundreds of clients, so it is still a powerful computer. By offloading some of the application's functions to the workstations, however, its processing burden is nowhere near what it would be on a mainframe system.

Operating Systems and Applications

Clients and servers are actually software components, although some people associate them with specific hardware elements. This confusion is due to the fact that some network operating systems (such as Novell NetWare) require that a computer be dedicated to the role of server and that other computers function solely as clients. This is a client/server operating system, as opposed to a peer-to-peer operating system, in which every computer can function as both a client and a server. The most basic client/server functionality provided by a network operating system (NOS) is the ability to share file system drives and printers, and this is what usually defines the client and server roles. For example, Windows 2000 comes in both Server and Professional (workstation) versions, but it's still a peer-to-peer operating system, because any Windows 2000 computer can access resources on another Windows 2000 computer and share its own resources, regardless of whether it is running Server or Professional. A Novell NetWare server, however, can share its drives and printers, but it can't access shared resources on other computers, nor can NetWare clients share their own resources. NetWare clients can communicate with servers only, not with other clients (see Figure 1-2).

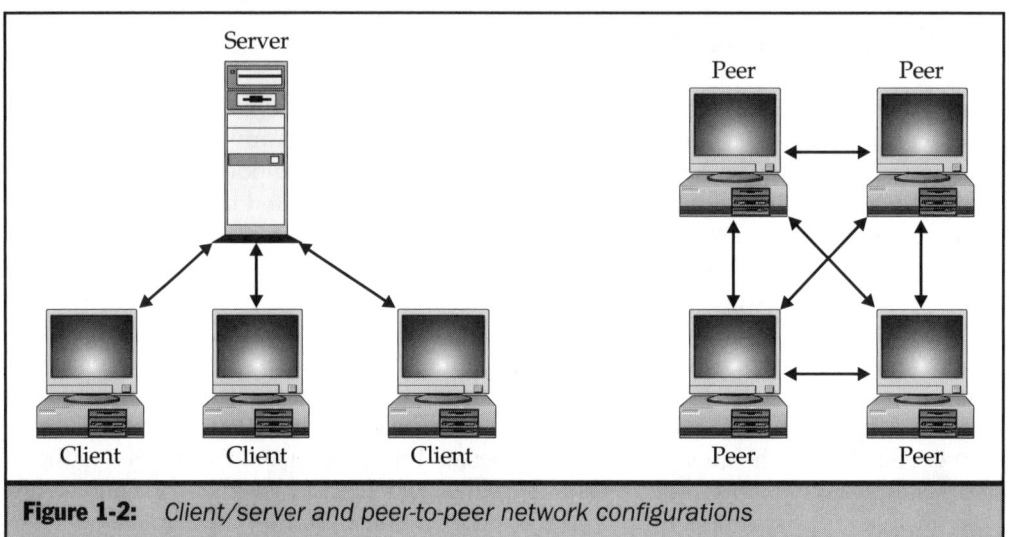

Figure 1-2:   *Client/server and peer-to-peer network configurations*

Note   *For more information on Windows 2000, see Chapter 16. For more information on Novell NetWare, see Chapter 19.*

Apart from the internal functions of network operating systems, many LAN applications and network services also operate using the client/server paradigm. Internet applications, such as the World Wide Web, consist of servers and clients, as do administrative services such as the Domain Name System (DNS).

Chapter 2

The OSI Reference Model

Network communications take place on many levels and can be difficult to understand, even for the knowledgeable network administrator. The Open Systems Interconnection (OSI) reference model is a theoretical construction that separates network communications into seven distinct layers, as shown in Figure 2-1. Each computer on the network uses a series of protocols to perform the functions assigned to each layer. The layers collectively form what is known as the *protocol stack* or *networking stack.* At the top of the stack is the application that makes a request for a resource located elsewhere on the network, and at the bottom is the medium that actually connects the computers and forms the network, such as a cable.

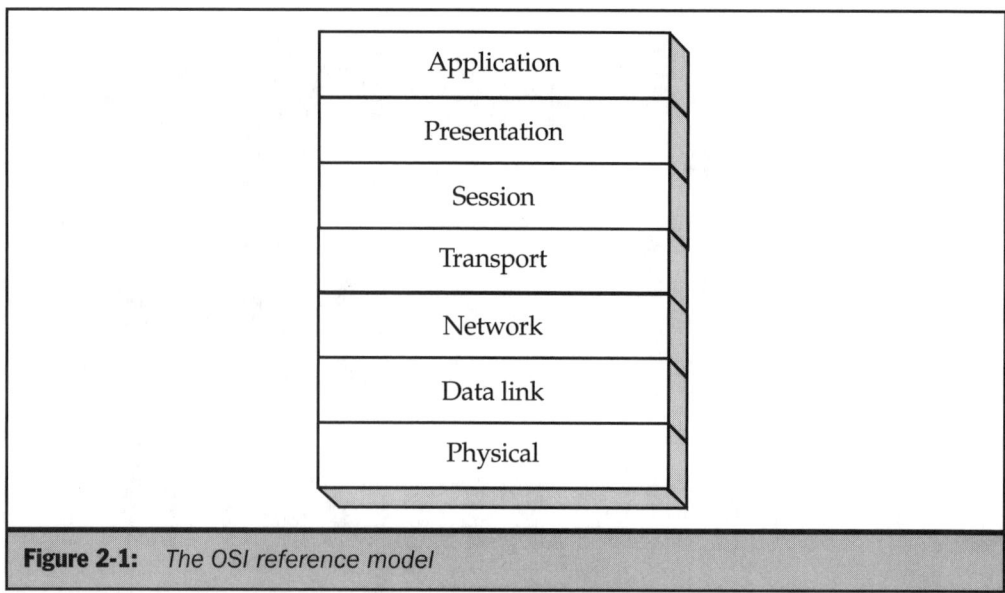

Figure 2-1:   *The OSI reference model*

The OSI reference model was developed in two separate projects by the International Organization for Standardization (ISO) and the Comité Consultatif International Téléphonique et Télégraphique (Consultative Committee for International Telephone and Telegraphy, or CCITT), which is now known as the Telecommunications Standardization Sector of the International Telecommunications Union (ITU-T). Each of these two bodies developed its own seven-layer model, but the two projects were combined in 1983, resulting in a document called "The Basic Reference Model for Open Systems Interconnection" that was published by the ISO as ISO 7498 and by the ITU-T as X.200.

The OSI stack was originally conceived to be the model for the creation of a protocol suite that would conform exactly to the seven layers. This suite never materialized in a commercial form, however, and the model has since been used as a teaching, reference, and communications tool. Networking professionals, educators, and authors frequently refer to protocols, devices, or applications as operating at a particular layer of the OSI model, because using this model breaks a complex process into manageable units that provide a common frame of reference. Many of the chapters in this book use the layers of the model to help define networking concepts. However, it is important to understand that none of the protocol stacks in common use today conform exactly to the layers of the OSI model. In many cases, protocols have functions that overlap two or more layers, such as Ethernet, which is considered a data link–layer protocol, but which also defines elements of the physical layer.

The primary reason why real protocol stacks differ from the OSI model is that many of the protocols used today (including Ethernet) were conceived before the OSI model documents were published. In fact, the TCP/IP protocols have their own layered model, which is similar to the OSI model in several ways, but uses only four layers (see Figure 2-2). In addition, developers are usually more concerned with practical functionality than with conforming to a preexisting model. The seven-layer model was designed to separate the functions of the protocol stack in such a way as to make it possible for separate development teams to work on the individual layers, thus streamlining the development process. However, if a single protocol can easily provide the functions that are defined as belonging in separate layers of the model, why divide it into two separate protocols just for the sake of conformity?

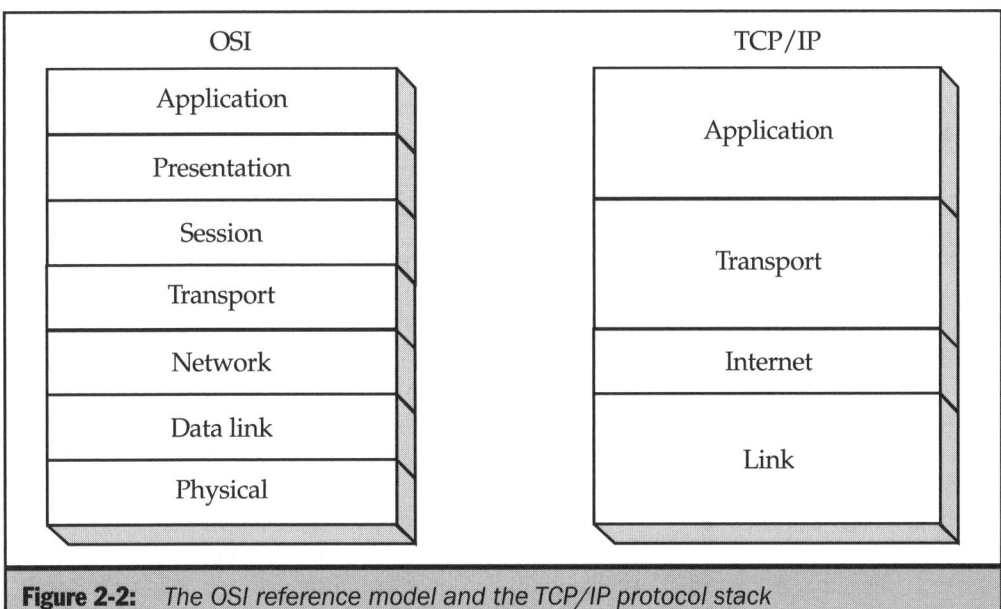

Figure 2-2:   *The OSI reference model and the TCP/IP protocol stack*

Interlayer Communications

Networking is the process of sending messages from one place to another, and the protocol stack illustrated in the OSI model defines the basic components needed to transmit messages to their destinations. The communication process is complex, because the applications that generate the messages have varying requirements. Some message exchanges consist of brief requests and replies that have to be exchanged as quickly as possible and with a minimum amount of overhead. Other network transactions, such as program file transfers, involve the transmission of larger amounts of data that must reach the destination in perfect condition, without a single bit value altered. Still other transmissions, such as streaming audio or video, consist of huge amounts of data that can survive the loss of an occasional bit, byte, or packet, but that must reach the destination in a timely manner.

The networking process also includes a number of conversions that ultimately take the API calls generated by applications and transform them into electrical charges, pulses of light, or other types of signals that can be transmitted across the network medium. Finally, the networking protocols must see to it that the transmissions reach the appropriate destinations in a timely manner. Just as you package a letter by placing it in an envelope and writing an address on it, the networking protocols package the data generated by an application and address it to another computer on the network.

Data Encapsulation

To satisfy all of the requirements just described, the protocols operating at the various layers work together to supply a unified quality of service. Each layer provides a service to the layers directly above and below it. Outgoing traffic travels down through the stack to the network medium, acquiring the control information needed to make the trip to the destination system as it goes. This control information takes the form of headers (and in one case a footer) that surround the data received from the layer above, in a process called *data encapsulation.* The headers and footer are composed of individual *fields* that contain control information used to get the packet to its destination. In a sense, the headers and footer form the envelope that carries the message received from the layer above.

In a typical transaction, shown in Figure 2-3, an application-layer protocol (which also includes presentation- and session-layer functions) generates a message that is passed down to a transport-layer protocol. The protocol at the transport layer has its own packet structure, called a *protocol data unit (PDU),* which includes specialized header fields and a Data field that carries the payload. In this case, the payload is the data received from the application-layer protocol. By packaging the data in its own

PDU, the transport layer *encapsulates* the application-layer data, and then passes it down to the next layer.

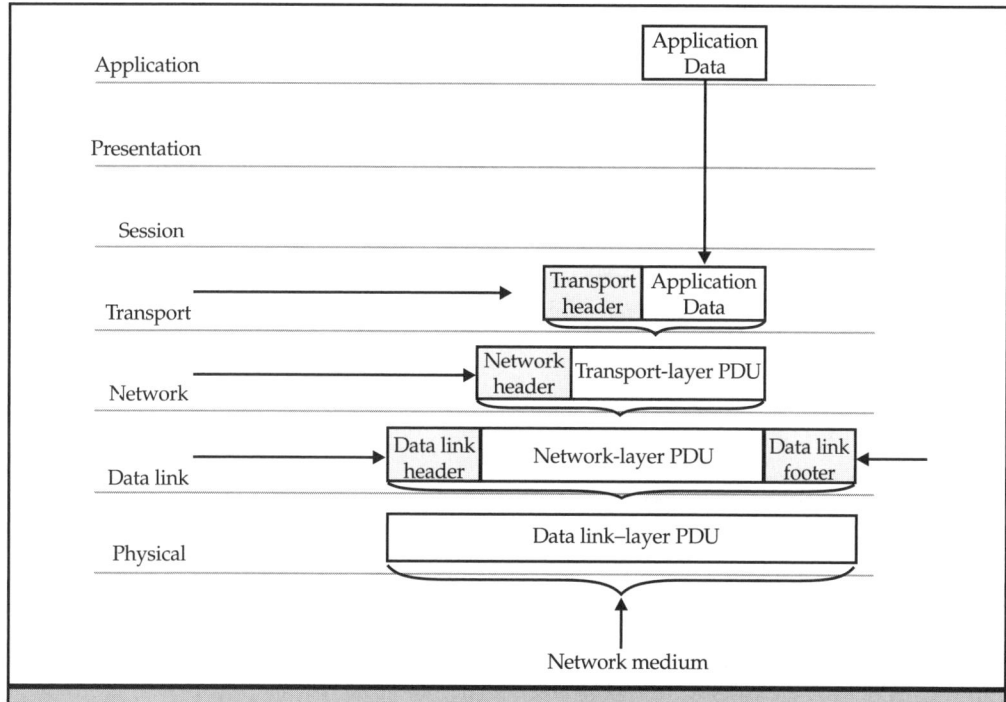

Figure 2-3:   *Application-layer data is encapsulated for transmission by the protocols operating at the lower layers in the stack.*

The network-layer protocol then receives the PDU from the transport layer and encapsulates it within its own PDU by adding a header and using the entire transport-layer PDU (including the application-layer data) as its payload. The same process occurs again when the network layer passes its PDU to the data link–layer protocol, which adds a header and a footer. To a data link–layer protocol, the data within the frame is treated as payload only, just as postal employees have no idea what is inside the envelopes they process. The only system that reads the information in the payload is the computer possessing the destination address. That computer then either passes the network-layer protocol data contained in the payload up through its protocol stack or uses that data to determine what the next destination of the packet should be. In the same way, the protocols operating at the other layers are conscious of their own header information, but are unaware of what data is being carried in the payload.

Once it is encapsulated by the data link–layer protocol, the completed packet (now called a *frame*) is then ready to be converted to the appropriate type of signal used by the network medium. Thus, the final packet, as transmitted over the network, consists of the original application-layer data plus several headers applied by the protocols at the succeeding layers, as shown in Figure 2-4.

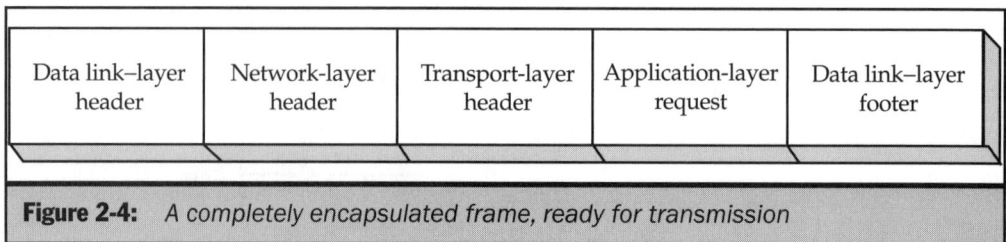

Figure 2-4:   *A completely encapsulated frame, ready for transmission*

Horizontal Communications

In order for two computers to communicate over a network, the protocols used at each layer of the OSI model in the transmitting system must be duplicated at the receiving system. When the packet arrives at its destination, the process by which the headers are applied at the source is repeated in reverse. The packet travels up through the protocol stack, and each successive header is stripped off by the appropriate protocol and processed. In essence, the protocols operating at the various layers communicate horizontally with their counterparts in the other system, as shown in Figure 2-5. The horizontal connections between the various layers are logical; there is no direct communication between them. The information included in each protocol header by the transmitting system is a message that will be carried to the same protocol in the destination system.

Vertical Communications

The headers applied by the various protocols implement the specific functions carried out by those protocols. In addition to communicating horizontally with the same protocol in the other system, the header information also enables each layer to communicate with the layers above and below it (see Figure 2-6). For example, when a system receives a packet and passes it up through the protocol stack, the data link–layer protocol header includes a field that identifies which network-layer protocol the system should use to process the packet. The network-layer protocol header in turn specifies one of the transport-layer protocols, and the transport-layer protocol identifies the application for which the data is ultimately destined. This vertical communication makes it possible for a computer to support multiple protocols at each of the layers simultaneously. As long as a packet has the correct information in its headers, it can be routed on the appropriate path through the stack to the intended destination.

Encapsulation Terminology

One of the most confusing aspects of the data encapsulation process is the terminology used to describe the PDUs generated by each layer. The term *packet* specifically refers to the complete unit transmitted over the network medium, although it also has become a generic term for the data unit at any stage in the process. Most data link–layer protocols are said to work with *frames,* because they include both a header and a footer that surround the data from the network-layer protocol. The term *frame* refers to a PDU of variable size, depending on the amount of data enclosed. A data link–layer protocol that uses PDUs of a uniform size, such as Asynchronous Transfer Mode (ATM), is said to deal in *cells*.

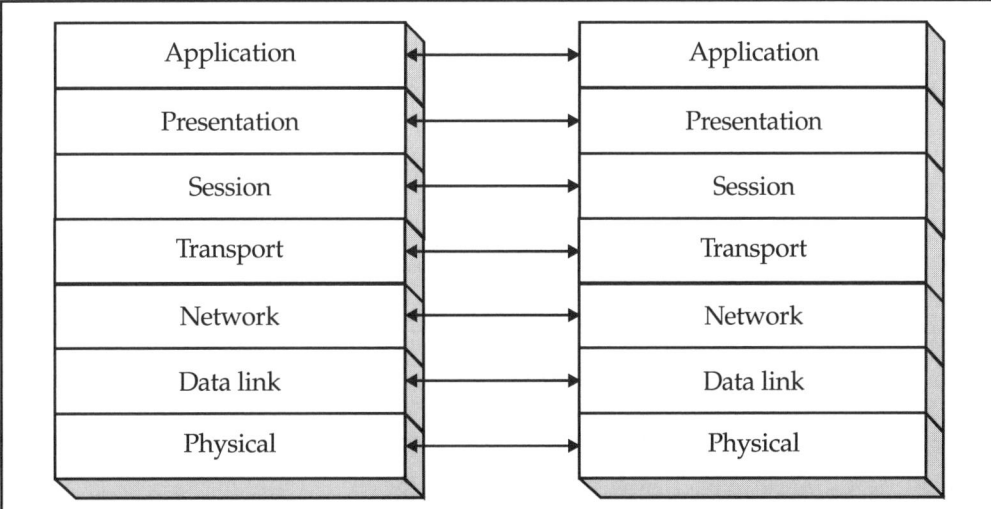

Figure 2-5:   *The protocols at the various layers have logical connections with their equivalents on other systems.*

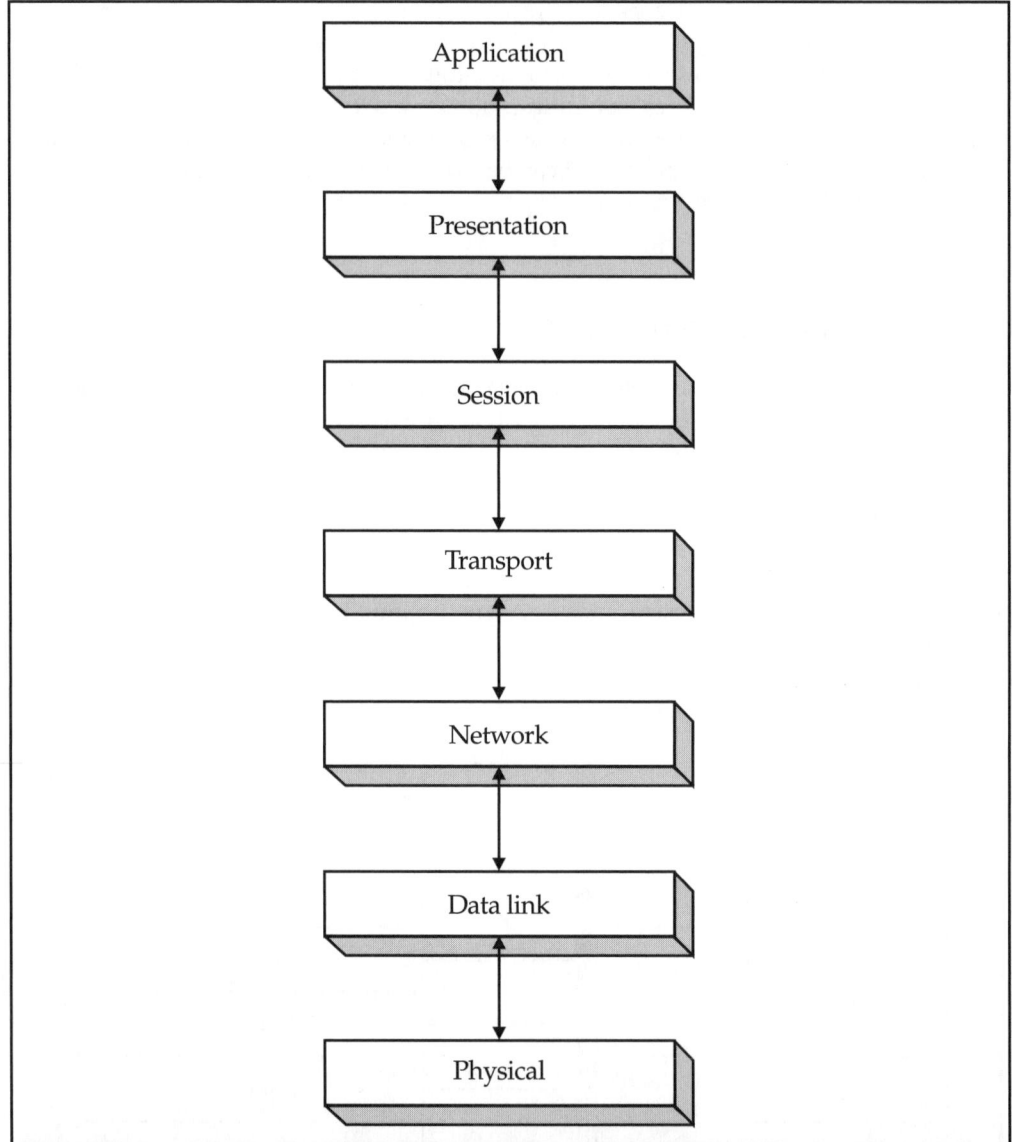

Figure 2-6: *Each layer in the OSI model communicates with the layers directly above and below it.*

NETWORK BASICS

When transport-layer data is encapsulated by a network-layer protocol, such as the Internet Protocol (IP) or Internetwork Packet Exchange (IPX), the resulting PDU is called a *datagram*. During the course of its transmission, a datagram might be split into *fragments*, each of which is sometimes incorrectly called a datagram. The terminology at the transport layer is more protocol specific than at the lower layers. TCP/IP, for example, has two transport-layer protocols. The first, called the *User Datagram Protocol (UDP)*, also refers to the PDUs it creates as *datagrams*, although these are not synonymous with the datagrams produced at the network layer. When the UDP protocol at the transport layer is encapsulated by the IP protocol at the network layer, the result is a datagram packaged within another datagram.

The difference between UDP and the Transmission Control Protocol (TCP), which also operates at the transport layer, is that UDP datagrams are self-contained units that were designed to contain the entirety of the data generated by the application-layer protocol. Therefore, UDP is traditionally used to transmit small amounts of data, while TCP, on the other hand, is used to transmit larger amounts of application-layer data that usually do not fit into a single packet. As a result, each of the PDUs produced by the TCP protocol is called a *segment*, and the collection of segments that carry the entirety of the application-layer protocol data is called a *sequence*. The PDU produced by an application-layer protocol is typically called a *message*. The session and presentation layers are usually not associated with individual protocols. Their functions are incorporated into other elements of the protocol stack, and they do not have their own headers or PDUs. All of these terms are frequently confused, and it is not surprising to see even authoritative documents use them incorrectly.

Note *Although UDP is still associated with brief request/response transactions, the roles of these protocols have evolved over the years, to some extent. Today, in addition to its original functions, UDP is sometimes used to transmit large amounts of data in the form of streaming audio or video. These applications use a great many UDP packets to carry their data, but they do not require the guaranteed delivery service provided by TCP. Unlike the transmission of a program file, which must be bit-for-bit accurate, an audio or video stream can suffer the loss of an occasional packet without damage to the overall transaction.*

The following sections examine each of the seven layers of the OSI reference model in turn, the functions that are associated with each, and the protocols that are most commonly used at those layers. As you proceed through this book, you will learn more about each of the individual protocols and their relationships to the other elements of the protocol stack.

The Physical Layer

The physical layer of the OSI model defines the actual medium that is used to carry data from one computer to another. The most common type of physical layer used in data networking is a copper-based electrical cable, although fiber-optic cable is becoming increasingly popular. There are also a number of wireless physical-layer implementations that use radio waves, infrared or laser light, microwaves, and other technologies. The physical layer includes the type of technology used to carry the data, the type of equipment used to implement that technology, the specifications of how the equipment should be installed, and the nature of the signals used to encode the data for transmission.

Note   *For more information on wireless networking, see Chapter 5.*

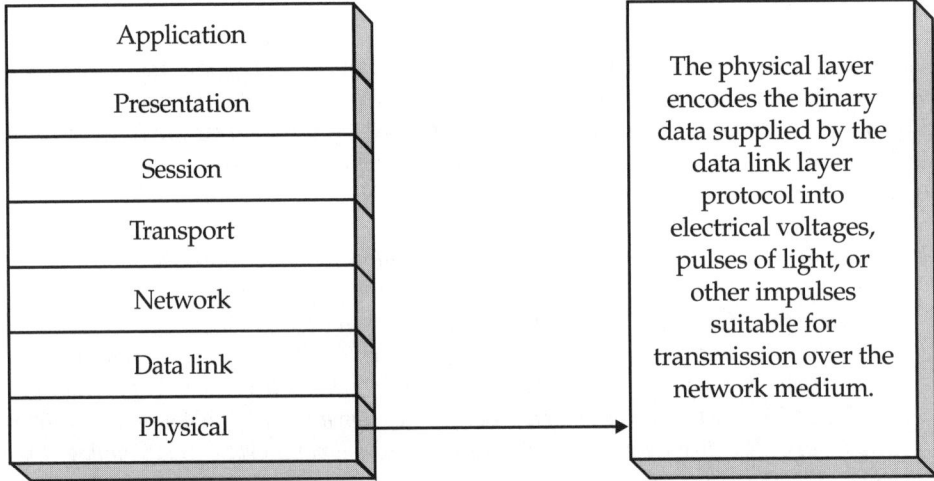

For example, one of the most popular physical-layer standards used for local area networking today is 10Base-T Ethernet. Ethernet is primarily thought of as a data link–layer protocol. However, as with most protocols functioning at the data link layer, Ethernet includes specific physical-layer implementations, and the standards for the protocol define the elements of the physical layer as well. 10Base-T refers to the type of cable used to form a particular type of Ethernet network. The Ethernet standard defines 10Base-T as an unshielded twisted-pair cable containing four pairs of copper wires enclosed in a single sheath.

However, the construction of the cable itself is not the only physical-layer element involved. The standards used to build an Ethernet network also define how to install the cable, including maximum segment lengths and distances from power sources. The standards specify what kind of connectors you use to join the cable, the type of network interface card (NIC) to install in the computer, and the type of hub you use to join the computers into a star topology. Finally, the standard specifies how the NIC should encode the data generated by the computer into electrical impulses that can be transmitted over the cable.

Thus, you can see that the physical layer encompasses much more than a type of cable. However, you generally don't have to know the details about every element of the physical-layer standard. When you buy Ethernet NICs, cables, and hubs, they are already constructed to the Ethernet specifications and designed to use the proper signaling scheme. Installing the equipment, however, is more complicated.

Physical-Layer Specifications

The installation of a network's physical layer is a job that is increasingly being outsourced to specialized contractors. While it is relatively easy to learn enough about a LAN technology to purchase the appropriate equipment, installing the cable (or other medium) is much more difficult, because you must be aware of all the specifications that affect the process. For example, the Ethernet standards published by the IEEE 802.3 working group specify the basic wiring configuration guidelines that pertain to the protocol's media access control (MAC) and collision detection mechanisms. These rules specify elements such as the maximum length of a cable segment, the distance between workstations, and the number of repeaters permitted on a network.

These guidelines are common knowledge to Ethernet network administrators, but these rules alone are not sufficient to perform a large cable installation. The American National Standards Institute/Electronic Industry Association/Telecommunication Industry Association (ANSI/EIA/TIA) 568, "Commercial Building Telecommunication Cabling Standard," defines in much greater detail how to install the cabling for data networks of various types. In addition, there are local building codes to consider, which might have a great effect on a cable installation. For these reasons, large physical-layer installations should in most cases be performed by professionals who are familiar with all of the standards that apply to the particular technology involved.

For more information on network cabling and its installation, see Chapter 4.

Physical-Layer Signaling

The primary operative component of a physical-layer installation is the transceiver found in NICs, repeating hubs, and other devices. The transceiver, as the name implies, is responsible for transmitting and receiving signals over the network medium. On networks using copper cable, the transceiver is an electrical device that takes the binary data it receives from the data link–layer protocol and converts it into signals of various voltages. Unlike all of the other layers in the protocol stack, the physical layer is not concerned in any way with the meaning of the data being transmitted. The transceiver simply converts zeros and ones into voltages, pulses of light, radio waves, or some other type of signal, but it is completely oblivious to packets, frames, addresses, and even the system receiving the signal.

The signals generated by a transceiver can be either analog or digital. Most data networks use digital signals, but some of the wireless technologies use analog radio transmissions to carry data. Analog signals transition between two values gradually, forming the sine wave pattern shown in Figure 2-7, while digital value transitions are immediate and absolute. The values of an analog signal can be determined by variations in amplitude, frequency, phase, or a combination of these elements.

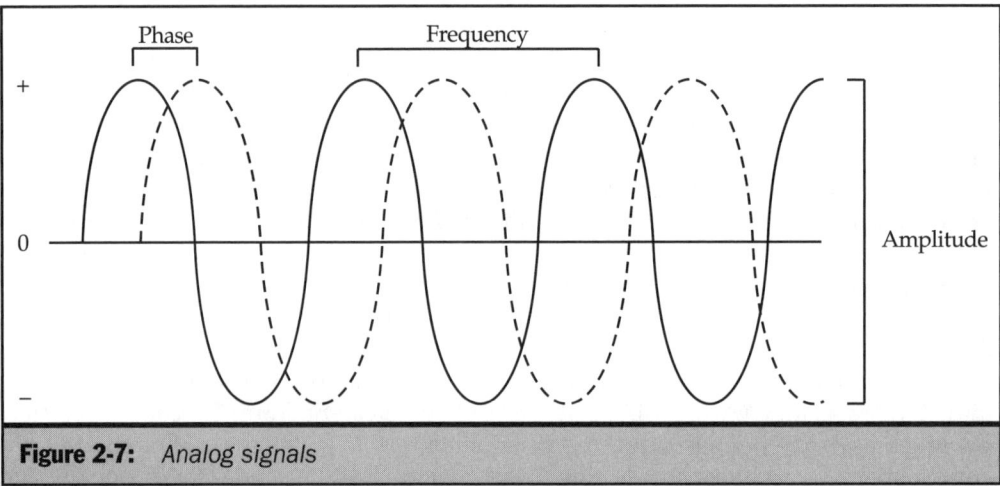

Figure 2-7:   *Analog signals*

The use of digital signals is much more common in data networking, however. All of the standard copper and fiber-optic media use various forms of digital signaling. The signaling scheme is determined by the data link–layer protocol being used. All Ethernet networks, for example, use the Manchester encoding scheme, whether they are running over twisted-pair, coaxial, or fiber-optic cable. Digital signals transition between values almost instantaneously, producing the square wave shown in Figure 2-8. Depending on the network medium, the values can represent electrical voltages, the presence or absence of a beam of light, or any other appropriate attribute of the medium. In most cases, the signal is produced with transitions between a positive and a negative voltage, although some use a zero value as well. The actual voltage is not relevant; the transitions create the signal.

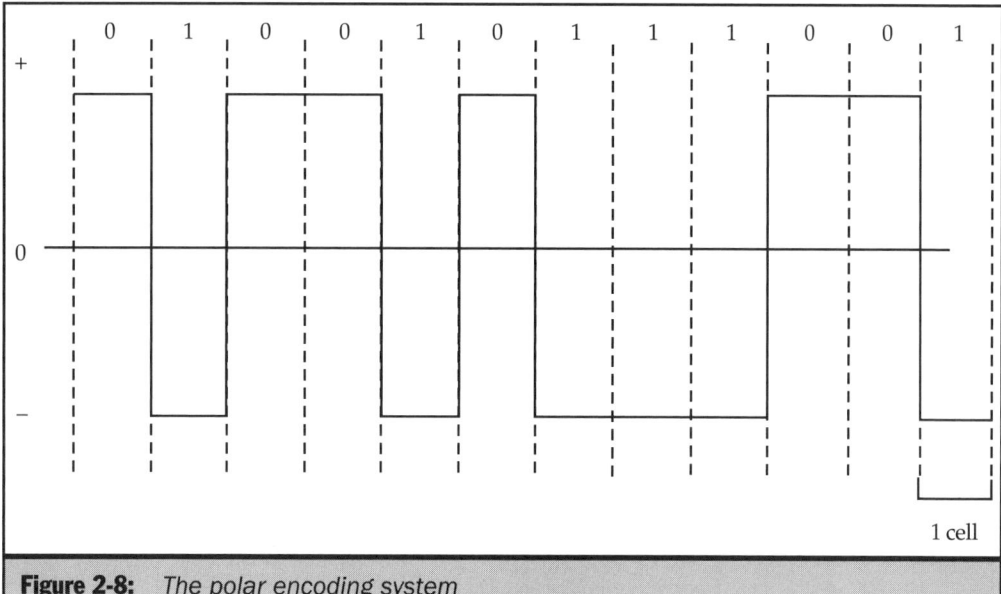

Figure 2-8:   *The polar encoding system*

Figure 2-8 illustrates a simple signaling scheme called *polar signaling.* In this scheme, the signal is broken up into units of time called *cells,* and the voltage of each cell denotes its binary value. A positive voltage is a zero, and a negative voltage is a one. This signaling code would seem to be a simple and logical method for transmitting binary information, but it has one crucial flaw, and that is timing. When the binary code consists of two or more consecutive zeros or ones, there is no voltage transition for the duration of two or more cells. Unless the two communicating systems have clocks that are precisely synchronized, it is impossible to tell for certain whether a voltage that remains continuous for a period of time represents two, three, or more cells with the same value. Remember that these communications occur at incredibly high rates of speed, so the timing intervals involved are extremely small.

Some systems can use this type of signal because they have an external timing signal that keeps the communicating systems synchronized. However, most data networks run over a baseband medium that permits the transmission of only one signal at a time. As a result, these networks use a different type of signaling scheme, one that is *self-timing.* In other words, the data signal itself contains a timing signal that enables the receiving system to correctly interpret the values and convert them into binary data.

The *Manchester encoding scheme* used on Ethernet networks is a self-timing signal, by virtue of the fact that every cell has a value transition at its midpoint. This delineates the boundaries of the cells to the receiving system. The binary values are specified by the direction of the value transition; a positive-to-negative transition indicates a value of zero, and a negative-to-positive transition indicates a value of one (see Figure 2-9). The value transitions at the beginnings of the cells have no function other than to set the voltage to the appropriate value for the midcell transition.

Token Ring networks use a different encoding scheme called *Differential Manchester,* which also has a value transition at the midpoint of each cell. However, in this scheme, the direction of the transition is irrelevant; it exists only to provide a timing signal. The value of each cell is determined by the presence or absence of a transition at the beginning of the cell. If the transition exists, the value of the cell is zero; if there is no transition, the value of the cell is one (see Figure 2-10). As with the midpoint transition, the direction of the transition is irrelevant.

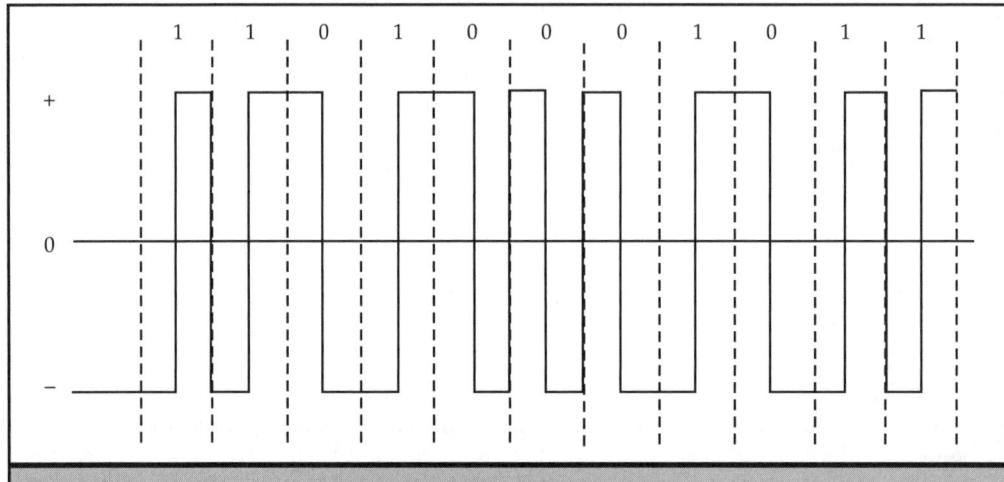

Figure 2-9: *The Manchester encoding scheme*

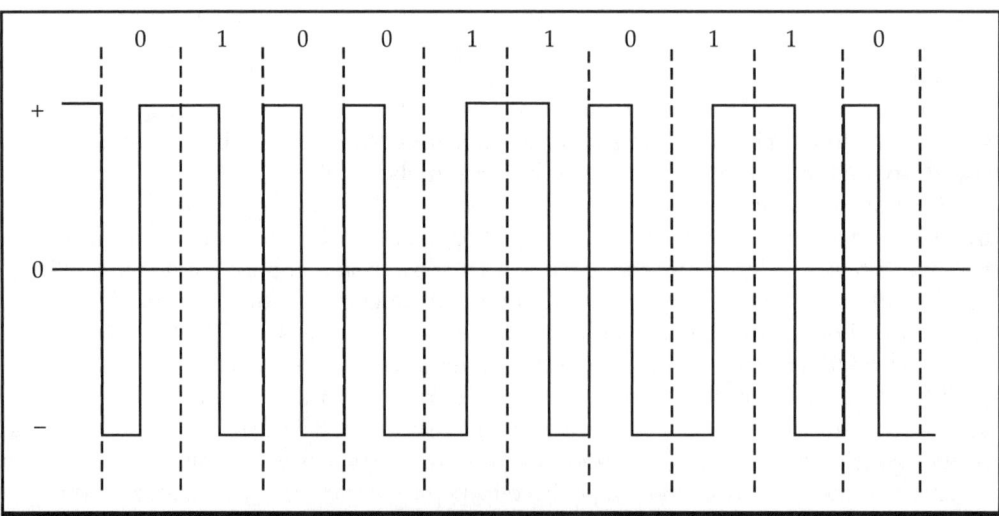

Figure 2-10: *The Differential Manchester encoding scheme*

The Data Link Layer

The data link–layer protocol provides the interface between the physical network and the protocol stack on the computer. A data link–layer protocol typically consists of three elements:

- The format for the frame that encapsulates the network-layer protocol data

- The mechanism that regulates access to the shared network medium

- The guidelines used to construct the network's physical layer

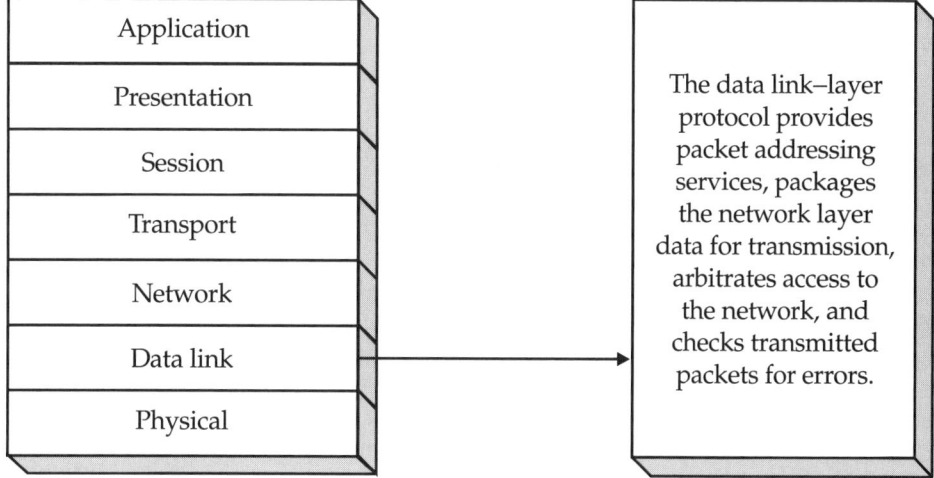

| Application |
| Presentation |
| Session |
| Transport |
| Network |
| Data link |
| Physical |

The data link–layer protocol provides packet addressing services, packages the network layer data for transmission, arbitrates access to the network, and checks transmitted packets for errors.

The header and footer applied to the network-layer protocol data by the data link–layer protocol are the outermost on the packet as it is transmitted across the network. This frame is, in essence, the envelope that carries the packet to its next destination and, therefore, provides the basic addressing information needed to get it there. In addition, data link–layer protocols usually include an error-detection facility and an indicator that specifies the network-layer protocol that the receiving system should use to process the data included in the packet.

On most LANs, multiple systems access a single shared baseband network medium. This means that only one computer can transmit data at any one time. If two or more systems transmit simultaneously, a collision occurs and the data is lost. The data link–layer protocol is responsible for controlling access to the shared medium and preventing an excess of collisions.

When speaking of the data link layer, the terms *protocol* and *topology* are often confused, but they are not synonymous. Ethernet is sometimes called a topology, when the topology actually refers to the way in which the computers on the network are cabled together. Some forms of Ethernet use a *bus topology,* in which each of the computers is cabled to the next one in a daisy-chain fashion, while the *star topology,* in which each computer is cabled to a central hub, is more prevalent today. A *ring topology* is a bus with the ends joined together, and a *mesh topology* is one in which each computer has a cable connection to every other computer on the network. These last two types are mainly theoretical; LANs today do not use them. Token Ring networks use a logical ring, but the computers are actually cabled using a star topology.

This confusion is understandable, since most data link–layer protocols include elements of the physical layer in their specifications. It is necessary for the data link–layer protocol to be intimately related to the physical layer, because media access control mechanisms are highly dependent on the size of the frames being transmitted and the lengths of the cable segments.

Addressing

The data link–layer protocol header contains the address of the computer sending the packet and the computer that is to receive it. The addresses used at this layer are the hardware (or MAC) addresses that in most cases are hardcoded into the network interface of each computer and router by the manufacturer. On Ethernet and Token Ring networks, the addresses are 6 bytes long, the first 3 bytes of which are assigned to the manufacturer by the Institute of Electrical and Electronic Engineers (IEEE), and the second 3 bytes of which are assigned by the manufacturer itself. A few older protocols, such as ARCnet, use addresses that are assigned by the network administrator, but the factory-assigned addresses are more efficient, insofar as they ensure that no duplication can occur.

Data link–layer protocols are not concerned with the delivery of the packet to its ultimate destination, unless that destination is on the same LAN as the source. When a packet passes through several networks on the way to its destination, the data link–layer protocol is only responsible for getting the packet to the router on the local network that provides access to the next network on its journey. Thus, the destination address in a data link–layer protocol header always references a device on the local network, even if the ultimate destination of the message is a computer on a network miles away.

The data link–layer protocols used on LANs rely on a shared network medium. Every packet is transmitted to all of the computers on the network segment, and only the system with the address specified as the destination reads the packet into its memory buffers and processes it. The other systems simply discard the packet without taking any further action.

Media Access Control

Media access control is the process by which the data link–layer protocol arbitrates access to the network medium. In order for the network to function efficiently, each of the workstations sharing the cable or other medium must have an opportunity to transmit its data on a regular basis. This is why the data to be transmitted is split into packets in the first place. If computers transmitted all of their data in a continuous stream, they could conceivably monopolize the network for extended periods of time.

Two basic forms of media access control are used on most of today's LANs. The *token passing* method, used by Token Ring and FDDI systems, uses a special frame called a *token* that is passed from one workstation to another. Only the system in possession of the token is permitted to transmit its data. A workstation, on receiving the token, transmits its data and then releases the token to the next workstation. Since there is only one token on the network at any time (assuming that the network is functioning properly), it isn't possible for two systems to transmit at the same time.

Note   *For more information on token passing, see Chapter 12.*

The other method, used on Ethernet networks, is called *Carrier Sense Multiple Access with Collision Detection (CSMA/CD).* In this method, when a workstation has data to send, it listens to the network cable and transmits if the network is not in use. On CSMA/CD networks, it is possible (and even expected) for workstations to transmit at the same time, resulting in packet collisions. To compensate for this, each system has a mechanism that enables it to detect collisions when they occur and retransmit the data that was lost.

Note   *For more information on CSMA/CD, see Chapter 10.*

Both of these MAC mechanisms rely on the physical-layer specifications for the network to function properly. For example, an Ethernet system can only detect collisions if they occur while the workstation is still transmitting a packet. If a network segment is too long, a collision may occur after the last bit of data has left the transmitting system, and thus may go undetected. The data in that packet is then lost and its absence can only be detected by the upper-layer protocols in the system that is the ultimate destination of the message. This process takes a relatively long time and significantly reduces the efficiency of the network. Thus, while the OSI reference model might create a neat division between the physical and data link layers, in the real world, the functionality of the two is more closely intertwined.

Protocol Indicator

Most data link–layer protocol implementations are designed to support the use of multiple network-layer protocols at the same time. This means that there are several possible paths through the protocol stack on each computer. To use multiple protocols at the network layer, the data link–layer protocol header must include a code that specifies the network-layer protocol that was used to generate the payload in the packet. This requirement is so that the receiving system can pass the data enclosed in the frame up to the appropriate network-layer process.

For example, on a network that uses both Windows 2000 and Novell NetWare servers, there might be some packets carrying IP datagrams and others using the IPX protocol at the network layer. To distinguish between the two, the DIX Ethernet II specification calls for a header field called the Ethertype, which contains a code identifying the network-layer protocol. The IEEE 802 specifications use a Sub-Network Access Protocol (SNAP) field to perform the same function, using the same codes as the Ethertype field.

Error Detection

Most data link–layer protocols are unlike all of the upper-layer protocols in that they include a footer that follows the payload field in addition to the header that precedes it. This footer contains a *frame check sequence (FCS)* field the receiving system uses to detect any errors that have occurred during the transmission. To do this, the system transmitting the packet computes a cyclical redundancy check (CRC) value on the entire frame and includes it in the FCS field. When the packet reaches its next destination, the receiving system performs the same computation and compares its results with the value in the FCS field. If the values do not match, the packet is assumed to have been damaged in transit and is silently discarded.

The receiving system takes no action to have discarded packets retransmitted; this is left up to the protocols operating at the upper layers of the OSI model. This error-detection process occurs at each hop in the packet's journey to its destination. Some upper-layer protocols have their own mechanisms for end-to-end error detection.

The Network Layer

The network-layer protocol is the primary end-to-end carrier for messages generated by the application layer. This means that, unlike the data link–layer protocol, which is concerned only with getting the packet to its next destination on the local network, the network-layer protocol is responsible for the packet's entire journey from the source system to its ultimate destination. A network-layer protocol accepts data from the transport layer and packages it into a datagram by adding its own header. Like a data link–layer protocol header, the header at the network layer contains the address of the destination system, but this address identifies the packet's final destination. Thus, the destination addresses in the data link–layer and network-layer protocol headers may actually refer to two different computers. The network-layer protocol datagram is essentially an envelope within the data link–layer envelope, and while the data link–layer envelope is opened by every system that processes the packet, the network-layer envelope remains sealed until the packet reaches its final destination.

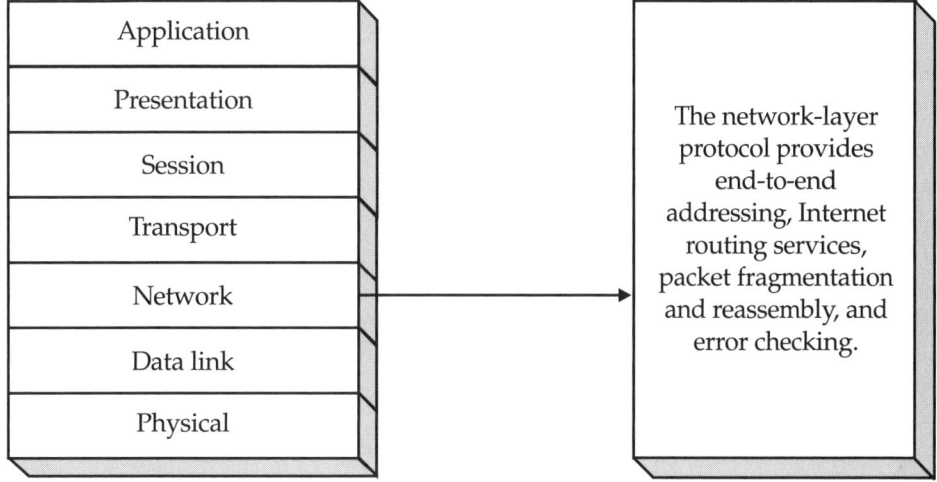

In addition to addressing, the network-layer protocol also performs some or all of the following functions:

- Routing
- Fragmenting
- Error checking
- Transport-layer protocol identification

Routing

Network-layer protocols use different types of addressing systems to identify the ultimate destination of a packet. The most popular network-layer protocol, the Internet Protocol (IP), provides its own 32-bit address space that identifies both the network on which the destination system resides and the system itself. NetWare's Internetwork

Packet Exchange (IPX) protocol uses a separate 32-bit network address and relies on the 48-bit hardware address coded into each network interface to identify particular systems.

An address by which individual networks can be uniquely identified is vital to the performance of the network-layer protocol's primary function, which is *routing*. When a packet travels through a large corporate internetwork or the Internet, it is passed from router to router until it reaches the network on which the destination system is located. A properly designed network will have more than one possible route to a particular destination, for fault-tolerance reasons, and the Internet has literally thousands of possible routes. Each router is responsible for determining the next router that the packet should use to take the most efficient path to its destination. Because data link–layer protocols are completely ignorant of conditions outside of the local network, it is left up to the network-layer protocol to choose an appropriate route with an eye on the end-to-end journey of the packet, not just the next interim hop.

The network layer defines two types of computers that can be involved in a packet transmission: end systems and intermediate systems. An *end system* is either the computer generating and transmitting the packet or the computer that is the ultimate recipient of the packet. An *intermediate system* is a router or switch that connects two or more networks and forwards packets on the way to their destinations. On end systems, all seven layers of the protocol stack are involved in either the creation or the reception of the packet. On intermediate systems, packets arrive and travel up through the stack only as high as the network layer (see Figure 2-11). The network-layer protocol chooses a route for the packet and sends it back down to a data link–layer protocol for packaging and transmission at the physical layer.

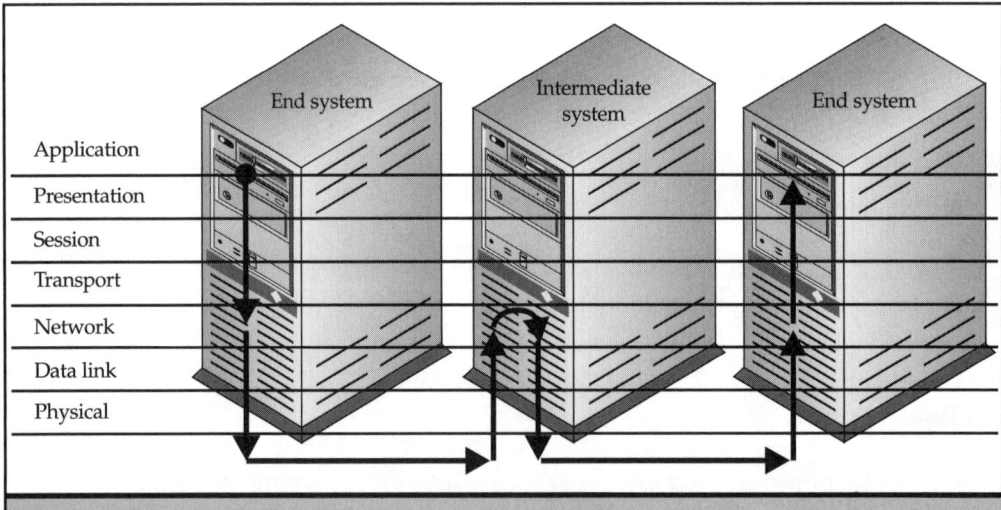

Figure 2-11:   *On intermediate systems, packets travel no higher than the network layer.*

When an intermediate system receives a packet, the data link–layer protocol checks it for errors and for the correct hardware address, and then strips off the data link header and footer and passes it up to the network-layer protocol identified by the Ethertype field or its equivalent. At this point, the packet consists of a datagram—that is, a network-layer protocol header and a payload that was generated by the transport-layer protocol on the source system. The network-layer protocol then reads the destination address in the header and determines what the packet's next destination should be. If the destination is a workstation on a local network, the intermediate system transmits the packet directly to that workstation. If the destination is on a distant network, the intermediate system consults its routing table to select the router that provides the most efficient path to that destination.

The compilation and storage of routing information in a reference table is a separate network-layer process that is performed either manually by an administrator or automatically by specialized network-layer protocols that routers use to exchange information about the networks to which they are connected. Once it has determined the next destination for the packet, the network-layer protocol passes the information down to the data link–layer protocol with the datagram, so that it can be packaged in a new frame and transmitted. When the IP protocol is running at the network layer, an additional process is required in which the IP address of the next destination is converted into a hardware address that the data link–layer protocol can use.

Fragmenting

Because routers can connect networks that use different data link–layer protocols, it is sometimes necessary for intermediate systems to split datagrams into fragments to transmit them. If, for example, a workstation on a Token Ring network generates a packet containing 4,500 bytes of data, an intermediate system that joins the Token Ring network to an Ethernet network must split the data into fragments no larger than 1,500 bytes, because that is the largest amount of data that an Ethernet frame can carry.

Depending on the data link–layer protocols used by the various intermediate networks, the fragments of a datagram may be fragmented themselves. Datagrams or fragments that are fragmented by intermediate systems are not reassembled until they reach their final destinations.

Connection-Oriented and Connectionless Protocols

There are two types of end-to-end protocols that operate at the network and transport layers: connection-oriented and connectionless. The type of protocol used helps to determine what other functions are performed at each layer. A *connection-oriented* protocol is one in which a logical connection between the source and the destination system is established before any upper-layer data is transmitted. Once the connection is established, the source system transmits the data, and the destination system acknowledges its receipt. A failure to receive the appropriate acknowledgments serves as a signal to the sender that packets have to be retransmitted. When the data transmission is completed successfully, the systems terminate the connection. By using this type of protocol, the sending system is certain that the data has arrived at the destination successfully. The cost of this guaranteed service is the additional network traffic generated by the connection establishment, acknowledgment, and termination messages, as well as a substantially larger protocol header on each data packet.

A *connectionless protocol* simply packages data and transmits it to the destination address without checking to see whether the destination system is available and without expecting packet acknowledgments. In most cases, connectionless protocols are used when a protocol higher up in the networking stack provides connection-oriented services, such as guaranteed delivery. These additional services can also include flow control (a mechanism for regulating the speed at which data is transmitted over the network), error detection, and error correction.

Most of the LAN protocols operating at the network layer, such as IP and IPX, are connectionless. In both cases, various protocols are available at the transport layer to provide both connectionless and connection-oriented services. If you are running a connection-oriented protocol at one layer, there is usually no reason to use one at another layer. The object of the protocol stack is to provide only the services that an application needs, and no more. There have been a few connection-oriented network-layer protocols, such as X.25, but the widespread adoption of TCP/IP, which supports only connectionless communications at the network layer, has all but eliminated this type of protocol. Despite its being connectionless, the IP protocol has an error-detection mechanism, but it checks only the IP header fields for errors, leaving the data in the payload to be checked by the protocols at the other layers.

The Transport Layer

Once you reach the transport layer, the process of getting packets from their source to their destination is no longer a concern. The transport-layer protocols and all the layers above them rely completely on the network and data link layers for addressing and transmission services. As discussed earlier, packets being processed by intermediate systems travel only as high as the network layer, so the transport-layer protocols only operate on the two end systems. The transport-layer PDU consists of a header and the data it has received from the application layer above, which is encapsulated into a datagram by the network layer below.

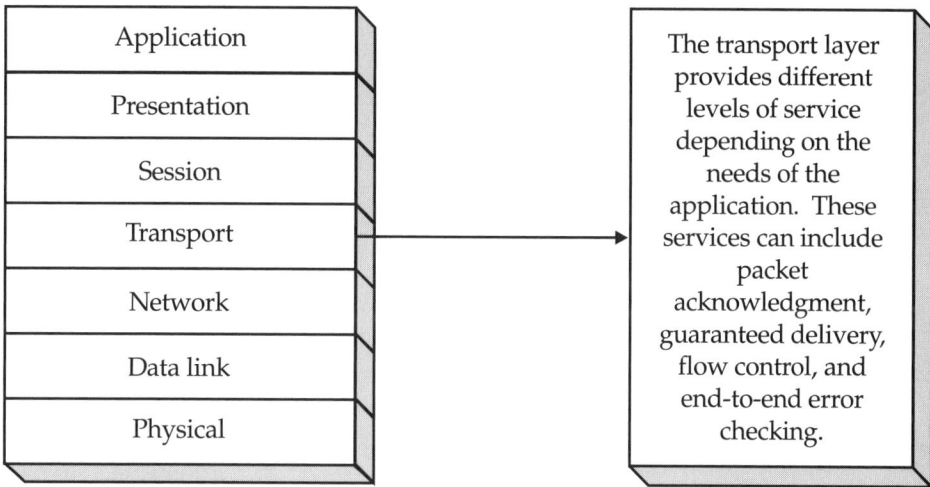

| Application |
| Presentation |
| Session |
| Transport |
| Network |
| Data link |
| Physical |

The transport layer provides different levels of service depending on the needs of the application. These services can include packet acknowledgment, guaranteed delivery, flow control, and end-to-end error checking.

One of the main functions of the transport-layer protocol is to identify the upper-layer processes that generated the message at the source system and that will receive the message at the destination system. The transport-layer protocols in the TCP/IP suite, for example, use port numbers in their headers to identify upper-layer services. Other functions that can be performed at the transport layer include error detection and correction, flow control, packet acknowledgment, and other connection-oriented services.

Protocol Service Combinations

Data link– and network-layer protocols operate together interchangeably; you can use almost any data link–layer protocol with any network-layer protocol. However, transport-layer protocols are closely related to a particular network-layer protocol and cannot be interchanged. The combination of a network-layer protocol and a transport-layer protocol provides a complementary set of services that is suitable for a specific application. As at the network layer, transport-layer protocols can be connection oriented (CO) or connectionless (CL). The OSI model document defines four possible combinations of CO and CL protocols at these two layers, depending on the services required, as shown in Figure 2-12. The process of selecting a combination of protocols for a particular task is called *mapping* a transport-layer service onto a network-layer service.

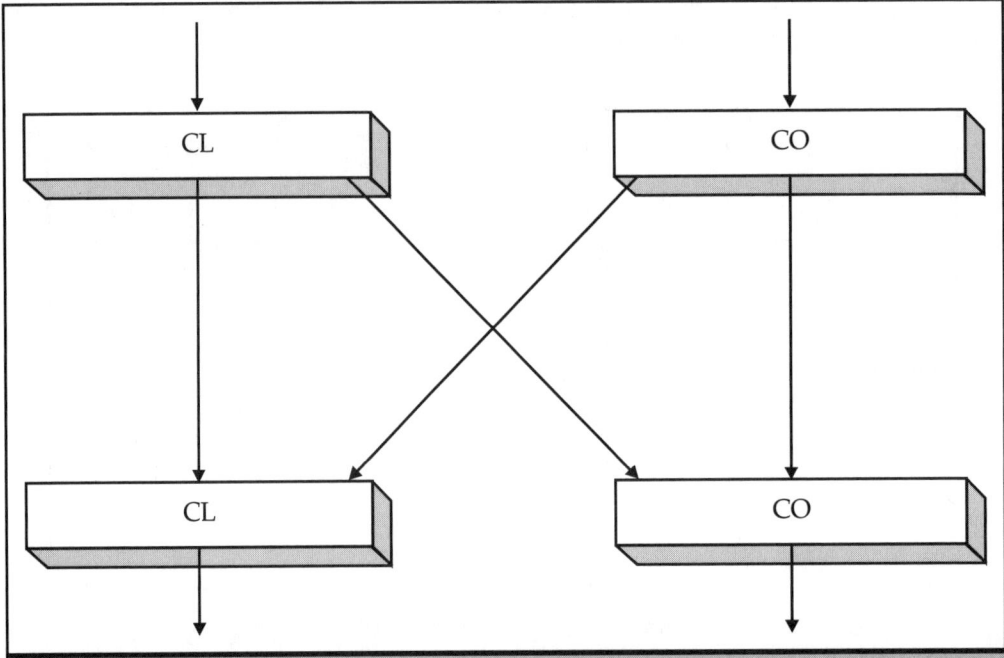

Figure 2-12: *A computer can use any combination of connection-oriented and connectionless protocols.*

The selection of a protocol at the transport layer is based on the needs of the application generating the message and the services already provided by the protocols at the lower layers. The OSI document defines five theoretical classes of transport-layer protocol, which are as follows:

TP0   No additional functionality. Assumes that the protocols at the lower layers already provide all of the services needed by the application.

TP1   Signaled error recovery. Provides the capability to correct errors that have been detected by the protocols operating at the lower layers.

TP2   Multiplexing. Includes codes that identify the process that generated the packet and that will process it at the destination, thus enabling the traffic from multiple applications to be carried over a single network medium.

TP3   Signaled error recovery and multiplexing. Combines the services provided by TP1 and TP2.

TP4   Complete connection-oriented service. Includes error detection and correction, flow control, and other services. Assumes the use of a connectionless protocol at the lower layers that provides none of these services.

This classification of transport-layer services is another place where the theoretical constructs of the OSI model differ substantially from reality. None of the protocol suites in common use have five different transport-layer protocols conforming to these classes. Most of the suites, like TCP/IP, have two protocols that basically conform to the TP0 and TP4 classes, providing connectionless and connection-oriented services, respectively.

Transport-Layer Protocol Functions

The UDP protocol is a connectionless service that, together with IP at the network layer, provides minimal services for brief transactions that do not need the services of a connection-oriented protocol. Domain Name System (DNS) transactions, for example, generally consist of short messages that can fit into a single packet, so no flow control is needed. A typical transaction consists of a request and a reply, with the reply functioning as an acknowledgment, so no other guaranteed delivery mechanism is needed. UDP does have an optional error-detection mechanism in the form of a checksum computation performed on both the source and destination systems. Because the UDP protocol provides a minimum of additional services, its header is only 8 bytes long, adding very little additional control overhead to the packet.

The TCP protocol, on the other hand, is a connection-oriented protocol that provides a full range of services, but at the cost of much higher overhead. The TCP header is 20 bytes long, and the protocol also generates a large number of additional packets solely for control procedures, such as connection establishment, termination, and packet acknowledgment.

Segmentation and Reassembly

Connection-oriented transport-layer protocols are designed to carry large amounts of data, but the data must be split into segments to fit into individual packets. The segmentation of the data and the numbering of the segments is a critical element in the transmission process and also makes functions such as error recovery possible. The routing process performed at the network layer is dynamic; in the course of a transmission, it is possible for the segments to take different routes to the destination and arrive in a different order from that in which they were sent. It is the numbering of the segments that makes it possible for the receiving system to reassemble them back into their original order. This numbering also makes it possible for the receiving system to notify the sender that specific packets have been lost or corrupted. As a result, the sender can retransmit only the missing segments and not have to repeat the entire transmission.

Flow Control

One of the functions commonly provided by connection-oriented transport-layer protocols is *flow control*, which is a mechanism by which the system receiving the data can notify the sender that it must decrease its transmission rate or risk overwhelming the receiver and losing data. The TCP header, for example, includes a Window field in which the receiver specifies the number of bytes that it can receive from the sender. If this value decreases in succeeding packets, the sender knows that it has to slow down its transmission rate. When the value begins to rise again, the sender can increase its speed.

Error Detection and Recovery

The OSI model document defines two forms of error recovery that can be performed by connection-oriented transport-layer protocols. One is a response to *signaled errors* detected by other protocols in the stack. In this mechanism, the transport-layer protocol does not have to detect the transmission errors themselves. Instead, it receives notification from a protocol at the network or data link layer that an error has occurred and that specific packets have been lost or corrupted. The transport-layer protocol only has to send a message back to the source system listing the packets and requesting their retransmission.

The more commonly implemented form of error recovery at the transport layer is a complete process of error detection and correction that is used to cope with *unsignaled errors*, which are errors that have not yet been detected by other means. Even though most data link–layer protocols have their own error-detection and correction mechanisms, they only function over the individual hops between two systems. A transport-layer error-detection mechanism provides error checking between the two end systems and includes the capability to recover from the errors by informing the sender which packets have to be resent. To do this, the checksum included in the transport-layer protocol header is computed only on the fields that are not modified during the journey to the destination. Fields that routinely change, such as the Time-to-Live indicator that is incremented by every router processing the packet, are omitted from the calculation.

The Session Layer

When you reach the session layer, the boundaries between the layers and their functions start to become more obscure. There are no discrete protocols that operate exclusively at the session layer. Rather, the session-layer functionality is incorporated into other protocols, with functions that fall into the provinces of the presentation and application layers, as well. NetBIOS (Network Basic Input/Output System) and

NetBEUI (NetBIOS Extended User Interface) are two of the best examples of these protocols.

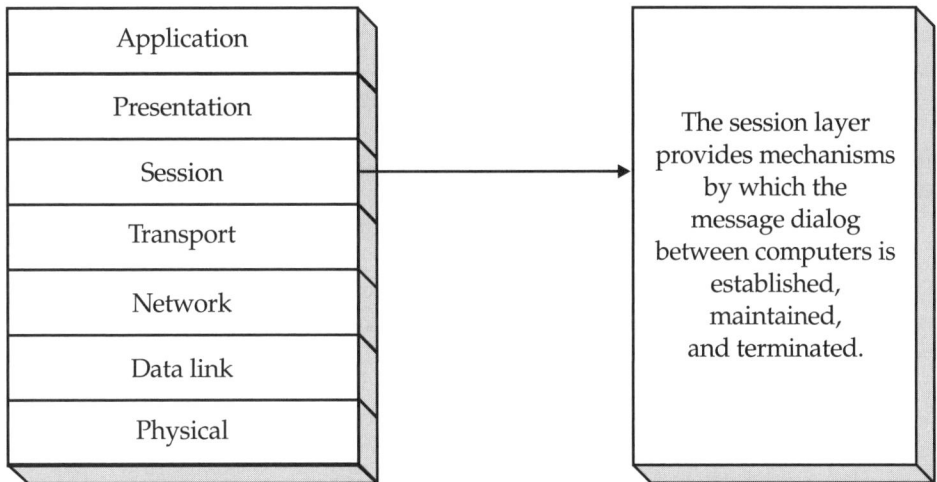

The boundary to the session layer is also the point at which all concern for the transmission of data between two systems is transcended. Questions of packet acknowledgment, error detection, and flow control are all left behind at this point, because everything that can be done has been done by the protocols at the transport layer and below.

The session layer is also not inherently concerned with security and the network logon process, as the name seems to imply. Rather, the primary functions of this layer concern the exchange of messages between the two connected end systems, called a *dialog.* There are also numerous other functions provided at this layer, which really serves as a multipurpose "toolkit" for application developers.

The services provided by the session layer are widely misunderstood, and even at the time of the OSI model's development, there was some question concerning whether they should be allotted a layer of their own. In fact, 22 different services are provided by the session layer, grouped into subsets such as the Kernel Function Unit, the Basic Activity Subset, and the Basic Synchronization Subset. Most of these services are of interest only to application developers, and some are even duplicated as a result of a compromise that occurred when the two committees creating OSI model standards were combined.

Communications between the layers of the OSI reference model are facilitated through the use of *service request primitives,* which are the tools in the toolkit. Each layer provides services to the layer immediately above it. A process at a given layer takes advantage of a service provided by the layer below by issuing a command using the appropriate service request primitive, plus any additional parameters that may be required. Thus, an application-layer process issues a request for a network resource using a primitive provided by the presentation layer. The request is then passed down through the layers, with each layer using the proper primitive provided by the layer below, until the message is ready for transmission over the network. Once the packet arrives at its destination, it is decoded into *indication primitives* that are passed upward through the layers of the stack to the receiving application process.

The two most important services attributed to the session layer are dialog control and dialog separation. *Dialog control* is the means by which two systems initiate a dialog, exchange messages, and finally end the dialog while ensuring that each system has received the messages intended for it. While this may seem to be a simple task, consider the fact that one system might transmit a message to the other and then receive a message back without knowing for certain when the response was generated. Is the other system responding to the message just sent or was its response transmitted before that message was received? This sort of *collision case* can cause serious problems, especially when one of the systems is attempting to terminate the dialog or create a checkpoint. *Dialog separation* is the process of inserting a reference marker called a *checkpoint* into the data stream passing between the two systems so that the status of the two machines can be assessed at the same point in time.

Dialog Control

When two end systems initiate a session-layer dialog, they choose one of two modes that controls the way they will exchange messages for the duration of the session: either *two-way alternate (TWA) mode* or *two-way simultaneous (TWS) mode*. Each session connection is uniquely identified by a 196-byte value consisting of the following four elements:

- Initiator SS-USER reference

- Responder SS-USER reference

- Common reference

- Additional reference

Once made, the choice of mode is irrevocable; the connection must be severed and reestablished in order to switch to the other mode.

In TWA mode, only one of the systems can transmit messages at any one time. Permission to transmit is arbitrated by the possession of a *data token*. Each system, at the conclusion of a transmission, sends the token to the other system using the S-TOKEN-GIVE primitive. On receipt of the token, the other system can transmit its message. There is also an S-TOKEN-PLEASE primitive that a system can use to request the token from the other system. The use of TWS mode complicates the communication process enormously. As the name implies, in a TWS mode connection, there is no token and both systems can transmit messages at the same time.

Note	*It is important to understand that the references to tokens and connections at the session layer have nothing to do with the similarly named elements in lower-layer protocols. A session-layer token is not the equivalent of the token frame used by the Token Ring protocol, nor is a session-layer connection the equivalent of a transport-layer connection such as that used by TCP. It is possible for end systems to terminate the session-layer connection while leaving the transport-layer connection open for further communications.*

The use of the token prevents problems resulting from crossed messages and provides a mechanism for the *orderly termination* of the connection between the systems. An orderly termination begins with one system signaling its desire to terminate the connection and transmitting the token. The other system, on receiving the token, transmits any data remaining in its buffers and uses the S-RELEASE primitive to acknowledge the termination request. On receiving the S-RELEASE primitive, the original system knows that it has received all of the data pending from the other system, and can then use the S-DISCONNECT primitive to terminate the connection.

There is also a *negotiated release* feature that enables one system to refuse the release request of another, which can be used in cases in which a collision occurs because both systems have issued a release request at the same time, and a *release token* that prevents the occurrence of these collisions in the first place by enabling only one system at a time to request a release.

All of these mechanisms are "tools" in the kit that the session layer provides to application developers; they are not automatic processes working behind the scenes. When designing an application, the developer must make an explicit decision to use the S-TOKEN-GIVE primitive instead of S-TOKEN-PLEASE, for example, or to use a negotiated release instead of an orderly termination.

Dialog Separation

Applications create checkpoints in order to save their current status to disk, in case of a system failure. This was a much more common occurrence at the time that the OSI model was developed than it is now. As with the dialog control processes discussed earlier, checkpointing is a procedure that must be explicitly implemented by an application developer as needed.

When the application involves communication between two systems connected by a network, the checkpoint must save the status of both systems at the same point in the data stream. Performing any activity at precisely the same moment on two different computers is nearly impossible. The systems might be performing thousands of activities per second, and their timing is nowhere near as precise as would be needed to execute a specific task simultaneously. In addition, the problem again arises of messages that may be in transit at the time the checkpoint is created. As a result, dialog separation is performed by saving a checkpoint at a particular point in the data stream passing between the two systems, rather than at a particular moment in time.

When the connection uses TWA mode, the checkpointing process is relatively simple. One system creates a checkpoint and issues a primitive called S-SYNC-MINOR. The other system, on receiving this primitive, creates its own checkpoint, secure in the knowledge that no data is left in transit at the time of synchronization. This is called a *minor synchronization* because it works with data flowing in only one direction at a time and requires only a single exchange of control messages.

It is still possible to perform a minor synchronization in TWS mode, using a special *minor synchronization token* that prevents both systems from issuing the S-SYNC-MINOR primitive at the same time. If it was possible to switch from TWS to TWA mode in midconnection, the use of an additional token would not be necessary, but mode switching is not possible. This is something that many people feel to be a major shortcoming in the session-layer specification.

In most cases, systems using TWS mode communications must perform a *major synchronization,* which accounts not only for traffic that can be running in both directions, but also for expedited traffic. A primitive called S-EXPEDITED enables one system to transmit to the other using what amounts to a high-speed pipeline that is separate from the normal communications channel. To perform a major synchronization, the system in possession of yet another token called the *major/activity token* issues a primitive called S-SYNC-MAJOR and then stops transmitting until it receives a response. However, the system issuing this primitive cannot create its checkpoint yet, as in a minor synchronization, because there may be traffic from the other system currently in transit.

On receiving the primitive, the other system is able to create its own checkpoint, because all of the data in transit has been received, including expedited data, which has to have arrived before the primitive. The receiving system then transmits a confirmation response over the normal channel and transmits a special PREPARE message over the expedited channel. The system that initiated the synchronization procedure receives the PREPARE message first and then the confirmation, at which time it can create its own checkpoint.

The Presentation Layer

Unlike the session layer, which provides many different functions, the presentation layer has only one. In fact, most of the time, the presentation layer functions primarily as a *pass-through service,* meaning that it receives primitives from the application layer and issues duplicate primitives to the session layer below, using the Presentation Service Access Point (PSAP) and the Session Service Access Point (SSAP). All of the discussion in the previous sections about applications utilizing session-layer services actually involves the use of the pass-through service at the presentation layer, because it is impossible for a process at any layer of the OSI model to communicate directly with any layer other than the one immediately above or beneath it.

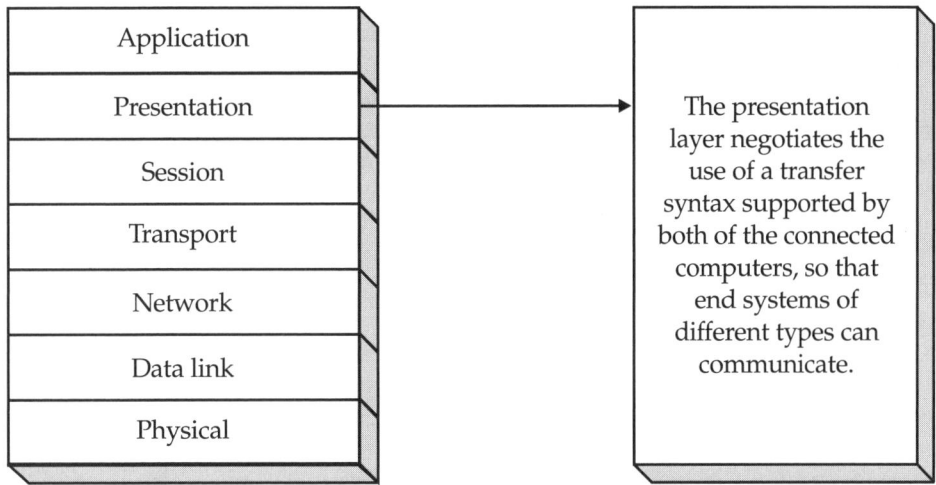

While the basic functions of the primitives are not changed as they are passed down through the presentation layer, they can undergo a crucial translation process that is the primary function of the layer. Applications generate requests for network resources using their own native syntax, but the syntax of the application at the destination system receiving the request may be different in several ways. For example, a connection between a PC and a mainframe may require a translation between the ASCII and EBCDIC bit-coding formats. The systems might also implement encryption and/or compression on the data to be transmitted over the network.

This translation process occurs in two phases, one of which runs at the presentation layer on each system. Each computer maintains an *abstract syntax*, which is the native syntax for the application running on that system, and a *transfer syntax*, which is a common syntax used to transmit the data over the network. The presentation layer on the system sending a message converts the data from the abstract syntax to the transfer syntax and then passes it down to the session layer. When the message arrives at the destination system, the presentation layer converts the data from the transfer syntax to the abstract syntax of the application receiving the message. The transfer syntax chosen for each abstract syntax is based on a negotiation that occurs when a presentation-layer connection is established between two systems. Depending on the application's requirements and the nature of the connection between the systems, the transfer context may provide data encryption, data compression, or a simple translation.

> **Note**
>
> *As with the session layer, the presentation-layer connection is not synonymous with the connections that occur at the lower layers, nor is there direct communication between the presentation layers of the two systems. Messages travel down through the protocol stack to the physical medium and up through the stack on the receiver to the presentation layer there.*

The syntax negotiation process begins when one system uses the P-CONNECT primitive to transmit a set of *presentation contexts,* which are pairs of associated abstract contexts and transfer contexts supported by that system. Each presentation context is numbered using a unique odd-numbered integer called a *presentation context identifier.* With this message, one system is essentially informing the other of its presentation-layer capabilities. The message may contain multiple transfer contexts for each abstract context, to give the receiving system a choice.

Once the other system receives the P-CONNECT message, it passes the presentation contexts up to the application-layer processes, which decide which of the transfer contexts supported by each abstract context they want to use. The receiver then returns a list of contexts to the sender with either a single transfer context or an error message specified for each abstract context. On receipt by the original sender, this list becomes the *defined context set.* Error messages indicate that the receiving system does not support any of the transfer contexts specified for a specific abstract context. Once the negotiation process is completed, the systems can propose new presentation contexts for addition to the defined context set or remove contexts from the set using a primitive called P-ALTER-CONTEXT.

The Application Layer

As the top layer in the protocol stack, the application layer is the ultimate source and destination for all messages transmitted over the network. All of the processes discussed in the previous sections are triggered by an application that requests access to a resource located on a network system. Application-layer processes are not necessarily synonymous with the applications themselves, however. For example, if you use a word processor to open a document stored on a network server, you are redirecting a local function to the network. The word processor itself does not provide the application-layer process needed to access the file. In most cases, it is an element of the operating system that distinguishes between requests for files on the local drive and those on the network. Other applications, however, are designed specifically for accessing network resources. When you run a dedicated FTP client, for example, the application itself is inseparable from the application-layer protocol it uses to communicate with the network.

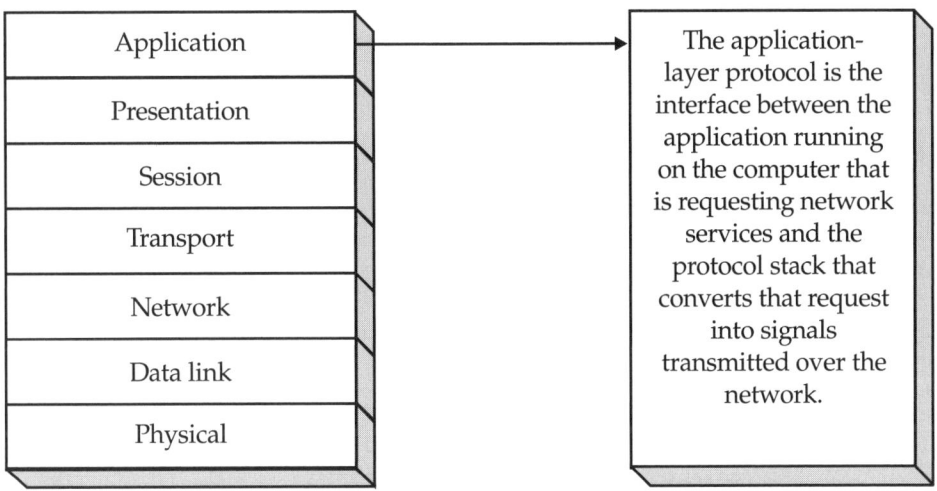

Application	The application-layer protocol is the interface between the application running on the computer that is requesting network services and the protocol stack that converts that request into signals transmitted over the network.
Presentation	
Session	
Transport	
Network	
Data link	
Physical	

Some of the other protocols that are closely tied to the applications that use them are as follows:

DHCP   Dynamic Host Configuration Protocol

TFTP   Trivial File Transfer Protocol

DNS   Domain Name System

NFS   Network File System

RIP   Routing Information Protocol

BGP   Border Gateway Protocol

In between these two extremes are numerous application types that access network resources in different ways and for different reasons. The tools that make that access possible are located in the application layer. Some applications use protocols that are dedicated to specific types of network requests, such as the Simple Mail Transport Protocol (SMTP) and Post Office Protocol (POP3) used for e-mail, the Simple Network Management Protocol (SNMP) used for remote network administration, the Hypertext Transfer Protocol (HTTP) used for World Wide Web communications, and the Network News Transfer Protocol (NNTP) used for Usenet news transfers. As you have seen in this chapter, the bottom four layers of the OSI reference model perform functions that are easily differentiated, while the functions of the session, presentation, and application layers tend to bleed together. Many of the application-layer protocols listed here contain functions that rightly belong at the presentation or session layers, but it is important not to let the OSI model assert itself too forcibly into your perception of data networking. The model is at this point a tool for understanding how networks function, not a guide for the creation of networking technologies. Manufacturers are not interested in designing their products to conform to the arbitrary divisions of a theoretical model, and you should not expect them to.

Protocol Stacks

As should be plain from this chapter, the interaction between the protocols operating at the various layers of the OSI reference model can be extremely complex. Figure 2-13 illustrates the interaction between the most commonly used protocols. As you progress through the chapters of this book, you'll learn more about each one and how it fits into the overall network communications picture.

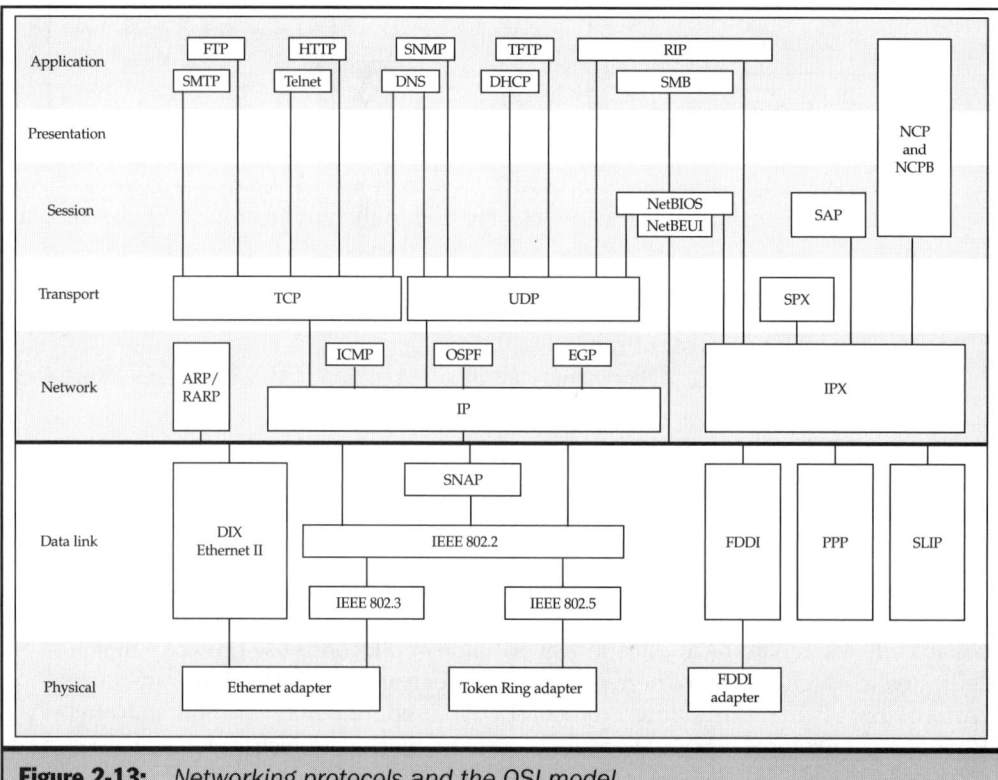

Figure 2-13: *Networking protocols and the OSI model*

The Complete Reference

Networking

Part II

Network Hardware

The
Complete
Reference

Networking

Chapter 3

Network Interface Adapters

very computer that participates on a network must have an interface to that network, using either a cable or some form of wireless signal that enables it to transmit data to the other devices on the network. The most common form of network interface is an adapter card that connects to the computer's expansion bus and to a network cable, typically referred to as a *network interface card* or *NIC* (see Figure 3-1). The NIC is usually a separate product that you can insert and remove from the computer, but quite a few systems today integrate the network adapter into the motherboard design. Modems are also a form of network interface, even when the network is nothing more than two computers joined together. The Windows operating systems, for example, treat modems as a part of their networking architecture, just like a NIC, but with a different set of hardware features.

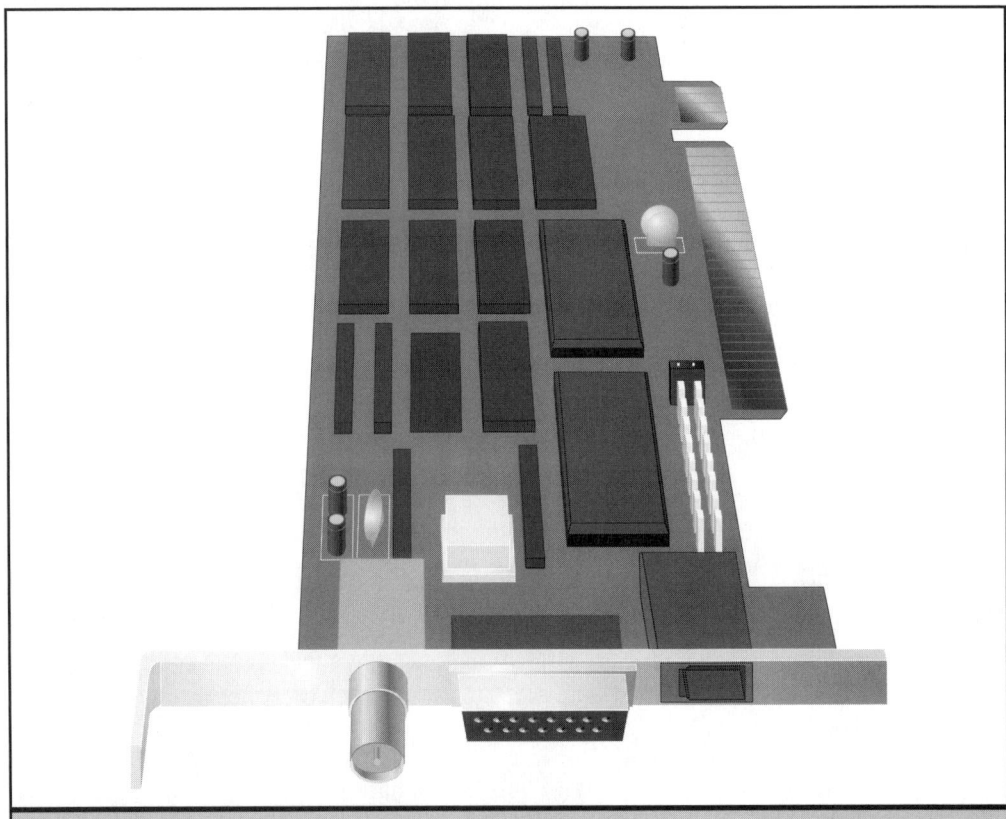

Figure 3-1: *A typical network interface adapter card*

NIC Functions

The *network interface adapter,* in combination with the network adapter driver, implements the data link–layer protocol used on the computer, such as Ethernet or Token Ring, as well as part of the physical layer. The NIC also provides the link between the network-layer protocol, which is implemented completely in the operating system, and the network medium, which is usually a cable connected to the NIC. When you purchase a NIC, you must select one that supports your network's data link–layer protocol, the type of cable or other medium used to build your network, and the type of expansion bus slot available in the computer. For example, one of the most common NIC configurations today is the PCI 100Base-TX Fast Ethernet NIC, which is a card that plugs into a computer's PCI bus and connects to a Fast Ethernet network using Category 5 unshielded twisted-pair (UTP) cable and RJ-45 connectors.

The NIC and its driver perform the basic functions needed for the computer to access the network. The process of transmitting data consists of the following steps (which, naturally, are reversed during packet reception):

1. **Data transfer**   The data stored in the computer's memory is transferred to the NIC across the system bus using one of the following technologies: *direct memory access (DMA),* shared memory, or programmed I/O.

2. **Data buffering**   The rate at which the PC processes data is different from the transmission rate of the network. The NIC includes memory buffers that it uses to store data so it can process an entire frame at once. A typical Ethernet NIC has 4KB of buffer space, divided into separate transmit and receive buffers of 2KB each, while Token Ring and higher-end Ethernet cards can have 64KB of buffer space or more, which may be split between the transmit and receive buffers in several configurations.

3. **Frame construction**   The NIC receives data that has been packaged by the network-layer protocol and encapsulates it in a frame that consists of its own data link–layer protocol header and footer. Depending on the size of the packet and the data link–layer protocol used, the NIC may also have to split the data into segments of the appropriate size for transmission over the network. Ethernet frames, for example, carry up to 1,500 bytes of data, while Token Ring frames can carry up to 4,500 bytes. For incoming traffic, the NIC reads the information in the data link–layer frame, verifies that the packet has been transmitted without error, and determines whether the packet should be passed up to the next layer in the networking stack. If so, the NIC strips off the data link–layer frame and passes the enclosed data to the network-layer protocol.

4. **Media access control**   The NIC is responsible for arbitrating the system's access to the shared network medium, using an appropriate *media access control (MAC)* mechanism. This is necessary to prevent multiple systems on the network from transmitting at the same time and losing data because of a packet collision. The MAC mechanism is the single most defining element of a data link–layer protocol. Ethernet's *Carrier Sense Multiple Access with Collision Detection (CSMA/CD)* system is radically different from the token-passing system used on Token Ring networks, but their basic purposes are ultimately the same. (The MAC mechanism is not needed for incoming traffic.)

5. **Parallel/serial conversion**   The system bus connecting the NIC to the computer's main memory array transmits data 16 or 32 bits at a time in parallel fashion, while the NIC transmits and receives data from the network serially—that is, one bit at a time. The NIC is responsible for taking the parallel data transmission that it receives over the system bus into its buffers and converting it to a serial bit stream for transmission out over the network medium. For incoming data from the network, the process is reversed.

6. **Data encoding/decoding**   The data generated by the computer in binary form must be encoded in a matter suitable for the network medium before it can be transmitted, and in the same way, incoming signals must be decoded on receipt. This and the following step are the physical-layer processes implemented by the NIC. For a copper cable, the data is encoded into electrical impulses; for fiber-optic cable, the data is encoded into pulses of light. Other media may use radio waves, infrared light, or other technologies. The encoding scheme is determined by the data link–layer protocol being used. For example, Ethernet uses Manchester encoding and Token Ring uses Differential Manchester.

7. **Data transmission/reception**   The NIC takes the data it has encoded, amplifies the signal to the appropriate amplitude, and transmits it over the network medium. This process is entirely physical and depends wholly on the nature of the signal used on the network medium.

Note

For more information on the protocol-specific elements of the previous procedure, see Chapter 2 for more information about physical layer signaling, Chapter 10 for more information about the Ethernet protocol and the CSMA/CD mechanism, and Chapter 12 for Token Ring and the token-passing MAC method.

The NIC also provides the data link–layer hardware (or MAC) address that is used to identify the system on the local network. Most data link–layer protocols, including Ethernet and Token Ring, rely on addresses that are hardcoded into the NIC by the manufacturer. In actuality, the MAC address identifies a particular network interface, not necessarily the whole system. In the case of a computer with two NICs installed and connected to two different networks, each NIC has its own MAC address that identifies it on the network to which it is attached.

There are a few exceptions to this rule. Some Ethernet adapters use addresses that are specified by the network administrator, as in the case of some IBM adapters. Other computing platforms (such as those made by Sun Microsystems) use a single address for all of the NICs installed in the computer. These examples are relatively rare, however, and are generally isolated to proprietary platforms designed to run a Unix variant manufactured by the same company.

Some older protocols, such as ARCnet, required the network administrator to set the hardware address manually on each NIC. If systems with duplicate addresses were on the network, communications problems resulted. Today, MAC addresses are assigned in two parts, much like IP addresses and domain names. The IEEE (Institute of Electrical and Electronic Engineers) maintains a registry of NIC manufacturers and assigns 3-byte address codes called *organizationally unique identifiers (OUIs)* to them as needed. Manufacturers use these codes as the first 3 bytes of the 6-byte MAC address for each NIC they produce. Then it is up to the manufacturer to see that the remaining 3 bytes are unique to each NIC they build.

The IEEE maintains a searchable database of OUIs on the Web at http://standards.ieee.org/ regauth/oui/index.html.

NIC Features

In addition to the basic functionality described thus far, NICs can have a variety of other features, depending on the manufacturer, protocol, price point, and the type of computer in which the device is to be used. Some of these features are discussed in the following sections.

Full Duplex

Most of the data link–layer protocols that use twisted-pair cable separate the transmitted and received signals onto different wire pairs. Even when this is the case, however, the NIC typically operates in *half-duplex mode,* meaning that at any given time, it can be transmitting or receiving data, but not both simultaneously. NICs that can operate in *full-duplex mode* can transmit and receive at the same time, effectively doubling the throughput of the network (see Figure 3-2).

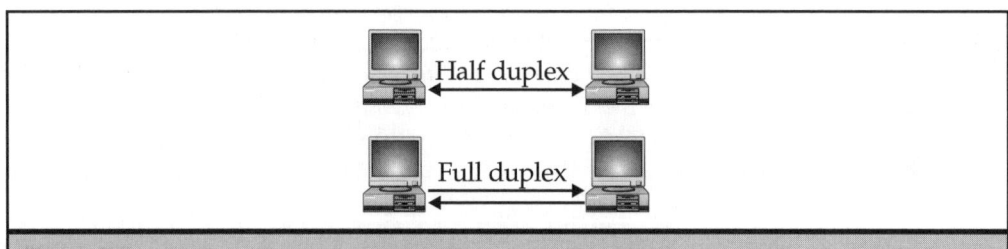

Figure 3-2: *Half-duplex systems transfer data in one direction at a time, while full-duplex systems can transfer data in both directions simultaneously.*

When a NIC is operating in full-duplex mode, it can transmit and receive data at any time, eliminating the need for a media access control mechanism. This also eliminates collisions, which increases the overall efficiency of the network. Running a full-duplex network requires more than just NICs that support this feature, however. The hub, switch, router, or other device to which each computer connects must also support full-duplex operation.

 *Full-duplex operation is a feature that is usually associated with switched Fast Ethernet networks. For more information on full-duplex networking, see Chapter 10.*

Bus Mastering

Normally, when data is transmitted between the computer's memory and an expansion card over the system bus, the processor functions as the middleman, reading data from the source and transmitting it to the destination. This utilizes processor clock cycles that could otherwise be running applications or performing other important tasks. An expansion card capable of *bus mastering* has a chipset that can arbitrate the card's access to the bus, eliminating the need for the system processor's involvement in the transfer of data to and from memory. Bus mastering NICs enable the computer to operate more efficiently because they conserve the processor clock cycles that would otherwise be expended in data transfers. Most of the bus mastering NICs on the market use the PCI bus. This is one reason why a PCI NIC may be beneficial, even for a computer connected to a network that does not require the additional bus speed, such as a 10 Mbps Ethernet LAN.

Parallel Tasking

Parallel Tasking is a feature that was developed by 3Com Corporation and subsequently implemented by other NIC manufacturers, using different names. The term describes a process by which the NIC can begin to transmit a packet over the network while the data is still being transferred to the NIC over the system bus. A NIC without this capability must wait until an entire packet is stored in its buffers before it can transmit.

3Com's next innovation is called *Parallel Tasking II*, which improves bus-mastering communications over the PCI bus. Previously, a PCI NIC could transfer only 64 bytes at a time during a single bus master operation, which required dozens of operations to transfer each packet. Parallel Tasking II enables the NIC to stream up to an entire Ethernet packet's worth of data (1,518 bytes) during one single bus master operation.

Wake on LAN

Some of the PCI NICs available today support a remote wake-up feature that enables an administrator to power up the machine from a remote system on the network. There are many products available that enable network support personnel to administer a computer from a remote location, performing tasks such as virus scans, backups, and software updates during non-working hours without having to physically travel to the location of the computer. One of the problems with this technology, however, is that the remote administration tools are of no use if users turn off their computers before leaving at the end of the day. A group called the Intel-IBM Advanced Management Alliance has addressed this problem with a technology called *Wake on LAN (WOL)*, which is part of their Wired for Management (WfM) specification.

Wake on LAN is an enhancement built into some network interface adapters and computer motherboards that enables an administrator to turn a computer on from a remote location. Once turned on, the administrator can perform any necessary maintenance tasks. For this feature to function, both the computer's motherboard and the NIC must have a three-pin remote wake-up connector, which are connected with a cable. When the computer is turned off, it actually switches to a low-power sleep state, instead of being completely powered off. Power management in the form of the Advanced Configuration and Power Interface (ACPI) standard, is an integral part of the WfM specification and makes this low-power state possible. While in this state, the NIC continuously monitors the network for a special wake-up packet that can be delivered to it by a desktop management application running on an administrator's computer. When the NIC receives the packet, it signals the motherboard, which in turn switches the power supply back into its full power state, effectively turning on the computer. Once the computer is up and running, the administrator can take control of the system using whatever tools are available.

NETWORK HARDWARE

IEEE 802.1p

The 802.1p standard published by the IEEE defines a method for assigning priorities to network packets, so the data for specific types of applications can be transmitted on a timely basis. This enables streaming audio and video applications, for example, to transmit their data across the network in real time, without being affected by other traffic on the network. For priorities to be implemented, both the NIC and the operating system must support a standard like IEEE 802.1p. Also called *quality of service (QoS)*, this is relatively new technology, and other standards exist that provide similar services.

Note   *For more information, see "Quality of Service" in Chapter 16.*

Selecting a NIC

The selection of a NIC for a particular computer is based on several different factors:

- The data link–layer protocol used by the network
- The transmission speed of the network
- The type of interface that connects the NIC to the network
- The type of system bus into which you will install the NIC
- The hardware resources the NIC requires
- The electric power the NIC requires
- The role of the computer using the NIC: server versus workstation and home versus office
- The availability of appropriate drivers

The following sections examine these criteria and how they can affect the performance of the NIC and your network.

Protocol

The data link–layer protocol is the single most defining characteristic of a network interface adapter. The most popular protocol used at the data link layer is Ethernet, but NICs are also available that support Token Ring, FDDI, ATM, and others, as well as variations on these protocols. There are also specialized NICs available that support dedicated server technologies, such as Fibre Channel. Other NIC characteristics, such as the bus type and hardware resources, relate only to the specific computer in which the device will be installed, but the protocol it uses is relevant to many other aspects of the network configuration.

All of the computers on the network must, of course, be using the same data link–layer protocol, and the selection of that protocol should be a decision made long before you're ready to purchase NICs. This is because all of the other network hardware, such as cables, hubs, and other devices, are also protocol specific. The NIC you select must also support the type of cable or other medium the network uses, as well as the transmission speed of the network. If, for example, you're running an Ethernet network, you can get NICs that support either the standard 10 Mbps speed, or 100 Mbps Fast Ethernet as well. You can also select Ethernet NICs that support the use of unshielded twisted-pair (UTP), two types of coaxial, or fiber-optic cable, as well as various types of wireless transmissions. These are all aspects of the network configuration that you must consider before making NIC purchases.

Manufacturers and vendors of NICs typically categorize their products by protocol first because consumers are interested only in comparing products that use a specific protocol. The protocol is also the characteristic that has the greatest effect on the price of a NIC. Different protocols communicate over the network in different ways and require different types of NIC components. The price of a NIC is also affected by a protocol's popularity. Because the vast majority of networked computers use Ethernet as their data link–layer protocol, these NICs are typically the cheapest. You can get a no-name brand Ethernet NIC for less than $20, with well-known brands starting at about $50. Token Ring NICs are more expensive ($125 and up) because fewer networks use that protocol, and relatively new technologies, such as ATM and Gigabit Ethernet, can command far higher prices for NICs (from $175 to well over $1,000) and other hardware devices.

Transmission Speed

Some data link–layer protocols can run at different speeds, and the capability of a NIC to support these speeds can be an important part of selecting the correct product for your network. In some protocols, an increase in speed has been fully assimilated into the technology, while in others, the faster version is still an optional feature. The Token Ring protocol, for example, originally ran at 4 Mbps, but now virtually all Token Ring networks run at 16 Mbps, and it might be difficult to find a NIC that does not support the higher speed.

The higher-speed Ethernet protocols, by contrast, are relatively new. Fast Ethernet (running at 100 Mbps) is rapidly replacing traditional 10 Mbps Ethernet, but it is still an optional feature when it comes to network interface adapters. Most of the Fast Ethernet NICs manufactured today are combination devices that support both 10 and 100 Mbps operation, making it possible to gradually upgrade a traditional Ethernet network to

Fast Ethernet. When the connection is established between the NIC and the hub, the devices negotiate the highest possible speed they have in common.

The popularity of Fast Ethernet has reached the point at which these dual-speed NICs are only slightly more expensive than the 10 Mbps devices; sometimes there is as little as $10 difference between comparable products. Eventually, all of the Ethernet NICs in production will support both speeds, and it will become difficult and expensive to buy cards that support only traditional 10 Mbps Ethernet. If there is even the remotest possibility that you will upgrade your 10 Mbps Ethernet network to Fast Ethernet sometime in the future, it is a good idea to make any NICs you purchase now dual-speed, so you don't have to replace them later.

The 100VG AnyLAN standard, which was introduced at about the same time as Fast Ethernet, has failed to achieve a large market share, but a few products are still available. There are combination NICs on the market that support both standard Ethernet at 10 Mbps and 100VG AnyLAN at 100 Mbps. These tend to be a good deal more expensive than Ethernet/Fast Ethernet cards, however, because 100VG AnyLAN is substantially different from Ethernet and requires different components. An Ethernet/100VG AnyLAN NIC is essentially two separate adapters on a single card.

The latest Ethernet innovation, Gigabit Ethernet, runs at 1,000 Mbps. Manufacturers and dealers generally place this technology into a different category than traditional and Fast Ethernet for several reasons. Because it is a relatively new technology, Gigabit Ethernet is still in the early stages of product development and marketing. This means that there are relatively few products on the market, and their prices are high. The current cost of a Gigabit Ethernet NIC can run anywhere from $125 to over $1,000. The lower priced NICs are the first attempts to make Gigabit Ethernet technology affordable to the average user, and the expensive models are NICs designed for use in servers that require optimum network performance. Gigabit Ethernet was designed with backbone networks and server connections in mind, because there are not many applications that require 1,000 Mbps service to the desktop. Nevertheless, these lower-priced Gigabit Ethernet NICs are clearly being positioned for the desktop market, and there are also some multispeed cards available that support speeds of 10, 100, and 1,000 Mbps.

Network Interface

The type of cable (or other medium) that forms the fabric of the network determines the network interface used on the NIC. The network cable type is typically selected at the same time as the data link–layer protocol, and the NICs you purchase must support that medium. Some data link–layer protocols support different types of cables, and NICs are available for each one, while other protocols are designed to use only one type of cable.

Ethernet supports more cabling options than any of the other commonly used data link–layer protocols. Most of the Ethernet networks installed today use UTP, which requires an RJ-45 jack on the NIC. However, NICs are also readily available with BNC and AUI connectors, for Thin Ethernet and Thick Ethernet, respectively (see Figure 3-3). This is rather surprising because, while networks running Thin or Thick Ethernet are rare these days, the NICs that support these media are quite common. Because these three media all use electrical signals running over copper cable, the components of the NIC used to generate those signals are the same, regardless of the cable type. This makes it easy for manufacturers to produce combination devices with multiple jacks that support two or even all three cable types. The inclusion of multiple jacks on a NIC can add greatly to its cost, however, almost doubling it in many cases.

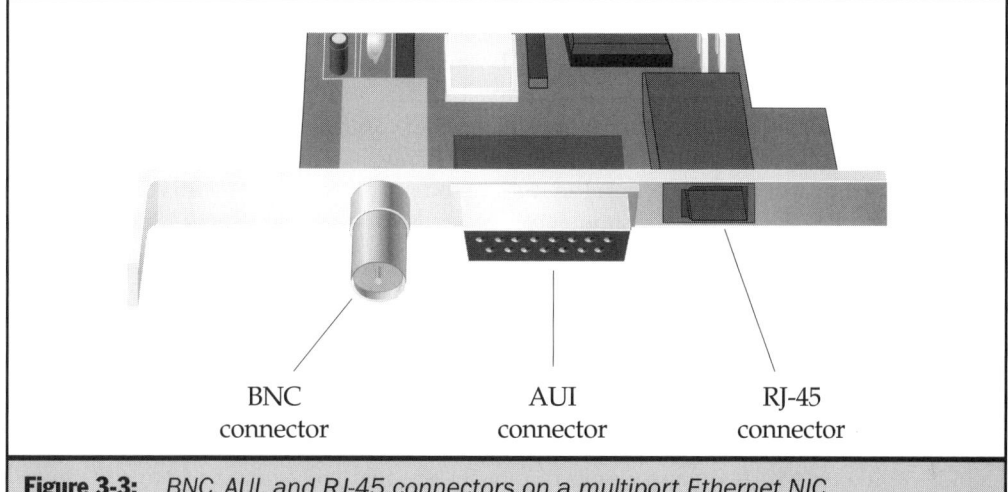

BNC   AUI   RJ-45
connector   connector   connector

Figure 3-3:   *BNC, AUI, and RJ-45 connectors on a multiport Ethernet NIC*

Note   *Dual-speed Ethernet NICs with multiple jacks are common, but Fast Ethernet cannot run over coaxial cable of any type. Thus, on a NIC of this type, only the RJ-45 connector supports both 10 and 100 Mbps speeds. The AUI and/or BNC connectors run only at 10 Mbps. For more information on cables and connectors, see Chapter 4.*

The purchase of NICs with multiple jacks is justified when a network is going to be upgraded from coaxial to UTP cable at some future time. Considering that relatively few coaxial networks are left, however, there is usually no good reason to spend the extra money for this type of combination NIC. One possible scenario when it might be justified is when an organization is running a complex network that consists of both coaxial and UTP cable segments. Even if the coaxial cable is not going to be upgraded, it may be economically wise to purchase a large quantity of combination NICs, rather than smaller quantities of NICs supporting each type of medium. This is because the prices of network adapter cards are often greatly discounted when purchased in multiple unit packs. As an example, the current prices for a 3Com Fast Etherlink 10/100 PCI NIC at one major vendor reflect a 10 percent discount for five units and almost a 40 percent discount for 100 units. In some cases, the discount may more than offset the additional cost for combination NICs.

Ethernet also supports the use of fiber-optic cable, which is fundamentally different from the copper media, in that it carries data coded into light pulses rather than electric voltages. The components on a fiber-optic NIC are therefore substantially different in form (if not function) from those on a copper-based Ethernet NIC, including the network interface, which is usually an *ST (straight tip)* connector. While there is a standard that defines the use of fiber-optic cable with 10 Mbps Ethernet, it is not often used. Fast Ethernet, however, can use fiber-optic cable to run at 100 Mbps over far longer distances than any copper medium. Because of these technological differences, fiber-optic Fast Ethernet NICs are not usually combined with other technologies. Fiber-optic network hardware is also more expensive than comparable copper-based products, with fiber-optic Ethernet NICs starting at about $100.

Most of the Token Ring NICs on the market today use standard RJ-45 connectors for UTP cable, which is referred to as the *Type 3 cable system* in Token Ring parlance. Older Token Ring networks used the Type 1 cable system; NICs for this system have a DB9 connector like that used for a PC's serial ports. Token Ring hardware is also more expensive than Ethernet hardware, with Token Ring NICs starting at about $125.

The use of various types of cables and connectors on Ethernet and Token Ring networks can be attributed mostly to the fact that these are the oldest data link–layer protocols still in common use today. Ethernet, for example, has evolved from the use of coaxial cable to twisted pair, and the connectors have changed accordingly. Other protocols, like ATM, FDDI, and Gigabit Ethernet, are relatively new and have fewer physical layer options.

Bus Interface

The network interface adapter enables a network system to transmit data from its main memory array to an outside destination, just as a parallel or serial port does. The data travels from the memory to the network adapter across the system bus, in the same manner as with any other expansion card, like a graphics or audio adapter. The type of bus the NIC uses to communicate with the computer can affect the performance of the network connection, but the selection of a bus type for the NIC is unique to each computer. In other words, you needn't use the same bus type for all of the NICs in your network workstations; one PC can use an ISA NIC, while others use PCI.

Integrated Systems Architecture (ISA) and *Peripheral Component Interconnect (PCI)* are the two bus types used in virtually all of the desktop computers sold today. Laptops and other portables use the PC Card bus (formerly known as the Personal Computer

Memory Card International Association, or PCMCIA bus). Older systems use various other types of expansion buses, such as *VLB (VESA Local Bus), MCA (Micro Channel Architecture),* or *EISA (Extended Industry Standard Architecture).* The latest innovation in the bus interface used for network adapters is the Universal Serial Bus (USB). USB adapters require no internal installation. You simply plug the adapter into a computer's USB port, plug the network cable into the adapter, and install the appropriate driver for the new device. No external power connection is needed; the adapter derives power from the bus itself. This makes for an extremely simple installation, but the performance of a USB network adapter is substantially inferior to even an ISA NIC. Table 3-1 lists the characteristics of these buses and their respective throughput speeds.

Bus Type	Bus Width	Bus Speed	Theoretical Maximum Throughput
ISA	16 bits	8.33 MHz	66.64 Mbps (8.33 MBps)
MCA	32 bits	10 MHz	320 Mbps (40 MBps)
EISA	32 bits	8.33 MHz	266.56 Mbps (33.32 MBps)
VLB	32 bits	33.33 MHz	1,066.56 Mbps (133.33 MBps)
PCI	32 bits	33.33 MHz	1,066.56 Mbps (133.33 MBps)
PC Card (CardBus)	32 bits	33.33 MHz	1,066.56 Mbps (133.33 MBps)
USB	1 bit	N/A	12 Mbps (1.5 MBps)

Table 3-1: *PC Bus Types, Widths, and Speeds*

Bottlenecks

The bus type selection can affect network performance if the selected bus is slow enough to cause a bottleneck in the network. In networking, a *bottleneck* occurs when one element of a network connection runs at a significantly slower speed than all of the others. This can cause the entire network to slow down to the speed of its weakest component, resulting in wasted bandwidth and needless expense.

As an exaggerated example, consider a network that consists of all top-of-the-line PCs with the fastest processors and hard drives available, connected by a Fast Ethernet network running at 100 Mbps. All of the workstations on the network have NICs that use the PCI bus except for the main database server, which has an ISA NIC. The PCI bus is twice as wide as the ISA bus (32 bits as opposed to 16) and runs at 1,066.56 Mbps, far faster than the 100 Mbps network itself, while the ISA bus runs at only 66.64 Mbps. The result of this is that the ISA NIC will probably be the slowest component in all of the workstation/server connections, and will be a bottleneck that prevents the rest of the equipment from achieving its full potential.

The process of identifying actual bottlenecks is rarely this clean-cut. Just because a network protocol runs at 100 Mbps doesn't mean that data is continuously traveling over the cable at that speed, and the raw speed of a particular bus type is not indicative of that actual throughput rate for the data generated by the system. However, it is a good idea to use common sense when purchasing NICs and to try to maximize the performance of your network.

ISA or PCI?

Although you may have to deal with the older bus types if you are using legacy PCs on your network, the choice for most desktop systems manufactured after about 1995 is between ISA and PCI. For a traditional Ethernet network running at 10 Mbps or a Token Ring network running at 4 or 16 Mbps, an ISA NIC is more than sufficient. In fact, ISA NICs can be perfectly serviceable on 100 Mbps networks as well, at least for workstations, because the average network user does not require anything approaching 100 Mbps of bandwidth on a continuous basis. The main reason for the ISA NIC being the bottleneck in the scenario described earlier is that it is installed in the server. A server PC that is handling data requests generated by dozens or hundreds of workstations simultaneously naturally requires more bandwidth than any single workstation. In a server, therefore, the use of the fastest bus available (usually PCI) is nearly always recommended.

Note   *If you are working with older PCs, you should be able to find a few EISA NICs still on the market, but MCA and VLB have all but disappeared. Fortunately, most of the systems that use these bus types support ISA as well, so you should be able to make them into functional workstations.*

The introduction of Fast Ethernet and the gradual elimination of the ISA bus have led to the dominance of PCI in the NIC market (and in expansion cards in general). However, there is another element to the bus type decision that you must consider, and that is the availability of expansion bus slots in your computers.

Obviously, to install a network interface card into a PC, it must have a free bus slot. PCs have varying numbers of PCI and ISA slots, and the hardware configuration of the machine determines how many of those slots (if any) are free. Today's full-featured computers often have peripheral devices installed that occupy many of the bus slots, such as audio adapters, SCSI adapters, MPEG decoder cards, and modems. Since it's possible for a card to occupy a slot without protruding through the back of the computer, simply looking at the outside of a system is not sufficient to determine how many free slots there are. You must open the machine to check for free slots and to determine which types of slots are available. If no slots are available, an external network adapter using the USB port may be your only recourse.

Administrators of large networks often purchase workstations that do not have all the state-of-the-art features found in many home systems, which may leave more slots free for additional components such as a NIC. In addition, PCs targeted at the corporate market are more likely to have peripheral devices like audio and video adapters integrated into the motherboard, which also can leave more slots free. However, an office computer may also use a slimline or low-profile case design that reduces the number of slots, to minimize the computer's footprint.

Wherever possible, the selection of the bus type for the NIC should be based on the network bandwidth requirements of the user and not on the type of bus slot the computer has free. Sometimes, though, you may have no other choice than to put an ISA NIC in a computer that could benefit from a PCI card or to install a PCI NIC in a computer that doesn't require the additional performance, but that has no ISA slots free. The latter situation, which only incurs an extra expense, is certainly less serious than the former, which can negatively affect the performance of the network.

Integrated Adapters

As mentioned earlier, some PCs, and particularly those intended for the corporate market, may have peripheral devices integrated into the motherboard. One of these devices may be the network interface adapter. Integrated network adapters are less common than integrated graphics and audio adapters, but they do exist, and are particularly popular in laptops and other portable systems, as well as dedicated network appliances, such as network attached storage units. Because an integrated network adapter is not a separate card, it cannot rightfully be called a NIC, but it does perform exactly the same function as a network adapter that installs into the system's expansion bus.

Although they reduce the distance the signals have to travel to reach the adapter and avoid the electrical interference that occurs during a bus transfer, the problem with integrated network adapters is that they are not upgradable. When 10 Mbps Ethernet was the only Ethernet game in town, integrating a 10Base-T network adapter onto the motherboard was a safe and practical alternative. However, the introduction of Fast Ethernet has complicated things by providing multiple cabling options, even for UTP cable, such as 100Base-TX and 100Base-T4. An integrated, dual-speed 10Base-T/100Base-TX adapter would be suitable for most consumers, but the industry has, instead, backed away from providing integrated network capabilities in desktop systems at all.

A system that has an integrated network adapter is under no obligation to use it. You can nearly always disable the adapter through the system BIOS, or by manipulating a switch or jumper on the motherboard, or simply by installing a NIC into a bus slot. You might find a deal on workstations with the wrong type of integrated network adapter that is good enough to be worth buying NICs for the computers as well.

Portable Systems

Network interface adapters for laptops and other portable systems all take the form of PC Cards, so there is no choice of bus type for these machines. However, there is a choice between PC Card NICs that support CardBus and those that do not. *CardBus* is an improved version of the PC Card bus that doubles the bandwidth from 16 to 32 bits and provides speeds up to 33 MHz, which yields performance comparable to the PCI bus on desktop systems. If the computer supports CardBus and your network is running at 100 Mbps or faster, it is a good idea to buy a CardBus NIC.

Hardware Resource Requirements

In addition to a bus slot, a computer must have the appropriate hardware resources free to support a NIC. A network interface adapter requires a free *interrupt request line (IRQ)* and usually either an I/O port address, a memory address, or both. When evaluating NICs, you must take into account both the resource requirements of the NIC and the resources available on the computer.

On a PC with a lot of peripheral devices already installed, most of the IRQs may already be in use, and adding a NIC may be difficult. This is because a NIC may only be able to use a select few of the system's IRQs, and if all of those IRQs are occupied, the card cannot function. Two devices configured to use the same resource will sometimes conflict, causing both to malfunction. In some cases, however, it's possible for two devices to share an IRQ. To free up one of the IRQs usable by the NIC, you may have to configure another device to use a different IRQ. Thus, you have to consider not only the number of available IRQs on the computer, but also which ones are available. The same is true for the other resources required by the card.

Many older NICs supported only two or three IRQs and other resources, and configuring the devices in the computer was a manual, trial-and-error process. System administrators could spend hours trying different combinations of hardware settings for the components in a single computer before finding one that enabled all of the devices to function simultaneously. Today, however, NICs are generally more flexible and support a wider range of resource settings. In addition, the BIOS and the operating system of a modern PC have features that simplify the process of configuring peripheral devices to work together.

Plug-and-play, when it functions properly, eliminates the need to worry about hardware resource configuration for peripheral devices. When a system has a BIOS, an operating system, and hardware that all support the plug-and-play standard, the computer assigns hardware resources to each device dynamically when the system starts. When plug-and-play is not supported for a particular device such as a NIC, operating systems (such as Microsoft Windows) provide tools that can identify the free resources in the machine and indicate whether the NIC's current configuration conflicts with any other devices in the system.

Thus, when selecting NICs, you should be conscious of the hardware resources in use on the computers that will use them. When using NICs and computers of recent manufacture, this is rarely a problem. However, a computer with a lot of installed peripherals may be unable to support an additional card without removing one of the existing components. In other cases, you may have to reconfigure other devices to support the addition of a NIC. Most NIC manufacturers publish specification sheets (often available on their Web sites) that list the hardware resources their NICs can use. By comparing this information to the current configuration of a PC, you can determine whether the computer has the resources to support the NIC.

Power Requirements

The power supplies in today's computers usually supply more than enough voltage to support a full load of expansion cards and other internal peripherals. However, if you're running a system with a large number of internal devices, such as hard disk, CD-ROM, or other drives, you may want to compare the power load incurred by these devices with the voltage furnished by the computer's power supply before you install a NIC. Because the power drain of mechanical drives varies depending on how often and how heavily they're used, a system putting out insufficient power to support its hardware load may experience intermittent problems that are difficult to diagnose. What may seem to be a faulty drive may, in fact, be the effect of an insufficient power supply for the hardware.

Server versus Workstation NICs

The NICs in servers and workstations perform the same basic functions, and yet there are cards on the market that are targeted specifically for use in servers. Some of these NICs use protocols, like Gigabit Ethernet, that are intended primarily for servers because their cost and capabilities make them impractical for use in desktop workstations. Others, however, are NICs that use standard protocols, like Ethernet and Fast Ethernet, but that contain additional features to make them more useful in servers. Naturally, these extra features drive the price of the NIC up considerably, and it is up to you to decide whether they are worth the extra expense. Because the basic functionality of all NICs is the same, there is no reason you can't use a standard NIC in a server.

The following sections examine some of the additional capabilities included in the server NICs marketed by various manufacturers. These features often have different names trademarked by their makers, and they may not function in exactly the same way, but the general principles are the same.

NETWORK HARDWARE

Multiport NICs

Servers must often be connected to two or more networks simultaneously, either to service clients on both networks or route traffic between them. To do this using standard NICs, you must install multiple cards in the computer, which can raise problems of slot, power, and hardware resource availability. Several manufacturers have NICs available that are designed for servers like these, which have two or more RJ-45 ports on them, to support multiple network connections. These devices are essentially two (or more) network interface adapters on a single card, that are able to use the hardware resources of the computer more effectively than separate cards. Multiport NICs also facilitate the implementation of other server-specific features, such as load balancing and fault tolerance.

Load Balancing

Because a server must support multiple clients, it is often the location of network bottlenecks. A single system running at 100 Mbps cannot conceivably keep up with dozens of clients, all requesting data at the same speed simultaneously. However, client systems usually do not all access the server at the same time, which is what makes LAN communications practical. When a network gets to the point at which the number of clients accessing a particular server at the same time gets too large, the server NIC becomes a bottleneck and the performance of the client systems degrades as a result.

One possible solution to this problem is to install another NIC into the server and divide the network into two segments. Half of the clients connect to one server NIC and half to the other, thus lessening the amount of traffic passing through each NIC. This process sounds easier than it is, however. The task requires that you take the server offline to install the new NIC, reassign IP addresses (on a TCP/IP network) for the new subnet, and possibly migrate clients from one segment to the other to properly balance the traffic load. If the network traffic continues to grow, the entire process must be performed again.

Several manufacturers, such as Intel and 3Com, have addressed this problem by creating NICs that can work together in a server to balance the network traffic load between them. After installing multiple NICs of the same type in a server and connecting them to a switch (or switches), you can configure the NICs to function as a group. Normally, individual NICs in one machine each have their own IP addresses, but the NICs in a load-balancing group share one IP address, even though they retain their individual MAC addresses (see Figure 3-4). The group thus forms a *virtual NIC*, which can process the aggregate amount of traffic support by the entire group combined. Different products support different numbers of NICs in a group. 3Com, for example, supports up to eight NICs, for a total bandwidth of 800 Mbps, while Intel supports up to four NICs and 400 Mbps. With this capability, administrators can compensate for increasing traffic on the network simply by installing another NIC in the server and adding it to the group.

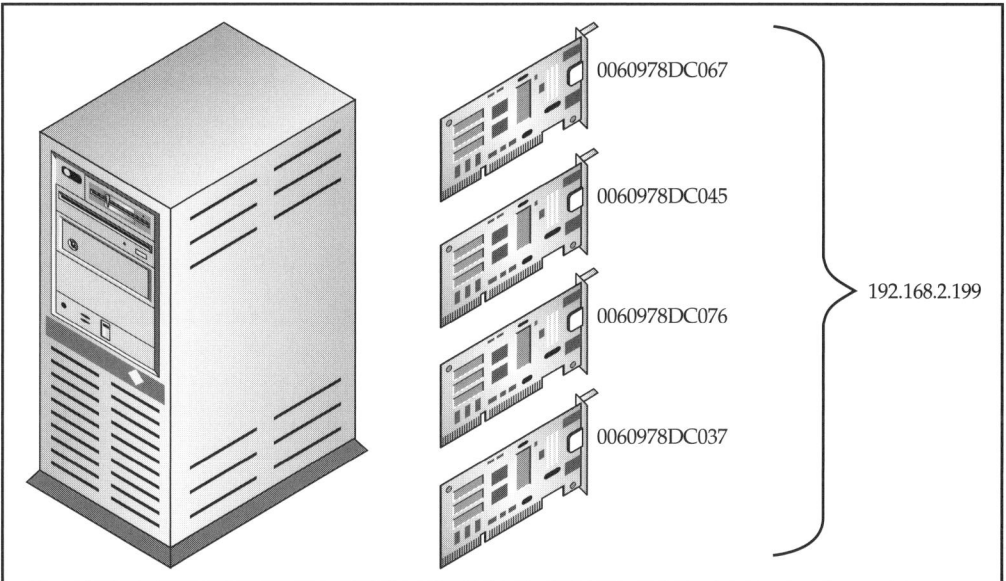

Figure 3-4:   *A virtual NIC consists of multiple network interface cards that have individual hardware addresses, but share a single IP address.*

When the server transmits data to clients, the traffic is distributed evenly among the NICs in the group, based on the IP address of the client system. In most cases, a TCP connection between the server and a particular client is assigned to one NIC in the group, which is used for the duration of that connection. When a client has multiple connections to the server running simultaneously, it may use different NICs. Incoming traffic is treated differently by various load-balancing implementations. Because the server's incoming traffic consists largely of relatively small request messages, some products use one NIC in the group to process all of the incoming packets, while others distribute the incoming packets among the NICs in the group in a round-robin fashion.

Another form of load balancing, called *link aggregation* by Intel, can balance both incoming and outgoing traffic. Intel's standard load-balancing feature always uses one NIC for incoming traffic, even when the outgoing traffic is distributed among up to four NICs. Link aggregation balances the load in both directions and supports full-duplex operation, for an aggregated bandwidth of up to 800 Mbps using Fast Ethernet and 8 Gbps using Gigabit Ethernet. This technology, however, requires hardware support in the switches as well as the NICs, unlike standard load balancing, which can use any type or brand of switch.

While different manufacturers may have load balancing products that appear similar, you usually cannot mix NICs from different makers in a server and expect them to work together in this way. Several vendors have gotten together to launch the IEEE 802.3ad Link Aggregation Task Force, and a standard for this technology was ratified in March 2000. However, many of the products that provide load balancing and link aggregation still use proprietary technologies and are not compatible with other manufacturers' products.

Fault Tolerance

There is also a valuable byproduct of the load-balancing concept, and that is the fault tolerance provided by the NIC group. If one NIC should fail, the traffic is automatically distributed evenly among the remaining ones until it is replaced. Even if you don't use load balancing, you can install a redundant NIC in your servers that the system automatically uses if the primary connection fails for any reason, including a malfunction in the cable, hub, or switch to which it is connected.

Some products even enable you to use a more inexpensive NIC as a backup. For example, if you have a server with a Gigabit Ethernet connection to the network, it may not be worth the expense to install a second Gigabit Ethernet NIC into the server, just for fault tolerance purposes. However, it may be possible to use a relatively inexpensive Fast Ethernet card as a backup instead. In the event that the Gigabit Ethernet connection fails, the Fast Ethernet connection will take over and provide at least nominal service until you restore the higher-speed connection.

Another fault tolerance mechanism, called PCI Hot Plug capability, was developed by Compaq and is being implemented in many servers and NICs. This feature enables administrators to replace a malfunctioning NIC without powering the server down. Also called *hot swapping*, this PCI Hot Plug works in conjunction with the NIC's failover capability. When a NIC fails, the connection switches to a redundant NIC or another NIC in the same group. The administrator can then remove the malfunctioning device from the server while it is running and insert a new one. This eliminates any of the down time usually associated with a NIC failure.

Actual NIC hardware failures are not common occurrences, but for some networks, even the remotest chance of an outage that causes an interruption of service is unacceptable. In addition, these fault tolerance features can also protect against failures that are more common, such as severed cables or power failures to a switch or hub.

Remote Management

Other features often included in NICs intended for servers provide administrators with the ability to configure a NIC from a remote location and receive information about its status. In a case in which a hardware failure causes a server to switch to a backup NIC, for example, these features can generate an alert and send it to network administrators to inform them of the event. Many server NICs support standards for this type of interaction between network hardware and an administrative console application, such as the Simple Network Management Protocol (SNMP) and the Desktop Management Interface (DMI) 2.0.

Home and Office NICs

Arguably the most rapidly growing segment of the NIC market is the one for homes and small offices. Many families and small business owners, even those with only two or three computers, are beginning to see the advantages of networking their systems together. Most of the major NIC manufacturers now have a line of products targeted at these markets. The basic premise behind these product lines is to provide users that have little or no networking experience with an inexpensive yet functional NIC that is easy to install and run.

With the PCI bus and plug-and-play support built into the Windows operating systems, installation of a NIC is easier than it has ever been before. In many cases, all that's involved is simply inserting the card into a bus slot. The operating system configures the card and installs the drivers. All that's left is to connect the computers to a hub (marketed as part of the same product line), and you have a basic network.

The NICs provided in these product lines don't have the advanced features found in server NICs, and may even lack some of the features directed at larger-scale networking operations, such as remote management. Their documentation is also simpler and reflects the type of user that is expected to be working with the card. However, in their basic functions the NICs are usually comparable to the other products targeted at the corporate market, for a substantially lower price.

Network Adapter Drivers

The final concern when evaluating NICs is the availability of drivers for the operating system that is running on the computer. In most cases, this is not an issue, because most, if not all, of the NICs on the market today include the various NDIS drivers required for the Windows operating systems, ODI drivers for Novell NetWare, and a packet driver that provides low-level access to basic NIC functions. In fact, operating systems usually include drivers for most of the major manufacturers' NICs. The only time you should be concerned about driver support is when you are working with very old NICs, with those by marginal manufacturers, or with an operating system that has limited driver support.

There are, however, some proprietary NICs on the market that are designed for use with specific computing platforms running specific operating systems. In the PC world, you can generally buy any NIC and install it in any computer, as long as the bus connectors match. However, if you buy a Unix workstation from Sun Microsystems, you will likely have to purchase NICs from them.

The
Complete
Reference

Networking

Chapter 4

Cabling a Network

Although there are networks that use radio transmissions and other wireless technologies to transmit data, the vast majority of local area networks use cable as the network medium. Most of the cables used for data networking use a copper conductor to carry electrical signals, but *fiber-optic,* a spun glass cable that carries pulses of light, is an increasingly popular alternative.

Cabling issues have, in recent years, become separated from the typical network administrator's training and experience. Many veteran administrators have never installed (or "pulled") cable themselves and are less than familiar with the technology that forms the basis for the network. In many cases, the use of twisted-pair cable has resulted in telephone system contractors being responsible for the network cabling. Network consultants typically outsource all but the smallest cabling jobs to outside companies.

However, although the cabling represents only a small part of a network's total cost (as little as 6 percent), it has been estimated to be responsible for as much as 75 percent of network down time. The cabling is also usually the longest-lived element of a network. You may replace servers and other components more than once before you replace the cable. For these reasons, spending a bit extra on good quality cable, properly installed, is a worthwhile investment. This chapter examines the types of cables used for LANs, their composition, the connectors they use, and their installation processes.

Cable Properties

Data link–layer protocols are associated with specific cable types and include guidelines for the installation of the cable, such as maximum segment lengths. In some cases, such as Ethernet, you have a choice as to what kind of cable you want to use with the protocol, while in others you do not. Part of the process of evaluating and selecting a protocol involves an examination of the cable types and their suitability for your network site. For example, a connection between two adjacent buildings is better served by fiber-optic than copper, so with that requirement in mind you should proceed to evaluate the data link–layer protocols that support the use of fiber-optic cable.

Your cable installation may also be governed, in part, by the layout of the site and the local building codes. Cables generally are available in both nonplenum and plenum types. A *plenum* is an air space within a building, created by the components of the building themselves, that is designed to provide ventilation, such as a space between floors or walls. Buildings that use plenums to move air usually do not have a ducted ventilation system. In most communities, to run cable through a plenum, you must use a plenum-rated cable that does not give off toxic gases when it burns, because the air in the plenum is distributed throughout the building. The outer covering of a plenum cable is usually some sort of Teflon product, while nonplenum cables have a PVC (polyvinyl chloride) sheath, which does produce toxic gases when it burns. Not surprisingly, plenum cable costs more than nonplenum, sometimes twice as much or more, and it is also less flexible, making it more difficult to install. However, it is important to use the correct type of cable in any installation. If you violate the building codes, the local authorities can force you to replace the offending cable and possibly make you pay fines as well.

Cost is certainly an element that should affect your cable selection process, not only of the cable itself, but also of the ancillary components such as connectors and mounting hardware, the NICs for the computers, and the labor required for the cable installation. The qualities of fiber-optic cable might make it seem an ideal choice for your entire LAN, but when you see the costs of purchasing, installing, and maintaining it, your opinion may change.

Finally, the quality of the cable itself is an important part of the evaluation and selection process. When you walk into your local computer center to buy a prefabricated 10Base-T cable, you won't have much of a selection, except for cable length and possibly color. Vendors that provide a full cable selection, however (many of whom sell online or by mail order), have a variety of cable types that differ in their construction, their capabilities, and, of course, their prices.

Depending on the cable type, a good vendor may have both bulk cable and prefabricated cables. *Bulk cable* (that is, unfinished cable without connectors) should be available in various grades, in both plenum and nonplenum types. The grade of the cable itself can depend on several features, including the following:

Conductor gauge   The gauge is the diameter of the actual conductor within a cable, which in the case of copper cables is measured using the American Wire Gauge (AWG) scale. The lower the AWG rating, the thicker the conductor. A 24 AWG cable, therefore, is thinner than a 22 AWG cable. A thicker conductor provides better conductivity and more resistance against attenuation.

Category rating   Some types of cables are assigned ratings by a standards body, like the EIA/TIA. Twisted-pair cable, for example, is given a category rating that defines its capabilities. Most of the twisted-pair cable installed today is Category 5.

Shielded or unshielded   Some cables are available with casings that provide different levels of shielding against electromagnetic interference. The shielding usually takes the form of foil or copper braid, the latter of which provides better protection. Twisted-pair cabling, for example, is available in shielded and unshielded varieties. For a typical network environment, unshielded twisted pair provides sufficient protection against interference, because the twisting of the wire pairs itself is a preventative measure.

Solid or stranded conductor   A cable with a solid metal conductor provides better protection against attenuation, which means it can span longer distances. However, the solid conductor hampers the flexibility of the cable. If flexed or bent repeatedly, the conductor inside the cable can break. Solid conductor cables, therefore, are intended for permanent cable runs that will not be moved, such as those inside walls or ceilings. (Note that the cable can be flexed around corners and other obstacles during the installation; it is repeated flexing that can damage it.) Cables with conductors composed of multiple copper strands can be flexed repeatedly without breaking, but are subject to greater amounts of attenuation. Stranded cables, therefore, should be used for shorter runs that are likely to be moved, such as for patch cables running from wall plates to computers.

Note   Attenuation *refers to the tendency of signals to weaken as they travel along a cable, due to the resistance inherent in the medium. The longer a cable, the more the signals attenuate before reaching the other end. Attenuation is one of the primary factors that limits the size of a data network. Different types of cable have different attenuation rates, with copper cable being far more susceptible to the effect than fiber-optic cable.*

These features naturally affect the price of the cable. A lower gauge is more expensive than a higher one, a higher category is more expensive than a lower, shielded is more expensive than unshielded, and solid is more expensive than stranded. This is not to say, however, that the more expensive product is preferable in every situation. In addition to the cable itself, a good vendor should have all of the equipment you need to attach the appropriate connectors, including the connector components and the tools for attaching them.

Prefabricated cables have the connectors already attached and should be available in various lengths and colors, using cable with the features already listed, and with various grades of connectors. The highest quality prefabricated cables, for example, usually have a rubber boot around the connector that seals it to the cable end, prevents it from loosening or pulling out, protects the connector pins from bending, and reduces signal interference between the wires (called *crosstalk)*. On lower-cost cables, the connector is simply attached to the end, without any extra protection.

Cabling Standards

Prior to 1991, the cabling used for LANs was specified by the manufacturers of specific networking products. This resulted in the incompatibilities that are common in proprietary systems, and the need was recognized for a standard to define a cabling system that could support a multitude of different networking technologies. To address this need, the American National Standards Institute (ANSI), the Electronic Industry Association (EIA), and the Telecommunications Industry Association (TIA), along with a consortium of telecommunications companies, developed the ANSI/EIA/ TIA-568-1991 Commercial Building Telecommunications Cabling Standard. This document was revised in 1995 and is now known as ANSI/TIA/EIA-T568-A.

ANSI/TIA/EIA-T568-A

The *T568-A standard* defines a structured cabling system for voice and data communications in office environments that has a usable lifespan of at least ten years, will support products of multiple technology vendors, and uses any of the following cable types for various applications:

- Unshielded twisted pair (UTP) (100 ohm, 22 or 24 AWG)

- Shielded twisted pair (STP) (150 ohm)

- Multimode optical fiber (62.5/125 micron)

- Single-mode optical fiber (8.3/125 micron)

For each cable type, the standard defines the following elements:

- Cable characteristics and technical criteria that determine its performance level

- Topology and cable segment length specifications

- Connector specifications and pinouts

The document also includes specifications for the installation of the cable within the building space. Toward this end, the building is divided into the following subsystems:

Building entrance   The location at which the building's internal cabling interfaces with outside cabling.

Equipment room   The location of equipment that can provide the same functions as that in a telecommunications closet, but which may be more complex.

Telecommunications closet   The location of localized telecommunications equipment, such as the interface between the horizontal cabling and the backbone.

Backbone cabling   The cabling that connects the building's various equipment rooms, telecommunications closets, and the building entrance, as well as connections between buildings in a campus network environment.

Horizontal cabling   The cabling and other hardware used to connect the telecommunications closet to the work area.

Work area   The components used to connect the telecommunications outlet to the workstation.

Thus, the cable installation for a modern building might look something like the diagram shown in Figure 4-1. The connections to external telephone and other services arrive at the building entrance and lead to the equipment room, which contains the PBX, network servers, and other equipment. A backbone network connects the equipment room to various telecommunications closets throughout the building, which contain network interface equipment, such as switches, bridges, routers, or hubs. From the

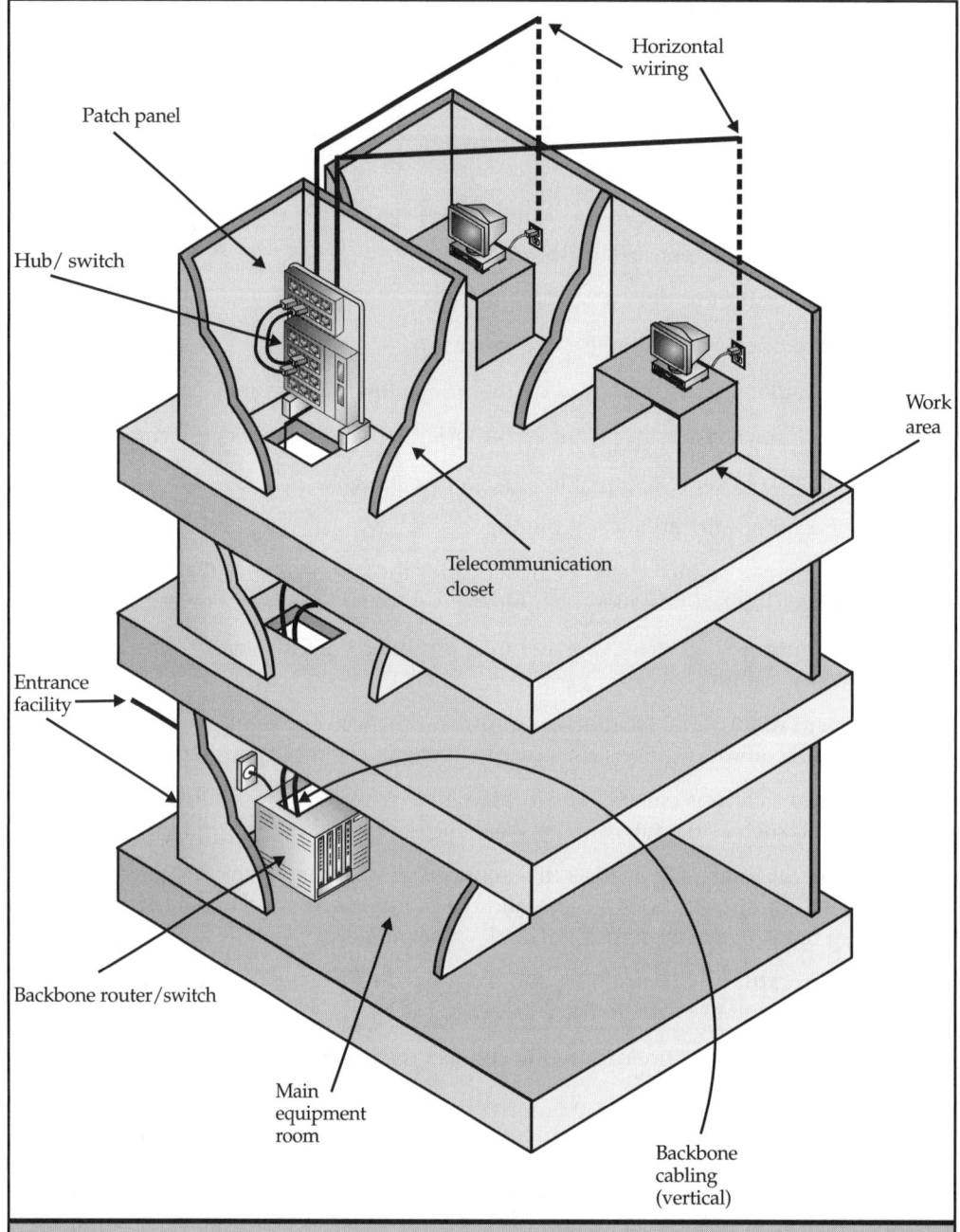

Figure 4-1: *A generic building cabling system as defined by ANSI/TIA/EIA-T568-A*

telecommunications closets, the horizontal cabling branches out into the work areas, terminating at wall plates. The work area then consists of the patch cables that connect the computers and other equipment to the wall plates.

This is, of course, a simplified and generalized plan. The T568-A standard, in coordination with other TIA/EIA standards, provides guidelines for the types of cabling within and between these subsystems that you can use to create a wiring plan customized to your site and your equipment. Some of these other standards are as follows:

TIA/EIA-569   Commercial Building Standard for Telecommunications Pathways and Spaces

TIA/EIA-606   Administration Standard for the Telecommunications Infrastructure of Commercial Buildings

TIA/EIA-607   Ground and Bonding Requirements for Telecommunications in Commercial Buildings

Contractors you hire to perform an office cable installation should be familiar with these standards and should be willing to certify in writing that their work conforms to the guidelines they contain.

ISO 11801E 1995

In addition to ANSI/TIA/EIA-T568-A, which defines cabling specifications used in the United States, the International Organization for Standardization (ISO) has published the ISO 11801E 1995 standard, which is the cabling standard most often used in Europe. Based on T568-A, this standard extends the cable types to include 100 ohm STP and 120 ohm UTP cabling, which are more popular in France and other European countries.

Data Link–Layer Protocol Standards

The protocols traditionally associated with the data link layer of the OSI reference model, such as Ethernet, Token Ring, and FDDI, also overlap into the physical layer in that they contain specifications for the network cabling. Thus, Ethernet and Token Ring standards, like those produced by the IEEE 802 working group and the ANSI X3T9.5 standard that defines FDDI, can also be said to be cabling standards. However, these documents do not go as deeply into the details of the cable properties and enterprise cable system design as T568-A. For more information on these standards, see Chapters 10, 11, and 12.

Coaxial Cable

The first commercially viable LAN technologies introduced in the 1970s used coaxial cable as the network medium. *Coaxial cable* is named for the two conductors that share the same axis running through the cable's center. Many types of copper cable have two separate conductors, such as a standard electrical cord. In most of these, the two conductors run side by side within an insulating sheath that protects and separates them. A coaxial cable, on the other hand, is round, with a copper core at its center that forms the first conductor. It is this core that carries the actual signals. A layer of dielectric foam insulation surrounds the core, separating it from the second conductor, which is made of braided wire mesh and functions as a ground. As with any electrical cable, the signal conductor and the ground must always be separated or a short will occur, producing noise on the cable. This entire assembly is then enclosed within an insulating sheath (see Figure 4-2).

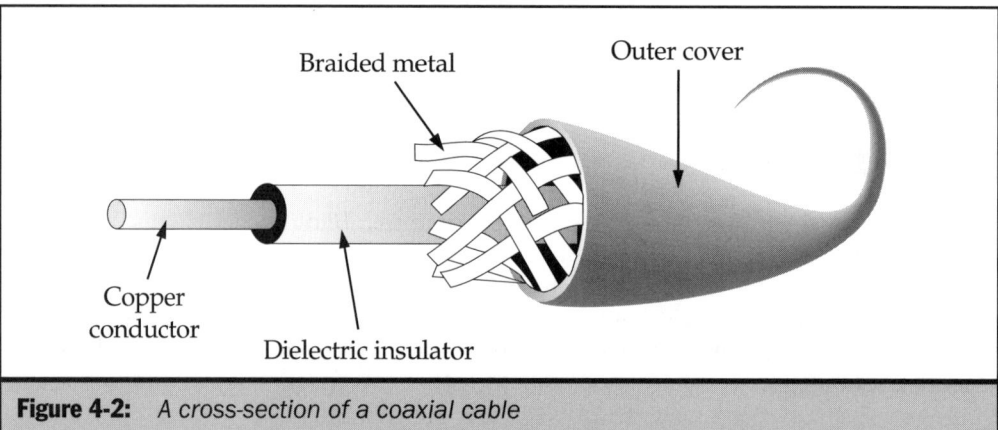

Figure 4-2: *A cross-section of a coaxial cable*

Note
Coaxial cables can have either a solid or a stranded copper core, and their designations reflect this difference. The suffix /U indicates a solid core, while A/U indicates a stranded core. Thin Ethernet, for example, can use either an RG-58/U or an RG-58A/U cable.

Several types of coaxial cables are used for networking, and they have different properties, even if they are similar in appearance. Table 4-1 lists the various types of coaxial cable. Data link–layer protocols call for specific types of cable, the properties of which determine the guidelines and limitations for the cable installation. The cable's attenuation, for example, determines how long a cable segment can be. The Attenuation column in the table specifies how much of a 100 MHz signal's strength (in decibels) is lost for every hundred feet of cable. A lower value indicates less signal loss, meaning the cable can span a longer distance before the signal is no longer viable.

Cable Designation	Cable Diameter	Impedance	Attenuation (dB/100' @ 100 MHz)	Connectors Used	Protocols Supported
RG-8/U	.405 inches	50 ohms	1.9	N	Thick Ethernet
RG-58/U or RG-58A/U	.195 inches	50 ohms	4.5	BNC	Thin Ethernet
RG-62A/U	.242 inches	93 ohms	2.7	BNC	ARCnet
RG-59/U	.242 inches	75 ohms	3.4	F	Cable TV

Table 4-1:   *Coaxial Cable Specifications*

The thickness of the cable also has a great effect on the nature of the installation. The layers of copper and foam insulation inside coaxial cable form a solid mass, unlike twisted-pair cables, for example, which contain separate wires and air space between and around them. Thus, coaxial cable is relatively heavy and relatively inflexible. The thicker the cable, the heavier and more inflexible it is. This inflexibility makes the cable difficult to install and to conceal.

Coaxial networks are cabled using a bus topology, in which the cable forms a segment with two ends and computers connected along its length. Each signal transmitted on the cable by a workstation travels on the bus in both directions to all of the other computers and, eventually, to the two cable ends. Each end of the bus must have a terminating resistor on it that removes the signals it receives by nullifying the voltages. Without terminators, the signals can reach the end of the cable and reflect back, causing data corruption.

Compared to other cable types, coaxial is also relatively inefficient for data networking. An Ethernet network constructed with coaxial cable is limited to a speed of 10 Mbps. There is no upgrade path to any faster technology as with twisted-pair or fiber-optic cable and Fast Ethernet. While you may encounter coaxial cable in networks that were installed years ago, virtually no new Ethernet LANs are being installed with it today. The following sections examine the applications for the various cable types, as well as the restrictions and advantages imposed by the cable itself.

Thick Ethernet

RG-8/U cable, when it is available at all, is usually referred to as *Thick Ethernet trunk cable*, because that is its primary use. The RG-8/U cable used for Thick Ethernet networks has the least amount of attenuation of the coaxial cables, due in no small part to its being much thicker than the other types. This is why a Thick Ethernet network can have cable segments up to 500 meters long, while Thin Ethernet is limited to 185 meters.

RG-8/U cable, at .405 inches in diameter, is similar in size to a garden hose, but much heavier and less flexible, which makes it difficult to bend around corners. For these reasons, the cable is typically installed along the floor of the site. The Ethernet specification calls for separate *Attachment Unit Interface (AUI)* cables that connect the NIC in each computer to the RG-8/U cable. By contrast, the RG-58A/U cable used by Thin Ethernet is thinner, lighter, and flexible enough to run directly to the NIC. RG-8/U is also far more expensive than the other coaxial cable types, which today may be partially due to its scarcity. One vendor currently sells a 500-foot spool of nonplenum RG-8/U cable for $399, as opposed to $129 for the same length of RG-58A/U (Thin Ethernet). Plenum cable is even more expensive: $1,049 for 500 feet of RG-8/U, as opposed to $259 for RG-58A/U.

Thick Ethernet cable is usually yellow and is marked every 2.5 meters for the taps to which the workstations connect. To connect a workstation to the cable, you apply what is known as a vampire tap. A *vampire tap* is a clamp that you connect to the cable after drilling a hole in the sheath. The clamp has metal "fangs" that penetrate into the core to send and receive signals (see Figure 4-3). The vampire tap also includes the transceiver (external to the computer on a Thick Ethernet network), which connects to the NIC with an AUI cable that has 15-pin D-shell connectors at both ends.

> *The fact that AUI cables can be up to 50 meters long is a major factor in planning a network layout. In most cases, a Thick Ethernet trunk cable can run through a room along one wall, and all of the computers in the room can connect to it using AUI cables.*

Because of this connection method, there is no need for breaks in the Thick Ethernet cable for every workstation. In fact, the Ethernet specification recommends using a single, unbroken cable segment whenever possible and even supplies the ideal places where breaks should be located, if they have to exist (see "Thick Ethernet" in Chapter 10). When you do have breaks in a Thick Ethernet cable, you use connectors, known as *N connectors,* to join the ends. You also use special N connectors with resistors in them to terminate the bus at both ends (see Figure 4-4).

As a result of the inconvenience caused by its expense and rigidity, and despite its better performance than Thin Ethernet, Thick Ethernet is never used for new Ethernet installations anymore and is rarely seen even on legacy networks.

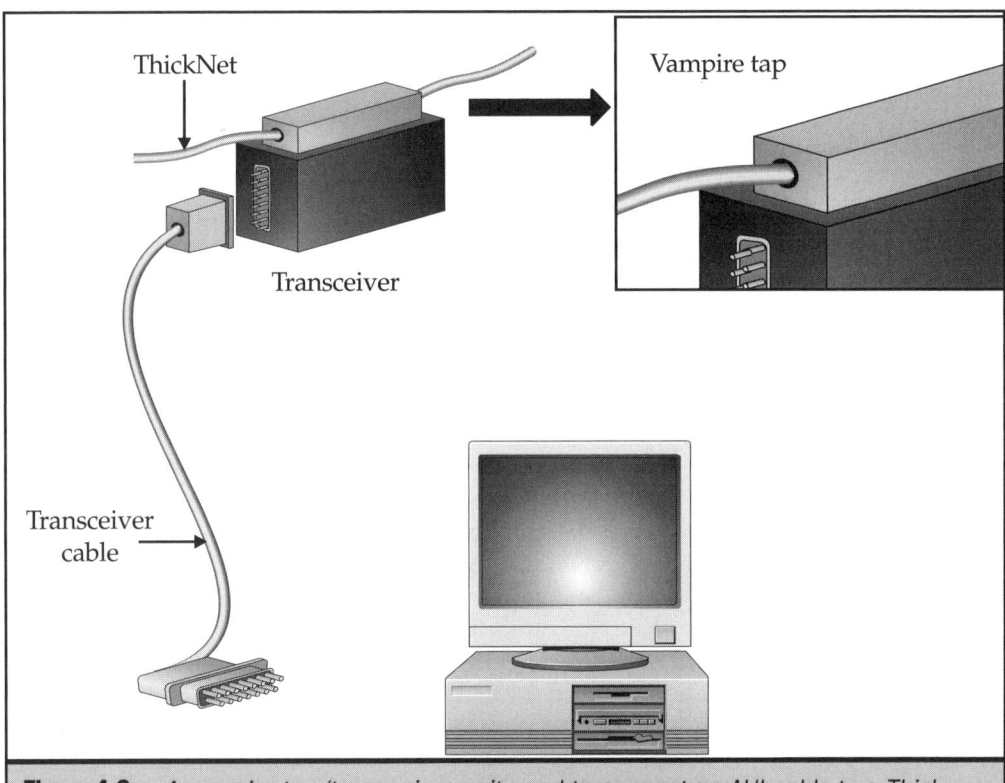

Figure 4-3: *A vampire tap/transceiver unit used to connect an AUI cable to a Thick Ethernet trunk*

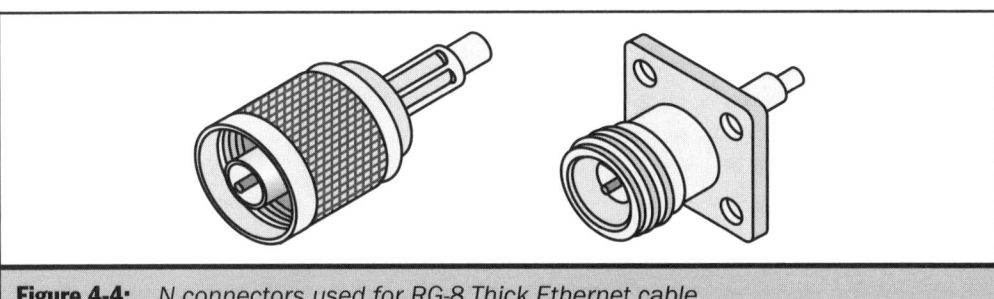

Figure 4-4: *N connectors used for RG-8 Thick Ethernet cable*

NETWORK HARDWARE

Thin Ethernet

The main advantage of the RG-58 cable used for Thin Ethernet networks over RG-8 is its relative flexibility, which simplifies the installation process and makes it possible to run the cable directly to the computer, rather than using a separate AUI cable. Compared to twisted pair, however, Thin Ethernet is still ungainly and difficult to conceal because every workstation must have two cables connected to its NIC using a T fitting. Instead of neat wall plates with modular jacks for patch cables, an internal Thin Ethernet installation has two thick, semi-rigid cables protruding from the wall for every computer.

As a result of this installation method, the bus is actually broken into separate lengths of cable that connect each computer to the next, unlike a Thick Ethernet bus, which ideally is one long cable segment pierced with taps along its length. This makes a big difference in the functionality of the network because if one of the two connections to each computer is broken for any reason, the bus is severed. When this happens, network communications fail between systems on different sides of the break, and the loss of termination on one end of each fragment jeopardizes all of the network's traffic.

RG-58 cable uses BNC (Bayonet Neil-Concelman) connectors to connect to the T and to connect the T to the NIC in the computer (see Figure 4-5). Even at the height of its popularity, Thin Ethernet cable was typically purchased in bulk and the connectors attached by the installer or administrator; prefabricated cables were relatively rare. The process of attaching a BNC connector involves stripping the insulation off the cable end to expose both the copper core and the ground, applying the connector as separate components (a socket that the cable threads through and a post that slips over the core), and then compressing the socket so it grips the cable and holds the post in place, using a pliers-like tool called a *crimper* (see Figure 4-6).

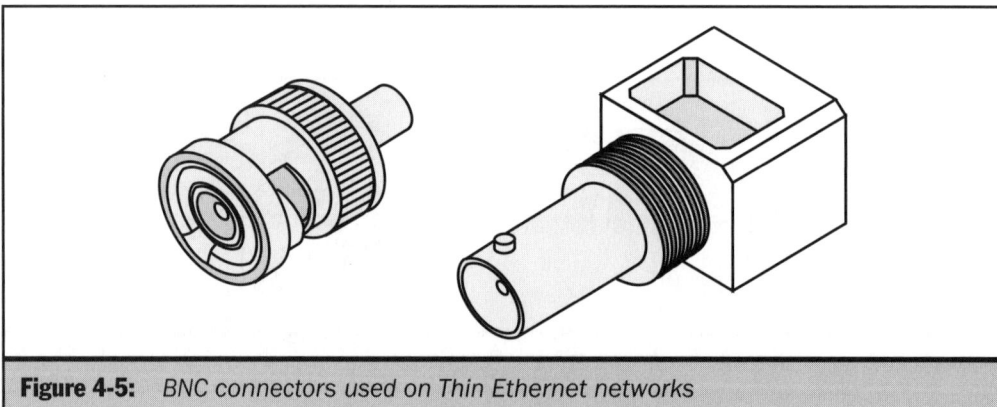

Figure 4-5:   *BNC connectors used on Thin Ethernet networks*

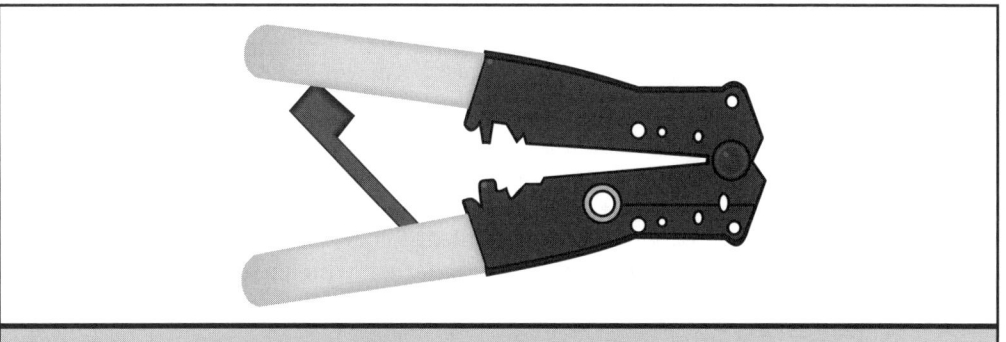

Figure 4-6:   *A crimper tool used to attach cable connectors*

Attaching BNC connectors properly is a skill that is rarely taught in the formal education of a network technician and requires some practice to master. Connectors that are not tightly crimped are easily pulled off the cable, or worse, loosened so that the electrical connection is partially severed. The result of this is a network with uneven performance and occasional outages that are difficult to track down without the proper cable-testing equipment. Largely because of faulty connections like these, Thin Ethernet sometimes gained a reputation for being quirky and, at times, unreliable.

RG-58 cable is much cheaper than RG-8 and is available in more varieties, but Thin Ethernet is still an all-but-dead technology. The bus topology, relatively difficult installation, and limited speed of coaxial cables has made them impractical for today's LANs.

ARCnet

The *Attached Resource Computing Network (ARCnet)* is the only other major LAN technology that used coaxial cable. Although similar in appearance to Thin Ethernet, the cable used in ARCnet networks is 93 ohm RG-62A/U, and the two are not interchangeable. ARCnet is a token-passing network that runs at only 2.5 Mbps and can use a mixture of star and bus technologies. ARCnet products are no longer available, but at one time this was a reasonably serviceable and very inexpensive networking solution.

Cable Television

Just because coaxial cable is hardly ever used for LANs anymore does not mean that it has totally outlived its usefulness. Antennas, radios, and particularly the cable television industry still use it extensively. The cable delivering TV service to your home is RG-59 75 ohm coaxial, used in this case for broadband rather than baseband transmission (meaning that the single cable carries multiple, discrete signals simultaneously). This cable is also similar in appearance to Thin Ethernet, but it has different properties and uses different connectors. The F connector used for cable TV connections screws into the jack, while BNC connectors use a bayonet lock coupling.

Many cable TV providers use this same coaxial cable to supply Internet access to subscribers, as well as television signals. In these installations, the coaxial cable connects to a device typically referred to as a cable modem, which then is connected to a computer using a 10Base-T Ethernet cable. However, while the coaxial cable may be part of an Ethernet network, do not confuse it with Thin Ethernet, which uses a different type of coaxial cable and uses baseband transmissions only.

Twisted-Pair Cable

Twisted-pair cable is the current standard for LAN communications. When compared to coaxial, it is easier to install, suitable for many different applications, and provides far better performance. Perhaps the biggest advantage of twisted-pair cable, however, is that it is already used in countless telephone system installations throughout the world. This means that a great many contractors are familiar with the installation procedures and that in a newly constructed office, it is possible to install the LAN cables at the same time as the telephone cables. In fact, many private homes now being built include twisted-pair network cabling as part of the basic service infrastructure.

Unlike coaxial cable, which has only one signal-carrying conductor and one ground, the twisted-pair cable used in most data networks has four pairs of insulated copper wires within a single sheath. Each wire pair is twisted with a different number of twists per inch to avoid electromagnetic interference from the other pairs and from outside sources (see Figure 4-7).

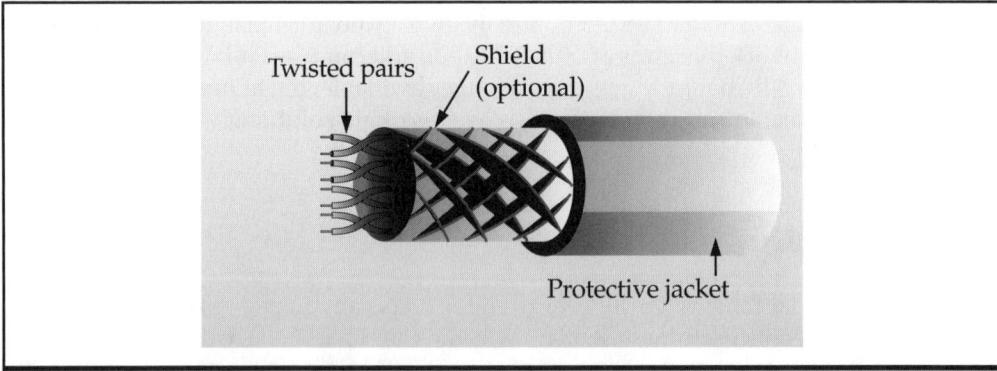

Figure 4-7: *A cross-section of a twisted-pair cable*

Each pair of wires in a twisted-pair cable is color-coded, using colors defined in the TIA/EIA-T568-A standard, which are as follows. In each pair, the solid-colored wire carries the signals, while the striped wire acts as a ground.

Pair 1   Solid blue, and white with blue stripe

Pair 2   Solid orange, and white with orange stripe

Pair 3   Solid green, and white with green stripe

Pair 4   Solid brown, and white with brown stripe

Unshielded Twisted Pair (UTP)

The outer sheathing of a twisted-pair cable can either be relatively thin, as in *unshielded twisted-pair (UTP)* cable, or thick, as in *shielded twisted pair (STP)*. UTP cable is the more commonly used of the two; most office Ethernet networks are more than adequately served by UTP cable. The UTP cable itself uses 22 or 24 AWG copper conductors and has an impedance of 100 ohms. The insulation can be plenum rated or nonplenum.

Beyond these specifications, the TIA/EIA-T568-A standard defines levels of performance for UTP cable that are referred to as categories. A higher category rating means that a cable is more efficient and able to transmit data at greater speeds. The major difference between the different cable categories is the tightness of each wire pair's twisting. Table 4-2 lists the categories defined by the T568-A standard, their speed ratings, and their applications.

Category	Frequency	Applications
1	Up to 0 MHz	Voice-grade telephone; POTS (plain old telephone service); alarm systems
2	Up to 1 MHz	Voice-grade telephone; IBM minicomputer and mainframe terminals; ARCnet; LocalTalk
3	Up to 16 MHz	Voice-grade telephone; 10Base-T Ethernet; 4 Mbps Token Ring; 100Base-T4; 100VG-AnyLAN
4	Up to 20 MHz	16 Mbps Token Ring
5	Up to 100 MHz	100Base-TX; OC-3 (ATM); SONet
5e	Up to 100 MHz	1000Base-T (Gigabit Ethernet)

Table 4-2:   *TIA/EIA Category Ratings for UTP Cable*

   *The TIA/EIA-T568-A standard recognizes Categories 3, 4, and 5, from Table 4-2, as viable choices for a telecommunications network.*

Category 3 cable was traditionally used for telephone system installations and is also suitable for 10Base-T Ethernet networks, which run at 10 Mbps. Category 3 is not suitable for the 100 Mbps speed used by Fast Ethernet, except in the case of 100Base-T4, which is specifically designed to run on Category 3 cable. 100Base-T4 (and also the less-than-successful 100VG-AnyLAN protocol) are only able to function on this cable because they use all four of the wire pairs to carry data, while the standard technologies use only two pairs.

Category 4 cable provides a marginal increase in performance over Category 3 and was, for a time, used in Token Ring networks. Since its ratification in 1995, however, most of the UTP cable installed for LANs (and telephone networks as well) is Category 5. Category 5 UTP cable (often known simply as Cat5) provides a substantial performance increase, supporting transmissions at up to 100 MHz. Even if you only intend to run 10Base-T today, it's a good idea to install Category 5 cable in anticipation of an upgrade to Fast Ethernet or another high-speed technology in the future.

   *Although the TIA/EIA category ratings are primarily used in relation to the cable itself, the other components that comprise the network medium are rated as well. To build a cabling system that is completely compliant with the Category 5 rating, for example, all connectors, wall plates, patch panels, and other components must also be rated Category 5.*

Anixter Cabling Standards

A company called Anixter, Inc., which played a prominent part in the development of the TIA/EIA standards, maintains its own cable ratings, in which it refers to *levels*, as opposed to categories. Table 4-3 lists the Anixter levels that go beyond the current Category 5 rating.

Level	Frequency
5	200 MHz
6	350 MHz
7	400 MHz

Table 4-3:   *Anixter Post-Category 5 UTP Cable Ratings*

Level 5 doubles the bandwidth of the Category 5 specification to 200 MHz, to conform to the ISO 11801 international standard. This cable supports throughput of up to 1.2 Gbps, which makes it suitable for Gigabit Ethernet communications. Level 6 increases the bandwidth of the cable to 350 MHz and Level 7 increases it to 400 MHz. These "unofficial" cable ratings are often used as guidelines for the manufacture of high-performance cables that exceed the TIA/EIA specifications.

Category 5E and Beyond

While Category 5 cable is sufficient for use on 100 Mbps networks such as Fast Ethernet, technology continues to advance, and Gigabit Ethernet products are now available, running at 1 Gbps (1,000 Mbps). To accommodate these ultra-high speeds, UTP cable ratings have continued to advance as well. However, the process by which the TIA/EIA standards are defined and ratified is much slower than the pace of technology, and many high-performance cable products arrived on the market that exceeded the Category 5 specifications to varying degrees. In 1999, after a surprisingly accelerated development period of less than two years, the TIA/EIA ratified the Category 5e (or Enhanced Category 5) standard.

The Category 5e standard was revised more than 14 times during its development, because there was a great deal of conflict among the concerned parties as to how far the standard should go. Category 5e is intended primarily to support the IEEE 802.3ab Gigabit Ethernet standard, also known as 1000Base-T, which is a version of the 1,000 Mbps networking technology designed to run on the standard 100-meter copper cable segments also used by Fast Ethernet. As you can see in Table 4-2, the Category 5e standard only calls for a maximum frequency rating of 100 MHz, the same as that of Category 5 cable. However, Gigabit Ethernet uses frequencies up to 125 MHz, and Asynchronous Transfer Mode (ATM) networks, which are also expected to use this cable, can run at frequencies of up to 155 MHz. As a result, there has been a good deal of criticism leveled at the new standard, saying that it doesn't go far enough to ensure adequate performance of Gigabit Ethernet networks.

It's important to understand that the TIA/EIA UTP cable standards consist of many different performance requirements, but the frequency rating is the one that is most commonly used to judge the transmission quality of the cable. In fact, the Category 5e standard is basically the Category 5 standard with slightly elevated requirements for some of its testing parameters, such as near end crosstalk (NEXT), the attenuation-to-crosstalk ratio (ACR), return loss, and differential impedance, as well as requirements for several new testing parameters, including power sum near end crosstalk (PSNEXT), far end crosstalk (FEXT), and equal level far end crosstalk (ELFEXT). These new tests are designed to ensure the cable's performance in full-duplex mode.

The result of this conflict is that there are now many different UTP cables on the market that are said to conform to the Category 5e standard, but many of them have actually been tested to levels that far exceed the standard. Some cables are tested to frequencies of 200, 250, and even 500 MHz, and are specified as conforming to the corresponding Anixter level, as well as to the Cat5e standard. Manufacturers frequently produce several different grades of Cat5e cable (obviously at different prices).

A cable that conforms strictly to the Cat5e standard will perform adequately with Gigabit Ethernet equipment, as long as it's installed precisely to the specifications. Like the Ethernet standards themselves, the Category 5 and earlier cable specifications have a built-in "fudge factor" that enables them to accommodate a less-than-perfect installation, which the Category 5e standard largely lacks. If you're building a network that you anticipate might be used at speeds faster than Fast Ethernet, either now or in the future, selecting a cable that exceeds the Cat5e performance levels by a substantial margin is a good idea.

A Category 6 standard calling for frequencies up to 250 MHz is currently in development, and a Category 7 standard pushing performance levels to 600 MHz is planned. While the Category 6 standard currently retains the basic UTP cable architecture (four wire pairs and RJ-45 connectors), Category 7 is expected to be a fully shielded cable system using an entirely new type of connector. While some manufacturers are producing cables that they claim to conform to the current draft standard of Category 6, the performance parameters are likely to change substantially before the standard is finally ratified.

Connector Pinouts

Twisted-pair cables use RJ-45 modular connectors at both ends (see Figure 4-8). An RJ-45 (RJ is the acronym for *registered jack*) is an 8-pin version of the 4-pin (or sometimes 6-pin) RJ-11 connector used on standard satin telephone cables. The pinouts for the connector, which are also defined in the TIA/EIA-T568-A standard, are shown in Figure 4-9, and have come to be known as the 568A pin assignments. However, other standards that predate TIA/EIA-T568-A provide alternate connector pinouts.

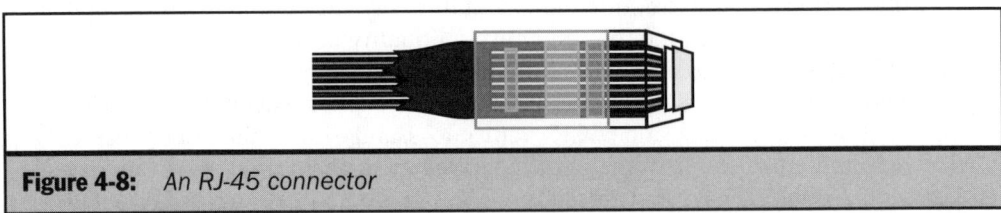

Figure 4-8:   *An RJ-45 connector*

The USOC standard (as shown on the right in Figure 4-10) was the traditional pinout originated for voice communications in the United States, but this configuration is not suitable for data. This is because, while pins 3 and 6 do connect to a single-wire pair, pins 1 and 2 are connected to separate pairs. AT&T discovered this shortcoming when it began doing research into computer networks that would run over the existing telecommunications infrastructure. In 1985, AT&T published its own standard, called 258A, which defined a new pinout in which the proper pins used the same wire pairs. The TIA/EIA, which was established in 1985 after the breakup of AT&T, then published this standard as an adjunct to TIA/EIA-T568-A in 1995, giving it the name T568-B (as shown on the left in Figure 4-10). Thus, while the pinout now known as 568B would seem to be newer than 568A, it is actually older. Pinout 568B began to be used widely in the United States before the TIA/EIA-T568-A standard was even published.

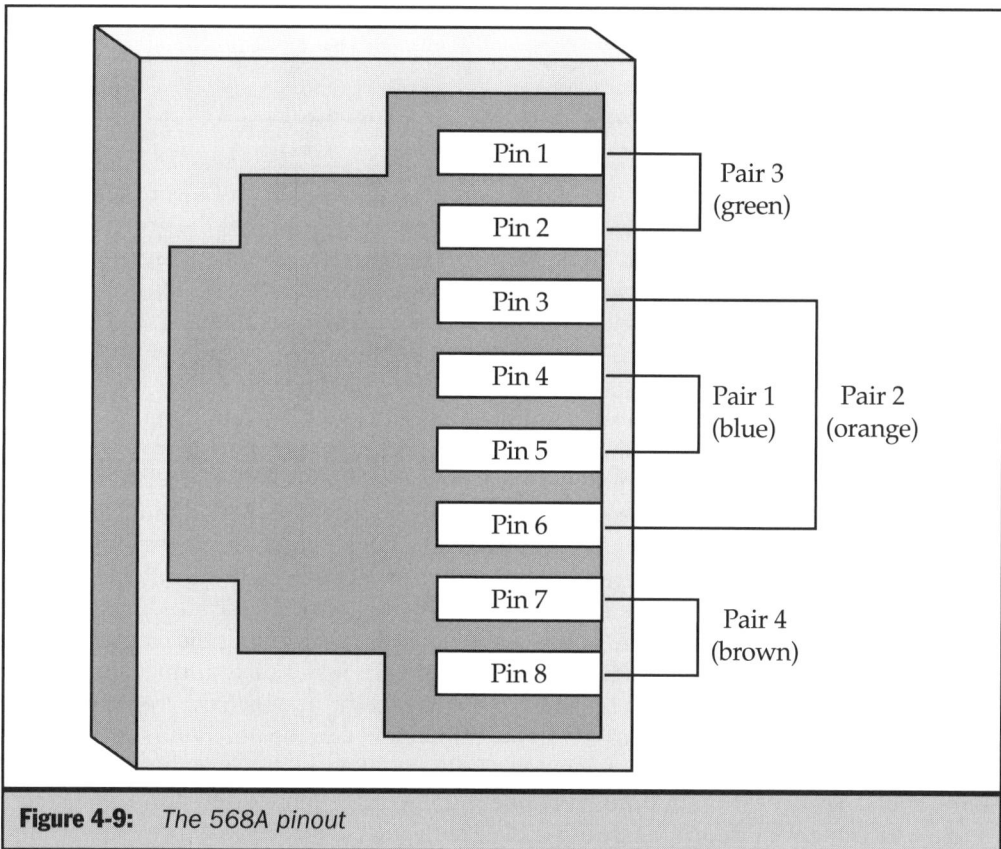

Figure 4-9:   *The 568A pinout*

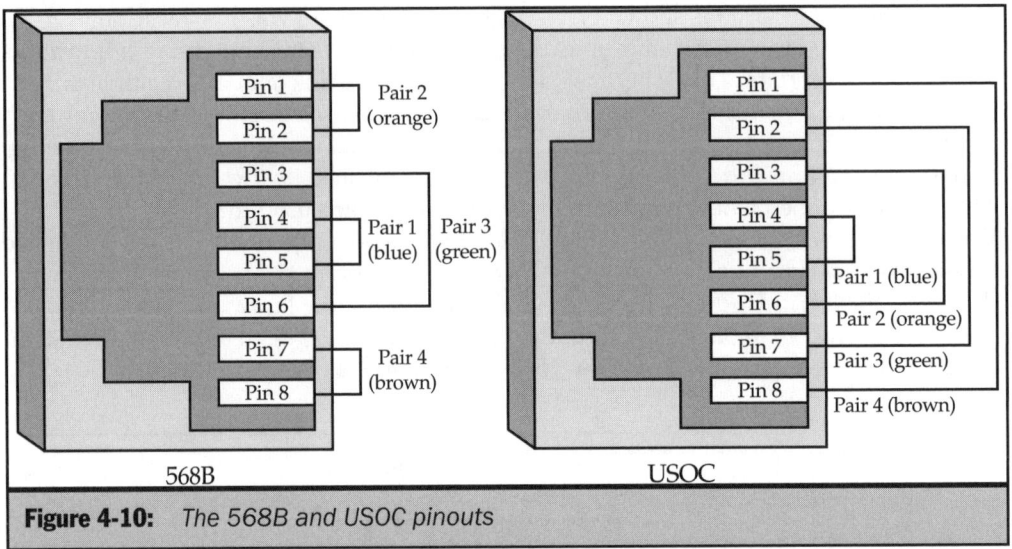

Figure 4-10: *The 568B and USOC pinouts*

As you can see in Figure 4-10, the USOC standard uses a different layout for the wire pairs, while the 568A and 568B pinouts are identical except that the green and orange wire pairs are transposed. Thus, the two TIA/EIA standards are functionally identical; neither one offers a performance advantage over the other, as long as both ends of the cable use the same pinout. Prefabricated cables are available that conform to either one of these standards. USOC cables are available as well, but under no circumstances should you use these on a LAN or other data network.

In most cases, twisted-pair cable is wired straight through, meaning that each of the pins on one connector is wired to its corresponding pin on the other connector, as shown in Figure 4-11. On a typical network, however, computers use separate wire pairs for transmitting and receiving data. For two machines to communicate, the transmitted signal generated at each computer must be delivered to the receive pins on the other, meaning that a signal crossover must occur between the transmit and receive wire pairs. The cables are wired straight through (that is, without the crossover) on a normal Ethernet LAN because the hub is responsible for performing the crossover.

If you want to connect one computer to another without a hub to form a simple two-node Ethernet network, you must use a *crossover cable*, in which the transmit pins on each end of the cable are connected to the receive pins on the other end, as shown in Figure 4-12. Crossover cables are a comparatively rare item in the average computer store. You can purchase them from an online or mail order vendor, but don't expect to find a wide range of options regarding cable grade and construction. Since crossover cables are only used for the most rudimentary networks, peak performance is usually not an issue.

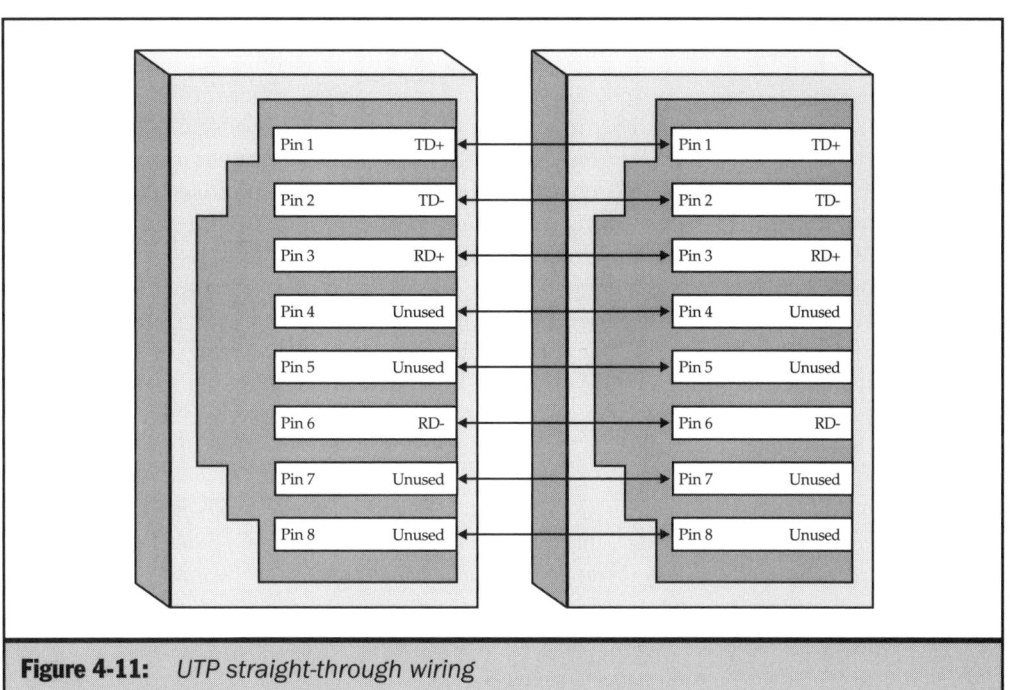

Figure 4-11: *UTP straight-through wiring*

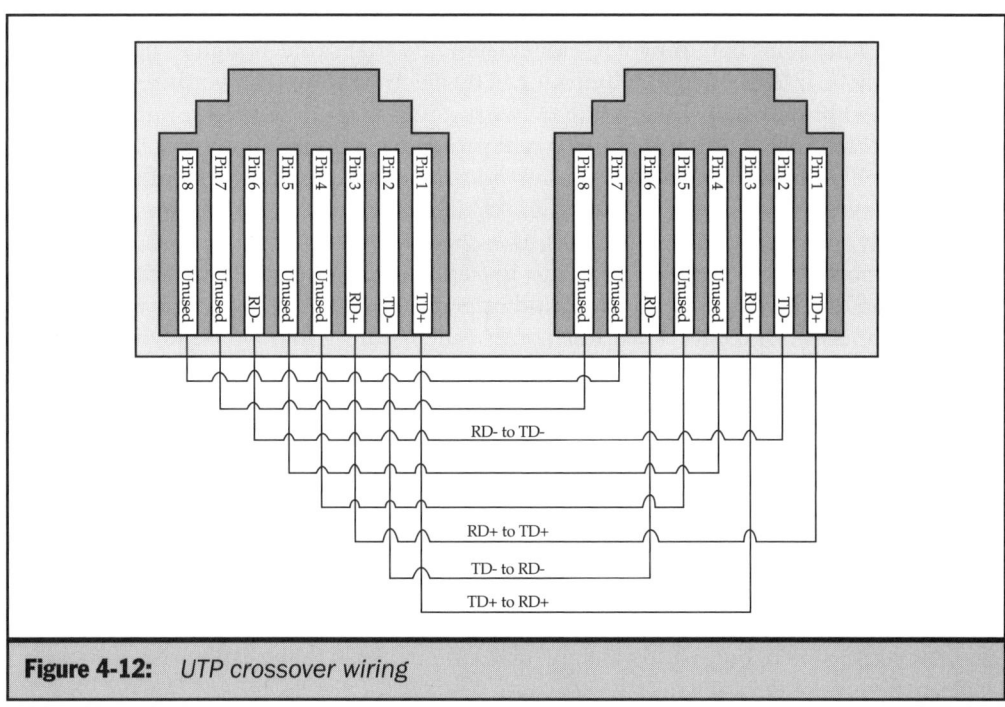

Figure 4-12: *UTP crossover wiring*

Because each pin on a straight-through cable is connected to the corresponding pin at the other end, it doesn't matter what colors the wires are, as long as the pairs are properly oriented. So, when purchasing prefabricated cables, either the 568A or 568B pinouts will function properly. The time when you must make a conscious decision to use one standard or the other is when you install bulk cable (or have it installed). You must connect the same colors on each end of the cable to the same pins, so you get a straight-through connection. Selecting one standard and sticking to it is the best way to avoid confusion that can result in nonfunctioning connections.

Attaching the connectors to a cable requires a crimper tool, much like the one used for coaxial cable, except that the process is complicated by having eight conductors to deal with, instead of only two. However, prefabricated twisted-pair cables are much more readily available than prefabricated Thin Ethernet cables. A network administrator who is not handy with a crimper can easily purchase twisted-pair cables with connectors attached in a wide variety of grades, lengths, and colors.

The crimper used for attaching RJ-45 connectors to UTP cable differs from the Thin Ethernet crimper only in the configuration of the jaws themselves. Some crimpers have modular dies that you can replace to support various types of cables and connectors.

Shielded Twisted Pair (STP)

STP is 150 ohm cable containing additional shielding that protects signals against the electromagnetic interference (EMI) produced by electric motors, power lines, and other sources. Used primarily in Token Ring networks, STP is also intended for installations where UTP cable would provide insufficient protection against interference.

The shielding in STP cable is not just an additional layer of inert insulation, as many people believe. Rather, the wires within the cable are encased in a metallic sheath that is as conductive as the copper in the wires. This sheath, when properly grounded, converts ambient noise into a current, just like an antenna. This current is carried to the wires within, where it creates an equal and opposite current flowing in the twisted pairs. The opposite currents cancel each other out, resulting in no noise to disturb the signals passing over the wires.

This balance between the opposite currents is delicate. If they are not exactly equal, the current can be interpreted as noise and can disturb the signals being transmitted over the cable. To keep the shield currents balanced, the entire end-to-end connection must be shielded and properly grounded. This means that all of the components involved in the connection, such as connectors and wall plates, must also be shielded. It is also vital to install the cable correctly, so that it is grounded properly and the shielding is not ripped or otherwise disturbed at any point.

The shielding in an STP cable can be either foil or braided metal. The metal braid is a more effective shield, but it adds weight, size, and expense to the cable. Foil-shielded cable, sometimes referred to as *screened twisted pair (ScTP)* or *foil twisted pair (FTP)*, is thinner, lighter, and cheaper, but is also less effective and more easily damaged. In both cases, the installation is difficult when compared to UTP because the installers must be careful not to flex and bend the cable too much, or they could risk damaging the shielding. The cable may also suffer from increased attenuation and other problems because the effectiveness of the shielding is highly dependent on a multitude of factors, including the composition and thickness of the shielding, the type and location of the EMI in the area, and the nature of the grounding structure.

The properties of the STP cable itself were defined by IBM during the development of the Token Ring protocol. These STP cable types are as follows:

Type 1A   Two pairs of 22 AWG wires, each pair wrapped in foil, with a shield layer (foil or braid) around both pairs, and an outer sheath of either PVC or plenum-rated material.

Type 2A   Two pairs of 22 AWG wires, each pair wrapped in foil, with a shield layer (foil or braid) around both pairs, plus four additional pairs of 22 AWG wires for voice communications, within an outer sheath of either PVC or plenum-rated material.

Type 6A   Two pairs of 22 AWG wires, with a shield layer (foil or braid) around both pairs, and an outer sheath of either PVC or plenum-rated material.

Type 9A   Two pairs of 26 AWG wires, with a shield layer (foil or braid) around both pairs, and an outer sheath of either PVC or plenum-rated material.

Note   *The TIA/EIA-T568-A standard recognizes only two of these STP cable types: Type 1A, for use in backbones and horizontal wiring, and Type 6A, for patch cables.*

Token Ring networks running on STP use large, proprietary connectors called IBM Data Connectors (IDCs). However, due to the bulkiness of the cable and the difficult installation process, most of today's Token Ring networks run on standard 4-pair UTP cable instead of STP.

Fiber-Optic Cable

Fiber-optic cable is completely different from all of the other cables covered thus far in this chapter because it is not based on electrical signals transmitted through copper conductors. Instead, fiber-optic cable uses pulses of light (photons) to transmit the binary signals generated by computers. Because fiber-optic cable uses light instead of electricity, nearly all of the problems inherent in copper cable, such as electromagnetic interference, crosstalk, and the need for grounding, are completely eliminated. In addition, attenuation is reduced enormously, enabling fiber-optic links to span much greater distances than copper—up to 120 kilometers in some cases.

NETWORK HARDWARE

Fiber-optic cable is ideal for use in network backbones, and especially for connections between buildings, because it is immune to moisture and other outdoor conditions. Fiber cable is also inherently more secure than copper because it does not radiate detectable electromagnetic energy like copper, and it is extremely difficult to tap.

The drawbacks of fiber optic mainly center around its installation and maintenance costs, which are usually thought of as being much higher than those for copper media. What used to be a great difference, however, has come closer to evening out in recent years. The fiber-optic medium itself is at this point only slightly more expensive than Category 5 UTP. Even so, the use of fiber does present some problems, such as in the installation process. Pulling the cable is basically the same as with copper, but attaching the connectors requires completely different tools and techniques—you can essentially throw everything you may have learned about electric wiring out the window.

Fiber optics has been around for a long time; even the early 10 Mbps Ethernet standards supported its use, calling it FOIRL, and later 10BaseF. Fiber optics came into its own, however, as a high-speed LAN technology, and today virtually all of the data link–layer protocols currently in use support it in some form, including the following:

■ Fast Ethernet (100BaseFX)

■ Gigabit Ethernet (1000BaseFX)

■ Token Ring

■ Fiber Distributed Data Interface (FDDI)

■ 100VG-AnyLAN

■ Asynchronous Transfer Mode (ATM)

■ Fibre Channel

As with copper, fiber-optic cable is typically installed using a star or a ring topology, although the FDDI protocol has popularized the *double ring*, which consists of two redundant rings with traffic traveling in opposite directions, for fault-tolerance purposes.

Fiber-Optic Cable Construction

A fiber-optic cable consists of a core made of glass or plastic, and a cladding that surrounds the core; then a plastic spacer layer; a layer of Kevlar fiber for protection; and an outer sheath of Teflon or PVC, as shown in Figure 4-13. The relationship between the core and the cladding enables fiber-optic cable to carry signals such long distances. The transparent qualities of the core are slightly greater than those of the cladding, which makes the inside surface of the cladding reflective. As the light pulses travel through the core, they reflect back and forth off of the cladding. This reflection enables you to bend the cable around corners and still have the signals pass through it without obstruction.

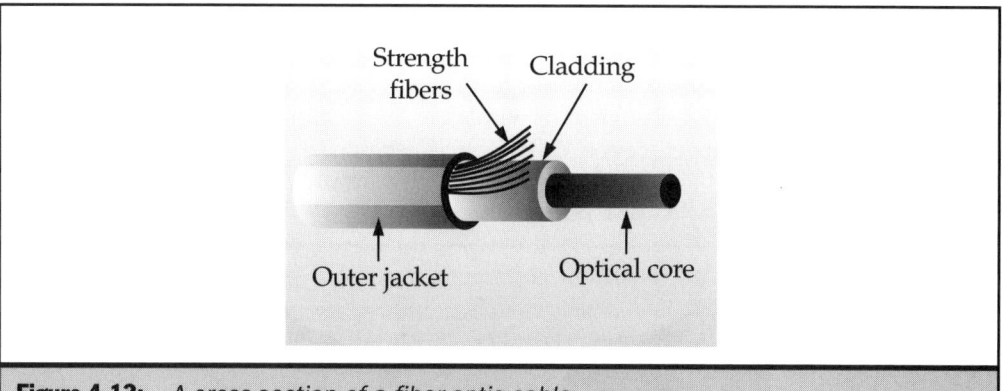

Figure 4-13: *A cross-section of a fiber-optic cable*

There are two main types of fiber-optic cable, called *singlemode* and *multimode*, that differ in several ways. The most important difference is in the thickness of the core and the cladding. Singlemode fiber is typically rated at 8.3/125 microns and multimode fiber at 62.5/125 microns. These measurements refer to the thickness of the core and the thickness of the cladding and the core together. Light travels down the relatively thin core of singlemode cable without reflecting off of the cladding as much as in multimode fiber's thicker core. The signal carried by a singlemode cable is generated by a laser and consists of only a single wavelength, while multimode signals are generated by a light-emitting diode (LED) and carry multiple wavelengths. Together, these qualities enable singlemode cable to operate at higher bandwidths than multimode and traverse distances up to 50 times longer.

However, singlemode cable is much more expensive and has a relatively high bend radius compared to multimode, which makes it more difficult to work with. Most fiber-optic LANs use multimode cable, which, although inferior in performance to singlemode, is still vastly superior to copper. Telephone and cable television companies tend to use singlemode fiber because they have to carry more data and span longer distances.

Fiber-optic cables are available in a variety of configurations, because the cable can be used for many different applications. Simplex cables contain a single fiber strand, while duplex cables contain two strands running side by side in a single sheath. Breakout cables can contain as many as 24 fiber strands in a single sheath, which you can divide to serve various uses at each end. Because fiber-optic cable is immune to copper cable problems such as EMI and crosstalk, it's possible to bundle large numbers of strands together without twisting them or worrying about signal degradation, as with UTP cable.

Fiber-Optic Connectors

The traditional connector used on fiber-optic cables is called an *ST (straight tip) connector*. It is a barrel-shaped connector with a bayonet locking system, as shown in Figure 4-14. A newer connector type, called the *SC (subscriber connector)*, is gaining in popularity, though. It has a square body and locks by simply pushing it into the socket.

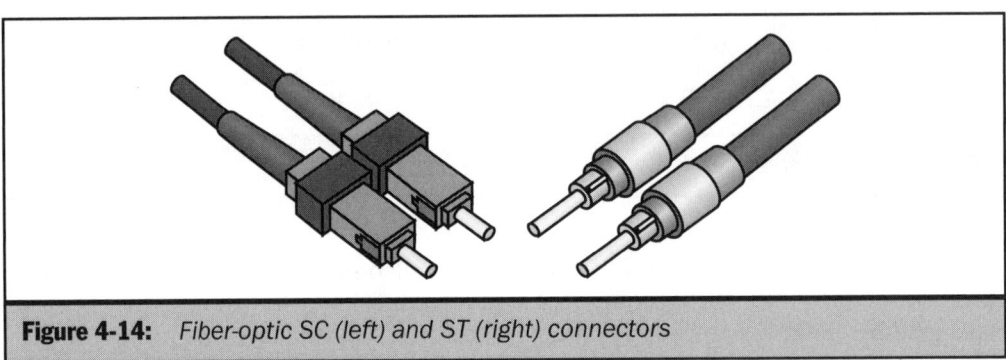

Figure 4-14:   *Fiber-optic SC (left) and ST (right) connectors*

Fiber-optic connectors can attach to the cable in several ways, using either a crimped compression fitting or an epoxy glue. Unlike the tools for crimping copper cables, a complete kit of which you can buy for under $100, a comparable fiber-optic toolkit costs well over $1,000, and requires considerably more skill to use.

Fiber-Optic Cable and Network Design

At this time, fiber-optic cable is primarily limited to backbones and is not often used for horizontal wiring because of its higher installation and maintenance costs. The technology has great potential in this area, however. Using fiber-optic cable imparts a freedom to the network designer that could never be realized with copper media. Because fiber optic permits segment lengths much greater than UTP's 100 meters, having telecommunications closets containing switches or hubs scattered about a large installation is no longer necessary.

Instead, horizontal cable runs can extend all the way from wall plates down to a central equipment room that contains all of the network's patch panels, hubs, switches, routers, and other such devices. This is known as a *collapsed backbone*. Rather than traveling constantly to remote areas of the installation, the majority of the infrastructure maintenance can be performed at this one location. For more information about network design, see Chapter 9.

Cable Installations

Installing network cable can be as simple as buying a few prefabricated cables at the computer center and stapling them to the baseboard, or as complicated as connecting a thousand workstations to a corporate backbone in a multifloored office building. As mentioned earlier in this chapter, cabling is a part of the LAN installation process that is frequently outsourced, not because it is especially difficult technically, but because it tends to be a dirty, time-consuming job. As with most professionals, however, cable installers with the proper tools, techniques, and experience can make the entire job look quick and easy.

There are two basic types of cable installations: external and internal, as described in the following sections.

External Installations

An external installation uses all prefabricated cables, and the cables typically run along the walls of the room near the floor and behind the furniture. This type of network is best suited to a single room, but it is possible to run the cables to adjacent rooms through doorways. Running cables to other rooms through walls is possible also, but remember that the connectors on the prefabricated cables force you to drill larger holes than bulk cable requires. The benefits of an external installation are its relatively low price, the fact that you can install the cables yourself, and its portability. If you ever need to move the network, you can simply roll up the cables and go. For a home or small office network, you can usually find all of the components you need (NICs, cables, and hubs) at your local computer store. Many hardware manufacturers have product lines that are specifically directed at this market, complete with instructions for the amateur networker.

The main disadvantage is the external network's appearance. Exposed cables are not nearly as neat and professional looking as an internal installation, and leaving the cables exposed does subject them to a greater risk of damage from everyday foot and furniture traffic. However, depending on the layout of your site, this may not be a major issue, and there are also steps you can take to disguise the installation and make it look more professional without punching holes in your walls.

The process of performing an external cable installation can be very informal, but to do the job right, you should follow a procedure like the following. This procedure assumes that you're going to be installing a Fast Ethernet network using UTP cable.

1. Select locations for each of your computers. Whenever possible, they should be located next to a wall, so that the cable doesn't have to run across the floor.

2. Select a location central to all of the computers for the hub. You will also need access to a power outlet to plug it in. Try to find a place where the hub will be protected, so that there is no chance of the cables being accidentally knocked out of the ports or the power cable being unplugged.

3. Measure the length of the cable path between each computer and the hub, taking into account the additional length needed to route the cable around any obstacles, such as doorways. Add a few extra yards of slack to each cable length as a safety factor.

4. Purchase prefabricated Category 5 or better UTP cables of the appropriate length for each of your computers.

5. Run the cables from the hub to the computers along the paths you've selected without securing the cables in place. Make sure that all of your cables are long enough and leave some slack at each end, so that you can move the computer and hub if necessary.

6. Starting at one end of each cable run, secure the cables in place using staples, cable ties, raceways, or some other type of fastener. Make sure that the cables are not kinked and that they're protected from damage by foot traffic or furniture.

7. Plug the cables into the ports of your hub at one end and into the network's adapters in your computers at the other end. When you plug in the hub's power cable and turn on the computers, the link pulse LEDs in both devices should light up, indicating that the connections are active.

A typical computer store might carry cables in only a few popular lengths, such as 10 and 20 feet. You might have to order from a catalog or Web site to get any other lengths you need. In addition, a well-stocked supplier should have cables available in a variety of colors, so that you can match the cables to the colors of your walls or furniture. This is one way to make an external installation appear more professional. Another method to improve the appearance of the installation is to secure the cables properly. There are several different ways you can do this, which are as follows:

Staples   Stapling your cables to the walls or floor in inconspicuous locations is a good way to protect the cables, as well as improving their appearance. However, you should not use a staple gun that shoots standard square-headed staples. The staples can compress the cables, possibly damaging the wires inside. Staple guns are available (often from cable suppliers) that shoot round-headed staples that don't crush the cables. A cable stapler is usually adjustable, so that you can set the depths that the staples are driven into the wall. The ideal setting is one that holds securely without compressing the cable in any way. You should be able to pull the cable back and forth inside the staple. If you don't want to buy a staple gun, you can also buy individual staples that you hammer into the wall manually.

Cable ties   Cable ties are strips of nylon or fabric with ratchet connectors or Velcro that hold the cables in place and a grommet for attaching them to a wall or floor.

—

These products are better suited to holding bundles of cables together, but you can use them to secure single cables as well.

Raceways   A raceway is a conduit made of plastic or metal that completely encloses your cables and secures to a wall or a floor. Raceways hide the cables completely, and protect them as well, but they are more expensive to purchase and more difficult to install. Because they're rigid, raceways are modular, and you must buy the appropriate fittings (such as corners, straightaways, as so on) to accommodate your site. Raceways are also used in buildings where the construction makes an internal installation impossible.

Internal Installations

While a relatively small network can be composed of all prefabricated cables, a good internal installation (that is, with the cables hidden from view in the walls and ceilings) is performed using bulk cable. A cable run in an internal installation consists of three elements: a permanent cable installation running from a port in a central wiring nexus called a *patch panel* to a wall plate in the vicinity of the computer to be connected, a patch cable connecting the patch panel port to a port in a hub or switch, and a second patch cable connecting the computer to the wall plate. A patch cable is a length of cable with standard male RJ-45 jacks at both ends. You can buy prefabricated patch cables or make them yourself by crimping connectors onto lengths of bulk cable.

The basic steps for this type of installation, assuming the use of UTP cable, are as follows:

1. Create a plan specifying the locations of the patch panel and the wall plates and the exact route that each cable will take.

2. Starting at the patch panel location, pull the cables through the walls and ceilings to where each of the workstations will be located.

3. Mount a wall plate near the site of each workstation and wire the cable end into the plate's connector.

4. At the location of the patch panel, mount the panel on the wall and "punch down" each of the cables to a connector in the panel.

5. Using the appropriate equipment, test each connection.

6. Using patch cables, connect the ports on the patch panel to the appropriate hub and connect the computer to the jack in the wall plate.

Of course, this description simplifies the process greatly, so it is worth devoting some detail to the individual steps.

Creating a Plan

Planning is the single most important part of the entire network construction process. You must know the exact location for each cable drop, preferably marked on a floor plan, so that you can keep track of the various cables during the installation. Remember that in a large cable installation, you may have hundreds of identical cables passing through walls and ceilings. Unless you are organized, confusion is sure to result. Your plan must also be designed with the data link–layer protocol requirements, local building codes, and fire laws in mind, so that you are not forced to pull all your cables out later. Of course, depending on the physical layout of your site and the construction of your building, you may be in for a few surprises that cause you to change your plans in midstream. This is why leaving a 10 percent reserve in your budget is always a good idea.

Even if you are outsourcing the cable installation, you should insist that your contractor provide you with a written diagram showing all of the cable routes. If you should ever have to work on the cabling at a later time, or on another service that might conflict with the cabling, you'll need this diagram.

Pulling Cables

The actual process of pulling cable begins at the server closet or Datacenter where you have decided to locate your patch panel. A *patch panel,* also called a *punchdown block* (see Figure 4-15), is a wall-mounted frame containing jacks for all of the cables you intend to pull. This will be your wiring nexus and the starting point for all of your cable runs.

Figure 4-15:   *A patch panel or punchdown block*

Bulk cable typically comes on a large spool. Locating this spool near the patch panel, you can start to strip cable off and pull it to the site of the first drop. How you do this depends on the construction of the building. In a modern office with hollow walls and a drop ceiling, you would typically run the cable through the ceiling to the approximate location of the drop, and then down through the wall to a hole you've cut for the wall plate.

Before you push that first cable end into the wall or ceiling, however, be sure to label it. Most installers use adhesive labels of some kind and some sort of code representing that cable's ultimate location. You may also want to attach a pull string to the leading edge of the cable with tape. A *pull string* is simply a length of string or twine you pull along with the cable, so that if you later want to pull additional cable all or part of the way to that same location, you can attach it to the string and pull it through from the other end.

Another tool used by cable installers working inside drop ceilings is called a telepole. A *telepole* is a thin, telescoping pole, sometimes 10 or 15 feet long when extended, with a clip on the end to hold a cable. With the pole in its closed state, you attach the leading cable end to it and then extend the pole inside the ceiling as far as it will reach. This tool is good for situations in which you have to pull cable by yourself because it prevents you from having to climb one ladder, throw the looped cable as far as possible inside the ceiling, and then climb another ladder across the room to retrieve it. Many professional installers also improvise their own tools for specific jobs.

When you are pulling cable in a drop ceiling, you must push each cable down inside the wall to the location of the wall plate. Installers have another tool called a *fish tape*, which is similar to a plumber's snake with a clamp on the end to hold the cable. The tape is flexible enough (like a tape measure) to push the cable either down inside a wall from the ceiling or up from the floor. Once the cable segment has reached the location of the wall plate, you pull through an extra few yards for slack and cut the cable at the spool, remembering to label that end with the same code you put on the leading end.

You may encounter obstacles at any stage of this process. You may find that the tops of the walls are capped with wood or metal studs, or that they have horizontal barriers halfway down the wall. Inside a drop ceiling, you may have to route your cable around fluorescent light fixtures or other sources of interference. Once the cables are in place, you should secure them using staples or cable ties, so that other maintenance workers cannot move them closer to a light fixture or other source of interference.

Mounting Hardware

Once you have your cable runs in place, it's time to terminate them at both ends by attaching them to the appropriate fixtures. At the workstation side, the cable usually terminates at a wall plate. A *wall plate* contains the jacks to which you connect the cables and is mounted flush with the wall, or sometimes in a box mounted on the wall surface. Wall plates are usually modular and can hold up to four jacks. You can install jacks of different types in the plate to support multiple voice and data connections as needed. Assuming a UTP installation, you attach the individual wires in the cable to the connector in the jack (using the pinout standard you decided on earlier), and then snap it into the plate and screw the plate into the wall.

At the other end of the run, the cable connects to the patch panel you mounted. The patch panel performs the same function as the wall plate, except that it contains more jacks. The patch panel is not a hub; no communication occurs between the ports. It is simply a convenient way of keeping track of the cable runs. To connect a cable to a port in the patch panel or a wall plate, you lay the individual wires in the appropriate slots in the connector and use a punchdown tool (see Figure 4-16) to push them into the slots. The tool strips the insulation off of the wire, firmly seats the wire in the slot, making a connection, and then cuts off the excess. Be sure to label the port in the patch panel so that you know the location of its terminus.

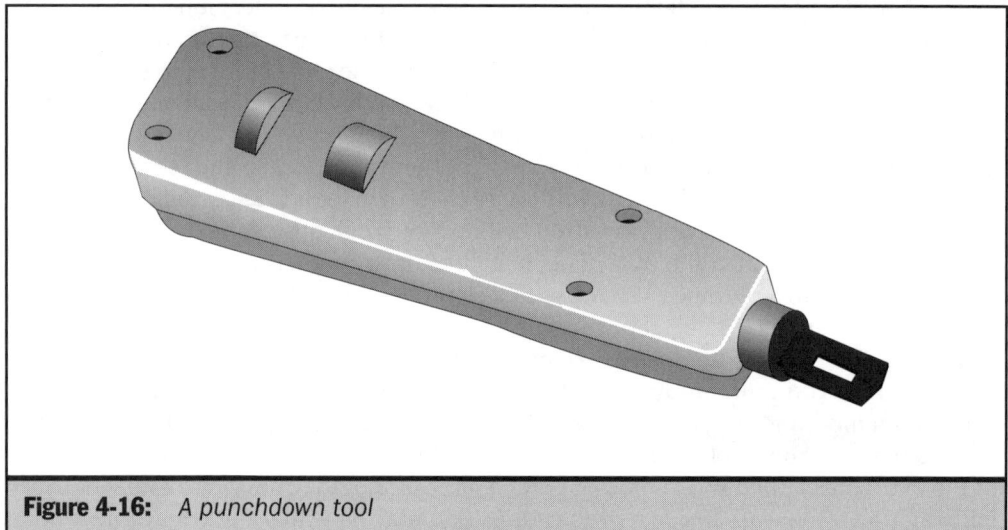

Figure 4-16: *A punchdown tool*

Testing Connections

Once you terminate both ends of a cable run, you should test the run to make sure it is functioning properly. You can do this by actually connecting a computer to the network using the cable run, but professional installers use cable testing equipment, which is easier and more accurate.

Connecting Computers

At this point, you are ready to connect both ends of the cable run to the appropriate devices. Using patch cables that you have either purchased or made yourself by attaching connectors to bulk cable, you connect the wall plate to the NIC in a computer and the patch panel port to a hub, switch, or other device in the server closet or Datacenter. Once the devices at both ends are powered, the link pulse LEDs at both ends should light, indicating a successful connection.

The Larger Picture

The process of installing the cable system for a large enterprise network can be considerably more complex than described here. This example covers only the installation of the horizontal cabling in one area, and doesn't attempt to anticipate the many problems that can arise or the different circumstances you'll encounter when installing cable types other than UTP. A large enterprise network can consist of many such horizontal cable installations, connected by a backbone that may have to run between the floors of a building or even between buildings. Except for small projects in limited areas, internal cable installations are better left to professionals.

NETWORK HARDWARE

The Complete Reference

Networking

Chapter 5

Wireless LANs

C omputer networks are traditionally thought of as using cables for their communications medium, but there have also been wireless networking solutions available for many years. Wireless networking products typically use some form of radio or light waves; these are called *unbounded media* (as opposed to *bounded media*, which refers to cabled networks). These media enable users with properly equipped computers to interact with other networked computers, just as if they were connected to them with cables. Wireless networking products have long had a reputation for poor performance and unreliability. It is only recently that these technologies have developed to the point at which they can be considered to be serious tools for business users.

Transmitting data over wireless media has recently become an extremely popular trend in the cellular world. There are a great many cellular phones on the market that can also connect to a data network run by the telephone service provider, enabling users to retrieve e-mail and access rudimentary Web sites that have been specially designed for small cellular phone displays. These data networks run at slow speeds, typically less than 14.4 Kbps, but the specialized nature of the data they carry lessens the frustration level. These are not data networks that you use to transmit document files and programs; the content is typically limited to brief text messages. The primary advantage of cellular-based data networking is its range. Users can access the network from any location supported by the cellular network.

Another emerging wireless technology that has great promise is called Bluetooth. Named for a tenth-century Danish king, Bluetooth provides short-range wireless communications between devices such as cellular phones and personal digital assistants (PDAs), or between desktop computers and peripherals, at a very low cost. Unlike the infrared communications used by laptops and other portables in recent years, Bluetooth uses radio frequency signals, which are not limited to line-of-sight transmissions. Bluetooth data communications are somewhat faster than cellular technologies. The symmetric data service runs at 433.9 Kbps in both directions, and the asymmetric data service runs at 723.2 Kbps in one direction and 57 Kbps in the other. This is a more useful speed for tasks like synchronizing schedules and contact lists, but it is still a long way from what is generally considered to be LAN speed.

The wireless technology that comes closest to emulating the cabled computer networks that are the primary focus of this book is the wireless LAN (WLAN), as defined in the 802.11 standard published by the Institute for Electrical and Electronics Engineers (IEEE). This standard defines the physical and data link–layer specifications for a wireless network that can use any one of several media and transmit data at speeds of up to 11 Mbps. This is not Fast Ethernet speed by any means, but it is more than sufficient for standard networking tasks, such as access to files, printers, and server applications.

Wireless Applications

The most immediate application for wireless local area networking that comes to mind is a situation where it is impractical or impossible to install a cabled network. In some cases, the construction of a building may prevent the installation of network cables, while in others cosmetic concerns may be the problem. For example, a kiosk containing a computer that provides information to guests might be a worthwhile addition to a luxury hotel, but not at the expense of running unsightly cables across the floor or walls of a meticulously decorated lobby. The same might be the case for a small two- or three-node network in a private home, where installing cables inside walls would be difficult and using external cables would be unacceptable in appearance.

The other primary application for wireless LANs is to support mobile client computers. These mobile clients can range from laptop-equipped technical support personnel for a corporate internetwork to roving customer service representatives with specialized handheld devices, such as rental car and baggage check workers in airports. With today's handheld computers and a wireless LAN protocol that is reliable and reasonably fast, the possibilities for its use are endless. For example:

- Hospitals can store patient records in a database and permit doctors and nurses to continually update them by entering new information into a mobile computer.

- Workers in retail stores can dynamically update inventory figures by scanning the items on the shelves.

- A traveling salesperson can walk into the home office with a laptop in hand, and as soon as the computer is within range of the wireless network, it connects to the LAN, downloads new e-mail, and synchronizes the user's files with copies stored on a network server.

The IEEE 802.11 Standards

In 1997, the IEEE published the first version of a standard that defined the physical and data link–layer specifications for a wireless networking protocol that would meet the following requirements:

- The protocol would support stations that are fixed, portable, or mobile, within a local area. The difference between portable and mobile is that a portable station can access the network from various fixed locations, while a mobile station can access the network while it is actually in motion.

NETWORK HARDWARE

■ The protocol would provide wireless connectivity to automatic machinery, equipment, or stations that require rapid deployment—that is, rapid establishment of communications.

■ The protocol would be deployable on a global basis.

This document, in its current form, is now known as IEEE 802.11, 1999 Edition, "Wireless LAN Medium Access Control (MAC) and Physical Layer (PHY) Specifications." In addition to this document, there are two supplements that add to the standard, called IEEE 802.11a-1999, "Wireless LAN Medium Access Control (MAC) and Physical Layer (PHY) Specifications: High-Speed Physical Layer in the 5 GHz Band," and IEEE 802.11b-1999, "Wireless LAN Medium Access Control (MAC) and Physical Layer (PHY) Specifications: High-Speed Physical Layer in the 2.4 GHz Band." The two supplements are not free-standing documents in themselves; they consist of amendments and edits that are to be applied to the original standard.

Because 802.11 was developed by the same the IEEE 802 committee responsible for the 802.3 (Ethernet) and 802.5 (Token Ring) protocols, it fits into the same physical- and data link–layer stack arrangement. The data link layer is divided into the logical link control (LLC) and media access control (MAC) sublayers. The 802.11 documents define the physical layer and MAC sublayer specifications for the wireless LAN protocol, and the systems use the standard LLC sublayer defined in IEEE 802.2. From the network layer up, the systems can use any standard set of protocols, such as TCP/IP or IPX.

Note   *For more information on LLC, see Chapter 10.*

Despite the inclusion of 802.11 in the same company as Ethernet and Token Ring, the use of wireless media calls for certain fundamental changes in the way you think about a local area network and its use. Some of these changes are as follows:

Unbounded media   A wireless network does not have readily observable connections to the network or boundaries beyond which network communication ceases.

Dynamic topology   Unlike cabled networks, in which the LAN topology is meticulously planned out before the installation and remains static until deliberate changes are made, the topology of a wireless LAN changes frequently, if not continuously.

Unprotected media   The stations on a wireless network are not protected from outside signals as cabled networks are. On a cabled network, outside interference can affect signal quality, but there is no way for the signals from two separate but adjacent networks to be confused. On a wireless network, roving stations can conceivably wander into a different network's operational perimeter, compromising security.

Unreliable media   Unlike a cabled network, a protocol cannot work under the assumption that every station on the network receives every packet and can communicate with every other station.

Asymmetric media   The propagation of data to all of the stations on a wireless network does not necessarily occur at the same rate. There can be differences in the transmission rates of individual stations that change as the device moves or the environment in which it is operating changes.

As a result of these changes, the traditional elements of a data link–layer LAN protocol (the MAC mechanism, the frame format, and the physical-layer specifications) have to be designed with different operational criteria in mind.

The Physical Layer

The 802.11 physical layer defines two possible topologies and three types of wireless media, operating at four possible speeds.

Physical-Layer Topologies

As you learned in Chapter 1, the term "topology" usually refers to the way in which the computers on a network are connected together. A bus topology, for example, means that each computer is connected to the next one, in daisy-chain fashion, while in a star topology, each computer is connected to a central hub. These examples apply to cabled networks, however. Wireless networks don't have a concrete topology as cabled ones do. Unbounded media, by definition, enable wireless network devices to transmit signals to all of the other devices on the network simultaneously. However, this does not equate to a mesh topology, as described in Chapter 1. Although each device theoretically can transmit signals to all of the other wireless devices on the network at any time, this does not necessarily mean that it will. Mobility is an integral part of the wireless network design, and a wireless LAN protocol must be able to compensate for systems that enter and leave the area in which the medium can operate. The result is that the topologies used by wireless networks are basic rules that they use to communicate, and not static arrangements of devices at specific locations. IEEE 802.11 supports two types of wireless network topologies: the ad hoc topology and the infrastructure topology.

The fundamental building block of an 802.11 wireless LAN is the *basic service set (BSS)*. A BSS is a geographical area in which properly equipped wireless stations can communicate. The configuration and area of the BSS are dependent on the type of wireless medium being used and the nature of the environment in which it's being used, among other things. A network using a radio frequency–based medium might have a BSS that is roughly spherical, for example, while an infrared network would deal more in straight lines. The boundaries of the BSS can be affected by environmental conditions, architectural elements of the site, and many other factors, but when a station moves within the basic service set's sphere of influence, it can communicate with other stations in the same BSS. When it moves outside of the BSS, communication ceases.

The simplest type of BSS consists of two or more wireless computers or other devices that have come within transmission range of each other, as shown in Figure 5-1. The process by which the devices enter into a BSS is called *association*. Each wireless device has an operational range dictated by its equipment, and as the two devices approach each other, the area of overlap between their ranges becomes the BSS. This arrangement, in which all of the network devices in the BSS are mobile or portable, is called an *ad hoc topology* or an *independent BSS (IBSS)*. The term ad hoc topology refers to the fact that a network of this type may often come together without prior planning, and exist only as long as the devices need to communicate. This type of topology operates as a peer-to-peer network, because every device in the BSS can communicate with every other device.

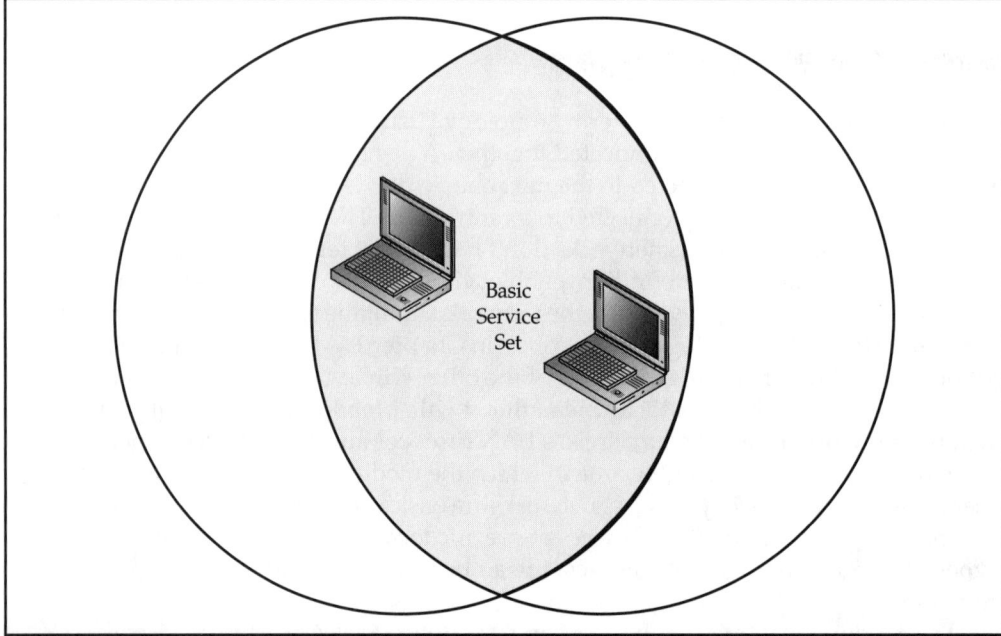

Figure 5-1:   *A basic service set can be as simple as two wireless stations within communication range of each other.*

 *While this illustration depicts the BSS as roughly ovular, and the convergence of the communicating devices as being caused by their physically approaching each other, the actual shape of the BSS is likely to be far less regular and much more ephemeral. The ranges of the devices can change instantaneously due to many different factors, and the BSS can grow, shrink, or even disappear entirely at a moment's notice.*

While an ad hoc network uses basic service sets that are transient and constantly mutable, it's also possible to build a wireless network with basic service sets that are more permanent. This is the basis of a network that uses an *infrastructure topology*. An infrastructure network consists of at least one wireless *access point (AP)*, which is either a standalone device or a wireless-equipped computer that is also connected to a standard bounded network using a cable. The access point has an operational range that is relatively fixed (when compared to an IBSS), and functions as the *base station* for a BSS. Any mobile station that moves within the AP's sphere of influence is associated into the BSS and becomes able to communicate with the cabled network (see Figure 5-2). Note that this is more of a client/server arrangement than a peer-to-peer one. The AP enables multiple wireless stations to communicate with the systems on the cabled network, but not with each other. However, the use of an AP does not prevent mobile stations from communicating with each other independently of the AP.

It is because the AP is permanently connected to the cabled network and not mobile that this type of network is said to use an infrastructure topology. This arrangement is typically used for corporate installations that have a permanent cabled network that also must support wireless devices that access resources on the cabled network. An infrastructure network can have any number of access points, and therefore any number of basic service sets. The architectural element that connects basic service sets together is called a *distribution system (DS)*. Together, the basic service sets and the DS that connects them are called the *extended services set (ESS)*. In practice, the DS is typically a cabled network using IEEE 802.3 (Ethernet) or another standard data link–layer protocol, but the network can conceivably use a *wireless distribution system (WDS)* also. Technically, the AP in a network of this type is also called a *portal*, because it provides access to a network using another data link–layer protocol. It's possible for the DS to function solely as a means of connecting APs together and not provide access to resources on a cabled network, but this is relatively rare. Whether the media used to form the BSS and the DS are the same or different (the standard takes no stance either way), 802.11 logically separates the wireless medium from the distribution system medium.

The basic service sets connected by a distribution system can be physically configured in almost any way. The basic service sets can be widely distant from each other, to provide wireless network connectivity in specific remote areas, or they can overlap, to provide a large area of contiguous wireless connectivity. It's also possible for an infrastructure BSS to be concurrent with an IBSS. The 802.11 standard makes no distinction between the two topologies, because both must present the exact same appearance to the LLC sublayer operating at the upper half of the data link layer.

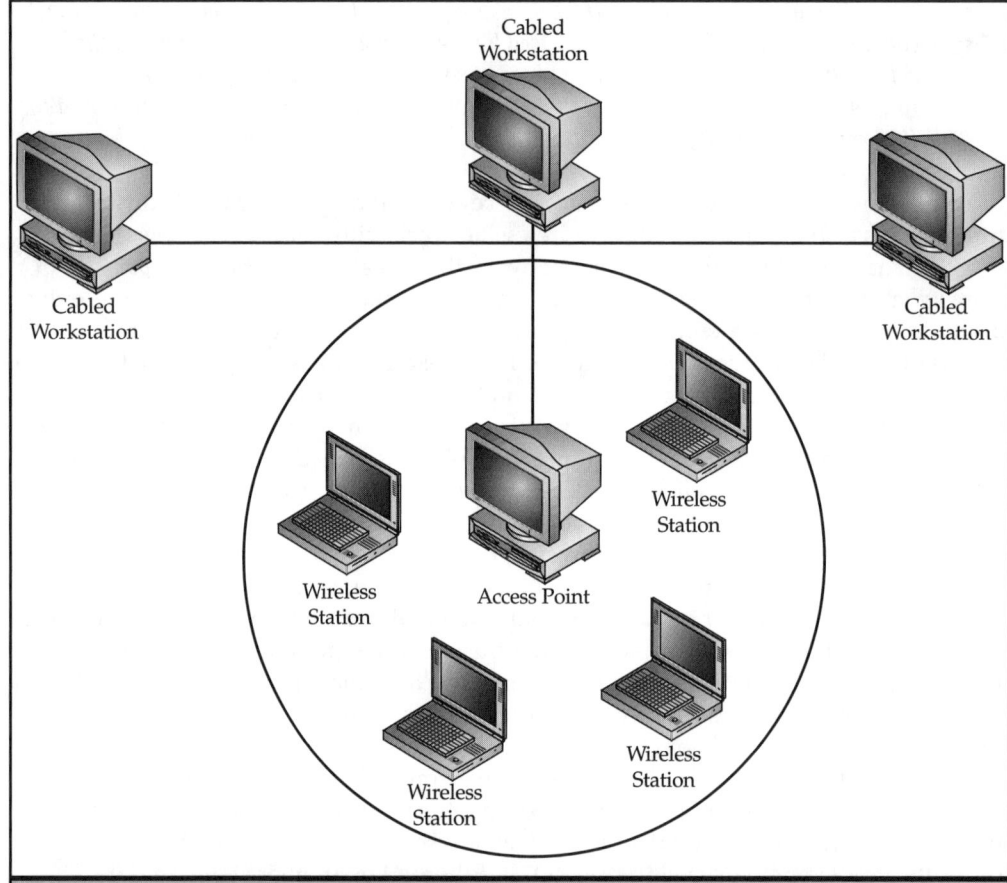

Figure 5-2: *An access point enables wireless stations to access resources on a cabled network.*

Physical-Layer Media

The IEEE 802.11 standard defines three physical-layer media, two that use radio frequency (RF) signals and one that uses infrared light signals. A wireless LAN can use any one of the three media, all of which interface with the same MAC layer. These three media are as follows:

- Frequency-Hopping Spread Spectrum (FHSS)
- Direct-Sequence Spread Spectrum (DSSS)
- Infrared

The two RF media both use spread spectrum communication, which is a common form of radio transmission used in many wireless applications. Invented during the 1940s, spread spectrum technology takes an existing narrowband radio signal and divides it among a range of frequencies in any one of several ways. The result is a signal that utilizes more bandwidth, but is louder and easier for a receiver to detect. At the same time, the signal is difficult to intercept, because attempts to locate it by scanning through the frequency bands turn up only isolated fragments. It is also difficult to jam, because you would have to block a wider range of frequencies for the jamming to be effective.

The 802.11 RF media operate in the 2.4 GHz frequency band, occupying the 83 MHz of bandwidth between 2.400 and 2.483 GHz. These frequencies are unlicensed in most countries, although there are varying limitations on the signal strength imposed by different governments.

The difference between the various types of spread spectrum communications lies in the method by which the signals are distributed among the frequencies. *Frequency-hopping spread spectrum*, for example, uses a predetermined code or algorithm to dictate frequency shifts that occur continually, in discrete increments, over a wide band of frequencies. The 802.11 FHSS implementation calls for 79 1-MHz channels, although some countries impose smaller limits. Obviously, the receiving device must be equipped with the same algorithm in order to read the signal properly. The rate at which the frequency changes (that is, the amount of time that the signal remains at each frequency before hopping to the next one) is independent of the bit rate of the data transmission. If the frequency-hopping rate is faster than the signal's bit rate, the technology is called a *fast hop system*. If the frequency-hopping rate is slower than the bit rate, you have a *slow hop system*. The 802.11 FHSS implementation runs at 1 Mbps, with an optional 2 Mbps rate.

In *direct-sequence spread spectrum* communications, the signal to be transmitted is modulated by a digital code called a *chip* or *chipping code*, which has a bit rate larger than that of the data signal. The chipping code is a redundant bit pattern that essentially turns each bit in the data signal into several bits that are actually transmitted. The longer the chipping code, the more the original data signal is enlarged. This enlargement of the signal makes it easier for the receiver to recover the transmitted data if some bits are damaged. The more the signal is enlarged, the less significance attributed to each bit. As with FHSS, a receiver that doesn't possess the chipping code used by the transmitter can't interpret the DSSS signal, seeing it as just noise. The DSSS implementation in the original 802.11 document supports 1 and 2 Mbps transmission rates. IEEE 802.11b expands this capability by adding transmission rates of 5.5 and 11 Mbps. Only DSSS supports these faster rates, which is the primary reason why it is the most commonly used 802.11 physical-layer specification.

Infrared communications use frequencies in the 850 to 950 nanometer range, just below the visible light spectrum. This medium is rarely implemented on wireless LANs, because of its limited range. Unlike most infrared media, the IEEE 802.11 infrared implementation does not require direct line-of-sight communications; an infrared network can function using diffuse or reflected signals. However, the range of communications is limited when compared to FHSS and DSSS, about 10 to 20 meters, and can only function properly in an indoor environment with surfaces that provide adequate signal diffusion or reflection. This makes infrared unsuitable for mobile devices and places more constraints on the physical location of the wireless device than either FHSS or DHSS. Like FHSS, the 802.11 infrared medium supports a 1 Mbps transmission rate and an optional rate of 2 Mbps.

Physical-Layer Frames

Instead of a relatively simple signaling scheme such as the Manchester and Differential Manchester techniques used by Ethernet and Token Ring, respectively, the media operating at the 802.11 physical layer have their own frame formats that encapsulate the frames generated at the data link layer. This is necessary to support the complex nature of the media.

The Frequency-Hopping Spread Spectrum Frame

The FHSS frame is illustrated in Figure 5-3, and consists of the following fields:

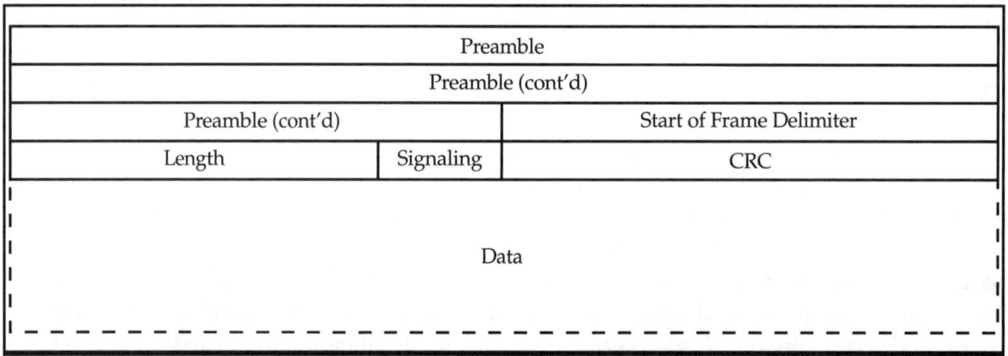

Preamble		
Preamble (cont'd)		
Preamble (cont'd)		Start of Frame Delimiter
Length	Signaling	CRC
Data		

Figure 5-3: *The FHSS frame format*

Preamble (10 bytes)   Contains 80 bits of alternating 0s and 1s that the receiving system uses to detect the signal and synchronize timing.

Start of Frame Delimiter (SFD) (2 bytes)   Indicates the beginning of the frame.

Length (12 bits)   Specifies the size of the data field.

Signaling (4 bits) Contains one bit that specifies whether the system is using the 1 or 2 Mbps transmission rate. The other three bits are reserved for future use. No matter which transmission rate the system is using, the preamble and header fields are always transmitted at 1 Mbps. Only the data field is transmitted at 2 Mbps.

CRC (2 bytes) Contains a cyclic redundancy check value, used by the receiving system to test for transmission errors.

Data (0 to 4,095 bytes) Contains the data link–layer frame to be transmitted to the receiving system.

The Direct-Sequence Spread Spectrum Frame

The DSSS frame is illustrated in Figure 5-4, and consists of the following fields:

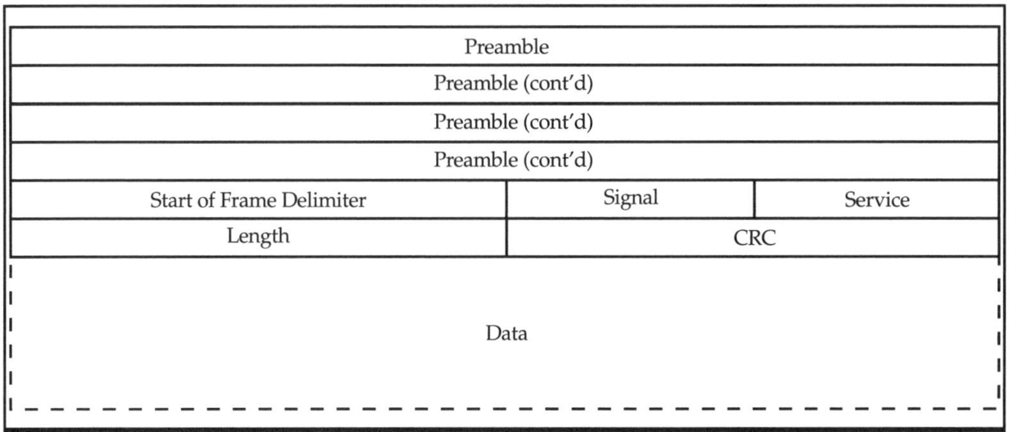

Preamble			
Preamble (cont'd)			
Preamble (cont'd)			
Preamble (cont'd)			
Start of Frame Delimiter		Signal	Service
Length		CRC	
Data			

Figure 5-4: *The DSSS frame format*

Preamble (16 bytes) Contains 128 bits that the receiving system uses to adjust itself to the incoming signal.

Start of Frame Delimiter (SFD) (2 bytes) Indicates the beginning of the frame.

Signal (1 byte) Specifies the transmission rate used by the system.

Service (1 byte) Contains the hexadecimal value 00, indicating that the system complies to the IEEE 802.11 standard.

Length (2 bytes) Specifies the size of the data field.

CRC (2 bytes)   Contains a cyclic redundancy check value, used by the receiving system to test for transmission errors.

Data (variable)   Contains the data link–layer frame to be transmitted to the receiving system.

The Infrared Frame

The frame used for infrared transmissions consists of the following fields:

Synchronization (SYNC) (57 to 73 slots)   Used by the receiving system to synchronize timing and, optionally, to estimate the signal-to-noise ratio and perform other preparatory functions.

Start of Frame Delimiter (SFD) (4 slots)   Indicates the beginning of the frame.

Data Rate (3 slots)   Specifies the transmission rate used by the system.

DC Level Adjustment (DCLA) (32 slots)   Used by the receiver to stabilize the DC level after the transmission of the preceding fields.

Length (2 bytes)   Specifies the size of the data field.

CRC (2 bytes)   Contains a cyclic redundancy check value, used by the receiving system to test for transmission errors.

Data (0 to 2,500 bytes)   Contains the data link–layer frame to be transmitted to the receiving system.

 The first four fields of the infrared frame are transmitted using pulse position modulation (PPM), which maps the data bits to be transmitted into symbols, using a unit of time called a slot, which is 250 nanoseconds long.

The Data Link Layer

As with IEEE 802.3 (Ethernet) and 802.5 (Token Ring), the 802.11 document defines only half of the functionality found at the data link layer. Like the other IEEE 802 protocols, the LLC sublayer forms the upper half of the data link layer and is defined in the IEEE 802.2 standard. The 802.11 document defines the MAC sublayer functionality, which consists of a connectionless transport service that carries LLC data to a destination on the network in the form of *MAC service data units (MSDUs)*. Like other data link–layer protocols, this service is defined by a frame format (actually several frame formats, in this case), and a media access control mechanism. The MAC sublayer also provides security services, such as authentication and encryption, and reordering of MSDUs.

Data Link–Layer Frames

The 802.11 standard defines three basic types of frames at the MAC layer, which are as follows:

Data frames Used to transmit upper-layer data between stations

Control frames Used to regulate access to the network medium and to acknowledge transmitted data frames

Management frames Used to exchange network management information to perform network functions such as association and authentication

The general MAC frame format is shown in Figure 5-5. The functions of the frame fields are as follows:

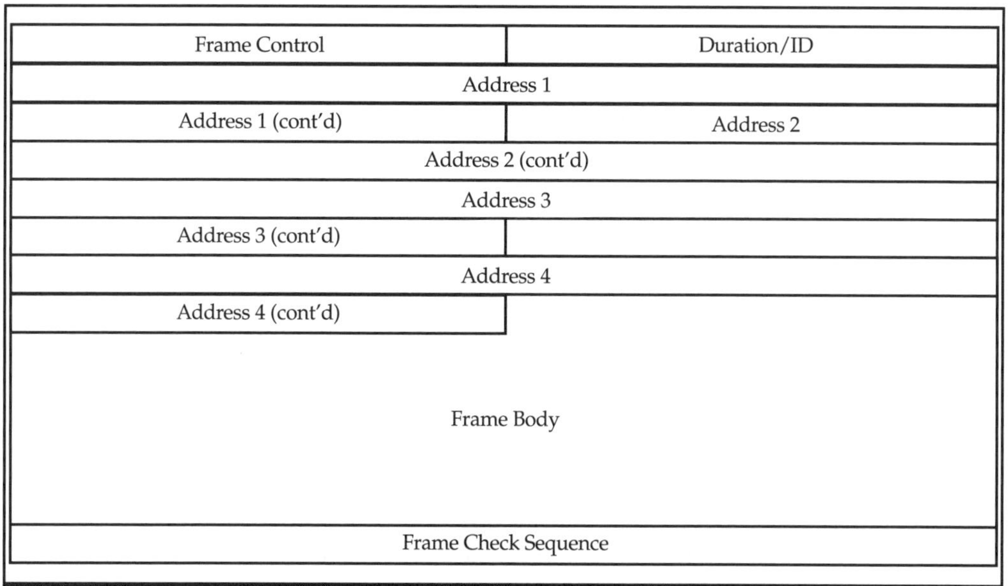

Frame Control	Duration/ID
Address 1	
Address 1 (cont'd)	Address 2
Address 2 (cont'd)	
Address 3	
Address 3 (cont'd)	
Address 4	
Address 4 (cont'd)	
Frame Body	
Frame Check Sequence	

Figure 5-5: *The IEEE 802.11 MAC sublayer frame format*

Frame Control (2 bytes)   Contains 11 subfields that enable various protocol functions. The subfields are as follows:

Protocol Version (2 bits)   Specifies the version of the 802.11 standard being used. At present, the only possible value for this field is 0.

Type (2 bits)   Specifies whether the packet contains a management frame (00), a control frame (01), or a data frame (10).

Subtype (4 bits)   Identifies the specific function of the frame.

To DS (1 bit)   A value of 1 in this field indicates that the frame is being transmitted to the distribution system (DS) via an access point (AP).

From DS (1 bit)   A value of 1 in this field indicates that the frame is being received from the DS.

More Frag (1 bit)   A value of 1 indicates that the packet contains a fragment of a frame, and that there are more fragments still to be transmitted. When fragmenting frames at the MAC layer, an 802.11 system must receive an acknowledgment for each fragment before transmitting the next one.

Retry (1 bit)   A value of 1 indicates that the packet contains a fragment of a frame that is being retransmitted after a failure to receive an acknowledgment. The receiving system uses this field to recognize duplicate packets.

Pwr Mgt (1 bit)   A value of 0 indicates that the station is operating in active mode; a value of 1 indicates that the station is operating in power-save mode. APs buffer packets for stations operating in power-save mode until they change to active mode or explicitly request that the buffered packets be transmitted.

More Data (1 bit)   A value of 1 indicates that an AP has more packets for the station that are buffered and awaiting transmission.

WEP (1 bit)   A value of 1 indicates that the Frame Body field has been encrypted using the *wired equivalent privacy (WEP)* algorithm, which is the security element of the 802.11 standard. WEP can only be used in management frames used to perform authentications.

Order (1 bit)   A value of 1 indicates that the packet contains a data frame (or fragment) that is being transmitted using the StrictlyOrdered service class, which is designed to support protocols that cannot process reordered frames.

Duration/ID (2 bytes)   In control frames used for power-save polling, this field contains the association identity (AID) of the station transmitting the frame. In all other frame types, the field indicates the amount of time (in microseconds) needed to transmit a frame and its short interframe space (SIFS) interval.

Address 1 (6 bytes)   Contains an address that identifies the recipient of the frame, using one of the five addresses defined in 802.11 MAC sublayer communications, depending on the values of the To DS and From DS fields.

Address 2 (6 bytes)   Contains one of the five addresses used in 802.11 MAC sublayer communications, depending on the values of the To DS and From DS fields.

Address 3 (6 bytes)   Contains one of the five addresses used in 802.11 MAC sublayer communications, depending on the values of the To DS and From DS fields.

Sequence Control (2 bytes)   Contains two fields used to associate the fragments of a particular sequence and assemble them into the right order at the destination system:

> **Fragment Number (4 bits)**   Contains a value that identifies a particular fragment in a sequence.

> **Sequence Number (12 bits)**   Contains a value that uniquely identifies the sequence of fragments that make up a data set.

Address 4 (6 bytes)   Contains one of the five addresses used in 802.11 MAC sublayer communications, depending on the values of the To DS and From DS fields. Not present in control and management frames and some data frames.

Frame Body (0 to 2,312 bytes)   Contains the actual information being transmitted to the receiving station.

Frame Check Sequence (4 bytes)   Contains a cyclic redundancy check (CRC) value used by the receiving system to verify that the frame was transmitted without errors.

The four address fields in the MAC frame identify different types of systems depending on the type of frame being transmitted and its destination in relation to the DS. You can determine the systems whose addresses are contained in the four address fields using the information in Table 5-1. The five different types of addresses referenced in the table are as follows:

Source Address (SA)   An IEEE MAC individual address that identifies the system that generated the information carried in the Frame Body field.

Destination Address (DA)   An IEEE MAC individual or group address that identifies the final recipient of an MSDU.

Transmitter Address (TA)   An IEEE MAC individual address that identifies the system that transmitted the information in the Frame Body field on the current wireless medium (typically an AP).

Receiver Address (RA)   An IEEE MAC individual or group address that identifies the immediate recipient of the information in the Frame Body field on the current wireless medium (typically an AP).

Basic Service Set ID (BSSID)   An IEEE MAC address that identifies a particular BSS. On an infrastructure network, the BSSID is the MAC address of the station functioning as the AP of the BSS. On an ad hoc network (IBSS), the BSSID is a randomly generated value generated during the creation of the IBSS.

To DS Value	From DS Value	Function	Address 1 Value	Address 2 Value	Address 3 Value	Address 4 Value
0	0	Data frames exchanged by stations in the same IBSS, and all control and management frames	DA	SA	BSSID	Not Used
0	1	Data frames transmitted to the DS	DA	BSSID	SA	Not Used
1	0	Data frames exiting the DS	BSSID	SA	DA	Not Used
1	1	Wireless distribution system (WDS) frames exchanged by APs in a DS	RA	TA	DA	SA

Table 5-1:   *MAC Sublayer Address Types*

Media Access Control

As with all data link–layer protocols that use a shared network medium, the media access control (MAC) mechanism is one of the protocol's primary defining elements. IEEE 802.11 defines the use of a MAC mechanism called Carrier Sense Multiple Access with Collision Avoidance (CSMA/CA), which is a variation of the Carrier Sense Multiple Access with Collision Detection (CSMA/CD) mechanism used by Ethernet.

The basic functional characteristics of wireless networks have a profound effect on the MAC mechanisms they can use. For example, the Ethernet CSMA/CD mechanism and the token-passing method used by Token Ring and FDDI networks both require every device on the network to receive every transmitted packet. An Ethernet system that doesn't receive every packet can't detect collisions reliably. In addition, the Ethernet collision detection mechanism requires full-duplex communications (because the indication that a collision has occurred is simultaneous transmit and receive signals), which is impractical in a wireless environment. If a token-passing system fails to receive a packet, the problem is even more severe, because the packet cannot then be passed on to the rest of the network, and network communication stops entirely. One of the characteristics of the wireless networks defined in 802.11, however, is that stations can repeatedly enter and leave the BSS because of their mobility and the vagaries of the wireless medium. Therefore, the MAC mechanism on a wireless network must be able to accommodate this behavior.

The carrier sense multiple access part of the CSMA/CD mechanism is the same as that of an Ethernet network. A computer with data to transmit listens to the network medium and, if it is free, begins transmitting its data. If the network is busy, the computer backs off for a randomly selected interval and begins the listening process again. Also like Ethernet, the CSMA part of the process can result in collisions. The difference in CSMA/CA is that systems attempt to avoid collisions in the first place by reserving bandwidth in advance. This is done by specifying a value in the Duration/ID field or using specialized control messages called request-to-send (RTS) and clear-to-send (CTS).

The carrier sense part of the transmission process occurs on two levels, the physical and the virtual. The physical carrier sense mechanism is specific to the physical-layer medium the network is using and is equivalent to the carrier sense performed by Ethernet systems. The virtual carrier sense mechanism, called a *network allocation vector (NAV)*, involves the transmission of an RTS frame by the system with data to transmit, and a response from the intended recipient in the form of a CTS frame. Both of these frames have a value in the Duration/ID field that specifies the amount of time needed for the sender to transmit the forthcoming data frame and receive an acknowledgment (ACK) frame in return. This message exchange essentially reserves the network medium for the life of this particular transaction, which is where the collision avoidance part of the mechanism comes in. Since both the RTS and CTS messages contain the Duration/ID value, any other system on the network receiving either one of the two observes the reservation and refrains from trying to transmit its own data during that time interval. This way, a station that is capable of receiving transmissions from one computer, but not the other, can still observe the CSMA/CA process.

In addition, the RTS/CTS exchange also enables a station to more easily determine if communication with the intended recipient is possible. If the sender of an RTS frame fails to receive a CTS frame from the recipient in return, it retransmits the RTS frame repeatedly, until a pre-established timeout is reached. Retransmitting the brief RTS message is much quicker than retransmitting large data frames, which shortens the entire process.

To detect collisions, IEEE 802.11 uses a positive acknowledgment system at the MAC sublayer. Each data frame that a station transmits must followed by an ACK frame from the recipient, which is generated after a CRC check of the incoming data. If the frame's CRC check fails, the recipient considers the packet to have been corrupted by a collision (or other phenomenon) and silently discards it. The station that transmitted the original data frame then retransmits it as many times as needed to receive an ACK, up to a predetermined limit. Note that the failure of the sender to receive an ACK frame could be due to the corruption or non-delivery of the original data frame or the non-delivery of an ACK frame that the recipient did send in return. The 802.11 protocol does not distinguish between the two.

IEEE 802.11 Products

Although wireless networking has been around for some years, the products on the market have gravitated toward either the low-cost home networking market (with low 1 or 2 Mbps transmission rates and spotty reliability) or the expensive corporate networking market, which was somewhat more reliable, but not much faster. The ratification of the IEEE 802.11b standard, however, has resulted in the release of many new 11 Mbps wireless LAN products that provide the speed and reliability needed to actually be useful on many types of networks. For example, you can now purchase an 802.11 "starter set" that consists of an access point and a PC Card NIC for less than $300. By connecting the access point to an existing Ethernet network and plugging the PC Card into a laptop, the computer can access the network and all of its resources from locations as far as 1,500 feet away from the access point (depending on the conditions where the network is operating). To add up to 127 other wireless computers to this network, you only have to purchase additional NICs, at prices that start at approximately $80 each. To enlarge your network even further, you can install additional access points. To build a simple ad hoc, peer-to-peer network without an access point or an Ethernet component, you simply install 802.11b NICs in two computers and begin communicating.

Chapter 6

Network Connection Devices

O
riginally, LANs consisted of nothing more than computers and cables, but as the technology evolved, more equipment was required. As the early coaxial cable networks grew to span longer distances, devices called repeaters were added to boost the signals. Later, when the dominant medium for Ethernet networks shifted from coaxial to unshielded twisted-pair (UTP) cable, hubs became an essential network component. As networks grew from tools for localized workgroups to company-wide resources, components such as bridges, switches, and routers were developed in order to create larger networks. Using these devices makes it possible to build networks that span longer distances, support more computers, and provide increased bandwidth for each system on the network. This chapter examines the functions of these devices and how you can integrate them into your network infrastructure.

Repeaters

As a signal travels over a cable, the natural resistance of the medium causes it to gradually weaken until it is no longer viable. The longer the cable, the weaker the signal gets. This weakening is called *attenuation,* and it is a problem that affects all types of cable to some degree. The effect of attenuation is dependent on the type of cable. Copper cable, for example, is much more prone to attenuation than fiber-optic. This is one reason why fiber-optic cable segments can be much longer than copper ones.

When building a LAN, the standard for the data link–layer protocol you intend to use contains specifications for the types of cable you can use and guidelines for installing them. These guidelines include, among other things, the minimum and maximum lengths for the cables connecting the computers. The cable's attenuation rate is one of the most important factors affecting the maximum cable length. When you have to run a cable across a longer distance than is specified in the standard, you can use a device called a *repeater* to amplify the signal, enabling it to travel greater distances without attenuating to the point of being unreadable by the destination system. In its simplest form, a repeater is an electrical device used on a copper-based network that receives a signal through one cable connection, amplifies it, and transmits it out through another connection.

Repeaters were first used in data networking to expand the length of coaxial cable segments on Ethernet networks. On a coaxial network, such as a Thin or Thick Ethernet LAN, a standalone repeater enables you to extend the maximum bus length past 185 meters (for Thin Ethernet) or 500 meters (for Thick Ethernet). This type of repeater is simply a small box with two BNC connectors on it and a power cable. Using T connectors and terminators as shown in Figure 6-1, you connect two cable segments to the repeater and the repeater to a power source. Signals entering either one of the two connectors are immediately amplified and transmitted out through the other connector. On a modern network, it is rare to see a standalone repeater. In most cases, the repeating function is built into another device, such as a hub or a switch.

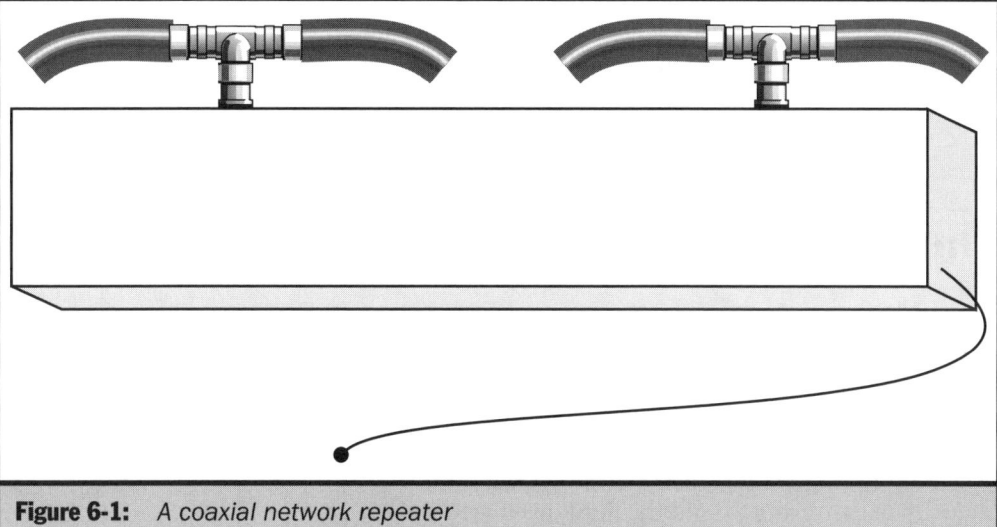

Figure 6-1: *A coaxial network repeater*

Collision Domains and Broadcast Domains

A *collision domain* is a group of computers connected by a network so that if any two computers transmit at the same time, a collision between the transmitted packets occurs, causing the data in the packets to be damaged. This is in contrast to a *broadcast domain*, which is a group of computers networked together in such a way that if one computer generates a broadcast transmission, all of the other computers in the group receive it. These two concepts are the tests used to define the functionality of network connection devices (such as repeaters, hubs, bridges, switches, and routers) and are used repeatedly in this chapter.

Other factors besides attenuation limit the maximum distance a network signal can travel. On an Ethernet network, for example, the first bit of a packet being transmitted by one computer must reach all the other computers on the local network before the last bit is transmitted. Therefore, you cannot extend a network segment without limit by adding multiple repeaters. A 10 Mbps Ethernet network can have up to five cable segments connected by four repeaters. Fast Ethernet networks are more limited, allowing a maximum of only two repeaters.

Because its function is purely electrical, this type of repeater functions at the network's physical layer only. The repeater cannot read the contents of the packets traveling over the network or even know that they are packets. The device simply amplifies the incoming electrical signals and passes them on. Repeaters are also incapable of performing any sort of filtration on the data traveling over the network. As a result, two cable segments joined by a repeater form a single collision domain, and therefore a single network.

Hubs

A *hub* is a device that functions as the cabling nexus for a network that uses the star topology. Each computer has its own cable that connects to the central hub. The responsibility of the hub is to see to it that traffic arriving over any of its ports is propagated out through the other ports. Depending on the network medium, a hub might use electrical circuitry, optical components, or other technologies to disseminate the incoming signal out among the outgoing ports. A fiber-optic hub, for example, actually uses mirrors to split the light impulses.

The hub itself is a box, either freestanding or rack-mounted, with a number of ports to which the cables connect. The ports can be the standard RJ-45 connectors used by twisted-pair networks, ST connectors for fiber-optic cable, or any other type of connector used on a star network. In many cases, hubs also have one or more LEDs for each port that light up to indicate when a device is connected to it, when traffic is passing through the port, or when a collision occurs.

The term *hub* or *concentrator* is used primarily in reference to Ethernet networks; the equivalent device on a Token Ring network is called a *multistation access unit (MAU)*. Other protocols typically use one or the other of these terms, depending on the media access control (MAC) mechanism the protocol uses. The internal functions of hubs and MAUs are very different, but they serve the same basic purpose: to connect a collection of computers and other devices into a single collision domain.

Passive Hubs

Unlike standalone repeaters, which are all essentially the same, many different types of hubs exist with different capabilities. At its simplest, a hub supplies cable connections by passing all the signals entering the device through any port out through all the other ports. This is known as a *passive hub* because it operates only at the physical layer, has no intelligence, and does not amplify or modify the signal in any way. This type of hub was at one time used on ARCnet networks, but it is almost never used on networks today.

Repeating Hubs

The hubs used on today's Ethernet networks propagate the signals they receive through any of their ports out through all of the other ports in the device simultaneously. This creates a shared network medium and joins the networked computers into a single collision and broadcast domain, just as if they were connected to the same cable, as on a coaxial Ethernet network. Ethernet hubs also supply repeating functionality by amplifying the incoming signals as they propagate them to the other ports. In fact, Ethernet hubs are sometimes referred to as *multiport repeaters*. Unlike a passive hub, a repeating (or active) hub requires a power source to boost the signal. The device still operates at the physical layer, however, because it deals only with the raw signals traveling over the cables.

Some hubs go beyond repeating and can repair and retime the signals as well, to synchronize the transmissions through the outgoing ports. These hubs use a technique called *store and forward*, which involves actually reading the contents of the packets to retransmit them over individual ports as needed. A hub with these capabilities can lower the network performance for the systems connected to it because of processing delays. At the same time, packet loss is diminished and the number of collisions reduced.

For a small workgroup network, such as that used in a home or small business, you can purchase a basic 10 Mbps Ethernet repeating hub with up to eight ports for well under $100. For a bit more money, you can get a hub with more ports, or a Fast Ethernet hub that supports 100 Mbps transfers, like the one shown in Figure 6-2. Other small network hubs incorporate modems into their design, enabling you to connect your computers into a LAN and connect the entire LAN to the Internet, using a single device. There is now a large market for small network hubs like this, so there are many products to choose from and prices are competitive. The hubs used on large networks are usually designed to be stacked or mounted in standard 19-inch racks and typically provide many more ports, as well as other features such as manageability. However, the basic functionality of the hub is the same as that of the small models.

An Ethernet hub connects all of your computers into a single collision domain, which is not a problem on a small network. Larger networks consist of multiple network segments connected by other types of devices, such as bridges, switches, or routers. Because an Ethernet hub also functions as a repeater, each of the cables connecting the hub to a computer can be the maximum length allowed by the protocol standard. For Ethernet running on UTP cable, the maximum length is 100 meters.

Using multiple hubs on a single LAN is possible, by connecting them together to form a hierarchical star network, as shown in Figure 6-3. When you do this using standard repeating hubs, all the computers remain in the same collision domain, and you must observe the configuration guidelines for the data link–layer protocol used on the network. Just as with the standalone repeaters discussed earlier in this chapter, the path between any two machines on a 10 Mbps Ethernet network cannot include more than four repeaters (hubs). Fast Ethernet networks typically support only two hubs.

NETWORK HARDWARE

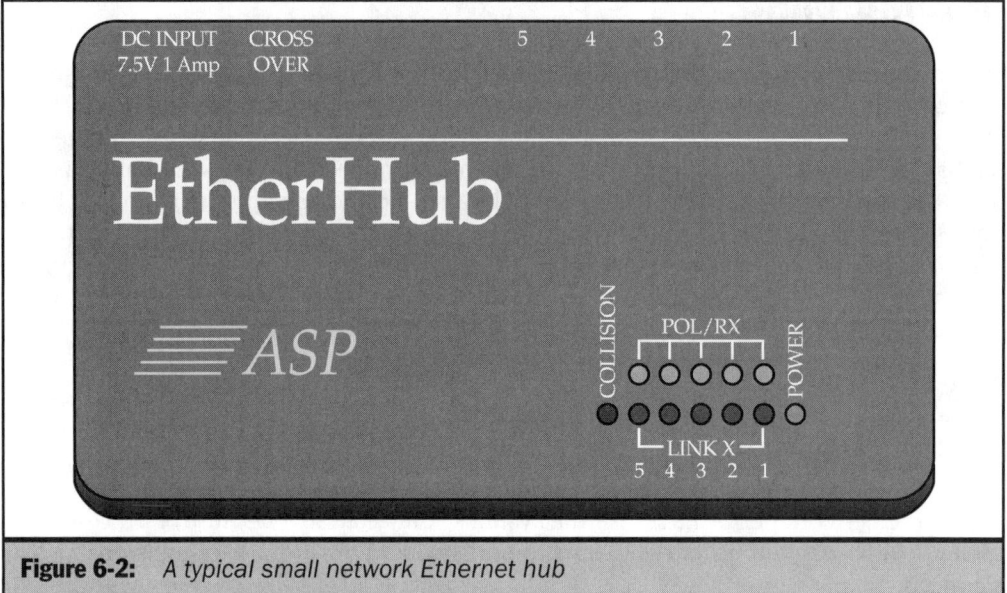

Figure 6-2:   *A typical small network Ethernet hub*

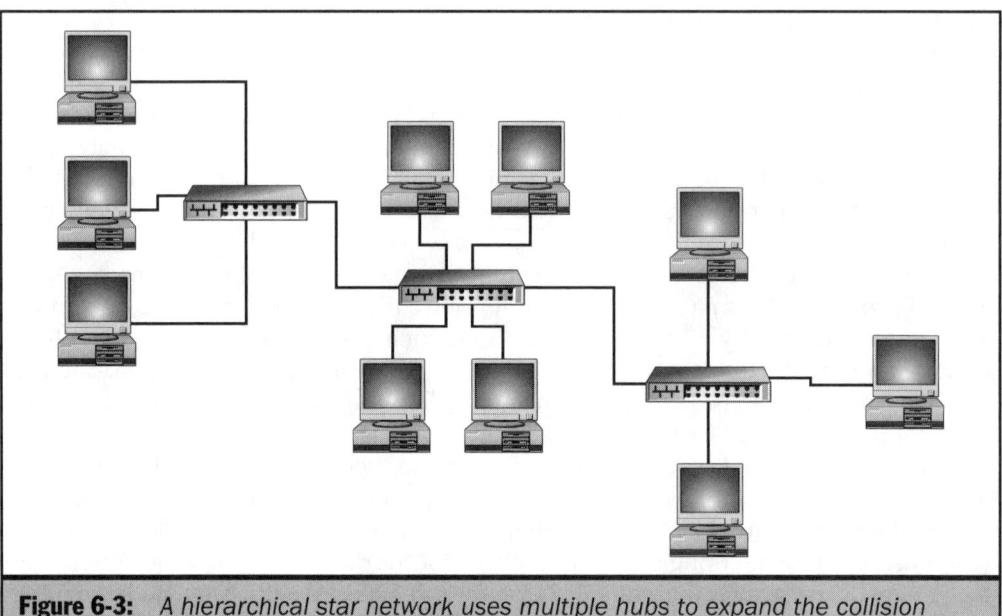

Figure 6-3:   *A hierarchical star network uses multiple hubs to expand the collision domain.*

For example, if you have a small 10Base-T workgroup network that eventually outgrows your existing hub, you can add another one by using a standard UTP cable to connect the uplink port on either one of the hubs to a standard port on the other. All the computers will still be in the same collision domain and they can receive all the packets transmitted over the network. This growth can continue until you have a maximum of four hubs. If your network grows even more, you can replace your hubs with models having a greater number of ports, or you can split the network into two LANs (or collision domains) by using a different type of device, such as a bridge, router, or switch.

 *For more information on the uplink port and the internal functioning of hubs, see "Hub Configurations," later in this chapter.*

Token Ring MAUs

Token Ring networks use hubs as well, although they call them *multistation access units, or MAUs*. While a MAU, to all external appearances, performs the same function as an Ethernet hub, its internal workings are quite different. Instead of passing incoming traffic to all of the other ports at one time, as in an Ethernet hub, a MAU transmits an incoming packet out through each port in turn, one at a time. After transmitting a packet to a workstation, the MAU waits until that packet returns through the same port before it transmits it out the next port. This implements the logical ring topology from which the protocol gets its name.

MAUs contain switches that enable specific ports to be excluded from the ring, in the event of a failure of some kind. This prevents a malfunctioning workstation from disturbing the functionality of the entire ring. MAUs also have ring-in and ring-out ports that you can use to enlarge the ring network by connecting several MAUs together.

 *For more information on MAUs and the other physical layer elements of a Token Ring network, see Chapter 12.*

Intelligent Hubs

Intelligent hubs are units that have some form of integrated management capability. A basic repeating hub is essentially an electrical device that propagates incoming packets to all available ports without discrimination. Intelligent hubs do the same thing, but they also monitor the operation of each port. The management capabilities vary widely between products, but many intelligent hubs use the Simple Network Management Protocol (SNMP) to send information to a centralized network management console, such as Hewlett-Packard's OpenView. Other devices might use a terminal directly connected to the hub or an HTML interface you can access from a Web browser anywhere on the network.

The object of the management capability is to provide the network administrator with a centralized source of information about the hubs and the systems connected to them. This eliminates the need for the staff supporting a large network to go running to each wiring closet looking for the hub or system causing a problem. The management console typically displays a graphical model of the network and alerts the administrator when a problem or failure occurs on any system connected to the hub.

On smaller networks, this capability isn't needed, but when you're managing an enterprise network with hundreds or thousands of nodes, a technology that can tell you exactly which one of the hub ports is malfunctioning can be helpful. The degree of intelligence built into a hub varies greatly with the product. Many hybrid devices are on the market today that have sufficient intelligence to go beyond the definition of a hub and provide bridging, switching, or routing functions as well.

Hub Configurations

Hubs are available in a wide variety of sizes and with many different features, ranging from small, simple devices designed to service a handful of computers to huge rack-mounted affairs for large enterprise networks. The range of hub designs fall into three categories, as follows:

- Standalone hubs
- Stackable hubs
- Modular hubs

Standalone Hubs

A *standalone hub* is a small box about the size of a paperback book that has anywhere from 4 to 16 ports in it. As the name implies, the device is freestanding, has its own power source, and can easily fit on or under a desk. Four- or five-port *minihubs* are good for home and small workgroup networks, or for providing quick, ad hoc expansions to a larger network. Larger units can support more connections and may have LEDs that indicate the presence of a link pulse signal on the connected cable and, possibly, the occurrence of a collision on the network.

Despite the name, a standalone hub usually has some mechanism for connecting with other hubs to expand the network within the same collision domain. The following sections examine how the most common mechanisms are used for this purpose.

The Uplink Port   The cables used on a twisted-pair network are wired *straight through,* meaning that each of the eight pins on the RJ-45 connector on one end of the cable is wired to the corresponding pin on the other end. UTP networks use separate wire pairs within the cable for transmitting and receiving data. For a UTP connection between two computers to function, however, the transmit contacts on each system must be connected to the receive contacts on the other. Therefore, a *crossover* must exist somewhere in the connection and, traditionally, this occurs in the hub, as shown in Figure 6-4. The pins in each of a hub's ports are connected to those of every other port using crossover circuits that transpose the transport data (TD) and receive data (RD) signals. Without this crossover circuit, the transmit contacts on the two systems are connected, as are the receive contacts, preventing any communication from taking place.

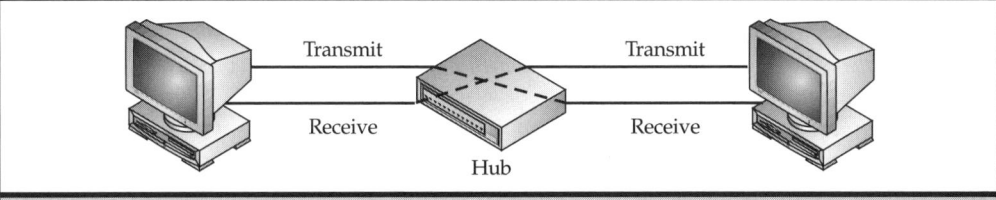

Figure 6-4:   *Ethernet hubs contain a crossover circuit so that the network cables can be wired straight through.*

Note   *For more information on network cabling and signal crossovers, see Chapter 4.*

Many hubs have a port that bypasses the crossover circuit, which you can use to connect to another hub. This port is typically labeled *Uplink* and may or may not have a switch that enables you to specify whether the port should be crossed over or wired straight through. If you have more than one hub on your system, you connect them using the uplink port on one hub only and a standard port on the other. If you connect two hubs using the uplink ports on both devices, the two crossovers would cancel each other out, and the connection between a computer attached to one hub and a computer attached to the other would be the equivalent of a straight-through connection. If a hub does not have an uplink port, you can still connect it to another hub using a standard port and a *crossover cable,* which is a cable that has the transmit pins on each end wired directly to the receive pins on the other end. You typically use the uplink port to connect hubs when they're located some distance away from each other and you want to use the same cable medium throughout the network. When you are evaluating hubs, being aware of just how many hub ports are available for workstation connections is important. A device advertised as an eight-port hub may have seven standard ports and one unswitched uplink port, leaving only seven connections for computers. No matter what the size of the network, purchasing hubs with a few ports more than you need right now—for expansion purposes—is always a good idea.

Backbone Connections   Some standalone hubs have an *attachment unit interface (AUI)* port you can use to create a *backbone,* that is, a separate cable segment that carries the traffic between hubs. This configuration is called a *star-bus topology.* Many inexpensive 10Base-T hubs, for example, have a BNC connector on them that you can use to connect the hub to a Thin Ethernet coaxial cable segment, although the AUI connector may also support Thick Ethernet, fiber-optic cable, or any other medium. Thin Ethernet is rarely, if ever, used now in new network installations, and you can't use the star-bus topology on Fast Ethernet networks, because coaxial cable is limited to speeds of 10 Mbps.

> **Note**   *This is only a basic example of the backbone concept, in which a dedicated cable segment is used to connect other individual segments. In this type of backbone, all of the cable segments are part of the same LAN, which means that all of the connected computers are in one collision domain. Larger networks, and even huge ones like the Internet, use separate backbone networks (often running at high speeds) to connect multiple LANs together using routers, which provides more efficient internetwork traffic paths. For more information on using backbones, see Chapter 9.*

When you have several 10Base-T Ethernet hubs connected in a hierarchical star topology using their uplink ports, each length of cable is a separate segment. Because the Ethernet guidelines allow the path from one system to another to travel across only five segments, connected by four repeaters, you are limited to four hubs on any particular LAN. When you connect the hubs using a Thin Ethernet bus, as shown in Figure 6-5, only one segment is added to the equation. The path between two systems involves, at most, three segments because the data travels over the bus to all the hubs at the same time.

As you expand this type of network further, you may run into another Ethernet limitation not yet mentioned. The bus connecting the hubs is called a *mixing segment* because it has more than two devices connected to it. A segment that connects only two devices, such as the UTP cable connecting hubs through the uplink port, is called a *link segment.* Of the five segments permitted on a 10Base-T LAN, only three of these can be mixing segments. This guideline, stating that you can connect up to five segments using four repeaters and that no more than three of the segments can be mixing segments, is known as the *Ethernet 5-4-3 rule.*

> **Note**   *For more information on Ethernet cabling guidelines, see Chapters 10 and 11.*

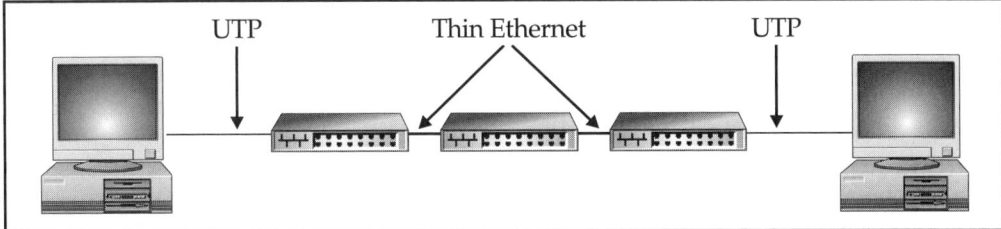

Figure 6-5: *Because the two lengths of Thin Ethernet coaxial cable in this illustration form a single bus, the path between the two workstations consists of three segments, not four.*

Stackable Hubs

As you move up the scale of hub size and complexity (as well as expense), you find units called *stackable hubs* that provide greater expandability. As the name implies, these hubs have cases designed to stack one on top of the other, but this is not the only difference. Unlike standalone hubs, which can be located in different rooms or floors and still connected together, stackable hubs are typically located in a data center or wiring closet and are connected together with short cables.

When you connect stackable hubs together, they form what is functionally a single larger hub. The cables connecting the units do not form separate segments, so you can have more than four hubs interconnected. In addition, these devices can share their capabilities. A single intelligent hub unit can manage its own ports, as well as those of all the other units in the array.

Stackable hubs have their own power supplies and can function independently, thus providing a much more expandable environment than standalone hubs. You can start with a single unit, without incurring the major expense of a chassis (as used by modular hubs), and connect additional units as the network grows.

Modular Hubs

Modular hubs are designed to support the largest networks, and provide the greatest amount of expandability and flexibility. A modular hub consists of a chassis (sometimes called a *card cage*) that is nearly always mounted in a standard 19-inch equipment rack and contains several slots into which you plug individual communications modules. The chassis provides a common power source for all the modules, as well as a backplane that enables them to communicate with each other. The modules contain the ports to which you connect the computer cables. When you plug multiple modules into the chassis, they become, in effect, a single large hub.

Modular hubs nearly always include management capabilities and are extremely flexible because you can insert modules supporting different technologies into the same chassis. By using various modules, you can mix media like 100Base-TX and 100Base-FX in the same hub; mix protocols, such as Ethernet and Token Ring; or insert cards that provide bridging, switching, or routing capabilities. Some modular products include additional fault-tolerance features such as redundant fans, power supplies, and backup battery power, as well as the capability to hot swap modules in the event of a malfunction. As you might imagine, modular hubs are the most expensive type and are intended for large, permanent network installations.

In some cases, the dividing line between stackable and modular hub products becomes rather indistinct. There are stackable hubs with expansion slots that accept modules providing additional ports, management capabilities, or even support for other media, such as a fiber-optic backbone.

Selecting a Hub

When evaluating hubs for your network, planning for future upgrades and expansion is arguably the most important element of your decision. You should always purchase a hub with a few more ports than you need, remembering that you may eventually want to connect printers, as well as workstations, directly to the hub. In addition, you should have a plan for how you are going to expand your network beyond the limits of a single hub.

Planning for Network Growth

For small networks, standalone hubs are sufficient, and you can expand your network by connecting another hub to the uplink port (or to a standard port using a crossover cable). This is also a good solution if you want to place hubs at different locations, as the cable connecting the two hubs can be up to 100 meters long on a UTP Ethernet network. You must, however, be conscious of the limits on the number of repeaters imposed by the Ethernet standards, as well as the maximum cable segment lengths. Four 10Base-T hubs and two Class II 100Base-T hubs are all that are permitted on an Ethernet LAN. If you suspect your network may grow larger than this arrangement can support, you may be better served by purchasing a stackable hub that you can expand later.

Stackable hubs are an excellent choice for mid-sized networks because they can provide much of the flexibility of modular systems with a relatively small initial investment. You can purchase one unit to start and later add more to provide additional ports or more features, such as network management or support for other protocols.

Because stackable hubs connect to form a single unit, you needn't worry about the Ethernet repeater limits unless you have several hub arrays in different locations. But be aware that stackable hub products are not infinitely expandable. Make sure you find out the maximum number of units you can interconnect, as this differs greatly between products.

Stackable hubs use short cables to connect to each other, so all the units must be together, typically in a wiring closet or other safe location. This type of arrangement requires more planning than standalone hubs, which you can deploy at will anywhere on the network where you need additional ports. You should choose a central location for the stacked hubs that is roughly equidistant from all the workstations you want to connect, and make sure the hubs have a reliable source of power. Remember, if the hubs go down, the network goes down, even if the PCs are protected by uninterruptible power supplies (UPSes). Stackable hubs also have the advantage of being easy to move around the office or even to a different site entirely.

Modular hubs require the most planning of all and the greatest initial expense because you have to purchase a chassis, as well as the cards containing the ports that will be mounted in it. Modular hubs are usually permanently mounted in a data center or wiring closet, which means that you also have to consider environmental controls and physical security for the unit's location. Investing in modular hubs is usually a long-term commitment because they are best suited to large networks that require the greatest amount of flexibility and expandability. The assumption is that your network will grow extensively over time and may require a variety of different technologies to keep up with that growth. Consider scenarios such as the possibility of your company merging with another company and having to combine different network types. You would want to buy modular hubs from a reputable vendor that has all the features you can anticipate needing in the next five years and one that you also expect to still be in business in five years.

Planning for Network Upgrades

The other primary concern when it comes to hub purchases today is the question of whether you will be performing a speed upgrade to your network in the near future. For example, if you're currently running 10Base-T, you may want to upgrade at least some of your workstations to Fast Ethernet (100Base-T) at some point.

Ethernet hubs can support 10Base-T, Fast Ethernet, or both. Unlike network interface cards, not every Fast Ethernet hub is capable of running at both 10 Mbps and 100 Mbps speeds. If you opt for single-speed hubs, you need to have separate network segments for the equipment running at each of the two speeds and two NICs in computers that must be accessible to both networks. If you decide to spend a bit more and purchase dual-speed hubs, each port in the hub automatically negotiates the best possible speed with the computer connected to it.

NETWORK HARDWARE

Dual-speed hubs work by maintaining a separate logical segment within the unit for each speed and may use a two-port switch to pass data between the segments (so that only the data destined for the other segment gets passed to that segment). In some stackable hub arrangements, the switch is built into one "master" hub that can service all the other "client" hubs in the stack. This saves you money by making it possible to switch between the segments on the interconnected hubs without having to pay for redundant switching circuitry in each unit.

Dual-speed hubs also simplify the migration path between standard and Fast Ethernet. There is no need for two NICs in servers and other shared systems because the hub provides a path between the two network segments. You also don't have to change any of your workstation NICs when you install the hub. Once the device is in place, your 10Base-T systems continue to function normally. At any subsequent time, you can replace a workstation's 10Base-T NIC with a 100Base-T model. When you boot the workstation, the new NIC connects with the hub at 100 Mbps and is added to the high-speed segment.

Not surprisingly, dual-speed hubs are significantly more expensive than the single-speed variety. Depending on where in the migration process you are when you buy the hubs, you may find sticking with single-speed units more economical. For example, if a large part of your network is already running Fast Ethernet and you have only a few 10 Mbps workstations left, spending a lot of extra money on dual-speed hubs is probably not worth the expense when you could just spend a long weekend upgrading the remaining systems to Fast Ethernet instead. If, however, you are still running 10Base-T throughout the network and your migration to Fast Ethernet is still in the planning stages, dual-speed hubs can be an ideal solution.

Bridges

A *bridge* is another device used to connect LAN cable segments together, but unlike hubs, bridges operate at the data link layer of the OSI model and are selective about the packets that pass through them. Repeaters and hubs are designed to propagate all the network traffic they receive to all of the connected cable segments.

A bridge has two or more network interfaces (complete with their own MAC addresses) with their ports connected to different cable segments and operating in *promiscuous mode*. Promiscuous mode means that the interfaces receive all of the packets transmitted on the connected segments. As each packet enters the bridge, the device reads its destination address in the data link–layer protocol header and, if the packet is destined for a system on another segment, forwards the packet to that segment. If the packet is destined for a system on the segment from which it arrived, the bridge discards the packet, as it has already reached its destination. This process is called *packet filtering*. Packet filtering is one of the fundamental principles used by network connection devices to regulate network traffic. In this case, the packet filtering is occurring at the data link layer, but it can also occur at the network and transport layers.

Just the ability to read the contents of a packet header elevates a bridge above the level of a hub or repeater, both of which deal only with individual signals. However, as with a hub or repeater, the bridge makes no changes in the packet whatsoever and is completely unaware of the contents within the data link–layer frame. In Chapter 2, the protocol operating at the OSI model's data link layer was compared to a postal system, in which each packet is a piece of mail and the data link–layer frame functions as the envelope containing the data generated by the upper layers. To extend that analogy, the bridge is able to read the addresses on the packet envelopes, but it cannot read the letters inside. As a result, you don't have to consider the protocols running at the network layer and above at all when evaluating or installing bridges.

By using packet filtering, the bridge reduces the amount of excess traffic on the network by not propagating packets needlessly. Broadcast messages are forwarded to all of the connected segments, however, making it possible to use protocols that rely on broadcasts, such as NetBEUI, without manual system configuration. Unlike a repeater or hub, however, a bridge does not relay data to the connected segments until it has received the entire packet. (Remember, hubs and repeaters work with signals, while bridges work with packets.) Because of this, two systems on bridged segments can transmit simultaneously without incurring a collision. Thus, a bridge connects network segments in such a way as to keep them in the same broadcast domain, but in different collision domains. The segments are still considered to be part of the same LAN, however.

If, for example, you have a LAN that is experiencing diminished performance due to high levels of traffic, you can split it into two segments by inserting a bridge at the midpoint. This will keep the local traffic generated on each segment local and still permit broadcasts and other traffic intended for the other segment to pass through. On an Ethernet network, reducing traffic in this way also reduces the number of collisions, which further increases the network's efficiency. Bridges also provide the same repeating functions as a hub, enabling you to extend the cable length accordingly.

Bridges are available in three basic types, as follows:

Local   A local bridge provides packet filtering and repeating services for network segments of the same type. This type of device is also called a *MAC-layer bridge* because the data arriving at the bridge only has to travel up the protocol stack as high as the media access control, or MAC, sublayer (that is, the lower of the two sublayers that comprise the data link layer, the other being the logical link control, or LLC, sublayer). This is the simplest type of bridge because it has no need for packet translation or buffering. The device simply propagates the incoming packets to the appropriate ports or discards them.

Translation   A translation bridge provides the same functions as a local bridge, except that it connects segments running at different speeds or using different protocols. You can, for example, use a translation bridge to connect Ethernet to Token

Ring, 10Base-T to 100Base-T, or 100Base-TX to 100Base-T4. In this type of bridge, the incoming packets travel up the protocol stack to the MAC sublayer, where they are stripped of their data link–layer protocol headers and passed to the LLC sublayer. The data is then encapsulated by the appropriate protocol for each of the ports over which the bridge will transmit the outgoing packets. This translation adds a measure of complexity (and expense) to the bridge itself and a delay to the propagation of data over the entire network, but it remains an effective solution for joining disparate networks together into a single broadcast domain.

Remote   A remote bridge connects network segments at different locations, using a wide area network (WAN) link such as a modem or leased line. WAN links are usually slower and more expensive than LAN connections and a bridge conserves their valuable bandwidth by minimizing the amount of traffic passing over the link while giving both segments full access to the entire network. Because of the difference in speed between the local and wide area links, a remote bridge usually has an internal buffer it uses to store the data it receives from the LAN while it is waiting for transmission to the remote site.

On most networks constructed today, bridging has largely been replaced by routing and switching technologies, which provide a more comprehensive service at what has become a competitive price. Routers and switches are covered in more depth later in this chapter.

Transparent Bridging

To filter the packets reaching it effectively, a bridge has to know which systems are located on which network segments, so it can determine which packets to forward and which to discard. The bridge stores this information in an address table that is internal to the unit. Originally, network administrators had to create the address table for a bridge manually, but today's bridges compile the address table automatically, a process called *transparent bridging*.

As soon as a transparent bridge (also known as a *learning bridge*) is connected to the network segments, it begins to compile its address table. By reading the source addresses in the arriving packets and noting the interface over which they arrived, the bridge can build a table of node addresses for each segment connected to it.

To illustrate, picture a network composed of three segments (A, B, and C), all connected to a local bridge, as shown in Figure 6-6. When the bridge is first activated, it receives a packet from Node 1 over the interface to Network A that is destined for Node 2 on Network B. Because the bridge now knows Node 1 is located on Network A, it creates an entry in its table for Network A that contains Node 1's MAC address.

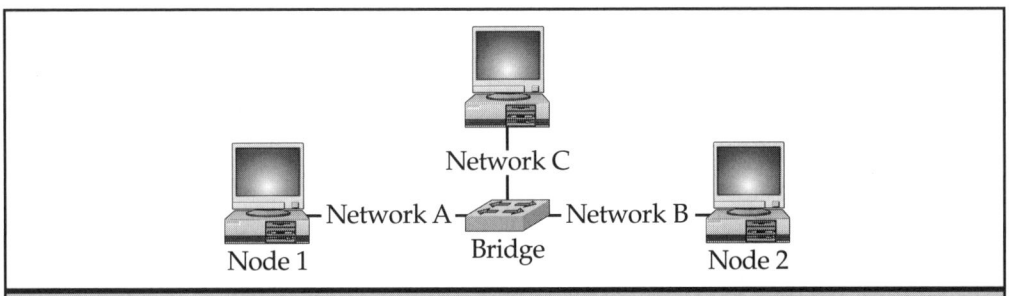

Figure 6-6: *A transparent bridge forwards packets based on address tables that it compiles from previously transmitted packets.*

At this time, the bridge has no information about Node 2 and the segment on which it's located, so it transmits the packet out to Networks B and C—that is, all of the connected segments except the one from which the packet arrived. This is the default behavior of a bridge whenever it receives a packet destined for a system not in its tables. It transmits the packet over all of the other segments to ensure that it reaches its destination.

Once Node 2 receives the packet, it transmits a reply to Node 1. Because Node 2 is located on Network B, its reply packet arrives at the bridge over a different interface. Now the bridge can add an entry to its table for Network B containing Node 2's address. On examining the packet, the bridge looks for the destination address in its tables and discovers that the address belongs to Node 1, on Network A. The bridge then transmits the packet over the interface to Node A only.

From this point on, when any other system on Network A transmits a packet to Node 1, the bridge knows to discard it because there is no need to pass it along to the other segments. However, the bridge still uses those packets to add the transmitting stations to its address table for Network A.

Eventually, the bridge will have address table entries for all of the nodes on the network, and it can direct all of the incoming packets to the appropriate outgoing ports.

One of the statistics often cited in the specifications for bridge products is the number of addresses the device can store in its table. Usually, the amount of storage space is far more than anyone would need, but it's good to make sure the product you purchase can support your network properly.

Bridge Loops

When the segments of a network are connected using bridges, the failure or malfunction of a bridge can be catastrophic. For this reason, administrators often connect network segments with redundant bridges, to ensure that every node can access the entire network, even if a bridge should fail.

In Figure 6-7, three segments are connected by two bridges. If one of the bridges fails, one of the segments is cut off from the rest of the network. To remedy this problem and to provide fault tolerance, you can add a third bridge connecting the two end segments, as shown in Figure 6-8. This way, each system always has two possible paths to the other segments.

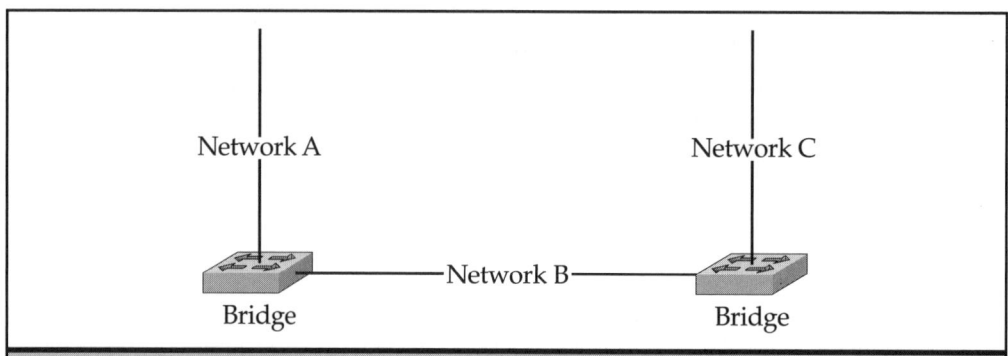

Figure 6-7:   *When each segment is connected to the others using one bridge, a single point of failure is created.*

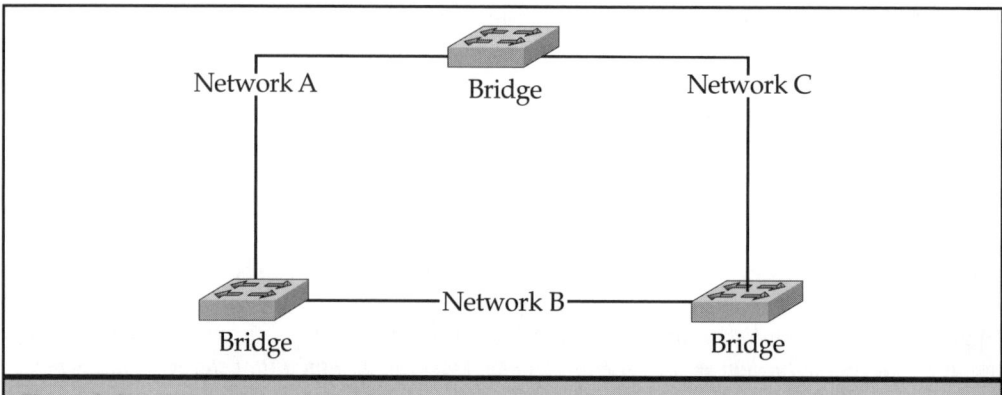

Figure 6-8:   *Connecting each segment to two bridges provides fault tolerance.*

Installing redundant bridges is a good idea, but it also produces what can be a serious problem. When a computer (Node 1) is located on a segment connected to two bridges, as shown in Figure 6-9, both of the bridges will receive the first packet the system transmits and add the machine's address to their tables for that segment, Network A. Both bridges will then transmit the same packet onto the other segment, Network B. As a result, each bridge will then receive the packet forwarded by the other bridge. The packet headers will still show the address of Node 1 as the source, but both bridges will have received the packet over the Network B interface. As a result, the bridges may (or may not) modify their address tables to show Node 1 as being on Network B, not A. If this occurs, any subsequent transmissions from Node 2 on Network B that are directed to Node 1 will be dropped because the bridges think Node 1 is on Network B, when it is, in fact, on A.

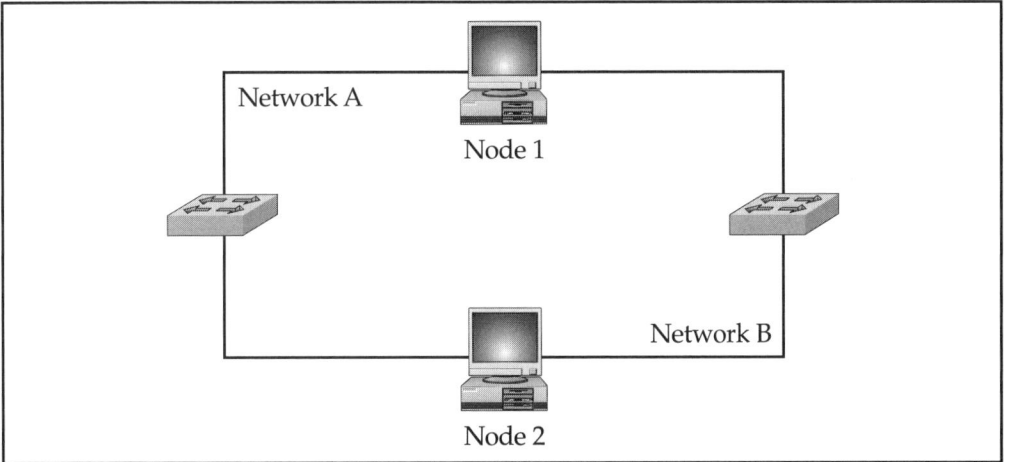

Network A

Node 1

Network B

Node 2

Figure 6-9:   *Redundant bridges provide fault tolerance, but they can also create bridging loops and broadcast storms.*

The result of this occurrence is lost data (because the bridges are improperly dropping frames) and degraded network performance. Eventually, the incorrect entries in the bridges' address tables will expire or be modified, but in the interim, Node 1 is cut off from the systems on the other network segments.

If this problem isn't bad enough, what happens when Node 1 transmits a broadcast message is worse. Both of the bridges forward the packet to Network B, where it is received by the other bridge, which forwards it again. Because bridges always forward broadcast packets without filtering them, multiple copies of the same message circulate endlessly between the two segments, constantly being forwarded by both bridges. This is called a *broadcast storm* and it can effectively prevent all other traffic on the network from reaching its destination.

The Spanning Tree Algorithm

To address the problem of endless loops and broadcast storms on networks with redundant bridging, the Digital Equipment Corporation devised the *spanning tree algorithm (SPA),* which preserves the fault tolerance provided by the additional bridges, while preventing the endless loops. SPA was later revised by the Institute of Electrical and Electronic Engineers (IEEE) and standardized as the 802.1d specification.

The spanning tree algorithm works by selecting one bridge for each network segment that has multiple bridges available. This designated bridge takes care of all of the packet filtering and forwarding tasks for the segment. The others remain idle, but stand ready to take over should the designated bridge fail.

During this selection process, each bridge is assigned a unique identifier (using one of the bridge's MAC addresses, plus a priority value) as is each individual port on each bridge (using the port's MAC address). Each port is also associated with a path cost, which specifies the cost of transmitting a packet onto the LAN using that port. Path costs typically can be specified by an administrator when a reason exists to prefer one port over another, or they can be left to default values.

Once all the components have been identified, the bridge with the lowest identifier becomes the *root bridge* for the entire network. Each of the other bridges then determines which of its ports can reach the root bridge with the lowest cost (called the *root path cost*) and designates it as the root port for that bridge.

Finally, for each network segment, a *designated bridge* is selected, as well as a *designated port* on that bridge. Only the designated port on the designated bridge is permitted to filter and forward the packets for that network segment. The other (redundant) bridges on that segment remain operative—in case the designated bridge should fail—but are inactive until they are needed. Now that only one bridge is operating on each segment, packets can be forwarded without loops forming.

To perform these calculations, bridges must exchange messages among themselves, using a message format defined in the 802.1d standard (see Figure 6-10). These messages are called *bridge protocol data units (BPDUs)* and contain the following fields:

Protocol Identifier (2 bytes)   Always contains the value 0

Version (1 byte)   Always contains the value 0

Message Type (1 byte)   Always contains the value 0

Flags (1 byte)   Contains two 1-bit flags, using the following values:

> **Bit**   Topology change—indicates the message is being sent to signal a change in the network topology
>
> **Bit 2**   Topology change acknowledgment—used to acknowledge receipt of a message with the topology change bit set

Root ID (8 bytes)   Identifies the root bridge by specifying its 2-byte priority value followed by its 6-byte MAC address

Root Path Cost (4 bytes)   Specifies the cost of the path from the bridge sending the BPDU message to the root bridge

Bridge ID (8 bytes)   Identifies the bridge sending the message by specifying its 2-byte priority value followed by its 6-byte MAC address

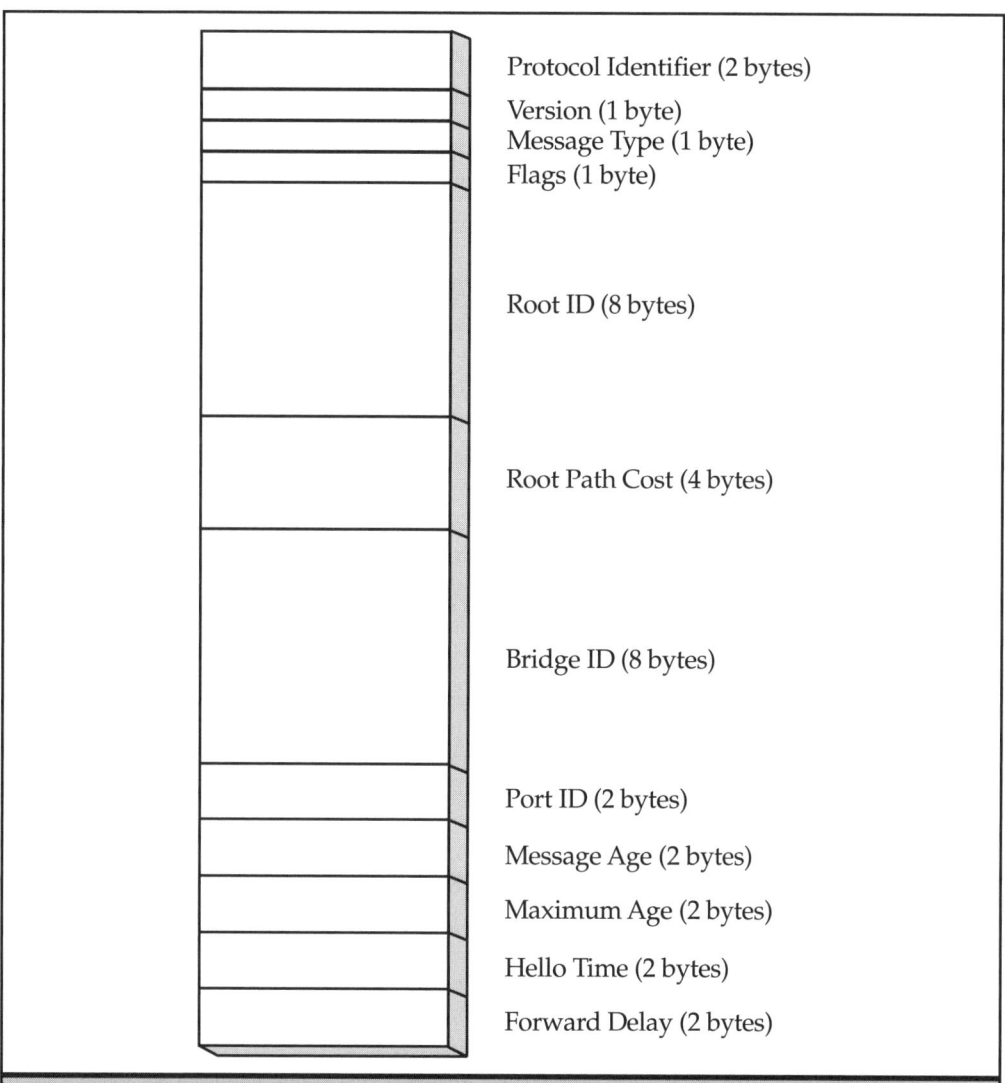

Figure 6-10:   *The format of the bridge protocol data unit message used during bridges' spanning tree algorithm computations*

Port ID (2 bytes)   Identifies the port over which the message is being sent

Message Age (2 bytes)   Specifies the elapsed time since the root bridge transmitted the message that triggered the generation of this message

Maximum Age (2 bytes)   Specifies the age at which this message should be deleted

Hello Time (2 bytes)   Specifies the time interval between root bridge configuration messages

Forward Delay (2 bytes)   Specifies the time interval bridges should wait to complete the spanning tree algorithm after a change in the network topology. Premature state transitions can cause loops to form if all the bridges have not completed the algorithm.

BPDU messages are encapsulated with standard data link–layer protocol frames using the SAP value 01000010 and addressed to the "all bridges" multicast address. Bridges generate the messages autonomously, but they do not forward them to other networks. Therefore, all of the BPDUs exchanged by bridges remain on the network segments to which the bridges are directly connected.

As with the learning process, the spanning tree algorithm begins as soon as the bridges are connected to the network and powered up. Initially, each bridge assumes it will be the root bridge and uses a path cost of 0, but as the bridge receives BPDU messages from the other bridges on the segment, it compares the information in the messages and determines which bridge is better suited to performing the bridging tasks for the segment. The algorithm for making this determination is based on the values for the following criteria, in order:

- Root ID

- Root path cost

- Bridge ID

- Port ID

For each criterion, a lower value is better than a higher one. If a bridge receives a BPDU message with better values than those in its own messages, it stops transmitting BPDUs over the port through which it arrived—in effect, relinquishing its duties to the bridge better suited for the job. The bridge also uses the values in that incoming BPDU to recalculate the fields of the messages it will send through the other ports.

The spanning tree algorithm must complete before the bridges begin forwarding any network traffic. Depending on the bridge implementation, the algorithm might even be completed before the bridges begin compiling their address tables.

Once the spanning tree algorithm has designated a bridge for each network segment, it must also continue to monitor the network so that the process can begin again when a bridge fails or goes offline. All of the bridges on the network store the BPDUs they've received from the other bridges and track their ages. Once a message exceeds the maximum allowable age, it is discarded and the spanning tree message exchanges begin again.

In addition, at periodic intervals (specified by the value of the Hello Time field), the root bridge transmits a new BPDU with a message age value of 0. This causes the other bridges on the network to follow suit. If one of the bridges on a network segment fails to transmit BPDU messages, the other bridges perform the entire algorithm again to select a new designated bridge for the segment. A 4-byte topology change message signals the other bridges to begin the algorithm again. This message consists of only the Protocol Identifier, Version, and Message Type fields from the BPDU format, with the first two fields having a value of 0 and the Message Type field having a value of 128.

Load-Sharing Bridges

In the case of remote bridges that connect network segments using WAN links, it makes no sense to pay for a redundant leased line or other expensive telecommunications link and to allow it to remain unused as a result of the spanning tree algorithm. To address this problem, load-sharing bridges are on the market that can use the backup WAN link to carry data without causing endless loops or broadcast storms.

Source Route Bridging

Source route bridging is an alternative to transparent bridging that was developed by IBM for use on multisegment Token Ring networks and is standardized in IEEE 802.5. On a network that uses transparent bridging, the path a packet takes to a destination on another segment is determined by the designated bridges selected by the spanning tree algorithm. In *source route bridging,* the path to the destination system is determined by the workstation and contained in each individual packet.

To discover the possible routes through the network to a given destination, a Token Ring system transmits an *All Rings Broadcast (ARB)* frame that all the bridges forward to all connected rings. As each bridge processes the frame, it adds its *route designator (RD),* identifying the bridge and port, to the packet. By reading the list of RDs, bridges prevent loops by not sending the packet to the same bridge twice.

If more than one route exists to the destination system, multiple ARBs will arrive there, containing information about the various routes they took. The destination system then transmits a reply to each of the ARBs it receives, using the list of RDs to route the packet back to the sender.

When the original sender of the ARBs receives the responses, it selects one of the routes to the destination as the best one, based on one or more of the following criteria:

- The amount of time required for the explorer frame to return to the sender

- The number of hops between the source and the destination

- The size of the frame the system can use

After selecting one of the routes, the system generates its data packets and includes the routing information in the Token Ring frame header.

The format for the ARB packet and for a data packet containing routing information is the same as a standard IEEE 802.5 frame, except that the first bit of the source address field, called the *routing information indicator (RII)* bit, is set to a value of 1, indicating that the packet contains routing information. The routing information itself, which is nothing more than a list of the bridges the packet will use when traveling through the network, is carried through the *routing information field (RIF)* that appears as part of the information field, just after the frame's source address field (see Figure 6-11).

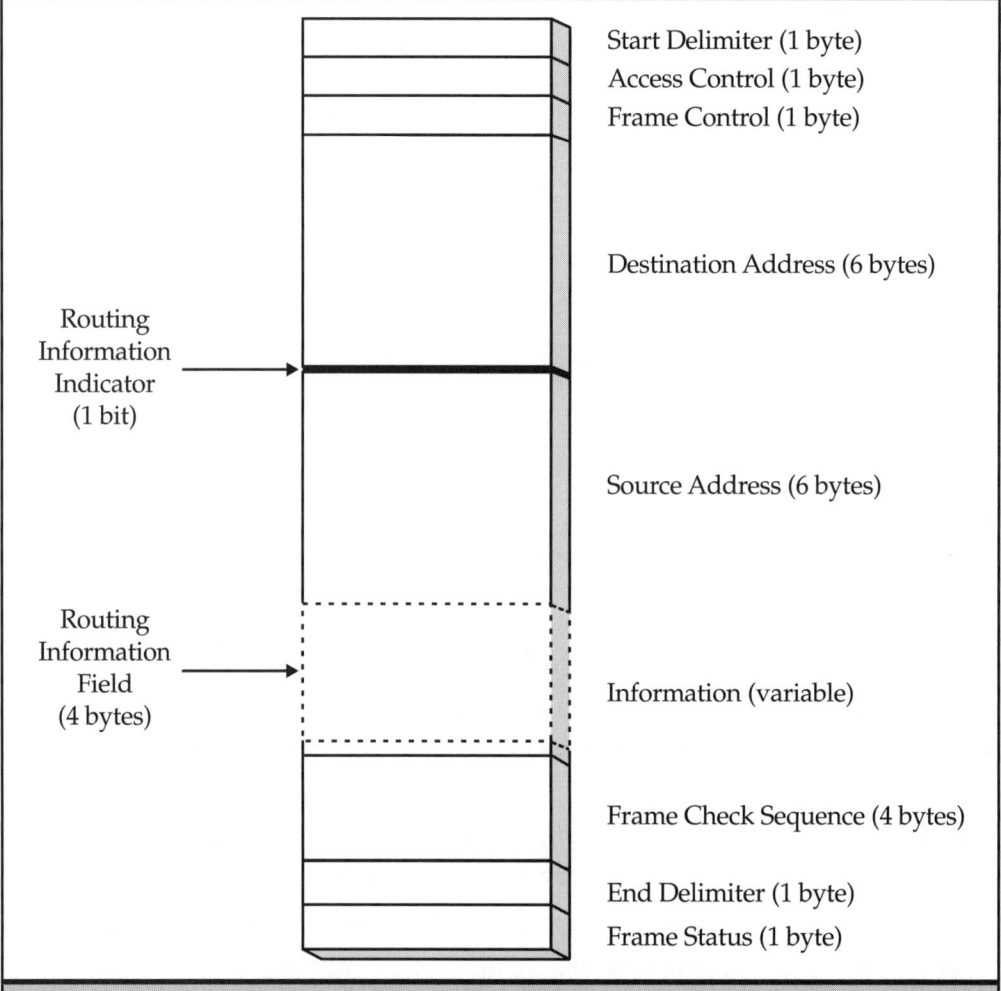

Routing Information Indicator (1 bit)

Routing Information Field (4 bytes)

Start Delimiter (1 byte)
Access Control (1 byte)
Frame Control (1 byte)

Destination Address (6 bytes)

Source Address (6 bytes)

Information (variable)

Frame Check Sequence (4 bytes)

End Delimiter (1 byte)
Frame Status (1 byte)

Figure 6-11: *The routing information indicator and the routing information field used in source bridge routing are carried within standard Token Ring frames.*

The RIF consists of a 2-byte routing control section and a number of 2-byte route designator sections, as shown in Figure 6-12. The routing control section contains the following fields:

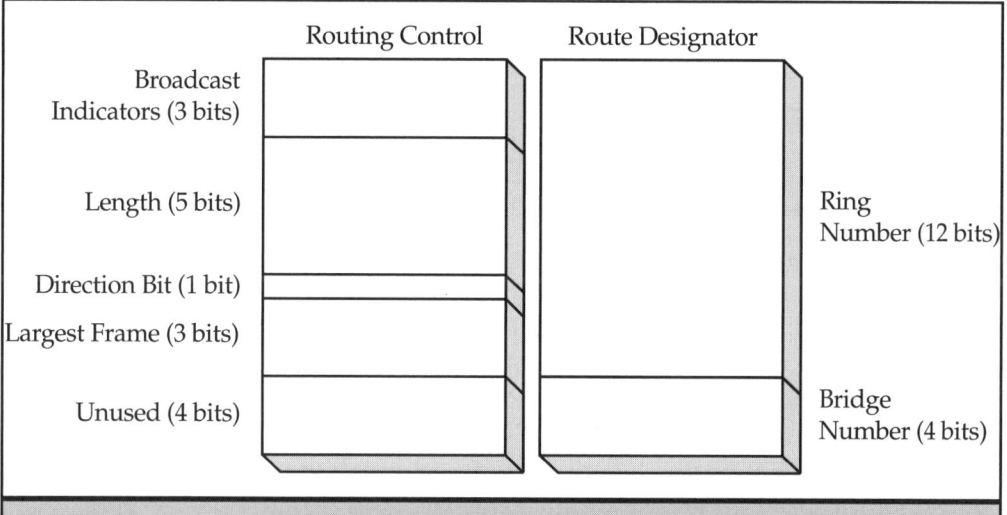

Figure 6-12: *The routing information field identifies the bridges the packet will use when traveling across the network.*

Broadcast indicators (3 bits)   Specifies the type of routing to be used by the frame, according to the following values:

000 – Non-broadcast   Indicates that the packet contains a specific route to the destination in the route designator sections of the RIF field.

100 – All routes broadcast   Indicates that the packet should be routed through all the bridges on the network (without traversing the same bridge twice) and that each bridge should add a route designator section to the RIF field identifying the bridge and the port onto which it is being forwarded.

110 – Single route broadcast   Indicates that the packet should be routed only through the bridges designated by the spanning tree algorithm and that each bridge should add a route designator section to the RIF field identifying the bridge and the port onto which it is being forwarded.

Length (5 bits)   Indicates the total length of the RIF field, from 2 to 30 bytes.

Direction bit (1 bit)   Specifies the direction in which the packet is traveling. The value of this bit indicates whether the transmitting node should read the route designator sections in the RIF field from left to right (0) or right to left (1).

Largest frame (3 bits)   Indicates the largest frame size that can be accommodated by the route, called the *maximum transfer unit (MTU)*. Initially set by the transmitting system, a bridge lowers this value if it forwards the packet onto a segment that only supports smaller frames. The permitted values are as follows:

000   Indicates a MAC MTU of 552 bytes

001   Indicates a MAC MTU of 1,064 bytes

010   Indicates a MAC MTU of 2,088 bytes

011   Indicates a MAC MTU of 4,136 bytes

100   Indicates a MAC MTU of 8,232 bytes

Unused (4 bits)

The IBM standard for source route bridging originally specified a maximum of 8 route designator sections in a single packet, but the IEEE 802.5 standard allows up to 14. Most bridge manufacturers have adhered to the IBM standard; however, newer IBM bridge implementations support up to 14 RDs as well. Each of the route designator sections in the RIF consist of the following two fields:

Ring number (12 bits)   Uniquely identifies the network segment (ring)

Bridge number (4 bits)   Identifies a specific bridge on the network, using a value that only has to be unique among the bridges connected to the network segment (ring)

Source route bridging is a relatively inefficient method because it relies heavily on broadcast transmissions that are propagated throughout all the segments on the network. Each workstation must maintain its own routing information to each of the systems with which it communicates. This can result in a large number of ARB frames being processed by a destination system before it even sees the first byte of application data.

Bridging Ethernet and Token Ring Networks

Generally speaking, Ethernet networks use transparent bridging and Token Ring networks use source route bridging. So what happens when you want to connect an Ethernet segment to a Token Ring using a bridge? The answer is complicated, both because the task presents a number of significant obstacles and because a well-defined standard providing a solution does not yet exist.

Some of the fundamental incompatibilities of the two data link–layer protocols are as follows:

Bit ordering   Ethernet systems consider the first bit of a MAC address to be the low-order bit, while Token Ring systems treat the first bit as the high-order bit.

MTU sizes   Ethernet frames have a maximum transfer unit size of 1,500 bytes, while Token Ring frames can be much larger. Bridges are not capable of fragmenting packets for transfer over a segment with a lower MTU and then reassembling them at the destination, as routers are. A too-large packet arriving at a bridge to a segment with a smaller MTU can only be discarded.

Exclusive Token Ring features   Token Ring networks use frame status bits, priority indicators, and other features that have no equivalent in Ethernet.

In addition, the two bridging methods have their own incompatibilities. Transparent bridges neither understand the special function of the ARB messages used in source route bridging nor can they make use of the RIF field in Token Ring packets. Conversely, source route bridges do not understand the spanning tree algorithm messages generated by transparent bridges and they do not know what to do when they receive frames with no routing information.

Two primary methods exist for overcoming these incompatibilities, neither of which is an ideal solution. These methods are:

- Translational bridging

- Source route transparent bridging

Translational Bridging

In *translational bridging*, a special bridge translates the data link–layer frames between the Ethernet and Token Ring formats. No standard at all exists for this process, so the methods used by individual product manufacturers can vary widely. Some compromise is needed in the translation process because no way exists to implement all the features fully in each of the protocols and to bridge those features to its counterpart. Some of the techniques used in various translational bridges to overcome the incompatibilities are described in the following paragraphs.

One of the basic functions of the bridge is to map the fields of the Ethernet frame onto the Token Ring frame and vice versa. The bridge reverses the bit order of the source and destination addresses for the packets passing between the segments and may or may not take action based on the values of a Token Ring packet's frame status, priority, reservation, and monitor bits. Bridges may simply discard these bits when translating from Token Ring to Ethernet and set predetermined values for them when translating from Ethernet to Token Ring.

To deal with the different MTU sizes of the network segments, a translation bridge can set the largest frame value in the Token Ring packet's RIF field to the MTU for the Ethernet network (1,500 bytes). As long as the Token Ring implementations on the workstations read this field and adjust their frame sizes accordingly, no problem should occur, but any frames larger than the MTU on the Ethernet segments will be dropped by the bridge connecting the two networks.

The biggest difference between the two types of bridging is that, on Ethernet networks, the routing information is stored in the bridges, while on Token Ring networks, it's stored at the workstations. For the translational bridge to support both network types, it must appear as a transparent bridge to the Ethernet side and a source route bridge to the Token Ring side.

To the Token Ring network, the translational bridge has a ring number and bridge number, just like a standard source route bridge. The ring number, however, represents the entire Ethernet domain, not just the segment connected to the bridge. As packets from the Token Ring network pass through the bridge, the information from their RIF fields is removed and cached in the bridge. From that point on, standard transparent bridging gets the packets to their destinations on the Ethernet network.

When a packet generated by an Ethernet workstation is destined for a system on the Token Ring network, the translational bridge looks up the system in its cache of RIF information and adds an RIF field to the packet containing a route to the network, if possible. If no route is available in the cache or if the packet is a broadcast or multicast, the bridge transmits it as a single-route broadcast.

Source Route Transparent Bridging

IBM has also come up with a proposed standard that combines the two primary bridging technologies, called *source route transparent (SRT)* bridging. This technology is standardized in Appendix C of the IEEE 802.1d document. SRT bridges can forward packets originating on either source route bridging or transparent bridging networks, using a spanning tree algorithm common to both. The standard spanning tree algorithm used by Token Ring networks for single-route broadcast messages is incompatible with the algorithm used by Ethernet, as defined in the 802.1d specification. This appendix reconciles the two.

SRT bridges use the value of the RII bit to determine whether a packet contains RIF information and, consequently, whether it should use source route or transparent bridging. The mixing of the two technologies is not perfect, however, and network administrators may find it easier to connect Ethernet and Token Ring segments with a switch or a router rather than either a translational or SRT bridge.

Routers

In the previous sections, you learned how repeaters, hubs, and bridges can connect network segments at the physical and data link layers of the OSI model, creating a larger LAN with a single collision domain. The next step up in the network expansion process is to connect two completely separate LANs together at the network layer. This is the job of a router. Routers are more selective than bridges in the traffic they pass between the networks, and they are capable of intelligently selecting the most efficient path to a specific destination. Because they function at the network layer, routers can also connect dissimilar networks. You can, for example, connect an Ethernet network to a Token Ring network because packets entering a router are stripped of their data link–layer protocol headers as they pass up the protocol stack to the network layer. This leaves a *protocol data unit (PDU)* encapsulated using whatever network-layer protocol is running on the computer (see Figure 6-13). After processing, the router then encapsulates the PDU in a new data link–layer header using whatever protocol is running on the other network to which the router is connected.

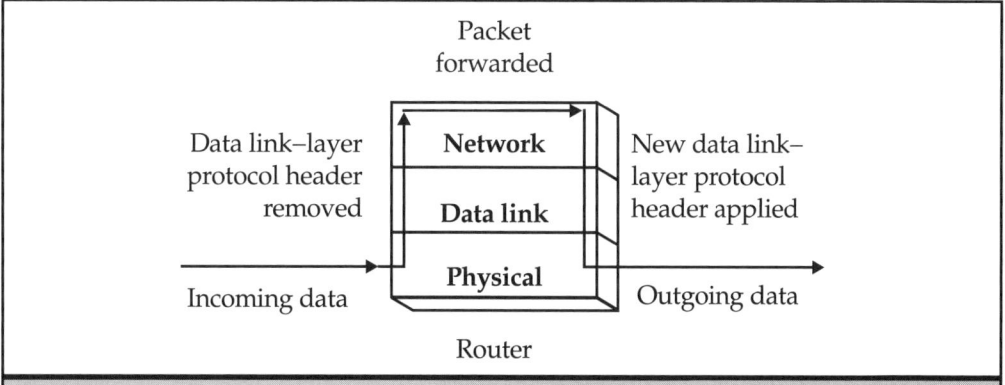

Figure 6-13: *Routers can connect networks of different types because they strip the data link–layer protocol header from a packet before processing and apply a new one before transmitting it.*

The general impression most people have of routers is that they are expensive, dedicated hardware devices used in large enterprise network installations. While this is certainly true in many cases, routing can also operate on a much smaller scale. If, for example, you use your home computer to dial into your system at work and access resources on the office network, your work computer is functioning as a router. In the same way, if you share an Internet connection with systems on a LAN, the machine connected to the Internet is a router. A router, therefore, can be either a hardware or a software entity, and it can range from the simple to the extraordinarily complex.

Routers are protocol specific; they must support the network-layer protocol used by each packet. By far, the most common network-layer protocol in use today is the Internet Protocol (IP), which is the basis for the Internet and for most private networks. In most cases, a reference to a router means an IP router. Some private networks, however, use Novell's *Internetwork Packet Exchange (IPX)* protocol at the network layer. A computer that is connected to two or more networks is said to be a *multihomed system*.

Novell NetWare servers with two or more network interface cards (NICs) installed have always been able to function as IPX routers, and now the product includes multiprotocol routing software that supports IP as well. Multihomed Windows systems can function as routers, too. Windows XP, 2000, and NT server systems also have multiprotocol routing capabilities that support IP and IPX. Windows 95, 98, Me, and XP workstations can route IPX by default, but not IP. However, installing the Internet Connection Sharing (ICS) feature now included in several versions of Windows provides IP routing services between a LAN and a dial-up Internet connection. The NetBEUI protocol, strictly speaking, is not routable, but Windows systems all support dial-up network access using NetBEUI.

Most of the routers used on large networks, though, are standalone devices that are essentially computers dedicated to routing functions. Routers come in various sizes, from small units that connect a workgroup network to a backbone to large, modular, rack-mounted devices costing well into six figures. However, while routers vary in their capabilities, such as the number of networks to which they connect, the protocols they support, and the amount of traffic they can handle, their basic functions are essentially the same.

Router Applications

Although the primary function of a router is to connect networks together and pass traffic between them, routers can fulfill several different roles in network designs. The type of router used for a specific function determines its size, cost, and capabilities. The simplest type of routing architecture is when a LAN must be connected to another LAN some distance away, using a *wide area network* (*WAN*) connection. A branch office for a large corporation, for example, might have a WAN connection to the corporate headquarters in another city (see Figure 6-14).

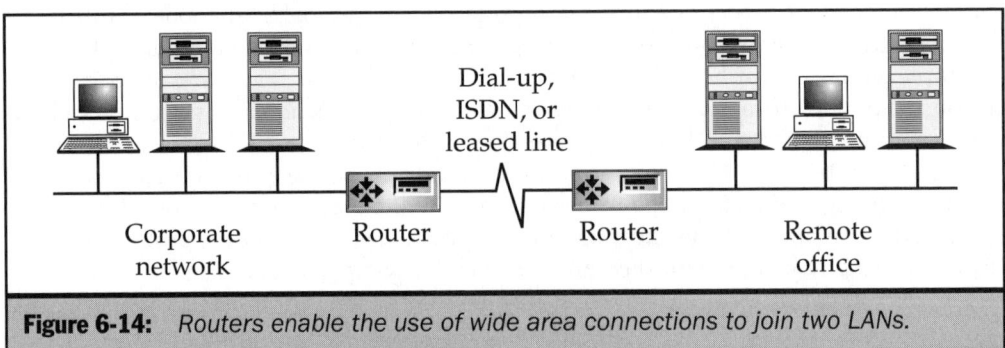

Figure 6-14:   *Routers enable the use of wide area connections to join two LANs.*

To make communications between the networks in the two offices possible, each must connect its LAN to a router, and the two routers are linked by the WAN connection. The WAN connection may take the form of a leased telephone line, an ISDN or DSL connection, or even a dial-up modem connection. The technology used to connect the two networks is irrelevant, as long as the routers in both offices are connected. Routers are required in this example because the LAN and WAN technologies are fundamentally incompatible. You can't run an Ethernet connection between two cities nor can you use leased telephone lines to connect each workstation to the file server in the next room.

In a slightly more complicated arrangement, a site with a larger internetwork may have several LANs, each of which is connected to a backbone network using a router (see Figure 6-15). Here, routers are needed because one single LAN may be unable to support the number of workstations required. In addition, the individual LANs may be located in other parts of a building or in separate buildings on the same campus, and may require a different type of network to connect them. Connections between campus buildings, for example, require a network medium that is suitable for outdoor use, such as fiber-optic cable, while the LANs in each building can use more inexpensive copper cabling. Routers are available that can connect these different network types, no matter what protocols they use.

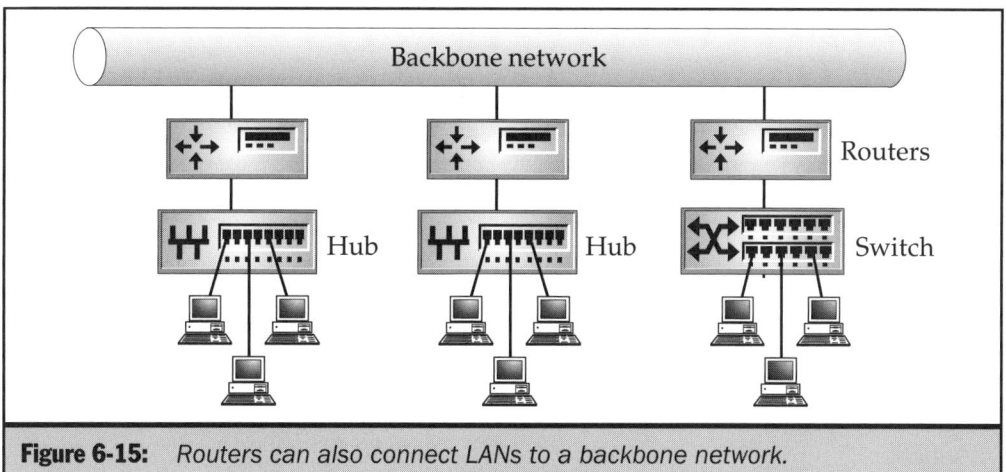

Figure 6-15:   *Routers can also connect LANs to a backbone network.*

These two examples of router use are often combined. A large corporate internetwork using a backbone to connect multiple LANs will almost certainly want to be connected to the Internet. This means that another router is needed to support some type of WAN connection to an Internet service provider (ISP). Users anywhere on the corporate network can then access Internet services.

Both of these scenarios use routers to connect a relatively small number of networks, and they are dwarfed by the Internet, which is a routed internetwork comprised of thousands of networks all over the world. To make it possible for packets to travel across this maze of routers with reasonable efficiency, a hierarchy of routers leads from smaller, local ISPs to regional providers, which in turn get their service from large national services (see Figure 6-16). Traffic originating from a system using a small ISP travels up through this virtual tree to one of the main backbones, across the upper levels of the network, and back down again to the destination.

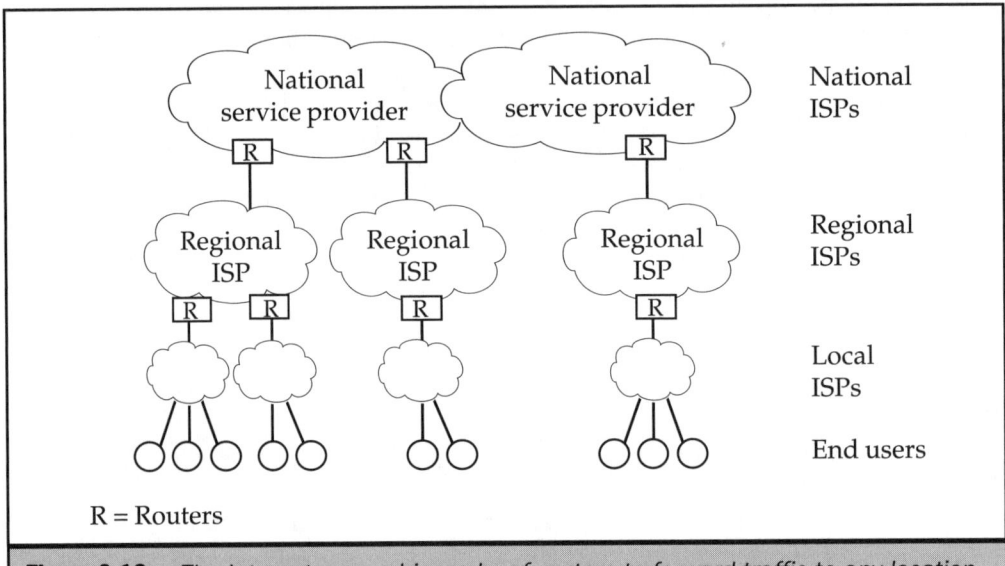

Figure 6-16: *The Internet uses a hierarchy of routers to forward traffic to any location.*

You can see the route that packets take from your computer through the Internet to a specific destination by using the traceroute utility. Called *traceroute* on Unix systems and *Tracert.exe* on Windows systems, this command-line utility takes the IP address or DNS name you specify and uses Internet Control Message Protocol (ICMP) messages to display the names and addresses of all the intermediate routers on the path to the destination. A typical traceroute display (here generated by a Windows 2000 system) appears as follows:

```
Tracing route to zacker.com [192.41.15.74] over a maximum of 30 hops:

  1    213 ms    226 ms    230 ms   vty254.as.wcom.net [206.175.104.254]

  2    212 ms    208 ms    205 ms   fas2.wan.wcom.net [209.154.35.35]
```

```
 3    250 ms    263 ms    219 ms   205.156.214.145

 4    242 ms    218 ms    214 ms   atm1.wan.wcom.net [205.156.223.134]

 5    290 ms    269 ms    263 ms   fdd4.wan.wcom.net [205.156.223.68]

 6    238 ms    369 ms    251 ms   pos3.wan.wcom.net [205.156.223.98]

 7    392 ms    370 ms    448 ms   f0.iad0.verio.net [192.41.177.121]

 8    326 ms      *       239 ms   iad0.iad3.verio.net [129.250.2.178]

 9    498 ms    341 ms      *      iad3.dfw2.verio.net [129.250.2.209]

10      *       342 ms    289 ms   dfw2.dfw3.verio.net [129.250.3.74]

11    327 ms    389 ms    359 ms   dfw3.pvu1.verio.net [129.250.2.41]

12    360 ms    376 ms    355 ms   vwhpvu1.verio.net [129.250.16.118]

13    372 ms    379 ms    325 ms   zacker.com [192.41.15.74]

Trace complete.
```

Note *For more information on traceroute and how it works, see Chapter 31.*

Router Functions

The basic function of a router is to evaluate each packet arriving on one of the networks to which it is connected and send it on to its destination through another network. The goal is for the router to select the network that provides the best path to the destination for each packet. A packet can pass through several different routers on the way to its destination. Each router on a packet's path is referred to as a *hop,* and the object is to get the packet where it's going with the smallest number of hops. On a private network, a packet may need three or four (or more) hops to get to its destination. On the Internet, a packet can easily pass through 20 or more routers along its path.

A router, by definition, is connected to two or more networks. The router has direct knowledge about those networks for the protocols that it supports. If, for example, a workstation on Network 1 (see Figure 6-17) transmits a packet to a system on Network 2, the router connecting Networks 1, 2, and 3 can directly determine which of the two networks (2 or 3) contains the destination system and forward the packet appropriately.

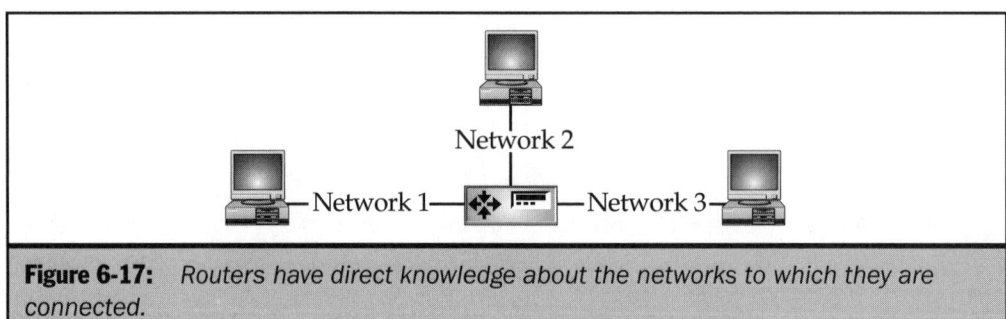

Figure 6-17:   *Routers have direct knowledge about the networks to which they are connected.*

Routing Tables

The router forwards packets by maintaining a list of networks and hosts, called a *routing table*. For computers to communicate over a network, each machine must have its own address. In addition to identifying the specific computer, however, its address must also identify the network on which it's located. On TCP/IP networks, for example, the standard 32-bit IP address consists of a network identifier and a host identifier. A routing table consists of entries that contain the network identifier for each connected network (or in some cases the network and host identifiers for specific computers). When the router receives a packet addressed to a workstation on Network 3, it looks at the network identifier in the packet's destination address, compares it to the routing table, and forwards it to the network with the same identifier.

This is a rather simple task, as long as the router is connected to all of the LANs on the internetwork. When an internetwork is larger and uses multiple routers, however, no single router has direct knowledge of all of the LANs. In Figure 6-18, Router A is connected to Networks 1, 2, and 3 as before, and has the identifiers for those networks in its routing table, but it has no direct knowledge of Network 4, which is connected using another router.

How then does Router A know where to send packets that are addressed to a workstation on a distant network? The answer is that routers maintain information in their routing tables about other networks besides those to which they are directly attached. A routing table may contain information about many different networks all over the enterprise. On a private internetwork, it is not uncommon for every router to have entries for all of the connected networks. On the Internet, however, there are so many networks and so many routers, that no single routing table can contain all of them and function efficiently. Thus, a router connected to the Internet sends packets to another router that it thinks has better information about the network to which the packet is ultimately destined.

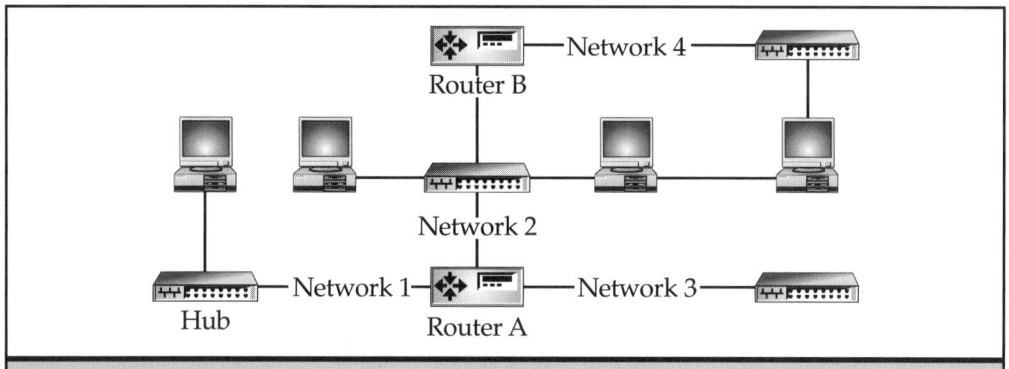

Figure 6-18:   *Router A has no direct knowledge of Network 4, because it is connected to a different router.*

Windows Routing Tables   Every computer on a TCP/IP network has a routing table, even if it is connected to only one network. At the very least, the routing table identifies the system's default gateway and instructs it how to handle traffic sent to the local network and the loopback network address (127.0.0.0). A typical routing table for a Windows system appears as follows:

Network Address	Netmask	Gateway Address	Interface	Metric
0.0.0.0	0.0.0.0	192.168.2.100	192.168.2.5	1
127.0.0.0	255.0.0.0	127.0.0.1	127.0.0.1	1
192.168.2.0	255.255.255.0	192.168.2.5	192.168.2.5	1
192.168.2.5	255.255.255.255	127.0.0.1	127.0.0.1	1
192.168.2.255	255.255.255.255	192.168.2.5	192.168.2.5	1
224.0.0.0	224.0.0.0	192.168.2.5	192.168.2.5	1
255.255.255.255	255.255.255.255	192.168.2.5	0.0.0.0	1

To display the routing table on a Windows system and on most Unix systems, type **netstat -nr** *at the command prompt.*

The entries in the table run horizontally. The function of the information in each column is as follows:

Network Address   Specifies the network address for which routing information is to be provided. While most entries have network addresses in this field, it's also possible to supply routing information for a specific host address. This is called a *host route*.

Netmask   Specifies the subnet mask used to determine which bits of the network address function as the network identifier.

Gateway Address   Specifies the IP address of the gateway (router) the system should use to send packets to the network address. When the entry is for a network to which the system is directly attached, this field contains the address of the system's network interface.

Interface   Specifies the IP address of the network interface the system should use to send traffic to the gateway address.

Metric   Specifies the distance between the system and the destination network, usually in terms of the number of hops needed for traffic to reach the network address.

Note   *TCP/IP and Internet terminology often use the term "gateway" synonymously with "router." In general networking parlance, a gateway is an application-layer interface between networks that involves some form of high-level protocol translation, such as an e-mail gateway or a gateway between a LAN and a mainframe. When a Windows system refers to its "default gateway," however, it is referring to a standard router, operating at the network layer.*

The system using this routing table has only one NIC, with the IP address 192.168.2.5. You can tell this from the fourth entry, which contains a host route directing that address to the loopback adapter (127.0.0.1). The system is connected to a LAN with 192.168.2 as its network address. You can see from the third entry in the table that all traffic destined for this network uses the system's own NIC as the gateway because no router is needed to access systems on the local network. This is called a *direct route* because the destination address in the IP header represents the same machine as the destination address in the data link–layer protocol header. The 0.0.0.0 entry represents the system's default gateway, which it uses for traffic addressed to networks not listed in the table. This entry instructs the system to send this traffic over the NIC to 192.168.2.100, which is the IP address of the router that connects the network to the Internet.

If the system was connected to the Internet using a modem, both the Gateway and the Interface fields of the 0.0.0.0 entry would contain the address assigned to the modem connection by the server on the ISP's network. In this case, the modem functions as a network interface, just like a NIC, and has its own IP address.

The last three entries in the table define routes for broadcast and multicast messages. The "Assigned Numbers" RFC contains Class D network addresses that have been assigned to specific multicast groups, all of which fall in the 224.0.0.0 network. 255.255.255.255 is the standard broadcast address. The 192.168.2.255 entry is for broadcasts to the local network.

For more information on TCP/IP concepts such as IP addresses, the loopback adapter, multicasting, and Requests for Comments (RFCs), see Chapter 13.

Unix Routing Tables   Other operating systems display the routing table slightly differently and may include other information, but the basic elements and functions of the table are the same. The following is a sample routing table from a System V–based Unix system:

```
Destination       Gateway           Flags   Refcnt    Use       Interface

127.0.0.1         127.0.0.1         UH      1         298       lo0

default           192.168.2.76      UG      2         50360     le0

192.168.2.0       192.168.2.21      U       40        111379    le0

192.168.4.0       192.168.2.1       UG      4         5678      le0

192.168.5.0       192.168.2.1       UG      10        8765      le0

192.168.3.0       192.168.2.1       UG      2         1187      le0
```

The functions of the columns in this table are as follows:

Destination   Specifies the address of the network or host for which routing information is being provided

Gateway   Specifies the address of the gateway (router) the system should use to send traffic to the specified network or host

Flags   Specify the special characteristics of each routing table entry, using the following values:

   U   Indicates the route is up and functioning

   H   Indicates this is a route to a host rather than a network

   G   Indicates the route uses a gateway to reach the specified network address (as opposed to being directly connected to that network)

   D   Indicates the entry was added to the table as a result of an ICMP Redirect message

NETWORK HARDWARE

Refcnt   Specifies the number of times the system has used the route to connect to another system

Use   Specifies the number of packets transmitted by the system using this route

Interface   Identifies the network interface in the computer that the system should use to access the specified gateway

Thus, in this table, you can see that the system is connected directly to the 192.168.2.0 network because the entry does not have a G flag, which indicates that the system needs a gateway to access the network. The value in the Gateway field is therefore the IP address of the computer's own network interface. The last three entries in the table are for other networks at the same site, which are accessible through the same router (192.168.2.1). The default entry specifies the address of a different router (in this case, one that provides access to the Internet) for all packets other than those destined for the networks listed elsewhere in the table.

Therefore, to return to the example network shown in Figure 6-18, Router A has entries in its routing table for all of the LANs in the internetwork that specify how it should transmit packets to each of those networks. The entries for the networks to which the router is directly connected specify the interface that the router should use to access those networks, and the entries for distant networks specify the address of another router. When packets reach the specified router, the same process occurs again and the data may be transmitted to still another router. On the Internet, this process may be repeated dozens of times. No one router knows the complete path that a packet will take from source to destination; each one is only responsible for the next hop. In fact, when a file transfer or other operation consists of multiple packets, constantly changing network conditions may cause the individual packets to take different routes to the same destination.

Routing Table Parsing   Whether a system is functioning as a router or not, the responsibility of a network-layer protocol like IP is to determine where each packet should be transmitted next. The IP header in each packet contains the address of the system that is to be its ultimate destination, but before passing each packet down to the data link–layer protocol, IP uses the routing table to determine what the data link–layer destination address should be for the packet's next hop. This is because a data link–layer protocol like Ethernet can only address a packet to a system on the local network, which may or may not be its final destination. To make this determination, IP reads the destination address for each packet it processes from the IP header and searches for a matching entry in the routing table, using the following procedure:

1. IP first scans the routing table looking for a host route that exactly matches the destination IP address in the packet. If one exists, the packet is transmitted to the gateway specified in the routing table entry.

2. If no matching host route exists, IP uses the subnet mask to determine the network address for the packet and scans the routing table for an entry that

matches that address. If IP finds a match, the packet is transmitted either to the specified gateway (if the system is not directly connected to the destination network) or out the specified network interface (if the destination is on the local network).

3. If no matching network address is in the routing table, IP scans for a default (or 0.0.0.0) route and transmits the packet to the specified gateway.

4. If no default route is in the table, IP returns a destination unreachable message to the source of the packet (either the application that generated it or the system that transmitted it).

Static and Dynamic Routing

The next logical question concerning the routing process is, how do the entries get into the routing table? A system can generate entries for the default gateway, the local network, and the broadcast and multicast addresses because it possesses all of the information needed to create them. For networks to which the router is not directly connected, however, routing table entries must be created by an outside process. The two basic methods for creating entries in the routing table are called *static routing*, which is the manual creation of entries, and *dynamic routing*, which uses an external protocol to gather information about the network.

On a relatively small, stable network, static routing is a practical alternative because you only have to create the entries in your routers' tables once. Manually configuring the routing table on workstations isn't necessary because they typically have only one network interface and can access the entire network through one default gateway. Routers, however, have multiple network interfaces and usually have access to multiple gateways. They must, therefore, know which route to use when trying to transmit to a specific network.

To create static entries in a computer's routing table, you use a program supplied with the operating system. The standard tool for this on Unix and Windows systems is a character-based utility called *route* (in Unix) or *Route.exe* (in Windows). To create a new entry in the routing table on a Windows computer, for example, you use a command like the following:

```
ROUTE ADD 192.168.5.0 MASK 255.255.255.0  192.168.2.1 METRIC 2
```

This command informs the system that to reach a network with the address 192.168.5.0, the system must send packets to a gateway (router) with the address 192.168.2.1, and that the destination network is two hops away.

Note   *For more information on using route and Route.exe, see Chapter 31.*

In some cases, graphical utilities are available that can perform the same task. A Windows 2000 Server system with the Routing and Remote Access Server service running, for example, enables you to create static routes using the interface shown in Figure 6-19. On a Novell NetWare server with the TCP/IP protocol installed, you can use either the Inetcfg.nlm or Tcpcon.nlm program to create static routes using a menu-based interface.

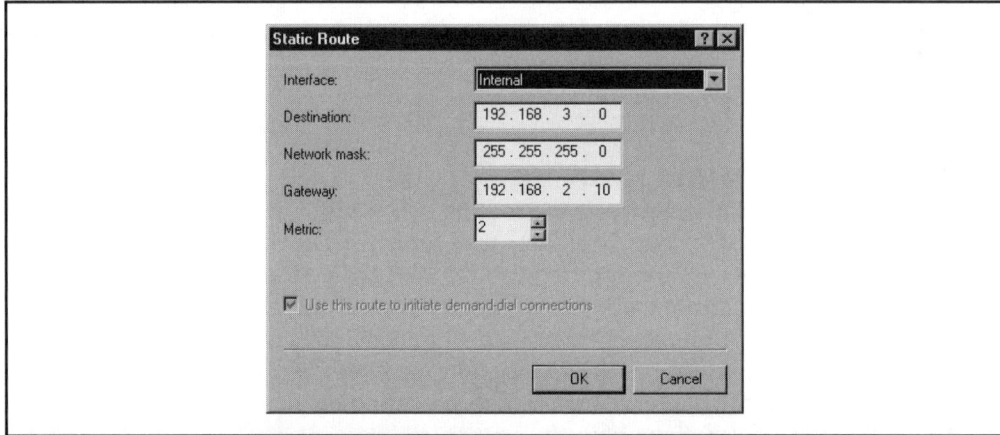

Figure 6-19: *The Routing and Remote Access Server Manager program enables you to create static routes using a standard dialog box.*

Static routes created this way remain in the routing table until you manually change or remove them, and this can be a problem. If a gateway specified in a static route should fail, the system continues to send packets to it, to no avail. You must either repair the gateway or modify the static routes that reference it throughout the network before the systems can function normally again.

On larger networks, static routing becomes increasingly impractical, not only because of the sheer number of routing table entries involved, but also because network conditions can change too often and too quickly for administrators to keep the routing tables on every system current. Instead, these networks use dynamic routing, in which specialized routing protocols share information about the other routers in the network and modify the routing tables accordingly. Once configured, dynamic routing needs little or no maintenance from network administrators because the protocols can create, modify, or remove routing table entries as needed to accommodate changing network conditions. The Internet is totally dependent on dynamic routing because it is constantly mutating, and no manual process could possibly keep up with the changes.

Selecting the Most Efficient Route

Many internetworks, even relatively small ones, are designed with multiple routers that provide redundant paths to a given destination. Thus, while creating an internetwork that consists of several LANs joined in a series by routers would be possible, most use something approaching a mesh topology instead, as shown in Figure 6-20. This way, if any one router should fail, all of the systems can still send traffic to any other system on any network.

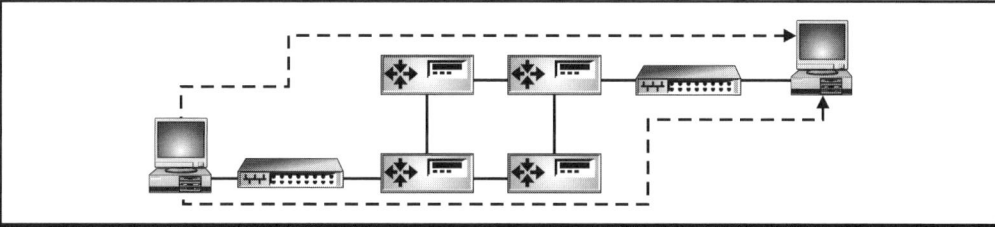

Figure 6-20: *By interconnecting routers, packets from one workstation can travel to a destination on another network by different routes.*

When an internetwork is designed in this way, another important part of the routing process is selecting the best path to a given destination. The use of dynamic routing on the network typically results in all possible routes to a given network being entered in the routing tables, each of which includes a metric that specifies how many hops are required to reach that network. Most of the time, the efficiency of a particular route is measured by the metric value because each hop involves processing by another router, which introduces a slight delay. When a router has to forward a packet to a network represented by multiple entries in the routing table, it chooses the one with the lower metric.

Discarding Packets

The goal of a router is to transmit packets to their destinations using the path that incurs the smallest number of hops. Routers also track the number of hops that packets take on the way to their destinations for another reason. When a malfunction or misconfiguration occurs in one or more routers, it is possible for packets to get caught in a router loop and be passed endlessly from one router to another.

To prevent this, the IP header contains a *Time-to-Live* (*TTL*) field that the source system gives a certain numerical value when a packet is created. (On Windows systems, the value is 128, by default.) As a packet travels through the network, each router that processes it decrements the value of this field by 1. If, for any reason, the packet passes through routers enough times to bring the value of this field down to 0, the last router removes it from the network and discards it. The router then returns an ICMP Time to Live Exceeded in Transit message to the source system to inform it of the problem.

For more information on the functions of the IP header fields and ICMP, see Chapter 13.

Packet Fragmentation

Routers can connect networks of vastly different types, and the process of transferring datagrams from one data link–layer protocol to another can require more than simply stripping off one header and applying a new one. The biggest problem that can occur during this translation process is when one protocol supports frames that are larger than the other protocol.

If, for example, a router connects a Token Ring network to an Ethernet one, it may have to accept 4,500-byte datagrams from one network and then transmit them over a network that can carry only 1,500-byte datagrams. Routers determine the maximum transfer unit (MTU) of a particular network by querying the interface to that network. To make this possible, the router has to break up the datagram into fragments of the appropriate size and then encapsulate each fragment in the correct data link–layer protocol frame. This fragmentation process may occur several times during a packet's journey from the source to its destination, depending on the number and types of networks involved.

For example, a packet originating on a Token Ring network may be divided into 1,500-byte fragments to accommodate a route through an Ethernet network, and then each of those fragments may themselves be divided into 576-byte fragments for transmission over the Internet. Note, however, that while routers fragment packets, they never defragment them. Even if the 576-byte datagrams are passed to an Ethernet network as they approach their destination, the router does not reassemble them back into 1,500-byte datagrams. All reassembly is performed at the network layer of the final destination system.

For more information on the IP fragmentation process, see "Fragmenting" in Chapter 13.

Routing and ICMP

The Internet Control Message Protocol (ICMP) provides several important functions to routers and the systems that use them. Chief among these is the capability of routers to use ICMP messages to provide routing information to other routers. Routers send ICMP Redirect messages to source systems when they know of a better route than the system is currently using. If, for example, a workstation on Network A sends a packet to Router A that is destined for a computer on Network B, and Router A determines that the next hop should be to Router B, which is on the same network as the transmitting workstation, Router A will use an ICMP message to inform the workstation that it should use Router B to access Network B instead (see Figure 6-21). The workstation then modifies the entry in its routing table accordingly.

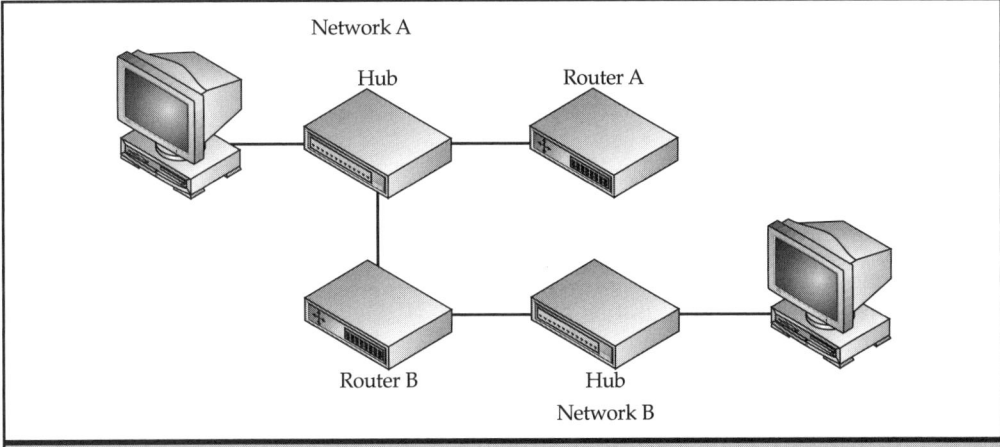

Figure 6-21:   *ICMP Redirect messages provide simple routing information to transmitting systems.*

Routers also generate ICMP Destination Unreachable messages of various types when they are unable to forward packets. If a router receives a packet that is destined for a workstation on a locally attached network, and it can't deliver the packet because the workstation is offline, the router generates a Host Unreachable message and transmits it to the system that originated the packet. If the router is unable to forward the packet to another router that provides access to the destination, it generates a Network Unreachable message instead. Network-layer protocols provide end-to-end communications, meaning it is usually the end systems that are involved in a dialog. ICMP is therefore a mechanism that enables intermediate systems (routers) to communicate with a source end system (the transmitter) in the event that the packets can't reach the destination end system.

Other ICMP packets, called Router Solicitation and Advertisement messages, can enable workstations to discover the routers on the local network. A host system generates a Router Solicitation message and transmits it either as a broadcast or a multicast to the All Routers on This Subnet address (224.0.0.2). Routers receiving the message respond with Router Advertisement messages that the host system uses to update its routing table. The routers then generate periodic updates to inform the host of their continued operational status. Windows XP, 2000, and NT 4.0 (with the Routing and Remote Access update installed) can both update their routing tables with information from ICMP Router Advertisement messages. Support for these messages in hardware router implementations varies from product to product.

The ICMP Redirect and Router Solicitation/Advertisement messages do not constitute a routing protocol per se because they do not provide systems with information about the comparative efficiency of various routes. Routing table entries created or modified as a result of these messages are still considered to be static routes.

For more information on ICMP messages, their formats, and their functions, see "ICMP" in Chapter 13.

Routing Protocols

Routers that support dynamic routing use specialized protocols to exchange information about themselves with other routers on the network. Dynamic routing doesn't alter the actual routing process; it's just a different method of creating entries in the routing table. There are two types of routing protocols: interior gateway protocols and exterior gateway protocols. Private internetworks typically use only *interior gateway protocols* because they have a relatively small number of routers and it is practical for all of them to exchange messages with each other.

On the Internet, the situation is different. Having every one of the Internet's thousands of routers exchange messages with every other router would be impossible. The amount of traffic involved would be enormous and the routers would have little time to do anything else. Instead, as is usual with the Internet, a two-level system was devised that splits the gigantic network into discrete units called *autonomous systems* (sometimes called *administrative domains,* or simply *domains*).

An autonomous system (AS) is usually a private internetwork administered by a single authority, such as those run by corporations, educational institutions, and government agencies. The routers within an AS use an interior gateway protocol, such as the Routing Information Protocol (RIP) or the Open Shortest Path First (OSPF) protocol, to exchange routing information among themselves. At the edges of an AS are routers that communicate with the other autonomous systems on the Internet, using an exterior gateway protocol (as shown in Figure 6-22), the most common of which on the Internet are the Border Gateway Protocol (BGP) and the Exterior Gateway Protocol (EGP).

The term "exterior gateway protocol" is a generic name for the routing protocols used between autonomous systems. It is also the name of a specific protocol used between autonomous systems. In the latter case, the phrase is capitalized (Exterior Gateway Protocol).

By splitting the routing chores into a two-level hierarchy, packets traveling across the Internet pass through routers that contain only the information needed to get them to the right AS. Once the packets arrive at the edge of the AS in which the destination system is located, the routers there contain more specific information about the networks within the AS. The concept is much like the way that IP addresses and domain names are assigned on the Internet. Outside entities track only the various network addresses or domains. The individual administrators of each network are responsible for maintaining the host addresses and host names within the network or domain.

The following sections examine the most common routing protocols in use today.

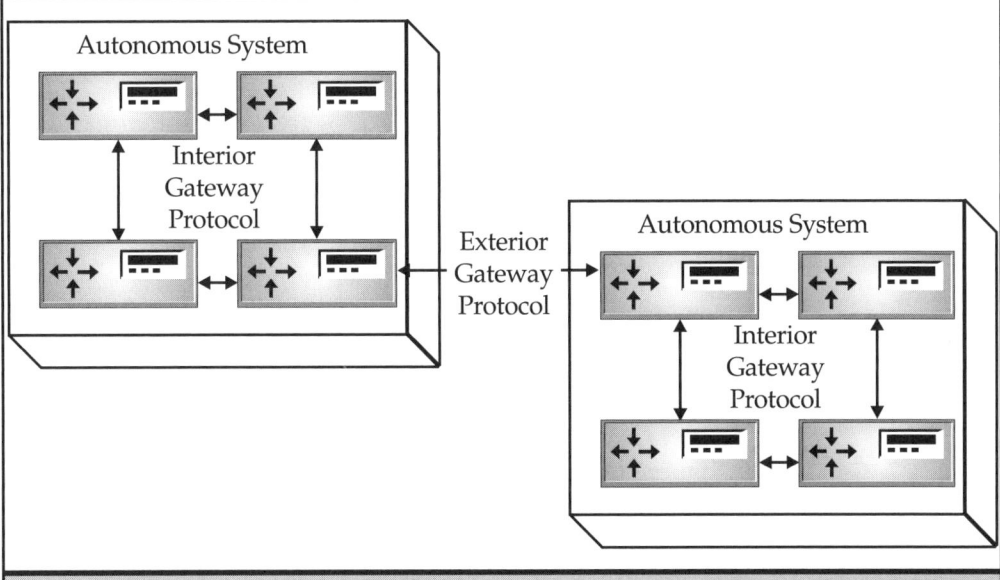

Figure 6-22: *Autonomous systems exchange routing information using an exterior gateway protocol.*

Routing Information Protocol

Routing Information Protocol (RIP) is the most commonly used of the interior gateway protocols, largely because it is supported by many operating systems and is easy to set up and use. In fact, RIP often requires no configuration at all. Originally conceived for Xerox Network Services (XNS) and included in Berkeley Unix (BSD 4.2 and later versions), RIP takes the form of a daemon called *routed* (pronounced *rout-dee*) on most Unix systems. In 1988, the Internet Engineering Task Force (IETF) standardized RIP as RFC 1058. Since then, the protocol has been implemented all but universally in hardware router products, as well as in the Microsoft Windows and Novell NetWare server operating systems.

NetWare has long used RIP to exchange information about IPX routers, but now uses it for IP routing as well. The Windows XP, 2000, and NT implementations also support both IP and IPX.

Routers that use RIP exchange request and reply messages using the User Datagram Protocol (UDP) and port 520, as specified in the "Assigned Numbers" RFC. When a router starts up, it sends a RIP request message to all the other routers on the network, using either a broadcast or a multicast transmission (depending on the RIP version). The other routers respond by transmitting their entire routing tables in RIP reply messages and repeat the advertisement every 30 seconds. Routers can also use RIP to request information on a specific route.

RIP always uses a hop count as the metric in a routing table entry and enforces a maximum hop count of 15. Networks or hosts more than 15 hops away are considered unreachable. This demonstrates that the protocol was designed for use on private internetworks and not the Internet, as Internet routes often require more than 15 hops. This limitation on the number of hops is independent of the IP header's Time-to-Live field, although RIP routers generate the same ICMP Destination Unreachable messages when the maximum number of hops is exceeded.

RIP routing table entries also have a timeout value of three minutes. If an entry is not updated by an incoming RIP message for three minutes, the router increases its metric to 16, which for RIP is infinity. One minute later, the entry is purged from the table completely.

The RIP Message Format   RIP messages consist of a 4-byte header and one or more 20-byte routes. A single message can contain up to 25 routes, for a total UDP datagram size of 512 bytes (including the 8-byte UDP header). If more than 25 entries are in a routing table, the router generates additional messages until the entire table has been transmitted.

The format of the RIP message is shown in Figure 6-23. The functions of the header fields are listed on the following page.

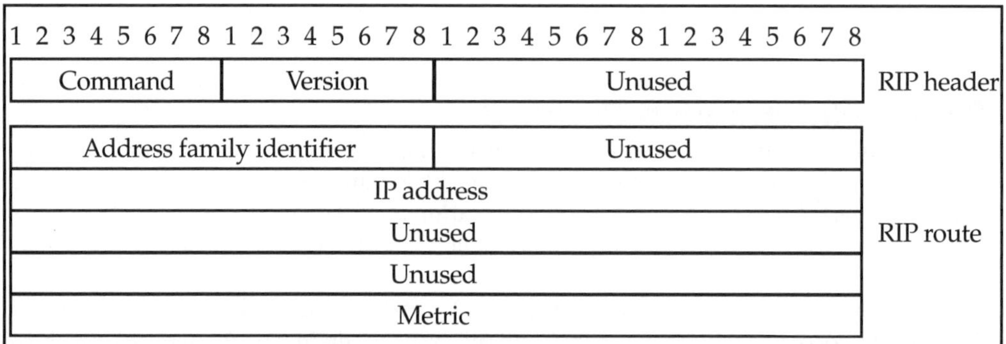

Figure 6-23:   *The RIP header and route format*

Command (1 byte)   Specifies the function of the message, using the following values:

1 – Request   Requests transmission of the entire routing table or a specific route from all routers on the local network.

2 – Reply   Transmits routing table entries.

Version (1 byte)   Specifies the version of RIP running on the system that generated the packet. Possible values are 1 and 2.

Unused (2 bytes)

The functions of the fields in each 20-byte route are as follows:

Address family identifier (2 bytes)   Identifies the network-layer protocol for which the message is carrying routing information. The value for IP is 2.

Unused (2 bytes)

IP address (4 bytes)   Specifies the address of a network or a host that is accessible through the router generating the message.

Unused (4 bytes)

Unused (4 bytes)

Metric (4 bytes)   Specifies the number of hops between the system generating the message and the network or host identified by the IP address field value.

RIP Problems   RIP is what is known as a *distance vector* routing protocol. This means that every router on the network advertises its routing table to its neighboring routers. Each router then examines the information supplied by the other routers, chooses the best route to each destination network, and adds it to its own routing table. The process of updating the routing tables on all of a network's routers in response to a change in the network (such as the failure or addition of a router) is called *convergence*. Distance vector routing is relatively simple and reasonably efficient in terms of locating the best route to a given network. It has some fundamental problems, however.

Distance vector protocols like RIP have a rather slow convergence rate because updates are generated by each router asynchronously—that is, without synchronization or acknowledgment. They are, therefore, prone to a condition known as the *count-to-infinity problem*. The count-to-infinity problem occurs when a router detects a failure in the network, modifies the appropriate entry in its routing table accordingly, and then has that entry updated by an advertisement from another router before it can broadcast it in its own advertisements. The routers then proceed to bounce their updates back and forth, increasing the metric for the same entry each time until it reaches infinity (16). The process eventually corrects itself, but the delay incurred each time a change occurs in the network slows down the entire routing process.

RIP is also widely criticized for the amount of broadcast traffic it produces. Every RIP router on an internetwork broadcasts its entire routing table every 30 seconds.

Depending on the size of the network, this may involve several RIP messages per server. One advantage of the use of broadcasts, however, is that it is possible for systems to process the advertisement messages without advertising their own routing tables. This is called *silent RIP*, and it is more likely to be implemented in host systems that are not routers.

RIP also does not include a subnet mask with each route in an advertisement message. The protocol is designed for use with network addresses that conform to the standard IP address classes, which can be identified by the first three bits of the address. If the network address for a routing table entry fits the address classes, the protocol uses the subnet mask associated with its class. When this is not the case, the protocol uses the subnet mask of the network interface over which the RIP message was received. If this mask does not fit, the protocol assumes that the table entry contains a host route and uses a subnet mask of 255.255.255.255. These assumptions can cause traffic on certain types of networks, such as those that use variable-length subnets or disjointed subnets, to be forwarded incorrectly.

RIP also does not support any form of authentication for participating routers. A RIP router accepts and processes messages from any source, making it possible for the entire network's routing tables to be corrupted with incorrect information supplied (either accidentally or deliberately) by a rogue router.

RIP v2   Other interior gateway protocols, such as OSPF, were developed as a result of the shortcomings of the original RIP standard, but the RIP protocol itself was also upgraded. RIP version 2 was initially published as RFC 1388, proposed as a draft standard in RFC 1723, and finally ratified as an IETF standard and published in November 1998 as RFC 2453. The original Windows NT Server 4.0 release supports RIP v1, but the Routing and Remote Access Server (RRAS) update adds support for RIP v2, as defined in RFC 1723. The Windows XP and 2000 server products also support RIP v2, as do many of the hardware routers on the market today.

RIP v2 addresses many of the problems inherent in version 1, including the following:

Broadcast traffic   RIP v2 supports the use of multicast transmissions for router advertisements, as well as broadcasts. The "Assigned Numbers" RFC assigns RIP v2 routers a multicast address of 224.0.0.9. Transmissions sent to that address are processed only by the routers and do not affect other systems. The use of multicasts is optional on all RIP v2 routers; broadcasts are still supported. The only possible drawback to the use of multicasts is if the network contains systems that are using silent RIP, which cannot monitor the multicast address for RIP traffic. *Silent RIP* is when a network device is configured to process the RIP broadcasts generated by other systems, but doesn't generate its own RIP broadcasts.

Subnet masks   Unlike RIP v1, RIP v2 includes a subnet mask for every route it advertises. This makes it possible for the protocol to support networks that use variable-length or disjointed subnets.

Authentication   RIP v2 supports the use of authentication to ensure that incoming RIP messages originate from authorized routers. Windows NT's RRAS and Windows 2000 support the use of simple passwords only, but some hardware routers can use more advanced authentication mechanisms such as Message Digest 5 (MD5) for this purpose.

The RIP v2 Message Format   The message format for RIP v2 is the same as that for RIP v1, except that the Version field has a value of 2, and the fields that were unused in the original format are now used to carry additional information. The format of the RIP v2 message is shown in Figure 6-24. The functions of the header fields are listed below.

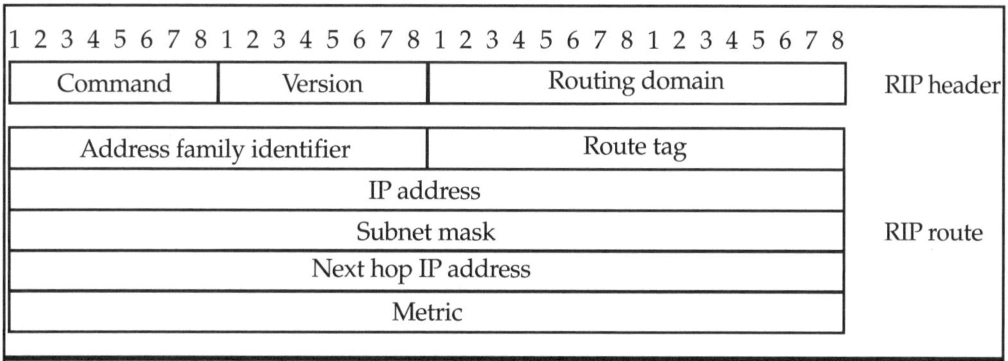

Figure 6-24:   *The RIP v2 header and route format*

Command (1 byte)   Specifies the function of the message, using the following values:

    1 – Request   Requests transmission of the entire routing table or a specific route from all routers on the local network.

    2 – Reply   Transmits routing table entries.

Version (1 byte)   Specifies the version of RIP running on the system that generated the packet. Possible values are 1 and 2.

Routing domain (2 bytes)   Identifies the routing process for which this message is intended. By using various values in this field, administrators can create independent routing domains and separate the routing information in each one. The default value is 0.

  The functions of the fields in each 20-byte route are as follows:

Address family identifier (2 bytes)   Identifies the network-layer protocol for which the message is carrying routing information. The value for IP is 2.

Route tag (2 bytes)   Contains a value that makes it possible to distinguish routes that have originated within the current autonomous system from those supplied by an exterior gateway protocol or a different autonomous system. Usually, the value is a number that uniquely identifies the autonomous system.

IP address (4 bytes)   Specifies the address of a network or a host that is accessible through the router generating the message.

Subnet mask (4 bytes)   Contains a mask that is used to distinguish the network identifier bits from the host identifier bits in the value of the IP address field.

Next hop IP address (4 bytes)   Identifies the gateway that the router should use to send traffic to the network or host specified in the IP address field. In most cases, the router should use the gateway from which it received the route, but this field is intended to prevent the propagation of less than optimal routing information. A host route, for example, should instruct systems on the same network as the host to send traffic directly to the host, not to a router on the network, and this field can be used to provide that host address. In another example, when a router runs both OSPF and RIP, it may use RIP to propagate routing information that it obtained from OSPF, in which case the next hop IP address field may contain the address of the OSPF router that was the source of the information.

Metric (4 bytes)   Specifies the number of hops between the system generating the message and the network or host identified by the IP address field value.

To provide authentication information, RIP v2 uses the first 20-byte route in a message, with the format shown in Figure 6-25. The functions of the fields in this section of the RIPv2 packet are listed below.

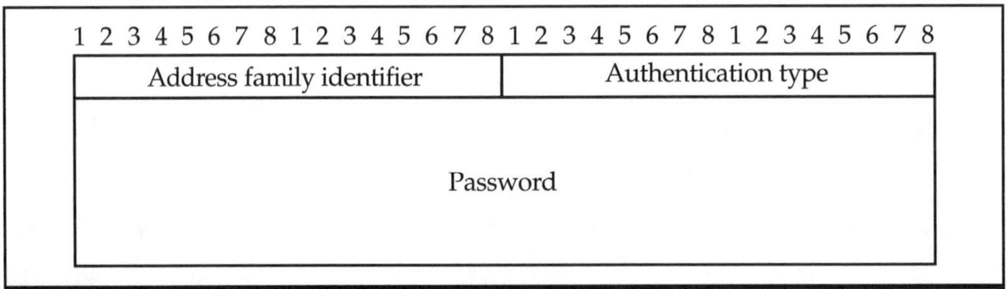

1 2 3 4 5 6 7 8 1 2 3 4 5 6 7 8 1 2 3 4 5 6 7 8 1 2 3 4 5 6 7 8	
Address family identifier	Authentication type
Password	

Figure 6-25:   *The RIP v2 authentication section*

Address family identifier (2 bytes)   Contains the hexadecimal value FF FF, indicating that this route contains authentication data. RIP v1 routers do not recognize this value and, therefore, ignore the route.

Authentication type (2 bytes) Specifies the type of authentication the router is using. Simple password authentication uses a value of 2.

Password (16 bytes) Contains the authentication password in a format specified by the value of the authentication type field.

Open Shortest Path First Protocol (OSPF)

Distance vector routing has a fundamental flaw: it bases its routing metrics solely on the number of hops between two networks. When an internetwork consists of multiple LANs in the same location, all connected using the same data link–layer protocol, the hop count is a valid indicator. When WAN links are involved, however, a single hop can refer to routers in two adjacent rooms or a transatlantic link, and there is a vast difference in the time needed to traverse the two.

The alternative to distance vector routing is called *link state routing*, most commonly used in the Open Shortest Path First (OSPF) protocol. OSPF is an interior gateway protocol that was documented by the IETF in 1989 and published as RFC 1131. The current specification, which has been ratified as an IETF standard, was published in April 1998 as RFC 2328. Most router products now support OSPF in addition to RIP, including Windows XP and 2000, Windows NT's RRAS, and Novell NetWare.

Unlike RIP and most other TCP/IP protocols, OSPF is not carried within a transport protocol such as UDP or TCP. The OSPF messages are encapsulated directly in IP datagrams using protocol number 89.

Link state routing, as implemented in OSPF, uses a formula called the Dijkstra algorithm to judge the efficiency of a route, based on several criteria, including the following:

Hop count While link state routing protocols still use the hop count to judge a route's efficiency, it is only part of the equation.

Transmission speed The speed at which the various links operate is an important part of a route's efficiency. Faster links obviously take precedence over slow ones.

Congestion delays Link state routing protocols consider the network congestion caused by the current traffic pattern when evaluating a route and bypass links that are overly congested.

Route cost The route cost is a metric assigned by the network administrator used to rate the relative usability of various routes. The cost can refer to the literal financial expense incurred by the link, or any other pertinent factor.

Link state routing is more complex than RIP and requires more processing by the router, but it judges the relative efficiency of routes more precisely and has a better convergence rate than RIP. OSPF also reduces the amount of bandwidth used by the routing protocol because it only transmits updates to other routers when changes in the network configuration take place, unlike RIP, which continually transmits the entire routing table.

Several of the advantages of OSPF are clearly the inspiration for the improvements made in the RIP version 2 specification. For example, all OSPF routes include a subnet mask and OSPF messages are all authenticated by the receiving router before they are processed. The protocol can also use routing information obtained from outside sources, such as exterior gateway protocols. In addition, OSPF provides the capability to create discrete areas within an autonomous system that exchange routing information among themselves. Only certain routers, called *area border routers,* exchange information with other areas. This reduces the amount of network traffic generated by the routing protocol.

Unlike RIP, OSPF can maintain multiple routes to a specific destination. When two routes to a single network address have the same metric, OSPF balances the traffic load between them.

Version 2 of RIP, therefore, is comparable to OSPF in its features and is definitely the preferable alternative on a relatively small internetwork that does not have severe traffic problems. However, on an internetwork that relies heavily on WAN connections or has many routers with large routing tables that would generate a lot of network traffic, OSPF is the preferable alternative.

Switches

The traditional internetwork configuration uses multiple LANs connected by routers to form a network that is larger than would be possible with a single LAN. This is necessary because each LAN is based on a network medium that is shared by multiple computers, and there is a limit to the number of systems that can share the medium before the network is overwhelmed by traffic. Routers segregate the traffic on the individual LANs, forwarding only those packets addressed to systems on other LANs.

Routers have been around for decades, but a newer type of device, called a *LAN switch,* has revolutionized network design and made it possible to create LANs of almost unlimited size. A *switch* is essentially a multiport bridging device in which each port is a separate network segment. Similar in appearance to a hub, a switch receives incoming traffic through its ports. Unlike a hub, which forwards the traffic out through all of its other ports, a switch forwards the traffic only to the single port needed to reach the destination (see Figure 6-26). If, for example, you have a small workgroup network with each computer connected to a port in the same switching hub, each system has what amounts to a dedicated, full-bandwidth connection to every other system. No shared network medium exists, and consequently, no collisions or traffic congestion. As an added bonus, you also get increased security because, without a shared medium, an unauthorized workstation cannot monitor and capture the traffic not intended for it.

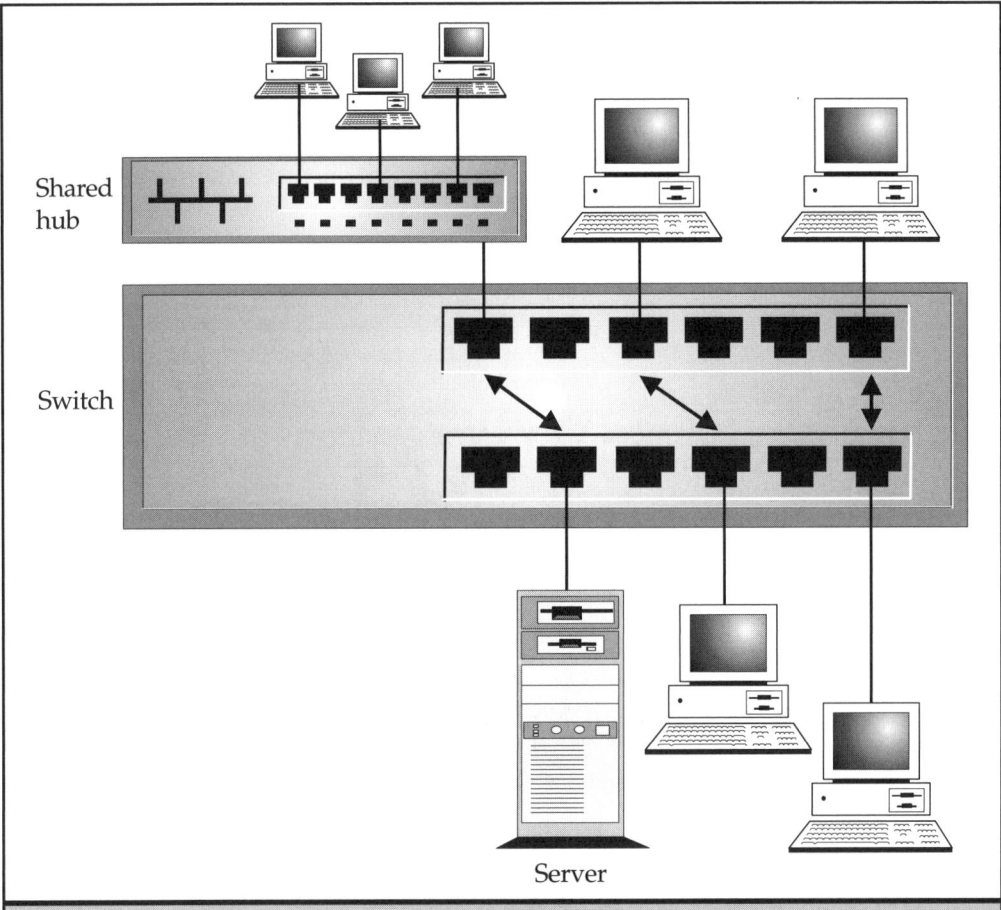

Figure 6-26: *Switches repeat incoming traffic, but only to the specific port for which it is intended.*

Switches operate at layer 2 of the OSI reference model, the data link layer, so consequently, they are used to create a single large network, instead of a series of smaller networks connected by routers. This also means that switches can support any network-layer protocol. Like transparent bridges, switches can learn the topology of a network and perform functions such as forwarding and packet filtering. Some switches are also capable of full-duplex communications and automatic speed adjustment.

In the traditional arrangement for a larger internetwork, multiple LANs are connected to a backbone network with routers. The backbone network is a shared-medium LAN like all of the others, however, and must therefore carry all of the internetwork traffic generated by the horizontal networks. This is why the backbone network traditionally uses a faster protocol. On a switched network, workstations are connected to individual workgroup switches, which in turn are connected to a single, high-performance switch, thus enabling any system on the network to open a dedicated connection to any other system (see Figure 6-27). This arrangement can be expanded further to include an intermediate layer of departmental switches as well. Servers accessed by all users can then be connected directly to a departmental switch or to the top-level switch, for better performance.

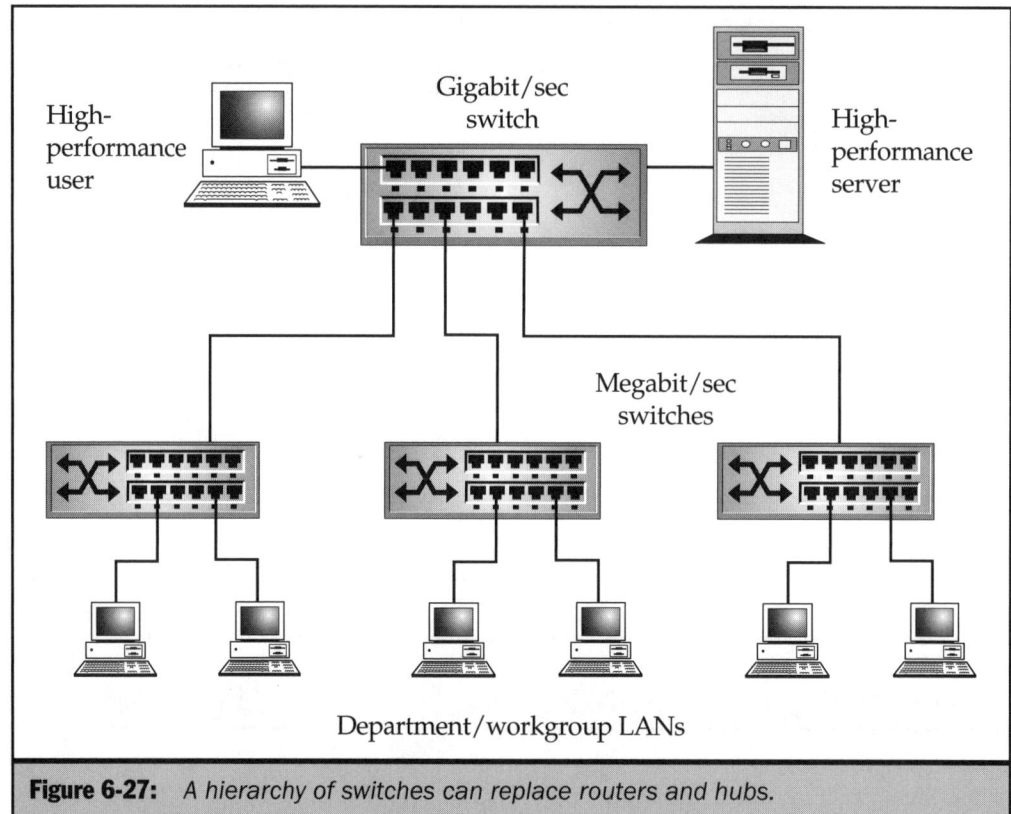

Figure 6-27: *A hierarchy of switches can replace routers and hubs.*

Replacing hubs with switches is an excellent way to improve the performance of a network without changing protocols or modifying individual workstations. Even a standard Ethernet network exhibits a dramatic improvement when each workstation is given a full 10 Mbps of bandwidth, rather than sharing it with 20 or 30 other systems. Full-duplex switches can double the effective bandwidth to 20 Mbps. While most of the LAN switches on the market are designed for Ethernet (including Fast and Gigabit Ethernet) networks, Token Ring and FDDI switches are available.

 *Asynchronous Transfer Mode (ATM) networks also rely on switching, but ATM is a connection-oriented, circuit-switched networking technology, and its switches are not interchangeable with those made for standard LANs. For more information on ATM, see Chapter 7.*

While a fully switched network provides an ideal level of performance, switches are far more expensive than standard repeating hubs, and most networks combine these two technologies to reach a happy medium. You can, for example, connect standard hubs to the ports of a switch and share the bandwidth of a switched connection among a handful of machines, rather than several dozen.

Switch Types

There are two basic types of switching: cut-through and store-and-forward. A *cut-through switch* reads only the MAC address of an incoming packet, looks up the address in its forwarding table, and immediately begins to transmit it out through the port providing access to the destination. The switch forwards the packet without any additional processing, such as error checking, and before it has even received the entire packet. This type of switch is relatively inexpensive and more commonly used at the workgroup or department level, where the lack of error checking will not affect the performance of the entire network. The immediate forwarding of incoming packets reduces the latency (that is, the delay) that results from error checking and other processing. If the destination port is in use, however, the switch buffers incoming data in memory, incurring a latency delay anyway, without the added benefit of error checking.

A *store-and-forward switch,* as the name implies, stores an entire incoming packet in buffer memory before forwarding it out the destination port. While in memory, the switch checks the packet for CRC errors and other conditions, such as runts, giants, and jabber. The switch immediately discards any packets with errors; those without errors are forwarded out through the correct port. These switching methods are not necessarily exclusive of each other. Some switches can work in cut-through mode until a preset error threshold is reached, and then switch to store-and-forward operation. Once the errors drop below the threshold, the switch reverts back to cut-through mode.

For more information on runts, giants, jabber, and other transmission problems, see Chapter 10.

LAN switches implement these functions using one of three hardware configurations. *Matrix switching,* also called *crossbar switching*, uses a grid of input and output connections, such as that shown in Figure 6-28. Data entering through any port's input can be forwarded to any port for output. Because this solution is hardware based, there is no CPU or software involvement in the switching process. In cases where data can't be forwarded immediately, the switch buffers it until the output port is unblocked.

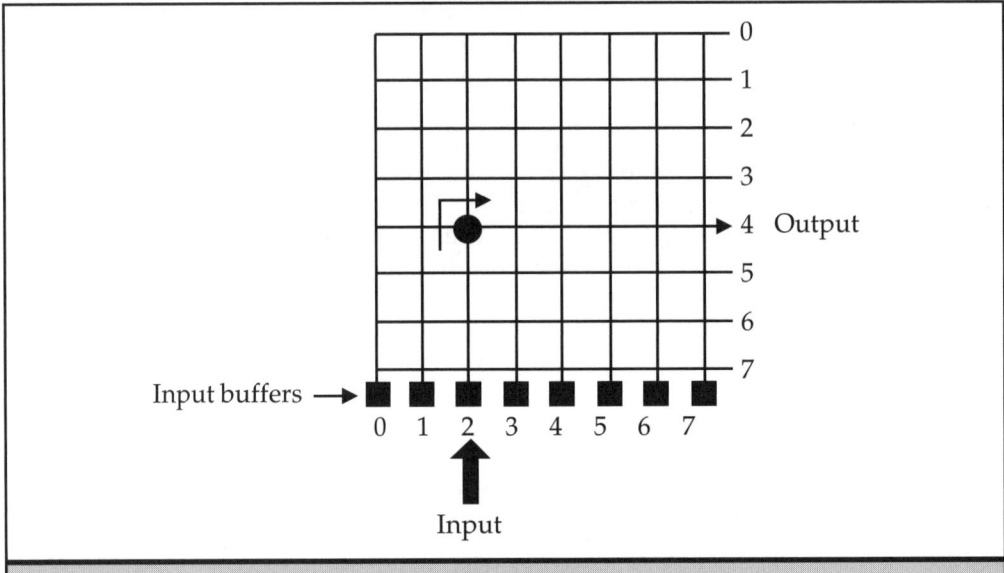

Figure 6-28:   *Matrix switching uses a grid of input and output circuits.*

In a *shared-memory switch*, all incoming data is stored in a memory buffer that is shared by all of the switch's ports and then forwarded to an output port (see Figure 6-29). A more commonly used technology (shown in Figure 6-30), called *bus-architecture switching*, forwards all traffic across a common bus, using time-division multiplexing to ensure that each port has equal access to the bus. In this model, each port has its own individual buffer and is controlled by an ASIC (application-specific integrated circuit).

Like hubs, switches are available for any size network, from inexpensive workgroup switches designed for small office networks to stackable and modular units with much higher prices.

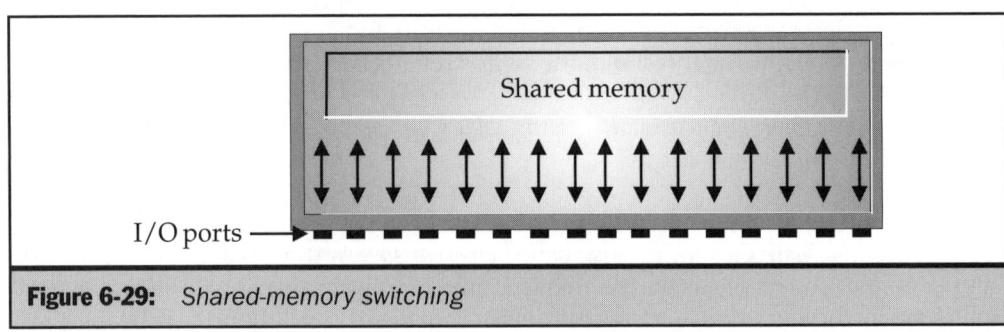

Figure 6-29:   *Shared-memory switching*

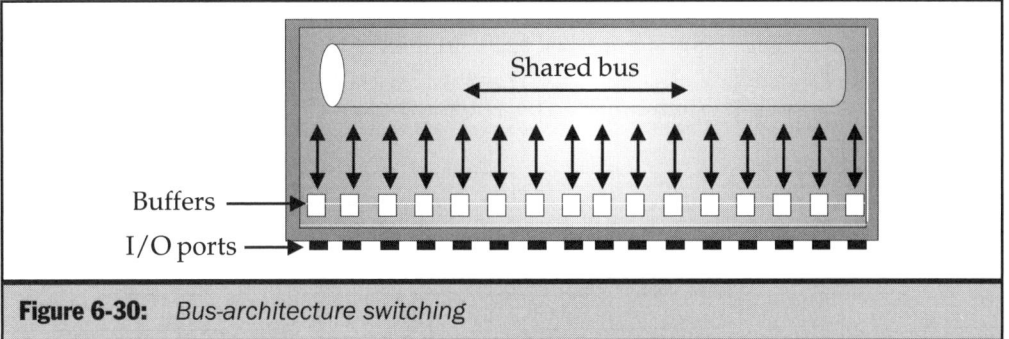

Figure 6-30:   *Bus-architecture switching*

Routing versus Switching

The question of whether to route or switch on a network is a difficult one. Switching is faster and cheaper than routing, but it raises some problems in most network configurations. By using switches, you eliminate subnets and create a single flat network segment that hosts all of your computers. Any two systems can communicate using a dedicated link that is essentially a temporary two-node network. The problems arise when workstations generate broadcast messages. Because a switched network forms a single broadcast domain, broadcast messages are propagated throughout the whole network and every system must process them, which can waste enormous amounts of bandwidth.

One of the advantages of creating multiple LANs and connecting them with routers is that broadcasts are limited to the individual networks. Routers also provide security by limiting transmissions to a single subnet. To avoid the wasted bandwidth caused by broadcasts, it has become necessary to implement certain routing concepts on switched networks. This has led to a number of new technologies that integrate routing and switching to varying degrees. Some of these technologies are examined in the following sections.

Virtual LANs

A *virtual LAN* or *VLAN* is a group of systems on a switched network that functions as a subnet and communicates with other VLANs through routers. The physical network is still switched, however; the VLANs exist as an overlay to the switching fabric, as shown in Figure 6-31. Network administrators create VLANs by specifying the MAC, port, or

IP addresses of the systems that are to be part of each subnet. Messages that are broadcast on a VLAN are limited to the subnet, just as in a routed network. Because VLANs are independent of the physical network, the systems in a particular subnet can be located anywhere, and a single system can even be a member of more than one VLAN.

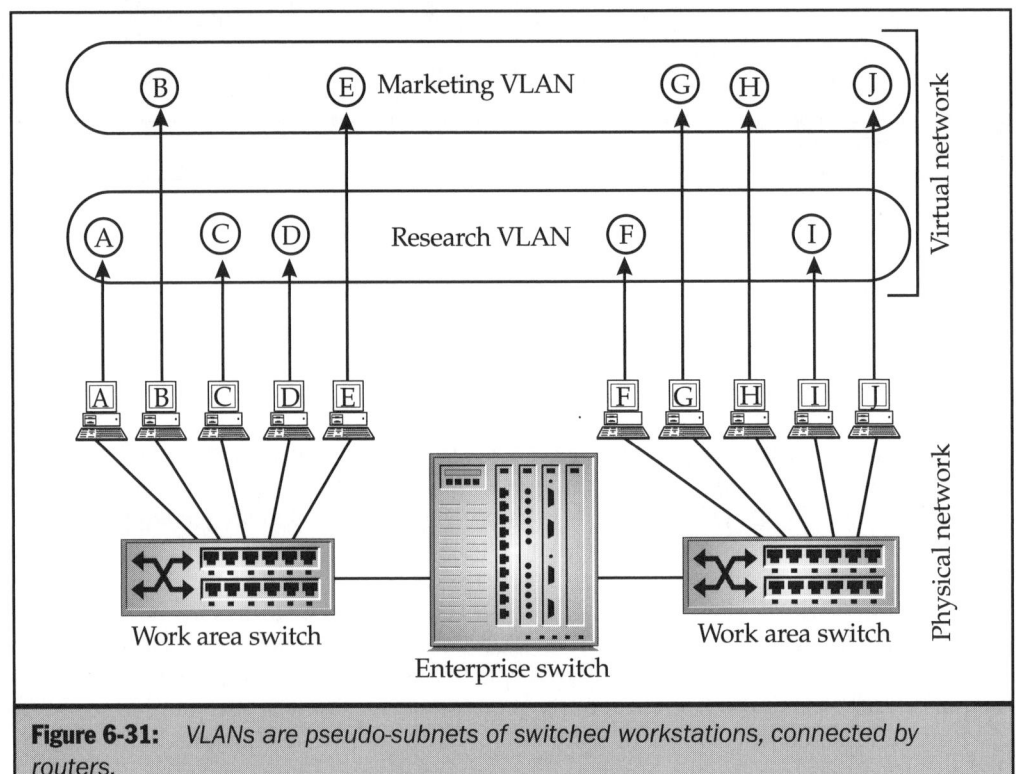

Figure 6-31:   *VLANs are pseudo-subnets of switched workstations, connected by routers.*

Despite the fact that all of the computers are connected by switches, routers are still necessary for systems in different VLANs to communicate. VLANs that are based solely on layer-2 technology, such as those that use port configuration or MAC addresses to define the member systems, must have a port dedicated to a router connection. In this type of VLAN, the network administrator either selects certain switch ports to designate the members of a VLAN or creates a list of the workstations' MAC addresses.

Because of the additional processing involved, routing is slower than switching. This particular arrangement is sometimes referred to as "switch where you can, route where you must," because routing is only used for communication between VLANs; all communication within a VLAN is switched. This is an efficient arrangement, as long as the majority of the network traffic (70 to 80 percent) is between systems in the same

VLAN. Communication speed within a VLAN is maximized, at the expense of the inter-VLAN communication. When too much traffic occurs between systems in different subnets, the routing slows down the process too much and the speed of the switches is largely wasted.

Layer-3 Switching

Layer-3 switching also uses VLANs, but it mixes routing and switching functions to make communication between VLANs more efficient. This technology is known by several different names—depending on the vendor of the equipment—including *IP switching, multilayer routing, cut-through routing*, and *Fast IP*. The essence of the concept is described as "route once, switch afterward." A router is still required to establish connections between systems in different VLANs, but once the connection has been established, subsequent traffic travels over the layer-2 switching fabric, which is much faster.

Most of the hardware devices called layer-3 switches that are being produced by the major manufacturers combine the functions of a switch and a router into one unit. The device is capable of performing all of a router's standard functions, but is also able to transmit data using high-speed switches, all at a substantially lower cost than a standard router. Layer-3 switches are optimized for use on LAN and MAN (metropolitan area network) connections, not WANs.

Layer-3 switching has not yet matured to the point at which vendors are manufacturing completely interoperable solutions, but this technology could potentially represent the ideal upgrade path for internetworks that are now based on routers and repeating hubs. By replacing the routers that connect workgroup or department networks to the backbone with layer-3 switches, you retain all of the router functionality, while increasing the overall speed at which data is forwarded. Eventually, by moving workstation connections from repeating hubs to layer-2 switches, you can migrate to a network with no shared media at all, except for wide area links, still connected with traditional routers.

NETWORK HARDWARE

Chapter 7

Wide Area Networking

T
he physical- and data link–layer protocols used to build local area networks (LANs) are quite efficient over relatively short distances. Even for campus connections between buildings, fiber-optic solutions enable you to use a LAN protocol such as Ethernet or FDDI throughout your whole internetwork. However, when you want to make a connection over a long distance, you move into an entirely different world of data communications called *wide area networking*. A wide area network (WAN) is a communications link that spans a long distance and connects two or more LANs together.

WAN connections make it possible to connect networks in different cities or countries, enabling users to access resources at remote locations. Many companies use WAN links between office locations to exchange e-mail, groupware, and database information, or even just to access files and printers on remote servers. Banks and airlines, for example, use WANs because they must be in continual communication with all of their branch offices to keep their databases updated, but WAN connections can also function on a much smaller scale, such as a system that periodically dials in to a remote network to send and retrieve the latest e-mail messages.

WAN Connections

A WAN connection requires a router or a bridge at each end to provide the interface to the individual LANs, as shown in Figure 7-1. This reduces the amount of traffic that passes across the link. *Remote link bridges* connect LANs running the same data link–layer protocol at different locations using an analog or digital WAN link. The bridges prevent unnecessary traffic from traversing the link by filtering packets according to their data link–layer MAC addresses. However, bridges do pass broadcast traffic across the WAN link. Depending on the speed of the link and applications for which it is intended, this may be a huge waste of bandwidth.

Note   *It's possible to make a good case that using remote link bridges to connect networks at two sites is technically not a WAN, because you are actually joining the two sites into a single network, instead of creating an internetwork. However, whether the final result is a network or an internetwork, the technologies used to join the two sites are the same, and are commonly called WAN links. For more information on bridges and their functions, see Chapter 6.*

If the WAN link is intended only for highly specific uses, such as e-mail access, data link–layer bridges can be wasteful, because they provide less control over the traffic that is permitted to pass over the link. Routers, on the other hand, keep the two LANs completely separate. In fact, the WAN link is a network in itself that connects only two systems, the routers at each end of the connection. Routers pass no broadcasts over the WAN link (except in exceptional cases, such as when you use DHCP or BOOTP relay agents), and administrators can exercise greater control over the traffic passing between the LANs. Routers also enable you to use different data link–layer protocols on each of the LANs, because they operate at the network layer of the OSI model.

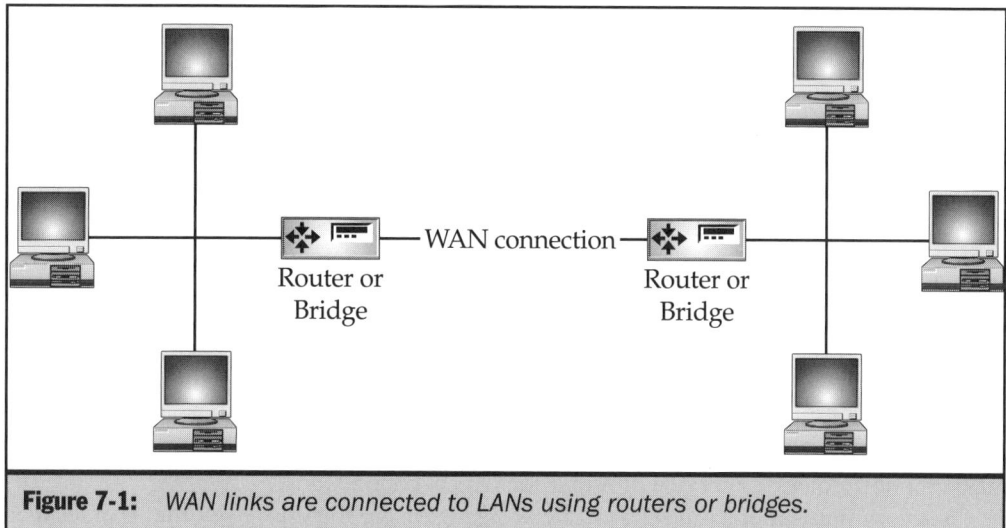

Figure 7-1: *WAN links are connected to LANs using routers or bridges.*

While bridges are always separate units, the routers used to connect two networks with a WAN link can take the form of either a computer or a dedicated hardware device. When a remote user connects to a host PC with a dial-up modem connection and accesses other systems on the network, the host PC is functioning as a router. For technologies other than dial-up connections, however, most sites use dedicated routers. The router or bridge located at each terminus of the WAN link is connected to the local LAN and to whatever hardware is used to make the physical-layer connection to the WAN, such as a modem, CSU/DSU, or NT-1.

 For more information on the types of routers and their functions, see Chapter 6.

Introduction to Telecommunications

When you enter the world of wide area networking, you experience a major paradigm shift from the local area networking world. When you design, build, and maintain a LAN, you are working with equipment that you (or your organization) owns and controls completely. Once you pay for the equipment itself, the network and its bandwidth are

yours to do with as you please. When you connect networks using WAN links, however, you almost never own all of the technology used to make the connections. Unless your organization has the means to run its own long distance fiber-optic cables or launch its own satellite (and we're talking millions, if not billions of dollars needed to do this in most cases), you have to deal with a third-party telecommunications service provider that makes it possible for you to send your data signals over long distances.

The need to rely on an outside service provider for WAN communications can enormously complicate the process of designing, installing, and maintaining the network. LAN technicians are often tinkerers by trade. When problems with the network occur, they have their own procedures for investigating, diagnosing, and resolving them, knowing that the cause is somewhere nearby if they can only find it. Problems with WAN connections can conceivably be caused by the equipment located at one of the connected sites, but it's more likely for the trouble to be somewhere in the service provider's network infrastructure. A heavy equipment operator a thousand miles away in Akron, Ohio, can sever a trunk cable while digging a trench, causing your WAN link to go down. Solar flares on the surface of the sun 93 million miles away can disturb satellite communications, causing your WAN link to go down. In either case, there is nothing that you can do about it except call your service provider and complain. Because of this reliance on outside parties, many network administrators maintain backup WAN links that use a different technology or service provider for critical connections.

Telecommunications is a separate networking discipline unto itself that is at least as complicated as data networking, if not more so. (If you think that local area networking has a lot of cryptic acronyms, wait till you start studying telecommunications.) A large organization relies at least as much on telecommunications technology as on their data networking technology. If the computer network goes down, people complain loudly; if the phone system goes down, people quickly begin to panic. In many large organizations, the people who manage the telecommunications infrastructure are different from those who administer the data network. However, it is in the area of WAN communications that these two disciplines come together. It isn't common to find technical people who are equally adept at data networking and telecommunications; most technicians tend to specialize in one or the other. However, a LAN administrator has to know something about telecommunications if the organization has offices at multiple locations that are to be connected using WANs.

All data networking is about bandwidth, the ability to transmit signals between systems at a given rate of speed. On a LAN, when you want to increase the bandwidth available to users, you can upgrade to a faster protocol or add network connection components such as bridges, switches, and routers. After the initial outlay for the new equipment and its installation, the network has more bandwidth—forever. In the world of telecommunications, bandwidth costs money, often lots of it. If you want to increase the speed of a WAN link between two networks, you not only have to purchase new equipment, but you probably also have to pay additional fees to your service provider. Depending on the technology you've chosen and your service provider, you may have to pay a fee to have the equipment installed, a fee to set up the new service, and permanent monthly subscriber fees based on the amount of bandwidth you want. Combined, these fees can be substantial, and they're ongoing; you continue to pay as long as you use the service.

The result of this expense is that WAN bandwidth is far more expensive than LAN bandwidth. In nearly every case, your LANs will run at speeds far exceeding those of your WAN connections, as shown in Figure 7-2. Today, a 10 Mbps Ethernet LAN is old technology, bordering on obsolescence. Most of the new LANs installed today use Fast Ethernet, running at 100 Mbps, and backbone networks are starting to use Gigabit Ethernet, at 1,000 Mbps. In telecommunications, 10 Mbps is a large amount of bandwidth. Every home Internet user with a dial-up modem connection dreams, while waiting for Web pages to download, of having a personal T-1 Internet connection. A T-1 is a leased digital telephone line running at 1.544 Mbps; Web pages that require up to a minute to load over a dial-up connection load in seconds with a T-1. Installing a T-1 in your home can easily cost you $2,000 per month or more, and that does not include the hardware, installation, or the Internet access fees. However, even the most basic Ethernet LAN that you can build using $100 worth of hardware from the computer store runs almost six and a half times faster than a T-1. A Fast Ethernet LAN runs more than 64 times faster. This gives you some idea why it's so important to minimize the amount of traffic passing over a WAN link joining two networks. That bandwidth costs real money, and you don't want to waste it transmitting needless broadcast messages or other "junk" traffic.

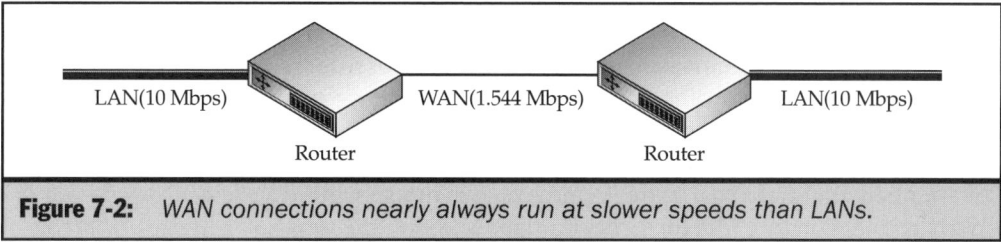

Figure 7-2:   *WAN connections nearly always run at slower speeds than LANs.*

WAN Utilization

WAN technologies vary in the way they're structured, the way you pay for them, and the way you use them. A T-1, for example, is a telephone connection that you lease from a telephone service provider on a permanent basis. You are essentially paying for a permanent connection between two specific points. A T-1 is not a dial-up service, so if you use one to connect to your Internet service provider (ISP) and you later decide to change ISPs, you cannot simply "call" someone else on your T-1. Instead, you have to have the old line removed and a new one installed, with all of the expenses that entails. Other types of telecommunications services that you can use for WAN links, such as ISDN, offer dial-up services with bandwidth comparable to that of a leased line.

Because the T-1 is a permanent connection, you are paying for a fixed amount of bandwidth, 24 hours a day. If your company only uses the T-1 from 9:00 A.M. to 5:00 P.M., it remains idle for 16 hours out of the day and two-thirds of the bandwidth you're paying for is going to waste. In addition, the amount of bandwidth provided by your T-1 is fixed. If you need more bandwidth during peak usage hours, you have no choice but to install another connection. There are telecommunications technologies, however, that provide more flexible amounts of bandwidth, enabling you to use as much as you need and pay only for what you use. This is called bandwidth on demand, meaning that the capacity of the connection increases with your needs.

Sometimes, as with a frame-relay connection, the bandwidth allocation process is automatic. A sudden burst of heavy traffic immediately triggers an increase in the bandwidth provided by the connection, and although there will probably be an additional charge, the application is served. In other cases, you might have to request additional bandwidth from the service provider as you notice a trend of increasing utilization, but this is still likely to be preferable to installing another leased line or taking other permanent action to accommodate what may be a temporary condition.

Suppose, for example, that a large accounting firm experiences a dramatic increase in WAN bandwidth utilization each year during tax season, only to have it die down again after April 15. Installing additional leased lines just for a few months would be totally impractical because of their high installation and equipment fees. On the other hand, a frame-relay connection or even primary rate ISDN can be configured to accommodate temporarily increased bandwidth requirements and then revert back to the original service as needed.

There are also alternatives to a T-1 that prevent you from paying for bandwidth you don't use in other ways. When you install a T-1, you agree to a flat monthly subscriber fee that covers the full-time connection. A digital dial-up service such as ISDN is more likely to charge a per-minute rate in addition to a more modest monthly fee. This type of arrangement might cost more if your WAN is in use around the clock, but if your WAN connections are idle for a substantial portion of the day, you can simply "hang up" the ISDN connection, avoiding the per-minute fee and lowering your monthly bill.

Another big difference between a leased line like a T-1 and some other services is that the price of a T-1 is based on the length of the connection. A T-1 connection between you and your ISP in the next town will cost a lot less than a transcontinental connection between your New York and San Francisco offices. The costs of specific technologies are also dependent on your location. In Europe, for example, leased lines are rarely used and are astronomically expensive (one quote for a short-distance link in France in 1996 was more than $35,000 per month!), but ISDN is much more prevalent there than in the United States for business WAN connections, and it's priced quite reasonably.

Selecting a WAN Technology

The selection of a WAN connection for a specific purpose is generally a tradeoff between speed and expense. Because your WAN links will almost certainly run more slowly than the networks that they connect, and cost more as well, it's important to determine just how much bandwidth you need and when you need it as you design your network.

It usually is not practical to use a WAN link in the same way that you would use a LAN connection. You might have to limit the amount of traffic that passes over the link in ways other than just using routers at each end. One way is to schedule certain tasks that require WAN communications to run at off-peak hours. For example, database replication tasks can easily monopolize a WAN link for extended periods of time, delaying normal user activities. Many applications that require periodic data replication, including directory services such as Active Directory, Windows NT domains, and NDS, enable you to specify when these activities should take place. Active Directory, for example, enables you to split your internetwork into units called *sites* and regulate the time and frequency of the replication that occurs between domain controllers at different sites.

Before you select a WAN technology, you should consider the applications for which it will be used. Different functions require different amounts of bandwidth, and different types as well. E-mail, for example, not only requires relatively little bandwidth, but also is intermittent in its traffic. High-end applications, such as full-motion video, not only require enormous amounts of bandwidth, but also require that the bandwidth be continuously available, to avoid dropouts in service. The needs of most organizations fall somewhere between these two extremes, but it is important to remember that the continuity of the bandwidth can sometimes be as important as the transmission rate.

Table 7-1 lists some common WAN applications and the approximate amount of bandwidth they require. When you examine these figures, remember to consider the number of people using the application. A hundred users querying a database at another site can use as much bandwidth as a single full-motion video stream.

Application	Transmission Rate
Personal communications	300 to 9,600 bps or higher
E-mail transmission	2,400 to 9,600 bps or higher
Remote control programs	9,600 bps to 56 Kbps or higher
Digitized voice phone call	64 Kbps
Database text query	Up to 1 Mbps
Digital audio	1 to 2 Mbps
Access images	1 to 8 Mbps
Compressed video	2 to 10 Mbps
Medical transmissions	Up to 50 Mbps
Document imaging	10 to 100 Mbps
Scientific imaging	Up to 1 Gbps
Full-motion video	1 to 2 Gbps

Table 7-1:   *WAN Applications and Their Approximate Required Bandwidth*

Table 7-2 lists the most popular technologies used for WAN connections and their transmission speeds. The sections following the table examine some of the technologies that are most commonly used for WAN connectivity.

Note *The transmission rates listed in this table represent the maximum rated throughput for these technologies and, for a variety of reasons, usually do not necessarily reflect the actual throughput realized by applications using them. In the real world, the throughput is generally lower.*

Connection Type	Transmission Rate
Dial-up modem connection	Up to 56 Kbps (53 Kbps in U.S., by FCC restriction)
X.25	64 Kbps to 2 Mbps
ISDN	Up to 128 Kbps or 1.544 Mbps
Fractional T-1	64 Kbps
T-1	1.544 Mbps
T-3	44.736 Mbps
Frame relay	56 kbps to 44.736 Mbps
DSL	Up to 51.84 Mbps
SONET	51.9 Mbps to 2.5 Gbps
ATM	25.6 Mbps to 2.46 Gbps

Table 7-2: *WAN Technologies and Their Transmission Rates*

PSTN Connections

A WAN connection does not necessarily require a major investment in hardware and installation fees. A standard asynchronous modem connection using dial-up telephone lines to connect your computer to a network (such as that of an ISP) is technically a wide area link, and for some purposes, this is all that is needed. For example, an employee working at home or on the road can dial in to a server at the office and connect to the LAN to access e-mail and other network resources. In the same way, a dial-up connection may be sufficient for a small branch office to connect to the corporate headquarters for the same purposes. The connection can be scheduled to occur at regular intervals or to dial in to the network whenever a user requests a remote resource. In telecommunications parlance, the standard dial-up telephone service is called the Public Switched Telephone Network (PSTN) or, believe it or not, the Plain Old Telephone Service (POTS).

The speed of a dial-up modem link is limited to 33.6 Kbps unless one side of the link uses a digital connection to the PSTN, in which case the maximum possible speed is 56 Kbps (for digital-to-analog traffic only; analog-to-digital traffic is limited to 31.2 Kbps). However, in the United States and Canada, the Federal Communications Commission imposes a 53 Kbps limit on transmission speed over standard telephone lines. In most cases, only ISPs and large corporations use the digital equipment necessary to operate at the higher speed. On a smaller scale, the increase in speed often isn't great enough to justify the equipment expense. Analog modem communications are also dependent on the quality of the lines involved. Most telephone companies certify their lines for voice communications only, and will not perform repairs to improve the quality of data connections. As a result, it is rare for a dial-up connection to achieve a full 53/31.2 Kbps throughput.

In most cases, a dial-up WAN connection uses a computer as a router, although there are standalone devices that perform the same function. The most basic arrangement uses the dial-up connection for remote network access. To do this, you configure a modem-equipped Windows computer on a LAN to function as a dial-up server. A user at a remote location, also running a Windows computer with a modem, dials in to the server's modem, connects to the server, and accesses the network through the server's LAN connection. The remote computer can be running an e-mail client, a Web browser, or another application designed to access network resources, or simply access the file system on the network's servers. This simple arrangement is best suited to users who want to dial in to their office computers while at home or traveling.

A computer can also host multiple dial-up connections, with the right software and equipment. The Windows workstation operating systems (Windows 2000 Professional, Windows NT Workstation, and Windows 9*x*/Me), for example, can host only a single remote user, but Windows 2000 Server and Windows NT Server can host up to 256 remote users simultaneously. There are a number of hardware products available that enable a single computer to host a large number of modems, making it possible for network administrators to set up a single *remote access server* that can host all of the network's remote users. For remote access on a larger scale, the dedicated hardware platforms used by service providers and large corporations can split a single leased line, such as a T-1 or T-3, into separate 64-Kbps channels supporting multiple dial-up or other connections. The process of splitting a single connection into multiple discrete channels is called *multiplexing.*

In addition to linking a single computer to a network, dial-up connections can connect two networks at remote locations. The Routing and Remote Access Service for

Windows NT Server, for example, enables you to configure a connection for *dial-on-demand*. When a user on one LAN performs an operation that requires access to the other LAN, the server automatically dials in to a server on the other network, establishes the connection, and begins routing traffic. When the link remains idle for a preset period of time, the connection terminates. There are also standalone routers that perform in the same way, enabling users to connect to a remote LAN or the Internet as needed. This arrangement minimizes the connection time, thus reducing the telephone charges, and provides WAN access to users without them having to establish the connection manually. The only evidence that the requested resource is at a remote location is the delay incurred while the modem dials and connects.

Obviously, the chief drawback to using the PSTN for WAN connections is the limited bandwidth, but the low cost of the hardware and services required make dial-up connections compelling, and many network administrators make use of them in interesting and creative ways.

Inverse Multiplexing

Some enterprise networks use PSTN connections as inexpensive backups for their higher-speed WAN links, in the event of a hardware or service failure, or to augment the bandwidth of a leased line during periods of high traffic. The process of combining the bandwidth of multiple connections into a single conduit is called *inverse multiplexing* (see Figure 7-3). Basic rate ISDN connections routinely use inverse multiplexing to combine two 64-Kbps B-channels into a single 128-Kbps channel, and there are routers that can combine the bandwidth of leased lines, as well. More recently, however, inverse multiplexing has been introduced in lower-end products intended for home and small business users.

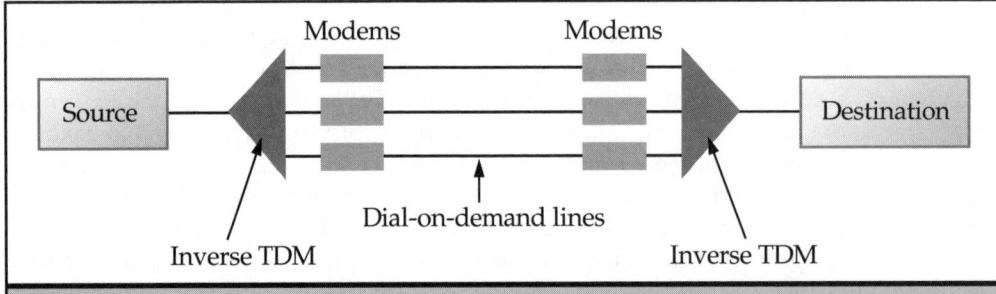

Figure 7-3: *Inverse multiplexing splits a data stream into channels for transmission across separate media.*

Several of the Windows operating systems, for example, include a feature called *multilink* that enables you to combine the bandwidth of multiple modems to form a single higher-speed connection. However, as with all inverse multiplexing arrangements, the implementation at one end that separates the signal into multiple channels for transmission over separate connections must be matched by an equivalent process at the other end that combines the connections back into a single signal. This means that if you want to use this feature to connect to the Internet, for example, you must find an ISP that supports it on the server end.

Some of the router products with inverse multiplexing features also have the capability to open additional dial-up connections automatically when traffic passing over the link reaches a certain threshold. This feature goes by different names, such as channel aggregation, bonding, and (amusingly) rubber bandwidth.

NETWORK HARDWARE

Leased Lines

A *leased line* is a dedicated, permanent connection between two sites that runs through the telephone network. The line is said to be *dedicated* because the connection is active 24 hours a day and does not compete for bandwidth with any other processes. The line is *permanent* because there are no telephone numbers or dialing involved in the connection, nor is it possible to connect to a different location without modifying the hardware installation. While this book is naturally more interested in leased lines as WAN technologies, it's important to understand that they are also a vital element of the voice telecommunications network infrastructure. When a large organization installs its own private branch exchange (PBX) to handle its telephone traffic, the switchboard is typically connected to one or more T-1 lines, which are split into individual channels with enough bandwidth to handle a single voice-grade connection (56 to 64 Kbps). Each of these channels becomes a standard voice "telephone line," which is allocated by the PBX to users' telephones as needed.

You install a leased line by contacting a telephone service provider, either local or long-distance, and agreeing to a contract that specifies a line granting a certain amount of bandwidth between two locations, for a specified cost. The price typically involves an installation fee, hardware costs, and a monthly subscription fee, and depends both on the bandwidth of the line and the distance between the two sites being connected.

As mentioned earlier, the advantages of a leased line are that the connection delivers the specified bandwidth at all times, and that the line is as inherently secure as any telephone line, because it is private. While the service functions as a dedicated line between the two connected sites, there is not really a dedicated physical connection, such as a separate wire running the entire distance. The service provider installs a dedicated line between each of the two sites and the provider's nearest *point of presence (POP)*, but from there, the connection uses the provider's standard switching facilities to make the connection. The provider guarantees that its facilities can provide a specific bandwidth and quality of service.

A leased line contract typically quantifies the quality of service using two criteria: service performance and availability. The performance of the service is based on the percentage of error-free seconds per day and its availability is computed in terms of the time that the service is functioning at full capacity during a three-month period, also expressed as a percentage. The contract will specify thresholds for these statistics, such as a guarantee of 99.99 percent error-free seconds per day (which computes out to about nine errored seconds per day), and 99.96 percent service availability over the course of a year. If the provider fails to meet the guarantees specified in the contract, the customer receives a financial remuneration in the form of service credits.

Leased-Line Types

Leased lines can be analog or digital, but digital lines are more common. An analog line is simply a normal telephone line that is continuously open. When used for a WAN connection, modems are required at both ends to convert the digital signals of the data network to analog form for transmission, and back to digital at the other end. In some cases, the line may have a greater service quality than a standard PSTN line. This type of leased line is relatively rare, because unless the connection is continuously transmitting data, it is often cheaper to use a dial-up connection with some form of dial-on-demand technology, so that you are not paying for the connection during nonproduction hours.

Digital leased lines are more common, because no analog-to-digital conversion is required for data network connections, and the signal quality of a digital line is usually superior to that of an analog line, whether leased or dial-up. Digital leased lines are based on a hierarchy of digital signal (DS) speeds used to classify the speeds of carrier links. These levels take different forms in different parts of the world. In North America, the DS levels are used to create the T-carrier (for "trunk-carrier") service. Europe and most of the rest of the world uses the E-carrier service, which is standardized by the Telecommunications sector of the International Telecommunications Union (ITU-T), except for Japan, which has its own J-carrier service. Each of these services names the various levels by replacing the DS prefix with that of the particular carrier. For example, the DS-1 level is known as a T-1 in North America, an E-1 in Europe, and a J-1 in Japan. The only exception to this is the DS-0 level, which represents a standard 64-Kbps voice-grade channel and is known by this name throughout the world. As you go beyond the DS-1 service, bandwidth levels rise steeply, as do the costs. In North America, most enterprise networks use multiple T-1 lines, for both voice and data. T-3s are used mainly by ISPs and other service providers with high bandwidth needs. The use of T-4s is quite rare.

NETWORK HARDWARE

Because of the different ways in which the signals are encoded, the bandwidth for the levels is not always consistent across all of the carriers, and even when it is, the services are not compatible. The T-1 lines used in North America, for example, run at 1.544 Mbps, while an E-1 in Europe is 2.048 Mbps. Table 7-3 lists the levels of the DS hierarchy and the characteristics of the North American T-carrier service. Table 7-4 lists the DS levels and their E-carrier and J-carrier equivalents.

DS Level	T-Carrier Service Name	Bandwidth	Number of 64-Kbps Voice-Grade Channels
DS-0	DS-0	64 Kbps	1
DS-1	Trunk Level 1 (T-1)	1.544 Mbps	24
DS-1C	Trunk Level 1 Combined (T-1C)	3.152 Mbps	48 (or 2 T-1s)
DS-2	Trunk Level 2 (T-2)	6.312 Mbps	96 (or 4 T-1s)
DS-3	Trunk Level 3 (T-3)	44.736 Mbps	672 (or 28 T-1s)
DS-4	Trunk Level 4 (T-4)	274.176 Mbps	4,032 (or 168 T-1s)

Table 7-3:   *North American T-Carrier Service Characteristics*

DS Level	E-Carrier Service Name	E-Carrier Bandwidth	J-Carrier Service Name	J-Carrier Bandwidth
DS-0	DS-0	64 Kbps	DS-0	64 Kbps
DS-1	E-1	2.048 Mbps	J-1	1.544 Mbps
DS-1C	No equivalent	No equivalent	No equivalent	No equivalent
DS-2	E-2	8.448 Mbps	J-2	6.312 Mbps
DS-3	E-3	34.368 Mbps	J-3	32.064 Mbps
DS-4	E-4	139.264 Mbps	J-4	97.728 Mbps

Table 7-4:   *E-Carrier and J-Carrier Service Characteristics*

While it's possible to install a leased line using any of the service levels listed for your geographical location, you are not limited to the amounts of bandwidth provided by these services. Because the bandwidth of each service is based on multiples of 64 Kbps, you can split a digital link into individual 64 Kbps channels and use each one for voice or data traffic. Service providers frequently take advantage of this capability to offer leased lines that consist of any number of these 64 Kbps channels that the subscriber needs, combined into a single data pipe. This is called *fractional T-1* service.

Leased-Line Hardware

A T-1 line requires two twisted pairs of wires, and originally the line was *conditioned,* meaning that a repeater was installed 3,000 feet from each endpoint and every 6,000 feet in between. Today, however, a signaling scheme called *High-bit-rate Digital Subscriber Line (HDSL)* makes it possible to transmit digital signals at T-1 speeds over longer distances without the need for repeating hardware.

The hardware required at each end of a digital leased line is called a *channel service unit/data service unit (CSU/DSU),* which is actually two devices that are usually combined into a single unit. The CSU provides the terminus for the digital link and keeps the connection active even when the connected bridge, router, private branch exchange (PBX), or other device isn't actually using it. The CSU also provides testing and diagnostic functions for the line. The DSU is the device that converts the signals it receives from the bridge, router, or PBX to the bipolar digital signals carried by the line. In appearance, a CSU/DSU looks something like a modem, and as a result, they are sometimes incorrectly called digital modems. (Since a modem, by definition, is a device that converts between analog and digital signals, the term "digital modem" is actually something of an oxymoron. However, just about any device used to connect a computer or network to a telephone or Internet service has been incorrectly called a modem, including ISDN and cable network equipment.) A single, basic CSU/DSU unit costs somewhere in the neighborhood of $1,000, but identical units are required at each end of the connection.

The CSU/DSU is connected to the leased line on one side using an RJ connector, and to a device (or devices) on the other side that provides the interface to the local network (see Figure 7-4), using a V.35 or RS-232 connector. This interface can be a bridge or a router for data networking, or a PBX for voice services. The line can be either *unchanneled,* meaning that it is used as a single data pipe, or *channeled*, meaning that a multiplexor is located in between the CSU/DSU and the interface, to break up the line into separate channels for multiple uses.

Digital leased lines use *time division multiplexing (TDM)* to create the individual channels in which the entire data stream is divided into time segments that are allocated to each channel in turn (see Figure 7-5). Each time division is dedicated to a particular channel, whether it is used or not. Thus, when one of the 64-Kbps voice lines that are part of a T-1 is idle, that bandwidth is wasted, no matter how busy the other channels are.

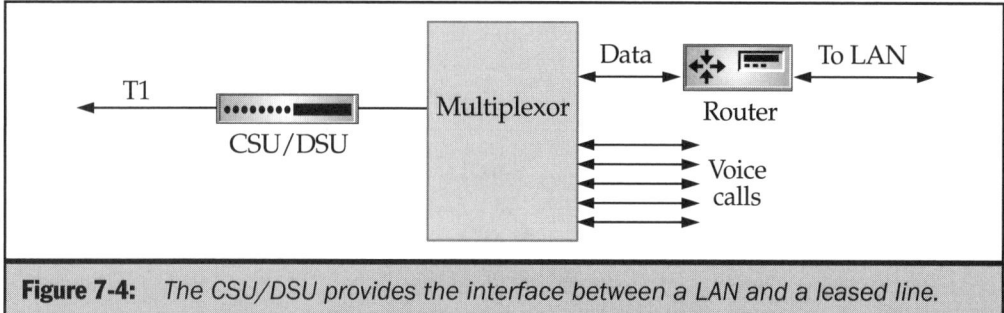

Figure 7-4:   *The CSU/DSU provides the interface between a LAN and a leased line.*

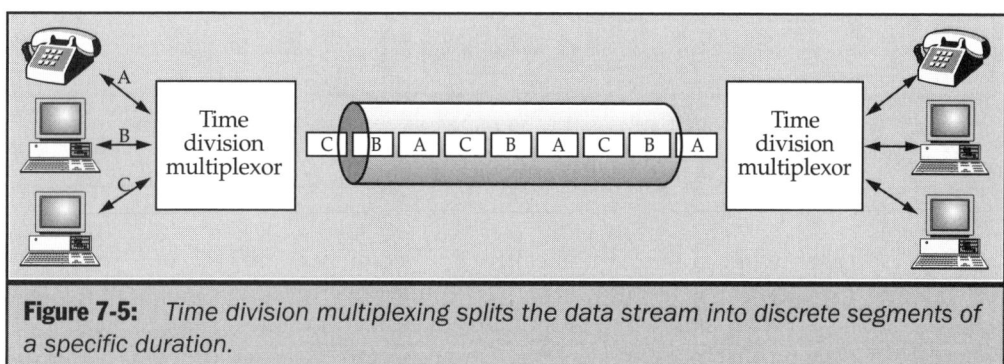

Figure 7-5:   *Time division multiplexing splits the data stream into discrete segments of a specific duration.*

Leased-Line Applications

T-1s and other leased lines are used for many different purposes. As mentioned earlier, T-1s are commonly used to provide telephone services to large organizations. On the WAN front, organizations with offices in several locations can use leased lines to build a *private network* for both voice and data traffic. With such a network in place, users can access network resources in any of the sites at will, and telephone calls can be transferred to users in the different offices. The problem with building a network in this manner is that it requires a true mesh topology of leased lines—that is, a separate leased line connecting each office to every other office—to be reliable. An organization with four sites, for example, would need 6 leased lines, as shown in Figure 7-6, and eight sites would require 28 leased lines! It would be possible for the sites to be connected in series, using seven links to connect eight sites, but then the failure of any one link or router would split the network in two.

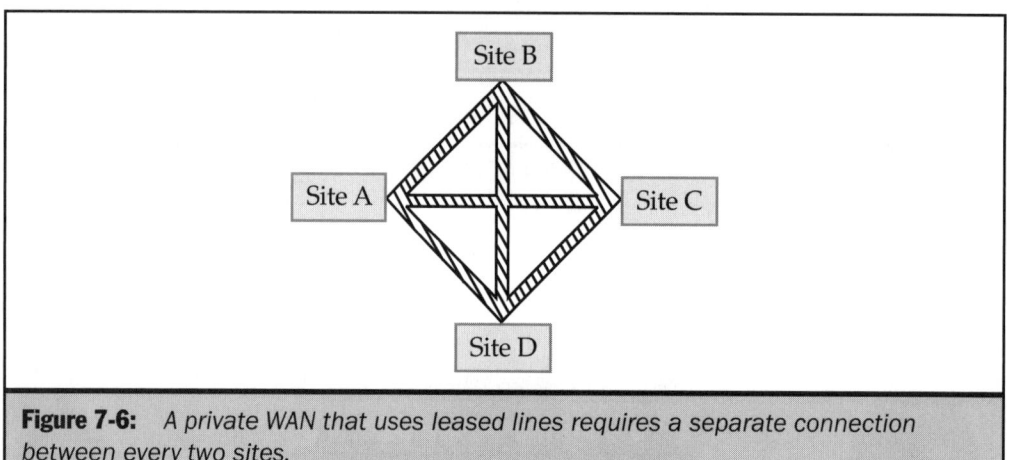

Figure 7-6: *A private WAN that uses leased lines requires a separate connection between every two sites.*

This type of network can provide excellent and secure performance, but it can be abominably expensive. The monthly cost of a T-1 line depends on the length of the connection. A short hop within a metropolitan area can cost in the area of $1,500 per month, while a long-distance connection running coast to coast in the United States can be $20,000 per month or more, and these prices do not include the installation and equipment fees. Thus, if an organization has eight sites scattered around the country, the monthly bill for the leased lines could reach $250,000 or more.

These days, most organizations use a less expensive technology to create WAN links between their various offices. One alternative to a private network would be to use leased lines at each site to connect to a public carrier network using a technology such as frame relay or ATM to provide the required bandwidth. Each site would require only a single, relatively short-distance leased line to a local service provider, instead of a separate line to each site. For more information on this alternative, see "Packet-Switching Services," later in this chapter. The most common application for T-1 lines in WANs today, however, is to use them to connect a private network to an ISP, in order to provide Internet access to its users and to host Internet services, such as Web and e-mail servers. T-1s are well-suited for providing Internet access to corporate networks, because services like e-mail have to be connected around the clock. ISPs also usually have a local point of presence, so the leased line does not have to span a tremendously long distance and is not too terribly expensive. A single T-1 connection to the Internet can serve the needs of hundreds of average users simultaneously.

ISDN

Integrated Service Digital Network (ISDN) and *Digital Subscriber Line (DSL)* are both services that utilize the existing copper POTS cable at an installation to carry data at much higher transmission rates. In both cases, the site must be relatively close to the telephone company's nearest point of presence (POP), a location containing telephone switching equipment. Basic rate ISDN, for example, requires a location no farther than 18,000 feet (3.4 miles) from the POP; DSL distances vary with the data rate. ISDN and DSL are sometimes called *last-mile technologies,* because they are designed to get data from the user site to the POP at high speed.

The copper cable running from the POP to the individual user site is traditionally the weakest link in the phone system. Once a signal reaches the POP, it moves through the telephone company's switches at high speed. By eliminating the bottlenecks at both ends of the link, traffic can maintain that speed from end to end. While these technologies are currently being marketed in the United States primarily as Internet connectivity solutions for home users, they both are usable for office-to-office WAN connections as well.

ISDN is a digital point-to-point telephone system that has been around for many years, but that has not been adopted as widely in the United States as its proponents had hoped. Originally, ISDN was designed to completely replace the current phone system with all-digital service, but it is now positioned as an alternative technology for home users who require high-bandwidth network connections and for links between business networks. In this country, ISDN technology has garnered a reputation for being overly complicated, difficult to install, and not particularly reliable—and to some extent, this reputation is justified. Up until a very few years ago, inquiries to most local phone companies about ISDN service would be met only with puzzlement, and horror stories from consumers about installation difficulties were common. The situation has now improved somewhat, and many ISPs offer a turnkey service for home ISDN Internet access, in which the ISP coordinates the entire installation process, including the negotiations with the telephone company. For business use, ISDN is not commonly used in North America, but it is a popular solution in Europe, where leased lines can be prohibitively expensive.

ISDN is a digital service that can provide a good deal more bandwidth than standard telephone service, but unlike a leased line, it is not permanent. ISDN devices dial a number to establish a connection, like a standard telephone, meaning that users can

NETWORK HARDWARE

connect to different sites as needed. For this reason, ISDN is known as a *circuit-switching service,* because it creates a temporary point-to-point circuit between two sites. For the home or business user connecting to the Internet, this means you can change ISPs without any modifications to the ISDN service by the telephone company. For organizations using ISDN for WAN connections between offices, this means you can dial in to different office networks when you need access to their resources, rather than maintain separate leased line connections to each site.

> **Note**
> *ISDN numbers are not the same as standard telephone numbers, however. You can only establish a digital connection by dialing another ISDN number from your ISDN number. However, it is possible to purchase a special ISDN telephone that enables you to dial standard analog telephone numbers from your ISDN service.*

This impermanence also means that, in many cases, you only pay for the bandwidth you use. Most ISDN service agreements include a per-minute charge, as well as installation fees and monthly subscriber fees. Unlike a leased line, you can disconnect an ISDN WAN link during off hours. For home users wanting a full-time Internet access solution, the per-minute charges make ISDN a rather costly alternative when compared to other options available today. However, for business use, a PRI ISDN connection can, in many cases, be far cheaper than a leased line and provide virtually the same service.

Unlike a modem connection using standard dial-up telephone lines, however, ISDN connections are nearly instantaneous because they're digital, like the computer network, and no analog-to-digital conversion is required. Picture a dial-up connection to your ISP without all of the chirping and scratching you hear during the connection process for a standard modem. When configured to dial on demand, you experience virtually no delay while the ISDN service connects. ISDN can carry different kinds of traffic, such as voice, fax, and data; but for these purposes, special hardware is required to convert the analog signals of the device into the digital format needed for transmission. This is exactly the opposite of the traditional arrangement, where a computer's digital signals are converted to analog format by a modem before transmission. Multiple devices can utilize the same channel using TDM to split the bandwidth. This side of the technology has not caught on; a few vendors sell ISDN telephones and other devices, but not nearly as many as those who sell the ISDN terminal adapter used to connect to a computer or router.

ISDN Services

There are two main types of ISDN service, which are based on units of bandwidth called *B channels,* running at 64 Kbps, and *D channels,* running at 16 or 64 Kbps.

B channels carry voice and data traffic, and D channels carry control traffic only. The service types are as follows:

BRI (Basic Rate Interface)   Also called *2B+D*, because it consists of two 64-Kbps B channels and one 16-Kbps D channel. BRI is targeted primarily at home users for connections to business networks or the Internet.

PRI (Primary Rate Interface)   Consists of up to 23 B channels and one 64-Kbps D channel, for a total bandwidth equivalent to a T-1 leased line. PRI is aimed more at the business community, as an alternative to leased lines that can provide the same bandwidth and signal quality with greater flexibility.

Another type of service, called Broadband ISDN (B-ISDN), is not directly comparable to BRI and PRI, because it defines a cell-switching service that uses SONET at the physical layer and ATM at the data link layer to carry voice, data, and video traffic at speeds in excess of 155 Mbps.

One of the primary advantages of ISDN is the ability to combine the bandwidth of multiple channels as needed, using inverse multiplexing. Each B channel has its own separate ten-digit number. For the home user, one of the B channels of the BRI service can carry voice traffic while the other B channel is used for data, or both B channels can be combined to form a single 128-Kbps connection to the Internet or to a private network. The PRI service can combine any number of the B channels in any combination, to form connections of various bandwidths.

In addition, the ISDN service supports bandwidth-on-demand, which can supplement a connection with additional B channels to support a temporary increase in bandwidth requirements. Depending on the equipment used, it's possible to add bandwidth according to a predetermined schedule of usage needs or to dynamically augment a connection when the traffic rises above a particular level. If your bandwidth needs fluctuate, an ISDN connection can be far more economical than a leased line because you only pay for the channels that are currently in use. With a leased line, you must pay whether it's being used or not.

ISDN Communications

The ISDN B channels carry user traffic only, whether in the form of voice or data. The D channel is responsible for carrying all of the control traffic needed to establish and terminate connections between sites. The traffic on these channels consists of protocols that span the bottom three layers of the OSI reference model. The physical layer establishes a circuit-switched connection between the user equipment and the telephone company's switching office that operates at 64 Kbps and also provides diagnostic functions such as loopback testing and signal monitoring. This layer is also responsible for the multiplexing that enables devices to share the same channel.

At the data link layer, bridges and PBXs using an ISDN connection employ the *LAPD (Link Access Procedure for D Channel)* protocol, as defined by the International Telecommunications Union (ITU-T) documents Q.920 through Q.923, to provide frame-relay and frame-switching services. This protocol (which is similar to the LAP-B protocol used by X.25) uses the address information provided by the ISDN equipment to create virtual paths through the switching fabric of the telephone company's network to the intended destination. The end result is a private network connection much like that of a leased line.

The network layer is responsible for the establishment, maintenance, and termination of connections between ISDN devices. Unlike leased lines and similar technologies, which maintain a permanently open connection, ISDN must use a handshake procedure to establish a connection between two points. The process of establishing an ISDN connection involves messages exchanged between three entities: the caller, the switch (at the POP), and the receiver. As usual, network-layer messages are encapsulated within data link–layer protocol frames. The connection procedure is as follows:

1. The caller transmits a SETUP message to the switch.

2. If the SETUP message is acceptable, the switch returns a CALL PROC (call proceeding) message to the caller and forwards the SETUP message to the receiver.

3. If the receiver accepts the SETUP message, it rings the phone (either literally or figuratively) and sends an ALERTING message back to the switch, which forwards it to the caller.

4. When the receiver answers the call (again, either literally or figuratively), it sends a CONNECT message to the switch, which forwards it to the caller.

5. The caller then sends a CONNECT ACK (connection acknowledgment) message to the switch, which forwards it to the receiver. The connection is now established.

ISDN Hardware

ISDN does not require any modifications to the standard copper POTS wiring. As long as your site is within 18,000 feet of a POP, you can convert an existing telephone line to ISDN just by adding the appropriate hardware at each end. The telephone company uses special data-encoding schemes (called *2B1Q* in North America and *4B3T* in Europe) to provide higher data transmission rates over the standard cable. All ISDN installations must have a device called an *NT1 (Network Termination 1)* connected to the telephone line at each end (see Figure 7-7). The service from the telephone company provides what is known as a *U interface* operating over one twisted pair of wires. The NT1 connects to the U interface and converts the signals to the four-wire *S/T interface* used by ISDN terminal equipment (that is, the devices that use the connection).

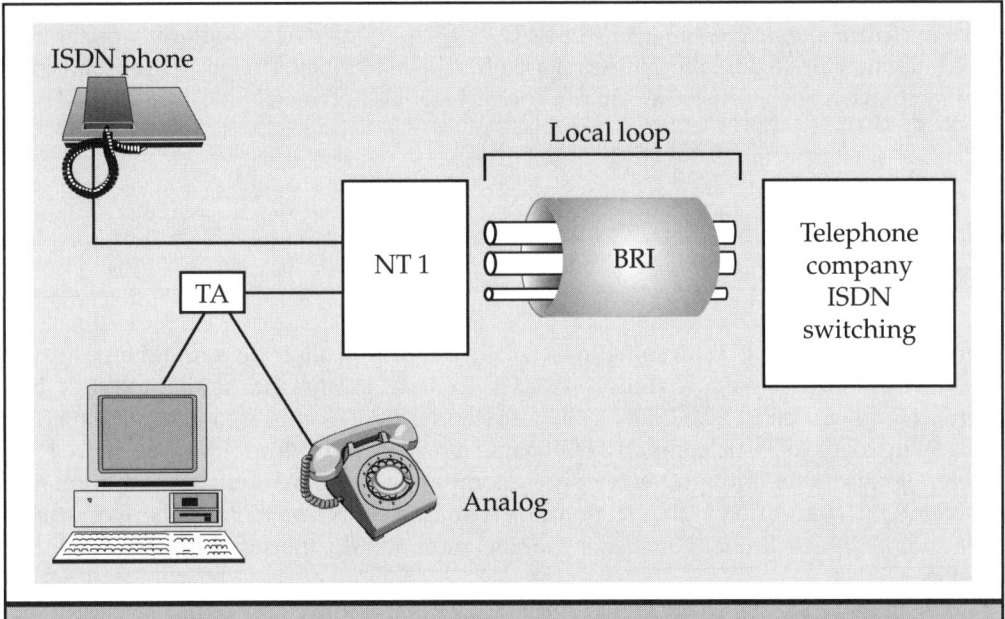

Figure 7-7: *ISDN connections use an NT1 to convert the U interface to the S/T interface.*

Note *In some cases, the S/T interface can use six or eight wires, with the additional wires providing emergency power to telephones in the event that local power fails. Power may also be delivered through the standard four wires in some instances.*

Devices that connect directly to the S/T interface, such as ISDN telephones and ISDN fax machines, are referred to as *terminal equipment 1 (TE1)*. Devices that are not ISDN capable, such as standard analog phones and fax machines, as well as computers, are called *terminal equipment 2 (TE2)*. To connect a TE2 device to the S/T interface, you must have an intervening *terminal adapter (TA)*. You can connect up to seven devices to an NT1, both TE1 and TE2.

In North America, it is up to the consumer to provide the NT1, which is available in several forms as a commercial product. In Europe and Japan, where ISDN is much more prevalent, the NT1 is owned and provided by the telephone company; users only have to provide the terminal equipment. For the BRI service, a separate NT1 is required if you are going to use more than one type of terminal equipment, such as a terminal adapter for a computer and an ISDN telephone. If the service is going to be used only for data networking, as is often the case in the United States, there are single devices available that combine the NT1 with a terminal adapter. These combination devices can take the form of an expansion card for a PC, or a separate device, similar in appearance to a modem. Once again, the units that are often called "ISDN modems" are technically not modems at all, because they do not convert signals between analog and digital formats.

DSL

Digital Subscriber Line (DSL), sometimes called *xDSL*, is a collective term for a group of related technologies that provide a WAN service that is somewhat similar to ISDN, but at much higher speeds. Like ISDN, DSL uses standard POTS wiring to transmit data from a user site to a telephone company POP using a private point-to-point connection. From there, signals travel through the telephone company's standard switching equipment to another DSL connection at the destination. Also like ISDN, the distance between the site and the POP is limited; the faster the transmission rate, the shorter the operable distance.

The transmission rates for DSL services vary greatly, and many of the services function *asymmetrically*, meaning that they have different upload and download speeds. This speed variance occurs because the bundle of wires at the POP is more susceptible to a type of interference called *near-end crosstalk* when data is arriving from the user site than when it is being transmitted out to the user site. The increased signal loss rate resulting from the crosstalk requires that the transmission rate be lower when traveling in that direction.

Standard telephone communications only use a small amount of the bandwidth provided by the POTS cable. DSL works by utilizing frequencies above the standard telephone bandwidth (300 to 3,200 Hz) and by using advanced signal encoding methods to transmit data at higher rates of speed. Some of the DSL services use only frequencies that are out of the range of standard voice communications, which makes it possible for the line to be used for normal voice traffic while it is carrying digital data. In other words, imagine a high-speed Internet connection that enables you to talk on the phone while you're surfing the Web, all using a single standard telephone line.

DSL is a relatively new technology, and is being promoted heavily as an Internet access solution. However, the higher-speed services like High-bit-rate Digital

Subscriber Line (HDSL) have been deployed heavily by local telephone carriers for several years. Asymmetrical operation is not much of a problem for services like Asymmetrical Digital Subscriber Line (ADSL), which are used for Internet access, because the average Internet users download far more data than they upload. For WAN connections, however, symmetrical services like HDSL are standard. DSL differs from ISDN in that it uses permanent connections; it has dial-up service, no numbers assigned to the connections, and no session-establishment procedures. The connection is continuously active and private, much like that of a leased line. In fact, it's estimated that as many as 70 percent of the T-1 circuits in the United States use HDSL as their underlying technology. As an Internet access solution, DSL has grown quickly, due to its relatively low prices and high transmission rates, and has all but eclipsed ISDN in this market. DSL and CATV (cable television network) connections are now the two biggest competing technologies in the end-user, high-speed Internet connection market. It remains to be seen if the telephone company legislation currently underway in Washington and the high prices of the routers and switches used to build the fiber-optic infrastructure that supports DSL communications don't cause consumer prices to rise steeply in the near future.

The various DSL services have abbreviations with different first letters, which is why the technology is sometimes called xDSL, with the x acting as a placeholder. These services and their properties are shown in Table 7-5.

The hardware required for a DSL connection is a standard POTS line and a DSL "modem" at both ends of the link. For services that provide simultaneous voice and data traffic, a POTS splitter is needed to separate the lower frequencies used by voice traffic from the higher frequencies used by the DSL service. In addition, the telephone line cannot use *loading coils,* inductors that extend the range of the POTS line at the expense of the higher frequencies that DSL uses to transmit data.

At this point, the two most commonly used DSL services are HDSL and ADSL, the latter of which has been ratified as ANSI standard T1.413. Telephone companies are using HDSL for their own feeder lines, and it can fulfill the same functions as the conventional T-1 and E-1 infrastructure, without the need for specially conditioned cable and with half as many repeaters along the length of the connection. For home users, however, ADSL is being marketed as an Internet access solution. The technology is still developing, though, as evidenced by the fact that two competing encoding schemes currently are in use. One is called *DMT (discrete multitone),* which splits the available frequency range into 256 channels of 4.3125 KHz each, and the other is called *CAP (carrierless amplitude and phase),* which uses a single channel and a modulation technique called *QAM (quadrature amplitude modulation).* As with the competing 56-Kbps modem standards of a few years ago, the only real concern for the user is that both ends of the link are using the same technology. Most of the service providers offering ADSL Internet access supply the necessary modem (for sale or lease), so you are certain of having compatible devices at either end.

Service	Transmission Rate	Link Length	Simultaneous Voice Capability	Applications
HDSL (High-bit-rate Digital Subscriber Line)	1.544 Mbps full duplex (two wire pairs) or 2.048 Mbps full duplex (three wire pairs)	12,000 to 15,000 feet, more with repeaters	No	T-1/E-1 substitute for Internet connections, LAN and PBX interconnections, and frame-relay traffic aggregation
SDSL (Symmetrical Digital Subscriber Line)	1.544 Mbps full duplex or 2.048 Mbps full duplex (one wire pair)	10,000 feet	Yes	T-1/E-1 substitute for Internet connections, LAN and PBX interconnections, and traffic aggregation
ADSL (Asymmetrical Digital Subscriber Line)	1.544 to 8.448 Mbps downstream; 640 Kbps to 1.544 Mbps upstream	10,000 to 18,000 feet, depending on speed	Yes	Internet/intranet access, remote LAN access, virtual private networking, video-on-demand, voice-over-IP
RADSL (Rate-Adaptive Digital Subscriber Line)	1.544 to 8.448 Mbps downstream; 640 Kbps to 1.544 Mbps upstream	10,000 to 18,000 feet	Yes	Same as ADSL, except that the transmission speed is dynamically adjusted to accommodate the link length and signal quality
ADSL Lite	Up to 1 Mbps downstream; up to 512 Kbps upstream	18,000 feet	Yes	Internet/intranet access, remote LAN access, IP telephony, videoconferencing

Table 7-5:   *DSL Services and Their Properties*

Service	Transmission Rate	Link Length	Simultaneous Voice Capability	Applications
VDSL (Very-high-bit-rate Digital Subscriber Line)	12.96 to 51.84 Mbps downstream; 1.6 to 2.3 Mbps upstream	1,000 to 4,500 feet, depending on speed	Yes	Multimedia Internet access, high-definition television delivery
ISDL (ISDN Digital Subscriber Line)	Up to 144 Kbps full duplex	18,000 feet, more with repeaters	No	Internet/ intranet access, remote LAN access, IP telephony, video-conferencing

Table 7-5:   *DSL Services and Their Properties (continued)*

Packet-Switching Services

A *packet-switching service* transmits data between two points by routing packets through the switching network owned by a carrier such as AT&T, MCI, Sprint, or another telephone company. The end result is a high-bandwidth connection similar in performance to a leased line, but the advantage of this type of service is that a single WAN connection at a network site can provide access to multiple remote sites, simply by using different routes through the network. Frame relay and ATM are the two most popular packet-switching services used today. Frame relay is based on variable-length packets called frames, while ATM uses 53-byte cells instead.

The packet-switching service consists of a network of high-speed connections that is sometimes referred to as a *cloud*. Once data arrives at the cloud, the service can route it to a specific destination at high speeds. It is up to the consumers to get their data to the nearest POP connected to the cloud, after which all switching is performed by the carrier. Therefore, an organization setting up WAN connections between remote sites installs a link to an *edge switch* at a local POP using whatever technology provides suitable performance. This local link can take the form of a leased line, ISDN, DSL, or even a dial-up connection. Once the data arrives at the edge switch, it is transmitted through the cloud to an edge switch at another POP, where it is routed on to a private link connecting the cloud to the destination site (see Figure 7-8).

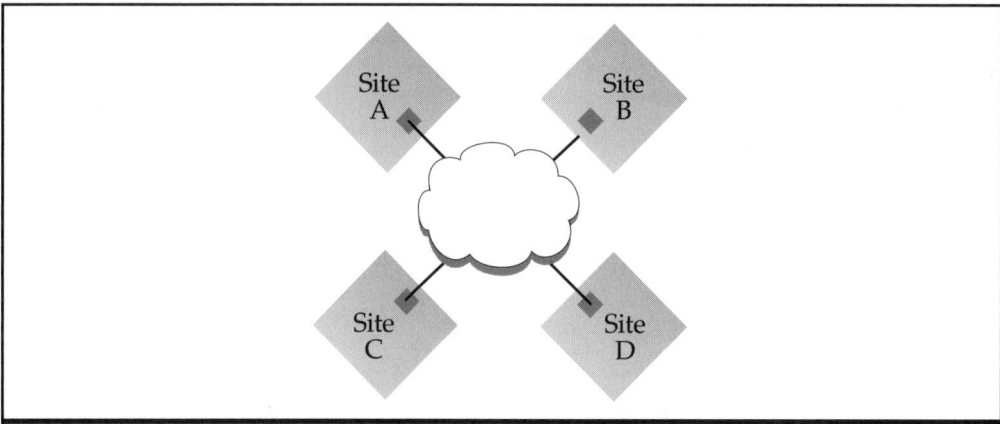

Figure 7-8: *Packet-switching networks use a network cloud to route data between remote sites.*

As mentioned earlier, an organization with eight offices scattered around the country would need 28 leased lines to interconnect all of the sites, some of which may have to span long distances. In this arrangement, the organization does all of its own switching. Using a packet-switching service instead requires one leased line connecting each site to the service's local POP. Eight leased lines is far cheaper than 28, especially when they span relatively short distances. To get the data where it's going, the carrier programs *virtual circuits (VCs)* from the POP used by each site to each of the seven other POPs. Thus, there are still 28 routes connecting each location to every other location, but the service maintains them, and the client only pays for the bandwidth used.

Unlike a leased line, however, a packet-switching service shares its network among many users. The link between two sites is not permanently assigned a specific bandwidth. In some instances, this can be a drawback, because your links are competing with those of other clients for the same bandwidth. However, you can now contract for a specific bandwidth over a frame-relay network, and ATM is built around a quality of service (QoS) feature that allocates bandwidth for certain types of traffic. In addition, these technologies enable you to alter the bandwidth allotted to your links. Unlike a leased line with a specific bandwidth that you can't exceed, and which you pay for whether you're using it or not, you contract with a packet-switching service to provide a certain amount of bandwidth, which you can exceed during periods of heavy traffic (possibly with an additional charge), and which you can increase as your network grows.

Frame Relay

Frame relay has come to be one of the most popular WAN technologies used today, because it provides the high-speed transmission of leased lines with greater flexibility and lower costs. Frame-relay service operates at the data-link layer of the OSI reference model and runs at bandwidths from 56 Kbps to 44.736 Mbps (T-3 speed). You negotiate a *committed information rate (CIR)* with a carrier that guarantees you a specific amount of bandwidth, even though you are sharing the network medium with other users. It is possible to exceed the CIR, however, during periods of heavy use, called *bursts.* A burst can be a momentary increase in traffic or a temporary increase of longer duration. Usually, bursts up to a certain bandwidth or duration carry no extra charge, but eventually, additional charges will accrue.

The contract with the service provider also includes a *committed burst information rate (CBIR),* which specifies the maximum bandwidth that is guaranteed to be available during bursts. If you exceed the CBIR, there is a chance that data will be lost. The additional bandwidth provided during a burst may be "borrowed" from your other virtual circuits that aren't operating at full capacity, or even from other clients' circuits. One of the primary advantages of frame relay is that the carrier can dynamically allocate bandwidth to its client connections as needed. In many cases, it is the leased line to the carrier's nearest POP that is the factor limiting bandwidth.

Frame-Relay Hardware

Each site connected to a frame-relay cloud must have a *frame relay access device (FRAD),* which functions as the interface between the local network and the leased line (or other connection) to the cloud (see Figure 7-9). The FRAD is something like a router, in that it operates at the network layer. The FRAD accepts packets from the LAN that are destined for other networks, strips off the data link–layer protocol header, and packages the datagrams in frames for transmission through the cloud. In the same way, the FRAD processes frames arriving through the cloud and packages them for transmission over the LAN. The difference between a FRAD and a standard router, however, is that the FRAD takes no part in the routing of packets through the cloud; it simply forwards all of the packets from the LAN to the edge switch at the carrier's POP.

The only other hardware element involved in a frame-relay installation is the connection to the nearest POP. In frame relay, the leased line is the most commonly used type of connection. When selecting a carrier, it is important to consider the locations of their POPs in relation to the sites that you want to connect, because the cost of the leased lines (which is not included in the frame-relay contract) depends on their length. The large long-distance carriers usually have the most POPs, scattered over the widest areas, but it is also possible to use different carriers for your sites and create frame-relay links between them.

NETWORK HARDWARE

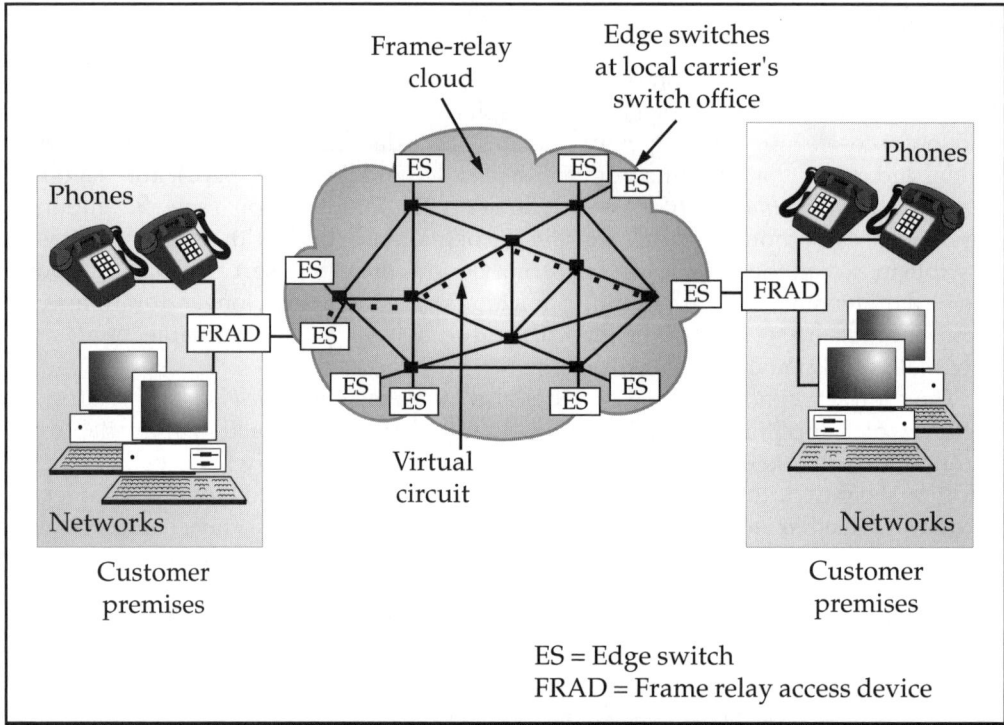

Figure 7-9: *Frame-relay connections use a FRAD to connect a LAN to the cloud.*

When installing leased lines, it is important to take into account the number of virtual circuits that will run from the FRAD to your various sites. Unlike the private network composed of separate leased lines to every site, the single leased-line connection between the FRAD and the carrier's edge server will carry all of the WAN data to and from the local network. Multiple VCs will be running from the edge server through the cloud to the other sites, and the leased line from the FRAD will essentially multiplex the traffic from all of those VCs to the LAN, as shown in Figure 7-10. Thus, if you are connecting eight remote sites together with frame-relay WAN links, the leased line at each location should be capable of handling the combined bandwidth of all seven VCs to the other locations.

In most cases, the actual traffic moving across a WAN link does not utilize all of the bandwidth allotted to it at all times. Therefore, it may be possible to create a serviceable WAN by contracting for T-1-speed VCs between all eight offices and using T-1 leased lines to connect all of the sites to the cloud. Be aware, however, that the leased lines are the only elements of the WAN that are not flexible in their bandwidth. If you find that

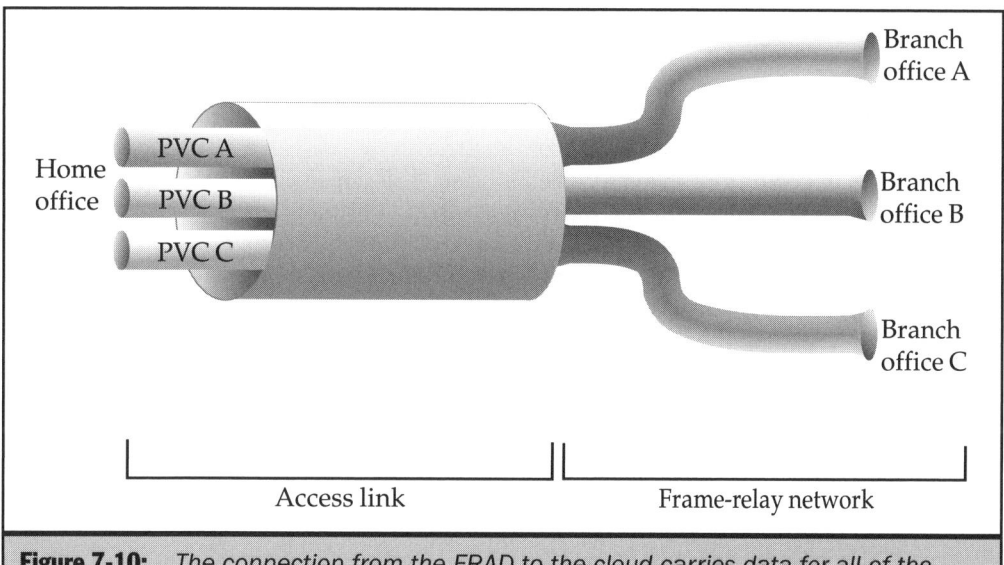

Figure 7-10: *The connection from the FRAD to the cloud carries data for all of the virtual circuits.*

your WAN traffic exceeds the capacity of the leased line, the only recourse is to augment its bandwidth by installing another connection. This does not necessarily mean installing another T-1, however. You can augment the bandwidth connecting the FRAD to the edge server by adding a fractional T-1 or even a dial-up connection that activates during periods of high traffic.

Virtual Circuits

The virtual circuits that are the basis for frame-relay communications come in two types: *permanent virtual circuits (PVCs)* and *switched virtual circuits (SVCs)*. PVCs are routes through the carrier's cloud that are used for the WAN connections between client sites. Unlike standard internetwork routing, PVCs are not dynamic. The frame-relay carrier creates a route through its cloud for a connection between sites, assigns it a unique 10-bit number called a *data link connection identifier (DLCI)*, and programs it into its switches. Programming a FRAD consists of providing it with the DLCIs for all of the PVCs leading to other FRADS. DLCIs are locally significant only; each FRAD has its own DLCI for a particular virtual circuit. Frames passing between two sites always take the same route through the cloud and use the DLCI as a data link–layer address. This is one of the reasons why frame relay is so fast; there is no need to dynamically route the packets through the cloud or establish a new connection before transmitting data.

Each PVC can have its own CIR and CBIR, and despite the description of the VC as permanent, the carrier can modify the route within a matter of hours if one of the sites moves. It is also possible to have the carrier create a PVC for temporary use, such as for a meeting in which a special videoconferencing session is required.

Although it was originally created for data transfers, you can also use frame relay to carry other types of traffic, such as voice or video. This capability is starting to become more popular, but the use of PVCs makes it difficult to implement practically. To set up a voice call or a videoconference between two sites, there has to be a virtual circuit between them. This is easy if the communications are between two of an organization's own sites, which are already connected by a PVC; but to conference with a client or other outside user requires a call to the carrier to set up a new PVC, which can take hours or days. To make applications of this type more practical, carriers are beginning to implement SVCs, as defined in the ITU-T Q.933 document, which are temporary routes through a cloud that are created dynamically as needed.

Frame-Relay Messaging

Frame relay uses two protocols at the data link layer: LAPD for control traffic, and *LAPF (Link Access Procedure for Frame-mode Bearer Services)* for the transfer of user data. The LAPD protocol, the same one used by ISDN (ITU-T Q.921), is used to establish VCs and prepare for the transmission of data. LAPF is used to carry data and for other processes, such as multiplexing and demultiplexing, error detection, and flow control.

The format of the frame used to carry data across a frame-relay cloud is shown in Figure 7-11. The functions of the fields are as follows:

Flag, 1 byte   Contains the binary value 01111110 (or 7E in hexadecimal form) that serves as a delimiter for the frame.

Link Info, 2 bytes   Contains the frame's address and control fields, as follows:

Upper DLCI, 6 bits   Contains the first 6 bits of the 10-bit DLCI identifying the virtual circuit that the frame will use to reach its destination.

C/R (Command/Response), 1 bit   Undefined.

EA (Extended Address), 1 bit   Indicates whether the current byte contains the last bit of the DLCI. The eighth bit of every byte in the Link Info field is an EA bit. When the frames use standard 10-bit DLCIs, the value of this bit will always be 0.

Lower DLCI, 4 bits   Contains the last 4 bits of the 10-bit DLCI identifying the virtual circuit that the frame will use to reach its destination.

NETWORK HARDWARE

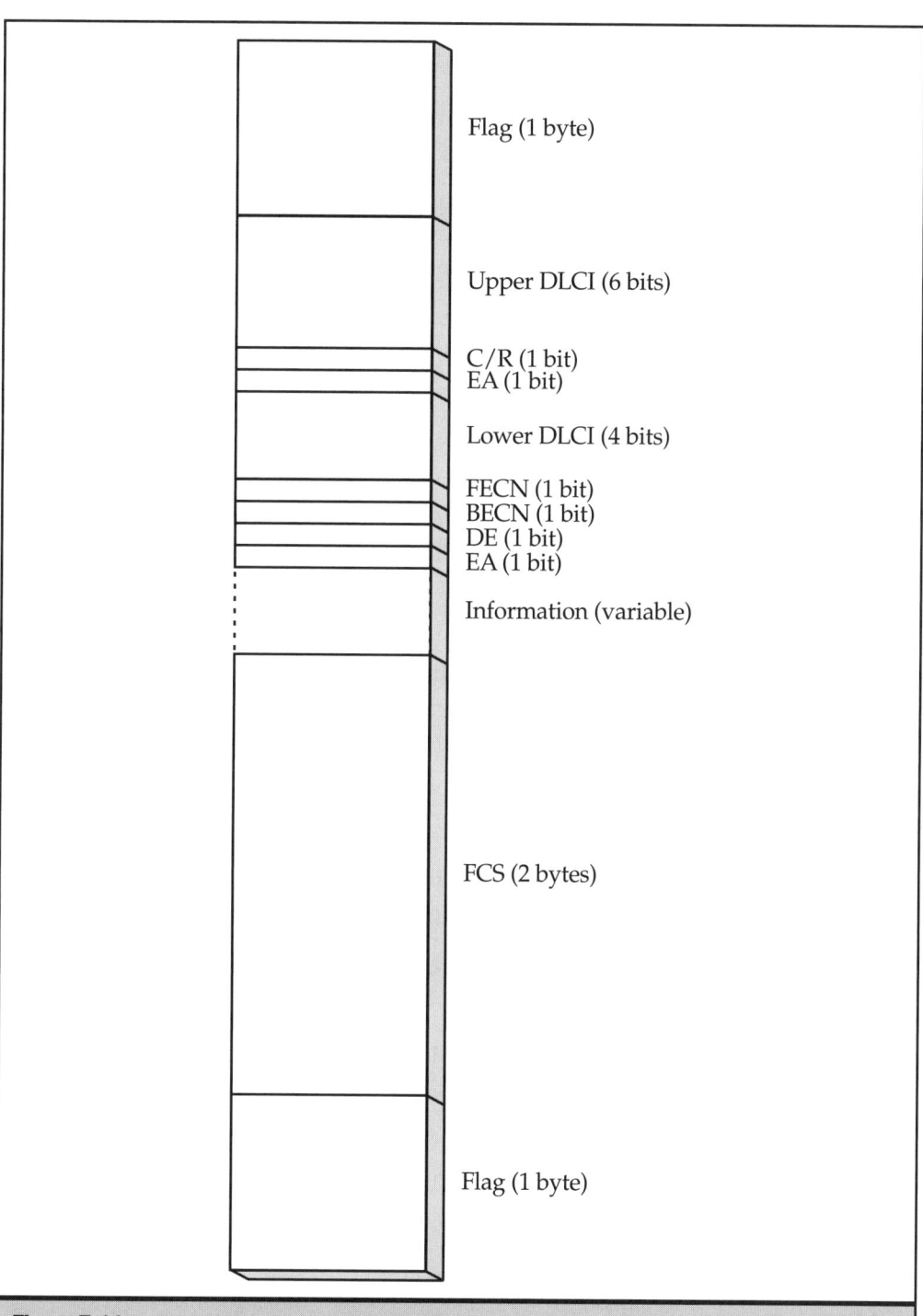

Flag (1 byte)

Upper DLCI (6 bits)

C/R (1 bit)
EA (1 bit)

Lower DLCI (4 bits)

FECN (1 bit)
BECN (1 bit)
DE (1 bit)
EA (1 bit)

Information (variable)

FCS (2 bytes)

Flag (1 byte)

Figure 7-11:   *The frame-relay frame format*

FECN (Forward Explicit Congestion Notification), 1 bit   Indicates that network congestion was encountered in the direction from source to destination.

BECN (Backward Explicit Congestion Notification), 1 bit   Indicates that network congestion was encountered in the direction from destination to source.

DE (Discard Eligibility), 1 bit   Indicates that a frame is of lesser importance than the other frames being transmitted and that it can be discarded in the event of network congestion.

EA (Extended Address), 1 bit   Indicates whether the current byte contains the last bit of the DLCI. When the frames use standard 10-bit DLCIs, the value of this bit will always be 1. The EA field is intended to support the future expansion of frame-relay clouds in which DLCIs longer than 10 bits are needed.

Information, variable   Contains a protocol data unit (PDU) generated by a network-layer protocol, such as an IP datagram. The frame-relay protocols do not modify the contents of this field in any way.

FCS (Frame Check Sequence), 2 bytes   Contains a value computed by the source FRAD that is checked at each switch during the frame's journey through the cloud. Frames in which this value does not match the newly computed value are silently discarded. Detection of the missing frame and retransmission are left to the upper-layer protocols at the end systems.

Flag, 1 byte   Contains the binary value 01111110 (or 7E in hexadecimal form) that serves as a delimiter for the frame.

ATM

Asynchronous Transfer Mode (ATM) has, since the early 1990s, been the holy grail of the networking industry. Fabled as the ultimate networking technology, ATM is designed to carry voice, data, and video over various network media, using a high-speed, cell-switched, connection-oriented, full-duplex, point-to-point protocol. Unfortunately, as with the holy grail, the quest is taking far longer than anyone expected and the ultimate goal continues to be elusive.

The theory behind ATM is perfectly sound. Instead of using variable-length frames like Ethernet, frame relay, and other protocols, all ATM traffic is broken down into 53-byte *cells*. This makes it easier to regulate and meter the bandwidth passing over a connection, because by using data structures of a predetermined size, network traffic becomes more readily quantifiable, predictable, and manageable. With ATM, it's possible to guarantee that a certain quantity of data will be delivered within a given time. This makes the technology more suitable for a unified voice/data/video network than a nondeterministic protocol like Ethernet, no matter how fast it runs. In addition, ATM has quality of service (QoS) features built into the protocol that enable administrators to reserve a certain amount of bandwidth for a specific application.

ATM is both a LAN and WAN protocol, and is a radical departure from the other lower-layer protocols examined in this book. All ATM communication is point-to-point. There are no broadcasts, which means that switching, and not routing, is an integral part of this technology. ATM can also be deployed on public networks, as well as private ones. Public carriers can provide ATM services that enable clients to connect LANs at remote locations. On private networks, ATM implementations at various speeds can run throughout the network, from the backbone to the desktop. Thus, the same cells generated by a workstation can travel to a switch that connects the LAN to an ATM carrier service, through the carrier's ATM cloud, and then to a workstation on the destination network. At no point do the cells have to reach higher than the data link layer of an intermediate system, and transmission speeds through the cloud can reach as high as 2.46 Gbps.

This dramatic potential has not been realized in the real world on a large scale, however. The reality of the situation is that ATM is being used as a high-speed backbone protocol and for WAN connections, but the 25.6 Mbps ATM LAN solution intended for desktop use has been eclipsed by Fast Ethernet, which runs at 100 Mbps and is far more familiar to the majority of network administrators. Approximately 20 percent of installed enterprise backbones run over ATM, largely because administrators find that its QoS capabilities and support for voice, data, and video make it a better performer than traditional LAN protocols.

You can use an ATM packet-switching service for your WAN links in roughly the same way as you would use frame relay, by installing a router at your sites and connecting them to the carrier's POPs using leased lines. This process transmits the LAN data to the POP first and then repackages it into cells. It's also possible, however, to install an ATM switch at each remote site, either as part of an ATM backbone or as a separate device providing an interface to the carrier's network (see Figure 7-12). This way, the LAN data is converted to ATM cells at each site before it is transmitted over the WAN.

Like frame relay, ATM supports both PVCs and SVCs, but ATM was designed from the beginning to support voice and video using SVCs, while in frame relay, PVCs and SVCs were a later addition. It is expected that the need to transmit alternative data types like voice and video will become more prevalent in the future, and if this happens, ATM will have an advantage over frame relay because of its greater speed and manageability. However, at this time, frame relay provides a more economical WAN solution running at speeds that are sufficient for standard data networking tasks.

NETWORK HARDWARE

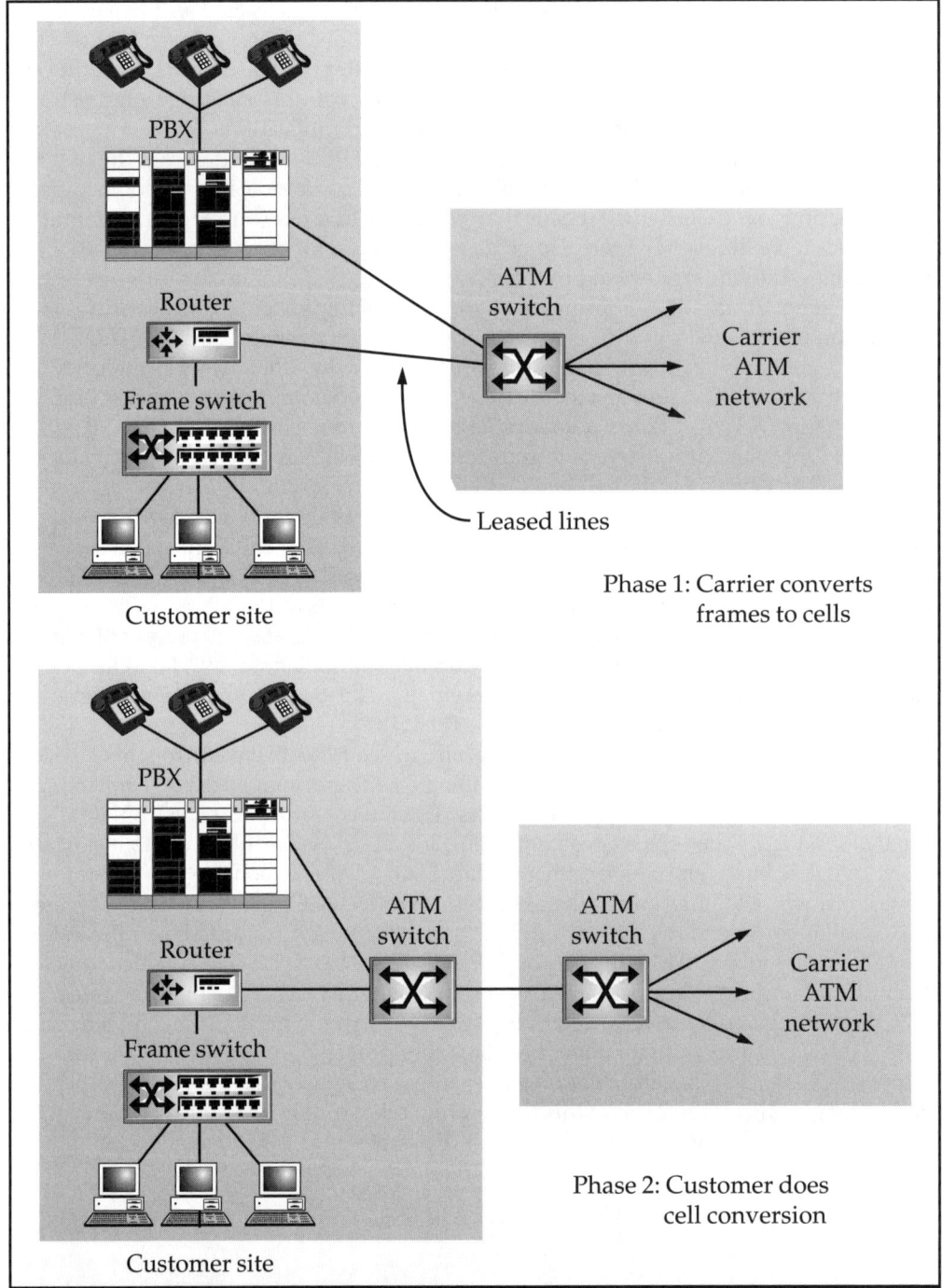

Figure 7-12:   *The conversion of LAN data into ATM cells can occur at either the LAN site or the ATM carrier's site.*

ATM Architecture

Many of the familiar concepts of other protocols, such as media access control and variable-length frames, are not applicable to ATM. Because ATM does not share bandwidth among systems, there is no need for a MAC mechanism such as CSMA/CD or token passing. Switches provide a dedicated connection to every device on the ATM network. Because all ATM transmissions are composed of fixed-length cells, the switching process is simpler and predictable. All ATM switching is hardware based because there is no need for software-managed flow control and other such technologies. References to ATM systems and devices refer to switches and routers, as well as actual computers.

The bandwidth delivered by an ATM network is also readily quantifiable, making it easier to designate the appropriate amount of bandwidth for a specific application. On an Ethernet network, for example, it may be necessary to provide much more bandwidth than is actually needed to ensure good performance from a videoconferencing application. This is because you must account for the bandwidth required for videoconferencing on top of the maximum bandwidth used by all other applications combined. The network, therefore, is designed to accommodate the peak traffic condition that occurs only a small fraction of the time. On an ATM network, bandwidth can be more precisely calculated.

Like Ethernet and Token Ring, ATM encompasses the physical and data link layers of the OSI reference model, but is itself divided into three layers (see Figure 7-13), which are as follows:

- Physical layer
- ATM layer
- ATM adaptation layer

The following sections examine the functions performed at each of these layers.

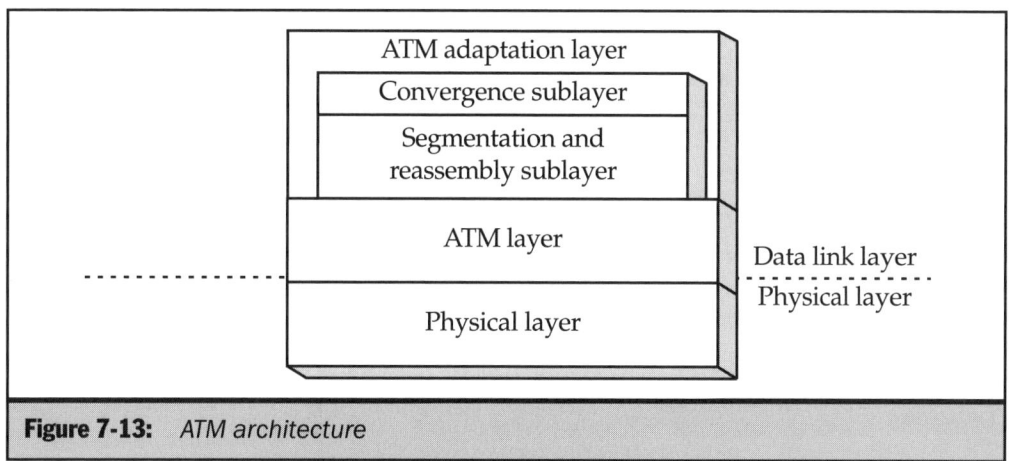

Figure 7-13: *ATM architecture*

The Physical Layer

The ATM standards do not specify precise physical-layer technologies as most other data link–layer protocols do. This media independence is one of the guiding design principles behind the technology. ATM can run at various speeds over SONET (Synchronous Optical Network) and DS-3 connections, and locally over multimode fiber-optic and shielded twisted-pair (STP) cable, among others. Speeds range from 25.6 Mbps for desktop connections to 2.46 Gbps, although the most common implementations run at 155 or 625 Mbps. It seems unlikely that the 25.6 Mbps implementation will ever become popular on the desktop, as Fast Ethernet provides four times the speed at far less cost. The higher speeds are commonly used for backbones and WAN connections.

SONET is a fiber-optic standard that defines a series of optical carrier (OC) services that range from OC-1, operating at 51.84 Mbps, to OC-192 at 9,952 Mbps.

The ATM physical layer itself is divided into two sublayers, called the *physical medium dependent (PMD)* sublayer and the *transmission convergence (TC)* sublayer. The PMD sublayer defines the actual medium used by the network, including the type of cable and other hardware, such as connectors, and the signaling scheme used. This sublayer is also responsible for maintaining the synchronization of all the clocks in the network systems, which it does by continuously transmitting and receiving clock bits from the other systems.

The TC sublayer is responsible for the following four functions:

Cell delineation   Maintains the boundaries between cells, enabling systems to isolate cells within a bit stream.

Header error control (HEC) sequence generation and verification   Ensures the validity of the data in the cells by checking the error-control code in the cell headers.

Cell rate decoupling   Inserts or removes idle cells to synchronize the transmission rate to the capacity of the receiving system.

Transmission frame adaptation   Packages cells into the appropriate frame for transmission over a particular network medium.

The ATM Layer

The ATM layer specifies the format of the cell, constructs the header, implements the error-control mechanism, and creates and destroys virtual circuits. There are two versions of the cell header, one for the *User Network Interface (UNI)*, which is used for communications between user systems or between user systems and switches, and the *Network-to-Network Interface (NNI)*, which is used for communications between switches.

In each case, the 53 bytes of the cell are divided into a 5-byte header and a 48-byte payload. Compared to an Ethernet header, which is 18 bytes, the ATM header seems quite small, but remember that an Ethernet frame can carry up to 1,500 bytes of data. Thus, for a full-sized Ethernet frame, the header is less than 2 percent of the packet, while an ATM header is almost 10 percent of the cell. This makes ATM considerably less efficient than Ethernet, as far as the amount of control data transmitted across the wire is concerned.

The format of the ATM cell is shown in Figure 7-14. The functions of the fields are as follows:

Generic flow control (GFC), 4 bits   Provides local functions in the UNI cell that are not currently used and are not included in the NNI cell.

Virtual path identifier (VPI), 8 bits   Specifies the next destination of the cell on its path through the ATM network to its destination.

Virtual channel identifier (VCI), 16 bits   Specifies the channel within the virtual path that the cell will use on its path through the ATM network to its destination.

Payload type indicator (PTI), 3 bits   Specifies the nature of the data carried in the cell's payload, using the following bit values:

Bit 1   Specifies whether the cell contains user data or control data.

Bit 2   When the cell contains user data, specifies whether congestion is present on the network.

Bit 3   When the cell contains user data, specifies whether the payload contains the last segment of an AAL-5 PDU.

Cell loss priority (CLP), 1 bit   Specifies a priority for the cell, which is used when a network is forced to discard cells because of congestion. A value of 0 indicates a high priority for the cell, while a value of 1 indicates that the cell may be discarded.

Header error control (EC), 8 bits   Contains a code computed on the preceding four bits of the header, which is used to detect multiple-bit header errors and correct single-bit errors. This feature detects errors in the ATM header only; there is no error control of the payload at this layer.

Payload, 48 bytes   Contains the user, network, or management data to be transported in the cell.

Note   *The only difference between the UNI header and the NNI header is the GFC field, which is omitted from NNI cells. The four bits from the GFC field are, in this case, added to the VPI field, making it 12 bits long instead of 8.*

NETWORK HARDWARE

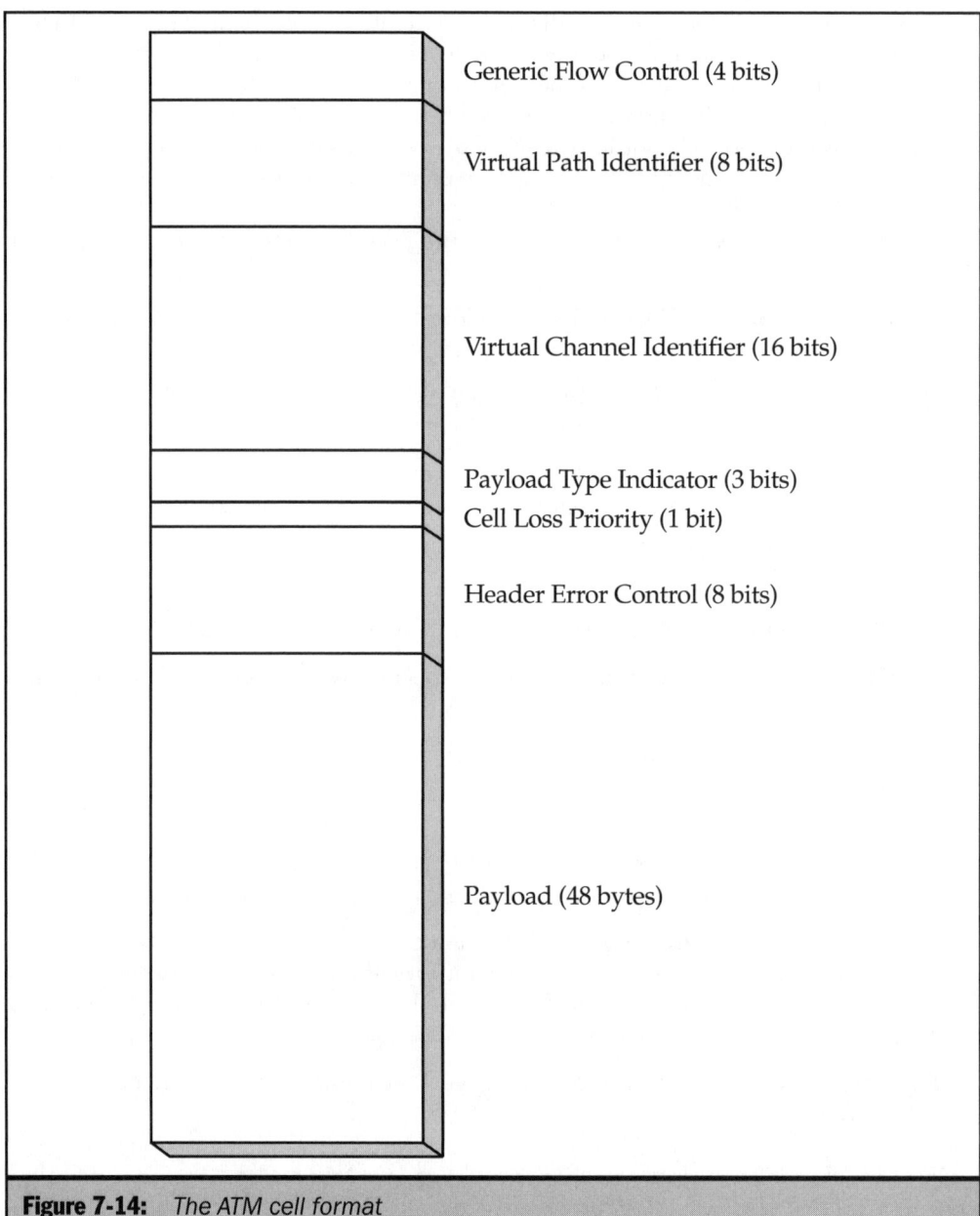

Generic Flow Control (4 bits)

Virtual Path Identifier (8 bits)

Virtual Channel Identifier (16 bits)

Payload Type Indicator (3 bits)
Cell Loss Priority (1 bit)

Header Error Control (8 bits)

Payload (48 bytes)

Figure 7-14:   *The ATM cell format*

Virtual Circuits   A connection between two ATM systems takes the form of a *virtual circuit*. Like frame relay, ATM uses two types of virtual circuits: *permanent virtual circuits (PVCs)*, which network administrators manually create and which are always available, and *switched virtual circuits (SVCs)*, which systems dynamically create as needed and then terminate after use.

Establishing a virtual circuit through the network to a destination enables the transmission of cells through that circuit without extensive processing by intermediate systems along the way. A virtual circuit is composed of a *virtual path (VP)* and a *virtual channel (VC)*. A virtual path is a logical connection between two systems that is comprised of multiple virtual circuits, much as a cable between two points can contain multiple wires, each carrying a separate signal. Once a VP is established between two points, creating an additional VC for a new connection within that VP is a relatively simple matter.

In addition, managing the VP is an easy way of modifying the properties of all of the VCs it contains. When a switch fails, for example, the VP can be rerouted to use another path, and all of its VCs are rerouted with it. Every ATM cell header contains a virtual path identifier and a virtual channel identifier, which specify the VP that the cell is using and the VC within that VP.

ATM Addressing   ATM networks have their own addresses for each device, in addition to any upper-layer addresses they might possess. The addresses are 20 bytes long and hierarchical, much like telephone numbers, enabling them to support extremely large networks. Unlike protocols that share network bandwidth, it isn't necessary to include source and destination addresses in each cell because ATM transmissions use dedicated point-to-point links. Instead, the addresses are used by the ATM switches to establish the VPIs and VCIs for a connection.

The ATM Adaptation Layer

The primary function of the *ATM adaptation layer (AAL)* is to prepare the data received from the network-layer protocol for transmission and segment it into 48-byte units that the ATM layer will package as cells by applying the header. The AAL consists of two sublayers, called the *convergence sublayer (CS)* and the *segmentation and reassembly sublayer (SAR)*. The CS prepares the network-layer data for segmentation by applying various fields that are specific to the type of service that will transmit the data, creating CS-PDUs (convergence sublayer protocol data units). The SAR then splits the CS-PDUs into segments of the appropriate size for packaging in cells.

Several AAL protocols are available at this sublayer, which provide different types of service to support various applications. The AAL protocols are as follows:

   AAL-1   A connection-oriented service intended for applications that require circuit emulation, such as voice and videoconferencing. This service requires clock synchronization, so a network medium that supports clocking, such as SONET, is required. For this service, the CS sublayer adds Sequence Number (SN) and Sequence Number Protection (SNP) fields to the data that enable the receiving system to assemble the cells in the proper order.

AAL-3/4   Supports both connection-oriented and connectionless data transfers with cell-by-cell error-checking and multiplexing. The CS creates a PDU by adding a beginning/end tag to the data as a header and a length field as a footer. After the SAR layer splits the CS-PDU into cell-sized segments, it adds a CRC value to each segment for error-detection purposes.

AAL-5   Also called *SEAL (Simple and Efficient Adaptation Layer)*, AAL-5 provides both connection-oriented and connectionless services, and is most commonly used for LAN traffic. The CS takes a block of network-layer data up to 64KB in size and adds a variable-length pad and an 8-byte trailer to it. The pad ensures that the data block falls on a cell boundary, and the trailer includes a block length field and a CRC value for the entire PDU. The SAR splits the PDU into 48-byte segments for packaging into cells. The third bit of the PTI field in the ATM header is then set to a value of 0 for all of the segments of the data block except the last one, in which it is set to 1.

ATM Drawbacks

There are serious drawbacks to consider before you even think about implementing ATM on a backbone (or any) network. The first is the cost: ATM hardware is much more expensive than that for Gigabit Ethernet or virtually any other high-speed protocol. ATM server NICs start at $500, and the prices for ATM switches run quickly into four and five figures.

Support

The other big problem is the cost and complexity of installing and supporting an ATM network. While a competent Ethernet LAN administrator should be able to install the components of a Gigabit Ethernet backbone with little trouble, an ATM backbone is a completely different story. ATM networks are a hybrid of telecommunications and data networking technologies. These are two separate types of networks, but in the case of ATM, both can use the same cables and switches. An ATM backbone, therefore, may be connected not only to data networking components such as routers, switches, and servers, but also to PBXs and other telecommunications devices.

The fact that these two types of networks are traditionally separated means that the people who support them are often separated as well. People with the advanced knowledge of both disciplines needed to master the intricacies of ATM are relatively rare and therefore command high salaries. As a result, part of the danger in using ATM is that you are likely to be trusting your network to a technology that only one or two highly trained people in your company fully understand. In today's volatile job market, this is a bad thing.

LANE

Another problem with using ATM as a network backbone is the need to make it run with other protocols. Unless you use ATM throughout your network, including runs to the desktops, the backbone must connect individual LANs running Ethernet or other data link–layer protocols. This is a problem because ATM is a connection-oriented protocol and Ethernet and Token Ring are connectionless. Connection-oriented protocols have no broadcast capabilities, so there is no inherent means for a system on an Ethernet (or Token Ring) LAN to discover the address or even the existence of a server on the ATM backbone.

Originally, the solution to this problem was for network administrators to map PVCs between the ATM/Ethernet switches and the servers and other resources on the ATM network. This way, the switches provide the address needed for the Ethernet systems to communicate with the ATM systems. This is a functional solution, but it requires a good deal of administrative effort that grows with the number of systems on the ATM network.

In 1995, *LANE (LAN Emulation)* 1.0 was introduced, which eliminates the need for this manual PVC configuration. LANE automatically creates and deletes SVCs between switches and ATM network systems. This makes the backbone transparent to the Ethernet LANs and creates a virtual broadcast domain that includes the ATM systems. LANE 2.0, ratified in 1997, adds support for ATM's QoS classes and multicasting.

LANE, unfortunately, requires many different software modules to do its job, including the following:

LAN Emulation Client (LEC)   A module on each ATM device that has both a LAN MAC address and an ATM address. These devices together form an *emulated LAN (ELAN)*. Ethernet systems contact the LECs on the switches joining the networks to discover the addresses of ATM systems. If the LEC is currently communicating with the requested ATM system, it furnishes the address to the Ethernet system. If not, the LEC contacts the LES and the BUS using the *LAN Emulation Address Resolution Protocol (LE_ARP)* to discover the address. Once the address is resolved, the LEC also performs data-forwarding services.

LAN Emulation Server (LES)   Maintains a database containing all of the addresses known to the LECs on an ELAN. All LECs have a *virtual channel connection (VCC)* to the LES that they use to request the addresses of specific systems on the ATM network. If the LES does not have the requested address in its database, it forwards the request to the other LECs on the network.

Broadcast and Unknown Server (BUS)   If the LES is unable to discover the proper address for the requested system, the BUS establishes VCCs to all of the LECs on the network and floods them with cells, which the LECs forward to the devices on the Ethernet LANs to which they are attached. The BUS also provides a broadcast emulation service by forwarding messages to all of the LECs on the network.

LAN Emulation Configuration Server (LECS)   Maintains a database of configuration information for each ELAN, coordinates the activities of the other modules, and inserts new clients into ELANs.

This is a lot of complex software just to make an ATM network visible to Ethernet or other LAN protocols, when a Gigabit Ethernet backbone can do same thing transparently. Many network administrators are staying away from ATM because it requires so much effort to perform what should be simple tasks. Many of the organizations that find ATM most suitable for their backbones are those that require support for voice and video, as well as data, such as campus networks servicing hospitals and universities. Medical imaging and remote learning applications need the dedicated bandwidth that ATM can provide more efficiently than other technologies. However, if you're upgrading an existing Ethernet LAN, already have a telephone network installed, and high-bandwidth applications like videoconferencing are not in your organization's immediate future, then Gigabit Ethernet is probably a more practical and economical choice.

The
Complete
Reference

Networking

Chapter 8

Server Technologies

Ⅱ of the computers on a local area network contain roughly the same components, such as a microprocessor, memory modules, mass storage devices, keyboards, video adapters, and other input/output mechanisms. However, you can still divide the computers into two basic categories: servers and client workstations. When Novell NetWare was the predominant network operating system on the market, it was easy to differentiate between servers and clients, because servers functioned only as servers and clients only as clients. Servers in those days were essentially computers with more of everything: faster processors, more memory, and larger hard drives, for example. Now that peer-to-peer network operating systems like Windows and Unix are the industry standard, and most computers can function as both servers and clients simultaneously, the boundary between the server and client functions has been obscured somewhat. Recent years have seen great developments in the features and technologies that make a server different from a workstation. This chapter examines some of these features and technologies and explains how they can enhance the performance of your network.

Purchasing a Server

When building a local area network (LAN), you can purchase virtually any computer and use it as a server. The primary attributes that make a computer a server are determined by the network operating system's hardware requirements. For example, the Windows 2000 Server requirements call for 256 megabytes of memory, but you can actually run the operating system on a standard workstation computer with as little as 128 megabytes. It won't run as well, but it will run. When shopping for computers, you'll see that some products are specifically designed to be servers, and not just because of the operating system installed on them or the amount of memory or disk space they contain. For a small network consisting of only a handful of nodes, it may not be practical for you to spend the extra money on a computer that is designed to be a server. Instead, you can purchase a high-end workstation with sufficient resources to run the server operating system, and use that instead. When you do need the features of a real server, it's important to understand how a server can differ from a workstation, and which features you need for your network.

When you look at the description of a server computer in a catalog or on a Web site, it may seem at first as though you're paying more money for less. Servers often do not come with monitors, and they generally do not include the high-performance video adapters and audio systems you find in nearly every home or office computer package. You also are not likely to get the free peripherals, such as printers and scanners, that are so often bundled with home computers these days.

The video adapter in a server is in many cases integrated into the computer's motherboard and includes sufficient memory to power a display at a variety of resolutions. However, the video subsystem in a server usually does not include the 3-D accelerator and other components found on a separate adapter card used in a workstation for more video-intensive tasks, such as game-playing and multimedia applications. A video adapter in a server also tends not to use the Accelerated Graphics Port (AGP) for its interface to the computer, because AGP uses system memory for some of its functions, and in a server, you want as much system memory as possible to be devoted to your server applications.

As for audio, most servers include no audio adapter at all, or at most, a rudimentary one that is also integrated into the motherboard. Speakers are usually not included. The only purpose for having any audio capabilities in a server is to provide audible feedback alerting the administrator of particular system conditions. However, since servers are often kept in a locked closet or data center, even this basic audio capability often isn't necessary. Servers can have other means of sending alerts to an administrator, such as the Simple Network Management Protocol (SNMP), as covered in Chapter 31.

Note *Although servers generally do not come equipped with high-end video and audio adapters, there is usually no reason why you can't add them later and use the computer for tasks more traditionally associated with client workstations.*

The question then remains, what you do get when you purchase a server for more money than you would spend on a workstation with the same processor and a comparable amount of memory and disk space? The following list examines the ways in which the basic components in a server differ from their counterparts in a workstation:

Case   A server case is typically larger than that of a workstation, in order to provide room for greater expansion. Server cases are usually either freestanding towers or specially designed to be mounted in a standard 19-inch equipment rack. Expandability is an important quality in a server, and the cases typically have a large number and variety of bays to support the installation of additional drives, such as hard disks, CD-ROMs, and tape drives. Since a server doesn't usually take up space on a user's desk, maintaining a small footprint is not a concern, and server cases tend not to have their components shoehorned into them, in the interest of saving space. The result is that there is more room to work inside the case and easier access to the components. A server case might also have greater physical security than a standard computer case, such as a key-lockable cover that prevents any access to the server controls and drives.

Power supply   To support the greater number of drives and other devices frequently found in a server, the power supply is typically more robust, often supplying 300 watts of power or more. The power supply usually also has more internal power connectors available to attach to installed devices. In some cases, a server's power supply might have its own internal surge protection circuitry. Some servers also have redundant power supplies, providing fault tolerance in the event of a power supply failure.

Fans   The possibility of having many more drives and multiple processors in a server means that the computer can potentially generate a lot more heat than a workstation. Server cases typically have multiple fans in them, aside from the one in the power supply. A well-designed case will also have a carefully planned ventilation path that blows the cooler air from the outside directly across the components that most need to be kept cool. In some cases, servers use a sealed case design in which all of the air entering the case runs through a filter, enabling the server to function in an industrial environment without contaminating the internal components with dust and other particles. Some high-end servers designed for mission critical applications also have hot-swappable modular fan assemblies, meaning that should a fan fail, it's possible to replace the unit without shutting down the server.

Processor   Servers use the same model processors as workstations, and given the computer industry's dedication to aggressively marketing the newest and fastest processors to home users, you may find that a server's processor is not any faster than a workstation's. In fact, because servers are designed with an emphasis on expandability, and because they cost more, they tend to have longer lives than workstations, meaning that they might have a processor that is slower than the "latest and greatest." Where servers do differ from workstation in this area is that they often have more than one processor. For more information, see "Using Multiple Processors," later in this chapter.

Memory   Servers are typically capable of supporting more memory than workstations, sometimes a lot more. Workstations today often come with 64 or 128 megabytes of memory standard, and can support additional memory up to 512 megabytes and sometimes more. Servers may also have only 128MB in their standard configurations, but they can usually handle up to two gigabytes of memory or more. Examining the inside of the server and a workstation, you may not see any difference, as a server may have the same number of memory slots as a workstation and use the same basic type of memory modules. The server will support modules containing more memory, however, in a greater variety of configurations.

Drive interface   The Small Computer System Interface (SCSI) is the interface of choice for server hard drives and tapes drives, while workstations are more likely to use the Integrated Drive Electronics (IDE) interface. For more information on the differences between SCSI and IDE, see "Server Storage Technologies," later in this chapter. Some servers have the SCSI host adapter integrated into the motherboard, while others come with a SCSI adapter card installed into a bus slot. The server's motherboard may also have an IDE interface built into it and may use it for the CD-ROM drive, but the server's hard drives are SCSI.

Expansion bus   The Peripheral Component Interconnect (PCI) bus is the industry standard for all personal computers today, and servers tend to differ from workstations in this respect mainly in the number of PCI bus slots they provide. Some of the servers being manufactured today have eliminated the venerable Industry Standard Architecture (ISA) bus, and contain only PCI bus slots, usually more than are found in the typical workstation.

In addition to these differences in a server's basic components, there are other more advanced technologies that can have an even greater impact on the computer's performance, as discussed in the following sections.

Using Multiple Processors

Even though the processor designs used in computers today are continually being enhanced and upgraded to run at ever faster speeds, servers often require more processing power than any single processor can provide. This is because a server application such as a database engine may have to service requests from dozens or even hundreds of users at the same time. To increase the processing power available to the application, you can add additional processors. You can multiply the processing power of a server in two ways: by installing multiple processors into the computer or by connecting multiple computers together using a hardware or software product that joins them into a cluster or a *system area network (SAN)*.

Parallel Processing

The use of multiple processors in a single computer is not a new idea, although it has only become common in the PC industry in the last few years. The two biggest advantages of using multiple processors are economy and expandability. When a processor manufacturer releases a new product, its price compared to the previous models is always disproportionately high for the performance increase it provides. As that new processor is superceded by the next model, its price drops quickly. By purchasing a server with multiple processors in it, you can realize nearly the same processing power as the latest chip on the market for much less money. With all other things being equal, a server with two 500 MHz processors is going to cost significantly less than the same server with a single 1 GHz processor, and, depending on the capabilities of the software, provide nearly the same performance. Multiple processor support can also extend the life of a server by enabling the owner to upgrade it as needed. You can buy a single-processor server containing a motherboard that supports up to four processors for only slightly more than a computer with a standard single processor motherboard. Later, as the burden on the server is increased by the addition of more users or applications, you can buy additional processors and install them into the empty motherboard sockets.

The method by which a computer makes use of multiple processors is known as *parallel processing*. This consists of distributing computing tasks among the available processors so that they are all continuously active. There are various methods in which computers with multiple processors can implement parallel processing. Supercomputer systems, for example, can combine the capabilities of hundreds of processors to perform complex tasks that require enormous numbers of computations, such as weather forecasting. In most cases, these supercomputers use a technique called *massively parallel processing (MPP)*, in which the processors are grouped into nodes and connected by a high-speed switch. In this arrangement, each node has its own memory array and its own bus connecting the processors to the memory. There is no sharing of resources between nodes, and communication between them is restricted to a dedicated messaging system.

Symmetric Multiprocessing

The servers with multiple processors used on LANs today employ a different method, called *symmetrical multiprocessing (SMP)*. In an SMP system, the processors share a single memory array, input/output (I/O) system, and interrupts, as shown in Figure 8-1. Processing tasks are distributed evenly between all of the processors, so it isn't possible for one processor to be overloaded while another sits idle. This is in contrast to another system, called *asymmetrical multiprocessing,* in which tasks are assigned to each processor individually and the workload may not be balanced.

Sharing a single memory array eliminates the need for the messaging system found in MPP. The processors in an SMP computer can communicate and synchronize their activities more quickly than most other parallel processing technologies. One resource that isn't shared in an SMP computer, however, is level 2 cache memory, which in the case of today's Pentiums, is built into the processor package. This causes a problem, because one processor may at times have to access data stored in another processor's cache. This problem is resolved by the motherboard hardware, which enables a processor to read from an address in another processor's cache, write it in its own cache, and invalidate the original cache entry.

This cache manipulation generates additional overhead, and is one of the contributing reasons why the aggregate processing power of an SMP computer is not equivalent to that of a computer with one processor running at a faster speed. In other words, a server with two 500 MHz processors will provide slightly less throughput than one with a single 1 GHz processor. However, when you're talking about a relatively small number of processors, the additional overhead generated by SMP is nowhere near that of MPP and other types of parallel processing. As you move beyond about eight processors in an SMP computer, the contention of the multiple chips for the memory and bus resources begins to be more of a problem.

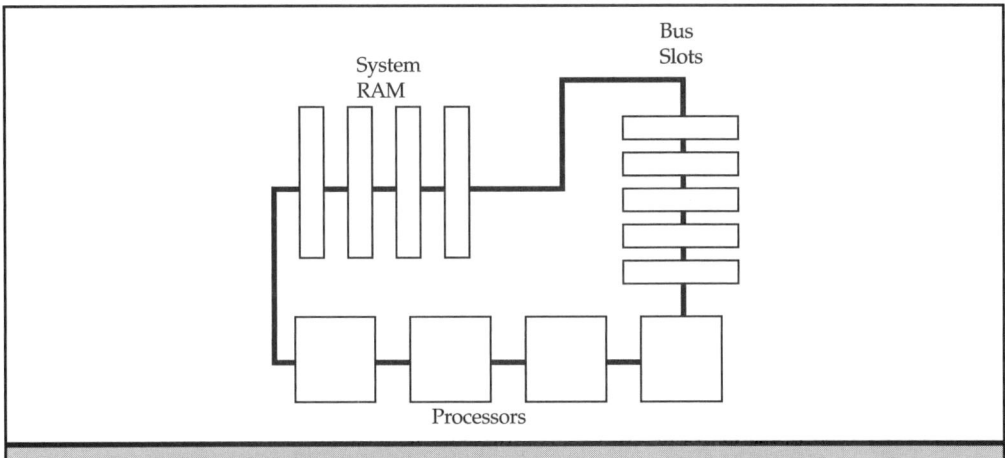

Figure 8-1:   *SMP computers have a single memory array and I/O bus, which are shared by all of the processors.*

It is important to note that having multiple processors in a computer is not considered to be a fault tolerance mechanism. If one of the processors should fail while the system is running, the coherency of the cached operating system and application information are likely to be affected, eventually causing a crash. Failure or removal of a processor while the computer is shut down, however, will not have a deleterious effect, since the operating system detects the number of available processors during the startup sequence and configures itself accordingly.

Hardware and Software Requirements

In order to use multiple processors in a LAN server, SMP must be supported by the processors themselves, the computer's motherboard, the operating system, and the applications running on the server. If you install an operating system or an application that doesn't support SMP on a server with multiple processors, the software functions in the normal manner using only one of the processors. SMP is supported by all of the Intel processors 80486 and above, as well as the SPARC, Alpha, and PowerPC platforms. The Intel clone processors, Via's Cyrix 6x86 and AMD's K6, can't use the Advanced Programmable Interrupt Controller (APIC) standard to implement SMP, because

NETWORK HARDWARE

it's owned by Intel. Instead, they use another standard called OpenPIC. Motherboards supporting multiple processors, either on the board itself or using plug-in daughterboards, are available from many manufacturers.

Upgrading to SMP

When you install Windows NT or Windows 2000 on a server, the operating system detects the number of actual processors in the computer and installs the appropriate Hardware Application Layer (HAL) driver to support them. If the computer has a motherboard that supports multiple processors, but there is only one processor chip installed, the setup program installs the single processor version of the HAL driver. If you add another processor to the computer at a later time, you must upgrade the HAL driver in order for the operating system to recognize and use the additional processor(s).

In Windows NT, you upgrade the HAL driver by running a command-line utility included with the operating system, called Uptomp.exe. In Windows 2000, you upgrade the driver in the Device Manager application, using the following procedure:

1. Open the Control Panel from the Start/Settings group and double-click the System icon.

2. Select the Hardware tab in the System Properties dialog box and click the Device Manager button.

3. In the Device Manager window (see Figure 8-2), expand the Computer entry and double-click the computer type listed beneath it to display the Properties dialog box for the currently installed HAL driver.

4. Select the Driver tab in the Properties dialog box and click the Update Driver button. This launches the Upgrade Device Driver Wizard.

5. After clicking the Next button to bypass the wizard's Welcome page, select the Display A List Of Known Drivers For This Device radio button and then click the Show All Hardware Of This Device Class radio button.

6. Select the appropriate HAL driver from the list shown (see Figure 8-3). Click the Next and Finish buttons to complete the driver upgrade. Depending on the server you're using, you may have to use a HAL driver supplied by the computer's manufacturer instead of those provided in Windows 2000. If this is the case, click the Have Disk button and browse to the location of the driver.

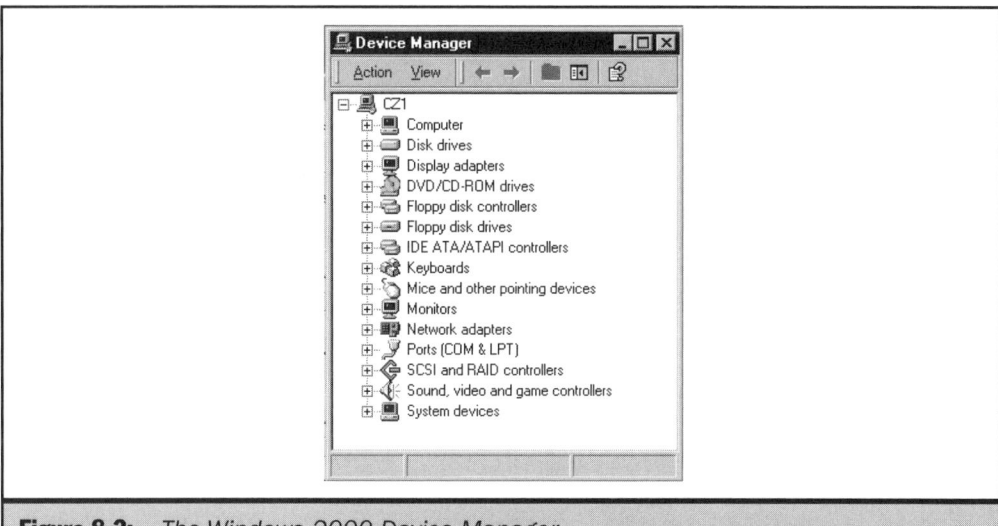

Figure 8-2:   *The Windows 2000 Device Manager*

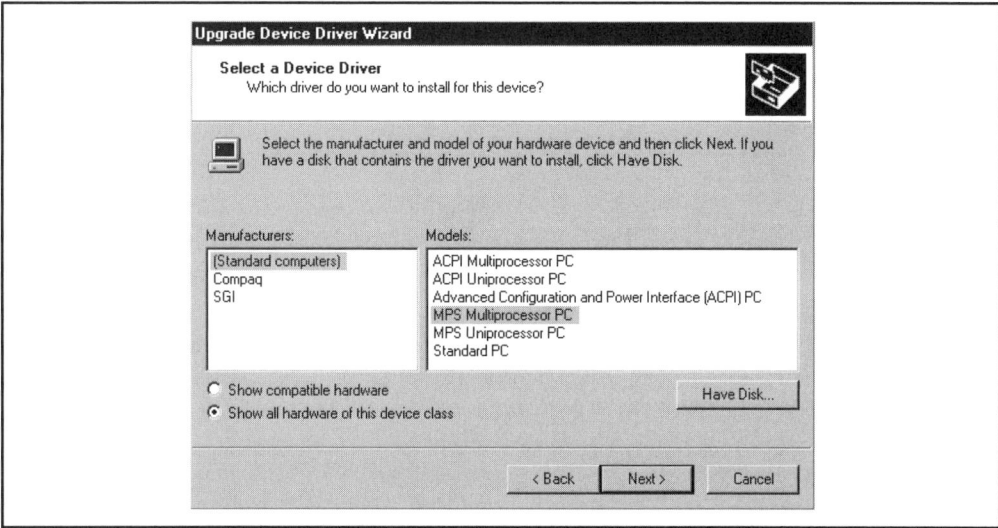

Figure 8-3:   *The Select A Device Driver page of the Upgrade Device Driver Wizard*

Most of the operating systems intended for use on servers support SMP. There are three Windows 2000 Server products, which differ mainly in the number of processors they can support. Windows 2000 Server supports up to 4 processors, Windows 2000 Advanced Server supports up to 4, and Windows 2000 Datacenter Server supports from 8 to 32 processors. Windows NT also supports SMP, but Windows Me, 98, 95, and MS-DOS do not. Most of the Unix operating systems support SMP, including Linux version 2.0 and higher, as well as MacOS 9 and OS/2 Warp Server. In some cases, such as FreeBSD, you have to substitute a multiprocessor kernel for the standard one supplied with the operating system.

Applications that support SMP include the Microsoft BackOffice servers, such as SQL and Exchange, as well as Lotus Notes and other database managers, such as the SQL products from Oracle, Informix, and Sybase. Interestingly, although it is not considered a server application, Adobe Photoshop also supports SMP, making it possible for graphic designers working with large image files and complex functions to take advantage of a computer with multiple processors.

Server Clustering

A *cluster* is a group of servers that are connected by cables and that function as a single entity. To a client on the network, the cluster appears to be a single server, even though it consists of two or more computers. Clustering can provide the same advantage as having multiple processors in a single server, since it is possible to divide the server's workload between the processors in the various computers that make up the cluster. However, clustering can also provide fault tolerance in ways that SMP cannot.

The computers that make up a cluster are connected programmatically as well as physically. In some cases, operating systems provide direct support for clustering, while in others, a separate application is required. For example Windows NT 4.0 Enterprise Edition supports clustering through the use of a separate product, called Microsoft Cluster Server. However, clustering is built into the Windows 2000 Advanced Server and Windows 2000 Datacenter Server operating systems, in the form of a Clustering Service. Clustering is supported by several other server operating systems, including Novell NetWare and some Unix variants.

Clustering can provide two basic advantages over a single server: load balancing and fault tolerance. Load balancing is the process by which the tasks assigned to the server are distributed evenly among the computers in the cluster. This concept can work in different ways, depending on the application involved. For example, a cluster of Web servers can balance its load by sending each of the incoming requests from Web browser clients to a different server. When you connect to a hugely popular Internet Web site like www.microsoft.com, you can be sure that all of its thousands of concurrent users are not being served by a single computer. Instead, the site uses a "server farm" that consists of many identically configured computers. Each time you connect to the site with your Web browser, you are probably accessing a different server. A clustered terminal server works in the same way; each new client connecting to the server is directed to the computer that is currently carrying the lightest load. Other applications that split the processing into threads can distribute those threads equally among the computers in the cluster.

This load balancing capability greatly enhances the expandability of the server. If you reach a point where the server is overburdened by the application traffic it must handle, you can simply add another computer to the cluster, and the workload will automatically be balanced among the available systems, thus reducing the load on each one. You can also upgrade the server by installing additional processors to SMP computers in the cluster, or by replacing a computer with one that is faster and more capable.

Load balancing also provides fault tolerance. If one of the computers in the cluster should fail, the others continue to function with the load redistributed between them. However, it's also possible to construct a cluster with more extensive failover capabilities. A failover cluster is one on which connected computers are configured so that when one fails, the other takes over all of its functions. This type of cluster is better suited to database and e-mail servers that must be continuously available. E-commerce is one of the few technologies that can require both load balancing and failover technologies in one cluster.

The first commercially successful server failover product (designed and marketed long before the term cluster came into general use in the PC networking industry) was Novell NetWare SFT III. SFT III was a version of NetWare that included two network interface adapter cards and a cable used to connect two identical computers together. The operating system, installed on both computers, continuously replicated its disk and memory contents, so that if the primary computer failed, the secondary computer could assume all of its functions instantaneously. However, the SFT III product used the second computer solely for fault tolerance; the second computer remained idle (except for its replication activities) until the first computer failed, as shown in Figure 8-4.

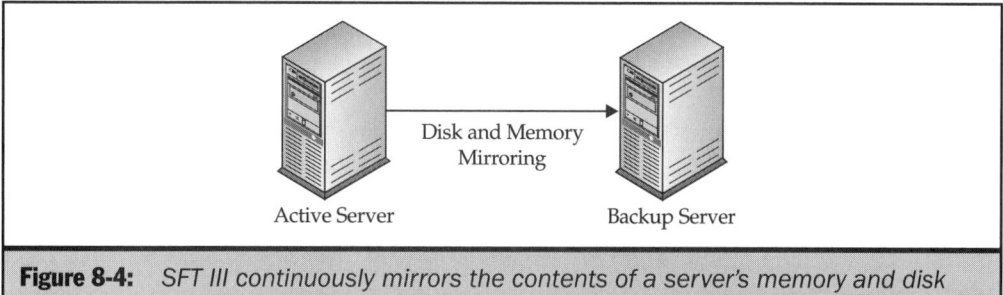

Figure 8-4: *SFT III continuously mirrors the contents of a server's memory and disk drives to a backup server.*

In today's clustering products, a group of computers can be clustered in a failover configuration without leaving some of the machines idle. For example, the Clustering Service included with Windows 2000 can create a failover cluster out of t wo (with Advanced Server) or four (with Datacenter Server) computers, in which each computer is running its own applications and performing its own tasks. If one of the computers fails, its applications are migrated to another computer in the cluster, which takes over its functions, as shown in Figure 8-5. (In order for this to occur, all of the computers in the cluster must have access to the applications and data used by the other computers.)

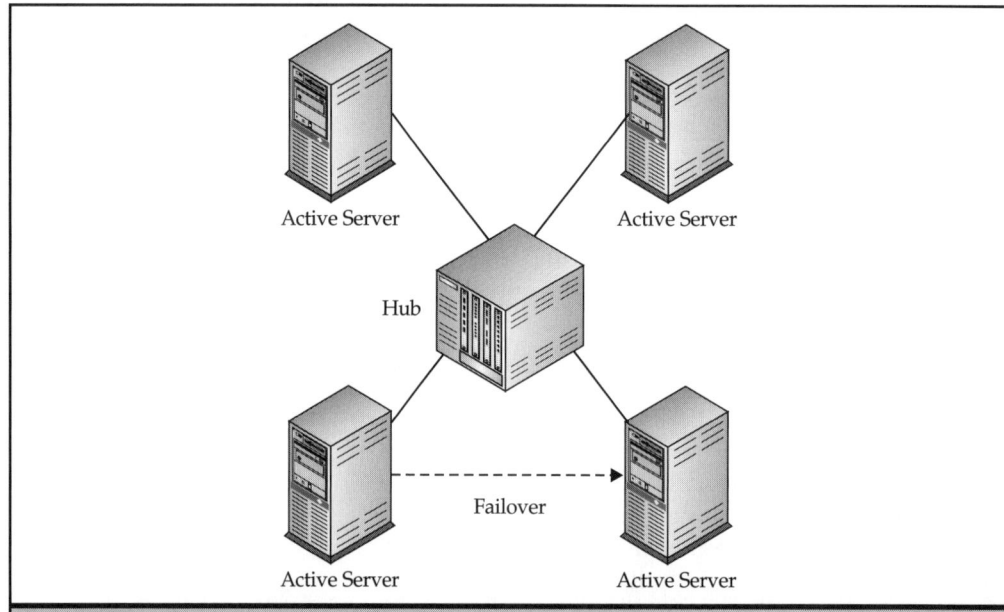

Figure 8-5:   *In a server cluster, all of the servers are active, with functions ready to fail over to other servers.*

While this failover arrangement in a Windows 2000 cluster does not leave some of the computers in the clusters idle, as SFT III does, you still have to consider the hardware requirements for the cluster carefully in order for the failover mechanism to function. If you have a failover cluster that consists of two Windows 2000 Advanced Server computers, you must make sure that each of the computers has sufficient resources to run all of the applications and perform all of the functions handled by both machines under normal conditions. For example, if you are using one of the computers in the cluster as a database server and the other as an e-mail server, each of the computers must have sufficient memory, disk space, and processing power to run both the database and e-mail server applications, in case the other computer fails. If each of the computers is operating at full capacity running just one of the server applications, adding the other application in a failover situation can cause a crash or a severe performance degradation. Thus, while entire computers do not sit idle in a failover cluster, the systems must be underused to some extent.

System Area Networks

There is a new communications technology for clustered computers that has just arrived on the market, which is designed to provide data throughput rates exceeding 1 Gbps. Called a *system area network* (or SAN, not to be confused with a storage area network, also abbreviated SAN), this technology is essentially a dedicated, switched network that connects a group of computers that are in the same administrative domain and located relatively close to each other. The network achieves greater transmission speeds by implementing a reliable transport service (much like TCP, the Transmission Control Protocol) in hardware instead of software. The SAN hardware consists of 32/64-bit PCI network interface adapter cards that use Fibre Channel connections to a central switch. A SAN network interface adapter makes individual transport endpoints (much like the ports used in a TCP software implementation) available to the connected computers. These endpoints are memory-based registers that are shared by the SAN network adapter and the computer's processor. The processor can therefore pass the incoming traffic directed at a particular endpoint immediately to the appropriate application running on the computer. In a sense, a SAN operates much like a distributed memory array, rather than a standard networking technology.

Cluster Networking Hardware

There are two areas in which the use of server clustering can affect the hardware used to construct a network: the network connections themselves and the server's mass storage hardware. The computers in a cluster use standard network connections to communicate with each other. In fact, it is possible to build a server cluster with no additional networking hardware other than each computer's normal connection to the enterprise network. In a failover configuration, the servers in the cluster communicate by exchanging signals at regular intervals called *heartbeats*. These heartbeats serve as an indication to each computer that the other computers in the cluster are up and running properly. If a computer fails to transmit a predetermined number of consecutive heartbeats, the other computers in the cluster assume that it has failed and take action to assume its functions. This same heartbeat method also functions at the application level. If a single application fails on one of the computers in the cluster, the Cluster Service attempts to restart it on the same computer. If this should fail, the service then migrates the application to another computer in the cluster.

The heartbeats can be exchanged over the normal network connection, but if the cluster is on a shared network with other systems, the additional traffic generated by the heartbeats can be a problem. In addition, the network connection provides a single point of failure. If a cable break or a failure in a hub or other network component should occur, the heartbeats can fail to reach all of the computers in the cluster, resulting in a condition in which both computers attempt to take on the functions of the other.

To address these problems, it's a good idea to build a separate, private network that is dedicated to the computers in the cluster, as shown in Figure 8-6. Fast Ethernet is typically the protocol of choice for this arrangement, with Gigabit Ethernet an option for installations that can benefit from greater speeds. Not only does this private network ensure that the heartbeats generated by each computer reach the others in a timely fashion, it also provides a backup for the intracluster communications. The Windows 2000 computers in a cluster, for example, use the private network to exchange their heartbeats by default. If the private network should fail, the computers switch to the standard network shared by the rest of the enterprise. Later in this chapter, you will see how this separate network can also be used with a higher-speed protocol such as Fibre Channel to connect the servers to external drive arrays and other storage devices. This is called a storage area network (SAN).

Microsoft has developed a revised version of the Windows Sockets (Winsock) protocol called the Windows Sockets Direct Path (WSD) for system area networks that is included in the Windows 2000 Datacenter Server product, and which enables the computer to distinguish between resource requests intended for the standard network connection and those intended for a computer on the SAN. For the latter, the traffic bypasses the TCP/IP protocol stack and is transmitted directly to the destination on the SAN, using a hardware-based transport service that provides the same reliability as TCP, but with far less latency. This enables applications that use standard Winsock programming calls to operate with the SAN technology without modification.

Cluster Storage Hardware

One of the elements that complicates the implementation of a clustering solution in a failover configuration is that each of the computers in the cluster requires access to the applications and data running on the other computers. There are three ways to

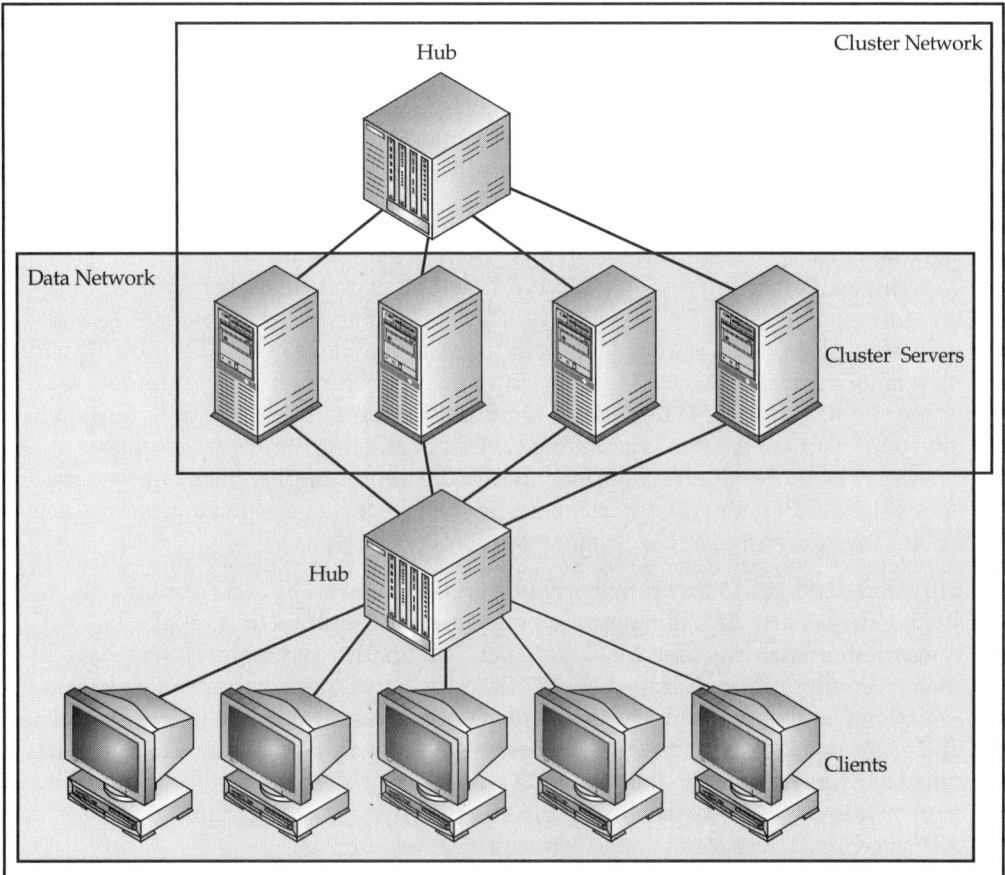

Figure 8-6: *A cluster network is a separate network that joins clustered servers together.*

accomplish this, which have come to define the three basic hardware configurations you can use in a computer that is part of a cluster. These three hardware configurations are as follows:

Shared disk In a shared disk configuration, the computers in the cluster are all connected to the same disk array using a common I/O bus, so that all of the computers can access the same applications and data simultaneously, as shown in Figure 8-7. The disk array typically uses some form of SCSI, Fibre Channel, or Serial Storage Architecture (SSA) to connect to the computers. Because this arrangement makes it possible for two computers to update files on the shared drives at the same time, an additional software component called a *distributed lock manager* is needed, to prevent files from being corrupted and new data from being overwritten. IBM's High Availability Cluster Multi-Processing (HACMP) product for its AIX operating system and Oracle Corporation's Oracle Parallel Server (OPS) are two products that enable you to use this type of configuration.

Shared nothing A shared nothing configuration is one in which there is no simultaneous access of the same data stores by different computers in the cluster. The Cluster Service in Windows 2000 and the Microsoft Cluster Server used by Windows NT are both considered to be shared nothing clustering technologies. Both of these products may at first seem to use a shared disk configuration, because they call for a hardware configuration in which the computers in the cluster are all connected to a single I/O bus. However, unlike a true shared disk configuration, only one of the computers is accessing the bus at any one time. The redundant connection is so that if one computer should fail and its applications fail over to another computer, the substitute can immediately access the same data stores as the original system and continue where it left off.

Mirrored disk In a mirrored disk configuration, each computer maintains its own storage drives, and data is replicated between the computers on a regular basis. This is the method that was used by Novell NetWare SFT III, and the products that still use this configuration, such as Legato's Standby Server, tend to resemble NetWare SFT III in that they use a cluster of only two computers and one of the two stands idle, ready to take over if the first computer fails. Legato also has another product, called Mirroring for MCS, that enables you to use a disk mirroring configuration with Windows NT and Microsoft Cluster Server, instead of the shared nothing array that Microsoft advocates.

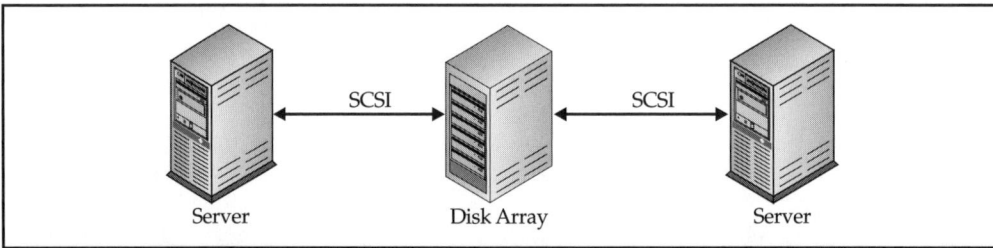

Figure 8-7: *Clustered servers can share a disk array so that they all have access to essential data.*

Server Storage Technologies

Data storage is an important element of all computers, but for servers, an efficient storage subsystem is even more important. Servers not only have to routinely store large amounts of data (even by today's standards), they have to handle data access requests from multiple clients simultaneously. In some cases, there might be dozens of users trying to access a single hard drive at one time, and servers typically use storage technologies that are better suited to this kind of traffic.

A hard disk drive consists of one or more spinning platters stacked on top of each other, and a head assembly that moves laterally back and forth across the platter surfaces to place the heads at the exact location where a particular piece of data is located. The heads can only be in one position at any given time, and there is a slight delay whenever the head assembly has to move the heads to a different location. Therefore, if ten network users are all requesting different files on the same server hard drive, there is no way for the drive to access them all at once. The drive must process each request sequentially. A server drive is one that is capable of processing large numbers of requests efficiently and accessing the requested data as quickly as possible.

Servers also tend to store data that is vitally important to businesses, organizations, and clients, data which if lost can cost huge amounts of money, time, and even lives. For this reason, many servers also use storage technologies that provide an added measure of protection against data loss and corruption. The following sections examine some of the storage technologies that are commonly found in network servers.

SCSI or IDE?

The two mass storage interfaces most commonly used in personal computers are Integrated Drive Electronics (IDE) and the Small Computer System Interface (SCSI). The general rule of thumb is that you use IDE for workstations and standalone computers and SCSI for servers and other high-performance computers. This rule still holds true in most cases. If you're trying to decide whether to use IDE or SCSI drives in your servers, the answer is SCSI without a doubt, although there are a few people who maintain that IDE is a viable alternative. Except in the case of a standard computer being used as a server on a very small network (such as a three-node home network), SCSI is always the preferable alternative.

The IDE interface was originally devised for hard drives only, its advantage over the other interfaces at the time being that the controller was built into the drive itself. Integrating the controller into the drive unit eliminated the need for a separate controller card in the computer and the need to maintain compatibility in communications between the drive and the controller. Today, the name IDE is something of an anachronism, because all hard drives have built-in controllers, including SCSI drives.

In its original conception, the IDE interface had many limitations when compared to SCSI, including the following:

- Limited drive capacity
- Support for internal devices only
- Support for hard drives only
- Support for only two devices

By comparison, SCSI, at that time, supported up to seven internal or external devices of various types, including hard drives with much greater capacities. Over the years, both the IDE and SCSI standards were updated and enhanced to improve their capabilities and performance. Their basic performance parameters as they now stand are shown in Table 8-1. The IDE interface can now support other types of devices, including CD-ROM and tape drives, as well as hard drives with enormous capacities. However, a standard IDE interface supports only four devices, and they must be internal ones. SCSI has always been an interface that was designed to support a variety of devices, and as the SCSI standards progressed, support for many different types of hardware was added, including optical drives, tape autochangers, and scanners, in addition to the traditional hard disk, CD-ROM, and tape drives. Many of the advanced mass storage devices intended for use on servers, such as RAID arrays and autochangers, are available only with the SCSI interface.

Note *Although SCSI was designed to support different types of hardware from the outset, you have to be careful when selecting the devices that you intend to connect to a single SCSI bus. If you have a SCSI bus populated with the latest and fastest Ultra3 drives, and connect an outdated CD-ROM drive to the same bus, the CD-ROM will cause the other devices to slow down to its speed.*

Interface	Maximum Transfer Rate	Number of Devices Supported	Internal/External Device Support	Maximum Bus Speed
IDE	100 MBps	2–4	Internal only	80 MHz
SCSI	160 MBps	7–15	Internal and external	80 MHz

Table 8-1: *IDE and SCSI Performance Parameters*

Both the IDE and SCSI interfaces require a host adapter, which is the conduit between the storage devices and the rest of the computer. The operating system sends file access requests to the host adapter, which then transmits the appropriate commands to the attached devices. Virtually all of the computers sold today (including servers) have an IDE host adapter built into the motherboard and support for IDE devices integrated into the system BIOS. SCSI host adapters, however, usually take the form of a separate expansion card that plugs into the computer's bus, although some servers have integrated SCSI adapters on the motherboard. Therefore, to use SCSI devices, you usually have to purchase a host adapter card. This, combined with the cost of the SCSI devices themselves, which is substantially more than equivalent IDE devices, is what makes SCSI a more expensive proposition than IDE.

In performance, SCSI also has the edge over IDE. The fastest IDE transfer mode is currently UltraATA/100, with a maximum data transfer rate of 100 MBps. The fastest SCSI transfer mode is Ultra3 SCSI, which has a maximum transfer rate of 160 MBps. These figures, like all speed ratings for data storage devices, assume the existence of perfect conditions that are almost never achieved in the real world, but you can still see from these figures that SCSI outperforms IDE by a substantial margin, at least in theory.

Sheer speed, however, is not the only factor that makes SCSI the preferable mass storage interface for a network server. The need to handle multiple concurrent requests is what distinguishes a server storage system from a workstation one, and SCSI excels in this respect as well. An IDE interface consists of two channels, each of which supports two devices, a master and a slave, as shown in Figure 8-8. IDE is a single-threaded interface, which means the host adapter can only issue a command to one of the devices on a particular channel at a time. If, for example, you're copying data from a master hard drive to a slave hard drive on the same channel, the host adapter must issue a read request to the master and wait for the drive to complete that request before it can issue a write request to the slave. For a hard drive to hard drive copy operation, this is not a noticeable drawback. However, if you have a hard drive and a CD-R drive on the same channel, your ability to successfully burn a CD with data from that hard drive can be affected. This is because the CD-R drive requires a consistent stream of data in order to function properly, and the continual stopping and starting caused by the interface can interrupt the data stream. If you compound this phenomenon by adding a dozen or more network users all trying to access data stored on IDE hard drives, the cumulative delays that result from the natural latency of the interface can add up to a significant amount of performance degradation.

SCSI, on the other hand, takes the form of a parallel bus in which each device is connected to the previous device in the chain, as shown in Figure 8-9. The SCSI host adapter is installed in a computer, the first device is connected to the adapter, the second device is connected to the first, and so on. Each device on the bus has its own unique SCSI ID, and terminating resistors at both ends of the bus prevent signals from reflecting back in the other direction. The result is what amounts to a "mini-network" of hardware devices, all taking orders from the SCSI host adapter. Unlike IDE, however, SCSI devices are capable of independently queuing and executing commands received from the host adapter. Once the adapter sends a command to a device on the bus, it can immediately move on to processing another command; it doesn't have to wait for the first command to complete, as an IDE adapter does. This is the key to more efficient server drive operations. Disk access requests can come in from all over the network, and the SCSI host adapter simply has to issue the appropriate commands to the devices on the bus and move on. The latency generated by IDE drives is all but eliminated, and the performances of the SCSI devices, the server hosting them, and the network itself become the primary determining factors in the speed at which network clients can access data.

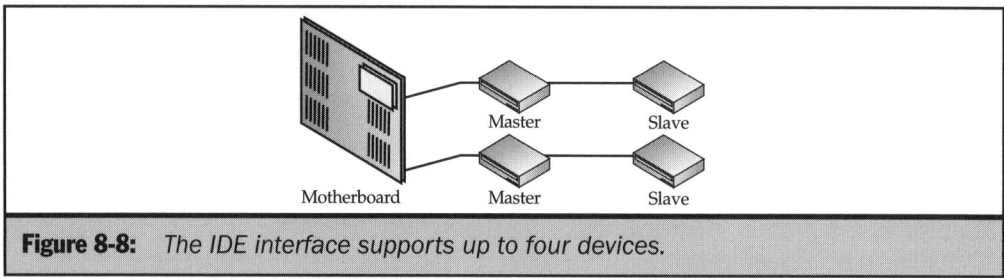

Figure 8-8: *The IDE interface supports up to four devices.*

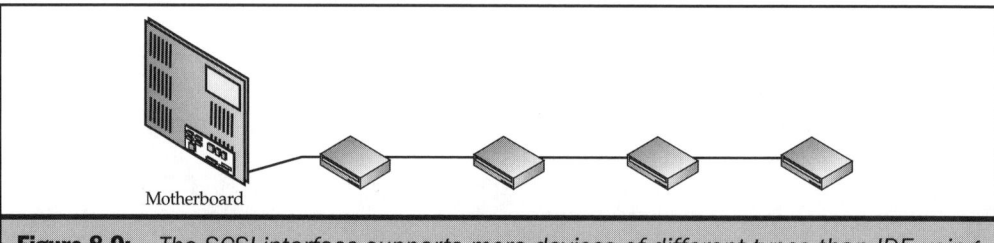

Figure 8-9: *The SCSI interface supports more devices of different types than IDE using daisy chain connections.*

Using RAID

The mass storage subsystems used in network servers frequently go beyond just having greater capacities, faster drives, and a SCSI interface. There are also more advanced storage technologies that provide better performance, reliability, and fault tolerance.

The most common of these technologies is called a Redundant Array of Independent (or sometimes Inexpensive) Disks, or RAID. A RAID array is a group of hard drives that function together in any one of various ways, called levels. There are six basic RAID levels, numbered from 0 to 5, plus several other RAID standards that are proprietary or variations on one of the other levels. The different RAID levels provide varying degrees of data protection and performance enhancement.

RAID can be implemented in hardware or software, in whole or in part. For example, network operating systems such as Windows 2000, Windows NT, and Novell NetWare all can implement some RAID functions, such as disk striping (RAID Level 0) and disk mirroring/duplexing (RAID Level 1), without any special hardware. Third-party software products can provide other RAID levels. Generally speaking, however, the best RAID performance comes from a hardware RAID implementation.

Hardware RAID solutions can range from dedicated RAID controller cards, which you install into a server like any other PCI expansion card and connect to your hard drives, to standalone RAID drive arrays. A RAID controller card typically contains a coprocessor and a large memory cache (often 32MB or more). This hardware enables the controller itself to coordinate the RAID activity, unlike a software solution that utilizes the computer's own memory and processor. When you use a hardware RAID solution, the drive array appears to the computer as a single drive. All of the processing that maintains the data stored array is invisible.

A RAID drive array is a unit, either separate or integrated into a server, that contains a RAID controller and slots into which you insert hard disk drives, like those shown in Figure 8-10. In some cases, the slots are merely containers for the drives, and you use standard SCSI and power cables to connect them to the RAID controller and to the computer's power supply. In higher-end arrays, the drives plug directly into a *backplane*, which connects all of the devices to the SCSI bus, supplies them with power, and eliminates the need for separate cables. In some cases, the drives are *hot swappable*, meaning that you can replace a malfunctioning drive without powering down the whole array. Some arrays also include a hot standby drive, which is an extra drive that remains idle until one of the other drives in the array fails, at which time the standby drive immediately takes its place. Some servers are built around an array of this type, while in other cases the array is a separate unit, either standing alone or mounted in a rack. These separate drive arrays are what you use when you want to build a server cluster with shared drives.

Standalone RAID drive arrays connect to a SCSI host adapter in a server like any other SCSI device, and this can be one of their drawbacks. A single SCSI channel can be flooded by as few as four hard drives, and the single SCSI channel used to carry all of the traffic to and from the server can end up being a bottleneck. RAID controller cards that you install into a server, on the other hand, are available in models that use two or more separate SCSI channels, which eliminates this potential bottleneck.

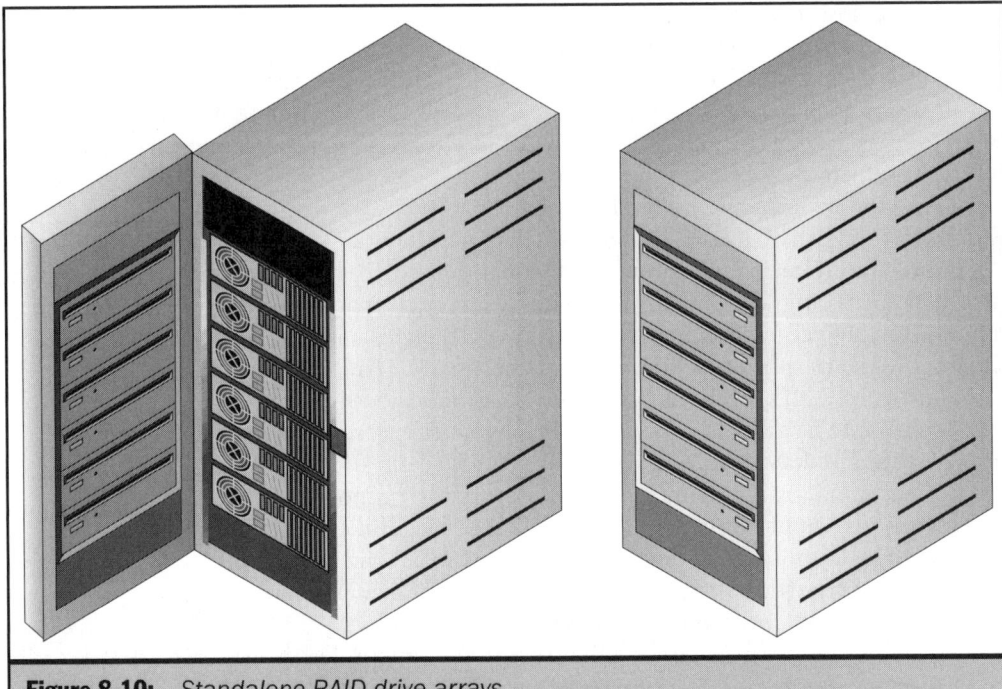

Figure 8-10: *Standalone RAID drive arrays*

> **Tip** *It's important to carefully investigate the capabilities of the RAID controller card you plan to buy. In some cases, relatively inexpensive products sold as RAID controllers are actually SCSI adapters that use software drivers to perform all of the RAID functions. These software drives use the computer's processor and memory, and can place a substantial additional burden on the computer. If you want a hardware RAID solution, make sure that the controller you select has a built-in coprocessor and memory cache.*

Whether you implement RAID using software or hardware, you choose the RAID level that best suits your installation. Although the various RAID levels are numbered consecutively, the higher levels are not always "better" than the lower ones in every case. In some cases, for example, you are trading off speed or disk space in return for added protection, which may be warranted in one installation, but not in another. The various levels of RAID are described in the following sections.

RAID 0: Disk Striping

Disk striping is a method for enhancing the performance of two or more drives by using them concurrently, rather than individually. Technically, disk striping is not RAID at all, because it provides no redundancy and therefore no data protection or fault tolerance. In a striped array, the blocks of data that make up each file are written to different drives in succession. In a four-drive array like that shown in Figure 8-11, for example, the first block (A) is written to the first drive, the second block (B) to the second drive, and so on through the fourth block (D). Then the fifth block (E) is written to the first drive, the sixth (F) to the second drive, and the pattern continues until all of the blocks have been written. Operating the drives in parallel increases the overall I/O performance of the drives during both reads and writes, because while the first drive is reading or writing block A, the second drive is moving its heads into position to read or write block B. This reduces the latency period caused by the need to move the heads between each block in a single drive arrangement. To reduce the latency even further, you can use a separate controller for each drive.

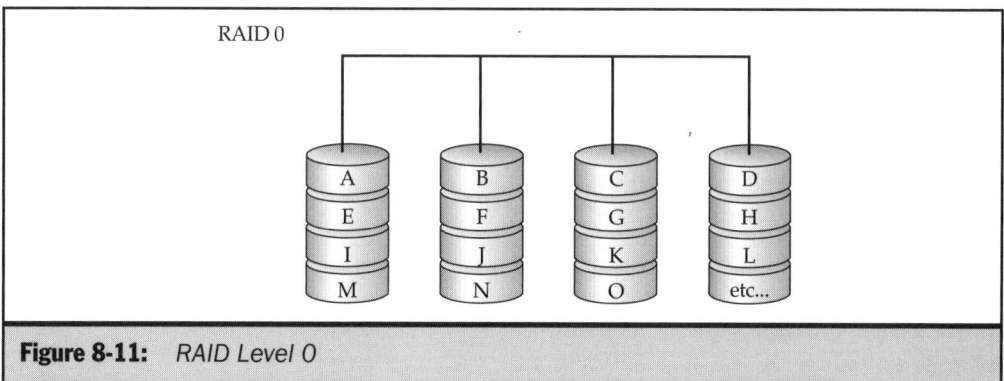

Figure 8-11: *RAID Level 0*

As mentioned earlier, disk striping provides no additional protection to the data, and indeed even adds an additional element of danger. If one of the drives in a RAID 0 array should fail, the entire volume is lost, and recovering the data directly from the disk platters is much more difficult, if not impossible. However, disk striping provides the greatest performance enhancement of any of the RAID levels, largely because it adds the least amount of processing overhead. RAID 0 is suitable for applications in which large amounts of data must be retrieved on a regular basis, such as video and high-resolution image editing, but you must be careful to back up your data regularly.

Note *It's possible to stripe data across a series of hard drives either at the byte level or the block level (one block typically equals 512 bytes). Byte-level striping is better suited to the storage of large data records, because the contents of a record can be read in parallel from the stripes on different drives, thus improving the data transfer rate. Block-level striping is better suited for the storage of small data records in an environment where multiple concurrent requests are common. A single stripe is more likely to contain an entire record, which enables the various drives in the array to process individual requests independently and simultaneously.*

RAID 1: Disk Mirroring and Duplexing

Disk mirroring and disk duplexing are the simplest arrangements that truly fit the definition of RAID. *Disk mirroring* is a technique where two identical drives are connected to the same host adapter, and all data is written to both of the drives simultaneously, as shown in Figure 8-12. This way, there is always a backup (or mirror) copy of every file immediately available. If one of the drives should fail, the other continues to operate with no interruption whatsoever. When you replace or repair the malfunctioning drive, all of the data from the mirror is copied to it, thus re-establishing the redundancy. *Disk duplexing* is an identical arrangement, except that the two drives are connected to separate controllers. This enables the array to survive a failure of one of the disks or one of the controllers.

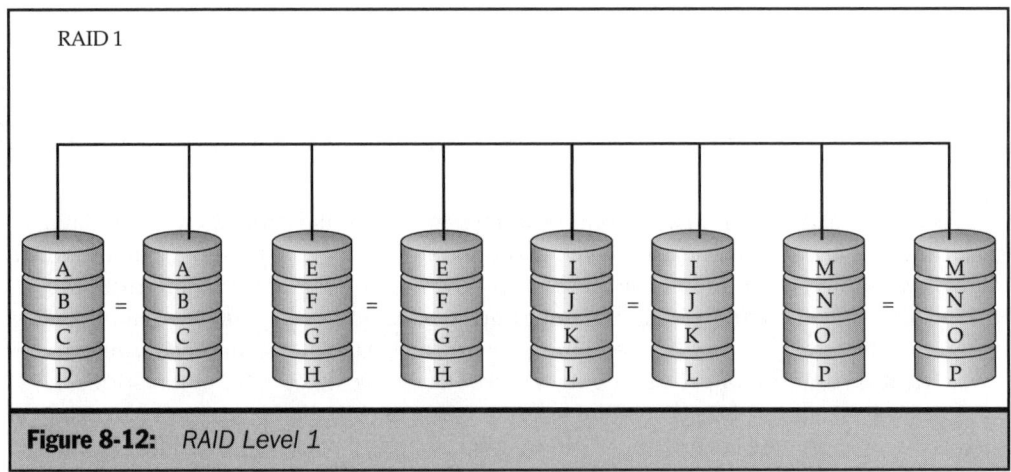

Figure 8-12: *RAID Level 1*

Obviously, disk mirroring provides complete hard drive fault tolerance, and disk duplexing provides both drive and controller fault tolerance, because a complete copy of every file is always available for immediate access. However, mirroring and duplexing do this with the least possible efficiency, because you realize only half of the disk space that you are paying for. Two 10GB drives that are mirrored yield only a 10GB volume. As you will see, other RAID levels provide their fault tolerance with greater efficiency, as far as available disk space is concerned.

Disk mirroring and duplexing do enhance disk performance as well, but only during read operations. During write operations, the files are written to both drives simultaneously, resulting in the same speed as a single drive. When reading, however, the array can alternate between the drives, doubling the transaction rate of a single drive. In short, write operations are said to be expensive and read operations efficient. Like disk striping, mirroring and duplexing are typically implemented by software, and are common features in server operating systems like Windows 2000. However, as mentioned earlier, using the system processor and memory for this purpose can degrade the performance of the server when disk I/O is heavy.

RAID 2: Hamming ECC

RAID 2 is a seldom used arrangement where each of the disks in a drive array are dedicated to either the storage of data or of error correcting code (ECC), as shown in Figure 8-13. As the system writes files to the data disks, it also writes the ECC to drives dedicated to that purpose. When reading from the data drives, the system verifies the data as correct using the error correction information from the ECC drives. The ECC in this case is hamming code, which is the same type of ECC used on SCSI hard drives that support error correction. Because all SCSI hard drives made today already support ECC, and because a relatively large number of ECC drives are required for the data drives, RAID 2 is an inefficient method that has almost never been implemented commercially.

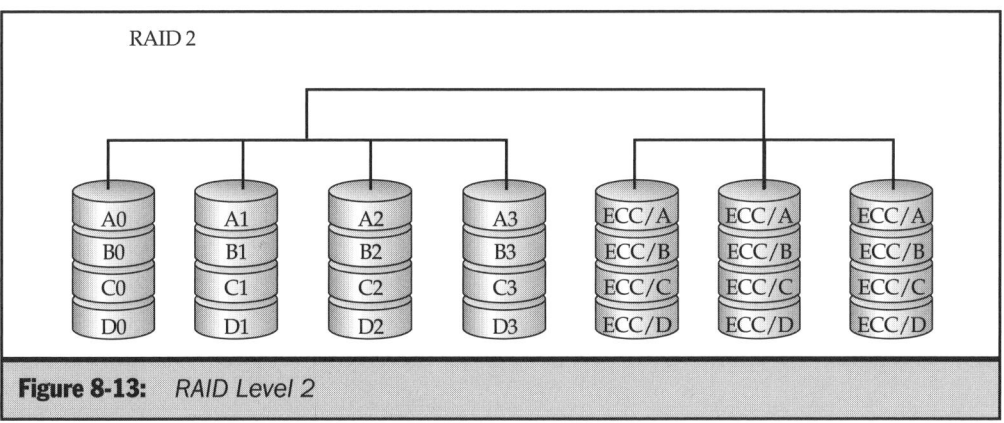

Figure 8-13:   *RAID Level 2*

RAID 3: Parallel Transfer with Shared Parity

A RAID 3 array is a combination of data striping and the storage of a type of ECC called *parity* on a separate drive. RAID 3 requires a minimum of three drives, with two or more of the drives holding data striped at the byte level and one drive dedicated to parity information, as shown in Figure 8-14. The use of striping on the data drives enhances I/O performance, just as in RAID 0, and using one drive in the array for parity information adds fault tolerance. Whenever the array performs a read operation, it uses the information on the parity drive to verify the data stored on the striped drives. Because only one of the drives holds the parity information, you realize a greater amount of usable disk space from your array than you do with RAID 2. If one of the striped drives should fail, the data it contains can be reconstructed using the parity information. However, this reconstruction takes longer than that of RAID 1 (which is immediate) and can degrade performance of the array while it is occurring.

When you hit RAID 3 and the levels above it, the resources required by the technology make them much more difficult to implement in software only. Most servers that use RAID 3 or higher use a hardware product, which could even be just a RAID controller card connected to standard SCSI hard drives.

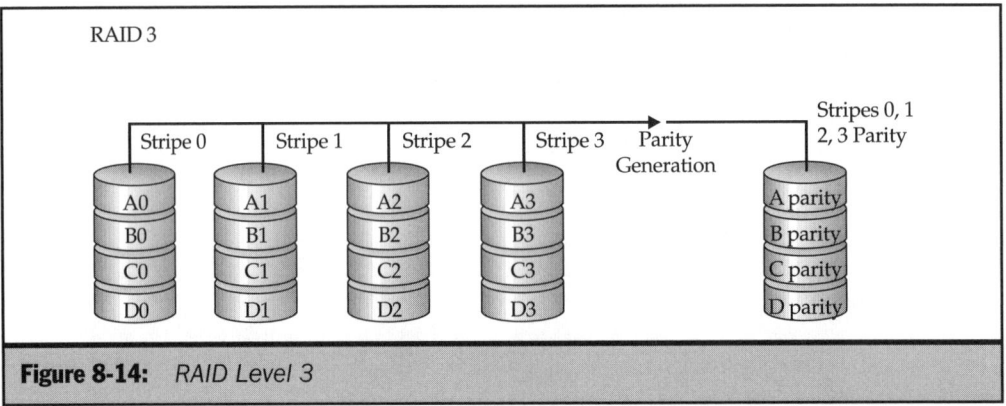

Figure 8-14: *RAID Level 3*

RAID 4: Independent Data Disks with Shared Parity

RAID 4 is similar to RAID 3, except that the drives are striped at the block level, rather than at the byte level, as shown in Figure 8-15. There is still a single drive devoted to parity information, which enables the array to recover the data from a failed drive if needed. The performance of RAID 4 in comparison to RAID 3 is comparable during read operations, but write performance suffers because of the need to continually update the information on the parity drive. RAID 4 is also rarely used, because it offers few advantages over RAID 5.

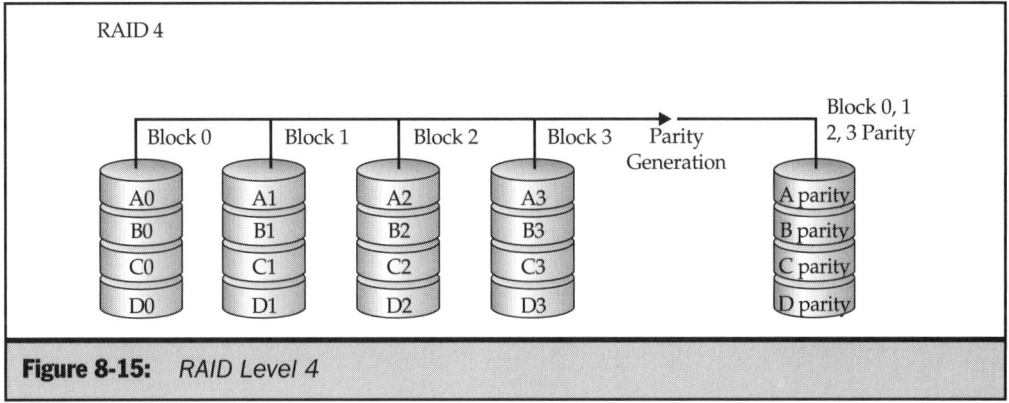

Figure 8-15: *RAID Level 4*

RAID 5: Independent Data Disks with Distributed Parity

RAID 5 is the same as RAID 4, except that the parity information is distributed among all of the drives in the array, as shown in Figure 8-16, instead of being stored on a drive dedicated to that purpose. Because of this arrangement, there is no parity drive to function as a bottleneck during write operations, and RAID 5 provides significantly better write performance than RAID 4, along with the same degree of fault tolerance. The rebuild process in the event of a drive failure is also made more efficient by the distributed parity information. Read performance suffers slightly in RAID 5, however, because the drive heads must skip over the parity information stored on all of the drives.

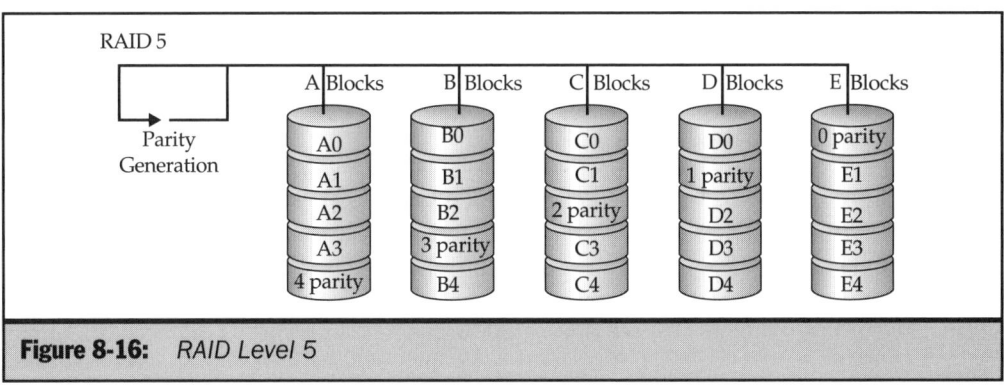

Figure 8-16: *RAID Level 5*

RAID 5 is the level that is usually implied when someone refers to a RAID array, because it provides a good combination of performance and protection. In a four-disk array, only 25 percent of the disk space is devoted to parity information, as opposed to 50 percent in a RAID 1 array.

RAID 6: Independent Data Disks with Two-Dimensional Parity

RAID 6 is a variation on RAID 5 that provides additional fault tolerance by maintaining two independent copies of the parity information, both of which are distributed among the drives in the array, as shown in Figure 8-17. The two-dimensional parity scheme greatly increases the controller overhead, since the parity calculations are doubled, and the array's write performance is also degraded, because of the need to save twice as much parity information. However, a RAID 6 array can sustain multiple simultaneous drive failures without data loss, and is an excellent solution for read-intensive environments working with mission-critical data.

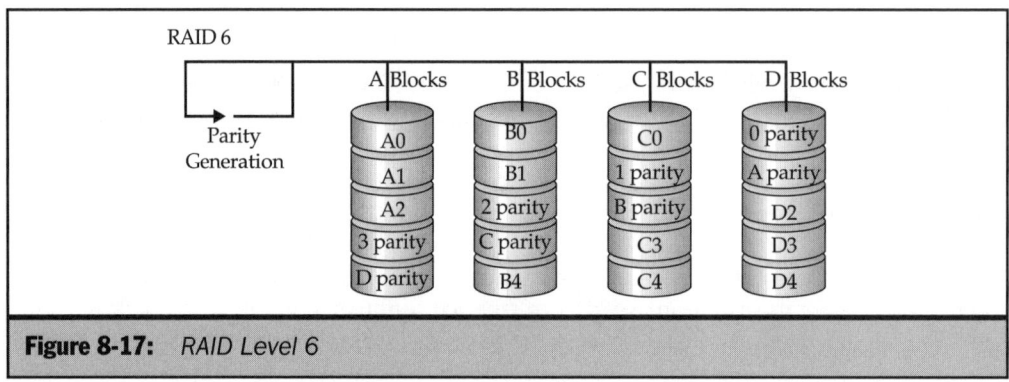

Figure 8-17:   *RAID Level 6*

RAID 7: Asynchronous RAID

RAID 7 is a proprietary solution marketed by Storage Computer Corporation, which consists of a striped data array and a dedicated parity drive, as shown in Figure 8-18. The difference in RAID 7 is that the storage array includes its own embedded operating system, which coordinates the asynchronous communications with each of the drives. Asynchronous communication, in this context, means that each drive in the array has its own dedicated high-speed bus and its own control and data I/O paths, as well as a separate cache. The result is increased write performance over other RAID levels and very high cache hit rates under certain conditions. The disadvantages of RAID 7 are its high cost and the danger resulting from any investment in a proprietary technology.

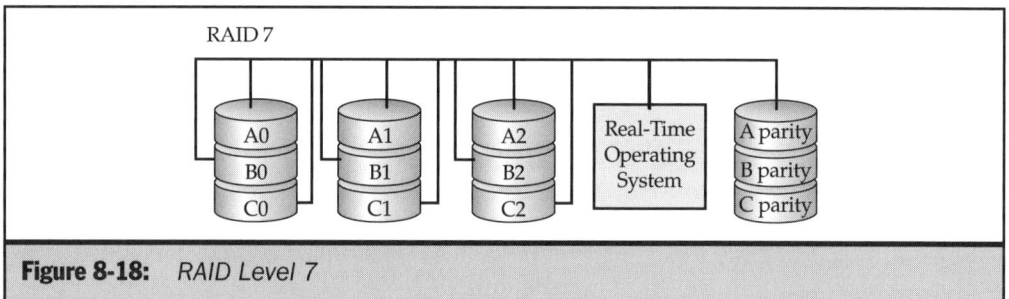

Figure 8-18:   *RAID Level 7*

RAID 10: Striping of Mirrored Disks

RAID 10 is a combination of the disk striping used in RAID 0 and the disk mirroring used in RAID 1. The drives in the array are arranged in mirrored pairs and data is striped across them, as shown in Figure 8-19. The mirroring provides complete data redundancy while the striping provides enhanced performance. The disadvantage of RAID 10 is the high cost (at least four drives are required), and the same low data storage efficiency as RAID 1.

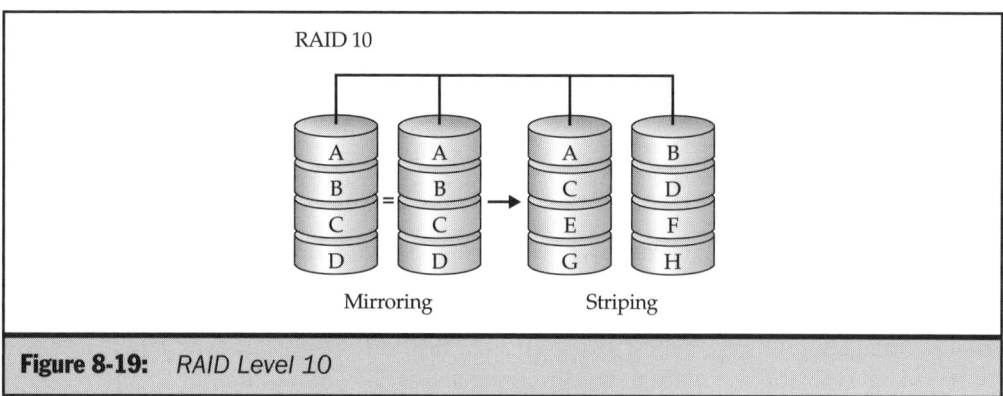

Figure 8-19:   *RAID Level 10*

RAID 53: Striped Array of Arrays

RAID 53 is a combination of RAID technologies that consists of multiple RAID 5 (or sometimes RAID 3) arrays that are used as the segments of a striped array. In other words, the array stripes data across several disks, just as in RAID 0, but each of the disks is actually a RAID 5 array in itself. This arrangement is very expensive to implement, but it provides the same fault tolerance as RAID 5 with the added performance of striping.

RAID 0+1: Mirroring of Striped Disks

RAID 0+1 is the opposite of RAID 10. Instead of striping data across mirrored pairs of disks, RAID 0+1 takes an array of striped disks and mirrors it, as shown in Figure 8-20. The resulting performance is similar to that of RAID 10, but a single drive failure turns the array back to a simple RAID 0 installation.

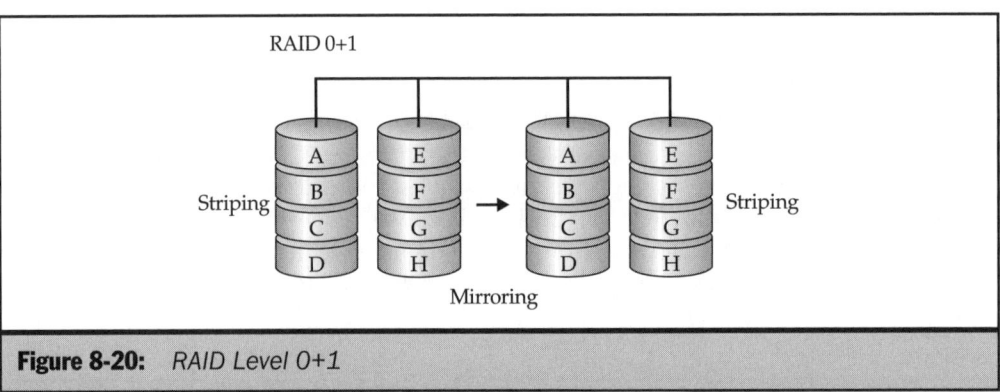

Figure 8-20: *RAID Level 0+1*

Using Hierarchical Storage Management (HSM)

Hierarchical Storage Management (HSM) is a technique for storing data on a variety of device types in order to minimize storage costs while providing easy accessibility. Hard disk drive costs are continually dropping, but they are still more expensive (in terms of cost per megabyte) than other media such as optical disks and magnetic tape. However, as a general rule, the cheaper the medium, the slower its access time. By installing these various types of drives in a server, you can minimize your storage costs by putting the most frequently used files on hard drives, occasionally used files on optical disks, and seldom used files on magnetic tape.

The problem with this arrangement is keeping track of which files are stored on which device, and this is where HSM provides a solution. HSM is a software product that automatically migrates files between the various media, depending on how often they're accessed. A typical HSM installation consists of a server with one or more hard drives and an optical disk jukebox or a tape autochanger, or both. Jukeboxes and autochangers are mechanical devices that can automatically select a disk or tape and insert it into a drive, eliminating the need for a person to manually insert the proper medium. These devices enable you to maintain large amounts of storage and still access it without human intervention. This is known as *nearline storage*.

When a file on a hard drive goes a certain number of days without being accessed, the HSM software migrates it to the secondary medium, such as an optical disk. After copying the file to the optical disk, the software creates a tiny key file in its place on the hard drive. The key file specifies the location of the actual file and provides a placeholder for network users. If the file goes even longer without being accessed, HSM migrates it to a tertiary medium (such as tape) and updates the key file. To a user on the network, the files that have been migrated to other media appear to still be on the hard drive. When the user attempts to access the file, HSM reads the contents of the key file, loads the appropriate disk or tape into the drive, reads the file, and supplies it to the user. The only sign to the user that the file is not stored on the hard drive is the additional time it takes for HSM to supply the file. Everything else is completely invisible. If the user modifies the file, HSM migrates it back to the hard drive, where it remains until it reaches the migration interval once again.

HSM software products are usually highly configurable, enabling you to use various combinations of media and specify whatever migration intervals you wish. An HSM installation is not cheap, mainly because jukeboxes and autochangers are expensive devices, but for a network that must store vast amounts of data while keeping it all available at a few minutes notice, HSM is a viable solution.

Fibre Channel Networking

SCSI is a parallel signaling technology, meaning that a SCSI connection consists of multiple signals transmitted over separate conductors at the same time. Parallel signaling is one of the factors that enables SCSI devices to transmit data at high rates of speed. However, parallel signaling does impose limitations on the length of the SCSI bus. Just like a parallel printer cable is limited in length when compared to a serial modem cable, the length of a parallel SCSI bus is limited to 25 meters. The development of new network storage technologies, such as Network Attached Storage (NAS) and storage area networks (SANs), that call for storage hardware external to the server, has resulted in the need for a means to transmit large amounts of data between relatively distant devices at high speeds. Several networking technologies have been adapted for this purpose, by mapping SCSI transmissions onto a serial connection, making greater distances between nodes possible.

The three technologies most often associated with serialized SCSI communications are IEEE 1394, also known as FireWire, Serial Storage Architecture (SSA), and Fibre Channel. IEEE 1394, as pioneered by Apple Computer, is being deployed primarily as a technology for connecting video equipment to computers. SSA is a protocol, developed by IBM, that uses bidirectional cabling (signals can travel in either direction) for fault tolerance, and is available in IBM's RS/6000 product line. Neither of these technologies has had the market acceptance of Fibre Channel in the area of network storage, however.

Fibre Channel was conceived in 1988 as a high-speed networking technology that its advocates hoped would be the successor to Fast Ethernet and Fiber Distributed Data Interface (FDDI) on backbone networks that required large amounts of bandwidth. Ratified in a series of American National Standards Institute (ANSI) standards in 1994, Fibre Channel never found acceptance as a general local area networking protocol, although Gigabit Ethernet, an extension of the Ethernet standard using the Fibre Channel physical layer options, did. Instead, Fibre Channel has become the protocol of choice for high-end network storage technologies, and has particularly become associated with SANs. A Fibre Channel connection can transfer data at the rate of 100 MBps, or 200 MBps when operating in full-duplex mode.

Note	*The unusual spelling of "fibre" is deliberate, and intended to distinguish the term Fibre Channel from fiber optic.*

Unlike standard SCSI, which connects storage devices and servers together using a bus, Fibre Channel is essentially a separate network that can connect various types of storage devices with the servers on a network. Fibre Channel uses standard networking hardware components, such as cables, hubs, and ports, to form the network medium, and the connected nodes transmit and receive data using any one of several services, providing various levels of performance. Fibre Channel differs from standard networking protocols like the Internet Protocol (IP) in that much of its "intelligence" is implemented in hardware, rather than in software running on a host computer.

The Fibre Channel protocol stack consists of five layers that perform the functions attributed to the physical and data link layers of the Open Systems Interconnection (OSI) reference model. These layers are as follows:

FC-0   This layer defines the physical components that make up the Fibre Channel network, including the cables, connectors, transmitters, and receivers, and their properties.

FC-1   This layer defines the encoding scheme used to transmit the data over the network, as well as the timing signals and error detection mechanism. Fibre Channel uses an encoding scheme called 8B/10B, in which 10 bits are used to represent 8 bits of data, thus yielding a 25 percent overhead.

FC-2   This layer defines the structure of the frame in which the data to be transmitted is encapsulated and the sequence of the data transfer.

FC-3   This layer defines additional services such as the striping of data across multiple signal lines to increase bandwidth and the use of multiple ports with a single alias address.

FC-4   This layer maps the Fibre Channel network to the upper layer protocols running over it. While it's possible to map Fibre Channel to standard networking protocols, such as the IP, the Fibre Channel Protocol (FCP) is the protocol used to adapt the standard parallel SCSI commands to the serial SCSI-3 communications used by storage devices on a Fibre Channel network.

The Fibre Channel Physical Layer

Fibre Channel supports both fiber-optic and copper cables, with fiber optic providing greater segment lengths. Fibre Channel can operate at 133 Mbps, 266 Mbps, 532 Mbps, or 1.0625 Gbps, with the latter providing the maximum transmission speed of 100 MBps. Future Fibre Channel standards are expected to increase the bandwidth to 2 and eventually 4 Gbps.

The three physical layer cable options are as follows:

Singlemode fiber optic   Nine-micron singlemode fiber-optic cable, using standard SC connectors, with a maximum cable length of 10,000 meters.

Multimode fiber optic   Fifty or 62.5-micron multimode fiber-optic cable with SC connectors, with a maximum cable length of 500 meters.

Shielded twisted pair (STP)   Type 1 STP cable with DB-9 connectors, with a maximum cable length of 30 meters.

Using any of these cable types, you can build a Fibre Channel network with any one of three following topologies:

Point-to-point   The point-to-point topology links a Fibre Channel host bus adapter installed into a computer to a single external storage device or subsystem.

Loop   The loop topology, also called a continuous arbitrated loop, can contain an unlimited number of nodes, although only 127 can be active at any one time. You can connect the nodes to each other using a physical loop, or you can implement the loop logically using a hub and a physical star topology, as in a Token Ring network. Traffic travels only one direction on the loop, unlike SSA and FDDI, which have redundant loops that permit bidirectional communications. Therefore, in the case of a physical loop, a cable break or node failure can take down the whole loop, while the hub in a logical loop can remove the malfunctioning node and continue operating. Each of the nodes in a Fibre Channel loop acts as a repeater, which prevents signal degradation due to attenuation, but a loop is still a shared network with multiple devices utilizing the same bandwidth, which can limit the performance of each device.

Fabric   The fabric topology consists of nodes connected to switches with point-to-point connections. Just as on an Ethernet network, switching enables each device to use the full bandwidth of the network technology in its transmissions. Fibre Channel uses nonblocking switches, which enable multiple devices to send traffic through the switch simultaneously. A switched Fibre Channel network has the benefit of almost unlimited expandability while maintaining excellent performance.

Fibre Channel Communications

Communications over a Fibre Channel network are broken down into three hierarchical structures. The highest level structure is called an *exchange*, which is a bidirectional, application-oriented communication between two nodes on the network. In the context of a storage operation, an exchange would be the process of reading from or writing to a file. A single device can maintain multiple exchanges simultaneously, with communications running in both directions, if needed.

An exchange is made up of unidirectional transmissions between ports called *sequences*, which in the context of a read or write operation are the individual blocks transmitted over the network. Each sequence must be completed before the next one can begin. Sequences are comprised of *frames*; the frame is the smallest protocol data unit transmitted over a Fibre Channel network. Fibre Channel frames are constructed much like the frames used in other networking protocols, such as Ethernet and IP. The frame consists of discrete fields that contain addressing and error detection information, as well as the actual data to be transmitted. In the storage context, a frame is the equivalent of a SCSI command.

Fibre Channel provides three classes of service, with different resource requirements and levels of performance provided by each. These service classes are as follows:

Class 1   Class 1 is a reliable, connection-oriented, circuit-switched service in which two ports on the network reserve a path through the network switches to establish a connection for as long as they need it. The result is the functional equivalent of a point-to-point connection that can remain open for any length of time, even permanently. Because a virtual circuit exists between the two nodes, frames are always transmitted and received in the same order, eliminating the additional processing required to reorder the packets, as on an IP network. The Class 1 service tends to waste bandwidth when the connection is not in use all of the time, but for applications that require a connection with the ultimate in reliability and performance, the expenditure can be worthwhile.

Class 2   Class 2 is a connectionless service that provides the same reliability as Class 1, through the use of message delivery and non-delivery notifications. Since Class 2 is not a circuit-switched service, frames may arrive at the destination port in the wrong order. However, it is the port in the receiving node that reorders the frames, not the processor inside the server or storage subsystem containing the port. By placing the responsibility for ordered delivery of frames on the port rather than on the switch, as in the Class 1 service, the switches are better able to provide the maximum amount of bandwidth to all of the nodes on the network. The Class 2 service can therefore provide performance and reliability that is nearly that of the Class 1 service, with greater overall efficiency. Most storage network implementations use Class 2 rather than Class 1 for this reason.

Class 3   Class 3 is an unreliable connectionless service that does not provide notification of delivery and non-delivery like Class 2. Removing the processing overhead required to implement the notifications reduces port latency and therefore greatly increases the efficiency of the network. This is particularly true in the case of a loop network, which uses a shared medium. In the case of a storage network, the FCP protocol provides frame acknowledgment and reordering services, making it unnecessary to implement them in the network hardware.

Note   *There is also an extension to the Class 1 service called* Intermix, *which enables other processes to utilize the unused bandwidth of a Class 1 connection for the transmission of Class 2 and Class 3 traffic. In this arrangement, however, the Class 1 traffic maintains absolute priority over the connection, which can cause the nodes to buffer or discard Class 2 and 3 frames, if necessary.*

Network Storage Subsystems

In the original client/server network design, the server was a computer constructed very much like a client, except with more storage capacity, more memory, a faster processor, and so on. As the years have passed and data storage requirements have increased at an exponential level, it has become unwieldy for a personal computer to contain enough space and power for the many drives used in modern storage arrays. The resulting solution is a separate drive array that connects to the server using a standard external SCSI connection. Moving the storage management tasks away from the server and into a dedicated device also reduces the processing burden on the server. Since the length of a parallel SCSI bus is limited, the drive array has usually been located near the server hosting its drives. Today, with server clusters and other advanced server technologies becoming more popular, there is a drive toward storage arrays with greater capabilities. Connecting multiple servers to a single storage array using SCSI connections, for example, isn't possible unless the servers are all located in the same place, and part of the reason for having multiple servers is often to provide backups in the event of fire or theft.

One of the solutions to this problem is to integrate the standard storage I/O architecture with the networking architecture used for other communications between systems. Combining I/O and networking makes it possible to locate the servers and the storage arrays virtually anywhere, build a more flexible and expandable storage solution, and enable any server on the network to work with any storage device. There are two technologies that are leading the way in this new area of development: Network Attached Storage (NAS) and storage area networks (SANs). These technologies are not mutually exclusive; in fact, the future network is likely to encompass both, to some degree.

Network Attached Storage (NAS)

Network Attached Storage is a term that is generally applied to a standalone storage subsystem that connects to a network and contains everything needed for clients and servers to access the data stored there. A NAS device, sometimes called a *network storage appliance*, is not just a box with a power supply and an I/O bus with hard drives installed in it. The unit also has a self-contained file system and a stripped-down, proprietary operating system that is optimized for the task of serving files. The NAS appliance is essentially a standalone file server that can be accessed by any computer on the network. For a network that currently has Windows, Unix, or NetWare servers dedicated primarily to file-serving tasks, NAS appliances can reduce costs and simplify the deployment and ongoing management processes. Because the appliance is a complete turnkey solution, there is no need to integrate separate hardware and operating system products or be concerned about compatibility issues.

NAS appliances can connect to networks in different ways, and it is here that the definition of the technology becomes confusing. A NAS server is a device that can respond to file access requests generated by any other computer on the network, including clients and servers. The device typically uses a standard file system protocol like the Network File System (NFS) or the Common Internet File System (CIFS) for its application layer communications. There are two distinct methods for deploying a NAS server, however. You can connect the appliance directly to the LAN, using a standard Ethernet connection, enabling clients and servers alike to access its file system directly, or you can build a dedicated storage network, using Ethernet or Fibre Channel, enabling your servers to access the NAS and share files with network clients.

The latter solution places an additional burden on the servers, but it also moves the I/O traffic from the LAN to a dedicated storage network, thus reducing network traffic congestion. Which option you choose largely depends on the type of data to be stored on the NAS server. If you use the NAS to store users' own work files, for example, it can be advantageous to connect the device to the LAN and let users access their files directly. However, if the NAS server contains databases or e-mail stores, a separate application server is required to process the data and supply it to clients. In this case, you may benefit more by creating a dedicated storage network that enables the application server to access the NAS server without flooding the client network with I/O traffic.

Storage Area Networks (SAN)

A storage area network is simply a separate network with an enterprise that is used to connect storage devices and the computers that use them. In practice, SANs are usually associated with Fibre Channel networks, but actually, you can use any type of network for this purpose, including SSA or Ethernet (usually Gigabit Ethernet). The reasons for building a SAN have been repeated throughout this chapter. Server technologies such as clustering and remote disk arrays require high-bandwidth connections, and using the same data network as the client computers for this purpose could easily result in massive amounts of traffic. In addition, the bandwidth requirements of a storage I/O network far exceed those of a typical data network. Constructing a separate SAN using Fibre Channel or Gigabit Ethernet is far cheaper than equipping all of the computers on your network with ultra–high-speed network interface adapters.

In a typical enterprise network containing a SAN, the servers have interfaces to both the data network (the LAN) and the storage network (the SAN). The LAN, therefore, is completely ordinary, containing client and server computers, and the storage devices are connected only to the SAN. Where the servers store their data is of no consequence to the clients, which do not even have to know of the SAN's existence.

A typical SAN using Fibre Channel to connect servers to the storage devices can take many forms. The simplest possible SAN consists of a single server connected to a drive array using a point-to-point Fibre Channel connection, as shown in Figure 8-21. The server accesses the data stored on the array, which would typically use RAID to provide added performance and fault tolerance. One of the primary differences between a SAN and a NAS device is that SANs provide block-level access to data, while NAS appliances provide file-level access.

A more complicated SAN would consist of several servers and several storage arrays, all connected to the same network, as shown in Figure 8-22. If the SAN uses Fibre Channel for its communications, the network's topology can take the form of a loop or a fabric, depending on whether the devices are all connected to a hub or a switch. This enables the servers to communicate with each other and with all of the storage devices on the SAN. The storage devices can be drive arrays using RAID, NAS servers, or any other technology that may evolve, as long as it supports Fibre Channel or whatever networking protocol the SAN uses.

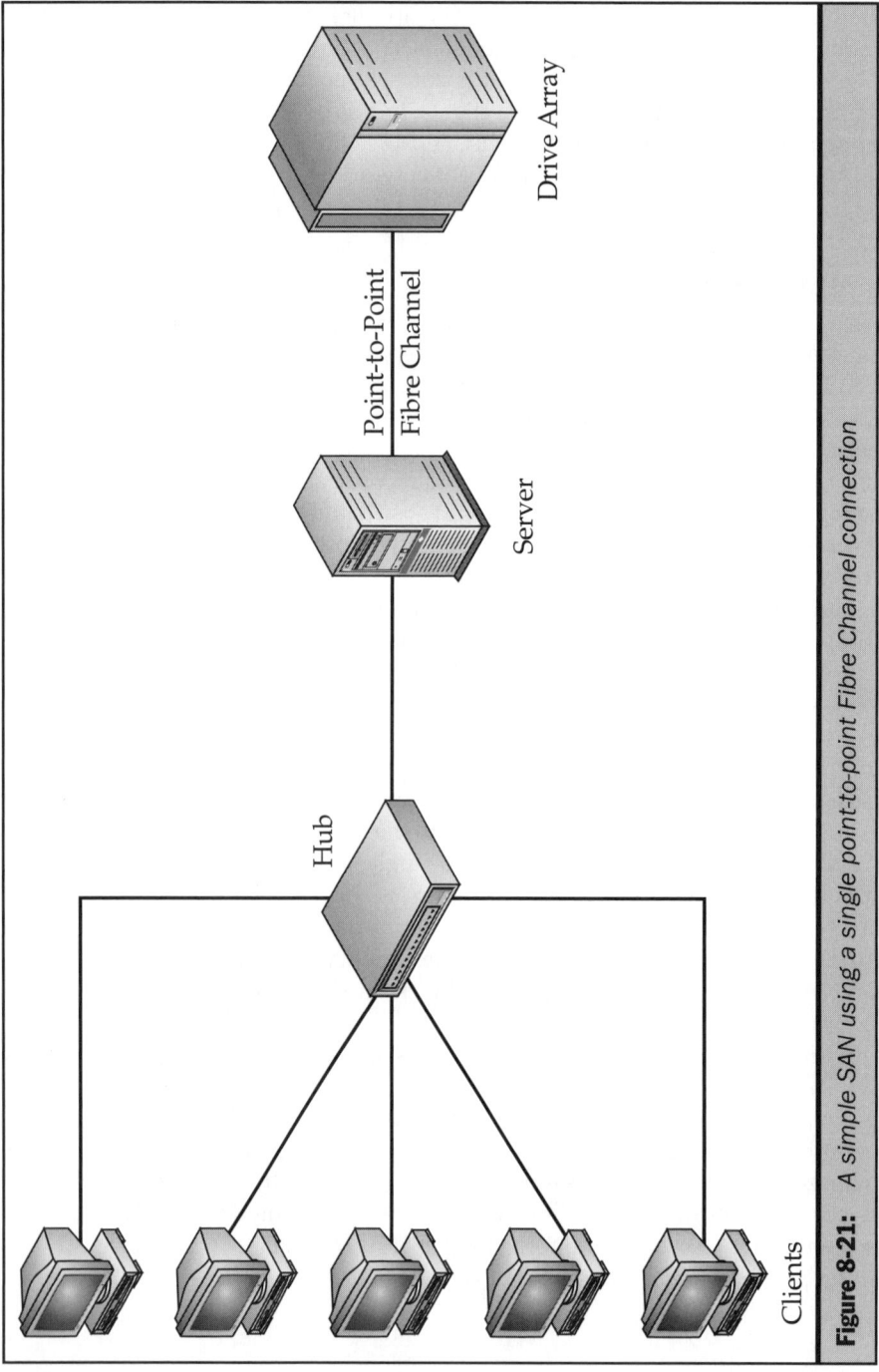

Figure 8-21: A simple SAN using a single point-to-point Fibre Channel connection

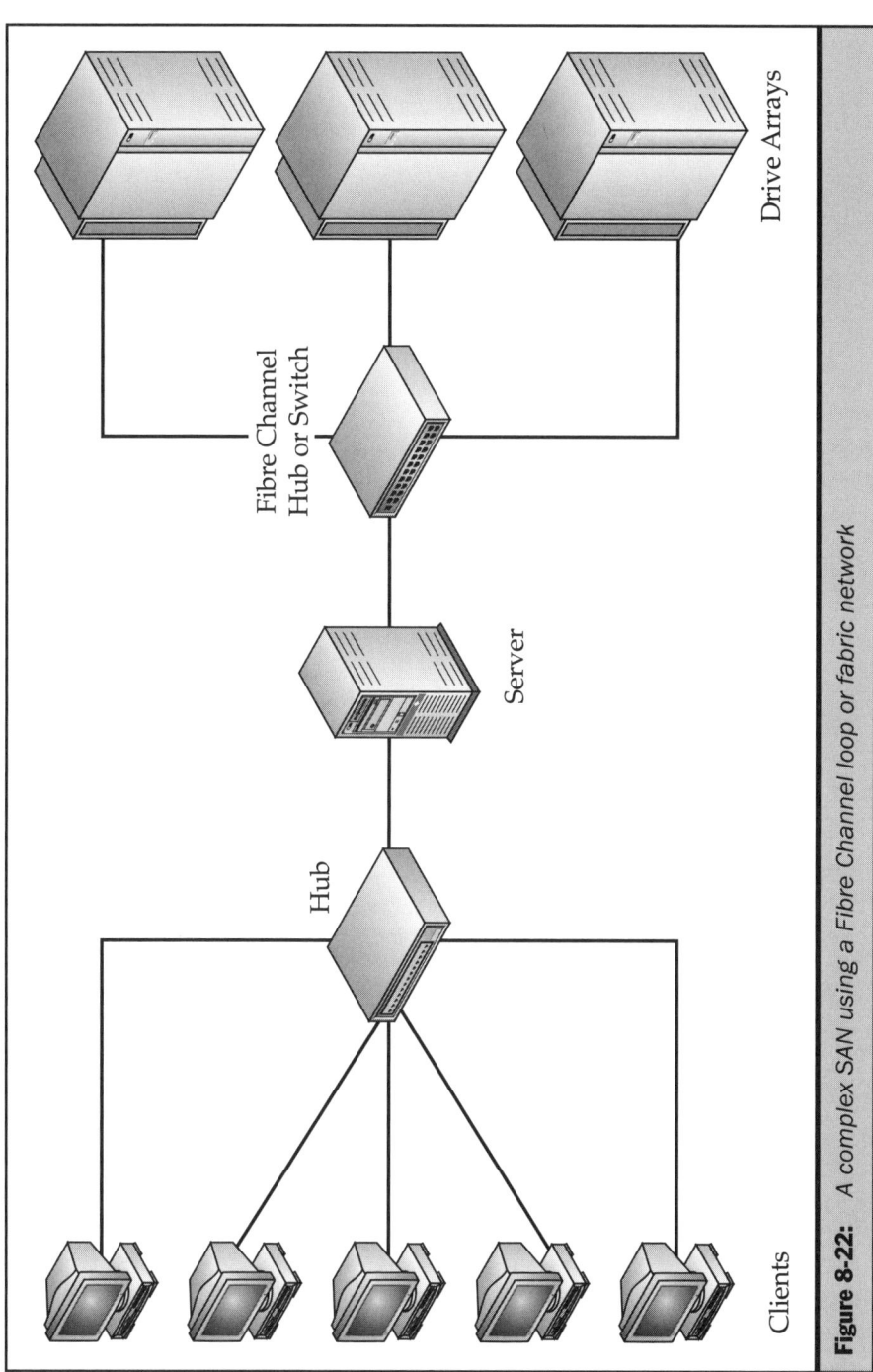

NETWORK HARDWARE

Figure 8-22: *A complex SAN using a Fibre Channel loop or fabric network*

Chapter 9

Designing a Network

P lanning is an essential part of any network deployment, and the design of the network is a crucial element of the planning process. Depending on its size and location, the process of designing your network can be simple or extremely complex. This chapter examines some of the concepts involved in designing networks that range from small home LANs to large enterprise internetworks.

Network Design Overview

A network design can encompass decisions made at many levels. At a minimum, the design should include what hardware you intend to purchase, how much it costs, where you're going to locate it at your site, and how you're going to connect it all together. For a home or small business network, this can be as easy as taking a few computers, choosing a network interface card (NIC) for each one, and buying some cables and a hub. You can make all of the other decisions involved in setting up and configuring the network as you proceed. For a large enterprise internetwork, the design process is considerably more complicated.

As you've learned, an internetwork is a collection of LANs that have been connected together so that each computer can communicate with any other computer on any of the LANs. You can design each LAN separately, using standard elements like the cables and hubs already mentioned, but then you must consider how you are going to connect the LANs into an internetwork and regulate the communications between them. You also have to consider all of the services that you must provide to your users, and how you intend to provide them. This means that the network design might include software products and configurations, outside services provided by third parties, and operating procedures, as well as a hardware list and a network diagram.

In addition to purely technical issues, designing a large internetwork involves a number of important business decisions as well. Generally speaking, the early phases of the internetwork design process tend to proceed as follows:

1. Identify the business needs that the network is intended to satisfy.

2. Create an ideal network design that satisfies all of the previously defined needs.

3. Estimate the cost of building the network as designed.

4. Determine whether or not the benefits of building the network rationalize the expense.

5. Revise the network design to bring the expense in line with the benefits.

This is a high-level overview of the network design process as a business decision, and while economic issues will probably not be the primary concern of the people involved in the technical side of the process, the cost of the project will certainly have a profound effect on the design. This chapter is more involved with the technical side of the design process than the business side, but having some idea of the budget allotted for the network and the cost of implementing the technologies you select can streamline the whole design and approval process considerably.

Reasoning the Need

The first step in designing a network is always to list the reasons for building it in the first place. For a home or small business network, the list is often short and simple, containing items such as the desire to share one printer among several computers and to access the Internet using a single connection. In most cases, the economic decision is equally simple. Weigh the price of a few NICs, cables, and a hub against the cost of supplying each computer with its own printer or Internet connection, and the conclusion is obvious.

For a large internetwork installation, the list of requirements is usually much longer, and the decision-making process far more complex. Some of the questions that you should ask yourself as you're first conceiving the network are as follows:

- What business needs will the network satisfy?

- What services do you expect the network to provide now? In the future?

- What applications must the network run now? In the future?

- What are the different types of users that you expect the network to support now?

- What types of users (and how many of them) do you expect the network to support in the future?

- What level of service do you expect the network to provide, in terms of speed, availability, and security?

- What environmental factors at the site can possibly affect the network?

- What is the geographic layout of the business? Are there remote offices to connect?

- What network maintenance skills and resources are available to the organization?

By answering questions like these, you should be able to come up with a basic, high-level conception of the type of network you need. This conception should include a sketch of the network indicating the number of levels in the hierarchy. For example, a network at a single site might consist of a number of LANs connected by a backbone, while a network encompassing multiple sites might consist of several LANs, connected by a backbone at each location, all of which are then connected by WAN links. This plan may also include decisions regarding the network media and protocols to use, a routing strategy, and other technical elements.

NETWORK HARDWARE

Seeking Approval

The next step is to start making generic technology and equipment selections in order to develop an estimate of the costs of building and maintaining the network. For example, you might at this point decide that you are going to build an internetwork consisting of ten Category 5 UTP LANs running Fast Ethernet, connected by a fiber-optic backbone running Gigabit Ethernet and using a T-1 line for access to the Internet. With this information, you can start to figure out the general costs of purchasing and installing the necessary equipment.

With a rough cost estimate in hand, it's generally time to decide whether building the network as conceived is economically feasible. In many cases, this requires an evaluation by non-technical people, so a layman's summary of the project and its cost is usually in order. At this point, some of the following questions may be considered:

- Does the network design satisfy all of the business needs listed earlier?

- Do the business needs that the network will satisfy justify the cost expenditures?

- Can the costs of the network be reduced while still providing a minimum standard of performance?

- How will reducing the quality of the network (in regard to elements such as speed, reliability, and/or security) affect the business needs it is able to satisfy?

- Can the network be reconceived to lower the initial costs while still providing sufficient capability for expansion at a later time?

This review process may involve individuals at several management layers, each with their own concerns. In many cases, business and economic factors force a redesign of the network plan at this point, either to better address business needs not considered earlier or to reduce costs. It's better for these modifications to occur now, while the network design plan is still in its preliminary stages. Once the elements of the plan are developed in greater detail, it will become more difficult and inefficient to drastically change them.

When the economic and business factors of the network design have been reconciled with the technical factors, you can begin to flesh out the plan in detail. The following sections examine some of the specific elements that should be included in your network design plan.

Designing a Home or Small Office Network

A network for a home or small office typically consists of a single LAN connecting anywhere from 2 to 16 computers together. The LAN might also have additional network devices attached to it, such as a network printer or a router providing a connection to the Internet or another office. For this kind of network, the design process consists mostly of selecting products that are suitable for your users' needs and for the physical layout of the site.

Selecting Computers

Virtually all of the computers on the market today can be connected to a network, so compatibility in this area is not as much of a problem as it used to be. However, for the sake of convenience, it's easier to design, build, and maintain a small network in which all of the computers use the same platform. If most of your users are accustomed to using Windows PCs, then make the network all Windows PCs. If you're more comfortable with Macintosh or Unix systems, then use those. It's not impossible to connect computers running different platforms to the same network by any means, but if you're planning a small network and you want to have as easy a time of it as possible, stick to one platform.

Standardizing on a single platform may be difficult in some situations, however. For a home network, for example, you may have kids who use Macs in school and adults who use PCs at work. In a small business environment, you are more likely to be able to impose one platform on your employees, unless they have special requirements such as different types of machines. If you do feel compelled to mix platforms, you must be careful to select products that are compatible with every type of computer you plan to use. It's generally not too difficult to configure different types of computers to access shared network resources such as printers and Internet connections. However, file sharing is more of a problem, because the computers may use different file formats.

The other important consideration when selecting the computers to be connected to a network is whether they have the resources needed for networking. For the most part, this just means that you must determine what type of network interface adapter the computer can use. Some computers come equipped with a network adapter, but in most, you must add one yourself. Most computers have PCI and/or ISA bus slots, into which you can plug a network interface card, but you must make sure that each computer you plan to network has a slot free, and that you purchase a NIC of the appropriate type for that slot. For computers without free slots, you can use a Universal Serial Bus (USB) network interface adapter.

Selecting a Networking Protocol

The protocol that your network uses at the data-link layer of the OSI reference model is the single most defining element of the network design. The data link–layer protocol determines, among other things, what network medium you will use, what networking hardware you will buy, how you will connect the computers together, and how fast the network can transfer data. The most common choice in data link–layer protocols is

IEEE 802.3 (Ethernet), and there are few situations where you can go wrong in installing an Ethernet network, particularly in a home or small business. The following sections examine some of the further decisions you have to make after deciding to use Ethernet on your network.

Note   *For more information on the Ethernet protocol, see Chapters 10 and 11.*

Choosing a Network Medium

The Ethernet protocol supports a variety of network media, but when installing a new network today, the choice for a bounded (cabled) network comes down to unshielded twisted-pair (UTP) or fiber-optic cable. The other alternative is a wireless (unbounded) medium. UTP cable is perfectly suitable for most home and small business networks. To use UTP, you have to purchase an Ethernet hub (unless you are networking only two computers together), and each of your network devices must be connected to the hub using a cable no more than 100 meters long. Category 5 UTP is the industry standard for Ethernet cabling and is more than sufficient for networks running at speeds up to 100 Mbps. Category 5E or higher is preferable only if you are planning to run Gigabit Ethernet at 1,000 Mbps, which is unlikely on a small network at this time.

The 100-meter segments supported by UTP cabling are more than sufficient for most networks in general and for virtually all home or small business networks. If you are in a situation where the locations of your computers call for longer segments, however, or the network must operate in an environment with extreme amounts of electromagnetic interference (EMI) present, you can opt to use fiber-optic cable. Fiber-optic cable is immune to EMI and supports longer segments, but it is also more expensive than UTP and more difficult to install.

For a small network, the ease of installation is often a major factor in the selection of a network medium. An Ethernet network using UTP is the simplest type of cabled network to install. UTP Ethernet NICs, hubs, and prefabricated cables are available in almost any computer store; all you have to do to install the network hardware is insert the NICs into the computers and use the cables to connect the computers to the hub. The same is not true for fiber-optic cables, which are generally purchased as components (bulk cable, connectors, and so on) from professional suppliers. Unless you are willing to spend a good deal of money, time, and effort on learning about fiber-optic cabling, you are not going to install it yourself.

It's possible to install UTP cable from components also, and this is usually how professional, internal installations are performed. An internal cable installation is one in which the cables are installed inside wall cavities and drop ceilings. The only elements of the installation that are visible to the network user are the wall plates to which their computers are attached. This type of installation is neater than an external one that uses prefabricated cables that are usually left exposed, but it requires more expertise to perform correctly, as well as additional tools and access to internal wall cavities. For a small business network in a traditionally designed office space, a small-scale internal installation is feasible, but homeowners are less likely to want to drill holes in their walls, floors, and ceilings for the installation of cables, despite a greater concern for the installation's cosmetic appearance.

 *For more information on network cables and their installation, see Chapter 4.*

For network installations where cables are impractical or undesirable, you can also elect to install a wireless LAN, using the IEEE 802.11b protocol that was ratified in 1999. Although IEEE 802.11b is not a form of Ethernet, it is the first practical wireless protocol to achieve traditional LAN speeds, up to 11 Mbps. There are many 802.11b products now on the market, at competitive prices, and for home users wanting to network their computers without leaving cables exposed or performing a major cable installation, this solution can be ideal.

 *For more information on wireless LANs, see Chapter 5.*

Choosing a Network Speed

Another consideration when designing an Ethernet LAN is the speed at which the network will run. Standard Ethernet runs at 10 Mbps, Fast Ethernet runs at 100 Mbps, and Gigabit Ethernet runs at 1,000 Mbps. For a small network, Gigabit Ethernet speeds generally aren't necessary and the hardware is still too expensive to be a practical choice. As a result, the average consumer has to choose between regular and Fast Ethernet. Most of the Ethernet NICs on the market today are dual-speed, supporting both regular and Fast Ethernet. The NIC auto-detects the speed of the hub to which it's attached and configures itself accordingly. It's still possible to find 10 Mbps-only NICs, but in some cases, you'll pay more for the older technology, because it is no longer the industry standard.

Hubs are available in standard Ethernet, Fast Ethernet, and dual-speed varieties. A dual-speed hub is only required if you have devices on your network running at both speeds. If you're designing a new network, you can make sure all of your selected equipment runs at one speed and avoid the need for a dual-speed hub. Be sure to consider devices other than computers in this respect, however. If, for example, you decide to run Fast Ethernet on the entire network and purchase a single-speed hub, the print server you use to connect your printer directly to the network must also support Fast Ethernet.

Generally speaking, you're better off running Fast Ethernet on a new network, even if you don't have an immediate need for the faster transmission rate. The additional expense of purchasing Fast Ethernet equipment, as opposed to standard Ethernet, is minimal if there is any at all.

Expanding the Network

Planning for future expansion is one of the most important elements of the network design process. For a small network, this may mean nothing more than buying Fast Ethernet equipment now, instead of upgrading to it later, and choosing a hub with a few more ports than you currently need. One of the important network design factors if you elect to use Fast Ethernet is that a LAN can only have two hubs, as opposed to standard Ethernet, which permits as many as four. Fast Ethernet also requires a shorter total length of the network cable than standard Ethernet when you use two hubs. Once you reach the limit of your network's capacity, because you're using the maximum number of hubs allowed and all of the hub ports are in use, you can only expand further by replacing your hubs with models having more ports or by creating a second LAN and connecting the two together.

Designing an Internetwork

The design elements discussed thus far apply to large internetworks as well as to small, single-segment LANs. Even the largest internetwork consists of individual LANs that require the same components as a standalone LAN, such as computers, NICs, cables, and hubs. For a large internetwork with more varied requirements, you can design each LAN separately, selecting protocols and hardware that best suit the physical environment and the requirements of the users, or create a uniform design suitable for all of the LANs.

Once you get beyond the individual LANs, however, you are faced with the problem of connecting them together to form the internetwork. The following sections examine the technologies you can use to do this.

Segments and Backbones

The traditional configuration for a private internetwork is to have a series of LANs (called *network segments* or sometimes *horizontal networks*) connected together using another, separate network called a backbone. A *backbone* is nothing more than a network that connects other networks together, forming an internetwork. The individual segments can be networks that service workgroups, departments, floors of a building, or even whole buildings. Each of the segments is then connected to a backbone network, using a router or a switch, as shown in Figure 9-1. This enables a workstation on any of the networks to communicate with any other workstation. The term backbone can refer to a LAN that connects other LANs (usually in the same building or campus) together, or to a network of wide area links that connect networks or internetworks at remote locations together.

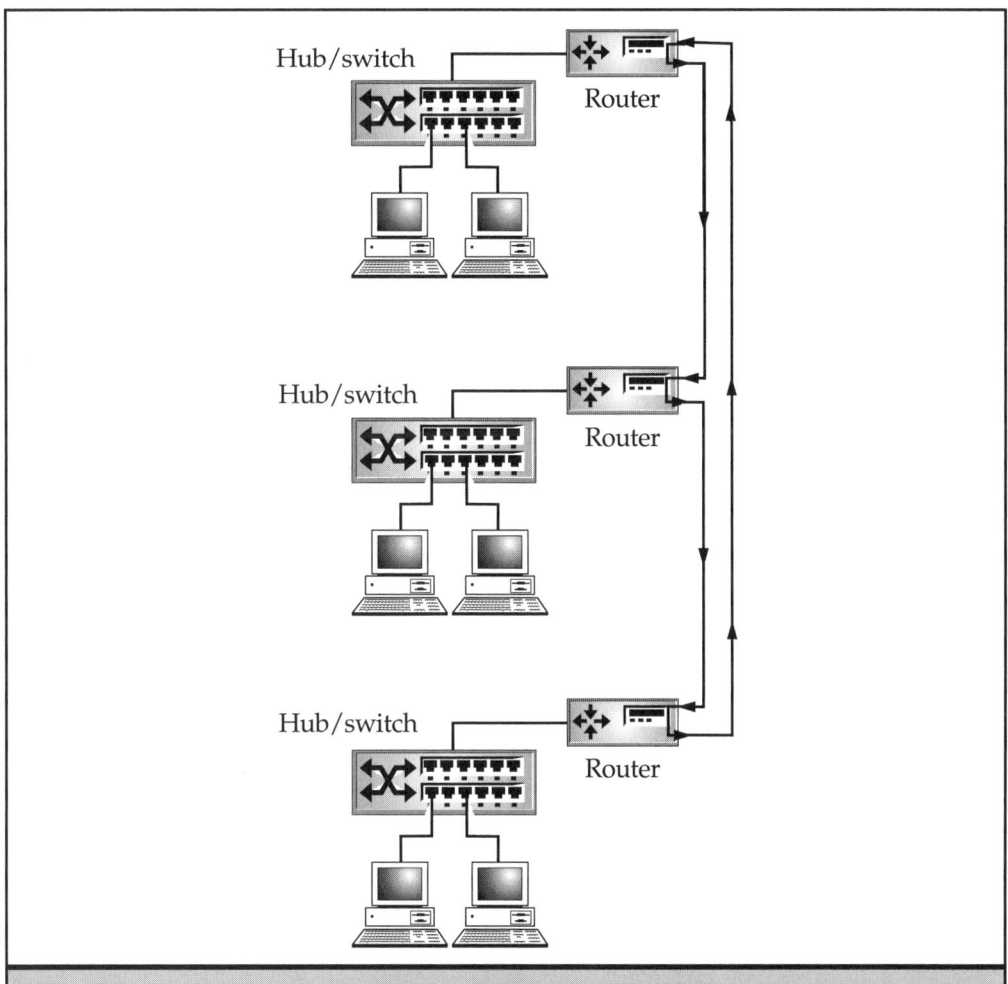

Figure 9-1: *An enterprise network, consisting of multiple LANs connected by a backbone*

One of the most common configurations for a large internetwork that encompasses an entire building with multiple floors is to have a separate LAN connecting all of the network devices on each floor (which is the origin of the term "horizontal network") and a backbone network running vertically between the floors, connecting all of the LANs together. Of course, the configuration you use must depend on the building in which the internetwork is installed. If your entire organization is housed in an enormous building with only two floors, you will probably have to create several LANs on each floor and connect them with a backbone that runs throughout the building.

When two computers on the same LAN communicate with each other, the traffic stays on that local network. However, when the communicating computers are on different LANs, the traffic goes through the router connecting the source computer to the backbone and then to the LAN on which the destination computer is located. It is also common practice to connect network resources required by all of the internetwork's users directly to the backbone, instead of to one of the horizontal networks. For example, if you have a single e-mail server for your entire organization, connecting it to one of the horizontal networks forces all of the e-mail client traffic from the entire internetwork to travel to that segment, possibly overburdening it. Connecting the server to the backbone network enables the traffic from all of the horizontal segments to reach it equitably.

Because the backbone is shared by the horizontal networks, it carries all of the internetwork traffic generated by each of the computers on every LAN. This can be a great deal of traffic, and for this reason, the backbone typically runs at a higher speed than the horizontal networks. Backbones may also have to traverse greater distances than horizontal networks, so it is common for them to use fiber-optic cable, which can span much longer distances than copper.

When the concept of the backbone network originated, the typical departmental LAN was relatively slow, running 10 Mbps Ethernet. The first backbones were Thick Ethernet trunks, selected because the RG-8 coaxial cable could be installed in segments up to 500 meters long. These backbones ran at the same speed as the horizontal networks, however. To support all of the internetwork traffic, a distributed backbone running at a higher speed was needed. This led to the use of data link–layer protocols like *Fiber Distributed Data Interface (FDDI)*. FDDI runs at 100 Mbps, which was faster than anything else at the time, and it uses fiber-optic cable, which can span much greater distances than Thick Ethernet.

Once Fast Ethernet products arrived on the market, the situation changed by an order of magnitude. 100 Mbps horizontal networks have become common, and an even faster backbone technology is needed to keep up with the traffic load they generate. This led to the development of protocols like Asynchronous Transfer Mode (ATM), running at speeds up to 655 Mbps, and Gigabit Ethernet, at 1,000 Mbps.

Distributed and Collapsed Backbones

There are two basic types of backbone LAN in general use: the distributed backbone and the collapsed backbone. In a *distributed backbone*, the backbone takes the form of a separate cable segment that runs throughout the enterprise and is connected to each of the horizontal networks using a router or switch. In a *collapsed backbone*, the hub on each of the horizontal networks is connected to a centrally located modular router or switch (see Figure 9-2). This router or switch functions as the backbone for the entire internetwork by passing traffic between the horizontal networks. This type of backbone uses no additional cable segment because the central router/switch has individual modules for each network, connected by a backplane. The *backplane* is an internal communications bus that takes the place of the backbone cable segment in a distributed backbone network.

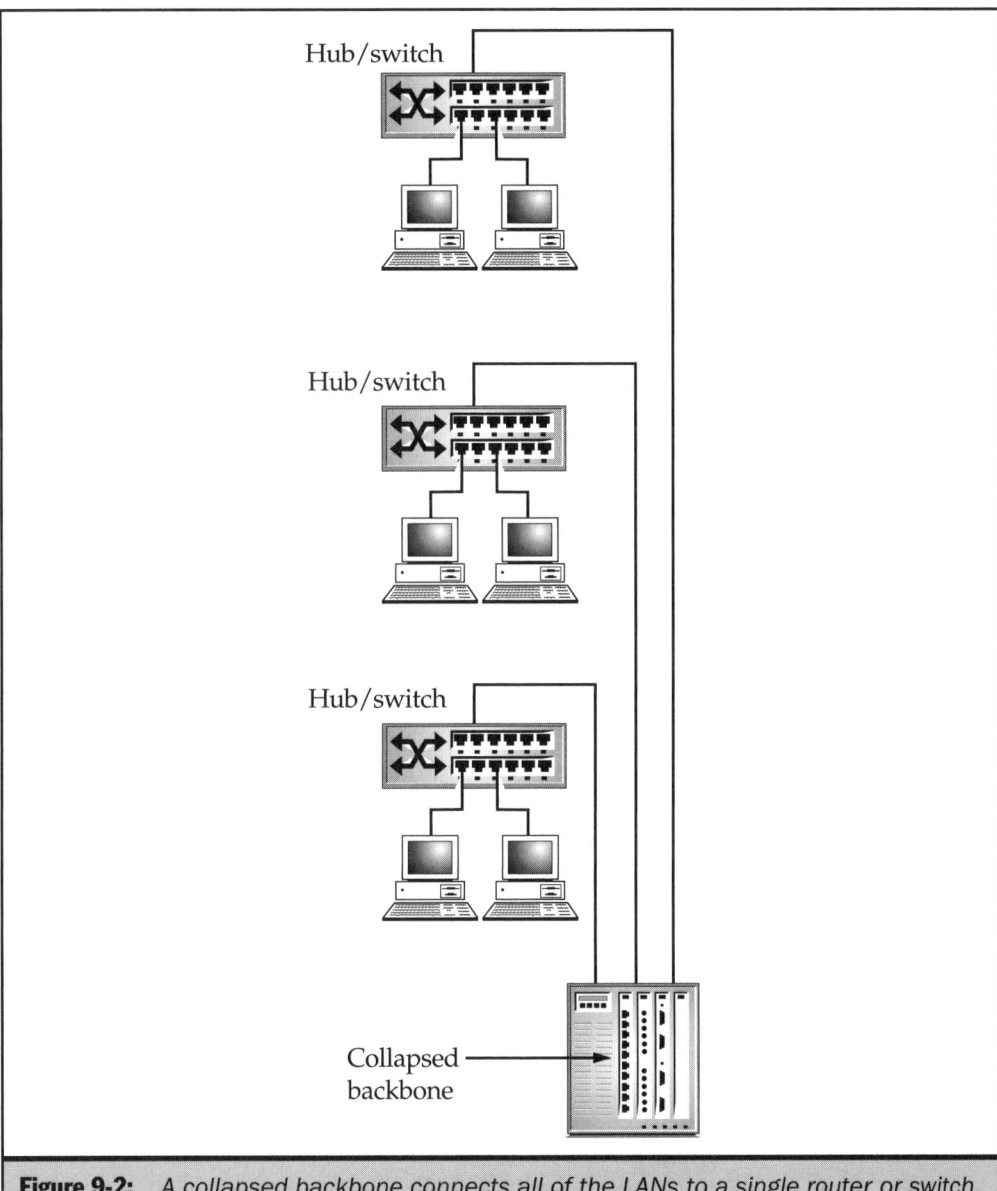

Figure 9-2:   *A collapsed backbone connects all of the LANs to a single router or switch.*

The advantage of a collapsed backbone is that internetwork traffic only has to pass through one router on the way to its destination, unlike a distributed backbone, which has separate routers connecting each network to the backbone. The disadvantage of a collapsed backbone is that the hub on each network must connect to the central router with one cable segment. Depending on the layout of the site and the location of the router, this distance may be too long for copper cable.

Because a collapsed backbone does not use a separate cable segment to connect the horizontal networks, it does not need its own protocol. Today's Fast Ethernet technology has made the collapsed backbone a practical solution. As an example, consider an enterprise that consists of horizontal networks running 100Base-TX and using modular hubs, which can connect different media types. While copper cable connects the individual workstations to the hub, a single 100Base-FX fiber-optic segment runs to the organization's data center, where a large multinetwork switch is installed. This switch routes the traffic between the networks, and the use of fiber-optic cable means it can be located virtually anywhere in the building. The entire network runs at 100 Mbps, and no traffic has to pass through more than two networks to get from one workstation to another.

While this may be an ideal solution for a new network being constructed today, there are thousands of existing networks that still use 10 Mbps Ethernet or other relatively slow protocols on their horizontal networks, and can't easily adapt to the collapsed backbone concept. Some or all of the horizontal networks might be using older media, such as Category 3 UTP or even Thin Ethernet, and can't support the long cable runs to a central router. The horizontal networks might even be in separate buildings on a campus, in which case a collapsed backbone would require each building to have a cable run to the location of the router. In cases like these, a distributed backbone is necessary.

Backbone Fault Tolerance

Because it provides all internetwork communications, the backbone network is a vitally important part of the overall design. A horizontal network that can't access the backbone is isolated. Computers on that LAN can communicate with each other, but not with the computers on other LANs, which can cut them off from vital network services. To ensure continuous access to the backbone, some internetworks design redundant elements into the plan for fault-tolerance purposes. You can, for example, use two routers on each LAN, both of which are connected to the backbone network hub, so that if one router fails, the other provides continued access to the rest of the network.

Some designs go so far as to include two separate distributed backbone networks. This plan also calls for two routers on each horizontal network, but in this case, the routers are connected to two different backbone networks, as shown in Figure 9-3. This way, the internetwork can continue to function despite the failure of a router, a backbone hub, or any backbone cable segment. Another benefit of this design is the ability to balance the internetwork traffic load among the two backbones. By configuring half of the computers to use one backbone and half the other (by varying their default gateway addresses), you split the internetwork traffic between the two. This can make the use of Fast Ethernet on both the horizontal and backbone networks a practical proposition, even on a highly trafficked network. With a single backbone connecting Fast Ethernet LANs, you may find that you need to use Gigabit Ethernet or another high-speed protocol to support the internetwork traffic.

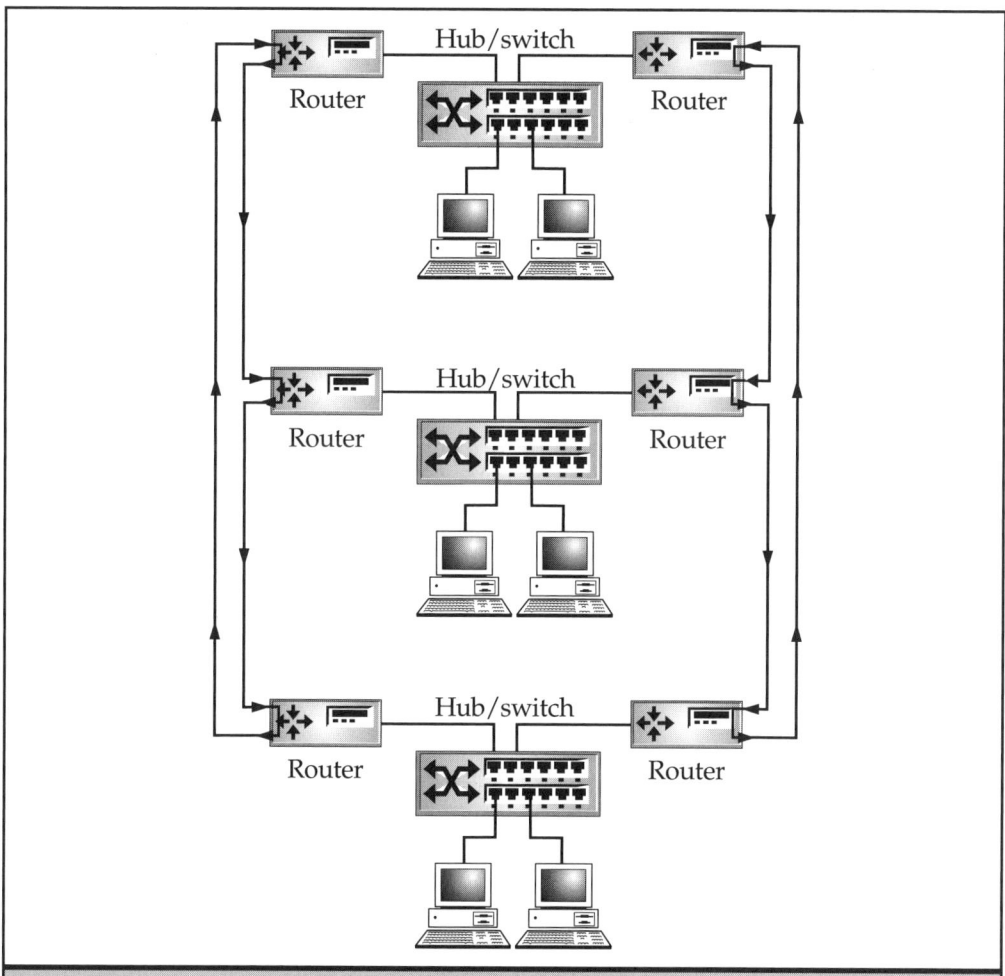

Figure 9-3:   *Redundant backbones can provide fault tolerance and load balancing.*

Selecting a Backbone LAN Protocol

The protocol that you use on the backbone connecting your horizontal networks together should depend on the amount of traffic it has to carry and the distance it has to span. In some organizations, most of the network communications are limited to the individual horizontal LANs. If, for example, your company consists of several departments that are largely autonomous, each with their own servers on a separate horizontal LAN, all of the intradepartmental traffic remains on the horizontal network and never reaches

NETWORK HARDWARE

the backbone. In a case like this, you can probably use the same technology on the backbone as the horizontal LANs, such as Fast Ethernet throughout. If, on the other hand, your company consists of departments that all rely on the same resources to do their work, such as a central database, it makes sense to connect the database servers directly to the backbone. When you do this, however, the backbone must be able to support the traffic generated by all of the horizontal networks combined. If the horizontal networks are running Fast Ethernet, the backbone should usually use a faster technology, such as Gigabit Ethernet, in order to keep up.

The distance that the backbone LAN must span and the environment in which it's used can also affect the protocol selection. If your site is large enough that the backbone cable runs are likely to exceed the 100-meter limit for unshielded twisted-pair (UTP) cable, you should consider using fiber-optic cable. Fiber optic is also the preferred solution if you have to connect horizontal LANs that are located in different buildings on the same campus. Fiber optic is more expensive to purchase and install than UTP, but it is interoperable with copper cable in most cases. For example, you can purchase Fast Ethernet hubs and routers that support both cable types, so that you can use UTP on your horizontal networks and fiber optic on the backbone.

Connecting to Remote Networks

In addition to connecting LANs together at the same site, many internetworks also use a backbone to connect to remote networks. In some cases, the organization consists of multiple offices in different cities or countries that must communicate with each other. If each office has its own internetwork, connecting the offices together with WAN links forms another backbone that adds a third level to the network hierarchy and creates a single, enterprise internetwork. However, even an organization with one internetwork at a single location is likely to need a WAN connection to an Internet service provider so that users can access e-mail and other Internet services.

The technology you select for your WAN connections depends on factors such as the amount of bandwidth your network needs, when it needs it, and as always, your budget. You can use anything from dial-on-demand telephone connections to high-speed leased lines, to flexible bandwidth solutions, such as frame relay. For more information on WAN technologies, see Chapter 7.

Selecting a WAN Topology

Another factor in selecting a WAN technology is the topology you will use to connect your various sites together. WAN topologies are more flexible than those on LANs, which are dictated by the data link– and physical-layer protocols you elect to use. You can use WAN links to build an internetwork in many different ways. For example, the *full mesh* topology, when used on a WAN, consists of a separate, dedicated link (such as a leased line) between each two sites in your organization. If you have five offices in different cities, each office has four separate WAN links connecting it to the other offices, for a total of ten links (see Figure 9-4). If you have eight offices, a total of 28 separate WAN links are required. This arrangement provides the greatest amount of fault tolerance, since a single link failure only affects the two sites involved, as well as the most efficient network, since each site can communicate directly with each of the other sites. However, this solution can also be outrageously expensive, as well as wasteful, unless your network generates sufficient WAN traffic between each pair of sites to fill all of these links most of the time.

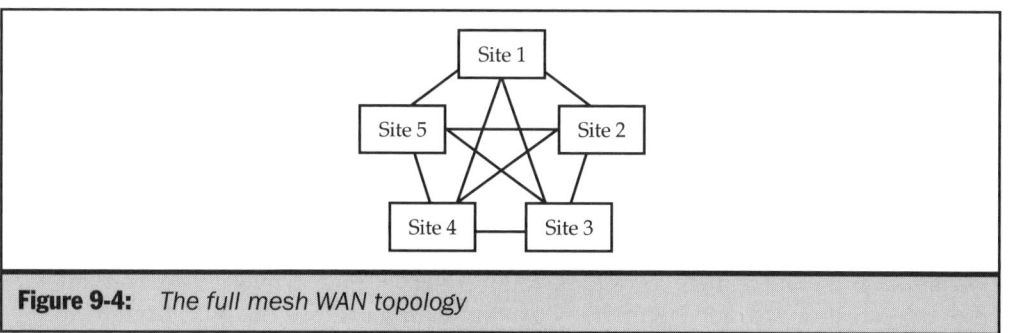

Figure 9-4:   *The full mesh WAN topology*

A full mesh topology consisting of individual links between the sites assumes the use of dedicated, point-to-point WAN connections such as leased lines. However, there are alternatives to this type of link that can provide what amounts to a full mesh topology at much less expense. Frame relay uses a single leased line at each site to connect to a service provider's network, called a cloud. With all of the sites connected to the same cloud (using access points local to each location), each site can establish a virtual circuit to every other site as needed.

At the other end of the spectrum from the full mesh topology is the *star* topology, which designates one site as the main office (or hub) and consists of a separate, dedicated connection between the hub and each of the other branch sites. This topology uses the fewest number of WAN links to connect all of the sites together, providing the greatest economy, and enables the main office to communicate directly with each of the branch sites. However, when two of the branch sites have to communicate, they must do so by going through the hub. Whether the star topology is suitable for your network depends on whether the branch sites frequently need to communicate with each other.

A *ring* topology has each site connected to two other sites, as shown in Figure 9-5. This topology uses only one link more than a star, but it provides a greater degree of fault tolerance. If any one link fails, it is still possible for any two sites to communicate by sending traffic around the ring in the other direction. By contrast, a link failure in a star internetwork disconnects one of the sites from the others completely. The disadvantage of the ring is the delay introduced by the need for traffic to pass through multiple sites in order to reach its destination, in most cases. A site on a star internetwork is never more than two hops from any other site, while ring sites may have to pass through several hops.

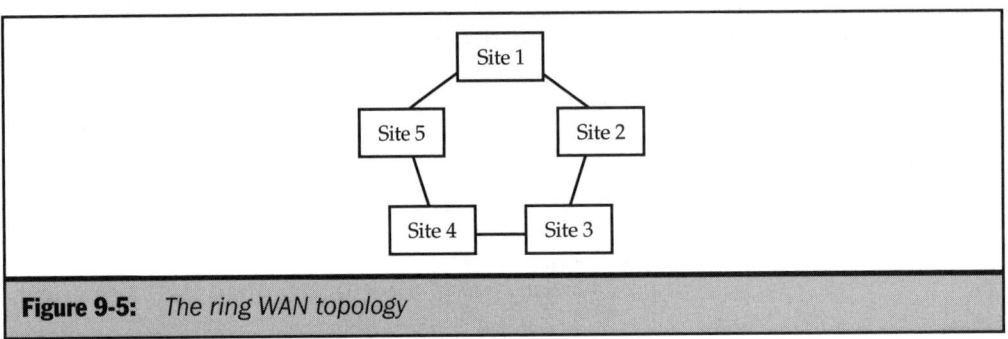

Figure 9-5: *The ring WAN topology*

Each of these topologies represents an extreme example of a network communication technique, but none of them have to be followed absolutely in every case. You can, for example, create a *partial mesh* topology by eliminating some of the links from the full mesh design. Not all of your sites may require a dedicated link to every other site, so you can eliminate the extraneous links, thus reducing the cost of the network. When a site has to communicate with another site to which it does not have a direct connection, it can go through one of its connected sites instead. In the same way, you can build more fault tolerance into a star network by having two hub sites instead of one, and connecting each of the other sites to both hubs. This requires twice as many links as a standard star topology, but still fewer than a full mesh.

Planning Internet Access

Connecting a network to the Internet is usually far less complicated than connecting multiple sites with WAN links. Even if your internetwork consists of several sites, it is more common to equip each one with its own Internet connection, rather than connect one site and have the other sites access the Internet through the inter-site WAN. The WAN technology you use to connect each site to the Internet should once again depend on the bandwidth you require and your budget. For a small office that requires only e-mail connectivity, a dial-up line may suffice, while larger sites often use leased lines, such as a T-1.

Note *For more information on connecting a network to the Internet, see Chapter 28.*

Locating Equipment

Designing the individual LANs that make up the internetwork is similar to designing a single, standalone LAN, except that you must work the backbone connections into the design. Large internetworks are more likely to use internal bulk cable installations for the network segments, rather than the prefabricated, external cables commonly used on home and small business networks. In an internal installation, cables run inside walls and ceilings and terminate at wall plates and patch panels. This type of installation is much more complicated than an external one where the cables are left exposed, and is frequently outsourced to a contractor that specializes in premises wiring. For these reasons, a detailed network plan showing the route of each cable and the location of each wall plate and patch panel is essential. You don't want to have to call the contractor back in after the installation is finished to pull additional cables.

Designing such a network and creating the plan are tasks that require an intimate knowledge of the building in which the network is to be located. As with a home or small business LAN, you must decide where all of the computers and other network devices are going to be located, and then work out how you are going to run the cables that connect them to the hub. For an internetwork design, you also have to decide where you're going to put the router that connects each LAN to the backbone (in the case of a distributed backbone network) or how you're going to connect each LAN to the main router/switch (in the case of a collapsed backbone network).

Wiring Closets

In the classic example of a multifloored office building with a horizontal network on each floor and a distributed backbone connecting them vertically, it is common practice to have a wiring closet on each floor. This closet can serve as the location for the patch panel where all of the cable runs for the floor terminate, as well as the hub that connects all of the devices on the floor into a LAN and the router that connects the LAN to the backbone network. It's also possible to install workgroup or even enterprise servers in these closets. To facilitate the backbone cabling, the best arrangement is for the wiring closets on each floor to be on top of each other, with a chase or wiring conduit running vertically through them, connecting all of the closets in the building.

To some people, the term "wiring closet" might invoke visions of hubs and routers shoved into a dark little space along with mops and buckets, but this should definitely not be the case. Wiring closets may already exist, even in a building not already cabled for a data network, to support telephone equipment and other building services. The closet may indeed be a small space, but it should be well lit, and have room enough to work in, if necessary. The room is called a closet because there is typically no room (or need) for desks and workstations inside. Most of the routers, servers, and other networking equipment available today can be equipped with remote administration capabilities, which minimizes the need to actually open the closet to physically access the equipment. Unlike an equipment storage closet, a wiring or server closet must also maintain an appropriate environment for the equipment inside. A space that is not heated in the winter and air-conditioned in the summer can greatly shorten the life of delicate electronics. Wiring closets must also be kept locked, of course, to protect the valuable equipment from theft and "experimentation" by unauthorized personnel.

Data Centers

Wiring closets are eminently suitable for distributed backbone networks, because this type of network requires that a relatively large amount of expensive equipment be scattered throughout the building. Another organizational option, better suited for a collapsed backbone network, is to have a single data center containing all of the networking equipment for the entire enterprise. In this context, a data center is really just a larger, more elaborate wiring closet. Typically, a data center is a secured room or suite that has been outfitted to support large amounts of electronic equipment. This usually includes special air conditioning, extra power lines, power conditioning and backup, additional fixtures such as a modular floor with a wiring space beneath it, and extra security to prevent unauthorized access.

The data center typically contains the network's enterprise servers and the routers that join the LANs together and provide Internet and WAN access. If the building housing the network is not too large, you can place all of the hubs for the individual LANs in the data center as well. This means that every wall plate in the building to which a computer is connected has a cable connecting it to a hub in the data center. This arrangement is only feasible if the length of the cable runs is under 100 meters, assuming that the horizontal networks are using UTP cable. If the distance between any of your wall plate locations and the data center exceeds 100 meters, you must either use fiber-optic cable (which supports longer segments) or place the hubs at the location of each LAN. If you choose to do the latter, you only have to find a relatively secure place for each hub.

When the hubs are distributed around the building, you only need one cable run from each hub to the data center. If you use centralized hubs, each of your cable runs extends all the way from the computer to the data center. Not only can this use much more cable, but the sheer bulk of the cables might exceed the size of the wiring spaces available in the building. However, the advantage of having centralized hubs is that network support personnel can easily service them and monitor their status, and connecting them to the hub or switch that joins the LANs into an internetwork is simply a matter of running a cable across the room.

The equipment in a data center is typically mounted in racks, which are 19 inches wide and can extend from floor to ceiling. Virtually all manufacturers of servers, hubs, routers, and other network devices intended for large enterprise networks have products designed to bolt into these standard-sized racks, which makes it easier to organize and access the equipment in the data center.

Finalizing the Design

As you flesh out the network design in greater detail, you can begin to select specific vendors, products, and contractors. This process can include shopping for the best hardware prices in catalogues and on Web sites, evaluating software products, interviewing and obtaining estimates from cable installation contractors, and investigating service providers for WAN technologies. This is the most critical part of the design process, for several reasons. First, this is the point at which you'll be able to determine the actual cost of building the network, not just an estimate. Second, it is at this phase that you must make sure all the components you select are actually capable of performing as your preliminary plan expects them to. If, for example, you discover that the router model with all of the features you need is no longer available, you may have to modify the plan to use a different type of router or to implement the feature you need in another way. Third, the concrete information you develop at this stage enables you to create a deployment schedule.

A network design plan can never have too much detail. Documenting your network as completely as possible both before, during, and after construction can only help you to maintain and repair it later. The planning process for a large network can be long and complicated, but it is rare for any of the time spent to be wasted.

Part III

Network Protocols

The
Complete
Reference

Networking

Chapter 10

Ethernet Basics

E thernet is the data link–layer protocol used by the vast majority of the local area networks operating today. In the course of more than 20 years, the Ethernet standards have been revised and updated to support many different types of network media, and to provide dramatic speed increases over the original protocol. Because all of the Ethernet variants operate using the same basic principles and because the high-speed Ethernet technologies were designed with backward compatibility in mind, upgrading a standard 10 Mbps network to 100 Mbps or more is usually relatively easy. This is in marked contrast to other high-speed technologies like FDDI and ATM, for which upgrades can require extensive infrastructure modifications, such as new cabling, as well as training and acclimation for the personnel supporting the new technology.

This chapter examines the fundamental Ethernet mechanisms and how they provide a unified interface between the physical layer of the OSI reference model and multiple protocols operating at the network layer. Then you'll learn how newer technologies like Fast Ethernet and Gigabit Ethernet improve on the older standards and provide sufficient bandwidth for the needs of virtually any network application. Finally, there will be a discussion of upgrade strategies and real-world troubleshooting techniques to help you improve the performance of your own network.

Ethernet Defined

The Ethernet protocol provides a unified interface to the network medium that enables an operating system to transmit and receive multiple network-layer protocols simultaneously. Like most of the data link–layer protocols used on LANs, Ethernet is, in technical terms, connectionless and unreliable. Ethernet makes its best effort to transmit data to the appointed destination, but no mechanism exists to guarantee a successful delivery. Instead, services such as guaranteed delivery are left up to the protocols operating at the higher layers of the OSI model, depending on whether or not the data warrants it.

Note *In this context, the term "unreliable" means only that the protocol lacks a means of acknowledging that packets have been successfully received.*

As defined by the Ethernet standards, the protocol consists of three essential components:

- A series of physical-layer guidelines that specify the cable types, wiring restrictions, and signaling methods for Ethernet networks

- A frame format the defines the order and functions of the bits transmitted in an Ethernet packet

- A media access control (MAC) mechanism called Carrier Sense Multiple Access with Collision Detection that enables all of the computers on the LAN equal access to the network medium

From a product perspective, the Ethernet protocol consists of the network interface adapters installed in the network's computers (usually in the form of network interface cards, or NICs), the network adapter drivers the operating system uses to communicate with the network adapters, and the hubs and cables you use to connect the computers together. When you purchase network adapters and hubs, you must be sure they all support the same Ethernet standards for them to be able to work together.

Ethernet Standards

As it was first designed in the 1970s, Ethernet carried data over a baseband connection using coaxial cable running at 10 Mbps (megabits per second) and a signaling system called Manchester encoding. This eventually came to be known as *Thick Ethernet* because the cable itself was approximately one centimeter wide, about the thickness of a garden hose (indeed, its color and rigidity led to its being called the "frozen yellow garden hose" by whimsical network administrators). The first Ethernet standard, which was titled "The Ethernet, a Local Area Network: Data Link Layer and Physical Layer Specifications," was published in 1980 by a consortium of companies that included DEC, Intel, and Xerox, giving rise to the acronym DIX, thus, the document became known as the DIX Ethernet standard.

Ethernet II

The DIX V2.0 standard, commonly known as DIX Ethernet II, was published in 1982 and expanded the physical-layer options to include a thinner type of coaxial cable, which came to be called *Thin Ethernet, ThinNet,* or *Cheapernet,* because it was less expensive than the original thick coaxial cable.

IEEE 802.3

During this time, a desire arose to build an international standard around the Ethernet protocol. In 1980, a working group was formed by a standards-making body called the Institute of Electrical and Electronics Engineers (IEEE), under the supervision of their Local and Metropolitan Area Networks (LAN/MAN) Standards Committee, for the purpose of developing an "Ethernet-like" standard. This committee is known by the number 802, and the working group was given the designation IEEE 802.3. The resulting standard, published in 1985, was called the "IEEE 802.3 Carrier Sense Multiple Access with Collision Detection (CSMA/CD) Access Method and Physical Layer Specifications." The term Ethernet was (and still is) scrupulously avoided by the IEEE 802.3 group, because they wanted to avoid creating any impression that the standard was based on a commercial product that had been registered as a trademark by Xerox. However, with a few minor differences, this document essentially defines an Ethernet network under another name and, to this day, the products conforming to the IEEE 802.3 standard are called by the name Ethernet.

After the release of the IEEE 802.3 document, a series of supplements were published, which are as follows:

IEEE 802.3a-1985   10Base2 Thin Ethernet

IEEE 802.3c-1985   10 Mbps Repeater Specifications

IEEE 802.3d-1987   Fiber-Optic Inter-Repeater Link (FOIRL)

IEEE 802.3i-1990   10Base-T Twisted Pair Ethernet

IEEE 802.3j-1993   10Base-F Fiber Optic Ethernet

IEEE 802.3u-1995   100Base-T Fast Ethernet and Auto-Negotiation

IEEE 802.3x-1997   Full Duplex Ethernet

IEEE 802.3z-1998   1000Base-X Gigabit Ethernet

IEEE 802.3ab-1999   1000Base-T Gigabit Ethernet (Twisted Pair)

IEEE 802.3ac-1998   Frame Size Extension to 1522 Bytes for VLAN Tag

IEEE 802.3ad-2000   Link Aggregation for Parallel Llinks

Believe it or not, the current version of the IEEE 802.3 standard, including all supplements, is now called "802.3, 1998 Edition Information Technology–Telecommunications and information exchange between systems–Local and metropolitan area networks–Specific requirements–Part 3: Carrier sense multiple access with collision detection (CSMA/CD) access method and physical layer specifications."

Since their inception, the IEEE 802 standards have only been available to the public in a costly printed edition. However, the IEEE has recently made all of the 802 standards, including the entire 1,555-page 802.3 document, available on the Web as PDF (Portable Document Format) files, at http://standards.ieee.org/getieee802.

DIX Ethernet and IEEE 802.3 Differences

While the DIX Ethernet II standard treats the data-link layer as a single entity, the IEEE standards divide the layer into two sublayers, called *logical link control (LLC)* and *media access control (MAC)*. The LLC sublayer isolates the functions that occur beneath it from those above it, and is defined by a separate standard: IEEE 802.2. The IEEE committee uses the same abstraction layer with the network types defined by other 802 standards, such as the 802.5 Token Ring network. The use of the LLC sublayer with the 802.3 protocol also led to a small but important change in the protocol's frame format, as described in "The Ethernet Frame," later in this chapter. The MAC sublayer defines the mechanism by which Ethernet systems arbitrate access to the network medium, as discussed in the forthcoming section "CSMA/CD."

By 1990, the IEEE 802.3 standard had been developed further and now included other physical-layer options that made coaxial cable all but obsolete, such as the twisted-pair cable commonly used in telephone installations and fiber-optic cable. Because it is easy to work with, inexpensive, and reliable, twisted-pair (or 10Base-T) Ethernet quickly became the most popular medium for this protocol. Most of the Ethernet networks installed today use twisted-pair cable, which continues to be supported by the new, higher-speed standards. Fiber-optic technology enables network connections to span much longer distances than copper and is immune from electromagnetic interference.

The primary differences between the IEEE 802.3 standard and the DIX Ethernet II standard are listed in the following table:

	IEEE 802.3	DIX Ethernet II
Physical-Layer Options	Coaxial, UTP, fiber optic	Coaxial only
Bits 13–14 of the Frame Header	Length of the data field	Ethertype
External Transceiver Test	SQE Test	Collision presence test (heartbeat)

IEEE Shorthand Identifiers

The IEEE is also responsible for the shorthand identifiers that are commonly used when referring to specific physical-layer Ethernet implementations, such as 10Base5 for a Thick Ethernet network. In this identifier, the "10" refers to the speed of the network, which is 10 Mbps. All of the Ethernet identifiers begin with 10, 100, or 1000.

The "Base" refers to the fact that the network uses baseband transmissions. As explained in Chapter 1, a baseband network is one in which the network medium carries only one signal at a time, as opposed to a broadband network, which can carry many signals simultaneously. All of the Ethernet variants are baseband, except for one broadband version, which is rarely, if ever, used.

The "5" in 10Base5 refers to the maximum possible length of a Thick Ethernet cable segment, which is 500 meters. Most of the other Ethernet variants use one or more initials instead of a number in this position, which in most cases specifies the type of medium the network uses. For example, the "T" in 10Base-T stands for twisted-pair cable.

The shorthand identifiers for the various types of Ethernet networks are listed in Table 10-1.

Beginning with the 10Base-T specification, the IEEE began including a hyphen after the "base" designator, to prevent people from pronouncing 10Base-T as "ten bassett."

NETWORK PROTOCOLS

Identifier	Common Name	Network Type
10Base5	Thick Ethernet	10 Mbps bus network using RG-8 coaxial cable in segments up to 500 meters long
10Base2	Thin Ethernet	10 Mbps bus network using RG-58 coaxial cable in segments up to 185 meters long
FOIRL	Fiber-Optic Inter-Repeater Link	10 Mbps point-to-point network using fiber-optic cable to connect two repeaters
10Broad36	Broadband Ethernet	10 Mbps broadband network with segments having up to 3,600 meters between stations
10Base-T	Twisted-Pair Ethernet	10 Mbps star network using twisted-pair cable
10Base-F	Fiber-Optic Ethernet	Generic term for three types of 10 Mbps fiber-optic network: 10Base-FB, 10Base-FP, and 10Base-FL
10Base-FB	Ethernet Fiber Backbone	10 Mbps fiber-optic network using active fiber hubs to extend a backbone network; rarely implemented
10Base-FP	Passive Fiber Ethernet	10 Mbps fiber-optic network using passive fiber hubs to connect workstations; rarely implemented
10Base-FL	Ethernet Fiber Link	10 Mbps point-to-point fiber-optic network; update of the FOIRL standard
100Base-T	Fast Ethernet	Generic term for all of the Fast Ethernet physical layer options, including twisted-pair and fiber-optic cables

Table 10-1:   *IEEE Shorthand Identifiers for Ethernet Networks*

Identifier	Common Name	Network Type
100Base-X	Fast Ethernet	Generic term for the 100Base-TX and 100Base-FX standards, both of which use the same 4B/5B block-encoding method.
100Base-TX	Fast Ethernet	100 Mbps network using two of the wire pairs in a Category 5 twisted-pair cable
100Base-FX	Fast Ethernet	100 Mbps network using multimode fiber-optic cable
100Base-T4	Fast Ethernet	100 Mbps network using all four of the pairs in a Category 3 twisted-pair cable
1000Base-X	Gigabit Ethernet	Generic term for the Gigabit Ethernet standards based on the Fibre Channel 8B/10B block-encoding scheme, including 1000Base-SX, 1000Base-LX, and 1000 Base-CX
1000Base-SX	Gigabit Ethernet	1,000 Mbps network using short wavelength fiber-optic cable
1000Base-LX	Gigabit Ethernet	1,000 Mbps network using long wavelength fiber-optic cable
1000Base-CX	Gigabit Ethernet	1,000 Mbps network using short copper cables like those defined in the Fibre Channel standard
1000Base-LH	Gigabit Ethernet	1,000 Mbps "long haul" network using singlemode fiber-optic cable
1000Base-ZX	Gigabit Ethernet	1,000 Mbps network using singlemode fiber-optic cable
1000Base-T	Gigabit Ethernet	1,000 Mbps network using four of the wire pairs in a Category 5 twisted-pair cable

Table 10-1:   *IEEE Shorthand Identifiers for Ethernet Networks (continued)*

NETWORK PROTOCOLS

CSMA/CD

The most definitive property of an Ethernet network is its media access control mechanism, which is called *Carrier Sense Multiple Access with Collision Detection (CSMA/CD)*. Like any MAC method, CSMA/CD enables the computers on the network to share a single baseband medium without data loss. There are no priorities on an Ethernet network, as far as media access is concerned; the protocol is designed so that every node has equal access rights to the network medium. The process by which CSMA/CD arbitrates access to the network medium on an Ethernet network is illustrated in Figure 10-1.

When a node on an Ethernet network wants to transmit data, it first monitors the network medium to see if it is currently in use. This is the *carrier sense* phase of the process. If the node detects traffic on the network, it pauses for a short interval and then listens to the network again. Once the network is clear, any of the nodes on the network may use it to transmit their data. This is the *multiple access* phase. This mechanism in itself arbitrates access to the medium, but it is not without fault.

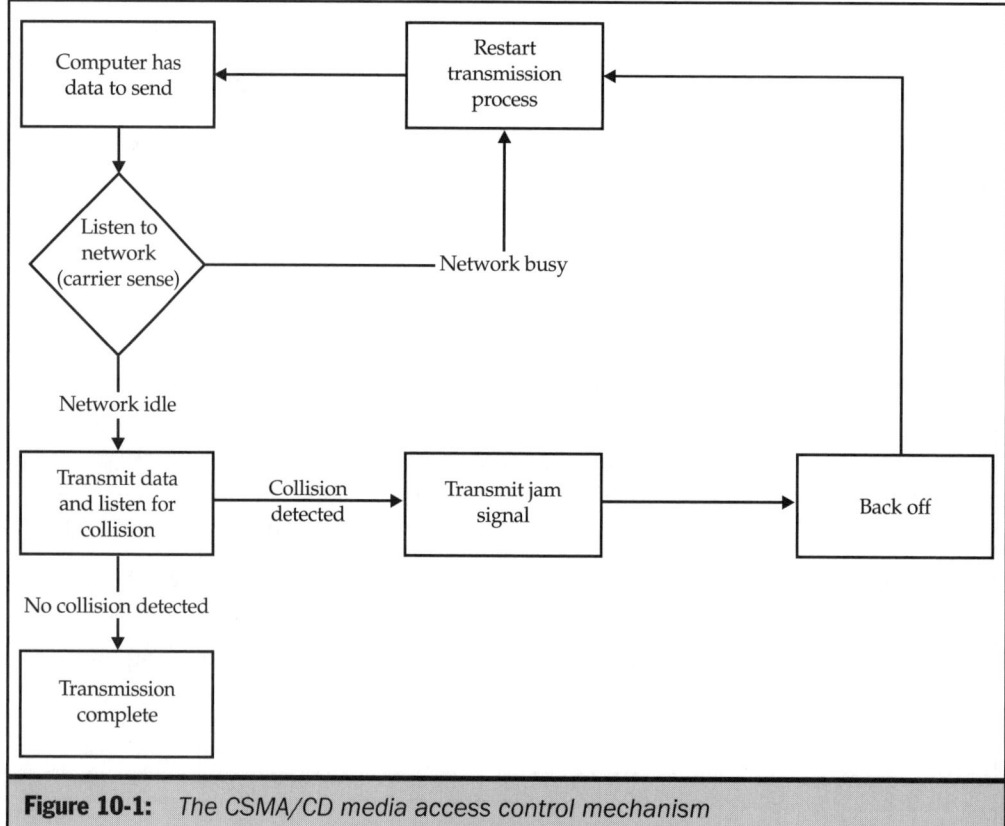

Figure 10-1:   *The CSMA/CD media access control mechanism*

It is entirely possible for two (or more) systems to detect a clear network and then transmit their data at nearly the same moment. This results in what the 802.3 standard calls a *signal quality error (SQE)* or, as the condition is more commonly known, a *packet collision*. Collisions occur when one system begins transmitting its data and another system performs its carrier sense during the brief interval before the first bit in the transmitted packet reaches it (see Figure 10-2). This interval is known as the *contention time* (or *slot time*), because each of the systems involved believes it has begun to transmit first. Every node on the network is, therefore, always in one of three possible states: transmission, contention, or idle.

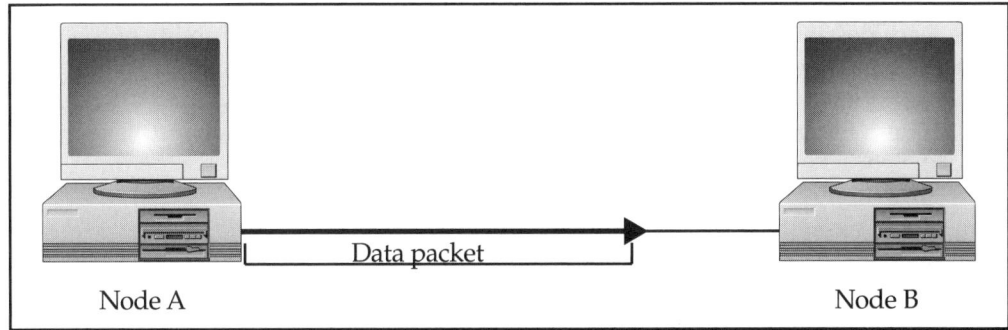

Data packet

Node A   Node B

Figure 10-2:   *Node A has begun to transmit its data, but since the beginning of the packet has not yet reached Node B, Node B senses the network as being clear. If Node B begins to transmit at this moment, a collision will occur.*

When packets from two different nodes collide, an abnormal condition is created on the cable that travels on toward both systems. On a coaxial network, the voltage level spikes to the point at which it is the same or greater than the combined levels of the two transmitters (+/− 0.85v). On a twisted-pair or fiber-optic network, the anomaly takes the form of signal activity on both the transmit and receive circuits at the same time.

When each transmitting system detects the abnormality, it recognizes that a collision has taken place, immediately stops sending data, and begins taking action to correct the problem. This is the *collision detection* phase of the process. Because the packets that collided are considered to be corrupted, both the systems involved transmit a *jam pattern* that fills the entire network cable with voltage, informing the other systems on the network of the collision and preventing them from initiating their own transmissions.

The jam pattern is a sequence of 32 bits that can have any value, as long as it does not equal the value of the cyclical redundancy check (CRC) calculation in the damaged packet's frame check sequence (FCS) field. A system receiving an Ethernet packet uses the FCS field to determine if the data in the packet has been received without error. As long as the jam pattern differs from the correct CRC value, all receiving nodes will discard the packet. In most cases, network adapters simply transmit 32 bits with the value 1. The odds of this also being the value of the CRC for the packet are 1 in 2^{32}—in other words, not likely.

After transmitting the jam pattern, the nodes involved in the collision both reschedule their transmissions, using a randomized delay interval they calculate with an algorithm that uses their MAC addresses as a unique factor. This process is called *backing off*. Because both nodes perform their own independent backoff calculations, the chances of them both retransmitting at the same time are substantially diminished. This is a possibility, however, and if another collision occurs between the same two nodes, they both increase the possible length of their delay intervals and back off again. As the number of possible values for the backoff interval increases, the probability of the systems again selecting the same interval diminishes. The Ethernet specifications call this process *truncated binary exponential backoff* (or *truncated BEB*). An Ethernet system will attempt to transmit a packet as many as 16 times and, if a collision results each time, the packet is discarded.

Collisions

Every system on an Ethernet network uses the CSMA/CD MAC mechanism for every packet it transmits, so the entire process obviously occurs quickly. Most of the collisions that occur on a typical Ethernet network are resolved in microseconds (millionths of a second). The most important thing to understand when it comes to Ethernet media arbitration is that *packet collisions are natural and expected occurrences on this type of network, and they do not necessarily signify a problem.* If you use a protocol analyzer or other network monitoring tool to analyze the traffic on an Ethernet network, you will see that a certain number of collisions always occur.

Note
The type of packet collision described here is normal and expected, but there is a different type, called a "late collision," that signifies a serious network problem. The difference between the two types of collisions is that normal collisions are detectable and late collisions are not. See the next section, "Late Collisions," for more information.

Normal packet collisions only become a problem when there are too many of them and significant network delays begin to accumulate. The combination of the backoff intervals and the retransmission of the packets themselves (sometimes more than once) incurs delays that are multiplied by the number of packets transmitted by each computer and by the number of computers on the network.

The fundamental fault of the CSMA/CD mechanism is that the more traffic there is on the network, the more collisions there are likely to be. The utilization of a network is based on the number of systems connected to it and the amount of data they send and receive over the network. When expressed as a percentage, the network utilization represents the proportion of the time the network is actually in use—that is, the amount of time that data is actually in transit. On an average Ethernet network, the utilization is likely to be somewhere in the 30 to 40 percent range. When the utilization increases to approximately 80 percent, the number of collisions increases to the point at which the performance of the network noticeably degrades. In the most extreme case, known as a *collapse*, the network is so heavily trafficked, it is almost perpetually in a state of contention, waiting for collisions to be resolved. This condition can conceivably be caused by the coincidental occurrence of repeated collisions, but it is more likely to result from a malfunctioning network interface that is continuously transmitting bad frames without pausing for carrier sense or collision detection. An adapter in this state is said to be *jabbering*.

Note *Data link–layer protocols that use a token-passing media access control mechanism, like Token Ring and FDDI, are not subject to performance degradation caused by high-network traffic levels. This is because these protocols use a mechanism that makes it impossible for more than one system on the network to transmit at any one time. On networks like these, collisions are not normal occurrences and signify a serious problem. For more information on token passing, see Chapter 12.*

Because of the possible performance degradation at high traffic levels, it is important to plan the expansion of Ethernet networks carefully, to prevent the network utilization from getting too high. When traffic levels become excessive, the use of bridges, switches, or other network connection devices that separate a LAN into multiple collision domains is preferable to expansion using simple repeating hubs that propagate all the incoming traffic to the entire network.

Note *For more information on expanding a network using bridges and switches, see Chapter 6.*

Late Collisions

The physical-layer specifications for the Ethernet protocol are designed so that the first 64 bytes of every packet transmission completely fill the entire aggregate length of cable in the collision domain. Thus, by the time a node has transmitted the first 64 bytes of a packet, every other node on the network has received at least the first bit of that packet. At this point, the other nodes will not transmit their own data because their carrier sense mechanism has detected traffic on the network.

It is essential for the first bit of each transmitted packet to arrive at every node on the network before the last bit leaves the sender. This is because the transmitting system can only detect a collision while it is still transmitting data. (Remember, on a twisted-pair or fiber-optic network, it is the presence of signals on the transmit and receive wires at the same time that indicates a collision.) Once the last bit has left the sending node, the sender considers the transmission to have completed successfully and erases the packet from the network adapter's memory buffer. It is because of this collision detection mechanism that every packet transmitted on an Ethernet network must be at least 64 bytes in length, even if the sending system has to pad it with useless (0) bits to reach that length.

If a collision should occur after the last bit has left the sending node, it is called a *late collision,* or sometimes an *out-of-window collision* (see Figure 10-3). (To distinguish between the two types of collisions, the normally occurring type is sometimes called an *early collision.*) Because the sending system has no way of detecting a late collision, it considers the packet to have been transmitted successfully, even though the data has actually been destroyed. Any data lost as a result of a late transmission cannot be retransmitted by a data link–layer process. It is up to the protocols operating at higher layers of the OSI model to detect the data loss and to use their own mechanisms to force a retransmission. This process can take up to 100 times longer than an Ethernet retransmission, which is one reason why this type of collision is a problem.

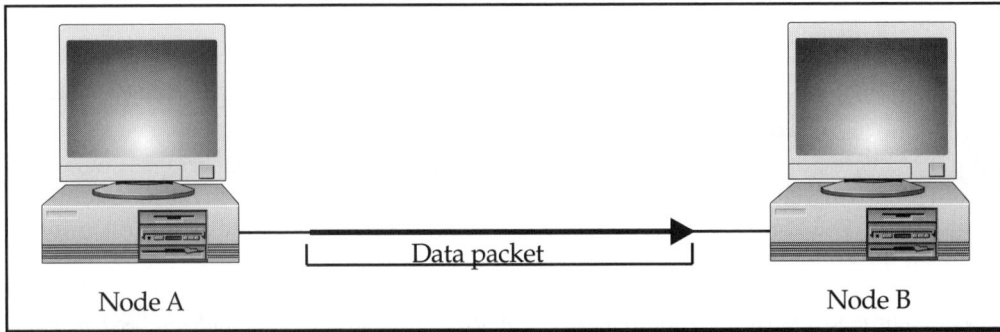

Node A   Node B

Figure 10-3: *Because of a network fault, such as an overlong cable segment, Node A has finished transmitting its packet before the first bit reaches Node B. When Node B, not sensing any traffic, begins to transmit, a collision occurs, but Node A cannot detect it or retransmit its packet. This is a called a late collision.*

Late collisions can result from several different causes. If a network interface adapter should malfunction and transmit a packet less than 64 bytes long (called a *runt*), the last bit could leave the sender before the packet has fully propagated around the net. In other cases, the adapter's carrier sense mechanism might fail, causing it to transmit at the wrong time. In both instances, you should replace the malfunctioning adapter.

Another possible cause of late collisions is a network that does not fall within the Ethernet cabling guidelines. If cable segments are too long or if there are too many repeaters on the network, the signal propagation delay can increase beyond the 600 nanoseconds specified in the Ethernet specification as the maximum time allowed for a transmission between two systems. See "Cabling Guidelines," later in this chapter, for more information. If this is the cause of the problem, the solution may be considerably more complex, as modifications to the cable installation or the network design may be necessary.

Late collisions are not an ordinary occurrence on an Ethernet network. They are, rather, a sign that a serious problem exists that should be addressed immediately.

The Capture Effect

The existence of collisions as a regular occurrence on Ethernet networks can have deep and complex repercussions on the way in which the network operates. Theoretically, every system on an Ethernet network has equal access to the network medium at any given moment. In practice, however, this has been found not to be so at certain times. When two nodes with a series of packets to transmit experience a collision, it is possible for one of the two to monopolize the network medium for the duration of its transmissions. This is known as the *capture effect*.

After the first collision, one of the two nodes wins the contention and successfully retransmits its packet. This system then tries to transmit the second packet in its sequence, while the other node is still trying to transmit its first. If a second collision occurs, one system backs off for the first time, while the other performs its second backoff using the truncated BEB mechanism. Statistically, the second system is more likely to lose this contention because it is selecting a backoff interval from a larger group of delay periods.

For a simplified example, the first system selects a backoff interval of either 1 or 2 milliseconds because it is backing off for the first time. However, the system backing off for the second time must select 1, 2, 3, or 4 seconds because the truncated BEB mechanism expands the pool of possible intervals with each successive backoff. The laws of probability dictate that the second system is likely to select a longer backoff interval than the first system and lose the second contention. If the same system does lose the second contention, its pool of possible backoff intervals will increase even more, as will the likelihood of its losing yet another contention. Thus, the first system, because it is continually transmitting new packets for the first time, has captured the network medium and prevented the second system from transmitting.

The ramifications of the capture effect on the network are, in most cases, not even detectable. The likelihood of this phenomenon occurring to the degree that it has a palpable effect on network performance is minimal, but the theory behind the capture effect is an excellent illustration of how complex the system interactions can be on an Ethernet network. To the members of the IEEE 802 committee, however, the problem was worth exploring. They convened a working group (IEEE 802.3w) to develop a specification for an alternate backoff algorithm called the *Binary Logarithmic Arbitration Method (BLAM)*.

BLAM resolves the capture effect problem by incrementing the collision counters on every network node symmetrically. Whenever a collision occurs, all the systems on the network modify their backoff interval selection algorithm in the same way, unlike the truncated BEB method, which is asymmetrical. This and other modifications resulted in an adequate solution to the problem, but the committee eventually decided to disband the group and declined to submit the BLAM document for standardization.

This decision was as much political as it was technical. The committee felt that an actual occurrence of the capture effect problem was only a remote possibility to begin with and that updating the Ethernet standard to address it implied that the problem was more serious than it actually was. In addition, the increasing popularity of switched and full-duplex Ethernet solutions had largely rendered the issue moot.

100Base-FX Physical-Layer Guidelines

The Ethernet specifications define not only the types of cable you can use with the protocol, but also the installation guidelines for the cable, such as the maximum length of cable segments and the number of hubs or repeaters permitted. As explained earlier, the configuration of the physical-layer medium is a crucial element of the CSMA/CD media access control mechanism. If the overall distance between two systems on the network is too long, or there are too many repeaters, diminished performance can result, which is quite difficult to diagnose and troubleshoot.

Standard 10 Mbps Ethernet, as defined in IEEE 802.3, can run on four different media configurations, as shown in Table 10-2. The cabling guidelines vary for each of the media to compensate for the performance characteristics of the different cable types.

	Thick Ethernet	Thin Ethernet	Twisted Pair	Fiber Optic
Designation	10Base5	10Base2	10Base-T	10Base-FL
Maximum Segment Length (meters)	500 meters	185 meters	100 meters	1,000/2,000
Maximum Nodes per Cable Segment	100	30	2	2
Cable Type	RG-8 coaxial	RG-58 coaxial	Category 3 unshielded twisted pair	62.5/125 multimode fiber optic
Connector Type	N	BNC	RJ-45	ST

Table 10-2: *Physical-Layer Options for 10 Mbps Ethernet*

Thick Ethernet

Thick Ethernet, or *ThickNet,* uses RG-8 coaxial cable in a bus topology to connect up to 100 nodes to a single segment no more than 500 meters long. Because it can span long distances and is well shielded, Thick Ethernet was commonly used for backbone networks in the early days of Ethernet. However, RG-8 cable, like all of the coaxial cables used in Ethernet networks, cannot support transmission rates faster than 10 Mbps, which limits its utility as a backbone medium. As soon as a faster alternative was available (such as FDDI), most network administrators abandoned Thick Ethernet. However, although it is hardly ever used anymore, the components of a Thick Ethernet network are a good illustration of the various components involved in the physical layer of an Ethernet network.

The coaxial cable segment on a Thick Ethernet network should, whenever possible, be a single unbroken length of cable, or at least be pieced together from the same spool or cable lot using N connectors on each cable end and an N barrel connector between them. There should be as few breaks as possible in the cable, and if you must use cable from different lots, the individual pieces should be 23.4, 70.2, or 117 meters long, to minimize the signal reflections that may occur. Both ends of the bus must be terminated with a 50-ohm resistor built into an N terminator, and the cable should be grounded at one (and only one) end, using a grounding connector attached to the N terminator.

> **Note**   *For more information on RG-8 and all of the cables used to build Ethernet networks, see Chapter 4.*

Unlike all of the other Ethernet physical-layer options, the Thick Ethernet cable does not run directly to the network interface card in the PC. This is because the coaxial cable itself is large, heavy, and comparatively inflexible. Instead, the NIC is connected to the RG-8 trunk cable with another cable, called the *attachment unit interface (AUI)* cable (see Figure 10-4). The AUI cable has 15-pin D-shell connectors at both ends, one of which plugs directly into the NIC, and the other into a *medium attachment unit (MAU),* also known as a transceiver. The MAU connects to the coaxial cable using a device called the *medium dependent interface (MDI),* which clamps to the cable and makes an electrical connection through holes cut into the insulating sheath. Because of the fang-like appearance of the connector, this device is commonly referred to as a *vampire tap.*

> **Note**   *Do not confuse the MAUs used on Thick Ethernet networks with the multistation access units (MAUs) used as hubs on Token Ring networks. The maximum of 100 nodes on a Thick Ethernet cable segment (and 30 nodes on a ThinNet segment) is based on the number of MAUs present on the network. Because repeaters include their own MAUs, they count toward the maximum.*

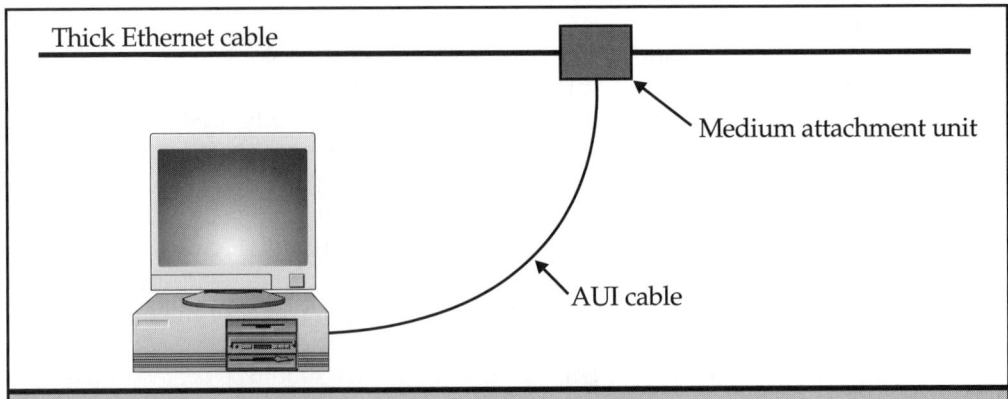

Figure 10-4: *Workstations on a Thick Ethernet network are connected to the main network cable using smaller AUI cables.*

If for no other reason, the DIX Ethernet standard should be fondly remembered for using more sensible names for many of Ethernet's technical concepts, such as "collision" rather than "signal quality error." The DIX Ethernet name for the "medium attachment unit" is the "transceiver" (because it both transmits and receives) and its name for the "attachment unit interface cable" is "transceiver cable."

Each standard AUI cable on a Thick Ethernet network can be up to 50 meters long, which provides for an added degree of flexibility in the installation. Standard AUI cables are the same thickness as the Thick Ethernet coaxial and similarly hard to work with. There are also thinner and more flexible "office grade" AUI cables, but these are limited to a maximum length of 12.5 meters, which is usually sufficient.

Thick Ethernet is the only form of Ethernet that does not incorporate the MAU (transceiver) into the network interface adapter. When you purchase combination NICs with two or three cable connectors, one of which is an AUI, this connector bypasses the MAU built into the card, which the other connectors use. If you have network interface adapters with AUI connectors only, it is still possible to connect the computers to a Thin Ethernet or even a 10Base-T network using an external MAU with a BNC or RJ-45 connector. This is rarely, if ever, necessary these days, however, because Thick Ethernet has been obsolete for several years and AUI-only NICs are not often seen.

The 500-meter maximum length for the Thick Ethernet cable makes it possible to connect systems at comparatively long distances and provides excellent protection against interference and attenuation. Unfortunately, the cable is difficult to work with and even harder to hide. Virtually no new Thick Ethernet networks are being installed today. Sites that require long cable segments or better insulation are now more likely to use fiber optic, which exceeds the performance of Thick Ethernet in almost every way. In addition, coaxial cable (both thick and thin) is limited to the 10 Mbps speed of the standard Ethernet specification.

Thin Ethernet

Thin Ethernet, or *ThinNet,* is similar in functionality to Thick Ethernet, except that the cable itself is RG-58 coaxial, about 5 millimeters in diameter, and much more flexible. For Thin Ethernet (and all other Ethernet physical-layer options except Thick Ethernet), the MAU (transceiver) is integrated into the network interface card and no AUI cable is needed.

Thin Ethernet uses BNC (Bayonet Neil-Concelman) connectors and a fitting called a *T-connector* that attaches to the network card in the PC. You create the network bus by running a cable to one end of the T-connector's crossbar and then using another cable on the other end of the crossbar to connect to the next system (see Figure 10-5). Like Thick Ethernet, a Thin Ethernet network must be terminated and grounded. The two systems at the ends of the bus must have a terminator containing a 50-ohm resistor on one end of their Ts to terminate the bus and one end (only) should be connected to a ground.

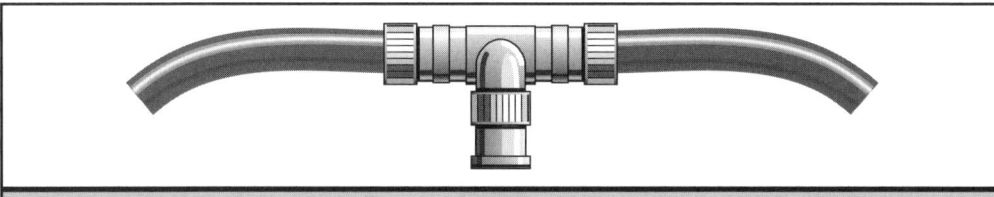

Figure 10-5: *Thin Ethernet networks use T-connectors to form a single cable segment connecting up to 30 computers in a bus topology.*

Note *The T-connectors on an Ethernet network must be directly connected to the network interface cards in the computers. Having a length of cable joining the T-connector to the computer is not permitted. The Ethernet standard calls for the distance from the MDI (built into the NIC) to the coaxial cable to be no more than 4 centimeters. Exceeding this length can cause signal reflections that damage packets, forcing them to be retransmitted by the higher-layer protocols. If you use these illegal "stub" cables, your network may seem to function properly, but the constant need for retransmissions can seriously degrade the performance of the network.*

Because the cable is thinner, Thin Ethernet is more prone to interference and attenuation, and is, therefore, limited to a segment length of 185 meters and a maximum of 30 nodes. Each piece of cable forming the segment must be at least 0.5 meters long.

Thin Ethernet is easier to work with than the thick variety and upon its introduction, it rapidly became the medium of choice for business Ethernet LANs. Both Thick and Thin Ethernet suffer from a failing common to all bus networks, however. If there is a break or a faulty connection anywhere on the bus, the network is effectively split into two segments that cannot access each other, and the lack of termination on one end of each segment produces signal reflections that make even communications between nodes on the same segment troublesome. This is similar to the "Christmas light effect," where one blown bulb causes an entire string of lights to fail.

Connector faults are not an uncommon occurrence on Thin Ethernet networks because prefabricated cables are relatively rare (compared to twisted pair) and the BNC connectors are usually crimped onto the RG-58 cables by network administrators, which can be a tricky process. Also, some cheap connectors are prone to a condition in which an oxide layer builds up between the conductors resulting in a serious degradation in the network connectivity. These connectors are notoriously sensitive to improper treatment. An accidental tug or a person tripping over one of the two cables connected to each machine can easily weaken the connection and cause intermittent transmission problems that are difficult to isolate and diagnose.

Note, however, that for a network failure to result, the problem must be in the cable or one of the T-connectors, not in the computer. You can safely shut off a system that lies in the middle of the bus without splitting the network or disturbing network communications between the other systems.

Twisted-Pair Ethernet

Most of the Ethernet networks today use *unshielded twisted-pair (UTP) cable,* originally known in the Ethernet world as *10Base-T,* which solves several of the problems that plague coaxial cables. Among other things, UTP Ethernet networks are:

Easily hidden UTP cables can be installed inside walls, floors, and ceilings with standard wall plates providing access to the network. Only a single, thin cable has to run to the computer. Pulling too hard on a UTP cable installed in this manner damages only an easily replaceable patch cable connecting the computer to the wall plate.

Fault tolerant UTP networks use a star topology in which each computer has its own dedicated cable run to the hub. A break in a cable or a loose connection affects only the single machine to which it is connected.

Upgradeable A UTP cable installation running 10 Mbps Ethernet now can be upgraded to 100 Mbps Fast Ethernet, or possibly even Gigabit Ethernet, at a later time.

Unshielded twisted-pair cable consists of four pairs of wires in a single sheath, with each pair twisted together at regular intervals to protect against crosstalk, and 8-pin RJ-45 connectors at both ends. Since this isn't a bus network, no termination or grounding is necessary. 10Base-T Ethernet uses only two of the four wire pairs in the cable, however, one pair for transmitting data signals (TD) and one for receiving them (RD), with one wire in each pair having a positive polarity and one a negative. The pin assignments for the connectors are shown in Table 10-3.

Pin	Pair	Polarity	Signal	Designation
1	1	Positive	Transmit	TD+
2	1	Negative	Transmit	TD-
3	2	Positive	Receive	RD+
4	3		Unused	
5	3		Unused	
6	2	Negative	Receive Data	RD-
7	4		Unused	
8	4		Unused	

Table 10-3:   *RJ-45 Pin Assignments for 10Base-T Networks*

NETWORK PROTOCOLS

Unlike coaxial networks, 10Base-T calls for the use of a *hub*, which is a device that functions both as a wiring nexus and as a signal repeater, to which each of the nodes on the network has an individual connection (see Figure 10-6). The maximum length for each cable segment is 100 meters, but because there is nearly always an intervening hub that repeats the signals, the total distance between two nodes can be as much as 200 meters.

UTP cables are typically wired *straight through,* meaning the wire for each pin is connected to the corresponding pin at the other end of the cable. For two nodes to communicate, however, the TD signals generated by each machine must be delivered to the RD connections in the other machine. In most cases, this is accomplished by a crossover circuit within the hub. You can connect two computers directly together without a hub by using a *crossover cable,* though, which connects the TD signals at each end to the RD signals at the other end.

For more information on network cables and their installation, see Chapter 4. For more information on hubs and repeaters, see Chapter 6.

Fiber-Optic Ethernet

Fiber-optic cable is a radical departure from the copper-based, physical-layer options discussed so far. Because it uses pulses of light instead of electric current, fiber optic is immune to electromagnetic interference and is much more resistant to attenuation than copper. As a result, fiber-optic cable can span much longer distances and, because of the electric isolation it provides, it is suitable for network links between buildings. Fiber-optic

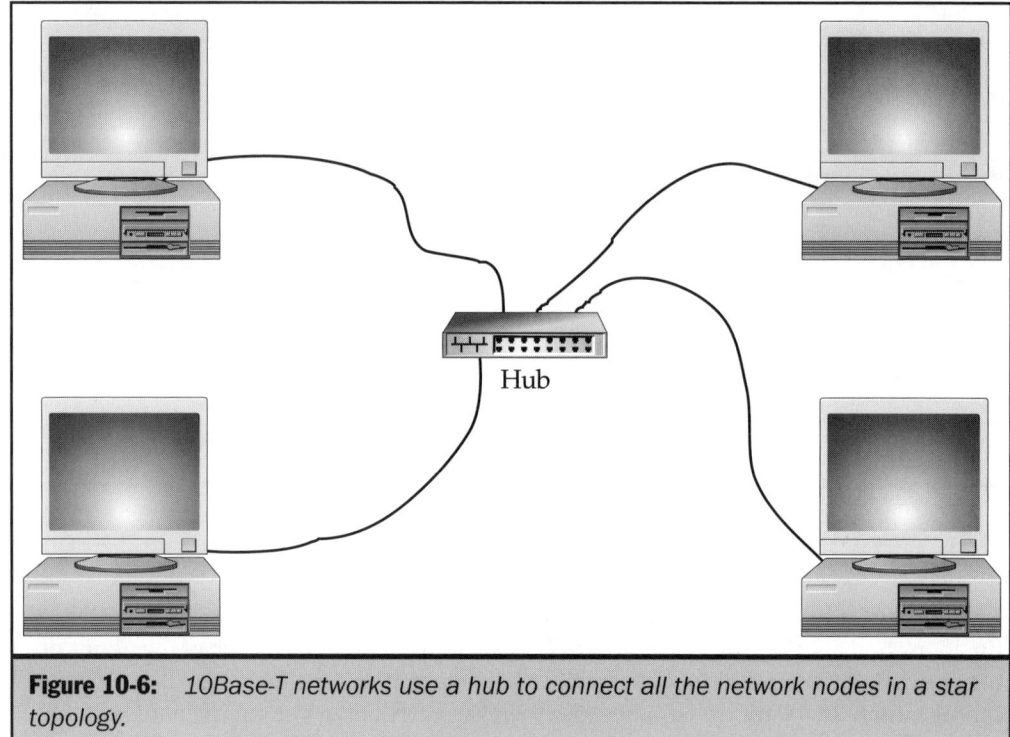

Figure 10-6: *10Base-T networks use a hub to connect all the network nodes in a star topology.*

cable is an excellent medium for data communications, but installing and maintaining it is somewhat more expensive than copper, and it requires completely different tools and skills.

The medium itself on a fiber-optic Ethernet network is two strands of 62.5/125 multimode fiber cable with one strand used to transmit signals and one to receive them.

There are two main fiber-optic standards for 10 Mbps Ethernet: the original FOIRL standard and 10Base-F, which defines three different fiber-optic configurations called 10Base-FL, 10Base-FB, and 10Base-FP. Of all these standards, 10Base-FL has always been the most popular, but running fiber-optic cable at 10 Mbps is an underuse of the medium's potential that borders on the criminal. Now that 100 Mbps data link–layer protocols, such as Fast Ethernet and FDDI, run on the same fiber-optic cable, there is no reason to use any of these slower solutions in a new installation.

FOIRL The original fiber-optic standard for Ethernet from the early 1980s was called the *Fiber-Optic Inter-Repeater Link (FOIRL)*. It was designed to function as a link between two repeaters up to 1,000 meters away. Intended for use in campus networks, FOIRL can join two distant networks, particularly those in adjacent buildings, using a fiber-optic cable.

10Base-FL   The 10Base-F supplement was developed by the IEEE 802.3 committee to provide a greater variety of fiber-optic alternatives for Ethernet networks. Designed with backward compatibility in mind, 10Base-FL is the IEEE counterpart to FOIRL. It increases the maximum length of a fiber-optic link to 2,000 meters, and permits connections between two repeaters, two computers, or a computer and a repeater.

> **Note**   *If you are using any old FOIRL hardware on a 10Base-FL network, you should limit the maximum segment length to 1,000 meters.*

As in all of the 10Base-F specifications, a computer connects to the network using an external fiber-optic MAU (or FOMAU) and an AUI cable up to 25 meters long. The other end of the cable connects to a fiber-optic repeating hub that provides the same basic functions as a hub for copper segments.

10Base-FB   10Base-FB is intended as a backbone cabling solution that connects hubs over distances up to 2,000 meters. By using synchronous signaling 10Base-FB hubs, you can safely exceed the number of repeaters permitted on an Ethernet network.

The other Ethernet standards used for links between repeaters (such as 10Base-FL and 10Base-T) keep their connections active through the use of an idle signal that is asynchronous with the normal packet transmissions. 10Base-FB's 2.5 MHz square wave idle signal, on the other hand, uses the same clock as the packet transmissions and, therefore, is said to be synchronous with them. Because the receiver of a communication is continuously locked to the transmitter's signal, no bits are stripped off at the beginning of the packet by the asynchronous squelching circuit in the receiver's MAU (transceiver).

Synchronous signaling means the *inter-packet gap* (that is, the brief interval between packets traveling over the network) is not varied nearly as much by a 10Base-FB repeater as it can be by other repeater types. A standard repeater can shrink the inter-packet gap by as much as 8 bits, while with a synchronous signaling repeater that variability is reduced to only 2 bits. Because of this reduction in the input to output variability, there can be up to 12 10Base-FB hubs on the transmission path between two nodes, rather than the normal 10 Mbps Ethernet maximum of 4.

10Base-FB also has a remote data link–layer diagnostics capability, provided by a special RF (remote fault) signal that a hub uses in place of the standard idle signal when it detects a problem. This ensures that the hubs on both sides of a transmission are informed when a fault occurs. Without the RF signal, a cable or interface problem that disrupts communications on one of the two fiber strands will be undetected by the hub that transmits data using that strand. The transmitting hub will send its data and receive nothing in return but the idle signal. The other hub will detect the fault, however, because it receives neither data nor the idle signal from the other end of the link. When this happens, the hub realizing the problem changes its idle signal to the RF signal, informing the other hub of the problem.

Because 10Base-FB is used only for backbone connections between hubs, there is no need for external MAUs or AUI cables, and hubs can connect directly together using standard fiber-optic cable.

10Base-FB is an excellent long-distance backbone technology, but it suffers from the same 10 Mbps speed limitation as all of the 10Base-F standards. By replacing the hubs, it is possible to upgrade a 10Base-FB connection to 100Base-FX and gain a tenfold speed increase.

Note

For more information on upgrading Ethernet networks to Fast Ethernet, see Chapter 11.

10Base-FP   10Base-FP defines a fiber network that uses a passive (that is, nonrepeating) star coupler to connect up to 33 workstations with segments up to 500 meters long. Intended as the 10Base-F element that would deliver fiber optic to the desktop, this part of the specification never gained commercial acceptance, most likely due to the added expense of installing fiber-optic cable to the desktop. A few products were produced, but they are no longer available.

Cabling Guidelines

In addition to the minimum and maximum segment lengths for the various types of Ethernet media, the standards also impose limits on the number of repeaters you can use in a single collision domain. This is necessary to ensure that every packet transmitted by an Ethernet node begins to reach its destination before the last bit leaves the sender. If the distance traveled by a packet is too long, the sender is unable to detect collisions reliably and data loss can occur.

Link Segments and Mixing Segments   When defining the limits on the number of repeaters allowed on the network, the 802.3 standard distinguishes between two types of cable segments, called link segments and mixing segments. A *link segment* is a length of cable that joins only two nodes, while a *mixing segment* joins more than two.

In the real world, this distinction has largely disappeared because the vast majority of Ethernet networks today use only link segments. All Ethernet twisted-pair and fiber-optic networks, for example, use only link segments because each node has its own dedicated cable leading to the hub. The same is true of 10Base-FL and 10Base-FB. The only types of Ethernet networks that employ mixing segments are the coaxial-based media, Thick Ethernet and Thin Ethernet, and 10Base-FP, which uses a hub that is passive and provides no repeater function. Because all three of these network types are seldom used today except on legacy installations, you can consider most Ethernet networks as composed only of link segments.

The 5-4-3 Rule   The Ethernet standards state that, in a single Ethernet collision domain, the route taken between any two nodes on the network can consist of no more than *five* cable segments, joined by *four* repeaters, and only *three* of the segments can be mixing segments. This is known as the *Ethernet 5-4-3 rule*. This rule is manifested in different ways, depending on the type of cable used for the network medium.

Note   *A collision domain is defined as a network configuration on which two nodes transmitting data at the same time will cause a collision. The use of bridges, switches, or intelligent hubs, instead of standard repeaters, does not extend the collision domain and does not fall under the Ethernet 5-4-3 rule. If you have a network that has reached its maximum size because of this rule, you should consider using one of these devices to create separate collision domains. See Chapter 6 for more information.*

On a coaxial network, whether it is Thick or Thin Ethernet, you can have five cable segments joined by four repeaters. On a coaxial network, a repeater has only two ports and does nothing but amplify the signal as it travels over the cable. A segment is the length of cable between two repeaters, even though in the case of Thin Ethernet the segment can consist of many separate lengths of cable. This rule means that the overall length of a Thick Ethernet bus (called the *maximum collision domain diameter*) can be 2,500 meters (500 × 5) while a Thin Ethernet bus can be up to 925 meters (185 × 5) long.

On either of these networks, however, only three of the cable segments can actually have nodes connected to them (see Figure 10-7).You can use the two link segments to join mixing segments located at some distance from each other, but you cannot populate them with computers or other devices.

On a UTP network, the situation is different. Because the repeaters on this type of network are actually multiport hubs, every cable segment connecting a node to the hub is a link segment. You can, therefore, have four hubs in a collision domain that are connected to each other and each of which can be connected to as many nodes as the hub can support (see Figure 10-8). Because data traveling from one node to any other node passes through a maximum of only four hubs and because all the segments are link segments, the network is in compliance with the Ethernet standards.

One potentially complicating factor to this arrangement is when you connect 10Base-T hubs using Thin Ethernet coaxial cable. Some 10Base-T hubs include BNC connectors that enable you to use a bus to chain multiple hubs together. When you do this with more than two hubs connected by a single coaxial segment, you are actually creating a mixing segment and you must count this toward the maximum of three mixing segments permitted on the network.

On a network that uses coaxial cable only to connect 10Base-T hubs together, this won't be a problem because you can't have more than one mixing segment when you only have four hubs. If part of your network uses coaxial cable (either Thick or Thin Ethernet) to connect to nodes on its own mixing segments, however, you must count the bus connecting the 10Base-T hubs as one of the maximum of three allowed by the standard.

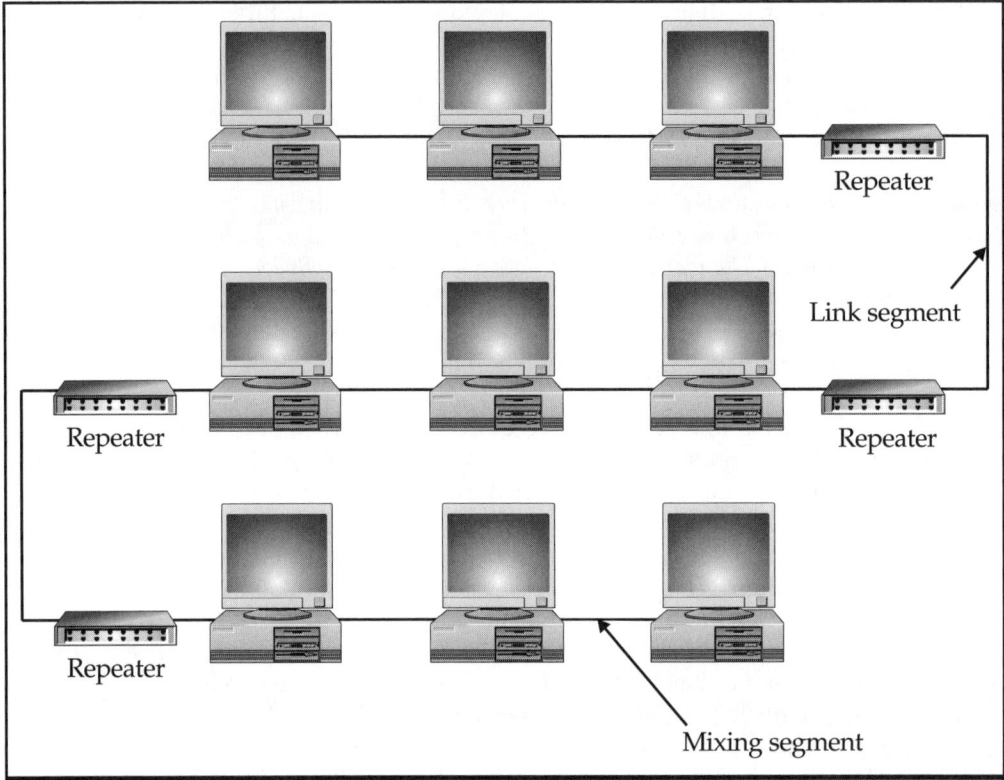

Figure 10-7:   *Coaxial networks can consist of up to five cable segments, with only three of the five connected to computers or other devices.*

The 10Base-F specifications include some modifications to the 5-4-3 rule. When five cable segments are present on a 10Base-F network connected by four repeaters, FOIRL, 10Base-FL, and 10Base-FB segments can be no more than 500 meters long. 10Base-FP segments can be no more than 300 meters long.

When four cable segments are connected by three repeaters, FOIRL, 10Base-FL, and 10Base-FB segments can be no more than 1,000 meters long and 10Base-FP segments can be no more than 700 meters long. Cable segments connecting a node to a repeater can be no more than 400 meters for 10Base-FL and 300 meters for 10Base-FP. Also, there is no limitation to the number of mixing segments when there are only a total of four cable segments on the network.

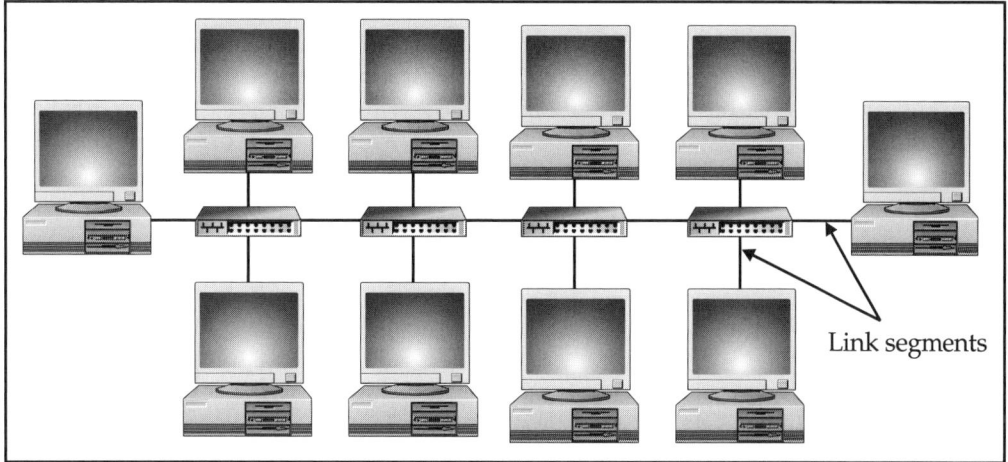

Link segments

Figure 10-8:   *Twisted-pair networks use link segments to connect to the computers, making it possible to have four populated hubs.*

Ethernet Timing Calculations   The 5-4-3 rule is a general guideline that is usually accurate enough to ensure your network will perform properly. However, it is also possible to assess the compliance of a network with the Ethernet cabling specifications more precisely by calculating two measurements: the *round trip signal delay time* and the *interframe gap shrinkage* for the worst case path through your network.

The round trip signal delay time is the amount of time it takes a bit to travel between the two most distant nodes on the network and back again. The interframe gap shrinkage is the amount the normal 96-bit delay between packets is reduced by network conditions, such as the time required for repeaters to reconstruct a signal before sending it on its way.

In most cases, these calculations are unnecessary; as long as you comply with the 5-4-3 rule, your network should function properly. If you are planning to expand a complex network to the point at which it pushes the limits of the Ethernet guidelines, however, it might be a good idea to get a precise measurement to ensure that everything functions as it should. If you end up with a severe late collision problem that requires an expensive network upgrade to remedy, your boss isn't likely to want to hear about how reliable the 5-4-3 rule usually is.

Note

Calculating the round trip signal delay time and the interframe gap shrinkage for your network is not part of a remedy for excessive numbers of early collisions. The collisions that result from networks not in compliance with these Ethernet specifications are late collisions, which do not register on elementary tools like Windows 2000's Performance Console and NetWare's Monitor.nlm. You must use a high-end protocol analyzer tool to detect late collisions at the data link–layer level.

Finding the Worst Case Path   The *worst case path* is the route data takes when traveling between the two most distant nodes on the network, both in terms of segment length and number of repeaters. On a relatively simple network, you can find the worst case path by choosing the two nodes on the two outermost network segments that have either the longest link segments connecting them to the repeater or are at the far ends of the cable bus, as shown in Figure 10-9.

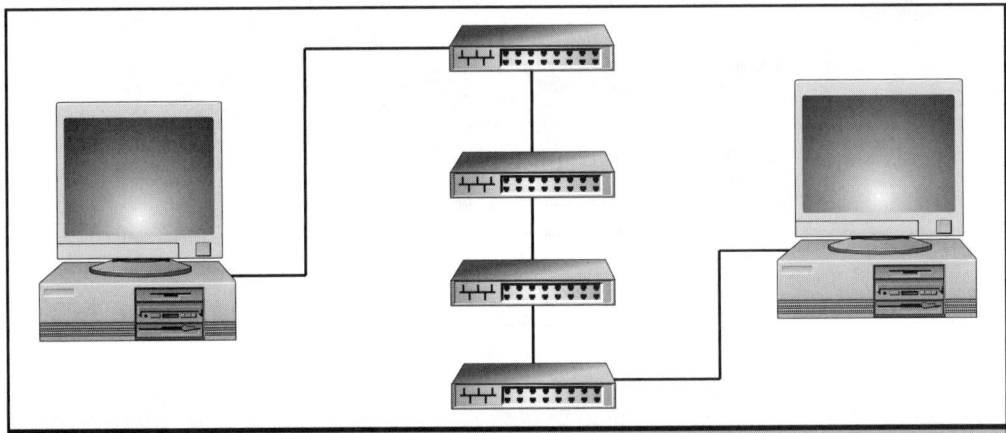

Figure 10-9: *On a simple network with all 10Base-T segments, the worst case path runs between the nodes with the longest cables on both end segments.*

On more complex networks using various types of cable segments, you may have to select several paths to test your network. In addition, you may have to account for additional distance generated by AUI cables and the variations caused by having different cable segment types at the left and right ends of the path.

If your network is well documented, you should have a schematic containing the precise distances of all your cable runs. You need these figures to make your calculations. If you don't have a schematic, determining the exact distances may be the most difficult part of the whole process. The most accurate method for determining the length of a cable run is to use a multifunction cable tester, which utilizes a technique called *time domain reflectometry (TDR)*. TDR is similar to radar, in that the unit transmits a test signal, precisely measures the time it takes the signal to travel to the other end of the cable and back again, and then uses this information to compute the cable's length. If you don't have a cable tester with TDR capabilities, you can measure the cable lengths manually by estimating the distances between the connectors. This can be particularly difficult when cables are installed inside walls and ceilings, because there may be unseen obstacles that extend the length of the cable. If you use this method, you should err on the side of caution and include an additional distance factor to account for possible errors. Alternatively, you can simply use the maximum allowable cable distances for the various cable segments, as long as you are sure the cable runs do not exceed the Ethernet standard's maximum segment length specifications.

Once you have determined the worst case path (or paths) you will use for your calculations, it's a good idea to create a simple diagram of each path with the cable distances involved, as shown in Figure 10-10. Each path will have left and right end segments and may have one or more middle segments. You will then perform your calculations on the individual segments and combine the results to test the entire path.

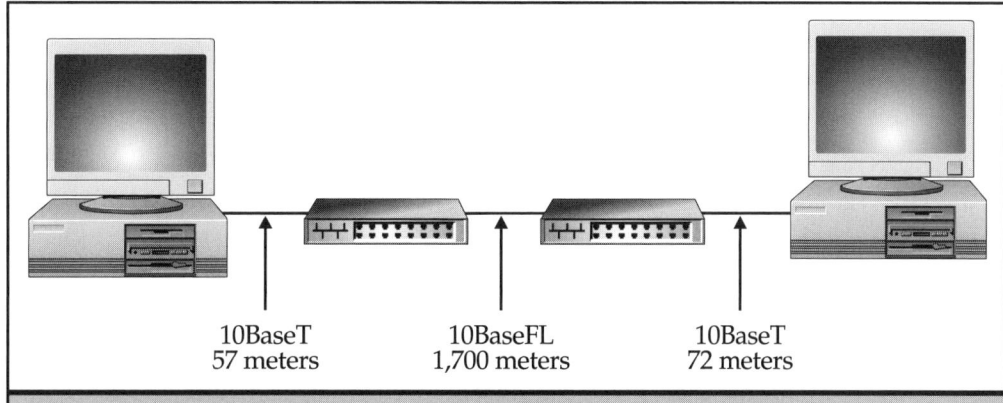

| 10BaseT | 10BaseFL | 10BaseT |
| 57 meters | 1,700 meters | 72 meters |

Figure 10-10: *Complex networks using several different media may require you to perform calculations for several different paths to test the whole network adequately.*

Calculating the Round Trip Signal Delay Time   To determine whether the most distant nodes on your network are capable of properly detecting collisions, you determine the total path delay by using the segment delay values in Table 10-4 to calculate the delay for each segment in the path and then add them together. The table contains base and maximum bit-time values for each of the Ethernet cable types. There are separate values for the segment at the left and right ends of your worst case path, while all the segments in the middle use the same value. A *bit time* is the amount of time required to send one bit of data over the network.

Cable Type	Maximum Length (Meters)	Left End Base	Middle Segment Base	Right End Base	Round Trip Delay/Meter
10Base5	500	11.75	46.5	169.5	0.0866
10Base2	185	11.75	46.5	169.5	0.1026
10Base-T	100	15.25	42	165	0.113
FOIRL	1,000	7.75	29	152	0.1
10Base-FL	2,000	12.25	33.5	156.5	0.1
10Base-FB	2,000	*	24	*	0.1
10Base-FP	1,000	11.25	61	183.5	0.1
Excess AUI	48	0	0	0	0.1026

*End connections not supported

Table 10-4:   *Round Trip Delay Values for Ethernet Cable Types (in Bit Times)*

To calculate the delay for a particular segment, use the following formula:

segment delay = (segment length × round trip delay/meter) + segment base

Thus, for a 50-meter 10Base-T segment at the left end of your worst case path, you would multiply 50 × 0.113 (the round trip delay/meter for 10Base-T) and add 15.25 (the base delay for a left-end 10Base-T segment) to get 20.9:

(50 × 0.113) + 15.25 = 20.9

If you want to use the delay for the maximum segment size permitted for each cable type, rather than measure the actual lengths of your network segments, you can use the values in Table 10-5, which displays the results of the formula using the maximum segment length.

When you have calculated the delay values for all the segments on your network, you add them together and include an additional 5 bit times for a margin of error. This provides you with the total round trip signal delay time for the worst case path on your network. If this figure is less than or equal to 575, your network conforms to the Ethernet specifications for this parameter. The value of 575 is derived from the 64 bytes (512 bits minus 1) required to fill the entire length of cable in the collision domain, plus the 64 bits that form the preamble and start of frame delimiter in the Ethernet frame. If the delay time is less than 575 bit times, this means the node at one end of the worst case path will be unable to send more than 511 bits of the frame plus the preamble and start of frame delimiter before it is notified of a collision.

Cable Type	Maximum Length (Meters)	Left End Maximum	Middle Segment Maximum	Right End Maximum
10Base5	500	55.05	89.8	212.8
10Base2	185	30.731	65.48	188.48
10Base-T	100	26.55	53.3	176.3
FOIRL	1,000	107.75	129	252
10Base-FL	2,000	212.25	233.5	356.5
10Base-FB	2,000	*	224	*
10Base-FP	1,000	111.25	161	284
External AUI	48	4.88	4.88	4.88

* Not supported

Table 10-5: *Round Trip Delay Values for Maximum Ethernet Segment Lengths*

Because the delay values for the left and right end segments are different, you must perform these calculations twice if your network uses a different cable type at each end. After calculating the total delay in one direction, reverse the path and perform the same calculations using the other end as the left segment.

If your network uses Thick Ethernet or 10Base-F segments with separate AUI cables connecting nodes to the network, you must figure these cable lengths into your calculations. The values for the standard cable types in Table 10-5 include a two-meter allowance for the total length of the AUI connections inside repeaters and network interfaces, but you can use the values shown in the External AUI row of the table to calculate the additional delay for AUI cables. The table also includes External AUI values for the maximum allowable AUI cable length, if you want to use this figure instead of measuring your cables. Once you determine the delay time for all the AUI cables in your path, you add it to the total for the rest of the network and compare it to the 575 bit-time maximum delay.

Calculating the Interframe Gap Shrinkage   The interframe gap shrinkage test ensures that a sufficient delay exists between packet transmissions to make certain that network interfaces have enough time to cycle between transmit mode and receive mode. If the variable timing delays in the network components and the signal reconstruction delays in the repeaters cause this gap to become too small, the frames may arrive too quickly and overwhelm the interface of the receiving node.

Note

The interframe gap of 9.6 microseconds specified by the Ethernet standard was ratified almost 20 years ago, and the technology used to design and manufacture network interface adapters has certainly advanced considerably since then. Many of the NICs manufactured today require a good deal less than 9.6 microseconds to cycle between modes. Some of them even deliberately use shorter interframe gaps to speed network transmissions. When all the network adapters on a network have this capability, no problem should occur. If you use cards like these on a network along with older network interfaces, you may find the older devices cannot keep up with the new ones.

To calculate the interframe gap shrinkage, you use the same worst case path through the network you used in the round trip signal delay time calculation, except that only the transmitting segment and the middle segments are pertinent here. This is because the shrinkage values for the transmitting and middle segments include the repeaters to which they deliver the packets. The final segment leading to the destination node begins after the last repeater in the path and does not contribute any more shrinkage. The interframe gap shrinkage values for the various Ethernet cable types in both transmitting and middle segments are shown in Table 10-6.

Cable Type	Transmitting Segment	Middle Segment
10Base5, 10Base2	16	11
10Base-T, FOIRL, 10Base-FL	10.5	8
10Base-FB	End connections not supported	2
10Base-FP	11	8

Table 10-6:   *Interframe Gap Shrinkage for Ethernet Cable Types (in Bit Times)*

The interframe gap shrinkage for the entire network is the sum of the values from the table for transmitting segment, plus all of the middle segments. If the total is less than 49 bit times, the network passes the test. If your worst case path uses different cable types at both ends, you should calculate the shrinkage twice, using first the value for one end and then the value for the other. The highest total of the two should be considered the interframe gap shrinkage measurement for the network.

Exceeding Ethernet Cabling Specifications

The Ethernet specifications have a certain amount of leeway built into them that makes it possible to exceed the cabling limitations, within reason. If a network has an extra repeater or a cable that's a little too long, it will probably continue to function without causing the late collisions that occur when the specifications are grossly exceeded. You can see how this is so by calculating the actual amount of copper cable filled by an Ethernet signal.

Electrical signals passing through a copper cable travel at approximately 200,000,000 meters/second (2/3 of the speed of light). Ethernet transmits at 10 Mbps, or 10,000,000 bits/second. By dividing 200,000,000 by 10,000,000, you arrive at a figure of 20 meters of cable for every transmitted bit. Thus, the smallest possible Ethernet frame, which is 512 bits (64 bytes) long, occupies 10,240 meters of copper cable.

If you take the longest possible length of copper cable permitted by the Ethernet standards, a 500-meter Thick Ethernet segment, you can see that the entire 500 meters would be filled by only 25 bits of data (at 20 meters/bit). Two nodes at the far ends of the segment would have a round trip distance of 1,000 meters.

When one of the two nodes transmits, a collision can only occur if the other node also begins transmitting before the signal reaches it. If you grant that the second node begins transmitting at the last possible moment before the first transmission reaches it, then the first node can send no more than 50 bits (occupying 1,000 meters of cable, 500 down and 500 back) before it detects the collision and ceases transmitting. Obviously, this 50 bits is well below the 512-bit barrier that separates early from late collisions.

Of course, this example involves only one segment. But even if you extend a Thick Ethernet network to its maximum collision domain diameter—five segments of 500 meters each, or 2,500 meters—a node would still only transmit 250 bits (occupying 5,000 meters of cable, 2,500 down and 2,500 back) before detecting a collision.

Thus, you can see that the Ethernet specifications for the round trip signal delay time are fully twice as strict as they need to be in the case of a Thick Ethernet network. For the other copper media, Thin Ethernet and 10Base-T, the specifications are even more lax because the maximum segment lengths are smaller, while the signaling speed remains the same. For a full-length five-segment 10Base-T network, only 500 meters long, the specification is ten times stricter than it needs to be.

This is not to say that you can safely double the maximum cable lengths on your network across the board or install a dozen repeaters (although it is possible to safely lengthen the segments on a 10Base-T network up to 150 meters if you use Category 5 UTP cable instead of Category 3). Other factors can affect the conditions on your network to bring it closer to the limits defined by the specifications. In fact, the signal timing is not as much of a restricting factor on 10 Mbps Ethernet installations as is the signal strength. The weakening of the signal due to attenuation is far more likely to cause performance problems on an overextended network than are excess signal delay times. The point here is to demonstrate that the designers of the Ethernet protocol built a safety factor into the network from the beginning, perhaps partially explaining why it continues to work so well more than 20 years later.

The Ethernet Frame

The *Ethernet frame* is the sequence of bits that begins and ends every Ethernet packet transmitted over a network. The frame consists of a header and footer that surround and encapsulate the data generated by the protocols operating at higher layers of the OSI model. The information in the header and footer specifies the addresses of the system sending the packet and the system that is to receive it, and also performs several other functions that are important to the delivery of the packet.

The IEEE 802.3 Frame

The basic Ethernet frame format, as defined by the IEEE 802.3 standard, appears as shown in Figure 10-11. The functions of the individual fields are discussed in the following sections.

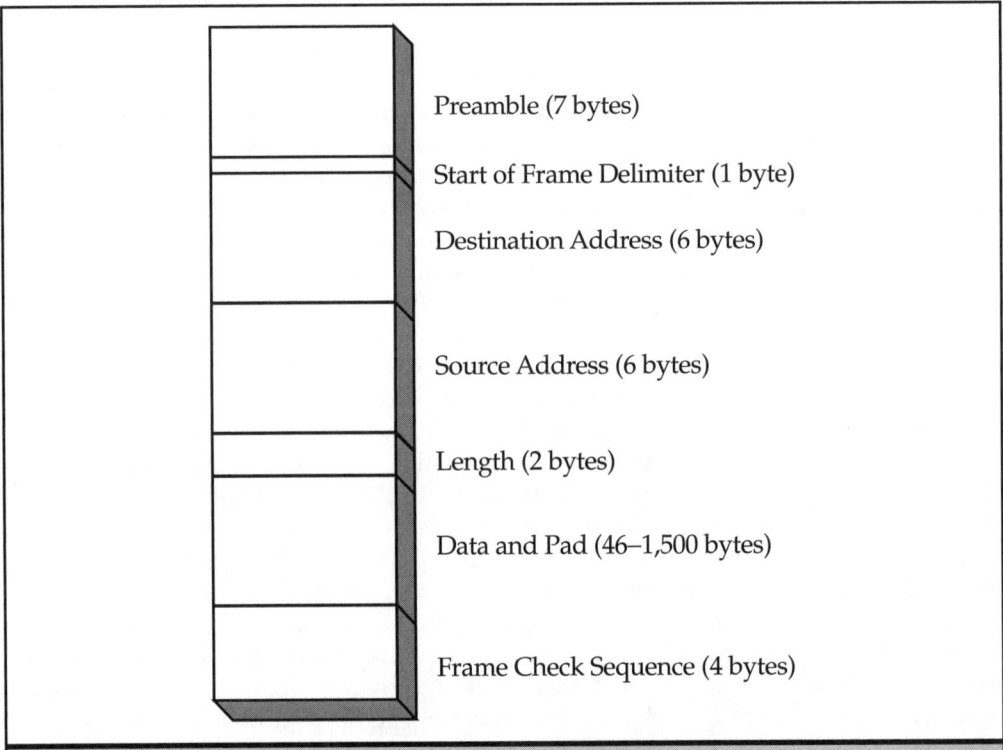

Preamble (7 bytes)

Start of Frame Delimiter (1 byte)

Destination Address (6 bytes)

Source Address (6 bytes)

Length (2 bytes)

Data and Pad (46–1,500 bytes)

Frame Check Sequence (4 bytes)

Figure 10-11: *The Ethernet frame encloses the data passed down the protocol stack from the network layer and prepares it for transmission.*

Preamble and Start of Frame Delimiter (SFD)   The preamble consists of 7 bytes of alternating 0s and 1s, which the systems on the network use to synchronize their clocks and then discard. The Manchester encoding scheme Ethernet uses requires the clocks on communicating systems to be in synch, so that they both agree on how long a bit time is. Systems in idle mode (that is, not currently transmitting and not in the process of rectifying a collision) are incapable of receiving any data until they use the signals generated by the alternating bit values of the preamble to prepare for the forthcoming data transmission.

> **Note**   *For more information on Manchester encoding and the signaling that occurs at the physical layer, see Chapter 2.*

By the time the 7 bytes of the preamble have been transmitted, the receiving system has synchronized its clock with that of the sender, but the receiver is also unaware of how much of the 7 bytes have elapsed before it fell into synch. (Most of the adapters made today are designed to synch up within 11 bit times, but this is not an absolutely reliable figure.) To signal the commencement of the actual packet transmission, the sender transmits a 1-byte start of frame delimiter, which continues the alternating 0s and 1s, except for the last two bits, which are both 1s. This is the signal to the receiver that any data following is part of a data packet and should be read into the network adapter's memory buffer for processing.

Destination Address and Source Address   *Addressing* is the most basic function of the Ethernet frame. Because the frame can be said to form an envelope for the network-layer data carried inside it, it is only fitting that the envelope have an address. The addresses the Ethernet protocol uses to identify the systems on the network are 6 bytes long and hardcoded into the network interface adapters in each machine. These addresses are referred to as *hardware addresses* or *MAC addresses*. The hardware address on every Ethernet adapter made is unique. The IEEE assigns 3-byte prefixes to NIC manufacturers that it calls *organizationally unique identifiers (OUIs)* and the manufacturers themselves supply the remaining 3 bytes. When transmitting a packet, it is the network adapter driver on the system that generates the values for the destination address and source address fields.

The destination address field identifies the system to which the packet is being sent. The address may identify the ultimate destination of the packet if it's on the local network, or the address may belong to a device that provides access to another network, such as a router. Addresses at the data-link layer always identify the packet's next stop on the local network. It is up to the network layer to control end-to-end transmission and to provide the address of the packet's ultimate destination.

> **Note**   *For more information on the interaction between data link–layer addresses and network-layer addresses, see "ARP" in Chapter 13.*

Every node on a shared Ethernet network reads the destination address from the header of every packet transmitted by every system on the network, to determine if the header contains its own address. A system reading the frame header and recognizing its own address then reads the entire packet into its memory buffers and processes it accordingly. A destination address of all 1s signifies that the packet is a *broadcast*, meaning it is intended for all of the systems on the network. Certain addresses can also be designated as *multicast* addresses by the networking software on the system. A multicast address identifies a group of systems on the network, all of which are to receive certain messages.

The source address field contains the 6-byte MAC address of the system sending the packet.

Length   The length field in an 802.3 frame is 2 bytes long and specifies how much data is being carried as the packet's payload in bytes. This figure includes only the actual upper-layer data in the packet. It does not include the frame fields from the header or footer or any padding that might have been added to the data field to reach the minimum size for an Ethernet packet (64 bytes). The maximum size for an Ethernet packet, including the frame, is 1,518 bytes. Because the frame consists of 18 bytes, the maximum value for the length field is 1,500.

> **Note**   *The frame format for the Ethernet II standard uses this field for a different purpose. See "The Ethernet II Frame," later in this chapter, for more information.*

Data and Pad   The data field contains the payload of the packet—that is, the "contents" of the envelope. As passed down from the network-layer protocol, the data will include an original message generated by an upper-layer application or process, plus any header information added by the protocols in the intervening layers. In addition, an 802.3 packet will contain the 3-byte logical link control header in the data field as well.

For example, the payload of a packet containing an Internet host name to be resolved into an IP address by a DNS server consists of the original DNS message generated at the application layer, a header applied by the UDP protocol at the transport layer, a header applied by the IP protocol at the network layer, and the LLC header. Although these three additional headers are not a part of the original message, to the Ethernet protocol they are just payload that is carried in the data field like any other information. Just as postal workers are not concerned with the contents of the envelopes they carry, the Ethernet protocol has no knowledge of the data within the frame.

The entire Ethernet packet (excluding the preamble and the start of frame delimiter) must be a minimum of 64 bytes in length, for the protocol's collision detection mechanism to function. Therefore, subtracting 18 bytes for the frame, the data field must be at least 46 bytes long. If the payload passed down from the network-layer protocol is too short, the Ethernet adapter adds a string of meaningless bits to pad the data field out to the requisite length.

The maximum allowable length for an Ethernet packet is 1,518 bytes, meaning the data field can be no larger than 1,500 bytes (including the LLC header).

Frame Check Sequence (FCS)   The last 4 bytes of the frame, following the data field (and the pad, if any), carry a checksum value the receiving node uses to determine if the packet has arrived intact. Just before transmission, the network adapter at the sending node computes a cyclical redundancy check (CRC) on all of the packet's other fields (except for the preamble and the start of frame delimiter) using an algorithm called the AUTODIN II polynomial. The value of the CRC is uniquely based on the data used to compute it.

When the packet arrives at its destination, the network adapter in the receiving system reads the contents of the frame and performs the same computation. By comparing the newly computed value with the one in the FCS field, the system can verify that none of the packet's bit values have changed. If the values match, the system accepts the packet and writes it to the memory buffers for processing. If the values don't match, the system declares an *alignment error* and discards the frame. The system will also discard the frame if the number of bits in the packet is not a multiple of 8. Once a frame is discarded, it is up to the higher-layer protocols to recognize its absence and arrange for retransmission.

The Ethernet II Frame

The function of the 2-byte field following the source address is different in the frame formats of the two predominant Ethernet standards. While the 802.3 frame uses this field to specify the length of the data in the packet, the Ethernet II standard uses it to specify the frame type, also called the *Ethertype*. The Ethertype specifies the memory buffer in which the frame should be stored. The location of the memory buffer specified in this field identifies the network-layer protocol for which the data carried in the frame is intended.

This is a crucial element of every protocol operating in the data link, network, and transport layers of a system's networking stack. The data in the packet must be delivered not only to the proper system on the network, but also to the proper application or process on that system. Because the destination computer can be running multiple protocols at the network layer at the same time, such as IP, NetBEUI, and IPX, the Ethertype field informs the Ethernet adapter driver which of these protocols should receive the data.

When a system reads the header of an Ethernet packet, the only way to tell an Ethernet II frame from an 802.3 frame is by the value of the length/Ethertype field. Because the value of the 802.3 length field can be no higher than 1,500 (0x05DC, in hexadecimal notation), the Ethertype values assigned to the developers of the various network-layer protocols are all higher than 1,500.

Xerox continues to function as the registrar of the Ethertype assignments. Some of the possible values for the Ethertype field are shown in Table 10-7.

Ethertype	Protocol	Ethertype	Protocol
0600	Xerox NS IDP	6002	DEC MOP (Remote Console)
0800	Internet Protocol (IP)	6003	DECNET Phase 4
0801	X.75	6004	DEC LAT
0802	NBS	6005	DEC
0803	ECMA	6006	DEC
0804	Chaosnet	8005	HP Probe
0805	X.25 Packet (Level 3)	8010	Excelan
0806	Address Resolution Protocol (ARP)	8035	Reverse ARP
0807	XNS Compatibility	8038	DEC LANBridge
1000	Berkeley Trailer	809B	AppleTalk
5208	BBN Simnet	80F3	AppleTalk ARP
6001	DEC MOP (Dump/Load)	8137	NetWare IPX/SPX

Table 10-7: *Ethertype Values for Network-Layer Protocols (in Hexadecimal Notation)*

The Logical Link Control Sublayer

As mentioned earlier, the IEEE splits the functionality of the data-link layer into two sublayers: media access control (MAC) and logical link control (LLC). On an Ethernet network, the MAC sublayer includes elements of the 802.3 standard: the physical-layer specifications, the CSMA/CD mechanism, and the 802.3 frame. The functions of the LLC sublayer are defined in the 802.2 standard, which is also used with the other 802 MAC standards.

The LLC sublayer is capable of providing a variety of communications services to network-layer protocols, including the following:

Unacknowledged connectionless service A simple service that provides no flow control or error control, and does not guarantee accurate delivery of data.

Connection-oriented service   A fully reliable service that guarantees accurate data delivery by establishing a connection with the destination before transmitting data, and by using error and flow control mechanisms.

Acknowledged connectionless service   A midrange service that uses acknowledgment messages to provide reliable delivery, but which does not establish a connection before transmitting data.

On a transmitting system, the data passed down from the network-layer protocol is encapsulated first by the LLC sublayer into what the standard calls a *protocol data unit (PDU)*. Then the PDU is passed down to the MAC sublayer, where it is encapsulated again in a header and footer, at which point it can technically be called a *frame*. In an Ethernet packet, this means the data field of the 802.3 frame contains a 3- or 4-byte LLC header, in addition to the network-layer data, thus reducing the maximum amount of data in each packet from 1,500 to 1,497 bytes.

The LLC header consists of three fields (see Figure 10-12), the functions of which are described in the following sections.

DSAP and SSAP   The Destination Service Access Point (DSAP) field identifies a location in the memory buffers on the destination system where the data in the packet should be stored. The Source Service Access Point (SSAP) field does the same for the source of the packet data on the transmitting system. Both of these 1-byte fields use values assigned by the IEEE, which functions as the registrar for the protocol. Some of the possible values are shown in Table 10-8.

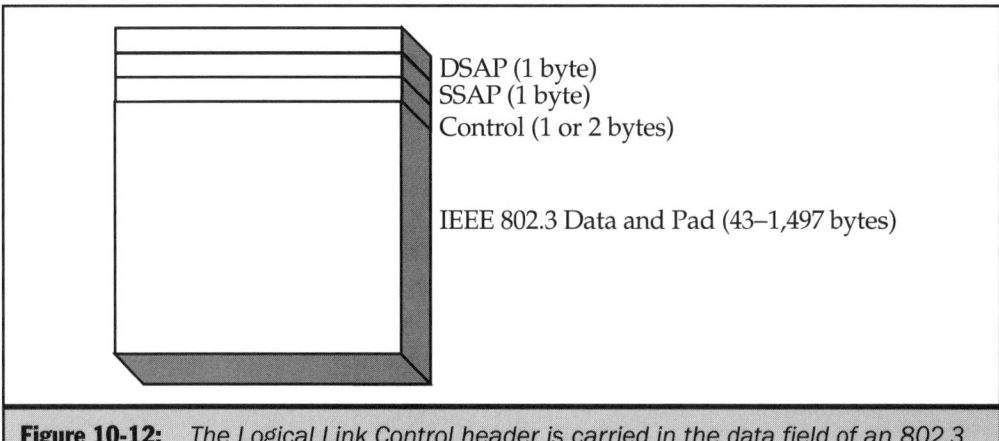

DSAP (1 byte)
SSAP (1 byte)
Control (1 or 2 bytes)

IEEE 802.3 Data and Pad (43–1,497 bytes)

Figure 10-12:   *The Logical Link Control header is carried in the data field of an 802.3 packet.*

NETWORK PROTOCOLS

DSAP/SSAP Value	Description
0	Null LSAP
2	Indiv LLC Sublayer Mgt
3	Group LLC Sublayer Mgt
4	SNA Path Control
6	Reserved (DOD IP)
14	PROWAY-LAN
78	EIA-RS 511
94	ISI IP
142	PROWAY-LAN
170	SNAP
254	ISO CLNS IS 8473
255	Global DSAP

Table 10-8: *IEEE DSAP and SSAP Values*

In an Ethernet SNAP packet, the value for both the DSAP and SSAP fields is 170 (or 0xAA, in hexadecimal form). This value indicates that the contents of the LLC PDU begins with a Sub-Network Access Protocol (SNAP) header. The SNAP header provides the same functionality as the Ethertype field to the 802.3 frame.

Control The control field of the LLC header specifies the type of service needed for the data in the PDU and the function of the packet. Depending on which of the services is required, the control field can be either 1 or 2 bytes long. In an Ethernet SNAP frame, for example, the LLC uses the unacknowledged, connectionless service, which has a 1-byte control field value using what the standard calls the *unnumbered format*. The value for the control field is 3, which is defined as an *unnumbered information frame*—that is, a frame containing data. Unnumbered information frames are quite simple and signify either that the packet contains a noncritical message or that a higher-layer protocol is somehow guaranteeing delivery and providing other high-level services.

The other two types of control field (which are 2 bytes each) are the *information format* and the *supervisory format*. The three control field formats are distinguished by their first bits, as follows:

■ The information format begins with a 0 bit.

■ The supervisory format begins with a 1 bit and a 0 bit.

■ The unnumbered format begins with two 1 bits.

The remainder of the bits specify the precise function of the PDU. In a more complex exchange involving the connection-oriented service, unnumbered frames contain commands, such as those used to establish a connection with the other system and terminate it at the end of the transmission. The commands transmitted in unnumbered frames are as follows:

UI (Unnumbered Information)   Used to send data frames by the unacknowledged, connectionless service

XID (Exchange Identification)   Used as both a command and a response in the connection-oriented and connectionless services

TEST   Used as both a command and a response when performing an LLC loopback test

FRMR (Frame Reject)   Used as a response when a protocol violation occurs

SABME (Set Asynchronous Balanced Mode Extended)   Used to request that a connection be established

UA (Unnumbered Acknowledgment)   Used as the positive response to the SABME message

DM (Disconnect Mode)   Used as a negative response to the SABME message

DISC (Disconnect)   Used to request that a connection be closed; a response of either UA or DM is expected

Information frames contain the actual data transmitted during connection-oriented and acknowledged connectionless sessions, as well as the acknowledgment messages returned by the receiving system. Only two types of messages are sent in information frames: N(S) and N(R) for the send and receive packets, respectively. Both systems track the sequence numbers of the frames they receive. An N(S) message lets the receiver know how many packets in the sequence have been sent and the N(R) message lets the sender know what packet in the sequence it expects to receive.

Supervisory frames are used only by the connection-oriented service and provide connection maintenance in the form of flow control and error-correction services. The types of supervisory messages are as follows:

RR (Receiver Ready)   Used to inform the sender that the receiver is ready for the next frame and to keep a connection alive

RNR (Receiver Not Ready)   Used to instruct the sender not to send any more packets until the receiver transmits an RR message

REJ (Frame Reject)   Used to inform the sender of an error and request retransmission of all frames sent after a certain point

NETWORK PROTOCOLS

LLC Applications   In some cases, the LLC frame plays only a minor role in the network communications process. On a network running TCP/IP along with other protocols, for example, the only function of LLC may be to enable 802.3 frames to contain a SNAP header, which specifies the network-layer protocol the frame should go to, just like the Ethertype in an Ethernet II frame. In this scenario, the LLC PDUs all use the unnumbered information format. Other high-level protocols, however, require more extensive services from LLC. NetBIOS sessions, for example, and several of the NetWare protocols, use LLC's connection-oriented services more extensively.

The SNAP Header

Because the IEEE 802.3 frame header does not have an Ethertype field, it would normally be impossible for a receiving system to determine which network-layer protocol should receive the incoming data. This would not be a problem if you ran only one network-layer protocol, but with multiple protocols installed, it becomes a serious problem. 802.3 packets address this problem by using yet another protocol within the LLC PDU, called the *Sub-Network Access Protocol (SNAP)*.

The SNAP header is 5 bytes long and found directly after the LLC header in the data field of an 802.3 frame, as shown in Figure 10-13. The functions of the fields are as follows:

Organization code   The organization code, or vendor code, is a 3-byte field that takes the same value as the first 3 bytes of the source address in the 802.3 header.

Local code   The local code is a 2-byte field that is the functional equivalent of the Ethertype field in the Ethernet II header, using the same values as assigned by Xerox.

Note   *Many, if not all, of the registered values for the NIC hardware address prefixes, the Ethertype field, and the DSAP/SSAP fields are listed in the "Assigned Numbers" document published as a Request for Comments (RFC) by the Internet Engineering Task Force (IETF). The current version number for this document is RFC1700 and is available at http://www.ietf.org/rfc.html.*

Full-Duplex Ethernet

The CSMA/CD media access control mechanism is the defining element of the Ethernet protocol, but it is also the source of many of its limitations. The fundamental shortcoming of the Ethernet protocol is that data can travel only in one direction at a time. This is known as *half-duplex* operation. With special hardware, it is also possible to run Ethernet connections in *full-duplex* mode, meaning that the device can transmit and receive data simultaneously. This effectively doubles the bandwidth of the network. Full-duplex capability for Ethernet networks was standardized in the 802.3x supplement to the 802.3 standard in 1997.

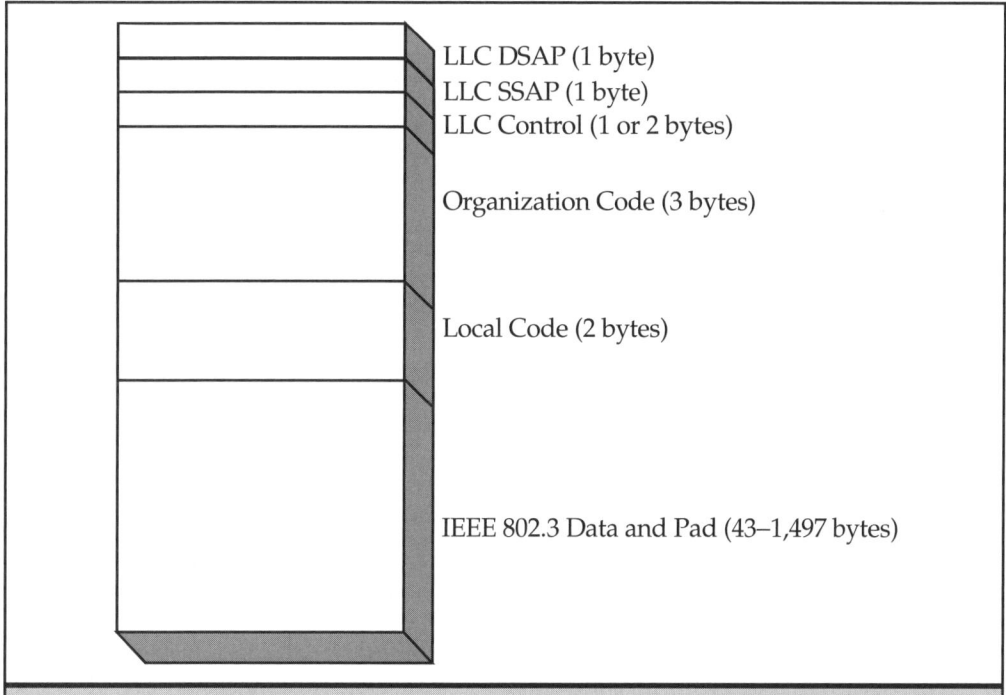

LLC DSAP (1 byte)
LLC SSAP (1 byte)
LLC Control (1 or 2 bytes)

Organization Code (3 bytes)

Local Code (2 bytes)

IEEE 802.3 Data and Pad (43–1,497 bytes)

Figure 10-13: *The SNAP header is carried with an LLC protocol data unit and provides the same functionality as the Ethertype field.*

When operating in full-duplex mode, the CSMA/CD MAC mechanism is ignored. Systems do not listen to the network before transmitting; they simply send their data whenever they want to. Because both of the systems in a full-duplex link can transmit and receive data at the same time, there is no possibility of collisions occurring. Because no collisions occur, the cabling restrictions intended to support the collision detection mechanism are not needed. This means that you can have longer cable segments on a full-duplex network. The only limitation is the signal transmitting capability (that is, the resistance to attenuation) of the network medium itself.

This is a particularly important point on a Fast Ethernet network using fiber-optic cable because the collision detection mechanism is responsible for its relatively short maximum segment lengths. While a half-duplex 100Base-FX link between two devices can only be a maximum of 412 meters long, the same link operating in full-duplex mode can be up to 2,000 meters (2 km) long because it is restricted only by the strength of the signal. A 100Base-FX link using singlemode fiber-optic cable can span distances of 20 km or more. The signal attenuation on twisted-pair networks, however, makes 10Base-T, 100Base-TX, and 1000Base-T networks still subject to the 100-meter segment length restriction.

Full-Duplex Requirements

There are three requirements for full-duplex Ethernet operation:

■ A network medium with separate transmit and receive channels

■ A dedicated link between each two systems

■ Network interface adapters and switches that support full-duplex operation

Full-duplex Ethernet is possible only on link segments that have separate channels for the communications in each direction. This means that twisted-pair and fiber-optic networks can support full-duplex communications using regular, Fast, and Gigabit Ethernet, but coaxial cable cannot. Of the Ethernet variants using twisted-pair and fiber-optic cables, 10Base-FB and 10Base-FP do not support full-duplex (which is not a great loss, since no one uses them), nor does 100Base-T4 (which is also rarely used). All of the other network types support full-duplex communications.

Full-duplex Ethernet also requires that every two computers have a dedicated link between them. This means you can't use repeating hubs on a full-duplex network, because these devices operate in half-duplex mode by definition and create a shared network medium. Instead, you must use switches, also known as switching hubs, which effectively isolate each pair of communicating computers on its own network segment and provide the packet-buffering capabilities needed to support bidirectional communications.

Finally, each of the devices on a full-duplex Ethernet network must support full-duplex communications and be configured to use it. Switches that support full-duplex are readily available, as are Fast Ethernet NICs. Full-duplex operation is an essential component of 1000Base-T Gigabit Ethernet, and many 1000Base-X Gigabit Ethernet adapters support full-duplex as well. Ensuring that your full-duplex equipment is actually operating in full-duplex mode can sometimes be tricky. Auto-negotiation is definitely the easiest way of doing this; dual-speed Fast Ethernet equipment automatically gives full-duplex operation priority over half-duplex at the same speed. However, adapters and switches that do not support multiple speeds may not include auto-negotiation. For example, virtually all 100Base-TX NICs are dual speed, supporting both 10 and 100 Mbps transmissions. Auto-negotiation is always supported by these NICs, which means that simply connecting the NIC to a full-duplex switch will enable full-duplex communications. Fast Ethernet NICs that use fiber-optic cables, however, are usually single-speed devices, and may or may not include auto-negotiation capability. You may have to manually configure the NIC before it will use full-duplex communications.

Full-Duplex Flow Control

The switching hubs on full-duplex Ethernet networks have to be able to buffer packets as they read the destination address in each one and perform the internal switching needed to send it on its way. The amount of buffer memory in a switch is, of course, finite, and as a result, it's possible for a switch to be overwhelmed by the constant input of data from freely transmitting full-duplex systems. Therefore, the 802.3x supplement defines an optional flow control mechanism that full-duplex systems can use to make the system at the other end of a link pause its transmissions temporarily, enabling the other device to catch up.

The full-duplex flow control mechanism is called the MAC Control protocol, which takes the form of a specialized frame that contains a PAUSE command and a parameter specifying the length of the pause. The MAC Control frame is a standard Ethernet frame of minimum length (64 bytes) with the hexadecimal value 8808 in the Ethertype or SNAP Local Code field. The frame is transmitted to a special multicast address (01-80-C2-00-00-01) designated for use by PAUSE frames. The data field of the MAC Control frame contains a 2-byte operational code (opcode) with a hexadecimal value of 0001, indicating that it is a PAUSE frame. At this time, this is the only valid MAC Control opcode value. A 2-byte *pause-time* parameter follows the opcode, which is an integer specifying the amount of time the receiving systems should pause their transmissions, measured in units called *quanta*, each of which is equal to 512 bit times. The range of possible values for the pause-time parameter is 0 to 65,535.

Full-Duplex Applications

Full-duplex Ethernet capabilities are most often provided in Fast Ethernet and Gigabit Ethernet adapters and switches. It's also possible to run standard Ethernet networks, such as 10Base-T, in full-duplex mode, but it generally is not worth upgrading a 10Base-T network to full-duplex when you can upgrade it to Fast Ethernet at the same time, without spending a lot more money.

While full-duplex operation theoretically doubles the bandwidth of a network, the actual performance improvement that you realize depends on the nature of the communications involved. Upgrading a desktop workstation to full duplex will probably not provide a dramatic improvement in performance. This is because desktop communications typically consist of request/response transactions that are themselves half-duplex in nature, and providing a full-duplex medium won't change that. Full-duplex operation is better suited to the communications between switches on a backbone, which are continually carrying large amounts of traffic generated by computers all over the network.

NETWORK PROTOCOLS

Networking

Chapter 11

Fast Ethernet and Gigabit Ethernet

F ast Ethernet and Gigabit Ethernet are the 100 and 1,000 Mbps variants of the Ethernet protocol, respectively. Although similar to standard Ethernet in many ways, these faster protocols have some configuration issues that you must be aware of in order to design, install, and administer the networks that use them.

Fast Ethernet

The IEEE 802.3u specification, ratified in 1995, defines what is commonly known as Fast Ethernet or 100Base-T, a data link–layer protocol running at 100 Mbps, ten times the speed of the original Ethernet protocol. Fast Ethernet is now the industry standard for new LAN installations, largely because it improves network performance so much while changing so little.

Fast Ethernet leaves two of the three defining elements of an Ethernet network unchanged. The new protocol uses the same frame format as IEEE 802.3 and the same CSMA/CD media access control mechanism. The changes that enable the increase in speed are in several elements of the physical-layer configuration, including the types of cable used, the length of cable segments, and the number of hubs permitted.

Physical-Layer Options

The first difference between regular and Fast Ethernet is that coaxial cable is no longer supported. Fast Ethernet runs only on UTP or fiber-optic cable, although shielded twisted-pair (STP) is an option as well. Gone also is the Manchester signaling scheme, which is replaced by the 4B/5B system developed for the FDDI protocol. The physical-layer options defined in 802.3u are intended to provide the most flexible possible installation parameters. Virtually every aspect of the Fast Ethernet protocol's physical layer specifications is designed to facilitate upgrades from earlier technologies and, particularly, from 10Base-T. In many cases, existing UTP networks can upgrade to Fast Ethernet without pulling new cable. The only exception to this would be in the case of a network that spanned longer distances than Fast Ethernet can support with copper cabling. See "Upgrading an Ethernet Network," later in this chapter, for more information on the upgrade process.

Fast Ethernet defines three physical-layer specifications, as shown in Table 11-1.

In addition to the connectors shown for each of the cable types, 802.3u standard describes a *medium-independent interface (MII)* that uses a 40-pin D-shell connector. Taking from the design of the original Thick Ethernet standard, the MII connects to an external transceiver called a *physical-layer device (PHY)*, which, in turn, connects to the network medium. The MII makes it possible to build devices such as hubs and computers that have integrated Fast Ethernet adapters, but are not committed to a particular media type. By supplying different PHY units, you can connect the device to a Fast Ethernet network using any supported cable type. Some PHY devices connect directly to the MII, while others use a cable not unlike the AUI cable arrangement in Thick Ethernet. If this is the case, the MII cable can be no more than 0.5 meters long.

	100Base-TX	**100Base-T4**	**100Base-FX**
Maximum Segment Length	100 meters	100 meters	412 meters
Cable Type	Category 5 UTP or Type 1 STP (two wire pairs)	Category 3 UTP (four wire pairs)	62.5/125 multimode fiber
Connector Type	RJ-45	RJ-45	SC, MIC, or ST

Table 11-1: *IEEE 802.3u Physical-Layer Specifications*

Most of the Fast Ethernet hardware on the market today uses internal transceivers and does not need an MII connector or cable, but a few products do take advantage of this interface.

100Base-TX

Using standards for physical media developed by ANSI (the American National Standards Institute), 100Base-TX and its fiber-optic counterpart 100Base-FX are known collectively as 100Base-X. They provide the core physical-layer guidelines for new cable installations. Like 10Base-T, 100Base-TX calls for the use of unshielded twisted-pair cable segments up to 100 meters in length. The only difference from a 10Base-T segment is in the quality and capabilities of the cable itself.

100Base-TX is based on the ANSI TP-PMD specification and calls for the use of Category 5 UTP cable for all network segments. The ratings for UTP cable are defined by the TIA/EIA (Telecommunications Industry Association/Electronic Industry Association), and are shown in Table 4-2 in Chapter 4.

As you can see in the table, the Category 5 cable specification provides the potential for much greater bandwidth than the Category 3 cable specified for 10Base-T networks. As an alternative, using Type 1 shielded twisted-pair cable (STP) is also possible for installations where the operating environment presents a greater danger of electromagnetic interference.

For the sake of compatibility, 100Base-TX (as well as 100Base-T4) uses the same type of RJ-45 connectors as 10Base-T and the pin assignments are also the same (see Table 10-3). The pin assignments are the one area in which the cable specifications differ from ANSI TP-PMD, to maintain backward compatibility with 10Base-T networks.

100Base-T4

100Base-T4 is intended for use on networks that already have UTP cable installed, but the cable is not rated as Category 5. The 10Base-T specification allows for the use of standard voice-grade (Category 3) cable, and there are many networks that are already wired for 10Base-T Ethernet (or even for telephone systems). 100Base-T4 runs at 100 Mbps on Category 3 cable by using all four pairs of wires in the cable, instead of just two, as 10Base-T and 100Base-TX do.

The transmit and receive data pair in a 100Base-T4 circuit are the same as that of 100Base-TX (and 10Base-T). The remaining four wires function as bidirectional pairs. The pin assignments for the RJ-45 connectors in a 100Base-T4 network are shown in Table 11-2.

Pin	Pair	Polarity	Signal	Designation
1	1	Positive	Transmit	TX_D1+
2	1	Negative	Transmit	TX_D1-
3	2	Positive	Receive	RX_D2+
4	3	Positive	Bidirectional	BI_D3+
5	3	Negative	Bidirectional	BI_D3-
6	2	Negative	Receive Data	RX_D2-
7	4	Positive	Bidirectional	BI_D4+
8	4	Negative	Bidirectional	BI_D4-

Table 11-2: *RJ-45 Pin Assignments for 100Base-T4 Networks*

As on a 10Base-T network, the transmit and receive pairs must be crossed over for traffic to flow. The crossover circuits in a Fast Ethernet hub connect the transmit pair to the receive pair, as always. In a 100Base-T4 hub, the two bidirectional pairs are crossed as well, so that pair 3 connects to pair 4, and vice versa.

Virtually all of the UTP cable installed today is Category 5 or better, which makes the market for 100Base-T4 equipment marginal at best. There are a few 100Base-T4 products available, but the vast majority of UTP Fast Ethernet networks use 100Base-TX.

100Base-FX

The 100Base-FX specification calls for exactly the same hardware as the 10Base-FL specification, except that the maximum length of a cable segment can be no more than 412 meters. As with the other Fast Ethernet physical-layer options, the medium is capable of transmitting a signal over longer distances, but the limitation is imposed to ensure the proper operation of the collision-detection mechanism. As mentioned earlier, when you eliminate the CSMA/CD MAC mechanism, as on a full-duplex Ethernet network, 100Base-FX segments can be much longer.

Cable Length Restrictions

Because the network is operating at ten times the speed of regular Ethernet, Fast Ethernet cable installations are more restricted than standard Ethernet. In effect, the Fast Ethernet standard uses up a good deal of the latitude built into the original Ethernet standards to achieve greater performance levels. In 10 Mbps Ethernet, the signal timing specifications are at least twice as strict as they have to be for systems to detect early collisions properly on the network. The lengths of the network segments are dictated more by the need to maintain the signal strength than the signal timing.

On 100Base-T networks, however, signal strength is not as much of an issue as signal timing. The CSMA/CD mechanism on a Fast Ethernet network functions exactly like that of a 10 Mbps Ethernet network and the packets are the same size, but they travel over the medium at ten times the speed. Because the collision detection mechanism is the same, a system still must be able to detect the presence of a collision before the slot time expires (that is, before it transmits 64 bytes of data). Because the traffic is moving faster, though, the duration of that slot time is reduced and the maximum length of the network must be reduced as well to sense collisions accurately. For this reason, the maximum overall length of a 100Base-TX network is approximately 205 meters. This is a figure you should observe much more stringently than the 500 meter maximum for a 10Base-T network.

Note

When you plan your network, be sure to remain conscious that the 100-meter maximum cable segment length specification in the Fast Ethernet standard includes the entire length of cable connecting a computer to the hub. If you have an internal cable installation that terminates at wall plates at the computer site and a patch panel at the hub site, you must include the lengths of the patch cables connecting the wall plate to the computer and the patch panel to the hub in your total measurement. The specification recommends that the maximum length for an internal cable segment be 90 meters, leaving 10 meters for the patch cables.

NETWORK PROTOCOLS

Hub Configurations

Because the maximum length for a 100Base-TX segment is 100 meters, the same as that for 10Base-T, the restrictions on the overall length of the network are found in the configuration of the repeating hubs used to connect the segments. The 802.3u supplement describes two types of hubs for all 100Base-T networks: Class I and Class II. Every Fast Ethernet hub must have a circled Roman numeral I or II identifying its class.

Class I hubs are intended to support cable segments with different types of signaling. 100Base-TX and 100Base-FX use the same signaling type, while 100Base-T4 is different (because of the presence of the two bidirectional pairs). A Class I hub contains circuitry that translates incoming 100Base-TX, 100Base-FX, and 100Base-T4 signals to a common digital format and then translates them again to the appropriate signal for each outgoing hub port. These translation activities cause comparatively long timing delays in the hub, so you can only have one Class I hub on the path between any two nodes on the network.

Class II hubs can only support cable segments of the same signaling type. Because no translation is involved, the hub passes the incoming data rapidly to the outgoing ports. Because the timing delays are shorter, you can have up to two Class II hubs on the path between two network nodes, but all the segments must use the same signaling type. This means a Class II hub can support either 100Base-TX and 100Base-FX together, or 100Base-T4 alone.

Additional segment length restrictions are also based on the combination of segments and hubs used on the network. The more complex the network configuration gets, the shorter its maximum collision domain diameter can be. These restrictions are summarized in Table 11-3.

	One Class I Hub	One Class II Hub	Two Class II Hubs
All copper segments (100Base-TX or 100Base-T4)	200 meters	200 meters	205 meters
All fiber segments (100Base-FX)	272 meters	320 meters	228 meters
One 100Base-T4 segment and one 100Base-FX segment	231 meters	Not applicable	Not applicable
One 100Base-TX segment and one 100Base-FX segment	260.8 meters	308.8 meters	216.2 meters

Table 11-3:   *Fast Ethernet Multisegment Configuration Guidelines*

Note that a network configuration that uses two Class II hubs actually uses three lengths of cable to establish the longest connection between two nodes: two cables to connect the nodes to their respective hubs, and one cable to connect the two hubs together. For example, the assumption of the standard is that the additional 5 meters added to the length limit for an all-copper network will account for the cable connecting the two hubs (see Figure 11-1). But in practice, the three cables can be of any length as long as their total length does not exceed 205 meters.

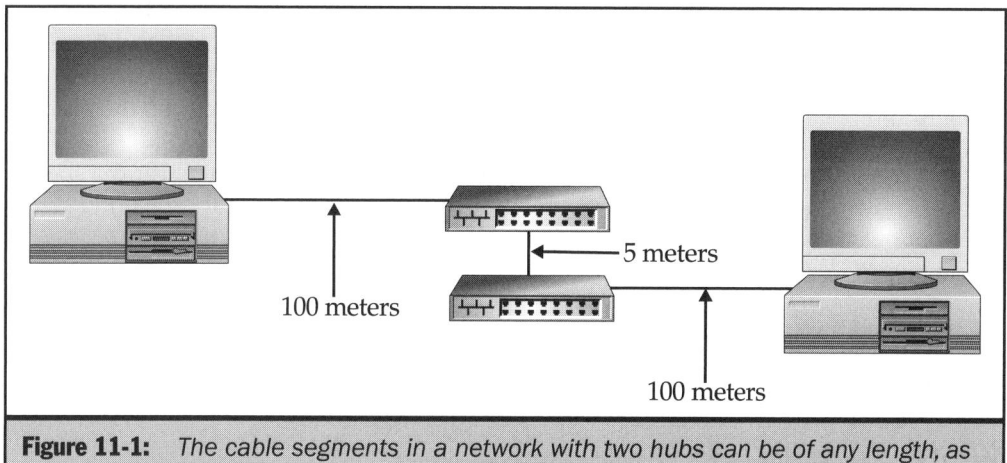

Figure 11-1:   *The cable segments in a network with two hubs can be of any length, as long as you observe the maximum collision domain diameter.*

What these restrictions mean to 100Base-FX networks is that the only fiber segment that can be 412 meters long is one that connects two computers directly together. Once you add a hub to the network, the total distance between computers drops drastically. This largely negates one of the major benefits of using fiber-optic cable. You saw earlier in this chapter that the original Ethernet standards allow for fiber-optic segments up to 2 kilometers (2,000 meters) long. The closer tolerances of the collision-detection mechanism on a Fast Ethernet network makes it impossible to duplicate the collision domain diameter of standards like 10Base-FL. Considering that other high-speed protocols such as FDDI use the same type of cable and can support distances up to 200 kilometers, Fast Ethernet might not be the optimal fiber-optic solution, unless you use the full-duplex option to increase the segment length.

Fast Ethernet Timing Calculations

As with the original Ethernet standards, the cabling guidelines in the previous sections are no more than rules of thumb that provide general size limitations for a Fast Ethernet network. Making more precise calculations to determine if your network is fully compliant with the specifications is also possible. For Fast Ethernet, these calculations consist only of determining the round trip delay time for the network. No interframe gap shrinkage calculation exists for Fast Ethernet because the limited number of repeaters permitted on the network all but eliminates this as a possible problem.

Calculating the Round Trip Delay Time   The process of calculating the round trip delay time begins with determining the worst case path through your network, just as in the calculations for standard Ethernet networks. As before, if you have different types of cable segments on your network, you may have more than one path to calculate. There is no need to perform separate calculations for each direction of a complex path, however, because the formula makes no distinction between the order of the segments.

The round trip delay time consists of a delay per meter measurement for the specific type of cable your network uses, plus an additional delay constant for each node and repeater on the path. Table 11-4 lists the delay factors for the various network components.

Component	Delay (in Bit Times)
Category 3 UTP cable segment	1.14/meter
Category 4 UTP cable segment	1.14/meter
Category 5 UTP cable segment	1.112/meter
STP cable segment	1.112/meter
Fiber-optic cable segment	1.0/meter
Two 100Base-TX/100Base-FX nodes	100
Two 100Base-T4 nodes	138
One 100Base-TX/100Base-FX node and one 100Base-T4 node	127
Class I hub	140
Class II 100Base-TX/100Base-FX hub	92
Class II 100Base-T4 hub	67

Table 11-4:   *Delay Times for Fast Ethernet Network Components*

To calculate the round trip delay time for the worst case path through your network, you multiply the lengths of your various cable segments by the delay factors listed in the table and add them together, along with the appropriate factors for the nodes and hubs and a safety buffer of 4 bit times. If the total is less than 512, the path is compliant with the Fast Ethernet specification. Thus, the calculations for the network shown in Figure 11-2 would be as follows:

(150 meters × 1.112 bit times/meter) + 100 bit times + (2 × 92 bit times) + 4 bit times = 454.8 bit times

150 meters of Category 5 cable multiplied by a delay factor of 1.112 bit times per meter yields a delay of 166.8 bit times, plus 100 bit times for two 100Base-TX nodes, two hubs at 92 bit times each, and an extra 4 for safety yields a total round trip delay time of 454.8 bit times, well within the 512 limit.

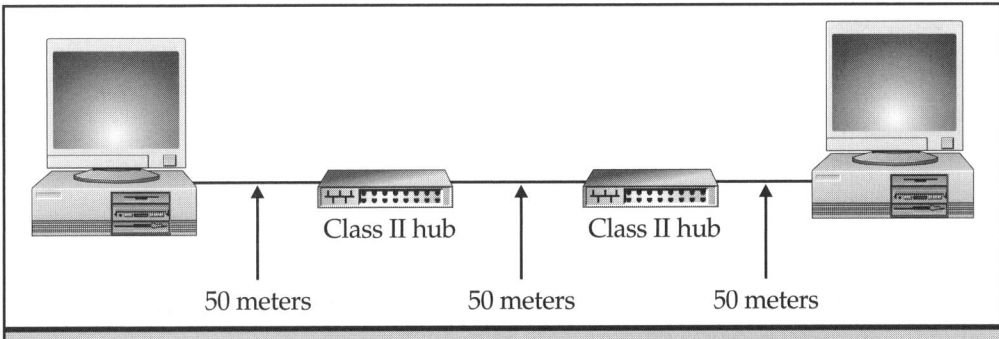

Figure 11-2: *This worst case path is compliant with the round trip delay time limitations defined in the Ethernet standard.*

> **Note**
> *As with the calculations for standard Ethernet networks, you may be able to avoid having to measure your cable segments by using the maximum permitted segment length in your calculations. Only if the result of this calculation exceeds the specification do you have to consider the actual lengths of your cables.*

Calculating Cable Delay Values   The delay values for the cable types shown in Table 11-4 are general estimations based only on the cable's rating. If you want to achieve even greater accuracy in your calculations and you have the necessary specifications from the cable manufacturer, you can also compute the rate of delay for your specific cables.

To do this, you must find out the cable's speed relative to the speed of light, usually referred to as the *Nominal Velocity of Propagation (NVP)* in the specifications for the cable. Once you have this figure, you can use the values in Table 11-5 to determine the delay factor of your cable in bit times/meter. Use this figure in place of the more general delay factors supplied in Table 11-4.

NVP	Nanoseconds/ Meter	Bit Times/ Meter	NVP	Nanoseconds/ Meter	Bit Times/ Meter
0.4	8.34	0.834	0.62	5.38	0.538
0.5	6.67	0.667	0.63	5.29	0.529
0.51	6.54	0.654	0.64	5.21	0.521
0.52	6.41	0.641	0.65	5.13	0.513
0.53	6.29	0.629	0.654	5.10	0.510
0.54	6.18	0.618	0.66	5.05	0.505
0.55	6.06	0.606	0.666	5.01	0.501
0.56	5.96	0.596	0.67	4.98	0.498
0.57	5.85	0.585	0.68	4.91	0.491
0.58	5.75	0.575	0.69	4.83	0.483
0.5852	5.70	0.570	0.7	4.77	0.477
0.59	5.65	0.565	0.8	4.17	0.417
0.6	5.56	0.556	0.9	3.71	0.371
0.61	5.47	0.547			

Table 11-5: *Nominal Velocity of Propagation (NVP) Figures and Their Equivalent Delay Times*

Note *Some manufacturers furnish the NVP as a percentage, in which case you divide the percentage by 100 and find the corresponding value in the table's NVP column.*

Autonegotiation

Virtually all of the Fast Ethernet adapters on the market today are dual-speed devices, meaning that they can operate at either 10 or 100 Mbps. This helps simplify the process of upgrading a 10Base-T network to Fast Ethernet. The Fast Ethernet standard also defines an autonegotiation system that enables a dual-speed device to sense the capabilities of the network to which it is connected and to adjust its speed accordingly. The autonegotiation mechanism in Fast Ethernet is based on *fast link pulse (FLP)* signals, which are themselves a variation on the *normal link pulse (NLP)* signals used by 10Base-T and 10Base-FL networks.

Standard Ethernet networks use NLP signals to verify the integrity of a link between two devices. Most Ethernet hubs and network interface adapters have a link pulse LED that lights when the device is connected to another active device. For example, when you take a UTP cable that is connected to a hub and plug it into a computer's NIC and turn the computer on, the LEDs on both the NIC and the hub port to which it's connected should light. This is the result of the two devices transmitting NLP signals to each other. When each device receives the NLP signals from the other device, it lights the link pulse LED. If the network is wired incorrectly, because of a cable fault or improper use of a crossover cable or hub uplink port, the LEDs will not light. These signals do not interfere with data communications, because the devices transmit them only when the network is idle.

Note	*The link pulse LED indicates only that the network is wired correctly, not that it's capable of carrying data. If you use the wrong cable for the protocol, you will still experience network communications problems, even though the devices passed the link integrity test.*

Fast Ethernet devices capable of transmitting at multiple speeds elaborate on this technique by transmitting FLP signals instead of NLP signals. FLP signals include a 16-bit data packet within a burst of link pulses, producing what is called an *FLP burst*. The data packet contains a *link code word (LCW)* with two fields: the *selector field* and the *technology ability* field. Together, these fields identify the capabilities of the transmitting device, such as its maximum speed and whether it is capable of full-duplex communications.

Because the FLP burst has the same duration (2 nanoseconds) and interval (16.8 nanoseconds) as an NLP burst, a standard Ethernet system can simply ignore the LCW and treat the transmission as a normal link integrity test. When it responds to the sender, the multiple-speed system sets itself to operate at 10Base-T speed, using a technique called *parallel detection*. This same method applies also to Fast Ethernet devices incapable of multiple speeds.

NETWORK PROTOCOLS

When two Fast Ethernet devices capable of operating at multiple speeds autonegotiate, they determine the best performance level they have in common and configure themselves accordingly. The systems use the following list of priorities when comparing their capabilities, with full-duplex 1000Base-T providing the best performance and half-duplex 10Base-T providing the worst:

1. 1000Base-T (full-duplex)

2. 1000Base-T

3. 100Base-TX (full-duplex)

4. 100Base-T4

5. 100Base-TX

6. 10Base-T (full-duplex)

7. 10Base-T

FLP signals account only for the capabilities of the devices generating them, not the connecting cable. If you connect a dual-speed 100Base-TX computer with a 100Base-TX hub using a Category 3 cable network, autonegotiation will still configure the devices to operate at 100 Mbps, even though the cable can't support transmissions at this speed.

The benefit of autonegotiation is that it permits administrators to upgrade a network gradually to Fast Ethernet with a minimum of reconfiguration. If, for example, you have 10/100 dual-speed NICs in all your workstations, you can run the network at 10 Mbps using 10Base-T hubs. Later, you can simply replace the hubs with models supporting Fast Ethernet, and the NICs will automatically reconfigure themselves to operate at the higher speed during the next system reboot. No manual configuration at the workstation is necessary.

Gigabit Ethernet

When 100 Mbps networking technologies like FDDI were first introduced, most horizontal networks used 10 Mbps Ethernet. These new protocols were used primarily on backbones. Now that Fast Ethernet has taken over the horizontal network market, a 100 Mbps backbone is, in many cases, insufficient to support the connections between switches that have to accommodate multiple Fast Ethernet networks. Gigabit Ethernet was developed to be the next generation of Ethernet network, running at 1 Gbps, or 1,000 Mbps, ten times the speed of Fast Ethernet.

Although it is still a relatively new technology, Gigabit Ethernet is virtually assured of a place in the market because, like Fast Ethernet before it, it uses the same frame format, frame size, and media access control method as standard 10 Mbps Ethernet. Fast Ethernet has overtaken FDDI as the dominant 100 Mbps solution because it prevented network administrators from having to use a different protocol on the backbone. In the same way, Gigabit Ethernet prevents administrators from having to use a different protocol like ATM for their backbones.

To connect an ATM or FDDI network to an Ethernet network requires that the data be converted at the network layer from one frame format to another. Connecting two Ethernet networks together, even when they're running at different speeds, is a data link–layer operation because the frames remain unchanged. In addition, using Ethernet throughout your network eliminates the need to train administrators to work with a new protocol and purchase new testing and diagnostic equipment. The bottom line is that in most cases, it is possible to upgrade a Fast Ethernet backbone to Gigabit Ethernet without completely replacing hubs, switches, and cables. This is not to say, however, that some hardware upgrades will not be necessary. Hubs and switches will need modules supporting the new protocol, and networking monitoring and testing products may also have to be upgraded to support the higher speed.

Gigabit Ethernet Architecture

Gigabit Ethernet was first defined in the 802.3z supplement to the 802.3 standard, which was published in June 1998. The 802.3z defines a network running at 1,000 Mbps in either half-duplex or full-duplex mode, over a variety of different network media. The frame used to encapsulate the packets is identical to that of 802.3 Ethernet, and the protocol (in half-duplex mode) uses the same Carrier Sense Multiple Access with Collision Detection (CSMA/CD) MAC mechanism as the other Ethernet incarnations.

As with standard and Fast Ethernet, the Gigabit Ethernet standard contains both physical and data link–layer elements, as shown in Figure 11-3. The data link–layer consists of the logical link control (LLC) and media access control (MAC) sublayers that are common to all of the IEEE 802 protocols. The LLC sublayer is identical to that used by the other Ethernet standards, as defined in the IEEE 802.2 document. The underlying concept of the MAC sublayer, the CSMA/CD mechanism, is fundamentally the same as on a standard Ethernet or Fast Ethernet network, but with a few changes in the way that it's implemented.

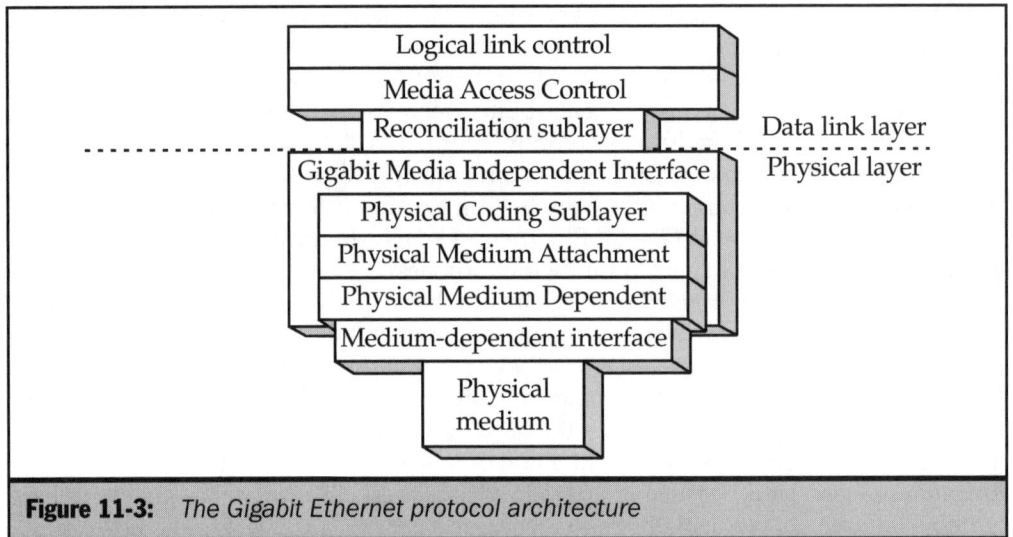

Figure 11-3: *The Gigabit Ethernet protocol architecture*

Media Access Control

Gigabit Ethernet is designed to support full-duplex operation as its primary signaling mode. As mentioned earlier, when systems can transmit and receive data simultaneously, there is no need for a media access control mechanism like CSMA/CD. For systems on a 1000Base-X network to operate in half-duplex mode, however, some modifications to the CSMA/CD mechanism were necessary. Ethernet's collision-detection mechanism only works properly when collisions are detected while a packet is still being transmitted. Once the source system finishes transmitting a packet, the data is purged from its buffers and it is no longer possible to retransmit that packet in the event of a collision.

When the speed at which systems transmit data increases, the round trip signal delay time during which a collision can be detected decreases. When Fast Ethernet increased the speed of an Ethernet network by ten times, the standard compensated by reducing the maximum diameter of the network. This enabled the protocol to use the same 64-byte minimum packet size as the original Ethernet standard and still be able to detect collisions effectively.

Gigabit Ethernet increases the transmission speed another ten times, but reducing the maximum diameter of the network again was impractical because it would result in networks no longer than 20 meters or so. As a result, the 802.3z supplement increases the size of the CSMA/CD carrier signal from 64 bytes to 512 bytes. This means that while the 64-byte minimum packet size is retained, the MAC sublayer of a Gigabit Ethernet system appends a carrier extension signal to small packets that pads them out to 512 bytes. This ensures that the minimum time required to transmit each packet is sufficient for the collision detection mechanism to operate properly, even on a network with the same diameter as Fast Ethernet.

The carrier extension bits are added to the Ethernet frame after the Frame Check Sequence, so that while they are a valid part of the frame for collision-detection purposes, the carrier extension bits are stripped away at the destination system before the FCS is computed and the results compared with the value in the packet. This padding, however, can greatly reduce the efficiency of the network. A small packet may consist of up to 448 bytes of padding (512 minus 64), the result of which is a throughput only slightly faster than Fast Ethernet. To address this problem, 802.3z introduces a packet-bursting capability along with the carrier extension. *Packet bursting* works by transmitting several packets back to back until a 1,500-byte burst timer is reached. This compensates for the loss incurred by the carrier extension bits and brings the network back up to speed.

When Gigabit Ethernet is used for backbone networks (as it still is almost exclusively), full-duplex connections between switches and servers are the more practical choice. The additional expenditure in equipment is minimal, and aside from eliminating this collision-detection problem, it increases the theoretical throughput of the network to 2 Gbps.

The Gigabit Media-Independent Interface

The interface between the data link and physical layers, called the *Gigabit Medium-independent Interface (GMII),* enables any of the physical-layer standards to use the MAC and LLC sublayers. The GMII is an extension of the medium-independent interface (MII) in Fast Ethernet, which supports transmission speeds of 10, 100, and 1,000 Mbps and has separate 8-bit transmit and receive data paths, for full-duplex communications. The GMII also includes two signals that are readable by the MAC sublayer, called *carrier sense* and *collision detect*. One of the signals specifies that a carrier is present, and the other that a collision is currently occurring. These signals are carried to the data link layer by way of the *reconciliation sublayer* located between the GMII and the MAC sublayer.

The GMII is broken up into three sublayers of its own, which are as follows:

- Physical Coding Sublayer (PCS)

- Physical Medium Attachment (PMA)

- Physical Medium-Dependent (PMD)

The following sections discuss the functions of these sublayers.

The Physical Coding Sublayer

The Physical Coding Sublayer (PCS) is responsible for encoding and decoding the signals on the way to and from the PMA. The physical-layer options defined in the 802.3z document all use the 8B/10B coding system, which was adopted from the ANSI Fibre Channel standards. In this system, each 8-bit data symbol is represented by an 11-bit code. There are also codes that represent control symbols, such as those used in the MAC carrier extension mechanism. Each code is formed by breaking down the 8 data bits into two groups consisting of the 3 most significant bits (y) and the 5 remaining bits (x). The code is then named using the following notation:

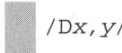

/Dx,y/

where x and y equal the decimal values of the two groups. The control codes are named the same way, except that the letter D is replaced by a K, as follows:

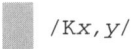

/Kx,y/

Note *Fibre Channel is a networking protocol that is designed to provide computers with high-speed access to peripheral devices such as RAID arrays and other mass storage systems, forming what is known as a storage area network (SAN). For more information on Fibre Channel and SANs, see Chapter 8.*

The idea behind this type of coding is to minimize the occurrence of consecutive 1s and 0s, which make it difficult for systems to synchronize their clocks. To help do this, each of the code groups must be composed of one of the following:

- Five 0s and five 1s
- Six 0s and four 1s
- Four 0s and six 1s

Note *The 1000Base-T physical-layer option does not use the 8B/10B coding system. See "1000Base-T," later in this chapter, for more information.*

The PCS is also responsible for generating the carrier sense and collision-detect signals, and for managing the autonegotiation process used to determine what speed the network interface card should use (10, 100, or 1,000 Mbps) and whether it should run in half-duplex or full-duplex mode.

The Physical Medium Attachment Sublayer

The Physical Medium Attachment (PMA) sublayer is responsible for converting the code groups generated by the PCS into a serialized form that can be transmitted over the network medium, and converting the serial bit stream arriving over the network into code groups for use by the upper layers.

The Physical Medium-Dependent Sublayer

The Physical Medium-Dependent (PMD) sublayer provides the interface between the coded signals generated by the PCS and the actual physical network medium. This is where the actual optical or electric signals that are transmitted over the cable are generated and passed on to the cable through the medium-dependent interface (MDI).

The Physical Layer

Collectively called 1000Base-X, there are three physical-layer options for Gigabit Ethernet defined in the 802.3z document, two for fiber-optic cable and one for copper. Another copper option is defined in a separate document, IEEE 802.3ab, which was ratified in June 1999. These three physical-layer options in 802.3z were adopted from the ANSI X3T11 Fibre Channel specifications. The use of an existing standard for this crucial element of the technology has greatly accelerated the development process, both of the Gigabit Ethernet standards and of the hardware products. In general, 1000Base-X calls for the use of the same types of fiber-optic cables as FDDI and 100Base-FX, but at shorter distances. The longest possible Gigabit Ethernet segment, using singlemode fiber cable, is 5 kilometers.

For its multimode cable options, the 802.3z standard pioneers the use of laser light sources at high speeds. Most fiber-optic applications use lasers only with singlemode cable, while the signals on multimode cables are produced by light-emitting diodes (LEDs). The jitter effect, which was a problem with previous efforts to use lasers with multimode cable, was resolved by redefining the properties of the laser transmitters used to generate the signals.

Unlike standard and Fast Ethernet, the fiber-optic physical-layer standards for 1000Base-X are not based on the properties of specific cable types, but rather on the properties of the optical transceivers that generate the signal on the cable. Each of the fiber-optic standards supports several grades of cable, using short or long wavelength laser transmitters. The physical-layer options for 1000Base-X are described in the following sections.

1000Base-LX

1000Base-LX is intended for use in backbones spanning relatively long distances, using long wavelength laser transmissions in the 1,270 to 1,355 nanometer range with either multimode fiber cable within a building or singlemode fiber for longer links, such as those between buildings on a campus network. Multimode fiber cable with a core diameter of 50 or 62.5 microns supports links of up to 550 meters, while 9-micron singlemode fiber supports links of up to 5,000 meters (5 km). Both fiber types use standard SC connectors. The cable types supported by 1000Base-LX are shown in Table 11-6.

Cable Type	Core Diameter	Bandwidth	Maximum Link Length
Singlemode	9 microns	N/A	5,000 meters
Multimode	50 microns	400 MHz/km	550 meters
Multimode	50 microns	500 MHz/km	550 meters
Multimode	62.5 microns	500 MHz/km	550 meters

Table 11-6:   *1000Base-LX Cable Specifications*

1000Base-SX

1000Base-SX uses short wavelength laser transmissions ranging from 770 to 860 nanometers and is intended for use on shorter backbones and horizontal wiring. This option is more economical than 1000Base-LX because it uses only the relatively inexpensive multimode fiber cable, in several grades, and the lasers that produce the short wavelength transmissions are the same as those commonly used in CD and CD-ROM players. As of this writing, most of the fiber-optic Gigabit Ethernet products on the market support the 1000Base-SX standard.

The cable types supported by 1000Base-SX are shown in Table 11-7.

Cable Type	Core Diameter	Bandwidth	Maximum Link Length
Multimode	50 microns	400 MHz/km	500 meters
Multimode	50 microns	500 MHz/km	550 meters
Multimode	62.5 microns	160 MHz/km	220 meters
Multimode	62.5 microns	200 MHz/km	275 meters

Table 11-7:   *1000Base-SX Cable Specifications*

1000Base-LH

1000Base-LH (LH stands for *long haul*) is not a standard that has been ratified by the IEEE, nor is it in the process of being ratified. This is a physical-layer specification that has been developed by a group of hardware vendors, including 3Com and Cisco, that are seeking a longer-distance Gigabit Ethernet solution suitable for metropolitan area network (MAN) applications. Specific cable options have not been solidified yet, as there are several manufacturers working on different implementations. 3Com, for example, has defined two cable options, both using 9-micron singlemode fiber. One, with a wavelength of 1,310 nanometers, is rated for distances of 1 to 49 kilometers, and the other uses a 1,550 nm wavelength for distances of 50 to 100 km.

1000Base-CX

There is only one standard for copper cabling in the original 802.3z document. 1000Base-CX is intended for links that span only short distances (under 25 meters), such as for connections within the same telecommunications closet or data center. These connections require the use of a special 150-ohm shielded twisted-pair cable. The standard specifically mentions that the use of UTP or IBM Type 1 STP is not recommended. 1000Base-CX is intended for equipment connections such as server clusters and links between switches, because it is less expensive and easier to install than fiber optic. The connections are often located within a controlled environment that doesn't need the long lengths and resistance to interference provided by fiber. There has not been a great deal of interest from hardware manufacturers in producing 1000Base-CX equipment, presumably because of the limits of its market and the subsequent publication of the 1000Base-T supplement, which supports standard UTP cable. There are no 1000Base-CX products on the market.

1000Base-T

Although it is not included in the 802.3z standard, one of the original goals of the Gigabit Ethernet development team was for it to run on standard Category 5 UTP cable and support connections up to 100 meters long. This enables existing Fast Ethernet networks to be upgraded to Gigabit Ethernet without pulling new cable or changing the network topology. 1000Base-T is defined in a separate document called 802.3ab, which was unanimously ratified by the IEEE in June 1999.

To achieve these high speeds over copper, 1000Base-T modifies the way that the protocol uses the UTP cable. While designed to use the same cable installations as 100Base-TX, 1000Base-T uses all four of the wire pairs in the cable, while 100Base-TX uses only two pairs. In addition, all four pairs can carry signals in either direction. The 1000Base-T signal assignments are shown in Table 11-8. This effectively doubles the throughput of 100Base-TX, but it still doesn't approach speeds of 1,000 Mbps. However, 1000Base-T also uses a different signaling scheme to transmit data over the cable than the other 1000Base-X standards. This makes it possible for each of the four wire pairs to carry 250 Mbps, for a total of 1,000 Mbps or 1 Gbps. This signaling scheme is called *Pulse Amplitude Modulation 5 (PAM-5)*.

Pin	Pair	Polarity	Signal	Designation
1	1	Positive	Bidirectional	BI_DA+
2	1	Negative	Bidirectional	BI_DA-
3	2	Positive	Bidirectional	BI_DB+
4	3	Positive	Bidirectional	BI_DC+
5	3	Negative	Bidirectional	BI_DC-
6	2	Negative	Bidirectional	BI_DB-
7	4	Positive	Bidirectional	BI_DD+
8	4	Negative	Bidirectional	BI_D4D

Table 11-8: *1000Base-T Gigabit Ethernet UTP Pin Assignments*

While designed to run over standard Category 5 cable, as defined in the TIA/EIA standards, the standard recommends that new 1000Base-T networks use at least Category 5E (or Enhanced Category 5) cable. Category 5E cable is tested for its resistance to return loss and equal-level far-end crosstalk (ELFEXT). As with Fast Ethernet, 1000Base-T NICs and other equipment are available that can run at multiple speeds, either 100/1000 or 10/100/1000 Mbps, to facilitate gradual upgrades to Gigabit Ethernet. Autonegotiation, optional in Fast Ethernet, is mandatory in Gigabit Ethernet.

While networks that run Gigabit Ethernet to the desktop are not likely to be commonplace for some time, it will eventually happen, if history is any indicator.

Upgrading an Ethernet Network

The two most typical upgrades performed on an Ethernet network are the addition of new computers and the migration to Fast Ethernet. The following sections examine the procedures involved for these upgrades and present some of the more common problems you may encounter.

Adding Workstations

For the most part, the addition of new workstations on an Ethernet network is purely a mechanical exercise. Depending on the network medium, you do one of the following:

- On a Thick Ethernet network, attach new vampire taps to the coaxial cable and connect them to the computers with an external MAU (transceiver) and an AUI cable.

- On a Thin Ethernet network, connect a new length of coaxial cable to the T-connector of one of the existing workstations and to the T-connector of the new workstation, and then connect the old cable to the other end of the T-connector on the new workstation.

- On a UTP or fiber-optic network, plug one end of a new cable into a free port on the hub and the other end into the new computer. For an internal cable installation, you may have to use patch cables to connect the hub to a port on the patch panel and the computer to the wall plate corresponding to the patch panel port.

When difficulties arise, this is usually because you have reached the maximum number of nodes permitted on a coaxial segment, run out of ports on your hub, or exceeded the segment length restrictions for the cable type. In some cases, you can work around these problems without major renovations; but if you're approaching the limits of your network configuration, it may be time to consider a more comprehensive upgrade.

Bus Network Expansion

When adding workstations to a Thick or Thin Ethernet network, the primary danger is that you will have too many nodes on the cable segment. You learned earlier in this chapter that the Ethernet cabling specifications have a built-in buffer, which usually enables you to exceed the configuration guidelines to some degree. As long as your maximum segment lengths comply with the specifications, having as many as 110 nodes on a Thick Ethernet segment or 35 nodes on Thin Ethernet (instead of 100 and 30, respectively) should not be a problem.

Exceeding the maximum segment length is a problem limited primarily to Thin Ethernet networks because every workstation you add means attaching another length of cable. On Thick Ethernet networks, you are only adding more taps to the already-installed cable. Again, if you find that the new additions bring your Thin Ethernet segment to 210 or 220 meters, the network should continue to function normally, as long as you have 30 nodes or less. Exceeding both the maximum length and maximum number of nodes is not recommended, as this compounds the risk of malfunctions significantly.

If your expansion sends you well over the maximum recommended segment length, you can add a repeater at some point near the middle of the cable to prevent the attenuation that is a major source of problems on overlong coaxial networks. If you do have to add a repeater, you must consider also the guideline governing the maximum number of repeaters allowed on your network. Again, you can probably get away with one too many repeaters, as long as your network is compliant in every other way. Only when you exceed the specifications in two dimensions should you begin to worry.

When you add more nodes to a coaxial network, observing the guidelines regarding the minimum amount of cable required between workstations is important. In fact, it is less dangerous to exceed the segment length than to save length by placing the node connections too close together. On a Thick Ethernet network, the RG-8 coaxial cable is usually marked with black stripes every 2.5 meters. This is the minimum safe distance between cable taps. On a Thin Ethernet network, every length of coaxial cable must be at least 0.5 meters long. This is usually not a problem, but even if you have two computers arranged back to back so that they are nearly touching, you should use a piece of cable at least a meter long to be safe.

Star Network Expansion

Adding nodes to your twisted-pair Ethernet network couldn't be easier, as long as you have the hub ports and cable drops you need. However, when you run short of either of these things, there are sometimes relatively easy solutions.

On a 10Base-T network, you can add a minihub to a location that has run short of cable drops or when you run out of ports on your other hubs. Just as on a coaxial network, if you have one too many hubs on your worst case path through the network, this will probably not be a problem. Also, remember that because 10Base-T hubs also function as repeaters, you can use a minihub to connect a workstation more than 100 meters away from your main hub.

Fast Ethernet networks, however, are a different matter entirely. The tolerances built into the Fast Ethernet specifications are much tighter than those for standard Ethernet. Exceeding the recommended cable lengths, and especially the maximum number of repeaters, is strongly discouraged.

For either standard or Fast Ethernet, when you expand beyond the point at which you can add more nodes, you must consider splitting the collision domain in two. To do this, you add a device that filters the network traffic that passes through it, such as a switching hub. Unlike repeating hubs, which are purely electrical devices that manage and amplify all the signals they receive indiscriminately, switches actually read the headers of the packets and propagate incoming traffic only to the ports for which it is destined.

Note

For more information on expanding your network with switches, see Chapter 6.

Upgrading to Fast Ethernet

Because Fast Ethernet is so similar to standard Ethernet, upgrading networks from 10 to 100 Mbps can often be quite easy. The key to simplifying the procedure in a production environment is the use of the dual-speed devices available today. Originally, Fast Ethernet devices were all single speed and you had to have separate hubs for the 10Base-T and 100Base-T workstations.

In addition, any servers you wanted accessible at both speeds had to have separate 10 and 100 Mbps NICs in them. Each NIC was connected to the appropriate speed hub along with the workstations running at that speed. Upgrading a workstation to Fast Ethernet meant installing a new NIC into the computer and plugging the other end of the cable into the other hub.

Upgrading NICs and Hubs

Today, virtually all Fast Ethernet NICs are dual speed, and many dual-speed hubs are available as well. This provides the network administrator with complete flexibility over the upgrade process. When faced with economic or scheduling constraints, you can gradually upgrade the network to Fast Ethernet at any desired pace.

If you are running 10Base-T and have even the remotest plans of upgrading to Fast Ethernet in the future, you should install 10/100 NICs in all your new workstations and think about replacing the old NICs as time permits. Dual-speed NICs are only slightly more expensive than 10Base-T-only cards, and can be cheaper when bought in bulk. In fact, as Fast Ethernet becomes more ubiquitous, 10Base-T-only cards are becoming harder to find than dual-speed NICs.

The other half of the equation is the Fast Ethernet hub. If you purchase Fast Ethernet–only hubs, you should either wait until all of your dual-speed NICs are installed to deploy them, or plan on maintaining your 10Base-T hubs for the standard Ethernet workstations until they are upgraded. As you upgrade a workstation to a Fast Ethernet NIC, you will also have to plug it into the new hub.

If, however, you buy dual-speed hubs, you can replace the old hubs completely and plug all of your workstations in, regardless of speed. The Fast Ethernet autonegotiation mechanism enables each NIC and hub port to function at the fastest possible speed. As you upgrade the workstation NICs, they will automatically negotiate the higher speed, without any adjustments at the hub end.

Upgrading the Cable Plant

In most cases, Fast Ethernet is designed to use the UTP cable already installed at a site. If you have Category 3 or 4 cable installed, you can run 100Base-T4; with Category 5 cable, you can run 100Base-TX. 100Base-T4 and 100Base-TX require different NICs and

NETWORK PROTOCOLS

hubs, so be sure that you purchase the correct hardware for your network. Due to the relative scarcity of 100Base-T4 equipment, you may want to consider installing Category 5 cable instead.

On a relatively simple network, migrating to Fast Ethernet can be as easy as replacing the NICs and hubs, with no modifications to the cable installation. Because the maximum segment length is the same (100 meters), there should be no problem with the locations of your hubs relative to the workstations. However, the big additional restriction in the Fast Ethernet standard is the limitation on the number of hubs.

If you have a network in which the paths between workstations routinely run through three or four hubs, you need to rethink your network architecture to use Fast Ethernet because the maximum of two Class II hubs is a restriction you should not ignore. To remain within the Fast Ethernet guidelines, you must eliminate some of the hubs on your network paths. You do this by dividing the network into two or more collision domains, so no more than two hubs are in any path between two workstations, using a switch or a device operating higher in the OSI model, such as a router. (Remember, mixing cable types using Class I hubs limits you to only one hub.)

Dividing a network into multiple collision domains is easier when the hubs are all located in the same place, such as a data center or wiring closet. If you have a more informal network with ad hoc upgrades in the form of minihubs scattered about in different locations, you may have a difficult time making the network compliant with the Fast Ethernet specifications.

Ethernet Troubleshooting

Troubleshooting an Ethernet network often means dealing with a problem in the physical layer, such as a faulty cable or connection, or possibly a malfunctioning NIC or hub. When a network connection completely fails, you should immediately start examining the cabling and other hardware for faults. If you find that the performance of the network is degrading, however, or if a problem is affecting specific workstations, you can sometimes get an idea of what is going wrong by examining the Ethernet errors occurring on the network.

Ethernet Errors

Following are some of the errors that can occur on an Ethernet network. Some are relatively common, while others are rare. Detecting these errors usually requires special tools designed to analyze network traffic. Basic software applications like Windows 2000's Network Monitor and NetWare's Monitor.nlm can detect some of these conditions, such as the number of early collisions and FCS errors. Others, like late collisions, are much more difficult to detect and may require high-end software or hardware tools to diagnose.

Early collisions   Strictly speaking, not an error, because collisions occur normally on an Ethernet network. But too many collisions (more than approximately 5 percent of the total packets) is a sign that network traffic is approaching critical levels. It is a good idea to keep a record of the number of collisions occurring on the network at regular intervals (such as weekly). If you notice a marked increase in the number of collisions, you might consider trying to decrease the amount of traffic, either by splitting the network into two collision domains or moving some of the nodes to another network.

Late collisions   Late collisions are always a cause for concern and are difficult to detect. They usually indicate that data is taking too long to traverse the network, either because the cable segments are too long or there are too many repeaters. A NIC with a malfunctioning carrier sense mechanism could also be at fault. Network analyzer products that can track late collisions can be extremely expensive, but are well worth the investment for a large enterprise network. Because late collisions force lost packets to be retransmitted by higher-layer protocols, you can sometimes detect a trend of network-layer retransmissions (by the IP protocol, for example) caused by late collisions, using a basic protocol analyzer like Network Monitor.

Runts   A *runt* is a packet less than 64 bytes long, caused either by a malfunctioning NIC or hub port, or by a node that ceases transmitting in the middle of a packet because of a detected collision. A certain number of runt packets occur naturally as a result of normal collisions, but a condition where more runts occur than collisions indicates a faulty hardware device.

Giants   A *giant* is a packet that is larger than the Ethernet maximum of 1,518 bytes. The problem is usually caused by a NIC that is *jabbering*, or transmitting improperly or continuously, or (less likely) by the corruption of the header's length indicator during transmission. Giants never occur normally. They are an indication of a malfunctioning hardware device or a cable fault.

Alignment errors   A packet that contains a *partial byte* (that is, a packet with a size in bits that is not a multiple of 8) is said to be *misaligned*. This can be the result of an error in the formation of the packet (in the originating NIC) or evidence of corruption occurring during the packet's transmission. Most misaligned packets also have CRC errors.

CRC errors   A packet in which the frame check sequence generated at the transmitting node does not equal the value computed at the destination is said to have experienced a CRC error. The problem can be caused by data corruption occurring during transmission (because of a faulty cable or other connecting device) or conceivably by a malfunction in the FCS computation mechanism in either the sending or receiving node.

NETWORK PROTOCOLS

Broadcast storms   When a malformed broadcast transmission causes the other nodes on the network to generate their own broadcasts for a total traffic rate of 126 packets per second or more, the result is a self-sustaining condition, known as a *broadcast storm*. Because broadcast transmissions are processed before other frames, the storm effectively prevents any other data from being successfully transmitted.

Isolating the Problem

Whenever you exceed any of the Ethernet specifications (or the specifications for any protocol, for that matter), the place where you're pushing the envelope should be the first place you check when a problem arises. If you have exceeded the maximum length for a segment, for example, try to eliminate some of the excess length to see if the problem continues. On a Thin Ethernet network, this usually means cross-cabling to eliminate some of the workstations from the segment. On a UTP network, connect the same computer to the same hub port using a shorter cable run. If you have too many workstations running on a coaxial bus (Thick or Thin Ethernet), you can determine if overpopulation is the problem simply by shutting down some of the machines.

Excessive repeaters on a UTP network is a condition that you can test for by checking to see if problems occur more often on paths with a larger number of hubs. You can also try to cross-cable the hubs to eliminate some of them from a particular path. This is relatively easy to do in an environment in which all the hubs are located in the same wiring closet or data center, but if the hubs are scattered all over the site, you may have to disconnect some of the hubs temporarily to reduce the size of the collision domain to perform your tests. The same is true of a coaxial network on which the primary function of the repeaters is to extend the collision domain diameter. You may have to disconnect the cable from each of the repeaters in turn (remembering to terminate the bus properly each time) to isolate the problem.

Reducing the size of the collision domain is also a good way to narrow down the location of a cable fault. In a UTP network, the star topology means that a cable break will only affect one system. On a coaxial network using a bus topology, however, a single cable fault can bring down the entire network. On a multisegment network, terminating the bus at each repeater in turn can tell you which segment has the fault.

A better, albeit more expensive, method for locating cable problems is to use a multifunction cable tester. These devices can pinpoint the exact location of many different types of cable faults.

Tip

Once you locate a malfunctioning cable, it's a good idea to dispose of it immediately. Leaving a bad cable lying around can result in someone else trying to use it, thus the need for another troubleshooting session.

100VG-AnyLAN

100VG-AnyLAN is a 100 Mbps desktop networking protocol that is usually grouped with Fast Ethernet because the two were created at the same time and briefly competed for the same market. However, this protocol cannot strictly be called an Ethernet variant because it does not use the CSMA/CD media access control mechanism.

100VG-AnyLAN is defined in the IEEE 802.12 specification, while all of the Ethernet variants are documented by the 802.3 working group. Originally touted by Hewlett-Packard and AT&T as a 100 Mbps UTP networking solution that is superior to Fast Ethernet, the market has not upheld that belief. While a few 100VG products are still available, Fast Ethernet has clearly become the dominant 100 Mbps networking technology.

As with Fast Ethernet, the intention behind the 100VG standard is to use existing 10Base-T cable installations and to provide a clear, gradual upgrade path to the faster technology. Originally intended to support all the same physical-layer options as Fast Ethernet, only the first 100VG cabling option has actually materialized, using all four wire pairs in a UTP cable rated Category 3 or better. The maximum cable segment length is 100 meters for Category 3 and 4 cables, and 200 meters for Category 5. Up to 1,024 nodes are permitted on a single-collision domain. 100VG-AnyLAN uses a technique called *quartet signaling* to use the four wire pairs in the cable.

100VG uses the same frame format as either 802.3 Ethernet or 802.5 Token Ring, making it possible for the traffic to coexist on a network with these other protocols. This is an essential point that provides a clear upgrade path from the older, slower technologies. As with Fast Ethernet, dual-speed NICs are available to make it possible to perform upgrades gradually, one component at a time.

A 10Base-T/100VG-AnyLAN NIC, however, is a substantially more complex device than a 10/100 Fast Ethernet card. While the similarity between standard and Fast Ethernet enables the adapter to use many of the same components for both protocols, 100VG is sufficiently different from 10Base-T to force the device to be essentially two network interface adapters on a single card, which share little else but the cable and bus connectors. This, and the relative lack of acceptance for 100VG-AnyLAN, has led the prices of the hardware to be substantially higher than those for Fast Ethernet.

The one area in which 100VG-AnyLAN differs most substantially from Ethernet is in its media access control mechanism. 100VG networks use a technique called *demand priority*, which eliminates the normally occurring collisions from the network and also provides a means to differentiate between normal and high-priority traffic. The introduction of priority levels is intended to support applications that require consistent streams of high bandwidth, such as real-time audio and video.

NETWORK PROTOCOLS

The 100VG-AnyLAN specification subdivides its functionality into several sublayers. Like the other IEEE 802 standards, the LLC sublayer is at the top of a node's data-link layer's functionality, followed by the MAC sublayer. On a repeater (hub), the *repeater media access control (RMAC)* sublayer is directly below the LLC. Beneath the MAC or RMAC sublayer, the specification calls for a physical medium–independent (PMI) sublayer, a medium-independent interface (MII), and a physical medium–dependent (PMD) sublayer. Finally, the medium-dependent interface provides the actual connection to the network medium. The following sections examine the activities at each of these layers.

The Logical Link Control Sublayer

The LLC-sublayer functionality is defined by the IEEE 802.2 standard and is the same as that used with 802.3 (Ethernet) and 802.5 (Token Ring) networks.

The MAC and RMAC Sublayers

100VG's demand priority mechanism replaces the CSMA/CD mechanism in Ethernet and Fast Ethernet networks. Unlike most other MAC mechanisms, access to the medium on a demand priority network is controlled by the hub. Each node on the network, in its default state, transmits an *Idle_Up* signal to its hub, indicating that it is available to receive data. When a node has data to transmit, it sends either a *Request_Normal* or *Request_High* signal to the hub. The signal the node uses for each packet is determined by the upper-layer protocols, which assign priorities based on the application generating the data.

The hub continuously scans all of its ports in a round-robin fashion, waiting to receive request signals from the nodes. After each scan, the hub selects the node with the lowest port number that has a high-priority request pending and sends it the *Grant* signal, which is the permission for the node to transmit. After sending the *Grant* signal to the selected node, the hub sends the *Incoming* signal to all of the other ports, which informs the nodes of a possible transmission. As each node receives the incoming signal, it stops transmitting requests and awaits the incoming transmission.

When the hub receives the packet from the sending node, it reads the destination address from the frame header and sends the packet out the appropriate port. All the other ports receive the *Idle_Down* signal. After receiving either the data packet or the *Idle_Down* signal, the nodes return to their original state and begin transmitting either a request or an *Idle_Up* signal. The hub then processes the next high-priority request.

When all the high-priority requests have been satisfied, the hub then permits the nodes to transmit normal priority traffic, in port number order. This exchange of signals is illustrated in Figure 11-4.

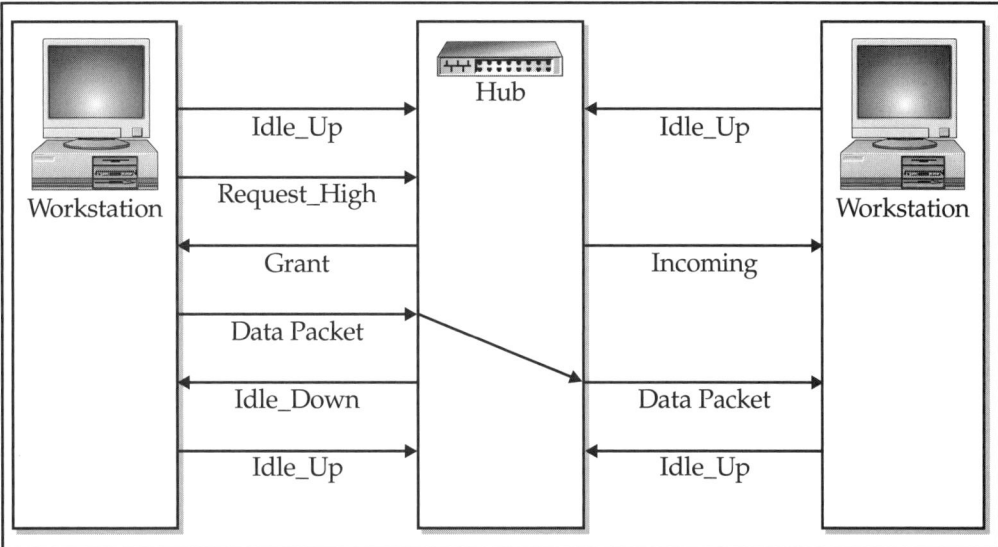

Figure 11-4: *100VG hubs arbitrate access to the network medium by granting nodes permission to transmit.*

Note

By default, a 100VG hub only transmits incoming packets out to the port (or ports) identified in the packet's destination address. This is known as operating in private mode. Configuring specific nodes to operate in promiscuous mode is possible, however, in which case they receive every packet transmitted over the network.

The processing of high-priority requests first enables applications that require timely access to the network to receive it, but a mechanism also exists to protect normal priority traffic from excessive delays. If the time needed to process a normal priority request exceeds a specified interval, the request is upgraded to high priority.

On a network with multiple hubs, one *root hub* always exists, to which all the others are ultimately connected. When the root hub receives a request through a port to which another hub is connected, it enables the subordinate hub to perform its own port scan and process one request from each of its own ports. In this way, permission to access the media is propagated down the network tree and all nodes have an equal opportunity to transmit.

NETWORK PROTOCOLS

MAC Frame Preparation

In addition to controlling access to the network medium, the MAC sublayer also assembles the packet frame for transmission across the network. Four possible types of frames exist on a 100VG-AnyLAN network:

- 802.3

- 802.5

- Void

- Link training

802.3 and 802.5 Frames   100VG-AnyLAN is capable of using either 802.3 (Ethernet) or 802.5 (Token Ring) frames, so that the 100VG protocol can coexist with the other network types during a gradual deployment process. Using both frame types at once is impossible, however. You must configure all of the hubs on the network to use one or the other frame type.

All 100VG frames are encapsulated within a Start of Stream field and an End of Stream field by the physical medium–independent sublayer, which informs the PMI sublayer on the receiving station when a packet is being sent and when the transmission is completed. Inside these fields, the 802.3 and 802.5 frames use the exact same formats defined in their respective specifications.

The MAC sublayer supplies the system's own hardware address for each packet's source address field and also performs the CRC calculations for the packet, storing them in the FCS field.

On incoming packets, the MAC sublayer performs the CRC calculations and compares the results with the contents of the FCS field. If the packet passes the frame check, the MAC sublayer strips off the two addresses and the FCS fields and passes the remaining data to the next layer.

Void Frames   *Void frames* are generated only by repeaters when a node fails to transmit a packet within a given time period after the repeater has acknowledged it.

Link Training Frames   Every time a node is restarted or reconnected to the network, it initiates a link training procedure with its hub by transmitting a series of specialized link training packets. This procedure serves several purposes, as follows:

Connection testing   For a node to connect to the network, it must exchange 24 consecutive training packets with the hub without corruption or loss. This ensures that the physical connection is viable and that the NIC and hub port are functioning properly.

Port configuration   The data in the training packets specifies whether the node will use 802.3 or 802.5 frames, operate in private or promiscuous mode, and whether it is an end node (computer) or a repeater (hub).

Address registration   The hub reads the node's hardware address from the training packets and adds it to the table it maintains of all the connected nodes' addresses.

Training packets contain two-byte *requested configuration* and *allowed configuration* fields that enable nodes and repeaters to negotiate the port configuration settings for the connection. The training packets the node generates contain its settings in the requested configuration field and nothing in the allowed configuration field. The repeater, on receiving the packets, adds the settings it can provide to the allowed configuration field and transmits the packets back to the node.

The packets also contain between 594 and 675 bytes of padding in the data field, to ensure that the connection between the node and the repeater is functioning properly and can transmit data without error.

The Physical Medium–Independent Sublayer

As the name implies, the *physical medium–independent (PMI) sublayer* performs the same functions for all 100VG packets, regardless of the network medium. When the PMI sublayer receives a frame from the MAC sublayer, it prepares the data for transmission using a technique called *quartet signaling*. The quartet refers to the four pairs of wires in a UTP cable, all of which the protocol uses to transmit each packet. Quartet signaling includes four separate processes, as follows:

1. Each packet is divided into a sequence of 5-bit segments (called *quintets*) and assigned sequentially to four channels that represent the four wire pairs. Thus, the first, fifth, and ninth quintets will be transmitted over the first pair, the second, sixth, and tenth over the second pair, and so on.

2. The quintets are scrambled using a different algorithm for each channel, to randomize the bit patterns for each pair and eliminate strings of bits with equal values. Scrambling the data in this way minimizes the amount of interference and crosstalk on the cable.

3. The scrambled quintets are converted to sextets (6-bit units) using a process called *5B6B encoding*, which relies on a predefined table of equivalent 5-bit and 6-bit values. Because the sextets contain an equal number of 0s and 1s, the voltage on the cable remains even and errors (which take the form of more than three consecutive 0s or 1s) are more easily detected. The regular voltage transitions also enable the communicating stations to synchronize their clocks more accurately.

4. Finally, the preamble, Start of Frame, and End of Frame fields are added to the encoded sextets and, if necessary, padding is added to the data field to bring it up to the minimum length.

NETWORK PROTOCOLS

The Medium-Independent Interface Sublayer

The *medium-independent interface (MII) sublayer* is a logical connection between the PMI and PMD layers. As with Fast Ethernet, the MII can also take the form of a physical hardware element that functions as a unified interface to any of the media supported by 100VG-AnyLAN.

The Physical Medium–Dependent Sublayer

The *physical medium–dependent (PMD) sublayer* is responsible for generating the actual electrical signals transmitted over the network cable. This includes the following functions:

Link status control signal generation   Nodes and repeaters exchange link status information using control tones transmitted over all four wire pairs in full-duplex mode (two pairs transmitting and two pairs receiving). Normal data transmissions are transmitted in half-duplex mode.

Data stream signal conditioning   The PMD sublayer uses a system called *NRZ* (non-return to zero) encoding to generate the signals transmitted over the cable. NRZ minimizes the effects of crosstalk and external noise that can damage packets during transmission

Clock recovery   NRZ encoding transmits one bit of data for every clock cycle, at 30 MHz per wire pair, for a total of 120 MHz. Because the 5B6B encoding scheme uses 6 bits to carry 5 bits of data, the net transmission rate is 100 MHz.

The Medium-Dependent Interface

The *medium-dependent interface (MDI)* is the actual hardware that provides access to the network medium, as realized in a network interface card or a hub.

Working with 100VG-AnyLAN

When compared to the success of Fast Ethernet products in the marketplace, 100VG-AnyLAN obviously has not been accepted as an industry standard, but a few networks still use it. The problem is not so much one of performance, because 100VG certainly rivals Fast Ethernet in that respect, but, instead, of marketing and support.

Despite using the same physical-layer specifications and frame formats, 100VG-AnyLAN is sufficiently different from Ethernet to cause hesitation on the part of network administrators who have invested large amounts of time and money in learning to support CSMA/CD networks. Deploying a new 100VG-AnyLAN would not be a wise business decision at this point, and even trying to preserve an existing investment in this technology is a doubtful course of action.

Mixing 100VG-AnyLAN and Fast Ethernet nodes on the same collision domain is impossible, but you can continue to use your existing 100VG segments and to add new Fast Ethernet systems, as long as you use a switch to create a separate collision domain. The most practical method for doing this is to install a modular switch into which you can plug transceivers supporting different data link–layer protocols.

Hewlett-Packard, for example (one of the original supporters of 100VG-AnyLAN), still has 100VG transceiver modules available for its AdvanceStack Switch 2000, which can also support standard, Fast, and Gigabit Ethernet segments, as well as FDDI and ATM. This arrangement provides your 100VG systems with full access to the rest of the network while enabling you to expand using any of these protocols, without making further investments in 100VG-AnyLAN technology.

It's probably only a matter of time before 100VG-AnyLAN is abandoned entirely and, at this time, you can think about replacing the transceiver and the NICs in your 100VG systems with Fast Ethernet or another technology. Switches of this type are an effective stopgap, but they are only a temporary solution to what has turned out to be an unsuccessful product.

NETWORK PROTOCOLS

Networking

Chapter 12

Token-Passing Protocols

Although the vast majority of local area networks use one of the Ethernet variants, other data link–layer protocols provide their own unique advantages. Chief among these advantages is the use of media access control mechanisms (MACs) other than CSMA/CD. Token Ring and Fiber Distributed Data Interface (FDDI) are both viable LAN protocols that approach the problem of sharing a network cable in a wholly different way.

Token Ring

Token Ring is the traditional alternative to the Ethernet protocol at the data-link layer. The supporters of Token Ring are stalwart and, while it is not likely ever to overtake Ethernet in popularity, it is far from being out of the race. Token Ring was originally developed by IBM and later standardized in the IEEE 802.5 document, so, like Ethernet, there are slightly divergent protocol standards. The products on the market today, however, usually conform to the IEEE specifications.

The biggest difference between Token Ring and Ethernet is the media access control mechanism. To transmit its data, a workstation must be the holder of the *token*, a special packet circulated to each node on the network in turn. Only the system in possession of the token can transmit, after which it passes the token to the next system. This eliminates all possibility of collisions in a properly functioning network, as well as the need for a collision-detection mechanism.

Token Ring is also faster than standard Ethernet. Although the technology originally ran at 4 Mbps, most implementations now run at 16 Mbps. A Fast Token Ring standard also exists that pushes the speed to 100 Mbps and provides full-duplex capability, making it the equal of Fast Ethernet in this respect, but this standard has not caught on in the marketplace the way that Fast Ethernet has.

The Token Ring Physical Layer

As the name implies, the nodes on a Token Ring network are connected in a ring topology. This is, in essence, a bus with the two ends connected to each other, so that systems can pass data to the next node on the network until it arrives back at its source. This is exactly how the protocol functions: the system that transmits a packet is also responsible for removing it from the network after it has traversed the ring.

This ring, however, is logical, not physical. That is, the network to all appearances takes the form of a star topology, with the workstations connected to a central hub called a *multistation access unit* (*MAU*, or sometimes *MSAU*). The *logical ring* (sometimes called a *collapsed ring*) is actually a function of the MAU, which accepts packets transmitted by one system and directs them out each successive port in turn, waiting for them to return over the same cable before proceeding to the next port (see Figure 12-1). In this arrangement, therefore, the transmit and receive circuits in each workstation are actually separate ports that just happen to use the same cable because the system always transmits data to the next downstream system and receives data from the next upstream.

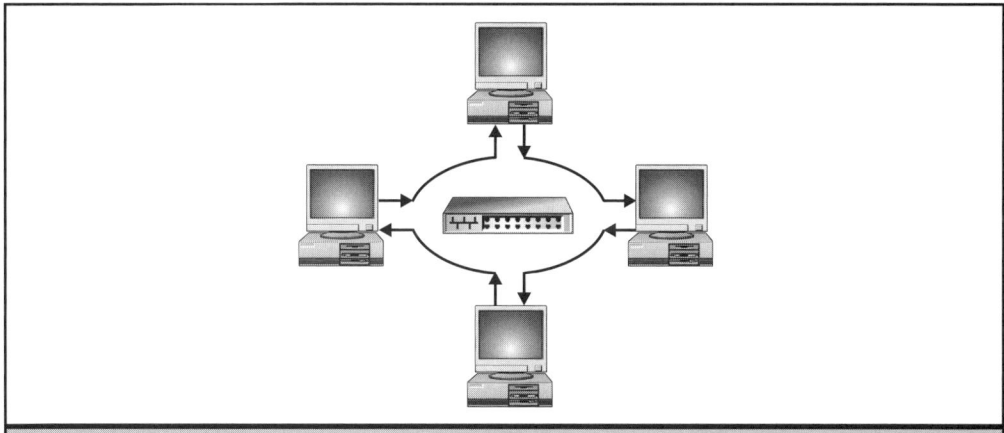

Figure 12-1:   *Token Ring networks appear to use a star topology, but data travels in the form of a ring.*

Cable Types

The original IBM Token Ring implementations use a proprietary cable system designed by IBM, which they refer to as their Type 1, or the IBM Cabling System (ICS). Type 1 is a 150-ohm shielded twisted-pair (STP) cable containing two wire pairs. The ports of a Type 1 MAU use proprietary connectors called IBM Data Connectors (IDCs) or Universal Data Connectors (UDCs), and the network interface cards use standard DB9 connectors. A cable with IDCs at each end, used to connect MAUs together, is called a *patch cable.* A cable with one IDC and one DB9, used to connect a workstation to the MAU, is called a *lobe cable*.

The other cabling system used on Token Ring networks, called Type 3 by IBM, uses standard unshielded twisted-pair (UTP) cable, with Category 5 recommended. Like Ethernet, Token Ring uses only two of the wire pairs in the cable, one pair to transmit data and one to receive it. Type 3 cable systems also use standard RJ-45 connectors for both the patch cables and the lobe cables. The signaling system used by Token Ring networks at the physical layer is different from that of Ethernet, however. Token Ring uses Differential Manchester signaling, while Ethernet uses Manchester.

 *For more information on the cables used for Token Ring networks and physical-layer signaling, see Chapter 4.*

Type 3 UTP cabling has largely supplanted Type 1 in the Token Ring world, mainly because it is much easier to install. Type 1 cable is thick and relatively inflexible when compared to Type 3, and the IDC connectors are large, making internal cable installations difficult. Type 1 cable can span longer distances than Type 3, however. Table 12-1 lists the general cabling guidelines for a Token Ring network.

	Type 1 Cable	Type 3 Cable
Maximum Lobe Cable Length	300 meters	150 meters
Maximum Number of Workstations	260	72
Maximum Ring Length @ 16 Mbps	160 meters	60 meters
Maximum Ring Length @ 4 Mbps	360 meters	150 meters
Maximum 8-Port MAUs	32	9

Table 12-1: *Token Ring Cabling Guidelines*

Note *The physical-layer standards for Token Ring networks are not as precisely specified as those for Ethernet. In fact, the IEEE 802.5 standard is quite a brief document that contains no physical-layer specifications at all. The cable types and wiring standards for Token Ring are derived from the practices used in products manufactured by IBM, the original developer and supporter of the Token Ring protocol. As a result, products made by other manufacturers may differ in their recommendations for physical-layer elements such as cable lengths and the maximum number of workstations allowed on a network.*

Token Ring NICs

The network interface cards for Token Ring systems are similar to Ethernet NICs in appearance. Most of the cards on the market today use RJ-45 connectors for UTP cable, although DB9 connectors are also available and the internal connectors support all of the major system buses, including PCI and ISA. Every Token Ring adapter has a VLSI (Very Large Scale Integration) chipset that consists of five separate CPUs, each of which has its own separate executable code, data storage area, and memory space. Each CPU corresponds to a particular state or function of the adapter. This complexity is one of the main reasons why Token Ring NICs are substantially more expensive than Ethernet NICs.

Token Ring MAUs

To maintain the ring topology, all of the MAUs on a Token Ring network must be interconnected using the Ring In and Ring Out ports intended for this purpose. Figure 12-2 illustrates how the MAUs themselves are cabled in a ring that is extended by the lobe cables connecting each of the workstations. It's also possible to build a Token Ring network using a *control access unit (CAU)*, which is essentially an intelligent MAU that supports a number of *lobe attachment modules (LAMs)*. To increase the number of workstations connected to a Token Ring network without adding a new MAU, you can use lobe access units (LAUs) that enable you to connect several workstations to a single lobe.

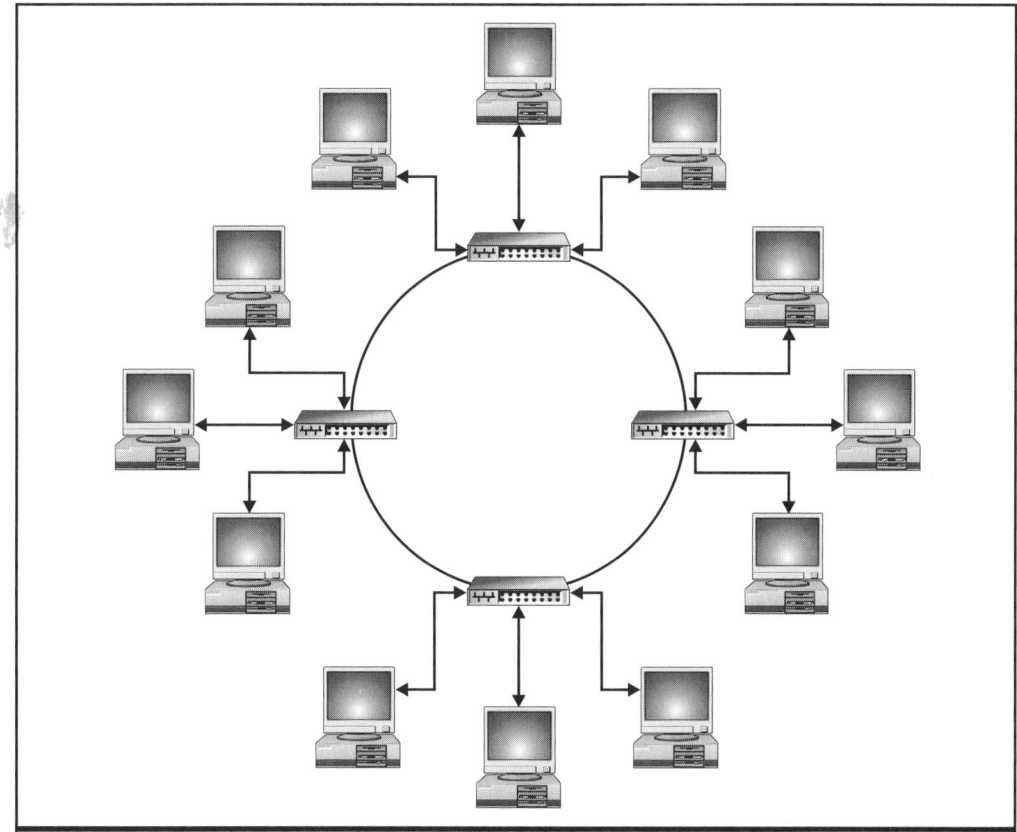

Figure 12-2: *The MAUs in a Token Ring network form the basic ring, which is extended with each workstation added to the network.*

Token Ring MAUs (not to be confused with an Ethernet hub, which is occasionally called a MAU, or medium access unit) are quite different from Ethernet hubs in several ways. First, the typical MAU is a *passive device,* meaning that it does not function as a

NETWORK PROTOCOLS

repeater. The cabling guidelines for Token Ring networks are based on the use of passive MAUs. There are repeating MAUs on the market, however, that enable you to extend the network cable lengths beyond the published standards.

Second, the ports on all MAUs remain in a loopback state until they are initialized by the workstation connected to them. In the *loopback state,* the MAU passes signals it receives from the previous port directly to the next port without sending them out over the lobe cable. When the workstation boots, it transmits what is known as a *phantom voltage* to the MAU. Phantom voltage does not carry data, it just informs the MAU of the presence of the workstation, causing the MAU to add it to the ring. On older Type 1 Token Ring networks, an administrator has to manually initialize each port in the MAU with a special "key" plug before attaching a lobe cable to it. This initialization is essential in Token Ring because of the network's reliance on each workstation to send each packet it receives from the MAU right back. The MAU can't send the packet to the next workstation until it receives it back from the previous one. If a MAU were to transmit a packet out through a port to a workstation that was turned off or nonexistent, the packet would never return, the ring would be broken, and the network would cease functioning. Because of the need for this initialization process, it is impossible to connect two Token Ring networks together without a MAU, as you can with Ethernet and a crossover cable.

Finally, MAUs always have two ports for connecting to the other MAUs in the network. Ethernet systems using a star topology connect their hubs in a hierarchical star configuration (also called a branching tree), in which one hub can be connected to several others, each of which, in turn, is connected to other hubs, as shown in Figure 12-3. Token Ring MAUs are always connected in a ring, with the Ring In port connected to the next upstream MAU and the Ring Out port to the next downstream MAU. Even if your network only has two MAUs, you must connect the Ring In port on each one to the Ring Out port on the other using two patch cables.

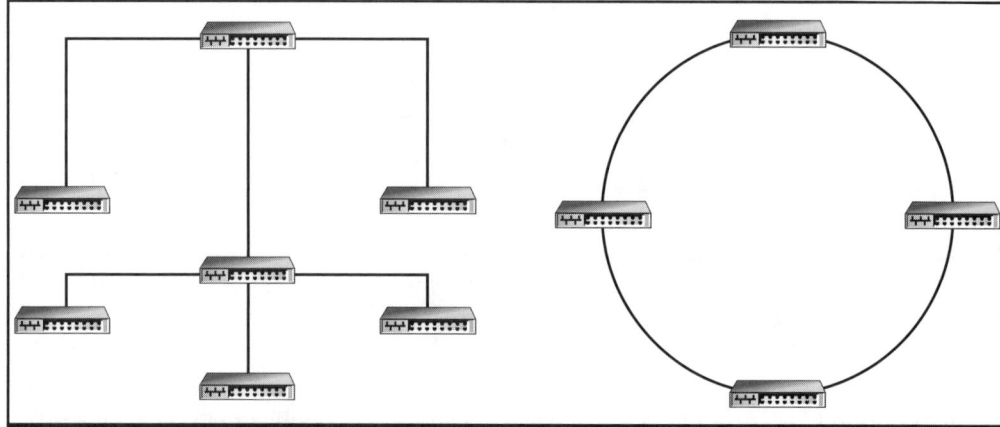

Figure 12-3:   *Ethernet hubs (at left) are connected using a branching tree arrangement, while Token Ring MAUs (at right) are connected in a ring.*

The connections between Token Ring MAUs are redundant. That is, if a cable or connector failure causes a break between two of the MAUs, the adjacent MAUs will transmit any data reaching them back in the other direction, so the packets will always reach all of the workstations connected to the network. The Token Ring standards use a specification called the *adjusted ring length (ARL)* to determine the total length of the data path in the event of this type of failure.

Calculating the ARL

To calculate the ARL for a network, you take the sum of all the patch cable lengths between wiring closets minus the length of the shortest patch cable connecting two wiring closets and make the following adjustments:

- Add 3 meters for every punchdown connection involved in the path between two MAUs.

- Add 30 meters for every surge protector used on the network.

- Add 16 meters for every 8-port MAU.

Because MAUs are often stored in wiring closets, the standard refers to the number of wiring closets used on the network to refer to MAUs more than 3 meters apart. Whether the MAUs are physically located in different closets is not relevant; any two MAUs connected by a cable more than 3 meters long are said to be in different wiring closets. Patch cables shorter than 3 meters should not be included in the ARL calculations.

Thus, the ARL is essentially the longest possible path (also called the *worst case distance*) between two MAUs on the network. The total ARL multiplied by 2 should not exceed 366 meters for the network to be within specifications. The ARL plus twice the length of the longest lobe cable is known as the *worst case maximum adapter signal drive distance* for the network.

 *All of the ring lengths discussed in reference to Token Ring networks refer to passive MAU networks. Unlike an Ethernet hub, a Token Ring MAU does not usually function as a repeater. When you use active MAUs that include signal-repeating capabilities, the cables can be much longer, depending on the capabilities of the individual MAU.*

Calculating the Maximum Main Ring Distance

The *main ring distance* is the total length of all the MAU-MAU connections on the network. The maximum main ring distance is determined by the number of MAUs on the network and the number of wiring closets used for the MAUs. Tables 12-2 through 12-5 list the maximum main ring distances for networks using Type 1 and Type 3 cable, running at 4 Mbps and 16 Mbps.

	1 WC	2 WC	3 WC	4 WC	5 WC
1 MAU	427/130				
2 MAU	416/216	410/124			
3 MAU	406/123	399/121	402/122		
4 MAU	396/120	389/118	382/116	375/114	
5 MAU	385/117	378/115	371/113	364/110	357/108
6 MAU	375/114	368/112	361/110	354/107	347/105
7 MAU	364/110	357/108	350/106	343/104	336/102
8 MAU	364/110	347/105	340/103	333/101	326/99

Table 12-2:   *Maximum Main Ring Distances in Feet and Meters for Type 1 Cable (16 Mbps)*

	1 WC	2 WC	3 WC	4 WC	5 WC
1 MAU	1220/371				
2 MAU	1190/362	1170/356			
3 MAU	1160/353	1140/347	1120/341		
4 MAU	1130/344	1110/338	1090/332	1070/326	
5 MAU	1100/335	1080/329	1060/323	1040/316	1010/310
6 MAU	1070/326	1050/320	1030/313	1010/307	990/301
7 MAU	1040/316	1020/310	1000/304	980/298	960/292
8 MAU	1010/307	990/301	970/295	950/289	930/283

Table 12-3:   *Maximum Main Ring Distances in Feet and Meters for Type 1 Cable (4 Mbps)*

	1 WC	2 WC	3 WC	4 WC	5 WC
1 MAU	210/64				
2 MAU	205/62	202/61			
3 MAU	200/61	197/60	195/59		
4 MAU	195/59	193/58	190/57	168/56	
5 MAU	190/58	188/57	185/56	164/55	179/54
6 MAU	186/56	183/55	180/54	160/54	174/52
7 MAU	181/55	178/54	175/53	155/52	169/51
8 MAU	176/53	173/52	170/51	151/50	165/50

Table 12-4: *Maximum Main Ring Distances in Feet and Meters for Type 3 Cable (16 Mbps)*

	1 WC	2 WC	3 WC	4 WC	5 WC
1 MAU	500/150				
2 MAU	485/147	475/144			
3 MAU	470/143	460/140	450/137		
4 MAU	455/138	445/135	435/132	425/129	
5 MAU	440/134	430/131	420/128	410/124	400/121
6 MAU	425/129	415/126	405/123	395/120	385/117
7 MAU	410/124	400/121	390/118	380/115	370/112
8 MAU	395/120	385/117	375/114	365/111	355/108

Table 12-5: *Maximum Main Ring Distances in Feet and Meters for Type 3 Cable (4 Mbps)*

Token Passing

Access to the network medium on a Token Ring network is arbitrated through the use of a 3-byte packet known as the *token*. When the network is idle, the workstations are said to be in *bit repeat mode*, awaiting an incoming transmission. The token circulates continuously around the ring, from node to node, until it reaches a workstation that has data to transmit. To transmit its data, the workstation modifies a single *monitor setting bit* in the token to reflect that the network is busy and sends it to the next workstation, followed immediately by its data packet.

The packet also circulates around the ring. Each node reads the destination address in the packet's frame header and either writes the packet to its memory buffers for processing before transmitting it to the next node or just transmits it without processing. (Compare this with Ethernet systems that simply discard packets that are not addressed to them.) In this way, the packet reaches every node on the network until it arrives back at the workstation that originally sent it.

On receipt of the packet after it has traversed the ring, the sending node compares the incoming data with the data it originally transmitted, to see if any errors have occurred during transmission. If errors have occurred, the computer retransmits the packet. If no errors have occurred, the computer removes the packet from the network and discards it, and then changes the monitor setting bit back to its free state and transmits it. The process is then repeated with each system having an equal chance to transmit.

Although it was not part of the original standard, most 16 Mbps Token Ring systems today include a feature called *early token release (ETR)*, which enables the transmitting system to send the "free" token immediately after the data packet (instead of the "busy" token before the data packet), without waiting for the data to traverse the network. This way, the next node on the network can receive the data packet, capture the free token, and transmit its own data packet, followed by another free token. This enables multiple data packets to exist on the network simultaneously, but there is still only one token. Early token release eliminates some of the latency delays on the network that occur while systems wait for the free token to arrive.

Note *Early token release is only possible on 16 Mbps Token Ring networks. Systems that use ETR can coexist on the same network with systems that do not.*

Because only the computer holding the token can transmit data, Token Ring networks do not experience collisions unless a serious malfunction occurs. This means that the network can operate up to its full capacity with no degradation of performance, as can happen in an Ethernet network. The token-passing system is also deterministic, which means that it can calculate the maximum amount of time that will elapse before a particular node can transmit.

Token Ring is not the only data link–layer protocol that uses token passing for its media access control method. FDDI uses token passing, as well as obsolete protocols like ARCnet. It isn't even necessary to use token passing in a ring topology. The IEEE

802.4 standard defines the specifications for a token-passing network that uses a bus topology, although it is rarely, if ever, used today.

System Insertion

Before it can join the ring, a workstation must complete a five-step insertion procedure that verifies the system's capability to function on the network (see Figure 12-4). The five steps are as follows:

1. **Media lobe check**   The media lobe check tests the network adapter's capability to transmit and receive data, and the cable's capability to carry the data to the MAU. With the MAU looping the incoming signal for the system right back out through the same cable, the workstation transmits a series of MAC Lobe Media Test frames to the broadcast address, with the system's own address as the source. Then the system transmits a MAC Duplication Address Test frame with its own address as both the source and the destination. To proceed to the next step, the system must successfully transmit 2,047 MAC Lobe Media Test frames and one MAC Duplication Address Test frame. The testing sequence can only be repeated two times before the adapter is considered to have failed.

2. **Physical insertion**   During the physical insertion process, the workstation sends a phantom voltage (a low-voltage DC signal invisible to any data signals on the cable) up the lobe cable to the MAU to trigger the relay that causes the MAU to add the system into the ring. After doing this, the workstation waits for a sign that an active monitor is present on the network, in the form of either an Active Monitor Present (AMP), Standby Monitor Present (SMP), or Ring Purge frame. If the system does not receive one of these frames within 18 seconds, it initiates a monitor contention process. If the contention process does not complete within one second, or if the workstation becomes the active monitor (see "Token Ring Monitors," later in this chapter) and initiates a ring purge that does not complete within one second, or if the workstation receives a MAC Beacon or Remove Station frame, the connection to the MAU fails to open and the insertion is unsuccessful.

3. **Address verification**   The address verification procedure checks to see if another workstation on the ring has the same address. Because Token Ring supports locally administered addresses (LAAs), it is possible for this to occur. The system generates a series of MAC Duplication Address Test frames like those in step 1, except that these are propagated over the entire network. If no other system is using the same address, the test frames should come back with their ARI and FCI bits set to 0, at which time the system proceeds to the next step. If the system receives two test frames with the ARI and FCI bits set to 1, or if the test frames do not return within 18 seconds, the insertion fails and the workstation is removed from the ring.

4. **Ring poll participation**   The system must successfully participate in a ring poll by receiving an AMP or SMP frame with the ARI and FCI bits set to 0, changing those bits to 1, and transmitting its own SMP frame. If the workstation does not receive an AMP or SMP frame within 18 seconds, the insertion fails and the workstation is removed from the ring.

5. **Request initialization**   The workstation transmits four MAC Request Initialization frames to the functional address of the network's ring parameter server (C0 00 00 00 00 02). If the system receives the frames back with the ARI and FCI bits set to 0, indicating that there is no functioning ring parameter server, the system's network adapter uses its default values and the initialization (as well as the entire system insertion) is deemed successful. If the system receives one of its frames back with the ARI and FCI bits set to 1 (indicating that a ring parameter server has received the frame), it waits two seconds for a response. If there is no response, the system retries up to four times, after which the initialization fails and the workstation is removed from the ring.

System States

During its normal functions, a Token Ring system enters three different operational states, which are as follows:

Repeat   While in the repeat state, the workstation transmits all the data arriving at the workstation through the receive port to the next downstream node. When the workstation has a packet of its own queued for transmission, it modifies the token bit in the frame's access control byte to a value of 1 and enters the transmit state. At the same time, the *token holding timer (THT)* that allows the system 8.9 ms of transmission time is reset to zero.

Transmit   Once in the transmit state, the workstation transmits a single frame onto the network and releases the token. After successfully transmitting the frame, the workstation transmits *idle fill* (a sequence of 1s) until it returns to the repeat state. If the system receives a Beacon, Ring Purge, or Claim Token MAC frame while it is transmitting, it interrupts the transmission and sends an Abort Delimiter frame to clear the ring.

Stripping   At the same time that a workstation's transmit port is in the transmit state, its receive port is in the stripping state. As the transmitted data returns to the workstation after traversing the ring, the system strips it from the network, so that it will not circulate endlessly. Once the system detects the end delimiter field on the receive port, it knows that the frame has been completely stripped and returns to the repeat state. If the 8.9 ms THT expires before the end delimiter arrives, the system records a *lost frame error* for later transmission in a *Soft Error Report frame* before returning to the repeat state.

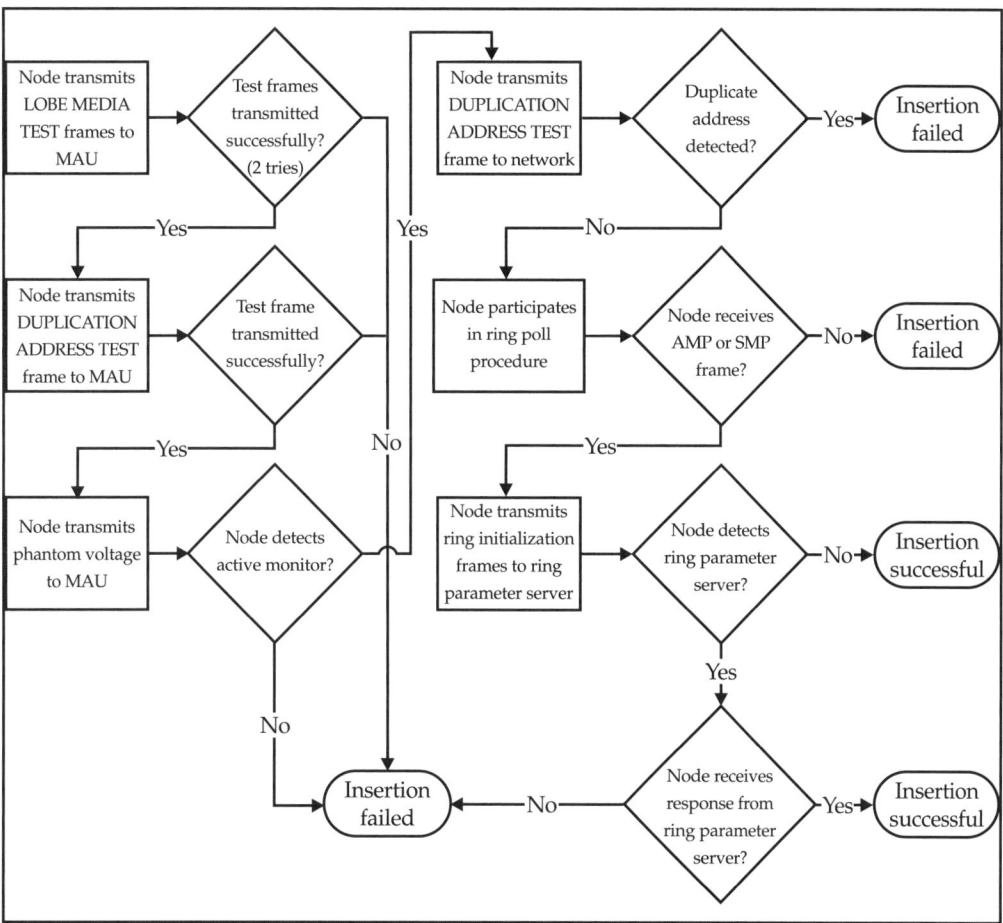

Figure 12-4: *The Token Ring workstation insertion procedure ensures that each system connected to the ring is functioning properly.*

Token Ring Monitors

Every Token Ring network has a system that functions as the *active monitor* that is responsible for ensuring the proper performance of the network. The active monitor does not have any special programming or hardware, it is simply elected to the role by a process called *monitor contention*. All of the other systems on the network then function as *standby monitors*, should the computer functioning as the active monitor fail. The functions of the active monitor are as follows:

Transmit Active Monitor Present frames   Every seven seconds, the active monitor (AM) transmits an Active Monitor Present MAC frame that initiates the ring polling process.

Monitor ring polling   The AM must receive either an Active Monitor Present or Standby Monitor Present frame from the node immediately upstream of it within seven seconds of initiating a ring polling procedure. If the required frame does not arrive, the AM records a ring polling error.

Provide master clocking   The AM generates a master clock signal that the other workstations on the network use to synchronize their clocks. This ensures that all the systems on the network know when each transmitted bit begins and ends. This also reduces network *jitter*, the small amount of phase shift that tends to occur on the network as the nodes repeat the transmitted data.

Provide a latency buffer   In the case of a small ring, it is possible for a workstation to begin transmitting a token and to receive the first bits back on its receive port before it has finished transmitting. The AM prevents this by introducing a propagation delay of at least 24 bits (called a *latency buffer*), which ensures that the token circulates around the network properly.

Monitor the token-passing process   The active monitor must receive a good token every 10 milliseconds, which ensures that the token-passing mechanism is functioning properly. If a workstation raises the token priority and fails to lower it, or fails to completely strip its packet from the ring, the AM detects the problem and remedies it by purging the ring and generating a new token. Every node, on receiving a Ring Purge MAC frame from the AM, stops what it's doing, resets its timers, and enters bit repeat mode, in preparation for the receipt of a new packet.

Ring Polling

Ring polling is the process by which each node on a Token Ring network identifies its nearest active upstream neighbor (NAUN). The workstations use this information during the beaconing process to isolate the location of a network fault.

The ring-polling process is initiated by the active monitor when it transmits an Active Monitor Present (AMP) MAC frame. This frame contains an Address Recognized (ARI) bit and a Frame Copied (FCI) bit, both of which have a value of 0. The first system downstream of the AM receives the frame and changes the ARI and FCI bits to 1. The receiving system also records the address of the sending system as its NAUN. This is because the first station that receives an AMP frame always changes the values of those two bits. Therefore, the system receiving a frame with zero-valued ARI and FCI bits knows the sender is its nearest active upstream neighbor.

After recording the address of its NAUN, the system then generates a MAC frame of the same type, except it is a Standby Monitor Present (SMP) frame instead of an Active Monitor Present. The system queues the SMP frame for transmission with a delay of 20 milliseconds, to give other systems a chance to send data. Without the delay, the ring would likely be clogged with ring-polling traffic, preventing the timely transmission of data packets.

After queuing the SMP frame, the system repeats the original AMP frame to the next system downstream. Because the ARI and FCI bits now have a value of 1, no further action is taken by any of the downstream systems except to pass the frame around the ring until it returns back to the active monitor, which strips it from the network.

When the 20-millisecond delay expires, the second system transmits the SMP packet and the entire process repeats with the next system downstream. Eventually, each system on the network will generate an SMP or AMP frame that identifies it as the NAUN of the next system downstream. When the active monitor receives an SMP packet with 0 values for the ARI and FCI bits, it knows the polling process has been completed.

The entire operation must take no more than seven seconds or the AM records a ring-poll error before initiating the whole process again. If any system on the network fails to receive an AMP packet for a 15-second interval, it assumes that the active monitor is not functioning properly and initiates the contention process that elects a new AM.

Beaconing

When a station on a Token Ring network fails to detect a signal on its receive port, it assumes that there is a fault in the network and initiates a process called *beaconing*. The system broadcasts MAC beacon frames to the entire network every 20 milliseconds (without capturing a token) until the receive signal commences again. Each station transmitting beacon frames is saying, in essence, that a problem exists with its nearest active upstream neighbor because it is not receiving a signal. If the NAUN begins beaconing also, this indicates that the problem lies farther upstream. By noting which stations on the network are beaconing, it is possible to isolate the malfunctioning system or cable segment. There are four types of MAC beacon frames, as follows:

Set Recovery Mode (Priority 1)   The Set Recovery Mode frame is rarely seen because it is not transmitted by a workstation's Token Ring adapter. This frame is used only during a recovery process initiated by an attached network management product.

Signal Loss (Priority 2)   The Signal Loss frame is generated when a monitor contention process fails due to a timeout and the system enters the contention transmit mode due to a failure to receive any signal from the active monitor. The presence of this frame on the network usually indicates that a cable break or a hardware failure has occurred.

Streaming Signal, Not Claim Token (Priority 3)   The Streaming Signal, Not Claim Token frame is generated when a monitor contention process fails due to a timeout and the system has received no MAC Claim Token frames during the contention period. The system has received a clock signal from the active monitor, however, or the Signal Loss frame would have been generated instead.

Streaming Signal, Claim Token (Priority 4)   The Streaming Signal, Claim Token frame is generated when a monitor contention process fails due to a timeout and the system has received MAC Claim Token frames during the contention period. This frame is usually an indication of a transient problem caused by a cable that is too long or by signal interference caused by environmental noise.

When a system suspects that it may be the cause of the network problem resulting in beaconing, it removes itself from the ring to see if the problem disappears. If the system transmits beacon frames for more than 26 seconds, it performs a *beacon transmit auto-removal test*. If the system receives 8 consecutive beacon frames that name it as the NAUN of a beaconing system downstream, it performs a *beacon receive auto-removal test*.

You can learn more about a Token Ring network problem by analyzing the packets transmitted during a beaconing event. For more information on protocol analysis, see Chapter 31.

Token Ring Frames

Four different types of frames are used on Token Ring networks, unlike Ethernet networks, which have one single frame format. The *data frame* type is the only one that actually carries the data generated by upper-layer protocols, while the *command frame* type performs ring maintenance and control procedures. The *token frame* type is a separate construction used only to arbitrate media access, and the *abort delimiter frame* type is only used when certain types of errors occur.

The Data Frame

Token Ring data frames carry the information generated by upper-layer protocols in a standard logical link control (LLC) protocol data unit (PDU), as defined in the IEEE 802.2 document. The fields that make up the frame format are shown in Figure 12-5; their functions are as follows:

Start Delimiter (SD), 1 byte   The start delimiter signals the beginning of the frame by deliberately violating the rules of the Differential Manchester encoding system. The bit pattern used is JK0JK000, where the *J*s are encoding violations of the value 0 and the *K*s are encoding violations of the value 1.

Access Control (AC), 1 byte The access control byte uses the bit pattern PPPTMRRR, where the *P*s are three *priority bits* and the *R*s are three *reservation bits* used to prioritize the data transmitted on Token Ring networks. Token Ring workstations can have priority levels from 0 to 7, where 7 is the highest priority. A system can only capture a free token and transmit data if that token has a priority lower than that of the workstation. When a node has a higher priority than that of the free token, it can raise the priority of the token by modifying the priority bits in order to transmit further packets more quickly. When the token returns to the system that raised its priority, it can transmit another packet at the same priority or return the token to its previous priority and change it to the "free" state. A system denied the token because its priority is too low can modify the reservation bits to request a token with a lower priority. Of the remaining two bits in the AC field, the *T* represents the *token bit,* the value of which indicates whether the frame is a data/command (1) or a token (0) frame. The *M* represents the *monitor bit,* which is changed from 0 to 1 by the system on the network designated as the active monitor. Because the active monitor is the only system able to change the value of this bit, the assumption is, if the active monitor ever receives a packet with a value of 1 there, the packet has, for some reason, not been removed from the network by the transmitting node and it is incorrectly traversing the ring a second time.

Frame Control (FC), 1 byte The frame control byte uses the bit pattern TT00AAAA, where the *T*s specify whether the packet contains a data or a command frame. The third and fourth bits are unused and always have a value of 0. The *A*s represent *attention code bits* that identify a specific type of MAC frame that should be written immediately to the receiving system's express buffer.

Destination Address (DA), 6 bytes The destination address field identifies the intended recipient of the packet, using either the hardware address coded into the network interface card or a broadcast or multicast address.

Source Address (SA), 6 bytes The source address field address identifies the sender of the packet using the hardware address coded into the network interface card.

Information (INFO), variable In a data frame, the information field contains the *protocol data unit* passed down from a network-layer protocol, plus a standard LLC header consisting of DSAP, SSAP, and control fields. The size of the information field can be up to 4,500 bytes and is limited by the *ring token holding time*—that is, the maximum length of time a workstation can hold on to the token.

Frame Check Sequence (FCS), 4 bytes The frame check sequence field contains the 4-byte result of the CRC computation calculated from the frame control, destination address, source address, and information fields, for the purpose of verifying the successful transmission of the packet. The CRC value is calculated by the sending node and stored in the FCS field. At the destination, the same calculation is performed again and compared to the stored results. A match indicates a successful transmission.

End Delimiter (ED), 1 byte   The end delimiter field indicates the end of the packet by again violating the Differential Manchester signaling rules. The bit pattern is JK1JK1IE, where the *J*s and *K*s are encoding violations of 1s and 0s, respectively (as in the start delimiter field). The *I* is an *intermediate frame bit* that has a value of 1 if more packets in the current sequence are waiting to be transmitted. The *E* is an *error detection bit,* which a receiving system sets to 1 if it detects a CRC error in the transmission. This prevents the systems downstream from having to report the same error.

Frame Status (FS), 1 byte   The frame status field uses the bit pattern AF00AF00, in which the *A* is the Address Recognized Indicator (ARI) and the *F* is the Frame Copied Indicator (FRI). The values for these bits are repeated because the frame status field is not included in the frame check sequence's CRC check. The ARI and FCI are both set to 0 by the sending workstation. If the receiving node recognizes the frame, it sets the ARI value to 1. If the receiving node can copy the frame to the adapter's buffer memory, it sets the FCI value to 1. Failure to modify the FCI bits is an indication the packet has failed the CRC check or has been damaged in some other way and must be retransmitted.

The Token Ring specification lists a number of functional addresses that define specialized roles fulfilled by certain systems on the network. Using these addresses, it is possible for a node to send messages directly to the system performing a specific function, without having to know the machine's hardware address. The predefined addresses are as follows:

Active monitor	C0 00 00 00 00 01
Ring parameter server	C0 00 00 00 00 02
Ring error monitor	C0 00 00 00 00 08
Configuration report server	C0 00 00 00 00 10
Source route bridge	C0 00 00 00 01 00

The Command Frame

Command frames, also called *MAC frames,* differ from data frames only in the information field and sometimes the frame control field. MAC frames do not use an LLC header; instead, they contain a PDU consisting of two bytes that indicate the length of the control information to follow, a 2-byte major vector ID that specifies the control function of the frame, and a variable number of bytes containing the control information itself (see Figure 12-6).

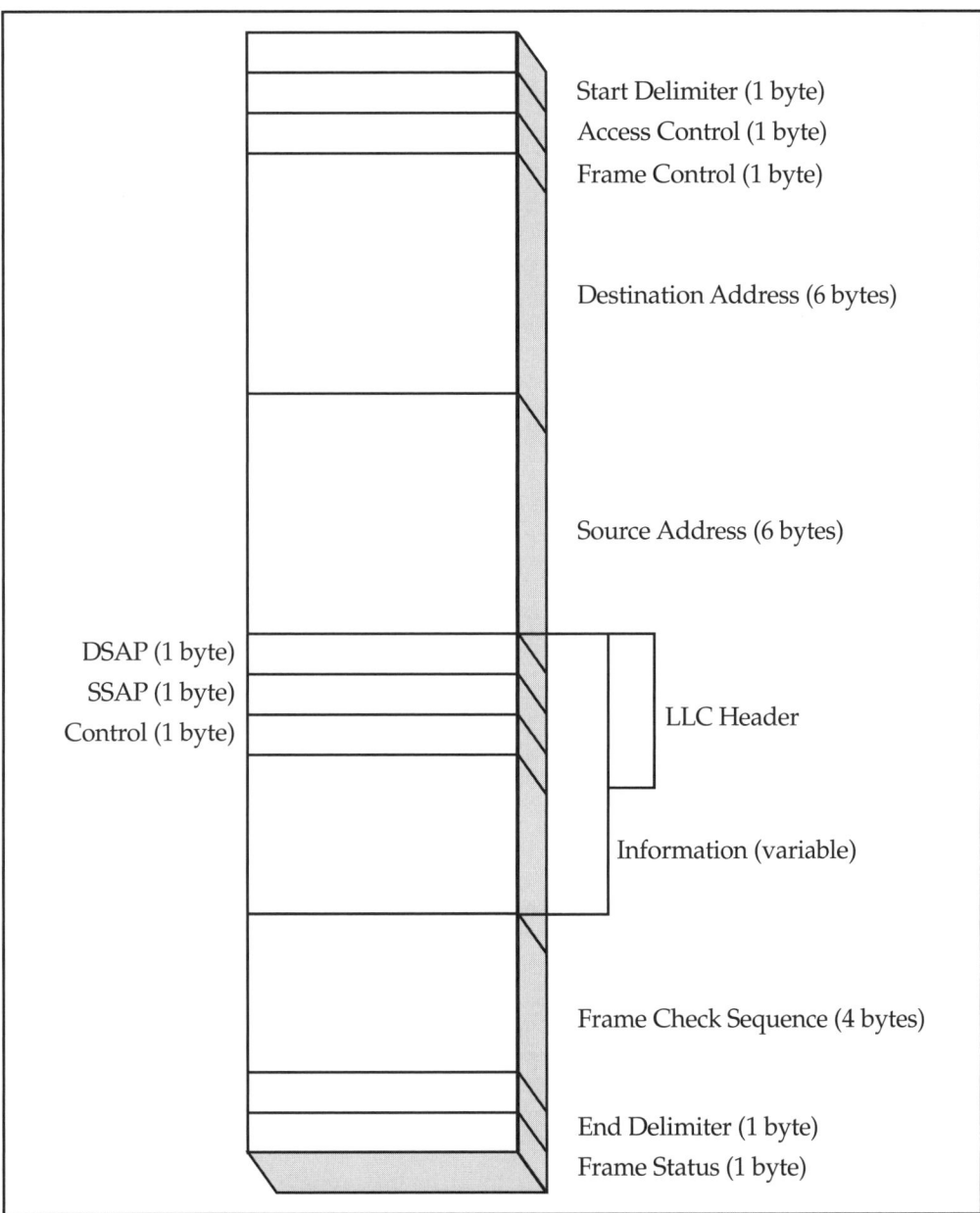

- Start Delimiter (1 byte)
- Access Control (1 byte)
- Frame Control (1 byte)
- Destination Address (6 bytes)
- Source Address (6 bytes)
- DSAP (1 byte)
- SSAP (1 byte)
- Control (1 byte)
- LLC Header
- Information (variable)
- Frame Check Sequence (4 bytes)
- End Delimiter (1 byte)
- Frame Status (1 byte)

Figure 12-5: *The Token Ring data frame carries information generated by upper-layer protocols.*

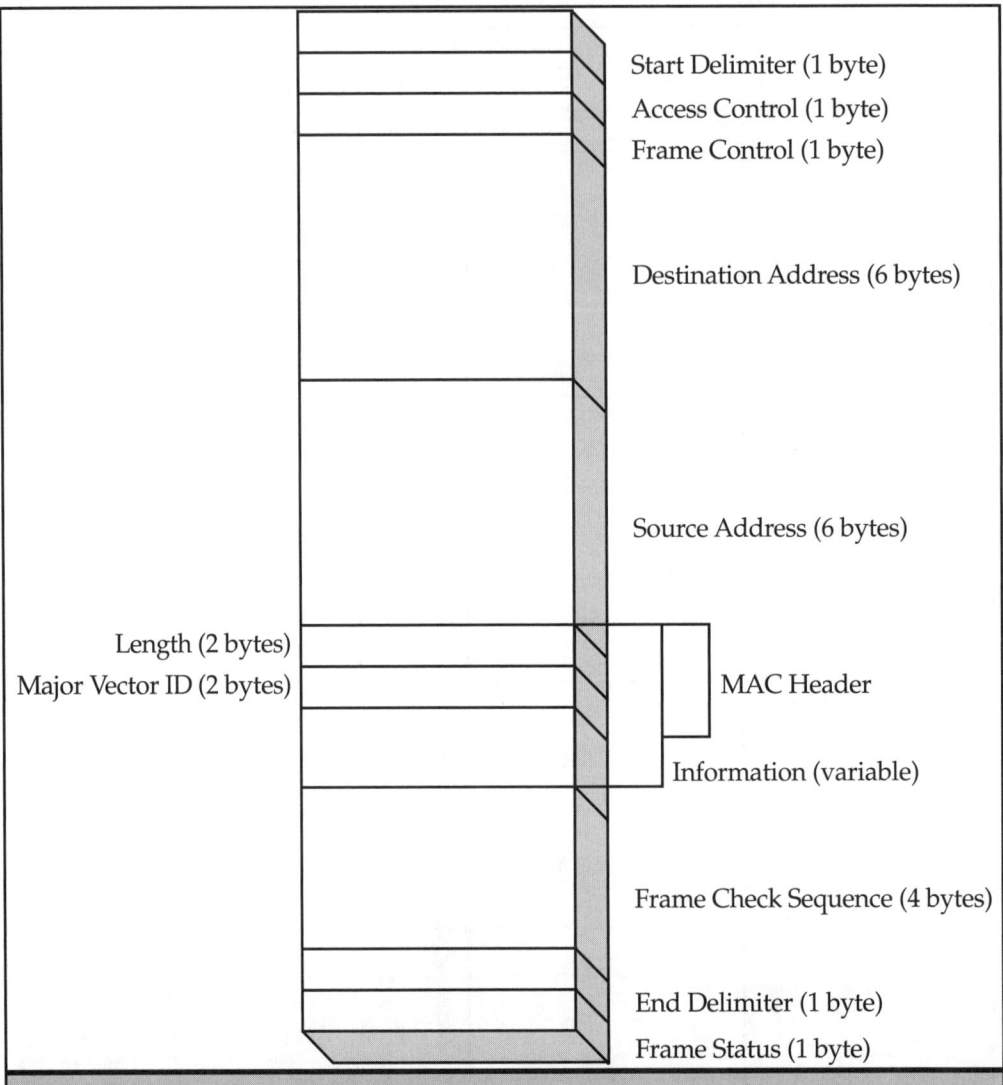

Start Delimiter (1 byte)

Access Control (1 byte)

Frame Control (1 byte)

Destination Address (6 bytes)

Source Address (6 bytes)

Length (2 bytes)

Major Vector ID (2 bytes)

MAC Header

Information (variable)

Frame Check Sequence (4 bytes)

End Delimiter (1 byte)

Frame Status (1 byte)

Figure 12-6: *The Token Ring command frame carries control messages used to perform ring maintenance procedures.*

MAC frames perform ring maintenance and control functions only. They never carry upper-layer data and they are never propagated to other collision domains by bridges, switches, or routers. Some of the most common functions are identified using only a 4-bit code in the frame control field, such as the following:

0010	Beacon
0011	Claim Token
0100	Ring Purge
0101	Active Monitor Present
0110	Standby Monitor Present

Some MAC frames with particular functions are processed by network adapters using a special memory area called the *express buffer*. This enables the node to process MAC frames containing important control commands at any time, even when it is busy receiving a large number of data frames.

The Token Frame

The *token frame* is extremely simple, consisting of only three 1-byte fields: the start delimiter, access control, and end delimiter fields (see Figure 12-7). The token bit in the access control field is always set to a value of one and the delimiter fields take the same form as in the data and command frames.

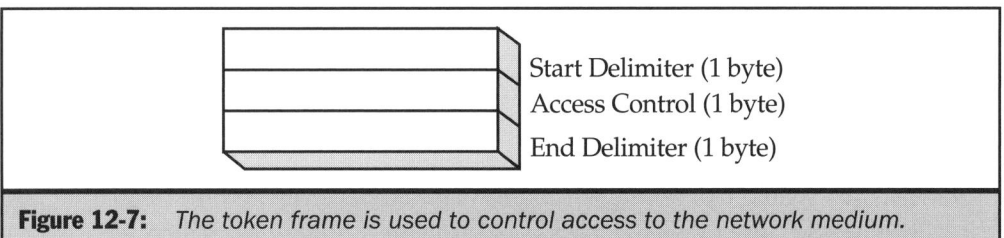

Start Delimiter (1 byte)
Access Control (1 byte)
End Delimiter (1 byte)

Figure 12-7:   *The token frame is used to control access to the network medium.*

The Abort Delimiter Frame

The *abort delimiter frame* consists only of the start delimiter and the end delimiter fields, using the same format as the equivalent fields in the data and command frames (see Figure 12-8). This frame type is used primarily when an unusual event occurs, such as when the transmission of a packet is interrupted and ends prematurely. When this happens, the active monitor transmits an abort delimiter frame that flushes out the ring, removing all the improperly transmitted data and preparing it for the next transmission.

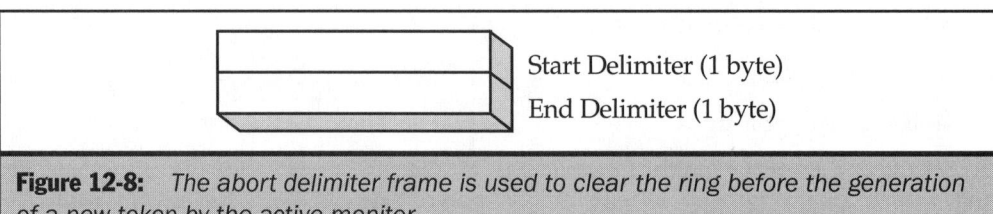

Start Delimiter (1 byte)

End Delimiter (1 byte)

Figure 12-8: *The abort delimiter frame is used to clear the ring before the generation of a new token by the active monitor.*

Token Ring Errors

The IEEE 802.5 standard defines a number of soft error types that systems on the network can report to the workstation functioning as the *ring error monitor* using MAC frames. When a Token Ring adapter detects a soft error, it begins a two-second countdown, during which it waits to see if other errors occur. After the two seconds, the system sends a soft error report message to the address of the ring error monitor (C0 00 00 00 00 08). The types of soft errors detectable by Token Ring systems are as follows:

Burst error   A burst error occurs when a system detects five half-bit times (that is, three transmitted bits) that lack the clock transition in the middle of the bit called for by the Differential Manchester encoding system. This type of error is typically caused by noise on the cable resulting from faulty hardware or some other environmental influence.

Line error   A line error occurs when a workstation receives a frame that has an error detection bit (in the end delimiter field) with a value of 1, either because of a CRC error in the frame check sequence or because a bit violating the Differential Manchester encoding system was detected in any fields other than the start delimiter and end delimiter. A network with noise problems will typically have one line error for every ten burst errors.

Lost frame error   A lost frame error occurs when a system transmits a frame and fails to receive it back within the 4 milliseconds allotted by the *return to repeat timer* (RRT). This error can be caused by excessive noise on the network.

Token error   A token error occurs when the active monitor's 10 millisecond *valid transmission timer (VTX)* expires without the receipt of a frame, and the AM must generate a new token. This error can be caused by excessive noise on the network.

Internal error   An internal error occurs when a system detects a parity error during direct memory access (DMA) between the network adapter and the computer. The problem could be with the adapter's own memory or with the computer's memory. If you install the adapter in another system and the error repeats, the problem is with the card itself.

Frequency error   A frequency error occurs when a standby monitor system receives a signal that differs from the expected frequency by more than a given amount. This error may mean that the active monitor is not supplying a proper clock signal. Shut down the current active monitor to force a new contention process. If no more frequency errors occur, then the network adapter on the original active monitor system is malfunctioning.

AC error   An AC error occurs when a system receives two consecutive ring-polling frames with ARI and FCI bits set to 0, in which the first frame is an AMP or an SMP, and the second frame is an SMP. Because the nearest downstream neighbor to the system transmitting an AMP or SMP frame should modify those bits, no system should ever receive two unmodified frames in this order. This error means that the system immediately upstream from the computer experiencing the error is failing to modify the ARI and FCI bits properly, probably due to a malfunctioning network adapter.

FC error   An FC (Frame Copied) error occurs when a system receives a unicast MAC frame with the ARI bit set to 1, indicating either a noise problem or a duplicate address on the network.

Abort delimiter transmitted error   An abort delimiter transmitted error occurs whenever a network condition causes a workstation to stop transmitting in the middle of a frame and to generate an abort delimiter frame. This occurs when the transmitting system receives a token with an invalid end delimiter or receives a claim token, beacon, or ring purge frame when it is expecting the start delimiter of its own transmitted frame.

Receive congestion error   A receive congestion error occurs when a system receives a unicast frame, but has no available buffer space to store the packet because it's being overwhelmed by incoming frames.

Note

The AC error is called an isolating error, because it points to a specific machine as the source of the problem. An error that does not indicate a specific source is called a nonisolating error.

FDDI

Appearing first in the late 1980s and defined in standards developed by the ANSI (American National Standards Institute) X3T9.5 committee, *Fiber Distributed Data Interface (FDDI*, pronounced *"fiddy")* was the first 100 Mbps data link–layer protocol to achieve popular use. In 1995, the committee became known as X3T12. The FDDI standards have also been approved by the International Organization for Standardization (ISO).

At the time of FDDI's introduction, 10 Mbps Thick and Thin Ethernet were the dominant LAN technologies, and FDDI represented a major step forward in speed. In addition, the use of fiber-optic cable provided dramatic increases in packet size, network segment length, and the number of workstations supported. FDDI packets can carry up to 4,500 bytes of data (compared to 1,500 for Ethernet) and, under certain conditions, a network can consist of up to 100 kilometers of cable, supporting up to 500 workstations. These improvements, in combination with fiber optics' complete resistance to the effects of electromagnetic interference, made it an excellent protocol for connecting distant workstations and networks, even those in different buildings. As a result, FDDI became known primarily as a backbone protocol, a role for which it is admirably suited.

Although it can run to the desktop, few networks use fiber-optic cable for this purpose because of its high hardware, installation, and maintenance costs. To address this problem, a standard was developed for running the same protocol over copper cable, called TP-PMD (Twisted Pair – Physical Media Dependent) or Copper Distributed Data Interface (CDDI), but this never achieved widespread acceptance.

Note *For more information on fiber-optic cable and its properties, see Chapter 4.*

Because of its common use as a backbone protocol, products like bridges and routers that connect Ethernet networks to FDDI backbones are common. FDDI is completely different from Ethernet, and the two network types can only be connected using a device like a router or a translation bridge that is designed to provide an interface between different networks. Today, the widespread acceptance of Fast Ethernet, which can run over the same fiber-optic cable, has resulted in a decline in the popularity of FDDI. A Fast Ethernet fiber network provides the same speed and comparable segment lengths and does not introduce a completely new frame format and media access control method onto the network. Therefore, fiber-optic Fast Ethernet segments can be joined to copper segments using relatively inexpensive hubs or switches, instead of routers.

FDDI Topology

FDDI is a token-passing protocol like Token Ring that uses either a double-ring or a star topology. Unlike Token Ring, in which the ring topology is logical and not physical, the original FDDI specification called for the systems to actually be cabled in a ring topology. In this case, it is a double ring, however. The *double ring,* also called a *trunk ring,* consists of two separate rings, a primary and a secondary, with traffic running in opposite directions to provide fault tolerance. The circumference of the double ring can be up to 100 km and workstations can be up to 2 km apart.

Workstations connected to both rings are called *dual attachment stations (DASs).* If a cable should break or a workstation malfunction, traffic is diverted to the secondary ring, running in the opposite direction, enabling it to access any other system on the network using the secondary path. A FDDI network operating in this state is called a *wrapped ring.* A properly functioning FDDI dual-ring network and a wrapped ring are depicted in Figure 12-9.

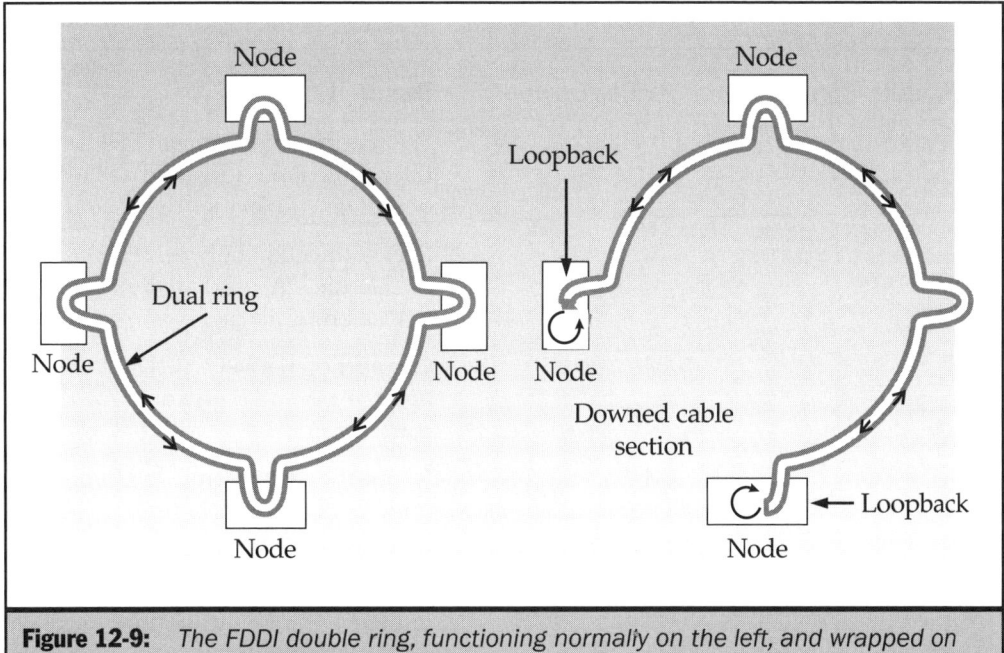

Figure 12-9: *The FDDI double ring, functioning normally on the left, and wrapped on the right*

If a second cable break should occur, the network is then divided into two separate rings, and network communications are interrupted. A wrapped ring is inherently less efficient than the fully functional double ring because of the additional distance that the traffic must travel and is, therefore, only meant to be a temporary measure until the fault is repaired.

FDDI can also use a star topology in which workstations are attached to a hub, called a *dual attachment concentrator (DAC).* The hub can either stand alone or be connected to a double ring, forming what is sometimes called a *dual ring of trees.* Workstations connected to the hub are *single-attachment stations (SASs);* they are connected only to the primary ring and cannot take advantage of the secondary ring's wrapping capabilities. The FDDI specifications define four types of ports used to connect workstations to the network. These are as follows:

A   DAS connection to secondary ring

B   DAS connection to primary ring

M   DAC port for connection to an SAS

S   SAS connection to M port in a concentrator

The various types of connections using the four types of FDDI ports are listed in Table 12-6.

Module Connection	Other Device	Description
A	B	Peer connection between FDDI fiber-optic DAS and another DAS device on trunk ring
B	A	Peer connection between FDDI fiber-optic DAS and another DAS device on trunk ring
S	M	Connection between FDDI fiber-optic SAS or UTP SAS and concentrator
A	M	Connection between FDDI fiber-optic DAS and concentrator; used for dual homing
B	M	Connection between FDDI fiber-optic DAS and concentrator; used for dual homing
S	S	Connection between FDDI fiber-optic SAS or UTP SAS and another SAS station

Table 12-6:   *FDDI Connection Types*

DASs and DACs have both A and B ports to connect them to a double ring. Signals from the primary ring enter through the B port and exit from the A port, while the signals from the secondary ring enter through A and exit through B. An SAS has a single S port, which connects it to the primary ring only through an M port on a DAC.

The 500 workstation and 100 km network-length limitations are based on the use of DAS computers. A FDDI network composed only of SAS machines can be up to 200 km long and support up to 1,000 workstations.

DAS computers that are attached directly to the double ring function as repeaters; they regenerate the signals as they pass each packet along to the rest of the network. When a system is turned off, however, it does not pass the packets along and the network wraps, unless the station is equipped with a bypass switch. A *bypass switch*, implemented either as part of the network interface adapter or as a separate device, enables incoming signals to pass through the station and on to the rest of the network, but it does not regenerate them. On a fiber-optic network, this is the equivalent of opening a window to let the sunlight into a room instead of turning on an electric light. As with any network medium, the signal has a tendency to attenuate if it is not regenerated. If too many adjacent systems are not repeating the packets, the signals can weaken to the point at which stations can't read them.

The DAC functions much like a Token Ring MAU in that it implements a logical ring while using a physical star topology. Connecting a DAC to a double ring extends the primary ring out to each connected workstation and back, as shown in Figure 12-10. Notice that while the DAC is connected to both the primary and secondary rings, the M ports connect only the primary ring to the workstations. Thus, while the DAC itself takes advantage of the double ring's fault tolerance, a break in the cable connecting a workstation to the DAC severs the workstation from the network. However, the DAC is capable of dynamically removing a malfunctioning station from the ring (again, like a Token Ring MAU), so that the problem affects only the single workstation and not the entire ring.

It is sometimes possible to connect a DAS to two DAC ports, to provide a standby link to the hub if the active link fails. This is called *dual homing*. However, this is different from connecting the DAS directly to the double ring, because both the A and B ports on the workstation are connected to M ports on the hub. M ports are connected only to the primary ring, so a dual-homed system simply has a backup connection to the primary ring, not a connection to both rings.

Cascading hubs are permitted on a FDDI network. This means that you can plug one DAC into an M port of another DAC to extend the network. There is no limit to the number of layers, as long as you observe the maximum number of workstations permitted on the ring. It is also possible to create a two-station ring by connecting the S ports on two SAS computers or by connecting an S port to either the A or B port of a DAC. Some FDDI adapters may require special configuration to do this.

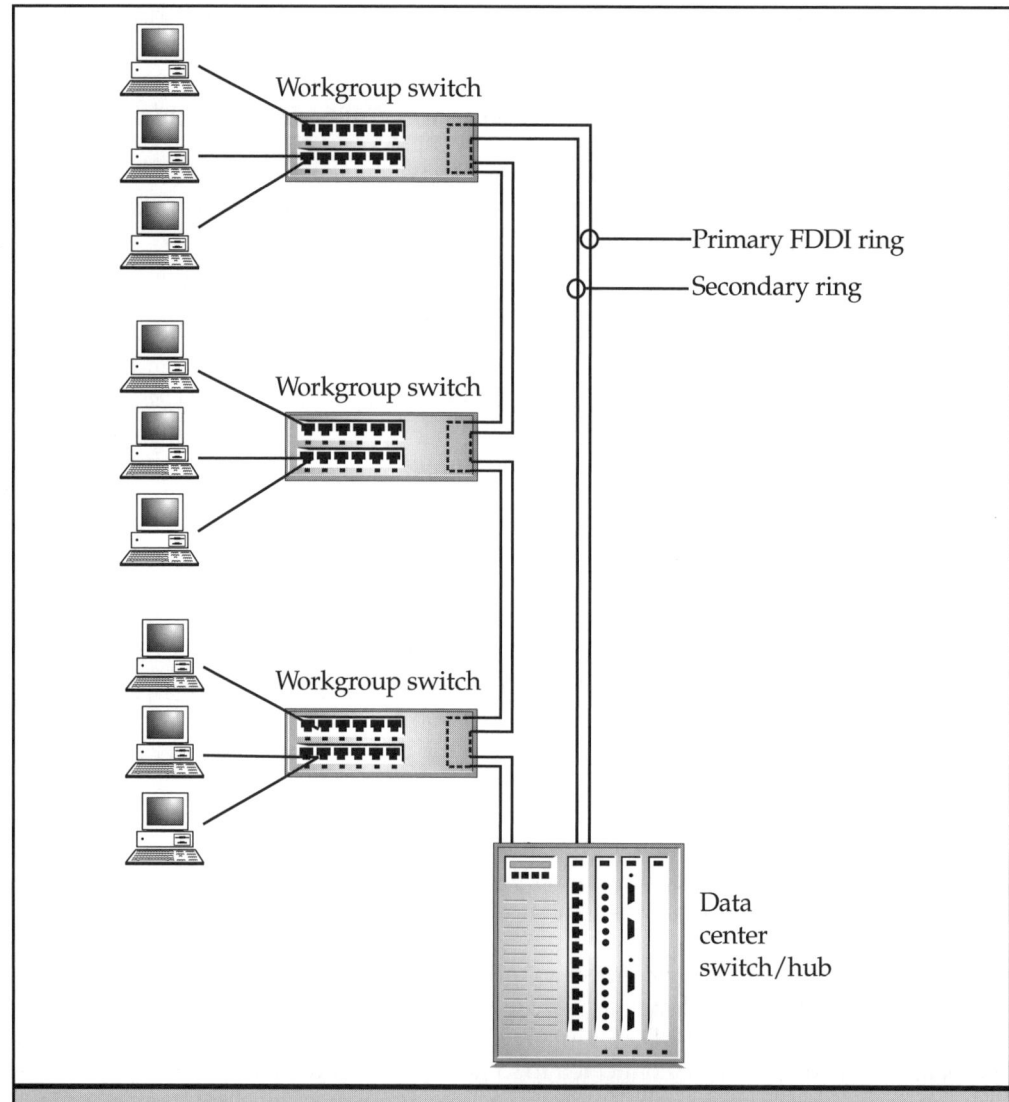

Figure 12-10:   *DACs connected to the double ring provide multiple SAS connections.*

FDDI Subsystems

The functionality of the FDDI protocol is broken down into four distinct layers, as follows:

Physical Media Dependent (PMD)   Prepares data for transmission over a specific type of network medium.

Physical (PHY)   Encodes and decodes the packet data into a format suitable for transmission over the network medium and is responsible for maintaining the clock synchronization on the ring.

Media Access Control (MAC)   Constructs FDDI packets by applying the frame containing addressing, scheduling, and routing data, and then negotiates access to the network medium.

Station Management (SMT)   Provides management functions for the FDDI ring, including insertion and removal of the workstation from the ring, fault detection and reconfiguration, neighbor identification, and statistics monitoring.

The FDDI standards consist of separate documents for each of these layers, as well as separate specifications for some of the options at certain layers. The operations performed at each layer are discussed in the following sections.

The FDDI standards, as with all ANSI and ISO standards, are not freely available, online or otherwise. You can purchase them, however, in either printed or digital form, from the American National Standards Institute at http://www.ansi.org/ or from Global Engineering Documents at http://global.ihs.com/.

The Physical Media Dependent Layer

The *Physical Media Dependent (PMD)* layer is responsible for the mechanics involved in transmitting data over a particular type of network medium. The FDDI standards define two physical layer options, as follows.

Fiber-Optic   The Fiber-PMD standards define the use of either singlemode or multimode fiber-optic cable, as well as the operating characteristics of the other components involved in producing the signals, including the optical power sources, photodetectors, transceivers, and medium interface connectors. For example, the optical power sources must be able to transmit a 25-microwatt signal, while the photodetectors must be capable of reading a 2 microwatt signal.

The original cable defined in the standard is 62.5/125 micron graded index multimode fiber, although cables with other core and cladding diameters are allowed, such as 50/125, 80/125, and 100/140. You can also use singlemode fiber cable, as defined in the SMF-PMD standard, with a core diameter from 8 to 10 microns and a 125-micron cladding diameter. Singlemode fiber provides significantly less signal attenuation than multimode, enabling it to span longer distances, but it is also more expensive and less flexible, making it more difficult to install.

The 2 km maximum distance between FDDI stations cited earlier is for multimode fiber; with singlemode cable, runs of 40 to 60 km between workstations are possible. There is also a low-cost multimode fiber cable standard, called LCF-PMD, that allows only 500 meters between workstations. All of these fiber cables use the same wavelength (1300 nm), so it's possible to mix them on the same network, as long as you adhere to the cabling guidelines of the least capable cable in use.

Note *For more information on the various types of fiber-optic cabling, see Chapter 4.*

Twisted-Pair The TP-PMD standard, published in 1995 and sometimes called the Copper Distributed Data Interface (CDDI, pronounced "*siddy*"), calls for the use of either standard Category 5 unshielded twisted-pair (UTP) or Type 1 shielded twisted-pair (STP) cable. In both cases, the maximum distance for a cable run is 100 meters. Twisted-pair cable is typically used for SAS connections to concentrators, while the backbone uses fiber optic. This makes it possible to use inexpensive copper cable for horizontal wiring to the workstations and retain the attributes of fiber optic on the backbone without the need to bridge or route between FDDI and Ethernet. CDDI never gained wide acceptance in the marketplace, probably because of the introduction of Fast Ethernet at approximately the same time.

The Physical Layer

While the PMD layer defines the characteristics of specific media types, the PHY layer is implemented in the network interface adapter's chipset and provides a media-independent interface to the MAC layer above it. In the original FDDI standards, the PHY layer is responsible for the encoding and decoding of the packets constructed by the MAC layer into the signals that are transmitted over the cable. FDDI uses a signaling scheme called *NRZI 4B/5B (Non-Return to Zero Inverted)*, which is substantially more efficient than the Manchester and Differential Manchester schemes used by Ethernet and Token Ring, respectively.

Note *For more information on physical-layer signaling techniques, see Chapter 2.*

The TP-PMD standard, however, calls for a different signaling scheme, called MLT-3 (Multi-Level Transition), which uses three signal values instead of the two used by NRZI 4B/5B. Both of these schemes provide the signal needed to synchronize the clocks of the transmitting and receiving workstations.

The Media Access Control Layer

The MAC layer accepts protocol data units (PDUs) of up to 9,000 bytes from the network-layer protocol and constructs packets up to 4,500 bytes in size by encapsulating the data within a FDDI frame. This layer is also responsible for negotiating access to the network medium by claiming and generating tokens.

Data Frames   Most of the packets transmitted by a FDDI station are data frames. A data frame can carry network-layer protocol data, MAC data used in the token claiming and beaconing processes, or station management data.

FDDI frames contain information encoded into symbols. A symbol is a 5-bit binary string that the NRZI 4B/5B signaling scheme uses to transmit a 4-bit value. Thus, two symbols are equivalent to one byte. This encoding provides values for the 16 hexadecimal data symbols, eight control symbols that are used for special functions (some of which are defined in the frame format that follows), plus eight violation symbols that FDDI does not use. Table 12-7 lists the symbols used by FDDI and the 5-bit binary sequences used to represent them.

Symbol	5-bit Binary Value
0 (binary 0000)	11110
1 (binary 0001)	01001
2 (binary 0010)	10100
3 (binary 0011)	10101
4 (binary 0100)	01010
5 (binary 0101)	01011
6 (binary 0110)	01110
7 (binary 0111)	01111

Table 12-7:   *FDDI Symbol Values*

Symbol	5-bit Binary Value
8 (binary 1000)	10010
9 (binary 1001)	10011
A (binary 1010)	10110
B (binary 1011)	10111
C (binary 1100)	11010
D (binary 1101)	11011
E (binary 1110)	11100
F (binary 1111)	11101
Q	00000
H	00100
I	11111
J	11000
K	10001
T	01101
R	00111
S	11001

Table 12-7: *FDDI Symbol Values* (continued)

The format of a FDDI data frame is shown in Figure 12-11. The functions of the frame fields are as follows:

Preamble (PA), 8 bytes   Contains a minimum of 16 symbols of idle, that is, alternating 0s and 1s, which the other systems on the network use to synchronize their clocks, after which they are discarded.

Starting Delimiter (SD), 1 byte   Contains the symbols J and K, which indicate the beginning of the frame.

Frame Control (FC), 1 byte   Contains two symbols that indicate what kind of data is found in the INFO field. Some of the most common values are as follows:

   40 (Void Frame)

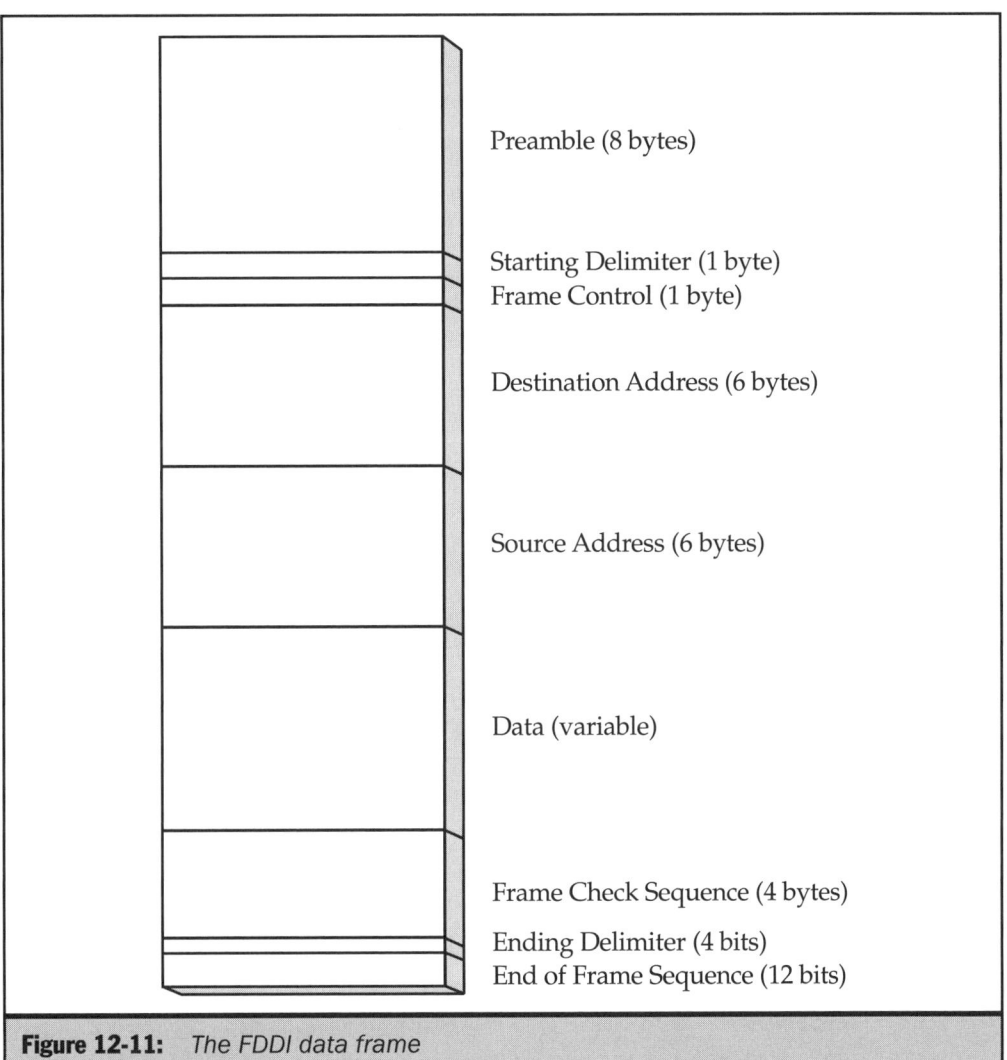

Preamble (8 bytes)

Starting Delimiter (1 byte)
Frame Control (1 byte)

Destination Address (6 bytes)

Source Address (6 bytes)

Data (variable)

Frame Check Sequence (4 bytes)

Ending Delimiter (4 bits)
End of Frame Sequence (12 bits)

Figure 12-11: *The FDDI data frame*

41, 4F (Station Management [SMT] Frame)   Indicates that the INFO field contains an SMT Protocol Data Unit, which is composed of an SMT header and SMT information.

C2, C3 (MAC Frame)   Indicates that the frame is either a MAC Claim frame (C2) or a MAC Beacon frame (C3). These frames are used to recover from abnormal occurrences in the token-passing process, such as failure to receive a token or failure to receive any data at all.

50, 51 (LLC Frame)   Indicates that the INFO field contains a standard IEEE 802.2 LLC frame. FDDI packets carrying application data use logical link control (LLC) frames.

60 (Implementer Frame)

70 (Reserved Frame)

Destination Address (DA), 6 bytes   Specifies the MAC address of the system on the network that will next receive the frame, or a group or broadcast address.

Source Address (SA), 6 bytes   Specifies the MAC address of the system sending the packet.

Data (INFO), variable   Contains network-layer protocol data, or an SMT header and data, or MAC data, depending on the function of the frame, as specified in the FC field.

Frame Check Sequence (FCS), 4 bytes   Contains a cyclic redundancy check value, generated by the sending system, that will be recomputed at the destination and compared with this value to verify that the packet has not been damaged in transit.

Ending Delimiter (ED), 4 bits   Contains a single T symbol indicating that the frame is complete.

End of Frame Sequence (FS), 12 bits   Contains three indicators that can have either the value R (Reset) or S (Set). All three have the value R when the frame is first transmitted, and may be modified by intermediate systems when they retransmit the packet. The functions of the three indicators are as follows:

E (Error)   Indicates that the system has detected an error, either in the FCS or in the frame format. Any system receiving a frame with a value of S for this indicator immediately discards the frame.

A (Acknowledge)   Indicates that the system has determined that the frame's destination address applies to itself, either because the DA field contains the MAC address of the system or a broadcast address.

C (Copy)   Indicates that the system has successfully copied the contents of the frame into its buffers. Under normal conditions, the A and the C indicators are set together; a frame in which the A indicator is set and C is not indicates that the frame could not be copied to the system's buffers. This is most likely due to the system's having been overwhelmed with traffic.

Token Passing   FDDI uses token passing as its media access control (MAC) mechanism, like the Token Ring protocol. A special packet called a *token* circulates around the network, and only the system in possession of the token is permitted to transmit its data. The optional feature called *early token release* on a Token Ring network, in which a system transmits a new token immediately after it finishes transmitting its last packet, is standard on a FDDI network. FDDI systems can also transmit multiple packets before releasing the token to the next station. When a packet has traversed the entire ring and returned to the system that originally created it, that system removes the token from the ring to prevent it from circulating endlessly.

The format of the token frame is shown in Figure 12-12. The functions of the fields are as follows:

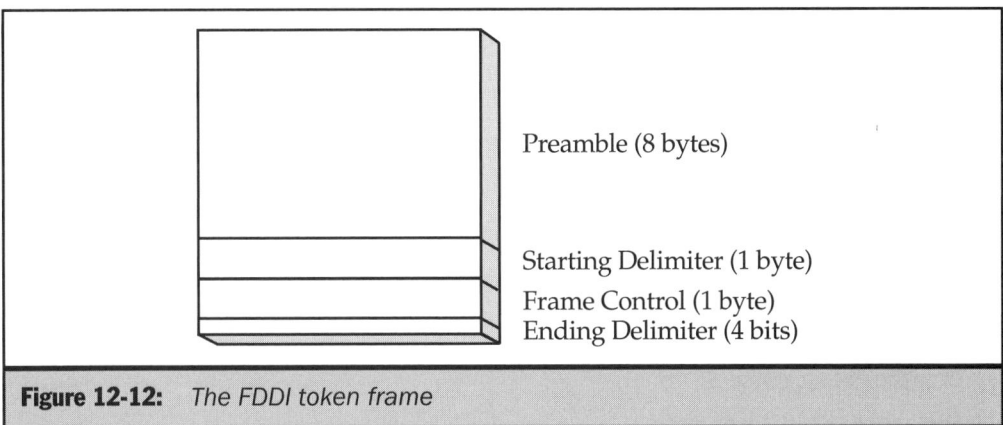

Figure 12-12: *The FDDI token frame*

Preamble (8 bytes)

Starting Delimiter (1 byte)
Frame Control (1 byte)
Ending Delimiter (4 bits)

Preamble (PA), 8 bytes   Contains a minimum of 16 symbols of idle, that is, alternating 0s and 1s, which the other systems on the network use to synchronize their clocks, after which they are discarded.

Starting Delimiter (SD), 1 byte   Contains the symbols J and K, which indicate the beginning of the frame.

Frame Control (FC), 1 byte   Contains two symbols that indicate the function of the frame, using the following hexadecimal values.

80 (Nonrestricted Token)

C0 (Restricted Token)

Ending Delimiter (ED), 1 byte   Contains two T symbols indicating that the frame is complete.

FDDI is a *deterministic* network protocol. By multiplying the number of systems on the network by the amount of time needed to transmit a packet, you can calculate the maximum amount of time it can take for a system to receive the token. This is called the *target token rotation time*. FDDI networks typically run in *asynchronous ring mode,* in which any computer can transmit data when it receives the token. Some FDDI products can also run in *synchronous ring mode*, which enables administrators to allocate a portion of the network's total bandwidth to a system or group of systems. All of the other computers on the network run asynchronously and contend for the remaining bandwidth in the normal manner.

The Station Management Layer

Unlike Ethernet and most other data link–layer protocols, FDDI has network management and monitoring capabilities integrated into it and was designed around these capabilities. The SMT layer is responsible for ring maintenance and diagnostics operations on the network, such as:

- Station initialization

- Station insertion and removal

- Connection management

- Configuration management

- Fault isolation and recovery

- Scheduling policies

- Statistics collection

A computer can contain more than one FDDI adapter, and each adapter has its own PMD, PHY, and MAC layer implementations, but there is only one SMT implementation for the entire system. SMT messages are carried within standard FDDI data frames with a value of 41 or 4F in the Frame Control field. In station management frames, the INFO field of the FDDI data frame contains an SMT PDU, which is composed of an SMT header and an SMT info field. The format of the SMT PDU is shown in Figure 12-13. The functions of the fields are as follows:

Frame Class, 1 byte   Specifies the function of the message, using the following values:

01 (Neighbor Information Frame [NIF])   FDDI stations transmit periodic announcements of their MAC addresses, which enable the systems on the network to determine their *upstream neighbor addresses (UNA)* and their *downstream neighbor addresses (DNA)*. This is known as the *Neighbor Notification Protocol*. Network monitoring products can also use these messages to create a map of the FDDI ring.

02 (Status Information Frame-Configuration [SIF-Cfg])   Used to request and provide a system's configuration information for purposes of fault isolation, ring mapping, and statistics monitoring.

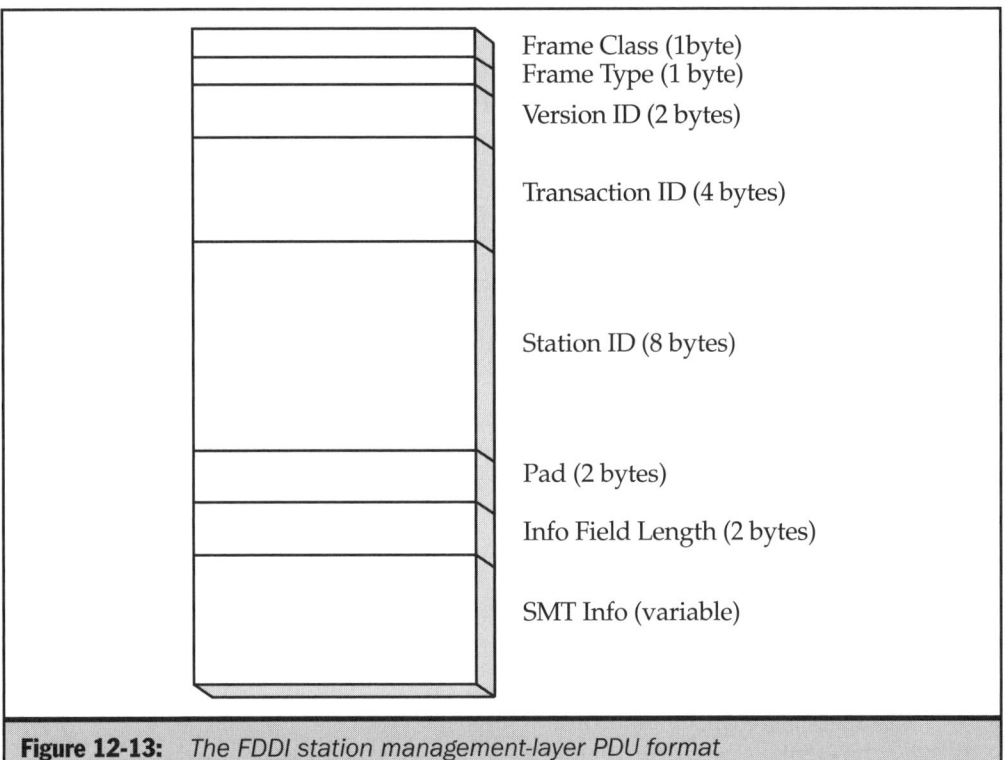

Frame Class (1byte)
Frame Type (1 byte)

Version ID (2 bytes)

Transaction ID (4 bytes)

Station ID (8 bytes)

Pad (2 bytes)

Info Field Length (2 bytes)

SMT Info (variable)

Figure 12-13:   *The FDDI station management-layer PDU format*

03 (Status Information Frame-Operation [SIF-Opr])   Used to request and
provide a system's operation information for purposes of fault isolation, ring
mapping, and statistics monitoring.

04 (Echo Frame)   Used for SMT-to-SMT loopback testing between FDDI
systems.

05 (Resource Allocation Frame [RAF])   Used to implement network policies,
such as the allocation of synchronous bandwidth.

06 (Request Denied Frame [RDF])   Used to deny a request issued by another
station because of an unsupported Version ID value or a length error.

07 (Status Report Frame [SRF])   Used to report a station's status to network
administrators when specific conditions occur, much like an SNMP trap. Some
of these conditions are as follows:

> **Frame Error Condition**   Indicates the occurrence of an unusually high
> number of frame errors.

NETWORK PROTOCOLS

LER Condition   Indicates the occurrence of link errors on a port above a specified limit.

Duplicate Address Condition   Indicates that the system or its upstream neighbor is using a duplicate address.

Peer Wrap Condition   Indicates that a DAS is operating in wrapped mode—in other words, that it is diverting data from the primary ring to the secondary due to a cable break or other error.

Hold Condition   Indicates that the system is in a holding-prm or holding-sec state.

NotCopied Condition   Indicates that the system's buffers are overwhelmed and that packets are being repeated without being copied into the buffers.

EB Error Condition   Indicates the presence of an elasticity buffer error on any port.

MAC Path Change   Indicates that the current path has changed for any of the system's MAC addresses.

Port Path Change   Indicates that the current path has changed for any of the system's ports.

MAC Neighbor Change   Indicates a change in either the upstream or downstream neighbor address.

Undesirable Connection   Indicates the occurrence of an undesirable connection to the system.

08 (Parameter Management Frame-Get [PMF-Get])   Provides the means to look at management information base (MIB) attributes on remote systems.

09 (Parameter Management Frame-Set [PMF-Set])   Provides the means to set values for certain MIB attributes on remote systems.

FF (Extended Service Frame [ESF])   Intended for use when defining new SMT services.

Frame Type, 1 byte   Indicates the type of message contained in the frame, using the following values:

01   Announcement

02   Request

03   Response

Version ID, 2 bytes   Specifies the structure of the SMT Info field, using the following values:

0001   Indicates the use of a version lower than 7.*x*

0002   Indicates the use of version 7.*x*

Transaction ID, 4 bytes   Contains a value used to associate request and response messages.

Station ID, 8 bytes   Contains a unique identifier for the station, consisting of two user-definable bytes and the 6-byte MAC address of the network interface adapter.

Pad, 2 bytes   Contains two bytes with a value of 00 that bring the overall size of the header to 32 bytes.

Info Field Length, 2 bytes   Specifies the length of the SMT Info field.

SMT Info, variable   Contains one or more parameters, each of which is composed of the following subfields:

Parameter Type, 2 bytes   Specifies the function of the parameter. The first of the two bytes indicates the parameter's class, using the following values:

00   General parameters

10   SMT parameters

20   MAC parameters

32   PATH parameters

40   PORT parameters

Parameter Length, 2 bytes   Specifies the total length of the Resource Index and Parameter Value fields.

Resource Index, 4 bytes   Identifies the MAC, PATH, or PORT object that the parameter is describing.

Parameter Value, variable   Contains the actual parameter information.

A FDDI system uses SMT messages to insert itself into the ring when it is powered up. The procedure consists of several steps, in which it initializes the ring and tests the link to the network. Then the system initiates its connection to the ring using a Claim Token, which determines whether a token already exists on the network. If a token frame already exists, the Claim Token configures it to include the newly initialized system in the token's path. If no token is detected, all of the systems on the network generate Claim frames, which enable the systems to determine the value for the token rotation time and determine which system should generate the token.

Because of the SMT header's size and the number of functions performed by SMT messages, the control overhead on a FDDI network is high, relative to other protocols.

FDDI-II

FDDI-II is a newer standard that is designed to provide a better form of bandwidth allocation than the original FDDI standard's synchronous ring mode. FDDI-II is intended for networks that require dedicated bandwidth for real-time applications, such as streaming audio and video. FDDI-II is essentially a circuit-switching technology, in which the existing bandwidth can be divided into 16 discrete channels of varying capacities. A specific application can then be assigned a dedicated circuit between a client and a server, providing consistent, continuous bandwidth.

FDDI-II never captured a significant market, largely because the technology requires that all of the systems on the network be running FDDI-II equipment. If there are any standard FDDI stations on the ring, all of the systems run in standard FDDI mode.

The
Complete
Reference

Chapter 13

TCP/IP

Since its inception in the 1970s, the TCP/IP protocol suite has evolved into the industry standard for data-transfer protocols at the network and transport layers of the OSI model. In addition, the suite includes myriad other protocols that operate as low as the data-link layer and as high as the application layer.

Operating systems tend to simplify the appearance of the network protocol stack to make it more comprehensible to the average user. On a Windows workstation, for example, you install TCP/IP by selecting a single module called a *protocol*, but this process actually installs support for a whole family of protocols, of which the Transmission Control Protocol (TCP) and the Internet Protocol (IP) are only two. The alternatives to TCP/IP function in much the same way: the IPX protocol suite consists of multiple protocols similar in function to those of TCP/IP, and NetBEUI, although much simpler, relies on other protocols as well, such as Server Message Blocks (SMB), for many of its operations. Understanding how the individual TCP/IP protocols function and how they work together to provide communication services is an essential part of administering a TCP/IP network.

TCP/IP Attributes

There are several reasons why TCP/IP has become the protocol suite of choice on the majority of data networks, not the least of which is that these are the protocols used on the Internet. TCP/IP was designed to support the fledgling Internet (then called the ARPANET) at a time before the introduction of the PC when interoperability between computing products made by different manufacturers was all but unheard of. The Internet was, and is, composed of many different types of computers and what was needed was a suite of protocols that would be common to all of them.

The main element that sets TCP/IP apart from the other suites of protocols that provide network and transport-layer services is its self-contained addressing mechanism. Every device on a TCP/IP network is assigned an IP address (or sometimes more than one) that uniquely identifies it to the other systems. Most of the PCs on networks today use Ethernet or Token Ring network interface adapters that have unique identifiers (MAC addresses) hardcoded into them, which makes the IP address redundant. Many other types of computers have identifiers assigned by network administrators, however, and no mechanism exists to ensure that another system on a worldwide internetwork like the Internet does not use the same identifier.

Because IP addresses are registered by a centralized body, you can be certain that no two (properly configured) machines on the Internet have the same address. Because of this addressing, the TCP/IP protocols can support virtually any hardware or software platform in use today. The IPX protocols will always be associated primarily with Novell NetWare, and NetBEUI is used almost exclusively on Microsoft Windows networks. TCP/IP, however, is truly universal in its platform interoperability, supported by all and dominated by none.

Another unique aspect of the TCP/IP protocols is the method by which their standards are designed, refined, and ratified. Rather than relying on an institutionalized standards-making body like the IEEE, the TCP/IP protocols are developed in a democratic manner by an ad hoc group of volunteers who communicate largely through the Internet itself. Anyone who is interested enough to contribute to the development of a protocol is welcome. In addition, the standards themselves are published by a body called the Internet Engineering Task Force (IETF) and are released to the public domain, making them accessible and reproducible by anyone. Standards like those published by the IEEE are available, but until very recently, you had to pay hundreds of dollars to purchase an official copy of an IEEE standard like the 802.3 document on which Ethernet is based. On the other hand, you can legally download any of the TCP/IP standards, called *Request for Comments (RFCs)*, from the IETF's Web site at http://www.ietf.org/ or from any number of other Internet sites.

The TCP/IP protocols are also extremely scaleable. As evidence of this, consider that these protocols were designed at a time when the ARPANET was essentially an exclusive club for scientists and academics and no one in their wildest dreams imagined that the protocols they were creating would be used on a network the size of the Internet as it exists today. The main factor limiting the growth of the Internet is the 32-bit size of the IP address space itself, and a new version of the IP protocol, called IPv6, is addressing that shortcoming with a 128-bit address space.

TCP/IP Architecture

TCP/IP is designed to support networks of almost any practical size. As a result, TCP/IP must be able to provide the services needed by the applications using it without being overly profligate in its expenditure of network bandwidth and other resources. For example, the NetBEUI protocol locates other systems by transmitting a broadcast message and expecting the desired system to respond. For this reason, NetBEUI is effective only on small networks comprised of a single broadcast domain. Imagine the state of the Internet today if every computer had to broadcast a message to all the millions of machines on the network each time it wanted to locate a single one! To accommodate the needs of specific applications and functions within those applications, TCP/IP uses multiple protocols in combination to provide the quality of service required for the task and no more.

The TCP/IP Protocol Stack

TCP/IP predates the OSI reference model, but its protocols break down into four layers that can be roughly equated to the seven-layer OSI stack, as shown in Figure 13-1.

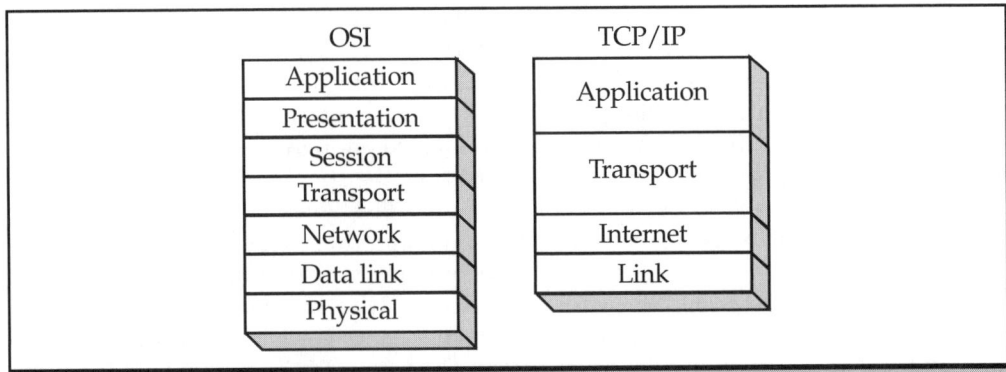

Figure 13-1: *The TCP/IP protocols have their own protocol stack, which is roughly analogous to the OSI reference model.*

On LANs, the link-layer functionality is not defined by a TCP/IP protocol, but by the standard data link–layer protocols, such as Ethernet and Token Ring. To reconcile the MAC address supplied by a network interface adapter with the IP address used at the network layer, systems use a TCP/IP protocol called the *Address Resolution Protocol (ARP)*. However, the TCP/IP standards do define the two protocols most commonly used to establish link-layer communications using modems and other direct connections. These are the *Point-to-Point Protocol (PPP)* and the *Serial Line Internet Protocol (SLIP)*.

At the internet layer is the *Internet Protocol (IP)*, which is the primary carrier for all of the protocols operating at the upper layers, and the *Internet Control Message Protocol (ICMP)*, which TCP/IP systems use for diagnostics and error reporting. IP, as a general carrier protocol, is connectionless and unreliable, because services like error-correction and guaranteed delivery are supplied at the transport layer when required.

Two protocols operate at the transport layer: the Transmission Control Protocol (TCP) and the User Datagram Protocol (UDP). TCP is connection-oriented and reliable, while UDP is connectionless and unreliable. An application uses one or the other, depending on its requirements and the services already provided for it at the other layers.

The transport layer can in some ways be said to encompass the OSI session layer as well as the transport layer in the OSI model, but not in every case. Windows systems, for example, can use TCP/IP to carry the NetBIOS messages they use for their file and printer-sharing activities, and NetBIOS still provides the same session-layer functionality as when a system uses NetBEUI or IPX instead of TCP/IP. This is just one illustration of how the layers of the TCP/IP protocol stack are roughly equivalent to those of the OSI model, but not definitively so. Both of these models are pedagogical and diagnostic tools more than they are guidelines for protocol development and deployment, and they do not hold up to strict comparisons of the various layers' functions with actual protocols.

The application layer is the most difficult to define because the protocols operating there can be fully realized, self-contained applications in themselves, such as the *File Transfer Protocol (FTP)*, or mechanisms used by other applications to perform a service, such as the *Domain Name System (DNS)* and the *Simple Mail Transfer Protocol (SMTP)*.

IP Addressing

The IP addresses used to identify systems on a TCP/IP network are the single most definitive feature of the protocol suite. The IP address is an absolute identifier of both the individual machine and the network on which it resides. Every IP datagram packet transmitted over a TCP/IP network contains the IP addresses of the source system that generated it and the destination system for which it's intended in its IP header. While Ethernet and Token Ring systems have a unique hardware address coded into the network interface card, there is no inherent method to effectively route traffic to an individual system on a large network using this address.

A NIC's hardware address is composed of a prefix that identifies the manufacturer of the card and a node address that is unique among all the cards built by that manufacturer. The manufacturer prefix is useless, as far as routing traffic is concerned, because any one manufacturer's cards can be scattered around the network virtually at random. To deliver network packets to a specific machine, a master list of all of the systems on the network and their hardware addresses would be needed. On a network the size of the Internet, this would obviously be impractical. In addition, all of the millions of computers that connect to the Internet using modems would not have addresses. By identifying the network on which a system is located, IP addresses can be routed to the proper location using a relatively manageable list of network addresses, not a list of individual system addresses.

IP addresses are 32 bits long and are notated as four 8-bit decimal numbers separated by periods, as in 192.168.2.45. This is known as *dotted decimal notation*; each of the 8-bit numbers is sometimes called an *octet* or a *quad*. (These terms were originally used because there are computers for which the more common term *byte* does not equal 8 bits.) Because each quad is the decimal equivalent of an 8-bit binary number, their possible values run from 0 to 255. Thus, the full range of possible IP addresses is 0.0.0.0 to 255.255.255.255.

IP addresses do not represent computers per se; rather, they represent network interfaces. A computer with two network interface cards, or one NIC and a modem connection to a TCP/IP server, has two IP addresses. A system with two or more interfaces is said to be *multihomed*. If the interfaces connect the computer to different networks and the system is configured to pass traffic between the networks, the system is said to function as a *router*.

> | Note |
>
> *A router can be a standard computer with two network interfaces and software that provides routing capabilities, or it can be a dedicated hardware device designed specifically for routing network traffic. At times, the TCP/IP standards refer to routers of any kind as gateways, while standard networking terminology defines a gateway as being an application layer device that forwards traffic between networks that use different protocols, as in an e-mail gateway. Do not confuse the two.*

Every IP address contains bits that identify a network and bits that identify an interface (called a *host*) on that network. To reference a network, systems use just the network bits, replacing the host bits with zeros. Routers use the network bits to forward packets to another router connected to the destination network, which then transmits the data to the destination host system.

Subnet Masking

IP addresses always dedicate some of their bits to the network identifier and some to the host identifier, but the number of bits used for each purpose is not always the same. Many common addresses use 24 bits for the network and eight for the host, but the split between the network and host bits can be anywhere in the address. To identify which bits are used for each purpose, every TCP/IP system has a subnet mask along with its IP address. A *subnet mask* is a 32-bit binary number in which the bits correspond to those of the IP address. A bit with a 1 value in the mask indicates that the corresponding bit in the IP address is part of the network identifier, while a 0 bit indicates that the corresponding address bit is part of the host identifier. As with an IP address, the subnet mask is expressed in dotted decimal notation, so although it may look something like an IP address, the mask has a completely different function.

As an example, consider a system with the following TCP/IP configuration:

```
IP address: 192.168.2.45
Subnet mask: 255.255.255.0
```

In this case, the 192.168.2 portion of the IP address identifies the network, while the 45 identifies the host. When expressed in decimal form, this may appear confusing, but the binary equivalents are as follows:

```
IP address:  11000000 10101000 00000010 00101101
Subnet mask: 11111111 11111111 11111111 00000000
```

As you can see in this example, the dividing line between the network and host bits lies between the third and fourth quads. The dividing line need not fall between quads, however. A subnet mask of 255.255.240.0 allocates 12 bits for the host address because the binary equivalent of the mask is as follows:

```
11111111 11111111 11110000 00000000
```

The dividing line between the network and host bits can fall anywhere in the 32 bits of the mask, but you never see network bits mixed up with host bits. A clear line always separates the network bits on the left from the host bits on the right.

IP Address Registration

For IP addresses to uniquely identify the systems on the network, it is essential that no two interfaces be assigned the same address. On a private network, the administrators must ensure that every address is unique. They can do this by manually tracking the addresses assigned to their networks and hosts, or they can use a service like DHCP (the Dynamic Host Configuration Protocol) to assign the addresses automatically.

Note *For more information on DHCP and automatic IP address assignment and TCP/IP configuration, see Chapter 23.*

On the Internet, however, this problem is considerably more complicated. With individual administrators controlling thousands of different networks, not only is it impractical to assume that they can get together and make sure that no addresses are duplicated, but no worldwide service exists that can assign addresses automatically. Instead, there must be a clearinghouse or registry for IP address assignments that ensures no addresses are duplicated.

Even this task is monumental, however, because literally millions of systems are connected to the Internet. In fact, such a registry exists, but instead of assigning individual host addresses to each system, it assigns network addresses to companies and organizations. The organization charged with registering network addresses for the Internet is called the *Internet Assigned Numbers Authority (IANA)*. After an organization obtains a network address, the administrator is solely responsible for assigning unique host addresses to the machines on that network.

Note *The IANA maintains a Web site at www.iana.org.*

This two-tiered system of administration is one of the basic organizational principles of the Internet. Domain name registration works the same way. A registry like Network Solutions registers domain names to organizations and individuals, and the individual administrators of those domains are responsible for assigning names in those domains to their hosts.

NETWORK PROTOCOLS

IP Address Classes

The IANA registers several different classes of network addresses, which differ in their subnet masks—that is, the number of bits used to represent the network and the host. These address classes are summarized in Table 13-1.

	Class A	Class B	Class C	Class D	Class E
Network Address Bits	8	16	24	N/A	N/A
Host Address Bits	24	16	8	N/A	N/A
Subnet Mask	255.0.0.0	255.255.0.0	255.255.255.0	N/A	N/A
Addresses Begin with: (Binary)	0	10	110	1110	1111
First Byte Values (Decimal)	0–127	128–191	192–223	224–239	240–255
Number of Networks	127	16,384	2,097,151	N/A	N/A
Number of Hosts	16,777,214	65,534	254	N/A	N/A

Table 13-1: *IP Address Classes*

The idea behind the different classes is to create networks of varying sizes suitable for different organizations and applications. A company building a relatively small network can register a Class C address which, because the addresses have only eight host bits, supports up to 254 systems, while larger organizations can use Class B or A addresses with 16 or 24 host bits and create subnets out of them. You create subnets by "borrowing" some of the host bits and using them to create subnetwork identifiers, essentially networks within a network.

The surest way to identify the class of a particular address is to look at the value of the first quad. Class A addresses always have a 0 as their first bit, which means that the binary values for the first quad range from 00000000 to 01111111, which translates into the decimal values 0 through 127. In the same way, Class B addresses always have 10 as their first two bits, providing first quad values of 10000000 to 10111111, or 128 to 191. Class C addresses have 110 as their first three bits, so the first quad can range from 11000000 to 11011111, or 192 to 223.

In practice, network addresses are not registered with the IANA directly by the companies and organizations running the individual networks. Instead, companies in the business of providing Internet access, called *Internet service providers (ISPs)*, register multiple networks and supply blocks of addresses to clients as needed.

Class D addresses are not intended for allocation in blocks like the other classes. This part of the address space is allocated for multicast addresses. *Multicast addresses* represent groups of systems that have a common attribute, but that are not necessarily located in the same place or even administered by the same organization. For example, packets sent to the multicast address 224.0.0.1 are processed by all of the routers on the local subnet. The block of addresses designated as Class E is reserved for future use.

Unregistered IP Addresses

IP address registration is designed for networks connected to the Internet with computers that must be accessible from other networks. When you register a network address, no one else is permitted to use it and the routers on the Internet have the information needed to forward packets to your network. For a private network that is not connected to the Internet, it is not necessary to register network addresses. In addition, most business networks connected to the Internet use some sort of firewall product to prevent intruders from accessing their networks from outside. In nearly all cases, there is no real need for every system on a network to be directly accessible from the Internet, and there is a genuine danger in doing so. Many firewall products, therefore, isolate the systems on the network, making registered IP addresses unnecessary.

For a network that is completely isolated from the Internet, administrators can use any IP addresses they wish, as long as there are no duplicates on the same network. If any of the network's computers connect to the Internet by any means, however, there is potential for a conflict between an internal address and the system on the Internet for which the address was registered. If, for example, you happened to assign one of your network systems the same address as a Microsoft Web server, a user on your network attempting to access Microsoft's site may reach the internal machine with the same address instead.

To prevent these conflicts, RFC 1918, "Address Allocation for Private Internets," specifies three address ranges intended for use on unregistered networks, as shown in Table 13-2. These addresses are not assigned to any registered network and can, therefore, be used by any organization, public or private.

Class A	10.0.0.0 through 10.255.255.255
Class B	172.16.0.0 through 172.31.255.255
Class C	192.168.0.0 through 192.168.255.255

Table 13-2: *Unregistered Network IP Addresses*

Using unregistered IP addresses not only simplifies the process of obtaining and assigning addresses to network systems, it also conserves the registered IP addresses for use by systems that actually need them for direct Internet communications. As with many design decisions in the computer field, no one expected at the time of its inception that the Internet would grow to be as enormous as it is now. The 32-bit address space for the IP protocol was thought to be big enough to support all future growth (as was the original 640KB memory limitation in PCs).

The Internet has now reached the point, however, where addresses are nearly always obtained from third parties and not directly from the IANA. In addition, the proliferation of other communications devices that use IP addresses, such as palmtop computers and cellular phones, could result in a severe shortage of addresses in the near future. The IPv6 protocol, currently in development, addresses this shortage by expanding the address space from 32 bits to 128.

Special IP Addresses

Aside from the blocks of addresses designated for use by unregistered networks, there are other addresses not allocated to registered networks because they are intended for special purposes. These addresses are listed in Table 13-3.

Address	Example	Function
All bits 0	0.0.0.0	Addresses the current host on the current network, such as during a DHCP transaction before a workstation is assigned an IP address.
All bits 1	255.255.255.255	Limited broadcast; addresses all the hosts on the local network.
Host bits all 0	192.168.2.0	Identifies a network.
Host bits all 1	192.168.2.255	Directed broadcast; addresses all the hosts on another network.
Network bits all 0	0.0.0.22	Addresses a specific host on the current network.
First quad 127	127.0.0.1	Internal host loopback address.

Table 13-3: *Special Purpose IP Addresses*

Subnetting

Theoretically, the IP addresses you assign to the systems on your network do not have to correlate exactly to the physical network segments, but in standard practice, it's a good idea if they do. Obviously, an organization that registers a Class B address does not have 65,534 nodes on a single network segment; they have an internetwork composed of many segments, joined by routers, switches, or other devices. To support a multisegment network with a single IP network address, you create subnets corresponding to the physical network segment.

A *subnet* is simply a subdivision of the network address that you create by taking some of the host identifier bits and using them as a subnet identifier. To do this, you modify the subnet mask on the machines to reflect the borrowed bits as part of the network identifier, instead of the host identifier.

For example, you can subnet a Class B network address by using the third quad, originally intended to be part of the host identifier, as a subnet identifier instead, as shown in Figure 13-2. By changing the subnet mask from 255.255.0.0 to 255.255.255.0, you divide the Class B address into 254 subnets of 254 hosts each. You then assign each of the physical segments on the network a different value for the third quad and number the individual systems using only the fourth quad. The result is that the routers on your network can use the value of the third quad to direct traffic to the appropriate segments.

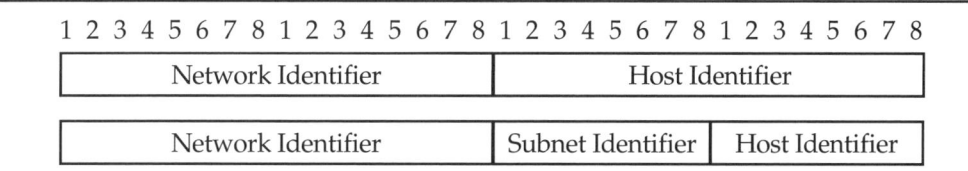

Figure 13-2:   *The top example shows a standard Class B address, split into 16-bit network and host identifiers. In the bottom example, the address has been subnetted by borrowing eight of the host bits for use as a subnet identifier.*

Note   *The subnet identifier is purely a theoretical construction. To routers and other network systems, an IP address consists only of network and host identifiers, with the subnet bits incorporated into the network identifier.*

The previous example demonstrates the most basic type of subnetting, in which the boundaries of the subnet identifier fall between the quads. However, you can use any number of host bits for the subnet identifier and adjust the subnet mask and IP address accordingly. This is called *variable mask subnetting*. If, for example, you have a Class B address and decide to use 4 host bits for the subnet identifier, you would use a subnet mask with the following binary value:

```
11111111  11111111  11110000  00000000
```

NETWORK PROTOCOLS

The first 4 bits of the third quad are changed from 0s to 1s, to indicate that these bits are now part of the network identifier. The decimal equivalent of this number is 255.255.240.0, which is the value you would use for the subnet mask in the system's TCP/IP configuration. By borrowing 4 bits in this way, you can create up to 14 subnets, consisting of 4,094 hosts each. The formula for determining the number of subnets and hosts is as follows:

$$2^x - 2$$

where x equals the number of bits used for the subnet identifier. You subtract two to account for identifiers consisting of all 0s and all 1s, which are traditionally not used, because the value 255 is used for broadcasts, and the value 0 to represent the network. For this example, therefore, you perform the following calculations:

$$2^4 - 2 = 14$$
$$2^{12} - 2 = 4,094$$

Note *Some TCP/IP implementations are capable of using 0 as a subnet identifier, but you should avoid this practice unless you are certain that all of your routers also support this feature.*

To determine the IP addresses you assign to particular systems, you increment the 4 bits of the subnet identifier separately from the 12 bits of the host identifier and convert the results into decimal form. Thus, assuming a Class B network address of 172.16.0.0 with a subnet mask of 255.255.240.0, the first IP address of the first subnet will have the following binary address:

```
10101100 00010000 00010000 00000001
```

The first two quads are the binary equivalents of 172 and 16. The third quad consists of the 4-bit subnet identifier, with the value 0001, and the first 4 bits of the 12-bit host identifier. Because this is the first address on this subnet, the value for the host identifier is 000000000001.

Although these 12 bits are incremented as a single unit, when converting the binary values to decimals, you treat each quad separately. Therefore, the value of the third quad (00010000) in decimal form is 16, and the value of the fourth quad (00000001) in decimal form is 1, yielding an IP address of 172.16.16.1.

Tip *You can use the Windows Calculator program to convert between binary and decimal numbers by switching it into scientific mode from the View menu. Once in scientific mode, click either the Dec or Bin radio button, enter the number you want to convert, then click the other radio button to perform the conversion.*

The last address in this subnet will have the following binary value:

```
10101100  00010000  00011111  11111110
```

which yields an IP address of 172.16.31.254.

For the next subnet, you increment the subnet identifier bits to 0010 and start again with 000000000001 as the first host in the new subnet. Thus, the first address in the second subnet is

```
10101100  00010000  00100000  00000001
```

or

```
172.16.32.1
```

Proceeding in this way, you can create all 14 subnets, using the following address ranges:

```
172.16.16.1   -   172.16.31.25
172.16.32.1   -   172.16.47.25
172.16.48.1   -   172.16.63.25
172.16.64.1   -   172.16.79.25
172.16.80.1   -   172.16.95.25
172.16.96.1   -   172.16.111.25
172.16.112.1  -   172.16.127.25
172.16.128.1  -   172.16.143.25
172.16.144.1  -   172.16.159.25
172.16.160.1  -   172.16.175.25
172.16.176.1  -   172.16.191.25
172.16.192.1  -   172.16.207.25
172.16.208.1  -   172.16.223.25
172.16.224.1  -   172.16.239.25
```

Fortunately, manually computing the values for your IP addresses isn't necessary when you subnet the network in this way. Utilities are available that enable you to specify a network address and class, and then select the number of bits to be used for the subnet identifier. The program then supplies you with the IP addresses for the machines in the individual subnets.

 *A freeware IP Subnet Calculator utility is available for download at http://www.wildpackets.com/products/ipsubnetcalculator/.*

Ports and Sockets

The IP address makes it possible to route network traffic to a particular system, but once packets arrive at the computer and begin traveling up the protocol stack, they still must be directed to the appropriate application. This is the job of the transport-layer protocol, either TCP or UDP. To identify specific processes running on the computer, TCP and UDP use port numbers that are included in every TCP and UDP header. Typically, the port number identifies the application-layer protocol that generated the data carried in the packet.

The port numbers permanently assigned to specific services, which are called *well-known ports*, are standardized by the Internet Assigned Numbers Authority (IANA) and published in the "Assigned Numbers" RFC (RFC 1700). Every TCP/IP system has a file called Services that contains a list of the most common well-known port numbers and the services to which they are assigned.

For example, the IP header of a DNS query message contains the IP address of a DNS server in its Destination Address field. Once the packet has arrived at the destination, the receiving computer sees that the UDP header's Destination Port field contains the well-known port value 53. The system then knows to pass the message to the service using port number 53, which is the DNS service.

 *The port number assignments for the TCP and UDP protocols are separate. Although not typical, it is possible for a service to use different port numbers for TCP and UDP, and for the same port number to be assigned to a different service for each protocol.*

The combination of an IP address and a port number is known as a *socket*. The Uniform Resource Locator (URL) format calls for a socket to be notated with the IP address followed by the port number, separated by a colon, as in 192.168.2.45:80.

Not all port numbers are well-known. When a client connects to a well-known service, such as a Web server, it uses the well-known port number for that service (which in the case of a Web server is 80), but selects the port number that it will use as its Source Port value at random. This is known as an *ephemeral port number*. The Web server, on receiving the packet from the client addressed to port 80, reads the Source Port value and knows to address its reply to the ephemeral port number the client has chosen. To prevent clients from selecting well-known ports for their ephemeral port numbers, all of the well-known port number assignments fall below 1,024, and all ephemeral port numbers must be over 1,024.

TCP/IP Naming

IP addresses are an efficient means of identifying networks and hosts, but when it comes to user interfaces, they are difficult to use and remember. Therefore, the Domain Name System (DNS) was devised to supply friendly names for TCP/IP systems. In a discussion of the network and transport-layer TCP/IP protocols, the most important information to remember about DNS names is that they have nothing to do with the actual transmission of data across the network.

Packets are addressed to their destinations using IP addresses only. Whenever a user supplies a DNS name in an application (such as a URL in a Web browser), the first thing the system does is initiate a transaction with a DNS server to resolve the name into an IP address. This occurs before the system transmits any traffic at all to the destination system. Once the system has discovered the IP address of the destination, it uses that address in the IP header to send packets to that destination; the DNS name is no longer used after that point.

Note	*The structure of DNS names and the functions of DNS servers are discussed more fully in Chapter 25.*

NETWORK PROTOCOLS

TCP/IP Protocols

The following sections examine the major protocols that make up the TCP/IP suite. There are literally dozens of TCP/IP protocols and standards, but only a few are commonly used by the systems on a TCP/IP network. Other chapters in this book discuss some of the more specialized protocols in the TCP/IP suite, such as the protocols used by routers to exchange routing data (see Chapter 6) and the application-layer protocols used by specific services (see Chapter 26).

SLIP and PPP

The Serial Line Internet Protocol (SLIP) and the Point-to-Point Protocol (PPP) are unique among the TCP/IP protocols because they provide full data link–layer functionality. Systems connected to a LAN rely on one of the standard data link–layer protocols, such as Ethernet and Token Ring, to control the actual connection to the network. This is because the systems are usually sharing a common medium and must have a MAC mechanism to regulate access to it.

SLIP and PPP are designed for use with modems and other direct connections in which there is no need for media access control. Because they connect only two systems, SLIP and PPP are called *point-to-point* or *end-to-end* protocols. On a system using SLIP or PPP, the TCP/IP protocols define the workings of the entire protocol stack, except for the physical layer itself, which relies on a hardware standard like that for the RS-232 serial port interface, which provides a connection to the modem.

In most cases, systems use SLIP or PPP to provide Internet or WAN connectivity, whether or not the system is connected to a LAN. Virtually every standalone PC that uses a modem to connect to an ISP for Internet access does so using a PPP connection, although a few system types still use SLIP. LANs also use SLIP or PPP connections in their routers to connect to an ISP to provide Internet access to the entire network or to connect to another LAN, forming a WAN connection. Although commonly associated with modem connections, other physical-layer technologies can also use SLIP and PPP, including leased lines, ISDN, frame relay, and ATM connections.

SLIP and PPP are connection-oriented protocols that provide a data link between two systems in the simplest sense of the term. They encapsulate IP datagrams for transport between computers, just as Ethernet and Token Ring do, but the frame they use is far simpler. This is because the protocols are not subject to the same problems as the LAN protocols. Because the link consists only of a connection between the two computers, there is no need for a medium access control mechanism like CSMA/CD or token passing. Also, there is no problem with addressing the packets to a specific destination—because only two computers are involved in the connection, the data can only go to one place.

SLIP

SLIP was created in the early 1980s to provide the simplest possible solution for transmitting data over serial connections. No official standard defines the protocol, mainly because there is nothing much to standardize and interoperability is not a problem. There is an IETF document, however, called "A Nonstandard for Transmission of IP Datagrams over Serial Lines" (RFC 1055), that defines the functionality of the protocol.

The SLIP frame is simplicity itself. A single 1-byte field with the hexadecimal value *c0* serves as an END delimiter, following every IP datagram transmitted over the link. The END character informs the receiving system that the packet currently being transmitted has ended. Some systems also precede each IP datagram with an END character. This way, if any line noise occurs between datagram transmissions, the receiving system treats it as a packet unto itself because it is delimited by two END characters (see Figure 13-3). When the upper-layer protocols attempt to process the noise "packet," they interpret it as gibberish and discard it.

| END | Data | END | Noise | END | Data | END |

Figure 13-3: *Data on a SLIP link can be surrounded by END characters to exclude line noise.*

If a datagram contains a byte with the value *c0*, the system alters it to the 2-byte string *db dc* before transmission, to avoid terminating the packet incorrectly. The *db* byte is referred to as the ESC (escape) character, which, when coupled with another character, serves a special purpose. If the datagram contains an actual ESC character as part of the data, the system substitutes the string *db dd* before transmission.

The ESC character defined by SLIP is not the equivalent of the ASCII ESC character.

SLIP Shortcomings

Because of its simplicity, SLIP is easy to implement and adds little overhead to data transmissions, but it also lacks features that could make it a more useful protocol. For example, SLIP lacks the capability to supply the IP address of each system to the other, meaning that both systems must be configured with the IP address of the other. SLIP also has no means of identifying the protocol it is carrying in its frame, which prevents it from multiplexing network-layer protocols (such as IP and IPX) over a single connection. SLIP also has no error-detection or correction capabilities, which leaves these tasks to the upper-layer protocols, causing greater delays than a data link–layer error-detection mechanism would.

Compressed SLIP (CSLIP)

When two systems communicate using SLIP, much of the control overhead contributed by the protocols at the network and transport layers becomes redundant, particularly in TCP connections. For example, every IP datagram contains 64 bits of data devoted to the IP addresses of the source and destination systems. Since there are only two computers on the network, repeating these addresses in every packet is unnecessary.

NETWORK PROTOCOLS

During the course of a typical SLIP connection, the two systems can exchange hundreds or thousands of packets with identical information in the network- and transport-layer protocol headers.

RFC 1144, "Compressing TCP/IP Headers for Low-Speed Serial Links," defines a mechanism by which the systems participating in a SLIP connection omit much of the redundant information from the headers, reducing the overhead from 40 bytes down to 5 bytes or less. This can speed up the performance of the connection considerably.

This type of header compression is also found in many PPP implementations as *Van Jacobson header compression*, named for the author of RFC 1144.

PPP

PPP was created as an alternative to SLIP that provides greater functionality, such as the capability to multiplex different network-layer protocols and support various authentication protocols. Naturally, the cost of these additional features is a larger header, but PPP still only adds a maximum of 8 bytes to a packet (as compared to the 16 bytes needed for an Ethernet frame). Most of the connections to Internet service providers, whether by standalone systems or routers, use PPP because it enables the ISP to implement access control measures that protect their networks from intrusion by unauthorized users.

A typical PPP session consists of several connection establishment and termination procedures, using other protocols in addition to the PPP itself. These procedures are as follows:

Connection establishment   The system initiating the connection uses the Link Control Protocol (LCP) to negotiate communication parameters that the two machines have in common.

Authentication   Although not required, the system may use an authentication protocol such as PAP (the Password Authentication Protocol) or CHAP (the Challenge Handshake Authentication Protocol) to negotiate access to the other system.

Network-layer protocol connection establishment   For each network-layer protocol that the systems use during the session, they perform a separate connection establishment procedure using a Network Control Protocol (NCP) such as IPCP (the Internet Protocol Control Protocol).

Unlike SLIP, PPP is standardized, but the specifications are divided among several different RFCs. The documents for each of the protocols are listed in Table 13-4.

Document	Title
RFC 1661	The Point-to-Point Protocol (PPP)
RFC 1662	PPP in HDLC-like Framing
RFC 1663	PPP Reliable Transmission
RFC 1332	The PPP Internet Protocol Control Protocol (IPCP)
RFC 1552	The PPP Internetworking Packet Exchange Control Protocol (IPXCP)
RFC 1334	PPP Authentication Protocols
RFC 1994	PPP Challenge Handshake Authentication Protocol (CHAP)
RFC 1989	PPP Link Quality Monitoring

Table 13-4: *PPP and Related Standards*

The PPP Frame

RFC 1661 defines the basic frame used by the PPP protocol to encapsulate other protocols and transmit them to the destination. The frame is small, only 8 (or sometimes 10) bytes, and is illustrated in Figure 13-4.

Figure 13-4: *The PPP frame format*

The functions of the fields are as follows:

Flag (1 byte)   Contains a hexadecimal value of *7e* and functions as a packet delimiter, like SLIP's END character.

Address (1 byte)   Contains a hexadecimal value of *ff*, indicating the packet is addressed to all stations.

Control (1 byte)   Contains a hexadecimal value of *03*, identifying the packet as containing an HDLC unnumbered information message.

Protocol (2 bytes)   Contains a code identifying the protocol that generated the information in the data field. Code values in the *0xxx* to *3xxx* range are used to identify network-layer protocols, values from *4xxx* to *7xxx* identify low-volume network-layer protocols with no corresponding NCP, values from *8xxx* to *bxxx* identify network-layer protocols with corresponding NCPs, and values from *cxxx* to *fxxx* identify link-layer control protocols like LCP and the authentication protocols. The permitted codes, specified in the TCP/IP "Assigned Numbers" document (RFC 1700), include the following:

0021   Uncompressed IP datagram (used when Van Jacobson compression is enabled)

002b   Novell IPX datagram

002d   IP datagrams with compressed IP and TCP headers (used when Van Jacobson compression is enabled)

002f   IP datagrams containing uncompressed TCP data (used when Van Jacobson compression is enabled)

8021   Internet Protocol Control Protocol (IPCP)

802b   Novell IPX Control Protocol (IPXIP)

c021   Link Control Protocol (LCP)

c023   Password Authentication Protocol (PAP)

c223   Challenge Handshake Authentication Protocol (CHAP)

Data and Pad (variable, up to 1,500 bytes)   Contains the payload of the packet, up to a default maximum length (called the *maximum receive unit*, or *MRU*) of 1,500 bytes. The field may contain meaningless bytes to bring its size up to the MRU.

Frame Check Sequence (FCS)(2 or 4 bytes)   Contains a CRC value calculated on the entire frame, excluding the flag and frame check sequence fields, for error-detection purposes.

Flag (1 byte)   Contains the same value as the flag field at the beginning of the frame. When a system transmits two packets consecutively, one of the flag fields is omitted, as two would be mistaken as an empty frame.

Several of the fields in the PPP frame can be modified as a result of LCP negotiations between the two systems, such as the length of the protocol and FCS fields and the MRU for the data field. The systems can agree to use a 1-byte protocol field or a 4-byte FCS field.

The LCP Frame

PPP systems use LCP to negotiate their capabilities during the connection establishment process, so they can achieve the most efficient possible connection. LCP messages are carried within PPP frames and contain configuration options for the connection. Once the two systems agree on a configuration they can both support, the link establishment process continues. By specifying the parameters for the connection during the link establishment process, the systems don't have to include redundant information in the header of every data packet.

The LCP message format is shown in Figure 13-5. The functions of the individual fields are listed below.

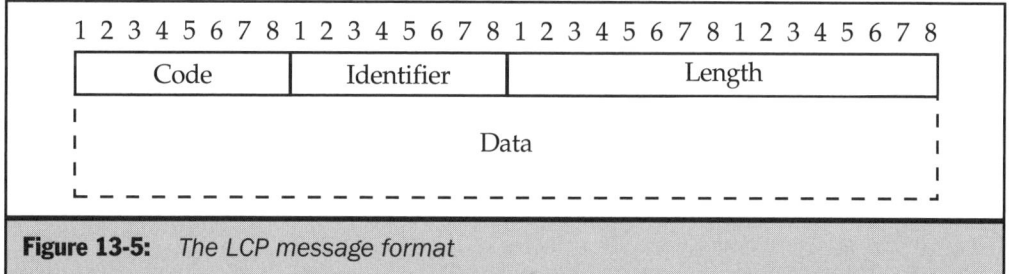

Figure 13-5:   *The LCP message format*

Code (1 byte)   Specifies the LCP message type, using the following codes:

1   Configure-Request

2   Configure-Ack

3   Configure-Nak

4   Configure-Reject

5   Terminate-Request

6   Terminate-Ack

7   Code-Reject

8   Protocol-Reject

9   Echo-Request

10   Echo-Reply

11   Discard-Request

Identifier (1 byte)   Contains a code used to associate the request and replies of a particular LCP transaction.

Length (2 bytes) Specifies the length of the LCP message, including the code, identifier, length, and data fields.

Data (variable) Contains multiple configuration options, each of which is composed of three subfields.

Each of the options in the LCP message's data field consists of the subfields shown in Figure 13-6. The functions of the subfields are listed below.

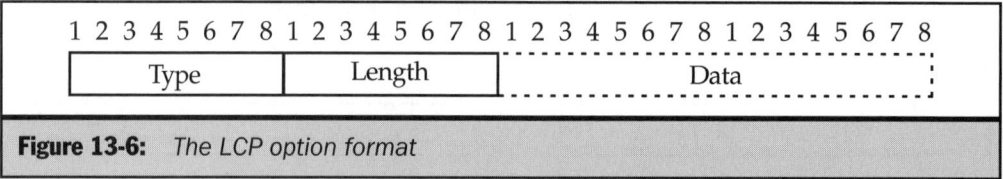

Figure 13-6: *The LCP option format*

Type (1 byte) Specifies the option to be configured, using a code from the "Assigned Numbers" RFC, as follows:

0 Vendor Specific

1 Maximum Receive Unit

2 Async Control Character Map

3 Authentication Protocol

4 Quality Protocol

5 Magic Number

6 Reserved

7 Protocol Field Compression

8 Address and Control Field Compression

9 FCS Alternatives

10 Self-Describing Pad

11 Numbered Mode

12 Multilink Procedure

13 Callback

14 Connect Time

15 Compound Frames

16 Nominal Data Encapsulation

17 Multilink MRRU

18 Multilink Short Sequence Number Header Format

19 Multilink Endpoint Discriminator

20 Proprietary

21 DCE Identifier

Length (1 byte) Specifies the length of the LCP message, including the code, identifier, length, and data fields.

Data (variable) Contains information pertinent to the specific LCP message type, as indicated by the code field.

The LCP protocol is also designed to be extensible. By using a code value of 0, vendors can supply their own options without standardizing them with the IANA, as documented in RFC 2153, "PPP Vendor Extensions."

Authentication Protocols

PPP connections can optionally require authentication to prevent unauthorized access, using an external protocol agreed on during the exchange of LCP configuration messages and encapsulated within PPP frames. Two of the most popular authentication protocols—PAP and CHAP—are defined by TCP/IP specifications, but systems can also use other proprietary protocols developed by individual vendors.

The PAP Frame PAP is the inherently weaker of the two primary authentication protocols because it uses only a two-way handshake and transmits account names and passwords over the link in clear text. Systems generally use PAP only when they have no other authentication protocols in common. PAP packets have a value of *c023* in the PPP header's protocol field and use a message format that is basically the same as LCP, except for the options. The functions of the message fields are as follows:

Code (1 byte) Specifies the type of PAP message, using the following values:

1 Authenticate Request

2 Authenticate Ack

3 Authenticate Nak

Identifier (1 byte) Contains a code used to associate the request and replies of a particular PAP transaction.

Length (2 bytes)   Specifies the length of the PAP message, including the code, identifier, length, and data fields.

Data (variable)   Contains a number of subfields, depending on the value in the code field, as follows:

Peer ID Length (1 byte)   Specifies the length of the peer ID field (Authenticate Request messages only).

Peer ID (variable)   Specifies the account the destination computer will use to authenticate the source system (Authenticate Request messages only).

Password Length (1 byte)   Specifies the length of the password field (Authenticate Request messages only).

Password (variable)   Specifies the password associated with the account name in the peer ID field (Authenticate Request messages only).

Message Length (1 byte)   Specifies the length of the message field (Authenticate Ack/Authenticate Nak messages only).

Message (variable)   Contains a text message that will be displayed on the user interface describing the success or failure of the authentication procedure (Authenticate Ack and Authenticate Nak messages only).

The CHAP Frame   The CHAP protocol is considerably more secure than PAP because it uses a three-way handshake and never transmits account names and passwords in clear text. CHAP packets have a value of *c223* in the PPP header's protocol field and use a message format almost identical to PAP's. The functions of the message fields are as follows:

Code (1 byte)   Specifies the type of CHAP message, using the following values:

1   Challenge

2   Response

3   Success

4   Failure

Identifier (1 byte)   Contains a code used to associate the request and replies of a particular CHAP transaction.

Length (2 bytes)   Specifies the length of the CHAP message, including the code, identifier, length, and data fields.

Data (variable)   Contains a number of subfields, depending on the value of the code field, as follows:

Value Size (1 byte)   Specifies the length of the value field (Challenge and Response messages only).

Value (variable)   In a Challenge message, contains a unique byte string that the recipient uses along with the contents of the identifier field and an encryption "secret" to generate the value field for the Response message (Challenge and Response messages only).

Name (variable)   Contains a string that identifies the transmitting system (Challenge and Response messages only).

Message (variable)   Contains a text message to be displayed on the user interface describing the success or failure of the authentication procedure (Success and Failure messages only).

The IPCP Frame

PPP systems use Network Control Protocols (NCPs) to negotiate connections for each of the network-layer protocols they will use during the session. Before a system can multiplex the traffic generated by different protocols over a single PPP connection, it must establish a connection for each protocol using the appropriate NCPs.

The *Internet Protocol Control Protocol (IPCP),* which is the NCP for IP, is a good example of the protocol structure. The message format of the NCPs is nearly identical to that of LCP, except that it supports only values 1 through 7 for the code field (the link configuration, link termination, and code reject values) and uses different options in the data field. Like LCP, the messages are carried in PPP frames, but with a value of *8021* in the PPP header's protocol field.

The options that can be included in the data field of an IPCP message use the following values in the type field:

2 (IP Compression Protocol)   Specifies the protocol the system should use to compress IP headers, for which the only valid option is Van Jacobson compression.

3 (IP Address)   Used by the transmitting system to request a particular IP address or, if the value is 0.0.0.0, to request that the receiving system supply an address (replaces the type 1 IP Addresses option, which is no longer used).

PPP Connection Establishment

Once the physical-layer connection between the two systems has been established (through a modem handshake or other procedure), the PPP connection establishment process begins. The two systems pass through several distinct phases during the course of the session, as illustrated in Figure 13-7 and discussed in the following sections.

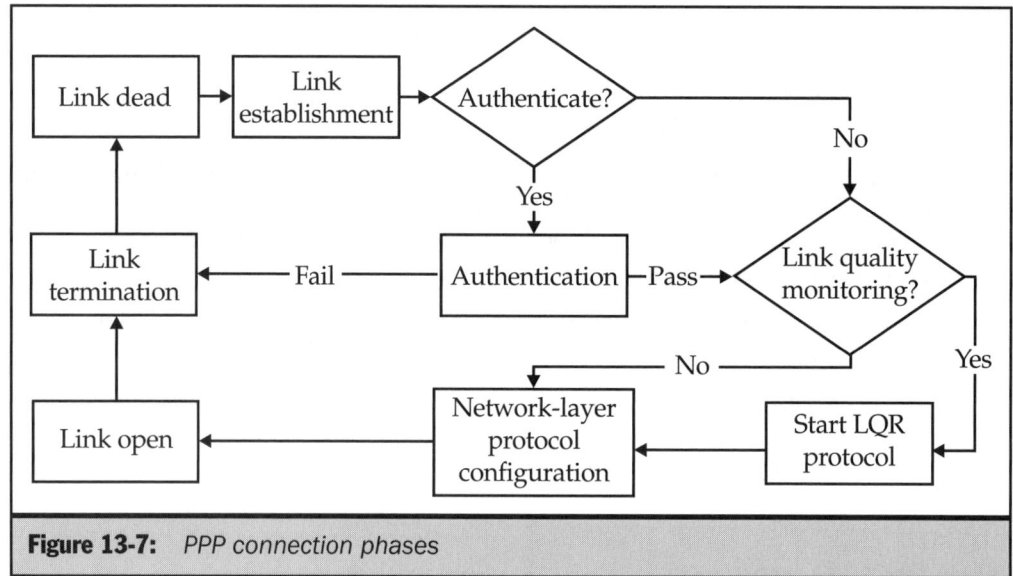

Figure 13-7: *PPP connection phases*

Link Dead   Both systems begin and end the session in the Link Dead phase, which indicates that no physical-layer connection exists between the two machines. On a typical session, an application or service on one system initiates the physical-layer connection by dialing the modem or using some other means. Once the hardware connection process is completed, the systems pass into the Link Establishment phase.

Link Establishment   In the Link Establishment phase, the system initiating the connection transmits an LCP Configure Request message to the destination containing the options it would like to enable, such as the use of specific authentication, link-quality monitoring, and network-layer protocols (if any), and whether the systems should modify standard features, such as the size of the FCS field or a different MRU value. If the receiving system can support all the specified options, it replies a Configure Ack message containing the same option values, and this phase of the connection process is completed.

If the receiving system recognizes the options in the request message, but cannot support the values for those options supplied by the sender (such as if the system supports authentication, but not with the protocol the sender has specified), it replies with a Configure Nak message containing the options with values it cannot support. With these options, the replying system supplies all the values it does support and also may include other options it would like to see enabled. Using this information, the connecting system generates another Configure Request message containing options it knows are supported, to which the receiver replies with a Configure Ack message.

If the receiving system fails to recognize any of the options in the request, it replies with a Configure Reject message containing only the unrecognized options. The sender then generates a new Configure Request message that does not contain the rejected options and the procedure continues as previously outlined. Eventually, the systems perform a successful request/acknowledgment exchange and the connection process moves on to the next phase.

Authentication   The Authentication phase of the connection process is optional and is triggered by the inclusion of the Authentication Protocol option in the LCP Configure Request message. During the LCP link establishment process, the two systems agree on an authentication protocol to use. Use of the PAP and CHAP protocols is common, but other proprietary protocols are available.

The message format and exchange procedures for the Authentication phase are dictated by the selected protocol. In a PAP authentication, for example, the sending system transmits an Authenticate Request message containing an account name and password and the receiver replies with either an Authenticate Ack or Authenticate Nak message.

CHAP is inherently more secure than PAP and requires a more complex message exchange. The sending system transmits a Challenge message containing data that the receiver uses with its encryption key to compute a value it returns to the sender in a Response message. Depending on whether the value in the response matches the sender's own computations, it transmits a Success or Failure message.

A successful transaction causes the connection procedure to proceed to the next phase, but the effect of a failure is dictated by the implementation of the protocol. Some systems proceed directly to the Link Termination phase in the event of an authentication failure, while others might permit retries or limited network access to a help subsystem.

Link Quality Monitoring   The use of a link quality monitoring protocol is also an optional element of the connection process, triggered by the inclusion of the Quality Protocol option in the LCP Configure Request message. Although the option enables the sending system to specify any protocol for this purpose, only one has been standardized, the Link Quality Report protocol. The negotiation process that occurs at this phase enables the systems to agree on an interval at which they should transmit messages containing link traffic and error statistics throughout the session.

Network-Layer Protocol Configuration   PPP supports the multiplexing of network layer protocols over a single connection, and during this phase, the systems perform a separate network-layer connection establishment procedure for each of the network-layer protocols that they have agreed to use during the Link Establishment phase. Each network-layer protocol has its own network control protocol (NCP) for this purpose, such as the Internet Protocol Control Protocol (IPCP) or the Internetworking Packet Exchange Control Protocol (IPXCP). The structure of an NCP message exchange is similar to that of LCP, except the options carried in the Configure Request message are unique to the requirements of the protocol. During an IPCP exchange, for example, the systems inform each other of their IP addresses and agree on whether or not to use Van Jacobson header compression. Other protocols have their own individual needs that the systems negotiate as needed. NCP initialization and termination procedures can also occur at any other time during the connection.

Link Open   Once the individual NCP exchanges are completed, the connection is fully established and the systems enter the Link Open phase. Network-layer protocol data can now travel over the link in either direction.

Link Termination   When one of the systems ends the session, or as a result of other conditions such as a physical-layer disconnection, an authentication failure, or an inactivity timeout, the systems enter the Link Termination phase. To sever the link, one system transmits an LCP Terminate Request message, to which the other system replies with a Terminate Ack. Both systems then return to the Link Dead phase.

NCPs also support the Terminate Request and Terminate Ack messages, but they are intended for use while the PPP connection remains intact. In fact, the PPP connection can remain active even if all of the network-layer protocol connections have been terminated. It is unnecessary for systems to terminate the network-layer protocol connections before terminating the PPP connection.

ARP

The Address Resolution Protocol (ARP) occupies an unusual place in the TCP/IP suite because it defies all attempts at categorization. Unlike most of the other TCP/IP protocols, ARP messages are not carried within IP datagrams. A separate protocol identifier is defined in the "Assigned Numbers" document that data link–layer protocols use to indicate that they contain ARP messages. Because of this, there is some difference of opinion about the layer of the protocol stack to which ARP belongs. Some say ARP is a link–layer protocol because it provides a service to IP, while others associate it with the internet layer because its messages are carried within link–layer protocols.

The function of the ARP protocol, as defined in RFC 826, "An Ethernet Address Resolution Protocol," is to reconcile the IP addresses used to identify systems at the upper layers with the hardware addresses at the data-link layer. When it requests network resources, a TCP/IP application supplies the destination IP address used in the IP protocol header. The system may discover the IP address using a DNS or NetBIOS name-resolution process, or it may use an address supplied by an operating system or application configuration parameter.

Data link–layer protocols like Ethernet, however, have no use for IP addresses and cannot read the contents of the IP datagram anyway. To transmit the packet to its destination, the data link–layer protocol must have the hardware address coded into the destination system's network interface adapter. ARP converts IP addresses into hardware addresses by broadcasting request packets containing the IP address on the local network and waiting for the holder of that IP address to respond with a reply containing the equivalent hardware address.

ARP was originally developed for use with DIX Ethernet networks, but has been generalized to allow its use with other data link–layer protocols.

The biggest difference between IP addresses and hardware addresses is that IP is responsible for the delivery of the packet to its ultimate destination, while an Ethernet implementation is only concerned with delivery to the next stop on the journey. If the packet's destination is on the same network segment as the source, the IP protocol uses ARP to resolve the IP address of the ultimate destination into a hardware address. If, however, the destination is located on another network, the IP protocol will not use ARP to resolve the ultimate destination address (that is, the destination address in the IP header). Instead, it will pass the IP address of the default gateway to the ARP protocol for address resolution.

This is because the data-link protocol header must contain the hardware address of the next intermediate stop as its destination, which may well be a router. It is up to that router to forward the packet on the next leg of its journey. Thus, in the course of a single internetwork transmission, many different machines may perform ARP resolutions on the same packet with different results.

ARP Message Format

ARP messages are carried directly within data link–layer frames, using *0806* as the Ethertype or SNAP Local Code value to identify the protocol being carried in the packet. There is one format for all of the ARP message types, which is illustrated in Figure 13-8.

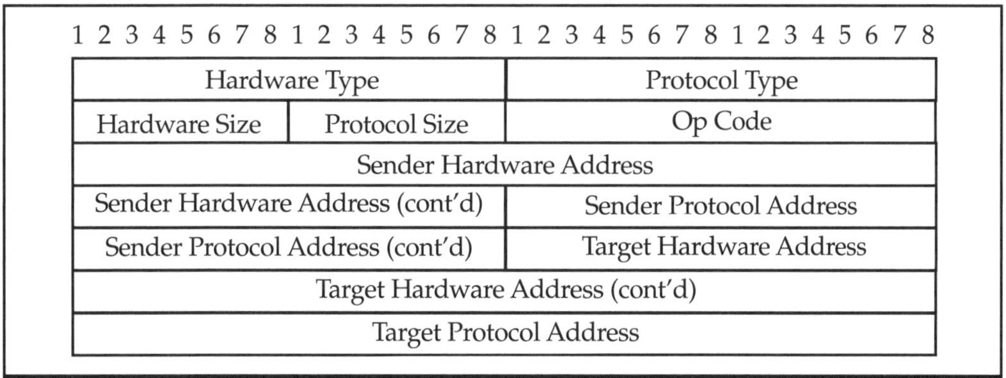

1 2 3 4 5 6 7 8	1 2 3 4 5 6 7 8	1 2 3 4 5 6 7 8	1 2 3 4 5 6 7 8
Hardware Type		Protocol Type	
Hardware Size	Protocol Size	Op Code	
Sender Hardware Address			
Sender Hardware Address (cont'd)		Sender Protocol Address	
Sender Protocol Address (cont'd)		Target Hardware Address	
Target Hardware Address (cont'd)			
Target Protocol Address			

Figure 13-8:   *The ARP message format*

The functions of the fields are as follows:

Hardware Type (2 bytes)   Specifies the type of hardware addresses found in the Sender Hardware Address and Target Hardware Address fields. The hexadecimal value for Ethernet is 0001.

Protocol Type (2 bytes)   Specifies the type of protocol addresses found in the Sender Protocol Address and Target Protocol Address fields. The hexadecimal value for IP addresses is *0800* (the same as the Ethertype value for IP).

Hardware Size (1 byte)   Specifies the size (in bytes) of the hardware addresses found in the Sender Hardware Address and Target Hardware Address fields. The value for Ethernet hardware addresses is 6.

Protocol Size (1 byte)   Specifies the size (in bytes) of the protocol addresses found in the Sender Protocol Address and Target Protocol Address fields. The value for IP addresses is 4.

Opcode (2 bytes)   Specifies the type of message contained in the packet, using the following values:

1   ARP Request

2   ARP Reply

3   RARP Request

4   RARP Reply

Sender Hardware Address (length specified by the value of the Hardware Size field)   Specifies the hardware (such as Ethernet) address of the system sending the message, in both requests and replies.

Sender Protocol Address (length specified by the value of the Protocol Size field)   Specifies the protocol (such as IP) address of the system sending the message, in both requests and replies.

Target Hardware Address (length specified by the value of the Hardware Size field)   Left blank in request messages; in replies, contains the value of the Sender Hardware Address field in the associated request.

Target Protocol Address (length specified by the value of the Protocol Size field)   Specifies the protocol (such as IP) address of the system to which the message is being sent, in both requests and replies.

Note   *The RARP Request and RARP Reply message types are not used in the course of standard TCP/IP network traffic. For more information on the Reverse Address Resolution Protocol (RARP), see Chapter 23.*

ARP Transactions

An ARP transaction occurs when the IP protocol in a TCP/IP system is ready to transmit a datagram over the network. The system knows its own hardware and IP addresses, as well as the IP address of the packet's intended destination. All it lacks is the hardware address of the system on the local network that is to receive the packet. The ARP message exchange proceeds according to the following steps:

1. The transmitting system generates an ARP Request packet containing its own addresses in the Sender Hardware Address and Sender Protocol Address fields (see the captured packet shown in Figure 13-9). The Target Protocol Address contains the IP address of the system on the local network that is to receive the datagram, while the Target Hardware Address is left blank. Some implementations insert a broadcast address or other value into the Target Hardware Address field of the ARP Request message, but this value is ignored by the recipient because this is the address the protocol is trying to ascertain.

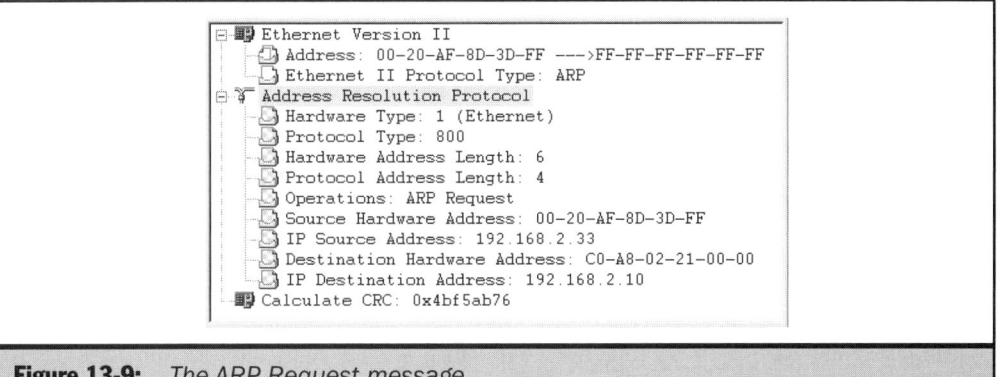

```
Ethernet Version II
    Address: 00-20-AF-8D-3D-FF ---)FF-FF-FF-FF-FF-FF
    Ethernet II Protocol Type: ARP
Address Resolution Protocol
    Hardware Type: 1 (Ethernet)
    Protocol Type: 800
    Hardware Address Length: 6
    Protocol Address Length: 4
    Operations: ARP Request
    Source Hardware Address: 00-20-AF-8D-3D-FF
    IP Source Address: 192.168.2.33
    Destination Hardware Address: C0-A8-02-21-00-00
    IP Destination Address: 192.168.2.10
Calculate CRC: 0x4bf5ab76
```

Figure 13-9: *The ARP Request message*

2. The system transmits the ARP Request message as a broadcast to the local network, asking in effect, "Who is using this IP address and what is your hardware address?"

3. Each TCP/IP system on the local network receives the ARP Request broadcast and examines the contents of the Target Protocol Address field. If the system does not use that address on one of its network interfaces, it silently discards the packet. If the system does use the address, it generates an ARP Reply message in response. The system uses the contents of the request message's Sender Hardware Address and Sender Protocol Address fields as the values for its reply message's Target Hardware Address and Target Protocol Address fields. The system then inserts its own hardware address and IP address into the Sender Hardware Address and Sender Protocol Address fields, respectively (see Figure 13-10).

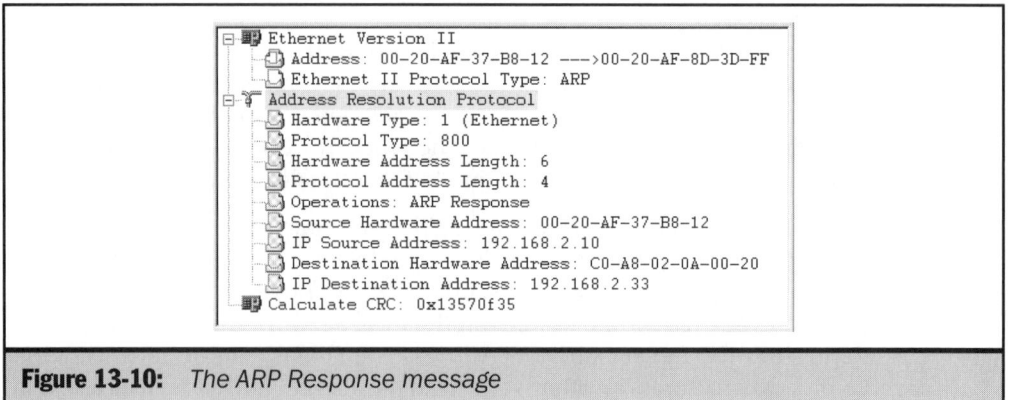

Figure 13-10: *The ARP Response message*

4. The system using the requested IP address transmits the reply message as a unicast back to the original sender. On receipt of the reply, the system that initiated the ARP exchange uses the contents of the Sender Hardware Address field as the Destination Address for the data link–layer transmission of the IP datagram.

ARP Caching

Because of its reliance on broadcast transmissions, ARP can generate a significant amount of network traffic. To lessen the burden of the protocol on the network, TCP/IP systems cache the hardware addresses discovered through ARP transactions in memory for a designated period of time. This way, a system transmitting a large string of datagrams to the same host doesn't have to generate individual ARP requests for each packet.

This is particularly helpful in an internetwork environment in which systems routinely transmit the majority of their packets to destinations on other networks. When a network segment has only a single router, all IP datagrams destined for other networks are sent through that router. When systems have the hardware address for that router in the ARP cache, they can transmit the majority of their datagrams without using ARP broadcasts.

The amount of time that entries remain in the ARP cache varies with different TCP/IP implementations. Windows systems purge entries after two minutes when they are not used to transmit additional datagrams.

Note
The Windows operating systems include a command-line utility called Arp.exe, which you can use to create entries manually in the ARP cache. Unlike dynamically created entries, manual ARP cache entries are permanent. If you have a stable network, you can reduce network traffic by adding ARP cache entries for the routers that provide access to other networks (and particularly to the Internet). Client applications (like Web browsers) are then able to access systems on other networks without generating repeated ARP Request broadcasts.

IP

The Internet Protocol (IP), as defined in RFC 791, is the primary carrier protocol for the TCP/IP suite. IP is essentially the envelope that carries the messages generated by most of the other TCP/IP protocols. Operating at the network layer of the OSI model, IP is a connectionless, unreliable protocol that performs several functions that are a critical part of getting packets from the source system to the destination. Among these functions are:

Addressing   Identifying the system that will be the ultimate recipient of the packet.

Packaging   Encapsulating transport-layer data in datagrams for transmission to the destination.

Fragmenting   Splitting datagrams into sections small enough for transmission over a network.

Routing   Determining the path of the packet through the internetwork to the destination.

The following sections examine these functions in more detail.

Addressing

IP is the protocol responsible for the delivery of TCP/IP packets to their ultimate destination. It is vital to understand how this differs from the addressing performed by a data link–layer protocol like Ethernet or Token Ring. Data link–layer protocols are only aware of the machines on the local network segment. No matter where the packet finally ends up, the destination address in the data link–layer protocol header is always that of a machine on a local network.

If the ultimate destination of the packet is a system on another network segment, the data link–layer protocol address will point to a router that provides access to that segment. On receipt of the packet, the router strips off the data link–layer protocol header and generates a new one containing the address of the packet's next intermediate destination, called a *hop*. Thus, throughout the packet's journey, the data-link protocol header will contain a different destination address for each hop.

The destination address in the IP header, however, always points to the final destination of the packet, regardless of the network on which it's located, and it never changes throughout the journey. IP is the first protocol in the stack (working up from the bottom) to be conscious of the packet's end-to-end journey from source to destination. Most of the protocol's functions revolve around the preparation of the transport-layer data for transmission across multiple networks to the destination.

Packaging

IP is also responsible for packaging transport-layer protocol data into structures called *datagrams* for its journey to the destination. During the journey, routers apply a new data link–layer protocol header to a datagram for each hop. Before reaching its final destination, a packet may pass through networks using several different data link–layer protocols, each of which requires a different header. The IP "envelope," on the other hand, remains intact throughout the entire journey, except for a few bits that are modified along the way, just as a mailing envelope is postmarked.

As it receives data from the transport-layer protocol, IP packages it into datagrams of a size suitable for transmission over the local network. A datagram (in most cases) consists of a 20-byte header plus the transport-layer data. The header is illustrated in Figure 13-11.

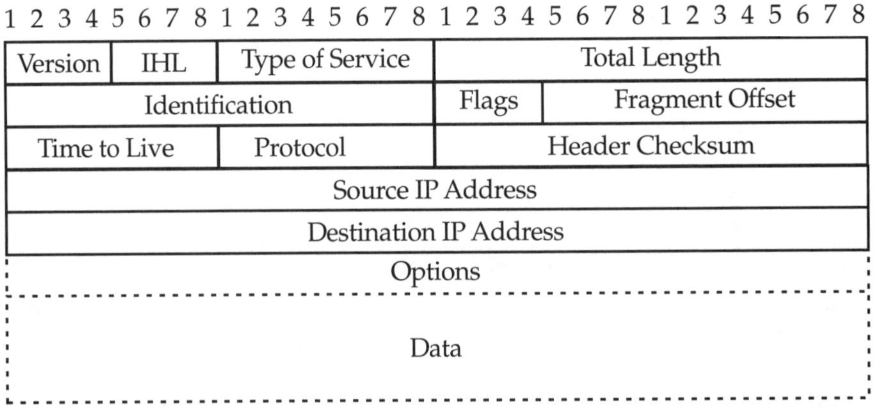

Figure 13-11: *The IP header format*

The functions of the header fields are as follows:

Version, 4 bits Specifies the version of the IP protocol in use. The value for the current implementation is 4.

IHL (Internet Header Length), 4 bits Specifies the length of the IP header, in 32-bit words. When the header contains no optional fields, the value is 5.

TOS (Type of Service), 1 byte Bits 1 through 3 and 8 are unused. Bits 4 through 7 specify the service priority desired for the datagram, using the following values:

0000 Default

0001 Minimize Monetary Cost

0010 Maximize Reliability

0100 Maximize Throughput

1000 Minimize Delay

1111 Maximize Security

Total Length, 2 bytes Specifies the length of the datagram, including all the header fields and the data.

Identification, 2 bytes Contains a unique value for each datagram, used by the destination system to reassemble fragments.

Flags, 3 bits Contains bits used during the datagram fragmentation process, with the following values:

Bit 1 Not used

Bit 2 (Don't Fragment) When set to a value of 1, prevents the datagram from being fragmented by any system.

Bit 3 (More Fragments) When set to a value of 0, indicates that the last fragment of the datagram has been transmitted; when set to 1, indicates that fragments still await transmission.

Fragment Offset, 13 bits Specifies the location (in 8-byte units) of the current fragment in the datagram.

TTL (Time to Live), 1 byte Specifies the number of routers the datagram should be permitted to pass through on its way to the destination. Each router that processes the packet decrements this field by 1. Once the value reaches 0, the packet is discarded, whether or not it has reached the destination.

Protocol, 1 byte Identifies the protocol that generated the information in the data field, using values found in the "Assigned Numbers" RFC (RFC 1700) and the PROTOCOL file found on every TCP/IP system, some of which are as follows:

1 Internet Control Message Protocol (ICMP)

2 Internet Group Management Protocol (IGMP)

3 Gateway-to-Gateway Protocol (GGP)

6 Transmission Control Protocol (TCP)

8 Exterior Gateway Protocol (EGP)

17 User Datagram Protocol (UDP)

Header Checksum, 2 bytes Contains a checksum value computer in the IP header fields only, for error-detection purposes.

NETWORK PROTOCOLS

Source IP Address, 4 bytes   Specifies the IP address of the system from which the datagram originated.

Destination IP Address, 4 bytes   Specifies the IP address of the system that will be the ultimate recipient of the datagram.

Options (variable)   Can contain any of 16 options defined in the "Assigned Numbers" RFC, described later in this section.

Data (variable, up to the MTU for the connected network)   Contains the payload of the datagram, consisting of data passed down from a transport-layer protocol.

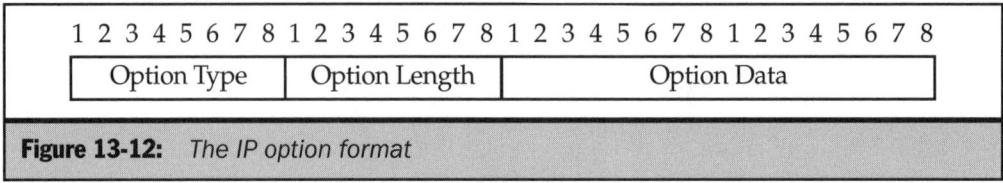

Figure 13-12:   *The IP option format*

Systems use the IP header options to carry additional information, either supplied by the sender or gathered as the packet travels to the destination. Each option is composed of the following fields (see Figure 13-12):

Option Type (1 byte)   Contains a value identifying the option that consists of the following three subfields:

> **Copy Flag (1 bit)**   When set to a value of 1, indicates the option should be copied to each of the fragments that comprise the datagram.
>
> **Option Class (2 bits)**   Contains a code that identifies the option's basic function, using the following values:
>
>> 0   Control
>>
>> 2   Debugging and measurement
>
> **Option Number (5 bits)**   Contains a unique identifier for the option, as specified in the "Assigned Numbers" RFC.

Option Length (1 byte)   Specifies the total length of the option, including the Option Type, Option Length, and Option Data fields.

Option Data (Option Length minus 2)   Contains the option-specific information being carried to the destination.

Table 13-5 lists some of the options systems can insert into IP datagrams, the values for the option subfields, and the RFCs that define the option's function. The functions of the options are described in the following sections.

Copy Flag	Option Class	Option Number	Option Value	RFC	Option Name
0	0	0	0	RFC 791	End of Options List
0	0	1	1	RFC 791	No Operation
1	0	3	131	RFC 791	Loose Source Route
0	2	4	68	RFC 791	Time Stamp
0	0	7	7	RFC 791	Record Route
1	0	9	137	RFC 791	Strict Source Route

Table 13-5:   *IP Header Options*

End of Options List   Consisting only of an Option Type field with the value 0, this option marks the end of all the options in an IP header.

No Operation   Consisting only of an Option Type field, systems can use this option to pad out the space between two other options, to force the following option to begin at the boundary between 32-bit words.

Loose Source Route and Strict Source Route   Systems use the Loose Source Route and Strict Source Route options to carry the IP addresses of routers the datagram must pass through on its way to the destination. When a system uses the Loose Source Route option, the datagram can pass through other routers in addition to those listed in the option. The Strict Source Route option defines the entire path of the datagram from the source to the destination.

Time Stamp   This option is designed to hold time stamps generated by one or more systems processing the packet as it travels to its destination. The sending system may supply the IP addresses of the systems that are to add time stamps to the header, or enable the systems to save their IP addresses to the header along with the time stamps, or omit the IP addresses of the time stamping systems entirely. The size of the option is variable to accommodate multiple time stamps, but must be specified when the sender creates the datagram and cannot be enlarged en route to the destination.

Record Route   This option provides the receiving system with a record of all the routers through which the datagram has passed during its journey to the destination. Each router adds its address to the option as it processes the packet.

Fragmenting

The size of the IP datagrams used to transmit the transport-layer data depends on the data link–layer protocol in use. Ethernet networks, for example, can carry datagrams up to 1,500 bytes in size, while Token Ring networks typically support packets as large as 4,500 bytes. The system transmitting the datagram uses the *maximum transfer unit (MTU)* of the connected network, that is, the largest possible frame that can be transmitted using that data link–layer protocol, as one factor in determining how large each datagram should be.

During the course of its journey from the source to the destination, packets may encounter networks with different MTUs. As long as the MTU of each network is larger than the packet, the datagram is transmitted without a problem. If a packet is larger than the MTU of a network, however, it cannot be transmitted in its current form. When this occurs, the IP protocol in the router providing access to the network is responsible for splitting the datagram into fragments smaller than the MTU. The router then transmits each fragment in a separate packet with its own IP header.

Depending on the number and nature of the networks it passes through, a datagram may be fragmented more than once before it reaches the destination. A system might split a datagram into fragments that are themselves too large for networks further along in the path. Another router, therefore, splits the fragments into still smaller fragments. Reassembly of a fragmented datagram takes place only at the destination system after it has received all of the packets containing the fragments, not at the intermediate routers.

| Note |

Technically speaking, the datagram is defined as the unit of data, packaged by the source system, containing a specific value on the IP header's Identification field. When a router fragments a datagram, it uses the same Identification value for each new packet it creates, meaning the individual fragments are collectively known as a datagram. Referring to a single fragment as a datagram is incorrect use of the term.

When a router receives a datagram that must be fragmented, it creates a series of new packets using the same value for the IP header's Identification field as the original datagram. The other fields of the header are the same as well, with three important exceptions, which are as follows:

- The value of the Total Length field is changed to reflect the size of the fragment, instead of the size of the entire datagram.

- Bit 3 of the Flags field, the More Fragments bit, is changed to a value of 1 to indicate that further fragments are to be transmitted, except in the case of the datagram's last fragment, in which this bit is set to a value of 0.

■ The value of the Fragment Offset field is changed to reflect each fragment's place in the datagram, based on the size of the fragments (which is, in turn, based on the MTU of the network across which the fragments are to be transmitted). The value for the first fragment is 0; the next is incremented by the size of the fragment, in bytes.

These changes to the IP header are needed for the fragments to be properly reassembled by the destination system. The router transmits the fragments like any other IP packets, and because IP is a connectionless protocol, the individual fragments may take different routes to the destination and arrive in a different order. The receiving system uses the More Fragments bit to determine when it should begin the reassembly process and the Fragment Offset field to assemble the fragments in the proper order.

Selecting the size of the fragments is left up to individual IP implementations. Typically, the size of each fragment is the MTU of the network over which it must be transmitted, minus the size of the data link and IP protocol headers, and rounded down to the nearest 8 bytes. Some systems, however, automatically create 576-byte fragments, as this is the default path MTU used by many routers.

Fragmentation is not desirable, but it is a necessary evil. Obviously, because fragmenting a datagram creates many packets out of one packet, it increases the control overhead incurred by the transmission process. Also, if one fragment of a datagram is lost or damaged, the entire datagram must be retransmitted. No means of reproducing and retransmitting a single fragment exists because the source system has no knowledge of the fragmentation performed by the intermediate routers. The IP implementation on the destination system does not pass the incoming data up to the transport layer until all the fragments have arrived and been reassembled. The transport-layer protocol must therefore detect the missing data and arrange for the retransmission of the datagram.

Routing

Because the IP protocol is responsible for the transmission of packets to their final destinations, IP determines the route the packets will take. A packet's route is the path it takes from one end system, the source, to another end system, the destination. The routers the packet passes through during the trip are called *intermediate systems*. The fundamental difference between end systems and intermediate systems is how high the packet data reaches in the protocol stack.

On the source computer, a request for access to a network resource begins at the application layer and wends its way down through the layers of the protocol stack, eventually arriving at the physical layer encapsulated in a packet, ready for transmission. When it reaches the destination, the reverse occurs, and the packet is passed up the stack to the application layer. On end systems, therefore, the entire protocol stack participates in the processing of the data. On intermediate systems, such as routers, the data arriving over the network is passed only as high as the network-layer protocol, which, in this case, is IP (see Figure 13-13).

NETWORK PROTOCOLS

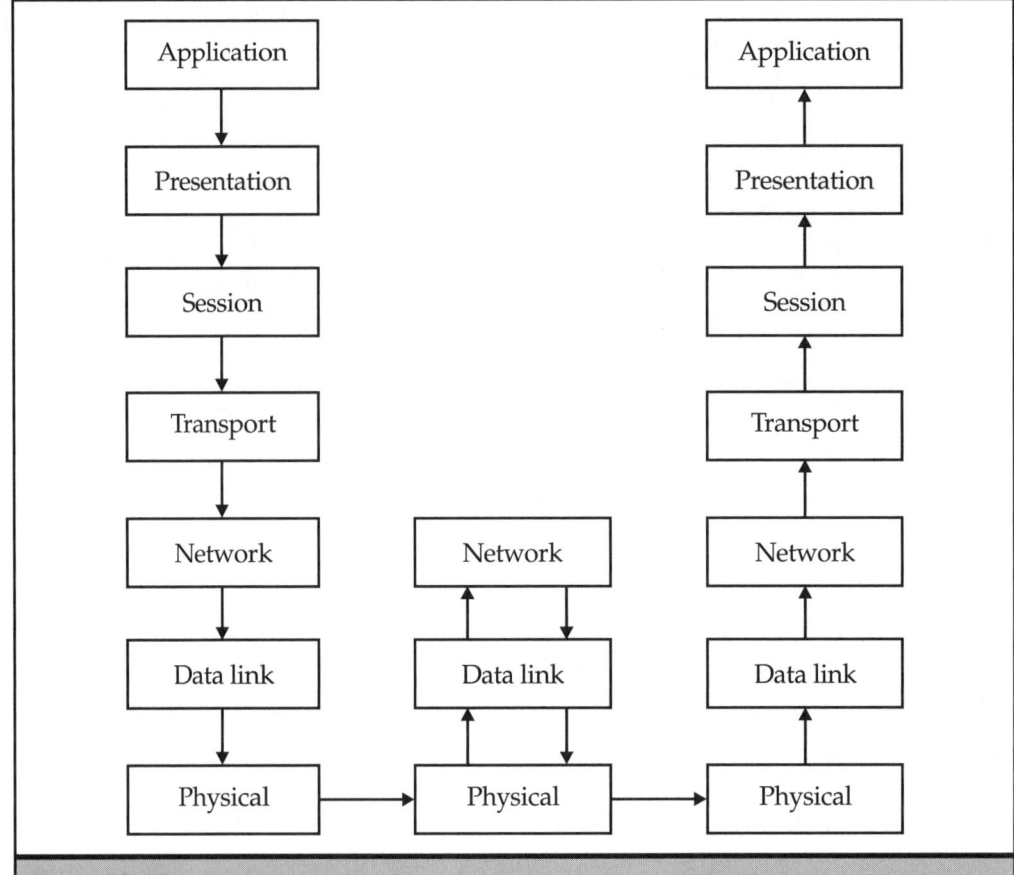

Figure 13-13: *Packets passing through routers travel no higher than the network layer of the protocol stack.*

IP strips off the data link–layer protocol header and, after determining where it should send the packet next, prepares it for packaging in a data link–layer protocol frame suitable for the outgoing network. This may involve using ARP to resolve the IP address of the packet's next stop into a hardware address and then furnishing that address to the data link–layer protocol.

Routing is a process that occurs one hop of a packet's journey at a time. The source system transmits the packet to its default gateway (router) and the router determines where to send the packet next. If the final destination is on a network segment to which the router is attached, it sends the packet there. If the destination is on another network, the router determines which of the other routers it should send the packet to, in order for it to reach its destination most efficiently. Thus, the next destination for the packet, identified by the destination address in the data link–layer protocol, may not be the same system as that specified in the IP header's Destination IP Address field.

Eventually, one of the routers will have access to the network on which the packet's final destination system is located, and will be able to send it directly to that machine. Using this method, the routing process is distributed among the network's routers. None of the computers involved in the process has complete knowledge of the packet's route through the network at any time. This distribution of labor makes huge networks like the Internet possible. No practical method exists for a single system to determine a viable path through the many thousands of routers on the Internet to a specific destination for each packet.

The most complex part of the routing process is the manner in which the router determines where to send each packet next. Routers have direct knowledge only of the network segments to which they are connected. They have no means of unilaterally determining the best route to a particular destination. In most cases, routers gain knowledge about other networks by communicating with other routers using specialized protocols designed for this purpose, such as the Routing Information Protocol (RIP). Each router passes information about itself to the other routers on the networks to which it is connected, those routers update their neighboring routers, and so on.

| **Note** | *For more information on routers, the routing process, and routing protocols, see Chapter 6.* |

Regular updates from the neighboring routers enable each system to keep up with changing conditions on the network. If a router should go down, for example, its neighbors will detect its absence and spread the word that the router is unavailable. The other routers will adjust their behavior as needed to ensure that their packets are not sent down a dead-end street.

Routing protocols enable each router to compile a table of networks with the information needed to send packets to that network. Essentially, the table says "send traffic to network x; use interface y," where y is one of the router's own network interfaces. Administrators can also manually configure routes through the network. This is called *static routing,* as opposed to protocol-based configuration, which is called *dynamic routing*.

On complex networks, there may be several viable routes from a source to a particular destination. Routers continually rate the possible paths through the network, so they can select the shortest, fastest, or easiest route for a packet.

IPv6

As mentioned earlier, no one involved in the original design and implementation of the Internet could have predicted the explosive growth it has undergone in the last five years. The TCP/IP protocols have held up remarkably well over the decades, proving that the scalability features incorporated into them were well designed. However, the single biggest problem with continuing to use these protocols is the rapid consumption of the address space provided by IP, version 4 (IPv4, the current version). IP addresses are no longer being used only by computers; other technologies have been devised that can also make use of the IP address space, including cellular phones, handheld computing devices, and global positioning systems. Anticipating the eventual depletion of the 32-bit address space, work commenced on an upgraded version of IP several years ago, which has resulted in several dozen RFCs, including RFC 2460, "Internet Protocol, Version 6 (IPv6) Specification," published in December 1998.

The primary improvement in IPv6 is the expansion of the address space from 32 to 128 bits. This should provide a sufficient number of IP addresses for all devices that can make use of them (which is probably what the designers of IPv4 said when they decided to use 32-bit addresses). In addition to the expanded address space, IPv6 also includes the following enhancements:

Simplified header format   IPv6 removes extraneous fields from the protocol header and makes other fields optional to reduce the network traffic overhead generated by the protocol.

Header extensions   IPv6 introduces the concept of extension headers, which are separate, optional headers located between the IP header and its payload. The extension headers contain information that is used only by the end system that is the packet's final destination. By moving them into extension headers, the intermediate systems don't have to expend the time and processor clock cycles needed to process them.

Flow labeling   IPv6 enables applications to apply a "flow label" to specific packets in order to request a nonstandard quality of service. This is intended to enable applications that require real-time communications, such as streaming audio and video, to request priority access to the network bandwidth.

Security extensions   IPv6 includes extensions that support authentication, data integrity, and data confidentiality.

IPv6 requires a number of fundamental changes to the hardware and software that make up the network infrastructure, apart from just the adaptation to 128-bit addresses. For example, the operating systems and applications that use IPv6 must also include the IPv6 version of ICMP, defined in RFC 2463. Also, networks that use IPv6 must support a maximum transfer unit value of at least 1,280 bytes. Issues like these complicate the process of transitioning the Internet from IPv4 to IPv6 enormously; the entire transition process is expected to take ten years. RFC 1933 defines mechanisms that are designed to facilitate the transition process, such as support for both IPv4 and IPv6 layers in the same system and the tunneling of IPv6 datagrams within IPv4 datagrams, enabling the existing IPv4 routing infrastructure to carry IPv6 information.

Although IPv6 support has not yet hit the mainstream, some operating systems, such as Sun Solaris 8 and IBM AIX, contain implementations, and for others there are test platforms available. Microsoft, for example, has released an IPv6 Technology Preview for Windows 2000 that you can use to test the functionality of IPv6. This product is available on the Web at http://msdn.microsoft.com/downloads/sdks/platform/tpipv6.asp.

ICMP

The Internet Control Message Protocol (ICMP) is a network-layer protocol that does not carry user data, although its messages are encapsulated in IP datagrams. ICMP fills two roles in the TCP/IP suite: it provides error-reporting functions, informing the sending system when a transmission cannot reach its destination, for example, and it carries query and response messages for diagnostic programs. The ping utility, for instance, which is included in every TCP/IP implementation, uses ICMP echo messages to determine if another system on the network can receive and send data.

The ICMP protocol, as defined in RFC 792, consists of messages carried in IP datagrams, with a value of 1 in the IP header's Protocol field and 0 in the Type of Service field. The ICMP message format is illustrated in Figure 13-14.

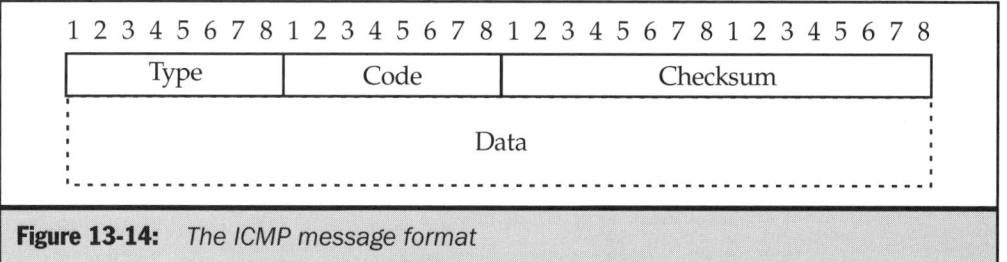

Figure 13-14:   *The ICMP message format*

The ICMP message format consists of the following fields:

Type (1 byte)   Contains a code identifying the basic function of the message.

Code (1 byte)   Contains a secondary code identifying the function of the message within a specific type.

Checksum (2 bytes)   Contains the results of a checksum computation on the entire ICMP message, including the Type, Code, Checksum, and Data fields (with a value of 0 in the Checksum field for computation purposes).

Data (variable)   Contains information specific to the function of the message.
The ICMP message types are listed in Table 13-6.

Type	Code	Query/Error	Function
0	0	Q	Echo Reply
3	0	E	Net Unreachable
3	1	E	Host Unreachable
3	2	E	Protocol Unreachable
3	3	E	Port Unreachable
3	4	E	Fragmentation Needed and Don't Fragment Was Set
3	5	E	Source Route Failed
3	6	E	Destination Network Unknown
3	7	E	Destination Host Unknown
3	8	E	Source Host Isolated
3	9	E	Communication with Destination Network is Administratively Prohibited
3	10	E	Communication with Destination Host is Administratively Prohibited
3	11	E	Destination Network Unreachable for Type of Service
3	12	E	Destination Host Unreachable for Type of Service
4	0	E	Source Quench
5	0	E	Redirect Datagram for the Network (or Subnet)
5	1	E	Redirect Datagram for the Host
5	2	E	Redirect Datagram for the Type of Service and Network

Table 13-6:   *ICMP Message Types*

Type	Code	Query/Error	Function
5	3	E	Redirect Datagram for the Type of Service and Host
8	0	Q	Echo Request
9	0	Q	Router Advertisement
10	0	Q	Router Solicitation
11	0	E	Time to Live Exceeded in Transit
11	1	E	Fragment Reassembly Time Exceeded
12	0	E	Pointer Indicates the Error
12	1	E	Missing a Required Option
12	2	E	Bad Length
13	0	Q	Time stamp
14	0	Q	Time stamp Reply
15	0	Q	Information Request
16	0	Q	Information Reply
17	0	Q	Address Mask Request
18	0	Q	Address Mask Reply
30	0	Q	Traceroute
31	0	E	Datagram Conversion Error
32	0	E	Mobile Host Redirect
33	0	Q	IPv6 Where-are-you
34	0	Q	IPv6 I-am-here
35	0	Q	Mobile Registration Request
36	0	Q	Mobile Registration Reply

Table 13-6:   *ICMP Message Types (continued)*

NETWORK PROTOCOLS

ICMP Error Messages

Because of the way TCP/IP networks distribute routing chores among various systems, there is no way for either of the end systems involved in a transmission to know what has happened during a packet's journey. IP is a connectionless protocol, so no acknowledgment messages are returned to the sender at that level. When using a connection-oriented protocol at the transport layer, like TCP, the destination system acknowledges transmissions, but only for the packets it receives. If something happens during the transmission process that prevents the packet from reaching the destination, there is no way for IP or TCP to inform the sender about what happened.

ICMP error messages are designed to fill this void. When an intermediate system, such as a router, has trouble processing a packet, the router typically discards the packet, leaving the upper-layer protocols to detect the packet's absence and arrange for a retransmission. ICMP messages enable the router to inform the sender of the exact nature of the problem. Destination systems can also generate ICMP messages when a packet arrives successfully, but cannot be processed.

The Data field of an ICMP error message always contains the IP header of the datagram the system could not process, plus the first 8 bytes of the datagram's own Data field. In most cases, these 8 bytes contain a UDP header or the beginning of a TCP header, including the source and destination ports and the sequence number (in the case of TCP). This enables the system receiving the error message to isolate the exact time the error occurred and the transmission that caused it.

However, ICMP error messages are informational only. The system receiving them does not respond nor does it necessarily take any action to correct the situation. The user or administrator may have to address the problem that is causing the failure.

In general, all TCP/IP systems are free to transmit ICMP error messages, except in certain specific situations. These exceptions are intended to prevent ICMP from generating too much traffic on the network by transmitting large numbers of identical messages. These exceptional situations are as follows:

■ TCP/IP systems do not generate ICMP error messages in response to other ICMP error messages. Without this exception, it would be possible for two systems to bounce error messages back and forth between them endlessly. Systems can generate ICMP errors in response to ICMP queries, however.

■ In the case of a fragmented datagram, a system only generates an ICMP error message for the first fragment.

■ TCP/IP systems never generate ICMP error messages in response to broadcast or multicast transmissions, transmissions with a source IP address of 0.0.0.0, or transmissions addressed to the loopback address.

The following sections examine the most common types of ICMP error messages and their functions.

Destination Unreachable Messages   Destination unreachable messages have a value of 3 in the ICMP Type field and any one of 13 values in the Code field. As the name implies, these messages indicate that a packet or the information in a packet could not be transmitted to its destination. The various messages specify exactly which component was unreachable and, in some cases, why. This type of message can be generated by a router when it cannot forward a packet to a certain network or to the destination system on one of the router's connected networks. Destination systems themselves can also generate these messages when they cannot deliver the contents of the packet to a specific protocol or host.

In most cases, the error is a result of some type of failure, either temporary or permanent, in a computer or the network medium. These errors could also possibly occur as a result of IP options that prevent the transmission of the packet, such as when datagrams must be fragmented for transmission over a specific network and the Don't Fragment flag in the IP header is set.

Source Quench Messages   The source quench message, with a Type value of 4 and a Code value of 0, functions as an elementary form of flow control by informing a transmitting system that it is sending packets too fast. When the receiver's buffers are in danger of being overfilled, the system can transmit a source quench message to the sender, which slows down its transmission rate as a result. The sender should continue to reduce the rate until it is no longer receiving the messages from the receiver.

This is a basic form of flow control that is reasonably effective for use between systems on the same network, but that generates too much additional traffic on routed internetworks. In most cases, this is unnecessary because TCP provides its own flow-control mechanism.

Redirect Messages   Redirect messages are generated only by routers to inform hosts or other routers of better routes to a particular destination. In the network diagram shown in Figure 13-15, a host on Network A transmits a packet to another host on Network B and uses Router 1 as the destination of its first hop. After consulting its routing table, Router 1 determines that the packet should be sent to Router 2, but also realizes that Router 2 is located on the same network as the original transmitting host.

Because having the host send the packets intended for that destination directly to Router 2 would be more efficient, Router 1 sends a Redirect Datagram for the Network message (Type 5, Code 0) to the transmitting host after it forwards the original packet to Router 2. The redirect message contains the usual IP header and partial data information, as well as the IP address of the router the host should use for its future transmissions to that network.

In this example, the redirect message indicates that the host should use the other router for the packets it will transmit to all hosts on Network B in the future. The other redirect messages (with Codes 1 through 3) enable the router to specify an alternative router for transmissions to the specific host, to the specific host with the same Type of Service value, and to the entire network with the same Type of Service value.

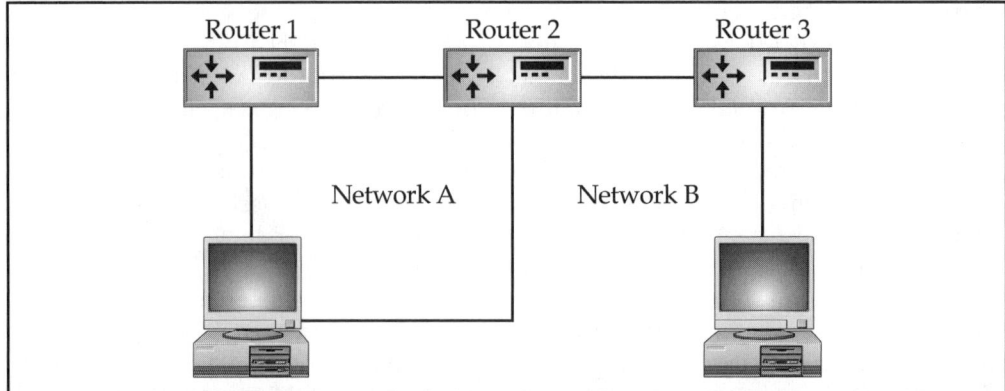

Figure 13-15: *Packets transmitted to a host on another network can often take any one of multiple routes to the destination.*

Time Exceeded Messages   Time exceeded messages are used to inform a transmitting system that a packet has been discarded because a timeout has elapsed. The Time to Live Exceeded in Transit message (Type 11, Code 0) indicates that the Time-to-Live value in a packet's IP header has reached zero before arriving at the destination, forcing the router to discard it.

This message enables the TCP/IP Traceroute program to display the route through the network that packets take to a given destination. By transmitting a series of packets with incremented values in the Time-to-Live field, each successive router on the path to the destination discards a packet and returns an ICMP time exceeded message to the source.

Note   *For more information on Traceroute, see Chapter 31.*

The Fragment Reassembly Time Exceeded message (Code 1) indicates that a destination system has not received all the fragments of a specific datagram within the time limit specified by the host. As a result, the system must discard all the fragments it has received and return the error message to the sender.

ICMP Query Messages

ICMP query messages are not generated in response to other activities, as are the error messages. Systems use them for self-contained request/reply transactions in which one computer requests information from another, which responds with a reply containing that information.

Because they are not associated with other IP transmissions, ICMP queries do not contain datagram information in their Data fields. The data they do carry is specific to the function of the message. The following sections examine some of the more common ICMP query messages and their functions.

Echo Requests and Replies   Echo Request and Echo Reply messages are the basis for the TCP/IP ping utility, which sends test messages to another host on the network to determine if it is capable of receiving and responding to messages. Each ping consists of an ICMP Echo Request message (Type 8, Code 0) that, in addition to the standard ICMP Type, Code, and Checksum fields, adds Identifier and Sequence Number fields the systems use to associate requests and replies. Packet captures of a typical request and reply exchange are shown in Figure 13-16 and Figure 13-17.

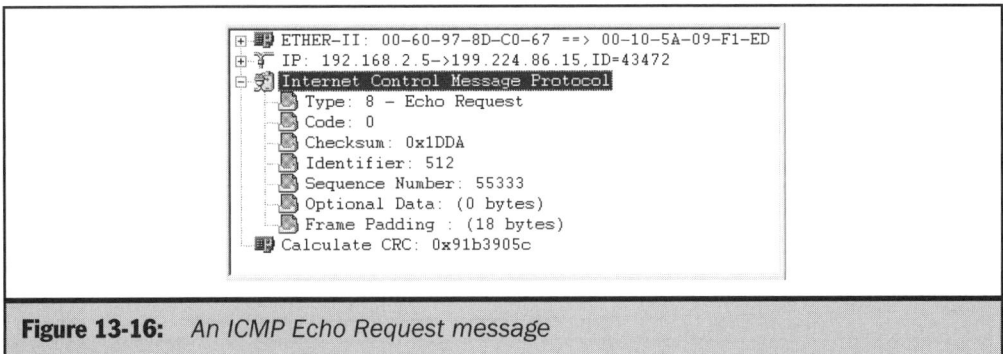

```
⊞ ▦ ETHER-II: 00-60-97-8D-C0-67 ==> 00-10-5A-09-F1-ED
⊞ ▼ IP: 192.168.2.5->199.224.86.15,ID=43472
⊟ 🗐 Internet Control Message Protocol
    🗐 Type: 8 - Echo Request
    🗐 Code: 0
    🗐 Checksum: 0x1DDA
    🗐 Identifier: 512
    🗐 Sequence Number: 55333
    🗐 Optional Data: (0 bytes)
    🗐 Frame Padding : (18 bytes)
  ▦ Calculate CRC: 0x91b3905c
```

Figure 13-16:   *An ICMP Echo Request message*

```
⊞ ▦ ETHER-II: 00-10-5A-09-F1-ED ==> 00-60-97-8D-C0-67
⊞ ▼ IP: 199.224.86.15->192.168.2.5,ID=35904
⊟ 🗐 Internet Control Message Protocol
    🗐 Type: 0 - Echo Reply
    🗐 Code: 0
    🗐 Checksum: 0x25DA
    🗐 Identifier: 512
    🗐 Sequence Number: 55333
    🗐 Optional Data: (0 bytes)
    🗐 Frame Padding : (18 bytes)
  ▦ Calculate CRC: 0xbb644cc9
```

Figure 13-17:   *An ICMP Echo Reply message*

If the system receiving the message is functioning normally, it reverses the Source and Destination IP Address fields in the IP header, changes the value of the ICMP Type field to 8 (Echo Reply), and recomputes the checksum before transmitting it back to the sender.

NETWORK PROTOCOLS

For more information on ping, see Chapter 31.

Router Solicitations and Advertisements   These messages make it possible for a host system to discover the addresses of the routers connected to the local network. Systems can use this information to configure the default gateway entry in their routing tables. When a host broadcasts or multicasts a Router Solicitation message (Type 10, Code 0), the routers on the network respond with Router Advertisement messages (Type 9, Code 0). Routers continue to advertise their availability at regular intervals (typically seven to ten minutes). A host may stop using a router as its default gateway if it fails to receive continued advertisements.

The Router Solicitation message consists only of the standard Type, Code, and Checksum fields, plus a 4-byte pad in the Data field. The Router Advertisement message format is shown in Figure 13-18.

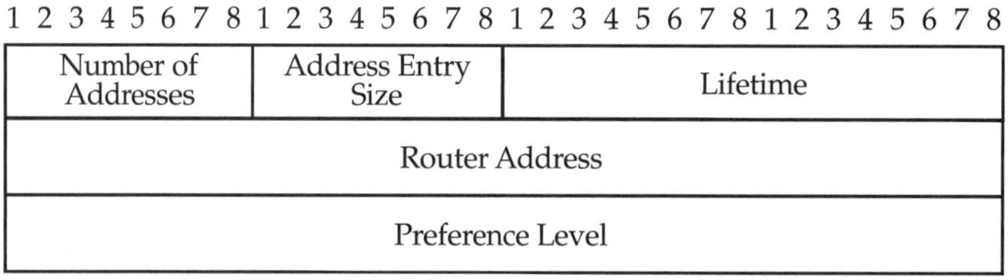

1 2 3 4 5 6 7 8	1 2 3 4 5 6 7 8	1 2 3 4 5 6 7 8 1 2 3 4 5 6 7 8
Number of Addresses	Address Entry Size	Lifetime
Router Address		
Preference Level		

Figure 13-18:   *The Router Advertisement message format*

The Router Advertisement message format contains the following additional fields:

Number of Addresses (1 byte)   Specifies the number of router addresses contained in the message. The format can support multiple addresses, each of which will have its own Router Address and Preference Level fields.

Address Entry Size (1 byte)   Specifies the number of 4-byte words devoted to each address in the message. The value is always 2.

Lifetime (2 bytes)   Specifies the time, in seconds, that can elapse between advertisements before a system assumes a router is no longer functioning. The default value is usually 1,800 seconds (30 minutes).

Router Address (4 bytes)   Specifies the IP address of the router generating the advertisement message.

Preference Level (4 bytes)   Contains a value specified by the network administrator that host systems can use to select one router over another.

UDP

Two TCP/IP protocols operate at the transport layer: TCP and UDP. The *User Datagram Protocol (UDP)*, defined in RFC 768, is a connectionless, unreliable protocol that provides minimal transport service to application-layer protocols with a minimum of control overhead. Thus, UDP provides no packet acknowledgment or flow-control services like TCP, although it does provide end-to-end checksum verification on the contents of the packet.

Although it provides a minimum of services of its own, UDP does function as a *pass-through protocol*, meaning that it provides applications with access to network-layer services, and vice versa. If, for example, a datagram containing UDP data cannot be delivered to the destination and a router returns an ICMP Destination Unreachable message, UDP always passes the ICMP message information up from the network layer to the application that generated the information in the original datagram. UDP also passes along any option information included in IP datagrams to the application layer and, in the opposite direction, information from applications that IP will use as values for the Time-to-Live and Type of Service header fields.

The nature of the UDP protocol makes it suitable only for brief transactions in which all the data to be sent to the destination fits into a single datagram. This is because no mechanism exists in UDP for splitting a data stream into segments and reassembling them, as in TCP. This does not mean that the datagram cannot be fragmented by IP in the course of transmission, however. This process is invisible to the transport layer, as the receiving system reassembles the fragments before passing the datagram up the stack.

In addition, because no packet acknowledgment exists in UDP, it is most often used for client/server transactions in which the client transmits a request and the server's reply message serves as an acknowledgment. If a system sends a request and no reply is forthcoming, the system assumes the destination system did not receive the message and retransmits. It is mostly TCP/IP support services like DNS and DHCP, services that don't carry actual user data, that use this type of transaction. Applications such as DHCP also use UDP when they have to send broadcast or multicast transmissions. Because the TCP protocol requires two systems to establish a connection before they transmit user data, it does not support broadcasts and multicasts.

The header for UDP messages (sometimes confusingly called datagrams, like IP messages) is small, only 8 bytes, as opposed to the 20 bytes of the TCP header. The format is illustrated in Figure 13-19.

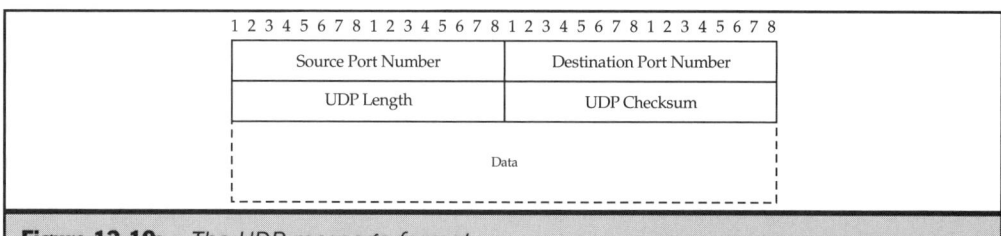

Figure 13-19:   *The UDP message format*

The functions of the fields are as follows:

Source Port Number (2 bytes)   Identifies the port number of the process in the transmitting system that generated the data carried in the UDP datagram. In some cases, this may be an ephemeral port number selected by the client for this transaction.

Destination Port Number (2 bytes)   Identifies the port number of the process on the destination system that will receive the data carried in the UDP datagram. Well-known port numbers are listed in the "Assigned Numbers" RFC and in the Services file on every TCP/IP system.

UDP Length (2 bytes)   Specifies the length of the entire UDP message, including the Header and Data fields, in bytes.

UDP Checksum (2 bytes)   Contains the results of a checksum computation computed from the UDP header and data, along with a pseudo-header comprised of the IP header's Source IP Address, Destination IP Address, and Protocol fields, plus the UDP Length field. This pseudo-header enables the UDP protocol at the receiving system to verify that the message has been delivered to the correct protocol on the correct destination system.

Data (variable, up to 65,507 bytes)   Contains the information supplied by the application-layer protocol.

TCP

The Transmission Control Protocol (TCP) is the connection-oriented, reliable alternative to UDP, which accounts for the majority of the user data transmitted across a TCP/IP network, as well as for giving the protocol suite its name. TCP, as defined in RFC 793, provides applications with a full range of transport services, including packet acknowledgment, error detection and correction, and flow control.

TCP is intended for the transfer of relatively large amounts of data that will not fit into a single packet. The data often takes the form of complete files that must be split up into multiple datagrams for transmission. In TCP terminology, the data supplied to the transport layer is referred to as a *sequence*, and the protocol splits the sequence into *segments* for transmission across the network. As with UDP, however, the segments are packaged in IP datagrams that may end up taking different routes to the destination. TCP, therefore, assigns sequence numbers to the segments, so the receiving system can reassemble them in the correct order.

Before any transfer of user data begins using TCP, the two systems exchange messages to establish a connection. This ensures that the receiver is operating and capable of receiving data. Once the connection is established and data transfer begins, the receiving system generates periodic acknowledgment messages. These messages inform the sender of lost packets and also provide the information used to control the rate of flow to the receiver.

The TCP Header

To provide these services, the header applied to TCP segments is necessarily larger than that for UDP. At 20 bytes (without options), it's the same size as the IP header. The header format is illustrated in Figure 13-20.

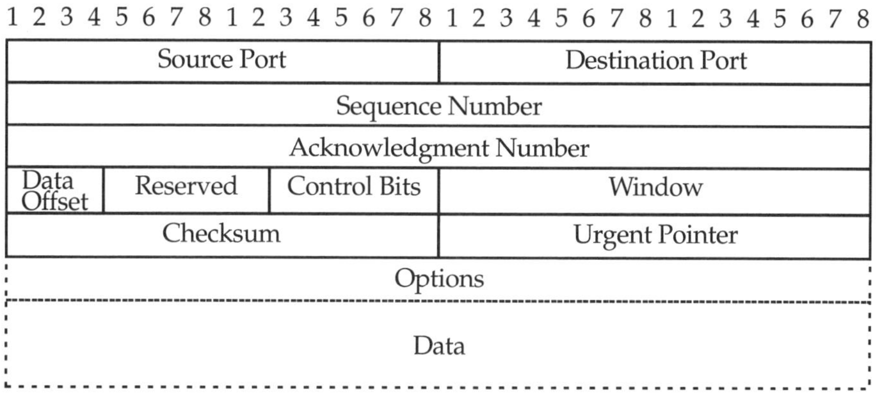

```
1 2 3 4 5 6 7 8 1 2 3 4 5 6 7 8 1 2 3 4 5 6 7 8 1 2 3 4 5 6 7 8
┌─────────────────────────────────┬─────────────────────────────────┐
│           Source Port            │        Destination Port          │
├─────────────────────────────────┴─────────────────────────────────┤
│                          Sequence Number                           │
├────────────────────────────────────────────────────────────────────┤
│                        Acknowledgment Number                       │
├───────┬───────────┬─────────────┬─────────────────────────────────┤
│ Data  │ Reserved  │ Control Bits │            Window               │
│Offset │           │              │                                 │
├───────┴───────────┴─────────────┼─────────────────────────────────┤
│            Checksum              │         Urgent Pointer          │
├─────────────────────────────────┴─────────────────────────────────┤
│                              Options                               │
├────────────────────────────────────────────────────────────────────┤
│                               Data                                 │
└────────────────────────────────────────────────────────────────────┘
```

Figure 13-20:   *The TCP message format*

The functions of the fields are as follows:

Source Port (2 bytes)   Identifies the port number of the process in the transmitting system that generated the data carried in the TCP segments. In some cases, this may be an ephemeral port number selected by the client for this transaction.

Destination Port (2 bytes)   Identifies the port number of the process on the destination system that will receive the data carried in the TCP segments. Well-known port numbers are listed in the "Assigned Numbers" RFC and in the Services file on every TCP/IP system.

Sequence Number (4 bytes)   Specifies the location of the data in this segment in relation to the entire data sequence.

Acknowledgment Number (4 bytes)   Specifies the sequence number of the next segment that the acknowledging system expects to receive from the sender. Active only when the ACK bit is set.

Data Offset (4 bits)   Specifies the length, in 4-byte words, of the TCP header (which may contain options expanding it to as much as 60 bytes).

Reserved (6 bits)   Unused.

Control Bits (6 bits)   Contains six 1-bit flags that perform the following functions:

    URG   Indicates that the sequence contains urgent data and activates the Urgent Pointer field.

ACK Indicates that the message is an acknowledgment of previously transmitted data and activates the Acknowledgment Number field.

PSH Instructs the receiving system to push all the data in the current sequence to the application identified by the port number without waiting for the rest.

RST Instructs the receiving system to discard all the segments in the sequence that have been transmitted thus far and resets the TCP connection.

SYN Used during the connection establishment process to synchronize the sequence numbers in the source and destination systems.

FIN Indicates to the other system that the data transmission has been completed and the connection is to be terminated.

Window (2 bytes) Implements the TCP flow-control mechanism by specifying the number of bytes the system can accept from the sender.

Checksum (2 bytes) Contains a checksum computation computed from the TCP header; data; and a pseudo-header composed of the Source IP Address, Destination IP Address, Protocol fields from the packet's IP header, plus the length of the entire TCP message.

Urgent Pointer (2 bytes) Activated by the URG bit, specifies the data in the sequence that should be treated by the receiver as urgent.

Options (variable) May contain additional configuration parameters for the TCP connection, along with padding to fill the field to the nearest 4-byte boundary. The available options are as follows:

Maximum Segment Size Specifies the size of the largest segments the current system can receive from the connected system.

Window Scale Factor Used to double the size of the Window Size field from 2 to 4 bytes.

Time stamp Used to carry time stamps in data packets that the receiving system returns in its acknowledgments, enabling the sender to measure the round-trip time.

Data (variable) May contain a segment of the information passed down from an application-layer protocol. In SYN, ACK, and FIN packets, this field is left empty.

Connection Establishment

Distinguishing TCP connections from the other types of connections commonly used in data networking is important. When you log on to a network, for example, you initiate a session that remains open until you log off. During that session, you may establish other connections to individual network resources like file servers that also remain open for extended lengths of time. TCP connections are much more transient, however, and typically remain open only for the duration of the data transmission. In addition, a system (or even a single application on that system) may open several TCP connections at once with the same destination.

As an example, consider a basic client/server transaction between a Web browser and a Web server. Whenever you type a URL in the browser, the program opens a TCP connection with the server to transfer the default HTML file that the browser uses to display the server's home page. The connection only lasts as long as it takes to transfer that one page. When the user clicks a hyperlink to open a new page, an entirely new TCP connection is needed. If there are any graphics on the Web pages, a separate TCP connection is needed to transmit each image file.

The additional messages required for the establishment of the connection, plus the size of the header, add considerably to the control overhead incurred by a TCP connection. This is the main reason why TCP/IP has UDP as a low-overhead transport-layer alternative.

The communication process between the client and the server begins when the client generates its first TCP message, beginning the three-way handshake that establishes the connection between the two machines. This message contains no application data, it simply signals to the server that the client wishes to establish a connection. The SYN bit is set, and the system supplies a value in the Sequence Number field, called the *initial sequence number (ISN)* (see Figure 13-21). In this packet capture, the sequence number is 119841003.

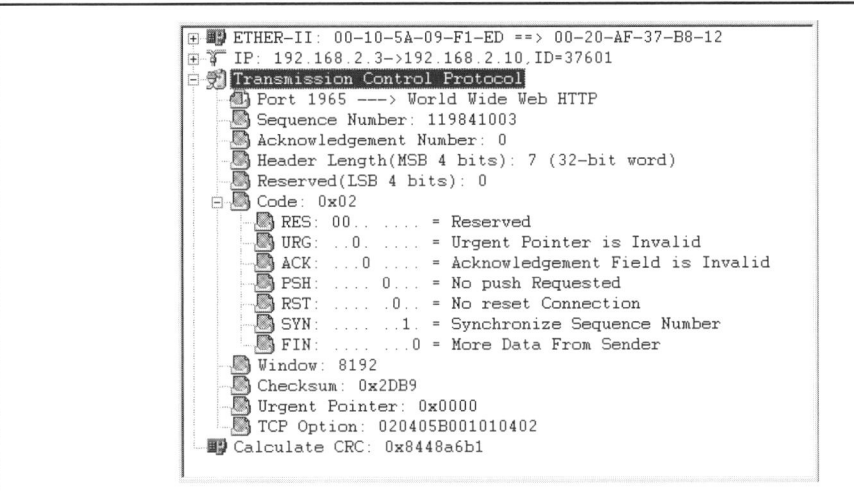

```
⊞ 🖳 ETHER-II: 00-10-5A-09-F1-ED ==> 00-20-AF-37-B8-12
⊞ 🖺 IP: 192.168.2.3->192.168.2.10,ID=37601
⊟ 🗐 Transmission Control Protocol
      🗐 Port 1965 ---> World Wide Web HTTP
      🗐 Sequence Number: 119841003
      🗐 Acknowledgement Number: 0
      🗐 Header Length(MSB 4 bits): 7 (32-bit word)
      🗐 Reserved(LSB 4 bits): 0
   ⊟ 🗐 Code: 0x02
      🗐 RES: 00.. .... = Reserved
      🗐 URG: ..0. .... = Urgent Pointer is Invalid
      🗐 ACK: ...0 .... = Acknowledgement Field is Invalid
      🗐 PSH: .... 0... = No push Requested
      🗐 RST: .... .0.. = No reset Connection
      🗐 SYN: .... ..1. = Synchronize Sequence Number
      🗐 FIN: .... ...0 = More Data From Sender
      🗐 Window: 8192
      🗐 Checksum: 0x2DB9
      🗐 Urgent Pointer: 0x0000
      🗐 TCP Option: 020405B001010402
   🖳 Calculate CRC: 0x8448a6b1
```

Figure 13-21:   *The client's SYN message initiates the connection establishment process.*

The system uses a continuously incrementing algorithm to determine the ISN it will use for each connection. The constant cycling of the sequence numbers makes it highly unlikely that multiple connections using the same sequence numbers will occur between the same two sockets. The client system then transmits the message as a unicast to the destination system and enters the SYN-SENT state, indicating that it has transmitted its connection request and is waiting for a matching request from the destination system.

The server, at this time, is in the LISTEN state, meaning that it is waiting to receive a connection request from a client. When the server receives the message from the client, it replies with its own TCP control message. This message serves two functions: it acknowledges the receipt of the client's message, as indicated by the ACK bit, and initiates its own connection, as indicated by the SYN bit (see Figure 13-22). The server then enters the SYN-RECEIVED state, indicating that it has received a connection request, issued a request of its own, and is waiting for an acknowledgment from the other system. Both the ACK and SYN bits are necessary because TCP is a *full-duplex* protocol, meaning that a separate connection is actually running in each direction. Both connections must be individually established, maintained, and terminated. The server's message also contains a value in the Sequence Number field (116270), as well as a value in the Acknowledgment Number field (119841004).

Figure 13-22: *The server acknowledges the client's SYN and sends a SYN of its own.*

Both systems maintain their own sequence numbers and are also conscious of the other system's sequence numbers. Notice that the server's value for the Acknowledgment Number field is the client's ISN plus 1, indicating that the server expects the client's next packet to have this value for its sequence number. Later, when the systems actually begin to send application data, these sequence numbers enable a receiver to assemble the individual segments transmitted in separate packets into the original sequence.

Remember, although the two systems must establish a connection before they send application data, the TCP messages are still transmitted within IP datagrams and are subject to the same treatment as any other datagram. Thus, the connection is actually a virtual one, and the datagrams may take different routes to the destination and arrive in a different order from that in which they were sent.

After the client receives the server's message, it transmits its own ACK message (see Figure 13-23), acknowledging the server's SYN bit and completing the bidirectional connection establishment process. This message has a value of 119841004 as its sequence number, which is the value the server expects, and an acknowledgment number of 116271, which is the sequence number it expects to see in the server's next transmission. Both systems now enter the ESTABLISHED state, indicating that they are ready to transmit and receive application data.

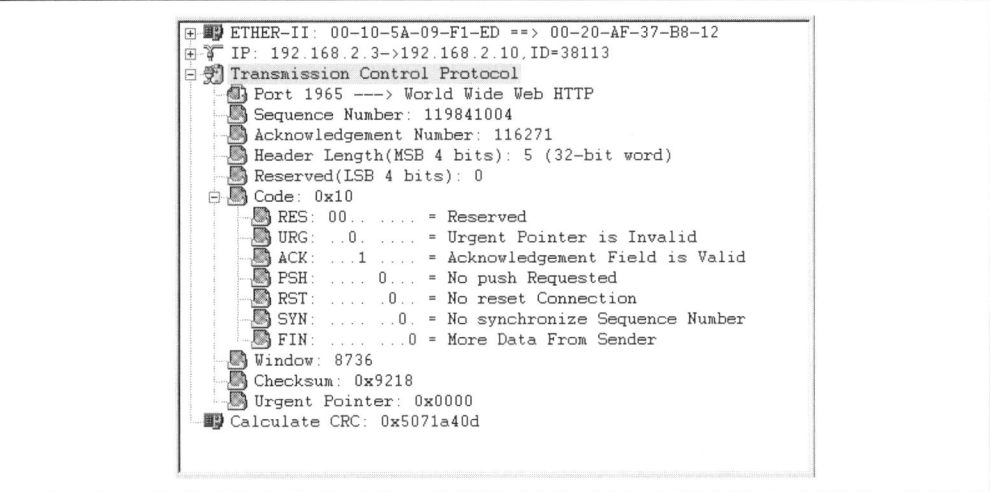

Figure 13-23:   *The client then acknowledges the server's SYN and the connection is established in both directions.*

Data Transfer

Once the TCP connection is established in both directions, the transmission of data can begin. The application-layer protocol determines whether the client or the server initiates the next exchange. In an FTP session, for example, the server sends a Ready message first. In an HTTP exchange, the client begins by sending the URL of the document it wants to receive.

The data to be sent is not packaged for transmission until the connection is established. This is because the systems use the SYN messages to inform the other system of the *maximum segment size (MSS)*. The MSS specifies the size of the largest segment each system is capable of receiving. The value of the MSS depends on the data link–layer protocol used to connect the two systems.

Each system supplies the other with an MSS value in the TCP message's Options field. As with the IP header, each option consists of multiple subfields, which for the Maximum Segment Size option, are as follows:

Kind (1 byte)   Identifies the function of the option. For the Maximum Segment Size option, the value is 2.

Length (1 byte)   Specifies the length of the entire option. For the Maximum Segment Size option, the value is 4.

Maximum Segment Size (2 bytes)   Specifies the size (in bytes) of the largest data segment the system can receive.

In the client system's first TCP message, shown earlier in Figure 13-21, the value of the Options field is (in hexadecimal notation) 020405B001010402. The first 4 bytes of this value constitute the MSS option. The Kind value is 02, the Length is 04, and the MSS is 05B0, which in decimal form is 1,456 bytes. This works out to the maximum frame size for an Ethernet II network (1,500 bytes) minus 20 bytes for the IP header and 24 bytes for the TCP header (20 bytes plus 4 option bytes). The server's own SYN packet contains the same value for this option because these two computers were located on the same Ethernet network.

Note   *The remaining 4 bytes in the Options field consist of 2 bytes of padding (0101) and the Kind (04) and Length (02) fields of the SACK-Permitted option, indicating that the system is capable of processing extended information as part of acknowledgment messages.*

When the two systems are located on different networks, their MSS values may also be different, and how the systems deal with this is left up to the individual TCP implementations. Some systems may just use the smaller of the two values, while others might revert to the default value of 536 bytes used when no MSS option is supplied. Windows 2000 systems use a special method of discovering the connection path's MTU (that is, the largest packet size permitted on an internetwork link between two systems). This method, as defined in RFC 1191, enables the systems to determine the packet sizes permitted on intermediate networks. Thus, even if the source and destination systems are both connected to Ethernet networks with 1,500-byte MTUs, they can detect an intermediate connection that only supports a 576-byte MTU.

Once the MSS for the connection is established, the systems can begin packaging data for transmission. In the case of an HTTP transaction, the Web browser client transmits the desired URL to the server in a single packet (see Figure 13-24). Notice that the sequence number of this packet (119841004) is the same as that for the previous packet it sent in acknowledgment to the server's SYN message. This is because TCP messages consisting only of an acknowledgment do not increment the sequence counter. The acknowledgment number is also the same as in the previous packet because the client has not yet received the next message from the server. Note also that the PSH bit is set, indicating that the server should send the enclosed data to the application immediately.

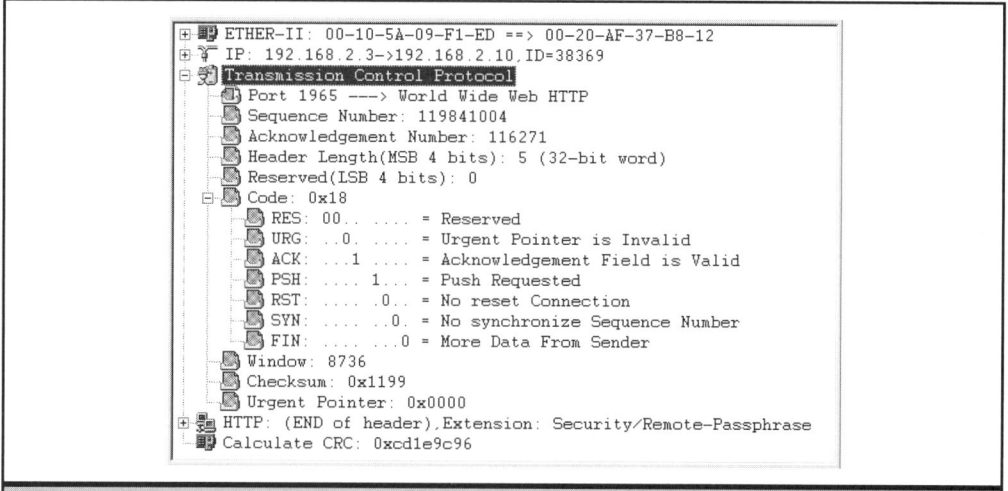

Figure 13-24: *The first data packet sent over the connection contains the URL requested by the Web browser.*

After receiving the client's message, the server returns an acknowledgment message, as shown in Figure 13-25, that uses the sequence number expected by the client (116271) and has an acknowledgment number of 119841363. The difference between this acknowledgment number and the sequence number of the client message previously sent is 359; this is correct because the datagram the client sent to the server was 399 bytes long. Subtracting 40 bytes for the IP and TCP headers leaves 359 bytes worth of data. The value in the server's acknowledgment message, therefore, indicates that it has successfully received 359 bytes of data from the client. As each system sends data to the other, they increment their sequence numbers for each byte transmitted.

The next step in the process is for the server to respond to the client's request by sending it the requested HTML file. Using the MSS value, the server creates segments small enough to be transmitted over the network and transmits the first one in the message, as shown in Figure 13-26. The sequence number is again the same as the server's previous message because the previous message contained only an acknowledgment. The acknowledgment number is also the same because the server is sending a second message without any intervening communication from the client.

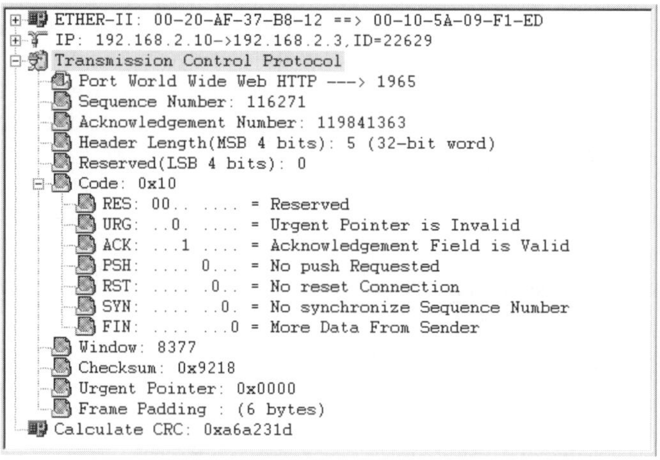

```
⊞ 📠 ETHER-II: 00-20-AF-37-B8-12 ==> 00-10-5A-09-F1-ED
⊞ 🖥 IP: 192.168.2.10->192.168.2.3,ID=22629
⊟ 🖥 Transmission Control Protocol
      📇 Port World Wide Web HTTP ---> 1965
      📇 Sequence Number: 116271
      📇 Acknowledgement Number: 119841363
      📇 Header Length(MSB 4 bits): 5 (32-bit word)
      📇 Reserved(LSB 4 bits): 0
   ⊟ 📇 Code: 0x10
      📇 RES: 00.. .... = Reserved
      📇 URG: ..0. .... = Urgent Pointer is Invalid
      📇 ACK: ...1 .... = Acknowledgement Field is Valid
      📇 PSH: .... 0... = No push Requested
      📇 RST: .... .0.. = No reset Connection
      📇 SYN: .... ..0. = No synchronize Sequence Number
      📇 FIN: .... ...0 = More Data From Sender
      📇 Window: 8377
      📇 Checksum: 0x9218
      📇 Urgent Pointer: 0x0000
      📇 Frame Padding : (6 bytes)
   📠 Calculate CRC: 0xa6a231d
```

Figure 13-25:   *The server acknowledges all of the data bytes transmitted by the client.*

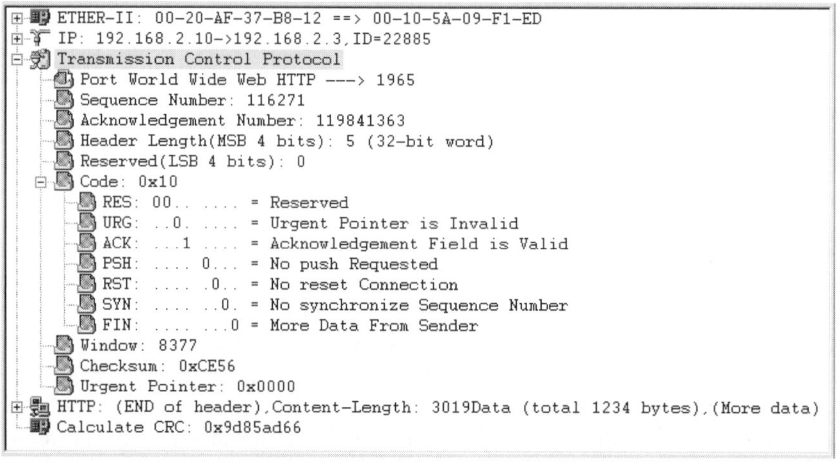

```
⊞ 📠 ETHER-II: 00-20-AF-37-B8-12 ==> 00-10-5A-09-F1-ED
⊞ 🖥 IP: 192.168.2.10->192.168.2.3,ID=22885
⊟ 🖥 Transmission Control Protocol
      📇 Port World Wide Web HTTP ---> 1965
      📇 Sequence Number: 116271
      📇 Acknowledgement Number: 119841363
      📇 Header Length(MSB 4 bits): 5 (32-bit word)
      📇 Reserved(LSB 4 bits): 0
   ⊟ 📇 Code: 0x10
      📇 RES: 00.. .... = Reserved
      📇 URG: ..0. .... = Urgent Pointer is Invalid
      📇 ACK: ...1 .... = Acknowledgement Field is Valid
      📇 PSH: .... 0... = No push Requested
      📇 RST: .... .0.. = No reset Connection
      📇 SYN: .... ..0. = No synchronize Sequence Number
      📇 FIN: .... ...0 = More Data From Sender
      📇 Window: 8377
      📇 Checksum: 0xCE56
      📇 Urgent Pointer: 0x0000
   ⊞ 📇 HTTP: (END of header),Content-Length: 3019Data (total 1234 bytes),(More data)
      📠 Calculate CRC: 0x9d85ad66
```

Figure 13-26:   *In response to the client's request, the server begins to transmit the Web page after splitting it into multiple segments.*

In addition to the acknowledgment service just described, the TCP header fields provide two additional services:

- Error correction
- Flow control

The following sections examine each of these functions.

Error Correction   You saw in the previous example how a receiving system uses the acknowledgment number in its ACK message to inform the sender that its data was received correctly. The systems also use this mechanism to indicate when an error has occurred and data is not received correctly.

TCP/IP systems use a system of *delayed acknowledgments,* meaning that they do not have to send an acknowledgment message for every packet they receive. The method used to determine when acknowledgments are sent is left up to the individual implementation, but each acknowledgment specifies that the data, up to a certain point in the sequence, has been received correctly. These are called *positive acknowledgments* because they indicate that data has been received. *Negative acknowledgments* or *selective acknowledgments,* which specify that data has not been received correctly, are not possible in TCP.

What if, for example, in the course of a single connection, a server transmits five data segments to a client and the third segment must be discarded because of a checksum error? The receiving system must then send an acknowledgment back to the sender indicating that all the messages up through the second segment have been received correctly. Even though the fourth and fifth segments were also received correctly, the third segment was not. Using positive acknowledgments means that the fourth and fifth segments must be retransmitted, in addition to the third.

The mechanism used by TCP is called *positive acknowledgment with retransmission* because the sending system automatically retransmits all of the unacknowledged segments after a certain time interval. The way this works is that the sending system maintains a queue containing all of the segments it has already transmitted. As acknowledgments arrive from the receiver, the sender deletes the segments that have been acknowledged from the queue. After a certain elapsed time, the sending system retransmits all of the unacknowledged segments remaining in the queue. The systems use algorithms documented in RFC 1122 to calculate the timeout values for a connection based on the amount of time it takes for a transmission to travel from one system to the other and back again, called the *round-trip time.*

Flow Control   *Flow control* is an important element of the TCP protocol because it is designed to transmit large amounts of data. Receiving systems have a buffer in which they store incoming segments waiting to be acknowledged. If a sending system transmits too many segments too quickly, the receiver's buffer fills up and any packets arriving at the system are discarded until space in the buffer is available. TCP uses a mechanism called a *sliding window* for its flow control, which is essentially a means for the receiving system to inform the sender of how much buffer space it has available.

Each acknowledgment message generated by a system receiving TCP data specifies the amount of buffer space it has available in its Window field. As packets arrive at the receiving system, they wait in the buffer until the system generates the message that acknowledges them. The sending system computes the amount of data it can send by taking the Window value from the most recently received acknowledgment and subtracting the number of bytes it has transmitted since it received that acknowledgment. If the result of this computation is zero, the system stops transmitting until it receives acknowledgment of outstanding packets.

Connection Termination

When the exchange of data between the two systems is complete, they terminate the TCP connection. Because two connections are actually involved—one in each direction—both must be individually terminated. The process begins when one machine sends a message in which the FIN control bit is set. This indicates that the system wants to terminate the connection it has been using to send data.

Which system initiates the termination process is dependent on the application generating the traffic. In an HTML transaction, the server can include the FIN bit in the message containing the last segment of data in the sequence, or it can take the form of a separate message. The client receiving the FIN from the server sends an acknowledgment, closing the server's connection, and then sends a FIN message of its own. Note that, unlike the three-way handshake that established the connection, the termination procedure requires four transmissions because the client sends its ACK and FIN bits in separate messages. When the server transmits its acknowledgment to the client's FIN, the connection is effectively terminated.

The Complete Reference

Networking

Chapter 14

NetWare Protocols

N ovell NetWare was originally designed at a time when proprietary computer networking products were commonplace. As a result, to provide transport services for the NetWare operating system, Novell created its own suite of protocols, usually referred to by the name of the network-layer protocol: IPX, or Internetwork Packet Exchange. As a parallel to TCP/IP, the suite is also sometimes called IPX/SPX, which adds a reference to the Sequenced Packet Exchange (SPX) protocol, which operates at the transport layer. Unlike the combination of TCP and IP, however, which usually accounts for a good deal of the traffic on a TCP/IP network, use of the IPX/SPX combination on a NetWare network is relatively rare.

As the networking industry developed, standardization and interoperability became the most important elements of networking product designs. The rise in popularity of the TCP/IP protocols and the Internet led most of the network operating system developers to adopt TCP/IP as their default protocols, if they were not using them already. Novell, however, held on to its proprietary protocols longer than anyone else, much to the detriment of its market share. It was only with the release of NetWare 5 in 1998 that TCP/IP was fully integrated into NetWare.

The IPX protocols are similar to TCP/IP in several ways. Both protocol suites use a connectionless, unreliable protocol at the network layer (IPX and IP, respectively) to carry datagrams containing the information generated by multiple upper-layer protocols that provide varying degrees of service for different applications. Like IP, IPX is responsible for addressing datagrams and routing them to their destinations on other networks.

Unlike TCP/IP, however, the IPX protocols were designed for use on LANs, and don't have the almost-unlimited scalability of the Internet protocols. IPX does not have a completely self-contained addressing system like IP. Systems on a NetWare network identify other systems using the node address (also called a hardware or firmware address) hardcoded into network interface adapters plus a network address assigned by the administrator (or the operating system) during the installation.

IPX also lacks the universality of TCP/IP because of Novell's policy of keeping many of the details regarding the inner workings of the protocols private. In retrospect, this policy can be seen as having worked against Novell. Microsoft seems to have had little difficulty in reverse-engineering IPX for its NWLink protocol, and by not releasing the protocol specifications to other companies, Novell squandered any chance for IPX to become an industry standard.

Data Link–Layer Protocols

IPX datagrams are carried within standard data link–layer protocol frames, just like IP. The IPX protocols have no data link–layer protocols of their own, although you can use them over a PPP connection. On most networks, however, Ethernet or Token Ring frames encapsulate IPX.

The only unusual aspect of configuring NetWare servers to use the Ethernet or Token Ring protocols is that you must specify the frame type (or types) for each network, using names that are less than intuitive. NetWare supports four Ethernet frame types, which differ only in certain aspects of the frame format. While all four are capable of carrying standard IPX traffic, the frame type you select can influence whether or not your network supports the use of other protocol suites (like TCP/IP) concurrently with IPX.

These four Ethernet frame types are as follows:

ETHERNET_802.3   Also called "raw Ethernet," this was the default frame type for NetWare versions through 3.11. This frame differs slightly from the exact format defined in the IEEE 802.3 document, which wasn't complete at the time of NetWare's initial release. The frame can be anywhere from 64 to 1,518 bytes long, and the field immediately following the source and destination addresses specifies the length of the packet, not the Ethertype value, as in the DIX Ethernet frame. Because of this, the Ethernet protocol has no means of identifying the network-layer protocol carried in the frame to the receiving system. Therefore, this frame type can only be used on a network running IPX exclusively at the network layer.

ETHERNET_802.2   The default frame type for NetWare versions 3.12 and later, the name is confusing because 802.2 refers to the Logical Link Control standard developed by the IEEE for use with all of the 802 protocols. Actually, this frame type uses the standard frame defined in the IEEE 802.3 standard, as well as the 802.2 header in the frame's Data field. Because this frame conforms to the standard, it can be used with other products that support IPX (such as Microsoft Windows). However, the frame still lacks the equivalent of an Ethertype field and can support only IPX traffic.

ETHERNET_II   As defined by the DIX Ethernet standard, this frame type differs from the IEEE frame mainly in that it has an Ethertype field that specifies which network-layer protocol generated the data carried in the frame. This, therefore, is the frame type you should use when you are running TCP/IP or other protocols on your network.

ETHERNET_SNAP   This frame type is identical to ETHERNET_802.2, except that it includes the SNAP header in the Data field in addition to the LLC header. The SNAP header also has a field that identifies the network-layer protocol, enabling you to use it on networks also running TCP/IP and/or AppleTalk.

NetWare enables you to select as many of these frame types for your Ethernet networks as you need, in order to support the various other systems connected to it. The main concern, apart from using either ETHERNET_II or ETHERNET_SNAP on networks running TCP/IP or other protocols, is that servers and workstations all have at least one frame type in common. On the Windows client systems today, this is not a problem, since the computer detects the frame types in use on the network and configures itself accordingly.

NETWORK PROTOCOLS

 **Note** *For more information on Ethernet frames and the standards that define them, see Chapters 10 and 11.*

The Internetwork Packet Exchange (IPX) Protocol

IPX is based on the Internetwork Datagram Packet (IDP) protocol designed for Xerox Network Services (XNS). IPX provides basic connectionless transport between systems on an internetwork, in either broadcast or unicast transmissions. Most of the standard traffic between NetWare servers and between clients and servers is carried within IPX datagrams.

The header on an IPX datagram is 30 bytes long (as compared to IP's 20-byte header). The format for the IPX header is illustrated in Figure 14-1, and the functions of the header fields are as follows:

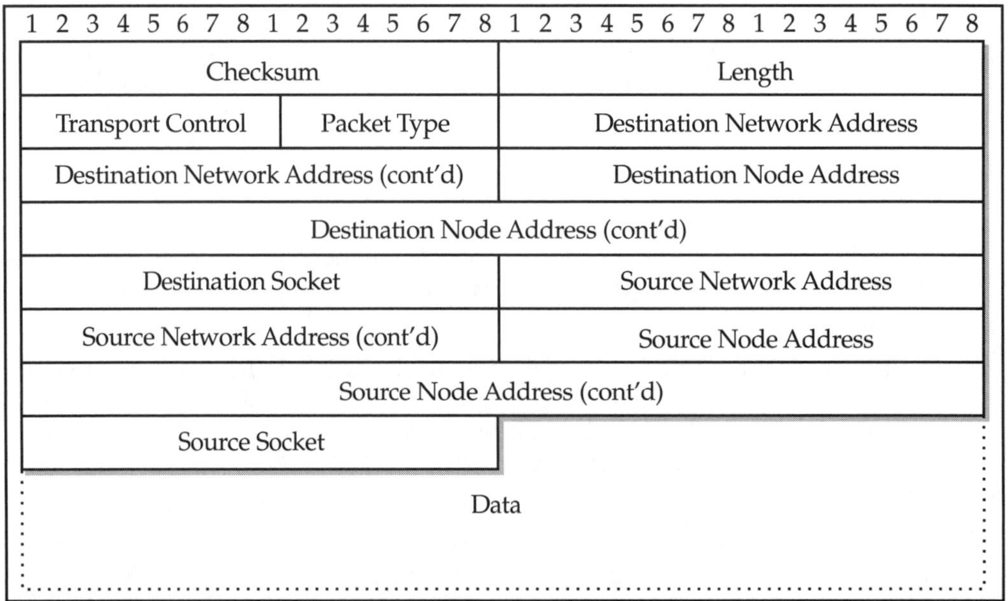

1 2 3 4 5 6 7 8	1 2 3 4 5 6 7 8	1 2 3 4 5 6 7 8	1 2 3 4 5 6 7 8
Checksum		Length	
Transport Control	Packet Type	Destination Network Address	
Destination Network Address (cont'd)		Destination Node Address	
Destination Node Address (cont'd)			
Destination Socket		Source Network Address	
Source Network Address (cont'd)		Source Node Address	
Source Node Address (cont'd)			
Source Socket			
Data			

Figure 14-1: *The IPX datagram format*

Checksum (2 bytes)   In the original IDP header, this field contained a CRC value for the datagram. Since the data link–layer protocol performs CRC checking, the function is disabled in IPX datagrams, and always contains the hexadecimal value ffff.

Length (2 bytes)   Specifies the length of the datagram, in bytes, including the IPX header and the data.

Transport Control (1 byte)   Also known as the *hop count*; specifies the number of routers the datagram has passed through on the way to its destination. Set to 0 by the transmitting system, each router increments the field by one as it processes the packet. If the value reaches 16, the router discards the datagram.

Packet Type (1 byte)   Identifies the service or upper-layer protocol that generated the data carried in the datagram, using the following values:

   0   Unknown Packet Type

   1   Routing Information Protocol

   4   Service Advertising Protocol

   5   Sequenced Packet Exchange

   17   NetWare Core Protocol

Destination Network Address (4 bytes)   Identifies the network on which the destination system is located using a value assigned by the administrator or the operating system during the NetWare installation.

Destination Node Address (6 bytes)   Identifies the network interface inside the computer to which the data is to be delivered, using the data link–layer protocol's hardware address. Broadcast messages use the hexadecimal address ff:ff:ff:ff:ff:ff.

Destination Socket (2 bytes)   Identifies the process on the destination system for which the data in the datagram is intended, using the following hexadecimal values:

   0451   NetWare Core Protocol (NCP)

   0452   Service Advertising Protocol (SAP)

   0453   Routing Information Protocol (RIP)

   0455   NetBIOS

   0456   Diagnostic Packet

   0457   Serialization Packet

   4000–6000   Custom sockets for server processes

   9000   NetWare Link Services Protocol

   9004   IPXWAN Protocol

NETWORK PROTOCOLS

Source Network Address (4 bytes)   Identifies the network on which the system sending the datagram is located, using a value assigned by the administrator or by the operating system during the NetWare installation.

Source Node Address (6 bytes)   Identifies the network interface inside the computer sending the datagram, using the data link–layer protocol's hardware address.

Source Socket (2 bytes)   Identifies the process on the local system from which the data originated, using the same hexadecimal values as the Destination Socket field.

Data (variable)   Contains the data generated by the upper-layer protocol.

To determine a viable route to the destination, NetWare servers use a routing protocol like RIP (Routing Information Protocol) or NLSP (NetWare Link Services Protocol). As with other network-layer protocols, an IPX router strips off the data link–layer protocol frame from each packet and adds a new frame for transmission over another network. The only modification the router makes to the IPX header is to increment the value of the Transport Control field.

Note   *For more information on routing, see Chapter 6.*

Because IPX is a connectionless protocol, it relies on the upper-layer protocols to affirm that data has been delivered correctly. For example, when a client sends a request to a server using a NetWare Core Protocol (NCP) message in an IPX datagram, it is the response from the server that functions as an acknowledgment of the request.

However, NetWare clients do maintain a response timeout clock that causes them to retransmit an IPX datagram if a response is not received in a given period of time. If there is a condition on the network that intermittently slows down traffic, a client may retransmit a datagram several times before it is received correctly at the destination. You can control the number of times that the system will attempt to retransmit a datagram by changing the client's IPX RETRY COUNT parameter from the default setting of 20. On a 32-bit Windows system with the Novell NetWare client installed, you find this parameter in the Properties dialog box for the IPX protocol module. On a DOS/Windows 3.1 system, you add the following line to the system's Net.cfg file:

```
IPX RETRY COUNT = 30
```

The Sequenced Packet Exchange (SPX) Protocol

Derived from the XNS Sequenced Packet Protocol (SPP), SPX operates at the transport layer and provides a connection-oriented, reliable service with flow control and packet sequencing, much like TCP in the TCP/IP protocol suite. However, NetWare systems use SPX far less often than TCP/IP systems use TCP. Typical file access procedures on a NetWare network use the NetWare Core Protocol, which accounts for the majority of the traffic produced. SPX is only used for tasks that require its services, such as communications between print servers, print queues, and remote printers; RCONSOLE sessions; and network backups.

The SPX header is illustrated in Figure 14-2, and the functions of the fields are as follows:

Connection Control (1 byte) Contains a code that regulates the bidirectional flow of data, using the following hexadecimal values:

10 End of Message

20 Attention

40 Acknowledgment required

80 System packet

Datastream Type (1 byte) Specifies the nature of the data in the message and the upper-layer process for which it is intended, using a value defined by the client or one of the following:

FE End-of-Connection

FF End-of-Connection Acknowledgment

Source Connection ID (2 bytes) Contains a unique value used to identify this connection, since a system can have multiple connections open to the same socket simultaneously.

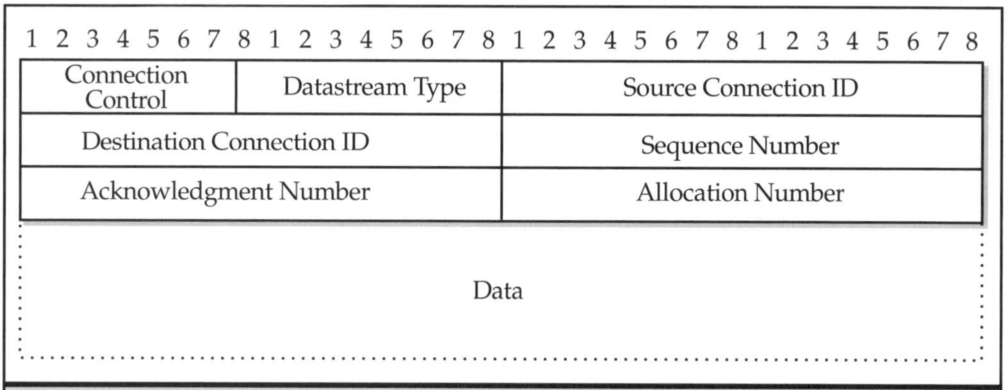

1 2 3 4 5 6 7 8 1 2 3 4 5 6 7 8	1 2 3 4 5 6 7 8 1 2 3 4 5 6 7 8
Connection Control	Datastream Type	Source Connection ID
Destination Connection ID	Sequence Number
Acknowledgment Number	Allocation Number
Data	

Figure 14-2: *The SPX message format*

NETWORK PROTOCOLS

Destination Connection ID (2 bytes)   Contains the unique value that the destination system uses to identify this connection. During the initial connection establishment process, the value of this field is ffff, since the other system's connection ID can't be known yet.

Sequence Number (2 bytes)   Contains a number, incremented for each message transmitted during the connection, that the receiving system uses to process the messages in the proper order.

Acknowledgment Number (2 bytes)   Contains the sequence number of the next message that the system expects to receive from the connected system, thereby acknowledging all of the packets having lower sequence numbers.

Allocation Number (2 bytes)   Implements the protocol's flow-control mechanism by specifying the number of packet receive buffers available on the system.

Data (variable)   Contains the data destined for an upper-layer process or protocol.

As with any connection-oriented protocol, the two systems exchange control messages in order to establish a connection before they transmit any application data. Once the connection is established, the systems send periodic keep-alive messages to maintain the connection when there is no activity. On a network with degraded performance due to heavy traffic or other problems, SPX connections may time out due to transmission delays.

Since SPX connections can provide crucial services (such as remote server console sessions), you may want to modify the timeout settings for the protocol if you are experiencing slow network performance. NetWare includes an SPX configuration utility for servers, called Spxconfg.nlm, that enables you to set the following parameters:

SPX Watchdog Abort Timeout   Specifies the maximum amount of time that an SPX connection can go unused before it's declared invalid. The default value is 540 ticks; the possible values range from 540 to 5,400 ticks (1 tick = 1/18 second).

SPX Watchdog Verify Timeout   Specifies the amount of time that must pass without a packet received before the system requests a keep-alive message from the connected system. The default value is 108 ticks; possible values range from 0 to 255 ticks.

SPX Ack Wait Timeout   Specifies the amount of time that a system waits for an acknowledgment before it retransmits an SPX message. The default value is 54 ticks; possible values range from 10 to 3,240 ticks.

SPX Default Retry Count   Specifies the number of times that a system will retransmit an SPX message without receiving an acknowledgment. The default value is 10; possible values range from 1 to 255.

Maximum Concurrent SPX Sessions   Specifies the maximum number of SPX sessions that an application can open. The default value is 1,000; possible values range from 100 to 2,000.

Maximum Open IPX Sockets   Specifies the maximum number of sockets that an application can have open. The default value is 1,200; possible values range from 60 to 65,520.

The NetWare Core Protocol (NCP)

As the name implies, the NetWare Core Protocol is responsible for the majority of the network traffic between NetWare clients and servers. Client systems use NCP to request files on server volumes and send print jobs to print queues. The server uses it to transfer the requested files back to the client. An NCP variant, called the NetWare Core Packet Burst (NCPB) protocol, enables servers to send large amounts of data to a client without the need to acknowledge every packet.

While SPX clearly operates at the transport layer, NCP's place in the OSI reference model is more nebulous. Because clients use NCP messages to log in to a server or NDS tree, the protocol is said to operate at the session layer, but its file-transfer and packet-acknowledgment capabilities place it in the transport layer. In addition, NCP provides file-locking and synchronization services and carries NDS messages, giving it attributes associated with the presentation and application layers. Like SPX, however, NCP messages are carried within standard IPX datagrams.

NCP communications typically follow a request/reply pattern, with the server generating a reply message for every client request. There are different message formats for NCP requests and replies, as outlined in the following sections.

The NCP Request Message

The NCP Request message format is illustrated in Figure 14-3. The functions of the fields are as follows:

Request Type (2 bytes)   Specifies the basic function of the message, using the following values:

1111 (Create a Service Connection)   Initiates a connection with a NetWare server.

2222 (File Server Request)   Used to request access to a NetWare server resource.

5555 (Connection Destroy)   Terminates a connection with a NetWare server.

7777 (Burst Mode Protocol Packet)   Used to request a burst mode transmission from a NetWare server.

Sequence Number (1 byte)   Contains a number that indicates the order in which the NCP messages have been transmitted, so that the receiver can process them in the correct order.

Connection Number Low (1 byte)   Specifies the number of the client's connection to the server, as displayed in the Monitor.nlm utility.

Task Number (1 byte)   Contains a unique value used to associate request messages with replies.

Connection Number High (1 byte)   Unused; contains a value of 00.

Function (1 byte)   Contains a code indicating the specific function of the message.

Subfunction (1 byte)   Contains a code that further defines the function of the message.

Subfunction Length (2 bytes)   Specifies the length of the Data field.

Data (variable)   Contains information pertinent to the processing of the request, such as the location of a file.

NCP is capable of a great many different functions; there are approximately 200 combinations of Function and Subfunction codes, providing services in the following categories:

Accounting Services   Retrieves account status, posts charges, and manages accounts.

Bindery Services   Accesses and modifies NetWare 3.x bindery objects and their properties.

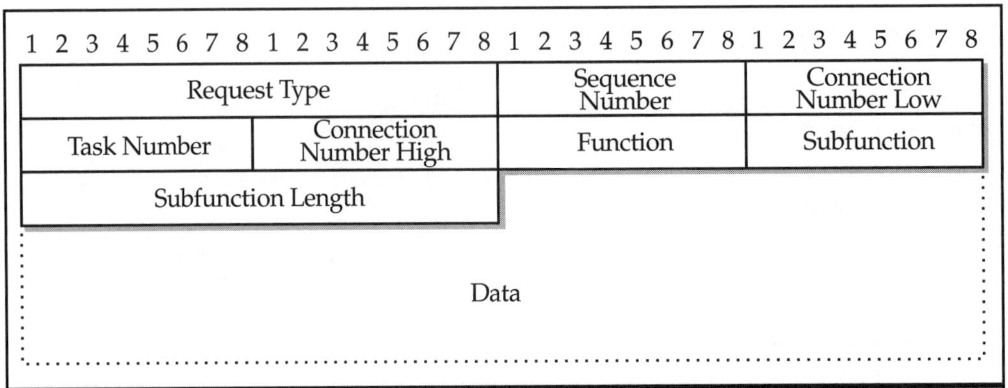

Figure 14-3:   *The NCP Request message format*

Connection Services   Creates, destroys, and retrieves information about connections to NetWare servers.

Directory Services   Views and manages directories on NetWare volumes and their trustees.

File Services   Accesses, views, and manages files and their attributes on NetWare volumes.

File Server Environment   Retrieves information about NetWare servers and manipulates their properties.

Message Services   Sends and receives broadcast messages.

Print Services   Sends print jobs to spool files in print queues.

Queue Services   Manipulates print queues and the jobs contained in them.

Synchronization Services   Manipulates records, file locks, and semaphores.

Transaction Tracking Services   Manages NetWare Transaction Tracking System (TTS) properties.

   The message format for NCP is not absolute. Some of the function codes trigger changes in the format in order to suit their specific purposes. For example, functions that do not take a value for the Subfunction field can eliminate that field from the message completely. Some functions also add their own specialized fields to the end of the message.

The NCP Reply Message

The NCP Reply message format is illustrated in Figure 14-4. The functions of the fields are as follows:

Reply/Response Type (2 bytes)   Specifies the nature of the reply, using one of the following values:

   3333 (File Server Reply)   Indicates that the message is a reply to a file server request, with a Request Type value of 2222.

   7777 (Burst Mode Protocol)   Indicates that a burst mode transmission process has been successfully initialized.

   9999 (Positive Acknowledgment)   Indicates that a request is being processed and that the reply is being sent only to prevent the client from timing out.

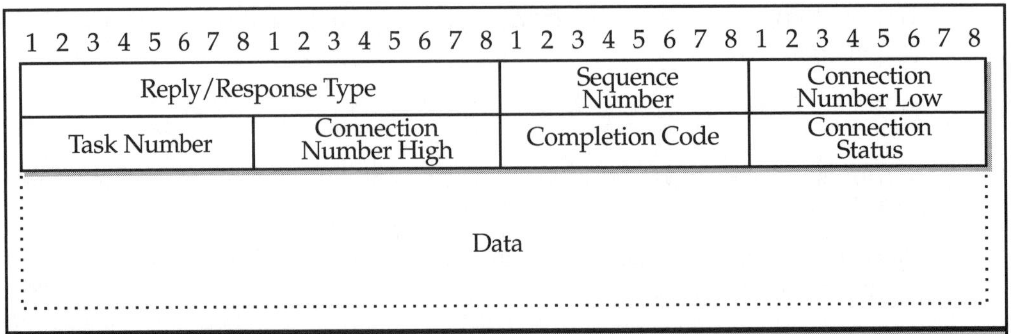

1 2 3 4 5 6 7 8	1 2 3 4 5 6 7 8	1 2 3 4 5 6 7 8	1 2 3 4 5 6 7 8
Reply/Response Type		Sequence Number	Connection Number Low
Task Number	Connection Number High	Completion Code	Connection Status
Data			

Figure 14-4: *The NCP Reply message format*

Sequence Number (1 byte) Contains a number that indicates the order in which the NCP messages have been transmitted, so that the receiver can process them in the correct order.

Connection Number Low (1 byte) Specifies the number of the client's connection to the server, as displayed in the Monitor.nlm utility.

Task Number (1 byte) Contains a unique value used to associate reply messages with requests.

Connection Number High (1 byte) Unused; contains a value of 00.

Completion Code (1 byte) Indicates the success or failure of the associated request. A value of 0 indicates successful completion of the request; nonzero values indicate that the request failed.

Connection Status (1 byte) Indicates whether or not the connection between the client and the server is still active. A value of 0 indicates that the connection is active; a value of 1 indicates that it is not.

Data (variable) Contains data sent by the server in response to the associated request.

The NetWare Core Packet Burst (NCPB) Protocol

The standard NCP protocol requires a response message for every request, which is sensible for some of its many functions, but not for others. For example, when a user logs in to a server, it makes sense for a message requesting the establishment of a connection to be answered immediately with a reply. However, for functions that involve the transfer of data sufficient to require multiple packets, this method is impractical.

When a client uses standard NCP messages to request a file from a server volume, the process is broken down into as many request/reply exchanges as are needed to transfer the file. The client begins by requesting the first part of the file, which it receives in a reply message. The client must then request the second part, receive it, request the third part, and so on. When transferring large files, the number of redundant request messages significantly erodes the efficiency of the protocol.

The NetWare Core Packet Burst Protocol was designed to address this shortcoming by making it possible for servers to send multiple data packets consecutively, without the need for individual replies or acknowledgments. Packet burst transmissions can send up to 64KB of data in a single burst, with only a single acknowledgment.

NCPB was first implemented as an add-on product to NetWare 3.11 that took the form of a server module called Pburst.nlm and a client shell called Bnetx.exe. Beginning with NetWare version 3.12 and the VLM client, NCPB was fully integrated into the protocol suite and is used automatically when a client accesses a file on a server, without any modification to the application generating the access request.

To provide this type of service, the NCPB protocol requires substantial changes to the NCP message format. The NCPB message format is illustrated in Figure 14-5, and the functions of the fields are as follows:

Request Type (2 bytes)   Specifies the basic function of the message, as in the NCP protocol. For packet burst messages, the value is always 7777.

Flags (1 byte)   Contains flags specifying the nature of the message or the data contained in it, using the following values:

Bit 1 (SYS)   Indicates that the packet contains a system message only and does not have any packet burst data associated with it.

Bit 2 (SAK)   Instructs the receiver to transmit its missing fragment list.

Bit 3 (Unused)

Bit 4 (EOB)   Indicates that the message contains the last data fragment in the burst.

Bit 5 (BSY)   Indicates that the server is busy and that the client should continue to wait for a response.

Bit 6 (ABT)   Indicates that the connection has been aborted and is no longer valid.

Bit 7 (Unused)

Bit 8 (Unused)

Stream Type (1 byte)   Specifies whether the server should respond to the request with a packet burst transmission. The only valid (hexadecimal) value is 02, signifying a "big send burst."

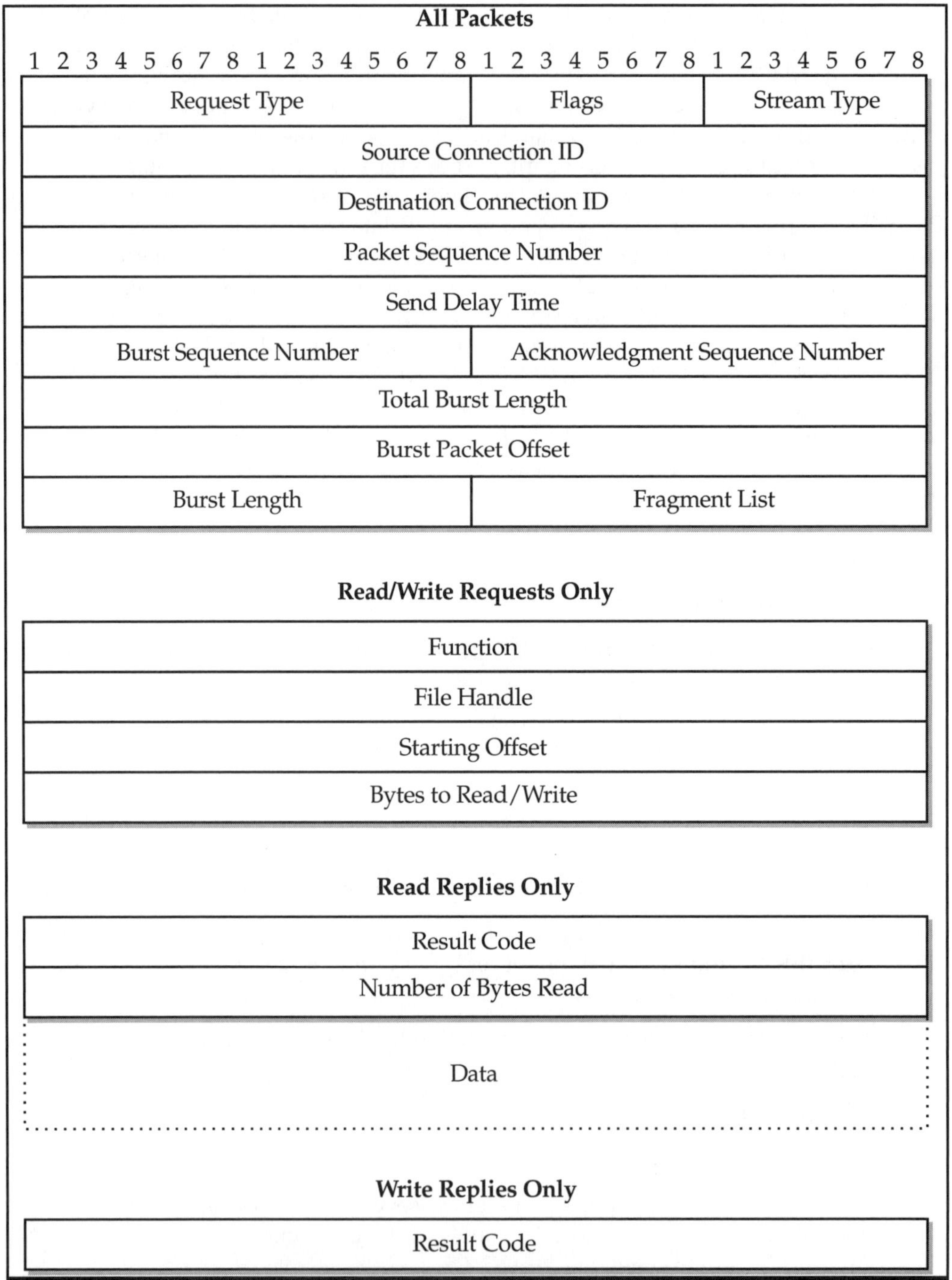

All Packets

1 2 3 4 5 6 7 8	1 2 3 4 5 6 7 8	1 2 3 4 5 6 7 8	1 2 3 4 5 6 7 8
Request Type		Flags	Stream Type
Source Connection ID			
Destination Connection ID			
Packet Sequence Number			
Send Delay Time			
Burst Sequence Number		Acknowledgment Sequence Number	
Total Burst Length			
Burst Packet Offset			
Burst Length		Fragment List	

Read/Write Requests Only

Function
File Handle
Starting Offset
Bytes to Read/Write

Read Replies Only

Result Code
Number of Bytes Read
Data

Write Replies Only

Result Code

Figure 14-5: *The NCPB message format*

Source Connection ID (4 bytes)   Contains a unique value (different from the NCP Connection ID value) derived by the sender from the current time of day, which is used to identify this packet burst connection.

Destination Connection ID (4 bytes)   Contains the connection identifier (equivalent to the Source Connection ID) generated by the destination system.

Packet Sequence Number (4 bytes)   Contains an incremental identifier for this individual packet (not to be confused with the Burst Sequence Number).

Send Delay Time (4 bytes)   Specifies the delay between the sender's packet transmissions (also called the *interpacket gap*), measured in units of 100 microseconds.

Burst Sequence Number (2 bytes)   Contains an incremental identifier for the packet burst (which consists of a sequence of packets containing a contiguous stream of data).

Acknowledgment Sequence Number (2 bytes)   Contains the Burst Sequence Number value that the system expects to see in the next burst, indicating that the previous burst was received successfully.

Total Burst Length (4 bytes)   Specifies the total length of the data being transmitted in the current burst (in bytes). A system can adjust this value to implement NCPB's sliding window flow control mechanism.

Burst Packet Offset (4 bytes)   Specifies the location of this packet within the current burst.

Burst Length (2 bytes)   Specifies how much of the Total Burst Length is included in this message.

Fragment List (2 bytes)   Contains a list of fragments still to be transmitted to complete the burst. The field initially contains a list of all of the fragments in the burst. As fragments are successfully transmitted, they are removed from the list. Any fragments remaining after the transmission is completed are considered to have been damaged or lost and must be retransmitted.

   In addition to the preceding fields, NCPB messages requesting a file read or write operation have the following fields:

Function (4 bytes)   Specifies whether the current transaction is a read or write operation.

File Handle (4 bytes)   Contains a code identifying the file to be read or written.

NETWORK PROTOCOLS

Starting Offset (4 bytes) Specifies the portion of the file indicated in the File Handle field that should be included in this packet.

Bytes to Read/Write (4 bytes) Specifies the number of bytes (beginning at the point specified in the Starting Offset field) that should be included in this packet.

NCPB Reply messages generated in response to a read request have the following fields, in addition to the basic message format:

Result Code (4 bytes) Specifies whether or not the associated request was successfully fulfilled, using the following values:

0 No error

1 Initial error

2 I/O error

3 No data read

Number of Bytes Read (4 bytes) Specifies the number of bytes that were successfully read.

Data (variable) Contains part of the data being transmitted in reply to the associated request.

NCPB Reply messages generated in response to a write request include the following field, in addition to the basic message format:

Result Code (4 bytes) Specifies whether or not the associated request was successfully fulfilled, using the following values:

0 No error

1 Write error

Packet Burst Transactions

NCPB is a connection-oriented protocol that is invoked when a client requests a burst-mode connection from a server using an NCP request with 101 as its function code and no subfunction code. The request also includes the client's maximum packet size and its maximum send and receive sizes, as shown in the captured packet in Figure 14-6. The server will use this information when packaging the data for transmission to the client.

802.2: Address: 00-60-97-B0-77-CA --->00-80-29-EB-F7-2C
LLC: Sap 0xE0 ---> 0xE0 (Command)
IPX: Packet=NCP, Net:31-FE-86-7F ---> 37-20-0D-D6
NetWare Core Protocol
NCP Request: Packet Burst Connection Request
Request Type: 0x2222 (Service Request)
Seq. number = 74
Conn. number low = 4
Task number = 48
Conn. number high = 0
Function code = 101
Local Connection ID: 0x3BB9652C
Local Maximum Packet Size: 0x05D2
Local Target Socket: 0xA40
Local Maximum Send Size: 65535
Local Maximum Recieve Size: 65535
Calculate CRC: 0x2df924b1

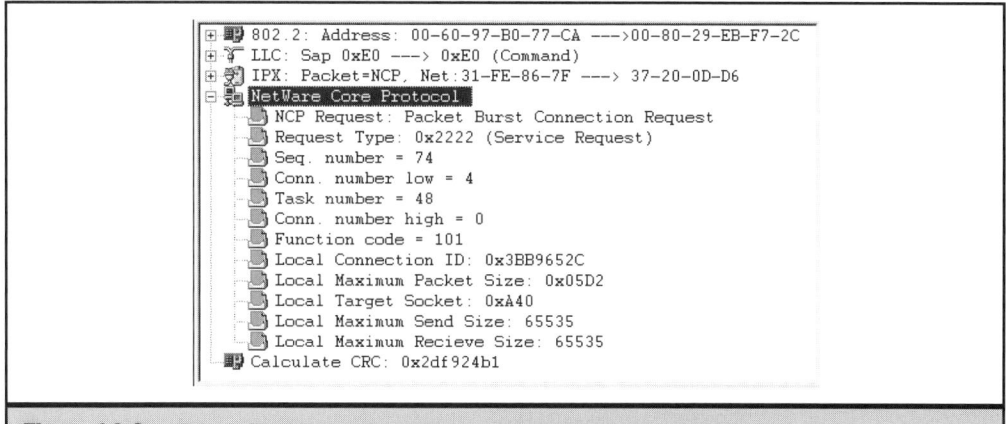

Figure 14-6: *The NCP Packet Burst Connection Request message*

When the server replies to the request, it transmits an NCP Service Reply message (Reply Type 3333) with a Completion Code value of 0 (see Figure 14-7).

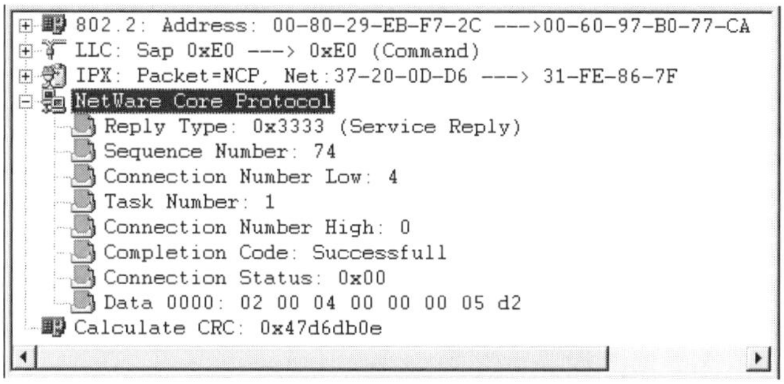
802.2: Address: 00-80-29-EB-F7-2C --->00-60-97-B0-77-CA
LLC: Sap 0xE0 ---> 0xE0 (Command)
IPX: Packet=NCP, Net:37-20-0D-D6 ---> 31-FE-86-7F
NetWare Core Protocol
Reply Type: 0x3333 (Service Reply)
Sequence Number: 74
Connection Number Low: 4
Task Number: 1
Connection Number High: 0
Completion Code: Successfull
Connection Status: 0x00
Data 0000: 02 00 04 00 00 00 05 d2
Calculate CRC: 0x47d6db0e

Figure 14-7: *The NCP Service Reply message completes the connection establishment process.*

At this time, the client switches to the NCPB message format (Request Type 7777) and sends a Read request for a particular file (see Figure 14-8).

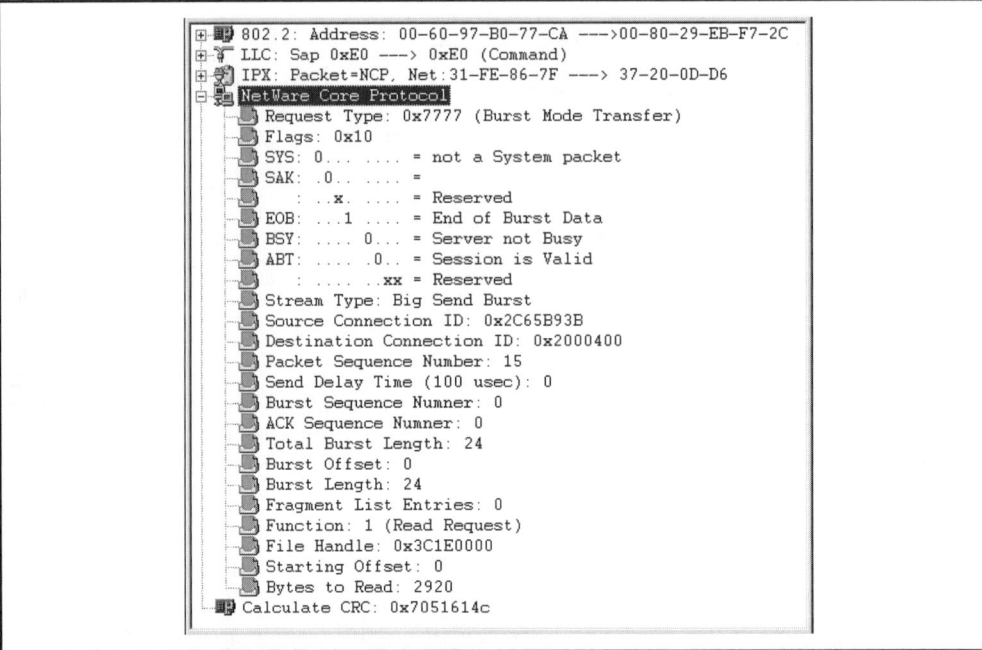

```
  802.2: Address: 00-60-97-B0-77-CA --->00-80-29-EB-F7-2C
  LLC: Sap 0xE0 ---> 0xE0 (Command)
  IPX: Packet=NCP, Net:31-FE-86-7F ---> 37-20-0D-D6
  NetWare Core Protocol
      Request Type: 0x7777 (Burst Mode Transfer)
      Flags: 0x10
      SYS: 0... .... = not a System packet
      SAK: .0.. .... =
         : ..x. .... = Reserved
      EOB: ...1 .... = End of Burst Data
      BSY: .... 0... = Server not Busy
      ABT: .... .0.. = Session is Valid
         : .... ..xx = Reserved
      Stream Type: Big Send Burst
      Source Connection ID: 0x2C65B93B
      Destination Connection ID: 0x2000400
      Packet Sequence Number: 15
      Send Delay Time (100 usec): 0
      Burst Sequence Numner: 0
      ACK Sequence Numner: 0
      Total Burst Length: 24
      Burst Offset: 0
      Burst Length: 24
      Fragment List Entries: 0
      Function: 1 (Read Request)
      File Handle: 0x3C1E0000
      Starting Offset: 0
      Bytes to Read: 2920
  Calculate CRC: 0x7051614c
```

Figure 14-8: *Once the connection is established, the client sends an NCPB Read request message.*

The server replies with a message containing the first fragment of the requested file (see Figure 14-9). In this case, the size of the requested file is shown in the Total Burst Length field as 2,928 bytes. Notice that the Burst Offset value is 0, indicating that this message contains the beginning of the file, and that the amount of data included in this packet, specified by the Burst Length field, is 1,424.

After transmitting a second message containing the next 1,424 bytes of the file, the server sends the message shown in Figure 14-10. Here, the Burst Offset value is 2,848 and the Burst Length is 80. Since 2,848 and 80 adds up to 2,928, the value of the Total Burst Length field, it is clear that this message contains the last 80 bytes of the requested file, even without noticing that the EOB (End of Burst Data) flag is set.

Once the packet burst transfer is complete, the client reverts back to NCP messages, and requests that the server close the file, as shown in Figure 14-11. Once the server returns a reply indicating that the request was completed successfully, the transaction is completed.

```
⊞ 🖳 802.2: Address: 00-80-29-EB-F7-2C --->00-60-97-B0-77-CA  ▲
⊞ ⌂ LLC: Sap 0xE0 ---> 0xE0 (Command)
⊞ 🗐 IPX: Packet=NCP, Net:37-20-0D-D6 ---> 31-FE-86-7F
⊟ 🖳 NetWare Core Protocol
    🗐 Request Type: 0x7777 (Burst Mode Transfer)
    🗐 Flags: 0x00
    🗐 SYS: 0... .... = not a System packet
    🗐 SAK: .0.. .... =
    🗐   : ..x. .... = Reserved
    🗐 EOB: ...0 .... = not End of Burst Data
    🗐 BSY: .... 0... = Server not Busy
    🗐 ABT: .... .0.. = Session is Valid
    🗐   : .... ..xx = Reserved
    🗐 Stream Type: Big Send Burst
    🗐 Source Connection ID: 0x2000400
    🗐 Destination Connection ID: 0x2C65B93B
    🗐 Packet Sequence Number: 15
    🗐 Send Delay Time (100 usec): 62553
    🗐 Burst Sequence Numner: 0
    🗐 ACK Sequence Numner: 1
    🗐 Total Burst Length: 2928
    🗐 Burst Offset: 0
    🗐 Burst Length: 1424
    🗐 Fragment List Entries: 0
    🗐 Reply Code: Successful
    🗐 Bytes Read: 2920
    🗐 Data 0000: 4d 5a bf 01 c2 01 00 00 20 00 64 19 ff ff a7 ▼
    🗐      0010: 80 00 00 00 10 00 cb 35 1e 00 00 00 01 00 00 00
◄                                                            ►
```

Figure 14-9:   *In response to the client's request, the server sends a reply message containing the first data fragment.*

Individual packet burst file transfers are typically integrated throughout a NetWare Core Protocol communications session. The packet burst connection to the server need not be terminated after every completed file transfer, as with TCP.

Packet Retransmission

One advantage of NCPB over TCP and most other connection-oriented protocols is its ability to carry a list of the fragments that need to be transmitted. Most protocols acknowledge packets by specifying a single point in the sequence, thus implying that all of the packets up to that point have been received successfully. When one packet in the sequence is lost, the whole sequence must be retransmitted from that point forward, even if subsequent packets were transmitted successfully. The NCPB fragment list enables the server to retransmit only the fragments that were lost.

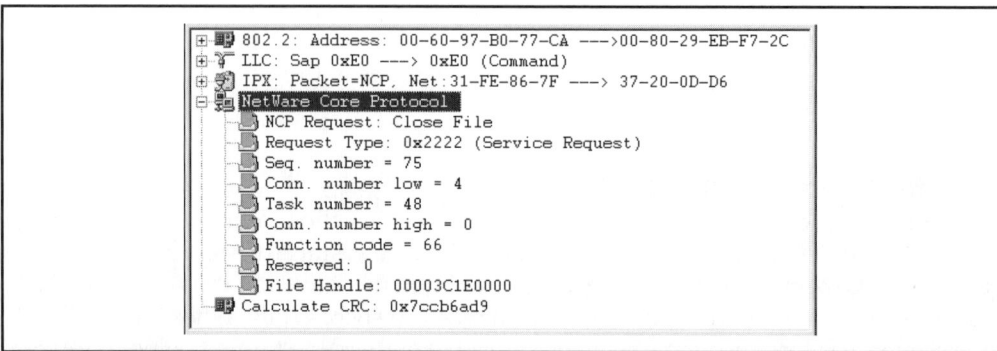

```
⊞ 🖳 802.2: Address: 00-80-29-EB-F7-2C --->00-60-97-B0-77-CA
⊞ 🖥 LLC: Sap 0xE0 ---> 0xE0 (Command)
⊞ 🖥 IPX: Packet=NCP, Net:37-20-0D-D6 ---> 31-FE-86-7F
⊟ 🖥 NetWare Core Protocol
     🖥 Request Type: 0x7777 (Burst Mode Transfer)
     🖥 Flags: 0x10
     🖥 SYS: 0... .... = not a System packet
     🖥 SAK: .0.. .... =
     🖥    : ..x. .... = Reserved
     🖥 EOB: ...1 .... = End of Burst Data
     🖥 BSY: .... 0... = Server not Busy
     🖥 ABT: .... .0.. = Session is Valid
     🖥    : .... ..xx = Reserved
     🖥 Stream Type: Big Send Burst
     🖥 Source Connection ID: 0x2000400
     🖥 Destination Connection ID: 0x2C65B93B
     🖥 Packet Sequence Number: 17
     🖥 Send Delay Time (100 usec): 62553
     🖥 Burst Sequence Numner: 0
     🖥 ACK Sequence Numner: 1
     🖥 Total Burst Length: 2928
     🖥 Burst Offset: 2848
     🖥 Burst Length: 80
     🖥 Fragment List Entries: 0
     🖥 Data 0000: 83 c4 02 52 50 b0 01 50 b1 00 51 9a 06 00 ce
     🖥      0010: 83 c4 0a b0 00 50 b9 8f 00 51 9a 36 03 09 1c
```

Figure 14-10: *The server transmits the last fragment of the transaction in a message with the EOB flag set.*

```
⊞ 🖳 802.2: Address: 00-60-97-B0-77-CA --->00-80-29-EB-F7-2C
⊞ 🖥 LLC: Sap 0xE0 ---> 0xE0 (Command)
⊞ 🖥 IPX: Packet=NCP, Net:31-FE-86-7F ---> 37-20-0D-D6
⊟ 🖥 NetWare Core Protocol
     🖥 NCP Request: Close File
     🖥 Request Type: 0x2222 (Service Request)
     🖥 Seq. number = 75
     🖥 Conn. number low = 4
     🖥 Task number = 48
     🖥 Conn. number high = 0
     🖥 Function code = 66
     🖥 Reserved: 0
     🖥 File Handle: 00003C1E0000
  🖳 Calculate CRC: 0x7ccb6ad9
```

Figure 14-11: *Once the file transfer is completed, the client switches back to NCP messages to close the file.*

The Service Advertising Protocol

NetWare systems use the *Service Advertising Protocol* (*SAP*) to compile and maintain a list of the file servers, print servers, gateway servers, and multiprotocol routers on the network, and the servers use SAP to inform the other systems of their presence. A NetWare client must learn about the servers on the network from SAP messages before it can send requests to them. Every server broadcasts a SAP message every 60 seconds, by default, containing its name, address, and the services that it provides. The other systems on the network, upon receiving a SAP message, create a temporary entry in its NDS or bindery database for each server listed in the message, in order to store the accompanying information.

In addition to these automatic service information broadcasts, servers can also generate SAP requests in order to solicit information from a particular server. NetWare uses this type of SAP exchange to implement the copy protection feature that prevents two servers with the same license number from running on the same network, and clients use it to locate the server nearest to them. For this type of transaction, there are separate Nearest Server Request and Nearest Server Reply packet types. The regular SAP broadcasts containing server information use the Standard Server Reply packet type. (The Standard Server Request message type is not used.)

SAP requests and replies use different message formats, but all SAP messages are carried in standard IPX datagrams with a Packet Type of 4 and a Destination Socket value of 4028, as shown in Figure 14-12. The message formats are covered in the following sections.

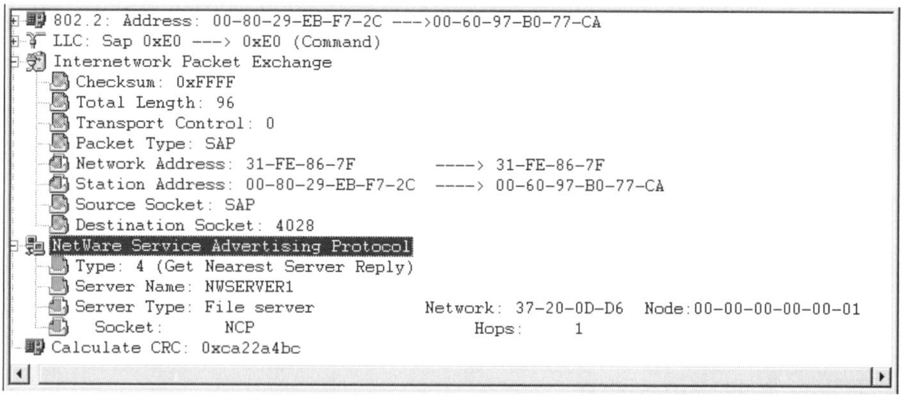

```
802.2: Address: 00-80-29-EB-F7-2C --->00-60-97-B0-77-CA
LLC: Sap 0xE0 ---> 0xE0 (Command)
Internetwork Packet Exchange
    Checksum: 0xFFFF
    Total Length: 96
    Transport Control: 0
    Packet Type: SAP
    Network Address: 31-FE-86-7F        ----> 31-FE-86-7F
    Station Address: 00-80-29-EB-F7-2C  ----> 00-60-97-B0-77-CA
    Source Socket: SAP
    Destination Socket: 4028
NetWare Service Advertising Protocol
    Type: 4 (Get Nearest Server Reply)
    Server Name: NWSERVER1
    Server Type: File server          Network: 37-20-0D-D6   Node:00-00-00-00-00-01
    Socket:      NCP                   Hops:    1
Calculate CRC: 0xca22a4bc
```

Figure 14-12:   *The SAP Get Nearest Server Reply message contains identification and routing information about the server.*

The SAP Request Frame

The SAP Request message format is used only when a system requests SAP information from a server, such as when client systems locate the nearest server. The messages are transmitted as broadcasts, and all servers receiving the message are expected to reply. The request message contains only two fields, which are as follows:

Packet Type (2 bytes)   Specifies the function of the message, using the following hexadecimal values:

1   Standard Server Request (unused)

3   Nearest Server Request

Server Type (2 bytes)   Specifies the type of service desired from a server, using the following hexadecimal values:

0000   Unknown

0003   Print Queue

0004   File Server

0005   Job Server

0007   Print Server

0009   Archive Server

000a   Job Queue

0021   NAS SNA Gateway

0024   Remote Bridge Server

002d   Time Synchronization VAP

002e   Dynamic SAP

0047   Advertising Print Server

004b   Btrieve VAP 5.0

004c   SQL VAP

007a   TES-NetWare VMS

0098   NetWare Access Server

009a   Named Pipes Server

009e   Portable NetWare-UNIX

0107 NetWare 386

0111 Test Server

0166 NetWare Management

026a NetWare Management

ffff Wildcard

The SAP Reply Frame

The SAP Reply message format is the same for both service information broadcasts and replies to Nearest Server Request messages. The difference between the two is that a Nearest Server Reply contains information about only one server, while the Standard Server Reply can contain information about up to seven servers. For the latter, the entire sequence of fields from Server Type to Intermediate Networks is repeated up to seven times, in sequence.

Since the Standard Server Reply messages are transmitted as broadcasts, they are limited to the local network segment. However, by sharing information about themselves as well as about all of the other servers on the local segment, every server on the network is able to build a complete list of all of the other servers.

The format for SAP Reply messages is illustrated in Figure 14-13, and the functions of the fields are as follows:

Packet Type (2 bytes)   Specifies the function of the message, using the following hexadecimal values:

2 Standard Server Reply

3 Nearest Server Reply

Server Type (2 bytes)   Specifies the type of service provided by the server, using the same values as the message format.

Server Name (48 bytes)   Specifies the name of the server.

Network Address (4 bytes)   Specifies the address of the network on which the server is located.

Node Address (6 bytes)   Specifies the node address of the server.

Socket (2 bytes)   Specifies the socket on which the server will receive service requests.

Intermediate Network (2 bytes)   Specifies the number of hops (that is, routers or network addresses) between the server and the destination.

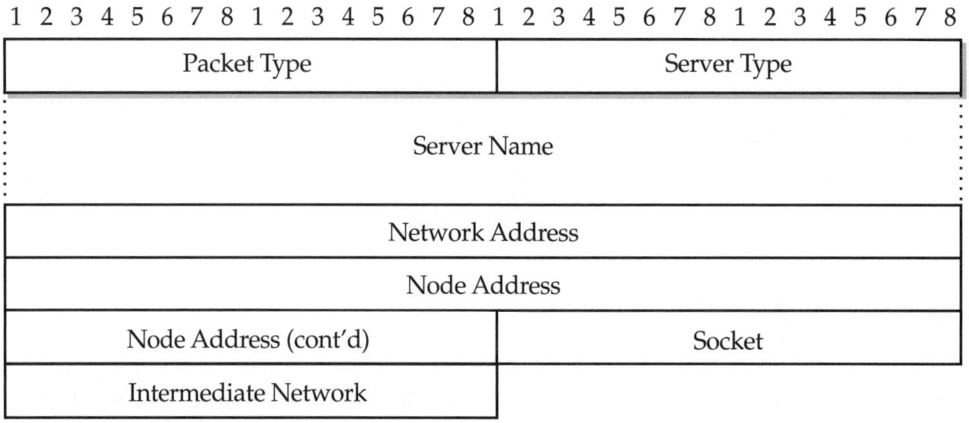

1 2 3 4 5 6 7 8	1 2 3 4 5 6 7 8	1 2 3 4 5 6 7 8	1 2 3 4 5 6 7 8
Packet Type		Server Type	
Server Name			
Network Address			
Node Address			
Node Address (cont'd)		Socket	
Intermediate Network			

Figure 14-13:   *The SAP Reply message format*

SAP Problems

One of the frequent criticisms of NetWare throughout its history has been its reliance on SAP and the amount of redundant broadcast traffic it creates on the network. NDS has reduced this traffic by storing the server information in the directory services database. In the course of the normal NDS replication process, the SAP data is replicated throughout the network using unicast messages between servers, rather than broadcasts. Network 5 addresses the problem further by including support for the Service Location Protocol (SLP), a service discovery protocol standardized by the Internet Engineering Task Force (IETF).

Chapter 15

NetBIOS, NetBEUI, and Server Message Blocks

E ven though TCP/IP has become the most popular protocol suite operating at the network and transport layers of the OSI reference model, alternatives still exist. *NetBEUI*, the *NetBIOS Extended User Interface*, is one of the oldest local area networking protocols still in use, yet it continues to be an excellent solution for relatively small networks because it requires less overhead than more comprehensive protocols.

NetBEUI was designed in the mid-1980s to provide network transport services for programs based on *NetBIOS (Network Basic Input/Output System)*. NetBEUI is just one method of transporting NetBIOS data across a network. Encapsulating the NetBIOS information using the TCP/IP or the IPX protocols is also possible.

When Microsoft began to introduce networking features into its operating systems, NetBEUI was their protocol of choice. Originally, both Windows for Workgroups and Windows NT used NetBEUI as their default protocols. Only later did Microsoft follow the lead of the rest of the networking industry and began relying on TCP/IP to carry NetBIOS data.

The reasons for the adoption of TCP/IP center around interoperability. The TCP/IP protocols were designed from the ground up to support communications between different computing platforms and operating systems. In addition, as local area networking began to take hold of the business world in a big way, NetBEUI's shortcomings made it clear that it would be unsuitable for anything other than small workgroup networks.

Today, NetBEUI is most commonly used on small Microsoft Windows networks because it provides good performance, requires little or no maintenance (because the protocol is self-configuring and self-tuning), and uses a relatively small amount of memory. Despite the criticisms leveled at it in networking circles, if you are running a home or small office network using Windows PCs, NetBEUI is still an excellent protocol solution.

NetBEUI's primary shortcoming is that it is not routable and should generally be used only on networks composed of a single collision domain. This is because the protocol relies on broadcast transmissions for some of its essential functions and has no means of identifying the network on which a system is located. The following sections examine the architecture of the NetBEUI protocol and how it works with NetBIOS and Server Message Blocks to provide basic networking services on a Windows LAN.

NetBIOS

NetBIOS was designed to provide a standardized programming interface between software applications and the networking hardware, so that applications would be more easily portable between systems. The interface includes a name space, which all Microsoft operating systems except Windows 2000 still use to identify the computers on the network. The computer name you assign to a Windows computer during the operating system installation is, in fact, a NetBIOS name, as are the names of domains and workgroups.

NetBIOS names are 16 bytes long, and on a Windows network, the sixteenth byte is reserved for a hexadecimal suffix that specifies the type of resource the name represents. The first 15 characters can be alphabetical or numeric. When you specify a name that is less than 15 characters, Windows left-justifies it and pads out the unused characters, so that the suffix always appears as the sixteenth byte. A list of the NetBIOS suffixes appears in Table 15-1. The NetBIOS name space performs the same function as the IP addresses used by the TCP/IP protocol suite and the network and node addresses used by the IPX/SPX protocols. NetBIOS names provide a unique identifier for every computer on the network, so that systems can send unicast transmissions directly to other systems. For this reason, individual system names are referred to as *unique names*, while NetBIOS names that represent a collection of systems, for purposes of multicasting, are called *group names*.

Suffix	Name Type	Usage
00	Unique	Workstation Service
01	Unique	Messenger Service
01	Group	Master Browser
03	Unique	Messenger Service
06	Unique	RAS Server Service
1F	Unique	NetDDE Service
20	Unique	File Server Service
21	Unique	RAS Client Service
22	Unique	Microsoft Exchange Interchange (MSMail Connector)
23	Unique	Microsoft Exchange Store
24	Unique	Microsoft Exchange Directory
30	Unique	Modem Sharing Server Service
31	Unique	Modem Sharing Client Service
43	Unique	SMS Clients Remote Control

Table 15-1:   *NetBIOS Suffixes Used in Microsoft Windows Networks*

NETWORK PROTOCOLS

Suffix	Name Type	Usage
44	Unique	SMS Administrators Remote Control Tool
45	Unique	SMS Clients Remote Chat
46	Unique	SMS Clients Remote Transfer
4C	Unique	DEC Pathworks TCP/IP Service on Windows NT
42	Unique	McAfee Anti-virus
52	Unique	DEC Pathworks TCP/IP service on Windows NT
87	Unique	Microsoft Exchange MTA
6A	Unique	Microsoft Exchange IMC
BE	Unique	Network Monitor Agent
BF	Unique	Network Monitor Application
03	Unique	Messenger Service
00	Group	Domain Name
1B	Unique	Domain Master Browser
1C	Group	Domain Controllers
1D	Unique	Master Browser
1E	Group	Browser Service Elections
1C	Group	IIS
00	Unique	IIS
[2B]	Unique	Lotus Notes Server Service
[2F]	Group	Lotus Notes
[33]	Group	Lotus Notes
[20]	Unique	DCA IrmaLan Gateway Server Service

Table 15-1: *NetBIOS Suffixes Used in Microsoft Windows Networks (continued)*

The main difference between the NetBIOS name space and the TCP/IP and IPX/SPX addresses is that the NetBIOS name space is flat. No naming hierarchy subdivides the network into individual subnetworks. The 32 bits that comprise an IP address are split between host address and network address bits, and the IPX/SPX address is naturally broken into network and node addresses. The NetBIOS name is a single name, however, and contains no identifying information about the network.

Because NetBEUI uses the NetBIOS name space to communicate with other systems and the name space has no inherent mechanism for identifying and addressing networks, NetBEUI cannot address communications to systems on other networks. This is one reason NetBEUI is not routable.

NetBEUI Frame

The point is often made that NetBIOS is an application programming interface (API) and not a protocol, so logically, you can make the same case for NetBEUI by saying an extended user interface for NetBIOS cannot be a protocol either. The Windows operating systems refer to it as such, however, and use the term *NetBEUI Frame* (or sometimes *NetBIOS Frame*, or more commonly *NBF*) to describe the actual protocol used to carry NetBEUI information over the network.

Note *Unlike TCP/IP and most other protocols, no official standard defines the architecture and functionality of NetBIOS and NetBEUI. Because NetBIOS was originally developed for use on early IBM PC networks, the "IBM LAN Technical Reference IEEE 802.2 and NetBIOS Application Program Interfaces" document is the closest thing there is to a standard. As a result, there have been many NetBIOS implementations over the years that were incompatible and limited to use with specific networking products.*

NBF operates at the session, transport, and network layers of the OSI reference model, although you can argue that NBF has no network layer because it lacks the routing functions that largely define this layer's functionality. On a Windows system, the protocol is used to register the system's NetBIOS names on the network, establish sessions between systems, and carry data generated by several different application-layer protocols and APIs. The most important API is *Server Message Blocks (SMB)*, the protocol used to carry shared file and print data.

In the OSI model, the functionality of the NBF protocol is bounded at the bottom by the NDIS interface, which provides a universal interface to the networking hardware. Data link–layer support is provided by the IEEE 802.2 Logical Link Control (LLC) frame, which encapsulates the NBF protocol message. The 802.2 frame for NBF packets uses (hexadecimal) values of F0 for the destination service access point (DSAP) and the source service access point (SSAP).

Note *For more information on the LLC frame, see "The Logical Link Control Sublayer" in Chapter 10.*

At the top of the OSI model, the protocol either interacts directly with the NetBIOS interface or on Windows XP, 2000, and NT systems with the *Transport Device Interface (TDI)*, an abstraction layer that lies between the NetBIOS interface and the transport-layer protocols.

The functions of the NBF protocol are divided into several different services, which are sometimes referred to as separate protocols. (The lack of a definitive standard makes the nomenclature difficult.) These services provide name registration and resolution, connectionless datagram delivery, diagnostic and monitoring functions, and session-based delivery, all using the same basic message format, which is shown in Figure 15-1 and consists of the following fields:

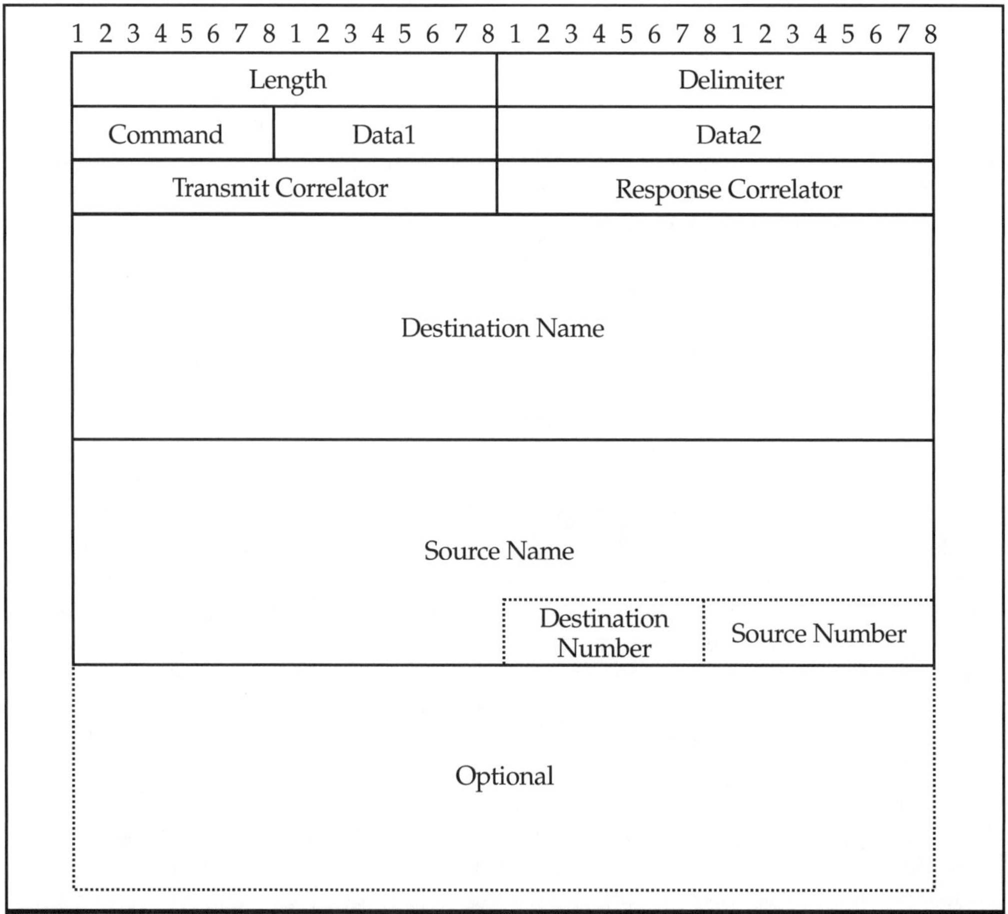

Figure 15-1: *The NetBEUI Frame message format*

Length (2 bytes)   Specifies the length of the NBF header field (including the Length field).

Delimiter (2 bytes)   Indicates that the data that follows is intended for the NetBIOS interface.

Command (1 byte)   Specifies the function of the message, using the following codes. Messages with command codes 00 through 0E are transmitted as unnumbered information (UI) frames, while command codes 0F through 1F are transmitted as information format LLC protocol data units (I-format LPDUs).

00	ADD GROUP NAME QUERY
01	ADD NAME QUERY
02	NAME IN CONFLICT
03	STATUS QUERY
07	TERMINATE TRACE
08	DATAGRAM
09	DATAGRAM BROADCAST
0A	NAME QUERY
0D	ADD NAME RESPONSE
0E	NAME RECOGNIZED
0F	STATUS RESPONSE
13	TERMINATE (local and remote) TRACE
14	DATA ACK
15	DATA FIRST MIDDLE
16	DATA ONLY LAST
17	SESSION CONFIRM
18	SESSION END
19	SESSION INITIALIZE
1A	NO RECEIVE
1B	RECEIVE OUTSTANDING
1C	RECEIVE CONTINUE
1F	SESSION ALIVE

Data1 (1 byte)   Contains optional data specific to the message type.

Data2 (2 bytes)   Contains optional data specific to the message type.

NETWORK PROTOCOLS

Transmit Correlator (2 bytes)   Contains a hexadecimal value from 0001 to FFFF used to associate requests and replies.

Response Correlator (2 bytes)   Contains a hexadecimal value from 0001 to FFFF that indicates the value expected in the Transmit Correlator field of the reply to the current message.

Destination Name (16 bytes)   Specifies the NetBIOS name of the intended destination system (not included in session service packets).

Source Name (16 bytes)   Specifies the NetBIOS name of the local system (not included in session service packets).

Destination Number (1 byte)   Specifies the session number on the destination system (not included in name, datagram, or diagnostic service packets).

Source Number (1 byte)   Specifies the session number on the destination system (not included in name, datagram, or diagnostic service packets).

Optional (variable)   Contains the actual data carried in session and datagram packets (not included in name or diagnostic service packets).

Four services use NBF messages: the name service, the datagram service, the diagnostic service, and the session service. These are sometimes referred to as separate protocols in themselves. These services are examined in the following sections.

Name Management Protocol

The name service, also called the Name Management Protocol (NMP), provides name registration and resolution services for network systems. When a computer on a Microsoft network boots, the system performs a name registration procedure designed to verify that the computer's NetBIOS name is unique on the network. A name resolution process occurs whenever a system tries to access another computer on the network. Because NetBIOS names have no permanent connection to the hardware addresses used to communicate on the LAN, a system trying to send unicast traffic directly to another system must first discover the hardware address of the intended destination.

Note   *Similar name registration and resolution procedures occur when NetBIOS traffic is encapsulated within the IP protocol, as defined in the standards for NetBIOS over TCP/IP, published as RFCs 1001 and 1002. The primary difference between using TCP/IP as the networking protocol on a Windows network and using NetBEUI is that TCP/IP inserts an intermediate step in the name registration and resolution procedures, in which the NetBIOS name is equated with a specific IP address, instead of the hardware address. A later TCP/IP process converts the IP address into the hardware address using the Address Resolution Protocol (ARP). For more information on NetBIOS name resolution on TCP/IP networks, see Chapter 24.*

Name Registration

The name registration process occurs when a system on a Windows network boots. To determine if another computer is on the network using the same NetBIOS name, the system transmits an ADD NAME QUERY message to the NetBIOS functional address (030000000001). The message contains a command code value of 01 and has the system's NetBIOS name in the Source Name field (see Figure 15-2).

Note   *Messages sent to the NetBIOS functional address are received by all NetBIOS systems on the network.*

```
⊟ ▒ IEEE 802.2
    🗋 Address: 00-60-97-B0-77-CA ---->03-00-00-00-00-01
    🗋 Length: 47
⊟ ⅄ Logical Link Control
    🗋 SSAP Address: 0xF0, CR bit = 0 (Command)
    🗋 DSAP Address: 0xF0, IG bit = 0 (Individual address)
    🗋 Unnumbered frame: UI
⊟ ▒ NetBIOS Protocol
    🗋 Length:    44  Delimiter: 0xEFFF
    🗋 Command: 1 (Add Name Query)
    🗋 Option 1: 0x00
    🗋 Option 2: 0x00
    🗋 Correlator Transmit: 0  Response: 1
    🗋 Name: CZ2              ---->
⊟ ▒ Calculate CRC: 0xefd82213
```

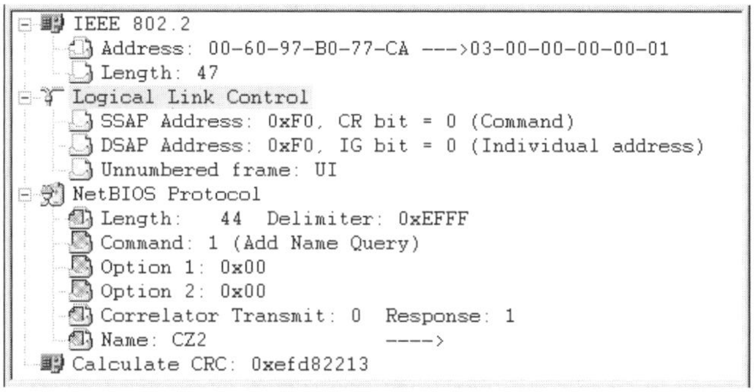

Figure 15-2:   *The NBF ADD NAME QUERY message*

The other NetBIOS systems on the network are required to respond if they possess the same name as that contained in the message. If the transmitting system receives no responses after repeated retries, the name is considered to be registered. If another machine has the same name, it transmits an ADD NAME RESPONSE message to the sender as a unicast (see Figure 15-3). This denies the name to the original system and forces the user to select a different one.

The ADD NAME RESPONSE message contains a Command code value of 0D and the name in question in both the Destination Name and Source Name fields. The Data1 field contains a binary flag with one of the following values:

0   Signifies that the Add Name procedure is in progress

1   Signifies that the Add Name procedure is not in progress

The Data2 field specifies whether the name in question is already in use on the network as a unique name or a group name, using the following values:

0   Unique

1   Group

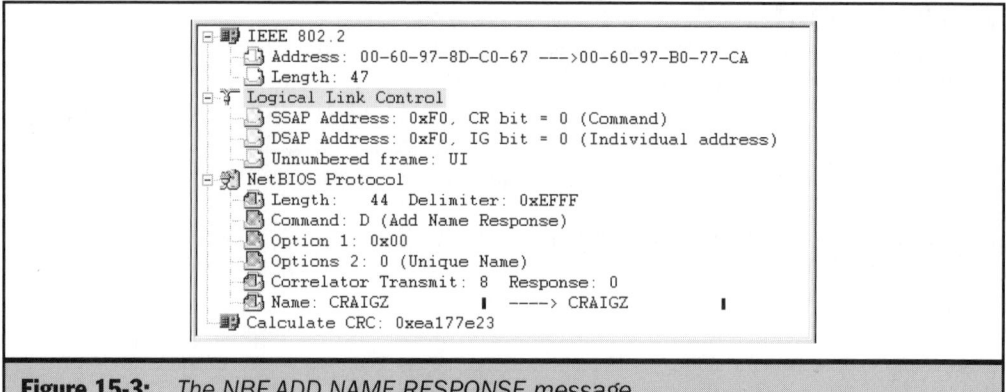

Figure 15-3: *The NBF ADD NAME RESPONSE message*

The Transmit Correlator field contains the same value as the Response Correlator field in the ADD NAME QUERY message. This is so the system receiving the message can associate it with the correct request.

If the system trying to register the name should receive ADD NAME RESPONSE messages from two or more other systems (or if the same name is registered both as a group name and a unique name), it generates a NAME IN CONFLICT message and transmits it to the NetBIOS functional address. The same message is generated when a system receives multiple ADD NAME RESPONSE replies to an ADD GROUP NAME QUERY message or NAME RECOGNIZED messages from two or more systems in response to a NAME QUERY.

The NAME IN CONFLICT message contains a Command code of 02 and has the name in question in the Destination Name field. The Source Name field contains the special NetBIOS *name number 1* for the transmitting system, which consists of 10 bytes worth of zeros followed by the system's 6-byte hardware address.

If the system is a member of a Windows domain, it also transmits an ADD GROUP NAME QUERY containing the name of the domain. This is intended to make sure that the group name is not being used by another system as a unique name, in which case the computer using the name generates an ADD NAME RESPONSE message.

Name Resolution

The name resolution process occurs whenever a system attempts to access another NetBIOS system on the network. Before a computer can transmit unicast packets, it must determine the hardware address of the destination system. To do this, the computer generates a NAME QUERY message that it transmits to the NetBIOS functional multicast address. This message contains the Command code 0A and the name of the system to be contacted in the Destination Name field (see Figure 15-4).

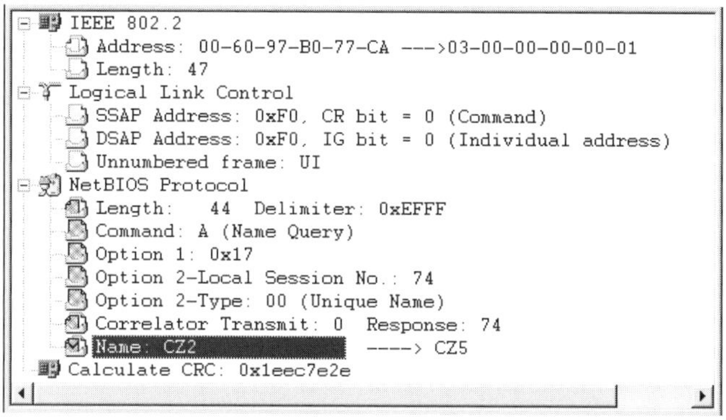

```
┌──────────────────────────────────────────────────────────┐
│ ⊟ 📑 IEEE 802.2                                            │
│    📄 Address: 00-60-97-B0-77-CA --->03-00-00-00-00-01    │
│    📄 Length: 47                                           │
│ ⊟ Ⴤ Logical Link Control                                  │
│    📄 SSAP Address: 0xF0, CR bit = 0 (Command)            │
│    📄 DSAP Address: 0xF0, IG bit = 0 (Individual address) │
│    📄 Unnumbered frame: UI                                 │
│ ⊟ 📑 NetBIOS Protocol                                      │
│    📑 Length:    44  Delimiter: 0xEFFF                     │
│    📑 Command: A (Name Query)                              │
│    📑 Option 1: 0x17                                       │
│    📑 Option 2-Local Session No.: 74                       │
│    📑 Option 2-Type: 00 (Unique Name)                      │
│    📑 Correlator Transmit: 0  Response: 74                 │
│    📑 Name: CZ2                       ----> CZ5            │
│    📑 Calculate CRC: 0x1eec7e2e                            │
│ ◄                                                      ►   │
└──────────────────────────────────────────────────────────┘
```

Figure 15-4: *The NBF NAME QUERY message*

The Data1 field is unused, but the Data2 field contains a 2-byte code specifying whether the name being queried is a unique or group name, using the following values:

00   Unique name

01   Group Name

NetBEUI systems use this NAME QUERY/NAME RECOGNIZED exchange for two purposes: to discover the address of another system or to initiate a session with another system. The second 2 bytes of the Data2 field contain either 00, indicating that the intent is only to determine the addresses of the system using the name, or a number from 01 to FE that acts as a local identifier for the session the system is trying to establish.

If the system receives no response to the NAME QUERY messages, it assumes the name doesn't exist on the network. Any computer using the name is required to respond with a NAME RECOGNIZED message for each NAME QUERY it receives, transmitted as a unicast to the sender. The NAME RECOGNIZED message contains 0E as its Command code (see Figure 15-5). The Destination Name field contains the name of the system that generated the NAME QUERY message and the Source Name field contains the name of the local system.

The Data1 field is again unused, and the Data field begins with the same 2-byte code identifying the name as unique or as that of group, just as in the NAME QUERY message. The second 2 bytes of the Data2 field specify the state of the name, using the following codes:

00   Indicates that the system is not listening for the SESSION INITIALIZE message from the sender that will initiate a session between the two machines

A value from 01 to FE   Specifies the value the local system will use to identify the session to be established

FF   Indicates that the system is listening for the SESSION INITIALIZE message from the sender, but that it cannot initiate a session between the two machines

NETWORK PROTOCOLS

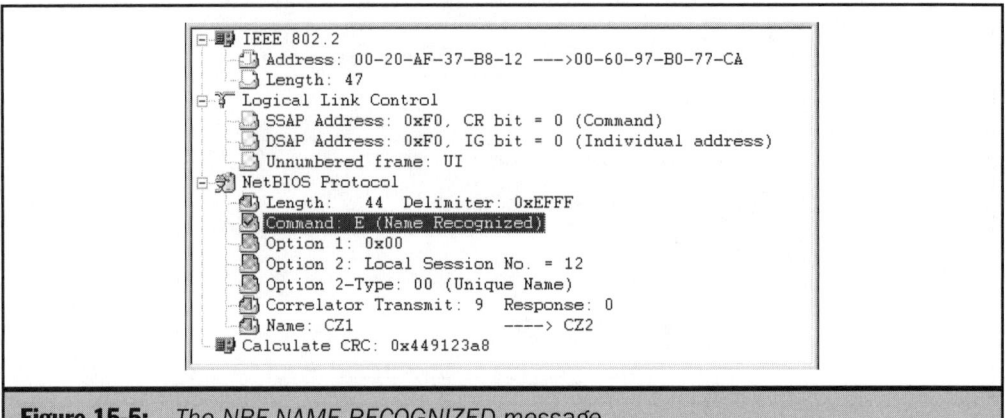

```
⊟ ▦ IEEE 802.2
    └─ 🗋 Address: 00-20-AF-37-B8-12 ---->00-60-97-B0-77-CA
    └─ 🗋 Length: 47
⊟ ¥ Logical Link Control
    └─ 🗋 SSAP Address: 0xF0, CR bit = 0 (Command)
    └─ 🗋 DSAP Address: 0xF0, IG bit = 0 (Individual address)
    └─ 🗋 Unnumbered frame: UI
⊟ 🗋 NetBIOS Protocol
    └─ 🗋 Length:    44   Delimiter: 0xEFFF
    └─ ☑ Command: E (Name Recognized)
    └─ 🗋 Option 1: 0x00
    └─ 🗋 Option 2: Local Session No. = 12
    └─ 🗋 Option 2-Type: 00 (Unique Name)
    └─ 🗋 Correlator Transmit: 9   Response: 0
    └─ 🗋 Name: CZ1            ----> CZ2
  └─ ▦ Calculate CRC: 0x449123a8
```

Figure 15-5:   *The NBF NAME RECOGNIZED message*

User Datagram Protocol

The messages used in the NetBIOS name service exchanges are transmitted as unnumbered information (UI) frames, sometimes called type 1 frames. This is NetBEUI's connectionless, unreliable service, used for brief exchanges in which retransmissions and expected responses eliminate the need for packet acknowledgments and guaranteed delivery. In addition to the name services messages, NBF also supports a datagram service that provides delivery of small amounts of data using the same connectionless, unreliable transmissions. The Server Message Block protocol often uses the datagram service for its request/reply transactions.

This service is sometimes called the User Datagram Protocol (UDP), which is unfortunate because TCP/IP has a transport-layer protocol of the same name (that provides basically the same service). In the vast majority of cases, documents that refer to UDP are referring to the TCP/IP protocol, not its NetBEUI equivalent.

The NetBEUI UDP is actually more comparable in function to the IP protocol in the TCP/IP suite, IPX in Novell's IPX/SPX, or AppleTalk's Datagram Delivery Protocol (DDP), except that UDP does not provide services for upper-layer protocols. IP, for example, is used to encapsulate various other protocols, including TCP, (the other) UDP, and ICMP, while NetBEUI's UDP carries only actual application data.

The DATAGRAM messages used to carry UDP data have a Command code of 08 and do not use either of the Data fields or the Correlator fields. The Destination Name field contains the NetBIOS name of the message's intended recipient, and the Source Name field contains the name of the transmitting system. The Optional field contains the data intended for the destination. There is also a DATAGRAM

BROADCAST message used to transmit to the entire network that is identical to the DATAGRAM message except that it has 09 for its Command code and contains no value in the Destination Name field.

Diagnostic and Monitoring Protocol

The *Diagnostic and Monitoring Protocol (DMP),* roughly analogous to the Simple Network Management Protocol (SNMP) in TCP/IP, is used to gather functional information about the systems on the network. A typical DMP exchange begins when a system generates a STATUS QUERY message (Command code 03) and transmits it to the NetBIOS functional address. The message contains a code in the Data1 field that indicates the type of request, using the following values:

00 NetBIOS 1.*x* or 2.0 type request

01 Initial NetBIOS 2.1 type request

Greater than 01 NetBIOS 2.1 type request for replies from more systems, where the value indicates the number of replies already received

The value in the Data2 field specifies the length of the system's status buffer. The Destination Name field contains the name of the system for which the status is being requested, and the Source Name field contains the name number 1 for the local system.

In reply to a STATUS QUERY transmission, a computer generates a STATUS RESPONSE message (Command code 0F) that it transmits as a unicast to the querying system. The Data1 field of the message indicates the status of the response, using the following codes:

00 NetBIOS 1.*x* or 2.0 type response

01 or greater NetBIOS 2.1 type response, where the value indicates the number of replies already received

The Data2 field contains two flags. The first bit is set to 1 if the length of the status data exceeds the frame size; the second bit is set to 1 if the length of the status data exceeds the size of the user's buffer. The remaining 14 bits in the field are used to specify the actual length of the status data. The Destination Name field contains the name of the system receiving the message and the Source Name field contains the NetBIOS name of the sender.

The DMP service also includes two messages used to end a network trace, both with the same name. The TERMINATE TRACE message with Command code 07 terminates the trace activity at a remote system, while the TERMINATE TRACE message with Command code 13 terminates the trace activity at both the local and remote systems. The latter message is never generated by the NetBIOS interface, but it is recognized when generated by another application.

Session Management Protocol

Much of the NetBEUI traffic generated by typical networking tasks on a Windows network is transmitted during a session between two machines. A session occurs when two systems establish a connection before they actually transmit any application data. The connection ensures that each system is prepared to communicate and enables the two machines to regulate the flow of data and acknowledge successful transmissions. The Session Management Protocol (SMP) provides this full-duplex, connection-oriented, reliable service between NetBIOS systems.

Session Establishment

The process of establishing a session between two machines begins with the name resolution procedure described earlier in this chapter. The client computer wishing to establish the session transmits a NAME QUERY message containing a session identifier (that is, a value other than 00) in the Data2 field to all of the NetBIOS systems on the network. The intended destination server responds with a NAME RECOGNIZED message that supplies its hardware address and indicates that it is listening for further session messages from the sender.

Note
As is often the case, the roles of client and server are tenuous on a Windows network because the computers can function as both clients and servers. The references here to clients and servers refer to the roles of the computers in this particular transaction. The two machines could just as easily reverse their roles and perform a session establishment originating from the other system.

Before the next exchange of NBF messages, the two systems perform a session establishment procedure at the LLC level, which consists of the client transmitting a SABME (Set Asynchronous Balance Mode Extended) message, to which the server replies with a UA (Unnumbered Acknowledgment) frame. The client then sends an RR (Receive Ready) message indicating that it is ready to receive data (see Figure 15-6).

Once the session is established at the LLC level, an NBF session establishment transaction is required before the system can begin to transmit actual application data. This procedure begins when the client system transmits a SESSION INITIALIZE message (with a Command code value of 19) as a unicast to the server (see Figure 15-7).

Note
After the initial NAME QUERY message, all the subsequent frames involved in session communications are I-format LPDU unicasts, which use the hardware address discovered during the name resolution process to direct the packets. Any timeouts and retries occurring during the transmissions are handled by the IEEE 802.2 LLC implementation.

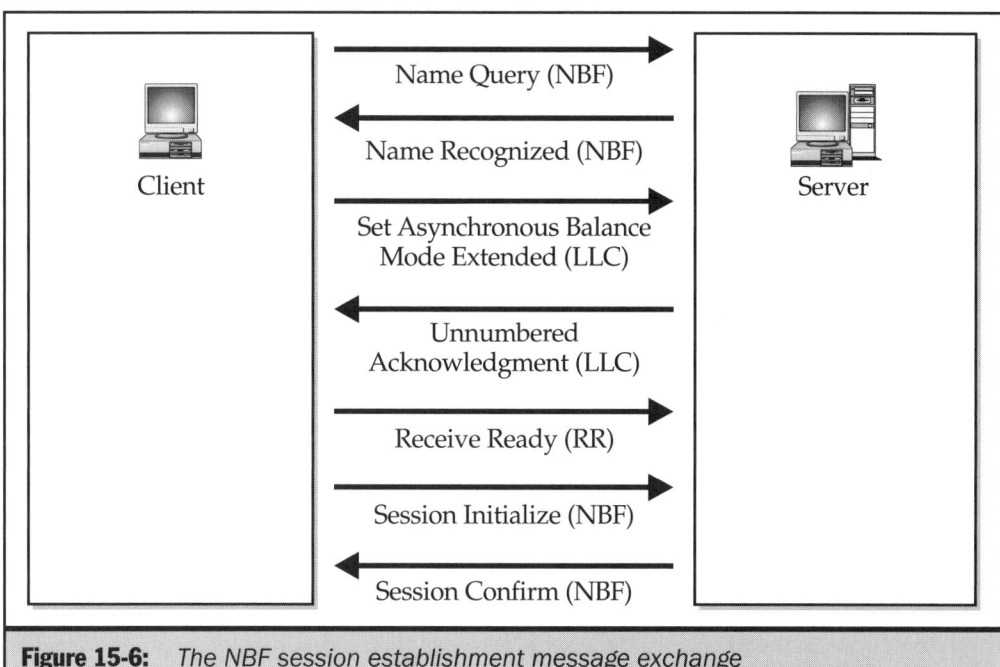

Figure 15-6:   *The NBF session establishment message exchange*

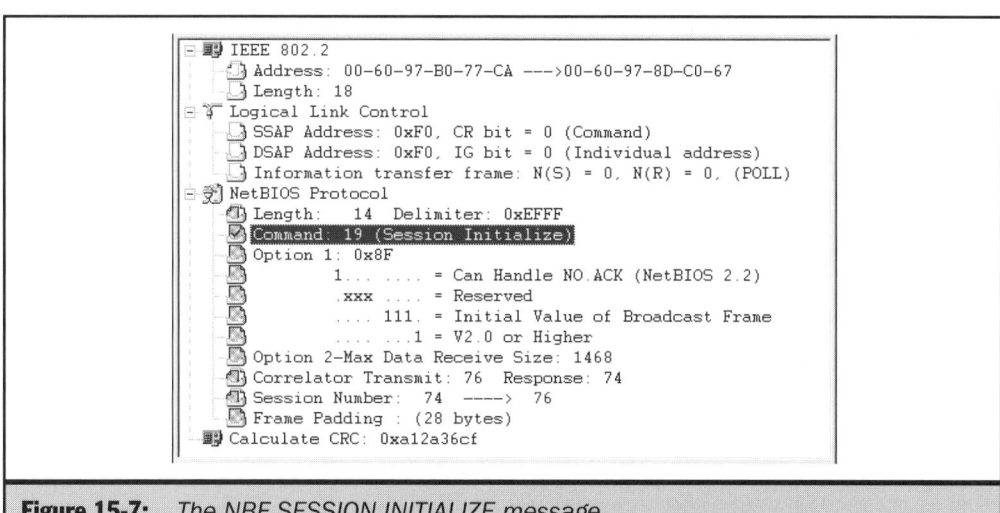

Figure 15-7:   *The NBF SESSION INITIALIZE message*

The Data1 field of this message uses the following bit values:

Bit 1   Flag specifying the NetBIOS version, using the following values:

   0   NetBIOS version 2.20 or lower

   1   NetBIOS version higher than 2.20

Bits 2–4   Unused

Bits 5–7   Specify the length of the largest frame value permitted by the MAC protocol

Bit 8   Flag specifying the NetBIOS version, using the following values:

   0   NetBIOS version 1.*x*

   1   NetBIOS version 2.0 or above. This indicates the system is capable of performing a certain type of data transfer that does not require acknowledgments

The Data2 field specifies the length of the user's receive buffer.

Unlike the messages for the other services, SMP messages do not have Destination Name and Source Name fields. Instead, they have a 1-byte Destination Number and Source Number fields that carry the unique identifiers the systems use to refer to the session. Each computer maintains its own session numbers.

In response to the SESSION INITIALIZE message, the second system generates a SESSION CONFIRM message that completes the initialization of the session (see Figure 15-8). This message has a Command code of 17 and is identical in format to the SESSION INITIALIZE message, except that bits 5 through 7 of the Data1 field are unused. The Session Number line of the display contains the identifying number for the session on the two computers involved (much like TCP/IP port numbers identify a process running on a computer).

Session Maintenance

During periods of inactivity, the computers involved in a session transmit SESSION ALIVE messages to confirm that the other system is still available and capable of receiving data (see Figure 15-9). The SESSION ALIVE message has a Command code value of 1F; all of the subsequent fields are unused.

Data Transfer

Once the session is established, the transfer of data can begin, using NBF messages that may or may not carry data generated by upper-layer protocols (such as SMB). When one computer connects to another to copy files from the server to a local drive, for example, the systems use NBF frames to carry the actual data. When you use a Windows application to open a file on a network drive, the system uses SMB messages (carried with NBF frames) to access the drive and transfer the file.

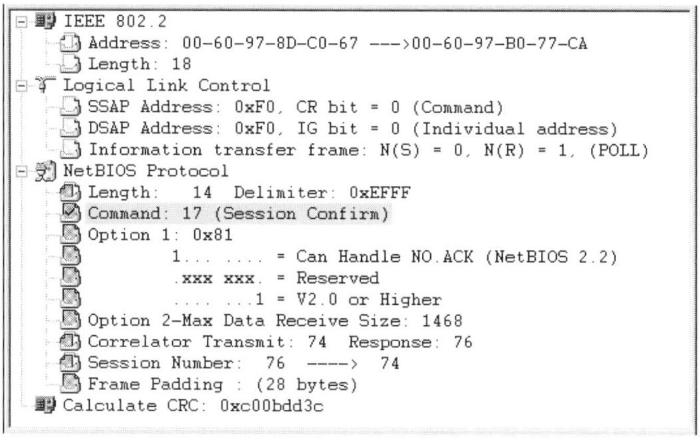

```
IEEE 802.2
   Address: 00-60-97-8D-C0-67 --->00-60-97-B0-77-CA
   Length: 18
Logical Link Control
   SSAP Address: 0xF0, CR bit = 0 (Command)
   DSAP Address: 0xF0, IG bit = 0 (Individual address)
   Information transfer frame: N(S) = 0, N(R) = 1, (POLL)
NetBIOS Protocol
   Length:   14   Delimiter: 0xEFFF
   Command: 17 (Session Confirm)
   Option 1: 0x81
            1... .... = Can Handle NO.ACK (NetBIOS 2.2)
            .xxx xxx. = Reserved
            .... ...1 = V2.0 or Higher
   Option 2-Max Data Receive Size: 1468
   Correlator Transmit: 74  Response: 76
   Session Number:   76  ---->  74
   Frame Padding : (28 bytes)
Calculate CRC: 0xc00bdd3c
```

Figure 15-8: *The NBF SESSION CONFIRM message*

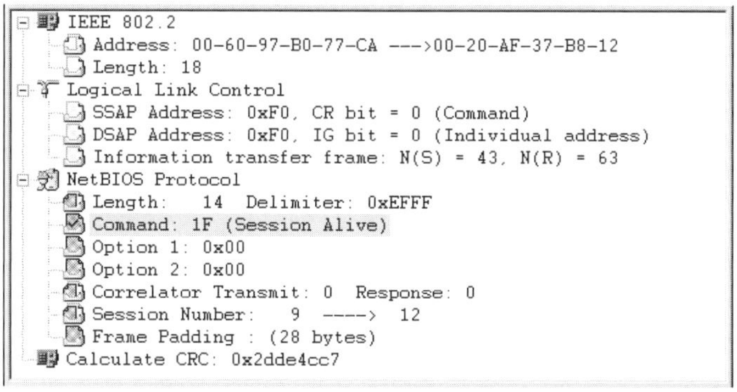

```
IEEE 802.2
   Address: 00-60-97-B0-77-CA --->00-20-AF-37-B8-12
   Length: 18
Logical Link Control
   SSAP Address: 0xF0, CR bit = 0 (Command)
   DSAP Address: 0xF0, IG bit = 0 (Individual address)
   Information transfer frame: N(S) = 43, N(R) = 63
NetBIOS Protocol
   Length:   14   Delimiter: 0xEFFF
   Command: 1F (Session Alive)
   Option 1: 0x00
   Option 2: 0x00
   Correlator Transmit: 0  Response: 0
   Session Number:   9  ---->  12
   Frame Padding : (28 bytes)
Calculate CRC: 0x2dde4cc7
```

Figure 15-9: *The NBF SESSION ALIVE message*

The NBF frames used to transmit the data depend on the amount of data to be transferred. When copying a file small enough to fit in a single packet, the sending system sends the data in a DATA ONLY LAST message. When the file spans multiple packets because it is too large for either the frame size or the transmit buffer size of the receiving system, all the segments are transmitted in DATA FIRST MIDDLE frames except for the last one, which goes in a DATA ONLY LAST frame.

> **Note** The term "message" refers to an entire data sequence, even if it is segmented into multiple packets. All the frames used in the name, datagram, and status services are self-contained messages, but SMP can require many frames to transmit a single message.

The DATA FIRST MIDDLE frame has a Command code of 15 (see Figure 15-10). The Data1 field contains flags that use the following bit values:

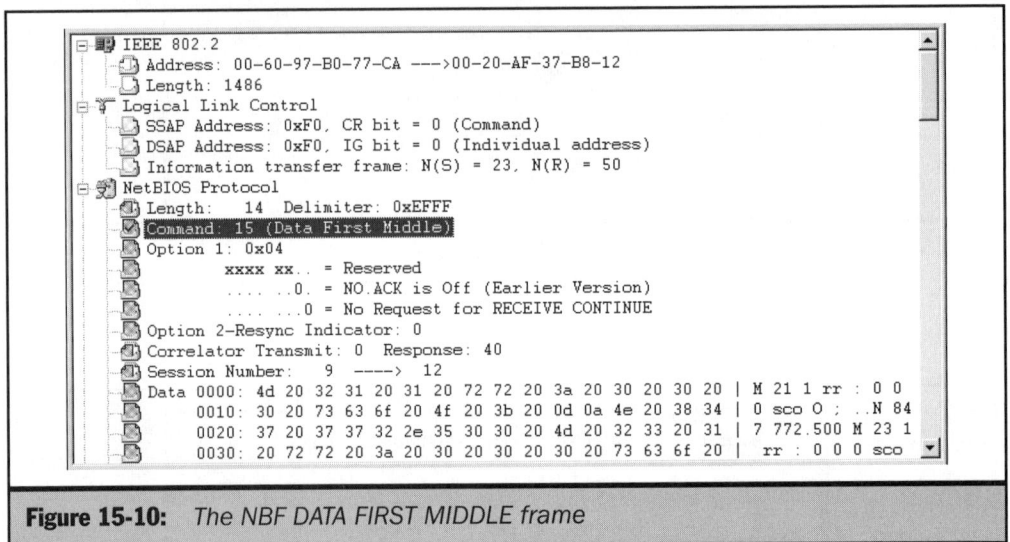

Figure 15-10: *The NBF DATA FIRST MIDDLE frame*

Bits 1–4 Unused

Bit 5 Specifies whether an acknowledgment is included with the frame, using the following values:

  0 Acknowledgment not included

  1 Acknowledgment included

Bit 6 Unused

Bit 7 Indicates the version of NetBIOS and whether an acknowledgment is expected from the receiver, using the following values:

  0 NetBIOS version prior to 2.20 (acknowledgment is expected)

  1 NetBIOS version 2.20 or later (acknowledgment is not expected)

Bit 8   Specifies whether a RECEIVE CONTINUE message is requested from the receiver, using the following values:

0   RECEIVE CONTINUE not requested

1   RECEIVE CONTINUE requested

The sender requests a RECEIVE CONTINUE message from the receiver if the packet contains the first segment sent during the session, or if the sender has received any NO RECEIVE responses during the previous message transmission. If the previous message transmission was completed without any NO RECEIVE responses from the receiving system, the sender does not request a RECEIVE CONTINUE message.

The Data2 field contains a resynchronization indicator with a value of 0001 if this is the first DATA FIRST MIDDLE frame following the receipt of a RECEIPT OUTSTANDING message (which indicates the receiver's ability to accept more data following a NO RECEIVE message). This enables the receiver to resynchronize the transmission sequence with this frame should problems occur during subsequent packet transfers.

The DATA ONLY LAST frame (see Figure 15-11) has a Command code value of 16 and is nearly identical in format to the DATA FIRST MIDDLE frame, except for the flag bits in the Data1 field, which are as follows:

Bits 1–4   Unused

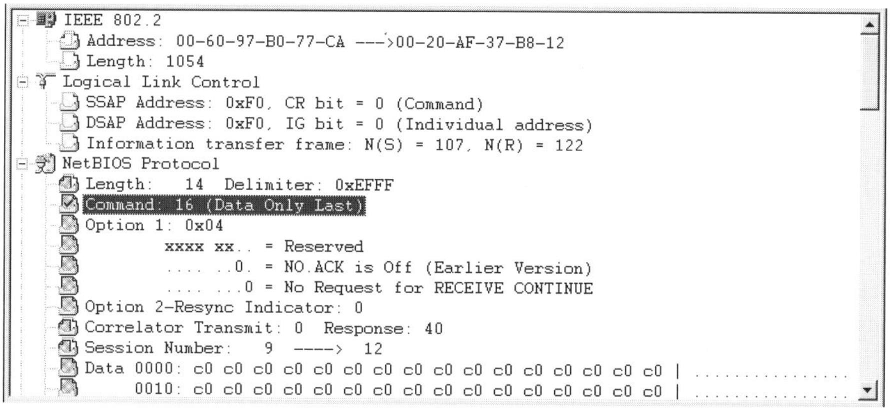

```
IEEE 802.2
    Address: 00-60-97-B0-77-CA ---->00-20-AF-37-B8-12
    Length: 1054
Logical Link Control
    SSAP Address: 0xF0, CR bit = 0 (Command)
    DSAP Address: 0xF0, IG bit = 0 (Individual address)
    Information transfer frame: N(S) = 107, N(R) = 122
NetBIOS Protocol
    Length:   14  Delimiter: 0xEFFF
    Command: 16 (Data Only Last)
    Option 1: 0x04
            xxxx xx.. = Reserved
            .... ..0. = NO.ACK is Off (Earlier Version)
            .... ...0 = No Request for RECEIVE CONTINUE
    Option 2-Resync Indicator: 0
    Correlator Transmit: 0  Response: 40
    Session Number:   9  ----> 12
    Data 0000: c0 c0 c0 c0 c0 c0 c0 c0 c0 c0 c0 c0 c0 c0 c0 | ................
        0010: c0 c0 c0 c0 c0 c0 c0 c0 c0 c0 c0 c0 c0 c0 c0 | ................
```

Figure 15-11:   *The NBF DATA ONLY LAST frame*

Bit 5   Specifies whether an acknowledgment is included with the frame, using the following values:

0   Acknowledgment not included

1   Acknowledgment included

Bit 6   Specifies whether the receiving system can acknowledge the transmission by sending either a DATA ACK message or a data frame in which bit 5 of the Data1 field has a value of 1, indicating that an acknowledgment is included with the frame. The *acknowledgment with data* feature must be supported by the NBF implementations on both systems to be used. The flag takes the following values:

0   Acknowledgment with data not allowed

1   Acknowledgment with data allowed

Bit 7   Indicates the version of NetBIOS and whether an acknowledgment is expected from the receiver, using the following values:

0   NetBIOS version prior to 2.20 (acknowledgment is expected)

1   NetBIOS version 2.20 or later (acknowledgment is not expected)

Bit 8   Unused

The receiver of a DATA ONLY LAST frame must acknowledge it by responding to the sender with one of the following frames:

- DATA ACK

- NO RECEIVE

- RECEIVE OUTSTANDING

- DATA FIRST MIDDLE or DATA ONLY LAST (with the acknowledgment with data feature enabled)

The DATA ACK message (Command code 14) is a simple frame that does nothing but acknowledge that a DATA ONLY LAST frame was received correctly. Both of the Data fields are unused.

The RECEIVE CONTINUE message is generated by a system that receives a DATA FIRST MIDDLE frame, in which the eighth bit of the Data1 field has a value of 1, indicating that the sender is requesting the response. The RECEIVE CONTINUE message serves as an acknowledgment of the data received thus far and indicates there is more data to transmit. The message itself has a Command code value of 1C and nothing in either of the Data fields.

When a system receives a DATA FIRST MIDDLE or DATA ONLY LAST frame that fills its receive buffer, it generates a NO RECEIVE message with a Command code of 1A. This message contains one flag in the Data1 field, using the following bit values:

Bits 1–6 Unused

Bit 7 Indicates the version of NetBIOS and that the acknowledgment included with the previously transmitted data was either partially received or not received at all, using the following values:

    0 NetBIOS version prior to 2.20

    1 NetBIOS version 2.20 or later (acknowledgment not received)

Bit 8 Unused

The Data2 field specifies how many bytes of data from the last frame were received before the buffer was filled. The sender uses this information when it resumes transmitting to restart the sequence at the point where the receiver left off.

Once the sender receives a NO RECEIVE message, it stops transmitting until it receives a RECEIVE OUTSTANDING message (Command code 1B) from the receiving system. This indicates that room now exists in the receive buffer for more data and that the sender should resume transmitting beginning with the byte immediately after the last byte acknowledged, as specified in the Data2 field.

Session Termination

When the client system wants to terminate the session with the server, it transmits a SESSION END message, with a Command code of 18 (see Figure 15-12). The Data1 field of this message is unused, but the Data2 field specifies the reason for the termination of the session, using the following values:

```
⊟ ▦ IEEE 802.2
    ⬚ Address: 00-60-97-B0-77-CA --->00-10-5A-10-6E-93
    ⬚ Length: 18
⊟ ⌐ Logical Link Control
    ⬚ SSAP Address: 0xF0, CR bit = 0 (Command)
    ⬚ DSAP Address: 0xF0, IG bit = 0 (Individual address)
    ⬚ Information transfer frame: N(S) = 7, N(R) = 6, (POLL)
⊟ ▩ NetBIOS Protocol
    ⬚ Length:    14  Delimiter: 0xEFFF
    ⬚ Command: 18 (Session End)
    ⬚ Option 1: 0x00
    ⬚ Option 2—Termination Indicator: 0 (Normal Session End)
    ⬚ Correlator Transmit: 0  Response: 0
    ⬚ Session Number:  10  ----> 44
    ⬚ Frame Padding : (28 bytes)
  ▦ Calculate CRC: 0xa9775d16
```

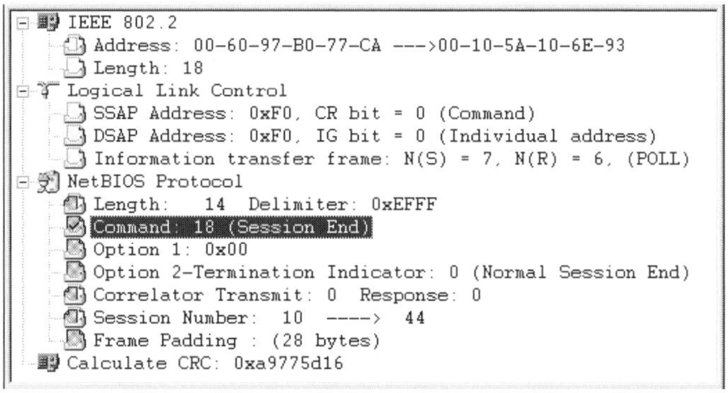

Figure 15-12: *The NBF SESSION END message*

00 Normal session end (due to an application command, for example)

01 Abnormal session end (due to a timeout, for example)

Server Message Blocks

In some cases, the NBF frame is the primary payload of a packet. For example, when a Windows system accesses a file on another system, the file itself is transmitted in NBF data frames. However, NBF messages can also carry upper-layer protocol messages as well. *Server Message Blocks (SMB)* is an application-layer protocol that the Windows redirector (the module responsible for sending application requests to the appropriate network resource) uses to perform many different file management and authentication tasks on remote systems. For example, before copying a file from a shared network drive to a local drive, the two systems involved exchange SMB messages that authenticate the user's access to the resource and establish a session with the share.

Note *The session established at the application layer by the SMB protocol is independent of the other sessions discussed earlier in this chapter: the NBF session and the LLC session. All three session establishment processes must be completed before the two Windows network systems can transfer application data.*

SMB Messages

SMB messages are not restricted to use with NetBEUI, but they are intimately connected with NetBIOS. When a Windows network uses TCP/IP as its network protocol, NetBT (NetBIOS over TCP/IP) frames carry the SMB messages. On a NetBEUI network, SMB messages are carried within the following NBF message types:

- DATAGRAM
- DATAGRAM BROADCAST
- DATA FIRST MIDDLE
- DATA ONLY LAST

Several dozen SMB message types exist, falling into four basic categories:

Session control messages   Used to establish and terminate a connection to a shared resource on a server

File system messages   Used to access and manipulate the file system on a remote server's shared drive

Printer messages   Used to send print jobs generated by local applications to a print queue on a remote server

Message messages   Used to carry messages with another system on the network

Each SMB message includes a 1-byte command field that identifies the function of the message, using the values shown in Table 15-2.

Command	Command Code
CREATE_DIRECTORY	00
DELETE_DIRECTORY	01
OPEN	02
CREATE	03
CLOSE	04
FLUSH	05
DELETE	06
RENAME	07
QUERY_INFORMATION	08
SET_INFORMATION	09
READ	0A
WRITE	0B
LOCK_BYTE_RANGE	0C
UNLOCK_BYTE_RANGE	0D
CREATE_TEMPORARY	0E
CREATE_NEW	0F
CHECK_DIRECTORY	10
PROCESS_EXIT	11

Table 15-2:   *Server Message Block Protocol Command Codes*

NETWORK PROTOCOLS

Command	Command Code
SEEK	12
LOCK_AND_READ	13
WRITE_AND_UNLOCK	14
READ_RAW	1A
READ_MPX	1B
READ_MPX_SECONDARY	1C
WRITE_RAW	1D
WRITE_MPX	1E
WRITE_COMPLETE	20
SET_INFORMATION2	22
QUERY_INFORMATION2	23
LOCKING_ANDX	24
TRANSACTION	25
TRANSACTION_SECONDARY	26
IOCTL	27
IOCTL_SECONDARY	28
COPY	29
MOVE	2A
ECHO	2B
WRITE_AND_CLOSE	2C
OPEN_ANDX	2D
READ_ANDX	2E
WRITE_ANDX	2F
CLOSE_AND_TREE_DISC	31
TRANSACTION2	32

Table 15-2: *Server Message Block Protocol Command Codes (continued)*

Command	Command Code
TRANSACTION2_SECONDARY	33
FIND_CLOSE2	34
FIND_NOTIFY_CLOSE	35
TREE_CONNECT	70
TREE_DISCONNECT	71
NEGOTIATE	72
SESSION_SETUP_ANDX	73
LOGOFF_ANDX	74
TREE_CONNECT_ANDX	75
QUERY_INFORMATION_DISK	80
SEARCH	81
FIND	82
FIND_UNIQUE	83
NT_TRANSACT	A0
NT_TRANSACT_SECONDARY	A1
NT_CREATE_ANDX	A2
NT_CANCEL	A4
OPEN_PRINT_FILE	C0
WRITE_PRINT_FILE	C1
CLOSE_PRINT_FILE	C2
GET_PRINT_QUEUE	C3
SEND_MESSAGE	D0
SEND_BROADCAST	D1
FORWARD_USER_NAME	D2
CANCEL_FORWARD	D3

Table 15-2: *Server Message Block Protocol Command Codes (continued)*

Command	Command Code
GET_MACHINE_NAME	D4
SEND_MULTIBLOCK_MESSAGE	D5
END_MULTIBLOCK_MESSAGE	D6
MULTIBLOCK_MESSAGE_TEXT	D7
READ_BULK	D8
WRITE_BULK	D9
WRITE_BULK_DATA	DA

Table 15-2: *Server Message Block Protocol Command Codes (continued)*

In addition to the Command code, each SMB message contains a 1-byte Flags field and a 2-byte Flags2 field that specify information about the message and the capabilities of the system generating it, such as whether the system supports long filenames and extended attributes, whether path names should be case sensitive, and whether the message is being sent by a server in response to a client request.

This last flag is included because separate request and reply messages do not exist for each SMB command. A system receiving an SMB message containing a command to perform an action typically responds with a reply that uses the same command code and contains some indication of the success or failure of the procedure. The response flag is set in the reply message to ensure that the receiver associates the reply with its previous request.

The rest of the message fields vary depending on the type and function of the message.

SMB Communications

SMB messages provide networking support services for Windows systems; they do not perform entire transactions by themselves. During a typical client/server process on a NetBEUI network, such as a workstation accessing a file on a shared drive, the communications intersperse LLC, NBF, and SMB messages during the various parts of the procedure. This is illustrated is Figure 15-13, which shows a sequence of packets captured during a simple transaction in which one Windows system opens the Autoexec.bat file on another system's shared drive using the WordPad text editor.

```
No.  Source.. Dest Address Layer   Summary
  1  cz2      cz5          NetBIOS NetBIOS Cmd: A (Name Query), CZ2             --->CZ5
  2  cz5      cz2          NetBIOS Cmd: E (Name Recognized), CZ5                --->CZ2
  3  cz2      cz5          LLC     Sap 0xF0 ---> 0xF0 (Command)
  4  cz5      cz2          LLC     Sap 0xF0 ---> 0xF0 (Response)
  5  cz2      cz5          LLC     Sap 0xF0 ---> 0xF0 (Command)
  6  cz5      cz2          LLC     Sap 0xF0 ---> 0xF0 (Response)
  7  cz2      cz5          NetBIOS Cmd: 19 (Session Initialize), 74--->76
  8  cz5      cz2          NetBIOS Cmd: 17 (Session Confirm), 76--->74
  9  cz5      cz2          LLC     Sap 0xF0 ---> 0xF0 (Response)
 10  cz2      cz5          LLC     Sap 0xF0 ---> 0xF0 (Response)
 11  cz2      cz5          SMB     C=Negotiate,Dialect[6]=NT LM 0.12
 12  cz5      cz2          LLC     Sap 0xF0 ---> 0xF0 (Response)
 13  cz5      cz2          SMB     R=Negotiate,Selected Dialect#=5
 14  cz2      cz5          LLC     Sap 0xF0 ---> 0xF0 (Response)
 15  cz2      cz5          SMB     C=Session_Setup+X,Account=CRAIGZ,XCmd=Tree_Connect+X,Server=\\CZ5\C
 16  cz5      cz2          LLC     Sap 0xF0 ---> 0xF0 (Response)
 17  cz5      cz2          NetBIOS Cmd: 14 (Data Ack), 76--->74
 18  cz2      cz5          LLC     Sap 0xF0 ---> 0xF0 (Response)
 19  cz5      cz2          SMB     R=Session_Setup+X, XCmd=Tree_Connect+X,Type=A:
 20  cz2      cz5          LLC     Sap 0xF0 ---> 0xF0 (Response)
 21  cz2      cz5          SMB     C=Open+X,Name=autoexec.bat
 22  cz5      cz2          LLC     Sap 0xF0 ---> 0xF0 (Response)
 23  cz5      cz2          SMB     R=Open+X,FID=0x41,File Size=0x20
 24  cz2      cz5          LLC     Sap 0xF0 ---> 0xF0 (Response)
 25  cz2      cz5          SMB     C=Read_Raw,FID=0x41,Read 4096 at 0x0
 26  cz5      cz2          LLC     Sap 0xF0 ---> 0xF0 (Response)
 27  cz5      cz2          NetBIOS Cmd: 16 (Data Only Last), 76--->74
 28  cz2      cz5          NetBIOS Cmd: 14 (Data Ack), 74--->76
 29  cz5      cz2          LLC     Sap 0xF0 ---> 0xF0 (Response)
 30  cz2      cz5          SMB     C=Close_File,FID=0x41
 31  cz5      cz2          LLC     Sap 0xF0 ---> 0xF0 (Response)
 32  cz5      cz2          SMB     R=Close_File
 33  cz2      cz5          LLC     Sap 0xF0 ---> 0xF0 (Response)
 34  cz2      cz5          SMB     C=Tree_Disconnect
 35  cz5      cz2          SMB     R=Tree_Disconnect
 36  cz2      cz5          NetBIOS Cmd: 14 (Data Ack), 74--->76
 37  cz5      cz2          LLC     Sap 0xF0 ---> 0xF0 (Response)
```

Figure 15-13: *The LLC, NBF, and SMB protocols working together to provide Windows systems with their networking capability*

The sequence proceeds as follows:

Packets 1–2   The machine with the NetBIOS name CZ2 sends an NBF NAME QUERY message to the network to locate the machine called CZ5 and resolve its name into a hardware address. CZ5 responds with a NAME RECOGNIZED message containing the address.

Packets 3–6   CZ2 initiates an LLC session with CZ5 at the data-link layer.

Packets 7–10   CZ2 establishes an NBF session with CZ5, and both systems transmit LLC Receive Ready messages to acknowledge that they are prepared for the next transmission. All subsequent messages will trigger LLC Receive Ready messages in response.

Packets 11–14   CZ2 sends an SMB NEGOTIATE message (Command code 72) to CZ5, containing the protocol dialects it understands (see Figure 15-14). CZ5 replies with the index number of the dialect it has selected (see Figure 15-15).

```
⊞ 🖫 802.2: Address: 00-60-97-B0-77-CA --->00-60-97-8D-C0-67
⊞ 🖳 LLC: Sap 0xF0 ---> 0xF0 (Command)
⊞ 🖳 NetBIOS: Cmd: 16 (Data Only Last), 74--->76
⊟ 🖳 Server Message Block Protocol
   ☑ Command: 114 - Negotiate
   ⊞ 🖳 Error Class:0 - Success;  Error  Code: Success
   ⊞ 🖳 Flags: Case Sens,Request
   ⊞ 🖳 2nd Flags:Naming=DOS8.3,Err=DOS,Str=ASCII
   ⊟ 🖳 IDs in SMB:
       🖳 Tree              ID: 0 (0x0)
       🖳 Caller's  Process ID: 5597 (0x15DD)
       🖳 Unauthenticated User ID: 0 (0x0)
       🖳 Multiplex        ID: 64770 (0xFD02)
   ⊟ 🖳 Multiplex          ID: 64770 (0xFD02)
       🖳 Word Count of Parameter: 0
   ⊟ 🖳 Negotiate BYTE LENGTH(2 bytes)+PARAMETERS(119 bytes):
       🖳 Byte Count of Data: 119
   ⊞ 🖳 Dialect Entry [1]: PC NETWORK PROGRAM 1.0
   ⊞ 🖳 Dialect Entry [2]: MICROSOFT NETWORKS 3.0
   ⊞ 🖳 Dialect Entry [3]: DOS LM1.2X002
   ⊞ 🖳 Dialect Entry [4]: DOS LANMAN2.1
   ⊞ 🖳 Dialect Entry [5]: Windows for Workgroups 3.1a
   ⊞ 🖳 Dialect Entry [6]: NT LM 0.12
   🖫 Calculate CRC: 0x988566d6
```

Figure 15-14: *The SMB NEGOTIATE request message*

```
⊞ 🖫 802.2: Address: 00-60-97-8D-C0-67 --->00-60-97-B0-77-CA
⊞ 🖳 LLC: Sap 0xF0 ---> 0xF0 (Command)
⊞ 🖳 NetBIOS: Cmd: 16 (Data Only Last), 76--->74
⊟ 🖳 Server Message Block Protocol
     🖳 Command: 114 - Negotiate
   ⊞ 🖳 Error Class:0 - Success;  Error  Code: Success
   ⊞ 🖳 Flags: Case Sens,Response
   ⊞ 🖳 2nd Flags:Naming=DOS8.3,Err=DOS,Str=ASCII
   ⊟ 🖳 IDs in SMB:
       🖳 Tree              ID: 0 (0x0)
       🖳 Caller's  Process ID: 5597 (0x15DD)
       🖳 Unauthenticated User ID: 0 (0x0)
       🖳 Multiplex        ID: 64770 (0xFD02)
   ⊟ 🖳 Multiplex          ID: 64770 (0xFD02)
       🖳 Word Count of Parameter: 17
     ☑ Index of Selected Dialect: 5
   ⊞ 🖳 Security Mode: User + Encrypt passwords
       🖳 Max  Pending  Mpx  Requests: 2
       🖳 Max  VCs in Client & Server: 1
       🖳 Max  Transmit  Buffer  Size: 2920 (0xB68)
       🖳 Max   Raw   Buffer   Size: 65536 (0x10000)
       🖳 Unique    Session    Key: 2147616588 (0x8002074C)
   ⊞ 🖳 Server Capabilities: Raw,Mpx,C_FIND,
       🖳 System Time of the Server: 01-Jul-1999 21:30:44
       🖳 Current Time Zone at Server: 240
       🖳 Encryption Key Length: 8
   ⊟ 🖳 Negotiate BYTE LENGTH(2 bytes)+PARAMETERS(8 bytes):
       🖳 Byte Count of Data: 8
       🖳 Encryption Key: (8 bytes in highligh area)
   🖫 Calculate CRC: 0x82d3dc64
```

Figure 15-15: *The SMB NEGOTIATE reply message*

Packets 15–20 CZ2 sends an SMB SESSION_SETUP_ANDX message (Command code 73) to CZ5, containing a user name, domain name, and password for authentication to the server. SMB's ANDX feature enables a system to batch multiple commands in the same message. Here, the packet contains a secondary TREE_CONNECT_ANDX command (code 75) that specifies the share on CZ5 to which CZ2 wants to connect (see Figure 15-16). CZ5 sends an NBF DATA ACK message acknowledging the transmission and an SMB reply indicating the success of the session establishment and tree connection.

```
⊞ 🖳 802.2: Address: 00-60-97-B0-77-CA --->00-60-97-8D-C0-67
⊞ 👎 LLC: Sap 0xF0 ---> 0xF0 (Command)
⊞ 🖳 NetBIOS: Cmd: 16 (Data Only Last), 74--->76
⊟ 🖳 Server Message Block Protocol
    🗋 Command: 115 - Session_Setup+X
  ⊞ 🗋 Error Class:0 - Success;  Error  Code: Success
  ⊞ 🗋 Flags: Case Sens,Canoni,Request
  ⊞ 🗋 2nd Flags:Naming=DOS8.3,Err=DOS,Str=ASCII
  ⊞ 🗋 TID = 0x0000, PID = 0x15dd, UID = 0x0001, MID = 0xfd02
  ⊟ 🗋 Multiplex            ID: 64770 (0xFD02)
        🗋 Word Count of Parameter: 13
        🗋 Secondary Command: 117 - Tree_Connect+X
        🗋 Reserved     (MSB): 0
        🗋 Offset to Next Command: 125 (0x7D)
        🗋 Consumer's Max Buffer Size: 2920
        🗋 Max  Mpx  pending Requests: 2
        🗋 Vc Number(0=1st,Non0=more): 0
        🗋 Unique Session Key: 2147616588 (0x8002074C)
        🗋 Case Insensitive Password Size: 24
        🗋 Case  Sensitive  Password Size: 0
        🗋 Reserved (Must be Zero): 0 (0x0)
    ⊞ 🗋 Client Capabilities:
  ⊟ 🗋 Session_Setup+X BYTE LENGTH(2 bytes)+PARAMETERS(64 bytes):
        🗋 Byte Count of Data: 64
        🗋 Case Insensitive Password: (24 bytes in highligh area)
        🗋 Case  Sensitive  Password: (0 bytes in highligh area)
        🗋 Account          Name: CRAIGZ
        🗋 Client's Primary Domain: NTDOMAIN
        🗋 Client's Native     OS: Windows 4.0
        🗋 Client's Native LAN Mgr: Windows 4.0
    ⊞ 🗋 Tree_Connect+X WORD LENGTH(1 byte)+PARAMETERS(4 words):
  ⊟ 🗋 Tree_Connect+X BYTE LENGTH(2 bytes)+PARAMETERS(15 bytes):
        🗋 Byte Count of Data: 15
        🗋 Password: (1 bytes in highligh area)
        🗹 Server&Share Name: \\CZ5\C
        🗋 Service Name: ?????
    🗋 No More Secondary Command.
    🖳 Calculate CRC: 0x78da5819
```

Figure 15-16: *The SMB SESSION_SETUP_ANDX message, including a TREE_CONNECT_ANDX command*

Packets 21–24 CZ2 sends an SMB OPEN_ANDX command (code 2d) to CZ5, specifying the name of the file it wants to open: Autoexec.bat (see Figure 15-17). CZ5 sends a reply indicating the successful completion of the command, assigning a file handle (FID) to Autoexec.bat, and specifying information about the requested file, such as its size and date last modified (see Figure 15-18). CZ2 uses the FID to reference the file in subsequent messages.

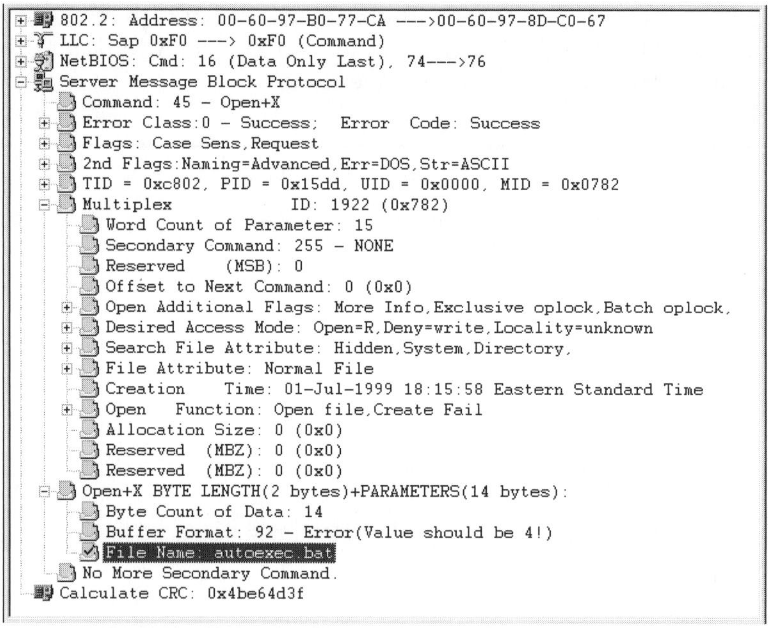

```
802.2: Address: 00-60-97-B0-77-CA --->00-60-97-8D-C0-67
LLC: Sap 0xF0 ---> 0xF0 (Command)
NetBIOS: Cmd: 16 (Data Only Last), 74--->76
Server Message Block Protocol
    Command: 45 - Open+X
    Error Class:0 - Success;  Error  Code: Success
    Flags: Case Sens,Request
    2nd Flags:Naming=Advanced,Err=DOS,Str=ASCII
    TID = 0xc802, PID = 0x15dd, UID = 0x0000, MID = 0x0782
    Multiplex           ID: 1922 (0x782)
        Word Count of Parameter: 15
        Secondary Command: 255 - NONE
        Reserved     (MSB): 0
        Offset to Next Command: 0 (0x0)
        Open Additional Flags: More Info,Exclusive oplock,Batch oplock,
        Desired Access Mode: Open=R,Deny=write,Locality=unknown
        Search File Attribute: Hidden,System,Directory,
        File Attribute: Normal File
        Creation     Time: 01-Jul-1999 18:15:58 Eastern Standard Time
        Open   Function: Open file,Create Fail
        Allocation Size: 0 (0x0)
        Reserved  (MBZ): 0 (0x0)
        Reserved  (MBZ): 0 (0x0)
    Open+X BYTE LENGTH(2 bytes)+PARAMETERS(14 bytes):
        Byte Count of Data: 14
        Buffer Format: 92 - Error(Value should be 4!)
        File Name: autoexec.bat
    No More Secondary Command.
Calculate CRC: 0x4be64d3f
```

Figure 15-17:   The SMB OPEN_ANDX request message, with the name of the requested file, Autoexec.bat, highlighted

Packets 25–26   CZ2 sends an SMB READ_RAW message (command code 1A) to CZ5 containing the FID of Autoexec.bat, the location in the file where the read should begin (in this case 0, the beginning of the file), and the maximum number of bytes to be returned (see Figure 15-19).

Packets 27–29   CZ5 reads the file on its local drive as directed and transmits it to CZ2 in a single NBF DATA ONLY LAST frame (see Figure 15-20). If the file was too large to fit in a single frame, the system would use as many DATA FIRST MIDDLE frames as needed, followed by a DATA ONLY LAST frame with the final bits of the file. CZ2 replies with an NBF DATA ACK message.

Packets 30–33   CZ2 sends an SMB CLOSE message (Command code 04) requesting that CZ5 close the file, using the same FID to reference Autoexec.bat (see Figure 15-21). CZ5 responds, indicating the successful completion of the command.

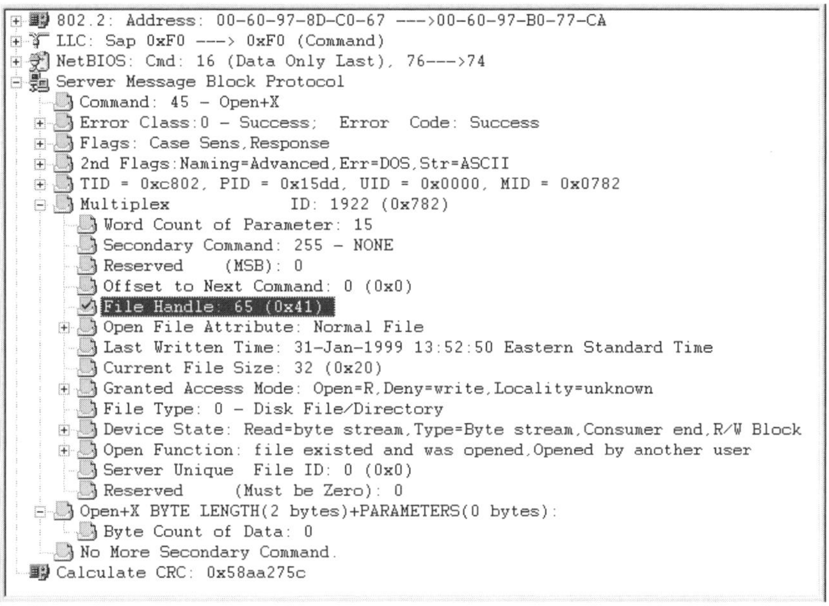

```
⊞ ▦ 802.2: Address: 00-60-97-8D-C0-67 --->00-60-97-B0-77-CA
⊞ ⅄ LLC: Sap 0xF0 ---> 0xF0 (Command)
⊞ 🗐 NetBIOS: Cmd: 16 (Data Only Last), 76--->74
⊟ 🖳 Server Message Block Protocol
   ─🗐 Command: 45 - Open+X
   ⊞ 🗐 Error Class:0 - Success;  Error  Code: Success
   ⊞ 🗐 Flags: Case Sens,Response
   ⊞ 🗐 2nd Flags:Naming=Advanced,Err=DOS,Str=ASCII
   ⊞ 🗐 TID = 0xc802, PID = 0x15dd, UID = 0x0000, MID = 0x0782
   ⊟ 🗐 Multiplex          ID: 1922 (0x782)
      ─🗐 Word Count of Parameter: 15
      ─🗐 Secondary Command: 255 - NONE
      ─🗐 Reserved     (MSB): 0
      ─🗐 Offset to Next Command: 0 (0x0)
      ─☑ File Handle: 65 (0x41)
      ⊞ 🗐 Open File Attribute: Normal File
      ─🗐 Last Written Time: 31-Jan-1999 13:52:50 Eastern Standard Time
      ─🗐 Current File Size: 32 (0x20)
      ⊞ 🗐 Granted Access Mode: Open=R,Deny=write,Locality=unknown
      ─🗐 File Type: 0 - Disk File/Directory
      ⊞ 🗐 Device State: Read=byte stream,Type=Byte stream,Consumer end,R/W Block
      ⊞ 🗐 Open Function: file existed and was opened,Opened by another user
      ─🗐 Server Unique  File ID: 0 (0x0)
      ─🗐 Reserved     (Must be Zero): 0
   ⊟ 🗐 Open+X BYTE LENGTH(2 bytes)+PARAMETERS(0 bytes):
      ─🗐 Byte Count of Data: 0
   ─🗐 No More Secondary Command.
   ─▦ Calculate CRC: 0x58aa275c
```

Figure 15-18: *The SMB OPEN_ANDX reply message, with the file handle assigned to Autoexec.bat highlighted*

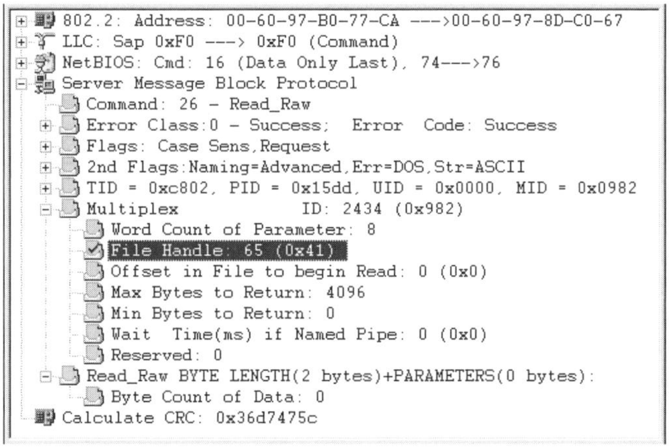

```
⊞ ▦ 802.2: Address: 00-60-97-B0-77-CA --->00-60-97-8D-C0-67
⊞ ⅄ LLC: Sap 0xF0 ---> 0xF0 (Command)
⊞ 🗐 NetBIOS: Cmd: 16 (Data Only Last), 74--->76
⊟ 🖳 Server Message Block Protocol
   ─🗐 Command: 26 - Read_Raw
   ⊞ 🗐 Error Class:0 - Success;  Error  Code: Success
   ⊞ 🗐 Flags: Case Sens,Request
   ⊞ 🗐 2nd Flags:Naming=Advanced,Err=DOS,Str=ASCII
   ⊞ 🗐 TID = 0xc802, PID = 0x15dd, UID = 0x0000, MID = 0x0982
   ⊟ 🗐 Multiplex          ID: 2434 (0x982)
      ─🗐 Word Count of Parameter: 8
      ─☑ File Handle: 65 (0x41)
      ─🗐 Offset in File to begin Read: 0 (0x0)
      ─🗐 Max Bytes to Return: 4096
      ─🗐 Min Bytes to Return: 0
      ─🗐 Wait  Time(ms) if Named Pipe: 0 (0x0)
      ─🗐 Reserved: 0
   ⊟ 🗐 Read_Raw BYTE LENGTH(2 bytes)+PARAMETERS(0 bytes):
      ─🗐 Byte Count of Data: 0
   ─▦ Calculate CRC: 0x36d7475c
```

Figure 15-19: *The SMB READ_RAW message, which uses the previously assigned file handle to refer to Autoexec.bat*

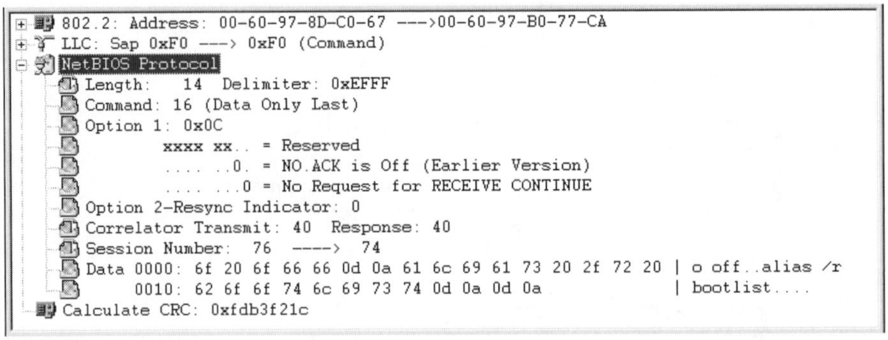

```
⊞ 🖳 802.2: Address: 00-60-97-8D-C0-67 --->00-60-97-B0-77-CA
⊞ 𝗬 LLC: Sap 0xF0 ---> 0xF0 (Command)
⊟ 🖳 NetBIOS Protocol
   🖳 Length:    14  Delimiter: 0xEFFF
   🖳 Command: 16 (Data Only Last)
   🖳 Option 1: 0x0C
   🖳        xxxx xx.. = Reserved
   🖳        .... ..0. = NO.ACK is Off (Earlier Version)
   🖳        .... ...0 = No Request for RECEIVE CONTINUE
   🖳 Option 2-Resync Indicator: 0
   🖳 Correlator Transmit: 40  Response: 40
   🖳 Session Number:  76  ----> 74
   🖳 Data 0000: 6f 20 6f 66 66 0d 0a 61 6c 69 61 73 20 2f 72 20 | o off..alias /r
   🖳      0010: 62 6f 6f 74 6c 69 73 74 0d 0a 0d 0a             | bootlist....
⊞ 🖳 Calculate CRC: 0xfdb3f21c
```

Figure 15-20:   *The NBF DATA ONLY LAST message carries the file requested by the SMB sequence.*

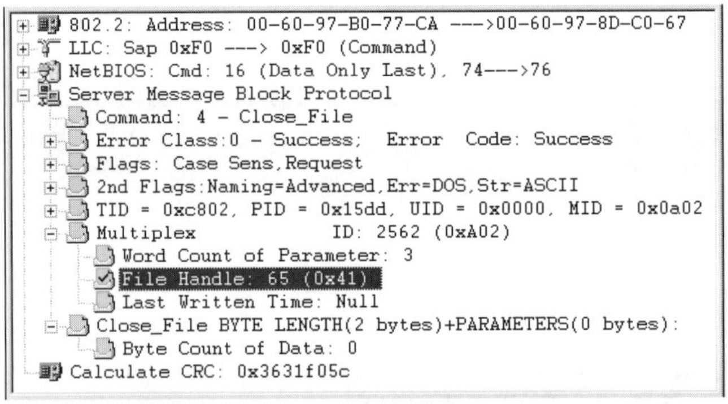

```
⊞ 🖳 802.2: Address: 00-60-97-B0-77-CA ---->00-60-97-8D-C0-67
⊞ 𝗬 LLC: Sap 0xF0 ---> 0xF0 (Command)
⊞ 🖳 NetBIOS: Cmd: 16 (Data Only Last), 74--->76
⊟ 🖳 Server Message Block Protocol
   🖳 Command: 4 - Close_File
   ⊞ 🖳 Error Class:0 - Success;  Error  Code: Success
   ⊞ 🖳 Flags: Case Sens,Request
   ⊞ 🖳 2nd Flags:Naming=Advanced,Err=DOS,Str=ASCII
   ⊞ 🖳 TID = 0xc802, PID = 0x15dd, UID = 0x0000, MID = 0x0a02
   ⊟ 🖳 Multiplex          ID: 2562 (0xA02)
      🖳 Word Count of Parameter: 3
      ✓ File Handle: 65 (0x41)
      🖳 Last Written Time: Null
   ⊟ 🖳 Close_File BYTE LENGTH(2 bytes)+PARAMETERS(0 bytes):
      🖳 Byte Count of Data: 0
⊞ 🖳 Calculate CRC: 0x3631f05c
```

Figure 15-21:   *The SMB CLOSE message*

Packets 34–37   CZ2 sends an SMB TREE_DISCONNECT message (Command code 71) to CZ5, requesting that the connection to the share be terminated (see Figure 15-22). CZ5 responds, indicating a successful disconnection, and transmits a final NBF DATA ACK message.

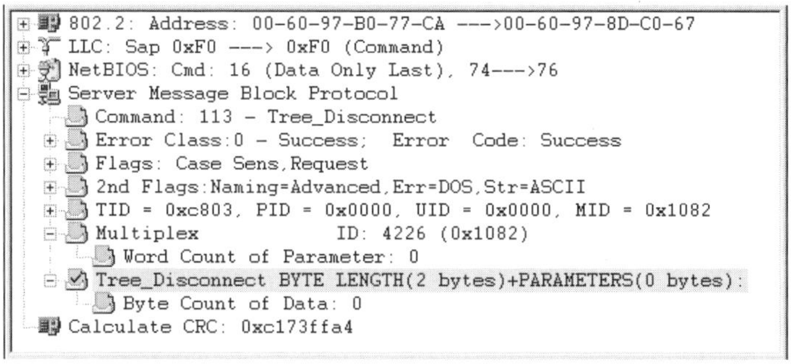

```
⊞ 📇 802.2: Address: 00-60-97-B0-77-CA --->00-60-97-8D-C0-67
⊞ 🔲 LLC: Sap 0xF0 ---> 0xF0 (Command)
⊞ 📇 NetBIOS: Cmd: 16 (Data Only Last), 74--->76
⊟ 📇 Server Message Block Protocol
   📄 Command: 113 - Tree_Disconnect
   ⊞ 📄 Error Class:0 - Success;  Error  Code: Success
   ⊞ 📄 Flags: Case Sens,Request
   ⊞ 📄 2nd Flags:Naming=Advanced,Err=DOS,Str=ASCII
   ⊞ 📄 TID = 0xc803, PID = 0x0000, UID = 0x0000, MID = 0x1082
   ⊟ 📄 Multiplex          ID: 4226 (0x1082)
      📄 Word Count of Parameter: 0
   ⊟ ☑ Tree_Disconnect BYTE LENGTH(2 bytes)+PARAMETERS(0 bytes):
      📄 Byte Count of Data: 0
   📇 Calculate CRC: 0xc173ffa4
```

Figure 15-22: *The SMB TREE_DISCONNECT message*

The Complete Reference

Networking

Part IV

Network Operating Systems

The
Complete
Reference

Networking

Chapter 16

Windows 2000 and Windows NT

I n the years since its initial release in 1993, Microsoft's Windows NT operating system has become the most popular network operating system on the market, taking the place of Novell NetWare. NT's familiar interface and ease of use enabled relatively unsophisticated users to install and maintain local area networks, making LAN technology an all but ubiquitous part of doing business today. Windows 2000, the latest incarnation of the NT operating system, addresses some of the shortcomings of Windows NT, with the intention of creating one operating system family suitable for use in all PCs, from standalone workstations to the most powerful servers.

The Role of Windows in the Enterprise

The strength of Novell NetWare is traditionally its file and print services, which were the original reason for the development of PC networks. Windows 2000 and NT servers still provide these services (though arguably not as well as NetWare), but they also place a much greater emphasis on being an effective application server platform.

Unlike NetWare, which is strictly a client/server platform and uses a proprietary OS at the server, Windows operates on a peer-to-peer model, in which each system can function both as a client and as a server. As a result, the same familiar interface is used in all Windows computers, both clients and servers, simplifying the learning curve for users as well as the development effort for software designers. Although applications for Windows NT were slow in coming at first, Windows is now the dominant software development platform in the networking industry. Windows applications tend to be easier to install and use than NetWare or Unix server applications.

At the time of NT's introduction, installing a NetWare server was largely a manual process in which you had to modify the server's configuration files in order to load the appropriate drivers. Windows, on the other hand, had an automated installation program much like those of most applications. While the process of setting up a NetWare network required considerable expertise, many people discovered that a reasonably savvy PC user could install the Windows OS and Windows applications with little difficulty. In fact, it was very likely these Windows qualities that led Novell to begin working on a more automated installation process for NetWare.

Although it took several years for the NT operating system to mature, and for large numbers of third-party developers to begin writing applications for it, many administrators began deploying it on their NetWare networks. Its favorable pricing and its ability to coexist with NetWare made Windows easy to experiment with and evaluate. In the ensuing years, as Windows replaced DOS on the desktop, Windows NT and 2000 became the natural choices for high-end workstations and servers. With the release of the 32-bit DOS-based Windows operating systems, starting with Windows 95 and proceeding through Windows 98 and ME, software developers were able to design 32-bit applications that ran on any Windows operating system.

Another major factor that contributed to Windows' rise in popularity was its adoption of TCP/IP as its default protocols. As the Internet grew in popularity, a market developed for a platform that was easier to use than Unix that would run Internet and intranet server applications, and Windows fit the bill nicely. Eventually, major database engines were running on Windows servers, and the similarity of the client and server platforms streamlined the development process. As it gained a reputation as an application server, the popularity of Windows grew to the point at which it largely replaced NetWare on business networks.

Today, most of the servers installed on new LANs run Windows 2000 or NT, and the OS is also making inroads into the desktop workstation market. Until the release of Windows 2000, Microsoft intended Windows 95 and 98 for the average network workstation, and Windows NT Workstation for higher-end applications. These roles are now filled by Windows 98 and Windows 2000 Professional. Windows ME, the final version of the Windows 9x OS, is targeted more at home users.

Versions

The first version of Windows NT (which was given the version number 3.1, to conform with the then-current version of Windows) was introduced in 1993. The motivation behind it was to create a new 32-bit OS from the ground up that left all vestiges of DOS behind. Although the interface was nearly identical in appearance to that of a Windows 3.1 system, NT was a completely new OS in many fundamental ways. Backward-compatibility with existing applications is a factor that has always hindered advances in operating system design, and once Microsoft decided that running legacy programs was not to be a priority with Windows NT, it was free to implement radical changes.

The various versions of Windows NT can be said to fall into three distinct generations, based on the user interface. The first generation consists of Windows NT 3.1, 3.5, and 3.51, all three of which use the same Windows 3.1–style interface. Version 3.1 used NetBEUI as its default protocol, which immediately limited it to use only on relatively small networks. TCP/IP and IPX support were available, but only through the STREAMS interface. The intention at that time was to build small workgroup networks using Windows NT servers and Windows for Workgroups on client computers. Windows NT versions 3.5 and 3.51 shifted the emphasis from NetBEUI to TCP/IP, and introduced some of the services that have come to be closely related with Windows NT, such as WINS and DHCP. At this time, Windows NT 3.1, 3.5, and 3.51 are considered to be obsolete, and any computers still running them should be upgraded.

The second generation consists of Windows NT 4.0, which was released in 1996 as an interim upgrade leading toward the major innovation that Microsoft began promising in 1993. NT 4 uses the same interface introduced in Windows 95, and positioned the OS more positively as an Internet platform with the inclusion of the Internet Explorer Web browser and Internet Information Server—a combination World Wide Web, FTP, and Gopher server.

The third generation is Windows 2000, which is the long-awaited release of the operating system that was originally code-named Cairo. The Windows 2000 interface is a refined version of the NT 4/Win95 GUI, but the biggest improvement is the inclusion of Active Directory, an enterprise directory service that represents a quantum leap over the domain-based directory service included in Windows NT. Windows XP is the next-generation operating system that brings the DOS-based world of Windows 95, 98, and ME together with the Windows NT/2000 design to form a single product line that is suitable for both home and office computers. However, this release is more of a comprehensive upgrade for Windows 98 and ME than it is for Windows 2000.

Windows NT/2000 Products

Every version of Windows NT and 2000 has been available in separate editions designed for use on servers and workstations. The editions for Windows NT 4.0 are as follows:

- Windows NT Workstation 4.0

- Windows NT Server 4.0

- Windows NT Server 4.0 Enterprise Edition

- Windows NT Server 4.0 Terminal Server Edition

These are the editions for Windows 2000:

- Windows 2000 Professional

- Windows 2000 Server

- Windows 2000 Advanced Server

- Windows 2000 Datacenter Server

Windows XP, formerly known by the code name "Whistler," is available in the following editions:

- Windows XP Personal

- Windows XP Professional

- Windows XP Server

- Windows XP Advanced Server

The core operating systems used to build the variants in the Windows NT and 2000 families are identical in almost every way. The differences between the server and workstation operating systems are found mostly in the additional services included, the number of users supported, and, of course, the price. The Windows 2000 server products, for example, include the Active Directory directory service and the ability to function as a domain controller (DC), as well as services like the DNS, WINS, and DHCP servers. Windows 2000 Professional lacks all of these features, but is still able to interact with other workstations on a peer-to-peer basis when servers are not present. The three Windows 2000 server products differ from each primarily in their support for computers with multiple processors. Windows 2000 Server is intended for computers with up to 4 processors, while the Advanced Server and Datacenter Server versions are for computers with up to 8 and up to 32 processors, respectively. The Windows 2000 Professional product is the updated equivalent of Windows NT Workstation, and is intended for client computers.

In addition to the standard Windows NT Server 4.0 product, there are also two special-purpose NT server packages. Windows NT Server Terminal Server Edition is used to deliver the Windows NT desktop to Windows-based terminals and thin clients, and Windows NT Server Enterprise Edition provides support for large-scale, distributed applications using specialized memory allocation techniques, symmetric multiprocessing (SMP), and server clustering.

The Windows XP server products are positioned much like those of Windows 2000, but the workstation editions are split into two products. Windows XP Personal is intended for the home user, while Windows XP Professional is positioned for business or higher-end client workstations.

Service Packs

Microsoft releases regular updates to the Windows products in the form of Service Packs, which contain numerous fixes and upgrades in one package, using a single installation routine. Microsoft was one of the first software companies to adopt this update release method, which is a vast improvement over dozens of small patch releases (sometimes called *hot fixes*) that address single, specific issues. Apart from the inconvenience of downloading and installing many small patches, this update method is a technical support nightmare, because it's difficult for both the user and the technician to know exactly which patches have been installed. Service Packs are designed to detect the components installed on a Windows computer and install only the updates needed by those components.

Service Packs consist of a single release for all of the various editions of an operating system. You use the same Service Pack for Windows 2000 Professional, Server, and Advanced Server, for example. Service Packs often consist of more than just bug fixes. They may include upgraded versions of operating system utilities, new features, or entirely new programs. All of the components are installed at the same time by the Service Pack's setup program. Service Packs are also cumulative, meaning that each successive Service Pack for a particular product contains the contents of all of the previous Service Packs for that product. This simplifies the process of installing Windows on a new computer or updating one that hasn't been patched in some time, but it also causes the Service Pack releases to grow very large. Microsoft makes its Service Packs available as free downloads or on CD-ROMs, for which you must pay postage, handling, and media fees.

When downloading Service Packs from the Microsoft Web site, there are several options you can use to limit the size of the file. If you run an Express installation from the computer you want to update, you can download only the files that are actually needed to update that computer. For Windows 2000 Service Pack 2, for example, the Express download option triggers a download ranging from 10 to 23 MB, depending on whether you are running Windows 2000 Server or Professional and what OS components you have installed. This option minimizes the size of the download, but it is only suitable for the single computer where the download is running. If you want to download Service Pack 2 only once and install it on a variety of Windows 2000 computers, you must select the Network installation option, which requires you to download a single 101MB file. This is obviously not a terribly practical option if you're limited to a dial-up modem connection.

For the network administrator who is heavily committed to the use of Microsoft products, Microsoft TechNet is a subscription-based CD-ROM product that is an invaluable resource for technical information and product updates. The monthly releases typically include six or more CD-ROMs containing Resource Kits, documentation, the entire Knowledge Base for all of the Microsoft products, and a lot of other material as well. TechNet Plus is an additional service that, for an additional charge, includes the latest beta releases of the products currently under development.

Generally speaking, it's a good idea to keep the Windows systems on your network up to date with the latest Service Pack. However, it's also a good idea to either test out a new Service Pack release in a controlled environment or wait a month or two after its release before you install it, to see if any problems arise. Once you've decided to install a Service Pack, you should update all of the computers on your network that are running that operating system, so that they're all using the same software. The Service Pack installation program includes an option to save copies of the files it's replacing, so that you can uninstall the Service Pack if necessary. However, the prudent administrator will still perform a full backup of all operating systems files, applications, and data before installing a Service Pack or any software update, just in case.

When you add new operating system components to a Windows NT computer that has already had a Service Pack installed, you must reinstall the most recent Service Pack to ensure that all of the OS files are up to date. Windows 2000 Service Packs, however, include a feature called *slipstreaming*, which copies the Service Pack files to an install share and automatically substitutes the latest versions of any modules being installed with a new OS component. This enables you to install the operating system on a new computer or add Windows components to an existing one without performing a separate Service Pack installation.

Between Service Pack releases, Microsoft releases *hot fixes,* which are small, interim patch releases intended to address specific issues. You should not install every hot fix that comes along, because they will all be included in the next Service Pack release. The rule of thumb is to install a hot fix only when you have a system that is suffering from the exact problem that the patch addresses. Many of the hot fixes address security issues, however, and you should generally install these as they become available, to maintain the security of your computers.

Operating System Overview

Windows 2000 and NT are modular operating systems that are designed to take advantage of the advanced capabilities built into the latest Intel and Alpha processors, while leaving behind the memory and storage constraints imposed by DOS-based operating systems. Early operating systems such as DOS were *monolithic*—that is, the entire OS consisted of a single functional unit, which made it difficult to upgrade and modify. By creating an OS composed of many separate components, Microsoft has made it easier to upgrade and modify parts of the operating system without affecting other elements in the overall functionality of the whole.

Note *Despite the large number of new features introduced in Windows 2000, its fundamental architecture is quite similar to that of Windows NT.*

Kernel Mode Components

The Windows 2000 and NT operating systems are composed of components that run in one of two modes: *kernel mode* and *user mode* (see Figure 16-1). A component running in kernel mode has full access to the system's hardware resources via the *hardware abstraction layer (HAL),* which is a virtual interface that isolates the kernel from the computer hardware. Abstracting the kernel from the hardware makes it far easier to port the OS to different hardware platforms. While Windows NT was at one time available in versions for four processor types, the MIPS and PowerPC versions have been discontinued, leaving Windows NT and 2000 with only the Alpha and Intel versions—the latter of which is by far the most popular.

The OS kernel itself is responsible for delegating specific tasks to the system processor or processors and other hardware. Tasks consist of *processes,* broken down into *threads,* which are the smallest units that the kernel can schedule for execution by a processor. A thread is a sequence of instructions to which the kernel assigns a priority level that determines when it will be executed. When the computer has multiple processors, the kernel runs on all of them simultaneously, sharing access to specific memory areas and allocating threads to specific processors according to their priorities.

In addition to the HAL and the kernel, Windows' *executive services* also run in kernel mode. These executive services consist of the following components.

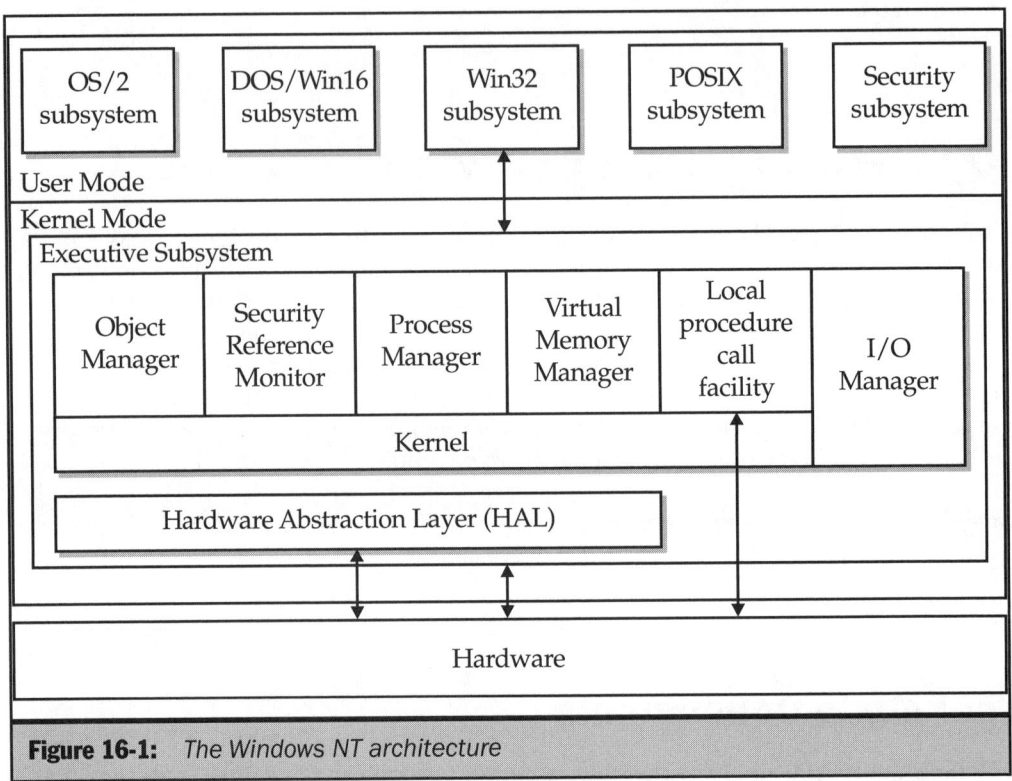

Figure 16-1: *The Windows NT architecture*

Object Manager

Windows creates objects that function as abstract representations of operating system resources, such as hardware devices and file system entities. An *object* consists of information about the resource it represents and a list of *methods*, which are procedures used to access the object. A file object, for example, consists of information such as the file's name and methods describing the operations that can be performed on the file, such as open, close, and delete.

The Windows *Object Manager* maintains a hierarchical, global name space in which the objects are stored. For example, when the system loads a kernel mode device driver, it registers a device name with the Object Manager, such as \Device\CDRom0 for a

CD-ROM drive or \Device\Serial0 for a serial port. The objects themselves are stored in directories similar to those in a file system, but they are not a part of any Windows file system. In addition to hardware devices, objects can reference both abstract and concrete entities, including the following:

- Files
- Directories
- Processes
- Threads
- Memory segments
- Semaphores

By using a standard format for all objects, regardless of the type of entities they represent, the Object Manager provides a unified interface for object creation, security, monitoring, and auditing. Access to objects in the name space is provided to system processes using *object handles,* which contain pointers to the objects and to access control information.

Note   *The kernel mode objects discussed here are not equivalent to the objects in the Active Directory database. They are two completely different hierarchies. Active Directory runs in user mode within the Windows security subsystem.*

Usually, the only places that you see devices referred to by these object names are entries in the registry's HKEY_LOCAL_MACHINE\HARDWARE key and error messages such as those displayed in the infamous "blue screen of death." Applications typically run in the Win32 subsystem, which is a user mode component that cannot use internal Windows device names. Instead, the Win32 subsystem references devices using standard MS-DOS device names, like drive letters and port designations such as COM1. These MS-DOS names exist as objects in the Object Manager's name space, in a directory called \??, but they do not have the same properties as the original resources; they are actually only *symbolic links* to the equivalent Windows device names.

Security Reference Monitor

Every Windows object has an *access control list (ACL)* that contains *access control entries (ACEs)* that specify the *security identifiers (SIDs)* of users or groups that are to be permitted access to the object, as well as the specific actions that the user or group can perform. When a user successfully logs on to the computer, Windows creates a *security access token (SAT)* that contains the SIDs of the user and all the groups of which the user is a member. Whenever the user attempts to access an object, the *Security Reference Monitor* is responsible for comparing the SAT with the ACL to determine whether the user should be granted that access.

Process Manager

The *Process Manager* is responsible for creating and deleting the process objects that enable software to run on a Windows system. A *process object* includes a virtual address space and a collection of resources allocated to the process, as well as threads containing the instructions that will be assigned to the system processor(s).

Virtual Memory Manager

The ability to use virtual memory is one of the major PC computing advancements introduced in the Intel 80386 processor, and Windows NT and 2000 were designed around this capability. *Virtual memory* is the ability to use the computer's disk space as an extension to the physical memory installed in the machine.

Every process created on a Windows computer by the Process Manager is assigned a virtual address space that appears to be 4GB in size. The *Virtual Memory Manager (VMM)* is responsible for mapping that virtual address space to actual system memory, as needed, in 4KB units called *pages*. When there is not enough physical memory in the computer to hold all of the pages allocated by the running processes, the VMM swaps the least recently used pages to a file on the system's hard disk drive called Pagefile.sys. This swapping process is known as *memory paging*.

Local Procedure Call Facility

The environmental subsystems that run in Windows' user mode (such as the Win32, DOS, Win16, OS/2, and POSIX subsystems) are utilized by applications (also running in user mode) in a server/client relationship. The messages between the clients and servers are carried by the *local procedure call (LPC)* facility. Local procedure calls are essentially an internalized version of the remote procedure calls used for messaging between systems connected by a network.

When an application (functioning as a client) makes a call for a function that is provided by one of the environmental subsystems, a message containing that call is transmitted to the appropriate subsystem using LPCs. The subsystem (functioning as the server) receives the message and replies using the same type of message. The process is completely transparent to the application, which is not aware that the function is not implemented in its own code.

I/O Manager

The I/O Manager handles all of a Windows computer's input/output functions by providing a uniform environment for communication between the various drivers loaded on the machine. Using the layered architecture shown in Figure 16-2, the I/O Manager enables each driver to utilize the services of the drivers in the lower layers.

For example, when an application needs to access a file on a drive, the I/O Manager passes an *I/O request packet (IRP)* generated by a file system driver down to a disk driver. Since the I/O Manager communicates with all of the drivers in the same way, the request can be satisfied without the file system having any direct knowledge of the disk device where the file is stored.

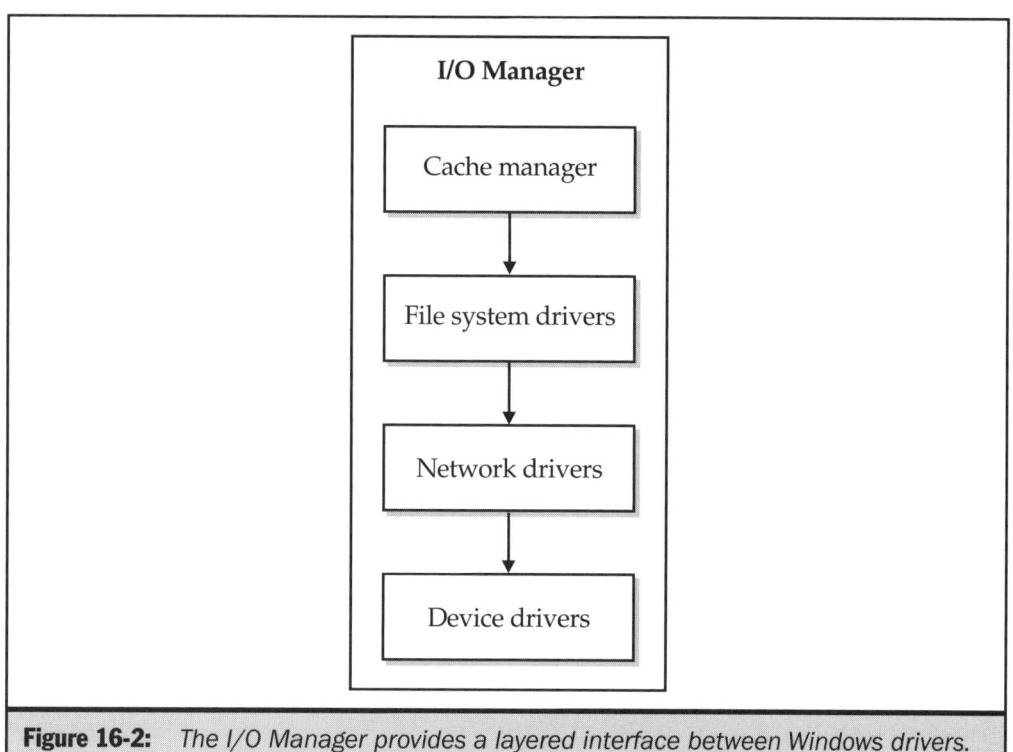

Figure 16-2: *The I/O Manager provides a layered interface between Windows drivers.*

Window Manager

The Window Manager, along with the *Graphical Device Interface (GDI)*, is responsible for creating the graphical user interface used by Windows applications. Applications make calls to Window Manager functions in order to create architectural elements on the screen, such as buttons and windows. In the same way, the Window Manager informs the application when the user manipulates screen elements by moving the cursor, clicking buttons, or resizing a window.

NETWORK OPERATING SYSTEMS

Prior to NT 4.0, the Window Manager was a user mode process, but it is now implemented as a single, kernel mode driver called Win32k.sys. This change is invisible to application developers, but it improves graphical performance while reducing the amount of memory required.

User Mode Components

In addition to the kernel mode services, Windows has two types of protected subsystems that run in user mode: *environment subsystems* and *integral subsystems.* The environment subsystems enable Windows to run applications that were designed for various OS environments, such as Win32, OS/2, and POSIX. Integral subsystems, like the security system, perform vital OS functions. User mode subsystems are isolated from each other and from the Windows executive services, so that modifications to the subsystem code do not affect the fundamental operability of the OS. If a user mode component like a subsystem or application should crash, the other subsystems and the Windows executive services are not affected.

The Win32 Subsystem

Win32 is the primary environment subsystem in Windows 2000 and NT that provides support for all native Windows applications. All of the other environment subsystems included with Windows are optional and loaded only when a client application needs them, but Win32 is required and runs at all times. This is because it is responsible for handling the keyboard and mouse inputs and the display output for all of the other subsystems. Since they rely on Win32 API calls, the other environment subsystems can all be said to be clients of Win32.

The DOS/Win16 Subsystem

Unlike Windows 95 and 98, Windows 2000 and NT do not run a DOS kernel, and as a result, they cannot shell out to a DOS session. Instead, 2000 and NT emulate DOS using a subsystem that creates *virtual DOS machines (VDMs).* Every DOS application that you run uses a separate VDM that emulates an Intel *x*86 processor in Virtual 86 mode (even on a non-Intel system). All of the application's instructions run natively within the VDM except for I/O functions, which are emulated using *virtual device drivers (VDDs).* VDDs convert the DOS I/O functions into standard Windows API calls and feed them to the I/O Manager, which satisfies the calls using the standard Windows device drivers.

The use of separate VDMs isolates the DOS applications from the rest of the system, and from any other DOS applications that may be running. If a DOS application halts, nothing other than that particular VDM will be affected. Windows can run as many

VDMs as the installed hardware will support, and once created, VDMs are not destroyed—even when the DOS application terminates. When you execute a DOS application, the system uses an idle VDM if one is available or creates a new VDM if one is not.

To run 16-bit applications designed for Windows 3.1, Windows 2000 and NT use a single VDM with a single additional software layer called *Win16 on Win32*, or *WOW*. The WOW layer emulates the Windows 3.1 environment in standard mode only, not enhanced mode. If you run multiple Win16 applications, they all execute in the same VDM, and Windows provides nonpreemptive multitasking between them. (The VDM itself, however, is preemptively multitasked with the other Windows components.) This means that Win16 applications running on Windows 2000 and NT are subject to the same problems as on Windows 3.1. It is possible for one application running on the VDM to use memory allocated for another application, and when one Win16 application crashes, all of the other Win16 apps go down with it (just like on a real Windows 3.1 system).

The OS/2 Subsystem

Windows 2000 and NT include an OS/2 subsystem that supports OS/2 1.*x* character-based applications only. It does not support the Presentation Manager or Warp applications. Unlike the other environment subsystems, OS/2 applications are supported on the Intel processor versions of Windows 2000 and NT only. However, a real mode OS/2 app that can run in a standard DOS session will run in a VDM on a non-Intel Windows machine.

The POSIX Subsystem

Windows 2000 and NT include support for applications that comply to the POSIX.1 standard as defined by the IEEE Std. 1003.1-1990 document. POSIX stands for *Portable Operating System Interface for Computing Environments,* and is an attempt to create a series of APIs that facilitate the porting of applications between Unix environments.

Services

A *service* is a program or other component that Windows loads with the OS, before a user logs on or sees the desktop interface. Services usually load automatically and permit no interference from the system user as they're loading. This is in contrast to other mechanisms that load programs automatically, such as the Startup program group. A user with appropriate rights can start, stop, and pause services using the Services console (see Figure 16-3) or the NET command, and also specify whether a particular service should load when the system starts, not load at all, or require a manual startup.

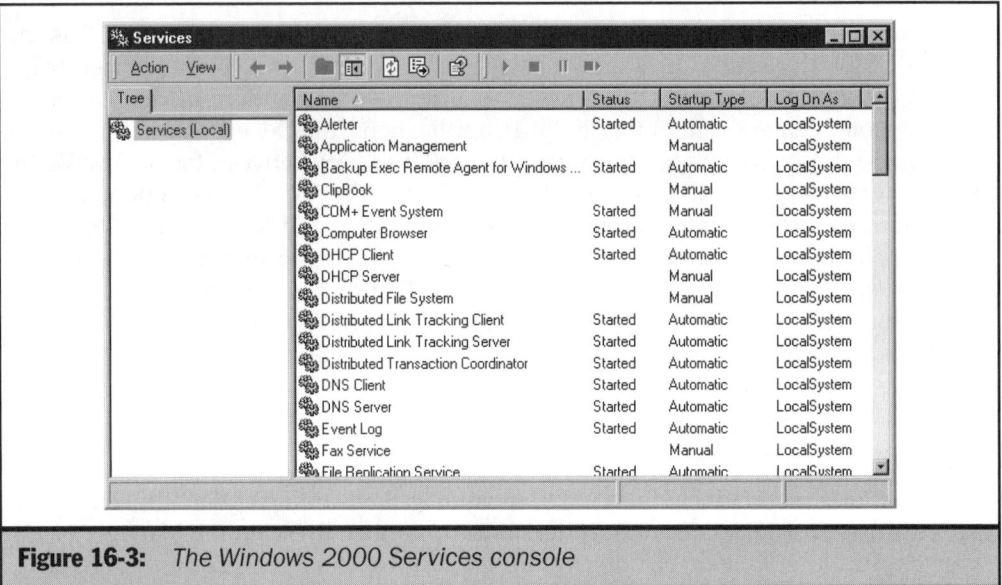

Figure 16-3: *The Windows 2000 Services console*

Users without administrative rights cannot control the services at all, which makes the services a useful tool for network administrators. You can, for example, configure a workstation to load a particular service at startup, and it will run whether a user logs on or not. The Server service, for example, which enables network users to access the computer's shares, loads automatically by default. Even if no one logs on to the computer, it is possible to access its shares from the network.

Windows 2000 and NT include a large number of services, several of which are required for basic system functions. Some of the fundamental Windows services are as follows:

Server   Enables the system to share drives and printers with other network systems.

Workstation   Enables applications running on the system to access network resources.

Browser   Maintains a list of the domains and the Windows computers on the network that the system uses to locate resources.

Messenger   Enables administrators or the Alerter service to send and receive messages, either manually or as a result of a system event.

Netlogon   Enables Windows computers to locate a DC and log on to the domain. If the computer does not participate in a domain, the Netlogon service does not start.

In addition to these essential services, Windows 2000 and NT include many other services that are optional. The Windows 2000 and NT Server products include a great many more services than Windows 2000 Professional or NT Workstation. All of the network servers, such as WINS, DHCP, DNS, and IIS, run as services. Selecting these components during the Windows installation process configures the computer to load the services automatically. Many third-party server applications also take the form of services that run continuously in the background.

The Windows 2000 and Windows NT Server 4.0 Resource Kits include a utility called Srvany.exe, which enables you to run any Windows application as a service. When you do this, the application continues to run even while the current user logs off the system and a new user logs on. In addition, a service can run using the access permissions of a user other than the user who is currently logged on to the system. Srvany.exe itself is actually a service that you install in Windows using another Resource Kit utility called Instsrv.exe. Once you've installed Srvany.exe, you specify the path to the application you want to run as a service in a registry entry, which loads that application when the system starts.

The Windows Networking Architecture

Networking is an integral part of Windows 2000 and NT, and the operating systems use a modular networking architecture that provides a great deal of flexibility for the network administrator. While not perfectly analogous to the OSI reference model, the Windows networking architecture is structured in layers that provide interchangeability of modules such as network adapter drivers and protocols. The basic structure of the networking stack is shown in Figure 16-4.

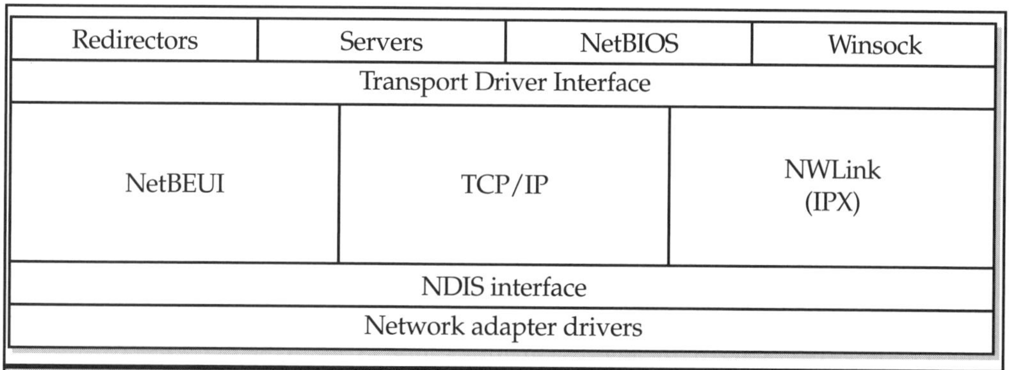

Redirectors	Servers	NetBIOS	Winsock
Transport Driver Interface			
NetBEUI	TCP/IP		NWLink (IPX)
NDIS interface			
Network adapter drivers			

Figure 16-4: *The Windows 2000/NT networking architecture*

Windows relies on two primary interfaces to separate the basic networking functions, called the *NDIS interface* and *Transport Driver Interface (TDI)*. Between these two interfaces are the protocol suites that provide transport services between computers on the network: TCP/IP, NetBEUI, and IPX. Although they have different features, these three sets of protocols are interchangeable when it comes to basic networking services. A Windows computer can use any of these protocols or all of them simultaneously. The TDI and NDIS interfaces enable the components operating above and below them to address whichever protocol is needed to perform a particular task.

For example, on a system with all three protocols loaded, a request for access to an Internet server generated by a Web browser utilizes the TCP/IP protocol, while an application trying to access a Novell NetWare server will use IPX. For applications that can use any of the protocols, the order in which the protocols are bound to the service determines which one the application uses.

For more information on the protocols used by Windows 2000 and NT, see Chapters 13, 14, and 15 for coverage of TCP/IP, IPX, and NetBEUI, respectively.

The NDIS Interface

The Network Driver Interface Specification (NDIS) is a standard developed jointly by Microsoft and 3Com that defines an interface between the network-layer protocols and the media access control (MAC) sublayer of the data link–layer protocol. On a Windows 2000 or NT system, the NDIS interface lies between the network adapter drivers and the protocol drivers. Protocols do not communicate directly with the network adapter; instead, they go through the NDIS interface. This enables a Windows computer to have any number of network adapters and any number of protocols installed, and any protocol can communicate with any adapter.

NDIS is implemented on a Windows 2000/NT system in two parts: the *NDIS wrapper* (Ndis.sys) and the *NDIS MAC driver.* The NDIS wrapper is not device specific; it contains common code that surrounds the MAC drivers and provides the interface between the network adapter drivers and the protocol drivers installed in the computer. This replaces the Protocol Manager (PROTMAN) used by other NDIS versions to regulate access to the network adapter.

The NDIS MAC driver is device specific, and provides the code needed for the system to communicate with the network interface adapter. This includes the mechanism for selecting the hardware resources the device uses, such as the IRQ and I/O port address. All of the network interface adapters in a Windows system must have an NDIS driver, which is provided by virtually all of the manufacturers producing NICs today.

The NDIS implementation in Windows NT 4.0 provides support for the connectionless communications used on most of the common data link–layer protocols, such as

Ethernet and Token Ring. Windows 2000 NDIS continues to provide this capability, but also adds support for connection-oriented media, such as Asynchronous Transfer Mode (ATM) and Integrated Services Digital Network (ISDN). This connection-oriented service uses a call manager to establish virtual circuits between two network endpoints. A virtual circuit (VC) is a data conduit that remains open for the life of the connection, enabling the protocol to regulate the bandwidth between the two points. The Windows 2000 NDIS drivers also support advanced network adapter features such as Wake On LAN, which enables an administrator to turn a computer on from a remote location, as well as TCP/IP segmentation and IPsec offload capabilities, which enable the network adapter to assume transport-layer processing and encryption tasks that are normally performed by the computer's processor.

The Transport Driver Interface

The TDI performs roughly the same basic function as the NDIS wrapper, but higher up in the networking stack. The TDI functions as the interface between the protocol drivers and the components operating above them, such as the server and the redirectors. Traffic moving up and down the stack passes through the interface and can be directed to any of the installed protocols or other components.

Above the TDI, Windows has several more components that applications use to access network resources in various ways, using the TDI as the interface to the protocol drivers. Because Windows is a peer-to-peer operating system, there are components that handle traffic running in both directions. The most basic of these components are the *Workstation* and *Server* services, which enable the system to access network resources and provide network clients with access to local resources (respectively). Also at this layer are *application programming interfaces (APIs),* such as NetBIOS and Windows Sockets, which provide applications running on the system special access to certain network resources.

The Workstation Service

When you open a file or print a document in an application, the process is the same whether the file or printer is part of the local system or on the network, as far as the user and the application are concerned. The Workstation service determines whether the requested file or printer is local or on the network and sends the request to the appropriate driver. By providing access to network resources in this way, the Workstation service is essentially the client half of Windows' client/server capability.

The Workstation service consists of two modules: Services.exe, the Service Control Manager, which functions as the user mode interface for all services; and the Windows network redirector. When an application requests access to a file, the request goes to the I/O Manager, which passes it to the appropriate file system driver. Depending on what file systems the computer's drives use, there may be NTFS or FAT file system drivers installed on the machine, or both. The *redirector* is also a file system driver, but instead of providing access to a local drive, the redirector transmits the request down through the protocol stack to the appropriate network resource. The I/O Manager treats a redirector no differently from any other file system drivers. Windows installs a redirector for the Microsoft Windows network by default, but client software for other network operating systems (such as Novell NetWare) can include additional redirectors.

The Multiple UNC Provider

In the case of a system with multiple network clients (and multiple redirectors), Windows uses one of two mechanisms for determining which redirector it should use, depending on how an application formats its requests for network resources. The Multiple UNC Provider (MUP) is used for applications that use Uniform Naming Convention (UNC) names to specify the desired resource, and the Multi-Provider Router (MPR) is used for applications that use Win32 network APIs.

The UNC defines the format that Windows uses for identifying network resources. UNC names take the following form:

*server**share*

On a Windows network, the server can be any Windows system, and the share can be any shared drive. On a NetWare network, the *server* and *share* variables correspond to a NetWare server and one of its volumes. When you browse through the shares displayed in Windows Explorer and select a file, the system returns a UNC path name whether the file is located on a Windows network share, a Novell NetWare server, or even a local drive. The function of the MUP is to determine which type of resource hosts the file and send the request to the appropriate redirector.

The MUP is implemented as a file called Mup.sys and functions using a trial-and-error method in which it sends the requested UNC name to each of the redirectors on the system in turn and awaits a response from each one. Once the replies are received, the MUP selects the appropriate redirector and transmits the request. If there is only one redirector on the system, obviously this process is simple. If there are two or more redirectors, it's possible for a resource of the same name to exist on both server types. For example, while you can't have two Windows systems with the same NetBIOS name on one network, you can have Windows and NetWare servers with the same name. Because of the confusion it can cause, this is a practice that you should generally try to avoid. If the MUP receives positive responses from two (or more) redirectors, indicating that the requested resource exists on both networks, it sends the request to the redirector with the highest priority. The MUP maintains a cache of the UNC names that it has processed within the last 15 minutes, which it consults before sending requested names to any of the redirectors. If it finds the name in the cache, it uses the redirector associated with that name. If the name is not in the cache, the MUP sends the request to the redirectors *synchronously*, meaning that the MUP sends the request to the redirector with the highest priority and waits for a response before it sends the request to the redirector with the next highest priority.

The Multi-Provider Router

For applications that request access to network resources using the Win32 network APIs (also known as the WNet APIs), the Multi-Provider Router (MPR) determines which redirector should process the requests. In addition to a redirector, a network client installed on a Windows computer includes a *provider DLL* that functions as an interface between the MPR and the redirector. The MPR passes the requests that it receives from applications to the appropriate provider DLLs, which pass them to the redirectors.

The Server Service

Just as the Workstation service provides network client capabilities to Windows 2000 and NT, the Server service enables other clients on the network to access the computer's local resources. When the redirector on a client system transmits a request for access to a file on a server, the receiving system passes the request up the protocol stack to the Server service. The Server service is a file system driver (called Srv.sys) that is started by the Service Control Manager, just like the Workstation service, that operates just above the TDI. When the Server service receives a request for access to a file, it generates a read request and sends it to the appropriate local file system driver (such as the NTFS or FAT driver) through the I/O Manager. The local file system driver accesses the requested file in the usual manner and returns it to the Server service, which transmits it across the network to the client. The Server service also provides support for printer sharing, as well as remote procedure calls (RPCs) and named pipes, which are other mechanisms used by applications to communicate over the network.

Bindings

For traffic to pass up and down the protocol stack, the components operating at the various layers must be bound together to form an unbroken path through the stack. In Windows, adapter drivers must be bound to protocol drivers, and protocol drivers must be bound to services, for communications to occur. By default, all of the networking components installed in Windows are bound to each other, meaning that all of the installed network adapters are bound to all of the installed protocols, and likewise with the protocols and the services. This makes it possible for a message traveling up or down the stack to take a variety of different paths.

Windows also recognizes the order in which the components are bound, and establishes priorities based on that order. If, for example, all of the systems on your network have both the TCP/IP and NetBEUI protocols installed, a request for access to a file on a network share can conceivably use either protocol. The redirector handling the request sends it to all of the available protocol drivers and then selects the protocol with the highest priority that was able to process the request successfully.

In Windows 2000, when you select Advanced Settings from the Advanced menu in the Network and Dial-up Connections control panel, you can use the Adapters and Bindings page of the Advanced Settings dialog box (see Figure 16-5) to manage the bindings between the installed networking components, creating and deleting bindings and modifying their priorities. The Provider Order page lets you specify which network should receive the highest priority.

Using the arrow buttons, you can control the priority of the bindings. Moving the TCP/IP protocol above the NetBEUI protocol for a particular service, for example, will ensure that the service always uses TCP/IP when both protocols are available. NetBEUI is used only when the attempt to connect via TCP/IP fails.

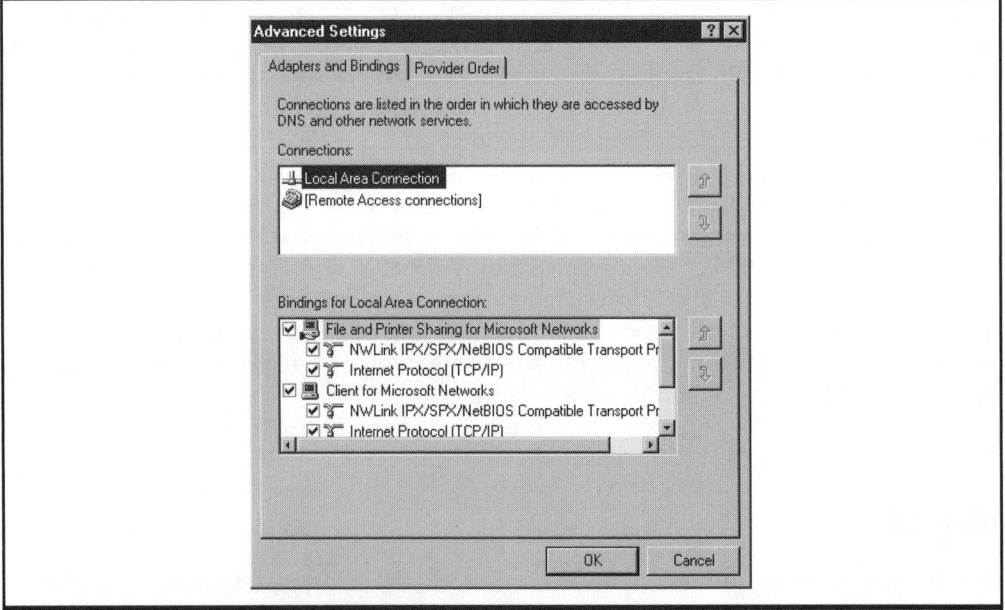

Figure 16-5:   *The Adapters and Bindings page in the Windows 2000 Network and Dial-up Connections Advanced Settings dialog box*

APIs

Services are not the only components that interact with the TDI on a Windows system. Application Programming Interfaces (APIs), such as NetBIOS and Windows Sockets, also send and receive data through the TDI, enabling certain types of applications to communicate with other network systems without using the Server and Workstation services. Windows also supports other APIs that operate higher up in the stack and use the standard services to reach the TDI.

NetBIOS

NetBIOS is an integral component of Microsoft Windows networking, because it provides the name space used to identify the domains, computers, and shares on the network. Because of its dependence on NetBIOS, Windows supports it in all of its protocols. NetBEUI is inherently designed for use with NetBIOS communications, and the NetBIOS over TCP/IP (NetBT) standards defined by the Internet Engineering Task Force (IETF) enable its use with the TCP/IP protocols. NWLink, the Windows implementation of the Novell IPX protocols, also supports NetBIOS. Although Windows 2000 relies on DNS names to identify network resources, it remains backward compatible with NetBIOS names, to support networks also using computers running earlier Windows versions.

 **Note**   *For more information on Windows NT's use of NetBIOS and NetBT, see Chapter 24. For more information on NetBEUI and NetBIOS messaging, see Chapter 15.*

Windows Sockets

The Windows Sockets specification defines one of the APIs that is most commonly used by applications, because it is the accepted standard for Internet network access. Web browsers, FTP clients, and other Internet client and server applications all use Windows Sockets (Winsock) to gain access to network resources. Unlike NetBIOS, Winsock does not support all of the Windows protocols. While it can be used with NWLink (IPX), the overwhelming majority of Winsock applications use TCP/IP exclusively. As with NetBIOS, Winsock is implemented in Windows as a kernel mode emulator just above the TDI, and a user mode driver, called Wsock32.dll.

File Systems

The FAT file system is a holdover from the DOS days that the developers of the original Windows NT product were seeking to transcend. While an adequate solution for a workstation, the 16-bit FAT file system used by DOS cannot support the large volumes typically required on servers, and it lacks any sort of access control mechanism. As a result, the NT developers again came up with a completely new solution to their problems, without any concern for backward-compatibility. The NT file system (NTFS) provides the advanced features needed on network servers, at the price of its complete invisibility to DOS. As a result, on Windows NT versions through 4.0, you can select either FAT16 or NTFS as the file system for each drive in the computer. Windows 2000 adds support for FAT32, the next-generation FAT file system first introduced in the OSR2 release of Windows 95, and upgrades NTFS to version 5.

FAT16

The traditional DOS file system divides a hard disk drive into volumes that are composed of uniformly sized clusters and uses a file allocation table (FAT) to keep track of the data stored in each cluster. Each directory on the drive contains a list of the files in that directory and, in addition to the file name and other attributes, specifies the entry in the FAT that represents the cluster containing the beginning of the file. That first FAT entry contains a reference to another entry that references the file's second cluster, the second entry references the third, and so on until enough clusters are allocated to store the entire file. This is known as a *FAT chain.*

Early DOS versions used FAT entries that were 12 bytes long, but DOS version 4.0 increased the size of a FAT entry to 16 bytes, and all versions of Windows NT retain this limitation. Because each cluster in a volume must be referenced by a FAT entry, a volume can have no more than 65,536 (2^{16}) clusters. Therefore, as a volume grows larger, the cluster size must increase as well, because the maximum number of clusters remains constant. The largest cluster size supported by DOS is 32KB, meaning that a volume with the maximum of 65,536 clusters at 32,768 bytes each can be no larger than 2,147,482,648 bytes or 2GB. Windows NT, however, supports clusters up to 64KB in size, meaning that a FAT volume can be up to 4GB.

 *It was only with the introduction of the FAT32 file system that the traditional FAT file system came to be called FAT16. In most cases, references to a FAT drive without a numerical identifier refer to a FAT16 drive.*

The other limiting factor of the FAT file system is that, as clusters grow larger, more drive space is wasted due to slack. *Slack* is the fraction of a cluster left empty when the last bit of data in a file fails to completely fill the last cluster in the chain. When 3KB of data from a file is left to store, for example, a volume with 4KB clusters will contain 1KB of slack, while a volume with 64KB clusters will waste 61KB. Windows NT is designed to be a server OS, as well as a workstation OS, and servers are naturally expected to have much larger drives. The amount of slack space and the 4GB limit on volume size are not acceptable for a server OS.

The other major shortcoming of the FAT file system is the amount of information about each file that is stored on the disk drive. In addition to the data itself, a FAT drive maintains the following information about each file:

File name   Limited to an eight-character name plus a three-character extension.

Attributes   Contains four usable file attributes: Read-only, Hidden, System, and Archive.

Date/time   Specifies the date and time that the file was created or last modified.

Size   Specifies the size of the file, in bytes.

Unlike DOS, Windows NT was designed to be a network operating system, and any drive or directory on an NT computer can be shared with network users. To share files effectively, there must be a way to grant specific users access to them, deny access to others, and control the degree of access that is granted. The FAT file system lacks all of these features. A network server is also expected to be able to store the files generated by workstations running other operating systems, files that may use different formats, attributes, and naming conventions. Windows NT improves on the FAT system by enabling files and directories to have names up to 255 characters long, but it cannot provide additional attributes and other elements.

If you boot a Windows NT system with a DOS boot disk, you can still access the drives, as long as they use the FAT file system. You won't see the long file and directory names, but you can still work with the directories and files using their 8.3 equivalents. You can also use standard FAT disk repair and maintenance tools on Windows NT FAT drives, like Norton Disk Doctor for disk repair and Norton Speed Disk for defragmentation. NTFS drives, on the other hand, are completely invisible to other operating systems, such as DOS and Windows 9x. When you boot an NT system with NTFS drives using a DOS boot disk, it is as if there are no drives installed in the machine at all. You cannot manipulate the partitions using FDISK or use FAT file system repair tools.

FAT32

As hard disk drive capacities grew over the years, the limitations of the FAT file system became more of a problem. To address the problem, Microsoft created a file system that uses 32-bit FAT entries instead of 16-bit ones. The larger entries mean that there can be more clusters on a drive. The results are that the maximum size of a FAT32 volume is 2 terabytes (or 2,048GB) instead of 2GB for a FAT16 drive, and the clusters can be much smaller, thus reducing the waste due to slack space.

The FAT32 file system was introduced in the Windows 95 OSR2 release, and is also included in Windows 98, Windows ME, and Windows 2000. However, Windows NT 4.0

does not support it. You cannot access a FAT32 drive from Windows NT, nor can you boot an NT system from a FAT32 drive. If you want to be able to boot multiple operating systems on a single PC, and one of them is Windows NT, you must format the boot drive using FAT16. The FDISK utility included in Windows 95 OSR2, 98, and ME prompts you to specify whether or not you want to enable large disk support whenever you load the program. When you answer Yes, FDISK uses FAT32 to create volumes on the system's drives; answering No causes the program to create FAT16 drives, with the size limitations mentioned earlier.

FAT32 supports larger volumes and smaller clusters, but it does not provide any appreciable change in performance, and it still does not have the access control capabilities needed for network servers as NTFS does. For any Windows NT or 2000 system with drives that you plan to share with network users, you should use NTFS to protect the files against unauthorized access or accidental damage.

NTFS

NTFS is the file system intended for use on Windows 2000 and NT computers. Without it, you cannot install Active Directory or implement the file and directory-based permissions needed to secure a drive for network use. Because it uses a completely different structure than FAT drives, you cannot create NTFS drives using the FDISK utility, but the version of FDISK included with Windows 9*x* is able to identify NTFS drives and delete them (as it can other types of non-DOS partitions).

In the NTFS file system, files take the form of *objects* that consist of a number of *attributes.* Unlike DOS, in which the term "attribute" typically refers only to the Read-only, System, Hidden, and Archive flags, NTFS treats all of the information regarding the file as an attribute, including the flags, the dates, the size, the file name, and even the file data itself. NTFS also differs from FAT in that the attributes are stored with the file, instead of in a separate directory listing.

The equivalent structure to the FAT on an NTFS drive is called the *Master File Table (MFT).* Unlike FAT, however, the MFT contains more than just pointers to other locations on the disk. In the case of relatively small files (up to approximately 1,500 bytes), all of the attributes are included in the MFT, including the file data. When larger amounts of data need to be stored, additional disk clusters called *extents* are allocated, and pointers are included with the file's attributes in the MFT. The attributes stored in the MFT are called *resident attributes*; those stored in extents are *nonresident attributes.*

In addition to the four standard DOS file attributes, an NTFS file includes a Compression flag; two dates/times specifying when the file was created and when it was last modified; and a security descriptor that identifies the owner of the file, lists the users and groups that are permitted to access it, and specifies what access they are to be granted.

Windows 2000 includes the NTFS 5.0 file system, which improves on the file system included with Windows NT by adding such features as support for file encryption and the ability to enlarge volumes without rebooting. Windows 2000 requires NTFS 5.0 volumes to host the Active Directory database. New NTFS volumes created with Windows 2000 use NTFS 5.0, and NTFS volumes that already exist prior to an OS upgrade are converted to the new file system.

One of the most frequent complaints of network administrators about Windows NT's file system concerns its inability to restrict the amount of disk space utilized by a user. NetWare has had this capability for many years, and Microsoft has finally incorporated it into Windows 2000. On the Quota tab of an NTFS drive's Properties dialog box (see Figure 16-6), you can specify the maximum amount of space to be allotted to a new user, as well as the behavior of the system when the user approaches and exceeds the limit. As users access the volume, the system lists them in the Quota Entries window (see Figure 16-7). From here, you can modify the quotas for individual users and monitor the amount of space they have in use.

Windows 2000 also includes a disk defragmentation utility. Like disk quotas, this is a tool that has been lacking ever since the initial Windows NT release. Windows 2000 Server and Professional both can improve disk efficiency by defragmenting volumes that use the NTFS, FAT16, or FAT32 file system.

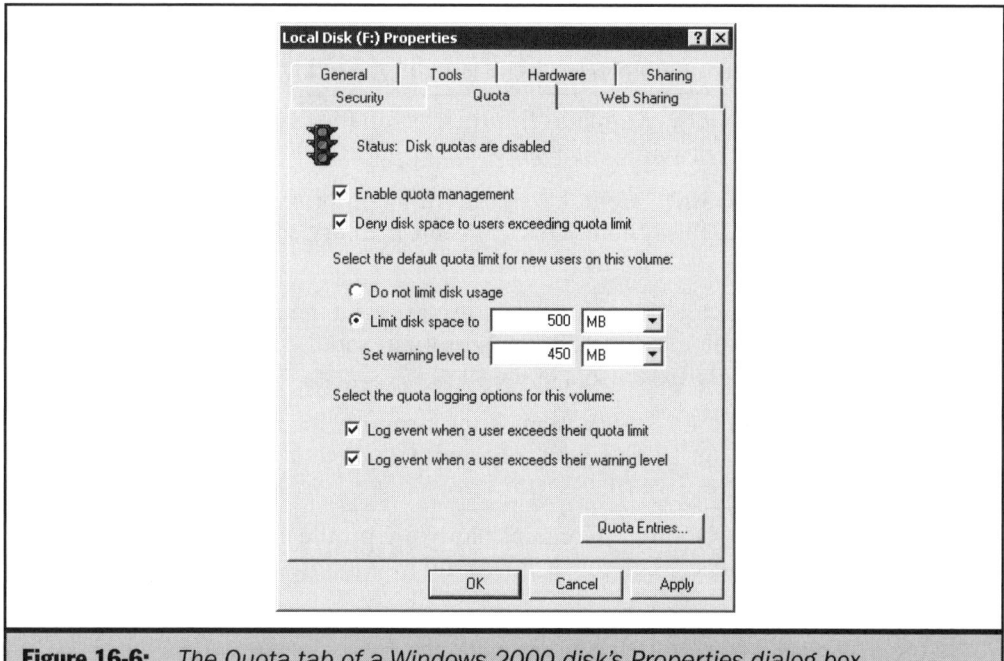

Figure 16-6:   *The Quota tab of a Windows 2000 disk's Properties dialog box*

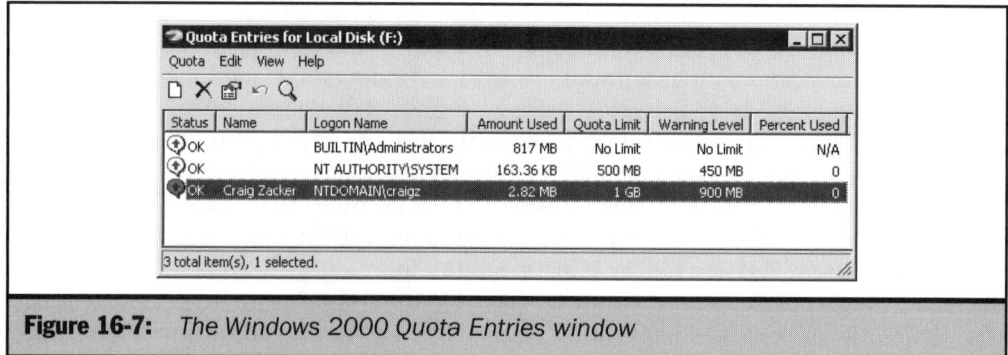

Figure 16-7:   *The Windows 2000 Quota Entries window*

The Windows Registry

The registry is the database where Windows stores nearly all of its system configuration data. As a system or network administrator, you'll be working with the registry in a variety of ways, since many of the Windows 2000 and NT configuration tools function by modifying entries in the registry. The registry is a hierarchical database that is displayed in most registry editor applications as an expandable tree, not unlike a directory tree. At the root of the tree are five containers, called *keys*, with the following names:

HKEY_CLASSES_ROOT   Contains information on file associations—that is, associations between file name extensions and applications.

HKEY_CURRENT_USER   Contains configuration information specific to the user currently logged on to the system. This key is the primary component of a user profile.

HKEY_LOCAL_MACHINE   Contains information on the hardware and software installed in the computer, the system configuration, and the SAM database. The entries in this key apply to all users of the system.

HKEY_USERS   Contains information on the currently loaded user profiles, including the profile for the user who is currently logged on and the default user profile.

HKEY_CURRENT_CONFIG   Contains hardware profile information used during the system boot sequence. This key is not found in Windows NT versions 3.51 and earlier.

In most cases, you work with the entries in the HKEY_LOCAL_MACHINE and HKEY_CURRENT_USER keys (often abbreviated as the HKLM and HKCU, respectively) when you configure a Windows system, whether you are aware of it or not. When the keys are saved as files, as in the case of user profiles, they're often referred to as *hives.* When you expand one of these keys, you see a series of *subkeys,* often in several layers. The keys and subkeys function as organizational containers for the registry *entries,* which contain the actual configuration data for the system. A registry entry consists of three components: the *value name,* the *value type,* and the *value* itself.

The value name identifies the entry for which a value is specified. The value type specifies the nature of the data stored in the entry, such as whether it contains a binary value, an alphanumeric string of a given size, or multiple values. The value types found in the registry are as follows:

REG_SZ   Indicates that the value consists of a string of alphanumeric characters. Many of the user-configurable values in the registry are of this type.

REG_DWORD   Indicates that the value consists of a 4-byte numerical value used to specify information such as device parameters, service values, and other numeric configuration parameters.

REG_MULTI_SZ   Same as the REG_SZ value type, except that the entry contains multiple string values.

REG_EXPAND_SZ   Same as the REG_SZ value type, except that the entry contains a variable (such as *%SystemRoot%*) that must be replaced when the value is accessed by an application.

REG_BINARY   Indicates that the value consists of raw binary data, usually used for hardware configuration information. You should not modify these entries manually unless you are familiar with the function of every binary bit in the value.

REG_FULL_RESOURCE_DESCRIPTOR   Indicates that the value holds configuration data for hardware devices in the form of an information record with multiple fields.

The registry hierarchy is large and complex, and the names of its keys and entries are often cryptic. Locating the correct entry can be difficult, and the values are often less than intuitive. When you edit the registry manually, you must be careful to supply the correct value for the correct entry or the results can be catastrophic. An incorrect registry modification can halt the computer or prevent it from booting, forcing you to reinstall Windows from scratch.

Because of the registry's sensitivity to improper handling, selecting the proper tool to modify it is crucial. The tradeoff in Windows' registry editing tools is between a safe, easy-to-use interface with limited registry access, and comprehensive access using a less intuitive interface. The following sections examine the various registry editing tools included with Windows 2000 and NT.

The Control Panel

Although it isn't evident from the interface, most of the functions in the Windows Control Panel work by modifying settings in the registry. The Control Panel's graphical interface provides users with simplified access to the registry and prevents them from introducing incorrect values due to typographical errors. You can also use Windows' security mechanisms to prevent unauthorized access to certain registry settings through the Control Panel. The main disadvantage of using the Control Panel to modify the registry is that it provides user access to only a small fraction of the registry's settings.

The System Policy Editor

System policies are collections of registry settings saved in a *policy file* that you can configure a Windows computer to load whenever a user logs on to the system or the network. You can create different sets of policies for each of your network users, so that when John Doe logs on to a workstation, his customized registry settings are downloaded to the computer and loaded automatically. Windows includes a tool called the System Policy Editor that you can use to create policy files; you can also use it to modify the registry directly. Like the Control Panel, the System Policy Editor uses a graphical interface to set registry values (see Figure 16-8), but it is far more configurable than the Control Panel and can provide access to a great many more registry entries.

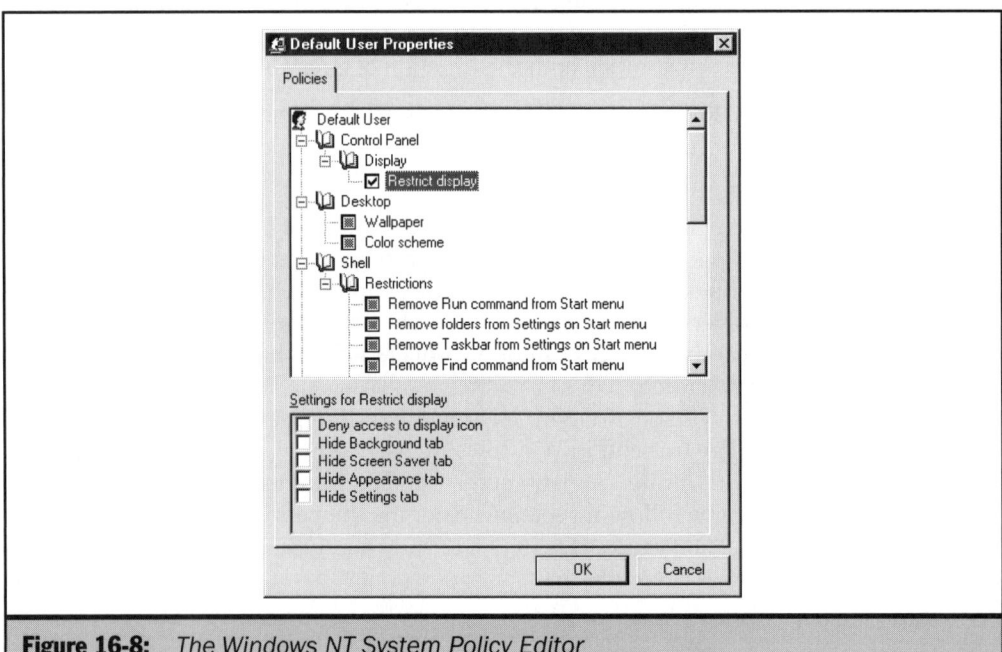

Figure 16-8:   *The Windows NT System Policy Editor*

The system policies that the System Policy Editor lists in its hierarchical display are derived from a file called a *policy template.* The template is an ASCII text file with an .adm extension that uses a special format to define how each policy should appear in the System Policy Editor and which registry settings each policy should modify. Windows includes several template files that define policies for a wide range of system settings, some of which are also configurable through the Control Panel. Because creating a new system policy is simply a matter of creating a new template, software developers can include with their products template files that define application-specific system policies. You can also create your own templates to modify other registry settings.

The process of setting values for a system policy by using the System Policy Editor consists of navigating through the hierarchical display and selecting a policy. Some policies consist of a single feature that you can toggle on and off, while others have additional controls in the form of check boxes, pull-down menus, or data entry fields. To create a policy file, you select the policies you want to set, specify values for them, and then save them to a file with a .pol extension.

The System Policy Editor can also directly modify the Windows registry, however. When you select File | Open Registry, the program connects to the registry on the local machine. When you configure a policy, the program applies the necessary changes directly to the registry. In addition, when you choose File | Connect, you can select another Windows NT, 2000, or 9x computer on the network and modify its registry from your remote location.

Note   *To access the registry on a Windows 95 or 98 machine from a remote location, the system must be running the Remote Registry service and be configured for user-level access control.*

The use of customizable template files makes the System Policy Editor a far more comprehensive registry-editing tool than the Control Panel. You can specify values for a wider range of registry entries, while still retaining the advantages of the graphical interface. Because the changes that the System Policy Editor makes to the registry are controlled by the policy template, the possibility of a misspelled value in a data entry field still exists, but the chances of an incorrect value damaging the system is far less than when editing the registry manually.

Group Policies

Windows 2000 group policies are the next step in the evolution of the system policies found in Windows NT and 98. Group policies include all of the registry modification capabilities found in NT system policies, plus a great deal more, such as the ability to install and update software, implement disk quotas, and redirect folders on user workstations to network shares. While NT system policies are associated with domain

NETWORK OPERATING SYSTEMS

users and groups, Windows 2000 group policies are associated with Active Directory objects, such as sites, domains, and organizational units.

The Registry Editors

Windows 2000 and NT include two Registry Editors, called Regedt32.exe and Regedit.exe, that provide direct access to the entire registry. There are many Windows features you can configure using the Registry Editors that are not accessible by any other administrative interface. These programs are the most powerful and comprehensive means of modifying registry settings in Windows, and also the most dangerous. These editors do not supply friendly names for the registry entries, and they do not use pull-down menus or check boxes to specify values. You must locate (or create) the correct entry and supply the correct value in the proper format, or the results can be wildly unpredictable. Windows installs both of these Registry Editors with the OS, but it does not create shortcuts for them in the Start menu or on the desktop. You must launch the Registry Editors from the Run dialog box, from Windows Explorer, or by creating your own shortcuts. Like the System Policy Editor, both Registry Editors enable you to connect to another Windows system on the network and access its registry.

The Regedit.exe editor (see Figure 16-9) was designed for use with Windows 95, but it has been included with all subsequent versions of Windows NT and 2000 because it has two important features that Regedt32.exe inexplicably lacks. The program uses an Explorer-like interface that displays an expandable tree of registry keys and subkeys in the left pane. When you select a subkey, the entries it contains are displayed in the right pane. Regedit.exe enables you to make most standard modifications to registry settings as well as create and delete keys and entries, but it lacks certain features specific to Windows 2000/NT that are found in Regedt32.exe, such as the ability to manage registry permissions and support for the REG_MULTI_SZ and REG_EXPAND_SZ value types.

The advantages of using Regedit.exe are that you can search for a string anywhere in the registry, including keys, entries, and their values, while Regedt32.exe can search the keys only. This is an important feature, because it enables you to locate a particular key by searching for its current value. If, for example, you want to know where the system's IP address is stored, you can search for the current address in Regedit.exe to find the correct key. In Regedt32.exe, you have to either find the key by manually browsing through the registry or use a registry reference, such as a book or a help file.

The other feature found exclusively in Regedit.exe is the ability to import and export registry information using text files with .reg extensions, called *registry scripts*. These scripts contain registry entries along with all of the information associated with them, including their locations in the key hierarchy, their value types, and the values themselves. You can select a key at any level in the hierarchy and create a script containing all of the entries and subkeys it contains. You can then apply those registry settings to another machine by running the script with the Regedit.exe program from the Windows command prompt. Because you can run scripts from the command line, this capability makes Regedit.exe an excellent tool for network administrators wanting to configure workstations using batch files.

Regedt32.exe can save the contents of a key to a text file, but the output file is not formatted in such a way that you can apply the settings automatically to another machine, nor can you run Regedt32.exe from the command line.

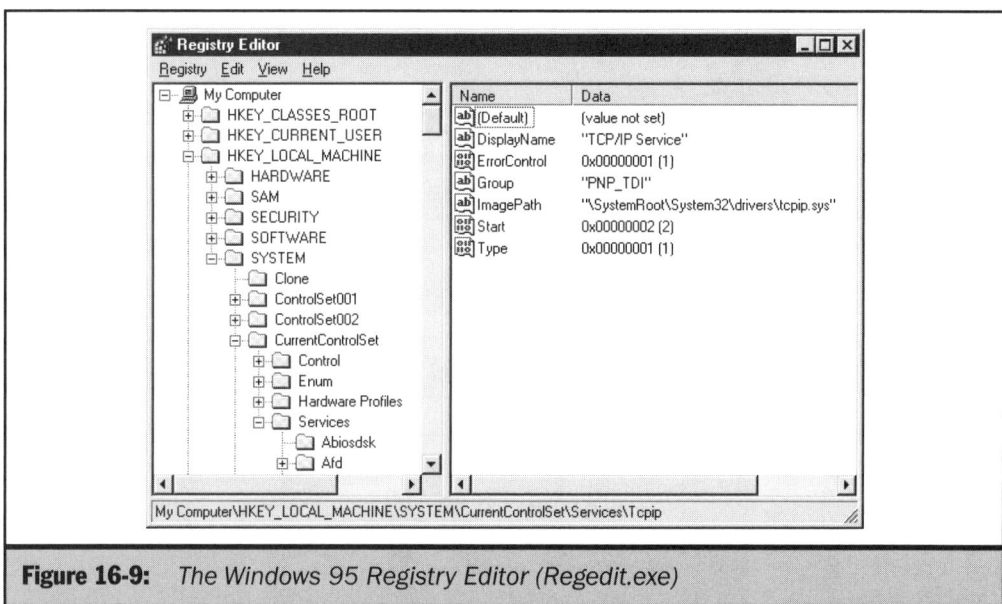

Figure 16-9:   *The Windows 95 Registry Editor (Regedit.exe)*

Regedt32.exe (see Figure 16-10) is the Registry Editor designed for use with Windows NT 3.*x*. The program opens a separate window for each of the top-level keys in the registry that enables you to navigate through each key independently to locate the desired entries. Regedt32.exe provides more display options than Regedit.exe. You can view values in binary, decimal, or hexadecimal notation, and choose to see the registry tree, the data contained in it, or both. When Windows is installed on an NTFS drive, Regedt32.exe also enables you to manage permissions for individual registry keys. Just as with file system permissions, you can grant specific users and groups various degrees of access to registry keys and their subkeys.

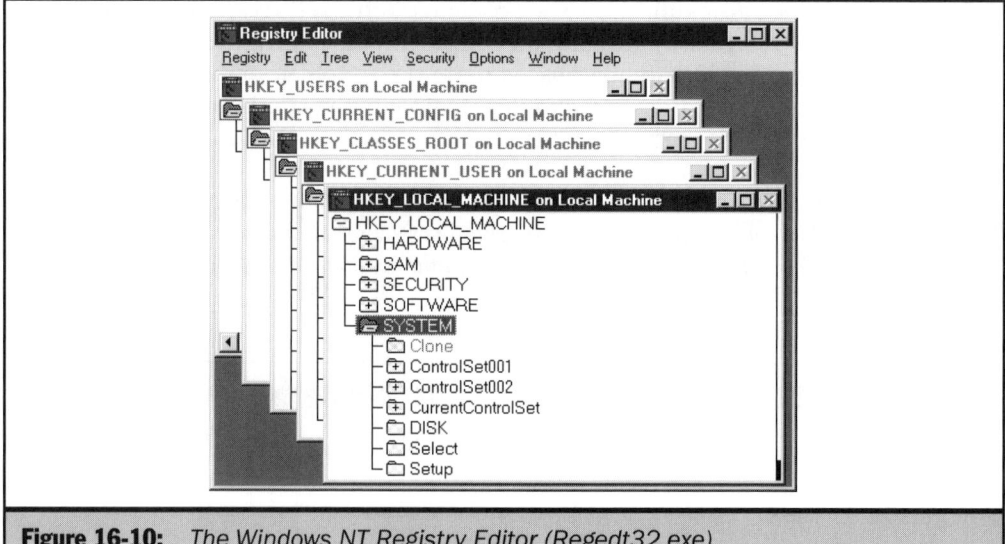

Figure 16-10:   *The Windows NT Registry Editor (Regedt32.exe)*

Optional Windows Networking Services

In addition to its core services, Windows 2000 and NT, particularly in the Server versions, include a large collection of optional services that you can choose to install either with the OS or at any time afterward. Some of these services are discussed in the following sections.

Active Directory

Active Directory, the enterprise directory service included with the Window 2000 Server products, is the single most revolutionary feature in Windows 2000. Active Directory is a hierarchical, replicated directory service designed to support networks of virtually unlimited size. For more information on Active Directory, see Chapter 17.

Microsoft DHCP Server

Unlike NetBEUI and IPX, using the TCP/IP protocols on a network requires that each computer be configured with a unique IP address, as well as other important settings. A

Dynamic Host Configuration Protocol (DHCP) server is an application designed to automatically supply client systems with TCP/IP configuration settings as needed, thus eliminating a tedious manual network administration chore. For more information about DHCP, see Chapter 23.

Microsoft DNS Server

The Domain Name System (DNS) facilitates the use of familiar names for computers on a TCP/IP network instead of the IP addresses they use to communicate. Designed for use on the Internet, DNS servers resolve domain names (Internet domain names, not NT domain names) into IP addresses, either by consulting their own records or by forwarding the request to another DNS server. The DNS server included with Windows 2000 and NT 4.0 enables an NT server to function on the Internet in this capacity. The role of the DNS is greatly expanded in Windows 2000, however. Windows 2000's Active Directory stores information about domain controllers in Service (SRV) resource records on DNS servers, and clients generate DNS queries to locate the domain controllers. Windows NT, by contrast, uses WINS to track the locations of network resources. To accommodate the needs of Active Directory, the DNS server included with Windows 2000 supports the new dynamic DNS (DDNS) standard, which enables the DCs to automatically update their resource records. In addition, the DHCP server included with Windows 2000 now works with the DNS server to dynamically update the records for network systems as they are assigned new IP addresses. For more information about DNS and its role in Active Directory, see Chapters 25 and 17.

Windows Internet Naming Service

Windows Internet Naming Service (WINS) is another service that supports the use of TCP/IP on a Windows network. Windows 9x and NT identify systems using NetBIOS names, but in order to transmit a packet to a machine with a given name using TCP/IP, the sender must first discover the IP address associated with that name. WINS is essentially a database server that stores the NetBIOS names of the systems on the network and their associated IP addresses. When a system wants to transmit, it sends a query to a WINS server containing the NetBIOS name of the destination system, and the WINS server replies with its IP address. Although Windows 2000 uses DNS for this purpose instead of WINS, the Windows 2000 Server products still include the WINS server application, for reasons of backward compatibility. For more information on NetBIOS naming and NetBEUI, see Chapter 24.

NETWORK OPERATING SYSTEMS

Gateway Services for NetWare

In addition to the default Microsoft Windows network client, Windows 2000 and NT include a client that enables the system to connect to Novell NetWare servers. NetWare connectivity was a sore point throughout the history of Windows NT. Both Microsoft and Novell have provided NetWare clients for NT since soon after its release, but for several years, both clients were limited in their capabilities as well as their performance. Microsoft's own NetWare client provided adequate connectivity, but early versions did not support Novell Directory Services (NDS) connections. The NetWare clients in Windows 2000 and Windows NT 4.0 do support NDS, but they do not provide the full NetWare administration capabilities of Novell's own client for Windows 2000 and NT.

 *Novell's Client for Windows NT/2000 provided NDS support long before Microsoft's client did, and to this day, you must use the Novell client to administer an NDS tree using the NetWare Administrator utility. For more information on the Novell Client for Windows NT/2000, see Chapter 22.*

There are two variants of the Microsoft client for NetWare: Windows 2000 Professional and Windows NT Workstation include Client Services for NetWare (CSNW), and Windows 2000 Server and Windows NT Server 4.0 include Gateway Services for NetWare (GSNW). Both of these services provide basic NetWare connectivity in the form of a redirector that enables you to log in to an NDS tree, connect to NetWare servers, and access NetWare volumes and printers in the same manner as Windows network resources. Once connected, you can map drive letters to NetWare volumes or reference them directly using either NetWare notation (*server:volume*) or the standard UNC notation used by Windows (*server**volume*).

GSNW is a superset of CSNW that includes the ability to provide Windows network clients with access to NetWare volumes without their having to run a NetWare client themselves. In the GSNW Control Panel (see Figure 16-11), you can use the NetWare client to connect to specific volumes on NetWare servers and publish them as Windows network shares. Other systems on the network see these volumes as shares on the Windows server, and can access them as such. These client systems do not have to be running a NetWare client themselves, because the Windows server functions as a gateway to the NetWare server.

The use of a Windows 2000 or NT server as a gateway to NetWare resources is a convenience, but is generally not a solution that is suitable for heavy use. Unless both the Windows server and the network connection between the Windows and NetWare servers are fast enough to support the combined traffic generated by a large number of users, the Windows server can end up being a bottleneck that slows down network communications.

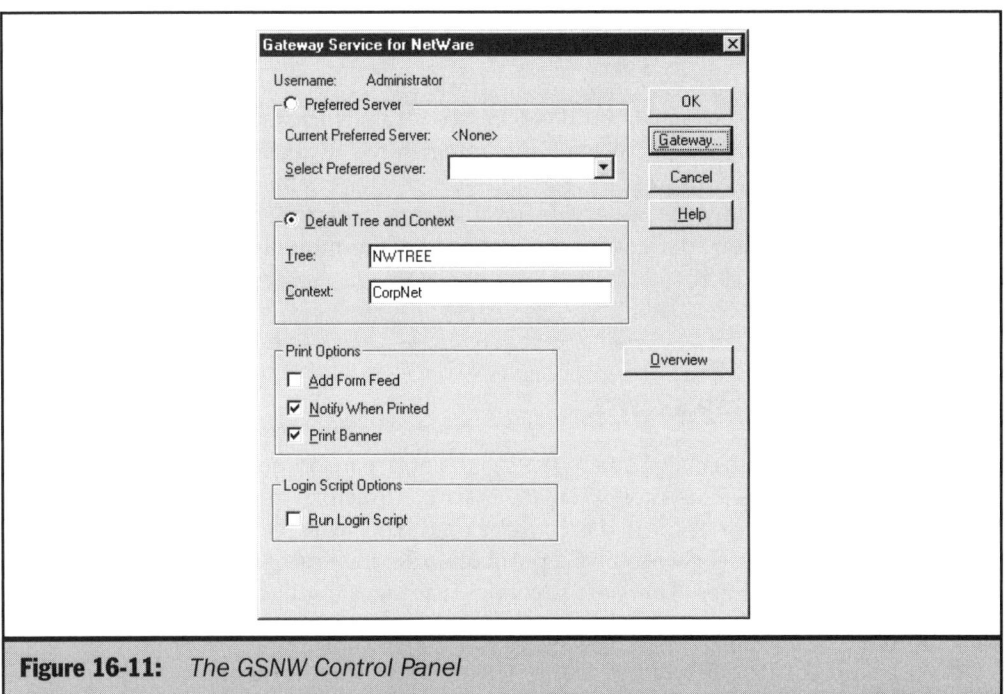

Figure 16-11:   *The GSNW Control Panel*

Routing and Remote Access Server

Windows 2000 and NT have always included multiprotocol routing capabilities as part of the OS. When two network interfaces are installed in the computer, in the form of either NICs or modem connections, traffic using any of the supported protocols can pass through the system, thus connecting two network segments to form an internetwork. Microsoft released an additional service for Windows NT Server 4.0 that enhances the operating system's routing capabilities, called the *Routing and Remote Access Server (RRAS)*. This service is also included with the Windows 2000 Server products.

RRAS provides Windows 2000 and NT with the ability to support more complex routing configurations, such as those using WAN and dial-up connections. The service includes dial-on-demand capability, the Open Shortest Path First (OSPF) routing protocol, virtual private networking support using the Point-to-Point Tunneling Protocol (PPTP), and a graphical utility for managing the routing table and network connections. For more information on the routing principles supported by the RRAS service, see Chapter 6.

Network Address Translation

Windows 2000 includes a Network Address Translation (NAT) server that enables a network to use private, unregistered IP addresses for its workstations. When a system on the network connects to the Internet, the unregistered IP address in its packets is translated to a registered address by the NAT server. This enables the NAT server to use a few registered addresses to service many clients, and protects the systems on the network from unauthorized access from the Internet. For more information on NAT, see Chapter 28.

Distributed File System

The Microsoft Distributed File System (DFS) is a service for Windows 2000 and Windows NT 4.0 that enables you to create a virtual directory tree that is comprised of shared drives or directories located anywhere on the network. Instead of users having to browse to the correct server to find a particular file, they can access all of the share's files from a single DFS tree. There are several advantages to this arrangement, as follows:

- Users are insulated from the actual, physical locations of the files.

- Administrators can relocate files to other volumes or servers without altering the DFS directory structure seen by users.

- The file server traffic load can be distributed among several servers, improving overall network performance.

- A DFS tree can provide fault tolerance by accessing alternate paths to duplicate files on another server in the event of an equipment failure.

When you install DFS, you create a *root volume* on a system running Windows 2000 Server or Windows NT Server 4.0, and then use the Distributed File System snap-in for the Microsoft Management Console (in Windows 2000) or the DFS Administrator graphical utility in Windows NT (see Figure 16-12). There is also a command-line utility called Dfscmd.exe that creates the *leaf volumes* that stem from the root. The leaf volumes can be located on a computer running any OS that is accessible by Windows 2000 or NT, including Windows 9*x*, Windows for Workgroups, and Novell NetWare. Users see the root volume as a share on the Windows server, and see the leaf volumes as subdirectories on that share. To a user browsing through the DFS tree, the leaf volumes appear just like subdirectories of the root, even though they are located on other drives or other systems.

Note   *Microsoft DFS is included with the three Windows 2000 Server packages, but it is an add-on product for Windows NT Server 4.0 that is available from Microsoft's Web site as a free download at http://www.microsoft.com/ntserver/nts/downloads/winfeatures/ NTSDistrFile.*

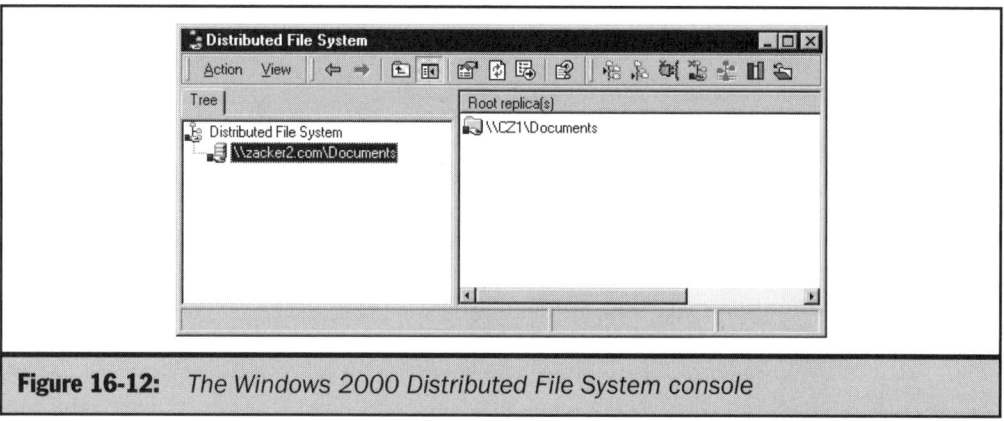

Figure 16-12:   *The Windows 2000 Distributed File System console*

DFS provides no additional security of its own. Users who have permission to access a volume, directory, or file in its original location can also access it through the DFS tree. Permission to access the root volume is not needed for a user to access the leaf volumes beneath it.

Internet Information Server

Microsoft Internet Information Services (IIS) is the Internet server package included with Windows 2000 and Windows NT 4.0, which includes Web and FTP servers, as well as other tools for building Web sites and Web-based applications. The Windows 2000 products included IIS 5.0. The original Windows NT 4.0 release included IIS version 2.0, but the current version for NT, IIS 4.0, is part of Windows NT Option Pack 4.0, now included with all Windows NT 4.0 Server products. For more information about IIS, see Chapter 26.

Load Balancing Service

The Network Load Balancing Service is included as part of the Windows 2000 Advanced Server product. The Windows NT equivalent, called the Windows Load Balancing Service (WLBS) is included with Windows NT Server 4.0, Enterprise Edition. WLBS enables up to 32 Windows servers to function as a cluster, providing identical services to a large number of TCP/IP clients.

WLBS is designed primarily for use on highly trafficked Web sites in which a single server is not sufficient to support the expected number of users. Since a Web server's DNS name can only be resolved to a single IP address, all client traffic must use that address as its destination. WLBS works by creating a virtual IP address that appears to the clients as though it represents only a single server. In actuality, there are multiple Web servers hosting the same content, all of which communicate through WLBS. The service distributes the incoming traffic evenly among all of the servers in the cluster, enabling them to function as though they were one machine.

The servers in a WLBS cluster continually exchange messages (called *heartbeats*) to inform each other of their continued operation. As the traffic to the Web site increases, administrators can simply add more servers to the cluster. If a server in the cluster malfunctions, the interruption of its heartbeat informs the other systems, and they modify their configurations (using a process called *convergence*) to take up the workload.

Microsoft Cluster Server

Microsoft Cluster Server is another service included with Windows 2000 Advanced Server and Windows NT Server 4.0, Enterprise Edition. In this context, a *cluster* is a pair of Windows servers that duplicate all of their functions, not just a single application. To client systems, the cluster appears as a single entity, and if one server malfunctions, the other continues to support the clients. However, the relationship between the servers is not just a fail-over model in which one system sits idle, in case the other one fails. Both servers can run applications independently, and if one server fails, the other launches the applications that the failed server was running and takes over its functions. In addition, applications that are *cluster-aware* can spread their functions across the servers in the cluster to balance the load. Although Cluster Server currently supports only two servers in a cluster, future versions will be able to cluster more machines.

IntelliMirror

Microsoft's Zero Administration Initiative for Windows is intended to simplify the administration of network systems and lower their total cost of ownership by limiting user access to the OS and maximizing remote administration capabilities. The Zero Administration Kit defines a methodology for automating the installation of the workstation OS and applications, and for using NTFS permissions and system policies to lock down the Windows desktop.

One byproduct of these techniques is the ability to store user profile information on a network drive, thus enabling users to log on to the network from any workstation and retrieve basic configuration settings, such as desktop icons and Start menu program groups. IntelliMirror takes this concept further by enabling more of a user's computing environment to follow him or her around the network, including applications and data files, as well as system and application configuration settings.

IntelliMirror consists of three basic features:

User Data Management   Provides users with access to their data files from any workstation on the network, or even when disconnected from the network, using the Windows 2000 Synchronization Manager to replicate folders on the local drive.

Software Installation and Maintenance   Installs applications and other software to any workstation on which they're needed.

User Settings Management   Provides users with access to their desktop environment, application configuration settings, and personal preferences on any network workstation.

In addition to providing users with access to their workstation resources from any computer, IntelliMirror also simplifies the process of building a new workstation. In the event of a hardware failure that puts a user's PC out of commission, an administrator can use IntelliMirror's capabilities to configure a new system quickly, with the user's exact same working configuration.

Quality of Service

Quality of service (QoS) is essentially the result of assigning priorities to various types of network traffic and adapting systems to transmit packets based on the priority they've been assigned, thus guaranteeing the timely delivery of information. Certain applications can benefit from the ability to transmit their data immediately, without delays incurred by traffic volume. Multimedia applications are the typical example used in the design of QoS protocols, because delivering audio and video over a network in real time requires a continuous data stream to avoid interruptions. However, other applications, such as e-mail, can also take advantage of QoS prioritizing. QoS also makes it possible to reserve a certain amount of bandwidth for use by a particular application, whether it is currently using it or not.

There are several different QoS standards and technologies available. Windows 2000 supports the following:

Differentiated Quality of Service   IETF standards (published as RFCs 2474 and 2475) that define a *differentiated services field* (also called the *diff-serve* or *DS* field) for the IPv4 and IPv6 headers, which provide *classes of service* that can be assigned to specific types of applications.

Admission Control Service (ACS)   Enables administrators to reserve bandwidth for specific users, applications, or sites, using the QoS Admission Control MMC plug-in (see Figure 16-13) to create QoS policies stored in Active Directory. Clients that support the IETF's Subnet Bandwidth Management (SBM) standard can request priority bandwidth from the service.

NETWORK OPERATING SYSTEMS

IEEE 802.1p Defines a frame format extension for Ethernet, Token Ring, FDDI, and other data link–layer protocols that carries the extra bits used to specify the priority of the packet. There are eight service classes, ranging from 0, which has the lowest priority, to 7, which has the highest.

Resource Reservation Protocol (RSVP) IETF standard (published as RFC 2205) that provides guaranteed, on-time delivery of data over an IP network in cooperation with the diff-serve standard by reserving bandwidth in response to client requests.

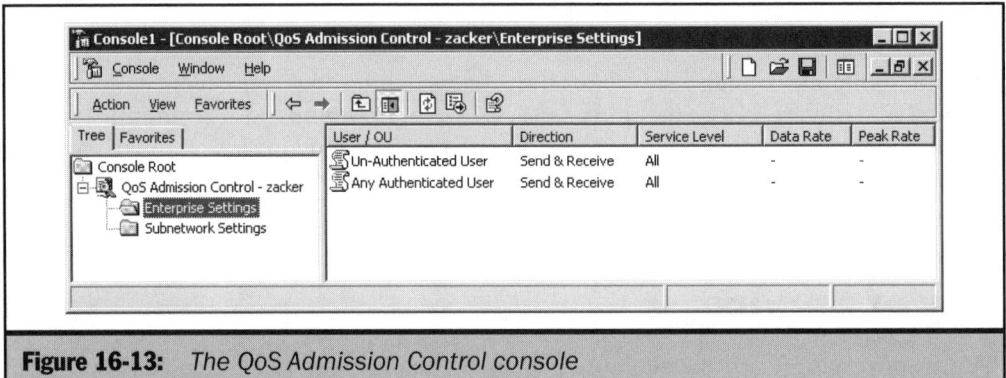

Figure 16-13: *The QoS Admission Control console*

Windows Terminal Support

Windows 2000 Server includes terminal services like those provided by Windows NT Server 4.0, Terminal Server Edition as part of the base OS. With this technology, you can use the server to run the 32-bit Windows interface on legacy desktop systems (such as 16-bit Windows, Macintosh, or Unix) and on Windows-based terminals.

Chapter 17

Active Directory

Т he domain-based directory service used by Windows NT has long come under fire for its inability to scale up to support larger networks. An enterprise network that consists of multiple domains is limited in its communication between those domains to the trust relationships that administrators must manually establish between them. In addition, because each domain must be maintained individually, the account administration process is complicated enormously. Since the original Windows NT 3.1 release in 1993, Microsoft has promised to deliver a more robust directory service that is better suited for use on large networks, and in Windows 2000's Active Directory, it has finally done that.

Active Directory (AD) is an object-oriented, hierarchical, distributed directory services database system that provides a central storehouse for information about the hardware, software, and human resources of an entire enterprise network. Based on the general principles of the X.500 global directory standard and similar to Novell Directory Services (NDS) in many ways, network users are represented by objects in the Active Directory tree. Administrators can use those objects to grant users access to resources anywhere on the network, which are also represented by objects in the tree. Unlike Windows NT, which uses a flat, domain-based structure for its directory, Active Directory expands the structure into multiple levels. The fundamental unit of organization in the Active Directory database is still the domain, but a group of domains can now be consolidated into a tree, and a group of trees into a forest. Administrators can manage multiple domains simultaneously by manipulating the tree, and manage multiple trees simultaneously by manipulating a forest.

A directory service is not only a database for the storage of information, however. It also includes the services that make that information available to users, applications, and other services. Active Directory includes a global catalog that makes it possible to search the entire directory for particular objects using the value of a particular attribute. Applications can use the directory to control access to network resources, and other directory services can interact with AD using a standardized interface and the Lightweight Directory Access Protocol (LDAP).

Active Directory Architecture

Active Directory is composed of objects, which represent the various resources on a network, such as users, user groups, servers, printers, and applications. An *object* is a collection of attributes that define the resource, give it a name, list its capabilities, and specify who should be permitted to use it. Some of an object's attributes are assigned automatically when they're created, such as the globally unique identifier (GUID) assigned to each one, while others are supplied by the network administrator. A user object, for example, has attributes that store information about the user it represents, such as an account name, password, telephone number, and e-mail address. Attributes also contain information about the other objects with which the user interacts, such as the groups of which the user is a member. There are many different types of objects, each of which has different attributes, depending on its functions.

The primary difference between Active Directory and Windows NT 4.0 domains is that Active Directory provides administrators and users with a global view of the network. Both directory services can use multiple domains, but instead of managing the users of each domain separately, for example, as in Windows NT 4.0, AD administrators create a single object for each user and can use it to grant that user access to resources in any domain.

Each type of object is defined by an object class stored in the *directory schema*. The schema specifies the attributes that each object must have, the optional attributes it may have, the type of data associated with each attribute, and the object's place in the directory tree. The schema are themselves stored as objects in the Active Directory called *class schema objects* and *attribute schema objects*. A class schema object contains references to the attribute schema objects that together form the *object class*. This way, an attribute is only defined once, although it can be used in many different object classes.

The schema is extensible, so that applications and services developed by Microsoft or third parties can create new object classes or add new attributes to existing object classes. This enables applications to use Active Directory to store information specific to their functions and provide that information to other applications as needed. For example, rather than maintain its own directory, an e-mail server application like Microsoft Exchange 2000 can modify the Active Directory schema so that it can use AD to authenticate users and store their e-mail information. When you install Exchange 2000 on a Windows 2000 server, the application adds attribute schema objects to the user object class, causing additional pages to appear in the user object's Properties dialog box (see Figure 17-1).

Object Types

There are two basic types of objects in Active Directory, called container objects and leaf objects. A *container object* is simply an object that stores other objects, while a *leaf object* stands alone and cannot store other objects. Container objects essentially function as the branches of the tree, and leaf objects grow off of the branches. Active Directory uses container objects called organizational units (OUs) to store other objects. Containers can store other containers or leaf objects, such as users and computers. The guiding rule of directory tree design is that rights and permissions flow downward through the tree. Assigning a permission to a container object means that, by default, all of the objects in the container inherit that permission. This enables administrators to control access to network resources by assigning rights and permissions to a single container rather than to many individual users.

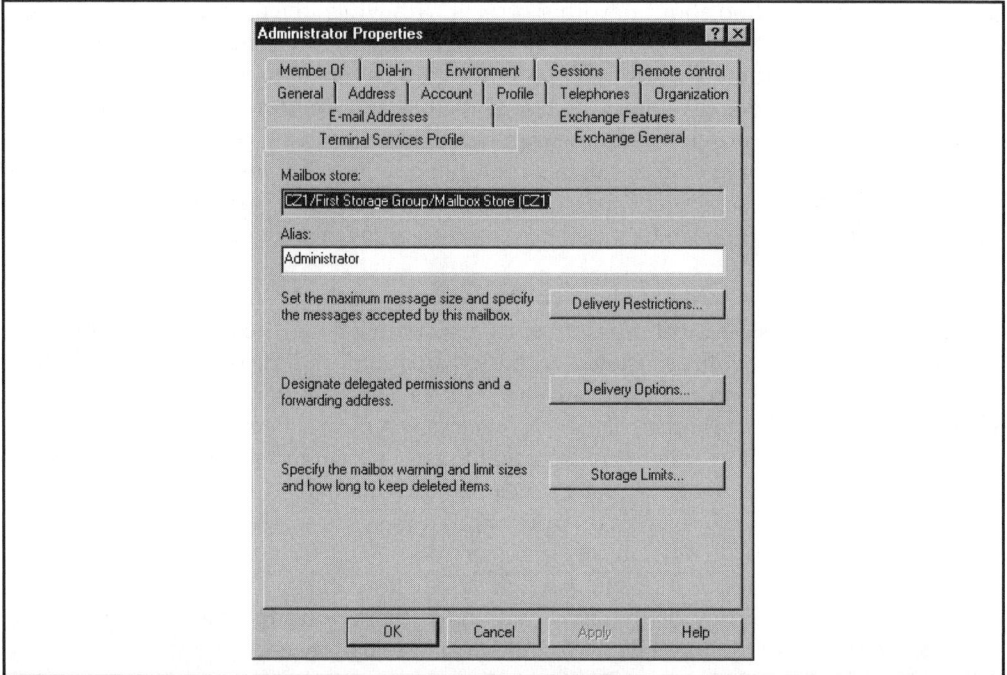

Figure 17-1: *Applications can extend the Active Directory schema to add new attributes to object classes, such as the Exchange-specific attributes shown here.*

By default, an Active Directory tree is composed of objects that represent the users and computers on the network, the logical entities used to organize them, and the folders and printers they regularly access. These objects, their functions, and the icons used to represent them in tools such as Active Directory Users and Computers are listed in Table 17-1.

Object Naming

Every object in the Active Directory database is uniquely identified by a name that can be expressed in several forms. The naming conventions are based on the Lightweight Directory Access Protocol (LDAP) standard defined in RFC 2251, published by the Internet Engineering Task Force (IETF). The *distinguished name (DN)* of an object consists of the name of the domain in which the object is located, plus the path down the domain tree through the container objects to the object itself. The part of an object's name that is stored in the object itself is called its *relative distinguished name (RDN)*.

Icon	Object Type	Function
	Domain	Container object that stores organizational unit objects and their contents.
	Organizational unit	Container object that stores computer, user, and group objects within the tree structure.
	User	Leaf object that represents a network user and stores identification and authentication data about that user.
	Computer	Leaf object that represents a computer on the network, stores information about the computer, and provides the machine account needed for the system to log on to the domain.
	Contact	Leaf object that represents a user outside the domain for specific purposes, such as e-mail delivery; does not enable the user to log on to the domain.
	Group	Container objects that represent logical groupings of users, computers, and/or other groups that are independent of the AD tree structure. Group members can be located in any organizational unit or domain in the tree.
	Shared folder	Represents a shared folder on a Windows 2000 system.
	Shared printer	Represents a shared printer on a Windows 2000 system.

Table 17-1: *Active Directory Object Types*

Note *The Lightweight Directory Access Protocol is an adaptation of the Directory Access Protocol (DAP) designed for use by X.500 directories. Active Directory domain controllers and several other directory services use LDAP to communicate with each other.*

NETWORK OPERATING SYSTEMS

By specifying the name of the object and the names of its parent containers up to the root of the domain, the object is uniquely identified within the domain, even if the object has the same name as another object in a different container. Thus, if you have two users, called John Doe and Jane Doe, you can use the RDN jdoe for both of them. As long as they are located in different containers, they will have different DNs.

Canonical Names

Most Active Directory applications refer to objects using their canonical names. A *canonical name* is a DN in which the domain name comes first, followed by the names of the object's parent containers working down from the root of the domain and separated by forward slashes, followed by the object's RDN, as follows:

```
zacker.com/sales/inside/jdoe
```

In this example, jdoe is a user object in the inside container, which is in the sales container, which is in the zacker.com domain.

LDAP Notation

The same DN can also be expressed in LDAP notation, which would appear as follows:

```
cn=jdoe,ou=inside,ou=sales,dc=zacker,dc=com
```

This notation reverses the order of the object names, starting with the RDN on the left and the domain name on the right. The elements are separated by commas and include the LDAP abbreviations that define each type of element. These abbreviations are as follows:

cn   Common name

ou   Organizational unit

dc   Domain component

In most cases, LDAP names do not include the abbreviations, and they can be omitted without altering the uniqueness or the functionality of the name. It is also possible to express an LDAP name in a URL format, as defined in RFC 1959, which appears as follows:

```
ldap://cz1.zacker.com/cn=jdoe,ou=inside,ou=sales,dc=zacker,dc=com
```

This format differs in that the name of a server hosting the directory service must appear immediately following the ldap:// identifier, followed by the same LDAP name as shown earlier. This notation enables users to access Active Directory information using a standard Web browser.

Globally Unique Identifiers

In addition to its DN, every object in the tree has a *globally unique identifier (GUID)*, which is a 128-bit number that is automatically assigned by the Directory System Agent when the object is created. Unlike the DN, which changes if you move the object to a different container or rename it, the GUID is permanent and serves as the ultimate identifier for an object.

User Principal Names

Distinguished names are used by applications and services when they communicate with Active Directory, but they are not easy for users to understand, type, or remember. Therefore, each user object has a *user principle name (UPN)* that consists of a user name and a suffix, separated by an @ symbol, just like the standard Internet e-mail address format defined in RFC 822. This name provides users with a simplified identity on the network and insulates them from the need to know their place in the domain tree hierarchy.

In most cases, the user name part of the UPN is the user object's RDN, and the suffix is the DNS name of the domain in which the user object is located. However, if your network consists of multiple domains, you can opt to use a single domain name as the suffix for all of your users' UPNs. This way, the UPN can remain unchanged even if you move the user object to a different domain.

The UPN is an internal name that is used only on the Windows 2000 network, so it doesn't have to conform to the user's Internet e-mail address. However, using your network's e-mail domain name as the suffix is a good idea, so that users only have to remember one address for accessing e-mail and logging on to the network.

 *You can use the Active Directory Domains and Trusts console to specify alternate UPN suffixes, so that all of your users can log on to the network using the same suffix.*

Domains, Trees, and Forests

Windows NT has always based its networking paradigm on domains, and all but small networks require multiple domains to support their users. Windows 2000 makes it easier to manage multiple domains by combining them into larger units called trees and forests. When you create a new Active Directory database by promoting a server to domain controller, you create the first domain in the first tree of a new forest. If you create additional domains in the same tree, they all share the same schema, configuration, and global catalog server (GCS, a master list directory of Active Directory objects that provides users with an overall view of the entire directory), and are connected by transitive trust relationships.

Trust relationships are how domains interact with each other to provide a unified network directory. If Domain A trusts Domain B, the users in Domain B can access the resources in Domain A. In Windows NT domains, trust relationships operate in one direction only and must be explicitly created by network administrators. If you want to create a full network of trusts between three domains, for example, you must create six separate trust relationships, so that each domain trusts every other domain. Active Directory automatically creates trust relationships between domains in the same tree. These trust relationships flow in both directions, are authenticated using the Kerberos security protocol, and are *transitive*, meaning that if Domain A trusts Domain B and Domain B trusts Domain C, then Domain A automatically trusts Domain C. A *tree*, therefore, is a single administrative unit that encompasses a number of domains. The administrative nightmare of manually creating trust relationships between large numbers of domains is diminished, and users are able to access resources on other domains.

The domains in a tree share a contiguous name space. Unlike a Windows NT domain, which has a single, flat name, an Active Directory domain has a hierarchical name that is based on the DNS name space, such as mycorp.com. Sharing a contiguous name space means that if the first domain in a tree is given the name mycorp.com, the subsequent domains in that tree will have names that build on the parent domain's name, such as sales.mycorp.com and mis.mycorp.com (see Figure 17-2).

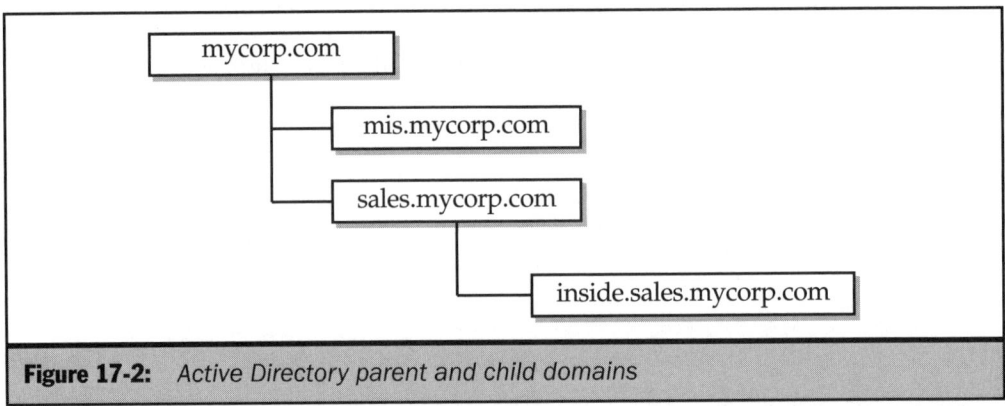

Figure 17-2: *Active Directory parent and child domains*

The parent/child relationships in the domain hierarchy are limited solely to the sharing of a name space and the trust relationships between them. Unlike the container hierarchy within a domain, rights and permissions do not flow down the tree from domain to domain.

In most cases, a single tree is sufficient for a network of almost any size. However, it is possible to create multiple trees and join them together in a unit known as a *forest*. All of the domains in a forest, including those in separate trees, share the same schema, configuration, and GCS. Every domain in a forest has a transitive trust relationship with the other domains, regardless of the trees they are in. The only difference between the trees in a forest is that they have separate name spaces. Each tree has its own root domain and child domains that build off of its name. The first domain created in a forest is known as the *forest root domain.*

The most common reason for having multiple trees is the merging of two organizations, both of which already have established domain names that cannot be readily assimilated into one tree. Users are able to access resources in other trees, because the trust relationships between domains in different trees are the same as those within a single tree. It is also possible to create multiple forests on your network, but the need for this is rare.

Different forests do not share the same schema, configuration, and GCS, nor are trust relationships automatically created between forests. It is possible to manually create unidirectional trusts between domains in different forests, just as you would on a Windows NT network. In most cases, though, the primary reason for creating multiple forests is to completely isolate two areas of the network and prevent interaction between them.

DNS and Active Directory

Windows NT is based on NetBIOS, and uses a NetBIOS name server called *Windows Internet Naming Service (WINS)* to locate computers on the network and resolve their names into IP addresses. The primary limitation of NetBIOS and WINS is that they use a flat name space, whereas Active Directory's name space is hierarchical. The AD name space is based on that of the Domain Name System (DNS), so the directory uses DNS servers instead of WINS to resolve names and locate domain controllers. You must have at least one DNS server running on your network in order for Active Directory to function properly.

The domains in an Active Directory are named using standard DNS domain names, which may or may not be the same as the names your organization uses on the Internet. If, for example, you have already registered the domain name mycorp.com for use with your Internet servers, you can choose to use that same name as the parent domain in your AD tree or create a new name for internal use. The new name doesn't have to be registered for Internet use, because its use will be limited to your Windows 2000 network only.

DNS is based on resource records (RRs) that contain information about specific machines on the network. Traditionally, administrators must create these records manually, but on a Windows 2000 network, this causes problems. The task of manually creating records for hundreds of computers is long and difficult, and it is compounded by the use of the Dynamic Host Configuration Protocol (DHCP) to automatically assign IP addresses to network systems. Because the IP addresses on DHCP-managed systems can change, there must be a way for the DNS records to be updated to reflect those changes.

Microsoft DNS Server included with Windows 2000 supports the new SRV resource record type that enables client systems to use DNS queries to locate Windows 2000 domain controllers. The Microsoft DNS server also supports dynamic DNS (DDNS), which works together with Microsoft DHCP Server to dynamically update the resource records for specific systems as their IP addresses change. Many of the older DNS server products used today (such as BIND version 4.*xx*) do not support these new features and will not work with Active Directory. However, these new DNS features have been standardized and are being implemented in other manufacturers' products, such as newer versions of BIND. The only one of these new features that is absolutely required by Active Directory is support for the SRV resource record. Other features, such as DDNS, secure DDNS, and incremental zone transfers, are recommended but not essential.

Active Directory is still a relatively new product, and is bound to undergo some revisions as it matures, which may affect the relationship between the directory service and DNS. If you are deploying AD on a production network, it is a good idea to stick with Microsoft DNS Server, because it will no doubt be upgraded to conform to any changes made to the directory service itself.

 *For more information on the Domain Name System standards and functions, see Chapter 25.*

Global Catalog Server

To support large enterprise networks, Active Directory can be both partitioned and replicated, meaning that the directory can be split into sections stored on different servers, and copies of each section can be maintained on separate servers. Splitting up the directory in this way, however, makes it more difficult for applications to locate specific information. Therefore, Active Directory maintains the *global catalog*, which provides an overall picture of the directory structure. While a domain controller contains the Active Directory information for one domain only, the global catalog is a replica of the entire Active Directory, except that it includes only the essential attributes of each object, known as *binding data*.

Because the global catalog consists of a substantially smaller amount of data than the entire directory, it can be stored on a single server and accessed more quickly by users and applications. The global catalog makes it easy for applications to search for specific objects in Active Directory using any of the attributes included in the binding data.

Deploying Active Directory

All of the architectural elements of Active Directory that have been described thus far, such as domains, trees, and forests, are logical components that do not necessarily have any effect on the physical network. In most cases, network administrators create domains, trees, and forests based on the political divisions within an organization, such as workgroups and departments, although geographical elements can come into play as well. Physically, however, an Active Directory installation is manifested as a collection of domain controllers, split into subdivisions called *sites*.

Creating Domain Controllers

A *domain controller (DC)* is a Windows 2000 Server system that hosts all or part of the Active Directory database and provides the services to the rest of the network through which applications access that database. When a user logs on to the network or requests access to a specific network resource, the workstation contacts a domain controller, which authenticates the user and grants access to the network.

Unlike Windows NT 4.0, Active Directory has only one type of domain controller. When installing an NT 4 server, you have to specify whether it should be a primary domain controller (PDC), a backup domain controller (BDC), or a member server. Once a system is installed as a domain controller for a specific domain, there is no way to move it to another domain or change it back to a member server. All Windows 2000 servers start out as standalone or member servers; you can then promote them to domain controllers and later demote them back to member servers. Active Directory has no PDCs or BDCs; all domain controllers function as peers.

Anyone who has worked with Windows NT 4.0 domains will understand how the ability to promote servers to domain controllers and demote them at will is a major boon to the network administrator. This capability enables you to add domain controllers when necessary or reassign a domain controller to another domain as needed. With this type of flexible deployment, you can modify Active Directory as the network grows and business conditions change. When a corporation reorganizes its divisions, for example, the network administrators can modify the AD structure by reassigning domain controllers as needed, while Windows NT 4.0 servers would probably have to be completely reinstalled to conform to the new organization.

To promote a Windows 2000 server into a domain controller, you use the Active Directory Installation Wizard utility (see Figure 17-3), which you launch by running the Dcpromo.exe program from the command line. To do this, you must be logged on to the local machine with administrator privileges. The wizard guides you through the process of either creating a new domain or a replica of an existing domain, and of deciding whether the new domain should be the first in a new tree or forest or be the child of an existing domain. If the server is already a domain controller, the wizard lets you demote it to a member server.

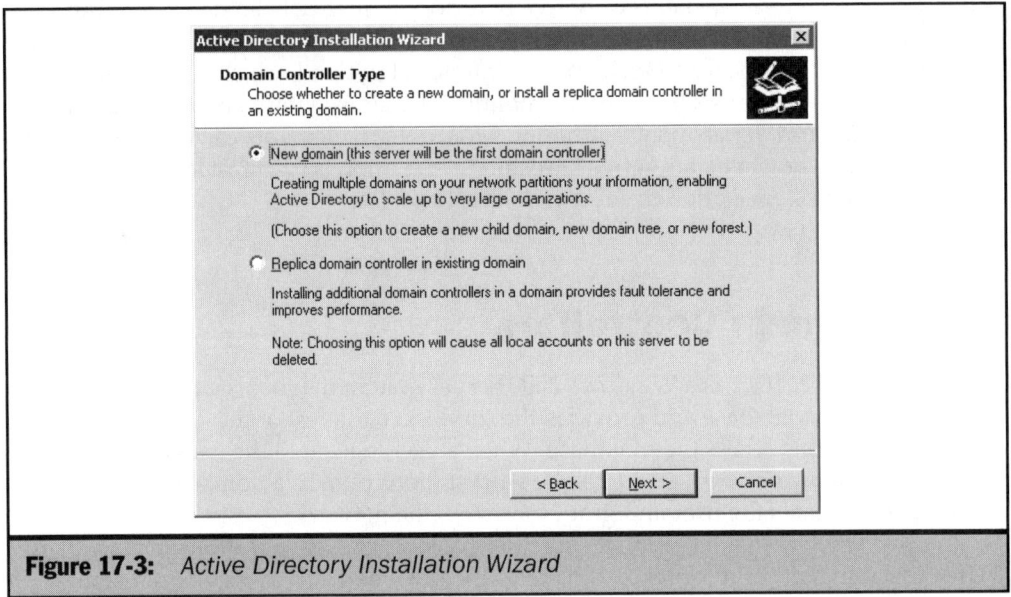

Figure 17-3: *Active Directory Installation Wizard*

A server that is to function as a domain controller must have at least one NTFS 5.0 drive to hold the Active Directory database, log files, and the system volume, and must have access to a DNS server that supports the SRV resource record and (optionally) dynamic updates. If the computer cannot locate a DNS server that provides these features, it offers to install and configure the Microsoft DNS Server software on the Windows 2000 system.

Windows 2000 includes an upgraded version of the NT file system (NTFS 5.0), which you must use to store Active Directory data. You cannot use FAT drives or NTFS drives created by Windows NT, unless you upgrade them during the Windows 2000 installation. The amount of disk space required by Active Directory varies, of course, depending on the size of the network and how you intend to partition the database. A 1GB partition should be the absolute minimum for an Active Directory domain controller.

Directory Replication

Every domain on your network should be represented by at least two domain controllers, for reasons of fault tolerance. Once your network is reliant on Active

Directory for authentication and other services, inaccessible domain controllers would be a major problem. Therefore, each domain should be replicated on at least two domain controllers, so that one is always available. Directory service replication is nothing new, but Active Directory replicates its domain data differently from Windows NT.

Windows NT domains are replicated using a technique called *single master replication,* in which a single PDC with read/write capabilities replicates its data to one or more BDCs that are read-only. In this method, replication traffic always travels in one direction, from the PDC to the BDCs. If the PDC fails, one of the BDCs can be promoted to PDC. The drawback of this arrangement is that changes to the directory can only be made to the PDC. When an administrator creates a new user account or modifies an existing one, for example, the User Manager for Domains utility must communicate with the PDC, even if it is located at a distant site connected by a slow WAN link.

Active Directory uses *multiple master replication,* which enables administrators to make changes on any of a domain's replicas. This is why there are no longer PDCs or BDCs. The use of multiple masters makes the replication process far more difficult, however. Instead of simply copying the directory data from one domain controller to another, the information on each domain controller must be compared with that on all of the others, so that the changes made to each replica are propagated to every other replica. In addition, it's possible for two administrators to modify the same attribute of the same object on two different replicas at virtually the same time. The replication process must be able to reconcile conflicts like these and see to it that each replica contains the most up-to-date information.

Multimaster Data Synchronization

Some directory services, such as NDS, base their data synchronization algorithms on time stamps assigned to each database modification. Whichever change has the later time stamp is the one that becomes operative when the replication process is completed. The problem with this method is that the use of time stamps requires the clocks on all of the network's domain controllers to be precisely synchronized, which is difficult to arrange. The Active Directory replication process only relies on time stamps in certain situations. Instead, AD uses *Update Sequence Numbers (USNs),* which are 64-bit values assigned to all modifications written to the directory. Whenever an attribute changes, the domain controller increments the USN and stores it with the attribute, whether the change results from direct action by an administrator or replication traffic received from another domain controller.

In addition to the USN stored with the attributes on the local system, each domain controller maintains a table containing the highest USN value it has received from each of the other domain controllers with which it replicates its data. When the replication process begins, the domain controller requests that its replication partners transmit all modifications with USNs greater than the highest value in the table. Domain controllers also use the USNs to recover from failures or down periods. When the server restarts, it transmits the highest value in its USN table to its replication partners, and the other domain controllers use this information to supply all updates that have occurred since the system went offline.

The only problem with this method is when the same attribute is modified on two different domain controllers. If an administrator changes the value of a specific attribute on Server B before a change made to the same attribute on Server A is fully propagated to all of the replicas, then a *collision* is said to have occurred. To resolve the collision, the domain controllers use property version numbers to determine which value should take precedence. Unlike USNs, which are a single numerical sequence maintained separately by each domain controller, there is only one property version number for each object attribute.

When a domain controller modifies an attribute as a result of direct action by a network administrator, it increments the property version number. However, when a domain controller receives an attribute modification in the replication traffic from another domain controller, it does not modify the property version number. A domain controller detects collisions by comparing the attribute values and property version numbers received during a replication event with those stored in its own database. If an attribute arriving from another domain controller has the same property version number as the local copy of that attribute, but the values don't match, a collision has occurred. In this case, and only in this case, the system uses the time stamps included with each of the attributes to determine which value is newer and should take precedence over the other.

Sites

A single domain can have any number of domain controllers, all of which contain the same information, thanks to the AD replication system. In addition to providing fault tolerance, you can create additional domain controllers to provide users with local access to the directory. In an organization with offices in multiple locations, connected by WAN links, it would be impractical to have only one or two domain controllers, because workstations would have to communicate with the AD database over a relatively slow, expensive WAN connection. Therefore, administrators often create a domain controller at each location where there are resources in the domain.

The relatively slow speed and high cost of the average WAN connection also affects the replication process between domain controllers, and for this reason, Active Directory can break a domain up into sites. A *site* is a collection of domain controllers that are assumed to be well-connected, meaning that all of the systems are connected using the same relatively high-speed LAN technology. The connections between sites are assumed to be WANs that are slower and possibly more expensive.

The actual speed of the intra-site and inter-site connections is not an issue. Your LANs can run at 10 to 100 Mbps or more, and the WAN connections can be anything from dial-ups to T-1s. The issue is the relative speed between the domain controllers at the same site and those at different sites. The reason for dividing a domain into logical units that reflect the physical layout of the network is to control the replication traffic that passes over the slower and more expensive WAN links. Active Directory also uses sites to determine which domain controller a workstation should access when authenticating a user. Whenever possible, authentication procedures use a domain controller located on the same site.

Intra-Site Replication

The replication of data between domain controllers located at the same site is completely automatic and self-regulating. A component called the Knowledge Consistency Checker (KCC) dynamically creates connections between the domain controllers as needed to create a replication topology that minimizes latency. *Latency* is the period of time during which the information stored on the domain controllers for a single domain is different—that is, the interval between the modification of an attribute on one domain controller and the propagation of that change to the other domain controllers. The KCC triggers a replication event whenever a change is made to the AD database on any of the site's replicas.

The KCC maintains at least two connections to each domain controller at the site. This way, if a controller goes offline, replication between all of the other domain controllers is still possible. The KCC may create additional connections to maintain timely contact between the remaining domain controllers while the system is unavailable, and then remove them when the system comes back online. In the same way, if you add a new domain controller, the KCC modifies the replication topology to include it in the data synchronization process. As a rule, the KCC creates a replication topology in which each domain controller is no more than three hops away from any other domain controller. Because the domain controllers are all located on the same site, they are assumed to be well-connected, and the KCC is willing to expend network bandwidth in the interest of replication speed. All updates are transmitted in uncompressed form because, even though this requires the transmission of more data, it minimizes the amount of processing needed at each domain controller.

Replication occurs primarily within domains, but when multiple domains are located at the same site, the KCC also creates connections between the global catalog servers for each domain, so that they can exchange information and create a replica of the entire Active Directory containing the subset of attributes that form the binding data.

Inter-Site Replication

By default, a domain consists of a single site, called Default-First-Site-Name, and any additional domains you create are placed within that site. You can, however, use the Active Directory Sites and Services console to create additional sites and move domains into them. Just as with domains in the same site, Active Directory creates a replication topology between domains in different sites, but with several key differences.

Because the WAN links between sites are assumed to be slower and more expensive, Active Directory attempts to minimize the amount of replication traffic that passes between them. First, there are fewer connections between domain controllers at different sites than with a site; the three-hop rule is not observed for the inter-site replication topology. Second, all replication data transmitted over inter-site connections is compressed, to minimize the amount of bandwidth utilized by the replication process. Finally, replication events between sites are not automatically triggered by modifications to the Active Directory database. Instead, replication can be scheduled to occur at specified times and intervals, to minimize the effect on standard user traffic and to take advantage of lower bandwidth costs during off hours.

Microsoft Management Console

Microsoft Management Console (MMC) is an application that provides a centralized administration interface for many of the services included in Windows 2000, including those used to manage the Active Directory. Windows NT relies on separate management applications for many of its services, such as the DHCP Manager, WINS Manager, and Disk Administrator. Windows 2000 consolidates all of these applications, and many others, into MMC. Most of the system administration tasks for the OS are now performed through MMC. In many cases, the exact same functions found in the standalone Windows NT utilities are ported to a substantially different MMC interface. The initial Windows 2000 experience for system administrators used to Windows NT is typically one of frustration as they search for familiar functions that are now found in very different places.

MMC has no administrative capabilities of its own; it is, essentially, a shell for application modules called *snap-ins* that provide the administrative functions for many of Windows 2000's applications and services. Snap-ins take the form of files with an .msc extension that you load either from the command line or interactively through the MMC menus. Windows 2000 supplies snap-in files for all of its tools, but the interface is designed so that third-party software developers can use the MMC architecture to create administration tools for their own applications.

MMC can load multiple snap-ins simultaneously using the Windows multiple-document interface (MDI). You can use this capability to create a customized management interface containing all of the snap-ins you use on a regular basis. When you run MMC (by launching the Mmc.exe file from the Run dialog box) and select Console | New, you get an empty Console Root window. By selecting Console | Add/Remove Snap-in, you can build a list of the installed snap-ins and load selected ones into the console. The various snap-ins appear in an expandable, Explorer-like display in the left pane of MMC's main screen, as shown here. By selecting Console | Save As, you can save your console configuration as an MSC file and load it again as needed.

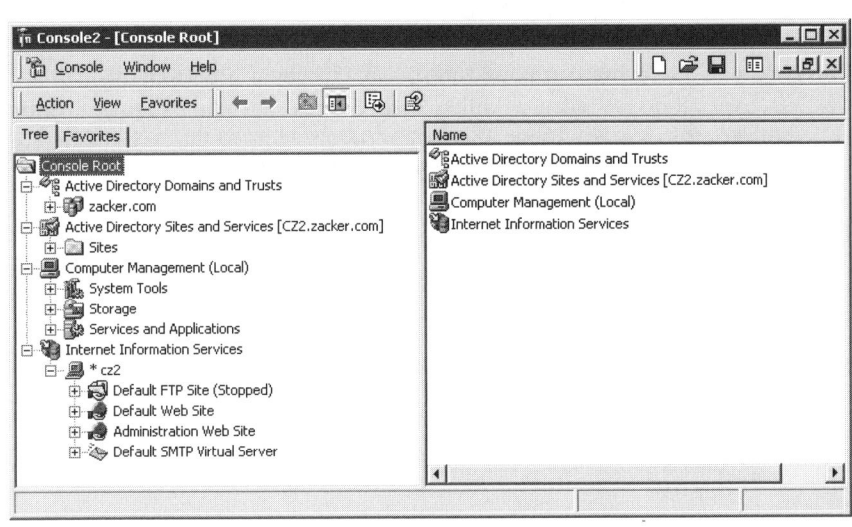

Many of Windows 2000's administrative tools, such as Active Directory Sites and Services, are actually preconfigured MMC consoles. Selecting Computer Management from the Programs/Administrative Tools group in the Start menu displays a console that contains a collection of the basic administration tools for a Windows 2000 system, as shown here. By default, the Computer Management console administers the local system, but you can use all of its tools to manage a remote network system by selecting Action | Connect to Another Computer.

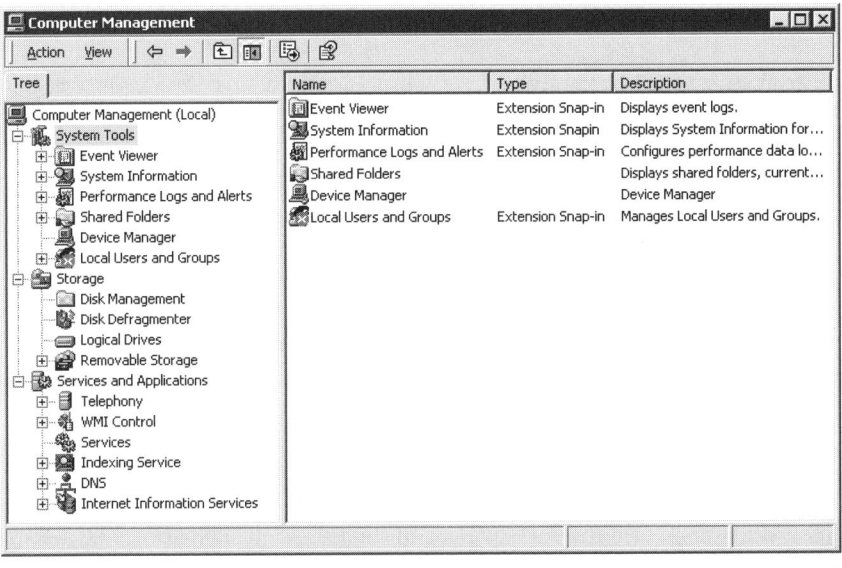

Creating and Configuring Sites

Splitting a network into sites has no effect on the hierarchy of domains, trees, and forests that you have created to represent your enterprise. However, sites still appear as objects in Active Directory, along with several other object types that you use to configure your network's replication topology. These objects are only visible in the Active Directory Sites and Services tool, as shown in Figure 17-4. The object called Default-First-Site-Name is created automatically when you promote the first server on your network to a domain controller, along with a server object that appears in the Servers folder beneath it. Server objects are always subordinate to site objects, and represent the domain controllers operating at that site. A site can contain server objects for domain controllers in any number of domains, located in any tree or forest. You can move server objects between sites as needed.

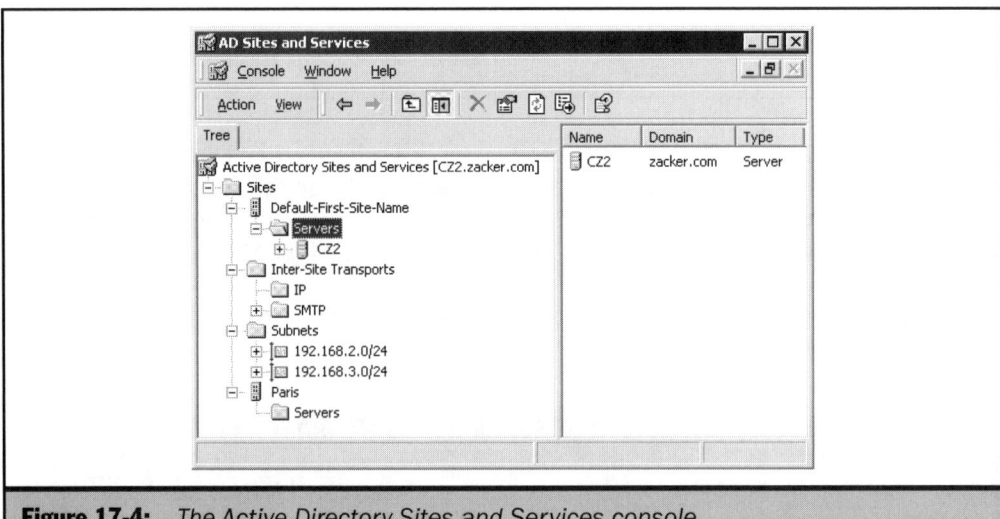

Figure 17-4:   *The Active Directory Sites and Services console*

The other two important object types associated with sites and servers are subnet and site link objects. Subnet objects represent the particular IP subnets that you use at your various sites, and are used to define the boundaries of the site. When you create a subnet object, you specify a network address and subnet mask. When you associate a site with a subnet object, server objects for any new domain controllers that you create on that subnet are automatically created in that site. You can associate multiple subnet objects with a particular site to create a complete picture of your network.

Site link objects represent the WAN links on your network that Active Directory will use to create connections between domain controllers at different sites. Active Directory supports the use of the Internet Protocol (IP) and the Simple Mail Transport Protocol (SMTP) for site links, both of which appear in the Inter-Site Transports folder in Active Directory Sites and Services. IP site links can use any type of WAN connection, such as a dial-up modem connection or a T-1. An SMTP site link can take the form of any applications you use to send e-mail using the SMTP protocol. When you create a site link object, you select the sites that are connected by the WAN link the object represents. The attributes of site link objects include various mechanisms for determining when and how often Active Directory should use the link to transmit replication traffic between sites (see Figure 17-5), including the following:

Cost   The cost of a site link can reflect either the monetary cost of the WAN technology involved or the cost in terms of the bandwidth needed for other purposes. For example, the monthly bills for a T-1 connection may be astronomical, but if you installed the T-1 primarily for the purpose of supporting replication traffic, you don't want to assign it a high cost that results in its not being used. The higher the value of the cost setting, the less frequently AD uses the link for replication traffic.

Schedule   Specifies the hours of the day during each day of the week that the link can be used to carry replication traffic. If you want to minimize the impact of replication on a link used for other types of traffic during work hours, you can create a schedule that enables replication to occur on this link only during non-production times.

Replication period   Specifies the interval between replication procedures that use this link, subject to the schedule described previously.

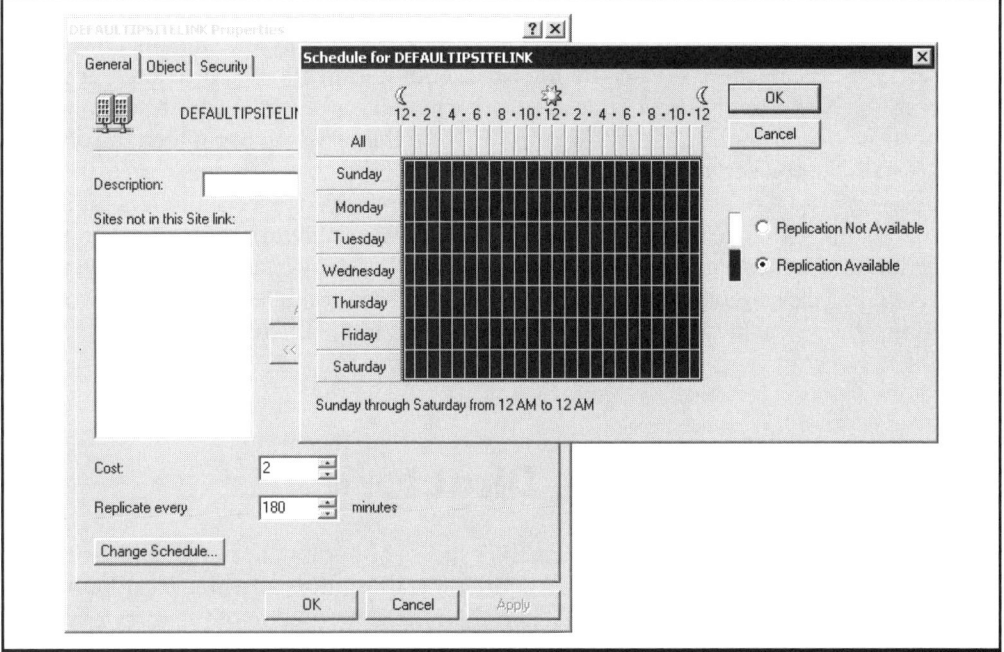

Figure 17-5:   *A site link object's Properties Schedule dialog boxes*

By default, Active Directory creates an IP site link object, DEFAULTIPSITELINK, that you can use as-is or modify to reflect the type of link used to connect your sites. If all of your sites are connected by WAN links of the same type, you don't have to create additional site link objects, because a single set of scheduling attributes should be applicable for all of your inter-site connections. If you use various types of WAN connections, however, you can create a separate site link object for each type and configure its attributes to reflect how you want it to be used.

There is another type of object that you can create in the Inter-Site Transports container, called a *site link bridge object,* that is designed to make it possible to route replication traffic through one remote site to others. By default, the site links that you create are transitive, meaning that they are bridged together, enabling them to route replication traffic. For example, if you have a site link object connecting Site A to Site B, and another one connecting Site B to Site C, then Site A can send replication traffic to Site C. If you want to, you can disable the default bridging by opening the Properties dialog box for the IP folder and clearing the Bridge All Site Links check box. If you do this, you must manually create site link bridge objects in order to route replication traffic in this way. A site link bridge object generally represents a router on the network. While a site link object groups two site objects together, a site link bridge object groups two site link objects together, making it possible for replication traffic to be routed between them.

Once you have created objects representing the sites that form your network and the links that connect them, the KCC can create connections that form the replication topology for the entire internetwork, subject to the limitations imposed by the site link object attributes. The connections created by the KCC, both within and between sites, appear as objects in the NTDS Settings container beneath each server object. A connection object is unidirectional, representing the traffic running from the server under which the object appears to the target server specified as an attribute of the object. In most cases, there should be no need to manually create or configure connection objects, but it is possible to do so. You can customize the replication topology of your network by creating your own connections and scheduling the times during which they may be used. Manually created connection objects cannot be deleted by the KCC to accommodate changing network conditions; they remain in place until you manually remove them.

Designing an Active Directory

As with any enterprise directory service, the process of deploying Active Directory on your network involves much more than simply installing the software. The planning process is, in many cases, more complicated than the construction of the directory itself.

Naturally, the larger your network, the more complicated the planning process will be. You should have a clear idea of the form that your AD structure will take and who will maintain each part of it, before you actually begin to deploy domain controllers and create objects.

In many cases, the planning process will require some hands-on testing before you deploy Active Directory on your production network. You may want to set up a test network and try out some forest designs before you commit yourself to any one plan. Although a test network can't fully simulate the effects of hundreds of users working at once, the time that you spend familiarizing yourself with the Active Directory tools and procedures can only help you later when you're building the live directory service.

Planning Domains, Trees, and Forests

To a Windows NT administrator, Windows 2000 and Active Directory may seem like an embarrassment of riches. While NT networks are based solely on domains, Active Directory expands the scope of the directory service by two orders of magnitude by providing trees and forests that you can use to organize multiple domains. In addition, the domains themselves can be subdivided into smaller administrative entities called organizational units. To use these capabilities effectively, you must evaluate your network in light of both its physical layout and the needs of the organization that it serves.

Creating Multiple Trees

In most cases, a single tree with one or more domains is sufficient to support an enterprise network. The main reason for creating multiple trees is if you have two or more existing DNS name spaces that you want to reflect in Active Directory. For example, a corporation that consists of several different companies that operate independently can use multiple trees to create a separate name space for each company. Although there are transitive trust relationships between all of the domains in a tree, separate trees are connected only by trusts between their root domains, as shown in Figure 17-6.

If you have several levels of child domains in each tree, the process of accessing a resource in a different tree involves the passing of authentication traffic up from the domain containing the requesting system to the root of the tree, across to the root of the other tree, and down to the domain containing the requested resource. If the trees operate autonomously, and access requests for resources in other trees are rare, this may not be much of a problem. If the trust relationships in a directory design like this do cause delays on a regular basis, you can manually create what are known as *shortcut trusts* between child domains lower down in both trees, as shown in Figure 17-7.

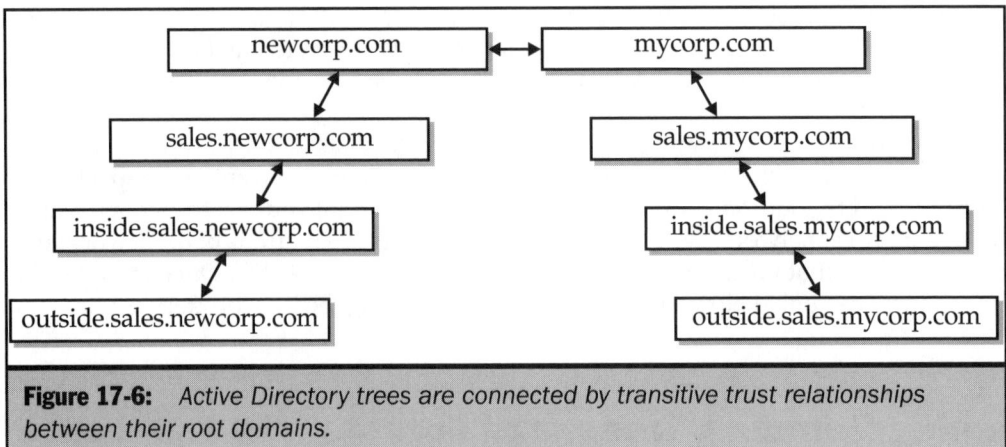

Figure 17-6:   *Active Directory trees are connected by transitive trust relationships between their root domains.*

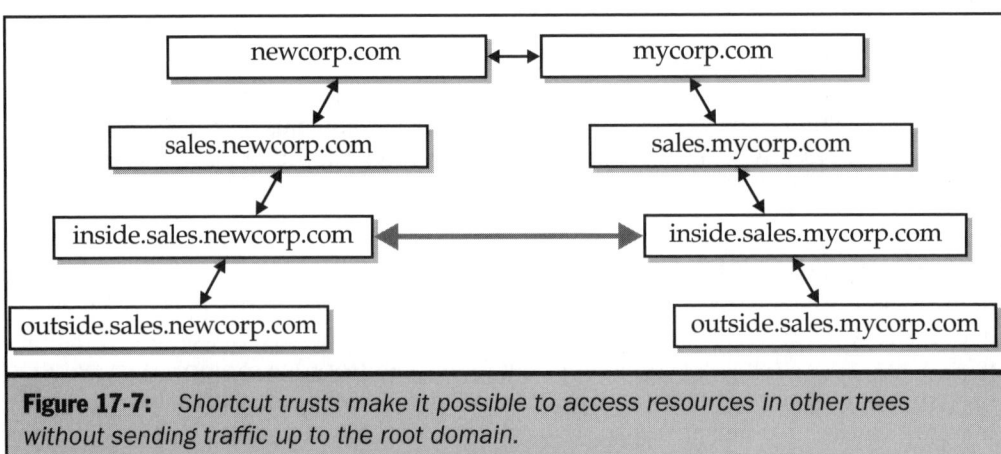

Figure 17-7:   *Shortcut trusts make it possible to access resources in other trees without sending traffic up to the root domain.*

Just as you can create multiple trees in a forest, you can create multiple forests in the Active Directory database. Scenarios in which the use of multiple forests is necessary are even rarer than those calling for multiple trees, because forests have no inherent trust relationships between them at all and use a different global catalog, making it more difficult for users even to locate resources. You may want to use a separate forest for a lab-based test network or for a project that you don't want other network users to know even exists.

Creating Multiple Domains

In Windows NT, the domain is the only administrative unit available in the directory service. Many administrators have become accustomed to creating separate domains to reflect the department or workgroup divisions in their organizations. Active Directory provides this same capability, but the hierarchical nature of the tree should affect the number of domains that you create. The compound domain names used in Active Directory make it impractical to create a tree that runs to too many layers; the domain names become too long and difficult to use and remember. In many cases, instead of creating a separate domain, it is preferable to create an organizational unit within a domain.

OUs provide administrative boundaries that are similar to those of domains, without extending the domain name space and complicating the network of trust relationships. OUs enable you to assign permissions and create policies for a group of users without modifying their individual user objects. In addition, using OUs instead of domains can yield significant hardware savings, because OUs don't need their own servers to function as domain controllers.

The first element that you should consider when planning your domains is the physical layout of your network, and particularly the WAN links. Splitting a domain among several sites may be desirable in some cases, but you must weigh this against the cost of the bandwidth connecting those sites, the frequency of the changes made to Active Directory, and the location of the people who will maintain the database. If, for example, your AD database is relatively static and your remote sites are relatively small, you can create domain controllers at the different locations, manage them from the central office, and configure the site link objects to perform infrequent database replications. The bandwidth utilized by the replication process will be minimal, and all of the sites will have local access to the resources in the domain.

If you make frequent changes to the AD database, you may be better off creating separate domains at each site, to prevent the need for frequent replication events. Even if you do this, however, there is still some replication traffic that passes across the WAN links, since all of the domains in the forest share a common global catalog, as well as the traffic generated by users accessing resources in domains at other sites.

Another factor that can influence the creation of separate domains is the language used at each site. If your organization has sites in different countries and uses the language indigenous to each country, creating separate domains for each site may be recommended, especially if the sites are located in countries that use non-Roman alphabets, such as Japan or Russia.

Object Naming

One of the most important elements of a directory service plan is the conventions you will use when naming the objects in the database and the domains that form the network. AD tree designs can be based on several different criteria, including the geographical locations of network resources and the political divisions within an organization. However you choose to represent the network in the AD design, you should be consistent in the names that you assign objects.

User object names, for example, should consist of a standardized pattern derived from the users' names, such as the first initial plus the surname. In the same way, domains and containers named for geographical locations or departmental divisions should use standardized abbreviations. This type of consistency makes it easier to search for specific objects in the global catalog.

Locating resources on a Windows 2000 network is more complicated than on Windows NT, because the directory service is designed to support networks of almost unlimited size. In Windows NT, each domain is a separate entity and there is no mechanism for performing a global search for a particular resource. Object naming can be a more informal process, because users and administrators are more likely to be intimately familiar with the domain configuration. Active Directory's global catalog, on the other hand, can contain information about thousands of objects located anywhere in the world, and the ability to search successfully for the data needed can be greatly enhanced by the implementation of consistent naming policies for the entire directory.

When it comes to naming domains and trees, the biggest decision you have to make is whether you should use the same name space for both your internal network resources and the external resources that are visible to the Internet. Registering one domain name for your organization with an Internet registrar like Network Solutions and using it both for your Internet servers and as the root domain name for your AD tree creates a unified name space for your network that has both advantages and disadvantages.

It is important to remember that the average network user is far less acquainted with the infrastructure than the average administrator. Using a single root domain name provides a consistency that makes the network easier for users to understand and also enables them to use their Internet e-mail addresses as their UPNs. The disadvantages of using the same name space are that it can be difficult to distinguish between internal and external resources, and the system is inherently less secure if your domain names are publicly known on the Internet.

Upgrading from Windows NT 4.0

In many cases, Windows 2000 will be installed on an existing Windows NT network, and the domain-based directory service will be upgraded to Active Directory. Windows 2000 and AD are capable of interacting with NT systems, either on a temporary or permanent basis, so there is no need to upgrade the entire network in one weekend. However, as with the installation of a new Windows 2000 network, the upgrade process must be carefully planned in order to proceed smoothly.

This section assumes that the NT systems being upgraded to Windows 2000 are running Windows NT 4.0. There are serious incompatibilities in the authentication methods used by Windows 2000 and Windows NT versions 3.51 and earlier that have led to a recommendation by Microsoft that these two operating systems not be used together.

Whenever you create an Active Directory domain, the Installation Wizard prompts you for a NetBIOS name for the domain, in addition to the DNS name. By default, the wizard creates the NetBIOS name simply by dropping the .com or other suffix from the DNS name you've supplied for the domain. Downlevel client systems (that is, clients running Windows versions prior to Windows 2000) will see the domain using the NetBIOS name rather than the full DNS name.

The basic plan when upgrading an NT network to Windows 2000 is to upgrade the PDCs first, then the BDCs, then the member servers, and finally the clients. You can spread the process out as long as you want, and run the NT and 2000 servers together indefinitely. When you upgrade a Windows NT 4.0 PDC to Windows 2000 and promote it to a domain controller, the system functions in *mixed mode*, meaning that it can use the BDCs for the domain you've upgraded as Active Directory domain controllers. While in mixed mode, the Windows NT 4.0 BDCs operate as fully functional Active Directory domain controllers, capable of multiple master replication and all other AD functions. Once you've upgraded the BDCs to Windows 2000 domain controllers, you can switch the servers to *native mode*, from the domain's Properties dialog box in the Active Directory Domains and Trusts tool (see Figure 17-8). The drawbacks to running in mixed mode are that you can't nest groups within groups in an Active Directory tree and you can't have groups with members in different domains. Once you've switched to native mode, you can do both of these things.

Switching from mixed mode to native mode is a one-way process. Once you've switched a server, you cannot switch it back without completely reinstalling it. Be sure that all of your upgraded domain controllers are functioning properly before you make this change.

If your NT network is relatively simple and based on the single domain model, migrating to Active Directory is simply a matter of upgrading the systems to Windows 2000 in the order previously specified. However, many networks are considerably more complicated, and further planning is required. If your network uses the single master domain model, you have one account domain containing all of your users and groups and one or more resource domains that contain all of your network's shared resources. This model fits into a single Active Directory tree that may or may not eventually include all of your original domains. For example, you might want to consolidate the tree into a single domain in which the former resource domains are converted to OUs in the original account domain, or even migrate the resource objects directly into the account domain without segregating them in OUs. If you eliminate some domains in this way, you can redeploy some of your servers by creating additional replicas of other domains or demoting them to member servers.

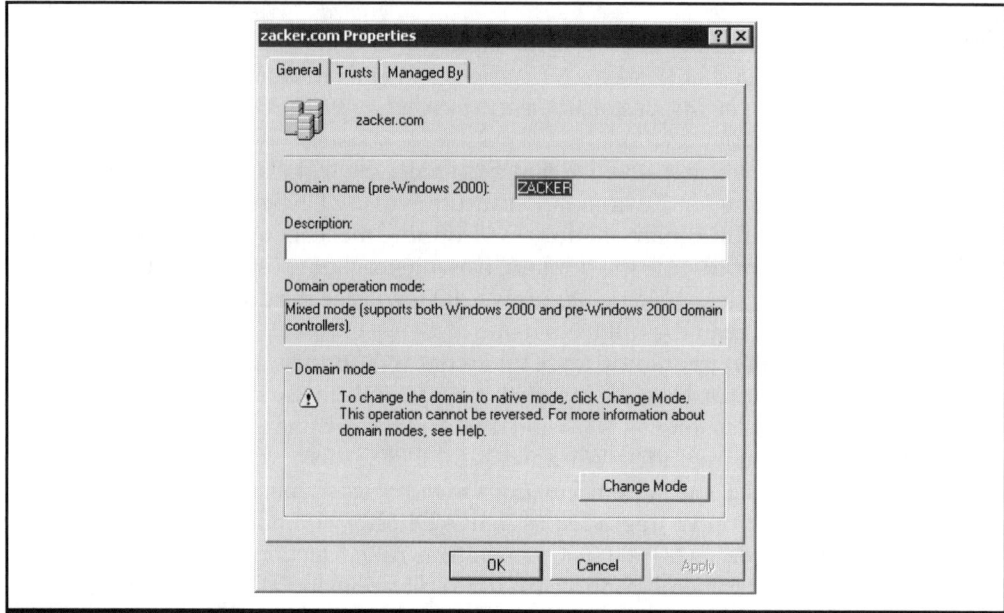

Figure 17-8: *A domain's Properties dialog box in the Active Directory Domains and Trusts utility*

One of the original purposes for using the single master domain model in Windows NT is to limit the number of trust relationships that administrators have to create and maintain. In that model, each of the resource domains has a one-way trust relationship with the account domain. No trusts between resource domains are needed. In Active Directory, the trust relationships between domains are created automatically, so this is not an issue. You can consolidate the NT domains before you perform the upgrade to AD or after, but you will probably find that the tools for manipulating Active Directory make it easier to consolidate the NT domains after you upgrade the servers.

To start the migration to Active Directory, you must upgrade the PDC in the account domain first, so that it becomes the root domain of the new AD tree. Then you should upgrade the PDCs of the resource domains, making them children of the root domain and establishing the tree structure. Once the PDC upgrades are completed, you can proceed to upgrade the BDCs for all of the domains, and then modify the tree design to take advantage of Active Directory's capabilities.

If your NT network uses the multiple master domain model, you must examine the reasons for using this model in the first place before you decide on an upgrade strategy. In the multiple master domain model, there are two or more account domains, connected by two-way trust relationships, and a number of resource domains, each of which trusts all of the account domains. In some cases, this model is used simply to support a larger number of users than can comfortably fit in a single account domain. In other cases, multiple masters are needed because the organization uses a decentralized support system in which the network is divided into sites or divisions, each of which has its own network administrators.

If the deciding factor is the number of users, an Active Directory domain can support many more users than an NT domain, which may lead you to consolidate all of the account domains into one or create a new directory service design entirely. If you have separate support teams for your domains, you may want to consider creating an Active Directory that uses multiple trees, with each account domain becoming the root of a separate tree. Again, because the trust relationships in AD are created automatically, there is no need to design the directory around the task of manually creating trust relationships.

Once you have made your design decisions, the upgrade process is roughly the same as that for the single master domain model. Upgrade the PDCs of the account domains first, either by selecting one to be the root domain of your tree or by creating a new tree for each account domain, and then upgrade the BDCs.

Note   *For more information on the Windows NT 4.0 domain models, see Chapter 18.*

Managing Active Directory

Once you've designed and built your Active Directory installation, you're ready to populate it with objects. Creating and managing user and group accounts is an everyday task for the typical network administrator, and in Windows 2000, you do this with the Active Directory Users and Computers console, as shown in Figure 17-9. Like most Microsoft Management Console (MMC) snap-ins, the Users and Computers utility consists of a scope pane (on the left) and a result pane (on the right). In this case, the scope pane displays the domains in your tree and the OUs in each domain, and the result pane displays the objects in the highlighted container, such as users and groups.

By default, Active Directory domains contain the following four container objects:

Builtin   Contains various built-in security groups used to delegate system administration tasks.

Computers   Contains computer objects representing the accounts for the machines that are members of the domain.

Domain Controllers   Contains computer objects representing the servers that are functioning as domain controllers for the selected domain.

Users   Contains the default user objects for the domain, such as the Administrator and Guest accounts, and group objects used for delegating domain administrative tasks.

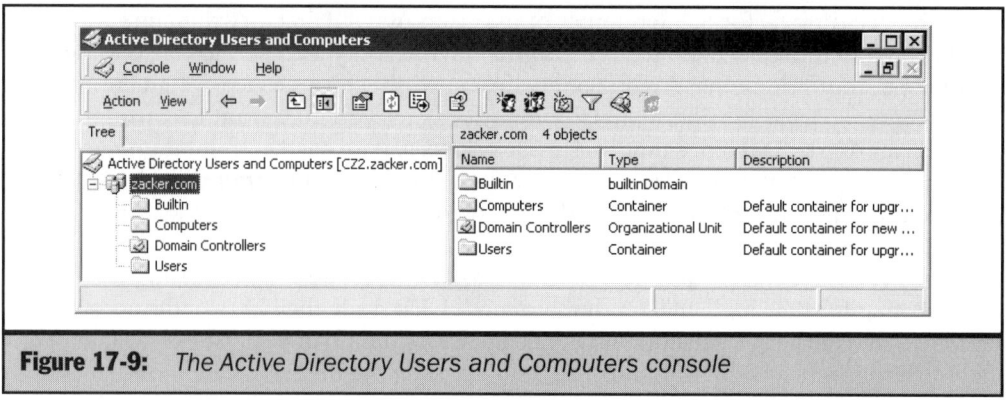

Figure 17-9:   *The Active Directory Users and Computers console*

You can use these existing containers by adding your own user accounts to the Users container and so forth, or you can create your own hierarchy of organizational units and populate it with objects as needed. One of the big differences between creating user objects in Active Directory and in Windows NT is the amount of information that an AD object can store. As you can see in Figure 17-10, the Properties dialog box for each user object has 12 screens that you can use to supply information about the user. For most administrators, the first inclination is to skip all of the informational screens and fields and configure only the attributes that are absolutely needed for the object to function. You should think twice before you do this, however, especially if you are working on a large network. One of the major advantages of Active Directory is the ability to search the entire database for objects, based on the values of specific attributes.

Consider a large corporation with offices in many different cities, for example. There is probably someone in the organization who is responsible for maintaining and publishing a corporate directory that lists all of the employees and vital information about them, such as telephone numbers and e-mail addresses. If you take the time to include just the phone number and e-mail address for each of the user objects that you create, you can eventually eliminate the need to maintain this directory by making it possible for users to search Active Directory for an employee's phone number instead.

This is just one example of how Active Directory can become the primary information resource about your network and the people who use it.

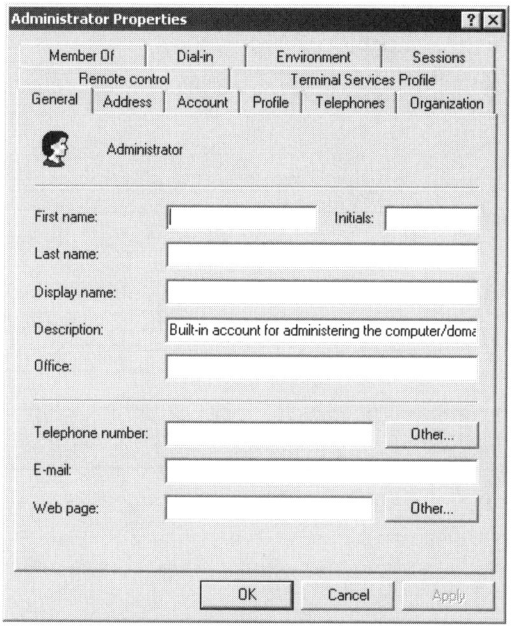

Figure 17-10:   *A user object's Properties dialog box*

Because Active Directory is designed to support networks of almost any size, it is not expected that one administrator will be responsible for the entire directory. There are likely to be situations in which certain administration tasks are delegated to a variety of individuals, not all of whom should have administrative rights to the entire directory service. To make this possible, Windows 2000 includes a Delegation of Control Wizard that enables you to grant users certain administrative rights to specific users and groups. This way, you can allow managers or supervisors to maintain the accounts of the people who work for them, without endangering the other parts of the directory tree. Once you select the users and groups for whom you want to delegate control and the user who will control them, you specify the tasks that the user can perform, using the dialog box shown in Figure 17-11.

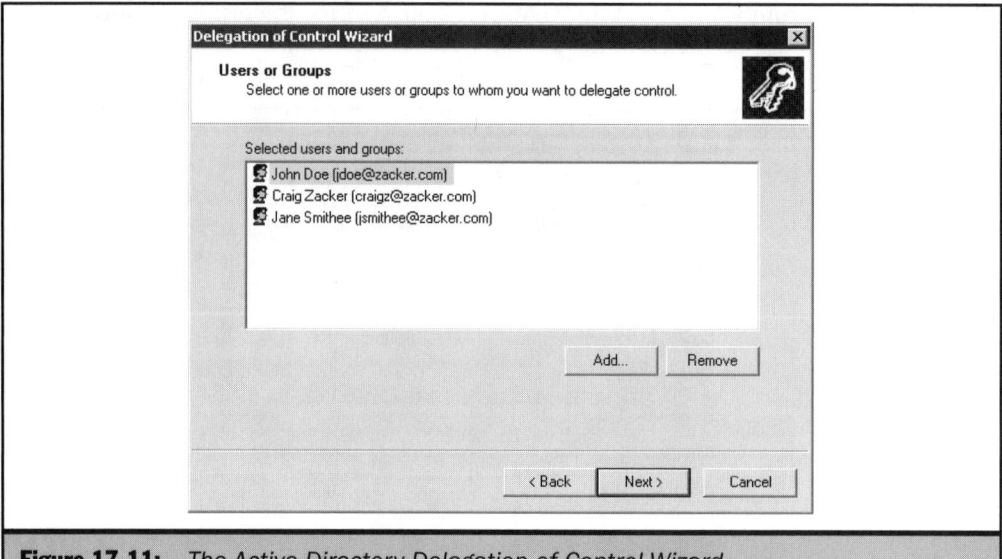

Figure 17-11:   *The Active Directory Delegation of Control Wizard*

The Complete Reference

Networking

Chapter 18

Windows NT Domains

Probably the most obvious shortcoming in the Windows NT operating system is its lack of a full-featured directory service. The domains that NT uses to organize and manage the computers on a network are serviceable, but lack the scalability and expandability of products like Novell Directory Services. For a small- to medium-sized network, the NT directory service is sufficient. As you move into larger enterprises, NT domains become difficult to administer and customize for specific purposes. Windows 2000 addresses these shortcomings with Active Directory, a hierarchical directory service that can support networks of virtually any size, but until Windows 2000 completely replaces NT, there will still be a great many NT domain networks operating for years to come.

Note

For more information on Active Directory, see Chapter 17.

The use of the term *domain* by Windows NT to describe a group of computers managed as a unit is unfortunate. At the time the concept was developed, the Internet was nowhere near as ubiquitous an entity as it is today, and the use of a term previously adopted for the Internet infrastructure was not expected to be a problem. Now, however, even people who have never used a computer are familiar with the phrase "dot com," and there can be some confusion at times between an Internet domain and a Windows NT domain.

The primary difference between the two types of domains is that whereas the Internet's Domain Name Service (DNS) is hierarchical, Windows NT domains are not. An NT domain is simply a group of computers on a network that share a security model and use a common *Security Accounts Manager (SAM)* database of user and group information stored on one or more systems that have been designated as *domain controllers.* A diagram of the computers in a Windows NT domain consists of two layers only: the domains and the computers in those domains. The domains used on the Internet can have any number of layers, with a minimum of three: the top-level domain (such as *com*), the second-level domain (such as *microsoft*, in *microsoft.com*), and the computers in the second-level domain. The owner of a second-level domain is free to create any number of subdomain layers, while an NT network is limited to a single domain layer.

The alternative to the use of domains in Windows NT is a *workgroup,* an informal grouping of computers in which each system maintains its own SAM database, containing user and group accounts that are valid on that system only. For a very small network (up to 20 workstations) that does not require stringent security, a workgroup is sufficient. You can run a workgroup network with any combination of Windows systems, including Windows 2000 Professional, Windows NT Workstation, Windows 95, 98, and ME, and even Windows for Workgroups. Although you can have Windows NT Server and Windows 2000 Server machines on a workgroup network, they are not needed for administration purposes as they are on a domain network.

From the standpoint of the network administrator, a workgroup network requires that users have individual accounts on each machine whose shares they will access. Either each user must be responsible for maintaining the accounts on his or her machine, or an administrator has to travel to each workstation to maintain it. On a domain network, a single domain user account provides the ability to access the shared resources on any computer in the domain. An administrator can perform account maintenance tasks for the entire domain from any workstation on the network.

Workgroups also function as an organizational resource only, whereas domains can provide security boundaries that restrict user access. In other words, there is nothing to prevent a user logged on to a system in one workgroup from accessing the resources on a system in another workgroup. A member of one domain, however, cannot access resources in another domain, unless an administrator has implemented a trust relationship between the domains.

Domain Controllers

The core of a Windows NT domain is its domain controllers. These are Windows NT Server computers that have been designated as domain controllers during the operating system installation. You cannot make an NT system into a domain controller after the OS installation is complete, nor can you change an existing domain controller back into a regular server. In either case, you must reinstall the operating system to change its domain controller status.

Note

One of the major improvements in Windows 2000's Active Directory is that you can convert a server into a domain controller at any time and then convert it back to a regular server as needed.

Windows NT domains can have two types of domain controllers, called primary and backup. Every domain must have one (and only one) *primary domain controller (PDC)*. This is the computer that contains the only read-write copy of the SAM database for the domain, including its users and groups and their password information. Whenever you create a new user or group (or modify the properties of an existing one), you are modifying the SAM database on the PDC.

A *backup domain controller (BDC)*, as the name implies, functions as a backup to the PDC in the event of a system failure or a break in network communications. If the PDC should become unavailable, you can promote a BDC to a PDC to take its place, using the Windows NT Server Manager utility. Promoting any BDC to a PDC causes the existing PDC to be demoted to a BDC. Once the original PDC server is back in operation, you can promote it back to its original role—which causes the temporary PDC to revert back to a BDC. A domain can have any number of backup domain controllers, or none at all, but at least one is recommended for every domain. Without a domain controller, a domain is useless, because no one can log on to the network or access network resources.

Replication

Every BDC on the network contains a replica of the master database stored on the PDC, which is updated at regular intervals by Windows NT's Netlogon service. In NT, the replication of the domain database always flows in one direction, from the PDC to each of the BDCs. This is called *single master replication*, because there is only one master copy of the database, to which administrators make all modifications, and which is replicated to all of the other copies. The alternative, as used in NDS and Active Directory, is *multiple master replication*, in which changes can be made to any copy of the database, and these changes are replicated to all of the other copies (see Figure 18-1).

The use of single master replication is one of the limiting factors of Windows NT domains that makes using the directory service on large enterprise networks connected by WAN links inconvenient. As a general rule, a large network that consists of multiple sites connected by WAN links should have at least one domain controller at each site, so that users can log on to the network locally (that is, without having to traverse a WAN connection to reach a domain controller). However, even though a BDC can authenticate users and grant them access to network resources, its copy of the SAM database is read-only. All modifications to the SAM database are made to the PDC, and an administrator at a remote location may have to connect to the PDC using a relatively slow, expensive WAN connection. With multiple master replication, an administrator can make changes to any copy of the SAM database, and the system will eventually propagate those changes to all of the other replicas automatically.

Understanding the Replication Process

Because Windows NT uses single master replication, the process of synchronizing the SAM databases on the domain controllers is relatively simple. During the replication process, SAM data travels only in one direction, from the PDC to the BDCs. By default, a domain's PDC sends pulses to the BDCs every five minutes, indicating that the BDCs should transmit a request for a database update to the PDC. The BDCs' requests are staggered, so that too many BDCs are not synchronizing at the same time. The PDC then responds to the requests with the changes that have been made to the database since each BDC's last update. This is known as a *partial synchronization*.

The PDC sends no pulses when there are no changes made to the SAM database during the time since the last BDC synchronization.

Most of the domain replication events that occur on the network are partial synchronizations. The PDC maintains a change log that lists all of the modifications to the SAM database, including password changes, the addition of new users and groups,

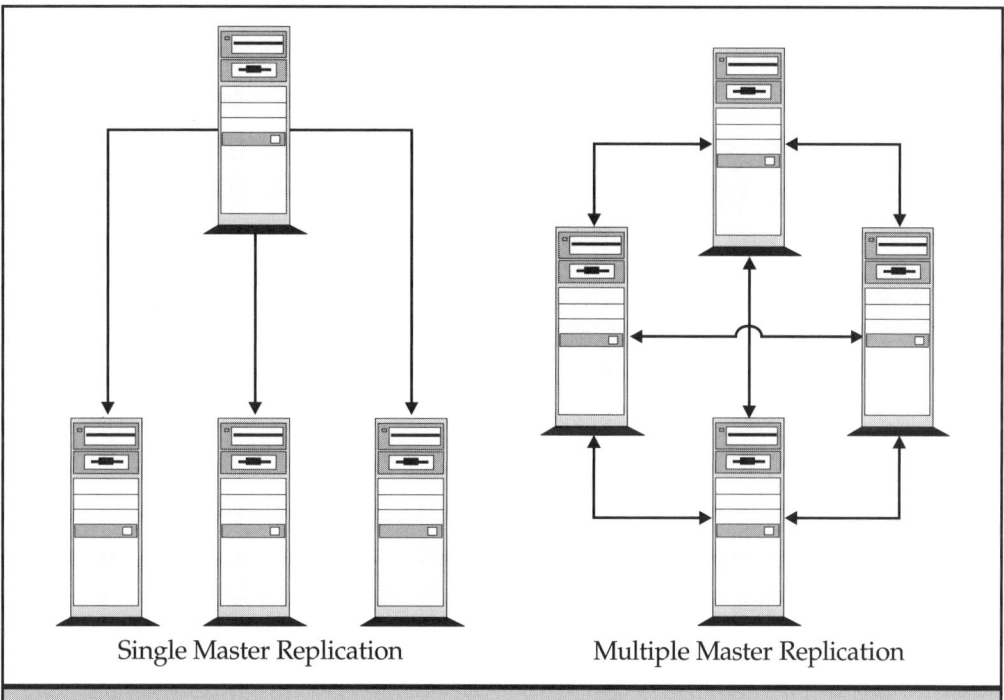

Single Master Replication   Multiple Master Replication

Figure 18-1:   *Single master replication traffic flows in one direction, from PDCs to BDCs, while in multiple master replication all domain controllers update each other.*

and the modification of existing ones. The change log has a fixed size (64KB, by default); the oldest entries are deleted as new ones are added. The update request sent by a BDC to the PDC specifies the last change that it received, and the PDC's reply includes all of the modifications made since that time.

If the BDC's request specifies a last change received that has been purged from the change log, the PDC must perform a *full synchronization* and transmit the entire SAM database to the BDC. This typically occurs when a BDC has been down for an extended period of time or when a new BDC is added to the network. Depending on the size of the network, a full synchronization can require the transmission of much more data than a partial one and should be avoided whenever possible.

Modifying Replication Parameters

The frequency of the synchronization events and the size of the change log maintained by the PDC are controlled by registry entries located in the following key:

```
HKEY_LOCAL_MACHINE\SYSTEM\CurrentControlSet\Services\Netlogon\Parameters
```

If the following entries do not already exist in the registry, you can create them to modify the default settings for the Windows NT Server.

ChangeLogSize is a REG_DWORD entry that defines the size of the PDC's change log. The log exists in memory, as well as on the hard drive as a file called Netlogon.chg in the \%SystemRoot% directory (C:\Winnt, by default). Each entry in the log is typically 32 bytes long (with some being longer, but none shorter), which enables the default 64KB log file to hold approximately 2,000 entries. However, you can modify the ChangeLogSize registry entry in order to maintain a log up to 4MB in size (by changing the hexadecimal value of the entry to 4000000). Increasing the size of the log does not degrade the performance of the system in any way, and it greatly reduces the chance that the PDC will have to perform a full synchronization because entries needed by a BDC have aged out of the log.

Note	*When modifying the value of the ChangeLogSize registry entry on the PDC, be sure to change the entry on all of the BDCs to the same value. Only the log on the PDC is actually used, but if a BDC ever has to be promoted to a PDC, it should have the same size change log.*

The Pulse REG_DWORD entry specifies the interval at which the PDC should transmit the pulses to the BDCs that trigger their synchronization requests to the PDC. By default, the value is 300 seconds (5 minutes), with possible values ranging from 60 to 172,800 seconds (48 hours). If you increase the size of the change log, you can conceivably increase the Pulse value as well, reducing the amount of network traffic generated by the replication process without forcing the PDC to perform full synchronizations. However, this also means that it will take longer for new user and group accounts to be propagated to all of the domain's BDCs.

Trust Relationships

Because an NT domain is not a hierarchical directory service, there is a point at which an NT domain can grow too large to be practically managed. Therefore, on larger networks, administrators create multiple domains. Theoretically, a single domain can support up to 26,000 users and approximately 250 groups, but you may find that splitting the network into multiple domains is advantageous for several reasons, including the following:

Delegating administration tasks By creating separate domains, you can grant support personnel administrative access to one domain while preventing them from accessing the other domains.

Creating security boundaries Domains are inherently secure from access by users outside the domain, unless administrators explicitly create trust relationships between the domains.

Improving performance   On enterprise networks connected using WAN links, separate domains can help keep network traffic within a site and help minimize the WAN traffic.

When users log on to the network from a workstation, they must log on using an account in a specific domain, which makes it possible for them to access resources in that domain, but not the resources in other domains. In order for users to access resources outside of their domain, the network administrator must create *trust relationships* between the various domains.

When one domain trusts another domain, the authentication process that grants users access to their home domain also grants them access to the other domain. If Domain A trusts Domain B, a user logging on to Domain A can access resources in Domain B. However, this does not mean that users logging on to Domain B can automatically access resources in Domain A. Trust relationships operate in one direction; for two domains to mutually trust each other, administrators must create trust relationships running in both directions.

In Windows NT (unlike in Windows 2000), trust relationships are also not transitive. If Domain A trusts Domain B, and Domain B trusts Domain C, it does not follow that Domain A trusts Domain C. Again, explicit trust relationships must exist between A and C for them to trust each other.

Even when a trust relationship exists between domains, users' access to shared resources on other domains is still subject to the same access permissions as on their local domain. To access a shared drive on a server in Domain B, a user in Domain A must still be granted permission to access that share, even when Domain B already trusts Domain A. When you set the access permissions for a share, you can select users and groups from any domain visible on the network, as shown in Figure 18-2. The domain to which the computer is currently attached is marked with an asterisk.

Creating Trust Relationships

To create trust relationships between domains, you launch the User Manager for Domains utility and select Trust Relationships from the Policies menu to display the dialog box shown in Figure 18-3. The upper box lists the trusted domains (the domains that the current domain trusts), and the bottom box lists the trusting domains (the domains that will be permitted to trust the current domain).

Note *You can also create trust relationships from the Windows NT command prompt, and perform many other domain-related maintenance tasks, using the Netdom.exe utility included in the Windows NT Server 4.0 Resource Kit. The latest version of Netdom, included in the Supplement 3 release of the Resource Kit, also enables you to create and manage the computer accounts for workstations and BDCs and reset secure channels.*

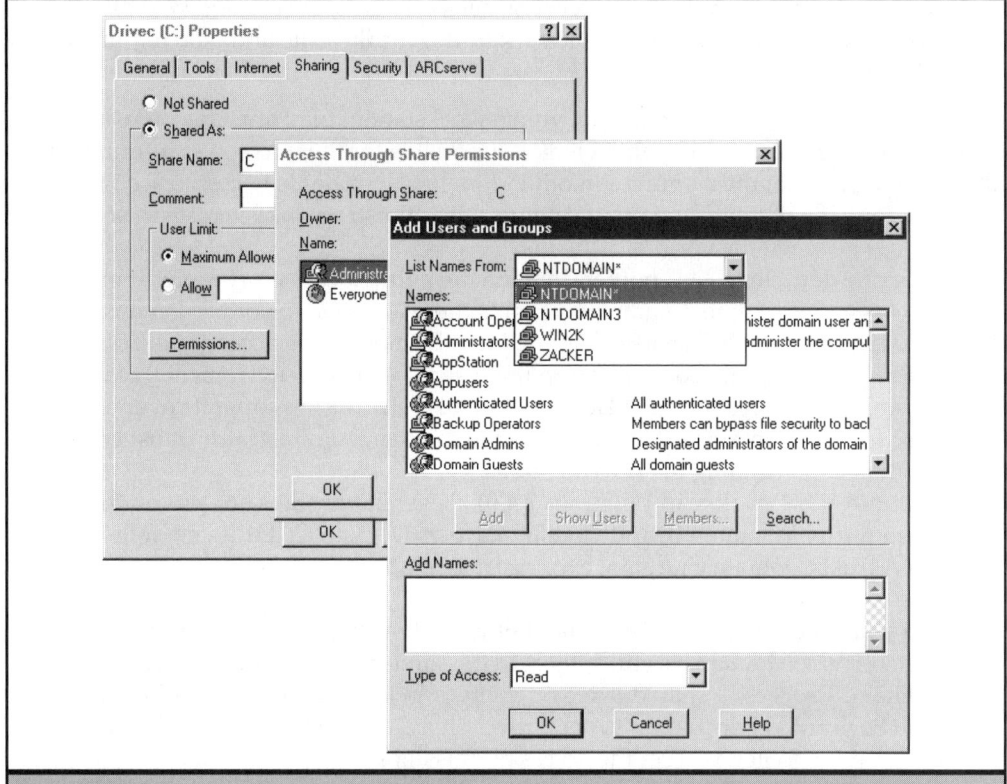

Figure 18-2: *You can grant shares access to users in any domain on the network.*

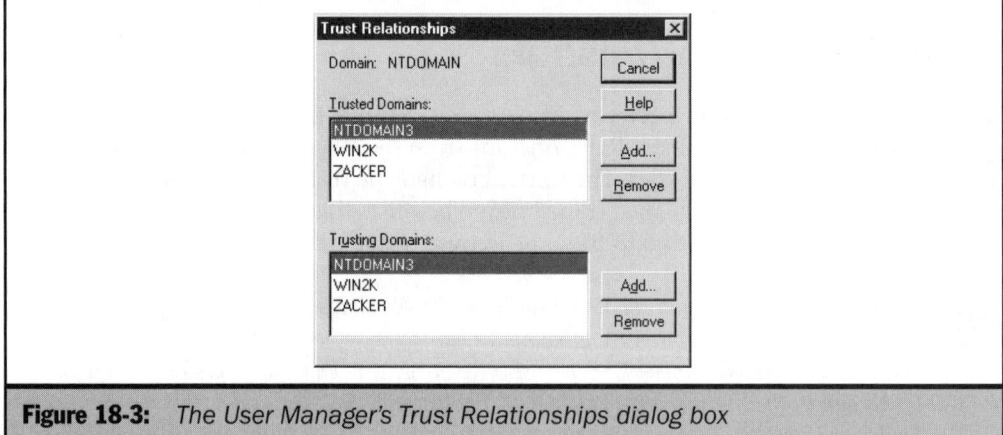

Figure 18-3: *The User Manager's Trust Relationships dialog box*

In order for a trust relationship to be established, both of the domains involved must approve. For example, when you, as the administrator of Domain A, add Domain B to the Trusting Domains list, you specify a password to be used for that trust relationship. An administrator of Domain B must then add Domain A to the Trusted Domains list in the same dialog box for their domain, and specify the same password. This establishes a one-way trust relationship in which Domain B trusts Domain A. In other words, users logging on to a machine in Domain A can access resources in Domain B. Domain B users cannot access Domain A resources unless the administrators perform the same process in the other direction.

Note

As a general rule, administrators should configure the Trusting Domain side of the relationship first, and then the Trusted Domain side. This enables the trusted domain to recognize the password and immediately establish the relationship. If the order is reversed, an error message will appear stating that the password could not yet be verified and the relationship will not be established until it is.

During the establishment of the trust relationship, the trusting domain creates a *trusted domain object* on its domain controllers, which contains the name of the trusted domain and its security identifier (SID). Each of the domain controllers in the trusting domain then creates an *LSA (Local Security Authority) secret object* containing the password for the trust relationship.

On the trusted domain, the User Manager program creates an *interdomain trust account* that also stores the password supplied by the administrator. After the trusted domain object, LSA secret object, and interdomain trust account have been created, the Netlogon service on the trusting domain's PDC begins the process of establishing an encrypted communications link called a *secure channel* between the two domains. This secure channel is authenticated by the PDC of the trusted domain, and the trust relationship is established. Once the PDCs on both domains replicate all of these elements to the BDCs, it is possible for any domain controller in the trusting domain to establish a secure channel to any domain controller in the trusted domain.

Each time a domain controller is restarted, it attempts to discover the domain controllers for all of its trusted domains, to reestablish the secure channel. The controller makes three attempts for each trusted domain, at five-second intervals, before the discovery is said to have failed. After a failure, additional discovery attempts occur every 15 minutes, or when a client attempts to access a resource in a trusted domain.

Organizing Trust Relationships

There are a number of different organizational paradigms you can use when creating multiple domains on a network and establishing trust relationships between them. On a relatively small network, you can establish trust relationships on an ad hoc basis. When users need access to a resource on another domain, you configure it to trust the domain in which those users reside. However, on larger networks, you should consider a more organized methodology that will enable you to use the domains to manage your network more efficiently. Some of the basic types of domain/trust models for an enterprise network are examined in the following sections.

The Single Domain Model

The single domain model requires no trust relationships, since only one domain is involved. This can be a viable alternative, even for a fairly large network, when the entire network is managed by a single group of administrators. Many networks begin as a single domain, and later evolve into one of the other models discussed in the following sections.

The main advantage of this model is that administration is quite simple. There are no trusts to create or maintain, defining groups is easier, and all of the user accounts are located in the one (and only) domain. The disadvantages become increasingly obvious as the network grows in size. Browsing the domain tree slows down as the number of computers and users increases and the lack of groupings provided by separate domains that define departmental or geographical boundaries becomes more of a problem.

The Complete Trust Domain Model

The complete trust domain model consists of multiple domains, each of which has a trust relationship with every other domain, running in both directions (see Figure 18-4). Unlike the master and multiple master domain models, every domain in the complete trust is a peer to all of the other domains, with each one containing its own user accounts. In many cases, this type of network model is the eventual result of the ad hoc trust establishment method described earlier, taken to the nth degree. From a user access standpoint, this arrangement is not much different from having a single, large domain, since every user can be granted access to any resource on the entire network.

As a network grows from a single domain into multiple domains, the complete trust model may be viable at first, but it can rapidly become unmanageable. Obviously, this model requires the least amount of planning, and it is best suited to an environment where political or other factors require the use of multiple domains, but users frequently require access to resources all over the network. The main drawbacks of this arrangement are the sheer number of trust relationships that must be established and the fact that user and group accounts are scattered in domains all over the network.

Since you create a separate trust relationship in each direction between every pair of domains on the network, the formula for computing the total number of trust relationships needed is $(n - 1) \times n$, where n is the number of domains on the network. Thus, a network consisting of 10 domains would require you to create 81 separate trust relationships $(10 - 1) \times 10 = 90$ to form a complete trust domain model, and, as a result of all of this effort, you achieve little more than you would with one large domain.

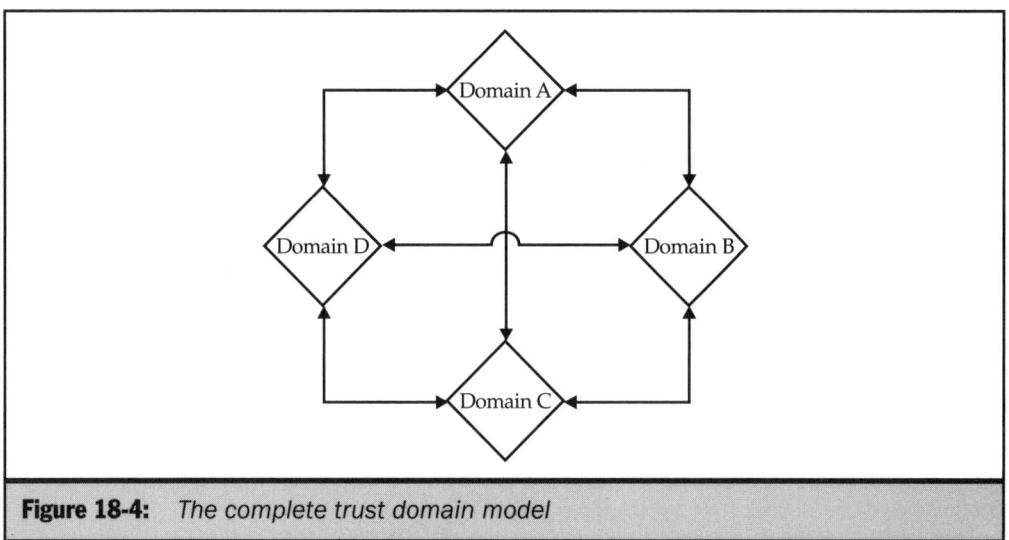

Figure 18-4:   *The complete trust domain model*

Managing permissions in this model is difficult, because there is no ready way of knowing which users are members of which domain. If you must grant a specific user rights to a share in the domain for which you are responsible, you either have to know which domain the user account resides in or go searching for it in each of the network's domains.

The Single Master Domain Model

The single master domain model uses standard NT domains in two different roles, called *master domains* and *resource domains.* The arrangement calls for a hierarchy in which there is one master domain that contains the user and group accounts for the entire network (also called the *account domain*), and a series of resource domains that contain the network's workstations, file servers, printers, and other resources, thus breaking up the network into more easily manageable units (see Figure 18-5).

In this model, all of the resource domains have a one-way relationship in which they trust the master domain. Every user logging on to the master domain, therefore, can access resources anywhere on the network (with the appropriate permissions). This eliminates the need to create trust relationships between the resource domains, and all of the user and group accounts are stored in the single master domain for easy centralized administration. Because the trust relationships run in only one direction, from the resource domains to the master domain, any user logging on to an account in one of the resource domains cannot access resources outside of that domain.

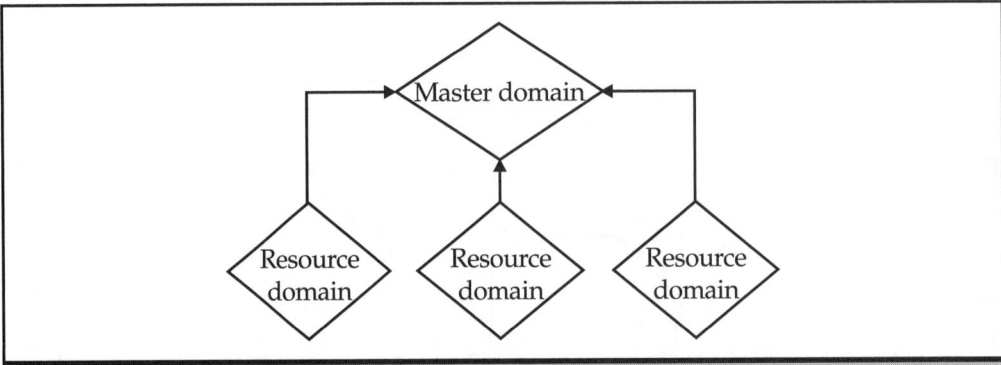

Figure 18-5: *The single master domain model; the arrows signify the direction of the trust relationships.*

Note, however, that despite the fact that all of the user and group accounts are located in the master domain, the computer accounts should be located in the individual resource domains. When you install a workstation, you specify the NetBIOS name for the computer and join it to the local resource domain, not the master domain. This enables the system to use the domain controllers for the resource domain to provide pass-through authentication during the logon process, and eliminates the need for additional BDCs in the master domain.

Note *See "Logging On to the Network," later in this chapter, for more information on pass-through authentication and the domain logon process.*

In the single master domain model, the master domain is responsible for authenticating all of the users as they log on. However, once the logon process is complete, most of a user's activity will be confined to one or another of the resource domains. This effectively splits the domain processing burden between the master and the resource domains. In addition, the master domain is ultimately responsible for network security, since all of the other domains trust it. Administration of the resource domains can, therefore, be delegated to personnel in the departments hosting the individual domains, without compromising the overall security of the network.

On the down side, however, is the fact that the entire network is reliant on the master domain for authentication services. This single master domain model does not provide the ability to support more users than a single domain, since all of the accounts are still located in the one master domain. Instead, it enables you to use multiple domains for organizational purposes and still maintain all of the user and group accounts in one place. If the network is growing on a regular basis, it is a good idea to have a plan ready for expanding it into a multiple master domain network. Otherwise, it is easy for a well-organized web of trust relationships to degenerate into the confusion of the complete trust domain model.

To further minimize the account administration required at the resource-domain level, you can create local groups in the resource domains and grant them access to the shared resources in those domains. Then, by creating global groups and making them members of the local groups, you can manage access to the resources by adding user accounts to the global groups instead of to the local ones. This makes it possible to perform all of the day-to-day account administration tasks in the master domain, without accessing resource domain accounts at all.

The Multiple Master Domain Model

The multiple master domain model is an extension of the single master domain model that is designed to support larger networks that may have individual administration teams for the organization's divisions or locations. In this model, there are at least two master domains and a number of resource domains. As in the single master model, the master domains contain the user and group accounts for the division, and the resource domains host the shared resources. In this model, however, each of the resource domains trusts all of the master domains, and the master domains all have two-way trust relationships between them (see Figure 18-6). It's also possible for trust relationships to exist between resource domains, but this should not be necessary, unless physical circumstances (such as slow WAN connections) warrant it.

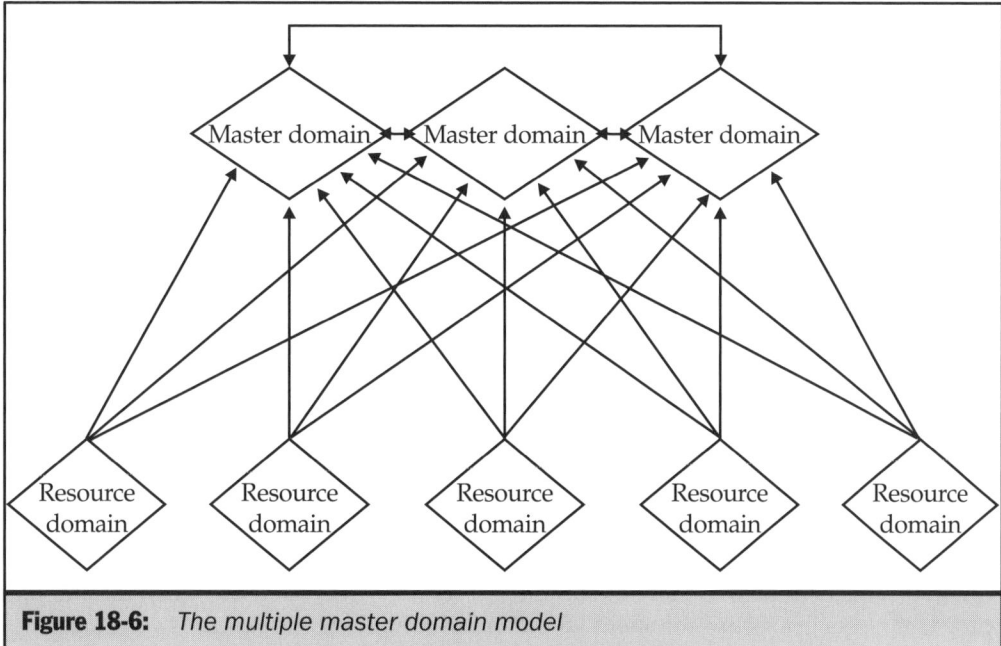

Figure 18-6: *The multiple master domain model*

In this arrangement, a user with an account in one of the master domains can access any resource on the network, because all of the resource domains trust all of the master domains. The two-way trusts between the master domains enable the users logging directly on to the master domain to access the resources in the other master domains.

The object of this model is to enable each division to maintain its own user and group accounts in its master domain, while still enabling users to access the resources in other divisions. For companies with several branch locations, for example, having a master domain at each site enables the administrators at each location to manage their own accounts.

The main advantage of the multiple master domain model is its potential for almost unlimited growth. By creating additional master domains, you can expand the network in an organized fashion or even join networks together (in the case of a company merger, for example). The number of trust relationships that administrators must create and maintain to support this model can be seen as a disadvantage, but the multiple master domain model requires fewer trusts than the complete trust domain model.

Viewing Trust Relationships

The Windows NT 4.0 Resource Kit includes a graphical utility called Domain Monitor (Dommon.exe) which you can use to examine the secure channels between the domain controllers on your network. When you run the program, you see a list of the domains found on the network and their trusted domains. Double-clicking an entry in the list produces a display like that shown in Figure 18-7, which displays the status of the domain controllers in the selected domain and the secure channels to the trusted domains.

Domain Monitor is a good way to ensure that all of the secure channels on your network are functioning properly. If the secure channel between a particular resource domain controller and a master domain controller is not functioning, local workstations may be unable to use that resource domain controller for pass-through authentication to the master domain controller. As a result, you may find that your workstations are connecting to another, more distant resource domain controller. If this other resource domain controller is located at another site, the workstation logon process can be delayed considerably, while generating needless traffic over a WAN link.

If you find there are secure channels on your network that are not functioning properly, you can reset them using the Nltest.exe command-line utility, also included with the Resource Kit. The following command resets the secure channel between the system on which you execute the command and the domain specified by the *domain* variable:

```
NLTEST /SC_RESET:domain
```

If you run the command on a workstation or a server that is not a domain controller, the program resets the secure channel to the domain containing the system's computer account. If you run Nltest.exe on a domain controller, you can reset the secure channel created by the trust relationship with another domain.

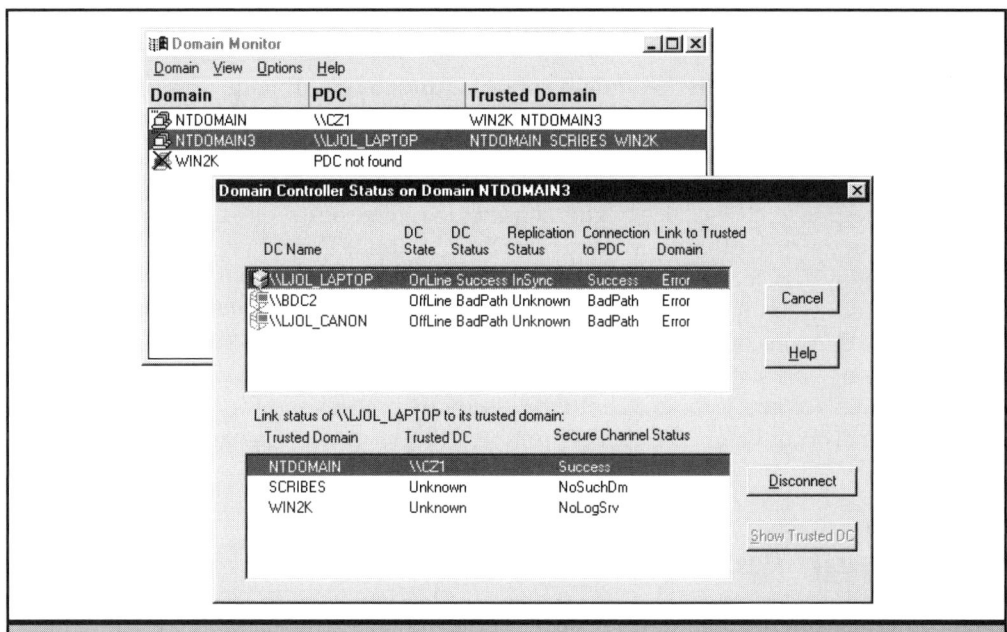

Figure 18-7: *The Domain Monitor utility displays the status of the secure channels between trusted domains.*

Note

Nltest.exe is a powerful utility that you can also use to list the domain controllers in a domain, control domain replication, and display the contents of the change log, among other things. For more information, see the documentation included with the Windows NT Server 4.0 Resource Kit.

Logging On to the Network

When you log on to a domain, the Netlogon service on the workstation performs a series of different processes, depending on how the network's domains are organized. During the boot process of a Windows NT workstation, before you see the screen prompting you to press CTRL-ALT-DEL to log on, the system attempts to locate a domain controller using a process called *discovery*. If the workstation is not configured to log on to a domain, the Netlogon service terminates and no discovery takes place. When it is logging on to a domain, the Netlogon service attempts to locate a domain controller in the domain of which it is a member, and in each of its trusted domains.

Domain Controller Discovery

The Netlogon service locates a domain controller first by sending queries to its designated WINS server for NetBIOS name entries with *1C* as the value of the 16th bit, which identifies them as resource domain controllers. The system then transmits local broadcasts for each of the entries returned by the WINS server. If those broadcasts do not elicit a response, the service begins sending logon requests, one at a time, to each of the entries, in the order that they were furnished by the WINS server.

When the system locates a domain controller for the domain in which the workstation has a computer account, the workstation establishes a secure channel to it. In the same way that the domain controllers involved in trust relationships have secure channels between them, workstations establish secure channels to the domain controller in which their computer account is located. Both the discovery process and the establishment of the secure channel occur before the user is prompted to supply a logon name and password.

Pass-Through Authentication

In the case of a single domain network, all of the domain controllers contain both the workstation's computer account and the user account that will be used to log on to the domain. However, when you build a domain infrastructure using master and resource domains, users logging on to the network can be authenticated by a domain controller in a resource domain, using a process called *pass-through authentication,* even though their user accounts are in the master domain.

On a single or multiple master domain network, the computer accounts for the workstations are located in the resource domains, while the user accounts are located in the master domain. This means that user authentication must ultimately be performed by the domain controller in the master domain, but the workstation will not establish a secure channel directly with the master domain controller.

When you log on to the network from a Windows NT workstation (or from a server that is not functioning as a domain controller), the pull-down menu for the Domain field in the Logon dialog box contains the domain in which the workstation's computer account is located, and all of the domains trusted by that domain. Since your user account is located in the master domain, you select the master domain in this dialog box, even though this is not the domain in which the computer account for the workstation is located.

However, it is the computer account that determines the domain to which the workstation will create a secure channel. Once the secure channel is created between the workstation and the resource domain (where the computer account is located), the resource domain controller passes the authentication information it has received from the workstation (that is, the username and password) to the master domain controller, using the secure channel that it has already created when establishing the trust relationship between the two domains (see Figure 18-8). The master domain controller then attempts to authenticate the user and returns the results to the workstation, via the resource domain controller. In this arrangement, it is the master domain controller that supplies the system policy file and logon script to be loaded by the workstation (if any), not the resource domain controller.

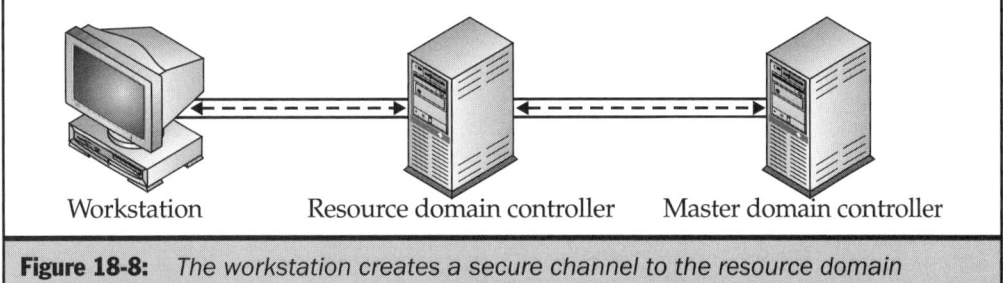

Workstation   Resource domain controller   Master domain controller

Figure 18-8:   *The workstation creates a secure channel to the resource domain controller, which then uses the existing secure channel to communicate with the master domain controller.*

The end result of this arrangement is that the workstation can log on using a local resource domain controller, instead of having to establish a secure connection to the master domain controller. In the case of a network that is split into remote sites connected by WAN links, there is no need to have a BDC for the master domain at each location just so users can authenticate locally. The resource domain controller performs this function and uses a secure channel that already exists to communicate with the master domain controller.

Selecting a Directory Service

The domain-based directory service is one of the most frequently cited shortcomings of the operating system, particularly in an enterprise network environment. While the NT directory service is quite serviceable for a small- to medium-sized network in which no more than a handful of domains are needed, it becomes increasingly unwieldy on large networks, especially those with multiple sites connected by a WAN. The multiple master domain model is really a makeshift solution for a directory service that is just not suited for large enterprise networks. For large networks that are expanding on a regular basis, or those that are in need of an overhaul to address out-of-control domain expansion, a hierarchical directory service that is designed to support large networks, such as Active Directory or Novell Directory Services, is recommended.

The
Complete
Reference

Networking

Chapter 19

Novell NetWare

Novell NetWare was one of the first commercial network operating systems designed for use on PC LANs and is certainly the oldest one still in general use today. In the early 1980s, PC networks were concerned primarily with providing basic functions, such as file and printer sharing, and NetWare has always done this very well. For many years, NetWare was the leader in a market with little commercial competition that could touch it.

As a platform for third-party application development, and particularly as a platform for Internet services, NetWare failed to keep up with its competition in the 1990s and has lost much of the market share it once enjoyed. NetWare is still used on many legacy networks, often in combination with other network operating systems such as Windows or Unix. Its deployment on new networks has reduced considerably, however, despite its obvious strength in certain areas, such as directory services.

Unlike Windows and Unix, NetWare is a dedicated client/server operating system, meaning that NetWare servers contain no client capabilities, and NetWare clients communicate only with servers, not with each other (unless another NOS client is also installed). A NetWare server runs a proprietary OS that launches from a DOS prompt, but runs independently from DOS once loaded. NetWare clients are usually Windows or DOS systems, with a client package installed that provides server connectivity. Microsoft supplies a basic NetWare client with the Windows operating systems, but Novell's own client software provides more complete NetWare functionality. Additional software packages provide client connectivity for other operating systems, such as Macintosh and Unix.

The NetWare server OS provides the ability to run applications developed by Novell and third parties, using executable program files called *NetWare Loadable Modules (NLMs)*. The server console uses a character-based interface, with some utilities driven by ASCII-character menus, as shown in Figure 19-1. The latest version, NetWare 5.1, includes a Java-based console called ConsoleOne that for the first time provides NetWare servers with a graphical interface.

NetWare's Role in the Enterprise

NetWare is best known for providing users with the basic file and print services that are the core of LAN functionality. To this day, these services are what NetWare does best, better than operating systems like Windows 2000 and NT that have taken its place as the market leader. The traditional NetWare file system is scaleable and secure, and the printing subsystem is simple to use and administer. Current versions of NetWare include updated versions of these elements, such as Novell Storage Services (NSS) and the Novell Distributed Print System (NDPS), but support for the older versions remains in place, because many sites continue to use NetWare as part of an "if it ain't broke, don't fix it" philosophy.

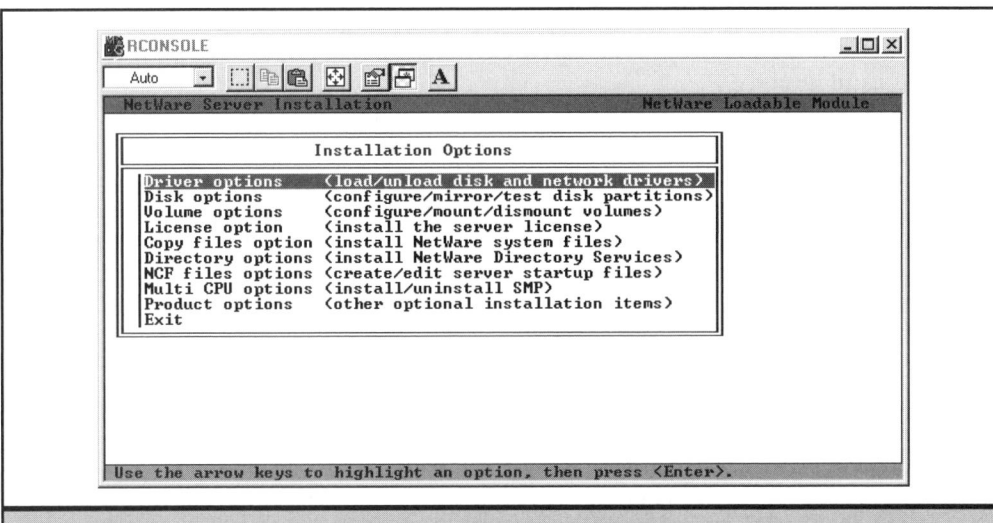

Figure 19-1: *The NetWare Install.nlm utility uses a character-based menu interface.*

As a platform for application development, NetWare is necessarily more problematic than Windows, which has made its name as an application server platform. Network software development for Windows operating systems uses the same types of executables on both servers and clients, whereas NetWare servers require the use of the proprietary NLM executable. In fact, its tendency to rely on proprietary elements is one of the primary causes of NetWare's decline in the network marketplace. The server NLMs and the IPX protocols are both Novell developments that it has retained long after the rest of the industry has moved to other, more publicly documented alternatives.

The single most redeeming factor in the NetWare camp today is Novell Directory Services (NDS), a hierarchical, X.500-based directory service that was originally deployed in 1993 and is now far more mature than the directory services available for Windows or Unix. NDS is an extremely flexible repository for all sorts of information about network elements, including hardware and software, as well as users, groups, and other incorporeal entities. In the years since its introduction, NDS has been tested and deployed on more networks than any other directory service, and there are now many third-party products that utilize its capabilities. As good as Microsoft's Active Directory is, Novell has had a seven-year head start in this respect.

Note   *For more information on NDS, see Chapter 20.*

As an Internet platform, NetWare was the last of the major network operating systems to adopt the TCP/IP protocols. It was only with the release of NetWare version 5 that the operating system was able to use TCP/IP for its core file and print services. Previous versions supported the protocols and included Internet applications, such as Web and FTP servers, but client/server LAN communications relied on the proprietary IPX protocols until recently.

The NetWare product, as it stands now, contains all the features that are expected in an enterprise NOS, as well as some exciting new ideas, but the continued viability of Novell in the marketplace is definitely in question. An organization that is already committed to NetWare might well benefit from an upgrade to the latest version, but large-scale deployment of the operating system on a new network or migration from Windows or Unix to NetWare would not, in most cases, be a wise decision.

NetWare Versions

Novell began producing networking products in the early 1980s, and was involved in PC networking almost from the introduction of the first IBM PC. The product has developed steadily over the years, culminating in the current NetWare 5.1 release. However, many NetWare shops continue to rely on older versions of the product. Table 19-1 lists the versions that are currently available and compares their features and capabilities. The following sections examine in more detail the differences between the major releases of the NetWare product.

	NetWare 4.2	NetWare 5.1
System Requirements	Intel 80386 or better processor; 16MB RAM; 55MB hard disk space (minimum)	Intel Pentium II or better processor; 128MB RAM; 1.3GB hard disk space
Directory Service	NDS	NDS
File System	NetWare file system	NetWare file system; Novell Storage Services (NSS)
Maximum Number of Connections Supported per Server	Thousands	Thousands
Maximum Number of Volumes per Server	64	Unlimited

Table 19-1: *Comparison of NetWare Versions and Features*

	NetWare 4.2	**NetWare 5.1**
Maximum Disk Storage Capacity	32TB	Unlimited
Maximum File Size	4GB	8TB
Maximum Volume Size	32GB	8TB on 32-bit systems; 8 exabytes (EB) on 64-bit systems
Maximum Concurrent Open Files per Server	100,000	1,000,000
Maximum Directory Entries per Volume	16,000,000	264
Maximum Number of Server Processors Supported	8	32
Core Protocols Supported	IPX only; IP(using NetWare/IP encapsulation only)	IPX and/or IP
Internet Services Included	Netscape FastTrack Server; Netscape Communicator; DHCP Server (v2.10); DNS Server (v2); IPX/IP Gateway (free download); FTP Services; CGI Scripting and NetBasic support; Multiprotocol WAN Router	NetWare Enterprise Web Server; NetWare FTP Server; NetWare News Server; NetWare Web Search Server; NetWare Multimedia Server; Novell ScriptPages; Novell Certificate Server; WAN Traffic Manager; DHCP Server; DNS Server; LDAP 3 Service; Network Address Translation
Retail Price (as of Summer 2001)	Server plus 5-Connections: $1,195; 25-Connection Additive License: $3,155	Server plus 5-Connections: $995; 25-Connection Additive License: $2,750

Table 19-1:   *Comparison of NetWare Versions and Features (continued)*

NetWare 2.x

The NetWare 2.x operating systems culminated in version 2.2, which is no longer sold or supported by Novell. Any network running a 2.x version of NetWare must purchase a full version of NetWare 3.12, 4.2, or 5 to upgrade, as there is no longer an upgrade path from this early product.

NetWare 2.x was designed to support PCs built around Intel processors up to the 80286 and is subject to limitations that seem absurd by today's standards, but which represented the state of the art back then. A NetWare 2.x server could have as much as 12MB of RAM installed and could support up to 2GB of hard disk space, which had to be divided into volumes no larger than 256MB.

Designed primarily to provide file and print services, NetWare 2.x included limited support for server applications in the form of *value-added processes (VAPs)*. VAPs enabled servers to run relatively simple applications, such as server backups. NetWare 2.x was also the last version of the operating system that enabled the server to be installed in nondedicated mode, meaning that the server computer could be used as a DOS workstation while performing server tasks in the background. This arrangement resulted in unsatisfactory performance of both workstation and server, however, and all subsequent versions of NetWare required a dedicated server.

NetWare 3.x

In 1989, Novell released NetWare 386, version 3, which was a major upgrade of the OS designed to take advantage of the (then) new Intel 80386 processor. Several maintenance releases followed, culminating in the NetWare 3.12 release in September 1993. Version 3.12 was planned to be the ultimate version of the NetWare 3.x generation, but Novell later decided that a 3.2 version was warranted, primarily to make the operating system Y2K compliant. NetWare 3.x requires a server with at least a 386 processor, 6MB of RAM, and 30MB of disk space. Development of the 3.x product has ceased as of February, 2001, but it will continue to be supported by Novell until February 2002.

Unlike version 2.x, NetWare 3.x is a 32-bit OS that introduced many of the elements that have come to be associated with NetWare, including multitasking, support for large amounts of RAM (up to 4GB) and disk space (up to 32TB), and NetWare Loadable Modules (NLMs), the application development platform that is still used in the latest NetWare versions. In the course of its release history, NetWare 3.x has added support for industry-standard technologies as they arrived, including CD-ROM drives and the TCP/IP protocols. Version 3.2 adds a Windows-based interface for server administration, but the fundamental structure of the operating system has remained intact throughout its history.

Designed for use on LANs, NetWare 3.x is not intended to be an enterprise NOS (as NetWare 4.x and 5.x are). The OS is organized around a nonhierarchical database called the *bindery*, which consists of *objects* representing users and groups, as well as hardware and software entities found on the network. As in most directory services, the objects are database records that contain attributes called *properties,* which can have one or more *values.* For example, an object representing a user will have a property that consists of a list of the groups of which the user is a member.

Every server on a NetWare network contains its own bindery, and there is no communication between the binderies on different servers. This is one of the reasons why NetWare 3.*x* is not a suitable OS for large networks that must rely on centralized management techniques. A new user who requires access to five different servers must have a separate bindery account on each server.

For its intended purposes, though, NetWare 3.*x* is an excellent solution that has continued to retain a staunch following throughout the years. Many of its devotees have shunned later releases, such as NetWare 4.*x* and 5.*x*, in favor of what they have determined to be the right tool for the job. The product has now reached the stage in its life, however, where users should consider upgrading, or risk being stuck with an orphaned product with no upgrade path.

NetWare 4.*x*

The next generation of NetWare was introduced in April 1993 with version 4. Several maintenance versions ensued during the next three years, and NetWare 4.11 was released in October 1996. This version persisted until early 1999, when version 4.2 was released to provide Y2K revisions and add updated versions of the intraNetWare utilities. Like version 3.*x*, NetWare 4.*x* will run on any server with at least a 386 processor, but requires 16MB or more of RAM and at least 55MB of disk space.

The primary innovation in NetWare 4 was NetWare Directory Services (NDS), which later was renamed Novell Directory Services to reflect its newly implemented cross-platform compatibility. NDS is a hierarchical, partitioned, replicated directory service that is intended to support enterprise networks of virtually any size. Like the bindery, NDS consists of objects, properties, and values, but the objects here are organized in a tree-like display that can represent a very large organization. An enterprise typically maintains a single NDS database that is split among several servers and replicated to provide fault tolerance and load balancing.

It took Novell several years to refine and mature NDS to the point that it could be considered reliable on mission-critical networks, and several years more for third-party developers to take advantage of its services and write applications that store information in the NDS database.

NetWare 4.*x* also introduced server administration tools for the Windows platform for the first time. The NetWare Administrator application provides GUI access to the NDS tree and the NetWare file system, although a similar utility for DOS, called NETADMIN.EXE, is included as well.

 *For more information about NetWare Administrator and NDS, see Chapter 20.*

intraNetWare

The intraNetWare release (formerly known as IntranetWare—go figure) was a bundle that combined NetWare 4.11 with a collection of Internet/intranet utilities at the same price as NetWare 4.11 alone. In addition to NetWare 4.11 itself, the package contained the following:

- Novell Web Server 2.51
- Novell FTP Server
- Novell DHCP Server
- Netscape Navigator
- NetWare Internet Access Server
- Novell Multi-Protocol Route 3.1
- NetWare/IP
- Novell IPX/IP Gateway

NetWare 4.2

Instead of continuing the intraNetWare brand, Novell elected to release its next operating system upgrade simply as NetWare 4.2. However, this release includes the latest versions of all the bundled intraNetWare components, making it the functional equivalent of intraNetWare, using the NetWare name. NetWare 4.2 is the ultimate release of the NetWare 4.x generation, and is intended for use by organizations that are not yet motivated to upgrade to NetWare 5, but that want the latest revisions of the bundled utilities, as well as the assurance that the OS is Y2K compliant.

NetWare 4.2 includes the following components:

Netscape Fast-Track Server   Beginning in NetWare 4.2 and in NetWare 5, Novell began bundling Netscape's entry-level Web server with NetWare, instead of its own Web server. A license for the Netscape Enterprise Web server (available as a free download) is also included with the product.

Novell DHCP Server v2.10   Dynamically allocates TCP/IP configuration settings to network clients.

Novell DNS Server v2.0   Enables client systems to resolve DNS names into the IP addresses required for TCP/IP communications.

Novell Multi-Protocol WAN Router   Enables a NetWare server to function as a WAN router using multiple transport and routing protocols on a number of industry-standard WAN communications technologies.

Novell FTP Server   Enables a NetWare server to function as an FTP server, providing TCP/IP file transfer services to local and remote clients.

Netscape Communicator   Includes a license (equivalent to the NetWare user license) for the Netscape Web browser and e-mail client package.

Symmetrical multiprocessing (SMP) support   Supports up to eight server processors using a separate multiprocessing NLM.

Oracle8   Includes a five-user version of the Oracle database server.

Perl v5.1 and NetBasic v6   Provide scripting capabilities for Web site development.

ZENworks Starter Pack   Subset of the full ZENworks product, that provides workstation application distribution and centralized management services.

NetWare 5.1

NetWare 5.1 is the current release of the operating system and represents a major step forward in many ways. Most importantly, 5.*x* is the first generation of NetWare that can use TCP/IP as its native protocols. Previous versions supported TCP/IP, but the core file and print services still relied on Novell's own IPX protocols. NetWare 5.*x* can eliminate IPX from the network completely, which is something that users have been clamoring for for years.

   NetWare 5.1 also expands the role of NDS in the enterprise, by including DNS and DHCP servers that store their data in the NDS database. The DNS server also supports the dynamic DNS standard (RFC 2136) that enables DHCP-assigned IP addresses to be automatically reflected in DNS resource records. This new version of NetWare also reflects Novell's commitment to the Java language by including, for the first time, a Java-based GUI installation program and a network administration utility called ConsoleOne.

NetWare Installation

NetWare versions 3.*x* and higher are launched from a single DOS executable file called Server.exe. Installing the OS requires a DOS partition on the server that is large enough to hold this file and a collection of drivers and other support files used during the OS loading process. For troubleshooting purposes, it is also a good idea to leave enough free space on the DOS partition to hold a core dump of the OS, which means having as much free space as there is memory in the computer. A *core dump* is an exact copy of the server's memory content at the time that a fault (called an *abend,* or abnormal end) occurs. In the event of a serious problem, Novell technicians can determine from the core dump exactly what was happening at the time the error occurred.

Many people place the blame for a failure to resolve a problem on the software manufacturer's technical support staff when, in fact, the real cause lies with the user who fails to provide the information needed to determine the cause of the problem. The information in a NetWare core dump is meaningless to all but a small percentage of network administrators, but it is an invaluable diagnostic tool for the Novell engineers.

The Server.exe file alone loads the NetWare OS, and the rest of the installation process consists of the selection and configuration of the drivers to support the server's storage subsystem, network interface cards (NICs), and protocols. The server console provides a command-line interface through which you can interactively perform these tasks using NetWare's internal console commands. For example, all NetWare drivers and executables must be loaded into memory using the LOAD command. In addition to the server console command line, various menu-driven utilities enable you to configure the properties of the server, its drivers, and applications. These utilities, such as MONITOR and INETCFG, are not graphical; they use ASCII characters only to create a cursor-driven menu system known as the *C-worthy interface* (see Figure 19-2).

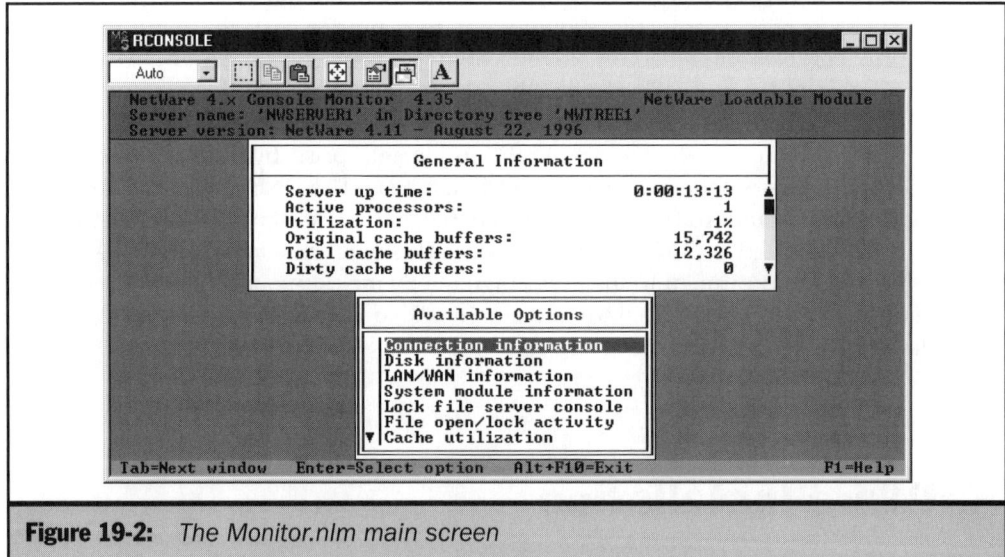

Figure 19-2: *The Monitor.nlm main screen*

Disk Drivers

Until NetWare version 4.11 was released, support for disk drives and other storage devices in NetWare was provided by separate, monolithic drivers with a .dsk extension. Development of the DSK drivers ceased in January 1997, however. NetWare 4.11 introduced the NetWare Peripheral Architecture (NPA) that consists of modular drivers called HAMs (host adapter modules) and CDMs (custom device modules). During the installation process, you install a HAM for the host adapter installed in the server and install individual CDMs for each of the devices connected to that host adapter.

For example, when the installation program detects a SCSI adapter in the system, it loads a HAM driver such as Scsi154x.ham. Loading this driver *spawns* (that is, autoloads) the main NPA program: Nwpa.nlm. Once the HAM is loaded, the system loads CDMs for each of the storage devices connected to the adapter, such as Scsihd.cdm for hard disk drives and Scsicd.cdm for CD-ROMs. If you are running NetWare 4.11 or higher with the old-style DSK drivers, it is recommended that you switch to NPA if you plan to upgrade the hardware in your server.

Note   *The NetWare installation program encounters a conflict whenever it attempts to install OS support for the computer's CD-ROM drive. This is because you must load DOS CD-ROM drivers in order to install NetWare from the distribution CD-ROM disk. In most cases, the installation program will generate an error while trying to load the NetWare CD-ROM drivers because the DOS drivers are already addressing the device. Be sure to remove the commands loading the DOS drivers from the computer's boot device once the installation process is completed, and the NetWare drivers for the CD-ROM should load without a problem the next time you start NetWare.*

NIC and Protocol Drivers

In order for a server to be able to communicate on the network, you must install drivers for the network interface card (or cards) in the computer and for the protocols the server will use to communicate with other systems. The NIC drivers have a LAN extension and use parameters on the LOAD command line to specify the hardware resources that the card requires. When you have multiple NICs in a NetWare server, the operating system routes traffic between the different networks by default.

In addition to the NIC drivers that address the network hardware, you must also install drivers for the protocols that you will run on the network. With the exception of NetWare 5.*x*, all NetWare versions require Novell's proprietary IPX protocols to communicate with clients. IPX is a connectionless datagram protocol that carries the messages generated by several upper-layer protocols, including SPX and the NetWare

Core Protocol. These are the protocols that clients use to access files and printers on servers and perform other standard NetWare functions.

Note *For more information on the IPX protocol suite, see Chapter 14.*

NetWare 3.*x* and 4.*x* support other protocols, such as TCP/IP, but not for NetWare client/server communications. If, for example, you run a Web server or other Internet service on a NetWare server, you must install the TCP/IP driver in order for Web browsers or other clients to communicate with the server.

You load protocols in the same manner as other drivers, using the LOAD command with appropriate parameters. To use a loaded protocol, you must bind it to one or all of the NIC drivers installed in the system. In the case of a server with multiple NICs, the use of individual bindings lets you control which protocols are routed between the connected networks. To bind a protocol to a NIC driver, you use the BIND console command to associate a particular protocol driver with a particular NIC driver. In the case of TCP/IP, the BIND command contains the IP address and subnet mask for the NIC (since these will be different for each interface bound to the TCP/IP protocol driver).

Building NCF Files

As on a DOS system, a NetWare server runs batch files when it starts, to configure the OS and load drivers and applications. Instead of using Config.sys and Autoexec.bat, however, NetWare servers use two files called Startup.ncf and Autoexec.ncf. An NCF (NetWare Command File) is simply a text file containing a series of console commands with parameters, just like those that you can issue from the server command prompt. Startup.ncf is stored on the server's DOS partition, because it contains the commands that load the disk drivers. These drivers must load before the server can mount its SYS volume, which contains the other batch file, Autoexec.ncf.

One of the functions of the installation program is to detect the hardware in the computer and build these batch files with commands appropriate to the hardware and to the environment specified by the installer. As NetWare progressed through its various versions, the installation program became increasingly sophisticated. Early 3.*x* versions did virtually no hardware detection and left much of the system configuration up to the installer. By the time version 4.11 was released, the installation process was much easier, because the program was capable of detecting most of the common hardware in modern systems and contained default values for most of the configuration parameters that worked on a majority of systems.

The installation program also enables you to create partitions and volumes on the server's hard disk drives, and then it creates the directory service (either bindery or NDS) on the default SYS volume. Once the installation program has completed all of these tasks, the server restarts and executes all of the commands in the NCF files. At this point, the server is functional and ready to accept connections from clients.

Once the NetWare server is installed, you can modify its startup configuration by editing the Startup.ncf and Autoexec.ncf files directly, using a text editor such as the Edit.nlm program included with NetWare, or by using NetWare's Install.nlm utility and the server console. Alternatively, NetWare versions 4.*x* and higher include an Internetworking Configuration utility (Inetcfg.nlm) that enables you to create more complex multinetwork configurations using WAN connections and routing protocols. The first time that you run Inetcfg.nlm on the server console, it removes the commands from Autoexec.ncf that load and bind the NIC and protocol drivers, and imports them into its databases. From this point on, you must use Inetcfg to modify your server's networking configuration. Inetcfg provides a menu-driven interface that makes it easier to reconfigure your server and explore NetWare's WAN and routing capabilities (see Figure 19-3).

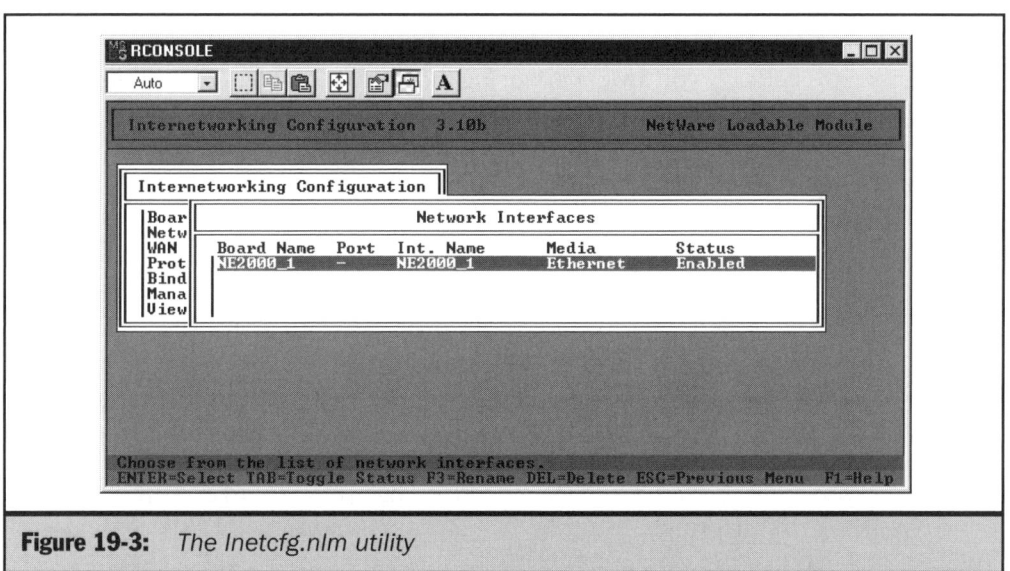

Figure 19-3:   *The Inetcfg.nlm utility*

NetWare Updates

As with most other operating systems, Novell issues patches between major releases to correct problems and sometimes to provide new features. The original model that Novell used for providing these fixes was to release each patch individually, sometimes resulting in dozens of individual patches that administrators had to download, evaluate, and install. This method is not only inconvenient; it can also result in problems due to incompatible modules. The more fixes there are to install, the greater the number of combinations that may be found on a particular server. If a server has only some of the patch releases installed, it may react differently from a server running all of the patches or a different combination of patches.

This is the model that was used through NetWare 3.2, although some of the fixes have been bundled into packages that simplify the update process. For NetWare 4.*x* and 5.*x*, Novell adopted the method used by Microsoft for its Windows (and other product) updates, releasing what it calls Support Packs. Like Microsoft Service Packs, Support Packs are numbered releases containing all of the patches recommended for the OS at that time. The patches are all installed at once, simplifying the process for the administrator, and have all been tested together, simplifying the development process for Novell. Support Pack releases are cumulative, meaning that each release includes all of the material in the previous releases.

Support Pack releases are intended for all servers running that NetWare version, but there are other patch releases intended to address only specific problems. In most cases, it is not a good idea to install every patch that comes along; just because it's newer doesn't mean it's better. If you are experiencing the specific problem that a patch addresses, install it, but otherwise your best bet is to leave well enough alone.

The NetWare Storage Subsystem

Unlike peer-to-peer network operating systems such as Windows, NetWare only shares the files and directories on designated servers, not on client systems. Because the server is largely dedicated to tasks like this, instead of running a GUI and user applications, the file-sharing performance is often better on a NetWare server than on a comparably equipped Windows machine.

NetWare uses a proprietary file system that can coexist on a hard disk drive with other types of partition formats, like the DOS FAT and Windows NTFS file systems, but it uses no part of these other formats itself. You do not use any of the standard tools (such as FDISK and FORMAT) to create and manage NetWare partitions, nor can you use diagnostic and repair tools like Microsoft's ScanDisk and Norton Disk Doctor on them.

The NetWare file system consists of *partitions,* of which there can be no more than one on a hard disk drive, and *volumes,* into which the partition is divided. A NetWare partition uses a standard partition table entry in the master boot record of a hard disk drive, just as FAT and NTFS partitions do. You create NetWare partitions on drives using the NetWare Install.nlm utility at the server console. When you install NetWare on a server, you must first have a standard FAT partition on the computer's primary drive to boot the system. The area of the disk that will be used for the NetWare partition should be left as free space.

During the installation process, the NetWare installation program will create a partition on the drive. On that partition, you must create at least one volume, called SYS, which is where the operating system files will be located. You should avoid storing your user data on this volume, so it doesn't have to be enormous, but remember also that this is where NetWare stores the NDS database (even though it doesn't appear as visible files), so be sure to leave enough space for its potential growth.

You can then use any remaining space in the partition to create up to seven additional *segments*, for a total of eight in any one partition. Each segment can be a self-contained volume or part of a volume that spans several segments, either in the same or different partitions. If you have other hard disk drives in the server, you can create partitions and volumes on them in the same way. While it is possible to create a single volume using segments in partitions on separate drives, be aware that if one of the drives fails, you will lose the entire volume. NetWare also includes the ability to create mirrored volumes on separate partitions, which provides fault tolerance at the expense of disk space.

Note *If you choose to create volumes that are composed of segments on different disk drives or mirrored volumes, be sure to use drives with comparable capabilities. If one segment of a volume or one mirror image is hosted by a drive that is demonstrably slower than the others, the performance of the entire volume will be degraded by the slower device.*

Once you have created a volume, you must mount it in before server processes or clients on the network can access it. A server's Autoexec.ncf file typically includes the MOUNT ALL command, which takes care of this task automatically each time the server starts. However, you can manually mount and dismount volumes at any time by using the MOUNT and DISMOUNT commands at the server console prompt.

Disk Allocation Blocks

When you create a volume, you must select the size for the blocks it will use: 4, 8, 16, 32, or 64 KB. The volume is divided into *disk allocation blocks* (much like a FAT drive). A disk allocation block was originally the smallest unit of disk space that the server was able to allocate when storing a file. For example, if you specify a 4KB block size, a 9KB file would require three blocks, or 12KB, with the remaining 3KB going to waste. NetWare 4.*x* introduced a feature called *block suballocation*, which mitigated some of this waste.

Unlike FAT drives, which automatically determine the block size based on the size of the volume, NetWare enables you to specify the block size for each volume you create. Selecting the appropriate block size can be an important element of building a high-performance storage subsystem. To use block sizes most efficiently, you have to consider the type of data that you intend to store on the server, and then organize your volumes accordingly.

If, for example, you intend to store large database files on a server, the best course of action is to dedicate a volume to those large files and use a larger block size for it. This will increase file access efficiency in several ways, such as the following:

Fewer blocks read   Using smaller blocks means a larger number of disk read operations are required to access a file. Using larger blocks reduces the mechanical overhead required for the hard disk drive to access the file.

Less server memory required   Each block requires an entry in the FAT for that drive. Larger blocks mean that fewer entries are required for a given file and that less memory is needed to read them.

Faster read-aheads   NetWare's file system attempts to anticipate users' needs when they request a file, by reading the subsequent files on the disk into memory before they are actually requested. Larger blocks enable the server to read more of this data into memory faster than small blocks will.

The drawback to a large block size is that if you store small files on the volume, the amount of wasted space will increase. To use the previous example, storing a 9KB file on a volume with a 64KB block size will waste 55KB of space. Multiply this by a large number of small files, and a great deal of the volume's capacity can end up being wasted.

By default, NetWare 3.*x* sets a block size of 4KB on all new volumes. NetWare versions 4.*x* and higher select a block size based on the volume's size, much like the FAT file system. For all volumes larger than 500MB, the default block size is 64KB, because the NetWare 4.*x* block suballocation feature enables the server to allocate space in increments smaller than the block size. You can select any valid block size for your volumes as you create them, even different sizes for volumes on the same hard disk. However, you cannot change the block size once the volume has been created without destroying the volume (erasing the data) and recreating it.

DETs and FATs

The *directory entry table (DET)* is where NetWare volumes store all of the information about the files stored on the volume, except for the file data itself. The directory structure you see when you look at a NetWare volume in Windows Explorer or another utility is not actually reflected in the storage of the data on the drive. The directory tree hierarchy is actually a virtual construction that exists physically only as information in the DET.

Each NetWare volume contains two copies of its DET, which is composed of 4KB blocks (no matter what block size you choose for the volume). Every file and directory on the volume has an entry in the DET that contains the following information:

- Whether the element is a file or directory

- The name of the file or directory

- The owner, attributes, and dates (created, last accessed, and last modified) of the file or directory

- The name of the file or directory's parent directory

- If a file, the location of the FAT entry for the first data segment in that file

- The location of name space information associated with the file or directory

- The trustee list for the file or directory

NetWare servers use the DET for all file management functions that do not access the actual file data. For example, when you view the contents of a directory, the server is actually scanning the DET for all entries specifying that directory as its parent.

NetWare volumes use file allocation tables (FATs) to keep track of the blocks used to store a particular file, just like DOS volumes do. Since the blocks containing the data for a single file are usually not stored contiguously on the disk, the FAT maintains the record of which blocks contain the data for that file. There is a numbered entry in the FAT for each block in the volume, and the DET specifies the number of the block containing the first data segment for each file. The first FAT entry for a file contains a reference to the FAT entry for the second data segment, the second entry contains a reference to the entry for the third segment, and so on, until the *FAT chain* reaches the entry representing the file's last data segment.

Name Spaces

NetWare uses the standard DOS 8.3 file and directory naming convention by default, but it also supports other file systems using *name spaces*, which take the form of additional DET entries that contain further information about the file or directory. For example, to support the long file and directory names created by Windows systems, you must install long name space support by loading a module called Long.nam at the server console. Since a standard DET entry cannot hold a file or directory name that is 255 characters long, the server uses additional entries to store the long name and includes references to the additional entries in the file's original DET entry.

When you load a name space module for the first time, you must execute the ADD NAME SPACE command at the server console prompt with the name space to be added and the volume to which it should be added, as in the following example:

```
ADD NAME SPACE LONG TO SYS
```

NetWare includes name space modules (all of which have a .nam extension) that support the following file systems:

- Windows VFAT
- Macintosh
- OS/2 HPFS
- NFS
- FTAM

Long file and directory names are only one feature that can be provided by name space modules. The capabilities that an individual name space adds to NetWare volumes depends on the file system that the name space supports.

While it may be tempting to add all of the available name spaces to your volumes, to provide maximum storage flexibility, you should avoid this practice. Each name space you install on a volume adds an entry to the DET for each file and directory on the volume. One name space, therefore, doubles the size of the DET. Adding many name spaces drastically increases the size of the table, causing it to occupy more disk space, and reducing the number of DET records cached in server memory.

NetWare servers store the most recently used DET entries in a memory cache for rapid access should they be needed again. Adding a name space cuts the number of cached files and directories in half, since two DET entries are required for each file or directory. Fewer entries in the cache means the server is more likely to have to access the DET information on the volume, instead of finding it in the cache. Installing too many name spaces on a volume can palpably degrade the performance of the volume. Therefore, it is a good idea to only install the name spaces you actually need on a NetWare volume.

In addition, it is a good idea to avoid using multiple name spaces on one volume by designating specific volumes for storage of certain types of data. For example, allocating one volume for Macintosh file storage is a better solution than adding the MAC name space to all of your volumes, in case someone wants to store Macintosh files there.

It is also possible to adjust the amount of memory that a server uses for caching DET entries. The SET MINIMUM DIRECTORY CACHE BUFFERS command specifies the number of buffers that are automatically allocated for DET caching when the server is started. The default value for this parameter is 20, but you can set it to any value from 10 to 8000. As the server runs, it allocates additional buffers to the DET cache as needed, up to the maximum specified by the SET MAXIMUM DIRECTORY CACHE BUFFERS command, which defaults to 500. You can set this parameter to any value from 20 to 20000.

Assuming that you have sufficient memory in the server, you can increase the maximum in cases where you must install name spaces, in order to maintain (or perhaps improve) file system performance. Each 4KB memory buffer that you add to the cache can hold eight DET entries. Thus, to double the size both of the initial cache and the maximum cache, you would issue the following commands from the server console prompt (or add them to the Autoexec.ncf file):

```
SET MINIMUM DIRECTORY CACHE BUFFERS = 40
SET MAXIMUM DIRECTORY CACHE BUFFERS = 1000
```

File System Improvements

The basic NetWare file system as described in the preceding sections persists to this day, but the NetWare 4 release introduced several new features. For example, NetWare 4.*x* and 5.*x* volumes can optionally use on-the-fly compression to increase their disk capacity. At a specified time each day or night, the server compresses all of the data on a volume in place. The next time a user or application requests access to that file, the server automatically decompresses the file before delivering it to the requestor. The file then remains in the decompressed state, ready for further use, until the next compression cycle. This type of compression does have an effect on the performance of the volume, as well as the server's processor utilization, but in some cases, the additional storage space provided is worth the expenditure in other areas.

Another feature of the storage system in NetWare 4.*x* and 5.*x* is *block suballocation,* a mechanism that addresses the problem of wasted storage space caused by blocks that are not completely filled. Using small 4KB blocks minimizes the waste, but the ever-increasing capacity of today's hard disk drives makes the use of small blocks increasingly impractical. By dividing blocks into smaller, 512-byte units, it is possible to store files more efficiently, with a minimum of waste.

When a volume's block suballocation feature is activated, the server creates a number of *suballocation reserved blocks (SRBs).* These are blocks that are divided into 512-byte segments and allocated for the storage of file fragments of various sizes. When writing a file to the volume, the server fills as many full blocks as it can and then uses the size of the leftover fragment to determine which SRB it should use to store the remaining data. As a result of this technique, the amount of space wasted in the worst possible case is no more than 512 bytes per file, which is a significant improvement over a volume that is not suballocated, even when it uses small 4KB blocks.

Block suballocation is an optional feature that is enabled by default and that you can control for each volume individually from the Install.nlm utility. There is rarely a good reason to disable it, because it functions completely invisibly and causes no obvious degeneration of system performance. If you do decide not to use block suballocation for a particular volume, you will probably want to modify the default 64KB block size for drives over 500MB. Without suballocation, a block size this large can lead to a great deal of wasted space.

Novell Storage Services

NetWare 5.*x* includes another improvement to the file system, called Novell Storage Services (NSS). NSS is a 64-bit, indexed storage service that uses the free space on multiple storage devices to create a single virtual partition you can use to create an unlimited number of volumes. On today's 32-bit servers, volumes can be up to 8TB in size and up to 8 exabytes (that's 8×10^{64} bytes) on 64-bit servers. An NSS volume can store billions of files, with individual files up to 8TB in size. NSS also greatly increases the speed at which volumes mount, the slowness of which was a frequent complaint of the standard NetWare file system (which Novell has dubbed NWFS, now that it has something to compare it to).

Other NetWare Information

Novell Directory Services is one of the most important features of the NetWare 4.*x* and 5.*x* operating systems, and it is discussed in detail in Chapter 20. The IPX protocols used by NetWare for its client/server communications are covered in Chapter 14. The NetWare printing subsystem is discussed in Chapter 27.

Chapter 20

Novell Directory Services

637

As originally conceived, Novell NetWare was an operating system designed to provide multiple users with access to file and print resources on a common server. To control access to the server, NetWare included a flat file database called the *bindery,* which consisted of user accounts, user groups, passwords, and other basic properties. Administrators granted access to specific resources on the server by selecting users and groups from the bindery.

As NetWare grew in popularity, it became common for an organization to have multiple servers, each of which had its own bindery. To add a new user, administrators had to create an account in the bindery of each server to which the user needed access. When the network had a handful of servers, this repetitive chore could be irritating, but it was still possible. When large networks grew to the point of having dozens or hundreds of NetWare servers, maintaining individual binderies on each machine became increasingly impractical.

To address this problem, Novell created a directory service that would function as a central repository for information about an entire enterprise network. Administrators create a single account for each user in the directory service, and then use that account to grant the user access to resources anywhere in the enterprise. Based largely on the X.500 standard developed by the International Organization for Standardization (ISO), NetWare Directory Services (NDS) first appeared as part of the NetWare 4 operating system in 1993. While somewhat unstable at first, NDS has had a long time to mature, and has also undergone a name change from NetWare Directory Services to Novell Directory Services, which reflects the utility of the product for applications and operating systems other than NetWare.

In the commercial networking industry, NDS has become the pattern for the development, deployment, and adaptation of a directory service into a general-purpose networking tool. Microsoft's Active Directory service, first released in Windows 2000, clearly builds on the foundation created by NDS. The use of a single storehouse for information about all of a network's hardware, software, human, and organizational resources greatly simplifies the job of the network administrator, and Novell was the first company to create a commercial enterprise directory service and deploy it on a large scale.

NDS, in the form of the Novell eDirectory product, can now run on Unix and Windows servers as well as on NetWare. The popularity of NetWare has certainly suffered in the face of competition from Windows 2000 and Windows NT, but Novell has had a huge head start in its directory service, and NDS is ahead of the competition in its flexibility and application support. Even though Active Directory has a great deal of potential, the fact remains that Novell and its partners have had seven years to exploit the capabilities of NDS, while AD is still a relatively new product.

NDS Architecture

NDS is essentially a database that is comprised of objects arranged in a hierarchical tree, just like that of a file system (see Figure 20-1). At the top of the tree is a theoretical object called [Root], from which all the other objects stem. *Objects* are logical entities that can represent users, hardware, software, and organizational components. Each object consists of a collection of *properties* that contain information about the object. For example, an object representing a user called John Doe might be called jdoe, and might contain a property called Telephone Number, and the value of that property would be John Doe's telephone number. The other properties for a user object contain identification information about the user, as well as a list of the groups to which the user belongs, the user's permissions to other objects in the NDS tree, and many other account restrictions.

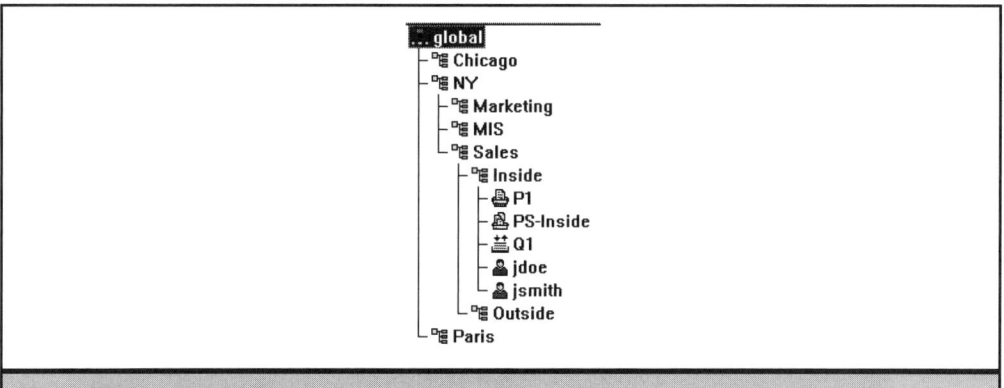

Figure 20-1:   *The NDS database takes the form of a hierarchical tree.*

Administrators can use any one of three tools to create objects and specify values for their properties. NetWare Administrator is a Windows application, available for the various versions of Windows, that provides a graphical view of the NDS tree and its elements (see Figure 20-2). NetWare Administrator is the easiest and most versatile tool to use for these and other NDS maintenance tasks. Each object in the tree has its own Details dialog box that lets you modify the object's property values.

Netadmin.exe is a menu-driven, character-based utility that you can run from the command prompt in any DOS-based operating system (see Figure 20-3). The interface is not as intuitive as that of NetWare Administrator and does not perform all of the same functions, but the program has the advantage of being very fast.

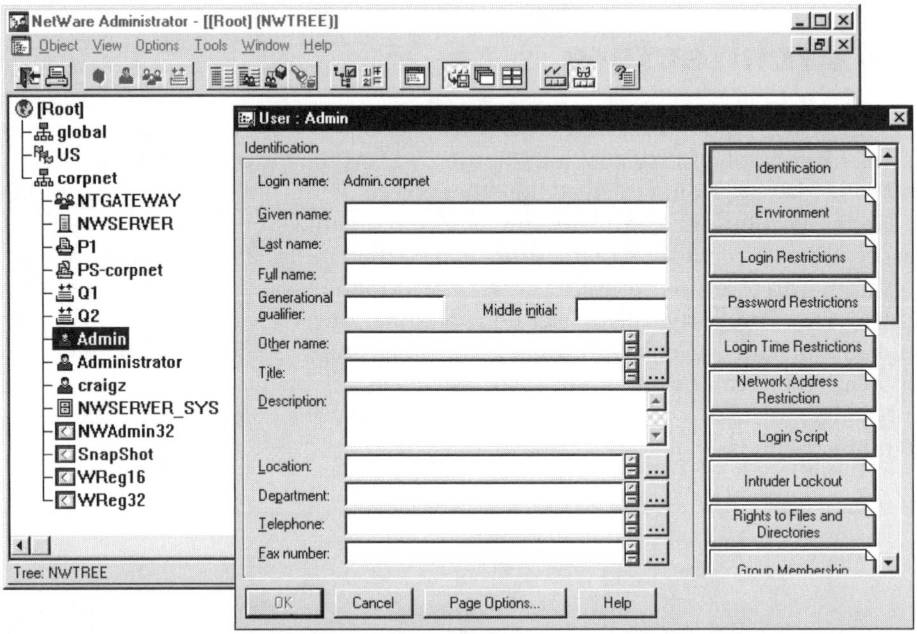

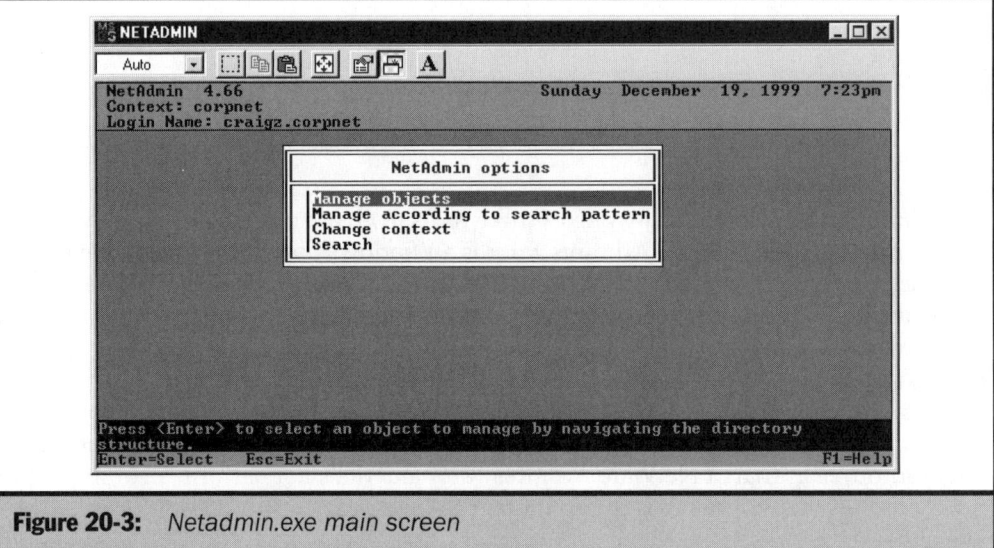

Figure 20-2: *NetWare Administrator tree display and Details dialog box*

Figure 20-3: *Netadmin.exe main screen*

ConsoleOne is a Java-based administration console (see Figure 20-4) first introduced in NetWare 5 that is designed for use with large enterprise networks. ConsoleOne provides NetWare and NDS administration capabilities on any platform that supports Java, including NetWare servers. A Web browser version of ConsoleOne enables any user with a Java-capable browser to work with the NDS database, even if no other NDS client is installed on the workstation.

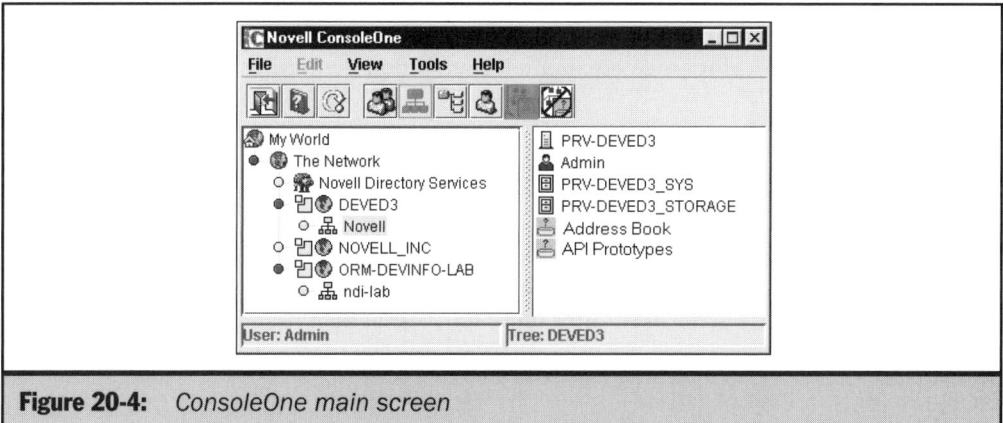

Figure 20-4:   *ConsoleOne main screen*

Containers and Leaves

There are two types of objects used to build the NDS tree, called *container objects* and *leaf objects.* As the name implies, a container object is one that has other objects subordinate to it, forming a branch of the tree. A leaf object cannot contain other objects, just as the leaves of a tree represent the endpoints of a branch.

Most of the container objects in an NDS tree are purely theoretical constructs that need not have any relationship to the physical objects they contain. For example, country, organization, and organizational unit objects exist only to form the hierarchy of the tree and to contain other objects. You can build the tree using these container objects to represent geographical locations, political divisions in your organization (such as departments), or any other paradigm. These objects simplify the process of assigning the same property values to multiple objects. When you modify a property of a container object, all of the objects in that container inherit the value for that property. For example, you can grant a container object the right to access a specific directory on a server, and every user object in that container will inherit that right.

Apart from user objects, there are other types of leaf objects in an NDS tree, representing NetWare servers, printers, applications, and other elements. Each object type has properties that are specific to the entity that the object represents. In addition to the default objects that you can create in the NDS database, it is possible for software developers to create new object types, or new properties for existing object types. The capabilities of NDS are defined by its *schema*, which specifies what kind of objects can exist in the database, the properties of those objects, and the relationships between the different object types. Applications can modify the schema by adding new object types that are specifically designed for use by that application. For example, Novell's ZENworks product modifies the schema to enable the creation of objects representing network workstations. Novell has also published software developer's kits (SDKs) that enable third-party developers to create their own schema modifications.

Objects and Properties

The object types included by default with NetWare, their functions, and their icons in NetWare Administrator are listed in Table 20-1. You can create objects of these types using the NDS management utilities immediately upon installing NetWare.

Object Icon	Object Name	Description
	AFP Server	Represents an AppleTalk Filing Protocol server.
	Alias	Functions as a duplicate of another object in the tree; enables the same object to exist in multiple contexts.
	Country	Represents a country in which the network has resources.
	Directory Map	References a particular directory on a NetWare file server; enables NetWare-aware utilities to reference the literal directory using the directory map object name.
	Group	Represents a group of users located anywhere in the NDS tree.
	Locality	Represents a geographical location in the NDS tree.

Table 20-1:   *Default NetWare 4.11 Object Types*

Object Icon	Object Name	Description
	NetWare Server	Represents a server running the NetWare operating system.
	Organization	Container object that represents an organization or company.
	Organizational Role	Represents a job responsibility that requires specific access rights to network resources.
	Organizational Unit	Container object that represents a department or division of an organization or company.
	Print Queue	Represents a NetWare print queue.
	Print Server (Non NDPS)	Represents a NetWare print server.
	Printer (Non NDPS)	Represents a NetWare printer.
	Profile	Represents a login script that can be assigned to users.
	Template	A leaf object that enables the rapid creation of multiple leaf objects with similar characteristics.
	User	Represents a network user.
	Volume	Represents a NetWare server volume.

Table 20-1:   *Default NetWare 4.11 Object Types (continued)*

As an example of NDS's expandability, the current Novell client software packages ship with the ZENworks Starter Pack, which includes the Novell Application Launcher and Workstation Manager. Both of these applications include schema extensions that add the new object types shown in Table 20-2.

NETWORK OPERATING SYSTEMS

Object Icon	Object Name	Description
	Application	Represents a networked application.
	Application Folder	Container for application objects displayed in Novell Application Launcher.
	Computer	Represents a nonserver computer on the network.
	Policy Package	Contains a collection of policies related to specific object types.
	Workstation	Represents a client workstation on the network.
	Workstation Group	Container holding a collection of workstation objects.

Table 20-2: *Object Types Added by the ZENworks Starter Pack*

The schema also defines the properties for each type of object in the NDS database, which are dependent on the function of the object. Some properties are mandatory, meaning that the object must have a value for them, while others are optional and can be left blank. Object properties carry information that falls into any of four categories, which are as follows:

Names Every object must have a name, of course, by which it is referenced in the NDS tree, but other properties can carry more complete naming information, such as a user's full name (for example, John Doe), as well as the object name (for example, jdoe).

Addresses Objects that represent hardware components, such as servers and printers, can have network addresses. Mailing addresses, telephone numbers, and other contact information are optional properties, often overlooked by administrators, that can be useful both to other users and to support personnel.

Descriptions Informational properties may not have technical functions, but taking the time to provide information about the type of equipment that an object represents, for example, or its exact location, can greatly reduce the support burden for network resources.

Memberships   The relationships of an object to other objects in the tree are among the most important types of property information. Group objects, for example, contain a list of their members, and virtually all objects contain a list of their trustees.

NDS Object Naming

The hierarchical structure of the NDS database functions not only as an organizational paradigm, but also as a means of uniquely identifying each object in the tree. In a flat file database like the bindery in older NetWare versions, every object has a single name that must be unique. This is practical for a single server solution like the bindery, but for an enterprise directory service like NDS, it is far more likely that you will run into a situation in which you want to have two objects with the same name. If, for example, there is a Joanne Smith who works in the Marketing department and a Joe Smith in Sales, your naming convention for user objects might require that both people have the user name jsmith.

In an NDS tree, you can have two objects with the same name, as long as they are in different contexts. The *context* of an object is simply its location in the NDS tree, as identified by the names of the containers in which it resides, stretching all the way up to the root, and separated by periods. Each object in the NDS tree is uniquely identified by a combination of the object name and its context. This combination is called the object's *distinguished name (DN)*. Thus, these two users can both exist in the tree with the same object name, using the following DNs:

```
jsmith.NY.Sales.corpnet

jsmith.Chicago.Marketing.corpnet
```

In these names, the context specifies the path from the [Root] object at the right (theoretically, since [Root] never actually appears in a distinguished name) down through the NDS tree to the object. The corpnet container is an organization object at the top layer of the tree, closest to the [Root], and is therefore said to be the *most significant* object in the name. The object that is farthest away from the root is the *least significant* object.

Using the Context

All applications and operating system functions that request information from the NDS database do so using the distinguished name of the object needed. However, when you perform an operation that triggers an NDS call, like entering your user name into a login dialog box, you never type the entire distinguished name of your user object, because the application automatically appends your current context to whatever object name you specify.

The user's default context is somehow specified in the client software that provides access to NDS, such as in a Windows registry or Net.cfg file entry, depending on the client operating system used. With the default context in place, the user can reference any other object in the same context using only its object name, which in this case would be called a *relative distinguished name* or *partial name*.

This automatic use of the context by applications and other software is why it is best to design your NDS tree so that users tend to be in the same container as the resources they access most often. To access files on a server in the same context, for example, a user can simply specify the server name. If the server is in a different context, however, the user has to specify either the complete distinguished name of the server or a context qualifier to indicate the name of the context in which the server is located.

It is possible to change your current context after logging in to the NDS tree using the NetWare Administrator, Netadmin.exe, or Cx.exe utility, enabling you to identify resources in the new context by using only their object names. However, most network users have no conception of what the NDS tree is and how contexts work, and will not usually be able to do this. The most common method used today to reference objects in other contexts is to browse through the containers in the NDS tree using a graphical dialog box like that shown in Figure 20-5. Selecting an object in another container automatically returns the distinguished name of that object to the application.

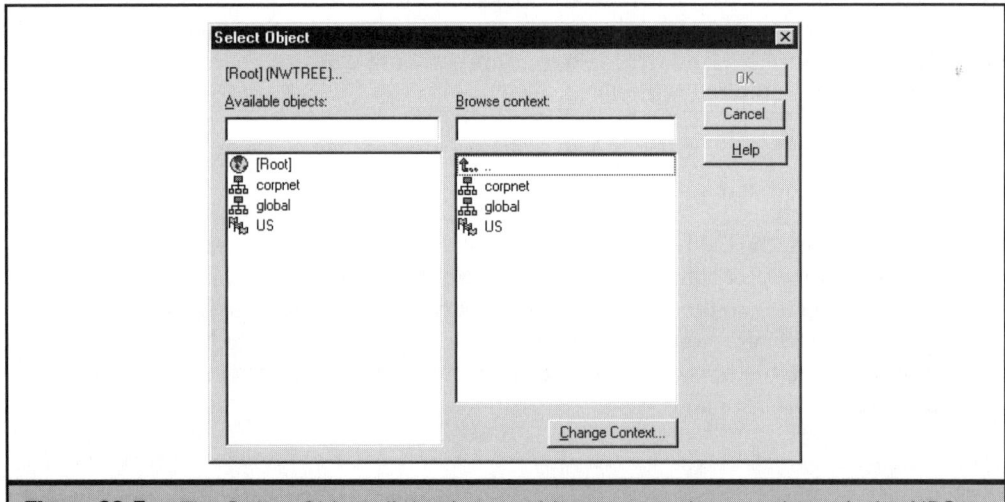

Figure 20-5: *The Select Object dialog box enables users to browse through the NDS tree to locate a specific object.*

Thus, for clients logged in to the NDS tree, the primary function of the context is to simplify the process of referring to other objects in the same container. Contexts also serve a function on NetWare servers, however. When clients who do not support NDS log in to a network, NetWare 4.*x* and 5.*x* servers can emulate the bindery used by NetWare 3.*x* and earlier versions. In order to support *bindery emulation,* you have to create one or more bindery contexts on a NetWare server using SET commands. These commands identify the only contexts that bindery clients can access. To the client, it appears as though the system is logging in to a bindery server, when actually it is logging in to a limited area of the NDS tree.

Typeful and Typeless Names

There are two ways of notating the distinguished name of any object in an NDS tree. The names used as examples so far are called *typeless* names, because they do not specify the object type for each name. A *typeful* name includes an abbreviation for each object name that specifies its type, in the following format:

```
CN=jsmith.OU=NY.OU=Sales.O=corpnet
```

```
CN=jsmith.OU=Chicago.OU=Marketing.O=corpnet
```

The object type abbreviations that can be used in typeful names are as follows:

C   Country

O   Organization

OU   Organizational unit

CN (Common Name)   All leaf objects

The distinguished names supplied by applications and operating systems making NDS calls are always typeful names. However, a user almost never has to supply a typeful name manually because clients use the *default typing rules*, which usually function correctly, to add the abbreviations to distinguished names. The default typing rules are as follows:

- The least significant partial name in a distinguished name should be designated as a leaf object and given the CN abbreviation.

- The most significant partial name in a distinguished name should be designated as an organization object and given the O abbreviation.

- All of the partial names between the least significant name and the most significant name should be designated as organizational unit objects and given the OU abbreviation.

Virtually the only time when these rules do not function properly is when the NDS tree uses country objects at its top layer. Country objects are a vestige from X.500, which was designed to be a true global directory service. It is because of the default typing rules that the use of country objects in an NDS tree is generally discouraged, even for multinational organizations.

When applying the default typing rules, an NDS client makes no attempt to check whether an object of the type it assigns actually exists. If, for example, you use country objects in your NDS tree, clients will submit names typed using these rules, and the names will all fail, because even though the name of the most significant container object will be correct, the type will be wrong.

Using Context Qualifiers

If, for any reason, you want to enter the name of an object in another context into an application, you can usually do so without typing the object's entire distinguished name, but you must be conscious of your current context when you do so. When you specify the name of a leaf object, your NDS client appends your current context to it. Thus, suppose a user is logged in to an NDS tree using the following default context:

```
OU=NY.OU=Sales.O=corpnet
```

When the user attempts to access a server called NW1 and manually types that name into an application dialog box, the NDS client uses the following distinguished name for the server in its call to the NDS database:

```
CN=NW1.OU=NY.OU=Sales.O=corpnet
```

If there is a second server, called NW2, that is located in the Chicago.Sales.corpnet container, the user can avoid typing the server's entire distinguished name by supplying only the name of the server object itself and that of its least significant container, which is the only part that is different from the user's current context. However, if the user supplies the name NW2.Chicago in the application, the client still appends the user's context, resulting in the following nonexistent name:

```
CN=NW2.OU=Chicago.OU=NY.OU=Sales.O=corpnet
```

The user wants to replace the NY container in the name with Chicago, not just add the Chicago container. To do this, you use a *context qualifier*, which essentially is a signal to the client instructing it to suppress a certain part of the context when it appends it to the name supplied by the user.

Trimmed Masking   The most commonly used type of context qualifier is called *trimmed masking,* which consists of adding a period to the end of the partial name supplied to the application for each container that you want omitted from the context when it is appended to that name. Thus, if you supply the name NW2.Chicago. (including the trailing period), the client will omit the NY container from the context and append only OU=Sales.O=corpnet, resulting in the following name, which is correct:

```
CN=NW2.OU=Chicago.OU=Sales.O=corpnet
```

If the user were to supply the name NW2.Chicago.. (including two trailing periods), the client would omit both the NY and Sales containers from the context, resulting in the following incorrect name:

```
CN=NW2.OU=Chicago.O=corpnet
```

Preceding Periods   Another form of context qualifier involves the insertion of a period before the name supplied to an application. When you do this, the client applies the object types of the user's current context to the object names supplied after the period. Thus, if a user with the context OU=NY.OU=Sales.O=corpnet.C=US enters the name .NW2.Paris.Marketing.corpnet.FR into an application, the client will take the object type from each of the four containers in the context and apply them to the names supplied, resulting in the following distinguished name:

```
CN=NW2.OU=Paris.OU=Marketing.O=corpnet.C=FR
```

The only time this technique is necessary is when the NDS tree uses country objects, and the default typing rules would result in an incorrect name. When you use the preceding period, the name you supply to the application must include the same number of container names as in the context. In other words, you supply a typeless distinguished name, and the preceding period enables the client to type it correctly.

Note   *Context qualifiers are not often needed anymore, because most operating systems enable users to browse the NDS tree in a GUI and select an object, rather than type the object's name. One of the instances in which you might use these techniques is when changing your current context using the NetWare CX utility. Cx.exe is a command-line program that enables you to navigate the NDS tree, much as the DOS CD command enables you to move around in a computer's file system.*

Partitions and Replicas

For NDS to be an effective repository for information about an entire enterprise network, the database can't be located on a single server. Not only would this be inefficient for access by systems that are separated from that server by relatively slow WAN links, but the failure of that one server could bring the entire network to a halt. To protect the database and make it accessible to users and administrators all over the network, NDS can be both partitioned and replicated.

Partitioning refers to splitting the database into segments, each of which is stored on a different server. Each partition is, in effect, a branch of the NDS tree. Creating partitions makes it possible to keep the objects in the NDS tree near to the physical entities they represent. For example, a company with four offices in remote cities connected by WAN links can split its NDS database into four partitions, each containing the users, servers, and other objects for one site and stored on a server at that site. This way, users logging in to the network do not have to access their user objects on a server at another site over a slow WAN link.

Even when WAN links are not involved, however, creating partitions on several servers helps to spread the NDS traffic around the network. When the entire database is stored on one server, every process that requires access to NDS objects must send traffic to that server. This may not be a problem if the server is located on a backbone or other high-speed network that can handle the traffic, but creating partitions on servers connected to different LANs balances the traffic load among several different network segments.

Replica Types

Partitioning the NDS tree creates a measure of fault tolerance, because the failure of one server does not render the entire database unavailable. However, it's possible to make the database even more fault-tolerant by replicating it. A *replica* is an exact duplicate of an NDS partition or the entire database that is stored on another server. You can create as many replicas of a partition as you want, with each one on a different server. The replicas of a specific partition are known collectively as a *replica ring*. If a server containing part of the NDS tree should fail, users and applications can still access the database from one of its replicas.

NDS recognizes four different types of replicas, as follows:

Master replica   The primary copy of a particular partition. There can only be one master replica of a partition. All partition management tasks, such as the creation of new replicas, must be performed on this replica, during which time all other replicas are locked.

Read-write replica   A copy of a partition that users and applications can access to process login and authentication requests, and that administrators can access to make changes to the database.

Read-only replica   A copy of a partition that users and applications can access to process login and authentication requests, but which cannot be modified by an administration utility. This type of replica is updated only by the NDS synchronization process.

Subordinate reference replica   Special-purpose replicas of subordinate partitions created on a server to point to the locations of those subordinate partitions on other servers. When a read-write or read-only replica of a subordinate partition is created on the same server as the parent partition, the subordinate reference replica for that partition is deleted.

Creating Partitions and Replicas

Administrators create partitions using either NDS Manager or Partition Manager. NDS Manager is a Windows utility that provides a graphical view of all of an NDS database's partitions and replicas (see Figure 20-6). Partition Manager is a DOS-based alternative that provides roughly the same functions without the comprehensive overview.

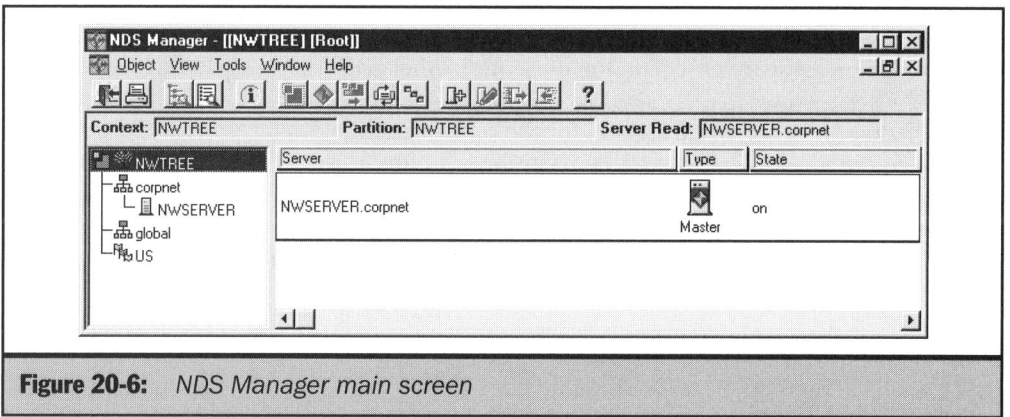

Figure 20-6:   *NDS Manager main screen*

Creating a partition is simply a matter of selecting a container in the tree. That container plus all of the objects in it (both containers and leaves) become part of the new partition. The selected container is then known as the *partition root object.* The partition root object tracks both the replicas of the partition that exist elsewhere on the network and the status of the replica synchronization process. Partitions with root objects that are closer to the [Root] of the NDS tree are said to be *superior* to the partition, and those with root objects farther from the [Root] are said to be *subordinate*.

Once you have created a partition, you can at any time select a container object within that partition and split it off into a partition of its own. You can also create new replicas of a partition at any time. The NDS tree can, therefore, grow to accommodate your network.

The number of partitions and replicas that you create should be based on the number of servers you have available, the capabilities of those servers, the needs of the network's users, and the layout of the network itself. In the case of the network mentioned earlier, consisting of four WAN-connected sites, creating a separate partition for each site is a logical solution. In addition, creating read-write replicas of the other three partitions on each of the four servers is a good idea (assuming that the servers are capable of hosting them), because then the users at each site can access the entire NDS database locally. The only directory service traffic that passes over the WAN links is for the synchronization of the database.

Walking the Tree

No matter how many partitions and replicas an NDS tree has and no matter where they're located, the NDS tree always appears as a single functional entity to the applications that utilize it. Every server that hosts any replica of an NDS partition functions as a *name server* and can locate the resource referenced by any NDS object name. Object name requests always specify the full distinguished name of the requested object. When a name server receives the request, it first checks its own partitions for the object. When the requested object does not exist in the server's own partitions, it must search for it on the network's other name servers. The process by which the server does this is called *walking the tree*, which involves searching the partitions on the other name servers for the requested name in a systematic fashion.

The partition root object of every partition contains a list of the other partitions that are directly superior and directly subordinate to it. If the server recognizes any of the containers in the requested name as being part of a subordinate partition, it can pass the request downward through the tree to the appropriate name server. If the name server makes no such recognition, it must pass the request farther up the tree toward the [Root] object.

Any object in the tree can be located by searching from the [Root] object, but this is not a practical method, because it would overburden the server containing the [Root] partition and possibly create multiple delays if that server is connected using a slow WAN link. Instead, a name server processing a request passes it to the name server that is immediately superior to it in the tree hierarchy. That server then searches its own partitions and satisfies the request, or passes it down to a subordinate name server of its own, or passes it up to the next superior server. Eventually, a name server will recognize one of the containers in the request name and be able to route the request down to the server containing the correct partition. This way, the request only gets passed up as far as is needed to find the name.

For example, consider the name server hierarchy shown in Figure 20-7. Suppose that an application sends a request for a user object called jdoe.Chicago.Sales to Server D. The partition on Server D contains the objects in the NY.Sales container, but it has no information about Chicago.Sales. Since the partition subordinate to Server D (Inside.NY.Sales on Server F) must have NY.Sales in its context, Server D knows that it has to pass the request upward to a superior partition. Server B contains the Sales partition and recognizes that part of the requested object name, but it does not contain the Chicago.Sales partition. However, Server B does recognize that Chicago.Sales must be subordinate to Sales, so it passes the request to Server E, which does contain the Chicago.Sales partition. This server can therefore locate the jdoe.Chicago.Sales object and supply the requested information.

In this particular case, the difference between using the tree-walking process and simply sending the request to the [Root] partition is only one layer in the tree. In a larger network with more layers of partitions, walking the tree can be much faster than querying the [Root] partition. This scenario also assumes that the name servers each host only a single partition. Creating replicas of several partitions on a name server reduces the number of other servers involved in the tree-walking process.

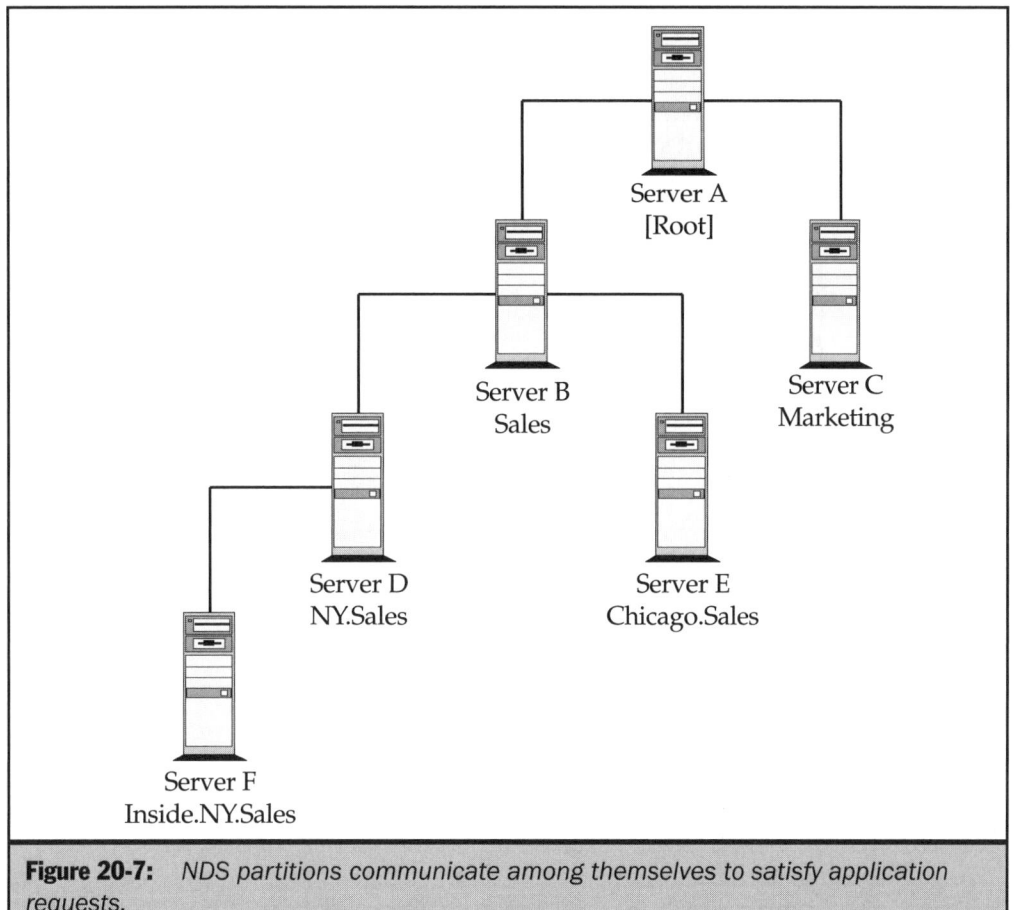

Figure 20-7: *NDS partitions communicate among themselves to satisfy application requests.*

NDS Synchronization

The messages exchanged by the name servers during the tree-walking process are only one of the ways in which the various NDS replicas communicate with each other. In the same way that an application can use any name server to access the entire NDS database, an administrator modifying the database can make changes to any replica that permits writes (which means the master replica or any read-write replica). Once the changes have been made, however, they must be propagated to all of the other replicas of that partition, so that applications accessing the NDS database retrieve the same information, no matter which replica they use. This process is called *synchronization,* and it is responsible for most of the traffic between NDS name servers.

A directory service that functions this way is said to use *multiple master replication,* meaning that you can make changes to any one of several replicas. In a *single master replication* system, such as that used by Windows NT domains, all changes to the database must be made to one specific replica (the primary domain controller, in Windows NT), and then that system propagates the changes to the other replicas. The pattern of replication traffic in the two systems is shown in Figure 20-8.

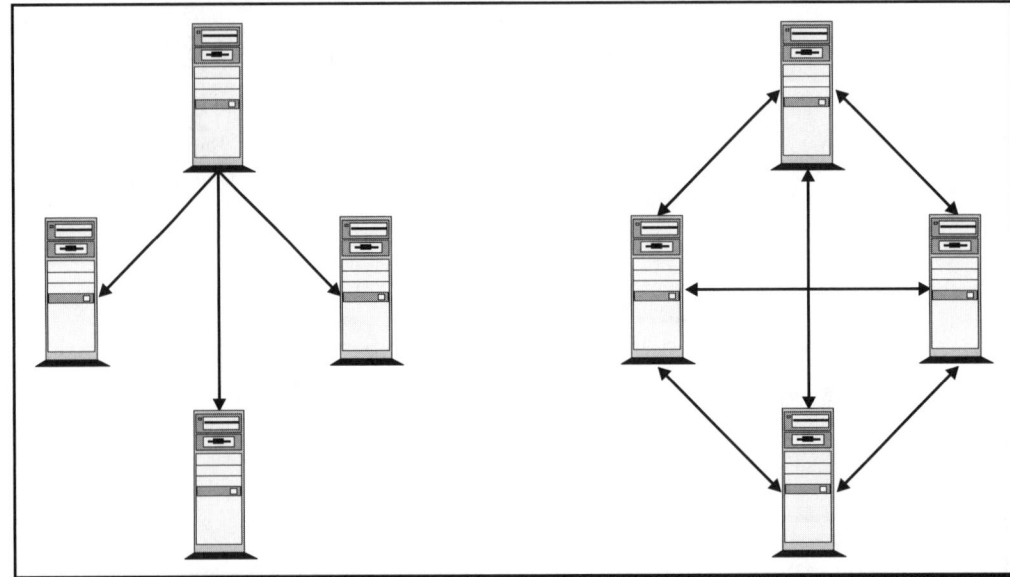

Figure 20-8: *Single master replication traffic moves in one direction, while in multiple master replication, servers communicate bidirectionally.*

The advantage of multiple master replication is that administrators can perform their database maintenance activities on a local replica, rather than having to access a name server that may be a long distance away or on the other side of a WAN link. The disadvantage is that the synchronization process is much more complicated. In a single master replication system, the synchronization traffic travels in only one direction, and there is no possibility of conflicting information. In a system that uses multiple master replication, it is possible for two administrators to make changes to different replicas of the same partition at nearly the same time. Each of the replicas must then send the modifications to all of the other replicas. If both administrators have modified the same datum, the synchronization processes will conflict, and there must be a mechanism for determining which version takes priority.

For example, suppose that two NDS administrators modify the same user object on two different replicas at roughly the same time. One administrator changes the user's telephone number, and the other administrator modifies the user's account password.

The synchronization system must function so that, at the end of the synchronization process, the object in both replicas has the new telephone number and the new password. Thus, in this case, the process is not simply a matter of overwriting the object on one replica with the object from another. The data in both copies of the object must be combined.

To complicate matters further, it is also possible for the modifications themselves to conflict. Suppose, for example, that a user calls a network administrator to give the administrator the user's new telephone number, realizes soon after that he has given the administrator the wrong number, and then calls again to remedy the error, but talks to a different administrator this time. The first administrator modifies the user's object on one replica by entering the new (incorrect) number, and a few seconds later, before the synchronization process can take place, the second administrator modifies the same object on another replica, entering the correct phone number this time. When the synchronization process begins, the systems must have a way of determining which version of the data should take precedence over the other.

The NDS synchronization process can overcome both of these problems, but although it is quite efficient, it isn't perfect, because it can't update all of the replicas in real time. It is still possible for an application to access outdated information from the database during the brief period before the synchronization process has completed. For this reason, the NDS database is said to be *loosely synchronized*. Replicas that are in the process of synchronizing are said to be *converging*. When the convergence process is completed and the replicas of each partition are identical, the database is said to be *fully synchronized*.

Time Synchronization

The NDS synchronization process is based on *time stamps* that are assigned to each modification made to the database. Name servers use these time stamps to determine which data should take precedence when a conflict occurs. In order for the time stamps to be accurate, it is essential that all of the name servers on an NDS network have their clocks synchronized. Keeping the clocks in synch on a collection of computers is more complicated than it may seem. The first problem is that the servers may be located at different sites, in different time zones, and with different daylight saving time policies. The second problem is that PC clocks are notorious for being wildly inaccurate.

To keep their clocks synchronized, NetWare servers use a program called Timesync.nlm. Each of the servers uses SAP (Service Advertising Protocol) messages to transmit messages containing a UTC (Universal Time Coordinated) signal, which is the same as Greenwich Mean Time. Every NetWare NDS server on the network also includes a series of SET commands in its Autoexec.ncf file that specify how the server should keep time. These commands appear as follows:

```
SET TIME ZONE = EST5EDT

SET DAYLIGHT SAVINGS TIME OFFSET = 1
```

```
SET START OF DAYLIGHT SAVINGS TIME = (APRIL SUNDAY FIRST  2:00:00 AM)

SET END OF DAYLIGHT SAVINGS TIME = (OCTOBER SUNDAY LAST  2:00:00 AM)

SET DEFAULT TIME SERVER TYPE = REFERENCE
```

The functions of the first four commands are self-explanatory, and can contain various values depending on the geographical location of the server. The last command specifies how the server should interact with the other servers on the network when processing time signals. The possible values are as follows:

Primary   Exchanges time signal messages with the other primary and reference time servers on the network to compute the average time and adjust its clock.

Secondary   Periodically requests a time signal from a primary or reference server and adjusts its clock accordingly.

Single Reference   Maintains a unilateral time setting without communicating with other servers. Used on a network with only one time server.

Reference   Participates in the computation of average server time, but unlike primary and secondary time servers, does not adjust its clock, because the server is assumed to be calibrated by an external time source—such as a radio clock or modem connection.

Note   *It is not necessarily imperative that all of the name servers participating in an NDS tree keep the correct time, only that they are synchronized to exactly the same time.*

Creating an effective time-synchronization strategy on a medium-to-large network is a tradeoff between the amount of network traffic generated by the process and the fault tolerance of the system. You can designate all of your NDS servers as primary or reference time servers, but then these systems will constantly be exchanging messages in order to compute the average time. At the other extreme, you can create one primary or reference server and designate all of the others as secondaries, but then if the primary time server should fail, you lose calibration for the entire network. A reasonable medium between these two extremes is the best solution.

NDS Tree Design

One of the most important elements of using NDS effectively is designing a tree that is suitable for your enterprise. This is not a task to be taken lightly or improvised as you create the objects in the tree. For all but the smallest networks, the tree design will have an immediate effect on the efficiency of the network. A poorly designed tree will slow down user logins and access to network resources and make the task of administering the NDS database more difficult.

It isn't easy to sit down with a blank sheet of paper and design an NDS tree. It takes experience with the administration tools and with the effects of design changes on network performance to become an NDS expert. None of the tree design decisions that you make are irrevocable. In fact, just the opposite is the case. The tools provided with NDS enable you to create a tree that can evolve both with your increasing expertise and with the changes in your network. You can move objects to different containers and even shift whole branches to other locations in the tree. Just as with the file system on a network server, though, making drastic changes like these can confuse the other people who access the tree.

When you install the first NDS name server on your network, the installation program enables you to create a simple tree hierarchy that consists of an organization object and (optionally) a few organizational unit objects. An object representing the new server is then created in one of these containers. For a simple, one-server network, this is all the tree structure you need. Just create your user objects in the same container as the server object, and you are ready to go. In this capacity, the way NDS functions isn't much different from how the old NetWare bindery functioned. However, NDS was designed for use on larger networks, even truly huge ones, and the tree-design process naturally gets more complex when this is the case.

Tree Design Rules

An effective NDS tree must be logical in its construction to make it easy for users and administrators to locate particular objects. You can, for example, create container objects using the colors of the rainbow as names and randomly distribute your network's user and server objects among them. But this artificial construct would make it very difficult to find anything, and the relationships between the objects would be difficult to manage.

Two fundamental concepts should guide virtually all of your NDS tree design decisions: *rights inheritance* and *ease of access.* In an NDS tree, rights flow downward through the tree, and this policy greatly simplifies the administration process. When an administrator grants access rights to a container, all the objects in that container inherit those rights. Assigning rights to a single container object is far easier than assigning them individually to many different user objects.

Rights inheritance is one of the main reasons why you should try to group together users with similar access requirements. You could conceivably create a tree with containers named for the letters of the alphabet and then create each user object in the container with the same name as the user's initial, but then you could not use the container objects to grant the users access to specific resources. You would have to create individual rights for each user instead, complicating the process enormously.

Ease of access means that users should be located in containers with the network resources they access most often. Most users remain blissfully unaware of the intricacies of the network they're using and are not the least bit interested in learning about NDS tree design. They simply want to access the printer across the room as easily as possible. Placing all of the printer objects in their own separate container for the sake of organizational consistency might look nice on paper, but it won't work well in real life. The users in a workgroup or department who routinely access specific resources, such as servers and printers, should be represented by objects in a container with those resources whenever possible.

Instead of using arbitrary container names and object groupings, NDS trees are typically designed using one or more of the organizational paradigms examined in the following sections. In many cases, the best solution is to use a combination of criteria to design an effective tree. Strict adherence to any one of these paradigms can force you to make impractical design choices in order to conform to a preexisting plan. Practicality is the cornerstone of any tree design, and you should feel free to violate your established design rules if the result is easier access or administration.

Partitions and WAN Links

The communication capabilities of your network are of principle importance when designing an NDS tree. If your network consists of multiple sites connected by WAN links that are slower and/or more expensive than your LANs, the tree design should reflect those WAN divisions. The general rule when dealing with WAN-connected sites is to create a separate NDS partition at each site containing objects representing the resources at that site. The intention behind this is to keep the majority of NDS communications internal to each site. A user wanting to connect to a server in the next room should not have to exchange messages with a name server a thousand miles away in order to be authenticated.

Thus, when you have multiple sites connected by WAN links, you should create a container object representing each one of those sites as high up as possible in the NDS tree hierarchy. The usual course of action is to create a single organization object and then an organizational unit object named for each site, as shown in Figure 20-9. The organization object enables you to assign property values that are inherited by the entire enterprise, and the organizational units enable you to create a separate partition and specify separate property values for each site.

Note *NetWare 5.x includes a utility called the WAN Traffic Manager, which enables you to exercise precise control over the NDS traffic that passes over WAN links. The utility works by extending the NDS schema to include an object type called a LAN area object. A snap-in for NetWare Administrator enables you to create LAN area objects and define WAN traffic policies, which exist as properties of either server or LAN area objects. These policies specify the conditions under which NDS traffic should be transmitted over a link connecting LAN areas, and every NDS name server on the network runs a program called Wtm.nlm, which applies the policies.*

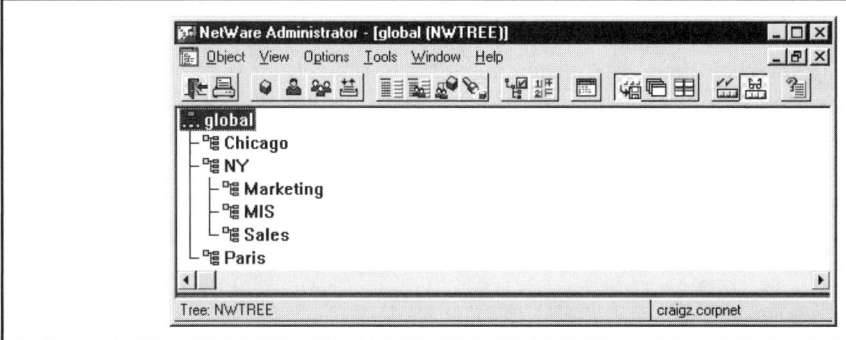

Figure 20-9:   *Sites connected by WAN links should be reflected by organizational unit objects placed as high as possible in the tree.*

Partitioning the NDS database is often a good idea (for load balancing and fault tolerance) even when WAN links aren't involved, and the tree design can center around those partitions as well, even though the network traffic issues are not as critical. Creating a separate partition for each of your network segments, for example, and storing it in a server on that segment, is a good way to maximize the efficiency of your network and reduce the burden on the routers or switches connecting the networks. If you choose to do this, the distribution of the partitions in your tree design should take precedence over any other organizational method you plan to use. If, for example, your tree design is based primarily around geographical locations, such as the rooms in which your users work, you should let the partitions dictate the arrangement of the upper-layer containers, and then apply your geographical method as you work your way down through the tree.

Geographical Divisions

Creating a tree based on the physical locations of the resources the NDS objects represent can be an effective way of grouping users with the objects they need to access. You can create a hierarchy of organizational units that corresponds to the layout of the buildings, floors, wings, or rooms that your network services, as shown in Figure 20-10. Navigating a geographical tree is easy for anyone who is familiar with the layout of the facility, and since users often work near the resources they access regularly, such as servers and printers, administrators can readily place them in the same container.

However, a strict adherence to geographical locations is not necessarily good. If, for example, all of the network's servers are located in a data center, creating a single container for all of the server objects may not be a good idea, because all of your users would have to access servers in another context. It would be better to locate the server for each department or workgroup with the users who actually access it. The same holds true if some users associated with particular resources are not physically near them. For example, even though all of the vice presidents in a corporation might have offices in a separate executive wing, it may be better to place their user objects in the containers associated with their respective departments, rather than in a single container with the other vice presidents.

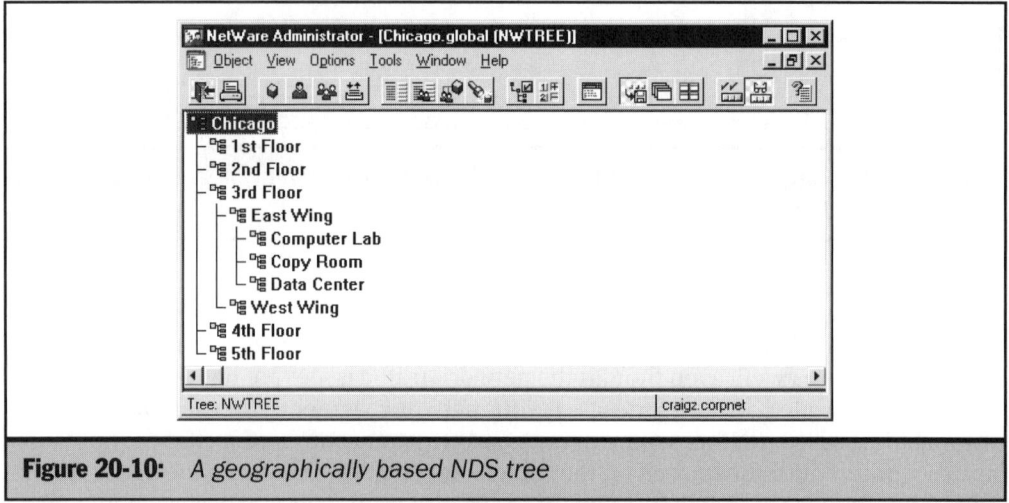

Figure 20-10: *A geographically based NDS tree*

A geographically designed tree can also be a problem if the organization moves to a new facility or if departments are shifted around within the current facilities. The question then arises of whether you should upset the entire tree design to make it correspond with the new locations or keep it the way it is despite its increasing discrepancy from the physical layout of the network.

Most of the problems that arise as a result of using a geographical tree design can be addressed through the judicious use of alias objects, as described in "Using Aliases," later in this chapter.

Departmental Divisions

Another solution that you can use when designing your tree is to take advantage of the divisions that already exist in your organization, such as departments and workgroups. Containers in the higher layers of the tree would represent corporate divisions or departments, such as Sales, with lower-layer containers subdividing each department into smaller units, such as Inside Sales and Outside Sales. This arrangement provides natural groupings of users with the resources they access, regardless of their respective locations, making it easily possible to assign rights using container objects.

A departmental tree could be as easy or even easier to navigate than a geographical one. In a very large organization, users might be more aware of other people's functions than of their exact locations. In addition, the modifications needed to keep the tree up to date would naturally correspond to the practical changes needed. When users move to other departments, they are likely to need access to different network resources, and simply moving their user objects to containers representing their new departments can provide them with that access, if the appropriate rights have been granted to the container. A purely geographical tree would likely require modifications that reflect only the movement of people and equipment from one place to another, and that serves no practical purpose other than to maintain the fidelity of the tree.

Physical Network Divisions

Another method for designing a tree, and one that in some ways functions as a lesson in what not to do, is to follow the physical layout of your network. You can create containers that represent the network segments and place objects in them that represent the equipment connected to those segments. This concept may be intuitive for the network administrator who has designed the network from the ground up, but not for the average network user.

However, while not a suitable model for the whole tree, the network layout can have an effect on the tree design. As mentioned earlier, creating partitions around WAN links must take priority over other more arbitrary design criteria. Therefore, on a network with WAN links, the top layer of the tree may appear to use the geographical method, by having containers named for cities, when it is actually conforming to the layout of the network. Beneath this layer, you can revert to a departmental or geographical design for the rest of the tree.

There is no reason why you can't "mix your metaphors" while designing a tree, as long as the result is intuitive to both users and administrators. You can have your upper-layer containers conform to sites separated by WAN links, use departmental divisions for the next layer of containers, and geographical designations for the layers below that, as shown in Figure 20-11.

Balancing the Tree

In order to make the NDS database function as efficiently as possible, Novell recommends building what it calls a *balanced tree,* using a cone as the optimal shape for the design, as shown in Figure 20-12. When the tree isn't balanced, such as when there are too many containers in the top layer, or just too many layers, its performance can suffer. If your organization has a large number of departments, branch offices, or other elements that lead you toward the creation of too many top-layer containers, try to find some logical criterion that you can use to group them, to keep the top layer relatively small.

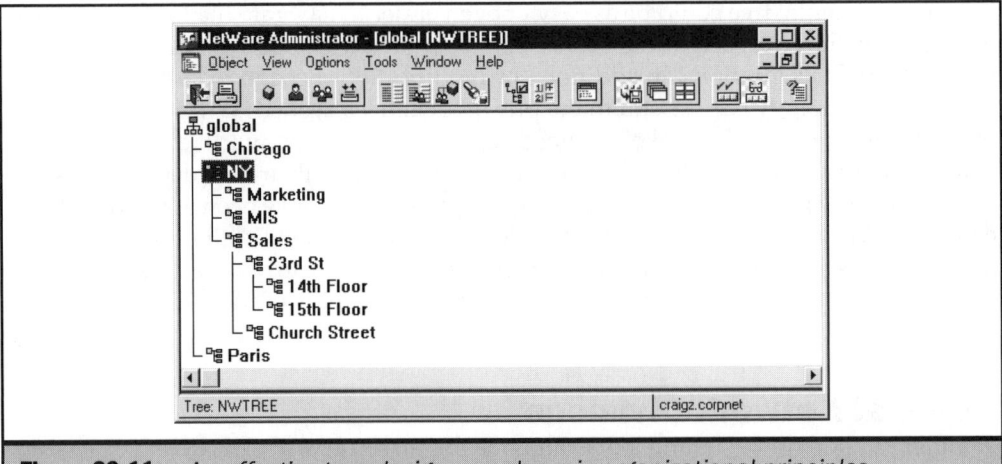

Figure 20-11:   *An effective tree design can also mix organizational principles.*

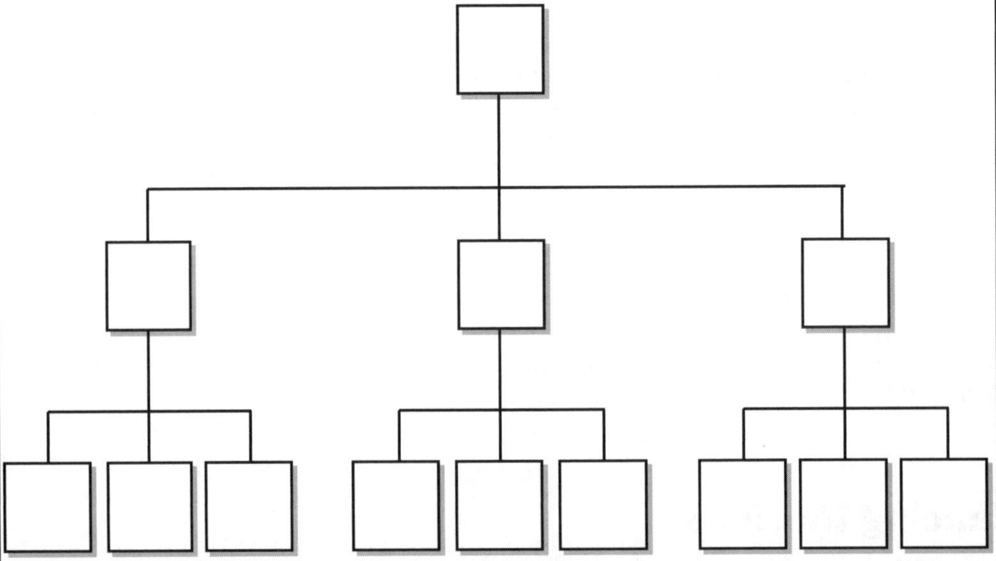

Figure 20-12:   *The ideal NDS tree configuration is a balanced, cone-like shape.*

Having too many layers in the tree can also have a negative effect on NDS performance. Aside from making the distinguished names of objects longer than necessary and complicating the tree-navigation process, the need to pass inherited rights down through many layers increases the processing time for each object. It is important not to let the organizational paradigm you've selected for your tree design take precedence over the tree's performance. Just because your workforce is ultimately broken down into five-person teams doesn't mean that you have to create dozens of tiny container objects to represent each one.

Building the Tree

After you decide how you will structure your NDS tree, it's time to think about the actual construction process. Creating objects is not difficult in itself, but you must consider the future maintenance of the tree as well. For example, part of the task of designing your tree should involve the ways in which you can assign trustee rights using containers instead of individual objects. It may seem easy to create individual trustee rights as part of the user objects, but later, when it comes time to modify the rights, you'll wish you could change a single container object's rights rather than those of multiple user objects.

The following sections examine some of the elements you should consider, as well as some of the techniques you can use to simplify the object-creation process.

Object Naming Conventions

The names that you select for the objects in your NDS tree can be as important as the structure of the tree itself. Objects that are logically named can more easily be identified by both users and administrators. The primary rule to follow in this respect is *consistency*. User object names, for example, should follow a specific formula, such as the user's first initial and surname. Allowing people to select their own user names only makes the process of working with the user objects more difficult later on, especially when there are several people maintaining the tree.

In the same way, the names for objects representing hardware resources should be in some way informational. For a small network, calling your servers SERVER1 and SERVER2 can be acceptable, but in a large enterprise, a department name, location, or function, such as SALES1 or DBASE1, is more functional.

Note also that while object names can be up to 64 characters in length, excessively long names are not recommended. Originally, the reason for keeping names short was because people had to type them. This is still a possibility, but most users and administrators today navigate the tree using a graphical interface. Even in a GUI program, however, an object name like NETWARE_SERVER_THIRD_FLOOR_EAST_ WING_BY_THE_WINDOW, while descriptive, will probably not display properly in a dialog box.

Object Relationships

The NDS schema determines the relationships between the various types of objects in the tree—that is, which objects can be superior and subordinate to other objects. For example, you cannot have a container object subordinate to a leaf object, or even a leaf subordinate to a leaf. The top layer of containers in the tree must be either country or organization objects. For the reasons given earlier, you should avoid using country objects, so you must create organizations directly beneath the [Root]. Beneath the organization objects, you create organizational units for the rest of the layers.

Fortunately, applications like NetWare Administrator enforce these relationships for you. When you highlight the [Root] and attempt to create an object, for example, the program only enables you to create a country, organization, or alias object.

Using Aliases

One method for resolving many tree design problems is through the use of Alias objects. *Aliases* are objects that function as copies of other objects located elsewhere in the tree, enabling them to appear as though they are in another container. For example, if your organization keeps all of the network servers in a data center, you can put all of your server objects into a single container representing the data center and then create an alias object for each server in the same container as the users who regularly access it.

You can create an alias for any object in the NDS tree, and use them for many different purposes. If you have users (such as technical support personnel) who log in to the network from many different workstations, you can create aliases of their user objects and place them in various contexts, so that the users can log in using any workstation's default context. Printer aliases permit users in multiple containers to access the same printer without having to browse the tree. When you use NetWare Administrator to move a container object to another location in the tree, the program offers to leave an alias object behind in the original location, so that any references to the container in its original location will not be orphaned.

When you open the Details dialog box for an alias object in NetWare Administrator, the properties the program displays are those of the original object, not the alias itself (with one exception). The process of displaying the original alias's properties is known as *dereferencing.* The only exception in which NetWare Administrator displays the properties of the alias object itself is for the list of trustees maintained by the alias object separately from the original object's trustees. You can control which trustee list NetWare Administrator displays by selecting either Get Alias Trustees or Get Aliased Object Trustees from the Options menu.

An alias of an object that you've created in a different container has exactly the same property values as the original. This includes values inherited from the container objects in which the original object resides. An alias of a user object, for example, will run the login script of the original object's container, but will not inherit rights to printers from the alias's own container. Therefore, to enable a user to access the local resources when logging in from a different context using an alias, you must explicitly grant the original user object rights to those resources.

Aliases must maintain the same relationships as the objects they reflect; you can only create an alias of an organization object directly beneath the [Root] or a country object, for example. Aliases are also dynamically connected with the objects they reflect. When you change the properties of the original object, the properties of the alias change as well, and when you delete the original object, all of its aliases are deleted.

Using Templates

When building a large NDS tree, setting property values on a large number of user objects can slow down the creation process. There are several methods you can use to simplify the user object creation process, chief of which is the use of property values inherited from container objects instead of user object property values. However, when you must assign the same property values to multiple user objects, you can do so using a template.

A *template* is an object type that functions as a pattern for the creation of new user objects. You can create a template and configure it with property values like those for a user object. Then, when you use the template to create new user objects, each one is given the same property values as the template.

The template was first introduced as a separate object type in the NetWare 4.11 release. Prior NetWare 4.x versions enabled you to create a USER_TEMPLATE entity from an existing user object, and then use it to create additional user objects with the same property values.

You can use template objects to create any number of new user objects, but you cannot apply a template to a user object that already exists. In addition, you must be aware that NDS utilities access the properties of the template object only when creating a new user object. Modifying the template object's property values has no effect on the user objects that have already been created using the template.

Using Groups

The NetWare bindery relied on groups as the sole mechanism for applying trustee rights to multiple users at once. In NDS, the various container objects, such as organization and organizational units, would seem to make groups superfluous, because they perform roughly the same function. However, group objects in NDS differ from the containers in that they enable you to assign rights to a collection of users who are not all located in the same context.

However, because group members can be located anywhere in the tree, including opposite sides of a WAN link, processing the trustee rights assigned to a group object is much more complicated than processing container object rights. For this reason, you should minimize the use of group objects in your tree design wherever possible and assign rights using containers instead. For example, placing the user objects for vice presidents into the containers with the rest of their departments enables them to access the same departmental resources as the other workers. However, you may want to create a group object with the vice presidents of all the departments as members, in order to grant them rights that only executives should have.

Security Equivalents and Organizational Roles

Another method for assigning a user object specific property values is to make it the *security equivalent* of another user object. When you do this, the new object inherits all of the rights granted to its equivalent. The problem with this method, however, is that it's dynamic. Changing property values in the original object causes the same changes to be reflected in its equivalent. If, for example, you assign access rights to a user object called jdoe, and then make all of the other users in the department security equivalents to jdoe, what happens when jdoe is fired or transferred to another department? The communication required between the security equivalents can also slow down NDS performance, particularly when the two objects are separated by a WAN link.

As an alternative to security equivalents, NDS includes an object called an *organizational role,* which defines a particular job, rather than a person performing that job. You can create an organizational role object, assign it access rights, and then specify a list of users who will occupy that role. Changing the property values of the organizational role will dynamically change the properties of the users occupying it.

Bindery Migration

If you are upgrading to NetWare 4.x or 5.x from a bindery-based NetWare version, you can get a head start on constructing your NDS tree by importing the binderies from your existing NetWare servers. The DS Migrate tool included in NetWare versions 4.11

and above is a separate program that reads the information from your NetWare binderies and stores it in a temporary database. Once there, you can model the bindery information into an NDS tree structure by creating container objects, moving user and group objects to other containers, and modifying object properties. After creating the tree structure you want, you can commit it to the real NDS database. Then, using the NetWare File Migration utility, you can migrate the files from the volumes on your bindery servers to NetWare 4.*x* or 5.*x* volumes, keeping intact the trustee rights you migrated with the bindery objects.

Merging Trees

Dsmerge.nlm is a server console utility that enables you to combine two separate NDS trees into a single database. You can use this tool when companies or divisions with separate NDS trees merge, or to incorporate an experimental database into your enterprise network tree. DSMERGE simply combines the roots of the two trees you specify. It does not combine objects or reconcile conflicts between similarly named objects. Therefore, you must be sure that there are no duplicate container names in the top layer of each tree before you perform the merge.

NDS Security

Just like a file system, a directory services database contains valuable information that must be secured against intrusion, unauthorized modification, and inadvertent deletion. NDS includes its own system of trustee rights that is independent from the trustee rights you assign to the files and directories on NetWare volumes. You can use these trustee rights to specify which users should be permitted access to the various objects in the tree and their properties. This way, you can delegate NDS maintenance tasks for a portion of the tree to certain administrators without giving them access to the entire tree, and grant users the ability to read information in the NDS database without giving them the ability to modify it. However, because NDS rights and file system rights are separate, it is possible for users to have access to the files and directories on a particular NetWare volume without having access to the volume object in the NDS tree.

You manage NDS security rights using the same tools that you use to create and configure objects. NetWare Administrator uses a dialog box like that shown in Figure 20-13 to manage all aspects of the object and property rights for a particular object, as well as its inherited rights filters and effective rights. Netadmin.exe provides the same capabilities, using a character-based menu interface.

NETWORK OPERATING
SYSTEMS

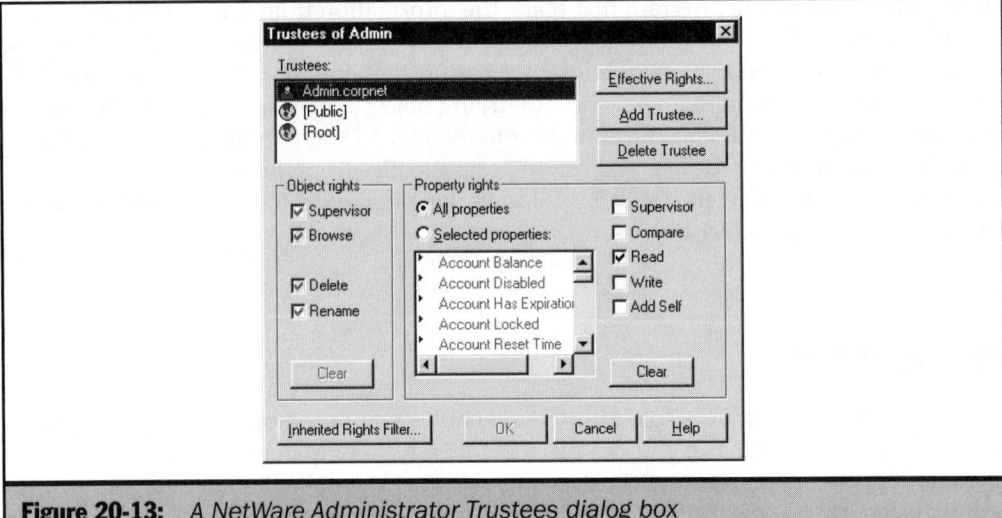

Figure 20-13:　*A NetWare Administrator Trustees dialog box*

Object and Property Rights

NDS uses both *object rights* and *property rights,* which enables you to grant users control over an entire object and all of its properties, or just specific properties. Like file system rights, the various object rights grant specific types of access to the trustees that possess them. The types of object rights are as follows:

Supervisor　Grants full control over an object, including the other four rights, all property rights, and the ability to grant rights to other users.

Browse　Enables the user to view the object in the directory tree. To a user who lacks this right, the object and all of its subordinate objects are invisible. By default, all users receive the Browse right to the [Root] object, enabling them to see the entire tree.

Create　Enables the user to create new objects in the tree, subordinate to the current object.

Delete　Enables the user to remove the current object from the tree.

Rename　Enables the user to change the name of the current object.

Object rights enable users to manipulate the objects themselves, but not the properties of those objects. The one exception to this is the Supervisor object right, which grants the trustee full control over all of the object's properties as well. When users do not have the Supervisor object right, they must be granted separate property rights in order to modify the value of specific properties.

Each of an object's properties has its own individual rights. You can grant a user rights to all of an object's properties at once, or to specific properties individually. The rights that you can grant to properties are as follows:

Supervisor   Grants full control over the value for a particular property.

Compare   Enables the user to check a property value for equality with another property value.

Read   Enables the user to see the value of a property.

Write   Enables the user to add, modify, or delete the value for a property.

Add Self   Enables users to add their own user objects to (or remove them from) a property value that accepts a list of object names.

Rights Inheritance

As with file system rights, NDS object and property rights flow downward through the tree. Any rights that you grant to a container object are inherited by the objects in that container. For example, the Admin user object created by default when you install your first NDS server is automatically granted the Supervisor object right to the tree's [Root]. This gives the user full control over all of the objects in the tree and their properties.

In addition, by inheriting the Supervisor object right from the [Root] object, the Admin user also receives full control over the file system on all NetWare server volumes. This is one of the few areas in which the NDS rights and file system rights converge. If you grant any user Supervisor object rights to a server object, the user gets not only the rights to manipulate the object, but rights to the server's volumes as well.

The Admin user is not inherently different from any other user object in the NDS tree, except for the object rights that it has been granted. Unlike the Supervisor in the NetWare bindery, you can delete the Admin user and modify its object rights. However, before you do this, be sure that every object in the tree has at least one user with the Supervisor object right to it. Otherwise, it is possible to orphan parts of the NDS tree by having no object with the rights needed to administer them.

As mentioned earlier, every user object in an NDS tree is granted the Browse right to every other object in the tree. This is not a right that is individually granted to each user on its creation. Instead, this right is granted through the use of a special entity called [Public]. This object doesn't actually appear in the tree, but it is listed in utilities like NetWare Administrator as an object to which you can grant rights. Any rights that you grant to the [Public] object are inherited by all of the objects in the tree, as well as all

NDS clients that have not yet logged in to the tree. The Browse object right to the [Root] object that is granted to [Public] is what makes the entire tree visible to all users and enables clients to browse the tree (such as when selecting a context) before they log in.

Inherited Rights Filters

In some cases, you may want to prevent rights from flowing down through the tree, and you can do this by applying an *inherited rights filter (IRF)* to a container object. An IRF functions like a dam, preventing the flow of specific object and property rights. When you create an IRF in a container to filter the Create object right, for example, all the other rights the container possesses will flow down to the objects in the container, but the Create right will not.

IRFs only prevent the inheritance of rights from superior objects; they do not block rights that are explicitly granted to an object below the filter. Thus, if you want to create a secret branch of your tree for experimental purposes or any other reason, you can create an IRF that filters out all inherited rights. The denial of the Browse object right will even make the tree branch invisible to other users. For those users who require access to the hidden branch, you can grant the necessary rights to their user objects or create a group object for that purpose.

Effective Rights

As you have seen, the object and property rights that a particular object possesses can come from several different sources, and all of those rights are combined in the final object. For example, a user can have some property rights to an object granted by inheritance from a container, while other rights are explicitly granted by an administrator. The end result is a combination of those rights. Add the [Public] object, group objects, inherited rights filters, and several layers of container objects, and it can sometimes be quite difficult to determine what the user's effective rights actually are.

The *effective rights* of an object are the net result of all of the mechanisms that can grant or revoke object or property rights. The rights explicitly granted to an object or granted through security equivalences or group memberships take precedence over those inherited from containers or the [Public] object. Specifically, effective rights are the combination of all of the following influences:

- Rights explicitly granted to the object

- Rights received through security equivalences and group memberships

- Rights received through the [Public] object

- Rights inherited from container objects

- Inherited rights blocked by inherited rights filters

While it is possible to ascertain the effective rights of an object by comparing the effects of all of these influences, it is fortunate that NDS tools like NetWare Administrator and Netadmin.exe can display the effective rights for a particular object, as shown in Figure 20-14.

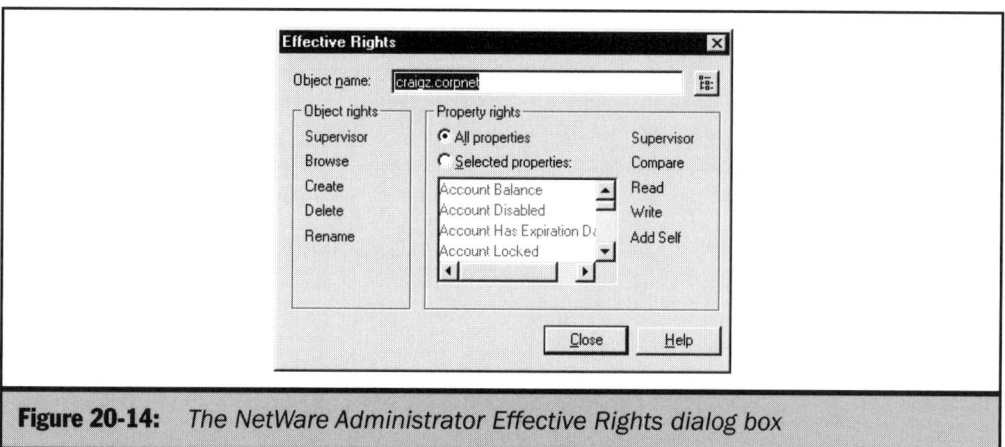

Figure 20-14: *The NetWare Administrator Effective Rights dialog box*

It is important for everyone who takes part in the maintenance of your NDS tree to understand the concept of effective rights and the various mechanisms you are using to assign object and property rights. Many administrators limit the mechanisms they use to assign rights in order to simplify maintenance tasks, and there must be specific policies in effect so that everyone knows which mechanisms they should use. One new staff member who is fond of using inherited rights filters can create a great deal of confusion if no one else is familiar with them.

Chapter 21

Unix

U nix is a multiuser, multitasking operating system (OS) with roots that date back to the late 1960s. It was developed throughout the 1970s by researchers at AT&T's Bell Labs, finally culminating in Unix System V Release 1 in 1983. During this time, and since then, many other organizations have built their own variants on the Unix formula, and now dozens of different operating systems function using the same basic Unix components. This was possible because, from the beginning, Unix has been more of a collaborative research project than a commercial product. Unlike companies such as Microsoft and Novell, which jealously guard the source code to their operating systems, many Unix developers make their code freely available. This enables anyone with the appropriate skills to modify the OS to their own specifications.

Unix is not a user-friendly OS, nor is it commonly found on the desktop of the average personal computer user. To its detractors, Unix is an outdated OS that relies primarily on an archaic, character-based interface. To its proponents, however, Unix is the most powerful, flexible, and stable OS available. As is usually the case, both opinions are correct, to some degree.

Although its popularity is growing tremendously, you are not going to see racks of Unix-based games and other recreational software at the computer store any time soon, nor are you likely to see offices full of employees running productivity applications, such as word processors and spreadsheets, on Unix systems. However, when you use a browser to connect to a site on the Web, there's a good chance that the server hosting the site is running some form of Unix. In addition, many of the vertical applications designed for specific industries, such as those used when you book a hotel room or rent a car, run on Unix systems.

Despite all this talk of its power and complexity, it is not impossible to provide a user with a Unix system that has a GUI and runs applications such as word processors and Web browsers that function very much like their Windows counterparts. GUIs are not as completely integrated into the Unix operating systems as they are in the various versions of Windows. Many of the most powerful Unix features require the use of the command prompt, and in many cases, the GUI is just another program running on the computer that provides a simpler method executing these commands, rather than being an integral part of the OS.

As a server operating system, Unix has a reputation for being stable enough to support mission-critical applications, portable enough to run on many different hardware platforms, and scaleable enough to support a user base of almost any size. All Unix systems use TCP/IP as their native protocols, so they are naturally suited for use on the Internet and for networking with other operating systems. In fact, Unix systems were instrumental in the development of the Internet from an experiment in decentralized, packet-switched networking to the worldwide phenomenon it is today.

Unix Principles

More than other operating systems, Unix is based on a principle of simplicity that makes it highly adaptable to many different needs. This is not to say that Unix is simple to use, because generally it isn't. Rather, it means that the OS is based on guiding principles that treat the various elements of the computer in a simple and consistent way. For example, a Unix system treats physical devices in the computer, such as the printer, the keyboard, and the display, in exactly the same way as it treats the files and directories on its drives. You can copy a file to the display or to a printer just as you would copy it to another directory, and use the devices with any other appropriate file-based tools.

Another fundamental principle of Unix is the use of small, simple tools that perform specific functions and that can easily work together with other tools to provide more complex functions. Instead of large applications with many built-in features, Unix operating systems are far more likely to utilize a small tool that provides a basic service to other tools. A good example is the sort command, which takes the contents of a text file, sorts it according to user-supplied parameters, and sends the results to an output device, such as the display or a printer. In addition to applying the command to an existing text file, you can use it to sort the output of other commands before displaying or printing it.

The element that lets you join tools together in this way is called a *pipe* (" | "), which enables you to use one tool to provide input to or accept output from another tool. DOS can use pipes to redirect standard input and output in various ways, but Unix includes a much wider variety of tools and commands that can be combined to provide elaborate and powerful functions.

Thus, Unix is based on relatively simple elements, but its ability to combine those elements makes it quite complex. While a large application attempts to anticipate the needs of the user by combining its functions in various predetermined ways, Unix supplies users with the tools that provide the basic functions and lets them combine the tools to suit their own needs. The result is an OS with great flexibility and extensibility, but that requires an operator with more than the average computer user's skills to take full advantage of it.

Because of this guiding principle, Unix is in many ways a "programmer's operating system." If a tool to perform a certain task is not included, you usually have the resources available to fashion one yourself. This is not to say that you have to be a programmer to use Unix, but many of the techniques that programmers use when writing code are instrumental to the use of multiple tools on the Unix command line.

If all of this talk of programming and command-line computing is intimidating, be assured that it is quite possible to install, maintain, and use a Unix system without a substantial investment in learning command-line syntax. Some of the Unix operating systems are being geared more and more to the average computer user, with most of the common system functions available through the GUI. You can perform most of your daily computing tasks on these operating systems without ever seeing a command prompt.

The various Unix operating systems are built around basic elements that are fundamentally the same, but they include various collections of tools and programs. Depending on which variant you choose and whether it is a commercial product or a free download, you may find that the OS comes complete with modules such as Web and DNS servers and other programs, or you may have to obtain these yourself. However, one of the other principles of Unix development that has endured through the years is the custom of making the source code for Unix software freely available to everyone. The result of this open source movement is a wealth of Unix tools, applications, and other software that is freely available for download from the Internet.

In some cases, programmers modify existing Unix modules for their own purposes and then release those modifications to the public domain so that they can be of help to others. Some programmers collaborate on Unix software projects as something of a hobby and release the results to the public. One of the best examples of this is the Linux operating system, which was designed from the beginning to be a free product and which has now become one of the most popular Unix variants in use today.

Unix Architecture

Because Unix is available in so many variants, Unix operating systems can run on a variety of hardware platforms, from simple Intel 80386-based PCs to dedicated workstations costing hundreds of thousands of dollars. Many of the Unix variants are proprietary versions created by specific manufacturers to run on their own hardware platforms. Sun Microsystems' Solaris, Hewlett-Packard's HP-UX, IBM's AIX, and Silicon Graphics' IRIX are all examples of operating systems that are packaged with a workstation built by the same manufacturer. Most of the software-only Unix solutions run on Intel-based PCs, and some are available in versions for multiple platforms. Solaris, for example, comes in a version for Sun's proprietary SPARC platform and in an Intel version.

The hardware requirements for the various Unix platforms vary greatly, depending on the functions required of the machine. You can run Linux on an old 386, for example, as long as you don't expect to use a GUI or run a server supporting a large number of users. At the high end, some of the proprietary workstation platforms, like those from Silicon Graphics, are massively powerful (and massively expensive) machines that the movie industry uses to create state-of-the-art special effects.

No matter what hardware a Unix system uses, the basic software components are the same (see Figure 21-1). The *kernel* is the core module that insulates the programs running on the computer from the hardware. The kernel uses device drivers that interact with the specific hardware devices installed in the computer, to perform basic functions like memory management, input/output, interrupt handling, and access control.

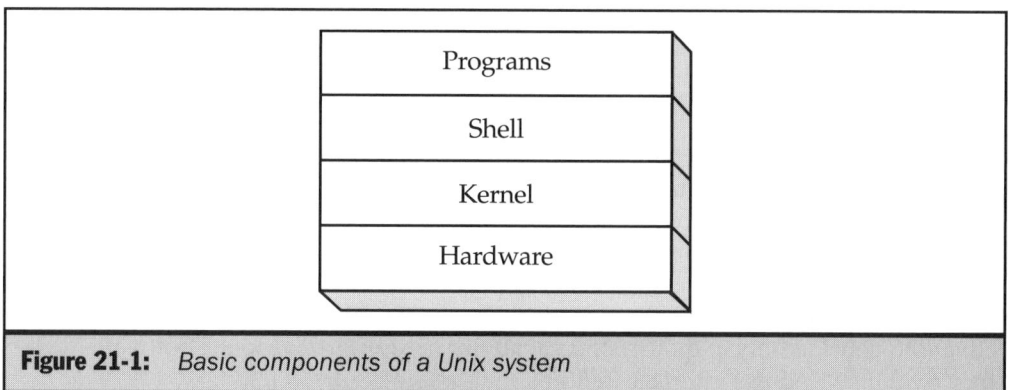

Figure 21-1: *Basic components of a Unix system*

The Unix kernel provides approximately 100 system calls that programs can use to execute certain tasks, such as opening a file, executing a program, and terminating a process. These are the building blocks that programmers use to integrate hardware-related functions into their applications' more complex tasks. The system calls can vary between the different Unix versions to some extent, particularly in the way that the system internals perform the different functions.

Above the kernel is the shell, which provides the interface you use to issue commands and execute programs. The shell is a command interpreter, much like Command.com in DOS and Cmd.exe in Windows 2000 and NT, which provides a character-based command prompt that you use to interact with the system. The shell also functions as a programming language you can use to create scripts, which are functionally similar to DOS batch files but much more versatile and powerful.

Unlike DOS and Windows, which limit you to a single command interpreter, Unix traditionally has several shells you can choose from, with different capabilities. The shells that are included with particular Unix operating systems vary, and others are available as free downloads. Often, the selection of a shell is a matter of personal preference, guided by the user's previous experience. The basic commands used for file management and other standard system tasks are the same in all of the shells. The differences become more evident when you run more complex commands and create scripts.

The original Unix shell is a program called *sh* that was created by Steve Bourne and is commonly known as the *Bourne shell.* Some of the other common shells are as follows:

csh   Known as the C shell, and originally created for use with BSD Unix; utilizes a syntax similar to that of the C language and introduces features such as a command history list, job control, and aliases. Scripts written for the Bourne shell usually need some modification to run in the C shell.

ksh   Known as the *Korn shell*; builds on the Bourne shell and adds elements of the C shell, as well as other improvements. Scripts written for the Bourne shell usually can run in the Korn shell without modification.

bash   The default shell used by Linux; closely related to the Korn shell, with elements of the C shell.

In addition to the character-based shells, it is also possible to use a GUI to interact with a Unix system. Many different GUIs are available for the various Unix versions, such as Motif and Open Look. Some Unix variants, such as Solaris, include their own proprietary GUI implementation, while others, such as Linux, use an X Window server that enables you to run any one of many window managers, such as the K Desktop Environment (KDE). In response to the popularity of Windows, Unix GUIs have advanced considerably in quality and capability in recent years, to the point at which they rival the Windows interface in functionality (see Figure 21-2).

Figure 21-2:   *The K Desktop Environment can be configured to look a lot like Windows.*

Above the shell are the commands that you use to perform tasks on the system. Unix includes hundreds of small programs, usually called *tools* or *commands*, which you can combine on the command line to perform complex tasks. Hundreds of other tools are available on the Internet that you can combine with those provided with the OS. Unix command-line tools are programs, but don't confuse them with the complex applications used by other operating systems, such as Windows. Unix has full-blown applications as well, but its real power lies in these small programs. Adding a new tool on a Unix system does not require an installation procedure; you simply have to specify the appropriate location of the tool in the file system in order for the shell to run it.

Unix Versions

The sheer number of Unix variants can be bewildering to anyone trying to find the appropriate operating system for a particular application. However, apart from systems intended for special purposes, virtually any Unix OS can perform well in a variety of roles, and the selection you make may be based more on economic factors, hardware platform, or personal taste than on anything else. If, for example, you decide to purchase proprietary Unix workstations, you'll be using the version of the OS intended for the machine. If you intend to run Unix on Intel-based computers, you might choose the OS based on the GUI that you feel most comfortable with, or you might be looking for the best bargain you can find and limit yourself to the versions available as free downloads. The following sections discuss some of the major Unix versions available.

Unix System V

Unix System V is the culmination of the original Unix work begun by AT&T's Bell Labs in the 1970s. Up until release 3.2, the project was wholly developed by AT&T, even while other Unix work was ongoing at the University of California at Berkeley and other places. Unix System V Release 4 (SVR4), released in the late 1980s, consolidated the benefits of the SVR operating system with those of Berkeley's BSD, Sun's SunOS, and Microsoft's Xenix. This release brought together some of the most important elements that are now indelibly associated with the name Unix, including networking elements like the TCP/IP Internet Package from BSD, which includes file transfer, remote login, and remote program execution capabilities, and the Network File System (NFS) from SunOS.

AT&T eventually split its Unix development project off into a subsidiary called Unix System Laboratories (USL), which released System V Release 4.2. In 1993, AT&T sold USL to Novell, which released its own version of SVR4 under the name UnixWare. In light of pressure from the other companies involved in Unix development, Novell transferred the Unix trademark to a consortium called X/Open, thus enabling any manufacturer to describe its product as a Unix OS. In 1995, Novell sold all of its interest in Unix SVR4 and UnixWare to the Santa Cruz Operation (SCO), which owns it to this day. In 1997, SCO released Unix System V Release 5 (SVR5) under the name OpenServer, as well as version 7 of its UnixWare product. These are the descendents of the original AT&T products, and they are still on the market.

BSD Unix

In 1975, one of the original developers of Unix, Ken Thompson, took a sabbatical at the University of California at Berkeley, and while there, he ported his current Unix version to a PDP-11/70 system. The seed he planted took root, and Berkeley became a major developer of Unix in its own right. BSD Unix introduced several of the major features associated with most Unix versions, including the C shell and the vi text editor. Several versions of BSD Unix appeared throughout the 1970s, culminating in 3BSD. In 1979, the U.S. Department of Defense's Advanced Research Projects Agency (DARPA) funded the development of 4BSD, which coincided with the development and adoption of the TCP/IP networking protocols.

Eventually, BSD Unix came to be the OS that many other organizations used as the basis for their own Unix products, including Sun Microsystems' SunOS. The result is that many of the programs written for one BSD-based Unix version are binary-compatible with other versions. Once the SVR4 release consolidated the best features of BSD and several other Unix versions into one product, the BSD product became less influential and culminated in the 4.4BSD version in 1992.

Although many of the Unix variants that are popular today owe a great debt to the BSD development project, the versions of BSD that are still commonly used are public domain operating systems, such as FreeBSD, NetBSD, and Open BSD. All of these operating systems are based on Berkeley's 4.4BSD release and can be downloaded from the Internet free of charge and used for private and commercial applications at no cost.

FreeBSD

FreeBSD, available at http://www.freebsd.org/ in versions for the Intel and Alpha platforms, is based on the Berkeley 4.4BSD-Lite2 release, and is binary-compatible with Linux, SCO, SVR4, and NetBSD applications. The FreeBSD development project is divided into two branches: the STABLE branch, which includes only well-tested bug fixes and incremental enhancements, and the CURRENT branch, which includes all of the latest code and is intended primarily for developers, testers, and enthusiasts. The current stable version as of April 2001 is 4.3.

NetBSD

NetBSD, available at http://www.netbsd.org/, is derived from the same sources as FreeBSD, but places portability as one of its highest priorities. NetBSD is available in formal releases for 15 different hardware platforms, ranging from Intel and Alpha to Macintosh, SPARC, and MIPS processors, including those designed for handheld Windows CE devices. Many other ports are in the developmental and experimental stages. NetBSD's binary compatibility enables it to support applications written for many other Unix variants, including BSD, FreeBSD, HP/UX, Linux, SVR4, Solaris, SunOS, and others. Networking capabilities supported directly by the kernel include NFS, IPv6, Network Address Translation (NAT), and packet filtering. The latest version of NetBSD, released in December 2000, is 1.5.

OpenBSD

OpenBSD is available at http://www.openbsd.org/; the current version is 2.9, released in June 2001. Like the other BSD-derived operating systems, OpenBSD is binary-compatible with most of its peers, including FreeBSD, SVR4, Solaris, SunOS, and HP/UX, and currently supports 11 hardware platforms, including Intel, Alpha, SPARC, PowerPC, and others. However, the top priority of OpenBSD's developers is security and cryptography. Because OpenBSD is a noncommercial product, its developers feel they can take a more uncompromising stance on security issues and disclose more information about security than commercial software developers. Also, because it is developed in and distributed from Canada, OpenBSD is not subject to the American laws that prohibit the export of cryptographic software to other countries. The developers are, therefore, more likely to take a cryptographic approach to security solutions than American-based companies.

Sun Solaris

Sun Microsystems (http://www.sun.com/) has been involved in Unix development since the early 1980s, when its operating system was known as SunOS. In 1991, Sun created a subsidiary called SunSoft that began work on a new Unix version based on SVR4, which it called Solaris. Several versions later, Solaris 8 is one of the most popular (if not the most popular) commercial Unix versions on the market.

Aside from the SunOS kernel, the Solaris OS includes the OpenWindows 3.x GUI, which is X Window-based. Sun also manufactures its own proprietary computer hardware, based around a processor called SPARC, and Solaris is naturally designed to take full advantage of the platform, although it is available in a version for Intel systems as well. Although Solaris is a commercial product, Sun also supplies the OS to noncommercial users both as a free download or on CD-ROM for a relatively low cost.

Linux

Developed as a college project by Linus Torvalds of Sweden, Linux has emerged as one of the most popular Unix variants in recent years. Like FreeBSD, NetBSD, and OpenBSD, Linux is available as a free download from the Internet in versions for most of the standard hardware platforms, and is continually refined by an ad hoc group of programmers that communicate mainly through Internet mailing lists and newsgroups. Because of its popularity, a lot of people are working on the development of Linux modules and applications. Many new features and capabilities are the result of programmers adapting the existing software for their own uses and then posting their code for others to use. As the product increases in popularity, more people work on it in this way, and the development process accelerates. This flurry of activity has also led to the fragmentation of the Linux development process. Many different Linux versions are now available, which are similar in their kernel functions but vary in the features they include. As with the various BSD operating systems now available, some of these Linux packages are available for download on the Internet, but the growth in the popularity of the OS has led to some commercial distribution releases as well.

The most popular version of Linux right now is called Red Hat Linux, distributed by a company whose success is based on the packaging and sales of a free product for a more general audience. The popularity of Linux has reached the point at which it is expanding beyond Unix's traditional market of computer professionals and technical hobbyists. In part, this is due to a backlash against Microsoft, which some people believe is close to holding a monopoly on operating systems. When you pay for a "commercial" Linux release like Red Hat, you get not only the OS and source code on CD-ROM, but also a variety of applications, product documentation, and technical support, which are often lacking in the free download releases. Other distributors, such as Slackware, Debian, and Caldera, provide similar products and services, but this does not necessarily mean that these Linux versions are binary-compatible. In some cases, software written for one distribution will not run on another one.

The free Linux distributions provide much of the same functionality as the commercial ones, but in a less convenient package. The downloads can be very large and time-consuming, and you may find yourself interrupting the installation process frequently to track down some essential piece of information or to download an additional module you didn't know you needed. Generally speaking, the commercial Linux products offer a slicker package with a smoother installation process and the knowledge that you have on the CD-ROMs all of the software you need to get a system up and running. You also have better documentation and someone to contact for help when you need it, all for a minimal cost (about $50 to $100). There are also commercial Linux distributions designed for use on large servers that are considerably more expensive.

One of the biggest advantages of Linux over other Unix variants is its excellent driver support. Device drivers are an integral part of any operating system, and if Unix is ever going to become a rival to Windows in the personal computer mainstream, it's going to have to run on the same computers that run Windows, using the same peripherals. Many of the other Unix variants have relatively limited device driver support. If you are trying to install a Unix product on an Intel-based computer with the latest and greatest video adapter, for example, you may not be able to find a driver that takes full advantage of its capabilities.

Device drivers, even those included with operating systems, are generally written by the device manufacturer. Not surprisingly, the driver development effort of most hardware manufacturers devotes most of its attention to Windows, with Unix getting only perfunctory support, if any at all. The fans of Linux are legion, however, and the OS's development model has led the operating system's supporters to develop their own drivers for many of the devices commonly found in Intel-based computers. If you are having trouble finding appropriate drivers for your hardware that run on other Unix variants, you are more likely to have success with Linux.

Unix Networking

Unix is a peer-to-peer network operating system, in that every computer is capable of both accessing resources on other systems and sharing its own resources. These networking capabilities take three basic forms, as follows:

■ The ability to open a session on another machine and execute commands on its shell.

■ The ability to access the file system on another machine, using a service like NFS.

■ The ability to run a service (called a *daemon*) on one system and access it using a client on another system.

The TCP/IP protocols are an integral part of all Unix operating systems, and many of the TCP/IP programs and services that may be familiar to you from working with the Internet are also implemented on Unix networks. For example, Unix networks can use DNS servers to resolve host names into IP addresses and use BOOTP or DHCP servers to automatically configure TCP/IP clients. Standard Internet services like FTP and Telnet have long been a vital element of Unix networking, as are utilities like ping and traceroute.

The following sections examine the types of network access used on Unix systems and the tools involved in implementing them.

NETWORK OPERATING SYSTEMS

Note   *For more information on the TCP/IP protocols, see Chapter 13. For more information on DHCP and DNS, see Chapters 23 and 25.*

Using Remote Commands

One form of network access that is far more commonly used on Unix than on other network operating systems is the remote console session, in which a user connects to another computer on the network and executes commands on that system. Once the connection is established, commands entered by the user at the client system are executed by the remote server and the output is redirected over the network back to the client's display. It's important to understand that this is not the equivalent of accessing a shared network drive on a Windows computer and executing a file. In the latter case, the program runs using the client computer's processor and memory. When you execute a command on a Unix computer using a remote console session, the program actually runs on the other computer, using its resources.

Because Unix relies heavily on the command prompt, character-based remote sessions are more useful than they are in a more graphically oriented environment like that of Windows. Although a remote console application is available for Windows 2000 and NT, its command-line capabilities are relatively limited when compared to those of Unix.

Berkeley Remote Commands

The Berkeley Remote Commands were originally a part of BSD Unix, and have since been adopted by virtually every other Unix OS. Sometimes known as the r* commands, these tools are intended primarily for use on LANs, rather than over WAN or Internet links. These commands enable you not only to open a session on a remote system, but also to perform specific tasks on a remote system without logging in and without working interactively with a shell prompt.

rlogin

The rlogin command establishes a connection to another system on the network and provides access to its shell. Once connected, any commands you enter are executed by the other computer using its processor, file system, and other components. To connect to another machine on the network, you use a command like the following:

```
rlogin [-l username] hostname
```

where the *hostname* variable specifies the name of the system to which you want to connect.

Authentication is required for the target system to establish the connection, which can happen using either host-level or user-level security. To use host-level security, the client system must be trusted by the server by having its host name listed in the /etc/host.equiv file on the server. When this is the case, the client logs in without a user name or password, because it is automatically trusted by the server no matter who's using the system.

User-level security requires the use of a user name and sometimes a password, in addition to the host name. By default, rlogin supplies the name of the user currently logged in on the client system to the remote system, as well as information about the type of terminal used to connect, which is taken from the value of the TERM variable. The named user must have an account in the remote system's password database, and if the client system is not trusted by the remote system, the remote system may then prompt the client for the password associated with that user name. It's also possible to log in using a different user name by specifying it on the rlogin command line with the -l switch.

For the user name to be authenticated by the remote system without using a password, it must be defined as an equivalent user by being listed in a .rhosts file located in the user's home directory on that system. The .rhosts file contains a list of host names and user names that specify whether a user working on a specific machine should be granted immediate access to the command prompt. Depending on the security requirements for the remote system, the .rhosts files can be owned either by the remote users themselves or the root account on the system. Adding users to your .rhosts file is a simple way of giving them access to your account on that machine without giving them the password.

Note
The root *account on a Unix computer is a built-in superuser that has full access to the entire system, much like the Administrator account in Windows 2000/NT and the Admin account in NetWare.*

Once you have successfully established a connection to a remote system, you can execute any command in its shell that you would on your local system, except for those that launch graphical applications. You can also use rlogin from the remote shell to connect to a third computer, giving you simultaneous access to all three. To terminate the connection to a remote system, you can use the exit command, press the CTRL-D key combination, or type a tilde followed by a period (~.).

rsh

In some instances, you may want to execute a single command on a remote system and view the resulting output without actually logging in. You can do this with the rsh command, using the following syntax:

```
rsh hostname command
```

NETWORK OPERATING SYSTEMS

where the *hostname* variable specifies the system on which you want to open a remote shell, and the *command* variable is the command to be executed on the remote system. Unlike rlogin, interactive authentication is not possible with rsh. For the command to work, the user must have either a properly configured .rhosts file on the remote system or an entry in the /etc/host.equiv file. The rsh command provides essentially the same command-line capabilities as rlogin, except that it works for only a single command and does not maintain an open session.

 *The rsh command is called remsh on HP-UX systems. There are many cases in which commands providing identical functions have different names on various Unix operating systems.*

rcp

The rcp command is used to copy files to or from a remote system across a network without performing an interactive login. The rcp functions much like the cp command used to copy files on the local system, using the following syntax:

```
rcp [-r] sourcehost:filename desthost:filename
```

where the *sourcehost:filename* variable specifies the host name of the source system and the name of the file to be copied, and the *desthost:filename* variable specifies the host name of the destination system and the name that the file should be given on that system. You can also copy entire directories by adding the -r parameter to the command and specifying directory names instead of file names. As with rsh, there is no login procedure, so to use rcp, either the client system must be trusted by the remote system or the user must be listed in the .rhosts file.

Secure Shell Commands

The downside of the Berkeley Remote Commands is that they are inherently insecure. Passwords are transmitted over the network in clear text, making it possible for intruders to intercept them. Because of this susceptibility to compromise, some administrators prohibit the use of these commands. To address this problem, there is a Secure Shell program that provides the same functions as rlogin, rsh, and rcp, but with greater security. The equivalent programs in the Secure Shell are called slogin, ssh, and scp. The primary differences in using these commands are that the connection is authenticated on both sides and all passwords and other data are transmitted in encrypted form.

 *For more information about the Secure Shell, see http://www.ssh.fi/.*

DARPA Commands

The Berkeley Remote Commands are designed for use on like Unix systems, but the DARPA commands were designed as part of the TCP/IP protocol suite, and can be used by any two systems that support TCP/IP. Virtually all Unix operating systems include both the client and server programs for telnet, ftp, and tftp, and install them by default, although some administrators may choose to disable them later.

telnet

The telnet program is similar in its functionality to rlogin, except that telnet does not send any information about the user on the client system to the server. You must always supply a user name and password to be authenticated. As with all of the DARPA commands, you can use a Telnet client to connect to any computer running a Telnet server, even if it is running a different version of Unix or a non-Unix OS. The commands you can use while connected, however, are wholly dependent on the OS running the Telnet server. If, for example, you install a Telnet server on a Windows system, you can connect to it from a Unix client, but once connected, you can only use commands recognized by Windows. Since Windows is not primarily a character-based OS, its command-line capabilities are relatively limited, unless you install outside programs.

ftp

The ftp command provides more comprehensive file transfer capabilities than rcp, and enables a client to access the file system on any computer running an FTP server. However, instead of accessing files in place on the other system, ftp provides only the ability to transfer files to and from the remote system. For example, you cannot edit a file on a remote system, but you can download it to your own system, edit it there, and then upload the new version back to the original location. As with Telnet, users must authenticate themselves to an FTP server before they are granted access to the file system. Many systems running FTP, such as those on the Internet, support anonymous access, but even this requires an authentication process of sorts in which the user supplies the name "anonymous" and the server is configured to accept any password.

 *For more information on the FTP protocol, see Chapter 26.*

tftp

The tftp command uses the Trivial File Transfer Protocol to copy files to or from a remote system. Whereas ftp relies on the Transmission Control Protocol (TCP) at the transport layer, tftp uses the User Datagram Protocol (UDP) instead. Because UDP is a connectionless protocol, no authentication by the remote system is needed. However, this limits the command to copying only files that are publicly available on the remote system. The TFTP protocol was designed primarily for use by diskless workstations that have to download an executable operating system file from a server during the boot process.

Network File System (NFS)

Sharing files is an essential part of computer networking, and Unix systems use several mechanisms to access files on other systems without first transferring them to a local drive, as with ftp and rcp. The most commonly used of these mechanisms is the Network File System (NFS), which was developed by Sun Microsystems in the 1980s and has now been standardized by the Internet Engineering Task Force (IETF) as RFC 1094 (NFS Version 2) and RFC 1813 (NFS Version 3). By allowing NFS to be published as an open standard, Sun made it possible for anyone to implement the service, and the result is that NFS support is available for virtually every OS in use today.

Practically every Unix variant available includes support for NFS, which makes it possible to share files among systems running different Unix versions. Non-Unix operating systems, such as Windows and NetWare, can also support NFS, but a separate product (marketed by either the manufacturer or a third party) is required. Since Windows and NetWare have their own internal file-sharing mechanisms, these other operating systems mostly require NFS only to integrate Unix systems into their networks.

NFS is a client/server application in which a server makes all or part of its file system available to clients (using a process called *exporting* or *sharing*), and a client accesses the remote file system by *mounting* it, which makes it appear just like part of the local file system. NFS does not communicate directly with the kernel on the local computer, but rather relies on the remote procedure calls (RPC) service, also developed by Sun, to handle communications with the remote system. RPC has also been released as an open standard by Sun, and published as an IETF document called RFC 1057. The data transmitted by NFS is encoded using a method called *External Data Representation (XDR)*, as defined in RFC 1014. In most cases, the service uses the UDP protocol for network transport and listens on port 2049.

NFS is designed to keep the server side of the application as simple as possible. NFS servers are *stateless*, meaning they do not have to maintain information about the state of a client to function properly. In other words, the server does not maintain information about which clients have files open. In the event that a server crashes, clients simply continue to send their requests until the server responds. If a client crashes, the server continues to operate normally. There is no need for a complicated reconnection sequence. Because repeated iterations of the same activities can be the consequence of this statelessness, NFS is also designed to be as *idempotent* as possible, meaning that the repeated performance of the same task will not have a deleterious effect on the performance of the system. NFS servers also take no part in the adaptation of the exported file system to the client's requirements. The server supplies file system information in a generalized form, and it is up to the client to integrate it into its own file system so that applications can make use of it.

The communication between NFS clients and servers is based on a series of RPC procedures defined in the NFS standard and listed in Table 21-1. These basic functions enable the client to interact with the file system on the server in all of the ways expected by a typical application.

Procedure Number	Procedure Name	Function
0	NULL	Does not do any work; used for server response testing and timing.
1	GETATTR	Retrieves attributes for a specified file system object.
2	SETATTR	Changes one or more of the attributes of a file system object on the server.
3	LOOKUP	Searches a directory for a specific name and returns the file handle for the corresponding file system object.
4	ACCESS	Determines the access rights that a user has with respect to a file system object.
5	READLINK	Reads the data associated with a symbolic link.
6	READ	Reads data from a file.
7	WRITE	Writes data to a file.
8	CREATE	Creates a regular file.
9	MKDIR	Creates a new subdirectory.

Table 21-1: *RPC Procedures Supplied by an NFS Version 3 Protocol Server*

Procedure Number	Procedure Name	Function
10	SYMLINK	Creates a new symbolic link.
11	MKNOD	Creates a new special file.
12	REMOVE	Deletes a file from a directory.
13	RMDIR	Deletes a subdirectory from a directory.
14	RENAME	Renames a file or directory.
15	LINK	Creates a link to an object.
16	READDIR	Retrieves a variable number of entries from a directory and returns the name and file identifier for each.
17	READDIRPLUS	Retrieves a variable number of entries from a file system directory and returns complete information about each.
18	FSSTAT	Retrieves volatile file system state information.
19	FSINFO	Retrieves nonvolatile file system state information and general information about the NFS Version 3 protocol server implementation.
20	PATHCONF	Retrieves POSIX information for a file or directory.
21	COMMIT	Forces or flushes data to stable storage that was previously written with a WRITE procedure call with the stable field set to UNSTABLE.

Table 21-1: *RPC Procedures Supplied by an NFS Version 3 Protocol Server (continued)*

On a system configured to function as an NFS server, you can control which parts of the file system are accessible to clients by using commands such as share on Solaris and SVR4 systems, and exportfs on Linux and HP-UX. Using these commands, you specify which directories clients can access and what degree of access they are provided. You can choose to share a directory on a read-only basis, for example, or grant read/write access, and you can also designate different access permissions for specific users.

Client systems access the directories that have been shared by a server by using the mount command to integrate them into the local file system. The mount command specifies a directory shared by a server, the access that client applications should have to the remote directory (such as read/write or read-only), and the mount point for the remote files. The *mount point* is a directory on the local system in which the shared files and directories will appear. Applications and commands running on the client system can reference the remote files just as if they were located on a local drive.

Client/Server Networking

Client/server computing is the basis for networking on Unix systems, as it is on many other computing platforms. Unix is a popular application server platform largely because its relative simplicity and flexibility enables the computer to devote more of its resources toward its primary function. On a Windows 2000 server, for example, a significant amount of system resources are devoted to running the GUI and other subsystems that may have little or nothing to do with the server applications that are its primary functions. When you dedicate a computer to functioning as a Web server, for example, and you want it to be able to service as many clients as possible, it makes sense to disable all extraneous functions, which is something that is far easier to do on a Unix system than in Windows.

Server applications on Unix systems typically run as *daemons*, which are background processes that run continuously, regardless of the system's other activities. There are many commercial server products available for various Unix versions, and also a great many that are available free of charge. Because the TCP/IP protocols were largely developed on the Unix platform, Unix server software is available for every TCP/IP application in existence. For example, a computer running Linux as its OS and Apache as its Web server software is a powerful combination that is easily equal or superior to most of the commercial products on the market—and the software is completely free.

Chapter 22

Network Clients

A lthough network administrators frequently spend a lot of time installing and configuring servers, the primary reason for the servers' existence is the clients. The choice of applications and operating systems for your servers should be based in part on the client platforms and operating systems that have to access them. Generally speaking, it is possible for any client platform to connect to any server, one way or another, but this doesn't mean that you should choose client and server platforms freely and expect them all to work well together in every combination.

For ease of administration, it's a good idea to use the same operating system on all of your client workstations wherever possible. Most network installations use standard Intel-based PCs running some version of Microsoft Windows, but even if you choose to standardize on Windows, you may have some users with special needs that require a different platform. Graphic artists, for example, are often accustomed to working on Macintosh systems, and other users may need Unix workstations. When selecting server platforms, you should consider what is needed to enable users on various client platforms to access them. Table 22-1 lists the most common server and workstation platforms and the clients most often used to connect them.

When you run various server platforms along with multiple clients, the process becomes even more complicated, because each workstation might require multiple clients. The impact of multiple network clients on the performance of the computer depends on exactly which clients are involved, but in some instances, the effect can be significant in terms of reduced speed, increased resource utilization, and more complicated troubleshooting. This chapter examines the client platforms commonly used on networks today, and the software used to connect them to various servers.

Windows Network Clients

Although Microsoft Windows began as a standalone operating system, networking soon became a ubiquitous part of Windows, and all versions now include a client that enables them to connect to any other Windows computer. Windows networking was first introduced in the Windows NT 3.1 and Windows for Workgroups releases in 1993. The Windows networking architecture is based on network adapter drivers written to the Network Device Interface Specification (NDIS) standard and, originally, on the NetBEUI protocol. Later, TCP/IP became the default networking protocol.

Windows networking is a peer-to-peer system that enables any computer on the network to access resources on any other computer. When Microsoft introduced networking into Windows, the predominant network operating system was Novell NetWare, which uses the client/server model that enables clients to access server resources only. Adding peer-to-peer networking to an already popular, user-friendly operating system like Windows led to its rapid growth in the business LAN industry and its eventual encroachment into NetWare's market share.

	Windows 2000/ NT Servers	Novell NetWare Servers	Unix Servers
DOS Clients	Microsoft Client 3.0 for MS-DOS	Novell Client for DOS/Windows	Third-party clients only
Windows for Workgroups Clients	Microsoft Windows Network internal client; TCP/IP-32	Novell Client for DOS/Windows	DARPA Commands (using Winsock and Telnet/FTP clients supplied in TCP/IP-32)
Windows 95/98 Clients	Microsoft Client for Microsoft Networks	Microsoft Client for NetWare Networks; Novell Client for Windows 95/98	DARPA Commands (FTP and Telnet)
Windows 2000 Professional/ Windows NT Workstation Clients	Microsoft Windows Network internal client	Microsoft Client Service for NetWare; Novell Client for Windows NT/2000	DARPA Commands (FTP and Telnet); Microsoft Windows Services for Unix
MacOS Clients	Microsoft Services for Macintosh	Novell Client for MacOS; Novell NetWare AppleTalk; Novell Native File Access for Macintosh	DARPA Commands (FTP and Telnet)
Unix Clients	Microsoft Windows Services for Unix (SAMBA, at http://www.samba.org/)	NetWare NFS Services	Berkeley Remote Commands; DARPA Commands (FTP and Telnet); NFS

Table 22-1: *Network Client/Server Connectivity Matrix*

Windows Networking Architecture

Windows 3.1 and 3.11 were the only major versions of the operating environment that lacked a networking stack of their own, but it is possible to use Microsoft Client 3.0 for MS-DOS to connect them to a Windows network. All of the other Windows versions, including Windows for Workgroups, Windows 95/98/Me, and Windows NT/2000, have built-in networking capabilities that enable the computer to participate on a Windows network.

The basic architecture of the Windows network client is the same in all of the operating systems, although the implementations differ substantially. In its simplest form, the client functionality uses the modules shown in Figure 22-1. At the bottom of the protocol stack is an NDIS network adapter driver that provides access to the network interface card (NIC) installed in the computer. Above the network adapter driver are drivers for the individual protocols running on the system. At the top of the stack is the client itself, which takes the form of one or more services on the 32-bit operating systems.

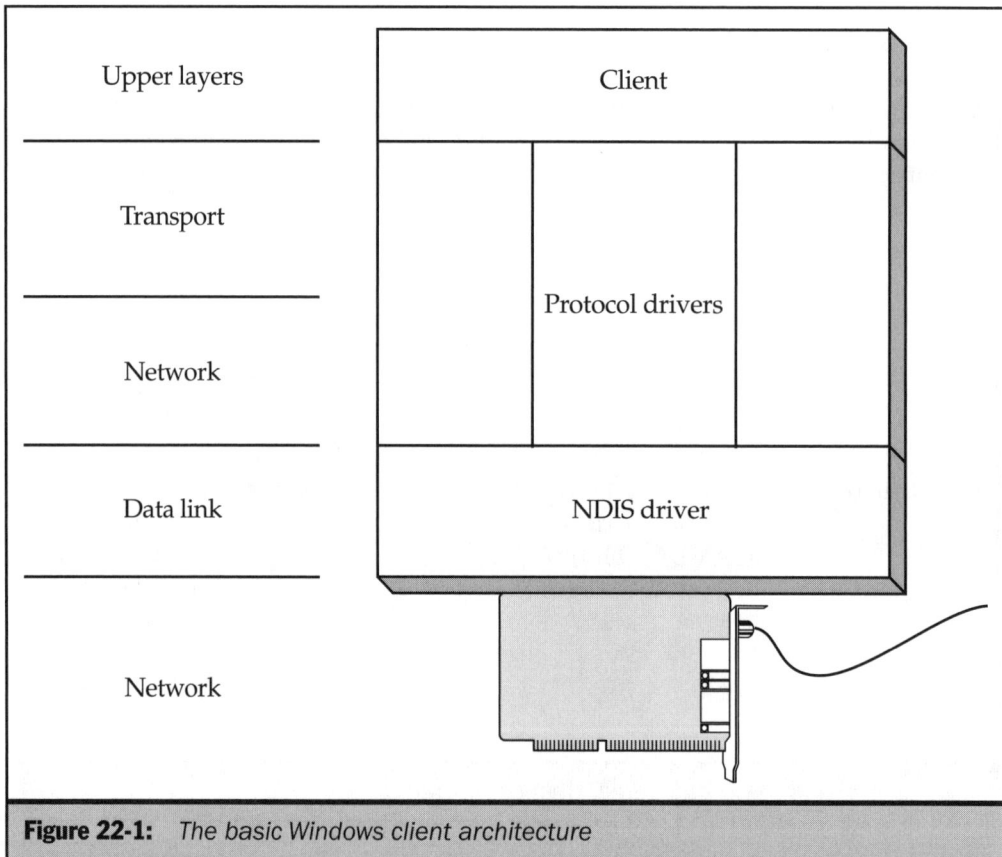

Figure 22-1: *The basic Windows client architecture*

 *Because Windows uses peer-to-peer networking, all of the operating systems can function as clients as well as servers. This means that all of the information about the Windows network clients pertains equally to the Windows 2000 and NT Server operating systems.*

These three layers form a complete protocol stack running from the application layer of the OSI model down to the physical layer. Applications generate requests for specific resources that pass through a mechanism that determines whether the resource is located on a local device or on the network. Requests for network resources are redirected down through the networking stack to the NIC, which transmits them to the appropriate devices. The following sections examine these elements in more detail.

NDIS Drivers

The Network Device Interface Specification was designed by Microsoft and 3Com to provide an interface between the data-link and network layers of the OSI model that would enable a single NIC installed in a computer to carry traffic generated by multiple protocols. This interface insulates the protocol drivers and other components at the upper layers of the protocol stack, so that the process of accessing network resources is always the same, no matter what NIC is installed in the machine. As long as there is an NDIS-compatible NIC driver available, the interface can pass the requests from the various protocol drivers to the card, as needed, for transmission over the network.

The various Windows network clients use different versions of NDIS for their adapter drivers, as shown in Table 22-2. NDIS 2 is the only version of the interface that runs in the Intel processor's real mode, using conventional rather than extended memory, and a driver file with a .dos extension. Microsoft Client 3.0 for MS-DOS relies on this version of the specification for network access, but the primary job of NDIS 2 is to function as a real-mode backup for Windows for Workgroups, Windows 95, 98, and Me. All four of these operating systems include later versions of the NDIS specification that run in protected mode, but the real-mode driver is included for situations in which it is impossible to load the protected-mode driver.

NDIS Version	Operating Systems
2	Client 3.0 for MS-DOS; Windows 95/98 (real mode)
3	Windows NT 3.1–3.51; Windows for Workgroups 3.11
3.1	Windows 95
4	Windows NT 4.0; Windows 95 OSR2
5	Windows 2000; Windows 98

Table 22-2: *NDIS Versions and the Operating Systems that Use Them*

NETWORK OPERATING SYSTEMS

For example, when you boot Windows 98 to the command prompt and load networking support using the NET START command, the system loads the NDIS 2 real-mode driver instead of the default protected-mode driver used by the GUI. The real-mode driver may not perform as efficiently as the standard driver, but it provides the workstation with basic network access using a minimum of system resources.

The primary advantage of the NDIS 3 drivers included with Windows for Workgroups and the first Windows NT releases is their ability to run in protected mode, which can use both extended and virtual memory. The driver takes the form of an NDIS wrapper, which is generic, and a miniport driver that is device specific. Because most of the interface code is part of the wrapper, the development of miniport drivers by individual NIC manufacturers is relatively simple.

NDIS 3.1, first used in Windows 95, introduced plug-and-play capabilities to the interface, which greatly simplifies the process of installing NICs. NDIS 4 provides additional functionality, such as support for infrared and other new media and power-management capabilities. NDIS 5 adds connection-oriented service that supports the ATM protocol in its native mode, as well as its quality of service functions. In addition, *TCP/IP task offloading* enables enhanced NICs to perform functions normally implemented by the transport-layer protocol, such as checksum computations and data segmenting, which reduces the load on the system processor.

All of the Windows network clients ship with NDIS drivers for an assortment of the most popular NICs that are in use at the time of the product's release. This means, of course, that older clients, like those for DOS and Windows for Workgroups, do not include support for the latest NICs on the market, but the NIC manufacturers all supply NDIS drivers for their products. Even real-mode NDIS 2 drivers are usually included with NICs, because Windows 9x can use them in safe mode.

Protocol Drivers

The Windows network clients all support the use of three protocols—NetBEUI, TCP/IP, and IPX—either alone or in combination. When Microsoft first added networking to Windows, NetBEUI was the default protocol, because it is closely related to the NetBIOS interface that Windows uses to name the computers on the network. NetBEUI is self-adjusting and requires no configuration or maintenance at all, but its lack of routing capabilities makes it unsuitable for large networks that consist of multiple segments. This shortcoming, plus the rise in the popularity of the Internet, led to TCP/IP being adopted as the protocol of choice on most networks, despite its need for individual client configuration.

The IPX protocol suite was developed by Novell for its NetWare operating system, which was the most popular networking solution at the time that Windows networking was introduced. As a result, all of the Windows clients include support for the IPX protocol, for compatibility reasons. The Windows IPX implementation is not the actual Novell IPX product, but rather an IPX-compatible protocol called NWLink that was reverse-engineered to be compatible with NetWare. Therefore, on networks already running NetWare, administrators can implement Windows networking without adding another protocol. Some administrators routinely run two or all three protocols on their networks, but this is usually not necessary on a properly organized network. Additional protocol modules consume system resources and complicate the process of administering the network.

The protocol drivers take different forms depending on the operating system. The DOS and Windows for Workgroups clients use real- and protected-mode protocol drivers that correspond to the NDIS version, while Windows 95/98/Me and NT/2000 use 32-bit drivers that conform to their particular architectures.

Client Services

The upper layers of the networking stack in a Windows client take different names and forms, depending on the operating system. On Windows 98, for example, Client for Microsoft Networks is a service that you (or the Windows setup program) must explicitly install along with at least one protocol and an adapter. A service is a program that runs continuously in the background while the operating system is loaded, the equivalent of a daemon in Unix. In Windows 2000 and NT, however, the client takes the form of a number of separate services that the setup program installs with the operating system.

In most cases, the Windows networking architecture enables you to install additional client services that can take advantage of the same protocol and adapter modules as the Windows network client. For example, the Client for NetWare Networks and the Client Services for NetWare modules in Windows 9x and Windows NT/2000, respectively, utilize the same protocol and adapter modules as the Windows network client. The exceptions to this are the Microsoft Windows Network client in Windows for Workgroups and Microsoft Client 3.0 for MS-DOS, which are both self-contained clients.

Windows Client Versions

The various Windows operating systems provide most of the same networking capabilities. You can log on to a domain or a workgroup, browse the other computers on the network, map drive letters to specific shares, and redirect print jobs to network printers. However, the clients are slightly different in their implementations. The following sections examine each of the clients and their peculiarities.

Microsoft Client 3.0 for MS-DOS

Windows NT Server 4.0 includes a real-mode client for MS-DOS systems that provides access to Windows networks. Microsoft Client 3.0 for MS-DOS is located in the \Clients\ Msclient directory on the Windows NT 4.0 Server CD-ROM, and includes a set of drivers for NICs and support for the three standard protocols (including TCP/IP). In addition to supporting DOS, you can use this client to enable Windows 3.1 systems to access a network. However, Windows for Workgroups uses virtually the same environment as Windows 3.1, except that it includes a protected-mode network client.

If you are committed to a 16-bit Windows environment for your network clients, you should at least consider upgrading Windows 3.1 systems to Windows for Workgroups. Client for MS-DOS runs in conventional and upper memory only, and therefore occupies a significant amount of the RAM needed to load other programs.

Windows NT Server 4.0 includes a tool called Network Client Administrator that enables you to create installation disks for the MS-DOS client (see Figure 22-2). Using this program, you can create a set of installation disks for the client or create a boot disk that contains a fully configured client. When you boot an MS-DOS system with the disk, the client loads, connects to the network, and runs the client installation program from a shared network drive.

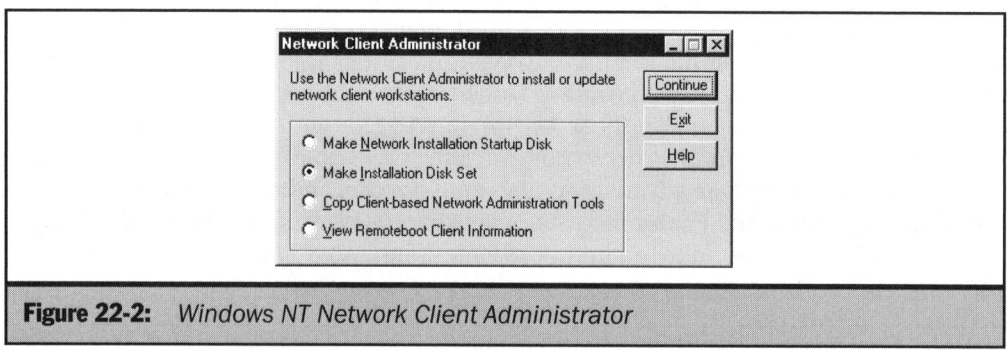

Figure 22-2: *Windows NT Network Client Administrator*

   *Of all the Windows network clients, Client for MS-DOS is the only one that operates exclusively in access-only mode. A workstation running the client can access resources of other Windows systems, but it cannot share its own drives and printers.*

Because it is intended for use on DOS systems, this client does not include any graphical configuration or user tools, but the installation program has a character-based menu interface, as shown in Figure 22-3. From these screens, you choose the adapter and protocol drivers that the program should install, and configure them with the appropriate settings. The client does include a Winsock driver, however, that enables the workstation to run Windows-based Internet applications like Web browsers. The client also includes a standard TCP/IP ping utility.

The primary tool for controlling the network environment on a DOS client is a command-line utility called Net.exe. The Net.exe program is also included with all of the other Windows network clients, even when the operating system provides graphical tools that perform the same functions.

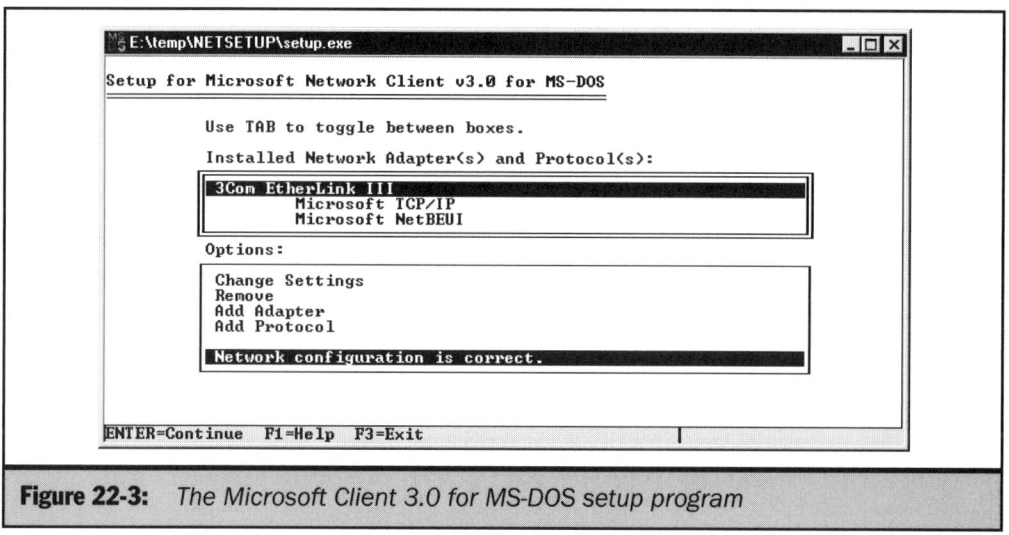

Figure 22-3: *The Microsoft Client 3.0 for MS-DOS setup program*

Note *For more information on using Net.exe, see Chapter 31.*

Windows for Workgroups Client

As mentioned earlier, the Microsoft Windows Network client runs in protected mode, which enables it to use mostly extended memory, freeing up all but 4KB of conventional memory for use by other programs. Unlike Client for MS-DOS, however, the network client loads with Windows, preventing you from accessing network resources with DOS programs except through a DOS window.

The Microsoft Windows Network client, as it ships with Windows for Workgroups, supports the NetBEUI and IPX protocols only. Support for the TCP/IP protocols is also available, but you must download the TCP/IP-32 protocol package from Microsoft's FTP site and install it after installing Windows for Workgroups.

Note *The TCP/IP-32 protocol stack for Windows for Workgroups is available as Wfwt32.exe at ftp://ftp.microsoft.com/peropsys/windows/public/tcpip/.*

As with Client for MS-DOS, the Windows for Workgroups client enables the workstation to connect to either a workgroup or a domain. The client incorporates networking functions into standard Windows utilities such as File Manager and Print Manager, and includes command-line networking tools such as Net.exe, Netstat.exe, and Nbtstat.exe. In addition, the TCP/IP-32 package provides a selection of TCP/IP tools, including the Ping, Arp, and Route utilities and a character-based FTP client.

NETWORK OPERATING SYSTEMS

Windows 95/98/Me Client

Windows 95, 98, and Me include a 32-bit, protected-mode Windows network client that includes support for all three of the major networking protocols and an extensive set of network adapter drivers. The Windows 95/98/Me networking architecture is based on virtual device drivers that enable multiple clients to share the NIC and protocol drivers. For example, after installing the Client for Microsoft Networks service with appropriate protocol and network adapter drivers, you can simply add Client for NetWare Networks to provide access to NetWare servers using the same drivers.

Networking functionality is completely integrated into the Windows 95, 98, and Me operating systems. The Network Neighborhood program enables you to browse network domains and workgroups and access their resources right alongside those on the local machine. In addition to what are strictly defined as client functions, the Windows operating systems (with the exception of Windows 3.1) include server capabilities that enable you to share the system's drives and printers with other users on the network. These elements are installed along with the client and are also integrated into Windows Explorer and other Windows utilities.

Windows 2000/NT Client

Functionally, Windows 2000 and NT are virtually identical to Windows 95, 98, and Me, as far as networking is concerned. However, the networking architecture of Windows 2000 and NT differs in that the upper layers of the networking stack are not implemented as a single service. The operating systems have separate Workstation and Server services that must both be running for the system to be able to access network resources and share its own. Other services, such as the Computer Browser, Messenger, and Netlogon services, provide additional networking functions that contribute to the overall capabilities of the operating system.

Note   *For more information on Windows 2000 and NT networking, see Chapter 16.*

NetWare Clients

Novell NetWare dominated the network operating system market when networking was being integrated into the Windows operating systems, so the ability to access legacy NetWare resources while running a Windows network was a priority for Microsoft's development team. The original arrangement was that Novell would supply the NetWare client functionality for Windows NT, but when delays and some unsatisfactory beta releases made this seem unlikely, Microsoft took it upon itself to create its own NetWare client. The original Windows NT release, version 3.1, included the Microsoft version of the IPX protocol suite (called NWLink), but no NetWare client. Microsoft eventually released NetWare Workstation Compatible Service (NWCS) as an add-on to Windows NT 3.1 and incorporated it into NT version 3.5. Eventually, Novell released its own NetWare client for Windows NT, and ever since, NetWare clients for the Windows operating systems have been available from both Microsoft and Novell.

NetWare and 16-Bit Windows

Neither Windows 3.1 nor Windows for Workgroups includes a NetWare client, but both of them can function with the clients supplied by Novell. In addition, Windows for Workgroups can run a Novell client along with the Microsoft Windows Network client, to provide access to both NetWare and Windows resources. At the time that the 16-bit versions of Windows were released, NetWare clients used either the NetWare shell (NETX) or the NetWare DOS Requestor (VLM) client for the upper-layer functionality, and used either a monolithic or Open Datalink Interface (ODI) driver for the NIC. A monolithic driver is a single executable (called Ipx.com) that includes the driver support for a particular NIC, while ODI is the Novell equivalent of NDIS, a modular interface that permits the use of multiple protocols with a single network card. The combination of an ODI driver and the VLM requestor was the most advanced NetWare client available at that time.

All of these client options loaded from the DOS command line, which meant that they provided network access to DOS applications outside of Windows, but also meant that they utilized large amounts of conventional and upper memory. In fact, without a carefully configured boot sequence or an automated memory management program, it was difficult to keep enough conventional memory free to load applications.

Today, Novell Client for DOS/Windows eliminates this memory management problem and provides full access to all NetWare resources (including Novell Directory Services) while using only 4KB of conventional memory. The main component of the client is NetWare I/O Subsystem (Nios.exe), which works with the DOS extended memory manager (Himem.sys) to provide an area of contiguous protected memory for the rest of the client components. Once Nios.exe is loaded, the ODI network adapter drivers and other client modules are loaded into that memory area. These modules take the form of NetWare Loadable Modules (NLMs), which are traditionally the program files used on NetWare servers. Nios.exe, however, enables client systems to use the same NIC drivers as servers do, and executes them from the command prompt using the LOAD command.

The NLMs needed for the client to function are as follows:

Network adapter driver The client uses 16- or 32-bit drivers with a .lan extension (the same as those used on NetWare servers) to provide communications with the network adapter.

Cmsm.nlm   The *media support module (MSM)* provides media-specific support in cooperation with *topology support modules (TSMs)* such as Ethertsm.nlm for Ethernet networks and Tokentsm.nlm for Token Ring networks.

Lslc32.nlm   The *link support layer (LSL)* module functions as the interface between the network adapter driver and the protocol drivers operating above it, enabling multiple protocols to use a single adapter.

Protocol driver   The client includes drivers for the IPX protocol suite as well as a full TCP/IP stack, which includes a Winsock driver that enables the workstation to run Web browsers and other Internet applications.

Client 32 Requestor   The Client32.nlm module provides the requestor functions that redirect the resource requests generated by applications to the networking stack. This single module takes the place of the multiple VLM modules used in the earlier NetWare DOS Requestor client.

NetWare and Windows 95/98/Me

With Windows 95, 98, and Me (as well as Windows 2000 and NT), you have a choice between running a NetWare client furnished by Microsoft and one provided by Novell. Generally speaking, the Microsoft client provides better performance, while the Novell client provides more features. The client you choose should be based on what type of access your users need and the resources available in your workstations. For a Windows network that still maintains some NetWare resources, the Microsoft client provides good access and works well together with Client for Microsoft Networks. The Novell client is preferable (or required) in the following client situations:

- Networks that rely exclusively (or heavily) on NetWare for shared network resources

- NetWare administrators that require access to the NetWare Administrator and other NDS applications

- Networks that use NetWare 5.x servers running TCP/IP as their only protocol

Microsoft Client for NetWare Networks

Windows 95, 98, and Me all include Client for NetWare Networks, which provides basic connectivity to NetWare resources. To log in to a Novell Directory Services (NDS) tree, you must also install Microsoft Service for NDS, which is also included with the operating system. These services work with the same network adapter and IPX/SPX-compatible protocol modules as Client for Microsoft Networks, and can coexist with the Windows network client as well.

However, although the NDS service enables the user to log in to an NDS tree and access NetWare resources, it does not supply the modules needed to run NDS applications like NetWare Administrator used to manage the NDS tree. To run these applications, the system must have access to the following files:

- Nwcalls.dll

- Nwlocale.dll

- Nwipxspx.dll

- Nwnet.dll

- Nwgdi.dll

- Nwpsrv.dll

These files are distributed by Novell in its client software and are not included with the Windows 95/98/Me operating system. You can copy the files to a workstation from a NetWare distribution disk or client download and use them with the Microsoft client to run NDS applications, as long as you have the appropriate license for the files. Place the files in a directory on the workstation's path, such as C:\Windows.

The Microsoft NetWare client also does not support the use of any protocol other than the IPX/SPX-compatible protocol included with Windows 95/98/Me. You can install TCP/IP on a system running the NetWare client, but you cannot bind the client to the TCP/IP protocol to access NetWare 5.x servers. To do this, you must use Novell Client for Windows 95/98.

Windows 95, 98, and Me also include a real-mode NetWare client that, like the real-mode Windows network client, is intended for use in situations in which the protected-mode client cannot load. This real-mode client lacks some of the more advanced features of the standard client, such as support for long file names, automatic server reconnection, and the NetWare Core Packet Burst protocol (NCPB).

Another service included with Windows 95/98/Me, called File and Printer Sharing for NetWare Networks, enables you to configure a Windows machine to share its drives and printers with other NetWare clients on the network. When the service is running, the Windows 95/98/Me computer appears to the network as a NetWare 3.12 server with its shared drives appearing as volumes on that server. NetWare users can log in and access the shared resources in the normal manner, unaware that the system they are accessing is actually running Windows.

Novell Client for Windows 95/98

Novell Client for Windows 95/98 (which also supports Windows Me) includes all the components needed to create a networking stack in the computer, but it also works together with the existing Windows networking architecture. If, for example, you have no networking software components installed on the computer, the Novell client's setup program will install an ODI network adapter driver, Novell's own 32-bit IPX protocol for the Novell NetWare Client, and the Novell NetWare Client service itself. If you already have Windows' own networking components installed, the Novell client can use the Windows NDIS driver for the NIC, but will still install its own protocols and client service.

Note *Novell Client for Windows 95/98, as well as the Novell clients for the other Windows operating systems, is available free of charge from Novell's Web site at http://www.novell.com/download/.*

The protocol drivers that ship with the Novell client are different than Windows' own. Microsoft supplies an IPX/SPX-compatible protocol of its own design, while Novell provides its own version. If you use Microsoft's IPX protocol for your Windows network communications and install the Novell client, there will be two separate IPX implementations running on the system, which is not a problem. NetWare 5.x also supports the use of TCP/IP as a native protocol. Previous versions of NetWare could use TCP/IP for Internet services and for standard LAN communications only when IPX data was encapsulated within UDP datagrams (a technique called *tunneling*). To support NetWare 5.x, though, the Novell client now supports native TCP/IP communications with NetWare servers (which the Microsoft NetWare client does not), using Windows 95/98/Me's own TCP/IP stack.

With the Novell client installed, you can still log in to an NDS tree or NetWare server and a Windows domain or workgroup using a single user name and password, as long as the account exists with the same password in both directories. By default, the Novell client makes itself the primary network client and uses the account name and password you supply for the NetWare login to perform the Windows network logon. If an account with the same name does not exist in the Windows domain, or if the same account exists but uses a different password, the system generates a second authentication dialog box that you can use to log on to the Windows network. If you change the primary network client to Client for Microsoft Networks, the process functions the same way but in reverse; the NetWare client logs in using the Windows network credentials.

The Novell client provides support for all of the latest NetWare features, including Workstation Manager and NetWare Distributed Print Services, as well as other NetWare-related utilities. You can, for example, copy files directly between two NetWare server volumes rather than have them go through the workstation where you are performing the copy, manage NetWare connections through a dedicated interface, log in to two NDS trees simultaneously, and run NetWare utilities like NetWare Administrator and NDS Manager. For network administrators and power NetWare users, these abilities can be useful, but for the average user who simply stores files on NetWare servers and prints to NetWare printers, they are not really necessary.

NetWare and Windows NT/2000

Like Windows 95, 98, and Me, Windows NT and 2000 include Microsoft clients that provide access to NetWare resources, or you can choose to run Novell Client for Windows NT/2000. The development of the NetWare clients for Windows NT was a long and difficult process for both Microsoft and Novell. Early versions of the clients functioned poorly and were lacking in features that users and administrators felt were vital. Today, however, the status of the Windows NT/2000 clients is roughly the same as those of Windows 95/98/Me. The Microsoft clients provide basic NetWare connectivity, while the Novell client implements the full set of NetWare features.

Microsoft Client Service for NetWare

The Windows 2000 Professional and Windows NT Workstation products include Microsoft Client Service for NetWare (CSNW), which enables the computer to log in to an NDS tree or a NetWare server and access its resources. Unlike Windows 95/98/Me, no additional service is required for NDS connectivity. CSNW (and GSNW, the Gateway Service for NetWare client included with the Windows 2000 and NT Server products) is *NDS-aware,* but not *NDS compatible.* This means that while the services can log in to an NDS tree, they do not provide full support for NDS applications. As with Windows 95/98/Me, if you want to use NetWare 5.x with TCP/IP, NDPS, ZENworks, or other NDS applications, you must install Novell Client for Windows NT/2000.

CSNW is based on NWLink, the Windows protocol that provides services compatible with Novell's IPX. NWLink includes modules that emulate the most important protocols in the IPX suite, including IPX itself, SPX, RIP, SAP, and NetBIOS. As with the Windows 95/98/Me client, the service uses the same NDIS adapter driver as the Windows network client, and can coexist with the Windows network, providing connectivity to both NetWare and Windows resources simultaneously.

 *For more information on the IPX protocol suite used by NetWare, see Chapter 14.*

Microsoft Gateway Service for NetWare

The NetWare client included with Windows 2000 Server and Windows NT Server is called Gateway Service for NetWare (GSNW). GSNW provides the same client services as CSNW, but adds a gateway function that enables Windows client systems to access NetWare resources through the 2000 or NT server. After connecting to the NDS tree or NetWare server using the same elements as CSNW, you can use the gateway to create shares on the Windows server that are associated with particular volumes on NetWare servers (see Figure 22-4). All Windows clients on the network can see the shares and access them in the normal way, thus enabling them to access NetWare resources without running a NetWare client. The Windows server accesses the associated NetWare server volumes using its client capabilities and then furnishes the files and directories to the Windows clients. Using GSNW is not a replacement for running NetWare clients on your Windows workstations, but it does provide a quick solution for users who need occasional access to NetWare volumes and reduces the administrative overhead for computers that would otherwise require the installation of multiple clients.

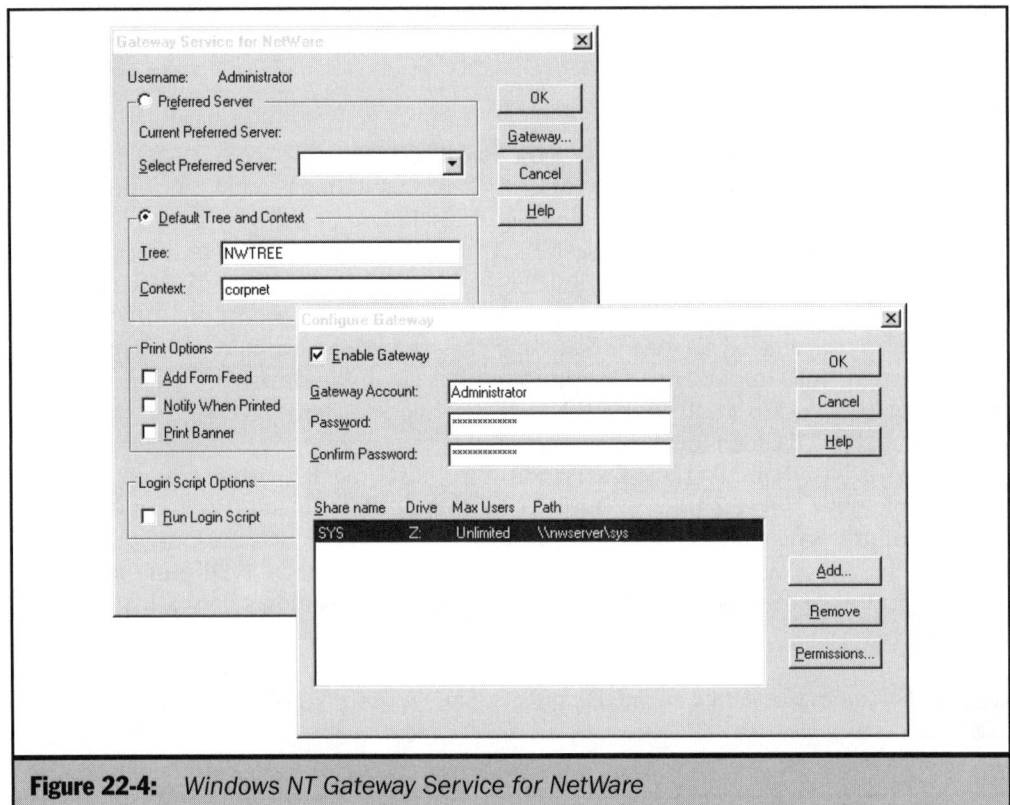

Figure 22-4:   *Windows NT Gateway Service for NetWare*

Novell Client for Windows NT/2000

Although the early versions of Novell Client for Windows NT were limited in their functionality, the current release provides full access to all NetWare features, just as the DOS and Windows 95/98 clients do. However, Client for Windows NT/2000 does not utilize NetWare I/O Subsystem to run NLMs on the workstation, as the other clients do.

All of the Novell clients have a setup program that simplifies the process of installing all of the individual components. You can also perform the installation in the traditional manner, from the Network Control Panel in Windows. The Novell NT/2000 client also includes the capability to integrate the client installation process into the operating system installation, so that an individual client installation procedure is not necessary. Further, the client includes an Automatic Client Upgrade (ACU) feature that checks the version number of a workstation's currently installed client against the version stored on a server. If the server has a newer version, the workstation automatically performs an upgrade. Features like these are designed to simplify the workstation maintenance process for administrators who may be responsible for hundreds or thousands of computers.

Macintosh Clients

While not nearly as popular as the PC, Macintosh systems have their place in the personal computer world, and you may find yourself needing to connect Mac workstations to your network. All Macintosh systems include an integrated network interface, and this has long been touted as evidence of the platform's simplicity and superiority. However, Macintosh workstations require special treatment to connect them to a network running other platforms, such as Windows 2000/NT, NetWare, or Unix. In most cases, however, you can configure your network to handle Macintosh clients, enabling Mac users to share files with Windows and other clients. If you select applications that are available in compatible versions for the different client platforms you're running, Mac users can even work on the same files as Windows users.

Connecting Macintosh workstations to a network running another platform is usually a matter of either configuring your existing servers to use AppleTalk, the native Macintosh networking protocol, or configuring the Macintosh systems to use a protocol supported by the servers, by installing an additional client. The following sections examine the basic procedures for connecting Macs to various types of networks, and the products you may need to implement the connections.

Connecting Macintosh Systems to Windows Networks

The Windows 2000 and NT Server products include Microsoft Services for Macintosh, which implements the AppleTalk protocol on the Windows computer, enabling Macintosh systems to access file and printer shares on the server. Unlike Windows clients,

Mac systems do not participate as peers on the Windows network. The relationship between the Mac workstation and the Windows NT or 2000 server is strictly client/server, meaning that the Mac can access the server's files and printers, but it can't share its own drives or access the shares on other network workstations, such as Windows 95/98/Me machines. It is possible for a Macintosh system to function as a peer with workstations running Windows 95/98/Me or other operating systems, but a third-party client software product is required.

Using Microsoft Services for Macintosh

Microsoft Services for Macintosh makes it possible for Macintosh systems to access Windows 2000 and NT Server shares without modifying the configuration of the workstations. When you install the Services for Macintosh package on a Windows 2000 or NT server from the Network Control Panel, three new modules are added to the machine, as follows:

AppleTalk Protocol   Implements the AppleTalk Phase 2 protocols that enable Macintosh systems to communicate with the server.

File Server for Macintosh (MacFile)   Enables you to create Macintosh-accessible volumes out of directories on the server's NTFS drives.

Print Server for Macintosh (MacPrint)   Enables Macintosh systems to send print jobs to spoolers on the Windows server.

Once the installation is complete, you must specify the name of the AppleTalk zone in which the server will participate. The service then creates a directory called \Microsoft UAM Volume on the server drive. By default, this is the only directory that Macintosh systems can access on the server. The mechanism by which you create and configure Macintosh volumes on the server is wholly separate from the system's standard drive-sharing capability. You can share the \Microsoft UAM Volume directory in the usual manner, making it available to standard Windows clients, but to manage its Macintosh-related properties or create new Macintosh volumes on the server, you use the Server Manager program.

Installing Services for Macintosh adds a MacFile menu to Server Manager, as shown in Figure 22-5. Selecting Volumes from this menu displays the Properties of Macintosh-Accessible Volume dialog box, from which you can create new volumes and manage the properties of existing ones. For example, Macintosh users are granted read-only privileges to \Microsoft UAM Volume, by default. From the Properties dialog box for the volume, you can grant users write access, assign a password to the volume, or create access permissions for specific users and groups, with the optional *user authentication module (UAM)*. Installing the UAM on the Macintosh system enables the Mac Chooser to authenticate the system to the 2000/NT server using encrypted Microsoft authentication rather than the clear text authentication the system uses natively.

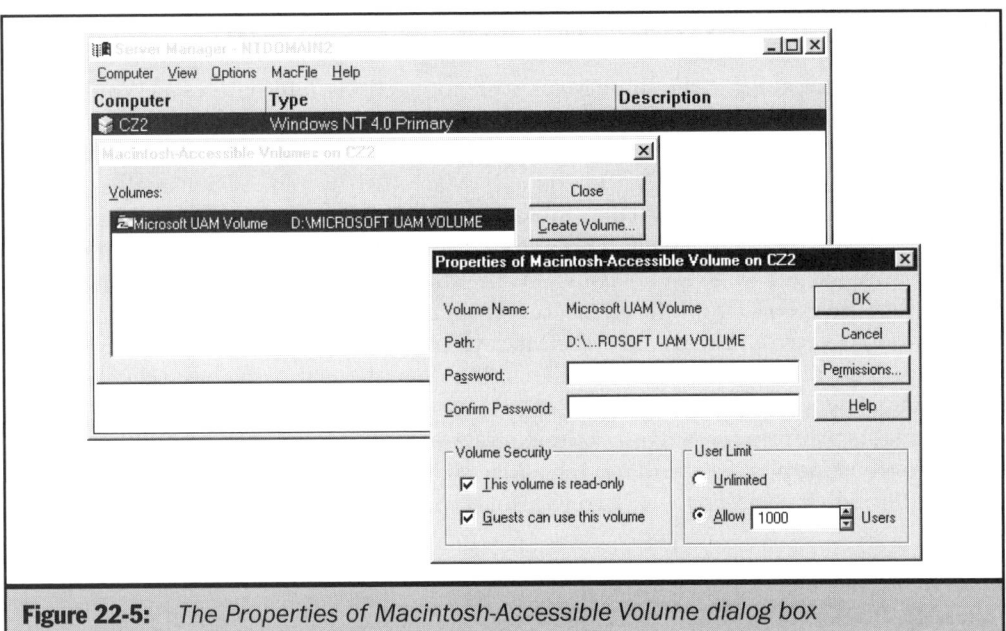

Figure 22-5:   *The Properties of Macintosh-Accessible Volume dialog box*

Third-Party Macintosh Clients

One of the primary advantages of Windows Services for Macintosh is that the additional networking software needed runs on the network. There is no need to purchase, install, and configure new software on each Macintosh workstation (although you must have a standard Windows 2000 or NT client license for each Mac system that accesses the server). To approach the Mac/Windows networking problem from the other side—that is, by altering the client—you must purchase a third-party product such as Thursby Software Systems's DAVE (http://www.thursby.com/products/dave.html) or Open Door Networks's ShareWay IP (www2.opendoor.com/shareway/).

DAVE is a peer-to-peer networking solution for Macintosh systems that enables the Mac to access Windows network shares and share its own drives with Windows clients. Unlike Microsoft Services for Macintosh, DAVE enables Macs to access all Windows shares, and not just those on Windows 2000 and NT Server systems. DAVE also utilizes the TCP/IP protocols instead of AppleTalk and, therefore, requires no changes to the Windows computers on the network. As with any client-based solution, you must purchase a copy of DAVE for each Mac client and install it individually. For a relatively small network or a larger network with only a few Macintosh clients, products like DAVE can be an ideal solution.

Connecting Macintosh Systems to NetWare

Novell NetWare provides the same Macintosh connectivity alternatives as Windows. You can configure your NetWare servers to run the AppleTalk protocol, enabling the Macintosh workstations to communicate with the servers in their native mode, or you can install NetWare Client for MacOS on each Macintosh system, which enables the Mac to use the IPX protocols to communicate with NetWare servers. In addition, a new product called Novell Native File Access for Macintosh makes it possible for Macintosh clients to access NetWare servers using AppleTalk Filing Protocol 3.0 over TCP/IP.

NetWare runs strictly on the client/server model, so both solutions enable workstations to access servers only. The Mac systems do not share their own drives nor can they access the resources of other clients. These solutions provide basic client services to Macintosh workstations, such as access to server volumes and printers, but a Mac is not a suitable platform for NetWare administration. There is no NetWare Administrator utility for the Mac platform, and though a Mac client can log in to an NDS tree and browse through its objects, there is no way for it to create or modify them.

Using NetWare for Macintosh

NetWare for Macintosh is included with the NetWare operating system and installs the AppleTalk protocols on the server, including the AppleTalk Filing Protocol (AFP) and AppleTalk Print Services (ATPS), which provide file and printer sharing services, respectively. You install the product onto an existing server using the Install.nlm utility in the normal manner. Like any other protocol on a NetWare server, AppleTalk must be bound to one or more of the installed NICs, and the Autoexec.ncf file must be modified to load it along with the AFP and ATPS modules automatically with the operating system. You must also add the MAC name space to every NetWare volume on which your clients will store Macintosh files.

Once NetWare for Macintosh is installed and configured on a NetWare server, Mac workstations can connect to it without further modification. However, because no NDS client capability has been added to the Mac system, it can log in to a NetWare 4.x or 5.x server using bindery emulation only. Security for Mac clients is no different than for any other NetWare client. You must create a user object for the person logging in from the Mac system in the server's bindery emulation context and grant the user rights to files and directories in the usual manner.

Using Native File Access for Macintosh

Novell Native File Access for Macintosh is a separate product that is similar to NetWare for Macintosh in that it provides Mac clients with access to NetWare resources with no modifications to the client computer at all. Native File Access provides TCP/IP connectivity for Mac clients with encrypted authentication, enabling administrators to avoid adding AppleTalk traffic to the network. Native File Access also makes it possible to create Macintosh user objects in the NDS tree, just like those for any other user. Network administrators can then use the objects to manage Mac users' access to NetWare resources in the same way as they would other NetWare users.

Using NetWare Client for MacOS

Installing NetWare Client for MacOS enables a Mac workstation to log in to an NDS tree (rather than perform a bindery emulation login) and use NetWare's own IPX protocols rather than AppleTalk. The client also supports NetWare/IP, which is a product that packages IPX traffic within UDP datagrams, a process called *tunneling*. There is not yet a Mac client that supports the use of the TCP/IP protocols in native mode with NetWare 5.*x* servers.

Novell no longer provides NetWare Client for MacOS itself. The product is no longer being developed, but Novell has granted an exclusive license for the distribution of the client to Prosoft Engineering (http://www.prosofteng.com/netware.asp).

Client for MacOS consists of both server and workstation components. Installing the client on the server adds the Macintosh name space to the volumes you select, enables you to specify the names by which the volumes will appear to the Macintosh systems, and installs Mac OS file system support in the form of a server module called Macfile.nlm, which provides AppleTalk Filing Protocol support.

The workstation components of the client include MacIPX and NW/IP, which provide support for the IPX and NetWare/IP protocols, and an RCONSOLE program that enables Mac systems using IPX to access the server console. The installation process also creates a NetWare Client Utilities folder on the workstation drive that contains the following tools:

NetWare Directory Browser   Enables users to browse (but not modify) the NDS tree.

NetWare Print Chooser   Enables users to select the NetWare printers to which they will send print jobs.

NetWare Volume Mounter   Enables users to mount specific volumes by dragging them from the NetWare Directory Browser window.

Once the client is installed on the workstation, you can log in to NetWare using the Chooser or tree icon on the right side of the menu bar. The tree icon provides NDS login capabilities, and the Chooser lets you select either an NDS or NetWare server login. If you choose to log in to a specific server, the system performs a bindery or bindery emulation login, depending on the NetWare version the server is running.

Unix Clients

Three primary mechanisms provide client/server access between Unix systems. Two of these have been ported to many other computing platforms, and you can use them to access Unix systems from workstations running other operating systems. These three mechanisms are as follows:

Berkeley Remote Commands   Designed for Unix-to-Unix networking, these commands provide functions such as remote login (rlogin), remote shell execution (rsh), and remote file copying (rcp).

DARPA Commands   Designed to provide basic remote networking tasks, like file transfers (ftp) and terminal emulation (telnet), the DARPA commands operate independently of the operating system and have been ported to virtually every platform that supports the TCP/IP protocols.

Network File System (NFS)   Designed by Sun Microsystems in the 1980s to provide transparent file sharing between network systems, NFS has since been published as RFC 1813, an informational RFC (Request for Comments), by the Internet Engineering Task Force (IETF). NFS is available on a wide range of computing platforms, enabling most client workstations to access the files on Unix systems.

In most cases, the TCP/IP stacks on client computers include applications providing the DARPA ftp and telnet commands. Since all Unix versions run FTP and Telnet server services by default, you can use these client applications to access any Unix system available on the network. These server applications have been ported to other operating systems as well. For example, Windows 2000 and NT include an FTP server as part of the Internet Information Server package.

All of the Windows TCP/IP clients include FTP and Telnet client applications, with the exception of Microsoft Client 3.0 for MS-DOS. Installing this client provides a TCP/IP stack and the Winsock driver needed to run Internet applications, but the FTP and Telnet programs are not included. You can, however, use third-party FTP and Telnet clients to access Unix and other server systems, or install Novell Client for DOS/Windows, which also includes FTP and Telnet clients as part of its TCP/IP stack.

While FTP and Telnet provide basic access to a Unix system, they are not the equivalent of full client capabilities. For example, FTP provides only basic file transfer and file management capabilities. To open a document on a Unix system using FTP, you must download the file to a local drive and use your application to open it from there. NFS, on the other hand, enables the client system to access a server volume as though it were available locally. NFS downloads only the blocks that the client application needs, instead of the whole file.

Thus, while FTP and Telnet are provided for free and are nearly always available, clients that need regular access to Unix file systems are better off using NFS. There are NFS products for both Windows NT/2000 and Novell NetWare that make file system communications with Unix systems possible, but both of them are products that must be purchased and installed separately.

Windows Services for Unix is an add-on package that enables Windows computers to access Unix volumes using NFS and to publish their drives as NFS volumes for Unix clients. The product also includes a Telnet server for Windows and a character-based Telnet client that improves on the version included with 2000/NT, as well as a password synchronization daemon for Unix systems. With the services in place, the Windows computer system can map a drive letter to an NFS volume on a Unix system or reference it using either standard UNC (Universal Naming Convention) names or the Unix *server:/export* format. Unix systems can access Windows drives just as they would any other NFS volume.

The NFS server feature creates a separate NFS Sharing tab in the Properties dialog box of each drive and directory on the Windows system. You create NFS shares and assign permissions to them independently of the standard Windows shares. Windows 2000 and NT use their own access control permissions to approximate the security model used on Unix systems. The password synchronization daemon enables administrators to manage passwords on both Unix and Windows systems from Windows.

In addition to the NFS client/server capabilities, Windows Services for Unix includes a toolkit that provides a working environment on the Windows system similar to that of a Unix system, including a Korn shell and an assortment of standard Unix utilities, such as cp, for copying files, mkdir, for creating directories, and vi, the command-line text editor.

Although Windows Services for Unix provides multiple Unix clients access to Windows volumes, it only enables a single Windows client to access NFS drives on Unix systems. Novell's NFS Services for NetWare is a server-based solution that enables Unix clients to access NetWare volumes and enables NetWare clients to access Unix NFS volumes using the NetWare server as a gateway. The product also provides Unix-to-NetWare and NetWare-to-Unix printing services that enable clients on each platform to access printers on the other.

Both of these products provide basic connectivity between Unix workstations and PC-based servers running Windows NT/2000 or NetWare. Third-party NFS products also are available that provide similar services. While the interoperability between different platforms is never completely transparent, you should be able to use whatever client operating systems your network users require, and still provide them with access to any resources anywhere on the network.

NETWORK OPERATING SYSTEMS

Networking

Part V

Network Connection Services

The Complete Reference

Chapter 23

DHCP

B ecause of their use on the Internet and their compatibility with virtually every network operating system in use today, the TCP/IP family of protocols are nearly ubiquitous on all but the smallest LANs. The chief administrative problem with deploying and maintaining a TCP/IP network is the need to assign each node a unique IP address and to configure the various other TCP/IP parameters with appropriate values. On a large network deployment, performing these tasks manually on individual workstations is not only labor intensive, it also requires careful planning to ensure that no IP addresses are duplicated.

The Dynamic Host Configuration Protocol (DHCP) was developed to address this problem. DHCP takes the form of a service that network administrators configure with ranges of IP addresses and other settings. Workstations configured to run as DHCP clients contact the service at boot time and are assigned a set of appropriate TCP/IP parameters, including a unique IP address. The workstation uses these parameters to configure its TCP/IP client, and network communication commences with no manual configuration necessary at the workstation.

DHCP is a platform-independent service that can configure the TCP/IP parameters of any operating system with DHCP client capabilities. DHCP server software is available for many platforms, including Microsoft Windows, Novell NetWare, and various flavors of Unix.

Origins

DHCP was developed by Microsoft in the early 1990s as a workstation configuration solution for enterprise networks and, particularly, for its own 35,000-node network rollout. After determining that TCP/IP was the optimum protocol for their needs, Microsoft realized that the task of manually assigning IP addresses to thousands of machines located at various sites in 50 countries was enormous, as was the continued tracking of those addresses as computers were added to and removed from the network.

The concept of server-based IP address assignment was not a new one, however. DHCP is based on two earlier protocols, called RARP and BOOTP.

RARP

RARP, the Reverse Address Resolution Protocol, does the opposite of ARP, the Address Resolution Protocol used on every TCP/IP system. While ARP converts network-layer IP addresses into data link–layer hardware addresses, RARP works by broadcasting a system's hardware address and receiving an IP address in return from a RARP server.

Note   *For more information on ARP, see Chapter 13.*

Designed for use with diskless workstations that have no means to store their own TCP/IP configuration information, a RARP server can supply IP addresses to all of the systems on a network segment. However, the concept, as defined in the Internet Engineering Task Force's RFC 903 document, was not sufficient for Microsoft's needs for the following reasons:

- RARP relies on broadcast messages generated by the client because no means exists for the diskless workstation to store the address of the RARP server. Therefore, because broadcasts are limited to the local network, there must be a RARP server on every network segment, to service the workstations on that segment.

- RARP is only capable of supplying client systems with IP addresses. To be useful on today's networks, a protocol must provide values for other configuration parameters as well, such as name servers and default gateways.

- RARP is only a mechanism for the storage and delivery of IP addresses. Administrators must still manually assign addresses to clients by creating a lookup table on the RARP server.

BOOTP

The *Bootstrap Protocol (BOOTP)* is an improvement over RARP and is still in use today, particularly on routers made by Cisco and other manufacturers. DHCP takes much of its functionality from the BOOTP standards (published as RFC 951 with extensions in RFC 1533 and RFC 1542). Also designed for use with diskless workstations, BOOTP is capable of delivering more than just IP addresses; it uses standard UDP/IP datagrams instead of a specialized data link–layer protocol like RARP.

BOOTP servers can use the *Trivial File Transfer Protocol (TFTP)* to deliver an executable boot file to a diskless client system, in addition to an IP address and other TCP/IP configuration parameters. Like RARP, BOOTP clients use broadcast transmissions to contact a server, but the standard calls for the use of BOOTP relay agents to make it possible for one BOOTP server to service clients on multiple network segments.

A *BOOTP relay agent* detects the BOOTP broadcasts on a network segment and transmits them to a server on another segment. Many of the routers on the market support BOOTP relay and, because DHCP uses exactly the same relay system, the DHCP relay feature built into Windows 2000 and Windows NT works with BOOTP traffic as well.

The primary shortcoming of BOOTP is that, like RARP, administrators must manually create a look-up table on the server containing the IP addresses and other configuration parameters to be assigned to the clients. This makes the system subject to many of the same errors and administrative problems as RARP and, for that matter, as manual client configuration.

NETWORK CONNECTION
SERVICES

DHCP Objectives

BOOTP eliminates the need for administrators to travel to every workstation so they can configure the TCP/IP client manually, but the possibility still exists for systems to be assigned duplicate IP addresses because of typographical errors in the look-up table. What was needed was a service that would automatically allocate addresses to systems on demand and keep track of which addresses have been assigned and which are available for use. DHCP improves on the BOOTP concept by enabling administrators to create a pool of IP addresses. As a client system boots, it requests an address from the DHCP server, which assigns one from the pool, along with the other static configuration parameters the client needs.

There are problems inherent in this concept. One problem is the possibility of a shortage of IP addresses. Once a client system is assigned an address on a particular subnet, what happens to that address if the machine is moved to another department on a different subnet? DHCP will assign the system a different address on the new subnet, but a mechanism must exist to reclaim the old address.

The solution to this problem lies in DHCP's mechanism for leasing IP addresses to client systems. Each time the server assigns an address, it starts a clock that will eventually run out if the system does not renew the lease. Each time the client system is restarted, the lease is renewed. If the lease runs out, the server releases the assigned IP address and returns it to the pool for reassignment.

The other problem with automatic address assignment is the possibility for a client's IP address to change periodically. If a lease expires for any reason (such as a user going on vacation), the server is likely to assign the system a different address the next time it connects to the network. The reason this can be a problem concerns the name resolution processes on the network.

Windows client systems are often assigned permanent names in either the DNS or the NetBIOS name service, or in both. In either case, a change of IP address can render the name resolution information invalid. In the NetBIOS name space, the problem arises only if the network relies on LMHOSTS files for name resolution. Networks using broadcasts to resolve names will have no problem with the new IP address because a new resolution is performed each time another system accesses the client. The best solution, however, is WINS, which registers each client system's NetBIOS name and new IP address each time it connects to the network.

Note *For more information on WINS and NetBIOS names, see Chapter 24.*

For DNS names, the problem is more complex because most DNS servers traditionally do not dynamically update the resource records containing host names and their equivalent IP addresses. If you have a system on your network that functions as an Internet server (such as a Web or FTP server), its IP address must be permanently assigned. Because of the distributed nature of DNS on the Internet, it takes too long for IP address changes to propagate to all the name servers involved. To address this problem, DHCP makes it possible to create permanent address assignments, as well as dynamic ones.

Note   *The DNS server included with the Windows 2000 Server products supports a new feature called Dynamic Update, which provides automatic modification of resource records. Addresses assigned using the Windows 2000 version of Microsoft DHCP Server can be automatically added to DNS records. For more information on DNS, see Chapter 25.*

The overall objectives Microsoft used when designing DHCP are as follows:

- The DHCP server should be able to provide a workstation with all the settings needed to configure the TCP/IP client so that no manual configuration is needed.

- The DHCP server should be able to function as a repository for the TCP/IP configuration parameters for all of a network's clients.

- The DHCP server should assign IP addresses in such a way as to prevent the duplication of addresses on the network.

- The DHCP server should be able to configure clients on other subnets through the use of relay agents.

- DHCP servers should support the assignment of specific IP addresses to specific client systems.

- DHCP clients should be able to retain their TCP/IP configuration parameters despite a reboot of either the client or the server system.

IP Address Assignment

The primary function of DHCP is to assign IP addresses and to accommodate the needs of all types of client systems. The standard defines three types of address assignments, which are as follows:

Manual allocation   The administrator configures the DHCP server to assign a specific IP address to a given system, which will never change unless it is manually modified. This is equivalent in functionality to RARP and BOOTP.

Automatic allocation   The DHCP server assigns permanent IP addresses from a pool (called a *scope*), which do not change unless they are manually modified by the user or the administrator.

Dynamic allocation The DHCP server assigns IP addresses from a pool using a limited-time lease, so the address can be reassigned if the client system does not periodically renew it.

> **Note** *Most DHCP server implementations support all three types of address allocation, but they usually do not let you select them using these names. For example, with the Microsoft DHCP Server, dynamic allocation is the default. If you want to use automatic allocation, you change the Lease Duration setting for a scope to unlimited. For manual allocation, you create what the Microsoft server calls reservations.*

Manual allocation is suitable for Internet servers and other machines that require static IP addresses because they rely on DNS name resolution for user access. This form of address allocation is nothing more than a remote configuration solution because the end result is no different than if the administrator manually configured the TCP/IP client. As an organizational aid, however, this method of address assignment is recommended over manual system configuration on a network that uses DHCP for its other machines. Keeping all the address assignments in one database makes it easier to track the assignments and reduces the likelihood of duplication. Duplicate IP addresses can occur accidentally if you configure a TCP/IP client by hand with an address that is included in a DHCP scope. Using DHCP for all address assignments, even manual ones, helps to minimize the chance that this can occur.

> **Note** *DHCP does not completely eliminate the possibility of IP address duplication. Since each DHCP server operates independently, it's possible to configure two servers with scopes containing addresses that overlap.*

Automatic allocation is useful on stable, single-segment networks or multisegment networks where machines are not routinely moved to other segments. This method reduces the network traffic by eliminating the address lease renewal procedures. In most cases, the savings are minimal. Automatic allocation is also not recommended if your organization is working with a limited supply of registered IP addresses.

Once configured, dynamic allocation provides the greatest amount of flexibility with the least amount of administrative intervention. The DHCP server assigns IP addresses to systems on any subnet and automatically reclaims the addresses no longer in use for reassignment. Also, there is no possibility of duplicate addresses being on the network (as long as DHCP manages all of the network's addresses).

TCP/IP Client Configuration

In addition to IP addresses, DHCP also can provide clients with values for the other parameters needed to configure a TCP/IP client, including a subnet mask, default gateway, and name server address. The object is to eliminate the need for any manual

TCP/IP configuration on a client system. For example, the Microsoft DHCP server includes more than 50 configuration parameters, which it can deliver along with the IP address, even though Windows clients can only use 11 of those parameters. The RFC 2132 document, "DHCP Options and BOOTP Vendor Extensions," defines an extensive list of parameters that compliant servers should support, and most of the major DHCP server packages adhere closely to this list. Many of these parameters are designed for use by specific system configurations and are submitted by vendors for inclusion in the standard document.

DHCP Architecture

The architecture of the DHCP system is defined by a public standard published by the IETF as RFC 2131, "Dynamic Host Configuration Protocol," and consists of two basic elements:

- A service that assigns TCP/IP configuration settings to client systems

- A protocol used for communications between DHCP clients and servers

The document defines the message format for the protocol and the sequence of message exchanges that take place between the DHCP client and server.

DHCP Packet Structure

DHCP communications use eight different types of messages, all of which use the same basic packet format. DHCP traffic is carried within standard UDP/IP datagrams, using port 67 at the server and port 68 at the client (the same ports as BOOTP). The packet format is shown in Figure 23-1 and contains the following fields:

op (op code, 1 byte)   Specifies whether the message is a request or a reply, using the following codes:

1   BOOTREQUEST

2   BOOTREPLY

htype (hardware type, 1 byte)   Specifies the type of hardware address used in the chaddr field, using codes from the ARP section of the IETF "Assigned Numbers" document (RFC 1700), some of which are as follows:

1   Ethernet (10MB)

4   Proteon ProNET Token Ring

5 Chaos

6 IEEE 802 Networks

7 ARCnet

11 LocalTalk

14 SMDS

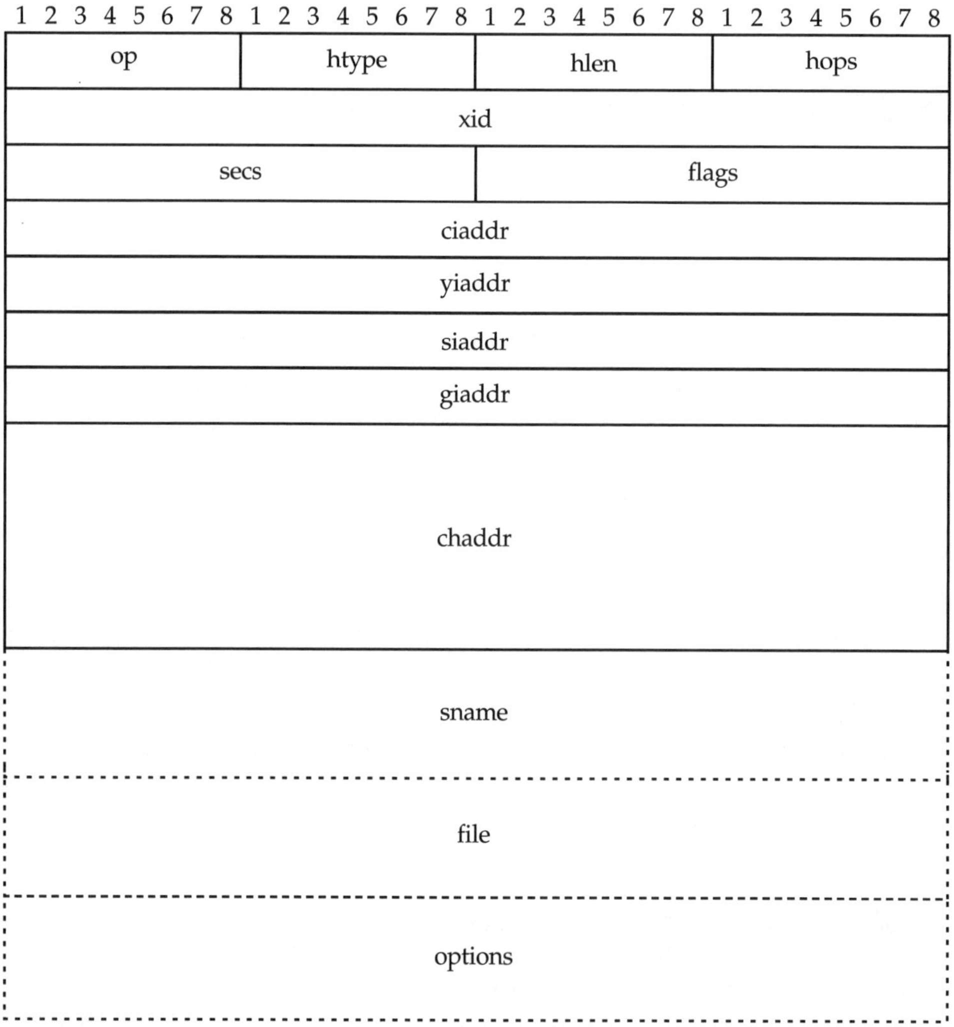
1 2 3 4 5 6 7 8 1 2 3 4 5 6 7 8 1 2 3 4 5 6 7 8 1 2 3 4 5 6 7 8
op
htype
hlen
hops
xid
secs
flags
ciaddr
yiaddr
siaddr
giaddr
chaddr
sname
file
options

Figure 23-1: *All DHCP messages use the same basic packet format.*

15   Frame Relay

16   Asynchronous Transfer Mode (ATM)

17   HDLC

18   Fibre Channel

19   Asynchronous Transfer Mode (ATM)

20   Serial Line

21   Asynchronous Transfer Mode (ATM)

hlen (hardware address length, 1 byte)   Specifies the length (in bytes) of the hardware address found in the chaddr field, according to the value of the htype field (for example, if htype equals 1, indicating an Ethernet hardware address, the value of hlen will be 6 bytes).

hops (1 byte)   Specifies the number of network segments between the client and the server. The client sets the value to 0 and each DHCP relay system increments it by 1 during the journey to the server.

xid (transaction ID, 4 bytes)   Contains a transaction identifier that systems use to associate the request and response messages of a single DHCP transaction.

secs (seconds, 2 bytes)   Specifies the number of seconds elapsed since the IP address was assigned or the lease last renewed. This enables the systems to distinguish between messages of the same type generated during a single DHCP transaction.

flags (2 bytes)   Contains the broadcast flag as the first bit, which, when set to a value of 1, specifies that DHCP servers and relay agents should use broadcasts to transmit to the client and not unicasts. The remaining bits in the field are unused and must have a value of 0.

ciaddr (client IP address, 4 bytes)   Specifies the client's IP address in DHCPREQUEST messages transmitted while in the bound, renewal, or rebinding state. At all other times, the value must be 0.

yiaddr (your IP address, 4 bytes)   Specifies the IP address being offered or assigned by a server in DHCPOFFER or DHCPACK messages. At all other times, the value must be 0.

siaddr (server IP address, 4 bytes)   Specifies the IP address of the next server in a bootstrap sequence. Servers include this information in DHCPOFFER and DHCPACK messages only when DHCP is configured to supply an executable boot file to clients and the boot files for various client platforms are stored on different servers.

giaddr (gateway IP address, 4 bytes)   Specifies the IP address of the DHCP relay agent to which a server should send its replies when the client and server are located on different subnets. When the client and server are on the same segment, the value must be 0.

chaddr (client hardware address, 16 bytes)   Specifies the hardware address of the client system in DHCPDISCOVER and DHCPREQUEST messages, which the server uses to address its unicast responses to the client. The format of the hardware address is specified by the values of the htype and hlen fields.

sname (server host name, 64 bytes)   Specifies the (optional) host name of the DHCP server. The field is more commonly used to hold overflow data from the options field.

file (boot file name, 128 bytes)   Specifies the name of an executable boot file for diskless client workstations in DHCPDISCOVER messages (in which case a generic file name is supplied) or DHCPOFFER messages (in which the field contains a full path and file name. The field is more commonly used to hold overflow data from the options field.

options (variable size, minimum 312 bytes)   Contains the magic cookie that specifies how the rest of the field should be interpreted and the DHCP Message Type option that defines the function of the message, as well as other options, defined in RFC 2132, that contain configuration data for other TCP/IP client parameters.

DHCP Options

The DHCP message format is almost identical to the BOOTP message defined in RFC 951. The primary difference is the options field, which replaces the 64-byte vend field in the BOOTP message. The options field in a DHCP message is a catchall area designed to carry the various parameters (other than the IP address) used to configure the client system's TCP/IP stack. Because you can configure a DHCP server to deliver many options to clients, defining separate fields for each one would be impractical.

The Magic Cookie

The options field always begins with the so-called *magic cookie,* which informs the server about what is contained in the rest of the field. The magic cookie is a 4-byte subfield containing the dotted decimal value 99.130.83.99.

The Option Format

The individual options in the options field contain various types and amounts of data, but most of them use the same basic structure, which consists of three subfields, as shown in Figure 23-2. The functions of the subfields are as follows:

code (1 byte)   Contains a code specifying the function of the option, as defined in RFC 2132

length (1 byte) Specifies the length of the data field associated with the option, making it possible for systems that do not support a particular option to skip directly to the next one

data (variable) Contains information used by the client in various ways depending on the code value and the message type

For example, in the Subnet Mask option, the code subfield has a value of 1, the length subfield has a value of 4, and the data field contains the 4-byte mask associated with the IP address assigned to the client.

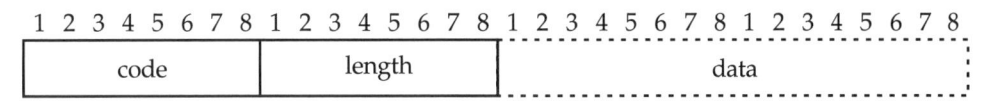

Figure 23-2: *A DHCP packet's options field contains multiple substructures for individual options, each of which is composed of three fields.*

The DHCP Message Type Option

The DHCP Message Type option identifies the overall function of the DHCP message and is required in all DHCP packets. The code subfield for the option is 53 and the length is 1. The data subfield contains one of the following codes:

1 – DHCPDISCOVER Used by client systems to locate DHCP servers and request an IP address

2 – DHCPOFFER Used by servers to offer IP addresses to clients

3 – DHCPREQUEST Used by clients to request specific IP address assignments or to renew leases

4 – DHCPDECLINE Used by clients to reject an IP address offered by a server

5 – DHCPACK Used by servers to acknowledge a client's acceptance of an offered IP address

6 – DHCPNAK Used by servers to reject a client's acceptance of an offered IP address

7 – DHCPRELEASE Used by clients to terminate a lease

8 – DHCPINFORM Used by clients that have already been assigned an IP address to request additional configuration parameters

The Pad Option

The *Pad option* is not really an option at all, but is instead a filler used to pad out fields so their boundaries fall between 8-byte words. Unlike most other options, the Pad option has no length or data field, and consists only of a 1-byte code field with a value of 0.

The Option Overload Option

Because DHCP messages are carried within UDP datagrams, the packets are limited to a maximum size of 576 bytes and the inclusion of a large number of options can test this limit. Because the DHCP message's sname and file fields are carryovers from the BOOTP protocol that is rarely used today, the DHCP standard allows these fields to be used to contain options that do not fit in the standard options field.

To include options in the sname and/or file fields, the packet's option field must contain the Option Overload option. This option has a value of 52 in the code subfield and a length of 1. The data subfield specifies which of the two auxiliary fields will carry additional options, using the following codes:

1   The file field will carry additional options

2   The sname field will carry additional options

3   Both the file and sname fields will carry additional options

The Vendor-Specific Information Option

An option is defined in the standard as code 43, which is specifically intended for use by vendors to supply information required for the operation of their products. This Vendor-Specific Information option can itself contain multiple options for use by a vendor's products. To identify the vendor of the products for which the information in the Vendor-Specific Information option is intended, a message uses the Vendor Class Identifier option, which has a code value of 60 and a variable length with a minimum of 1 byte.

The Vendor-Specific Information option can contain encapsulated vendor-specific options, which are essentially options within an option. The structure of the encapsulated options is the same as that of the standard DHCP options, with code, length, and data fields. No magic cookie is used, however, and the appearance of the End option (code 255) signals the end of the encapsulated options only—not the end of the entire options field. The codes used in the encapsulated options are not defined by the DHCP standards because they only need to be understandable to systems using the vendor's products.

The End Option

The *End option* signifies the end of the option field. Any bytes in the option field coming after the End option must contain nothing but 0 (Pad option) bytes. Like the Pad option, the End option consists only of a 1-byte code, with no length or data fields. The code has a value of 255.

Other Configuration Options

The DHCP options defined in RFC 2132 fall into several functional categories, which are discussed in the following sections. Each category includes a list containing some of the available options, as well as their code field values.

BOOTP Vendor Information Extensions   These options are included in the DHCP standard exactly as defined in RFC 1497 for use with BOOTP and include many of the basic TCP/IP configuration parameters used by most client systems, such as the following:

Pad (code 0)   See "The Pad Option," earlier in this chapter

End (code 255)   See "The End Option," earlier in this chapter

Subnet Mask (code 1)   Specifies which bits of the IP address identify the host system and which bits identify the network where the host system resides

Router (code 3)   Specifies the IP address of the router (or default gateway) on the local network segment the client should use to transmit to systems on other network segments

Domain Name Server (code 6)   Specifies the IP addresses of the servers the client will use for DNS name resolution

Host Name (code 12)   Specifies the DNS host name the client system will use

Domain name (code 15)   Specifies the name of the DNS domain on which the system will reside

Host-Specific IP Layer Parameters   These options affect the overall functionality of the IP protocol on the client system:

IP Forwarding Enable/Disable (code 19)   Specifies whether IP forwarding (that is, routing) should be enabled on the client system

Maximum Datagram Reassembly Size (code 22)   Specifies the largest size datagram the client should reassemble

Default IP Time-to-Live (code 23)   Specifies the time-to-live value the client should use in its outgoing IP datagrams

Interface-Specific IP Layer Parameters   These options affect the IP protocol functionality of individual network interfaces on the client system. On multihomed systems (that is, systems with two or more network interfaces) these options can have different values for each interface:

Interface MTU (code 26)   Specifies the maximum transfer unit to be used by the Internet Protocol on this network interface only

Broadcast Address (code 28)   Specifies the address to be used for broadcast messages on this network interface only

Static route (code 33)   Specifies a list of static routes to be added to the system's routing table

Link-Layer Parameters   These options affect the interface-specific functionality of the data link–layer protocol:

ARP Cache Timeout (code 35)   Specifies the amount of time that entries should remain in the system's Address Resolution Protocol cache

Ethernet Encapsulation (code 36)   Specifies the Ethernet frame type the client will use when transmitting IP traffic

TCP Parameters   These options affect the interface-specific functionality of the TCP protocol:

TCP Default TTL (code 37)   Specifies the time-to-live value the client should use in its outgoing TCP segments

TCP Keepalive Interval (code 38)   Specifies the amount of time that should elapse before the client sends a keepalive signal over an idle TCP connection

Application and Service Parameters   These options configure various application-layer functions:

Network Information Service Domain (code 40)   Specifies the name of the NIS domain to which the client belongs. The equivalent option for NIS+ uses code 64.

Network Information Servers (code 41)   Specifies the IP addresses of the NIS servers the client will use. The equivalent option for NIS+ uses code 65.

Vendor-Specific Information (code 43)   See "The Vendor-Specific Information Option," earlier in this chapter.

NetBIOS over TCP/IP Name Server (code 44)   Specifies the IP addresses of the servers (usually Windows 2000 or NT WINS servers) the client will use for NetBIOS name resolution.

NetBIOS over TCP/IP Node Type (code 46)   Specifies the NetBIOS name resolution mechanisms the client will use and the order in which it will use them.

Simple Mail Transport Protocol (SMTP) Server (code 69)   Specifies the IP addresses of the SMTP servers the client will use.

Post Office Protocol (POP3) Server (code 70)   Specifies the IP addresses of the POP servers the client will use.

Network News Transport Protocol (NNTP) Server (code 71)   Specifies the IP addresses of the NNTP servers the client will use.

DHCP Extensions   These options are used to provide parameters that govern the DHCP lease negotiation and renewal processes, as well as performing basic tasks, such as specifying the function of a message:

Requested IP Address (code 50)   Used by the client to request a particular IP address from the server.

IP Address Lease Time (code 51)   Specifies the duration of a dynamically allocated IP address lease.

Option Overload (code 52)   See "The Option Overload Option," earlier in this chapter.

DHCP Message Type (code 53)   See "The DHCP Message Type Option," earlier in this chapter.

Server Identifier (code 54)   Specifies the IP address of the server involved in a DHCP transaction; used by the client to address unicasts to the server.

Parameter Request List (code 55)   Used by the client to send a list of requested configuration options (identified by their code numbers) to the server.

Message (code 56)   Used to carry an error message from the server to the client in a DHCPNAK message.

Renewal (T1) time value (code 58)   Specifies the time period that must elapse before an IP address lease enters the renewing state.

Rebinding (T2) time value (code 59)   Specifies the time period that must elapse before an IP address lease enters the rebinding state.

Vendor Class Identifier (code 60)   Identifies the vendor and configuration of the client. See "The Vendor-Specific Information Option," earlier in this chapter.

Client Identifier (code 61)   Unique identifier for the client delivered to the server for use in its index of client systems.

DHCP Communications

When you configure a workstation to be a DHCP client, the system initiates an exchange of messages with a DHCP server. Whether you are using dynamic, automatic, or manual address allocation, the first exchange of messages, resulting in an IP address assignment for the client, is the same.

IP Address Assignment

The entire IP address negotiation process is illustrated in Figure 23-3. Before the initial client/server exchange can begin, the question rises of how the client is to find the server and communicate with it when its TCP/IP stack has not yet been configured. A DHCP client that does not yet have an IP address is said to be in the *init state*. In this state, even though the workstation has no information about the servers on the network and no IP address of its own, it is still capable of sending broadcast transmissions.

The client begins the exchange by broadcasting a series of DHCPDISCOVER messages. In this packet, the DHCP Message Type option has a value of 1 in its data field and the hardware address of the client is included in the chaddr (or client hardware address) field, as shown in Figure 23-4. In most cases, this will be the MAC address hardcoded into the network interface adapter, as identified by the values of the htype and hlen fields. Because the client has no IP address of its own yet, the source address field in the IP header contains the value 0.0.0.0. The message may also contain other options requesting specific information from the server, such as the Requested IP Address option.

Because the basic DHCP message format is the same as that of BOOTP, some protocol analyzer programs interpret DHCP traffic using BOOTP labels, as in the packet captures shown in Figures 23-4 through 23-7.

Each of the servers receiving the DHCPDISCOVER packet responds to the client with a DHCPOFFER message that contains an IP address in the yiaddr (or your IP address) field. This is the address the server is offering for the client's use. Whether the server transmits its DHCPOFFER and other messages to the client as unicasts or broadcasts is determined by the status of the Broadcast flag in the client's DHCPDISCOVER message (which the observant reader will note is not displayed in the packet captures shown here), as well as the capabilities of the server. The message will also contain options specifying values for the other TCP/IP parameters that the server is configured to deliver, as well as the DHCP extension options, as shown in Figure 23-5. Because each DHCP server on the network operates independently, the client may receive several offers, each with a different IP address.

Note   *If systems are functioning as DHCP (or BOOTP) relay agents on the same network segment as the client, the agent systems will propagate the broadcasts to other segments, resulting in additional DHCPOFFER messages from remote servers being transmitted back to the client through the agent. See "Relay Agents," later in this chapter, for more information.*

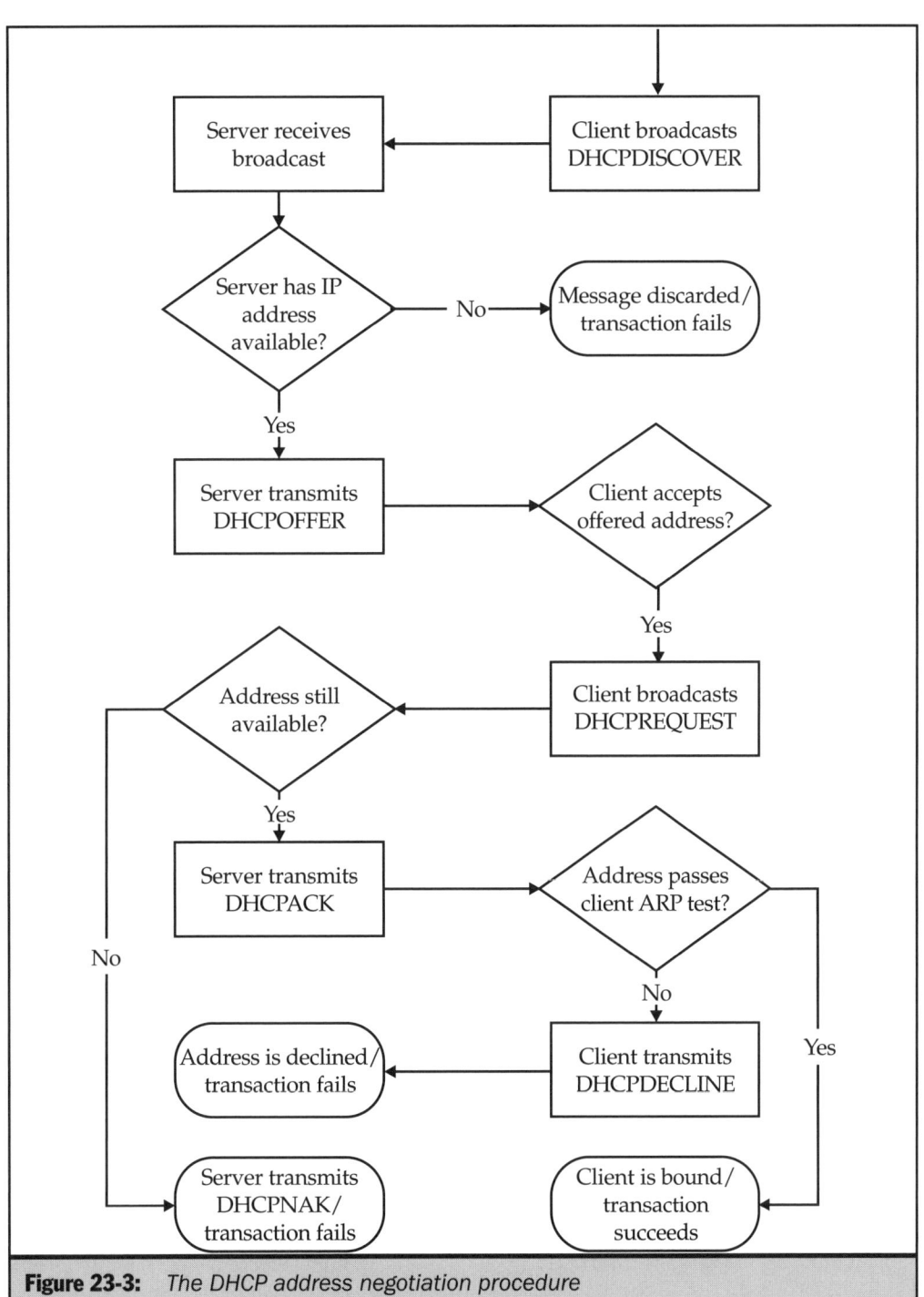

Figure 23-3:   *The DHCP address negotiation procedure*

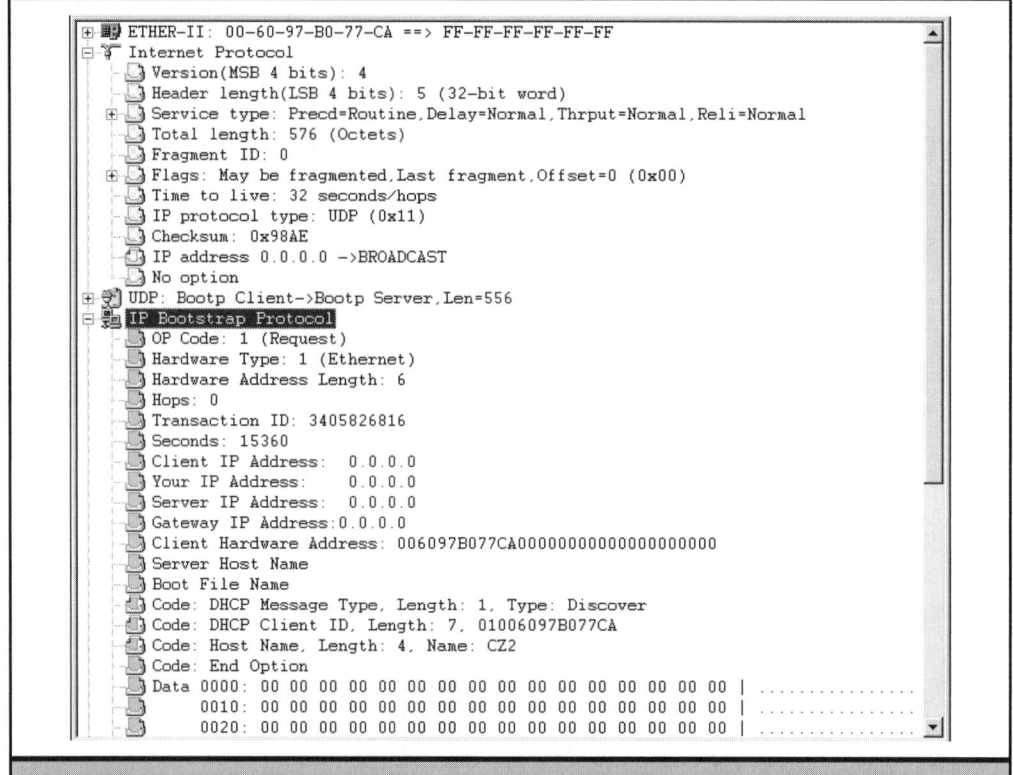

```
⊞ ■▓ ETHER-II: 00-60-97-B0-77-CA ==> FF-FF-FF-FF-FF-FF
⊟ ▼ Internet Protocol
    ⤷ □ Version(MSB 4 bits): 4
    ⤷ □ Header length(LSB 4 bits): 5 (32-bit word)
   ⊞ □ Service type: Precd=Routine,Delay=Normal,Thrput=Normal,Reli=Normal
    ⤷ □ Total length: 576 (Octets)
    ⤷ □ Fragment ID: 0
   ⊞ □ Flags: May be fragmented,Last fragment,Offset=0 (0x00)
    ⤷ □ Time to live: 32 seconds/hops
    ⤷ □ IP protocol type: UDP (0x11)
    ⤷ □ Checksum: 0x98AE
    ⤷ ◻ IP address 0.0.0.0 ->BROADCAST
    ⤷ □ No option
  ⊞ ▓ UDP: Bootp Client->Bootp Server,Len=556
  ⊟ ▓ IP Bootstrap Protocol
    ⤷ ▓ OP Code: 1 (Request)
    ⤷ ▓ Hardware Type: 1 (Ethernet)
    ⤷ ▓ Hardware Address Length: 6
    ⤷ ▓ Hops: 0
    ⤷ ▓ Transaction ID: 3405826816
    ⤷ ▓ Seconds: 15360
    ⤷ ▓ Client IP Address:  0.0.0.0
    ⤷ ▓ Your IP Address:    0.0.0.0
    ⤷ ▓ Server IP Address:  0.0.0.0
    ⤷ ▓ Gateway IP Address:0.0.0.0
    ⤷ ▓ Client Hardware Address: 006097B077CA00000000000000000000
    ⤷ ▓ Server Host Name
    ⤷ ▓ Boot File Name
    ⤷ ◻ Code: DHCP Message Type, Length: 1, Type: Discover
    ⤷ ◻ Code: DHCP Client ID, Length: 7, 01006097B077CA
    ⤷ ◻ Code: Host Name, Length: 4, Name: CZ2
    ⤷ ◻ Code: End Option
    ⤷ □ Data 0000: 00 00 00 00 00 00 00 00 00 00 00 00 00 00 00 00 | ...............
    ⤷ □      0010: 00 00 00 00 00 00 00 00 00 00 00 00 00 00 00 00 | ...............
    ⤷ □      0020: 00 00 00 00 00 00 00 00 00 00 00 00 00 00 00 00 | ...............
```

Figure 23-4: *The DHCPDISCOVER message, as captured in a network analyzer program*

After a predetermined interval, the client stops broadcasting DHCPDISCOVER messages and selects one of the offers it has received. If the client receives no DHCPOFFER messages, it retries its broadcasts and eventually times out with an error message. No other TCP/IP communications are possible while the machine is in this state.

When the client accepts an IP address offered by a server, it generates a DHCPREQUEST message that contains the name of the selected server in the Server Identifier option and the offered IP address (taken from the yiaddr field of the DHCPOFFER message) in the Requested IP Address option (see Figure 23-6). The client also uses the DHCP Parameter Request List option to request additional parameters from the server.

The client transmits the DHCPREQUEST message as a broadcast because the message not only informs the selected server that the client has accepted its offer, it also informs

```
⊞ ⬛ ETHER-II: 00-20-AF-37-B8-12 ==> 00-60-97-B0-77-CA
⊞ ⬛ IP: 192.168.2.10->192.168.2.22,ID=40705
⊞ ⬛ UDP: Bootp Server->Bootp Client,Len=308
⊟ ⬛ IP Bootstrap Protocol
    ⬛ OP Code: 2 (Reply)
    ⬛ Hardware Type: 1 (Ethernet)
    ⬛ Hardware Address Length: 6
    ⬛ Hops: 0
    ⬛ Transaction ID: 3982093657
    ⬛ Seconds: 0
    ⬛ Client IP Address:  0.0.0.0
    ⬛ Your IP Address:    192.168.2.22
    ⬛ Server IP Address:  0.0.0.0
    ⬛ Gateway IP Address:0.0.0.0
    ⬛ Client Hardware Address: 006097B077CA00000000000000000000
    ⬛ Server Host Name
    ⬛ Boot File Name
    ⬛ Code: DHCP Message Type, Length: 1, Type: Offer
    ⬛ Code: Subnet Mask, Length: 4       Address255.255.255.0
    ⬛ Code: DHCP Renewal (T1) Time, Length: 4, Value:129600
    ⬛ Code: DHCP Rebinding  (T2) Time, Length: 4, Value:226800
    ⬛ Code: DHCP IP Address Lease Time, Length: 4, Value:259200
    ⬛ Code: DHCP Server ID, Length: 4
    ⬛       Address: 192.168.2.10
    ⬛ Code: Router, Length: 4
    ⬛       Address: 192.168.2.100
    ⬛ Code: Domain Name Server, Length: 8
    ⬛       Address: 199.224.86.15
    ⬛       Address: 199.224.86.16
    ⬛ Code: NetBIOS Name Server, Length: 4
    ⬛       Address: 192.168.2.10
    ⬛ Code: NetBIOS over TCP/IP, Length: 1, Node Type:0x8 H-node
    ⬛ Code: End Option
    ⬛ Data 0000: 00                                |  .
  ⬛ Calculate CRC: 0xfab71c9d
```

Figure 23-5:   *The DHCPOFFER message, returned to the client by a server*

the other servers that their offers have been declined. The message must have the same value in the secs field and use the same broadcast address as the DHCPDISCOVER messages the client previously transmitted. This is so that the broadcast is certain to reach all the servers that responded with offers (including those on other segments that require the assistance of relay agents).

At the time a server generates a DHCPOFFER message, the IP address it offers is not yet exclusively allocated to that client. If addresses are in short supply or if the client takes too long to respond, the server may offer that address to another client in the interim. When the server receives the DHCPREQUEST message saying its offer has been accepted, it generates either a DHCPACK message indicating that the IP address assignment has been completed (see Figure 23-7) or a DHCPNAK message indicating that the offered address is no longer available. The DHCPACK message contains all the options requested by the client in the DHCPREQUEST message and, as with the DHCPOFFER message, can be unicast or broadcast depending on the value of the Broadcast flag in the client's messages.

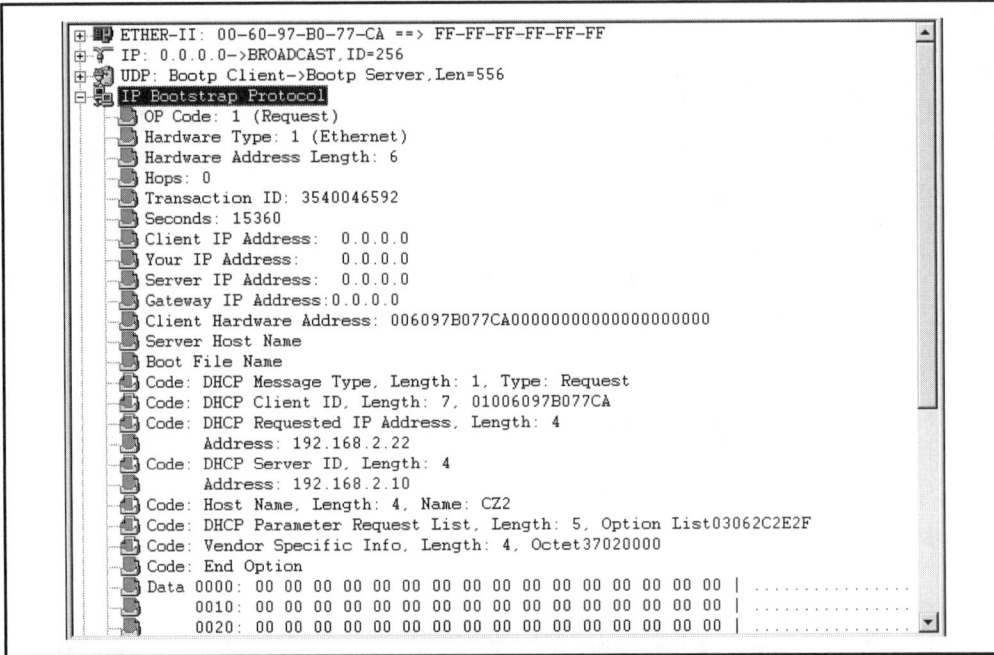

```
⊞ 🖳 ETHER-II: 00-60-97-B0-77-CA ==> FF-FF-FF-FF-FF-FF
⊞ 🖵 IP: 0.0.0.0->BROADCAST,ID=256
⊞ 📱 UDP: Bootp Client->Bootp Server,Len=556
⊟ 📱 IP Bootstrap Protocol
      📄 OP Code: 1 (Request)
      📄 Hardware Type: 1 (Ethernet)
      📄 Hardware Address Length: 6
      📄 Hops: 0
      📄 Transaction ID: 3540046592
      📄 Seconds: 15360
      📄 Client IP Address:  0.0.0.0
      📄 Your IP Address:    0.0.0.0
      📄 Server IP Address:  0.0.0.0
      📄 Gateway IP Address:0.0.0.0
      📄 Client Hardware Address: 006097B077CA00000000000000000000
      📄 Server Host Name
      📄 Boot File Name
      📄 Code: DHCP Message Type, Length: 1, Type: Request
      📄 Code: DHCP Client ID, Length: 7, 01006097B077CA
      📄 Code: DHCP Requested IP Address, Length: 4
      📄        Address: 192.168.2.22
      📄 Code: DHCP Server ID, Length: 4
      📄        Address: 192.168.2.10
      📄 Code: Host Name, Length: 4, Name: CZ2
      📄 Code: DHCP Parameter Request List, Length: 5, Option List03062C2E2F
      📄 Code: Vendor Specific Info, Length: 4, Octet37020000
      📄 Code: End Option
      📄 Data 0000: 00 00 00 00 00 00 00 00 00 00 00 00 00 00 00 00 | ................
      📄      0010: 00 00 00 00 00 00 00 00 00 00 00 00 00 00 00 00 | ................
      📄      0020: 00 00 00 00 00 00 00 00 00 00 00 00 00 00 00 00 | ................
```

Figure 23-6:   *The DHCPREQUEST message, sent by a client to a server to accept its offered address*

When the server generates a DHCPACK message, it creates an entry in its database that commits the offered IP address to the client's hardware address. The combination of these two addresses will, from this point until the address is released, function as a unique identifier for that client, called the *lease identification cookie*. If the server sends a DHCPNAK message to the client, the entire transaction is nullified and the client must begin the whole process again by generating new DHCPDISCOVER messages.

As a final test of its newly assigned address, the client can (but is not required to) use the ARP protocol to make sure no other system on the network is using the IP address furnished to it by the server. If the address is in use, the client sends a DHCPDECLINE message to the server, nullifying the transaction. If the address is not in use, the address assignment process is completed and the client enters what is known as the *bound state*.

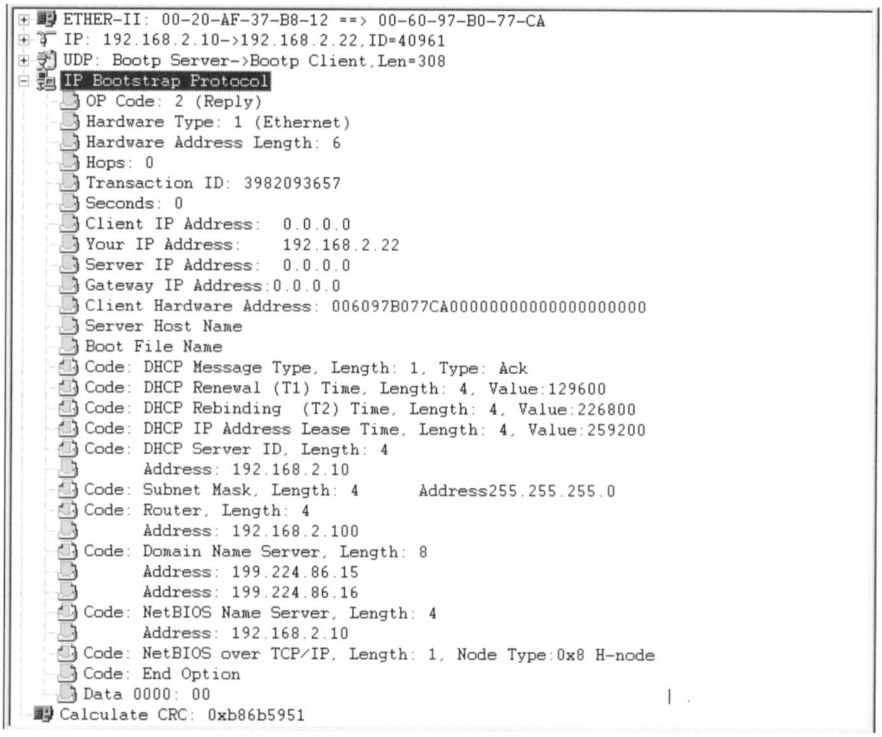

```
⊞ ▦ ETHER-II: 00-20-AF-37-B8-12 ==> 00-60-97-B0-77-CA
⊞ ⊻ IP: 192.168.2.10->192.168.2.22,ID=40961
⊞ ⊛ UDP: Bootp Server->Bootp Client,Len=308
⊟ ▦ IP Bootstrap Protocol
      ⌐ OP Code: 2 (Reply)
      ⌐ Hardware Type: 1 (Ethernet)
      ⌐ Hardware Address Length: 6
      ⌐ Hops: 0
      ⌐ Transaction ID: 3982093657
      ⌐ Seconds: 0
      ⌐ Client IP Address:  0.0.0.0
      ⌐ Your IP Address:    192.168.2.22
      ⌐ Server IP Address:  0.0.0.0
      ⌐ Gateway IP Address:0.0.0.0
      ⌐ Client Hardware Address: 006097B077CA00000000000000000000
      ⌐ Server Host Name
      ⌐ Boot File Name
      ⌐ Code: DHCP Message Type, Length: 1, Type: Ack
      ⌐ Code: DHCP Renewal (T1) Time, Length: 4, Value:129600
      ⌐ Code: DHCP Rebinding  (T2) Time, Length: 4, Value:226800
      ⌐ Code: DHCP IP Address Lease Time, Length: 4, Value:259200
      ⌐ Code: DHCP Server ID, Length: 4
      ⌐       Address: 192.168.2.10
      ⌐ Code: Subnet Mask, Length: 4        Address255.255.255.0
      ⌐ Code: Router, Length: 4
      ⌐       Address: 192.168.2.100
      ⌐ Code: Domain Name Server, Length: 8
      ⌐       Address: 199.224.86.15
      ⌐       Address: 199.224.86.16
      ⌐ Code: NetBIOS Name Server, Length: 4
      ⌐       Address: 192.168.2.10
      ⌐ Code: NetBIOS over TCP/IP, Length: 1, Node Type:0x8 H-node
      ⌐ Code: End Option
      ⌐ Data 0000: 00                                    |  .
   ▦ Calculate CRC: 0xb86b5951
```

Figure 23-7: *The DHCPACK message, confirming that the server has bound the offered address to the client*

DHCPINFORM

The addition of a new message type to the DHCP standard makes it possible for clients to request TCP/IP configuration parameters without being assigned an IP address. When a server receives a DHCPINFORM message from a client, it generates a DHCPACK message containing the appropriate options for the client without including the lease time options or an IP address in the yiaddr field. The assumption is the client has already been manually configured with an IP address and does not require regular renewals or any other maintenance beyond the initial parameter assignment.

The most obvious application for the DHCPINFORM message is to configure DHCP servers themselves, which typically cannot use a DHCP-supplied IP address. The server must have a manually configured IP address, but administrators can use DHCPINFORM messages to request values for the other required TCP/IP configuration parameters. This eliminates the need for administrators to configure any of the client parameters manually, apart from the IP address.

Lease Renewal

When a DHCP server is configured to use manual or automatic address allocation, no further contact occurs with the client unless (or until) the client manually releases the address. When using dynamic allocation, however, the DHCPOFFER messages sent to the client contain options that specify the nature of the address lease agreement. These options include the IP Address Lease Time, the Renewal (T1) Time Value, and the Rebinding (T2) Time Value.

These time values are supplied to the client in units of seconds and do not include specific clock times (to account for possible discrepancies between the client and server system clocks). Administrators can specify values for these intervals in the DHCP server configuration. As an example, the Microsoft DHCP Server has a default IP Address Lease Time of three days. The Renewal (T1) Time Value defaults to 50 percent of the lease time and the Rebinding (T2) Time Value to 87.5 percent of the lease time. Because the IP Address Lease Time option uses a 4-byte data subfield to specify the number of seconds, the maximum possible absolute value is approximately 136 years. A hexadecimal value of 0xffffffff (or a binary value of 32 1s) indicates infinity.

Once a client system with a leased address has entered the bound state, it has no further communications with the server until the system restarts or it reaches the T1, or renewal time. The lease renewal transaction that begins at this time is illustrated in Figure 23-8.

When the client's lease reaches the T1 time, it enters the *renewing* state and begins transmitting DHCPREQUEST messages to the server that assigned its IP address. The messages contain the client's lease identification cookie and are transmitted to the server as unicasts (unlike the DHCPREQUEST messages in the initial lease negotiation, which are broadcasts). If the server receives the message and is capable of renewing the lease, it responds with a DHCPACK message and the client returns to the bound state with a reset lease time. No further communication is necessary until the next renewal.

If the server cannot renew the lease, it responds with a DHCPNAK message, which terminates the transaction and the lease. The client must then restart the entire lease negotiation process with a new sequence of DHCPDISCOVER broadcasts.

Note   *This is virtually the same message exchange that occurs each time a DHCP client system reboots, except the client remains in the bound state. Usually, the server will respond to the DHCPREQUEST message with a DHCPNAK only when the client system has been moved to a different subnet and requires an IP address with a different network identifier. If the server fails to respond at all after repeated retransmissions, the client continues to use the address until the lease reaches the T1 time, at which point it enters the renewing state and begins the lease renewal process.*

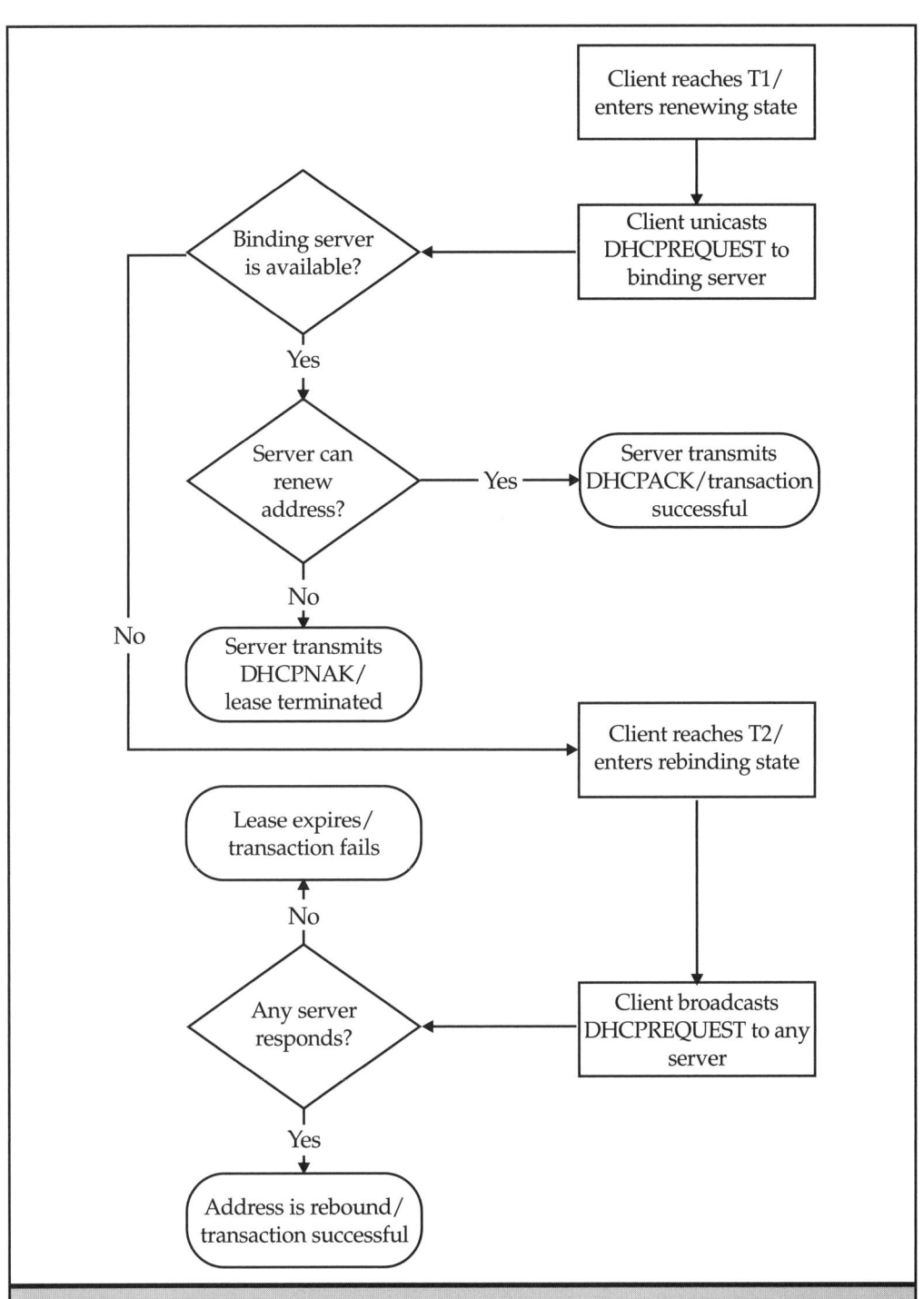

Figure 23-8:   *The DHCP lease renewal procedure*

If the client receives no response to its unicast DHCPREQUEST, it retransmits the message each time half the interval between the current time and the T2 time has expired. Thus, using the default time values for the Microsoft server, the lease duration is 72 hours (three days), the T1 time is 36 hours (50 percent of 72), and the T2 time is 63 hours (87.5 percent of 72). The client will send its first DHCPREQUEST message to the server at 36 hours into the lease (the T1 time) and then retransmit at 49.5 hours (half the time until T2), 55.75 hours (half the remaining time until T2), 59.375 hours, and so on until it reaches the T2 time.

Once the lease time hits the T2 point, the client enters the *rebinding state* and begins transmitting its DHCPREQUEST messages as broadcasts to solicit an IP address assignment from any available server. Once again, the client awaits either a DHCPACK or DHCPNAK reply from a server. If no replies occur, the client continues retransmitting whenever half the remaining time in the lease expires. If the lease time does expire with no response from the server, the client releases the IP address and returns to the init state. It cannot send any further TCP/IP transmissions with the exception of DHCPDISCOVER broadcasts.

Address Release

While in the bound state, a DHCP client can relinquish its possession of an IP address (whether leased or permanent) by transmitting a unicast DHCPRELEASE message to the server, containing the client's lease identification cookie. This returns the client to the unbound state, preventing any further TCP/IP transmissions, except for DHCPDISCOVER broadcasts, and causes the server to place the IP address back into its pool of available addresses. In most cases, an address release like this only occurs when the user of a client workstation explicitly requests it, using a utility such as Ipconfig.exe in Windows 2000 or Winipcfg.exe in Windows 98.

Relay Agents

Because of its reliance on broadcast transmissions—at least from the client side—it would seem that DHCP clients and servers must be located on the same network segment in order to communicate. If this were the case, however, DHCP would not be a practical solution for enterprise networks, because there would have to be a DHCP server on every LAN. To resolve this problem, DHCP takes its cue from BOOTP and uses relay agents as intermediaries between clients and servers. In fact, DHCP uses BOOTP relay agents exactly as defined in RFC 1542, "Clarifications and Extensions for the Bootstrap Protocol."

A *DHCP relay agent* (or *BOOTP relay agent*—the names are interchangeable) is a module located in a workstation or router on a particular network segment, which enables the other systems on that segment to be serviced by a DHCP server located on a remote segment. The relay agent works by monitoring UDP port 67 for DHCP messages being broadcast by clients on the local network. Normally, a workstation or router would ignore these messages because they do not contain a valid source address but, like the DHCP server, the relay agent is designed to accept a source address of 0.0.0.0.

When the relay agent receives these messages, it inserts its own IP address into the giaddr field of the DHCP message, increments the value of the hops field, and retransmits the packets to a DHCP server located on another segment. Depending on the location of the relay agent, this retransmission can take two forms. For an agent built into a router, the device may be able to broadcast the message using a different interface than the one over which it received the message. For a relay agent running on a workstation, it is necessary to send the message to the DHCP server on another segment as a unicast.

A message can be passed along by more than one relay agent on the way to the DHCP server. An agent only inserts its IP address into the giaddr field if this field has a value of 0. In addition, the agent must silently discard messages that have a value greater than 16 in the hops field (unless this limit is configurable and has been adjusted by the network administrator). This prevents DHCP messages from cycling endlessly around the network.

When a DHCP server receives messages from a relay agent, it processes them in the normal manner, but transmits them back to the address in the giaddr field, rather than to the client. The relay agent may then use either a broadcast or a unicast (depending on the state of the Broadcast flag) to transmit the reply, unchanged, to the client.

For an entire internetwork to be serviced by DHCP, every segment must have either a DHCP server or a relay agent on it. Most of the routers on the market today have DHCP/BOOTP relay agent functionality built into them but for those that don't, the server versions of Windows 2000 and Windows NT 4.0 also include a relay agent service.

DHCP Implementations

Most of the operating systems in use today are capable of functioning as DHCP clients, and DHCP servers are available on many different platforms. Novell includes a DHCP server in all of its NetWare products, and DHCP server products (many of which are open source) exist for several flavors of Unix, including Linux. By far, the most popular implementation is the Microsoft DHCP Server.

Most of the DHCP products conform to the IETF standards, although you should be aware that some implementations support the older version of the DHCP standard,

RFC 1541. The latest standard, RFC 2131, was published in March 1997. The most important modifications to the standard are:

- The inclusion of a new message type—DHCPINFORM—that clients already having an IP address can use to request other local configuration parameters from a server.

- The addition of vendor-specific options that make it possible for a DHCP server to supply nonstandardized information to clients based on vendor class identifiers.

When you are evaluating DHCP servers for use on your network, you will find that while the basic functionality in most of the server implementations is the same, the options supported by each one can differ greatly. All DHCP servers will certainly support the basic parameters required by all TCP/IP clients, such as the Subnet Mask, Router, and Domain Name Server options, but the other options they support may be limited. Certain products may also include nonstandardized options intended for specific clients. For example, the Novell DHCP server includes options to support older NetWare/IP clients.

If you are running a variety of different clients on your network, you should list the options you need to supply to each one and try to find a server to satisfy them all. Mixing DHCP servers from different manufacturers on the same internetwork is also possible but because the clients use broadcast transmissions for their DHCPDISCOVER messages, no sure way exists to predict which server will configure which client.

Microsoft DHCP Server

Microsoft's DHCP Server is supplied with the Windows 2000 Server and Windows NT 4.0 Server operating systems. While obviously intended for use with the Microsoft client operating systems, the server also supports all the options defined in the DHCP standards, many of which are not used by Microsoft clients. The DHCP server included in Windows NT 4.0 is designed to support the RFC 1533, 1534, 1541, and 1542 standards because the operating system was released in 1996, before RFC 2131 and 2132 were published. The Windows 2000 version supports the newer standards.

Creating Scopes

Microsoft DHCP Server enables you to define a range of IP addresses, called a *scope*, for each subnet on the network, and to specify the lease duration for that scope (see Figure 23-9). When a client on another subnet solicits an address from the server through a relay agent, the server uses the information added to the message by the agent to assign an IP address on the proper subnet. You then specify values for the options you want to deliver with each IP address the server assigns. You can assign options for individual scopes, global options that are applied to all scopes, and default options.

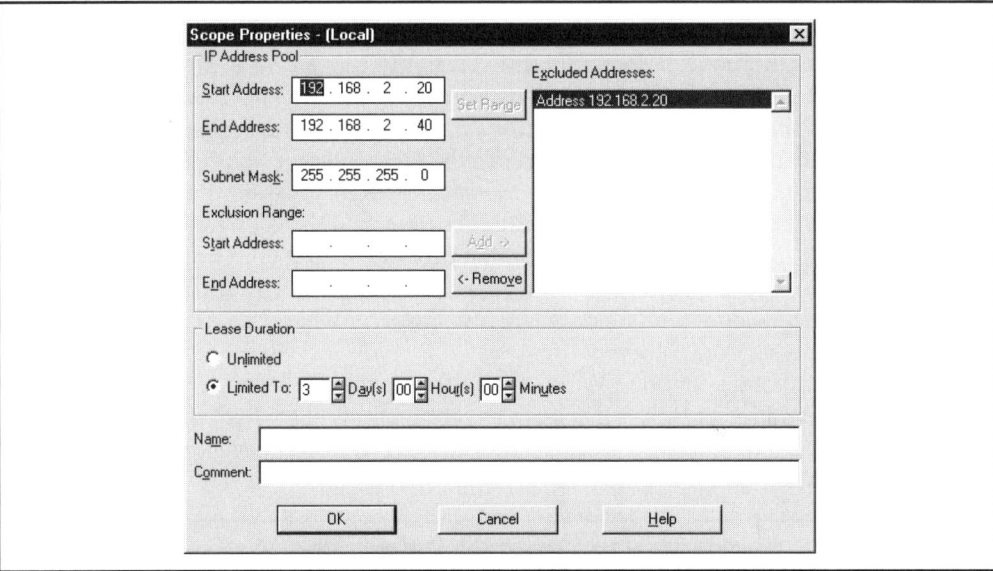

Figure 23-9: *The Microsoft DHCP Server enables you to define a range of addresses to be excluded from a scope, as well as leases of varying duration.*

One of the persistent problems with DHCP has always been that no communication exists between servers. Each server operates independently and assigning IP addresses from a single, common pool shared by multiple servers can result in workstations with duplicate addresses. Thus, if you intend to allocate the addresses from 192.168.10.100 to 192.168.10.200 using two DHCP servers, you traditionally must split the addresses between two scopes, one on each server. If one server should fail, the addresses allocated by that server become unavailable for renewal or reallocation unless an administrator modifies the scope on the remaining server.

Microsoft DHCP Server enables you to specify multiple address ranges to exclude from a scope. When using two or more DHCP servers to service a single subnet for fault tolerance purposes, Microsoft recommends you create a single identical scope for each subnet on each server and then exclude different address ranges on each one in an 80/20 split. Thus, to continue the example, you would configure both DHCP servers with a scope of addresses running from 192.168.10.100 to 192.168.10.200, but one server would exclude addresses 192.168.10.181 to 192.168.10.200, while the other would exclude 192.168.10.100 to 192.168.10.180. This technique makes it easy for an administrator to modify the configuration by altering the excluded range, should one of the servers become unavailable.

Since the release of Service Pack 3 for Windows NT 4.0, there is another solution to this problem. The DHCP standards state that a client may use a gratuitous ARP request to determine if the IP address it has just been assigned is already in use on the network. If the address is in use, the client sends a DHCPDECLINE message to the server, which nullifies the entire transaction. If all clients did this, it would be possible to create identical scopes on different DHCP servers without having identically configured clients as a result. Some DHCP clients (most notably those in Windows 95 and 98), however, do not perform this gratuitous ARP test.

The Service Pack 3 (and later) releases for Windows NT 4.0 (as well as Windows 2000) improve on this concept. This update modifies the DHCP server to include a *conflict detection* mechanism that uses ICMP Echo messages (pings) to verify that an address is not in use before it is assigned to a client. This is better than the client-side ARP check because it occurs earlier in the lease negotiation process, saving both time and network traffic. When you enable conflict detection on all your DHCP servers, you can configure them with identical scopes and the only negative result will be some additional "bad address" error messages in the event logs.

To enable conflict detection in Windows 2000, you highlight the server in the DHCP console display and select Properties from the context menu to display the dialog box shown in Figure 23-10. You can select the number of pings the server uses to test each address before assigning it.

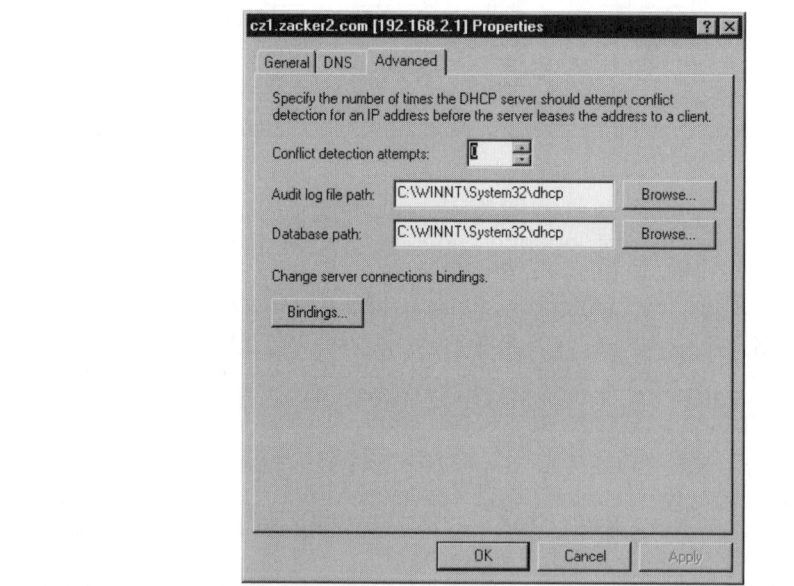

Figure 23-10:   *In the server Properties dialog box you can configure the Microsoft DHCP Server to detect address conflicts by sending a number pings to each IP address before allocating it to a client.*

Server to Client Transmissions

By default, the Microsoft DHCP Server included with Windows NT 4.0 ignores the state of the Broadcast flag in the DHCPDISCOVER and other messages transmitted by the client and broadcasts all its responses. This significantly increases the overall amount of network traffic generated by the DHCP address allocation process. In prior versions of Windows NT, this behavior was not controllable, but Windows NT 4.0 supports a registry entry that forces the DHCP server to use the Broadcast flag to determine how it will transmit its responses to DHCP clients. The entry is called IgnoreBroadcastFlag and you create it in the following registry key:

```
HKEY_LOCAL_MACHINE\System\CurrentControlSet\Services\DHCPServer\

Parameters
```

Create IgnoreBroadcastFlag as a REG_DWORD entry and configure it using one of the following values:

0   DHCP transmissions from the server are controlled by the status of the Broadcast flag in the previous client messages

1   All DHCP transmissions from the server are sent as broadcasts

Microsoft Client Utilities

Little client-side control is available for the Microsoft DHCP clients, mainly because little is needed. Simply enabling DHCP instead of manually configuring the TCP/IP stack is all you have to do to use the client. The operating systems do include a utility you can use to view the settings assigned to the client by the DHCP server, as well as manually release and renew the current configuration. This is handy, because without these utilities, there is no easy way for the user on a DHCP client workstation to find out its IP address.

Microsoft Client 3.0 for MS-DOS, Windows for Workgroups (with TCP/IP-32 installed), Windows NT, and Windows 2000 include a utility called Ipconfig.exe, which runs from the command line. The syntax for Ipconfig.exe is as follows:

```
Ipconfig [/all] [/release {adaptername}] [/renew {adaptername}]
```

/all   Displays a complete listing of the TCP/IP parameters for all interfaces installed in the system

/release {*adaptername*}   Terminates the DHCP lease for a specific adapter installed in the system, or if no adapter name is supplied, for all of the adapters installed in the system

/renew {*adaptername*}   Initiates a new DHCP lease negotiation procedure for a specific adapter installed in the system, or if no adapter name is supplied, for all of the adapters installed in the system

   Running Ipconfig from the command line with no parameters produces a basic TCP/IP client configuration display. Running the program with the /all parameter produces a more detailed display, as follows:

```
Windows NT IP Configuration

        Host Name . . . . . . . . . : cz1

        DNS Servers . . . . . . . . : 199.204.86.15

                                      199.204.86.16

        Node Type . . . . . . . . . : Hybrid

        NetBIOS Scope ID. . . . . . :

        IP Routing Enabled. . . . . : No

        WINS Proxy Enabled. . . . . : No

        NetBIOS Resolution Uses DNS : Yes

Ethernet adapter Elnk31:

        Description . . . . . . . . : ELNK3 Ethernet Adapter.

        Physical Address. . . . . . : 00-20-AF-37-B8-12

        DHCP Enabled. . . . . . . . : Yes

        IP Address. . . . . . . . . : 192.168.2.10

        Subnet Mask . . . . . . . . : 255.255.255.0

        Default Gateway . . . . . . : 192.168.2.100

        Primary WINS Server . . . . : 192.168.2.10

        Lease Obtained. . . . . . . : Wednesday, June 27, 2001 4:32:59 PM

        Lease Expires . . . . . . . : Saturday, June 30, 2001 4:32:59 PM
```

This display provides the network adapter names (such as Elnk31) you can use when you manipulate the DHCP client using the /release and /renew command-line parameters.

   Windows 95 and Windows 98 have a GUI version of the same utility called Winipcfg.exe. Executing this program from the Run dialog box produces a display like that shown in Figure 23-11. You can use the buttons to release and renew the DHCP client, just as you would with Ipconfig.exe's command-line parameters.

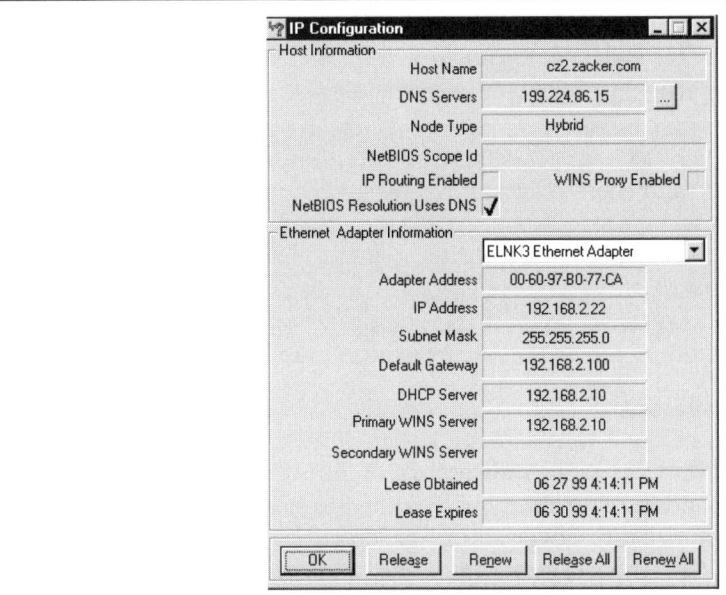

Figure 23-11: Winipcfg.exe *provides the same functions as* Ipconfig.exe, *using a graphical interface.*

Client Option Support

Even though the Microsoft DHCP Server supports all the options defined in the IETF standards, this does not mean the DHCP client in the Microsoft operating systems supports them all as well. In fact, the clients support only a few of the available options, even though they could conceivably make good use of them. In addition to the DHCP extension options required for control of the lease negotiation process, the options supported by the DHCP clients in Microsoft Client 3.0 for MS-DOS, Windows for Workgroups (with TCP/IP-32 installed), Windows 95, Windows 98, and Windows NT are as follows:

- Subnet Mask
- Router
- Domain Name Server
- Domain Name

- NetBIOS over TCP/IP Name Server

- NetBIOS over TCP/IP Node Type

- NetBIOS over TCP/IP Scope

The Windows 2000 client also supports these and adds support for vendor-specific information using vendor and client class identifiers.

 *For more information on the DHCP support in Windows 2000, see "Windows 2000 and DHCP," later in this chapter.*

DHCP Lease Duration

Once configured, DHCP servers require little or no monitoring. Depending on the volatility of the network and the availability of IP addresses, administrators can modify the lease duration to minimize the network traffic generated by DHCP communications. On an unregistered TCP/IP network, no shortage of IP addresses should occur, so you can safely increase the duration of the lease to make renewals less frequent.

By increasing the lease from the default of 3 days to 30 days, for example, a system would have to run continuously for 15 days before attempting to renew its address. If some addresses were to remain allocated for a few weeks after a machine was moved to another subnet, no harm would be done.

On a network that uses registered IP addresses and that has a limited supply of them, however, you can decrease the duration of the lease to prevent addresses from being orphaned for too long a period. Of course, on a stable network—that is, one on which computers are rarely moved around—it may not be necessary to have the leases expire at all.

 If the users of a network turn their computers off at the end of the day's work, the lease will be rebound each time the machine is turned on again. Be aware that DHCP activity occurs during the system startup process, not during the network logon. Simply logging off the network at the end of the day, instead of shutting the machine down, will not trigger a renewal at the next logon.

Moving a DHCP Server

If at any time you want to use a different Windows computer as your DHCP server, it is possible to migrate the database and configuration settings to another system without interrupting service to the clients. To do this, you must move the DHCP database files themselves as well as the registry entries that govern the DHCP server's operations.

The DHCP database is located in the *\%SystemRoot%\System32\dhcp* directory on the server, which, by default, would be *C:\Winnt\System32\dhcp*. After installing the Microsoft DHCP Server service on the new machine, you must stop the service (from the Services console or Control Panel) on both machines and copy the entire contents of this directory from the existing DHCP server (including all subdirectories and their contents) to the same location on the new system.

The next step is to launch the Regedit.exe program to export the entire contents of the following registry key to a registry script file:

```
HKEY_LOCAL_MACHINE\SYSTEM\CurrentControlSet\Services\DHCPServer\

Configuration
```

Once you have created a registry script file (with a .reg extension) containing these settings, you import them on the new machine by double-clicking the file name in Windows Explorer (or by using the Import Registry Key command on the File menu). You can then restart the service on the new server and it will service the clients from that point forward.

Note   *Do not allow the DHCP service to restart on the old system or you may end up with clients on your network that have duplicate IP addresses. To prevent a service duplication, modify the Startup Type of the Microsoft DHCP Server on the old machine in the Services console to prevent it from starting automatically when the server boots, deactivate the scopes duplicated on the new server, or modify the scopes to assign different IP addresses.*

DHCP Maintenance

DHCP address allocation on a large network is a rather disk-intensive process. This is particularly true on networks where a large number of users all turn on their computers at the same time each day. The PCs to be used as DHCP servers should be selected with an eye toward disk efficiency—RAID arrays, high-speed disk drives, and other such disk technologies are a good idea.

Another good idea is to compact the DHCP database on a regular basis, to keep it running at maximum efficiency. Windows 2000 and NT include a command-line utility called Jetpack.exe, which enables you to create a batch file that performs the compression. You can execute the batch file manually on a regular basis or schedule it to automate the process. The batch file used to compress the DHCP database should appear as follows:

```
cd %SystemRoot%\System32\dhcp

net stop dhcpserver
```

```
jetpack Dhcp.mdb Tmp.mdb

net start dhcpserver
```

This same batch file, run from the \%SystemRoot%\System32\wins directory with the Wins.mdb file name instead of Dhcp.mdb, will compress the WINS database.

The Jetpack.exe program reads the first file name on the command line and compacts it to the second file name. After the process completes, the program deletes the original database file and renames the new, compacted version to the original file name.

The Jetpack.exe program creates another temporary file called Temp.mdb during the database compression process. Do not use this file name on the command line and be sure no file of this name exists in the database directory before you launch Jetpack.exe.

Windows 2000 and DHCP

The DHCP server included with the Windows 2000 Server operating systems adds several features not found in the Windows NT 4.0 version. The server is now compliant with the newer DHCP standards, published as RFC 2131 and 2132, and is also committed to compliance with emerging standards, such as those defining interaction between DHCP and dynamic DNS and DHCP server-to-server communications.

DHCP and DNS

The Windows NT 4.0 version of the Microsoft DHCP Server can interact with WINS by entering its IP address assignments into the WINS database as they are assigned. In Windows 2000, DNS largely replaces the functionality of WINS, but DNS servers must traditionally be manually updated with new information. The Microsoft DNS Server included with Windows 2000 Server supports dynamic updates of the DNS, as defined in RFC 2136.

As a result of this technology, it is also possible for a DHCP server to update address (A) and pointer (PTR) records on a DNS server with new IP addresses as they are assigned. The draft standard that defines this interaction also calls for a new DHCP option (assigned code number 81), which enables a client to supply its fully qualified domain name to a DHCP server. This capability makes it possible for a Windows 2000 DHCP server to function as a proxy for clients that do not themselves support dynamic DNS updates, like those running Windows 95, 98, and NT 4.0.

The interaction between DHCP and DNS requires all the DNS servers on the network to support the dynamic update feature. This makes the technology suitable for a Windows network that uses only Microsoft DNS servers, but not for the Internet, which uses many different DNS servers. Using DHCP to assign IP addresses to Web servers and other machines accessible from the Internet can result in periodic address changes that may be reflected immediately in the records of the local DNS server, but still take hours or days to propagate throughout the Internet.

Superscopes

Windows 2000's DHCP server also includes a new feature, called *superscopes*, that resolves one of the problems that occurs when you divide the IP addresses to be assigned on a particular subnet between two servers, for fault tolerance purposes. If you divide 100 IP addresses between two separate scopes, a client in the rebinding state (that is, one that has reached the T2 time interval) that broadcasts DHCPREQUEST messages in an attempt to renew its address lease may receive a DHCPNAK message from a server that does not have that address in its scope. This message would terminate the lease and force a renegotiation for a new address. This can occur even when you create identical scopes on both servers and exclude different address ranges on each one, as recommended in "Creating Scopes," earlier in this chapter.

Deactivating a scope also causes a server to issue DHCPNAK messages when a client attempts to renew an address in that scope. If you want to disable the allocation of certain addresses temporarily, it is better to exclude the address ranges from the active scope than to deactivate the scope entirely.

You can prevent the generation of these DHCPNAK messages by using Windows 2000 to run your DHCP servers and create superscopes on both that contain all the IP addresses for the subnet (including the excluded ones). Because both servers are aware of all the addresses assigned to the subnet (even those assigned by the other server), they are able to determine which addresses should receive DHCPNAK messages and which should not.

New DHCP Options

The DHCP implementation in Windows 2000 includes support for assignment of vendor-specific options, as well as TCP/IP configurations based on user classes. *Vendor-specific options* are defined in RFC 2132, and they are essentially a mechanism for supplying product-specific configuration data to certain clients without the need to put new options through the standardization process. For example, the Windows 2000

DHCP server defines a vendor class for Microsoft clients that includes three new options, as follows:

Microsoft Disable NetBIOS option   Disables NetBIOS support on the client

Microsoft Release DHCP Lease on Shutdown option   Causes the client to release its IP address lease each time it shuts down

Microsoft Default Router Metric Base   Specifies the default metric to be used for new entries to the client's routing table

These options are useful only to Microsoft DHCP clients, so rather than go through the time and effort of submitting them for inclusion in the DHCP standard, the server delivers them only to clients identifying themselves as Microsoft systems with an appropriate Vendor Class Identifier. The options themselves are delivered as suboptions within the Vendor-Specific Options option.

Another method Windows 2000 uses to supply customized configurations to clients is through user classes. A *user class* is a category of users defined by the administrator and associated with a set of options and their values. In the past, DHCP servers treated all clients equally; every system on a particular subnet received the same combination of options with its IP address. With user classes, it's possible to assign different options or different option values to systems based on any criteria. If, for example, you want to balance the traffic load generated by your clients between two DNS servers, you can create two separate classes, each specifying different values for the Domain Name Server option.

DHCPINFORM

Windows 2000 supports the new DHCPINFORM message type defined in the RFC 2131 standard for supplying configuration parameters to clients without including an IP address. Windows 2000 also uses this message for communications between DHCP servers to verify all the servers operating on the network are authorized.

Windows 2000 includes a DhcpServer object type in Active Directory that you can use to register the DHCP servers operating on the network and to authorize them for use. Whenever the DHCP service starts on a Windows 2000 machine, it broadcasts DHCPINFORM messages, looking for the other DHCP servers on the network. These servers respond with DHCPACK messages containing their DS enterprise root in the Vendor-Specific Options field. The initializing server builds a list of the other DHCP servers on the network, all of which should have the same root. The server also communicates with the directory service by requesting a list of the machines that have been authorized to function as DHCP servers. As long as the server finds its own address on the list, it provides DHCP services to the clients on the network. If its address is not on the list, the DHCP service fails to initialize. This prevents unauthorized Microsoft DHCP servers, called rogue servers, from operating on the network.

The
Complete
Reference

Networking

Chapter 24

WINS and NetBIOS Name Resolution

T he *Windows Internet Naming Service (WINS)* is a NetBIOS name server (NBNS) as defined in the RFC 1001 and 1002 standards published by the Internet Engineering Task Force (IETF). These standards define the use of NetBIOS over a TCP/IP network, called *NetBT* for short. Windows NT and Windows 2000 servers include WINS as a means of registering the NetBIOS names of the computers on a network and resolving those names into their equivalent IP addresses. Using WINS can significantly reduce the amount of broadcast traffic on your network and simplify the administration of Windows networks that span multiple segments.

Note *The inclusion of the Active Directory directory service in Windows 2000 represents a major shift in Microsoft's networking philosophy. Windows 2000 no longer relies on WINS for computer name registration and resolution; it uses the Domain Name System (DNS) instead. However, the Windows 2000 Server products include WINS to provide support for workstations running earlier versions of Windows. If you are running an Active Directory network that consists of only Windows 2000 computers, there's no need for you to run WINS, but if you have computers running older Windows versions, such as NT or 98, you should consider using WINS. For more information on how Windows 2000 and Active Directory use DNS, see Chapter 25.*

NetBIOS Names

Windows networks use computer (NetBIOS) names to uniquely identify each of the systems on the network. When you browse through the network on a Windows system, the domain, workgroup, and computer names you see in the display are actually NetBIOS names. The NetBIOS name space is flat (not hierarchical), meaning that each name can be used only once on a network. The NetBIOS name space also calls for names up to 16 characters in length, the last character of which Windows systems use for a resource identifier that identifies the function of the system, leaving 15 characters for the actual name supplied during the installation.

For example, when a system is sharing a drive or directory with another computer on the network, its NetBIOS name consists of the computer name assigned during the installation of the operating system, plus a sufficient number of spaces to pad the name out to 15 characters (if necessary), followed by the resource identifier. In this case, the resource identifier has a hexadecimal value of 20, indicating that the system is functioning as a file server. The possible values for the NetBIOS name resource identifier are shown in Table 24-1.

Resource Name	Resource Identifier (hex)	Resource Type	Function
<computername>	00	U	Workstation Service
<computername>	01	U	Messenger Service
<\\—__MSBROWSE__>	01	G	Master Browser
<computername>	03	U	Messenger Service
<computername>	06	U	RAS Server Service
<computername>	1F	U	NetDDE Service
<computername>	20	U	File Server Service
<computername>	21	U	RAS Client Service
<computername>	22	U	Microsoft Exchange Interchange (MSMail Connector)
<computername>	23	U	Microsoft Exchange Store
<computername>	24	U	Microsoft Exchange Directory
<computername>	30	U	Modem Sharing Server Service
<computername>	31	U	Modem Sharing Client Service
<computername>	43	U	SMS Clients Remote Control
<computername>	44	U	SMS Administrators Remote Control Tool
<computername>	45	U	SMS Clients Remote Chat
<computername>	46	U	SMS Clients Remote Transfer
<computername>	4C	U	DEC Pathworks TCP/IP service on Windows NT

Table 24-1: *Windows NetBIOS Resource Identifiers (16th Character Codes)*

NETWORK CONNECTION SERVICES

Resource Name	Resource Identifier (hex)	Resource Type	Function
<computername>	52	U	DEC Pathworks TCP/IP service on Windows NT
<computername>	87	U	Microsoft Exchange MTA
<computername>	6A	U	Microsoft Exchange IMC
<computername>	BE	U	Network Monitor Agent
<computername>	BF	U	Network Monitor Application
<username>	03	U	Messenger Service
<domain>	00	G	Domain Name
<domain>	1B	U	Domain Master Browser
<domain>	1C	G	Domain Controllers
<domain>	1D	U	Master Browser
<domain>	1E	G	Browser Service Elections
<INet~Services>	1C	G	IIS
<IS~computer name>	00	U	IIS
<computername>	[2B]	U	Lotus Notes Server Service
IRISMULTICAST	[2F]	G	Lotus Notes
IRISNAMESERVER	[33]	G	Lotus Notes
Forte_$ND800ZA	[20]	U	DCA IrmaLan Gateway Server Service

Table 24-1: *Windows NetBIOS Resource Identifiers (16th Character Codes) (continued)*

When Windows systems communicate using the NetBEUI protocol at the network and transport layers, they use the NetBIOS names to determine the hardware address to which they send their packets. To communicate using TCP/IP, a system requires the IP address of the intended destination. The process of converting a name into an IP address is known as *name resolution*. Windows systems on a TCP/IP network resolve NetBIOS names by using a lookup table that lists the names of the computers on the network and their equivalent IP addresses.

DNS servers also perform name resolutions, except that they resolve the DNS names of computers into IP addresses, not the NetBIOS names. For more information on DNS, see Chapter 25.

NetBIOS names must be unique on the network, so that traffic is always sent to the correct machine. To make sure that no two computers on a network have the same NetBIOS name, Windows systems use a *name registration* mechanism to check for duplicates during the network logon process. The name registration mechanism is how the system compiles the lookup table it will use for name resolutions.

The NetBT standards define the message exchanges that systems use during their name registration and resolution activities, as well as the formats of the messages themselves. Windows systems can use three different mechanisms to perform name registrations and resolutions, which are as follows:

LMHOSTS A manually created text file containing a lookup table of NetBIOS names and IP addresses

Broadcasts A process by which systems broadcast NetBT messages to the entire network segment and rely on the responses from other systems for NetBIOS name information

WINS A service running on a Windows 2000 or NT server that maintains a dynamic database of NetBIOS names and IP addresses

The following sections examine the advantages and drawbacks of these methods and explain how WINS, in most cases, is the best NetBIOS name registration and resolution method for Windows networks.

Name Registration Methods

During name registration, a system "claims" a NetBIOS name for its own use and, in some cases, prevents other systems from taking the same name. The name may also be added to a lookup table that systems will use later for name resolution.

LMHOSTS Name Registration

LMHOSTS is a simple text file that contains the NetBIOS names of the systems on the network and their equivalent IP addresses, much like the HOSTS file performs the same function for DNS names on TCP/IP systems. Because the LMHOSTS file must be manually created, the name registration process for a system occurs when a user or administrator edits the file and adds the computer's NetBIOS name and IP address.

The only mechanism in an LMHOSTS file that prevents the existence of duplicate NetBIOS names on the network is that systems parse the file from top to bottom and use only the first entry they find for a particular NetBIOS name. All subsequent entries for the same name in the file are ignored.

Broadcast Name Registration

When using the broadcast name registration method, a Windows computer transmits a series of broadcast messages to the local network that contain the NetBIOS name assigned to the computer during the installation of the operating system. These NAME REGISTRATION REQUEST messages are carried within a User Datagram Protocol (UDP) packet using a format defined in the NetBT standards. The system repeats the broadcast three times at 250 millisecond intervals.

Any computers on the local network receiving the broadcasts are required to reply if they are using the same NetBIOS name as the broadcasting system. If this is the case, the computer with the duplicate name transmits a NEGATIVE NAME REGISTRATION RESPONSE message back to the original system as a unicast transmission. When this occurs, the registration of the name fails and the system attempting the registration prompts the user for a different NetBIOS name. This prevents any duplicate NetBIOS names from existing on the same network segment.

If the system receives no replies to its NAME REGISTRATION REQUEST broadcasts, it broadcasts a NAME OVERWRITE DEMAND message that declares its possession of the NetBIOS name in question. No replies are expected to demand-type messages, as they are unequivocal declarations of an operating condition. This entire broadcast name registration process is illustrated in Figure 24-1.

Strictly speaking, no permanent lookup table is associated with the broadcast name registration method. Each Windows system on the network does maintain a NetBIOS name cache, however, in which names and their equivalent addresses are stored temporarily to prevent repetitive name resolution procedures.

Because broadcast transmissions are propagated only within the LAN on which they originate, this method of name registration does nothing to prevent systems on other network segments from using duplicate NetBIOS names. Also, on a properly configured network (that is, one without duplicate NetBIOS names), these name registration broadcasts generate a lot of network traffic for no good purpose. Overall, the broadcast name registration method is effective, but not particularly efficient.

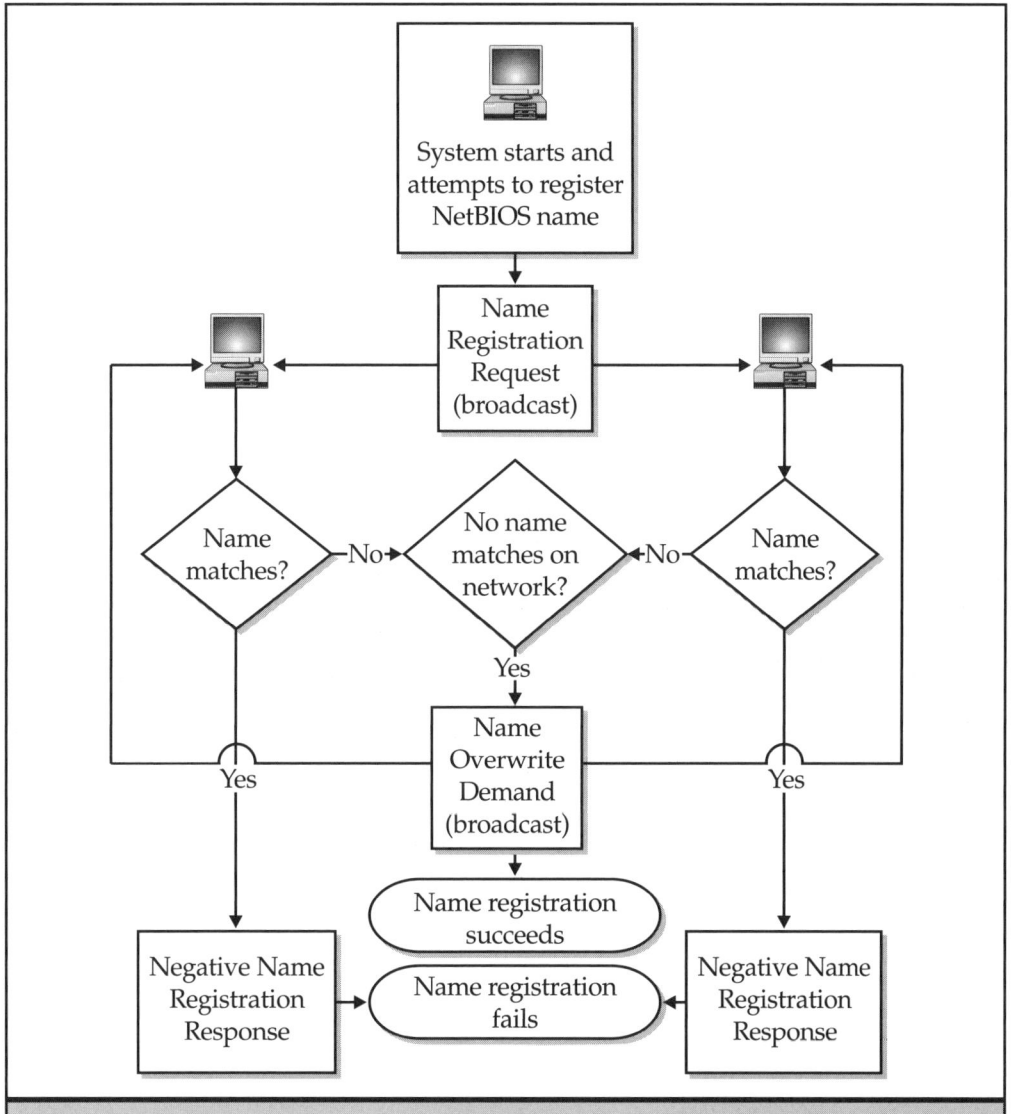

Figure 24-1:   *A system can successfully register its NetBIOS name using broadcasts, as long as no other system on the network segment is using the same name.*

WINS Name Registration

When a system configured to use WINS logs on to the network, it also generates a NAME REGISTRATION REQUEST message, but instead of broadcasting it to the entire network, the computer sends it as a unicast to the WINS server specified in the computer's TCP/IP client configuration using UDP port 137. The only differences in the message itself are the values of a few minor fields (see "NetBT Message Formats," later in this chapter).

The WINS server maintains a database of the NetBIOS names assigned to other systems on the network along with their IP addresses. When it receives a NAME REGISTRATION REQUEST unicast, it checks to see if the NetBIOS name specified in the request has already been registered by another system. If the name is not in use, the WINS server adds it to its database and returns a POSITIVE NAME REGISTRATION RESPONSE back to the sender. This message contains a *time-to-live (TTL)* value that specifies how long the name registration will remain in the database without being renewed by the client. This prevents a NetBIOS name from being perpetually assigned to a system that is no longer running.

WINS Name Challenges

If the WINS server already has the requested NetBIOS name in its database, it initiates a name challenge procedure. This is to ensure that the name in question is actually in use. If, for example, you were to move a computer physically to another location, it might be connected to another subnet, which would give it a different IP address. As far as WINS is concerned, however, the NetBIOS name is still in use by the original IP address. The name challenge determines whether the system to which the name is registered is actually using it. In this example, because the registered system no longer exists at the old IP address, it does not respond to the challenge and the name is released.

The name challenge process begins when the WINS server transmits a series of unicast NAME QUERY REQUEST messages to the IP address of the system registered as using the NetBIOS name. If the system is still using the name, it replies to the server with a POSITIVE NAME QUERY RESPONSE message. The server then transmits a NEGATIVE NAME REGISTRATION RESPONSE message to the original client, denying the client the registration and forcing the selection of a new NetBIOS name.

If the registered client does not respond to the server's NAME QUERY REQUEST message, the server retransmits up to three times at 500 millisecond intervals. If no responses occur or if the registered client returns a NEGATIVE NAME QUERY RESPONSE, indicating that it is no longer using the name, the server purges the record from the WINS server's database and assigns the name to the new client system. The WINS name registration and name challenge process is illustrated in Figure 24-2.

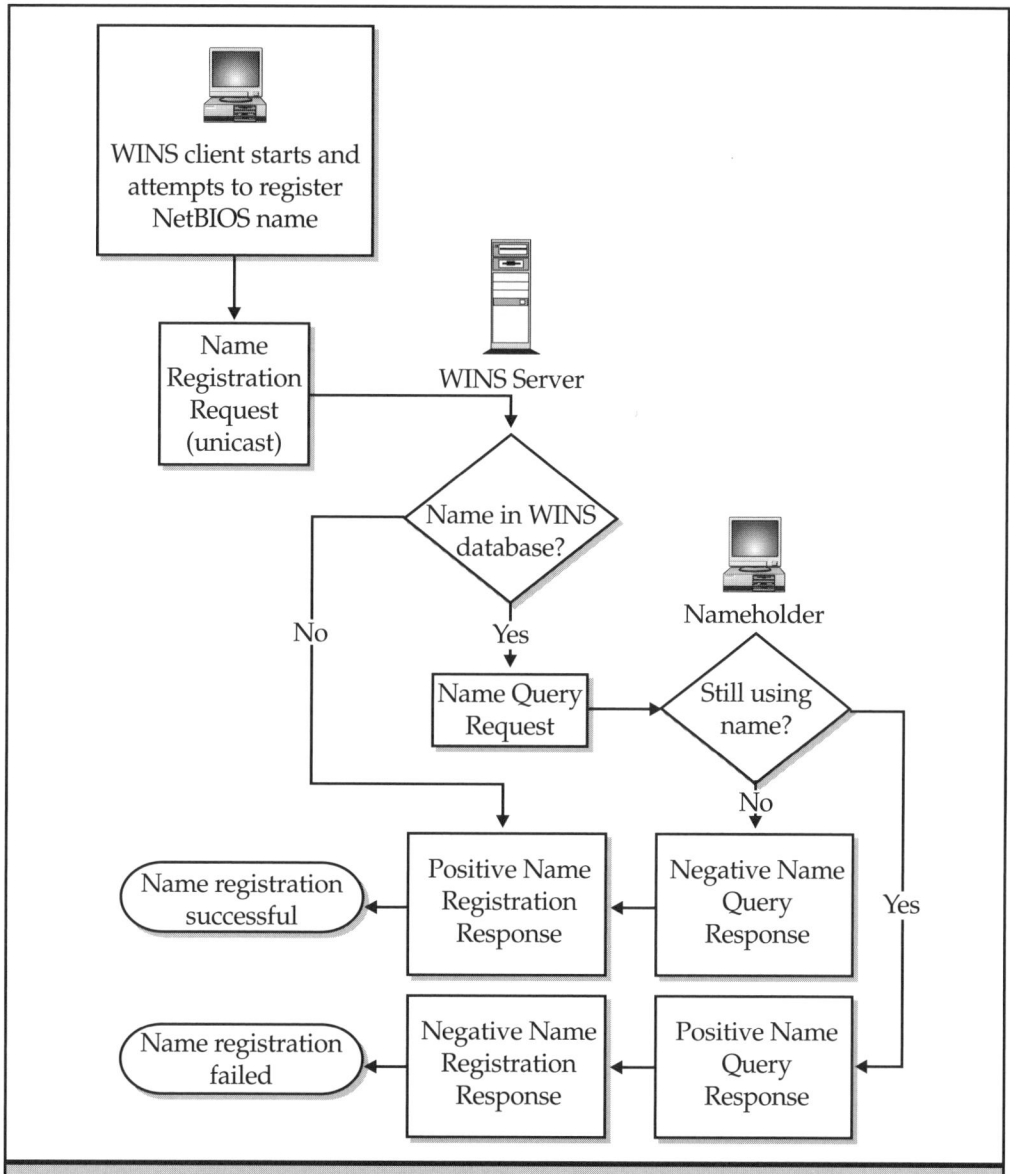

Figure 24-2: *When a WINS server receives a request to register a NetBIOS name that is already in use, the server checks with the nameholder to verify its status.*

WINS name registration is a great improvement over the broadcast method because all of its communications are in the form of unicast messages instead of broadcasts. This greatly reduces the amount of network traffic generated by the name registration process. In addition, because it uses unicast transmissions, WINS can handle name registration and resolution for an entire internetwork, not just for a single segment.

WINS Name Renewal

By default, the TTL interval assigned to each WINS name registration is six days, although network administrators can modify this value. The TTL clock for a given system resets itself to the default interval each time it logs on to the network. When a system remains logged on continuously for half the TTL interval (three days, by default), it begins attempting to renew the name registration.

The name renewal process begins when the client transmits a NAME REFRESH REQUEST message to the WINS server. The server then replies with either a POSITIVE NAME REFRESH RESPONSE message containing a new TTL interval or a NEGATIVE NAME REFRESH RESPONSE that forces the client to register a different name.

If the server fails to respond to the request, the client retransmits at two-minute intervals until half the remaining TTL interval (1.5 days, by default) remains. At this point, the client begins sending NAME REFRESH REQUEST messages to the secondary WINS server specified in the client's TCP/IP configuration. If the client receives no response, it continues transmitting requests to the secondary server until half the remaining TTL interval is left and then switches back to the primary server.

This process continues, with the client switching servers each time it reaches half the remaining time, until it either receives a response from one of the servers or the TTL interval expires. Once the TTL interval expires, the client reverts to the broadcast name registration method.

WINS Name Release

As part of its normal shutdown sequence, a client system transmits a NAME RELEASE REQUEST message to the WINS server. The server replies with either a POSITIVE NAME RELEASE RESPONSE message, in which case the client continues the shutdown process, or a NEGATIVE NAME RELEASE RESPONSE, which only occurs when the IP address in the server's record for that NetBIOS name contains an IP address different from that in the message.

Once the release sequence is complete, the WINS server can register the NetBIOS name to any other system that requests it. If a client system's TTL interval expires without the WINS server receiving any renewal messages from the client, the server purges the associated record from its database and releases the NetBIOS name.

Name Resolution Methods

Name resolution is the process by which a client system discovers the IP address of a computer on the network using a particular NetBIOS name. This process must occur before a computer can transmit TCP/IP traffic to another specific computer on the network. The methods for resolving names largely correspond to the methods for registering them.

NetBIOS Name Cache Resolution

By far, the fastest method of resolving NetBIOS names uses a lookup table stored in the client system's memory, called the *NetBIOS name cache*. Whenever the client system resolves a name using one of the other methods, the information is stored in the cache for a limited amount of time to prevent the need for repeated resolutions of the same name during a single network transaction. Because accessing the cache requires no network communications or even disk drive access, it is much faster than any other method. Windows systems always check the cache before they attempt any other name resolution method.

You can view the current contents of the NetBIOS name cache on a Windows system at any time by running the Nbtstat.exe program at the command line with the -c switch, as shown:

```
c:\>nbtstat -c

Node IpAddress: [192.168.2.5] Scope Id: []

            NetBIOS Remote Cache Name Table

    Name              Type      Host Address      Life [sec]

    ----------------------------------------------------------

    CZ3            <20>  UNIQUE     192.168.2.3          360

    CZ1            <20>  UNIQUE     192.168.2.10         360

    CZ1            <00>  UNIQUE     192.168.2.10         360

    CZ1            <03>  UNIQUE     192.168.2.10         360
```

The number in angle brackets that follows the NetBIOS name is the value assigned to the NetBIOS name's resource identifier, the 16th character that defines the function of the machine. Because a Windows system can perform multiple functions at the same time (such as client and server), you may see multiple entries in the cache for the same computer, with different resource identifiers.

Note *For more information on using the Windows Nbtstat.exe utility, see Chapter 31.*

Entries in the NetBIOS name cache have a limited life, so outdated information does not remain cached indefinitely. However, it is possible to preload the cache by including entries in the LMHOSTS file that are marked with the #PRE tag. When a Windows system boots, it reads the LMHOSTS file and loads the #PRE-tagged entries into the cache without assigning them a time limit. This makes it possible to bypass the other name resolution methods for specific systems and provide virtually instantaneous access to the resolution information.

LMHOSTS Name Resolution

The LMHOSTS file is the simplest method of name resolution because it does not require any additional network communications. To resolve a name, a system simply opens the LMHOSTS text file on the local drive, searches for the desired NetBIOS name, and reads its equivalent IP address. The process is fast because there are no network traffic delays, but it also suffers from a major problem.

To function properly, an LMHOSTS file must contain the NetBIOS names and IP addresses of all the computers on the network, and no mechanism exists to update the file automatically. A user or administrator must manually update the file on each computer whenever a change occurs in the network configuration. Obviously, once you get beyond a handful of systems on a network, this method becomes far too labor intensive.

In most cases, the LMHOSTS file is only used as a companion to the broadcast name resolution method. Because broadcasts are limited to the local network segment, administrators can use an LMHOSTS file to resolve the names of servers and other key systems on other networks. Even for this purpose, however, the maintenance required is usually more than most administrators find convenient.

Broadcast Name Resolution

In the broadcast name resolution method, a computer resolves a given NetBIOS name by generating a series of NAME QUERY REQUEST messages and broadcasting them to the local network segment. Each computer on the network examines the request, and if the message contains that computer's NetBIOS name, it replies with a POSITIVE NAME QUERY RESPONSE (see Figure 24-3). If the names do not match, the system discards the packet.

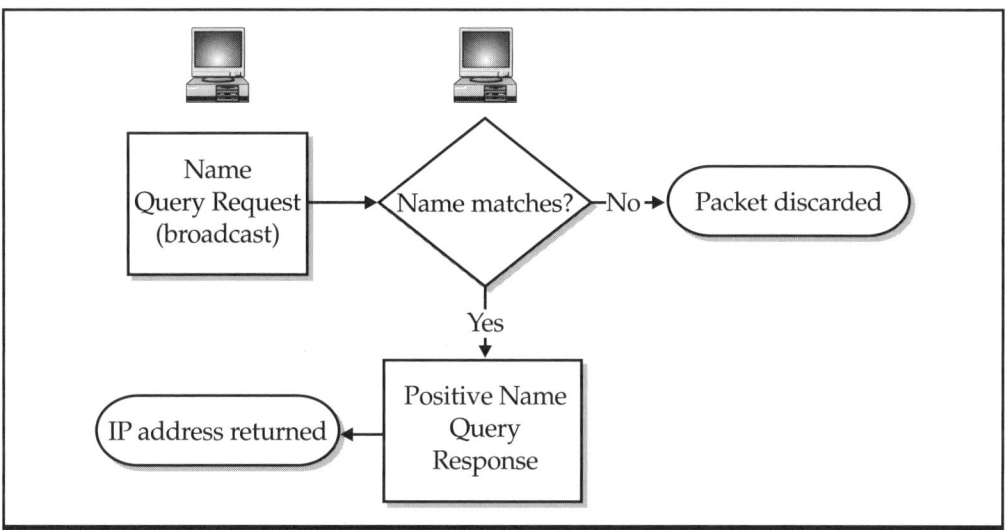

Figure 24-3:   *When a node configured to use the broadcast name resolution method receives a NAME QUERY REQUEST containing its own name, it must respond with its IP address.*

The broadcast method works quite well on a small network, but its main problem lies in the excess traffic it generates. Just as with broadcast name registration, every successful name resolution also results in a failed resolution by every other computer on the network segment. However, name registration only occurs once for each computer when it logs on to the network; name resolution occurs continually, every time one computer initiates communications with another. Also, the impact of these failures is not limited to the excess traffic volume transmitted over the network.

The data link–layer protocol of each system receiving a NAME QUERY REQUEST must pass the message up through the layers of the protocol stack. Not until the message reaches the NetBIOS interface, roughly corresponding to the session layer in the OSI reference model, does the system read the NetBIOS name in the request and either respond to it or discard it. This requires a certain number of processor cycles on the client computer, which is multiplied by the total number of broadcast messages generated by all the systems on the network. The result is that all of the computers on the network using broadcast name resolution must constantly devote a small but significant percentage of their resources to processing broadcasts that are not intended for them.

One factor that does reduce this burden, to some degree, is the NetBIOS name cache on every Windows computer. After successfully resolving a name into an IP address, a computer adds the information to the cache it keeps in memory. This prevents the machine from having to perform repeated resolutions of the same name while transmitting a large number of packets to a single destination. This cache is volatile, however, meaning the information is purged whenever the system is restarted. This prevents the computer from using outdated name resolution information when transmitting data.

The other drawback of the broadcast method is its limitation to the local network segment. Data link–layer broadcasts are not propagated to other network segments by routers. While configuring a router to do this would be possible, the amount of traffic generated by having all the broadcasts from all the networks propagated throughout the enterprise would be enormous. When a Windows network that uses the broadcast name resolution method consists of more than one LAN, administrators must make special arrangements for the computers to resolve the names of systems on other LANs, usually in the form of LMHOSTS file entries. It's far easier, in most cases, to use WINS instead.

WINS Name Resolution

As with name registration, WINS makes the name resolution process more efficient by using only unicast transmissions. A computer trying to resolve a NetBIOS name transmits a NAME QUERY REQUEST message, just as in the broadcast method, except that it transmits the message as a unicast to the WINS server specified in its TCP/IP configuration. The server consults its database and replies with either a POSITIVE NAME QUERY RESPONSE containing the equivalent IP address for the requested name or a NEGATIVE NAME QUERY RESPONSE informing the system that the name does not exist in the database.

The server may also send interim WAIT FOR ACKNOWLEDGMENT RESPONSE (WACK) messages back to the client if some delay occurs in satisfying the request, to prevent the client from timing out. If the client receives a negative response or no response at all from the server, it sends the same request to the secondary WINS server specified in its TCP/IP configuration (see Figure 24-4). If the secondary server also fails to reply with a positive response, the client switches to a different name resolution method, according to its node type (as discussed in "Node Types," later in this chapter).

WINS, therefore, reduces the amount of network traffic devoted to NetBIOS name resolution and eliminates the need for systems to process extraneous broadcast messages. Because unicast messages (unlike broadcasts) can be transmitted to any location in an internetwork, WINS can provide name resolution services for the entire enterprise. In addition, because the WINS name registration process automatically compiles its database of NetBIOS names and IP addresses, there is no need for administrators to edit a lookup table manually.

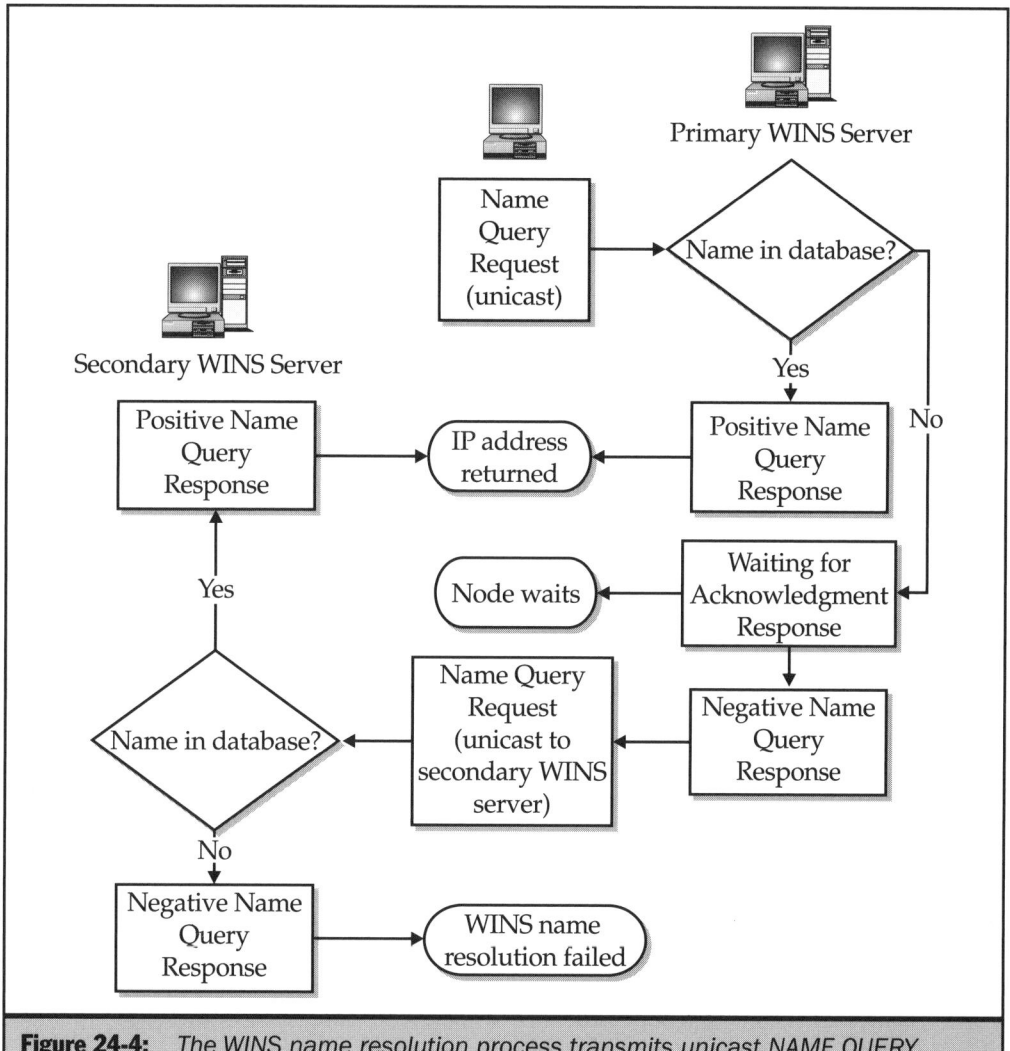

Figure 24-4: *The WINS name resolution process transmits unicast NAME QUERY REQUEST messages to both WINS servers, if necessary.*

WINS and Internetwork Browsing

In addition to name registration and resolution services, WINS also makes it possible for client systems to browse the shares on other network segments, without any manual configuration by the administrator. Browsing on a Windows network is the ability to see the domains, workgroups, computers, and shares in the Windows My Network Places or Network Neighborhood display. Browsing the shares on the network is entirely separate from actually accessing the shares. A user may, for example, be unable to see the shares in My Network Places, but still have the ability to access those shares by mapping a drive letter directly to a UNC name.

Do not confuse the concept of the Windows network browser with Web browsers like Internet Explorer and Netscape Navigator. These are applications, while the Windows network browser is a service performed by Windows machines.

On every segment of a Windows network, one system is elected to the role of master browser. The *master browser* is responsible for compiling a definitive list of all the computers and shares on the network and replicating it to other systems that function as *backup browsers*. In the event that the master browser fails or is shut down, a new election delegates a system to take its place.

When WINS servers are present on a network, the master browser gets its information about the computers on the network from WINS, instead of from the computers directly. WINS also simplifies the communications process between browsers on different network segments. Without WINS, the computers functioning as browsers must be listed in the LMHOSTS file of the browsers on other network segments for their names to be resolved during the browser replication process.

Node Types

The NetBIOS name registration and resolution methods that a computer uses and the order in which it uses them are specified by the computer's *node type*, as defined in the NetBT standards. Each node type defines a name registration method and a primary name resolution method and, in some cases, a series of fallbacks to use if the primary resolution method should fail. The three possible node types defined in the standards are as follows:

b-node (broadcast node)   Uses the broadcast method for name registration and resolution exclusively

p-node (point-to-point node)   Uses NetBIOS name servers (that is, WINS servers) for name registration and resolution exclusively

m-node (mixed mode node)   Uses the broadcast method for name registration exclusively; uses broadcasts for name resolution and NetBIOS name servers if broadcasts fail to resolve a name

Microsoft Node Types

The node types as defined in the NetBT standards are not particularly well suited to the capabilities of Windows systems. On a b-node system, a name resolution fails completely if the requested name is located on another network segment. On a p-node system, the name resolutions fail completely if the NetBIOS name servers are not functioning. An m-node system is intended for a situation in which a NetBIOS name server is used only for resolving names on other segments. This is impractical on a Windows network because WINS is designed to completely replace broadcasts. As a result of these inadequacies, Microsoft created three additional node types for use with its operating systems, which are as follows:

Modified b-node   Uses the broadcast method for name registration exclusively; uses broadcasts for name resolution and the LMHOSTS file if broadcasts fail to resolve a name. This is the default node type for a Windows computer that is not configured to use WINS.

h-node (hybrid node)   Uses NetBIOS name servers for name registration exclusively; uses NetBIOS name servers for name resolution and the broadcast method if the NetBIOS name servers fail. The system then reverts back to using the name servers as soon as they become available. This is the default node type for a WINS-enabled client.

Microsoft-Enhanced h-node   Windows NT systems include options that can supplement an h-node system with LMHOSTS name resolution, as well as Windows Sockets calls to a DNS server, and a HOSTS file, all to be used if both WINS servers and broadcasts fail to resolve a name.

The DNS option, which appears as an Enable DNS For Windows Resolution check box on the WINS Address page of the Microsoft TCP/IP Properties dialog box on Windows NT systems, enables users to specify a DNS name in a UNC path. The DNS name of a system need not be identical to its NetBIOS name, but because both are associated with the same IP address, the end result is the same.

With all of its options enabled, a Microsoft-Enhanced h-node uses all the name resolution mechanisms available to it, in the manner shown in Figure 24-5.

Windows computers configured to use WINS do not reply to the NAME REGISTRATION REQUEST broadcasts generated by b-node, modified b-node, and m-node systems. For this reason, it is recommended that you do not mix node types on your network.

NETWORK CONNECTION SERVICES

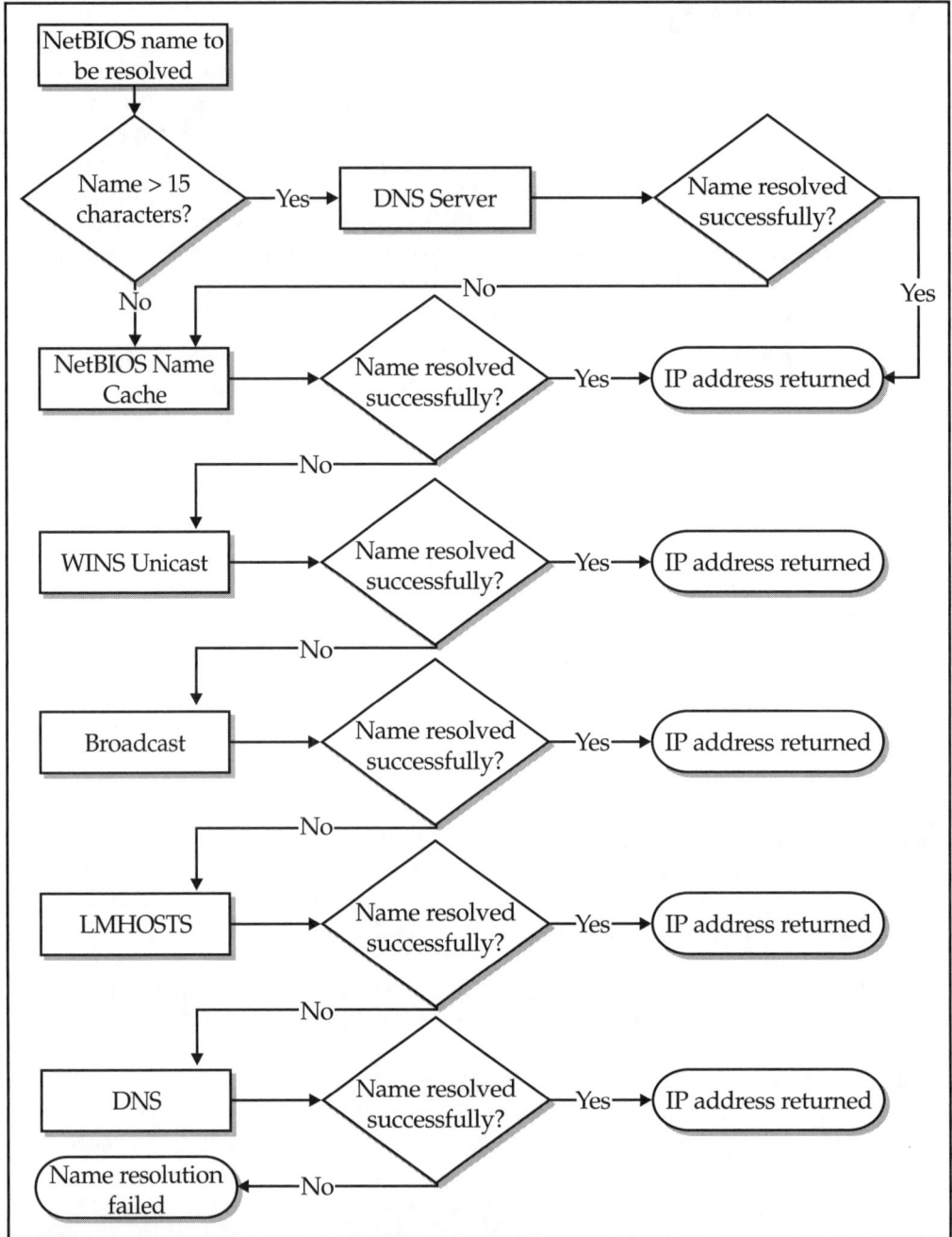

Figure 24-5: *Windows computers can use up to six different mechanisms in an attempt to resolve a NetBIOS name.*

Setting Node Types

The node type on a Windows system is determined by the status of the WINS client and whether the WINS client is activated manually or by using DHCP. Generally speaking, systems that use WINS are h-nodes, while systems that do not use WINS are b-nodes.

When you use DHCP to activate a system's WINS client by supplying a value for the WINS/NBNS Servers option (option 044), the node type is set to h-node by default. When you don't activate WINS using DHCP, the system is a b-node. You can set the node type independently of the WINS client by modifying the DHCP WINS/NBT Node Type option (option 046). When you enable the WINS client manually by supplying WINS server addresses in the system's TCP/IP Properties dialog box, the settings override the DHCP options for WINS and the node type, and make the client an h-node.

You can also manually set the node type on a workstation that is not using DHCP by modifying the registry directly. On a Windows 2000 or NT computer, you create a entry called **NodeType** in the following registry key:

```
HKEY_LOCAL_MACHINE\System\CurrentControlSet\Services\NetBT\Parameters
```

Assign the new entry one of the following REG_DWORD values:

- 0x00000001 for b-node
- 0x00000002 for p-node
- 0x00000004 for m-node
- 0x00000008 for h-node

On a Windows 95 or 98 system, create the NodeType entry (if it does not already exist) as a string value in the following registry key:

```
HKEY_LOCAL_MACHINE\System\CurrentControlSet\Services\VxD\MSTCP
```

Use the following values for the NodeType entry:

- 1 for b-node
- 2 for p-node
- 4 for m-node
- 8 for h-node

The mechanism for the incorporation of the Microsoft enhancements to these node type settings depends on which operating system the client is running. Windows 2000 and NT, for example, have a check box on the WINS Address page of the Microsoft TCP/IP Properties dialog box that enables LMHOSTS lookups. You can also import an LMHOSTS file from a network drive in this dialog box, which simplifies the process of deploying the same LMHOSTS file on systems throughout a network. Windows 95 and 98 b-node systems, on the other hand, parse the LMHOSTS file automatically during the system startup and do not have the import option (although you can specify an alternate location for the LMHOSTS file by modifying the value of the LMHostFile entry in the registry's \MSTCP key).

A problem currently exists with the TCP/IP client in Windows 95 and 98 that prevents them from using the LMHOSTS file for NetBIOS name resolution when the DNS client is enabled. However, the system does preload the LMHOSTS entries tagged with the #PRE option into the NetBIOS name cache during the system startup process.

In most cases, WINS clients on a properly configured network should rarely, if ever, have to use any name resolution mechanism other than WINS. A properly configured WINS implementation consists of multiple WINS servers that replicate their database information on a regular basis, so that even if one server fails, clients can access another.

NetBT Message Formats

The name service messages used during the Windows name registration and resolution procedures have the basic form of the message packets for the Domain Name Service (DNS). As with DNS messages, NetBT messages are typically carried within UDP datagrams, which limits their length to 576 bytes. The packet consists of header fields that contain codes defining the function of the message, plus Question and/or Resource Record sections to carry the query and response data (that is, NetBIOS names and IP addresses), as shown in Figure 24-6. The message format provides the capability to carry multiple resource records in response packets, in separate sections called the Answer, Authority, and Additional sections, but this is not necessary in the NetBIOS name server messages used in Windows networking.

The functions of the name service message fields are explained in the following sections.

The Header Section

Every name service message has a header section that defines the basic type of message carried in the packet and whether it is a request or a response. The header fields are as follows:

NAME_TRN_ID (16 bits)   Contains a transaction ID used to match requests with responses.

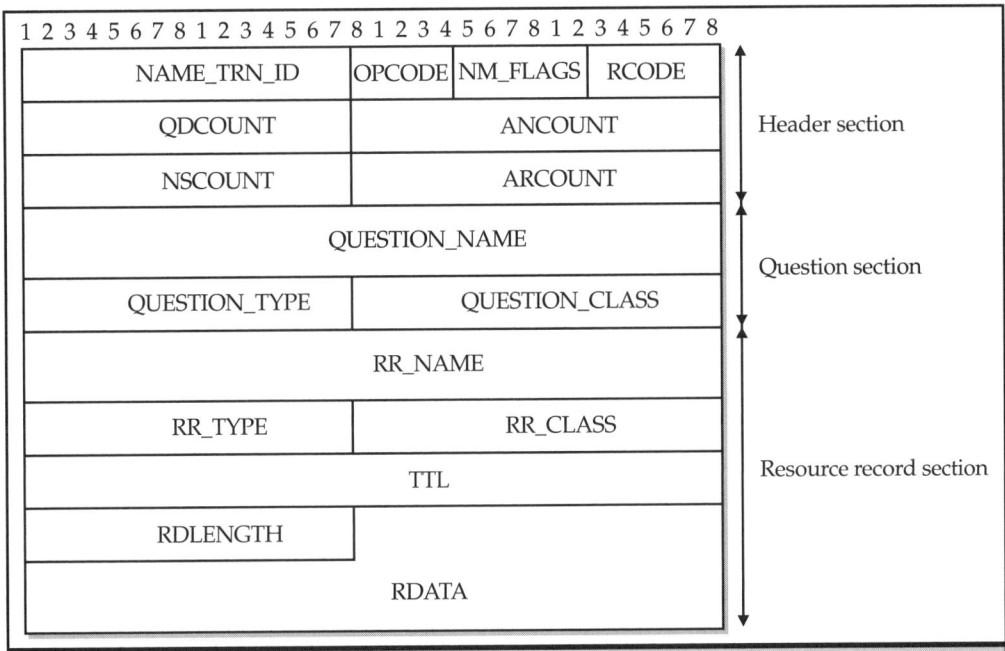

Figure 24-6: *Windows systems use the same message format for all of their name registration and resolution procedures.*

OPCODE (5 bits)   The first bit is the R flag, which specifies whether the message is a request or a response, using the following values:

0   Request

1   Response

The remaining 4 bits specify the type of message contained in the packet, using the following values:

0   Query

5   Registration

6   Release

7   WACK

8   Refresh

NM_FLAGS (7 bits) Contains five 1-bit flags (and 2 null bits) with the following functions:

AA – Authoritative Answer Specifies whether the response is being furnished by an authoritative source. In request messages (where the R flag in the OPCODE field is 0), the value of this flag is always 0. WINS servers always set the value of this flag to 1.

TC – Truncation Specifies whether the message had to be truncated to fit into a UDP datagram; serves as a signal to retransmit using TCP. Because packet length is not a problem in NetBIOS Name Service messaging, the value of this flag is always 0.

RD – Recursion Desired Used in NBNS request messages to obtain information about the name server's recursive capabilities. Windows clients always set this flag to a value of 1, indicating that recursion is desired.

RA – Recursion Available Used in NBNS response messages to indicate the name server supports recursive queries, registrations, and releases. For most types of messages (WACK messages are the exception), WINS servers set this value to 1, indicating that recursion is available.

0 Null bit

0 Null bit

B – Broadcast Specifies whether the packet was broadcast or unicast, using the following values:

  0 Unicast

  1 Broadcast or multicast

RCODE (4 bits) Used in response packets to specify the results of a particular request, using the following codes:

0 No Error

1 – FMT_ERR Format Error (request was formatted incorrectly)

2 – SRV_ERR Server Failure (request could not be processed due to NBNS malfunction)

3 – NAM_ERR Name Error (requested name does not exist in the name server)

4 – IMP_ERR Unsupported Request Error (used only when an NBNS is challenged by an update type registration request)

5 – RFS_ERR   Refused Error (policy prevents the server from registering the requested name to this host)

6 – ACT_ERR   Active Error (requested name is already owned by another node)

7 – CFT_ERR   Name in Conflict Error (a unique NetBIOS name is already owned by another node)

QDCOUNT (16 bits)   Specifies the number of entries in the Question section of the message. Windows NetBIOS Name Service messages always contain only one question entry.

ANCOUNT (16 bits)   Specifies the number of resource records in the Answer section of the message. Windows NetBIOS Name Service messages always contain only one answer section entry.

NSCOUNT (16 bits)   Specifies the number of resource records in the Authority section of the message. The Authority section is not used in Windows NetBIOS Name Service messages.

NRCOUNT (16 bits)   Specifies the number of resource records in the additional section of the message. The additional section is not used in Windows NetBIOS Name Service messages.

Note

The name service messages used in name registration and resolution procedures are nearly identical, whether the client system uses the broadcast or the WINS method. The primary difference between the two is the state of the Broadcast bit in the NM_FLAGS field.

From the values of these header fields, you can determine exactly what kind of message is included in the packet. For example, a message with an R flag value of 0, an OPCODE value of 0, a B flag value of 0, and a QDCOUNT value of 1 would be a NAME QUERY REQUEST packet sent to a WINS server with a NetBIOS name in the Question section. If the WINS server replies with a message that changes the R flag to 1, the QDCOUNT field to 0, and the ANCOUNT field to 1, this would be a POSITIVE NAME QUERY RESPONSE with the IP address for the requested NetBIOS name in the Resource Record section. If the server did not have the name in its database, the resulting NEGATIVE NAME QUERY RESPONSE message would have QDCOUNT and ANCOUNT fields with values of 0 and an RCODE field with a value of 3.

The Question Section

The Question section contains the NetBIOS name to be registered or resolved, and appears only in request and demand packets, such as the following:

- NAME REGISTRATION REQUEST
- NAME OVERWRITE DEMAND
- NAME QUERY REQUEST
- NAME REFRESH REQUEST
- NAME RELEASE REQUEST

The Question section fields are as follows:

QUESTION_NAME (variable)   Contains the NetBIOS name to be registered or resolved

QUESTION_TYPE (16 bits)   Specifies the type of request, using the following values:

0x0020   NB (NetBIOS Name Service resource record)

0x0021   NBSTAT (NetBIOS Node Status resource record)

QUESTION_CLASS (16 bits)   Specifies the class of the request, for which only one possible value exists, as follows:

0x0001   Internet Class

The Resource Record Section

A Resource Record section appears in positive response packets, like the following, as well as in WACK packets:

- POSITIVE NAME REGISTRATION RESPONSE
- POSITIVE NAME QUERY RESPONSE
- POSITIVE NAME REFRESH RESPONSE
- POSITIVE NAME RELEASE RESPONSE
- WAIT FOR ACKNOWLEDGMENT RESPONSE

Although resource records can appear in any of three sections—Answer, Authority, and Additional—Windows NetBIOS Name Service messages only carry a single resource record, typically in the Answer or Additional section. The Resource Record section fields are as follows:

RR_NAME (variable)   Contains the NetBIOS name from the request message to which this is the response.

RR_TYPE (16 bits)   Specifies the type of resource record, using the following values:

0x0001   A (IP Address resource record)

0x0002   NS (name server resource record)

0x000A   NULL (null resource record)

0x0020   NB (NetBIOS Name Service resource record)

0x0021   NBSTAT (NetBIOS Node Status resource record)

RR_CLASS (16 bits)   Specifies the class of the resource record, for which only one possible value exists, as follows:

0x0001   Internet Class

TTL (32 bits)   Specifies the time to live for the information in the resource record.

RDLENGTH (16 bits)   Specifies the number of bytes in the RDATA field.

RDATA (variable)   Specifies the IP address of the system identified by the value of the RR_NAME field. When the RR_TYPE field contains the NB code (as WINS server responses do), the RDATA field begins with 32 NB_FLAGS bits, broken down as follows:

G (1 bit) – Group Name Flag   Specifies whether the name supplied in the RR_NAME field is a unique or group NetBIOS name, using the following values:

0   Unique

1   Group

ONT (2 bits) – Owner Node Type   Specifies the node type of the system identified in the RR_NAME field, using the following values:

00   b-node

01   p-node

10   m-node

11   Reserved for future use

Reserved for Future Use (13 bits)

Sample Transactions

The following figures show the contents of packets captured with a protocol analyzer during some of the basic NetBIOS Name Service transactions that occur regularly on a Windows network. Figure 24-7 shows the NAME REGISTRATION REQUEST being sent to a WINS server.

```
NETBIOS Name Service
  HEADER SECTION:
  Transaction Identifier: 4
  Flags:
    0... .... = Request packet
    .010 1... = OP Code is 0x05 – Registration
    .... .0.. = Non-Authoritative Answer
    .... ..0. = No Truncation Packet
    .... ...1 = Recursion Desired
    0... .... = Recursion Not Available
    .00. .... = Reserved Bits
    ...0 .... = unicast packet
    .... 0000 = Response Code is 0 – No Error
  Section Entries:
    Question   Section: 1 Entrie(s)
    Answer     Section: 0 Entrie(s)
    Authority  Section: 0 Entrie(s)
    Additional Section: 1 Entrie(s)
  QUESTION SECTION[1]:
    NetBios Name: CZ2                     ▌
    Question  Type: 0x0020 = NB – NetBios General Name Service Resource Record
    Question Class: 0x0001 = IN – Internet
  ADDITIONAL SECTION[1]:
    NetBios Name (w/Pointer): CZ2         ▌
    RR  Type: 0x0020 = NB – NetBios General Name Service Resource Record
    RR Class: 0x0001 = IN – Internet class
    RR Time To Live: 300000 second(s)
    RR Data  Length: 6 Octet(s)
    0... .... = Group Name Flag – Unique NetBios Name
    .00. .... = Owner Node Type – B node
    ...0 0000 = Reserved Bits
    0000 0000 = Reserved Bits
    A NetBios address: 192.168.2.20
  Calculate CRC: 0x4e00a566
```

Figure 24-7: *A NAME REGISTRATION REQUEST message sent to a WINS server*

Notice that the desired name (CZ2) is supplied in the Question section, but that the Additional section also contains the name and the IP address the client is requesting to be added to the WINS name service. Figure 24-8 shows a POSITIVE NAME

REGISTRATION RESPONSE message from the server containing the same information in the Answer section, indicating that it has been successfully added to the WINS database.

```
⊞ 🖳 ETHER-II: 00-20-AF-37-B8-12 ==> 00-60-97-B0-77-CA
⊞ 📄 IP: 192.168.2.10->192.168.2.20,ID=11950
⊞ 📄 UDP: NETBIOS Name Service->NETBIOS Name Service,Len=70
⊟ 📄 NETBIOS Name Service
   📄 HEADER SECTION:
   📄 Transaction Identifier: 4
   ⊟ 📄 Flags:
      📄 1... .... = Response packet
      📄 .010 1... = OP Code is 0x05 - Registration
      📄 .... .1.. = Authoritative Answer
      📄 .... ..0. = No Truncation Packet
      📄 .... ...1 = Recursion Desired
      📄 1... .... = Recursion Available
      📄 .00. .... = Reserved Bits
      📄 ...0 .... = unicast packet
      📄 .... 0000 = Response Code is 0 - No Error
   ⊟ 📄 Section Entries:
      📄 Question   Section: 0 Entrie(s)
      📄 Answer     Section: 1 Entrie(s)
      📄 Authority  Section: 0 Entrie(s)
      📄 Additional Section: 0 Entrie(s)
   ⊟ 📄 ANSWER SECTION[1]:
      📄 NetBios Name: CZ2                    ▌
      📄 RR  Type: 0x0020 = NB - NetBios General Name Service Resource Record
      📄 RR Class: 0x0001 = IN - Internet class
      📄 RR Time To Live: 518400 second(s)
      📄 RR Data  Length: 6 Octet(s)
      📄 0... .... = Group Name Flag - Unique NetBios Name
      📄 .00. .... = Owner Node Type - B node
      📄 ...0 0000 = Reserved Bits
      📄 0000 0000 = Reserved Bits
      📄 A NetBios address: 192.168.2.20
   🖳 Calculate CRC: 0x757da172
```

Figure 24-8:   *The POSITIVE NAME REGISTRATION RESPONSE message returned to a client by the WINS server*

The next two packet captures show the responses from the WINS server during a failed name registration attempt. While trying to process the request, the server sends a WAIT FOR ACKNOWLEDGMENT RESPONSE, or WACK, packet to the client, as shown in Figure 24-9. The Answer section of this message contains a five-second time-to-live value, so the client does not time out and revert to the broadcast name registration method.

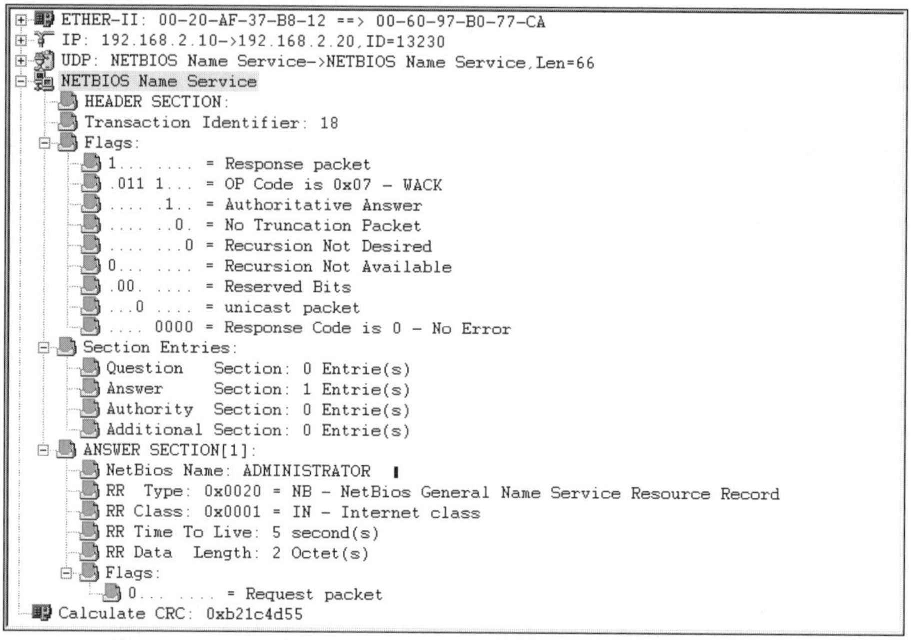

```
⊞ ▦ ETHER-II: 00-20-AF-37-B8-12 ==> 00-60-97-B0-77-CA
⊞ 〒 IP: 192.168.2.10->192.168.2.20,ID=13230
⊞ 🖳 UDP: NETBIOS Name Service->NETBIOS Name Service,Len=66
⊟ 🖳 NETBIOS Name Service
    ─🗋 HEADER SECTION:
    ─🗋 Transaction Identifier: 18
  ⊟ 🗋 Flags:
      ─🗋 1... .... = Response packet
      ─🗋 .011 1... = OP Code is 0x07 - WACK
      ─🗋 .... .1.. = Authoritative Answer
      ─🗋 .... ..0. = No Truncation Packet
      ─🗋 .... ...0 = Recursion Not Desired
      ─🗋 0... .... = Recursion Not Available
      ─🗋 .00. .... = Reserved Bits
      ─🗋 ...0 .... = unicast packet
      ─🗋 .... 0000 = Response Code is 0 - No Error
  ⊟ 🗋 Section Entries:
      ─🗋 Question    Section: 0 Entrie(s)
      ─🗋 Answer      Section: 1 Entrie(s)
      ─🗋 Authority   Section: 0 Entrie(s)
      ─🗋 Additional Section: 0 Entrie(s)
  ⊟ 🗋 ANSWER SECTION[1]:
      ─🗋 NetBios Name: ADMINISTRATOR ▌
      ─🗋 RR  Type: 0x0020 = NB - NetBios General Name Service Resource Record
      ─🗋 RR Class: 0x0001 = IN - Internet class
      ─🗋 RR Time To Live: 5 second(s)
      ─🗋 RR Data  Length: 2 Octet(s)
    ⊟ 🗋 Flags:
        ─🗋 0... .... = Request packet
  ─▦ Calculate CRC: 0xb21c4d55
```

Figure 24-9: *The WAIT FOR ACKNOWLEDGMENT message generated by a WINS server experiencing a delay*

When the server finally generates a reply, as shown in Figure 24-10, the message includes the Answer section that a POSITIVE NAME REGISTRATION RESPONSE would have, but, in fact, this is a NEGATIVE NAME REGISTRATION RESPONSE, as demonstrated by the value of the response code (RCODE) field. This value indicates that an Active Error has occurred because the requested name is already in use by another system.

Once it has been registered, a WINS client system can begin using the WINS server to resolve the NetBIOS names of other systems on the network. A NAME QUERY REQUEST sent to a WINS server, as shown in Figure 24-11, contains a Question section that specifies the name the client seeks to resolve.

The POSITIVE NAME QUERY RESPONSE message (see Figure 24-12) contains an Answer section supplying the requested IP address.

As you can see in Figure 24-13, a broadcast NAME QUERY REQUEST is almost exactly the same as the unicast version sent to a WINS server. The B flag indicates that the message is a broadcast, but the messages are otherwise identical.

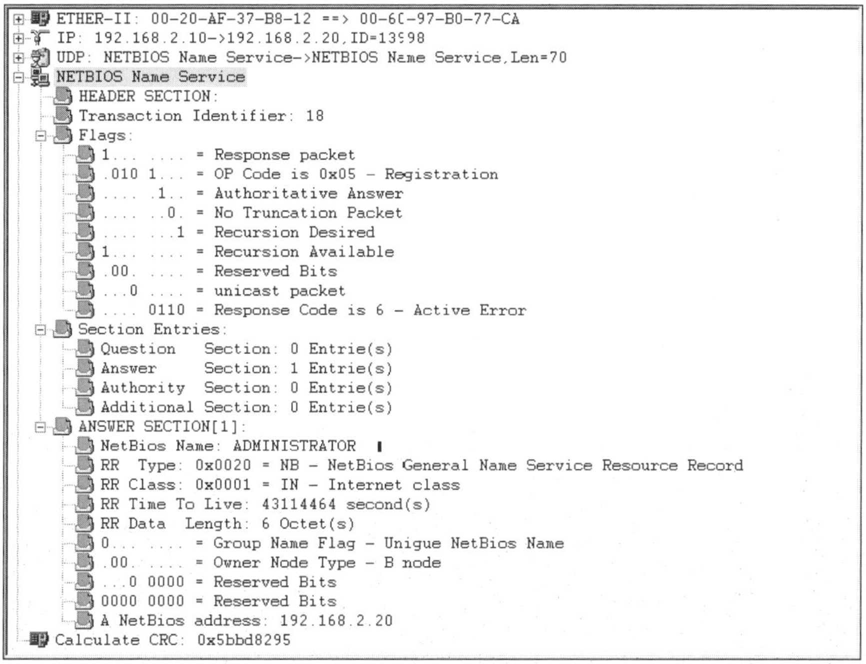

```
ETHER-II: 00-20-AF-37-B8-12 ==> 00-6C-97-B0-77-CA
IP: 192.168.2.10->192.168.2.20,ID=13598
UDP: NETBIOS Name Service->NETBIOS Name Service,Len=70
NETBIOS Name Service
   HEADER SECTION:
   Transaction Identifier: 18
   Flags:
       1... .... = Response packet
       .010 1... = OP Code is 0x05 - Registration
       .... .1.. = Authoritative Answer
       .... ..0. = No Truncation Packet
       .... ...1 = Recursion Desired
       1... .... = Recursion Available
       .00. .... = Reserved Bits
       ...0 .... = unicast packet
       .... 0110 = Response Code is 6 - Active Error
   Section Entries:
       Question   Section: 0 Entrie(s)
       Answer     Section: 1 Entrie(s)
       Authority  Section: 0 Entrie(s)
       Additional Section: 0 Entrie(s)
   ANSWER SECTION[1]:
       NetBios Name: ADMINISTRATOR  ▌
       RR  Type: 0x0020 = NB - NetBios General Name Service Resource Record
       RR Class: 0x0001 = IN - Internet class
       RR Time To Live: 43114464 second(s)
       RR Data  Length: 6 Octet(s)
       0... .... = Group Name Flag - Unique NetBios Name
       .00. .... = Owner Node Type - B node
       ...0 0000 = Reserved Bits
       0000 0000 = Reserved Bits
       A NetBios address: 192.168.2.20
Calculate CRC: 0x5bbd8295
```

Figure 24-10:   *The NEGATIVE NAME REGISTRATION RESPONSE message generated by a WINS server when a requested NetBIOS name is already in use*

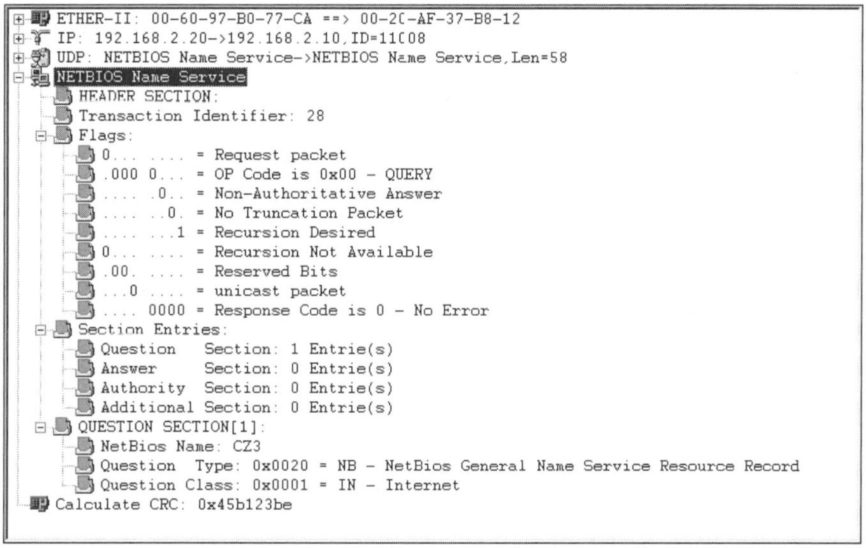

```
ETHER-II: 00-60-97-B0-77-CA ==> 00-2C-AF-37-B8-12
IP: 192.168.2.20->192.168.2.10,ID=11C08
UDP: NETBIOS Name Service->NETBIOS Name Service,Len=58
NETBIOS Name Service
   HEADER SECTION:
   Transaction Identifier: 28
   Flags:
       0... .... = Request packet
       .000 0... = OP Code is 0x00 - QUERY
       .... .0.. = Non-Authoritative Answer
       .... ..0. = No Truncation Packet
       .... ...1 = Recursion Desired
       0... .... = Recursion Not Available
       .00. .... = Reserved Bits
       ...0 .... = unicast packet
       .... 0000 = Response Code is 0 - No Error
   Section Entries:
       Question   Section: 1 Entrie(s)
       Answer     Section: 0 Entrie(s)
       Authority  Section: 0 Entrie(s)
       Additional Section: 0 Entrie(s)
   QUESTION SECTION[1]:
       NetBios Name: CZ3
       Question  Type: 0x0020 = NB - NetBios General Name Service Resource Record
       Question Class: 0x0001 = IN - Internet
Calculate CRC: 0x45b123be
```

Figure 24-11:   *A NAME QUERY REQUEST message sent to a WINS server*

NETWORK CONNECTION SERVICES

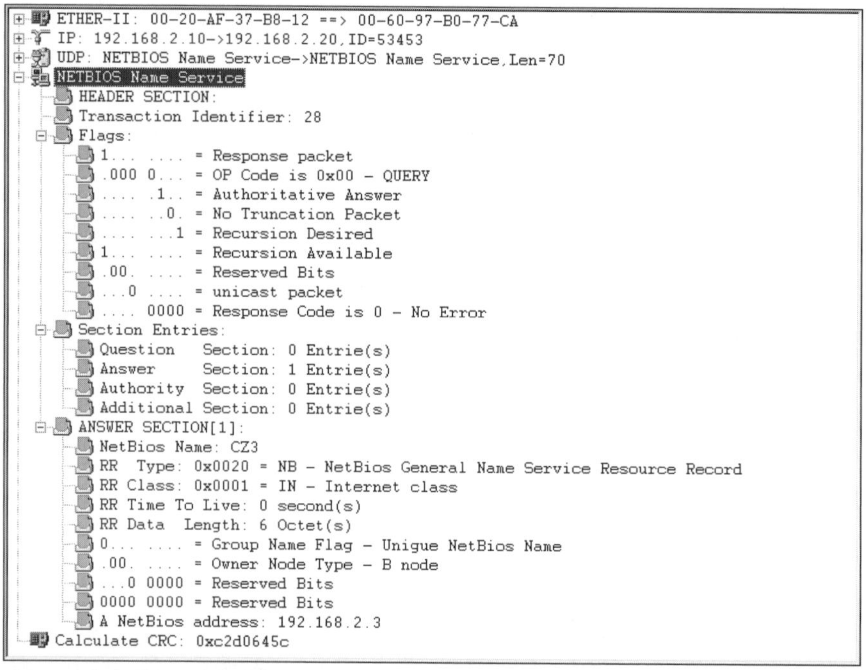

```
⊞·📠 ETHER-II: 00-20-AF-37-B8-12 ==> 00-60-97-B0-77-CA
⊞·📶 IP: 192.168.2.10->192.168.2.20,ID=53453
⊞·📱 UDP: NETBIOS Name Service->NETBIOS Name Service,Len=70
⊟·📠 NETBIOS Name Service
   ·📄 HEADER SECTION:
   ·📄 Transaction Identifier: 28
   ⊟·📄 Flags:
      ·📄 1... .... = Response packet
      ·📄 .000 0... = OP Code is 0x00 - QUERY
      ·📄 .... .1.. = Authoritative Answer
      ·📄 .... ..0. = No Truncation Packet
      ·📄 .... ...1 = Recursion Desired
      ·📄 1... .... = Recursion Available
      ·📄 .00. .... = Reserved Bits
      ·📄 ...0 .... = unicast packet
      ·📄 .... 0000 = Response Code is 0 - No Error
   ⊟·📄 Section Entries:
      ·📄 Question   Section: 0 Entrie(s)
      ·📄 Answer     Section: 1 Entrie(s)
      ·📄 Authority  Section: 0 Entrie(s)
      ·📄 Additional Section: 0 Entrie(s)
   ⊟·📄 ANSWER SECTION[1]:
      ·📄 NetBios Name: CZ3
      ·📄 RR  Type: 0x0020 = NB - NetBios General Name Service Resource Record
      ·📄 RR Class: 0x0001 = IN - Internet class
      ·📄 RR Time To Live: 0 second(s)
      ·📄 RR Data  Length: 6 Octet(s)
      ·📄 0... .... = Group Name Flag - Unique NetBios Name
      ·📄 .00. .... = Owner Node Type - B node
      ·📄 ...0 0000 = Reserved Bits
      ·📄 0000 0000 = Reserved Bits
      ·📄 A NetBios address: 192.168.2.3
   ·📠 Calculate CRC: 0xc2d0645c
```

Figure 24-12: *The POSITIVE NAME QUERY RESPONSE returned by the WINS server when it has successfully resolved a NetBIOS name*

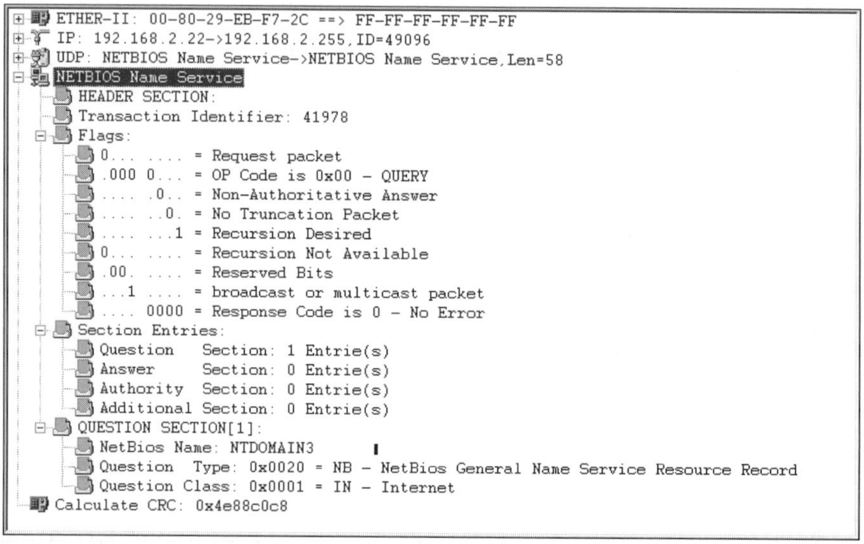

```
⊞·📠 ETHER-II: 00-80-29-EB-F7-2C ==> FF-FF-FF-FF-FF-FF
⊞·📶 IP: 192.168.2.22->192.168.2.255,ID=49096
⊞·📱 UDP: NETBIOS Name Service->NETBIOS Name Service,Len=58
⊟·📠 NETBIOS Name Service
   ·📄 HEADER SECTION:
   ·📄 Transaction Identifier: 41978
   ⊟·📄 Flags:
      ·📄 0... .... = Request packet
      ·📄 .000 0... = OP Code is 0x00 - QUERY
      ·📄 .... .0.. = Non-Authoritative Answer
      ·📄 .... ..0. = No Truncation Packet
      ·📄 .... ...1 = Recursion Desired
      ·📄 0... .... = Recursion Not Available
      ·📄 .00. .... = Reserved Bits
      ·📄 ...1 .... = broadcast or multicast packet
      ·📄 .... 0000 = Response Code is 0 - No Error
   ⊟·📄 Section Entries:
      ·📄 Question   Section: 1 Entrie(s)
      ·📄 Answer     Section: 0 Entrie(s)
      ·📄 Authority  Section: 0 Entrie(s)
      ·📄 Additional Section: 0 Entrie(s)
   ⊟·📄 QUESTION SECTION[1]:
      ·📄 NetBios Name: NTDOMAIN3     ▌
      ·📄 Question  Type: 0x0020 = NB - NetBios General Name Service Resource Record
      ·📄 Question Class: 0x0001 = IN - Internet
   ·📠 Calculate CRC: 0x4e88c0c8
```

Figure 24-13: *The header of a broadcast NAME QUERY REQUEST message differs from the unicast version in the value of only 1 bit.*

Using LMHOSTS

To use an LMHOSTS file for NetBIOS name resolution, you must first create the file by adding the names you want to resolve and their IP addresses. The basic format of the file is simple; each IP address should appear in a separate line along with its NetBIOS name, separated by at least one space, as shown in the following listing:

```
192.168.2.2        CZ2
192.168.2.3        CZ3
192.168.2.10       CZ1
```

LMHOSTS also supports the use of tags that provide special functions for specific entries. In most cases, you use a tag by adding it to an LMHOSTS entry after the NetBIOS name, separated by at least one space. The tags recognized by Windows systems are as follows:

#PRE   Adding the #PRE tag to an LMHOSTS entry causes the computer to preload the entry into the NetBIOS name cache during system startup. Name resolutions performed using cached information are faster than any other method. Because Windows systems parse the LMHOSTS file from top to bottom, you should place #PRE-tagged entries at the bottom to provide the best possible speed for standard name resolutions.

#DOM:*domainname*   Use the #DOM tag to identify Windows NT domain controllers located on other network segments, where *domainname* is the name of the domain. This tag adds the entry to the system's domain name cache, making it possible for a non-WINS system to send unicasts directly to domain controllers on other segments. Use #DOM in combination with the #PRE tag to preload the domain controller entries into the cache, as follows:

```
    192.168.2.10       CZ1        #PRE#DOM:NTDOMAIN
```

#MH   Use the #MH tag for multihomed systems (that is, computers with more than one IP address). You can have up to 25 separate #MH entries containing different IP addresses for a single NetBIOS name.

#SG:*groupname*   Use the #SG tag to create a NetBIOS group with the name specified by the *groupname* variable. Each group can have up to 25 members, each of which must have its own entry in the LMHOSTS file.

\0x##   Use the \0x## tag to specify the hexadecimal value for a nonprinting character as part of a NetBIOS name. To use the tag, you enclose the entire NetBIOS name in double quotation marks and insert \0x## in place of the desired character, where ## is the hexadecimal value for that character. Note that when you enclose the

NetBIOS name in quotation marks, you must account for all 16 characters of the name. For example, to specify a nonprinting character as the 16th character of the name, you must pad out the preceding 15 characters with spaces or other characters.

#INCLUDE *filename*   Use the #INCLUDE tag to specify the location of an alternate LMHOSTS file on a shared network drive. This enables network administrators to maintain a single LMHOSTS file and to have multiple client systems access it. Place the #INCLUDE tag on its own line of the LMHOSTS file and replace *filename* with the full UNC path to the desired alternate file, as follows. If the UNC name uses a NetBIOS name to identify the system, be sure to preload it into the name cache by creating a separate entry with the #PRE tag.

```
#INCLUDE \\CZ1\C\WINNT\LMHOSTS
```

#BEGIN_ALTERNATE/#END_ALTERNATE   Use the #BEGIN_ALTERNATE tag on a separate line to mark the beginning of a subroutine that contains multiple #INCLUDE commands, followed by the #END_ALTERNATE tag, also on a separate line, as follows. When parsing the LMHOSTS file, the system will attempt to access each of the alternative LMHOSTS files specified in the #INCLUDE commands until it successfully locates one. This provides fault tolerance if the system hosting the LMHOSTS file is unavailable.

```
#BEGIN_ALTERNATE

#INCLUDE \\CZ1\C\WINNT\LMHOSTS

#INCLUDE \\CZ2\D\WINNT\LMHOSTS

#INCLUDE \\CZ3\C\WINDOWS\LMHOSTS

#END_ALTERNATE
```

Once you've created the LMHOSTS file, put it in the \Winnt\System32\Drivers\etc folder on a Windows 2000 or NT computer, or in the \Windows folder on a Windows 95, 98, or Me computer.

Using Broadcasts

The broadcast method of NetBIOS name resolution is the default for Windows systems that are not configured to use WINS. On a relatively small, single-segment network, broadcasts are an effective solution that should not have an adverse effect on network performance (unless your bandwidth is already being stretched to the limit). For a network with more than one segment, you can use broadcasts in combination with LMHOSTS files (to resolve names on other segments), but this rapidly becomes an administrative problem best resolved by the addition of one or more WINS servers.

Using WINS

Deploying a single WINS server is a simple matter. Once you install the service itself on a Windows 2000 or NT server, the default settings are sufficient for most network environments. It is recommended that a network have at least two WINS servers to provide fault tolerance in case one fails.

On a single-segment network, having two servers is not strictly necessary because the client systems will fall back to broadcast name resolution if they can't contact the WINS server. When multiple segments are involved, however, WINS quickly becomes an essential element of the network configuration, unless the administrators continue to maintain the LMHOSTS files that will be needed if the WINS server fails.

WINS Replication

To provide true fault tolerance, WINS servers on the same network must replicate their data. This makes it possible for a client that has registered its NetBIOS name with one WINS server to appear in the databases of all WINS servers on the network. Replication between WINS servers is not automatic. The network administrator must devise a proper replication strategy that keeps all the servers updated at reasonable intervals without overloading the network with traffic.

WINS server replication traffic is unidirectional. You can configure a server to push its data to another server or to pull data from another server. To replicate in both directions, you must configure the traffic in each direction separately. This capability makes it possible to deploy a series of WINS servers and create a *replication ring*.

On a WINS replication ring, each server pushes its data to its nearest downstream neighbor and pulls data from its nearest upstream neighbor, as shown in Figure 24-14. The relationship between two WINS servers is known as a *partnership*, with one machine being referenced as the *pull partner* or *push partner* of the other. The ring of partnerships, however, is purely a logical construction that has no correspondence to the physical location of the servers.

WINS Architecture

On a network in which all the segments are connected by LAN links, the number of WINS servers required is governed solely by the number of clients to support. Microsoft states that a single WINS server can support as many as 1,500 name registrations per minute and 4,500 queries per minute. Assuming the existence of traditional peak use times (such as a specific time of day in which many users arrive at the office and turn on their computers, triggering a large number of name registration requests), this works out to approximately 10,000 users who can be supported by a single WINS server.

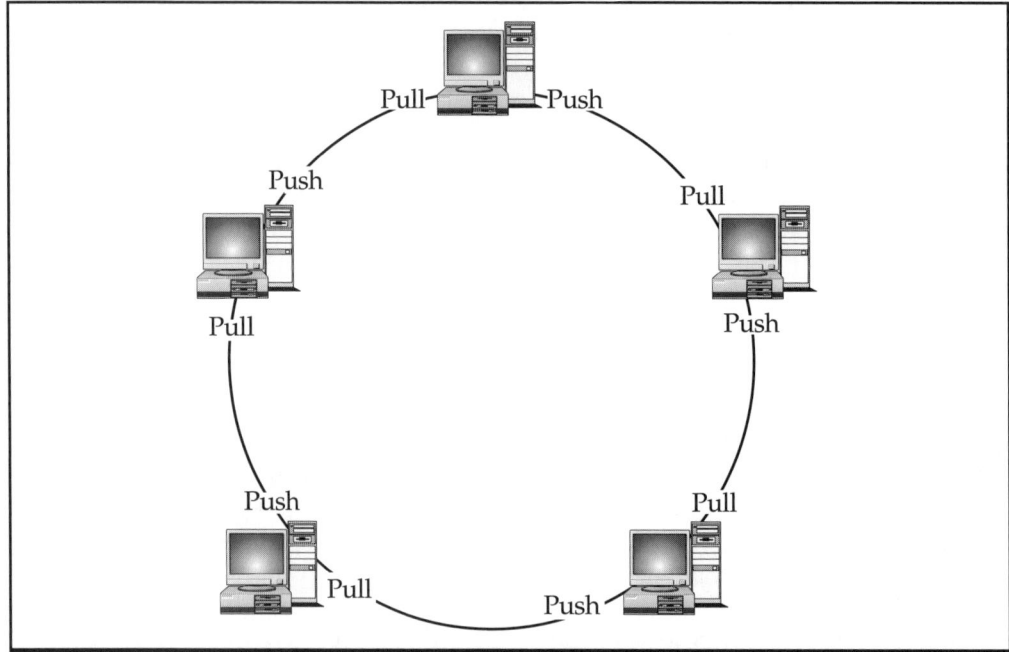

Figure 24-14: *A WINS replication ring uses push and pull partnerships to see to it that every server has identical information in its database.*

Thus, two WINS servers (and two are required only for fault-tolerance reasons) can support all but the largest networks in a single location. Because the communications among WINS servers and among clients and servers are all unicasts, the servers can be located on different segments anywhere on the internetwork.

When an internetwork spans several locations connected by WAN links, the design of the WINS architecture becomes considerably more complex. Because WAN links are usually slower and more expensive than LAN connections (and sometimes much slower and more expensive), having at least two WINS servers at every site makes it possible for every client system on the network to register and resolve NetBIOS names without sending traffic over a WAN link.

In an enterprise that consists of a large number of small offices, however, this may not be practical. In this case, it's important to consider how much network traffic actually travels between the sites. In an environment where the majority of the traffic remains within the boundaries of each site, using remote WINS servers and allowing small, individual LANs to fall back to broadcast name registration and resolution if the WAN link fails may be perfectly feasible.

Replication traffic, however, must go over the WAN links and you should configure your servers to replicate themselves in such a way as to provide backup connections if one or more WAN links should go down. A ring arrangement, where each server replicates its data to one other server down the line, doesn't work well in this case because one nonfunctioning WAN link between two sites breaks the chain and prevents the database information from being identical on all servers.

One way to maintain the database replication in a failure is to create a double-ring configuration, as shown in Figure 24-15. With data traveling in both directions, all the servers will continue to be updated, even if a WAN link fails.

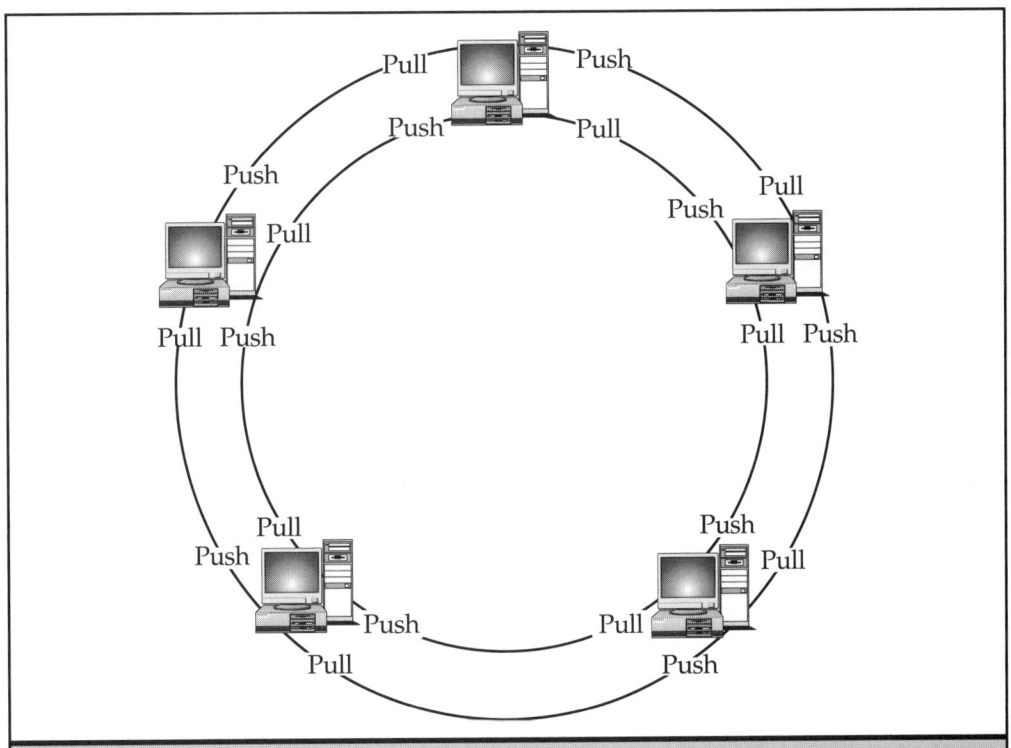

Figure 24-15: *With a double ring that has traffic flowing in both directions, no single connection failure can prevent the servers from being replicated.*

When you have more than two WINS servers at a particular site, there is no need to make all the servers part of the ring (or double ring) connecting the WAN sites. For large networks, creating a replication ring within each site and then joining the sites with a separate ring is best. Every site should have more than one machine that replicates with the other sites, however. You don't want to have the WINS servers at an entire site cut off from the rest of the enterprise due to the failure of one computer.

WINS Proxies

Mixing b-node and h-node systems on the same network segment is not a good idea, but sometimes a situation arises when a separate segment of b-nodes is connected to a WINS network. Rather than use LMHOSTS files on the b-node systems to resolve NetBIOS names on other segments, it is possible to configure a Windows machine on the b-node segment to function as a WINS proxy.

A *WINS proxy* is a WINS client system configured to listen and respond to broadcast NAME QUERY REQUEST messages on the local network. When a b-node broadcasts a request, the WINS proxy system tries to satisfy the request using the information in its name cache. If this is not possible, the proxy system transmits the request as a unicast to the WINS server (presumably on another segment) specified in its TCP/IP configuration. The WINS server replies to the proxy, which then relays the POSITIVE NAME QUERY RESPONSE message to the original b-node.

WINS proxies, in essence, enable a large number of b-node systems to take advantage of WINS without configuring each individual machine to be a WINS client. Naturally, the name resolution process takes longer when an intervening proxy is involved, but as a temporary measure, WINS proxies can be quite useful.

To configure a Windows 2000 or NT system to function as a WINS proxy, you must configure it as a normal WINS client by specifying the IP address of one or more WINS servers in its TCP/IP configuration dialog box. Then you modify the registry by creating a REG_DWORD entry called **EnableProxy** with a value of 1 in the following key:

```
HKEY_LOCAL_MACHINE\SYSTEM\CurrentControlSet\Services\NetBT\Parameters
```

The
Complete
Reference

Networking

Chapter 25

The Domain
Name System

C omputers are designed to work with numbers, while humans are more comfortable working with words. This fundamental dichotomy is the reason why the Domain Name System (DNS) came to be. Back in the dark days of the 1970s, when the Internet was the ARPANET and the entire experimental network consisted of only a few hundred systems, a need was recognized for a mechanism that would permit users to refer to the network's computers by name, rather than by address. The introduction of the TCP/IP protocols in the early 1980s led to the use of 32-bit IP addresses, which even in dotted decimal form were difficult to remember.

Host Tables

The first mechanism for assigning human-friendly names to addresses was called a *host table*, which took the form of a file called /etc/hosts on Unix systems. The host table was a simple ASCII file that contained a list of network system addresses and their equivalent host names. When users wanted to access resources on other network systems, they would specify a host name in the application, and the system would resolve the name into the appropriate address by looking it up in the host table. This host table still exists on all TCP/IP systems today, usually in the form of a file called Hosts somewhere on the local disk drive. If nothing else, the host table contains the following entry, which assigns to the standard IP loopback address the host name localhost:

```
127.0.0.1       localhost
```

Today, the Domain Name System has replaced the host table almost universally, but when TCP/IP systems attempt to resolve a host name into an IP address, it is still possible to configure them to check the Hosts file first before using DNS. If you have a small network of TCP/IP systems that is not connected to the Internet, you can use host tables on your machines to maintain friendly host names for your computers. The name resolution process will be very fast, because no network communications are necessary, and you will not need a DNS server.

Host Table Problems

The use of host tables on TCP/IP systems caused several problems, all of which were exacerbated as the fledgling Internet grew from a small "family" of networked computers into a much larger entity. The most fundamental problem was that each computer had to have its own host table, which listed the names and addresses of all of the other computers on the network. When a new computer was connected to the network, you could not access it until an entry for it was added to your computer's host table.

In order for everyone to keep their host tables updated, it was necessary for administrators to be informed when a system was added to the network or a name or address change occurred. Having every administrator of an ARPANET system e-mail every other administrator each time they made a change was obviously not a practical solution, so it was necessary to designate a registrar that would maintain a master list of the systems on the network, their addresses, and their host names.

The task of maintaining this registry was given to the Network Information Center (NIC) at the Stanford Research Institute (SRI), in Menlo Park, California. The master list was stored in a file called Hosts.txt on a computer with the host name SRI-NIC. Administrators of ARPANET systems would e-mail their modifications to the NIC, who would update the Hosts.txt file periodically. To keep their systems updated, the administrators would use FTP to download the latest Hosts.txt file from SRI-NIC and compile it into a new Hosts file for their systems.

Initially, this was an adequate solution, but as the network continued to grow, it became increasingly unworkable. As more systems were added to the network, the Hosts.txt file grew larger and more people were accessing SRI-NIC to download it on a regular basis. The amount of network traffic generated by this simple maintenance task became excessive, and changes started occurring so fast that it was difficult for administrators to keep their systems updated.

Another serious problem was that there was no control over the host names used to represent the systems on the network. Once TCP/IP came into general use, the NIC was responsible for assigning network addresses, but administrators chose their own host names for the computers on their networks. The accidental use of duplicate host names resulted in misrouted traffic and disruption of communications. Imagine the chaos that would result today if anyone on the Internet was allowed to set up a Web server and use the name www.microsoft.com for it. Clearly, a better solution was needed, and this led to the development of the Domain Name System.

DNS Objectives

To address the problems resulting from the use of host tables for name registration and resolution, the people responsible for the ARPANET decided to design a completely new mechanism. Their primary objectives at first seemed to be contradictory: to design a mechanism that would enable administrators to assign host names to their own systems without creating duplicate names, and to make that host name information globally available to other administrators without relying on a single access point that could become a traffic bottleneck and a single point of failure. In addition, the mechanism had to be able to support information about systems that use various protocols with different types of addresses, and it had to be adaptable for use by multiple applications.

The solution was the Domain Name System, designed by Peter Mockapetris and published in 1983 as two IETF (Internet Engineering Task Force) documents called RFC 882, "Domain Names: Concepts and Facilities," and RFC 883, "Domain Names: Implementation Specification." These documents were updated in 1987, published as RFC 1034 and RFC 1035, respectively, and ratified as an IETF standard. Since that time, numerous other RFCs have updated the information in the standard to address current networking issues. Some of these additional documents are proposed standards, while others are experimental. These documents include the following:

RFC 1101   "DNS Encoding of Network Names and Other Types"

RFC 1183   "New DNS RR Definitions"

RFC 1348   "DNS NSAP RRs"

RFC 1794   "DNS Support for Load Balancing"

RFC 1876   "A Means for Expressing Location Information in the Domain Name System"

RFC 1982   "Serial Number Arithmetic"

RFC 1995   "Incremental Zone Transfer in DNS"

RFC 1996   "A Mechanism for Prompt Notification of Zone Changes (DNS NOTIFY)"

RFC 2052   "A DNS RR for Specifying the Location of Services (DNS SRV)"

RFC 2181   "Clarifications to the DNS Specification"

RFC 2136   "Dynamic Updates in the Domain Name System (DNS UPDATE)"

RFC 2137   "Secure Domain Name System Dynamic Update"

RFC 2308   "Negative Caching of DNS Queries (DNS NCACHE)"

RFC 2535   "Domain Name System Security Extensions"

The DNS, as designed by Mockapetris, consists of three basic elements:

- A hierarchical name space that divides the host system database into discrete elements called *domains*

- Domain name servers that contain information about the host and subdomains within a given domain

- Resolvers that generate requests for information from domain name servers

These elements are discussed in the following sections.

Domain Naming

The Domain Name System achieves the designated objectives by using a hierarchical system, both in the name space used to name the hosts and in the database that contains the host name information. Before the DNS was developed, administrators assigned simple host names to the computers on their networks. The names sometimes reflected the computer's function or its location, as with SRI-NIC, but there was no policy in place that required this. At that time, there were few enough computers on the network to make this a practical solution.

To support the network as it grew larger, Mockapetris developed a hierarchical name space that made it possible for individual network administrators to name their systems, while identifying the organization that owns the systems and preventing the duplication of names on the Internet. The DNS name space is based on domains, which exist in a hierarchical structure much like the directory tree in a file system. A *domain* is the equivalent of a directory, in that it can contain either subdomains (subdirectories) or hosts (files), forming a structure called the DNS tree (see Figure 25-1). By delegating the responsibility for specific domains to network administrators all over the Internet, the result is a *distributed database* scattered on systems all over the network.

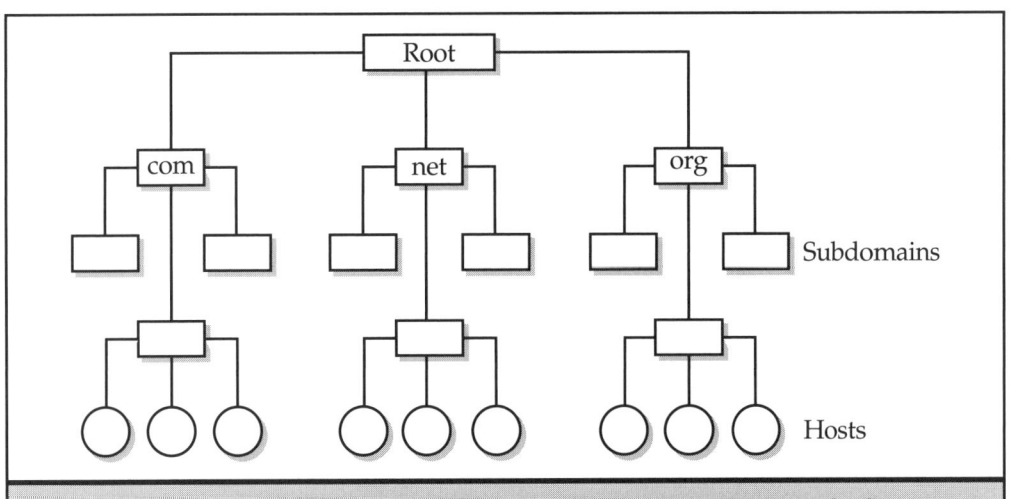

Figure 25-1: *The Domain Name System uses a tree structure like that of a file system.*

Note   *The term "domain" has more than one meaning in the networking industry. Windows NT, for example, uses the term to refer to an administrative grouping of computers on a private network, identified by a NetBIOS name. This type of domain is completely independent of, and has nothing to do with, a DNS domain. A system on a Windows network can be a member of either a Windows NT domain or a DNS domain, or both, but the domains perform different functions and can have different names. Sun Microsystems' Network Information Service (NIS) also uses the term "domain" to refer to a group of hosts, but this, too, is not synonymous with a DNS domain.*

To assign unique IP addresses to computers all over the Internet, a two-tiered system was devised in which administrators receive the network identifiers that form the first part of the IP addresses, and then assign host identifiers to individual computers themselves, to form the second part of the addresses. This distributes the address assignment tasks among thousands of network administrators all over the world. The DNS name space functions in the same way: administrators are assigned domain names and are then responsible for specifying host names to systems within that domain.

The result is that every computer on the Internet is uniquely identifiable by a *DNS name* that consists of a host name plus the names of all of its parent domains, stretching up to the root of the DNS tree, separated by periods. Each of the names between the periods can be up to 63 characters long, with a total length of 255 characters for a complete DNS name, including the host and all of its parent domains. Domain and host names are not case sensitive, and can take any value except the null value (no characters), which represents the root of the DNS tree. Domain and host names also cannot contain any of the following symbols:

```
_ : , / \ ? . @ # ! $ % ^ & * ( ) { } [ ] | ; " < > ~ `
```

In Figure 25-2, a computer in the mycorp domain functions as a Web server, and the administrator has therefore given it the host name www. This administrator is responsible for the mycorp domain, and can therefore assign systems in that domain any host name he wants to. Because mycorp is a subdomain of com, the full DNS name for that Web server is www.mycorp.com. Thus, a DNS name is something like a postal address, in which the top-level domain is the equivalent of the state, the second-level domain is the city, and the host name is the street address.

Because a complete DNS name traces the domain path all the way up the tree structure to the root, it should theoretically end with a period, indicating the division between the top-level domain and the root. However, this trailing period is nearly always omitted in common use, except in cases in which it serves to distinguish an absolute domain name from a relative domain name. An *absolute domain name* (also called a *fully qualified domain name,* or *FQDN*) does specify the path all the way to the root, while a *relative domain name* specifies only the subdomain relative to a specific domain context. For example, when working on a complex network called zacker.com that uses several levels of subdomains, you might refer to a system using a relative domain name of mail.paris without a period, because it's understood by your colleagues that you're actually referring to a system with an absolute name of mail.paris.zacker.com. (with a period).

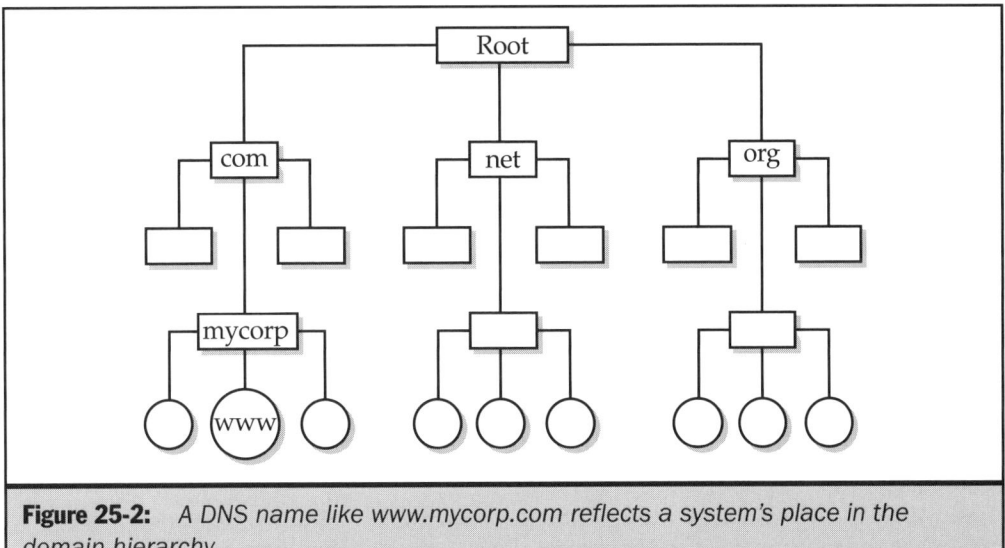

Figure 25-2:   *A DNS name like www.mycorp.com reflects a system's place in the domain hierarchy.*

It's also important to understand that DNS names have no inherent connection to IP addresses or any other type of address. Theoretically, the host systems in a particular domain can be located on different networks, thousands of miles apart.

Top-Level Domains

In every DNS name, the first word on the right represents the domain at the highest level in the DNS tree, called a *top-level domain*. These top-level domains essentially function as registrars for the domains at the second level. For example, the administrator of zacker.com went to the com top-level domain and registered the name zacker. In return for a fee, that administrator now has exclusive use of the name zacker.com and can create any host or subdomain names in that domain that he wishes. It doesn't matter that thousands of other network administrators have named their Web servers www, because they all have their own individual domain names. The host name www may be duplicated anywhere, as long as the DNS name www.zacker.com is unique.

The original DNS name space called for seven top-level domains, dedicated to specific purposes, as follows:

com   Commercial organizations

edu   Four-year, degree-granting educational institutions in North America

gov   United States government institutions

int   Organizations established by international treaty

mil   United States military applications

net   Networking organizations

org   Noncommercial organizations

The edu, gov, int, and mil domains are reserved for use by certified organizations, but the com, org, and net domains are called *global domains,* because organizations anywhere in the world can register second-level domains within them. Originally, these top-level domains were managed by a company called Network Solutions, Inc. (NSI, formerly known as InterNIC, the Internet Network Information Center) as a result of cooperative agreement with the United States government. You can still go to its Web site at http://www.networksolutions.com/ and register names in these top-level domains.

In 1998, the agreement with the U.S. government was changed to permit other organizations to compete with NSI in providing domain registrations. An organization called the Internet Corporation for Assigned Names and Numbers (ICANN) is responsible for the accreditation of domain name registrars. Under this new policy, the procedures and fees for registering names in the com, net, and org domains may vary, but there will be no difference in the functionality of the domain names nor will duplicate names be permitted. The complete list of registrars that have been accredited by ICANN is available at http://www.icann.org/registrars/accredited-list.html.

com Domain Conflicts

The com top-level domain is the one most closely associated with commercial Internet interests, and names of certain types in the com domain are becoming scarce. For example, it is difficult at this time to come up with a snappy name for an Internet technology company that includes the word "net" that has not already been registered in the com domain.

There have also been conflicts between organizations that feel they have a right to a particular domain name. Trademark law permits two companies to have the same name, as long as they are not directly competitive in the marketplace. However, A1 Auto Parts Company and A1 Software may both feel that they have a right to the a1.com domain, and lawsuits have arisen in some cases. In other instances, forward-thinking private individuals who registered domains using their own names several years ago have later been confronted by corporations with the same name who want to jump on the Internet bandwagon and feel that they have a right to that name. If a certain individual of Scottish extraction registers his domain, only to find out some years later that a fast-food company (for example) is very anxious to acquire that domain name, the end result can be either a profitable settlement for the individual or a nasty court case.

This phenomenon has also given rise to a particular breed of Internet bottom-feeder known as *domain name speculators*. These are people who register large numbers of domain names that they think some company might want someday, hoping that they can receive a large fee in return for selling them the domain name. Another unscrupulous practice is for a company in a particular business to register domains using the names of their competitors. Thus, when Internet users go to www.pizzaman.com, expecting to find Ray the Pizza Man's Web site, they instead find themselves redirected to the site for Bob's Pizza Palace, which is located across the street from Ray's. Both of these practices have resulted in civil suits, but neither one is illegal.

Generic Top-Level Domains

ICANN is also responsible for the ratification of new top-level domains to address this perceived depletion of available names. There have been numerous proposals for the establishment of various new top-level domains; those selected for negotiation of agreements in November 2000 are aero, biz, coop, info, museum, name, and pro. Of these, only biz and info are scheduled to become operational in the near future.

Country-Code Domains

There are 239 *country-code domains* (also called *international domains*), named for specific countries using the ISO designations, such as *fr* for France and *de* for Deutschland (Germany). Many of these 239 countries allow free registration of second-level domains to anyone, without restrictions. For the other countries, an organization must conform to some sort of local presence, tax, or trademark guidelines in order to register a second-level domain. Each of these country-code domains is managed by an organization in that country, which establishes its own domain name registration policies.

For a list of the country codes, as maintained by the ISO (the International Organization for Standardization), see http://www.din.de/gremien/nas/nabd/ iso3166ma/codlstp1/en_listp1.html.

Some of the countries that permit free registration of second-level domains have been more aggressive than others in pursuing registrations of company domains, which has resulted in the fairly common appearance of top-level domains from obscure island countries, such as *nu* (Niue), *to* (Tonga), and *cc* (Cocos-Keeling Islands).

There is also a *us* top-level domain that is a viable alternative for organizations unable to obtain a satisfactory name in the com domain. The us domain is administered by the Information Sciences Institute of the University of Southern California, which registers second-level domains to businesses and individuals, as well as to government agencies, educational institutions, and other organizations. The only restriction is that all us domains must conform to a naming hierarchy that uses two-letter state abbreviations at the third level and uses local city or county names at the fourth level. Thus, an example of a valid domain name would be something like zacker.chicago.il.us.

Second-Level Domains

The registrars of the top-level domains are responsible for registering second-level domain names, in return for a subscription fee. As long as an organization continues to pay the fees for its domain name, it has exclusive rights to that name. The domain registrar maintains records that identify the owner of each second-level domain and specify three contacts within the registrant's organization—an administrative contact, a billing contact, and a technical contact. In addition, the registrar must have the IP addresses of two DNS servers that function as the source for further information about the domain. This is the only information maintained by the top-level domain. The administrators of the registrant's network can create as many hosts and subdomains within the second-level domain as they want without informing the registrars at all.

To host a second-level domain, an organization must have two DNS servers. A DNS server is a software program that runs on a computer. DNS server products are available for all of the major network operating systems. The DNS servers do not have to be located on the registrant's network; many companies outsource their Internet server hosting chores and use their service provider's DNS servers. The DNS servers identified in the top-level domain's record are the *authority* for the second-level domain. This means that these servers are the ultimate source for information about that domain. When network administrators want to add a host to the network or create a new subdomain, they do so in their own DNS servers. In addition, whenever a user application somewhere on the Internet has to discover the IP address associated with a particular host name, the request eventually ends up at one of the domain's authoritative servers.

Thus, in its simplest form, the Domain Name System works by referring requests for the address of a particular host name to a top-level domain server, which in turn passes the request to the authoritative server for the second-level domain, which responds with the requested information. This is why the DNS is described as a *distributed database*. The information about the hosts in specific domains is stored on their authoritative servers, which can be located anywhere. There is no single list of all the host names on the entire Internet, which is actually a good thing, because at the time that the DNS was developed, no one would have predicted that the Internet would grow as large as it has.

This distributed nature of the DNS database eliminates the traffic-congestion problem caused by the use of a host table maintained on a single computer. The top-level domain server handles millions of requests a day, but they are requests only for the DNS servers associated with second-level domains. If the top-level domains had to maintain records for every host in every second-level domain they have registered, the resulting traffic would bring the entire system to its knees.

Distributing the database in this way also splits the chores of administering the database among thousands of network administrators around the world. Domain name registrants are each responsible for their own area of the name space and can maintain it as they wish with complete autonomy.

Subdomains

Many of the domains on the Internet stop at two levels, meaning that the second-level domain contains only host systems. However, it is possible for the administrators of a second-level domain to create subdomains that form additional levels. The us top-level domain, for example, requires a minimum of three levels: the country code, the state code, and the local city or county code. There is no limit on the number of levels you can create within a domain, except for those imposed by practicality and the 255-character maximum DNS name length.

In some cases, large organizations use subdomains to subdivide their networks according to geographical or organizational boundaries. A large corporation might create a third-level domain for each city or country in which it has an office, such as paris.zacker.com and newyork.zacker.com, or for each of several departments, such as sales.zacker.com and mis.zacker.com. The organizational paradigm for each domain is left completely up to its administrators.

The use of subdomains can make it easier to identify hosts on a large network, but many organizations also use them to delegate domain maintenance chores. The DNS servers for a top-level domain contain the addresses for each second-level domain's authoritative servers. In the same way, a second-level domain's servers can refer to authoritative servers for third-level administrators at each site to maintain their own DNS servers.

To make this delegation possible, DNS servers can break up a domain's name space into administrative units called *zones*. A domain with only two levels consists of only a single zone, which is synonymous with the domain. A three-level domain, however, can be divided into multiple zones. A zone can be any contiguous branch of a DNS tree, and can include domains on multiple levels. For example, in the diagram shown in Figure 25-3, the paris.zacker.com domain, including all of its subdomains and hosts, is one zone, represented by its own DNS servers. The rest of the zacker.com domain, including newyork.zacker.com, chicago.zacker.com, and zacker.com itself, is another zone. Thus, a *zone* can be defined as any part of a domain, including its subdomains, that is not designated as part of another zone.

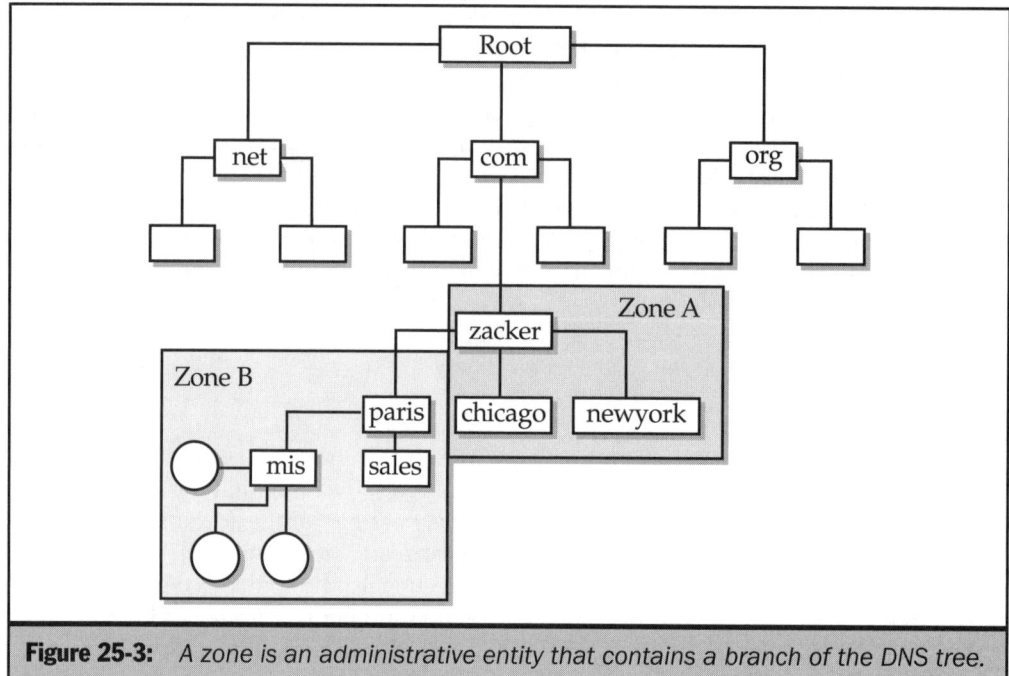

Figure 25-3: *A zone is an administrative entity that contains a branch of the DNS tree.*

Each zone must be represented by DNS servers that are the authority for that zone. A single DNS server can be authoritative for multiple zones, so you could conceivably create a separate zone for each of the third-level domains in zacker.com and still only have two sets of DNS servers.

DNS Functions

DNS servers are a ubiquitous part of most TCP/IP networks, even if you aren't aware of it. If you connect to the Internet, you use a DNS server each time you enter a server name or URL into a Web browser or other application to resolve the name of the system you specified into an IP address. When a standalone computer connects to an Internet service provider (ISP), the ISP's server usually supplies the addresses of the DNS servers that the system will use. On a TCP/IP network, administrators or users configure clients with the addresses of the DNS servers they will use. This can be a manual process performed for each workstation, or an automatic process performed using a service such as DHCP (Dynamic Host Configuration Protocol).

TCP/IP communications are based solely on IP addresses. Before one system can communicate with another, it must know its IP address. Often, the user supplies a friendly name (such as a DNS name) for a desired server to a client application. The application must then resolve that server name into an IP address before it can transmit a message to it. If the name resolution mechanism fails to function, no communication with the server is possible.

Virtually all TCP/IP networks use some form of friendly name for host systems and include a mechanism for resolving those names into the IP addresses needed to initiate communications between systems. If the network is connected to the Internet, DNS name resolution is a necessity. Private networks do not necessarily need it, however. Microsoft Windows NT networks, for example, use NetBIOS names to identify their systems and have their own mechanisms for resolving those names into IP addresses. These mechanisms include WINS, the Windows Internet Naming System, and also the transmission of broadcast messages to every system on the network. NetBIOS names and name resolution mechanisms do not replace the DNS; they are intended for use on relatively small, private networks and would not be practical on the Internet. A computer can have both a NetBIOS name and a DNS host name, and use both types of name resolution.

For more information on NetBIOS naming and Windows network name resolution, see Chapter 24.

Resource Records

DNS servers are basically database servers that store information about the hosts and subdomain for which they are responsible in *resource records (RRs)*. When you run your own DNS server, you create a resource record for each host name that you want to be accessible by the rest of the network. There are several different types of resource records used by DNS servers, the most important of which are as follows:

SOA (Start of Authority)   Indicates that the server is the best authoritative source for data concerning the zone. Each zone must have an SOA record, and only one SOA record can be in a zone.

NS (Name Server)   Identifies a DNS server functioning as an authority for the zone. Each DNS server in the zone (whether primary master or slave) must be represented by an NS record.

A (Address)   Provides a name-to-address mapping that supplies an IP address for a specific DNS name. This record type performs the primary function of the DNS, converting names to addresses.

PTR Provides an address-to-name mapping that supplies a DNS name for a specific address in the in-addr.arpa domain. This is the functional opposite of an A record, used for reverse lookups only.

CNAME (Canonical Name) Creates an alias that points to the *canonical* name (that is, the "real" name) of a host identified by an A record. CNAME records are used to provide alternative names by which systems can be identified. For example, you may have a system with the name server1.zacker.com on your network that you use as a Web server. Changing the host name of the computer would confuse your users, but you want to use the traditional name of www to identify the Web server in your domain. Once you create a CNAME record for the name www.zacker.com that points to server1.zacker.com, the system is addressable using either name.

MX (Mail Exchanger) Identifies a system that will direct e-mail traffic sent to an address in the domain to the individual recipient, a mail gateway, or another mail server.

In addition to functioning as the authority for a small section of the DNS name space, servers process client name resolution requests by either consulting their own resource records or forwarding the request to another DNS server on the network. The process of forwarding a request is called a *referral,* and this is how all of the DNS servers on the Internet work together to provide a unified information resource for the entire domain name space.

DNS Name Resolution

Although all Internet applications use DNS to resolve host names into IP addresses, this name resolution process is particularly easy to see when you're using the Microsoft Internet Explorer browser to access a Web site. When you type a URL containing a DNS name (such as www.microsoft.com) into the browser's Address field and press the ENTER key, if you look quickly at the status bar in the lower-left corner, you'll see a message that says "Finding Site: www.microsoft.com." In a few seconds, you'll then see a message that says "Connecting to," followed by an IP address. It is during this interval between the Finding Site message and the Connecting to message that the DNS name resolution process occurs.

From the client's perspective, the procedure that occurs during these few seconds consists of the application sending a query message to its designated DNS server that contains the name to be resolved. The server then replies with a message containing the IP address corresponding to that name. Using the supplied address, the application can then transmit a message to the intended destination. It is only when you examine the DNS server's role in the process that you see how complex the procedure really is.

Resolvers

The component in the client system that generates the DNS query is called a *resolver.* In most cases, the resolver is a simple set of library routines in the operating system that generates the queries to be sent to the DNS server, reads the response information from the server's replies, and feeds the response to the application that originally requested it. In addition, a resolver can resend a query if no reply is forthcoming after a given timeout period, and can process error messages returned by the server, such as when it fails to resolve a given name.

DNS Requests

A TCP/IP client usually is configured with the addresses of two DNS servers to which it can send queries. A client can send a query to any DNS server; it does not have to use the authoritative server for the domain in which it belongs nor does the server have to be on the local network. Using the DNS server that is closest to the client is best, however, because it minimizes the time needed for messages to travel between the two systems. A client only needs access to one DNS server, but two are usually specified, to provide a backup in case one server is unavailable.

There are two types of DNS queries: recursive and iterative. When a server receives a *recursive query*, it is responsible for trying to resolve the requested name and for transmitting a reply back to the requestor. Even if the server does not possess the required information itself, it must send its own queries to other DNS servers until it obtains the requested information or an error message stating why the information was unavailable, and must then relay the information back to the requestor. The system that generated the query, therefore, receives a reply only from the original server to which it sent the query. The resolvers in client systems nearly always send recursive queries to DNS servers.

When a server receives an *iterative query* (also called a *nonrecursive query)*, it can either respond with information from its own database or refer the requestor to another DNS server. The recipient of the query responds with the best answer it currently possesses, but is not responsible for searching for the information, as with a recursive query. DNS servers processing a recursive query from a client typically use iterative queries to request information from other servers. It is possible for a DNS server to send a recursive query to another server, thus in effect "passing the buck" and forcing the other server to search for the requested information, but this is considered to be bad form and is rarely done without permission.

One of the scenarios in which DNS servers do send recursive queries to other servers is when you configure a server to function as a *forwarder.* On a network running several DNS servers, you may not want all of the servers sending queries to other DNS servers on the Internet. If the network has a relatively slow connection to the Internet, for example, several servers transmitting repeated queries may use too much of the available bandwidth.

To prevent this, some DNS implementations enable you to configure one server to function as the forwarder for all Internet queries generated by the other servers on the network. Any time that a server has to resolve the DNS name of an Internet system, and fails to find the needed information in its cache, it transmits a recursive query to the forwarder, which is then responsible for sending its own iterative queries over the Internet connection. Once the forwarder resolves the name, it sends a reply back to the original DNS server, which relays it to the client.

This request-forwarding behavior is a function of the original server only. The forwarder simply receives standard recursive queries from the original server and processes them normally. A server can be configured to use a forwarder in either exclusive or nonexclusive mode. In *exclusive mode*, the server relies completely on the forwarder to resolve the requested name. If the forwarder's resolution attempt fails, the server relays a failure message back to the client. A server that uses a forwarder in exclusive mode is called a *slave*. In *nonexclusive mode*, if the forwarder fails to resolve the name and transmits an error message to the original server, that server makes its own resolution attempt before responding to the client.

Root Name Servers

In most cases, DNS servers that do not possess the information needed to resolve a name requested by a client send their first iterative query to one of the Internet's root name servers. The *root name servers* possess information about all of the top-level domains in the DNS name space. When you first install a DNS server, the only addresses that it needs to process client requests are those of the root name servers, because these servers can send a request for a name in any domain on its way to the appropriate authority.

The root name servers contain the addresses of the authoritative servers for all of the top-level domains on the Internet. In fact, the root name servers are the authorities for certain top-level domains, but they can also refer queries to the appropriate server for any of the other top-level domains, including the country-code domains, which are scattered all over the world. There are currently 13 root name servers, and they process millions of requests each day. The servers are also scattered widely and connected to different network trunks, so the chances of all of them being unavailable are minimal. If this were to occur, virtually all DNS name resolution would cease and the Internet would be crippled.

Resolving a Domain Name

With the preceding pieces in place, you are now ready to see how the DNS servers work together to resolve the name of a server on the Internet (see Figure 25-4). The process is as follows:

1. A user on a client system specifies the DNS name of an Internet server in an application such as a Web browser or FTP client.

2. The application generates an API call to the resolver on the client system, and the resolver creates a DNS recursive query message containing the server name.

3. The client system transmits the recursive query message to the DNS server identified in its TCP/IP configuration.

4. The client's DNS server, after receiving the query, checks its resource records to see if it is the authoritative source for the zone containing the requested server name. If it is the authority, it generates a reply message and transmits it back to the client. If the DNS server is not the authority for the domain in which the requested server is located, it generates an iterative query and submits it to one of the root name servers.

5. The root name server examines the name requested by the original DNS server and consults its resource records to identify the authoritative servers for the name's top-level domain. Because the root name server received an iterative request, it does not send its own request to the top-level domain server. Instead, it transmits a reply to the original DNS server that contains a referral to the top-level domain server addresses.

6. The original DNS server then generates a new iterative query and transmits it to the top-level domain server. The top-level domain server examines the second-level domain in the requested name and transmits to the original server a referral containing the addresses of authoritative servers for that second-level domain.

7. The original server generates yet another iterative query and transmits it to the second-level domain server. If the requested name contains additional domain names, the second-level domain server replies with another referral to the third-level domain servers. The second-level domain server may also refer the original server to the authorities for a different zone. This process continues until the original server receives a referral to the domain server that is the authority for the domain or zone containing the requested host.

8. Once the authoritative server for the domain or zone containing the host receives a query from the original server, it consults its resource records to determine the IP address of the requested system and transmits it in a reply message back to that original server.

9. The original server receives the reply from the authoritative server and transmits the IP address back to the resolver on the client system. The resolver relays the address to the application, which can then initiate communications with the system specified by the user.

This procedure assumes a successful completion of the name resolution procedure. If any of the authoritative DNS servers queried returns an error message to the original server stating, for example, that one of the domains in the name does not exist, this error message is relayed back to the client and the name resolution process is said to have failed.

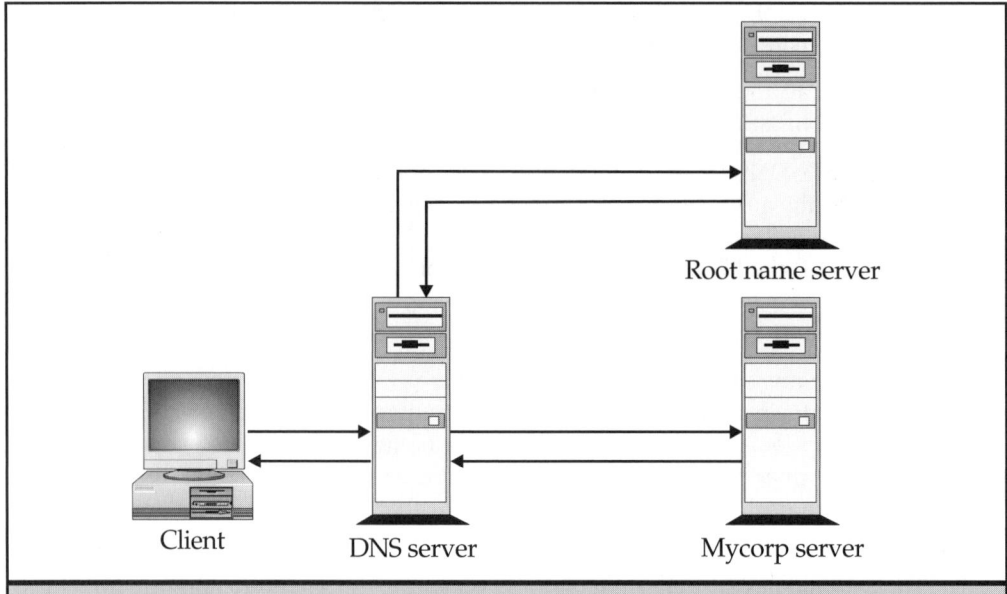

Figure 25-4: *DNS servers communicate among themselves to locate the information requested by a client.*

DNS Server Caching

This process may seem extremely long and complex, but in many cases, it isn't necessary for the client's DNS server to send queries to the servers for each domain specified in the requested DNS name. DNS servers are capable of retaining the information they learn about the DNS name space in the course of their name resolution procedures and storing it in a cache on the local drive.

A DNS server that receives requests from clients, for example, caches the addresses of the requested systems, as well as the addresses for particular domains' authoritative servers. The next time that a client transmits a request for a previously resolved name, the server can respond immediately with the cached information. In addition, if a client requests another name in one of the same domains, the server can send a query directly to an authoritative server for that domain, and not to a root name server. Thus, users should generally find that names in commonly accessed domains resolve more quickly, because one of the servers along the line has information about the domain in its cache, while names in obscure domains take longer, because the entire request/referral process is needed.

Negative Caching   In addition to storing information that aids in the name resolution process, most modern DNS server implementations are also capable of negative caching. *Negative caching* occurs when a DNS server retains information about names that do not exist in a domain. If, for example, a client sends a query to its DNS server containing a name in which the second-level domain does not exist, the top-level domain server will return a reply containing an error message to that effect. The client's DNS server will then retain the error message information in its cache. The next time a client requests a name in that domain, the DNS server will be able to respond immediately with its own error message, without consulting the top-level domain.

Cache Data Persistence   Caching is a vital element of the DNS architecture, because it reduces the number of requests sent to the root name and top-level domain servers, which, being at the top of the DNS tree, are the most likely to act as a bottleneck for the whole system. However, caches must be purged eventually, and there is a fine line between effective and ineffective caching. Because DNS servers retain resource records in their caches, it can take hours or even days for changes made in an authoritative server to be propagated around the Internet. During this period, users may receive incorrect information in response to a query. If information remains in server caches too long, the changes that administrators make to the data in their DNS servers take too long to propagate around the Internet. If caches are purged too quickly, the number of requests sent to the root name and top-level domain servers increases precipitously.

The amount of time that DNS data remains cached on a server is called its *time to live*. Unlike most data caches, the time to live is not specified by the administrator of the server where the cache is stored. Instead, the administrators of each authoritative DNS server specify how long the data for the resource records in their domains or zones should be retained in the servers where it is cached. This enables administrators to specify a time-to-live value based on the volatility of their server data. On a network where changes in IP addresses or the addition of new resource records is frequent, a lower time-to-live value increases the likelihood that clients will receive current data. On a network that rarely changes, you can use a longer time-to-live value, and minimize the number of requests sent to the parent servers of your domain or zone.

DNS Load Balancing

In most cases, DNS servers maintain one IP address for each host name. However, there are situations in which more than one IP address is required. In the case of a highly trafficked Web site, for example, one server may not be sufficient to support all of the clients. To have multiple, identical servers with their own IP addresses hosting the same site, some mechanism is needed to ensure that client requests are balanced among the machines.

One way of doing this is to control how the authoritative servers for the domain on which the site is located resolve the DNS name of the Web server. Some DNS server implementations enable you to create multiple resource records with different IP addresses for the same host name. As the server responds to queries requesting resolution of that name, it uses the resource records in a rotational fashion to supply the IP address of a different machine to each client.

DNS caching tends to defeat the effectiveness of this rotational system, because servers use the cached information about the site, rather than issuing a new query and possibly receiving the address for another system. As a result, it is generally recommended that you use a relatively short time-to-live value for the duplicated resource records.

Reverse Name Resolution

The Domain Name System is designed to facilitate the resolution of DNS names into IP addresses, but there are also instances in which IP addresses have to be resolved into DNS names. These instances are relatively rare. In log files, for example, some systems convert IP addresses to DNS names to make the data more readily accessible to human readers. Certain systems also use reverse name resolution in the course of authentication procedures.

The structure of the DNS name space and the method by which it's distributed among various servers is based on the domain name hierarchy. When the entire database is located on one system, such as in the case of a host table, searching for a particular address to find out its associated name is no different from searching for a name to find an address. However, locating a particular address in the DNS name space would seem to require a search of all of the Internet's DNS servers, which is obviously impractical.

To make reverse name resolution possible without performing a massive search across the entire Internet, the DNS tree includes a special branch that uses the dotted decimal values of IP addresses as domain names. This branch stems from a domain called in-addr.arpa, which is located just beneath the root of the DNS tree, as shown in Figure 25-5. Just beneath the in-addr domain, there are 256 subdomains named using the numbers 0 to 255, to represent the possible values of an IP address's first byte. Each of these subdomains contains another 256 subdomains representing the possible values of the second byte. The next level has another 256 domains, each of which can have up to 256 numbered hosts, which represent the third and fourth bytes of the address.

Using the in-addr.arpa domain structure, each of the hosts represented by a standard name on a DNS server also has an equivalent DNS name constructed using its IP address. Therefore, if a system with the IP address 192.168.214.23 is listed in the DNS server for the zacker.com domain with the host name www, there is also a resource record for that system with the DNS name 23.214.168.192.in-addr.arpa, meaning that there is a host with the name 23 in a domain called 214.168.192.in-addr.arpa, as shown in Figure 25-6. This domain structure makes it possible for a system to search for the IP address of a host in a domain (or zone) without having to consult other servers in the DNS tree. In most cases, you can configure a DNS server to automatically create an equivalent resource record in the in-addr.arpa domain for every host you add to the standard domain name space.

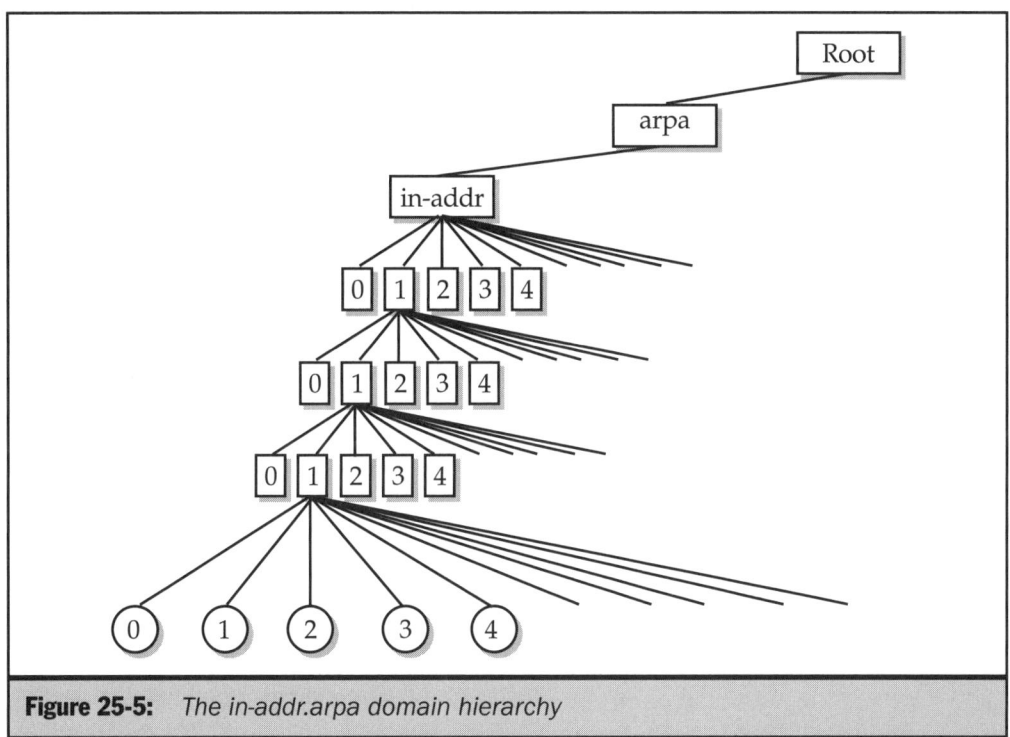

Figure 25-5:   *The in-addr.arpa domain hierarchy*

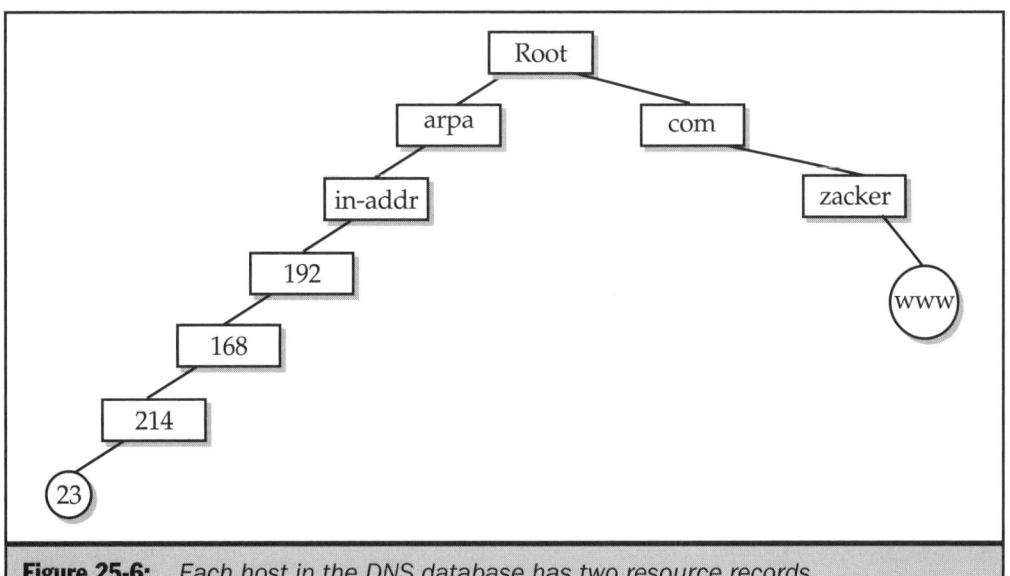

Figure 25-6:   *Each host in the DNS database has two resource records.*

The byte values of IP addresses are reversed in the in-addr.arpa domain, because in a DNS name, the least significant word comes first, whereas in IP addresses, the least significant byte comes last. In other words, a DNS name is structured with the root of the DNS tree on the right side and the host name on the left. In an IP address, the host identifier is on the right and the network identifier is on the left. It would be possible to create a domain structure using the IP address bytes in their regular order, but this would complicate the administration process by making it harder to delegate maintenance tasks based on network addresses.

If, for example, a corporate internetwork consists of three branch offices, each with its own Class C IP address (192.168.1, 192.168.2, and 192.168.3), it is possible to have a different person administer the reverse lookup domain for each site, because the domains would have the following names:

```
1.168.192.in-addr.arpa

2.168.192.in-addr.arpa

3.168.192.in-addr.arpa
```

If the in-addr.arpa name space was constructed using the byte values for the IP addresses in their traditional order, the third-level domain would be named using the host identifier (that is, the fourth byte) from the IP address. You might need up to 256 third-level domains to represent these three networks instead of 1. Delegating administration chores would be all but impossible, because the 72.in-addr.arpa domain (for example) could contain a host from all three networks.

DNS Name Registration

As you have already learned, name resolution is the process by which IP address information for a host name is extracted from the DNS database. The process by which host names and their addresses are added to the database is called *name registration*. Name registration refers to the process of creating new resource records on a DNS server, thus making them accessible to all of the other DNS servers on the network.

The name registration process on a traditional DNS server is decidedly low-tech. There is no mechanism by which the server can detect the systems on the network and enter their host names and IP addresses into resource records. In fact, a computer may not even be aware of its host name, because it receives all of its communications using IP addresses and never has to answer to its name.

To register a host in the DNS name space, an administrator has to manually create a resource record on the server. The method for creating resource records varies depending on the DNS server implementation. Unix-based servers require you to edit a text file, while Microsoft DNS Server uses a graphical interface.

Manual Name Registration

The manual name registration process is basically an adaptation of the host table for use on a DNS server. It is easy to see how, in the early days, administrators were able to implement DNS servers on their network by using their host tables with slight modifications. Today, however, the manual name registration process can be problematic on some networks.

If you have a large number of hosts, manually creating resource records for all of them can be a tedious affair, even with a graphical interface. However, depending on the nature of the network, it may not be necessary to register every system in the DNS. If, for example, you are running a Windows NT network using unregistered IP addresses, you may not need your own DNS server at all, except possibly to process client name resolution requests. Windows NT networks have their own NetBIOS naming system and name resolution mechanisms, and you generally don't need to refer to them using DNS names.

The exceptions to this would be systems with registered IP addresses that you use as Web servers or other types of Internet servers. These must be visible to Internet users and, therefore, must have a host name in a registered DNS domain. In most cases, the number of systems like this on a network is small, so manually creating the resource records is not much of a problem. If you have Unix systems on your network, however, you are more likely to use DNS to identify them using names and in this case, you must create resource records for them.

Dynamic Updates

As networks grow larger and more complex, the biggest problem arising from manual name registration stems from the increasing use of DHCP servers to dynamically assign IP addresses to network workstations. The manual configuration of TCP/IP clients is another long-standing network administration chore that is gradually being phased out in favor of an automated solution. Assigning IP addresses dynamically means that workstations can have different addresses from one day to the next, and the original DNS standard has no way of keeping up with the changes.

On networks where only a few servers have to be visible to the Internet, it isn't too great an inconvenience to configure them manually with static IP addresses and use DHCP for the unregistered systems. This situation has changed with the advent of Windows 2000 and Active Directory, its new enterprise directory service. Windows NT networks use WINS to resolve NetBIOS names into IP addresses, but name registration is automatic with WINS. WINS automatically updates its database record for a workstation that is assigned a new IP address by a DHCP server, so that no administrator intervention is required. Active Directory, however, relies heavily on DNS instead of WINS to resolve the names of systems on the network and to keep track of the domain controllers that are available for use by client workstations.

To make the use of DNS practical with technologies like Active Directory that require regular updates to resource records, members of the IETF have developed a new specification, published as RFC 2136, "Dynamic Updates in the Domain Name System," that is currently a proposed standard. This document defines a new DNS message type, called an *Update*, that systems like domain controllers and DHCP servers can generate and transmit to a DNS server. These Update messages can modify or delete existing resource records or create new ones, based on prerequisites specified by the administrator. For Active Directory, the Update message can contain the information for the new SVR resource record type, as defined in RFC 2052, that specifies the locations of servers that perform particular functions.

Note   *For more information on the role of DNS on an Active Directory network, see Chapter 17.*

Zone Transfers

Most networks use at least two DNS servers to provide fault tolerance and to give clients access to a nearby server. Because the resource records (in most cases) have to be created and updated manually by administrators, the DNS standards define a mechanism that replicates the DNS data among the servers, thus enabling administrators to make the changes only once.

The standards define two DNS server roles: the primary master and the secondary master, or slave. The *primary master* server loads its resource records and other information from the database files on the local drive. The *slave* (or *secondary master*) server receives its data from another server in a process called a *zone transfer,* which the slave performs each time it starts and periodically thereafter. The server from which the slave receives its data is called its *master server,* but it need not be the primary master. A slave can receive data from the primary master or another slave.

Zone transfers are performed for individual zones, and because a single server can be the authority for multiple zones, more than one transfer may be needed to update all of a slave server's data. In addition, the primary master and slave roles are also zone specific. A server can be the primary master for one zone and the slave for another, although this practice generally should not be necessary and is likely to generate some confusion.

Although slave servers receive periodic zone transfers from their primaries, they are also able to load database files from their local drives. When a slave server receives a zone transfer, it updates the local database files. Each time the slave server starts up, it loads the most current resource records it has from the database files and then checks this data with the primary master to see whether an update is needed. This prevents zone transfers from being performed needlessly.

DNS Messaging

DNS name resolution transactions use UDP (User Datagram Protocol) datagrams on port 53 for servers and on an ephemeral port number for clients. Communication between two servers uses port 53 on both machines. In cases in which the data to be transmitted does not fit in a single UDP datagram, in the case of zone transfers, the two systems establish a standard TCP connection, also using port 53 on both machines, and transmit the data using as many packets as needed.

The Domain Name System uses a single message format for all of its communications that consists of the following five sections:

Header   Contains information about the nature of the message.

Question   Contains the information requested from the destination server.

Answer   Contains RRs supplying the information requested in the Question section.

Authority   Contains RRs pointing to an authority for the information requested in the Question section.

Additional   Contains RRs containing additional information in response to the Question section.

Every DNS message has a Header section, and the other four sections are included only if they contain data. For example, a query message contains the DNS name to be resolved in the Question section, but the Answer, Authority, and Additional sections aren't needed. When the server receiving the query constructs its reply, it makes some changes to the Header section, leaves the Question section intact, and adds entries to one or more of the remaining three sections. Each section can have multiple entries, so that a server can send more than one resource record in a single message.

The DNS Header Section

The Header section of the DNS message contains codes and flags that specify the function of the message and the type of service requested from or supplied by a server. The format of the Header section is shown in Figure 25-7.

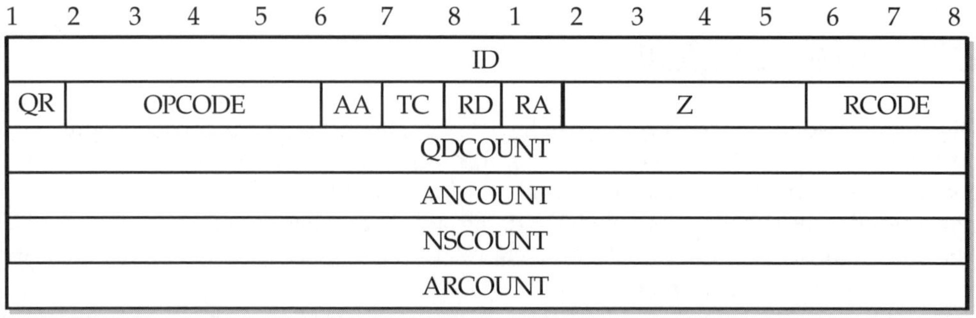

1	2	3	4	5	6	7	8	1	2	3	4	5	6	7	8
ID															
QR	OPCODE				AA	TC	RD	RA	Z				RCODE		
QDCOUNT															
ANCOUNT															
NSCOUNT															
ARCOUNT															

Figure 25-7: *The DNS Header section format*

The functions of the Header fields are as follows:

ID, 2 bytes Contains an identifier value used to associate queries with replies.

Flags, 2 bytes Contains flag bits used to identify the functions and properties of the message, as follows:

QR, 1 bit Specifies whether the message is a query (value 0) or a response (value 1).

OPCODE, 4 bits Specifies the type of query that generated the message. Response messages retain the same value for this field as the query to which they are responding. Possible values are as follows:

0   Standard query (QUERY).

1   Inverse query (IQUERY).

2   Server status request (STATUS).

3–15   Unused.

AA (Authoritative Answer), 1 bit Indicates that a response message has been generated by a server that is the authority for the domain or zone in which the requested name is located.

TC (Truncation), 1 bit Indicates that the message has been truncated because the amount of data exceeds the maximum size for the current transport mechanism. In most DNS implementations, this bit functions as a signal that the message should be transmitted using a TCP connection rather than a UDP datagram.

RD (Recursion Desired), 1 bit   In a query, indicates that the destination server should treat the message as a recursive query. In a response, indicates that the message is the response to a recursive query. The absence of this flag indicates that the query is iterative.

RA (Recursion Available), 1 bit   Specifies whether a server is configured to process recursive queries.

Z, 3 bits   Unused.

RCODE (Response Code), 4 bits   Specifies the nature of a response message, indicating when an error has occurred and what type of error, using the following values:

   0   No error has occurred.

   1 – Format Error   Indicates that the server was unable to understand the query.

   2 – Server Failure   Indicates that the server was unable to process the query.

   3 – Name Error   Used by authoritative servers only to indicate that a requested name or subdomain does not exist in the domain.

   4 – Not Implemented   Indicates that the server does not support the type of query received.

   5 – Refused   Indicates that server policies (such as security policies) have prevented the processing of the query.

   6–15   Unused.

QDCOUNT, 2 bytes   Specifies the number of entries in the Question section.

ANCOUNT, 2 bytes   Specifies the number of entries in the Answer section.

NSCOUNT, 2 bytes   Specifies the number of name server RRs in the Authority section.

ARCOUNT, 2 bytes   Specifies the number of entries in the Additional section.

The DNS Question Section

The Question section of a DNS message contains the number of entries specified in the header's QDCOUNT field. In most cases, there is only one entry. Each entry is formatted as shown in Figure 25-8.

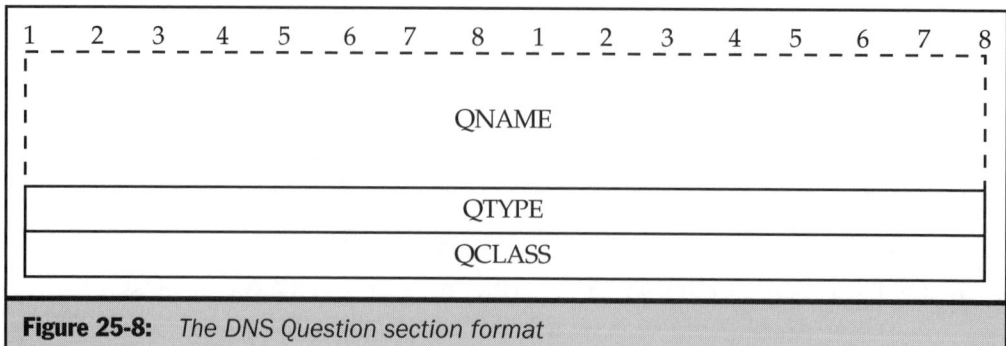

Figure 25-8: *The DNS Question section format*

The functions of the fields are as follows:

QNAME, variable   Contains the DNS, domain, or zone name about which information is being requested.

QTYPE, 2 bytes   Contains a code that specifies the type of RR the query is requesting.

QCLASS, 2 bytes   Contains a code that specifies the class of the RR being requested.

DNS Resource Record Sections

The three remaining sections of a DNS message, the Answer, Authority, and Additional sections, each contain resource records that use the format shown in Figure 25-9. The number of resource records in each section is specified in the header's ANCOUNT, NSCOUNT, and RCOUNT fields.

The functions of the fields are as follows:

NAME, variable   Contains the DNS, domain, or zone name about which information is being supplied.

TYPE, 2 bytes   Contains a code that specifies the type of RR the entry contains.

CLASS, 2 bytes   Contains a code that specifies the class of the RR.

TTL, 4 bytes   Specifies the amount of time (in seconds) that the RR should be cached in the server to which it is being supplied.

RDLENGTH, 2 bytes   Specifies the length (in bytes) of the RDATA field.

RDATA, variable   Contains RR data, the nature of which is dependent on its TYPE and CLASS. For an A-type record in the IN class, for example, this field contains the IP address associated with the DNS name supplied in the NAME field.

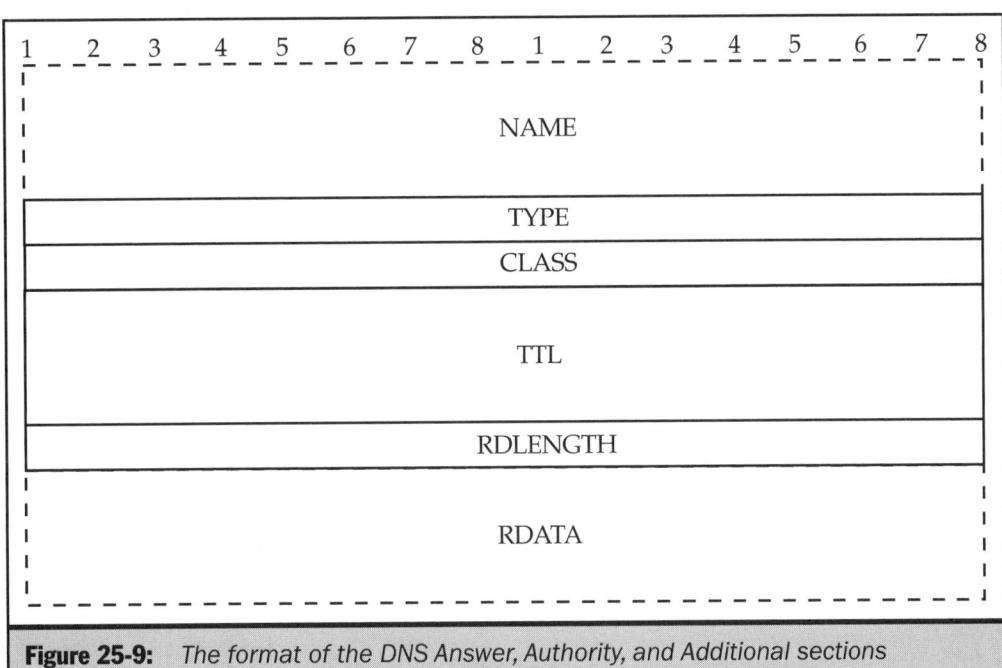

Figure 25-9:   *The format of the DNS Answer, Authority, and Additional sections*

Different types of resource records have different functions and, therefore, may contain different types of information in the RDATA field. Most resource records, such as the NS, A, PTR, and CNAME types, have only a single name or address in this field, while others have multiple subfields. The SOA resource record is the most complex in the Domain Name System. For this record, the RDATA field is broken up into seven subfields, as shown in Figure 25-10.

The functions of the SOA resource record subfields are as follows:

MNAME, variable   Specifies the DNS name of the primary master server that was the source for the information about the zone.

RNAME, variable   Specifies the e-mail address of the administrator responsible for the zone data. This field has no actual purpose as far as the server is concerned; it is strictly informational. The value for this field takes the form of a DNS name. Standard practice calls for the period after the first word to be converted to the @ symbol, in order to use the value as an e-mail address. For example, a value of administrator.zacker.com is equivalent to the e-mail address administrator@zacker.com.

SERIAL, 4 bytes   Contains a serial number that is used to track modifications to the zone data on the primary master server. The value of this field is incremented (either manually or automatically) on the primary master server each time the zone data is modified, and the slave compares its value to the one supplied by the primary master, to determine if a zone transfer is necessary.

NETWORK CONNECTION SERVICES

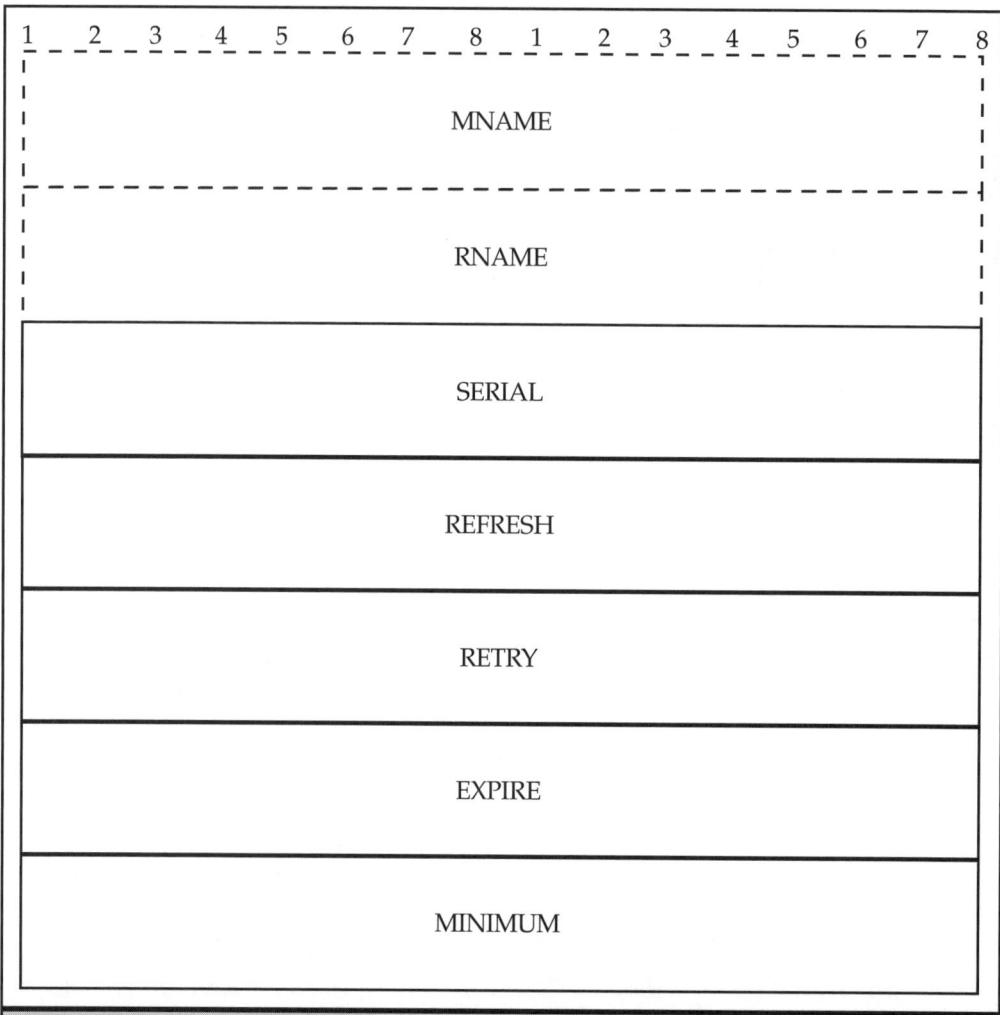

Figure 25-10: *The RDATA subfields in the SOA resource record*

REFRESH, 4 bytes Specifies the time interval (in seconds) at which the slave should transmit an SOA query to the primary master to determine if a zone transfer is needed.

RETRY, 4 bytes Specifies the time interval (in seconds) at which the slave should make repeat attempts to connect to the primary master after its initial attempt fails.

EXPIRE, 4 bytes   Specifies the time interval (in seconds) after which the slave server's data should expire, in the event that it cannot contact the primary master server. Once the data has expired, the slave server stops responding to queries.

MINIMUM, 4 bytes   Specifies the time-to-live interval (in seconds) that the server should supply for all of the resource records in its responses to queries.

DNS Message Notation

The latter four sections of the DNS message are largely consistent in how they notate the information in their fields. DNS, domain, and zone names are all expressed in the same way, and the sections all use the same values for the resource record type and class codes. The only exceptions are a few additional codes that are used only in the Question section, called QTYPES and QCLASSES, respectively. The following sections describe how these values are expressed in the DNS message.

DNS Name Notation

Depending on the function of the message, any or all of the four sections can contain the fully qualified name of a host system, the name of a domain, or the name of a zone on a server. These names are expressed as a series of units, called *labels*, each of which represents a single word in the name. The periods between the words are not included, so to delineate the words, each label begins with a single byte that specifies the length of the word (in bytes), after which the specified number of bytes follows. This is repeated for each word in the name. After the final word of a fully qualified name, a byte with the value of 0 is included to represent the null value of the root domain. Thus, to express the DNS name www.zacker.com, the value of the QNAME or NAME field would appear as follows, in decimal form:

```
3 w w w 5 m y c o r p 3 c o m 0
```

Resource Record Types

All of the data distributed by the Domain Name System is stored in resource records. Query messages request certain resource records from servers, and the servers reply with those resource records. The QTYPE field in a Question section entry specifies the type of resource record being requested from the server, and the TYPE fields in the Answer, Authority, and Additional section entries specify the type of resource record supplied by the server in each entry. Table 25-1 contains the resource record types and the codes used to represent them in these fields. All of the values in this table are valid for both the QTYPE and TYPE fields. Table 25-2 contains four additional values that represent sets of resource records that are valid for the QTYPE field in Question section entries only.

Type	Type Code	Function
A	1	Host address
NS	2	Authoritative name server
MD	3	Mail destination (obsolete)
MF	4	Mail forwarder (obsolete)
CNAME	5	Canonical name for an alias
SOA	6	Start of a zone of authority
MB	7	Mailbox domain name (experimental)
MG	8	Mail group member (experimental)
MR	9	Mail rename domain name (experimental)
NULL	10	Null RR (experimental)
WKS	11	Well-known service description
PTR	12	Domain name pointer
HINFO	13	Host information
MINFO	14	Mailbox or mail list information
MX	15	Mail exchange
TXT	16	Text strings
SVR	33	Network server

Table 25-1: *DNS Resource Record Types and Values for Use in the TYPE or QTYPE Field*

QTYPE	QTYPE Code	Function
AXFR	252	Request for transfer of an entire zone
MAILB	253	Request for mailbox-related records (MB, MG, or MR)
MAILA	254	Request for mail agent RRs (obsolete)
*	255	Request for all records

Table 25-2: *Additional Values Representing Sets of Resource Records for Use in the QTYPE Field Only*

Class Types

The QCLASS field in the Question section and the CLASS field in the Answer, Authority, and Additional sections specify the type of network for which information is being requested or supplied. Although they performed a valid function at one time, these fields are now essentially meaningless, as virtually all DNS messages use the IN class. CSNET and CHAOS class networks are obsolete, and the Hesiod class is used only for a few experimental networks at MIT. For academic purposes only, the values for the CLASS and QCLASS values are shown in Tables 25-3 and 25-4.

Class	Class Code	Function
IN	1	Internet
CS	2	CSNET
CH	3	CHAOS
HS	4	Hesiod

Table 25-3: *Values for the Resource Record CLASS and QCLASS Fields*

QCLASS	QCLASS Code	Function
*	255	Any Class

Table 25-4: *Additional Value for the Resource Record QCLASS Field Only*

Name Resolution Messages

The process of resolving a DNS name into an IP address begins with the generation of a query by the resolver on the client system. Figure 25-11 shows a query message, captured in a network monitor program, generated by a Web browser trying to connect to the URL http://www.zacker.com/. The value of the message's OPCODE flag is 0, indicating that this is a regular query, and the RD flag has a value of 1, indicating that this is a recursive query. As a result, the DNS server receiving the query (which is called CZ1) will be responsible for resolving the DNS name and returning the results to the client. The QDCOUNT field indicates that there is one entry in the Question section, and no entries in the three resource record sections, which is standard for a query message. The Question section specifies the DNS name to be resolved (www.zacker.com) and the type (1 = A) and class (1 = IN) of the resource record being requested.

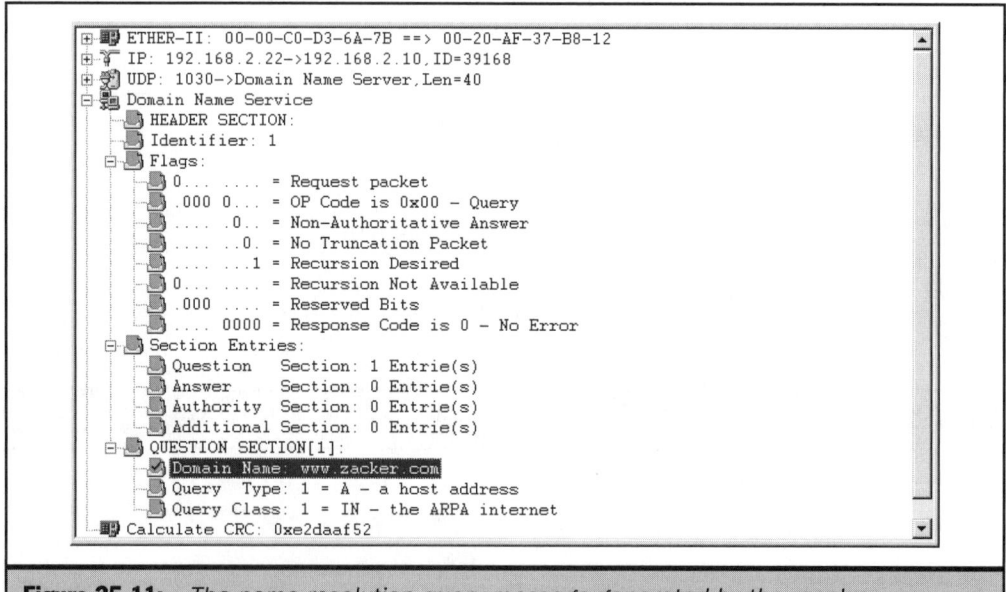

Figure 25-11: *The name resolution query message generated by the resolver*

CZ1 is not the authoritative server for the zacker.com domain, nor does it have the requested information in its cache, so it must generate its own queries. CZ1 first generates a query message (see Figure 25-12) and transmits it to one of the root name servers (198.41.0.4) configured into the server software. The entry in the Question section is identical to that of the client's query message. The only differences in this query are that the server has included a different value in the ID field (4114) and has changed the value of the RD flag to 0, indicating that this is an iterative query.

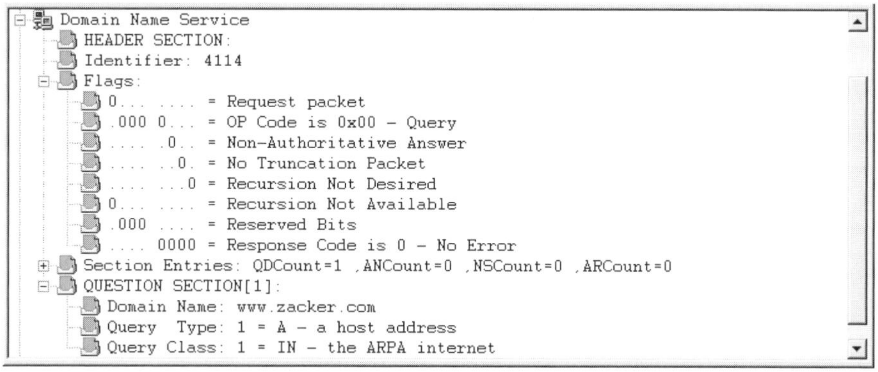

```
⊟ 🖳 Domain Name Service
   ├─ 🗐 HEADER SECTION:
   ├─ 🗐 Identifier: 4114
   ⊟ 🗐 Flags:
      ├─ 🗐 0... .... = Request packet
      ├─ 🗐 .000 0... = OP Code is 0x00 - Query
      ├─ 🗐 .... .0.. = Non-Authoritative Answer
      ├─ 🗐 .... ..0. = No Truncation Packet
      ├─ 🗐 .... ...0 = Recursion Not Desired
      ├─ 🗐 0... .... = Recursion Not Available
      ├─ 🗐 .000 .... = Reserved Bits
      └─ 🗐 .... 0000 = Response Code is 0 - No Error
   ⊞ 🗐 Section Entries: QDCount=1 ,ANCount=0 ,NSCount=0 ,ARCount=0
   ⊟ 🗐 QUESTION SECTION[1]:
      ├─ 🗐 Domain Name: www.zacker.com
      ├─ 🗐 Query  Type: 1 = A - a host address
      └─ 🗐 Query Class: 1 = IN - the ARPA internet
```

Figure 25-12:   *The query message sent to the root name server*

The response that CZ1 receives from the root name server bypasses one step of the process, because this root name server is also the authoritative server for the com top-level domain. As a result, the response contains the resource record that identifies the authoritative server for the zacker.com domain. If the requested DNS name had been in a top-level domain for which the root name server was not authoritative, such as one of the country-code domains, the response would contain a resource record identifying the proper authoritative servers.

The response message from the root domain server (see Figure 25-13) has a QR bit that has a value of 1, indicating that this is a response message, and the same ID value as the request, enabling CZ1 to associate the two messages. The QDCOUNT field again has a value of 1, because the response retains the Question section, unmodified, from the query message. The NSCOUNT and ARCOUNT fields indicate that there are two entries each in the Authority and Additional sections. The first entry in the Authority section contains the NS resource record for one of the authoritative servers for zacker.com known to the root name/top-level domain server, and the second entry contains the NS record for the other. The type and class values are the same as those requested in the query message; the time-to-live value assigned to both records is 172,800 seconds (48 hours). The RDATA field in the first entry is 16 bytes long and contains the DNS name of the first authoritative server (ns1.secure.net). The RDATA field in the second entry is only 6 bytes long, and contains only the host name (ns2) for the other authoritative server, since it's in the same domain as the first one.

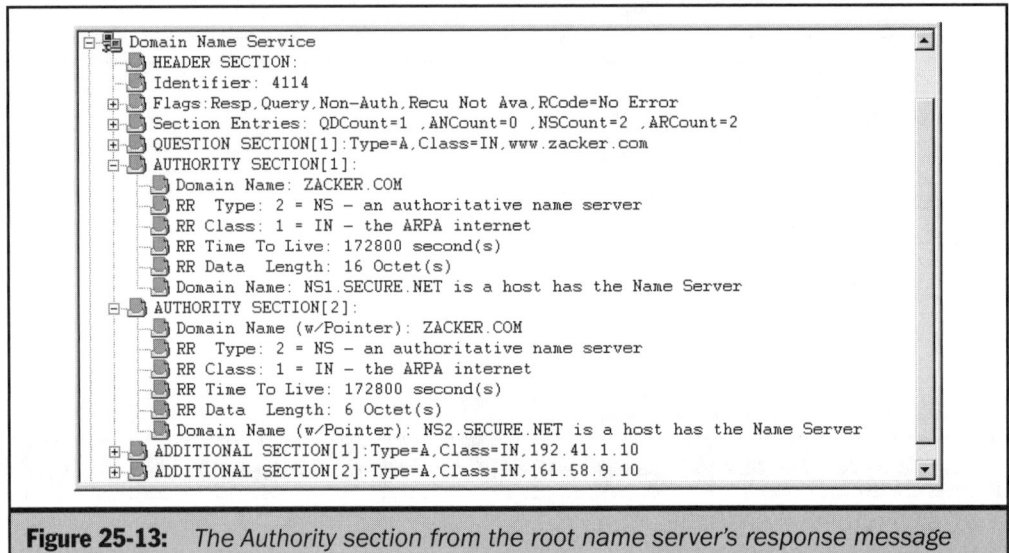

```
⊟ 🖳 Domain Name Service
   📄 HEADER SECTION:
   📄 Identifier: 4114
 ⊞ 📄 Flags:Resp,Query,Non-Auth,Recu Not Ava,RCode=No Error
 ⊞ 📄 Section Entries: QDCount=1 ,ANCount=0 ,NSCount=2 ,ARCount=2
 ⊞ 📄 QUESTION SECTION[1]:Type=A,Class=IN,www.zacker.com
 ⊟ 📄 AUTHORITY SECTION[1]:
   📄 Domain Name: ZACKER.COM
   📄 RR  Type: 2 = NS - an authoritative name server
   📄 RR Class: 1 = IN - the ARPA internet
   📄 RR Time To Live: 172800 second(s)
   📄 RR Data  Length: 16 Octet(s)
   📄 Domain Name: NS1.SECURE.NET is a host has the Name Server
 ⊟ 📄 AUTHORITY SECTION[2]:
   📄 Domain Name (w/Pointer): ZACKER.COM
   📄 RR  Type: 2 = NS - an authoritative name server
   📄 RR Class: 1 = IN - the ARPA internet
   📄 RR Time To Live: 172800 second(s)
   📄 RR Data  Length: 6 Octet(s)
   📄 Domain Name (w/Pointer): NS2.SECURE.NET is a host has the Name Server
 ⊞ 📄 ADDITIONAL SECTION[1]:Type=A,Class=IN,192.41.1.10
 ⊞ 📄 ADDITIONAL SECTION[2]:Type=A,Class=IN,161.58.9.10
```

Figure 25-13:   *The Authority section from the root name server's response message*

These Authority section entries identify the servers that CZ1 needs to contact to resolve the www.zacker.com domain name, but it does so using DNS names. In order to prevent CZ1 from having to go through this whole process again to resolve ns1.secure.net and ns2.secure.net into IP addresses, there are two entries in the Additional section that contain the A resource records for these two servers, which include their IP addresses, as shown in Figure 25-14.

Using the information contained in the previous response, CZ1 transmits a query to the first authoritative server for the zacker.com domain (ns1.secure.net – 192.41.1.10). Except for the destination address, this query is identical to the one that CZ1 sent to the root name server. The response message, shown in Figure 25-15, that CZ1 receives from the ns1.secure.net server (finally) contains the information that the client originally requested. This message contains the original Question section entry and two entries each in the Answer, Authority, and Additional sections.

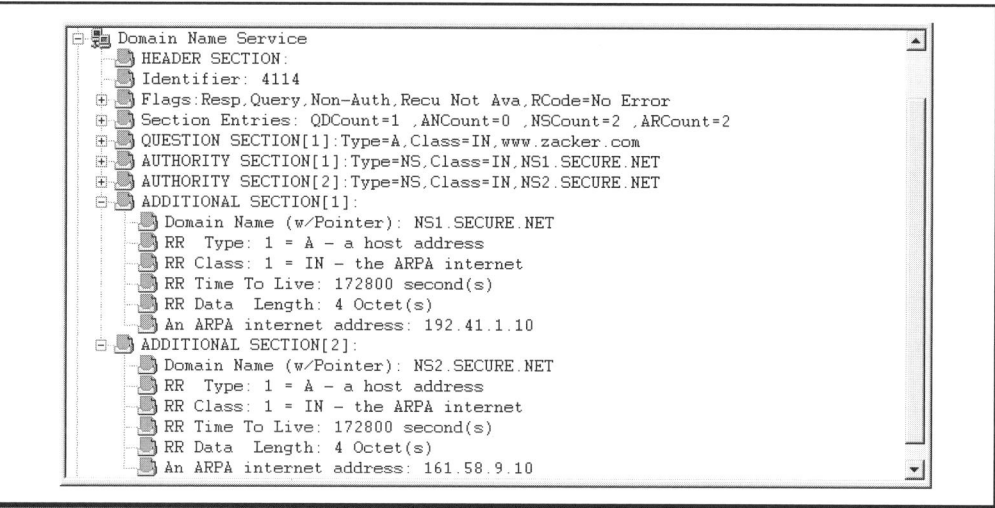

Figure 25-14:   *The Additional section from the root name server's response message*

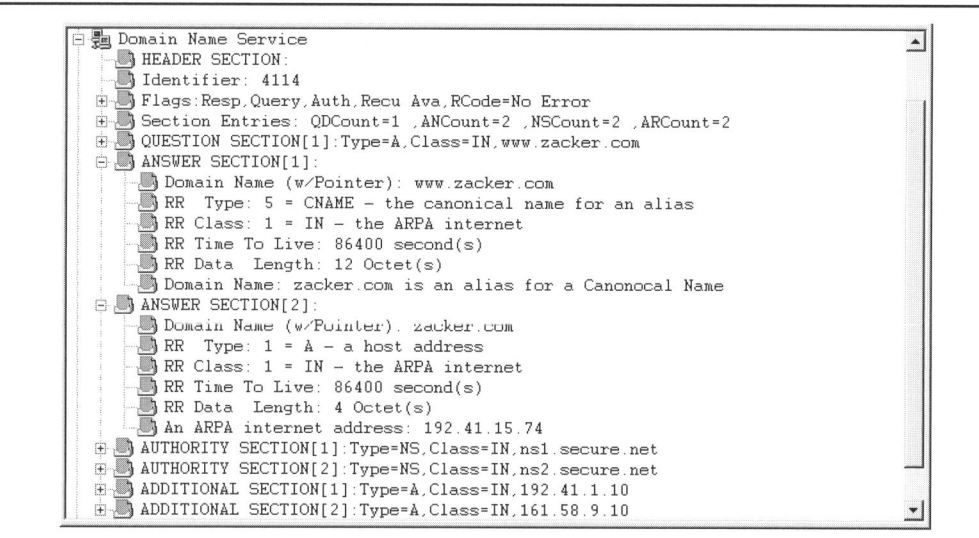

Figure 25-15:   *The response message from the authoritative server for the requested domain*

The first entry in the Answer section contains a resource record with a TYPE value of 5 (CNAME) and a time-to-live value of 86,400 seconds (24 hours). The inclusion of a CNAME resource record in a response to a query requesting an A record indicates that the host name www exists in the zacker.com domain only as a canonical name (that is, an alias for another name), which is specified in the RDATA field as zacker.com. The second entry in the Answer section contains the A resource record for the name zacker.com, which specifies the IP address 192.41.15.74 in the RDATA field. This is the IP address that the client system must use to reach the www.zacker.com Web server. The entries in the Authority and Additional sections specify the names and addresses of the authoritative server for zacker.com, and are identical to the equivalent entries in the response message from the root name server.

Root Name Server Discovery

Each time the DNS server starts, it loads the information stored in its database files. One of these files contains root name server hints. Actually, this file contains the names and addresses of all of the root name servers, but the DNS server, instead of relying on this data, uses it to send a query to the first of the root name servers, requesting that it identify the authoritative servers for the root domain. This is to ensure that the server is using the most current information. The query (see Figure 25-16) is just like that for a name resolution request, except that there is no value in the NAME field.

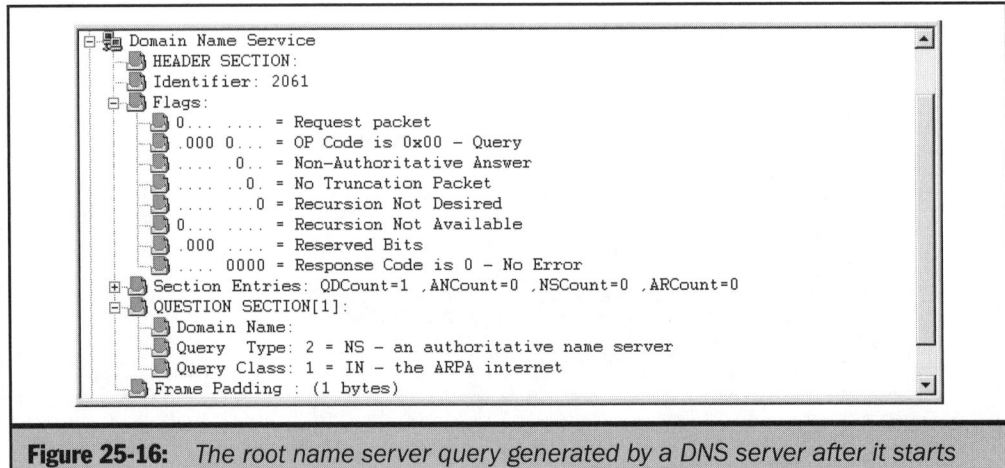

Figure 25-16:   *The root name server query generated by a DNS server after it starts*

The reply returned by the root name server contains 13 entries in both the Answer and Additional sections, corresponding to the 13 root name servers currently in operation (see Figure 25-17). Each entry in the Answer section contains the NS resource record for one of the root name servers, which specifies its DNS name, and the corresponding entry in the Additional section contains the A record for that server, which specifies its IP address. All of these servers are located in a domain called root-server.net and have incremental host names from *a* to *m*. Because the information about these servers does not change often, if at all, their resource records can have a long time-to-live value: 518,400 seconds (144 hours or 6 days) for the NS records and 3,600,000 (1,000 hours or 41.67 days) for the A records.

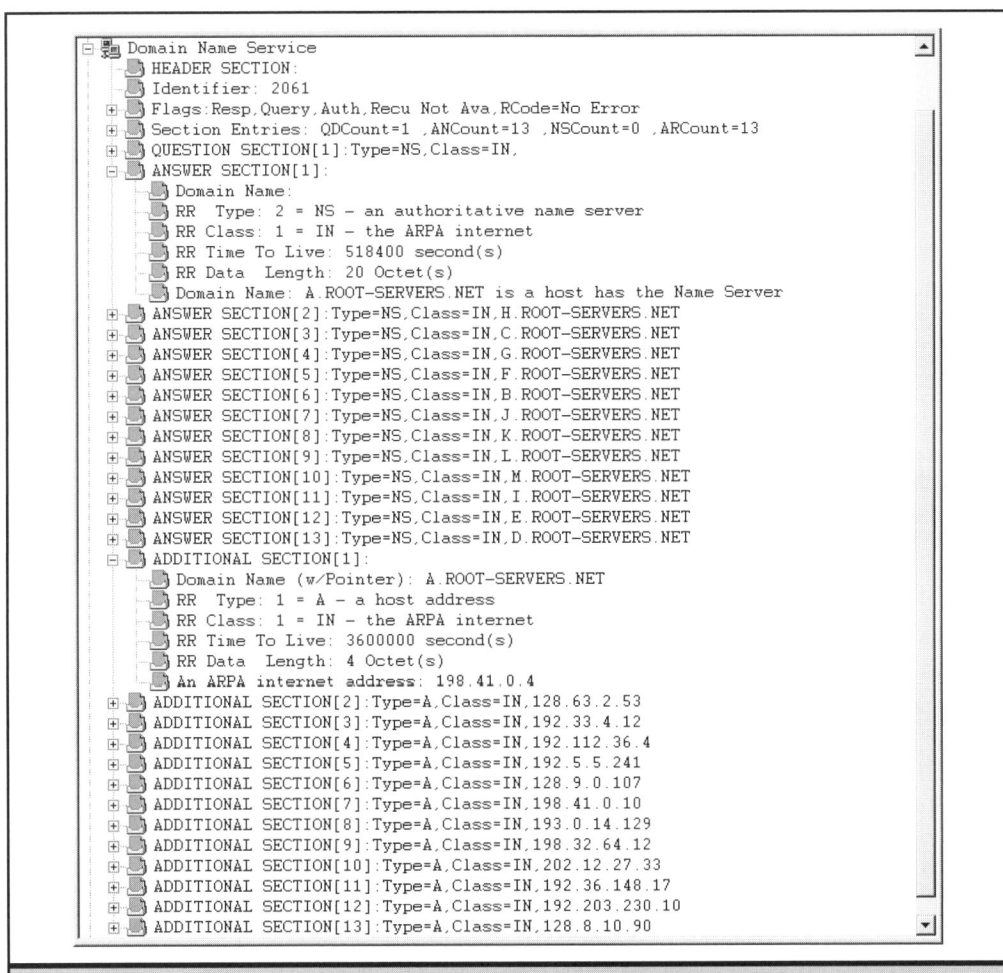

```
⊟ 🔳 Domain Name Service
   📄 HEADER SECTION:
      📄 Identifier: 2061
   ⊞ 📄 Flags:Resp,Query,Auth,Recu Not Ava,RCode=No Error
   ⊞ 📄 Section Entries: QDCount=1 ,ANCount=13 ,NSCount=0 ,ARCount=13
   ⊞ 📄 QUESTION SECTION[1]:Type=NS,Class=IN,
   ⊟ 📄 ANSWER SECTION[1]:
      📄 Domain Name:
      📄 RR  Type: 2 = NS - an authoritative name server
      📄 RR Class: 1 = IN - the ARPA internet
      📄 RR Time To Live: 518400 second(s)
      📄 RR Data  Length: 20 Octet(s)
      📄 Domain Name: A.ROOT-SERVERS.NET is a host has the Name Server
   ⊞ 📄 ANSWER SECTION[2]:Type=NS,Class=IN,H.ROOT-SERVERS.NET
   ⊞ 📄 ANSWER SECTION[3]:Type=NS,Class=IN,C.ROOT-SERVERS.NET
   ⊞ 📄 ANSWER SECTION[4]:Type=NS,Class=IN,G.ROOT-SERVERS.NET
   ⊞ 📄 ANSWER SECTION[5]:Type=NS,Class=IN,F.ROOT-SERVERS.NET
   ⊞ 📄 ANSWER SECTION[6]:Type=NS,Class=IN,B.ROOT-SERVERS.NET
   ⊞ 📄 ANSWER SECTION[7]:Type=NS,Class=IN,J.ROOT-SERVERS.NET
   ⊞ 📄 ANSWER SECTION[8]:Type=NS,Class=IN,K.ROOT-SERVERS.NET
   ⊞ 📄 ANSWER SECTION[9]:Type=NS,Class=IN,L.ROOT-SERVERS.NET
   ⊞ 📄 ANSWER SECTION[10]:Type=NS,Class=IN,M.ROOT-SERVERS.NET
   ⊞ 📄 ANSWER SECTION[11]:Type=NS,Class=IN,I.ROOT-SERVERS.NET
   ⊞ 📄 ANSWER SECTION[12]:Type=NS,Class=IN,E.ROOT-SERVERS.NET
   ⊞ 📄 ANSWER SECTION[13]:Type=NS,Class=IN,D.ROOT-SERVERS.NET
   ⊟ 📄 ADDITIONAL SECTION[1]:
      📄 Domain Name (w/Pointer): A.ROOT-SERVERS.NET
      📄 RR  Type: 1 = A - a host address
      📄 RR Class: 1 = IN - the ARPA internet
      📄 RR Time To Live: 3600000 second(s)
      📄 RR Data  Length: 4 Octet(s)
      📄 An ARPA internet address: 198.41.0.4
   ⊞ 📄 ADDITIONAL SECTION[2]:Type=A,Class=IN,128.63.2.53
   ⊞ 📄 ADDITIONAL SECTION[3]:Type=A,Class=IN,192.33.4.12
   ⊞ 📄 ADDITIONAL SECTION[4]:Type=A,Class=IN,192.112.36.4
   ⊞ 📄 ADDITIONAL SECTION[5]:Type=A,Class=IN,192.5.5.241
   ⊞ 📄 ADDITIONAL SECTION[6]:Type=A,Class=IN,128.9.0.107
   ⊞ 📄 ADDITIONAL SECTION[7]:Type=A,Class=IN,198.41.0.10
   ⊞ 📄 ADDITIONAL SECTION[8]:Type=A,Class=IN,193.0.14.129
   ⊞ 📄 ADDITIONAL SECTION[9]:Type=A,Class=IN,198.32.64.12
   ⊞ 📄 ADDITIONAL SECTION[10]:Type=A,Class=IN,202.12.27.33
   ⊞ 📄 ADDITIONAL SECTION[11]:Type=A,Class=IN,192.36.148.17
   ⊞ 📄 ADDITIONAL SECTION[12]:Type=A,Class=IN,192.203.230.10
   ⊞ 📄 ADDITIONAL SECTION[13]:Type=A,Class=IN,128.8.10.90
```

Figure 25-17:   *The root name server's response message, containing the RRs for all 13 root name servers*

Zone Transfer Messages

A zone transfer is initiated by a DNS server that functions as a slave for one or more zones whenever the server software is started. The process begins with an iterative query for an SOA resource record that the slave sends to the primary master to ensure that it is the best source for information about the zone (see Figure 25-18). The single Question section entry contains the name of the zone in the QNAME field and a value of 6 for the QTYPE field, indicating that the server is requesting the SOA resource record.

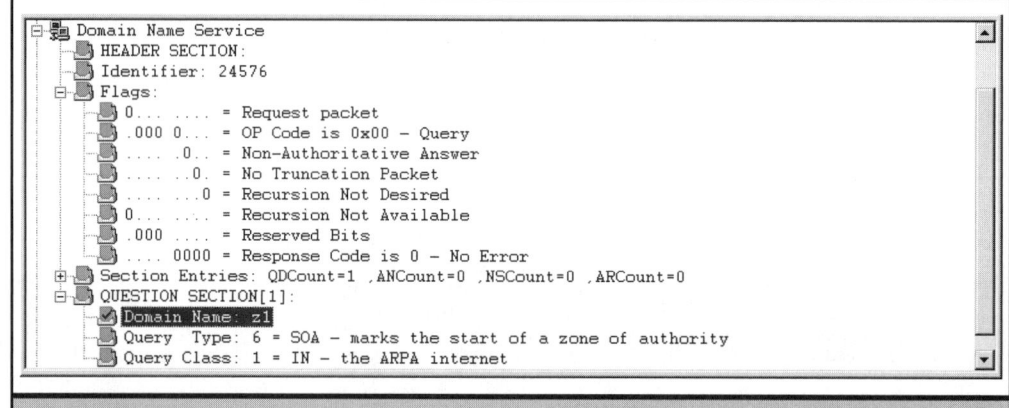

```
⊟ 🖳 Domain Name Service
    🗐 HEADER SECTION:
    🗐 Identifier: 24576
  ⊟ 🗐 Flags:
      🗐 0... .... = Request packet
      🗐 .000 0... = OP Code is 0x00 - Query
      🗐 .... .0.. = Non-Authoritative Answer
      🗐 .... ..0. = No Truncation Packet
      🗐 .... ...0 = Recursion Not Desired
      🗐 0... .... = Recursion Not Available
      🗐 .000 .... = Reserved Bits
      🗐 .... 0000 = Response Code is 0 - No Error
  ⊞ 🗐 Section Entries: QDCount=1 ,ANCount=0 ,NSCount=0 ,ARCount=0
  ⊟ 🗐 QUESTION SECTION[1]:
      ☑ Domain Name: z1
      🗐 Query  Type: 6 = SOA - marks the start of a zone of authority
      🗐 Query Class: 1 = IN - the ARPA internet
```

Figure 25-18: *The SOA query message generated by a slave server to determine if a zone transfer is warranted*

The primary master then replies to the slave with a response that includes the original Question section and a single Answer section containing the SOA resource record for the zone (see Figure 25-19). The slave uses the information in the response to verify the primary master's authority and to determine whether a zone transfer is needed. If the value of the SOA record's SERIAL field, as furnished by the primary master, is greater than the equivalent field on the slave server, then a zone transfer is required.

A zone transfer request is a standard DNS query message with a QTYPE value of 252, which corresponds to the AXFR type. AXFR is the abbreviation for a resource record set that consists of all of the records in the zone. However, in most cases, all of the resource records in the zone will not fit into a single UDP datagram. UDP is a connectionless, unreliable protocol in which there can be only one response message for each query, because the response message functions as the acknowledgment of the query. Because the primary master will almost certainly have to use multiple packets in order to send all of the resource records in the zone to the slave, a different protocol is needed. Therefore, before it transmits the zone transfer request message, the slave server initiates a TCP connection with the primary master using the standard three-way handshake. Once the connection is established, the slave transmits the AXFR query in a TCP packet, using port 53 (see Figure 25-20).

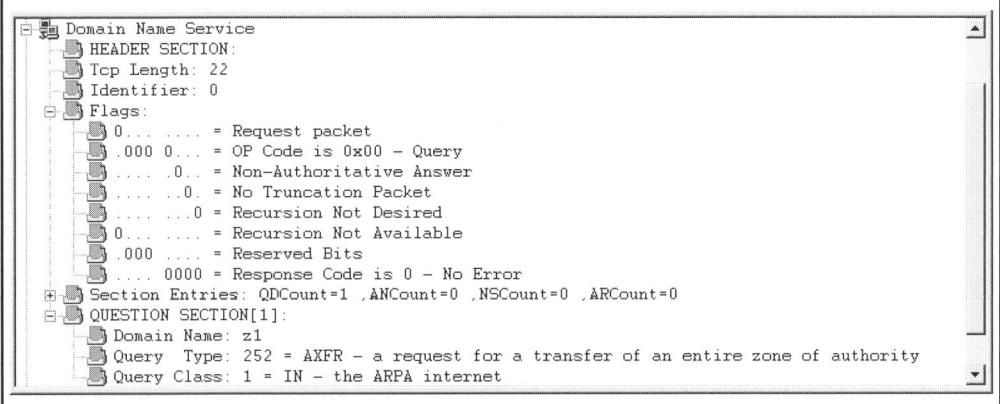

```
Domain Name Service
    HEADER SECTION:
    Identifier: 24576
    Flags:
        1... ... = Response packet
        .000 0... = OP Code is 0x00 - Query
        .... .1.. = Authoritative Answer
        .... ..0. = No Truncation Packet
        .... ...0 = Recursion Not Desired
        1... ... = Recursion Available
        .000 .... = Reserved Bits
        .... 0000 = Response Code is 0 - No Error
    Section Entries: QDCount=1 ,ANCount=1 ,NSCount=0 ,ARCount=0
    QUESTION SECTION[1]:Type=SOA,Class=IN,z1
    ANSWER SECTION[1]:
        Domain Name (w/Pointer): z1
        RR  Type: 6 = SOA - marks the start of a zone of authority
        RR Class: 1 = IN - the ARPA internet
        RR Time To Live: 3600 second(s)
        RR Data  Length: 52 Octet(s)
        Domain Name: cz1.zacker.com is a Name Server of original source
        Domain Name (w/Pointer): administrator.zacker.com is a mailbox of responsible person
        Serial Number: 0
        Refresh Time: 131072 second(s)
        Retry   Time: 235929600 second(s)
        Expire  Time: 39321601 second(s)
        Minimum  TTL: 20864 second(s)
```

Figure 25-19:   *The response message from the primary master server containing the SOA resource record*

```
Domain Name Service
    HEADER SECTION:
    Tcp Length: 22
    Identifier: 0
    Flags:
        0... ... = Request packet
        .000 0... = OP Code is 0x00 - Query
        .... .0.. = Non-Authoritative Answer
        .... ..0. = No Truncation Packet
        .... ...0 = Recursion Not Desired
        0... ... = Recursion Not Available
        .000 .... = Reserved Bits
        .... 0000 = Response Code is 0 - No Error
    Section Entries: QDCount=1 ,ANCount=0 ,NSCount=0 ,ARCount=0
    QUESTION SECTION[1]:
        Domain Name: z1
        Query  Type: 252 = AXFR - a request for a transfer of an entire zone of authority
        Query Class: 1 = IN - the ARPA internet
```

Figure 25-20:   *The AXFR query requesting a zone transfer, transmitted to the primary master server using a TCP connection*

In response to the query, the primary master server transmits all of the resource records in the requested zone as entries in the Answer section, as shown in Figure 25-21. Once all of the data has been transmitted, the two systems terminate the TCP connection in the usual manner, and the zone transfer is completed.

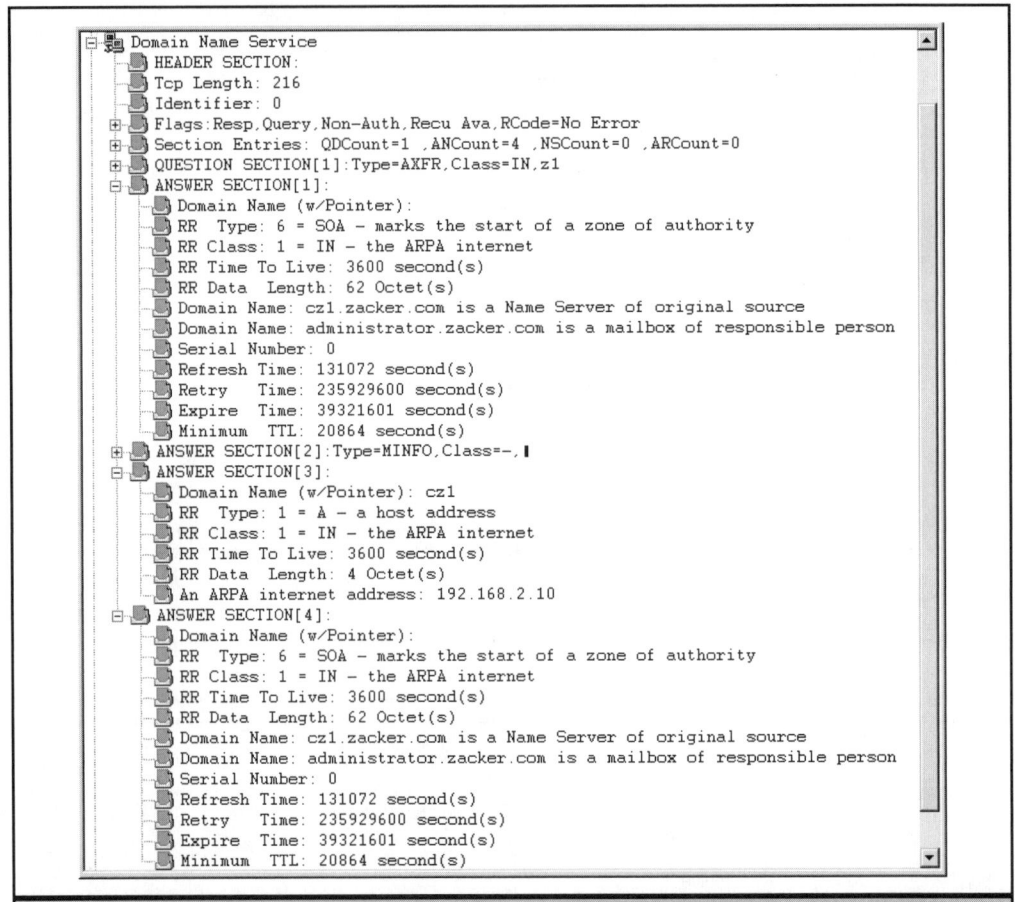

```
Domain Name Service
  HEADER SECTION:
    Tcp Length: 216
    Identifier: 0
  Flags:Resp,Query,Non-Auth,Recu Ava,RCode=No Error
  Section Entries: QDCount=1 ,ANCount=4 ,NSCount=0 ,ARCount=0
  QUESTION SECTION[1]:Type=AXFR,Class=IN,z1
  ANSWER SECTION[1]:
    Domain Name (w/Pointer):
    RR  Type: 6 = SOA - marks the start of a zone of authority
    RR Class: 1 = IN - the ARPA internet
    RR Time To Live: 3600 second(s)
    RR Data  Length: 62 Octet(s)
    Domain Name: cz1.zacker.com is a Name Server of original source
    Domain Name: administrator.zacker.com is a mailbox of responsible person
    Serial Number: 0
    Refresh Time: 131072 second(s)
    Retry   Time: 235929600 second(s)
    Expire  Time: 39321601 second(s)
    Minimum  TTL: 20864 second(s)
  ANSWER SECTION[2]:Type=MINFO,Class=-, ▌
  ANSWER SECTION[3]:
    Domain Name (w/Pointer): cz1
    RR  Type: 1 = A - a host address
    RR Class: 1 = IN - the ARPA internet
    RR Time To Live: 3600 second(s)
    RR Data  Length: 4 Octet(s)
    An ARPA internet address: 192.168.2.10
  ANSWER SECTION[4]:
    Domain Name (w/Pointer):
    RR  Type: 6 = SOA - marks the start of a zone of authority
    RR Class: 1 = IN - the ARPA internet
    RR Time To Live: 3600 second(s)
    RR Data  Length: 62 Octet(s)
    Domain Name: cz1.zacker.com is a Name Server of original source
    Domain Name: administrator.zacker.com is a mailbox of responsible person
    Serial Number: 0
    Refresh Time: 131072 second(s)
    Retry   Time: 235929600 second(s)
    Expire  Time: 39321601 second(s)
    Minimum  TTL: 20864 second(s)
```

Figure 25-21: *One packet from a zone transfer transmitted by the primary master server*

Obtaining DNS Services

On a private network that is connected to the Internet, you still can use your ISP's DNS servers, or you can run your own. If you are not running your own domain, the network only needs the client capabilities of the DNS servers to resolve names into addresses. An organization that is hosting an Internet domain must have DNS servers to function as the authority for that domain. NIS and other domain registrars require addresses for two DNS servers, for fault-tolerance purposes. Again, even if you are hosting a domain, you can still use your ISP's DNS servers. However, there will usually be an additional fee for domain hosting, whereas client access to the servers is virtually always included in the Internet access fee.

Outsourcing DNS Services

The advantages of using the ISP's DNS servers are convenience, accessibility, and fault tolerance. Having the ISP maintain the DNS records for your domain is certainly easier than doing it yourself, and a reliable ISP will have redundant, high-speed Internet connections that ensure your domain's continuous availability to Internet users. A good ISP should also have fault-tolerance mechanisms available to keep its servers running, such as backup power supplies and redundant disk arrays.

The disadvantages of outsourcing your domain hosting chores are the expense and maintenance delays that can result. To make a change in your DNS configuration, you have to call your ISP, who may or may not perform the requested task in a timely manner. In addition, some ISPs charge you for each change that you make to the DNS records, in addition to a monthly hosting fee. This means that every new host you want to add to your domain will cost you money.

Another thing to watch out for are ISPs that offer complete domain hosting packages in which they fill out the forms and send them to the domain registrar, all for one price that includes the registration fee. This can be a nice convenience, but some ISPs supply themselves as all three of the contacts required by the registrar. Only the three people designated as the billing, technical, and administrative contacts are permitted to modify the domain registration information. If you ever decide to change ISPs, the domain records must be changed to reflect different DNS servers, and your old provider is not likely to do this for you after you've dumped them for a competitor. You may, therefore, lose control of the domain name that you've paid for.

To prevent this from happening, you should always register the domain name yourself. NSI provides online forms at its Web site that make the process quite easy, and you can charge the fee to a credit card. Be sure to supply contact names of people who you know will be involved in the organization for some time to come.

Running Your Own DNS Servers

The practicality of running DNS servers on your own network depends on several factors, including which DNS server implementation you will use, how large your network is, and how much maintenance is required. On a Unix network, for example, all of the workstations usually have to be registered in the DNS server's database, while a Windows or NetWare network might need only a few registered hosts.

Most of the DNS servers on the Internet today run a variation of Unix, Windows NT, or Windows 2000. Novell also has a DNS server product that runs on a NetWare server. All of these servers are compliant with the core DNS standards and provide basically the same functions in the same way, but with different interfaces. Windows NT and 2000 are by far the easiest to install, configure, and maintain, because they provide a graphical interface to the server's database files. The DNS Manager in Windows NT Server 4.0 and the DNS snap-in for Microsoft Management Console in Windows 2000 Server both enable you to view the information in the DNS database and in the cache (see Figure 25-22). Creating and modifying resource records and configuring server properties is a simple matter of navigating standard Windows dialog boxes and controls.

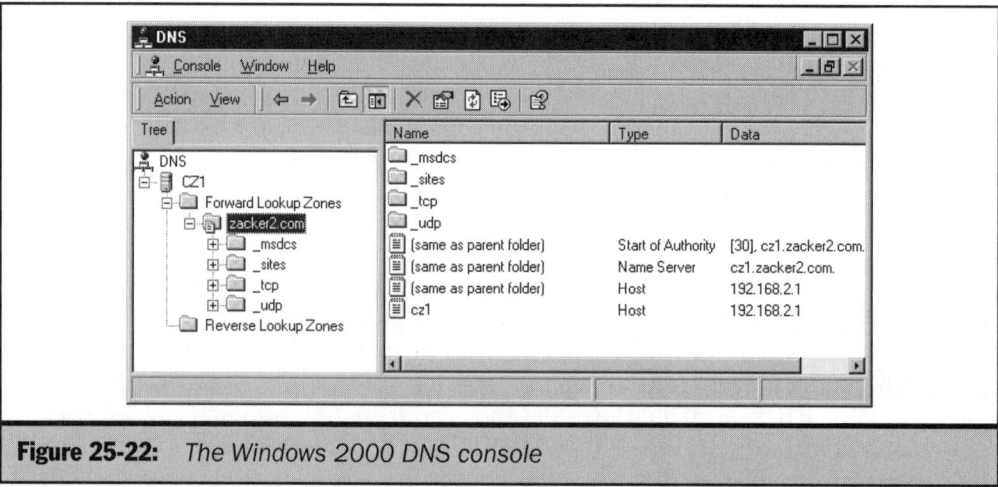

Figure 25-22:   *The Windows 2000 DNS console*

As is usually the case, the DNS servers included with most Unix variants are powerful, flexible, and far less user friendly than their Windows counterparts. The most popular of these is BIND (Berkeley Internet Name Domain), which was written for use with BSD Unix and has since been ported to many other operating systems, including Windows. As with most Unix programs, BIND does not have a fancy interface. To configure the server or add and modify resource records, you must edit the program's database files directly, the syntax for which can be rather cryptic. Virtually anyone can create a new resource record in a Windows DNS server, but working with BIND is more difficult and less forgiving of mistakes.

Despite its relative ease of use, however, the NT/2000 DNS servers remain compatible with the Unix implementations because they store most of their data in the same database files that BIND uses. In fact, the primary function of the Microsoft DNS Manager utility and the Windows 2000 DNS console is to provide a graphical editor for these database files.

The usability of NetWare's DNS server falls somewhere between that of Windows and Unix. You configure the program and create resource records using UNICON, a character-based, menu-driven utility that runs on the NetWare server.

With one important exception, you can use any of these DNS server implementations on your network, or even a combination of them. As long as a DNS server is compliant with the RFC 1034 and RFC 1035 standards, it will be able to interact with another compliant server. Therefore, you can run a primary master server on Windows and a slave on Unix if you wish, or any other combination.

The exception is when you are running a Windows 2000 network that uses Active Directory. Active Directory relies heavily on DNS, and the DNS server included with the Windows 2000 Server products is compliant with RFC 2052, which implements the SVR resource record type, and RFC 2136, which provides the Update message. Active Directory needs these additional features in order to register its domain controllers in the DNS database and keep the information current. Therefore, you should stick to Windows 2000 DNS servers if you are using Active Directory, unless you are certain that the implementation you plan to use supports these features.

The
Complete
Reference

Networking

Part VI

Network Services

The Complete Reference

Networking

Chapter 26

Internet Services

839

A t one time, the term "server" in computer networking was nearly always used in the phrase "file server," referring to a PC running a network operating system (NOS), such as Novell NetWare, that enables users to access shared files and printers. However, the rapid growth of the Internet and the rise of Windows NT/2000 as the most popular NOSes on the market have changed the common meaning of the term. To average Internet users, servers are the invisible systems that host World Wide Web and FTP sites, or that enable them to send and receive e-mail. For LAN users, servers still fill the traditional file and printer sharing roles, but also provide application-related functions, such as access to databases. Thus, people are gradually learning that a *server* is a software, not a hardware, entity, and that a single computer can actually function in multiple server roles simultaneously.

Internet servers are software products that provide traditional Internet services to clients, whether or not they are actually connected through the Internet. Web, FTP, and e-mail are all services that can be as useful on a LAN as on the Internet, and many sites are deploying them for a variety of reasons. This chapter examines the technology behind these services and the procedures for implementing them on your network.

Web Servers

The Web is rapidly becoming a ubiquitous tool for business, education, and recreation. It seems as though virtually every business worth speaking of has a Web site, and the components needed to host and access Web sites are integrated into most operating systems. The basic building blocks of the Web are as follows:

Web servers   Computers running a software program that processes resource requests from clients

Browsers   Client software that generates resource requests and sends them to Web servers

Hypertext Transfer Protocol (HTTP)   The TCP/IP application-layer protocol that servers and browsers use to communicate

Hypertext Markup Language (HTML)   The markup language used to create Web pages

Why Run Your Own Web Servers?

Most organizations and individuals with Web sites use an Internet service provider (ISP) or an outside service to host their sites. Many ISPs offer subscribers a free site with a subscription to their service, and organizations with more complex sites often do business with a Web hosting company that can provide a wider range of services. However, there are several good reasons why you might want to run one or more Web servers on your own network.

The first of these reasons is for Web site development. If you create your own sites, there is no substitute for testing and debugging on a live Web server. Even though you can open the HTML files containing your Web pages directly from a local drive, getting the links between pages to function properly requires a server. In addition, if you intend to use scripts, applications, or other advanced technologies on your site, you need a server to test them.

Another reason for running your own Web server is to build an intranet. The term *intranet* is defined as a TCP/IP-based network with access restricted to a specific group of users, but in more common usage, it refers to a Web site running on a private network. An intranet is a tool with which you can publish information for your users in many forms. Standard Web pages can reproduce documents in an easily readable form, but you can also use a Web server to create a library of the files stored on your network.

Shared network drives are an excellent way to provide users with access to files, but it can often be difficult for users to locate the exact information they need. A simple text-based intranet Web site can list the files that are available, along with brief descriptions, and enable users to download them to their workstations using standard hyperlinks. Because an intranet is a read-only medium when used in this way, there is no danger of the files being accidentally modified or deleted. In addition to basic document and file access, an intranet can serve as a development platform for Web-based applications that provide access to databases and other resources using a browser as a client.

This discussion also leads to the question of whether, if you have a Web server on your network, you should host your own site on the Web. The Web server software included with your operating system is certainly capable of hosting an Internet Web site, but there are several other factors that you must consider before you do so. The Web server is only a small part of running a Web site. In addition to the server itself, you must consider the bandwidth that the Web traffic will consume, the ability of your network infrastructure to tolerate hardware or service faults, and the security of your network. Depending on the nature of your site and your business, the first two of these considerations may be minimized, but security is always a concern.

To run a successful and professional Internet Web site, your network must be capable of coping with the traffic generated by constant access from users, and your server must be continuously available. If, for example, the users on your network access the Internet using the same connection as the Web server, a period of heavy internal use can reduce the bandwidth available to the server. You should make sure that you have either a separate Internet connection for the Web server or sufficient bandwidth to support both your internal users and the Internet server traffic at the same time.

Web hosting services typically have large amounts of bandwidth available for their servers, and often provide redundant Internet connections as well. Nothing turns a potential customer off more than an exceedingly slow or unavailable Web site. The impression given is that the company has not made a sufficient commitment to the Internet or that it doesn't care, either of which is worse than having no Web site at all. Hosting services often have fault-tolerance mechanisms for their Web servers as well, such as mirrored drives or RAID arrays, redundant power supplies, and other technologies. There's no reason why you can't have these features as well, but many businesses are not ready to spend an exorbitant sum of money to run a Web server.

Security is another major issue. A computer that is accessible from the Internet also provides a gateway to your network for potential intruders. Most LANs use unregistered IP addresses for their workstations, which makes them invisible to the Internet, but the address of an Internet server must be registered in order for outside users to connect to it. Most administrators who run their own Web servers place them on the "edge" of the network—that is, on a LAN separate from the unregistered workstations—and use a firewall product to prevent outsiders from accessing the servers using anything other than the prescribed protocol and port.

Selecting a Web Server

A Web server is actually a rather simple device. When you see complex pages full of fancy text and graphics on your monitor, you're actually seeing something that is more the product of the page designer and the browser technology than of the Web server. In its simplest form, a Web server is a software program that processes requests for specific files from browsers and delivers those files back to the browser. The server does not read the contents of the files nor does it participate in the rendering process that controls how a Web page is displayed in the browser. The differences between Web server products are in the additional features they provide and their ability to handle large numbers of requests.

Web Server Platforms

A Web server does not have to be a big, expensive machine. Depending on your requirements and those of your users, you can run a Web server on any PC, from a basic Windows 98 system to a high-end Unix workstation. Many of the most popular Web server software products are free, such as Microsoft's Internet Information Server (IIS), which is included with Windows 2000 and NT, and the Apache Web server for Unix and Windows, while others can be quite expensive. The computer platform on which you choose to run your server generally does not affect connectivity with clients. Any browser can connect to any server, although certain site-development technologies are platform specific.

Active Server Pages, for example, is a technology created by Microsoft that (not surprisingly) requires the use of a Windows 2000 or NT system to host the site, although any browser can display the resulting pages. Microsoft Front Page is a WYSIWYG Web page editor that requires a Windows system to run, but that can upload pages to a Web server running on virtually any platform, using server extension modules provided with the product.

Web server products are available for all PC platforms, including Windows 9*x*, Windows 2000/NT, NetWare, and Unix. Windows 9*x* as a Web server platform is suitable for development purposes and light internal use on a LAN, but generally isn't suitable for use as an Internet server. The Windows 98 operating system includes the Personal Web Server product, which provides basic capabilities that are quite serviceable for personal use. In some organizations, individual users or departments use this combination to create small intranet sites that improve communications between colleagues.

Internet Information Server is Microsoft's full-featured Web server product, which, like all Microsoft Web servers, is free. Windows 2000 (Server and Professional) include IIS version 5. The original Windows NT Server 4.0 release includes IIS version 2, but the latest version, IIS 4, is available as part of the Windows NT Server Option Pack (see http://www.microsoft.com/ntserver/nts/downloads/recommended/NT4OptPk/default.asp). IIS on Windows 2000 or NT is one of the most popular Web server platforms in use today. Windows provides the flexibility and the security needed to support a large number of users, and IIS includes a large collection of additional features and services that enable you to create complex, cutting-edge Web sites.

Windows has grown rapidly as a Web server platform, largely due to its ease of installation and administration. All of the Microsoft Web server products include a graphical configuration utility. However, Unix was the original Web server platform, and it is still used by the majority of sites on the Web. By far the most popular Unix Web server is Apache, a public-domain product based on the original httpd server created by NCSA (National Computational Science Alliance). The name "Apache" is derived from the fact that the server uses the httpd code plus a collection of patches, thus making it *a patchy server*.

Apache is one of the best examples of open source software. The Apache source code is freely available to anyone (see http://www.apache.org/) and maintained by an informal group of programmers, most of whom have never met in person, who communicate using e-mail and newsgroups. Many add-on modules are available for the core server, providing more advanced features, such as support for various types of scripts and authentication options. Programmers who create new modules or modify existing ones routinely upload their work to public servers for use by others.

Because Apache is not a commercial product and its source code is freely available, it is far more flexible than any Windows Web server. In the Unix world, the general rule is that if you don't like how a piece of software runs, go ahead and change it, and then make your modifications available to others. However, there are also compiled versions of the server available for over 20 Unix variants, as well as Windows and Macintosh, so even a nonprogrammer can get the software up and running. Although not as user friendly as Windows, Unix is generally thought to be a more stable server platform with excellent security, and the various Unix operating systems can run on everything from standard Intel PCs to hugely powerful workstations.

Although Unix and Windows systems account for the vast majority of Web servers, there are products for other platforms, such as NetWare and Macintosh. Novell's intraNetWare was the first version of the operating system to include the Novell Web Server product, but this was later replaced by Netscape FastTrack Web Server, which is included with NetWare version 4.2, and NetWare Enterprise Web Server, included with version 5.*x*. Macintosh systems do not include a Web server of their own, but there are third-party products available, like StarNine Software's WebStar.

Web Server Functions

A Web server is a program that runs in the background on a computer and listens on a particular TCP port for incoming requests. A program of this type has different names on various operating systems. In Windows, it's a *service*; on a Unix system, it's a *daemon*; on NetWare, it's a *NetWare Loadable Module (NLM)*. The standard TCP port for an HTTP server is 80, although most servers enable you to specify a different port number for a site and may use a second port number for the server's administrative interface. To access a Web server using a different port, you must specify that port number as part of the URL.

Uniform Resource Locators   The format of the Uniform Resource Locator (URL) that you type into a browser's Address field to access a particular Web site is defined in the RFC 1738 document published by the Internet Engineering Task Force (IETF). A URL consists of four elements that identify the resource that you want to access:

Protocol   Specifies the application-layer protocol that the browser will use to connect to the server. The values defined in the URL standard are as follows (others have been defined by additional standards published since RFC 1738):

http   Hypertext Transfer Protocol

ftp   File Transfer Protocol

gopher   Gopher protocol

mailto   E-mail address

news   Usenet news

nntp   Usenet news using NNTP access

telnet   Reference to interactive sessions

wais   Wide Area Information Servers

file   Host-specific file names

prospero   Prospero Directory Service

Server name Specifies the DNS name or IP address of the server.

Port number Specifies the port number that the server is monitoring for incoming traffic.

Directory and file Identifies the location of the file that the server should send to the browser.

The format of a URL is as follows:

protocol://name:port/directory/file.html

Most of the time, users do not specify the protocol, port, directory, and file in their URLs, and the browser uses its default values. When you enter just a DNS name, such as www.zacker.com, the browser assumes the use of the HTTP protocol, port 80, and the Web server's home directory. Fully expanded, this URL would appear something like the following:

```
http://www.zacker.com:80/index.html
```

The only element that could vary among different servers is the file name of the default Web page, here shown as index.html. The default file name is configured on each server, and specifies the file that the server will send to a client when no file name is specified in the URL. The traditional default file name for Unix systems is index.html; for Microsoft Web servers, it is default.htm.

If you configure a Web server to use a port other than 80 to host a site, users must specify the port number as part of the URL. Most Web users don't even know that port numbers exist, so the use of nonstandard ports is relatively rare. The main exception to this is when the administrator wants to create a site that is hidden from the average user. Some Web server products, for example, are configurable using a Web browser, and the server creates a separate administrative site containing the configuration controls for the program. During the software installation, the program prompts the administrator for a port number that it should use for the administrative site. Thus, specifying the name of the server on a browser opens the default site on port 80, but specifying the server name with the selected port accesses the administrative site.

The use of a nonstandard port is not really a security measure, because there are programs available that can identify the ports that a Web server is using. The administrative site for a server usually has security in the form of user authentication as well; the port number is just a means of keeping the site hidden from curious users.

CGI Most of the traffic generated by the Web travels from the Web server to the browser. The upstream traffic from browser to server consists mainly of HTTP requests for specific files. However, there are mechanisms by which browsers can send other types of information to servers. The server can then feed the information to an application for processing. The Common Gateway Interface (CGI) is the most widely supported mechanism of this type. In most cases, the user supplies information in a form built into a Web page using standard HTML tags, and then submits the form to a server. The server, on receiving the data from the browser, executes a CGI script that defines how the information should be used. The server might feed the information as a query to a database server, use it to perform an online financial transaction, or use it for any other purpose.

Logging   Virtually all Web servers have the capability to maintain logs that track all client access to the site and any errors that have occurred. The logs typically take the form of a text file, with each server access request or error appearing on a separate line. Each line contains multiple fields, separated by spaces or commas. The information logged by the server identifies who accessed the site and when, as well as the exact documents sent to the client by the server. While it is possible for administrators to examine the logs in their raw format and learn some things about the site's usage, such as which pages receive the most hits, there are a number of third-party statistics programs, such as WebTrends (http://www.webtrends.com/), that take the log files as input and produce detailed reports illustrating trends and traffic patterns.

Most Web servers enable the administrator to choose among several formats for the logs they keep. Some servers use proprietary log formats, which generally are not supported by the statistics programs, while other servers may also be able to log server information to an external database using an interface like ODBC. Most servers, however, support the Common Log File format defined by NCSA. This format consists of nothing but one-line entries with fields separated by spaces. The format for each Common Log File entry and the functions of each field are as follows:

```
remotehost logname username date request status bytes
```

remotehost   Specifies the IP address of the remote client system. Some servers also include a DNS reverse lookup feature that resolves the address into a DNS name for logging purposes.

logname   Specifies the remote log name of the user at the client system. Most of today's browsers do not supply this information, so the field in the log is filled with a placeholder, such as a dash.

username   Specifies the user name with which the client was authenticated to the server.

date   Specifies the date and time that the request was received by the server. Most servers use the local date and time by default, but may include a Greenwich Mean Time differential, such as –0500 for U.S. Eastern Standard Time.

request   Specifies the text of the request received by the server.

status   Contains one of the status codes defined in the HTTP standard that specifies whether the request was processed successfully, and if not, why.

bytes   Specifies the size (in bytes) of the file transmitted to the client by the server in response to the request.

There is also a log file format created by the World Wide Web Consortium (W3C), called the Extended Log File format, that addresses some of the inherent problems of the Common Log File format, such as difficulties in interpreting logged data due to spaces within fields. The Extended Log File provides an extendable format with which administrators can specify the information to be logged or information that shouldn't be logged. The format for the Extended Log File consists of *fields,* as well as *entries.* Fields appear on separate lines, beginning with the # symbol, and specify information about the data contained in the log. The valid field entries are as follows:

#Version: *integer.integer*   Specifies the version of the log file format. This field is required in every log file.

#Fields: [*specifiers*]   Identifies the type of data carried in each field of a log entry, using abbreviations specified in the Extended Log File format specification. This field is required in every log file.

#Software *string*   Identifies the server software that created the log.

#Start-Date: *date time*   Specifies the date and time that logging started.

#End-Date: *date time*   Specifies the date and time that logging ceased.

#Date: *date time*   Specifies the date and time at which a particular entry was added to the log file.

#Remark: *text*   Contains comment information that should be ignored by all processes.

These fields enable administrators to specify the information to be recorded in the log while making it possible for statistics programs to correctly parse the data in the log entries.

Remote Administration   All Web servers need some sort of administrative interface that you can use to configure their operational parameters. Even a no-frills server lets you define a home directory that should function as the root of the site, and other basic features. Some server products include a program that you can run on the computer that provides this interface, but many products have taken the opportunity to include an administrative Web site with the product. With a site like this, you can configure the server from any computer, using a standard Web browser (see Figure 26-1). This is a convenient tool for the network administrator, especially when the Web server system is located in a server closet or other remote location, or when one person is responsible for maintaining several different servers.

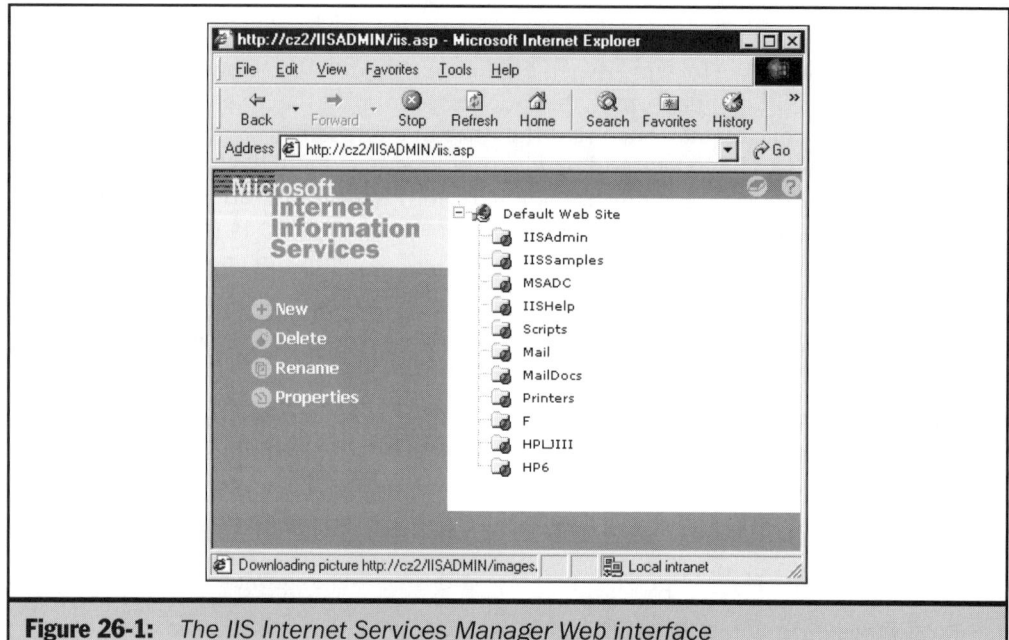

Figure 26-1:   *The IIS Internet Services Manager Web interface*

The biggest problem with this form of remote administration is security, but there are several mechanisms that can prevent unauthorized users from modifying the server configuration. The most basic of these mechanisms, as mentioned earlier, is the use of a nonstandard port number for the administrative site. Servers that use nonstandard ports typically require that you specify the port number during the server installation.

A second method is to include a means by which you can specify the IP addresses of the only systems that are to be permitted access to the administrative interface. IIS includes this method, and, by default, the only system that can access the Web-based interface is the one on which the server is installed (see Figure 26-2). However, you can open up the server to remote administration by specifying the addresses of other workstations to be granted access or by opening up the server to free access and specifying the addresses of systems that are to be denied.

The third method is the use of a directory service to specify which users are permitted to configure the system. IIS uses the accounts in standard Windows NT domains or in Active Directory for this purpose. Netscape provides its own directory server for access control, and Unix systems can use any one of several authentication protocols.

NETWORK SERVICES

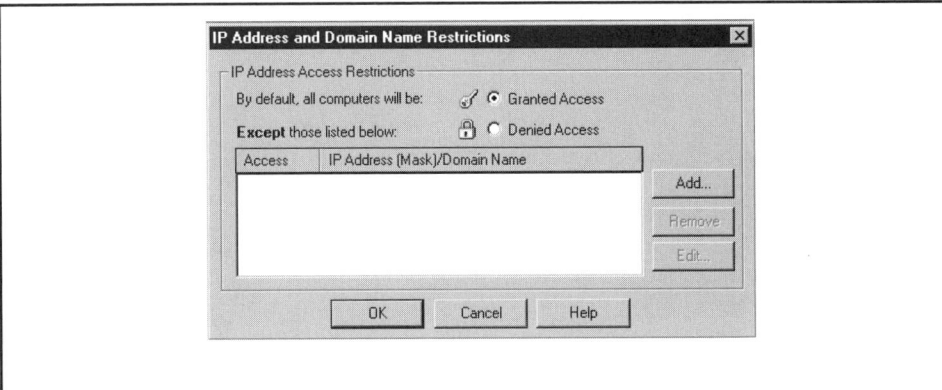

Figure 26-2:   *IIS's Internet Services Manager enables you to grant or deny Web site access to specific computers.*

Virtual Directories   A Web server utilizes a directory on the computer's local drive as the home directory for the Web site it hosts. The server transmits the default file name in that directory to clients when they access the site using a URL that consists only of a DNS name or IP address. Subdirectories beneath that directory also appear as subdirectories on the Web site. IIS, for example, uses the C:\InetPub\wwwroot directory as the default home directory for its Web site. If that Web server is registered in the DNS with the name www.zacker.com, the default page displayed by a browser accessing that site will be the default.htm file in the wwwroot directory. A file in the C:\InetPub\wwwroot\docs directory on the server will, therefore, appear on the site in http://www.zacker.com/docs.

Using this system, all the files and directories that are to appear on the Web site must be located beneath the home directory. However, this is not a convenient arrangement for every site. On an intranet, for example, administrators may want to publish documents in existing directories using a Web server without moving them to the home directory. To make this possible, some server products enable you to create virtual directories on the site. A *virtual directory* is a directory at another location—elsewhere on the drive, on another drive, or sometimes even on another computer's shared drive—that is published on a Web site using an alias. The administrator specifies the location of the directory and the alias under which it will appear on the site. The alias functions as a subdirectory on the site that users can access in the normal manner, and contains the files and subdirectories from the other drive.

Multiple Site Support   A basic Web server product like Microsoft's Personal Web Server enables a computer to host a single Web site, but more advanced products like IIS can host multiple sites at the same time without running multiple copies of the server program. There are several different methods that a single computer can use to host multiple sites, and high-end servers usually give you a choice between them. These methods are as follows:

Different port numbers   When hosting sites using the same IP address but different ports, users must specify a port number in the URL, except for the default site using port 80. This is rarely done, except in cases where an administrator wants to keep a site hidden, because most Internet users are unaware of the use of port numbers, and even if users are aware, there is no easy way of knowing which one to use.

Different IP addresses   It is possible to assign multiple IP addresses to a single network interface, and once you do this, some Web servers enable you to select a different address for each Web site. You can then register each address in the DNS with a different host name, making the sites appear to be completely independent to users, even though they are running on the same computer. This is a common method for hosting multiple sites, the only drawback being the need for a separate registered IP address for each site.

Different DNS names   Some servers can host multiple sites with different DNS names but only one IP address, using a technique called *virtual hosting*. In this method, each site has its own DNS name, but all of the names are registered in the DNS with the same IP address. When the Web server receives HTTP requests at that address, it uses the contents of the Host field in the message header to determine which site should receive the request. Even though the DNS name specified by the client in the URL is resolved into an IP address before the request is transmitted, the DNS name is retained in the Host field. The advantage of this method is that only one registered IP address is required for the server, but it is also impossible to reach a site by specifying only the IP address in the browser.

Security   Many Web sites have no need for security measures at all beyond those that protect the computer running the server software from outside intrusion. However, two forms of security are needed in some cases: the need to restrict site access to selected users, and the need to protect the data being transmitted between clients and servers.

The first form of security usually takes the form of an authentication protocol that restricts access to specific users with valid passwords. Just as you can protect your server's administration interface by restricting access to selected users, you can protect your Web sites as well. This capability is used more on intranets than on Internet sites, because access to certain documents (such as financial data) may have to be restricted to certain employees. However, some organizations are implementing *extranets,* intranet Web sites that are accessible to selected outside entities via the Internet. Companies use extranets to grant their clients access to information about products and services and to enable them to perform tasks like checking inventory and placing orders. Some Web servers, such as Apache, maintain their own directories of user accounts, while others use the directory service provided by the operating system or one running as a separate service.

The other form of security implemented by Web servers enables clients to send and receive data over the Internet using a secure channel that can't easily be intercepted or penetrated by outsiders. The most common application for this type of security is *e-commerce*, the purchasing of products and services over the Web. Securing communications between a client browser and Web server is a matter of authenticating the systems so that the client can be sure of communicating with the correct server, and encrypting the data transmitted between the systems so that anyone intercepting the transmissions will not be able to read them.

The standard protocol for Web server authentication and encryption is Secure Sockets Layer (SSL), a public key encryption protocol developed by Netscape and later submitted to the IETF for standardization. To initiate an SSL connection, the Web site developer creates a hyperlink that uses https as the protocol identifier instead of http, as follows:

```
https://www.zacker.com/
```

Specifying https as the protocol initiates the use of the SSL Handshake Protocol between the server and the client using port 443 instead of the standard HTTP port 80. The two systems first perform a certificate negotiation to discover the strongest encryption method that they have in common, and the server sends a *digital certificate* to the client to verify its identity. Digital certificates are encrypted data files provided by trusted third-party companies, called *certificate authorities (CAs)*, that uniquely identify a particular company or organization. A client receiving such a certificate from a server can be reasonably sure that the server is actually affiliated with that company or organization. It's also possible for a client to accept an "untrusted" certificate and initiate an SSL session without a certificate signed by a CA.

To read the certificate, the client must possess the public key for the certificate authority that generated the server's digital certificate. The two most common CAs, VeriSign and Thawte, are supported by most browsers. Using the CA's public key, the browser decrypts the certificate and extracts the public key for the server. The server, meanwhile, encrypts the file requested by the server using the SSL Record Protocol and transmits it to the client. The client can then transmit information (such as an order form containing a credit card number, for example) to the server, using the same key. Only a system possessing that key can decrypt the data.

Support for the SSL protocol is provided in the Internet Explorer and Netscape Navigator Web browsers and in most servers, including IIS. Apache servers can use OpenSSL, an open source SSL toolkit available at http://www.openssl.org/.

HTML

HTML, the Hypertext Markup Language, is the *lingua franca* of the Web, but it actually has very little to do with the functions of a Web server. Web servers are programs that deliver requested files to clients. The fact that most of these files contain HTML code is immaterial, because the server does not read them. The only way in which they affect the server's functions is when the client parses the HTML code and requests additional files from the server that are needed to display the Web page in the browser, such as image files. Even in this case, however, the image file requests are just additional requests to the server.

HTTP

Communication between Web servers and their browser clients is provided by an application-layer protocol called the Hypertext Transfer Protocol. The most current version of the HTTP specification, version 1.1, was published by the IETF as RFC 2616 in June 1999. HTTP is a relatively simple protocol that takes advantage of the services provided by the TCP protocol at the transport layer to transfer files from servers to clients. When a client connects to a Web server by typing a URL in a browser or clicking a hyperlink, the system generates an HTTP request message and transmits it to the server. This is an application-layer process, but before it can happen, communication at the lower layers must be established.

Unless the user or the hyperlink specifies the IP address of the Web server, the first step in establishing the connection between the two systems is to discover the address by sending a name resolution request to a DNS server. This address makes it possible for the IP protocol to address traffic to the server. Once the client system knows the address, it establishes a TCP connection with the server's port 80, using the standard three-way handshake process defined by that protocol.

Note *For more information on DNS name resolution, see Chapter 25. For more information on the TCP three-way handshake, see Chapter 13.*

Once the TCP connection is established, the browser and the server can exchange HTTP messages. HTTP consists of only two message types, requests and responses. Unlike the messages of most other protocols, HTTP messages take the form of ASCII text strings, not the typical headers with discrete coded fields. In fact, you can connect to a Web server with a Telnet client and request a file by feeding an HTTP command directly to the server. The server will reply with the file you requested in its raw ASCII form.

Each HTTP message consists of the following elements:

Start line   Contains a request command or a reply status indicator, plus a series of variables

Headers [optional]   Contains a series of zero or more fields containing information about the message or the system sending it

Empty line   Contains a blank line that identifies the end of the header section

Message body [optional]   Contains the payload being transmitted to the other system

HTTP Requests

The start line for all HTTP requests is structured as follows:

```
RequestType RequestURI HTTPVersion
```

Version 1.1 of the HTTP standard defines seven types of request messages, which use the following values for the *RequestType* variable:

GET   Contains a request for information specified by the *RequestURI* variable. This type of request accounts for the vast majority of request messages.

HEAD   Functionally identical to the GET request, except that the reply should contain only a start line and headers; no message body should be included.

POST   Requests that the information included in the message body be accepted by the destination system as a new subordinate to the resource specified by the *RequestURI* variable.

OPTIONS   Contains a request for information about the communication options available on the request/response chain specified by the *RequestURI* variable.

PUT   Requests that the information included in the message body be stored at the destination system in the location specified by the *RequestURI* variable.

DELETE   Requests that the destination system delete the resource identified by the *RequestURI* variable.

TRACE   Requests that the destination system perform an application-layer loopback of the incoming message and return it to the sender.

CONNECT   Reserved for use with proxy servers that provide SSL tunneling.

The HTTP 1.0 specification (RFC 1945) defines only the GET, HEAD, and POST request types. In addition to the five new request types, version 1.1 also enables a client to establish a persistent connection to a server, so that multiple files can be transmitted during one TCP session. Version 1.0 requires a separate TCP connection for each file. The new specification also provides better cache management at the browser, which improves the overall client response time. Both the client and the server must support the 1.1 standard in order to use these new features.

The *RequestURI* variable contains a *Uniform Resource Identifier (URI)*, a text string that uniquely identifies a particular resource on the destination system. In most cases, this variable contains the name of a file on a Web server that the client wants the server to send to it, or the name of a directory from which the server should send the default file. The *HTTPVersion* variable identifies the version of the HTTP protocol that is supported by the system generating the request. Currently, the three possible values for this variable are as follows:

- HTTP/0.9
- HTTP/1.0
- HTTP/1.1

Thus, when a user types the name of a Web site into a browser, the request message generated contains a start line that appears as follows:

```
GET / HTTP/1.1
```

The GET command requests that the server send a file. The use of the forward slash as the value for the *RequestURI* variable represents the root of the Web site, so the server will respond by sending the default file located in the server's home directory.

HTTP Headers

Following the start line, any HTTP message can include a series of headers, which are text strings formatted in the following manner:

```
FieldName: FieldValue
```

where the *FieldName* variable identifies the type of information carried in the header, and the *FieldValue* variable contains the information itself. The various headers mostly provide information about the system sending the message and the nature of the request, which the server may or may not use when formatting the reply. The number, choice, and order of the headers included in a message are left to the client implementation, but the HTTP specification recommends that they be ordered using four basic categories. The possible values for the *FieldName* variable defined in the HTTP 1.1 specification are as listed in the following sections, by category.

General Header Fields   General headers apply to both request and response messages, but do not apply to the entity (that is, the file or other information in the body of the message). The general header *FieldName* values are as follows:

Cache-Control   Contains directives to be obeyed by caching mechanisms at the destination system

Connection   Specifies options desired for the current connection, such that it be kept alive for use with multiple requests

Date   Specifies the date and time that the message was generated

Pragma   Specifies directives that are specific to the client or server implementation

Trailer   Indicates that specific header fields are present in the trailer of a message encoded with chunked transfer-coding

Transfer-Encoding   Specifies what type of transformation (if any) has been applied to the message body in order to safely transmit it to the destination

Upgrade   Specifies additional communication protocols supported by the client

Via   Identifies the gateway and proxy servers between the client and the server and the protocols they use

Warning   Contains additional information about the status or transformation of a message

Request Header Fields   Request headers apply only to request messages, and supply information about the request and the system making the request. The request header *FieldName* values are as follows:

Accept   Specifies the media types that are acceptable in the response message

Accept-Charset   Specifies the character sets that are acceptable in the response message

Accept-Encoding   Specifies the content codings that are acceptable in the response message

Accept-Language   Specifies the languages that are acceptable in the response message

Authorization   Contains credentials with which the client will be authenticated to the server

Expect   Specifies the behavior that the client expects from the server

From   Contains an e-mail address for the user generating the request

Host   Specifies the Internet host name of the resource being requested (usually a URL), plus a port number if different from the default port (80)

If-Match   Used to make a particular request conditional by matching particular entity tags

If-Modified-Since   Used to make a particular request conditional by specifying the modification date of the client cache entry containing the resource, which the server compares to the actual resource and replies with either the resource or a cache referral

If-None-Match   Used to make a particular request conditional by not matching particular entity tags

If-Range   Requests that the server transmit the parts of an entity that the client is missing

If-Unmodified-Since   Used to make a particular request conditional by specifying a date that the server should use to determine whether to supply the requested resource

Max-Forwards   Limits the number of proxies or gateways that can forward the request to another server

Proxy-Authorization   Contains credentials with which the client will authenticate itself to a proxy server

Range   Contains one or more byte ranges representing parts of the resource specified by the *ResourceURI* variable that the client is requesting be sent by the server

Referer   Specifies the resource from which the *ResourceURI* value was obtained

TE   Specifies which extension transfer-codings the client can accept in the response and whether the client will accept trailer fields in a chunked transfer-coding

User-Agent   Contains information about the browser generating the request

Response Header Fields   The response headers apply only to response messages and provide additional information about the message and the server generating the message. The response header *FieldName* values are as follows:

Accept-Ranges   Enables a server to indicate its acceptance of range requests for a resource (used in responses only)

Age   Specifies the elapsed time since a cached response was generated at a server

Etag   Specifies the current value of the entity tag for the requested variant

Location   Directs the destination system to a location for the requested resource other than that specified by the *RequestURI* variable

Proxy-Authenticate   Specifies the authentication scheme used by a proxy server

Retry-After   Specifies how long a requested resource will be unavailable to the client

Server   Identifies the Web server software used to process the request

Vary   Specifies the header fields used to determine whether a client can use a cached response to a request without revalidation by the server

WWW-Authenticate   Specifies the type of authentication required in order for the client to access the requested resource

Entity Header Fields   The term *entity* is used to describe the data included in the message body of a response message, and the entity headers provide additional information about that data. The entity header *FieldName* values are as follows:

Allow   Specifies the request types supported by a resource identified by a particular *RequestURI* value

Content-Encoding   Specifies additional content-coding mechanisms (such as gzip) that have been applied to the data in the body of the message

Content-Language   Specifies the language of the message body

Content-Length   Specifies the length of the message body, in bytes

Content-Location   Specifies the location from which the information in the message body was derived, when it is separate from the location specified by the *ResourceURI* variable

Content-MD5   Contains an MD5 digest of the message body (as defined in RFC 1864) that will be used to verify its integrity at the destination

Content-Range   Identifies the location of the data in the message body within the whole of the requested resource when the message contains only a part of the resource

Content-Type   Specifies the media type of the data in the message body

Expires   Specifies the date and time after which the cached response is to be considered stale

Last-Modified   Specifies the date and time at which the server believes the requested resource was last modified

Extension-Header   Enables the use of additional entity header fields that must be recognized by both the client and the server

HTTP Responses

The HTTP responses generated by Web servers use many of the same basic elements as the requests. The start line also consists of three elements, as follows:

```
HTTPVersion StatusCode StatusPhrase
```

The *HTTPVersion* variable specifies the standard supported by the server, using the same values listed earlier. The *StatusCode* and *StatusPhrase* variables indicate whether the request has been processed successfully by the server, and if it hasn't, why not. The code is a three-digit number and the phrase is a text string. The code values are defined in the HTTP specification and are used consistently by all Web server implementations. The first digit of the code specifies the general nature of the response, and the second two digits give more specific information. The status phrases are defined by the standard as well, but some Web server products enable you to modify the text strings in order to supply more information to the client. The codes and phrases defined by the standard are listed in the following sections.

Informational Codes   Informational codes are used only in responses with no message bodies, and have the numeral 1 as their first digit. No informational codes were defined in HTTP version 1 and earlier. The only code of this type in the current standard is as follows:

100 – Continue   Indicates that the request message has been received by the server and that the client should either send another message completing the request or continue to wait for a response. A response using this code must be followed by another response containing a code indicating completion of the request.

Successful Codes   Successful codes have a 2 as their first digit and indicate that the client's request message has been successfully received, understood, and accepted. The valid codes are as follows:

200 – OK   Indicates that the request has been processed successfully and that the response contains the data appropriate for the type of request.

201 – Created   Indicates that the request has been processed successfully and that a new resource has been created.

202 – Accepted   Indicates that the request has been accepted for processing, but that the processing has not yet been completed.

203 – Nonauthoritative Information   Indicates that the information in the headers is not the definitive information supplied by the server, but is gathered from a local or a third-party copy.

204 – No Content   Indicates that the request has been processed successfully, but that the response contains no message body. It may contain header information.

205 – Reset Content   Indicates that the request has been processed successfully and that the client browser user should reset the document view. This message typically means that the data from a form has been received and that the browser should reset the display by clearing the form fields.

206 – Partial Content   Indicates that the request has been processed successfully and that the server has fulfilled a request that uses the Range header to specify part of a resource.

Redirection Codes   Redirection codes have a 3 as their first digit and indicate that further action from the client (either the browser or the user) is required to successfully process the request. The valid codes are as follows:

300 – Multiple Choices   Indicates that the response contains a list of resources that can be used to satisfy the request, from which the user should select one.

301 – Moved Permanently   Indicates that the requested resource has been assigned a new permanent URI and that all future references to this resource should use one of the new URIs supplied in the response.

302 – Found   Indicates that the requested resource resides temporarily under a different URI, but that the client should continue to use the same *RequestURI* value for future requests, since the location may change again.

303 – See Other   Indicates that the response to the request can be found under a different URI and that the client should generate another request pointing to the new URI.

304 – Not Modified   Indicates that the version of the requested resource in the client cache is identical to that on the server and that retransmission of the resource is not necessary.

305 – Use Proxy   Indicates that the requested resource must be accessed through the proxy specified in the Location header.

306 – Unused

307 – Temporary Redirect   Indicates that the requested resource resides temporarily under a different URI, but that the client should continue to use the same *RequestURI* value for future requests, since the location may change again.

Client Error Codes   Client error codes have a 4 as their first digit and indicate that the request could not be processed due to an error by the client. The valid codes are as follows:

400 – Bad Request   Indicates that the server could not understand the request due to malformed syntax.

401 – Unauthorized   Indicates that the server could not process the request because user authentication is required.

402 – Payment Required   Reserved for future use.

403 – Forbidden   Indicates that the server is refusing to process the request and that it should not be repeated.

404 – Not Found   Indicates that the server could not locate the resource specified by the *RequestURI* variable.

405 – Method Not Allowed   Indicates that the request type cannot be used for the specified *RequestURI*.

406 – Not Acceptable   Indicates that the resource specified by the *RequestURI* variable does not conform to any of the data types specified in the request message's Accept header.

407 – Proxy Authentication Required   Indicates that the client must authenticate itself to a proxy server before it can access the requested resource.

408 – Request Timeout   Indicates that the client did not produce a request within the server's timeout period.

409 – Conflict   Indicates that the request could not be processed because of a conflict with the current state of the requested resource, such as when a PUT command attempts to write data to a resource that is already in use.

410 – Gone   Indicates that the requested resource is no longer available at the server, and that the server is not aware of an alternate location.

411 – Length Required   Indicates that the server has refused to process a request that does not have a Content-Length header.

412 – Precondition Failed   Indicates that the server has failed to satisfy one of the preconditions specified in the request headers.

413 – Request Entity Too Large   Indicates that the server is refusing to process the request because the message is too large.

414 – RequestURI Too Long   Indicates that the server is refusing to process the request because the *RequestURI* value is longer than the server is willing to interpret.

415 – Unsupported Media Type   Indicates that the server is refusing to process the request because the request is in a format not supported by the requested resource for the requested method.

416 – Requested Range Not Satisfiable   Indicates that the server cannot process the request because the data specified by the Range header in the request message does not exist in the requested resource.

417 – Expectation Failed   Indicates that the server could not satisfy the requirements specified in the request message's Expect header.

Server Error Codes   Server error codes have a 5 as their first digit and indicate that the request could not be processed due to an error by the server. The valid codes are as follows:

500 – Internal Server Error   Indicates that the server encountered an unexpected condition that prevented it from fulfilling the request.

501 – Not Implemented   Indicates that the server does not support the functionality required to satisfy the request.

502 – Bad Gateway   Indicates that a gateway or proxy server has received an invalid response from the upstream server it accessed while attempting to process the request.

503 – Service Unavailable   Indicates that the server cannot process the request due to it being temporarily overloaded or under maintenance.

504 – Gateway Timeout   Indicates that a gateway or proxy server did not receive a timely response from the upstream server specified by the URI or some other auxiliary server needed to complete the request.

505 – HTTP Version Not Supported   Indicates that the server does not support, or refuses to support, the HTTP protocol version used in the request message.

After the start line, a response message can contain a series of headers, just like those in a request, that provide information about the server and the response message. The header section concludes with a blank line, after which comes the body of the message, typically containing the contents of the file requested by the client. If the file is larger than what can fit in a single packet, the server generates additional response messages containing message bodies, but no start lines or headers.

HTTP Message Exchanges

In the most basic form of HTTP message exchange, the client browser establishes a TCP connection to a server and then transmits an HTTP request message, like that shown in Figure 26-3.

The start line for the message indicates that it's a GET command, that the *RequestURI* value identifies the default file in the Web site's root directory, and that the client is using HTTP version 1.1. The Accept header lists the media types acceptable in the response, including the */* value, which enables the client to accept any media type. The Accept-Language and Accept-Encoding headers indicate, respectively, that the response should be in U.S. English and that the gzip and deflate compression formats are acceptable. The User-Agent header identifies the browser used by the client (Internet Explorer 5, in this case), and the Host header provides the URL supplied by the user in the browser's Address field. The Connection header contains the Keep-Alive value, indicating that the same TCP connection will be used to transmit multiple files, and the Extension header contains proprietary information about the authentication used by the client to access the server.

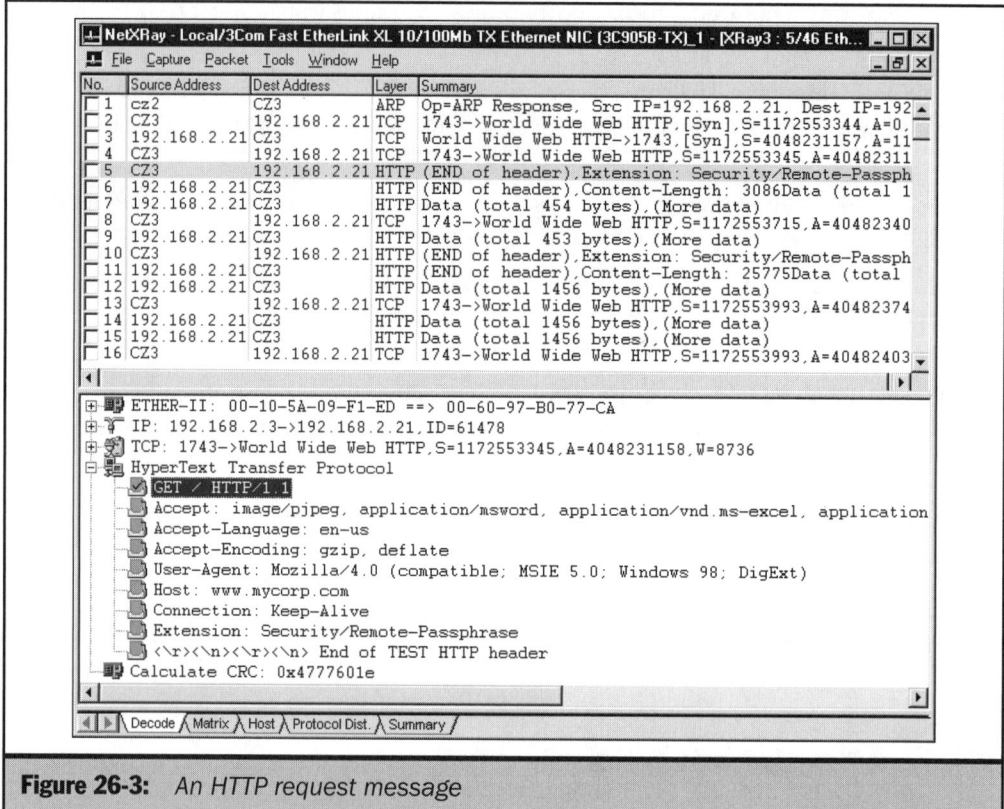

Figure 26-3: *An HTTP request message*

Before issuing a response, the server transmits a standard TCP acknowledgment message back to the client, indicating that the request has been received intact. This message is not strictly necessary, since the response also serves as an indication that the request was received, but all HTTP messages are transmitted within a TCP connection and must be acknowledged.

The server's response to the request (shown in Figure 26-4) specifies that the server is also using HTTP version 1.1, and the StatusCode value 200 indicates that the request was processed successfully. The Server header identifies the server as running IIS 5, and other headers specify the date and time that the request was processed and the media type of the requested file. The Content-Location header specifies the name of the file included in the response, which also specifies the default file name that the server is configured to use (that is, the request had only a forward slash for the *RequestURI* value). The Last-Modified header indicates that the requested file has not been modified since October 2, 1999, and the Content-Length header provides the total length of the file. The Etag header provides an entity tag for the file, which has no function here, but can conceivably be used with other headers such as If-Match and If-None-Match for cache checking.

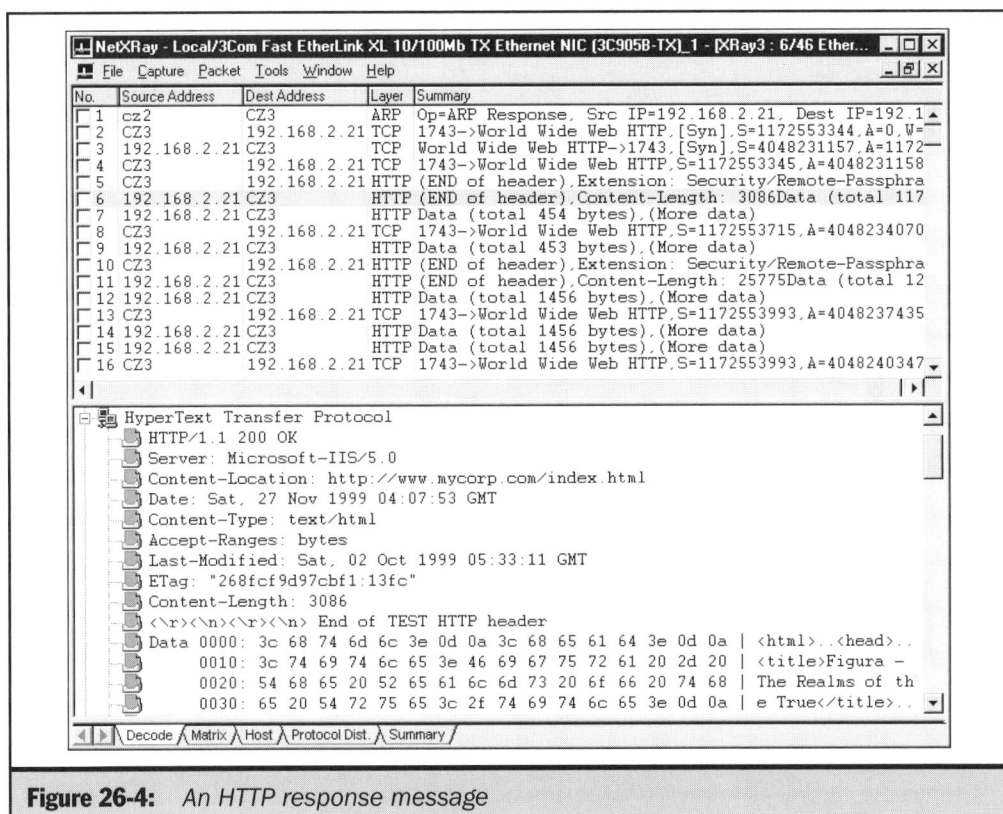

Figure 26-4: *An HTTP response message*

Because of its size, two additional response messages are required before the entire file is transmitted to the client. Once it has received the file, the client parses its HTML code. Encountering an image tag, the browser generates another request message for a file called 5.gif and transmits it to the server (see Figure 26-5). Because both the browser and the server support persistent connections, the request and response messages for the 5.gif file are both transmitted as part of the same TCP connection. If this was not the case, the server would have begun the connection termination process immediately after sending the last part of the index.html file, and the browser would have had to establish a new connection in order to send the 5.gif request.

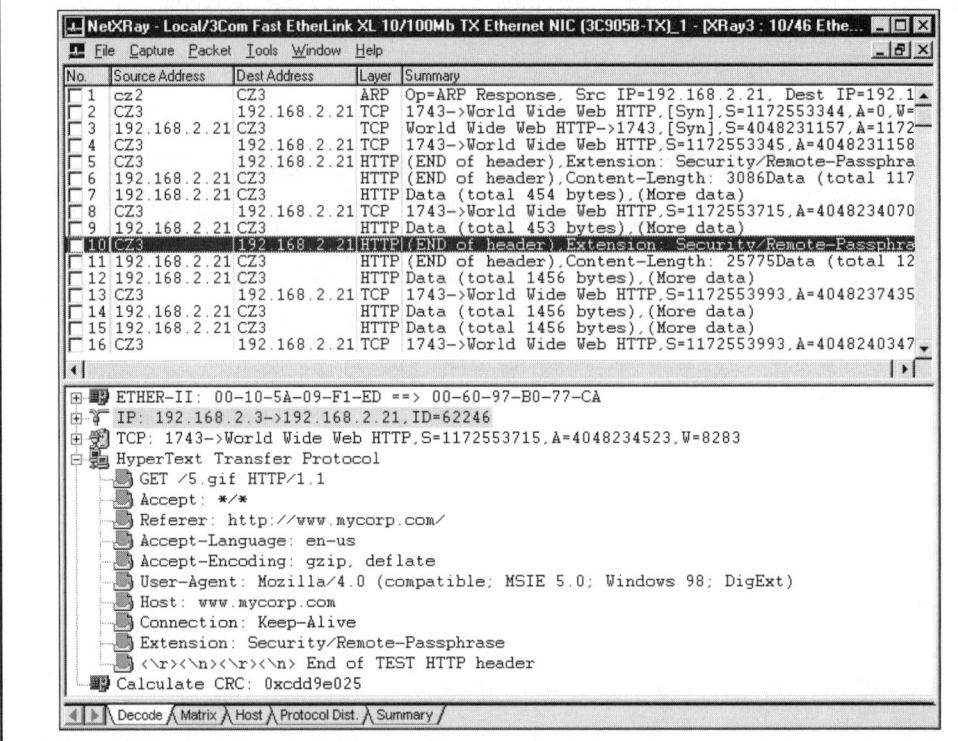

Figure 26-5: *An HTML file can trigger requests for additional files, such as graphic images.*

Once the client has requested and the server has transmitted all of the files needed to display the home page for the Web site, the server begins the process of terminating the TCP connection with the client. Once this process is completed, there is no further communication between client and server until the user initiates another request by clicking a hyperlink or typing a URL.

FTP Servers

FTP, the File Transfer Protocol, is an application-layer TCP/IP protocol that enables an authenticated client to connect to a server and transfer files to and from the other machine. FTP is not the same as sharing a drive with another system on the network.

Access is limited to a few basic file management commands, and the primary function of the protocol is to copy files to your local system, not to access them in place on the server.

Defined by the IETF in RFC 959, FTP has been a common fixture on Unix systems for many years. All Unix workstations typically run an FTP server daemon and have an FTP server client, and many users rely on the protocol for basic LAN file transfers. FTP is also a staple utility on the Internet, with thousands of public servers available from which users can download files. Although not as ubiquitous as on Unix, every Windows operating system with a TCP/IP stack has a character-based FTP client, and Web browsers can access FTP servers as well. FTP server capabilities are provided as a part of Web server packages such as IIS.

Like HTTP, FTP uses the TCP protocol for its transport services and relies on ASCII text commands for its user interface. All of the original FTP implementations on Unix are character based, as is the FTP client included with Windows. However, there are now many graphical FTP clients available that automate the generation and transmission of the appropriate text commands to a server.

The big difference between FTP and HTTP (as well as most other protocols) is that FTP uses two port numbers in the course of its operations. When an FTP client connects to a server, it uses port 21 to establish a control connection. This connection remains open during the life of the session; the client and server use it to exchange commands and replies. When the client requests a file transfer, the server establishes a second connection on port 20, which it uses to transfer the file and then terminates immediately afterward.

While Unix systems include FTP client and server capabilities as part of the default operating system installation, running an FTP server on Windows requires a separate installation and configuration process. Microsoft bundles its FTP server software for Windows 2000/NT with its Web server as part of the IIS product. FTP runs as a service that administrators can configure through the same Internet Service Manager application as the Web server. There are also several popular and robust FTP server products available as freeware/shareware.

The IIS FTP server uses Windows NT domains or Active Directory to authenticate client connections, while Unix systems typically use a list of approved user names on the local system for access control. Most FTP access on the Internet is anonymous, but on LANs, more security is often required. While you can protect FTP servers from unauthorized access with passwords, the FTP messages themselves are utterly unprotected. As with HTTP, the communications exchanged by clients and servers over the FTP control connection take the form of ASCII strings, which are transmitted over the control connection in clear text. If someone is using a network monitor program to capture packets as they travel over the network, the account names and passwords used to authenticate FTP client connections are easily visible.

For this reason, administrators should not use accounts with access to sensitive materials when connecting to an FTP server. For example, when running the IIS FTP server on Windows, you can conceivably use the Administrator account for your domain, but this would compromise the security of your entire network. Instead, create a new account that has only the permissions you need when using FTP, one that you can easily change or delete if the password is intercepted.

FTP Commands

An FTP client consists of a user interface, which may be text based or graphical, and a *user protocol interpreter.* The user protocol interpreter communicates with the *server protocol interpreter* using text commands that are passed over the control connection (see Figure 26-6). When the commands call for a data transfer, one of the protocol interpreters triggers a *data transfer process,* which communicates with a like process on the other machine using the data connection. The commands issued by the user protocol interpreter do not necessarily correspond to the traditional text-based user interface commands. For example, to retrieve a file from a server, the traditional user interface command is GET plus the file name, but after the user protocol interpreter receives this command, it sends a RETR command to the server with the same file name. Thus, the user interface can be modified for purposes of language localization or other reasons, but the commands used by the protocol interpreters remain consistent.

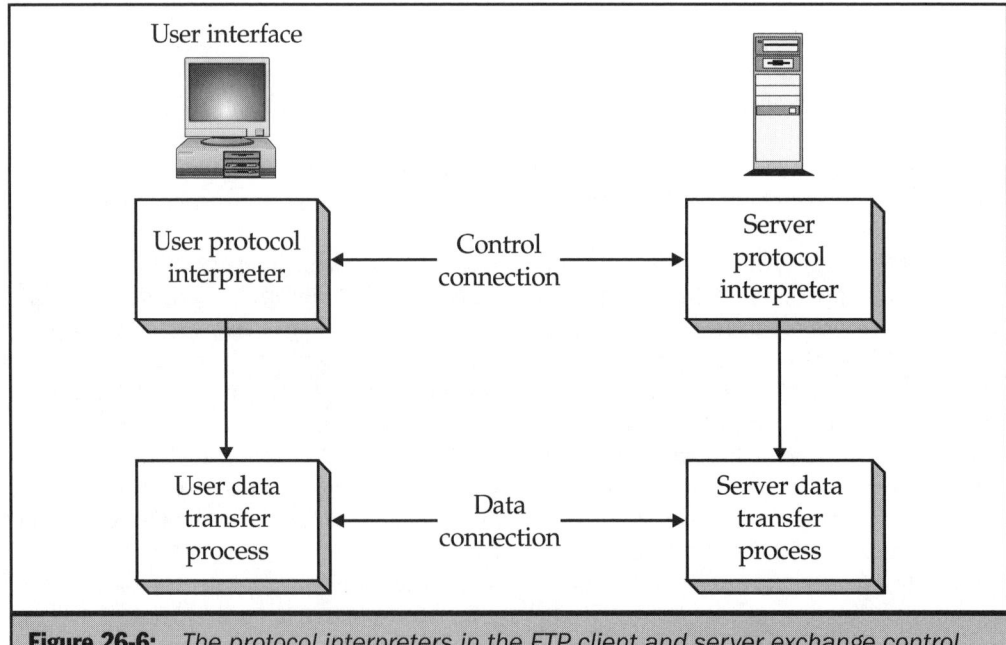

Figure 26-6: *The protocol interpreters in the FTP client and server exchange control messages.*

The following sections list the commands used by the FTP protocol interpreters.

Access Control Commands

FTP clients use the access control commands to log in to a server, authenticate the user, and terminate the control connection at the end of the session. These commands are as follows:

USER *username*   Specifies the account name used to authenticate the client to the server.

PASS *password*   Specifies the password associated with the previously furnished user name.

ACCT *account*   Specifies an account used for access to specific features of the server file system. The ACCT command can be issued at any time during the session, and not just during the login sequence, as with USER.

CWD *pathname*   Changes the working directory in the server file system to that specified by the *pathname* variable.

CDUP   Shifts the working directory in the server file system one level up to the parent directory.

SMNT *pathname*   Mounts a different file system data structure on the server, without altering the user account authentication.

REIN   Terminates the current session, leaving the control connection open and completing any data connection transfer in progress. A new USER command is expected to follow immediately.

QUIT   Terminates the current session and closes the control connection after completing any data connection transfer in progress.

Transfer Parameter Commands

The transfer parameter commands prepare the systems to initiate a data connection and identify the type of file that is to be transferred. These commands are as follows:

PORT *host/port*   Notifies the server of the IP address and ephemeral port number that it expects a data connection to use. The *host/port* variable consists of six integers, separated by commas, representing the four bytes of the IP address and two bytes for the port number.

PASV   Instructs the server to specify a port number that the client will use to establish a data connection. The reply from the server contains a *host/port* variable, like PORT.

TYPE *typecode*   Specifies the type of file to be transferred over a data connection. Currently used options are as follows:

A   ASCII plain text file

I   Binary file

STRU *structurecode*   Specifies the structure of a file. The default setting, F (for File), indicates that the file is a contiguous byte stream. Two other options, R (for Record) and P (for Page), are no longer used.

MODE *modecode*   Specifies the transfer mode for a data connection. The default setting, S (for Stream), indicates that the file will be transferred as a byte stream. Two other options, B (for Block) and C (for Compressed), are no longer used.

FTP Service Commands

The FTP service commands enable the client to manage the file system on the server and initiate file transfers. These commands are as follows:

RETR *filename*   Instructs the server to transfer the specified file to the client.

STOR *filename*   Instructs the server to receive the specified file from the client, overwriting an identically named file in the server directory if necessary.

STOU   Instructs the server to receive the file from the client and give it a unique name in the server directory. The reply from the server must contain the unique name.

APPE *pathname*   Instructs the server to receive the specified file from the client and append it to the identically named file in the server directory. If no file of that name exists, the server creates a new file.

ALLO *bytes*   Allocates a specified number of bytes on the server before the client actually transmits the data.

REST *marker*   Specifies the point in a file at which the file transfer should be restarted.

RNFR *filename*   Specifies the name of a file to be renamed; must be followed by an RNTO command.

RNTO *filename*   Specifies the new name for the file previously referenced in an RNFR command.

ABOR   Aborts the command currently being processed by the server, closing any open data connections.

DELE *filename*   Deletes the specified file on the server.

RMD *pathname*   Deletes the specified directory on the server.

MKD *pathname*   Creates the specified directory on the server.

PWD   Returns the name of the server's current working directory.

LIST *pathname*   Instructs the server to transmit an ASCII file containing a list of the specified directory's contents, including attributes.

NLST *pathname*   Instructs the server to transmit an ASCII file containing a list of the specified directory's contents, with no attributes.

SITE *string*   Carries nonstandard, implementation-specific commands to the server.

SYST   Returns the name of the operating system running on the server.

STAT *filename*   When used during a file transfer, the server returns a status indicator for the current operation. When used with a *filename* argument, the server returns the LIST information for the specified file.

HELP *string*   Returns help information specific to the server implementation.

NOOP   Instructs the server to return an OK response. Used as a session keep-alive mechanism; the command performs no other actions.

FTP Reply Codes

An FTP server responds to each command sent by a client with a three-digit reply code and a text string. As with HTTP, these reply codes must be implemented as defined in the FTP standard on all servers, so that the client can determine its next action, but some products enable you to modify the text that is delivered with the code and displayed to the user.

The first digit of the reply code indicates whether the command was completed successfully, unsuccessfully, or not at all. The possible values for this digit are as follows:

1## – Positive preliminary reply   Indicates that the server is initiating the requested action and that the client should wait for another reply before sending any further commands.

2## – Positive completion reply   Indicates that the server has successfully completed the requested action.

3## – Positive intermediate reply   Indicates that the server has accepted the command, but that more information is needed before it can execute it, and that the client should send another command containing the required information.

4## – Transient negative completion reply Indicates that the server has not accepted the command or executed the requested action due to a temporary condition, and that the client should send the command again.

5## – Permanent negative completion reply Indicates that the server has not accepted the command or executed the requested action, and that the client is discouraged (but not forbidden) from resending the command.

The second digit of the reply code provides more specific information about the nature of the message. The possible values for this digit are as follows:

#0# – Syntax Indicates that the command contains a syntax error that has prevented it from being executed.

#1# – Information Indicates that the reply contains information that the command requested, such as status or help.

#2# – Connections Indicates that the reply refers to the control or data connection.

#3# – Authentication and accounting Indicates that the reply refers to the login process or the accounting procedure.

#4# – Unused

#5# – File system Indicates the status of the server file system as a result of the command.

The actual error codes defined by the FTP standard are as follows:

- 110 Restart marker reply
- 120 Service ready in *nnn* minutes
- 125 Data connection already open; transfer starting
- 150 File status okay; about to open data connection
- 200 Command okay
- 202 Command not implemented, superfluous at this site
- 211 System status, or system help reply
- 212 Directory status
- 213 File status
- 214 Help message
- 215 NAME system type
- 220 Service ready for new user

- 221 Service closing control connection

- 225 Data connection open; no transfer in progress

- 226 Closing data connection

- 227 Entering Passive Mode (h1,h2,h3,h4,p1,p2)

- 230 User logged in, proceed

- 250 Requested file action okay, completed

- 257 "PATHNAME" created

- 331 User name okay, need password

- 332 Need account for login

- 350 Requested file action pending further information

- 421 Service not available; closing control connection

- 425 Can't open data connection

- 426 Connection closed; transfer aborted

- 450 Requested file action not taken

- 451 Requested action aborted; local error in processing

- 452 Requested action not taken; insufficient storage space in system

- 500 Syntax error, command unrecognized

- 501 Syntax error in parameters or arguments

- 502 Command not implemented

- 503 Bad sequence of commands

- 504 Command not implemented for that parameter

- 530 Not logged in

- 532 Need account for storing files

- 550 Requested action not taken; file unavailable (e.g., file not found, no access)

- 551 Requested action aborted; page type unknown

- 552 Requested file action aborted; exceeded storage allocation (for current directory or dataset)

- 553 Requested action not taken; file name not allowed

FTP Messaging

An FTP session begins with a client establishing a connection with a server by using either a GUI or the command line to specify the server's DNS name or IP address. The first order of business is to establish a TCP connection using the standard three-way handshake. The FTP server is listening on port 21 for incoming messages, and this new TCP connection becomes the FTP control connection that will remain open for the life of the session. The first FTP message is transmitted by the server, announcing and identifying itself, as follows:

```
220 CZ2 Microsoft FTP Service (Version 5.0)
```

As with all messages transmitted over a TCP connection, acknowledgment is required. During the course of the session, the message exchanges will be punctuated by TCP ACK packets from both systems, as needed. After it sends the initial acknowledgment, the client prompts the user for an account name and password and performs the user login sequence, as follows:

```
USER anonymous

331 Anonymous access allowed, send identity (e-mail name) as password.

PASS jdoe@zacker.com

230 Anonymous user logged in.
```

The client then informs the server of its IP address and the port that it will use for data connections on the client system, as follows:

```
PORT 192,168,2,3,7,233

200 PORT command successful.
```

The values 192, 168, 2, and 3 are the four decimal byte values of the IP address, and the 7 and 233 are the 2 bytes of the port number value, which translates as 2025. By converting these 2 port bytes to binary form (00000111 11101001) and then converting the whole 2-byte value to a decimal, you get 2025.

At this point, the client can send commands to the server requesting file transfers or file system procedures, such as the creation and deletion of directories. One typical client command is to request a listing of the files in the server's default directory, as follows:

```
NLST -l
```

In response to this command, the server informs the client that it is going to open a data connection, because the list is transmitted as an ASCII file:

```
150 Opening ASCII mode data connection for /bin/ls.
```

The server then commences the establishment of the second TCP connection, using its own port 20 and the client port 2025 specified earlier in the PORT command. Once the connection is established, the server transmits the file it has created containing the listing for the directory. Depending on the number of files in the directory, the transfer may require the transmission of multiple packets and acknowledgments, after which the server immediately sends the first message in the sequence that terminates the data connection. Once the data connection is closed, the server reverts to the control connection and finishes the file transfer with the following positive completion reply message:

```
226 Transfer complete.
```

At this point, the client is ready to issue another command, such as a request for another file transfer, which repeats the entire process beginning with the PORT command, or some other function that uses only the control connection. When the client is ready to terminate the session by closing the control connection, it sends a QUIT command, and the server responds with an acknowledgment like the following:

```
221
```

E-mail

While Internet services such as the Web and FTP are wildly popular, the service that is the closest to being a ubiquitous business and personal communications tool is e-mail. E-mail is a unique communications medium that combines the immediacy of the telephone with the precision of the written word, and no Internet service is more valuable to the network user. Until the mid-1990s, the e-mail systems you were likely to encounter were self-contained, proprietary solutions designed to provide an organization with internal communications. As the value of e-mail as a business tool began to be recognized by the general public, businesspeople began swapping the e-mail addresses supplied to them by specific online services. However, if you subscribed to a different service than your intended correspondent, you were out of luck. The rise of the Internet revolutionized the e-mail concept by providing a single, worldwide standard for mail communications that was independent of any single service provider. Today, e-mail addresses are almost as common as telephone numbers, and virtually every network with an Internet connection supplies its users with e-mail addresses.

E-mail Addressing

The e-mail address format soon becomes second nature to beginning e-mail users. An Internet e-mail address consists of a user name and a domain name, separated by an "at" symbol (@), as in *jdoe@mydomain.com*. As in the URLs used to identify Web and FTP sites, the domain name in an e-mail address (which is everything following the @ symbol) identifies the organization hosting the e-mail services for a particular user. For individual users, the domain is typically that of an ISP, which nearly always supplies one or more e-mail addresses with an Internet access account. For corporate users, the domain name is usually registered to the organization, and is usually the same domain used for their Web sites and other Internet services.

The user name part of an e-mail address (which is everything before the @ symbol) represents the name of a mailbox that has been created on the mail server servicing the domain. The user name often consists of a combination of names and/or initials identifying an individual user at the organization, but it's also common to have mailboxes for specific roles and functions in the domain. For example, most domains running a Web site have a *webmaster@mydomain.com* mailbox for communications concerning the functionality of the Web site.

Because Internet e-mail relies on standard domain names to identify mail servers, the Domain Name System (DNS) is an essential part of the Internet e-mail architecture. As you learned in Chapter 25, DNS servers store information in units of various types called resource records. The MX resource record is the one used to identify an e-mail server in a particular domain. When a mail server receives an outgoing message from an e-mail client, it reads the address of the intended recipient and performs a DNS lookup of the domain name in that address. The server generates a DNS message requesting the MX resource record for the specified domain, and the DNS server (after performing the standard iterative process that may involve relating the request to other domain servers) replies with the IP address of the e-mail server for the destination domain. The server with the outgoing message then opens a connection to the destination domain's mail server using the *Simple Mail Transfer Protocol (SMTP)*. It is the destination mail server that processes the user name part of the e-mail address by placing the message in the appropriate mailbox, where it waits until the client picks it up.

E-mail Clients and Servers

Like HTTP and FTP, Internet e-mail is a client/server application. However, in this case, there are several types of servers involved in the e-mail communication process. SMTP servers are responsible for receiving outgoing mail from clients and transmitting the mail messages to their destination servers. The other type of server is the one that maintains the mailboxes, and which the e-mail clients use to retrieve their incoming mail. The two predominant protocols for this type of server are the *Post Office Protocol, version 3 (POP3)* and the *Internet Message Access Protocol (IMAP)*. This is another case where it's important to understand that the term "server" refers to an application, and not necessarily to a separate computer. In many cases, the SMTP and either the POP3 or IMAP server run on the same computer.

E-mail server products generally fall into two categories, those that are designed solely for Internet e-mail and those that provide more comprehensive internal e-mail services as well. The former are relatively simple applications that typically provide SMTP support, and may or may not include either POP3 or IMAP as well. If not, you have to purchase and install a POP3 or IMAP server also, so that your users can access their mail. One of the most common SMTP servers used on the Internet is a free Unix program called sendmail, but there are many other products, both open source and commercial, that run on a variety of computing platforms.

After installing the mail server applications, the administrator creates a mailbox for each user and registers the server's IP address in a DNS MX resource record for the domain. This enables other SMTP servers on the Internet to send mail to the users' mailboxes. Clients access the POP3 or IMAP server to download mail from their mailboxes and send outgoing messages using the SMTP server. ISPs typically use mail servers of this type, because their users are strictly concerned with Internet e-mail. The server may provide other convenience services for users as well, such as Web-based client access, which enables users to access their mailboxes from any Web browser.

The more comprehensive e-mail servers are products that evolved from internal e-mail systems. Products like Microsoft Exchange started out as servers that a corporation would install to provide private e-mail service to users within the company, as well as other services such as calendars, personal information managers, and group scheduling. As Internet e-mail became more prevalent, these products were enhanced to include the standard Internet e-mail connectivity protocols as well. Today, a single product like Exchange provides a wealth of communications services for private network users. On this type of e-mail product, the mail messages and other personal data are stored permanently on the mail servers, and users run a special client to access their mail. Storing the mail on the server makes it easier for administrators to back it up, and enables users to access their mail from any computer. E-mail applications like Exchange are much more expensive than Internet-only mail servers, and administering them is more complicated.

An e-mail client is any program that can access a user's mailbox on a mail server. Some e-mail client programs are designed strictly for Internet e-mail, and can therefore access only SMTP, POP3, and/or IMAP servers. The Microsoft Outlook Express client is one of these, but there are many other products, both commercial and free, that perform the same basic functions. In many cases, e-mail client functionality is integrated into other programs, such as personal information managers (PIMs). Because the Internet e-mail protocols are standardized, users can run any Internet e-mail client with any SMTP/POP3/IMAP servers. Configuring an Internet e-mail client to send and retrieve mail is simply a matter of supplying the program with the IP addresses of an SMTP server (for outgoing mail) and a POP3 or IMAP server (for incoming mail), as well as the name of a mailbox on the POP3/IMAP server and its accompanying password.

The more comprehensive e-mail server products, such as Microsoft Exchange, require a proprietary client to access all of their features. In the case of Exchange, the client is the Microsoft Outlook program included as part of the Office product. Outlook is an unusual e-mail client in that you can configure it to operate in corporate/workgroup mode, in which the client connects to an Exchange server, or in Internet-only mode. Both modes enable you to access SMTP and POP3/IMAP services, but corporate/workgroup mode provides access to all of the Exchange features, such as group scheduling, and stores the user's mail on the server. Internet-only mode stores the mail on the computer's local drive.

Simple Mail Transfer Protocol (SMTP)

SMTP is an applications-layer protocol that is standardized in the IETF's RFC 821 document. SMTP messages can be carried by any reliable transport protocol, but on the Internet and most private networks, they are carried by the TCP protocol, using well-known port number 25 at the server. Like HTTP and FTP, SMTP messages are based on ASCII text commands, rather than the headers and fields used by the protocols at the lower layers of the protocol stack. SMTP communications can take place between e-mail clients and servers or between servers. In each case, the basic communication model is the same. One computer (called the sender-SMTP) initiates communication with the other (the receiver-SMTP) by establishing a TCP connection using the standard three-way handshake.

SMTP Commands

Once the TCP connection is established, the sender-SMTP computer begins transmitting SMTP commands to the receiver-SMTP, which responds with a reply message and a numeric code for each command it receives. The commands consist of a keyword and an argument field containing other parameters in the form of a text string, followed by a carriage return/line feed (CR/LF).

The SMTP standard uses the terms sender-SMTP and receiver-SMTP to distinguish the sender and the receiver of the SMTP messages from the sender and the receiver of an actual mail message. The two are not necessarily synonymous.

The commands used by the sender-SMTP and their functions are as follows (the parentheses contain the actual text strings transmitted by the sending computer):

HELLO (HELO)   Used by the sender-SMTP to identify itself to the receiver-SMTP by transmitting its host name as the argument. The receiver-SMTP responds by transmitting its own host name.

MAIL (MAIL)   Used to initiate a transaction in which a mail message is to be delivered to a mailbox by specifying the address of the mail sender as the argument and, optionally, a list of hosts through which the mail message has been routed (called a *source route*). The receiver-SMTP uses this list in the event it has to return a non-delivery notice to the mail sender.

RECIPIENT (RCPT) Identifies the recipient of a mail message, using the recipient's mailbox address as the argument. If the message is addressed to multiple recipients, the sender-SMTP generates a separate RCPT command for each address.

DATA (DATA) Contains the actual e-mail message data, followed by a CRLF, a period, and another CRLF (<CRLF>.<CRLF>), which indicates the end of the message string.

SEND (SEND) Used to initiate a transaction in which mail is to be delivered to a user's terminal (instead of to a mailbox). Like the MAIL command, the argument contains the sender's mailbox address and the source route.

SEND OR MAIL (SOML) Used to initiate a transaction in which a mail message is to be delivered to a user's terminal, if they are currently active and configured to receive messages, or to the user's mailbox, if they are not. The argument contains the same sender address and source route as the MAIL command.

SEND AND MAIL (SAML) Used to initiate a transaction in which a mail message is to be delivered to a user's terminal, if they are currently active and configured to receive messages, and to the user's mailbox. The argument contains the same sender address and source route as the MAIL command.

RESET (RSET) Instructs the receiver-SMTP to abort the current mail transaction and discard all sender, recipient, and mail data information from that transaction.

VERIFY (VRFY) Used by the sender-SMTP to confirm that the argument identifies a valid user. If the user exists, the receiver-SMTP responds with the user's full name and mailbox address.

EXPAND (EXPN) Used by the sender-SMTP to confirm that the argument identifies a valid mailing list. If the list exists, the receiver-SMTP responds with the full names and mailbox addresses of the list's members.

HELP (HELP) Used by the sender-SMTP (presumably a client) to request help information from the receiver-SMTP. An optional argument may specify the subject for which the sender-SMTP needs help.

NOOP (NOOP) Performs no function other than to request that the receiver-SMTP generate an OK reply.

QUIT (QUIT) Used by the sender-SMTP to request the termination of the communications channel to the receiver-SMTP. The sender-SMTP should not close the channel until it has received an OK reply to its QUIT command from the receiver-SMTP, and the receiver-SMTP should not close the channel until it has received and replied to a QUIT command from the sender-SMTP.

TURN (TURN) Used by the sender-SMTP to request that it and the receiver-SMTP should switch roles, with the sender-SMTP becoming the receiver-SMTP and the receiver-SMTP the sender-SMTP. The actual role switch does not occur until the receiver-SMTP returns an OK response to the TURN command.

Not all SMTP implementations include support for all of the commands listed here. The only commands that are required to be included in all SMTP implementations are HELO, MAIL, RCPT, DATA, RSET, NOOP, and QUIT.

SMTP Replies

The receiver-SMTP is required to generate a reply for each of the commands it receives from the sender-SMTP. The sender-SMTP is not permitted to send a new command until it receives a reply to the previous one. This prevents any confusion of requests and replies. The reply messages generated by the receiver-SMTP consist of a three-digit numerical value plus an explanatory text string. The number and the text string are essentially redundant; the number is intended for use by automated systems that take action based on the reply, while the text string is intended for humans. The text messages can vary from implementation to implementation, but the reply numbers must remain consistent.

The reply codes generated by the receiver-SMTP are as follows (italicized values represent variables that the receiver-SMTP replaces with an appropriate text string):

- 211 System status, or system help reply
- 214 Help message
- 220 *Domain* service ready
- 221 *Domain* service closing transmission channel
- 250 Requested mail action okay, completed
- 251 User not local; will forward to *forward-path*
- 354 Start mail input; end with <CRLF>.<CRLF>
- 421 *Domain* service not available, closing transmission channel
- 450 Requested mail action not taken: mailbox unavailable
- 451 Requested action aborted: local error in processing
- 452 Requested action not taken: insufficient system storage
- 500 Syntax error, command unrecognized
- 501 Syntax error in parameters or arguments
- 502 Command not implemented
- 503 Bad sequence of commands

- 504 Command parameter not implemented

- 550 Requested action not taken: mailbox unavailable

- 551 User not local; please try *forward-path*

- 552 Requested mail action aborted: exceeded storage allocation

- 553 Requested action not taken: mailbox name not allowed

- 554 Transaction failed

SMTP Transactions

A typical SMTP mail transaction begins (after a TCP connection is established) with the sender-SMTP transmitting a HELO command, to identify itself to the receiver-SMTP by including its host name as the command argument. If the receiver-SMTP is operational, it responds with a 250 reply. Next, the sender-SMTP initiates the mail transaction by transmitting a MAIL command. This command contains the mailbox address of the message sender as the argument on the command line. Note that this sender address refers to the person that generated the e-mail message, and not necessarily to the SMTP server currently sending commands.

Note
In the case where the SMTP transaction is between an e-mail client and an SMTP server, the sender of the e-mail and the sender-SMTP refer to the same computer, but the receiver-SMTP is not the same as the intended receiver (that is, the addressee) of the e-mail. In the case of two SMTP servers communicating, such as when a local SMTP server forwards the mail messages it has just received from clients to their destination servers, neither the sender-SMTP nor the receiver-SMTP refer to the ultimate sender and receiver of the e-mail message.

If the receiver-SMTP is ready to receive and process a mail message, it returns a 250 response to the MAIL message generated by the sender-SMTP. After receiving a positive response to its MAIL command, the sender-SMTP proceeds by sending at least one RCPT message that contains as its argument the mailbox address of the e-mail message's intended recipient. If there are multiple recipients for the message, the sender-SMTP sends a separate RCPT command for each mailbox address. The receiver-SMTP, on receiving a RCPT command, checks to see if it has a mailbox for that address, and if so, acknowledges the command with a 250 reply. If the mailbox does not exist, the receiver-SMTP can take one of several actions, such as generating a 251 User Not Local; Will Forward response and transmitting the message to the proper server, or rejecting the message with a failure response, such as 550 Requested Action Not Taken: Mailbox Unavailable or 551 User Not Local. If the sender-SMTP generates multiple RCPT messages, the receiver-SMTP must reply separately to each one before the next can be sent.

The next step in the procedure is the transmission of a DATA command by the sender-SMTP. The DATA command has no argument, and is followed simply by a CRLF. On receiving the DATA command, the receiver-SMTP returns a 354 response and assumes that all of the lines that follow are the text of the e-mail message itself. The sender-SMTP then transmits the test of the message, one line at a time, ending with a period on a separate line (in other words, a CRLF.CRLF sequence). On receipt of this final sequence, the receiver-SMTP responds with a 250 reply and proceeds to process the mail message by storing it in the proper mailbox and clearing its buffers.

MIME

SMTP is designed to carry text messages using 7-bit ASCII codes and lines no more than 1,000 characters long. This excludes foreign characters and 8-bit binary data from being carried in e-mail messages. To make it possible to send these types of data in SMTP e-mail, another standard called the *Multipurpose Internet Mail Extension (MIME)* was published in five RFC documents, numbered 2045 through 2049. MIME is essentially a method for encoding various types of data for inclusion in an e-mail message.

The typical SMTP e-mail message transmitted after the DATA command begins with a header containing the familiar elements of the message itself, such as the To, From, and Subject fields. MIME adds two additional fields to this initial header, a MIME-Version indicator that specifies which version of MIME the message is using and a Content-Type field that specifies the format of the MIME-encoded data included in the message. The Content-Type field can specify any one of several predetermined MIME formats, or it can indicate that the message consists of multiple body parts, each of which uses a different format.

For example, the header of a multipart message might appear as follows:

```
MIME-Version: 1.0

From: John Doe jdoe@anycorp.com

To: Tim Jones timj@anothercorp.com

Subject: Network diagrams

Content-Type: multipart/mixed;boundary=gc0p4Jq0M2Yt08j34c0p
```

The Content-Type field in this example indicates that the message consists of multiple parts, in different formats. The *boundary* parameter specifies a text string that is used to delimit the parts. The value specified in the boundary parameter can be any text string, just as long as it does not appear in the message text itself. After this header comes the separate parts of the message, each of which begins with the boundary value on a separate line and a Content-Type field that specifies the format for the data in that part of the message, as follows:

```
—gc0p4Jq0M2Yt08j34c0p
Content-Type: image/jpeg
```

The actual message content then appears, in the format specified by the Content-Type value.

The header for each part of the message can also contain any of the following fields:

Content-Transfer-Encoding   Specifies the method used to encode the data in that part of the message, using values such as 7-bit, 8-bit, Base64, and Binary

Content-ID   Optional field that specifies an identifier for that part of the message that can be used to reference it in other places

Content-Description   Optional field that contains a description of the data in that part of the message

The most commonly recognizable elements of MIME are the content types used to describe the nature of the data included as part of an e-mail message. A MIME content type consists of a type and a subtype, separated by a forward slash, as in image/jpeg. The type indicates the general type of data and the subtype indicates a specific format for that data type. The *image* type, for example, has several possible subtypes, including *jpeg* and *gif*, which are both common graphics formats. Systems interpreting the data use the MIME types to determine how they should handle the data, even if they do not recognize the format. For example, an application receiving data with the *text/richtext* content type might display the content to the user, even if it cannot handle the *richtext* format. Because the basic type is *text*, the application can be reasonably sure that the data will be recognizable to the user. If the application receives a message containing *image/gif* data, however, and is incapable of interpreting the *gif* format, it can be equally sure, because the message part is of the *image* type, that the raw, uninterpreted data would be meaningless to the user, and as a result would not display it in its raw form.

The seven MIME content types defined in the RFC 2046 document are as follows:

Text   Contains textual information, either unformatted (subtype: *plain*) or enriched by formatting commands

Image   Contains image data that requires a device such as a graphical display or graphical printer to view the information

Audio   Contains audio information that requires an audio output device (such as a speaker) to present the information

Video   Contains video information that requires the hardware/software needed to display moving images

Application   Contains uninterpreted binary data, such as a program file, or information to be processed by a particular application

Multipart   Contains at least two separate entities using independent data types

Message   Contains an encapsulated message, such as those defined by RFC 822, which may themselves contain multiple parts of different types

Post Office Protocol (POP3)

The Post Office Protocol, version 3 (POP3), as defined in the RFC 1939 document, is a service designed to provide mailbox services for client computers that are themselves not capable of performing transactions with SMTP servers. For the most part, the reason for the clients requiring a mailbox service is that they may not be continuously connected to the Internet, and are therefore not capable of receiving messages any time a remote SMTP server wants to send them. A POP3 server is continuously connected and is always available to receive messages for offline users. The server then retains the messages in an electronic mailbox until the user connects to the server and requests them.

POP3 is similar to SMTP in that it relies on the TCP protocol for transport services (using well-known port 110) and communicates with clients using text-based commands and responses. As with SMTP, the client transmits commands to the server, but in POP3, there are only two possible response codes, +OK, indicating the successful completion of the command, and –ERR, indicating that an error has occurred to prevent the command from being executed. In the case of POP3, the server also sends the requested e-mail message data to the client, rather than the client sending outgoing messages to the server as in SMTP.

A POP3 client/server session consists of three distinct states: the *authorization* state, the *transaction* state, and the *update* state. These states are described in the following sections.

The Authorization State

The POP3 session begins when the client establishes a TCP connection with an active server. Once the TCP three-way handshake is complete, the server transmits a greeting to the client, usually in the form of a +OK reply. At this point, the session enters the authorization state, during which the client must identify itself to the server and perform an authentication process before it can access its mailbox. The POP3 standard defines two possible authentication mechanisms. One of these utilizes the USER and PASS commands, which the client uses to transmit a mailbox name and the password associated with it to the server in clear text. Another, more secure, mechanism uses the APOP command, which performs an encrypted authentication. Other authentication mechanisms are defined in RFC 1734, "POP3 AUTHentication Command." Although POP3 doesn't require that servers use one of the authentication mechanisms described in these documents, it does require that servers use some type of authentication mechanism.

While in the authorization state, the only command permitted to the client other than authentication-related commands is QUIT, to which the server responds with a +OK reply before terminating the session without entering the transaction or update states.

Once the authentication process has been completed and the client granted access to its mailbox, the session enters the transaction state.

The Transaction State

Once the session has entered the transaction state, the client can begin to transmit the commands to the server with which it retrieves the mail messages waiting in its mailbox. When the server enters the transaction state, it assigns a number to each of the messages in the client's mailbox and takes note of each message's size. The transaction state commands use these message numbers to refer to the messages in the mailbox. The commands permitted while the session is in the transaction state are as follows. With the exception of the QUIT command, all of the following commands can only be used during the transaction state.

STAT   Causes the server to transmit a *drop listing* of the mailbox contents to the client. The server responds with a single line containing a +OK reply, followed on the same line by the number of messages in the mailbox and the total size of all the messages, in bytes.

LIST   Causes the server to transmit a *scan listing* of the mailbox contents to the client. The server responds with a multiliine reply consisting of a +OK on the first line, followed by an additional line for each message in the mailbox, containing its message number and its size, in bytes, followed by a line containing only a period, which indicates the end of the listing. A client can also issue the LIST command with a parameter specifying a particular message number, which causes the server to reply with a scan listing of that message only.

RETR   Causes the server to transmit a multiline reply containing a +OK reply, followed by the full contents of the message number specified as a parameter on the RETR command line. A separate line containing only a period serves as a delimiter, indicating the end of the message.

DELE   Causes the server to mark the message represented by the message number specified as a parameter on the DELE command line as deleted. Once marked, clients can no longer retrieve the message, nor does it appear in drop listings and scan listings. However, the server does not actually delete the message until it enters the update state.

NOOP   Performs no function other than to cause the server to generate a +OK reply.

RSET Causes the server to unmark any messages that have been previously marked as deleted during the session.

QUIT Causes the session to enter the update state prior to the termination of the connection.

The Update State

Once the client has finished retrieving messages from the mailbox and performing other transaction state activities, it transmits the QUIT command to the server, causing the session to transition to the update state. After entering the update state, the server deletes all of the messages that have been marked for deletion and releases its exclusive hold on the client's mailbox. If the server successfully deletes all of the marked messages, it transmits a +OK reply to the client and proceeds to terminate the TCP connection.

Internet Message Access Protocol (IMAP)

POP3 is a relatively simple protocol that provides clients with only the most basic mailbox service. In nearly all cases, the POP3 server is used only as a temporary storage medium; e-mail clients download their messages from the POP3 server and delete them from the server immediately afterward. It is possible to configure a client not to delete the messages after downloading them, but the client must then download them again during the next session. The Internet Message Access Protocol (IMAP), version 4 of which is defined in RFC 2060, is a mailbox service that is designed to improve upon POP3's capabilities.

IMAP functions similarly to POP3 in that it uses text-based commands and responses, but the IMAP server provides considerably more functions than POP3. The biggest difference between IMAP and POP3 is that IMAP is designed to store e-mail messages on the server permanently, and provides a wider selection of commands that enable clients to access and manipulate their messages. Storing the mail on the server enables users to easily access their mail from any computer or from different computers.

Take, for example, an office worker who normally downloads her e-mail messages to her work computer using a POP3 server. She can check her mail from her home computer if she wants to by accessing the POP3 server from there, but any messages that she downloads to her home computer are normally deleted from the POP3 server, meaning that she will have no record of them on her office computer, where most of her mail is stored. Using IMAP, she can access all of her mail from either her home or office computer at any time, including all of the messages she has already read at both locations.

To make the storage of clients' e-mail on the server practical, IMAP includes a number of organizational and performance features, including the following:

- Users can create folders in their mailboxes and move their e-mail messages among the folders to create an organized storage hierarchy.

- Users can display a list of the messages in their mailboxes that contains only the header information, and then select the messages they want to download in their entirety.

- Users can search for messages based on the contents of the header fields, the message subject, or the body of the message.

While IMAP can be a sensible solution for a corporate e-mail system in which users might benefit from its features, it is important to realize that IMAP requires considerably more in the way of network and system resources than POP3. In addition to the disk space required to store mail on the server indefinitely, IMAP also requires more processing power to execute its many commands, and consumes more network bandwidth because users remain connected to the server for much longer periods of time. For these reasons, POP3 remains the mailbox server of choice for the largest consumers of these server products: Internet service providers.

The Complete Reference

Networking

Chapter 27

Network Printing

S haring printers was one of the original motivations for networking computers, and now, decades later, it is still one of the primary reasons to install a LAN. In most cases, users have to print documents on a regular basis, but not continuously, so it is not worth the expense of devoting a separate printer to each user who needs one. In addition to the expense, individual printers occupy valuable desk space and represent an additional support burden for the system administrators.

Network Printing Issues

Sharing printers among multiple users presents several technical and administrative issues that administrators must resolve in the planning phase, preferably before they purchase or install the printers. The most obvious issue is that sharing the printer makes it possible for two or more users to print jobs at the same time. A network printing solution, therefore, must include some means to store pending jobs in a *queue* until the printer is ready to process them. The process of temporarily storing print jobs on a disk drive is called *spooling*.

Print Job Spooling

Depending on the print architecture used, print jobs may be spooled on the machine where they were generated or on a network server directory dedicated to that purpose. The location of the spool file determines how long the user's workstation is involved in the printing process. When print jobs are queued on the local machine, the system processor must continue to expend clock cycles and utilize network bandwidth in order to send the job to the printer when it's ready. Using a network print queue provides better performance for the user, because the job is transmitted immediately to the network server, and the workstation is no longer involved in the printing process.

The location of the queued files can also determine how much control network administrators can exercise over the printing process. When the queue is stored on a network server, administrators can usually manage the jobs by reordering, pausing, and canceling them. When the queued files are stored on individual workstations, it's more difficult for administrators to exercise control over the entire printing process from a centralized location.

Printer Connections

Another important issue is the location of the printers themselves. Finding locations that are convenient to the network users is certainly important, but there are also limitations imposed by the type of connection used to attach the printer to the network. There are three basic types of network printer connections, which are as follows:

Server connections   On a client/server network, such as one using Novell NetWare, connecting the printer to a server can minimize the amount of network traffic generated by the printing process, because the print queue is typically located on the same server. The drawbacks of this method are that the use of parallel or serial connections limits both the maximum distance between the printers and the server and the number of printers you can conveniently connect to the server. If the server is located in a wiring closet or data center, access to the printers by users may be limited.

Workstation connections   Connecting a printer to a workstation is possible on either a client/server or peer-to-peer network. Although workstations use the same parallel or serial connections as servers, with the same limitations on the distance between the printer and the computer, workstations are nearer to the users and are more numerous than servers, which provides greater flexibility in finding convenient printer locations. However, workstation-based printers can generate more network traffic if the queue is located on a different machine. In addition, the printing process imposes an additional burden on the workstation's processor. The same is true for server connections, but servers are usually faster machines that can better sustain the additional load.

Direct network connections   One of the most popular network printing solutions is to connect the printers directly to the network cable using a standalone print server that takes the form of either a network interface card that you install into the printer or a dedicated device that connects to the printer with a parallel cable. This method enables you to locate the printer anywhere a network connection is available. Print jobs must always be queued on a computer somewhere on the network, but this can be a server or a workstation, and the administrator can select a system that is powerful enough to service the printer.

The type of connections you use influences the locations that you select for your printers, but you should also consider their proximity to the users. It's inconvenient to have to walk down the hall every time you print a document, but it can be equally inconvenient to have a big laser printer right next to your desk with a constant stream of people using it. In addition, printers make noise that can be intrusive and expel gases that some people think may be harmful. The ideal location for a printer is one that is convenient to users but away from work areas.

Selecting Printers

Virtually any printer can be connected to a network, because it is the operating system that is responsible for tasks like spooling print jobs, not the printer itself. Laser, inkjet, and dot matrix printers all have their uses. Lasers are the most popular type of printer, especially in business environments, because they provide the best print quality. They tend to be the most expensive, but in the past five years, their prices have dropped to the point at which a wide selection is affordable, ranging from small personal printers to large business machines.

Inkjets are less expensive than lasers, but generally print at lower resolutions and with less quality. However, inkjet printers can print in color much more easily and economically. The output quality is far from professional, but good enough for proofs and home use. Inkjet technology is also used in home office devices that combine faxing, printing, and scanning functions in one unit.

Dot matrix printers are the cheapest of the three main printer types, both to purchase and to run. But their poor-quality output, low speed, and noisy operation have relegated them to specialized uses such as forms and receipts.

There are printers that are specifically intended for use on networks, because they have special features that make them more suitable for the network environment. These features can include the following:

Higher print speed A printer intended for use on a network typically runs faster than a personal printer, to keep up with jobs generated by multiple users. Manufacturers usually produce printers at several speeds (and price points) that are marketed as personal printers, workgroup printers, high-volume printers, and so forth.

Higher usage rating Printers have a usage rating that specifies the recommended maximum number of pages that should be printed per month. It's a good idea to use this rating when evaluating network printers. Even a high-quality printer is likely to have a shorter operational life if you consistently push it beyond its capabilities.

Integrated print server Some printers have a built-in print server with a network interface that enables you to plug the network cable directly into the printer. In most cases, the print server takes the form of a card that is installed in an expansion slot inside the printer, so that the device can be removed to accommodate cards for different networks. This is not an essential feature, because you can always purchase a print server separately, but it is a convenience.

Multiple paper trays A printer with multiple paper trays and/or high-capacity trays can support a variety of print jobs for longer periods without refilling. When users print jobs that call for different-sized pages, a printer with multiple trays and a driver that recognizes them can service the jobs with no manual intervention. If a printer has only one tray, a job requiring a different page size can cause all print job processing to halt until someone inserts the right size paper.

Remote printer administration   Most printers today have bidirectional communication capabilities that enable administrators to install, configure, monitor, and manage the functions of multiple network printers from a central interface. Using a dedicated program or a Web interface, it is possible to check the printer's operational status and manipulate its controls over the network, just as you would from the control panel on the printer. Note that this is a capability of the printer and its accompanying software, whereas managing a print queue is a function of the operating system.

Internal hard disk drive   Some printers can use internal hard drives to store frequently accessed data, such as font files and fax cover sheets, which speeds up the printing process and reduces the traffic sent over the network.

Combination devices   Combination devices are a rapidly growing part of the printer market. A copier, for example, is essentially a scanner and a laser printing engine, so it is only logical for manufacturers to add a printer's data processing components and a network interface, creating a hybrid device called a *mopier*. A mopier provides all of the standard copier functions, but also enables users to generate documents directly from their computers, bypassing the scanner. Unlike most standard laser printers, a mopier can produce multiple copies of a document without having to process the data for each page multiple times. Some combination devices also enable you to use the scanner for faxing and digitizing documents, in addition to making copies.

None of these features are essential. You can buy any small personal printer with only the most basic features and it will function perfectly well on a network. Your printer selection should be based more on the specific needs of your users than on a set of "network-ready" features that you may or may not need.

Selecting an Operating System

You can use Windows, NetWare, or Unix systems as servers to host your network printers. Many networks run more than one or even all of these operating systems, and the question will arise of which you should use. As with all network planning questions, the answer should be based primarily on the needs of your clients and the capabilities of their workstations.

Windows 2000, NT, and NetWare provide comparable printer hosting capabilities. Any of these operating systems enables you to implement complex printing strategies that support heavy use and a wide variety of printer types. If your network uses mostly Windows workstations, you can configure them to access either Windows or NetWare network printers, or both. However, you must install a NetWare client on every Windows system that is to use NetWare printers (unless you are using Gateway Services for

NetWare, supplied in Windows 2000/NT and described in Chapter 22). Generally speaking, running a NetWare client on a Windows workstation just to provide printer access is not recommended, because the client tends to slow down the system in several ways. Therefore, if your workstations don't need NetWare access for other reasons, Windows-based printers are the better choice.

Windows 2000 and NT provide better printer hosting capabilities than Windows 9x. If the choice is between using NetWare servers or Windows 9x workstations to host your printers, NetWare is probably the better solution, particularly if you want to use advanced capabilities such as printer pooling.

Unix workstations complicate the network printing problem further. Windows 2000 includes an LPD (line printer daemon) implementation called the Print Services for Unix that enables RFC 1179-compliant Unix clients to send print jobs to Windows 2000 printers. The NetWare Print Services for Unix, which is included with NetWare 4.2 and NetWare 5.x, provides a bidirectional solution that enables Unix clients to access NetWare printers, and NetWare clients to access Unix printers.

Selecting Print Servers

Virtually any computer on your network can host a network printer, but it is important to consider the effect that the role of print server can have on a working system. As mentioned earlier, print jobs can be huge files, and it is important that the system functioning as print server have sufficient disk space and processing power to handle the burden. This is particularly true when the drive where the spooled files are stored also contains the operating system.

On a Windows 2000 system, for example, the spooling directory is C:\Winnt\ System32\Spool by default. If the printer is offline for a long period of time, such as when it runs out of paper, a large number of print jobs can build up in the queue and fill up the disk. If the system drive (that is, the drive on which the operating system is installed) runs out of free space, Windows may be unable to write to the memory paging file or the registry, causing serious problems. If the machine happens to be a domain controller, Active Directory performance may be affected, which can be a serious problem that affects the whole network.

The same can be true for NetWare if print queues are located on the SYS volume and the server contains part of the NDS (Novell Directory Services) database. Administrators may be unable to make additions or changes to the NDS tree, and the automated database replication events might fail. On NetWare, you can select the volume on which a print queue is to be located when you create it. It is recommended that you not use the SYS volume for this purpose. On Windows 2000, you can change the default spooling directory by opening the Printers window in the Control Panel and selecting Server Properties from the File menu. On the Advanced page, you can change the Spool Folder value by specifying a path to a directory on a different drive.

There are also Windows registry settings that you can use to specify the location of an alternative spooling directory that the system uses when no disk space is available on the drive on which the main spooling directory is located. To create an alternative spooling directory for all of the printers hosted by the system, create a REG_DZ registry entry called DefaultSpoolDirectory that contains the directory path in the following key:

```
HKEY_LOCAL_MACHINE\SYSTEM\CurrentControlSet\Control\Print\Printers
```

To create an alternative spooling directory for an individual printer, create an entry of the same type called SpoolDirectory in this key:

```
HKEY_LOCAL_MACHINE\SYSTEM\CurrentControlSet\Control\Print\Printers\
PrinterName
```

Replace the *PrinterName* variable with the name of the printer that should use the specified directory.

The PCs that you use for your Windows or NetWare servers should have the hardware needed to support the processing and I/O burden of a print server's functionality. However, if you plan to share printers that are connected to workstations, this may not be the case. Print server functions can seriously debilitate the performance of an average workstation, particularly if the print jobs are spooled on the workstation, as in the case of a Windows network printer share. A continual influx of print jobs to the workstation can affect application responses and frustrate the user working on that machine.

The effects of printer processing on a NetWare client with a shared local printer are less debilitating, because the print queue and print server functions are located on other computers, but there can still be a significant I/O burden. If you must use workstations to host your network printers, you should select PCs that have sufficient resources to support the additional functions.

Printer Administration

Printer administration is often more of an organizational issue than a technical one. Often, when a single printer is shared by multiple users, no one wants to take responsibility for it. This can result in a large backlog of queued print jobs because no one bothered to fill the paper tray. There should be someone who is responsible for performing everyday services to the printer, such as filling the paper trays, clearing jams, and replacing the toner or ink cartridge. Depending on the location of the printer and the size of the organization, this could be either an end user or a network support person.

Even if the network users are responsible for basic printer maintenance tasks, there is still a need for a knowledgeable administrator to handle more complex problems. For example, it is important for someone to be available to manage the jobs in the print queue. It is not uncommon for a print job to be garbled on its way to the printer, and if the initial characters of the print job do not follow the correct format, the printer may end up producing hundreds of pages of gibberish. Someone with the knowledge and the permissions needed to manage the jobs in the queue can delete the offending job before too much time and paper is wasted. Depending on the print architecture in use, it may also be possible for an administrator to modify the order in which jobs are printed, or put certain jobs on hold because they would take too long or require a paper change. Apart from operational problems, there are also the inevitable hardware breakdowns and malfunctions, which require an experienced person to troubleshoot.

Windows Network Printing

Printing on a Windows network is a matter of installing a printer on the system that is to function as the print server and then creating a printer share out of it. Users can access the printer by configuring their workstations to send print jobs to the share, rather than to a locally attached printer. Windows printing is based on the concept of a logical printer, which is realized by the installation of a printer driver on both client and server systems. A *logical printer* is the software entity created when you install a printer driver on a Windows computer using the Printers window in the Control Panel. Applications send print jobs to a logical printer, which in turn relays them to the appropriate physical printing device, either on the local machine or a print server on the network.

Logical printers make it possible to create multiple print configurations that are serviced by a single physical printer. If you have a printer that uses both the PCL and PostScript page description languages (PDLs), you can create a separate logical printer for each PDL. You can also create logical printers with different printer configuration settings. For example, one logical printer can use separator pages while another doesn't, and another printer can be configured to print large, complex jobs only during the night hours.

The Windows Printing Process

In the Windows network printing architecture, both the client and server systems must have a logical printer installed. The logical printer on the server points to the physical printer using either a parallel or serial port or a custom port created by an external print server device. A user working at the server can then print jobs locally using that logical printer. After creating a share out of the server's logical printer, clients can create their own logical printers that use a printer driver to create jobs and send them to the share.

Note

External network print servers, like the Hewlett-Packard JetDirect devices, are not print servers in the same sense as a Windows system. They do not store the print jobs internally and do not themselves appear as shares on the network. There must be a computer on the network with a logical printer that is configured to send its jobs to the port created by the print server device.

The networking printing process as performed by Windows 2000 systems is illustrated in Figure 27-1. On other Windows operating systems, the file and directory names may be different, but the basic concepts are the same. The printing procedure is as follows:

1. The client loads the printer driver, from either the local drive or the print server. The printer driver consists of three components: a *print graphics driver DLL,* a *printer interface driver DLL,* and either a *minidriver* or a *PostScript Printer Description (PPD)* file. The print graphics driver module provides image rendering and management services and the API calls used by the Windows GDI (Graphical Device Interface) when an application prints a document. The printer interface driver module provides the configuration interface (the printer's Properties dialog box). The minidriver or PPD provides the device-specific configuration parameters for the printer.

2. Through the GDI, the application running on the client system creates an output file containing API calls to the printer driver using the *Device Driver Interface (DDI).* This output file is called a *DDI journal file.*

3. The spooler on the client system receives the DDI journal file and stores it in the C:\Winnt\System32\Spool directory until the print processor can service it. The spooler can also perform a limited amount of rendering on the file, which is completed by the print processor.

4. The print processor (Winprint.dll) receives the DDI journal file from the spooler and processes it to create a print job file using the format specified in the logical printer's Properties dialog box. The RAW format means that the print job is rendered by the client system and the output is sent to the server. The *EMF (enhanced metafile)* format sends the journal file data to the server for rendering there. After processing, the print job file is returned to the spooler.

5. The print router (Winspool.drv) retrieves the job from the client spooler, locates the printer for which it is intended, and transmits it over the network to the spooler on the print server. The server's spooler assigns a priority to the job and tracks its progress. The print router is also responsible for copying the printer driver from the print server. After the initial driver installation, the print router compares the version of the driver installed on the client system with the version on the print server at regular intervals and updates the client if necessary. This way, network administrators can install new drivers on the server and automatically update all of the clients.

6. The print monitor (Localmon.dll) on the server retrieves the print job file from the spooler and sends it to the parallel, serial, or other port associated with the printer share.

7. Once the print job file is processed by the printer, the print monitor sends a notification message to the client system, informing it of the job's completion. The spooler then deletes the job from the queue. The print monitor is also responsible for handling errors generated by the printer and resubmitting spooled jobs that have to be reprinted due to an error.

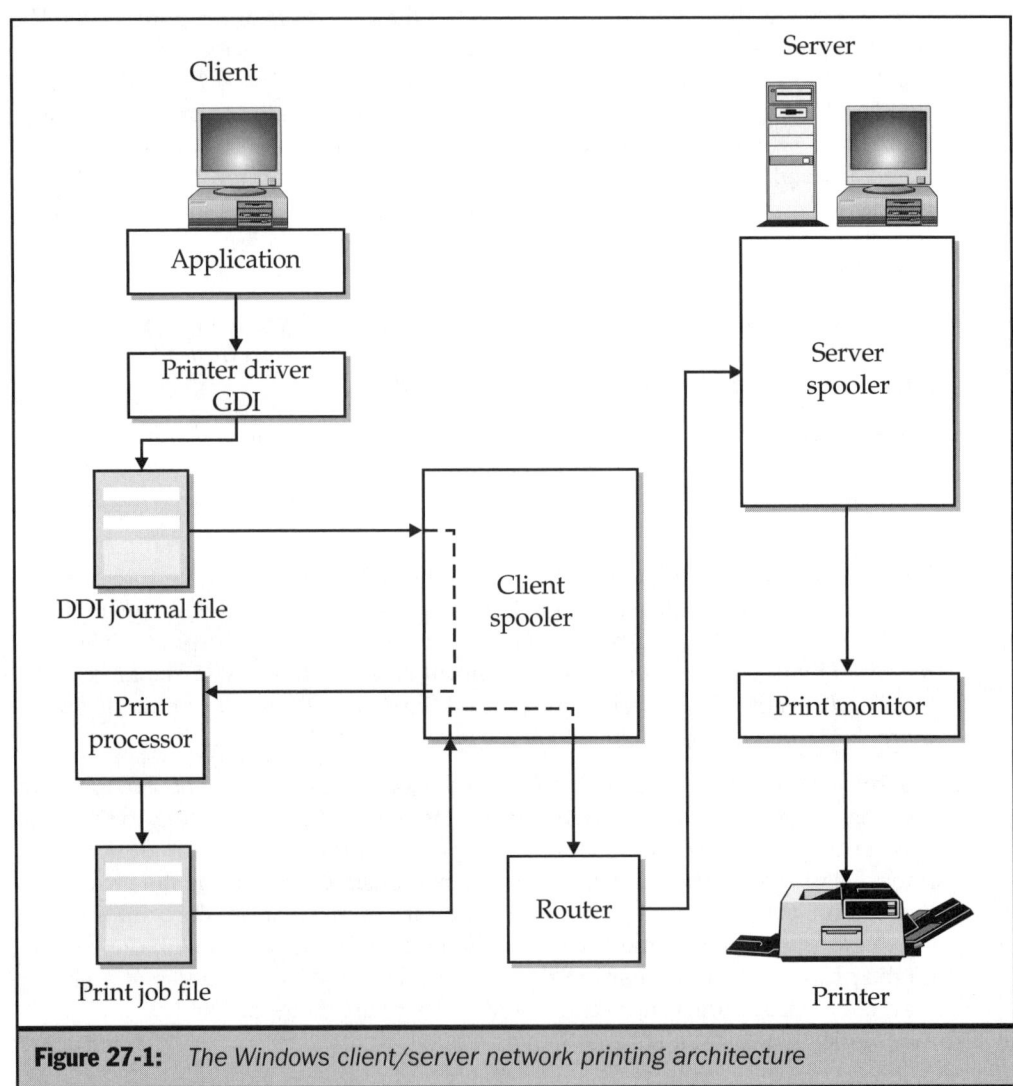

Figure 27-1: *The Windows client/server network printing architecture*

Windows Printer Configuration

Virtually all administrative access to the printing architecture in Windows 95, 98, NT, and 2000 is through the Printers Control Panel. The Add Printer Wizard walks you through the steps of creating a logical printer, either on a machine that will function as a print server or on a client system. On a Windows 2000 or NT system, you are given the option to create a share whenever you install a local printer; on the other operating systems, you have to share the printer manually after the installation is completed. When installing a client, you select a printer share from an expandable network tree. If the share is hosted by a 2000/NT server with the appropriate drivers installed, the client automatically downloads the drivers and installs them. Otherwise, you must select the appropriate driver for the printer.

Once you've created a logical printer, you can modify its configuration from its Properties dialog box, accessible from the File menu in the Printers Control Panel. The controls available in this dialog box vary depending on the operating system running on the machine. A Windows 98 system, for example, always displays the Windows 98 dialog box, even if the logical printer is pointing to a Windows 2000 print server.

Any Windows system can function as a print server, but Windows 2000 and NT systems provide more advanced features with the greatest amount of flexibility. Note that you can only control these options from a Windows 2000 or NT system; the Properties dialog box on a Windows 9*x* system does not have the required controls. Some of the advanced printing functions that Windows 2000 and NT provide are examined in the following sections.

Using Printer Pools

While any Windows system can have multiple logical printers that are serviced by one physical printer, a Windows 2000/NT system can have one logical printer that is serviced by multiple physical printers. This is called a *printer pool*, and it enables you to cope with an increasing number of print jobs by adding physical printers to the logical printer, instead of reconfiguring some of the workstations to use a different logical printer.

The printers in a pool can be connected to the local machine or to remote systems, which makes it possible to create a pool of almost any size. The only limitation is that all of the printers in the pool must be the same model, because the same driver is used to process all of the print jobs. For the convenience of your users, the printers should also be located in the same general area, because there is no way to know which of the printers in the pool will actually produce a particular document.

To create a printer pool, you select the Enable Printer Pooling check box on the Ports page of the printer's Properties dialog box (see Figure 27-2). Also on this page, each of the computer's LPT and COM ports is listed with a check box, and you can select all of the ports to which a printer is connected. When you use an external print server to connect a printer to the network, its software creates additional ports that you can add to the pool as well.

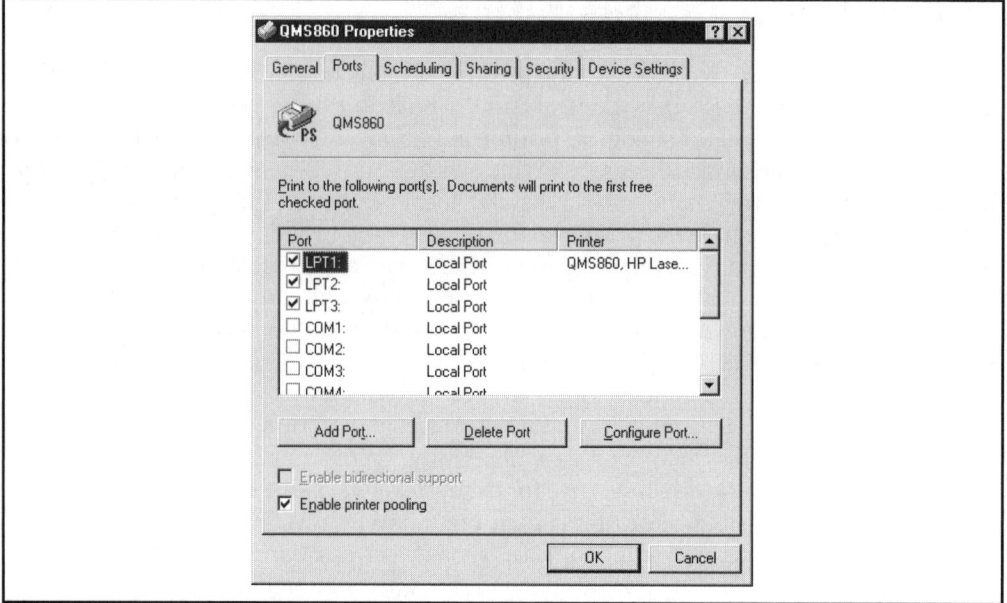

Figure 27-2: *The printer pooling feature in Windows 2000 and NT enables multiple printers to service a single print queue.*

Installing Additional Printer Drivers

As part of the process of creating a logical printer on a Windows 2000 or NT system that represents a locally attached device, you install the printer driver for the operating system. It is also possible to install additional drivers for other Windows operating systems, including Windows 9*x*, previous versions of Windows NT, and the Windows 2000/NT versions for other platforms, such as Alpha.

Although the system hosting the printer does not run these additional drivers itself, it does make them available to clients. When a user on a client system creates a logical printer that points to a share on the 2000/NT system, the Add Printer Wizard contacts the server and specifies what operating system version it is running. If the driver for that operating system is installed on the server, the client automatically copies it to its local drive and installs it. If the proper driver is not available on the server, the user must select a driver from the standard Windows list of manufacturers and printer models.

To install additional drivers for an existing logical printer on a Windows 2000 computer, you select the Sharing tab in the printer's Properties dialog box and click the Additional Drivers button, to display the page shown in Figure 27-3. After you select the operating systems for which you want to install drivers, the system prompts you for the location of the various distribution files needed and copies them to the local drive.

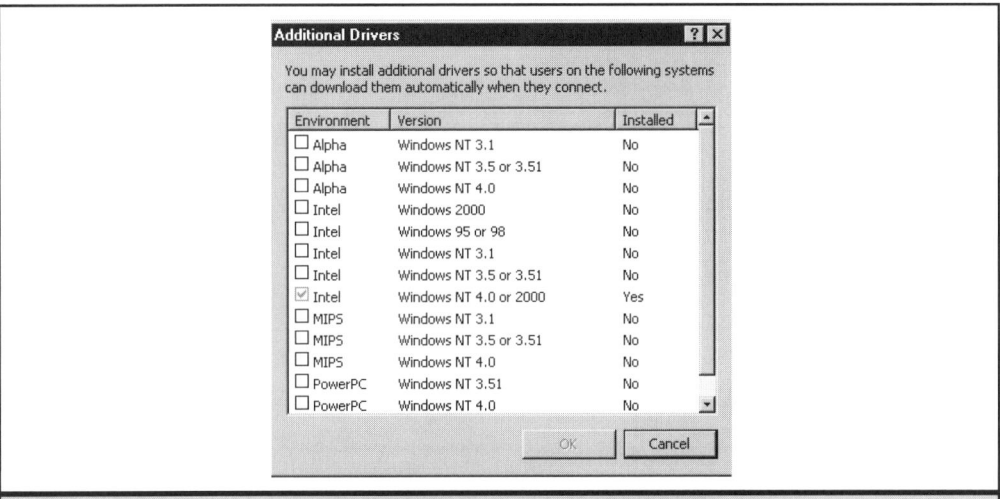

Figure 27-3:   *A Windows 2000 or NT print server can provide printer drivers to multiple client operating systems.*

Scheduling and Prioritizing Printers

Some print jobs can monopolize a printer for a long period of time, because they consist of many pages or contain elaborate graphics that take a long time to render. In cases like this, you can configure a logical printer to process jobs only during specific hours of the day. On the Advanced page of the Properties dialog box (see Figure 27-4), you can specify a range of hours during which the printer is available.

If you can count on your users to understand the difference between them, you can create two shared logical printers on the print server pointing to the same physical printer and configure one of them with access only after business hours. This assumes that the users will be conscious of the size of their print jobs and conscientious enough to defer them by using the alternate printer share when necessary.

You can also use this feature to grant users limited access to a special printer. For example, you may not want certain users to send jobs to a color printer during business hours. You can create one logical printer for selected users that permits access to the printer at any time, secure it using share permissions, and then create a second logical printer for all other users that grants access only after business hours.

Another way to control access to a physical printer is to create multiple logical printers with different priority ratings and share permissions. The Priority control on the Advanced page enables you to assign a priority value between 1 and 99 to the logical printer. The default value for this setting is 1, which is the lowest priority. You

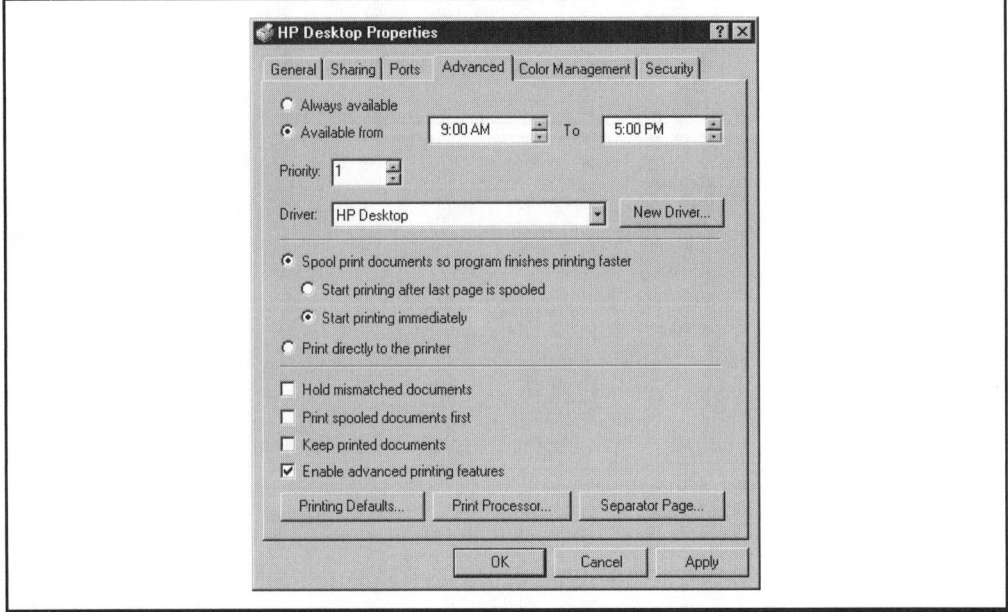

Figure 27-4: *Windows 2000 and NT can schedule printer use hours and assign printing priority ratings to various clients.*

can create several logical printers, assign them escalating priority values, and then use permissions to specify which users should have priority access to the printer. When the print queue on the server contains multiple jobs, the jobs with the highest priority are always processed first. Once a job begins printing, it is not interrupted if a job with a higher priority arrives in the queue.

Securing Printers

All Windows operating systems enable you to restrict access to printer shares with whatever form of access control the system is configured to use. Windows 95 and 98 computers, for example, can use share-level access control to authenticate users with a single share password before granting them access, or use user-level access control to grant specific users and groups access to the printer. In both cases, access is granted using an "all or nothing" model. Either the user has full access to the printer or no access at all.

Windows 2000 and NT systems provide a more flexible form of access control that enables you to specify what activities users are permitted to perform. This capability makes it easier to delegate printer administration chores to specific users without giving them complete control over a printer. To do this on a Windows 2000 system, click

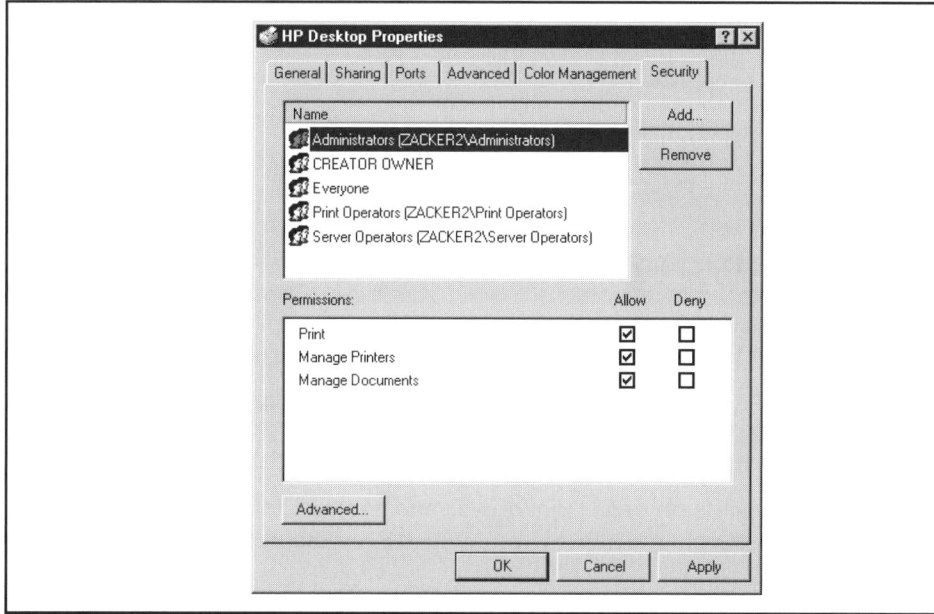

Figure 27-5: *A standard Windows 2000 Security dialog box controls access to a printer.*

the Security tab on the Properties dialog box, to display the page shown in Figure 27-5. Here you can select Active Directory objects and allow or deny them any one of the following three permissions:

Print Enables users and groups to print documents, but denies them any other access to the printer

Manage Printers Enables users to pause, resume, and purge the printer; reorder jobs in the print queue; modify the printer properties and permissions; and delete the printer

Manage Documents Enables users to pause, resume, restart, and delete jobs from the print queue, as well as control the properties of individual documents

Clicking the Advanced button expands this flexibility further by enabling you to allow or deny users and groups the following six permissions for the current printer, for documents only, or for both the printer and the documents:

■ Print

■ Manage Printers

■ Manage Documents

- Read Permissions
- Change Permissions
- Take Ownership

NetWare Printing

Because NetWare is a client/server network operating system, its printing architecture relies more heavily on servers. It is still possible to connect a printer to a server, a workstation, or directly to the network cable, but the jobs in the print queue must be stored on a server. The traditional NetWare printing architecture consists of the following three elements:

Printer   Represents the physical printer itself and contains information about how the printer is connected to the network and which print queues it will service

Print queue   Represents a directory on a NetWare server volume where print jobs are stored while waiting to be serviced by a printer

Print server   A hardware or software module that accesses the print jobs stored in a queue and submits them to the appropriate printer

These elements are represented by NDS objects on a NetWare 4 or 5 network, which makes them configurable using the NetWare Administrator or Pconsole.exe utility. In its simplest form, the NetWare print architecture consists of a printer connected to a server using a standard parallel or serial port, a print queue directory on that same server, and the Pserver.nlm print server, also running on that server, as shown in Figure 27-6. In this case, the printing process would proceed as follows:

1. A NetWare client produces a print job file using its native processes. On a Windows system, this involves the use of the standard drivers for the printer type and the GDI. NetWare print servers do not provide any rendering capabilities; the print job file produced by the client must be ready to be submitted to the printer.

2. The print job file is routed to the print queue on the NetWare server in one of two ways. Either the application is configured to submit jobs directly to the queue, or the print output is captured from a printer port (such as LPT1) and redirected to the queue. Once the job is submitted to the queue, the client's role in the printing process is completed. Once queued, print jobs can be paused, reordered, or scheduled by an administrator.

3. The print server reads the print job from the queue and sends it to the printer using the appropriate port. Once the printing is complete, the job is purged from the queue.

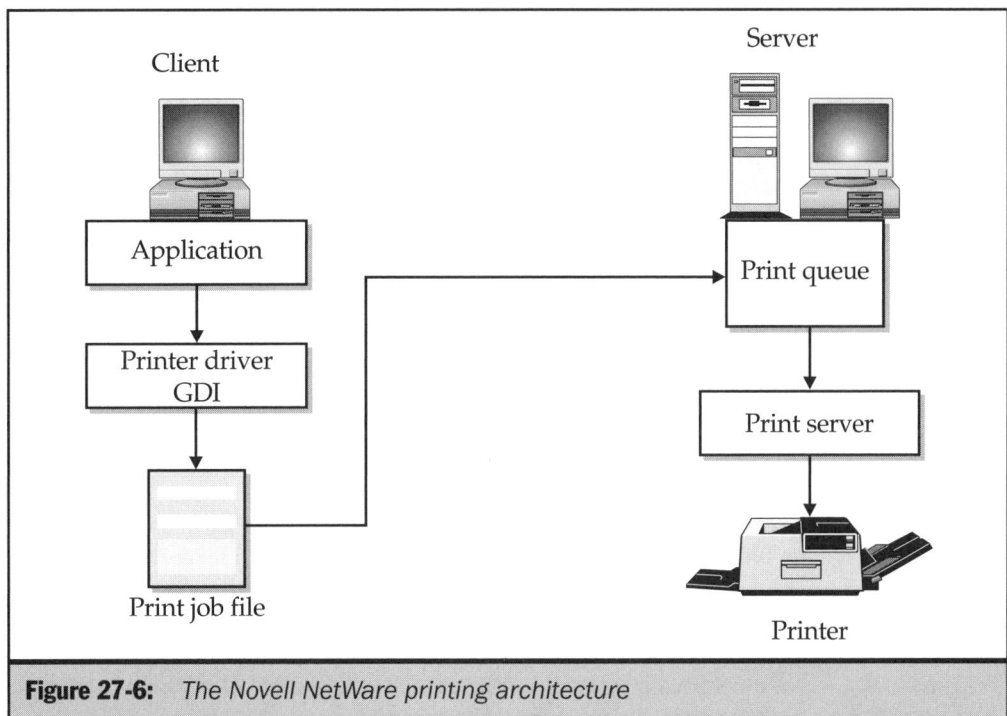

Figure 27-6: *The Novell NetWare printing architecture*

The relationships between the print objects are configurable through the NetWare Administrator or Pconsole utility. Print server objects are associated with specific printers, and printers, in turn, are associated with specific print queues. In this scenario, all of the print objects are located on the same server, but they can just as easily be separated. You can connect a printer to another server or a workstation using a standard parallel or serial port, but you must then run on that system a program called a *port driver* that enables it to receive jobs from the print server and send them out the appropriate port to the printer. In much the same way, the print queue can be located on any volume of any server on the network. As with Windows printing, you can have multiple print queues serviced by one printer, or have multiple printers service one queue. The print server must take the form of either an NLM running on a NetWare server or a standalone device like a HP JetDirect print server. The JetDirect print server enables you to connect the printer directly to the network cable, and services only the printer (or printers) directly attached to it. The Pserver.nlm program running on any NetWare server version can manage up to 256 printers connected to servers and workstations all over the network.

While separating the print elements by placing them on different machines can be convenient, you should also consider the impact that it has on your network traffic situation. Print job files can be enormous, because they can contain graphics, fonts, and other data in a raw, uncompressed form. The print output file for a modest word processor document can be several megabytes long, and files with intensive graphics and other elements can be 100MB or more. When all of the print elements are on one server, these files only have to be transmitted over the network once, from the client to the server. However, if you have the printer, print queue, and print server on separate systems, the file must be transmitted three times.

Novell Distributed Print Services

NetWare 5 supports the traditional queue-based NetWare printing system, but the preferred network printing architecture for this new NetWare version is called *Novell Distributed Print Services (NDPS)*. The original NetWare printing architecture was developed long before NDS, and when the directory service was introduced, objects representing the three print elements were included in the directory. NDPS replaces the older print architecture with a single entity called a *Printer Agent (PA)* that performs all the tasks of the printer, print server, and print queue objects.

You can connect NDPS printers to servers, workstations, or the network cable, just as before, and the PA takes the form of either a software program running on a server or a hardware device embedded in the printer. The PA manages the processing of print jobs, controls bidirectional communications between the printer and network clients, and notifies users of job status and printer errors. To create a PA on a server, you must first create an NDPS Manager object on the NDS tree.

An NDPS printer can be configured as either a public access printer or a controlled access printer. A *public access printer* is available to all users on the network, with limited administration and notification capabilities. A *controlled access printer* is represented in the NDS database by a Printer object that not only enables you to use NDS permissions to limit access to the printer, but also provides more comprehensive administration and notification options. For example, an administrator can configure a Printer object to send a notification by pop-up message or e-mail whenever the printer goes offline or experiences a specific error such as a paper jam. The administration architecture is also extensible so third-party developers can create interfaces that provide beeper notification and other features.

NDPS is a protocol-independent print architecture, unlike queue-based printing, which requires the use of IPX. NDPS is also more completely integrated into NDS than the old system. NetWare Administrator functions as the interface to a wide range of printer configuration parameters and status indicators. The functionality is limited only by the bidirectional communication capabilities of the printer itself. You can gather printers together into groups in order to regulate access to them and, by assigning properties to the printer objects, search for printers with certain capabilities.

Novell Print Manager is an NDPS utility for users that enables them to locate printers on the network and install them, as well as modify printer configuration parameters (such as page size) and check the current status of a printer, including the number of jobs currently waiting to be processed. NDPS also provides an automatic

NETWORK SERVICES

driver-download capability, much like that of Windows 2000/NT. When a user installs an NDPS printer, the workstation downloads the driver from a database and installs it without the user having to determine the exact printer model.

NetWare Printer Configuration

When the NetWare print architecture was first adapted to NDS, you had to create the Printer, Print Queue, and Print Server objects separately, and then manually configure the associations between them so that the print server could manage the printer and the printer could service the print queue. The NetWare Administrator and Pconsole utilities now have a Print Services Quick Setup feature that creates all three objects at once and automatically configures the relationships between them (see Figure 27-7). Once the utility has created the three basic objects, you can specify values for their properties or add objects to modify your printing strategy.

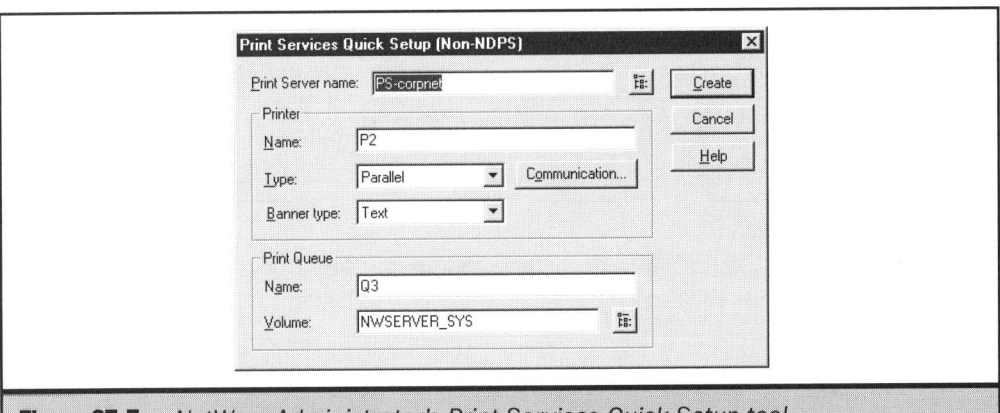

Figure 27-7:   *NetWare Administrator's Print Services Quick Setup tool*

The Printer object's details dialog box (see Figure 27-8) enables you to both specify which print queues the printer should service and assign a priority value (from 1 to 10) for each queue. As with the Windows print server, you can use this feature to grant certain users priority access to the printer. In the Print Queue object's details dialog box, you can view the Printer and Print Server objects that are associated with the queue, but you can't modify them. However, this dialog box is where you view the list of jobs currently in the queue and select both the users who are to be allowed to submit jobs to the queue and the users who are to be designated as queue operators. Queue users are permitted to manipulate their own queued jobs, by deleting them or putting them on hold. A queue operator can manage any jobs in the queue by reordering, pausing, or deleting them.

The Print Server object's dialog box enables you to select the printers that the server will manage, grant print server user and print server operator privileges, and display a print layout that diagrams all of the objects associated with that server, as shown in Figure 27-9. On networks with a complex printing infrastructure, this feature provides an excellent high-level view of print operations.

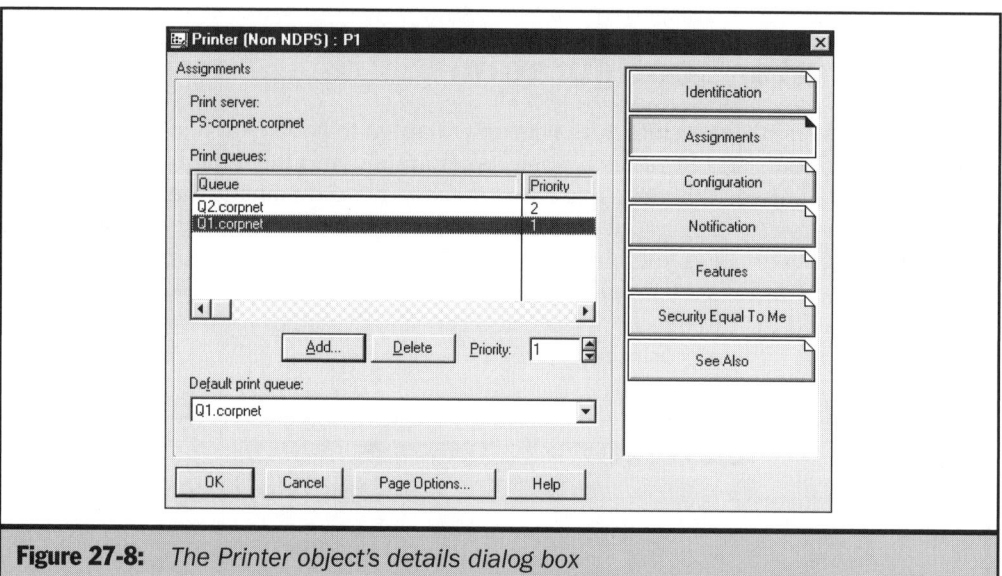

Figure 27-8:   *The Printer object's details dialog box*

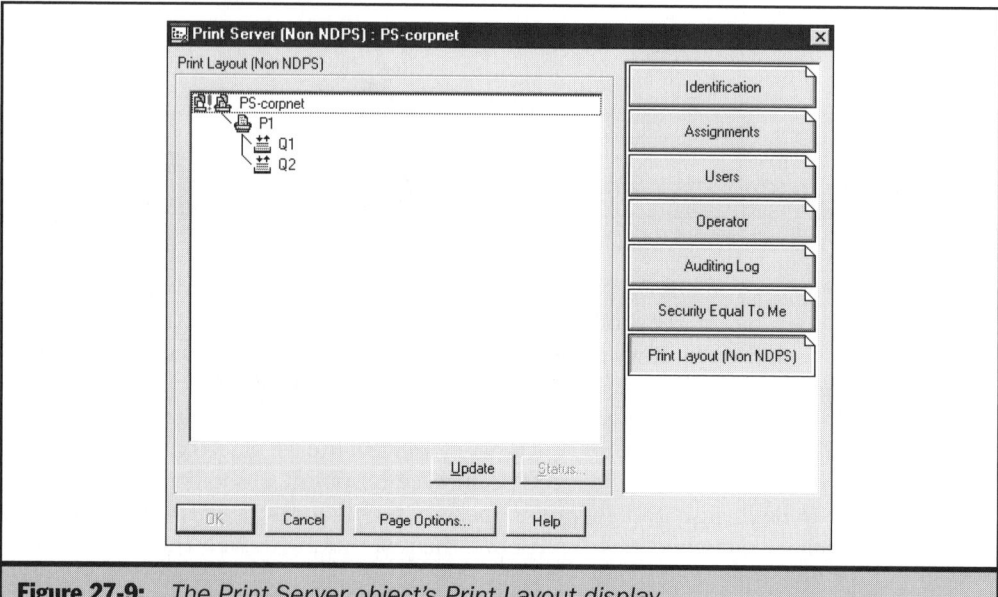

Figure 27-9:   *The Print Server object's Print Layout display*

Unix Printing

The dozens of different Unix variations use a wide variety of printing solutions, but one of the most common is the *line printer daemon (LPD)* running on a workstation. A *daemon* on a Unix system is a program that runs continuously in the background, much like a service in Windows. LPD is a daemon that causes the system on which it's running to function as a print server. Client systems can send their print jobs to the LPD system, where they're spooled until the printer can service them. Thus, the basic architecture of Unix network printing consists of clients, servers, and jobs spooled into a print queue, much like on Windows and NetWare networks.

Once a printer is set up for local use, configuring the system to function as a print server generally consists of creating a spool directory, where the print jobs will be stored while they await processing, and a printer capability database, called a *printcap* file, that specifies the names of the printers the server manages and their capabilities. The daemon accesses the printcap file every time a file is submitted to the spooler for printing. Administrators control access to the printer by listing the names of authorized users in a hosts.lpd file.

The protocol for communications between the line printer daemon and client systems is defined in the RFC 1179 document, "Line Printer Daemon Protocol," published by the Internet Engineering Task Force (IETF) in August 1990. RFC 1179 is an informational document, not a TCP/IP standard, that specifies the print server functions created for the BSD Unix operating system in the 1980s. The LPD protocol calls for the print server running the daemon to listen on port 515 for incoming TCP connections from clients. The clients must use a port number ranging from 721 to 731 (inclusive) and, once connected, can send print jobs to the server as well as commands that enable them to check the status of and manage queued jobs.

Most Unix applications provide direct access to print functions, just like on any other operating system, but client systems can also use a variety of commands to submit print jobs to the LPD and manage existing jobs, the names of which can vary depending on the Unix implementation. The primary command for submitting jobs is lp (or lpr), the syntax of which is as follows:

```
lp [options] filename
```

The command has a wide array of options that enable users to specify which printer should receive the job, specify how the job should print, and control the use of banner pages, among other things. The lp command is intended for sending ASCII print jobs to a printer, possibly with a few basic formatting commands. It is also possible to send PostScript files to a printer using the lp command. Other print-related commands are as follows:

lprm (or cancel)   Removes jobs from the print queue

lpq (or lpstat)   Displays the contents of the print queue

An LPD print job consists of two files, a data file that contains the actual data to be printed, and a control file that contains information about the data file, such as the name of the document being printed and its attributes. You can send either file to the print server first, but there are some LPD implementations (such as routers that support printers connecting to asynchronous ports) that ignore the contents of the control file because they have no facility for storing large data files, and instead feed them directly to the printer, even if the control file hasn't yet arrived.

The
Complete
Reference

Networking

Chapter 28

Connecting to
the Internet

909

Internet access has become an all but ubiquitous part of computer networking. Even if employers don't want the staff surfing the Web on company time, they probably do provide Internet e-mail and may rely on the Internet for other vital business services as well. This chapter examines the process of connecting a network to the Internet, providing users with the services they need, and protecting the network from outside intrusion.

Providing network users with Internet access is a matter of establishing an Internet connection and sharing it with the rest of the network. The basic steps of the procedure are as follows:

1. Contract with an Internet service provider (ISP) for a connection.

2. Install the hardware needed for the connection.

3. Install and configure a router that will connect your LAN to the ISP's network.

4. Configure your client systems to access the Internet using the router.

5. Arrange for the clients to have access to the Internet services they need.

6. Protect your network from unauthorized access by Internet intruders.

Selecting an ISP

The Internet is essentially a pyramid of service providers, with large backbone networks at the top, connected to regional networks that sell bandwidth to local ISPs, which in turn split the bandwidth into smaller units and resell it to other providers or end users, as shown in Figure 28-1. The providers higher up in the pyramid use high-bandwidth connections to a backbone (sometimes called *fat data pipes*) to handle all the traffic generated by their client ISPs, which use smaller pipes. The place that your organization occupies in the food chain depends on how much bandwidth you need, what type of connection you'll use to get it, and which ISP provides it for you. The bandwidth you obtain from your provider might have been resold several times before it gets to you, but this is not necessarily a bad thing, as long as you obtain your service from a reputable and well-equipped ISP.

In a business environment, Internet access can be crucial. When your users can't exchange e-mail with business partners, and customers can't access your Web site, business suffers. Therefore, it is important that you select an ISP that can be relied on to supply you with a consistent stream of bandwidth. The more bandwidth you require, the better the ISP's facilities have to be to support your connection reliably.

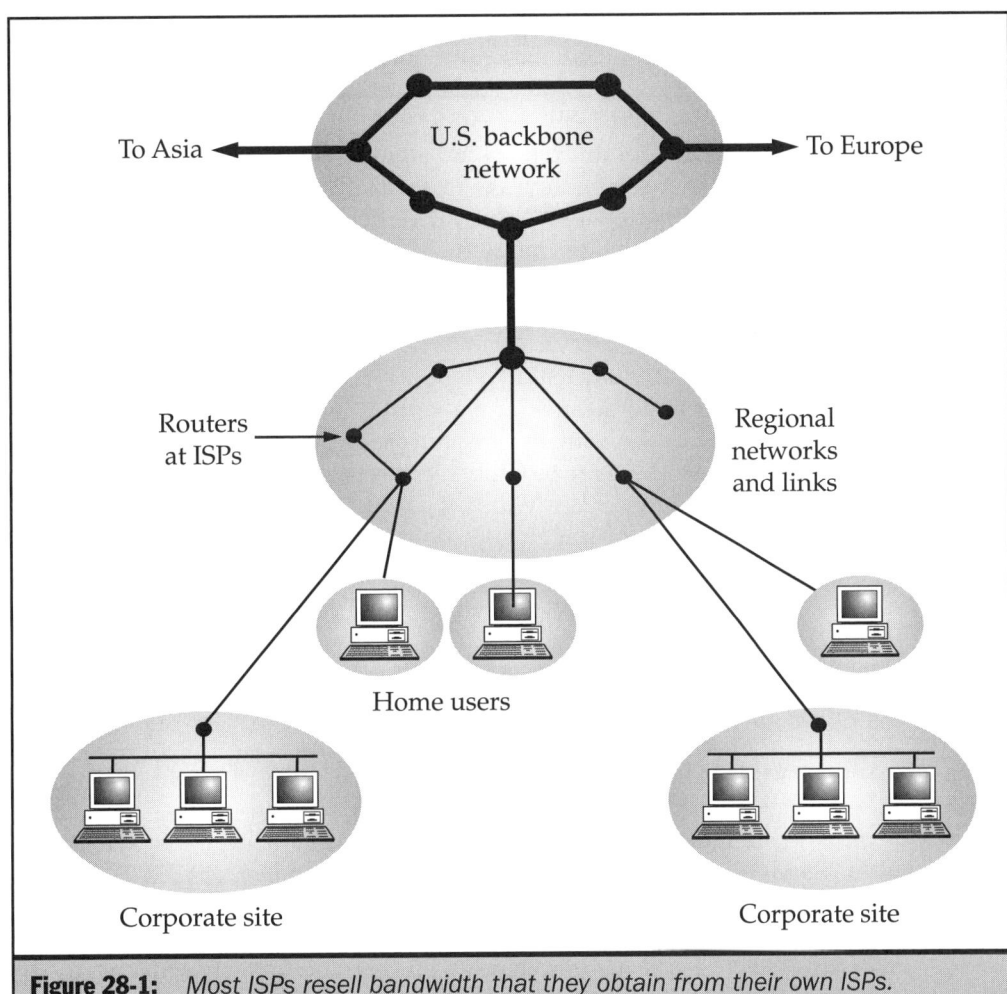

Figure 28-1: *Most ISPs resell bandwidth that they obtain from their own ISPs.*

For example, a small, local ISP that services dial-up users might have a T-1 connection (running at 1.544 Mbps) to its own provider, which it subdivides into a number of 56 Kbps connections. A single T-1 can support 25–30 user connections at this speed, and the ISP might sell 50 or more subscriptions, because it knows that not every user is connected all the time. This also means that it is possible for some users to be unable to connect at times, because all of the lines are in use. The reliability of the ISP's service is based on its ratio of subscribers to available connections. A failure to connect may not be a major problem for a home user, who can simply try again later, but a business has more stringent requirements.

In most cases, businesses do not use dial-up Internet connections, but it is possible to provide a small number of users with basic e-mail access using a shared dial-up that connects on demand or at regular intervals. For this arrangement to work, the ISP must have a client/connection ratio that enables the user to connect reliably. There is often no way to determine the reliability of this type of ISP other than to use its service for a while. The size of the provider's company is not necessarily indicative of the quality of its service. One of the largest national ISPs recently came under harsh criticism for signing up far more subscribers than its infrastructure could handle. However, since this type of connection doesn't require specialized equipment or large installation fees, changing providers is easy.

Connecting a large number of users to the Internet requires more than a dial-up connection. For large organizations, a leased telephone line is a more common solution. A T-1 line can provide Internet connectivity for about 100 average users, and fractional T-1 services are available in 64 Kbps increments. A leased-line connection to the Internet is an expensive undertaking. There are monthly fees for both the telephone company providing the connection and the ISP providing the service, which can easily run to $2,000 or more. There are also significant installation and equipment fees, and because a leased line is a permanent connection between two sites, changing providers is a major production. Therefore, you must be sure to select an ISP that can adequately and reliably support your needs.

| Note |

For more information on leased lines and other WAN technologies, see Chapter 7.

Many of the ISPs that cater primarily to dial-up users also offer T-1 connections and other business services, but they may or may not be properly equipped to support them. A provider that offers T-1 service should have at least a T-3 line of its own (running at 44.736 Mbps) that connects it to the Internet. Remember, any ISP that you deal with has its own ISP, from whom it purchases the bandwidth that it resells to you. If its service goes down, yours does too. That's why the best providers offering business services have redundant T-3s with multiple providers high up in the Internet hierarchy, to ensure that your service is not interrupted. About 90 percent of all Internet traffic is carried by the networks owned by Sprint, Cable & Wireless, and UUnet Worldcom. A provider connected directly to a backbone on one or more of these networks is more likely to provide better service than one that is connected farther down in the pyramid.

ISPs are also just as liable as any other type of business to experience failures of other kinds, such as those caused by power failures and natural disasters. The best providers also have facilities that can keep their services running in spite of these problems, such as backup power supplies in the form of battery arrays and/or generators. ISPs that have advanced capabilities like these are usually quite proud of them and should be more than willing to tell you all about them, or even give you a tour of their facilities. This type of investigation is well worth the effort when you are planning to commit your organization to a contract that will cost you many thousands of dollars a year.

In between the dial-up service and the leased line are other types of Internet connections, such as ISDN, DSL, and cable modems, that are popular in the home market, but that also have potential for widespread business use. All of these connections are constrained by the limitations of the technology. ISDN and DSL connections require that the telephone company providing the line have a switching office relatively close to your site, and cable modem connections are usually provided by the cable television company that services your area. Of the three connection types, only ISDN offers the possibility of changing service providers without an additional installation fee, as it is a dialed, circuit-switching service. Switching to a new DSL or cable provider requires an entirely new installation, if a competing service is even available.

Connection Types

The following sections examine the technologies that you can use to connect a network to the Internet, and the advantages and drawbacks of each. These various types of Internet connections are essentially wide area network (WAN) links, all of which are discussed in more technical detail in Chapter 7.

Dial-Up Connections

Most users with standalone systems connect to the Internet with a standard asynchronous modem and a dial-up connection using a POTS (Plain Old Telephone Service) line. Although intended for use by a single computer, you can share a connection of this type with a network. The primary drawback is, of course, the relatively low bandwidth. Most ISPs today have the equipment to support 56 Kbps modem connections, but in the real world, the connections are not nearly this fast. The Federal Communications Commission limits the data transfer speed over POTS lines to 53 Kbps, but even this speed is rarely achieved. Connections of 40 to 50 Kbps are more common, due primarily to the condition of the lines connecting the user site to the telephone company's switching office. In addition, this higher speed is for downstream traffic only—that is, traffic from the ISP to the user. The maximum upstream speed is 33.6 Kbps, which is also the maximum speed for ISPs that do not have the digital links and other equipment needed for the faster connections.

Dial-up connections are the most economical way to connect to the Internet. Although the client is responsible for buying the modem, the market is quite competitive and modems are heavily discounted. A standard telephone line is needed, but there are thousands of ISPs available that provide dial-up service. It should be no problem finding one within your local calling area, thus keeping the telephone charges to a minimum. A typical unlimited dial-up account with an ISP costs about $20 per month, possibly with a small setup fee. But if you plan to share a dial-up connection with a network, you should be aware that ISPs can have different definitions of the term "unlimited." In almost every case, an unlimited account does not enable you to remain connected to the Internet 24 hours a day. Most ISPs impose a limit on their "unlimited" accounts, either in the form of a timeout that disconnects you after a certain number of hours or a maximum number of connected hours you are permitted per month. Therefore, a dial-up connection to the Internet cannot provide network users with continuous access.

There are several methods you can use to overcome these limitations. Some ISPs offer a dedicated dial-up service. This is a dial-up connection with a port and telephone line at the ISP's site that are dedicated solely to your use. You can remain connected continuously for a single monthly fee. The cost of this type of connection is significantly more than that of a standard dial-up account, typically from $100 to $150 per month, but as long as you have local unlimited service for the phone line, there are no additional charges.

Another method is *dial-on-demand*, which is a feature provided by most of the router products that enable you to share a dial-up connection. Dial-on-demand causes the modem to connect to the ISP whenever a client generates a request to an Internet service. After a specified period of inactivity, the connection is then terminated. This enables the modem to reconnect as needed without manual intervention, while also conforming to the ISP's policies. This method saves money by using the ISP's standard dial-up account, although the savings over a dedicated dial-up are slightly mitigated by the additional telephone calls needed to reestablish the connection.

In addition to these solutions to the problem, there is also the low-tech option, in which you use a standard dial-up account and a user is responsible for periodically disconnecting and reconnecting to the ISP. This may seem like a viable alternative, unless you are the person with that responsibility.

The downside of sharing a dial-up connection is obviously the limited bandwidth. A dial-up connection, though intended for a single user, can conceivably serve two or three users on a network, as long as they don't require superior performance. Shared dial-ups are more common in residential settings than in business settings for this reason. However, it is possible to combine the bandwidth of two or more dial-up connections into a single data pipe. The technique by which this is done is called *inverse multiplexing*; Windows 98, NT, and 2000 call this feature *multilink*.

A multilink installation requires a separate COM port, modem, telephone line, and ISP port for each of the connections to be combined. In addition, the ISP must support the multilink feature, and this is one of the factors that has prevented this technology from achieving widespread use. Another limiting factor is that there are now other, faster connection types available, such as cable modems and DSL, that provide far better service at a lower cost.

ISDN

Integrated Services Digital Network is a dial-up service that operates over standard POTS lines, but that uses digital connections at higher speeds than standard analog modems. There are two ISDN services: *Basic Rate Interface (BRI)*, which provides

128 Kbps of bandwidth, and *Primary Rate Interface (PRI)*, which runs at up to 1.544 Mbps. BRI service is directed primarily at the home market, whereas PRI is directed at businesses. Both have a far greater installed user base in Europe than in North America.

ISDN was designed as a general-purpose digital replacement for the current analog phone system and has been around since the mid-1980s, but it received very little commercial attention in the United States until the Internet boom sent people searching for faster connections. There are many ISPs that offer ISDN Internet connections, mostly using the BRI service, but the technology has a reputation for being difficult to install and cranky in its operation.

The primary advantages of ISDN are that it provides dedicated digital service at the specified bandwidth and that it is a dial-up service, just like the analog phone system. Once you have the ISDN service installed, if you ever want to change ISPs, all you have to do is change the numbers dialed; no hardware or service modifications are necessary. The PRI service provides the same bandwidth as a T-1, but it actually consists of 24 channels of 64 Kbps each that can be combined into a single data pipe or used independently for voice or data traffic.

The ISDN service itself must be installed by a telephone company. No special wiring is required for the BRI service, but the line plugs into a hardware device called an *NT1* at the client site, which in turn is connected to a *terminal adapter,* which provides the interface for the computer or router. In most cases, you must purchase the NT1 and the terminal adapter yourself, although several manufacturers integrate the two into a single device. In addition to the hardware, there is an installation fee from the phone company, a monthly charge, and usually a connection charge of about one cent per minute. Because it is a dial-up service, you can disconnect from your ISP during off hours to minimize the connection charges.

These fees are all in addition to the ISP charges for Internet access. Some providers offer a turnkey solution in which they arrange for the ISDN installation by the phone company, while with other providers, you must arrange for the installation yourself. The end result is that ISDN can be an expensive proposition, and compared to newer technologies like DSL and cable modems, the increase in bandwidth for the average home user (from 40 or 50 Kbps to 128 Kbps) is not worth the difficulty and the expense. You can usually share a BRI ISDN connection with a network by using the same hardware or software routers as you would use for a standard dial-up, but the additional bandwidth supports only a few more users.

DSL

Digital Subscriber Line connections provide high-speed Internet access over standard POTS lines, much like ISDN, except that the connection is a permanent link between the client site and the ISP and usually runs at much higher speeds. The most popular DSL service for Internet access is Asymmetrical Digital Subscriber Line (ADSL). An *asymmetrical* connection is one with a downstream speed that differs from the upstream speed. ADSL providers typically offer the service at a variety of speeds and prices, which can run as high as 8.448 Mbps downstream and 1.544 Mbps upstream. The difference in upstream and downstream speeds makes the connection a good choice for Internet client access, which consists mainly of downstream traffic, but not for hosting Internet servers, which requires more upstream bandwidth.

In many cases, Symmetrical Digital Subscriber Line (SDSL) service is more suitable for business Internet access, because it provides various bandwidths running at the same speed downstream and upstream. Compared to leased lines and other types of connections providing similar transmission speeds, DSL is an economical alternative that requires a relatively small initial investment.

DSL is the latest Internet access technology to hit the market, and it is being marketed primarily to home users, but there is no reason why DSL connections could not eventually become a popular business solution as well. The only hardware required is a DSL "modem" and a standard Ethernet card for the computer hosting the connection. Most ISPs supply the hardware, on either a leased or purchase basis, as part of a subscription package that includes a monthly fee and an installation charge. No additional wiring is needed, since DSL enables you to continue to use the phone line for voice traffic while you're connected to the Internet.

Cable Modems

Many cable television providers are taking advantage of the broadband fiber-optic networks they have installed in neighborhoods all over North America to supply Internet access in addition to the TV signals. Since most potential subscribers already have the cable wired into their homes, and the cable company already has a fleet of service technicians, the installation process consists of a cable technician putting a splitter on the provider's coaxial cable and connecting it to a modem-like device. This unit functions as the interface to the cable network and also connects to a standard Ethernet card installed in a computer. Since the same company is providing the hardware, the cable, and the Internet access, there is only one bill to think about, which is generally approximately $40 per month (not including the cable TV service). The hardware may be leased, with the cost included in the monthly fee, or you may have the option to buy it from the cable company in return for a lower monthly charge. Be aware, however, that this is relatively new technology that has not yet been standardized in the same way as analog modems. You will very likely not be able to use the same hardware if you move or switch to another cable system.

Cable modems deliver excellent bandwidth, as much as 512 Kbps, but they are asymmetrical, like DSL, so this speed is restricted to downstream traffic. This asymmetry is due to the fact that cable television networks are designed primarily to send signals from the provider to the subscriber. Relatively little upstream bandwidth is available, so the upstream speed is typically capped at 128 Kbps or less. Even this speed is as fast as a BRI ISDN connection and far faster than any dial-up connection. Therefore, while a cable modem connection is an excellent solution for Web surfing and other Internet client applications, even when shared among many network users, you probably would not want to use one to run a heavily trafficked Web server.

The potential downside to cable access is that, unlike all of the other connection types discussed here, the cable network uses a shared network medium. The fiber-optic network owned by the cable system is essentially a *metropolitan area network (MAN)* that joins you to your neighbors on a large Ethernet network. The greater the number of other users accessing the Internet, the less bandwidth that is available to you. As a result, you may notice a slowdown in the service during peak usage hours. In addition, you must be careful to secure your network against intrusion from other users on the same MAN.

Cable companies are used to dealing primarily with residential users, and many of them intend their service for use only on standalone computers, not networks. However, there is no technological reason why you can't share a cable modem connection with a network, as long as the cable company permits it. Some cable systems are new to the Internet-access side of the business, and they might not yet have a policy in place regarding the sharing of the connection. Others might specify a maximum number of users that you can connect to the system for the standard subscription fee, or might charge a much higher rate for a network connection. For the amount of bandwidth provided, cable modem access to the Internet is currently the best bargain in the industry.

Leased Lines

Most of the connection types discussed thus far are primarily marketed at home Internet users, but also can be shared with a network in a business environment. Leased lines are the Internet access standard for the business world because they provide a large amount of consistent, continuous bandwidth. The most common type of leased line is the T-1, running at 1.544 Mbps, which can function as a single data pipe or be split into 64 Kbps voice channels. Individual 64 Kbps channels are also available from some providers; this is called *fractional T-1* service. A leased line is a permanent connection between two sites, installed by a telephone carrier, that runs continuously at a specific bandwidth. The cost of the line itself is based on the distance between the two sites. Since there are many ISPs that offer T-1 service, it should not be difficult to find one nearby. However, even a short-distance T-1 line can cost $1,500 per month or more, plus a hefty installation fee and the cost of the required hardware, called a *channel service unit/data service unit (CSU/DSU)*, which can be over $1,000. This is in addition to the ISP charges, which will also run at least $1,000 per month. A leased line, therefore, is an expensive proposition, but it can service a large number of Internet users and enables you to host your own Internet servers.

Because a leased line is a permanent connection, you cannot easily switch ISPs. You must have the old line removed and a new one installed, which can take a great deal of time and money. Therefore, you should be careful to choose a provider that is going to be around for a while and that can provide all of the services you are liable to need in the future.

Frame Relay

Frame relay is an alternative to a leased line that lets you pay only for the bandwidth you use. When you install a T-1 line for Internet access, you pay for the full amount of bandwidth at all times, even when you're not using it. Frame relay is a type of WAN link that uses a leased line to connect to a local provider, which routes your traffic through its network to another provider, which in turn is connected to your destination site using another leased line. The concept is a little different for an ISP connection. You still must install a leased line between your site and the ISP, and pay all of the telephone company charges for that line, but the ISP's fees are substantially lower because you are not paying for bandwidth that goes unused when your company is closed. When you contract with an ISP for frame relay service, you agree to a *committed information rate (CIR)* that specifies the amount of bandwidth that will always be available to you. At times of increased traffic (called *bursts*), you can exceed the CIR, for an additional charge. If your needs increase over time, you can negotiate a new CIR.

Thus, in a typical installation, you would install a T-1 line between your site and the ISP, and contract for a CIR of 768 Kbps, for example. You are guaranteed to have 768 Kbps available to you at all times, but you are not paying for bandwidth during times when your offices are closed or not in use. During high-traffic periods, your transmission rate may burst up to the full capacity of the leased line, as long as the ISP has the additional bandwidth available. Frame relay does not affect the cost of the leased line or the hardware needed to connect it to your network, but it can reduce ISP fees from 20 to 40 percent in cases where you do not utilize all of the bandwidth of the T-1 (or other leased-line connection).

Bandwidth Requirements

Although the fundamental principles of connecting a LAN to the Internet are always the same, the amount of bandwidth you need for your Internet connection will influence much of the process, from purchasing equipment to selecting an ISP. Table 28-1 lists the most common Internet connection types used today, the bandwidth they supply, and the approximate number of clients they can support for particular applications.

The estimates provided in the table refer to the approximate number of clients of each type that can use the given connection at any one time. They do not indicate that the connection will support the number of all three clients given simultaneously. For example, a 33.6 Kbps dial-up connection will support up to six e-mail users, two Web browsers, or two FTP clients, but not all three at the same time. TCP/IP traffic tends to be "bursty," meaning that the amount of data traveling over the cable is more likely to come in spurts rather than a continuous stream.

NETWORK SERVICES

Connection Type	Approximate Actual Speed	Applications
Basic dial-up	28.8–33.6 Kbps	E-mail for up to 6 users Web browsing for 1 or 2 simultaneous users Large FTP downloads for 1 or 2 simultaneous users
High-speed dial-up	Up to 53 Kbps	E-mail for up to 10 users Web browsing for 2 to 3 simultaneous users Large FTP downloads for 1 or 2 simultaneous users
ISDN	128 Kbps	E-mail for up to 20 users Web browsing for 6 to 8 simultaneous users Large FTP downloads for 3 or 4 simultaneous users
Cable modem	Up to 512 Kbps downstream; up to 128 Kbps upstream	E-mail for 50 or more users Web browsing for 25 to 30 simultaneous users Large FTP downloads for 12 to 15 simultaneous users
DSL	Up to 640 Kbps downstream; up to 160 Kbps upstream	E-mail for 60 or more users Web browsing for 30 to 35 simultaneous users. Large FTP downloads for 15 to 18 simultaneous users
T-1	1.544 Mbps	E-mail for 120 or more users Web browsing for 75 to 100 simultaneous users Large FTP downloads for 40 to 50 simultaneous users

Table 28-1:   *Internet Connection Types and Estimated Bandwidth Utilization*

The table also refers to users who are actually working in the client programs specified. You may budget a T-1 connection as supporting 100 Web users, but you can use that connection to provide Web access to several hundred users, as long as you don't expect too many more than 100 users to be surfing the Web at any one time. In most cases, network users have access to all three of the specified clients, and possibly others as well.

When you are evaluating Internet connection types and the needs of your users, be sure to account for the fact that a single user might be running more than one client at the same time, such as surfing the Web while downloading a file with FTP. You should also try to account for the growth both of your network and of the reliance of your business on the Internet.

The following sections examine the most common Internet applications and the issues that affect their bandwidth utilization.

E-mail

E-mail is one of the most basic Internet applications, and the one that in most cases utilizes the least amount of bandwidth. E-mail messages themselves consist of ASCII text, which takes up relatively little bandwidth. Therefore, you can support more e-mail users with a given amount of bandwidth than you can support any other type of Internet client.

The factor that complicates the equation is the ability to attach files of almost any size to e-mail messages. If you use a relatively low-bandwidth connection to provide network users with e-mail access, it is possible for a single large attachment to monopolize the connection for a long period of time. Therefore, you may want to impose a limit at the mail server on the size of the attachments that you permit e-mail users to send and receive. This is particularly true when you are using an asymmetrical connection, such as DSL or a cable modem. In these cases, you should calculate your attachment size limit in relationship to the upstream transmit rate rather than the downstream one, unless you are confident that an internal policy against sending e-mails with large attachments will be observed by your users.

For more information on e-mail and other Internet services, see Chapter 26.

Web Surfing

Web pages are primarily made up of HTML files, which are all ASCII text, and relatively small image files, in most cases. Therefore, the amount of data that makes up a page is fairly modest. However, there are several factors that make Web surfing a more bandwidth-intensive application than e-mail.

E-mail is an application in which a delay between the transmission of a message and the receipt of a reply is expected, and delays caused by a busy Internet connection are generally acceptable, within reason. Users accessing the World Wide Web expect a more immediate response. Delays in the receipt of the files that make up a Web page are more frustrating to users, especially in a business environment. A large download or other form of congestion on a low-bandwidth connection can bring Web performance to a virtual halt. Another factor is the ability of Web pages to link to other file types that require more bandwidth, such as binary program files, large images, and sound files.

Because of these factors, you should count on an Internet connection supporting fewer Web users than e-mail users. In a home environment where people use the Internet primarily for recreation, two or three users can conceivably share a connection intended for one, because they can more easily wait until later if the delays become too great. In a business environment where users are (presumably) using the connection for business purposes, delays can result in reduced productivity and other negative effects.

It's also important to consider the amount of time that you expect your network clients to spend using the Internet connection. The Web is the application most prone to abuse, so you should calculate the amount of time that you would expect the average user to spend surfing for business purposes and double it to account for their other interests, unless you plan on implementing a mechanism that monitors or restricts their access.

Web traffic utilizes much more downstream bandwidth than upstream. The messages generated by Web browsers are mostly small requests for specific URLs transmitted to Web servers. The servers respond by supplying the requested files, which, while usually not enormous in size, are far bigger than the requests. Therefore, asymmetrical Internet connections like those provided by cable modems and DSL are eminently suited for Web browsing.

FTP

Once the mainstay of Internet communications, the File Transfer Protocol has been eclipsed by the Web in popularity, but it is still widely used. Many of the hyperlinks on Web pages that trigger file downloads use the FTP protocol instead of the Hypertext Transfer Protocol (HTTP), the standard for Web communications. The files downloaded using FTP are usually binaries, such as program and image files, and are often larger than the files that make up a home page.

When triggered from Web links, FTP communications tend to be intermittent. A typical user might perform one or two FTP downloads in the midst of an hour of Web surfing. Dedicated FTP clients spend the majority of their time performing file transfers and can consume a great deal more bandwidth than Web browsers. Serious FTP users also tend to transfer larger files, some of which can be enormous. Advanced users involved in software beta testing might have to download dozens or even hundreds of megabytes worth of program files in the course of the testing process.

Because their bandwidth needs are greater and more continuous, you should count on FTP clients utilizing more bandwidth than virtually any other Internet application. Another difference between Web-triggered FTP and a dedicated FTP client is that Web links produce downstream file transfers only, while an FTP client can transfer files in either direction. Asymmetrical connections are suitable for Web-based FTP transfers, but not for dedicated FTP clients if upstream transfers will be frequent.

Usenet News

Usenet is the Internet news service, but it is not "news" in the traditional sense of the word. Usenet is actually a collection of messaging forums numbering in the tens of thousands, called *newsgroups,* on every topic under the sun, from computer-related subjects to entertainment, that are maintained on news servers located all over the Internet. Users access these forums with a client program called a *newsreader* that downloads the latest messages for particular forums and posts their replies back to the server.

Much of the traffic generated by Usenet is plain ASCII text, much like e-mail, so the burden on an Internet connection is similar to that of an e-mail client. Some of the newsgroups are dedicated to the posting of binaries, such as images or program files, encoded into ASCII text. These binaries are the functional equivalent of files attached to e-mails, as far as bandwidth utilization is concerned.

As long as users are accessing a news server on the Internet or one supplied by your ISP, you can consider a Usenet client as the equivalent of an e-mail client. If you plan to run your own news server, the bandwidth requirements are greater than for virtually any other Internet application. A full news feed consists of several gigabytes of data every day, which can tax even the fastest connection.

Web Site and FTP Hosting

The bandwidth needed for client access to the Internet is relatively quantifiable, but when you run your own Web and FTP servers, the traffic is determined by the number of outside users that access them. In most cases, businesses make a concerted effort to draw as much traffic to their Web sites as possible, so you must be sure that you have sufficient bandwidth to support the maximum number of users you can realistically expect.

When running a business Web site, the consequences of having insufficient bandwidth can be far worse than when supporting only internal Internet clients. Temporarily inconveniencing your employees is nothing compared to alienating potential customers and portraying your business in an unprofessional light. If you plan on using promotional activities to draw Internet users to your site, you must be prepared to handle them.

Unlike the traffic generated by Internet client programs, which mostly run downstream, Web and FTP servers generate mostly upstream traffic, so asymmetrical technologies like ADSL and cable modems are unsuitable for this purpose.

Internet Services

At the most basic level, an ISP supplies access to the Internet and nothing more. Additional services are required to use Internet applications, and you may have to obtain those services from your ISP if you don't plan on running them internally. For example, your clients will need access to DNS servers; you can use your ISP's servers or run your own servers on your local network. ISPs usually can also provide access to mail and news servers, and can host your Web site for you, for additional charges. All of these are services that you can provide yourself, however. For a large network with hundreds of users, running these services in-house is a practical solution. For small networks, it's generally better to obtain them from the ISP. The following sections examine these services and what's involved in running them.

IP Addresses

One of the few things that you can only get from an ISP, apart from the Internet connection itself, is the registered IP address (or addresses) that Internet systems will use. Every client system connected to the Internet must have an IP address, but only systems with registered IP addresses are visible from the Internet. ISPs obtain registered addresses either from the Internet Assigned Numbers Authority (IANA) or from their own service providers, and sublet them to their clients.

Note   *For more information about registered and unregistered IP addresses, see Chapter 13.*

Depending on both the type of connection you use to access the Internet and the configuration of the ISP's own network, registered IP addresses may or may not be available. Cable systems typically run their own private, unregistered networks and assign unregistered addresses to the computers connected to it. The computers then access the Internet using a network address translation (NAT) or proxy server, which makes them appear to have a registered address. When this is the case, your clients must all use unregistered addresses, meaning that they can access the Internet, but Internet users cannot access their machines. This is perfectly acceptable for Internet client applications, but you cannot run a Web or FTP server using this arrangement.

If you have a T-1, ISDN, frame relay, or even a dial-up connection, your ISP should be able to provide you with as many registered IP addresses as you need, although there may be an additional charge for them. Generally speaking, you should only use registered IP addresses on Web servers and other computers that must be accessible from the Internet. Every system with this type of address is vulnerable to unauthorized access from the Internet and must be protected against intruders.

For more information about protecting your systems from unauthorized access and using unregistered IP addresses to access the Internet, see "Firewalls," in Chapter 29.

DNS

The Domain Name System is the service that converts the friendly names by which Internet servers are known into the IP addresses needed to communicate with them. Internet clients need access to a DNS server in order to process the URLs, server names, and e-mail addresses they supply in their applications. All ISPs have DNS servers, and in many cases a client system is automatically configured to use them when connecting to the provider. A system using a dial-up Internet connection receives an IP address and DNS server addresses from the ISP's server during the network logon sequence. Whenever the user of the client system types a URL or clicks a hyperlink, the computer sends a name resolution request to the ISP's DNS server, which returns the IP address of the specified destination.

A system that is directly connected to the ISP through a modem or other connection can be automatically configured with the addresses of DNS servers, but when you share the connection with other network users, you must configure the client systems with the DNS server addresses either manually or by using some other means, such as a DHCP (Dynamic Host Configuration Protocol) server.

Internet client systems use the Domain Name System only to resolve the names of the servers on the Internet that they want to access. If you run Web or FTP servers of your own, you must have a DNS server that is configured with resource records containing your servers' names and their equivalent IP addresses. This is necessary in order for other users on the Internet to be able to access your servers by specifying a name, such as www.zacker.com. When your ISP or a hosting service runs your Web and FTP servers for you, adding the appropriate resource records to the provider's DNS servers is usually part of the package.

To register a domain name, you must contact one of the registrars that are charged with maintaining the records for specific top-level domains, such as com, org, and net. At one time, a company called Network Solutions (http://www.networksolutions.com/) was the sole registrar for these domains, but there are now many organizations that function as domain name registrars.

For more information on the Domain Name System, running DNS servers, and registering your own domain name, see Chapter 25.

You can, however, run your own DNS servers, both for use by your clients and for hosting your own Web and FTP sites. Most server operating systems, such as Windows 2000, NT, NetWare, and most forms of Unix, include DNS server software. The computer functioning as a DNS server must have a registered IP address so that it is visible from the Internet. If you are hosting your own Internet servers, you must register the addresses of the DNS servers with the authority that provides you with your domain name.

Mail Servers

In order for your users to send and receive Internet e-mail, they must have their own e-mail addresses and access to mail servers. Internet e-mail is based on the Simple Mail Transfer Protocol (SMTP), which carries e-mail traffic between servers and defines the well-known addressing format *username@domain*.com. Most ISPs provide e-mail service to their clients by using two types of mail servers: an SMTP server for sending outgoing mail, and a Post Office Protocol (POP3) or Internet Message Access Protocol (IMAP) server for receiving incoming mail. Both of these servers (which may or may not run on the same computer) have registered IP addresses, so that they are visible from the Internet.

Most of the Internet access accounts provided by ISPs, especially those intended for home users, include at least one e-mail address and access to the provider's mail servers. Some ISPs provide multiple e-mail addresses with the account, either for free or for a nominal additional fee. Unlike DNS servers, you must manually configure each e-mail client to access the appropriate servers, whether the computer is connected directly to the ISP or accesses it through the network.

When you share an Internet connection with a network and use the ISP's mail servers, your provider must either supply an e-mail address and POP3 or IMAP server account for each of your users, or give you a shell account that enables you to connect to the server and create the accounts yourself.

If you choose to run your own mail servers, you can perform all of the account maintenance tasks yourself and create as many accounts as you wish, but there are other factors you must consider. As with DNS servers, the computers running the e-mail server software must have registered IP addresses, and the SMTP server must be connected to the Internet at all times, in order for e-mail messages from outside users to find it. If the server is down or disconnected, mail messages directed to any of the accounts bounce back to the sender. E-mail server software is not included with server operating systems, and prices for e-mail server products can vary greatly. Full-featured e-mail products for Windows, like Microsoft Exchange, can be quite expensive, while sendmail, the most popular mail server on the Internet, is free. You must also have your own registered domain name and the appropriate resource records in the DNS, so that e-mail messages sent by users elsewhere on the Internet can find their way to your server.

Hosting your own e-mail servers is a major commitment in both time and resources, and thus is recommended only for network administrators with a large number of users and a full-time, high-speed connection to the Internet, such as a T-1.

News Servers

Most ISPs provide access to a Usenet news server, and in most cases, there is no persuasive reason to run your own, because of the enormous amount of data required for a full news feed. Some companies run news servers in order to host their own newsgroups for technical support purposes or product announcements. A private news server does not have to receive the Usenet news feed, so its bandwidth requirements are no more than that of a Web server.

Web Hosting

Many ISPs offer Web site hosting services and include a small site with an Internet access account. This level of service is usually only suitable for a personal Web site, and if you want a professional site, you will have to pay an additional monthly fee for the disk space and other services. As with e-mail and DNS servers, you can have your ISP or an outside service host your Web site, or you can run it yourself.

As with e-mail servers, Web servers must have registered IP addresses and be visible from the Internet. You also need a permanent connection with sufficient bandwidth for the traffic your site will generate, so that it is continuously available. Web server software is included with most server operating systems, but there are many reasons why it is better to use a service to host your Web sites instead. Many Web hosting services have redundant Internet connections that provide all the bandwidth you need and fault-tolerant systems that can compensate for potential disasters like power outages and disk failures. Some services can also provide access to advanced features such as database servers and e-commerce tools that enable you to build a state-of-the-art Web site. These are all features that you can conceivably provide yourself, and for a large enough company, it may be economical to do so. But for most Web sites, a hosting service can provide a more reliable and consistent presence on the Web.

Internet Routers

To share an Internet connection with a network, you must have a router that provides the interface between your internal LAN and your ISP's network. The ISP network, in turn, is connected to another ISP or to an Internet backbone. Most people envision a router as an expensive standalone device used only on large networks, but an Internet router can also take the form of a software program running on a PC or a small hardware device designed to provide Internet access to a small network.

The router, in its basic form, receives the packets generated by the workstations on the network that are intended for the Internet, repackages them for transmission over the connection to the ISP, and sends them on to the ISP's server. The ISP then forwards them to the appropriate Internet destination. The format of the packets generated by the router depends on the type of Internet connection you're using. Since the router is a network-layer device and the Internet uses TCP/IP communications exclusively, the packets all contain IP datagrams that may be repackaged several times by different data link–layer protocols on the way to their destination.

At one time, the high-end standalone routers intended for use on large networks performed only the standard functions associated with routers. Today, the routers designed for connecting networks to the Internet often include other features as well, such as firewall capabilities, Network Address Translation (NAT), and DHCP servers. These features are designed to make the process of connecting your network to the Internet easier and safer.

Software Routers

A software router is a program that enables a computer to share its connection to the Internet with other systems on the LAN. Some operating systems, such as Windows 2000, can route the IP traffic to the Internet themselves, but others, such as Windows 98, cannot. A software router program provides this IP routing capability, as well as other functions that simplify the process of sharing an Internet connection. Windows 98 Second Edition includes a software routing service called Internet Connection Sharing (ICS), and third-party products, such as ACT Software's NAT32, and ITWIN Technology's WinRoute, provide roughly the same features. These products are designed to work on a small scale, providing Internet access to users on a small LAN. You would not use them to provide access to dozens or hundreds of Internet clients.

Note	*For information on NAT32 and to download a trial version of the program, see http://www.nat32.com/. For information on WinRoute, see http://www.itwin.com.my/winroute.php.*

Like any router, the computer providing the Internet access is connected to two networks: the LAN, using a standard network interface card, and the ISP's network, using a modem, ISDN, DSL, or cable modem connection. The Internet connection itself is not modified in any way by the router software. To the ISP and to Internet systems, all of the traffic appears to be coming from the router. While performing its router functions, it is still possible to use the connected system as a normal workstation.

Each of the other workstations on the LAN accesses the Internet by using the router system's IP address as its default gateway. Most Internet client programs can access the Internet through the router with no modification.

For more information on routers and routing, see Chapter 6.

Network Address Translation

Software routers are typically used to share connections that are normally intended for single users, such as dial-up, BRI ISDN, and cable modem connections. The providers of these types of connections typically do not furnish registered IP addresses for multiple systems as part of the account, so the router software uses a technique called *Network Address Translation (NAT)* to enable the network clients to use unregistered IP addresses.

Windows 2000 is capable of routing IP traffic, but to use an Internet-connected Windows 2000 system as a router, you must use registered IP addresses for all of the client workstations on the network. Not only can this be an expensive proposition, since some ISPs charge $10 to $20 per month for each additional registered address, but it is also a dangerous one, because it can open up your network to outside intruders. Registered IP addresses make your workstations visible to any user on the Internet who cares to access them, and installations using connection-sharing on a small scale are rarely willing to invest in elaborate firewalls to protect the network from intrusion.

With NAT, you assign unregistered IP addresses to your workstations. The IANA has designated three ranges of IP addresses for use on unregistered networks, one for each address class. These addresses are not assigned to any specific network. By using them on your private network, you can be sure that there is no conflict between the addresses of your systems and those of public machines on the Internet. These addresses are listed in Table 28-2.

Class	Private Network Addresses
Class A	10.0.0.0 through 10.255.255.255
Class B	172.16.0.0 through 172.31.255.255
Class C	192.168.0.0 through 192.168.255.255

Table 28-2: *Unregistered IP Addresses for Use on Private Networks*

When a client system on your network sends a packet to a server on the Internet, the packet's IP header contains the IP address of the sender, which is unregistered. If this packet was to reach the destination server unaltered, the server would transmit its reply to the unregistered IP address. This attempt would fail, because Internet routers do not forward packets with unregistered destination addresses. Therefore, the router on your LAN translates the IP address of the sending system in each packet from the unregistered address of the client to the registered address of the router itself. The destination server can then send its response to the router, which forwards it to the appropriate client.

Thus, NAT enables any number of workstations to access the Internet using a single registered IP address. At the rate that the Internet is currently growing, the 32-bit IP address space is rapidly becoming depleted, and the use of NAT helps to conserve the registered IP addresses by enabling networks to use them only for the systems that must be directly accessible from the Internet. At the same time, the router protects the workstations from direct access by intruders on the Internet. Only the router itself has an address that is visible from outside, making it impossible to address IP traffic directly to a workstation on the LAN.

DHCP

Software router programs are directed at users who have little technical knowledge about networking, so they try to make the process of configuring the workstations to access the router as simple as possible. The program typically includes a DHCP service that automatically configures the TCP/IP client on the LAN workstations with an unregistered IP address and the address of the router as the default gateway. As a result, no manual configuration is required for the client workstations other than to specify that they use DHCP to obtain their TCP/IP settings.

Hardware Routers

There are also hardware-based Internet routers designed for use on networks of various sizes. This type of unit is essentially a special-purpose computer with its own processor, software, and IP addresses that performs all the functions of a software router and typically has other capabilities, such as packet filtering and support for multiple protocols. The primary difference is that, in this case, the router is a completely separate device, not reliant on any single workstation on the network and not as vulnerable as a workstation to interference. When you use a software router, all the clients on the LAN rely on that machine for their Internet access. If the computer goes down or the user must reboot, all Internet communications cease. A hardware router is wholly dedicated to providing Internet access and, unlike a regular workstation, is not subject to failure because of user error or an errant application.

Hardware routers connect both to the local network and to an ISP. Some devices have an Ethernet hub integrated into the unit, enabling you to plug the LAN workstations into it, while others just have an Ethernet interface, which you plug into a port on a standard hub. This latter type provides more versatility, because you can use any size hub, rather than being limited to the number of ports integrated into the former unit. The router also supports a connection to an ISP, with different products supporting various types of connections. Devices intended for use on home or small business networks might have a serial port or PC card slot for a standard modem connection, or they might support BRI ISDN or DSL. Higher-end units might support PRI ISDN, frame relay, or leased lines. Like software routers, these devices typically use NAT to provide client access to the Internet, and might include a DHCP server. Most units also have an integrated Web server that lets you configure the unit using a standard browser.

Hardware routers are significantly more expensive than software ones, ranging from a few hundred dollars for small business units to thousands for more advanced models, but they generally provide a more reliable connection-sharing solution.

Client Requirements

Part of the process of providing your network users with Internet access is the configuration of the client systems themselves. The Internet is a TCP/IP-based network, so if you have not already done so, you must configure all of the client workstations that need Internet access to use the TCP/IP protocols. Most LANs today use TCP/IP as their primary networking protocol; unless you have Novell NetWare servers that require IPX, no other protocols are necessary.

Every workstation that will access the Internet must be configured with the following TCP/IP parameters:

- IP address

- Subnet mask

- Default gateway address

- DNS server addresses

The primary consideration when you are assigning IP addresses is whether they are registered or unregistered. Some network administrators, when designing a private network that will not be connected to the Internet, choose IP addresses arbitrarily because no conflict with the Internet can exist. If they decide to connect the network to the Internet later, they must change all of the addresses.

The type of addresses you use is determined by the type of connection you will have to the Internet and the type of firewall you will use to protect your network from intrusion. If you're going to use NAT, for example, you must use unregistered addresses on your workstations. Other types of firewalls, such as those that filter packets, permit the use of registered IP addresses.

The default gateway address for each workstation is the internal IP address of the router that provides access to the Internet. If the workstation is on the same LAN as the router that is actually connected to the Internet, that router is the default gateway. On a multisegment network, the default gateway for some of your workstations will be a router that provides access to the segment on which the Internet router is located. On this type of network, you must configure your internetwork routers so they know the address of the Internet router, using either a static route or a routing protocol.

Each workstation must also be configured with the IP address of at least one DNS server, either on the local network or the ISP's network. Although it is still possible for a TCP/IP workstation to communicate with another system without a DNS server, the inability to resolve names means that the user must specify the IP address of the target system, instead of a friendly name, which is usually not practical. The addresses of two DNS servers are preferable, but the second one is only used if the first is unavailable.

The
Complete
Reference

Networking

Chapter 29

Network Security

933

S ecurity is an essential element of any network, and many of the daily maintenance tasks performed by the network administrator are security related. Simply put, all of the security mechanisms provided by the various components of a network are designed to protect a system's hardware, software, and data from accidental damage and unauthorized access. The goal of the security administration process is to provide users with access to all of the resources they need, while insulating them from those they don't need. This can be a fine line for the administrator to draw and a difficult one to maintain. Proper use of all the security administration tools provided by the network components is essential to maintaining a secure and productive network. There are many different security mechanisms on the average network, some of which are all but invisible to users and even at times to administrators, while others require attention on a daily basis. This one chapter cannot hope to provide anything close to a comprehensive treatise on network security, but it does examine some of the major components you can use to protect your network and your data from unauthorized access.

Securing the File System

All of your data is stored in files on your computers, and protecting the file system is one of the most basic forms of network security. File system security not only prevents unauthorized access to your files, it also enables you to protect your data from being modified or deleted, either accidentally or deliberately. There are two basic forms of security that you can apply to the file system on your computers: access permissions and data encryption.

File system permissions are the most commonly used security element on network servers. All of the major server operating systems have file systems that support the use of permissions to regulate access to specific files and directories. File system permissions typically take the form of an *access control list (ACL)*, which is a list of users (or groups of users), maintained by each file and directory, that have been granted a specific form of access to that file or directory. Each entry in the ACL contains a user or group name, plus a series of bits that define the specific permissions granted to that user or group.

Note *Not every file system supports the use of permissions. Novell NetWare and Unix both use permissions in their default file systems, but in Windows 2000 and Windows NT, you have a choice. The FAT16 and FAT32 file systems supported by all of the current versions of Windows do not have ACLs for each file, and therefore cannot use permissions. To protect files and directories on Windows 2000 and Windows NT computers with permissions, you must use the NTFS file system.*

It is standard practice for a file system to break down access permissions into individual tasks, such as read and write, and to assign them to users separately. This enables the network administrator to specify exactly what access each user should have. For example, you may want to grant certain users the read permission only, enabling them to read the contents of a file but not modify it. Manipulating permission assignments is an everyday task for the administrator of a properly protected network.

The following sections examine the file system permissions, as implemented by each of the major server operating system platforms.

The Windows 2000/NT Security Model

Security is an integral part of the Windows 2000 and Windows NT operating system design, and to fully understand the use of permissions in these operating systems, it helps to have some knowledge of the overall security model they use. The security subsystem in Windows 2000 and NT is integrated throughout the OS and is implemented by a number of different components, as shown in Figure 29-1. Unlike other Windows environmental subsystems running in user mode, such as Win32, the security subsystem is known as an *integral subsystem,* because it is used by the entire OS. All of the security subsystem components interact with Security Reference Monitor, the kernel mode security arbitrator that compares requests for access to a resource to that resource's ACL.

Note   *All of the information in this section pertains to Windows 2000 and Windows NT, but not to other products such as Windows 95, 98, and Me. These operating systems do not support the use of permissions and have a much simpler security architecture.*

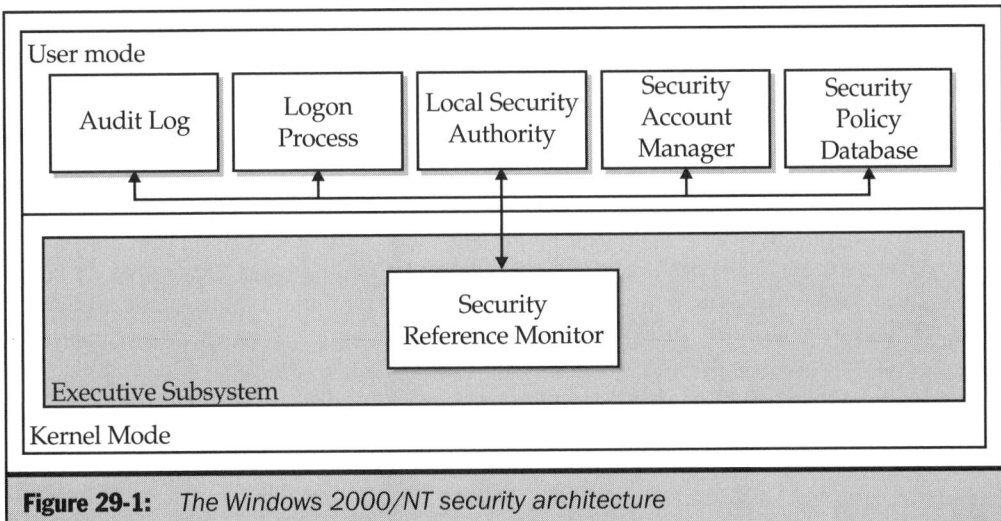

Figure 29-1:   *The Windows 2000/NT security architecture*

The user mode security subsystem components and their functions are as follows:

Logon Process The mechanism that accepts logon information from the user and initiates the authentication process.

Local Security Authority (LSA) Functions as the central clearinghouse for the security subsystem by initiating the logon process, calling the authentication package, generating access tokens, managing the local security policy, and logging audit messages.

Security Accounts Manager (SAM) Database containing the user and group accounts for the local system.

Security Policy Database Contains policy information on user rights, auditing, and trust relationships.

Audit Log Contains a record of security-related events and changes made to security policies.

During a typical user logon to the local machine, these components interact as follows:

1. The logon process appears in the form of the Logon dialog box produced when the user presses CTRL-ALT-DELETE after the system boots. Other logon processes may be substituted by other network clients (such as the Novell Client for Windows NT/2000). The user then supplies a user name and password.

2. The logon process calls the LSA that runs the authentication package.

3. The authentication package checks the user name and password against the local SAM database.

4. When the user name and password are verified, the SAM replies to the authentication package with the security IDs (SIDs) of the user and all the groups of which the user is a member.

5. The authentication package creates a logon session and returns it to the LSA with the SIDs.

6. The LSA creates a *security access token* containing the SIDs and the user rights associated with the SIDs, as well as the name of the user and the groups to which the user belongs, and sends it to the logon process, signaling a successful logon. The system will use the SIDs in this token to authenticate the user whenever he or she attempts to access any object on the system.

7. The logon session supplies the access token to the Win32 subsystem, which initiates the process of loading the user's desktop configuration.

Note *This procedure occurs when a user logs on using an account on the local machine only, not when logging on to an Active Directory domain. Active Directory logons are more complex, and are examined later in this chapter in the "Kerberos" section.*

Much of the Windows security subsystem's work is transparent to users and administrators. The security components that are most conspicuous in day-to-day activities are the SAM database, which holds all the local Windows user, group, and computer accounts (plus Windows NT domain accounts), and Active Directory. Every Windows 2000/NT system has a SAM database for its local accounts, and NT domains have their own separate SAM database, a copy of which is stored on each DC. Active Directory is a separate service that has its own security architecture, but for the purpose of assigning permissions, Active Directory objects function in the same way as accounts in the SAM database. Every object on the system that is protected by Windows security includes a security descriptor that contains an ACL. The ACL consists of *access control entries (ACEs)* that specify which users and groups are to be granted access to the object and what access they are to receive. When you specify the permissions for an object, such as a file, directory, share, or registry key, you are modifying the entries in that object's ACL. Clicking the Add button on the Security page in the Properties dialog box for a specific folder, for example (see Figure 29-2), displays a list of the users and groups in the SAM database or the objects in the Active Directory. Selecting users and granting them permission to access the share adds the users to the ACL for that share.

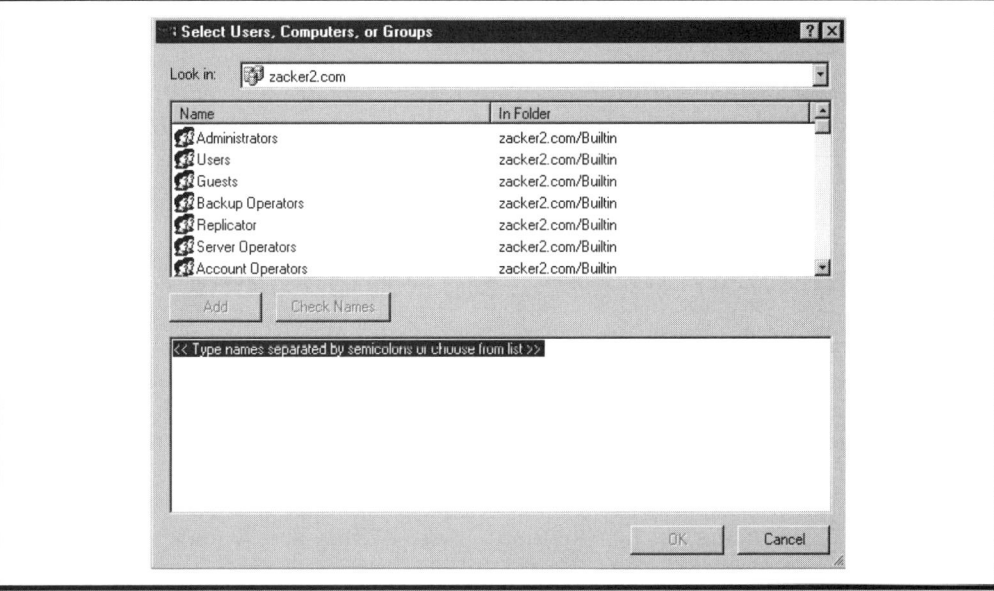

Figure 29-2: *You use a Select Users, Computers, or Groups dialog box like this one to create ACEs for Windows 2000/NT objects.*

When you log on to an Active Directory or Windows NT domain, the system accesses an account database that is located on one of the network's domain controllers (DCs) for authentication. The user, group, and computer accounts for the domain are stored in the DCs and are accessed whenever you use a utility that modifies the ACLs of system objects. During a domain session, you use the same Security page shown in Figure 29-2 to select the users and groups in the domain as you would those in the local SAM. You can also select users and groups from other domains on the network, as long as those other domains are trusted by the domain in which the system is currently participating.

Note *For more information on Active Directory, see Chapter 17. For more information on Windows NT domains, see Chapter 18.*

When a Windows 2000 or NT computer is a member of a domain, the local SAM database still exists. The Log On To Windows dialog box lets you select a domain or the local system for the current session. Note that a domain and a local SAM database can have user and group accounts with the same name. There is, for example, an Administrator account in the domain and an Administrator account for the local system, both of which are automatically created by default. These two accounts are not interchangeable. They can have different passwords and different rights and permissions. To install a network adapter driver, you must be logged on as the Administrator of the local system (or an equivalent). By default, a domain Administrator account does not have the rights to modify the hardware configuration on the local system.

Windows 2000/NT File System Permissions

Granting a user or group permissions to access a Windows resource adds them as an ACE to the resource's ACL. The degree of access that the user or group is granted depends on what permissions they are assigned. NTFS defines six standard permissions for files and folders—read, read and execute, modify, write, list folder contents, and full control—plus one extra for folders only. The standard permissions for NTFS files and folders are actually combinations of individual permissions.

The following lists the functions of the standard permissions when applied to a folder:

Read Enables a user/group to

■ See the files and subfolders contained in the folder

■ View the ownership, permissions, and attributes of the folder

Read and Execute   Enables a user/group to

- Navigate through restricted folders to reach other files and folders
- Perform all actions associated with the Read and List Folder Contents permissions

Modify   Enables a user/group to

- Delete the folder
- Perform all actions associated with the Write and the Read & Execute permissions

Write   Enables a user/group to

- Create new files and subfolders inside the folder
- Modify the folder attributes
- View the ownership and permissions of the folder

List Folder Contents   Enables a user/group to

- View the names of the files and subfolders contained in the folder

Full Control   Enables a user/group to

- Modify the folder permissions
- Take ownership of the folder
- Delete subfolders and files contained in the folder
- Perform all actions associated with all of the other NTFS folder permissions

The following lists the functions of the standard permissions when applied to a file:

Read   Enables a user/group to

- Read the contents of the file
- View the ownership, permissions, and attributes of the file

Read and Execute   Enables a user/group to

- Perform all actions associated with the Read permission
- Run applications

Modify   Enables a user/group to

- Modify the file
- Delete the file
- Perform all actions associated with the Write and the Read & Execute permissions

Write   Enables a user/group to

■ Overwrite the file

■ Modify the file attributes

■ View the ownership and permissions of the file

Full Control   Enables a user/group to

■ Modify the file permissions

■ Take ownership of the file

■ Perform all actions associated with all of the other NTFS file permissions

The following specifies the individual permissions that make up each of the standard permissions:

Read   Enables a user/group to

■ List Folder/Read Data

■ Read Attributes

■ Read Extended Attributes

■ Read Permissions

■ Synchronize

Read and Execute   Enables a user/group to

■ List Folder/Read Data

■ Read Attributes

■ Read Extended Attributes

■ Read Permissions

■ Synchronize

■ Traverse Folder/Execute File

Modify   Enables a user/group to

■ Create Files/Write Data

■ Create Folders/Append Data

■ Delete

■ List Folder/Read Data

- Read Attributes
- Read Extended Attributes
- Read Permissions
- Synchronize
- Write Attributes
- Write Extended Attributes

Write   Enables a user/group to

- Create Files/Write Data
- Create Folders/Append Data
- Read Permissions
- Synchronize
- Write Attributes
- Write Extended Attributes

List Folder Contents   Enables a user/group to

- List Folder/Read Data
- Read Attributes
- Read Extended Attributes
- Read Permissions
- Synchronize
- Traverse Folder/Execute File

Full Control   Enables a user/group to

- Change Permissions
- Create Files/Write Data
- Create Folders/Append Data
- Delete
- Delete Subfolders and Files
- List Folder/Read Data
- Read Attributes

- Read Extended Attributes
- Read Permissions
- Synchronize
- Take Ownership
- Write Attributes
- Write Extended Attributes

The functions of the individual permissions are as follows:

Traverse Folder/Execute File   The Traverse Folder permission allows or denies users the ability to move through folders that they do not have permission to access, so as to reach files or folders that they do have permission to access (applies to folders only). The Execute File permission allows or denies users the ability to run program files (applies to files only).

List Folder/Read Data   The List Folder permission allows or denies users the ability to view the file and subfolder names within a folder (applies to folders only). The Read Data permission allows or denies users the ability to view the contents of a file (applies to files only).

Read Attributes   Allows or denies users the ability to view the NTFS attributes of a file or folder.

Read Extended Attributes   Allows or denies users the ability to view the extended attributes of a file or folder.

Create Files/Write Data   The Create Files permission allows or denies users the ability to create files within the folder (applies to folders only). The Write Data permission allows or denies users the ability to modify the file and overwrite existing content (applies to files only).

Create Folders/Append Data   The Create Folders permission allows or denies users the ability to create subfolders within a folder (applies to folders only). The Append Data permission allows or denies users the ability to add data to the end of the file but not to modify, delete, or overwrite existing data in the file (applies to files only).

Write Attributes   Allows or denies users the ability to modify the NTFS attributes of a file or folder.

Write Extended Attributes   Allows or denies users the ability to modify the extended attributes of a file or folder.

Delete Subfolders and Files   Allows or denies users the ability to delete subfolders and files, even if the Delete permission has not been granted on the subfolder or file.

Delete   Allows or denies users the ability to delete the file or folder.

Read Permissions   Allows or denies users the ability to read the permissions for the file or folder.

Change Permissions   Allows or denies users the ability to modify the permissions for the file or folder.

Take Ownership   Allows or denies users the ability to take ownership of the file or folder.

Synchronize   Allows or denies different threads of multithreaded, multiprocessor programs to wait on the handle for the file or folder and synchronize with another thread that may signal it.

Permissions are stored as part of the NTFS file system, not in the Active Directory or the SAM database. To modify the permissions for a file or directory, you select the Security tab in the Properties dialog box of a file or folder, to display controls like those shown in Figure 29-3. Here you can add users and groups from the local SAM, from the current domain, and from other trusted domains, and specify the standard permissions that each one is to be allowed or denied.

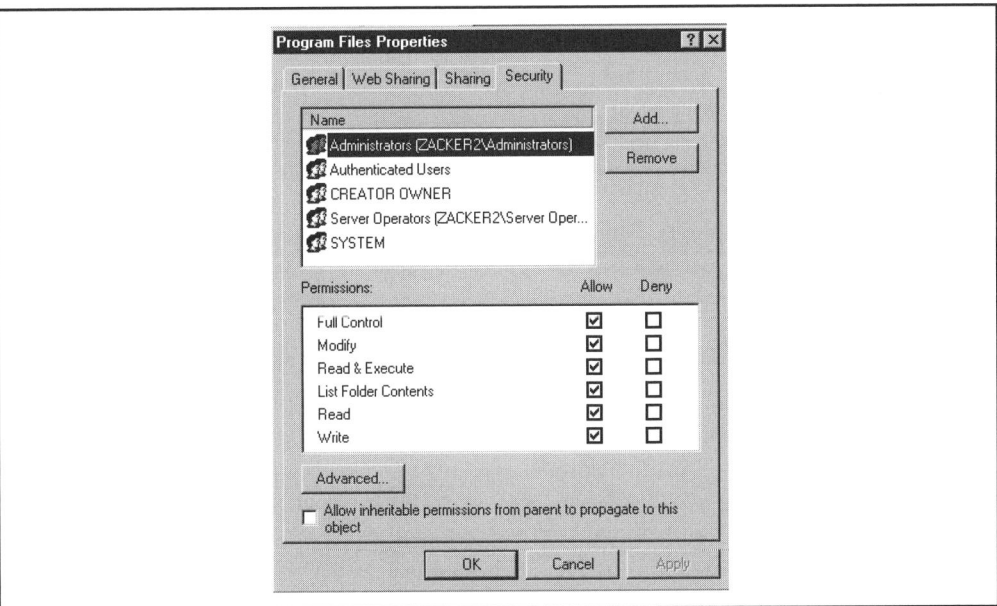

Figure 29-3:   *From the Properties dialog box for NTFS file system objects in Windows 2000 and NT use the Security dialog box to assign permissions.*

As with all file systems, the permissions that you assign to a folder are inherited by all of the files and subfolders contained in that folder. By judiciously assigning permissions throughout the file system, you can regulate user access to files and folders with great precision.

If the standard NTFS permissions do not provide you with the exact degree of access control you need, you can work directly with the individual permissions by clicking the Advanced button to display an Access Control Settings dialog box for the file or folder, like the one in Figure 29-4.

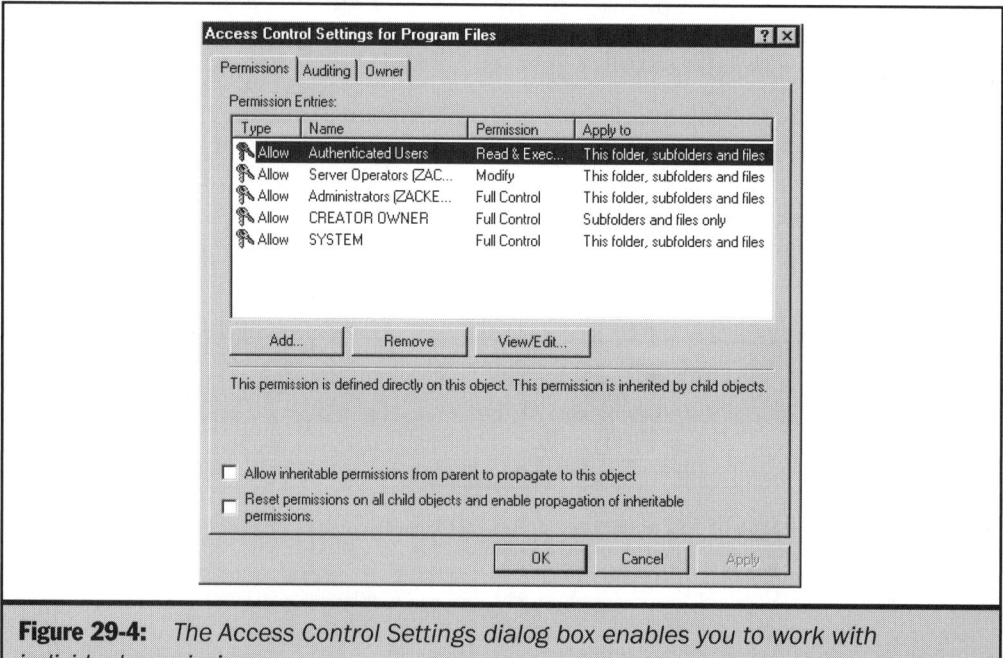

Figure 29-4:   *The Access Control Settings dialog box enables you to work with individual permissions.*

This dialog box lists all of the entries in the ACL for the selected file or folder. By selecting one of the entries and clicking the View/Edit button, you display a Permission Entry dialog box (see Figure 29-5) that lists the individual permissions allowed or denied the specified user or group. You can modify these permissions at will to customize the user or group's access to the file system resource.

The file and directory permissions apply to everyone who accesses the object, either on the local system or through the network. It is also possible to control network access to the file system by using *share permissions*. To make an NTFS drive or directory available for access over the network, you have to create a share out of it, and shares have access control lists just like files and directories do. To set share permissions, you open a drive or folder's Properties dialog box, select the Sharing tab, and click the Permissions button to display a dialog box like that shown in Figure 29-6. To access the files on a share, a network user must have permissions for both the share and the files and directories in the share.

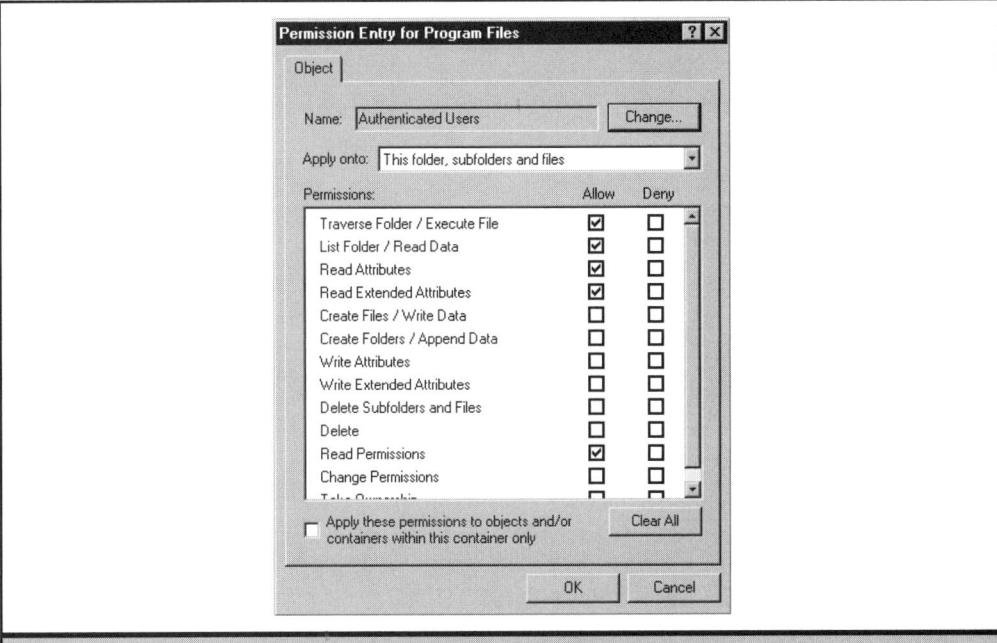

Figure 29-5:   *Working with individual permissions enables you to customize the NTFS access control mechanism.*

The permissions that you can grant to specific users and groups for shares are different than those used for files and directories. For a share, you can only allow or deny the Full Control, Change, and Read permissions.

Note   *In Windows 2000 and NT, it's important to understand that permissions are not the same thing as rights.* Rights *are rules that identify specific actions a user is allowed to perform on the local system, such as Access This Computer From The Network and Back Up Files And Directories. Many people use the term "rights" incorrectly when they mean "permissions," as in "the user has the rights to access the directory."*

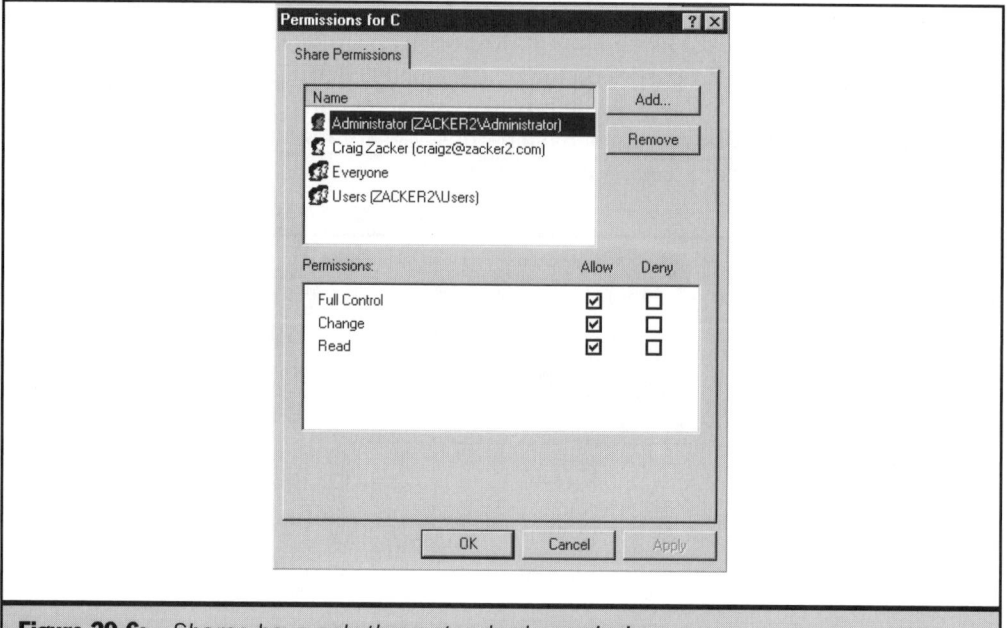

Figure 29-6: *Shares have only three standard permissions.*

NetWare File System Permissions

Novell NetWare file system permissions are usually referred to as *rights*, which can be confusing to people accustomed to Windows. Unlike Windows, NetWare does not use the terms rights and permissions to refer to two separate security elements; the terms are synonymous. NetWare file system permissions function in much the same way as the NTFS permissions used by Windows servers. Like NTFS, every file and directory on a NetWare volume has an ACL, but the entries in a NetWare ACL are called *trustees* instead of ACEs. Access control for the NetWare file system (and for the NDS database) is based on the creation of trustees, which are NDS objects that have been granted rights to a specific resource. A NetWare user or group (as represented by an NDS object or a bindery account) can be a trustee of any file or directory on a NetWare volume. As in NTFS, the ACL is stored with its file or directory, not in the NDS database or the bindery.

Distinguishing Between File System and NDS Permissions

It is extremely important to understand the difference between the permissions that NetWare uses to protect the file system and those that protect the NDS tree. Both NDS objects and the files and directories on a NetWare volume have ACLs that list the trustees of that resource. On a NetWare network using NDS, you add NDS objects (such as users and groups) to the ACL of a file or directory to grant users access to that file or directory. You can also add NDS objects to the ACL of another NDS object to grant users permission to access and modify that object.

The file system and NDS use many of the same names for their permissions, which is what can lead to confusion between the two. You can, for example, grant a user the Supervisor permission to a container object in the NDS tree, meaning that the user is permitted to modify the properties of the NDS objects in that container. You can also grant the same user the Supervisor permission to a NetWare directory, providing full access to the files and subdirectories that it contains. In the first case, the user with the Supervisor permission to the container object is a trustee of that object. In the second case, the user is a trustee of a file system directory, which is not an NDS object, but still uses a Supervisor permission. It is important to distinguish between these two permission systems because, in many cases, the access they provide is used for completely different purposes.

On the average NetWare network, administrators do not have to grant users NDS object rights on a regular basis. However, every user on the network requires file system permissions, and most network administrators work with these permissions on a regular basis.

In the NetWare file system, every file and directory has its own list of trustees. A trustee can be a user, a group of users, or any one of several other NDS objects that you can grant permission to access the file or directory. Many of the objects that you can add to a file system trustee list (such as organizational units and other container objects) do not actually require access to the files themselves. You grant the objects access solely for the purpose of passing along the rights to other objects through inheritance.

As with NTFS, you can assign each trustee in an ACL a number of different permissions to a file or directory, each of which grants the trustee a different kind of access. In NetWare, there is no distinction between individual and standard permissions; administrators always work with the individual permissions shown in the following list:

Read Enables the trustee to read the contents of, or execute, the file or any file within the directory.

Write   Enables the trustee to modify the contents of the file, or any file within the directory.

Create   Enables the trustee to create a new file or subdirectory in the directory.

Erase   Enables the trustee to delete the file, or any file or subdirectory from the directory.

Access Control   Enables the trustee to grant other objects rights to the file or directory, and to modify its inherited rights filter.

Modify   Enables the trustee to rename or change the attributes of the file or directory, or any file or subdirectory contained within the directory.

File Scan   Enables the trustee to list the name and other attributes of the file or directory and of the files and subdirectories contained within the directory. Withholding the File Scan permission effectively hides a file or directory from a user or group.

Supervisor   Provides the trustee with the equivalent of all the other rights to the file or directory, and all of the files and subdirectories contained within the directory. The Supervisor right also overrides the effect of all other rights assignments and inherited rights filters applied to the file, the directory, or any of the files and subdirectories contained within the directory.

Inheriting Rights

Rights inheritance is a crucial element of access control maintenance in all file systems. Without it, network administrators would be forced to grant each individual user the appropriate rights to every individual file they need to access. The task would be monumental, even on small networks, and agonizingly tedious. Instead, NetWare's access control systems enable trustee rights to be passed down through the file system and NDS object hierarchies. This rights inheritance is manifested in two ways:

- NDS objects with trustee rights to file system directories inherit rights to all of the files and subdirectories contained in those directories.

- Trustee rights granted to NDS container objects are inherited by all of the container and leaf objects below them in the NDS tree.

In its simplest form, the inheritance of rights to file system directories means that when you make a user a trustee of a directory, the rights that you assign to the user apply equally to all of the files and subdirectories in that directory as well. This, in effect, establishes a default rights assignment for that directory. You can then create additional trustee rights assignments for specific files or subdirectories in the directory that add or subtract certain rights. This leads to one of the most basic rules of NetWare trustee assignments: trustee rights granted to a specific file or subdirectory override the rights granted to the parent directory. Even when you create a file trustee that grants fewer rights than the parent directory's trustee assignment, the file rights take precedence, effectively blocking the inheritance of rights from the directory. The only exception to this rule is when you assign the Supervisor right. Once you grant a user the Supervisor right to a directory, the user inherits the same right to all of the files and subdirectories below, which cannot be blocked by any means.

NDS objects can have trustee rights to other objects, as well as to files and directories. Because file and directory rights become a property of an object, those property values flow down the branches of the NDS tree, just like any other properties. Just as you can grant a user rights to a directory, so that the user will inherit rights to the files in the directory, you can grant file system rights to a container object, so that the leaf objects in the container will inherit those rights. Thus, while a container object itself has no use for file system rights, you can assign it rights to a volume's files and directories, just as though it was a user. Then all of the user objects in that container, as well as its child containers and their user objects, receive the same rights.

Using Inherited Rights Filters

Many network administrators arrange the directory structure of their volumes to accommodate NetWare's file system rights inheritance behavior. There are times, however, when it is necessary to prevent rights from being passed down from a directory to its files or subdirectories. To prevent rights from flowing downward, you create an *inherited rights filter (IRF)*, which functions something like a dam, blocking selected rights from flowing downstream. You don't apply IRFs to specific trustees of a directory or file; they are general barriers that prevent all rights from being passed to a certain file or directory, and everything below it in the tree.

When you create an IRF at a specific point in a directory structure, you select which rights are to be blocked. You can, for example, have a directory to which users have full rights, and create a filter to block only the Write and Erase rights to a certain file in that directory. Users will still inherit other rights from the directory, but will be prevented from changing or deleting the file.

Note *The only file system right that you cannot block with an inherited rights filter is Supervisor. Once users are granted Supervisor rights to a particular directory, their access to all of the files and subdirectories below cannot be blocked by any means.*

As the name implies, inherited rights filters can only prevent rights to a file or directory from being inherited. They have no effect on rights that are explicitly assigned to a particular element of the file system. You can, therefore, grant your users Read and File Scan rights to an entire volume, and place IRFs on certain directories. The users will not be able to see that those directories even exist. Then, for users who require access to those filtered directories, you can make them trustees of the filtered directories, in any of the usual ways. The rights that you explicitly grant are not filtered, and those users will have the access they need.

It should be noted that inherited rights filters introduce an entirely new twist on the issue of access control procedures. IRFs are the only subtractive access control mechanism used by NetWare and are not commonly found in other file systems. IRFs can make it difficult to keep track of users' effective rights to particular files and directories. For this reason, administrators tend to use IRFs less frequently than the other NetWare access control mechanisms, and many avoid them entirely.

Understanding Effective Rights

There are many methods of providing users with permission to access the NetWare file system resources they need. You can choose any of these methods, or all of them, to suit your needs. However, using various forms of file system access control at once can result in a confusing array of mechanisms, all of which are influencing a single user's rights at the same time. The phrase that NetWare uses to describe the cumulative effect of all of these access control mechanisms is *effective rights*. A user's effective rights to a particular resource are determined by evaluating the combined influences of the following factors:

- Rights inherited from container objects
- Rights obtained through security equivalences
- Rights granted by group memberships
- Rights provided by organizational role occupancies
- Rights inherited from parent directories
- Rights blocked by inherited rights filters
- Rights explicitly granted to the user

It is recommended that you establish a regimen of access control procedures for your network that make use of only a few of these techniques and stick to them. If you have a staff of network support personnel with different access control philosophies, you could end up with one person applying IRFs, while another uses group memberships. This can result in extreme confusion. As with most aspects of network administration, prior planning and the development of a working standard are the keys to efficient maintenance.

Unix File System Permissions

Unix also uses permissions to control access to its file system, but the system is substantially different from those of Windows and NetWare. In Unix, there are only three permissions: read, write, and execute.

The following lists the access provided by each permission when applied to a directory:

Read   Enables a user to list the contents of the directory.

Write   Enables a user to create or remove files and subdirectories in the directory.

Execute   Enables a user to change to the directory using the cd command.

The following lists the access provided by each permission when applied to a file:

Read   Enables the user to view the contents of the file.

Write   Enables the user to alter the contents of the file.

Execute   Enables a user to run the file as a program.

Each of these three permissions can be applied to three separate entities: the file's owner, the group to which the file belongs, and all other users. When you list the contents of a directory using the ls -l command, you see a display for each file and directory like the following:

```
-rwxr-xr-    1    jdoe    sales    776    Sep 15 09:34    readme
```

The first character in the display identifies the file system element, using the following values:

- - File
- d   Directory
- b   Special block file
- c   Special character file
- l   Symbolic link
- P   Named pipe special file

The next three characters (*rwx*) indicate the permissions granted to the owner of the file (jdoe). In this case, the owner has all three permissions. The next three characters indicate the permissions granted to the file's group and the following three indicate the permissions granted to all other users. In this example, the *r-x* value indicates that the file's group (sales) has been granted the read and execute permissions only, and the *r—* value indicates that the other users have been granted only the read permission. To change the permissions, you use the chmod command.

This access control mechanism is common to all Unix variants, but it doesn't provide anywhere near the granularity of the NTFS and NetWare file systems. The system only recognizes three basic classes of users (users, groups, and others), making it impossible to grant permissions to several users in different groups, while blocking access by everyone else. To address this shortcoming, some Unix operating systems include more advanced access control mechanisms. Hewlett-Packard's HP-UX supports the use of ACLs that, as in NTFS, consist of ACEs that specify a specific user account and the permissions granted to that user.

Verifying Identities

User authentication is another one of the important security mechanisms on a data network. Assigning file system permissions to specific users is pointless unless the system can verify the user's identity and prevent unauthorized people from assuming that identity. Authentication is an exchange of information that occurs before a user is permitted to access secured network resources. In most cases, the authentication process consists of the user supplying an account name and an accompanying password to the system hosting the resources the user wants to access. The system receiving the name and password checks them against an account directory and, if the password supplied is the correct one for that account, grants the user access to the requested resource.

Applications and services use different types of authentication mechanisms, ranging from the simple to the extremely complex. The following sections examine some of these mechanisms.

FTP User Authentication

The File Transfer Protocol (FTP) is a basic TCP/IP service that enables users to upload files to and download them from another computer on the network, and to perform basic file management tasks. However, before an FTP client can do any of this, it must authenticate itself to the FTP server. FTP is an example of the simplest possible type of authentication mechanism and one of the most insecure. After the FTP client establishes a standard TCP connection with the server, it employs the USER and PASS commands to transmit an account name and password. The server checks the credentials of the user and either grants or denies access to the service.

Note *In many cases, the authentication sequence remains invisible to the user operating the FTP client. This is because, on the Internet, access to many FTP servers is unrestricted. The server accepts any account name and password, and the tradition is to use* anonymous *as the account name and the user's e-mail address as the password. Many FTP client programs automatically supply this information when connecting to a server, to save the user from having to supply it manually.*

The FTP authentication process is inherently insecure because it transmits the user's account name and password over the network in cleartext. Anyone running a protocol analyzer or other program that is capable of capturing the packets transmitted over the network and displaying their contents can view the name and password and use them to gain access to the FTP server. If the user should happen to be a network administrator who is thoughtless enough to use an account that also provides high-level access to other network resources, the security compromise could be severe.

Clearly, while FTP may be suitable for basic file transfer tasks, you should not count on its access control mechanism to secure sensitive data, because it is too easy for the account passwords to be intercepted.

Kerberos

At the other end of the spectrum of authentication mechanisms is a security protocol called Kerberos, developed by MIT and defined in the RFC 1510 document published by the Internet Engineering Task Force (IETF). Windows 2000/Active Directory networks use Kerberos to authenticate users logging on to the network. Because Kerberos relies on the public key infrastructure when exchanging data with the clients and servers involved in the authentication process, all passwords and other sensitive information are always transmitted in encrypted form, and never in cleartext. This ensures that even if an unauthorized individual were to capture the packets exchanged during the authentication procedure, no security compromise would result.

One of the fundamental principles of Active Directory is that it provides users with a single network logon capability, meaning that one authentication procedure can grant a user access to resources all over the network. Kerberos is a perfect solution for this type of arrangement, because it is designed to function as an authentication service that is separate from the servers hosting the resources that the client needs to access. For example, during an FTP authentication, only two parties are involved, the client and the server. The server has access to the directory containing the account names and password information for authorized users, checks the credentials supplied by each connecting client, and either grants or denies access to the server on that basis. If the client wants to connect to a different FTP server, it must perform the entire authentication process all over again.

By contrast, during an Active Directory logon, the client sends its credentials to the Kerberos *Key Distribution Center (KDC)* service running on a domain controller, which in Kerberos terminology is called an *authentication server (AS)*. Once the AS checks the client's credentials and completes the authentication, the client can access resources on servers all over the network, without performing additional authentications. For this reason, Kerberos is called a *trusted third-party authentication protocol*.

Public Key Infrastructure

Windows 2000 uses a *public key infrastructure (PKI)* that strengthens its protection against hacking and other forms of unauthorized access. In traditional cryptography, also called *secret key cryptography,* a single key is used to encrypt and decrypt data. For two entities to communicate, they must both possess the key, which implies the need for some previous communication during which the key is exchanged. If the key is intercepted or compromised, the entire encryption system is compromised.

The fundamental principle of a PKI is that the keys used to encrypt and decrypt data are different. Each system has a *public key* used to encrypt data and a *private key* used to decrypt it. By supplying your public key to other systems, you enable them to encrypt data before sending it to you, so that you can decrypt it using your private key. However, the public key cannot decrypt the data once it has been encrypted. Thus, while intruders may intercept public keys as they are transmitted across the network, they can't access any encrypted data unless they have the private keys as well, and private keys are never transmitted over the network.

The use of a PKI makes it possible to transmit authentication data across a Windows 2000/Active Directory network with greater security than clear text authentication mechanisms like that of FTP or even other secret key cryptography mechanisms. A PKI also provides the capability to use digital signatures to positively identify the sender of a message. A *digital signature* is a method for encrypting data with a particular user's private key. Other users receiving the transmission can verify the signature with the user's public key. Changing even one bit of the data invalidates the signature. When the transmission arrives intact, the valid signature proves not only that the transmission has not been changed in any way, but also that it unquestionably originated from the sending user. Thus, the potential exists for a digitally signed transmission to eventually carry as much legal and ethical weight as a signed paper document.

Note *It is entirely possible for the KDC service to run on the same server that hosts the network resources clients need to access. The point is not that the AS must be a separate computer, but that it is a separate service, independent of the network resources it protects.*

Kerberos authentication is based on the exchange of *tickets* that contain an encrypted password that verifies a user's identity. When a user on a Windows client system logs on to an Active Directory domain, it transmits a logon request containing the user's account name to an AS, which is an Active Directory domain controller. The KDC service on the domain controller then issues a *ticket-granting ticket (TGT)* to the client that includes the user's SID, the network address of the client system, a time stamp that helps to prevent unauthorized access, and the session key that is used to encrypt the data. The AS encrypts the response containing the TGT using a key that is based on the password associated with the user's account (which the AS already has in its directory). When the client receives the response from the AS, it decrypts the message by prompting the user for the password, which is the decryption key. Thus, the user's identity is authenticated without the password being transmitted over the network.

The TGT is retained by the client system, to be used as a license for future authentication events. It is essentially a pass affirming that the user has been authenticated and is authorized to access network resources. Once a client has a TGT, it can use it to identify the user, eliminating the need to repeatedly supply a password when accessing various network resources.

When the user wants to access a resource on a network server, the client sends a request to a *ticket-granting service (TGS)* on the domain controller, which identifies the user and the resource server and includes a copy of the TGT. The TGS, which shares the session key for the TGT with the AS, decrypts the TGT to affirm that the user is authorized to access the requested resource. The TGS then returns a *service ticket* to the client that grants the user access to that particular resource only. The client sends an access request to the resource server that contains the user's ID and the service ticket. The resource server decrypts the service ticket, and as long as the user ID matches the ID in the ticket, grants the user access to the requested resource. A client system can retain multiple service tickets to provide future access to various network resources. This system protects both the server and the user because it provides mutual authentication; the client is authenticated to the server and the server to the client.

Digital Certificates

For the PKI to operate, computers must exchange the public keys that enable their correspondents to encrypt data before transmitting it to them over the network. However, the distribution of the public keys presents a problem. For the transmission to be truly secure, there must be some way to verify that the public keys being distributed actually came from the party they purport to identify. For example, if your employer sends you an e-mail encrypted with your public key, you can decrypt the message using your private key, sure in the knowledge that no one could have intercepted the message and read its contents. But how do you know the message did indeed come from your boss, when it's possible for someone else to have obtained your public key? Also, what would stop someone from pretending to be you and distributing a public key that others can use to send encrypted information intended for you?

One answer to these questions is the use of digital certificates. A *certificate* is a digitally signed statement, issued by a third party called a *certificate authority (CA)*, that binds a user, computer, or service holding a private key with its corresponding public key. Because both correspondents trust the CA, they can be assured that the certificates they issue contain valid information. A certificate typically contains the following:

Subject identifier information   Name, e-mail address, or other data identifying the user or computer to which the certificate is being issued.

Subject public key value   The public key associated with the user or computer to which the certificate is being issued.

Validity period   Specifies how long the certificate will remain valid.

Issuer identifier information   Identifies the system issuing the certificate.

Issuer digital signature   Ensures the validity of the certificate by positively identifying its source.

On the Internet, certificates are used primarily for software distribution. For example, when your Web browser downloads a plug-in created by Big Graphics Corporation that is required to display a particular type of Web page, a certificate supplied by the server verifies that the software you are downloading did actually come from Big Graphics. This prevents anyone else from modifying or replacing the software and distributing it as Big Graphics' own.

The certificates used on the Internet are typically defined by the ITU-T X.509 standard and issued by a separate company that functions as the CA. The most commonly known public CA is called VeriSign. It's also possible to create your own certificates for internal use in your organization. Windows 2000 Server includes a component called Certificate Services, which can function as the CA for your network. You can use certificates to authenticate users to Web servers, send secure e-mail, and (optionally) authenticate users to domains. For the most part, the use of certificates is transparent to users, but administrators can manage them manually using the Certificates snap-in for the Microsoft Management Console.

Token-Based and Biometric Authentication

All of the authentication mechanisms described thus far rely on the transmission of passwords between clients and servers. Passwords are a reasonably secure method of protecting data that is somewhat sensitive, but not extremely so. When data must remain truly secret, passwords are insufficient for several reasons. Most network users have a tendency to be sloppy about the passwords they select and how they protect them. Many people choose passwords that are easy for them to remember and type, unaware that they can easily be penetrated. Names of spouses, children, or pets, as well as birthdays and other such common-knowledge information does not provide much security. In addition, some users compromise their own passwords by writing them down in obvious places or giving them to other users for the sake of convenience.

A carefully planned regimen of password length and composition requirements, rotations, and maintenance policies can help make your passwords more secure. There are also mechanisms that you can use in addition to passwords that can greatly enhance the security of your network.

To address the inherent weakness of password-based authentication and provide greater security, it's possible for each user to employ a separate hardware device as part of the authentication process. *Token-based authentication* is a technique in which the user supplies a unique token for each logon, as well as a password. The token is a one-time value that is generated by an easily portable device, such as a smart card. A *smart card* is a credit card-sized device with a microprocessor in it that supplies a token each time the user runs it through a card reader connected to a computer. The idea behind the use of a token is that a password, even in encrypted form, can be captured by a protocol analyzer and "replayed" over the network to gain access to protected resources. Because a user's token changes for each logon, it can't be reused, so capturing it is pointless. Token-based authentication also requires the user to supply a personal identification number (PIN) or a password to complete the logon, so that if the smart card is lost or stolen, it can't be used by itself to gain access to the network. Because this type of authentication is based on something you have, the token, and something you know, the PIN or password, the technique is also called *two-factor authentication*.

Smart cards can also contain other information about their users, including their private keys. The security of Windows 2000's PKI relies on the private encryption keys remaining private. Typically, the private key is stored on the workstation, which makes it susceptible to both physical and digital intrusion. Storing the private key on the card instead of on the computer protects it against theft or compromise and also enables the user to utilize the key on any computer.

Another tool that can be used to authenticate users is a biometric scanner. A *biometric scanner* is a device that reads a person's fingerprints, retinal patterns, or some other unique characteristic, and compares the information it gathers against a database of known values. Clearly, we are getting well into James Bond territory here, but these devices do exist, and they provide excellent security, since the user's "credentials" cannot easily be misplaced or stolen. The down side to this technology is its great expense, and it is used only in installations requiring extraordinary security.

Securing Network Communications

Authentication is a means for verifying a users' identities to ensure that they are authorized to access specific resources. Many authentication systems use encryption to prevent passwords from being intercepted and compromised by third parties. However, authorization protocols like Kerberos only use encryption during the authentication process. Once the user has been granted access to a resource, the participation of the authentication protocol and the encryption it provides ends. Thus, you may have data that is secured by permissions (or even by file system encryption) while it is stored on the server, but once an authorized client accesses that data, the server usually transmits it over the network in an unprotected form. Just as with the FTP passwords discussed earlier, an intruder could conceivably capture the packets as they travel over the network and view the data carried inside.

In many cases, the danger presented by unprotected network transmissions is minor. For instances when extra protection is warranted, it is possible to encrypt data as it travels over the network. The following sections examine the IP Security (IPsec) protocol and the Secure Sockets Layer (SSL) protocol, both of which are capable of encrypting data before it is transmitted over the network and decrypting it on receipt at the destination.

IPsec

Virtually all TCP/IP communication uses the Internet Protocol (IP) at the network layer to carry the data generated by the protocols operating at the upper layers. IPsec is a series of standards that define a method for securing IP communications using a variety of techniques, including authentication and encryption. Windows 2000 supports the use of IPsec, as do many Unix variants. Unlike many other TCP/IP protocols, IPsec is defined by many different documents, all published as RFCs by the IETF. The primary IPsec standard documents are as follows:

RFC 2411   IP Security Document Roadmap

RFC 2401   Security Architecture for the Internet Protocol

RFC 2402   IP Authentication Header

RFC 2403   The Use of HMAC-MD5-96 Within ESP and AH

RFC 2404   The Use of HMAC-SHA-1-96 Within ESP and AH

RFC 2405   The ESP DES-CBC Cipher Algorithm with Explicit IV

RFC 2406   IP Encapsulating Security Payload (ESP)

RFC 2407   The Internet IP Security Domain of Interpretation for ISAKMP

RFC 2408   Internet Security Association and Key Management Protocol (ISAKMP)

RFC 2409   The Internet Key Exchange (IKE)

RFC 2410   The NULL Encryption Algorithm and Its Use with IPsec

RFC 2412   The OAKLEY Key Determination Protocol

Although IPsec is usually thought of primarily as an encryption protocol, it actually provides several data protection services, including the following:

Encryption   The IPsec standards allow for the use of various forms of encryption. Windows 2000, for example, can use the Data Encryption Standard (DES) algorithm or the Triple Data Encryption Standard (3DES) algorithm. DES uses a 56-bit key to encrypt each 64-bit block, while 3DES encrypts each block three times with a different key, for 168-bit encryption. Both DES and 3DES are *symmetrical encryption algorithms*, meaning that they use the same key to encrypt and decrypt the data.

Authentication   IPsec supports a variety of authentication mechanisms, including Kerberos, Internet Key Exchange (IKE), digital certificates, and preshared keys. This enables different IPsec implementations to work together, despite using different methods of authentication.

Nonrepudiation   By employing public key technology, IPsec can affix digital signatures to datagrams, enabling the recipient to be certain that the datagram was generated by the signer. The sending computer creates the digital signatures using its private key, and the receiver decrypts them using the sender's public key. Since no one but the sender has access to the private key, a message that can be decrypted using the public key must have originated with the holder of the private key. The sender, therefore, cannot deny having sent the message.

Replay prevention   It is sometimes possible for an unauthorized user to capture an encrypted message and use it to gain access to protected resources without actually decrypting it, by simply replaying the message in its encrypted form. IPsec uses a technique called *cipher block chaining (CBC)* that adds a unique initialization vector to the data encryption process. The result is that each encrypted datagram is different, even when they contain exactly the same data.

Data integrity   IPsec can add a *cryptographic checksum* to each datagram that is based on a key possessed only by the sending and receiving systems. This special type of signature, also called a *hash message authentication code (HMAC)*, is essentially a summary of the packet's contents created using a secret, shared key, which the receiving system can compute using the same algorithm and compare to the signature supplied by the sender. If the two signatures match, the receiver can be certain that the contents of the packet have not been modified.

Encrypting network transmissions at the network layer provides several advantages over doing it at any other layer. First, network-layer encryption protects the data generated by all of the protocols operating at the upper layers of the protocol stack. Some other security protocols, such as SSL, operate at the application layer and therefore can only protect specific types of data. IPsec protects the data generated by any application or protocol that uses IP, which is virtually all of them.

Second, network-layer encryption provides data security over the entire journey of the packet, from source to destination. The computer that originates the packet encrypts it, and it remains encrypted until it reaches its final destination. This not only provides excellent security, but also means that the intermediate systems involved in the transmission of the packet do not have to support IPsec. A router, for example, receives packets, strips off the data link–layer protocol headers, and repackages the datagrams for transmission over another network. Throughout this process, the datagram remains intact and unmodified, so there is no need to decrypt it.

IPsec is composed of two separate protocols: the *IP Authentication Header (AH)* protocol and the *IP Encapsulating Security Payload (ESP)* protocol. Together, these two protocols provide the data protection services just listed. IPsec can use the two protocols together, to provide the maximum amount of security possible, or just one of the two.

IP Authentication Header

The IP Authentication Header protocol provides the authentication, nonrepudiation, replay prevention, and data integrity services listed earlier, in other words, all of the services IPsec provides except data encryption. This means that when AH is used alone, it is possible for unauthorized users to read the contents of the protected datagrams, but they cannot modify the data or reuse it without detection.

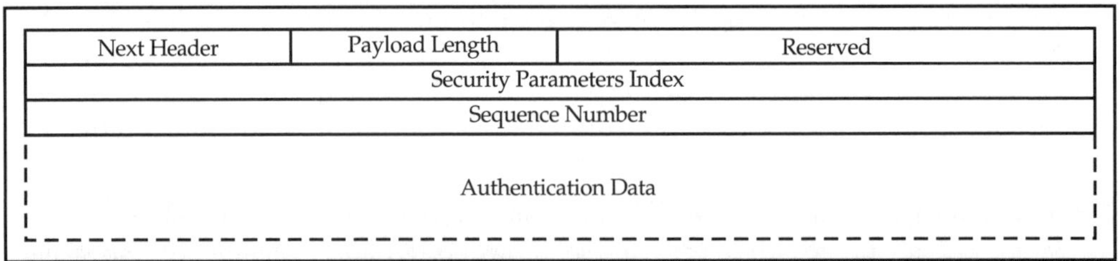

Next Header	Payload Length	Reserved
Security Parameters Index		
Sequence Number		
Authentication Data		

Figure 29-7: *The Authentication Header protocol header*

AH adds an extra header to each packet, immediately following the IP header and preceding the transport layer or other header encapsulated within the IP datagram. The fields of the AH header are illustrated in Figure 29-7. The functions of the fields are as follows:

Next Header (1 byte)   Identifies the protocol that generated the header immediately following the AH header, using values defined in the "Assigned Numbers" RFC (RFC 1700).

Payload Length (1 byte)   Specifies the length of the AH header.

Reserved (2 bytes)   Reserved for future use.

Security Parameters Index (4 bytes)   Contains a value that, in combination with the IP address of the destination system and the security protocol being used (AH or ESP), forms a security association for the datagram. A *security association* is a combination of parameters (such as the encryption key and security protocols to be used) that the sending and receiving systems agree upon before they begin to exchange data. The systems use the SPI value to uniquely identify this security association among others that may exist between the same two computers.

Sequence Number (4 bytes)   Implements the IPsec replay prevention service by containing a unique, incrementing value for each packet transmitted by a security association. The receiving system expects every datagram it receives in the course of a particular security association to have a different value in this field. Packets with duplicate values are discarded.

Authentication Data (variable)   Contains an *Integrity Check Value (ICV)* that the sending computer calculates for the entire AH header, including the Authentication Data field (which is set to a value of zero for this purpose), and the encapsulated protocol header (or headers) and data that follow the AH header. The receiving system performs the same ICV calculation and compares the results to this value to verify the packet's integrity.

The IP standard dictates that the Protocol field in the IP header must identify the protocol that generated the first header found in the datagram's payload. Normally, the first header in the payload is a TCP or UDP header, so the Protocol value is 6 or 17, respectively. ICMP data can also be carried in IP datagrams, with a Protocol value of 1. When IPsec adds an AH header, it becomes the first header found in the datagram's payload, so the value of the Protocol field is changed to 51. To maintain the integrity of the protocol stack, the Next Header field in the AH header identifies the protocol that follows AH in the datagram. In the case of datagrams that use AH alone, the Next Header field contains the value for the TCP, UDP, or ICMP protocol formerly found in the IP header's Protocol field. If IPsec is using both AH and ESP, the AH Next Header field contains a value of 50, which identifies the ESP protocol, and ESP's own Next Header field identifies the TCP, UD, or ICMP protocol data encapsulated within.

IP Encapsulating Security Payload

Unlike AH, the ESP protocol completely encapsulates the payload contained in each datagram, using both header and footer fields, as shown in Figure 29-8. The functions of the ESP fields are as follows:

Security Parameters Index (4 bytes)   Contains a value that, in combination with the IP address of the destination system and the security protocol being used (AH or ESP), forms a security association for the datagram. A *security association* is a combination of parameters (such as the encryption key and security protocols to be used) that the sending and receiving systems agree upon before they begin to exchange data. The systems use the SPI value to uniquely identify this security association among others that may exist between the same two computers.

Sequence Number (4 bytes)   Implements the IPsec replay prevention service by containing a unique, incrementing value for each packet transmitted by a security association. The receiving system expects every datagram it receives in the course of a particular security association to have a different value in this field. Packets with duplicate values are discarded.

Payload Data (variable)   Contains the original TCP, UDP, or ICMP header and data from the datagram.

Padding (0–255 bytes)   Some algorithms are only capable of encrypting data in blocks of a specific length. This field contains padding to expand the size of the payload data to the boundary of the next 4-byte word.

Pad Length (1 byte)   Specifies the size of the Padding field, in bytes.

Next Header (1 byte)   Identifies the protocol that generated the header immediately following the ESP header, using values defined in the "Assigned Numbers" RFC.

Authentication Data (variable)   Optional field that contains an *Integrity Check Value (ICV)* that the sending computer calculates for all of the fields from the beginning of the ESP header to the end of the ESP trailer (excluding the original IP header and the ESP Authentication Data field itself). The receiving system performs the same ICV calculation and compares the results to this value to verify the packet's integrity.

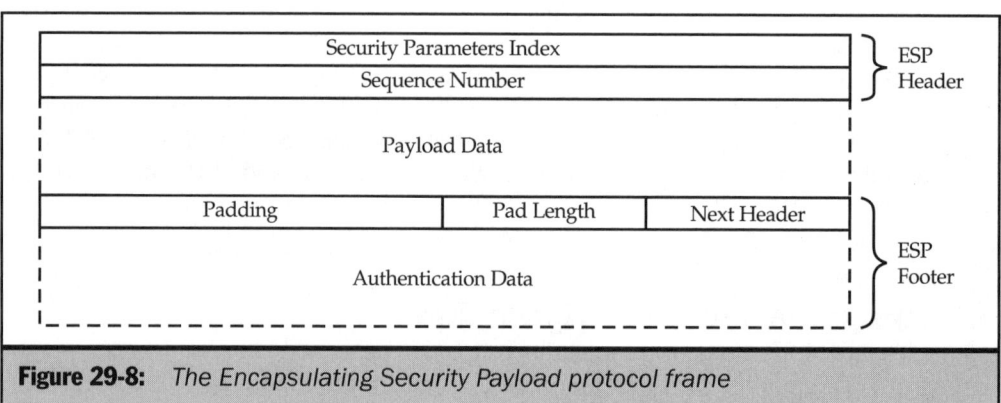

Figure 29-8:   *The Encapsulating Security Payload protocol frame*

ESP encrypts the data beginning at the end of the ESP header (that is, the end of the Sequence Number field) and proceeding up to the end of the Next Header field in the ESP footer. ESP is also capable of providing its own authentication, replay prevention, and data integrity services, in addition to those of AH. The information that ESP uses to compute the integrity signature runs from the beginning of the ESP header to the end of the ESP trailer. The original IP header from the datagram is not included in the signature (although it is in the AH signature). This means that, when IPsec uses ESP alone, it's possible for someone to modify the IP header contents without the changes being detected by the recipient. Avoiding this possibility is why the use of both AH and ESP is recommended, for maximum protection. A packet using both the AH and ESP protocols, and showing the signed and encrypted fields, is shown in Figure 29-9.

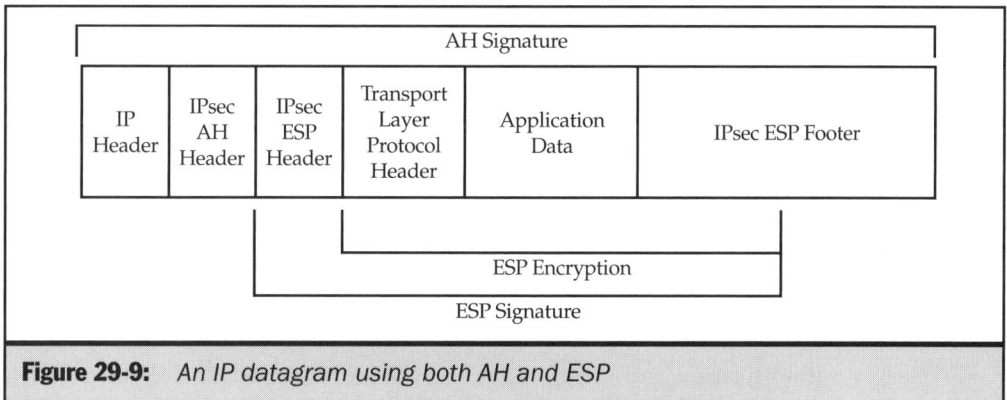

Figure 29-9: *An IP datagram using both AH and ESP*

Note   *The formats for the IPsec protocols shown here are for when IPsec runs in* transport mode, *which is the mode used for end-to-end security on a LAN or WAN connection. IPsec is also capable of running in* tunnel mode, *in which case an entirely new IP header is created. In this mode, the entire original datagram is encrypted by ESP and the entire new datagram (including the new IP header) is signed by AH. Tunnel mode is used for virtual private network (VPN) connections and is implemented only by the gateways providing the connection through the Internet, instead of by the end systems that are the ultimate source and destination of the packets.*

SSL

Secure Sockets Layer (SSL) is a series of protocols that provide many of the same services as IPsec, but in a more specialized role. Instead of protecting all TCP/IP traffic by signing and encrypting network-layer datagrams, SSL is designed to protect only the TCP traffic generated by specific applications, most notably the Hypertext Transfer Protocol (HTTP) traffic generated by Web servers and browsers. In most cases, when you use a Web browser to connect to a secured site (for the purpose of conducting a credit card or other transaction), the client and server open a connection that is secured by SSL, usually evidenced by an icon on the browser's status bar. The major Web servers and browsers all support SSL, with the result that its use is virtually transparent to the client.

SSL consists of two primary protocols: the SSL Record Protocol (SSLRP) and the SSL Handshake Protocol (SSLHP). SSLRP is responsible for encrypting the application-layer data and verifying its integrity, while SSLHP negotiates the security parameters used during an SSL session, such as the keys used to encrypt and digitally sign the data.

SSL Handshake Protocol

Clients and servers that use SSL exchange a complex series of SSLHP messages before they transmit any application data. This message exchange consists of four phases, which are as follows:

1. **Establish security capabilities**   During this phase, the client and the server exchange information about the versions of SSL they use and the encryption and compression algorithms they support. The systems need this information in order to negotiate a set of parameters supported by both parties.

2. **Server authentication and key exchange**   If the server needs to be authenticated, it sends its certificate to the client, along with the algorithms and keys that it will use to encrypt the application data.

3. **Client authentication and key exchange**   After verifying the server's certificate as valid, the client responds with its own certificate, if the server has requested one, plus its own encryption algorithm and key information.

4. **Finish**   The client and server use a special protocol called the SSL Change Cipher Spec Protocol to modify their communications to use the parameters they have agreed upon in the earlier phases. The two systems send handshake completion messages to each other using the new parameters, which completes the establishment of the secure connection between the two computers. The transmission of application data using SSLRP can now begin.

SSL Record Protocol

The process by which SSLRP prepares application-layer data for transmission over the network consists of five steps, which are as follows:

1. **Fragmentation**   SSLRP splits the message generated by the application-layer protocol into blocks no more than 2 kilobytes long.

2. **Compression**   Optionally, SSLRP can compress each fragment, but the current implementations do not do this.

3. **Signature**   SSLRP generates a message authentication code (MAC) for each fragment, using a secret key exchanged by the transmitting and receiving systems during the SSLHP negotiation, and appends it to the end of the fragment.

4. **Encryption**   SSLRP encrypts each fragment with any one of several algorithms using keys of various sizes. The encryption is symmetrical, with a key that is also exchanged during the SSLHP negotiation.

5. **Encapsulation**   SSLRP adds a header to each fragment before passing it down to the TCP protocol for further encapsulation.

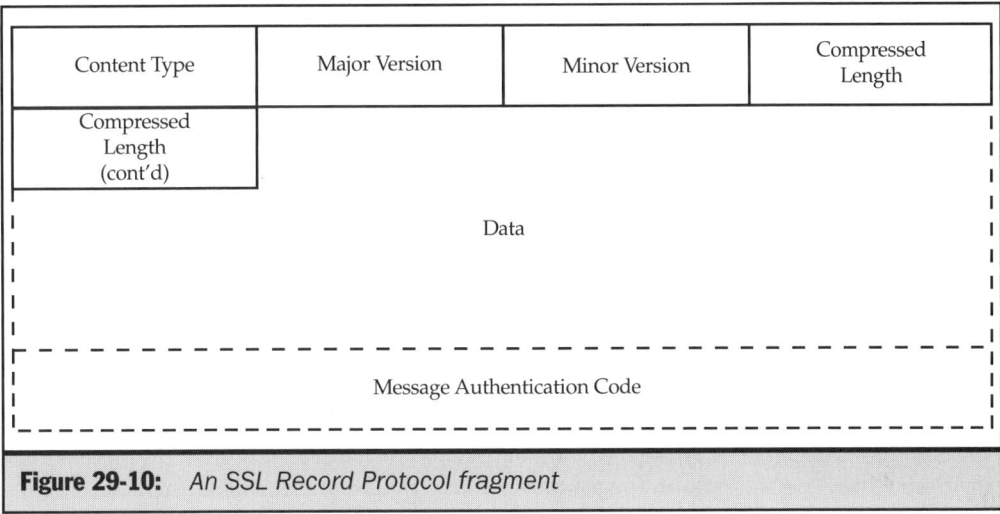

Content Type	Major Version	Minor Version	Compressed Length
Compressed Length (cont'd)			
Data			
Message Authentication Code			

Figure 29-10:   *An SSL Record Protocol fragment*

After this entire process is completed, each SSLRP fragment consists of the following fields, as shown in Figure 29-10:

Content Type (1 byte)   Identifies the application-layer protocol that generated the data fragment.

Major Version (1 byte)   Specifies the major version of SSL in use.

Minor Version (1 byte)   Specifies the minor version of SSL in use.

Compressed Length (2 bytes)   Specifies the length of the Data field.

Data (up to 2 kilobytes)   Contains a fragment of (possibly compressed) application-layer data.

Message Authentication Code (0, 16, or 20 bytes)   Contains the digital signature for the fragment, which the receiving system uses to verify its integrity.

Firewalls

A *firewall* is a hardware or software entity that protects a network from intrusion by outside users by regulating the traffic that can pass through a router connecting it to another network. The term is most often used in relation to protection from unauthorized users on the Internet, but a firewall can also protect a LAN from users on other LANs, either local or WAN connected. Without some sort of a firewall in place, outside users can access the files on your network, plant viruses, use your servers for their own purposes, or even delete your drives entirely.

Completely isolating a network from communication with other networks is not difficult, but this is not the function of a firewall. A firewall is designed to permit certain types of traffic to pass over the router between the networks, while denying access to all other traffic. You want your client workstations to be able to send HTTP requests from their Web browsers to servers on the Internet, and for the servers to be able to reply, but you don't want outside users on the Internet to be able to access those clients. Firewalls use several different methods to provide varying degrees of protection to network systems. A client workstation has different protection requirements than a Web server, for example.

Depending on the size of your network, the functions of your computers, and the degree of risk, firewalls can take many forms. The term has come to be used to refer to any sort of protection from outside influences. In fact, a true firewall is really a set of security policies that may be implemented by several different network components that work together to regulate not only the traffic that is permitted into the network, but possibly also the traffic that is permitted out. In addition to preventing Internet users from accessing the systems on your network, you can also use a firewall to prevent certain internal users from surfing the Web, while allowing them the use of Internet e-mail.

An inexpensive software router program can use network address translation (NAT) to enable client workstations on a small network to use unregistered IP addresses, and in a loose sense of the term, this is a form of a firewall. A large corporation with multiple T-1 connections to the Internet is more likely to have a system between the internal network and the Internet routers that is running software dedicated to firewall functions. Some firewall capabilities are integrated into a router, while other firewalls are separate software products that you must install on a computer.

Firewall protection can stem from either one of the following two basic policies, the choice of which is generally dependent on the security risks inherent in the network and the needs of the network users:

- Everything not specifically permitted is denied.
- Everything not specifically denied is permitted.

These two policies are essentially a reflection of seeing a glass as being either half full or half empty. You can start with a network that is completely secured in every way and open up portals permitting the passage of specific types of traffic, or you can start with a completely open network and block the types of traffic considered to be intrusive. The former method is much more secure, and is generally recommended in all environments. However, it tends to emphasize security over ease of use. The latter method is less secure but makes the network easier to use. This method also forces the administrator to try to anticipate the techniques by which the firewall can be penetrated. If there is one thing that is known for certain about the digital vandals that inhabit the Internet, it is that they are endlessly inventive, and keeping up with their diabolical activities can be difficult.

Network administrators can use a variety of techniques to implement these policies and protect the different types of systems on the network. The following sections examine some of these techniques and the applications for which they're used.

Packet Filters

Packet filtering is a feature implemented on routers and firewalls that uses rules specified by the administrator to determine whether a packet should be permitted to pass through the firewall. The rules are based on the information provided in the protocol headers of each packet, including the following:

- IP source and destination addresses

- Encapsulated protocol

- Source and destination port

- ICMP message type

- Incoming and outgoing interface

By using combinations of values for these criteria, you can specify precise conditions under which packets should be admitted through the firewall. For example, you can specify the IP addresses of certain computers on the Internet that should be permitted to use the Telnet protocol to communicate with a specific machine on the local network. As a result, all packets directed to the system with the specified destination IP address and using port 23 (the well-known port for the Telnet protocol) are discarded, except for those with the source IP addresses specified in the rule. Using this rule, the network administrators can permit certain remote users (such as other administrators) to Telnet into network systems, while all others are denied access. This is known as *service-dependent filtering,* because it is designed to control the traffic for a particular service, such as Telnet.

Service-independent filtering is used to prevent specific types of intrusion that are not based on a particular service. For example, a hacker may attempt to access a computer on a private network by generating packets that appear as though they originated from an internal system. This is called *spoofing*. Although the packets might have the IP address of an internal system, they arrive at the router through the interface that is connected to the Internet. A properly configured filter can associate the IP addresses of internal systems with the interface to the internal network, so that packets arriving from the Internet with those source IP addresses can be detected and discarded.

Packet filtering is a feature integrated into many routers, so no extra monetary cost is involved in implementing protection in this way, and no modification to client software or procedures is required. However, creating a collection of filters that provides adequate protection for a network against most types of attack requires a detailed knowledge of the way in which the various protocols and services work, and even then the filters may not be sufficient to prevent some types of intrusion. Packet filtering also creates an additional processing burden on the router, which increases as the filters become more numerous and complex.

Network Address Translation

Network address translation (NAT) is a technique that enables a LAN to use private, unregistered IP addresses to access the Internet. A NAT server or a router with NAT capabilities modifies the IP datagrams generated by clients to make them appear as though they were created by the NAT server. The NAT server (which has a registered IP address) then communicates with the Internet, and relays the responses back to the original client. Because the clients do not have valid Internet IP addresses, they are invisible to outside Internet users. For more information on NAT, see Chapter 28.

Proxy Servers

Proxy servers, also known as *application-level gateways,* provide a much stricter form of security than packet filters, but they are designed to regulate access only for a particular application. In essence, a proxy server functions as the middleman between the client and the server for a particular service. Packet filtering is used to deny all direct communication between the clients and servers for that service; all traffic goes to the proxy server instead (see Figure 29-11).

Because the proxy server has a much more detailed knowledge of the specific application and its functions, it can more precisely regulate the communications generated by that application. A firewall might run individual proxy servers for each of the applications needed by client systems, as shown in Figure 29-12.

Proxy
Internet
Internal network

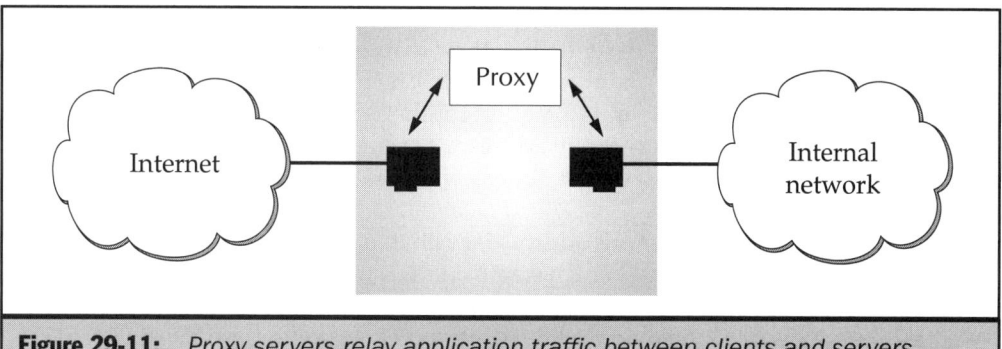

Figure 29-11: *Proxy servers relay application traffic between clients and servers.*

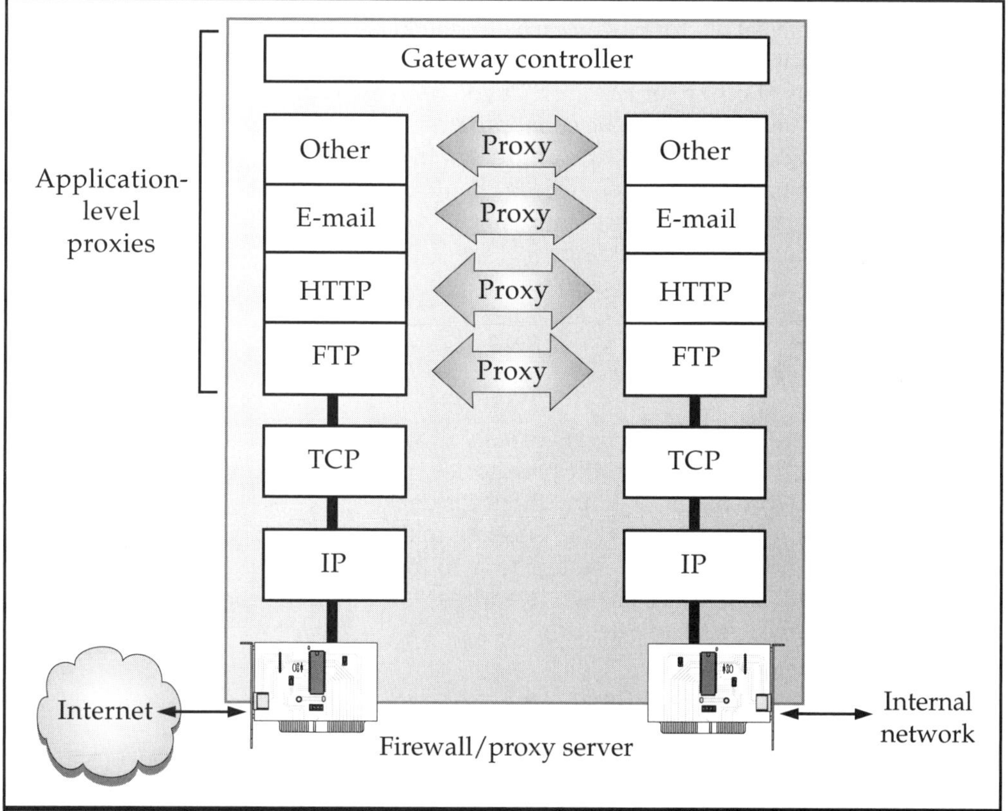

Gateway controller
Application-level proxies
Other
E-mail
HTTP
FTP
Proxy
Proxy
Proxy
Proxy
Other
E-mail
HTTP
FTP
TCP
IP
TCP
IP
Internet
Firewall/proxy server
Internal network

Figure 29-12: *A single proxy server product may provide individual gateways for several applications.*

The most common form of proxy server used today is for the Web. The client browsers on the network are configured to send all of their requests to the proxy server, instead of to the actual Internet server they want to reach. The proxy server (which does have access to the Internet) then transmits a request for the same document to the appropriate server on the Internet, using its own IP address as the source of the request, receives the reply from the server, and passes the response on to the client that originally generated the request.

Because only the proxy server's address is visible to the Internet, there is no way for Internet intruders to access the client systems on the network. In addition, the server analyzes each packet arriving from the Internet. Only packets that are responses to a specific request are admitted, and the server may even examine the data itself for dangerous code or content. The proxy server is in a unique position to regulate user traffic with great precision. A typical Web proxy server, for example, enables the network administrator to keep a log of users' Web activities, restrict access to certain sites or certain times of day, and even cache frequently accessed sites on the proxy server itself, enabling other clients to access the same information much more quickly.

The drawbacks of proxy servers are that you need an individual server for every application and modifications to the client program are required. A Web browser, for example, must be configured with the address of the proxy server before it can use it. Traditionally, manual configuration of each client browser was needed to do this, but there are now proxy server products that can enable the browser to automatically detect a server and configure itself accordingly.

Circuit-Level Gateways

A *circuit-level gateway*, a function that is usually provided by application-level gateway products, enables trusted users on the private network to access Internet services with all the security of a proxy server, but without the packet processing and filtering. The gateway creates a conduit between the interface to the private network and the Internet interface, which enables the client system to send traffic through the firewall. The gateway server still substitutes its own IP address for that of the client system, so that the client is still invisible to Internet users.

Combining Firewall Technologies

There are various ways in which these firewall technologies can be combined to protect a network. For a relatively simple installation in which only client access to the Internet is required, packet filtering or NAT alone—or packet filtering in combination with a proxy server—can provide a sufficient firewall. Adding the proxy server increases the security of the network beyond what packet filtering provides, because a potential intruder has to penetrate two levels of protection. However, if you run servers that must be visible to the Internet, the problem becomes more complicated.

One of the most secure firewall arrangements you can use for this type of environment is called a *screened subnet firewall.* This consists of a *demilitarized zone (DMZ)* network between the private network and the Internet. Using two routers with packet-filtering capabilities, you create a DMZ network that contains your proxy server, as well as your Web, e-mail, and FTP servers, and any other machines that must be visible to the Internet, as shown in Figure 29-13.

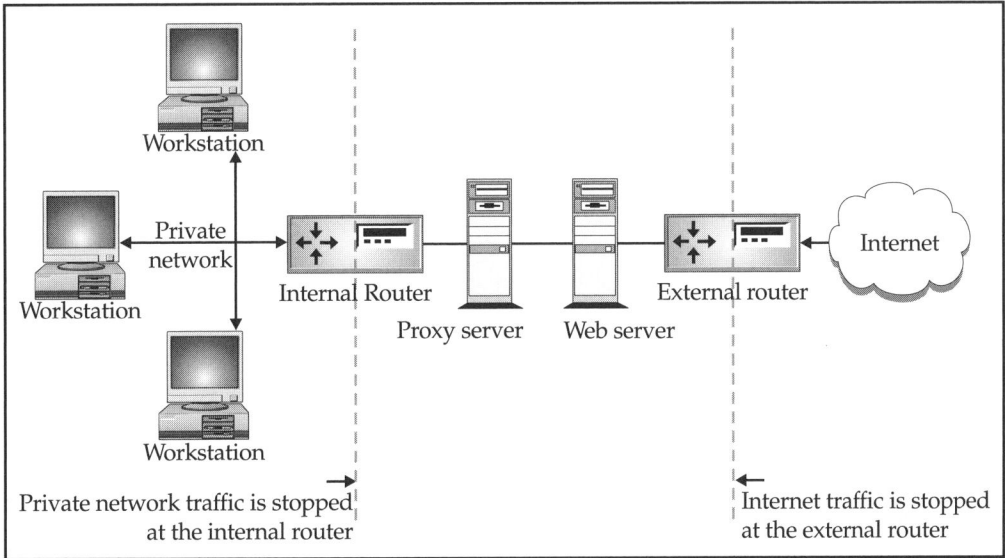

Figure 29-13:   *A screened subnet firewall provides a private network with three layers of protection.*

The two routers are configured to provide systems on the private network and the Internet with a certain degree of access to certain systems on the DMZ network, but no traffic passes directly through the DMZ. Users from the Internet must then pass through three separate layers of security (router, proxy, and router) before they can access a system on the private network.

Firewalls of this type are complex mechanisms that must be configured specifically for a particular installation, and can require a great deal of time, money, and expertise to implement. The prices of comprehensive firewall software products for enterprise networks can run well into five figures, and deploying them is not simply a matter of running an installation program. However, compared to the potential cost in lost data and productivity of a hacker intrusion, the effort taken to protect your network is not wasted.

Part VII

Network Administration

Chapter 30

Windows Network Administration

lthough business networks often run a variety of operating systems, particularly on their servers, the majority of user workstations run some form of Windows. Whether or not you agree with Microsoft that the Windows interface is user friendly and intuitive, there is no question that administering a fleet of hundreds or thousands of Windows workstations is an extremely formidable task. Windows 95, Windows 98, Windows NT Workstation, and Windows 2000 Professional all include tools that network administrators can use to simplify the process of installing, managing, and maintaining the operating system on a large number of workstations. This chapter examines some of these tools and how you can use them to configure workstations en masse, rather than working on them one at a time.

Note   *Windows Me is deliberately omitted here, as this operating system is intended for the home user and has had many of the administration tools removed.*

One of the primary goals of any network administrator should be to create workstation configurations that are standardized and consistent, so that when problems occur, the support staff is fully acquainted with the user's working environment. Failure to do this can greatly increase the time and effort needed to troubleshoot problems, thus increasing the overall cost of operating the computer. Unfortunately, many users have a tendency to experiment with their computers, such as by modifying the configuration settings or installing unauthorized software. This can make the system unstable and can interfere with the maintenance and troubleshooting processes. Therefore, it is advisable that administrators impose some form of restraints on network workstations to prevent this unauthorized experimentation.

Features such as user profiles and system policies are basic tools you can use to do this on Windows systems, to whatever degree you judge is necessary for your users. Using these tools, you can limit the programs that a system is able to run, deny access to certain elements of the operating system, and control access to network resources. These tools, and several others, are all part of what Microsoft calls its Zero Administration Initiative (ZAI). The ZAI is misunderstood by many people who think that its goal is a workstation that needs no maintenance or administration of any kind. In actuality, the ZAI is designed to prevent users from performing system administration tasks for which they are not qualified. By installing a workstation environment that has been carefully planned and tested, and then preventing users from changing it, the cost of maintaining workstations can be greatly reduced.

Imposing restrictive policies and limiting users' access to their workstations can be sensitive undertakings, and network administrators should carefully consider the capabilities of their users before making decisions like these. Unsophisticated computer users can benefit and may even appreciate a restricted environment that insulates them from the more confusing elements of the operating system. However, users with more experience might take offense at being limited to a small subset of the computer's features, and their productivity may even be impaired by it.

Locating Applications and Data

One of the basic tasks of the network administrator is to decide where data should be stored on the network. Network workstations require access to operating system files, applications, and data, and the locations where these elements are stored is an important part of creating a safe and stable network environment. Some administrators actually exercise no control over where users store files. Fortunately, most Windows applications install themselves to a default directory located in the C:\Program Files folder on the local system, which provides a measure of consistency if nothing else. Some applications even create default data directories on the local drive, but leaving users to their own devices when it comes to storing their data files is an inherently dangerous practice. Many users have little or no knowledge of their computer's directory structure and little or no training in file management. This can result in files for different applications all being dumped into a single common directory and left unprotected from accidental damage or erasure.

Server-Based Operating Systems

In the early days of Windows, running the operating system from a server drive was a practical alternative to having individual installations on every workstation. Storing the operating system files on a server enabled the network administrator not only to prevent them from being tampered with or accidentally deleted, but also to upgrade all the workstations at once. The technique also saved disk space on the workstation's local drive. However, as the years passed, the capacity of a typical hard drive on a network workstation grew enormously, as did the size of the Windows operating system itself.

Today, the practice of installing an operating system onto a mapped server drive is not practical. A workstation running Windows 95/98 or NT/2000 must load many megabytes of files just to boot the system, and when you multiply this by hundreds of computers, the amount of network traffic created by this practice could saturate even the fastest network. In addition, disk space shortages are not a big problem now that workstations routinely ship with drives that hold anywhere from 10 to 20 GB or more. Installing the operating system onto the local drive is, in most cases, the obvious solution.

However, there are newer technologies available today that are once again making it practical to run a Windows operating system from a server. This time, the workstations do not download the entire operating system from the server drive. Instead, the workstations function as client terminals that connect to a terminal server. The workstation operating system and applications actually run on the server, while the terminal functions solely as an input/output device. As a result, the workstations require only minimal resources, because the server takes most of the burden.

The terminals in this arrangement are either relatively modest computers, such as 486s, running a terminal emulation program, or dedicated Windows terminals that are designed solely to run the client software. In either case, the cost of a workstation is far less than that of a new PC with the hardware needed to run an individual copy of Windows 2000 or NT. Windows 2000 Server includes the terminal server capability with the operating system, and Windows NT 4.0 is available in a special version that supports this technology, called Windows NT Server 4.0 Terminal Server Edition. Running a terminal-based Windows network is a complete departure from a standard LAN, and is not a practical alternative for a network already running full Windows versions on its workstations. However, if you're building a new network or performing a major expansion, using Windows terminals is an option you may want to consider.

Server-Based Applications

Running applications from a server drive rather than individual workstation installations is another way to provide a consistent environment for your users and minimize the network's administrative burden. At its simplest, you do this by installing an application in the usual manner and specifying a directory on a network drive instead of a local directory as the location for the program files. Windows applications are rarely simple, however, and the process is usually more complicated.

Running applications from server drives has both advantages and disadvantages. On the plus side, as with server-based operating systems, are the disk space savings on the local drives, the ability to protect the application files against damage or deletion, and the ability to upgrade and maintain a single copy of the application files rather than individual copies on each workstation. The disadvantages are that server-based applications nearly always run more slowly than local ones, generate a substantial amount of network traffic, and do not function when the server is malfunctioning or otherwise unavailable.

In the days of DOS, applications were self-contained and usually consisted of no more than a single program directory that contained all of the application's files. You could install the application to a server drive and then let other systems use it simply by running the executable file. Today's Windows applications are much more complex, and the installation program is more than just a means of copying files. In addition to the program files, a Windows application installation may include registry settings and Windows DLLs that must be installed on the local machine, as well as a procedure for creating the Start menu entries and icons needed to launch the application.

When you want to share a server-based application with multiple workstations, you usually still have to perform a complete installation on each computer. This is to ensure that each workstation has all of the DLL files, registry settings, and icons needed to run the application. One way to implement a server-based application is to perform a complete installation of the program on each workstation, specifying the same directory on a server drive as the destination for the program files in each case. This way, each workstation receives all of the necessary files and modifications, and only one copy of the application files is stored on the server.

However, another important issue is the ability to maintain individual configuration settings for each of the computers accessing the application. When one user modifies the interface of a shared application, you don't want those modifications to affect every other user. As a result, each of the application's users must maintain their own copies of the application configuration settings. Whether this is an easy task, or even a possible one, depends on how each individual application stores its configuration settings. If, for example, the settings are stored in the registry or a Windows INI file, the installation process will create a separate configuration on each workstation. However, if the settings are stored with the program files on the server by default, you must take steps to prevent each user's changes from overwriting those of the other users.

In some cases, it is possible to configure an application to store its configuration settings in an alternate location, enabling you to redirect them to each workstation's local drive or to each user's home directory on a server. If this is not possible, the application may not be suitable for use in a shared environment. In many cases, the most practical way to run applications from a server is to select applications that have their own networking capabilities. Microsoft Office, for example, lets you create an administrative installation point on a server that you can use to install the application on your workstations. When you perform each installation, you can select whether the application files should be copied to the local drive, run from the server drive, or split between the two.

Storing Data Files

On most of today's Windows networks, both the operating system and the applications are installed on local workstation drives, but it is still up to the network administrator to decide where the data files generated and accessed by users should be stored. The two primary concerns that you must evaluate when making this decision are accessibility and security. Users must certainly have access to their own data files, but there are also files that have to be shared by many users. Important data files also have to be protected from modification and deletion by unauthorized personnel, and backed up to an alternative medium to guard against a disaster, such as a fire or disk failure.

Data files come in various types and formats that can affect the way in which you store them. Individual user documents, such as those created in word processor or spreadsheet applications, are designed for use by one person at a time, while databases can support simultaneous access by multiple users. In most cases, database files are stored on the computer running the database server application, so administrators can regulate access to them with file system permissions and protect them with regular backups. Other types of files may require additional planning.

Since Windows 95/98 and NT/2000 are all peer-to-peer network operating systems, you can allow users to store their document files on either their local drives or a server and still share them with other users on the network. However, there are several compelling reasons why it is better for all data files to be stored on servers. The first and most important reason is to protect the files from loss due to a workstation or disk failure. Servers are more likely to have protective measures in place, such as RAID arrays or mirrored drives, and are more easily backed up. Servers also make the data available at all times, while a workstation might be turned off when the user is absent.

The second reason is access control. Although Windows workstations and servers both have the same capabilities when it comes to granting access permissions to specific users, users rarely have the skills or the inclination to protect their own files effectively, and it is far easier for network administrators to manage the permissions on a single server than on many individual workstations. Another important reason for storing data on servers is that sharing the drives on every workstation can make it much more difficult to locate information on the network. To look at a Windows 2000 or NT domain and see dozens or hundreds of computers, each with its own shares, makes the task of locating a specific file much more complicated. Limiting the shares to a relatively few servers simplifies the process.

As a result, the best strategy for most Windows networks is to install the operating system and applications on local drives and implement a strategy for storing all data files on network servers. The most common practice is to create a home directory for each user on a server, to which they have full access permissions. You should then configure all applications to store their files in that directory, by default, so that no valuable data is stored on local drives. Depending on the needs of your users, you can make the home directories private, so that only the user that owns the directory can access it, or grant all users read-only access to all of the home directories. This makes it possible for users to share files at will simply by giving another user the file name or location.

When you create a user object in the Windows 2000 Active Directory or a user account in a Windows NT domain, you have the option of creating a home directory for the user at the same time in a dialog box like that shown in Figure 30-1. By default, users are given full control over their home directories, and no one else is given any access at all. You may want to modify these permissions to grant access to the directory to the other users on the network or, at the very least, to administrators.

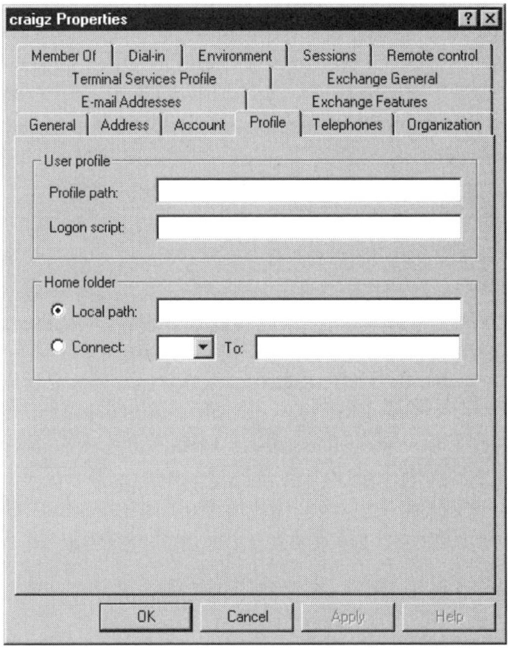

Figure 30-1: *The User Environment Profile dialog box in Windows NT's User Manager for Domains*

Controlling the Workstation Environment

In an organization comprised of expert computer users, you can leave everyone to their own devices when it comes to managing their Windows desktops. Experienced users can create their own desktop icons, manage their own Start menu shortcuts, and map their own drive letters. However, not many networks have only power users; in most cases, it is better for the network administrator to create a viable and consistent workstation environment.

Drive Mappings

Many less sophisticated computer users don't fully understand the concept of a network and how a server drive can be mapped to a drive letter on a local machine. A user may have the drive letter F mapped to a particular server drive and assume that other users' systems are configured the same way. If workstation drive mappings are inconsistent, confusion results when one user tells another that a file is located on the F drive, and the other user's F drive refers to a different share. To avoid problems like these, administrators should create a consistent drive-mapping strategy for users who will be sharing the same resources.

As an example, in many cases users will have a departmental or workgroup server that is their "home" server, and it's a good idea for every workstation to have the same drive letter mapped to that home server. If there are application servers that provide resources to everyone on the network, such as a company database server, then every system should use the same drive letter to reference that server, if a drive letter is needed. Implementing minor policies like these can significantly reduce the number of nuisance calls to the network help desk generated by puzzled users.

To implement a set of consistent drive mappings for your users, you can create logon script files containing NET USE commands that map drives to the appropriate servers each time the user logs on to the network. By structuring the commands properly, you should be able to create a single logon script for multiple users. To map a drive letter to each user's own home directory, you use a command like the following:

```
NET USE X: /home
```

where *home* is the name of the directory.

To designate a command file as a user's logon script, you add it to the Profile page in the user object's Properties dialog box in the Windows 2000 Active Directory Users and Computers console (see Figure 30-2), or to the User Environment Profile dialog box in Windows NT 4.0's User Manager for Domains utility.

Note *For more information on the many uses of the Windows NET command, see Chapter 31.*

User Profiles

Creating user profiles is a method of storing the shortcuts and desktop configuration settings for individual users in a directory, where a computer can access them during the system startup sequence. By creating separate profiles for different users, each person can retrieve their own settings when they log on. When you store multiple profiles on a local machine, you make it possible for users to share the same workstation without overwriting each other's settings. When you store the profiles on a network server, users can access their settings from any network workstation; this is called a *roaming profile*. In addition, you can force users to load a specific profile each time they log on to a system and prevent them from changing it; this is called a *mandatory profile*.

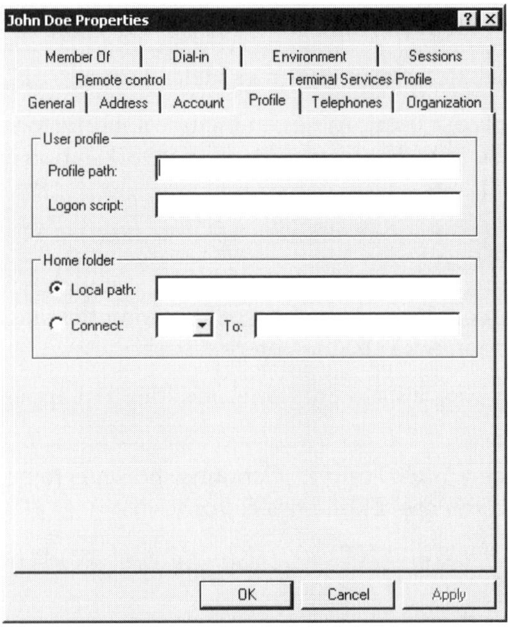

Figure 30-2:   *The Profile page in the Properties dialog box of an Active Directory user object*

The registry on a Windows 95 or 98 computer consists of two files on the local drive, called System.dat and User.dat. User.dat corresponds to the HKEY_CURRENT_USER key in the registry, which contains all of the environmental settings that apply to the user who is currently logged on. On a Windows 2000 or NT computer, the corresponding file is called Ntuser.dat. This file, called a *registry hive*, forms the basis of a user profile. By loading a User.dat or Ntuser.dat file during the logon sequence, the computer writes the settings contained in the file to the registry and they then become active on the system.

The user hive contains the following types of system configuration settings:

■   All user-definable settings for Windows Explorer

■   Persistent network drive connections

■   Network printer connections

■   All user-definable settings in the Control Panel, such as the Display settings

■   All taskbar settings

- All user-definable settings for Windows accessories, such as Calculator, Notepad, Clock, Paint, and HyperTerminal

- All bookmarks created in the Windows Help system

In addition to the hive, a user profile can include subdirectories that contain shortcuts and other elements that form parts of the workstation environment. These subdirectories are as follows:

Application Data   Contains application-specific data, such as custom dictionary files

Cookies   Contains cookies used by Internet Explorer to store information about the system's interaction with specific Internet sites

Desktop   Contains shortcuts to programs and files that appear on the Windows desktop

Favorites (Windows NT/2000 only)   Contains shortcuts to programs, files, and URLs that appear in Internet Explorer's Favorites list

History (Windows 95/98 and NT only)   Contains shortcuts to the URLs previously visited by the user in Internet Explorer

My Documents (Windows 2000 only)   Contains shortcuts to personal documents and other files

NetHood   Contains shortcuts that appear in the Network Neighborhood window

Personal (Windows NT only)   Contains shortcuts to personal documents and other files

PrintHood (Windows NT/2000 only)   Contains shortcuts that appear in the Printers window

Recent   Contains shortcuts to files that appear in the Documents folder in the Start menu

SendTo (Windows NT/2000 only)   Contains shortcuts to programs and file system locations that appear in the context menu's Send To folder

Start menu   Contains folders and shortcuts to programs and files that appear in the Start menu

Templates (Windows NT/2000 only)   Contains shortcuts to document templates

Note   *The NetHood, PrintHood, and Templates directories are hidden by default. To view them, you must configure Windows Explorer to display hidden files.*

Between the hive and the subdirectories, the user profile configures most of a user's workstation environment—including cosmetic elements, such as screen colors and wallpaper, and operational elements, such as desktop icons and Start menu shortcuts. The more concrete elements of the system configuration, such as hardware device drivers and settings, are not included in the user profile. If, for example, you install a new piece of hardware on a system, all users will have access to it, regardless of which profile is in use.

By default, Windows 2000 creates a user profile for each different user who logs on to the machine, and stores them in the \Documents and Settings folder directory on the system drive. Windows NT stores the profiles in the \Winnt\Profiles folder. The system also creates a default user profile during the operating system installation process that functions as a template for the creation of new profiles. If there are elements that you want included in all of the new profiles created on a computer, you can make changes to the profile in the \Default User subdirectory before any of the users log on. The system will then copy the default profile to a new subdirectory each time a new user logs on. Changing the \Default User subdirectory does not affect the user profiles that have already been created, however.

In Windows 95 and 98, the use of user profiles is optional and controlled by the User Profiles page in the Passwords Control Panel (see Figure 30-3). In this dialog box, you can specify whether the computer should create a new profile for each user who logs on and whether the profile should include desktop icons, Network Neighborhood settings, the Start menu, and program groups. The profiles are created in the \Windows\Profiles directory on the system drive. Unlike Windows 2000 and NT, there is no default user profile on a Windows 95/98 system.

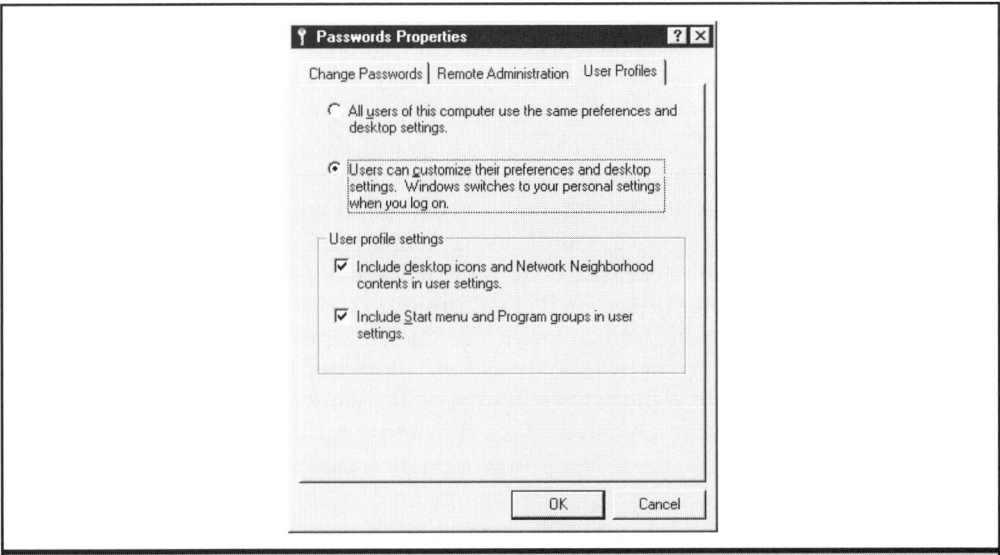

Figure 30-3: *The User Profiles page in Windows 98's Passwords Control Panel*

Creating Roaming Profiles

Both Windows NT/2000 and 95/98 store user profiles on the local machine by default. You can modify this behavior by specifying a location on a network server for a particular user's profile in the same Windows 2000 Profile page or NT User Environment Profile dialog box in which you specified the location of the user's home directory. Windows 2000 and NT systems use the path specified in the Profile field, while Windows 95 and 98 systems use the home directory. The profile server can be any system that is accessible by the workstation, running any version of Windows 95/98 or NT/2000, or even Novell NetWare. Once you specify the location for the profile, the operating system on the workstation copies the active profile to the server drive the next time the user logs off the network.

The best way to organize user profiles on the network is to designate a single machine as a profile server and create subdirectories named for your users, in which the profiles will be stored. When you specify the location of the profile directory for each user, you can use the *%UserName%* variable as part of the path, as follows:

```
\\Ntserver\Profiles\%UserName%
```

The system then replaces the *%UserName%* variable with the user's logon name, as long as the variable appears only once in the path and the variable is the last subdirectory in the path. In other words, the path \\Ntserver\Users\ *%UserName%*\ Profile would not be acceptable. However, the system does recognize an extension added to the variable, making \\Ntserver\Profiles\ *%UserName%*.man an acceptable path.

Storing user profiles on a server does not delete them from the workstation from which they originated. Once the server-based profile is created, each logon by the user triggers the following process. The workstation compares the profile on the server with the profile on the workstation. If the profile on the server is newer than that on the workstation, the system copies the server profile to the workstation drive and loads it from there into memory. If the two profiles are identical, the workstation loads the profile on the local drive into memory without copying from the server. When the user logs off, the workstation writes to both the local drive and the server any changes that have been made to the registry keys and shortcut directories that make up the profile.

Because the profile is always loaded from the workstation's local drive, even when a new version is copied from the server, it is important to consider the ramifications of making changes to the profile from another machine. If, for example, an administrator modifies a profile on the server by deleting certain shortcuts, these changes will likely have no effect, because those shortcuts still exist on the workstation, and copying the server profile to the workstation drive does not delete them. To modify a profile, you must make changes on both the server and workstation copies.

Once you create a server-based profile for a given user, that user can log on to the network from any workstation and load the profile, with one exception: the user profiles created by Windows NT/2000 and Windows 95/98 are not interchangeable, because the registries of the two operating systems are fundamentally different. A user with a Windows NT/2000 profile on the server cannot log on to a Windows 95 or 98 system and load that same profile or use that same server directory to store a 95/98 profile as well.

One of the potential drawbacks of storing user profiles on a network server is the amount of data that must be transferred on a regular basis. The registry hive and the various shortcut subdirectories are usually not a problem. But if, for example, a Windows 2000 Professional user stores many megabytes worth of files in the \My Documents directory, the time needed to copy that directory to the server and read it back again can produce a noticeable delay during the logoff and logon processes. The reason for including directories such as \My Documents and \Personal in the user profile is to enable users to access their personal documents if they log on to the network using another machine. If you store your users' document files on server drives already, as recommended earlier, this is not necessary and you should instruct your users not to store large amounts of data in these directories.

Creating Mandatory Profiles

When users modify elements of their Windows environment, the workstation writes those changes to their user profiles so that the next time they log on, the changes take effect. However, it's possible for a network administrator to create mandatory profiles that the users are not permitted to change, so that the same workstation environment loads each time they log on, regardless of the changes they made during the last session. To prevent users from modifying their profiles when logging off the system, you simply change the name of the registry hive in the server profile directory from Ntuser.dat to Ntuser.man, or from User.dat to User.man. When the workstation detects the MAN file in the profile directory, it loads that instead of the DAT file and does not write anything back to the profile directory during the logoff procedure.

 *When creating a mandatory profile, be sure that the user is not logged on to the workstation when you change the registry hive file extension from .dat to .man. Otherwise, the hive will be written back to the profile with a .dat extension during the logoff.*

Another modification you can make to enforce the use of the profile is to add a .man extension to the directory in which the profile is stored. This prevents the user from logging on to the network without loading the profile. If the server on which the profile is stored is unavailable, the user can't log on. If you choose to do this, be sure to add the .man extension both to the directory name itself and to the path specifying the name of the profile directory in the user object's Properties dialog box (for Windows 2000) or the User Environment Profile dialog box (for Windows NT 4.0).

It's important to note that making profiles mandatory does not prevent users from modifying their workstation environments; it just prevents them from saving those modifications back to the profile. Also, making a profile mandatory does not in itself prevent the user from manually modifying the profile by adding or deleting shortcuts or accessing the registry hive. If you want to exercise greater control over the workstation to prevent users from making any changes to the interface at all, you must use another mechanism, such as system policies, and be sure to protect the profile directories on the server using file system permissions.

Replicating Profiles

If you intend to rely on server-based user profiles to create workstation environments for your users, you should take pains to ensure that those profiles are always available to your users when they log on. This is particularly true if you intend to use mandatory profiles with .man extensions on the directory names, because if the server on which the profiles are stored is malfunctioning or unavailable, the users cannot log on. One way of doing this is to create your profile directories on a domain controller and then use the Directory Replicator service in Windows 2000 or NT to copy the profile directories to the other domain controllers on the network on a regular basis.

Once you have arranged for the profile directories to be replicated to all of your domain controllers, you can use the *%LogonServer%* variable in each user's profile path to make sure that they can always access the profile when logging on, as in the following example:

```
\\%LogonServer%\users\%UserName%
```

During the logon process, the workstation replaces the *%LogonServer%* variable with the name of the domain controller that authenticated the user. Since the profile directories have been copied to all of the domain controllers, the workstation always has access to the profile as long as it has access to a domain controller. If none of your domain controllers are available, you have much bigger problems to worry about than user profiles.

Creating a Network Default User Profile

Windows 2000 and NT systems have a default user profile they use as a template for the creation of new profiles. As mentioned earlier, you can modify this default profile so that all of the new profiles created on that machine have certain characteristics. It is also possible to create a default user profile on your network to provide the same service for all new profiles created on the network. To do this, you create a directory called \Default User in the root of the Netlogon share on your Windows 2000 or NT domain controllers. By default, the Netlogon share is located in the \Winnt\System32\ Repl\Import\Scripts directory on the server's system drive. Then you copy the entire profile you want to use, including the registry hive and the subdirectories, to the \Default User directory.

Whenever new users log on to a Windows 2000 or NT system, their profiles are created by copying the default profile from the domain controller.

The network default user profile works with Windows 2000 and Windows NT client systems only. Windows 95 and 98 systems do not create new profiles from the network drive.

Controlling the Workstation Registry

The registry is the central repository for configuration data in Windows 95, 98, NT, and 2000 systems, and exercising control over the registry is a major part of a system administrator's job. The ability to access a workstation's registry in either a remote or automated fashion enables you to control virtually any aspect of the system's functionality and also protect the registry from damage due to unauthorized modifications.

Using System Policies

All of the 32-bit Windows operating systems include *system policies,* which enable you to exercise a great deal of control over a workstation's environment. By defining a set of policies and enforcing them, you can control what elements of the operating system your users are able to access, what applications they can run, and the appearance of the desktop. System policies are really nothing more than collections of registry settings that are packaged into a system policy file and stored on a server drive. When a user logs on to the network, the workstation downloads the system policy file from the server and applies the appropriate settings to the workstation's registry. Because workstations load the policy file automatically during the logon process, users can't evade them. This makes system policies an excellent tool for limiting users' access to the Windows interface.

Using system policies is an alternative to modifying registry keys directly, and reduces the possibility of system malfunctions due to typographical or other errors. Instead of browsing through the registry tree, searching for cryptic keys and value names, and entering coded values, you create system policy files using a graphical utility called System Policy Editor (SPE). SPE displays registry settings in the form of *policies,* plain-English phrases with standard Windows dialog box elements arranged in a tree-like hierarchy (see Figure 30-4).

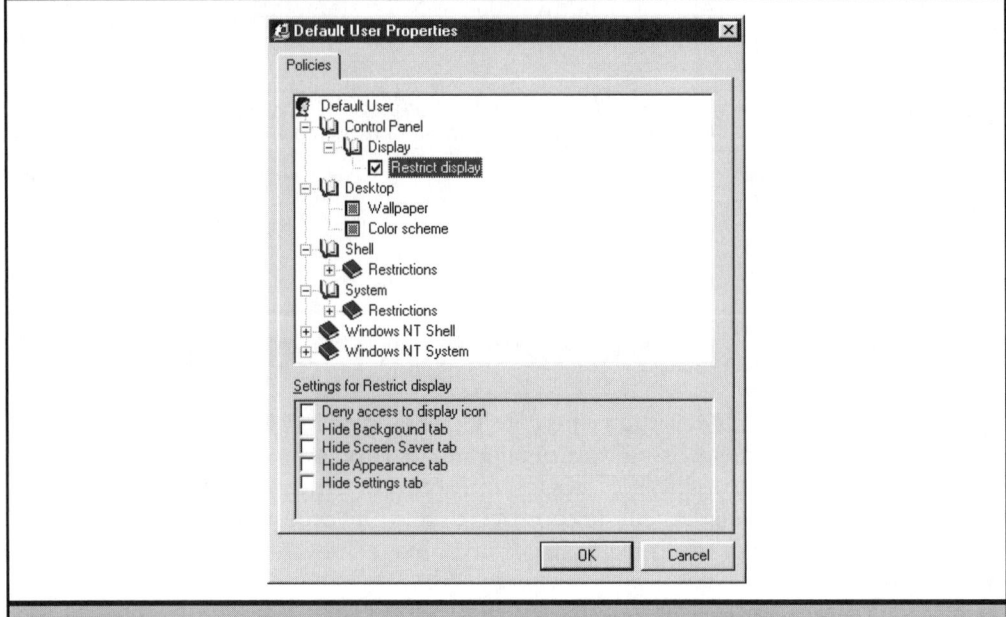

Figure 30-4:   *The System Policy Editor Default User Properties dialog box*

System Policy Templates

System Policy Editor is simply a tool for creating policy files; it has no control over the policies it creates. The policies themselves come from system policy templates, which are ASCII files that contain the registry keys, possible values, and explanatory text that make up the policies displayed in SPE. For example, the following excerpt from the Common.adm policy template creates the Remote Update policy shown in Figure 30-5:

```
CATEGORY !!Network

    CATEGORY !!Update

        POLICY !!RemoteUpdate

        KEYNAME System\CurrentControlSet\Control\Update

        ACTIONLISTOFF

            VALUENAME "UpdateMode"        VALUE NUMERIC 0

        END ACTIONLISTOFF

            PART !!UpdateMode          DROPDOWNLIST REQUIRED
```

```
                VALUENAME "UpdateMode"

                ITEMLIST

                    NAME !!UM_Automatic      VALUE NUMERIC 1

                    NAME !!UM_Manual         VALUE NUMERIC 2

                END ITEMLIST

                END PART

                PART !!UM_Manual_Path        EDITTEXT

                VALUENAME "NetworkPath"

                END PART

                PART !!DisplayErrors         CHECKBOX

                VALUENAME "Verbose"

                END PART

                PART !!LoadBalance           CHECKBOX

                VALUENAME "LoadBalance"

                END PART

            END POLICY

        END CATEGORY      ; Update

    END CATEGORY      ; Network
```

All of the Windows operating systems include a variety of template files in addition to the SPE program itself. The three main templates are Winnt.adm, which contains Windows NT/2000 policies; Windows.adm, which contains Windows 95/98 policies; and Common.adm, which contains policies that apply to both Windows 95/98 and NT/2000. Other applications, such as Microsoft Office and Internet Explorer, include their own template files containing policies specific to those applications, and you can even create your own custom templates, to modify other registry settings.

By selecting Options | Policy Template, you can load the templates that SPE will use to create policy files. You can load multiple templates into SPE, and the policies in them will be combined in the program's interface. Whenever you launch SPE, it loads the templates that it was using when it was last shut down, as long as the files are still in the same locations. When you use multiple policy templates in SPE, it is possible for policies defined in two different templates to configure the same registry setting. If this type of duplication occurs, the policy closest to the bottom of the hierarchy in the object's Properties dialog box takes precedence.

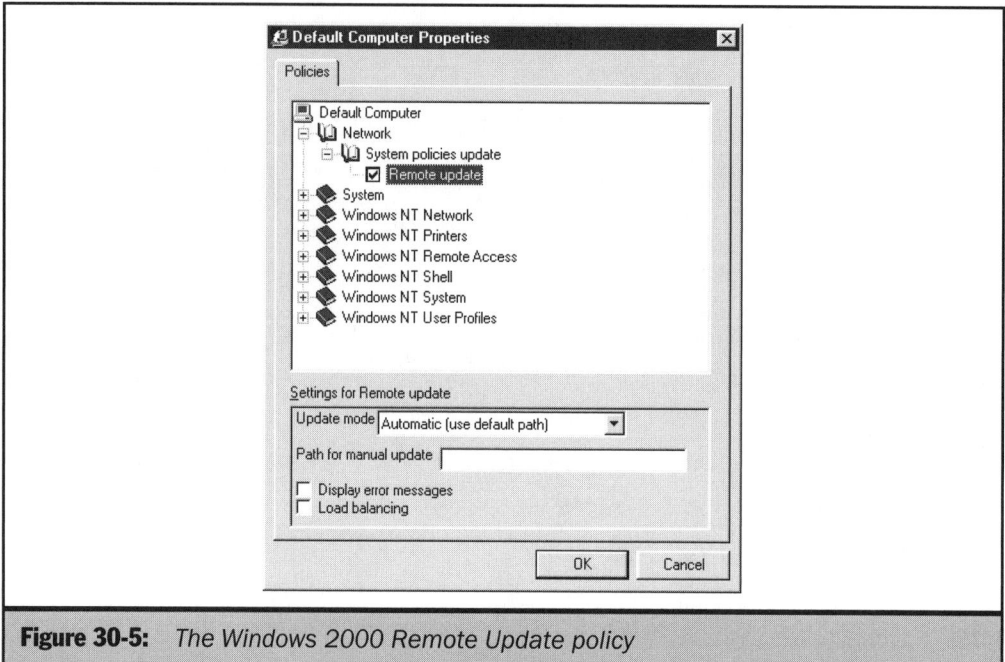

Figure 30-5:   *The Windows 2000 Remote Update policy*

System Policy Files

Using SPE, you can create policies that apply only to specific users, groups, and computers, as well as create Default User and Default Computer policies. Policies for multiple network users and computers are stored in a single file that every computer downloads from a server as it logs on to the network. The policy file for Windows NT/2000 systems is called Ntconfig.pol, and the file for Windows 95/98 systems is called Config.pol. Windows 95, 98, NT, and 2000 all ship with their own versions of System Policy Editor, and the versions all use the same interface. Windows 95/98 policy files are different from NT/2000 policy files, so they must be created separately.

The Windows NT and 2000 versions of SPE are installed with the operating system by default, and you can run them on any Windows NT, 2000, 95, or 98 system simply by copying the files to a drive on another computer. However, the type of policy files that SPE creates depends on the operating system on which it's running. You can create NT/2000 policy files by running SPE on an NT or 2000 system, and create 95/98 policy files by running SPE on a Windows 95 or 98 system, but you can't create a 95/98 policy file on an NT or 2000 system, and vice versa. The versions of SPE included with Windows 95 and 98 are not installed by default, must be manually installed (and not just copied) from the distribution CD-ROM, and can create Windows 95/98 policy files only.

Creating Policy Files   Creating a new policy file is a matter of opening SPE and selecting File | New Policy. By default, the program creates Default Computer and Default User objects. Opening an object displays the policies that you can configure for it. Policies that you configure within the default objects will be applied to all workstations logging on to the network. You can also create additional computer and user objects, as well as group objects, that correspond, respectively, to the NetBIOS names of the computers, and to the user and group account names in your Windows NT or 2000 domain. Computer policies modify registry settings in the HKEY_LOCAL_MACHINE key, while group and user policies affect settings in the HKEY_CURRENT_USER key. When you create a new computer, user, or group object, SPE copies the contents of the corresponding default object and creates a new icon, as shown in Figure 30-6. With these tools, you can implement different policies for the various types of users on your network.

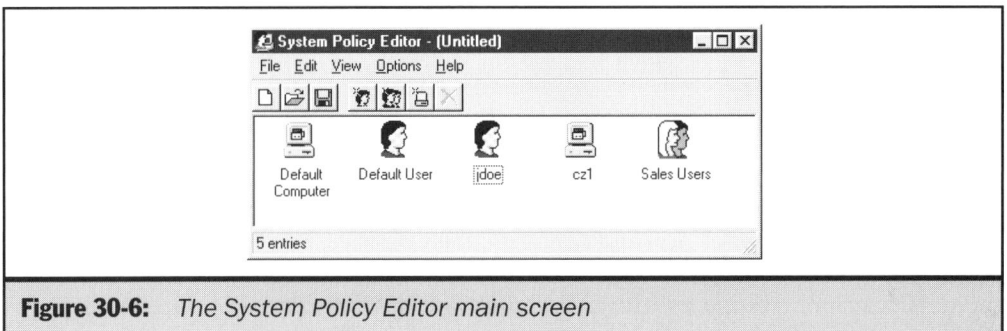

Figure 30-6:   *The System Policy Editor main screen*

After saving the policy file with the appropriate name, deploying it on the network typically is as simple as copying it to the Netlogon share on your domain controllers, which by default is the \Winnt\System32\Repl\Import\Scripts directory on the system drive. Workstations automatically check this share for a policy file during the logon process and, if one is present, download and process it. System policy files are always processed after user profiles are loaded, so the registry settings in the policy file always take precedence over those in the registry hive of a user profile.

Policy Priorities   It is up to you to create a policy strategy that is easy to maintain and that exercises sufficient control over your workstations. As with most network administration tasks, it is easier to implement policies on a group level than it is to create individual user policies. When creating group policies in SPE, you can specify a priority for each group (by selecting Options | Group Priority) that controls the order in which the policies are applied to a system. When a workstation processes a policy file and the user is a member of more than one group, the system applies the group policies in order, from the lowest priority to the highest, so that the policies with higher priorities can overwrite the registry settings of the lower-priority policies.

When you create individual computer objects in SPE, they take priority over the Default Computer object. When a user logs on to the network from a workstation that is represented by a computer object, the system processes the policies for the individual computer and ignores the Default Computer policies. In the same way, when a user is represented by an individual user object in the policy file, the system loads the policies for that user object and ignores all of the groups to which the user belongs, as well as the Default User policies. User and group policies also take precedence over computer policies in the case of the few registry keys that can exist in either the HKEY_LOCAL_MACHINE or HKEY_CURRENT_USER key. When a setting exists in both keys, the value in HKEY_CURRENT_USER takes precedence.

System policies apply to all users, including those with administrative access. If you plan to impose severe restrictions on your users' workstations, be sure to create user or group objects that provide exceptional access for administrators.

Setting Policy Values

Once you have loaded the policy templates into SPE and created a new policy file, you can begin to create new objects and configure their policies. Each computer, user, and group object in a policy file contains a hierarchy of categories that contains the various policies. Each policy appears with a check box that has three possible states, which are as follows:

Enabled   Applies the policy to the registry using the given value

Disabled   Does not apply the policy to the registry and removes it from the registry if it already exists

Undefined   Ignores the policy, making no changes to the registry, whether it exists or not

In addition to the check box, a policy can have a number of other controls associated with it, which appear in the settings area at the bottom of the dialog box when the policy is enabled. These controls can take several forms, including additional check boxes, data fields, and selectors. The way that the controls affect the registry depends on the requirements of the individual setting and the structure of the policy template. Some policies simply create a value name in the registry, while others assign a binary, alphanumeric, or hexadecimal value to a particular setting.

Restricting Workstation Access with System Policies

One of the primary functions of system policies is to prevent users from accessing certain elements of the operating system. There are several reasons for doing this, such as these:

■ Prohibiting users from running unauthorized software

■ Preventing users from adjusting cosmetic elements of the interface

■ Insulating users from features they cannot use safely

By doing these things, you can prevent users from wasting time on nonproductive activities and causing workstation malfunctions through misguided experimentation that require technical support to fix. The following sections describe how you can use specific system policies to control the workstation environment.

Restricting Applications   One of the primary causes of instability on Windows workstations is the installation of incompatible applications. Most Windows software packages include dynamic link library modules (DLLs) that get installed to the Windows system directories, and many times these modules overwrite existing files with new versions designed to support that application. The problem with this type of software design is that installing a new version of a particular DLL may affect other applications already installed in the system.

The way to avoid problems stemming from this type of version conflict is to assemble a group of applications that supplies the users' needs, and then test the applications thoroughly together. Once you have determined that the applications are compatible, you install them on your workstations and prevent users from installing other software that can introduce incompatible elements. Restricting the workstation software also prevents users from installing nonproductive applications, such as games, that can occupy large amounts of time, disk space, and even network bandwidth.

There are several techniques that you can employ to prevent users from installing unauthorized software on their workstations. One is the brute-force method, in which you simply deny them access to the media with which they would install new software. By purchasing computers without CD-ROM drives, you cut off the primary source of unauthorized software. It is also possible to prevent access to the system's floppy drive, either by installing a physical lock or by running a program such as the FloppyLocker service included with the Microsoft Zero Administration Kit.

The third potential source of unauthorized software is the Internet. If you are going to provide your users with access to services such as the Web, you may want to take steps to prevent them from installing downloaded software. One way of doing this, and of preventing all unauthorized software installations, is to use system policies that prevent users from running the setup program needed to install the software. Some of the policies that can help you do this are as follows:

Remove Run Command from Start menu   Prevents the user from launching application installation programs by preventing access to the Run dialog box.

Run Only Allowed Windows Applications   Enables the administrator to specify a list of executable files that are the only programs the user is permitted to execute. When using this policy, be sure to include executables that are needed for normal Windows operation, such as Systray.exe and Explorer.exe.

Disable MS-DOS Prompt   Prevents Windows 95/98 users from launching programs from the DOS prompt.

Locking Down the Interface   There are many elements of the Windows interface that unsophisticated users do not need to access, and suppressing these elements can prevent the more curious users from exploring things they don't understand and possibly damaging the system. Some of the policies you can use to do this are as follows:

Remove Folders from Settings on Start menu   Suppresses the appearance of the Control Panel and Printers folders in the Start menu's Settings folder. This policy does not prevent users from accessing the Control Panel in other ways, but it makes the user far less likely to explore it out of idle curiosity. You can also suppress specific Control Panel icons on Windows 95/98 systems using policies such as the following:

- Restrict Network Control Panel
- Restrict Printer Settings
- Restrict Passwords Control Panel
- Restrict System Control Panel

Remove Taskbar from Settings on Start menu   Prevents users from modifying the Start menu and taskbar configuration settings.

Remove Run Command from Start menu   Prevents users from launching programs or executing commands using the Run dialog box. This policy also provides users with additional insulation from elements such as the Control Panel and the command prompt, both of which can be accessed with Run commands.

Hide All Items on Desktop   Suppresses the display of all icons on the Windows desktop. If you want your users to rely on the Start menu to launch programs, you can use this policy to remove the distraction of the desktop icons.

Disable Registry Editing Tools   Direct access to the Windows registry should be limited to people who know what they're doing. This policy prevents users from running the registry editing tools included with the operating system.

Disable Context Menus for the Taskbar   Prevents the system from displaying a context menu when you click the secondary mouse button on a taskbar icon.

You can also use system policies to secure the cosmetic elements of the interface, preventing users from wasting time adjusting the screen colors and desktop wallpaper.

In addition, you can configure these items yourself to create a standardized desktop for all of your network's workstations. These policies are as follows:

Deny Access to Display Icon   Removes the Display icon from the Control Panel window, preventing users from accessing all display configuration parameters.

Hide Background Tab   Suppresses the Background tab in the Display Properties dialog box.

Hide Screen Saver Tab   Suppresses the Screen Saver tab in the Display Properties dialog box.

Hide Appearance Tab   Suppresses the Appearance tab in the Display Properties dialog box.

Hide Settings Tab   Suppresses the Settings tab in the Display Properties dialog box.

Wallpaper Name   Enables you to specify the path and file name of a bitmap image that the system will use as desktop wallpaper.

Color Scheme   Enables you to specify the colors that the system should use for the various elements of the desktop.

As an alternative to user profiles, system policies enable you to configure with greater precision the shortcuts found on the Windows desktop and in the Start menu. Instead of accessing an entire user profile as a whole, you can specify the locations of individual shortcut directories for various elements of the interface, using policies such as the following:

Custom Programs Folder/Custom Shared Programs Folder   Specifies the location of a directory containing shortcuts that will appear in the Start menu's Programs folder.

Custom Desktop Icons/Custom Shared Desktop Icons   Specifies the location of a directory containing shortcuts that will appear on the Windows desktop.

Custom Startup Folder/Custom Shared Startup Folder   Specifies the location of a directory containing shortcuts that will appear in the Start menu's Startup folder.

Custom Start Menu/Custom Shared Start Menu   Specifies the location of a directory containing shortcuts that will appear in the Start menu.

Hide Start Menu Subfolders   Suppresses the display of the Start menu subfolders included in a user profile to prevent duplication with the folders specified in the previous policies.

These system policies have different names, depending on whether they apply to Windows 95/98 or Windows NT/2000. The first policy listed in each of these bullets is for Windows 95/98 and the second is for Windows NT/2000.

NETWORK ADMINISTRATION

Protecting the File System Limiting access to the file system is another way of protecting your workstations against user tampering. If you preconfigure the operating system and applications on your network workstations and force your users to store all of the data files on server drives, there is no compelling reason why users should have direct access to the local file system. By blocking this access with system policies, you can prevent users from moving, modifying, or deleting files that are crucial to the operation of the workstation. You can limit users' access to the network also, using policies such as the following:

Hide Drives in My Computer Suppresses the display of all drive letters in the My Computer window, including both local and network drives.

Hide Network Neighborhood Suppresses the display of the Network Neighborhood icon on the Windows desktop and disables UNC connectivity. For example, when this policy is enabled, users can't open access network drives by opening a window with a UNC name in the Run dialog box.

No Entire Network in Network Neighborhood Suppresses the Entire Network icon in the Network Neighborhood window, preventing users from browsing network resources outside the domain or workgroup.

No Workgroup Contents in Network Neighborhood Suppresses the icons representing the systems in the current domain or workgroup in the Network Neighborhood window.

Remove Find Command from Start Menu Suppresses the Find command, preventing users from accessing drives that may be restricted in other ways. If, for example, you use the Hidden attribute to protect the local file system, the Find command can still search the local drive and display the hidden files.

Locking down the file system is a drastic step, one that you should consider and plan for carefully. Only certain types of users will benefit from this restricted access, and others may severely resent it. In addition to system policies, you should be prepared to use file system permissions and attributes to prevent specific types of user access.

Above all, you must make sure that the system policies you use to restrict access to your workstations do not inhibit the functionality your users need to perform their jobs, and that the features you plan to restrict are not accessible by other methods. For example, you might prevent access to the Control Panel by removing the folder from the Settings group in the Start menu, but users will still be able to access it from the My Computer window or the Run dialog box, unless you restrict access to those as well.

Deploying System Policies

The use of system policies by a Windows computer is itself controlled by a policy called Remote Update, which is applicable to all of the Windows operating systems. This policy has three possible settings:

Off The system does not use system policies at all.

Automatic   The system checks the root directory of the Netlogon share on the authenticating domain controller for a policy file called Ntconfig.pol (for Windows NT/2000 systems) or Config.pol (for Windows 95/98 systems).

Manual   The system checks for a policy file in a directory specified as the value of another policy called Path for Manual Update.

Using the Remote Update policy, you can configure your systems to access policy files from the default location or from any location you name. In order for workstations to have access to the policy files at all times, it is a good idea to replicate them to all of your domain controllers, either manually or automatically, just as you can do with user profiles.

Remote Registry Editing

In addition to preconfigured methods, such as user profiles and system policies, it is also possible to manage the registry on Windows workstations interactively, even from a remote location, using any one of several different tools. Windows 2000 and NT systems are capable of remote registry access by default, but on Windows 95 and 98 systems, you must install the Remote Registry Service from the distribution CD-ROM and configure the system for user-based access control.

Both the Windows 95/98 registry editor (Regedit.exe) and the Windows NT registry editor (Regedt32.exe) are able to connect to another system on the network and access its registry. In addition, you can use SPE to interactively modify the registry of the local or a remote system. However, the registry access available through SPE is limited to the registry settings that have been defined in the system policy templates that are currently loaded.

Windows 2000 Group Policies

While Windows 2000 systems support the use of Windows NT 4.0 system policies, the operating system also introduces an expanded feature called *group policies* that works in conjunction with Active Directory to create more comprehensive desktop environments. In addition to the registry-based capabilities of NT system policies, group policies can include the following types of policies:

Security policies   Policies containing local computer, domain, and network security settings.

Software installation and maintenance policies   Policies that enable administrators to remotely install, update, repair, and remove workstation software.

Script policies   Policies that can implement specific user logon and logoff scripts using the variety of different scripting languages supported by the Windows Scripting Host.

Folder redirection policies   Policies that redirect users' special folders to network drives, where they can be accessed by any system.

You implement group policies by creating a group policy object in one of the Active Directory consoles included with Windows 2000, such as Active Directory Users and Computers (see Figure 30-7). Once created, you can associate the group policy object with any other object in Active Directory to apply the policies to it.

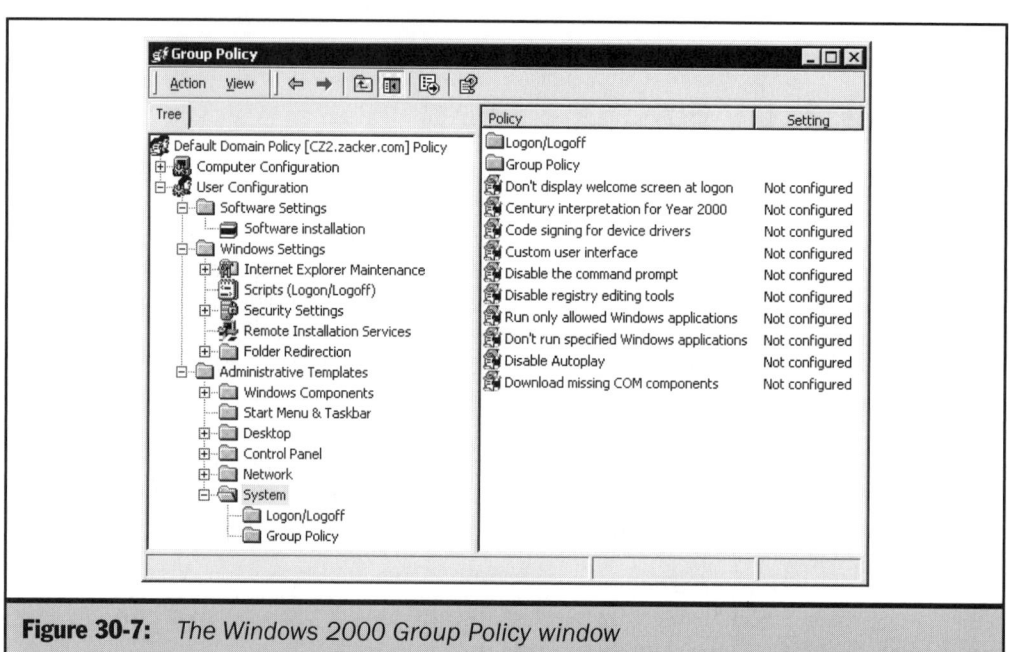

Figure 30-7:   *The Windows 2000 Group Policy window*

The Complete Reference

Chapter 31

Network Management and Troubleshooting Tools

No matter how well designed and well constructed your network is, there are going to be times when it does not function properly. Part of the job of a network administrator is to monitor the day-to-day performance of the network and cope with any problems that arise. To do this, you must have the appropriate tools. In Chapter 2, you learned about the seven layers of the networking stack as defined in the OSI reference model. Breakdowns can occur at virtually any layer, and the tools used to diagnose problems at the various layers are quite different. Knowing what resources are available to you is a large part of the troubleshooting battle; knowing how to use them properly is another large part.

Operating System Utilities

Many administrators are unaware of the network troubleshooting capabilities that are built into their standard operating systems, and as a result, they sometimes spend money needlessly on third-party products and outside consultants. The following sections examine some of the network troubleshooting tools that are provided with the operating systems commonly used on today's networks.

Windows Utilities

The Windows operating systems include a variety of tools that you can use to manage and troubleshoot network connections. Most of these tools are included in both the 95/98 and 2000/NT packages, although they may take slightly different forms.

NET

The NET command is the primary command-line control for the Windows network client. You can use NET to perform many of the same networking functions that you can perform with graphical utilities, such as Windows Explorer. Because NET is a command-line utility, you can include the commands in logon scripts and batch files. For example, you can use NET to log on and off of the network, map drive letters to specific network shares, start and stop services, and locate shared resources on the network.

The NET command is implemented as a file called Net.exe, which is installed to the system directory (C:\Windows or C:\Winnt) during the operating system installation. To use the program, you execute the file from the command line with a subcommand, which may take additional parameters. Although Windows 95/98 and Windows 2000/NT share some of these subcommands, others are unique to each operating system. These subcommands and their functions are listed in Table 31-1, and some of the key functions are examined in the following sections.

NET Subcommand	Operating Systems Supported	Function
NET ACCOUNTS	Windows 2000/NT	Configures settings and policies for all of the accounts on a particular computer or domain
NET COMPUTER	Windows 2000/NT	Adds or removes computers from the current domain
NET CONFIG	Windows 95/98	Displays network client information
NET CONFIG SERVER	Windows 2000/NT	Configures Server service parameters
NET CONFIG WORKSTATION	Windows 2000/NT	Configures Workstation service parameters
NET CONTINUE	Windows 2000/NT	Resumes a service that has been paused
NET DIAG	Windows 95/98	Exchanges diagnostic messages with another system to test the network connection
NET FILE	Windows 2000/NT	Displays and closes files shared with network users and removes file locks
NET GROUP	Windows 2000/NT	Creates or deletes global groups and adds users to or deletes them from those groups
NET HELP	Windows 95/98; Windows 2000/NT	Displays help information for specific NET subcommands
NET HELPMSG	Windows 2000/NT	Displays additional information about a specific four-digit error code
NET INIT	Windows 95/98	Loads network adapter and protocol drivers without binding them to the Protocol Manager

Table 31-1:   *Windows NET Subcommands*

NETWORK ADMINISTRATION

NET Subcommand	Operating Systems Supported	Function
NET LOCALGROUP	Windows 2000/NT	Creates or deletes local groups and adds users to or deletes them from those groups
NET LOGOFF	Windows 95/98	Logs the user off of the network and severs connections with all shared network resources
NET LOGON	Windows 95/98	Logs a user on to a workgroup or domain
NET NAME	Windows 2000/NT	Administers the list of names used by the Messenger service to send messages
NET PASSWORD	Windows 95/98	Changes the current user's logon password
NET PAUSE	Windows 2000/NT	Pauses a specific service without unloading it until resumed by the NET CONTINUE command
NET PRINT	Windows 95/98; Windows 2000/NT	Administers print queues and the print jobs in them
NET SEND	Windows 2000/NT	Transmits a text message to another user or computer using the Messenger service
NET SESSION	Windows 2000/NT	Displays information about, and disconnects currently active sessions with, other network users
NET SHARE	Windows 2000/NT	Displays, creates, and deletes shares on the current system
NET START	Windows 95/98; Windows 2000/NT	Starts a specific network service

Table 31-1:   *Windows NET Subcommands (continued)*

NET Subcommand	Operating Systems Supported	Function
NET STATISTICS	Windows 2000/NT	Displays statistics for the Server or Workstation service
NET STOP	Windows 95/98; Windows 2000/NT	Stops a specific network service
NET TIME	Windows 95/98; Windows 2000/NT	Displays the time on the current system or synchronizes the time with another system
NET USE	Windows 95/98; Windows 2000/NT	Displays information about and administers connections to shared network resources
NET USER	Windows 2000/NT	Creates, modifies, and deletes user accounts
NET VER	Windows 95/98	Displays the type and version number of the workgroup redirector currently in use
NET VIEW	Windows 95/98; Windows 2000/NT	Displays available resources on the network

Table 31-1: *Windows NET Subcommands (continued)*

NET CONFIG The NET CONFIG command displays network client information about the current system, such as the following:

```
Computer name              \\CZ5

User name                  CRAIGZ

Workgroup                  NTDOMAIN

Workstation root directory C:\WINDOWS

Software version           4.00.950

Redirector version         4.00

The command was completed successfully.
```

NET DIAG   The NET DIAG command initiates a low-level diagnostics test between two computers on the network. For this purpose, NET can function as either a diagnostic client or server. When you run the NET DIAG command, the system first attempts to detect a diagnostic server on the network. If it fails to detect a server, it starts functioning as one itself. If the system has both the IPX protocol and a NetBIOS protocol (either NetBIOS over TCP/IP or NetBEUI) installed, the program prompts you for the protocol it should use for the diagnostic test.

When the system successfully locates a server, it transmits a series of NetBIOS session messages or IPX Service Advertising Protocol (SAP) messages and examines the replies it receives. During a NetBIOS test, the client establishes a TCP connection or a NetBEUI session with the server and then begins sending session messages containing test data. In an IPX test, the client transmits a SAP broadcast and receives a reply from the server. After that, the client can transmit unicast SAP packets containing test messages to the server.

The NET DIAG command tests the networking functionality of the entire protocol stack, and when used in combination with other diagnostic utilities, you can try to isolate a problem to a particular service or protocol. If, for example, two systems can successfully exchange Ping messages, but the NET DIAG test between the two machines fails, you can deduce that a problem exists somewhere above the network layer of the OSI model.

NET START and NET STOP   The NET START and NET STOP commands are used to start and stop network services on the current system. On a Windows 95, 98, or Me system, you can use these commands to select which redirector you want to load on the system to provide access to network resources, using the following syntax, the individual elements of which are described next:

```
NET START [BASIC | NWREDIR | WORKSTATION | NETBIND | NETBEUI |
NWLINK] [/LIST] [/YES] [/VERBOSE]
```

BASIC   Starts the basic redirector

NWREDIR   Starts the Microsoft Client for NetWare Networks redirector

WORKSTATION   Starts the default redirector

NETBIND   Binds protocols and network-adapter drivers

NETBEUI   Starts the NetBIOS interface

NWLINK   Starts the IPX/SPX-compatible interface

/LIST   Displays a list of the services that are running

/YES   Executes the command without user input

/VERBOSE   Displays information about device drivers and services as they are loaded

By default, Windows 95, 98, and Me load the full workstation redirector, which provides complete access to domains and workgroups on the network. The operating system also includes a basic redirector that you can use for basic network connectivity, using a minimum of system resources. The basic redirector enables you to connect to a workgroup system and access shared resources, but you cannot use it to log on to a Windows 2000 or NT domain.

In situations where elements of the system are not functioning properly, such as when you can't load the GUI, you can load the real-mode network client included with Windows 95/98/Me, using either the NET START BASIC or NET START WORKSTATION command. These commands load the NDIS 2.0 driver for your NIC and the drivers for your installed protocols, and then bind the drivers to the protocol manager.

Once the redirector is fully loaded, the system logs you on to the default workgroup or domain and prompts you for your password. You can then map drive letters to shared network drives using the NET USE command, and access those drives, or execute other commands that perform network functions, such as NET VIEW to display the resources available on the network, or NET DIAG to test network communications. After they're started, you can stop specific services by using the NET STOP command with the same parameters that started the services, such as NET STOP WORKSTATION, but only outside of the Windows GUI. You cannot execute the NET STOP command from a DOS session within Windows.

On Windows 2000 and NT systems, you can use the NET START and NET STOP commands to start and stop any service running on the machine, or use NET PAUSE and NET CONTINUE to temporarily suspend a service. Typing NET START at the command prompt on a 2000/NT system displays a list of the services currently running, such as the following:

```
These Windows 2000 services are started:

   COM+ Event System

   Computer Browser

   DHCP Client

   Distributed Link Tracking Client

   DNS Client

   Event Log

   Infrared Monitor

   IPSEC Policy Agent

   Logical Disk Manager

   Logical Disk Manager Administrative Service
```

```
       Messenger

       Net Logon

       Network Connections

       Plug and Play

       Print Spooler

       Protected Storage

       Remote Access Connection Manager

       Remote Procedure Call (RPC)

       Remote Registry Service

       Removable Storage

       RunAs Service

       Security Accounts Manager

       Server

       System Event Notification

       Task Scheduler

       TCP/IP NetBIOS Helper Service

       Telephony

       Windows Management Instrumentation

       Windows Management Instrumentation Driver Extensions

       Windows Time

       WMDM PMSP Service

       Workstation

The command completed successfully.
```

The behavior of the NET START command, without any further parameters, is different on Windows 95/98 and 2000/NT systems. While the command is purely informational on a 2000/NT system, on Windows 95/98 it starts the default redirector.

NET SESSION   It is possible, in both Windows 2000 and NT, to disable a particular user account in the user object's Properties dialog box or in the User Manager, thus preventing the user from logging on to the network. However, this action doesn't take effect until the next time the user tries to log on. If you want to disconnect a user from a system immediately, you can use the Computer Management Console in Windows 2000 or the Windows NT Server Manager, or you can use NET SESSION from the command line.

Running NET SESSION with no parameters displays a list of the current active sessions on the system, like the following:

```
Computer      User name              Client Type       Opens    Idle time

--------------------------------------------------------------------

\\CZ2         Administrator          Windows 2000 2195   5       00:14:51

\\CZ3         JDOE                   Windows 4.0         0       00:00:08

\\CZ5         CRAIGZ                 Windows 2000 2195   0       06:02:51

The command completed successfully.
```

To disconnect a session immediately, you use NET SESSION with the following syntax:

```
NET SESSION [\\computername] /delete
```

When you specify the NetBIOS name of a computer on the command line, NET SESSION immediately disconnects all of the sessions from that computer to the current system and closes all open files. If you omit a computer name, NET SESSION terminates all sessions from all computers.

Net Watcher

Net Watcher is a utility included in Windows 95/98 and in Windows NT Server 4.0 Resource Kit that enables you to monitor the network users connected to your computer, the shares that are currently being accessed, and the files that the remote users have open. You can also disconnect a user from a share, forcibly close a file that a user has open, and create or remove shares. Net Watcher is a useful tool for determining who is accessing the files or shares on your system at any given time. However, from a network administration standpoint, the best feature of this application is that you can connect to other computers on the network and perform these actions remotely for those systems.

Connecting to a Remote System   Net Watcher takes the form of an executable file called Netwatch.exe that Windows 95 and 98 install with the operating system by default. When you launch Net Watcher, the program displays the connections to your own machine that are currently open. To monitor the activity of another Windows 95 or 98 workstation on the network, you choose Select Server from the Administer menu and either browse to the computer you want to monitor or specify its NetBIOS name or IP address.

To connect to another Windows 95 or 98 system with Net Watcher, the other computer must have Remote Administration enabled. To do this, a user at the other machine must open the Passwords Control Panel to the Remote Administration page, select the Enable Remote Administration of This Server check box, and specify a password that you will use when connecting to the workstation.

> **Note** *It is advisable that you use a strong Remote Administration password for systems that contain sensitive data, since a user with remote administration privileges can access all of the system's hard drives and create new shares without limitation.*

When you enable Remote Administration, Windows 98 creates two administrative shares, as follows:

ADMIN$ Provides administrators with access to the file system, even when drives are not shared

IPC$ Provides an interprocess communication (IPC) channel between the user's computer and the administrator's computer

These shares enable you to interact with the remote system and observe its networking activities.

> **Note** *The Windows 98 Remote Administration feature is not the same as the Remote Registry service that enables you to modify registry settings on other network systems. You can enable Remote Administration on any Windows 98 system, whereas the Remote Registry service requires user-level access control and a Windows 2000 or NT system on the network.*

Using the Connections Window When you first connect to another system with the Windows 95/98 version of Net Watcher, the program displays the Connections window, which contains a list of the users and computers currently accessing the system's shares, as shown in Figure 31-1. The left pane displays the number of shares and files that each user has open, while the right pane lists the open files in each share. You can disconnect a user from the computer (as well as any shares and files they are accessing) by highlighting a name and selecting Disconnect User from the Administer menu.

As a security tool, Net Watcher enables you to monitor the network for unauthorized access to specific systems and shares, and take steps to prevent continued intrusion. When you discover someone accessing a share without authorization, you can immediately disconnect that user from the machine, and then switch to the Shared Folders window to change the share's permissions or password.

> **Note** *Disconnecting a user from a system is a drastic step when the user has files open. The connection is severed with no warning to the user, and the interruption can result in data loss.*

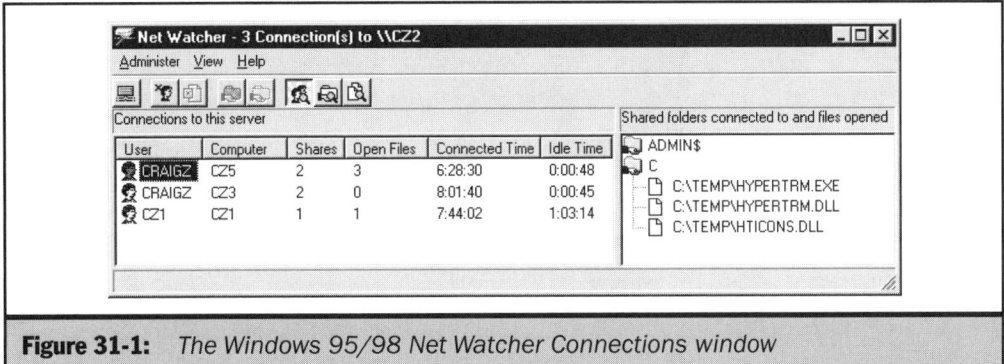

Figure 31-1: *The Windows 95/98 Net Watcher Connections window*

Using the Shared Folders Window   The other two Net Watcher windows display the same information in a different format. The Shared Folders window lists the drive shares on the system, the computers connected to them, and the files opened by each computer (see Figure 31-2). From this window, you can create and delete shares on the remote system and modify the properties of existing shares.

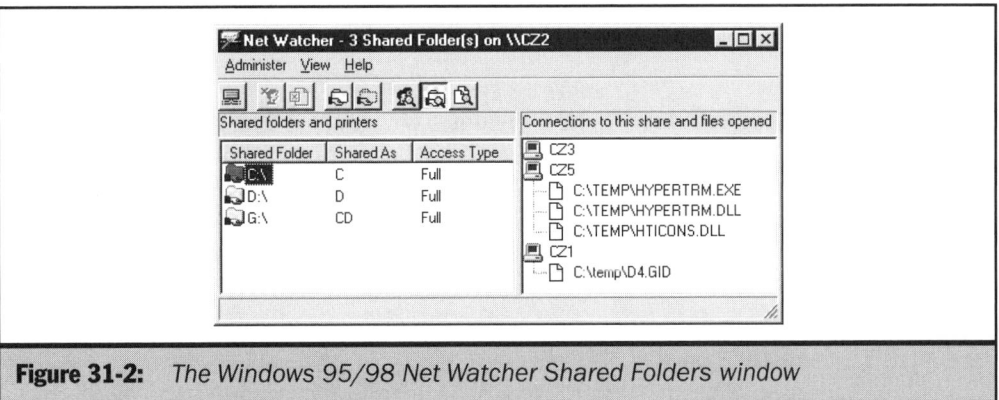

Figure 31-2: *The Windows 95/98 Net Watcher Shared Folders window*

To create a new share, you select Add Shared Folder from the Administer menu and select the desired drive or directory from the Browse for Folder dialog box, shown in Figure 31-3. This dialog box displays the existing shares on the computer, as well as the administrative shares, represented by a drive letter followed by a dollar sign (such as C$). You must select one of these administrative shares or a subdirectory of one of these shares as the root directory for your new share. After you make your selection, you see the standard Sharing dialog box in which you specify a name for the share, the type of access you want to grant to users, and a password.

NETWORK ADMINISTRATION

Figure 31-3: *The administrative shares for a Windows 95/98 system appear as drive letters followed by a dollar sign.*

You can also modify the password for a share, or the share name itself, by selecting a share from the list and choosing Shared Folder Properties from the Administer menu. In the event of a security breach, you can even delete the share entirely to prevent all users from accessing it.

Using the Open Files Window   The Open Files window lists the files that are in use, and identifies who is using them (see Figure 31-4). From this window, you can close an individual file (instead of disconnecting a user completely from the system) by selecting Administer | Close File. For example, if a file is inaccessible because another user has left it open and walked away from their computer, you can close it without disturbing the user's other work.

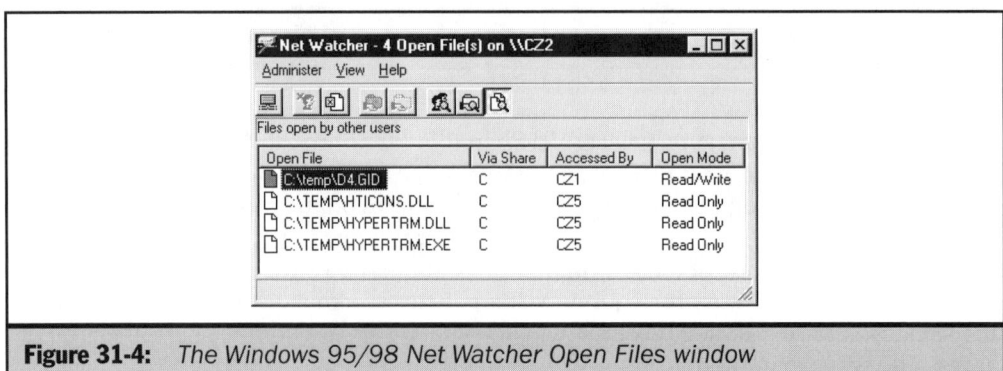

Figure 31-4: *The Windows 95/98 Net Watcher Open Files window*

Monitoring Network Activity on Windows 2000 and Windows NT   Windows 2000 and Windows NT both include utilities that provide the same functions as Net Watcher. The Windows NT Server 4.0 Resource Kit includes Net Watch, its own version of the Net Watcher program, which uses a different interface (see Figure 31-5), but can perform all of the same tasks as Net Watcher. The NT version of the Net Watcher program can monitor other NT and 2000 systems, and the Windows 95/98 version can monitor other 95/98 systems, but the two cannot interact.

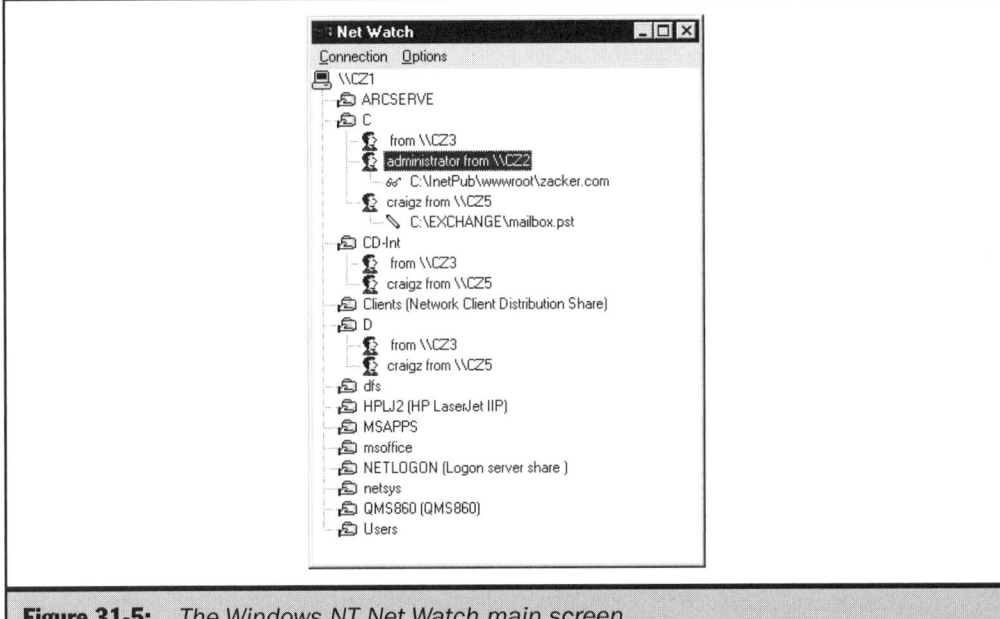

Figure 31-5:   *The Windows NT Net Watch main screen*

Instead of the three different windows that Net Watcher displays, Net Watch displays shares, connections, and open files in one hierarchical tree, and you can disconnect users and close files from their context menus. The same functions are provided by the Server Manager utility included with Windows NT, using screens like that shown in Figure 31-6.

In Windows 2000, you find the equivalent functionality in the Computer Management console (see Figure 31-7). The Shares, Sessions, and Open Files panels enable you to monitor the network connections of any system on the network, just as you can with Server Manager and Net Watcher.

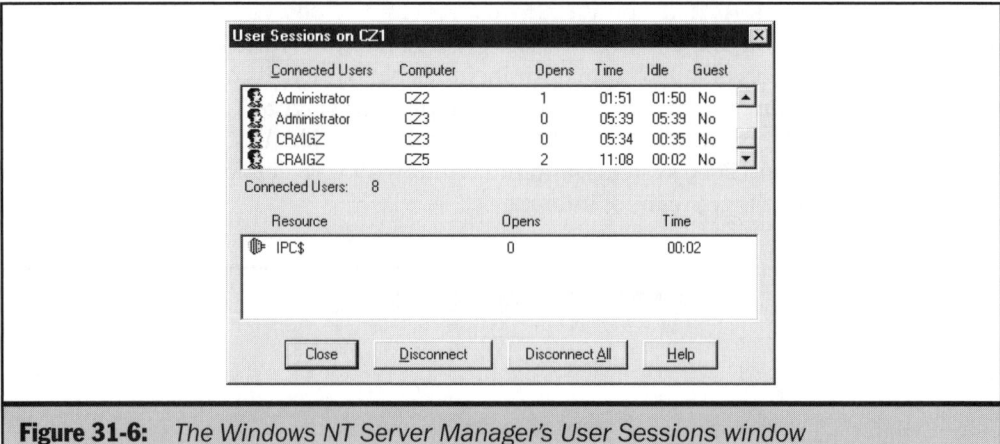

Figure 31-6:   *The Windows NT Server Manager's User Sessions window*

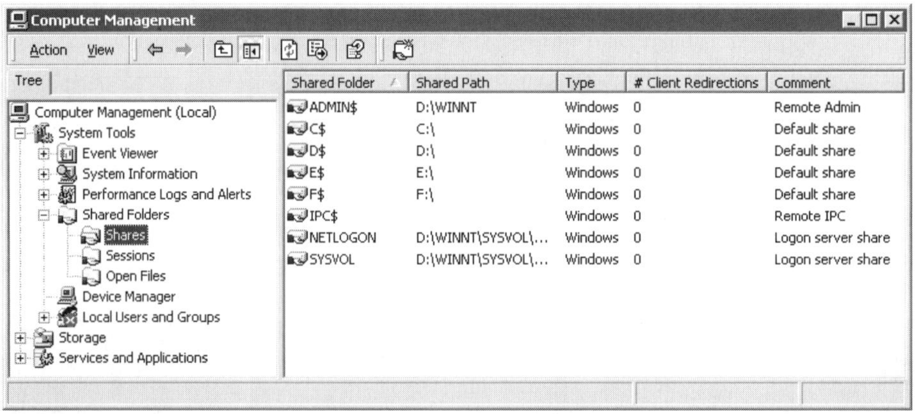

Figure 31-7:   *The Windows 2000 Computer Management console*

Web Administrator

Web Administrator is an add-on product for Microsoft's Internet Information Server (IIS) that enables you to manage many elements of a Windows NT 4.0 Server system using any Java-capable Web browser. Available free of charge from Microsoft's Web site at http://www.microsoft.com/ntserver/nts/downloads/management/NTSWebAdmin/, installing the product creates a new subdirectory called \NTADMIN on the Web site hosted by your NT Server. When you point a browser to that directory, the main Web Administrator page appears (see Figure 31-8).

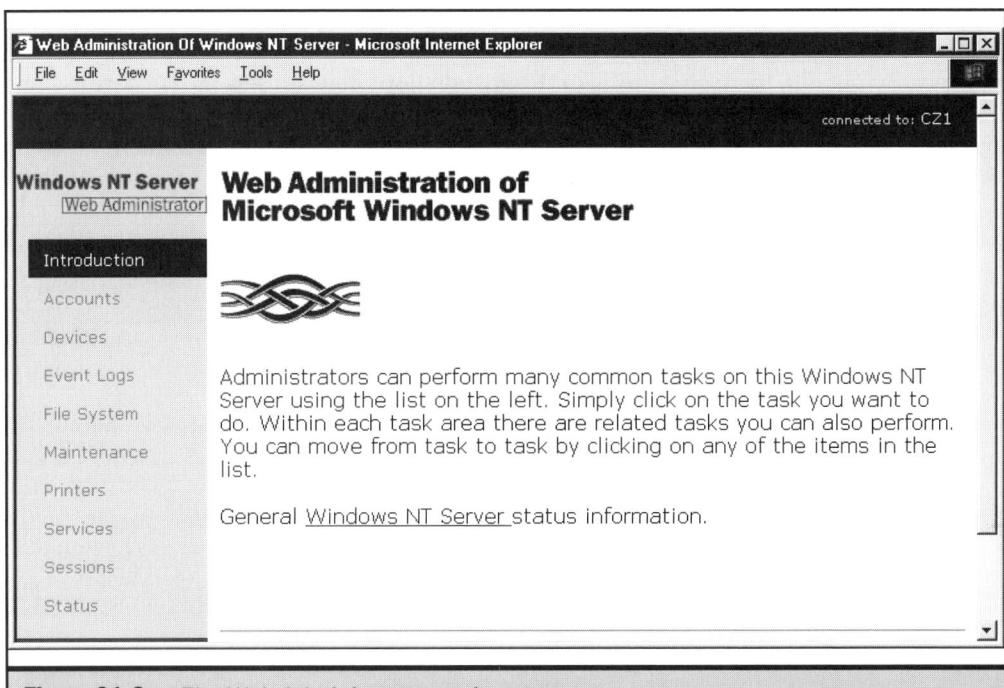

Figure 31-8: *The Web Administrator main screen*

From this screen, you can perform many of the same tasks as you can with the native NT administration programs—such as Control Panel, Server Manager, User Manager, Performance Monitor, and other utilities—including the following:

- Manage local and domain user, group, and computer accounts

- Manage the device drivers installed on the system

- View the NT event logs

- Manage shares and file system permissions

- Send messages to users logged on to the server

- Open a remote console session to the server

- Reboot the server from a remote location

- Manage print queues and their contents

- Start, stop, and pause services running on the server

- Monitor sessions and disconnect users

- View server status and performance statistics

Web Administrator uses Java to simulate the controls found in standard Windows applications, as shown in Figure 31-9. Most of the functionality of the original programs is intact. By default, the Web Administrator installation program sets the IIS permissions so that only a Web browser running on the server itself can access the site. You can modify the permissions so that only specific users or machines with specific IP addresses can access Web Administrator.

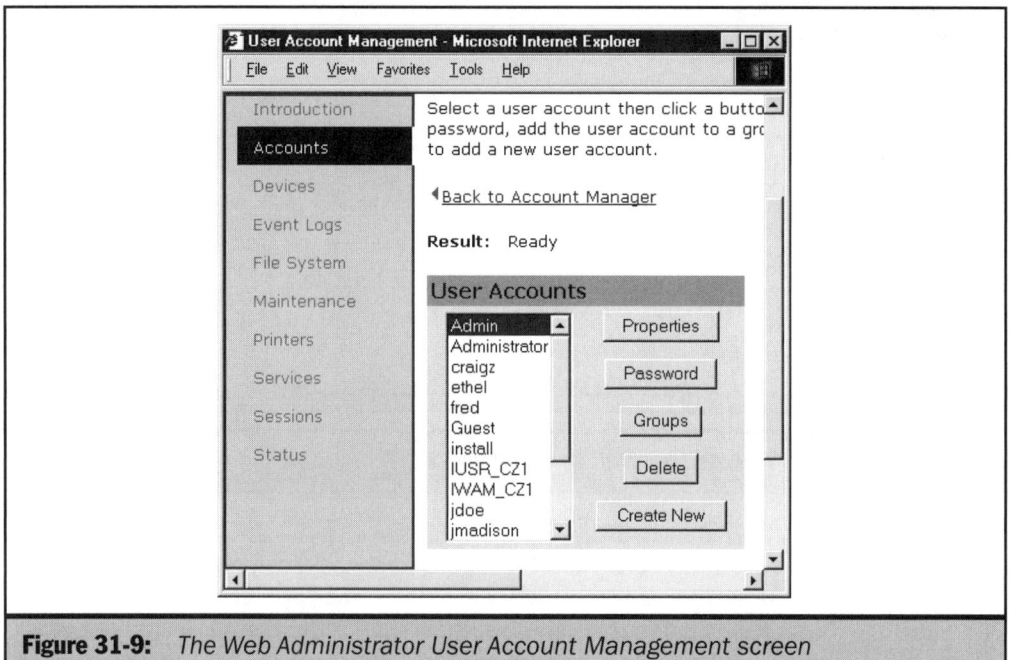

Figure 31-9:   *The Web Administrator User Account Management screen*

The Web Administrator utility is an excellent tool for networks running several different operating systems, because it provides administrative access to an NT server from any computer with a supported browser. Although many of the Windows NT tools can perform their functions for any NT system on the network, Web Administrator enables network support personnel working on other operating systems, such as Windows 95/98, Macintosh, or Unix, to access native NT functions. Web Administrator has not been ported to Windows 2000.

NetMeeting

Microsoft NetMeeting is a conferencing and collaboration tool that is intended for use over the Internet, but it can also be a valuable asset to the LAN administrator.

NetMeeting is a part of the full installation of Internet Explorer versions 4 and 5, but it is also available separately in versions for Windows 95/98 and 2000/NT (see www.microsoft.com/windows/netmeeting). Unless you have high-speed connections, NetMeeting's performance over the Internet is usually disappointing. However, the program runs much better on a LAN.

Although Internet users are primarily concerned with NetMeeting's audio- and videoconferencing capabilities, network administrators will find that its collaboration features are an excellent means of providing technical support without traveling to the user's location. In addition to audio- and videoconferencing, NetMeeting includes text-based chat and a whiteboard, as well as the ability to collaborate on a specific application.

For example, if a user is having trouble with a particular application, she can call the help desk and arrange to activate a NetMeeting connection with an administrator. When the user shares the application within NetMeeting, the administrator can take control over it to demonstrate a particular procedure. Only one user can have control of the application at a time, but the user is able to see the administrator's keystrokes and mouse clicks, and can learn to use the software interactively. In the same way, the administrator can watch the user try to perform a task, to see what she's doing wrong.

Another NetMeeting feature, called Remote Desktop Sharing, enables a remote user to take full control over a computer. Administrators can use this feature to remotely configure a system, install software, and even launch applications, all without any direct contact with the machine.

TCP/IP Utilities

TCP/IP has become the most commonly used protocol suite in the networking industry, and many network administration and troubleshooting tasks involve working with various elements of these protocols. Because virtually every computing platform supports TCP/IP, a number of basic tools have been ported to many different operating systems, some of which have also been adapted to specific needs. The following sections examine some of these tools, but do so more from the perspective of their basic functionality and usefulness to the network administrator than from the operational elements of specific implementations.

Ping

Ping is unquestionably the most common TCP/IP diagnostic tool, and is included in virtually every implementation of the TCP/IP protocols. In most cases, Ping is a command-line utility, although some graphical or menu-driven versions are available that use a different interface to perform the same tasks. The basic function of Ping is to send a message to another TCP/IP system on the network to determine if the protocol stack up to the network layer is functioning properly. Because the TCP/IP protocols function in the same way on all systems, you can use Ping to test the connection between any two computers, regardless of processor platform or operating system.

All Windows systems install a command-line Ping program called Ping.exe to the system directory (such as C:\Windows or C:\Winnt\System32) as part of their TCP/IP stacks. In the same way, all of the shells in the various Unix variants support the ping command. Novell NetWare includes a menu-driven Ping.nlm utility that runs on a server, as well as a server command-line implementation called Tping.nlm.

Ping works by transmitting a series of Echo Request messages to a specific IP address using the Internet Control Message Protocol (ICMP). When the computer using that IP address receives the messages, it generates an Echo Reply in response to each Echo Request and transmits it back to the sender. ICMP is a TCP/IP protocol that uses several dozen message types to perform various diagnostic and error-reporting functions. ICMP messages are carried directly within IP datagrams. No transport-layer protocol is involved, so a successful Ping test indicates that the protocol stack is functioning properly from the network layer down. If the sending system receives no replies to its Echo Requests, something is wrong with either the sending or the receiving system, or the network connection between them.

Note *For more information on the ICMP protocol and its various functions, refer to Chapter 13.*

When Ping is implemented as a command-line utility, you use the following syntax to perform a Ping test:

```
PING destination
```

where the *destination* variable is replaced by the name or address of another system on the network. The destination system can be identified by its IP address, or by a name, assuming that an appropriate mechanism is in place for resolving the name into an IP address. This means that you can use a host name for the destination, as long as you have a DNS server or HOSTS file to resolve the name. On Windows networks, you can also use NetBIOS names, along with any of the standard mechanisms for resolving them, such as WINS servers, broadcast transmissions, or an LMHOSTS file.

The screen output produced by a PING command on a Windows system appears as follows:

```
Pinging cz3 [192.168.2.3] with 32 bytes of data:


Reply from 192.168.2.3: bytes=32 time=1ms TTL=128

Reply from 192.168.2.3: bytes=32 time<10ms TTL=128

Reply from 192.168.2.3: bytes=32 time=1ms TTL=128
```

```
Reply from 192.168.2.3: bytes=32 time<10ms TTL=128


Ping statistics for 192.168.2.3:
    Packets: Sent = 4, Received = 4, Lost = 0 (0% loss),
Approximate round trip times in milli-seconds:
    Minimum = 0ms, Maximum =  1ms, Average =  0ms
```

Note *Because most Ping implementations display the IP address resolved from the system name specified on the command line, the program is also a quick and easy tool for determining a given system's IP address.*

The program displays a result line for each of the four Echo Request messages it sends by default, specifying the IP address of the recipient, the number of bytes of data transmitted in each message, the amount of time elapsed between the transmission of the request and the receipt of the reply, and the target system's time to live (TTL). The TTL is the number of routers that a packet can pass through before it is discarded.

Ping has other diagnostic uses apart from simply determining whether a system is up and running. If you can successfully ping a system using its IP address, but pings sent to the system's name fail, you know that a malfunction is occurring in the name resolution process. When you're trying to contact an Internet site, this indicates that there is a problem with either your workstation's DNS server configuration or the DNS server itself. If you can ping systems on the local network successfully, but not systems on the Internet, you know there is a problem with either your workstation's Default Gateway setting or the connection to the Internet.

Note *Sending a Ping command to a system's loopback address (127.0.0.1) tests the operability of the TCP/IP protocol stack, but it is not an adequate test of the network interface because traffic sent to the loopback address travels down the protocol stack only as far as the network transport layer and is redirected back up without ever leaving the computer through the network interface.*

In most Ping implementations, you can use additional command-line parameters to modify the size and number of the Echo Request messages transmitted by a single Ping command, as well as other operational characteristics. In the Windows Ping.exe program, for example, the parameters are as follows:

```
ping [-t] [-a] [-n count] [-l size] [-f] [-i TTL] [-v TOS] [-r count]
[-s count] [[-j host-list] | [-k host-list]] [-w timeout] destination
```

-t   Pings the specified destination until stopped by the user (with CTRL-C)

-a   Resolves destination IP addresses to host names

-n *count*   Specifies the number of Echo Requests to send

-l *size*   Specifies the size of the Echo Request messages to send

-f   Sets the IP Don't Fragment flag in each Echo Request packet

-i *TTL*   Specifies the IP TTL value for the Echo Request packets

-v *TOS*   Specifies the IP Type of Service (TOS) value for the Echo Request packets

-r *count*   Records the IP addresses of the routers for the specified number of hops

-s *count*   Records the time stamp from the routers for the specified number of hops

-j *host-list*   Specifies a partial list of routers that the packets should use

-k *host-list*   Specifies a complete list of routers that the packets should use

-w *timeout*   Specifies the time (in milliseconds) that the system should wait for each reply

There are many different applications for these parameters that can help you manage your network and troubleshoot problems. For example, by creating larger than normal Echo Requests and sending large numbers of them (or sending them continuously), you can simulate user traffic on your network to test its ability to stand up under heavy use. You can also compare the performance of various routes through your network (or through the Internet) by specifying the IP addresses of the routers that the Echo Request packets must use to reach their destinations. The -j parameter provides *loose source routing,* in which the packets must use the routers whose IP addresses you specify, but can use other routers also. The -k parameter provides *strict source routing,* in which you must specify the address of every router that packets will use to reach their destination.

Traceroute

Traceroute is another utility that is usually implemented as a command-line program and included in most TCP/IP protocol stacks, although it sometimes goes by a different name. On Unix systems, the command is called traceroute, but Windows implements the same functions in a program called Tracert.exe. The function of this tool is to display the route that IP packets are taking to reach a particular destination system. When you run the program with the name or IP address of a destination system as the command-line parameter, the result is a display like the following:

```
Tracing route to zacker.com [192.41.15.74] over a maximum of 30 hops:

  1    254 ms    194 ms    162 ms   qrv1-67.epoch.net [199.224.67.3]
```

```
 2    151 ms     135 ms     154 ms    qrvl.epoch.net [199.224.67.1]

 3    163 ms     150 ms     173 ms    svcr03-7b.epoch.net [199.224.103.125]

 4    136 ms     160 ms     164 ms    router05.epoch.net [216.37.155.162]

 5    161 ms     145 ms     170 ms    cpbg01-7.epoch.net [199.224.88.62]

 6    165 ms     149 ms     164 ms    Serial1.PH.ALTER.NET [157.130.7.213]

 7    182 ms     242 ms     169 ms    161.ATM2.ALTER.NET [146.188.162.118]

 8    178 ms     149 ms    1839 ms    294.ATM7.ALTER.NET [146.188.160.126]

 9    168 ms     147 ms     155 ms    192.ATM10.ALTER.NET [146.188.160.93]

10    260 ms     150 ms     176 ms    uu.iad1.verio.net [137.39.23.22]

11    163 ms     175 ms     166 ms    iad3.dca0.verio.net [129.250.2.62]

12    235 ms     243 ms     244 ms    dca0.pao5.verio.net [129.250.2.245]

13    224 ms     249 ms     255 ms    p4-01.us.bb.verio.net [129.250.2.74]

14    406 ms     272 ms     265 ms    pao6.pvu0.verio.net [129.250.3.26]

15    267 ms     250 ms     271 ms    pvu0.vwh.verio.net [129.250.16.14]

16    257 ms     270 ms     278 ms    zacker.com [192.41.15.74]

Trace complete.
```

Each of the entries in the trace represents a router that processed the packets generated by the traceroute program on the way to their destination. In this case, the packets required 16 hops to get to the zacker.com server. The three numerical figures in each entry specify the round-trip time to that router, in milliseconds, followed by the DNS name and IP address of the router. In a trace like this one, to a destination on the Internet, the round-trip times are relatively high and can provide you with information about the backbone networks your ISP uses (in this case, alter.net) and the geographical path that your traffic takes. For example, when you run a trace to a destination system on another continent, you can sometimes tell when the path crosses an ocean by a sudden increase in the round-trip times. On a private network, you can use traceroute to determine the path through your routers that local traffic typically takes, enabling you to get an idea of how traffic is distributed around your network.

Most traceroute implementations work by transmitting the same type of ICMP Echo Request messages used by Ping, while others use UDP packets by default. The only difference in the messages themselves is that the traceroute program modifies the TTL field for each sequence of three packets. The TTL field is a protective mechanism that prevents IP packets from circulating endlessly around a network. Each router that processes a packet decrements the TTL value by one. If the TTL value of a packet reaches 0, the router discards it and returns an ICMP Time to Live Exceeded in Transit error message to the system that originally transmitted it.

In the first traceroute sequence, the packets have a TTL value of 1, so that the first router receiving the packets discards them and returns error messages back to the source. By calculating the interval between a message's transmission and the arrival of the associated error, traceroute generates the round-trip time and then uses the source IP address in the error message to identify the router. In the second sequence of messages, the TTL value is 2, so the packets reach the second router in their journey before being discarded. The third sequence of packets has a TTL value of 3, and so on, until the messages reach the destination system.

It is important to understand that although traceroute can be a useful tool, a certain amount of imprecision is inherent in the information it provides. Just because a packet transmitted right now takes a certain path to a destination does not mean that a packet transmitted a minute from now to that same destination will take that same path. Networks (and especially those on the Internet) are mutable, and routers are designed to compensate automatically for the changes that occur. The route taken by traceroute packets to their destination can change, even in the midst of a trace, so it is entirely possible for the sequence of routers displayed by the program to be a composite of two or more different paths to the destination, because of changes that occurred in midstream. On a private network, this is less likely to be the case, but it is still possible.

Route

The routing table is a vital part of the networking stack on any TCP/IP system, even those that do not function as routers. The system uses the routing table to determine where it should transmit each packet. The Route.exe program in Windows and the route command included with most Unix versions enable you to view the routing table and add or delete entries to it. The syntax for the Windows Route.exe program is as follows:

```
ROUTE [-f] [-p] [command [destination] [MASK netmask] [gateway]
[METRIC metric] [IF interface]]
```

The *command* variable takes one of the following four values:

PRINT   Displays the contents of the routing table

ADD   Creates a new entry in the routing table

DELETE   Deletes an entry from the routing table

CHANGE   Modifies the parameters of a routing table entry

The other parameters used on the Route.exe command line are as follows:

–f   Deletes all of the entries from the routing table

–p   Creates a permanent entry in the routing table (called a *persistent route*) when used with the ADD command

destination   Specifies the network or host address of the routing table entry being added, deleted, or changed

MASK *netmask*   Specifies the subnet mask associated with the address specified by the *destination* variable

gateway   Specifies the address of the router used to access the host or network address specified by the *destination* variable

METRIC *metric*   Indicates the relative efficiency of the routing table entry

IF *interface*   Specifies the address of the network interface adapter used to reach the router specified by the *gateway* variable

For more information on routing tables and the principles of IP routing, refer to Chapter 6.

Netstat

Netstat is a command-line utility that displays network traffic statistics for the various TCP/IP protocols and, depending on the platform, may display other information as well. Most Unix variants support the netstat command, and Windows operating systems include a program called Netstat.exe that is installed by default with the TCP/IP stack. The command-line parameters for Netstat can vary in different implementations, but one of the most basic ones is the -s parameter, which displays the statistics for each of the major TCP/IP protocols, as follows:

```
IP Statistics

   Packets Received              = 130898

   Received Header Errors        = 0

   Received Address Errors       = 19

   Datagrams Forwarded           = 0

   Unknown Protocols Received    = 0

   Received Packets Discarded    = 0

   Received Packets Delivered    = 130898

   Output Requests               = 152294
```

```
Routing Discards                      = 0
Discarded Output Packets              = 0
Output Packet No Route                = 0
Reassembly Required                   = 0
Reassembly Successful                 = 0
Reassembly Failures                   = 0
Datagrams Successfully Fragmented     = 0
Datagrams Failing Fragmentation       = 0
Fragments Created                     = 0


ICMP Statistics

                          Received      Sent

   Messages               499           683
   Errors                 44            0
   Destination Unreachable 0            154
   Time Exceeded          414           0
   Parameter Problems     0             0
   Source Quenchs         0             0
   Redirects              0             0
   Echos                  1             522
   Echo Replies           27            1
   Timestamps             0             0
   Timestamp Replies      0             0
   Address Masks          0             0
   Address Mask Replies   0             0


TCP Statistics
   Active Opens                       = 1893
```

```
Passive Opens                    = 12

Failed Connection Attempts       = 37

Reset Connections                = 657

Current Connections              = 0

Segments Received                = 117508

Segments Sent                    = 142099

Segments Retransmitted           = 378


UDP Statistics

   Datagrams Received   = 12399

   No Ports             = 943

   Receive Errors       = 0

   Datagrams Sent       = 9129
```

Apart from the total number of packets transmitted and received by each protocol, Netstat provides valuable information about error conditions and other processes that can help you troubleshoot network communication problems at various layers of the OSI model. The Windows version of Netstat also can display Ethernet statistics (using the -e parameter), which can help to isolate network hardware problems, such as the following:

```
Interface Statistics

                            Received          Sent

Bytes                       44483612          20434045

Unicast packets             94653             92824

Non-unicast packets         4543              743

Discards                    0                 0

Errors                      0                 0

Unknown protocols           15452
```

When executed with the -a parameter, Netstat.exe displays information about the TCP connections currently active on the computer and the UDP services that are listening for input, as follows:

```
Active Connections

  Proto  Local Address          Foreign Address        State

  TCP    cz5:1044               CZ5:0                  LISTENING

  TCP    cz5:1025               CZ5:0                  LISTENING

  TCP    cz5:1025               CZ1:nbsession          ESTABLISHED

  TCP    cz5:137                CZ5:0                  LISTENING

  TCP    cz5:138                CZ5:0                  LISTENING

  TCP    cz5:nbsession          CZ5:0                  LISTENING

  TCP    cz5:nbsession          CZ3:1531               ESTABLISHED

  TCP    cz5:2521               netsurge.com:pop3      TIME_WAIT

  UDP    cz5:1044               *:*

  UDP    cz5:nbname             *:*

  UDP    cz5:nbdatagram         *:*
```

The State column indicates whether a connection is currently established or a program is listening on a particular port for messages from other computers, waiting to establish a new connection.

It is also possible to display network traffic statistics for upper-layer protocols, such as Server Message Blocks (SMBs), using the NET STATISTICS command.

Nslookup

Nslookup is a utility that enables you to send queries directly to a particular DNS server in order to resolve names into IP addresses or request other information. Unlike other name resolution methods, such as using Ping, Nslookup lets you specify which server you want to receive your commands, so that you can determine if a DNS server is functioning properly and if it is supplying the correct information. Originally designed for Unix systems, an Nslookup.exe program is also included with the Windows 2000 and NT and 2000 TCP/IP clients, but not with Windows 95 and 98. However, third-party Nslookup implementations are available, such as the graphical one included with Luc Neijens' CyberKit (available at http://www.cyberkit.net/).

Nslookup.exe can run in either interactive or noninteractive mode. To transmit a single query, you can use noninteractive mode, using the following syntax from the command prompt:

```
Nslookup hostname nameserver
```

Replace the *hostname* variable with the DNS name or IP address that you want to resolve, and replace the *nameserver* variable with the name or address of the DNS server that you want to receive the query. If you omit the *nameserver* value, the program uses the system's default DNS server. The output of the program in noninteractive mode on a Windows 2000 system is as follows:

```
Server:   ns1.secure.net

Address:   192.41.1.10


Name:     zacker.com

Address:   192.41.15.74

Aliases:   www.zacker.com
```

To run Nslookup in interactive mode, you execute the program from the command prompt with no parameters (to use the default DNS server) or with a hyphen in place of the *hostname* variable, followed by the DNS server name, as follows:

```
Nslookup - nameserver
```

The program produces a prompt in the form of an angle bracket (>), at which you can type the names or addresses you want to resolve, as well as a large number of commands that alter the parameters that Nslookup uses to query the name server. You can display the list of commands by typing **help** at the prompt. To exit the program, press CTRL-C.

Ipconfig

The Ipconfig program is a simple utility for displaying a system's TCP/IP configuration parameters. This is particularly useful when you are using Dynamic Host Configuration Protocol (DHCP) servers to automatically configure TCP/IP clients on your network, because there is no other simple way for users to see what settings have been assigned to their workstations. All Unix implementations include the ifconfig command (derived from *interface configuration*), and Windows 98, NT, and 2000 systems have a command-line program called Ipconfig.exe. Windows 95 and 98 also include a graphical utility that performs the same functions, called Winipcfg.exe (see Figure 31-10).

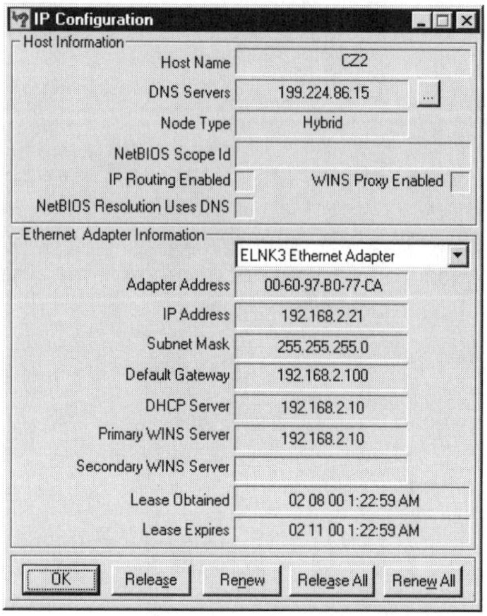

Figure 31-10:   *The Winipcfg.exe IP Configuration dialog box*

Running Ipconfig.exe on a Windows 2000 system with the /all parameter produces the following output:

```
Windows 2000 IP Configuration

        Host Name . . . . . . . . . . . . : cz2

        Primary DNS Suffix  . . . . . . . : zacker.com

        Node Type . . . . . . . . . . . . : Hybrid

        IP Routing Enabled. . . . . . . . : No

        WINS Proxy Enabled. . . . . . . . : No

        DNS Suffix Search List. . . . . . : zacker.com

Ethernet adapter Local Area Connection:

        Connection-specific DNS Suffix  . :

        Description . . . . . . : 3Com EtherLink III (3C509/3C509b)

        Physical Address. . . . : 00-60-97-B0-77-CA
```

```
        DHCP Enabled. . . . . .  : Yes

        Autoconfiguration Enabled: Yes

        IP Address. . . . . . .  : 192.168.2.21

        Subnet Mask . . . . . .  : 255.255.255.0

        Default Gateway . . . .  : 192.168.2.100

        DHCP Server . . . . . .  : 192.168.2.10

        DNS Servers . . . . . .  : 199.224.86.15

                                   199.224.86.16

        Primary WINS Server . .  : 192.168.2.10

        Lease Obtained. . . . .  : Sunday, Feb 06, 2000 10:08:23 PM

        Lease Expires . . . . .  : Wednesday, Feb 09, 2000 10:08:23 PM
```

You can also use Ipconfig to terminate or renew a workstation's lease with its DHCP server, using the /release and /renew parameters, respectively.

Network Analyzers

A *network analyzer,* sometimes called a *protocol analyzer,* is a device that captures the traffic transmitted over a network and analyzes its properties in a number of different ways. The primary function of the analyzer is to decode and display the contents of the packets captured from your network. For each packet, the software displays the information found in each field of each protocol header, as well as the original application data carried in the payload of the packet (see Figure 31-11). Analyzers often can provide statistics about the traffic carried by the network as well, such as the number of packets that use a particular protocol and the amount of traffic generated by each system on the network. A network analyzer is also an excellent learning tool. There is no better way to acquaint yourself with networking protocols and their functions than by seeing them in action.

There is a wide variety of network analyzer products, ranging from self-contained hardware devices costing thousands of dollars to software-only products that are relatively inexpensive or free. Windows 2000 Server and Windows NT Server 4.0, for example, include an application called Network Monitor (shown in Figure 31-11) that enables you to analyze the traffic on your network.

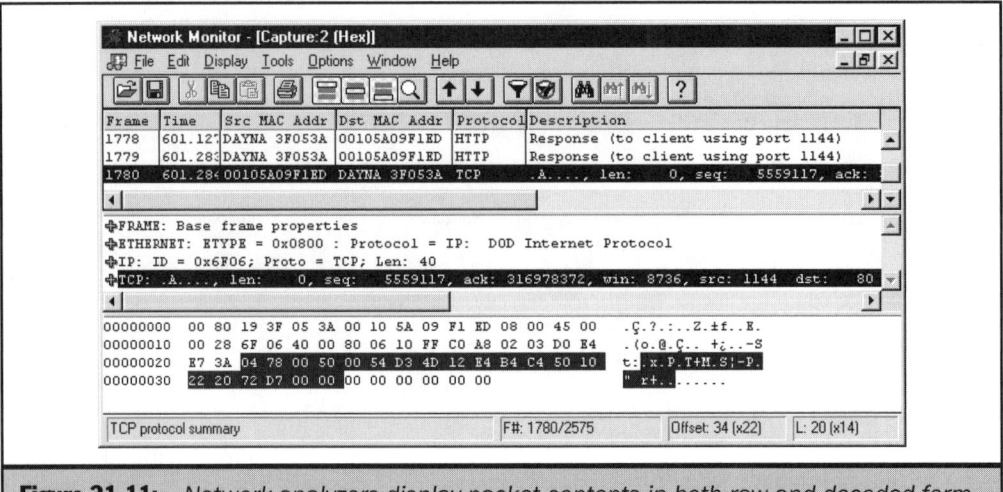

Figure 31-11:   *Network analyzers display packet contents in both raw and decoded form.*

A network analyzer is essentially a software application running on a computer with a network interface. This is why products can either include hardware or take the form of software only. A traveling network consultant might have a portable computer with comprehensive network analyzer software and a variety of NICs to support the different networks at various sites, while an administrator supporting a private network might be better served by a less-expensive software-based analyzer that supports only the type of network running at that site.

A network analyzer typically works by switching the NIC in the computer on which it runs into *promiscuous mode.* Normally, a NIC examines the destination address in the data link–layer protocol header of each packet arriving at the computer, and if the packet is not addressed to that computer, the NIC discards it. This prevents the CPU in the system from having to process thousands of extraneous packets. When the NIC is switched into promiscuous mode, however, it accepts all of the packets arriving over the network, regardless of their addresses, and passes them to the network analyzer software for processing. This enables the system to analyze not only the traffic generated by and destined for the system on which the software is running, but also the traffic exchanged by other systems on the network.

Once the application captures the traffic from the network, it stores the entire packets in a buffer from which it can access them later during the analysis. Depending on the size of your network and the amount of traffic it carries, this can be an enormous amount of data, so you can usually specify the size of the buffer to control the amount of data captured. You can also apply filters to limit the types of data the analyzer captures.

Filtering Data

Because of the sheer amount of data transmitted over most networks, controlling the amount of data captured and processed by a network analyzer is an important part of using the product. You exercise this control by applying *filters* either during the capture process or afterward. When you capture raw network data, the results can be bewildering, because all the packets generated by the various applications on many network systems are mixed together in a chronological display. To help make more sense out of the vast amount of data available, you can apply filters that cause the program to display only the data you need to see.

Two types of filters are provided by most network analyzers:

Capture filters   Limit the packets that the analyzer reads into its buffers

Display filters   Limit the captured packets that appear in the display

Usually, both types of filters function in the same way; the only difference is in when they are applied. You can choose to filter the packets as they are being read into the analyzer's buffers, or capture all of the data on the network and use filters to limit the display of that data (or both).

You can filter the data in a network analyzer in several different ways, depending on what you're trying to learn about your network. If you're concerned with the performance of a specific computer, for example, you can create a filter that captures only the packets generated by that machine, the packets destined for that machine, or both. You can also create filters based on the protocols used in the packets, making it possible to capture only the DNS traffic on your network, for example, or on pattern matches, enabling you to capture only packets containing a specific ASCII or hexadecimal string. By combining these capabilities, using Boolean operators such as AND and OR, you can create highly specific filters that display only the exact information you need. Figure 31-12 shows the Capture Filter dialog box from the Microsoft Network Monitor application. In this example, the application is configured to capture all IP packets traveling to or from the machine called LJOL_LAPTOP. Other network analyzers may use a different interface and offer additional features, such as the ability to filter packets based on their size or on specific error conditions, but the basic functionality is the same.

Agents

Hardware-based network analyzers are portable and designed to connect to a network at any point. Software-based products are not as portable, and often include a mechanism (sometimes called an agent) that enables you to capture network traffic using the NIC in a different computer. Using agents, you can install the analyzer product on one machine and use it to support your entire network. The agent is usually

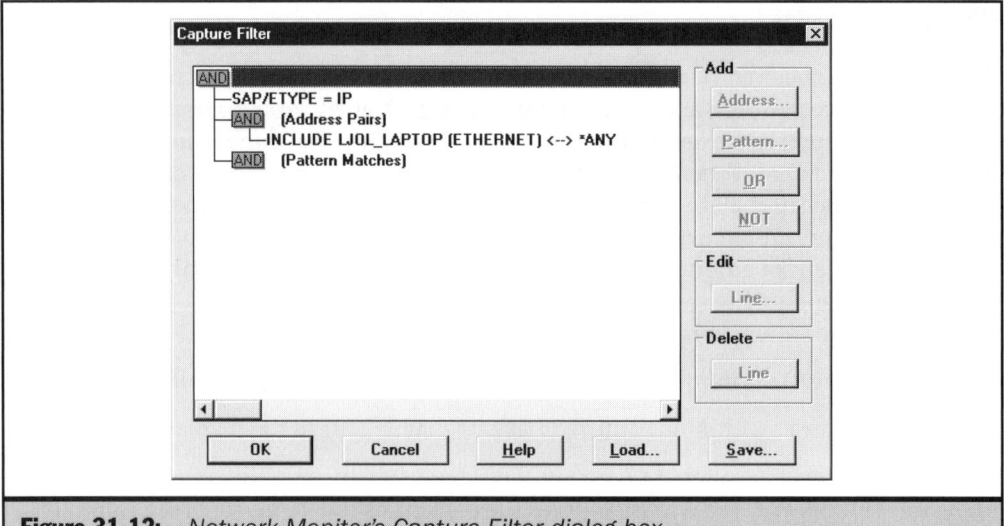

Figure 31-12: *Network Monitor's Capture Filter dialog box*

a driver or service that runs on a workstation elsewhere on the network. For example, all versions of Windows 95/98, NT, and 2000 include the Network Monitor Agent that provides remote capture capabilities for the Network Monitor application running on a Windows 2000 or NT server.

Note *The version of Network Monitor that is included with Windows 2000 Server and Windows NT 4.0 Server is limited to capturing only the network traffic sent to or from the computer on which it is running. In other words, it does not switch the NIC into promiscuous mode. To capture all the traffic on your network, you must use the full version of Network Monitor, which is included as part of the Microsoft Systems Management Server (SMS) product.*

When you run a network analyzer on a system with a single network interface, the application captures the data arriving over that interface by default. If the system has more than one interface, either in the form of a second NIC or a modem connection, you can select the interface from which you want to capture data. When the analyzer is capable of using agents, you can use the same dialog box to specify the name or address of another computer on which the agent is running. The application then connects to that computer, uses its NIC to capture network traffic, and transmits it to the buffers in the system running the analyzer. When you use an agent on another network segment, however, it's important to be aware that the transmissions from the agent to the analyzer themselves generate a significant amount of traffic.

Traffic Analysis

Some network analyzers can display statistics about the traffic on the network while it is being captured, such as the number of packets per second, broken down by workstation or protocol. Depending on the product, you may also be able to display these statistics in graphical form. You can use this information to determine how much traffic each network system or each protocol is generating. Network Associates' Sniffer Basic can display a matrix that graphically illustrates which computers on the network are communicating with each other, as shown in Figure 31-13.

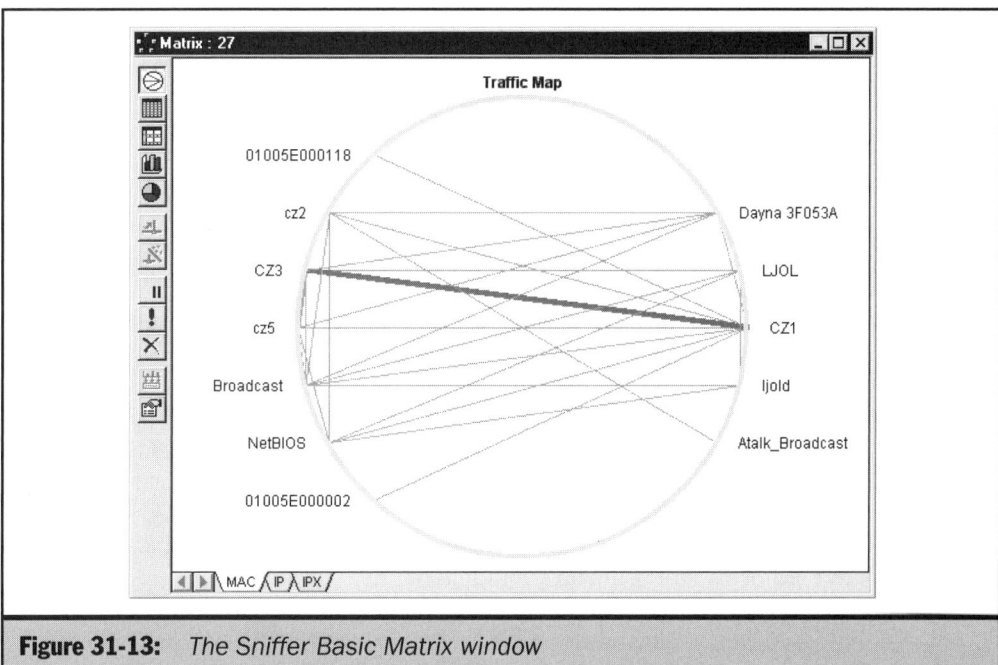

Figure 31-13:   *The Sniffer Basic Matrix window*

Using these capabilities, you can determine how much of your network bandwidth is being utilized by specific applications or specific users. If, for example, you notice that user John Doe's workstation is generating a disproportionate amount of HTTP traffic, you might conclude that he is spending too much company time surfing the Web when he should be doing other things. With careful application of capture filters, you can also configure a network analyzer to alert you of specific conditions on your network. Some products can generate alarms when traffic of a particular type reaches certain levels, such as when an Ethernet network experiences too many collisions.

In addition to capturing packets from the network, some analyzers can also generate them. You can use the analyzer to simulate traffic conditions at precise levels, to verify the operational status of the network, or to stress-test equipment.

Protocol Analysis

Once the analyzer has a network traffic sample in its buffers, you can examine the packets in great detail. In most cases, the packets captured during a sample period are displayed chronologically in a table that lists the most important characteristics of each one, such as the addresses of the source and destination systems, and the primary protocol used to create the packet. When you select a packet from the list, you see additional panes that display the contents of the protocol headers and the packet data, usually in both raw and decoded form.

The first application for a tool of this type is that you can see what kinds of traffic are present on your network. If, for example, you have a network that uses WAN links that are slower and more expensive than the LANs, you can use an analyzer to capture the traffic passing over the links, to make sure that their bandwidth is not being squandered on unnecessary communications.

One of the features that differentiates high-end network analyzer products from the more basic ones is the protocols that the program supports. To correctly decode a packet, the analyzer must support all the protocols used to create that packet at all layers of the OSI reference model. For example, a basic analyzer will support Ethernet and possibly Token Ring at the data link layer, but if you have a network that uses FDDI or ATM, you may have to buy a more elaborate and expensive product. The same is true at the upper layers. Virtually all analyzers support the TCP/IP protocols, and many also support IPX and NetBEUI, but be sure before you make a purchase that the product you select supports all the protocols you use. You should also consider the need for upgrades to support future protocol modifications, such as IPv6.

By decoding a packet, the analyzer is able to interpret the function of each bit and display the various protocol headers in a user-friendly, hierarchical format. In Figure 31-14, for example, you can see that the selected packet contains HTTP data generated by a Web server and transmitted to a Web browser on the network. The analyzer has decoded the protocol headers, and the display indicates that the HTTP data is carried in a TCP segment, which in turn is carried in an IP datagram, which in turn is carried in an Ethernet frame. You can expand each protocol to view the contents of the fields in its header.

A network analyzer is a powerful tool that can just as easily be used for illicit purposes as for network troubleshooting and support. When the program decodes a packet, it displays all of its contents, including what may be sensitive information. The FTP protocol, for example, transmits user passwords in clear text that is easily visible in a network analyzer when the packets are captured. An unauthorized user running an analyzer can intercept administrative passwords and gain access to protected servers.

This is one reason why the version of Network Monitor included with Windows 2000 and NT is limited to capturing the traffic sent to and from the local system.

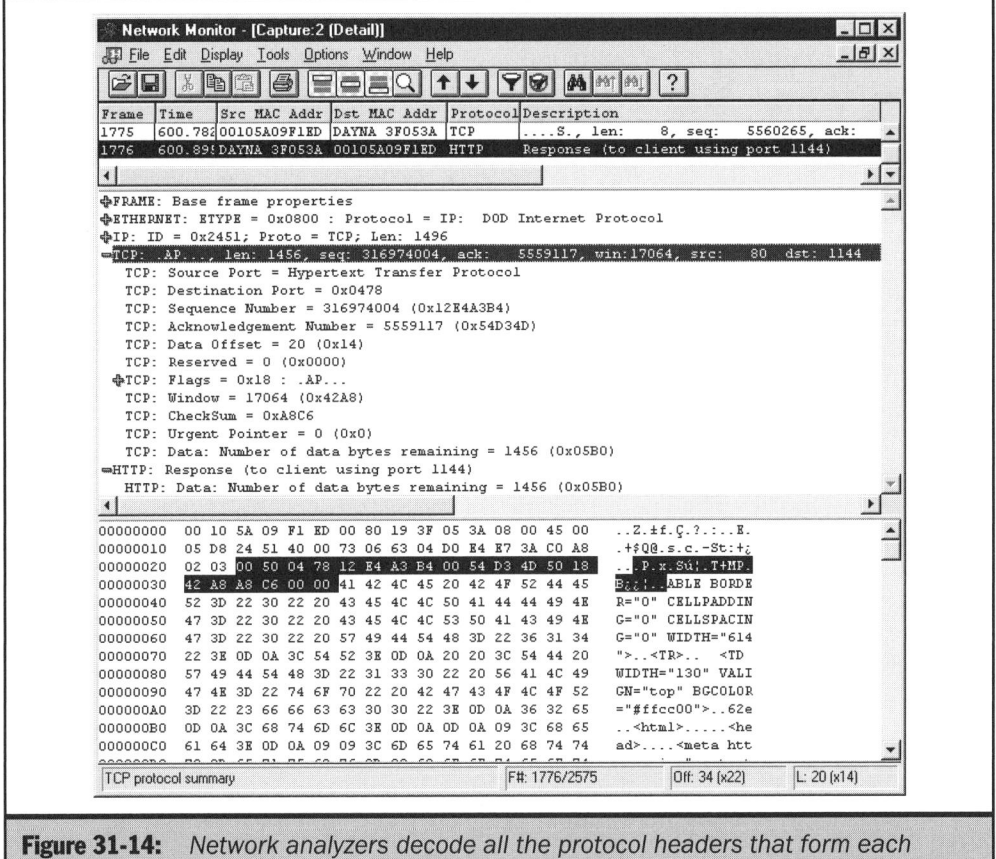

Figure 31-14: *Network analyzers decode all the protocol headers that form each packet.*

Cable Testers

Network analyzers can help you diagnose many types of network problems, but they assume that the physical network itself is functioning properly. When there is a problem with the cable installation that forms the network, a different type of tool, called a *cable tester*, is required. Cable testers are usually handheld devices that you connect to a network in order to perform a variety of diagnostic tests on the signal-conducting capabilities of the network cable. As usual, there is a wide range of devices to choose from that vary greatly in their prices and capabilities. Simple units are available for a few hundred dollars, while top-of-the-line models can cost several thousand dollars. Some combination testers can connect to various types of network cables, such as unshielded twisted-pair (UTP), shielded twisted-pair (STP), and coaxial, while others can test only a single cable type. For completely different signaling technologies, such as fiber-optic cable, you need a separate device.

Cable testers are rated for specific cable standards, such as Category 5, so that they can determine if a cable's performance is compliant with that standard. This is called *continuity testing*. During a cable installation, a competent technician tests each link to see if it is functioning properly, taking into account problems that can be caused by the quality of the cable itself or by the nature of the installation. For example, a good cable tester tests for electrical noise caused by proximity to fluorescent lights or other electrical equipment; crosstalk caused by signals traveling over an adjacent wire; attenuation caused by excessively long cable segments or improperly rated cable; and kinked or stretched cables, as indicated by specific levels of capacitance.

In addition to testing the viability of an installation, cable testers are good for troubleshooting cabling problems. For example, a tester that functions as a time-delay reflectometer can detect breaks or shorts in a cable by transmitting a high-frequency signal and measuring the amount of time it takes for the signal to reflect back to the source. Using this technique, you can determine that a cable has a break or other fault a certain distance away from the tester. Knowing that the problem is 20 feet away, for example, can prevent you from having to poke your head up into the ceiling every few feet to check the cables running through there. Some testers can also help you locate the route that a cable takes through walls or ceilings, using a tone generator that sends a strong signal over the cable that can be detected by the tester unit when it is nearby.

Network Management

Many of the tools examined thus far in this chapter are for use when a problem occurs, but many network administrators prefer to use a more proactive approach, by continuously monitoring and gathering information about the network using a network management console, such as Hewlett-Packard's OpenView or Microsoft's Systems Management Server. These products are designed to gather information about various devices located all over the network and display it at a central console. Some products are able to compile a graphical view of the network that enables administrators to select a device and check its status and statistics. When problems occur, causing a device to generate a special event called a *trap*, the console can be configured to notify administrators in a variety of ways, such as by e-mail and pages. Other network management functions include traffic monitoring, network diagnostics, software inventory and metering, and report generation.

There are two network management standards that are currently popular, called the Simple Network Management Protocol (SNMP) and Remote Monitoring (RMON). These standards are similar in their basic architecture, but handle information in different ways. The basic components of a network management system are as follows:

Network management console   Receives the information gathered by the agents and collates it for presentation to the administrator in the form of statistical readouts, graphical displays, and/or printed reports.

Agents   Programs running on network devices that gather information for eventual transmission to the network management console. Most network hardware devices, such as routers, switches, and hubs, have built-in agents that support SNMP, RMON, or both. On computers, the agent takes the form of a program or service that may be included with the operating system or furnished separately.

Management information bases (MIBs)   Store the *managed objects,* the individual pieces of information compiled by an agent.

SNMP   Includes the query language used between the agents and the console, and the UDP-based transport mechanism that carries the information collected by the agents.

While SNMP and RMON use the same basic components, RMON is the newer standard, developed to address the shortcomings of SNMP. In an SNMP installation, the agents are simple components that gather information and transmit it on request. The network management console is responsible for continually polling the agents and retrieving the information they've gathered. This places most of the burden for supporting the network management system on the console, which must constantly gather and process new data, and also generates a considerable amount of network traffic.

RMON also uses a central console and individual agents (which it calls *probes*), but instead of the agents functioning as the clients of the console, as in SNMP, the probes are the servers and the console is the client. RMON probes are more complex than SNMP agents, and are capable of gathering and maintaining information about the devices they serve by themselves. The console only has to retrieve the data from a probe when it is needed for display or processing. This reduces both the burden on the console and the network traffic generated by the process.

Deploying a network management application requires a considerable investment in time, money, and equipment. But on a larger network, it enables administrators to monitor and maintain hundreds or thousands of individual components without traveling to the far reaches of the network installation.

NETWORK ADMINISTRATION

The Complete Reference

Networking

Chapter 32

Backing Up

O ne of the primary functions of a computer network is to store, manipulate, and supply data, and protecting that data against damage or loss is a crucial part of the network administrator's job description. Hard disk drives contain most of the relatively few moving parts involved in the network data storage process, and are constructed to incredibly tight tolerances. As a result, they can and do fail on occasion, causing service interruptions and data loss, and server drives work the hardest of all. When you examine the inner workings of a hard drive, you may actually wonder why they don't fail more often. In addition to mechanical drive failures, data loss can occur for many other causes, including viruses, computer theft, natural disaster, or simple user error. To protect the data stored on your network, it is absolutely essential that you perform regular backups to an alternative storage medium.

While backing up data is an important maintenance task for all computers, it is particularly vital on a network, for several reasons. First, the data tends to be more important; a loss of crucial data can be a catastrophe for a business that results in lost time, money, business, reputation, and in some cases even lives. Second, network data is often more volatile than the data on a standalone computer, because many different users might access and modify it on a regular basis.

Network backups differ from standalone computer backups in four major ways: speed, capacity, automation, and price. A business network typically has data stored on many different computers, and that combined with the ever-increasing drive capacities in today's computers means that a network backup solution may have to protect hundreds or even thousands of gigabytes of data. In order to practically back up this much data, backup drives that are capable of unprecedented speeds are required. The big advantage of backing up multiple computers that are all connected to a network is that you can use one backup drive to protect many computers, using the network itself to transfer the data (as shown in Figure 32-1), rather than a separate drive on each computer. In order for this to be practical, the network administrator must be able to control the backup process for all of the computers from a central location. Without this type of automation, the administrator would have to travel to each computer to create an individual backup job. By installing the backup drive and backup software on one of the network's computers, you create a backup server that can protect all of the other computers on the network.

Automation also enables backups to occur during night or other non-working hours, when the network is idle. Backing up remote computers naturally entails transferring large amounts of data across the network, which generates a lot of traffic that can slow down normal network operations. In addition, data files that are being used by applications are frequently locked open, meaning that no other application can gain access to them. These files are skipped during a typical backup job and are therefore not protected. Network backup software programs enable you to schedule backup jobs to occur at any time of the day or night, when the files are available for access. With appropriate hardware, the entire backup process can run completely unattended.

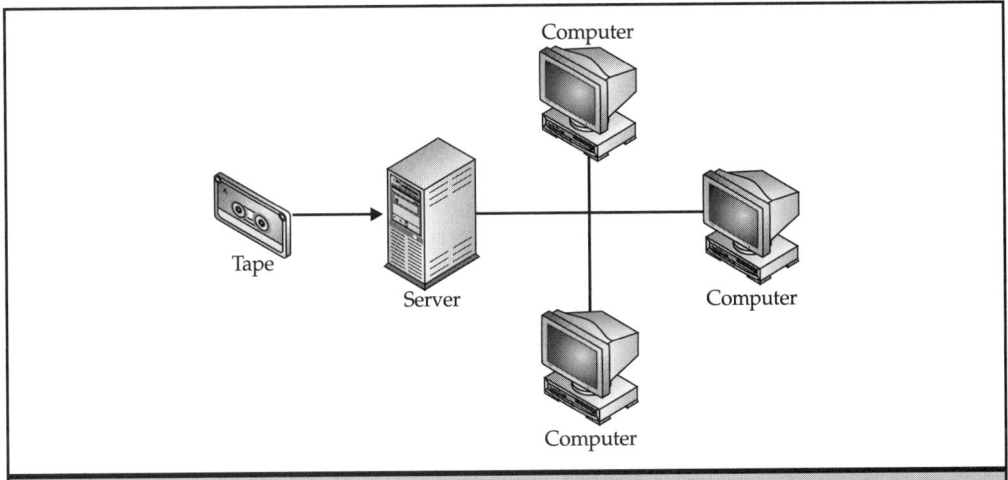

Figure 32-1:   *Computers all over a network can transmit data to a server for backup to tape.*

Network backup software costs a good deal more than a backup product intended for a standalone computer. Most of the major network backup software products are modular in nature, enabling you to purchase a base package and then add options providing other capabilities as needed. For a medium to large enterprise network, you can easily end up spending several thousand dollars on the backup software modules needed to protect all of the data resources on your network.

A network backup solution consists at the very least of a backup drive, backup media for the drive, and backup software. Depending on the amount and type of data to be backed up and the amount of time available to perform the backups, you may also need other equipment, such as multiple backup drives, an autochanger, or optional software components. Selecting appropriate hardware and software for your backup needs and learning to use them correctly are the essential elements of creating a viable network backup solution. In many cases, backup products are not cheap, but as the saying goes, you can pay now or you can pay later.

Backup Hardware

You can use virtually any type of drive that employs removable media as a backup drive. Writable CD-ROM drives like CD-Rs and CD-RWs are possible solutions, as are cartridge drives like the Iomega Zip and Jaz. However, while these drives may be suitable for a single system backup, they are almost never used as network backup drives, for two main reasons: insufficient capacity and excessive media cost. Writable CD disks and drives are inexpensive and reasonably fast, but they only hold about 670 megabytes of data. Backing up a single 10GB file server drive would require 15 CDs, which means that someone has to sit around while the network is otherwise idle and feed blank disks into the drive until the backup is completed. One of the main objectives of a network backup solution is to avoid the need for media changes during a job, so that the entire process can run unattended.

Zip cartridges hold even less data than CDs and are therefore unsuitable for the same reason. Iomega Jaz cartridges can hold up to 2GB of data, which would appear to make them a suitable backup medium for smaller networks. However, a single Jaz cartridge can cost $125 or more, which works out to a media cost of at least 6 cents per megabyte. By comparison, CD-R storage costs about 0.075 cents per megabyte and even the most expensive server hard drives cost less than 2 cents per megabyte, with common workstation hard drives costing 0.5 cents per megabyte or less. Another objective of a network backup solution is to retain a history of the protected data for a given period of time, so that it's possible to restore files that are several weeks or months old. Maintaining a backup archive like this requires a lot of storage, and the price of the medium is a major factor in the overall economy of the backup solution.

The result of this need for high media capacities and low media costs is that magnetic tape is nearly always the backup medium of choice in a network environment. Magnetic tapes can hold enormous amounts of data in a very small package, and the cost of the media is low, often less than one half cent per megabyte. In addition, magnetic tapes are durable and easy to store.

Unlike most of the other mass storage devices used in computers, magnetic tape drives do not provide random access to the stored data. Hard disk, floppy disk, and CD-ROM drives all have heads that move back and forth across a spinning medium, enabling them to place the head at any location on the disk almost instantaneously and read the data stored there. The magnetic tape drives used in computers work just like audio tape drives; the tape is pulled off of a spool and dragged across a head to read the data, as shown in Figure 32-2. This is called linear access. To read the data at a point near the end of a tape, the drive must unspool all of the preceding tape before accessing the desired information. Because they are linear access devices, magnetic tape drives are not mounted as volumes in the computer's file system. You can't assign a drive letter to a tape and access its files through a directory display, as you can with a CD-ROM or a floppy disk. Magnetic tape drives are used exclusively by backup software programs, which are specifically designed to access them.

Linear access devices like tape drives also cannot conveniently use a table containing information about the files they contain, as with a hard or floppy disk. When a backup system writes hard drive files to tape, it reads the information about each file from the hard drive's file allocation table (or whatever equivalent that particular drive's file system uses) and writes it to tape as a header before copying the file itself. The file is frequently followed by an error correction code (ECC) that ensures the validity of the file. This way, all of the information associated with each file is found at one location on the tape. However, some tape drive technologies, such as digital audio tape (DAT) and digital linear tape (DLT), do create an index on each tape of all of the files it contains, which facilitates the rapid restoration of individual files.

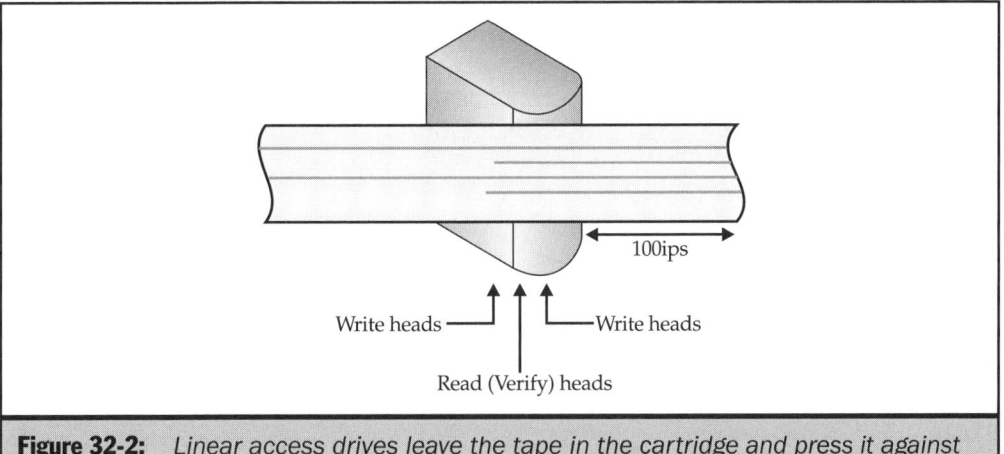

Figure 32-2: *Linear access drives leave the tape in the cartridge and press it against static heads.*

Caution *Many networks use data storage technologies such as RAID (Redundant Array of Independent Disks) to increase data availability and provide fault tolerance. However, despite the fact that these technologies can enable your network to survive a hard drive failure or similar problem, they are not a replacement for regular backups. Viruses, fires, and other catastrophes can still cause irretrievable data loss in hard drive-based storage arrays, while tape backups with offsite storage provide protection against these occurrences.*

Backup Capacity Planning

Magnetic tape has been a mainstay of data backup technology for many years, and as a result, there are many different tape formats and tape drives. In addition to the price and compatibility considerations important to every purchase, the criteria you should use to evaluate tape backup drives are capacity and media costs, as mentioned earlier, plus the speed at which the drive can copy data to the medium. Together, the tape capacity and the transfer speed dictate whether the drive is capable of backing up your data in the time you have available. Not surprisingly, the backup drives with greater capacity and faster speeds command higher prices. Depending on your situation, you may be able to trade off some speed for increased capacity or emphasize maximum speed over capacity.

In order to evaluate tape backup technologies, it's a good idea to first estimate the amount of data that you have to protect and the amount of time you will have for the backup jobs to run. The object is to select a drive (or drives) that can fit all of the data you need to protect during the average backup job on a single tape in the time available. Be sure to consider that it may not be necessary for you to back up all of the data on all of your computers during every backup job. Most of the files that make up a computer's operating system and applications do not change, so it isn't necessary to back them up every day. You can back these up once a week or even more seldom and still provide your computers with sufficient protection. The important files that you should back up every day are the data and system configuration files that change frequently, all of which might add up to far less data.

Consider, for example, a small network that consists of 20 computers with a total combined hard drive capacity of 200 gigabytes. An unattended tape backup solution that can protect the entire 200GB of data without a media change would require several high-end tape drives or an autochanger (a device that automatically moves tapes in and out of tape drives), either of which represents a serious investment in hardware. However, when you consider that the actual user data stored on the network is, in this case, about one tenth of its total storage capacity, you end up with about 20GB of data that need frequent backups, with the rest backed up only occasionally. This means that a single tape drive with a 20GB capacity is sufficient, which is a much more economical solution. Daily backups of the data only can run unattended, with only a weekly full backup requiring media changes.

Tape Drive Interfaces

In addition to the capabilities of the tape drive itself, you must also consider the interface that connects it to the computer that will host it. The process of writing data to a magnetic tape requires that the tape drive receive a consistent stream of data from the computer. Interruptions in the data stream force the tape drive to stop and start repeatedly, which wastes both time and tape capacity. The traditional interface for network backups, and the best one, is the Small Computer System Interface (SCSI), because SCSI tape drives can operate independently of the other devices connected to the interface.

For more information on SCSI and its use in network servers, see Chapter 8.

There are also tape drives available that connect to the Integrated Drive Electronics (IDE) interface used on most workstations today, but when two devices are connected to an IDE interface, only one can process a command at a time. This can lead to drive contention problems on a heavily used server, which negatively affect the backup process. Before the IDE interface was enhanced by the addition of the AT Attachment Packet Interface (ATAPI) standard, which permits devices other than hard drives to use the interface, there were some tape drives that connected to the floppy drive interface, which is also not a satisfactory solution for network backups.

> **Note**
>
> *There are also tape drives on the market that connect to computers' parallel or Universal Serial Bus (USB) ports. These are SCSI devices that simply use an alternative connection to the computer, which makes it easier to move the drive to a different location. On a network, however, this is not necessary, and neither the parallel nor the USB port provides the throughput of a dedicated SCSI host adapter.*

Magnetic Tape Capacities

The storage capacity of a magnetic tape is one of its most defining characteristics, and can also be one of the most puzzling aspects of the backup process. Many users purchase tape drives with rated capacities, and then are disappointed to find that the product does not store as much data on a tape as the manufacturer states. In most cases, this is not a matter of false claims on the part of the drive's maker.

There are three elements that can affect the data capacity of a magnetic tape, which are as follows:

■ Compression

■ Data stream

■ Write errors

Compression

Magnetic tape storage capacities are often supplied by manufacturers in terms of compressed data. A reputable manufacturer will always state in its literature whether the capacities it cites are compressed or uncompressed. Most of the tape drives designed for computer backups include hardware-based compression capabilities that use standard data compression algorithms to store the maximum amount of data on a tape. In cases where the drive does not support hardware compression, the backup software might implement its own compression algorithms. When you have a choice, you should always use hardware-based compression over software compression, because implementing the data compression process in the software places an additional processing burden on the computer. Hardware-based compression is performed by a processor in the tape drive itself and is inherently more efficient.

> **Note**   *Some manufacturers express tape drive capacities using the term "native." A drive's native capacity refers to its capacity without compression.*

The degree to which data can be compressed, and therefore the capacity of a tape, depends on the format of the files being backed up. A file in a format that is already compressed, such as a GIF image or a ZIP archive, cannot be compressed any further by the tape drive hardware or the backup software, and therefore has a compression ratio of 1:1. Other file types compress at different ratios, ranging from 2:1, which is typical for program files such as EXEs and DLLs, to 8:1 or greater, as with uncompressed image formats like BMP. It is standard practice for manufacturers to express the compressed storage capacity of a tape using a 2:1 compression ratio. However, your actual results might vary greatly, depending on the nature of your data.

Data Stream

In order to write data to the tape in the most efficient manner, the tape drive must receive the data from the computer in a consistent stream at an appropriate rate of speed. The rate at which the data arrives at the tape drive can be affected by many factors, including the interface used to connect the drive to the computer, the speed of the computer's processor and system bus, or the speed of the hard drive on which the data is stored. When you are backing up data from the network, you add the speed of the network itself into the equation. Even if you have a high-quality tape drive installed in a state-of-the-art server, slow network conditions caused by excessive traffic or faulty hardware can still affect the speed of the data stream reaching the tape drive. This is one of the reasons why network backups are often performed at night or during other periods when the network is not being used by other processes.

Tape drives write data to the tape in units called *frames* or sometimes *blocks*, which can vary in size depending on the drive technology and the manufacturer. The frame is the smallest unit of data that the drive can write to the tape at one time. The drive contains a buffer equal in size to the frames it uses, in which it stores the data to be backed up as it arrives from the computer. When the backup system is functioning properly, the data arrives at the tape drive, fills up the buffer, and then is written to the tape with no delay. This enables the tape drive to run continuously, drawing the tape across the heads, writing the buffered data to the tape, and then emptying the buffer for the next incoming frame's worth of data. This is called *streaming*.

> **Note**   *The frames used by tape drives do correspond in size or construction with the data link–layer protocol frames used in data networking.*

When the data arrives at the tape drive too slowly, the drive has to stop the tape while it waits for the buffer to fill up with data. This process of constantly stopping and starting the tape is called *shoe-shining*, and it is one of the main signals that the drive is not running properly. The buffer has a built-in data retention timeout, after which the drive flushes the buffer and writes its contents to tape, whether it's full of data or not. If the buffer is not full when the timeout period expires, the drive pads out the frame with nonsense data to fill it up and then writes the contents of the buffer (including the padding) to the tape. The end result is that each frame written to the tape contains only a fraction of the actual data that it can hold, thus reducing the amount of usable data stored on the tape.

The way to avoid having partially filled buffers flushed to tape is to ensure that there are no bottlenecks in the path from the sources of your data to the tape drive. The path is only as fast as its slowest component, and to speed up the data transfer rate, you may have to do any of the following:

- Replace hard drives with faster models

- Upgrade the network from standard Ethernet to Fast Ethernet

- Use a SCSI tape drive instead of an IDE model

- Install the tape drive in a faster computer

- Reduce the processing load on the computer hosting the tape drive

- Schedule backup jobs to occur during periods of low network traffic

Write Errors

Another possible reason for diminished tape capacity is an excess of recoverable write errors. A write error is considered to be recoverable when the tape drive detects a bad frame on the tape while the data is still in the buffer, making it possible for the drive to immediately write the same frame to the tape again. Drives typically detect these errors by positioning a read head right next to the write head, so that the drive can read each frame immediately after writing it.

When the drive rewrites a frame, it does not overwrite the bad frame by rewinding the tape; it simply writes the same frame to the tape again, immediately following the first one. This means that one frame's worth of data is occupying two frames' worth of tape, and if there are many errors of this type, a significant amount of the tape's storage capacity can be wasted. Recoverable write errors are most often caused by dirty heads in the tape drive or bad media. Most backup software products can keep track of and display the number of recoverable write errors that occur during a particular backup job. The first thing you should do when you notice that more than a handful of recoverable write errors have occurred during a backup job is to clean the drive heads using a proper cleaning tape, and then run a test job using a brand new, good quality tape. If the errors continue, this might be an indication of a more serious hardware problem.

**Tip** *Dirty drive heads are the single most common cause of tape drive problems. The importance of regular head cleaning cannot be overemphasized.*

Magnetic Tape Technologies

There are five main types of magnetic tape drives used for computer backups, as detailed in Table 32-1, some of which also have several subtypes. Generally speaking, the five types of drives are listed in the table in order from least to most expensive. These drives and their media are examined in more detail in the following sections.

Type	Tape Width	Cartridge Size	Maximum Capacity (uncompressed)	Maximum Transfer Rate
Quarter Inch Cartridge (QIC)	¼-inch	4 × 6 × 0.625 inches ; (data cartridge) 3.25 × 2.5 × 0.6 inches (minicartridge)	20 GB	120 MB/min
Digital Audio Tape (DAT)	4 mm	2.875 × 2.0625 × 0.375 inches	20 GB	144 MB/min
8 mm	8 mm	3.7 × 2.44 × 0.59 inches	60 GB	180 MB/min
Digital Linear Tape (DLT)	½-inch	4.16 × 4.15 × 1 inches	40 GB	360 MB/min
Linear Tape-Open(LTO), Ultrium media	½-inch	4.0 × 4.16 × 0.87 inches	100 GB	1920 MB/min

Table 32-1: *Magnetic Tape Drive Technologies*

Quarter-Inch Cartridge (QIC) Tape

The QIC (pronounced *quick*) tape standard was introduced in 1972 by the 3M Corporation, and soon became a common data storage system for personal computers. Named for the width of the tape as it was first designed, most but not all QIC tapes are one quarter inch wide; some of the later standards use wider tape as a means of increasing capacity. The QIC tape cartridge itself comes in two sizes, the data cartridge and the minicartridge, as shown in Figure 32-3. The data cartridge is the larger of the two (at 4 × 6 × 0.625 inches) and came first; the minicartridge is 3.25 × 2.5 × 0.6 inches, and rapidly overtook the data cartridge in popularity after its introduction. Both cartridges are similar in design, with two spools inside, one of which holds the tape and the other functioning as a take-up reel. A motor in the drive turns the spools via a belt in the cartridge, pulling the tape from one spool to the other. A metal rod built into the drive (called a *capstan*) protrudes through the tape cartridge and presses the tape against a rubber wheel (called a *pinch roller*) as the spools turn. This holds the tape in place against a head assembly that reads data from and writes it to the tape. Overall, the construction of the QIC drive and tape cartridge is similar in its basic principles to a standard audio cassette drive.

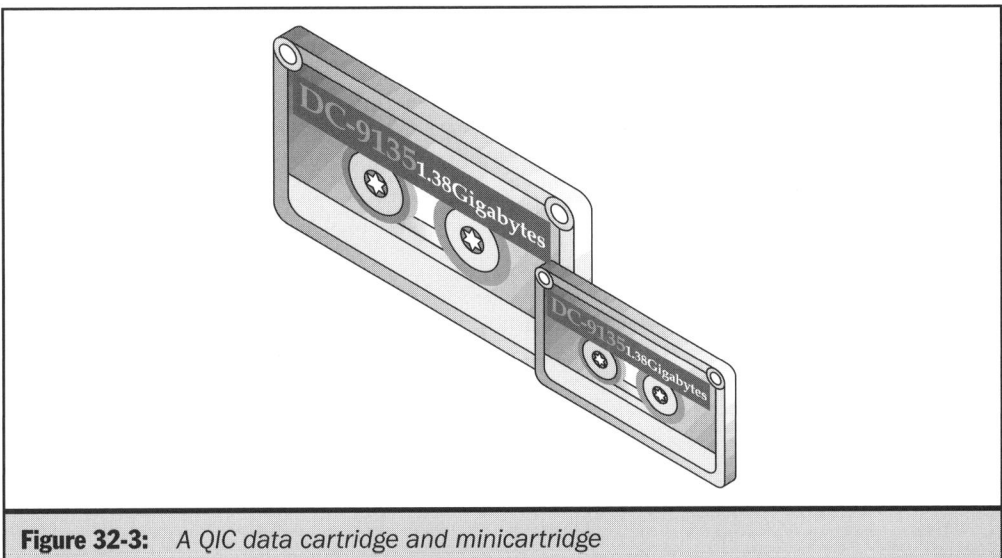

Figure 32-3:   *A QIC data cartridge and minicartridge*

QIC tape drives function by drawing the tape perpendicularly across an array of heads at a speed of 100 to 125 inches per second. The heads write data to the tape in straight, parallel lines along its length called *tracks*. When the drive reaches the end of the tape, it reverses direction and the heads begin writing to the next adjacent tracks on the tape. Some QIC drives use a single write head to write one track at a time, while others increase the rate at which they can write data by using multiple heads. Various QIC standards create different numbers of tracks on a tape; a greater number of tracks is in most cases an indication of greater capacity. As mentioned earlier, some QIC standards use wider 0.315-inch tape to increase capacity by squeezing a larger number of tracks onto a tape. Other standards achieve the same end by increasing the length of the tape in the cartridge.

QIC tape technology was introduced at a time when the standardization of computer hardware was not a great priority for manufacturers. As a result, there are more than 100 different QIC standards, many of which are incompatible with each other. The different standards use different tapes, different data coding formats, and different numbers of tracks. This incompatibility means that although drives built to different standards may in some cases use the same media, they cannot read the data from each others' tapes. This large number of standards has long been the biggest impediment to the acceptance of QIC drives as a large scale backup solution.

The early QIC standards, such as QIC-40 and QIC-80 (holding 40 and 80 uncompressed megabytes on a tape, respectively), reflect the data storage capacities that were common at the time of their inception, and over the years, a constant succession of new standards supporting greater capacities and higher speeds were introduced. The result is a hopeless jumble of alphanumeric QIC standards with a complex web of compatibilities. This long history also means that QIC drives have been manufactured for every possible interface, including the floppy drive, IDE, and parallel port interfaces, as well as SCSI. In an attempt to streamline the QIC technologies, a group of manufacturers led by 3M created a series of QIC standards that they called Travan. The Travan standards, as listed in Table 32-2, include drives with the highest speeds and greatest capacities available in the QIC format, as well as some lower-end drives that are more suitable for individual workstation backups.

Standard	Maximum Capacity (uncompressed)	Maximum Transfer Rate	Number of Tracks	Compatibility
TR-1	400 MB	125 KBps	36	QIC 80 (R/W); QIC 40 (R only)
TR-2	800 MB	125 KBps	50	QIC 3010 (R/W); QIC 80 (R only)
TR-3	1.6 GB	250 KBps	50	QIC 3010/QIC 3020 (R/W);QIC 80 (R only)
TR-4	4 GB	70 MB/min	72	QIC 3080/QIC 3095 (R/W);QIC 3020 (R only)
TR-5	10 GB	110 MB/min	108	QIC 3220 (R/W); TR-4; QIC 3095 (R only)

Table 32-2: *Travan Tape Drive Standards*

Despite these more easily comprehensible standards, however, QIC/Travan drives are not often used for network backups. As an inexpensive backup solution for a standalone computer, a lower-end QIC drive can be a practical and economical solution. However, the high-end QIC drives present no advantages in performance that make them preferable to industry standard network backup solutions such as DAT, and while the drives may be substantially less expensive than the alternatives, the media are not.

Digital Audio Tape (DAT)

Despite its name, digital audio tape is more commonly used for data backups than for audio recording. In fact, DAT is the most popular network backup solution today, largely because it occupies a central niche in the marketplace. DAT drives are not the fastest drives available, nor do they have the greatest capacity, nor are they the least expensive. However, the combination of speed, capacity, and price that DAT provides makes it an ideal backup solution for many network installations. All DAT drives use the same size tape cartridge, shown in Figure 32-4, which is slightly smaller than the QIC minicartridge and contains tape that is 4 mm wide. However, the way in which DAT drives handle the tape is completely different from QIC.

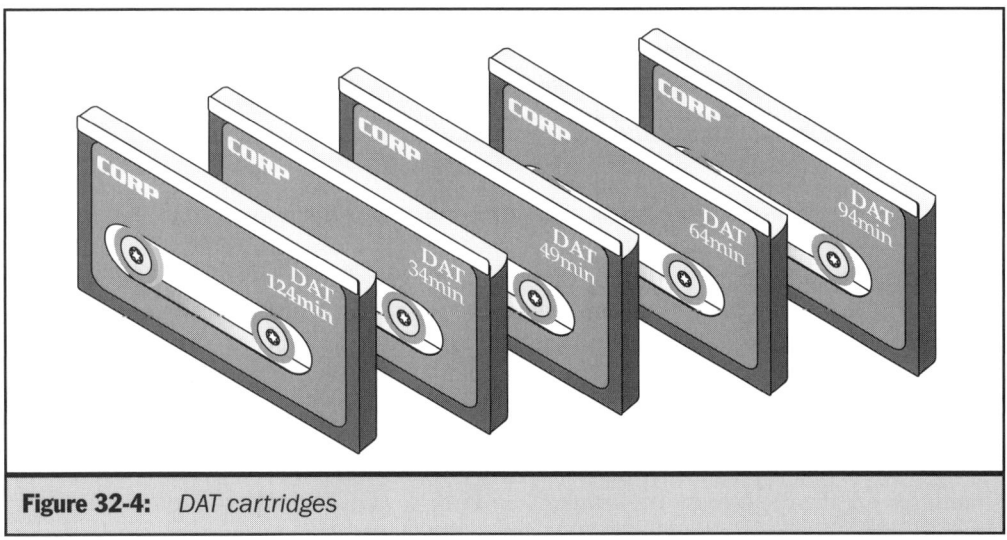

Figure 32-4:   *DAT cartridges*

In a QIC drive, the read and write heads are pressed up against the surface of the tape as it is transferred from one spool to the other. DAT tapes also have two spools, but instead of pressing the heads to the tape in the cartridge, the DAT drive actually pulls a loop of tape out of the cartridge and threads it around a rotating head assembly, as shown in Figure 32-5. This is very similar to the way in which a standard VCR handles video tapes. The head assembly in a DAT drive is a drum with two read heads and two write heads, which rotates at 2,000 rpm. The heads are positioned on the drum at 12, 3, 6, and 9 o'clock, and are mounted on the drum at an angle, so that as they spin, they write their tracks diagonally across the surface of the tape. This is called *helical scan recording*. Each track is approximately 32 mm long and contains 128 KB of data plus an error correction code (ECC) sequence. The drive pulls the tape in the direction opposite that in which the head assembly rotates, at the relatively slow speed of one inch per second.

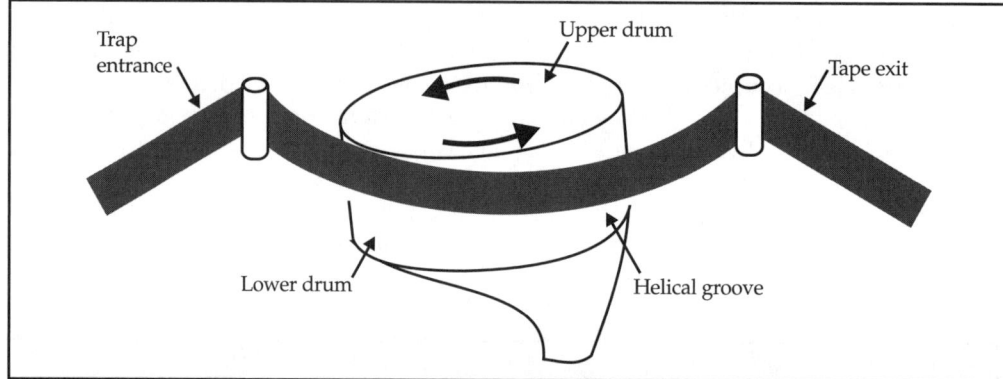

Figure 32-5: *Helical scan drives pull a tape loop out of the cartridge and wrap it around a spinning head assembly.*

The two write heads work at the same time, writing independent tracks at different angles about 40 degrees apart. This causes the tracks to cross over each other, but the heads use different magnetic polarities so the drive can distinguish the tracks created by each head. The two read heads are configured to use the same polarities, so that each write head is paired with a read head located adjacent to it. This enables the read heads to test the viability of each track as it is written. If a track is not written properly, the drive writes it again to the next track on the tape before the buffer is cleared.

Most DAT drives conform to one of the Digital Data Storage (DDS) standards, as developed by the DDS Manufacturers Group and listed in Table 32-3. The DDS standards are the model that the Travan standards for QIC drives attempt to emulate. Each of the DDS standards is backward compatible with all of the earlier standards, meaning that a drive can read from and write to tapes made by an older DDS drive. Each successive DDS standard provides greater tape capacity and higher transfer speeds than its predecessors, using various different techniques. For example, DDS-4, the most current standard, increases the uncompressed capacity of a single tape by 8 gigabytes by reducing the track pitch (that is, the distance between the tracks) from 9.1 microns to 6.8 microns, and by increasing the length of the tape from 125 to 150 meters.

Standard	Year Introduced	Maximum Capacity (Uncompressed)	Maximum Transfer Rate
DDS	1989	2 GB	55 KBps
DDS-2	1992	4 GB	1.1 MB/sec
DDS-3	1995	12 GB	2.2 MB/sec
DDS-4	1998	20 GB	4.8 MB/sec

Table 32-3: *DDS Standards for DAT Drives and Media*

All DAT drives use the SCSI interface, which means that they are available as both internal and external devices. External drives are always a bit more expensive than internal ones because they have their own power supply.

8 mm Tape

8 mm tape backup drives are another adaptation of a medium intended for a different application. The 8 mm tape cartridge was developed by Sony in the late 1980s as a compact video medium for camcorders. Soon after, a company called Exabyte began using the technology to manufacture tape backup drives using the Sony camcorder mechanisms. With the success of these drives, Exabyte began manufacturing their own 8 mm tape drive mechanisms, which they dubbed MammothTape technology. Today, their M2 drives represent the cutting edge of the MammothTape standards. There is also another 8 mm standard, called Advanced Intelligent Tape (AIT), which was developed by Sony and the Seagate Corporation. The capacities and data transfer speeds of the various 8 mm standards are listed in Table 32-4.

Because 8 mm tape is still commonly used in camcorders, users may be tempted to use 8 mm video tapes for data backups. This practice is strongly discouraged by 8 mm tape drive manufacturers, as video tapes are generally not manufactured to the same standards as tapes intended for data storage.

Like DAT, 8 mm tape uses a helical scan recording mechanism that pulls the tape out of the cartridge and wraps its around a rotating drum containing the read and write heads. The tape is twice as wide as DAT tape, however. This and other factors, such as a special Advanced Metal Evaporated (AME) medium, enables 8 mm tape drives to store enormous amounts of data on a single tape at speeds exceeding those of DAT. In addition, the Mammoth drives by Exabyte have innovative features such as a gentler tape transport system that replaces the traditional capstan and pinch roller, and an internal, automatic head-cleaning mechanism.

Standard	Maximum Capacity (Uncompressed)	Maximum Transfer Rate
Standard 8 mm	3.5 GB	0.53 MB/sec
Standard 8 mm	5 GB	1 MB/sec
Standard 8 mm	7 GB	1 MB/sec
Standard 8 mm	7 GB	2 MB/sec
Mammoth	20 GB	3 MB/sec
M2	60 GB	12 MB/sec
AIT-1	25 GB	6 MB/sec
AIT-2	50 GB	6 MB/sec

Table 32-4:   *8 mm Tape Standards*

This performance comes at a price, as 8 mm drives are significantly more expensive, with the high-end M2 drives topping $5,000. However, for networks with a large amount of data to protect, a single 8 mm drive can often do the same job as two or more DAT drives.

Digital Linear Tape (DLT)

Digital linear tape (DLT) is a mass storage technology that has been around since the late 1980s, and which also provides enormous amounts of storage on a single tape cartridge with high data transfer speeds. What makes DLT unique is its tape-handling mechanism. Unlike DAT and 8 mm tapes, a DLT cartridge contains only a single spool; the take-up spool is located in the drive and one end of the tape in the cartridge is left loose. While DAT and 8 mm drives pull a loop of tape out of the cartridge and wrap it around the rotating drum containing the heads, DLT drives pull the loose end of the tape out of the cartridge and thread it around an array of six rollers that form the drive's head guide assembly (HGA) and onto the take-up reel, as shown in Figure 32-6.

The HGA holds the tape up to the drive heads, which are static (not rotating as in helical scan drives). The mechanism is essentially a modernized version of the old open reel tape recorders, except that you don't have to manually thread the tape through the heads.

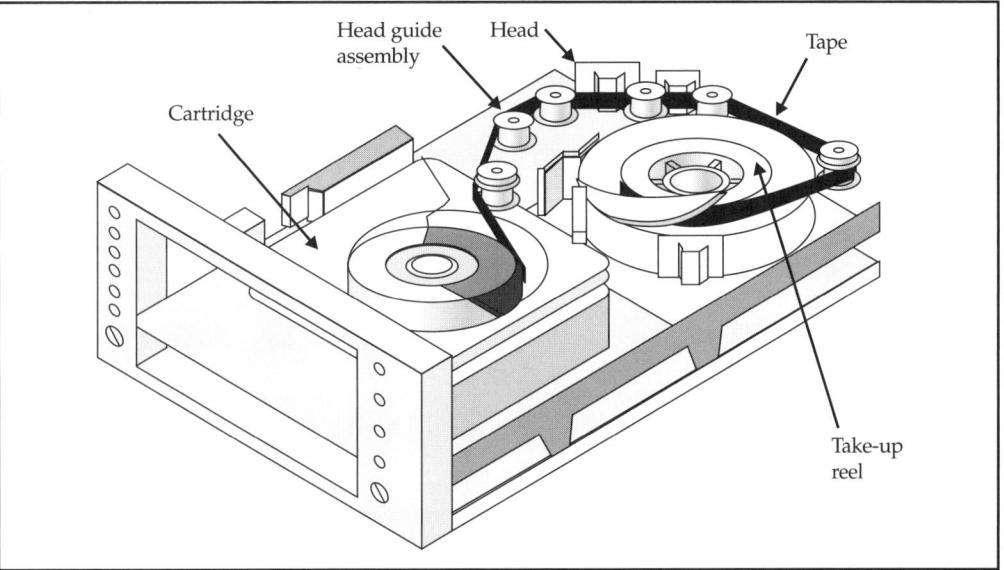

Figure 32-6: *DLT drives pull one end of the tape out of the cartridge and thread it around a head guide assembly.*

The tape in a DLT cartridge is one half inch wide, wider than QIC, DAT, or 8 mm tape, and the HGA is designed to provide head contact without pulling on the tape and bending it into acute angles, which can be a major cause of premature tape wear and damage. In addition, the entire tape-handling mechanism is constructed so that no component of the drive touches the side of the tape containing the data. DLT drives write data on the tape in pairs of parallel tracks from one end of the tape to the other. Then the drive reverses direction, shifts the heads, and writes a different pair of tracks in the other direction. A tape that is completely filled with data typically has either 128 or 208 tracks.

As with other magnetic tape technologies, DLT evolved over the years to increase tape capacity and the data transfer rate. The various DLT standards are shown in Table 32-5. While most of the DLT standards are comparable to those of 8 mm tape, the SuperDLT standards, of which the first drives are now on the market, promise to greatly exceed the capabilities of all other magnetic tape technologies. The SuperDLT drives that are currently available conform to the first of four SuperDLT standards. The ultimate goal of SuperDLT is a drive that holds 2 terabytes (2,000 gigabytes) of data on a single tape. The prices for DLT drives are high, approximately the same as 8 mm drives. DLT is clearly a backup technology for networks that require very large amounts of storage and high-speed backups.

Standard	Maximum Capacity (Uncompressed)	Maximum Transfer Rate
TK50	200 MB	1.6 MB/sec
TK70	588 MB	1.6 MB/sec
DLT260	5.2 GB	1.6 MB/sec
DLT700	12 GB	1.6 MB/sec
DLT2000	15 GB	2.5 MB/sec
DLT4000	20 GB	3 MB/sec
DLT7000	35 GB	10 MB/sec
DLT8000	40 GB	12 MB/sec
SuperDLT	110 GB	22 MB/sec

Table 32-5: *DLT Standards*

Linear Tape Open (LTO)

Linear Tape Open (LTO) is the newest of the high-speed/high-capacity magnetic tape standards to hit the market. Developed jointly as an open format by IBM, Hewlett-Packard, and Seagate, LTO consists of two standards, Accelis, which emphasizes fast access to data stored on tape, and Ultrium, which emphasizes high capacity. Ultrium is the standard best suited for network backups. Ultrium tape cartridges measure 4.1 × 4.0 × 0.8 inches and contain a single reel of tape (like DLT) that can store up to 100 GB of (uncompressed) data at speeds of 209 MB/sec. The roadmap for future Ultrium development calls for eventual capacities of 800 GB (uncompressed) on a single tape and speeds of up to 160 MB/sec. LTO is comparable to DLT and 8 mm tape, both in performance and price.

Tape Autochangers

In some cases, even the highest capacity tape drive is not sufficient to back up a large network. You can purchase more than one drive, but another device called an autochanger can automate the backup process even further. An *autochanger* (also called a *tape library* or a *jukebox*) is a combination device that contains at least one tape drive, a robotic mechanism for inserting tapes into the drive, and an array of slots into which you insert tapes. The robotic portion of the device can select any tape in the array, insert it into the drive, and remove the tape after the drive writes to it or reads from it. This arrangement enables the autochanger to execute backup jobs that span more than one tape or execute multiple jobs over the course of days or weeks, all without any user intervention.

Autochangers use a variety of mechanisms to manipulate the tapes. The simpler devices stack the tapes in a row and slide the entire row along a platform until the correct tape is lined up with the drive slot. Then the mechanism pushes the tape into the slot. This type of autochanger is typically a desktop unit that holds anywhere from 4 to 12 tapes. Larger autochangers often have a rack of tape slots and a robotic arm that pulls a tape out of a slot, carries it to the drive, and inserts it into the drive slot, as shown in Figure 32-7. These types of changers can have multiple tape drives, hold more than 100 tapes, and be as large as a refrigerator.

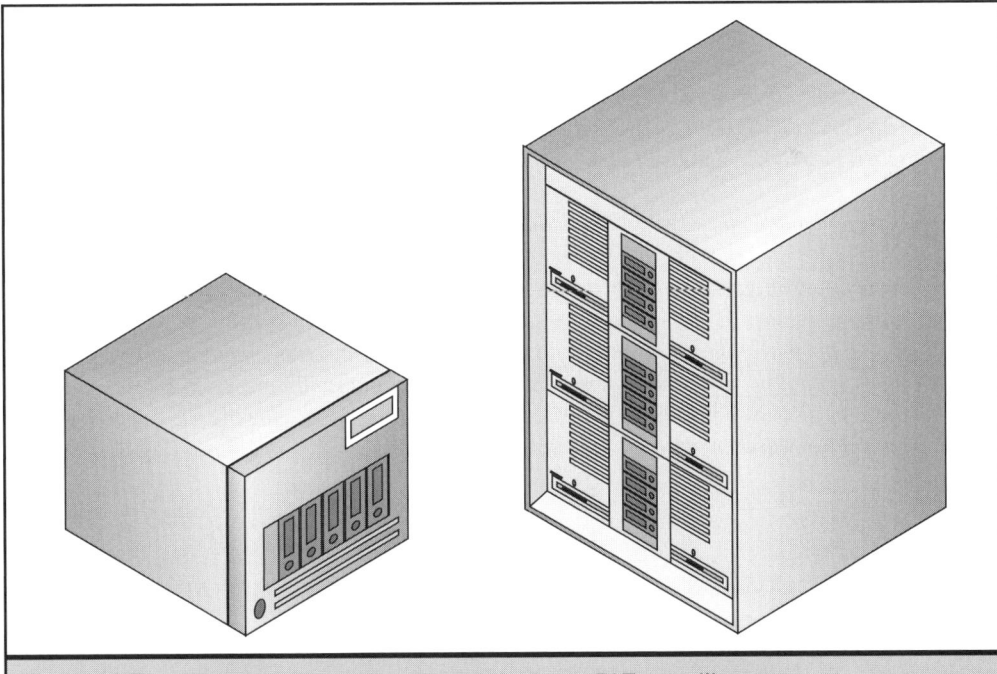

Figure 32-7: *A small DAT autochanger and a large DLT tape library*

For an autochanger to function properly, it has to know where each tape is located in the array. If an administrator creates a backup job that instructs the changer to load tape X, it has to know which of the tapes in the array is tape X before it can do so. Autochangers make this possible using one of two basic methods. The obvious method is for the autochanger to insert each tape in the array into the drive, take note of its contents, and store that information in memory. This is called performing an inventory. For a relatively small autochanger that holds only a few tapes, the inventory method is suitable. The robotic mechanisms in autochangers are not known for moving at great speed, however, and it may take several minutes for the inventory to complete.

For a refrigerator-sized autochanger with dozens of slots, the inventory process can take several hours. Because the inventory information is stored in volatile memory, it is lost when the device is unplugged or power fails. When this happens, the entire inventory process must start over again. To avoid this lengthy inventory process, some autochangers include a bar code reader in the robotic mechanism. By putting a label containing a bar code on each tape cartridge, administrators can identify the tapes in the array beforehand. The autochanger only has to move the robotic arm past each tape to scan its code, rather than take it from its slot, insert it into the drive, and remove it again.

Autochangers are available for all of the current magnetic tape technologies, in many different sizes. All are SCSI devices and usually use multiple SCSI IDs, one for each tape drive and one for the robotics. Smaller units costing around $2,000 make sense for networks on which administrators want to run unattended backup jobs that require two or more tapes. The largest autochanger units can have hundreds of tape slots, up to eight separate tape drives, total capacities of several terabytes, and price tags running as high as $200,000 or more. These autochangers can completely automate the backup process for a large enterprise network, so that administrators only have to change the tapes in the array in order to store copies offsite and replace expired tapes.

Tip *It is not a good idea to purchase a tape autochanger without seeing it first and examining how the robotic mechanism works. For expensive products, the designs of some autochangers are surprisingly like a Rube Goldberg contraption. One particular model uses normal standalone tape drives with buttons that a user can push to eject the tape. This particular autochanger, rather than use an internal command to eject the tapes, actually has a robotic arm with a finger that manually presses the eject buttons on the drives! This is not the type of solution you should expect to see on a device costing tens of thousands of dollars.*

Backup Software

In addition to the backup tape drive, the computer in which it's installed, and tape media, a backup system requires specialized software that can read the data from selected sources and transmit it to the drive in a controlled manner. As mentioned earlier in this chapter, magnetic tape drives are not directly addressable by a computer's file system. You can't simply drag and drop files to tape, or copy them from the command line as you would with a hard or floppy disk. Backup software products include device drivers for tape drives, so that the program can send commands to the drive and receive information from it. The software also provides an interface that you use to select the files, directories, or drives that you want to back up and to configure the various backup job parameters. These are the basic components, but most backup software products also include other options and features that simplify the process of performing backups.

When purchasing backup software, it's always a good idea to make sure that the software supports your tape drive. Most software manufacturers maintain a list of the drives (and the firmware revisions of the drives) that have been tested with the product and determined to function properly.

Network backup software is a more specialized type of backup program that enables administrators to control complex backup scenarios for computers located all over a network. Most of the backup programs designed for single computers can back up a network drive. Typically, you do this by mapping a drive letter to a drive on another computer and configure the software to back up that drive letter. While this is effective for backing up data files, it does not provide comprehensive backups of networked computers. Backing up a drive in this way does not protect operating system elements like the Windows registry or directory services databases, nor does it address the problem of open files. Network backup software is designed to provide more complete protection for networked computers by backing up operating system elements, commonly referred to as the *system state*.

Most tape drives come with backup software, and some operating systems have a backup utility as well. It is up to you to determine if this software is sufficient for your needs; if it isn't, you'll have to purchase a product from a third party. Operating system backup utilities typically provide only basic functionality and are often not suitable for network backups. Windows 2000's Backup program, for example (which is licensed from Veritas Software Corporation), provides an Explorer-like interface that you use to select what you want to back up, and lets you schedule the backup job to occur at a specific date and time. Backup can protect the registry on the computer where the software is running, and it can back up drives on other network computers, but it can't protect the registry on other computers. If you only want to back up data files from various computers on the network, Backup is a good solution, far better than some of the operating system backup utilities of the past, which lacked even basic job scheduling capabilities.

If you decide that you must purchase a network backup software package, it's a good idea to familiarize yourself with the capabilities of the various products on the market, and then compare them with your needs. In some cases, you can obtain evaluation versions of backup software products and test them on your network. This can help you identify potential problems you may encounter while backing up your network. The following sections examine some of the basic functions of a backup software package and how they apply to a typical network backup situation.

Selecting Backup Targets

The simplest type of backup job is a full backup, in which you back up the entire contents of a computer's drives. However, full backups usually aren't necessary on a daily basis, because many of the files stored on a computer do not change, and because full backups can take a lot of time and use a lot of tape. One of the best strategies when planning a backup solution for a network is to purchase a drive that can save all of your data files and the important system configuration files on a single tape. This enables you to purchase a less expensive drive and still provide your network with complete protection.

Being selective about what you want to back up complicates the process of creating a backup job, and a good backup software program provides several different ways to select the computers, drives, directories, or files (collectively called *targets*) that you want to back up. Most backup programs provide an expandable directory interface like that used in Windows Explorer, with check boxes that enable you to make your selections, as shown in Figure 32-8. As with Explorer, you can browse the network as well and select targets on other computers.

Selecting a drive or directory for backup includes all of the files and subdirectories it contains as well. You can then deselect certain files or subdirectories that you want to exclude from the backup. Some backup software programs can also list the targets for a backup job in text form, as shown in Figure 32-9. When you're creating a large, complex job involving many computers, this format can sometimes be easier to comprehend and modify.

Using Filters

The expandable display is good for selecting backup targets based on the directory structure, but it isn't practical for other types of target selection. Many applications and operating systems create temporary files as they're running, and these files are

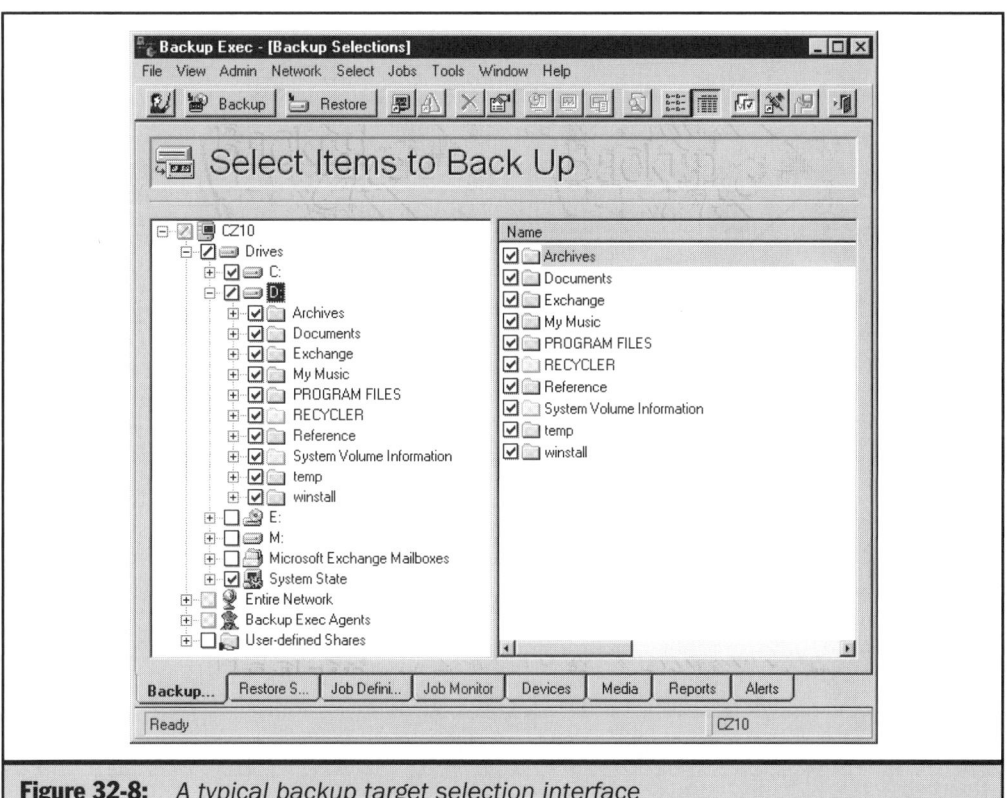

Figure 32-8:   *A typical backup target selection interface*

frequently named using a specific pattern, such as a TMP extension. In most cases, you can safely exclude these files from a backup, because they would only be automatically deleted at a later time anyway. However, manually deselecting all of the files with a TMP extension in a directory display would be very time-consuming, and you also have no assurance that there might not be other TMP files on your drives when the backup job actually runs.

To select (or deselect) files based on characteristics like extension, file name, date, size, and attributes, most backup software programs include filters. A filter is a mechanism that is applied to all or part of a backup target that instructs the software to include or exclude files with certain characteristics. For example, to exclude all files with a TMP extension from a backup job, you would apply an exclude filter to the drives that specified the file mask *.*tmp,* using an interface like that shown in Figure 32-10.

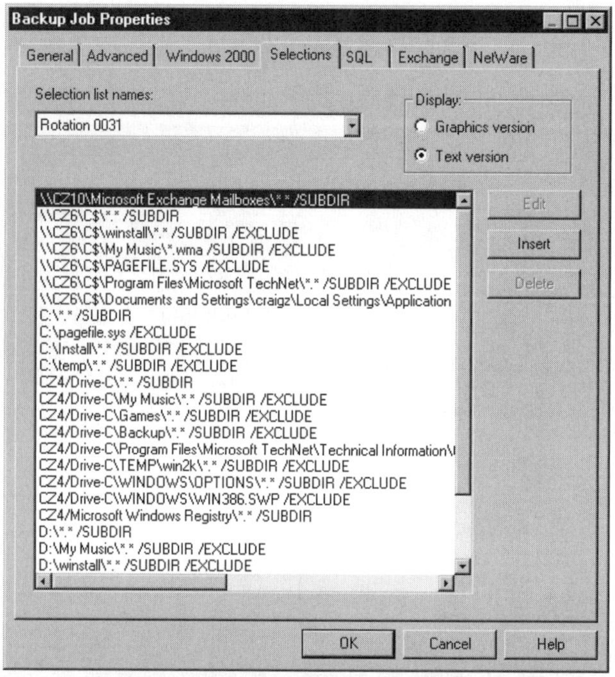

Figure 32-9: *A text-based backup target selection interface*

You can use filters in many ways to limit the scope of a backup job, such as the following:

■ Create an include filter specifying a modification date to back up all the files that have changed since a particular day.

■ Create exclude filters based on file extensions to avoid backing up program files, such as EXEs and DLLs.

■ Create a filter based on access dates to exclude all files from a backup that haven't been accessed in the last 30 days.

Incremental and Differential Backups

The most common type of filter used in backups is one that is based on the Archive attribute. This is the filter that backup software products use to perform incremental and differential backups. File attributes are single bits included with every file on a disk drive that are dedicated to particular functions. Different file systems have various

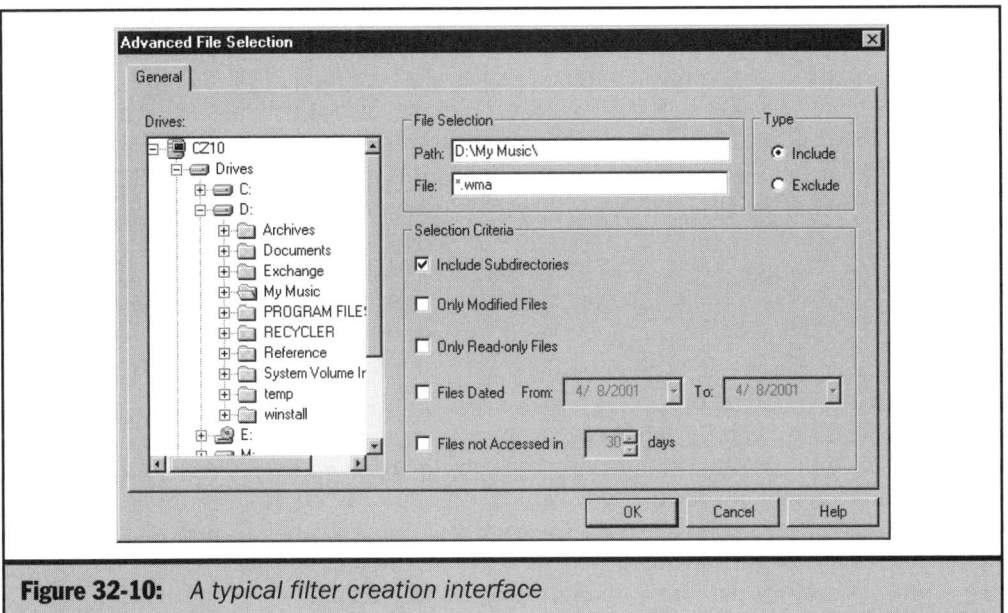

Figure 32-10:   *A typical filter creation interface*

attributes, but the most common ones found in almost all file systems are Read-only, Hidden, and Archive. The Read-only and Hidden attributes affect how specific files are manipulated and displayed by file management applications. Under normal conditions, a file with the Hidden attribute activated is invisible to the user and a Read-only file can't be modified. The Archive attribute has no effect in a normal file management application, but backup programs use it to determine whether files should be backed up.

A typical backup strategy for a network consists of a full backup job that is repeated every week with daily incremental or differential jobs in between. When you configure a backup software program to perform a full backup of a drive, the software typically resets the Archive attribute on each file, meaning that it changes the value of all of the Archive bits to 0. After the full backup, whenever an application or process modifies a file on the drive, the file system automatically changes its Archive bit to a value of 1. It is then possible to create a backup job that uses an attribute filter to copy to tape only the files with Archive bit values of 1, which are the files that have changed since the last full backup. The result is a backup job that uses far less tape and takes far less time than a full backup.

An *incremental backup* job is one that copies only the files that have been modified since the last backup and then resets the Archive bits of the backed-up files to 0. This means that each incremental job you perform copies only the files that have changed since the last job. If you perform your full backups on Sunday, Monday's incremental job consists of the files that have changed since Sunday's full backup. Tuesday's incremental job consists of the files that have changed since Monday's incremental, Wednesday's job consists of the files changed since Tuesday, and so forth. Files that are modified frequently might be included in each of the incremental jobs, while occasionally modified files might only be backed up once or twice a week.

The advantage of performing incremental jobs is that you use the absolute minimum amount of time and tape, because you never back up any files that haven't changed. The drawback of using incremental jobs is that in order to perform a complete restoration of a drive or directory, you have to restore the copy from the last full backup, and then repeat the same restore job from each of the incrementals performed since that full backup, in order. This is because each of the incremental jobs may contain files that don't exist on the other incrementals, and because they might contain newer versions of files on the previous incrementals. By the time you complete the restore process, you have restored all of the unique files on all of the incrementals, and overwritten all of the older versions of the files with the latest ones.

If you have a lot of data to back up and want the most economical solution, performing incremental jobs in the way to go. The restore process is more complex, but performing a full restore of a drive is (hopefully) a relatively rare occurrence. When you have to restore a single file, you just have to make sure that you restore the most recent copy from the appropriate full or incremental backup tape.

A *differential backup* job differs from an incremental only in that it does not reset the Archive bits of the files it backs up. This means that each differential job backs up all of the files that have changed since the last full backup. If a file is modified on Monday, the differential jobs back it up on Monday, Tuesday, Wednesday, and so on. The advantage of using differential jobs is that to perform a complete restore, you only have to restore from the last full backup and the most recent differential, because each differential has all of the files that have changed since the last full backup. The disadvantage of differentials is that they require more time and tape, because each job includes all of the files from the previous differential jobs. If your tape drive has sufficient capacity to store all of your modified data for a full week on a single tape, differentials are preferable to incrementals, because they simplify the restoration process.

In most cases, the incremental and differential backup options are built into the software, so you don't have to use filters to manipulate the Archive attributes. The software typically provides a means of selecting from among basic backup types like the following, shown in Figure 32-11:

Normal   Performs a full backup of all selected files and resets their Archive bits

Copy   Performs a full backup of all selected files and does not reset their Archive bits

Incremental   Performs a backup only of the selected files that have changed and does not reset their Archive bits

Differential   Performs a backup only of the selected files that have changed and resets their Archive bits

Daily Performs a backup only of the selected files that have changed today

Working Set Performs a backup only of the selected files that have been accessed in a specified number of days

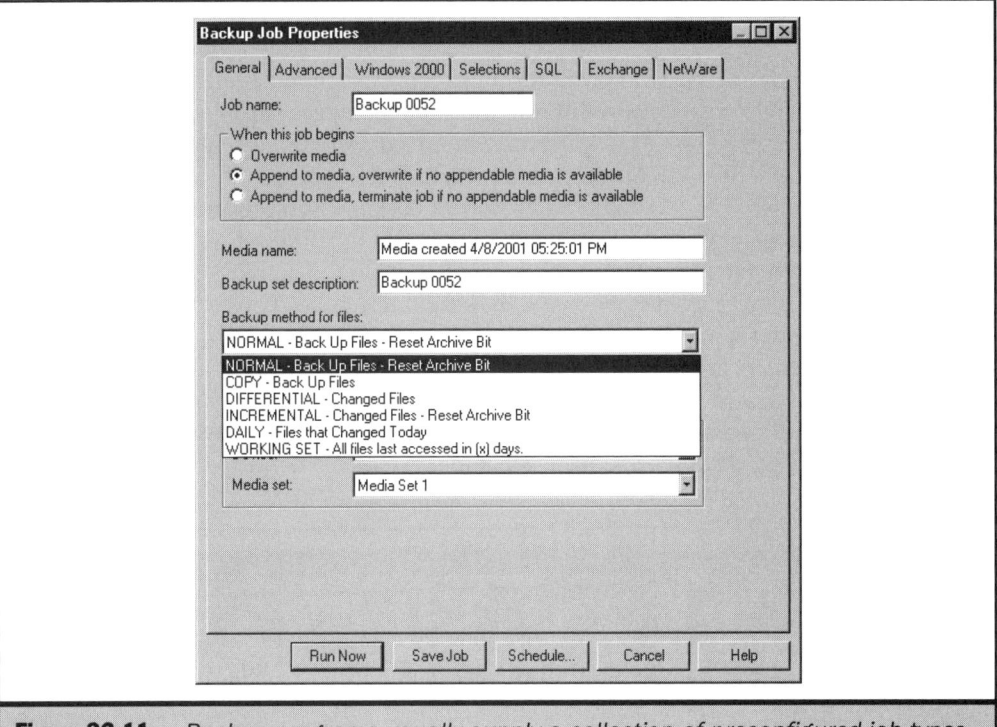

Figure 32-11: *Backup programs usually supply a collection of preconfigured job types.*

 *Different backup software products may not provide all of these options or may provide additional options. They may also refer to these options using different names.*

Using Backup Agents

As mentioned earlier, virtually all backup software programs enable you to select target drives and directories on other computers using the standard program interface. However, to perform comprehensive network backups, you must be able to back up operating system elements like the Windows registry and directory service databases.

The reason why backing up these elements is a problem is that the operating system accesses them continuously, and when a file is in use by one application, it often is not accessible by another. To back up remote computers fully, most network backup products use a special program that runs on the target computer, often called a backup agent. Agents can take different forms, depending on the operating system where they're running. On Windows computers, backup agents are usually services, which load automatically when the system boots and run continuously in the background. Unix computers have an equivalent program called a daemon.

Backup software packages also differ in the agents that are supplied with the base product. When you purchase a backup software product for a particular operating system, this typically refers to the computer in which the tape drive (or other backup hardware) is installed. When you buy a Windows 2000 backup product, for example, the main program runs on a Windows 2000 computer, and the product will usually include an agent that enables you to back up other computers on the network that are running the same operating system. In some cases, the software enables you to remotely install the agent on other computers over the network, while other products require you to travel to the other computers and install the agent manually. Once the agent is installed, the main backup software program typically enables you to select the computers for which agents are detected (as shown in Figure 32-12) and back them up completely, including the Windows registry or other operating system–specific elements.

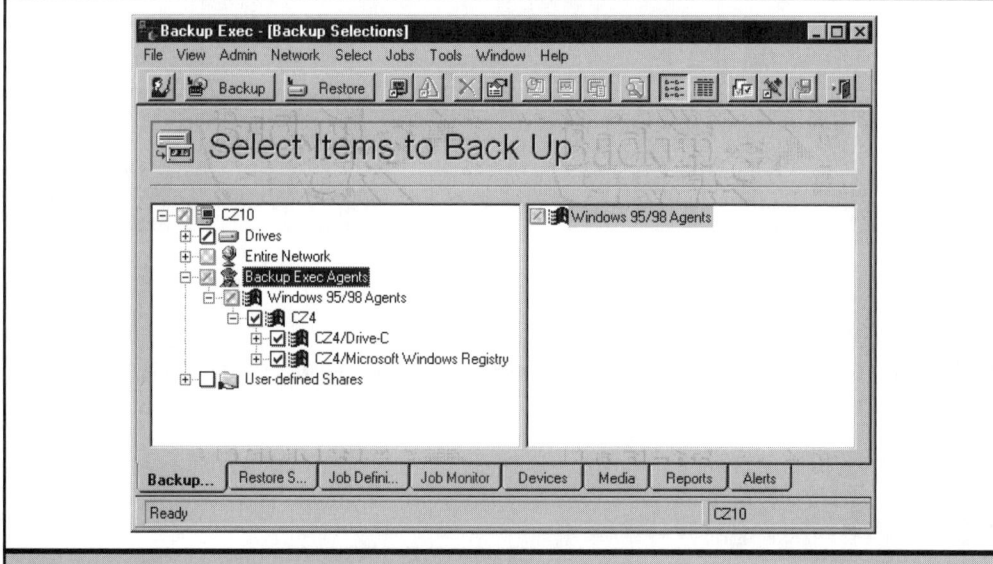

Figure 32-12: *When the backup software detects agents, they appear in the target selection interface.*

Backing up computers running other operating systems usually requires additional agents, which may or may not be included in the base software package. Backing up Windows 98, Unix, Macintosh, or other client computers with a Windows 2000 backup product may in some cases require you to purchase additional agents, and the cost of these agents can sometimes be a significant expense. When you're evaluating backup software products, be sure to determine what agents are supplied with the software and how much additional agents cost.

Another factor to consider when evaluating backup software is the licensing for the product. Network backup software can be licensed in several ways, the two most common being by client and by server. Licensing by client means that you purchase a package that supports a given number of clients. The software may be able to detect the actual number of clients it's backing up, or it may simply operate only with an operating system licensed for a specific number of users. For example, if you are running a 100-user version of Novell NetWare, some backup products require you to purchase a 100-user version of the backup software, even if you only want to back up the data on a handful of servers. Other products enable you to run a 10-user version of the backup software on a 100-user network if you only want to back up 10 computers. Upgrades for these products are usually available, enabling you to expand the number of supported clients.

Licensing by server typically means that you can back up an unlimited number of client computers, but only the number of servers specified by the license. The licensing terms for some backup products distinguish between Windows 2000 Server and Windows 2000 Professional computers. For example, you might be able to purchase a single server version of a Windows 2000 backup software package that enables you to install agents on an unlimited number of Windows 2000 Professional computers. If you want to back up other Windows 2000 servers, you may have to purchase additional licenses for a separate server agent.

Backing Up Open Files

The single biggest problem you are likely to encounter while performing backups in a network environment is that of open files. When a file is being used by an application, in most cases it is locked open, meaning that another application cannot open it at the same time. When a backup program with no special open file capabilities encounters a file that is locked, it simply skips it and proceeds to the next file. The activity log kept by the backup software typically lists the files that have been skipped and may declare a backup job as having failed when files are skipped (even when the vast majority of files were backed up successfully). Obviously, skipped files are not protected against damage or loss.

NETWORK ADMINISTRATION

Open files are one of the main reasons for performing backups during times when the network is not in use. Even during off hours, files can be left open for a variety of reasons. For example, users may leave their computers at the end of the day with files loaded into an application. The agents included with most network backup products are capable of backing up files left open in this way. This is one of the big advantages of using an agent, rather then simply accessing files through the network.

The most critical type of open file situation involves applications and data files that are left running continuously, such as database and e-mail servers. These applications often must run around the clock, and since their data files are constantly being accessed by the application, they are always locked open. A normal backup product can back up most of an application's program files in a case like this, but the most important files, containing the databases themselves or the e-mail stores, are skipped. This is a major omission that must be addressed in order to fully protect a network.

In most cases, network backup products are capable of backing up live databases and e-mail stores, but you must purchase extra software components to do so. Network backup software products usually have optional modules for each of the major database and e-mail products, which are sold separately. The optional component may consist of an upgrade to the main backup application, a program that runs on the database or e-mail server, or both. These options generally work by creating a temporary database file or e-mail store (sometimes called a *delta file*) which can process transactions with clients and other servers while the original data files in the server are being backed up. Once the backup is complete, the transactions stored in the delta file are applied to the original database and normal processing continues.

Recovering from a Disaster

Another add-on module available from many backup software manufacturers is a disaster recovery option. In this context, a disaster is defined as a catastrophic loss of data that renders a computer inoperable, such as a failure of the hard drive containing the operating system files in a server. This type of data loss can also result from a virus infection, theft, fire, or a natural disaster, such as a storm or earthquake. Assuming that you have been diligently performing your regular backups and storing copies offsite, your data should be safe if a disaster occurs. However, restoring the data to a new drive or a replacement server normally means that you must first reinstall the operating system and the backup software, which can be a lengthy process. A disaster recovery option is a means of expediting the restoration process in this type of scenario.

A disaster recovery option usually works by creating some form of boot medium that provides only the essential components needed to perform a restore job from a backup tape. In the event of a disaster, a network administrator only has to repair or replace any computer hardware that was lost or damaged, insert a CD-ROM (and possibly a floppy disk), and boot the computer. The disaster recovery disk supplies the files needed to bring the computer to a basic operational state from which you can perform a restore, using your most recent backup tapes.

Job Scheduling

Another important part of a network backup software product is its ability to schedule jobs to occur at particular times. Some rudimentary backup software products (such as those that come free with a tape drive) can only execute a backup job immediately. An effective network backup solution requires that you create a series of jobs that execute at regular intervals, preferably when the network is not otherwise in use. A good backup software product can be configured to execute jobs at any time of the day or night, and repeat them at specified intervals, such as daily, weekly, and monthly, using an interface like the one in Figure 32-13. More complicated scheduling options are also useful, such as the ability to execute a job on the last day of the month, the first Friday of the month, or every three weeks.

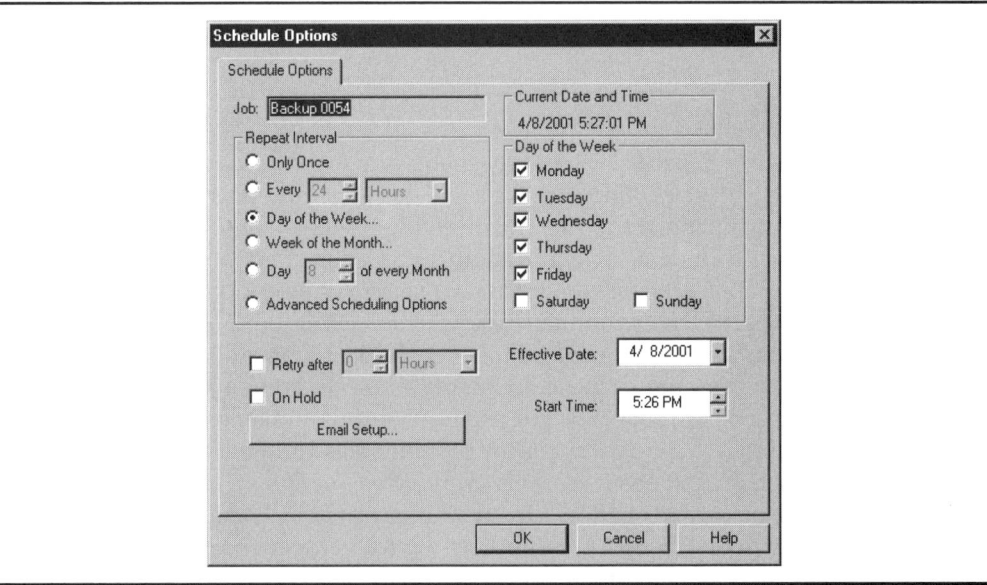

Figure 32-13: *A typical backup job scheduling interface*

Different software products use various methods to actually execute the jobs according to the schedule you specify. Some products utilize the capabilities of the operating system, such as the Scheduled Tasks window supplied with Windows 2000. Others extend the schema of a directory service, such as Microsoft's Active Directory or Novell NetWare, to create a job queue, and still others might supply their own program or service that runs in the background and executes the scheduled jobs at the right time.

The types of jobs you create and how often you run them should depend on the amount of data you have to back up, the amount of time you have to perform the backups, the capabilities of your hardware, and the importance of your data. For example, a typical network backup scenario would call for a full backup performed once a week, and incremental or differential jobs performed on the other days, with all of the jobs running during the night. An appropriate tape drive for this scenario would be able to fit all of the data from an incremental or differential job on a single tape, while the full backup might require a media change. This means that, with proper scheduling, an administrator would have to be present to change the tape during the full backup job, but the incrementals or differentials can run unattended, with someone changing the tape each day in preparation for that night's job. If it was important for all of the jobs to run unattended, a tape drive with a higher capacity or an autochanger would fit the bill.

Rotating Media

Network backup software products typically enable you to create your own backup strategy by creating and scheduling each job separately, but most also have preconfigured job scenarios that are suitable for most network configurations. These scenarios usually include a media rotation scheme, which is another part of an effective network backup strategy. A media rotation scheme is an organized pattern of tape labeling and allocation that enables you to fully protect your network using the minimum possible number of tapes. You can conceivably use a new tape for every backup job you run, but this can get very expensive. When you reuse tapes instead, you must be careful not to overwrite a tape you may still need in the event of a disaster. When you implement a media rotation scheme provided by the backup software, the program tells you which tape to put into the drive for each job, so that no vital data is overwritten.

The most common media rotation scheme implemented by backup software products is called Grandfather-Father-Son. These three generations refer to monthly, weekly, and daily backup jobs, respectively. The "Son" jobs run each day, and are typically incrementals or differentials. The scheme calls for four, five, or six daily tapes (depending on how many days per week you perform backups), which are reused each week. For example, you would have a tape designated for the Wednesday incremental job, which you overwrite every Wednesday. The "Father" jobs are the weekly full backups, which are overwritten each month. There will be four or five weekly jobs each month (depending on the day you perform the jobs), requiring one or more tapes per job. The tapes you use for the first full backup of the month, for example, will be overwritten during the first full backup of the next month. The "Grandfather" jobs are monthly full backups, the media for which are reused once every year.

The monthly tapes in the media rotation are often designated for offsite storage, which is an essential part of a good backup strategy. Diligently making backups will do you and your company no good if the building burns down, taking all of your tapes with it. Periodic full backups should be stored at a secured site, such as a fireproof tape vault or a bank safe deposit box. Some administrators simply bring the tapes home on a regular basis, which can be equally effective.

Backup Administration

When creating an automated network backup solution, proper planning and purchasing are the most important factors. Once the system is in place, there should be little user interaction required, except for making sure that the proper tape is inserted into the drive each day. It's also important for the administrator to make sure that the backup jobs are executing as designed.

Event Logging

Network backup software products nearly always have an indicator that specifies whether each backup job has completed successfully or has failed. However, simply checking this indicator does not necessarily give an adequate picture of the job's status. The criteria used to evaluate a job's success or failure can vary from product to product. A job failure can be an indication of a major problem, such as a hardware failure that has prevented any data from being written to the tape. With some products, a single file that is skipped because it is locked open can cause a job to be listed as having failed, even though all of the other files have been successfully written to the tape.

To check the status of the job in greater detail, you examine the event logs maintained by the software. Backup logs can contain a varying amount of detail, and many software products let you specify what information you want to be kept in the log, as shown in Figure 32-14. A full or complete log contains an exhaustive account of the backup job, including a list of all of the files copied to tape. This type of log contains everything you could ever want to know about a backup job, including which targets were backed up and which were skipped, as well as any errors that may have occurred. The complete file listing causes a log like this to be enormous in most cases, and the average administrator is less likely to check the logs regularly when it's necessary to scroll through hundreds of pages of file names to do so.

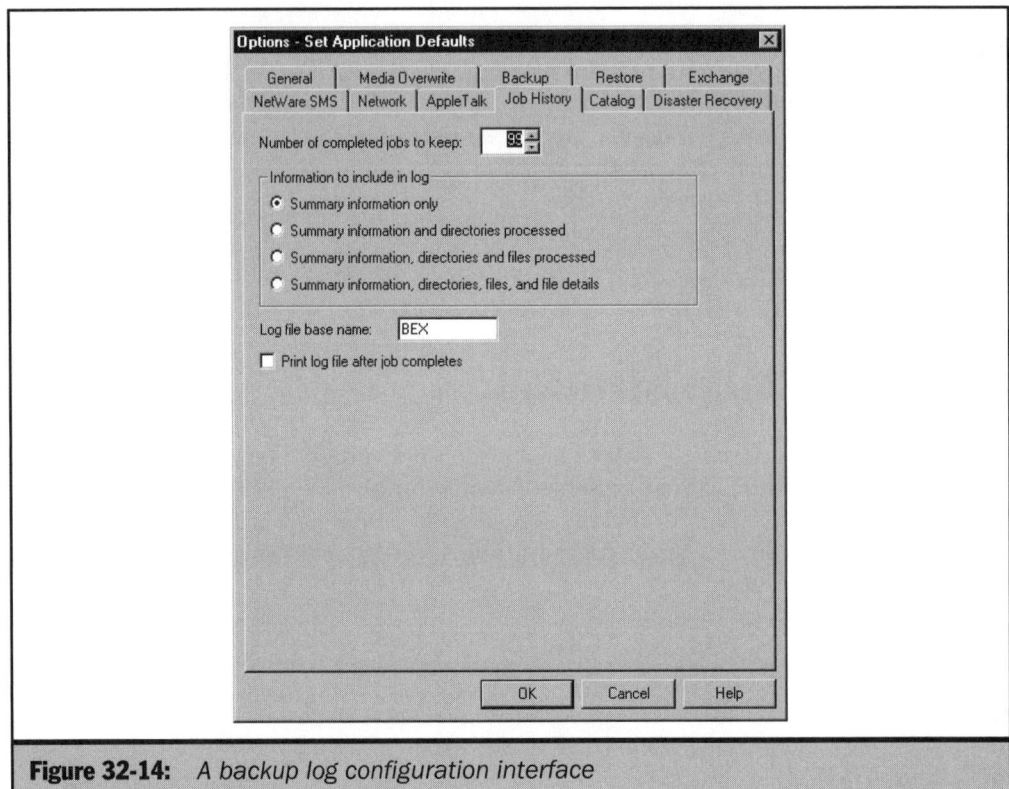

Figure 32-14:   *A backup log configuration interface*

Maintaining a full log might be a good idea as you are learning the intricacies of your backup software, but after the first few jobs, you'll probably want to reconfigure the software to keep a summary log containing only the details that you need to examine on a regular basis, such as whether target computers were backed up or not, the names of files that were skipped, and error messages. Administrators should examine the logs frequently to make sure that the backup jobs are running as planned.

Performing Restores

Logs and success indicators are usually reliable methods of confirming that your backups are completing successfully, but they are no substitute for performing a regular series of test restores. The whole reason for running backups in the first place is so you can restore data from the tapes when necessary. If you can't do this, then all of the time and money you've spent is wasted. It's entirely possible for a job to be listed as having completed successfully and for the logs to indicate that all of the targets have been backed up, only to find that it's impossible to restore any data from the tape. The reasons for this are manifold, but there are many horror stories told by network administrators about people who have diligently performed backups for months or years and have carefully labeled and stored the backup tapes, only to find that when they suffer a disaster, all of the tapes are blank. Performing test restores on a regular basis can prevent this sort of catastrophe.

Backup software products have a restore function that usually looks a lot like the interface you use to create backup jobs. You can browse through a directory structure to locate the files that you want to restore, as shown in Figure 32-15. When you browse in this way, you are looking at an index of all of the files stored on tape. Without the index, the software has no way of knowing what files are on each tape and where on the tape they're located. All backup software products create an index for each backup job they complete, but where they store the index can vary.

NETWORK ADMINISTRATION

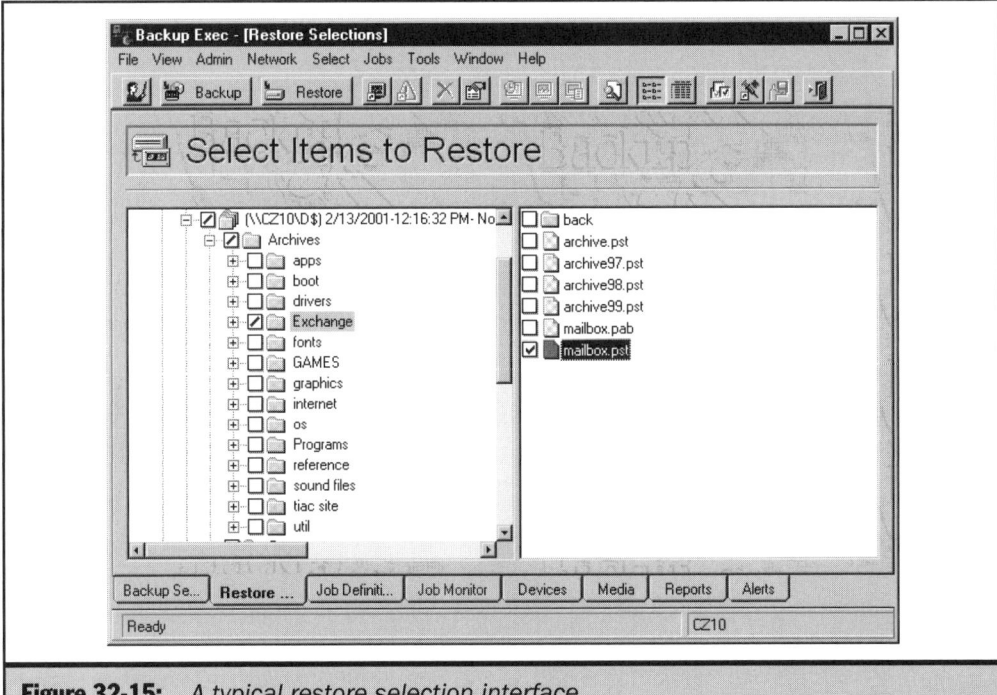

Figure 32-15: *A typical restore selection interface*

Lower-end backup software products, such as the Backup program included with Windows 2000, store the index for each tape on the tape itself. This ensures that you're never left with a tape that lacks an index, but you have to insert the tape into the drive in order to know what's stored on it. Higher-end network backup software products store the index on the tape as well, but they also maintain a composite index of all of your tapes on the computer's hard drive. This enables you to view your complete collection of backed-up files in one interface. The result is that you can select files in more than one way. A media view (shown in Figure 32-16) lists all of your tapes and lets you view the directories and files stored on each tape. If you know precisely which tape you want to restore a file from, this is the easiest way to locate it.

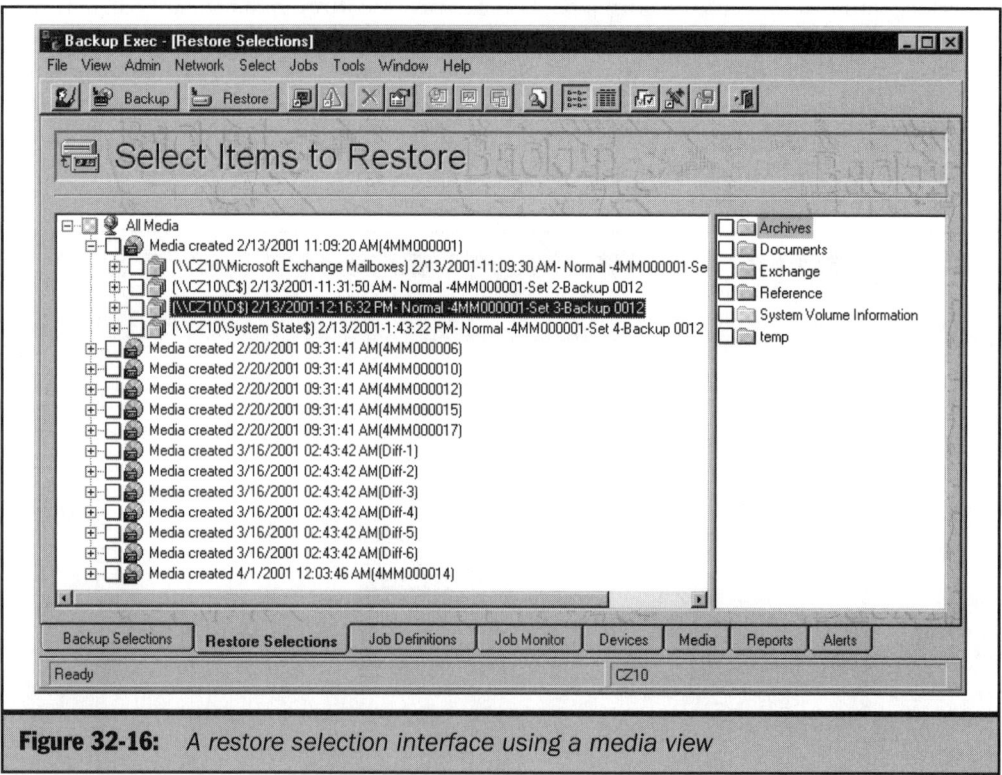

Figure 32-16:   *A restore selection interface using a media view*

A directory view (or volume view, as shown in Figure 32-17) lists the target computers and drives on your network and lets you select a backup tape from those containing backups of a particular target. If you know the file that you want to restore and want to see what versions of it are available on your various tapes, this is the restore mode you should use.

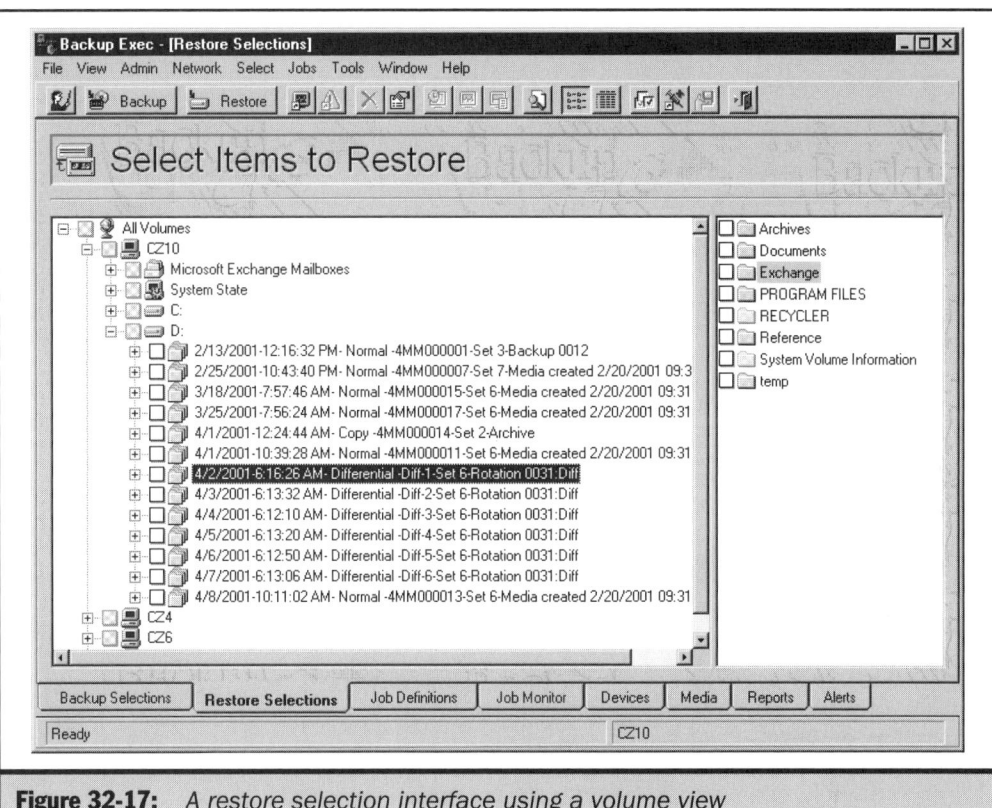

Figure 32-17: *A restore selection interface using a volume view*

In most cases, the process of selecting the files, directories, or drives to restore is very much like selecting those you want to back up; you click the check boxes next to the appropriate items. Once you've made your selections, you can then specify whether you want to restore them to their original locations or to an alternate location. You can also control the program's overwrite behavior when it encounters a file that already exists while performing a restore. Since restores are always performed on the spot, there is no need for the scheduling capabilities used in backup jobs. When you execute the job, the software uses the index created during the backup to locate the selected files on the tape and copies them to the destination you've selected.

Index

S

INTERNATIONAL CONTACT INFORMATION

AUSTRALIA
McGraw-Hill Book Company Australia Pty. Ltd.
TEL +61-2-9417-9899
FAX +61-2-9417-5687
http://www.mcgraw-hill.com.au
books-it_sydney@mcgraw-hill.com

CANADA
McGraw-Hill Ryerson Ltd.
TEL +905-430-5000
FAX +905-430-5020
http://www.mcgrawhill.ca

**GREECE, MIDDLE EAST,
NORTHERN AFRICA**
McGraw-Hill Hellas
TEL +30-1-656-0990-3-4
FAX +30-1-654-5525

MEXICO (Also serving Latin America)
McGraw-Hill Interamericana Editores S.A. de C.V.
TEL +525-117-1583
FAX +525-117-1589
http://www.mcgraw-hill.com.mx
fernando_castellanos@mcgraw-hill.com

SINGAPORE (Serving Asia)
McGraw-Hill Book Company
TEL +65-863-1580
FAX +65-862-3354
http://www.mcgraw-hill.com.sg
mghasia@mcgraw-hill.com

SOUTH AFRICA
McGraw-Hill South Africa
TEL +27-11-622-7512
FAX +27-11-622-9045
robyn_swanepoel@mcgraw-hill.com

**UNITED KINGDOM & EUROPE
(Excluding Southern Europe)**
McGraw-Hill Publishing Company
TEL +44-1-628-502500
FAX +44-1-628-770224
http://www.mcgraw-hill.co.uk
computing_neurope@mcgraw-hill.com

ALL OTHER INQUIRIES Contact:
Osborne/McGraw-Hill
TEL +1-510-549-6600
FAX +1-510-883-7600
http://www.osborne.com
omg_international@mcgraw-hill.com